F. Vogel   A. G. Motulsky

Human Genetics

Problems and Approaches

Second, Completely Revised Edition

With 447 Figures and 217 Tables

Springer-Verlag
Berlin Heidelberg New York Tokyo

Professor Dr. Friedrich Vogel
Institut für Anthropologie und Humangenetik
Im Neuenheimer Feld 328, 6900 Heidelberg, FRG

Arno G. Motulsky, M.D.
Professor of Medicine and Genetics
University of Washington, Division of Medical Genetics
Seattle, WA 98195, USA

ISBN 3-540-16411-1 Springer-Verlag Berlin Heidelberg New York Tokyo
ISBN 0-387-16411-1 Springer-Verlag New York Heidelberg Berlin Tokyo

ISBN 3-540-09459-8 1. Auflage Springer-Verlag Berlin Heidelberg New York
ISBN 0-387-09459-8 1st edition Springer-Verlag New York Heidelberg Berlin

Library of Congress Cataloging-in-Publication Data.
Vogel, Friedrich, 1925 –. Human genetics. Bibliography: p. Includes indexes.
1. Human genetics. I. Motulsky, Arno, G., 1923 –. II. Title. [DNLM: 1. Genetics.
2. Genetics, Medical. QH 431 V878h] QH431.V59 1986 573.2'1 86-3957

Typesetting, Printing, and Binding: Appl, Wemding
2121/3140-543210

To our wives and children

Preface to the Second Edition

The first edition of this book, published in 1979, was found useful by many students and was well received by the scientific community. Since the book was first written, human genetics has undergone dramatic developments, mainly due to the introduction of new concepts and techniques from molecular biology. Concomitantly, "basic" scientists have become increasingly interested in problems of human genetics. More than 700 human genes have been mapped, genes of previously unsuspected complexity - such as the gene for factor VIII - have become known, and the structure of noncoding DNA sequences is being analyzed with the aim of understanding gene regulation. DNA diagnosis is being rapidly introduced into medical genetics. All this, as well as the extensive progress in most other fields of human and medical genetics, had to be considered in the preparation of this second edition.

The book has been extensively revised and rewritten. A substantial new section dealing with gene and chromosomal structure at the molecular level has been added. The newer knowledge of molecular genetics has been incorporated, and the conceptual and practical contribution of DNA methods (for example in the hemoglobinopathies and in some other diseases) is discussed. Many new figures and tables have been added, and some illustrative material has been replaced.

We have read carefully the many friendly and sometimes flattering reviews of the first edition. A frequently voiced regret was that we did not deal more extensively with the practical aspects of clinical genetics. In response, we have added more material of clinical interest and have significantly expanded the chapter dealing with genetic counseling and prenatal diagnosis. However, the increasing body of knowledge of human genetics spanning both the basic and the clinical sciences makes it impossible to cover all hereditary diseases and their clinical implications. Furthermore, much of this material is dealt with in recent books and compendia which are listed in the enlarged general reference section. This book's aim remains the presentation of a cohesive and up-to-date exposition of the concepts, data, and problems underlying the theory and practice of human and medical genetics.

Many colleagues helped us in the preparation of this volume, including W. Buselmaier, T. Cremer, S. Gartler, E. Giblett, D. Götze, Vesna Najfeld, P. Propping, T. M. Schroeder-Kurth, K. Sperling, V. Sybert, and R. Stern. They should not be held responsible for any errors which may have crept in. Again, J. Krüger proved to be extremely helpful in all statistical matters. He advised us in replacing old-fashioned pencil-and-paper methods by the use of computers. Our secretaries, Adelheid Fengler in Heidelberg, Ingrid Rudolph in Berlin, and Sylvia Waggoner in Seattle, typed the new parts of the manuscript and succeeded in the formidable task of keeping everything in order. The new figures were again drawn by Edda Schalt.

Most of this edition was prepared in 1984, when we spent several months together at the Institute for Advanced Study *(Wissenschaftskolleg)* in Berlin (West), and in 1985. Partial support for A. G. M.'s stay in Berlin was provided by an Alexander v. Humboldt Award.

We hope that this second edition will, like the first, fulfill its purpose: to provide the necessary background of genetic concepts and their applications to those interested in and fascinated by the scientifically challenging and practically important field of human genetics.

Spring 1986

Friedrich Vogel, Heidelberg
Arno G. Motulsky, Seattle

Preface to the First Edition

Human genetics provides a theoretical framework for understanding the biology of the human species. It is a rapidly growing branch of science. New insights into the biochemical basis of heredity and the development of human cytogenetics in the 1950s heightened interest in this field. The number of research workers and clinicians who define themselves as full-time or part-time human and medical geneticists has increases sharply, and detailed wellfounded knowledge has augmented exponentially. Many scientists and physicians are confronted with genetic problems and use concepts and methodology of human genetics in research and diagnosis. Methods developed in many different fields of the biologic, chemical, medical, and statistical sciences are being utilized toward the solution of genetic problems. The increasing number and sophistication of well-defined and elegantly solved problems helps to refine an extensive framework of genetic theory. These new conceptual insights in their turn lead to solutions of new questions. To mention only one example, the structure of hemoglobin genes has been elucidated using methods derived from protein chemistry and DNA technology. It is an exciting experience to participate in these developments!

Moreover, scientific progress in genetics has practical implications for human well-being. Improved knowledge of the genetic cause of an increasing number of human diseases helps to refine diagnosis, to find new therapeutic approaches, and above all, to prevent genetic diseases. So far, human genetics has had less of an impact on the behavioral and social sciences. It is possible that genetic differences involved in shaping personality structure, cognitive faculties, and possibly human social behavior may be at least as important as genetic variation affecting health and disease. The data, however, are less clear and more controversial. These problems are discussed in detail in the text. The rapid progress of human genetics in recent decades has attracted - and is still attracting - an increasing number of students and scientists from other fields. Various elementary textbooks, more advanced monographs of various branches of the field, and the original journal literature are the usual sources of introduction to human genetics. What seems to be lacking, however, is a fairly thorough and up-to-date treatise on the conceptual basis of the entire field of human genetics and its practical applications. Often, the absence of a broadly based background in the field leads to misunderstanding of its scope, unclear goals for research, improper selection of methods, and imbalanced theoretical discussions. Human genetics is based on a powerful theory, but this implicit conceptual foundation should be made explicit. This goal is the purpose of this book. It certainly is a formidable and possibly even too audacious task for two sole authors. However, both of us have been active in the field for more than 25 years. We have worked on various problems and with a variety of methods. Since the early years of our careers, we have met occasionally, followed each other's writings, and were often surprised by the similarity of our opinions and judgments despite quite different early medical and scientific backgrounds. Moreover, our knowledge of the literature turned out to be in part overlapping and in part complementary. Since we are working in different continents, AGM had a better knowledge of concepts and results in the USA, while FV knew more of the continetal European literature. Moreover, both of us have extensive experience as editors of journals in human genetics and one (FV) published a fairly comprehensive textbook in Germany some time ago (*Lehrbuch der allgemeinen Humangenetik,* Springer 1961), parts of which were still useful for the new book. We finally decided to take the risk, and, by writing an "advanced" text, to expose our deficiencies of knowledge, shortcomings of understanding, and biases of judgement.

A text endeavoring to expose the conceptual framework of human genetics cannot be dogmatic and has to be critical. Moreover, we could not confine ourselves to hard facts and well-proved statements. The cloud of conjectures and hypotheses surrounding a rapidly growing science had to be depicted. By doing so, we face the risk of being disproved by further results.

A number of colleagues helped by reading parts of the manuscript on which they had expert knowledge and by making useful suggestions: W. Buselmaier, U. Ehling, G. Flatz, W. Fuhrmann, S. Gartler, Eloise Giblett, P. Propping, Laureen Resnick, and Traute M. Schroeder. They should not be held responsible for possible errors. J. Krüger was of supreme help in the statistical parts. Our secretaries, Mrs. Adelheid Fengler and Mrs. Gabriele Bauer in Heidelberg, Mrs. Sylvia Waggoner in Seattle, and Mrs. Helena Smith in Stanford gave invaluable aid. The figures were drawn by Edda Schalt and Marianne Lebküchner. Miriam Gallaher and Susan Peters did an expert job of copy editing. The authors are especially grateful to Dr. Heinz Götze and Dr. Konrad F. Springer, of Springer Publishing Company, for the excellent production. The work could not have been achieved had the two authors not been invited to stay at the Center for Advanced Study in the Behavioral Sciences at Stanford (California) for the academic year of 1976/1977. The grant for AGM was kindly provided by the Kaiser Family Foundation, while the Spencer Foundation donated the grant for FV.

The cover of this book shows the mythical first human couple, Adam and Eve, as imagined by Albrecht Dürer (1504). They present themselves in the full beauty of their bodies, ennobled by the genius and skill of a great artist. The drawing should remind us of the uniqueness and dignity of the human individual. Human genetics can help us to understand humanity better and to make human life happier. This science is a cardinal example of Alexander Pope's statement. "The proper study of mankind is man."

Spring 1979

Friedrich Vogel, Heidelberg
Arno G. Motulsky, Seattle

Table of Contents

Introduction

Human Genetics as Fundamental and Applied Science. Human genetics is both a fundamental and an applied science. As a fundamental science, it is part of genetics - the branch of science that examines the laws of storage, transmission, and realization of information for development and function of living organisms. Within this framework, human genetics concerns itself with the most interesting organism - the human being. This concern with our own species makes us scrutinize scientific results in human genetics not only for their theoretical significance but also for their practical value for human welfare. Thus, human genetics is also an applied science. Its value for human welfare is bound to have repercussions for theoretical research as well, since it will influence the selection of problems by human geneticists, their training as well as financing of research. Because of its continued theoretical and practical interest, human genetics offers fascination and human fulfillment unparalleled by work in fields that are either primarily theoretical or entirely practical in subject matter.

Science of Genetics. Genetics has become a mature science. It is based on a powerful and penetrating theory. The profundity of a theory depends on the depth of the problems it sets out to solve and can be characterized by three attributes: the occurrence of high-level constructs, the presence of a mechanism, and high explanatory power [243]. In genetics, the high-level "construct" is the gene as a unit of storage, transmission, and realization of information. Since the rediscovery of Mendel's laws in 1900, genetic mechanisms have been worked out step by step to the molecular level - deciphering of the genetic code, analysis of transcription and translation, and function of gene-determined proteins. The fine structure of genes is now being elucidated. The problems of regulation of gene activity in the development and function of organisms are currently the main goal of fundamental research. So far, the explanatory power of the theory has not nearly been exhausted.

How Does a Science Develop? Kuhn (1962) [257] described the historical development of a science as follows: In the early, protoscientific stage, there is much competition between various attempts at theoretical foundation and empirical verification. Basic observations suggest a set of problems that, however, are not yet visualized clearly. Then, one "paradigm" unifies a group within the scientific community in the pursuit of a common goal, at the same time bringing into sharper focus one or a few aspects of the problem field, and suggesting a way for their solution. If the paradigm turns out to be successful, it is accepted by an increasing part of the scientific community, which now works under its guidance, exploring its possibilities, extending its range of application, and developing it into a scientific theory.

This concept of a paradigm has three main connotations:

1. It points to a piece of scientific work that serves as an "exemplar," which suggests ways in which a certain problem should be approached.
2. It describes a group of scientists who try to explore this approach, expand its applicability, deepen its theoretical basis by exploration of basic mechanisms, and enhance its explanatory power.
3. Finally, while an elaborate theory must not - and, in most cases, will not - exist when a paradigm is initiated, its germ will be there, and a successful paradigm will culminate in elaboration of this theory.

This process of development of a science within the framework of a paradigm has been described by Kuhn as "normal science": The basic theory is more and more taken for granted. It would be sterile at this stage to doubt and re-examine its very cornerstones; instead, it is applied to a variety of problems, expanded in a way that is comparable to puzzle solving. From time to time, however, results occur that, at first glance, defy explanation. First, this will lead to attempts at accommodating such results within the theoretical framework by additional ad hoc hypotheses. These attempts will often be successful; sometimes, however, they fail. If in such a situation an alternative paradigm is brought forward that explains most of the phenomena accounted for by the old theory as well as the new, hitherto unexplained

phenomena, a scientific revolution might occur: The new paradigm gains support from an increasing majority of the scientific community, it is soon developed into a new – more explanatory – theory, and the process of normal science begins anew.
This portrayal of scientific development has been criticized by some philosophers of science [258]. The concept of "normal" science as outlined above does not appeal to some theorists. Working within the framework of a given set of concepts has been denounced as dull, boring, and in any case not as science should be. According to these philosophers, scientists ought to live in a state of permanent revolution, always and again questioning the basic foundations of their field, always eager to put them to critical tests and, if possible, to refute them (Popper, [279, 280, 281]; Watkins [287]). Many scientists actively involved in research, on the other hand, have readily accepted Kuhn's view; apparently, he has helped them to recognize some important aspects in the development of their own fields.

Central Theory of Genetics Looked at as a Paradigm. While Kuhn's concepts were developed on the basis of the history of the physical sciences, his description fits the development of genetics well: Up to the second half of the nineteenth century, the phenomena of heredity eluded analysis. Obviously, children were sometimes – but by no means always – similar to their parents; some diseases were shown to run in families; it was possible to improve crops and domestic animals by selective breeding. Even low-level laws were discovered, as for example Nasse's law that hemophilia only affects boys but is transmitted by their mothers and sisters (Sect. 3.1.4). However, a convincing overall theory was missing, and attempts at developing such at theory were unsuccessful. In this situation, Mendel, in his work *Versuche über Pflanzenhybriden* (1865) [266], first improved a procedure: He complemented the breeding experiment by counting the offspring. Then, he interpreted the results in terms of random combination of basic units; by assuming these basic units, he founded the gene concept – the nuclear concept of genetic theory (Sect. 1.4).
Since then, Mendel's work has served as a paradigm in all three connotations: It provided an exemplar as to how breeding experiments should be designed and evaluated; it resulted in the establishment of a scientific community of geneticists; and it led to the development of a deep and fertile scientific theory.
A special problem that has not been answered satisfactorily in our opinion concerns the question of why Mendel's paradigm had to await acceptance for as long as 35 years after his experiments had been published. It would be too simplistic to blame academic arrogance and shortsightedness of contemporary biologists who did not want to accept the work of a "nonacademic" outsider, even if this factor might be one of the components for this neglect. We rather believe that the many new biologic discoveries in the 35 years following Mendel's discovery were of such a revolutionary nature as to qualify as a scientific crisis in the Kuhnian sense and therefore required a completely new approach.
Soon after the rediscovery of Mendel's laws in 1900, however, an initially small, but quickly growing group of scientists gathered who developed genetics in an interplay between theory and experiment and launched the major scientific revolution of the twentieth century in the field of biology.

Human Genetics and the Genetic Revolution. Meanwhile, the biologic revolution of the nineteenth century – evolutionary theory – had been accepted by the scientific community. One major consequence was the realization that human beings had evolved from other, more "primitive" primates, that humans are part of the animal kingdom, and that the laws of heredity that had been found to apply for all other living beings were also valid in our species. Hence, Mendel's laws were soon applied to traits that were found in human pedigrees – primarily hereditary malformations and diseases. Analyzing the mode of inheritance of alcaptonuria – a recessive disease – Garrod (1902) [249] clearly recognized the cardinal principle of gene action: genetic factors specify chemical reactions (Sect. 1.5). His insight also required 30 years before being incorporated into the body of "normal" science.
Elucidation of inheritance in humans did not begin with Mendel's paradigm. Many relevant observations had been reported before, especially on malformation and diseases. Moreover, another paradigm had been founded by F. Galton in his work on *Hereditary Talent and Character* (1865) [248] and in later works: to derive conclusions as to inheritance of certain traits such as high performance, intelligence and stature, one should measure these traits as accurately as possible and then compare the measurements between individuals of known degree of relationship (for example, parents and children, sibs, or twins) using statistical methods. This approach did not contain the potential for elucidating the mechanisms of heredity. On the other hand, it seemed to be much more generally applicable to human characters than Mendelian analysis; predigree analysis in terms of Mendel's laws was hampered by the fact that most human traits simply could not be classified as alternate characters such as round and shrunken peas. Human characters were usually graded and showed no alternative distribution in the

population. Moreover, the phenotypes were obviously determined not only by the genetic constitution but by external, environmental influences as well; they were the result of an interaction between "nature and nurture" (Galton). Therefore, naive attempts at applying Mendel's laws to such traits were doomed to failure. For traits that were regarded as important, such as intelligence and personality, but also for many diseases and mental retardation, there was only the choice between research along the lines suggested by Galton or no research at all. Investigations on genetic mechanisms would have to await elucidations of the genetics of other, more accessible organisms. Under these circumstances, scientists chose to follow Galton. This choice had not only theoretical reasons; it was strongly influenced by the desire to help individuals and families by calculating risk figures for certain diseases, thereby creating a sound basis for genetic counseling. More important, however, was the concern of some scientists about the biologic future of the human species, which they saw threatened by deterioration due to relaxation of natural selection. The motives for their research were largely eugenic: it seemed to provide a rational foundation for measures to curb reproduction of certain groups who were at high risk to develop disease.

History of Human Genetics: a Contest Between Two Paradigms. The two paradigms – Mendel's gene concept and Galton's biometric approach – developed side by side from 1900 up to the present; many present-day controversies, especially in the field of behavior genetics but also concerning strategies in the genetic elucidation of common diseases, are immediately understandable when the history of human genetics is conceived as a contest between these two paradigms. This does not mean that the two paradigms were mutually exclusive; in fact, correlations between relatives as demonstrated by biometric analysis were interpreted in terms of gene action by Fisher (1918) [664]. Some human geneticists have worked during some part of their career within the framework of the one, and during another, within the framework of the other paradigm. By and large, however, the two streams of research have little interconnections and may even become further polarized because of highly specialized training for both groups epitomized by the biochemical laboratories for one and the computer for the other group.

In the first decades of this century, the biometric paradigm of Galton appeared to be very successful: genetic variability within the human population was believed to be established for normal traits such as stature or intelligence as well as for a wide variety of pathologic conditions such as mental deficiency and psychosis, epilepsy, and common diseases such as diabetes, allergies, and even tuberculosis. Mendelian analysis, on the other hand, seemed to be confined to rare hereditary diseases; the ever repeated attempts at expanding Mendelian explanation into the fields of normal, physical characters and common diseases were usually undertaken without critical assessment of the inescapable limitations of Mendelian analysis. The first major breakthrough of Mendelian genetics was the establishment of the three-allele hypothesis for the ABO blood groups by Bernstein in the 1920s [240] (Sect. 3.2.2). Further progress, however, had to await the development of genetic theory by work on other organisms such as *Drosophila,* bacteria, and viruses, especially bacteriophages.

The advent of molecular biology in the late 1940s and 1950s had a strong influence on human genetics and, indeed, brought the final breakthrough of Mendel's paradigm. A major landmark was the discovery of Pauling et al. (1949) [1260] that sickle cell anemia was due to an abnormal hemoglobin molecule.

The foundation of human chromosome research in the late 1950s and early 1960s (Sect. 2.1) came as a second, important step. At present, most research in human genetics has become a part of mainstream research within the framework of genetic theory. The human species regarded by most early experimental geneticists as a poor tool for genetic research is now displaying definite advantages for attacking basic problems. Some of these advantages are the large size of available populations, the great number and variety of known mutants and chromosome anomalies, and the unparalleled detailed knowledge of human physiology and biochemistry in health and disease.

One would expect that these breakthroughs have led to the establishment of Mendel's paradigm as the only leading paradigm in human genetics. This, however, is not the case. In spite of the fact that genetic theory is now pervading many fields that seemed to be closed to it only a short time ago, the paradigm of Galton – biometric analysis – has recently attained an unsurpassed level of formal sophistication. The availability of computers has greatly facilitated the development and application of biometric techniques. Moreover, in some fields, such as behavior genetics, application of genetic theory – Mendel's paradigm – is still hampered by severe difficulties (Sect. 8), and here biometric methods still dominate. In the same field, however, they are most severely criticized.

Progress in Human Genetics and Practical Application. The two achievements mentioned above – molecular biology and chromosome research – have not only altered human genetics as a pure science, but they have also brought marked progress in its application for human welfare. At the beginning, this progress did not appear very conspicuous; the diagnosis of hereditary diseases was improved, and many, so far unexplained malformations were accounted for by chromosome aberrations. The first practical success came in the early 1950s when the knowledge of enzyme defects in phenylketonuria (Sect. 4.2.2.7) and galactosemia led to succesful preventive therapy by a specific diet. However, a breakthrough on a much larger scale was achieved when the methods of prenatal diagnosis for chromosome aberrations and for some metabolic defects were introduced in the late 1960s and early 1970s (Sect. 9.1.1). Now, suddenly, genetic counseling could be based not only on probability statements but, in an increasing number of cases, on certainty of individual diagnoses. This scientific development coincided with a growing awareness in large parts of the human population that unlimited human reproduction must not be accepted as a natural law but can - and should - be regulated in a rational way. Introduction of oral contraceptive agents signaled this awareness. The chance to avoid the births of severely handicapped children is being realized and accepted by a rapidly increasing fraction of the population. Knowledge in human genetics as a practical tool to prevent suffering and disease has found wide resonance and is now one of the most rewarding approaches in preventive medicine. In many countries, the politically responsible bodies have created, or are now creating the institutions for widespread application of the new tools.

Effects of Practical Application on Research. These practical applications have led to a marked increase in the number of research workers and the amount of work within the last 20-30 years. From the beginning of our century up to the early 1950s, human genetics had been the interest of a handful of scientists for most of whom it was not even a full-time occupation. Many of them were trained and worked much of their lifetime as physicians in special fields of medicine, such as Waardenburg and Franceschetti in ophthalmology, and Siemens in dermatology. Others were interested in theoretical problems of population genetics and evolution and chose problems in human genetics as the field of application for their theoretical concepts, most notably J. B. S. Haldane and R. A. Fisher. Yet others had their point of departure in physical anthropology. This heterogenous group of scientists could not form a coherent scientific community. For a long time, there was almost no social structure as required for the development of a scientific specialty. There were almost no special departments, journals, and international conferences. This lack of focus resulted in a marked heterogeneity in quality and content of scientific contributions.

All this has changed. There are now departments and units of human and medical genetics in many countries; universities and medical schools have introduced special curricula, many journals and a series of progress reports exist, and numerous congresses and conferences are being held. The overall impression is that of an active and vigorous field.

Dangers of Widespread Practical Application for Scientific Development. This development, however, satisfactory as it is, has also a number of potentially dangerous consequences:

a) Research is promoted primarily in the fields of immediate practical usefulness such as clinical cytogenetics and prenatal diagnosis of hereditary diseases; fields of less immediate practical importance may be neglected.
b) In the past, the contact with fundamental research in related fields such as molecular genetics and cell biology was not intensive enough. This may have led to a slowdown in the transfer of scientific concepts and experimental approaches from these fields. Fortunately, this has changed recently with the advent of "new genetics" (Sect. 2.3).
c) Areas of great importance for our understanding of human evolution (and possibly of human history), as well as for functioning of the human society and its institutions, are neglected if mainstream research in human genetics is directed exclusively to medical problems. Population and evolutionary genetics, on the one hand, and behavior genetics, on the other, are the two branches that suffer most. If they were to be excluded from mainstream research in human genetics, they will all too easily lose meaningful contact with human biology.
d) Much medical research applies established methods to answer straightforward questions. Many studies collect data with new techniques. Individual results often are not of great import, but the ensemble of such data is the essential building block for future progress of normal science. Much of such work is being done in human and medical genetics and is quite essential for many medical and anthropologic applications. However, there is continued need in human genetics to develop testable hypotheses and try to test their consequences from all viewpoints.

Human geneticists have to avoid the danger of neglecting the further development of genetic theory.

Basic research is needed in fields in which immediate practical application of results is not possible but, in the long run, might be at least quite as important for the future of the human species as current applications in preventive medicine.

Advantages of Practical Application for Research. Necessities of medical diagnosis and counseling have also given strong incentives to fundamental research. Many phenomena that fundamental research tries to explain would simply be unknown had they not been uncovered by study of diseases. We would be ignorant regarding the role of sex chromosomes in sex determination had there not been patients with sex chromosomal anomalies. Phenomena such as spontaneously enhanced chromosome instability in Fanconi's anemia or Bloom's syndrome with all its consequences for somatic mutation, cancer formation (Sect. 5.1.6), and even for understanding the normal interphase nucleus were discovered accidentally in the process of examining certain patients for diagnostic reasons. Genetic analysis of the "supergene" determining the major histocompatibility complex in humans contributes much to our fundamental understanding of how the genetic material above the level of a single gene locus is structured and how high genetic variability within the human population might be maintained (Sect. 3.5.5). Yet, research in this field would certainly be much less active had there not been the incentive of improving the chances of organ transplantation.

Whether we like it or not, our societies pay increasing amounts of money for research in human genetics because they want to have the practical benefits. Hence, to promote basic research, we have to push forward widespread practical applications. To guarantee progress in practical application for the future as well - and not only in the field of medicine - basic research needs to be promoted. This is also the only way to attract good research workers and to maintain - or even improve - scientific standards. This paradox creates priority problems for all those concerned with research planning.

Human Genetics and the Sociology of Science. From the foregoing discussion, it should have become abundantly clear that human genetics - like all other sciences - has not developed in a sociologic vacuum, following only the inherent logical laws of growth of theory and experimental testing. Human genetics is the work of social groups of human beings who are subject to the laws of group psychology and are influenced by the society at large in their attitudes toward research and their selection of problems. Unfortunately, sociologic investigations of group formation and structure in human genetics have not been carried out. Another group active in the foundation of molecular biology, the group that introduced the bacteriophages of *E. coli* into the analysis of genetic information, has been studied extensively [237]. We know from this and from other examples that, in a phase when a new paradigm is founded, the group that shares this paradigm establishes close within-group contacts. The normal channels of information exchange such as scientific journals and congresses are superseded by more informal information transfer through telephone calls, preprints, and personal visits. Within the group, influential personalities serve as intellectual and/or organizational leaders. Outside contacts, on the other hand, are often loose. When the acute phase of the scientific revolution is over, the bonds within the group are loosened, and information is, again, largely exchanged by normal channels of publication.

Similar developments can be observed in human genetics. In Sect. 2.1, we shall sketch the group structure of the British chromosome workers in the late 1950s when the first human chromosome aberrations were discovered and clinical cytogenetics was founded. Other, present-day examples are the groups active in the elucidation of the major histocompatibility complex (Sect. 3.5.5) and in the assignment of gene loci to chromosome segments by cell hybridization (Sect. 3.4).

Of similar influence on population genetics has been the one "big science" research project in human genetics - the Atomic Bomb Casualty Commission (ABCC) project that was launched in the late 1940s in Japan by American and Japanese research workers to examine the genetic consequences of the atomic bombs in Hiroshima and Nagasaki (Sect. 5.2.1.4). In later years, this project led to comprehensive studies of the genetic effects of parental consanguinity.

Many, if not most of the more interesting developments in the field were not initiated by investigators who would declare themselves human geneticists or who worked in human genetics departments. They were launched by research workers from other fields such as general cytogenetics, cell biology, molecular biology, biochemistry, immunology, but also from clinical specialities such as pediatrics, hematology, or psychiatry. A red thread running through many recent developments has been the application of nongenetic techniques from many different fields such as biochemistry and immunology to genetic concepts. In fact, most recent progress in human genetics comes from such interdisciplinary approaches. The number of research workers in the field has increased rapidly. Most did not start as human geneticists but as medical specialists, biochemists, statisticians, general cytogeneticists, etc. They were drawn into human genetics in the course of their research. This very variety of backgrounds makes discussions between human geneticists stimulating and is one of the intellectual assets of the present state of our field. However, such diversity is also a liability, since it might lead to an overrating of one's small specialty at the expense of a loss of an overview of the whole field. With increasing complexity of research methods, specialization within human genetics has become inevitable. However, it brings about the danger that the outlook of the scientist narrows, that whole fields are neglected, and promising research opportunities remain unexploited.

Human Genetics in Relation to Other Fields of Science and Medicine. The rapid development of human genetics during recent decades has created many interactions with other fields of science and medicine. Apart from general and molecular genetics and cytogenetics, these interactions are especially close with cell biology, biochemistry, immunology, and - in the clinical fields - with many specialties such as pediatrics, ophthalmology, and dermatology. On the other hand, there are few if any connections with physiology - probably a disadvantage for the development of both fields. One reason for this failure to establish fruitful interactions might be a difference in the basic approach: Mendelian genetic analysis is reductionistic and attempts to trace the causes of a trait to its most elementary components. The geneticist knows in principle that the phenotype is produced by a complex net of interactions between various genes, but he is interested more in the components than in the exact mechanism of such interactions. At present, genetic analysis has reached the level of gene structure and the genetic code; it would be a final goal to explain the properties of this code in terms of quantum physics. A malevolent observer could compare the geneticist with a man who, to understand a book, burns it and analyzes the ashes chemically.

The physiologist, on the other hand, tries to read the book. However, he often presupposes that every copy of the book should be exactly identical; variation is regarded as a nuisance. To put it differently, physiology is concerned not with the elements themselves but with their mode of interaction in complicated functional systems.[1] Physiologists are more concerned with integration of interacting systems than with the analysis of their components. The analysis of regulation of gene activities by feedback mechanisms, for example, the Jacob-Monod model in bacteria, and some approaches in developmental genetics of higher organisms are now teaching some geneticists the usefulness of the thinking in terms of systems. Therefore, it can be hoped that the gap between genetics and physiology will be closed in the near future. With the increasing interest of human genticists in the genetic basis of common diseases and individual genetic variation in response to influences such as nutrition and stress, genetic concepts will increasingly influence the many branches of medicine that, so far, have profited relatively little from genetic theory.

Future of Human Genetics. Research methods in science are becoming more and more complicated and expensive, and human genetics is no exception. As a necessary consequence, mastering of these methods increasingly requires specialization in a narrow field. Purchase of big instruments creates financial difficulties. Hence, selection of research problems is often directed not by the intrinsic scientific interest of the problems or the conviction that they could, in principle, be solved, but by availability of research methods, skilled co-workers, and instruments. The tendency toward specialization will inevitably continue, and it is very well possible that, in this process, important parts of human genetics will be resolved into fields that are mainly defined by research methods, such as biochemistry, chromosome research, or immunology. Already now, prenatal diagnosis - including cell culture and chromosome determination - has occasionally become the domain of the obstetrician; hereditary metabolic diseases are often studied and treated by pediatricians with little genetic training. Will human genetics in the future be reduced to population genetics, on the one hand, and cytogenetic diagnosis on the other?

Survival of an established field of science has no value in itself. If a field dies because its concepts and accomplishments have been accepted and are being successfully integrated into other fields, nothing much is lost. In human genetics, however, this state has not been reached. Many concepts of molecular biology are still awaiting their application to humans, and fields such as psychology and other social sciences have experienced but little impact from human genetics. The chances for such an impact would be slim if human genetics were reduced to biochemistry, diagnosis, and therapy.

Fields of Human and Medical Genetics. The field of human genetics is large and its borders are indistinct. The development of different techniques and methods has led to the development of many fields of subspecialization. Many of these overlap and are not mutually exclusive. The field of *human biochemi-*

[1] For a discussion cf. H. Mohr (1977) [267].

cal genetics deals with the biochemistry of nucleic acids, proteins, and enzymes in normal and mutant individuals. Laboratory methods of the biochemist and molecular biologist are being used (chromatography; enzyme assays; DNA digestion by restriction endonucleases). *Human cytogenetics* deals with the study of human chromosomes in health and disease. *Immunogenetics* concerns itself largely with the genetics of blood groups and of tissue antigens such as the HLA types. *Formal genetics* studies segregation of Mendelian genes and investigates more complex types of inheritance by statistical techniques.

Clinical genetics deals with diagnosis, prognosis, and to some extent with treatment of various genetic diseases. Diagnosis requires knowledge of etiologic heterogeneity and acquaintance with many disease syndromes. Genetic counseling is an important area of clinical genetics and requires skills in diagnosis, risk assessment, and interpersonal communication. *Population genetics* deals with the behavior of genes in large groups and is concerned with the evolutionary forces of drift, migration, mutation, and selection in human populations. The structure and gene pool of human populations are studied by considering gene frequencies of marker genes. In recent years, population geneticists have become interested in the epidemiology of complex genetic disease that require biometric techniques for their studies. *Behavioral genetics* is a science that studies the hereditary factors underlying behavior in health and disease. Behavior geneticists attempt to work out the genetic factors determining personality and cognitive skills in human beings. The genetics of mental retardation and various psychiatric diseases are also considered. The new field of sociobiology explains social behavior by using biologic and evolutionary concepts.

Somatic cell genetics is that branch of human genetics that studies the transmission of genes at the cellular level. Cell hybridization between different species has become an important tool for the study of the cartography of human genes. *Developmental genetics* studies genetic mechanism of normal and abnormal development. The field is relatively underdeveloped in its human applications and largely relies on experimental material in lower mammals such as mice. *Reproductive genetics* is that branch of genetics that studies details of gamete and early embryo formation by genetic techniques. This area is closely related to reproductive physiology and is developing rapidly. *Pharmacogenetics* deals with genetic factors governing the disposal and kinetics of drugs in the organism. Special interest in human pharmacogenetics relates to untoward drug reactions.

Clinical genetics has grown very rapidly in recent years because of the many practical applications of diagnosis and counseling, intrauterine diagnosis, and screening for genetic disease. Most research in human genetics is currently carried out in clinical genetics, cytogenetics, molecular and biochemical genetics, somatic cell genetics, and immunogenetics under medical auspices. Research in formal and population genetics has benefited enormously from the general availability of the computer.

Possible Function of a Textbook. In his essay on *"The Structure of Scientific Revolutions,"* Kuhn (1962) [257] described the function of textbooks not very flatteringly: they are "pedagogic vehicles for the perpetuation of normal science" that create the impression as if science would grow in a simple, cumulative manner. They tend to distort the true history of the field by only mentioning those contributions in the past that can be visualized as direct forerunners of present-day achievements. "They inevitably disguise not only the role but the very existence of ... revolutions ... ."

In the following text, we shall proceed in the same way: We shall describe the present-day problem situation in human genetics as we see it. The result will be a largely affirmative picture of normal science in a phase of growth and success. Anomalies and discrepancies might be there, but we shall very probably not identify them because we share the "blind spots" with most of the other members of our paradigm group. What we can do, however, is to alert the reader that human genetics, like all other branches of science, is by no way a completed and closed complex of theory and results that only needs to be complemented in a straightforward way and without major changes in conceptualization. It even has not developed - and will not develop in the future - as a self-contained system. Rather, it is an undertaking of human beings - social groups and single outsiders - who are motivated by a mixture of goals such as search for truth, ambition, desire to be acknowledged by one's peer group, the urge to convince the society at large to allocate resources in their field - but also the wish to help people and to do something useful for human society.

Therefore, we shall emphasize the history and development of problems and approaches. Occasionally, we shall ask the reader to step back, reflecting with us what might have been the reason why a certain development occurred at the time it did, why another development did not occur earlier, or why a certain branch of human genetics did not take the direction that one would have expected logically. Inevitably, this implies much more criticism than is usually found in textbooks. Such criticism will - at least partially - be subjective, reflecting the personal stance of the authors. Our goal is to convince the reader that a critical attitude improves one's own grasp of the problems and their possible solutions – it is not our intention to convince him that we are always right.

We would have liked to give more information on how sociologic conditions within the field and – still more important – the developments in the society at large have influenced the development of human genetics and how thinking on these problems influenced, in turn, the societies. The eugenics movement in the United States and the *Rassenhygiene* ideology in Germany have had a strong – and sometimes devastating – influence on human beings as well as the social structure of society at large. Too little systematic research has been done, however, to justify a more extended discussion than that presented in Sect. 1.8 [256]. Much more historical research along these lines is all the more urgent, as many of the ethical problems inherent – for example, in the sterilization laws of many countries during the first decades of our century – are now recurring with full force in connection with prenatal diagnosis, selective abortion and genetic engineering (Sect. 9). What has been the role of scientists active in human genetics in promoting ethically abhorrent measures in the past such as killing severely malformed newborns and mentally defectives in Nazi Germany – and how will future generations judge our own activities? These are intriguing questions. They show the Janus face of human genetics: It is fundamental science – guided by a fertile theory and full of fascinating problems. It is also an applied science, and its applications are bound to have a strong impact on society, leading to novel and difficult philosophical and ethical problems.

1 History of Human Genetics

The history of human genetics is particularly interesting since, unlike in many other natural sciences, concepts of human genetics have often influenced social and political events. At the same time, the development of human genetics as a science has been influenced by various political forces. Human genetics because of its concern with the causes of human variability has found it difficult to either remain a pure science or one of strictly medical application. Current concerns regarding the heritability of I. Q. and the existence of inherited patterns of behavior again have brought the field into public view. A consideration of the history of human genetics with some attention to the interaction of the field with societal forces will therefore be of interest. We will concentrate our attention on historical events of particular interest for human genetics and refer only to landmarks in general genetics insofar as such developments are essential for the understanding of the evolution of human genetics.

1.1 The Greeks

Prescientific knowledge regarding inherited differences between different human beings has probably existed since ancient times. Early Greek physicians and philosophers not only reported such observations but also developed some theoretical concepts and even proposed "eugenic" measures.

In the texts that are commonly ascribed to Hippocrates, the following sentence can be found:

> Of the semen, however, I assert that it is secreted by the whole body – by the solid as well as by the smooth parts, and by the entire humid matters of the body ... The semen is produced by the whole body, healthy by healthy parts, sick by sick parts. Hence, when as a rule, baldheaded beget baldheaded, blue-eyed beget blue-eyed, and squinting, squinting; and when for other maladies, the same law prevails, what should hinder that longheaded are begotten by longheaded?

This remarkable sentence not only contains observations on the inheritance of normal and pathologic traits but also a theory that explains inheritance on the assumption that the information carrier, the semen, will be produced by all parts of the body, healthy and sick. This theory became known later as the "pangenesis" theory. Anaxagoras, the Athenian philosopher, (500–428 *B. C.*) had similar views: "... in the same semen are contained hairs, nails, veins, arteries, tendons, and their bones, albeit invisible as their particles are so small. While growing, they gradually separate from each other." Because, he said, "how could hair come out of non-hair, and flesh out of non-flesh?"[1] In his opinion, the males produced the seed, the females the breeding ground.[2]

A comprehensive theory of inheritance was developed by Aristotle [7]. He also believed in a qualitatively different contribution of the male and of the female principle to procreation. The male gives the impulse to movement whereas the female contributes the matter, like the carpenter who forms a bedstead out of wood. When the male impact was stronger, a son will be born who, at the same time, is more like his father, and vice versa. This is the reason why sons will usually be similar to their fathers and daughters are similar to their mothers.

Barthelmess (1952) [7] writes: "Reading the texts from this culture, one gets the overall impression that the Greeks in their most mature minds came closer to the theoretical problems than to the phenomena of heredity."[3] Aristotle's assertion even provides an early example of how observation can be misled by a preconceived theoretical concept. Sons are not more similar to their fathers, nor daughters to their mothers.

Plato, in his *Statesman (Politikos),* explained in detail the task to carefully select spouses to produce children who will develop into bodily and ethically eminent personalities. He wrote:

> They do not act on any sound or self-consistent principle. See how they pursue the immediate satisfaction of their desire by hailing with delight those who are like themselves and by disliking those who are different. Thus they assign far too great an importance to their own likes and dislikes.

---

[1] Fragment 10 (see Capelle [244].

[2] Freeman, p. 272.

[3] English translation ours.

The moderate natures look for a partner like themselves, and so far as they can, they choose their wives from women of this quiet type. When they have daughters to bestow in marriage, once again they look for this type of character in the prospective husband. The courageous class does just the same thing and looks for others of the same type. All this goes on, though both types should be doing exactly the opposite...
Because if a courageous character is reproduced for many generations without any admixture of the moderate type, the natural course of development is that at first it becomes superlatively powerful but in the end it breaks out into sheer fury and madness...
But the character which is too full of modest reticence and untinged by valor and audacity, if reproduced after its kind for many generations, becomes too dull to respond to the challenges of life and in the end becomes quite incapable of acting at all.[4]

And in his *Republic,* Plato not only requires for the "guards" (one of the highest categories in the social hierarchy of his utopia) common property of women and public education of children, but the "best" of both sexes shall beget children who are to be educated with care. The children of the "inferior," on the other hand, are to be abandoned (Politeia, 459 a ff.). Democritus, on the other hand, writes: "More people become able by exercise than by their natural predisposition." Here (as in other places), the nature-nurture problem comes up already.

1.2 Scientists Before Mendel and Galton

The literature of the Middle Ages does not seem to contain many allusions to heredity. The new attitude of looking at natural phenomena from an empirical point of view created modern science and distinguishes modern man from others in earlier periods. This approach succeeded first in investigation of the inorganic world and only later in biology. In the work *De Morbis Hereditariis* by the Spanish physician Mercado (1605), the influence of Aristotle is still overwhelming, but there are some hints of a beginning emancipation of reasoning. One example is his contention that both parents, not only the father, contribute a seed to the future child. Malpighi (1628-1694) brought forward the hypothesis of "preformation," which implies that in the ovum the whole organism is preformed in complete shape, only to grow later. Even after the discovery of sperm (Leeuwenhoek, van Ham and Hartsoeker, 1677), the preformation hypothesis was not abandoned altogether, but it was believed by some that the individual was preformed in the sperm, only being nurtured by the mother. The long struggle between the "ovists" and the "spermatists" was brought to an end only when C. F. Wolff (1759) attacked both sides and stressed the necessity of further empirical research. A short time later, experimental research on heredity in plants was carried out by Gaertner (1772-1850) and Koelreuter (1733-1806). Their work prepared the ground for Mendel's experiments [7].

The medical literature of the eighteenth and early nineteenth centuries contains reports showing that men capable of clear observation were able to recognize correctly some phenomena relating to the inheritance of diseases. Maupertuis, for example, published in 1752 an account of a family with polydactyly in four generations and demonstrated that the trait could be equally transmitted by father and mother. He further showed, by probability calculation, that chance alone could not account for the familial concentration of the trait. Probably the most remarkable example, however, was Joseph Adams (1756-1818) [268], a British doctor who, in 1814, published a book with the title *A Treatise on the Supposed Hereditary Properties of Diseases.* This book was meant to provide a basis for genetic counseling. The following findings are remarkable:

a) Adams differentiated clearly between congenital "familial" (recessive) and "hereditary" (dominant) conditions.
b) He knew that in familial diseases the parents are frequently near relatives.
c) Hereditary diseases must not be present at birth; they may manifest themselves at different ages.
d) There are disease predispositions that lead to a manifest disease only under the additional influence of environmental factors. The progeny, however, is endangered also when the predisposed do not become ill themselves.
e) Intrafamilial correlations as to age of onset of a disease can be used in genetic counseling.
f) Clinically identical diseases may have a different genetic basis.
g) A higher frequency of familial diseases in isolated populations may be caused by inbreeding.
h) Reproduction of many patients with hereditary diseases is reduced. Hence, these diseases would disappear in the course of time, if they would not turn up from time to time among children of healthy parents (i.e. new mutations!)

Adams' attitude toward "negative" eugenic measures was critical. He proposed the establishment of registries for families with inherited diseases.

C. F. Nasse, a German professor of medicine, correctly recognized in 1820 one of the most important formal characteristics of the X-linked recessive mode of inheritance in hemophilia and presented a typical comprehensive pedigree. He wrote:

[4] Sect. 310, cf. translation by Skemp (1952).

All reports on families, in which a hereditary tendency towards bleeding was found, are in agreement that the bleeders are persons of male sex only in every case. All are explicit on this point. The women from those families transmit this tendency from their fathers to their children, even when they are married to husbands from other families who are not afflicted with this tendency. This tendency never manifests itself in these women. . . .[5]

Nasse also observed that some of the sons of these women remain completely free of the bleeding tendency.

In the medical literature of the nineteenth century, many more examples of observations and attempts to generalize and to find rules for the influence of heredity on disease can be found. The once very influential concept of "degeneration" should be mentioned, which is closely related to the concept of "anticipation," meaning that inherited diseases tend to have an earlier onset and a more severe course from generation to generation. We know now that "degeneration" has no biologic basis and that "anticipation" is a statistical artifact (see Sect. 3.1.7). Some features that older authors described as "signs of degeneration" in the external appearance of mentally deficient patients are now known to be characteristic of autosomal chromosomal aberrations or X-linked mental retardation.

In the work of most of the nineteenth century authors, true facts and wrong concepts were inextricably mixed, and there were few if any criteria for getting at the truth. This state of affairs was typical for the plight of a science in its prescientific state. Human genetics had no leading paradigm. The field as a science was to start with two paradigms in 1865: biometry, which was introduced by Galton, and Mendelism introduced by Mendel with his pea experiments. The biometric paradigm was very influential in the early decades of this century, and many examples and explanations in this book will utilize its framework. With the advent of molecular biology and insight into gene action, its use is on the decline. Nevertheless, many new applications in behavioral or social genetics where gene action cannot yet be studied rely on this paradigm and its modern elaborations. The laws that Mendel derived from his experiments, on the other hand, have been of almost unlimited fruitfulness and analytic power. The gene concept emerging from these experiments has become the central concept of all of genetics, including human genetics. Its possibilities have not been exhausted.

[5] Our translation.

1.3 F. Galton's Work: *Hereditary Talent and Character* [248]

In 1865, F. Galton published two short papers with the above-mentioned title. He wrote:

The power of man over animal life, in producing whatever varieties of form he pleases, is enormously great. It would seem as though the physical structure of future generations was almost as plastic as clay, under the control of the breeder's will. It is my desire to show, more pointedly than – so far as I am aware – has been attempted before, that mental qualities are equally under control.

A remarkable misapprehension appears to be current as to the fact of the transmission of talent by inheritance. It is commonly asserted that the children of eminent men are stupid; that, where great power of intellect seems to have been inherited, it has descended through the mother's side; and that one son commonly runs away with the talent of the whole family.

Then he stresses how little we know about the laws of heredity in man and mentions some reasons, such as long generation time, that make this study very difficult. However, he considers the conclusion to be justified that physical features of man are transmissible because resemblances between parents and offspring are obvious. Breeding experiments with animals, however, had not been carried out at that time, so that direct proof of hereditary transmission was lacking even with animals.

In humans, "we have . . . good reason to believe that every special talent or character depends on a variety of obscure conditions, the analysis of which has never yet been seriously attempted." For these reasons, he concluded that single observations must be misleading, and only a statistical approach can be adequate.

Galton evaluated collections of biographies of outstanding men as to how frequently persons included in these works were related to each other. The figures were much higher than expected, if the distribution would be random.

Galton himself was fully aware of the obvious sources of error of such biologic conclusions to which these data are subject. He stressed that "when a parent has achieved great eminence, his son will be placed in a more favorable position for advancement, than if he had been the son of an ordinary person. Social position is an especially important aid to success in statesmanship and generalship . . . ."

"In order to test the value of hereditary influence with greater precision, we should therefore extract from our biographical list the names of those that have achieved distinction in the more open fields of science and literature." Here, and in the law, which in his opinion was "the most open to fair competition," he found an equally high percentage of close relatives reaching eminence. This was especially obvious with Lord Chancellors, the most distinguished lawyers of Great Britain.

Galton concluded from his investigations that high talent and eminent achievement were strongly influenced by heredity. Having stressed the social obstacles that inhibit marriage and reproduction of the talented and successful, he proceeds to describe a utopic society,

> in which a system of competitive examination for girls, as well as for youths, had been so developed as to embrace every important quality of mind and body, and where a considerable sum was yearly allotted. ... to the endowment of such marriages as promised to yield children who would grow into eminent servants of the State. We may picture to ourselves an annual ceremony in that Utopia or Laputa, in which the Senior Trustee of the Endowment Fund would address ten deeply-blushing young men, all of twenty-five years old, in the following terms ....

In short, they were informed that the commission of the endowment fund had found them to be the best, had selected for each of them a suitable mate, would give them a substantial dowry, and promised to pay for the education of their children.

This short communication already shows human genetics as both a pure and an applied science: on the one hand, the introduction of statistical methods subjects general impressions to scientific scrutiny, thereby creating a new paradigm and turning prescience into science. Later, Galton and his student K. Pearson proceeded along these lines and founded biometric genetics. On the other hand, however, the philosophical motive of scientific work in this field is clearly shown: the object of reasearch is an important aspect of human behavior. The prime motive is the age-old "*γνῶθι σεαυτόν.*"[6]

Hence, with Galton, research in human genetics began with strong eugenic intentions. Later on, with increasing methodologic precision and increasing analytic success, such investigations were more and more removed from this prime philosophical motive. This motive helps to understand the second aspect of Galton's work: the utopian idea to improve the quality of the human species by conscious breeding. During the Nazi time in Germany (1933-1945), we saw how cruel the perverted consequences of such an idea may become (Sect. 1.8). Even these experiences, however, are sometimes forgotten, as testified by some aspects of recent discussions in genetic manipulation and engineering (Sect. 9.2). Nevertheless, the question first visualized by Galton is still, and now more than ever before, of pressing importance: What will be the biologic future of mankind?

[6] "Know yourself," inscription on the Apollo temple at Delphi.

1.4 Work of Gregor Mendel [266]

The other leading paradigm was provided by Mendel in his work *Experiments in Plant Hybridization*, which was read on February 8 and March 8, 1865 before the *Naturforschender Verein* (Natural Science Association) in Brünn (now Brno, Czechoslovakia) and subsequently published in its proceedings. It has frequently been told how this work went largely unnoticed for 35 years and was rediscovered independently by Correns, Tschermak, and de Vries in 1900 [7]. From then on, Mendel's insights triggered the development of modern genetics, including human genetics.

Mendel was stimulated to do his experiments by observations on ornamental plants, in which he had tried to breed new color variants by artificial insemination. Here he had been struck by certain regularities. He selected the pea for further experimentation. He crossed varieties with differences in single characters such as color (yellow or green) or form of seed (round or angular wrinkled) and counted all alternate types in the offspring of the first generation crosses and of crosses in later generations. Based on combinatorial reasoning, he gave a theoretical interpretation: The results pointed to free combination of specific sorts of egg and pollen cells. In fact, this concept must have occurred to him before, and he only verified and illustrated it by its "best" results since statistical correspondence between his figures and their expectations from the theoretical segregation ratios is much "too good" statistically. Mendel discovered three laws:

The law of uniformity, which states that after crossing of two homozygotes of different alleles the progeny of the first filial generation (F_1) are all identical and heterozygous; the law of segregation, which postulated 1:2:1 segregation in intercrosses of heterozygotes and 1:1 segregation in backcrosses of heterozygotes with homozygotes; and the law of independence, which states that different segregating traits are transmitted independently.

What is so extraordinary in Mendel's contribution that sets it apart from numerous other attempts in the nineteenth century to solve the problem of heredity? Three points are most important:

1. He simplified the experimental approach by selecting characters with clear alternative distributions, examining them one by one, and proceeding only then to more complicated combinations.
2. Evaluating his results, he did not satisfy himself with qualitative statements but counted the different types. This enabled him to recognize the statistical law governing these phenomena.
3. He hinted at the correct biologic interpretation for this statistical law: The germ cells represent the

constant forms that can be deduced from these experiments.

With this conclusion, Mendel founded the gene concept, which has proved so fertile ever since. The history of genetics since 1900 is governed by analysis of the gene. What had first been a formal concept derived from statistical evidence has emerged as the base pair sequence of DNA, which contains the information for protein synthesis and for life in all its forms [247 a].

1.5 Application to Man: Garrod's Inborn Errors of Metabolism

Only the first step of this development will be described in this historical introduction: A. Garrod's (1902) [249] paper on "The Incidence of Alkaptonuria: a Study in Chemical Individuality." There are two reasons for giving special attention to this paper. For the first time, Mendel's gene concept was applied to a human character and Mendel's paradigm was introduced into research on humans. Additionally, this work contains many new ideas set out in a most lucid way. Garrod was a physician and in his later life became the successor of Osler in the most prestigious chair of medicine at Oxford. His seminal contribution to human genetics remained unappreciated during his lifetime. Biologists paid little attention to the work of a physician. Their interest was concentrated more on the formal aspects of genetics rather than on gene action. The medical world did not understand the importance of his observations for medicine. Garrod mentioned first the isolation of homogentisic acid from the urine of patients with alkaptonuria by Walkow and Baumann (1891). Then he stated as the most important result of the investigation carried out so far:

> ... that, as far as our knowledge goes, an individual is either frankly alkaptonuric or conforms to the normal type, that is to say, excretes several grammes of homogentisic acid per diem or none at all. Its appearance in traces, or in gradually increasing or diminishing quantities, has never yet been observed ....

As a second important feature "the peculiarity is in the great majority of instances congenital ...." Thirdly: "The abnormality is apt to make its appearance in two or more brothers and sisters whose parents are normal and among whose forefathers there is no record of its having occurred." Fourthly, in six of ten reported families, the parents were first cousins, whereas the incidence of first-cousin marriages in contemporary England was estimated to be not higher than 3%. On the other hand, however, children with alkaptonuria are observed in a very small fraction only of all first-cousin marriages.

> There is no reason to suppose that mere consanguinity of parents can originate such a condition as alkaptonuria in their offspring, and we must rather seek an explanation in some peculiarity of the parents, which may remain latent for generations, but which has the best chance of asserting itself in the offspring of the union of two members of a family in which it is transmitted.

Then, Garrod mentioned the law of heredity discovered by Mendel, which "offers a reasonable account of such phenomena" that are compatible with a recessive mode of inheritance. He cited another remark of Bateson and Saunders[7] with whom he had discussed his data:

> ... we note that the mating of first cousins gives exactly the conditions most likely to enable a rare, and usually recessive, character to show itself. If the bearer of such a gamete mates with individuals not bearing it the character will hardly ever be seen; but first cousins will frequently be the bearers of similar gametes, which may in such unions meet each other and thus lead to the manifestation of the peculiar recessive characters in the zygote.

After having cited critically some opinions on the possible causes of alkaptonuria, Garrod proceeded:

> The view that alkaptonuria is a "sport" or an alternative mode of metabolism will obviously gain considerably in weight if it can be shown that it is not an isolated example of such a chemical abnormality, but that there are other conditions which may reasonably be placed in the same category.

Having mentioned albinism and cystinuria as possible examples, he went on: "May it not well be that there are other such chemical abnormalities which are attended by no obvious peculiarities [like the three mentioned above] and which could only be revealed by chemical analysis?" And further:

> If it be, indeed, the case that in alkaptonuria and the other conditions mentioned we are dealing with individualities of metabolism and not with the results of morbid processes the thought naturally presents itself that these are merely extreme examples of variations of chemical behaviour which are probably everywhere present in minor degrees and that just as no two individuals of a species are absolutely identical in bodily structure neither are their chemical processes carried out on exactly the same lines.

He suggested that differential responses toward drugs and infective agents could be the result of such chemical individualities. The paper brings forward the following new insights:

a) Whether a person has alkaptonuria or not is a matter of a clear alternative - there are no transitory forms. This is indeed a condition for straightforward recognition of simple modes of inheritance.
b) The abnormality is congenital.

---

7 Report to the Evolution Committee of the Royal Society, (1902).

c) It is observed in sibs and not in parents.
d) The parents are frequently first cousins.
The two last-mentioned features were explained by the hypothesis of a recessive mode of inheritance according to Mendel. The significance of first-cousin marriages is stressed especially for rare conditions; this may be a foreboding of population genetics.
e) Apart from alkaptonuria, several other similar "sports" may exist. Albinism and cystinuria are possible examples. This makes alkaptonuria the paradigm for the "inborn errors of metabolism." In 1908, Garrod published his classic monograph on this topic [75].
f) These sports may be extreme and therefore conspicuous examples only for a principle with *much more widespread applicability*. Lesser chemical differences between human beings are so frequent that no human being is identical chemically to anybody else.

All through this book, the principle of a genetically determined biochemical individuality will govern our discussions. Garrod's contribution may be contrasted with that of Adams. Apart from the "familial" occurrence of some hereditary diseases, Adams observed a number of phenomena that were not noted by Garrod, such as the late onset of some diseases, the intrafamilial correlation of age of onset, and the genetic predisposition leading to manifest illness only under certain environmental conditions. However, Adams did not have Mendel's paradigm. Therefore, his efforts could not lead to the development of an explanatory theory and of a coherent field of science. Garrod had this paradigm and used it, creating a new area: human biochemical genetics.

1.6 Visible Transmitters of Genetic Information: Early Work on Chromosomes

Galton's biometric analysis and Mendel's hybridization experiments both started with visible phenotypic differences between individuals. The gene concept was a construct derived from the phenotypic outcome of certain crossings. When Mendel carried out his experiments, nothing was known about a possible substantial bearer of the genetic information in the germ cells. During the decades to follow, however, up to the end of the nineteenth century, the chromosomes were identified, and mitosis and meiosis were analyzed. These processes were found to be highly regular and so obviously suited for orderly distribution of genetic information that only a very short time after the rediscovery of Mendel's laws in 1900 the parallelism of Mendelian segregation and chromosomal distribution during meiosis was realized, and chromosomes were identified as bearers of the genetic information. [244a]

Many research workers contributed to the development of cytogenetics [7; 239]. O. Hertwig (1875) first observed animal fertilization and established the continuity of cell nuclei: "omnis nucleus e nucleo." Flemming (1880–1882) found the separation of sister chromatids in mitosis; van Beneden (1883) established the equal and regular distribution of chromosomes to the daughter nuclei. Boveri (1888) found evidence for the individuality of each pair of chromosomes. Waldeyer (1888) coined the term "chromosomes." [244a]

Meanwhile, Naegeli (1885) had developed the concept of "idioplasma," a small part of the plasma that contains – to use a modern term – the "information" for the development of the next generation. He did not identify the idioplasma with any specific part of the cell. W. Roux seems to have been the first to set out by logical deduction which properties a carrier of genetic information was expected to have. He also concluded that the behavior of cell nuclei during division would perfectly fulfill these requirements. The most important specific property of meiotic divisions, the ordered reduction of genetic material, was first recognized by Weismann.

These results and speculations set the stage for the identification of chromosomes as carriers of the genetic information, which followed a short time after the rediscovery of Mendel's laws and apparently independently by different authors (Boveri; Sutton; Correns, 1902; de Vries, 1903).

Chromosome studies and genetic analysis have remained intimately connected in the field of cytogenetics ever since. Most basic facts were discovered and concepts were developed using insects and plants as the principal experimental tools.

The development of the field of human cytogenetics was delayed until 1956 when the correct number of chromosomes in man was established as 46 by use of rather simple methods. It should be stressed that this delay could not be explained by the introduction of new cytologic methods at that time. In fact, this discovery could have been made many years earlier. The delay was probably related to the lack of interest in human genetics by most laboratory-oriented medical scientists. Human genetics did not exist as a scientific discipline in medical schools since the field was not felt to be a basic science fundamental to medicine. Hereditary diseases were considered as oddities that could not be studied by the methodology of medical science as exemplified by the techniques of anatomy, biochemistry, physiology, microbiology, pathology, and pharmacology. Thus, most geneticists worked in biology departments of universities, colleges, or in agricultural stations.

They were usually not attuned to problems of man and his diseases, and there was little interest to study the human chromosomes. The discovery of trisomy 21 as the cause of Down's syndrome and the realization that many problems of sex differentiation owed their origin to sex chromosomal abnormalities established the central role of cytogenetics in medicine. Further details in the development of cytogenetics will be described in Sect. 2.

1.7 Early Achievements in Human Genetics

1.7.1 AB0 Blood Groups

The discovery of the AB0 blood group system by Landsteiner in 1900 [259] and the proof that these blood types are inherited (von Dungern and Hirschfeld, 1911 [245]) was one outstanding example of Mendelian inheritance applied to a human character. Bernstein in 1924 [240] demonstrated that A, B, and 0 blood group characters were multiple alleles at one locus. The combined efforts of Wiener, Levine, and Landsteiner 25–30 years later led to the discovery of the Rh factor and established that hemolytic disease of the newborn owed its origin to immunologic maternal-fetal incompatibility. The stage was set for the demonstration in the 1960s that Rh hemolytic disease of the newborn could be prevented by administration of anti-Rh antibodies to mothers at risk [278; 291].

1.7.2 Hardy-Weinberg Law

Hardy [252], a British mathematician and Weinberg [289], a German medical doctor, at about the same time (1908) set out the fundamental theorem of population genetics, which explained why a dominant gene did not increase in frequency from generation to generation. Hardy published his contribution in the United States in *Science*. He felt that this work would be considered as too trivial by his mathematical colleagues. Weinberg was a practicing physician who made many contributions to formal genetics. He developed a variety of methods in twin research [288] and first elaborated methods to correct for biased ascertainment in recessive inheritance [290].

1.7.3 Development Between 1910 and 1930

The years between 1910 and 1930 saw no major new paradigmatic discoveries in human genetics. Most of the data in formal genetics (such as linkage, nondisjunction, mutation rate) as well as the mapping of chromosomes were achieved by study of the fruit fly, largely in the United States, but also in many other parts of the world. Many scientists tried to apply the burgeoning insights of genetics to man. British scientists exemplified by Haldane excelled in the elaboration of a variety of statistical techniques required to deal with biased human data. The same period saw the development of the basic principles of population genetics by Haldane and Fisher in England and by Wright in the United States. This body of knowledge became the foundation of population genetics and is used by current workers in that field. In 1918, Fisher was able to resolve the bitter controversies in England between the Mendelians on the one hand, and followers of Galton, such as Pearson, on the other, by pointing out that correlations between relatives in metric traits could be explained by the combined action of many individual genes. Major steps in the development of medical genetics during this period were the establishment of empirical risk figures for mental and affective disorders by the Munich school of psychiatric genetics and introduction of sound criteria for such studies.

1.8 Human Genetics, the Eugenics Movement, and Politics

1.8.1 Great Britain and the United States [236, 246, 256, 263, 283]

The first decade of the century saw the developments of eugenics in Europe and in the United States. Many biologic scientists were impressed by their interpretation of an apparently all-pervasive influence of genetic factors on most normal physical and mental traits as well as on mental retardation, mental disease, alcoholism, criminality, and various other sociopathies. They became convinced that the human species should concern itself with encouragement of breeding between persons with desirable traits (positive eugenics) and discourage the sick, mentally retarded, and disabled from procreation (negative eugenics). Galton became the principal early proponent of such ideas. Various eugenic study units were established in the United States (Eugenics Record Office, Cold Spring Harbor) and in England. Much of the scientific work published by these institutions was of poor quality. Particularly, many different kinds of human traits such as "violent temper" and "wandering trait" were forced into Mendelian "straightjackets." Most serious geneticists became disenchanted and privately disassociated themselves from this work. For various reasons, including those of friendship and collegiality with the eugenicists, the scientific geneticists did not register their disagreement in public. Thus, the propagandists of eugenics continued their work with enthusiasm, and the field acquired a much better reputation among some of the public than it deserved. Thus, many college courses on eugenics were introduced in the United States.

These trends had several important political influences. Eugenics sterilization laws were passed in many states in the United States, which made it possible to sterilize a variety of persons for traits including criminality where no good scientific basis for inheritance existed. The attitude that led to the introduction of these laws could be epitomized by U. S. Supreme Court Justice Holmes' statement that "three generations of imbeciles are enough."

Eugenic influences also played an important role in the passing of restrictive laws on immigration to the United States. Using a variety of arguments, the proponents of eugenics claimed to show that Americans of northern and central European origin were more useful citizens than those of southern European origin or from Asia. Since such differences were claimed to be genetic in origin, immigration from southern and eastern European countries and from Asia was sharply curtailed. Similar trends were operative in England. While solid work in human genetics was carried out by a few statistical geneticists, there was also much eugenic propaganda including that by the distinguished statistician Pearson, the successor to Galton's academic chair in London.

Kevles [256] has recently published a wide-ranging and insightful history of eugenics and human genetics in the Anglo-Saxon countries. His book is the most carefully researched and exhaustive study of the uses and abuses of eugenic concepts.

1.8.2 Germany [250; 236 a]

In Germany, eugenics took the name of *Rassenhygiene* from a book of that name published in 1895 by Ploetz [277]. The *Rassenhygiene* movement became associated with mystical concepts of race, Nordic superiority, and the fear of degeneration of the human race in general and that of the German *Volk* in particular by alcoholism, syphilis, and increased reproduction of feebleminded or people from the lower strata of society. Some representatives of this movement became associated with a dangerous type of sociopolitical prejudice: anti-semitism. They warned the public against contamination of the German "blood" by foreign, especially Jewish, influences. Most followers of the racial hygiene concept were nationalistic and opposed to a development of an "open society" that allowed individual freedom and democratic participation. They shared this attitude with a significant segment of the educated classes in Germany. General eugenic ideas divorced from racism and other nationalist notions were often espoused by intellectuals who were concerned about the biologic future of mankind. Thus, socialists publicized such views in Germany [250]. In 1931, two years before Hitler's coming into power, the German society of racial hygiene added eugenics to its name. However, all efforts in that area soon became identified with the Nazi ideology.

Prominent German human geneticists identified themselves with the use of human genetics in the service of the Nazi State. Recognized scientists, such as Fischer, F. Lenz, Rüdin, and von Verschuer, accepted Nazi leadership and, at least in public, Nazi philosophy. While most of the propaganda for the new racial hygiene was not formulated by scientists but by representatives of the Nazi party, men such as Fischer and von Verschuer participated in spreading Nazi race ideology. Jews were declared foreign genetic material to be removed from the German *Volk* [286]. A eugenic sterilization law was already passed in 1933 and made forced sterilization obligatory for a variety of illnesses thought to be genetic in origin. Heredity courts were established to deal with appropriate interpretation of the sterilization law [273]. Sterilization laws for eugenic indications were also passed in various Scandinavian countries around the same time period but allowed voluntary (in contrast to forced) sterilization [273].

The exact role of the German human geneticists in the increasing radicalization and excesses of the application of Nazi philosophy has recently begun to be assessed by archival study [272 a]; von Verschuer's role in sponsoring twin and other genetic research by his former assistant Mengele in the Auschwitz concentration and extermination camp is clear. We have no record that any voices were raised in public by these men in protest against "mercy killings" of the mentally retarded and newborn children with severe congenital defects nor against the mass killings of Jews. The new historical evidence suggests that von Verschuer at least must have had some ideas of such events, since he had continued contact with Mengele when the mass killings at Auschwitz were at their height. The "final solution" to the "Jewish problem" resulted in the murder of about 6 million Jews in the early 1940s [282]. While we have no record that human geneticists favored this type of solution, their provision of so-called scientific evidence for a justification of Nazi anti-Semitism helped to create a climate in which these mass murders became possible. This episode is one of the most macabre and sorrowful chapters in the history of man's inhumanity to man in the name of pseudoscientific nationalism.

1.8.3 The Soviet Union [246, 250]

Eugenics was initiated in the Soviet Union in the 1920s by the establishment of eugenics departments, a eugenic society, and a eugenics journal. Eugenic ideals and the interpretation of Marxism-Leninism that came into official use during those years soon clashed, and these efforts were abandoned by the late 1920s. Scientists who had become identified with eugenics left the field to work with plant and animal material.
Interest in the medical application of human genetics persisted longer. A large institution of medical genetics was established in the Soviet Union in the 1920s. Its director, L. E. Levit, disappeared in the 1930s, and human genetics was officially declared a Nazi science. The later ascendance of Lysenko stifled all work in genetics, including that of human genetics, and no work whatsoever was carried out in this field until the early 1960s, when Lysenko's domination was over [255]. The reintroduction of human genetics into the Soviet Union occurred by way of medical genetics. A textbook of medical genetics was published by Efroimson in 1964 [142a]. A new institute of medical genetics was established in 1969 under the directorship of the cytogeneticist Bochkov. Work in many areas of medical genetics is carried out in that institution and in other places.

1.8.4 Human Behavior Genetics

Violent discussions continue regarding the role of genetic determinants in behavior, I.Q., and personality in both the East and the West. Some scientists deny the applicability of genetic influences on normal behavior or social characters such as personality and intellect entirely. This attitude toward genetics is shared more or less openly by some psychologists and social scientists and even a few geneticists who are concerned about the possible future political and social misuse of studies in human behavioral genetics and sociobiology that claim to show strong genetic determinants of intelligence and social behavior.
We do not agree with those who deny any genetic influence on behavior or social traits in man. However, we also caution about a too ready acceptance of results from biometric comparison of twins and other relatives, which claim very high heritabilities for many of these traits. Genetic data and pseudodata may be seriously misused by political bodies. Yet, as biologists and physicians impressed by biologic variation under genetic control, we would be surprised if the brain would not also show significant variation in structure and function. Such variation is expected to affect intellect, personality, behavior, and social interactions. The extent to which genetic variation contributes to such traits, and especially the biological nature of such variation, will have to await further studies. (See Sect. 8)

1.9 Development of Medical Genetics (1950-the Present)

1.9.1 Genetic Epidemiology

In the 1940s and 1950s, some institutions pioneered in research on epidemiology of genetic diseases. T. Kemp's institute in Copenhagen, J. V. Neel's department in Ann Arbor, Michigan, and A. C. Stevenson's in Northern Ireland and later in Oxford contributed much to our knowledge on prevalence, modes of inheritance, heterogeneity, and mutation rates of hereditary diseases. Recent years have seen a renaissance of this area with special attention to complex analysis of common diseases. At present, the time seems to be ripe for a new approach to genetic epidemiology or clinical population genetics with inclusion of the new laboratory methods together with more powerful methods of biometric analysis [270, 270a].

1.9.2 Biochemical Methods

The years after World War II brought a rapid renaissance of human genetics by the development of both biochemical and cytologic methods. Human genetics, which had largely been the concern of statistically oriented scientists, now entered the mainstream of medical research. The demonstration by Pauling that sickle cell anemia was a molecular disease [1260] was a key event that initiated this area. The hemoglobins allowed detailed study of the consequences of mutation. The genetic code was found to be valid for organisms as far apart as viruses and humans. Most detectable mutations were found to be single amino acid substitutions, but deletions of various sorts and frameshift mutations similar to those discovered in microorganisms were discovered. The actual nucleotide sequence of the hemoglobin gene was worked out using techniques developed in biochemistry and molecular genetics. Many inborn errors of metabolism were shown to owe their origin to various enzyme deficiencies each caused by a genetic mutation that changed enzyme structure. Methemoglobinemia due to diaphorase deficiency and glycogen storage disease were the first enzyme defects to be demonstrated (Sect. 4.1).

1.9.3 Biochemical Individuality

Work on variants of the enzyme glucose-6-phosphate-dehydrogenase (G6PD) helped to establish the concept of extensive mutational variation. Biochemical individuality explained some drug reactions and led to the development of the field of pharmacogenetics [269]. Marked biochemical heterogeneity of human enzymes and proteins was first shown in man by Harris and others [253]. The uniqueness of man, which is apparent by the physiognomonic singularity of each human being could be shown to apply at the biochemical and immunologic level as well. Here, as in several other fields (such as the hemoglobin variants and the mechanism of sex determination), studies in humans led the way to generally valid biologic rules. The significance of polymorphism for the structure of populations including that of man is being widely studied by population geneticists. The hypothesis that some polymorphisms may be the genetic substrate against which the environment acts to determine susceptibility and resistance to common disease led to the development of the field of ecogenetics [271]. The histocompatibility gene complex has become an important paradigm for the understanding of why several genes with related function might occur in closely linked clusters. This locus may be of great importance to understand susceptibility to many autoimmune and some other diseases. More recently, an enormous amount of apparently unexpressed genetic variation has been demonstrated at the DNA level [328].

1.9.4 Cytogenetics, Somatic Cell Genetics, Prenatal Diagnosis

After cytogenetic techniques became available, they were applied to many types of birth defects and intersex states. A specific type of malignancy, chronic myelogenous leukemia, was shown to be caused by a unique chromosomal aberration. Banding techniques developed by Caspersson in 1969 made it possible to visualize each human chromosome and gave cytogenetic methods added powers of resolution.

Soon biochemical and cytogenetic techniques were combined in somatic cell genetics. Specific enzyme defects were identified in single cells grown in tissue cultures. The development of methods to hybridize human with mouse cells by Henry Harris [254] and Ephrussi [247] soon allowed the assignment of many genes to specific chromosomes and the construction of a human linkage map that is already rivaling (Sect. 3.4.3) that of *Drosophila* and mice (App. 9).

The developments in somatic cell genetics led to the introduction of prenatal diagnosis in the late 1960s, when amniocentesis at the beginning of the 2nd trimester of pregnancy was developed. This allowed tissue cultures of amniotic cells of fetal origin, permitting both cytogenetic and biochemical characterization of fetal genotypes, assignment of sex, and the diagnosis of a variety of disorders in utero. In the early 1980s, chorion villus biopsy - a procedure that is already possible during the 1st trimester of pregnancy - was introduced, and is being widely tested. The discovery that neural tube defects were associated with increases in α-fetoprotein of the amniotic fluid permits intrauterine diagnosis of an important group of birth defects [242]. Fetoscopy allowed puncture of fetal vessels to obtain blood for diagnosis of the hemoglobinopathies and even made possible visualization of the fetus for certain types of external malformations. Ultrasound methods to visualize the placenta and to diagnose fetal abnormalities added to the diagnostic armamentarium. This method is being rapidly perfected and allows phenotypic diagnosis of the fetus more and more frequently. Since it is noninvasive, it is largely replacing fetoscopy for these purposes.

Clinical Genetics. Clinical genetics is growing rapidly [264]. Many hospitals are establishing special clinics where genetic diseases can be diagnosed and genetic counseling is given. The heterogeneity of genetic disease has been increasingly recognized. Genetic counseling is now intensified to provide patients and their families with information on natural history, recurrence risks and reproductive options [129]. Screening programs of the entire newborn populations for diseases such as phenylketonuria are being introduced in many countries and other screening programs such as those for Tay-Sachs disease have undergone extensive trials [238] (Sect. 9.1.2).

With these clinical developments, evolution of new scientific concepts in human genetics per se had become for a while somewhat less prominent since many human geneticists were involved in the many clinical problems to which they could make important practical contributions. With the advent of the new DNA techniques (Sect. 2.3), this has changed rapidly. Basic work in human genetics is now increasingly performed by a variety of scientists such as cell biologists, molecular biologists, biochemists, and others, who do not necessarily have training in human genetics [272]. Human genetics is now identified with medical genetics in most of its activities. The scientific developments of the past decades are widely applied in practical medicine.

1.9.5 DNA Technology in Medical Genetics [221]

Advances in molecular genetics and DNA technology are being rapidly applied to practical problems of medical genetics. Since understanding of the hemoglobin genes is more advanced than that of other genetic systems, the initial applications related to the diagnoses of the hemoglobinopathies. Several methodologies are being utilized. Inherited variation in DNA sequence that is phenotypically silent was found to be common, supplying a vast number of DNA polymorphisms for study. Just as everyone's physiognomy is unique, each person (except for identical twins) has a unique DNA pattern. DNA variants are being used in family studies as genetic "markers" to detect the presence of closely linked genes causing monogenic diseases. Direct detection of genetic disease has been achieved by utilizing specially constructed short stretches of nucleotides ("probes") that are homologous to the mutations that are searched for. Rarely, a specific restriction enzyme may detect the mutational lesion. Different DNA mutations at the same locus usually cause an identical phenotypic disease. This finding makes direct DNA diagnosis without family study difficult unless the specific mutation that causes the disease is known.

Efforts to construct a map of the human genome are under way. Several hundred DNA markers that are evenly spaced over all the chromosomes should provide the necessary landmarks for detection of all monogenic diseases and may provide aid in unravelling the contribution of specific genes to common diseases.

The use of DNA to treat genetic diseases is also under study. Current efforts are concerned with the insertion of the DNA of normal genes into somatic cells such as bone marrow (somatic gene therapy). In vitro and animal experiments are under way using retroviruses as vectors for these genes. No human studies had been done by spring of 1985. Early efforts of human somatic gene therapy for argininemia using Shope papilloma virus and for beta thalassemia using beta hemoglobin genes were premature and had no clinical effects. Germinal gene therapy, i.e., insertion of normal genes into defective germ cells, fertilized eggs or early embryos for treatment of genetic disease, is far in the future.

1.9.6 Unsolved Problems

Human genetics had been most successful by being able to guide work made possible by the development of techniques from various areas of biology using Mendelian concepts. Important basic frontiers that are still being extended concern problems of gene regulation, especially during embryonic development, control of the immune system and of brain function. Human genetics is likely to contribute to these problems by imaginative use of the study of genetic variation and disease by newer techniques. In medical genetics, the problem of common diseases including many birth defects requires many additional insights into the mechanisms of gene action during embryonic development and study of the specific genes involved in various common diseases.

At first glance, the history of human genetics of the last 30 years reads like a succession of victories. The reader could conclude that human geneticists of the last generation pursued noble science to the benefit of mankind. However, how will posterity judge current efforts to make use of our science for the benefit of mankind as we understand it? Will the ethical distinction between selective abortion of a fetus with Down's syndrome and infanticide of severely malformed newborns be recognized by our descendants? Are we again moving down the "slippery slope?"

2 Human Chromosomes

2.1 Human Cytogenetics, a Successful Late Arrival

The chromosome theory of Mendelian inheritance was launched in 1902 by Sutton and Boveri. In the same year, Garrod, establishing the autosomal-recessive mode of inheritance for alkaptonuria and commenting on metabolic individuality in general, created the paradigm of "inborn errors of metabolism." Simple modes of inheritance were soon established for many other human disorders. Not much more than 10 years later, Bridges (1916) [311] examined in *Drosophila* the first case of a disturbance in chromosome distribution during meiosis and named it "nondisjunction." Cytogenetics of animals and plants flourished during the first half of the century, and almost all important phenomena in the field of cytogenetics were discovered during this time period. Moreover, cytogenetic methods helped to elucidate many basic laws of mutation.

One might have expected that these results and concepts of general cytogenetics would soon flow into human genetics, helping to explain phenomena that might be genetic in origin but did not fit the expectations derived from Mendel's laws. This transfer, however, was not to occur until 50 years later. The age of human cytogenetics began only when Tjio and Levan (1956) [532] and Ford and Hamerton (1956) [351] established the diploid human chromosome number of 46. Lejeune (1959) [417] discovered trisomy 21 in Down's syndrome, and Ford et al. (1959) [352] and Jacobs and Strong (1959) [395] described the first cytogenetic analyses in the Turner and Klinefelter syndromes.

The late arrival of human cytogenetics is usually ascribed to shortcomings of the preparation methods of chromosomes. Indeed, the jumbled masses of chromosomes in old illustrations make us understand the difficulties encountered by the pioneers who tried to count human chromosomes. Still, it is hard to conceive that development of more adequate methods would have been delayed so long had the cytogeneticists realized there were human anomalies awaiting explanation. Some human geneticists did consider the possibility that certain anomalies could be due to chromosomal aberrations. For example. Waardenburg (1932) [537] remarked:

> The stereotyped recurrence of a whole group of symptoms among the Mongoloids offers an especially fascinating problem. I would like to suggest to the cytologists that they examine whether it may be possible that we are dealing with a human example of a certain chromosome aberration. Why should it not occur occasionally in humans, and why would it not be possible that – unless it is lethal – it would cause a radical anomaly of constitution? Somebody should examine in mongolism whether possibly a "chromosomal deficiency" or "nondisjunction" – or the opposite, "chromosomal duplication" – is involved. ... My hypothesis has at least the advantage of being testable. It would also explain the possible influence of maternal age.[1]

He then remarked that the very rare familial occurrence of Down's syndrome and the concordance of monozygotic twins would be compatible with this hypothesis. Waardenburg, being a practicing ophthalmologist who, in his spare time, became one of the foremost specialists for inherited eye disease, had no opportunity to put his suggestion into practice. The cytogeneticists of his time, however, did not carry through. The spark was there, but nothing caught fire.

2.1.1 *History and Development of Human Cytogenetics*

First Observations on Human Mitotic Chromosomes [522]. Research on human cytogenetics could be said to have begun with the work of Arnold (1879) [297] and Flemming (1882) [348], who for the first time examined human mitotic chromosomes. In the following years, a number of reports appeared with various estimates of the number of human chromosomes. Outstanding among these early contributions was the work of von Winiwarter (1912) [543]. He examined the testicular histology from four men aged 21, 23, 25, and 41, fixed and cut into sections of only 7.5 μ in diameter, which impeded chromosome counting. Thirty-two spermatogonial mitoses could be evaluated; for 29 he counted 47 elements, 46 for two others, and 49 for one (Fig. 2.1). Sixty diplotene plates were evaluated; 57 showed 24 elements, two seemed to have 25, and one 23. In diplotene, he even

[1] Translation ours.

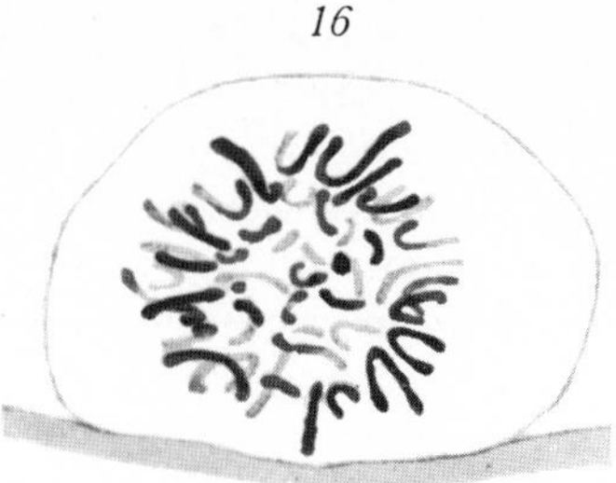

Fig. 2.1. An early picture of a spermatogonial mitosis. (Von Winiwarter, 1912 [543])

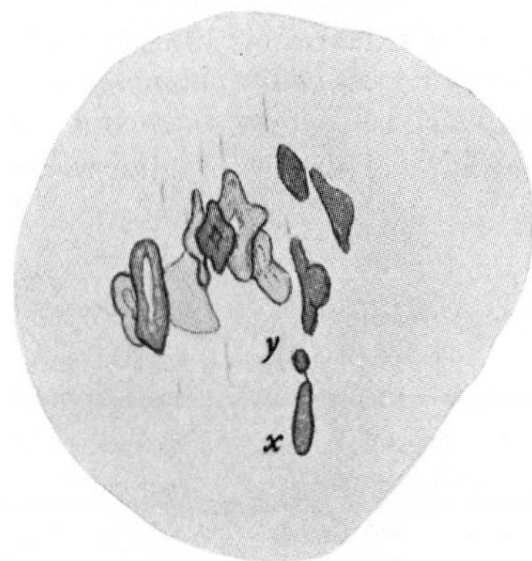

Fig. 2.2. The sex bivalent in a first meiotic anaphase. (Painter, 1923 [467])

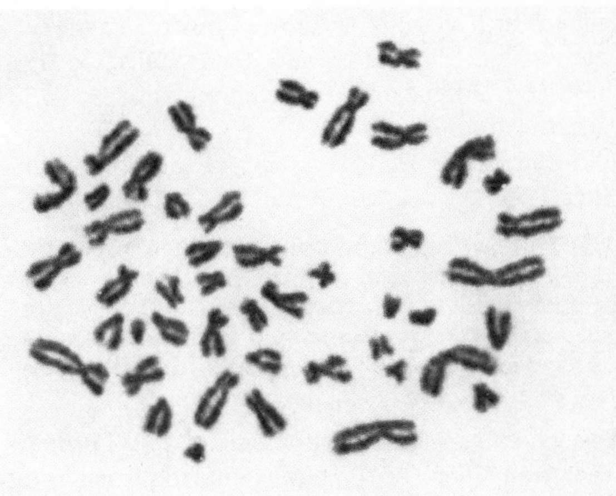

Fig. 2.3. A metaphase of a human embryonic lung fibroblast grown in vitro. From the first report in which the human chromosome number was established as 46. (Tjio and Levan, 1956 [532])

observed the sex chromosomes but explained them as one chromosome that was transported to one pole. He concluded that human males have 47 and females 48 chromosomes. His evidence for females was slender, as he only found three clear oogonial mitoses in a 4-month-old fetus. The results were compatible with the assumption that there were 48 chromosomes.

The report with the strongest impact was that of Painter (1921, 1923) [467]. He examined the testicles of three individuals from the Texas State Insane Asylum. In all three cases the cause for the removal of the testes was "excessive self-abuse coupled with certain phases of insanity." The results were based mainly on examinations from two of the three individuals. In a preliminary report (1921) he described the chromosome number as 46 or 48, but in the definitive report (1923) he had decided in favor of 48 chromosomes. In first meiotic divisions he was able to demonstrate the sex bivalent, consisting of the X and Y chromosomes, which at anaphase migrated to opposite poles (Fig. 2.2).

In the following years, a chromosome number of 48 in man was confirmed in a number of reports [379]. However, two technical difficulties impeded further progress:

1. Sectioning by the usual histologic techniques would often disturb mitoses.
2. The chromosomes tended to lie on top of each other and even to clump together.

These difficulties were ultimately overcome by:

a) Use of suspensions of intact cells that might be squashed or simply air-dried rather than of histologic sections.
b) Subjecting cells to a brief treatment with a hypotonic solution, causing them to swell and burst, thus spreading out the chromosomes for better definition.

The hypotonic shock technique paved the way for easy chromosome counting [384, 386].[2]

An Old Error Is Corrected and a New Era Begins [352]. In the summer of 1955, Levan (a Swedish cytogeneticist) visited Hsu in New York and learned the technique of squash preparation using hypotonic shock. He and Tjio then improved the technique by shortening the hypotonic treatment and adding colchicine, a chemical that arrests cells in metaphase to increase the number of countable cells. They examined lung fibroblasts of four human embryos. To their surprise, they found a chromosome number of 46 in most of 261 metaphases. Figure 2.3 shows one example. In discussing their findings, they mentioned three Swedish investigators who had studied

[2] Interestingly enough, even 30 years later Painter's estimate of 48 was so strongly imprinted on investigators' minds that in the first study on human chromosomes using the new technique the human chromosome number was reported as 48 [384].

mitoses in liver cells of aborted human embryos a year earlier. This study was discontinued because they were unable to find 48 chromosomes; in all cells, they found only 46.
Tjio and Levan were very careful in their conclusions:

Before a renewed, careful control has been made of the chromosome number in spermatogonial mitoses of man we do not wish to generalize our present findings into a statement that the chromosome number of man is $2n=46$, but it is hard to avoid the conclusion that this would be the most natural explanation of our observations.

This evidence was soon supplemented by Ford and Hamerton (1956) [351]. They examined testicular tissue from three men of relatively advanced age. In the great majority of all metaphase I cells, 23 bivalents were found, confirming the results of Tjio and Levan. Spermatogonial mitoses were difficult to find, but a few clear counts confirmed the chromosome number to be 46.
With these results, the stage was set for the development of clinical cytogenetics. Still, it was to be almost another three years before the first abnormal karyotypes in humans were reported.

Solution to an Old Riddle: Down's Syndrome (Mongolism) Is Due to Trisomy 21. In the spring of 1959, Waardenburg's suggestion finally was followed. Lejeune et al. [417] reported chromosome studies from fibroblast cultures in nine children with Down's syndrome. Fifty-seven diploid cells were regarded as technically perfect. In all of them, the chromosomes numbered 47. The supernumerary chromosome was described as small and "telocentric." Meiotic nondisjunction was suggested as the most likely explanation for the additional chromosome.

First Reports on Trisomies and Monosomies of Sex Chromosomes. Barr and Bertram [298] discovered the "X chromatin," an intranuclear body 0.8–1.1 μm in size, which is commonly located at the periphery of the interphase nuclei of females and is not present in males. The discovery was accidental since it originated in an investigation of the effects of fatigue on the central nervous system in cats. What seemed to be first a sex difference only in the neurons of cats turned out to be a normal finding characteristic of the nuclear inheritance of female mammals including human females. Corresponding structures, the drumsticks, were found by Davidson and Smith (1954) [331] in polymorphonuclear neutrophil leukocytes. The obvious next step was the examination of X chromatin in cells of patients with disturbances in sexual development. Here, most male patients with the Klinefelter syndrome (Sect. 2.2.3.2) turned out to be X chromatin-positive in spite of their predominantly male phenotype, whereas most female patients with the Turner syndrome (Sect. 2.2.3.3) were X chromatin-negative – again in contrast to their female phenotype. If the X chromatin was directly related to the X chromosomes, this finding pointed to X chromosome anomalies in these two syndromes.
In the absence of methods to examine sex chromosomes directly, examination of color vision was a promising approach, one that had been recommended as long ago as 1931 by Goldschmidt for elucidation of the genotypic sex in intersexes. In two series of altogether 89 cases of Klinefelter syndrome, three were detected with anomalies of red-green color vision – a common X-linked trait in males (Sect. 3.5.3) [462, 478]. This was not quite the number expected in males (about 7%–9%) but, on the other hand, this frequency was much higher than expected among females (<1%). Moreover, with a normal female XX chromosomal constitution, these three patients would have to have received an X chromosome from each parent. Since only homozygous females express the color vision defect, all their fathers were expected to have a color vision anomaly. In fact, two of the three fathers were available for study and were found to have normal color vision.
This atypical situation found its explanation when Jacobs and Strong (1959) [395], examining the chromosomes from bone marrow mitoses in Klinefelter patients, found 47 chromosomes, whereas both parents had a normal karyotype. The supernumerary chromosome belonged to the group of chromosomes including X chromosomes; the karyotype was tentatively identified as XXY. In the two Klinefelter patients with red-green color vision anomaly and the unaffected father, the two X chromosomes were obviously both of maternal origin, included in one germ cell due to meiotic nondisjunction (Sect. 5.1.2.3).
Soon after this first report, the XXY karyotype in the Klinefelter syndrome was confirmed in many more cases and is now known as the standard karyotype in this condition. At the same time, the result in the Klinefelter syndrome was complemented by chromosome examinations in another syndrome in which a discrepancy between phenotypic and nuclear sex seemed to exist: the Turner syndrome. Ford et al. (1959) [352] showed that the karyotype had only 45 chromosomes, obviously with one X and no Y chromosome. A third anomaly, with 47 chromosomes and three X chromosomes, was soon described in a slightly retarded woman with dysfunction of the sexual organs (Jacobs et al. 1959) [393]. The analytic possibilities afforded by human sex chromosome anomalies for sex determination in humans will be deferred to Sect. 2.2.3.

Birth of Human Cytogenetics 1956–1959: a Scientific Revolution. Kuhn [257] pointed to the difference between the progress of "normal science" and the occasional occurrence of "scientific revolutions" (see Introduction). In normal science, the inherent possibilities of existing paradigms are explored in a process that may be likened to puzzle solving. A scientific revolution is distinguished by the appearance of a new paradigm that, at first, is shared by only a small group of scientists. If it shows its superiority by offering solutions to so-far unsolved problems and – probably still more important – making it possible to ask new and more specific questions, it will be accepted by the scientific community, and everything in the whole field will now take on a different aspect.

From the standpoint of human genetics, the development of cytogenetics between 1956 and 1960, based on a new method and not a new concept, has been such a "revolution." The whole field attained a new dimension. Since only methods and not concepts changed, this advance did not make most of the earlier work obsolete but supplemented it in many directions. Any discussion of gene regulation, linkage, the structure of genetic material, spontaneous and induced mutations, population genetics, human evolution, and the practical use of genetic knowledge in the prevention of genetic disease would now be obsolete without due regard to human cytogentic data and concepts.

From the viewpoint of experimental geneticists, human cytogenetics had a much humbler appearance. The many advances were viewed as the belated application to man of concepts that had been known for many years, sometimes even half a century or still longer. More recently, human cytogenetics has reached the stage where the unique advantage of human material for the solution of more general biologic questions is now being explored. It is now at the forefront of basic cytogenetics.

What triggered this revolution? As observed frequently, it was a technical improvement: the hypotonic treatment for spreading chromosomes, accompanied by the examination of single isolated nuclei rather than tissue sections. Still, it was the scientists themselves who had to realize the inherent possibilities of this improvement.

Paradigm Group in Early Human Cytogenetics. These possibilities were apparently realized in two steps. Tjio and Levan [532] discovered the correct number of chromosomes but – apparently being too far away from medicine – did not see the potential application to human pathology. This step, taken by a group of British scientists, was one in which concepts of basic cytogenetics underwent a most lucky combination with experience in medicine. The same step, however, was taken by a scientist outside the academic structure of medicine in France, J. Lejeune [417]. During a scientific revolution, the group of scientists working on the new paradigm usually creates its own net of scientific interaction. The early phases of human cytogenetics offer an interesting object of research in the history of science, which may be useful for devising research strategies of the future. Such research should be carried out soon, while most of the protagonists are still active. One of them, D. G. Harnden, gave us the following interesting information.

The leading figures in the British group were Ch. Ford in Harwell and W. M. Court Brown in Edinburgh. Both worked in units sponsored by the Medical Research Council (MRC). Ford's interest in human chromosomes grew out of his work on mouse tumors and meiotic cells; Court Brown decided to work with human chromosomes because as an epidemiologist he felt it was necessary to combine epidemiologic with basic biologic studies. The two groups soon established relationships; for example, Patricia Jacobs, a nonmedical cytogeneticist, was sent by Court Brown to Ford. In his laboratory, she adapted the bone marrow culture technique developed by Lajtha for the examination of human chromosomes. Harnden for his part developed the technique of growing fibroblasts from skin biopsies, which he felt were more readily available than bone marrow.

The Edinburgh group was located in a hospital and had easy access to clinical material. Here, a physician, J. Strong, now a professor of medicine, seems to have had the idea to look at Klinefelter's syndrome. Harwell, where Ford worked, was an atomic energy biologic station and had no direct hospital connections; however, a connection with Guy's hospital in London soon developed, and P. Polani suggested looking at Turner's syndrome. Co-operation between the two groups was intensive; there was a great deal of interaction by letters, by telephone, by visits, and by exchange of material. Special conferences were not held. Human geneticists, such as P. Polani, L. S. Penrose, and J. Edwards, sent material to Harwell and advised the clinically inexperienced laboratory workers on medical matters.

The idea of examining Down's syndrome offered itself to the British workers as the obvious next choice after the search for aberrations in Klinefelter and Turner syndromes turned out to be successful. The idea seems to have come up independently in both Harwell and Edinburgh, where the workers were well advanced with their study before they knew of Lejeune's work.

The success of the two British groups was made possible by a lucky combination of people with different, but complementary backgrounds. Close co-operation developed for a few years during which the "paradigm" shared by the group revealed its explanatory power. Later, the co-operation slowly abated. At the same time, however, two other investigators had independently recognized the possibilities of the new methods. One was Lejeune in France; the other was the team of Fraccaro and Lindsten in Sweden, who started work on the Turner syndrome without knowledge of the investigations at Harwell.

Steps in the Development of Human Cytogenetics. The most important steps in the development of human cytogenetics were as follows:

1956 Tjio and Levan and Ford and Hamerton established the number of chromosomes in the diploid human cell (46).

1959 Lejeune discovered trisomy 21 in Down's syndrome; Ford et al. and Jacobs and Strong found the XXY karyotype for the Klinefelter syndrome and the XO karyotype for the Turner syndrome.

1960 Moorhead et al. [447] published the method for chromosome preparation from short-time lymphocyte cultures. Two autosomal trisomies, later identified as trisomies 13 and 18, were described by Patau et al. [472] and Edwards et al. [343]. Nowell and Hungerford described the Philadelphia chromosome in chronic myeloid leukemia [1584].

1963 The first deletion syndrome, the cri du chat syndrome, was observed by Lejeune et al. [418].

1964/65 Schroeder et al. (1964) [515] and German et al. (1965) [359] discovered a genetically determined increased chromosome instability in Fanconi's anemia and Bloom's syndrome, respectively. Jacobs et al. [394] associated criminal psychopathy with XYY males.

1968/70 Chromosome banding techniques were introduced. This permitted unequivocal identification of all human chromosomes [320].

Clinical Cytogenetics, the Most Popular Speciality of Human Genetics. In the years since 1960, human – and especially clinical – cytogenetics has developed into the most popular branch of human genetics. One reason, of course, is the fact that the cause of many previously unexplained malformation syndromes became apparent. Another likely reason is related to the fact that, after relatively simple manipulations, a "real" particulate appearance actually could be seen under the microscope. Visual images appeal to the medically trained and to many nonmedical biologists. In striking contrast, the more abstract concepts of formal genetics and population genetics do not attract the majority of physicians and biologists.

The surging popularity of clinical cytogenetics was all the more remarkable since during the first decade almost no practical significance of these results for medical therapy or prevention apart from diagnosis and genetic counseling seemed to be in sight. This changed dramatically when antenatal diagnosis became possible.

2.1.2 Normal Human Karyotype in Mitosis and Meiosis

2.1.2.1 Mitosis

Cell Cycle. Figure 2.4 diagrams the cell cycle of a dividing mammalian cell. The timing relates to rat hepatoma cells in vitro but is very similar in other cells. In the G_1 phase, RNA and proteins are synthesized and the cell prepares for DNA replication, which takes place in the S phase. The various parts of the chromosomes replicate asynchronously as evidenced when ^{3}H thymidine is offered at a certain time during the S phase. Then, only chromosomes that have not finished replication take up the labeled compound and can be identified by autoradiography. A certain amount of "unscheduled" or repair replication goes on during the G_2 phase, in which the cell prepares for mitosis (M). During the G_1 phase, the material of every chromosome of the diploid set (2 n) is present once. In the G_2 phase, on the other hand, every chromosome has doubled into two identical elements that are called sister chromatids.[3] The material of every chromosome is now present twice ($2 \times 2n = 4n$).

During or after replication, the two sister chromatids exchange segments repeatedly so that the two chromatid arms of a mitotic chromosome have parts of both chromatids (sister chromatid exchange, SCE). This can be made visible by using a specific staining technique after treatment with BrdU – a thymine analogue (Fig. 2.5) [411].

[3] Sister chromosomes would be a less confusing term but the terminology was developed at a time in which only the morphologic and not the biochemical aspects of mitosis were known.

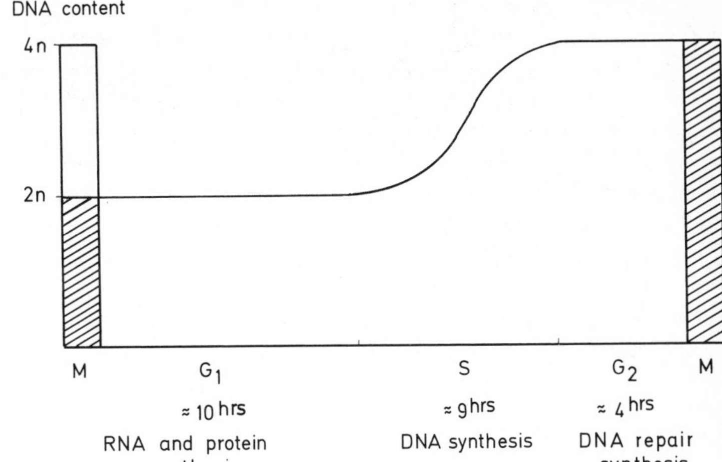

Fig. 2.4. Cell cycle of a dividing mammalian cell. In the G_1 phase, the diploid chromosome set (2 n) is present once. After DNA synthesis (S phase), the diploid chromosome set is present in duplicate (4 n). *M*, mitosis; ▨ DNA content during mitosis. Refer to text for details

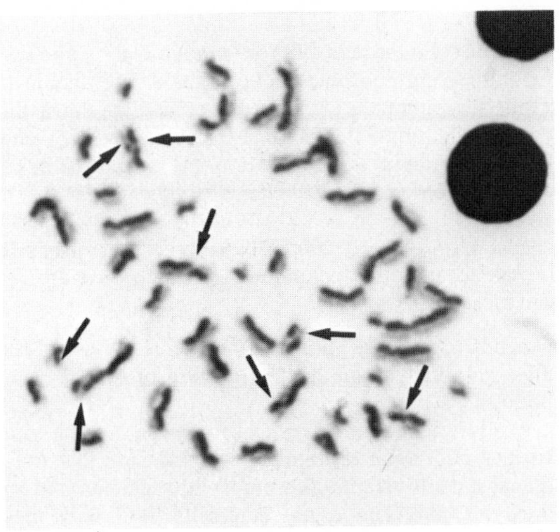

Fig. 2.5. Sister chromatid exchanges in a normal human metaphase. The locations of sister strand exchanges are indicated by arrows. Courtesy of Dr. T. M. Schroeder-Kurth

Mitosis. The phases of mitosis are set out in Fig. 2.6. The mitosis starts by condensation of chromatin (Fig. 2.6a; early prophase). At the end of prophase, the chromosomes are clearly visible; the two sister chromatids lie side by side. Meanwhile, the nuclear membrane has been dissolved, the nucleolus has disappeared, and the spindle is formed. The spindle consists of microtubules that are formed by a protein called tubulin and are visible under the microscope as spindle fibers. They connect the centromeric regions of the chromosomes with the centrioles. With dissolution of the nuclear membrane, prophase is finished, and the cell is now in metaphase. The centromeres are located in the equatorial plane between the two spindle poles. Now, the two chromatids of each chromosome start separating, until they are connected only at the centromeric region. Finally, the centromeres are also separated, to form half-chromosomes that are drawn to the opposite poles by the spindle fibers. The function of spindle microtubules can be demonstrated by colchicine treatment, which inhibits aggregation of tubulin and dissolves microtubules. It disturbs chromosome arrangement in the equatorial plane and inhibits their anaphase movement. Separation of chromatids occurs even in the presence of colchicine. In the last phase of mitosis, telophase, chromosomes are decondensed, spindle fibers disintegrate, the tubulin is stored in the cell, a new nuclear membrane is formed, and cell division begins. Chromosomes can most easily be examined in metaphase.

2.1.2.2 *Preparation and Staining of Mitotic Metaphase Chromosomes* [201, 88, 406]

Preparation. In principle, chromosome preparations can be made from all tissues and all suspensions that contain mitoses. In humans, direct preparations from bone marrow and preparations from short-term blood cultures or from long-term fibroblast cultures of other growing cells are used for most purposes. The most convenient technique is blood culture, as blood from patients is easily available, whereas bone marrow puncture or skin biopsy for fibroblast cultivation are more complicated and bone marrow aspiration is more unpleasant. Bone marrow preparations have the advantage that in vivo mitoses can be examined.

The blood of healthy, nonleukemic individuals contains no dividing cells. Therefore, cell divisions have to be stimulated artificially. This is possible for example by phytohemagglutinin (PHA). One hour after PHA incubation of a blood sample, the small (T-)lymphocytes show RNA synthesis, and about 24 h later DNA synthesis follows. The leukocyte suspension is grown in a culture medium for 72 hrs, and then chromosome preparations are made. To arrest as many cells as possible in the premetaphase or metaphase, formation of the spindle is prevented by a drug with col-

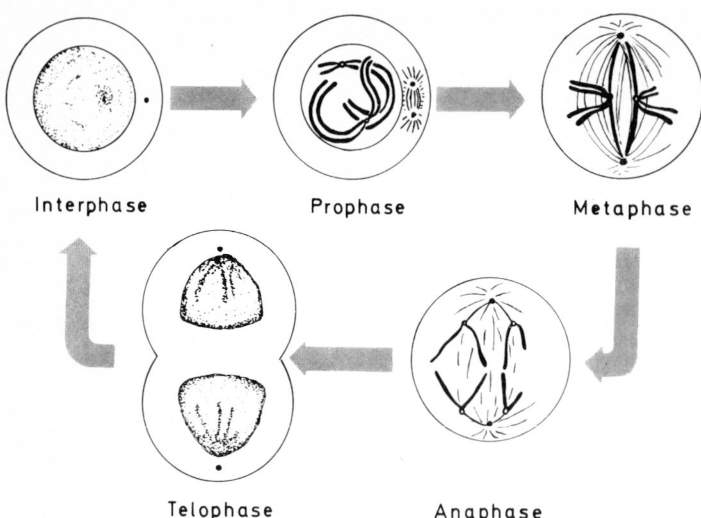

Fig. 2.6. Mitosis. Only 2 of the 46 chromosomes are drawn. (Buselmaier, 1976)

chicine-like effect, preferably colcemid. Under special conditions, culture time may be reduced to 48 hrs.
To obtain preparations in which the chromosomes are spread out in one plane, the cells are treated for a short time (10–30 min) with a hypotonic solution. The cells are then fixed with ethanol and acetic acid, a drop of the cell suspension is placed on the slide, air-dried, and stained.
Bone marrow preparations require sternal or iliac crest puncture. Cells are cultivated for only about 2 hrs with colcemid. Preparation differs in some details from that described above. Fibroblast cultures are prepared from skin biopsies, and the skin is minced into very small pieces and grown in culture medium in such a way that the tissue particles stick to the surfaces. After about 10 days, cells start growing on these surfaces and after about 21 days, are brought in suspension, prepared, stained, and examined.

Staining. The simplest staining methods are those using Giemsa solution, 2% acetic orcein, or 2% karmin solution. These dyes stain the entire chromosome uniformly and intensively. For some diagnostic purposes, such as diagnosis of the common numeric aberrations, this method is adequate. To obtain a more detailed picture of chromosome structure and to identify single chromosomes or chromosome segments, banding methods are used.

Banding Methods. Faint horizontal density differences may be seen occasionally in conventionally stained chromosomes. These gained little attention until Caspersson et al. (1968) [320] discovered that fluorescence of chromosomes after quinacrine mustard staining showed a distinctive sequence of bands for each chromosome. In the following year, Caspersson et al. demonstrated that each human chromosome could be identified by this staining method. Previously, identification of individual chromosomes was only possible for a few chromosomes and chromosome segments.
Soon afterward, it was shown that very similar banding patterns could be elicited by Giemsa staining with certain additional techniques. A number of investigators developed methods for staining of the pericentromeric regions, and partial heat denaturation was also shown to produce banded chromosomes.

At the Paris conference in 1971 on standardization and nomenclature of chromosomes [468], all the data available at that time were compared. It turned out that all of the methods revealed in principle the same structures, even if some techniques exposed some chromosome segments more clearly, while others worked better with other segments.

Available Methods [341; 200]. The various types of bands were named after the techniques by which they are revealed most clearly.

a) Q bands (quinacrine) are the fluorescing bands visible after staining with quinacrine mustard or similar compounds.
b) G bands (Giemsa) are revealed by Giemsa staining with various additional techniques, which insure that only the most readily staining chromosome segments take up the dye. Q and G bands are identical. Most laboratories prefer G banding for routine purposes since no fluorescence microscope is needed and the slides preserve the dye on storage. A specific advantage of the Q method is that it enables identification of the human Y chromosome by its strong fluorescence even in interphase nuclei.
c) R bands (reverse) are stained after controlled denaturation by heat. They are located between the Q or G bands, behaving like a photographic negative in relation to its positive image.
d) C bands (constitutive heterochromatin) are localized in the pericentromeric regions.
e) T bands (telomeric) mark the telomeric regions of chromosomes.

Detailed technical advice for use of these methods can be found in a number of publications. Many laboratories have worked out their own versions.

Chemical Differences Revealed by the Banding Methods. The chemical differences that are revealed by the banding methods are still under investigation. Two main hypotheses are usually discussed: the DNA hypothesis and the protein hypothesis. The DNA hypothesis is based on the observation that

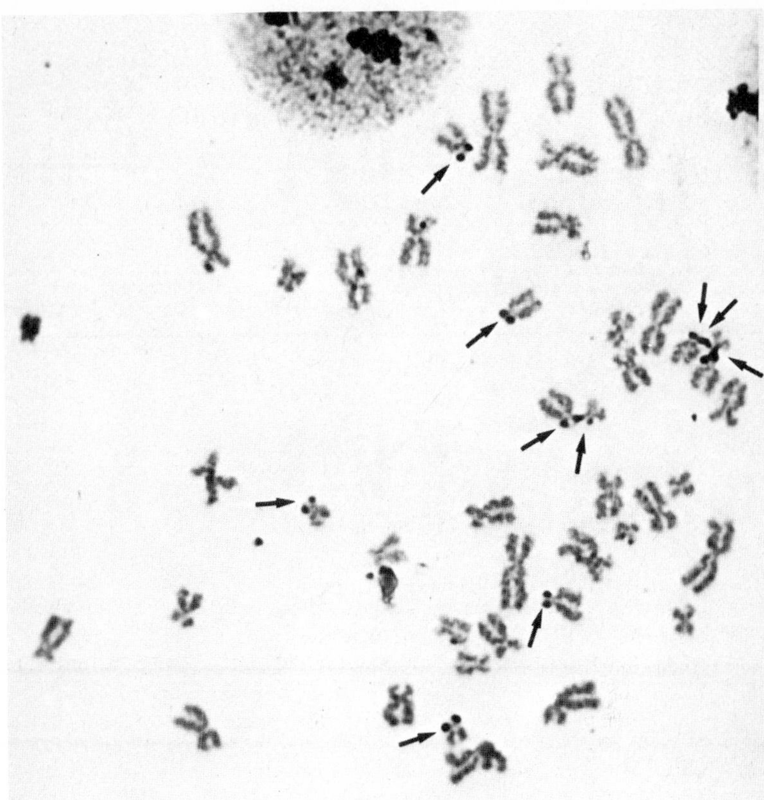

Fig. 2.7. Silver staining *(arrows)* of the nucleolus organizer regions of acrocentric chromosomes. Courtesy of Dr. T. M. Schroeder-Kurth

various parts of human chromosomes differ in their content of A-T (adenine-thymine) and G-C (guanine-cytosine) base pairs. Quinacrine fixes itself mainly to A-T rich segments [466; 341]. Acridine orange gives red fluorescence with single-stranded DNA. After controlled heat denaturation, the R bands are stained red. From these data, the following hypothesis may be derived:

a) Q bands indicate A-T rich chromosome segments.
b) R bands indicate G-C rich segments that are more resistant to heat denaturation than are A-T regions.

This hypothesis, however, does not explain all features of the banding patterns.
The protein hypothesis, on the other hand, is based on the observation that proteolytic treatment induces the appearance of G bands. Different kinds of DNA are linked in the chromosome to different protein species. The banding pattern probably depends in one way or another on properties of the whole DNA-protein complex.

Silver Staining of Nucleolus Organizer Regions [363; 511; 518]. A silver staining method is specific for the nucleolus organizer regions. They appear as black dots on the yellow-brownish background of the chromosomes (Fig. 2.7). Only those nucleolus organizers are stained that were functionally active during the preceding interphase [518].

Chromosomes from Human Spermatozoa. Some years ago, a method for making chromosome preparations directly from human spermatozoa by inducing mitoses through incubation of the semen together with zona pellucida-free golden hamster oocytes was described [489]. Such a method could be of great importance for direct investigation of chromosome anomalies in human spermatozoa. Replication of this method, however, turned out to be extremely difficult [431]. In one study the frequency of chromosomally abnormal sperm was 8.5% [432].

2.1.2.3 Normal Human Karyotype in Mitotic Metaphase Chromosomes

Conventional Staining. The chromosomes are ordered and numbered according to their length. The Denver convention (1960) had proposed numbering the pairs from 1 to 23. Patau (1960) [471] opposed this convention and showed that within some groups the chromosomes could not unequivocally be classified and attributed to a specific pair on the basis of their length and centromere position, which may be metacentric, submetacentric, or acrocentric. He proposed subdividing the 23 pairs into groups A–G. This was later accepted as an alternative procedure.

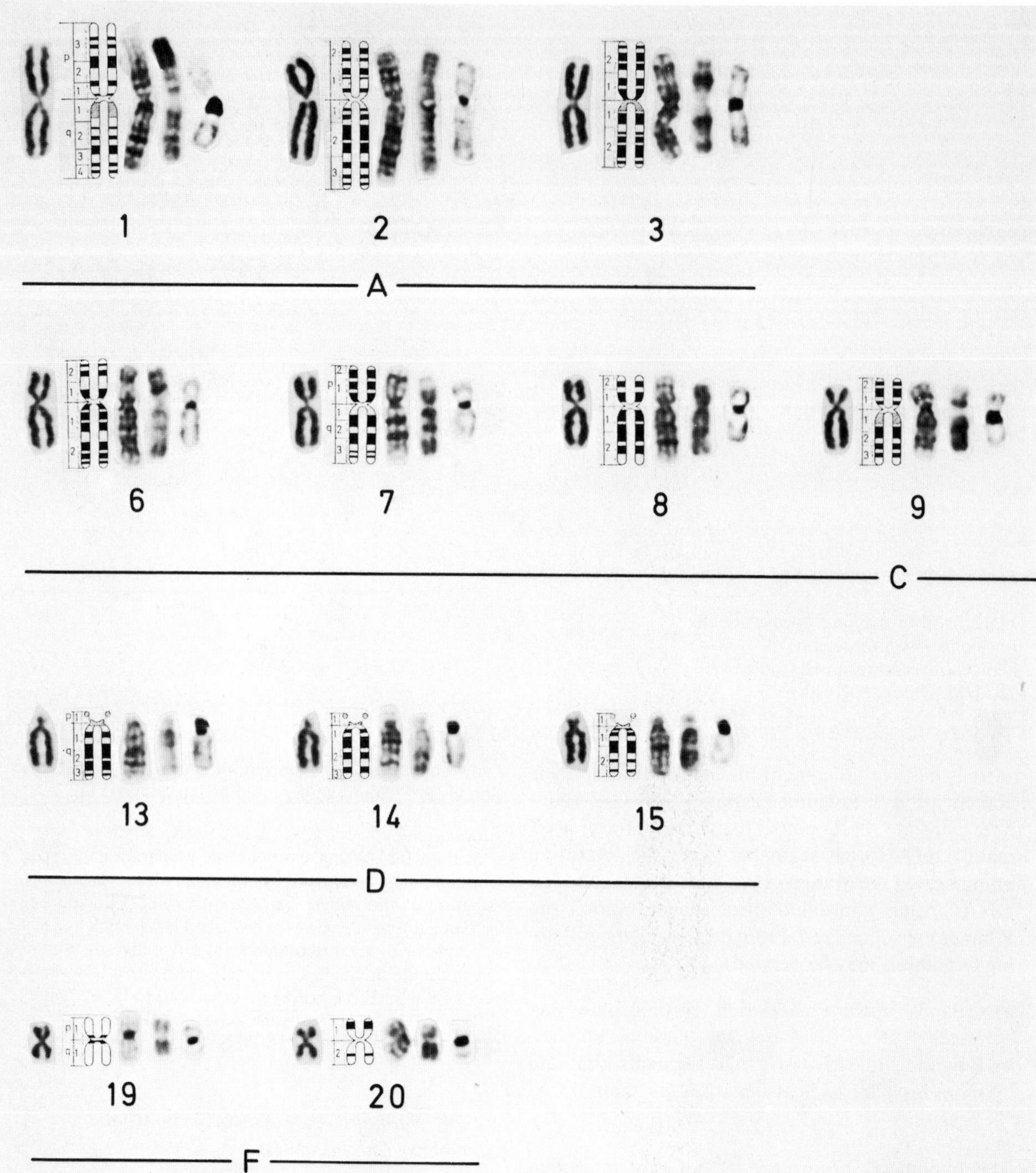

Fig. 2.8. Karyotype of a human male stained conventionally and using different banding techniques. From *left* to *right:* conventional staining; schematic representation of banding patterns; *G*, banding; *R*, banding; *C*, banding. Courtesy of Dr. T. M. Schroeder-Kurth

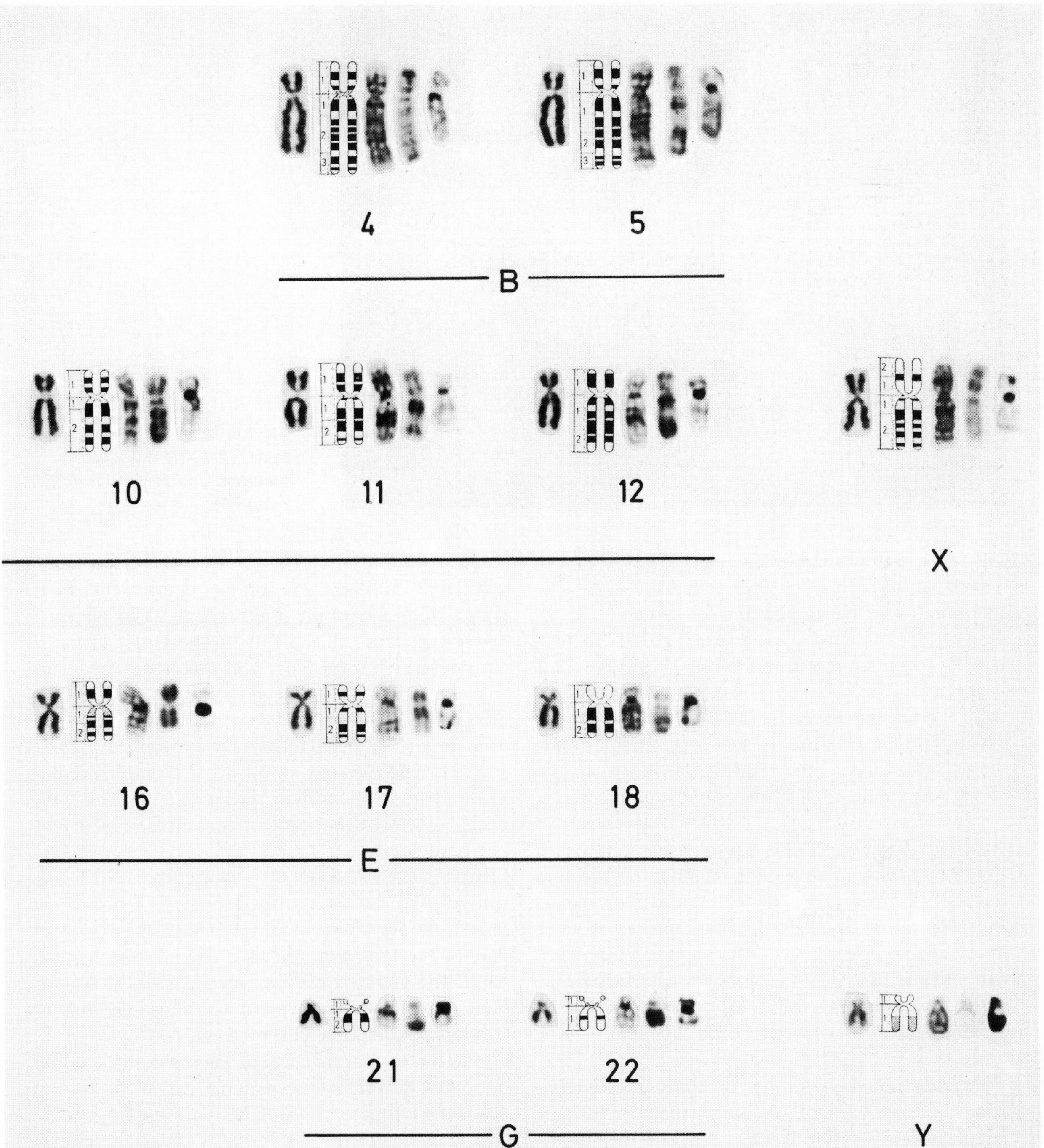

Fig. 2.8 *(continued)*

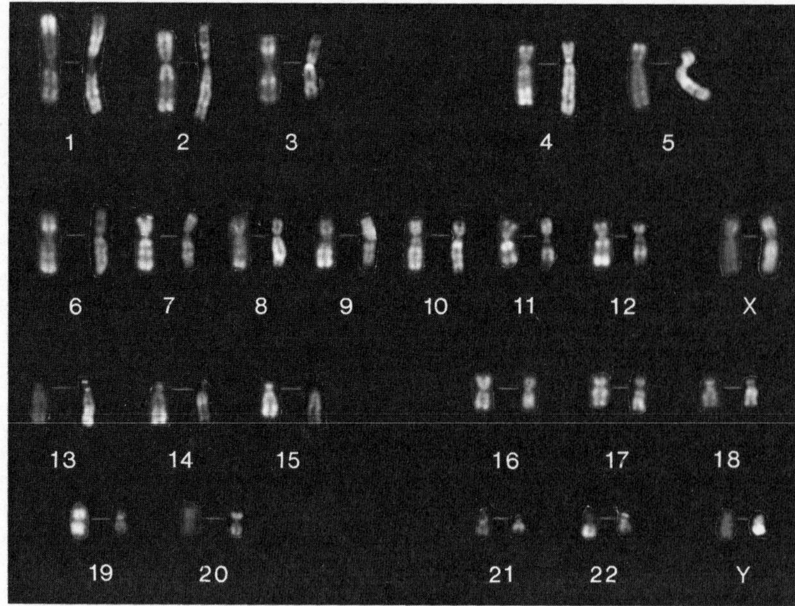

Fig. 2.9. Karyotype of a human male; Q (*right*) and R (*left*) banding. Courtesy of Dr. T. M. Schroeder-Kurth

The metacentric chromosomes of the A group can be identified. This is not possible for groups B, C including X, D, F, and G. Within the E group No. 16 is usually identifiable, and so are No. 17 and No. 18 if preparations are very good. The Y chromosome can usually be distinguished from others in the G group. An important measure is the centromere index, which is defined as the ratio of the length of the shorter arm to the total chromosome length.

Banding Techniques. The karyotype is shown in Figs. 2.8-2.10, stained with a number of banding techniques. Now, every chromosome can be identified. Figure 2.11 gives a schematic representation of G or Q bands, together with the number assigned to every band. The single chromosomes, together with their most frequently observed "normal" variants, may be described as follows [392]:

Individual Characterization of Human Chromosomes. Group A (No. 1-3). Large, metacentric and submetacentric chromosomes; No. 1 is the largest metacentric chromosome. The centromere is in the middle, the centromere index being 48-49. Close to the centromere, in the proximal part of the long arm, a "secondary constriction" is found fairly frequently. This constriction causes the occasional elongation of the long arm (Fig. 2.12). The extended segment may be very thin compared to the rest of the chromosome, suggesting "uncoiling" of the chromatid supercoil characterizing the metaphase chromosomes. This "uncoiler" phenomenon, like all individual variants of chromosome morphology, is transmitted to all cells, including about one-half of the germ cells, thus fulfilling the formal requirements of a simple dominant mode of inheritance. This "uncoiler-1 locus" was utilized to map the Duffy locus to chromosome 1 (Sect. 3.4). The secondary constriction shows little fluorescence with Q banding, but it does show a dark G band.

The largest submetacentric chromosome is No. 2. Its centromere index is 38-40. Autoradiographic studies with ^{3}H thymidine have shown that No. 2, especially the proximal parts of both arms, is relatively late-replicating.

Chromosome 3 is about 20% shorter than No. 1 and can therefore be distinguished easily. Its centromeric index is about 45-46. With Q banding, the proximal part of the long arm shows a brightly fluorescent band. The intensity of fluorescence varies strikingly between different individual chromosomes but is constant for the same chromosome.

Group B (No. 4 and 5). Large, submetacentric chromosomes. Chromosomes 4 and 5 cannot be distinguished without autoradiography or banding methods. Their centromeric index is 24-30. ^{3}H thymidine labeling shows No. 4 to be late-replicating over its whole length, while only the short arm of No. 5 is late-replicating over its whole length. R and G banding shows striking differences between these two chromosomes.

Group C (No. 6-12). Medium-sized, submetacentric chromosomes. After standard staining, the X chromosome cannot be distinguished from this group. Chromosomes 6, 7, 8, 11, and 12 are relatively submetacentric with a centromeric index of about 27-35; No. 9 frequently shows a secondary constriction in the proximal part of its long arm. All chromo-

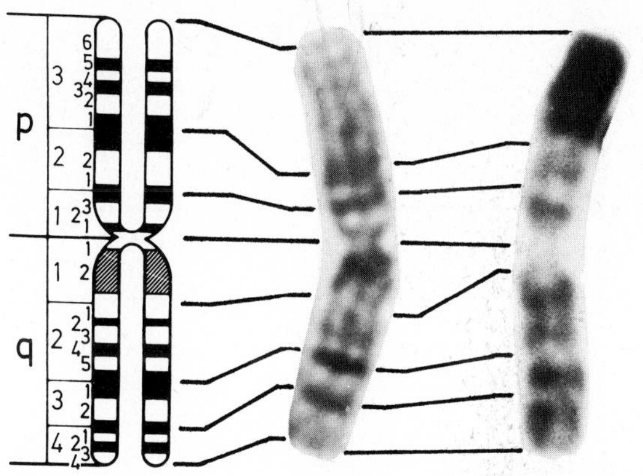

Fig. 2.10. Chromosome 1: comparison of G and R banding with schematic representation. Courtesy of Dr. T. M. Schroeder-Kurth

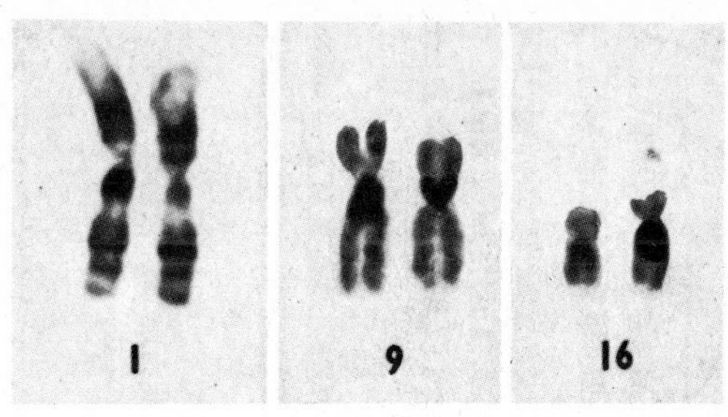

Fig. 2.12. Heteromorphism of constitutive heterochromatin at the secondary constrictions of chromosomes *1, 9,* and *16,* C banding. (Koske-Westphal and Passarge, 1974 [406])

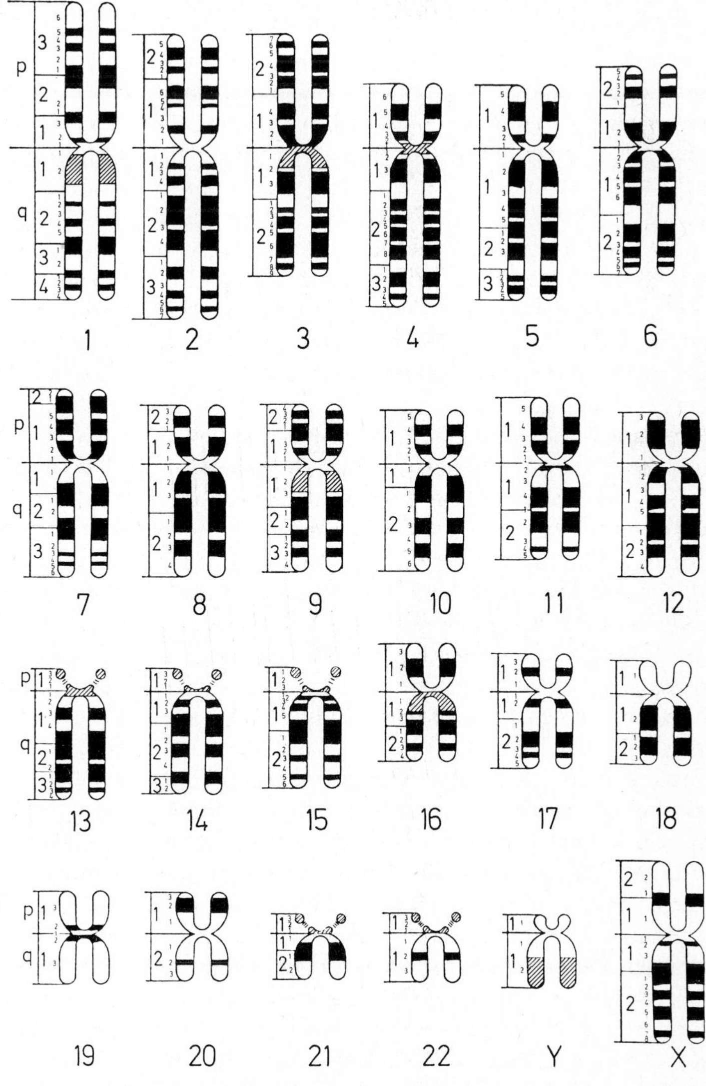

Fig. 2.11. Banding patterns according to the Paris nomenclature (G, Q, and R banding). Positive G and Q bands and negative R bands are *black,* variable regions are *hatched.* (Paris Conference, 1971 [468])

somes can definitely be identified by Q or G banding. The secondary constriction in No. 9 stains neither with quinacrine nor with Giemsa. Chromosomes 11 and 12 show very similar patterns, suggesting a possible common origin in evolution. (They contain loci for lactate dehydrogenase A and B, respectively, the common origin of which is suggested by biochemical evidence.) However, the centromeric index of No. 11 is somewhat greater than that of No. 12.

In contrast to the other chromosomes of this group, the X chromosome varies considerably in length. In general, it is similar to the longer C chromosomes. The centromeric index is high but somewhat variable. In female cells, one of the two X chromosomes still replicates in the late S phase, whereas replication of the other C chromosomes is complete except for short segments. The same late-replicating chromosome not only finishes but also starts replication later.

Group D (No. 13-15). These acrocentric chromosomes look quite different from other human chromosomes. The centromeric index is ≈ 15, being the lowest within the human karyotype. All three pairs may have satellites; their short-arm region shows strong interchromosomal variability. The proximal short arms are of varying length, satellites may be lacking or especially large; they may or may not show fluorescence; in some cases, double (tandem) satellites are observed. The long arms of all three D chromosomes are clearly distinguishable by Q and G banding. The following criteria are used for definition of variants in the D-G groups: The length of a short arm is compared with that of the short arm of No. 18 from the same cell. Normally it is shorter. It is called long (ph+), if it is as long as the short arm of No. 18, and very long, if it is longer. Large satellites are called (ps+), double satellites (pss), shortened short arms with or without satellites (ph−). The frequency of D group heteromorphism has been given as 3.7% (8 of 216) with banding techniques and 2.3% (411 of 24,440) without banding techniques [422] (Fig. 2.14).

Group E (No. 16-18). Relatively short, metacentric or submetacentric chromosomes. Chromosome 16 has a centromeric index of about 40. In general, its total length is normally somewhat more than one-third of No. 1 but shows striking variations. The long arm shows a secondary constriction in about 10% of all cases. The length of a proximal G band varies with this constriction. Chromosome 18 is about 5%-10% shorter than No. 17 and has shorter long arms (centromeric index 31 in No. 17 as compared with 26 in No. 18); No. 17 replicates early, No. 18 late.

Group F (No. 19 and 20). These two metacentric pairs have centromeric indices of between 36 and 46.

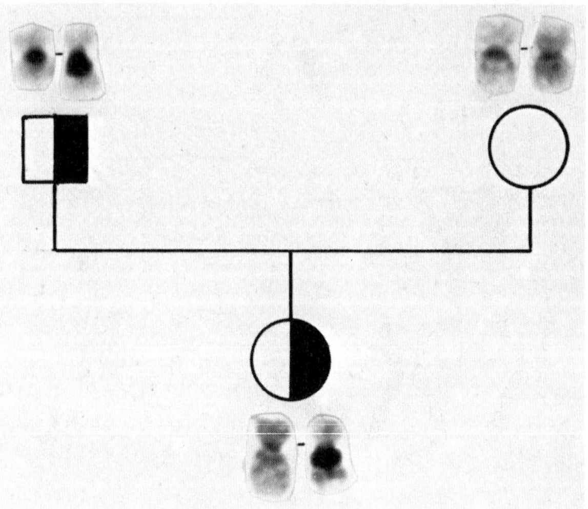

Fig. 2.13. Transmission of a C chromosome with especially large constitutive heterochromatin (C band) from father to child. Courtesy of Dr. T. M. Schroeder-Kurth

They cannot be distinguished in conventionally stained preparations, but their banding patterns are distinctively different.

Group G (No. 21 and 22). These small acrocentric chromosomes have centromeric indices of between 13 and 33. They can easily be distinguished by their banding patterns. The variability of their short-arm region is quite as large as that of D chromosomes. The same variants as in D chromosomes are usually distinguished (Fig. 2.14). Satellites and short arms may show weak, moderate, or strong fluorescence and G banding. Of 2,444 newborns, 3.5% had an elongated short arm. Other variants - such as giant satellites, elongated and shortened short arms - are much rarer. In another report, the overall frequency of G group variants was estimated as 1.8% with banding methods and 1.6% with standard methods. The short arms of D and G chromosomes contain the nucleolus organizer regions that are stained specifically by the silver method.

The Y chromosome is usually but not always larger than the G group chromosomes, and the two long arms are usually closely opposed to each other in contrast to G chromosomes where the angle subtending the long arms is usually wider. The centromere is less distinct, satellites are absent, and the long arm shows individual and transmissible variations in length. The centromeric index varies from 0 to 26 (average ≈ 16). The greatly variable long arm shows distal brilliant fluorescence after quinacrine staining. In most cases, two strongly fluorescing bands - in some rare cases even three - can be distinguished. In a population study, striking variants in length were found in 5.6% of 2,444 newborns. In

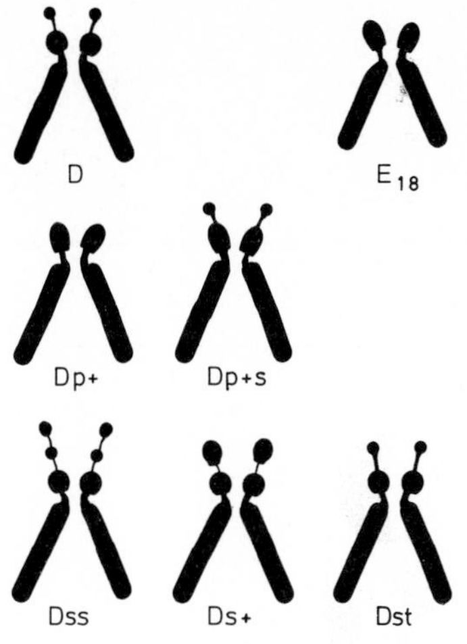

Fig. 2.14. Heteromorphism of acrocentric marker chromosomes of group D or G. *First row,* normal D and No. 18 chromosomes for reference. *Second row,* Dp+ : short arms at least as long as the short arms of a No. 18 chromosome; p+s: normal-sized satellites on elongated short arms, *third row, D,* chromosomes with structural variations of the satellite region. *ss,* double satellites; *s+,* enlarged satellites; *st,* enlarged satellite stalks. (Zankl and Zang, 1974 [1711])

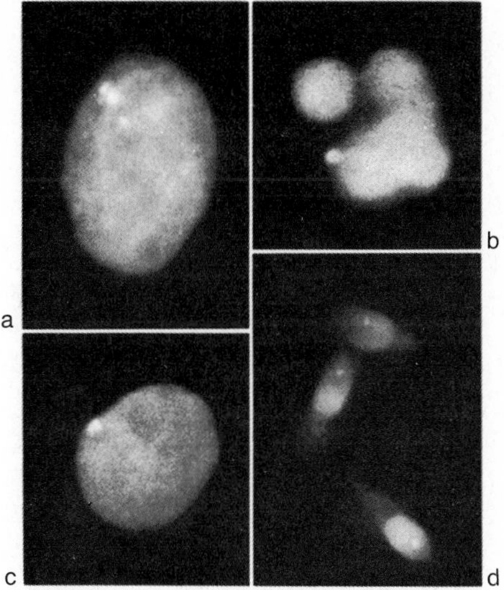

Fig. 2.15 a–d. Quinacrine mustard stainings of different cell nuclei of chromosomally normal men. **a** Buccal smear. The Y chromatin appears as a double structure. **b** Granulocyte from a blood smear. The Y chromatin is protruding as a small appendage. **c** Large lymphocyte from a blood smear. **d** Sperms. The Y chromatin is found near the border of the strongly fluorescent part of the sperm head (×2,400). (Schwarzacher and Wolf, 1974 [201])

most cases, the Y chromosome was elongated; in 5%, it was longer than an F chromosome, and in 0.33% longer than No. 18; 0.25% of the sample had very small Y chromosomes.

Chromatin [201, 516]. The distally brilliant fluorescent part of the long arm of the Y chromosome can be identified in the interphase nucleus as a bright dot about 0.3–1.0 μm in diameter. Figure 2.15 shows this "Y chromatin" in an epithelial cell, in a granulocyte, in a large lymphocyte, and in sperms.

Chromosome Measurements. Measurements of mitotic chromosomes meet with some difficulties, as the centromere position cannot always be determined accurately. A set of rules was laid down at the Paris conference in 1971 [468] Table 2.1 contains some typical measurements.

Chromosome Heteromorphisms. As mentioned in the description of single chromosomes, they do not always appear completely identical in all individual members of a population. Chromosome "heteromorphisms" are observed especially in the satellite regions of acrocentric chromosomes; the length of the Y, especially its heterochromatic part; and the "secondary constrictions" of chromosomes 1 and 9. But they also occur in heterochromatic segments of other chromosomes (For heterochromatin, see Sect. 2.3.14.). In aneuploidies, they may be used for identification of the parental origin of a chromosome (Sect. 5.1.2.3). In a number of chromosomes, fragile sites, i.e., chromosome sites showing an increased risk of chromosome or chromatid breaks (Sect. 2.2.2), have been discovered. Such breaks are especially easily induced by folic acid depletion of the culture medium [390 a]. A fragile site on the tip of the long arm of the X has recently aroused particular interest, since it is associated with a characteristic form of mental retardation (Sect. 8.2.1.2).

High-Resolution Banding [376]. Chromosomes in prophase and prometaphase are less tightly condensed than metaphase chromosomes. By suitable treatment of lymphocyte cultures with methotrexate for partial synchronization of cell cycles, a stage in which relatively many cells are in prophase or prometaphase can be picked out for preparation. Shortening of Colcemid treatment helps to decrease the degree of condensation. In such a preparation, single bands as revealed by standard methods can be resolved into subbands. The degree of resolution depends on the stage at which the cell has been picked. Some authors have described up to 2000 bands [550]; about 800–1200 bands can normally be seen in late prophase (Fig. 2.16). The method cannot replace standard methods in routine diagnosis; it is, however, useful for more precise identification of break-

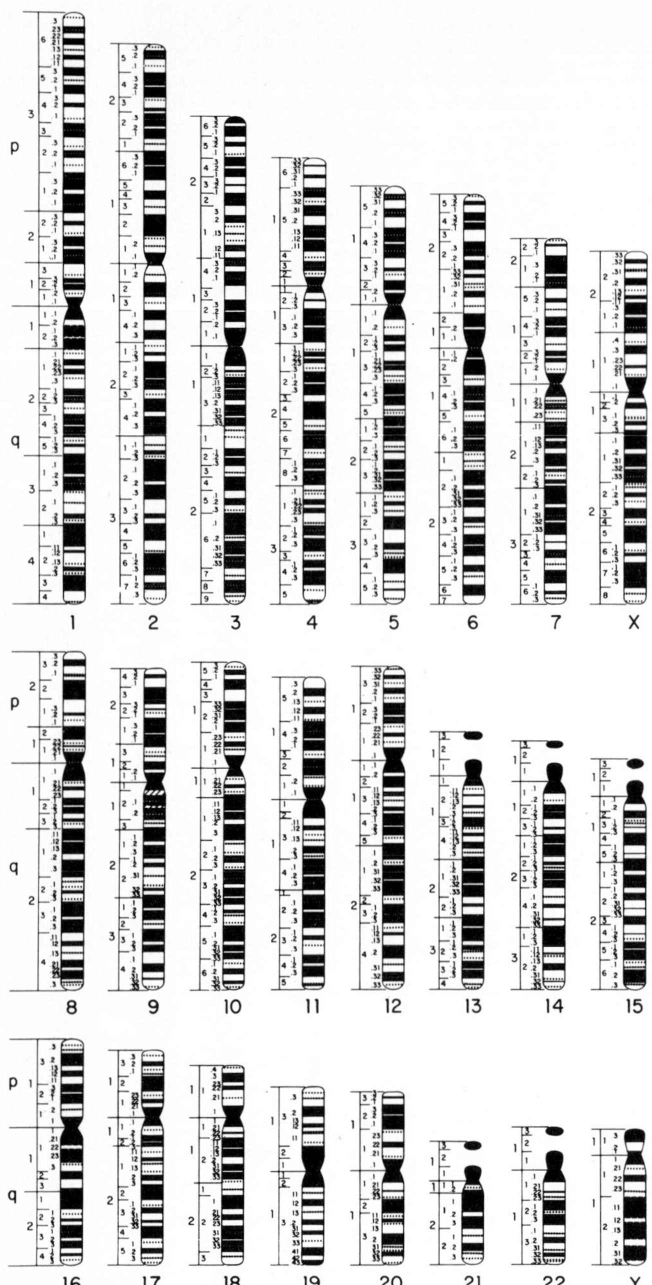

Fig. 2.16. Schematic representation of human chromosomes illustrating 1700 bands per haploid set. Solid black and white bands represent G-positive and G-negative bands seen at prometaphase or 850-band stage. Subbands visualized at late prophase (1300-band stage) and midprophase (1700-band stage) are represented with dashed bands. (From Yunis, Hum. Genet. 56, p. 296, 1980)

points and small aberrations, e.g., in families with balanced and unbalanced translocations, or particularly in tumor cytogenetics.

Electron-Microscopic Pictures from Human Chromosomes [490; 517]. A number of methods in electron microscopy have been used to gain insight into the overall structure of human chromosomes. Present models of the organization of genetic material in eukaryotes will be discussed in Sect. 2.3, at this point it will suffice to say that the evidence from electron microscopy does not contradict models assuming a chromatin thread that is supercoiled in several orders (Fig. 2.17). Three types of fibrils have been found: one type has a diameter of ~250 Å, a second measures ~100 Å, and a third only 30–50 Å. There seems to be good evidence that the latter fiber is the genetically active chromatin. A pure DNA double helix has a diameter of ~20 Å; hence 30–50 Å corresponds to a DNA fiber together with proteins (histone and nonhistone). The 100 Å fiber seems to be a secondary coil of the 30–50 Å fiber, and the 250 Å fiber may be a tertiary coil. These tertiary fibers may be arranged in a metaphase chromosome somewhat as in Fig. 2.17. Probably about nine of the 250 Å fibers are somehow bundled together, and two of these bundles seem to again form a coiled structure that can be discerned in appropriate electron-microscopic pictures and seems to be characteristic for each chromosome [490]. In some prepa-

Table 2.1. Measurements of relative length (in percent of the total haploid autosome length) and centromere index (length of short arm divided by total chromosome length × 100).
Chromosomes stained with orcein or the Giemsa 9 method and preidentified by Q band patterns. (Paris Conference, 1971) [468]

Chromosome No.	Relative length	Centromere index
1	8.44 ± 0.433	48.36 ± 1.166
2	8.02 ± 0.397	39.23 ± 1.824
3	6.83 ± 0.315	46.95 ± 1.557
4	6.30 ± 0.284	29.07 ± 1.867
5	6.08 ± 0.305	29.25 ± 1.739
6	5.90 ± 0.264	39.05 ± 1.665
7	5.36 ± 0.271	39.05 ± 1.771
X	5.12 ± 0.261	40.12 ± 2.117
8	4.93 ± 0.261	34.08 ± 1.975
9	4.80 ± 0.244	35.43 ± 2.559
10	4.59 ± 0.221	33.95 ± 2.243
11	4.61 ± 0.227	40.14 ± 2.328
12	4.66 ± 0.212	30.16 ± 2.339
13	3.74 ± 0.236	17.08 ± 3.227
14	3.56 ± 0.229	18.74 ± 3.596
15	3.46 ± 0.214	20.30 ± 3.702
16	3.36 ± 0.183	41.33 ± 2.74
17	3.25 ± 0.189	33.86 ± 2.771
18	2.93 ± 0.164	30.93 ± 3.044
19	2.67 ± 0.174	46.54 ± 2.299
20	2.56 ± 0.165	45.45 ± 2.526
21	1.90 ± 0.170	30.89 ± 5.002
22	2.04 ± 0.182	30.48 ± 4.932
Y	2.15 ± 0.137	27.17 ± 3.182

Data from 95 cells provided by Drs. H. A. Lubs, T. Hostetter, and L. Ewing from 11 normal subjects (six to ten cells per person). Average total length of chromosomes per cell: 176 µm. Standard deviations are an average of the standard deviations found in each of 11 subjects (six to ten cells per subject).

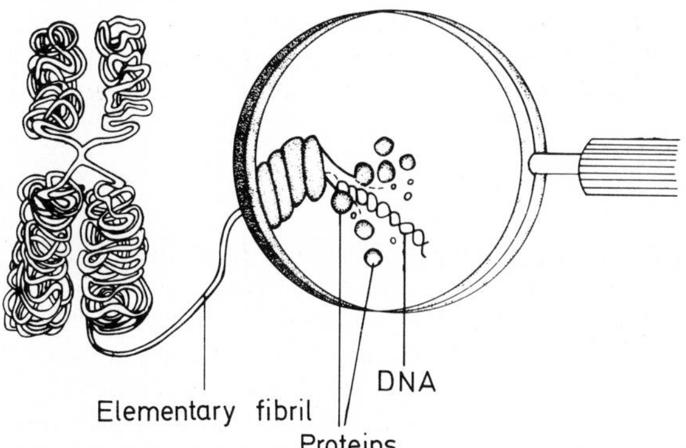

Fig. 2.17. Schematic representation of a metaphase chromosome. (Buselmaier, Biologie für Mediziner, 1985)

rations, relics of a membrane, presumably the nuclear membrane, are seen. Some investigators regard these as evidence that the interphase chromosomes are fixed to the membrane at various points. In view of numerous steps of preparation for electron microscopy, it is difficult to decide whether these – or other – details reflect in vivo structure or are merely preparative artifacts.

2.1.2.4 Meiosis

Biologic Function of Meiosis. While in the usual type of cell division, or mitosis, the number of divisions in daughter cells remains constant, the meiotic process is designed to reduce the number of chromosomes from the diploid number (46 in man) to one-half this number (23 in man). Fertilization of two germ cells, each with the haploid number, will reconstitute the diploid number of 46 in the zygote and in all of its descendant cells. Chance alone determines which of two homologous chromosomes ends up in a given germ cell. Genetic variability is thus enhanced. The somatic cell is diploid, containing both members of a pair of homologous chromosomes (2 n), whereas the germ cell is haploid, containing only one of each pair (n) The last regular DNA synthesis occurs during the interphase before the first meiotic division and precedes the meiotic phases shown in Fig. 2.18.

Meiotic Division I. Prophase I: Long chromosome threads become visible (leptotene) followed by pairing of homologous chromosomes, frequently from the chromosome ends (zytogene). The exact molecular mechanism of chromosome pairing is not yet known. The paired homologous chromosomes are connected by the so-called synaptonemal complex, a characteristic double structure (Fig. 2.19). After completion of pairing, the chromosomes become shorter through contraction (pachytene). A longitudinal cleft in each pair of chromosomes then becomes visible; four chromatids of each kind are seen side by side (diplotene). Non-sister chromatids are separated, while sister chromatids remain paired. In this phase, chromatin crossings – "chiasmata" – between non-sister chromatids become visible.

Metaphase I: The chromosomes are ordered in the metaphase plane as the centromeres are drawn to the poles. The homologous chromosomes are drawn somewhat apart but are still kept together by the chiasmata, frequently at the ends.

Anaphase I: The chiasmata are first "terminalized," i.e., they seem to migrate to the chromosome ends and are then resolved. The paired chromosomes separate and migrate to the opposite poles. The daughter nuclei are formed (interkinesis).

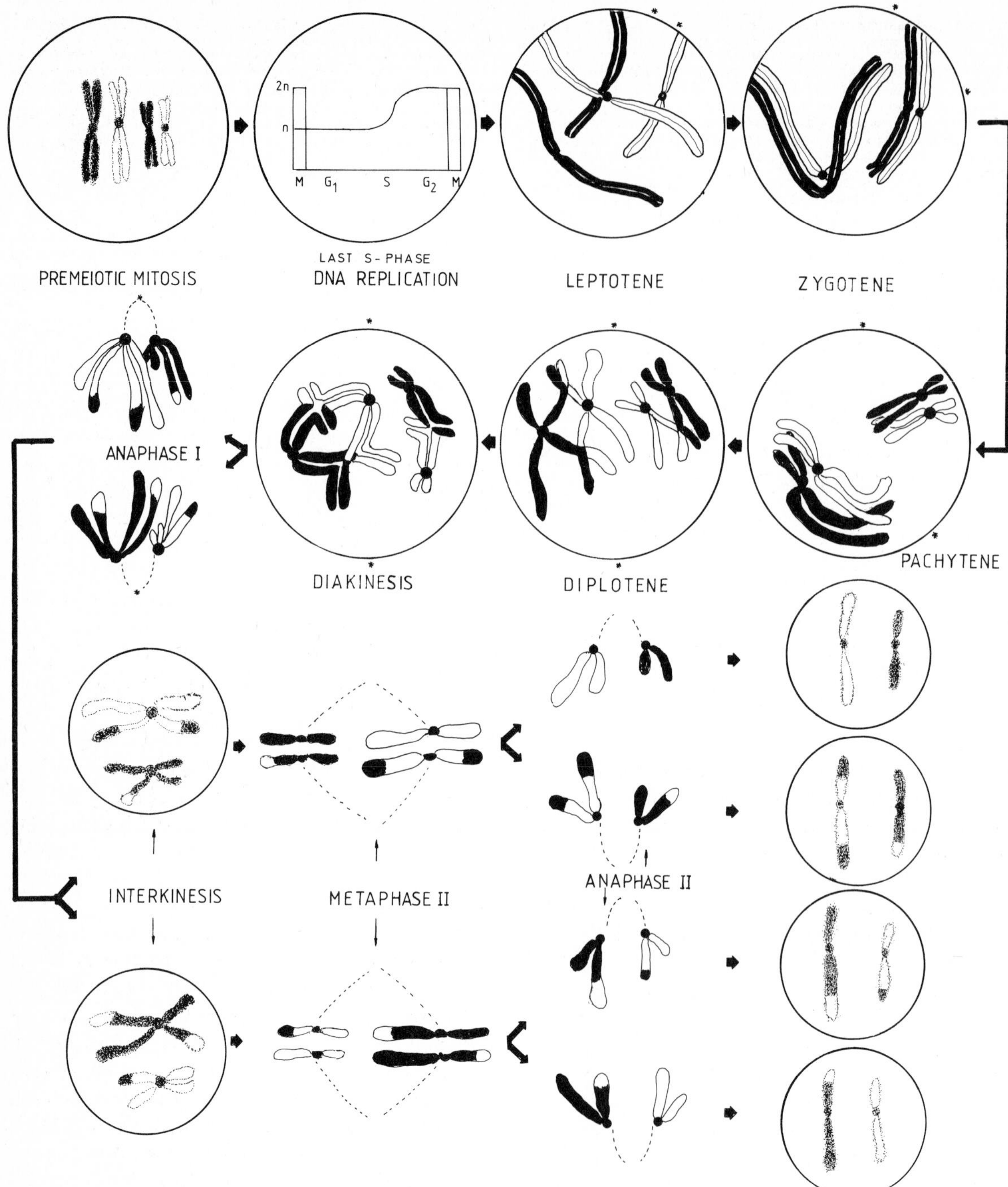

Fig. 2.18. The stages of meiosis. Paternal and maternal chromosomes black and white. The figure depicts male meiosis; in female meiosis, polar body formation occurs

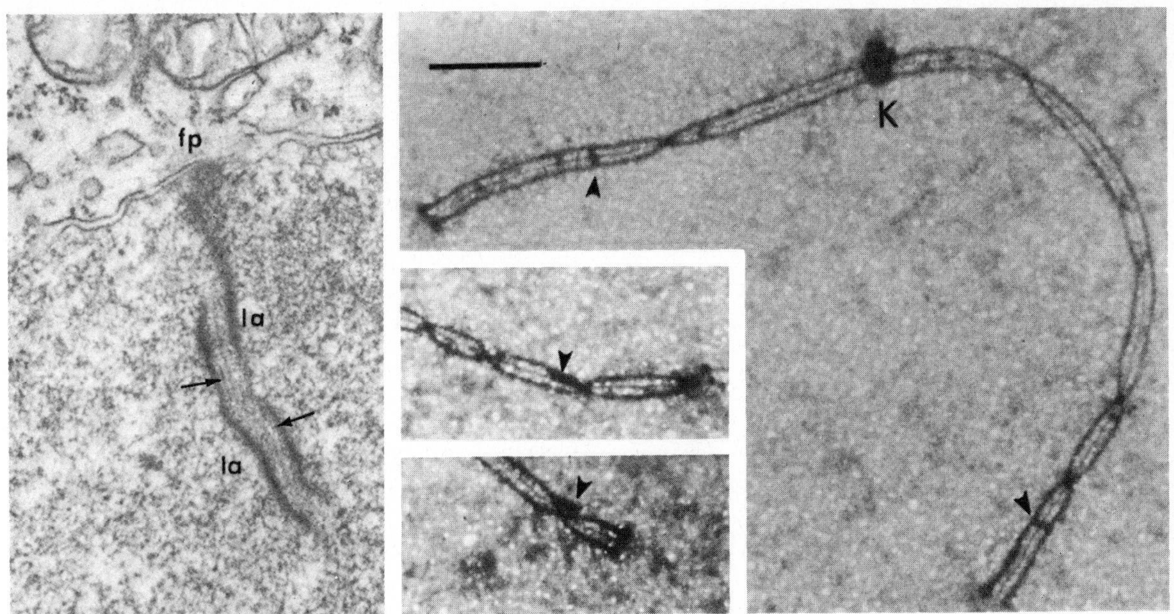

Fig. 2.19. *Left:* Electron micrograph of a synaptonemal complex with fixation point *(fp)* in the middle of pachytene in a mouse spermatocyte. Two electron-dense lateral arms *(la)* and a bright pairing space with a median density can be recognized. OsO_4, Vestopal, ×36,000. (Schleiermacher and Schmidt, 1973). *Right:* Synaptonemal complex of a human spermatocyte. *K,* centromer; *arrows* transverse dense bars. *Upper and lower insert:* Bars parallel to the axis of the synaptonemal comples. Magnification × 15,800; Bar = 1 μm. Bars may represent sites of recombination. (From Solari, Chromosoma 81, p. 330, 1980)

Meiotic Division II. This is in principle a mitotic division of the replicated haploid set of chromosomes. As noted above, meiosis begins after replication. The genetic material, which during division I becomes fourfold (2 × 2 homologous chromosomes) is, at the completion of division II, ordinarily distributed to four cells. A second, important aspect of meiosis is the random distribution of nonhomologous chromosomes, which leads to a very large number of possible combinations of possible germ cells. In humans with 23 chromosome pairs, the number of possible combinations in one germ cell is $2^{23} = 8{,}388{,}608$. The number of possible combinations of chromosomes in an offspring of a given pair of parents is $2^{23} \times 2^{23}$ and is further enhanced by crossing over during pairing of homologous chromosomes. The morphologic counterpart of crossing over is chiasma formation. Every chiasma corresponds to one crossing over event involving two non-sister chromatids (Fig. 2.20). For some time, it was disputed whether crossing over occurs during regular DNA synthesis - by "copy choice" - or after regular DNA synthesis - by breakage of non-sister chromatids at homologous sites and subsequent crosswise reunion (Fig. 2.21). The controversy seems now to be resolved in favor of the exchange hypothesis. Prophase I shows no regular, but much unscheduled DNA synthesis, which could easily indicate the reunion phase of crossing over.

The molecular mechanisms of recombination are not a specific problem of human genetics. They are discussed extensively in textbooks on molecular genetics.

Meiosis in the Human Male. From the beginning of puberty, human spermatocytes continuously undergo meiosis; the timing of spermatogenesis up to this stage will be described in Sect. 2.1.3.3. After the second meiotic division, DNA and mitochondria are densely packed during sperm development, and the sperm acquires the ability to move actively. Chromosomes of spermatogonial mitoses and second meiotic divisions may be prepared from testicular biopsy material or from surgically removed testicles.
Male diakinesis chromosomes are seen in Fig. 2.22. The homologous chromosomes are still close together at the ends, whereas the centromeric regions have already started on their way to the poles. The sex bivalent is clearly distinguished from all the others by end-to-end association of the X and Y chromosomes. No chiasma formation can be seen. During pachytene and prophase, the sex bivalent is already prematurely condensed and contained in a "sex vesicle." Part of the short - arm region of the X chromosome and the short arm of the Y are paired (Fig. 2.23), and hybridization studies with DNA probes have shown these regions to be structure-homologous [502].

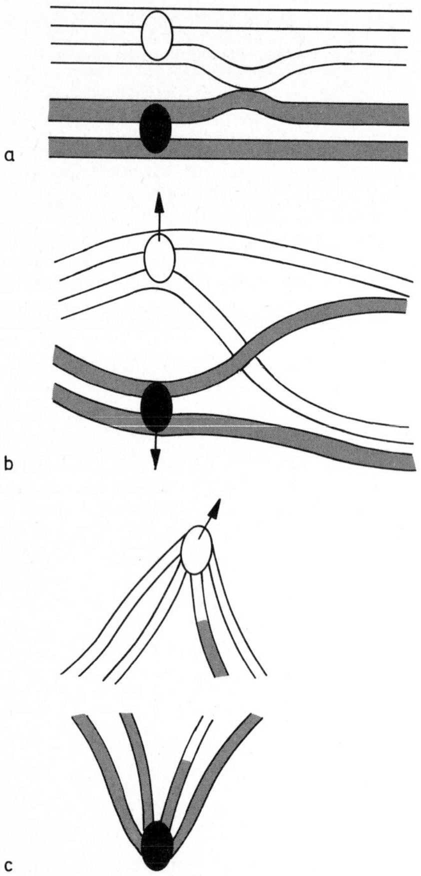

Genes located on a comon segment of the X and Y chromosome would be indistinguishable from autosomal genes if free recombination would occur between them. Such "pseudoautosomal" DNA sequences with X and Y linked alleles have been identified [315a; 488a]. Haldane (1936) [372] suggested the existence of incomplete sex linkage caused by occasional crossovers between X and Y for some human genes postulated to reside on a common X and Y chromosomal segment. However, no plausible evidence for such partial sex linkage in man exists. Furthermore, the human steroid sulfatase and Xg loci that are located very closely to the pseudoautosomal region of the X chromosome follow classic X linked transmission.

Table 2.2. Number of chiasmata in male meiosis (1st division)

No. individuals	Age range	No. of cells	Chiasmata/cell Range	Chiasmata/cell Mean	Chiasmata/bivalent, mean
48	15–79	817	39–64	54.4	2.36

Data from different authors; cf. Hamerton, 1971 [88].

Fig. 2.20 a–c. Crossing over and chiasma formation. **a** Homologous chromatids are attached to each other. **b** Crossing over with chiasma formation occurs. **c** Chromatid separation occurs

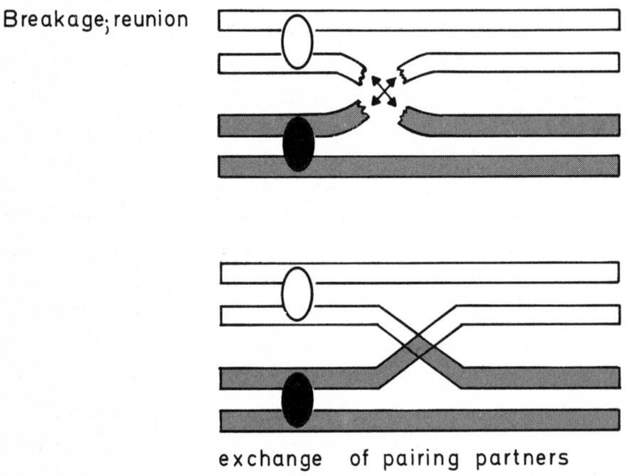

Fig. 221. Breakage and reunion of non-sister chromatids in crossing over

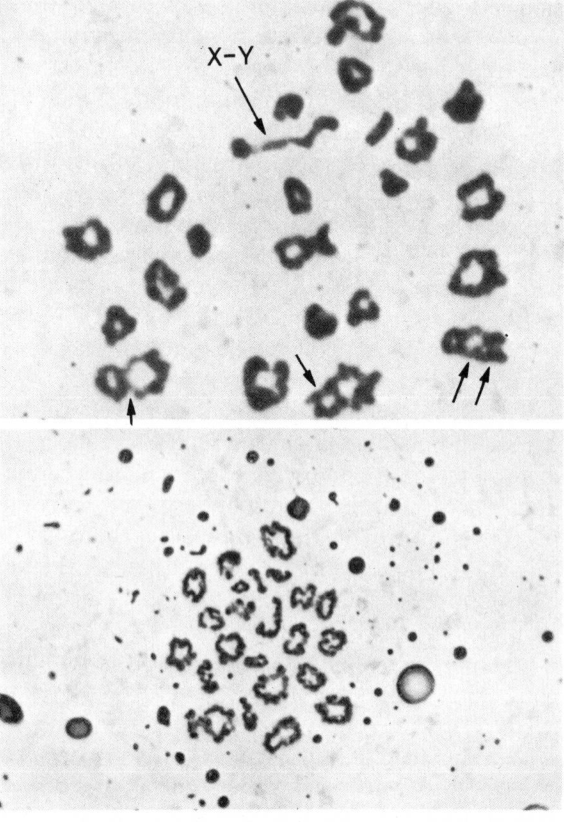

▷

Fig. 2.22. Meiotic metaphases I (diakinesis) of a male with clearly visible XY bivalent (Kjessler, 1966 [405]. *Arrows* indicate chiasmata

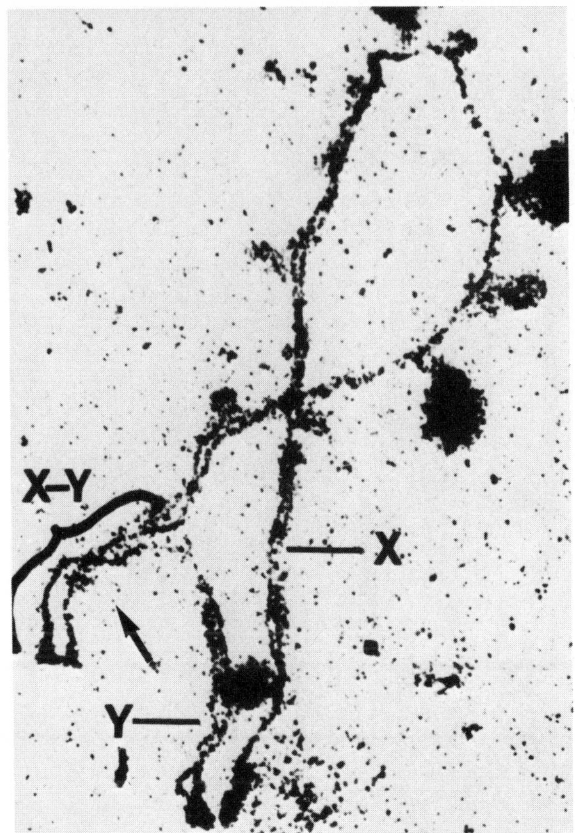

Fig. 2.23. Pairing of the short arms of human X and Y chromosomes in early meiosis. Courtesy of Dr. Goetz

The range of variation and the average number of chiasmata per cell are given in Table 2.2. Some bivalents may contain several, up to five or even six, chiasmata. From the number of chiasmata, the genetic map length (Sect. 3.4) of the human genome was estimated to be ~25.8 morgans in the male; it is longer but cannot be estimated in the female [450]. In the house mouse, the only other mammal for which an estimate is available, map length was estimated to be between 16.2. and 19.2 morgans [88].[4]

Meiosis in the Human Female. In all mammals, oogenesis is very much different from spermatogenesis; the overall timing is described in Fig. 5.13 (Sect. 5.1.3.3.). Figures 2.24 and 2.25 show the cell processes (see also [409]). The oocytes are already formed in the late embryonic stage; after the diplotene, the cell enters a stage in which the chromosomes have a lampbrush-like appearance and which is called dictyotene. Meiosis remains arrested at this stage for many years. After birth, most oocytes degenerate. After puberty, some oocytes start growing, finish the first meiotic division, and go into prophase II and metaphase II. At the same time, the egg is ovulated. Meiosis is finished only after fertilization. Nuclear membranes are formed around the haploid male and female chromosome sets and the zygote now contains two "pronuclei." At this stage, it is especially susceptible to disturbances, for example, by mutagenic agents (Sect. 5.2.1). Several hours later, the two pronuclei fuse to form a diploid nucleus, and the zygote starts dividing.

Examinations of female meiotic chromosomes are difficult, and few satisfying pictures have been published. Linkage analysis has shown crossing over in females to be more frequent than in males (Sect. 3.4). Hence, more chiasmata should be expected.

In females, only one of the four meiotic products develops into an oocyte, the others becoming polar bodies that under normal conditions are not fertilized. It is usually assumend that the risk of a chromosome being discarded in a polar body is unrelated to its genetic properties. This assumption is correct for most gene mutations, as shown by their undisturbed segregation ratios (50:50, 25:75, etc.). Even here, however, exceptions have been reported (Sect. 3.1.4). For structurally abnormal chromosomes, disturbances of their segregation into germ cells may be explained by nonrandom expulsion of normal and abnormal chromosomes into the polar body.

Sex Difference in Meiosis. There are two main aspects by which meiosis is different in males and females:

1. In males, all four division products develop into mature germ cells, whereas in females, only one of them becomes a mature oocyte, the others are lost.
2. In males, meiosis immediately follows a long series of mitotic divisions; it is completed when spermatids start developing into mature sperms. In females, meiosis begins at a very early stage of development, immediately after a much smaller series of mitotic divisions. It is then arrested for a long period and is finished only after fertilization.

These sex differences are important in human genetics. The fact that only one of the four division products develops into a mature oocyte and the three polar bodies contain little or no cytoplasm enables this oocyte to transmit to the new zygote a full set of cytoplasmic constituents such as mitochondria and messenger RNA (Sect. 4.7.1). These differences in cell kinetics probably are responsible for sex differences in mutations rates for trisomies, on the one hand, and point mutations, on the other (Sect. 5.1 and 5.2).

[4] One Morgan is the unit of map distance between linked genes; this unit measures chromosome length.

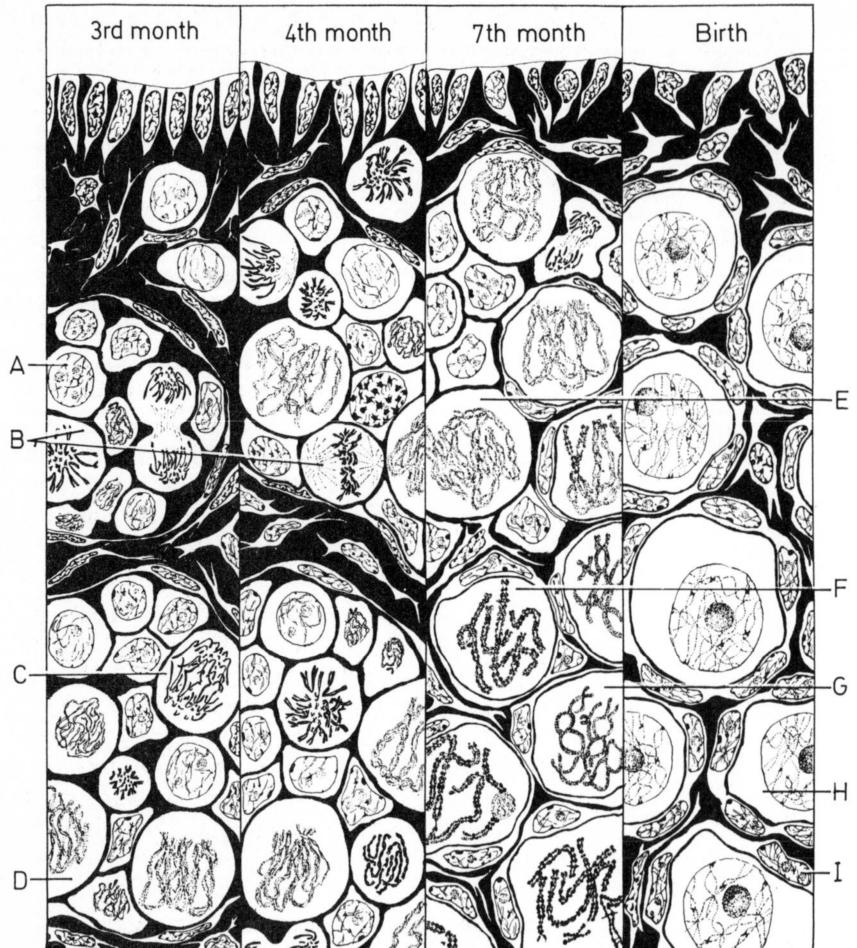

Fig. 2.24. Mitosis and meiosis in the human female. Up to the 3rd month, only mitotic divisions are to be seen (*A*, interphase; *B*, metaphase; *C*, anaphase). Then, the first meioses become visible (*D*, leptotene; *E*, zygotene). Up to the 7th month, new oocytes enter meiosis. The first pachytene *(F)* and diplotene *(G)* stages are observed in the 7th month. Meiosis does not proceed any further. Instead, the tetrads are again stretching, a nuclear membrane and a nucleolus are formed, and the cell enters a "resting phase," the dictyotene *(H)*. The function of the cells *(I)* around the oocytes is nutritional; they will later form follicles in which oocytes are embedded. (The semischematic drawings are taken from: Ohno et al., 1962; see also Bresch and Hausmann, klassische und molekulare Genetik, 1972)

2.2 Human Chromosome Pathology

2.2.1 Syndromes Due to Numeric Anomalies of Autosomes

Mechanisms Creating Anomalies in Chromosome Numbers (Genome Mutations) (Fig. 2.26). Anomalies in chromosome numbers may be caused by various mechanisms:

a) The most important mechanism is nondisjunction. Chromosomes that should normally be separated during cell division stick together and are transported in anaphase to one pole. This may occur at mitotic division but is observed more frequently during meiosis. The exact reasons are unknown, but in humans the acrocentric chromosomes run a higher risk of being involved (Sect. 5.1.2). Meiotic nondisjunction was discovered by Bridges (1916) in *Drosophila* [311]. For every gamete with one additional chromosome another one is formed with one chromosome less. After fertilization with a normal gamete, the zygote will be either trisomic or monosomic. Somatic nondisjunction in mitotic cell division during early development may lead to mosaics with normal, trisomic, and monosomic cells.

b) A second mechanism leading to numeric abnormalities is loss of single chromosomes, presumably due to "anaphase lagging": during anaphase movement, one chromosome may lag behind the others. Chromosome loss leads to mosaics with one euploid and one monosomic cell population.

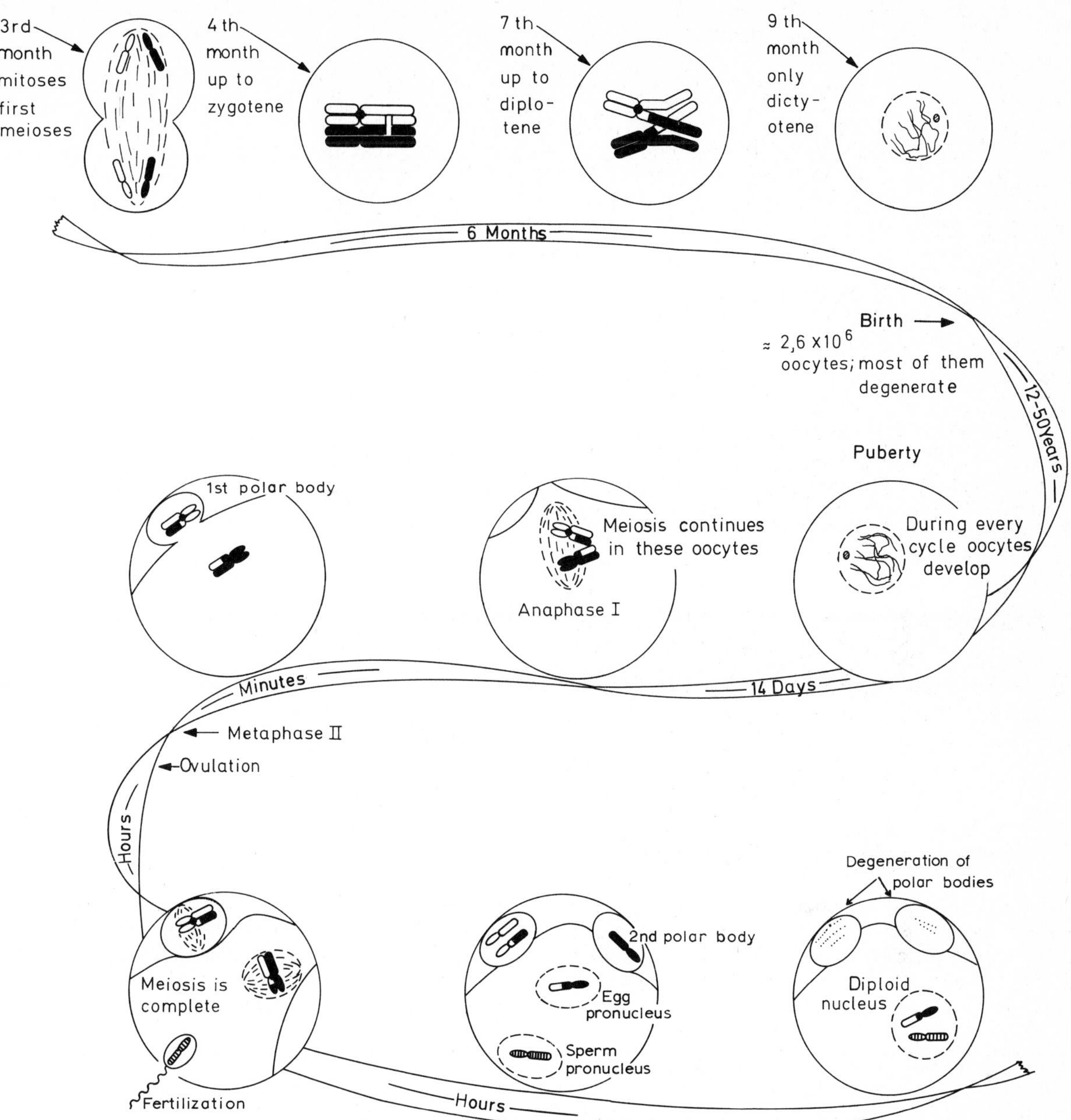

Fig. 2.25. Meiosis in the human female. Meiosis starts after 3 months of development. During childhood, the cytoplasm of oocytes increases in volume but the nucleus remains unchanged. About 90% of all oocytes degenerate at the onset of puberty. During the first half of every month the luteinizing hormone *(LH)* of the pituitary stimulates meiosis which is now almost completed (end of the prophase that began during embryonic age; metaphase I, anaphase I, telophase I and - within a few minutes - prophase II and metaphase II). Then, meiosis stops again. A few hours after metaphase I is reached, ovulation is induced by *LH*. Fertilization occurs in the Fallopian tube. Then, the second meiotic division is completed. Nuclear membranes are formed around the maternal and paternal chromosomes. After some hours, the two "pronuclei" fuse, and the first cleavage division begins. (Bresch and Hausmann, klassische und molekulare Genetik, 1972)

In the mouse, the pronucleus stage, i.e., the time between impregnation and fusion of the two haploid parental nuclei, is especially susceptible to loss of the paternal X chromosome. This phase, and possibly the first cleavage stages, could easily be likewise vulnerable in humans, since many mosaics are formed during these stages (Sect. 5.1.6).

c) A third mechanism is polyploidization. Here, the whole genome is present more than twice in every cell. In humans, only triploidy is observed. The chromosome number is $3n = 69$.

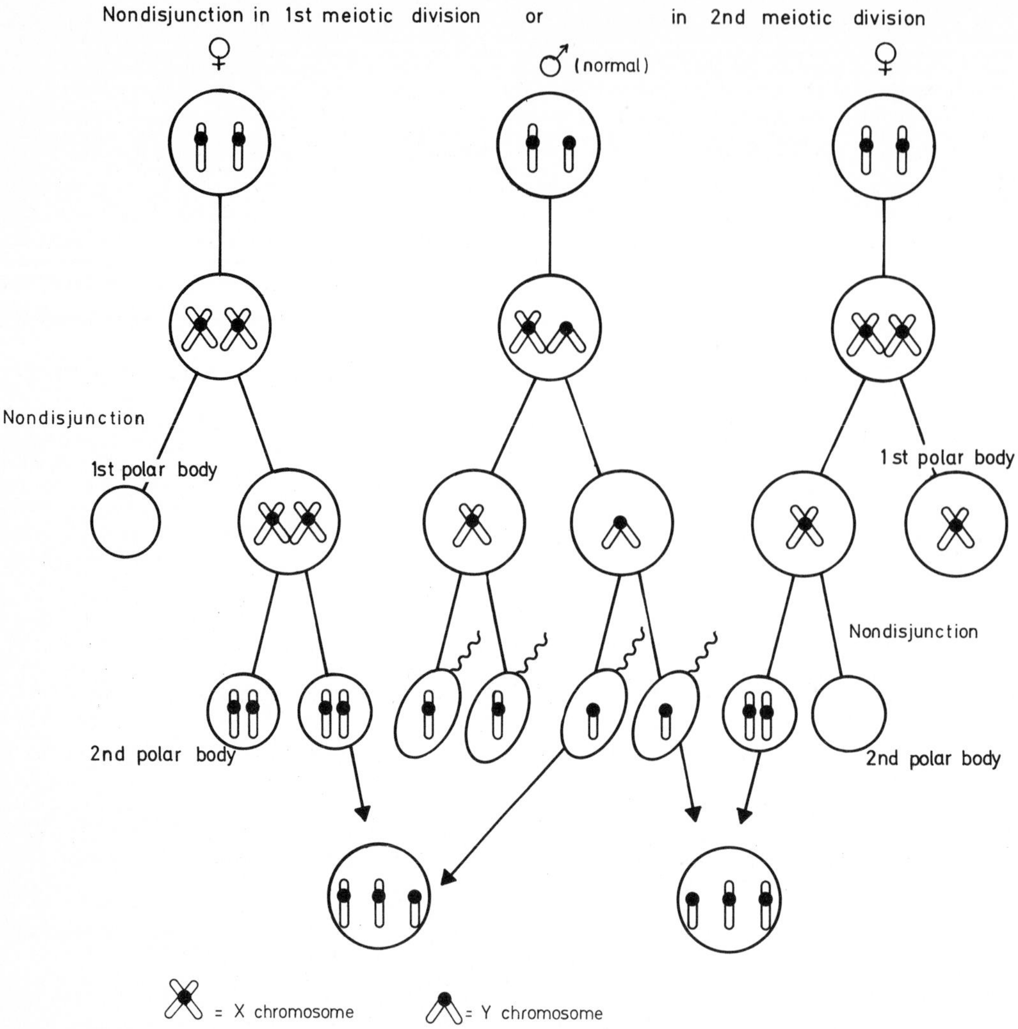

Nondisjunction in 1st meiotic division or in 2nd meiotic division

♀ ♂ (normal) ♀

Nondisjunction

1st polar body

1st polar body

Nondisjunction

2nd polar body

2nd polar body

= X chromosome = Y chromosome

Fig. 2.26. Nondisjunction of the X chromosome in the first *(left-hand side)* and 2nd *(right-hand side)* meiotic division in a woman. Fertilization by a normal Y sperm. An XXY individual can result from both 1st and 2nd meiotic division nondisjunction

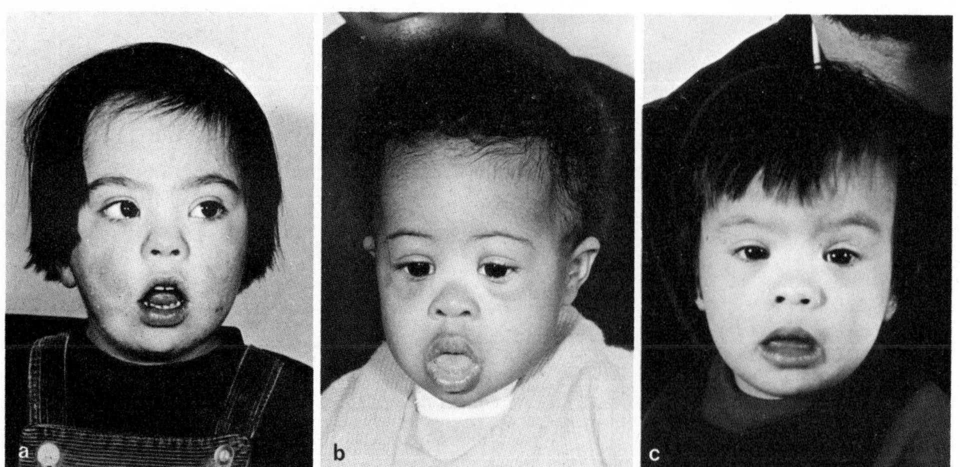

Fig. 2.27 a–c. Children with Down's syndrome. **a** European, **b** Afro-American, **c** Oriental-European. The common features of Down's syndrome are more impressive than the racial differences. Courtesy of Dr. T. M. Schroeder-Kurth

An abnormal number of chromosomes in a cell (aneuploidy) increases the risk of further irregularities, such as chromosome loss due to anaphase lagging in subsequent cell divisions. For many mosaics with two cell populations of equal proportions, one trisomic and one euploid, this is the most plausible explanation (Sect. 5.1.6). The partnerless chromosome seems to interfere in these cases with normal chromosome pairing.

Down's Syndrome. With an incidence at birth of 1-2/1000, Down's syndrome is the most frequent chromosome aberration syndrome in humans, and one of the most common conditions encountered in most genetic counseling services. Figure 2.27 demonstrates how physical differences between the three main racial groups are overshadowed by the similarity due to this syndrome. Figure 2.28 represents schematically the most frequent clinical symptoms. The following observations on Down's syndrome are important:

a) The condition is a well-defined syndrome. In spite of appreciable variability of signs, the diagnosis is rarely in doubt for experienced clinicians.
b) Its frequency increases with the age of the mother.
c) Most cases are the only ones affected in an otherwise healthy family; in a small minority of families, more than one case in observed.
d) Monozygotic (MZ) twins are usually concordant, while the great majority of dizygotic twins are discordant. This rule, however, has exceptions: Discordant MZ pairs occasionally occur [545]. This is probably caused by chromosome loss in the cells forming one of the twins.
e) Males with Down's syndrome have no children. However, at least 17 women with this syndrome did reproduce. Among their 19 children, including one pair of MZ twins, seven had Down's syndrome, eight were normal, two were retarded without Down's syndrome, one child was sillborn without signs of Down's syndrome, and two stillborn MZ twins with normal karyotypes were counted as one individual [443]. All mothers and affected children for which chromosome examinations were done showed the standard karyotype 47,G+; one of the retarded children without Down's syndrome had a normal karyotype 46,XY.
f) Life expectancy of the patients is reduced [160]. In an Australian sample published in 1963 [327], 31.1% had died by the end of their 1st year, and 46% by the end of the 3rd. Life expectancy is reduced in later life as well. In another sample [465], 37 of 73 patients died from respiratory disease, not including tuberculosis, a frequency 123 times higher than that of the general population of the same age. Another five patients died from nonrespiratory infections. These data suggest a defect of the immune system. Congenital cardiac anomalies were also increased. With antibiotic therapy and heart surgery, the patients are now surviving much longer (Fig. 2.29). Their ultimate life expectancy is unlikely to be in the normal range, since there are suggestions that they age more quickly than normal.
g) Expressivity of phenotypic features is variable. Congenital heart disease, for example, is present in some

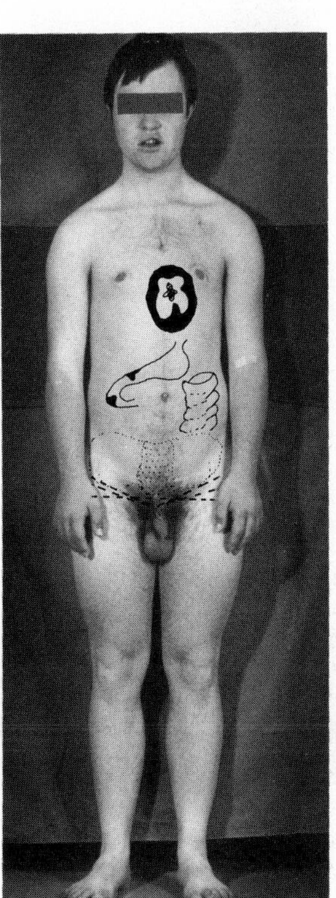

Fig. 2.28. The main clinical findings of Down's syndrome

but not all patients, and the same is true for many other clinical signs described above and in Fig. 2.28. This increased variability of phenotypic manifestation is characteristic for all chromosomal aberration syndromes in humans.

h) There is a 20-fold increased risk of dying from acute leukemia. The reasons are unknown. Three different hypotheses come to mind: a higher risk of aneuploidy due to mitotic disturbances in blood stem cells, a lower resistance against infection with a leukemogenic virus, a lower efficiency in the function of repair enzymes for which experimental evidence is available (Sect. 5.1.6).

Standard Karyotype in Down's Syndrome. The G chromosomes of a patient with Down's syndrome are seen in Fig. 2.30 (G and Q banding). The banding patterns of No. 21 and 22 can be distinguished easily, No. 21 being the stronger and broader fluorescent and having one to two dark G bands. Chromosome 22 has a dark G band at the proximal long arm and a fainter band more distally.

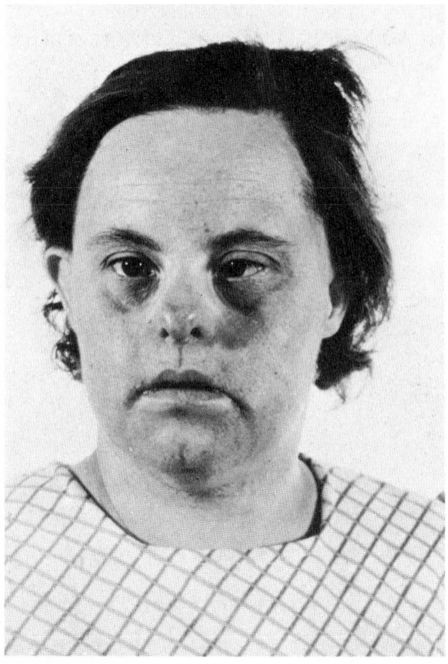

Fig. 2.29. Woman with Down's syndrome at the age of 38. Courtesy of Dr. T. M. Schroeder-Kurth

For a brief period after trisomy 21 was discovered, isolated cases allegedly without the additional No. 21 were described. However, it is now generally acknowledged that every patient with this syndrome has the supernumerary chromosome, either as free trisomy 21, or within a translocation chromosome formed from a chromosome 21 and another chromosome such as No. 21, 22, 13, 14, or 15. Observations on some rare cases with reciprocal translocations suggest that the distal part of the long arm of chromosome 21, and especially the band 21q22, is responsible for this phenotype [371]. For example, a girl with the tandem duplication of one chromosome 21 except for band 21q22 (Fig. 2.31) in addition to another free No. 21 was moderately mentally retarded but did not show most of the features of Down's syndrome Trisomy only of 21q22 led to mild manifestation of this syndrome [370].

Down's syndrome had been known as a clinical entity long before trisomy 21 was discovered. The other autosomal chromosome syndromes were buried in the great number of multiple malformations and could be singled out only after the abnormal chromosome complement had been discovered. In retrospect, some of these syndromes are so singular that they probably could have been delimited on purely clinical grounds.

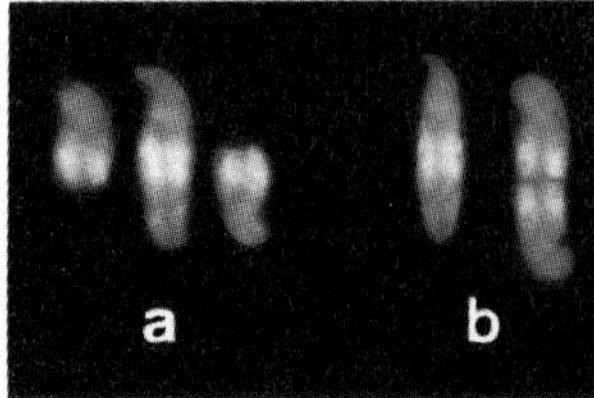

Fig. 2.31. a Tandem duplication of one chromosome 21 except for band 21q22 *(middle chromosome)* in a mildly mentally retarded child without most signs of Down's syndrome. (Hagemeijer and Smit, 1977 [370] **b** The abnormal chromosome *(left)* and, for comparison, two normal No. 21 placed closely opposite each other *(right)*. Here, in distinction to the duplication chromosome, the band 21q22 is visible twice

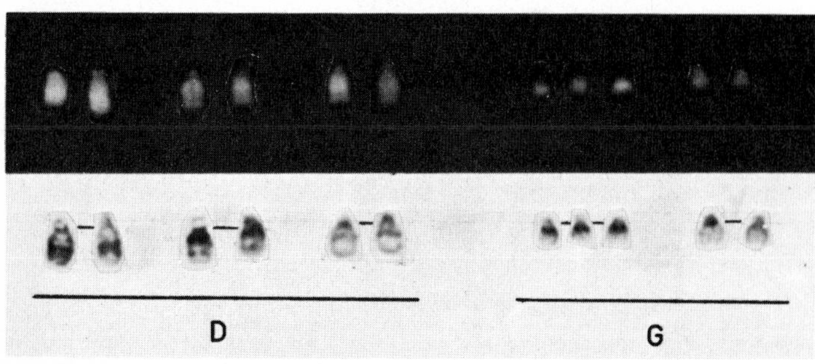

Fig. 2.30. D and G chromosomes of a patient with Down's syndrome. Q and G staining. Note the broad G band in the proximal region of 21p, which distinguishes No. 21 from No. 22. Courtesy of Dr. T. M. Schroeder-Kurth

Other Autosomal Trisomies. Patau et al. (1960) [472] described the first case of an autosomal trisomy other than trisomy 21. This discovery was the result of a deliberate search, which was guided by a hypothesis specified by the authors as follows:

> On genetic grounds it was not to be expected that the addition of an autosome to the normal complement would have a similarly restricted effect (as in X trisomy). Only one type of autosomal trisomic has been reported to date, and although the extra chromosome is one of the two smallest autosomes ..., its presence in triplicate results in mongolism.... It was to be expected that other autosomal trisomics, if they should be at all viable, would also display multiple congenital disturbances.

Their systematic search along these lines produced three cases in which trisomies were found: two involving trisomy 18 and one case with trisomy D. At the same time, trisomy 18 (first incorrectly labeled as trisomy 17) was also discovered by Edwards et al. [343]. Trisomy D was identified later as trisomy 13. The main signs and symptoms of both trisomies are seen in Figs. 2.32 and 2.33.

In the following years, all attempts to discover new autosomal trisomy syndromes among newborns failed, and these were assumed to be invariably lethal, especially as chromosome studies on spontaneous abortions revealed a variety of other trisomies. Discovery of three new syndromes, trisomies 8, 9 and 22, had to await development of the banding techniques [292; 396; 402]. Here, again, the children had severe and complex malformations. These three more recently discovered trisomy syndromes seem to be very rare [503; 195].

Triploidy. Apart from a doubtful mosaic [308] the first discovered cases of triploidy were two aborted fetuses [335; 475]. Later, examinations of spontaneous abortions confirmed that triploidy is not rare, but a small number of cases were also observed among children born alive [458]. In 1974, more or less detailed information was available from 275 cases of triploid abortuses with a gestation time of less than 20 weeks. Twenty-two fetuses survived past the age of 28 weeks; five died in utero, the others survived for only a few hours or days after birth. All children surviving for longer than a few days, eight in number up to 1974 were triploid-diploid mosaics.

The most characteristic feature among triploid abortions is the hydatidiform mole of the placenta. Some of the embryos showed local malformations such as myelomeningocele; most of them seemed normal.

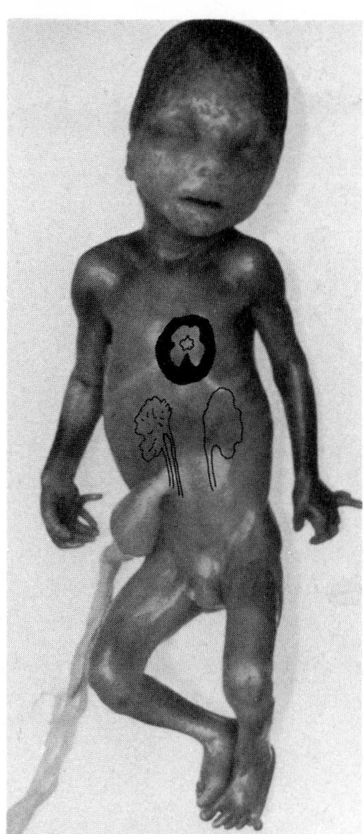

Fig. 2.32. Main clinical findings of trisomy 13

Growth failure
Mental retardation
Dolichocephaly with protruding occiput
Retroflexion of head
Arches on three or more finger tips
Absence of skin creases above distal joints
Simian crease
Short sternum
Horse shoe kidney
Abduction deformity of hips
Muscular hypertonus
Pes equinovarus
Prominent heel
Dorsal flexion of big toes

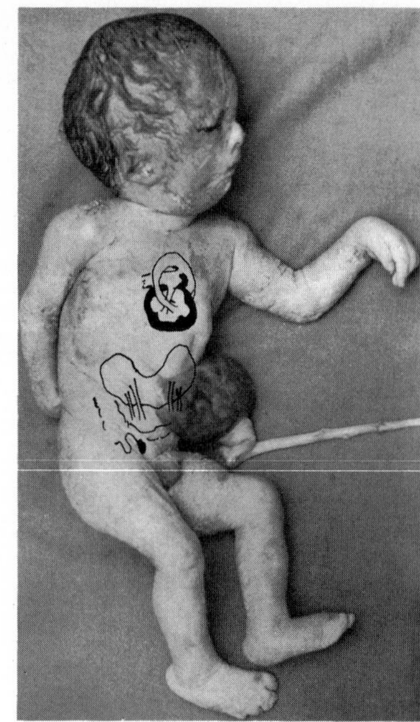

Open skull sutures and wide fontanelles at birth
Hypertelorism
High arched eyebrows
Low set and deformed ears
Micrognathia
Flexion deformities of fingers
Persistent ductus arteriosus
Ventricular septal defect
Meckel's diverticulum
Absence of labia majora
Prominent external genitalia
Hydramnios
Small placenta

Fig. 2.33. Main clinical findings of trisomy 18

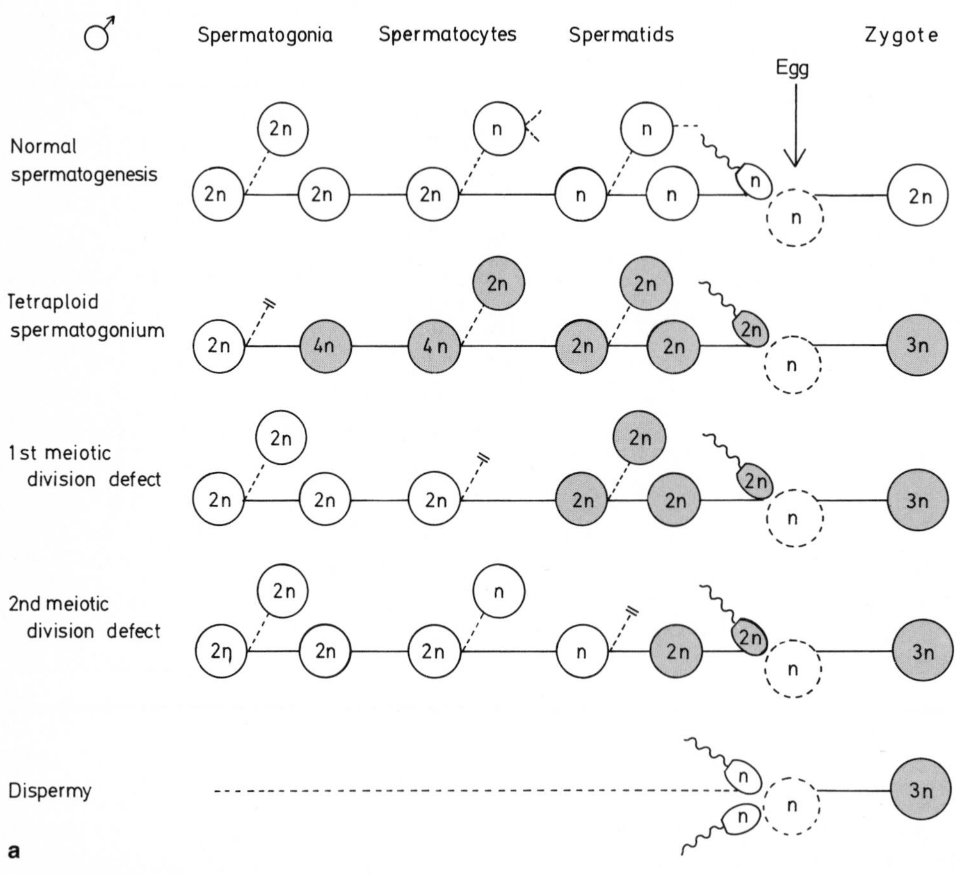

Fig. 2.34 a. Legend see p. 47

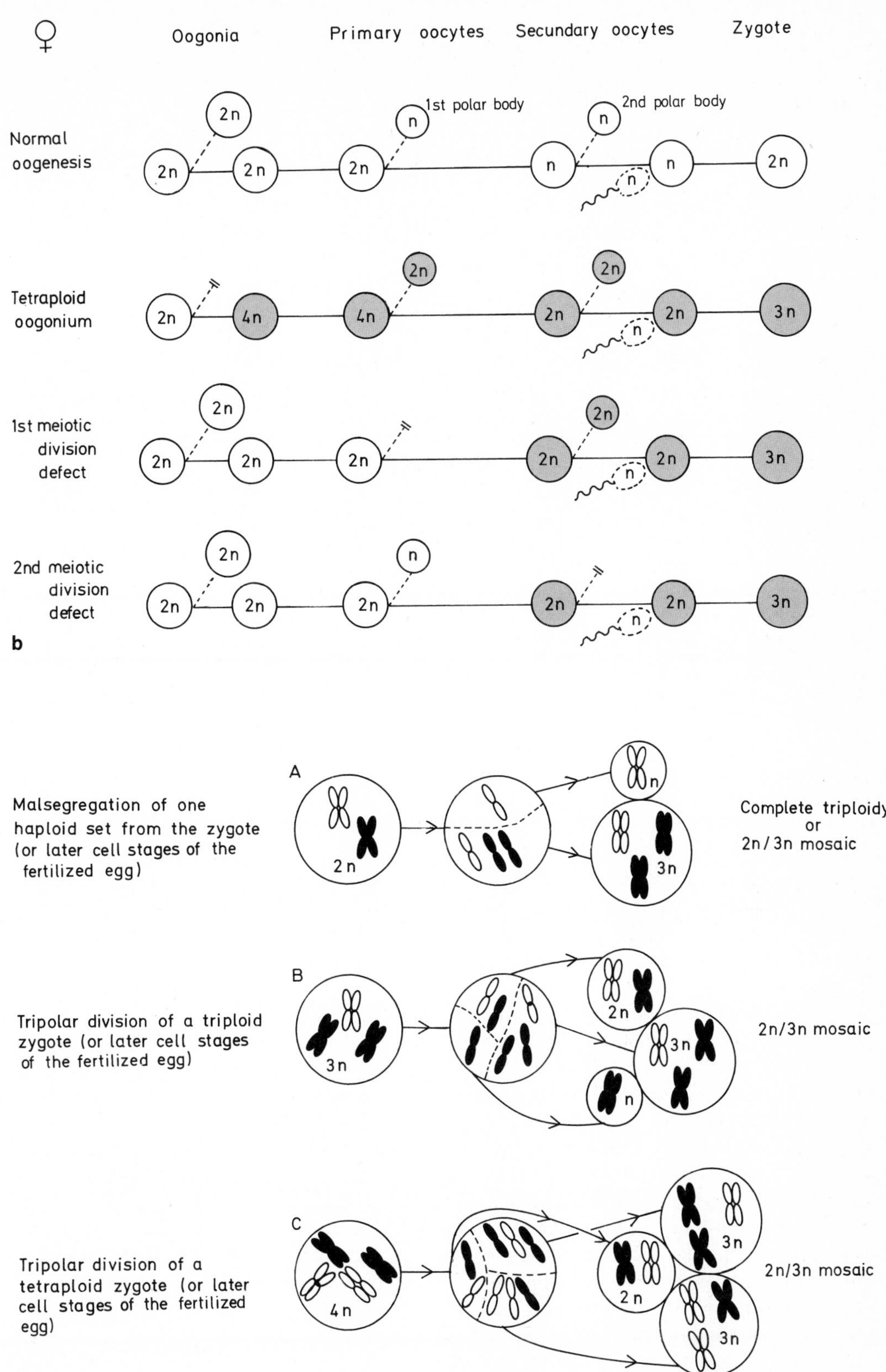

◁ **Fig. 2.34 a–c.** Anomalies in oogenesis, spermatogenesis or fertilization that may lead to triploidy. **a** In the male, triploidy may result from a tetraploid spermatogonium, from a disturbance in the first or second meiotic division (spermatocytes or spermatids), or from fertilization by two sperms. (Adapted from Niebuhr, Hum. Genet. 21, 1974) **b** In the female, the same mechanism may occur as in the male germ cells. **c** Abnormal division of the zygote or later cell stages of the fertilized egg giving rise to complete triploidy or mosaicism. (Niebuhr, 1974 [458])

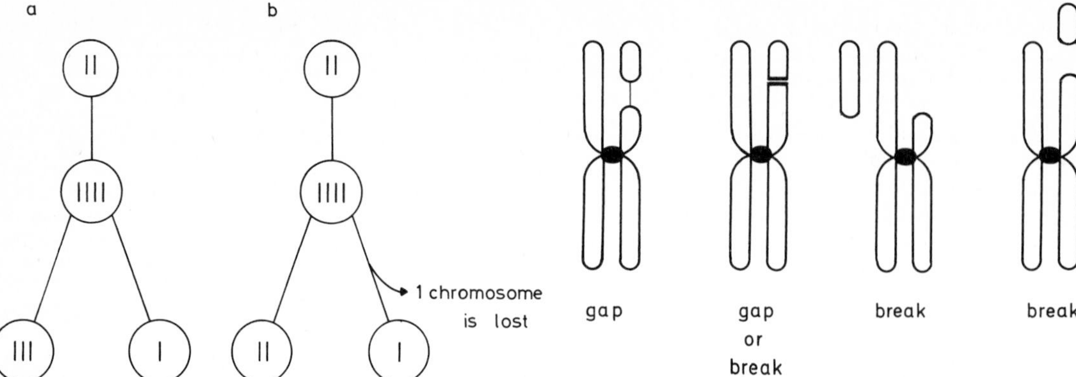

Fig. 2.35 a, b. Mitotic nondisjunction (**a**) and anaphase lagging (**b**). **a** After the homologous chromosomes have doubled to form four chromatids, three of them come into one division product in the next anaphase; the other division product only gets one chromosome. **b** One chromosome is lost during anaphase movement

Fig. 2.36. Definition of chromosome gaps and breaks. From *left to right,* in a gap, the separate segment is not dislocated; there might be a tiny connection. If this connection is lacking, it cannot be decided whether a gap or a break has occurred. The two *right-hand* figures clearly show breaks with different locations of the broken segment. Courtesy of Dr. T. M. Schroeder-Kurth

The triploids born alive have a low birth weight, large posterior fontanelles with underdevelopment of the occipital and parietal bones of the skull, and other nonspecific malformations also found in other autosomal aberrations. Those with 69,XXY karyotypes have grossly malformed genitals with a small penis and/or hypospadias, bifid scrotum, and nondescended testicles. Some of the mosaics have survived. Their clinical features are not very distinct, but the diagnosis may be suspected in mentally retarded children with abnormal placentas, syndactyly, abnormal genitals, and asymmetry.

Various errors in germ cell formation may lead to triploidy (Fig. 2.34). Some cause differences in the ratios of XXX, XXY, and XYY individuals among triploids. Evidence available so far indicates a high frequency of double fertilization or failure of the first meiotic division in the oocyte [1504; 414].

Mosaics. Mosaics are individuals with two or more genetically different cell populations. They are found relatively often in numerical chromosome aberrations of the sex chromosomes but also in autosomal aberrations. Chromosomal mosaics are sometimes called mixoploids. A mosaic may be formed either by mitotic nondisjunction or by loss of single chromosomes due to anaphase lagging (Fig. 2.35). Frequencies of these mitotic errors have been determined for Down's syndrome. The risk in trisomic zygotes was about 400 times higher than in euploid zygotes for anaphase lagging, and about 70 times higher for mitotic nondisjunction. These calculations were based on relative incidences of different types of mosaicism (Sect. 5.1.6) and on analysis of the maternal age effect.

Mosaics that result from meiotic nondisjunction with subsequent loss of the supernumerary chromosome due to anaphase lagging in one division product are expected to show the same increase with maternal age as do standard trisomies. Mosaics that result from mitotic nondisjunction, on the other hand, are not expected to show any increase with maternal age. Hence, the proportion of mosaics due to anaphase lagging may be estimated from comparison ot the maternal age effect between mosaic and standard trisomies. An exact estimate is made difficult by the fact that some mosaics escape diagnosis, as among the limited number of cells utilized for karyotyping aberrant cells present in a low percentage may easily escape observation. Moreover, cases with a small number of aberrant cells, and correspondingly, with little or no phenotypic abnormalities, will be detected only occasionally - mainly if there are trisomic cells in their germinal tissue and hence, cases with trisomies have occurred among their progeny. Still, among published mosaics, 17%-30% have been estimated to be due to mitotic nondisjunction. As expected, the maternal age was especially low in cases with 34% or fewer trisomic cells among those examined [483]. The overall frequency of mosaics among all cases with clinical symptoms of Down's syndrome is around 2%.

Statistical Problem in the Detection of Mosaics. What is the incidence of chromosome mosaics in the population compared with other chromosome aberrations? This question poses a statistical problem, the chance of detecting a mosaic depending on the percentage of aberrant cells in the tissue to be examined and the number of cells examined in a survey. In most populations surveys (Sect. 5.1.2), very few (three to five) cells per individual are usually examined,

and the proportion of mosaics is underestimated systematically. The problem is made still more difficult by artifacts, i. e., loss or - in some cases - gain of single chromosomes during preparation. On the other hand, it is not feasible, owing to lack of time and manpower, to examine several hundred cells per individual. Bochkov et al. (1974) [306] have proposed a workable compromise.
At the outset, a limit has to be set to the percent of cells with a deviating cell clone that one wishes to detect. Then, the number of cells needed for detecting at least one abnormal cell with a probability of 95% is caculated from the normal distribution. The probability that no abnormal cell will be present among n analyzed cells is $(1-p)^n$, where p is the given proportion of abnormal cells. It is reasonable to choose 25% as the lower limit for the diagnosis of mosaicism, since individuals with less than 25% abnormal cells will usually show few clinical abnormalities. Now, let $P_{1,n}$ be the probability of detecting at least one abnormal cell in a sample of n cells; $p=0.25$.
Then:

$$P_{1,10}=1-(1-0.25)^{10}=0.944$$

$$P_{1,11}=1-(1-0.25)^{11}=0.958$$

Therefore, the number of cells to be examined per individual should be 11. If no aneuploid cell is found, the case can be diagnosed as not being mosaic, i. e., having not more than 25% abnormal cells. If more than one abnormal cell with the same abnormality has been detected, the diagnosis of mosaicism is confirmed. With one abnormal cell, either the case may be a mosaic or the abnormal cell may be a preparatory artifact. Hence, the sample size has to be enhanced to the extent that at least two abnormal cells will be detected with $P_{2,n}=0.95$, $P_{2,17}=0.951$. In the second step, six additional cells should be examined, bringing the number of observed cells to 17. If no second abnormal cell of the same type is found, the first one may be regarded as an artifact. If a second cell is found, a third one should turn up with examination of 23 cells. An improved method with tables has been given by Hook, Amer. J. Hum. Genet. 29, 94-97 (1977).
In clinical diagnosis, a higher number of cells will frequently be examined, as a mosaic should be excluded with more confidence, or it may be desirable in some cases to detect mosaics having a lower proportion of abnormal cells.

2.2.2 *Syndromes Due to Structural Anomalies of Autosomes*

2.2.2.1 *Karyotypes and Clinical Syndromes*

First Observations of Down's Syndrome. Once trisomy 21 had been identified as the cause of Down's syndrome, the obvious next question was whether this trisomy was present in all cases. If not, the karyotypes of the exceptions would be of great interest. As the risk for meiotic nondisjunction was known to increase with the age of the mother, and as a single event could lead to only one affected offspring at a time, exceptions had to be searched for primarily among affected children of younger mothers and in exceptional families with more than one affected patient.

Polani et al. (1960) [479] examined three such schildren with Down's syndrome. In one girl, the first child of a 21-year-old mother and 23-year-old father, they found 46 chromosomes. There were four normal chromosomes of the G group. One chromosome of the D group had an elongated short arm. The authors suspected that one chromosome 21 was translocated to the short arm of the D chromosome. A short time afterward, this suspicion was confirmed in familial cases. Two healthy mothers or the three patients and the common grandmother had only 45 chromosomes and only three free G chromosomes. However, one of the D chromosomes - the investigators suspected it would be No. 15 - had an elongated short arm. If this arm included the missing chromosome 21, the karyotype of these women was "balanced" and the genetic material was completely present. In some of their offspring, on the other hand, the translocation chromosome included most of the chromosome 21 material occurring in their cells together with two normal chromosomes No. 21; these children were effectively trisomic for 21 and developed Down's syndrome in spite of their normal number of chromosomes. This karyotype was unbalanced.
At about the same time, the first G/G translocation was reported [354]. Shortly afterward, studies of the meiotic division I in a balanced heterozygote revealed a trivalent, i. e., a figure consisting of three chromosomes, giving conclusive evidence that the unusual chromosomes found in these families are, indeed, translocation chromosomes [373].

Frequency of Translocation Down's Syndrome. Translocation Down's syndrome explains a number of familial cases, but not all. Standard trisomies 21 may also occur repeatedly in the same families, pointing to constitutional factors or a mosaic status in the parents (Sect. 5.1.2). Table 2.3 shows the frequencies of translocation cases, inherited and noninherited, among affected children of younger as compared with older mothers. Most translocations are of the types described above, i. e., D/G or G/G.
There are, however, a few reciprocal translocations in which, apart from chromosome 21, other, nonacrocentric chromosomes are involved. Before details of the different structural aberrations are discussed, some remarks on the formation of structural chromosome aberrations are worthwhile.

Gaps and Breaks. For chromosomes to form any kind of rearrangements, they first have to be broken. Chromosome breakage involves breakage of the DNA sugar phosphate backbones, if the DNA is a single continuous thread through the chromosome. Under the light microscope, it may be difficult to distinguish chromosome breaks from achromatic regions called gaps. These gaps may be true breaks, but they may also be local despiralizations. Chromosome breakage is frequently analyzed in mutation research. Therefore, there has to be some agreement as to which aberrations are to be scored as breaks and which as gaps. One of the proposed agreements is illustrated in Fig. 2.36. The listed distinctions are very conservative and probably underestimate the number of breaks. Breaks and gaps may occur during interphase before and after replication. If occurring before replication, the lesion will be visible in the following metaphase in both daughter chromatids (isochromatid break). After replication, only one chromatid will be affected (chromatid break). These various types of breaks and gaps are explained in Fig. 2.37.

Fate of Broken Chromosomes. A break not affecting the centromere produces a shorter chromosome with a centromere and an acentric fragment. This fragment may or may not form a small ring but, lacking a centromere, runs a high risk of being lost during the subsequent mitosis. Hence, chromosome breakage will often leave behind a cell deficient in a chromosome segment. In some cases, however, chromosomes broken at two breaking points will rejoin under the influence of repair enzymes. The mechanisms involved are now being investigated [456]. If the broken ends happen to rejoin with each other, the chromosomes - and with them the cell - will be intact again. In fact, experience with repair-deficient human diseases (Sect. 5.1.6) suggests that this may happen over and over again in many human tissues. In other cases, broken ends will join with break points from other chromosomes, homologous or nonhomologous, one condition being that the two breaks have occurred within a reasonably short time period and reasonably close to each other. This leads to the various types of chromosome rearrangements.

Intrachromosomal Rearrangements (Intrachanges). A single chromosome may break at two different sites, and the intermediate part may rejoin upside down. This rearrangement does not lead to disturbances in mitosis, especially if the breaks have oc-

Table 2.3. Incidence of translocations among children with Down's syndrome

Maternal age under 30				Maternal age over 30			
Total No. of patients	No. of translocations			Total No. of patients	No. of translocations		
	Sporadic	Inherited	Parents not examined		Sporadic	Inherited	Parents not examined
1,431	69	32	14	1,058	7	5	4
Sum total	115 = 8.04%			Sum total	16 = 1.51%		

Data from Mikkelsen, 1971 [443].

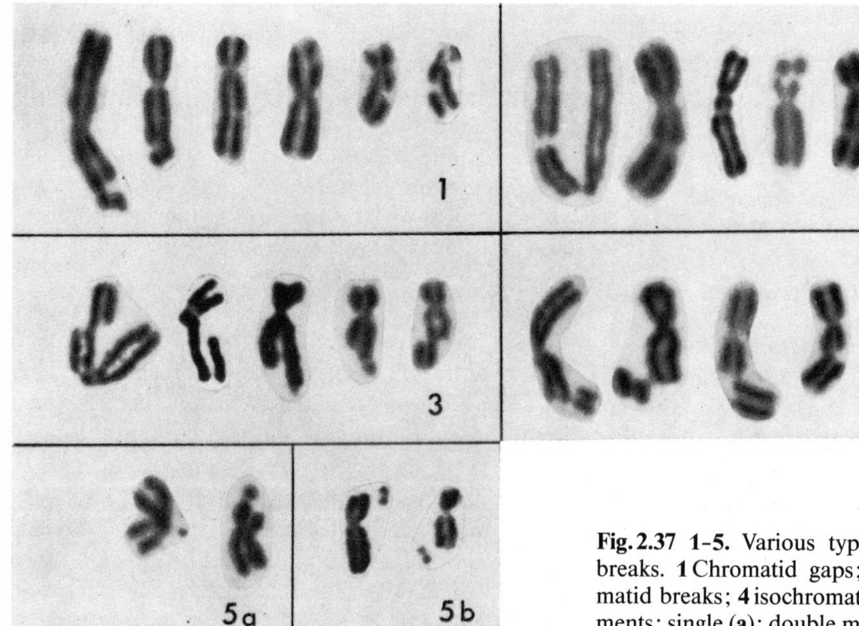

Fig. 2.37 1-5. Various types of chromosome gaps and breaks. **1** Chromatid gaps; **2** isochromatid gaps; **3** chromatid breaks; **4** isochromatid breaks; **5** minutes and fragments: single (**a**); double minute (**b**). (Gebhart [212])

curred in the G_1 phase. It can be diagnosed by use of banding methods, if the centromere is not included (paracentric inversion), and, in addition, by the location of the centromere, if it is included (pericentric inversion). Inversion heterozygotes are not particularly rare in human populations (Fig. 2.38). There may be difficulties in chromosome pairing at meiosis, which may lead to partial elimination of certain types of germ cells in inversion heterozygotes (Fig. 2.39). These difficulties do not occur in homozygotes. Inversions - and especially pericentric inversions - have played a major role in the phylogeny of higher primates (Sect. 7.2.1).

Another type of intrachange is the ring chromosome (Fig. 2.40). Here, two telomeres are usually lost as fragments and the open ends rejoin. A ring chromosome may or may not be able to go through mitosis, depending on whether the two chromatids happened to join in a crossed manner. If there is no sister strand exchange between the breakage points during DNA replication, the ring may replicate, forming two separate rings with one centromere each. They can pass the next mitosis without difficulty. One sister strand exchange will form one big ring with two centromeres. This structure normally will be destroyed in the next mitosis. Two sister strand exchanges may form two rings that are entwined with each other. Various possibilities are detailed in Fig. 2.40, some of which will be demonstrated with one case report (p. 59). Occasionally, chromatid breakage and ring formation occurs in the G_2 phase and patterns such as that shown in Fig. 2.41 are observed in single cells.

Interchromosomal Rearrangements (Interchanges). In many cases, joining will occur between different

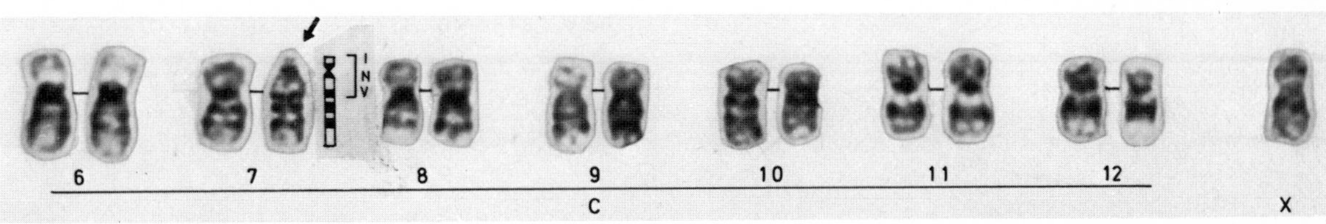

Fig. 2.38. A pericentric inversion of chromosome 7 in a healthy male, G banding. The inverted segment is shown diagrammatically *(INV)*. The other chromosomes are shown for comparison. Courtesy of Dr. T. M. Schroeder-Kurth

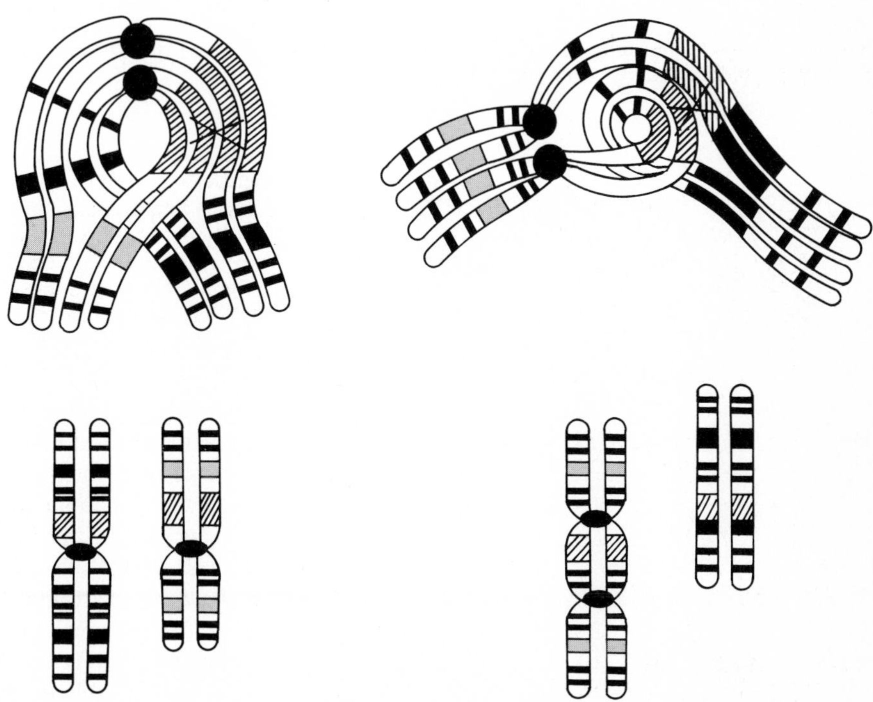

Fig. 2.39. Chromosome mispairing during meiosis in heterozygotes of pericentric *(left)* and paracentric *(right)* inversions. In the two figures, crossing over is assumed in the segments that are marked by an *X*. As a consequence, abnormal chromosomes are found that lead to aneuploidy of zygotes in the next generation

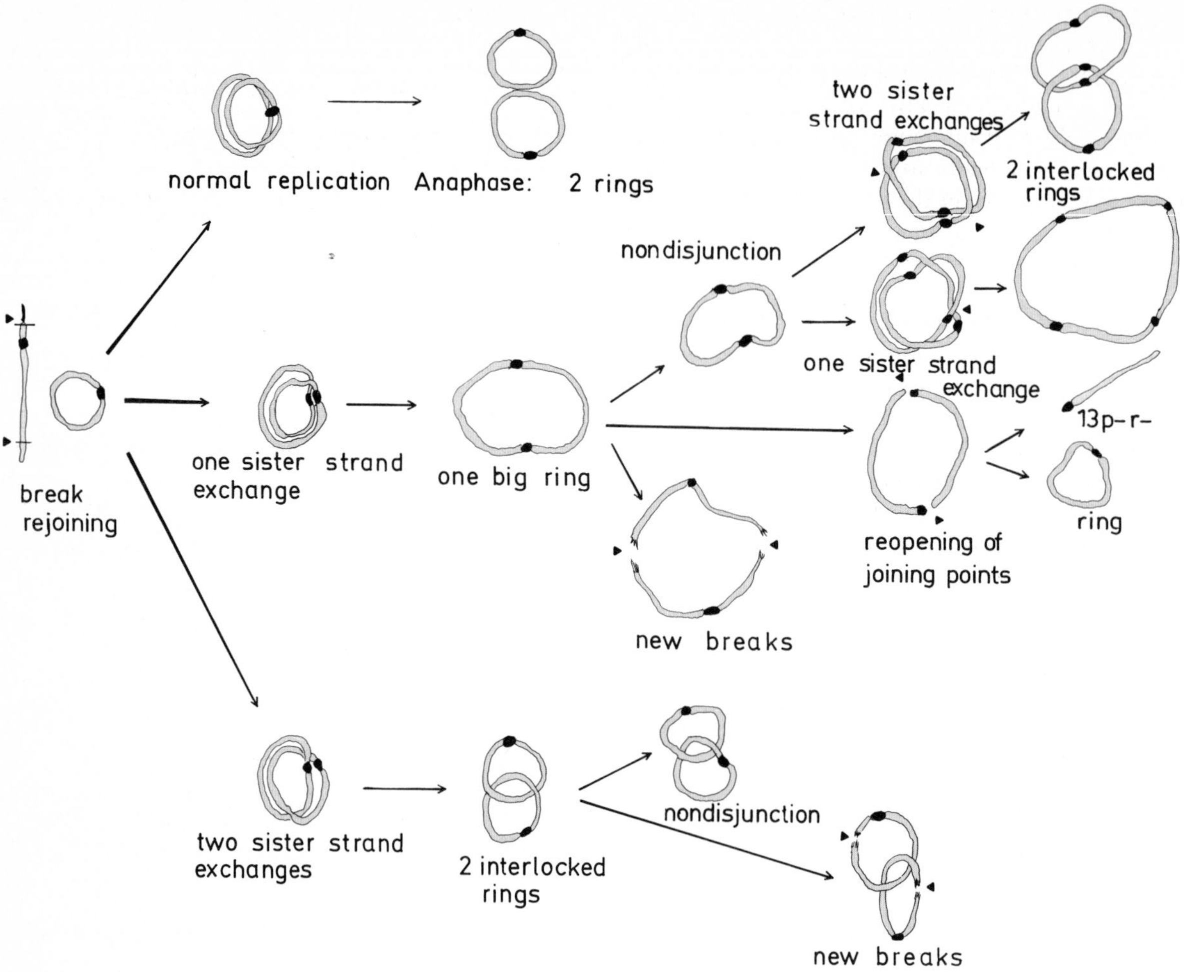

Fig. 2.40. Ring chromosome formation in the G_1 phase and fate of the ring chromosome in the following mitotic divisions. See text for details (*r-*, no ring)

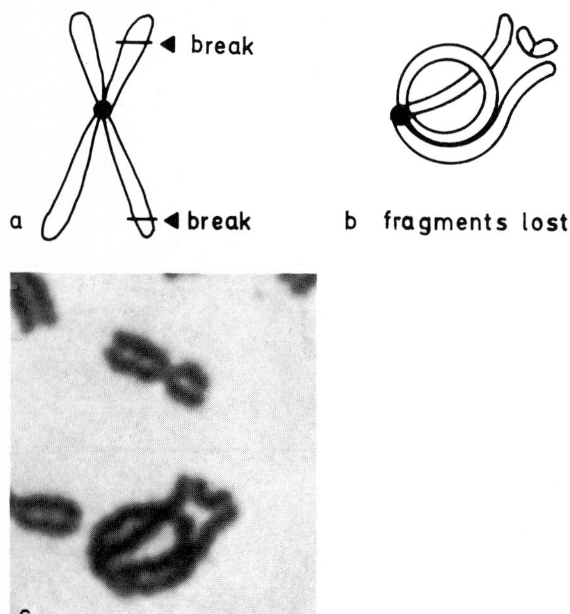

◁ **Fig. 2.41 a–c.** Formation of a ring chromosome in the G_2 phase. **a** Two breaks in one of two sister chromatids. **b** Rejoining of the broken ends; pairing of fragments with the homologous chromatid segments. **c** The same ring chromosome in a human metaphase. Courtesy of Dr. T. M. Schroeder-Kurth

chromosomes, homologous or nonhomologous. If breakage occurs in the G_1 phase, joining follows in the G_1 (or early S) phase before DNA replication. If each of the resulting chromosomes happens to have one centromere, the translocation chromosomes may pass through the next mitosis without difficulties. If one of the resulting chromosomes happens to get two centromeres, a dicentric chromosome will be formed. Depending on the exact mode of replication, it may be able to pass the next mitosis, under the following conditions: (1) the centromeres migrate to the same pole and (2) replication and sister chromatid exchange between the two centromeres has not led to interwining of the two chromatids (Fig. 2.42). If breakage and rejoining occur after DNA replication, only one sister chromatid of each chromosome will be affected. The rejoined sister chromatids are still paired with their unaffected partners. This leads to the interchange patterns shown in Fig. 2.43 in the first mitotic division after reunion. The mitotic anaphase will proceed without further difficulties if the two centromeres happen to be located on different elements (Fig. 2.43; classes I, III, and V). If the centromeres are located on the same element, the resulting daughter cell will in any case be aneuploid. Either the centromeres migrate to different poles in which case an "anaphase bridge" will be formed and the chromosome will finally break again, or the two centromeres migrate to the same pole, which can only happen with nonhomologous reunions (Fig. 2.43, classes V, VI and VII). In the latter case, the problem is postponed to the next mitosis, in which the chromosome appears as dicentric. It may or may not survive this mitosis. In any

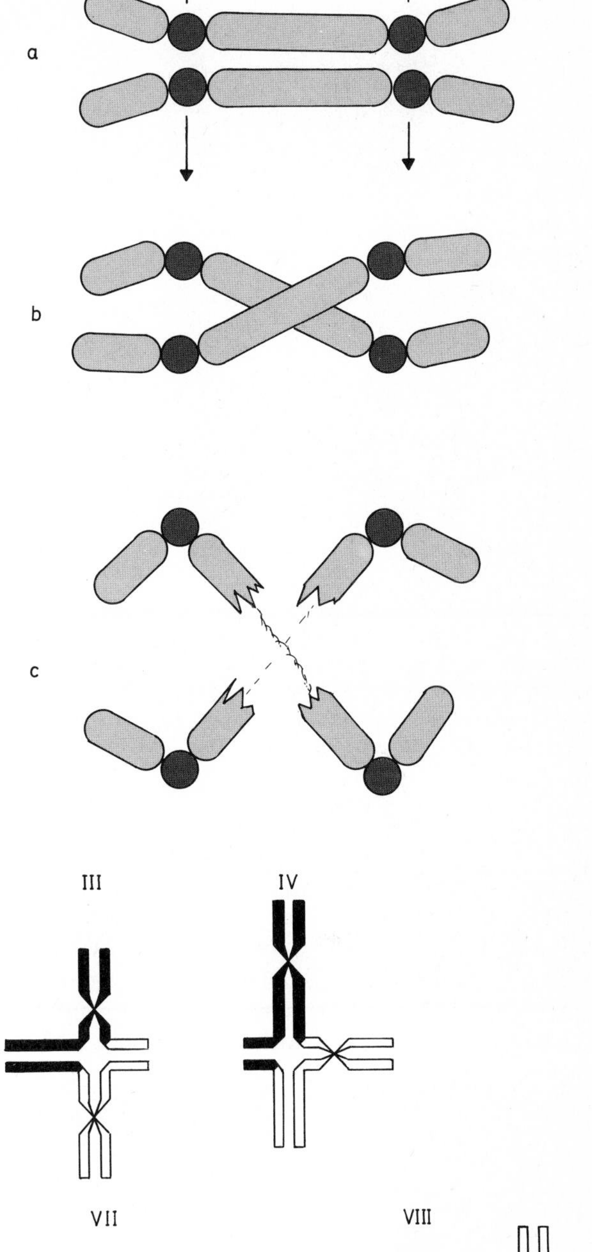

Fig. 2.42 a–c. Mitotic anaphase of a dicentric chromosome. **a** Both centromeres migrate to the same pole; chromosomes remain intact. **b** The centromeres migrate to opposite poles. Anaphase bridges are formed. **c** The chromosomes are broken

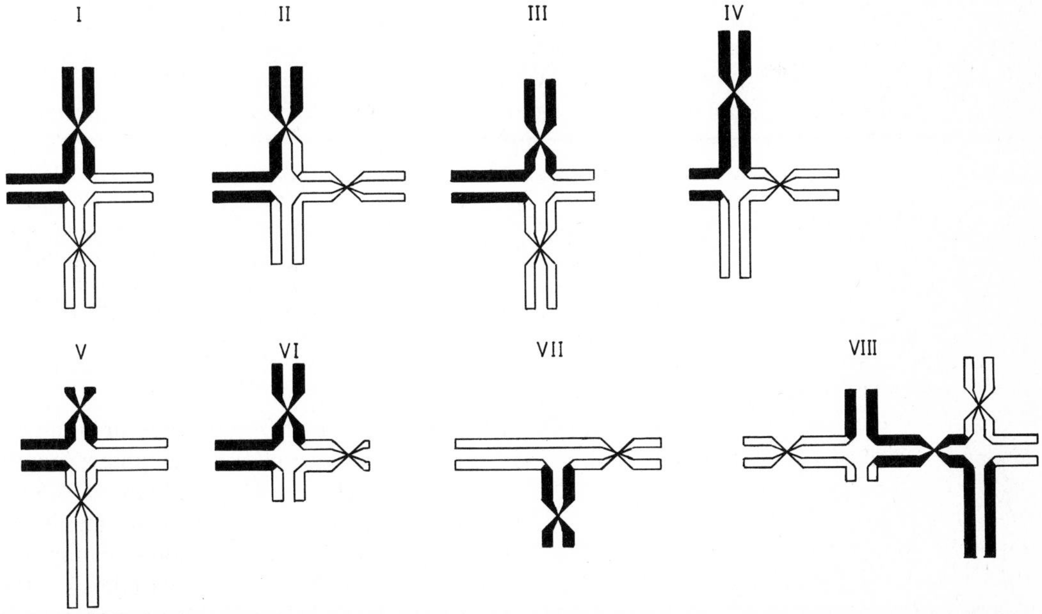

I II III IV V VI VII VIII

Fig. 2.43 I–VIII. Classes of interchanges found after translocation during the G_2 phase. Involvement of two homologous chromosomes. **I** Alternate position of the centromeres; exchange of fragments of equal length. **II** Adjacent position of the centromeres; exchange of fragments of equal length. **III** Alternate position of the centromeres; exchange of fragments of different length. **IV** Adjacent position of the centromeres; exchange of fragments of unequal length. Involvement of two nonhomologous chromosomes. **V** Alternate position of the centromeres. **VI** Adjacent positions of the centromeres. **VII** Triradial configuration (loss of fragments required). Complex of nonhomologous chromosomes (more than two). **VIII** One example of a figure with three chromosomes involved

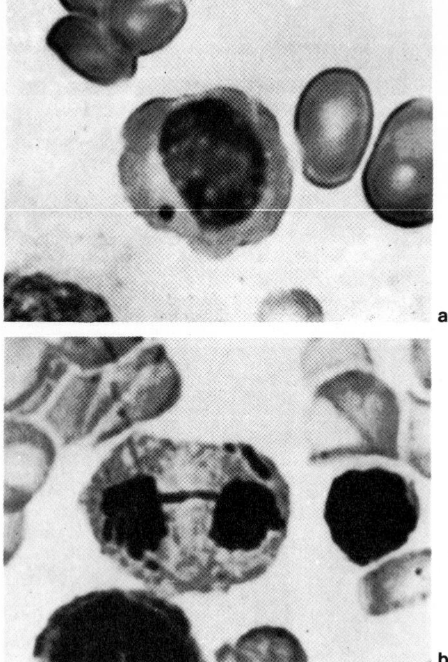

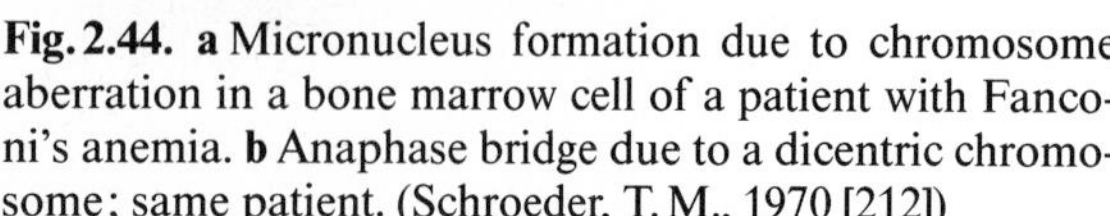

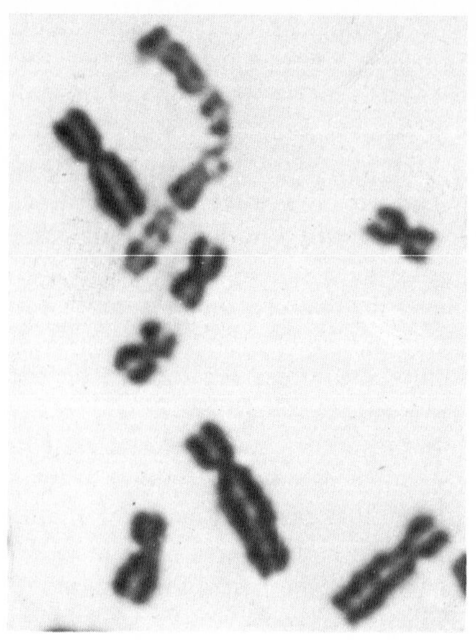

Fig. 2.44. a Micronucleus formation due to chromosome aberration in a bone marrow cell of a patient with Fanconi's anemia. **b** Anaphase bridge due to a dicentric chromosome; same patient. (Schroeder, T. M., 1970 [212])

Fig. 2.45. Premature chromosome condensation: Prophaselike chromosomes from a micronucleus together with several other, normal metaphase chromosomes. Courtesy of Dr. T. M. Schroeder-Kurth

case, under the conditions mentioned above, interchanges cause a great deal of cell loss due to aneuploidy or mitotic disturbance.

In human somatic tissue, many of these mitotic disturbances are visible even in conventionally stained cells, without special chromosome preparations. Figure 2.44 a and b shows an anaphase bridge and a so-called micronucleus in human bone marrow cells. Micronuclei are formed from chromosomes or chromosome fragments not taking part in normal mitosis at the same time. This leads to the phenomenon of premature chromosome condensation. The main nucleus is observed at metaphase with normally contracted chromatids, while the chromosomes of the micronucleus show a prophase-like condensation (Fig. 2.45). These cytologic anomalies have become important for quick evaluation of mutagenic agents (Sect. 5.2). Premature chromosome condensation can also be elicited in vitro by fusion of a cell in interphase with another cell that is just preparing for mitosis [500]. This method may be used, for example, for examining problems of chromosome structure in the interphase nucleus.

In meiosis, translocations may, again, lead to disturbances, as homologous chromosome segments tend to pair with each other. In metaphase I, they form so-called chains-of-three, if three chromosomes are involved, for example, in a balanced D/G or G/G translocation carrier. Figure 2.46 shows a chain-of-three in a man with a balanced D/D translocation. Figure 2.46 a and b shows the chromosome at pachytene with crossing over; Fig. 2.46 c shows the diakinesis as expected if each of the two free D chromosomes had one crossing over with the translocation chromosome, and for comparison the chain-of-three that was actually observed. If four chromosomes are involved, a chain-of-four may be formed. This event may or may not lead to further aneuploidy, depending on the anaphase movements of the four centromeres. If the two centromeres of one element happen to move to one pole - and if the chromatids happen not to be intertwined between the centromeres - normal anaphase may take place. Very often, however, additional chromosome breakage will occur. This is one of the reasons why meiosis is such a good filter for chromosome rearrangements.

Chromosome breakage together with its consequences may be observed in somatic cells and in germ cells. Breakage in somatic cells has become important for mutation research and will be discussed in Sect. 5.2. Breakage in germ cells may or may not be transmissible to the next generation. If transmitted, it often causes zygote death during the embryonic stage. In a certain number of cases, however, the aberration is compatible with postnatal life, leading to a chromosome aberration syndrome. Before some

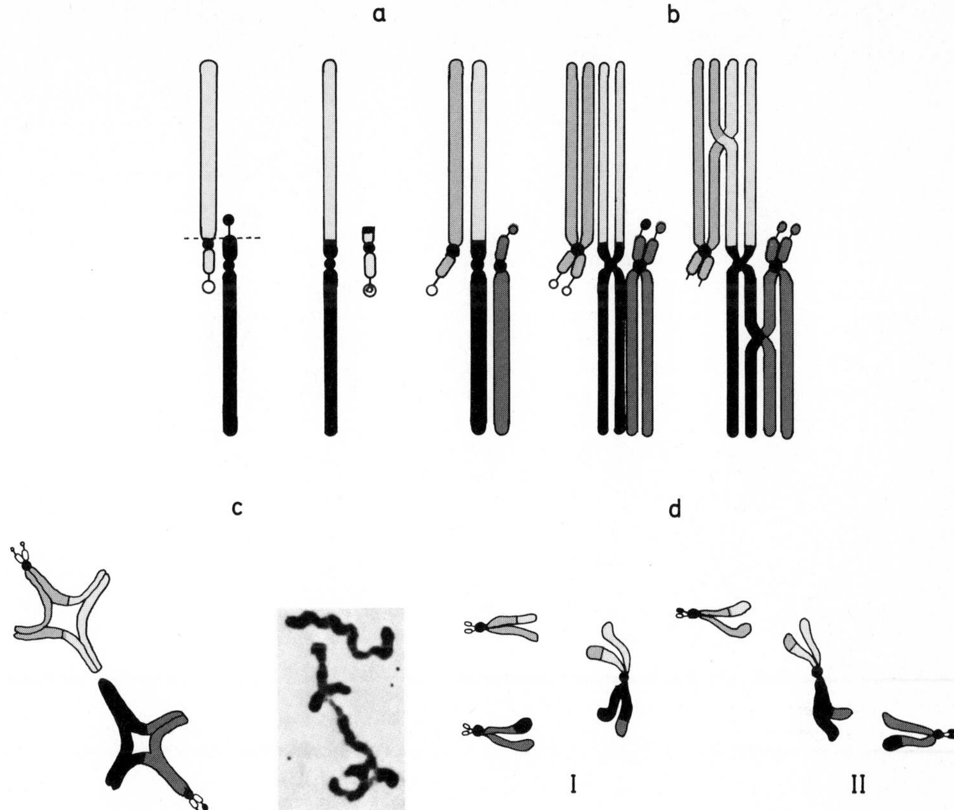

Fig. 2.46. **a** Possible origin and expected pairing of the D/D translocation chromosome with the two homologous chromosomes during meiotic prophase. **b** The translocation chromosome and its two homologous partners after replication without *(left)* and with *(right)* crossing over. **c** From the drawing (**b**), the expected shape of the translocation pattern, a trivalent, during diakinesis and metaphase I can be deduced *(left)* and compared with the actually observed pattern *(right)*. **d** The chromosomes involved in the trivalent are likely to be distributed to the daughter cells in two main classes giving rise to different types of balanced *(I)* and unbalanced gametes *(II)*. (Kjessler, 1966 [405])

of these syndromes are described, the universally accepted nomenclature of human karyotype description should be explained. This nomenclature was devised by a group of cytogeneticists and was last brought up to date in Paris in 1971 [468].

Description of Human Karyotypes. First, the total number of chromosomes is given, followed by the sex chromosome complement. Then, it is indicated which chromosomes are too many, or too few, or structurally altered. Some examples:

Table 2.4

46,XX	Normal female karyotype
46,XY	Normal male karyotype
47,XY,+G	Male karyotype with 47 chromosomes; one G chromosome too many
47,XY,+21	As above; the additional chromosome has been identified als No. 21
46,XY,1q +	Male karyotype with 46 chromosomes; the long arm (q) of one chromosome 1 is longer than normal
47,XY,+14p+	Male karyotype with 47 chromosomes, including an additional chromosome No. 14, which has en elongated short arm (p)[5]

[5] Petit (p) = small or short.

Table 2.4 *(continued)*

45,XX,-D,-G,+t(DqGq)	Female karyotype with a balanced Robertsonian translocation between a D and G group chromosome
46,XY,-5,-12,t(5p12p, t(5q12q)	Male karyotype with two translocations involving interchange of both whole arms of chromosomes 5 and 12. The breaks have occurred at or very near the centromere, and no information is available as to which centromere is included in either product

Changes in length of secondary constrictions, or negatively staining regions, should be distinguished from increases or decreases of arm length owing to other structural alterations by placing the symbol h between the symbol of the arm and the + or − sign. For example:

46,XY,16qh+	Male karyotype with 46 chromosomes, showing an increase in length of the secondary constriction on the long arm of chromosome 16

All symbols for rearrangements are to be placed before the designation of the chromosome or chromosomes involved, and the rearrangement chromosome or chromosomes always should be placed in parentheses:

46,XX,r(18)	Female karyotype with 46 chromosomes, including a ring (r) chromosome 18
46,X,i(Xq)	Female karyotype with 46 chromosomes, including one normal X and one isochromosome (i) for the long arm of X

The intensity of Q banding patterns is to be described as follows:

Negative	No or almost no fluorescence
Pale	As on distal 1p
Medium	As the two broad bands on 9q
Intense	As the distal half of 13q
Brilliant	As on distal Yq

Proposed Chromosome Band Nomenclature. Each chromosome is considered to consist of a continuous series of bands, with no unbanded areas. The bands are allocated to various regions along the chromosome arms and delimited by specific chromosome landmarks. The bands and the regions they belong to are identified by numbers, with the centromere serving as the point of reference for the numbering scheme. In designating a particular band, four items are required: the chromosome number, the arm symbol, the region number, and the band number within that region. For example, 1p33 indicates chromosome No. 1, short arm, region 3, band 3. The region and band numbers may be taken from Fig. 2.12, Table 2.5 gives the recommended abbreviations. Some examples will illustrate the principle of description:

Isochromosomes (always the abbreviated and the exact description):

46,X,i(Xq) 46,X,i(X)(qter →cen→pter)	Break points are at or close to the centromere and cannot be specified. The designation indicates that both entire long arms of the X chromosome are present and separated by the centromere

Terminal deletion:

46,XX,del(1)(q21) 46,XX,del(1)(pter→q21)	This indicates a break at band 1q21 and deletion of the long-arm segment distal to it. The remaining chromosome consists of the entire short arm and the part of the long arm lying between the centromere and band 1q21

Reciprocal translocations:

46,XY,t(2;5)(q21;q31) 46,XY,t(2;5)(2pter→2q21::5q31→4qter;5pter→5q31::2q21→2qter)	Breakage and reunion have occurred at bands 2q21 and 5q31 in the long arms of chromosomes 2 and 5 respectively. The segments distal to these bands have been exchanged between the two chromosomes. Note that the derivative chromosome with the lowest number (i.e., No. 2) is designated first

These examples should aid in understanding the symbols used in the following text and in cytogenetic publications. High-resolution banding requires logical extension of this nomenclature (see Fig. 2.16).

Table 2.5. Nomenclature symbols additional to those recommended by the Chicago Conference (1966) (Paris Conference, 1971 [468])

del	Deletion
der	Derivative chromosome
dup	Duplication
ins	Insertion
inv ins	Inverted insertion
rcp	Reciprocal translocation[a]
rec	Recombinant chromosome[a]
rob	Robertsonian translocation[a] ("centric fusion")
tan	Tandem translocation[a]
ter	Terminal or end (pter, end of short arm; qter, end of long arm)
:	Break (no reunion, as in a terminal deletion)
::	Break and join
→	From - to

[a] Optional, where it is desired to be more precise than provided by the use of *t* (translocation) as recommended by the Chicago Conference.

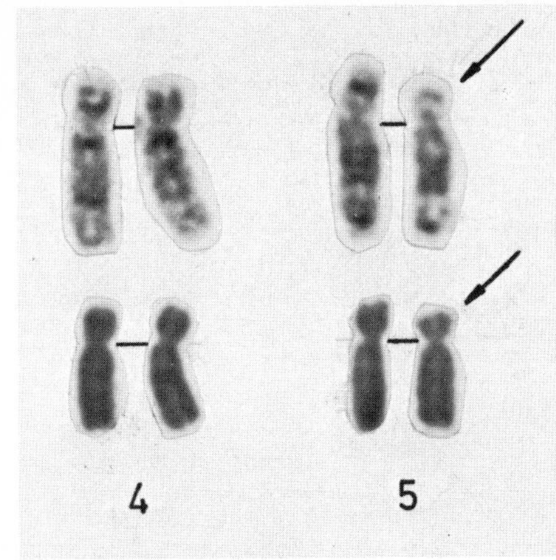

Fig. 2.47 Partial karyotype of a case with cri du chat syndrome - deletion 5p

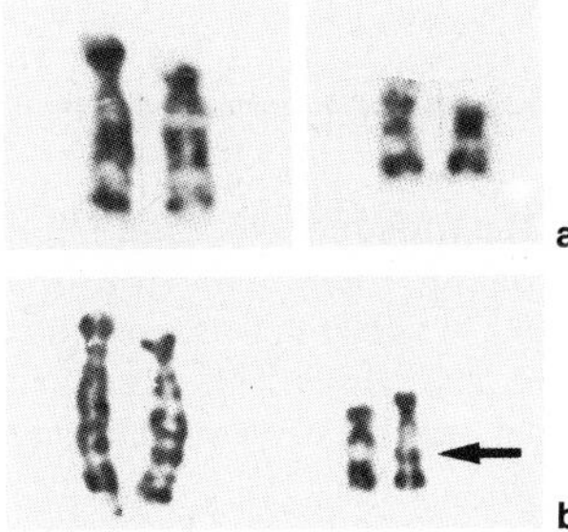

Fig. 2.48. a Partial karyotype of a patient with cri du chat syndrome and deletion of the terminal segment of chromosome 5. **b** Balanced translocation in a mother of this case. The terminal segment of chromosome 5 is inserted in chromosome 17. (Berger et al., 1974 [300])

Deletion Syndromes. An individual who is heterozygois for a deletion is monosomic for a part of the chromosome. De Grouchy et al. (1963) [367] were apparently the first to publish a case with del 18p −. The first deletion syndrome was established by Lejeune et al., also in 1963 [418]. They described three children with a deletion of the short arm of chromosome 5 (del 5 p−). In addition to the usual signs of autosomal chromosome aberration such as general retardation and low birth weight, the children showed a moon-like face with hypertelorism. Their appearance was not extraordinarily peculiar but they had a striking cry resembling that of a cat (cri du chat = cat cry) (Fig. 2.47).

There are various mechanisms by which a deletion may be formed: (1) true terminal deletion, (2) interstitial deletion, and (3) translocation. A number of reports have pointed to a translocation in the cri du chat syndrome. Figure 2.48 shows one of them - the partial karyotype of the proband with a chromosome 5p−. The deleted segment corresponds to the band 5p15 and part of 5p16. The phenotypically normal mother showed the same chromosome 5 but one chromosome 17 had an extra band on the long arm between 17q12 and 17q21. Hence, this band is an insertion of the terminal segment of a No. 5 into the long arm of No. 17. The case in Fig. 2.47, on the other hand, was proved by G banding to be a true terminal deleetion.

Intrachanges: Paracentric and Pericentric Inversions [401 a]. Paracentric inversions (i.e., those not including the centromere) in humans are difficult to diagnose. They will be discussed in the context of chromosome evolution (Sect. 7.2.1). Since the early 1960s, numerous reports on presumed pericentric inversions (i.e., with inclusion of the centromere) have been published. Some of the probands had various anomalies like mental retardation or malformation. Others were themselves phenotypically normal, but they or their wives had repeated abortions. Again, others did not show any abnormality at all. When only the conventional staining methods were available, pericentric inversions were thought to be rare.

After banding methods had been introduced, higher frequencies were reported in some populations. Chromosome 9 seemed to be affected frequently, and a relatively high prevalence was reported from Finland [323]. The material in this study consisted of karyotypes of 631 individuals who had been studied for various diagnostic purposes. In nine of them, pericentric inversions were detected, six of which involved No. 9. All these six were obviously identical. In the remaining three probands, No. 10 was involved: one inversion was present in two, the other in only one individual. The No. 9 inversion (Fig. 2.49) may readily be identified by the nonspecialist, as the secondary constric-

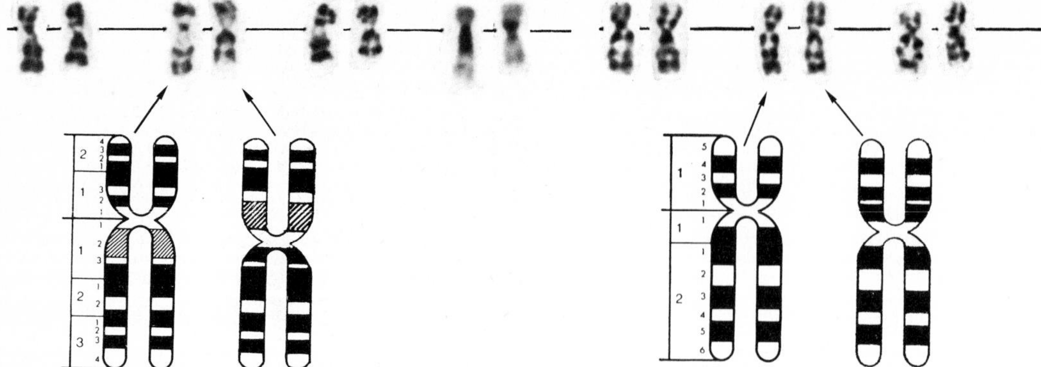

Fig. 2.49. No. 9 chromosomes from different heterozygotes for the same inv 9 (p1; q13). In each pair, the normal homologue is on the *left* and the inverted chromosome on the *right*. Three G-banded and one C-banded cell. The two *right-hand* pairs are from individuals with a secondary constriction in the normal homologue. (De la Chapelle et al., 1974 [323])

Fig. 2.50. No. 10 chromosomes from different individuals heterozygous for the same inv(10) (p1; q21). In each pair, the normal homologue is on the *left* and the inverted chromosome on the *right*. Three G-banded cells. (De la Chapelle et al., 1974 [323])

tion is now on the short arm. Identification of the No. 10 inversion (Fig. 2.50) needs considerable special experience.

In meiotic studies on two of the probands with the No. 9 inversion, the No. 9 bivalent could be identified in the meiotic division I. It looked normal. A chiasma was never observed close to the secondary constrictions. This may have been due to imperfect pairing; suppression of crossing over by inversion is well-known in other forms, for example, in *Drosophila*. For loci assigned to chromosome 9 (Sect. 3.4) this inversion can be used to test whether they are situated within the affected segment.

These inversions do not influence chromosome segregation during meiosis and do not bring about an increased risk of prenatal death for heterozygotes, as evidenced by the normal (corrected, cf. Sect. 3.3) segregation ratios: In matings between one normal homozygote and one heterozygote for inv(9), 25 offspring had a normal karyotype, 23 were heterozygous. For both inv(10) types together, this ratio was 10 : 11. In one mating between two remotely related heterozygotes, one homozygote was observed. Observations of this type may yield clues for mechanisms of chromosome evolution (Sect. 7.2.1).

As noted, the probands of this study had been examined for diagnostic purposes. Therefore, it is not surprising that they showed a variety of anomalies. These anomalies, however, did not show a common pattern; moreover, relatives with the inversion were clinically normal. Hence, these inversions very probably did not influence the phenotypes of their carriers nor their fecundity or the number of miscarriages.

Cromosome 2 also shows occasional pericentric inversions, as the same inversion has been described in three unrelated families (Fig. 2.51) [419]. Two of the families were examined because they had a malformed child; one was examined because of repeated abortions. Therefore, ascertainment was strongly biased. Yet, the possibility remains that the recurrent abortions could be caused by the inversion. Considering the origin of the probands' families in different countries, the authors regarded it as unlikely that the inversion had a common origin. They cite a case where this same inversion was observed as a new mutation [383]. They consider an increased vulnerability of the breakage points as a more probable explanation. However, the alleged new mutation was described before banding methods had been introduced.

Small inversions may be fairly frequent in particular populations. They may not influence health status or fecundity at all. If the inversions are larger, impairment of normal meiosis is more likely. But inversion carriers are euploid; therefore, phenotypic abnormalities are not to be expected.

Aneusomie de Recombination. Occasionally, families have been observed in which one parent seemed to have the same aberration as the child - for example, a pericentric inversion or a translocation. Yet, the parent was phenotypically normal, whereas the child showed a severe malformation syndrome. In some of these cases, a chance coincidence of a harmless chromosome variant with a malformation syndrome of different origin is the most likely explanation. However, crossing over between the abnormal chromosome and its normal counterpart in the displaced region might lead to unbalanced germ cells. This mechanism was suggested by Lejeune and Berger (1965) [416] but its confirmation had to await introduction of the banding techniques.

The first case in which this mechanism could be demonstrated was a boy with multiple malformations [340]. Figure 2.52 shows the No. 10 chromosomes of the proband and his mother. Apparently, the mother had a large pericentric inversion. Crossing over within this inversion led to

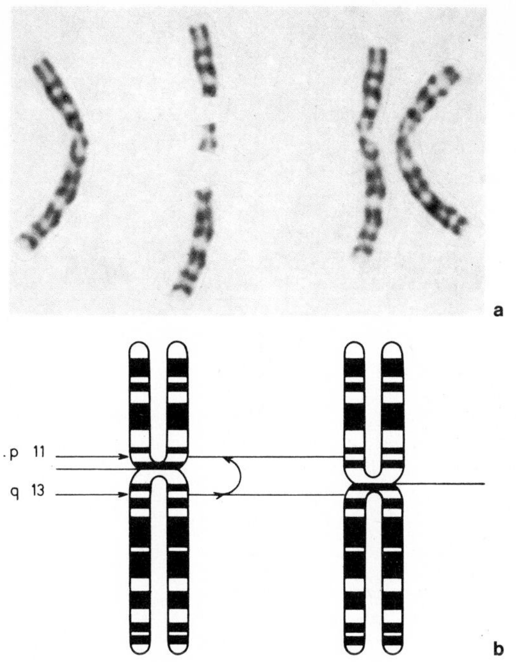

Fig. 2.51 a, b. Pericentric inversion in chromosome 2. **a** G banding. **b** Schematic representation. (Leonard et al., 1975 [419])

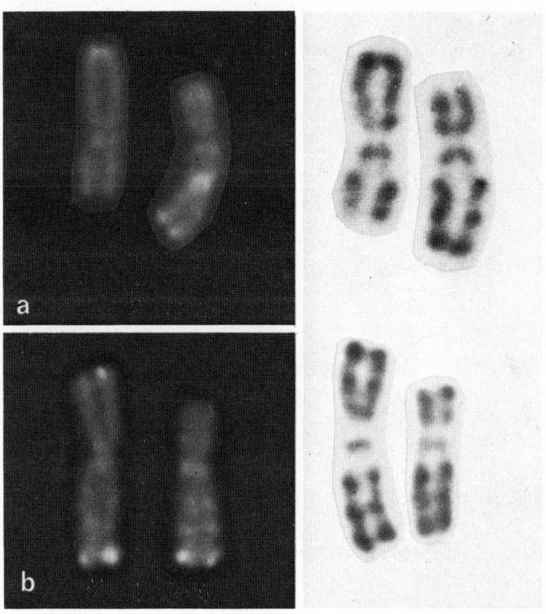

Fig. 2.52 a, b. *Aneusomie de recombinaison?* Karyotype of a malformed proband and his normal mother (Q and G banding). For explanation, cf. text. **a** Partial karyotype of No. 10 of the mother. **b** Partial karyotype of No. 10 of the son. (Dutrillaux et al., 1973 [340])

an abnormal chromosome, making the child trisomic for the segments q456. Without banding methods, all C chromosomes would probably have been classified differently, and the karyotypes of mother and child would have been regarded as identical. More such cases can be discovered by high-resolution banding.

Ring Chromosomes. The situation is different in ring chromosomes, as ring formation implies loss of usually telomeric chromosome segments. Individuals with ring chromosomes will resemble the corresponding deletion carriers. They may, for example, have the cri du chat syndrome, if 5p is affected [455]. In other cases, the symptoms may be less striking, depending on the size of the deleted segments.

A ring chromosome 13 was found in a 14-month-old child showing mental retardation, and symptoms such as microcephaly, epicanthus, broad nasal bridge, prominent ears, and micrognathia [382]. In 85% of blood cells and 82% of fibroblasts, a simple ring was present that could be identified as 13r(p11; q34). In 7% of blood cells and 6% of fibroblasts, a double-sized, dicentric ring was observed, which was composed of two chromosomes 13. In 5% of blood cells and 8% of fibroblasts, the ring was lost, and the rest of the cells contained miscellaneous anomalies, including one metaphase with two interlocked double-sized rings. Figure 2.40 shows the fate of a ring chromosome in mitosis; in most cases, the ring replicates and passes mitosis normally. Sometimes one sister strand exchange occurs, and a double ring with two centromeres is formed; in others, double sister strand exchange leads to two interlocked rings. In the next interphase, the double ring may again have one or two (or more) sister strand exchanges, leading either to interlocked double rings or to a fourfold ring. Thus, an indefinite number of combinations is possible. Figure 2.53 shows the intertwined double ring, Fig. 2.54 a fourfold ring. Many of these newly formed rings lead to metaphase disturbances and to more breakage and aneuploidy in the daughter cells. Figure 2.55 shows an anaphase with breaks of dicentric rings in equal and unequal parts. Most of the theoretically possible configurations (Fig. 2.40) were actually observed in this case.

Fragments. Chromosome fragments are usually lost in mitosis or meiosis, unless retaining a centromere (or a part of it), and they may segregate as supernumerary marker chromosomes. Such markers were not rare in a Danish random sample of newborns (Sect. 5.1.2.1); in some cases, phenotypic abnormalities were reported.

Isochromosomes. Occasionally, chromosomes that consist of two identical arms are found. Such chromosomes are known as isochromosomes and presumably originate by abnormal division of metaphase chromosomes, as shown in Fig. 2.56. If the chromosome involved has arms of unequal length, isochromosomes for the short arm or isochromosomes for the long arm may result.

Isochromosomes are observed relatively frequently for the X chromosome; an isochromosome of the long arm of the X, i (Xq), leads to the Turner syn-

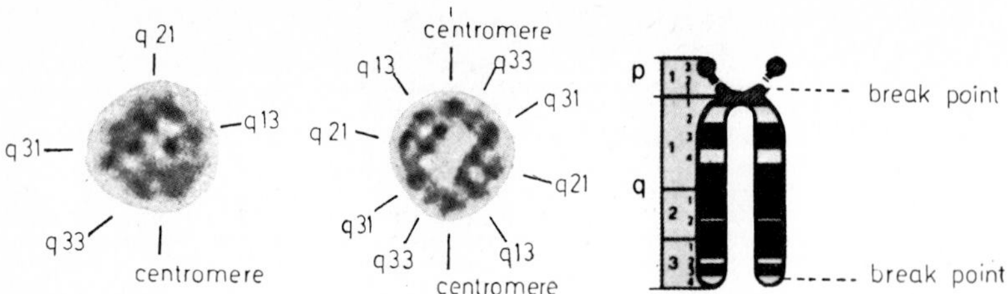

Fig. 2.53. The banding pattern of a monocentric *(left)* and a dicentric *(right)* ring chromosome 13. (Hoo et al., 1974 [382])

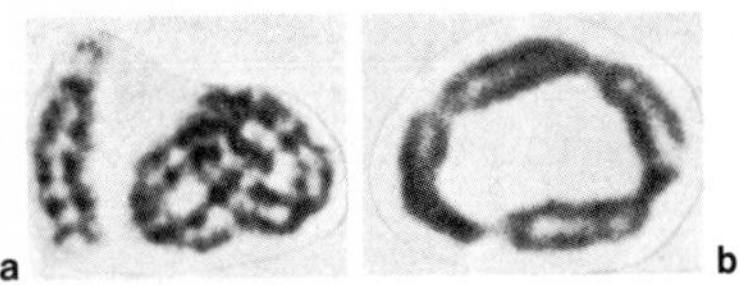

Fig. 2.54. a Two interlocked double-sized rings from a diploid cell with the other No. 13 in the neighborhood. **b** A tetracentric ring from a tetraploid cell. (Hoo et al., 1974 [382])

drome, since this chromosome is always inactivated and only the normal X is active (Sect. 2.2.3).

Interchanges: Centric Fusions (Robertsonian Translocations). Centric fusion is the most frequent type of chromosome rearrangement in human populations. The first reported cases of translocation Down's syndrome were due to centric fusion between the long arm of chromosome 21 and one D or G chromosome. Subsequently, similar cases have been observed repeatedly. Among all cases of Down's syndrome, centric fusion accounts for only a few percent, and many of them are due to new mutation. Only the five acrocentric pairs undergo centric fusion. In the interphase nucleus, the short arms and centromeric regions of these chromosomes are located close to the nucleolus, the short arms containing the nucleolus organizers, especially genes for rRNA (Sect. 2.3). After banding methods had been introduced, participation of single acrocentric pairs in centric fusions could be examined; it turned out to be nonrandom (Table 2.6). Since Table 2.6 is based on data from newborns, the data could be biased by unequal frequency of embryonic death in the different classes.

Centric fusion indicates that the short arms of the two participating chromosomes and possibly one of the centromeres will be lost (Fig. 2.57). This means that ribosomal RNA genes are lost, as well; and indeed, the average number of rRNA genes as estimated by DNA-RNA hybridization studies is lower in so-called balanced carriers of centric fusions than in the normal population [1020]; [1061]. Functionally, this does not make much difference, as, to the best of our knowledge, these carriers are perfectly healthy.

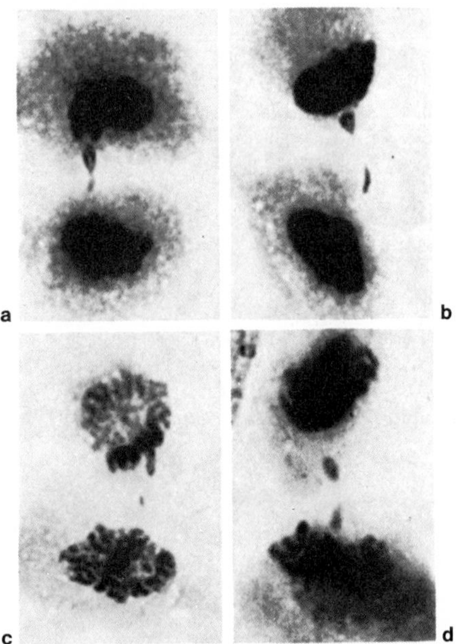

Fig. 2.55 a–d. Break of a dicentric ring that has formed an anaphase bridge could result in two equal parts (**b, d**) or in two unequal parts (**a, c**). (Hoo et al., 1974 [382])

Figure 2.58 shows the possible combinations of chromosomes in germ cells of a carrier for a D/G translocation, and for a 21/21 translocation. After fertilization with a normal sperm, there are six different possibilities. However, the first two, monosomy D and trisomy D, are never observed, and monosomy 21 is – at least in the great majority of all cases – lethal. Each of the remaining three – trisomy 21, balanced, and normal – might be expected to have a probability of ⅓. This expectation, however, is not confirmed by experience: If the mother is the carrier the probability is around 15%, and if the father is the carrier it is no more than 5%. However, the risk of a balanced zygote is the expected segregation rate of ≈50%. The problem will be discussed in greater detail in Sect. 2.2.2.

In a 21/21 translocation – or a 21/21 isochromosome – prospects are much more gloomy: Either the child will be trisomic and affected with Down's syndrome, or the aneuploidy will be lethal. Fortunately, it is now possible with banding techniques to distinguish 21/21 translocations from those between No. 21 and 22, for which the chances are much bet-

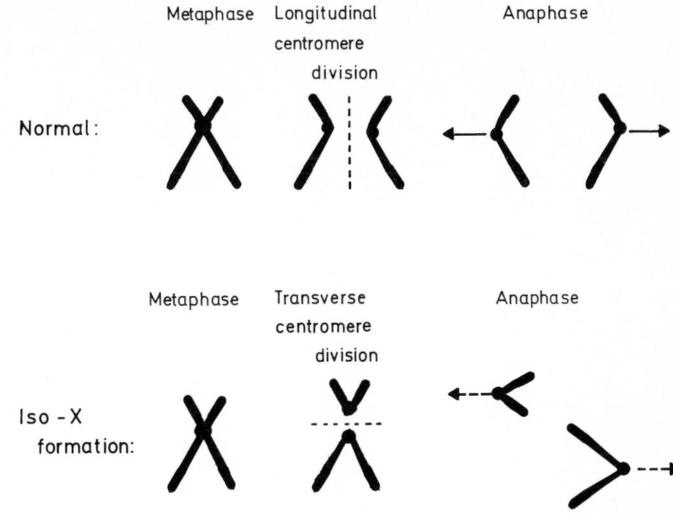

Fig. 2.56. Formation of an isochromosome by abnormal cleavage of the centromere

Table 2.6. Chromosomes involved in Robertsonian translocations. (Family material analyzed by Schäfer [501 a])

	14	15	21	22
13	101	8	10	6
14		8	45	3
15			17	3
21				17
22				

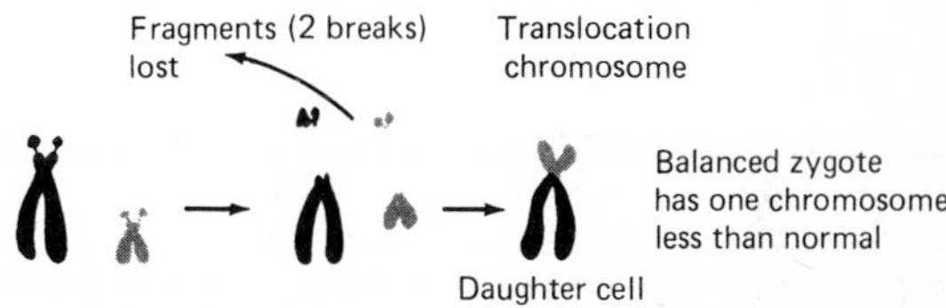

Fig. 2.57. Principle of a centric fusion (Robertsonian translocation). Two acrocentric chromosomes lose their short arms, and the long arms fuse. The translocation chromosome may have one or two centromeres; in the latter case, one centromere must be suppressed. In any case, the balanced zygote will have one chromosome less than normal (as compared with the reciprocal translocation, which leads to a balanced zygote with normal chromosome number)

ter, resembling those for translocations involving a D chromosome.

Interchanges: Reciprocal Translocations. Unlike centric fusions, reciprocal translocations do not necessarily entail loss of material. The broken parts are joined with other chromosomes. Therefore, the balanced zygote has 46 and not 45 chromosomes. Among the daughter cells, the types seen in Fig. 2.60 can be expected. Most often only the partially trisomic or the partially monosomic are found, the others presumably being lethal.

A typical case is the following [504]. Figure 2.61 shows two severely retarded sibs aged 11 (♀) and 9 (♂) years. They showed some concordant and some discordant symptoms (Table 2.7). In the conventionally stained karyotype, both children showed an elongated long arm of a C chromosome (Fig. 2.62); mother and maternal grandmother had the same chromosome, and additionally, another abnormal C chromosome, the short arm of which was almost completely lacking (Fig. 2.63). G banding in the mother revealed a reciprocal translocation between No. 7 and 10, the karyotype being 46,XX,t(7;10)(p22;p11). This resulted in partial trisomy 10p+ in both children. Remarkable are not only the concordant symptoms in the two children that establish a clinical syndrome, but also the discordances that show the variability in developmental disturbances caused by the same aberration.

Table 2.7. Concordant signs (Schleiermacher et al., 1974 [504])

Concordant signs	
Mental retardation	
Small body size	
Dolichocephaly	
Abnormalities of the teeth	
Hypoplastic and hypotonic skeletal muscles	
Club feet	
Brachymesophalangy of the third toe	
Discordant signs	
Girl	Boy
Low origin of hair growth	Seizures
High and narrow palate	Harelip and cleft palate
Low-set ears	Dysplasia of the ear
Peripheral neuropathy (N. tibialis and N. fibularis)	Scoliosis
	Camptodactyly

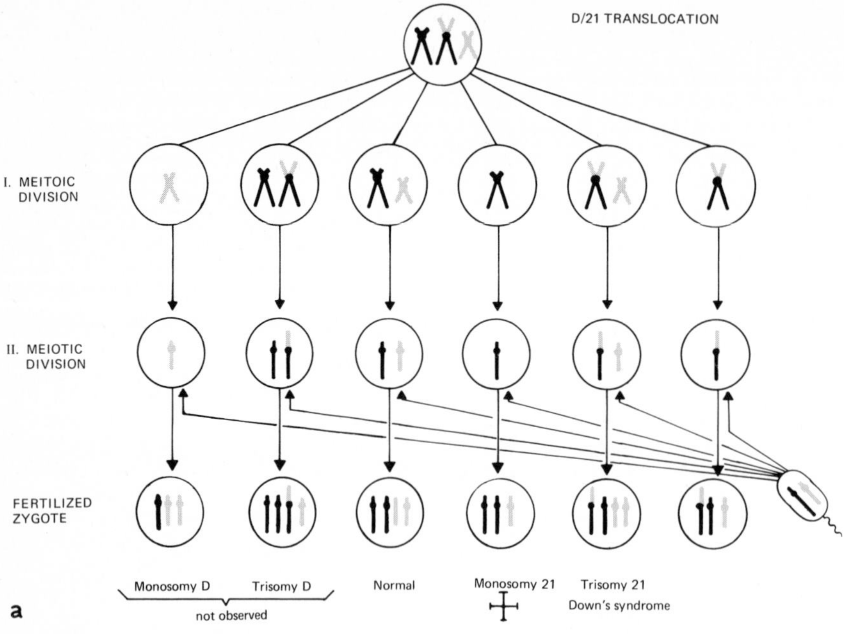

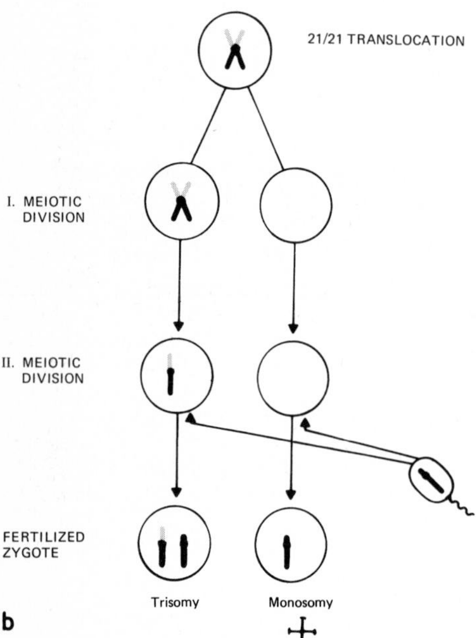

◁ **Fig. 2.58. a** Diagram of germ cell formation if the mother is the carrier of a D/21 translocation in a balanced state: the one D chromosome has acquired the translocated long arm of chromosome 21. As a result there is only one free chromosome 21. Since this free chromosome 21 and the two D-chromosomes combine according to the laws of chance, six different germ cells can theoretically be formed and it should be possible to find six different types of zygotes after fertilization with a normal sperm. However, three of the possible six have not been observed. The remainder are either normal, or balanced, or trisomic in a ratio that can only be established empirically. **b** Formation of germ cells with a 21/21 translocation and a 21-iso-chromosome. The possibilities are that either the translocation chromosome forms the germ cell – a resulting zygote would be functionally trisomic and the child would manifest Down's syndrome – or the translocation chromosome does not get into the germ cell, in which case the zygote is missing a chromosome 21 and dies

Phenotypes in Autosomal Chromosome Aberrations. The most conspicuous feature of the phenotypes in autosomal chromosome aberration syndromes is the considerable overlap of signs and symptoms. The main signs are:

a) General
   - Low birth weight (small for date)
   - Failure to thrive
   - Mental retardation (usually severe)
   - Short stature

b) Head and face
   - Microcephaly
   - Incomplete ossification
   - Micrognathia
   - Anomalous positioning of eyes
   - "Dysmorphic facies"
   - Low-set, deformed ears

c) Various anomalies of hands and feet
   - Anomalous dermatoglyphic patterns

d) Internal organs
   - Congenital defects of heart and/or great vessels
   - Cerebral malformations
   - Malformations of the genitourinary system

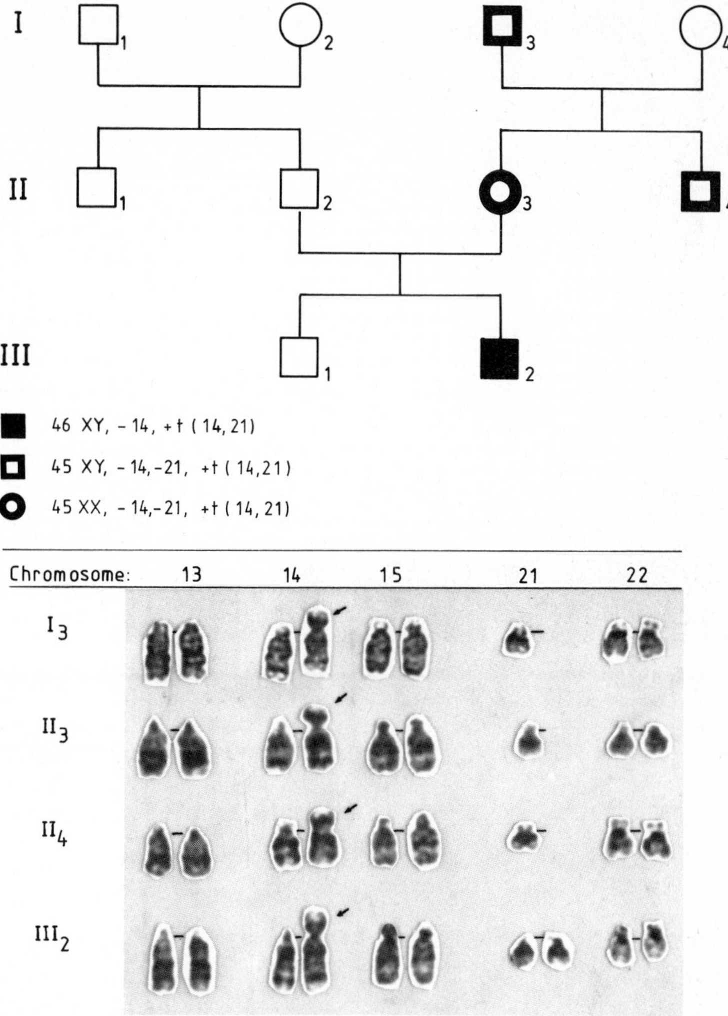

Fig. 2.59. Pedigree of a male patient with Down's syndrome and a 14/21 Robertsonian translocation *(III,2.)*. Mother *(II,3)*, her brother *(II,4)* and maternal grandfather *(I,3)* have a balanced karyotype. Courtesy of Dr. T. M. Schroeder-Kurth

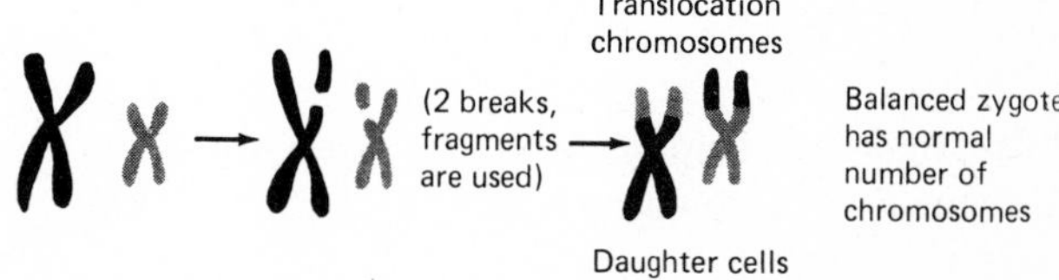

Fig. 2.60. Reciprocal translocation. The balanced zygote has 46 chromosomes; two chromosomes show complementary structural anomalies

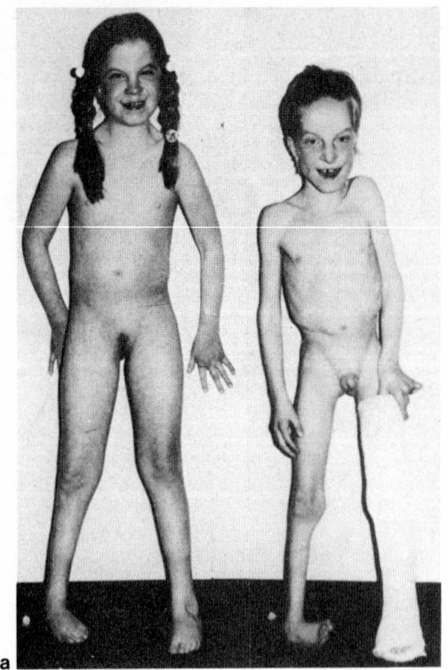

a

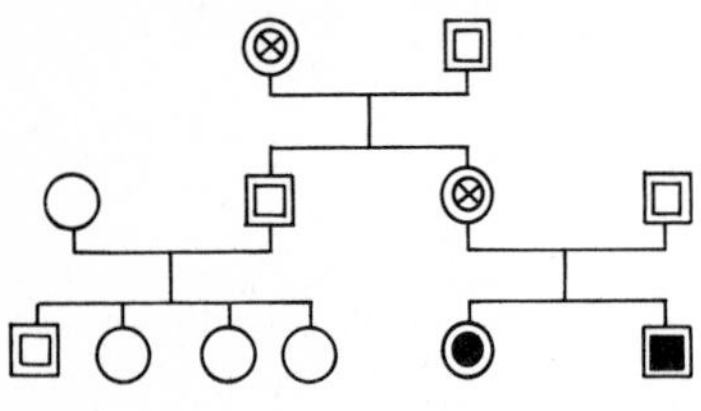

⊗ 46,t(7p+;10p−)

● ■ 46,t(7p+)

b ◎ ⧈ Examined persons

Fig. 2.61. a Brother and sister with severe multiple malformations and mental retardation. **b** The pedigree of the two siblings shown in Fig. 2.68 a (mother: balanced translocation; children: partial trisomy 10p+)

The following symptoms generally do not point to an autosomal chromosome anomaly, but exceptions occur:

Mental retardation without additional malformation
Malformations associated with normal mental capacity
Isolated, single malformations

Many, though not all autosomal aberrations show – apart from these common malformations – others that are more or less specific. Moreover, the common signs may be more or less severely expressed. The signs caused by a specific aberration usually form a pattern that is characteristic for this aberration and makes a preliminary impression possible on clinical grounds although such a diagnosis should always be checked by chromosome examination. With limited capacity of chromosome laboratories, these symptom complexes help in deciding when chromosome study is indicated.

Within the patient group showing a common aberration, the variability of many of the findings tends to be great. Some cases of Down's syndrome, for example, may be only mildly retarded, whereas most are severely mentally retarded; cardiac malformations are found in many, and intestinal atresia in only a few. The sibs in Fig. 2.61 with the same translocation show – apart from some common features – definite phenotypic differences. This variability is presumably due in part to the fact that the abnormal chromosome is superimposed on very different genotypes, i.e., the genetic background differs. Still, variability in development seems to be abnormally enhanced. Increased lability of embryonic development by yet unknown mechanisms has been postulated.

The most surprising observation about the phenotypes of chromosomal aberration syndromes is that trisomies cause any anomaly at all. Their carriers have a full set of genetic material, and no single gene function is altered or lost. Besides, we know from heterozygotes of autosomal recessive diseases (Sect. 4.2.2.8) that for most enzymes one-half of the normal production is sufficient to maintain normal function. It is hardly understandable why an amount of 3/2 of these gene products as present in trisomies should make so much difference. These problems will be discussed in more depth in Sect. 4.7.4.

Moreover, the common symptoms of all the autosomal syndromes are independent of the single chromosomes involved. One might note that the organ systems mainly involved are known to have a long and complicated embryonic development and that therefore many different genes will be needed for this development to proceed normally. But this explanation is also very general, and besides, the necessary genes are all there.

What is the nature of the genetic disturbance leading to these chromosome syndromes? The answer is: The syndromes are probably due chiefly not to surplus activity or a defect of single genes, but to failure in the regulation of genes during embryonic development. Analysis of autosomal chromosome aberrations should therefore be suitable to teach us much regarding the mechanisms of gene regulation in humans. This problem will be taken up again in Sect. 4.7.4. For a special problem – the development of sexual characters – studies of patients with numeric and structural chromosome aberrations have already been highly instructive. Before sex chromosomal aberrations are discussed, however, some remarks are necessary regarding segregation and pre-

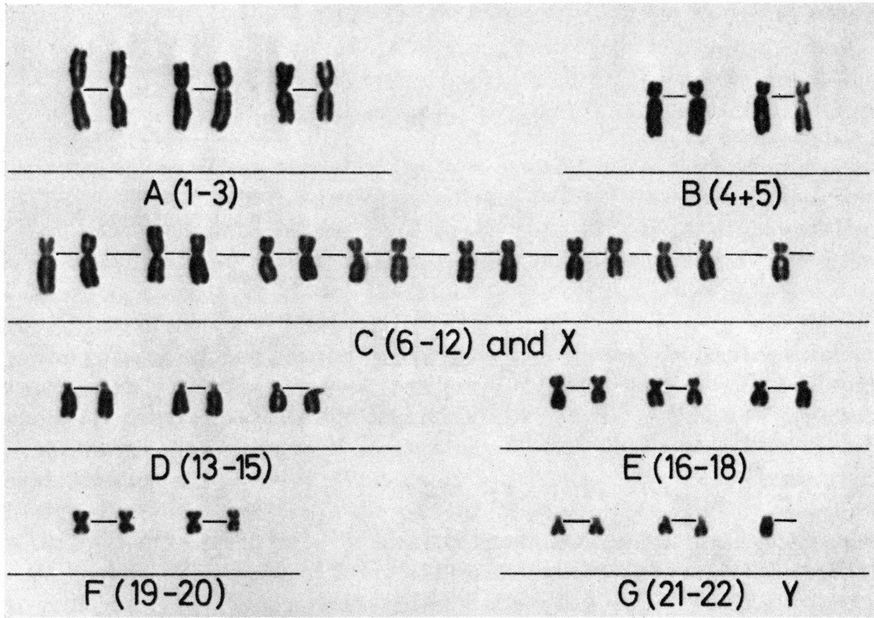

Fig. 2.62. Karyotype of the boy shown in Fig. 2.68. Conventional staining

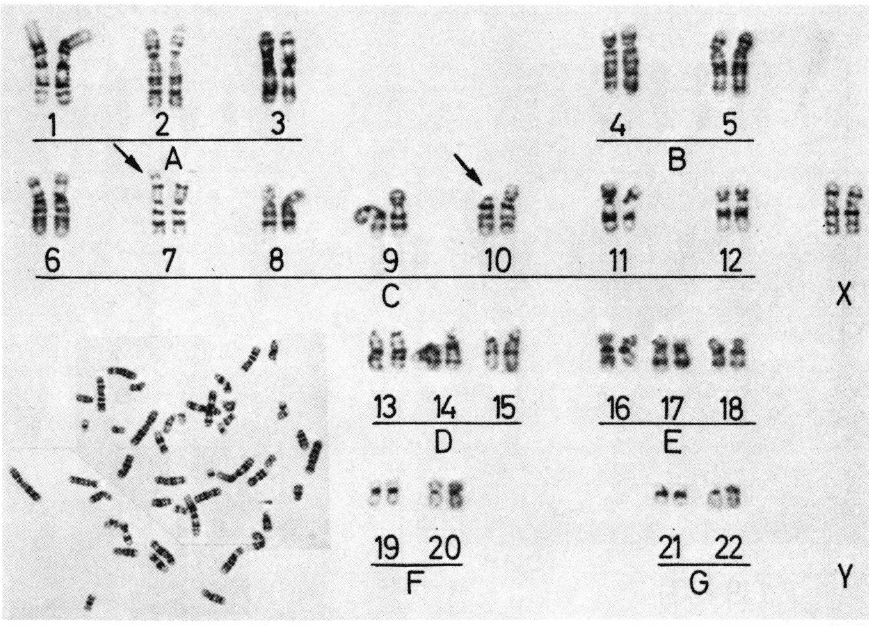

Fig. 2.63. G-banded karyotype of the mother of two children of Fig. 2.68 with a reciprocal translocation involving chromosome 7 and 10. These two chromosomes are marked by *arrows*. (Schleiermacher et al., 1974 [504])

natal selection of unbalanced, and possible clinical signs in "balanced" translocation carriers. In addition to their theoretical interest, these questions are important practically, - for risk assessment and genetic counseling.

2.2.2.2 *Segregation and Prenatal Selection of Translocations: Methodological Problems*

The problem of segregation and prenatal selection of translocations has often been studied, but there were many contradictions and the studies failed to give a clearcut picture. Recently, however, many of these problems have been solved by Schäfer [501 a]. We shall, in principle, follow his analysis.

Translocations are relatively rare. No single research group will therefore be able to collect sufficient data for definite conclusions. Hence, data published in the literature have to be analyzed. Such data, however, are subject to many biases. Some of these biases will be discussed later (Sect. 3.3.4. and Appendix 3); for example, sibships in which abnormal chromosomes segregate will, as a rule, be ascertained only if at least one sib is affected. It also makes a difference whether all sibships with at least one affected sib in the population are ascertained or a small fraction only. These problems can be handled relatively simply if hereditary diseases are involved, since in this case families are always ascertained through at least one "proband" who suffers from the disease in question. With translocations, however, families might be ascertained, for example, because of multiple abortions or through an unbalanced proband at birth or at prenatal diagnosis. Occasionally, balanced carriers are discovered in population surveys and a family study is initiated. It is impossible to correct completely for all biases - especially as in many published reports necessary information is lacking. The correction by Schäfer, however, appears to be the optimum attainable with the case material presently available.[6]

Data Used for This Analysis. The study was based on 1050 families with segregating translocations, with altogether 2109 pairs of parents and 4745 progeny. Moreover, 556 reported instances of pathologic effects in carriers of balanced translocations were collected, as were results from 814 prenatal diagnoses, and about 130000 individuals examined in various screening programs were evaluated. This statistical study led to detailed risk estimates for various kinds of pregnancy outcome, and to results on phenotypic effects in unbalanced and balanced translocation carriers. For an understanding of these results, it is necessary to visualize the consequences of translocations during meiosis.

Segregation of Translocations in the First Meiotic Division. In the first meiotic division, homologous chromosomes pair. This rule also applies to translocated chromosome segments: they pair with their *original* partners. This leads to complexes of *four* chromosomes in reciprocal translocations and of *three* chromosomes in Robertsonian translocations. As in normal meiosis, spindle fibers fix at the centromeres, and homologous chromosomes are moved to opposite poles. This regular anaphase movement may lead to four different division products with equal probabilities (Fig. 2.64):

1. and 2. Let the two normal chromosomes be A_1, B_1; they come into one division product (haploid cell), and the two translocation chromosomes, A_2, B_2, come into the other (=alternate disjunction).
3. and 4. One normal and one translocated chromosome come into one haploid cell (=adjacent-1 disjunction). Here, two possibilities exist: A_1B_2 or A_2B_1. Each of the four combinations has the probability 0.25.

A_1B_1 will be karyotypically normal, A_2B_2 will be balanced, since the two chromosomes A_2 and B_2 have exchanged segments. A_1B_2 and A_2B_1, however, will be unbalanced. In addition, other, abnormal types of segregation may occur as result of the chromosomal aberration. For example, homologous centromeres might occasionally come into the *same* division product (adjacent – 2, Fig. 2.65), or three centromeres may come into one, one centromere into the second division product (3:1 disjunction; Fig. 2.66).

Neglecting these abnormal situations, we conclude that, among offspring of individuals with balanced reciprocal translocations, we must expect 50% to be unbalanced and, therefore, phenotypically abnormal; 25% to be balanced and phenotypically normal; and 25% to have normal karyotype and normal phenotype. As has long been known, experience does not bear out these theoretical expectations. Fortunately, the number of children with unbalanced karyotypes is much *lower.* This might have a number of reasons, such as other anomalies of the first meiotic division; selection against germ cells with abnormal karyotypes in gametogenesis; preferential fertilization by normal and/or balanced germ cells; or selection against the aneuploid zygote in various stages of development. Statistical analysis has yielded evidence for some of these mechanisms.

Expectations for Unbalanced Zygotes. Using the abovementioned corrections for ascertainment biases (Appendix 3), the average risk for abnormal, unbalanced offspring at birth was estimated to be 7% for carrier mothers and 3% for carrier fathers. These risk figures were estimated for *all* balanced carriers. For carriers in families in which unbalanced zygote offspring had already been observed, the risk figures for sons and daughter together are higher (14% of all births for mothers and 8% for fathers); the risk for unbalanced zygote sons of carrier fathers is especially low (5%). When the mothers were carriers, 66% of all unbalanced zygote offspring were found to be of the adjacent-1 type, 3% adjacent-2, and 31% 3:1. With carrier fathers, 90% were adjacent-1, 3% adjacent-2, and 8% 3:1 (see above).

For translocation carriers ascertained through prenatal diagnosis, the estimated risk figures were: 11.7% unbalanced offspring for mothers and 12.1% for fathers.

The low *overall* estimate (7% for carrier mothers; 3% for fathers) is due to the fact that only about one half of all translocations can give rise to malformation syndromes at birth; the others invariably lead to fetal death.

For *Robertsonian* translocations involving chromosome 21, a 13% risk of unbalanced offspring has been estimated when the mother is a carrier; the risk is 3% with a

[6] The analysis will be set out in greater detail in Appendix 3.

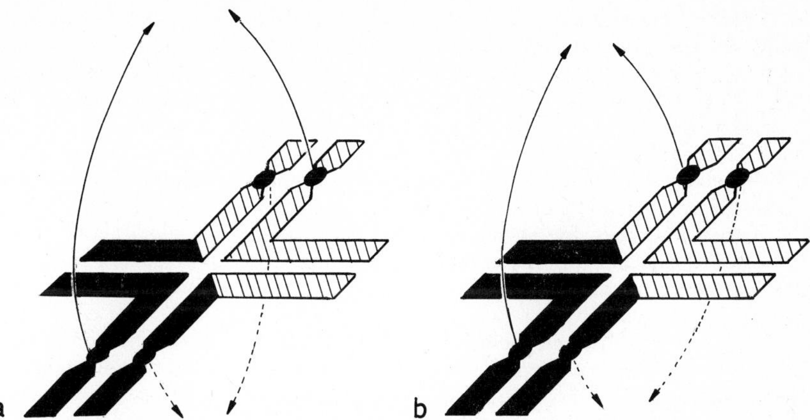

Fig. 2.64 a, b. Schematic representation of a translocation quadrivalent predisposed either to alternate (**a**) or to adjacent-1 (**b**) disjunction. The centric segments are longer than the translocated ones. (**a**) leads to normal and balanced zygotes, (**b**) to two types of unbalanced zygotes

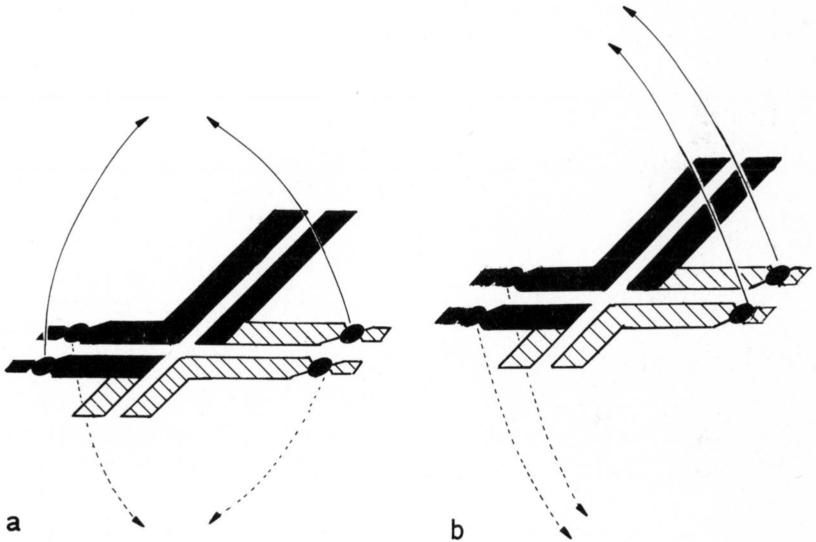

Fig. 2.65 a, b. Here, both chromosomes involved in a translocation are acrocentric. One of the paired arms that do *not* carry the centromere is much longer than the arms carrying the centromere. Here, it may (rarely) occur that homologous centromeres are transported into the same daughter cell (adjacent-2 disjunction). This type of disjunction is also possible if one of the two chromosomes is a chromosome No. 9

Fig. 2.66. 3:1 segregation with formation of tertiary trisomy (and monosomy). The length of the paired segments between the centromeres make chiasma formation likely; the chiasma on the right-hand side cannot be terminalized properly

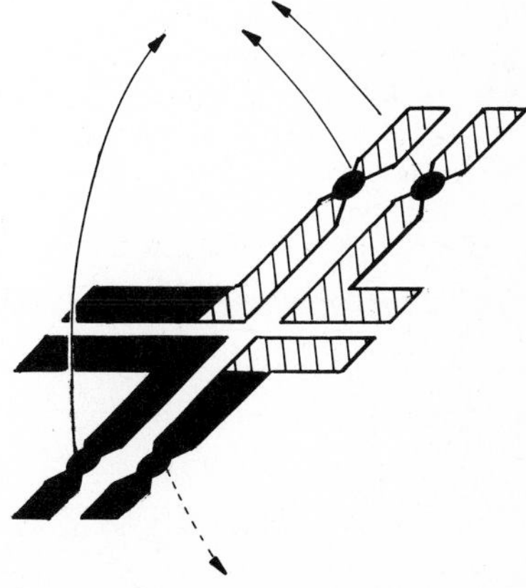

carrier father. DqDq carriers, on the other hand, run practically no risk of having unbalanced offspring. The same seems to apply for Dq22q translocations. For genetic counseling, the figures for Robertsonian translocations can be used with confidence, but those for reciprocals are too crude, since they are derived from many types of translocations involving a great number of different chromosomes; a refinement is possible. For example, the risk is higher if the carrier who asks for advice has been ascertained through an unbalanced proband rather than a balanced one. Moreover, special parameters, such as the length of the chromosomes involved in the translocation and, especially, the size of the trisomic (or monosomic) chromosome segments, should be considered: As a rule of thumb, the larger these segments, the stronger the prenatal selection against the aneuploid germ cell or zygote.
Increasing amounts of data are also available suggesting that the risks are higher when the total length of the two chromosomal segments involved in the imbalance is less than about 2% of total genome length. With longer chromosomal structures affected there is usually very early loss of the unbalanced zygote. Partial monosomies are more damaging than partial trisomies. Ideally, such a risk assessment should be based on empirical data about precisely the same translocation that is observed in the family in question [355]. This, however, is impossible in most cases, as too many different translocations exist.
Boué [310] has one of the largest single bodies of experience in this field. His data emphasize that the mode of ascertainment of a given Robertsonian translocation or reciprocal translocation determines the recurrence risks. Thus, if a balanced translocation is discovered during cytogenetic studies carried out because of multiple abortions the risk of an unbalanced fetus is small (~2%; Table 2.8). In contrast, the chance of an unbalanced offspring is almost 20% if it was ascertained via an abnormal child or fetus. Such analyses have allowed theoretically interesting insights into segregation of translocations in the first meiotic division of humans. In general, segregation appears to be very regular: homologous centromeres do indeed migrate to opposite poles, leading either to alternate, or to adjacent-1 segregation. Exceptions, such as adjacent-2 or 3:1 segregation occur only occasionally; 3:1 segregation occurs mainly (Fig. 2.66), if a small chromosome remains unpaired. In males, it often leads to a breakdown of meiosis. Selection against unbalanced *gametes* occurs in males but not in females. Selection against unbalanced *zygotes* depends on kind and extent of chromosomal imbalance; very severe imbalances will cause early abortions that go unnoticed. Moderately severe ones lead to recognizable abortions. The rate of abortions is increased especially when the *mother* is a carrier and may be due to the fact that there is no gamete selection in females.
It was suspected earlier that translocations might increase the risk for disturbances in the segregation of other chromosomes. For example, free trisomy 21 was thought by some authors to be more frequent when the mothers were translocation carriers. Careful analysis failed to substantiate this claim.

Table 2.8. Ascertainment of translocations [310]

Abnormalities	Ascertainment via unbalanced abnormalities		Ascertainment via abortions	
	No. of tests	Unbalanced aberration	No. of tests	Unbalanced aberration
Robertsonian translocations				
DqDq	7	0	30	0
DqGq	25	3	7	0
GqGq	4	1	0	0
Regional translocation	46	10	43	2
Inversions	3	3	12	0
Total	85	16	92	2

Segregation of Karyotypically Normal and Balanced Zygotes. As mentioned, alternate segregation should lead theoretically to karyotypically normal and balanced zygotes in equal frequencies (50:50). In practice, most reports record a *slight* excess of balanced zygotes. Schäfer's [501 a] analysis has shown that this excess might well be biologically real (and not caused by an ascertainment bias). It is observed especially in sons of carrier fathers, and is strong in sons of 13q14 carriers.

Phenotypic Deviations in Balanced Translocation Carriers. Balanced translocation carriers have the complete set of genetic material and should therefore be phenotypically normal. As a rule, this expectation is borne out by experience. However, there have been many reports claiming a somewhat higher incidence of malformations, mental retardation, and minor birth defects. Analysis of the available evidence has shown that multiple malformations and mental deficiency are rare among balanced translocation carriers, but more common than among karyotypically normal subjects (Table 2.9). As a rule, such clinical findings are encountered in *sporadic* or in familial Dq21q translocations: In this familial group, undetected mosaics might be suspected. In sporadic translocations, breaks within single genes or disconnection of genes from their functionally important location ("position effects") might explain the phenotype. Most carriers of familial translocations are clinically unaffected. Selection against unbalanced zygotes leads to an increase in the abortion rate; among couples with multiple miscarriages, translocation carriers are overrepresented. Infertility is common among male carriers; in females it only occurs in connection with X-autosomal translocations. However, these findings should not obscure the fact that the great majority of translocation carriers, familial and sporadic, have a normal phenotype.
These results on human translocations have implications for the mutation rate, as well as for fitness, selection, and theory of evolution. These aspects will be discussed in Sects. 5.1.2, 6.2.1, and 7.2.1.

Table 2.9. Prevalence (per 1000 individuals) of balanced translocations in various groups [501 a]

Class	Reciprocal translocations	DqDq	DqGq	GqGq	n (Examined)
Newborns	0.838	0.614	0.182	0.014	71603
Adults	0.639	0.930	0	0.058	17202
Abortions	1.455	0.727	0.242	0	4124
Stillbirths and premature births	0.734	0.734	1.468	0	1362
Multiple malformations	3.305	0.991	0.661	0.165	6052
Mental retardates	2.283	0.351	0.176	0	17083
Antisocial behavior	1.204	0.747	0.267	0	3463
Psychiatric patients (passive)		2.600			4014
Infertility (Male)	2.061	5.247	0.562	0.187	5336
Infertility (Female)	2.309	0.770	0.770	0	1299
Patients with multiple abortions (Male)	7.543	2.382	0.794	0	2519
(Female)	17.718	6.024	1.417	0	2822

2.2.3 Sex Chromosomes

2.2.3.1 First Observations

Nondisjunction of Sex Chromosomes and Sex Determination in Drosophila. Meiotic nondisjunction was discovered by Bridges (1916) [311] in the sex chromosome of *Drosophila melanogaster.* Morgan (1910) [448] had earlier described the X-linked mode of inheritance, and at the same time had elucidated the X-Y mechanism of sex determination in *Drosophila.* In his experiments, a few exceptions had occurred that did not conform to the predictions of X-linkage. Bridges explained them by an anomaly in the mechanism of meiosis.

Drosophila has four chromosome pairs, three pairs of autosomes, and two sex chromosomes. Just as in humans, the males have the complement XY, the females XX. Hence, each normal male germ cell will have either one X or one Y chromosme; all female germ cells will have an X. In corsses between an affected homozygote for the X-linked recessive trait white and a wild-type or normal male, all male offspring would be expected to have white eyes like their mothers. All daughters should be heterozygous and have normal red eyes. As a rule, this expectation was fulfilled. In exceptional cases, however, male offspring had normal red eyes, and some females were white-eyed. This was shown by Bridges to be due to nondisjunction of the maternal X chromosome leading to an oocyte with either two or no X chromosomes (Fig. 2.67). Fertilization with sperm from a wild-type male was expected to lead to four different types of zygotes: XXX; XXY; XO; YO. YO was not observed; apparently, zygotes without an X chromosome cannot survive. The other three types were observed and gave evidence regarding the mechanism of sex determination:

a) XXX } Female phenotype
b) XXY }
c) XO Male phenotype: sterile

Hence, the phenotypic sex in this fruit fly depends on the number of X chromosomes. One X chromosome makes a male, more than one X chromosome makes a female. The Y is also involved in sex determination, as XO males are sterile.

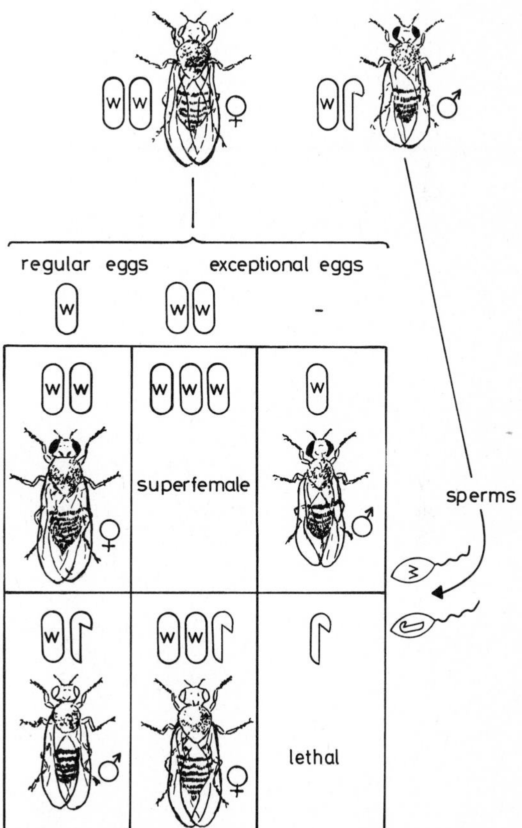

Fig. 2.67. Nondisjunction of X chromosomes in *Drosophila melanogaster,* mating of a white female with a wild-type male. (Sinnott-Dunn-Dobzhansky, Principles of Genetics, 1958)

XO Type in the Mouse. The X-linked mutation scurfy (sf) appeared first by spontaneous mutation. The animals have scurfy skin. The hemizygotes are sterile; therefore, the strain can be maintained only by crossing heterozygotes (X^{sf}/X^{+}) with normal males (X^{+}/Y). From this mating, scurfy and normal males are expected in a segregation ratio of 1:1; all females should be normal. From time to time, however, an exceptional sf female was observed. Like male hemizygotes, they are sterile. But their ovaries can be transplanted to normal females, which were mated with wild-type males. The sons were all sf; the daughters were all normal but fell into two groups, those transmitting sf and those not transmitting it. Further analysis showed that these daughters had two different karyotypes, X^+/O and X^+/X^{sf}; the first group did not transmit sf, the second did. This experiment showed that, contrary to the findings in *Drosophila,* XO is a fertile female in the mouse. Hence, in this animal, the Y and not the X chromosome is decisive for the phenotypic sex. Subsequently, the XO types of the mouse were found to be fairly frequent. In most cases, the condition is caused not by meiotic nondisjunction but by chromosome loss after fertilization. In mutation research, this chromosome loss has become an important tool for assessment of mutagenic activity (Sect. 5.2). Not long after the XO type, the XXY type was also discovered in the mouse. It is a sterile male unlike *Drosophila* where the XXY type was female.

First X Chromosomal Aneuploidies in Humans: XXY, XO, XXX. The X chromosomal aneuploidies were first discovered in humans (Sect. 2.1). When Jacobs and Strong (1959) [395] studied a 42-year-old male with the typical features of Klinefelter's syndrome (Fig. 2.68), including gynecomastia, small testicles, and hyalinized testicular tissue (Fig. 2.69), X chromatin in cells of buccal smears and drumsticks in granulocytes were found. Chromosome examination from bone marrow revealed an additional, submetacentric chromosome "in the medium size range." The authors felt that the patient very probably had the constitution XXY. However, "the possibility can not be excluded ... that the additional chromosome is an autosome carrying feminizing genes." The patient's parents both had normal karyotypes with 46 chromosome; hence, nondisjunction had occurred in one of their germ cells. Shortly afterward, the XXY status for Klinefelter's syndrome was confirmed in many other cases.

At the same time, the XO type was discovered by Ford et al. (1959) [352]. Their patient, a 14-year-old girl, presented clinically as the Turner syndrome (Fig. 2.70) and was X chromatin-negative. The modal number of chromosomes in bone marrow cells was 45; there were only 15 "medium length metacentric chromosomes" as in normal males. The evidence strongly suggested a chromosome constitution XO. The authors, comparing this result with that known from *Drosophila* (Fig. 2.67), concluded that, contrary to the fly, the XO type in humans may lead to an "agonadal" individual with female phenotype. Noting the XXX state of *Drosophila,* they remarked that it was still unknown in humans.

This gap was soon closed by the report of a 35-year-old woman with poor development of external sexual characteristics and secondary amenorrhea, who showed 47 chromosomes obviously with an additional X: 47,XX,+X. In this case, two tissues - bone

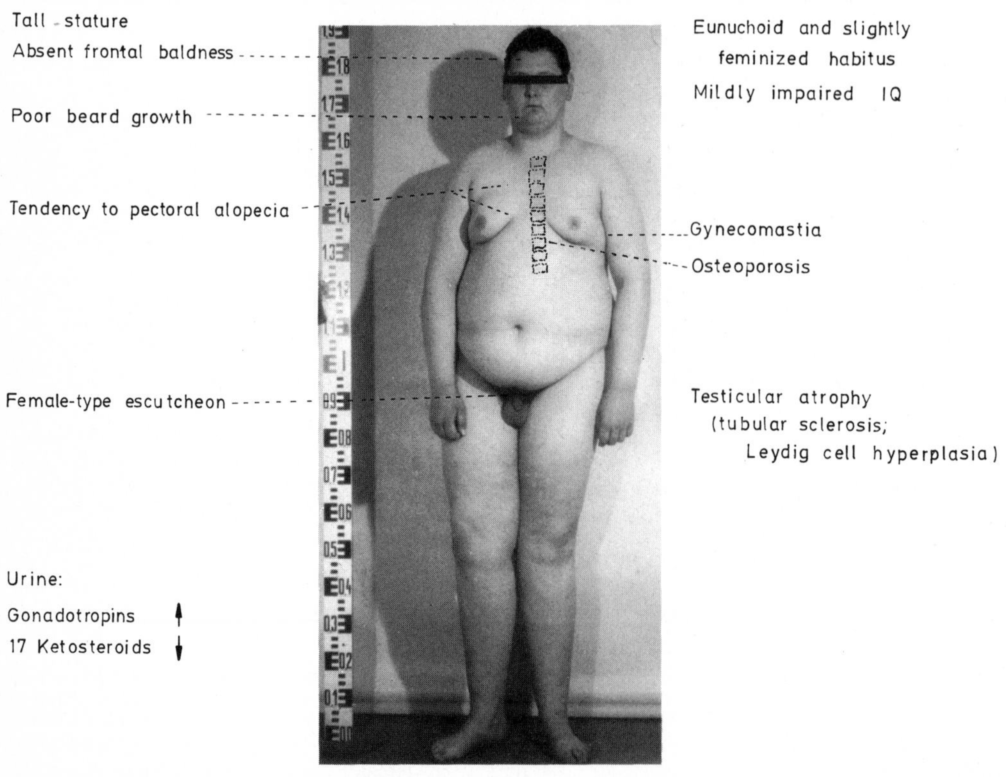

Fig. 2.68. Main clinical findings in Klinefelter's syndrome

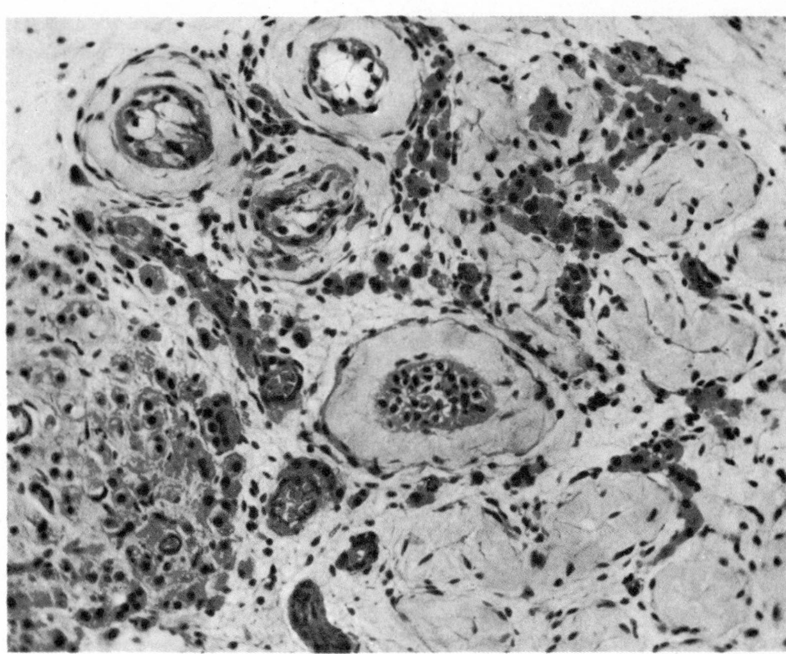

Fig. 2.69. Hyalinized testicular tissue in Klinefelter's syndrome. The normal tubules are lacking and are replaced by hyalinized tissue

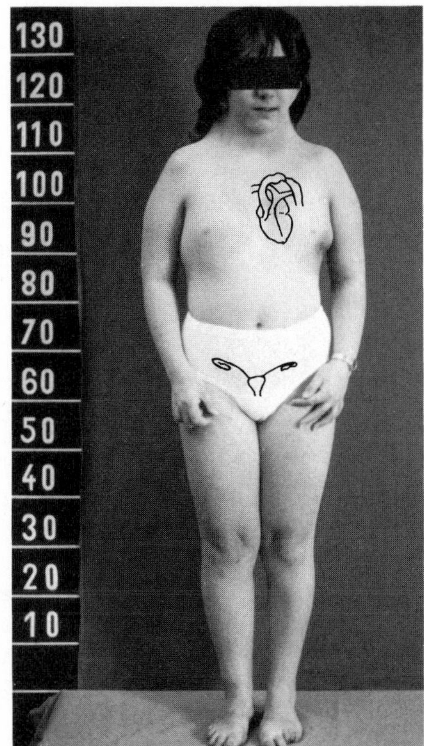

Fig. 2.70. Main clinical findings in Turner's syndrome

marrow and fibroblasts - were examined, and both showed the same aneuploidy. In many of her buccal smear cells, and in some granulocytes, the patient had two X chromatin bodies.

1. Contrary to *Drosophila,* the phenotypic sex in humans is determined by the presence or absence of the Y chromosome, not by the number of X chromosomes. In this aspect, humans are similar to mice, but the XO type in mice is a fertile female; in humans, it is a female with a nonfunctioning ovary.
2. The number of X chromatin bodies is one less than the number of X chromosomes.

These two observations were soon to become the cornerstones of our knowledge and hypotheses on sex determination and on the genetic activity of X chromosomes.

2.2.3.2 X Chromosomal Aneuploidies in Humans: Current Knowledge

Difference Between X Chromosomal and Autosomal Aneuploidies. Soon after these first discoveries, a great number of other aneuploidies of sex chromosomes were described. As a group they show some remarkable differences from the autosomal aneuploidies discussed before.

a) Mean intelligence is often below the norm, but the anomaly of brain development is not nearly as pronounced as in the autosomal conditions; many probands have normal intelligence, and in a few, it is even above average (Sect. 8.2.2.2).
b) The phenotypic disturbances most severely affect development of the sexual organs and sexual hormone-dependent growth. Other malformations do occur - mainly in the Turner syndrome - but they are less frequent and less severe.

In brief, X-chromosomal aneuploidy does not disturb embryonic development nearly as much as does autosomal aneuploidy. The reason is that normal women have two, normal men only one X chromosome. This difference led to the development in evolution of a powerful mechanism of gene dosage compensation that happened to benefit carriers of X aneuploidies.

Clinical Classification of X-Chromosomal Aneuploidies: Mosaics. The most important numeric and structural anomalies of the X chromosome are found in Table 2.10. In general, the number of additional X chromosomes enhances the degree of mental retardation. The number of X chromatin bodies is always one less than the number of X chromosomes.

Table 2.10. Numeric and structural X chromosomal aneuploidies in humans

Karyotype	Phenotype	Approxim. incidence
XXY	Klinefelter syndrome	1/700 ♂
XXXY	Klinefelter variant	≈1/2500 ♂
XXXXY	Low grade mental deficiency; severe sexual underdevelopment; radioúlnar synostosis	Very rare
XXX	Sometimes mild oligophrenia occasionally disturbances of gonadal function	1/1000 ♀
XXXX XXXXX	Physically normal; severe mental retardation	Rare
XXY/XY and XXY/XX mosaics	Klinefelter-like, sometimes with milder symptomatology	≈5–15% of all Klinefelter-like patients
XXX/XX mosaics	Like XXX	Rare
XO	Turner syndrome	≈1/2500 ♀ at birth
XO/XX and XO/XXX mosaics	(Turner); very different degrees of manifestation	Not uncommon
Various structural anomalies of X chromosomes		Not uncommon
XYY	Increased stature; occasional behavioral abnormalities	1/800 ♂
XXYY	Increased stature; otherwise resembling Klinfelter syndrome	Rare

Table 2.10 also lists the most frequent mosaics. It does not include interchanges involving the X chromosome. The same rules apply here as for reciprocal translocations of autosomes. A great variability in symptomatology has been observed in single families.

Some of these interchanges became important for our theoretical concepts of X inactivation. The most severely malformed phenotypes are found in the gonadal dysgenesis group that is, at the same time, clinically as well as cytogenetically the most heterogeneous. Some subcategorizations have been proposed on clinical grounds [88, 477], the most important being that between simple gonadal agenesis without additional symptoms and Turner's syndrome with the findings shown in Fig. 2.70. But the cytogenetic data show little if any correlation with these categorizations.

The XO type is much rarer than the XXY and XXX types. The expectations for the different types of zygotes can be derived from the distribution of germ cells after nondisjunction within first and second meiotic divisions in the two sexes (Fig. 2.71), with the additional assumption that the probability of being fertilized by an X or Y germ cell is ½ each for all female germ cells, and that all male nondisjunctional products will fertilize X oocytes. The events may be designated:

- NDF1 — Nondisjunction during the first meiotic diivision in females
- NDF2 — Nondisjunction during a second meiotic division in females
- NDM1 — Nondisjunction in first meiotic division in males
- NDM2X — Nondisjunction affecting X chromosomes in second meiotic division in males
- NDM2Y — Nondisjunction affecting Y chromosomes in second meiotic division in males

The relative frequencies (probabilities P) of the different types of zygotes to be expected are:

XXY: $\frac{1}{3} P_{NDF1} + \frac{1}{7} P_{NDF2} + \frac{1}{2} P_{NDM1}$
XXX: $\frac{1}{3} P_{NDF1} + \frac{1}{7} P_{NDF2} + \frac{1}{4} P_{NDM2X}$
XO: $\frac{1}{3} P_{NDF1} + \frac{1}{7} P_{NDF2} + \frac{1}{2} P_{NDM1} + \frac{1}{4} P_{NDM2X} + \frac{1}{4} P_{NDM2Y}$
XYY: $\frac{1}{4} P_{NDM2Y}$
XX : $\frac{2}{7} P_{NDF2} + \frac{1}{2} P_{NDM2Y}$
XY : $\frac{2}{7} P_{NDF2} + \frac{1}{2} P_{NDM2X}$

Hence, in the absence of other influences, XXY should be slightly more common than XXX if NDM1 and NDM2X have the same probability. There is some evidence in autosomes that nondisjunction is more common in the first than in the second meiotic division (Sect. 5.1.2.3).

Theoretically, XO zygotes should be somewhat more frequent than any other types. This expectation does not fit at all with the observed data, as all the karyotypes together that lead to the Turner syndrome are much rarer than XXX or XXY. This points to strong selection against germ cells without the X chromosome and/or to strong intrauterine selection against XO zygotes. The latter expectation is corroborated by observations on abortions, among which the XO type is, indeed, usually frequent. Another line of evidence points in the same direction: the risk of nondisjunction in general increases with the age of the mother (Sect. 5.1.2.2). For XXY and XXX karyotypes, this increase can be clearly demonstrated; but not for the XO karyotypes. Hence, it is now generally assumed that surviving XO zygotes are the result not of meiotic but of mitotic nondisjunction or of early chromosome loss. The relatively greater fraction of mosaics in this group compared with XXX and XXY points in the same direction.
XYY zygotes, on the other hand, can only be formed by nondisjunction during the second meiotic division in males. Nevertheless, they are about as frequent as XXY zygotes. Therefore, the probability for nondisjunction of Y chromosomes, P_{NDF1}, seems to be much higher than all probabilities for X chromosome nondisjunction. Mosaics have been observed for all types. The mechanisms for mosaic formation are discussed in Sect. 5.1.6.

Intersexes. From clinical observation three types of intersexes are distinguished:

1. True hermaphroditism: germ cells of both sexes are present
2. Male pseudohermaphroditism: only testicles are observed

Nondisjunction in:		♀		♂			
1st meiotic division	Products: of 1st meiotic division	XX	0	XY	0		
	Products of 2nd meiotic division (GERM CELLS)	XX XX	0 0	XY XY	0 0		
Nondisjunction in:				X		Y	
2nd meiotic division	Products of 1st meiotic division	X	X	X	Y	X	Y
	Products of 2nd meiotic division (GERM CELLS)	XX 0	X X	XX 0	Y Y	X X	YY 0

Fig. 2.71. Distribution of germ cells by nondisjunction of X and Y chromosomes in first and second meiotic divisions. For details see text

3. Female pseudohermaphroditism: only ovaries can be found

Unfortunately, this simple categorization is not supported by the cytogenetic evidence. Many different karyotypes can be found, even 46,XX males. Many intersexes are mosaics for cells with different sex chromosome complements in various combinations. The phenotype of 45,XX/46,XY mosaics, for example, may be ovarian dysgenesis, gonadal dysgenesis with male pseudohermaphroditism, or "mixed gonadal dysgenesis" - one gonad being a streak, the other a dysplastic testicle. Of the true hermaphrodites, some have a 46,XX karyotype. Others are 46,XX/46,XY mosaics, or XY, or 46,XX mosaics. This XX/XY state may originate from any of nine different mechanisms, such as fertilization of the oocyte by two different sperms, fusion of two fertilized eggs, mitotic errors during cleavage, or exchange of blood stem cells between dizygotic twins of different sex during embryonic life (Sect. 3.8.3).
The primary function of the sex-determining factors is induction of gonads. They, in turn, determine development of the other sexual organs and the secondary sex characters. Disturbances of gonad induction may be caused either by an abnormal sex chromosome complement or by other interfering factors not directly involving the sex chromosomes. In the latter case, the intersex may have a normal XX or XY constitution. Balanced structural changes involving the X often lead to infertility in both sexes [427].

Y Chromosome as the Male-Determining Unit. The development of testicles is determined by the Y chromosome. Bühler [315], collecting and evaluating the scattered reports on patients with abnormalities of sexual differentiation and structural anomalies of the Y, tried to localize various functions in

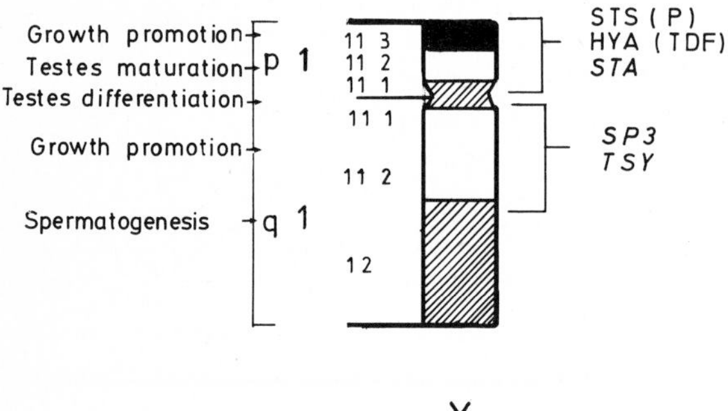

Fig. 2.72. *Left-hand side:* Functions, derived from anomalies in patients with aberrations involving the Y chromosome. (Bühler, 1980 [315]) *Right-hand side:* Localization of Y-linked DNA probes; from McKusick's (1985) tabulation

euchromatic regions of the human Y (Fig. 2.72). This figure, however, does not contain information gathered from Y-specific DNA probes. The various types of X and Y anomalies have taught us much about sexual development in humans. However, the available evidence on this problem also comes from metabolic anomalies with simple modes of inheritance, from animal experimentation with artificial chimeras, and from the HY antigen. Moreover, knowledge on the genetic regulatory activity of hormones, and specifically the concept of receptor diseases, is necessary for a full understanding. Therefore, the discussion is deferred to Sect. 4.7.5. Only one specific aspect will be discussed here - dosage compensation.

2.2.3.3 Dosage Compensation for Mammalian X Chromosomes [357]

Nature of the X Chromatin. Once the X chromatin had been discovered by Barr and Bertram (1949) [298], there were speculations regarding its nature. In analogy to *Drosophila,* it was first assumed to consist of heterochromatic parts of the two X chromosomes. Demonstration of its bipartite nature seemed to corroborate this conclusion. However, Ohno et al. (1959) [463; 464] showed it to represent one single X chromosome. In diploid prophase preparations of regenerating rat liver cells, the X chromatin body of the preceding interphase was not resolved as heterochromatic regions of two chromosomes. Instead, a rather large chromosome heavily condensed along its entire length was regularly observed. In sharp contrast, no such condensed chromosome was seen in male cells. It was concluded that each X chromatin body represents a single heteropyknotic X chromosome. Despite the single derivation of the sex chromatin body, a sexual dimorphism occurs because the X chromosome of the male, as well as one of the two Xs of the female, is euchromatic.
This conclusion was confirmed in other mammals, and Taylor (1960) [525] showed by labeling of the late S phase with ^{3}H thymidine that the female heterochromatic X shows DNA replication only near the end of the S phase in somatic cells of the Chinese hamster. Taylor's finding was confirmed in many other mammalian cells. The heterochromatization of the X occurs in the early embryonic stage. Cleaving mammalian zygotes have no X chromatin. The time of its first appearance in various species ranges from the blastocyst to early primitive streak stages, from a cell number of about 50 in the pig to probably thousands in humans and from before to after implantation. In the human trophoblast, it appears at the 12th day of development and in the embryo proper

at the 16th day. The X chromatin is formed rather suddenly in the entire embryo.
Evidence from aneuploid human individuals with more than two X chromosomes shows that only one X chromosome remains in the euchromatic stage, whereas all others are heterochromatic.

X Inactivation as the Mechanism of Gene Dosage Compensation: Lyon Hypothesis. M. Lyon (1961) [425] made the step from morphologic evidence to function, concluding that the heteropyknotic X chromosome may be either paternal or maternal in origin and is functionally inactive. With this, she created one of the most fertile hypotheses in mammalian genetics:

> The evidence had two parts. First, the normal phenotype of XO females in the mouse shows that only one active X chromosome is necessary for normal development, including sexual development. The second piece of evidence concerns the mosaic phenotype of female mice heterozygous for some sex-linked mutants. All sex-linked mutants so far known affecting coat color cause a "mottled" or "dappled" phenotype, with patches of normal and mutant color, in females heterozygous for them.
> It is here suggested that this mosaic phenotype is due to the inactivation of one or the other X chromosome in embryonic development ... This hypothesis predicts that for all sex-localized gene action the heterozygote will have a mosaic appearance, and that there will be a similar effect when autosomal genes are translocated to the X chromosome. When the phenotype is not due to localized gene action, various types of result are possible. Unless the gene action is restricted to the descendants of a very small number of cells at the time of inactivation, these original cells will ... include both types. Therefore, the phenotype will be intermediate between the normal and hemizygous types, or the presence of any normal cell may be enough to ensure a normal phenotype, or the observed expression may vary as the proportion of normal and mutant cells varies, leading to incomplete penetrance in heterozygotes.

In the same year, M. Lyon tentatively explained an observation on a human X-linked disease in the same way: In X-linked ocular albinism, the male hemizygotes lack retinal epithelial pigment and have a pale eye fundus. Heterozygous females have irregular retinal pigmentation, with patches of pigment and patches lacking pigment, so that the fundus has a stippled appearance. Figure 2.73 shows this condition. Lyon also predicted that mosaicism should be demonstrable in other X-linked genes, among them the variants of the enzyme glucose-6-phosphate dehydrogenase (G6PD).

Evidence from the Human G6PD Variants. Fur color in mice as affected by X-linked mutations or a stippled appearance of the eye fundus as shown in X-linked ocular albinism in man are phenotypic characteristics separated from primary gene action by the process of differentiation. Hence, interpretation of the origin of such phenotypes can always be disputed. These findings served to suggest the hypothesis of X-inactivation but were not sufficient to prove it. A critical test of such a hypothesis should utilize simpler and less ambiguous situations. X-linked gene products whose presence can be detected at the protein level provided the experimental material. The first X-linked gene for which such an analysis became possible was the human G6PD locus. And indeed, Beutler (1962, 1964) [302; 303], without knowledge of Lyon's hypothesis, had independently developed the concept of X-inactivation by observations on human G6PD variants. In spite of the fact that females have two and males only one copy of the G6PD gene, the average level of G6PD enzyme activity was found to be the same in both sexes as well as in individuals possessing additional numbers of X chromosomes (XXX, XXY). Hence a mechanism of dosage compensation must have been acting. If a female is heterozygous for an electrophoretic G6PD variant, the hypothesis of random inactivation predicts that in some cells the X chromosome with the normal allele, in others that with the variant allele will be active. Therefore a given single cell will be capable of determining only one of the two enzyme variants. Such mosaicism was, indeed, first characterized by Beutler in red cells by ingenious but indirect methods [304] and later confirmed by a number of authors with various techniques [358; 424]. One approach utilized cloning of fibroblasts in tissue culture. In the Black population, the *G6PD* gene is polymorphic, two frequent alleles, Gd A and

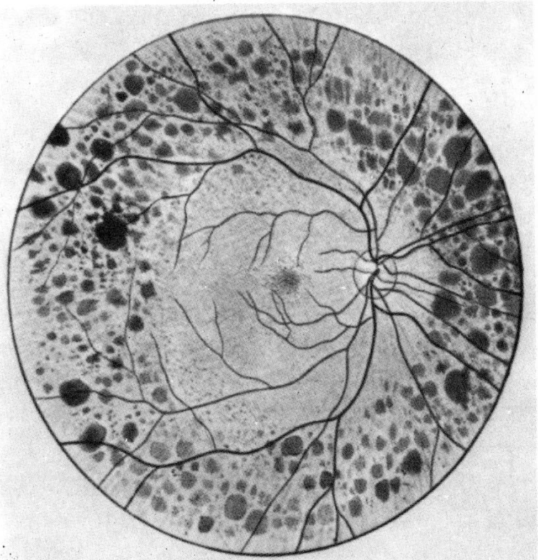

Fig. 2.73. Right fundus in a 6-year-old daughter of a male patient with X-linked ocular albinism. The distribution of pigmented patches in this heterozygote is clearly nonrandom. (Franceschetti and Klein, 1964, see [8])

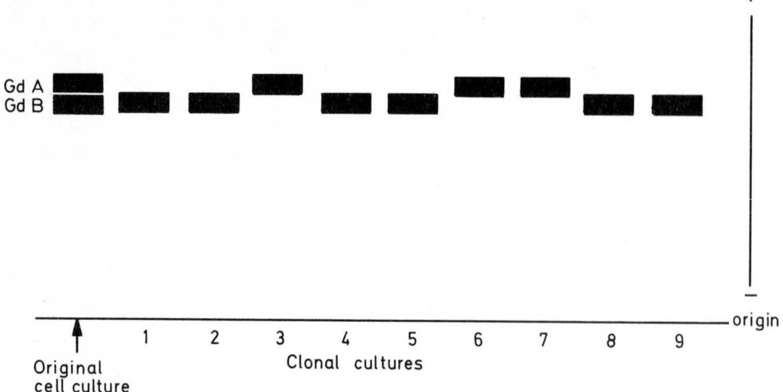

Fig. 2.74. Electrophoretic separation of G6PD components in ultrasound-treated cells from a tissue culture of a heterozygote woman with genotype Gd AB. The original cell culture shows the two G6PD components GdA and GdB. The clones derived from single cells show either GdA or GdB but not both. (Harris, H., 1980 [93])

Gd B, being present. Cloned cells of fibroblast cultures from black women heterozygous for these alleles showed either the Gd A or the Gd B variant (Fig. 2.74) but not both as found in their normal tissues. When women heterozygous for one of the *G6PD* deficiency variants were examined, the same phenomenon was observed: some cell clones had normal acitivity; in others, there was little activity.
Other evidence came from leiomyomata of the uterus in women heterozygous for the A and B *G6PD* variants. Tumor tissues invariably showed only one of the two mutant types while normal uterine tissues showed both types. This finding was possible only under three conditions:

1. Only one allele is active
2. The whole tumor was initiated by a single cell, i.e., it represents a single cell clone
3. The individual X chromosomes remain either active or inactive over the whole period of tumor growth

Hence, this experiment not only corroborated the Lyon hypothesis, but also helped to establish the formation of tumors from single cells.
Experiments with single cells were also carried out with another X-linked enzyme anomaly: the HPRT (hypoxanthine-guanine phosphoribosyl transferase) defect, and random inactivation was confirmed. This enzyme defect has been utilized for examining many problems of human gene action. Its discussion will therefore be deferred to Sect. 4.2.2.6.
Especially interesting were studies on a women heterozygous for both the G6PD locus and another X-linked enzyme, phosphoglycerate kinase (PGK) [303].
Fifty-six separate cell clones were examined, 33 of which showed, the GdA, PGK 1 phenotype, while 23 showed the GdB, PGK 2 phenotype. Had the inactivation of the two loci occurred independently of each other, clones with GdA, PGK 2 or GdB, PGK 1 would also have been expected.

Other Examples of X Inactivation in Humans. With a variety of methods, X inactivation has now been shown for a number of X-linked conditions in man. Especially interesting is the demonstration of retinal mosaicism in red-green color blindness [309]. By shining a very narrow beam of red or green light into the retinas of women heterozygous for color vision defects, patches of defective color perception could be found as would be expected if the retina were a mosaic consisting of normal and color-defective clones.
Anhidrotic ectodermal dysplasia is a rare X-linked condition. The affected males show absence of teeth, hypotrichosis, and absence of sweat glands. Patches with and without sweat glands can be recognized in many heterozygous females [470].
In chronic granulomatous disease with leukocyte malfunction (30640), [714], the bactericidal activity of granulocytes is very much reduced; they ingest staphylococci normally but are defective in their ability to digest them. Heterozygous females have two populations of leukocytes, normal and abnormal [480; 542]. In other X-linked diseases, observations have been made that are compatible with the predictions of the Lyon hypothesis [133].

Cells in Which the Second X Is Not Inactivated [424; 426]. The X chromatin body becomes visible around the 16th day, in the blastocyst stage. Functional inactivation is unlikely to occur much earlier. If one X were to be *always* inactive, the difference in phenotype between the normal male (XY) and Klinefelter's syndrome (XXY) as well as between normal females (XX) and Turner's patients (XO) would re-

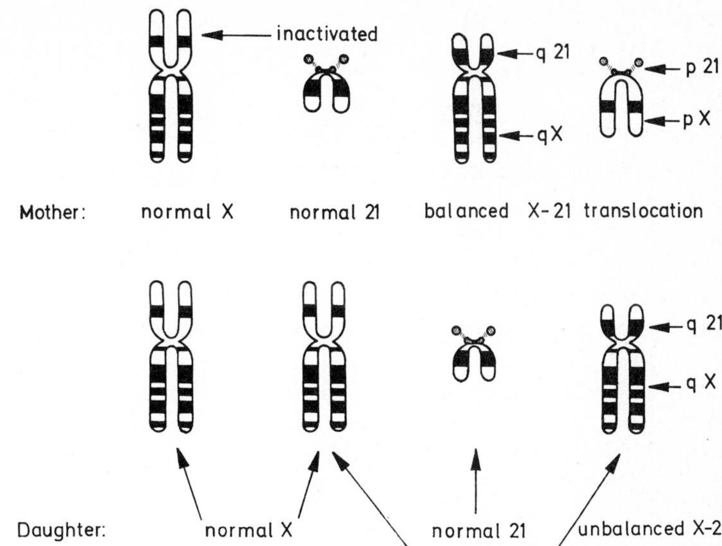

Fig. 2.75. X dosage compensation in a mother and a daughter with different sets of X-chromosomes. The mother has two X/21 translocation chromosomes and one normal X which is inactivated; the daughter has two normal X chromosomes, one of which is inactivated, and one translocation chromosome, which is also inactivated. (Summit, 1973 [501])

quire explanations other than the possibility of full gene action of the X chromosomes prior to inactivation, and there is good evidence that in oocytes as well as in male germ cells no inactivation occurs. In the mouse, the enzyme LDH is specified by an autosomal gene while G6PD, as in humans, by an X-linked gene. In fertilized XO oocytes. G6PD was found to be half as active as in XX oocytes, whereas LDH activity was the same in both [345]. This is a gene dosage effect as predicted if both X chromosomes were active.

One human blood group system, the Xg system, has an X-linked mode of inheritance. The Lyon hypothesis would predict that heterozygous females have two distinct types of red blood cells, each carrying only the antigen determined by the X chromosome active in its precursor cell. However, it became clear early in the study of X inactivation that this prediction was wrong: Two different erythrocyte populations could not be detected. One possibility was that the Xg antigenicity is not produced by the red cells themselves but is taken up from their environment, for example the serum. This, however, was disproved by observation of a blood chimera (Sect. 3.8.3) - a woman who had received blood stem cells from her dizygotic twin during embryonic life in addition to her "own" red cells. Some of her red blood cells were O and Xg^a+, others were AB and Xg^a-. Had the Xg come from the serum, all cells would have had the same Xg type - the genetically "own" type of the proband.

This riddle was solved when the Xg locus was shown to be located close to the tip of the short arm of the X chromosome and when at least one other locus that is closely linked with the Xg locus - that of the enzyme steroid sulfatase - was also shown to escape inactivation: The distal part of the short arm of the human X is *not* inactivated [486; 487]. This part seems to extend over the bands Xp 22.13 and Xp 22.3 [502], since, unlike the rest of the inactivated X, these regions replicate early.

Which Is Earlier, X Inactivation or X Chromatin Formation? At first glance, it is easy to visualize how the X chromosome first permanently condenses by forming X chromatin and then is unable to function properly. The real course of events, however, seems to be the opposite: Inactivation occurs first and is followed by X chromatin formation. The evidence leading to this conclusion is that X chromatin formation is never complete in all cells; for example, clonal fibroblast cultures show about 30% of the cells without X chromatin. The percentage seems to be independent of the cell cycle. Functional inactivation as measured by G6PD activity was almost complete. Neither was there any relation of enzyme activity to number of X chromatin bodies in fibroblasts from humans with different numbers of X chromosomes. The exact mechanism of inactivation is still under examination [357].

Genetic Differences in X Inactivation Patterns? Cattanach (1975) [321] described an X-linked gene in the mouse that controls X inactivation (X chromosome controlling element, Xce). Its "high variegation" mutant (O^{hv}) causes the X chromosome on which it is located to remain in the active state. From this observation in the mouse, it can be inferred that X inactivation may be under genetic control in humans as well. Nonrandom inactivation should result in occasional heterozygotes with clinical symptoms of an X-linked recessive disease. If inactivation occurs early enough during embryonic development - at a time when the number of stem cells for the tissue in question is still small enough - such affected heterozygotes should occasionally occur as well. They represent one of the extremes of the binominal distribution of all inactivation patterns. However, the hypothesis of random inactivation does not predict accumulation of such cases within the same sibships. Such an accumulation was observed, however, in the Duchenne type of muscular dystrophy [451], and in a family with Fabry's disease, a sphingolipidosis [488]. In this family, the eight heterozygous daughters of a male patient formed two classes: a "high group" of four with the 50% α-galactosidase A activity and a "low group" with an average activity of 20% in leukocytes. The authors discussed the hypothesis of a gene that causes preferential inactivation of the X chromosome carrying the normal allele. The above-mentioned observations in heterozygotes of the Duchenne type of muscular dystrophy could possibly be explained in the same way. Careful assessment of gene activity in heterozygotes of X-linked diseases may lead to other similar observations.

X Inactivation and Abnormal X Chromosomes [529]. When the first abnormal X chromosomes in humans - such as isochromosomes of the long arm, ring chromosomes, or deletions of parts of the long arm - were observed, the rules of inactivation appeared simple: The abnormal X was always inactivated, leaving the cell with one normal active X. To explain this specific inactivation pattern, two hypotheses have been brought forward. According to the selection hypothesis, the normal and the abnormal X were originally inactivated at random in the same way as two normal X. Cells with inactivation of the abnormal X chromosome would be genetically grossly unbalanced and for this reason would have a lower divsion rate than the effectively normal cells with inactivation of the normal X chromosome. The second hypothesis assumed inactivation to be an inherent property of the abnormal X [156].

In the meantime, a number of translocations with involvement of the X chromosome have been discovered, which show that neither of these two hypotheses can be entirely correct. There are three such groups of translocations: those with 46 chromosomes and a balanced reciprocal translocation, practically all of them the X-autosome type; those with 46 chromosomes and an unbalanced X-autosomal or X/X translocation; and those with 45 chromosomes and an unbalanced X-autosomal translocation. In the following, only the first group will be considered. Observations of the second and third basically confirm the conclusions from the first.

In most cases of such translocations, the normal X chromosome is inactivated; the phenotype is that of a multiple gonadal dysgenesis, sometimes combined with mild signs of Turner's syndrome. Some families were observed where in one member, the normal, and in another member, the abnormal X chromosome was inactivated. For example in one family, the mother had a balanced X/21 translocation (Fig. 2.75). One translocation chromosome was formed from the long arms, the other from the short arms of one X and one No. 21, the breaking point being close to the centromere - on which side could not be decided. In this woman, the normal X chromosome was inactivated, as evidenced by its late replication. Her cells showed one X chromatin body. Her daughter had the large translocation chromosome, but not the smaller one, and two normal X chromosomes. One of the latter was inactivated but - unlike in the mother's case - the translocation chromosome was also inactivated. Hence, dosage compensation was achieved in mother and daughter alike. In the daughter, however, inactivation extended beyond the boundaries of the X into the translocated long arm of no. 21, creating additional clinical findings similar to those occasionally described in monosomy 21.

This case shows that the hypothesis according to which inactivation is determined by the structure of the abnormal X cannot generally be true. In this family, however, and in others in which only one inactivation pattern was found, the pattern was one that left the cell genetically relatively balanced.

The abnormal phenotypes of most of these balanced translocation carriers involving the X-chromosome are remarkable, as in all balanced translocations between autosomes the carriers are normal. Two explanations are possible between which the observations do not discriminate. Either the continuity of a certain region of the long arm is needed for full female differentiation - in which case the defective phenotype would be due to a phenomenon called in experimental genetics a position effect - or the inactivation of the same X in all cells causes the effect, possibly due to functional hemizygosity of a recessive gene. Conceivably, one or the other of the mechanisms is at work in different cases.

In some patients with unbalanced X translocations, bipartite X chromatin bodies have been found that do not occur in normal individuals. Still another observation is relevant for the mechanism of X inactivation. Whereas many cases of isochromosomes i(Xq) have been observed, only very few cases of i(Xp) have become known in spite of the fact that these cases should be quite as frequently produced by abnormal centromere division [519]. On the other hand, Xq deletions are known to occur. All these observations have led to the hypothesis that an inactivation center exists on the proximal part of the long arm of the X chromosome. If this center is present in the abnormal chromosome, it can be inactivated. If two centers are present - as in some unbalanced X translocations - two X chromatin bodies may be formed. If no center is left - as in most i(Xp) chromosomes - inactivation cannot occur, and the zygote, being functionally trisomic for the short arm of the X chromosome, will, as a rule, be grossly unbalanced and incapable of normal development. In a more recent study of three cases with a partial Xq deletion, the inactivation center could be localized near the border between the proximal Q-dark and Q-bright region (~q13) [530].

Some observations even seem to indicate that the inactivation impulse created by this center may extend beyond the limits of the X chromosome in the direction of the short arm but not the long arm. Figure 2.76 shows the various types of abnormal X chromosomes, their inactivation patterns, and phenotypes.

Much more work on X-autosome translocation has been reported in the mouse than in humans. Some data in this species also support the hypothesis that an inactivation center may exist. It might even be possible to enhance or diminish the activity of this center as measured by the degree of inactivation of an autosomal segment translocated to an X by selection experiments.

There are many hypotheses and several studies on the molecular mechanism of X inactivation [357]. So far, however, no definite conclusions can be drawn.

X-Inactivation in Spermatogenesis? Liftschytz and Lindsley (1972) [420] brought forward the hypothesis that X inactivation occurs not only in individuals having more than one X chromosome, but also in primary spermatocytes during normal spermatogenesis, and that it might be necessary for normal sperm maturation. Male patients with Down's syndrome are known to be infertile due to spermatogenic arrest. Studies on spermatogenesis in a patient with Down's syndrome showed pairing of the supernumerary chromosome 21 with the X-Y complex in pachytene [399]. Some observations in mice might also be explained by the assumption that meiotic pairing of autosomes or autosome arms with the unpaired portion of the X is common, and that this might inhibit X inactivation and sperm maturation.

2.2.4 *Chromosome Aberrations and Spontaneous Miscarriage* [413]

Incidence of Prenatal Zygote Loss in Humans. About 15% of all pregnancies in humans end by recognizable spontaneous abortion, if abortion is defined as pregnancy termination before the 22nd week (body weight of the embryo: 500 g or less). However, there is good evidence in humans as well as in other mammals that many more zygotes are lost at an earlier stage of development; they are often severely malformed [316; 318; 413]. According to recent estimates, almost 50% of all conceptuses are lost within the first 2 weeks of development, - before the pregnancies are recognized [498]. In humans, this early zygote loss usually goes unnoticed. In earlier years, the amazingly high incidence of abortions was thought to be due to environmental factors such as a low-grade endometritis, which would interfere with normal nutrition of the embryo. However, a high frequency of developmental abnormalities suggested additional, endogenous causes. When chromosomal aberrations in humans were shown to cause multiple malformation syndromes with a reduced chance for survival, chromosomal studies of aborted embryos were performed.

Incidence of Chromosome Aberrations. Already in 1961, two abortuses with triploidy were described [335, 475] and in 1963 the first two series of cytogenetic studies on abortions [316, 324] showed a surprisingly high fraction of chromosomally abnormal abortuses. During the years to follow, a great number of series were published. In a recent survey, 3714 specimens from fairly unbiased series were mentioned, 1499 of which (40.4%) had chromosomal aberrations [413]. There was considerable variation between these series in the fraction of abnormal karyotypes, probably due to selective factors such as maternal age, failures of tissue culture, or gestational age. The latter appears to be the most important parameter. In Fig. 2.77, the available data are broken down according to gestational age. The highest frequency of miscarriages was found between the 8th and the 15th week of gestation. It goes down to about 5% in the last weeks of pregnancy. The relatively low frequency in the early weeks is explained by longer retention of aberrant embryos in the uterus and by the fact that such early pregnancies are often unrecognized. Assuming 15% as the incidence of perceptible spontaneous abortions among all recognized pregnancies, antenatal zygote loss due to chromosome aberrations can be estimated to be about 5%-6%. This is about ten times more than the incidence of chromosome aberrations among living newborns (about 0.5%-0.6%; Sect. 5.1.2.1). Moreover, this figure does not include loss of zygotes before implantation in the uterus. There is now good evidence from various sources, for example from mutation experiments (Sect. 5.2) that preimplantation loss may be even higher. Obviously, spontaneous abortion is a even higher. Obviously, spontaneous abortion is a powerful tool for early elimination of defective zygotes.

Types of Chromosome Aberrations in Aborted Fetuses. At the onset of abortion studies, it became obvious

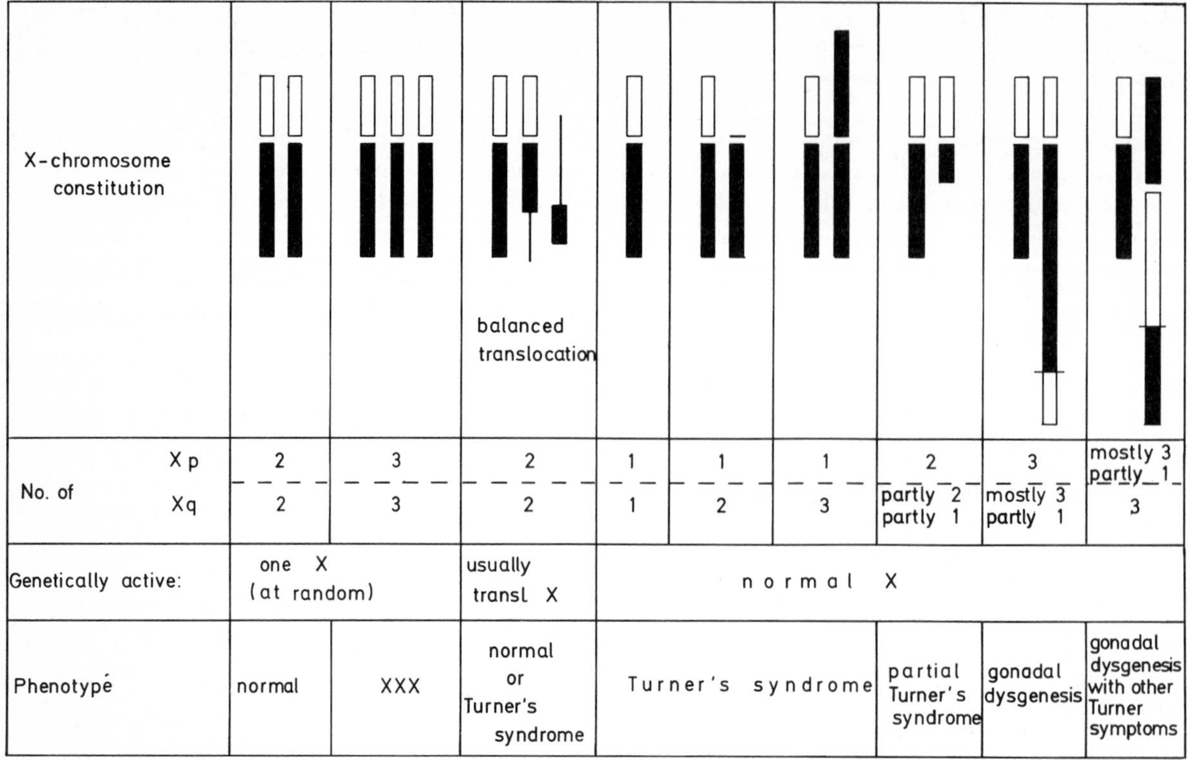

Fig. 2.76. Monosomy, disomy, and trisomy for different parts of the human X chromosome, their inactivation patterns, and phenotypic effects. (Therman and Pätau, 1974 [529]) The long arms are drawn in *black*, the short arms in *white*. This figure is based on evaluation of numerous reported cases

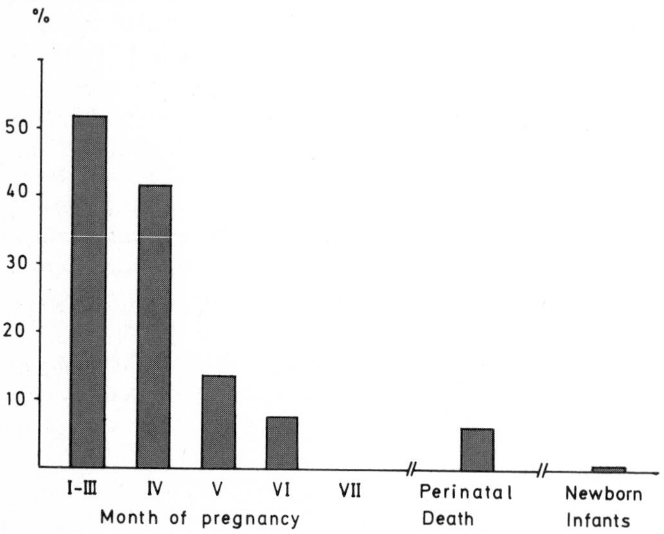

Fig. 2.77. Frequency of chromosome anomalies among 1,641 spontaneous abortuses in relation to duration of pregnancy. The frequency among 675 infants dying during the perinatal period and among 59,749 newborns is also shown. (Data of various authors; Lauritsen, 1976 [413])

that the distribution of types of chromosome anomalies observed in abortions differed from that in newborns. Some aberrations, such as the XO type, were present in newborns as well as in abortions; others, for example triploidies, led almost always to miscarriage and were compatible with birth of a living child only in exceptional cases (Sect. 2.2.1); others such as trisomy 16 were exclusively observed in aborted fetuses. A more detailed analysis had to await introduction of banding methods. Creasy et al. (1976) [329] have published the most comprehensive data:

Fifteen large hospitals in southeastern England collected all products of spontaneous abortion between September 1971 and April 1974: 2607 presumptive spontaneous miscarriages were studied.
A fetus or sac was identified in 1767 singleton and 36 twin pregnancies; in the remaining 804, neither an embryo nor a sac was obtained. Tissue culture was attempted for 1655 of the singleton specimens; the rest were rejected because they were extremely macerated, or had been incorrectly treated before they arrived in the laboratory; 513 cultures produced no growth, and a further 201 showed initial cell proliferation but no metaphases could be evaluated for chromosomes. Thus, the karyotype of 941 singleton spontaneous abortions could be determined. This demonstrates how many abortions are usually lost even in well-designed studies due to unavoidable technical reasons. This will inevitably bias the results as to incidence of chromosome anomalies among abortions in general and relative frequencies of specific aberrations.
Of the 941 singleton abortuses, 287 (30.5%) were chromosomally abnormal. Table 2.11 shows frequencies of the main groups of trisomies. One-half were primary autosomal trisomies, nearly one-quarter were X monosomies, and one-eighth were polyploid. The remainder were mostly monosomies or translocations. 89 of the 143 primary autosomal trisomies were identified by banding techniques. Of these, 35 had an extra chromosome 16. Extra chromosome 21 and 22 were each encountered in slightly more than 10% of all trisomies, while an additional chromosome 2 or chromosome 18 accounted for just over 5% each. No examples of an extra chromosome 1, 5, 6, 7, 11, 12, 17, or 19 were detected. Of the 36 twin pregnancies, karyotyping was possible in 26 cases in at least one twin. No chromosome anomalies were found.

Phenotypes of Abortuses. There were significant differences in phenotypes between conceptuses with the various chromosome complements. The presence of an extra chromosome 2 or 3, for example, may be incompatible with the formation of an embryo and leads to the production of an empty sac.
Trisomy 9 seems to result in limited and abnormal embryonic development; this is in accordance with the occasional observation of living though severely malformed newborns (Sect. 2.2.1). A fair amount of embryonic development, albeit with malformations, seems to be compatible with all types of trisomy D. Trisomy 16, on the other hand, leads to severe and early developmental disturbance; empty sacs and severely disorganized embryos were observed in most cases. In contrast, trisomy 18 causes much less disturbance, again fitting with the relatively more frequent survival into postnatal age. Of the two types of trisomy G, trisomy 21 is compatible with better development than trisomy 22. Yet, the authors estimated from their own as well as from literature data that more than 60% of all zygotes with trisomy 21 are aborted!
The widest variability in phenotypic manifestation has been encountered in this study - as well as in many others - among XO zygotes that, at the same time, constitute the most frequent single karyotype among all observed abortions. From apparently normal embryos to empty sacs, a wide range of phenotypes was observed. Characteristic are hygromata, i.e., edematous thickening of tissue, also observed in living XO newborns (Sect. 2.2.3).
The 12 triploids were mostly embryos and fetuses with various malformations (Sect. 2.2.1). In contrast to this, tetraploids were nearly all intact empty sacs; two of them with abnormal amniotic cavities. Hence, this abnormality seems not to be compatible with the development of an embryo.

Table 2.11. Frequency of different autosomal trisomies (Creasy et al., 1976 [329]) in aborted material (percentage in 183 cases)

1	–
2	4.48
3	1.12
4	1.90
5	–
6	0.53
7	1.60
8	3.72
9	3.72
10	2.13
11	–
12	–
13	2.36
14	6.50
15	10.04
16	32.11
17	–
18	5.58
19	–
20	1.90
21	12.54
22	9.76
Total	99.99

Referring again to the collection of 3714 abortions from recent studies [498], somewhat more than half the abnormal karyotypes were trisomies, ≈20% were monosomies, ≈18% polyploidies, ≈3% structural anomalies, and the rest others. All types of trisomies, excepting trisomy 1, were observed, albeit with greatly varying frequencies. These frequencies gave rise to a number of hypothetical calculations [498] on the overall frequency of numerical aberrations (trisomies and monosomies together). If monosomies and trisomies of all autosomes are assumed to occur in identical frequencies, and if early elimination accounts for the unequal frequencies in abortions, about 10%-30% of all human zygotes would be abnormal chromosomally at conception. To a certain degree, such speculations are corroborated by results from chromosomes studies in human sperm [431, 432]. In 1000 chromosome complements of sperm from 33 normal men, 8.5% were found to be abnormal chromosomally, among them 5.2% aneuploids. Sperms that would lead to monosomies and trisomies occurred in about equal frequencies, and all chromosome groups were present - with a slight tendency for G group chromosomes to be overrepresented. For aneuploids produced during oogenesis, no such data are available. It is well known, however, that nondisjunction during oogenesis is much more common (or leads to a fertilized zygote much more often) than nondisjunction during spermatogenesis (Sect. 5.1.2). On the other hand, it is at least doubtful whether observations on human oocytes artificially fertilized in vitro (⅔ with chromosomal aberrations; [294]) can be regarded as representative of the normal situation.

Some Conclusions. From studies on chromosomes in abortions, a number of conclusions can be derived: Various chromosomes participate in very unequal proportions in the recognizable overall zygote loss. This nonrandom pattern becomes especially obvious when the numbers and percentages of autosomal trisomies are compared. This does not necessarily indicate differences in nondisjunction rates during meiosis or during early cleavage divisions. A higher nondisjunction risk, however, is very likely for the five acrocentric pairs of the D and G groups. The apparent differences in frequency of trisomies for the remaining autosomes could easily be caused by the different times of zygote death. For example, if trisomy of No. 1 leads to zygote death already before or during morula formation, all No. 1 trisomies will go unnoticed. Phenotypic variability may be wide even among groups of cytogenetically uniform zygotes. It is especially striking between zygotes with different karyotypes. Some - such as trisomy 21 - are "near misses." Others, for example, trisomy 16, are not compatible even with early stages of embryonic development and are therefore invariably lethal. Comparison of aneuploid abortuses - together with tissue cultures of surviving carriers of aneuploidies at various levels of biochemical and morphologic analysis - may therefore become an important tool for elucidation of genetic regulation during embryonic development. This topic will be taken up again in Sect. 4 on gene action (Sect. 4.7.4).

2.3 Organization of Genetic Material in Human Chromosomes

In the first two decades of modern research on human chromosomes, many aspects of the organization of genetic material in chromosomes were analyzed; however, there was little concrete information on how this knowledge could be integrated with information from molecular biology into a molecular chromosome model. More recently, however, - and especially since the "new genetics" began to develop in the 1970s, - new information has accumulated rapidly. At present, answers are emerging for many questions thought to be unanswerable only a few years ago. In the following section, we shall try to draw a rough picture. The new genetics has had an impact on many aspects of human genetics; we shall return to these results in many of the later sections.

2.3.1 Chromatin Structure

2.3.1.1 Single-Copy and Repetitive DNA

Too Much DNA in a Human Genome? Shortly after the genetic code was deciphered - in the early 1960s -, scientists were impressed by the abundance of NA present in eukaryotic cells. According to various studies, the DNA content of a diploid human cell is in the order of $\approx 7.3 \times 10^{-12}$ g (range between 6.6 and 8.0). On the basis of molecular weights, it can be calculated that a nucleotide pair of adenine and thymine (A=T) has a weight of 1.025×10^{-21} g, whereas a nucleotide pair with guanine and cytosine (G=C) weighs 1.027×10^{-21} g. Hence, the total diploid set has about 7.1×10^9 nucleotide pairs:

$$\frac{7.3 \times 10^{-12}}{1.026 \times 10^{-21}} = 7.1 \times 10^9$$

If this DNA consisted of structural genes coding for proteins and if the average protein - like the hemoglobin genes - were comprised of about 150 amino acids, 6-7 million genes could be accomodated in the human genome [1338; 1339]. This figure is known to be too high by about two orders of magnitude; "informative" DNA sequences are intercalated with stretches that are *not* translated into an amino acid sequence. Some have some specific function; for others such a function has not been detected so far and is still a matter of speculation.
Some early evidence came from the giant chromosomes of insects such as *Drosophila* or *Chironomus*. In these, chromosome bands have an average length of about 20000-50000 nucleotide pairs (≈20-50 kilobases; kb). Genetic analysis, on the other hand, showed that one band (+interband) normally con-

tained only one gene [1042]. More direct analysis of the human genome, however, had to await the introduction of new methods.

Repetitive DNA [1317; 509; 499]. One major development was the discovery that DNA of higher organisms contains a large fraction of repetitive DNA sequences. When DNA is isolated and cut into fragments of about equal length, the double-stranded structure can be separated into single strands by heating in the presence of salt solutions. They move freely and meet one another at random. When the temperature is lowered, single strands meeting a complementary partner will connect to form DNA double strands. This offers a simple method of identifying complementary DNA strands.

When bacterial DNA is heat-denatured in this way and the fraction of newly reannealed double-strand DNA is registered in relation to concentration of molecules (C_0) and reaction time (t) the relationship is linear; logarithmic plotting gives a sigmoid curve, the C_0t curve (Fig. 2.78). A similar experiment with human DNA fragments ≈ 600 bases in length leads to an entirely different curve. Immediately after the experiment has started, a few percent of the DNA are double-stranded. The steep slope of the curve immediately afterwards shows that a further DNA fraction reanneals about 50000 times faster than bacterial DNA; still another fraction reanneals 10–1000 times as fast as bacterial DNA. The remaining 50+% of DNA shows kinetics similar to that found in bacteria. These data can be explained as follows: A few percent of human DNA consist of regions whose complementary sequences are located on the same strand but in reverse (= palindromic) order. This DNA can reanneal very fast by simply folding together. Another fraction contains repetitive sequences that reanneal to form DNA double strands; here, the speed of reannealing depends on the number of identical (or near-identical) repeats. Finally, there is also single-copy DNA with a reaction kinetics similar of that found in bacteria (Fig. 2.78).

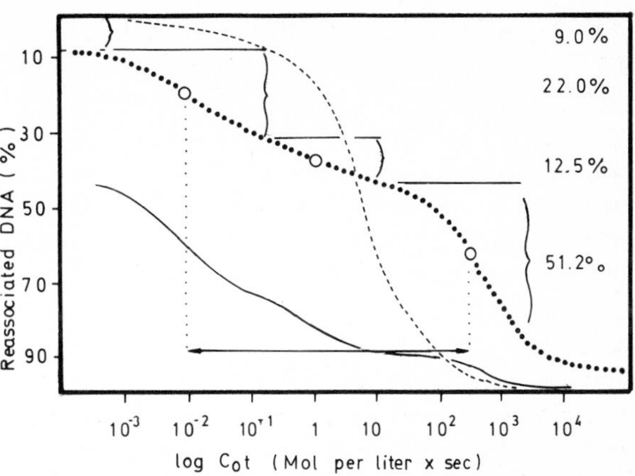

Fig. 2.78. Reannealing kinetics of bacterial DNA and human DNA of various fragment lengths. Percentage of double-stranded, reannealed DNA is plotted against the product of DNA concentration (C_0) and time (t). The dotted, sigmoid curve (-----) corresponds to bacterial DNA and is characteristic for single-copy DNA. The *curve in the middle* (....) shows the reassociation profile of human DNA fragments 600 bases in length. It can be subdivided into four classes: 9% have an unmeasurably fast $C_0t½$ value; 22% a $C_{ot}½$ value of 10^{-2}; 12,5% a $C_{ot}½$ value of 1,0 and 51,2% a $C_{ot}½$ value of 495. $C_{ot}½ = 10^{-2}$ means that reannealing is about 50000 times faster than with $C_{ot}½ = 495$. The *lower curve* (—) shows reaction kinetics with a fragment length of 1.3 kb. They will reanneal much faster. This means that most segments contain repetitive sequences. Only about 10% of the DNA behave like single-copy DNA. (Data from Schmid and Deininger, 1975; figure from Sperlig [499])

How Are Single-Copy and Repetitive DNA Located Relative to Each Other? Various studies have shown that somewhat more than 50% of human DNA consist of single-copy stretches about 2 kb long. They are interspersed with – mainly intermediary – repetitive sequences about 0.3 kb in length. Many of these repetitive sequences are very similar to each other. In addition, highly repetitive DNA sequences formed by a millionfold repetition of oligonucleotides were found in specific areas, such as the centromeric regions (cf. C-banding; Sect. 2.1.2.2) or the long arm of the Y chromosome. They often show conspicuous individual quantitative and qualitative differences with no apparent effects on the phenotype. The single-copy DNA comprises the structural genes, but these genes occupy only a minor portion of this DNA. The sequence pattern described is very widespread and occurs in species as widely separated from each other as mammals, amphibia, gastropods, and even flagellates (for details see [509]). This widespread occurrence of a relatively stable pattern suggests an important function, which, however, has not been identified so far. In some other species, such as *Drosophila melanogaster* or *Chironomus tentans* no such interspersion with short-period DNA sequence was found.

Repetitive DNA Sequences with Specific Functions. Some intermediate repetitive sequence contain genes necessary in all cells and in each phase of individual development (ribosomal RNA, histone, transfer RNA). In general, the genes for ribosomal RNA (rRNA) are part of the nucleolus organizer region; the nucleolus contains an rRNA pool. In humans, the nucleolus organizer regions comprise a part of

the short arms of acrocentric chromosomes (No. 13–15; 21; 22). In vitro RNA-DNA hybridization techniques have been used to estimate the number of rRNA genes in humans [1021; 1022]. By comparison with the fraction of human DNA that hybridizes with rRNA and the total amount of DNA in a human cell, the average total number of ribosomal genes was estimated to be of the order of 416–443.
The multigene family formed by the numerous genes for variable sequences of immunoglobulins (Sect. 4.4) has so many similar copies that the corresponding DNA sequences must be expected to be intermediate repetitive. Other multigene families, some of which may contribute to the repetitive fraction, are discussed in Sect. 2.3.6.7.

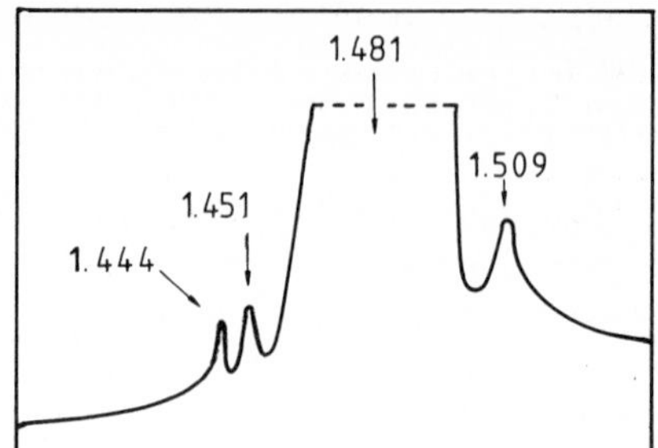

Fig. 2.79. Human satellite DNA: Analytical ultracentrifugation of total placental DNA in a cesium sulfate gradient in the presence of silver ions demonstrating the presence of satellites I (1,444), II (1,451), and III (1,509). (From Miklos and John, Amer. J. Hum. Genet. 31, p. 266, 1979)

Satellite DNA. Much of the DNA, especially in the highly repetitive sequences, has been characterized as *satellite DNA*. When fragmented DNA is centrifuged in a cesium chloride density gradient, a main band or peak is noted. In addition, however, some smaller peaks – or shoulders of the main peak – are often visible. The DNA of these smaller peaks is called satellite DNA; the number and location of satellite DNA peaks is characteristic for the species (Fig. 2.79). Their location in the cesium chloride gradient is determined by their base composition. A separate peak can be seen only if the base composition deviates from that of the main DNA fraction. Within the chromosomes, satellite DNA is usually confined to the constitutive heterochromatin; in humans, it has also been found outside the centromeric areas in the Y chromosome, and in chromosomes 1, 9, and 16. It consists of short but highly repetitive DNA sequences that might be present in several million copies.[7] Comparison of satellite DNA fractions in humans and other species – especially the great apes – has become important for our understanding of human evolution (Sect. 7.2.2).
The function of satellite DNA is unknown and has, therefore, been the subject of much speculation. An influence was suspected, for example, on recognition of homologous chromosomes during meiotic pairing or on some regulatory function in gene action. So far, no conclusive evidence has been brought forward in favor of any of these hypotheses. However, studies on *Drosophila* suggest an influence on crossing over [437]. Since the discovery of satellite DNA cytogeneticists have been struck by its location within parts of the chromatin known for many decades from microscopic analysis as heterochromatin.
For the newly discovered "minisatellite" regions, see Sect. 2.3.2.7.

2.3.1.2 Heterochromatin

Definitions and Properties. The name "heterochromatin" was coined by Heitz (1928) [469]. He wrote:[8] "In ... *P(ellia) epiphylla"* (a moss) "some parts of five of the nine chromosomes behave differently. In telophase, they do not become invisible as the remaining parts and the other four chromosomes but can be observed as such in young interphase nuclei and also in nuclei of grown-up cells ...". Maintenance of the condensed state in interphase has remained the main characteristic of heterochromatin [313]. Later, other peculiarities were discovered. DNA replication during S-phase, for example, mostly occurs somewhat later in heterochromatin than in euchromatic chromosome segments. Two subclasses are usually distinguished: Constitutive and facultative heterochromatin. In humans, the facultative fraction is represented by the inactivated X chromosome in females and in males having more than one X (Sect. 2.2.3.3).

Heteromorphisms: Function and Relation with Satellite DNA [410]. There is a large amount of interindividual variability in heterochromatin (Sect. 2.1.2.3), more than in the euchromatic parts of the genome. Such variants are called "heteromorphisms." In addition to the regions mentioned above (secondary constrictions of Nos. 1, 9, 16), heteromorphisms are found mainly in the centromeric and satellite regions of acrocentric chromosomes. In the practice of med-

[7] Satellite DNA must not be confused with the satellite regions of acrocentric chromosomes. The use of the same term is an unfortunate coincidence of nomenclature.

[8] Translation ours.

ical genetics they have been utilized, for example, for identifying the maternal vs paternal origin of chromosomes in patients with genome mutations such as Down's syndrome (Sect. 5.1.2.3), or in cases of disputed paternity. It has been known for many years that no classical genes can be assigned to constitutive heterochromatin, but most research workers are reluctant to assign no function at all to it. Many functions have been suspected. Examples include stabilization of chromatin structure and a "bodyguard" function for protection of the more valuable euchromatic DNA sequences against external impacts ([385]).

These considerations suggest that the phenomena observed by the classic cytogeneticists which led to the concept of heterochromatin are closely related to more recent data on highly repetitive DNA and satellite DNA, which are derived from entirely different experimental approaches. Satellite DNA, highly repetitive DNA, and heterochromatin are located mainly close to the centromeres, but can also be found in other regions of some chromosomes (1, 9, 16, Y). All are devoid of known structural genes, but they are not entirely identical: For example, small amounts of satellite DNA have been shown by in situ hybridization (Sects. 2.3.2.3 and 7.2.2) to be located outside centromeric constitutive heterochromatin. The reasons for these localizations are unknown.

2.3.1.3 The Nucleosome Structure of Chromatin [1172; 421]

Chemical Composition of Chromatin. In addition to DNA, chromosomes contain a number of proteins. Together with the DNA double helix, these proteins form chromatin. Most abundant are the histones, - positively charged alkaline proteins with a molecular weight of about 10000–20000. They can be subdivided into five classes (H1, H2A, H2B, H3, and H4). Other, so-called nonhistone, proteins are present in varying, but mostly smaller amounts. The nonhistone fraction is heterogeneous. It includes, for example, a number of enzymes.

Nucleosomes [1172]. The chromatin thread consists of repeat units, made up of a set of histone molecules in association with about 200 DNA base pairs. The set of histones consists of two each of the four types H2A, H2B, H3, and H4. They are folded in a globular fashion, forming a cyclinder. The DNA component of a nucleosome has two parts: a "core" of 140 base pairs, and a "linker" which varies in length from about 15 to about 100 base pairs, depending on the cell type. Such linkers apparently connect the nucleosomes with each other. The histone H1, which is about twice as long as the other histones, keeps the entire structure close together. When it is removed (which is not difficult experimentally) the chain becomes much looser (Fig. 2.80, 2.81). There is only one H1 molecule per nucleo-

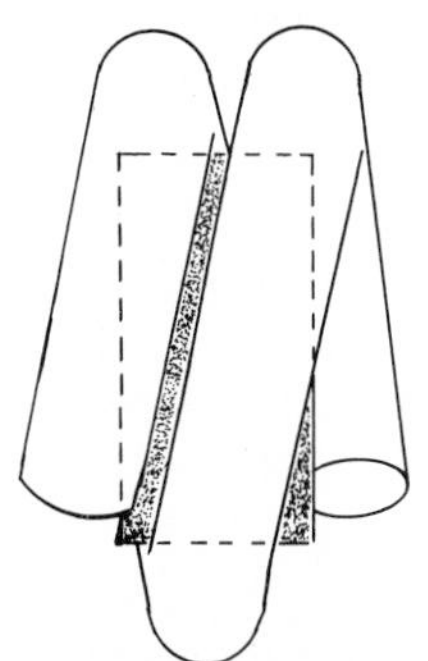

Fig. 2.80. Single nucleosome, schematic

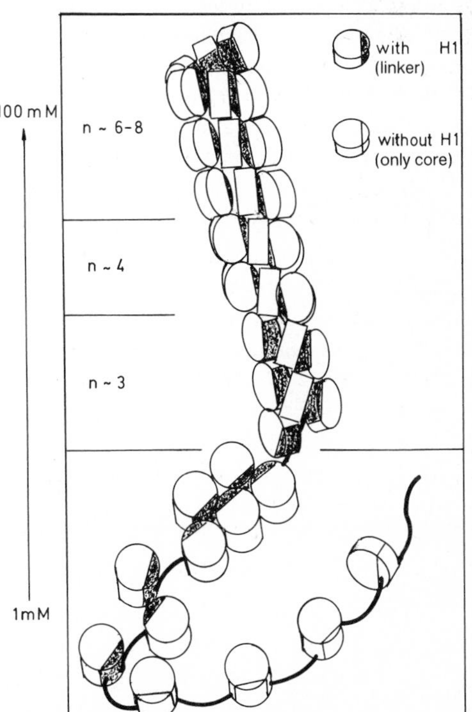

Fig. 2.81. Schematic representation of the nucleosome structure of chromatin: In in-vivo experiments, the precise structure depends on the salt concentration. At 100 mmol NaCl, 6–8 nucleosomes are combined in one turn of the chromatin thread *(above)*. At lower salt concentrations, only 3–4 nucleosomes/turn can be seen *(middle)*. In the absence of salt, nucleosomes have little contact with each other. (From Küppers, Molekulare Genetik, 3rd. ed., 1982) ▷

some. The DNA is wrapped around the set of eight histones, forming a roughly spherical particle with a diameter of about 100 Å. Such particles lie close to each other along the length of a chromatin fibre. The exact way in which the DNA is associated with histones is unknown; however, the double-helix structure is apparently undisturbed. Studies with DNA-RNA hybridization techniques (see Sect. 2.3.2.3) indicate that a wide variety of functionally different DNA stretches occur in the nucleosomes, from single-copy to repeated sequence DNA, and actively transcribed sequences as well as those belonging to constitutive heterochromatin. Probably, virtually all chromosomal DNA in eukaryotic cells is packaged in nucleosomes. The evidence for the nucleosome structure comes mainly from three lines of work: chains of particles were observed in electron micrographs of chromatin, X-ray diffraction studies suggested a repeat unit of chromatin, and enzymatic digestion with micrococcus nucleases allowed the isolation of single nucleosomes. Once the nucleosome concept had been established it stimulated additional experiments along various lines, which finally established the existence and properties of these structures.

2.3.1.4 Integration of the Chromatin Thread in Chromosome Structure

Interphase. The chromosome in interphase can be visualized as an elementary fibril that consists of a sequence of nucleosomes connected by linkers. This fibril does not extend through the entire nucleus but occupies certain domains; it is possible that chromosome segments that are being transcribed extend to the center of the nucleus. Normally, chromatin is highly spiralized. Some aspects of the precise order of this spiralization are still being disputed; the fibres representing increasing orders of spiralization in inter- and metaphase may be described as follows [1042]:

Fibril	Degree of shortening		Diameter
	to the next unit	From DNA	
DNA	1	1	10 Å
Nucleosome	7	7	100 Å
Nucleoprotein fibre (spheroid, superbead, elementary fibril)	6	42	200- 300 Å
Chromonema Interphase	40	1600	1000-2000 Å
Chromatid Metaphase	5	8000	5000-6000 Å

Mitotic and Meiotic Chromosomes. As seen in the above tabulation, chromosomes in mitosis and meiosis show a much higher degree of coiling than in interphase (Sect. 2.1.2). Their banding patterns have been discussed in Sect. 2.1.2.3. The number of subbands into which a band can be resolved varies with the degree of condensation (e.g. mitotic prophase or metaphase) and the quality of the staining method, as demonstrated especially clearly by the method of premature chromosome condensation (PCC; see Sect. 2.2.2). An upper limit appears to be given by the number of chromomeres ≈ 30000-100000 base pairs in length (see below [201 a]). Considering the number of base pairs/haploid genome ($\approx 3.5 \times 10^9$) and the number of bands seen even in the best preparations (up to ≈ 2000; Sect. 2.1.2.3), no-one has yet even come close to achieving this level of resolution. Chromosome bands can also be visualized in chromosomes during the early phases of meiosis.

Studies on replication patterns of mitotic chromosomes have shown that dark G bands (= light R bands and, as a rule, bright fluorescent Q bands) usually replicate in the second part of DNA synthesis (S phase). The single band in prometaphase chromosomes appears to be the unit of replication (which consists of a number of replicons; that is to say, replication appears to start at several points within this unit at about the same time). There have been speculations that the organization into units of replication might have some functional significance. These units contain many highly repetitive and nontranscribed DNA sequences. The number of visible bands depends on the degree of chromosome condensation, as outlined in Fig. 2.82. In a completely despiralized chromosome, every functional unit that consists of repetitive, noncoding parts and unique, transcribing portions would *ideally* be recognized as a G band together with an R band.

It follows that the R-banded (in most cases Q-darker) chromosome segments should have higher gene densities than the G-banded, Q-bright segments. In the human genome, such regions with prominent R-bands are found especially in chromosome regions 3p, 6p, 11q, 12q, 17q, and 19 (p or q). Linkage studies (Sect. 3.4.3) have indeed localized more genes in these areas than would have been expected with a random distribution of genes. Furthermore, the number of recognized abortions trisomic for these regions is lower than expected, indicating very early, and therefore undetectable, lethality of these trisomies [1481].

Many studies suggest a fine structure of the metaphase chromosome with certain "areas of constraint" (probably identical with G-dark areas) alternating with other areas in which loops may be formed under certain conditions [535].

Fig. 2.82. Production of chromosome banding (G-banding) patterns by coiling of the chromatid strand, which includes lightly stained (euchromatin-rich) as well as dark (heterochromatin-rich) areas! Note that the number of microscopically visible G bands decreases with increasing density of coiling. (Schwarzacher [201 a])

This fine structure appears to be different in areas named by classic cytogeneticists as "euchromatic" from that in areas labeled as "heterochromatic." (2.3.1.2)

2.3.1.5 Integrated Model of Chromosome Structure

These data, together with results from molecular biology (see below), suggest an integrated model of the human chromosome: It consists basically of a single DNA double helix combined with histones in nucleosomes over its whole length. In some regions, this double helix is mainly composed of repetitive sequences; highly repetitive satellite DNA stretches might be interspersed. In areas in which such repetitive sequences are abundant, - primarily the centromeric regions and secondary constrictions -, the characteristics of (constitutive) heterochromatin will ensue. In other parts, the double helix consists mainly of unique sequences, which are often about 2000 nucleotide pairs long but may in some cases be much longer, and which are interspersed with low or intermediate repetitive regions. These segments when examined with classical cytogenetic methods will have the properties of euchromatin. Under certain conditions, they may show more or less extended loops. Transcribing DNA sequences ("genes"; see below, Sect. 2.3.5) are located predominantly in these sequences, which tend to be light with G-banding and dark with R-banding. Special sequences coding for rRNA are located in the nucleolus organizer regions.

In especially well-suited cells (the relatively large oocytes of amphibia) this structure can be studied in detail, and even transcription can be observed [440] [535].

2.3.2 The Genetic Code

One of the principal achievements in the 1960s that made the new genetics possible was the breaking of the genetic code. Using synthetic trinucleotides it was shown that a given triplet of bases specified the ribosomal "translation" of a given amino acid. Soon, the codons (3 nucleotides coding for an amino acid) for all amino acids had been established (Table 2.12). All amino acids except tryptophan and methionine had more than one codon, i.e., the code was degenerate. The base in the third position of a codon had reduced specificity, since four codons that differ only in the third base are synonymous and represent the same amino acid. This feature and the tendency for similar amino acids (polar etc.) to be specified by related codons ensure that a random mutational alteration of a nucleotide has minimal effects. Three codons specified termination signals; wherever these three triplets appeared, translation was stopped. The AUG codon for methionine specifies the initiation of translation by *N*-formyl methionine at the beginning of a polypeptide chain.

The genetic code is universal and is used by organisms as far apart as viruses and humans - an impressive demonstration of the unity of life on planet earth. All mutations in human hemoglobins (Sect. 4.3) fit the coding rules worked out with lower organisms. In recent years some minor exceptions to codon usage have been noticed in mitochondria, in that UGA is a tryptophan rather than a termination codon.

2.3.3 Fine Structure of Human Genes: "New Genetics"

Around 1970, molecular biology seemed to have reached a certain degree of completeness. The structure of DNA [1347], the mechanisms of DNA replication, and the "central dogma" of gene action - transcription and translation -, as well as some ma-

Table 2.12. Genetic code

			Second base			
	DNA		A	G	T	C
		mRNA	U	C	A	G
	A	U	UUU, UUC Phe UUA, UUG Leu	UCU, UCC, UCA, UCG Ser	UAU, UAC Tyr UAA, UAG TERM	UGU, UGC Cys UGA TERM* UGG Trp
	G	C	CUU, CUC, CUA, CUG Leu	CCU, CCC, CCA, CCG Pro	CAU, CAC His CAA, CAG Gln	CGU, CGC, CGA, CGG Arg
First base	T	A	AUU, AUC, AUA Ile AUG Met	ACU, ACC, ACA, ACG Thr	AAU, AAC Asn AAA, AAG Lys	AGU, AGC Ser AGA, AGG Arg
	C	G	GUU, GUC, GUA, GUG Val	GCU, GCC, GCA, GCG Ala	GAU, GAC Asp GAA, GAG Glu	GGU, GGC, GGA, GGG Gly

* *TERM*, terminater (stop) codon

jor aspects of gene regulation were well established [220]. Since the basic structures and processes had been analyzed mainly in microorganisms, special features in eukaryotes (including humans) presented a number of additional problems; but entirely new results were not expected with the methods available at that time. In the early 1970s, however, a completely new development was triggered mainly by detection of a new research tool: the *restriction endonucleases*. The new genetics that has developed since that time has opened the way for large-scale industrial production of gene products such as biologically important proteins, and for genetic manipulation of various organisms. Our understanding of structure and function of the genetic material - especially in eukaryotes, including humans - has been deepened far beyond our keenest hopes. Completely unexpected facts were discovered, with implications for both theoretical and practical fields such as gene action, population genetics, evolution, and genetic counseling including prenatal diagnosis (Sects. 4.3 and 9.1). This enormous progress also gave rise to widespread public concern regarding possible unethical consequences of genetic engineering either by inadvertent production of dangerous germs or even by manipulation of human embryos. Many of these concerns were first articulated by the scientists actively involved in this work. At present, most scientists regard concerns about safety as largely unfounded; but many ethical problems remain, and new ones continue to arise.

Whereas in earlier decades human and medical genetics developed as a relatively separate branch of science, large parts of the field have now been incorporated into mainstream research in molecular genetics. This makes it more difficult to delineate the field. A textbook of human genetics cannot describe in detail all the techniques of molecular biology that have led to such enormous scientific progress in human genetics. More specific sources should be used [366, 493, 60]. However, the principles of the new approaches need to be understood by all medical geneticists, and also by students and research workers interested, for example, in evolution or behavior genetics. In the next section, the analysis of a group of human genes - the β-globin gene cluster (Sect. 4.3) - is described. In addition to restriction enzymes, techniques for nucleid acid hybridization, sequencing of DNA, and chromosome sorting by cytofluorometry will be discussed.

2.3.3.1 Analysis of a Human Gene

The β-Globin Gene. The hemoglobin molecule and its clinical and biochemical analysis will be described in detail in Sect. 4.3. Hemoglobin (Hb) A_1 of the human adult consists of four polypeptide chains - two α and two β chains. As was known earlier, the gene for the β-chain is closely linked to several other genes specifying hemoglobin chains; the genes for the γ chain found in fetal hemoglobin, and that for the δ chain – which is the characteristic sequence of Hb A_2, a small component found in adults. This Hb β gene cluster has now been identified and analyzed by molecular methods. What were the important steps of this analysis, and what methods were required?

Steps of the Analysis. The analysis of the hemoglobin protein (Sects. 4.3.1 and 4.3.2) was completed between the late 1940s and the early 1960s. The Hb β chain has 146 amino acids. The entire transcribing part of this gene therefore has 438 nucleotides ($3 \times 146 = 438$). Its length is ≈ 1500 Å. This small piece must be identified within a DNA thread of a total length of 2 m. The difficulty of this task becomes clearer when translated into another order of magnitude: In a thread 20 km long, a piece of 1.5 mm has to be found. Ideally, the entire Hb β gene cluster should be identified. To obtain precise understanding of this structure the following three conditions must be fulfilled:

1. The relevant DNA segments must be identified unequivocally.
2. They must be isolated and amplified so that enough material for a biochemical analysis becomes available.
3. The sequence of nucleotides must be determined.

The principles on which these three methods are based will be described briefly below. We will begin with the second, since progress made in isolation and amplification of genes has been decisive for the new genetics.

2.3.3.2 Restriction Endonucleases

The Germinal Observations. In the course of his work on infectivity of the λ phage of various strains of *E. coli*, Arber [296] discovered that the DNA of this phage was cut – and infectivity influenced – by passage of the phage through the bacterium; classical recombination processes or mutations were not involved. Moreover, this was not specific for phage DNA; any foreign DNA was cut by these bacteria in the same way. This cutting could be regarded as a defense mechanism of the cell against foreign DNA, and was performed, as further studies showed, by enzymes named restriction endonucleases. This immediately raised the question why these enzymes did not cut the DNA of their own cells. The answer found by Arber was that in order to act they only react with certain specific recognition sites of the DNA, and that these sites are protected by methylation. The restriction endonucleases that were discovered first did not cut DNA at their specific recognition sites, but at other random sites. The first restriction enzyme that cleaved DNA at a sequence specific site, the Hind enzyme, was discovered by Smith at the end of the 1960s [496]. It was used first by Nathans and his coworkers to construct a cleavage map of the genetic material of an organism – the SV40 virus [457]. Berg recognized the special advantages of DNA double strands in which the two strands were cut so that "sticky ends" were produced: One of the two strands is some bases longer than the other; these bases are now free to pair with other bases, for example, from another piece of DNA with sticky ends [299]. By this means, DNA from various sources and various species can be joined to produce recombinant DNA.

Principles of DNA Recombination Technology [2397, 60, 493]. A great number of such restriction enzymes (> 150) that cleave DNA at specific sites have now been discovered [117]. The enzyme Ri, for example, cuts DNA double strands at two different sites in such a way that two adhesive ends are produced:

```
 ↓
G–A–A–T–T–C
||| ||  ||  ||  ||  |||
C–T–T–A–A–G
          ↑
```

Adhesive ends of different DNA molecules split by this enzyme will connect by fourfold A=T pairing. A ligase is required for closing the gap. The various restriction endonucleases [117] differ in their sequence recognition sites. They can be used for a variety of purposes. A common application is the amplification of specific DNA, as required for determining the nucleotide sequence of a DNA segment or studying the conditions for gene expression. The latter problem is of great importance practically: Genes that produce functional proteins can now be introduced into bacteria and amplified; proteins otherwise available in only minute amounts can be produced in large quantities. The following principle is used: Apart from their single ring-shaped chromosome, bacteria often carry additional small, double-stranded DNA rings known as plasmids. Plasmids replicate autonomously and include genes for resistance against antibiotics and for substances that kill other bacteria, the colicins (Fig. 2.83). Such plasmids can be selected so that they are split by a restriction enzyme at one site only. Double-stranded DNA that has been cut by the same restriction enzyme may be introduced and replicated together with the rest of the plasmid within the bacterium (Fig. 2.84). The source of the exogenous DNA is immaterial. It could come, for example, from human

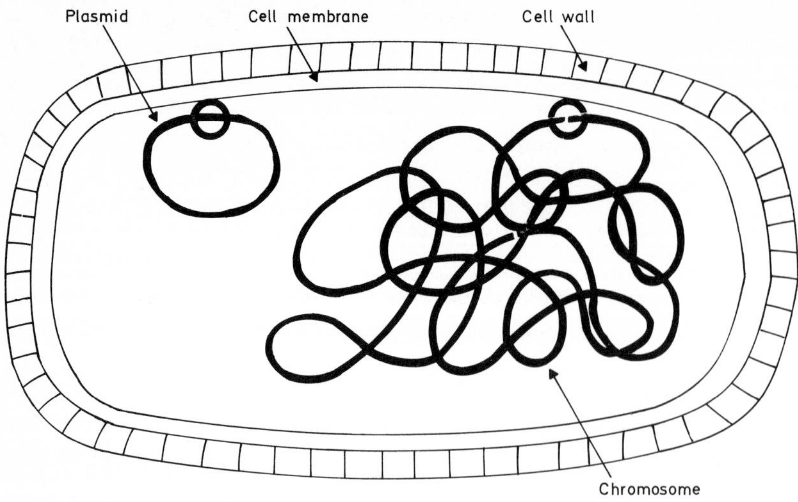

Fig. 2.83. *E. coli* cell with chromosome and plasmid. (Klingmüller, 1976 [2327])

cells, and might even consist of artificially synthesized genes.
Besides bacterial plasmids, λ phages (the objects of Arber's studies) are also being used as DNA vectors. Part of the λ genome is not essential for lytic growth of the phage; instead, the phage is able to take up foreign DNA and to propagate it together with its own genome after infection of a bacterium.
Once recombinant DNA has been replicated and amplified together with plasmid or phage DNA in transformed or transfected cells, two questions arise:

1. How can clones containing hybrid DNA be recognized among the progeny of transformed cells or viable bacteriophages?
2. How are specific DNA segments identified among the many cloned fragments?

Bacterial cells can be selected, for example, if the plasmids contain a resistance factor against an antibiotic and if the culture is grown in the presence of this antibiotic. A certain percentage of these plasmids will have no cloned DNA, and most plasmids will include DNA pieces that do not comprise the gene (or genes) to be investigated. A variety of selection methods have been developed in recent years that cannot be described in detail.
For genetic engineering of proteins it is not only necessary to select and amplify certain DNA regions, but also to induce them to express genetic activity and produce the required protein. This requires that the desired DNA sequences are combined with the machinery that promotes DNA transcription and translation, and that secondary processing at the transcriptional and translational levels is rectified.

Identification and Analysis of Genes. Another application of restriction enzymes is the analysis of presence and number of genes [332]. For this purpose a technique described by Southern (1975; [492]) is used. The total DNA of human cells is cut by a restriction enzyme into approximately 500000 fragments with a size range of 10^2 to 10^5 bp. The fragments are separated according to molecular weight by gel electrophoresis in agarose, and the double helices are separated by alkali treatment to produce single strands. The fragments are then blotted onto a microcellulose filter and fixed by baking at 80 °C. The resultant pattern represents a replica of the electrophoretically separated DNA from the original agarose gel. The DNA fragments can be identified by hybridization with radioactive DNA probes that are specific for genes or chromosomal regions. Any fragment that contains part or all of the probed gene sequence will be seen as a dark band after autoradiography. (Fig. 4.60)

Probes and Gene Libraries. An essential condition for such an analysis is availability of a gene-specific, radioactive DNA probe that can be used for hybridization (Table 2.13). In cases where messenger RNA is available, e.g., for β-hemoglobin, a specific probe can be produced by the enzyme reverse transcriptase. This enzyme catalyzes transcription of the mRNA nucleotide sequence into the complementary DNA sequence, the so-called cDNA (mRNA is not used directly for hybridization, since mRNA with sufficient radioactive label is too difficult to produce). mRNA obtained from various sources can be used to construct a cDNA gene library. Such cDNA libraries mainly contain single-sequence DNA specific for transcribing structural genes or parts of them, and DNA sequences from their immediate neighborhood. They are used mainly for finding and characterizing such genes. In addition, however, genomic libraries are available. They are produced by cutting DNA into segments using restriction enzymes, and amplifying the resulting DNA pieces in a vector. They are used, for example,

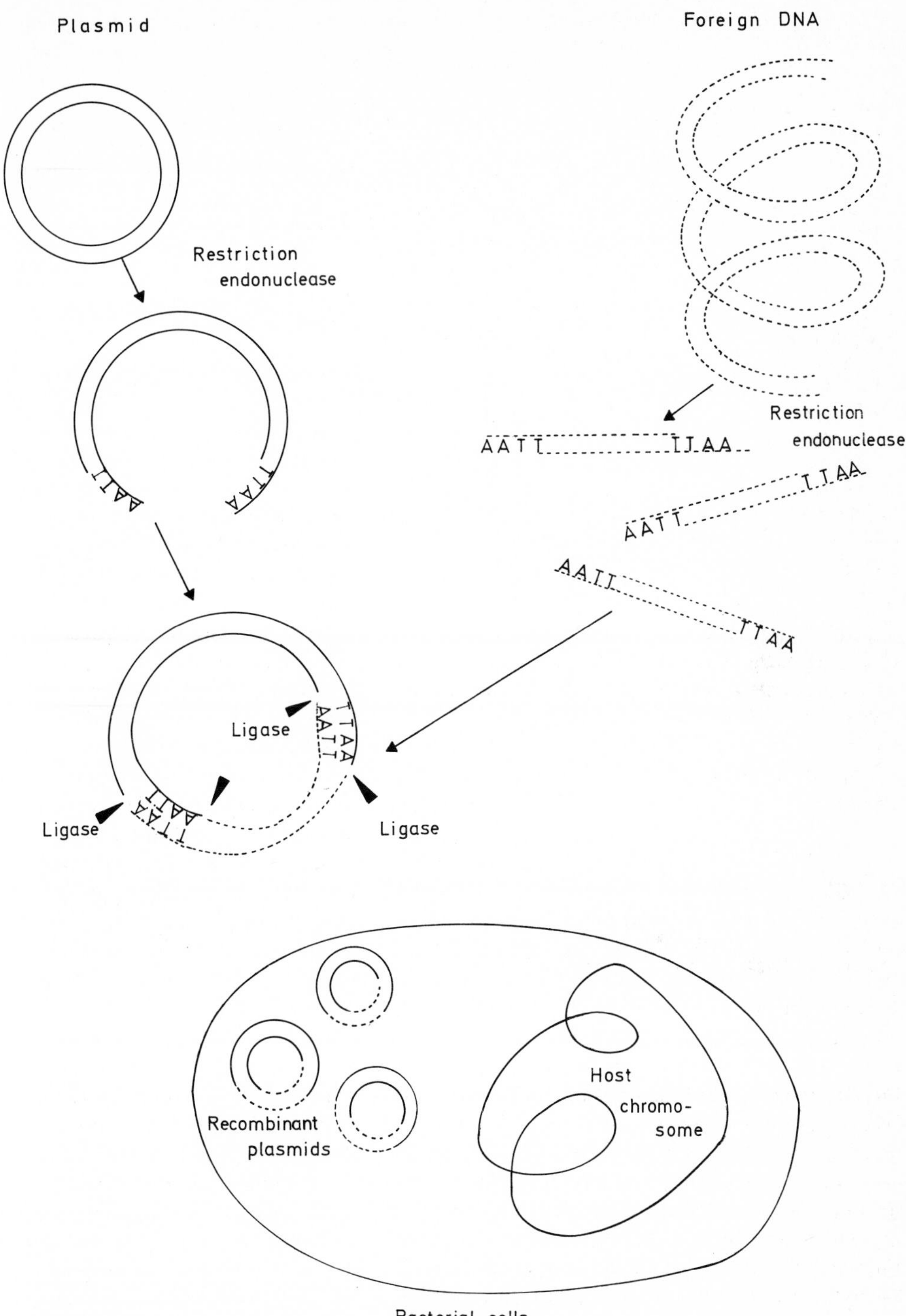

Fig. 2.84. Principle of introduction of foreign DNA into a bacterial plasmid using the RI endonuclease. (Adapted from Vosberg, 1977 [2397])

Table 2.13. DNA probes of potential medical interest (see also [328 a])

Gene specific
Globin gene cluster (α)
Globin gene cluster (gamma-delta-β)
Growth hormone
α-antitrypsin
Phenylalanine hydroxylase (PKU)
HGPRT locus (Lesch-Nyhan syndrome)
Prealbumin (amyloidosis)
Insulin
Immunoglobulin genes
Somatomammotropin
Collagen
G6PD
HLA genes
Clotting factor VIII (hemophilia A)
Clotting factor IX (hemophilia B)
Clotting factor VII
Antithrombin 3
LDL receptor (familial hypercholesterolemia)
Apolipoprotein AI
Apolipoprotein AII
Apolipoprotein B
Apolipoprotein CI
Apolipoprotein CII
Apolipoprotein E
HMG-Co A reductase
Nonspecific
Available for almost all chromosomes including X and Y chromosome. Many probes for chromosomal segments available and under development

for finding complementary sequences in the genome. In this way, distribution of homologous sequences in the genome may be investigated. Often, a restriction *polymorphism* is discovered inside a certain DNA sequence; that is to say, restriction sites vary between different individuals (Sect. 2.3.2.7). In these cases, such probes may be used for classical linkage studies in human families by methods described in Sect. 3.4.2.

Working with such genomic libraries is often unsatisfactory because of the sheer size of the human genome and the number of fragments from which the "interesting" ones have to be selected. For many problems, it is preferable to have *chromosome-specific* banks. Construction of such banks, however, requires that specific chromosomes can be isolated. Such preparations are now possible by cytofluorimetric chromosome sorting (Sect. 2.3.2.5).

Artificial oligonucleotides are another powerful tool, especially in cases where no mRNA is available for the construction of cDNA [390, 476, 527, 539]. Using our knowledge of the genetic code, they may be constructed, for example, for structural genes of protein for which amino acid sequences are available. Machines for construction of any desired nucleotide sequences are available.

When a gene has been identified – and especially when mRNA is available – its fine structure can be analyzed by combining various methods, which are described here in some detail. The ultimate goal of such studies is elucidation of the complete nucleotide sequence of the entire genetic region and assignment of specific functions in gene transcription and its control to identifiable groups of nucleotides.

2.3.3.3 Nucleic Acid Hybridization

The Principle. In Sect. 2.3.1.1, techniques for identification of repetitive DNA sequences have been described. They involve separation of DNA double strands by increasing the temperature and reannealing double strands when the temperature is lowered again. Thus, the tendency of complementary nucleic acid strands to hybridize to form double helices is exploited analytically. The same property is used for identification of electrophoretically separated DNA fragments by Southern blotting (Sect. 2.3.2.2). It occurs repeatedly in many analytical steps necessary in molecular biology. Therefore, it might be useful to describe this principle. Most naturally occurring DNA forms double helices. In a DNA double helix the pyrimidine base cytosine (C) pairs with the purine base guanine (G), whereas the pyrimidine base thymine (T) pairs with the purine base adenine (A). These complementary base pairs ($C \equiv G$; $T = A$) are connected by hydrogen bonds, which can be resolved relatively easily but have a strong tendency to reanneal, forming double helices again when single-stranded DNA chains are mixed. The origin of single-stranded DNA is not important for this reannealing; it does not even require complete complementarity of the single strands and it even works when some percentage of the bases in each strand does not fit (Fig. 2.85). Single-stranded DNA will pair (hybridize) even with RNA, provided only that complementarity of bases is maintained.

"Gene Walking". The hybridization technique may be used, for example, for analysis of a long gene when only part of its sequence is available: The probe is hybridized with sequences from a DNA library. A hybridizing library sequence will, as a rule, be longer than the probe; its ends will overlap with another library sequence and will hybridize with it in part; its free end will hybridize with the next, until a long portion, e.g., an entire structural gene, has been assembled. In this way, the structural gene for the human blood clotting factor VIII, an extraordinarily long gene with 180000 bases, has been reconstructed starting with an oligonucleotide probe only 36 bases in length. The above-mentioned way of first identifying a specific mRNA and then constructing a cDNA by reverse transcription was not feasible – mainly because of its low concentration. The DNA sequence of the oligonucleotide probe was therefore inferred from the amino acid sequence of a fac-

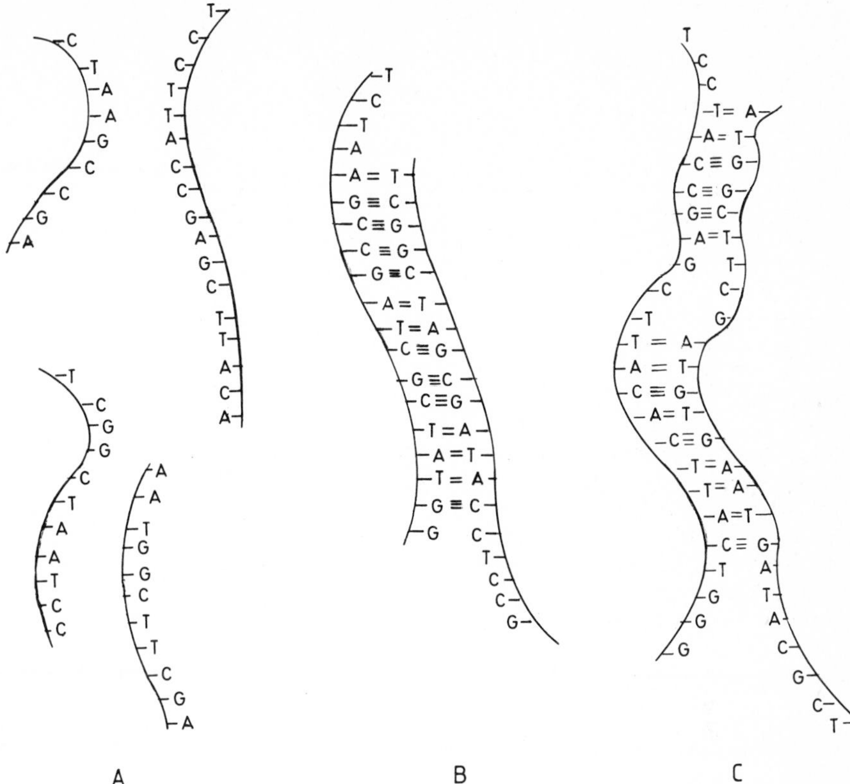

Fig. 2.85 A–C. Principle of DNA (or RNA) hybridization. **A** Nucleotide half-chains in solution. **B** The half-chains hybridize according to the pairing rules: Thymine pairs with adenine, cytosine pairs with guanine. **C** Hybridization may also occur, if matching of the two chains is not perfect, provided that the differences are not too large

tor VIII protein fragment, the precision achieved was sufficient for hybridization. For the entire analysis of the factor VIII gene, see Sect. 2.3.3.7.

In Situ Hybridization. A technique especially well-suited for analysis of eukaryotic genomes and for molecular cytogenetics is in situ hybridization. Here, metaphase chromosome preparations are treated in situ with a radioactive DNA probe under conditions permitting hybridization of this probe with the chromosomal DNA. In this way, the gene of interest can be localized on a specific chromosomal segment. When the method was introduced, it was so relatively insensitive that only highly repetitive DNA sequences such as satellite DNA could be localized. Even this led to interesting conclusions on human evolution (Sect. 7.2.2). The method has been refined to allow localization of single-copy genes such as the insulin gene ([377]; Sect. 3). In part, this was achieved by statistical evaluation of radioactivity in many metaphases in addition to the evaluation of single chromosome spreads. The essential steps are now discussed with reference to myosin heavy-chain gene as an example [482].

For this experiment, a cDNA probe for the myosin heavy-chain gene was available from the rabbit. Since homologous structural genes from various mammals are generally similar, DNA hybridization between DNA of mammalian species is not impaired. This probe was amplified in a plasmid and labeled with ^{3}H by *nick translation.* In this procedure, DNA is incubated with a small amount of DNAase I; this enzyme produces several single-strand breaks in the DNA. Radioactive nucleotide building blocks are then added, together with a polymerase; these blocks are introduced into the DNA structure.

Mitotic chromosome preparations were obtained from lymphocyte cultures. In the metaphase spreads, after chromosomal DNA had been denatured the ^{3}H-labeled DNA probe was incubated for 16–18 h at 40 °C. After removal of unbound or nonspecifically associated probe DNA, autoradiography was carried out for 3 weeks. Following staining with quinacrine mustard (Q-staining) the preparations were photographed. Figure 2.86 shows a typical metaphase plate, while Fig. 2.87 shows distribution of the label among human chromosome bands. The bulk of the label was localized to the short arm of chromosome 17, band 17 pl,2→pter. Hence, it was concluded that this myosin heavy-chain gene is located on this chromosome. Since this

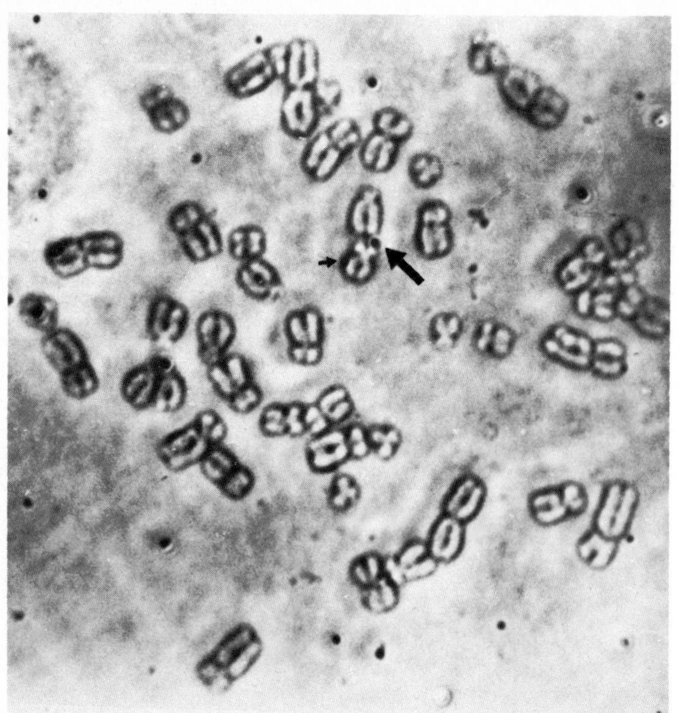

Fig. 2.86. In-situ hybridization of a myosin heavy-chain gene. Phase-contrast photograph of a representative metaphase spread after hybridization with a ^{3}H-labeled cDNA probe and exposure to autoradiographic emulsion for 20 days. The *arrow* points to a grain at the terminal end of the short arm of chromosome 17. (From [482])

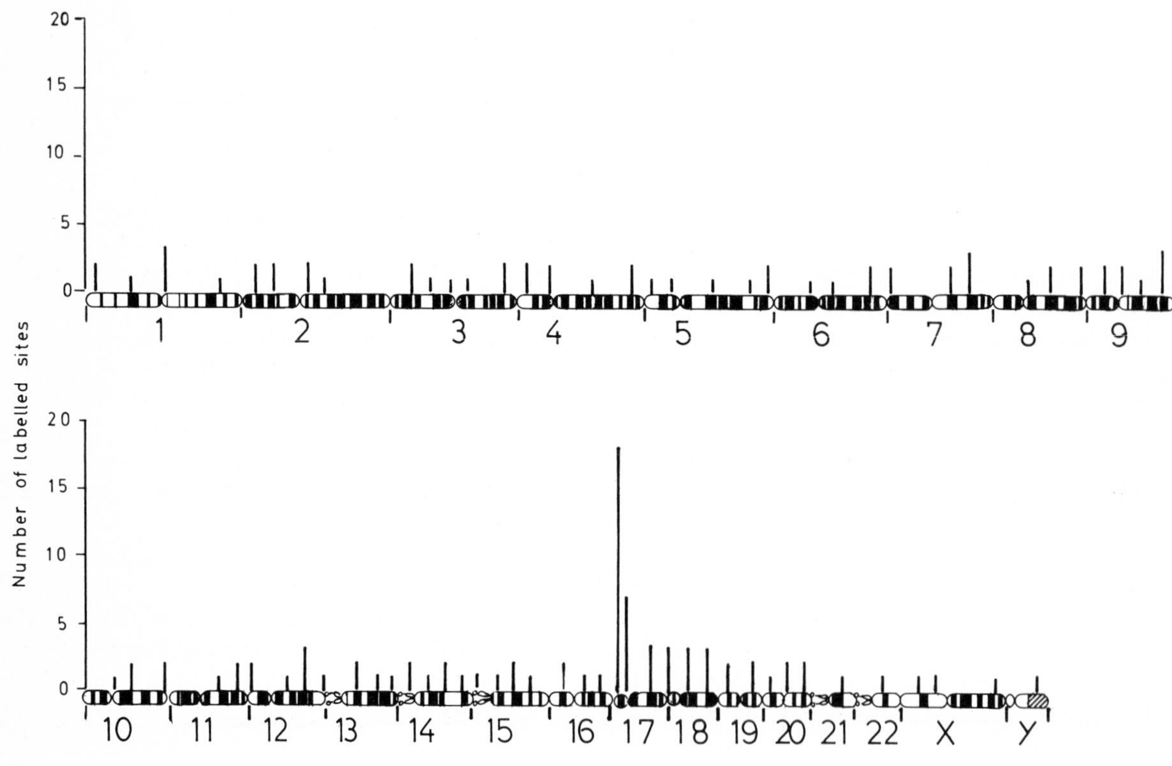

Fig. 2.87. Grain distribution in 36 metaphase spreads after in situ-hybridization with a myosin heavy-chain cDNA probe. This histogram results from an analysis based on the division of the human haploid karyotype into 110 equal segments (Chromosomes drawn in one quasi - continuous sequence). The number of labelled sites was plotted for each segment. Clustering of grains was found on the short arm of chromosome 17. (From [482])

experiment has been performed with a cDNA probe, it is theoretically not certain that the myosin gene identified in this way is, indeed, an active gene; it could also be a "pseudogene," i.e., a DNA sequence that is structure homologous to the active myosin gene but determines no myosin, having lost important flanking sequences outside the transcribed portion. Such pseudogenes have been discovered, for example, within the α- and β-globin regions (Sect. 4.3). An increasing number of human genes have been localized using this technique (Sect. 3.4) (Table 2.14).

Table 2.14. Some human genes identified by in situ hybridization

Gene, length of DNA sequence, and number of copies	Localization
β-globin (4400 bp)	11 p
α-globin (800 bp)	16 p
Insulin (900 bp)	11 p 15
LGH (550 bp) Placental lactogen growth hormone gene cluster	17 q 22–24
Interferon IFN $\alpha+\beta$ IFN γ	9 p 2.1-pter 12 q 24.1
c-myc-oncogene (2750 bp) c-mos-oncogene (2750 bp) (?)	8 q 24 8 q 22
Ig (6600 bp) heavy chain (γ4) genes (many)	14 q 32
Ig Kappa (10500 bp) light chain (VK) genes (50)	2 p 12
myb-oncogene (2000 bp) fes-oncogene (4000 bp)	6 q 22–24 15 q 24-qter
α-fetoprotein (380 bp) serum albumin (1600 bp)	4 q 11–22 4 q 11–22
c-myc-oncogene (−)	8 q
c-abl-oncogene (1100 + 600 bp) (?)	9 q 3.4
RFLP (5000 bp) D 14 S 1	14 q 31.2–3.2
Ig C lambda (203 bp) (gene family)	22 q 11
N-ras oncogene (4000 bp) (?)	1 cen – p 21
Myosin MHC (2200 bp) (?)	17 p 1.2-pter
Collagen-gene (COL1A2) (?)	7 q 22
ras (mil) oncogene (2500 bp) (?)	3 p 25

2.3.3.4 *Sequencing of DNA* [117, 122, 381]

Nucleotide Sequence and the Genetic Code. Methods for determining the sequence of amino acids in polypeptide chains have been known since the 1950s. The problem was relatively easy, since the 20 amino acids found in naturally occurring proteins all have different properties. The nucleotide sequence of DNA, on the other hand, is relatively monotonous, consisting only of the four bases guanine, cytosine, adenine, and thymine. Since the genetic code is encoded in the nucleotide sequence, and since this code had been deciphered as long ago as the early 1960s, the nucleotide sequence of the transcribed DNA segments could largely be inferred from the amino acid sequences of the corresponding proteins. However, the code is degenerate, i.e., several different triplets of nucleotides specify the same amino acid. The evidence from protein sequence is therefore ambiguous and furthermore cannot provide information regarding the sequence of noncoding DNA segments. Appropriate methods for DNA sequencing are now available [117].

The principle is the following: A long nucleic acid molecule is cut into smaller segments by agents cutting it at specific sites. Then, the nucleotide sequence is determined for these segments separately. The sequence of segments in the entire chain is ascertained by the principle of overlapping ends: Identical chains are cut a second time by an agent with different specificity, and the sequences of fragments produced by the two cuttings are compared. In this way, the entire sequence can be combined like a puzzle. Nucleotide sequences within these fragments can be identified by additional methods. Whereas in earlier times, sequencing of nucleic acids was a formidable job, it has now become relatively easy and very fast: The untractably long DNA chains can be cut into handy pieces by restriction enzymes and, if necessary, augmented by cloning (Sect. 2.3.2.2). Very long DNA chains have now been sequenced, for example the entire human mitochondrial DNA (Sect. 2.3.5) and the Hb β gene family (Sect. 4.3). More precise knowledge on non-transcribed DNA regions of importance for transcription control, such as operators or promoters, has also been achieved by DNA sequencing.

2.3.3.5 Chromosome Sorting by Cytofluorimetry

Why Do We Need Chromosome Sorting and Preparations from Single Chromosomes? Chromosome sorting by cytofluorimetry can be used for two different purposes: Characterization and measurement of fluorescence patterns of very many chromosomes within a very short time; and preparative separation of chromosomes. The first method has two advantages over more conventional methods of chromosome evaluation: It is automatic, so that the subjective element is eliminated; and it is very much faster. For example, some chromosome aberrations too small to be discovered by conventional techniques can be detected under certain conditions.

Still more important, however, the method permits preparative separation of chromosomes. Research on structure and properties of specific genes is relatively straight-forward if probes of these genes are available. The genes can then be localized by in situ hybridization, augmented by cloning, and sequenced. In addition to this extension of classical genetic analysis, however, the structure of the genetic material beyond the limits of the single gene can now be investigated. Genomic DNA libraries are a source of material for such studies. Their disadvantage is the difficulty or impossibility of selecting the few sequences of interest out of the huge number of fragments. Moreover, distribution of DNA on various chromosomes is often in itself an interesting topic for research. For such research, libraries of DNA for specific chromosomes (or even chromosome fragments) are needed.

The Physical Principle [364]. The principle can be seen in Fig. 2.88. The figure shows a single system; dual systems using two beams are also in use. For analysis by this dual system, chromosomes are stained with two dyes with fluorescence at different wave lengths. Then, the mitotic chromosomes are separated from the rest of the cell material and placed into the machine. During operation, chromosomes are forced to flow one by one across a water-filled measuring region where they are illuminated sequentially by two laser beams (e.g., UV and visible light of 458 nm wave length). The two fluorescence emissions resulting from the intersection point of the chromosome stream and the laser beam are collected by a lens and projected onto separate photomultipliers. They are then processed to give a two-dimensional representation (Fig. 2.89), one dimension for each dye and wave length. In the experiment illustrated in this Fig. 2.89, the two dyes were selected in such a way that one of them mainly stained A-T rich areas whereas the other stained predominantly G-C rich segments.

The machine can be supplemented to allow the chromosomes to be deflected from their track depending on their fluorescence patterns. In this way, relatively pure preparations of groups of similar chromosomes and even of single chromosomes can be made.

Sorting of X and Y Chromosomes. Preparative flow cytophotometry has been performed, for example, for human X and Y chromosomes [333, 330]. The X chromosome has been sorted out from an XXXX cell line by a one-beam chromosome sorter; this material has then been used to

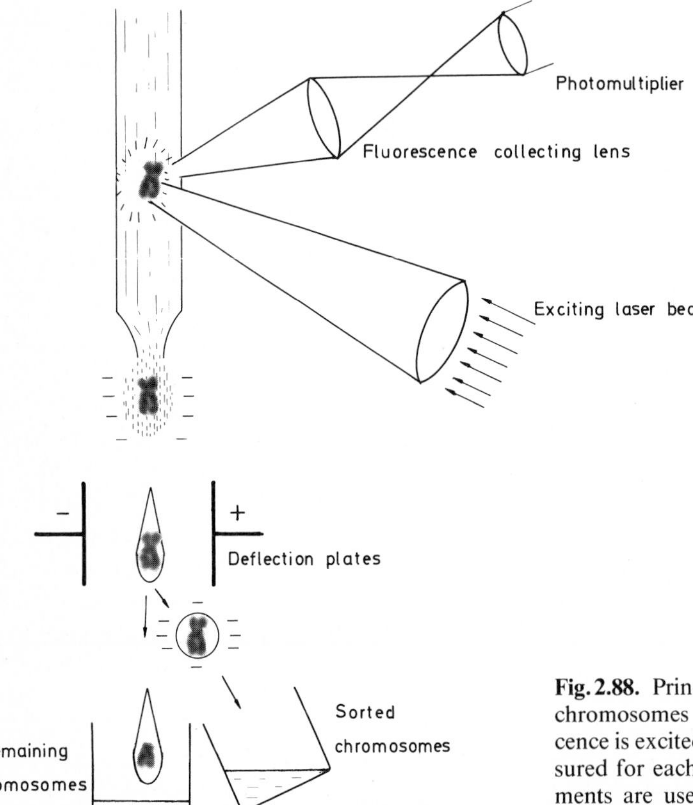

Fig. 2.88. Principle of chromosome sorting by Laser. The chromosomes are stained by a fluorescent stain. Fluorescence is excited by a laser beam. Fluorescence is then measured for each chromosome individually; these measurements are used to induce deflection and sorting out of chromosomes. Courtesy of Dr. C. Cremer

prepare a chromosomal DNA bank after restriction with the enzyme EcoRI and cloning in a λphage (see Sect. 2.3.2.2). The Y chromosome was sorted by dual-beam analysis from a Chinese hamster x man hybrid cell line. The production of cell hybrids will be described in Sect. 3.4.3 in the context of gene assignment to chromosomes. It is the most significant property of hybrids between human cells and those from mice or Chinese hamster that they loose human chromosomes at random. In this way a cell line was produced that had only kept the human Y in addition to the full hamster complement. Since all hamster chromosomes differ from human chromosomes more than human chromosomes from one another, these hybrid cells offered especially good conditions for cell sorting.
Figure 2.89 shows the peaks in the two-dimensional plane: The chromosome marked with L is the human Y. It is separated so clearly that preparation did not offer any specific difficulties. It could now be used to prepare a chromosomal DNA bank.

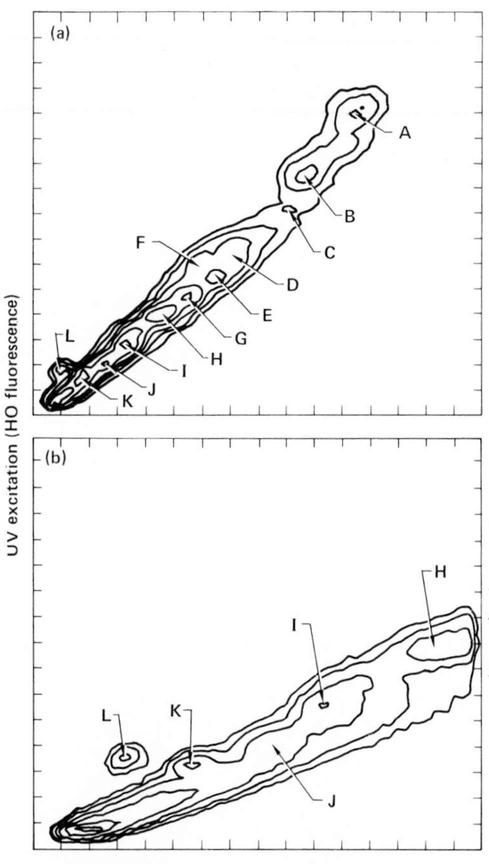

Fig. 2.89 a, b. Contour plot (intensity of fluorescence) after two-dimensional Laser chromosome sorting from Chinese hamster x man hybrid cells in which the Y chromosome was the only human chromosome retained. Letters *A* to *K* refer to various hamster chromosomes; *L* refers to the human Y chromosome. **a** All chromosomes; **b** small chromsomes only (increased gains). (From C. Cremer et al, Hum. Genet. 60, p. 264, 1982 [330])

2.3.3.6 Analysis of the β-Globin Gene and Generalizations with Experience from One Gene

The Paradigmatic Role of the β-Globin Gene. In Sect. 2.3.2.1, we began with the analysis of the β-globin gene using a number of new methods and approaches. The most important of these have been described in previous sections; therefore, we can return now to the analysis of this gene.
As mentioned, the first step must be to identify this gene in the large amount of human DNA. This has been done by isolation of DNA from human cells and treatment with a specific restriction enzyme. The many resultant DNA fragments are then separated according to their length by agarose gel electrophoresis. They are transferred to a nitrocellulose filter by Southern blotting and denatured. The DNA is now single-stranded and fixed to the filter. The next step is identification of DNA fragments containing the β-globin gene. This requires a radioactive probe which was made from β-globin mRNA using the enzyme reverse transcriptase (mRNA→cDNA). This cDNA probe is now used for hybridization with the DNA on the filter. Autoradiography shows the positions of the fragments containing globin genes.
This technique can also be used for localizing the β-gene on a human chromosome by in situ hybridization, as described in Sect. 2.3.2.3 for the myosin heavy-chain gene. The Hb β gene was localized to the short arm of chromosome 11 (11 p; see Sect. 4.3).
Further characterization of the β-globin gene family requires enrichment of the DNA specifying these genes by cloning in a vector such as a bacterial plasmid (see Sect. 2.3.2.2). The Hb β gene family was analyzed by comparing the DNA sequence of the transcribing region and the cDNA by electron microscopy; by sequencing DNA regions inside and outside transcribed sequences; and by identifying control sequences. One of the first and most conspicuous results of such studies was that hybrid molecules between the β-globin genomic DNA and cDNA showed peculiar loops in electron microscopy [1329]. They were caused by DNA regions not represented in the cDNA and obviously not transcribed, since cDNA is a true copy of mRNA. In the Hb β gene, two such intervening sequences (introns) were discovered that separated three distinct units (exons) at the DNA level. Meanwhile, studies on many other eukaryotic genes have shown such introns to be the rule rather than the exception, quite in contradistinction to bacteria and viruses, where genes are continuously transcribed. Often, exons represent functional subunits of the gene; they might have developed from separate genes during evolution (Sect. 7.2.3).

These and later studies confirmed inferences from family studies on abnormal hemoglobins (Sect. 4.3.2) that there is only one functional Hb β gene, whereas, for example, the genes for the α and γ chains are present in duplicate. In addition, however, pseudogenes were discovered: DNA regions very similar in DNA sequence to functional genes, which were not transcribed because of mutations in the coding or flanking areas.

Figure 4.44 shows the Hb β region. In addition to the gene itself and a pseudogene, there are two γ genes, one δ gene (for the Hb δ chain found in Ab A_2), and genes for an early embryonic hemoglobin. The molecular analysis confirmed inferences regarding the structure of this gene region from formal genetics and protein biochemistry (Sect. 4.3) but provided much entirely new information about gene structure and organization.

Additional studies on this and other genes have also shed some light on how transcription occurs and how mature mRNA is produced (Fig. 2.90). First, the entire gene is transcribed, including introns and flanking sequences proximal and distal to the structural information. Then, transcripts of introns are removed stepwise and a "cap" is added at the 5′ end and a poly-A sequence at the 3′ end. Finally, the processed mRNA leaves the nucleus to proceed to the ribosomes and acts as template for protein biosynthesis. The DNA sequences for the various globin genes are known, and many general problems of gene organization and gene action have been examined using these genes as models.

For further discussion of the hemoglobin genes, especially genetic variation at clinical, protein, and DNA levels, see Sect. 4.3. Aspects of the mutation process will be discussed in Sect. 5.1.4.3, and aspects of evolution in Sect. 7.2.3.

2.3.3.7 Structure of the Factor VIII (Antihemophilic Factor) Gene

The Antihemophilic Factor (Factor VIII). Hemophilia A is a "classical" hereditary disease with an X-linked recessive mode of inheritance (Sect. 3.1.4). Analysis of the blood clotting process led, in the 1950s, to identification of a plasma protein – the antihemophilic factor (factor VIII) – as the gene-determined protein that is lacking in this disease. Factor VIII is necessary for the first step of blood coagulation, – thromboplastin formation (see Sect. 4.2.2.9). Substitution with factor VIII concentrates is now the established therapy of hemophilia A; nowadays, it enables hemophilia patients to lead a near-normal life. The protein molecule, however, turned out to be large and complex; its synthetic production appeared to be impossible. Now, this situation has changed dramatically: The structure of the factor VIII gene has been elucidated, and expression of this gene has been observed in tissue culture [361, 362, 531, 536].

Research Strategy in Elucidation of the Factor VIII Gene. This has been achieved independently, but with similar research strategies, by two private gene technology companies: Genentech, San Francisco and Genetics Institute, Boston. The Genentech group proceeded as follows.

A DNA oligonucleotide probe only 36 bases long was synthesized as one codon choice of a tryptic peptide from human factor VIII. This very short DNA probe was used to screen a genomic DNA library in bacteriophage λ that had been derived from an individual with karyotpye 49, XXXXY. Hence, a higher concentration of X-specific probes was achieved not by chromosome sorting as described above (Sect. 2.3.2.5), but by utilization of a naturally occurring abnormality. Clones identified in this way by DNA hybridization (Sect. 2.3) had overlapping ends, allowing initial identification of part of the DNA sequence (Fig. 2.91). This part was then used to place factor VIII mRNA in a T-cell hybridoma cell line. This mRNA could then be used to produce cDNA of the entire coding part of the gene. This portion (~9 kb) was then sequenced (Sect. 2.3.2.4). The boundaries of exons were established by comparison of cDNA with genomic DNA. The complete gene consists of 186000 base pairs; it has 26 exons ranging in length from 69 to 3106 base pairs; introns may be as long as 32.4 kb. The factor VIII protein consists of 2351 amino acids (Fig. 2.92).

To achieve expression of the gene in a mammalian tissue, the complete 7-kb protein-coding sequence was assembled from portions of overlapping cDNA

Fig. 2.90. The DNA sequence which is depicted here as single nucleotide chain is characterized by a specific sequence of bases. At the 5′ side, where transcription begins, two characteristic sequences, CAATT and TATA were described 80 and 30 bp apart. From analogy with the bacterial genome, it was concluded that the CAATT sequence serves as recognition site for RNA polymerase whereas the TATA sequence serves as promoter region for polymerase-induced transcription. DNA is transcribed into the complementary RNA sequence – first including the introns. Then, the RNA is processed stepwise; the introns are removed; a 'cap' is added at the 5′ end, and a poly-A sequence at the 3′ end. Then, the processed mRNA passes through the nuclear membrane and attaches to the ribosomes, where the genetic information is translated into a protein sequence ▷

Fig. 2.91. See p. 100

Fig. 2.92. The factor VIII gene. *Open bar:* The gene. – *Filled in-bars:* the 26 exons. *Lower series of lines:* Location of the recognition sites of the 10 restriction endonucleases used for identification. *Grey boxes* represent the extent of human DNA contained in each λ phage (λ) and cosmid (p) clone. (From Gitschier et al., Nature 312, p. 327, 1984)

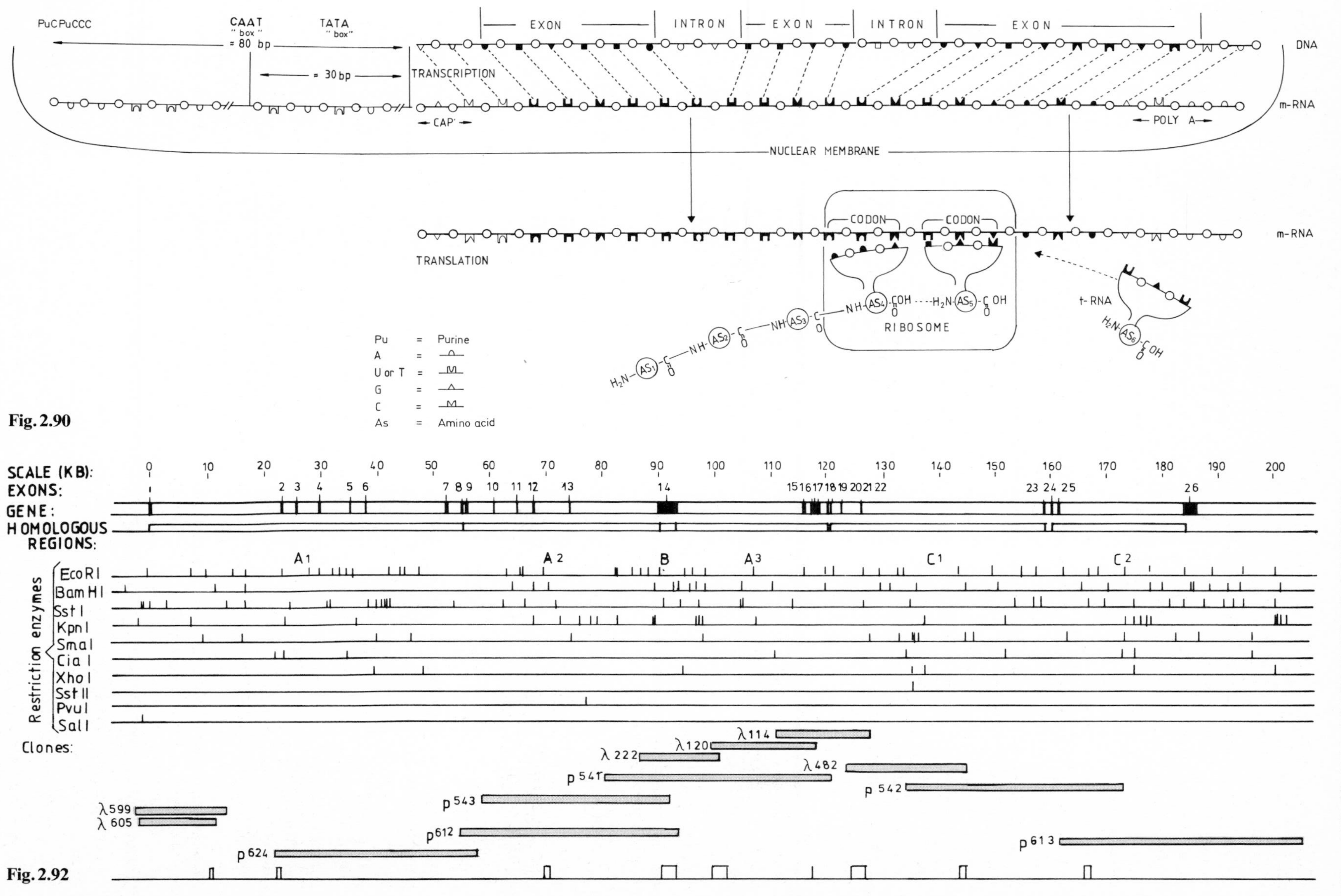

Fig. 2.90

Fig. 2.92

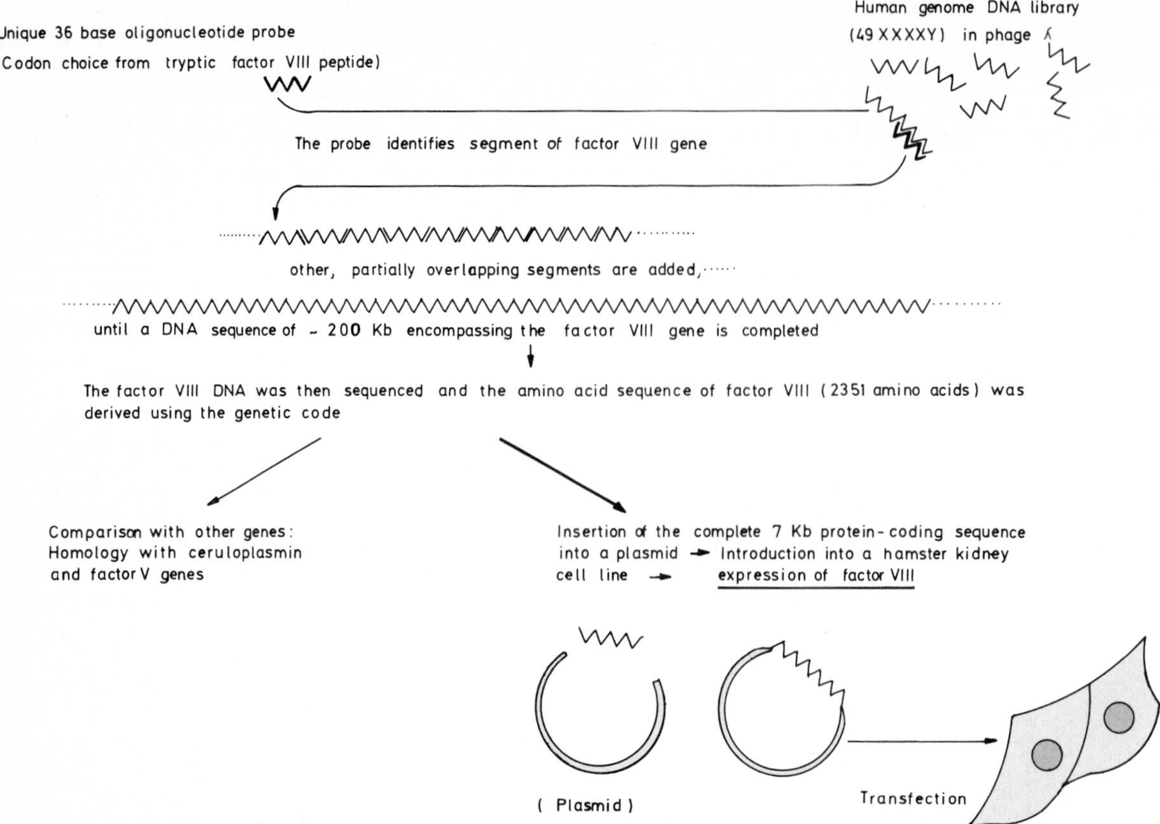

Fig. 2.91. Analysis of the human factor VIII gene, starting from a 36-base oligonucleotide probe (Codon choice from a tryptic factor VIII peptide)

and inserted into a plasmid; in this plasmid, factor VIII sequences were placed between promoter sequences and a polyadenylation sequence of viral origin. This plasmid was then introduced into a hamster cell line by a chemical method (calcium-phosphate coprecipitation), and gene expression was measured using monoclonal antibodies against part of the factor VIII protein. This indicated a 300-fold increase over cross-reacting activity in the control cells and ~7% of the normal plasma activity, as shown by additional biochemical studies.

The other group (Genetics Institute), working on similar lines, achieved comparable results.

Significance of These Studies. In these studies, which are described above only very briefly and in a much oversimplified way, methods from molecular biology were combined in an ingenious way to analyze an unusually complex gene. This result is significant for several reasons:

1. This is the first gene of such length and complexity to be analyzed in humans and, indeed, in any eukaryote. It can be predicted that many other genes coding for long and complex proteins might have a similar length and structure.
2. This structure permitted new conclusions regarding evolution of this gene [361], including an unexpected homology of ≈35% of the amino acid sequence with the copper-binding protein ceruloplasmin (see Sect. 7.2.3).
3. The result offers prospects for an improved therapy of hemophilia A (see also Sect. 4.2.2.9). Substitution treatment with factor VIII preparations is one of the success stories of therapy in hereditary diseases: Life expectation of hemophilics has increased dramatically, and many of them are now leading near-normal lives. This therapy, however, is not without some serious flaws. First, it is very expensive, since factor VIII must be prepared from human blood. Secondly, virus infections pose a serious problem – especially with hepatitis or, more dangerous with the AIDS virus [314].

At present, this infection can be avoided by careful virological screening of preparations, and heating of factor preparations.

After the decisive steps have been completed, some years will certainly elapse until a safe, effective and clinically tested factor VIII preparation will be available. We would not be surprised, however, if it were in our hands by the time this book appears.

An Exercise in Sociology of Science. This analysis has been achieved by two very large groups, as shown by the large number of authors of each paper. This effort approaches "big science," and is more customary in certain branches of physics such as high-energy physics. "In the application of molecular biology to practical affairs, it is for the time being clear that bigger battalions move faster" [428]. And these battalions have been set up not by universities or research institutes in basic research, but by two private companies and with the ultimate purpose of producing saleable products. This connection between research and commerce is new to human geneticists, but has become widespread in recent years in molecular biology. Should this development cause serious concern? Is there a danger that commercial interests will influence the development of science too heavily, diverting manpower and resources from scientifically important problems to those promising profits in the near future?

The danger may not be too great. After all, close relationships between basic research and industrial utilization has been customary not only in applied fields, such as engineering, but also in chemistry (for example) for a long time. To the best of our knowledge, it has not impaired the quality of basic research. For biologists, however, the situation is new. It requires careful observation.

2.3.3.8 Gene Families

Examples for Gene Families. The globin regions, α- and β-globin genes together, provide a good example of a gene family, i.e., a group of functionally related genes having similar structure and, presumably, a common origin in evolution (see Sect. 7.2.3). Other gene families include, for example, the immunoglobulin genes (Sect. 7.4); the genes for ribosomal RNA (Sect. 2.3.1.1); and the components of the major histocompatibility complex (MHC) (Sect. 3.5.5; see also [307]). There appears to be no general rule on location of gene families on chromosomes. Some are clustered in the same chromosome regions, showing close linkage with or without linkage disequilibrium (Sect. 3.5). The hemoglobin family forms two clusters, the Hb α cluster on chromosome 16 and the Hb β cluster on chromosome 11. Other gene families such as the muscle protein genes are scattered over many different chromosomes.

Genes for Actin and Myosin. It is the biological function of muscle to perform mechanical work by contracting. This problem, which involves transforming chemical energy into mechanical energy, has been solved by nature by creating extremely long, multinucleated cells, a large portion of which is occupied by contractile elements - the myofibrils - arranged as parallel bundles in the axis of contraction [120]. Mechanical labor is performed by interaction of two kinds of protein molecules - myosins and actins. In addition to muscle contracture, actins are involved in many other cell functions, such as maintenance of cytoskeletal structure, cell motion, and mitosis.

The genes determining both types of proteins - actin and myosin - are now being examined.

In one study, cDNA probes for actin genes were available from chicken and from *Drosophila* [344]. They were used for hybridization with human nuclear DNA from a single individual digested with restriction enzymes that cleave infrequently, i.e., produce relatively long fragments on average (Sect. 2.3.2.2). In this way, 20-30 bands were discovered. It was shown that at least 12 clones contained nonoverlapping sets of restriction fragments; nine of them hybridized well to human actin mRNA, while the remaining three were thought to encode slightly deviant smooth muscle actin. While formal linkage studies have not been performed, the authors concluded from various sources of evidence that there are *at least* ten different actin genes in humans that are not closely linked; and might even be located on different chromosomes. For example, each clone contained a unique set of fragments. Other groups [388, 403] estimated the number of human actin genes as ranging between 9 and 20.

Actin genes have been preserved well through evolution. In addition to mammals and *Drosophila,* they were also found, for example, in yeast and slime molds. Skeletal muscle α-actins from humans, rabbits and rats were found to be identical; untranslated regions were only partially identical. The split between skeletal muscle and heart muscle actin genes must have occurred a long time before the split between these various species of mammals [375].

Similar to actins, myosins exist in humans as multiple isozymes. These isozymes appear in a distinctive order during individual development [295]. Studies on DNA have again shown a multigene family which consists of many - probably more than ten - genes not located close to each other.

A New Principle in Genetic Analysis. These observations on muscle protein gene families introduce a new principle into genetic analysis. Until recently, genetic analysis always started with genetic variability. Such variability might be identified at the phenotypic level, e.g., by the presence of a hereditary disease, or at some intermediary levels, by lack of a functional protein, an electrophoretic protein variant, or a different antigenic site at the cell surface. This phenotypic variability is then traced by classical mendelian segregation analysis to a corresponding variability at the gene level. Basic mechanisms of gene action at all levels can often be elucidated using genetic variants as tools for research. For the actin and myosin gene families, however, no normal or pathologic genetic variants are known. Genetic analysis started at the level of the proteins and the genes themselves, irrespective of any interindividual variation. It became possible because mRNA was available in relatively large amounts. It will now be the task of medical geneticists to find hereditary diseases that might be caused by genetic variation in actin or myosin genes. It is possible, however, (though not very likely) that such diseases simply do not exist - either because any genetic defect would be lethal, or because gene expression of such a multigene family is so "elastic" that mutations at one gene locus are compensated by activity of other loci, so that no phenotypic change will ensue. Examination of the human gene map (Sect. 3.4.2) shows a number of genes that have been identified and located in the absence of any known genetic variability within the human species. These include genes for histones, rRNA, and for susceptibility to bacterial toxins or viruses. Analysis of structure and action of such genes will certainly become more important in the future.

2.3.3.9 Restriction Site Polymorphisms [548; 507; 508]

Genetic Variability Outside Coding Genes. In a classical analysis in 1978, Kan and Dozy [1157] detected a DNA polymorphism closely linked to the β-globin gene, which permitted prenatal diagnosis of sickle cell anemia in many cases following family studies. Since then, a great many DNA polymorphisms have been detected (see Sect. 6.1.3).

In publications on DNA polymorphisms, relevant information such as the number of individuals examined has often been omitted; in part probably because many molecular biologists are not oriented towards human population genetics. Such information is necessary, however, for estimation of how often a polymorphism may have been missed by chance. Often, authors specify the enzyme(s) by which polymorphisms were discovered but do not specify those enzymes that failed to detect polymorphisms. Such data are essential for assessment of the number of sites tested, and for estimation of the frequency of polymorphic sites (Sect. 6.1).

Why Is Knowledge of DNA Polymorphisms Useful for the Human Geneticist? Genetic variability at the DNA level - and especially outside transcribed DNA regions - appears to be much more common than anticipated on the basis of protein data (Sect. 6.1.2). Therefore, its analysis contributes to population history. It is also important for the genetic theory of evolution, for example, with respect to the much-discussed question of to what degree genetic differences between species and between population groups within species are brought about by natural selection or by random genetic drift (neutral hypothesis; see Sect. 7.2.3). Moreover, analysis of restriction polymorphisms throws new light on the molecular mechanisms of mutation (Sect. 5.1.4) and is likely to contribute to knowledge regarding the significance of noncoding DNA for gene regulation (Sect. 4.7). Here, preliminary data indicating that DNA polymorphisms seem to be rarer in the X-chromosome than in autosomes [328] extend Ohno's [156] observation that the X chromosome has been more strictly preserved during evolution. Possibly, functional constraints regarding the structure of the X apply not only to coding genes but to the entire genetic material.

Most important, however, is the contribution of such data for a detailed analysis of the arrangement and order of genes in the human genome. A rapidly increasing number of human mutations are being localized to specific chromosome regions by detection of close linkage with a polymorphic DNA site. These developments have far-reaching implications for genetic counseling and prenatal diagnosis (Sect. 9.1; for further applications see Sect. 6.1.3). Complete coverage of the human genome so that every newly detected mutation can be mapped immediatly to a specific chromosomal site by close linkage with a known DNA polymorphism is now a realistic goal for research. These aspects will be covered in the section on linkage (Sect. 3.4).

2.3.4 The Dynamic Genome

The research tools of new genetics have increased our knowledge of the structure of the genetic material. The simplistic model of a chromosome as a string of pearls - with the genes as pearls - fits the real data less than ever before. It is not even clear what we should call a "gene" (see below). One aspect of our increased knowledge is the realization that the genetic material appears much less static than perceived in the past. Some brief remarks on the dynamics of the genome are therefore necessary, even though these new data cannot yet be applied to human biology or pathology.

Movable Elements and Transposons. Maize (Indian corn) is distinguished by beautifully mottled corn cobs (Fig. 2.93). The genetics of this phenomenon was examined by B. McClintock [433]. She showed that "controlling elements" occur, which can be transposed in the genome from one location to the other, and may cause increased instability of genes leading to somatic mutations in the tissue of this plant. The beautifully variegated pattern of maize is caused by such somatic mutations. Their specific properties have been analyzed in a long series of elegant studies. For a long time, these controlling ele-

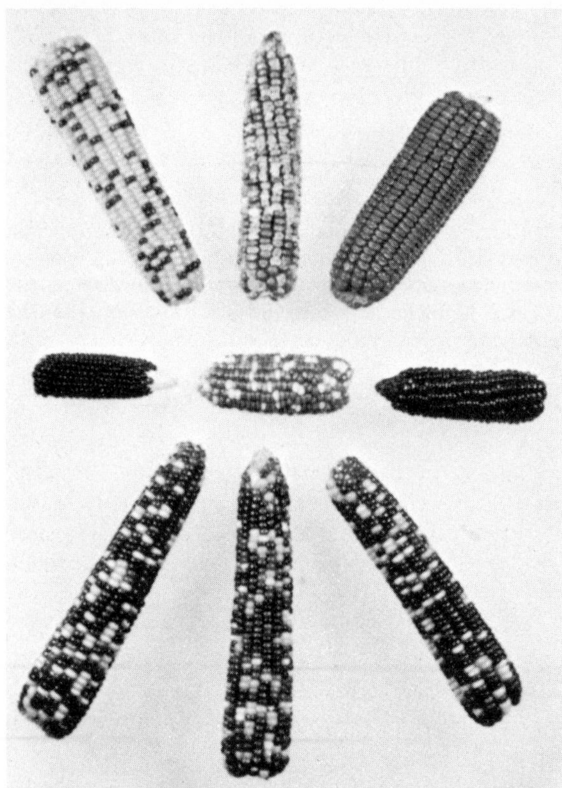

Fig. 2.93. Hybrid maize; different varieties; notice the mottled appearance of the cobs. (From Singleton, Elementary Genetics, Princeton etc.: Van Nostrand 1962)

ments seemed to be a unique exception in genetics, until Taylor (1963; [524]) described "phage-induced mutation in *E. coli*"; this phage is now called Mu (mutator). Shortly afterwards, Starlinger and Saedler [520] described the Is elements in bacteria.

Movable Elements in Bacteria [520, 358]. Transposable elements are now defined as specific DNA segments than can repeatedly insert into a few or many sites in a genome. Three classes of such elements are distinguished in prokaryotes:

1. Is elements (= simple insertion sequences) contain no known genes unrelated to insertion function. They are generally shorter than 2 kb.
2. Tn elements (transposons). These behave formally like Is elements, but are generally larger than 2 kb and contain additional genes unrelated to insertion function. They often contain two copies of an Is element.
3. Episomes. They are complex, self-replicating elements, often containing Is and Tn elements.

Such movable elements were shown by DNA sequencing and other techniques to share the following properties:

Is sequences terminate in perfect or nearly perfect inverted repeats of 20–40 bp; most Tn elements terminate in long (800–1500 bp) Is-like sequences.
Movable elements when integrated into the host genome are flanked by short duplications (4–12 bp) of host DNA. Elements generating host repeats of the same size share more or less pronounced homologies, which indicate some evolutionary relationship between them (Fig. 2.94).
Movable elements can be inserted at multiple sites of the host genome by nonhomologous recombination; sometimes they insert at specific sites. This insertion has often been shown to involve a copy of the element, whereas the original element stayed at the donor site (Fig. 2.95).
If a transposable element inserts into a structural gene, gene mutations result as a consequence of gene splitting. Moreover, chromosome aberrations such as deletions, duplications, inversions, and translocations may be induced by inserted transposable elements.

Transposable Elements in Eukaryotes. As mentioned, the controlling elements in maize were the first transposable elements detected. Later, such elements became known in other eukaryotes, such as *Drosophila* [365]. In this organism, insertion mutants are produced at high frequency at specific, preferred loci. Three sources of such insertions exist: (a) They may come from gene sequences dispersed over the genome. Elements of these gene sequences are known to transpose. (b) As a second possible source of transposing DNA, the repetitive sequences of constitutive, centromeric heterochromatin have been discussed. (c) RNA viruses known to be harbored by *Drosophila* have been implicated as possible sources; the enzyme reverse transcriptase (Sect. 2.3.2.2) may transcribe viral RNA into DNA that is then inserted.
In addition to important aspects of their structure, such as presence of inverse repeat units at their ends, transposable elements in *Drosophila* share some properties with those described in bacteria; mainly the ability to induce gene mutations at an unusually high rate; the instability and high reversion rate of these mutations; the independence of normal DNA replication; and the frequent induction of chromosomal aberrations. In yeast, transposable elements have also been described [358].

Significance of Movable Elements in Evolution? There have been speculations that gene transfer by movable elements might have played a role in evolution. Possibilities for genetic change would indeed be increased enormously if, in addition to the classical ways of (sexual and asexual) transmission of ge-

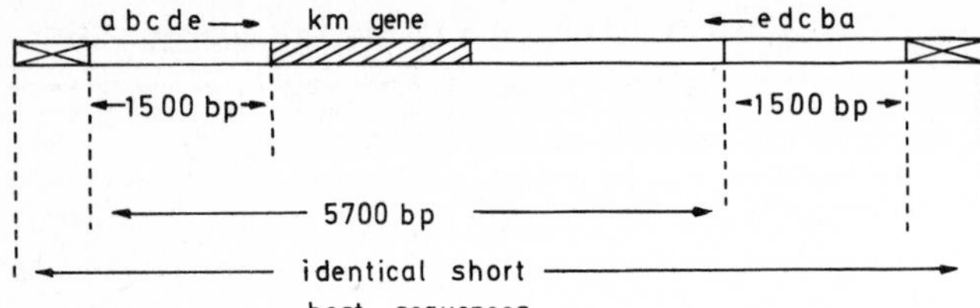

Fig. 2.94. Example for one transposon (Tn 5) of E. coli: This transposon carries on both ends an inverted sequence repeat of 1500 nucleotide pairs. These sequences are identical with one exception: The repeat to the left carries an A-T pair at a place where the right repeat has a GC pair. This creates a promoter sequence from which transcription of the Kanamycin-resistance gene (km) starts. The inverted repeats code for two proteins that belong to the enzyme system necessary for transposition of Tn. However, as the A-T pair in the left repeat includes a stop codon, only the right-hand repeat produces a functional gene product. (From Kueppers, 1982 [117]; modified)

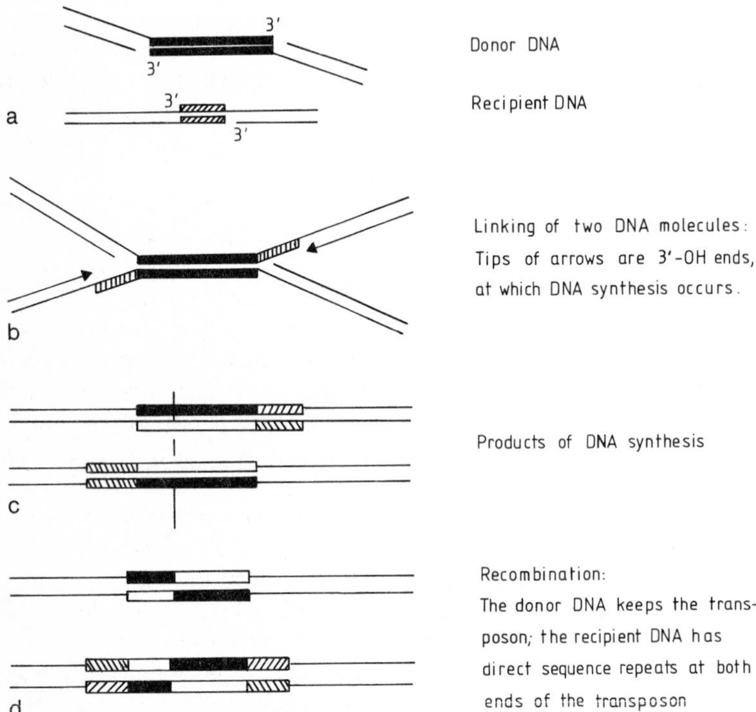

Fig. 2.95 a–d. A model, in which transposition is explained using known principles of molecular genetics. At the two ends of the transposon (or the Is elements; black segment of the upper DNA molecule in (**a**), one of the two DNA strands is cut by a restriction enzyme. In the same way, the recipient DNA strands are opened at opposite ends (**b**). Then, DNA replication leads to doubling of the transposon as well as the flanking sequences of the recipient (**c**). Finally, recombination takes place. (**d**) The donor DNA retains the transposon; the recipient DNA has the transposon as well as the directly repeated flanking sequences. (From: Shapiro; Proc. Note Acad. Sc. USA 76, 1933, 1979)

netic material from parent to offspring, "transverse" transmission often occurred – possibly even between quite different species. In this connection, it is good to remember that, for example, transduction of genes from one bacterium to the other by phages has been known for many years, and is now being used in eukaryotes, including mammalian cells for artificial gene transfer and genetic engineering (Sects. 2.3.2.2 and 9.2). It is possible that such processes might also occur in nature. Moreover, a sequence homologous to the human hemoglobin gene has been discovered in leguminous plants [312]. Its function might be "to ensure an adequate oxygen supply for bacteroid respiration in module tissue." The presence of this gene could possibly be explained by transfer from an insect or a mammal.

Movable Elements in the Human Genome? So far, no movable elements have been shown convincingly to occur in the human genome. However, like the ge-

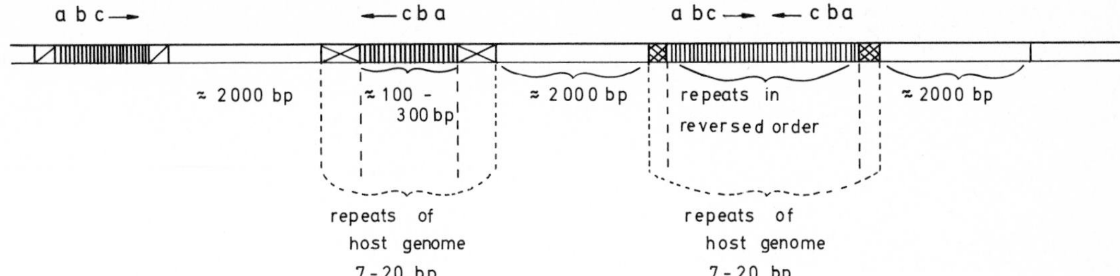

Fig. 2.96. Short-period interspersion pattern with Alu elements. Repeated sequences of about 100-300 base pairs in length alternate with single-copy sequences of ≈ 2000 base pairs. Repeats may occur in two directions (abc→; ←cba), or two repeated sequences in reversed order may immediately follow each other. Alu elements are flanked by short direct repeats of the 'host' genome (7-20 bp) that are different from one 'insertion' site to the next. About ⅓ of all 100-300 bp repeats are members of the Alu family

nome of *Drosophila*, the human genome also contains interspersed and centromeric repeated DNA segments (Sect. 2.3.1.1) - sometimes even with palindromic sequences - that could be good candidates for transposing elements by analogy. For example, the oncogenes have structural homologies with cellular RNA viruses (= retroviruses; Sect. 5.1.6); retrovirus-like repetitive elements have been identified in human DNA [429]; and DNA viruses have been shown to be mutagenic in mammalian cells [1463]. A special group of dispersed repetitive sequences has been discovered in the human genome - the Alu sequences [505].

It has been mentioned that much of the human DNA is organized according to the *Xenopus* pattern, i.e., single-copy sequences ≈ 1-2 kb in length, interspersed with repeated sequences ≈ 0.1-0.3 kb in length. We also mentioned that some of these sequences are palindromic, containing repeated sequences in inverse order (Sect. 2.3.1.1). Whereas in the *Xenopus* genome these repeat sequences belong to many different sequence families, much of this material shows strong homologies in mammals such as rodents or primates [505]. In humans, ≈ 3%-6% of the total DNA consists of such 300-bp repetitive sequences, and ≈ 60% of this material has been shown by restriction enzyme analysis to be homologous. The number of such Alu sequences is now estimated at ≈ 500000 for the haploid genome, which means that on average, such a sequence occurred about every 5000 base pairs; but they are unequally distributed. One in about 2½ interspersed repeat sequences is an Alu sequence. They might occur in straight, in reversed, or in palindromic order (Fig. 2.96). On both sides, they are flanked by direct repeats ranging in length from 7 to 20 bp. Unlike the Alu sequences themselves, these repeats are unique to each different Alu sequence. As mentioned above, such repeats were also found to flank bacterial transposons, as well as movable elements in eukaryotic DNA. Therefore, it was concluded that Alu sequences have also originated as movable elements, and that the flanking repeats may have resulted from duplication of the DNA sequence at the target site of Alu insertion.

Alu sequences are often found within primary RNA transcripts; they are often but not always removed by RNA processing (Sect. 2.3.3.6). It has been speculated that they might be distributed over the genome by their relatively short mRNA transcript, which might then be transcribed to DNA by reverse transcriptase, and inserted at various places. Since they have been conserved during mammalian evolution, as evidenced by their partial homology between primates (including man) and rodents, they might have an important function. By analogy with similar elements in other eukaryotes, such as *Zea mays* and *Drosophila*, they may be involved in gene expression, mutation (germ line and somatic), or recombination in germ and somatic cells.

A first step to transposition of certain DNA sequences could be the formation of extrachromosomal, circular DNA repeats, which has been observed for sequences normally localized between Alu repeat clusters in aging human fibroblasts in vitro [494; see also 400].

Gene Conversion. Another phenomenon observed in experimental genetics is *gene conversion* [122]. Various data from the hemoglobin gene suggest that this event also occurs in the human genome (Sect. 4.3; see also Fig. 2.97).

Gene conversion is the modification of one of two alleles by the other, which would alter, for example, a heterozygote Aa into a homozygote AA. Winkler, who was the first to discuss this concept more than 50 years ago, assumed a "physiological interaction" between alleles. Studies in yeast showed that atypical recombination is involved. The process is shown in Fig. 2.97. Crossing over always involves destruc-

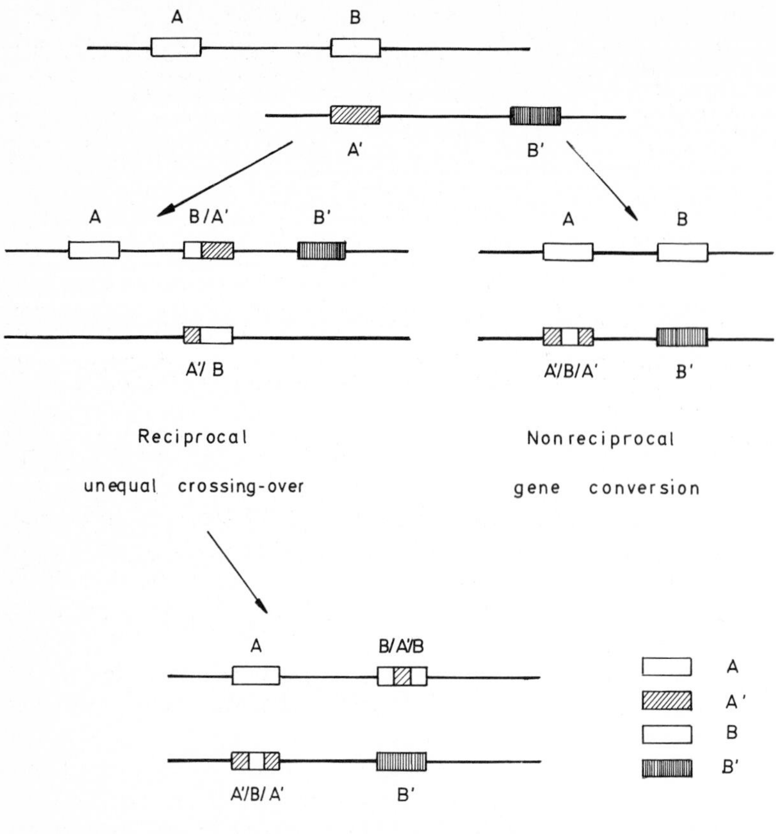

Fig. 2.97. Gene conversion vs. double unequal crossing over. A pair of homologous genes A, A^1 as well as B, B^1 are arranged in tandem. Because of homology, an "illegitimate" alignment between B and A^1 has occurred. Recombination within the gene produces a triplicated gene (A, B/A^1, B) on one strand and a deletion product (A^1/B) on the other strand. In a future generation, a person who is a double heterozygote for the triplicated and single gene (as shown) might undergo an additional unequal crossover creating the gene products as shown. In gene conversion, there is direct copying of part of the gene B into the middle of gene A^1 residing on the other DNA strand. The event is nonreciprocal and the strand carrying A and B remains unchanged. The haplotype created by conversion is identical to one produced by a double crossover as shown. Since both recombination products cannot be ascertained in humans, gene conversion would not be distinguished from such a double crossover. Statistically, however, a single conversion event is much more likely to occur than an event requiring two crossovers

tion of DNA sequences around the crossover site. Normally, the destroyed sequences are repaired using the sequences of the sister chromatid as template. In this way, the original double helix is restored. Sometimes, however, the double-strand structure is repaired using the strand of the homologous chromosome as template. In this case, the abnormal segregation ratio is observed.

Gene conversion has also been observed in somatic tissue, especially in plants. Probably, similar, deviant recombination processes are involved. This is not surprising, since somatic pairing of homologous chromosomes and somatic crossing over has been observed in many species (see, for example, [368]). A similar situation may exist in some cases of retinoblastoma, the malignant eye tumor of children (Sect. 5.1.6). There is some evidence that gene conversion might have occurred in the HLA region and in hemoglobin variants such as βthalassemia, Hb S or Hb E (Sect. 4.3).

Does the Genome Fluctuate? How Constant Is the Genetic Information and Its Transmission? When we observe inheritance of a monogenic disease or variants of a genetic polymorphism, such as the ABO or MN blood groups (Sect. 3.2), we are impressed by the regularity of genetic transmission indicating constancy of the genetic material over many generations. This regularity has been confirmed over and over again; we are justified in concluding that the few apparent exceptions are accounted for by other than biologic explanations, such as false paternity. Our confidence in the all-pervading dependability of heredity is somewhat attenuated by the fact that

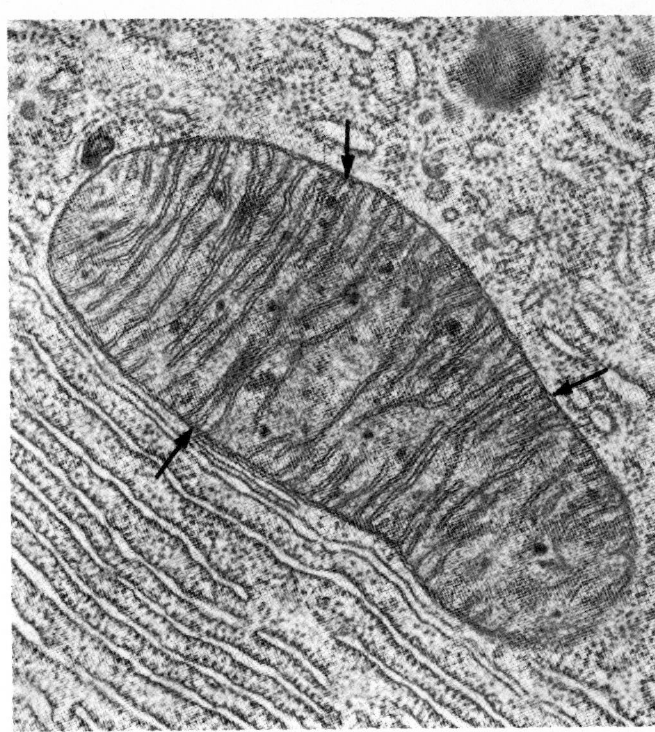

Fig. 2.98. Electron-microscopic photograph of a mitochondrium 53000× enlarged. *Arrows* point to the (outer and inner) membrane. (From Nielsen et al., Fundamental concepts of Biology, New York: Wiley 1970)

new mutations are observed occasionally (Sect. 5.1.); but mutation rates are usually very low. Moreover, mutations, once they have occurred, obey the rules of genetic transmission.

But now, the new findings of molecular biology raise problems - genes are cut into pieces, jump around in the genome, convert their alleles to the own structure; they might even be introduced into our genome by a vagrant virus and not by the time-honored and pleasant method practiced by our parents and forefathers. Must we forget our elementary genetics and be prepared to distrust all rules? Fortunately, the old rules can still be trusted in most instances. As Arber said in his Nobel lecture on genetic exchange: "In spite of possessing a multitude of natural mechanisms to promote exchange between genetic materials of unrelated origin, *E. coli* and other high organisms have succeeded in achieving a relatively high overall stability in their genetic makeup." The new results deepen our understanding of how the genetic material is structured and how it works. Conceivably, they might even help to prevent genetic disease. But the old rules are still valid in the overwhelming majority of situations.

2.3.5 *The Genome of Mitochondria*

Structure and Function of Mitochondria. Mitochondria are organelles found in the cytoplasm. Their number and form vary depending on the function of the cell; a mammalian liver cell, for example, contains ≈1000-1500 mitochondria. All of them have certain structural aspects in common (Fig. 2.98): A matrix, an internal and an external membrane. The internal membrane forms characteristic folds, sometimes "cristae," in other cases "tubuli." Important biochemical functions, such as aerobic oxidation, take place in the mitochondria, which are often called the power plants of the body. Energy is stored as ATP. Of the three energy sources in our food, amino acids and fats can be degraded only by aerobic oxidation; this oxidation takes place in mitochondria. Moreover, they harbor the citric acid cycle. Mitochondria form an ordered multienzyme system; distribution of enzymes in a functionally meaningful order guarantees ordered sequences of biochemical reactions.

Mitochondria multiply, like living organisms, by division; de novo synthesis is impossible. They also have ribosomes, which are, however, smaller than those found in the cytoplasm (70S type in distinction to the cytoplasmic 80S type). This and other evidence has led to the hypothesis that mitochondria originated as external microorganisms that underwent a symbiotic relationship with the cell some time early in evolution, have since been integrated, but still maintain some of their specific properties.

The Genome of Mitochondria. It has been known for some time that mitochondria contain DNA of their own, which also codes for genes, e.g., for tRNA. Genes necessary for mitochondrial enzymes, on the other hand, are often (not always) located on the chromosomes of nuclear DNA.

More recently, a well-organized effort by research groups at the MRC Laboratory of Molecular Biology, Cambridge, has led to full elucidation of the

DNA sequence and genic organization of the human mitochondrial genome (mtDNA; [293]; Fig. 2.99).
This genome has 16569 base pairs, which are arranged in a circular fashion. It contains genes for 12 S and 16 S rRNA, 22 different tRNAs, cytochrome C oxidase subunits I, II, and III, ATPase subunit 6, cytochrome b, and eight other protein-coding genes. In contrast to the nuclear genome in chromosomes (Sect. 2.3.1.1), the sequence of the mitochondrial genome shows extreme economy: The genes have none or only a few noncoding bases between them. In contrast to nuclear DNA, both DNA strands are transcribed and translated. In many cases the termination codons are not coded in the DNA but are created post-transcriptionally. Moreover, the genetic code of human mtDNA differs from the universal code in some important aspects: UGA[9] codes for tryptophan and not for termination, AUA codes for methionine, not isoleucine; AGA and AGG are termination rather than arginine codons. Moreover, the third codon positions - which are the main source of codon degeneracy - are more often A or C (and less often G or T) than in the nuclear genome.

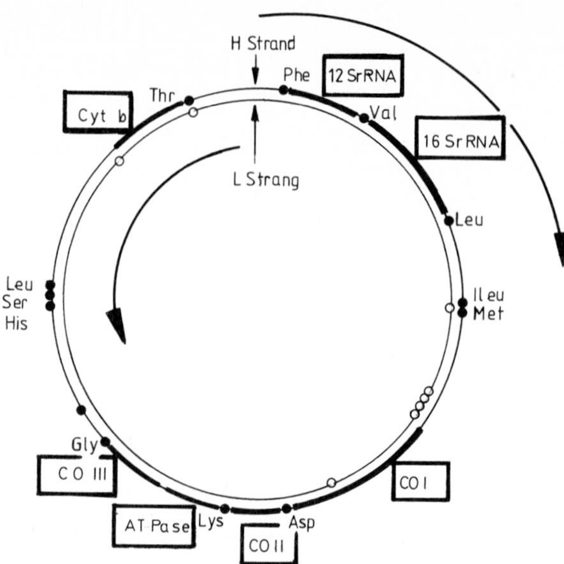

Fig. 2.99. The human mitochondrial genome is a double-stranded ring. The two single strands are distinguished based on their density in a CsCl gradient: A heavy *(H)*, and a light *(L)* strand. Arrows indicate direction of transcription. Arrows start at the promoter sites. Thick black areas indicate identified genes for two rRNA molecules; the genes Co I, Co II and Co III for cytochrome c-oxidase subunits, for subunit 6 of ATP synthase, and for cytochrome B. The tRNA genes for various aminoacids are indicated by points. The L strand contains 8 genes for tRNA. Empty areas very probably code for genes that have not yet been identified. (From Küppers, Molekulare Genetik, 4th ed. 1985, modified)

DNA Polymorphism and the Question of Hereditary Diseases Due to Mitochondrial Mutations. After the nucleotide sequence of human mitochondria had become known, several DNA cleavage site polymorphisms (Sect. 2.3.2.7) were discovered (see Sect. 6.1). In one study [305], 12 restriction enzymes were used to screen for cleavage sites at 441 locations in the mitochondrial DNA of 112 humans of different racial origin. The cleavage maps were constructed by comparison of DNA fragment sizes to those expected from the sequence of one human mtDNA [293]. Of the sites detected, 163 were polymorphic, i.e., present in some individuals and absent in others, while 278 did *not* show any variation. The polymorphisms occurred in all parts of the genome. Moreover, racial differences exist in the frequencies of certain polymorphic variants [336, 305].
So far, genetic recombination within human mtDNA has not been observed to the best of our knowledge; if it occurs, it is very probably very rare. Therefore, patterns of restriction site polymorphisms in populations probably give an undisturbed picture of the mutational history of a population. From comparison of such patterns between population groups, their origin and history might be derived in a much more unambiguous way than by comparison of the classical polymorphisms (Sect. 6.3.3).

Mitochondria are abundant in oocytes, but in sperm only four mitochondria (produced by fusion of a greater number) are found at the neck of the sperm head; they do not enter the oocyte at fertilization. Therefore, the entire set of mitochondria in all cells of any individual comes from the mother [360]. This immediately raises the question as to whether a mutation at a suitable site may cause a hereditary disease. Such a disease would be transmitted by the mother only, and to *all* her children. A probable example will be mentioned in Sect. 3.1.5.
One could argue that this mode of inheritance would be very unlikely on theoretical grounds, since every oocyte contains multiple mitochondria. Even if a mutation were to occur in one of them, the overwhelming majority would still have the unmutated DNA site (or gene); therefore, there could be no phenotypic effect. However, the same argument also holds for DNA polymorphisms of mtDNA. And still, these polymorphisms are transmitted to all children by maternal inheritance, and all mitochondria of one individual appear to be homogeneous genetically. What is the reason for this peculiar behavior? Do all mitochondria in the oocyte go back to one stem mitochondrium?

[9] Expressed, as usual, in terms of the mRNA code.

2.3.6 New Genetics and the Gene Concept

Molecular Cytogenetics. In addition to their significance for a deeper understanding of the structure of genes, the methods of the new genetics (Sect. 2.3) have an increasing impact on cytogenetic diagnosis. For example, minor chromosome aberrations such as deletions or translocations can be identified with much greater precision than has hitherto been possible. Small deletions that escape cytogenetic diagnosis by classical methods can now be diagnosed when a probe of the deleted segment is available. Prenatal sex diagnosis using Y-specific DNA probe is already possible very early in pregnancy (8–10 weeks) using chorionic biopsy (Sect. 9.1) and Southern blotting. Moreover, new and exciting results on structure, phylogenetic relationship, and evolution of specific chromosomes, e.g., the X and Y chromosomes, have been achieved (Sect. 3.4). Considering these cytogenetic results together with the new data on DNA polymorphisms and their significance for prenatal diagnosis of monogenic diseases, it can be concluded that DNA techniques should be available to institutions providing genetic counseling and prenatal diagnosis. Thus, up-to-date services in medical genetics have drastically increased the needs for equipment, personnel, and technical skill. In many countries these developments will create difficult problems. Will human societies be able and prepared to pay for such services on a large scale? Can a responsible consultant even recommend introduction of such expensive services in developing countries with many urgent economic problems? Completely new problems can arise. There are still cultural traditions in some populations that regard a son as more valuable than a daughter. Will the avail-

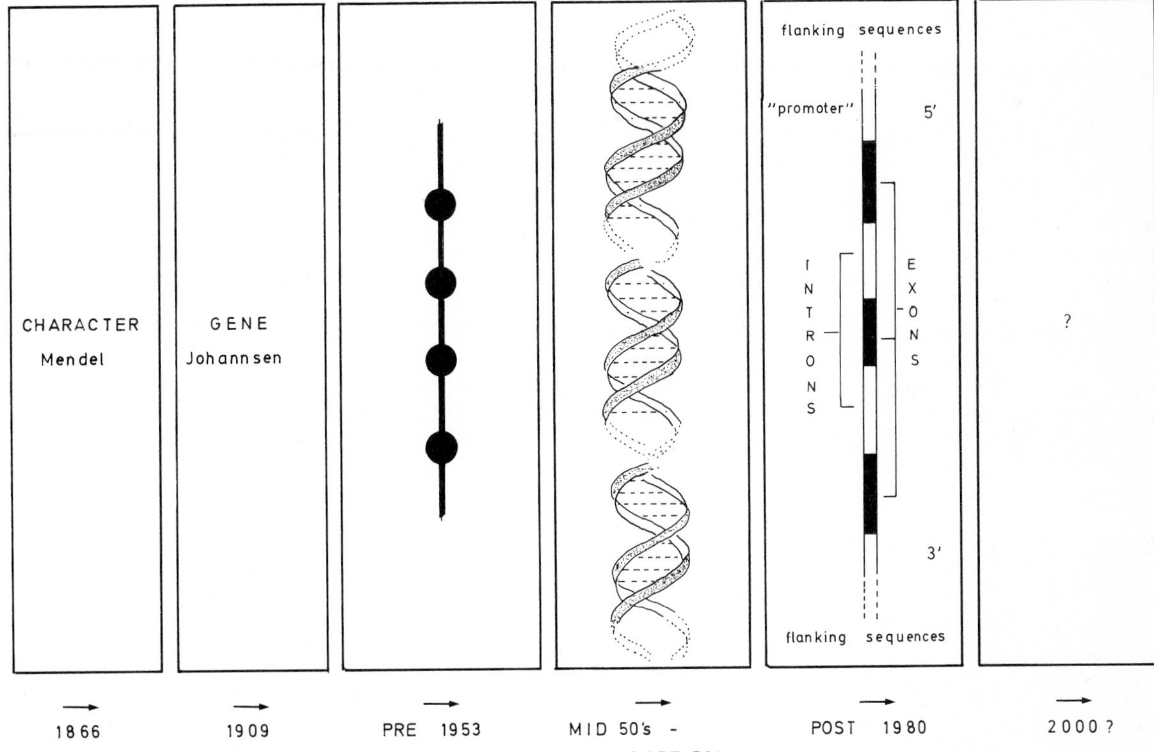

Fig. 2.100. Historical development of the gene concept. Genes were first postulated by Johannsen to account for the "characters" shown by Mendel to be responsible for hereditary transmission of traits. No material basis for the gene could be postulated until the demonstration of linkage of genes on chromosomes led to the "bead-on-a-string" model. The beads were considered to be the genes and the string reflected the chromosome. The material basis of the gene remained undefined. The demonstration of DNA as the genetic material led to the definition of a gene as a specified sequence of DNA that coded for a polypeptide sequence (3 nucleotide bases specify one amino acid). It was soon realized that the unit of polypeptide specification ("cistron") was different from the unit of recombination ("recon") that was different from the unit of mutation ("muton"). The "muton" could be as small as a single DNA base. It was then shown that vast stretches of DNA did not code for proteins, that some DNA sequences were regulatory, that structural genes were interrupted by noncoding intervening sequences (introns). Coding sequences of the structural genes are known as exons. The borders of the flanking and "regulatory" sequences upstream (5′) and downstream (3′) remain poorly defined. It is therefore impossible in the middle 1980's to define a gene precisely. The coding and intervening sequences of a gene can be accurately specified. (See also [247 a])

ability of modern biologic techniques of sex choice lead to complex and unforeseen consequences? (See also Sect. 9.2.)

What Is a Gene? (Fig. 2.100). In classical, formal genetics, a gene was the common unit of mutation, recombination, and action. Genes were initially regarded as arranged in a linear fashion on chromosomes, like pearls on a string. Detailed genetic analysis, however, showed this concept as too crude: For example, there were closely linked genes in *Drosophila,* where two mutations on the same chromosome side by side (in *cis* position) had a smaller phenotypic effect than the same two mutations on homologous chromosomes, opposite to each other (in *trans* position). Often, there was no phenotypic effect at all. Genes showing such *cis-trans* effects were called pseudoalleles (see also Sect. 3.5.1; Fig. 3.30). When biochemical analysis became possible such *cis-trans* effects were shown to occur when the two mutations involved two different sites within the same functional gene, i.e., the area determining one protein (e.g., an enzyme). When such mutations occur in *cis* position a functionally intact protein can be formed by the homologous gene; the two normal sites can complement each other. Mutations in *trans* position, on the other hand, do not show this complementation, since no intact protein is formed.

In microorganisms such complementation groups can easily be analyzed.

Immediately after the advent of molecular genetics in the 1950s new conceptualization became necessary. Benzer (1957; [569]) proposed partition of the gene concept into three aspects: The units of recombination (recon); mutation (muton); and action (cistron; from the *cis-trans* effect). In subsequent years, the recon and the muton were found to be as small as the single nucleotide, the smallest unit in the genetic material; the cistron was identified with that segment of DNA that codes for one polypeptide chain. Of these three terms, only the term cistron became popular among geneticists; it was synonymous with the functional gene."

With the increasing awareness of the complexities of the genetic material, i.e. introns; promoter sequences; pseudogenes, etc., and especially with the discovery of gene families such as the Hbβ family, the delimitation of "genes" became more and more blurred. At present, it is not clear at all which of the long DNA sequences that are interspersed between the coding parts, including repetitive sequences such as Alu elements, are necessary for gene function. While we know more about the Hbβ locus in humans than about any other mammalian gene (see also Sect. 4.3), it is difficult to provide a clear definition regarding the exact limits of this gene complex. It appears reasonable to keep the *cis-trans* or complementation criterion as the basis for definition of a gene; and in analysis of genetic heterogeneity at the biochemical level this criterion is indeed used. But the information it provides is generally not specific enough; additional information from DNA analysis will be necessary for a new definition of a "gene."

The New Results on Structure of Genes and Formal Genetics. The discussions in Sect. 2.3 on gene structure could create the impression that most results of classical genetic analysis are now outdated. This, however, is not the case. For the analysis of modes of inheritance in families, the presence or absence of linkage between nonallelic genes, and the basic properties of genes in populations the principles of formal genetics are still applicable and necessary. The situation might be compared to that encountered in physics: Quantum mechanics has helped us in understanding the nature of light much more completely than before. However, geometrical optics is not only as correct as it always was, but also necessary for many practical applications, such as construction of spectacles or microscopes. It is therefore a necessary part of every physics textbook.

3 Formal Genetics of Man

3.1 Mendel's Modes of Inheritance and Their Application to Humans

Mendel's fundamental discoveries are usually summarized in three laws:

1. Crosses between organisms homozygous for two different alleles at one gene locus lead to genetically identical offspring (F_1 generation), heterozygous for this allele. It of no importance which of the two homozygotes is male and which is female (law of uniformity and reciprocity). Such reciprocity applies only for genes not located on sex chromosomes.
2. When these F_1 heterozygotes are crossed with each other (intercross), different genotypes segregate out: One-half is heterozygous again and one-quarter is homozygous for each of the parental types. This segregation 1:2:1 is repeated after crossing of heterozygotes in the following generations, whereas the two types of homozygotes breed pure. As noted above (Sect. 1.4), Mendel interpreted this result correctly, assuming formation of two types of germ cells with a 1:1 ratio in heterozygotes (law of segregation and law of purity of gametes).
3. When organisms differing in more than one gene pair are crossed, every single gene pair segregates independently and the resulting segregation ratios follow the statistical law of independent segregation (law of free combination of genes).

This third law applies only when there is no linkage (Sect. 3.4). Human diploid cells have 46 chromosomes: the two sex chromosomes and 44 autosomes forming 22 pairs of two homologues each. The pairs of homologues are separated during meiosis, forming haploid germ cells or gametes. After impregna-

1. GENOTYPES of PARENTS:
AA
Gametes
AA
A
A
Gametes
A
A A
A A
A
A A
A A
GENOTYPES of CHILDREN:
A A

2. GENOTYPES of PARENTS:

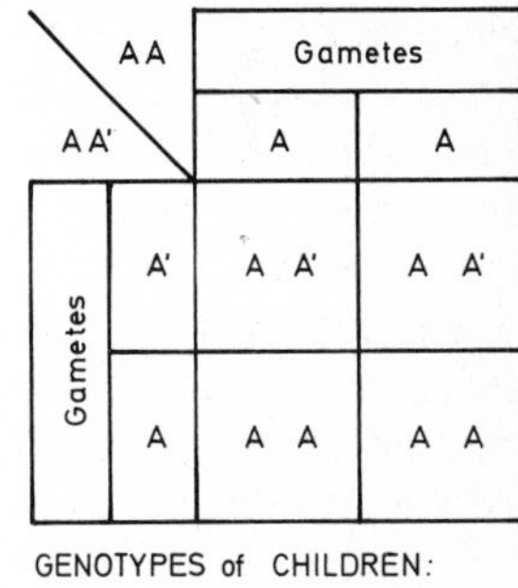

GENOTYPES of CHILDREN:
1 AA : 1AA'

3. GENOTYPES of PARENTS:
A A'
Gametes
A A'
A
A'
Gametes
A
A A
A A'
A'
A A'
A' A'
GENOTYPES of CHILDREN:
1AA : 2AA' : 1A'A'

4 GENOTYPES of PARENTS:

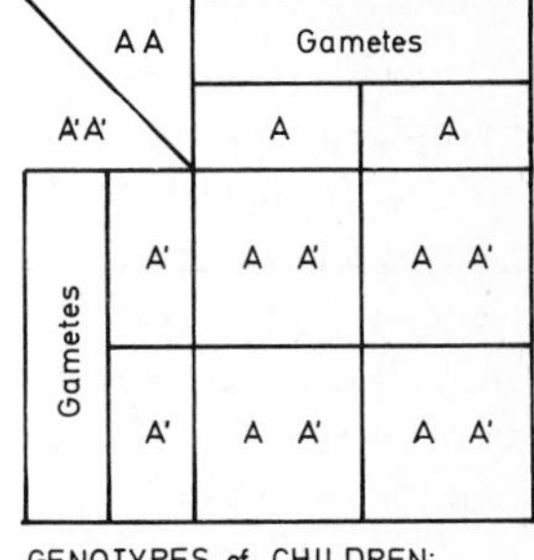

GENOTYPES of CHILDREN:
AA'

Fig. 3.1 Mating types with two alleles

tion, paternal and maternal germ cells unite to form the zygote, which is diploid again. Sex is determined genotypically; women normally have two X chromosomes, men have one X and one Y chromosome (Sect. 2.1.2).

For an understanding of the statistical character of segregation ratios in man, it is important to realize that the number of germ cells formed (Sect. 5.1.3) is particularly high among males. Only a very small sample comes to fertilization. When we look at single gene loci, this sampling process can in most cases be regarded as random. (For apparent exceptions, cf. Sect. 3.1.5.)

Two alleles may be termed A and A′. The combinations set out in Fig. 3.1 are possible. As noted above, these theoretical segregation ratios are probabilities; segregation ratios found empirically should be tested by statistical methods to determine whether they are compatible with the theoretical ratio implied by the genetic hypothesis.

The mating type of identical homozygotes (AA × AA or A′A′ × A′A′) is uninteresting except where it permits conclusions regarding genetic heterogeneity of a recessive condition (Sect. 3.1.3). Usually, mating between both homozygous types (AA × A′A′) is extremely rare and therefore of no practical importance. Matings between homozygotes and heterozygotes (AA′ × AA) and between two heterozygotes (A′A × A′A) are most important practically as explained below.

Mendel found that a genotype does not always determine one distinct phenotype. Frequently, heterozygotes resemble (more or less) one of the homozygotes. Mendel called the allele that determines the phenotype of the heterozygote dominant, the other recessive. With more penetrating analysis, some human geneticists believe that these terms have become useless and even misleading, especially in human genetics, and should be abandoned. In fact, at the level of gene action, genes are not dominant or recessive. At the phenotypic level, the distinction is important and useful. Biochemical mechanisms of dominant hereditary diseases (Sect. 4.6) are usually different from those of recessive conditions (Sect. 4.2). Hence, the mode of inheritance gives a hint regarding the biochemical mechanism likely to be involved.

In recent years, with the introduction of methods permitting analysis at a level closer to gene action, an increasing number of instances have become known in which every genotype determines a distinctive phenotype. This mode of inheritance is sometimes called codominant.

In addition, the somewhat outdated term "intermediate" is sometimes used when the phenotype of the heterozygote is in-between the phenotypes of the two homozygotes.

Table 3.1. Family studies of the genetics of MN blood types

Mating Type	Number of families	Types of children			Total children
		M	N	MN	
M × M	153	326	0	(1)	327
M × N	179	(1)	0	376	377
N × N	57	0	106	0	106
MN × M	463	499	(1)	473	973
MN × N	351	(3)	382	411	796
MN × MN	377	199	196	405	800
	1,580	1,028	685	1,666	3,379

Source: adapted from Wiener et al., 1953 [952]. ( ) indicates illegitimacy.

3.1.1 Codominant Mode of Inheritance

The first examples of codominance in man were found in the genetics of blood groups; the MN blood types (11130) may serve as one example (Table 3.1). When methods for genetic analysis at the protein level became available, many more examples were soon discovered (Sect. 6.1.2). The example in Table 3.1 clearly points toward a genetic model with two alleles, M and N, the phenotypes M and N being the two homozygotes, MN the heterozygote. This example will be used later for a statistical comparison between expected and observed segregation ratios. The "aberrant" cases in parentheses, which at first glance seem to contradict the genetic hypothesis, were the result of illegitimacy.

3.1.2 Autosomal Dominant Mode of Inheritance

The first description of a pedigree showing autosomal dominant inheritance of a human anomaly was Farabee's (1905) [656] paper on "Inheritance of Digital Malformations in Man." In textbooks, the condition is usually referred to as brachydactyly (short digits), but from the original paper it is clear that not only were the phalanges of hands and feet shortened, but the number of phalanges was also reduced (Fig. 3.2). Besides, stature was low (average of 159 cm in three males), apparently due to shortness of legs and inferentially also of arms. In every other aspect, Farabee wrote,

"the people appear perfectly normal ... and seem to suffer very little inconvenience on account of their malformation. The ladies complain of but one disadvantage in short fingers, and that is in playing the piano; they cannot reach the full octave and hence are not good players."

Figure 3.3 shows the pedigree. There are 36 affected in generations II through V, 13 of which are male, 23 female. Among the unaffected, 18 are male and 15 are female. The

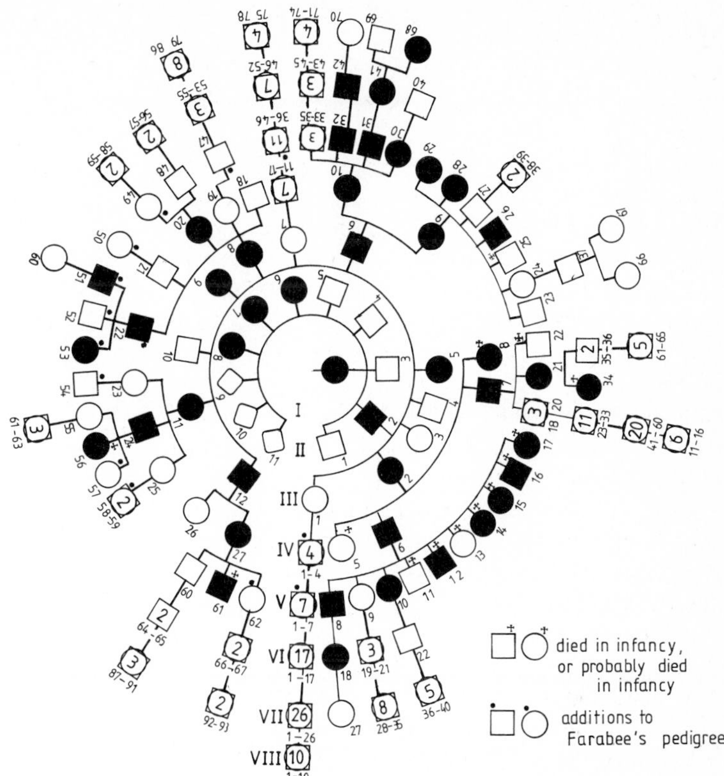

Fig. 3.2. The brachyphalangy pedigree of Farabee (1905) [656]. Black symbols refer to affected females (●) and males (■), numbers to their place in the pedigree

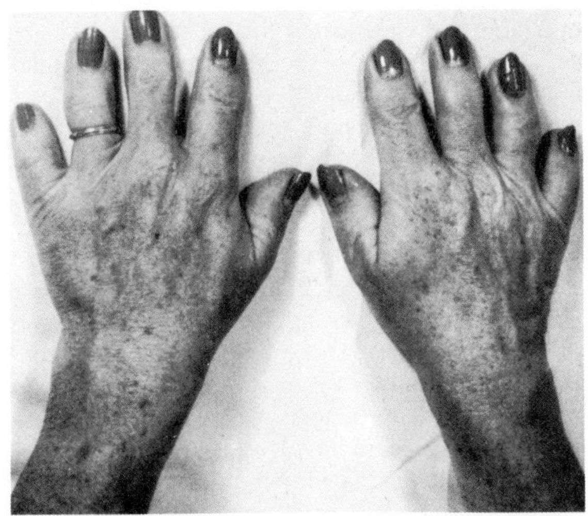

Fig. 3.3. Brachyphalangy in one member of a younger generation of Farabee's pedigree. (Haws and McKusick, 1963 [708])

trait is transmitted from one of the parents to about half the children; transmission is independent of sex. Unfortunately, Farabee did not take note of the children of the unaffected. Had he done so, he would have found them free from the anomaly. Many other pedigrees have shown absence of the trait among offspring of parents who do not carry the dominant gene. More recently, the family has been re-examined [708]. The children of the unaffected family members and some affected family members were added, and X-ray examination confirmed that not only hands and feet were affected but the distal limb bones as well. The basic defect is assumed to affect the epiphyseal cartilage. It is now called brachydactyly, type A1 (11250).[1]

Affected patients are heterozygous for an autosomal allele leading to a clear-cut and regular abnormality in the heterozygote. Therefore, the trait is, by definition, dominant. The family shows two other characteristics that were later found to be widespread:

1. The anomalies were described as being almost identical in all family members, and in each person appearing in all four extremities. This is a freguent finding in malformations with a regular mode of inheritance. The reason for the symmetry is evident considering that the same genes act on all four extremities.
2. The anomaly affected the well-being of its bearers very little. This lack of health impairment is typical for such extended pedigrees. Reproduction is normal. Otherwise, the trait would not be transmitted and would soon disappear. This is why, especially in the more serious dominant conditions, extended pedigrees are the exception rather than the rule. Most diseases caused by mutations ob-

[1] The numbers refer to identifying numbers of diseases listed in McKusick [133].

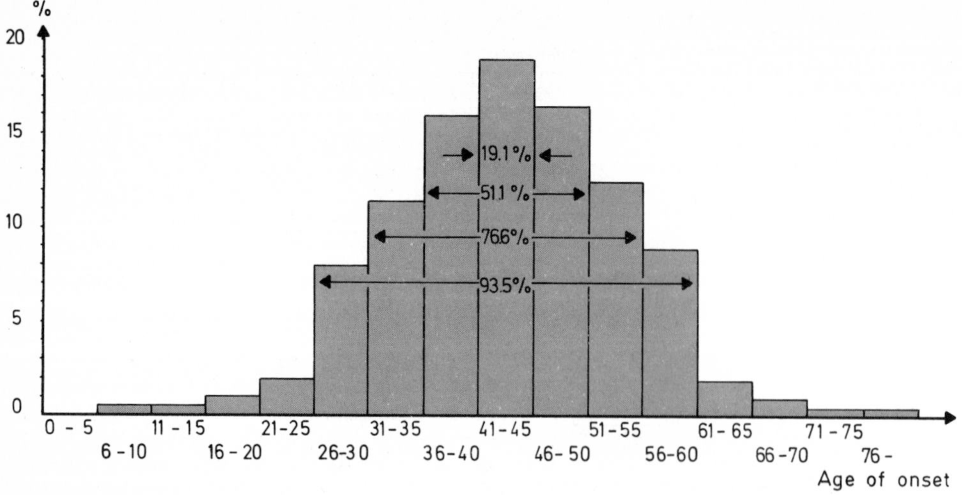

Fig. 3.4. Distribution of ages at onset in 802 cases of Huntington's chorea. (Wendt and Drohm, 1972 [941])

served in the present generation have originated rather recently, often even in the germ cell of one of the parents (cf. Sect. 5.1.3).

Late Manifestation, Incomplete Penetrance, and Variable Expressivity. Sometimes a severe dominant condition manifests only during or after the age of reproduction. Here, extended predigrees are usually observed in spite of the severity of the condition. The classic example is Huntington's chorea (14310), a degenerative disease of the nerve cells in the basal ganglia (caudate nucleus and putamen) leading to involuntary extrapyramidal movements, personality changes, and a slow deterioration of mental abilities.

Wendt and Drohm [941] carried out a comprehensive study of all cases of Huntington's chorea in West Germany. The distribution of ages at onset is seen in Fig. 3.4. The great majority of their patients were married when they developed clinical symptoms. Yet among thousands of patients, the authors were not able to locate even one case that could be ascribed with confidence to a new mutation. Similar results were obtained in another study from Michigan [850].

Another phenomenon occasionally encountered in dominant traits is incomplete penetrance.[2] Penetrance is a statistical concept and refers to the fraction of cases carrying a given gene that manifests a specified phenotype. The transmission seems occasionally to skip one generation, leaving out a person who judging from the pedigree must be heterozygous, or the fraction of those affected among sibs (after appropriate corrections, Sect. 3.3), turns out to be lower than the expected segregation ratio. An example is retinoblastoma (18020), a malignant eye tumor of children. Bilateral cases (and cases with more than one primary tumor) are always dominantly inherited, whereas most unilateral, single tumors are nonhereditary, probably being caused by somatic mutation (Sect. 5.1.6). Even in pedigrees otherwise showing regular dominant inheritance, however, apparent skipping of a generation is observed occasionally (Fig. 3.5). Estimation of the segregation ratio in a large sample showed that about 45% of sibs were affected instead of the 50% expected in regular dominant inheritance. The penetrance of all cases (unilateral and bilateral), therefore, is about 90%. Penetrance in families with bilateral cases is higher than in those with unilateral cases.

In many cases, penetrance is a function of the methods used for examination.

In many dominant conditions, the gene may manifest in all heterozygotes, but the *degree of manifestation* may be different. One example is neurofibromatosis (16220). Some cases may show the full-blown picture with many tumors of the skin, café-au-lait spots, and systemic involvement, whereas other cases – even in the same families – may show only a few café-au-lait spots. The term used to describe this phenomenon is "variable expressivity."[3] Expressivity relates to the variable manifestations of a given gene and is used interchangeably with the term "manifestation rate." Terms such as "incomplete penetrance" and "variable expressivity" are needed to convey quick understanding about certain phenomena. They may also become dangerous, once we forget that they do not explain a biologic mechanism, but rather are labels for our ignorance.

[2] Timoféeff-Ressovsky, 1931 [912].

[3] Timoféeff-Ressovsky, 1931 [912].

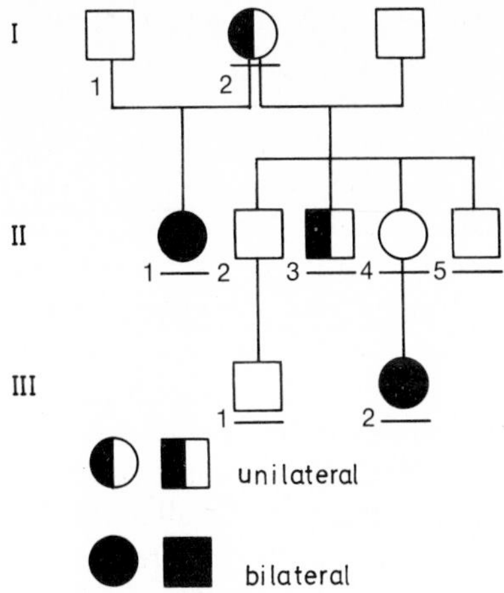

Fig. 3.5. Incomplete penetrance in retinoblastoma. The unaffected woman II, 4 must be heterozygous, her mother I, 2 and her daughter III, 2 being affected. □ Examined personally

It is, indeed, somewhat surprising that so many dominant conditions show such a large interindividual variability in age at onset and severity of manifestation. It would be more understandable if such variability were observed only between different families. From our knowledge of molecular biology (Sect. 5.1.4), we can infer that the mutational events leading to these conditions will almost always be slightly different between families. Indeed, there is usually intrafamilial correlation for age at onset and severity of manifestation. For Huntington's chorea, for example, Wendt and Drohm [941] calculated a correlation coefficient of +0.57 for age of on set. But there usually remains appreciable variability within families, in which the abnormal genes are identical by descent. It is again no more than a label for our ignorance when we invoke the "genetic background" or the action of all other genes for help. Analysis using the methods of formal genetics has contributed little to a better knowledge (cf. the discussion on allelic modification and sex-limiting modifying genes, Sect. 3.1.7).

Influence of Homozygosity on the Manifestation of Abnormal Dominant Genes. An abnormal gene is called dominant when the heterozygote deviates clearly from the normal. Indeed, almost all bearers of dominant conditions in the human population are heterozygotes. From time to time, however, it may happen that two bearers of the same anomaly marry and have children. One quarter of them will then be homozygous. This has been observed in several instances, especially when the spouses were relatives. Probably the first example was that described by Mohr and Wriedt (1919) [792]. In a consanguineous marriage between two bearers of a moderate brachydactyly (11260), a child was born who not only lacked fingers and toes but also showed multiple malformations of the skeleton and died at the age of one year. A sister, however, had only the moderate anomaly like her parents [792].

One other example is the Pelger-Huet anomaly of leukocytes (16940) [823]. This is a harmless anomaly of polymorphic leukocyte nuclei that, instead of the normal segmentation, show only two segments of approximately equal size. The chromatin looks coarse and pyknotic. This anomaly follows a regular, autosomal-dominant mode of inheritance. It is not very rare (about 1:1000 to 1:3000 in Central European populations).
The same anomaly was also discovered in the rabbit. Therefore, the possibility offered itself to breed the homozygote. But this turned out to be difficult: Homozygotes showed a tendency to die prenatally. Finally, however, a few living animals were obtained and kept alive for a short time. They showed some remarkable peculiarities: all granulocyte nuclei were round and segmentation was not observed. The chromatin looked extremely coarse (Fig. 3.6). This coarseness was apparent even in lymphocytes or plasma cells. Apart from the hematological anomalies, however, there were other symptoms such as chondrodysplasia of limb bones and ribs, and generally delayed development. The rib anomaly compressed the thoracic organs and eventually killed the animals. Nachtsheim predicted the same leukocyte anomaly and similar bone symptoms in homozygous humans. Fortunately, this prediction turned out to be only partially correct.
The first homozygous human Pelger case was described in 1952 in Holland [707]. The child came from a first-cousin mating between gypsies. Ninety-four percent of its granulocytes were similar to those of the homozygous rabbits (Fig. 3.6, 3.7). However, there were no signs of any skeletal anomaly in this or other homozygotes.

Further examples of homozygosity of dominant anomalies are known. In one family, two parents with hereditary hemorrhagic teleangiectasia had a child showing multiple, severe internal and external telangiectasias who died at the age of 2.5 months [885]. Similarly, a very severe form of epidermolysis bullosa was observed in two of eight children of a couple, both of whom were afflicted with a mild type of this disease.
Another couple, both having a myopathy affecting the distal limb muscles, had 16 children, three of whom showed atypical and especially severe symptoms: the long flexors and the proximal hip muscles were also afflicted, and onset was earlier in life [940].
Epithelioma adenoides cysticum (13270) is a dominant skin disease characterized by multiple nodular tumors. One female patient, whose parents were both affected, had especially severe symptoms and her eight children all showed this anomaly (Fig. 3.8) [677]. Further examples include achondroplasia (10080), the Ehlers-Danlos syndrome (13000), and others [832]. All these cases indicate that homozygotes of dominant anomalies are more severely affected than heterozygotes.

From what we know about gene action, this is not surprising. In familial hypercholesterolemia (14440)

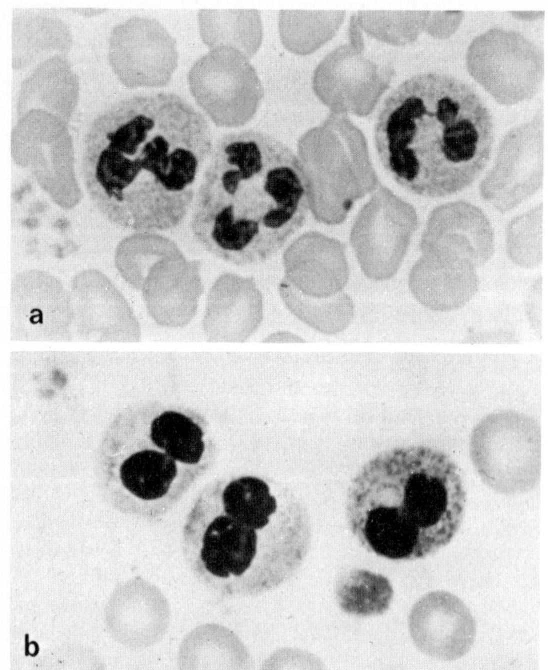

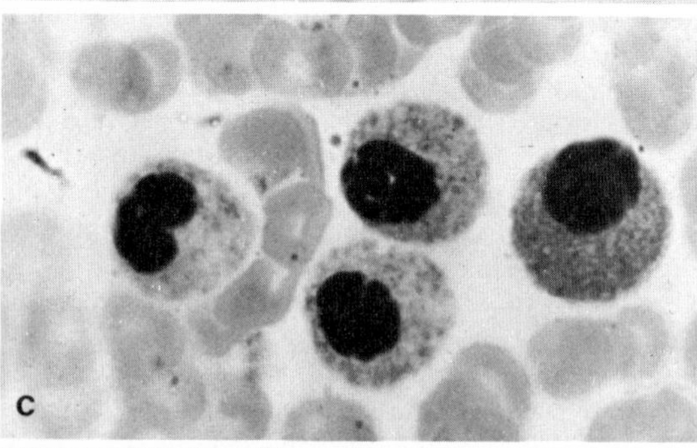

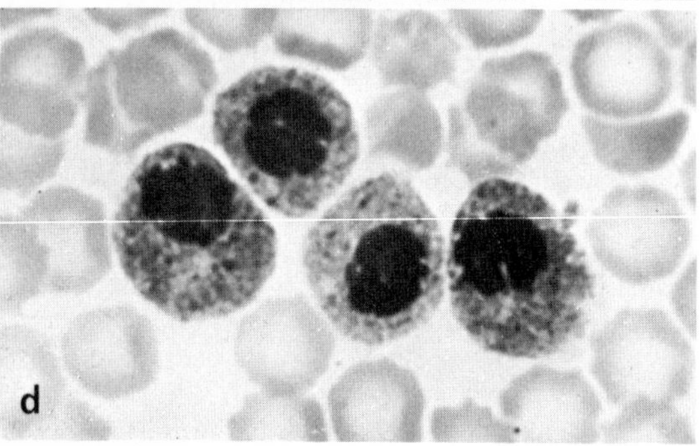

Fig. 3.6 a–d. The Pelger anomaly of polymorphnuclear leukocytes. **a** Normal granulocytes; **b** Pelger cells of a human heterozygote; **c, d** granulocytes of a human (**c**) and a rabbit (**d**) homozygote

for example, the mechanism for action of a dominant gene was discovered. A decreased number of receptors for a regulatory substance (low-density lipoprotein – LDL) showed the expected differences between heterozygotes and affected homozygotes: 50% decrease and complete absence of receptors, respectively (Sect. 4.6.4). Affected homozygotes have massive hypercholesteremia and usually die of myocardial infarction before the age of 30.

As noted before, Mendel called a gene dominant when the phenotype of the heterozygote resembled that of one homozygote. The examples of more severe manifestation of dominant genes in the homozygous than in the heterozygous state show that this strict definition is not maintained in human genetics. Here, all conditions are called dominant in which the heterozygote deviates consistently and perceptibly from the normal homozygote – irrespective of the phenotype of the anomalous homozygote. In Mendel's strict definition, most or even all dominant conditions in humans would be "intermediate". However, the more lenient connotation of "dominance" is now in general use.

3.1.3 Autosomal-Recessive Mode of Inheritance

The mode of inheritance is called recessive when the heterozygote does not phenotypically differ from the normal homozygote. In many cases, special methods will uncover slight detectable differences (Sect. 4.2.2.8). Contrary to dominant inheritance, in which almost all crosses are between heterozygotes × homozygous normals (Sect. 3.1), the great majority of all matings observed in recessive anomalies involve heterozygous and phenotypically normal parents. Since the three genotypes AA, Aa, and aa occur in the ratio 1 : 2 : 1 among the offspring, the probability of a child's being affected is 25%. At the turn of the century when Garrod wrote his paper on alkaptonuria (Sect. 1.5), the "familial" character of recessive diseases could easily be recognized, as family size was large. Today, however, two-children families are preferred in industrialized societies. This means that the patient with a recessive disease very often will be the only one affected in an otherwise healthy family. However, once one affected child has been born, the genetic risk for any further child of the same parents is 25%. This is important for genetic counseling.

A recessive disease that recently has attracted much attention by molecular biologists is xeroderma pigmentosum (27870). After exposure to ultraviolet light, erythema develops, especially in the face, and is followed by atrophy and telangiectases. Finally, skin cancers develop that, if untreated, lead to death. Figure 3.9 b shows a typical pedigree. Here the parents are first cousins. The rate of consanguinity among parents of patients with rare recessive diseases is well above the population average. Usually these parents have inherited this gene from a common ancestor (Sect. 6.3.1). In Garrod's time, this was a powerful tool for recognizing rare recessive dis-

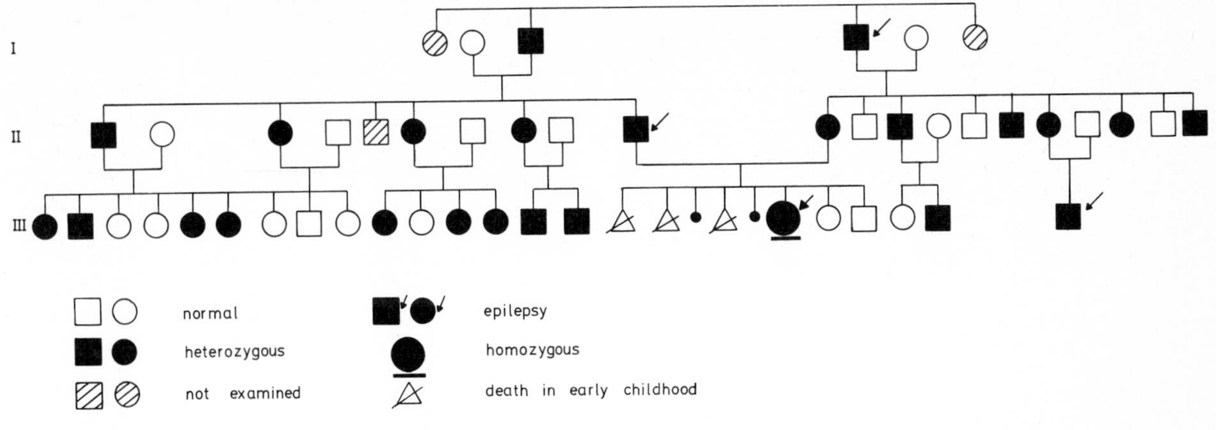

△ **Fig. 3.7.** Human pedigree with a homozygous Pelger girl. (Haverkamp-Begemann and van Lockeren-Campagne, 1952 [707])

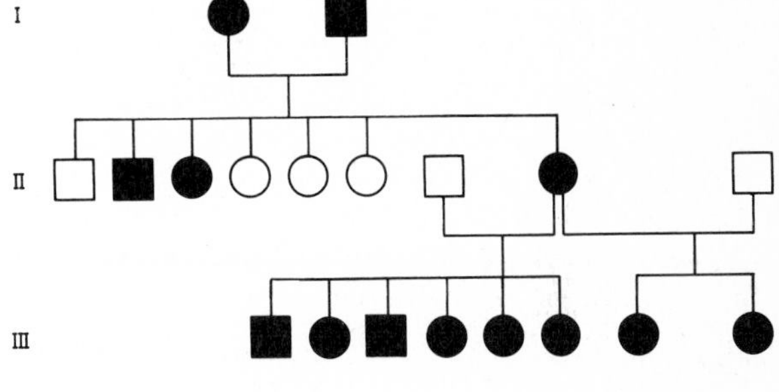

▷

Fig. 3.8. Woman homozygous for epithelioma adenoides cysticum and her progeny in two marriages. (Gaul, 1953 [677])

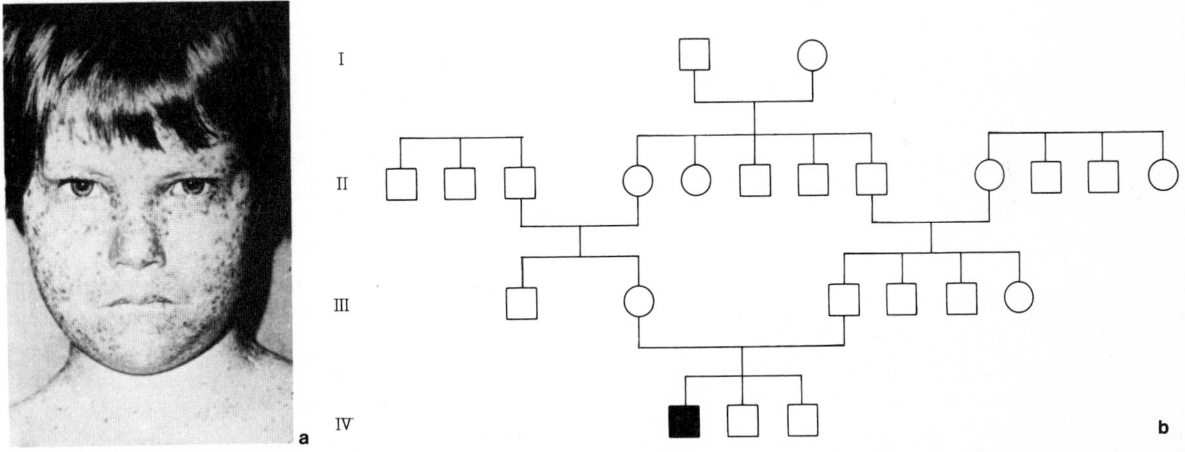

Fig. 3.9 a, b. Xeroderma pigmentosum. **a** Girl with this condition (Courtesy of Dr. U. W. Schnyder); **b** pedigree of single case with first-cousin marriage. (Dorn, 1959 [630]) See also Sect. 5.1.6.3.

eases; among ten families of alkaptonurics for which this information was available, the parents were first cousins in six cases (Sect. 1.5). Today, however, the consanguinity rate has decreased in most industrialized societies. Hence, even if the rate of consanguinity in families with affected children is substantially increased above the population average, this will not necessarily lead to the appearance of a consanguineous mating when a limited number of families are studied. This phenomenon together with the small average family size makes it increasingly difficult to recognize an autosomal-recessive mode of inheritance with certainty. Fortunately, however, we no longer need to depend solely on formal genetics. When a rare disease, especially in a child, shows signs of being an inborn error of metabolism, and especially when an enzyme defect can be demonstrated, a recessive mode of inheritance can be inferred in the absence of evidence to the contrary. For purposes of genetic counseling, it must be assumed.

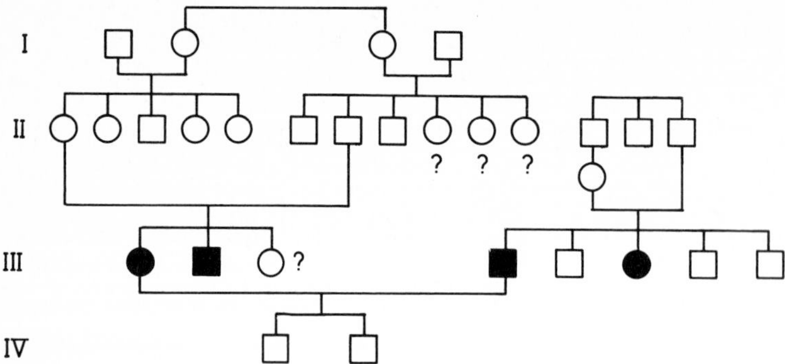

Fig. 3.10. Pedigree with deaf-mutism showing genetic heterogeneity (Mühlmann, 1930 [817]). Both parents are affected with a hereditary type of deaf mutism: they have affected sibs and come from consanguineous marriages. Yet, the two sons are not deaf. They are compound heterozygotes for different deaf mutism genes

As a rule, the vast majority of patients with autosomal-recessive diseases are children of two heterozygotes. Especially decisive for recessive inheritance are the rare matings of two homozygotes with the same anomaly. If both parents are homozygous for the same recessive gene, their mating should exclusively produce affected children. A number of such examples are reported in albinism (20310, 20320). Some marriages between albinos, on the other hand, have produced normally pigmented children [914]. Unless these children are all illegitimate, this proves that the parents must be homozygous for different albino mutations, i.e., more than one albino locus must exist in man. This is the kind of proof formal genetics can provide for genetic heterogeneity of diseases demonstrating an autosomal recessive mode of inheritance and the same (or a very similar) phenotype. In albinism, genetic heterogeneity has now been shown by biochemical methods [203].

Another condition for which genetic heterogeneity has been proved in this way is deaf mutism (Fig. 3.10). Since environmental causes can also cause deafness, it is remarkable that in the pedigree shown both spouses have an affected sibling, and both parents were consanguineous. The hypothesis of genetic heterogeneity has since been confirmed for this condition with a variety of methods (App. 3).

Pseudodominance in Autosomal Recessive Inheritance. Occasionally, matings between heterozygotes and affected homozygotes are observed. One parent is affected, and the expected segregation ratio among children is 1:1. Since this segregation pattern mimics that found with dominant inheritance, this situation is aptly named "pseudodominance." Fortunately for genetic analysis, such matings are very rare.

Garrod's alkaptonuria (20350) (Sect. 1.5) provides an example. In all families described after Garrod, the autosomal-recessive mode of inheritance was confirmed until 1956 when a family with a phenotypically similar but apparently dominant form was reported (Fig. 3.11) - a surprising finding. Some years later the authors had to disavow their conclusions: further family investigations had shown typical, recessive alkaptonuria. A number of marriages between relatives (homozygotes × heterozygotes) had led to pseudodominance. If an individual suffering from a recessive disease mates with a normal homozygote, all children are heterozygotes, and, hence, phenotypically normal. As soon as we learn to treat recessive diseases successfully, marriages of affected but treated homozygotes will increase.

Expressivity is, as a rule, more uniform within the same family in recessive than in dominant disorders. Incomplete penetrance seems to be rare. Variability between families, however, may be appreciable.

Compound Heterozygotes. When a more penetrating biochemical analysis becomes possible, alleles of different origin frequently have slightly different properties. Hence, if a homozygote is defined according to the specific base substitution within a certain codon, many patients with recessive disorders are probably not homozygotes but compound heterozygotes. As will be shown in Sect. 4.3, homozygotes for the hemoglobinopathies are defined on the basis of base substitutions. However, considering our present state of knowledge for most recessive diseases, it is not yet feasible to introduce this definition into general usage (Fig. 3.12). As a rule, we do not even know whether our present terminology relates to single genes or to more extensive chromosomal alterations which have not been detected so far. With the new DNA techniques (Sect. 2.3), this will change rapidly. It is good to remember that definitions are shifted with better understanding of the underlying mechanism.

We can be reasonably sure that an affected homozygote carries two copies of the same mutation if both copies have a common origin; for example, if his parents are first cousins and if the condition is very rare. In a relatively common condition, such as cystic fibrosis of the pancreas, the two mutations might have different sources. Another source of identity by descent are cases from an isolate in which

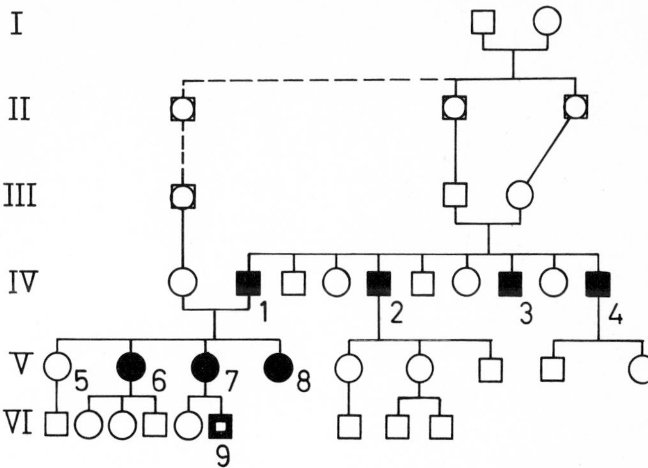

Fig. 3.11. Pedigree with pseudodominance of alkaptonuria, an autosomal-recessive condition. ▣ suspected alcaptonuric; ⊡ sex unknown. (Milch and Milch, 1959 [790])

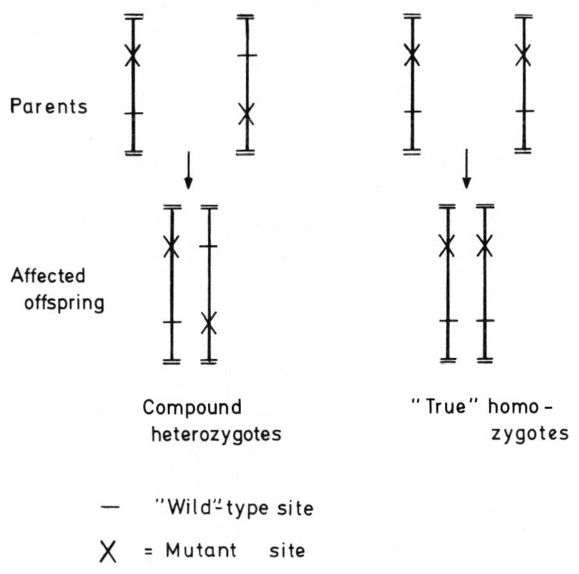

Fig. 3.12. Formation of a compound heterozygote. Each line represents the mutant locus on one chromosome in a parent. Among the many possibilities for mutation, two are shown. If parents are heterozygous for mutations that are at identical sites, the affected child will be a "true" homozygote. Otherwise, it will be a compound heterozygote

a single mutation – which has been introduced by one individual – became frequent, such as the skin disease called Mal de Meleda on the Yugoslavian island of Mljet (Sect. 6.4.2).

3.1.4 X-Linked Modes of Inheritance

In man, every mating is a Mendelian backcross with respect to the X and Y chromosomes:

		Paternal gametes	
		X	Y
Maternal gametes	X	¼ XX	¼ XY
	X	¼ XX	¼ XY
Total sum		½ XX ♀ +	½ XY ♂

This would imply that, on the average, female and male zygotes are formed in a 1:1 ratio. This, however, is not quite true. The sex ratio at birth (known as the secondary sex ratio in contrast to the primary sex ratio at conception) is slightly shifted in favor of boys (102–106 boys/100 girls). The primary sex ratio is not known exactly, but there are hints that it is also somewhat variable.[4] The formal characteristics of X-linked modes of inheritance can easily be derived from the mode of sex determination.

X-Linked Recessive Mode of Inheritance. If we use A for the dominant, normal wild-type and a for the recessive alleles, the following matings are possible:

a) AA ♀ × A ♂. All children will have the phenotype A. Neither this nor the analogous mating aa × a is useful for genetic analysis.
b) AA ♀ × a ♂. All sons have one of the mother's normal alleles. They are healthy. All daughters are heterozygous Aa. They are phenotypically

[4] Many studies on the (primary and secondary) sex ratio have been published. Chromosome studies on abortions should reflect the primary sex ratio and point to a value not too far from 100 (boys and girls in a ratio of 1:1). However, the primary and secondary sex ratio also depend on the interval between sexual intercourse and ovulation, frequency of intercourse, general cultural conditions, and even war and peace. After artificial insemination, the fraction of male offspring appears appreciably increased. [704].

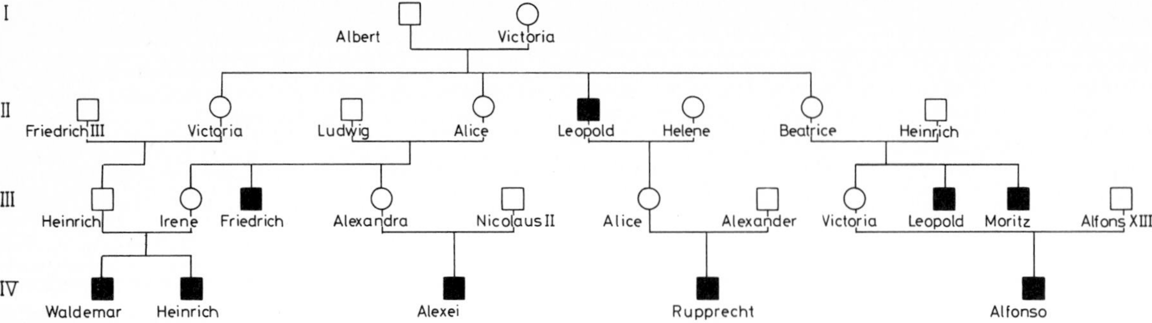

Fig. 3.13. Pedigree with X-linked recessive hemophilia A in the European Royal Houses. Queen Victoria (I, 2) was heterozygous; she transmitted the mutant gene to one hemophilic son and to three daughters

healthy, but carriers of the abnormal allele. In the analogous mating aa ♀ + A ♂, all sons are affected (a), and all daughters are heterozygous (Aa).

c) Aa ♀ + A ♂. This type is most important. All daughters are phenotypically normal; half are heterozygous carriers. Half of their sons are hemizygous a and affected. The analogous mating Aa ♀ × a ♂ is very rare. There is a 1:1 ratio of affected and heterozygotes among female children and an 1:1 ratio of affected and normals among males.

The principal formal characteristics of X-linked recessive inheritance can be summarized as follows: Males are predominantly – and in rare X-linked conditions almost exclusively – affected. All their phenotypically healthy but heterozygous daughters are carriers. If no new mutation has occurred and the mother of the affected male is heterozygous, then half of his sisters are heterozygous carriers. Among sons of heterozygous women, there will be a 1:1 ratio between affected and unaffected.

Strictly speaking, transmission from affected grandfathers via healthy mothers to affected grandsons is not altogether decisive for locating the gene on the X chromosome. An autosomal gene with manifestation limited to the male sex could show the same pattern. The fact that all sons of affected men are unaffected is decisive, however. This criterion can create difficulties in interpretation when a disease is so severe that the patients do not reproduce.

The two most famous and, from a practical standpoint, very important examples are hemophilia A and B (30670, 30690). Due to its alarming manifestations, hemophilia had been known to doctors for a long time and had given rise to the formulation of Nasse's rule (Sect. 1.2). Figure 3.13 shows the famous pedigree of Queen Victoria's descendants in the European royal houses. One of the hemophilics was the Czarevich Alexey of Russia, and in this case genetic disease influenced politics. Rasputin's power over the imperial couple was based at least partially on his ability to comfort the Czarevich when he was frightened by bleedings. Much larger pedigrees have been described, probably the most extensive being that of hemophilia B in Tenna (Switzerland). As a rule, however, the pedigrees observed in practice are much smaller. Frequently, there is only one sibship with affected brothers, or the patient is even the only one affected in an otherwise healthy family. Again, as in dominant conditions (Sect. 3.1.2), this is caused by the reduced reproductive capacity of the patients, which leads to the elimination of most hemophilia genes within one or a few generations after they have been produced by new mutation. As expected, almost all hemophilia patients are males. However, there are a few exceptions. Figure 3.14 shows a pedigree from Czechoslovakia in which a hemophilic had married a heterozygote (who was his double first cousin because in their parents' generation two brothers had married two sisters). The homozygous sisters both had moderately severe hemophilia similar to their affected male relatives.

Another X-linked recessive condition is the Lesch-Nyhan syndrome (30800), a rare anomaly of purine metabolism caused by deficiencies of the enzyme HGPRT leading to severe hyperuricemia, neurological anomalies and a compulsive tendency to self-destruction (Fig. 3.15). A number of basic biochemical and genetic problems are being examined using this model, and a selective system has been found to identify and isolate mutant cells with this defect in vitro (Sect. 4.2.2.6).

Some X-linked conditions have reached considerable frequencies. The most widespread are red-green color vision defects and variants of the enzyme glucose-6-phosphate dehydrogenase (G6PD) (Sect. 4.2) but X-linked mental retardation with marker X (Sect. 8.2.1.2) is also common.

X-Linked Dominant Mode of Inheritance. An X-linked dominant condition manifests itself in hemizygous men and heterozygous women. However, all sons of affected males are free of the trait unless their mothers are also affected, and their children are also unaffected. On the other hand, all daughters of affected males are affected. Among children of affected women, the segregation ratio is 1:1 regardless of the child's sex, quite like in autosomal-dominant inheritance. If affected individuals have a normal rate of reproduction, about twice as many affected females as males are found in the population.

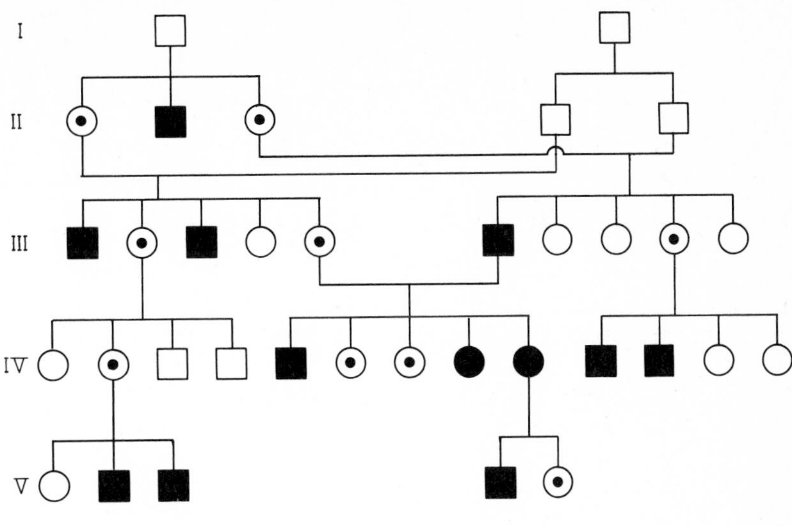

Fig. 3.14. Pedigree with two female homozygotes for X-linked hemophilia. The parents are double first cousins. ⊙ obligatory heterozygotes. (Pola and Svojitka, 1957 [840])

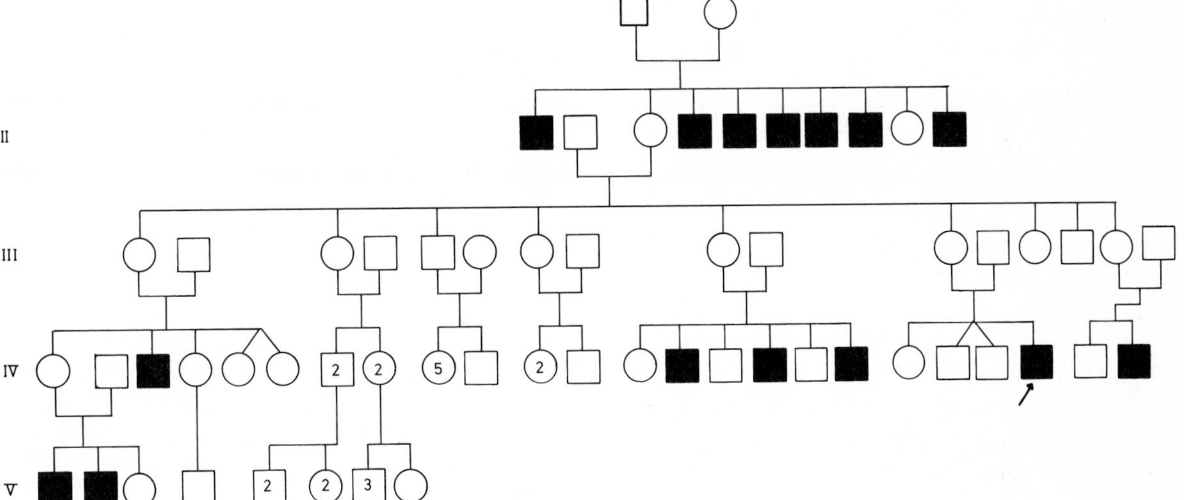

Fig. 3.15. Pedigree with Lesch-Nyhan syndrome. (Nyhan, see [203])

Since only children of affected males provide information in discriminating X-linked dominant from autosomaldominant inheritance, it is difficult or even impossible to distinguish between these modes of inheritance when the available data are scarce.

The first clear-cut example was described by Siemens (1925) [871] in a skin disease that he named "keratosis follicularis spinulosa decalvans cum ophiasi" (30880). The disease manifests follicular hyperkeratosis leading to partial or total loss of eyelashes, eyebrows, and head hair (Fig. 3.16). Severe manifestations were, however, confined to the male members of this family.

Since then it has been confirmed for all traits for which an X-linked dominant mode of inheritance can be established that males are on the average more severely affected than females. This finding was no surprise, since heterozygous women have a normal allele for compensation, but a satisfactory explanation was possible only when random inactivation of one of the X chromosomes in females was discovered (Sect. 2.2.3.3).

Another example of X-linked dominant inheritance is vitamin D-resistant rickets with hypophosphatemia (30780) [958]: In the pedigree shown in Fig. 3.17, all 11 daughters of the affected men suffered from rickets or had hypophosphatemia; all ten of their sons, however, were healthy. The affected women have both affected and healthy sons and daughters. The probability for the mode of inheritance to be autosomal-dominant and for the affected males to have only affected daughters and only healthy sons is less than 1:10000. Moreover, in this family male members also tended to be more severely affected than females.

X-Linked Dominant Inheritance with Lethality of the Male Hemizygotes [943]. Females with X-chromosomal diseases tend to have milder symptoms than males, as noted above. In some cases, the male

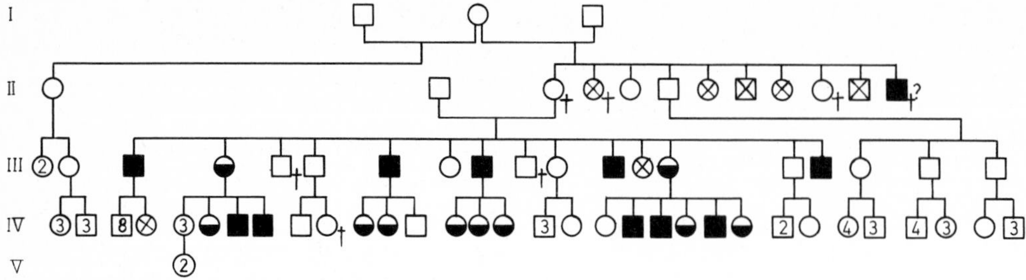

Fig. 3.16. Pedigree with keratosis follicularis spinulosa decalvans, the first example establishing X-linked dominant inheritance in humans. ■ severe; ○ mild manifestation; ⊗ dead before age of manifestation. (Siemens, 1925 [871])

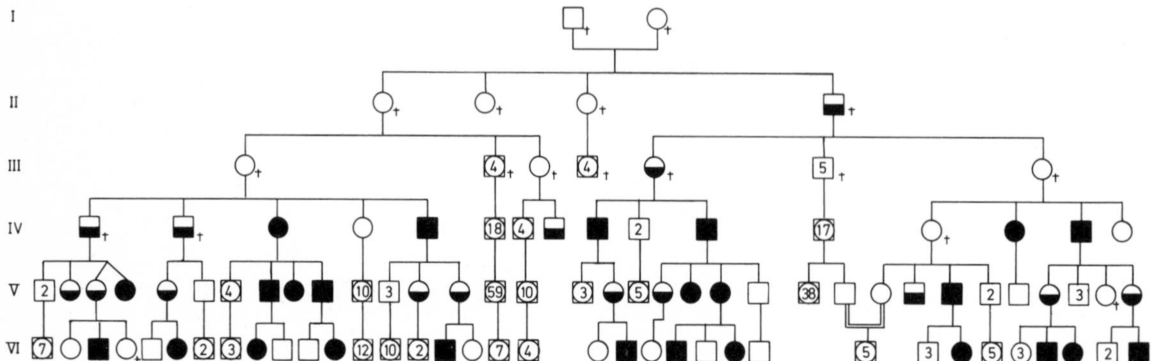

Fig. 3.17. Pedigree with X-linked dominant vitamin D-resistant rickets and hypophosphatemia. ■ hypophosphatemia and rickets; ⬓ hypophosphatemia without rickets. (Winters et al., 1957 [958])

zygotes may be so severely affected that they die before birth, and only the females survive. This would result in pedigrees containing only affected females, and among their children affected daughters, normal daughters, and normal sons would be formed in the ratio of 1:1:1. Among those male hemizygotes that did not die in very early pregnancy, spontaneous abortions (or male stillbirths) would be expected. W. Lenz (1961) [759] was the first to show that this mode of inheritance exists in humans in the condition known as incontinentia pigmenti (Bloch-Sulzberger) (30830).

Around the time of birth, the girls affected with this disease develop inflammatory erythematous and vesicular skin disorders. Later on, marble cake-like pigmentations appear (Fig. 3.18a). The syndrome additionally comprises tooth anomalies. Figure 3.18b shows a typical pedigree. The alternative hypothesis would be that of an autosomal-dominant mode of inheritance with manifestation limited to the female sex. The two hypotheses would have the following consequences:

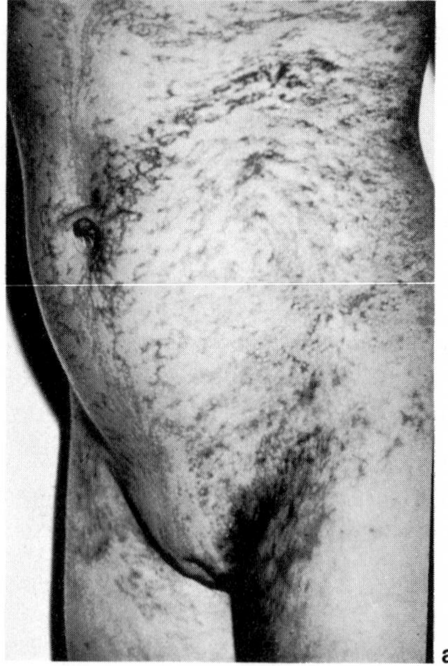

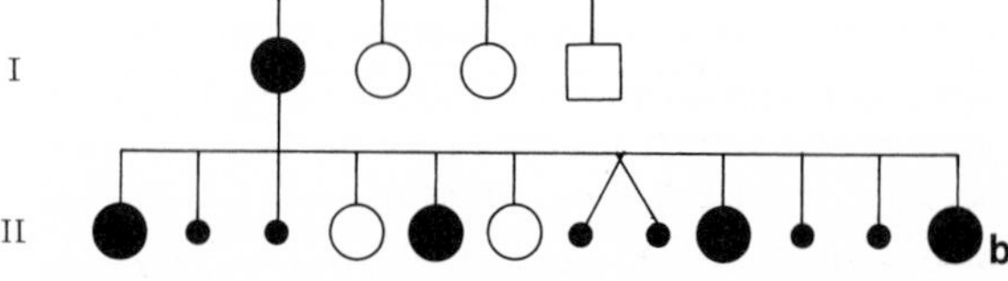

Fig. 3.18. a Incontinentia pigmenti (Bloch-Sulzberger). (Fuhrmann, 1974 [433]). Note marble cake appearance of skin. **b** Pedigree with incontinentia pigmenti. • spontaneous abortion; ● incontinentia pigmenti. (Lenz, 1961 [759]) ▷

a) With autosomal-dominant sex-limited inheritance, and after proper correction (Sect. 3.3.4), there would be a 1:1 ratio of affected to unaffected among sisters of propositae. All brothers would be healthy. If the population sex ratio is assumed to be 1:1, then a sex ratio of 2♂:1♀ would be expected among healthy sibs. With X-linked inheritance, on the other hand, the expected number of healthy brothers is much lower, because one-half of the male zygotes are expected to die before birth (possibly leading to an increased rate of spontaneous miscarriages). Among healthy sibs, a 1♂:1♀ ratio would be expected.
b) With autosomal-dominant inheritance, the abnormal gene may come from the father or from the mother. Therefore, more remotely related affected relatives are to be expected in paternal as well as in maternal relatives. With X-linked inheritance, on the other hand, the gene must come from the mother. Considering the rarity of the condition, additional cases would not occur in the father's families.
c) With autosomal-dominant inheritance, the loss of mutant genes per generation would be relatively small compared with the total number of these mutations in the population, as the male carriers, being free of symptoms, would reproduce normally. Therefore, assuming genetic equilibrium (Sect. 5.1.3.1), the number of new mutations would be small compared with the overall number of cases in the population. With X-linked inheritance, on the other hand, the loss of zygotes is high due to death of the hemizygote. Hence, many of the cases in the population will be caused ba recent mutation and extensive pedigrees will be rare [924].

The available statistical evidence has consistently supported the hypothesis of an X-linked dominant mode of inheritance with lethality of the male hemizygote. According to Carney (1976) [613], 593 female and 16 male cases were reported. 55,4% of the female patients had a positive family history. How can the sporadic males be explained? Of course, the phenomenon of *Durchbrenners* (Hadorn [696] used the term "escapers" - the occasional survival of individuals affected with a lethal genotype) is well-known, but Lenz (1975) [760] suggested a more specific explanation, assuming, on the basis of a suggestion made by Gartler and Francke (1975) [676], that a mutation in only one half-strand of the DNA double helix of either the sperm or the oocyte occurred.
Meanwhile, a few other conditions have been added to this category. One is the orofaciodigital (OFD) syndrome (31120) which consists of a number of malformations of mouth and tongue, a median hare lip, and a peculiar type of syndactyly [674]. Other examples may be focal dermal hypoplasia, X-linked chondrodysplasia punctata, ornithine transcarbamylase (OTC) deficiency (31125; lethality in the *neonatal* hemizygous male), and partial lipodystrophy with lipotrophic diabetes. Moreover, single pedigrees suggesting this mode of inheritance have been published for some other conditions, for example a special variant of muscular dystrophy of the limb girdle type [565, 943].

Genes on the Y Chromosome. Up to the 1950s, most geneticists were convinced that the human Y chromosome contained genes that occasionally mutated, giving rise to a Y-linked (or holandric) mode of inheritance with male-to-male transmission and males solely being affected. Stern (1957) [897] reviewed the evidence with the result that the time-honored textbook example of Y-linked inheritance of the porcupine man (severe ichthyosis) could no longer be maintained as valid. The only character for which Y-linked inheritance can still be discussed are hairy pinnae, i.e., hair on the outer rim of the ear. A number of extensive pedigrees have been published that show male-to-male transmission. However, the late onset mostly in the 3rd decade of life, as well as the extremely variable expressivity and the high prevalence in some populations (up to 30%), makes distinction from a multifactorial mode of inheritance with sex limitation very difficult. Y-linkage can therefore not be fully accepted for this trait.
Eichwald et al. (1955) [643] described a Y-determined transplantation antigen in the mouse, which they called HY. Its significance for sex determination and, specifically, for male gonad differentiation has been suggested. According to a hypothesis put forward by Ohno [1248], HY is expressed on all cells of a male organism but only gonadal cells have a HY receptor that binds the antigen. When bound to the receptor, this substance becomes active in organizing the development of testicular tissue. Extensive studies have been performed to test this hypothesis [1342]; since they are important for our understanding of sex determination. They will be discussed elsewhere (Sect. 4.7.5).
Anti-HY antisera of the mouse react with cells of all mammals tested so far, including man. Even birds and amphibia show a cross reaction. Hence, the antigen seems to have been conserved during evolution.

3.1.5 Pedigrees Not Fitting a Known, Simple Mode of Inheritance

From time to time, pedigrees have been reported that do not fit one of the usual modes of inheritance. Usually, errors of ascertainment or of documentation are the cause. Nevertheless, a few such pedigrees remain. One is the famous Cunier (1839) [898] pedigree of "color blindness" (Fig. 3.19).

The trait was transmitted through four generations to all 11 female descendants of a colorblind woman. A reappraisal of the original, very carefully reported ophthalmologie data showed that the condition was not one of the typically X-linked color blindness or any known autosomal type. Apparently, the findings were part of a syndrome of which the cardinal feature was a mild form of congenital cataract. Other symptoms were a greyish-brown iris and a weakness in distinguishing red and blue. Various hypotheses are possible to explain this pedigree, such as extranuclear inheritance of a cytoplasmatic element with manifestations only in females [898].

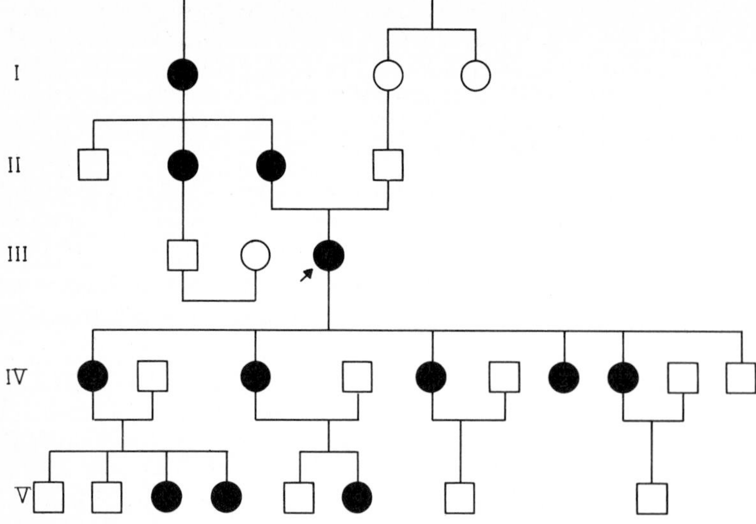

Fig. 3.19. The Cunier pedigree. All women – and only women – suffer from an unusual color vision anomaly [898]

In experimental animals, segregation ratios deviating from those expected from Mendelian expectations were occasionally reported, one example being the T locus of the mouse [568].

The Alport syndrome (10420) is sometimes mentioned as an example of aberrant segregation [673]. However, the evidence is not entirely convincing, especially, since genetic heterogeneity is very likely and autosomal as well as X-linked types may exist.
Another case of unusual inheritance is Leber's optical atrophy (30890). All reports agree that this condition is more frequent in men than in women [767]. Moreover, transmission from an affected grandfather via an unaffected mother to an affected son seems to be very rare; the trait is almost always transmitted by female family members. On the other hand, among daughters of heterozygous women, heterozygous affected women seem to be much more frequent than normal daughters. In this condition, a defect of a mitochondrial enzyme (thiosulfate-sulphur transferase) has been claimed [598]. As mentioned in Sect. 2.3.4, the DNA sequence of human mitochondria is now known, and some genes, e.g., those for several tRNA species and enzymes, have been found to be located here. The above-mentioned enzyme does not belong to this group, but since it is a mitochondrial enzyme it is certainly possible that a subunit is coded by the mitochondrial genome, as observed for other mitochondrial enzymes. The mode of inheritance is compatible with the well-known maternal transmission of mitochondria for all children if incomplete penetrance – for example due to an element of randomness in distribution of mitochondria to daughter cells – is assumed in addition. Another and still more convincing example of mitochondrial inheritance is mitochondrial cytopathy [639]. In this syndrome, structural mitochondrial anomalies are combined with deficiencies of multiple mitochondrial enzymes that apparently are the consequences of the structural defect. Clinical signs are variable; they might include progressive muscular weakness, ptosis, ophthalmoplegia, anomalies of the central and peripheral nervous system, and glomerular dysfunction of kidneys. As a rule, there was maternal transmission to the majority of children in the study cited above; expressivity, however, was very variable; again, as in Leber's disease, incomplete penetrance was observed. Occasional paternal transmission might be due to chromosomal localization of an enzyme subunit.
Other cases for which abnormal segregation has been asserted are less well-documented. Since families with many children have become the exception in most industrial societies, the prospect for tracking down and verifying abnormal segregation of pathologie genes is becoming more difficult.

3.1.6 *"Lethal Factors"* [696]

Animal Models. Mutations showing a simple mode of inheritance often lead to more or less severe impairment of their bearer's health. There is even evidence (Sect. 3.1.4) that some X-linked conditions prevent the male hemizygote from surviving to birth. It can be assumed that mutations exist that interfere with the development of their carriers so severely as to render them unable to survive.

The first reported case of a lethal mutation in mammalian genetics was the so-called yellow mouse. L. Cuénot (1905) [615] reported an apparent deviation from Mendel's law. A mutant mouse with yellow fur color did not breed true. When yellow animals were crossed with each other, normal grey mice would always segregate out. All yellow mice were heterozygous. They all had the same genetic constitution A^Y/A^+; A^Y is a dominant allele of the agouti series, the wild allele of which is termed A^+. When A^Y/A^+ heterozygotes were mated with A^+/A^+ homozygotes, the expected 1:1 ratio between yellow and grey mice was observed. In 1910, it was found that A^Y/A^Y homozygotes were formed but died in utero. Later on, abnormal embryos were discovered in the expected frequency of 25%.

In this case, the allele that is lethal in the homozygous state can be recognized in the heterozygotes by the yellow fur color. A similar case is the Pelger anomaly (Sect. 3.1.2). Contrary to the yellow mice, however, some of the homozygous rabbits survive. Cases of this sort are exceptional. As a rule, heterozygotes of lethals will not be readily recognizable; therefore, lethals occurring spontaneously are difficult to ascertain even in experimental animals and much more so in man. In mutation experiments with animals, however, an increase of lethals among the progeny of animals treated with a mutagen is one of the most convenient measures for mutagenicity.

Usually, a lethal mutation kills the embryo in a characteristic phase of its development ("effective lethal phase"; Hadorn [696]. This can easily be explained by the assumption that in this phase the action of the mutant gene would be required for further development.

Lethals in Humans. In man, many different types of lethals must occur since many metabolic pathways and their enzymes are essential for survival. It is likely that many enzyme defects that have not yet been detected do indeed occur but are not compatible with zygote survival. Moreover, many types of defects of inducer substances needed during embryonic development, and enzymes involved in nucleic acid and protein synthesis, may occur and add to the high incidence of zygote death, which has so far been unexplainable genetically. This problem will be discussed from a different standpoint in the context of population genetics (Sect. 6.3.2).

According to current estimates, about 15%-20% of all recognized human pregnancies end in spontaneous miscarriage. Experiments with other mammals suggest that an appreciable number of additional zygote losses go unnoticed, as death occurs during migration through the fallopian tubes. How much of this zygote wastage is due to genetic factors is unknown. A high fraction is caused by numeric or structural chromosome aberrations (Sect. 2.2.4). However, there are certainly other maternal causes for abortion as well. Whereas it seemed hopeless to try to relate any fraction of antenatal (or even postnatal) zygote loss to autosomal-dominant or recessive lethals, it appeared more reasonable to speculate about X-linked lethals, as they could influence the sex ratio.

In 1923, F. Lenz [757] advanced the hypothesis that the well-documented higher mortality rate of boys in their 1st year of life might be due to X-linked lethals. Subsequently, it was shown that in the period 1901-1935, the relative excess of male over female deaths increased with decreasing overall mortality, pointing to a "stable" (and possible genetic) component in male mortality [695]. Furthermore, years in which the general mortality increased showed a decrease of the male excess. A male excess is also found in stillbirths. However, during the period 1936-1964, the stillbirth rate decreased in England and Wales by more than one-half. Within the same time, the male excess, which had been appreciable at the beginning, gradually diminished and finally almost disappeared. This result would suggest that the number of X-linked lethals was rather low or that these mutations were lethal only under adverse environmental conditions [1650].

For fetal deaths during pregnancy, the situation is still more obscure than for stillbirths. The important parameter is always the sex ratio - either at fertilization (primary sex ratio) or at birth (secondary sex ratio) or sometime in between. The human sex ratio, however, varies with many factors. Stevenson and Bobrow (1967) [1650] have given a critical account of these problems. We agree with them that any conclusions as to frequency and mutation rates of lethals, especially X-linked lethals, in man have to be regarded with caution. This conclusion is unfortunate for human genetics since much work on spontaneous and induced mutation in *Drosophila melanogaster* and in the mouse has used lethal mutations. Direct comparisons with humans would be highly desirable.

It is possible that some case reports in the literature could be explained by the action of lethal genes. A rare disease of agenesis of the corpus callosum with flexor spasms, epileptic seizures, and chorioretinal abnormalities (30405) was observed in 19 infant girls; an exogenous cause could not be found. An X-linked mutation that kills the male hemizygote before birth is a possibility [121].

3.1.7 Modifying Genes

Up to now we have considered phenotypic traits depending on one gene only. However, the phenotypic expression of one gene is usually influenced by other genes. Experiments with animals, especially with mammals, show the importance of this "genetic background". One way to overcome analytic difficulties caused by such variation is the use of inbred strains.

The genetic background is a fairly diffuse concept, but in a number of cases it has been possible to show that penetrance or expressivity of a certain gene can be influenced by another gene. Such a gene is called a "modifier gene" when expressivity is influenced. When penetrance is suppressed altogether, the word "epistasis" (and "hypostasis" of the suppressed gene) is used. In experimental animals, cases have been analyzed in which the interaction of two mutations at different loci leads to a completely new phenotype. The classic example is the cross of chickens with "rose" combs and "pea" combs, which leads to the "walnut" comb in homozygotes for both of these mutations. To the best of our knowledge, a similar situation has not been described in man.

Modifier genes and epistasis, however, have been demonstrated.

Modifying Genes in the ABO Blood Group System. The best analyzed examples of modifying genes are offered by the ABO blood group systems. Occurrence of the ABH antigens in saliva (and other secretions) depends on the Se/se secretor gene. Homozygotes se/se are nonsecretors; heterozygotes Se/se and homozygotes Se/Se are secretors. Hence, se is a recessive suppressor gene. Other rare suppressor genes even prevent the expression of ABH antigens on the surface of erythrocytes.

Bhende et al. (1952) [576] discovered a phenotype, which they called "Bombay". The erythrocytes were not agglutinated either by anti-A or anti-B or anti-H. The serum contained all three of these agglutinins. Later, another family was discovered showing that the bearers of this unusual phenotype did have normal ABO alleles, but their manifestation was suppressed (Fig. 3.20; a woman, II, 6, has a Bombay phenotype but transmitted the B allele to one of her daughters). It was further shown that A can also be suppressed, and the available family data suggested an autosomal-recessive mode of inheritance. In the family shown in Fig. 3.20, the parents of the proposita are first cousins.

The locus is not linked to the ABO locus. The gene pair was named H, h, the Bombay phenotype representing the homozygote, h/h. Depending on the nature of the suppressed allele, the phenotype is designated O_hA_1, O_hA_2, or O_hB. The phenotype has a frequency of about 1 in 13000 among Maharati-speaking Indians in and around Bombay, and is common in the population isolate on Reunion island [679]. It is though to represent the defect of an enzyme that converts a precursor substance into the H antigen, which, in turn, is a precursor of the A and B antigens [93; 166; 931]. A second gene pair Yy, the rare homozygous conditions of which partially suppresses the A antigen, was postulated, and subsequently a number of additional families with this condition have been reported.

Sex-Limiting Modifying Genes. In other, less directly accessible traits, the action of modifying genes has been analyzed with statistical methods.

Haldane (1941) [697] tried to identify such genes in Huntington's chorea, using the family data assembled by J. Bell (1934) [565a]. H. Harris (1948) [702a] examined the problem in a condition called diaphyseal aclasis (13370), which is characterized by multiple exostosis near the cartilaginous epiphyses.

The mode of inheritance is dominant; however, the condition is about two times more common in males than in females. It may be transmitted in some families through unaffected females but not through unaffected males. Statistical analysis of the comprehensive pedigree data collected by Stocks and Barrington (1925) [901] suggests in part of the families independent segregation of a factor leading to incomplete penetrance only in females: a sex-limiting modifying gene.

Modification by the Other Allele: Anticipation. Phenotypic expression of a gene may be modified not only

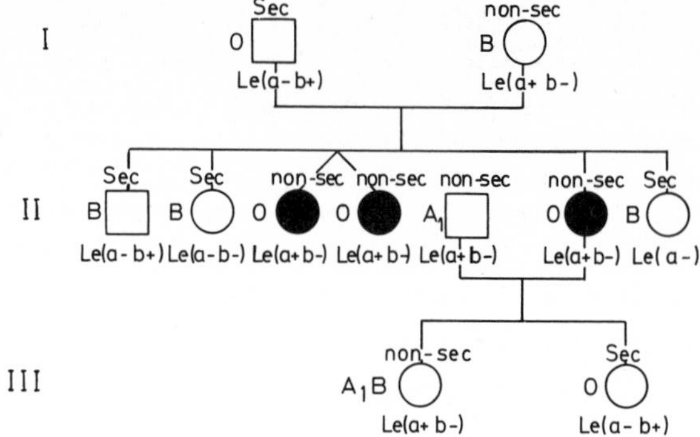

Fig. 3.20. The Bombay blood type. Manifestation of the B antigen is suppressed by a recessive gene X. (Bhende et al., 1952 [576]) Note that an O mother (II, 6) has an A_1B child

by genes at other loci but also by the "normal" allele. One example comes from the genetics of the Rh factor (Sect. 3.5.4). Occasional blood specimens, when tested with an anti-Rh D serum, do not give either a strong positive or a negative reaction, but an attenuated positive reaction. These are called D^u. In most cases, a special allele is responsible for this effect, but there are exceptions. In several families, the D^u reaction was observed only in family members having Cde as the homologous allele (Fig. 3.21). When such an apparent CD^ue allele segregates away from Cde, it is disclosed as a normal CDe [605].

By more indirect statistical analysis, allelic modification has been shown in a dominantly inherited disease, myotonic dystrophy (16090). In this slowly progressive condition, myotonia is associated with relatively mild muscular dystrophy and cataracts. This condition shows an unusual degree of variability in the age of onset. A comprehensive pedigree survey conducted by Bell (1947) [566] showed the following peculiarities. The cases in earlier generations exhibited cataracts in the middle of life that frequently remained the only symptom. In more recent generations, on the other hand, a more severe disease with earlier onset was frequently observed.

At first glance, these pedigrees seem to show a phenomenon considered important in the pre-Mendelian era of medical genetics: anticipation. Hereditary diseases were thought to begin earlier in life and to have a more severe course with their progress through the generations. Weinberg was the first to show that this phenomenon is a statistical artifact. The pedigrees are usually ascertained through a proband or probands in the youngest generations. Among patients in the parents' generation, however, only those having children will be ascertained. These, it turns out, will in most cases have a milder form of the disease than will those without children.

The problem in myotonic dystrophy was analyzed by Penrose (1948) [836]. He found a relatively low correlation for age at onset between parents and children ($r = 0.32$), but a much higher correlation between affected siblings ($r = 0.66$).

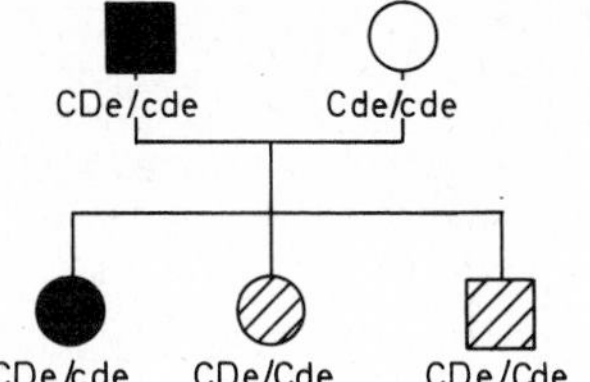

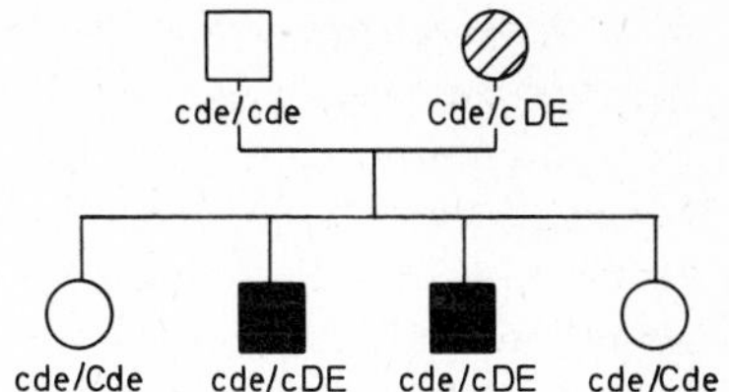

Fig. 3.21. Modification by the homologous allele in the Rh system. ● D-positive blood with normal reaction; ⊘ weak reaction (D^u variant); ○ D-negative blood. The haplotype Cde reduces expression of the D factor to D^u. (Ceppellini et al., 1955 [605])

Five factors had to be considered as possible "causes" of anticipation:

1. Selection of affected parents in whom the onset of the disease is late. Only affected persons who reach adulthood can have affected children.
2. Selection of affected offspring in whom the onset of the disease is early. This may occur since only the more severely affected propositi come to medical attention.
3. Selection of cases with simultaneous onset in parents and offspring, due to the fact that studies in clinical genetics cover only a limited time period. This bias will be very powerful.
4. Any of the above-mentioned biases may produce spurious anticipation, if correlation between parents and offspring is low.
5. General variability in age at onset.

The parent-child correlation Table for age of onset in myotonic dystrophy (Table 3.2) is not symmetric and points toward an age difference at onset in parent and child of between 20–40 years. However, such differences were likely to be caused by ascertainment bias in finding families with simultaneous onset in parents and children. Therefore, this table offers a plausible explanation for the apparent anticipation.

The data do not offer, however, an explanation of the difference in correlations for age at onset between parents and offspring, on the one hand, and between sibs, on the other. Here, the simplest explanation is that expressivity depends not only on the mutant allele but on the normal allele as well (Fig. 3.22). This allele always comes from the unaffected parent. Therefore, if the modification is entirely caused by the normal allele, the parent-child correlation is expected to be 0. Affected sibs have a 50% chance of sharing the normal allele. It was shown by Penrose that using plausible assumptions the observed correlations fit with the expectations of such allelic modification.

A more recent study of myotonic dystrophy was based on all families ascertained in Northern Ireland during a defined time period and not on a survey of the literature with many possible ascertainment biases. Here Penrose's results were only partially confirmed. The parent-offspring correlation for age at onset is absent only when cataract as a manifestation of the disease is omitted. Conceivably, modification by the normal allele influences all disease manifestation except for cataracts.

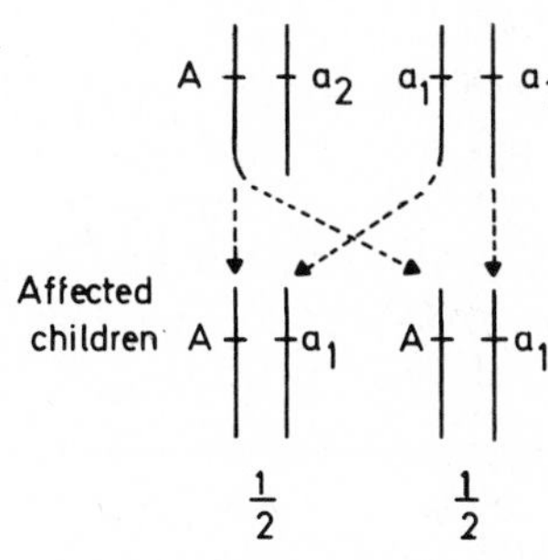

Fig. 3.22. Allelic modification. If manifestation of a dominant, abnormal gene A is modified by the normal allele and if the allele a_1 causes severe, a_2 milder manifestation of A, there is a correlation in the degree of manifestation between affected sibs but not between affected parent and child. An affected child cannot receive the modifying a_2 allele

Table 3.2. Age at onset in myotonic dystrophy. (Penrose, 1948 [836]. Heavy lines: age difference between parent and child at onset about 20–40 years

Children

Age	0	10	20	30	40	50	Total
50	–	4	3	2	–	–	9
40	8	5	5	–	–	–	18
30	3	9	1	1	–	–	14
20	3	1	–	–	–	–	4
10	2	4	–	–	–	–	6
0	–	–	–	–	–	–	–
Total	16	23	9	3	–	–	51

Parents

Another example of allelic modification is the nail-patella syndrome (16120) [854]. Altogether, the number of cases in man where interaction of well-defined genes has been analyzed is small. Some examples in which the analysis has been possible on the molecular level will be discussed under the topic of hemoglobin variants (Sect. 4.3). Analysis of the interaction between different genes in their relation to the phenotype will be one of the major tasks of human genetics in the future.

3.1.8 Number of Conditions with Simple Modes of Inheritance Known So Far in Humans

For many years, V. A- McKusick [133] has undertaken the task of collecting and documenting known conditions with simple modes of inheritance in man. Table 3.3 is based on the sixth edition with his later additions. Since the third edition was published (1971), the number of autosomal-dominant traits (confirmed and unconfirmed) increased from 943 to 2106, the autosomal-recessives from 783 to 1321, the X-linked from 150 to 267.

While genetic polymorphisms are included (Sect. 6.1.2), most conditions listed in this catalogue are rare. Many are rare hereditary diseases. At first glance, the list is impressive. However, more detailed scrutiny of the conditions shows that our knowledge of these rare diseases is not nearly as good as it should and could be. There are several reasons:

a) Most hereditary diseases have become known by occasional observation of affected patients and their families. For rare diseases, there is no other way to assess abnormalities whether they may or may not have a genetic basis.
b) Some recessive diseases have become known because they happened to be frequent in special populations, primarily in isolates. Isolate studies permit examination of the manifestation of recessive diseases caused by a single mutation. One problem with this approach is that chance determines the genes that can be studied.
c) Most human and medical geneticists are working in relatively few industrialized countries. However, genes for rare diseases show a very unequal distribution in different populations. This is particularly true for recessives but has also been shown for dominants with normal or only slightly lowered biologic fitness, i. e., when the incidence is not determined by the mutation rate. Hence, the developing countries, especially in Asia, Africa, and Latin America, can be expected to abound with hereditary anomalies and diseases that are unclassified to date. Any clinical geneticist who has ever walked through - say - Indian villages knows that this suggestion is not merely a theoretical speculation.
d) Genetic defects with simple modes of inheritance have a good chance of being detected when they show a clearcut phenotype that is readily recognizable. This is why the inherited conditions of the skin and eye are so relatively well-known. Other defects, however, may in some families cause anomalies or diseases that are precipitated by environmental factors. Most of these hidden defects are unknown at present.

Table 3.3. Number of known traits with simple modes of inheritance in man (McKusick, 1985 [133])

Mode of inheritance	Number of conditions	
	Total	Confirmed
Autosomal-dominant	2106	1096
Autosomal-recessive	1321	611
X-linked	267	119
Total	3694	1826

e) The real significance of hereditary disease and its total impact can be established only by studies in large populations, using epidemiologic methods. Such studies offer the opportunity to detect heterogeneity in etiology and aid in distinguishing genetic and nongenetic causes. They afford the only basis on which genetic parameters such as mutation rates, biologic fitness, and the relative incidence of mild and severe mutations of the same gene can be established. They also help in predicting the long-term and public health effects of medical therapy and of genetic counseling for future generations. Many epidemiologic studies in human genetics were carried out in the 1940s and 1950s. A few institutes played a leading role - most notably T. Kemp's Institute in Copenhagen. Here, a genetic register of the Danish population was established, and on this basis studies for several hereditary conditions were performed, such as achondroplasia (10080), polycystic kidneys (17390), limb malformations, and others.

Other active groups operated in Northern Ireland and in the United States in Michigan. Studies for single groups of diseases were also undertaken in Great Britain, the United States, Sweden, Finland, Switzerland, and Germany. Judged by present standards, many studies were necessarily imperfect. Yet, whatever we know about incidence, different genetic types, mutations rates, and biologic fitness we owe to these studies. With the detection of the first human chromosome anomalies in the late 1950s these efforts have been de-emphasized. We know of few groups that actively pursue this line of research. The time is again ripe for human geneticists to return to the field, combining various laboratory methods with the methods of formal genetics, mutation research, and epidemiology.

Difference in the Relative Frequencies of Dominant and Recessive Conditions in Man and Animals? At first glance, there appears to be a difference between humans and experimental animals in the relative frequencies of dominant and recessive conditions. Of the better known mutants of *Drosophila melanogas-*

ter, 200 are recessive and only 13 (6.1%) dominant. In the chicken, 40 recessive and 28 dominant mutations were reported. In the mouse, only 17 of 74 mutants were dominant (23%), the rest being recessive; in the rabbit 32 recessive and 6 dominant mutations have been found. (Multiple allelism were counted as one gene locus.) In man, on the other hand, more dominant than recessive conditions are known (Table 3.4). This discrepancy, however, is likely to be caused by diagnostic bias. Man observes himself most carefully; therefore, he detects defects that would probably escape observation when present in experimental animals. It would be difficult, for example, to detect brachydactyly in the mouse. This condition, however, leads to a much more severe defect when homozygous (Sect. 3.1.1). Hence, such a defect, dominant in man, would be counted as recessive in the mouse. Another reason might be that the population of the industrialized countries is not in equilibrium for recessive genes. The frequency of consanguineous matings has dropped sharply, and therefore the chance of a recessive gene meeting an identical partner and becoming homozygous in an individual is reduced. A new equilibrium will be reached only in the very distant future when recessive genes have become sufficiently frequent again (Sect. 6.3.1.2). In our opinion, there is no significant reason to assume that man is unique in regard to the ratio of dominant and recessive mutations.

3.2 Hardy-Weinberg Law and Its Applications

3.2.1 Formal Basis

So far, the application of Mendel's Laws in man has been considered from the standpoint of the single family. What, however, are the consequences for the genetic composition of the population? The field of research in which this problem is examined is called population genetics and will be discussed in Chap. 6. Some basic concepts will be introduced here.

These concepts have evolved around the so-called Hardy-Weinberg Law, discovered by these two authors independently in 1908 [252; 289], K. Pearson (1904) [833] – in the process of reconciling the consequences of Mendel's Laws for the population with biometric results – had already derived this law for the special case of equal gene frequencies of two alleles.

The law in its more general form may be formulated as follows: Let the gene frequencies of two alleles in a certain population be p for the allele A, q for the allele B; $(p+q=1)$. Let mating and reproduction be random with respect to this gene locus. Then the gene frequencies will remain the same, and the genotypes AA, AB, and BB in the F_1 generation will occur in the relative frequencies p^2, $2pq$, and q^2, the terms of the binomial expression $(p+q)^2$. In autosomal genes, and in the absence of disturbing influences, this proportion is maintained through all subsequent generations.

Derivation of the Hardy-Weinberg Law. We assume that at the beginning the proportions of genotypes AA, AB, and BB in the population of males as well as in the population of females are D, $2H$, and R, respectively. Symbolically, the distribution of genotypes in both sexes may be written as:

$$D \times \mathrm{AA} + 2H \times \mathrm{AB} + R \times \mathrm{BB} \tag{3.1}$$

From this, the distribution of mating types for random mating is obtained by formal squaring:

$$(D \times \mathrm{AA} + 2H \times \mathrm{AB} + R \times \mathrm{BB})^2 = D^2 \times \mathrm{AA} \times \mathrm{AA} + 4DH \times \mathrm{AA} \times \mathrm{AB} + 2DR \times \mathrm{AA} \times \mathrm{BB} + 4H^2 \times \mathrm{AB} \times \mathrm{AB} + 4HR \times \mathrm{AB} \times \mathrm{BB} + R^2 \times \mathrm{BB} \times \mathrm{BB}$$

The distribution of genotypes in the offspring of the different mating types is:

AA × AA	AA
AA × AB	$\frac{1}{2}$AA + $\frac{1}{2}$AB
AA × BB	AB
AB × AB	$\frac{1}{4}$AA + $\frac{1}{2}$AB + $\frac{1}{4}$BB
AB × BB	$\frac{1}{2}$AB + $\frac{1}{2}$BB
BB × BB	BB

Inserting these distributions for the mating types in Eq. (3.1) yields the distribution of genotypes in the F_1 generation:

$$(D^2 + 2DH + H^2)\mathrm{AA} + (2DH + 2DR + 2H^2 + 2HR)\mathrm{AB} + (H^2 + 2HR + R^2)\mathrm{BB} = p^2\mathrm{AA} + 2pq\mathrm{AB} + q^2\mathrm{BB}$$

where $p = D + H$, $q = H + R$ are the frequencies of the alleles A and B, respectively, in the parental generation. Thus, the distribution of genotypes in the offspring generation is uniquely determined by the gene frequencies in the parental population:

$$D' = p^2, \quad 2H' = 2pq, \quad R' = q^2.$$

As:

$$p' = D' + H' = p^2 + pq = p,$$
$$q' = H' + R' = pq + q^2 = q,$$

the gene frequencies in the F_1 generation are equal to those in the parental generation. Thus, also the genotype distribution in the next generation (F_2) is the same as in the F_1 generation, and this holds true for all following generations.

This means that in autosomal inheritance these proportions are expected in the first generation and will be maintained in the following generations. For X-linked genes, the situation is slightly more complicated (Sect. 6.1.1). At the same time, the concept of gene frequencies $p+q=1$ was created.

The Hardy-Weinberg Law can also be rephrased, indicating that random mating is equivalent to drawing random samples of size two from a pool of genes containing the two alleles A and a with relative frequencies p and q. One of the advantages of this law is that frequencies of genetic traits in different populations can be expressed and compared in terms of gene frequencies.

Apart from making it possible to simplify population descriptions, the Hardy-Weinberg Law can also help to elucidate modes of inheritance in cases where the straightforward approach through family studies would be too difficult. The classic examples are the AB0 blood types.

3.2.2 *Hardy-Weinberg Expectations Establish the Genetic Basis of AB0 Blood Group Alleles*

Table 3.4. Comparison of the two theories for inheritance of AB0 blood groups[a]

Parents	Children expected from the hypothesis of	
	Two gene pairs	Multiple alleles
0×0	0	0
0×A	0, A	0, A
0×B	0, B	0, B
A×A	0, A	0, A
A×B	0, A, B, AB	0, A, B, AB
B×B	0, B	0, B
0×AB	0, A, B, AB	A, B
A×AB	0, A, B, AB	A, B, AB
B×AB	0, A, B, AB	A, B, AB
AB×AB	0, A, B, AB	A, B, AB

[a] Adapted from A. S. Wiener, 1943.

Multiple Allelisms. So far, only two different alleles for each locus have been considered. Frequently, however, more than two different states for one gene locus, i.e., more than two alleles, are possible. Examples of such "multiple allelism" in humans and experimental animals abound. Two of the classics are the white series in *Drosophila melanogaster,* and the albino series in rabbits.

The formal characteristics can easily be derived:

a) In any one individual a maximum of two alleles can only be present (unless there are more than two homologous chromosomes, as in trisomics).
b) Between these alleles, crossing over can be disregarded as they are located at homologous loci. The qualifications of this second condition will be dealt with in Sect. 3.5, in connection with modern concepts of the gene. Here, the simplest formal model will be described, using the AB0 blood groups as an example.

Genetics of the AB0 Blood Groups. AB0 blood groups were discovered by Landsteiner (1900) [259]. Compared with other blood group systems, their most important property is the presence of isoantibodies that led to frequent transfusion accidents. These accidents helped in the discovery of the blood groups. The first relevant genetic theory was developed by v. Dungern and Hirszfeld (1911) [631]. To explain the four phenotypes A, B, 0, and AB, they assumed two independent pairs of alleles (A, 0; B, 0), with dominance of A and B. F. Bernstein (1925) [574] tested this hypothesis, using the Hardy-Weinberg expectations for the first time. He found their concept to be wrong and replaced it by the correct explanation – three alleles with six genotypes, leading to the four phenotypes due to the dominance of A and B over 0.

The most obvious method to discriminate between these two hypotheses is by family investigation. However, differences between them are to be expected only in matings in which at least one parent belongs to group AB (Table 3.4). The two-locus hypothesis allows for 0 children, the three-allele hypothesis does not. Although AB is the rarest group, the early literature contained some reports of supposedly 0 children with AB parents; these children were either misclassified or illegitimate. Bernstein, however, was not misled by these observations. His argument goes as follows. It may be assumed that the two-gene pair theory is correct; p may be the gene frequency of A, $1-p=p'$ of a; q the frequency of B, $1-q=q'$ of b. Then, the following frequencies have to be expected in the population:

Phenotype	Genotype	Frequency	
0	aabb	$(1-p)^2(1-q)^2=p'^2q'^2$	
B	aaBB	$(1-p)^2q^2$	$=p'^2(1-q'^2)$
	aaBb	$2(1-p)^2q(1-q)$	
A	AAbb	$p^2(1-q)^2$	$=(1-p'^2)q'^2$
	Aabb	$2p(1-p)(1-q)^2$	
AB	AABB	p^2q^2	$=(1-p'^2)(1-q'^2)$
	AaBB	$2p(1-p)q^2$	
	AABb	$2p^2q(1-q)$	
	AaBb	$2p(1-p)2q(1-q)$	

This leads to the following relationships ($\overline{A}$, $\overline{B}$: frequencies of phenotypes):

$$\overline{0}\times\overline{AB}=\overline{A}\times\overline{B}$$

and

$$\overline{A}+\overline{AB}=1-p'^2;\quad \overline{B}+\overline{AB}=1-q'^2$$

Thus, it follows:

$$(\overline{A}+\overline{AB})\times(\overline{B}+\overline{AB})=\overline{AB}$$

These identities can be tested. It turned out – and has turned out ever since – that $(\overline{A}+\overline{AB})\times(\overline{B}+\overline{AB})>\overline{AB}$, and $\overline{0}\times\overline{AB}<\overline{A}+\overline{B}$. The differences are so large – and so consistent – that an explanation by chance deviations is inadequate. The first alternative possibility considered by Bernstein was heterogeneity within the examined population.

This explanation, however, turned out to be insufficient. On the other hand, it could be shown that the distributions in all populations for which data were available are in perfect agreement with expectations derived from the multiple-allele hypothesis.

To understand Bernstein's argument, a fresh look at the Hardy-Weinberg Law is necessary. Up to now, it had been derived for the special case of two alleles only. However, it can be shown to apply for more than two alleles. Assuming n alleles $p_1, p_2, \ldots, p_n$, the relative frequencies of genotypes are given by the terms of the expansion of $(p_1+p_2+\ldots p_n)^2$. It follows for the special case of A, B, and O with the frequencies p, q, and r that the distribution of genotypes is:

$$p^2(\text{AA})+2pq(\text{AB})+2pr(\text{A0})+q^2(\text{BB})+2pr(\text{B0})+r^2(00).$$

Now, we follow Bernstein again[5] "For the classes" (phenotypes)

$\bar{0}=00 \quad \bar{\text{B}}=\text{B0}+\text{BB} \quad \bar{\text{A}}=\text{A0}+\text{AA} \quad \overline{\text{AB}}=\text{AB}$

the following probabilities can be derived:

$r^2 \qquad 2qr+q^2 \qquad 2pr+p^2 \qquad 2pq$

It follows:

$$\bar{0}+\bar{\text{A}}=(r+p)^2$$
$$\bar{0}+\bar{\text{B}}=(r+q)^2$$

and therefore:

$$q=1-\sqrt{\bar{0}+\bar{\text{A}}}$$
$$p=1-\sqrt{\bar{0}+\bar{\text{B}}}$$
$$r=\sqrt{\bar{0}}$$

and the relation:

$$1=p+q+r=1-\sqrt{\bar{0}+\bar{\text{B}}}+1-\sqrt{\bar{0}+\bar{\text{A}}}+\sqrt{\bar{0}}$$

This can be tested using the AB0 phenotype distributions in various populations of the world. The criterion is that the gene frequencies calculated with this formula must add to 1. Besides, expected genotype frequencies can be calculated from these gene frequencies and can be compared with all the observed frequencies. Apart from the correctness of the genetic hypothesis, however, this result is subject to still another condition. There must be random mating with regard to this character.

In the data analyzed by Bernstein, the agreement already was excellent, and this has turned out to hold true for the huge amount of data collected ever since. One example may help in understanding the principle of calculation. The following phenotype frequencies were reported from the city of Berlin:

Total persons:

n	A	B	0	AB
21,104	9,123 (43.23%)	2,987 (14.15%)	7,725 (36.60%)	1,269 (6.01%)

[5] Translation ours.

Using Bernstein's formula, the gene frequencies are:

$$p=1-\sqrt{(0.3660+0.1415)}=0.2876$$
$$q=1-\sqrt{(0.3600+0.4323)}=0.1065$$
$$r=\sqrt{0.3660} \qquad =\underline{0.6050}$$
$$0.9991$$

Thus:

$$p+q+r=0.9991$$

At first glance, this result agrees well with its expectation, i.e., 1. As a statistical test for examining whether the deviation is significant, the χ^2 method can be applied (Stevens, 1950 [625]):

$$\chi_1^2=2n\left(1+\frac{r}{pq}\right)D^2$$
$$D=1-(p+q+r)$$

In our example, the result is:

$$\chi_1^2=0.88$$

Hence, it is confirmed that the values found are in good agreement with the genetic hypothesis and with the assumptions of random mating for the AB0 system.

In a later paper, Bernstein showed how the difference D may be utilized to correct the calculated gene frequencies. The uncorrected gene frequencies may be named p', q', and r', and the following formulas may be used:

$$p=p'(1+D/2)$$
$$q=q'(1+D/2)$$
$$r=(r'+D/2)(1+D/2)$$

and for the example:

$$p=0.2876(1+0.00045)=0.2877$$
$$q=0.1065(1+0.00045)=0.1065$$
$$r=(0.6050-0.00045)(1+0.00045)=0.6057$$

In the process of testing the two genetic hypotheses for the AB0 system, Bernstein developed a method for calculating gene frequencies. Such methods have become important practically and will therefore be treated in a separate section (App. 1).

Meaning of a Hardy-Weinberg Equilibrium. Populations showing agreement of the observed genotype proportions with the expectations of the Hardy-Weinberg Law are said to be "in Hardy-Weinberg equilibrium". This equilibrium must be distinguished from that between alleles, which will be discussed in the contexts of selection (Sect. 6.2.1) and of mutation (Sect. 5.1.3). Hardy-Weinberg equilibrium is an equilibrium of the distribution of genes in the population ("gene pool") among the various genotypes. Under random mating, this equilibrium will be re-established after one generation, possibly with changed gene frequencies if it is disturbed by opposing forces.

It follows from our discussion, however, that the Hardy-Weinberg Law can be expected to be valid only when the following prerequisites are not violated:

a) The matings must be random with respect to the genotype in question. This can safely be assumed for such traits as blood groups or enzyme polymorphisms. It cannot be assumed for visible characters such as stature, and still less for behavioral characteristics such as intelligence. This should be kept in mind when measures used in quantitative genetics, for example, correlations between relatives, are interpreted in genetic terms.
b) A deviation from random mating is caused by consanguineous matings. If the consanguinity rate in a population is high, an increase in the number of homozygotes must be expected (Sect. 6.3.1). It is even possible to calculate the frequency of consanguinity in a population by means of the deviations from the Hardy-Weinberg proportions.
c) Recent migrations might disturb the Hardy-Weinberg proportions.
d) Occasionally, selection is mentioned as a factor leading to deviations. This can be true but need not necessarily apply. As a rule, selection tends to cause changes in gene frequencies; selection before reproductive age, for example, in the prenatal period, or during childhood and youth, does not influence the Hardy-Weinberg proportions in adults at all. A deviation may be visible among children, depending on the special type of selection involved. Besides, even assuming an appreciable amount of selection in a suitable age group, ascertainment of statistically significant deviations from Hardy-Weinberg proportions requires large sample sizes - larger than are usually available. Sometimes, absence of significant selection is inferred from the observation that Hardy-Weinberg proportions are preserved in a population. This conslusion, however, unless carefully qualified might easily be wrong. Considering all the theoretical possibilities for disturbance, it is indeed amazing how frequently the Hardy-Weinberg proportions are found to be preserved in the human population.
e) Formally, a deviation from the Hardy-Weinberg Law may be observed if the population is a mixture of subpopulations that do not completely interbreed (random mating only within subpopulations) and, consequently, the gene frequencies in these subpopulations are different. This was first described by Wahlund (1928) [929], who gave a formula for calculating the coefficient F of the apparent inbreeding from the variance of the gene frequencies between the subpopulations.
f) Another cause of deviation may be the existence of a hitherto undetected ("silent") allele, a heterozygous carrier of which cannot be distinguished from a homozygous carrier of the usual allele. C. A. B. Smith (1970) [879] however, has pointed out that genetically a silent allele will cause a significant deviation from the Hardy-Weinberg Law only when it occurs at a high enough frequency for the homozygous to be detected.

3.2.3 Gene Frequencies

One Gene Pair: Only Two Phenotypes Known. In rare autosomal-recessive diseases only one gene pair is present, and only two phenotypes are usually known when the heterozygotes cannot be identified or, as is usually the case, when direct data on population frequencies of heterozygotes are not available. This also applies for blood group systems for which only one type of serum is available. Here, the frequency of homozygotes aa being q^2, the gene frequency is simply $\sqrt{aa}$. There is no way to test the assumption of random mating.

One result of practical importance is the high frequency of heterozygotes even for relatively rare recessive diseases.

Table 3.5 (from Lenz, 1976 [121]) is slightly oversimplified; some of the frequencies given vary in different populations (Sect. 6.1.3). However, the data point out how much more frequent the heterozygotes are, especially for rare conditions. This is important for genetic counseling, and for the much-discussed problem of how many lethal or detrimental genes the average human being might be heterozygous for (Sect. 6.3.2). Methods for calculation of gene frequencies are given in App. 1.

3.3 Statistical Methods in Formal Genetics: Analysis of Segregation Ratios

3.3.1 Segregation Ratios as Probabilities

During meiosis - and in the absence of disturbances - germ cells are formed in exactly the relative frequencies expected from Mendel's laws. A diploid spermatocyte heterozygous for the alleles A and a produces two haploid sperms with A, and two with a. If all the sperms of a given male were to come to fertilization and if none of the zygotes should die before birth, the segregation ratio among his offspring would be exactly 1:1. There would be no place for any statistics.

Organisms in which such an analysis is, indeed, possible are yeast or the bread mould *Neurospora crassa,* which has become so important in biochemical genetics. In the development of such an organism, there is a phase in which the diploid state has just

Table 3.5. Different homozygote and heterozygote frequencies for different gene frequencies (with examples of recessive conditions)

Homozygote frequency q^2	Gene frequency q	Heterozygote frequency $2pq$	Approximate homozygote frequencies in European populations
0.64	0.8	0.32	Lp(a-) lipoprotein variant
0.49	0.7	0.42	Acetyl transferase, "slow" variant (Sect. 4.5)
0.36	0.6	0.48	Blood group 0
0.25	0.5	0.50	Nonsecretor (se/se)
0.16	0.4	0.48	Rh negative (dd)
0.09	0.3	0.42	
0.04	0.2	0.32	Le(a-b-) negative
0.01	0.1	0.18	
1:2,500	1:50	1:25	Pseudocholinesterase (dibucain-resistant variant), cystic fibrosis; α_1-antitrypsin deficiency
1:4,900	1:70	1:35	Adrenogenital syndrome (Kanton Zürich)
1:10,000	1:100	1:50	Phenylketonuria (Switzerland; USA)
1:22,500	1:150	1:75	Albinism; adrenogenital syndrome with loss of NaCl
1:40,000	1:200	1:100	Cystinosis
1:90,000	1:300	1:150	Mucopolysaccharidosis type 1
1:1,000,000	1:1,000	1:500	Afibrinogenemia

Source: adapted from Lenz, [121].

been reduced to the haploid, and all four meiotic products are lying in a regular sequence. They can be taken out separately, grown, and examined ("tetrad analysis"). Expected segregation ratios are found with precision.

In higher plants and animals, including man, only a minute sample of all germ cells comes to fertilization. In the human female, about 6.8×10^6 oogonia are formed; the number of spermatogonial stem cells in the male is estimated at about 1.2×10^9 (Sect. 5.1.3.3). The actual number of sperm is a multiple of this figure. Hence, any given germ cell has a very small probability of coming to fertilization. Besides, the sampling process is usually random with respect to a given gene pair A,a (for apparent exceptions, see Sect. 3.1.5). This means that, for the distribution of genotypes among germ cells coming to fertilization, the rules of probability theory apply, and empirically found segregation ratios may show deviations from their statistical expectations.

Modern man is more or less used to thinking in statistical terms when he solves his daily-life problems. These experiences help in the understanding of simple applications of probability theory. Everyone, for example, will readily recognize that the following rationale is wrong.

A young mother had always wished to have four children. After the third, however, there was a long intermission. The grandmother asked her daughter whether she had now decided differently. Answered the daughter: "Yes, in principle, I would still like four children. But I read in the newspaper that every fourth child to be born is Chinese. And a Chinese child ... there I am reluctant."

In another example, the mistake is not quite so obvious. The parents of two albino children visit a doctor for genetic counseling. They wish to know the risk of a third child also being albino. The doctor knows that albinism is an autosomal-recessive condition, with an expected segregation ratio of 1:3 among children of heterozygous parents. He also knows that sibships in which all sibs are affected are very rare. Hence, he informs the parents: "As you already have two affected children, the chance that the third child will also be affected is very small. The next child should be healthy." The actual risk, of course, remains ¼.

It cannot be the task of a textbook of human genetics to teach probability theory and basic statistics. Therefore, it will be assumed that the reader of this section has some knowledge of the basic concepts of probability theory, that he knows the most important distributions (binomial, normal, and Poisson distribution), and has some idea of standard statistical methods. In the following, some applications to problems in human genetics will be given. We are aware of the danger that the present section may be used as a kind of "cookbook", without understanding of the basic principles, and recommend that the reader familiarize himself with these principles referring, for example, to the opening chapters in W. Feller's *Probability Theory and Its Applications* [659].

3.3.2 Simple Probability Problems in Human Genetics

Independent Sampling and Prediction in Genetic Counseling. The doctor, who gave the wrong genetic counsel to the couple with two albino children, did not take into account that the fertilization events leading to the three children are independent of each other and that each child has the probability of ¼ to be affected, regardless of the genotypes of any other children. The probabilities for each child must be multiplied. He was right when he said that illness of all three children is rare in a recessive condition: The probability is $(1/4)^3 = 1/64$ for all three children to be affected. However, the family to be counseled already had two such children, the probability of this occurring being $(1/4)^2 = 1/16$. It needs only one event with the probability 1/4 to complete the three-child family with albinism, $1/16 \times 1/4 = 1/64$. It is also intuitively obvious that there is no way for a given zygote to influence the sampling of gametes of the same parents many years later. Chance has no memory!

All possible combinations of affected and unaffected siblings in three-child families can be enumerated as follows (A = affected; U = unaffected):

UUU, AUU, UAU, AAU, UUA, AUA, UAA, AAA

In recessive inheritance, the event U has the probability 3/4. Thus, the first of the eight combinations *(UUU)* has the probability $(3/4)^3 = 27/64$. This means that of all heterozygous couples having three children 27/64 for fewer than 50% will have only healthy children. On the other hand, all three children will be affected in $(1/4)^3 = 1/64$ of all such families. There remain the intermediate groups. Three-child families with one affected child and two healthy ones obviously have the probability $1/4 \times 3/4 \times 3/4 = 9/64$. However, we are not that much interested in the sequence of healthy and affected children. Therefore, the three cases of such families, *UUA, UAU,* and *AUU* can be treated as equivalent, giving $3 \times 9/64 = 27/64$. The group with two affected can be treated accordingly, giving $3 \times 1/4 \times 1/4 \times 3/4 = 9/64$. As a control, let us see whether the various probabilities add up to 1:

$$\frac{27+27+9+1}{64} = 1$$

This is a special case of the binomial distribution. For Mendelian segregation ratios, it has two consequences: one theoretical, the other extremely practical. First, it follows that among all families for which a certain segregation ratio must be expected, an appreciable part - 27 of 64 in a three-child family with recessive inheritance - cannot be observed because chance has favored them by not producing any affected homozygotes. Hence, the segregation ratio in the remainder is systematically distorted. Special methods have been devised to correct for this "ascertainment bias" (Sect. 3.3.3). Secondly, and this is a most practical conclusion, with limitation of the number of children to two or three, most parents both of whom are heterozygous for a recessive disease will not have more than one affected child. Since the probability of affected children occurring in another branch of the family is very low - and the rate of consanguinity in current populations of industrialized countries has likewise decreased - almost all affected children will be sporadic cases in an otherwise healthy family; there will be no distinct sign of recessive inheritance. Yet, a subsequent child will, again, run the risk of 1/4. The layman cannot know that the condition is inherited. Therefore, genetic counseling must be actively offered to these families.

Differentiation Between Different Modes of Inheritance. In Sect. 3.1.4., an X-linked dominant pedigree was shown (Fig. 3.17) for vitamin D-resistant rickets and hypophosphatemia. What would be the probability for this pedigree to occur if the gene were, in fact, located on one autosome? Only the children of affected males are informative, because among children of affected women, a 1:1 segregation irrespective of sex must be expected. The seven affected fathers have 11 daughters, all of whom are affected. The probability for this outcome with autosomal inheritance would be $(1/2)^{11}$. The same fathers have 10 sons who are all healthy, giving a probability of $(1/2)^{10}$. Hence, the combined probability of 11 affected daughters and 10 healthy sons is:

$$(1/2)^{21} = \frac{1}{2{,}097{,}152}$$

This probability is so extremely low that the alternative hypothesis of an autosomal dominant mode of inheritance is rejected convincingly. The only reasonable alternative is the X-linked dominant mode. This hypothesis is corroborated independently by the observation (Sec. 3.1.4) that, on the average, male patients are more severely affected than female.

This is different for a rare skin disease (keratoma dissipatum Brauer). For this condition, a Y-chromosomal mode of inheritance has been considered - and, indeed, all nine sons of affected fathers in a published pedigree show the trait, whereas five daughters in both generations are unaffected. This gives:

$$(1/2)^9 \times (1/2)^5 = (1/2)^{14} = \frac{1}{16{,}384}$$

Hence, the probability of this pedigree having occurred by chance as an autosomal-dominant trait is very low indeed. There is an important difference, however, from the example of vitamin D-resistant rickets. Further pedigrees showing autosomal-dominant inheritance are unknown for this type of rickets, and all observations confirm the location of this gene on the X chromosome. For keratoma dissipatum Brauer, on the other hand, some families have been observed exhibiting very similar phenotypes that showed clear-cut autosomal-dominant inheritance. Hence, it is likely that the described pedigree has been selected from an unknown number of observations because of its peculiar transmission. The calculation is misleading as the "sample space" from which the sample of observations has really been drawn (all pedigrees with the same phenotype) is much larger (and ill-defined), and the sample (the pedigree) is biased. The trait seems to be autosomal-dominant.

Another, more obvious example of an error in the definition of the sample space is the mother who did not want a Chinese baby.

3.3.3 Testing for Segregation Ratios Without Ascertainment Bias: Codominant Inheritance

Apart from these limiting cases, calculation of exact probabilities for certain families or groups of families would usually be impracticable. Therefore, statistical methods are used that are either based on the parameters of the "normal" distribution, which is a good approximation of the binomial distribution (parametric tests), or go directly back to probabilistic reasoning (nonparametric tests). One method especially well-suited for genetic comparisons is the χ^2 test. This enables us to compare frequencies of observations in two or more discrete classes with their expectations. The most usual form is:

$$\chi^2 = \Sigma \frac{(E-O)^2}{E}$$

(E = expected number; O = observed number). In Farabee's pedigree with dominant inheritance (Sect. 3.1.2), there were 36 affected and 33 unaffected children of affected parents. With dominant inheritance, E is ½ of all children, i.e., 34.5:

$$\chi_1^2 = \frac{(36-34.5)^2}{34.5} = \frac{(33-34.5)^2}{34.5} = 0.13$$

The probability P for an equal or greater deviation from expectation can be taken from a χ^2 table for 1 degree of freedom. The number of degrees of freedom indicates in how many different ways the frequencies in the different classes can be changed without altering the total number of observations. In this case, the content of class 2, unaffected, is unequivocally fixed by the content of class 1. Therefore, the number of degrees of freedom is 1. In general, the number of degrees of freedom is equal to the number of classes less 1.

A second example is taken from the codominant mode of inheritance (Sect. 3.1.2). In Table 3.1, Wiener's family data for the MN blood types were summarized. Are the resultant segregation ratios compatible with the genetic hypothesis? For this problem, matings MM × MM, MM × NN, and NN × NN give no information. Expectations in the matings MM × MN and NN × MN are 1:1, in the mating MN × MN 1:2:1. This leads to Table 3.6 for the χ^2 test: For four degrees of freedom, we find in the χ^2 table: $P = 0.75$. This is very good agreement with expectation.

Dominance. The situation becomes slightly more complicated when one allele is dominant, the other one recessive. This is the case, for example, in the AB0 blood group system. Here, the phenotype $\overline{\text{A}}$ consists of the genotypes AA and A0. The expected segregation ratios among their offsprings are different. Some of the heterozygous parents A0 can be recognized, for example, in matings with $\overline{0}$ partners by the observation of 0 children. Others will have only $\overline{\text{A}}$ children just by chance. Special statistical methods are necessary to calculate correct expectations and to compare empirical observations with these expectations (cf. App. 2).

3.3.4 Testing for Segregation Ratios: Rare Traits

Principal Biases. If the condition under examination is rare, families are usually not ascertained at random; one starts with a "proband" or "propositus", i.e., a person showing the condition. This leads to an *ascertainment bias,* which must be corrected. The bias can be of different kinds, depending on the way in which the material has been ascertained.

a) Family or truncate selection. All individuals suffering from a specific disease in a certain population at a certain time (or within certain time limits) are ascertained. The individual patients are ascertained independently of each other, i.e., the second case in a sibship would always have been found. Such truncate ascertainment is possible,

Table 3.6. Comparison between expected and observed segregation figures in the MN data of Wiener et al. (1953) Table 3.1 [952]

Mating type	MM	MN	NN	χ^2	Degrees of freedom
MM × MN	$\frac{(499-486)^2}{486}$	$\frac{(473-486)^2}{486}$	-	0.6955	1
MN × MN	$\frac{(199-200)^2}{200}$	$\frac{(405-400)^2}{400}$	$\frac{(196-200)^2}{200}$	0.1475	2
MN × NN	-	$\frac{(411-396,5)^2}{396,5}$	$\frac{(382-396,5)^2}{396,5}$	1.0605	1

for example, if the condition always leads to medical treatment and all doctors report every case to a certain registry – as when an institute carries out an epidemiologic study. As a rule, case collections approaching completeness have been possible only in ad hoc studies of research workers specializing in a condition or group of conditions.

Here, the ascertainment bias is caused exclusively by the fact that only those sibships are ascertained that contain at least one patient. As shown above, however, (Sect. 3.3.3), this leaves out all sibships in which no affected individual has occurred just by chance. Their expected number is:

$$\sum_{s} q^{s} n_{s} \quad (3.2)$$

(s = number of siblings/sibship; p = segregation ratio; $q = 1 - p$; n_s = number of sibships of size s). In recessive disorders, $p = 0.25$. The smaller the average sibship size, the stronger the deviation from the 3:1 ratio in the ascertained families.

b) Incomplete multiple (proband) selection; single selection as limiting case. In most studies, not all individuals in a population are ascertained; frequently, a study starts with a full cohort of military conscripts or with all patients in a hospital population who have the condition, etc. Here, an additional bias must be considered: The more affected members a sibship has, the higher is its chance to be represented in the sample. This bias causes a systematic excess of affected persons, which is added to the excess caused by truncate selection as explained above.

Koller (1940) [744] gave a simple example that demonstrates the nature of this excess. Let us assume that the probands are ascertained during examination of one year's group of conscripts. The population will contain a number of families with three children, one of whom belongs to the current year's group, and in which the condition occurs. Thus, all families with three affected siblings but only two-thirds of the families with two affected, and one-third of those with only one affected will be ascertained.

The methods of correction to be described below can be regarded as reliable only if the probability for ascertainment of consecutive siblings is independent of the ascertainment of the first one. In an examination of conscripts, as described above, this may be the case. Most studies, however, start from a hospital population or from another group of medically treated persons. Here, according to general experience, a subsequently affected child will much more frequently be brought to a hospital when another child has been treated successfully. The opposite trend, however, is also possible. P. E. Becker (1953) [564] for example, collected all cases of X-linked recessive Duchenne-type of muscular dystrophy in a restricted area of southwestern Germany (Suedbaden). He had good reason to assume that ascertainment was complete for this area. Nevertheless, brothers developing muscular dystrophy as the second or later cases in their sibships were, as a rule, not ascertained as probands (i. e., through hospitals and doctors), but through the first proband in the family. In his interviews with the parents, Becker found the reason for this unusual behavior. In the case of the first patient in the sibship, the parents would usually see a doctor. Then, however, they would find out that, in spite of examinations and therapeutic attempts, the course of the disease could not be influenced. Hence, they refrained from presenting a second child to the hospital or the doctor.

c) Apart from these biases, which can be corrected statistically to a certain degree, there are other biases that cannot be corrected. Frequently, for example, a genetic hypothesis is discussed on the basis of families sampled from the literature. Experience shows that such sampling usually leads to reasonable results in autosomal-dominant and X-linked recessive disorders. Autosomal recessive diseases, however, are more difficult to handle. Families with an impressive accumulation of affected sibs will have a higher chance of being reported than those with only one or two affected members. This selection for "interesting cases" was more important early in this century because families used to have more children. Furthermore, recessive conditions discovered today are usually interesting from a clinical and biochemical point of view as well.

This type of selection can be avoided only by publishing all cases and by critical interpretation of material from the literature. A statistically sound correction is impossible, as such bias has no simple and reproducible direction.

To summarize: the method of segregation analysis depends on the way in which the family material has been ascertained. It follows that the method of ascertainment should always be described carefully. Above all, the probands should always be fully indicated. It is also of interest whether the author, in the process of his case collection, has become aware of any ascertainment biases.

These considerations show that complete (truncate) ascertainment of cases in a population, and within defined time limits, is the optimal method of data collection.

Methods of Correction of Bias. Two different types of correction methods are possible: test methods and estimation methods.

In a test method, the observed values are compared with their expectations, which have previously been corrected for ascertainment bias. The appropriate test method was first published by Bernstein (1929) [744]; it corrected for truncate selection. The expected number of affected E_r is:

$$E_r = sn_s \frac{p}{1-q^s} \tag{3.3}$$

in all sibships of size n (Definition of symbols as for formula 3.2). A similar test method can also be used for proband selection.

These test methods answer a very specific question: Do the observed proportions fit the expectation according to a certain genetic hypothesis?

In many, if not in most, actual cases, the question will be much more general: What is the unbiased segregation ratio in the observed sibships? This is an estimation problem. The earliest method was published by Weinberg (1912) [936] and was called the sib method. Starting from every affected sib in the sibship, the number of affected and unaffected among the sibs is determined. This method is adequate for "truncate selection", i.e., when each affected person is, at the same time, a proband. The sib method is the limiting case of the "proband method" used when the families are ascertained by incomplete multiple proband selection. The number of affected and unaffected siblings is counted, starting from every proband. The limiting case to the lower side is single selection. Here, each sibship has only one proband, and the counting is done once among his sibs.

These estimates converge with increasing sample size to the parameter p, the true segregation ratio; they are *consistent.* It was realized early, however, that they were not fully *efficient,* except for the limiting case of single selection, i.e. they did not make optimal use of all available information. Therefore, improvements were devised by a number of authors. Here, the method of weighted scores by Finney [663] in a version by Kaelin (1955) [729] will be described, since application is easy with a small desk computer. A detailed description of the estimation method will be given in App. 3 for the two extreme cases: Truncate selection ($k=1$) and single selection ($k=0$). The assumption $k=1$ gives the highest estimate $\hat{p}$ of the segregation ratio p, the assumption $k=0$ the lowest. In addition, other statistical problems that occur in genetic analysis, such as genetic heterogeneity, admixture of sporadic cases, and testing for birth order effects will be discussed in Appendix 3. The various methods are applied to a collection of families with deaf mutism. More complicated problems arise when segregation is examined in chromosomal translocations, since here families might be ascertained through unbalanced or balanced probands or through population surveys; and in many other ways (see Sect. 2.2.2.2). Appropriate methods of analysis will be discussed in App. 3.

3.3.5 Discrimination of Genetic Entities: Genetic Heterogeneity

It is a common experience in clinical genetics that similar or identical phenotypes are often caused by a variety of genotypes. The splitting of a group of patients into smaller, but genetically more uniform subgroups has been a major topic of research in medical genetics during recent decades. Frequently such heterogeneity analysis is another aspect of the application of Mendel's paradigm (Sect. 3.6.1) and its consequences: carrying genetic analysis through different levels ever closer to gene action.

It would appear at first glance that with modern biologic methods discrimination of genetic entities on purely descriptive grounds, i.e., on the level of the clinical phenotype, would no longer hold interest. In our opinion, however, knowledge of the phenotypic variability of genetic disease in man is needed for many reasons:

a) Such knowledge provides heuristic hypotheses for systematic application of the more penetrating methods from biochemistry, molecular biology, immunology, micromorphology, and other fields.
b) We require insight into the genetic burden of the human population.
c) Better data are needed for many of our attempts to understand the problems of spontaneous and induced mutation.

Genetic Analysis of Muscular Dystrophy as One Example. One group of diseases in which analysis using the clinical phenotype together with the mode of inheritance proved to be successful are the muscular dystrophies. These conditions have in common a tendency to slow muscular degeneration, incapacitating affected patients who ultimately die from respiratory failure. There are major differences in age at onset, location of the first signs of muscular weakness, progression of clinical symptoms, and mode of inheritance. These criteria were utilized by medical geneticists to arrive at the classification of muscular dystrophies:

I) X-linked muscular dystrophies
   1. Infantile or severe type (Duchenne) (31020)
   2. Juvenile or benign type (Becker-Kiener) (31010)
   3. Benign type with early contractions (Cestan-Lejonne and Emery-Dreifuss) (31030)
   4. Late type (Heyck-Laudan)
   5. Hemizygous lethal type (Henson-Muller-de Myer) (30995)

II) Autosomal-dominant dystrophy
   Facio-scapulo-humeral type (Erb-Landouzy-Déjérine) (15890)

III) Autosomal-recessive muscular dystrophies
   1. Infantile type
   2. Juvenile type
   3. Adult type
   4. Shoulder girdle type

This classification is based on many reports from various populations and - for the rarer variants - on reports of pedigrees. It does not include pedigrees in which affected members showed involvement only of restricted parts of the muscular system, such as distal and ocular types. Congenital myopathies were also excluded. The main criteria for discrimination are obvious from the descriptive terms used in the tabulation; for details, see Becker (1972) [565].

Multivariate Statistics. The critical human mind is an excellent discriminator. However, statistical methods for identifying subgroups within a population on the basis of multiple characteristics are now available (multivariate statistics). Such methods can also be applied to the problem of making discrimination of genetic entities more objective. A first attempt at applying such methods to the muscular dystrophies, was not entirely successful [610]; a more recent attempt at analyzing genetic heterogeneity of myotonias led to exactly the same result that the clinical geneticist had achieved without these methods [9]. Still this approach deserves to be pursued further. However, the main emphasis in the analysis of genetic heterogeneity should be placed on the development of clinical and biochemical knowledge rather than on statistical analyses per se.

3.3.6 Conditions Without Simple Modes of Inheritance

The methods discussed so far are used mainly for genetic analysis of conditions thought to follow a simple mode of inheritance. In many diseases, however, and especially in some that are both serious and frequent, there are problems:

a) Diagnosis of the condition may be difficult. There are borderline cases. Expressed more formally: The distribution of affected and unaffected in the population is not outright alternative (examples: schizophrenia; hypertension; diabetes).
b) It is known from various investigations, including twin studies, that the condition is not entirely genetic but that certain environmental factors influence manifestation (example: decline of diabetes in European countries during World War II).
c) The condition is so frequent that clustering of affected patients in some families must be expected just by chance (examples: many types of cancer).
d) It can be concluded from our knowledge of pathogenic mechanisms that the condition is not a single disease but a complex of symptoms common to a number of different causes (example: epilepsy). In fact, it is becoming apparent that diagnosis such as hypertension or diabetes subsume groups of heterogeneous disease entities.

In all such cases, a genetic analysis that starts from the phenotype cannot be expected to lead to simple modes of inheritance (for more complete discussion, see Sect. 3.6). However, for many such conditions, two questions of practical importance arise:
1. What is the risk for relatives of various degrees to be affected? Is it higher than the population average?
2. What is the contribution of genetic factors to the disease? Under what conditions will the disease manifest?

Familial aggregation can be assessed by calculation of empirical risk figures. Twin studies and comparisons of incidences among close relatives of probands with that in the general population are required to answer the several questions discussed in Sect. 3.6.2. Here, we shall make some remarks on risk figures.

Empirical Risk Figures. The expression "empirical risk" is used in contrast to "theoretic risks" as expected by Mendelian rules in conditions with simple modes of inheritance. The early methodology was largely developed by the Munich school of psychiatric genetics in the 1920s with the goal of obtaining risk figures for psychiatric diseases.
The basic concept is to examine a sufficiently large sample of affected patients and their close relatives. From this material unbiased risk figures for defined classes of relatives are calculated. These figures are used to predict the risk for relatives in future cases. This approach makes the implicit assumption that, as a rule, risk figures are constants "in space and time", i.e., among various populations and under

changing conditions within the same population. Considering the environmental changes influencing the occurrence of diseases such as diabetes, this assumption is not necessarily true but is useful as a first approximation.

The approach can be extended, to include the question of whether two conditions A and B have a common genetic component, leading to increased occurrence of patients with disease A among close relatives of patients with disease B.

Selecting and Examining Probands and Their Families. In conditions that have simple modes of inheritance, selection of probands is usually straightforward. The modes of ascertainment have been discussed in Sect. 3.3.5. For empirical risk studies, the same rules apply. In fairly frequent conditions, complete ascertainment of cases in a population will rarely if ever be feasible and is also unnecessary for purposes of these investigations. In most situations, a defined sample of probands, such as all cases coming to a certain hospital, during a predefined time period can be used. The mode of ascertainment will be single selection, or very close to it. This approach simplifies correction of the ascertainment bias among sibs of probands. The empirical risk figures can be calculated by counting affected and unaffected among the sibs, excluding the proband. Risk figures among children ascertained through the parental generation are unbiased and need no correction.

Frequently, the diagnostic categories are not clear-cut. In these cases, criteria for accepting a person as a proband have to be defined unambiguously beforehand, and all possible biases of selection should be considered. Are more severe cases normally admitted to the hospital selected for study? Are patients selected from a particular social or ethnic group? Are there any other biases that might influence the comparability of the results? Really unbiased samples are hardly if ever available, but the biases should be known. Most important, such biases should be independent of the problem to be analyzed. For example, it would be a mistake to consider only cases where the case history shows that other family members have been similarly affected.

The goal of the examinations is to obtain maximal and precise information about the probands and their families as far as possible. Methods to achieve this goal, however, vary. Clinical experience and the study of publications on similar surveys are helpful.

Once the proband and his family are ascertained, the relatives should be noted as completely as possible, and information on their health must be collected. Here, personal examination by the investigator is of great importance, but in many instances case histories provided by the patients or their relatives are indispensable. Examination and case history should be backed by objective data such as hospital records and various laboratory and radiologic studies. Even results of clinical examinations should be regarded with scepticism since not all physicians are equally knowledgeable and careful, and official documents, such as death certificates, are often unreliable regarding diagnostic criteria.

In most cases, establishing genetic risk figures answers the question of whether the risk is higher than in the average population. Sometimes adequate incidence and/or prevalence data from a complete population in which the study is carried out or a very similar one are available. More often than not, however, a control series must be examined with the same criteria as used for the "test" populations. It is a good idea to use matched controls, i.e., to examine for every patient a control person matched in all criteria not connected with the condition to be investigated (such as age, sex, ethnic origin, etc.).

Statistical Evaluation, Age Correction. In conditions that manifest a birth, such as congenital malformations affecting the visible parts of the body, further calculations are straightforward: The empirical risk for children is given by the proportion of affected in the sample. In many cases, however, onset is some time during later life, and the period at risk may be extended – in schizophrenia, for example, between the 15th and the 45th years of life. Here, the question asked is: What is the risk of a person's becoming affected with the condition, provided he or she lives beyond the manifestation period? The appropriate methods of age correction have been discussed extensively in the older literature [744], one much-used is Weinberg's "shortened method". First, the period of manifestation is defined on the basis of a sufficiently large sample (usually larger than the sample of the study itself). Then, all relatives who dropped out of the study before the age of manifestation are discarded. Dropping out may be for any of a variety of reasons: death, loss of contact due to change of residence, or termination of the study. All persons dropping out during the age of manifestation are counted as one-half, and all who have survived the upper limit of manifestation age are counted full.

Example. Among children of schizophrenics, there are 50 affected and 200 unaffected. Of these, 100 have reached the age of 45 and 100 are between the age of 15 and 45 (i.e., the age of manifestation for the great majority of schizophrenic cases).

Thus, the corrected number of unaffected is: $200 - \frac{1}{2} \times 100 = 150$; the empirical risk is:

$$\frac{50}{150+50} = 25\%$$

Calculation of Risk Figures for Schizophrenia. The above procedure will be demonstrated using Kallmann's (1938) [731] study on the genetics of schizophrenia. This disorder has a number of properties

that are unfavorable for a genetic analysis. Diagnostic criteria differ from one school of psychiatrists to the other, and it is known from lack of complete concordance in identical twins that genetic liability is not the only pathogenic factor (Sect. 8.2.3.7).
The probands were all schizophrenic cases admitted to the Psychiatric State Hospital Berlin-Herzberge during the period 1893-1902. The only condition of inclusion in the sample was the diagnosis of schizophrenia according to the criteria established by Kraepelin. These criteria are more restrictive than those commonly used in American psychiatry. The diagnosis was confirmed by the investigator himself on the basis of all available data. There were 1087 cases, 647 female and 440 male.
The families (children, grandchildren, great-grandchildren) were contacted by an assistant and then examined by the investigator, either in the hospital or at their home. The diagnoses made on the basis of these examinations were corroborated by all available kinds of objective data. In total, 13851 persons were covered. In precomputer times, the evaluation must have been a prodigious job.
Table 3.7 shows the risk figures for schizophrenia and schizoid psychopathy for all degrees of relationship. These were calculated by means of the shortened method of Weinberg discussed above. More sophisticated correction methods were also used, but the results did not differ appreciably. The data were subsequently subdivided according to clinical characteristics of the disorder in spouses, parents, and siblings. The principal results of the study were:

a) The clinical picture of the disease is variable. This variability allows a subdivision of schizophrenia into a nuclear group (catatonics and hebephrenics) and a more marginal group (paranoid and simple forms). The risk is higher for relatives of patients with the nuclear forms.
b) The different forms may occur within the same families. These findings mean that there is a common genetic background to all types of schizophrenia. However, there is also intrafamilial correlation for subtypes showing that, apart from the general liability, there is also a more specific inherited predisposition.
c) The risk for children of schizophrenics to become affected is about 19 times higher than in the general population. For grandchildren, nephews, and nieces, it is about five times higher. Kallmann did not examine a control series since Panse (1935) had collected control material from the same population and had found the frequency of schizophrenia to be 0.89%.
d) Probands without affected sibs had about the same risk of schizophrenia in their children as probands with affected siblings. This finding suggests that there is no admixture of nonhereditary cases.
e) Children whose parents were both schizophrenic had a 68% risk of becoming affected.
f) Kallmann also examined whether other diseases tended to be more frequent in close relatives of schizophrenics. He found tuberculosis to be much more frequent than in the general population, and this liability paralleled the liability to schizophrenia.
g) Fertility of patients as well as that of their close relatives was lower than the population average.

Other studies have been conducted in various populations, and the data now available for empirical

Table 3.7. Empirical risk figures for schizophrenia (Kallmann, 1938 [731])

Degree of relationship with proband	Number of individuals above the age of 15	Risk for schizophrenia				Prevalence of schizoid psychopathy
		Total material		Clinical subgroups		
		Minimum	Age corrected	Nuclear group (age corrected)	Marginal group (age corrected)	
		%	%	%	%	%
Children	1,000	13.9	16.4	20.9	10.4	32.6
Grandchildren	543	2.9	4.3	5.1	2.9	22.8
Great-grandchildren	29	–	–	–	–	3.4
Siblings	2,581	7.5	11.5	12.9	8.9	10.5
Half-sibs	101	6.4	7.6	7.6	–	7.9
Nephews and nieces	1,654	1.9	3.9	4.7	3.4	6.2

Prevalence of schizophrenia in the general population: 0.9%. Note that the empirical risks for relatives have not essentially changed in many other studies done since 1938. There is less agreement regarding the categorization into nuclear and marginal groups and there is significant disagreement regarding the significance of a schizoid psychopathy, which is difficult to distinguish objectively.

risks in schizophrenia are considerable. The topic will be taken up again in Sect. 8.2.3.7.

Theoretical Risk Figures Derived from Heritability Estimates? There are suggestions [803] that empirical risk figures should be replaced by theoretical risk figures computed from heritability estimates for the multifactorial model (Sect. 3.6.2), after agreement of the data with expectations from this model as compared with a simple diallelic model has been established. Such heritability estimates can be achieved by comparison of incidence of the condition in the general population with the incidence in certain categories of relatives, for example, sibs, or, with precautions (Sect. 3.8.4), from twin data. Such theoretical risk figures have been derived, for example, for pyloric stenosis [752]. In theory, the method permits inclusion of environmental, for example, maternal, effects. This method may help in future to improve risk calculations especially for categories of relatives for which sufficient empirical risk figures are not available. However, the disadvantage of this approach is that it depends critically on the assumption that the genetic model fits the actual situation sufficiently well. Since the genetic model chosen may not apply to a given set of data, there is danger that the sophisticated statistical approach suggests a spuriously high degree of precision of the results. Empirical testing of risk figures predicted by such methods is desirable but may not always be practically possible.

3.4 Linkage: Localization of Genes on Chromosomes

Genes are located in a linear fashion on the chromosomes. This has the logical consequence that genes located on the same chromosome are transmitted together, i.e., that their segregation will not be independent. On the other hand, it is know from cytogenetics that, during the first meiotic division, chiasmata are formed and that certain chromosome segments are exchanged between homologous chromosomes (crossing over; cf. Sect. 2.1.2.4). Hence, even genes located on the same chromosome will not always be transmitted together; the probability of transmission of two linked genes will depend on the distance between them and how often they are separated by crossing over. If located on a fairly long chromosome and if the distance is large enough so that numerous crossing over events occur between them, genes located on the same chromosome may even seem to segregate independently. Such genes are syntenic but not linked. It has been the great achievement of Morgan and his school in the first two decades of this century to exploit linkage for localizing genes relative to each other on chromosomes and developing gene maps in the fruit fly *Drosophila melanogaster.*

Studies on linkage and gene mapping in man have lagged behind this development for many decades. Sophisticated statistical techniques were developed to get around the difficulty that directed breeding experiments are impossible in humans, so that information from naturally occurring families must be utilized. Application of such techniques, however, was only sparsely rewarded by detection of linkage [855]. A breakthrough occurred only when the new techniques of somatic cell genetics and especially cell fusion were introduced. These techniques permitted assignment of genes to chromosomes and even chromosome segments. Still later, methods taken from molecular biology, especially detection of restriction site polymorphisms and in situ hybridization with DNA probes, brought further progress. Today, the human gene map contains many genes (Table A 9,5), the number of successful assignments is growing quickly, and new insights into the organization of the genetic material are forthcoming. As more genes are assigned to the human gene map, there is a greater chance to localize a marker that is newly introduced into the analysis.

In the following, we shall first describe the principle of the classic approach to gene localization, as introduced by Morgan and his followers. This will give an opportunity to introduce some general concepts. Then, statistical methods for detecting and measuring linkage in humans will be discussed; calculated examples are given in App. 9. Finally, the principles of cell fusion and their application to linkage studies will be described.

3.4.1 Classic Approaches in Experimental Genetics: Breeding Experiments and Giant Chromosomes

According to Mendel's third law, segregation of two different pairs of alleles is independent; all possible zygotes of two pairs of alleles are formed by free recombination. In the mating between the double heterozygote AaBb and the double homozygote aabb, this leads to:

Paternal gametes		AB	Ab	aB	ab
Maternal gametes	ab	¼ AaBb	¼ Aabb	¼ aaBb	¼ aabb

The four genotypes are formed in equal proportions. Soon after Mendel's laws were rediscovered, Bateson, Saunders, and Punnett (1908) [561] found an exception from this rule in the vetch, *Lathyrus odora-*

tus. Certain combinations were observed more frequently, others less frequently than expected. In some cases, the two parental combinations – in our example AB (father) and ab (mother) – were increased among the progeny; in other cases, the two other types Ab or aB were more frequent.

Paternal gametes		AB	Ab	aB	ab
Maternal gametes	ab	AaBb	Aabb	aaBb	aabb
First case (coupling)		$\frac{1}{2}-\Theta$	Θ	Θ	$\frac{1}{2}-\Theta$
Second case (repulsion)		Θ	$\frac{1}{2}-\Theta$	$\frac{1}{2}-\Theta$	Θ

Θ = Recombination fraction

The alleles of the parental combination seemed either to attract one another or to repel one another. Bateson et al. [561] coined the terms "coupling" for the former phase and "repulsion" for the latter phase. Morgan (1910) [448] recognized that coupling and repulsion are two aspects of the same phenomenon, (i.e., location of two genes on the same or on homologous chromosomes). He coined the term "linkage". Coupling occurs when the genes A and B are localized in the doubly heterozygous parent on the same chromosome $\frac{AB}{ab}$, repulsion occurs when they are localized on homologous chromosomes $\frac{Ab}{aB}$. The terms *cis* and *trans* are more frequently used to refer to genes in coupling or repulsion, respectively. If linkage is complete, only two types of progeny can occur. In most cases, however, all four types are found, albeit two types in smaller numbers. Morgan explained this finding by exchange of chromosome pieces between homologous chromosomes during meiotic crossing over. He also recognized that frequency of crossing over depends on the distance between two gene loci in one chromosome. Using recombination analysis as an analytic tool, he and his co-workers succeeded in locating a great number of gene loci in *Drosophila* and in establishing chromosome maps. Their results were confirmed when, in the early 1930s, Heitz and Bauer and Painter discovered the giant chromosomes of some *Dipterae*. With this experimental tool, many gene localizations known from indirect evidence could be confirmed by direct inspection when they were accompanied by small structural chromosomal variation. In the meantime, linkage analyses have been carried out in a great number of species.

Linkage and Association. Sometimes, it is assumed that genes that are linked should always show a certain association in the population, i.e., that the chromosomal combinations AB or ab (coupling) occur more frequently than the combinations Ab or aB (repulsion). However, this is not the case in a randomly mating population. Even if the linkage is fairly close, repeated crossing over in many generations will cause, in the long run, all four combinations, AB, ab, Ab, and aB, to be randomly distributed. As a rule, association of genetic traits does not point to linkage; in most cases, it has other reasons.

This rule, however, has exceptions. Some combinations of closely linked genes do, indeed, occur more often than expected with random distribution. Such "linkage disequilibrium" had first been postulated in humans for the Rhesus blood types (Sect. 3.5.4) and has been proved for the major histocompatibility complex (MHC), especially the HLA system (Sect. 3.5.5), and for DNA polymorphisms. Linkage disequilibrium may have two reasons:

1. The population under examination originated from a mixture of two populations with different frequencies of the alleles A,a and B,b, and the time elapsed since the populations were mixed was not sufficient for complete randomization.
2. Certain combinations of alleles at linked gene loci are maintained in high frequency by natural selection.

These problems will be discussed in greater detail in connection with the MHC system (Sect. 3.5.5) and in the discussion on association between HLA and disease (Sect. 3.7.3).

3.4.2 Linkage Analysis in Humans: Classic Pedigree Method

Direct Observation of Pedigrees. In humans, linkage analysis by classic methods is difficult since no directed breeding occurs. However, in some cases, pedigree inspection gives some information. Linkage is excluded, for example, if one of the genes under scrutiny can be localized to the X chromosome, while the other is on an autosome. By the same token, there is a high probability of demonstrating formal linkage if both genes are X-linked. Even in this case, however, formal linkage may not be demonstrable since the loci may be so far from each other that crossing over separates them. Similar considerations hold for genes located on a given autosomal chromosome. Synteny is a term used to describe that two or more genes are situated on the same chromosome, whether formal linkage by family methodology can be demonstrated or not. Either a large pedigree or a number of smaller pedigrees must be screened to assess the extent of crossing over. Figure 3.23a shows a pedigree with red-green color blindness (30380, 30390) and hemophilia. The males

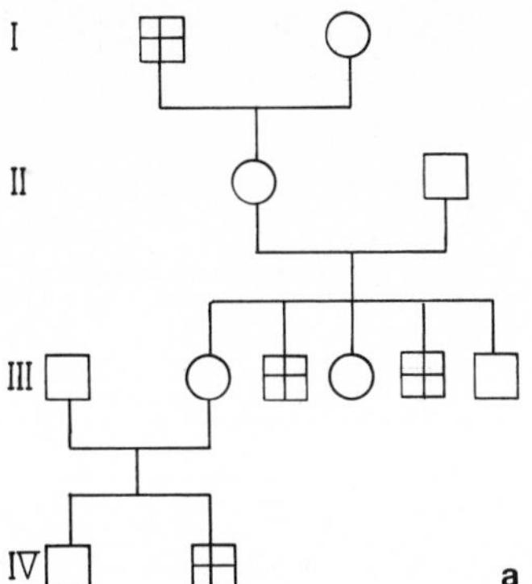

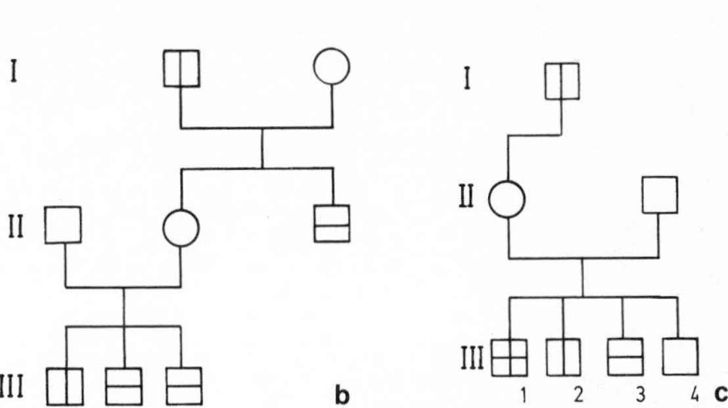

Fig. 3.23 a–c. Pedigrees with red-green color blindness (⊟), hemophilia (◫), and both conditions (⊞). **a** Both abnormal genes in coupling. **b** In repulsion. **c** In this family, crossing over between the two loci must have occurred twice – either in the oocyte leading to III, 1 and III, 4 or in the oocyte leading to III, 2 and III, 3. (**a** Madlener, 1928 [772]; **b** Birch, 1937; **c** Rath, 1938 [849]; Stern, 1973 [204])

in the sibships at risk either have both conditions or are normal. The genes are in the coupling (or *cis*) state. The pedigree in Fig. 3.23 b shows the opposite; here these genes are in the repulsion (or *trans*) phase. In the pedigree in Fig. 3.23 c, crossing over must have occurred twice in the mother's oocyte. She either has carried the two mutant alleles in cis and the second and third sons would be crossovers, or the two mutant alleles were in trans and the first and fourth sons would be crossovers. Unfortunately, there is no information on the color vision status of the maternal grandfather, which would have decided the issue. A fairly detailed map of the human X chromosome is now available (Sect. 3.4.3; Fig. 3.28).

In some cases, linkage between gene loci localized on an autosome can also be established by simple inspection of an extensive pedigree. Figure 3.24 a shows a large pedigree, in which Huntington's disease segregates together with a polymorphism of the G8 probe of Hind III – digested human genomic DNA [693]. Four allelic variants of this probe are observed in this pedigree, A, B, C, and D. The Huntington's gene invariably segregates together with the allele C. One individual, VI, 5 (arrow), has so far been unaffected by Huntington's disease. She will be affected later. The pedigree points to close linkage. Some crossovers have been detected, and the recombination fraction is 4% or less. Figure 3.24 b shows a pedigree with segregation for elliptocytosis (oval shape of the erythrocytes) and the Rh gene complex. Almost all family members with elliptocytosis also have the complex CDe; there are only two known exceptions, (II, 9; II, 11). Many of the unaffected siblings have other combinations. From this pedigree, linkage between the Rh locus and the elliptocytosis gene can be deduced. The deduction was confirmed by additional pedigrees. These examples show that the phase or alleles at two loci (*cis* or *trans*) can often be ascertained with great precision, and recombinants can be identified if (at least) three generations are available for analysis and if there are many sibs.

Statistical Analysis. In most cases, linkage analysis is more difficult. Extensive pedigrees like those in Fig. 3.24 are exceptional; most available families consist of two parents and their children. Here the problem is that the chromosomal phase is usually unknown: a double heterozygote may be AB/ab *(cis)* or Ab/aB *(trans)*. When the alleles are randomly distributed in the population, the two types are expected in about equal frequencies: an AB/ab person will form germ cells in the ratio:

AB	Ab	aB	ab
$\frac{1-\Theta}{2}$	$\frac{\Theta}{2}$	$\frac{\Theta}{2}$	$\frac{1-\Theta}{2}$

If these two types occur in about equal frequencies, the overall frequency of all four types of germ cells in the population will be:

AB	Ab	aB	ab
$\frac{1}{4}$	$\frac{1}{4}$	$\frac{1}{4}$	$\frac{1}{4}$

whereas a heterozygote Ab/aB will form germ cells in the ratio:

AB	Ab	aB	ab
$\frac{\Theta}{2}$	$\frac{1-\Theta}{2}$	$\frac{1-\Theta}{2}$	$\frac{\Theta}{2}$

All four types of germ cells occur with the same frequencies, regardless of the probability of recombination Θ. Linkage does not lead to any association of alleles A, B or a, b in the population. Another criterion for linkage has to be found, one that is independent of the phase of the double heterozygote.

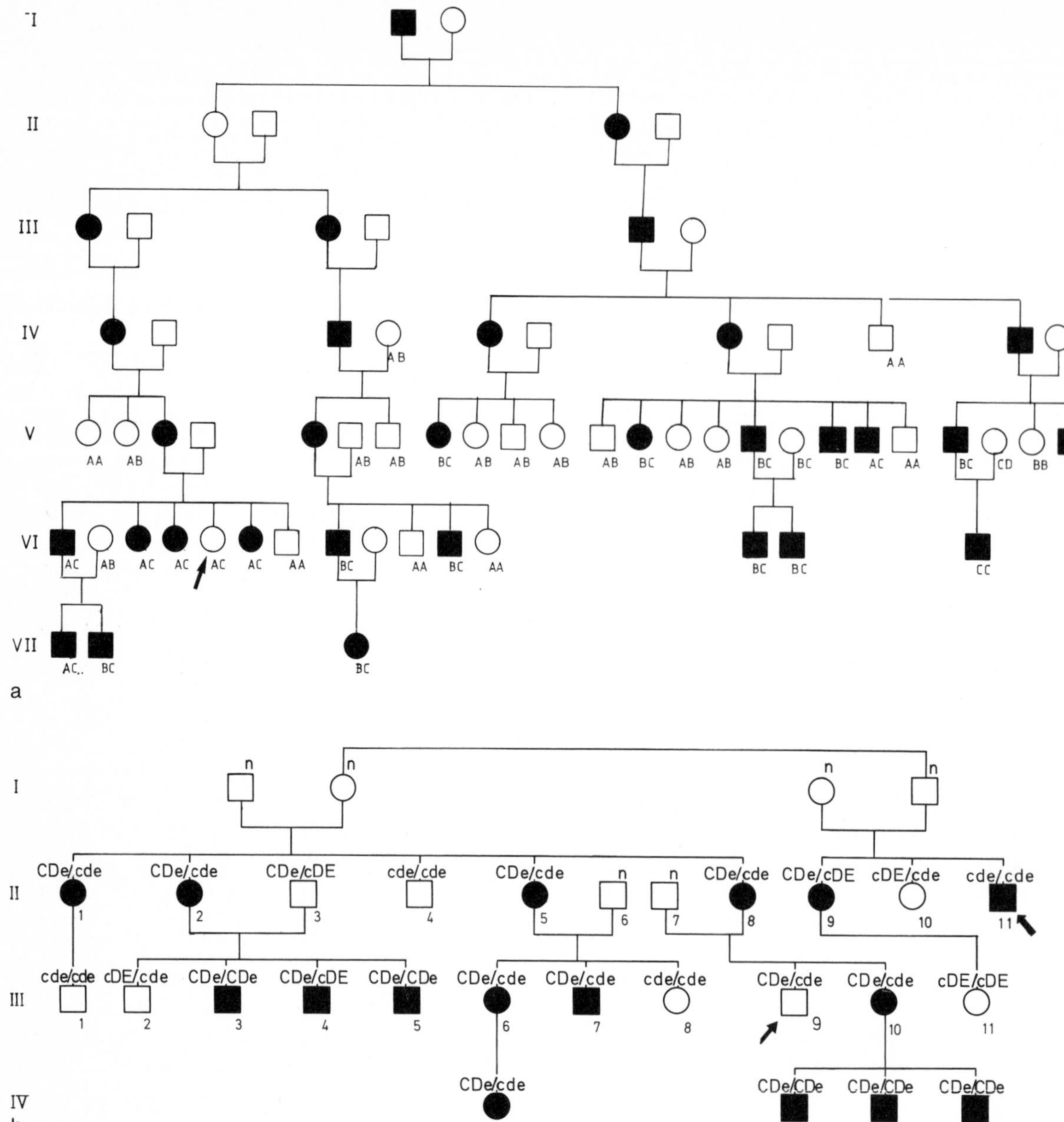

Fig. 3.24. a Large pedigree from Venezuela with Huntington's disease. A, B, C designate three different "alleles" of a DNA polymorphism. The Huntington gene is transmitted together with "allele" C. One individual, VI,5 (*arrow*) has so far been unaffected. She will most likely be affected later. (From Gusella et al. [694]) **b** Autosomal linkage between the Rh locus and dominant elliptocytosis ■. There are two crossovers (arrows): II, 11 and III, 9. In all other cases, the elliptocytosis gene is in coupling (cis) with CDe. *n*: not examined. (Lawler and Sandler, Ann. Eugen. (London) 1954)

Such a criterion would be the distribution of children within sibships. In mating of AB/ab persons (cis phase), most children will show the allele combinations of their parents; in matings Ab/aB (trans phase), most children will show these alleles in a new combination. How can these deviations from random distribution within sibships be measured and utilized for establishing linkage and determining the probability of recombination? Bernstein (1931) [571] was the first to develop such a method. It has now been replaced by the method of "lod scores" as developed by Haldane and Smith (1947) [699] and Morton (1955, etc.) [796; 797; 798; 799] and is generally used to assess linkage. Its principle can be described as follows:

The probability P_2 that the observed family data conform to the behavior of two loci under full recombination without any linkage is calculated. Simi-

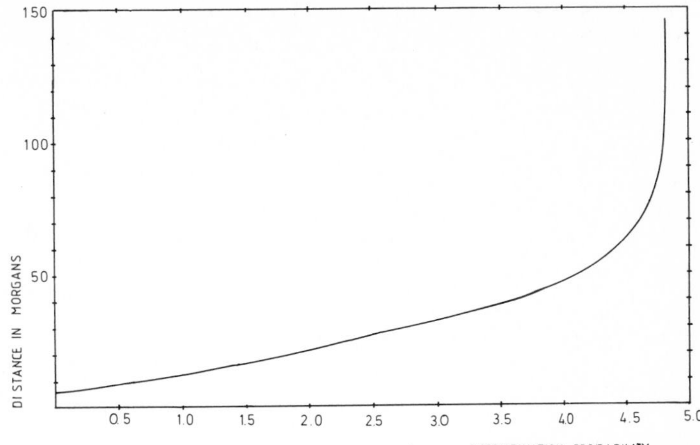

Fig. 3.25. Relationship between recombination probability and the distance (w) in Morgans. [612 a] The relationship is exponential because the number of double (and multiple) crossovers in creases with increasing map distance

larly, a probability P_1 that the identical family data are the result of two linked loci under a specified recombination fraction (Θ) is determined. The ratio of these two probabilities is the likelihood ratio and expresses the odds for and against linkage. This ratio $\frac{P_1(F/\Theta)}{P_2(F/(\frac{1}{2}))}$ must be calculated for each family F.

A man may be double heterozygous for the gene pairs A,a and B,b. His wife may be homozygous for the two recessive alleles aa, bb. Assume that his two sons, like the father, are double heterozygous, i.e., they inherited the alleles A and B from the father. This probability is $\frac{1}{2} \times \frac{1}{2} = \frac{1}{4}$, if the genes segregate independently. If the gene loci are closely linked without crossing over, the probability for occurrence of this pedigree may be calculated as follows. Either the genes occur in coupling state: AB/ab, then the possibility for common transmission to the two sons is ½ (transmission of the combination ab would also have a probability of ½) or the genes occur in repulsion state Ab/aB. Then, transmission of both dominant alleles to the same son requires crossing over; with close linkage, and in absence of crossing over, the probability of common transmission = 0. Hence, the total probability for transmission of the combination aB to both sons is ½ and the likelihood ratio is $P_1/P_2 = (\frac{1}{2})/(\frac{1}{4}) = 2$ in favor of close linkage. In the same way, likelihood ratios for the various degree of loose linkage can be calculated.

For convenience, the logarithm of the ratio is used, and a lod score z (meaning "log odds" or "log probability ratio") is used:

$$z = \log_{10} \frac{P(F \mid \Theta)}{P(F \mid (\frac{1}{2}))} \tag{3.3}$$

Here, $P(F \mid \Theta)$ denotes the probability of occurrence for a family F when the recombination fraction is Θ. Using the logarithms instead of the probabilities themselves has the advantage that the score of any newly found family can be added, giving a combined score $z = \Sigma z_i$ for all families examined. Equation 3.3 implies an identical recombination fraction for both sexes. Since sex differences in recombination rates have been described [855], the z score in actual data should be computed separately for the sexes:

$$z = \log_{10} \frac{P(F/\Theta, \Theta')}{P(F/(\frac{1}{2}, \frac{1}{2}))} \tag{3.4}$$

where Θ is the recombination fraction in females and Θ' in males.

It follows from the definition of the likelihood ratio that the higher its numerator, the stronger the deviation in the direction of linkage. In terms of logarithms, the higher the z score, the better the evidence for linkage. A lod score of 3 or higher is generally considered as proof of linkage. For setting up scores, minor corrections for dominance and for ascertainment of pedigrees showing rare traits had to be introduced but are not dealt with here [882].

The score $z(\Theta, \Theta')$ for the whole set of data is the sum of the scores of the separate families. For a first approach $\Theta = \Theta'$ is assumed to simplify the calculation. When linkage has been established, a possible sex difference can be examined.

Lod Scores. The lod scores were tabulated extensively and published with rules for their application. To extract full information from a more extensive pedigree, the procedure given by Ott [831 a, b, c; 612 a] should be followed. The ideal mating for linkage studies involves a double heterozygote, i.e. a person heterozygous for two different traits, with a person homozygous for the two genes. The following types of families do not contribute information regarding linkage:

a) Families in which neither parent is doubly heterozygous

b) Families in which there cannot be any observable segregation
c) Families in which the phases of the parents are unknown and there is only one examined child

Most linkage studies involve analysis of two common markers or of a common gene with a rare genetic disease. Opportunities to study linkage between two rare genes hardly ever present themselves. The ideal family for linkage studies is a kindred with three generations, many matings, and a large number of offspring [945; 946]. Such families are becoming rare in Western societies. An alternate approach involves testing of many small families. While most data obtained from such small sampling will be noncontributory for linkage, a large-scale data collection of this sort may yield occasional new linkage information.

The LIPED program is a computer program which provides the maximum likelihood of linkage between two genes considering all pedigree data. The program calculates the most likely genotype assignments and uses these data to provide the most likely recombination value for the data. Since the computer can make a very large number of calculations this program is superior to the usual manual calculations and has become the standard procedure in human linkage work.

Therefore, this computer programm LIPED (Ott [831 a, b, c]) is often used for practical analysis (App. 9). As mentioned in Sect. 2.1.2.4, the human male genome has a map length of ≈ 25.8 morgans. With $\approx 3.5 \times 10^9$ base pairs/haploid genome (Sect. 2.3), this means that 1 cM corresponds to $\approx 1.356 \times 10^6$ base pairs (or 1356 kb). However, crossing over sites are not distributed equally, as is discussed below.

When linkage has been established and a maximum likelihood estimate of Θ has been achieved, the question of heterogeneity should be examined. If, for example, linkage between the locus for a genetic polymorphism and a rare dominant condition has been established, linkage analysis can help to prove genetic heterogeneity if only part of the family material shows linkage.

Two calculated examples, one for moderately close linkage, and another one for loose linkage (or independence) are given in App. 9.

Recombination Probabilities and Map Distances. When a number of linkages have been established, the next step is to estimate the map distances between these loci. These distances are expressed in morgans and centimorgans, one centimorgan (map unit; cM) meaning 1% recombination, ($\Theta = 0.01$) for smaller map distances. For larger distances, this value must be corrected for double crossing over. Various methods have been proposed [612 a]. Given Θ, the map distance can be read directly from Fig. 3.2.5.

Results for Autosomal Linkage, Sex Difference, and Parental Age. The first autosomal linkage in man was found by J. Mohr [791] between the Lutheran blood groups and the AB0 secretor locus. Some years later, linkage between the Rh loci and elliptocytosis (16690) was established and utilized to detect genetic heterogeneity of elliptocytosis, since not all families with elliptocytosis showed this linkage. A short time later, linkage between the AB0 blood group locus and the dominant nail-patella syndrome (16120) locus was found. This linkage established for the first time a sex difference of recombination probabilities in man: map distance between these loci for males was 8 cM, for females 14 cM. An analogous sex difference was established for the Lu/Se pair (♂ : 10 cM; ♀ : 16 cM); for the AB0/Ak (adenylate kinase) pair (♂ : 12; ♀ : 19 cM), and for the HLA-PGM_3 pair (♂ : 15; ♀ : 36 cM). A sex difference was also found for linkage between the first and second loci of the HLA system and for the Bf locus, which is closely linked with the HLA loci. As already mentioned, many restriction fragment length polymorphisms (RFLPs) have been introduced into linkage analysis. In some cases, e.g., for the long arm of chromosome 13, the above-mentioned sex difference has been confirmed [945]. On the distal one third of the short arm of chromosome 11, however, the recombination rate has been reported to be higher in males [944].

A higher recombination frequency in females was also observed in the mouse [853]. These results confirm a rule set out by Haldane in 1922, according to which crossing over is more frequent in the homogametic (i.e., XX) than in the heterogametic sex (i.e., XY). In *Drosophila,* there is no crossing over at all in the male.

There has been considerable discussion as to whether recombination frequency is also influenced by parental age. In the mouse, the data are consistent with decreasing recombination rates with aging in females and increasing rates in males. Weitkamp (1972) [939] found a significantly increased incidence of recombinants with increasing birth order in humans for eight closely linked pairs of loci indicating a parental age effect. There was no difference between males and females for this effect. A similar parental age effect was found for the Lutheran/secretor and Lutheran/myotonic dystrophy (16090) pairs but not for AB0/nail-patella and Rh/PGD pairs. Possibly the age effect is different for various loci [855].

In a survey of cytogenetically determined chiasma frequencies from 204 males from the literature, little or no linear trend with age was found [754a]. No cytogenetic data are available for females. The discrepancy between formal recombination data and chiasma frequencies is unexplained.

Information from Chromosome Morphology. Pairs or clusters of autosomal loci found to be linked (linkage groups) could not be assigned to specific chromosomes by a formal methodology of family study. The first chromosomal localization was accomplished as follows [629; 855].

The long arms of chromosome 1 frequently show a secondary constriction close to the centromere. In about 0.5% of the population, this constriction appears much thinner and longer than normal. The variants is dominantly inherited. Presumably, an uncoiler locus (Un-l) appears to be localized on chromosome 1, as only one of the two homologous No. 1 chromosomes shows the abnormal phenotype. Linkage studies showed close linkage between the blood group Duffy locus and the Un-l trait; $\Theta = 0.05$. Linkage between Duffy and congenital zonular cataract (11620) had been discovered earlier. Hence, a linkage group with three loci, cataract, Duffy, Un-l could be assigned [629].

Another possibility to localize genes on specific chromosomes was afforded by deletions. If a gene locus whose mutation has a dominant effect is lost by deletion, the absence of that gene may occasionally have a phenotype similar to the dominant mutation. More extensive symptoms may also be present, since more genetic material than a single gene would be expected to be lost. In 1963, a retarded child with bilateral retinoblastoma was found to have a deletion of the long arm of one D chromosome [1531]. Later on, this chromosome was identified as No. 13, and deletion 13q14 was found in a number of other cases with retinoblastoma and additional anomalies. Patients with retinoblastoma without additional symptoms usually have no deletion. A locus sometimes involved in retinoblastoma could thus be assigned tentatively to this chromosome.

Another approach, thought to be more generally useful, was the quantitative examination of enzyme activities in cases with chromosome anomalies. Most enzymes show a clear-cut gene dosage effect in heterozygotes, i. e. heterozygotes for an enzyme deficiency have ≈50% of enzyme activity. Hence, a similar gene dosage effect might be expected when a gene locus is localized on a chromosome segment that had been lost by deletion. To exploit this possibility, a great number of genetic markers were examined in many cases. The result turned out to be negative in most cases, but ultimately such "exclusion mapping" may be expected to narrow down possible localizations of marker genes. However, several misassignments have been made with this approach since the presence of a "silent" gene, i. e., a mutation leading to absence of a gene product, may mimic a deletion effect.

However, if heterozygotes and by inference monosomics show a gene dosage effect, it was not too far-fetched to expect such an effect in trisomics as well. The first examinations of enzymes in Down's syndrome (trisomy 21) seemed to confirm this expectation. However, when it appeared that more and more enzyme loci were localized on chromosome 21 and increased enzyme activities of the X-linked enzyme G6PD were found, this approach was discredited somewhat. Quantitative deviations of enzyme activities in trisomics in vivo might be a problem of regulation.

But an increasing number of such gene dosage effects have been described in vitro, on trisomic and monosomic cells [1185] (Sect. 4.7.4.3). To mention only one example, the activity of the enzyme phosphoribosylglycinamide synthetase (GARS) was studied in several cases of partial monosomy and full and partial trisomy 21, as earlier studies had suggested a gene dosage effect for this enzyme. In regular trisomy 21 an excess was found with a trisomy 21-to-normal ratio of 1.55. A ratio of 0.99 was found in 21q21→21pter monosomy; of 0.54 in 21q22→21qter monosomy; of 0.88 in 21q21→21pter trisomy; and of 1.46 in 21q22.1 trisomy. Hence, the GARS gene locus could be localized in subband 21q22.1 [322]. For some other examples see Table 4.27 and Appendix 9, Table A 9.5. Utilization of variants in chromosome morphology (heteromorphisms), such as the secondary constriction on chromosome 1 mentioned above, along with gene dosage studies, opened a slow way to linkage and gene localization. A new method, utilizing cell fusion, has led to much more rapid progress.

3.4.3 Linkage Analysis in Humans: Cell Hybridization and DNA Techniques

First Observations on Cell Fusion. The history of cell fusion has been described by Henry Harris [702; 692]. Binucleate cells were observed by J. Mueller (1838) in tumors, and afterwards by Robin in bone marrow, by Rokitansky in tuberculous tissue, and by Virchow both in normal tissues and in inflammatory and neoplastic lesions. The view that some of these cells were produced by fusion of mononucleate cells derived from the work of de Bary (1859), who observed that the life cycle of certain myxomycetes involved the fusion of single cells to form multinucleated plasmodia. The earliest reports of multinucleated cells in lesions that can with certainty be identified as of viral origin appear to be those of Luginbuehl (1873) and Weigert (1874), who described such cells at the periphery of smallpox pustules.

Following the introduction of tissue culture methods, numerous observations were made on cell fusion in cultures of animal tissue, and 1927 there were 21 references to such observations. Enders and Peebles (1954) found that the measles virus induced cells in tissue culture to fuse together to form multinucleated syncytia. A number of similar observations were made by others, and Okada (1958) showed that animal tumor cells in suspension could be rapidly fused to form multinucleate giant cells using high concentrations of HVJ virus, a member of the parainfluenza group.

In 1960 Barski et al., identified cells generated by spontaneous fusion in a mixed culture of two different but related mouse tumor cell lines. These cells contained within a single nucleus the chromosome complements of both parent cells. This phenomenon was then examined by Ephrussi et al., who concluded that not only closely related mouse cells could be hybridized; even larger genetic differences did not exclude spontaneous cell fusion. However, it soon became obvious that the frequency of spontaneous cell fusion was very low and that many cell types never fused spontaneously. Fusion frequency had to be increased in some manner. Furthermore, isolation of hybrid cells was only possible when culture conditions were chosen to give these cells a selective advantage.

Both problems were soon solved: Littlefield (1964) isolated the rare products of spontaneous fusion in mixed cultures by a technique adopted from microbial genetics. Fusion of two cells deficient in two different enzymes resulted in hybrids that recovered the complete enzyme set by complementation. Only these cells survived selection against the deficient cells.

Harris and Watkins (1965) [254] enhanced the fusion rate of various cells by treatment with UV-inactivated Sendai virus. Along with introduction of this method, they showed that fusion could be induced between cells from widely different species and that the fused cells were viable. With this work, widespread use of the cell fusion method in various branches of cell biology began.

First Observation of Chromosome Loss in Human-Mouse Cell Hybrids and First Assignment of a Gene Locus. Weiss and Green (1967) [938] fused a stable, aneuploid mouse cell line, a subline of mouse L cells, with a diploid strain of human embryonic fibroblasts. The mouse cell line was deficient in the thymidine kinase (TK) locus and did not grow in HAT medium, a culture medium selective for cells containing the human TK locus (18830).

Cultures were initiated by mixing the two types of cells and growing them on standard medium. After 4 days, cultures were placed in the selective HAT medium. This led to degeneration of the mouse cells leaving a single layer of human cells. After 14–21 days, hybrid colonies could be detected growing on the human cell monolayer. A number of these colonies were then isolated, grown for a longer time period, and examined. They turned out to maintain the mouse chromosome complement but 75%–95% of the human chromosomes were lost. One human chromosome, however, was present in almost all cells growing in the HAT medium. This suggested the hypothesis that the locus for thymidine kinase was localized on this chromosome. Therefore, control experiments were carried out with a BrdU-containing medium. BrdU, a base analogue for thymine, is accepted by thymidine kinase in place of thymine, and selects against cells containing this enzyme. A special chromosome described as "having a distinctive appearance" was present in almost all HAT cultures but in none of the BrdU cultures. It was concluded that the TK locus is, indeed, localized to this chromosome. A short time later, the chromosome bearing the TK locus was found to be No. 17 ([788] Fig. 3.26).

This work led to two principles, which later were decisive for the use of cell hybridization in linkage work:

1. Hybrids between mouse and human cells tend to lose many human chromosomes. It was shown later that this loss is random, so that with a great number of hybrids one can expect to find a cell that has kept any one specified chromosome.
2. By using an appropriate selective system, it is possible to select cells with a certain enzyme activity and to localize the gene loci specifying this enzyme to a specific chromosome.

Whereas genetics historically has been the science of genetic variability within a species, the hybridization method permits localization of genes that do not show genetic variability in humans, provided only that the gene products of the human and nonhuman cells can be identified. One way of identification is the use of a selective system.

Since 1967 selective systems have been developed for several enzymes. One example uses the HPRT locus on the X chromosome (Sect. 4.2.2.6). This system can be used for selection not only of other X-linked loci but also of autosomal loci if a part of the autosome has undergone translocation with the X chromosome. It is also possible to assign loci for which no selective system exists, provided that enzymes produced by the two species have recognizable differences such as electrophoretic variation. This method, however, is more cumbersome, requiring thorough biochemical and cytogenetic examination of a great number of cell clones. The task would be hopeless had not the methods of chromosome identification by banding been developed in the meantime.

Influence of Banding Methods for Chromosome Identification. Further progress in gene localization depended on the development of new methods for individual identification of chromosomes. These banding methods were introduced in 1970 and developed further in the years thereafter (Sect. 2.1.2.2). They soon permitted identification not only of whole chromosomes but of chromosome parts. In hybrid cultures, chromosome breakage and rearrangement are relatively frequent events. This chromosomal behavior made possible the suitable selection of hybrid clones containing identifiable parts of chromosomes, thereby combining the advantages of deletion mapping and hybridization. In this way, a number of loci could be assigned to special chromosome bands (or small groups of bands).

Other Sources of Information for Gene Localization. Another method utilized for gene localization is the in situ DNA-RNA hybridization technique. In this method, labeled messenger RNA of the gene sought to be localized (or DNA produced from RNA by the enzyme reverse transcriptase) is brought into contact with chromosomal DNA, which is made single-stranded by appropriate techniques. The RNA or DNA sequences that match their DNA counterpart on the chromosome will hybridize at a specific chromosomal locus. This technique is currently available for genes with highly repetitive DNA sequences and for single-copy DNA, where m RNA or specific DNA probes are available. Whereas for highly repetitive DNA sequences radioactivity of the DNA probe is sufficient for demonstration of hybridization in one metaphase (see, for example, Sect. 7.2.2), a statistical approach must be used for identification of single-copy DNA sequences, e.g., genes whose mutation leads to hereditary diseases with simple modes of inheritance: Grains showing

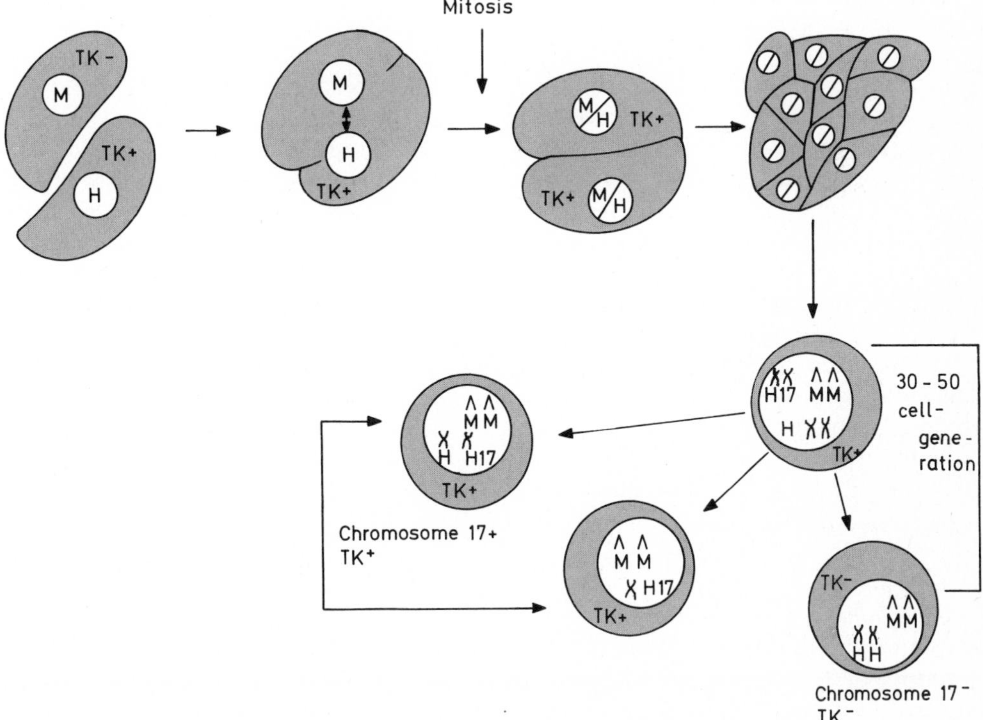

Fig. 3.26. The principle of gene localization on an autosome. Thymidine kinase-deficient mouse cells (M, TK^-) are grown in mixed cell culture with normal human cells (H, TK^+). The cells are fused spontaneously, chemically, or by Sendai virus. After 30-50 cell generations, the cells have lost part of their human chromosomes. Only cells having kept chromosome 17 show thymidine kinase activity (two cells at *left*). Cells without chromosome 17 show not *Tk* activity (cell at *bottom right*)

radioactive discharges must be counted in many metaphases (see Sect. 2.3.2.3).

DNA Polymorphisms and Gene Assignment. In recent years, detection of an increasing number of DNA restriction site polymorphisms (Sect. 2.3.2.7; 6.1.2) has opened up an additional approach to mapping of the human genome. Genes for important hereditary diseases could be localized on specific chromosomal segments when close linkage with a DNA restriction marker was established. In Fig. 3.24a a large pedigree with Huntington's disease is shown; the DNA marker, - and, hence, the chorea gene, - maps on chromosome 4. Model calculations have been performed [584; 754; 887] showing that only a few hundred DNA restriction sites distributed at random along the human genome are necessary for the entire human genome to be mapped and for at least one marker closely linked to the gene locus of a given hereditary disease to be found that can be used in genetic counseling and prenatal diagnosis (Sect. 9.1).
Recently, lymphoblastoid cell lines from large three-generation families with known genotypes for many RFLP loci have become available for the study of new markers [944]. With increasing numbers of available RFLP markers, analysis of linkage relationships not only between two gene loci (e.g., one disease gene and one RFLP marker) but also between a disease gene and a set of (more or less closely linked) markers, - a haplotype, - are becoming more and more important. Such haplotypes are being used increasingly in genetic counseling and prenatal diagnosis of thalassemia (Sect. 4.3). The fraction of kindreds in which the combination of genotypes is informative for linkage - and, hence, for prenatal diagnosis (see Sect. 3.4.2.) - can be enhanced appreciably by using such sets of markers.
In recent years, progress in establishing linkage - or synteny - and in assigning loci to certain chromosomes has been very rapid. Published conferences are held every two years and newsletters are circulated. In this rapidly evolving field of science centered around a special set of methods, scientists have created their own system of scientific interaction independent of "official" channels such as journals to keep pace with rapid progress.

Present Status of Gene Localization and Assignment to Autosomes. All the information noted above has been brought together in Figs. 3.27 and 3.28 and in Appendix 9, Table A9.5, which shows how the methods of mapping are symbolized.
Assignments and linkages not having any of these signs are regarded as confirmed.

Linkage of X-Linked Gene Loci. Assignment of loci to the X chromosome is no problem when the pedigrees show the typical pattern of X-linked inheritance. Assignment of genes to special segments of the X requires the newer techniques described above, supported by family studies. Most of the more than 115 X-linked loci have been assigned to this chromosome by pedigree analysis (and often confirmed by somatic cell techniques); a few only by somatic cell techniques. Many loci could be localized by these techniques to specific regions of the X; these studies were complemented and often confirmed by linkage studies in families. Two gene clusters were identified: The Xg cluster comprising the gene loci linked with the Xg blood group gene (31470), and the G6PD cluster. The Xg gene is located close to the tip of the short arm (band Xp 22.3) closely linked with the locus for X-linked ichthyosis (steroid sulfatase; 30810). The G6PD locus (30590) is located in band Xq28, not far from the tip of the long arm. The genes for hemophilia A (30670) and B (30690), for mental retardation with mar (X) chromosome (30955; Sect. 8.2.1.2), and for the enzyme HPRT (30800; Sect. 4.2.2.6) belong to this cluster, which also comprises the genes for protan (30390) and deutan (30380) color blindness. Figure 3.28 shows locations of gene loci. Loci at the tip of the short arm (Xg; X-linked ichthyosis) escape lyonization (Sect. 2.2.3.3); this segment of the X pairs with the Y chromosome during meiosis (Sect. 2.1.2.4). Moreover, many DNA probes have been isolated and identified in gene banks - mainly from sorted X chromosomes (Sect. 2.3.2.5) [624, 750]. It turned out from hybridization studies with such probes that the X has DNA sequences in common with the Y not only in the area on the short arm that pairs with the Y, but also in other areas [848]. These results are important for our understanding of the evolution of sex chromosomes and the mechanism of genotypic sex determination (Sect. 7.2.1).
In Fig. 3.28, the X chromosome is shown as an example for marking segments of a chromosome by restriction fragment length polymorphisms (RFLPs). Meanwhile, still more such polymorphisms have been discovered. Prenatal diagnosis using such RFLPs is being attempted for X-linked disorders (Sect. 9.1).

Unequal Distribution of Recombinational Events over the Length of Chromosome 1? With so much new information available, additional problems present themselves. The theoretically most important question posed by these data is whether genes with related functions show a tendency to cluster closely on the same chromosomes. This problem will be deferred to Sect. 3.5.

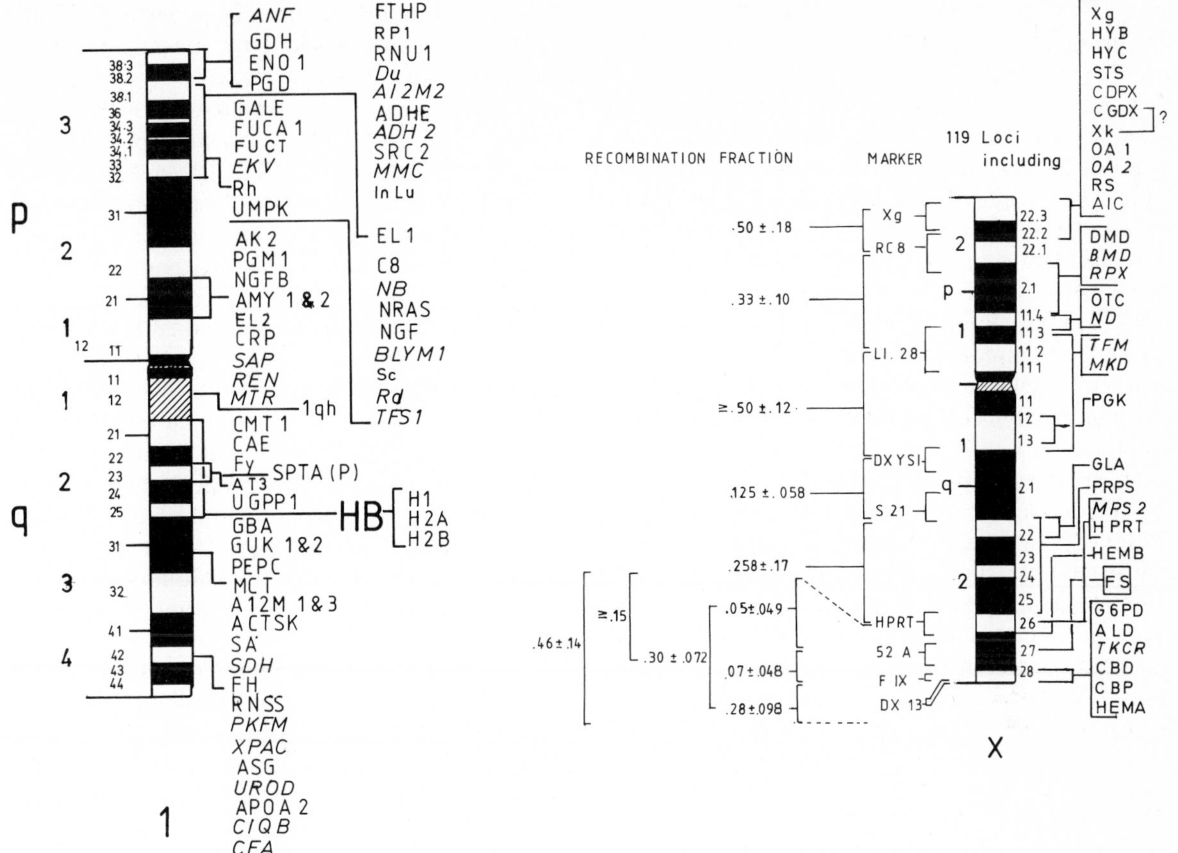

Fig. 3.27. Localization of genes on human chromosome 1. For explanation of abbreviations cf. Table A9.5. Provisional assignments are given in italics

Fig. 3.28. Gene loci on the human X chromosome. *Right:* Transcribed genes and their likely locations. *Left:* DNA restriction sites. For abbreviations, see Table A9.5

Another problem is whether map distances estimated from linkage studies in families conform with those estimated from physical assignment of gene loci to specific chromosome bands, for example by studies on man-mouse cell hybrids in which partially deleted human chromosomes have been kept. This problem was examined in chromosome No. 1, the most extensively mapped human chromosome. The evidence available so far points to a fairly good – though not perfect – agreement between localizations and distances based on such physical data and those based on family data [736].

Linkage Analysis with Genetically Ill-Defined, Quantitative Traits? In earlier linkage studies, quantitative traits with multifactorial inheritance were sometimes included. The idea behind these studies was that linkage would detect major genes influencing these phenotypes. Theoretically, this approach is certainly correct. If a measurable character shows fairly close linkage with a genetic marker, this indeed points to a major gene closely linked with this marker locus. If linkage can be shown for two measurable characters, both may be influenced by two major genes. Practically, however, great care is needed. The reason is twofold:

1. If many quantitative characters are included in such an analysis, chance alone will yield some statistical significance values that suggest linkage.
2. Linkage predicts correlations in families but not in the population. The prediction, however, depends on random mating. Some measurable graded characters sometimes are associated with assortative mating.

Results achieved so far, with such traits, are not encouraging. Such studies are conceivably marginally useful if a well-defined genetic marker is tested against a multifactorial trait. With the advent of many new DNA restriction markers, the chances of this approach are now increasing. Since linkage cannot discriminate between major and minor genes (i.e., the genetic background) that are involved in a given trait or disease (Sects. 3.6, 3.7), the analysis of data from such studies will be difficult.

DNA Variants in Linkage. The large number of DNA polymorphisms provides many new markers and most linkage work in the future will be carried out with DNA variants (Sects. 2.3.3.9, 6.1.2). When working with gene-specific probes (Table 2.13) such as hemoglobin as an example, the physical distance of a DNA polymorphism at the beta hemoglobin locus in relation to a beta hemoglobinopathy is so small that the possibility of a recombinational event can be ignored for practical purposes. On the other hand, when a linkage is found between a genetic disease locus and an "anonymous" probe developed from DNA of known or unknown chromosomal origin, it is unlikely that such linkage will be extremely tight. The same reasoning applies for linkage detected by a gene-specific probe and a disease locus that is biochemically unrelated to the probe. Thus, crossovers between the DNA marker and the disease gene will usually be found under these circumstances. Examples are Huntington's disease (G8 marker, 5 cM) and Duchenne muscular dystrophy markers (X-linked markers, 15 cM) [369, 667, 2306].

Linkage disequilibrium (i.e., failure of free assortment) has frequently been found between the sites of various RFLPs at a given locus. Since these sites are physically very close, crossovers between them will be rare and many generations must pass before linkage equilibrium will be reached. Furthermore, current data suggest that recombination rates at closely linked RFLP markers may vary considerably between different chromosomal locations. Thus, both "hot" and "cold" spots of recombination appear to exist [1097, 1959].

Practical Application of Results from Linkage Studies. Up to now, the main interest of linkage studies has been theoretical. Practical applications, however, are increasingly forthcoming. If, for example, gene A causes a rare hereditary disease manifesting itself later in life and B is a genetic marker closely linked to A and segregating in the same family, the disease can be predicted in a young individual, and this prediction may be utilized for genetic counseling. For example, the X-linked loci for hemophilia A and G6PD are closely linked. Therefore, information on a G6PD type can be utilized for calculating the risk to be heterozygous for the hemophilia A gene for the sister of a hemophilic patient. Let us assume that the mother is heterozygous G6PD A/B and that the proband is hemizygous A. If his sister is homozygous B, she must have inherited the allele B from her mother. Hence, she inherited the X chromosome carrying the allele B and very probably (in the absence of crossing over) the normal allele for factor VIII production. There is little or no risk for her sons to inherit the gene for hemophilia A. Prenatal diagnosis of hemophilia A follows similar principles. Since myotonic dystrophy and the secretor gene are linked, secretor typing has been used for detection of fetuses affected with myotonic dystrophy [895a] in intrauterine diagnosis. Fetal diagnosis of β-thalassemia has been achieved by identification of a nucleotide sequence polymorphisms closely linked to the Hb β gene [1253] in amniotic fluid cells. Such polymorphisms are used increasingly for prenatal diagnosis (Sect. 9.1).

This approach has several limitations, which are not always appreciated. Thus, there must be heterozygosity for the DNA marker. If in a given family, variation at the relevant DNA locus is not present, diagnosis becomes impossible. Fortunately, advances in this field are rapid, and an ever-increasing number of DNA variants is becoming identified. However, even if heterozygosity at the marker DNA locus exists, the family must be large enough to allow the necessary tests to be performed in several living family members, so that the appropriate inference of linkage of the marker gene with the disease gene can be made. Large kindreds are particularly favorable, but rarely exist in practice. X-linked diseases are more favorable for study than autosomal diseases, since males carry only a single X chromosome which simplifies the exact assignment of a linked marker to the disease gene (Fig. 3.28). In the X-linked lethal diseases such as Duchenne muscular dystrophy, and to a lesser extent in hemophilia (Sect. 9.1), there are many new mutations. Often one cannot be certain whether the new mutations occurred in the germ cells of the parental or of the grandparental generation. If the mutation happened in the parents, a patient's sister would not be at risk for the carrier state, while a 50% risk would apply if the mutation occurred in the grandparental generation. The resolution of this problem may be difficult, since biochemical carrier testing frequently does not give definite answers (see Appendix 8; Sect. 4.2.2.8).

It was shown in a large number of British families with Huntington's disease that only 15% had the appropriate family structure to detect the Huntington's disease gene in adults who were at 50% risk even if all families were informative for the DNA marker [2307]. This result was explained by the usual absence of living grandparents and the presence of few affected sibs. In contrast, a practically useful diagnosis in a fetus at risk (or in a child already born) in families where a grandparent had Huntington's disease was possible in 90% of cases. In about half such cases the disease could definitely be ruled out, while in the other half there would be a 50% risk of the disease. Definitive fetal diagnosis would be feasible only when definite prediction in the parent was possible (i.e., 15% of families, as above).

3.5 Gene Loci Localized Close to Each Other and Having Related Functions

3.5.1 Some Phenomena Observed in Experimental Genetics

Closely Linked Loci May Show a Cis-Trans Effect. When series of multiple alleles were analyzed in *Drosophila,* crossing over within these series was observed occasionally, indicating that what had been considered as one "gene" could be subdivided by genetic recombination. Such alleles were called "pseudoalleles" (McClintock, 1944 [782]). In some, a so-called *cis-trans* effect was shown. When two mutations were localized side by side on the same chromosome (*cis* position), the animal was phenotypically normal; when they were localized on homologous chromosomes (*trans* position), a phenotypic anomaly was seen [764].

Figure 3.29 shows one example. There are two mutations on the second chromosome of *Drosophila melanogaster,* the dominant S (star), and the recessive ast (asteroid). In the usual genetic analysis, these mutations behave like alleles. However, analysis of very large numbers shows a recombination frequency of 0.02% (0.02 cM). Genotypes S ast/+ + and S +/+ + are identical phenotypically. The eye in both is somewhat smaller than in the wild type and has a rough surface (Fig. 3.29 a). Genotype S +/+ ast, on the other hand, shows much more pronounced abnormalities. The eye is very small and rough, and the wings have abnormal veins (Fig. 3.29 b).

Explanation in Terms of Molecular Biology. The star-asteroid example has not been analyzed biochemically. In fungi, bacteria, and phages, however, genetic recombination is normally observed within functional genes, i.e., DNA regions carrying information for one polypeptide chain. A *cis-trans* effect is now considered to be typical for two mutations that are not able to complement each other functionally, i.e., that are localized within the same structural gene. Complementation between two mutations, by the same token, is regarded as an indication that these mutations are localized in different functional genes. This insight caused Benzer (1957) [569] to replace the "gene" as the unit of mutation, recombination, and function by three terms: the muton as the smallest unit of mutation; the recon as the smallest unit of recombination; and the cistron - derived from the *cis-trans* effect - as the smallest unit of function. In the usage of geneticists, these notions have not eliminated the word "gene," which is maintained with the meaning of "cistron" but has many mutational sites, and may be subdivided by recombination. Evidence is now accumulating that in higher organisms this functional gene may be much longer than the DNA sequence containing the specific information for a polypeptide. It includes long DNA stretches with regulatory function, which shows a *cis-trans* effect, as well as DNA segments at each end of the "structural" gene, which may also be involved in regulation (see Sect. 2.3.5).

A Number of Genes May Be Closely Linked. Close linkage has frequently been described between mutations affecting closely related functions, which are perfectly able to complement each other functionally and show no *cis-trans* effect. In bacteria like *E. coli,* gene loci for enzymes acting in one sequence were found to be closely linked and arranged in the sequence of their metabolic pathway. Their activity is subject to a regulating mechanism by a common operator and promoter [117].

3.5.2 Some Observations in the Human Linkage Map

Types of Gene Clusters That Have Been Observed. The first impression when scrutinizing the human

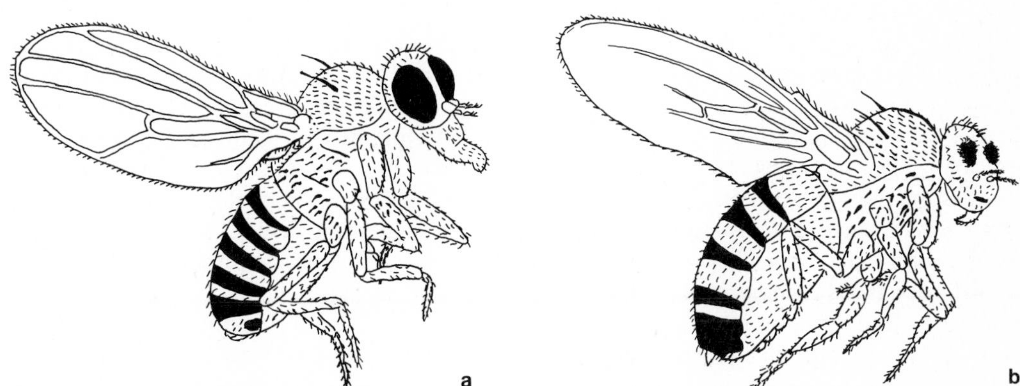

Fig. 3.29 a, b. *Cis-trans* effect of the mutations *S (star)* and ast *(asteroid)* in *Drosophila melanogaster.* **a** S ast/+ + (*Cis*-position); **b** S+/+ ast (*Trans*-position). Note the smaller eyes and the reduced veins on the wings in the animal carrying the two genes in *trans*-position. (Lewis, 1951 [764])

linkage map is that most loci are distributed pretty much at random. However, there are exceptions:

a) The loci for human hemoglobins γ, δ, and β, are closely linked.
b) The immune globulin region contains a number of loci responsible for synthesis of γ-globulin chains.
In both cases, incisive genetic analysis has been possible by establishing the amino acid sequence of proteins, and the differences between the closely linked genes have become known even at the level of base sequence within the transcribing strand of the DNA.
c) No less than four gene loci involved in the glycolytic pathway are located on chromosome 1.
d) A number of genes determining closely related enzymes seem to be closely linked, for example, pancreatic and salivary amylase on chromosome 1, and on the same chromosome, guanylate kinase 1 and 2.
e) While the protan and deutan loci for red-green color blindness are both located in the same cluster on the X chromosome, crossing over between them has been occasionally observed. Nevertheless, the data indicate that they are indeed located fairly close together.
f) Another quite distinctive group is represented by surface antigens mainly of the red blood cells and manifested by antigen-antibody reactions. Examples are the subtypes within the Rh blood group system.
g) Also exceptional is a cluster of genes, which are obviously related functionally, being involved in the immune response. This is the MHC (major histocompatibility complex) cluster and various components of complement on chromosome 6.

Clusters Not Observed So Far. As mentioned above, functionally related genes in bacteria are frequently closely linked; they are subject to common control within an operon. One would expect that, in humans, such operons might also occur. Linkage data known so far, however, do not give any positive evidence. Two genes linked in the same operon in bacteria are those for galactose-1-phosphate uridyl transferase and galactokinase. In humans, these genes are localized on chromosome 3 and chromosome 17, respectively. Similarly, the gene for G6PD is located on the X chromosome, that for 6-PGD, the following enzyme in the shunt pathway, is situated on autosome 1. Attempts in man to find the types of regulator mutations so frequently encountered in bacteria have failed so far. Does this mean that the Jacob-Monod model of gene regulation so successfully analyzed in bacteria does not apply to higher organisms like humans? The problem will be taken up in Sect. 4.7.

3.5.3 Why Do Gene Clusters Exist?

They Are Traces of Evolutionary History. In some cases, clustering has simply been left over from the evolutionary history of these genes. Early in evolution, there was one locus for a given gene. Then, gene duplication occurred and offered the opportunity of functional diversification. The first duplication paved the way for further duplications due to unequal crossing over (Sect. 3.5.8) and, hence, for further functional specialization.
With no further chromosomal rearrangements, the gene clusters remain closely linked. It is unknown whether, in these cases, close linkage is necessary for orderly function. While it may be so in some cases, this explanation is not needed to explain clustering. Explanation from evolution is sufficient. Examples are the closely linked isoenzyme loci or the various hemoglobin chains. For some isoenzymes, genes are localized at different chromosomes, for example, lactate dehydrogenase (LDH) A on No. 11, LDH B on No. 12. This may be due to polyploidization during early evolution or to a chromosomal rearrangement some time after gene duplication. The human Hb α genes are obviously related by descent to the δ, β, and γ genes, and yet they are not linked.

Color Vision Genes on the X Chromosome. The protan and deutan genes appear also to have arisen by gene duplication. Here, analysis at the protein level has not been possible so far, but ingenious experimentation with methods from sensory physiology, taking advantage of the unique possibilities offered by the human eye, has come very close to this level.
The Young-Helmholtz theory, developed in the nineteenth century, assumed three elementary mechanisms for color vision: one with maximum sensitivity for red, another for green and a third for blue-violet. The three main types of defect in color vision are explained by deficiencies in one of these mechanisms. Red and green deficiencies are fairly common; the blue-violet is extremely rare and does not concern us here [214]. New experiments mainly using reflection densitometry supplemented by microbeam experiments in retinas have shown that red and green sensitivity is mediated by two different pigments. These are localized in the cones of the retina, each cone containing only one type of pigment. Protan and deutan color vision defectives lack one of these two pigments completely; the pigments of intermediary types, protanomaly and deuteranoma-

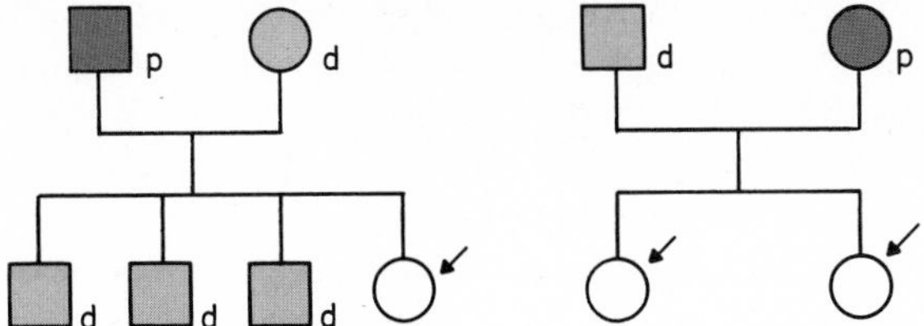

Fig. 3.30, 3.31. Two families in which normal daughters were born in a marriage of two individuals with different X-linked color vision defects p and d. Note that the three daughters indicated by *arrows* have normal color vision in spite of the fact that they inherited a color vision defect from each parent. There is complementation and such double heterozygotes have normal color vision. The fact that the mothers in both matings are homozygotes derives from X-linked inheritance of color vision defects. Only homozygous females manifest with color vision defects. (Franceschetti and Klein, 1957 [668])

ly, show altered absorption spectra of their pigments [863].

It is known from analysis of pedigree data that there are two series of alleles, one for the protan, the other for the deutan series. Pedigrees such as those in Figs. 3.30, 3.31 show that they are independent genetically. Occasional observations, however, seem to point to rare mutations not showing full complementation [668].

According to recent results of molecular genetics, the protan and deutan genes originated from one gene by duplication and mutations appear to be mostly due to unequal crossing over or gene conversion [825a].

Duplication and Clustering May Be Used for Improvement of Function. In the examples above, the clustering of genes was without obvious functional significance. It would be surprising, however, if evolution were not to take advantage of this situation, combining products of such gene clusters to form higher functional units. This might be the case for the hemoglobin molecule, since in the β cluster, the ε, γ, β, and δ genes are arranged in the sequence of their successive activation during individual development (Sect. 4.3). In the immunoglobulins close linkage of a number of genes, possibly a great many, has become important functionally (Sect. 4.4), as the gene products determined by these genes combine to form various classes of functional molecules.

3.5.4 Blood Groups: Rh Complex, Linkage Disequilibrium

History. Levine and Stetson (1939) [762] discovered a special antibody in the serum of a woman who had just delivered a macerated stillborn child and had received blood transfusions from her AB0-compatible husband. Of 101 0 bloods, only 21 showed a negative reaction with this antibody. There was no association with AB0, MN, or P systems.

Landsteiner and Wiener (1940) [753] immunizing rabbits with blood of Rhesus monkeys, obtained an immune serum that gave positive reactions with the erythrocytes of 39 of 45 individuals. Later on, the antibodies were compared with those of Levine and Stetson and thought to give reactions with the same antigens. Still much later, this was found to be not quite true, and now the antigen uncovered by the true anti-Rhesus antibody is called LW - in honor of Landsteiner and Wiener. Rh typing in man is always carried out with sera of human origin, i.e., according to Levine and Stetson's observation. The following discussion relates only to reactions with these human sera.

The great practical importance of the Rhesus system became apparent when transfusion accidents were traced to this antibody, and especially when erythroblastosis fetalis, a hemolytic disease of the newborn, could be explained by Rh-induced imcompatibilities between mother and fetus. The red blood cells of about 85% of all whites give positive reactions; family examinations showed that the Rh-positive individuals are homozygous Rh/Rh or heterozygous Rh/rh, whereas the rh-negative individuals were homozygous rh/rh.

In 1941, Wiener discovered a different antibody that reacted with the cells of 70% of all individuals and was independent of the basic Rh factor (Rh′, according to Wiener). A third related factor was discovered in 1943. These three factors were found in all possible combinations with one another, and the combinations were inherited together. Wiener brought forward the hypothesis that these serologic "factors" are properties of "agglutinogens" and that these agglutinogens are determined by one allele each of a series of multiple alleles. Hence, the agglutinogens were thought to determine the factors in different combinations. This descriptive hypothesis is so general that it indeed explains all the complexities discovered later. There are research workers who believe that more cannot be said about the internal order of the Rh locus until biochemical analysis is successful.

Fisher's Hypothesis of Two Closely Linked Loci. R. A. Fisher developed a more specific hypothesis. At that time, another antibody had been detected, anti-Hr. In 1943, Fisher examined a tabulation prepared by Race, containing the data accumulated so far. He recognized that Rh′ and Hr were complementary. All humans have either Rh′, Hr, or both antigens. Individuals having both antigens never transmit them together to the same child, and a child always receives one of the two. Fisher explained these findings by assuming one pair of alleles for the two antigens. The pair was named C/c. In analogy, an additional pair of alleles D/d was postulated for the original antigens Rh+ and rh−, and a third pair of alleles for the third factor that had been discovered. To explain the genetic data, close linkage between these three loci was assumed.

Fisher's hypothesis predicted discovery of the two missing complementary antigens d and e. This prediction was confirmed later on for e but not for d. Apparently, this area does not define a structure leading to antibody formation. In developing this hypothesis, Fisher went one step further. In the British population, three classes of frequency of the Rh gene complexes are observed (Figs. 3.32 and 3.33). Fisher explained this finding by assuming that the rare combinations could have originated from the more frequent ones by occasional crossing over. All four combinations of the less common class may have originated from occasional crossing over between the most frequent combinations; not however, CdE. This complex needs inclusion of a second-order chromosome. Hence, the hypothesis explains why CdE is so rare. Still another prediction is possible. In every crossing over leading, on the one hand, to Cde, CDE, or cdE, the complex cDe must also be produced. It follows that the frequency of the three first-mentioned combinations together should equal the frequency of cDe. Frequencies actually found were: cDe, 0.0257; Cde+cdE+CDE, 0.0241 (in blacks, however, cDe has a high frequency).

Furthermore, Fisher assumed that the sequence of the three loci is D-C-E, since cdE – which must have originated by crossing over between D and E from the genotype cDE/cde – is much more frequent in comparison with this genotype than Cde in relation to genotype CDe/cde (crossing over between C and D) and CDE in relation to CDe/cDE (crossing over between C and E).

Confirmation and Tentative Interpretation of the Sequential Order. In the 30 years following Fisher's hypothesis, many new observations were made, the most important for the question of sequence being the combined antigens, for example, ce. These compound antigens all seem to be compatible with the sequence D-C-E, whereas no such antigen suggesting close linkage between D/d and E/e has turned up. Fisher's hypothesis leads to two questions:

1. If the rare types have been formed by occasional crossing over from the more frequent ones, cases

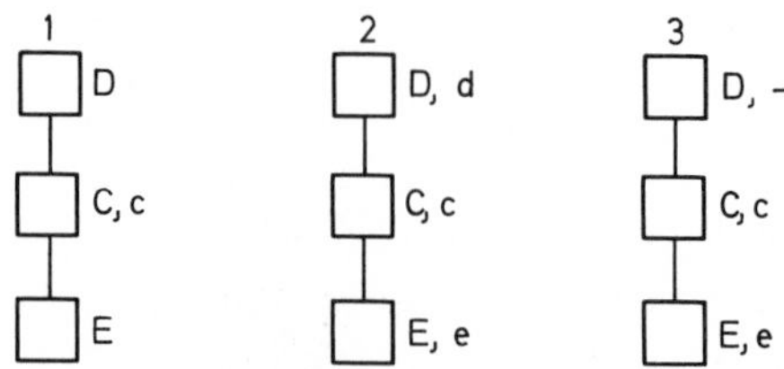

Fig. 3.32. A hypothetical structure of the Rh complex. *1,* On the basis of the evidence known in 1941; *2,* antigens predicted by Fisher and Race; *3,* antigens discovered; antigen *d* was not found

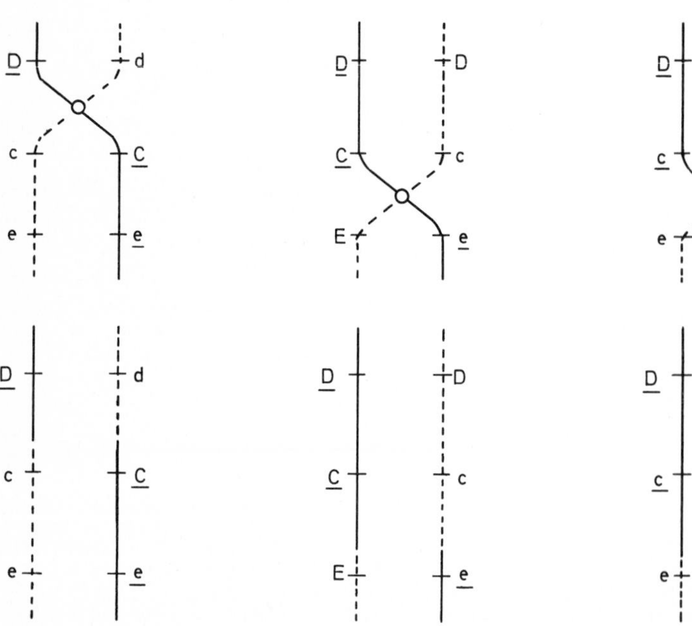

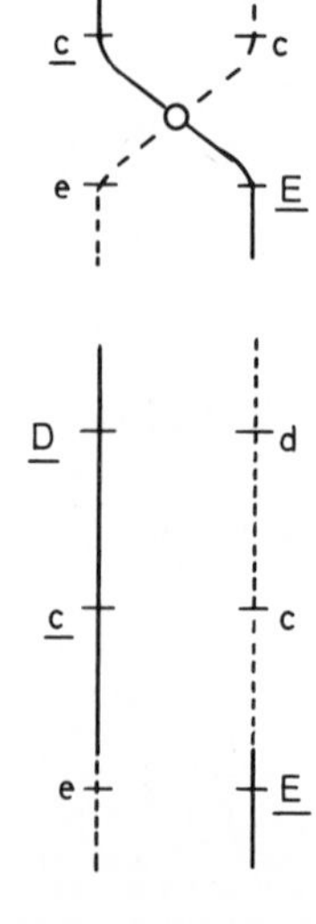

Fig. 3.33. Postulated production of three rare Rh haplotypes from the more common ones by crossing over. (Race and Sanger [166]) Each diagram refers to a different crossing over event

of crossing over should occasionally turn up in family studies. One such family has, indeed, been reported [896]: the father was CDe/cde, the mother cde/cde, four children were cde/cde, and three others CDe/cde, all in concordance with genetic theory. The sixth-born child, however, was Cde/cde. As the discrepancy involved father and child, it could be argued that the child was illegitimate. This, however, was made unlikely by blood and serum groups and by the fact that the family belonged to a sect with especially strict rules against adultery. It is surprising that no more families of this sort have been observed. Many routine laboratories, however, will possibly not follow up such atypical cases but will rather assume an occasional technical error.

2. How should we envision the structure of the Rh gene(s) in the light of evidence from molecular genetics? There are two possibilities in principle:

a) The Rh complex is one cistron with many mutational sites. Mutational changes are expressed as antigenic differences.

b) The Rh complex is composed of a number of closely linked cistrons, possibly three, and the main antigens reflect genetic variability at these cistrons. In absence of any conclusive biochemical evidence, this question cannot be answered. However, one important criterium is the *cis-trans* effect found in mutations affecting the same functional gene (Sect. 4.8). As the ce compound antigen can be found in *cis*-phase CE/ce but not in *trans*-phase Ce/cE, Race and Sanger (1969) [846] have tentatively concluded that C/c and E/e might be localized within the same functional gene.

During recent decades, many more observations have been added, and evidence has accumulated that not only qualitative expression of Rh cases but quantitative expression as well is under strict genetic control. Furthermore, while little or no evidence seems to be available on the biochemical basis of Rh specificity, we know much more on the structure of the cell membranes in which Rh specificity - together with other blood groups specificities - is located. Rosenfeld et al. (1973) [860] have attempted to combine this evidence in a new model on the structure of the Rh locus. According to this model, this locus consists of a number of areas carrying the information for membrane polypeptides (structural genes). These areas are under control of a common operator or promotor that regulates quantitative expression, possibly via a number of operator regions close to the single structural gene loci. This model combines Fisher's concept with more recent results from molecular biology. So far, however, it has no biochemical support.

Linkage Disequilibrium. In the process of developing his hypothesis on the Rh complex, Fisher also created another concept: linkage disequilibrium. Linkage normally does not lead to association between certain traits in the population (Sect. 3.4.1). Even if the population starts out with a nonrandom distribution of phases, repeated crossing over will randomize the linkage groups, and in the end the coupling and repulsion phases for two linked loci will be equally frequent. There is linkage equilibrium. However, when the population starts out with a deviation from this equilibrium, the time within which equilibrium will be reached depends on the closeness of linkage: the closer the linkage, the longer the time until equilibrium is reached. It will never be reached if certain types have a selective disadvantage.

A selective disadvantage for certain Rh complexes that could lead to their becoming less frequent has not been demonstrated so far; selection works against heterozygotes (Sect. 6.2), but this does not mean that a general disadvantage has never existed, neither has a conclusive explanation in terms of population history been brought forward. Fisher's hypothesis, by answering some questions, has posed a number of others. However, the concept of linkage disequilibrium was shown to be still more important in the genetic analysis of DNA polymorphisms (Sect. 6.1) and the major histocompatibility complex.

3.5.5 *Major Histocompatibility Complex (MHC)* [193; 188]

History. It had long been known that skin grafts from one individual to another (allotransplants) are usually rejected after a short time. K. H. Bauer (1927) [562] observed that rejection does not occur when skin is transplanted from one monozygotic twin to the other (isotransplant). Such a transplant is accepted just like a transplant in the same individual (autotransplant). Hence, the rejection reaction was shown to be genetically determined. In the years following, skin - and later on kidney - transplantations between monozygotic twins were occasionally reported. Research on histocompatibility antigens in humans started only when leukocytes were shown to be useful as test cells.

Dausset (1954) observed that some sera of polytransfused patients contained agglutinins against leukocytes. Later on, he showed that sera from seven such patients agglutinated leukocytes from about 60% of the French population, but not the leukocytes of the patients themselves. Twin and family investigations soon established that these isoantigens were genetically determined. Other isoantigens (now part of the HLA-B) were discovered by van Rood. Another important achievement was the microlymphocyte toxicity test introduced by Terasaki and McClelland (1964) [911], which is now the most frequently used method (Figs. 3.34 and 3.35). Subsequently, the number of leukocyte antigens increased rapidly, and in 1965 it was suggested that most of these antigens belong to one and the same genetic system. At the histocompatibility workshop in 1967, 16 different teams typed identical samples from Italian families. Here, the basic relationships among the different antigens were established. Finally, Kissmeyer-Nielsen [739] proposed the hypothesis of two closely linked loci (now A and B) controlling two series of alleles.

Social Phenomenon: Formation of a "Paradigm Group." Meanwhile, the same sociologic phenomenon noted for linkage studies by cell hybridization had become apparent in histocompatibility research as well. A group that maintained close contacts was formed. Special international meetings were organized, direct exchange of information was intensified, and the group's own journal was founded. In this process, the third histocompatibility workshop organized by Ceppellini in 1967 played a major role. Contacts among research workers had to be especially intensive, as HLA typing depends vitally on exchange of antisera between the groups. A study on

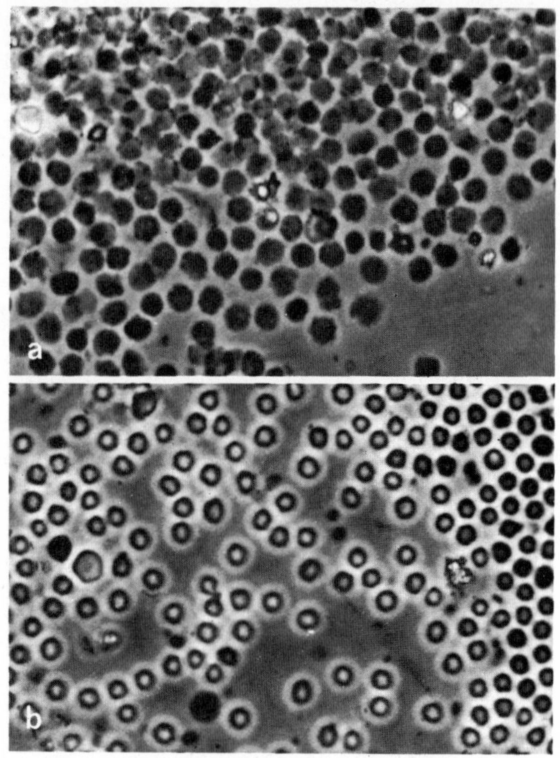

Fig. 3.35 a, b. Lymphotoxicity test **a** positive and **b** negative reaction. Positive reaction is indicated by staining of the cells. Courtesy of Dr. Greiner

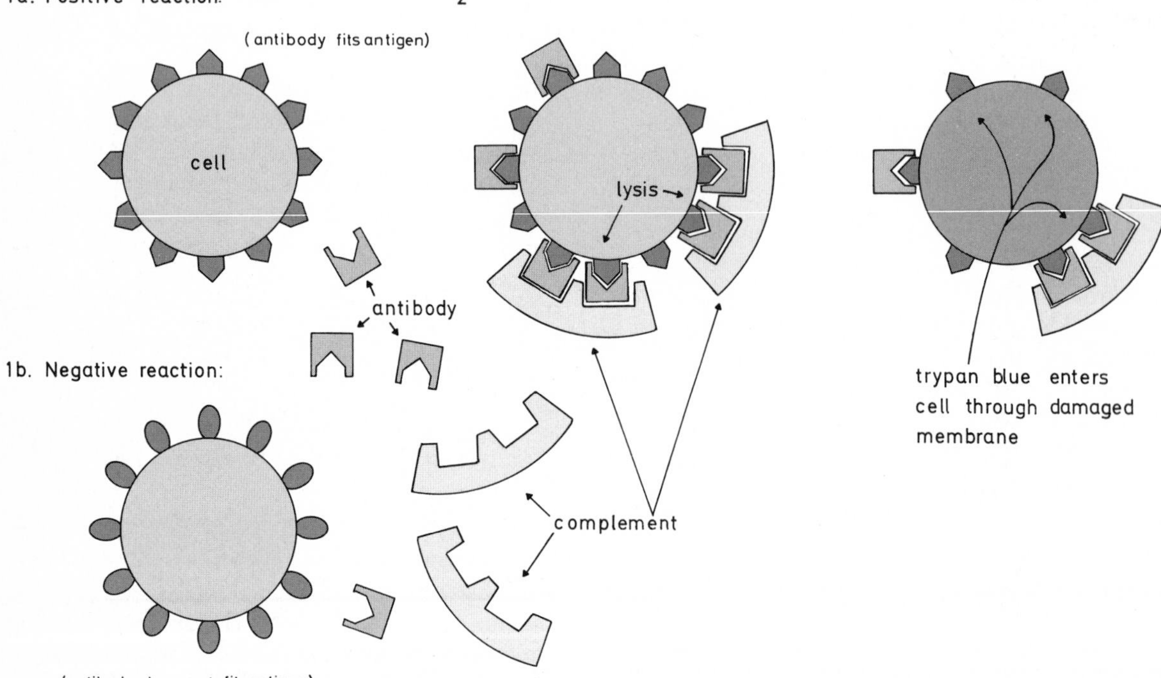

Fig. 3.34. Principle of the lymphocytotoxicity test: A cell having an appropriate antigen reacts with a specific antibody and complement. As a result, trypane blue enters the cell through the damaged membrane and indicates that the cell surface antigen has been recognized by a specific antibody

this "paradigm" group in the late 1960s and early 1970s and the role played by research workers such as Bodmer, Dausset, Ceppellini, Kissmeyer-Nielsen, van Rood, and Terasaki in the development of its activities will be of much interest for the history and sociology of modern biologic research. Rapid progress in this field was fueled not only by the intrinsic scientific interest but also by the hope that success of organ transplantation could be improved.

Main Components of the MHC on Chromosome 6. The linkage group of the MHC is seen in Fig. 3.36. There are now four loci of the basic HLA system: A, C, B, and D - in that order (for the D locus see next section). For each of them, we know many alleles that can be identified by specific antisera. A survey, together with allele frequencies is given in Table 3.8. For many other frequencies from various populations, see Terasaki (1980) [910]; Albert et al. (1984) [553]. The gene frequencies do not add up to 100%, as some antigens are still unknown (indicated as blank). The concept of four series of alleles is based on the following lines of evidence:

a) No individual possesses more than two antigens from each of the four series.
b) Recombination between these series has been observed, for example, between the loci for A and B series, 40 crossovers among 4614 meiotic divisions were described up to 1975, giving a combined (♀ + ♂) recombination frequency of 40/4614 = 0.0087 = 0,87 cM. Ten A-B recombinants informative for the C series have been reported. In eight of these, the C antigen followed B, in two, it followed A. Therefore, C is localized between A and B and probably a little closer to B.
c) When two antigens from the same series are present together in a parent, he or she always transmits one of them - never both or none - to the child. The segregation ratio is 0.5, corresponding to a simple codominant mode of inheritance.
d) Hardy-Weinberg proportions have been demonstrated for each of the three allele series separately in fairly large population samples.
e) Serologic cross reactions occur almost exclusively within the series, not between them. This points to a close biochemical relationship of the antigens within a given series. Figure 3.37 shows transmission en bloc to four of the five children examined and crossing over between A and C in a fifth child.

Mixed Lymphocyte Cultures: Typing for HLA-D Alleles. When lymphocytes from two individuals are mixed and cultured in vitro, they will usually stimulate each other to divide. This reaction is due to the fact that lymphocyte surfaces have antigens as well

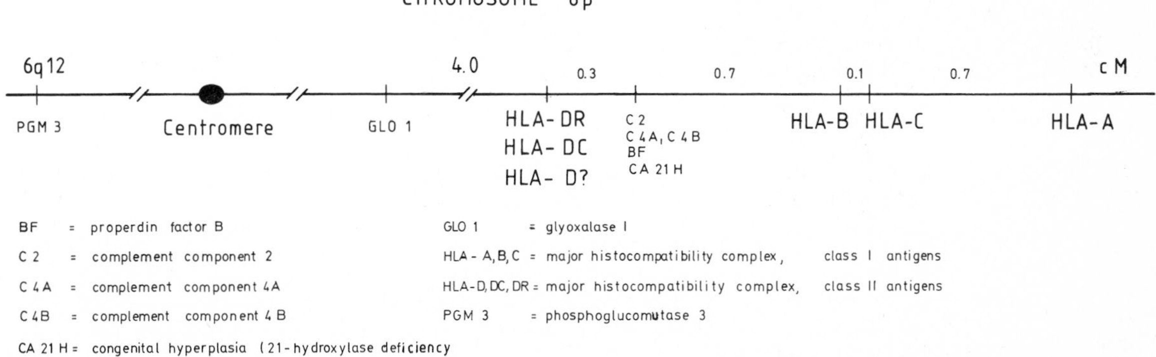

Fig. 3.36. Linkage group of MHC on chromosome 6. The HLA complex is ≈15 cM (centimorgans) away from the gene locus for PGM_3 and 10 cM from the locus for the enzyme glyoxalase (GLO). Within the HLA complex, the sequence D-B-C-A is the most likely. In the same region, other gene loci involved in the immune response are located, for example, those determining complement factors (C_2, C_4, C_8, Bf)

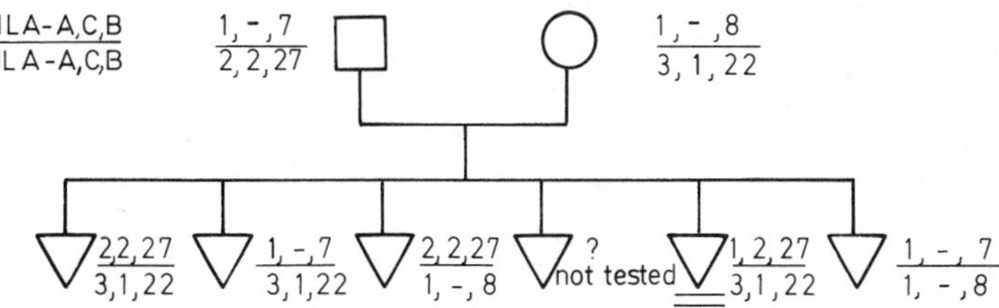

Fig. 3.37. Predigree with crossing over between the HLA-A and HLA-C genes within the MHC. Crossing over between A and C must have occurred in the father's germ line, making for the haplotype 1, 2, 27 in the fifth child. (Svejgaard et al., 1975 [193])

Table 3.8. Nomenclature for factors of the HLA system[d] and their gene frequencies among European Caucasoids[e]

HLA-A	Gene frequency	HLA-B	Gene frequency	HLA-C	Gene frequency	HLA-D	Gene frequency[f]	HLA-DR[c]	Gene frequency
A1	0.149	B5		Cw1	0.041	Dw1	0.069	DR1	0.069
A2	0.260			Cw2	0.051	Dw2	0.075	DR2	0.134
A3	0.116			Cw3	0.101	Dw3	0.082	DR3	0.108
A9		B7	0.088	Cw4	0.121	Dw4	0.053	DR4	0.096
A10		B8	0.082	Cw5	0.060	Dw5	0.056	DR5	0.103
A11	0.059	B12		Cw6	0.079	Dw6	0.103	DRw6	0.022
Aw19		B13	0.028	Cw7	0.023	Dw7	0.103	DR7	0.125
Aw 23(9)	0.023	B14	0.030	Cw8	0.019	Dw8	0.031	DRw8	0.027
Aw 24(9)	0.096	B15		C –[b]	0.506	Dw9	0.014	DRw9	0.011
A25	0.019	Bw16				Dw10	0.026	DRw10	0.007
A26	0.037	B17				Dw11	not tested	DR –[b]	0.298
A28	0.040	B18	0.058			Dw12	not tested		
A29	0.038	B21				D –[b]	0.388		
Aw30	0.024	Bw22							
Aw31	0.027	B27	0.039						
Aw32	0.045	Bw35	0.095						
Aw33	0.017	B37	0.015						
Aw34	0.006	Bw38[a](w16)	0.025						
Aw36	0.003	Bw39[a](w16)	0.021						
Aw43	0.000	B40							
A –[b]	0.043	Bw41	0.010						
		Bw42	0.003						
		Bw44[a](12)	0.110						
		Bw45[a](12)	0.011						
		Bw46	not tested						
		Bw47	0.004						
		Bw48	0.005						
		Bw49[a](w21)	0.023						
		Bw50[a](w21)	0.012						
		Bw51 (5)	0.072						
		Bw52 (5)	0.015						
		Bw53	0.009						
		Bw54[a]	0.000						
		Bw55[a](w22)	0.022						
		Bw56[a](w22)	0.006						
		Bw57[a](17)	0.031						
		Bw58[a](17)	0.011						
		Bw59	0.005						
		Bw60[a](40)	0.034						
		Bw61[a](40)	0.017						
		Bw62[a](15)	0.053						
		Bw63 (15)	0.005						
		B –[b]	0.061						
		Bw4	0.411						
		Bw6	0.589						

[a] These specificities are clear-cut splits of the antigens noted in parentheses.
[b] Alleles which are not yet detectable. These frequencies are calculated by subtracting the total frequencies from one.
[c] DR, D related. These specificities can be identified by serological techniques on B lymphocytes.
[d] Terasaki PI (1980) Histocompatibility testing 1980. UCLA Tissue Typing laboratory, Los Angeles, pp 18–20.
[e] Baur MP, Danilovs JA (1980) Population analysis of HLA-A, B, C, DR and other genetic markers: In: Terasaki PI (ed) Histocompatibility testing 1980. UCLA Tissue Typing Laboratory, Los Angeles, pp 955 ff.
[f] Dupont B, Braun DW Jr, Yunis EJ, Carpenter CB (1980) HLA-D by cellular typing. In: Terasaki PI (ed) Histocompatibility testing 1980. UCLA Tissue Typing Laboratory, Los Angeles, p 249.

as receptors for foreign antigens. In the one-way mixed lymphocyte culture (MLC) test, the stimulating lymphocytes are prevented from dividing by X irradiation or treatment with mitomycin C (Fig. 3.38). The phenotype of the "responder" can be assessed through stimulation by a number of different "stimulator" cells with known genotype.

Typing for antigens of the HLA-D locus is being done by this mixed lymphocyte culture (MLC) typing. New methods have been developed that are based on the phenomenon of immunologic memory, which can be created in vitro by mixed leukocyte culture. The test for the presence of HLA-D locus antigens is based on a rapid and strong anamnestic response upon restimulation with the antigen, which was used for the immunization in vivo. This test has been called "primed lymphocyte typing" (PLT).

Independently of these procedures, HLA-D antigens can also be typed by standard lymphotoxicity tests performed on B-lymphocyte enriched cell suspensions. In contrast to the antigens that are coded for the HLA-A, -B, and -C genes and are expressed on the surface of T and B cells, HLA-D antigens are predominantly expressed on B cells and macrophages. It remains open at this time whether HLA-D antigens detected by MLC typing are *completely* identical to those defined by serology. Figure 3.39 shows a model on the biochemical nature of HLA proteins and their location in the cell membrane. The HLA region has also been analyzed at the DNA level by recombinant DNA techniques (see Sect. 2.3); genes of the major HLA antigens and related genes and pseudogenes have been identified and, in part, sequenced, and restriction site polymorphisms have been identified [621, 652, 839].

Complement Components. Complement consists of a series of at least ten different factors present in fresh serum. The factors are called C1, C2, C3, etc., and the first of them is activated by antibodies that have reacted to their corresponding antigens. Then, C1 activates C4, this activates C2, and so on. The end result of this "complement cascade" is damage to the cell membrane carrying the antigen and often lysis of the cell. Moreover, activated complement components have a number of other biologic properties, such as chemotaxis or histamine release. They are important immune mediators in the body's defense against microbial infection.

The complement system can be activated not only via C1 (the classic pathway) but also via C3 through an alternative pathway involving the "properdin factors." The factor B(BF) acts as "proactivator" for C3.

For some of the complement factors, hereditary deficiencies have been described and polymorphisms are known. C2, C3, C4, and possibly C6 are polymorphic. The loci of C2 and C4 belong to the linkage group of the major histocompatibility complex discussed here as does the properdin factor B with the main alleles BF^F and BF^S. The locus for C3, on the other hand, with the alleles $C3^F$ and $C3^S$, is located elsewhere.

Immune Region-Associated Antigens. In Sect. 3.7.3, the evidence for immune region-associated antigens in the mouse and their possible role in human HLA disease associations will be discussed. Experimental evidence appears to place such genes in close proximity to the HLA-D/DR locus.

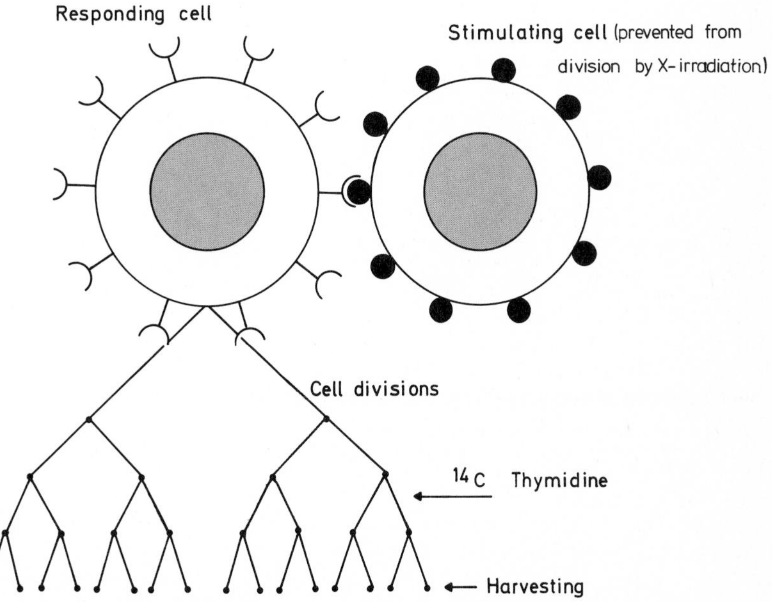

Fig. 3.38. Principle of the mixed lymphocyte culture technique. The stimulating cells, which have been prevented from dividing by X irradiation, react with the responding cells to start dividing. Division can be assessed by incorporation of thymidine. A negative reaction (no appropriate receptor on the cells to be examined) would mean no ^{14}C intake

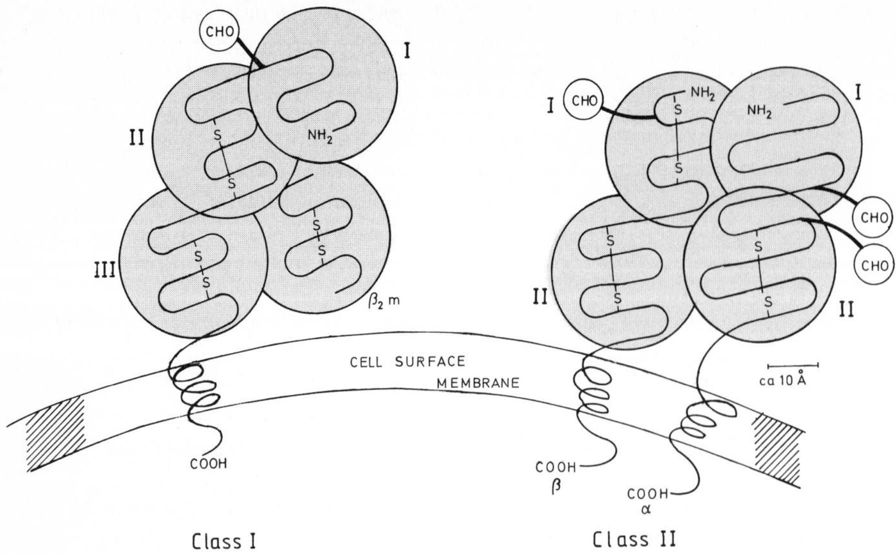

Fig. 3.39. Arrangement of class I (HLA-A, B, C) and II antigen (HLA-D/DR) domains which may yield similar overall structures. Roman numerals denote domains. *β_2m, β*-microglobulin; *CHO*, carbohydrate. (Kämpe et al., 1983 [729a])

Linkage Relationships with Other Markers. In 1971 linkage of the MHC with the gene for PGM3 (phosphoglucomutase-3) was shown. The map distance from HLA is about 15 cM in males and 30-45 cM in females. The locus for PGM3 is located on the long arm of chromosome 6 (Figure 3.36). PGM3 seems to be closer to the B than to the A locus. Another enzyme locus, glyoxalase I, is localized between PGM3 and HLA. The linkage with PGM3 made it possible to assign the whole linkage group to chromosome 6 by cell hybridization; probably within 75 cM from the centromere.

Svejgaard et al. (1975) [193] have noted an interesting parallel with the Rh system: the Rhesus complex and the PGM1 locus are localized at a distance of 35 cM (in the male) on chromosome 1. It is possible that these two linkage groups have one common ancestor in evolution. In this case, the Rh complex could have developed into a surface antigen system specialized to erythrocytes and the HLA system into a similar system not for erythrocytes, but for many other types of cells. Humans and cattle have two such systems; mice and chickens, on the other hand, have only one complex blood group system, which controls antigens on erythrocytes as well as on leukocytes.

Significance of HLA in Transplantation. One of the main motives for rapid development of our knowledge of HLA antigens has been the hope to improve the survival rate of transplanted organs, primarily kidneys. Indeed, kidneys from HLA-identical and AB0-compatible siblings have a survival rate in the recipient almost equaling that of monozygotic twins. Survival rate is worse in unrelated recipients even if HLA matching is as perfect as possible and AB0 compatibility is secured. This shows that, apart from the major histocompatibility system – the HLA system – there must be other systems of importance for graft survival. This is not surprising. A great number of such systems are known in the mouse. These systems lead to host-versus-graft reactions in almost all transplantations (Fig. 3.40). These reactions, however, can frequently be handled by immunosuppressive therapy. The chances for survival, and the survival times, of transplanted kidneys have increased substantially (Table 3.9).

Considering the high degree of polymorphism and the low gene frequencies of HLA alleles, successful matching of potential recipients with donor kidneys from others than sibs requires large-scale international organizations. Such organizations have now been founded. Still, the results of transplantations are not quite as good as expected, but further research on histocompatibility is likely to improve them.

Linkage Disequilibrium. One of the most conspicuous properties of the HLA system is that some HLA alleles tend to occur more frequently together than expected by chance. Table 3.10 shows some examples. The A1, B8 haplotype, for example, occurs about five times as often as expected.

Consider two alleles at two linked loci, with frequencies p_1 and p_2. With free recombination between them their combined frequency, i.e., the haplotype frequency h, should be $p_1 \times p_2$. If such a result is obtained, the two loci are said to be in linkage equilib-

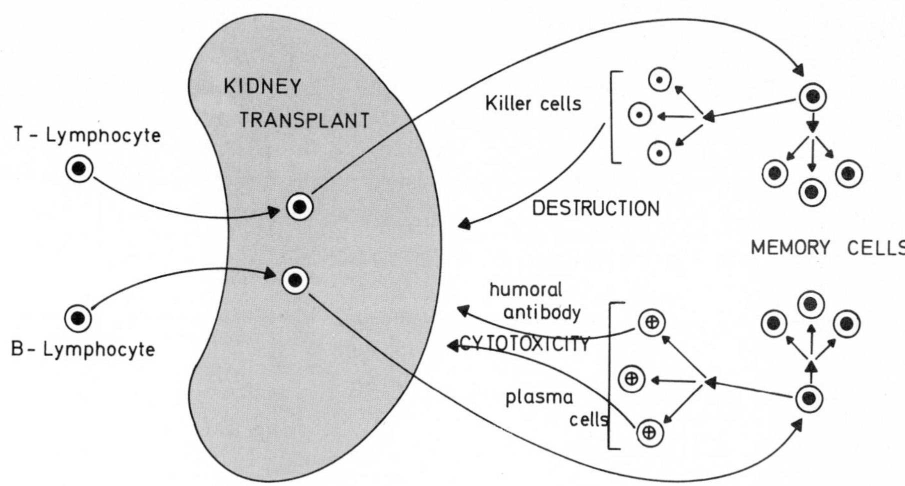

Fig. 3.40. Simplified diagram of the activation of the immune system by a kidney allotransplant. The transplant is recognized as foreign to the host organism by its T- and B-lymphocytes. This leads to activation of cellular and tumor immune response. (Svejgaard et al., 1975 [193])

Table 3.9. Influence of HLA in transplantation [889]

	Donor	Recipient	Skin graft (mean survival in days)	Kidney graft (1-year graft survival) %	Bone marrow graft
	a/c→	*a/c*	20.0	90	Often successful
Siblings	*a/d*→	*a/c*	13.8	70	Failure
	b/d→	*a/c*	12.5	60	Failure
Unrelated	*x/y*→	*a/c*	12.1	50	Failure

Letter *a, b, c, d, x,* and *y* indicate different HLA haplotypes. All transplants were ABO compatible. Skin graft data are from Ceppellini et al. and kidney data are approximate values based on Opelz et al. and Thorsby et al.

rium. If the haplotype frequency h is higher than expected with free recombination there is linkage disequilibrium (Δ), which is often symbolized as $\Delta = h - p_1p_2$. Haplotype and gene frequencies can be estimated from family and population data. In families, the haplotypes of the parents can in most cases be derived from their children. In the family shown in Fig. 3.37, for example, one of the maternal haplotypes should be 3, 1, 22, as she has transmitted it to three of her children. Frequencies of the single alleles can be deduced from the same sample of unrelated parents, and the extent of linkage disequilibrium Δ can be calculated. In a sample from a random mating population, deviation from linkage equilibrium can be tested in a 2×2 table χ^2 test as shown in Table 3.11 for a Danish sample (data and calculation from Svejgaard, 1975 [193].

In the HLA system, the deviations from linkage equilibrium are indeed striking. The situation is similar to that encountered with the Rh system (Sect. 3.5.4), but there is one important difference: In the Rh system only one case of recombination has been discovered, whereas many cases are known for the HLA system. Hence, linkage seems to be much closer in the Rh system. If the one case of crossing over can either be discarded or be explained by intracistronic crossing over in the Rh system, the possibility still remains that the Rh chromosomes are, indeed, true alleles, the specificities D, C, c, E, e representing characteristic sites within the same polypeptide. The HLA loci, on the other hand, are further apart from each other.

Table 3.10. Linkage disequilibrium (gametic association) [889]

Haplotype			Frequency, %	
A	B	D	Observed	Expected
A1	B8		9.8	2.1
A3	B7		5.4	2.1
	B8	Dw3	8.6	1.4
	B7	Dw2	3.9	1.8

The expected haplotype frequencies were calculated under the assumption of no association.

The observation of linkage disequilibrium - together with identification of immune response (Ir) genes in the mouse - initiated the investigations of HLA associations with diseases that proved to be successful in recent years (Sect. 3.7.3) as mentioned above.
Linkage disequilibrium may have two main causes:

1. Either two populations homozygous for different haplotypes have mixed a relatively short time ago, and so far repeated crossing over at a low rate has not been sufficient to lead to random distribution of alleles.
2. Certain combinations of alleles on closely linked gene loci have caused a selective advantage for their bearers and have therefore been preserved.

To be able to decide between these two possibilities, Bodmer (1972) [581] calculated how long a linkage disequilibrium would need to disappear in a random mating population.

Table 3.11. Association of HLA-A1 and B8 in unrelated Danes (2 × 2 table) [889]

	Number of individuals		Total
	B8-positive	B8-negative	
A1-positive	376	235	611
A1-negative	91	1,265	1,356
Total	467	1,500	1,967

The table is often given as follows:

First antigen	Second antigen	+/+ a	+/− b	−/+ c	−/− d	Total N
A1	B8	376	235	91	1,265	1,967

where, for example, +/− means number of individuals possessing the first character (A1) and lacking the second (B8). The *chi-square* is

$$\chi^2 = \frac{(\mathrm{ad}-\mathrm{bc})^2\, N}{(\mathrm{a}+\mathrm{b})(\mathrm{c}+\mathrm{d})(\mathrm{a}+\mathrm{c})(\mathrm{b}+\mathrm{d})} = 699.4$$

corresponding to a correlation coefficient

$$r = \sqrt{\chi^2/N} = \sqrt{699.4/1967} = 0.60.$$

Gene frequencies for A1 and B8 can be calculated by Bernstein's formula

$$p = 1 - \sqrt{1-\alpha}$$

(where α is the antigen frequency) as 0.170 and 0.127, respectively.
The Δ value can be calculated by the formula

$$\Delta = \sqrt{\frac{d}{N}} - \sqrt{\frac{(\mathrm{b}+\mathrm{d})(\mathrm{c}+\mathrm{d})}{N^2}} = 0.077.$$

Thus, the frequency of the HLA-A1, B8 haplotype is
$h_{A1,B8} = P_{A1}\, P_{B8} + \Delta_{A1,B8} = 0.170 \times 0.127 + 0.077 = 0.099.$

For these calculations, he utilized a result of Jennings (1917), according to which Δ decreases to zero at a rate of $1-\Theta$ per generation where Θ is the recombination fraction between the two loci. Between the HLA-A and HLA-B loci, Θ was found to be in the order of magnitude of 0.008. Taking linkage disequilibrium between HLA-A1 and B8 as an example, Δ values of ≈ 0.06–0.1 have been found in European populations. On the other hand, Δ values between 0.01–0.02 are not significant statistically with reasonable sample sizes. Therefore, it is meaningful to examine how many generations are needed to reduce Δ from 0.1 by a factor of 5 to 0.02.
Using the above-mentioned principle of Jennings, we get:

$$(1-\Theta)^n = (1-0.008)^n = 1/5; \quad n \approx 200$$

This means that Δ would be reduced to an insignificant value within ≈ 200 generations of random mating, corresponding to ≈ 5000 years, if a generation is assumed to last around 25 years.

This period is just about the length of time since agriculture first came to parts of Northern Europe and is certainly a very short time considering the evolutionary life span of the human species. The fact that such a significant Δ could be eroded in so short a time in the absence of selection suggests that, at least, this particular combination HLA-A1, B8, is being maintained at its comparatively high frequency by some sort of interactive selection [581]. We consider it likely that selection will also explain some of the other common cases of linkage disequilibrium and that the effect of recent population mixture may be of minor importance. Certain haplotypes seem to have a selective advantage that keeps them more frequent than others. This selective advantage, on the other hand, cannot be directly connected with the diseases for which associations have been shown so far, as they are too rare. Besides, the onset of most of them is usually delayed until after the age of reproduction. Factors that have influenced survival up to the age of reproduction in the past need to be examined. The topic will be discussed in Sect. 6.2.1.

The Normal Function of the System. The HLA determinants are localized at the surface of the cell and are strong antigens. They exhibit the most pronounced polymorphism known so far in humans. There is linkage disequilibrium not only between the HLA loci themselves, but probably also between them and closely linked immune-response genes. Disease associations have been shown between HLA antigens and diseases for which an autoimmune mechanism had been previously suspected. Furthermore, similar systems are known in all other mammals examined so far. Finally, there is close linkage with other loci concerned with the immune response. All this evidence together is very suggestive of a system that regulates contact of cells with their environment. In recent years, many details of this function have been elucidated. HLA identity is required for cooperation of different cells at the various stages of immune response, for example, when macrophages, which are the first cells to bind with an antigen, present this antigen to the T lymphocytes, and when T and B lymphocytes cooperate in the initiation of antibody formation (Sect. 4.4). Various haplotypes have been shown in vitro to differ in the responsiveness to antigens, e.g., the induction of T cell proliferation [623, 942]. Hence, these cell surface structures are important mediators of the immune reaction. Some investigators believe that cell recognition mechanisms may be important in embryonic development and differentiation, especially when they are present on only certain cell types. For example, Ia antigens are present in the mouse on B-lymphocytes, macrophages, and certain other cells, but not (or rarely) on T-lymphocytes or blood platelets; Ir genes act primarily on the collaboration of B- and T-lymphocytes. The HLA-A, -B, and -C antigens, on the other hand, seem to be present in all cells except erythrocytes. In a more specific version of this hypothesis, it is assumed that these antigens are important for the development of different clones of immunocompetent cells during embryonic age. The hypothesis of differentiation antigens would not explain the selective advantage of the high degree of polymorphism in this system.

Another possible function is protection against viral or bacterial infection. Antigenic material of human origin may be incorporated in the outer membrane of the virus, which is thereby made less recognizable to another human host. However, if the virus contains MHC material from a genetically different individual, it will be more readily inactivated by the immune system. Such a mechanism would also explain why the extreme polymorphism of the MHC system has a selective advantage. Another possible function of the MHC region is protection against "infection" with tumor cells from other individuals of the same species. This explanation would neatly account for what is already known of its major role in the transplantation system and for the high degree of polymorphism. On the other hand, it is difficult to understand how humans - especially those living in a hunting and gathering group or even in neolithic agricultural societies - could be endangered by infection with tumor cells. Further elucidation of the MHC will teach us a great deal about how the organism handles its interaction with the environment. This knowledge is important to our understanding of how natural selection has shaped our genetic constitution in the past and how recent changes in our environment may influence it in the future.

To broaden the empirical basis for such understanding, however, it may be useful to ask: Are there other examples in nature of such gene clusters with related functions? Can their analysis provide us with hints for a better understanding of the MHC cluster? There is, indeed, one such example that has already been analyzed very carefully - mimicry in butterflies.

3.5.6 Genetic Determination of Mimicry in Butterflies[6]

False Warning Coloration [866]. Some animals have developed certain protective devices against their predators, for example, cryptic or warning coloration. Warning coloration is typical for species that are relatively unpalatable for their potential predators. A relatively palatable species (the "mimic") can be protected from its enemies if it assumes the warning color of an inedible species (the "model"). H. W. Bates (1862) recognized that such a false warning coloration would be advantageous and might be evolved. Thus, the phenomenon is known as Batesian mimicry.

To serve its purpose, the protecting mechanism must fulfill certain conditions:

a) The model must be relatively inedible or otherwise protected.
b) It must have a conspicuous color pattern.
c) It must be common, usually but not necessarily more common than the mimic. Should the model be rarer than the mimic, the predator will have little or no chance of learning that it is protected and therefore will not avoid the mimic.
d) Model and mimic must usually be found together in the same area at the same time.
e) The mimic should resemble the model closely. However, this resemblance will only extend to visible structures, color pattern, or behavior. As Sheppard puts it, "it should deceive the artist (or predator hunting by sight) but not the anatomist."

"Batesian" mimicry is different from "Muellerian" mimicry, which means that two unpalatable species mimic each other, thus decreasing the number of individuals killed from each species during the learning process of the predators. Batesian mimicry has been observed fairly frequently. In our context, a few species of butterflies are of special interest.

Butterfly Especially Efficient in Mimicking Others. Papilio memnon is a swallowtail butterfly widely distributed in southeastern Asia, from India and Ceylon to the Philippines and the Moluccas. The males are monomorphic, nonmimetic, and tail-less in all regions except Palawan and Celebes, where they are tailed (Fig. 3.41). In Japan, the females are also monomorphic, nonmimetic, and tail-less,

---

6 This section is not absolutely necessary for understanding the principles of human genetics.

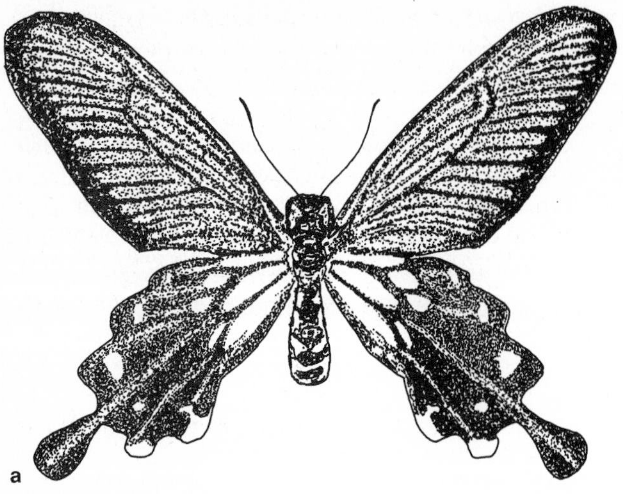

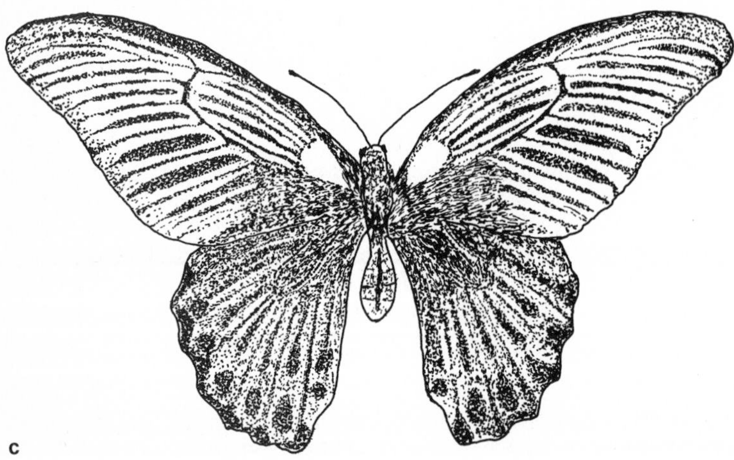

Fig. 3.41 a–c. Batesian mimicry in *Papilio memnon*. **a** the model, *Papilio coon* ♀; **b** the mimic, *P. memnon* ♀; **c** *Papilio memnon* ♂, from Japan. (Clarke et al., 1968, see [866])

though still different in appearance from the males. In other areas, either the females are mimetic, or if nonmimetic forms occur the species is polymorphic. A number of butterfly species serve as models, the resemblance between model and mimic often being very striking. Figure 3.41b shows one example: *P. coon* ♀ (from Java) as model, *P. memnon f. achates* ♀ as mimic. Figure 3.41c shows the nonmimicking forms of *P. memnon*, ♂ and ♀.
Very similar observations have also been made in an African species, *P. dardanus*. As with *P. memnon*, the females were found to be highly polymorphic within the same population; mimics for different models, as well as nonmimic forms, may be observed. This can be understood in terms of the conditions of Batesian mimicry discussed above. As soon as the mimic becomes too frequent in relation to the model, its selective advantage will disappear. The only way to maintain the selective advantage, and at the same time increase population size above a certain minimum, is to develop a polymorphism that will enable the species to mimic more than one model at the same time.

Genetic Determination. Genetic analysis has shown that the butterfly mimic pattern is controlled by a cluster of closely linked genes with rare crossing over between them; a "supergene." According to genetic analysis, the haplotypes behave like one set of alleles with a highly pleiotropic pattern of action, simultaneously influencing shape of the wing, body color, and wing pattern. However, there is conclusive evidence that crossing over between these loci occasionally occurs. The most likely sequence of these loci is that loci controlling body color (B) and the presence or absence of tails (T) are on opposite ends of the gene cluster; loci controlling hindwing pattern (W), epaulette color (E), and forewing pattern (F) seem to be localized in between. The probable sequence of loci has been deduced from the frequency of crossovers in relation to non-crossovers within the examined populations. Hence, the mode of reasoning is the same as that of Fisher when he proposed a sequence for the D-C-E determinants within the Rh loci. Moreover, it can be concluded from the available data that crossovers are very rare in the butterfly populations examined. There is marked linkage disequilibrium.

Similarities with the MHC Situation. There are two similarities between the analysis of butterfly mimicry and that of the major histocompatibility complex. Firstly, there is a cluster of genes with related functions. In the butterfly, these genes determine the mimic pattern; in man, they (probably) influence the handling of environmental agents by the cell. Secondly, in both systems there is marked polymorphism, together with marked linkage disequilibrium. How can evidence from the butterfly help in understanding how the MHC gene cluster has evolved?
The mode of natural selection for the butterfly patterns can be understood in terms of the special requirements for a selective advantage by Batesian mimicry. An individual happening to inherit a suitable combination of alleles for the mimicry pattern has a selective advantage. However, this advantage will disappear as soon as the pattern is disrupted by crossing over: only the pattern as a whole gives the advantage, not any parts of it. The advantage will be maintained in the next generation only if, by a chromosome rearrangement, more than one of these genes come together on the same chromosome, thereby increasing the chance for common transmission. In this way, gradual accumulation of these genes in a common cluster is favored. Up to now, however, selection has not been strong enough - or has not been working long enough - to prevent crossing over completely. Occasional occurrences tend to disrupt the mimicking pattern. But crossovers are devoured more frequently by predators. This selective disadvantage helps to maintain linkage disequilibrium.
This does not, however, explain the high degree of polymorphism that, in turn, is understandable in terms of the influence of population frequency. As noted, the mimic loses its advantage if it becomes too frequent, i.e., more frequent than the model (Frequency-dependent Selection; see Sect. 6.2.1.5). Hence, the advantage for the species is maintained if only part of the population has the mimic phenotype. The advantage will be still greater if different parts of the population manage to mimic different models occuring in their habitat. Genetic diversity as such gives the selective advantage.
Evolution of the MHC cluster can be accounted for if among the genes assembled within this cluster certain combinations are more suitable than others for dealing with the environment. Evolution would favor non-crossovers. This would at the same time explain linkage disequilibrium. The analogy with butterfly mimicry is not quite so straightforward toward helping to explain why such a high degree of polymorphism was maintained. However, the human species is exposed to a great variety of external influences - such as bacterial and viral infections - which are different throughout space and time. Too high a frequency of a human type adapted to one variety of a virus could easily lead to selection of other, mutant varieties of the same virus especially suited to attack this human type, thus obviating its adaptation. The human polymorphism could be a way to meet the challenge by different mutant types of the virus. This topic will be taken up again in the section on natural selection by infectious diseases.

3.5.7 Genes with Related Functions on the Human X Chromosome?

Unlike the autosomes, the X chromosome remained relatively unchanged during mammalian evolution. There are a number of hints pointing to homology between different species [156]. The X chromosome has the peculiar feature that it is present in one copy in the male and in two copies in the female - a difference that is not entirely compensated for by X-inactivation (Sect. 2.2.3.3). Is there any hint that X-linked genes are a nonrandom sample of all human genes? Do they show any peculiarities? This question has been examined by classification of all known X-linked and autosomal mutations in four categories [920]:

1. Those affecting the sense organs (eye, inner ear), the skin, and the teeth.
2. Those affecting the brain and the nervous system.
3. Structural anomalies of the skeleton, muscles and connective tissue, of inner organs systems like heart and digestive tract; cell surface antigens

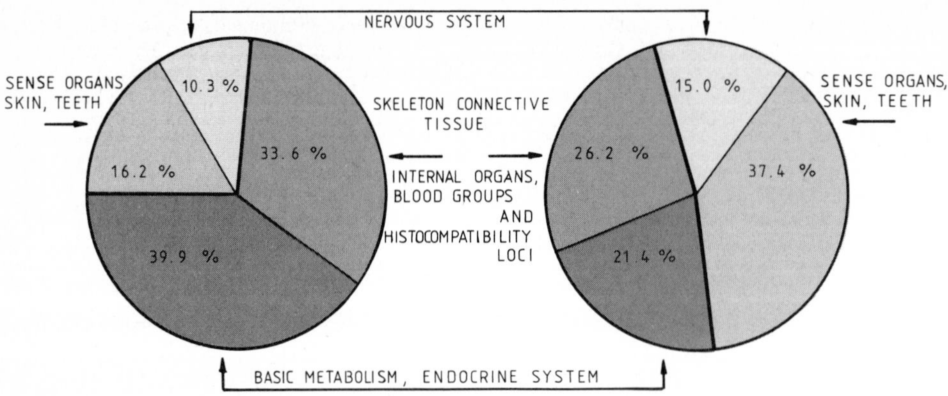

Autosomal mutations X-linked mutations

Fig. 3.42. Possible clustering of mutations affecting sense organs, skin, teeth, and the nervous system on the human X chromosome. Four groups of phenotypes in 1029 autosomal and 92 X-linked mutations. (Data from McKusick 1982 [133] only confirmed mutations)

(blood and histocompatibility antigens), and tumors.

4. Mutations affecting basic metabolism; blood clotting and other hematologic diseases; enzyme and serum and protein polymorphisms; endocrinologic disorders.

Figure 3.42 shows the result: categories 1 and 2 are much more frequent in X-linked mutations. Statistical comparison of categories 1+2 versus 3+4 between X-linked and autosomal mutations gives $\chi_1^2 = 17.4$, which is significant with $P < 0.01$. In addition, category 4 contains a number of X-linked mutations influencing endocrine functions of the neuropituitary that, with equal justification, could be included in category 2. Hence, the "higher" functions of the nervous system and sense organs seem to be overrepresented on the human X chromosome, whereas genetic representation of body structure, as well as of basic metabolism, is under-represented.

Ascertainment and classification of human mutations has never been very systematic. The conditions for recognizing and delimiting X-linked mutations are other than those for autosomal, and especially autosomal-recessive mutations. This bias could simulate a difference between the X chromosome and the autosomes. Yet, a real clustering of genes with related functions is very likely possible. In *Drosophila,* where ascertainment can be better controlled, significant nonrandomness of mutations affecting various organ systems has been described [648]. In case the difference between X chromosome and autosomes in humans turns out to be real on closer analysis, the question arises: Has this difference anything to do with the special properties of X-linked genes in the regulation of gene action? Do these genes run a lower risk of becoming recessive lethals by mutation, and has this been an important selective advantage in a genetic situation in which approximately every other individual is a male hemizygote, who would be killed by a recessive lethal? Or is this clustering just a trace of the evolutionary history of these genes?

3.5.8 Unequal Crossing Over

Discovery of Unequal Crossing Over. In the early years of work with *Drosophila,* some authors observed that the bar mutation, an X-linked dominant character, occasionally reverted to normal, whereas in other cases, homozygotes for the allele produced offspring with a new and more extreme allele, which was later called "double bar." Sturtevant (1925) [904] showed that this peculiar behavior was not due to mutations but to unequal crossing over, producing, on the one hand, a chromosome with two bar loci (double bar) and, on the other hand, a chromosome with no bar locus at all. When the giant salivary chromosomes of *Drosophila* permitted visual testing of genetic hypotheses, Bridges (1936) [588] was able to show that the simple, dominant bar mutation was caused by a duplication of some chromosomal bands. The reversion corresponds to the unduplicated state, whereas double bar is caused by a triplication of that band. Both reversion and triplication can be produced by one single event of unequal crossing over. Bridges did not yet formulate clearly the obvious reason for this event: the mispairing of "structure-homologous" but not "position homologous" chromosome sites (Fig. 3.43; see also Fig. 2.97).

Unequal Crossing Over in Human Genetics. Haptoglobin [584a], a transport protein for hemoglobin, is found in the blood serum and shows a polymor-

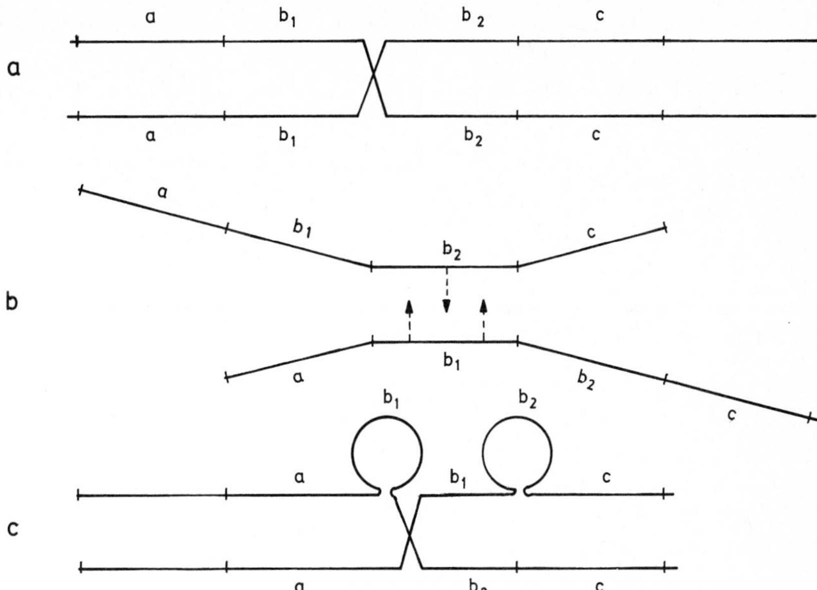

Fig. 3.43 a–c. The principle of unequal crossing over. **a** Normal pairing and crossing over. The two genes b_1 and b_2 are assumed to have very similar DNA sequences. **b** Genes b_1 and b_2 are pairing. This leads to a shift of the two homologous chromosomes relative to each other. **c** Such pairing requires formation of two loops in the upper chromosome

phism, the most common alleles being HP^{1F}, HP^{1S} and HP^2. Smithies et al. (1962) [884] discovered that the allele HP^2 is almost twice the length of each of the two alleles HP^{1F} and HP^{1S}, as evidenced by the composition of its polypeptide chain. In the HP^2 chain, the amino acid sequence of the HP^1 chain is repeated almost completely. They concluded that the HP^2 allele might have been produced by gene duplication. Moreover, they predicted that unequal crossing over might again occur with a relatively high probability between HP^2 alleles, producing, on the one hand, an allele similar to HP^1 and, on the other hand, an allele comprising the genetic information almost in triplicate. Repeated occurrence of this event might lead to still longer alleles and, hence, to a polymorphism of allele lengths in the population.

There is an essential difference between the first unique event that produces the almost double-sized gene (for example, HP^2) from a single gene HP^1, and the unequal but homologous crossing over that becomes possible as soon as the first duplicated allele is present in the population [748].

First Event. Given a pair of homologous chromosomes, both partner chromosomes consist of largely identical sequences of nucleotides. Normally, these partner chromosomes will pair at meiosis and there can be no unequal crossing over. To effect mispairing and, hence, unequal crossing over, an initial duplication of at least one gene is necessary. Mechanisms for such a duplication are known in cytogenetics, the simplest being two breaks at slightly different sites in adjacent homologous chromatids during meiosis and subsequent crosswise reunion. Another mechanism would be mispairing due to homology of short base sequences in nonhomologous positions. Our present knowledge of the structure of DNA sequences between the transcribing segments of structural genes (Sect. 2.3) suggests ample opportunities for such a mispairing.

If the sites of breakage are separated just by the length of one structural gene, this event results in two gametes that do not contain this gene at all, together with two others containing it in duplicate (Fig. 3.44). The gametes containing the deletion have a high risk of not being transmitted because of lethality of the ensuing embryo. On the other hand, a gamete with the duplication is likely to develop into a diploid individual, providing for the first time a chance for mispairing of homologous cistrons and, hence, for unequal crossing over.

Consequences of Unequal Crossing Over. The consequences are seen in Fig. 3.44. As long as the duplication remains heterozygous, all gametes will contain either one or two copies of the duplicated gene. When the duplication becomes homozygous, however, other types of gametes may be formed. Unequal crossing over may lead, on the one hand, to gametes with only one copy and, on the other hand, to gametes containing three, and in subsequent a Homologous unequal crossing over if the primary duplication is heterozygous

Crossover products

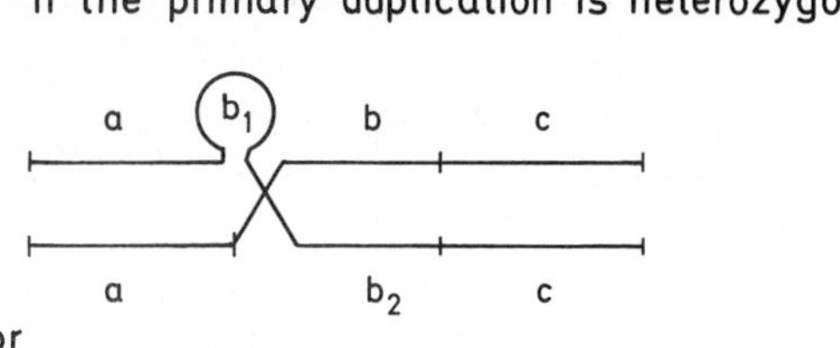

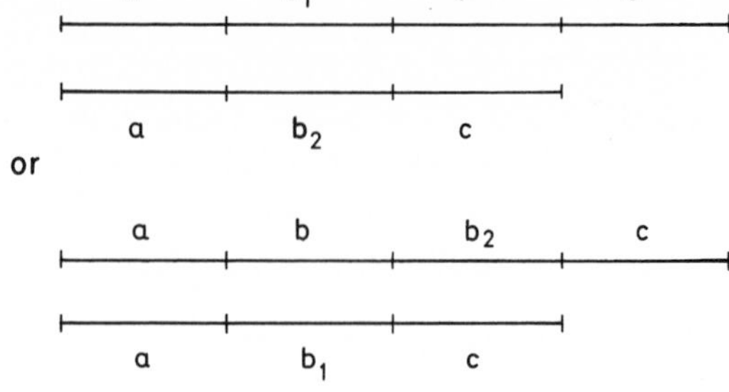

or

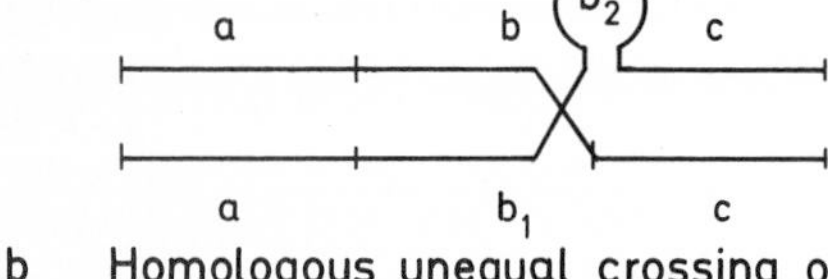

b Homologous unequal crossing over if the primary duplication is homozygous

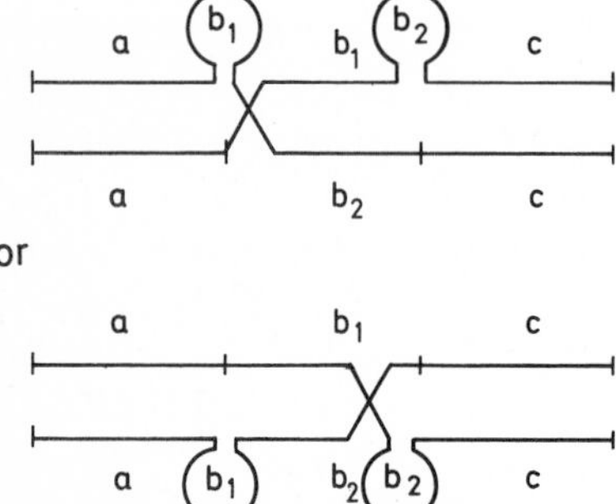

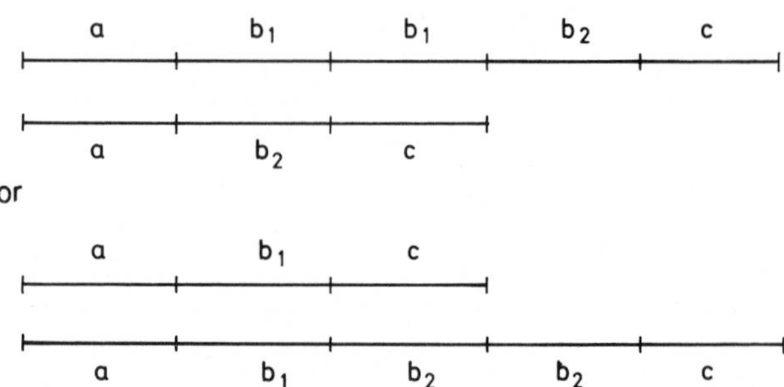

Fig. 3.44 a, b. Unequal crossing over between structure-homologous but not position-homologous genes. **a** Unequal crossing over will always lead to one crossover product with two genes b (b_1b or bb_2) and to another with only one gene (**b**). Formation of larger allele sequences becomes possible if the primary duplication is homozygous. In this case, a chromosome with three alleles b (b_1 b_1 b_2 or b_1 b_2 b_2) may be formed (Krüger and Vogel, 1975 [748])

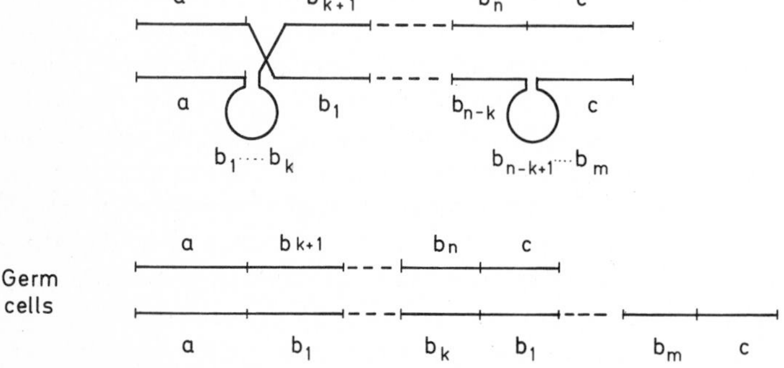

Fig. 3.45. The consequence of unequal crossing over. In subsequent generations, chromosomes with (theoretically) unlimited numbers of a alleles may be formed. Unequal crossing over between any of them may lead to still larger (or still shorter) haplotypes. $b_1 \dots b_k \dots b_n$ refer to homologous genes

generations, more than three copies (Fig. 3.44, 3.45).

If the probability of unequal crossing over is not too low, there will soon be a high variability in the number of homologous chromosomal segments that resemble each other in structure but not in position. If selection favors a certain number of such chromosomal segments that may be as small as a single gene, this number will soon become the most common one. Selection relaxation will lead to an increase in variability in both directions: The fraction of individuals with a very high number of such genes as well as those with a low gene number will gradually increase [748]. Another genetic mechanism resembling unequal crossing over in some aspects is gene conversion where non-reciprocal products result (Sect. 2.3; Fig. 2.97).

Possible Significance in Human Genetics. As mentioned, the haptoglobin HP^2 allele is almost a duplicate of the HP^1 alleles. Here, the unequal crossing over is occasionally expected to lead to alleles that contain the information in triplicate. Indeed, such alleles have occasionally been observed; they are known as Johnson-type alleles [883].

Other examples are the closely linked hemoglobin β- and δ-cistrons (Sect. 4.3). Here, the Lepore-type mutants may be caused by unequal crossing over (Fig. 4.51), and the X-linked color vision genes [825 a]. Moreover, there are many examples for moderately or highly repetitive DNA sequences within which unequal crossing over should be possible. The rRNA genes that are located within the nucleolus organizer regions have ca. 300–400 identical genes – with appreciable variation. At first glance, this situation seems to provide the best conditions for unequal crossing over. However, these genes are located close to the centromeres of acrocentric chromosomes where crossing over hardly ever occurs.

It could be questioned whether these acrocentric regions are at all able to undergo crossing over, and whether acrocentric location may be a mechanism protective against deviation from the optimum amount of rRNA genes by unequal crossing over. Other repeated DNA sequences are those coding for the immunoglobulins (Sect. 4.4). With increasing knowledge of the functional significance of repeated DNA sequences, we shall learn more about the significance of unequal crossing over.

Intrachromosomal Unequal Crossing Over. With structure-homologous but not position-homologous genes, such as those found in multigene families (Sect. 2.3.3.8.), unequal crossing over becomes possible not only between homologous chromosomes but also between sister chromatids (intrachromosomal unequal crossing over). Theoretical considerations have shown that this process could have played a role in molecular evolution [1941].

3.6 Conditions and Limitations of Genetic Analysis in Humans: Multifactorial Inheritance

3.6.1 Levels of Genetic Analysis

The paradigm created by Mendel with his pea crosses (Sect. 1.4) has since developed in several steps to the level of the gene in its definition as an information-carrying segment of the DNA double helix. The principal goal of genetic analysis is to pursue this path for the character under investigation. In the early part of this book, characters were selected as examples of basic principles in which the genotype-phenotype relationship was fairly straight-forward: in the "normal" range, blood groups; in the "abnormal" range, rare hereditary diseases.

However, there are many normal characters for which genetic variability obviously exists but no simple mode of inheritance can be found. These are, for example, stature and body proportion, physiognomic features and facial expression ("this child is the absolute image of his father"), skin color, blood pressure, and others. Many diseases may have a complex of various causes, but there are hints that the liability may differ among individuals and may be genetic in origin. In earlier years, a Mendelian framework was often superimposed naively on these data, without testing the formal requirements for simple modes of inheritance. Today, there is a tendency to describe observations in these traits in terms of biometric models developed on the basis of oversimplifying assumptions. Conclusions are then drawn from these models, which depend on these assumptions so heavily that their biologic significance is often questionable. It might, therefore, be useful to set out the logical background for a discussion of genetic hypotheses and to distinguish a number of levels at which genetic analysis has become possible.

3.6.1.1 Findings at the Gene: DNA Level

An analysis has occasionally been able to reach the DNA level and to identify the site of the mutation. The DNA sequence contains the information for sequence of amino acids in a polypeptide chain. Therefore, it is our goal to trace genetic differences to the DNA level or at least to corresponding differences in the amino acid sequence of proteins, from which the alteration at the DNA level can be inferred. This became possible first for the hemoglobin variants (Sect. 4.3) and later for some other proteins, inferring alterations in DNA from the observed changes in amino acids and knowledge of the genetic code. In most cases, one amino acid is replaced by another but other changes such as deletions, frameshifts and nonsense mutations also could be inferred (Sect. 4.3 and 5.1). Here, a genetic variant has been pursued down to its primary cause – a specific change at the DNA level: the "information carrier."

Techniques developed in recent years sometimes allow the direct demonstration of mutational alterations at the DNA level (Sect. 4.3.5). The increasing availability of such methods made it possible to elucidate novel types of mutations, such as those interfering with the regulation of transcription and others producing defective splicing during processing of gene transcripts. Most work of this type was done with the human β-hemoglobin system and the resulting phenotypic changes presented as so-called β-

thalassemias, which are characterized by absent (β°) or deficient β-chain production (β^{+}). As more DNA probes for different genes become available the elucidation of mutations at the DNA level is accelerating. We already know that different mutations analogous to those observed in the hemoglobinopathies occur in the hemophilias and in familial hypercholesterolemia. It can be expected that in the future most studies on mutational alterations in humans will be carried out directly on DNA rather than at the gene product level.

3.6.1.2 Analysis at the Gene Product: Biochemical Level

Here, identification of the mutant site within the gene is not possible, but the individual gene in which the mutation has occurred can be identified. There are several possibilities:

a) Specific proteins can be characterized by biochemical methods (Sect. 4.2). Genetic variability can be shown to reflect differences in proteins or enzymes. When a protein consists of more than one polypeptide chain, identification of the individual polypeptide chain involved may be possible. Examples are the numerous serum protein and enzyme polymorphisms.
b) Many proteins serve as enzymes, catalyzing specific metabolic steps. Hence, when a specific genetic block has been demonstrated, the enzyme defect identified, and all other biochemical explanations have been excluded, a mutation in a specific gene specifying the involved enzyme can be inferred. The next step will be characterization of the enzyme protein (see above).
c) Another subgroup within this category is comprised of cases in which an antigenic profile of the cell surface can be identified using specific antibodies. Examples are the blood groups and the HLA types (Sect. 3.5.5). This method permits identification not only of specific gene loci but also – within certain limits – of structural differences within these gene loci.

Only the results of an analysis at this level are strictly comparable to results from thoroughly examined species in experimental genetics, such as *Drosophila melanogaster,* the mouse, maize, silkworm, and others. In many of these mutations, no specific protein changes, enzyme defects, or aberrant antigen profiles have been demonstrated, but breeding experiments and recombination analysis offer an efficient alternative path toward identification of individual genes.

It might be interesting to examine how this purely formal gene concept, which dominated thinking of experimental geneticists for a long time, has influenced thinking and conceptualization in fundamental genetics. Human biochemical genetics in its early phases was further advanced than biochemical genetics dealing with other species. The synthesis of insights from biochemistry and genetics led to rapid progress in microbial and fungus genetics. Biochemical human genetics remains more advanced than that of other mammalian species. However, in spite of recent progress in DNA techniques, biochemical genetics and linkage analysis, many genetic analyses in man need to be carried out at a less sophisticated level.

3.6.1.3 Analysis at the Qualitative Phenotypic Level: Simple Modes of Inheritance

Conclusions often must be based on phenotypic differences far away from the primary gene action. Still, the relationship between genotype and phenotype is so straightforward that simple Mendelian modes of inheritance can be inferred with certainty. However, the involvement of a single gene cannot be identified with surety: The same phenotype, with the same mode of inheritance, may be caused by mutations at several different gene loci.

Rare Conditions Qualitatively Different from the Normal. This category comprises most inherited diseases. An individual either is normally pigmented or lacks skin pigment (albinism). When measurements are possible, they show a distribution in two different classes: affected and unaffected. When measurements of metabolites in blood or urine are possible, the values are distributed in two modes. Such findings frequently point to the enzyme defect and to identification of the gene involved. Examples are the increased excretion of homogentisic acid in the urine of alkaptonurics – Garrod's paradigm for inborn errors of metabolism (Sect. 1.5) – or the distribution of phenylalanine in the blood plasma of phenylketonurics in comparison to normals (Fig. 3.46). These conditions are usually rare. Whenever more discriminating analysis was done in such conditions, even diseases with similar phenotypes and identical modes fo inheritance turned out to be genetically heterogeneous. Criteria for such heterogeneity include:

a) Children from matings between two homozygotes for a recessive condition are phenotypically normal indicating that each homozygote parent carries a different recessive gene.
b) Linkage analysis shows close linkage with a marker gene in a number of the analyzed fami-

lies; in others, the two genes segregate independently. One example is linkage of one of the loci for dominant elliptocytosis with the Rh locus on chromosome 1.

c) Different protein or enzyme defects are demonstrated in various families by biochemical analysis. Conditions initially considered homogeneous turn out to have a different genetic basis (examples: hemophilia A and B, glycogen storage diseases, hereditary hemolytic anemias; Sect. 4.2).

Frequent Variants; Bimodal Distribution. In another group of traits analyzed at the phenotypic level, the phenotypes do not show a clear-cut distribution into two classes. Not every individual can be attributed to one specific class; there is overlap. Yet, as long as the area of overlap is relatively small, the overall distribution remains bimodal.

An example from pharmacogenetics illustrates this phenomenon. After administration of a single dose of the tuberculostatic drug isonicotinyl hydrazine (INH), the plasma level in various individuals was different, and the distribution showed two modes (Fig. 3.47). This suggested a simple mode of inheritance of drug biotransformation. This hypothesis was confirmed by family studies. Homozygotes Ac^s/Ac^s (Ac^s = allele for slow inactivation) showed high drug levels, whereas heterozygotes Ac^s/Ac^r (Ac^r = allele for rapid inactivation) and homozygotes Ac^r/Ac^r had low drug levels. This difference was inferred to be caused by a variant of the enzyme *N*-acetyltransferase. These findings confirmed the genetic hypothesis put forward when the bimodal distribution of INH concentrations was discovered.

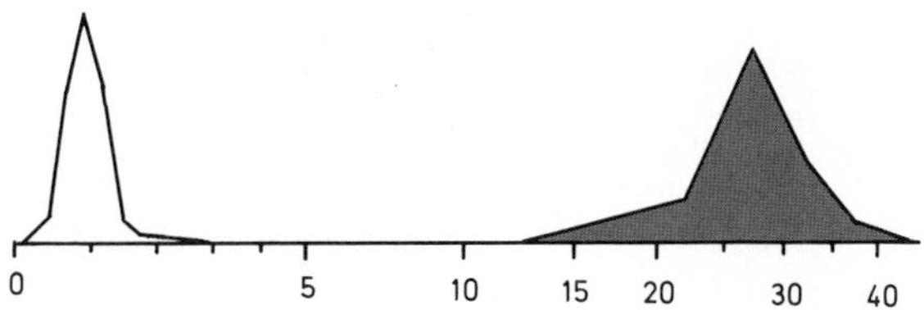

Fig. 3.46. Phenylalanine level in the blood plasma of healthy individuals and phenylketonurics *(shaded)* expressed in mg%. (Adapted from Penrose, 1951)

In the absence of other, more conclusive criteria, a simple mode of inheritance may be inferred but is not proved if the variable under investigation is measurable and shows a bimodal distribution. Exceptions to this rule occur in either of two directions:

1. A bimodal (or multimodal) distribution may be simulated by other mechanisms. This is especially likely when some threshold effect is involved (Sect. 3.6.2). However, even when a threshold effect seems to be absent, a bimodal distribution may arise as a secondary effect when the variable under investigation shows a tendency to self-enhancement once it has reached a certain level. Blood pressure is one example: above a certain level, the kidneys may be damaged, which leads to a further increase in blood pressure. Particular methods of ascertainment may simulate bimodality.
2. More frequently, a bimodal or trimodal distribution may be hidden. The means of the two distributions may be so similar that bimodality is obscured. Harris and Smith [703] examined the conditions under which a combination of two normal distributions can lead to a bimodal distribution:

a) Two normal distributions with identical variances combine to a bimodal distribution only, if the difference between the means is at least twice as large as the common standard deviation.

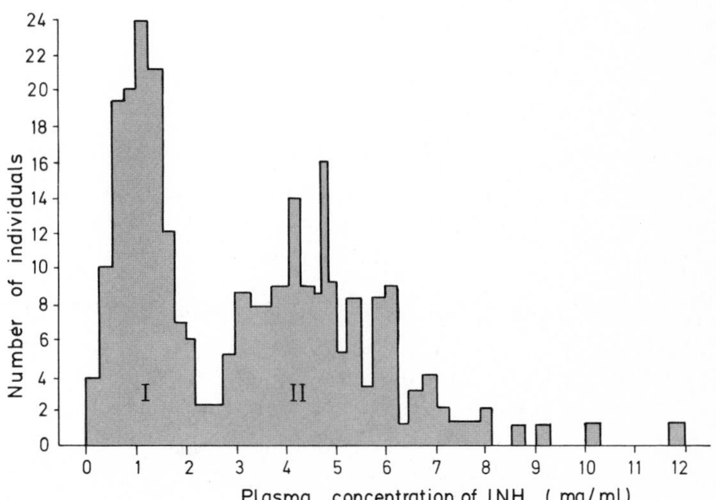

Fig. 3.47. Plasma concentration of isoniazid (*INH*) in 267 members of 53 families; bimodal distribution. The antimode is between 2–3 mg%. (Adapted from Evans et al., 1960 [196])

b) When the variances are different, the difference of the means must be at least equal to a certain multiple of the smaller standard deviation, which varies from 2 (when the variances are equal) to about 2.6 (when the variances are extremely different).
c) When the means are closer together, so that no bimodal distribution results and the number of individuals in the two distributions are not too different, then a "bitangential" distribution indicates that two different distributions are involved (Fig. 3.48).

In practice, such bitangential distributions are difficult to evaluate. Variables occurring in nature only rarely show an ideal normal distribution, and chance deviations must be taken into account. To see how chance deviations may distort distribution of empirically observed values with moderate sample size, compare Fig. 3.48 with Fig. 3.47.

To summarize: *A unimodal distribution may be compatible with a monogenic mode of inheritance. However, without additional data, discrimination from multifactorial models is impossible.*

Figure 3.49 shows the opposite example: The enzyme activity distributions of electrophoretically identified genetic variants of GPT (glutamate-pyruvate transaminase) are clearly different. However, a

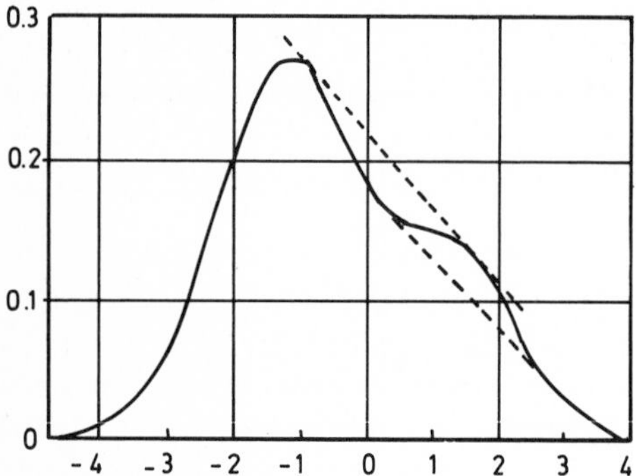

Fig. 3.48. Bitangential distribution. (Harris and Smith, 1951 [703])

combined distribution in the population is almost normal and would usually be interpreted as multifactorial in origin. Yet, the total distribution in the population owes its origin to only two alleles *(GPT1, GPT2)* and their corresponding phenotypes (GPT1, GPT2-1, GPT2). Many examples of this kind have been found when analyzing the variance of enzyme levels in various enzyme polymorphisms (Sect. 6.1).

Recognition of bimodality may be especially difficult if the two classes have very different frequencies, for example, if the number of individuals in one type is 10% or less than in the other (Fig. 3.50). Here, it may be doubtful whether the smaller mode is genuine, representing a genetically different group. It could as well be due to:

a) chance deviations.
b) a threshold: this possibility should be seriously considered where the values comprising the second mode are close to zero.

Chance deviations can be excluded by examining more individuals. Under such circumstances, a threshold effect obviously cannot be ruled out; however, family studies may help, especially if the mode of inheritance is autosomal dominant or X-linked. In such families, the two genotypes are expected in about equal frequencies, with the consequence that bimodality will be more clear cut than in a population sample. When families of probands from the lesser mode are examined, the following criteria confirm a simple mode of inheritance, if present in the majority of families:

a) Clear-cut bimodal distrubtion of the trait within sibships; the sibs who can be assigned to the mode reflecting normals are distributed in a manner similar to the general population.
b) Parents' values are also bimodal, with the additional condition that at least one of the two parents should fall into the second mode.

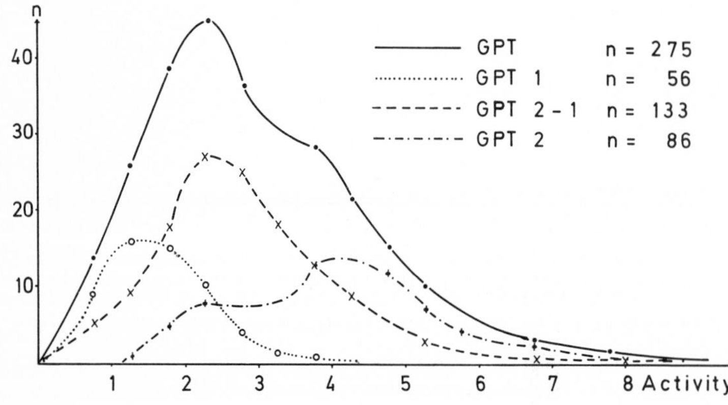

Fig. 3.49. Distribution of enzyme activities for three GPT genotypes, almost combining to a somewhat skewed normal distribution. (Data from Becker, P. E. (ed.), 1976 [8])

c) An approximate 1:1 ratio of unaffected to affected should be found in these sibships. An approximate 3:1 ratio should be found in sibships where both parents fall into the second mode, i.e., when both parents are heterozygous.

With multifactorial inheritance, spurious bimodality might result due to a threshold at 0. As a result, the distribution of sibs whose values can be assigned to the first mode will show a lower mean value than that of the population. Moreover, parents' values will more frequently be assignable to the first mode with a lower mean value than that of the population (Fig. 3.51).

One trait in which an autosomal-dominant mode of inheritance has been inferred using these criteria is the low voltage electroencephalogram (EEG) [921]. The human brain constantly produces certain voltage oscillations that, after passing through suitable amplifiers, can be drawn as curves on paper. As a rule, several leads (8-16) from different points of the head are taken simultaneously. The proband relaxes and rests with closed eyes but does not sleep. Sleeping patterns are different and are used for special diagnostic purposes.
The resting EEG of the healthy adult consists of a few wave types, the α waves being especially noticeable. In addition, β waves (> 13/s) and a few ϑ waves (4-8/s) may occur (Fig. 3.52). These few elements, however, may be formed, distributed, and combined in so many different ways that a comparison with handwriting is inviting. Almost every human being has his own, characteristic EEG, which remains constant over many years in the absence of diseases, such as epilepsy and brain tumors, and excepting such transitory physiologic states as severe fatigue and intoxication. During childhood and youth, the EEG develops from irregular forms with relatively slow waves to the final pattern, which will be reached by about age 19 at the latest and changes only very slowly with advancing age. Individual differences in speed of development are formidable, leading to a high variability during childhood. Twin studies have shown the normal, individual EEG pattern to be almost exclusively genetically determined.
In about 4% of the adult population, an EEG type is found with the following characteristics:

a) The occipital α waves are completely absent or can be seen for only a short time and with very low amplitude.
b) Hence, the EEG may look absolutely flat or show an irregular pattern with β or ϑ waves of low amplitude.
c) Contrary to the normal α EEG, there is no reaction when the eyes are opened. After the eyes are closed, a very few α waves may but do not necessarily appear.

The first requirement for a genetic analysis is a measurement to quantify the extent of occipital α formation. One such measurement is the so-called α index, defined as follows:

$$I = \frac{\text{number of } \alpha \text{ waves/10 s}}{\alpha \text{ frequency} \times 10\,\text{s}}$$

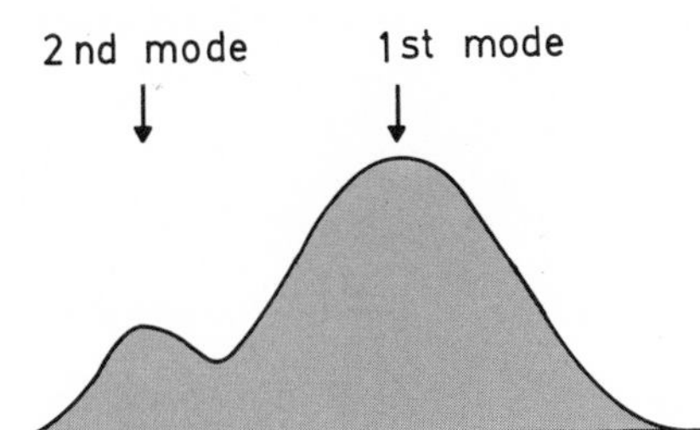

Fig. 3.50. Bimodal distribution of a quantitative trait in the human population. One of the two types is much more common than the other

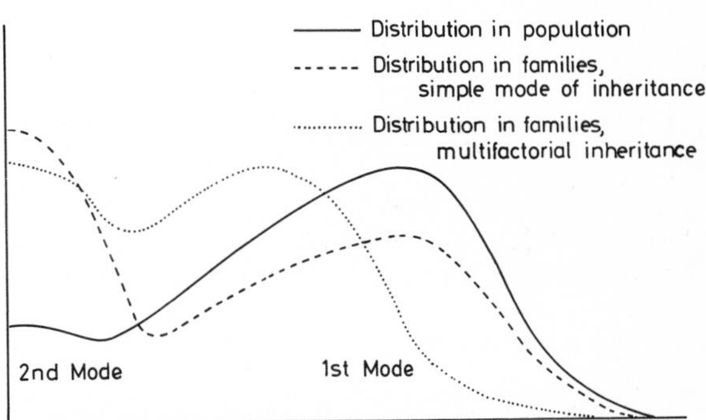

Fig. 3.51. Bimodal distribution when the second mode is close to 0. Discrimination between a simple diallelic mode of inheritance and a multifactorial model. Note that with multifactorial inheritance, the first mode is shifted to the *left,* whereas with diallelic inheritance, the first mode is identical in the population and in the families in which Segregation occurs in the two modes

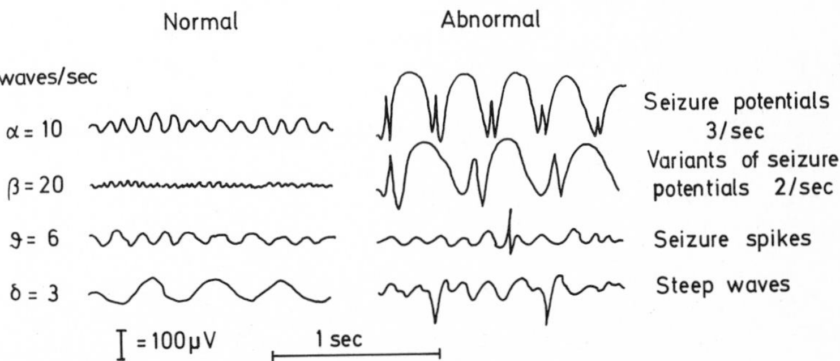

Fig. 3.52. Wave types of the human electroencephalogram

Figure 3.53 shows the distribution of this index in 30 sibships ascertained through parents of siblings with low-voltage EEG. This distribution has two maxima, one at about 70–80, the other at 0, the second maximum corresponding to the low-voltage type. At first glance, this distribution seems to favor the hypothesis of a simple mode of inheritance. Since the second maximum, however, is 0, a fictitious bimodal distribution could be due to a threshold at 0.

Are there additional arguments in favor of a monogenic inheritance? The distribution among the parents is, again, bimodal, the distribution around the first mode corresponds well with the distribution in the general population. Still more important, in all families ascertained by a child, at least one parent had a low-voltage EEG (Fig. 3.54). Analysis of the segregation ratio gave an estimate close to 75% for all families with two affected parents. For families with only one affected parent, the estimated overall segregation ratio turned out to be significantly below its expectation ($\approx 50\%$). This was found to be due to inclusion of teenagers between 10 and 19. The low-voltage EEG only gradually develops during this period. Limitation of analysis to persons 19 years old or older led to an excellent fit between expected and observed segregation ratios (Table 3.12). (The expectations are somewhat higher than 0.5 and 0.75, respectively, since because of the frequency of the character some homozygotes are expected among the parents.)

In this case, an autosomal-dominant mode of inheritance could be inferred by analysis of the distribution of a quantitative measure (the α index), together with family studies.

In principle, similar criteria can be used in studying X-linked recessive inheritance. Here, however, analysis of the distribution might be more difficult in females, since a trimodal distribution is expected: two homozygotes and one heterozygote. An example are glucose-6-phosphate dehydrogenase levels in G6PD deficiency. Since there is much overlap between normals and heterozygotes as well as between heterozygotes and hemizygotes, no clear trimodality could be discerned (see Sect. 4.2.2.2).

3.6.1.4 Genetic Analysis at the Quantitative Phenotypic-Biometric Level

Additive Model. In many cases, phenotypic variability is so complex that the action of single mutations can no longer be identified. Here, the genetic conclusions to be deduced from similarities among relatives are necessarily of a more general kind. Still, the "multifactorial" genetic models applied have certain characteristics in common – and more importantly, some predictions derived from these models are fulfilled when tested on observations.

In the simplest possible model, co-operation of a number of gene pairs is assumed. It is assumed that a capital letter allele (A and B but not a or b) contributes to the trait ("positive" allele), while the small letter allele (a and b but not A or B) is silent and has no effect on the trait ("negative" allele). The phenotypic character varies gradually, depending only on the relative number of positive and negative alleles, whose contribution is assumed to be equal and additive in this model. These gene pairs may be named A, a; B, b; C, c; D, d; etc. Thus, it makes no difference for the phenotype whether the genotype is AABBccdd ..., AaBbCcDc ..., or aabbCCDD ... (additive polygeny). This model will be used in the following explanation for a variety of concepts. It should be made clear that the model represents an abstraction and is oversimplified. In reality, the contributions of genes acting in a multifactorial system will almost always differ in quantity and quality. Some will be more important than others.

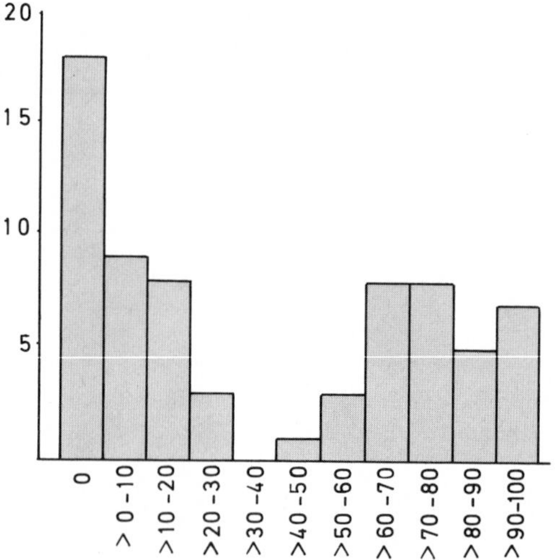

Fig. 3.53. Distribution of the α-index in 30 sibships from families with the low voltage EEG (see text)

Table 3.12. Low voltage EEG: Segregation analysis for 60 families with 117 children; low voltage EEG (+)

Mating types	+ × –	All children + × –	Only children > 19 years + × –
Expected (+)	$P=0.759$	$P=0.509$	$P=0.509$
Observed (+)	$\hat{P}=0.75 \pm 0.153$	$\hat{P}=0.364 \pm 0.052$	$\hat{P}=0.447 \pm 0.075$

P, expected; $\hat{P}$, estimated segregation ratio. Expectations were calculated assuming that the great majority of parents showing a low-voltage EEG (+) were heterozygous (expectations for children in matings + × + : 0.75; + × – : $P=0.05$), but that a fraction calculated with reference to the Hardy-Weinberg law was homozygous.

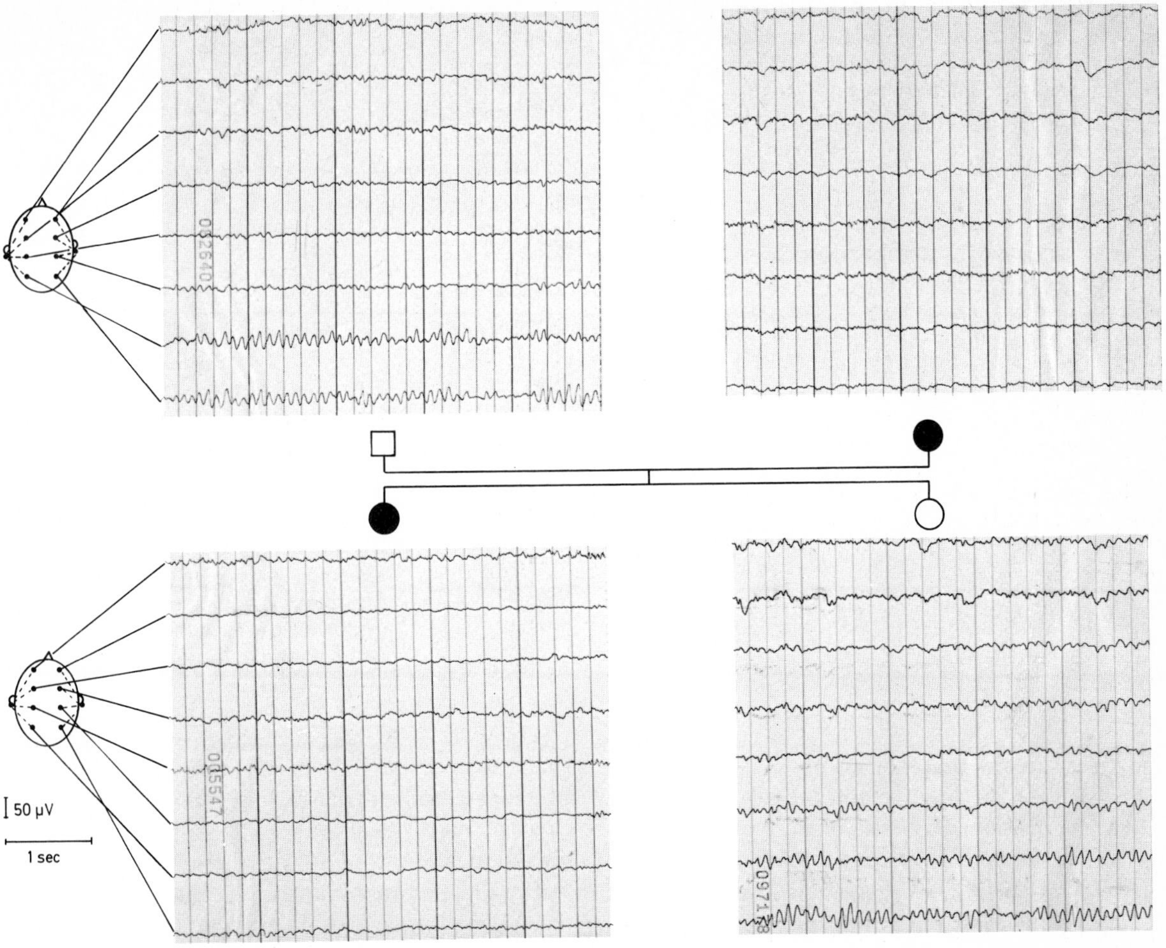

△ **Fig. 3.54.** Family observation: mother and first daughter have a low voltage EEG, father and second daughter have an α-EEG. Unipolar Leads

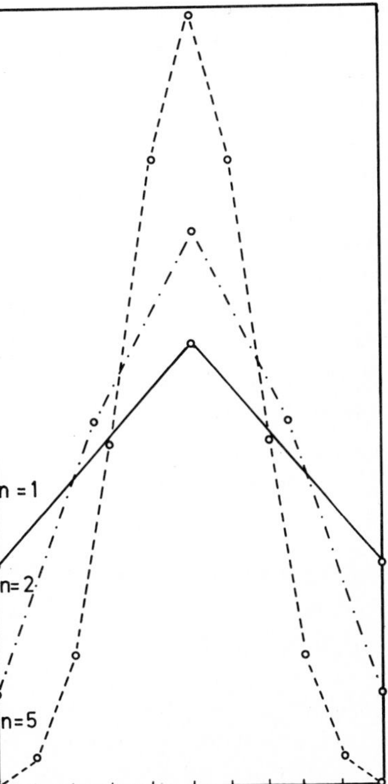

▷ **Fig. 3.55.** Distribution of genotypes according to the binomial distribution $(p+q)^{2n}$ with $p=q=0.5$ for 1, 2 and 5 gene pairs ($n=1, 2, 5$) *Abscissa*: Variation of measurements

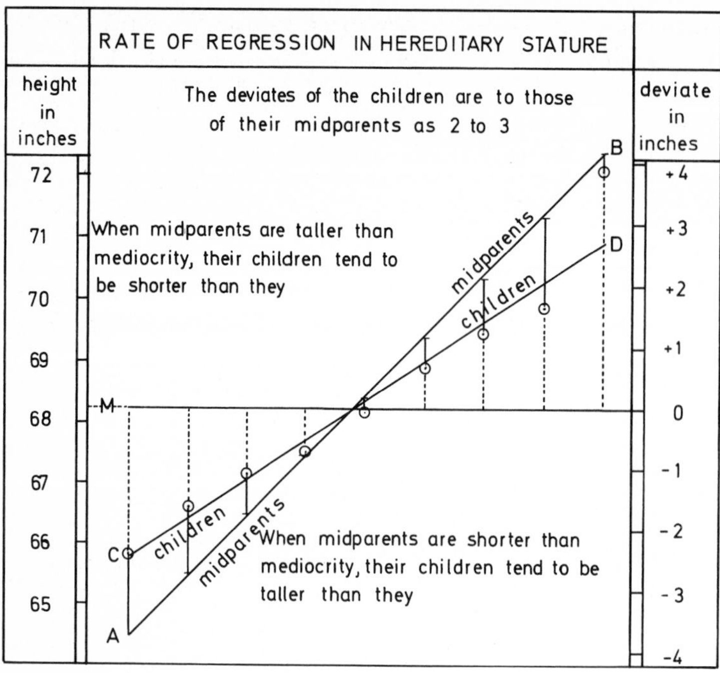

Fig. 3.56. Midparent-child correlation for stature; regression to the mean (original drawing and wording by Francis Galton)

Let us assume n gene pairs with gene frequencies $p=q=0.5$ for the positive and the negative alleles. Then, the distribution of the phenotypic classes on an arbitrary quantitative scale is given by the binomial formula $(p+q)^{2n}$ (Fig. 3.55). The higher the number of gene pairs, the more individuals found in the central (i.e., more average values) as distinguished from the peripheral part of the distribution (i.e., more extreme values). At first glance, the distinctions appear to provide a criterion for the number of gene pairs contributing to a character when the empirical distribution is compared with a number of theoretical ones. However, such an inference will only hold if each gene will contribute just as much to the variable at the extremes of the distribution as it does in the center. This assumption can be challenged by the general - and often biologically plausible - hypothesis that at the extremes further deviation in the same direction might be more difficult to achieve. For instance, for measurements of biologically active substances and enzymes, values of less than zero do not exist.

The argument has, in fact, been used to estimate the number of gene pairs involved in the genetic variability of skin pigmentation. In our opinion, it is likely that the number of skin pigmentation genes cannot be very high since segregation of completely "white" individuals on the one hand and completely "black" ones on the other is not infrequently observed from matings of hybrid persons (i.e. mulattoes).

The distributions in Fig. 3.55 have only one mode; they are unimodal. Furthermore, they are similar in shape to the "normal" distribution. This similarity becomes closer with increasing number of gene pairs (n): The normal distribution is the limiting case for the binomial distribution with increasing n. It can be shown that this approximation becomes just as good when gene frequencies of positive and negative alleles are not equal. Higher values of n are required to reach the same degree of approximation as in the symmetric case. In general, a unimodal distribution of a variable - with the shape of the distribution curve more or less aproximating the normal distribution - is typical for this genetic model of additive polygeny. However, neither a unimodal distribution nor the shape of the distribution nor the shape of the curve depend on the specific features of this model - equal and additive contributions of genes. They are good indicators for multifactorial inheritance in a more general sense.

On the other hand, as shown in Sect. 3.6.1.3, these properties do not exclude the possibility of a "major gene" with a simple mode of inheritance. In fact, it is biologically likely that only a few major genes may be the principal genetic factors in several diseases acting against a background of many genes of less pathophysiologic significance.

The first condition for establishing a unimodal and nearly normal distribution for the character is that the character in question can be measured and that a quantitative scale is used. It is, for example, possible to distribute all adult men into two alternative classes: those taller than 1.67 m and those shorter than 1.67 m. With such limited information, family investigations could easily lead to the conclusion that variability in human stature depends on a dominant gene with incomplete penetrance. The example seems trivial and the point it tries to make is self-evident, but the literature is still full of examples of this type of error.

A genetic hypothesis cannot exclusively be based on the population distribution of the variable. Family data are also needed. What type of family data does the model predict? This will be examined using the simplest possible case: two gene pairs, A, a and B, b, acting additively and equally. The gene frequencies may be p_1, p_2 and q_1, q_2, respectively. There are nine possible genotypes with five different phenotypes, occurring with the frequencies given in Table 3.13. The frequencies of the possible mating types and the distribution of the children's genotypes for each mating type can be calculated. For the special case of the last column (all gene frequencies equal and 0.5), the calculation is given in Table 3.14. From these distributions of genotypes, the corresponding phenotypic distributions of the children can be deduced (Table 3.15).

The properties of the examined model are:

a) In principle, all resulting distributions have about the same form: they are symmetric and unimodal.
b) If the parents' phenotypes are identical, the mean of the children equals the parents' phenotype. If the parents' phenotypes are different, the childrens' mean is exactly the mean of the parents (midparent value).
c) With increases of heterozygosity of the parents, the expected variance of the children becomes higher. It is largest with mating types 0×0 and zero in mating types $+4 \times +4$, -4×-4, as well as $+4 \times -4$.
d) The mean of the children of all persons with the same phenotype (for example, children of all persons with type +4) deviates half as much from the population mean as does the phenotype of these parents (for example, the mean phenotype of children of +4 parents is +2).

Table 3.13. Genotypes and phenotypes in additive polygenic inheritance

Phenotype	Genotype	Frequency	$p_1=p_2=q_1=q_2=0.5$	
+4	AA BB	$p_1^2p_2^2$	0.0625	
+2	AA Bb	$p_1^2 2p_2q_2$	0.125	0.25
	Aa BB	$2p_1q_1p_2^2$	0.125	
0	AA bb	$p_1^2q_2^2$	0.0625	0.357
	aa BB	$q_1^2p_2^2$	0.0625	
	Aa Bb	$2p_1q_1 2p_2q_2$	0.25	
−2	Aa bb	$2p_1q_1q_2^2$	0.125	0.25
	aa Bb	$q_1^2 2p_2q_2$	0.125	
−4	aa bb	$q_1^2q_2^2$	0.0625	
			1.000	

This model is highly particularized and simple. Yet the analysis proved to be fairly cumbersome. For examining the general case (n gene pairs; gene frequencies $p_1 \ldots p_n$; $q_1 \ldots q_n$), the method should be changed. First we assume a gene pair of heterozygotes with the phenotypic effect α, whereas one homozygote (AA) has the effect 2α and the other (aa) has the effect 0. Thus, we again assume that heterozygotes hold exactly the middle between the two homozygotes. The mean and variance of the corresponding character, x, can now derived as follows:

$$E_x = \frac{p^2(2\alpha)+2pq(\alpha)}{p^2+2pq+q^2} = \frac{2p\alpha(p+q)}{(p+q)^2} = 2p\alpha, \quad (3.5)$$

$$V_x = E(x^2)-(Ex)^2 = p^2(4\alpha^2)+2pq(\alpha^2)-4p^2\alpha^2 = 2pq\alpha^2 \quad (3.6)$$

(using $p+q=1$). α may be visualized as the contribution of the allele A to the character x. V_x is called the genetic variance in the population.

Now, the more general case of n gene pairs with gene frequencies $p_i = p_1, p_2 \ldots p_n$ for the genes $A_1, A_2 \ldots A_n$ and $q_i = q_1, q_2 \ldots q_n$ for the genes $a_1, a_2, \ldots, a_n$ can be examined:

$$E_x = 2\alpha \sum_{i=1}^{n} p_i \qquad V_x = 2\alpha^2 \sum_{i=1}^{n} p_iq_i$$

The following considerations, which, for the sake of simplicity, will be given for one gene pair only are valid for n gene pairs as well.

Now, the relationships between parents and children, as well as between siblings, are considered. To simplify the calculation, α will be assumed to be 1, making the phenotypic value of homozygotes AA = 2, of heterozygotes Aa = 1, of homozygotes aa = 0. Table 3.16 shows the frequencies of all possible mother-child combinations. They can be explained as follows. The frequency of AA mothers among all mothers is p^2. Each of their children gets one A gene. The probability that this gene meets another A gene in the zygote is p. This gives the overall frequency $p^2 \times p = p^3$. For the other maternal genotypes, an analogous calculation can be made. The overall distribution for the whole population (parents as well as children) is of course, $p^2+2pq+q^2$ (marginal sums in Table 3.16).

Now, the variable under investigation may be named x_1 in the parents, x_2 in the child. Then, Eqs. (3.5) and (3.6) derived above lead to:

$$\bar{x}_1 = \bar{x}_2 = 2p; \quad (3.7)$$

$$V_{x1} = V_{x2} = 2pq \quad (3.8)$$

The covariance between parent and child can be derived from Table 3.17. In general, the covariance of the two variables, x_1 and x_2, is defined as:

$$\text{Cov}(x_1, x_2) = E(x_1x_2) - x_1x_2$$

Here, $E(x_1x_2)$ is defined as: $\sum x_{1i};\, x_{2i}p(x_{1i}, x_{2i})$

The values $x_{1,i}$ and $x_{2,i}$ represent the phenotypic expression of the trait, i.e., in our example, 2, 1 and 0. The $p(x_{1i}, x_{2i})$ are the corresponding entries in Table 3.16; $p(2, 2)$, for example, has the value q^3. It follows:

Table 3.14. Mating types, their frequencies, and segregation ratio among children with two additive gene pairs and $p_1 = p_2 = q_1 = q_2 = 0.5$

		AABB	AABb	AaBB	AAbb	aaBB	AaBb	Aabb	aaBb	aabb
AABB × AABB	0.003906	1	—	—	—	—	—	—	—	—
× AABb	0.015625	½	½	—	—	—	—	—	—	—
× AaBB	0.015625	½	—	½	—	—	—	—	—	—
× AAbb	0.007813	—	1	—	—	—	—	—	—	—
× aaBB	0.007813	—	—	1	—	—	—	—	—	—
× AaBb	0.031250	¼	¼	¼	—	—	¼	—	—	—
× Aabb	0.015625	—	½	—	—	—	½	—	—	—
× aaBb	0.015625	—	—	½	—	—	½	—	—	—
× aabb	0.007813	—	—	—	—	—	1	—	—	—
AABb × AABb	0.015625	¼	½	—	¼	—	—	—	—	—
× AaBB	0.031250	¼	¼	¼	—	—	¼	—	—	—
× AAbb	0.015625	—	½	—	½	—	—	—	—	—
× aaBB	0.015625	—	—	½	—	—	½	—	—	—
× AaBb	0.062500	⅛	¼	⅛	⅛	—	¼	⅛	—	—
× Aabb	0.031250	—	¼	—	¼	—	¼	¼	—	—
× aaBb	0.031250	—	—	¼	—	—	½	¼	—	—
× aabb	0.015625	—	—	—	—	—	½	½	—	—
AaBB × AaBB	0.015625	¼	—	½	—	¼	—	—	—	—
× AAbb	0.015625	—	½	—	—	—	½	—	—	—
× aaBB	0.015625	—	—	½	—	½	—	—	—	—
× AaBb	0.062500	⅛	⅛	¼	—	⅛	¼	—	⅛	—
× Aabb	0.031250	—	¼	—	—	—	½	—	¼	—
× aaBb	0.031250	—	—	¼	—	¼	¼	—	¼	—
× aabb	0.015625	—	—	—	—	—	½	—	½	—
AAbb × AAbb	0.003906	—	—	—	1	—	—	—	—	—
× aaBB	0.007813	—	—	—	—	—	1	—	—	—
× AaBb	0.031250	—	¼	—	¼	—	¼	¼	—	—
× Aabb	0.015625	—	—	—	½	—	—	½	—	—
× aaBb	0.015625	—	—	—	—	—	½	½	—	—
× aabb	0.007813	—	—	—	—	—	—	1	—	—
aaBB × aaBB	0.003906	—	—	—	—	1	—	—	—	—
× AaBb	0.031250	—	—	¼	—	¼	¼	—	¼	—
× Aabb	0.015625	—	—	—	—	—	½	—	½	—
× aaBb	0.015625	—	—	—	—	½	—	—	½	—
× aabb	0.007813	—	—	—	—	—	—	—	1	—
AaBb × AaBb	0.062500	1/16	⅛	⅛	1/16	1/16	¼	⅛	⅛	1/16
× Aabb	0.062500	—	⅛	—	⅛	—	¼	¼	⅛	⅛
× aaBb	0.062500	—	—	⅛	—	⅛	¼	⅛	¼	⅛
× aabb	0.031250	—	—	—	—	—	¼	¼	¼	¼
Aabb × Aabb	0.015625	—	—	—	¼	—	—	½	—	¼
× aaBb	0.031250	—	—	—	—	—	¼	¼	¼	¼
× aabb	0.015625	—	—	—	—	—	—	½	—	½
aaBb × aaBb	0.015625	—	—	—	—	¼	—	—	½	¼
× aabb	0.015625	—	—	—	—	—	—	—	½	½
aabb × aabb	0.003906	—	—	—	—	—	—	—	—	1

Table 3.15. Distribution of children in additive polygenic inheritance (the lacking 5 categories, 0x−2, 0x−4, −2×−2, −2×−4, −4×−4, can be calculated according to the same rule)

Parents' Genotypes	+4 AABB	+2 AABb; AaBB	0 AAbb; aaBB; AaBb	−2 Aabb; aaBb	−4 aabb
+4×+4	1				
+4×+2	0.5	0.5			
+4×0	0.1666	0.6667	0.1666		
+4×−2		0.5	0.5		
+4×−4			1		
+2×+2	0.25	0.5	0.25		
+2×0	0.08333	0.41667	0.41667	0.08333	
+2×−2		0.25	0.5	0.25	
+2×−4			0.5	0.5	
0×0	0.02778	0.22222	0.50000	0.22222	0.02778

Table 3.16. Frequency of parent-child (father- or mother-child combinations) in the human population in a randomly mating population (see text)

Parent	Children			
	AA	Aa	aa	
AA	p^3	$p^2 q$	–	p^2
Aa	$p^2 q$	pq	pq^2	$2pq$
aa	–	pq^2	q^3	q^2

Table 3.17. Stature of parents and adult children (Galton from Johannsen, 1926 [726]). (Units used: inches, 1 inch = 2.54 cm)

Stature of mid-parent	Stature of children							
	60.7	62.7	64.7	66.7	68.7	70.7	72.7	74.7
64	2	7	10	14	4	–	–	–
66	1	15	19	56	41	11	1	–
68	1	15	56	130	148	69	11	–
70	1	2	21	48	83	66	22	8
72	–	–	1	7	11	17	20	6
74	–	–	–	–	–	–	4	–
	5	39	107	255	287	163	58	14

$$\mathrm{Cov}(x_1, x_2) = 4p^3 + 4p^2q + pq - (2p)^2 = pq$$

and for the correlation coefficient r_{PC} between parent and child:

$$r_{PC} = \frac{\mathrm{Cov}(x_1, x_2)}{\sigma x_1 \sigma x_2} = \frac{pq}{2pq} = 0.5$$

This important result was derived by Fisher (1918) [664]. In a random mating population and with additive gene action, the correlation coefficient between a parent and child is 0.5. It can be shown in a similar way that, under the same conditions, the correlation coefficient between full sibs is also 0.5. These correlation coefficients are independent of the gene frequencies p_1 and q_1 and state in a statistical way that parents and children as well as sibs have 50% of their genes in common.

The situation becomes more complicated when A is more or less dominant over a. In this case, the correlation coefficients are influenced by the gene frequencies. Parent-child correlation is no longer equal to the sib-sib correlation, but - apart from the case $q=1$ - is lower.

3.6.1.5 *Heritability Concept*

The concept of heritability is widely used in quantitative genetics. The graded character under examination, expressed in metric units, may be called the "value." The value measured in a given individual is its phenotypic value. This phenotypic value for most biologic characters owes its origin to both genetic and environmental factors. Environment is considered in a broad sense, i.e., comprises all nongenetic circumstances that influence the phenotypic value. The two components are usually called the genotypic value and the environmental value[7]:

$$P = G + E$$

where P = phenotypic value, G = genotypic value, and E = environmental value.

The phenotypic values of all individuals in a population have a mean, and a variance, that measures the variation around this mean. The variance is distinguished from other measures of variability by one mathematical property: Different variances can be added to give a common variance and, conversely, a common phenotypic variance V_P can be broken down into its components such as the genotypic variance V_G and the environmental variance V_E:

[7] Falconer [63] uses the term environmental deviation.

$$V_P = V_G + V_E$$

However, the addition rule for variances applies only if genotypic and environmental values are independent of each other, i.e., when they are not correlated. If there is a correlation between the two, the covariance of G and E must be added:

$$V_P = V_G + V_E + 2\,\mathrm{Cov}_{GE}$$

Let us take one example from that area of genetics that first introduced these concepts - agricultural studies (Falconer [63]). It is normal practice in dairy husbandry to feed cows according to their milk yield. Cows that produce more milk are given more food. Human society often behaves in a similar way toward its own members, as will be discussed in the section on behavior genetics.

Another assumption is that specific differences in environments have the same effect on different genotypes. When this is not so, there is an interaction between genotype and environment, giving an additional component to the variance V_I. Even in experimental animals, this component can be measured only under special conditions.

The genotypic value V_G can be subdivided into several components: an additive component (V_A) and a component (V_D) measuring the deviation due to dominance and epistasis from the expectation derived from the additive model. The dominance variance is contributed by heterozygotes (Aa) that are not exactly intermediate in value between the corresponding homozygotes (aa and AA). The variance contributed by epistasis refers to the action of genes that affect the expression of other genes. Hence, the concept of additive variance does not imply the assumption of purely additive action of the genes involved. Even the action of genes showing dominance or epistasis tends to have an additive component. The whole genotypic variance can be written:

$$V_P = \overbrace{V_A + V_D}^{\text{Genetic Variance}} + \overbrace{V_E + V_I}^{\text{Environmental variance}} + \overbrace{\mathrm{Cov}_{GE} + V_M}^{\text{Measurement variance}}$$

Here, a new component (V_M) has been introduced that relates to the variability in measurement of the same character at different times. The value may represent truly different values, such as different test results on different days, or measurement errors, such as differences in test results of the same blood specimen as well as differences on repeated testing of the same individual. If all these variables are known, they can be incorporated in the calculations. In the following, covariance between heredity and environment (Cov_{GE}) and interaction variance (V_I) will be assumed to be 0 and neglected. Measurement variance (V_M) will also be neglected but will be considered in the section on twin methods (App. 6).

For convenience, it is useful to introduce a new concept: Heritability, which is defined as:

$$h^2 = \frac{V_A}{V_P}$$

and provides a value ranging from 0 to 1 (0-100%) expressing the contribution of additive genetic elements to the phenotype under study. Otherwise stated, heritability is a population statistic parameter that expresses the (additive) genetic contribution to the trait under study in a single percentage value. A low value implies few contributions of additive genes to the trait, while a high value suggests a larger contribution. The concept was developed for purposes of selection in plant and animal breeding of economically useful traits such as milk production in cows and egg laying in chickens. The additive part of genetic variability was most important for these purposes. Any other genetic component, such as dominance, tends to impair accuracy of prediction. In humans, however, we are more interested in the total genetic variability, whether it is additive or not.

In human genetics, therefore, heritability as defined above is often called "heritability in the narrow sense" and is supplemented by another definition:

$$h^2 = \frac{V_G}{V_P}$$

where V_G and V_P refer to the total genotypic and phenotypic variance, respectively. This formulation is known as the heritability in the broad sense or degree of genetic determination.

There is a relationship between the heritability in the narrow sense (h^2_N) and the theoretical correlation coefficients between relatives as calculated above (Eq. 3.5). For the most important degrees of relationship, the following formulae apply:

Monozygotic twins:	$h^2 = r$
Sib-sib or dizygotic twins:	$h^2 = 2r$
One parent - one offspring:	$h^2 = 2r$
Midparent-offspring:	$h^2 = r/\sqrt{1/2} = r/\,0.7071$
First cousins:	$h^2 = 8r$
Uncle-nephew:	$h^2 = 4r$

Properties of h^2. In considering the biologic significance of heritability measurements, its properties need careful scrutiny:

a) Heritability is a ratio. A ratio changes when either the numerator or the denominator changes. h^2 will increase when the numerator (V_G, the genotypic, or V_A, additive variance) increases, or V_E, the environmental variance decreases.

b) The estimation of heritability is based on theoretical correlations between relatives. These correlations are valid only under the assumption of random mating. Assortative mating leads to other correlations and unless taken into consideration, produces systematic errors in the estimation of h^2. The correlations resulting from assortative mating were calculated first by Fisher (1918) [664] (cf. Cavalli-Sforza and Bodmer, 1971 [36]; for a more complete treatment, Wilson [955, 956]). These correlations can be utilized for adjustment of h^2.
c) An estimation of h^2 is possible only when the assumption is made that covariance and interaction between genotypic value and environmental deviation are 0.

Falconer tried to escape this dilemma for covariance by proposing the following convention. If the genetic constitution of an individual makes for environmental conditions that improve or worsen his phenotype, then this phenomenon can be included as part of the genotypic value. Formally, this is correct, even if it tends to obscure the problems involved in genotype-phenotype relationships. For animal breeding, this convention may be useful. Applied to humans, it leads to difficulties. Does the genotypic value of the American black imply that certain members of the society deny him full civil rights because of his skin color?
More difficulties arise when these concepts are applied to interpretation of heritability values from twin data (App. 6). The interaction term poses another difficulty in interpretation for which no solution has been proposed so far.

3.6.1.6 One Example: Stature

An example of a biometric study in which heritability can be estimated is F. Galton's classic work on inheritance of stature.[8] He measured 204 parental couples along with their 928 adult children. There was a methodologic difficulty due to the lower average stature of women. Galton overcame this by multiplying all measurements of females by 1.08, thus adjusting them to the male measurements: On the average, the stature of males in his sample was 1.08 times the stature of females. Having made this correction, he determined the midparent value for each couple: 1/2(♂ + ♀). The result of the study is seen in the correlation Table 3.19. A correlation is obvious at first glance. The correlation coefficient is:

$$r_{pc}=0.59 \quad p<0.01,$$

where r_{pc} symbolizes midparent-child correlation.

This value can be used to calculate h^2. The midparent-offspring correlation is:

$$h^2=\frac{r}{\sqrt{1/2}}$$

with random mating. This gives:

$$h^2=\frac{0.59}{0.7071}=0.834$$

Obviously, stature is predominantly genetically determined, but there is a component of 0.166 = 1 − 0.834 not accounted for by additive genetic variance. It may be due mainly to 'environmental factors.' Do these data offer any hints for environmental influences?
The same data may be arranged in a different way (Tables 3.17 and 3.18). Here, another divergence from expectation is obvious. With additive gene action, the children's mean is expected to be exactly one-half between parents' values, i. e., should be identical to the midparent value. This, however, is not the case. Instead, the data show something else: if the midparent value is higher than the population mean, the children's mean is lower than that of their parents. On the other hand, if the midparent value is lower than the population mean, the children's mean is higher. Hence, as a rule, the children's mean tends to deviate from the parents' mean in the direction of the population mean.
This phenomenon was observed by Galton and named "regression to the mean." It can also be shown in other, similarly continuously distributed characters (Fig. 3.56).

What is the reason for this divergence from genetic expectations? Individuals who can be ranged at the extremes of a distribution curve presumably obtain not only the genetic factors that make for the extreme phenotype but probably have benefited in addition from unusual environmental circumstances. Furthermore, specific gene-gene and gene-environmental interactions may have been operative in actualizing their extreme phenotypes. Their children on the average are less likely to have benefited from the special environmental influences and gene-environmental interaction that placed the parent in the extreme categories. Their phenotypic values are, therefore, more likely to be similar to the mean of the population – a regression to the mean.

3.6.1.7 Quantitative Genetics and the Paradigms of Mendel and Galton

How do the two paradigms on which human genetics was founded relate to each other? The gene

Table 3.18. (Johannsen, 1926 [726])

Stature of mid-parent	64.5	65.5	66.5	67.5	68.5	69.5	70.5	71.5	72.5
Mean stature of children	65.8	66.7	67.2	67.6	68.3	68.9	69.5	69.9	72.2

[8] Data from Johannsen, 1909, 1926 [726].

concept developed from Mendel's experiments (Sect. 1.4), while Galton's paradigm was based on concepts of correlation between human relatives and regression analysis. The two concepts can be theoretically linked to each other. Results of correlation studies in relatives can be interpreted in terms of action of individual genes, as was first extensively shown by Fisher (1918) [664]. Such correlation studies using biometric methods can complement genetic analysis.

As mentioned in the introduction, a paradigm comprises three main aspects: An exemplary approach, a group of scientists who follow this approach, and at least the germ for a scientific theory. The long-term success of a paradigm depends mainly on the value, i. e., the depth and explanatory power of this theory. Therefore, it might be useful to compare the two paradigms regarding values of their underlying theories, using some criteria that were developed in the philosophy of science [243]. According to Bunge (1967), "the basic desiderata of scientific theory construction are the following: (1) To *systematize knowledge* by establishing logical relations among previously disconnected items; in particular, to explain empirical generalizations by deriving them from higher-Level hypotheses. (2) To *explain facts* by means of systems of hypotheses entailing the propositions that express the facts concerned. (3) To *increase knowledge* by deriving new propositions (e. g., predictions) from the premises in conjunction with relevant information. (4) To *enhance the testability* of the hypotheses, by subjecting each of them to the control of the other hypotheses of the system ...".

"A few scientific theories comply not only with the basic desiderata (1)-(4) but also with the following additional goals: (5) To *guide* research either (a) by posing or reformulating fruitful problems, or (b) by suggesting the gathering of new data which would be unthinkable without the theory, or (c) by suggesting entire new lines of investigation. (6) *To offer a map of a chunk of reality,* i. e., a representation ... of real objects and not just a summary of actual data and a device for producing new data."

Bunge mentioned Darwin's theory of evolution as an example for a theory that fulfills all the above-mentioned criteria. In general, the ability of a theory to fulfill its task depends on its depth. Criteria for the depth of a theory are: "The occurrence *of high-level constructs;* the presence of a *mechanism,* and a *high explanatory power.* The three propositions are intimately linked: it is only by introducing high-brow (transempirical) concepts that unobservable 'mechanisms' can be hypothesized, and only what is hypothesized to occur in the depths can explain what is observed at the surface."

Less deep theories are closer to the phenomena; therefore, they are called "phenomenological"; in distinction to theories hypothesizing definite "mechanisms," which are often called "representational" or "mechanistic." Often, such deep mechanistic theories reward the scientists with an unexpected bonus: Their explanatory power extends beyond the range of phenomena for the explanation of which they had been created.

When the theories that developed within the paradigms of Galton and Mendel are compared using these criteria, Galton's approach reveals itself as having created a phenomenological theory. K. Pearson, Galton's master student, pointed out as long ago as 1904 that quantitative comparison of phenotypes between relatives with biometric methods leads to "a purely descriptive statistical theory." To a certain degree, it systematizes knowledge. But it only offers relatively low-level and nonspecific hypotheses, i. e., similarity between relatives can be explained by heredity. Or, more specifically, by additive gene action either with or without a contribution of dominance, or of environmental contributions. Such propositions are of a very general nature; and there is only occasional enhancement by additional hypotheses. As an example, see the Carter effect described in Sect. 3.6.2.3; here, the higher incidence of a birth defect in relatives of female probands was predicted and explained by the additional hypothesis of identical distribution of liability genes in both sexes, despite the unequal sex distribution among the probands. Conditions (5) and (6) are not fulfilled at all: Problems are not reformulated in a fruitful way, nor does the theory suggest gathering of new data. It only suggests the obvious: comparison of relatives.

Compare with this result the theory founded by Mendel's paradigm: Soon after its discovery a high-order construct – the unit of transmission, recombination, and function which is now called a *"gene"* – was introduced. It opened the way for investigating the *mechanisms* of gene replication, transmission, recombination, and action. Stepwise elucidation of these mechanisms constitutes the history and present problem situation of genetics. The *explanatory power* of this theory has been proven far beyond the range of phenomena for which it had been developed. At present, the theory explains not only transmission of traits from parent to offspring, but also between different cell generations of an organism.

Its explanatory power has not yet been exhausted. Coming back to our classification of genetic analysis (i. e., at the DNA-gene level, at the gene product-biochemical level, at the qualitative phenotypic level, and at the quantitative phenotypic-biometric level), the Galtonian biometric paradigm provides answers

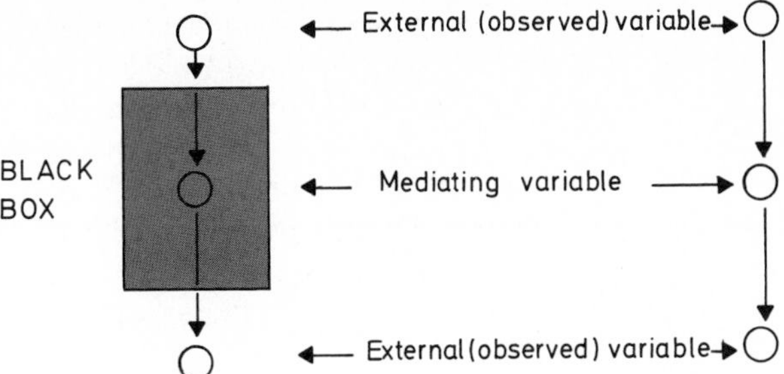

Fig. 3.57. Difference between a black box hypothesis and a hypothesis involving a mechanism. In a black box hypothesis *(left)*, the mediating variable by which one observable variable influences the other is unknown. In an explanatory hypothesis *(right)*, the mediating variable can be deduced from scientific theory; a mechanism for the influence of one observable variable to the second can be found

at the level most remote from gene action. Research with biometric genetic methods can be said to be guided by "black box" theory. Two external observable variables - the measurements of parents and children or other sets of relatives - are compared with each other, but the mediating biologic variables are unknown and remain in a black box (Fig. 3.57).

All of human development, structure, and function is ultimately controlled by genes. Differences between human beings can be demonstrated by unique physiologic, biochemical, and immunologic features of each individual. Genetic determination of these features can be shown by family resemblance. The degree of resemblance depends upon the closeness of the relationship. Monozygous twins share all their genes and will be more similar than any other relatives. Sibs share 50% of their genes, while more remote relatives share only a small fraction of their genes.

A comparison of relatives using biometric techniques for the measurement of any human phenotype is therefore likely to show genetic factors underlying that trait. A "heritability" or the fraction of the total variability assigned to genetic causation will often be estimated as higher than 0. Since the principal biologic basis of *human behavior* lies in the brain, and the brain is likely to show genetic variation as any other organ, some genetic factors are likely to determine behavior on a priori grounds. The teasing apart of shared genes and a shared environment in the family setting becomes particularly difficult in behavioral traits and causes problems in interpretation. Polemics with potentially explosive sociopolitical ramifications have resulted (see Sect. 8).

Analysis of any human trait and particularly of human behavior will yield more meaningful information if the phenotype under investigation can be studied by Mendelian techniques at the level of gene action. The black box is thus opened, and the unknown mediating variable is replaced by a known biologic mechanism.

Considering the marked differences in scientific value of both theories, one might ask why much work in human genetics is still being performed along Galtonian lines. The principal explanation is that analysis guided by advanced genetic theory is often impossible. Most human phenotypes - such as behavioral phenotypes and disease liabilities, - simply cannot be studied directly according to Mendelian principles. The underlying variables to be analyzed must first be isolated using additional, sometimes sophisticated, biologic techniques of all types. Finding such variables requires specialized knowledge in normal and abnormal human biology using the methodology of the various biomedical sciences. Counting and measurement of more simple and obvious phenotpyes, on the other hand, can often be done. This is why Galtonian techniques are often the first approach to further analysis and can often bring practically useful results despite their theoretical weakness.

Moreover, the Galtonian approach continues to be important for formulation of hypotheses, for selection of traits to be studied with more incisive methods, and for devising research strategies. Human traits that are under control of a large number of genes, each contributing to the total variability, are difficult to study by the Mendelian approach. For some of these traits, however, the conventional biometric paradigm, which presupposes a large number of genes of small effect, may not be true. One or several genes with major effect that are detectable individually with biologic techniques may be at work, with the remaining genes providing the "genetic background."

Consequently, we should use Galtonian techniques as long as no better alternatives are available. But we should refrain from making them a goal in themselves; the challenge that *in principle* a better alternative is at hand should always be kept in mind. It is all too easy for groups of geneticists trained mainly in statistical methods and the use of computers to develop highly sophisticated statistical methods for computing heritabilities, developing path analyses for contribution of various factors from heredity, family environment, economic status in a phenotype, or comparing genetic models with and without contributions for major genes to a certain phenotype. However, the final result is often disappointing to biologically oriented scientists, who demand more incisive data. Statistical methods are of great importance for analysis in human genetics. They should be used for testing of biologically well-founded hypoth-

eses that have been proposed under the guidance of a powerful biological theory. For a deeper understanding of human biology, sophisticated statistical methods that help in a genetic analysis at the quantitative phenotypic-biometric level are, as a rule, less useful than simpler methods applied in a genetic analysis at the gene or gene product levels.
This should be kept in mind when the following sections on more complex models of inheritance are studied.

3.6.2 Multifactorial Inheritance in Combination with a Threshold Effect

3.6.2.1 Description of the Model: Animal Experiments

In the previous section, genetic analysis at the quantitative phenotypic-biometric level was described for normal traits with a unimodal and nearly normal distribution in the population. It was shown that the simple model of additive polygenic inheritance accounts for these properties so that parent-offspring and sib-sib correlations can be used to estimate heritability h^2.
In many diseases and malformations, however, clear qualitative distributions apply: the individual either suffers from a given disease or is free from it. However, neither family investigations nor chromosome studies have been able to uncover a simple mode of inheritance or a visible chromosome abnormality. Family studies show an increased empirical risk for near relatives to be affected with the same condition (familial aggregation). In many cases, pathophysiologic considerations suggest a complicated etiology. Various biologic influences are often obvious, and environmental factors, such as malnutrition, infective organisms, and unknown agents, are additionally implicated. When all these genetic and environmental influences together exceed a certain threshold, the capacity of the organism to cope breaks down and the individual becomes ill or dies.
The terms "threshold" and "liability" are often used in discussion of multifactorial inheritance. A threshold implies a sharp qualitative difference beyond which individuals are affected. While the concept of a threshold is useful for models of multifactorial inheritance, it is unlikely to exist often. The concept of a liability implies a graded continuum of increasing susceptibility to the disease. It is more difficult to deal with analytically but biologically is more likely to apply in most situations.
In hereditary diseases with simple mode of inheritance, malfunction caused by a mutation at a single gene prevents normal function. In other conditions, the mutation leads to difficulties in special circumstances only, such as in the monogenically determined drug reaction. Most conditions, however, are so complex that direct analysis of all contributing factors becomes impossible. Obviously, many different genes are involved. We are left with the black box situation - genetic analysis can more readily be carried out by statistical rather than biologic methods.
To make genetic predictions on this complex level, several assumptions are necessary:

a) The genetic liability to disease is more or less normally distributed, and the distribution shows one mode.
b) The liability is caused by a great number of genes acting additively, each contributing equally to the liability.
c) The individual becomes sick or malformed when the liability exceeds a certain threshold. This threshold may be sharply defined; in most cases, however, there will be a threshold area within which additional environmental circumstances determine whether or not the individual falls ill. Or, to define it in the terms used above, heritability is less than 1 (Fig. 3.58).

Obviously, this model oversimplifies the actual situation but may be useful in understanding the occurrence of certain common diseases and malformations.

Animal Experiments. Certain observations in experimental genetics of mammals have been explained by threshold characters such as polydactyly in the guinea pig (Wright [961]). Two strains were crossed, one with the normal three toes on the hind feet, the other with four toes. Among the F_1 animals, only a few had four toes, whereas in the second generation of $F_1 \times F_1$ crosses, about one-fourth of all animals showed four toes. Genetic analysis suggested that the two strains differed in additive alleles at four gene loci: any animal could have a maximum of eight and a minimum of zero plus alleles. In matings of two homozygous animals (8×0) (Fig. 3.59) the F_1 generation being heterozygous should have four plus alleles. This genotype leads to four

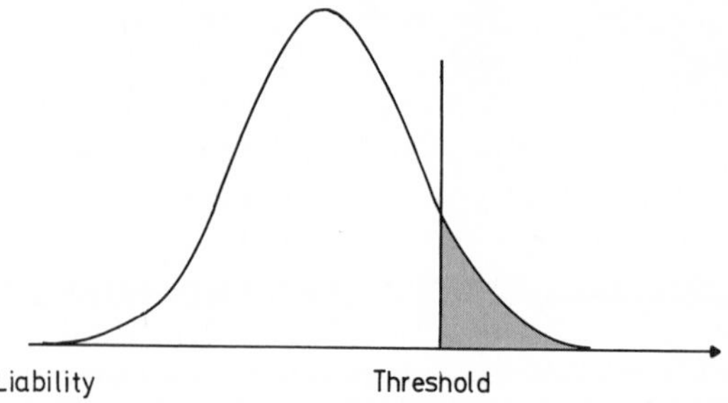

Fig. 3.58. Multifactorial inheritance in combination with a threshold effect - the simplest situation. The disease liability shows a normal distribution; individuals to the *right* of the threshold are affected with the disease

toes only in exceptional cases. In the F_2 generation ($F_1 \times F_1$), all combinations of plus alleles occur, giving a continuous distribution. In this case, it was shown in principle that additive gene action may, indeed, be associated with a threshold character (Fig. 3.59). In another example, not only the discontinuous, but also the continuous phase could be demonstrated, i.e., a phenotypic effect of the quantitatively varying liability. Grueneberg (1952) [690; 691] analyzed such a system in the mouse. In mice of the inbred CBA strain, a third molar tooth is frequently lacking. In 133 of 744 CBA animals, at least one of the four third molars was missing. In the C57 black strain, however, this molar is almost always present. Crossing between the two strains (CBA × C57) showed that the mode of inheritance is not simple, in spite of the fact that the character (tooth present or absent) is a clear-cut alternative trait. Even in animals of the CBA strain with the extra tooth, it was on the average much smaller than in the C57 black strain (Fig. 3.60). Hence, in the CBA strain, tooth size varies continuously down to a certain minimum size. Below this threshold, the tooth is not formed at all. Grueneberg named this phenomenon "quasi-continuous variation." The threshold is not absolutely sharp, and there seems to be a threshold area instead. The multifactorial genetic basis was revealed mainly by the strong difference between the two strains and by the interstrain crosses. Within the genetically uniform CBA strain, variability was caused by environmental influences.

Demonstration of continuously distributed liability as well as discontinuous thresholds has been tried repeatedly in humans (see, for example, 619), but in most cases only the discontinuous phase can be observed; the individual is either affected or unaffected. To find out what type of familial pattern can be expected with threshold characters, a theoretical model will be examined.

3.6.2.2 Simple Theoretical Model

Note the model set out in Sect. 3.6.1: two gene pairs A, a; B, b with equal and additive contributions to the phenotype and gene frequencies $p_1 = p_2 = q_1 = q_2 = 0.5$. This genotype is assumed to determine a liability, leading to a manifest disease when three or four plus alleles (A or B) are present (Fig. 3.61). The relative number of affected and unaffected children from the mating types, plus × plus, plus × minus, and minus × minus are seen in Fig. 3.62.

These expectations are alarmingly similar to those of a simple autosomal-dominant mode of inheritance: expectations for mating plus × plus are almost identical when a certain number of homozygotes among the plus parents are assumed. For matings plus × minus, the expectation is somewhat but certainly not very much lower with the additive model. Still, regular dominance with full penetrance in the heterozygote will always be clearly distinguishable, especially when more than two generations of a family can be investigated. With incomplete penetrance, however, the problem of discrimination from multifactorial inheritance with a threshold becomes very difficult; some sibships must now be expected in which both parents are unaffected and the segregation ratio is lower than 0.5. Here, however, a comparison between sibships from plus × minus and minus × minus matings may help. In the multifactorial model, a lower ratio of affected is expected when both parents are unaffected than when one parent is affected, whereas with simple autosomal dominance and incomplete penetrance, segregation ratios should be identical under such circumstances. This argument could be challenged by the assertion that penetrance might be influenced by the genetic background. Then, the problem becomes largely semantic; it is

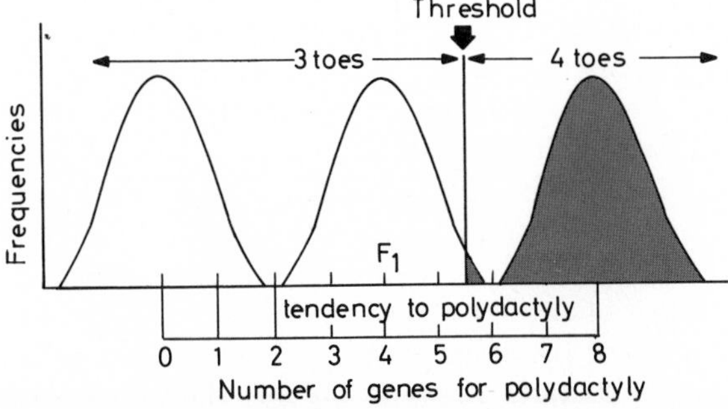

Fig. 3.59. Multifactorial inheritance in combination with a threshold: presence of an extra toe in guinea pigs. Two parental stocks, one showing three toes, the other one four toes. A fraction of the F_1 hybrids shows four toes. Genetic analysis showed eight genes to be responsible for this trait. The number of animals showing this extra toe depends on the number of "plus" genes. (Wright, 1931 [961])

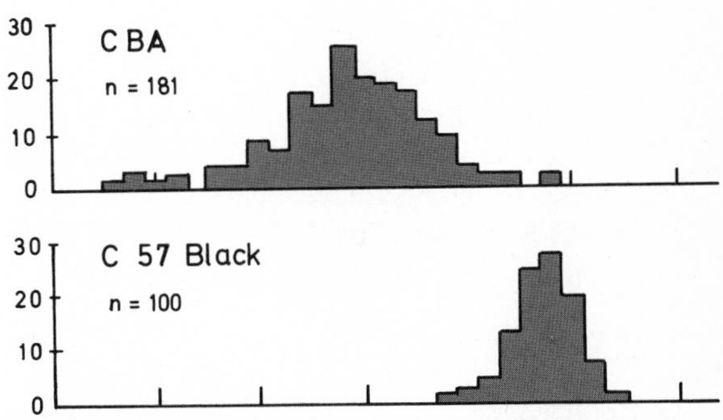

Fig. 3.60. Distribution of the sizes of the third lower molar in two inbred mouse strains, CBA *(above)* and C57 black *(below)*. (Grüneberg, 1952 [690])

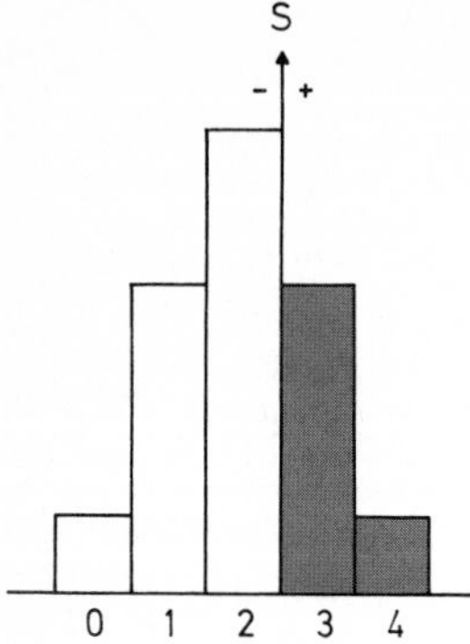

Fig. 3.61. Multifactorial inheritance of two gene pairs A, a, B, b in combination with a threshold: phenotypes −, □ and +, ■ depending on the number of genes A, B in a random mating population. Gene frequencies A = B = a = b = 0.5. Five phenotypes (0, 1, 2, 3, 4) are possible

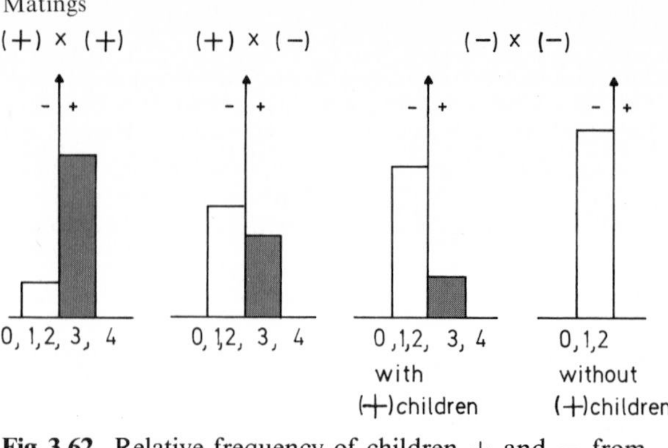

Fig. 3.62. Relative frequency of children + and − from four different mating types, if the genetic model described in Fig. 3.63 holds true

obvious from the onset that the assumption of equal contributions of all genes to the phenotypes is an oversimplification. However, if their contribution were to be unequal, at what degree of contribution of one locus to phenotypic variability should we start talking about a "major gene"?

The case examined above is a very special one. In App. 4, the model of multifactorial inheritance of an anomaly with a threshold effect will be treated in a more general way. The following criteria for multifactorial inheritance and against the simple diallelic mode of inheritance follow intuitively from the simple, special model set out above but can be derived in a more stringent way from the considerations in App. 4.

3.6.2.3 How Should the Model Be Used for Analysis of Data? [925]

These theoretical results have to be utilized critically in the analysis of actual data. As mentioned repeatedly, the multifactorial model is an abstraction and gives an oversimplified picture of the way in which multiple gene loci co-operate to create a liability. Besides, the data normally available are limited and, therefore, tend to have high sampling variances.

Qualitative (or Semiquantitative) Criteria for Multifactorial Inheritance. Four such criteria can be derived:

1. The twin criterion: if concordance in MZ twins is more than four times higher than in DZ twins, a multifactorial model is more adequate than a simple diallelic model (Table 3.19). The opposite does not apply: a concordance ratio of less than 4 does not necessarily rule out the multifactorial hypothesis.

Table 3.19. Twin concordance with various modes of inheritance

	MZ twin concordance	DZ twin concordance
Autosomal-dominant inheritance	100%	50%
Autosomal-recessive inheritance	100%	25%
Multifactorial inheritance with environmental influences	~40%–60%	~4%–8%

2. Segregation ratio of affected and unaffected sibs in matings plus × minus and minus × minus: if affected sibs are more than 2.5 times more frequent in matings with one affected parent than in matings with two unaffected parents, the multifactorial model should be preferred. Here, again, a ratio of less than 2.5 does not exclude multifactorial inheritance.
3. Sex ratio of affected persons: many anomalies for which multifactorial inheritance should be considered show a sex difference in incidence. In most cases, only a small part of this sex difference can be directly related to the sex chromosomes. Most of it will be due to physiologic differences between the sexes. Hence, it is reasonable to assume that the genotypic liability shows the same distribution in both sexes but that the threshold is different. Consequently, an affected person belonging to the sex with the lower incidence, will have, on the average, a higher personal lability than individuals of the other sex. This liability should show up by differences in the frequency of

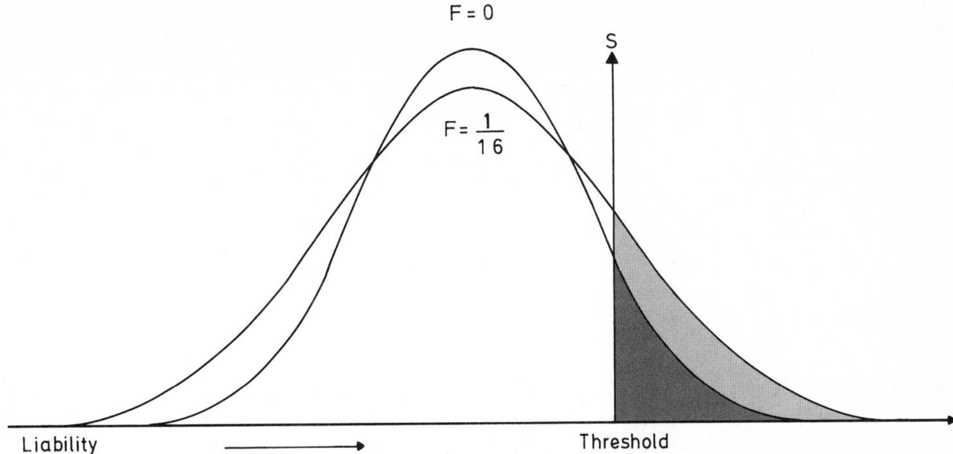

Fig. 3.63. Distribution of genetic liability with random mating and with $F=\frac{1}{16}$ (first-cousin marriage). The areas at the *right* of the threshold indicate the increase in frequency of a threshold character. S, threshold

Table 3.20. Pyloric stenosis: frequency among the close relatives of male and female probands (Fuhrmann and Vogel, 1983 [71])

Number and sex of the probands	Brother	Sister	Son	Daughter	Nephew	Niece	Male cousin	Female cousin
♂ 281	5/230 2.17%	5/242 2.07%	19/296 6.42%	7/274 2.55%	5/231 2.16%	1/213 0.47%	6/1061 0.57%	3/1043 0.29%
♀ 149	11/101 10.89%	9/101 8.91%	14/61 22.95%	7/62 11.48%	4/60 6.67%	1/78 1.28%	6/745 0.81%	2/694 0.29%

Source: adapted from Carter [601].

affected relatives. The sex with the lower incidence should have a higher proportion of affected relatives, when the same degrees of relationship are compared. This argument was first brought forward by C.O. Carter [601] and is sometimes called the Carter effect. Carter demonstrated it in pyloric stenosis, an anomaly in the newborn, in which the thickening of the pyloric muscle prevents release of stomach contents into the duodenum. While this defect is much more frequent in infant boys than in girls, a higher incidence was demonstrated among relatives of affected girls than among relatives of affected boys (Table 3.20 and Fig. 3.65). The Carter effect can also be demonstrated for the electroencephalogram (EEG) showing "diffuse" β-waves, a condition that is much more frequent in women than in men. Here, relatives of male probands show much higher incidences [921]. On the other hand, this effect could not be clearly demonstrated with family data on cleft lip and palate where one would expect more affected relatives of female probands – the lesser affected sex.

4. Consanguinity: in the models examined above, random mating was assumed. With consanguinity, however, the distribution of the liability in the population will have a higher variance:

$$V_F = V_O \times (1+F) \tag{3.9}$$

Here, F = the inbreeding coefficient, V_F = variance among all progeny from matings with inbreeding coefficient F, V_O = variance with random mating (Fig. 3.63). Figure 3.64 shows the increased incidence among children from first cousin matings ($F=1/16$) compared with the random mating population. The much larger increase observed with monogenic, autosomal recessive inheritance is given for comparison. In most cases, however, autosomal dominant inheritance with reduced penetrance rather than autosomal recessive inheritance will be the obvious alternative to the multifactorial model. Therefore, a moderate increase of the condition with inbreeding is an additional argument favoring a multifactorial model over an autosomal-dominant model, provided that admixture of families with a rare, autosomal-recessive type can be excluded.

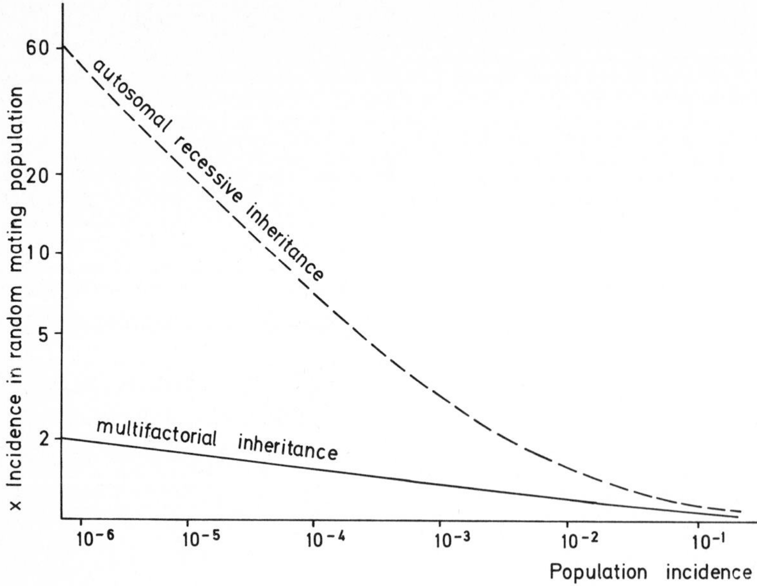

Fig. 3.64. Increased incidence of autosomal recessive and multifactorial characters among children from first-cousin matings compared with population incidence

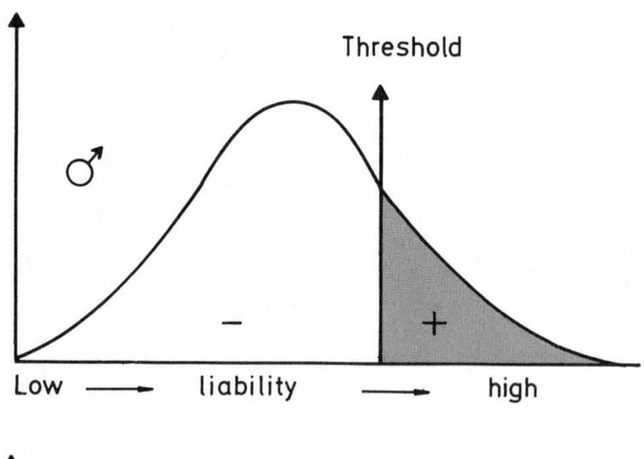

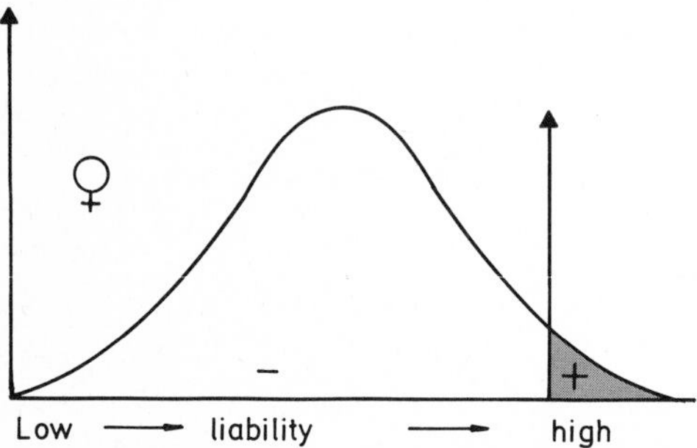

Fig. 3.65. A multifactorial condition may be more common in one sex than in the other. Pyloric stenosis, for example, is more common in males than in females. The genetic liability, on the other hand, can be assumed to be identical in both sexes. The position of the threshold differs. As a consequence, the average affected male expresses the trait with a lower genetic liability than the average affected female. Therefore, the incidence of this condition among relatives of male probands is expected to be lower than among relatives of affected female probands who carry more of the predisposing genes than affected males. This phenenomen is sometimes called the "Carter effect" [601]

Quantitative Criteria. It is not entirely satisfactory to have only semiquantitative criteria on which to base a genetic model. Thus, a method for quantitative comparison is needed. Here, it is possible to examine the 95% confidence intervals for the incidences Q_{11}, Q_{21}, and Q_{22} in children of mating types minus × minus, plus × minus, and plus × plus calculated from incidences in a sample to see whether the expectations fall within these confidence intervals (App. 4). However, a better way may be to test whether the overall pattern of the incidences in different types of relatives better fits the expectations derived from one or the other model. Such methods have been described by Morton et al. [804] and by Ch. Smith [875 a]. For an example see a study on infant glaucoma [628].

The incidence of the character in question are determined among the different degrees of relatives of the probands. Then, the joint probability of all these incidences together is compared with its theoretical expectation from the multifactorial model on the one hand and the diallelic model on the other. If one model fits the data within statistical limits while the other one shows significant deviations, the first model is accepted. The technical details of calculation cannot be described here (see Appendix 4). Similar methods for identification of "major genes" have been proposed by various authors [646; 852; 963].

3.6.2.4 If the Statistical Analysis Gives No Clear Answer, How Should We Decide?

The above discussion indicated that compatibility of a set of data with a genetic model does not mean that this model is true. A completely different model could fit the same data equally well. As shown above and in App. 4, there is considerable overlap in expectations between the special models chosen here as examples for a diallelic single locus and a multifactorial model. It is a general rule that scientific hypotheses can be refuted when the observations do not fit but cannot be accepted until all other possible and plausible hypotheses have been excluded. However, the human geneticist working with anomalies that have simple modes of inheritance tends to forget this rule, as the relationship between observation and genetic hypothesis is quite straightforward in the usual cases of simple modes of inheritance.

How should one proceed when the statistical data do not permit a decision in favor of either of these hypotheses? The most obvious answer would be to leave the problem open. However, the description of family data evading clear interpretation are a challenge and, furthermore, the diseases involved are frequent and practically important so that more thorough research into their causes is needed. Hence, some guidance for further studies might be desirable.

The hypotheses of a major gene has many advantages for research strategy. An anomaly showing a simple mode of inheritance must have a simple biochemical cause: the lack of an altered gene product of a structural or regulatory gene. Acceptance of a major gene hypothesis will naturally lead to a search for this cause, which for autosomal-dominant inheritance remains in its very beginnings (Sect. 4.6). In multifactorial conditions caused by a combination of various small physiologic deviations - possibly together with environmental influences - the major gene hypothesis leads usually nowhere and has caused therefore considerable frustration for research workers. The history of research into the neurophysiologic, biochemical, and immunologic basis of schizophrenia is one example (Sect. 8.2.3.7).

The genetic hypothesis of multifactorial inheritance is more cautious and conservative; adopting it as a preliminary description of the data, we remain conscious that it represents analysis at the level most remote from the gene action: the black box must still be opened up. In thinking about strategies, we are not guided into one direction by an overpowering genetic hypothesis but remain open-minded to various possibilities. If pursuing one of them should indeed lead to discovery of major gene action, we would be overjoyed, as this would bring our analysis down to a more genetic or biochemical plane. However, if the attempt should not succeed, we are still open to considering how a smaller deviation of a physiologic parameter - which might even be present in only a fraction of our probands - might interact with other small deviations to cause a truly multifactorial disease liability. Moreover, other research strategies intended to define precisely and on a level closer to gene action at least one component of the multifactorial system may be considered.

Therefore, if one cannot be reasonably certain that single gene action applies by a clear-cut genetic or biochemical criterion or both, acceptance of the more general multifactorial model is the wiser decision. However, it must be kept in mind that in many cases a major gene has not really been excluded. This may have consequences for our attitude toward genetic risks due to mutagenic agents (Sect. 5.2.1): adopting a multifactorial model without reservation may lead us to underestimate genetic threats. To avoid this error, it is good to remember some experiences from genetic research in experimental mammals.

3.6.2.5 Radiation-Induced Dominant Skeleton Mutations in the Mouse: Major Gene Mutations That Would Not Be Discovered in Humans

Experimental work with mammals whose developmental physiology is closer to man's than that of the more usual species of genetic research has shown how major gene action may be hidden in the phenotypic variability of the organism. Such major genes may be exposed by suitable breeding studies or by phenotypic analysis of induced mutations. One example will be discussed, which is also important for risk assessment of mutation induction in humans [865, 640].

Genetic damage induced by dominant mutations can be assessed by comparing first-generation descendants from treated and untreated animals, but for many characters it is difficult to distinguish between the effects of newly occurring genetic damage and the variation existing within a strain. This difficulty was overcome for a group of skeletal abnormalities of the mouse. In a mutation experiment, the abnormalities observed in the F_1 generation can be subdivided into those occurring only once in the entire experiment (class 1) and those occurring more frequently (class 2). It is reasonable working hypothesis to assume that most of the rarer abnormalities (class 1) will be due to mutation, whereas most of the more frequent abnormalities (class 2) represent the variability of the strain when examined in a study of several hundred animals. With this hypothesis, mutagenic influences, such as ionizing radiation, are expected to enhance primarily the number of rarer (class 1) abnormalities. This expectation has been confirmed in a fair number of experiments with ionizing radiation and chemical mutagens.

Phenotypically, the majority of rare (class 1) and more common (class 2) abnormalities are multiple, minor skeletal variants. Some of these, such as those affecting the vertebral column, are detrimental to various degrees. 31 of such skeletal variants have been confirmed as dominant mutations by breeding tests. Two characteristics of these dominant mutations stand out:

1. Some or all of the abnormalities caused by each mutation have low penetrance.
2. Only a small proportion of these mutations can be detected externally, and those that can be detected do not manifest in most gene carriers. Breeding experiments showed that among the offspring of the F_1 animals segregation ratios for abnormal phenotypes remained far below the expected 0.5.

Applying these data to inferences regarding dominant mutations in humans, difficulties in detection could possibly be explained away by arguing that because of closer medical supervision of humans, some or even most of these mutations would be detected. The finding of low penetrance of autosomal mutations is, however, very important for a consideration of dominant genes with incomplete penetrance as compared to multifactorial inheritance. Furthermore, the various anomalies usually found showed an amazing degree of variability between animals carrying mutant genes identical by descent. The patterns of abnormal findings in carriers of different mutations overlapped strongly; several mutations showed almost identical phenotypic effects.

In comparing these results with genetic situations in humans, one should look for human skeletal malformations that might be comparable to those observed in mice. This comparison has, in fact, been attempted in discussions between human and mouse geneticists but was made difficult by the imperfect state of genetic knowledge regarding human skeletal anomalies. The recent flourishing of new research in human skeletal anomalies [774] would make a second look worthwhile at this time. It is discouraging, however, that so far there are hardly any mouse mutations, which are identical to human mutations.

These experiments leave a number of questions open. In particular, the possibility of minor chromosome change has not been excluded so far. However, one general conclusion can be drawn tentatively: There may be a large number of dominant mutations in man causing a wide range of morphologic variability with possible effects on health. Since we are currently limited to methods of phenotypically oriented genetic analysis, few chances to uncover the genetic basis for this variability exist.

3.6.2.6 Isolation of Specific Genetic Types with Simple Diallelic Modes of Inheritance Using Additional, Phenotypic Criteria

It may sometimes be possible to define specific Mendelian subtypes of diseases from a large heterogeneous group of patients. The combination of careful clinical analysis and laboratory studies with genetic analysis may often be successful in isolating genetic from nongenetic entities. Early success was obtained with cases of mental retardation [2157], deafness [669], and blindness [670]. With the development of improved nosology in this field coupled with involvement of careful clinical observers, many patients with mental retardation previously thought to be unclassifiable can now be categorized into specific entities. X-linked mental retardation with marker-X may be mentioned as an example for a very common condition [2220]. Studies on blind and deaf children in residential institutions were also successful. In deafness and blindness, about 50% of all residential cases could be shown to be genetic in origin. Practically all cases were of Mendelian rather than multifactorial origin. Many different types were found.

It is almost a general rule that among the multifactorially caused cases of diseases, rare Mendelian variants can often be distinguished. Thus, X-linked HGPRT deficiency causes 1% of cases of gout. Some cases of hypertension are caused by the rare inherited pheochromocytoma. Peptic ulcer is part of the symptom complex of the Zollinger-Ellison syndrome. Cancer of the esophagus may be rarely caused by a gene producing keratomas of the palms and soles at the same time (Fig. 3.66).

There are a number of syndromes in which cancer occurs (see Sect. 5.1.6) as part of a more comprehensive pleiotropic pattern. Occasionally, families show dominant inheritance of a more or less common cancer. Here, early onset and multiple occurrence of cancers helps to delimit these major gene manifestations from the usual cancer type. In the pedigree (Fig. 3.66), the ages of the cancer patients are 34, 37, 38, 43, 44, 45, 46, 52, and 63, all but the last being very unusual for cancer of the esophagus. In dermatology, many benign and malignant tumors are observed as isolated cases as well as in families. Here, Schnyder [84] established a rule according to which single tumors in one patient favor a nongenetic origin, whereas multiple tumors tend to be inherited and frequently show an autosomal-dominant mode of inheritance (see also Sect. 5.1.6).

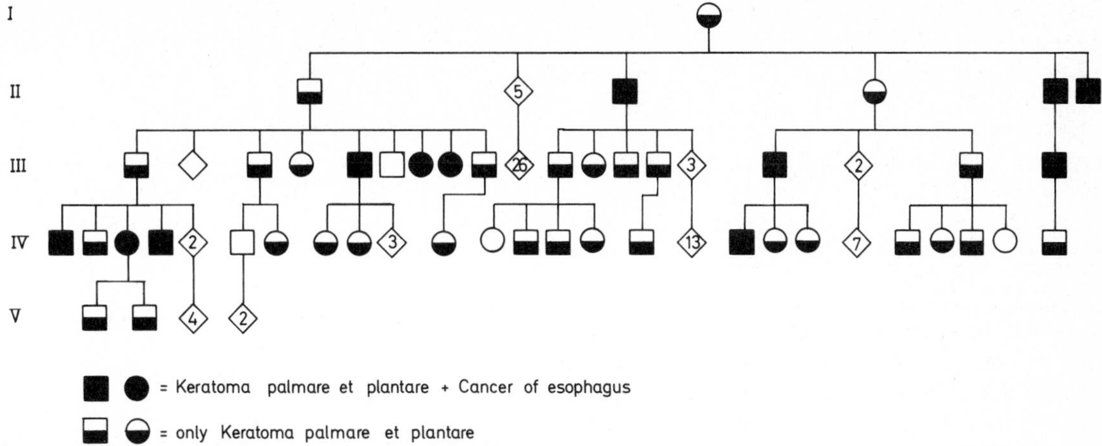

Fig. 3.66. Cancer of the esophagus as an additional symptom (■) in patients with a special, autosomal-dominant type of keratoma palmare et plantare (Howell-Evans et al., 1958 [716])

3.6.2.7 How Can an Apparently Multifactorial Condition Be Analyzed Further, when Special Types with Simple Modes of Inheritance Cannot Be Isolated?

A Complex Functional Defect Is Caused by a Combination of Small Aberrations. As mentioned above, the additive model used for a somewhat more quantitative understanding of multifactorial inheritance is an oversimplifying abstraction. In reality, the variability is not unidimensional, and a variety of different genetically determined physiologic influences may co-operate to give rise to a certain condition. It should be possible to isolate some of these influences.

In two series of children with strabismus examined by one investigator [856], the figures in Table 3.27 were found for parents and for siblings resulting from different mating types. Of 12 monozygotic twin pairs, 11 were concordant, whereas only 7 of 27 dizygotic pairs showed concordance. The data strongly point toward multifactorial inheritance. Incomplete dominance cannot be excluded but would require additional influence by the genetic background.
It is known that strabismus is the end result of a number of minor physiologic aberrations. Each of these alone can be overcome to achieve normal vision. When several of these aberrations are combined, the regulatory capacity of the visual system decompensates, and squinting results. Such aberrations were shown to occur more frequently among close relatives of the proband. In the pedigree in Fig. 3.67, three patients squint; two parents show isolated heterophoria (slight motoric weakness). One parent had an isolated anomaly of refraction, another shows heterophoria. The eyes of one parent were completely normal. The conclusions from this study - that strabismus is a multifactorial condition and that some physiologic factors can be isolated - have been confirmed later and were extended in a different population [709].
An attempt to analyze the genetic susceptibility to congenital dislocation of the hip was successful in demonstrating presumably polygenic factors affecting the concavity of the acetabulum together with a possibly monogenic factor affecting joint laxity [619].

Table 3.21. Frequency of manifest strabismus among sibs of children with strabismus (Richter, 1966) (Vogel and Krüger, 1967 [925]) (+ = manifest strabismus)

Mating type of parents	Number of propositi	Number of sibs	Manifest strabismus in sibs
Series of 697 patients (4-7 years old)			
+ × +	24	33	11 (33.3%)
+ × −	288	301	95 (30.6%)
− × −	385	478	98 (20.5%)
Series of 136 school children (12 years old)			
+ × +	6	6	3 (50.0%)
+ × −	61	120	52 (43.3%)
− × −	69	82	2 (29.3%)
Population frequency of strabismus: 3%-4%			

S. Richter's twin series on strabismus (1966)

	Concordant	Discordant	Total
Monozygotic twins	11	1	12
Dizygotic twins	7	20	27
Total	18	21	39

Family investigations that consist of careful examination of family members for related and associated abnormalities may help in understanding the relative importance of elements leading, when combined with each other, to a complex functional defect. This is possible even if single gene action cannot be identified.

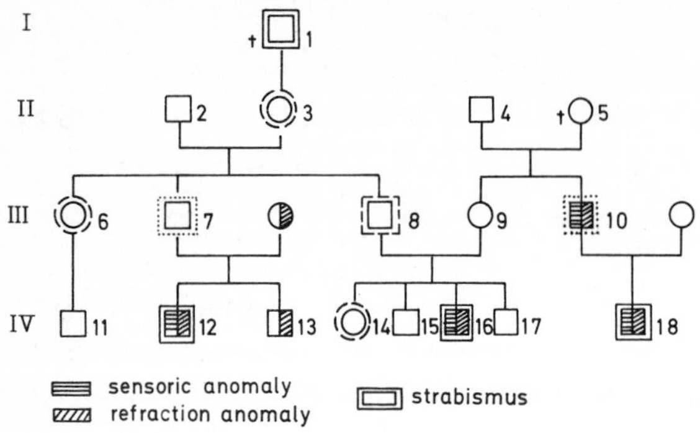

Fig. 3.67. Strabismus in three members of a family. Other relatives show different, minor anomalies. *Dotted and stippled line* represent various borderline results. Sensory anomalies observed in such pedigrees include, for example, amblyopia or imperfect binocular vision. (Richter, 1966 [856])

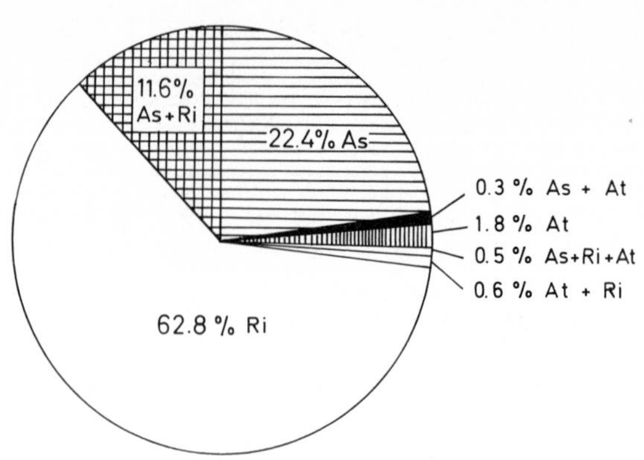

Fig. 3.68. Relative frequencies of probands manifesting one, two or even three atopic diseases. *As*, asthma; *Ri*, rhinitis; *At*, atopic dermatitis. (Data from Zurich, Switzerland; Schnyder, 1960 [894])

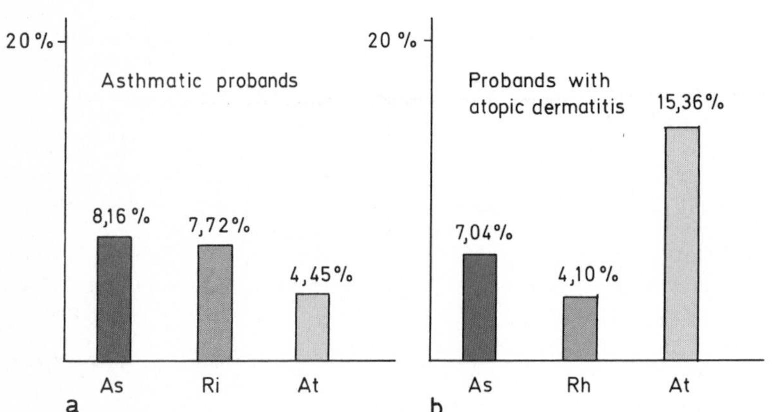

Fig. 3.69 a, b. Frequency of atopic dermatitis and respiratory atopies in relatives of probands with asthma (**a**), with atopic dermatitis (**b**). *As*, asthma; *Ri*, rhinitis; *At*, atopic dermatitis. (Data from Dorn and Schwarz [924])

A Multifactorial System Comprises a General Disposition That May Lead to a Group of Related Diseases; Specific Dispositions Influencing the Clinical Manifestation Pattern. The group of "atopic diseases" comprises atopic dermatitis, bronchial asthma, and hay fever. Figure 3.68 shows the relative frequencies of probands manifesting one, two, or three atopies in the population of Zurich [894].

Family investigations are most compatible with multifactorial inheritance. A further question can be posed. Is the influence of the genes on the liability to atopic diseases only one-dimensional and quantitative, or are there other genes influencing the organ specificity of disease manifestation?

If the liability shows a one-dimensional distribution, skin atopies (dermatitis) and respiratory atopies (asthma, hay fever) should occur in the same ratio among relatives of probands with either skin atopies or respiratory atopies. On the other hand, if organ-specific factors are involved, a certain accumulation of similar atopies among the probands' relatives should appear.

Figure 3.69 illustrates this comparison; among first-degree relatives of asthmatics, respiratory atopies are much more frequent, whereas among relatives of dermatitis patients, atopic dermatitis prevails. Thus, within the multifactorial genetic system determining the genetic liability to atopic diseases there are factors increasing the liability in general that act side by side with others influencing special organ manifestations.

Such an analysis is more satisfying than mere attempts to fit overall incidence figures with expectations derived from genetic models that are oversimplified from the outset. Despite this improvement, genetic analysis remains at the biometric level, remote from gene action. "Breaking open the black box" is now on the way. Thus, it could be shown that clinical ragweed pollinosis (hay fever) is influenced by interaction between two gene loci one of which regulates basic IgE production, whereas the other acts on IgE production in reaction of the allergen. The latter is identical with or closely linked to the HLA-A2 allele [775]. Genetic influences at other levels of the immune answer are very well possible [844]. A probable selective advantage of atopic genotypes under more primitive living conditions will be discussed in Sect. 6.2.1.

3.7 Genetic Polymorphism and Disease

3.7.1 New Research Strategy

To achieve deeper insight into the mechanisms involved in multifactorial inheritance, a change in research strategy is needed. If the direct path from the phenotype to the genotype fails, the opposite route – from the gene and its action to the phenotype – might possibly be more successful.

At first glance, this proposal sounds paradoxic; we have started from the phenotype because there is no other access to the genotype. Any other way seems barred by the nature of the genetic material. However, the multifactorial model assumes the co-operation of many genes; some of these are expected to be frequent. Work on human genetic polymorphisms, on the other hand, has succeeded in uncovering and analyzing genetic variability at many gene loci, which determine cell surface antigen structures, enzyme proteins, and serum proteins with many different – and in many cases unknown – functions. Hence, it is not farfetched to examine whether some of these polymorphisms might be parts of the multifactorial system influencing disease liabilities.

Harris and others [1787] showed that at least one-third of structural genes determining blood enzymes exist as polymorphisms. Thus, not every "normal" person will have the same gene product, and variant proteins and enzymes will be commonly found. It has been estimated that there are between 50,000 and 100,000 structural genes in humans. Therefore, thousands of polymorphisms presumably exist but only fewer than 150 have been discovered so far. At the present time, therefore, a search for diseases associated with existing polymorphisms will usually be fruitless unless a pathophysiologic relationship can be demonstrated. An important research strategy needs to document the existence of further polymorphisms to allow future search for individual genes sometimes involved in disease. The finding of additional polymorphisms will also be of great importance for completion of the human gene map.

3.7.2 Disease Association of the Blood Groups

3.7.2.1 AB0 Blood Groups

Shortly after the AB0 blood groups had been discovered, associations with certain diseases were suspected. During the 1920s, the first phase in examination of these associations reached its culmination. At that time, almost every common disease was thought by some authors to be associated with the blood groups. Most of these studies, however, were carried out with insufficient numbers and inadequate methods. The results were largely contradictory. Therefore, the reaction of most specialists during the following years in understandable: In their – basically justified – criticism, they threw out the "baby with the bath water," and for a long time the blood groups were regarded as not being associated with disease.

Wrong Hypothesis Leads to an Important Discovery. The first hint of blood group disease associations was the discovery of serologic mother-child incompatibility in the Rh system. A short time thereafter, associations with common diseases were discovered.

Aird et al. (1953) [552] described an association between group A and cancer of the stomach. Earlier, P. Stocks had shown in 1950 mortality of stomach cancer to be higher on the average in northern English than in southern English cities. In his opinion, this effect could be due to some substance more abundant in the north, which might irritate the mucous membrane of the stomach. In the search for such a substance, he found a relatively weak correlation with hardness (calcium content) of the water: Cities with low-calcium water tended to have a lower mortality rate from stomach cancer.

Aird et al., on the other hand, thought rather that the difference might be genetically determined. At that time, mouse strains with high and low cancer frequencies had been described. In their search for a possible genetic parameter, they came across an analogy with AB0 distribution. In northern England, 0 is more frequent, in the south A. Their working hypothesis was that group 0 would be associated with stomach cancer, causing its higher incidence in the north. To examine this hypothesis, they collected cancer cases in various English and Scottish cities and compared their AB0 distribution with that of carefully selected control samples, which were, as a rule, patients treated in the same hospitals for a number of different complaints.

Table 3.22 shows the result. Contrary to the working hypothesis, a significant association with group A and not 0 was found. Hence, the higher cancer mortality in the north could not be caused by the higher frequency of group 0 in this area. This study triggered a flood of investigations on blood group associations of disease.

Statistical Standard Method. [959] Before the most important results are presented, the standard statistical method used for their analysis must be explained. Incidence of two characters (or groups of characters, for example, A versus 0 or A+B+AB versus 0) are compared in two samples, the patient sample and the control sample. The ratio:

Table 3.22. Differences in relative frequencies of group A and 0 in patients with Cancer of the stomach and controls (Aird et al., 1953 [552])

	Cancer cases		Controls		$\frac{A}{A+0} \times 100$			χ^2
	0	A	0	A	Cancer	Controls	Differences	
Manchester	343	349	402	295	50.43	42.32	+ 8.11	9.183
Liverpool	85	97	108	86	53.30	44.33	+ 8.97	3.022
Leeds	92	104	102	87	53.06	46.03	+ 7.03	1.902
Birmingham	37	57	50	44	60.64	46.81	+13.08	3.616
Newcastle	44	44	53	37	50.00	41.11	+ 8.89	1.418
London	578	617	614	565	51.63	47.92	+ 3.71	3.267
Scotland	245	174	252	155	41.53	38.08	+ 3.45	1.022
	1,424	1,442	1,581	1,269	50.31	44.53	+ 5.78	19.198

The last column contains the single χ^2 values calculated for the 2×2 tables.

	Degrees of freedom	χ^2	P
Sum of single χ^2	7	23.430	0.0015
Overall χ^2 for England and Scotland	1	19.198	$10^{-4}-10^{-5}$
Heterogeneity χ^2	6	4.232	0.65

$$x=\frac{\text{A(Pat)} \times \text{0(contr)}}{\text{0(Pat)} \times \text{A(contr)}}; \quad y=\ln x \tag{3.10}$$

has the expectation 1, when the ratio A/0 is identical in the two samples, i. e., if there is no association.[9] Otherwise, the ratio x will be higher or lower than 1. This ratio x is commonly called the "relative incidence." In our example, its meaning is that the incidence of stomach cancer in persons of group A is x times higher than in persons of group 0. The deviation of x from 1 can be tested for significance as follows:

$$V=\frac{1}{w}=\frac{1}{\text{A(Pat)}}+\frac{1}{\text{0(Pat)}}+\frac{1}{\text{A(Contr)}}+\frac{1}{\text{0(Contr)}}$$

χ^2 of deviation $= y^2\, w$ (1 degree of freedom).

A number of estimates of x can be combined to a common estimate as follows:

$$Y=\frac{\sum wy}{\sum w} \quad Y=\ln X;$$

χ^2 of deviation: $Y^2 \sum w$ (1 degree of freedom)

χ^2 of heterogeneity: $\sum w\, y^2 - Y^2 \sum w$

(degrees of freedom = number of single comparisons - 1).

Standard deviation of Y: $\sigma=\frac{1}{\sqrt{\sum w}}$

[9] A (Pat) is the absolute number of A individuals in the patients' sample and A(Contr) the absolute number of A individuals among controls.

A Flood of Investigations and Their Results [211, 145]. Within a period of about 15 years, a number of associations for common diseases were detected (Table 3.23). Apart from stomach cancer, which was examined in at least 101 different samples, several other malignant neoplasias showed that the risk of being affected was somewhat higher for patients with blood group A. A tendency in this direction was also found in a number of nonmalignant diseases such as rheumatic diseases, pernicious anemia, and - with a stronger effect - in thrombotic and thromboembolic diseases. An association with group O, on the other hand, was shown for gastric and duodenal ulcers. These data indicate that if blood group A has a small but significant disadvantage in predisposing its carriers to some diseases comprising some of the most frequent causes of death in our society, a higher frequency of blood group 0 in healthy, old people as compared with the general population might be expected. This expectation was confirmed in one study of persons over 75, who, at the time of examination, were still in reasonable good health. The difference was especially striking when aged persons who had undergone major surgery were used as controls [728].

Possible Biases. Large-scale statistical investigations of this type are subject to certain biases:

a) Selection of appropriate controls. Human populations are not uniform in their blood group distributions. In spite of conformity of the data to Hardy-Weinberg proportions, there may be hidden stratification of subgroups differing in gene frequencies at the gene loci under examination. If the controls happen to be taken consistently from a subgroup other than the patients, a spurious association might result. For example, if blood group 0 confers especially good health on its bearers, the incidence of 0 in samples of blood donors who are more likely to be a particularly

Table 3.23. Significant associations between blood groups and noninfectious disease

Diagnosis	No. of series	No. of Patients	No. of Controls	Comparison	X	χ^2 for X (df=1)	Significance	χ^2 for heterogeneity	df	
Neoplasias of the intestinal tract										
Cancer, stomach	101	55,434	1,852,288	A:0	1.2238	386.267	c	2,178.127	100	c
Cancer of colon and rectum	17	7,435	183,286	A:0	1.1099	13.790	c	10.163	16	
Malignant tumors of salivary glands	2	285	12,968	A:0	1.6432	13.008	c	14.515	1	c
Cancer, pancreas	13	817	108,408	A:0	1.2359	7.549	b	15.048	12	
Cancer, mouth and pharynx	2	757	41,098	A:0	1.2478	7.703	b	1.084	1	
Other neoplasias										
Cancer, cervix	19	11,927	197,577	A:0	1.1334	30.959	c	29.362	18	a
Cancer, corpus uteri	14	2,598	160,602	A:0	1.1515	10.163	c	17.511	13	
Cancer, ovary	17	2,326	243,914	A:0	1.2789	26.630	c	19.362	15	
Cancer, breast	24	9,503	355,281	A:0	1.0827	11.183	c	31.042	23	
Multiple primary cancers	2	433	7,823	A:0	1.4340	10.401	c	1.396	1	
Nonmalignant tumors										
Nonmalignant salivary tumors	2	581	12,968	A:0	2.0153	54.874	c	23.183	1	b
Other internal diseases										
Duodeni ulcers	44	26,039	407,518	0:A	1.3492	394.710	c	80.977	43	b
				0:A+B+AB	1.3344	447.196	c	84.415	43	b
Gastric ulcers	41	22,052	448,354	0:A	1.1694	95.933	c	78.964	40	b
				0:A+B+AB	1.1774	125.107	c	62.978	40	a
Duodenal and gastric ulcers	6	957	120,544	0:A	1.5291	26.973	c	19.453	5	b
				0:A+B+AB	1.3561	18.722	c	24.120	5	b
Ulcus, without differentiation between stomach and duodenum	11	4,199	88,239	0:A	1.1462	14.864	c	8.833	10	
				0:A+B+AB	1.1765	24.988	c	16.828	10	
Bleeding ulcers (gastric and duodenal)	2	1,869	28,325	0:A	1.4640	52.973	c	0.457	1	
				0:A+B+AB	1.5076	72.879	c	0.567	1	
Rheumatic diseases	17	6,589	179,385	A:0	1.2350	49.765	c	28.575	16	b
				A+B+AB:0	1.2341	57.402	c	32.850	16	
Pernicious anemia	13	2,077	119,989	A:0	1.2453	20.149	c	11.904	12	
Diabetes mellitus	20	15,778	612,819	A:0	1.0710	13.719	c	37.543	19	b
				A+B+AB:0	1.0721	16.243	c	42.198	19	b
Ischemic heart disease	12	2,763	218,727	A:0	1.1817	13.906	c	22.808	11	a
				A+B+AB:0	1.1743	15.033	c	29.183	11	b

Table 3.23 *(continued)*

Diagnosis	No. of series	No. of Patients	No. of Controls	Comparison	X	χ^2 for X (df=1)	Significance	χ^2 for heterogeneity	df	
Cholecystitis and cholelithiasis	10	5,950	112,928	A:0	1.1734	25.746	c	9.637	9	
Eosinophilia	3	730	1,096	A:0	2.3792	45.757	c	0.597	2	
				A+B+AB:0	2.1315	48.920	c	0.961	2	
Thromboembolic disease	5	1,026	287,246	A:0	1.6135	45.500	c	23.364	4	c
				A+B+AB:0	1.6040	48.894	c	22.537	4	c

[a] $P \leqq 0.05$; [b] $P \leqq 0.01$; [c] $P \leqq 0.0027$. No letter is given, if the difference is not significant.

healthy subgroup of the population might be too high.

b) Publication only of positive results. It is an understandable wish of research workers to have their work rewarded by "positive" results, i.e., in this case, by discovering an association. It is therefore possible that only those who find a "significant" association (possibly by chance) will publish their data. Others, who have been less "lucky," will leave their data unpublished. The accumulation of positive publications leads to a spurious association.

It has been shown that these biases cannot be responsible for the associations found [211]. It is comforting that data collected for many other diseases gave consistently negative results despite the fact that patient and control samples were collected identically and evaluated so that the same biases should be at work. Congenital malformations are one example. The entire group that includes congenital heart disease, harelip and cleft palate, malformation of kidney and urinary tract, hydrocephalus, and others did not show any blood group association even though 4,762 patients and 156,716 controls were examined [211].

Failure to Find a Mechanism. In the early years of studies of blood groups and diseases, many authors tried to speculate on biologic reasons for these associations. The role of blood group substances, for example, in the secretions of stomach and duodenum, was thought to be responsible. Indeed, tumors of organs containing much of this substance, e.g., salivary glands or ovaries, showed especially strong associations. A more general hypothesis tried to connect these associations with stronger immune responses in group 0 carriers as compared with group A persons. This hypothesis led to studies in population genetics (Sect. 6.2.1.8) but was not pursued experimentally for the common diseases such as duodenal ulcer and cancer of the stomach. Better understanding will possibly have to await more thorough immunologic knowledge about the role of the cell surface, specifically its glycoproteins, in interaction with other cells and with environmental influences. The fact that attempts to demonstrate a convincing mechanism for the associations have failed so far may have contributed to a disappointment among research workers. In recent years, the flood of work on blood group associations has dried up almost completely.

It has become clear also that the total contribution of the AB0 genes to the genetic etiology of these diseases probably is small as shown in an analysis of the contribution of blood group 0 to peptic ulcer [637]. These studies, therefore, while statistically clear-cut in the case of peptic ulcer, cancer of the stomach, and some other conditions, have not aided in further understanding of the genetic and environmental etiology of these diseases.

3.7.2.2 The Kell System

Kell System Mutations, Acanthocytosis, and Chronic Granulomatous Disease. Apart from disease associations of common blood groups, some examples of hereditary anomalies due to rare or modifying genes of "blood group" genes are known. In Sect. 3.1.7, modifying genes in the AB0 system are described; to the best of our knowledge, they are without major significance for the health of their bearers. However, the Kell blood group system provides examples on direct associations between rare blood group alleles and disease. They are especially interesting, because the Kell substance is known to be involved in the structure of cell membranes. For a deeper understanding of blood group disease and HLA associations, better knowledge of membrane function will be essential, and such rare alleles might be especially useful for further understanding.

Various alleles at the autosomal Kell locus exist in populations of European origin; there are two alleles K, k, the rarer allele K has a frequency of 0.05.
An allele of the Kell system (Js) is found in 14%-20% of black American individuals but is extremely rare in other populations and therefore constitutes an excellent marker gene for African origin. Hemolytic disease of the newborn rarely is caused by anti-Kell antibodies; but if it occurs the basic mechanism is similar to that of Rh hemolytic disease.
In addition to the autosomal locus for the Kell antigen, an X-linked locus that codes for a precursor substance of Kell known as Kx has been identified. All normal persons have Kx antigenic activity on both red and white cells. Some individuals are homozygous for a "silent" allele (Ko) [609] of the autosomal Kell locus. In such cases, none of the usual Kell antigens but a strong Kx reaction can be detected [777]. This finding is compatible with the interpretation that unconverted Kx material specified by the X-linked locus is the only Kell-related antigen in homozygous carriers of the Ko or silent allele. Such persons are clinically and hematologically normal. Mutations at the Kx locus have been identified and may lead to phenotypic expression in red or white cells or in both types of cells [776].
The McLeod phenotype of red cells [957] is caused by an X-linked mutation causing lack of the Kx substance. Absence of the Kx antigen from the red cells causes a membrane abnormality associated with acanthocytosis ("spiny" red cells) and increased red cell destruction. The severity of hemolysis may range from compensated blood destruction to severe hemolytic anemia [776]. A-β-lipoproteinemia [592] - the usual cause of acanthocytosis (14595) - is not present. The red cell anomalies clearly are caused by absence of Kx since red cells lacking all Kell antigens except Kx (Ko) are morphologically normal. As expected, the fully expressed McLeod phenotype is only seen in males.
Mothers of males with the McLeod or mutant Kx phenotype are expected to be heterozygotes for both the normal Kx and the mutant Kx allele. The principle of X inactivation (Sect. 2.2.3.3) postulates that such women would have a mosaic population comprising cells expressing the normal Kx as well as the mutant Kx allele. In fact, red cell populations consisting of normal and abnormal cells have been observed [890]. Such mosaicism could be demonstrated by both immunologic and morphologic techniques since the Kx-negative cells were acanthocytes. The abnormal cells were outnumbered by the normal cells as explained by the shortened red cell survival of the Kx-negative cells. The X-linked Kx antigen presumably codes for a membrane protein. Mutations at that locus produce pathologic membrane alterations leading to the morphologic red cell abnormalities and hemolysis.
Effects of mutations affecting Kx are often observed in white cells. Giblett et al. [682] recognized poorly characterized Kell types in several boys with chronic granulomatous disease (CGD). These abnormalities later were shown to be mutations of Kx. Patients with CGD [589] have a high susceptibility to low-grade bacterial pathogens and develop granulomatous lesions filled with lipid-containing macrophages. Skin, lymph node, and pulmonary infections are common. Lymphadenopathy, hepatosplenomegaly, and hypergammaglobulinemia are often seen. The anomaly was known to be X-linked in many cases before the Kell blood group associations were known. The white cells in this disease have impaired killing power against many bacterial organisms with preservation of normal phagocytosis and normal lysosomal activity [715]. The exact relationship of the mutant Kx gene to the lack of bactericidal activity of white cells has not been determined but may be related to the involvement of the Kx membrane antigen in triggering activities of the NADH dehydrogenase required for bactericidal activity [776]. All other leukocyte antigens are normal. Some patients with granulomatous disease lack the Kx substance in both white cells and red cells. Such patients have this disease as well as acanthocytosis and hemolysis [776]. Most patients with Kx-associated CGD have normal red cells, while a few patients with X-linked acanthocytosis (McLeod phenotype) and normal leukocytes have been observed. At least six X-linked and three autosomal recessive types of CGD are known; the enzyme defects range from defective initiation of oxidative metabolism to defective NADPH supply and cytochrome b deficiency. Hence, the syndrome is highly heterogeneous; Kx deficiency is only one of several types [714].

3.7.3 The HLA System and Disease [888, 207 a]

As explained (Sect. 3.5.5), the major histocompatibility complex (MHC) on chromosome 6 is homologous with the H2 complex of the mouse [113]. Immunization of inbred mouse strains with a variety of apparently unrelated antigens (synthetic polypeptides, serum proteins, cell surface antigens) induces high levels of antibodies in some strains and low levels (or no response) in others. The quantity of antibodies induced is controlled by immune response loci (Ir), which are part of the H2 complex. Inoculation of mice with leukemia virus induces cancer in some strains more easily than in others [766]. These differences are controlled by genes that, like the Ir genes, are part of the H2 complex [741, 740, 765, 783]. Later linkage with the H2 complex was also demonstrated in mice for genetic factors predisposing to autoimmune thyroiditis [859] and to susceptibility with infection by the lymphocytic choriomeningitis virus.
No specific antibody response could be demonstrated for resistance to leukemogenesis and susceptibility to the choriomeningitis virus infection. There was, however, such a response in the case of autoimmune thyreoiditis. Here, the connection between a particular transplantation antigen type, a specific antithyroglobulin antibody response, and severity of the disease could be established. This was an important step toward elucidation of the mechanism of the association. (It may be mentioned in passing that in humans an association between autoimmune thyreoiditis and the antigen HLA-B8 has been described.)
The results suggested the hypothesis that in man immune response genes might also be closely linked with the HLA genes. As linkage disequilibrium had been demonstrated for the well-defined genes of the

Table 3.24. Associations between HLA and some diseases (cf. Svejgaard et al. 1983 [888])

Condition	HLA	Frequency (%)		Relative risk
		Patients	Controls	
Hodgkin's disease	A1	40	32.0	1.4
Idiopathic hemochromatosis	A3	76	28.2	8.2
	B14	16	3.8	4.7
Behcet's disease	B5	41	10.1	6.3
Congenital adrenal hyperplasia	Bw47	9	0.6	15.4
Ankylosing spondylitis	B27	90	9.4	87.4
Reiter's disease	B27	79	9.4	37.0
Acute anterior uveitis	B27	52	9.4	10.4
Subacute thyroiditis	Bw35	70	14.6	13.7
Psoriasis vulgaris	Cw6	87	33.1	13.3
Dermatitis herpetiformis	D/DR3	85	26.3	15.4
Celiac disease	D/DR3	79	26.3	10.8
	D/DR7	also increased		
Sicca syndrome	D/DR3	78	26.3	9.7
Idiopathic Addisons's disease	D/DR3	69	26.3	6.3
Graves' disease	D/DR3	56	26.3	3.7
Insulin-dependent diabetes	D/DR3	56	28.2	3.3
	D/DR4	75	32.2	6.4
	D/DR2	10	30.5	0.2
Myasthenia gravis	D/DR3	50	28.2	2.5
	B8	47	24.6	2.7
Systemic lupus erythematosus (SLE)	D/DR3	70	28.2	5.8
Idiopathic membranous nephropathy	D/DR3	75	20.0	12.0
Multiple sclerosis	D/DR2	59	25.8	4.1
Optic neuritis	D/DR2	46	25.8	2.4
Goodpasture's syndrome	D/DR2	88	32.0	15.9
Rheumatoid arthritis	D/DR4	50	19.4	4.2
Pemphigus (Jews)	D/DR4	87	32.1	14.4
IgA nephropathy	D/DR4	49	19.5	4.0
Hydralazine-induced SLE	D/DR4	73	32.7	5.6
Hashimoto's thyroiditis	D/DR5	19	6.9	3.2
Pernicious anemia	D/DR5	25	5.8	5.4
Juvenile rheumatoid arthritis:				
Pauciarticular	D/DR5	50	16.2	5.2
All cases	D/DRw8	23	7.5	3.6

human HLA complex, it might also be assumed for these as yet hypothetic immune response genes. Hence, disease associations with HLA types were anticipated.

The first anomaly to be examined in man was Hodgkin's disease, a malignant disorder of the lymphatic system. A collaborative study on 523 patients showed a significant association with HLA 1. Examinations of other malignant diseases such as acute lymphatic and myelogenous leukemia gave conflicting results. Much more striking associations, however, were found for a number of nonmalignant diseases, such as ankylosing spondylitis, gluten-sensitive enteropathy (sprue), Reiter's disease, multiple sclerosis, and psoriasis (Table 3.25). In some cases, the extent of the associations is enormous. In ankylosing spondylitis, a relative incidence X (Sect. 3.7.2) of 87 was given, i.e., the disease was 87 times more likely in bearers of the HLA type B 27 than in the rest of the population.

While almost all patients with ankylosing spondylitis have the HLA-B27 type, most carriers of HLA-B27 did not have ankylosing spondylitis. The frequency of HLA-B27 in the white population of the United States is about 5%, while the frequency of ankylosing spondylitis is about 1/2,000. However, careful clinical and radiologic investigations have shown that minor symptoms and X-ray findings suggestive of mild ankylosing spondylitis were found in about 20% of B27 carriers [48].

Several differences stand out when HLA associations are compared with AB0 associations. Most HLA associations are much stronger. For AB0 blood groups and disease, most relative incidences were less than twice the incidence of controls, whereas for the HLA associations the frequencies were usually much higher. However, the very high predisposition of HLA-B27 carriers to ankylosing spondylitis was exceptional, and most associations were significantly lower. Nevertheless, the data suggest that the contribution of HLA types to the multifactorial systems causing these diseases is more substantial than that of ABO genes to diseases found to be associated with them. Therefore, attempts to elucidate mechanisms have a better chance of success.

Are There, Indeed, HLA-Linked Immune-Response Genes in Man? And What Is Their Mode of Action? The important role played by HLA specificities in mediation of the immune response (Sect. 3.5.5; [623]) strongly suggests genetic variability of the immune response depending on the various HLA types. Regarding the precise nature of this influence, however, various possibilities exist. At least five mechanisms have been discussed:

1. The HLA specificity at the cell surface may act as a receptor for a virus or other pathogenic agent. This possibility should be explored, especially if the association is as strong as in ankylosing spondylitis.
2. A second possibility is a cross reaction of an HLA antigen with a viral or bacterial antigen leading to a weaker immune response because of immunologic tolerance or a stronger response because of reaction to a foreign antigen. These mechanisms have been discussed for associations between AB0 blood groups and infectious diseases and will be discribed further in the context of natural selection (Sect. 6.2.1).
3. The association might be produced by linkage disequilibrium with an HLA antigen of one of the other HLA loci. It has been shown, for example, that associations with autoimmune disease can mainly or exclusively be traced back to the HLA-D/DR locus. However, alleles at this locus show linkage disequilibrium with those of the B locus. This may cause a weaker association with B-locus antigens. An association was found, for example, between a type of juvenile diabetes mellitus (autoimmune form) and HLA-B8. When the D locus was included in the analysis, however, D3 showed a much stronger association with the same type of diabetes, and D3 was shown to be associated by linkage disequilibrium with B8. Therefore, the disease association with B8 was obviously caused by the D3 association together with linkage disequilibrium.
4. Association due to linkage disequilibrium may also occur if a rare mutation happened to affect a gene that is closely linked to the MHC without being functionally connected. One example seems to be hemochromatosis.
5. The fifth and most likely possibility is the hypothesis of immune-response (Ir) genes closely linked with the HLA genes and showing strong linkage disequilibrium. Analogous results in the mouse as discussed above offer a strong a priori argument in favor of this concept.

This hypothesis is not mutually exclusive with the idea that HLA (or Ir) specificities at the cell surface may act as receptors for pathogenic agents. This concept can be tested directly by two kinds of family studies. Finding the condition under study and a similar HLA type in affected members of a family would be compatible with the hypothesis of viral receptors and that of microbial cross reactions as well as with that of closely linked immune-response genes. The combination of Ir alleles in the *trans*-state with the HLA type commonly associated with them by linkage disequilibrium would yield families in which none of the patients manifesting the condition show haplotypes carrying the antigen normally associated with the disease. Such families have, indeed, been observed. However, with linkage disequilibrium, the HLA marker gene and the disease gene should be in *cis*-phase in the majority of cases in affected families and in the *trans*-phase in the minority of cases of affected families. Hence, for families carrying this gene in the *trans*-phase, the prediction of the hypothesis of an Ir locus would be contrary to that of other hypotheses.

If a disease association is due to closely linked Ir genes, it is also possible that these genes are predominantly linked with the allele for one specific HLA antigen and in another population with the allele for another antigen. Therefore, disease association studies, especially if carried out in racially different populations, may lead to apparently contradictory results.

The fact that a disease shows association with HLA may give hints as to its pathogenesis. In multiple sclerosis, for example, immunologic investigations guided by the HLA associations revealed a specifically decreased cellular immunity to measles and other paramyxoviruses [662, 689].

Multiple sclerosis is associated with HLA-B7. The same antigen also shows association with lepromatous leprosy, a type of leprosy infection with especially weak response of the T-lymphocytes, and hence, weak cellular immunity [688]. B7 cells also show reduced responsiveness and lymphotoxicity in vitro [838]. These and some other results have led to the conclusion that bearers of the B7 antigen might be "weak responders", i.e., their T-lymphocytes need more time to react to certain antigenic stimuli by cell proliferation. This may help to work out the mechanisms of HLA disease associations.

It was tentatively concluded that multiple sclerosis might be the result of a slow virus infection primarily affecting those individuals predisposed by an HLA-linked abnormal immune response. In diabetes, the observation that juvenile diabetes shows associations with HLA antigens, but mature onset diabetes does not, lends support to the contention that the etiology might be different and that in juvenile diabetes, an autoimmune or virus etiology might be involved. Much lower concordance for identical twins with juvenile diabetes as compared with mature onset diabetes as well as higher personal risk for family members of patients with mature onset diabetes lead to the same conclusion [906] (see Sect. 3.8.14).

Recent evidence even points to two types of juvenile diabetes mellitus: One type, which is associated with HLA-D3 (and B8) and seems to be caused by an autoimmune mechanism, and another one, associated with D4 in which the patients are often insulin antibody responders. Hence, analysis of HLA associations may help to refine nosologic classification of a group of diseases and to detect genetic heterogeneity.

Some hypotheses regarding the functional significance of the whole MHC area on chromosome 6 have been discussed in Sect. 3.5.5. This chromosomal area contains a remarkable cluster of closely linked gene loci – all with closely related functions. The investigation of disease associations promises insight not only into the etiology of certain diseases but also into normal functions.

Linkage and Association. Linkage and association should be carefully distinguished from each other. Linkage refers to two genes being located on the same chromosome within detectable distance of each other. The term "association" is often used when a higher frequency of a given gene is found in a certain disease or trait. Association does *not* imply that the gene involved in the disease and the marker gene are located on the same chromosome. Confusion regarding these concepts can arise in studies of HLA gene frequencies in disease [809]. The HLA complex is located on chromosome 6. The gene for 21 hydroxylase deficiency in the homozygous state produces congenital adrenal hyperplasia (20910) and is closely linked to the HLA complex [633]. Similarly, the gene of one type of dominant spinocerebellar ataxia (16440) is linked to the HLA complex [725]. Data on the iron storage disease hemochromatosis (14160) (presumably inherited as an autosomal-recessive trait with occasional heterozygote manifestations) can also be interpreted to mean that the gene for this disease also is linked to the HLA complex [872a, 872b, 745]. These diseases are monogenic conditions, and their respective loci are situated within measurable distance of the HLA complex on the 6th chromosome. There is no reason to believe that the genes for these diseases and HLA genes are physiologically related.

Table 3.25. α_1-antitrypsin concentration of different α_1-antitrypsin phenotypes

Pheno type	n	α_1-antitrypsin concentration (mg/100 ml)	Percent of normal (MM = 100%)	
			Kueppers	Fagerhol
M/M	21	212	100	100
S/S	2	80; 112		63
Z/Z	10	25	12	
M/S	19	167	79	83
M/Z	17	120	57	61

Source: adapted from Kueppers, 1975 [749].

The diseases with HLA associations (Table 3.25) are not simple monogenic conditions but have usually been shown to be of multifactorial origin. In one group of HLA-related diseases (chronic hepatitis, myasthenia gravis, rheumatoid arthritis, Addison's disease, thyrotoxicosis, juvenile diabetes, celiac disease, and multiple sclerosis), associations with D/DR specificities of the HLA system have been demonstrated [623]. The common factor in these conditions is the presence of autoantibodies, and they have been classified as autoimmune diseases or at least as immunologically associated diseases. Familial aggregation without clear Mendelian pattern has been detected when appropriate studies were done. The relative risks for each disease in carriers of the involved Dr specificity are modest and range between two to eight times the frequency observed in the control population (with the exception of celiac disease – seen 65 times more frequently than in controls [582]). Here, the explanations mentioned above for HLA disease associations appear pertinent. For example, immune response (Ir) genes closely linked to the D/DR alleles may be involved, or such alleles participate directly in the immune response – probably via the macrophage T cell system – or the entire cell surface pattern, which is specified by HLA-D/DR and Ir alleles (and, possibly, other closely linked genes), is involved [818].

Organ specific autoantibodies lead to the manifestations of the various autoimmune diseases. Whether additional genes on other chromosomes are implicated is unknown. The production of autoantibodies probably requires appropriate environmental stimuli which often are viral in origin as suggested for diabetes, hepatitis, and multiple sclerosis. Persons with certain D/DR specificities of HLA presumably are more susceptible to antibody formation than those who lack such genes; this explains at least part of the genetic susceptibility to autoimmune diseases. D/DR genes of the HLA complex and autoimmune diseases show association but no demonstrated linkage.

3.7.4 *α_1-Antitrypsin Polymorphism and Disease* [749; 653]

α_1-Antitrypsin (PI) Polymorphism. The AB0 blood groups show a weak association with a great number of diseases, but no convincing explanation for these associations exist. HLA antigens show a much stronger association with a smaller number of diseases. A convincing biologic explanation is not at hand, but there are reasonable and some experimentally testable hypotheses. The α_1-antitrypsin polymorphism is associated in adults mainly with one disease, chronic obstructive pulmonary disease. Its mechanism has been elucidated to some extent.

The antiproteolytic activity of the human serum was detected by Camus and Gley (1897) and by Hahn (1897). Landsteiner (1900) showed this activity to be located in the albumin fraction. Of the six antiproteases identified in the human serum, α_1-antitrypsin and α_2-macroglobulin have the highest concentrations. Both are able to inhibit a great number of proteases, including thrombin. Antiproteolytic activity is measured by hydrolysis of artificial substrates by trypsin in the presence of the serum to be tested. There is a close correlation between immunologically measured concentration and activity. The concentration increases quickly, for example, with bacterial infection, after injection of typhoid vaccine and during pregnancy. Synthesis occurs in the liver. Interindividual differences were first observed in 1963 [755]. A simple recessive mode of inheritance was proposed for low α_1-antitrypsin levels. Using electrophoretic techniques and isoelectric focusing, many phenotypes were observed, and at least 23 different phenotypes were known (Fig. 3.70; see also Table 3.25). The genetic basis of this heterogeneity is a series of multiple alleles. The locus was named PI, the various alleles PI^M, PI^Z, etc., the phenotypes M/M, M/Z, etc. In all populations examined so far, the PI^M alleles, M_1, M_2, and M_3, are the most frequent ones with a common gene frequency of 0.9 or higher. Other, rarer alleles were designated by letters. The position of these letters in the alphabet gives an approximation of electrophoretic mobility. Figure 3.70 illustrates the electrophoretic patterns. Nucleotide sequence analysis (Sect. 2.3.3.4) of cloned cDNA has demonstrated, for example, that the Z variant is caused by a single nucleotide substitution [751]. Two variants, Z and S, are especially important because the α_1-antitrypsin level is appreciably reduced. In another, very rare, allele PI^-, no activity of the enzyme at all is found in the homozygote (null allele). The heterozygote PI^M/PI^- has an M phenotype with only 50% of the normal concentration.

Intravenous injection of typhoid vaccine and diethylstilbestrol leads to a 100% increase of activity in subjects with the MM type. Heterozygotes of the MZ type show a moderate increase, whereas in homozygotes of the ZZ type hardly any increase can be seen.

Association with Chronic Obstructive Pulmonary Disease (COPD). Eriksson (1965) [651] reported 33 homozygotes of the ZZ type; at least 23 had definite symptoms of chronic obstructive pulmonary disease. On the basis of his family data, Eriksson estimated obstructive pulmonary disease to be at least 15 times more frequent among these homozygotes than in the general population. This observation was confirmed by many investigators in a great number of patients. In one group of 295 patients with this diagnosis, 20 PI^Z homozygotes were detected, whereas less than one would have been expected on the basis of the gene frequency. Usually, the first symptoms are recognized during the 3rd or 4th decade of life; patients with the usual variety of chronic obstructive pulmonary disease usually become affected during their 50s or 60s. Loss of lung tissue and blood vessels in the lower lobes is characteristic. Interestingly enough, even before this defect was discovered, a special group of patients with these symptoms had been delineated from the majority of patients with chronic obstructive pulmonary disease. The question whether heterozygotes also have a higher frequency of COPD has been much discussed. However, it has been claimed that heterozygotes have an approximately threefold risk of developing COPD compared with M homozygotes [653]. Pulmonary function tests in heterozygotes showed more frequent abnormalities. Their symptoms tend to appear later in life and heterozygotes.

Of the homozygotes, only 70%–80% develop obstructive emphysema, and in heterozygotes the frequency is much lower. Environmental factors appear to influence manifestation: α_1-antitrypsin also inhibits proteolytic enzymes released by granulocytes or macrophages. It is therefore likely that these enzymes, which normally are released during inflammatory process, are insufficiently inactivated. When a patient is exposed to recurrent bronchial irritation, such as caused by smoking or frequent infections, these enzymes will cause digestive damage of the lungs. Tobacco smoking enhances the danger of bronchial infections and hastens the progress of the disease [653]. Therefore, Z/Z homozygotes and

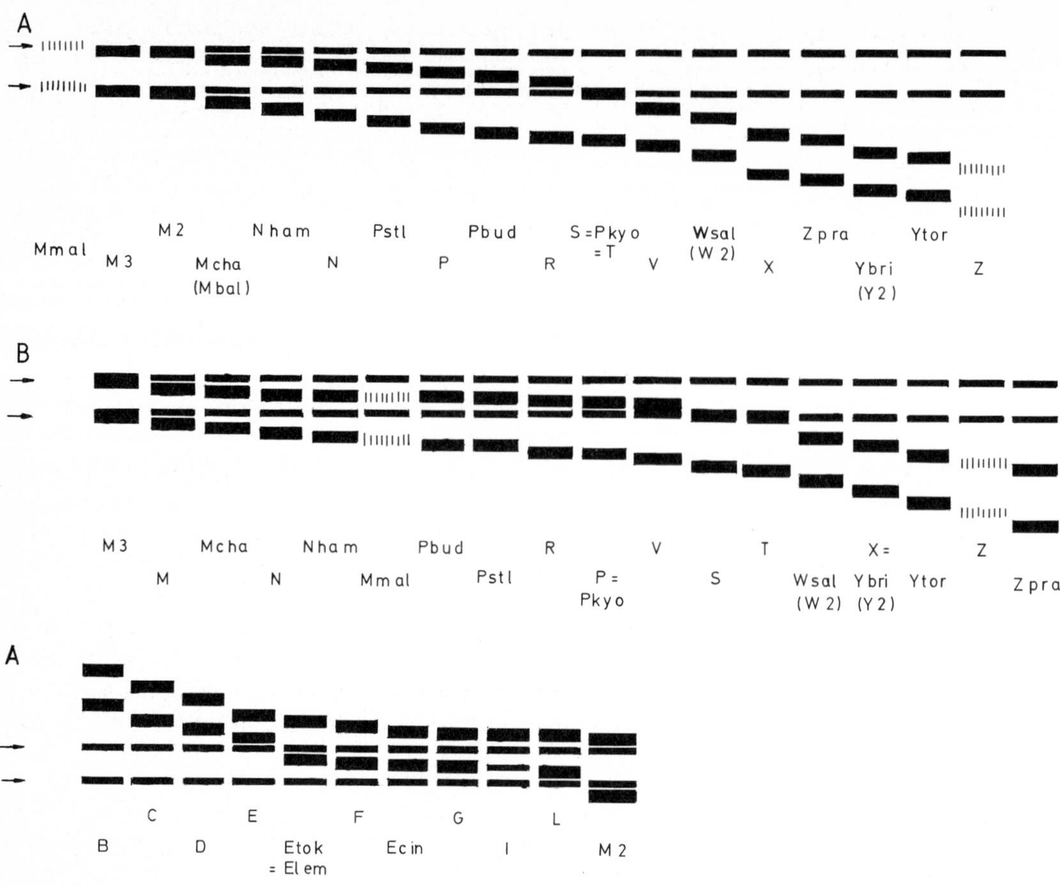

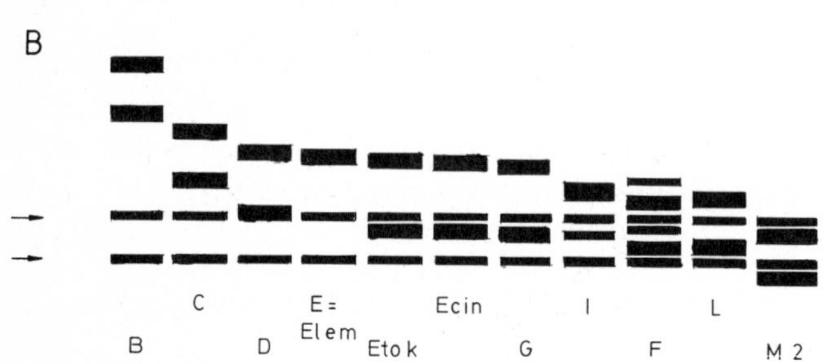

Fig. 3.70. Electrophoretic patterns of various α_1-antitrypsin-(Pi) proteins. Homozygotes of Z have the lowest activities

heterozygotes should be advised to refrain from smoking and to avoid jobs leading to bronchial irritation. Bronchitis should be treated early and intensively. "If we are able to change the smoking habits of a PI^Z individual, this may add 15 years to his life" [653].

Another disease associated with low α_1-antitrypsin values in homozygotes is childhood cirrhosis of the liver. This association is firmly established but is more rarely seen than chronic pulmonary disease. Cryptogenic cirrhosis in adults also appears to be more common in ZZ homozygotes.

Significance of the New Research Strategy. The α_1-antitrypsin polymorphism is remarkable, since the mechanism of lung damage can be explained. This result is in striking contrast to the associations described in the AB0 system and even in the HLA system. The situation is simpler than in these systems. One genotype is affected in not all, but in a high fraction of cases, and the disease is rather specific, so that it was identified even before the biochemical cause became known. The Z/Z state could also be regarded as a recessive disease with "incomplete penetrance." There are probably many other such recessive diseases still buried within the large groups of frequent multifactorial conditions, either because the phenotype is difficult to define or because all parameters for delineation have not yet been considered.

The contribution of the PI^Z allele to the disease liability of the heterozygotes seems to be quite similar to the AB0 and HLA situations; chronic obstructive pulmonary disease occurs only in a minority of cases, it resembles the more usual type of obstructive pulmonary disease, after chronic bronchitis, and the influence of additional environmental and genetic factors appears to be strong.

The example of this disease association shows the more recently developed indirect research strategy at its best. A genetic polymorphism is analyzed at the genetic level. The product of a single gene is identified using biochemical methods. The consequences of the polymorphism for gene action are sought. In this case, the quantity of protein is defective and enzyme activity is low so that the organism does not adequately respond to environmental challenges such as infections. This specific functional weakness leads to disease, especially with frequent exposure to certain environmental influences such as bronchial irritation.

It is likely that such a research strategy will become useful for genetic analysis in other areas, where genotype-phenotype relationships are so complex that a clear analysis from the phenotype down to the genes by Mendelian methods is barred, so that methods of quantitative genetics have to be relied upon. Or, to put it in a slightly different way: This strategy will help in applying Mendelian approaches to situations where Galtonian techniques now are the only applicable methods.

Disease Associations of Other Polymorphisms [145]. Apart from the three systems described above, disease associations have been examined (and in some cases made likely) for a number of other polymorphisms, for example, blood group systems [211]; haptoglobins, and tasting of phenylthiourea (PTC). Some of these will be described in the context of population genetics (Sect. 6.1.2). Especially interesting are the associations that have been reported between the Apo E lipoprotein polymorphism and atherosclerosis [916; 917] (Sect. 3.13 and variants of the third complement component and some diseases: The $C3^F$ allele seems to be associated with rheumatoid arthritis [590, 591, 657], hepatitis [657], the quantitative immune response. The C6 deficiency has been found in about half the patients with meningococcal meningitis. If confirmed, these associations would be of considerable interest because plausible hypotheses about biologic mechanisms and genetic consequences could then be discussed.

3.8 Nature-Nurture Concept: Twin Method

In discussing methods of quantitative genetics, the use of twin data for quantitative assessment of the degree of genetic determination was mentioned repeatedly. Indeed, twin investigations have played a major role in the history of human genetics. At one time, the twin method was even regarded as the "royal road" to genetic analysis in man. In one field of major importance, behavior genetics, many of our conclusions are still based on twin data. Therefore, critical assessment of the twin method, its advantages and limitations is important.

3.8.1 Historical Remarks

Introduction of the twin method is generally accredited to F. Galton (1876) [675], who also - consciously or unconsciously following Shakespeare - adopted the alternative terms "nature" and "nurture".[10] It is doubtful, however, whether Galton recognized the essence of the problem. Very probably he did not know that there are two types of twins, monozygotic and dizygotic. This distinction had been discovered only a short time previously by Camille Dareste, who reported it to the Société d'Anthropologie in 1874 [927]. Most likely, Galton intuitively had the right idea but no clear concept about the usefulness of the twin method. Such intuitive insights are frequently observed in the history of science when a new paradigm starts on its way and do not detract from the usefulness of the paradigm.

Poll (1914) [841] was the next to use the twin method in assessing genetic determination. But he failed since methods for distinguishing monozygotic from dizygotic twins were lacking. After Poll, single twin pairs continued to be described, but zygosity diagnosis remained ambiguous.

The method was given a sound foundation with the work of Siemens (1924) [869]. Siemens' achievement was threefold:

1. He showed that twin series of a size suitable for meaningful statistical evaluation can easily be found when schools are asked for help. Hence, investigation of normal variability became possible.
2. He developed a reliable method for zygosity diagnosis. Up to that time, research workers had tried in vain to achieve such a diagnosis using one single criterion. Siemens, however, showed that reliable separation of these groups was possible when a large number of criteria were examined. Each of these criteria may on the average show only somewhat more similarity in monozygotic twins, but all of them together will separate the two groups with a high degree of reliability.
3. Siemens proposed to examine not only monozygotic (MZ) but also dizygotic (DZ) twins. DZ twins are no more similar genetically than are other siblings, having on the average 50% of their

[10] In the *Tempest*, Prospero says of Caliban: "A devil, a born devil, on whose nature nurture can never stick".

genes in common by descent. However, like MZ, they are born at the same time, and exposed to similar environmental conditions.

3.8.2 Basic Concept

The twin method is founded on the biologic peculiarity that MZ twins originate from division of one zygote. Therefore, they must, as a rule, be identical genetically. A group of genetically identical individuals is called a clone. It follows that any phenotypic differences between MZ twins must be caused by environmental influences. Here, environment is defined in the widest possible sense: anything that is not fixed genetically.

Hence, to recognize whether and to what degree a character is determined genetically and to what degrees its variability might be modified by environmental influences, the degree of similarity between MZ twins must be measured. As DZ twins are thought to be influenced by the same environmental differences but have only one-half of their genes in common by descent, they are used as suitable controls.

In the following sections, it will be shown how this concept can be quantified, and its limitations will be discussed.

3.8.3 Biology of Twinning

Dizygotic Twins. In most mammals (rodents, carnivores, some ungulates), every birth is multiple. In every ovulation, the ovary discharges several ova, which may be fertilized by one sperm each. The marmoset monkey regularly gives birth to dizygotic twins. In higher ungulates, such as horses and cattle, and in the higher primates including humans, only one ovum is discharged as a rule at ovulation. There are occasional exceptions. If two oocytes are discharged at the same time and fertilized by different sperms, dizygotic (DZ) twins result. In the same way, polyovulation occasionally leads to trizygotic triplets and quadrizygotic quadruplets. However, not all triplets, quadruplets, and quintuplets arise in this manner.

It follows that DZ twins need not necessarily have the same father. The two oocytes may be fertilized by sperm from different men with whom the mother has had intercourse around the time of ovulation. One such possible case from Nazi Austria is of interest [681].

The fraternal twins were 25 at the time of examination. The legal father was a Jew, the mother was not Jewish. At that time Austria was incorporated into Nazi Germany, and racial laws discriminating against Jews were introduced. Since there was interest in exonerating the children from the "blemish" of being half Jewish, the mother reported an extramarital relationship at the time the twins were conceived. The sexual partner was still available for testing. The AB0 and MN blood group tests (the only systems routinely available at that time) gave the following results:

Legal father	B, M
Alleged father	A, MN
Mother	O, M
Twin brother	B, M
Twin sister	A, MN

If one accepts these tests as accurate, the results are:

1. Exclusion of the legal father for the girl - first, as she had to have inherited the A allele from one of her parents, and, secondly, as her N allele was not accounted for.
2. Exclusion of the alleged father for the boy, as neither he nor the mother had the allele B.

Theoretically, a third man other than the alleged sexual partner could have been father of both twins, but examination with additional anthropological methods gave highly suggestive evidence that the boy indeed resembled most the legal father and the daughter the alleged father. More recent dizygotic twin pairs with two fathers have been reported and were discovered in the course of a disputed paternity case. In one case, one of the two fathers was black, the other one white.

Anastomosis of blood vessels, which is quite normal between MZ twins during fetal development, may occur in exceptional cases in DZ twins as well. This may lead to mutual transfusion of blood stem cells since early embryos are immunologically tolerant to each other. The result are twins who are blood chimeras with two populations of genetically different blood cells [632, 828]. In cattle, a vascular connection between DZ twins is the rule and leads to partial sexual transformation in the female partner of fraternal twin pairs. Such a masculinized calf is known as a freemartin.

Monozygotic Twins. Much more interesting is the formation of monozygotic (MZ) twins. In a certain sense, they may be said to be the most extreme state of duplication. Less extreme duplications, such as Siamese twins, or doubleheaded monsters, are observed occasionally in man. Many of these duplications are lethal.

Some unusual types of twins, however, have survived and become famous, for example, the "Siamese twins" Chang and Eng (Fig. 3.71), who were born in Thailand in 1811. At the age of 18, the twins came to the United States and made a living by displaying themselves in curiosity shows. Later on, they married two sisters. Eng had 12, and Chang had ten children. They settled in the Carolinas and grew tobacco. At age 61, Chang had a stroke and died 2 years later from bronchitis. Eng who had been healthy up to that moment survived his brother by only 2 hours. Chang and Eng were connected by a tissue bridge about 10 cm wide

Fig. 3.71 The Siamese twins Chang and Eng. (From Lotze, 1937)

Table 3.26. Incidence of twin births per 10,000 births (Propping and Krüger, 1976 [842])

Country	Time period	DZ/10,000 births	MZ/10,000 births
Spain	1951-1953	59	32
Portugal	1955-1956	56	36
France	1946-1951	71	37
Austria	1952-1956	75	34
Switzerland	1943-1948	81	36
Fed. Rep. of Germany	1950-1955	82	33
Sweden	1946-1955	86	32
Italy	1949-1955	86	37
England and Wales	1946-1955	89	36
U.S.A. whites	?	67	39
U.S.A. blacks (California)	1905-1959	110	39
U.S.A. Chinese	?	22	48
U.S.A. Japanese	?	21	46
Japanese	1955-1962	24	40

reaching from the lower end of the sternum almost down to the navel. At postmortem examination, this bridge was discovered to contain liver tissue connecting the two livers. Hence, any attempt to separate the brothers surgically would hardly have been successful in 1872. Today, even more extensive connections between such twins have been severed.

The factors inducing division at an early cleavage stage and giving rise to MZ development in humans are unknown. But MZ twins were produced in experimental embryology - many decades ago in amphibians and, more recently, in mammals (for an example, see Sect. 4.7.1). Mirror-image similarity between human MZ twins has been discussed repeatedly. Since strong asymmetry could be experimentally produced in animals, true asymmetry in some MZ pairs would not be surprising.

Very rarely, twin pregnancies occur in which an oocyte and a polar body (see Sect. 2.1.24) are fertilized by different sperm. One such event led, in addition to birth of a normal child, to an acardic monster. The abnormal twin had been produced by fertilization of the first polar body (diploid) by a separate sperm, as shown by chromosome heteromorphisms and HLA haplotypes [577].

Frequency of Twinning [30]. Table 3.26 shows incidence figures in different populations for monozygotic (MZ) and dizygotic twin births. The proportion of MZ twins was calculated by Weinberg's difference method, which is based on the fact that MZ twins are always of the same sex. Among DZ twins, on the other hand, one-half are of the same sex, the other half are oppositely sexed. Hence:

Frequency of DZ twins = 2 × DZ of opposite sex
Frequency of MZ twins = Frequency of all twins − DZ twins

This method gives only approximate results, since somewhat more boys are born than girls. Moreover, there is some inconclusive evidence that DZ twins of the same sex are more frequent than expected. This is possibly caused by the fact that the time difference between ovulation and cohabitation influences the sex ratio at fertilization. It can be assumed with confidence that these deviations are small. Hence, Table 3.27 presents a fairly good approximation of the truth.

The frequency of MZ births shows little variability among different populations. Frequencies of DZ births, on the other hand, are different, the highest DZ frequency being found among African blacks with variability among tribes. The Yorubas in Nigeria have a twinning frequency of 4.5%; 4.2% of these are dizygous. Among American blacks, DZ twins are born more frequently than among American whites. In Europe, the rate of dizygosity is about 8/1000 births. But here higher frequencies have also been observed in some populations. On the Åland islands, the twinning rate was 15.2/1000 between 1900 and 1949. The lowest rates are found in Mongoloid populations, especially in the Japanese.

The differences in DZ twinning rates between the main racial groups are maintained when the data are corrected for maternal age and birth order.

Factors Influencing Frequency of Twin Births: Maternal Age and Birth Order. The probability of a twin

Table 3.27. Incidence of congenital malformations in twins and singletons per 1000 births [843]

Source	Approximate sample size	Incidence in singletons (%)	Incidence in twins (%)
Hendricks (1966)	~35,000	3.3	10.6
Stevenson et al. (1966)	421,000	12.7	14.4
Hay and Wehrung (1970)	10.2 million	5.8	6.2
Onyskowová et al. (1971)	240,000	13.2	26.4
Emanuel et al. (1972)	25,000	13.2	23.2

birth increases with maternal age. This increase affects exclusively DZ births, as recognized by Weinberg in 1901. Subsequent work confirmed a maternal age effect, showing that the DZ rate increases from 0 at puberty at a rate of 0.7%–0.8% a year up to the age of 35–39 and drops afterward [781, 842, 747]. The reason for the maternal age effect is probably an increase in the gonadotropin level (FSH). This hormone was shown generally to increase with maternal age, and it could easily cause an increased tendency to polyovulation. In the Yoruba, for example, women having had two twin births had the highest FSH levels and mothers of single-birth children had the lowest FSH levels. The hypothesis is corroborated by the fact that women treated with gonadotropic hormones for sterility due to anovulatory cycles frequently have multiple births. Discontinuation of birth control pills does not influence the twinning rate [787]. The reduced DZ twinning rates during the last years of reproductive age may be due to the ovaries being unable to discharge more ova in spite of high FSH levels. DZ twinning rate increases not only with maternal age but with birth order as well.

Genetic Factors. At the beginning of this century, Weinberg [934, 935] recognized that twin births show clustering in families. This familial aggregation is true for DZ twins only. After appropriate corrections for maternal age, the probability of future DZ twin births for the mother is about four times the frequency of DZ twins in the population. The chance for her female relatives is also increased; for her sisters, it equals her own chance. For male DZ twins, and fathers of DZ twins, on the other hand, this chance is not increased. The mode of inheritance seems to be multifactorial; gonadotropin levels could easily be the major genetically determined cause.
For MZ twins, there are no indications of any genetically influenced variability. The recurrence probability for MZ mothers does not exceed the population average. DZ mothers are, on the average, ~1–2 cm taller than either MZ, or singleton mothers [613].

Decrease in Twin Births in Industrialized Countries. In almost all industrialized countries, a decrease in the frequency of twinning has been observed during recent years. Hence, the old rule of one twin birth to 80 single births no longer holds. In present-day Germany, for example, there is less than one twin birth for 100 single births. Figure 3.72 shows the decrease for Germany. This decline was already apparent in the period after World War I (1914–1918) and, after a short peak in the late 1930s, became very pronounced after 1945. The decrease has commonly

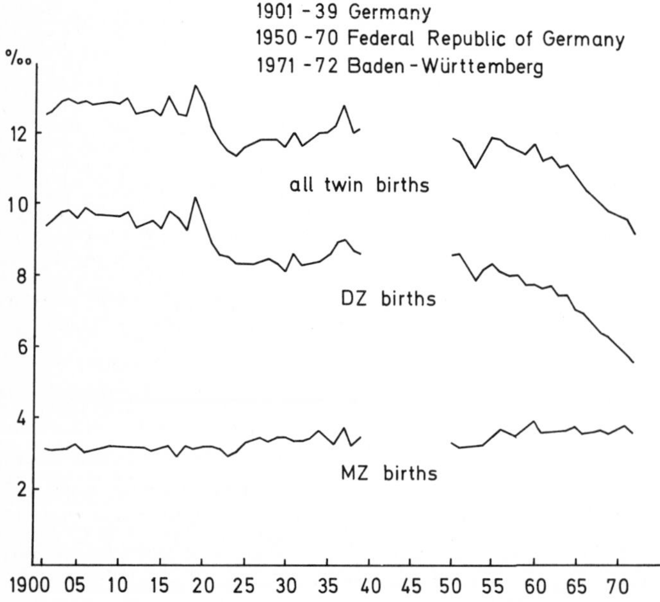

Fig. 3.72. Decrease of twin births in Germany during recent years. The decrease is entirely due to the DZ twins. (Propping and Krüger, 1976 [842])

been explained by a maternal age effect. The average number of pregnancies has decreased during this time period, and most pregnancies have occurred at an age at which the chance for twins is lower. However, this explanation is insufficient and explains only a small part of the decrease. Considering known physiologic and genetic data, the following hypothesis seems to be of interest [842].

Polyovulation is correlated with fecundability, i.e., the probability per cohabitation of conceiving a child. One common factor applying to both polyovulation and fecundability is the FSH level. In earlier years, women whose fecundability was high contributed more than their average share to the birth rate, thereby enhancing the number of DZ births. Currently, the number of children is largely being regulated by parents using birth control, and the significance of biologic fecundability for actual reproduction has decreased. Hence, the number of DZ births has decreased as well. This hypothesis is supported by statistical data. In 1946, for example, twin frequency increased steeply in the United States. One year before, many soldiers had come home and birth control presumably was less extensively practiced. As Fig. 3.72 shows, Germany had an increase in twin births in the late 1930s. This was at the time of Nazi propaganda in favor of large families, which led to an appreciable increase in the birth rate. Moreover, mothers of illegitimate children who may represent a subgroup of women of high fecundability, have a high twinning rate. Considered from this standpoint, the racial gradient for twin frequencies (Black-White-Mongoloid) could possibly be due to natural selection for fecundability. In Africa, high infant mortality made it necessary to exploit the reproductive capacity of women to the extreme, whereas in Japan, birth control had been practiced for centuries, probably diminishing the selective advantage of high fecundability.

Frequencies of Multiple Births of More than Two Children. Hellin's rule – where the frequency of twin births = t; triple births = t^2, etc. – holds true only very approximately. All combinations of mono-, di-, and trizygosity, etc. are possible. The famous Dionne quintuplets in Canada, for example, were monozygotic.

3.8.4 Limitations of the Twin Method

Systematic Differences Between Twins and Non-Twins. The purpose of twin studies is to obtain results that apply not only to twins but to the whole population. In any twin study, the following question has to be posed: Do twins differ from non-twins in the trait under study? Any differences might impair the general validity of any conclusions drawn from a twin sample.

Comparison of several physiologic characters has shown differences between twins and non-twins during embryonic development. Twins suffer from a higher frequency of abnormalities during pregnancy and at birth. Their lower birth weight can only partly be attributed to the shorter duration of gestation. The stillbirth rate and infant mortality in early life are considerably higher in multiple births than in single ones; in later years, twins run a higher risk than non-twins of becoming mentally retarded, presumably at least partly due to complications during pregnancy and at birth. Even the mean I.Q. of both MZ and DZ twins is slightly lower than that of control populations.

Do nongenetic factors act differently upon MZ and DZ twins? Could this difference alter the probability of manifestation of the condition under investigation? This is important because the basic concept (Sect. 3.8.2) of the twin method assumes that MZ and DZ twins are exposed to identical prenatal and postnatal environmental factors. Birth weight may serve as a simple and measurable indicator. In an extensive survey, Carney et al. (1972) [see 843] compared the average birth weights of 572 individuals from twin pairs of identical sex, classified by sex, placentation, and zygosity. The following mean birth weights were calculated:

Males (*n* = 304) 2659 g			*Females* (n = 268) 2547 g		
Dichorionic		Monochorionic	Dichorionic		Monochorionic
(196) 2703 g		(108) 2579 g	(162) 2577 g		(106) 2500 g
DZ	MZ	MZ	DZ	MZ	MZ
(160) 2728 g	(36) 2595 g	(108) 2579 g	(144) 2601 g	(18) 2385 g	(106) 2500 g

MZ twins of both sexes weigh less than DZ twins. The type of chorion and placenta has no effect on the mean birth weight of the surviving individuals. Therefore, it appears likely that zygosity rather than placentation is responsible for the difference in birth weight.

When individuals within monochorionic pairs (who are always monozygotic) were compared, greater differences come to light. These twins can differ in birth weight by even more than 1000 g. Such differences may be the result of arteriovenous anastomoses leading to a chronic "transfusion syndrome," consisting of chronic malnutrition with reduction of cytoplasmic mass of parenchymal organs and markedly reduced hemoglobin and serum protein in the donor twin. Since more than 20% of all MZ twins have only one chorion, the transfusion syndrome could account for considerable intrapair differences in birth weight that are not observed in DZ twins [567].

It follows immediately that birth weight is not one of the characters for which meaningful use of twin data can be made to estimate, for example, heritability. But intrauterine development influences other characters, as well. Table 3.27 shows the overall frequencies of congenital malformations in twins and in single children as reported by several authors. Although frequencies of congenital malformation vary greatly in the five series, presumably due to different definitions, on the whole these anomalies appear to be more frequent in twins in every analysis published. This tendency is more obvious when particular types of malformation are taken into consideration. The risk for twins is enhanced at least for congenital heart disease, anencephalus, hydrocephalus, and cleft lip and palate. In all four instances, the risk is higher in samesexed twins than in those of opposite sex. This indicates that MZ may be more often affected than DZ twins. The "transfusion syndrome" could easily account for this difference. If this explanation holds true, malformations should be expected in only one of the twins. This expectation is actually fulfilled. Unfortunately, no twin series with malformations seem to be available in which the type of placentation has been taken into account. Sirenomely is a rare condition, in which the lower body is underdeveloped, and the two legs are not separated. On the basis of data collected by Lenz (1973) [761], the incidence of that malformation can be estimated to be about 1:1000 MZ births, as compared with 1:60,000 in the general population.

Twin studies on congenital malformations using unselected case series show relatively low concordance rates in MZ twins (Table 3.31).

However, the twin method can only produce ambiguous results in regard to these malformations [843]. As for all other conditions, the possible influence of intrauterine factors in twin pregnancies should be considered before a twin study is undertaken.

Peculiarities of the Twin Situation in Postnatal Life. Can twins be regarded as "normal" children? Can the results of measurements be extrapolated to the non-twin population? The following considerations need mentioning.

I.Q. tests are lower in twins than in singletons especially in younger age groups as determined in 1947 for all Scottish school children of age 11, including 794 single twin individuals, and for 95,237 French school children between ages 6 and 13 including 808 twin individuals. Zazzo (1960) [27], recalculating these test results in I.Q. values, found a mean of 93 in the twins, compared to a population mean of 100. This difference in I.Q. tests can even be found at the end of the 2nd decade, as evidenced in the study by Husén (1959) [721] on all male military conscripts (1948–1952) in Sweden. The sample comprised 2,935 twin individuals including the male partners of differently sexed twins and those whose co-twin had died. The average difference amounted to about 4 I.Q. points (0.25 of a standard deviation). According to Husén, the I.Q. variance was also greater in twins, and the frequency of the mentally retarded was twice as high as in the general population.

The reasons for a lower intelligence quotient seem to be manifold. Premature birth with higher risk of minor brain damage may be one factor; the higher burden on the family that has to care for two young children at the same time could be another.

Twins form a social group. They depend less on exchange of information with the outside world because they have each other [587, 27]. Studies in central Europe have shown that they frequently develop a "private language". These findings probably explain why twins learn to speak later than other children. This process is more intensive in MZ than in DZ twins. MZ twins generally spend more time together; they frequently tend consciously toward uniformity, whereas DZ twins tend rather to stress the difference. The tendency toward uniformity, however, is stronger in female than in male MZ. *Protesting against identity* especially among male MZ twins occurs and may even lead to "twin hostility" [918].

Table 3.28. Incidence of selected congenital malformations in twins and singletons per 1000 births

Type of malformation	Source	Incidence in singletons	Incidence in twins		
			Total	Same sex	Different sex
Congenital heart disease	a	0.74	1.65	1.82	1.27
	b	2.8	6.3	–	–
	c	0.59	0.71	0.81	0.49
Anencephaly	a	0.92	1.24	1.52	0.64
	b	1.3	1.2	–	–
	c	0.23	0.37	0.45	0.22
Hydrocephalus	a	0.61	0.72	0.91	0.32
	b	1.0	3.1	–	–
	c	0.30	0.40	0.45	0.31
Cleft lip and/or cleft palate	a	1.21	0.34	1.68	0.64
	b	0.8	0.4	–	–
	c	1.11	1.07	1.10	1.01

Sources: *a*, Stevenson et al. (1966); *b*, Edwards (1968); *c*, Hay and Wehrung (1970) [see 843].

One twin may more frequently follow the father, the other his or her mother, and the identifications may lead to conspicuous behavioral differences between the twins. Another phenomenon frequently observed is role differentiation: one is the spokesman to the outside world who usually answers when both are addressed; the other may be the decision-maker for problems involving both twins. Or one may dominate, the other may be more submissive. This role differentiation occurs in both MZ and DZ twins but seems to be more frequent in MZ twins. This may lead to spurious discordance in MZ twins for behavioral traits.

Obviously, these peculiarities of the twin situation must be considered, especially when personality traits such as "extraversion" or "neuroticism" are investigated. Still more important, these attitudes are influenced by changing cultural patterns. In earlier years, identification of twins with each other was usually encouraged, for example, by identical clothing or by sending them to the same school. Today, many educators rather recommend an emphasis on differences. These special conditions of twin life will mainly influence personality characteristics, making the twin method especially controversial in behavior genetics. It is unfortunate that the method is so prominently used in these fields, not so much because its explanatory power might be overrated but rather because so few other methods are available (Sect. 8.2). In theory, these difficulties could be prevented by examining twins who were separated early in life and reared apart. In practice, however, such twins are rare, and the very fact that an ideal study from a scientific viewpoint would require adoption into homes with quite different environments make them exceptional.

These postnatal peculiarities of the twin situation may not be quite as biased for somatic diseases as for behavioral genetics. For a chronic infectious disease, such as tuberculosis or leprosy, for example, it may be important whether a twin has more contact with one of the parents if this parent is infectious for this disease. Fortunately, less biased results can be expected for multifactorially determined "constitutional" diseases of the adult.

3.8.5 Diagnosis of Zygosity

Every twin study requires a reliable method for zygosity diagnosis. Since Siemens (1924) [870] established the principle of the polysymptomatic similarity diagnosis, this problem had largely but not entirely been solved. More recently, study on genetic markers has made twin diagnosis more independent of the personal judgment and experience of the investigator. Details are given in App. 5.

3.8.6 Application of the Twin Method to Alternatively Distributed Characters

The twin method can serve three purposes in this area:

1. The difference in concordance between monozygotic and dizygotic twins can be used to determine whether genetic variability plays a role in a given disease.
2. Penetrance (i.e., the probability of manifestation of the disease) can be estimated.
3. The conditions of manifestation can be examined.

In the earlier phases of twin research, most of the work was centered around the first two aspects. Recently, however, the third aspect has often been emphasized. Four approaches have been used to achieve these purposes [769]:

1. *Case reports.* Descriptions of cases of concordant or discordant twin pairs, especially MZ pairs, are being published, most of them in nongenetic journals. Many are reported for the sake of their curiosity value. The scientific value of this approach lies in the possibility of a thorough analysis of discordant MZ pairs. In rare disorders, case reports may provide the only available information on whether the conditions have any genetic basis. A single carefully analyzed discordant MZ pair for any disease shows that the condition cannot be exclusively genetically determined.
2. *Accumulated case reports.* Even when a number of case reports are accumulated, the same limitations apply as for single case reports. Such a compilation of twin pairs is not representative, but systematic analysis of MZ twins discordant for a certain disease may provide valuable information on the environmental conditions favoring manifestation. A relatively small series of discordant twins could produce much better information on risk factors in the manifestation of disease than would surveys of large populations in which the problem of adequate control samples might be insuperable.
3. *"Limited representative" sampling.* This method is the most frequent approach for obtaining large, unbiased series of twin pairs. Luxenburger considered this approach of "limited representative value" because the sample is taken not from a defined region and time period but from a patient series. All twin individuals are ascertained within a population of affected patients of the disease under investigation. The co-twins are examined to find out whether they are affected as well. It is important, however, that *all* twins within the patient population be ascertained; otherwise concordant pairs will usually have a higher chance of being

included in the sample than will discordant ones. Successful ascertainment of all twins is achieved when the frequency of detected twins in the series equals that of the frequency of twins in the general population. The proportion of like-sexed and different-sexed twins and – after zygosity diagnosis – the proportion of MZ and DZ should agree with that in the general population. Practical application of this method would be much easier if the routine questionnaire for hospital patients were supplemented by the question: is the patient a twin?

4. *Unlimited representative sample.* Here every twin in a population is ascertained and examined whether he or she suffers from the condition under investigation. Usually the complete birth registers over a span of several years have to be screened. The number of individuals who must be examined is enormous in comparison to "limited representative sampling." If the condition has an incidence, say, of 0,5% and the number of twin individuals is 1:50, with the "limited representative" approach 10,000 persons have to be screened to find 200 twin individuals. For unlimited representative sampling, a population of 2 million must be screened. On the other hand, for some traits, especially mental illnesses, this approach – which was undertaken in Denmark, Norway, and Finland [2042, 2108, 2217] – has produced results showing interesting variations from those of investigations using limited representative sampling (Sect. 8.2.3.7). In Budapest (Hungary), all twins born since 1970 are being registered [618].

A recent large twin sample is a modified unlimited representative sample. All male twins who were registered in the U.S. Armed Services during World War II are being sampled by a twin registry maintained by the National Research Council in Washington D.C. A variety of investigations have been carried out with these data or are in progress.

3.8.7 *One Example: Leprosy in India*

A twin study on leprosy in India may serve as an example for application of the twin method [608]. Leprosy is caused by *Mycobacterium leprae* (Hansen's bacillus); however, not everybody exposed to the bacillus actually becomes infected, and not everybody who becomes infected develops clinical symptoms. Furthermore, the infection produces varying manifestations depending on the immunologic state of the organism. One patient may show only depigmented and anesthetic macules (tuberculoid leprosy), whereas another may have diffuse infiltrations (lepromatous leprosy).

The apparent differences in susceptibility may have many causes, but available information makes a genetic influence probable. Two such causes are the clustering of the same type of leprosy among near relatives and racial differences in relative frequencies of different leprosy types. In whites and blacks, the tuberculoid type is more frequent, in Orientals the lepromatous leprosy prevails. Moreover, twin investigations in tuberculosis have shown the importance of genetic factors. Other studies with smaller patient series, although not entirely satisfactory on methodologic grounds, have suggested genetic influences in leprosy as well [608].

The twin study discussed here was carried out in endemic leprosy areas in West Bengal and Andhra Pradesh, India, where at least 2%–4% of the total population is known to be affected. A determined effort was made to ascertain all twins suffering from leprosy within these districts. First, all those in permanent and mobile leprosy clinics were asked (a) are you a twin and (b) do you have any twin pairs in your family or in your village? The investigation was then extended to village surveys. One hundred and two twin pairs with at least one twin affected by leprosy were examined.

Table 3.29 shows that the concordance rate in MZ is significantly higher than in DZ. In addition, in many of the affected MZ pairs, the course of the disease, as well as the extent of the lesions, showed striking similarity. The intrapair differences at the age of onset of all concordant (MZ and DZ) twin pairs tended to be smaller in MZ than in DZ twins.

As leprosy may show different clinical manifestations, this disease allows analysis of concordance regarding the particular type of leprosy present. Table 3.30 shows that of the 37 MZ pairs concordant for leprosy, five were discordant as to leprosy type, one being tuberculoid, the other lepromatous. These pairs offered the opportunity for an additional finding. It has been suggested in the past that a sim-

Table 3.29. Concordance in 102 twin pairs with leprosy, 62 MZ and 40 DZ twins (Chakravartti and Vogel, 1973 [608])

Sex	MZ pairs	DZ pairs
	Concordant	Concordant
♂	24 = 60.0%	5 = 22.7%
♀	13 = 59.1%	1 = 16.7%
♂ ♀	–	2 = 16.7%
Total	37 = 59.7%	8 = 20.0%

$X^2 = 15.53$; $P \sim 0.003$.

Table 3.30. Concordance and discordance for type of leprosy in MZ and DZ leprous twins[a] [608]

	Concordant for type	Discordant for type	Sum total
MZ twins	32	5	37
DZ twins	6	2	8
Sum total	38	7	45[a]

[a] Only twin pairs concordant for leprosy are included.

ple mode of inheritance might be responsible for lepromatous leprosy. A possible candidate would be impaired function of T-lymphocytes. However, the discovery of five MZ twins concordant for leprosy but discordant for the type renders this possibility unlikely. Thus, twin studies, apart from providing evidence on how genetic variability in a population influences susceptibility to a disease, may help to illuminate more specific hypotheses on pathogenesis.

A possible bias must be considered. A determined effort was made to ascertain all twin individuals with leprosy in the regions under investigation. However, as the relative frequencies show, ascertainment of MZ twins was much more complete than that of DZ twins. (In the Indian population, the ratio MZ/DZ is quite similar to that in European populations). The reason for this bias will be found in certain living conditions in this part of India. Most of the individuals examined lived in rural areas where there was a high percentage of illiteracy; most people did not even know their exact age. Therefore, a twin pair was usually recognized as such only when the similarity could not be overlooked. Under these circumstances, DZ twins frequently would not even be noticed. Sometimes the sibs themselves did not realize they were twins.

How might incomplete ascertainment of DZ twins have influenced the result? As ascertainment of concordant pairs will for the most part be favored, the differences between MZ and DZ pairs may be underestimated. More important, however, is the question: Were the MZ pairs completely ascertained? Not likely. Some pairs may have successfully hidden their disease; many of the pairs lived in beggars' colonies and received no treatment; twins from high social classes may have escaped the survey due to treatment by private doctors; some patients may have given wrong answers to avoid the social stigma of leprosy for their afflicted twins or relatives. As most of these factors apply to both MZ and DZ twins, regardless of concordance or discordance, a stronger influence on concordance ratio in MZ due to proband selection is unlikely. Nevertheless, the concordance figure may still be too high.

As to environmental risk factors, analysis of discordant MZ pairs confirmed that continuous and intensive contact with infectious cases is most important. Therefore, concordance figures of the same high order of magnitude can only be expected in those areas where leprosy is highly endemic. Infection is almost ubiquitous, and contracting the disease depends mainly on inherited susceptibility. In populations with lower incidence of leprosy, infections depend more on chance. Therefore, low concordance rates between MZ may be expected.

Surveys on tuberculosis show similar results [919]. Earlier studies gave concordance rates of the order of magnitude encountered in this leprosy study. Patients included in these studies grew up at a time when almost everybody in an industrialized area such as central Europe or the United States was exposed to the infection as evidenced by positive tuberculin tests. In a more recent study, concordance was found to be lower [873]. In the meantime, the risk of infection had been reduced appreciably.

Even for a purely somatic condition such as leprosy, concordance rates found and hence conclusions as to the degree of genetic determination or liability are valid only for the environmental conditions prevalent in the population in which the twins are living. Generalization to other populations should be subject to careful consideration of living conditions. In central Europe, for example, leprosy disappeared during the seventeenth and eighteenth centuries without any therapy, only due to improvement of living conditions. There was probably little or no influence of genetic changes.

3.8.8 Twin Studies in Other Common Diseases

Table 3.31 lists diseases in which the twin method has helped to establish the significance of genetic factors in suceptibility. The first three entries are malformations, and therefore the transfusion syn-

Table 3.31. Twin series in multifactorial diseases (excluding mental diseases)

Condition	Twins	*n*	Concordant		MZ more frequently concordant than DZ
			n	%	
Club foot	MZ	35	8	22.9	10.0 times
	DZ	135	3	2.3	
Cong. dislocation of hip	MZ	29	12	41.4	14.8 times
	DZ	109	3	2.8	
Cleft lip & palate	MZ	125	37	29.6	6.4 times
	DZ	236	11	4.7	
Cancer	MZ	196	34	17.4	1.6 times
	DZ	546	59	10.8	
Coronary heart disease	MZ	21	4	19.0	2.4 times
	DZ	47	4	8.5	
Diabetes mellitus	MZ	181	101	55.8	4.9 times
	DZ	394	45	11.4	
Atopic diseases	MZ	12	6	50.0	11 times
	DZ	23	1	4.4	
Hyperthyroidism	MZ	49	23	47.0	15.1 times
	DZ	64	2	3.1	
Psoriasis	MZ	31	19	61.0	4.7 times
	DZ	46	6	13.0	
Cholelithiasis	MZ	49	13	26.6	4.1 times
	DZ	62	4	6.5	
Tuberculosis	MZ	381	202	51.6	2.3 times
	DZ	843	187	22.2	
Sarcoidosis	MZ	4	2	50.0	5.9 times
	DZ	11	1	8.5	

Source: adapted from von Verschuer, 1959 [919] and Jörgensen, 1974 [728].

Table 3.32. Twin concordances in four infectious diseases[a]

Disease	Twins	MZ			DZ			MZ higher than DZ
	n	*n*	Concord.	Conc. %	*n*	Concord.	Conc. %	
Measles	3,645	1,629	1,586	97.4	2,016	1,901	94.3	1.03
Scarlet fever	702	321	175	54.6	381	179	47.1	1.16
Pneumonia	800	328	106	32.3	412	86	18.2	1.77
Tuberculosis	1,316	386	204	52.8	930	192	20.6	2.56

[a] The data are from several different sources. (Jörgensen, 1974 [728])

drome may have influenced the concordance rate. In all diseases, higher concordance in MZ twins as compared to DZ twins is remarkable. The data fulfill the "twin criterion" (Sect. 3.6.2) for multifactorial inheritance.

Table 3.32 gives figures for four frequent infectious diseases. High concordance is not in itself sufficient to establish the existence of genetic factors in susceptibility; the difference between MZ and DZ must be considerable. For example, almost every child catches measles sooner or later. Therefore, concordance in both MZ and DZ will naturally be high, indicating that genetic factors are of no particular importance. The data of Table 3.33 had been collected at a time when these diseases were very common.

Analysis of discordance can shed some light on genetic versus environmental factors in disease. Studies by Lemser as early as 1938 [756] showed, for example, that pregnancies - and especially multiple pregnancies - could lead to manifestation of diabetes in predisposed mothers, as in a number of cases one co-twin became diabetic after several pregnancies, whereas the other one, who became pregnant less frequently, remained healthy (Fig. 3.73). However, the information supplied by twin studies on genetic aspects of disease susceptibility tends to be general and nonspecific. Hence, it is not surprising that twin studies for internal diseases have lost popularity in recent years. The feeling seems to prevail that, compared with the input of time and resources, the return in terms of new and specific knowledge is too meager. However, recent twin studies have helped to establish that genetic factors play a role in the biotransformation of all drugs examined so far (Sect. 4.5).

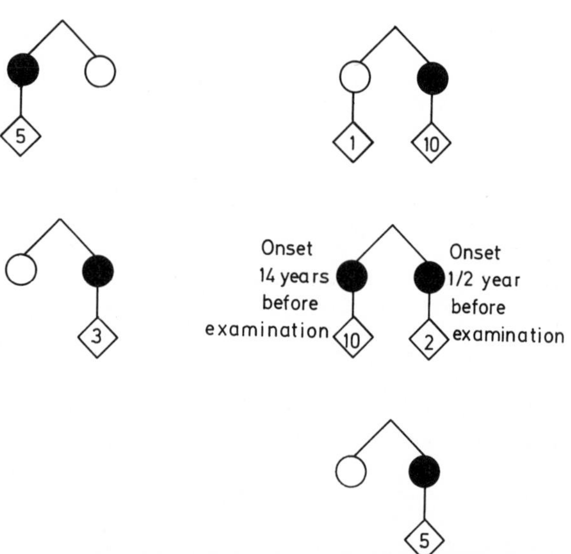

Fig. 3.73. Five MZ adult pairs discordant for diabetes mellitus. The twin sister with the higher number of pregnancies developed diabetes mellitus, whereas the sister with only a few or no pregnancies remained healthy or developed diabetes much later. (Data from Lemser, 1938 [756])

3.8.9 Twin Method in Investigating Continuously Distributed Characters

To what degree is the variability of a character in a population genetically determined? One condition for a meaningful investigation is that the character can be measured. This would seem to be self-evident, but in behavior genetics (Sect. 8.2) the problem of appropriate instruments for measurement is in fact a major problem. Once this problem has been resolved, the next problem would be: How do we get a twin sample? Usually schools, colleges, or military conscription provide good possibilities for locating an "unlimited representative sample" (Sect. 3.8.6). However, it is not easy to find a really unbiased sample. One bias at least is always present, the readiness to cooperate voluntarily. This is usually correlated with social class, people from the higher classes being more co-operative on the average. MZ twins are more likely to volunteer than DZ twins. It is plausible to assume that co-operation may also be correlated with personality variables. Hence, this source of selection may introduce a bias into many studies in behavior genetics.

When measuring the character in question, an important but much neglected aspect are errors of measurement. These can be dealt with inheritability estimates. For anthropologic measurements, the individual investigator can test his variability by measuring the same subjects repeatedly at different times.

Heritability Estimates from Twin Data. The concept of heritability has already been introduced in Sect. 3.6.1.5. In a continuously distributed character, such as stature, heritability was estimated from comparison between parents and children. Twin data can be utilized as an alternative way to get heritability estimates. The method will be discussed in App. 6. There, it is suggested that three alternative estimates of h^2 should be calculated:

h_1^2 from comparison of MZ with DZ pairs
h_2^2 from comparison of MZ pairs with age-matched, unrelated control pairs of the same series
h_3^2 from the intraclass correlation coefficients within the whole series of MZ, on the one hand, and DZ, on the other

It is also explained, that all these three estimates are subject to various biases, i.e., that an unbiased estimate of h^2 from twin data is impossible. This is especially true for the estimate h_3^2 from intraclass correlation coefficients, which has mostly been used in the literature.
Most heritability estimates from twin studies are based on unrealistic assumptions. Thus, twins in general are assumed to be an unbiased sample of the population, and the twins examined are taken for an unbiased sample of all twins. Among the total twin population, it is assumed that the environment of twins is identical to that of the general population and that identical environmental influences act on MZ and DZ twins. As explained earlier (Sect. 3.8.4), this assumption may well be the most troublesome, since MZ twins often search out and create more similar environments. However, interaction between heredity and environment as well as covariance between heredity and environment are usually neglected. Effects of dominance cannot be separated from additive genetic variance. The limitation of the heritability concept precludes any hints as to the number of genes operative and any clues on the genetic mechanism involved.
All these considerations should warn us not to take heritability estimates from twin data too literally. They are crude measures that may serve as a first orientation in the estimate of a genetic component in the phenotypic variability of a certain character. They pose questions rather than answering them.

3.8.10 Meaning of Heritability Estimates: Evidence from Stature

High heritability has been found for stature. This means that the environmental variations found in the population from which the twins came had little influence on the phenotypic value. One would tend to conclude that stature is in general a stable character that cannot be altered by any changes in the environment, except possibly in extreme situations such as by severe malnutrition. This conclusion has been shown to be wrong.

Increase in Stature During the Recent Century [758]. An appreciable increase in stature has been observed in Europe and the United States over the last century. A statistical analysis may help to find an explanation for this increase.
The average stature in central and western Europe had remained more or less constant from Neolithic times up to the middle of the nineteenth century. Since that time, it has increased steadily. Some examples are shown in Table 3.34, and many other data from various countries show the same tendency. However, the time at which the increase began is somewhat different from one population to another. In northern Holland, for example, the average stature of conscripts decreased between 1821 and 1858, and the same decrease was observed in Flanders during the 1840s. During this time, both countries were suffering from economic depression. In other countries as well, people tended to become shorter in times of economic difficulties and to grow taller when the economic situation improved. In Argentina, for example, a noticeable increase has been observed in the twentieth century only. A difference in stature between urban and rural areas has been reported frequently. In some regions, rural populations tended to be taller than city populations and in others vice versa. The difference between social classes, on the other hand, is consistent (Table 3.34). The strong class differences that were apparent around 1900 had almost disappeared by 1930. During this time period, stature of the lower classes caught up with that of the upper class.
Comparable differences between social classes have been found in age of onset of puberty and in increase of height and weight during childhood. There traits also have shown fewer or no social class differences in recent times.

More Detailed Analysis. In any attempt to explain the increase in stature, factors that have mainly changed among the lower class population need to be examined. It has been popular to think in terms of the migration from rural to urban areas. "Stimulation" of the nervous system with possible influ-

Table 3.33. Average stature of adult males (in centimeters) (Lundmann; cf. Lenz, 1959 [758])

	Sweden	Norway	Denmark
Stone Age	169.5	164	170.0
Bronze Age	166.5	–	166.5
Iron Age	167.0	167.0	168.0
Middle Ages	167.5	167.0	–
1855	167.5	168.0	165.5
1939	174.5	174.5	171.5

Table 3.34. Stature (in cm) of Swiss conscripts (Lenz, 1959 [758])

Kanton Luzern	1897–1902	1927–1932	Increase
Merchants and students	166.6	171.2	+ 4.4
Factory laborers	161.8	167.0	+ 5.2
Farmers	163.1	166.1	+ 3.0
Kanton Schwyz	1887	1935	
Intellectuals	167.0	170.6	+ 3.6
Heavy physical labor laborers	164.0	169.4	+ 5.5
Light physical labor laborers	163.2	168.0	+ 4.8
Farmers	162.9	168.7	+ 5.8
Factory laborers	155.9	169.6	+13.7
City of Zürich	1910	1930	
Merchants and students	169.6	172.7	+ 3.1
Tailors	166.5	169.5	+ 3.0
Factory workers	166.4	170.5	+ 4.1
Farmers	165.8	168.4	+ 2.6
Blacksmiths	165.7	168.8	+ 3.1

ences on growth hormone production by urbanization was brought forward as an explanatory hypothesis. This explanation, however, is not supported by the statistical evidence. Differences in stature between urban and rural populations were not found generally but only when the standard of living in rural areas was also lower. Earlier sexual maturation was observed in urban populations only when the general standard of living was higher as well.

Another factor to be considered is the age at which the secular trend in growth first becomes apparent. It turns out that even average birth weight is higher today than 100 years ago by some 100–300 g. This figure may overestimate the difference, as the sample of newborns examined 100 years ago was biased toward the lower social class since middle-class mothers usually delivered at home. At the end of their 1st year of life, however, children in the early 1950s tended to be 1.5–2 kg heavier than those in the same age group 100 years ago. This secular increase is much more pronounced than that of weight at birth. Besides, numerous investigations have shown that increases in growth and body weight during the 1st year of life are largely independent of birth weight. Hence, an appreciable part of the higher growth rate can be traced back to infancy.

The decisive factor for increased growth must be operative before school age. Menarche is reached 3–4 years earlier in more recent times than a century ago. In spite of this trend, however, the child at puberty is, on the average, about 10 cm taller than his age mates 100 years ago. The temporal increase of growth in recent years was more intense than the earlier onset of puberty. Hence, the increase in average stature of adults cannot be due to a greater growth rate during or after puberty.

Most Likely Explanation. Some authors have tried to explain the phenomenon genetically. One possibility would be heterosis, i.e., an enhancement of the growth rate due to increased heterozygosity. There is no doubt that average heterozygosity has indeed been increased in many populations. However, the inbreeding coefficient (Sect. 6.3.1.1) of most human populations up to the last century was low in comparison with that of non-human populations in which inbreeding affected most quantitative traits negatively (inbreeding depression). Besides, the increase in stature was also found in areas in which there was no breakup of isolates. Furthermore, examinations of inter-racial hybrids have shown that they are on the average not taller than the populations of the races they were derived from. Therefore, data claiming depressed stature in inbred human populations must be considered with caution.

Nor can selection be the decisive factor. On the contrary, the higher ranking social classes had at the same time a much lower average reproduction rate. We may conclude that the trend very probably has no genetic cause. There must be some environmental factor. Taking into account the differences between populations and those within the same population, i.e., between social classes, improved nutrition during infancy and childhood is the most plausible single factor in explaining the higher stature of all Western or Japanese populations studied in recent times. Improved protection against infectious diseases such as infant diarrhea might add to the effect. Even some of the differences in average stature among various racial groups can be explained by differences in nutrition and exposure to intestinal infection in early infancy. Walter (1976) [930], for example, showed that the so-called Bergmann rule – according to which varieties of the same species tend to be smaller and lighter in a warm climate and larger and heavier in a colder climate – applies to humans as well. However, he also showed that the same correlation can be found between amount of food available and stature.

Lesson to Be Learned from This Example. A higher heritability for a given trait as measured under certain environmental conditions does not prevent this trait from being influenced strongly by secular trends in environmental conditions to which the whole population or appreciable parts of it may be subjected.

This phenomenon is especially true for characters developing over a prolonged period of time, during which the human organism may be exposed to various changing external influences. However, it would be premature to conclude that every environmental change necessarily influences such a trait. Some - and even those that plausibly would have an effect - might still be without influence. A prediction in general terms is impossible; every situation may be different. This topic will be taken up again in the section on behavior genetics.

3.8.11 Twin-Family Method [768, 732]

It usually is burdensome to amass twin or family material for a proposed study. Hence, if one plans to investigate a certain condition by means of both methods, examination of the families of twins is an obvious possibility for reducing the effort involved.

The two approaches may be linked in a more special way. The fact that some of the MZ twins will be concordant and others discordant for a given disorder may be caused by one or both of two reasons: (1) manifestation of the disorder may be influenced by nongenetic factors and (2) there may be two different types - an inherited and a noninherited form. The two possible explanations can be separated by comparing empirical risk figures for the disorder in close relatives of concordant as compared with discordant MZ pairs. If there is heterogeneity and one type of the disease is nongenetic, the risk of relatives of discordant MZ pairs will be no higher than that in the general population, while discordance caused by nongenetic factors will lead to similar risks among relatives of concordant or discordant twins.

To the best of our knowledge, Luxenburger [769] was the first to use the twin-family method. He showed that in schizophrenia, risks are about equal in the groups of relatives of concordant and discordant MZ twin pairs, suggesting that sporadic nonhereditary cases are rare or nonexistent.

A good example of application of the twin-family method is a study on diabetes [906]. Ninety-six MZ twin pairs were examined, 65 pairs were concordant for diabetes. As emphasized by the authors, these cases did not constitute an unbiased sample but were biased for concordance. Therefore, they were not included in Table 3.35. However, family analysis of concordant and discordant pairs turned out to be interesting. The number of twin pairs with one affected parent was much higher in the concordant than in the discordant group. This suggests that there are two types of diabetes; one mainly inherited, the other mainly environmental. The conclusion is corroborated by a number of other differences. Concordance is much higher for probands diagnosed after the age of 40; in 75% of the concordant pairs, the interval between the diagnoses was less than 3 years, whereas half of the discordant pairs had been discordant for more than 10 years; most of the unaffected cotwins had normal glucose tolerance tests. Furthermore, the concordant diabetics showed the usual tendency to give birth to heavy babies before they manifested the disease, but the unaffected partners of discordant pairs did not. The largely nongenetic type comprises cases of juvenile diabetes. Specific environmental factors were not discovered in the study. Work done since this study was completed has established that maturity onset diabetes is not associated with HLA while juvenile diabetes usually is. Thus, genetic factors presumably related to immune responsiveness (Sect. 3.7.3) are operating in juvenile but not in maturity-onset diabetes.

Table 3.35. Number of twin pairs with a diabetic parent (Tattersall and Pyke, 1972 [906])

	Concordant	Discordant
All ages	21 of 65 (32%)	1 of 31 (3%)
Proband twin diagnosed under 40	6 of 30 (20%)	1 of 28 (3%)
Proband twin diagnosed after 40	15 of 35 (42%)	0 of 3

Note that the frequency of affected parents is much higher in concordant than in discordant pairs and especially in those pairs in which the diabetes had been discovered after the age of 40.

3.8.12 Co-Twin Control Method [680]

Since monozygotic twins are very similar or identical in a number of characters, this fact can be utilized to examine whether and to what degree certain environmental influences are capable of changing a given character. Often a character may change spontaneously over time; a disease may remit spontaneously, and this may be attributed erroneously to intervention or external influences.

MZ twins offer a good investigative opportunity by exposing one twin to the influence under study and the other not. This method allows the most perfect experimental device by complete control of possible host variability. The method can be compared to utilization of inbred animal strains in experimental medicine.

Although the method was developed for investigating educational influences on human behavioral characteristics, it can be utilized in a more general way. It can be used for testing, for example, whether certain therapeutic measures are useful.

In one study [742], 22 MZ and 28 DZ twins were examined by psychological test procedures to find out whether certain aspects of intelligence can be improved by "psychological exercise." First the tests were carried out without any prior training. Then the twin with the poorer performance received training once every week for 5 weeks. At the end of the 5 weeks, the twins were examined again, and an increase in performance was found in the trained twins but not in their co-twins.

In a Swedish study [824], two methods of teaching reading and writing were compared in ten MZ and eight DZ twins of the same sex. The advantages commonly claimed for the analytic method, which starts with the reading of whole words rather than single letters, could not be confirmed. Certain advantages of the more traditional methods in which individual letters are taught to be combined into words were, however, confirmed. Due to the different structure of languages, this result cannot immediately be generalized. Replication of the study in an English-speaking population would be interesting, as the sound of letters depends much more on their context within words than it does in other European languages.

3.8.13 Contribution of Human Genetics to a Theory of Disease [923]

Diseases with Simple Causes. Modern medicine tries to understand disease scientifically. Scientific understanding requires theoretical foundations. Requirements for a good scientific theory have been discussed in Sect. 3.6.1.7. A general theory underlying the explanation of all diseases does not exist and probably never will. However, theories are possible for certain aspects of disease. For example, the concept of a *disease* that is determined by one single, main *cause* turned out to have a high explanatory value. For example, the multiple and varied signs of tuberculosis will only occur following infections with mycobacterium tuberculosis. The specific development of the tubercular infection and the natural history of tuberculosis in an individual depend on many additional circumstances, including genetic factors. A theory of disease that is centered around the concept of a disease unit produced by a single cause is specific, requires an elucidation of mechanisms, and therefore has a high explanatory value. As a first choice, such a theory is preferable to a concept that is based on mere description of disease signs, such as coughing or hemoptysis, or on the low-order constructs on which the organ pathology of the 19th century was based, such as "chronic productive inflammation of the lungs." The goal of scientific research on disease is the replacement of descriptive pathology by more explanatory concepts.

Obviously this is easier when a distinct disease has a single cause. A hundred years ago, this concept was successfully applied to various infectious diseases. Understandably enough, this success encouraged scientists to apply the concept to conditions where a single cause did not exist, and where diagnostic criteria were fuzzy. A probably example is schizophrenia. Here, the search for the one major biological or psychological cause has been unsuccessful, although it is conceivable that a major cause has eluded research [2164].

Hereditary diseases with simple, monogenic inheritance are excellent examples for the successful application of the monocausal disease concept. Using mutations of the hemoglobin genes as examples, it can be shown how genetic analysis based on Mendel's paradigm and its extension into molecular biology not only permitted identification of the *causes* of disease, but paved the way for an elucidation of the *mechanisms* by which well-defined mutations cause impairment of function, i.e., disease (Sect. 4.3). It is noteworthy, however, that interaction with other genes (and possibly with the environment) determines the severity of monogenic diseases. Sickle cell anemia is a well-studied example. Higher Hb F levels make for milder clinical manifestations in sickle cell anemia, and various well-defined mutations that cause elevated levels of fetal hemoglobin (hereditary persistence of fetal hemoglobin) ameliorate the clinical picture. However, even more subtle alterations in the chromosomal environment around the Hb S mutation (as defined by DNA variant haplotypes) apparently affect critical Hb F regulatory sites [1344]. Thus, the "Senegalese" type of sickle cell anemia is associated with more Hb F, a preponderance of the Hb Gγ chains, and a lower proportion of irreversibly sickled cells compared with the mutationally identical "Benin" type of sickle cell anemia which differs in DNA haplotype [1233, 1232]. The simultaneous presence of α-thalassemia is another modifying factor associated with a less severe clinical pattern. Our increasing abilities to define specific genetic determinants that affect clinical severity in sickle cell anemia is an excellent model for elucidation of the pathogenesis and clinical severity of other genetic diseases by analysis of interacting genes.

The hemoglobin variants demonstrate another phenomenon. Mutations within the same gene may lead to quite different phenotypes. Methemoglobinemia, for example, is a different disease than sickle cell anemia. Other mutations at different sites of the same gene, with different effects have been observed. Conversely, genetic heterogeneity, i.e., causation of similar or even identical phenotypes by mutations at different gene loci, is also quite common, so that a variety of causes may lead to the same end effect.

In chromosomal aberrations, the *causes* of many birth defects have been identified. Chromosomal aberration syndromes are defined unequivocally by their abnormal chromosomal constitution. But the *mechanisms* by which these aberrations lead to abnormal phenotypes, i.e. the pathway from genotype to phenotype, remains poorly understood (Sect. 4.7.4).

The situation is *different* in many genetically *influenced* diseases and anomalies. In these – as in schizophrenia – a single cause cannot be identified and, in many cases, may not exist. The same pathogenetic process causing the disease might be triggered by a variety of causes – either alone, or in combination. Some of these causative factors might be genetic, while others might be "environmental" including somatic (e.g., allergens), behavioral (e.g., feeding and drinking habits), social (e.g., influences of parents, school, occupation), and other factors. Often, a preliminary description in terms of "multifactorial inheritance with or without threshold effect" (Sect. 3.6.2) allows some preliminary conclusions. However, the identification and analysis of the specific genetic and environmental components contributing to a disease risk is the *next* goal. A genetic liability might have different causes and components from one individual and family to the next, as shown in the discussion of hyperlipidemias and coronary heart disease (Sect. 3.8.13). The same is true for a disease that has been called "the nightmare of the medical geneticist" (Neel) – diabetes mellitus.

Genetics of Diabetes Mellitus [614, 831, 862]. The developments in our progressive understanding of diabetes mellitus illustrate how a common disease is gradually becoming better understood. Very early in medical history, the disease was diagnosed when there was thirst, polyuria, weight loss, weakness, coma, and death associated with sweet-tasting urine. Today, a quantitative blood glucose level with a cut-off point of 140 mg per 100 ml is used for diagnosis. This cut-off point is arbitrary, however, and therefore causes difficulties in classification and problems for genetic analysis. Diabetes is highly heterogeneous, i.e., different genetic and possibly nongenetic causes produce a clinical condition that is diagnosed as diabetes. Both common and uncommon varieties of diabetes exist. The two most common varieties are known as type I and II diabetes and can be differentiated by many different criteria (Table 3.36). Their etiology is different and they run true to form, i.e., there is familial aggregation limited to the type of diabetes observed in the index case. Although familial aggregation is less striking in the more severe type I diabetes, the pathophysiology is better understood [757b]. There is increasing evidence that this disease may be caused by a viral insult to the islets of Langerhans in the pancreas, followed by the production of anti-islet autoantibodies which gradually destroy the islets. This process leads to insulin deficiency and characteristic clinical findings. However, not every person will develop the disease. It appears that genetic determinants at the HLA locus associated with HLA DR3 and/or DR4 specificities determine the outcome [614]. The use of DNA probes for this HLA region is revealing restriction site differences between type I diabetics and controls [612], but further study is required to clarify the details. Type I diabetes appears to be caused by a virus that acts on a genetically susceptible host to form pancreatic islet autoantibodies. However, even among identical twins there is only 50% concordance, suggesting that various other factors, such as lack of equal exposure, random events, and environmental factors that are still poorly understood also play an important role in the etiology.

Table 3.36. Two common types of diabetes (after Olefsky, 1985 [831])

	Type I (insulin dependent) IDDM	Type II (insulin independent) NIDDM
Prevalence	0.2%–0.3%	2%–4%
Proportion of all diabetes[a]	7%–10%	90%–93%
Onset	< 30 years	> 40 years
Body fat	Lean	Obese
Ketoacidosis	Common	Rare
Insulin deficiency	Absolute	Rare
Therapy	Insulin	Diet
Complications	Vasculopathy, neuropathy, nephropathy	Infrequent and late
MZ twin concordance	40%–50%	100%
First-degree relatives affected	5%–10%	10%–15%
HLA D3/D4 association	Yes	No
Circulating pancreatic islet cell autoantibodies	Yes	No
Other autoimmune phenomenon	Occasional	No
Insulin resistance	Occasional; insulin antibodies	Usual: postreceptor defects?

[a] All other "diabetes" is quite rare: < 1%.

Type II diabetes is common in the middle-aged and elderly but is usually mild. Genetic factors have an important role, as evidenced by the high identical twin concordance rate. The nature of the genetic factors and their mode of transmission have not yet

been elucidated. Some studies using restriction mapping of the insulin gene have suggested that patients with type II diabetes have a higher frequency of the hypervariable restriction site variant immediately upstream of the insulin gene that may be involved with regulation of insulin synthesis [713, 743, 964]. These results could not always be replicated.
Further heterogeneity (based on a variety of autoimmune phenomena in type I diabetes and by various criteria such as obesity in type II diabetes) is likely, but has not yet found general acceptance.
In contrast to type I and type II diabetes, which do not follow Mendelian inheritance, a rare form of the disease with early onset and a mild course without complications is transmitted by autosomal dominant inheritance. The defect is unknown but may relate to subnormal insulin secretion. This condition is known as MODY (Maturity Onset Diabetes of the Young).
A variety of very rare types of diabetes have been differentiated. These involve insulins with amino acid substitutions that make for lessened activity of the insulin molecule [867]. Defective conversion of proinsulin to insulin owing to amino acid sustitution at critical sites has also been described as an autosomal dominant trait [857 a]. Most forms of diabetes, however, are not associated with structurally abnormal insulin.
Insulin receptor action has been extensively studied [861]. Various receptor abnormalities manifesting as a decreased number of insulin receptors or decreased binding have been found in some rare genetic conditions such as leprechaunism, lipodystrophy, ataxia telangiectasia, and acanthosis nigricans.
Receptor or postreceptor anomalies have also been suggested in type II diabetes but have not been definitely demonstrated.

Disease Concepts and Diagnosis [948, 949]. When a physician makes a diagnosis, a certain cluster of clinical signs and laboratory data is subsumed as a disease unit. Hence, it is implicitly assumed that a "natural system of diseases"exists. And, indeed, such an assumption is justified if a single major cause can be pinpointed, such as in infectious and monogenic genetic diseases. However, the great majority of all diseases are defined phenomenologically, or in more recent times by some metric characteristics, such as diabetes or hypertension.
Such a disease classification grew historically and now comprises disease units that have been defined sometimes arbitrarily in different ways, may have blurred borders, and often overlap. In medical practice, this procedure is often successful, since medical diagnosis is meant to serve as a guide for therapy. Therapy may not require the careful heterogeneity analysis necessary for genetic investigations. Therefore, it is reasonable and advisable to stop diagnostic procedures at a point at which no further benefit for the patient can be expected. If the management remains identical it is no longer important for practical purposes to differentiate between subtle diagnostic categories.
However, there may be pitfalls in that a given category may be too superficial to provide appropriate therapeutic guidance for all patients encompassed by that diagnosis. Thus, a diagnosis of fever 100 years ago comprised many different diseases, which today can be subclassified and demand different therapies. Similarly, a diagnosis of anemia 75 years ago is all that could be specified in pale patients with too little blood. Today, we know of different types of hereditary and acquired anemia, many of which require different treatments. Blood transfusion would be an inappropriate therapy for *all* anemias, since iron deficiency anemias can be treated specifically with iron, pernicious anemia with vitamin B_{12}, and hereditary spherocytosis with splenectomy. Another example: Even though we treat hypertension today empirically with many different drugs, it is likely that better understanding of the heterogeneous mechanisms of hypertension would lead to a more specific treatment appropriate for certain subgroups of patients. We know already that blacks with hypertension respond better to diuretics than to beta blockers (compared with whites), but do not yet know the reason for this variable response.
The medical geneticist always requires a highly specific diagnosis with special attention to heterogeneity, to allow him or her, to give appropriate genetic advice regarding recurrence risks and prenatal diagnosis (Sect. 9.1).
Diseases with similar manifestations may be inherited by different genetic mechanisms or may not be genetic at all.

Normal Variation and Disease. The distinction between a disease and the upper limits of normal variation is important. As an example, hypertension is not a disease (although often considered as such), since it represents the designation for a certain percentage of the population whose blood pressure level is higher than an arbitrary cut-off value. The risk of complications of hypertension increases with higher blood pressure levels, but there is no threshold value at which the risk disappears. The "diagnosis" of hypertension in a sense is inappropriate. Hypertension is a "risk faktor" for coronary heart disease, stroke, and renal failure, rather than an illness.
As we learn more about various genetic risk factors conferring susceptibility to certain diseases, similar problems arise. Many persons carry the HLA D3 or

D4 determinant, but only a small percentage will develop type I diabetes (Sect. 3.7). The relative risk is 8–10 times that of a person not carrying these HLA types. The absolute risk of HLA D3 and D4 varies for developing diabetes of HLA D3 and D4 types remains quite small. Homozygotes for the Pi Z phenotype will often develop chronic obstructive pulmonary disease, but not all gene carriers will get sick. Such Pi Z homozygotes are not diseased, but may become so in the future.

One of the goals of medical genetics is the elaboration of "marker profiles" that will aid in identifying those subgroups at high risk of developing certain diseases - particularly if measures can be taken to prevent, defer, or ameliorate the deleterious effects of the genetic predisposition by environmental manipulation. This approach is particularly promising, since the development of disease frequently requires interaction between genetic susceptibilities and environmental factors. Such preventive medicine of the future will be "tailor made" to the unique genotypes of an individual rather than directed at the entire population.

Creating the scientific basis for recommending such rules will be the specific goal of ecogenetics, to be discussed in Sect. 4.5.2. The theory of disease develops into a theory of health preservation. To explicate the aspects sketched in this section in greater detail will be one major pupose of the chapters to follow - especially our considerations on gene action (Sect. 4) and mutation (Sect. 5).

3.8.14 Current Status of the Genetics of Common Diseases [808, 810]

Genetic diseases caused by chromosomal aberrations and by Mendelian mutations affecting single genes are relatively well-understood. Their mechanisms can be approached by studies of individual gene action (see Sect. 4) or by consideration as to how gross chromosomal defects cause developmental damage (see Sect. 2). Various data show that familial aggregation is frequent in many other diseases. Appropriate studies (see Sect. 3.6) need to be carried out to prove that familial aggregation is caused by common genes rather than by a common family environment.

A variety of experimental designs to discriminate between the role of environment and heredity have been established. Such designs include study of identical twins reared apart in different environments and the comparisons of disease frequency in adopted children with that of their biologic and adopted relatives (see Sect. 8). When identical twins even in different environments show greater concordance than DZ twins in similar environments, genetic rather than environmental factors are suggested. Similarly, when adopted children resemble their biologic parents rather than their adoptive parents, genetic factors appear certain. The frequency of a trait or disease is also often studied in spouses who share the same environment in comparison with biologic relatives who share both heredity and environment. Absence of correlation in spouses when relatives have an increased disease frequency argues for genetic factors.

Based on various investigations of this type, it has been concluded that genetic factors are operative in the following groups of diseases:

a) Common birth defects (i.e., neural tube defects, cleft lip and palate, club foot, congenital heart disease, and others)
b) Common psychoses (schizophrenia and affective disorders).
c) Common diseases of middle life (diabetes, hypertension, coronary heart disease).

When family studies were done in these diseases, Mendelian inheritance could not be demonstrated. Based on models of polygenic gene action, it has been inferred that many unspecified genes acting together with environmental factors are operative in the etiology of these disorders. The biologic action of the involved genes remains largely unknown and is considered a "black box". It is usually assumed that the number of operative genes is relatively large and that the contribution of each of the postulated individual genes to the pathogenesis of the disease is relatively small. Additive gene actions of these genes is usually assumed. When the disease occurs as a qualitative phenomenon with a dichotomized class of "normals" and "sick", such as in the congenital malformations, a threshold is assumed. If the number of genes in a person falls above this threshold, the disease occurs. For some other diseases, the abnormality will occur in persons who do not have a sufficient number of genes and therefore fall below the threshold.

Our concept of multifactorial inheritance is summarized in Fig. 3.74. We would like to emphasize the potential role of one or a few major genes in many supposedly multifactorial traits. In the common diseases, particularly, a relatively small number of potentially identifiable major genes may contribute to the genetic etiology and explain most of the genetic variation. Such genes do not act in a vacuum. The ensemble of all other genes against which such major genes act are considered to be the various polygenes that constitute the "genetic background." It is well-known that the genetic background may modify and influence expression of major genes (Sect. 3.1.7).

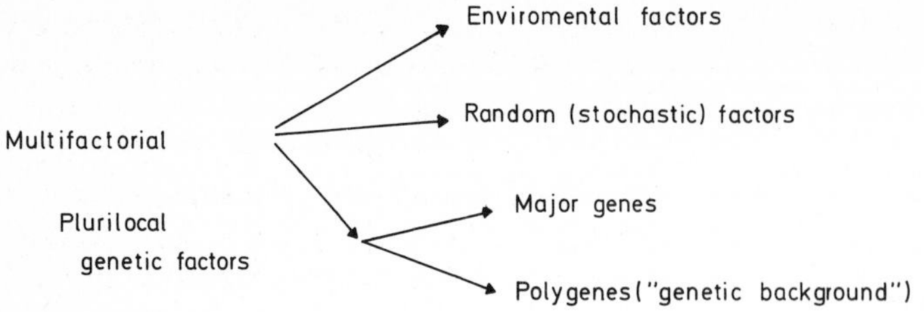

Fig. 3.74. Conceptual model for causation of multifactorial disease. In contrast to conventional models, the importance of major genes is emphasized

In birth defects, particularly, random factors may play a role [919]. We postulate factors that are not determined by either a genetic or environmental agent but act stochastically. Thus, one can visualize how some cardiac defects may occur, because the complex dynamic sequence of twisting and turning to form the normal heart may fail to synchronize by chance only. The relatively low concordance of genetically determined birth defects for MZ twins who share both heredity and intrauterine environment may be partially explained by such random factors. Another part of the explanation may lie in the transfusion syndrome sometimes seen in MZ twins.

Further research on the genetics of common diseases is most likely to be fruitful by attention to the action of individual genes to be studied by combined genetic, biochemical, immunologic, clinical and statistical methods. Biometric approaches alone are unlikely to provide new insights.

3.8.14.1 Biologic and Pathophysiologic Approaches to the Genetic Etiology of Common Diseases

Heterogeneity Analysis: Differentiation of Monogenic Subtypes from the Common Varieties. Often rare subvarieties of clear Mendelian inheritance such as X-linked HPRT deficiency in gout [737] and familial hypercholesterolemia in coronary heart disease [686] need to be differentiated from the multifactorial common diseases by appropriate clinical, laboratory, and genetic methods. Similar considerations apply to most other common diseases.

Clinical Population Genetics. Clinical, laboratory, and family studies of unselected cases of heterogeneous disease groupings such as mental retardation [835], deafness [669], blindness [670] and coronary heart disease [686] can achieve separation of familial from sporadic cases. Appropriate statistical and biochemical analyses of the sporadic cases can establish a certain number as caused by autosomal recessive inheritance. Family study will establish genetic heterogeneity as manifest by different modes of monogenic inheritance in some cases and multifactorial transmission in others. Such studies involving populations of patients and their families will be most informative if they incorporate the most advanced available laboratory techniques.

Search for Biological Heterogeneity. Geneticists need to be aware of biomedical research advances in the diseases in which they are interested. Incorporation of newer biologic and biochemical test methods in genetic studies is highly desirable. Similarly, researchers on disease can often be aided by the incorporation of a genetically oriented study design for the investigation of pathophysiology and biochemistry of a given disease.

Polymorphism and Disease. Some of the extensive polymorphisms of human genes may represent part of the genetic basis making for differential susceptibility to common diseases. Examples include the malaria dependent polymorphism Hb S, β-thalassemia and G6PD deficiency as well as the Duffy trait [1952]. Associations of HLA and disease are often of particular interest and may relate to differential responses to autogenous antigens [48]. Different disease associations with ABO types are less clear from a pathophysiological viewpoint [211]. In general, the approach of relating polymorphic marker genes will be most successful if markers which are pathophysiologically related to the disease can be studied. Random genetic markers investigated in random diseases are less likely to produce meaningful data.

Heterozygotes for Rare Diseases May Be More Susceptible to Develop a Functionally Related Common Disease. Even with rare autosomal-recessive disease, there are many heterozygotes in the population (Table 3.6). Such heterozygotes may be at higher risk for common diseases that are pathophysiologically related to the enzyme defect [1340]. As an example, heterozygotes for the rare autosomal recessive methemoglobinemia do not have methemoglobinemia unless they ingest methemoglobin-forming drugs. Normal homozygotes have sufficient methemoglobin reductase to reduce the methemoglobin formed by the drug while heterozygotes have sufficient amounts (50% of normal) under usual conditions but not enough when hemoglobin formation is excessive on drug administration.

Increased frequency of cancer among heterozygotes for ataxia-teleangiectasia and Bloom's syndrome (Sect. 5.1.6) and an increased frequency of heterozygotes for α_1-antitrypsin deficiency among patients with chronic obstructive pulmonary disease [749] (Sect. 3.7.4), are other examples of heterozygote susceptibility.

3.8.14.2 Genetics of Coronary Heart Disease

[847; 827; 570].

The prevalence of coronary heart disease varies widely in different parts of the world. In general, this condition has the highest frequency in "Western" countries and in subpopulations participating in "Western" life styles [580]. Its frequency in underdeveloped populations is usually quite low. Temporal trends in the USA with increasing mortality of coronary heart disease for many years but significant lowering of coronary death rate over the last 15 years show the strong effects of environmental factors [763]. The important role of environment is also shown by the increasing rate of coronary heart disease when migrants from low-frequency countries (i.e., Japan) move to high-frequency areas (i.e., USA) [730].

Genetically oriented studies in atherosclerosis aim to (a) detect genetic differences between individuals which predispose to atherosclerosis; (b) sort out genetic from environmental determinants; and (c) identify subpopulations at risk for preventive measures. With the development of appropriate genetic and other "markers", an increasing number of persons can be identified whose genetically determined internal makeup makes them more susceptible to certain environmental factors (including diet) that will cause coronary atherosclerosis. Twin studies have shown a significantly greater frequency of coronary heart disease among MZ twins than among DZ twins [2297]. However, such studies are difficult in view of problems with diagnostic endpoints. An ideal twin study for coronary heart disease would require angiography (or a noninvasive technique outlining coronary vessels) that delineates the extent of coronary atherosclerosis.

There is general agreement about familial aggregation of coronary atherosclerosis. The frequency of coronary heart disease is about 2–6 times higher in patients' families than in control families (for references see [570, 650, 701, 858, 902]). The following facts regarding familial aggregation are noteworthy:

1. Familial aggregation increases with decreasing age of affected patients, i.e., in premature coronary heart disease.
2. While women have lower frequency of coronary heart disease than men, affected women have a stronger familial aggregation than men. The less frequently affected sex has a higher extent of genetic "loading" (Sect. 3.6.2).
3. A family history of premature coronary heart disease (<55 years) emerges from recent work as the strongest risk factor for coronary heart disease and is stronger than all other risk factors (see below [830]).
4. Hyperlipidemia, hypertension, and diabetes represent risk factors for coronary heart disease that have strong genetic determinants. However, various studies suggest that familial aggregation cannot be entirely accounted for by these three well-known genetic risk factors [650, 830, 909]. Additional familial factors appear to contribute to familial aggregation.
5. Familial aggregation does not necessarily mean genetic determination. Families share similar environments that may include agents causing a higher frequency of coronary heart disease in family members. There is evidence for assortative mating for coronary risk factors. Spouses' families show the same extent of familial aggregation for coronary heart disease as the families of their affected husbands. Spouses of affected patients also have a higher frequency of coronary heart disease than the control population [908]. Assortative mating presumably occurs for social class, life style, smoking, and dietary habits. It is therefore likely that a significant component of familial aggregation is mediated by environmental factors shared by families. Furthermore, there may be complex genetic-environmental interaction.
6. A family history of premature coronary heart disease regardless of the nature of the various genetic and environmental factors involved serves to identify high-risk individuals and families for intervention schemes to prevent coronary heart disease.

Risk Factors. Extensive epidemiologic work has been done to identify a variety of risk factors for coronary (and cerebrovascular) atherosclerosis. Age, male sex, hypertension, hypercholesterolemia, low levels of high density lipoprotein (HDL), and diabetes carry a particularly strong risk [903].

Other risk factors that have been implicated include hypertriglyceridemia, high levels of apolipoprotein B, low levels of apolipoprotein A-1, sedentary life, obesity, and certain personality patterns. Inborn variation in thickness of the vascular intima and of musculoelastic layers has also been suggested [903].

These findings indicate a multifactorial etiology of coronary heart disease. Genetic factors can be elucidated by studying the extent of genetic contribution to the various risk factors.

Hyperlipidemias (See Table 3.37). Current data suggest that hypercholesterolemia and low high-density lipoprotein (HDL) levels are strong risk factors [723]. Hypertriglyceridemia is not generally accepted as an independent risk factor [718]. Among the inherited hyperlipidemias several conditions need to be differentiated.

Familial Hypercholesterolemia [671; 685; 875]. The best understood condition is autosomal dominant familial hypercholesterolemia (Sect. 4.6) with a heterozygote frequency of about 1/500 in the United States of America (see Section 5.2.1.5). Heterozygotes have an elevation of cholesterol and of LDL cholesterol in the upper 5% and usually in the upper 1% of the population range. Fifty percent of affected males have some manifestation of coronary heart disease by the age of 50. Tendon xanthomas and early corneal arcus may occur. Clinical manifestation in females occur 10-15 years later. Other risk factors, such as hypertension, cigarette smoking, and low HDL levels, interact with the gene for familial hypercholesterolemia in hastening the onset of atherosclerosis symptoms. The homozygote state is extremely rare, and coronary heart disease occurs in the teens or twenties. Familial hypercholesterolemia occurs in many countries and populations. Laboratory detection is difficult, since there is no generally available laboratory test for unequivocal diagnosis. Tests for receptor function can only be performed on a research basis. (Sect. 5.2.1.5)

Familial hypercholesterolemia needs to be distinguished by clinical, laboratory, and genetic criteria from various hypercholesterolemias caused by other acquired and genetic entities. Only about 1 in 25 persons with a cholesterol level in the upper 5 percentiles will carry the gene for familial hypercholesterolemia. About 3-8% of unselected male patients 60 years and below with myocardial infarct have heterozygote familial hypercholesterolemia (Table 3.37). With decreasing age at first myocardial infarction the frequency of this conditions increases.

Isoalleles for the LDL Receptor? The work on familial hypercholesterolemia may have implications for control of cholesterol levels and atherogenesis that goes beyond this specific entity. There are suggestions that alleles each specifying LDL receptors with different degrees of functional capacity for LDL binding exist [771, 933]. According to this, persons with a low-capacity allele for LDL binding would tend to have higher LDL cholesterol levels than those having a receptor that binds more LDL choles-

Table 3.37. Common hyperlipidemias associated with coronary heart disease

Name	Prevalence	Physiologic abnormality	Defect	Genetics	Frequency in unselected myocardial infarct	
					survivors < 60 years	Average age[a]
Familial hypercholesterolemia	1/500	Diminished LDL breakdown	Abnormal LDL	Autosomal dominant	3%-6%	46 years
Polygenic hypercholesterolemia	5%			"Polygenic"	Increased	58 years
Familial combined hyperlipidemia	0,3-1%	Increased apo B synthesis		Autosomal dominant	11%-20%	52 years
Familial hypertriglyceridemia	1%	Increased VLDL synthesis		Autosomal dominant	4%-5%	57 years
Type III hyperlipidemia (remnant removal disease)	1/10000	Diminished remnant catabolism	Abnormal Apo E binding and additional factors	Homozygote for Apo E_2	1%-2%	50's

LDL, Low density lipoproteins; *VLDL,* very low density lipoproteins.
[a] Average age of male patients with myocardial infarct.

terol. It is postulated that current levels of LDL cholesterol in Western countries exceed the saturation capacity of low-capacity receptors, thus causing atherosclerosis [684]. The normal range of cholesterol levels and the reported sib-sib and parent-child (but not spouse-spouse) correlation for cholesterol levels [727] would be at least partially explained by such an isoallelic mechanism. Polygenic hypercholesterolemia could largely be explained by such isoallelic variability.

Familial Combined Hyperlipidemia. Studies of probands with hyperlipidemia have resulted in the definition of a familial disorder characterized by elevation of both cholesterol and triglyceride (type IIb), of cholesterol alone (type II), or of triglyceride alone (type IV). This disorder has also been termed "multiple lipoprotein type hyperlipidemia" of familial combined hyperlipidemia and probably is common in the general population (1/100-1/300) [583]. The condition appears to segregate as a Mendelian trait, but does not reach full penetrance until age 30. Its detection therefore requires extensive family studies. Several large pedigrees with the disorder have been documented. It has recently been suggested that apolipoprotein B elevation may be a better marker than other lipid parameters and will allow more ready detection of familial combined hyperlipidemia [712]. The condition was seen in approximately 10% of unselected myocardial infarct survivors below the age of 60 years. When probands with familial combined hyperlipidemia without myocardial infarction were studied the familial aggregation of premature coronary heart disease was striking [594].

While apolipoprotein B elevations are helpful for diagnosis, an elevated apolipoprotein B level is not specific or pathognomic for familial combined hyperlipidemia.

Familial Hypertriglyceridemia. A common autosomal dominant condition associated with segregation of high triglyceride levels alone has been postulated. Children may not present with this entity. The association of the entity with coronary heart disease is contested. Work with this entity is rendered difficult by the considerable lability of triglyceride levels that are influenced by a variety of dietary and other factors including alcohol. Isolated hypertriglyceridemia of any sort as a risk factor for coronary heart disease has been questioned [718]. No basic defect for the claimed autosomal dominant variety of hypertriglyceridemia has been demonstrated, but increased VLDL-triglyceride synthesis has been suggested.

Broad Beta Disease of Type III Hyperlipoproteinemia (Remnant Removal Disease). Type III or dysbetalipoproteinemia occurs only in homozygotes of an apolipoprotein E polymorphism (E_2-E_2) that is found in about 1% of European populations. However, additional lipid-raising factors, such as familial combined hyperlipidemia or various secondary forms of hyperlipoproteinemia, are required to produce this disease. This disorder threfore is the result of the interaction of two genetic disorders or of a genetic and an acquired disease. While the frequency of the underlying polymorphism is common (1%), the actual condition is rare (1/10000).

Associations of Coronary Heart Disease with Genetic Markers [570, 801].

Protein Markers. Genetic polymorphisms that are biochemically and pathophysiologically related to a disease may represent the "genetic background" which makes certain individuals more likely to be affected. An analysis of such polymorphisms may assemble a group of markers that in the aggregate could contribute significantly to disease susceptibility. A variety of markers have been studied in coronary atherosclerosis. Most of them contribute only a small fraction to the total etiology. Individuals with blood group A of the AB0 system have a higher chance of thrombotic coronary heart disease and have a higher cholesterol level as well. Minor effects that raise cholesterol levels are also exerted by the nonsecretor gene, the haptoglobin^2 genes, and the Gm^a genes. A genetic polymorphism of lipoprotein (Lpa^+) (related to but different from LDL lipoprotein) is found with a higher frequency (2-3 times) among patients with coronary heart disease in the Scandinavian countries. Lpa^+ persons have higher cholesterol levels. Persons with a beta-lipoprotein polymorphism AgX^- have higher cholesterol levels than those with AgX^+.

A genetic polymorphism affecting apolipoprotein E (gene locus E on chromosome 19) has a significant effect on lipid levels [874]. E_2/E_2 homozygotes have lower cholesterol levels than others - an effect that can also be observed among heterozygotes for the E_2 polymorphism. The effect of the corresponding alleles on the frequency of hyperlipidemia and coronary heart disease is not yet completely clear [916, 786, 2372]. Convincing differences have not yet emerged, and interpretation is difficult because the E_2/E_2 genotype is associated with type III or remnant removal disease, which will be overrepresented among populations with coronary disease.

DNA Markers [560, 953]. Intensive research on the molecular biology of apolipoproteins, receptors, and enzymes involved in lipid metabolism has led to the development of DNA probes for many of these molecules. These probes have been used for a variety of purposes. The linked restriction enzyme fragment length polymorphism at the LDL locus [719] can be used for preclinical diagnosis in informative families where the phenotypic diagnosis of cholesterol elevation is not quite clear but where at least one definitely diagnosed patient is available. Isoallelic variation of the LDL receptor (see above) might also be demonstrated in this manner. Direct diagnostic probes for defective LDL receptos are not yet available and present problems in view of the mutational heterogeneity at the locus (Sect. 4.6.4).

Various apolipoprotein probes are used to study various hyperlipidemia and coronary heart disease populations [916, 917]. A DNA polymorphism affecting the AI-CIII locus (chromosome 11) has been claimed to be more frequent in undefined hypertriglyceridemia [851] and in myocardial infarct survivors [660], suggesting that a variant apolipoprotein gene (A_1 CIII) closely linked to this DNA marker might make for susceptibility to hyperlipidemia and coronary heart disease. An insertion at the AI-CIII locus produces severe coronary heart disease in homozygotes by markedly decreasing HDL [891]. Conversely, a DNA marker has been found to cosegregate with apolipoprotein AII (chromosome 1) [864]. Since this apolipoprotein is affected by dietary responses of HDL, this marker may define individuals who are more resistant to atherosclerosis. The very large apolipoprotein B molecule has been cloned recently [626] and many studies on the genetics of this major lipoprotein will now be possible. It is expected that certain "polygenic" hyperlipidemias will be elucidated by the variation existing in the different genes for apolipoproteins, enzyme, and receptors and their interactions.

Studies that are based on comparisons of RFLP frequencies between patients and controls must be interpreted cautiously unless there are major differences, since it is not easy to control the exact ethnic identity of the control population. Small differences in gene frequencies caused by different ancestry and random fluctuations may produce spurious results in such work.

HDL Levels (see [635] for references). Low plasma HDL levels predict an enhanced risk for coronary heart disease on a population level. Autosomal dominance of low HDL level was found in a large kindred with coronary heart disease. Pedigrees with high HDL levels and unusual longevity have also been reported but there is no agreement yet whether a single gene or a multifactorial mechanism applies [724]. Various twin and family studies suggest some genetic determination of HDL but its extent remains unclear.

Genetic Factors Other than Lipids. The familial aggregation of coronary heart disease in the absence of lipid elevation suggests the operation of genetic and environmental factors that do not affect lipids. Much work needs to be done. The response of blood vessels to atherogenic stimuli and the elucidation of the genes involved in hypertension which is a risk factor for coronary heart disease are only a few of the possible areas of future research.

Implications. Can premature coronary heart disease be prevented? The declining frequency in mortality of these diseases in the last 15 years in the USA suggests that various environmental changes can affect these conditions. A high priority must therefore be assigned to the identification of high-risk groups.

Hypertension is a strong risk factor for both coronary and cerebrovascular disease. The condition can be readily identified, and population screening is clearly justifiable. Since hypertension is familial, identification of a hypertensive person should lead to blood pressure screening of first-degree family members. The unknown long-term consequences of prolonged antihypertensive drug treatment is a potential source of problems, however [596].

Population screening for hyperlipidemia by cholesterol and triglyceride determinations is not indicated at this time. A vast number of hyperlipidemics would be identified if the upper 5% of triglyceride or cholesterollevels were to be used as cut-offs. No threshold for the identification of a specific risk group exists since the increased risk is gradual. (However, there are precedents in medical genetics with the use of arbitrary cut-offs, such as the recommendation of amniocentesis for expectant mothers older than 35 years for the detection of trisomy 21.) HDL levels also might be measured and some algorithm could be worked out to identify a logistically manageable group of persons at high risk for atherosclerosis. However, intervention with drugs raises problems of long-term safety. Nevertheless, careful large-scale long-term studies need to be set up to assess various preventive schemes.

A more general approach, with the advocation via the mass media (newspapers, TV, radio) of cessation of smoking, a sensible diet, exercise, and blood pressure monitoring has been validated in North Karelia (Finland) [845] and in the Stanford (California) area [658]. Such schemes address the whole population, but it is difficult to maintain momentum for long periods.

Targeted screening [575] of individuals with a family history of coronary heart disease is a possibility. For instance, school children could be given questionnaires asking their parents about familial coronary heart diseases with the aim of identifying high-risk families for appropriate intervention. The logistic problems would be surmountable, but this approach would raise problems of confidentiality and privacy.

Triggered screening following a clinical diagnosis of premature coronary heart disease is already feasible and strongly recommended [829]. Thus, following such a diagnosis family members would be studied for lipids and hypertension, and those found to be affected could be treated. This mode of ascertainment is "retrospective", but it can be instituted relatively simply by good physician education. Since many patients will be seen in hospitals, hospital medical staff should be sensitized to the desirability of initiating such procedures.

The goal of good public health practices will be the coordination of biochemical and genetic work with epidemiologically oriented population studies. As definite insights are gained, various schemes can be worked out for prophylaxis.

Occasionally, it may be worthwhile to "treat" everyone, regardless of genetic susceptibility, as in the case of fluoridation of water supplies to prevent caries or vaccination against various infectious diseases. However, as preventive schemes become more complex and cannot readily be applied or are not acceptable to the whole population, approaches that concentrate on smaller subpopulations at high genetic risk have much to offer.

4 Gene Action

4.1 Development of Mendel's Paradigm

Galton's and Mendel's Paradigm: Gene Action Is Well Understood. The discussion of the twin method has shown the possibilities, but still more clearly the limitations of genetic analysis involving comparison of phenotypes without reference to the action of single genes. The twin method is an extension of the paradigm created by F. Galton when he measured and compared quantitatively varying characters in close relatives. Heritability estimates whether based on twins or on family data indicate an underlying genetic variability in the population but do not provide any information as to the mechanisms of this variability. They pose questions but do not answer them.

Mendel's paradigm, on the other hand, proved to be fertile in answering questions about which genetic factors produce a certain phenotype and how they produce this phenotype. The black box was dissected step by step (Sect. 3.6).

The first step was accomplished by A. Garrod when he created the concept of inborn errors of metabolism (Sect. 3.6). Later, it was shown that genes specify enzymes and that genetic defects with simple modes of inheritance are frequently enzyme defects. Some genetic polymorphisms in the „normal" range were explained by qualitative differences in genes determining a certain range of enzyme activity. Introduction of advanced methods for protein analysis made it possible to analyze this genetic variability at the level of amino acid sequences in proteins. Once the structure of DNA had been elucidated by Watson and Crick (1953) [1347] and the genetic code had been deciphered, differences in amino acid sequences could be explained in terms of base substitutions within the DNA.

The mechanisms for regulation of gene action in higher organisms are still unknown, leaving open the question of how activities of the various genes are regulated and integrated in the development and functioning of the entire organism. Here, models were developed and experimentally tested that provided satisfactory mechanisms of regulation in bacteria. These models, however, proved to be much less satisfactory when applied to problems of regulation in higher organisms where the requirements for regulation are quite different from those in bacteria. Therefore, it comes as little surprise that the mechanisms that evolved to meet these requirements are also different.

Application to Human Genetics. In human genetics, a number of phenomena have been observed that can be satisfactorily explained by differences in action of single genes: in the normal range, the genetic polymorphisms, and in the pathologic range, the hereditary diseases with simple modes of inheritance. Another group of phenomena – the so-called multifactorial diseases – can be partially understood by action of single genes (Sect. 3.7). Still another group – the syndromes due to numeric or structural chromosome aberrations – are not understood at all in terms of gene action. These offer the most challenging problem.

The best-analzyed examples, the variants and anomalies with single modes of inheritance, have, at the same time, proved to be useful for analysis of certain aspects of normal function. By the same token, the chromosome aberration syndromes, once the genotype-phenotype relationship will be understood somewhat better, may help to gain deeper insight into gene regulation of normal embryonic development.

As mentioned in Sect. 3.6.1, genetic analysis can be performed at several levels. Methods of biochemistry and molecular biology have been applied to the analysis of genetic defects. Such studies helped in elucidating biological mechanisms at three levels: The DNA gene or the biochemical gene product or, in part, the qualitative phenotypical level. In this way, a variety of functional disturbances has been discovered, and much has been learned that is helpful for the understanding of normal function. Table 4.1 shows types of genetic defects and examples of hereditary diseases in which these defects were analzyed. A few diseases, such as hemoglobinopathies and the hemophilias (30670, 30690), have been analyzed at all three levels. For most conditions the basic mechanisms could only be elucidated at the second (enzymatic or protein function) or third level (disturbed cell and organ function).

Table 4.1. Etiology and pathogenesis of genetic diseases as elucidated by methods from molecular biology and biochemistry. (Modified from [203])

Level of analysis	Type of anomaly	Example
Altered DNA structure	1. Deletion mutations	α-Thalassemia; Lepore hemoglobins; hemophilias (Sect. 4.3.4)
	2. Missense mutations	Sickle-cell disease (Sect. 4.3.2)
	3. Splicing mutations	Some β-thalassemias (Sect. 4.3.4)
	4. Nonsense mutation	Some β-thalassemia variants (Sect. 4.3.4; 4.3.5)
	5. Frame-shift mutations	Hemoglobin Wayne (Sect. 4.3.3)
	6. Gene duplications	Hemoglobin Grady (Sect. 4.3)
	7. Regulatory mutations (For these types of mutation, see Sect. 5.1.4)	Some β-thalassemias
Disturbed protein function: Enzymes		
	1. Absent activity	
	a) Protein detectable immunologically	some variants of Lesch-Nyhan syndrome (Sect. 4.2.2.6)
	b) No protein detectable immunologically	most variants of Lesch-Nyhan syndrome, variants of homocystinuria (Sect. 4.2.2.9)
	2. Reduced activity	
	a) Decreased affinity for substrates	G-6-PD deficiency, Freiburg variant (Sect. 4.2.2.2)
	b) Decreased affinity for cofactors	homocystinuria (pyridoxine-responsive type) (Sect. 4.2.2.9)
	c) Unstable structures	G-6-PD deficiency, some variants (Sect. 4.2.2.2)
	3. Enhanced activity	G-6-PD Hektoen variant (Sect. 4.2.2.2)
	4. Defect of enzyme activator protein	AB variant of G_{M2} gangliosidosis [203]
	5. Reduced availability of cofactors	Pyridoxine (vitamin B 6) dependency (Sect. 5.2.2.5)
Disturbed protein function: Non-enzymic proteins		
	6. Defective posttranslational modification	α-1-Antitrypsine deficiency, ZZ variant (Sect. 3.7.4)
	7. Enhanced tendency to aggregation	Sickle-cell disease (Sect. 4.3.2)
	8. Defective receptor binding	Familial hypercholesterolemia (Sect. 4.6.4); testicular feminization (Sect. 4.7.5)
Disrupted cell and organ function	1. Altered flux through metabolic pathways	
	a) Accumulation of a toxic precursor (catabolic pathway)	Phenylketonuria (Sect. 4.2.2.7); mucopolysaccharidoses and other lysosomal defects (Sect. 4.2.2.3)
	b) Deficiency of product (anabolic pathway)	Various types of hypothyroidism with goiter (Sect. 4.2.2.7)
	c) Overproduction of product (anabolic pathway)	Rare form of gout due to altered PRPP synthetase [203, Chap. 50]
	2. Disordered feedback regulation of synthetic pathways	
	a) Overproduction of end product due to decreased synthesis or availability of feedback regulator	Acute intermittent porphyria (Sect. 4.6.3); familial hypercholesterolemia (Sect. 4.6.4)
	3. Disordered membrane function	
	a) Deficient transmembrane transport	Cystinuria (See [203, Chap. 80]) hereditary spherocytosis (Sect. 4.6.5)
	b) Deficient receptor-mediated endocytosis	Familial hypercholesterolemia, receptor-negative and receptor-defective variants (Sect. 4.6.4)
	c) Deficient generation of second messenger	Pseudohypoparathyroidism (See [203, Chap. 69])

Table 4.1 *(continued)*

Level of analysis	Type of anomaly	Example
	4. Disordered intracellular compartimentation	
	a) accumulation of unprocessed protein	α-l-Antitrypsin deficiency, ZZ variant (Sect. 3.7.4)
	b) Mislocation of protein	I-Cell disease (Sect. 4.2.2.3) Familial hypercholesterolemia (internalization-defective variant (Sect. 4.6.4)
	5. Distorted cellular tissue architecture	
	a) Alteration of cell shape	Sickle cell disease (Sect. 4.3.2); hereditary spherocytosis (Sect. 4.6.5)
	b) Alteration of organelle structure	Immotile cilia syndrome, especially Kartageners syndrome (See [203, Chap. 91])
	c) Alteration of extracellular matrix	Epidermolysis bullosa, Pasini type (Sect. 4.6.7); lysylhydroxylase deficiency (Ehlers-Danlos sydrome type VI)

In the following, the analysis will largely follow the path from gene action at the enzyme and protein level to identification of gene alteration at the level of DNA transcription. This account will be followed by the various mechanisms of dominant gene action and the problems of regulation. At all levels, the practical applications of theoretical insights for medical diagnosis and therapy will be considered.

4.2 Genes and Enzymes

4.2.1 One-Gene-One-Enzyme Hypothesis

Early Forerunners. Garrod (1902) [249] (Sect. 1.5) related the gene defect in alkaptonuria to a specific inability of the organism to degrade homogentisic acid. The obvious next problem was to determine the specific mechanism responsible for this inability. Metabolic steps are catalyzed by enzymes. Therefore, alterations of enzymes offered a plausible explanation. This had already been discussed by Driesch (in 1896), and was brought forward by Haldane (1920; see [1117]) and by Garrod at the end of his career (1923) [1091]. Important early analytic steps in biochemical genetics were the analyses of eye color mutants in the flour moth *Ephestia kühniella* by Kühn and Butenandt [1178, 1027] and in *Drosophila* by Beadle and Ephrussi (1936) [987]. In these first attempts, mutants in insects that had been analyzed by genetic methods were chosen to elucidate mechanisms of gene action. This approach, however, had only limited success, as the problem proved to be too complex for a direct attack. For a more successful approach, two conditions were required:

1. A simpler test organism had to be found that provided better opportunities for experimentation.
2. The problem needed to be examined by looking for genetic explanations of the biochemical phenotypes rather than to find biochemical explanations for genetically defined traits.

Both requirements were met by the work of Beadle and Tatum, 1941 [988]; cf. also Beadle, 1945 [986].

Beadle's and Tatum's Simple Organism and Method of Attack. The paper by these two investigators begins:

> From this standpoint of physiological genetics the development and functioning of an organism consists essentially of an integrated system of chemical reactions controlled in some manner by genes. It is entirely tenable to suppose that these genes ... control or regulate specific reactions in the system either by acting directly as enzymes or by determining the specificities of enzymes. Since the components of such a system are likely to be interrelated in complex ways, and since the synthesis of the parts of individual genes are presumably dependent on the functioning of other genes, it would appear that there must exist orders of directness of gene control ranging from simple one-to-one relations to relations of great complexity. In investigating the roles of genes, the physiological geneticist usually attempts the physiological and biochemical bases of already known hereditary traits. This approach ... has established that many biochemical reactions are in fact controlled in specific ways by specific genes. Furthermore, investigations of this type tend to support the assumptions that gene and enzyme specificities are of the same order. There are, however, a number of limitations inherent in this approach. Perhaps the most serious of these is that the investigator must in general confine himself to a study of nonlethal heritable characters. Such characters are likely to involve more or less non-essential so-called "terminal" reactions ... A second difficulty ... is that the standard approach to the problem implies the use of characters with

visible manifestations. Many such characters involve morphological variations, and these are likely to be based on systems of biochemical reactions so complex as to make analysis exceedingly difficult.

Considerations such as those just outlined, have led us to investigate the general problem of the genetic control of developmental and metabolic reactions by *reversing the ordinary procedure*[1] and, instead of attempting to work out the chemical basis of known genetic characters, to set out to determine *if and how genes control known biochemical reactions*.[1] The ascomycete neurospora offers many advantages for such an approach and is well suited for genetic studies. Accordingly, our program has been built around this organism. The procedure is based on the assumption that X-ray treatment will induce mutations in genes concerned with the control of known specific chemical reactions. If the organism must be able to carry out a certain chemical reaction to survive on a given medium, a mutant unable to do this will obviously be lethal on this medium. Such a mutant can be maintained and studied, however, if it will grow on a medium to which has been added the essential product of the genetically blocked reaction.

Then, Beadle and Tatum described their experimental design (Fig. 4.1). The complete medium contained agar, inorganic salts, malt extract, yeast extract, and glucose. The minimal medium, on the other hand, contained only agar, salts, biotin, a disaccharide and fat, or another carbon source. Mutants that grow on the complete but not on the minimal medium were tested systematically by the gradual adding of the complete medium components to ascertain which component mutants were unable to synthesize.

In this manner, mutants were isolated that were unable to synthesize growth factors such as pyridoxin, thiamine, and p-aminobenzoic acid. These defects were shown to be caused by mutations at specific gene loci. This work inaugurated an abundance of investigations on neurospora, bacteria, and yeast, in which such mutants were analyzed and "genetic blocks" in single metabolic steps were related to specific enzyme defects. Soon this approach became an important tool for better assessment of single steps within metabolic pathways.

The one-gene-one-enzyme hypothesis was established and had now a solid experimental foundation. This hypothesis proved to be highly fertile during the decades that followed. Analysis of enzyme defects and variants soon also provided evidence for genetic blocks in which only the function of the enzyme was impaired while an enzyme protein was still present that had kept its antigenic properties (cross-

[1] Italics ours.

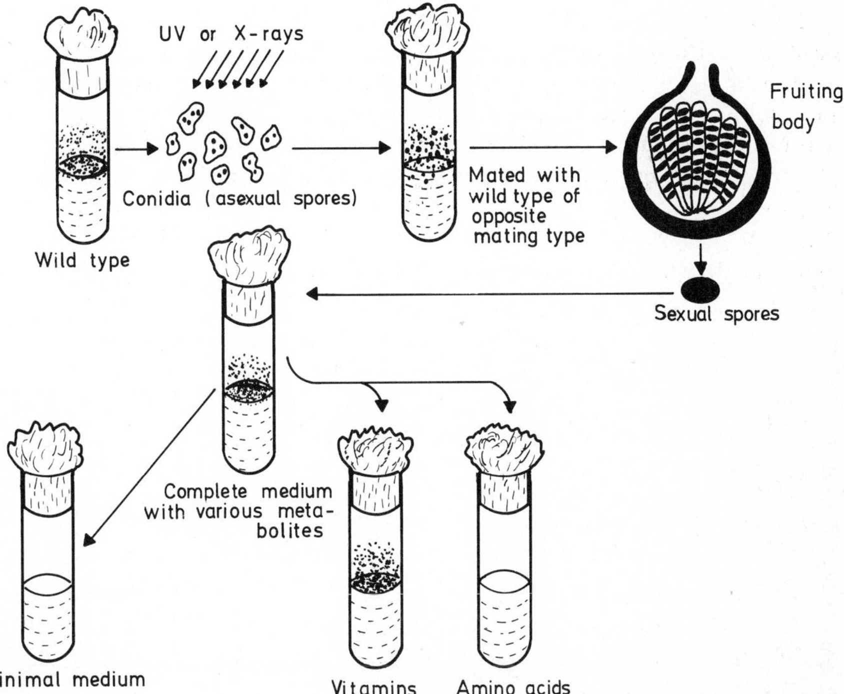

Fig. 4.1. The experimental design for discovering biochemical mutants in neurospora. The X-ray or UV-induced mutation does not impair fungus growth on complete medium. On minimal medium, however, the fungus cannot grow. Addition of vitamins restores growth capacity shown as "stippling". Addition of amino acids causes no growth. The figure suggests that the mutation has affected a gene, which influences vitamin metabolism. The next step in the experiment would be to find out which vitamin is able to restore normal function. The genetic block will be found in the metabolism of this vitamin. (Sinnott et al. [1303], modified)

reacting material; CRM). In other cases, the enzyme had an altered temperature optimum for its action. Some variants could be explained by a mutation causing altered activity of a series of enzymes by affecting a common control unit. From such studies, the concept of regulation of bacterial gene action, which included the operon concept (Sect. 4.7.3) emerged.

First Enzyme Defects in Humans. The first genetic disease in humans for which an enzyme defect could be shown was a recessively inherited type of methemoglobinemia (Gibson and Harrison, 1947 [1100]; Gibson, 1948 [1099]) (25080). The enzyme deficient in these cases is the NADH-dependent methemoglobin reductase. The first systematic attempt to elucidate a group of human metabolic diseases was made in 1951 by the Coris in glycogen storage disease [1044].

The Coris showed first that the structure of liver glycogen in ten cases of what was called at that time v. Gierke's disease (23220) was within the normal range of variation in eight and definitely abnormal in two cases. It was also obvious that liver glycogen, which accumulates in excessive amounts, is not readily available for blood sugar formation, as the patients show a tendency to hypoglycemia. Many enzymes are required for the conversion of glycogen to glucose in the liver. Two of them, amylo-1,6-glucosidase and glucose-6-phosphatase, were selected as possible candidates to be the deficient enzyme. Liberation of phosphate from glucose-6-phosphate was measured in liver homogenates at different pH. Figure 4.2 shows the results. In a normal liver, strong activity is measured with a maximum at pH 6–7. Severe liver damage due to cirrhosis leads only to a moderate decrease. In a fatal case of v. Gierke's disease, on the other hand, no activity at all can be detected; this was confirmed in a second case. Two patients with milder disease symptomatology showed markedly reduced activity.

It was concluded that there was an enzyme defect of glucose-6-phosphatase in these fatal cases of v. Gierke's disease. At the same time, however, in most of the milder cases, activity of this enzyme was not reduced below the level found in liver cirrhosis; only the two patients shown in Fig. 4.2 had moderately reduced values. The Coris offered no explanation for these last-mentioned results. They also noted that abnormal muscle storage of glycogen cannot be explained by a lack of glucose-6-phosphatase, since this enzyme is normally absent from muscle. For cases with muscular glycogenosis, they suggested a defect of amylo-1,6-glucocidase at the possible explanation. This prediction was soon confirmed, as Forbes [1081] discovered this defect in a clinical case of glycogen storage disease involving both heart and skeletal muscles.

Currently, many different enzyme defects have been discovered in glycogen storage disease. For details see [1133; 1244].

While the patterns of manifestation differ somewhat between the different types, there is much overlap in clinical manifestations. The mode of inheritance is with one exception autosomal recessive. Had the enzyme defects not been discovered, glycogen storage disease would be regarded as a single genetic disease with intrafamilial correlations as to severity, details of symptoms and time of death. Hence, we have here an example of how genetic heterogeneity, which can only be suspected at the phenotypic level (Sect. 3.3.5), is confirmed by analysis at the biochemical level: identification of specific genes by analysis of enzyme activities.

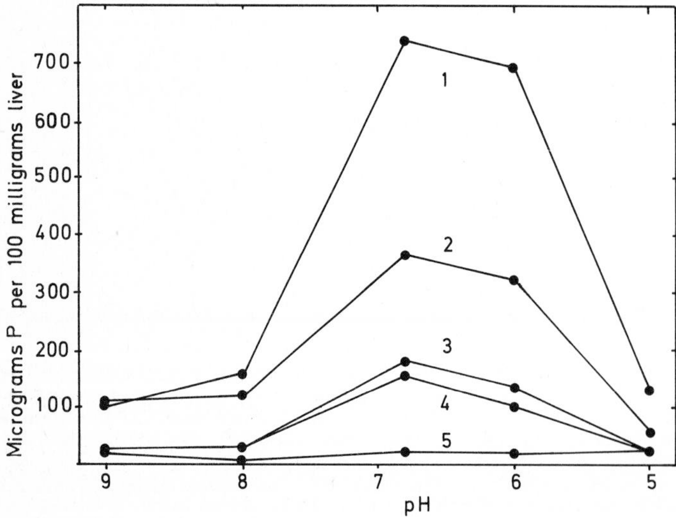

Fig. 4.2. Liberation of phosphate from glucose-6-phosphate in liver homogenates from various patients as a measure of glucose-6-phosphatase activity. *1* High phosphate liberation in a patient with normal liver function; *2* moderate reduction in cirrhosis of the liver; *3, 4* marked reduction in two patients with the milder form of glycogen storage disease; *5* complete absence of enzyme activity in a patient with severe v. Gierke's disease. (Cori and Cori, 1952 [1044])

In the following years, enzyme defects were discovered with increasing momentum, and in the sixth edition of his *Mendelian Inheritance in Man,* McKusick (1983) [133] could list among 588 confirmed autosomal recessive conditions more than 170 in which a specific enzyme defect had been located. Many achievements in methodology have contributed to this progress. Most are beyond the scope of this book, but some were connected with the development of concepts and methods in molecular genetics.

Some Steps in the Knowledge of Human Enzyme Defects. A number of steps were especially important for this continuing development:

1934 Phenylketonuria was detected by Følling [1080]
1941 Beadle and Tatum introduced the one-gene-one-enzyme hypothesis [988]
1948 Gibson described the first enzyme defect in a human disease (recessive methemoglobinemia) [1099]
1952 Glucose-6-phosphatase deficiency in v. Gierke's disease was discovered by the Coris [1044]
1953 Jervis demonstrated the lack of phenylalanine hydroxylase in phenylketonuria [1144]. Bickel reported the first attempt to alleviate an enzyme defect by a low-phenylalanine diet [1004]
1955 Smithies introduced starch gel electrophoresis [1307, 1308]
1956 Carson et al. discovered the defect of glucose-6-phosphate dehydrogenase (G6PD) in drug-induced hemolytic anemia [1030]
1957 The enzyme defect in galactosemia (P-gal-transferase deficiency) was described by Kalckar et al. showing that an identical enzyme deficiency existed in humans and bacteria [1150]
1961 Krooth and Weinberg demonstrated the enzyme defect of galactosemia in fibroblasts cultured in vitro [1177]
1967 Seegmiller et al. discovered the defect of hypoxanthine-guanine-phosphoribosyl transferase (HPRT) in Lesch-Nyhan syndrome [1295]
1968 Deficiency of excision repair in xeroderma pigmentosum was described by Cleaver [1035]
1970 Elucidation of the enzyme defects in the mucopolysaccharidoses by Neufeld [1240] allowing full understanding of the catabolic pathways of mucopolysaccharide metabolism
1974 Demonstration by Brown and Goldstein that genetically determined overproduction of an enzyme (HMG CoA reductase) in familial hypercholesterolemia is caused by a membrane receptor defect (low-density lipoprotein receptor), which modulates intracellular HMG CoA reductase activity [1023]
1977 Demonstration by Sly et al. [1306] that the mannose 6-phosphate components of lysosomal enzymes are recognized by fibroblast receptors. Genetic processing defect prevents binding of lysosomal enzymes with failure to enter cells and secretion into plasma (I-cell disease).
1980 Defect of receptor cyclase coupling protein in pseudohypoparathyroidism.

4.2.2 Genes and Enzymes in Humans: Present State of Knowledge

Scope and Limitations of This Review. Each enzyme defect presents special problems in methodology and interpretation. Limitations of space enforce a short and highly selective discussion of these problems, two groups of which were mainly selected.

1. Those of significance for an understanding of general principles of genetic determination and control in man;
2. Those of significance for diagnosis of enzyme defects and their contribution to our understanding of disease.

For special groups of disease, we refer the reader to more specialized monographs [203, 182] and to the many reviews on single diseases or groups of diseases.

4.2.2.1 Discovery and Analysis of Enzyme Defects

Difference in Research Strategy Between Humans and Neurospora. The progress in analysis of enzyme defects in neurospora and bacteria had been achieved by a novel research strategy. Instead of searching for biochemical explanations of known mutants, mutants were induced and screened as to whether they affected known metabolic steps. Such an approach can work only if mutations indeed cause genetic blocks due to enzyme defect. Moreover, recovery of mutants is limited to those actually leading to enzyme defects irrespective of whether they are a large or a small fraction of all occurring mutations. In practice, this limitation proved useful, since it helped to make the problem analytically accessible. At the same time, there were sufficient degrees of freedom for concepts of genetic control mechanisms of enzyme activity to emerge.
In humans, the approach via known metabolic pathways is barred, as we can neither induce mutations

artificially nor screen them in a system comparable to the selective systems for identification of auxotrophic mutants in neurospora. We have to start somehow with the phenotype and try to find and analyze the enzyme defect underlying the phenotype. The obvious disadvantage of this approach is its dependence on chance observations of individuals with rare diseases. But there are also advantages. In no experimental animal are so many individuals constantly being examined for their state of health as in man; moreover, the variety of methods available for analysis ranging from refined descriptions of clinical symptoms down to characterizations of enzyme proteins is only existing in man. As a consequence, a rich spectrum of phenotypes is offered for observation.

Clinical Symptoms Leading to the Detection of Enzyme Defects. How are enzyme defects discovered? Many ways exist. The defect of glucose-6-phosphatase in v. Gierke's disease offers the simplest example. The disorder had been known for a long time, and the clinical symptoms suggest an anomaly in a specific metabolic pathway. As soon as this pathway is sufficiently well-known and enzyme assays are available, the research worker has to find an affected patient to ascertain which of a limited number of enzymes is defective. However, difficulties may occur. These may be primarily technical. For example, many enzyme defects in humans are due not to complete lack of the enzyme but to mutationally altered enzyme properties causing anomalies such as reduced substrate affinity. Most in vitro assays use high substrate concentration and thus may enable even an altered enzyme to work properly [1166]. In vitro assays therefore may not reflect in vivo activity. Sometimes, the symptoms may point in the wrong direction. An example is glycogen storage disease type II (Pompe's disease). Here, all enzymes known to be involved in the main pathway of glycogen degradation are normal, and still, there is glycogen accumulation in most tissues of the body, especially marked in the heart muscle. Eventually, the enzyme defect was shown to affect α-1,4-glucosidase, which normally occurs along with other hydrolytic enzymes in organelles known as lysosomes (Sect. 4.2.2.3) and was previously not known to be involved in glycogen metabolism.

In other conditions, the clinical symptoms may be so unspecific that no clues as to the metabolic defect exist. Thus, a failure of infants to thrive is associated with many different inborn errors of metabolism affecting various metabolic enzymes.

A small proportion (~1%) of mentally retarded children resident in institutions have phenylketonuria (PKU). This condition was discovered by Følling (1934) [1080] in two sibs who had a peculiar mousy odor and excreted large amounts of phenylpyruvic acid in their urine. This discovery provided much hope that many other types of mental retardation might be caused by various other inborn errors. Many surveys for abnormal urinary metabolites were done among patients with mental retardation. Unfortunately, the yield was low and while some other conditions were discovered such as homocystinuria (see below), most mentally retarded persons were not affected with inborn errors that could be detected by this approach.

While widespread clinical findings involving bone and connective tissue and gross defects are usually not associated with inborn errors of metabolism, there are exceptions, such as homocystinuria, an anomaly in the metabolism of the sulfur-containing amino acid methionine, caused by a deficiency of the liver enzyme cystathionine synthase. The patients suffer from three groups of symptoms: (1) connective tissue and eye anomalies such as osteoporosis, knock knees, spider fingers and toes, and dislocated lenses; (2) anomalies in the function of the central nervous system such as mental retardation in about 50% of the cases; and (3) arterial and venous thrombosis. Some of the findings are similar to those found in Marfan's syndrome, a dominant condition that may occur as a new mutation, may therefore lead to sporadic nonfamilial cases, and consequently be confused with homocystinuria. But even without this parallel, no one familiar with the general symptomatology of recessive enzyme defects would have suspected this particular enzyme defect in a disorder with so many different, and mainly structural, symptoms. The disorder was discovered in a program for screening mentally defectives.

Clinical Diagnosis of Metabolic Defects. Metabolic defects are quite rare. This means that even busy pediatricians will see only a few of them during their careers, and those seen will be encountered only once or a very few times. Therefore, the complexity of diagnosis and particularly of therapy cannot be expected from every pediatrician. A few departments of pediatrics or medicine in the United States and in European countries are increasingly specializing in the diagnosis (including antenatal diagnosis) and therapy of single or small groups of enzyme defects. This specialization provides the highest possible level of medical care for these patients.

However, it is the task of every doctor, whether general practictioner, pediatrician, or medical geneticist, to arrange for appropriate diagnosis of these metabolic diseases. Early diagnosis is important not only for disorders where specific therapy is possible (Sect. 4.2.2.9), but where births of further affected siblings can be prevented by antenatal diagnosis. Careful diagnosis is therefore important for most inborn errors of metabolism, which usually will be manifested in infants as failure to thrive.

Methods Used for Analysis of Enzyme Defects. The appropriate methods for analysis of enzyme defects are generally those of enzymology. In the elucidation of the genetic basis of enzyme defects in inborn errors, not only quantitative assessments of enzyme activity but also qualitative differences in enzyme characteristics must be examined.

Temperature instability was found, for example, in the defective hypoxanthine-guanine-phosphoribosyl transferase (HPRT) of some children with the Lesch-Nyhan syndrome (30800) [1163]. Unusual thermoresistance, on the other hand, was found with the α-galactosidase in Fabry's disease, a lysosomal enzyme defect.

Frequently the difference between the normal and abnormal enzyme can be analyzed at the protein level, for example, by altered electrophoretic mobility. In these cases, the abnormal protein, while losing the catalytic capacity by which it is characterized as an enzyme, may have maintained its immunologic properties. It reacts with an antibody that has been produced against the normal enzyme. Such *cross-reacting material* (CRM) proteins were first discovered in bacteria, for example, for trypophan synthetase of *E. coli.* Such CRM proteins are frequent in human enzyme defects (Table 4.2) and are of special clinical significance in the detection of heterozygotes for hemophilia A (Sect. 4.2.2.8).

It is characteristic of enzyme deficiencies in humans that unlike many enzyme deficiencies in bacteria, qualitatively altered enzymes are more frequently observed than complete or nearly complete loss of an enzyme protein. This finding indicates that most if not all currently known enzyme defects in man are caused by structural mutations and not by regulatory mutations as are often found in bacteria. These facts are of great significance for the understanding of gene regulation in higher organisms including humans (Sect. 4.7). Among the many methods utilized for analysis of enzyme defects, one - described below - has gained special significance.

Examination of Enzyme Defects in Human Fibroblast Cultures. When the elucidation of the genetics of microorganisms proved to be successful in the 1940s and 1950s, many scientists believed that genetic analysis in higher organisms using individual cells would be a way of increasing the resolving power of analysis by several orders of magnitude (Pontecorvo, 1959 [162]. The technical conditions - the growing of cell lines in culture - had been available for several years. However, cell lines capable of growing in cultured medium for an indefinite time either had been derived from malignant tumors - such as the much-used HeLa cells - or had undergone a change in their growth characteristics in vitro, thereby losing their ability for contact inhibition; they were "transformed". These cells were genetically different from normal cells; above all, they were almost always aneuploid with a wide range of chromosome numbers within the same cell line and even in the same culture. Such cells cannot be used for genetic research; methods had to be developed for growing normal euploid cells in culture. Quantitative biochemical work, such as measurement of enzyme activities, is meaningful only when the growth of cells is carefully controlled. Some remarks on the principle of these methods, which also apply for studies of cells from amniotic fluid, may be useful.

Table 4.2. Human biochemical defects showing enzymes and proteins with cross-reacting material (CRM) indicating presence of a mutationally altered protein (Lenz, [121])

Pseudocholinesterase deficiency[a] ("silent" phenotype)
Metachromatic leukodystrophy
McArdle's glycogenosis[a]
Acatalasia
Sandhoff's Gangliosidosis
Lesch-Nyhan syndrome
Fructose intolerance, type I
Phenylketonuria
Galactosemia
Fibrinogen deficiency[a]
Mucopolysaccharidosis III B
Prothrombin deficiency
Proconvertin deficiency[a]
Hemophilia A[a]
Hemophilia B[a]
Stuart-Prower factor deficiency[a]
Deficiency of fibrin-stabilizing enzyme
C4 deficiency
Sucrase-isomaltase deficiency
Tay-Sachs disease

[a] CRM-negative cases have also been described.

Difficulties of the Method. The method of growing human fibroblasts in cell culture has a number of difficulties.

a) Growth of these cells is impossible or unsatisfactory in chemically defined media. Serum has to be added that contains a number of necessary nutrients. As a rule, fetal calf serum is used. Culture conditions cannot be standardized once and for ever; they always have to be controlled and adjusted. Variations in pH, glucose content, and other parameters influence cell growth and function.
b) Fibroblasts do not grow in suspension. Taking samples at various time intervals from one growing cell suspension is a routine procedure. Fibroblasts do not usually allow this method of sampling, because they grow attached to a surface as a monolayer. Multiple samplings have to be performed by setting up many subcultures. This introduces additional, and poorly controllable, sources of variation.
c) The amount of material available for study is small. Fibroblasts, unlike cells of established tumor cell lines, have a limited capacity for cell divisions, restricting the potentialities for mass cultivation.

Growth Characteristics of Fibroblasts. A skin biopsy is taken in the same way as when tissue culture cells are required for chromosome examinations. The pieces obtained after mincing the sample are implanted in such a way as to permit growth on a surface and are incubated with media. Culturing in Petri dishes in an incubator in which the air contains 5% CO_2 ensures a constant pH. After about 15 days, fibroblasts start growing at the surface, and eventually form a monolayer. The culture is then removed from the surface by trypsinization and the centrifuged cell suspension is reimplanted.

Cells in culture show a growth cycle consisting of an initial lag phase which is followed by a phase of logarithmic growth until the cell number per culture becomes stationary ("lag-log stationary cycle"). Enzyme activities and intracellular concentrations of metabolites fluctuate during this growth cycle. Therefore, comparative determinations of biochemical parameters in different cell strains require analysis of these fluctuations through all stages of the growth cycle.

To sum up, enzyme work with human fibroblast cultures requires an unusual amount of painstaking technical care [1208]. This approach has been increasingly rewarding since Krooth (1961) [1177] was able to demonstrate the galactose-1-phosphate-uridyltransferase defect in galactosemia in cultivated human fibroblasts. This defect had been known from in vivo studies since 1957. Many enzyme defects would never have been diagnosed without this method, and antenatal diagnosis - which is now so successful in detecting enzyme defects - would not be possible.

Some enzyme defects are not expressed in fibroblasts and therefore cannot be studied by the tissue culture method. Occasionally, use of other tissues such as lymphocytes or red blood cells may be successful in these instances. In general, enzyme defects that are not expressed in fibroblasts cannot be assayed in amniotic cells.

4.2.2.2 Typical Group of Enzyme Defects: Erythrocyte Enzymes

A well-examined group of enzyme deficiencies affects the enzymes of the red blood cell [993, 1345]. The human erythrocyte has no nucleus and is therefore unable to synthesize mRNA. Protein synthesis in nucleated red cell precursors supplies the erythrocyte with a number of enzyme systems that are active for a limited time only. They gradually lose their activity and the cell "dies" after circulating for 120 days. For many of these, enzyme deficiency syndromes are known; some of them cause nonspherocytic hemolytic anemia.

Enzyme Defects in Glycolysis. The most important catabolic pathway for obtaining energy-rich phosphates (ATP) in mature erythrocytes is glycolysis (Fig. 4.3) by the Embden-Meyerhof pathway. This anaerobic pathway leads to the formation of 2 mol of lactate from 1 mol of glucose. Moreover, 1 mol glucose generates 4 mol of ATP, 1 mol of which is needed for phosphorylation when glucose-6-phosphate is transformed to fructose-1,6-diphosphate and also when one mol of glucose is changed to one mol of glucose-6-phosphate. Hence, the net gain is 2 mol of ATP per 1 mol glucose. ATP is utilized for erythrocyte functions including maintenance of shape as a biconcave disk, for energy of the cation pump, as well as for the synthesis of such metabolites as glutathion (GSH) or AMP. The glycolytic pathway is catalyzed by 13 enzymes.

About 5%-10% of glucose-6-phosphate is degraded oxidatively via the hexose monophosphate cycle or the "shunt" pathway. In a number of steps, pentose phosphate is converted into fructose phosphate or glyceraldehyde-3-phosphate and recycled into glycolysis. The hexose monophosphate cycle is important, as it provides the NADPH needed for reduction of oxidized glutathions. This reduction is catalyzed by glutathione reductase.

Nonspherocytic Hemolytic Anemias. Dacie et al. (1953) [1049] delimited a group of hemolytic anemias, which they described as nonspherocytic, as contrasted to hereditary spherocytosis. The patients suffer from increased hemolysis and its consequences, e.g., jaundice of varying degrees, slight to moderate splenomegaly, and an increased occurrence of gall stones. Contrary to the findings in hereditary spherocytosis (18290), osmotic fragility of the red cells was not increased and there was no qualitative hemoglobin abnormality. On the basis of refined hematologic criteria, the condition was considered heterogeneous, but the findings in the various forms overlapped to some extent. Full analysis had to await development of enzymatic methods.

Enzyme Defects in the Glycolytic Pathway. Between 1961 and 1975, genetic defects were described for 11 of the 13 glycolytic enzymes. In at least eight of them, a causal relationship with hereditary nonspherocytic hemolytic anemia was shown - in some cases with additional disturbances in the central nervous system and muscles. In general, a critical reduction of enzyme activities leads to an accumulation of the metabolite prior to and diminution of the metabolite after the block. Secondary effects such as ATP reduction can be expected in some deficiencies. Due to the internal regulatory capacity of the system, however, direct inferences from metabolic, clinical and hematologic findings as to the nature and degree of the enzyme defect are often misleading. Besides, the examinations are usually carried out in an erythrocyte population, which comprises a larger fraction of young red cells. Since enzyme activities are often higher in young than in old cells, enzyme deficiency may be overlooked.

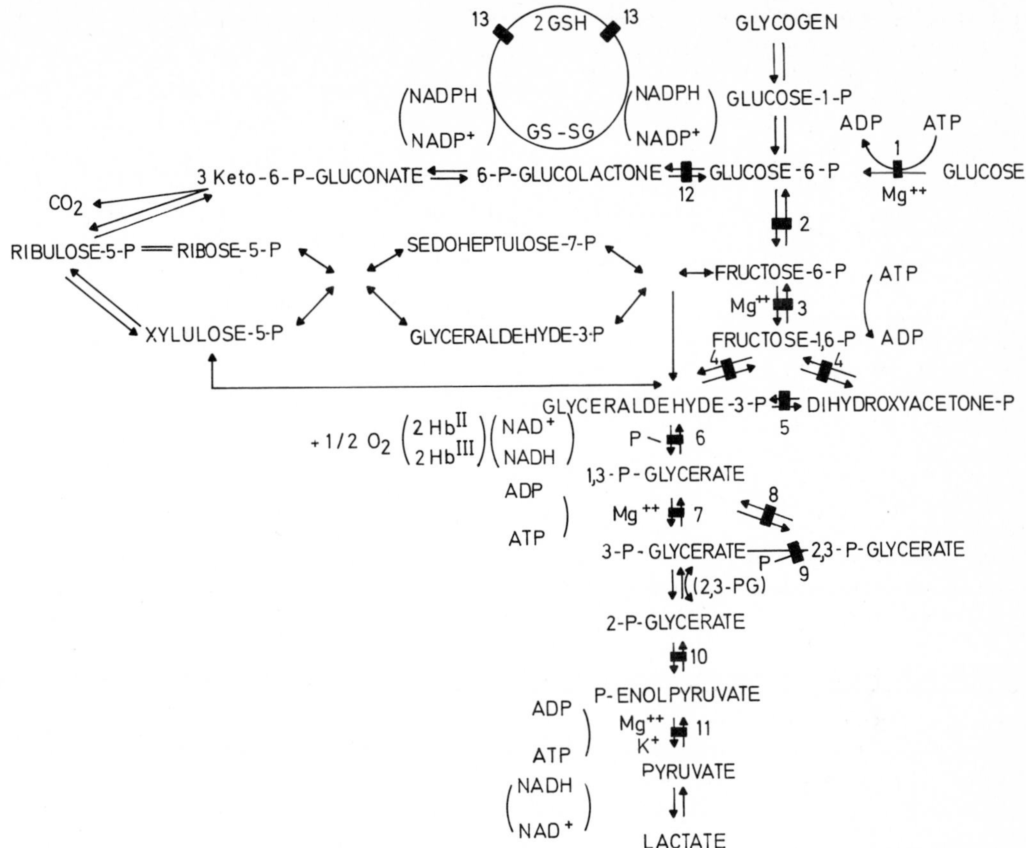

Fig. 4.3. The glycolytic pathway and its metabolic blocks in erythrocytes. This pathway is catalyzed by 11 enzymes. The rate-limiting enzyme is hexokinase, which converts glucose into glucose-6-phosphate. This is then metabolized in steps into 1,3-diphosphoglycerate. This compound may be converted directly by phosphoglycerate kinase into 3-phosphoglycerate and ATP. In an alternative pathway (Rapoport-Lübering cycle), however, 1,3-diphosphoglycerate may be converted into 2,3-diphosphoglycerate. 2,3-diphosphoglycerate is cleaved into inorganic phosphate and 3-phosphoglycerate, which is then recycled into the glycolytic pathway. The Rapoport-Lübering cycle produces no ATP. Hence, degradation of glucose may proceed with different net gains of ATP. However, the 2,3-diphosphoglycerate content of erythrocytes is important for normal oxygen dissociation of hemoglobin. Another requirement for a functional hemoglobin is availability of NADH by the glyceraldehyde-P-dehydrogenase reaction. NADH is needed not only for hydrogenation of pyruvate to lactate but also for reduction of methemoglobin. About 5%–10% of glucose-6-phosphate is degraded oxidatively via the hexose-monophosphate cycle. In a number of steps, pentose phosphate is converted into fructose phosphate or glyceraldehyde-3-phosphate and recycled into glycolysis. The hexose monophosphate cycle is important, as it provides NADPH which is needed for reduction of oxidized glutathion. This reduction is catalyzed by glutathion reductase. Glycolysis is controlled via a "multistep control" system in which hexokinase, phosphofructokinase, and the concentration of inorganic phosphates and Mg are important. The metabolic blocks that have been analyzed in humans are also shown. The numbers of the genetic blocks refer to the numbers in Table 4.3. Blocks at 6 (glyceraldehyde 3-phosphate dehydrogenase), and 10 (Enolase) are not listed in Table 4.3 since the exact relationship between the block and the enzyme deficiency is not proven

A number of enzyme defects are described in Table 4.3. The numbers refer to Fig. 4.3. Some observations of a more general significance for human enzyme defects can be made using these examples.

Material for Examination Is Readily Available. For almost every step of the glycolytic pathway in erythrocytes, enzyme deficiencies are now known. This is in striking contrast to other pathways, for which little if any evidence of such deficiencies is available. One obvious reason is that the material for examination – blood – is readily available. Repeated venepunctures are not unreasonable impositions on patients, as compared, for example, to skin, let alone brain biopsies. Besides, the erythrocytes are specialized cells that contain only a portion of the enzyme system found in other cells. This restriction in the enzymatic makeup reduces the number of possibly affected reactions and facilitates analysis. These advantages of blood – and especially the red blood cell – have

Table 4.3. Nonspherocytic hemolytic anemias *(NSHA)* caused by proven enzyme deficiencies in the red cell glycolytic pathway (numbers correspond to Fig. 4.4)

No.	Enzyme defect	Enzyme activity in defective individuals	Hematologic symptoms	Symptoms in other organ. systems (especially CNS)	Mode of inherit.	Remarks
1	Hexokinase	30-60%	NSHA, often severe	(skeletal abnormalities in one patient)	a.r.	In some cases only erythrocytes, in others leukocytes are also affected
2	Glucose phosphate isomerase (GPI)	15%-25%	NSHA, often severe; sometimes icterus neonatorum	No other symptoms	a.r.	Thermostability of enzyme reduced. A number of electrophoretically different variants is known. Enzyme activity is reduced in all examined tissues (no tissue-specific enzymes)
3	Phosphofructokinase (PFK)	8%-80%	Mild NSHA	Severe myopathy and myoglobinemia in some families; glycogen storage disease (without NSHA), type VII, in some cases	a.r.	Organ-specific enzymes; possibly two different enzymes even in erythrocytes, clinical and biochemical heterogeneity
4	Aldolase		NSHA		a.r.	Very few observations
5	Triosephosphate isomerase (TPI)	≈10%	NSHA	Neuromuscular anomalies; dementia; early death	a.r.	Enzyme defect also in leukocytes, skeletal muscles, and serum. It is not certain that muscular symptoms are directly caused by the enzyme defect
7	Phosphoglycerate kinase (PGK)	5%-30%	Severe NSHA	Oligophrenia, ataxia, and aphasia in some cases	X-linked	Enzyme activities in leukocytes in some cases reduced, in others normal
9	Diphosphoglycerate mutase/phosphatase	~3%	Moderate NSHA	No other signs	a.r.	
11	Pyruvate kinase (PK)	5%-20%	Very variable (normal to severe NSHA)	No constant other symptoms	a.r.	Variability of enzyme characteristics between various types (cf. text)

Numbers not referred to in Table 4.3 (6, 8, 10) relate to enzyme deficiencies in the glycolytic pathway not definitely proven to cause NSHA. For more biochemical details of these enzyme defects see [1148].

been widely exploited, as testified by the work on glucose-6-phosphate dehydrogenase and especially on the hemoglobin molecule, which provides us with the leading paradigms on molecular aspects of gene-determined proteins (Sect. 4.3) and on natural selection in human populations (Sect. 6.2.1.6).

Analysis at the Enzyme Level Reveals Genetic Heterogeneity. It was shown in Sect. 3.3 that analysis of genetic heterogeneity meets with severe limitations at the phenotypic level. If two conditions both have an autosomal recessive mode of inheritance and the overlap of phenotypic manifestations is fairly strong, the only clue to heterogeneity would be a mating of two affected homozygotes with only normal unaffected children (Sect. 3.1.3). Once the analysis is carried out at the enzyme level, genetic heterogeneity is obvious under the following conditions:

a) All the enzyme defects in the glycolytic pathway of red blood cells as described in Table 4.3 and Fig. 4.3 lead to very similar hemolytic anemias. One source of genetic heterogeneity is obviously the fact that mutations of different genes determining the various enzymes of a given pathway may have similar or identical phenotypes. This conclusion could also be made using the example of the glycogen storage diseases.
b) A second source of heterogeneity is provided by the many ways in which an enzyme that is determined by one gene may change its properties due to various mutations. The more methods for examining enzyme properties are applied, the more differences are exposed. Genetic heterogeneity at a given locus is of course expected since the number of mutations causing amino acid substitutions and deletions is very large (Sect. 4.3).

In Almost All Enzyme Defects, a Residual Activity Is Found Among Homozygotes. In the second column of Table 4.3, the enzyme activities of the homozygotes for glycolytic defects are shown: in all cases in which quantitative determinations were possible, residual activity sometimes of considerable magnitude was found. In some instances, this might have been caused by activity of another enzyme, which catalyzed the same metabolic step. Usually, however, the mutation does not change the protein so much that enzyme activity is completely lost. This maintenance of residual activity has been claimed by Kirkman [1166] to be a general property of many or most human enzyme deficiencies, quite contrary to bacteria, where many mutations lead to complete enzyme blocks. This phenomenon may be partially due to selection for survival: for example, it is easy to imagine that a complete genetic block of an enzyme in a key metabolic sequence would be lethal for the individual. In bacteria, on the other hand, mutants can mainly be observed if enzyme activity has been entirely or almost entirely lost. A mutant with an incomplete block ("leaky" mutant) will often manage to survive on a minimal medium.

The principal difference between human and bacterial mutations is their mode of ascertainment. Bacterial mutations are usually found by failure of bacterial cultures to grow under certain culture conditions. Most human mutations causing enzyme deficiency are detected in patients with disease. When appropriate techniques are used to screen all types of bacterial mutations, the expected range of different structural mutations similar to human structural enzyme mutations has been found. The difference between residual activity in man and microbes, particularly if one allows for lethality of mutations of key enzymes in man, may therefore be largely spurious.

Clinical Findings Caused by an Enzyme Defect Depend on the Normal Activity of This Enzyme in a Variety of Different Tissues. Multiple forms of a given enzyme may occur within a single organism or even in a single cell [120]. These are called isozymes. Isozymes may be generated by secondary alterations of the enzyme in the tissue and in such cases are of nongenetic origin. Genetically determined isozymes owe their origin to different genetic loci coding structurally distinguishable polypeptide chains that may have a common origin in evolution (Sect. 7.2.3). The term isozyme has also been applied to allelic variants at a single genetic locus that are detectable by electrophoresis. Isozymes are known for many enzymes. They catalyze the same overall reactions but are usually adapted to slightly different conditions under which these reactions occur in various tissues. Since the differences in intracellular milieu are small and largely unexplored, it cannot be predicted on theoretical grounds for which enzymes isozymes may exist.

Enzyme defects are due to mutations affecting single genes. Hence, they will usually only affect one of a series of isozymes. If more than one isozyme is affected, these variant isozymes may share a common polypeptide chain or secondary effects on enzyme structure have occurred.

An enzyme deficiency caused by a mutation of a gene that is active only in one tissue will affect the phenotype of its carrier in a different way from an enzyme deficiency affecting many tissues. Gene mutations may show pleiotropy, i.e., a single mutation may have many different consequences in a single individual. An enzyme deficiency affecting more than one tissue is expected to have a pleiotropic effect. This is one mechanism of pleiotropy, but certainly not the only one. Even if an enzyme - or any other protein - is active in only one tissue, its deficiency may affect other tissues by perturbations induced by the primary defect. However, an enzyme defect that is found in all tissues sometimes leads to a phenotypic anomaly in one tissue only - probably because the defect can be compensated for more easily in other tissues.

The enzyme deficiencies listed in Table 4.3, show examples for all these types of pleiotropism. For example, PFK deficiency leads in some cases to relatively mild nonspherocytic hemolytic anemia. Here, the pleiotropic pattern consists of mild anemia, slight jaundice, and mild splenomegaly. All these symptoms can be explained by a shortened life span of the red blood cells. Such a patient had normal PFK activities in leukocytes, thrombocytes, and skeletal muscles. The defect seems to be confined to the red cell [1147].

Patients in other families had very mild hemolytic anemia and severe myopathy [1191] associated with glycogen storage disease [1327]. Muscle and erythrocyte PFK are different from each other as shown by electrophoresis, chromatography, and immunology. The isozyme pattern seems to be fairly complicated; even within erythrocytes, at least two enzyme components have been found [1148]. The differences in patterns of pleiotropic action between families can in principle be explained by mutations in genes determining polypeptide chains that may be present or absent in tissue-specific isozymes.

However, an enzyme deficiency can be present in all examined tissues, and still the primary phenotypic effects may be confined to one tissue. Glucose phosphate isomerase (GPI) deficiency is one example. In probands ascertained via the erythrocyte defect, the activity of this enzyme is usually reduced similarly in leukocytes, thrombocytes, fibroblasts, muscles, and liver. In all these tissues, the enzyme appears to have the same biochemical properties. In spite of thorough examinations, there is no hint of tissue-specific enzymes. Still, the erythrocyte defect dominates the pattern of clinical manifestations. Patients suffer from severe hemolytic anemia which may present at birth as severe jaundice. Many different structural mutations affecting glucose phosphate isomerase (GPI) have been detected and compound heterozygotes for these defects are frequent. On the other hand, not all enzyme deficiencies cause any disease at all; they might not be associated with any detectable signs [1003].

Pyruvate Kinase (PK) Deficiency (26620). Pyruvate kinase (PK) deficiency is the most frequent defect of a glycolytic enzyme in red blood cells. Homozygotes may show a wide range of hematologic symptoms. Some patients have a fully compensated nonspherocytic hemolytic anemia, others suffer from severe and repeated hemolytic episodes. Some features seem to be undisputed:

a) Homozygotes usually have a residual activity of about 5%–20% of the normal enzyme activity. Heterozygotes have values around 50% and are clinically healthy.
b) Examination of qualitative enzyme characteristics such as kinetic properties, nucleotide specificity for ADP and UDP, temperature stability, urea stability, pH optimum, and electrofocusing have exposed the existence of many variants with different properties. Structural mutations therefore seem to be responsible for most if not all variants [1149; 1148]. It is very difficult from enzyme assays to draw conclusions whether affected patients are truly homozygous for the same variant or have two different defective alleles ("compound heterozygotes") (Sect. 3.1.3, Fig. 3.12).

Enzyme Activities and Clinical Symptoms in Heterozygotes. For most of the glycolytic defects listed in Table 4.3, determinations of enzyme activity have been performed in heterozygotes. As a rule, the activities are halfway between those of normal and defective homozygotes. This finding is an example of a rule that applies more generally: In most of human enzyme defects analyzed so far, heterozygotes show roughly 50% of normal activity. Usually, this amount of reduction does not lead to any obvious clinical manifestations; half the enzyme activity is sufficient to maintain function under normal circumstances.

It is of great interest that heterozygotes in fact only have 50% of enzyme activity in most enzyme deficiencies, which in man have so far all been shown to be structural mutants. This finding clearly establishes that the amount of enzyme activity is rigorously specified by the structural gene locus specifying the enzyme activity under its control. Thus, normal homozygotes who have two structural genes for the enzyme have 100% enzyme activity while heterozygotes have only 50%. The single normal gene in such heterozygotes is therefore unable to compensate for the mutant structural gene that produces an inactive gene product. This finding is of great importance for a consideration of gene regulation in mammals since it differs from regulatory phenomena in bacteria.

Aerobic Energy Production in the Red Cell: Hexose Monophosphate (HMP) Pathway [994]. The left-hand side of Fig. 4.3 shows the aerobic pathway via the so-called hexose monophosphate (HMP) cycle, also called the pentose phosphate or shunt pathway. Its main function is to generate reducing power in the form of NADPH. Glucose-6-phosphate is oxidized through the action of glucose-6-phosphate dehydrogenase [1030] to 6-phosphogluconate. Through various further steps, D-ribose 5-phosphate is formed.

Enzymatic reduction of oxidized glutathione (GSSG) oxidizes NADP. Reduced glutathione (GSH) keeps SH groups in the reduced stage, and may also help to protect the cell from damage by compounds such as H_2O_2. An inherited condition has become known in which GSH is completely lacking. The anomaly is caused by glutathione synthetase deficiency which, in some cases, also leads to oxiprolinemia [1148]. GSH reduction leads to a nonspherocytic hemolytic anemia, which is associated with drug sensitivity to oxidant drugs. Many drugs have an oxidizing effect and stress the reductive capacity of the cell [1009, 1269].

Deficiency of G6PD (Glucose-6-Phosphate Dehydrogenase) (30590) [1079, 999, 1002, 1225, 1234, 1224, 1146] (Fig. 4.3). A genetic block in the hexose monophosphate pathway that also manifests as increased drug sensitivity has gained special significance and has become one of the leading paradigms of pharmacogenetics (see Sect. 4.5.1). During the Korean War (1950–1952), American soldiers received prophylactic treatment with an antimalaria drug, primaquine [8-(4,amino-1-methylbutylamino)-6-methoxyquinoline]. In about 10% of black soldiers and in much fewer whites (usually of Mediterranean origin) (fewer than $\frac{1\text{-}2}{1000}$), an intravascular hemolytic reaction was observed. Similar hemolytic reactions had been observed earlier in darkly pigmented patients when drugs, such as sulfanilamide and pamaquine were administered. Hemolytic reactions had long been known in Mediterranean areas, for example, in Sardinia, to occur in some individuals after eating broad beans *(Vicia faba)*.

Critical studies of the life span of "primaquine-sensitive" red cells showed that the defect was intracellular and that the older red cells were more susceptible to hemolysis in black individuals. This fact explained the short duration of the hemolytic reaction (Fig. 4.4). Once the older cells are destroyed, hemolysis stops despite continuation of drug treatment.

At first an immune mechanism was suspected: later the sensitive cells were found to show glutathione instability on incubation with acetylphenylhydrazine. In 1956 Carson et al. [1030] demonstrated the specific enzyme defect. The following reactions were examined (cf. also Fig. 4.6):

a) $\text{GSSG} + \text{NADPH} + \text{H}^+ \xrightarrow{\text{GSSG reductase}} 2\,\text{GSH} + \text{NADP}^+$

b) $\text{Glucose-6-phosphate} + \text{NADP}^+ \xrightarrow[\text{dehydrogenase}]{\text{glucose-6-phosphate}} \text{6-phosphogluconate} + \text{NADPH} + \text{H}^+$

c) $\text{6-Phosphogluconate} + \text{NADP}^+ \xrightarrow[\text{dehydrogenase}]{\text{6-phosphogluconate}} \text{pentose phosphate} + \text{CO}_2 + \text{NADPH} + \text{H}^+$

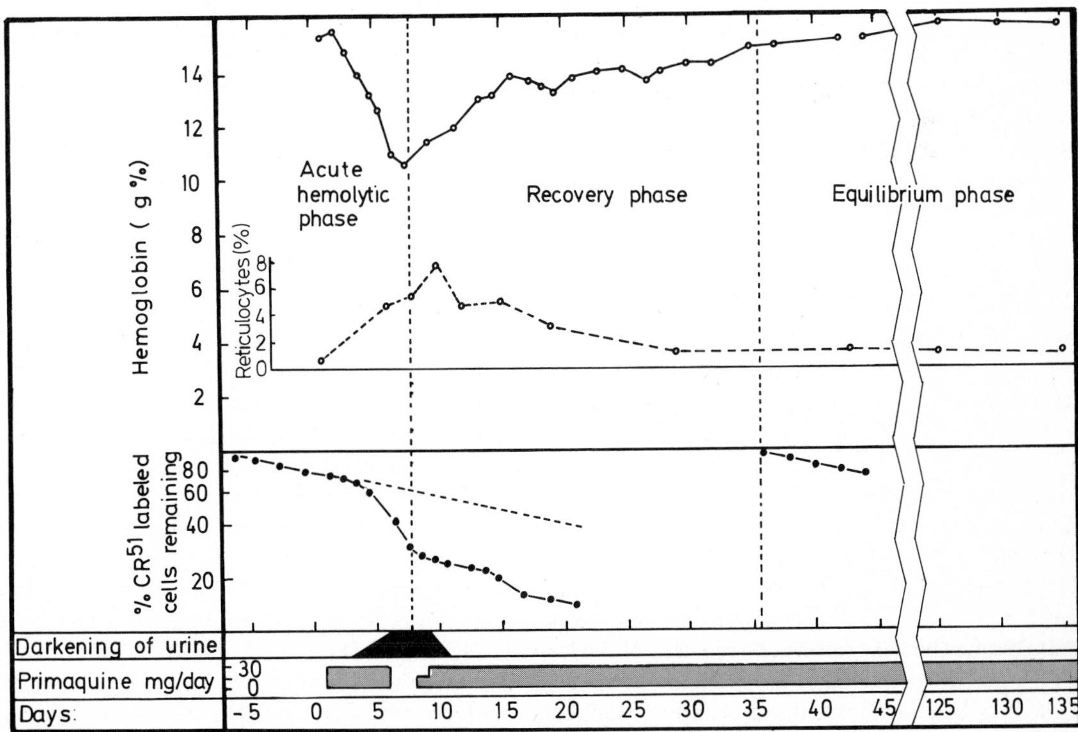

Fig. 4.4. Hemolytic reaction after treatment with primaquine. Within the first days after beginning of drug administration, many erythrocytes are destroyed by hemolysis. This leads to increased production of new erythrocytes; reticulocytes increase, and hemoglobin rises again. Hemolysis affects only red cells older than about 60 days; during therapy, the red cell population consists of younger cells. (Carson et al., 1956 [1030])

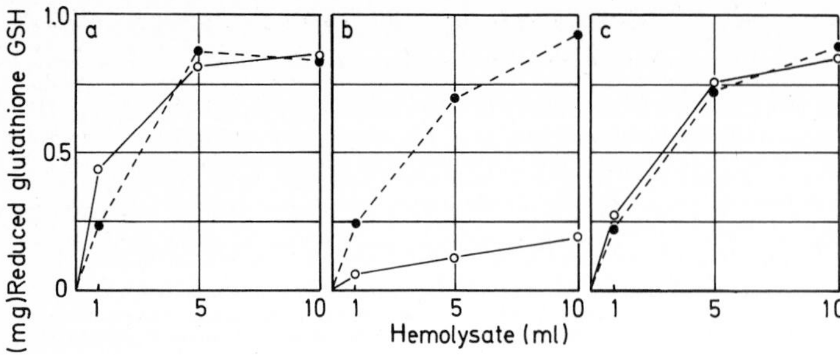

Fig. 4.5 a–c. Formation of reduced glutathione (GSH) from oxidized glutathione (GSSG) as function of the amount of dialyzed hemolyzate (in ml). *Solid line,* primaquine-sensitive individual; *dotted line,* normal individual. **a** Assay of GSSG reductase; **b** assay of G6P-dehydrogenase; **c** assay of 6PG-dehydrogenase. Only reaction 12 of Fig. 4.3. (G6PD) is abnormal (**b**) (Carson et al. 1956 [1030])

The critical defect was found to be G6PD deficiency. Activity of GSSG reductase and 6-phosphogluconate dehydrogenase were normal. These studies clearly established G6PD deficiency as the cause of the hemolytic reactions in primaquine-sensitive men.

It soon became obvious that men were more frequently affected with hemolytic reactions than women. Quantitative assay using the glutathione stability test,[2] in which GSH is measured before and after incubation of red cells with acetylphenylhydrazine gave a clearly bimodal distribution of postincubation glutathione values in 144 American black males, with an appreciable part of the population having very low values. In 184 black females, the distribution of glutathione values was shifted to the left, and the number of persons with very low values was much lower than among males. This sex difference suggested an X-linked mode of inheritance with very low values in male hemizygotes and female homozygotes; female heterozygotes had somewhat intermediate values. The hypothesis of X-linked inheritance was soon confirmed by family studies [1034]. In later years, when direct G6PD enzyme assays were carried out in population samples, a similar distribution was found, but the values of female heterozygotes were more nearly halfway between normal and abnormal homozygotes (Fig. 4.6) with much overlap with normals.

Males
Number of individuals
40
30
20
10
0
100
200
300
400
G-6-PD level

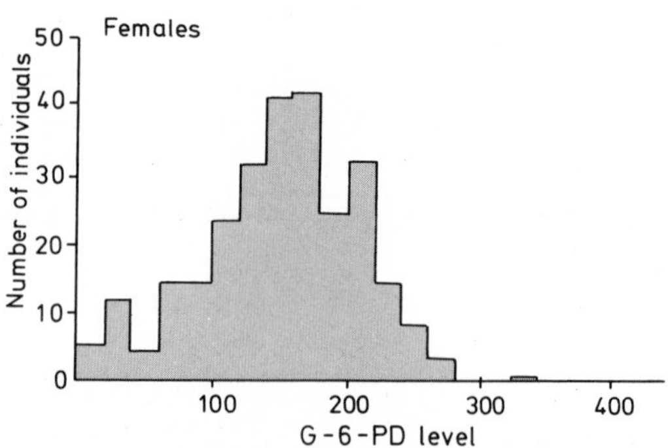

Fig. 4.6. Distribution of G6PD activities in males and females in a black population. Note almost perfect distinction between affected and normals in males. (Harris, [93])

Difference Between the African and Mediterranean Variants. Within a few years after discovery of G6PD deficiency, differences in severity of G6PD deficiency between African and Mediterranean male carriers became known. In red cells of Africans with G6PD deficiency, a residual activity of 10%–20% was regularly found, whereas Mediterranean carriers showed only minimal activity (below 5%). In addition, Africans had a nearly normal activity in their leukocytes, which was moderately or martedly reduced in Mediterraneans.

Electrophoretic methods for examination of affected enzyme proteins were developed. The mobility of the normal wild-type enzyme was designated as B. In blacks with normal enzyme activities, an electrophoretic variant with rapid migration was discovered in 20% of males, which was designated as A. Blacks with enzyme deficiencies always had a G6PD band with strongly reduced activity and the mobility of the A variant. This was called A^-. The variant of G6PD deficiency in the Mediterranean population (G6PD Med) migrated with mobility similar to nor-

[2] This test was used as an indicator of primaquine sensitivity before the G6PD defect was discovered.

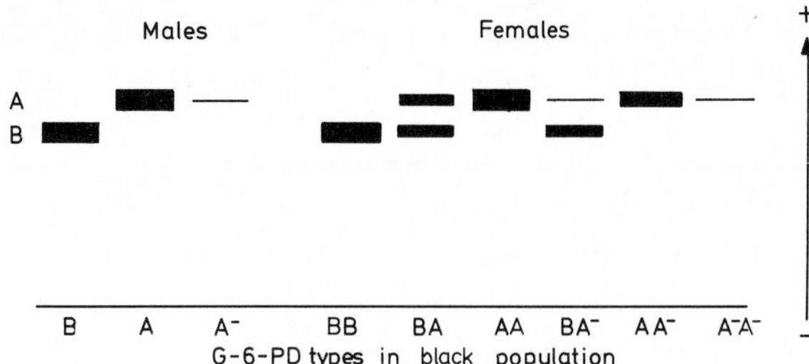

Fig. 4.7. G6PD electrophoretic phenotypes in a black population. Males are hemizygous A, A^-, or B; females may be homozygous for any of these alleles or heterozygous for all possible combinations. The A^- bounds are indicated by *thin lines* since little staining occurs because of enzyme deficiency. BB and BA^- as well as AA and AA^- types can usually not be distinguished. (Harris, [93])

mal G6PD and therefore was sometimes referred to as G6PD B^-. In normal white populations, G6PD migrated almost exclusively as the normal B component (Fig. 4.7).

More Detailed Characterization of G6PD Variants. Many rare G6PD variants were discovered in various populations, and standardization of methods for classification became necessary. A proposal by a group of specialists in this field was published in 1967 by the World Health Organization [996]. According to this proposal, characterization of a variant should include the following:

a) Determination of the enzyme activity by enzyme assay.
b) Electrophoresis of the enzyme (if needed, after partial purification) in various buffer systems.
c) Determination of substrate specificity (Michaelis-Menten constant K_m) for glucose-6-phosphate and NADP (and NAD).
d) Utilization of substrate analogues; these examinations were usually made with 2-desoxyglucose-6-phosphate, galactose-6-phosphate, and deamino-NADP. Utilization of such substrate analogues is a frequently used method for eliciting qualitative differences in enzyme properties.
e) Determination of thermostability.
f) Determination of pH dependency of enzyme activity.

Enzyme Variants Observed in Human Populations. With these methods, more than 200 G6PD variants have been described so far [1001] see also Sect. 7.5.8). They may be classified as follows (Table 4.4a and b):

a) Variants with enhanced enzyme activity. Only two such variants are known; G6PD Hektoen and G6PD Hartford.
b) Variants with almost normal activity. One of these is the variant A^+ mentioned above, which is

Table 4.4a. Classes of G6PD variants (See also McKusick [133], pp. 1110–1129)

Class	Examples
Enzyme activity increased	G6PD Hektoen
Enzyme activity nearly normal	G6PD A^+ (common in tropical Africa)
Enzyme activity moderately decreased (≃10%–60% in hemizygous males) (susceptibility to oxidizing drugs increased; no favism)	G6PD A^- (common in tropical Africa); G6PD Anant (Thailand); G6PD Canton (southern China)
Severe enzyme defect and compensated hemolysis; susceptibility to oxidizing drugs strongly increased; favism may occur	G6PD (Mediterranean type); G6PD El Fayoum (Egypt); G6PD Zähringen (Germany)
Severe enzyme defect and chronic hemolysis even without exposure to oxidizing drugs; nonspherocytic hemolytic anemia	G6PD Albuquerque; G6PD Beaujon; G6PD Freiburg

Table 4.4b. Some common G6PD variants

	Gd A^-	Gd Med	Gd Canton	Gd Mahidol
Population	African	Greek, Italian, Mediterran.	Chinese	Thai
G6PD activity (%)	10%–20%	0–5%	4%–25%	5%–16%
Hemolysis with drugs and infections	Yes	Yes	Yes	Yes
Favism	?	Yes	Yes	?
Neonatal jaundice	Rare	Yes	Yes	Yes
Drop in g/100 ml of Hb in hemolysis	2–5	4–10	4–10	2–8
Increased G6PD levels in red cells following hemolysis	Yes	Minimal	?	Yes

found in about 20%-25% of the male population of tropical Africa and their American descendants.

c) Variants with moderately reduced activity. These show activities between 10%-50% among hemizygous males. Sensitivity to hemolysis-provoking drugs may occur; favism is not observed.
d) Variants with severe enzyme deficiency and mild compensated hemolysis. A characteristic member of this group is the Mediterranean variant.
e) Variants with severe enzyme defect and chronic hemolysis even without additional exposure to oxidizing agents. These variants lead to a congenital nonspherocytic hemolytic anemia.

This subdivision does not consider important properties of some variants; nevertheless, it is useful as a first landmark for orientation.

More Incisive Biochemical and Molecular Analysis [1167, 1000, 1002]. An X-linked mode of inheritance was confirmed for all variants for which family investigations were carried out. This strongly suggested the hypothesis that all these mutations have, indeed, occurred within one gene locus and that no other autosomal loci for red cell G6PD exist. Furthermore, all variants seem to be caused by different mutations within this structural gene.

The active enzyme has a molecular weight of ~120,000 and consists of dimers. The subunits are polypeptide chains consisting of about 450 amino acids; their sequence has been analyzed [1368]. However, tryptic digestion (fingerprinting; cf. Sect. 4.3), revealed the molecular aberration in at least two variants to be caused by only one amino acid: G6PD A^+ differs from G6PD B^+ by one replacement of aspartic acid by asparagine. In G6PD Hektoen, histidine is replaced by tyrosine. Considering the genetic code, these substitutions can be related to single base replacements in the transcribing DNA strand. Thus, genetic analysis has been carried to the DNA-gene level (Sect. 3.6). This matter will be taken up again in Sect. 4.3.

G6PD activity has been shown to be present in most if not all tissue cells; there appear to be no tissue-specific isozymes and if there is a G6PD mutation, the altered G6PD can be demonstrated in all tissues.

Significance of G6PD Variants for Understanding of Human Enzyme Deficiencies. The G6PD system has been an excellent model because males with an X-linked structural mutation will only exhibit the mutant gene product. In contrast, heterozygotes for autosomal mutations will present a 50/50 ratio of normal and abnormal gene product, so that minor alterations in physico-chemical properties of the mutant gene product are difficult to detect. There are other characteristics of G6PD variants, such as a relatively high frequency of some variants in certain populations, which have permitted a more thorough genetic analysis than is possible with many other human enzyme deficiencies.

Among the more general findings that also apply to other human enzyme defects are:

a) With a sufficiently large spectrum of methods for characterization of the enzyme, the identifiable number of mutants with different characteristics is very large. Almost every mutational change in the enzyme seems to make for different physiochemical properties.
b) The phenotypic consequences of the molecular changes caused by these mutations form a continuum: from variants that do not alter the capacity for normal biologic function at all and can only be detected by special methods to others that are easily compensated under normal living conditions and lead to an abnormal reaction only under special environmental conditions to still others causing manifest disease even in the absence of harmful environmental factors. Most variants are benign and do not cause disease. There is a bias in detecting variants associated with illness since patients with hemolytic anemia are more frequently examined for enzyme deficiency than normals.
c) Most variants are quite rare, some having been identified only in single individuals and their family members. A few, however, have become frequent in some human populations (Table 4.4b); the reasons will be discussed in Sect. 6.2.1.6.
d) Practically all variants show some residual activity and, considering their qualitative differences in properties, all of them can be explained by structural mutations within an X-chromosomal gene locus that has been precisely localized (Sect. 3.4.3).

While there is evidence that these conclusions may hold true for most if not all human enzyme defects, there is a fifth conclusion related to localization of this gene on the X chromosome. In most cells of female heterozygotes, either one or the other of the two alleles is functionally active, the other having been inactivated. This inactivation can be used as a tool to examine problems of tumor formation and cell differentiation. It has been shown, for example, that uterine leiomyomata of women heterozygous for two electrophoretic types of G6PD [1002] exhibit either one or the other of these G6PD types in the tumor. This finding can be explained by the origin of all the cells of each tumor from a single cell. Similar findings suggesting a so-called monoclonal origin of tumors have been made for most neoplastic processes (see Sect. 5.1.6).

Another problem is the number of blood stem cells from which the erythrocyte population develops. If a woman is heterozygous for, say, G6PD A^+ and B^+ and inactivation occurs at random, then the relative number of stem cells with A^+ and B^+ follows a binomial distribution:

$$(\frac{1}{2}A^+ + \frac{1}{2}B^+)^n$$

n being the number of stem cells. Hence, the probability that a heterozygote will have only A^+ blood cells is $(\frac{1}{2})^n$. The probability of having only B^+ blood cells is identical. Practical application of this principle involves quantitative assessment of relative ratios of the A and B enzyme in the tissue to be tested. The preliminary data seem to point to a small number of blood stem cells at the time of inactivation (day 12-16; about 5,000 total cells; Sect. 2.2.3.3). Their number may not exceed three to five.

The method rests on the assumption that there is no selection for one or the other G6PD cell populations. There is some evidence for a selective advantage of G6PD normal cells in heterozygotes for G6PD deficiency. Under such conditions, the estimate of an embryologic or stem cell pool becomes hazardous.

Phenocopy of a Genetic Enzyme Defect: Glutathione Reductase Deficiency [1077, 1078]. An enzymatic reaction related with the shunt pathway is the reduction of GSSG to GSH. The enzyme involved is glutathione reductase (cf. Fig. 4.3; 13). The older literature contains a number of observations of families with alleged defects of this enzyme; various hematologic disorders were found in these families. However, the family data did not really fit the usual genetic hypotheses. More recently, it has been shown that deficiencies of glutathione reductase are almost always due to a nutritional lack of the coenzyme flavin adenine dinucleotide (FAD) [997]. The cause is usually riboflavin (vitamin B_2) deficiency; enzyme activity of glutathione reductase is normalized within a few days after administration of riboflavin to deficient individuals.

This condition has been found to be frequent in northern Thailand and is caused by riboflavin shortage in the customary diet. As attested by this example, not every enzyme anomaly found to be frequent in a particular population must be genetically determined.

Some other cases may reflect a genuine, genetically determined glutathione reductase defect. We do know of other hereditary anomalies in man caused by an abnormally high demand for a certain coenzyme, which has to be supplied as a therapeutic vitamin. X-linked vitamin D-resistant rickets [958] and pyridoxal dependency with epileptic seizures are two examples (cf. also Sect. 3.1.4). It may be that in some families, glutathione reductase deficiency will be related to genetically determined riboflavin deficiency.

Mimicry of a genetic defect by an environmental injury is called a *phenocopy* – a term coined by Goldschmidt (1935) [1106]. A phenocopy is defined as a simulation of an inherited character by external factors. Goldschmidt, treating wild-type *Drosophila* in various stages of development by heat shocks, succeeded in producing numerous phenotypic abnormalities that were similar to variants usually produced by mutation.

Phenocopy experiments have been carried out in many species and appeared to offer insights into the mechanisms of normal embryonic development and the production of malformation. However, their explanatory power has been overrated. Still, in human metabolic diseases, the possibility that a phenocopy may be present should not be too readily discarded especially since effective therapy may be available.

From a genetic point of view, a common phenocopy in the field of a human metabolic disease is hypothyroidism due to lack of inorganic iodide – a frequent condition in the Alpine countries of Europe and in some other parts of the world. Here, deficiency of an essential inorganic ion leads to the same end result as defective synthesis of the thyroid hormone observed in some families as a consequence of any one of a number of enzyme defects in thyroid hormone synthesis (Fig. 4.24). Nevertheless, while the clinical end results are similar, the pathophysiology of nutritional iodine lack and of genetically defective enzymes in thyroxin biosynthesis are different.

Deficiencies in Nucleotide Metabolism. The necessary level of ATP is maintained in the red cell by the glycolytic pathway (Fig. 4.3). Another possible reaction for ATP production is the following:

$$1\,\text{AMP} + 1\,\text{ATP} \xrightleftharpoons{\text{adenylate kinase}} 2\,\text{ADP}$$

ADP is then converted to ATP by glycolysis. Cation transport is one of the energy-requiring cell functions for which ATP is needed. It is directly related to the enzymatic activity of ATPase. This enzyme is located in the cell membrane and is present in two fractions, which may be activated by K^+ and Na^+ or by Mg^{++}. The enzyme hydrolyzes ATP to AMP. ATPase deficiency may also cause nonspherocytic hemolytic anemia. A number of cases with slightly different symptoms and hematologic characteristics have been described [1046, 1118, 1120, 1203]. Adenylate kinase (AK) deficiency was observed (20160) in a 13-year-old boy with NSHA. The AK activity was only 1%-13% of normal in erythrocytes. ATP and ADP content of the red cells was normal. AMP content was increased. Again, the mode of inheritance appeared to be autosomal recessive. Pyrimidine 5'nucleotidase deficiency was observed first in a patient suffering from moderate, chronic hemolysis [1333]; and shortly afterwards, in a few other patients [1148].

Mature erythrocytes have the capacity to synthesize large amounts of NAD from nicotinic acid, ATP and PRPP (5-phospho-D-ribose-l-pyrophosphate). Glycolytic activity is required for this pathway, but the hexose monophosphate pathway does not appear to be essential as a source of ribose phosphate, as the NAD content of G6PD-deficient red cells is normal. Synthesis of NADP from NAD requires kinase for these enzymes; no defects have become known. It is possible that the apparently relatively low number of observed deficiencies in the nucleotide metabolism group is not real but may be caused by the complicated methods of enzyme assay.

Analysis of enzymes not present in blood cells and of their deficiencies poses more difficulties. Many such enzymes are found in fibroblasts grown, for example, from skin biopsies. In contrast to red cells, the fibroblasts contain nuclei. They are able to divide and to carry out all stages of protein synthesis. Their enzymatic endowment is much more complete than that of red cells. Fibroblasts sometimes lack only those enzymes that are confined to single groups of specialized cells, such as liver cells (i. e. phenylalanine hydroxylase deficient in phenylketonuria). Deficiencies of enzymes for many different metabolic pathways will manifest in fibroblasts. This is why enzyme assays from cultured fibroblasts have brought such progress to our knowledge of enzyme defects.

In the following, only one group of diseases will be discussed, which also permits some more general conclusions regarding properties of human enzyme deficiencies. These are the mucopolysaccharidoses, which are classified with the larger group of conditions due to deficiencies of lysosomal enzymes, other subgroups being the sphingolipidoses and mucolipidoses.

4.2.2.3 Mucopolysaccharidoses

Deficiencies of Lysosomal Enzymes. Enzymes or enzyme systems usually are located in one characteristic cell compartment. For example, the enzymes of electron transport and oxidative phosphorylation of ADP as well as other oxidative enzymes for pyruvate, fatty acids, and some amino acids are located in the mitochondria. A number of hydrolytic enzymes are concentrated in organelles called lysosomes [1128]. If these enzymes are liberated by destruction of the lysosomal membrane, the entire cell will be destroyed by self-digestion. Normally, this digestion will proceed within the lysosomes, which degrade not only defective cell particles and material from intercellular connective tissue but also external material taken up by the cell. Among the cell elements degraded by the lysosomes are mucopolysaccharides, mucolipids, and sphingolipids (Fig. 4.8). Deficiencies are now known for many of the enzymatic pathways involved in this degradation [1029, 1240, 1241].

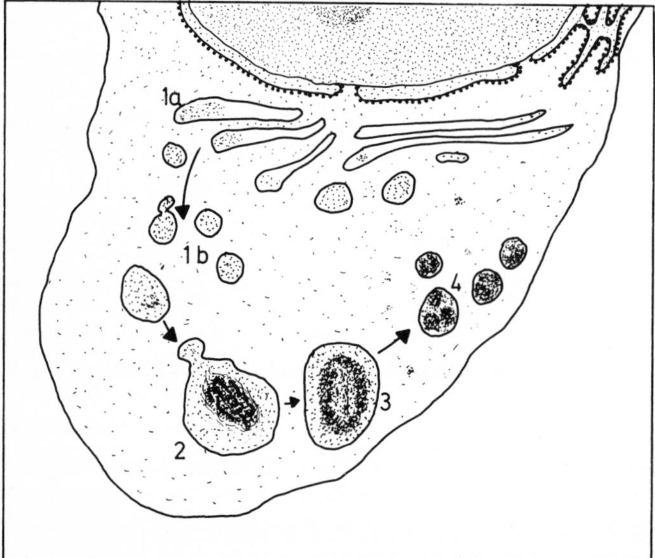

Fig. 4.8. Schematic representation of a lysosome and its functional apparatus. The figure shows the functional cycle of a normal lysosome. *1a,* Golgi apparatus; *1b,* primary lysosome; *2,* phagolysosome; *3,* secondary lysosome; *4,* residual body. Courtesy of Dr. W. Buselmaier

The metabolic pathways causing red cell defects manifesting as hemolytic anemia were already known in biochemistry, and analysis of enzyme deficiencies involved testing the known metabolic steps in these pathways. The situation was different with the mucopolysaccharidoses. Here, the genetic diseases were known first, and analysis of the enzyme defects led to elucidation of the enzymatic pathway. Therefore, we will start with a description of these diseases and proceed stepwise to the biochemical defect, the enzyme defects, and the reconstruction of the normal pathway. This example shows how genetic diseases as experiments of nature contribute to the understanding of normal biochemistry and physiology.

Mucopolysaccharidoses: Clinical Picture. All mucopolysaccharidoses are rare diseases with complex and in most cases severe manifestations ranging from abnormalities of the skeletal and vascular systems to impairment and deterioration of mental functions. The clinical symptoms result from excessive and progressive storage of sulfated polysaccharides in various tissues.

These disorders have been classified into six distinct categories on the basis of clinical, genetic, and biochemical evidence. Table 4.5 shows this classification together with the major clinical findings. With the exception of type II (Hunter), the mode of inheritance is autosomal recessive. The clinical symptoms range from very severe to relatively mild forms. For type II (Hunter) (30990), two forms have been described that can be distinguished after the age of 4 years; a severe "juvenile" form in which death occurs before puberty, and a mild "late" form with mild or no mental retardation and a generally better prognosis. Type VI (Maroteaux-Lamy) (25320) differs from type I (Hurler) (25280) in that it manifests normal intelligence and less pronounced facial dysmorphism. Here two different subtypes have also been observed - one with a fairly rapid course and more severe clinical features, the other with a relatively slow progression. In both types, cardiac impairment may lead to death in the 2nd or 3rd decade. For type IV (Morquio-) two subtypes have also been discovered - IV A with very severe, IV B with milder manifestations.

Table 4.5. Major clinical findings in the mucopolysaccharidoses (Cantz and Gehler, 1976 [1029]; Spranger 1983 [1312]; modified)

Mucopolysaccharidosis		Mental and motor retardation	Growth retardation	Coarse facies	Bone dysplasia	Joint contractures	Hepatosplenomegaly	Corneal opacities	Mode of inheritance
Type[a]	Name								
I H	Hurler (1)	+++	++	+++	+++	++	++	+	Autosomal-recessive
I S	Scheie (1)	–	–	±	+	+	±	+	
I H/S	Hurler-Scheie compound (1)	±	+	++	++	++	+	+	
II A	Hunter, severe (2)	+	+	+	++	+	+	±	X-linked recessive
II B	Hunter, mild (2)	±	+	+	++	+	+		
III A	Sanfilippo A (3 A)	+++	–	+	+	±	++	–	Autosomal-recessive
III B	Sanfilippo B (3 B)	+++	–	+	+	±	++	–	
III C	Sanfilippo C (3 C)	+++	–	+	+	±	++	–	Autosomal-recessive
III D	Sanfilippo D (3 D)	+++	–	+	+	±	++	–	
IV A	Morquio A (4)	–	+++	+	+++	+	+	+	Autosomal-recessive
IV B	Morquio B	–	++	+	++	+	+	+	Autosomal-recessive
V		Vacant							
VI A	Maroteaux-Lamy, classic form (6)	–	++	++	++	+	+	+	Autosomal-recessive
VI B	Maroteaux-Lamy, mild form (6)	–	+	+	+	+	+	+	
VII	Sly (7)	+	±	±	+	–	++	±	Autosomal-recessive

[a] Classification of McKusick (1972); supplemented.
– absent; ± occasionally present; + mild; ++ less severe; +++ severe. The numbers in parentheses refer to the genetic blocks shown in Fig. 4.11 and Fig. 4.12.

Lysosome Storage and Urinary Excretion. Histochemical studies have shown that these conditions are storage diseases. In many cells - including fibroblasts, liver cells, Kupffer cells, reticulum cells of spleen and lymph nodes, leukocytes, epithelial cells of kidney glomeruli, and nerve cells - enlargement and vacuolization is observed due to large quantities of stored material. The main storage compounds have been identified as sulfated glycosaminoglycans.

Further clarification has come from electron-microscopic studies. The compounds were found to be stored in rounded vacuoles, similar to lysosomes seen in experimental animals that had been injected with nonmetabolizable substances. Therefore, it was concluded that these vacuoles were lysosomes engorged with undigested or only partially degraded glycosaminoglycans [1129]. This conclusion was confirmed and extended to other tissues. The overloading of the lysosomal system led to classifying these disorders as lysosomal diseases even at a time when the metabolic defects were still unknown. Since the most important function of lysosomes lies in the hydrolytic degradation of macromolecules, it was probable that the storage resulted from deficiencies of lysosomal hydrolytic enzymes. Other evidence for a disturbed glycosaminoglycan metabolism came from the detection of an excessive excretion of these compounds in the urine. The approximate urinary excretion of various glycosaminoglycans in the different syndromes can be seen in Table 4.6. These excretion patterns reflect the basic metabolic lesions (Table 4.7).

Biochemistry of Sulfated Glycosaminoglycans. The sulfated glycosaminoglycans are complex heterosaccharides consisting of long polysaccharide chains covalently linked to a protein core. In dermatan sulfate, heparan sulfate, and chondroitin 4- and 6-sulfates, the polysaccharide chains are composed of alternating residues of uronic acid and sulfated hexosamine. Keratan sulfate differs from the other glycosaminoglycans in that the uronic acid residues are replaced by galactose. The polymeric chains may be about 100 residues long and are bound to specific proteins via a distinct linkage region; several sugar chains are attached to the same polypeptide backbone. Such proteoglycans may form even larger

Table 4.6. Urinary excretion patterns of sulfated glycosaminoglycans in the mucopolysaccharidoses (Cantz and Gehler, 1976 [1029])

Mucopolysaccharidosis	Glycosaminoglycans excreted[a]
I-H and I-S	Dermatan sulfate and heparan sulfate; proportion about 3:1
II	Dermatan sulfate and heparan sulfate; proportion about 1:1
III A, III B, III C, III D	Heparan sulfate
IV A, IV B	Keratan sulfate and chondroitin-6-sulfate
VI	Dermatan sulfate; sometimes also some heparan sulfate (?)
VII	Dermatan sulfate, and possibly heparan sulfate

[a] Adapted from Spranger (1972).

Table 4.7. The metabolic defects in the mucopolysaccharidoses (Cantz and Gehler, 1976 [1029])

Mucopolysaccharidosis[a]		Major storage substance	Enzymatic defect
I H	Hurler	Dermatan sulfate und heparan sulfate	α-L-iduronidase
I S	Scheie		
I H/S	Hurler-Scheie compound		
II A	Hunter, severe form	Dermatan sulfate und heparan sulfate	Iduronate sulfatase
II B	Hunter, mild form		
III A	Sanfilippo A	Heparan sulfate	Heparan N-sulfatase
III B	Sanfilippo B	Heparan sulfate	α-*N*-acetylglucosaminidase
III C	Sanfilippo C	Heparan sulfate	α-Glucosaminidase (?)
III D	Sanfilippo D	heparan sulfate	*N*-ac-glucosamine-6-sulfate sulfatase
IV A	Morquio A	Keratan sulfate	*N*-acetylgalactosamine 6-sulfatase
IV R	Morquio B	keratane sulfate	β-galactosidase
VI A	Maroteaux-Lamy, classic form	Dermatan sulfate	*N*-acetylgalactosamine 4-sulfatase (arylsulfatase B)
VI B	Maroteaux-Lamy, mild form	Dermatan sulfate	
VII	Sly	Dermatan sulfate and heparan sulfate	β-glucuronidase

[a] Classification of McKusick (1972); supplemented.

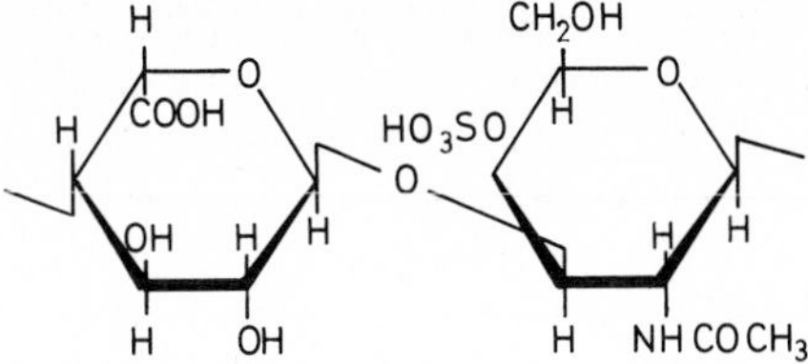

Fig. 4.9. Dimer of L-iduronic acid and 4-sulfated *N*-acetyl-galactosamine as found in dermatan sulfate

complexes through noncovalent bonds. Within the carbohydrate chains, there is considerable variation of the constituent sugars, as well as in the degree of sulfation.

For example, the major part of the polysaccharide chain in dermatan sulfate is composed of repeating dimers of L-iduronic acid linked to 4-sulfated *N*-acetyl-galactosamine (Fig. 4.9). The other glycosaminoglycans have similar structures. They are constituents of connective tissue and the ground substance.

In patients with mucopolysaccharidoses, the connective tissue glycosaminoglycans consist of the same large proteoglycan entities that are found in normal individuals, indicating that the enzyme deficiency does not affect their synthesis. In those tissues where abnormal storage occurs, and also in the urine, the molecules are smaller and of varying length. This suggests that the cell has cleaved as many linkages as possible before stopping at a residue for which the specific degradative enzyme is deficient.

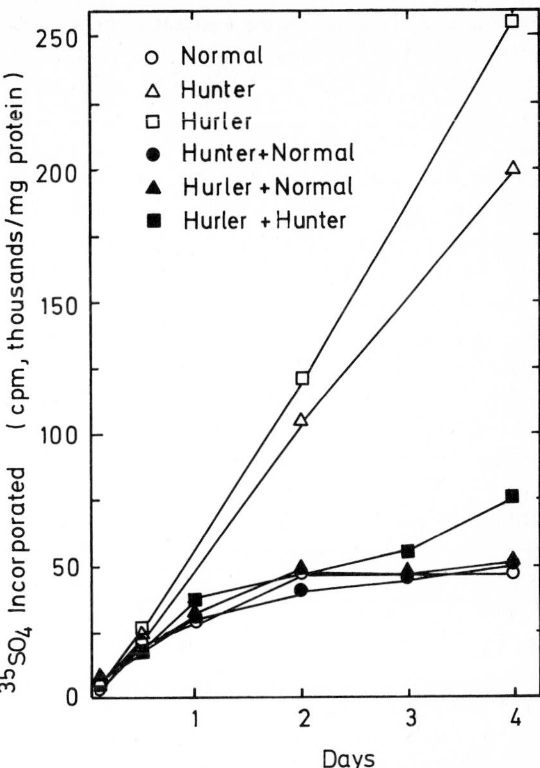

Fig. 4.10. Abnormal incorporation of $^{35}SO_4$ in cells from Hunter and Hurler patients. Mixing of Hunter and Hurler cells with normal cells as well as mixing of Hunter cells with those of a Hurler patient leads to reduction of incorporated $^{35}SO_4$, and the results are similar to those obtained with cells from a normal individual. (Fratantoni et al., 1968 [1086])

Enzyme Deficiencies. The most direct approach to investigating the deficiency in a metabolic disorder is to define the compound whose metabolism is at fault and to measure the activities of the enzymes participating in its turnover. This approach was followed in the analysis of the red cell defects described above. The investigation of the mucopolysaccharidoses was not simple, however, since neither the precise chemical structure of the relevant glycosaminoglycans nor their normal catabolic enzymes were known. A systematic study became possible when fibroblasts cultured from the skin of Hunter or Hurler patients were found to accumulate glycosaminoglycans. The most important step in further analysis was the demonstration that the accumulation could be reduced to normal levels in vitro by a correcting factor from tissue fluids in culture. This was first shown by Neufeld et al. in 1968 [1240]. It could be assumed that type I (Hurler) and type II (Hunter) are genetically different since the inheritance of Hurler's syndrome was autosomally recessive, while Hunter's syndrome was X-linked. Hence, the enzyme defect could be assumed to occur at different points in the pathway of mucopolysaccharide degradation. If it were possible to fuse the nucleus of a cell from a patient with Hurler's syndrome with that of a patient with Hunter's syndrome (as had been accomplished by somatic cell geneticists for many different cell types) complementation of each defect by the other should have resulted. In the process of these experiments, the problem turned out to be much simpler. Cell fusion was not necessary. Hurler and Hunter cells were found to compensate each other when simply mixed in culture or even when cells of one genotype were exposed to culture medium preincubated with the other. The accumulation of mucopolysaccharides was measured by $^{35}SO_4$. Figure 4.10 shows these experiments. Compensation of one defect (Hurler's syndrome) could be achieved by cells with the other defect (Hunter's syndrome) as well as by normal cells [1086].

In the following years, such experiments were carried out with the other clinical types, and the correcting factors were characterized by means of biochemical methods. Fibroblasts from patients with Sanfilippo's syndrome were found to fall in at least two groups, A and B, each deficient in a different factor and thus able to cross-correct each other. Hence, Sanfilippo disease was genetically heterogeneous.

On the other hand, fibroblasts from patients with Scheie's syndrome were deficient in the same factor as Hurler patients - in spite of the wide difference in clinical symptoms.

All corrective factors were soon shown to be specific proteins and upon more thorough analysis were identified as lysosomal enzymes involved in the degradation of sulfated glycosaminoglycans. In the elucidation of these enzyme defects, recent knowledge on the structure of the compounds was combined with analysis of the function of corrective factors and with direct attempts at identifying the enzymes.

The hypothesis to be tested was that the enzymes that were deficient in these disorders were specific for the different types of bonds occurring in the glycosaminoglycans. The predictions derived from this hypothesis were confirmed. For example, when Sanfilippo A (25290) corrective factor was incubated in vitro with ^{35}S sulfate - labeled heparan sulfate isolated from Sanfilippo fibroblasts, a release of inorganic sulfate was observed. Further investigations revealed that the factor was acting on the *N*-sulfate bond of heparan sulfate [1175]. The Hunter corrective factor, when incubated in vitro with ^{35}S-labeled dermatan sulfate or heparan sulfate isolated from Hunter fibroblasts, also catalyzed the release of inorganic sulfate. The Hunter gene is X-linked, and therefore the enzyme defect had to be different from that found in Sanfilippo A disease. Since both glycosaminoglycans have in common occasional sulfated iduronyl residues, it was tentatively assumed that the Hunter correction factor might be a sulfatase. This was confirmed using an artificial substrate.

A different approach for elucidating the specific type of the enzyme block was to determine the nature of the terminal residues in the polysaccharide chains stored in these diseases. In Hunter's disease, for example, the terminal residue of the stored dermatan sulfate was shown to be sulfated iduronic acid. This was expected from the sulfatase defect suggested by the experiments with an artificial substrate. Therefore, the experiment confirmed this defect. Moreover, it suggested that the glycosaminoglycans are normally degraded stepwise and that this degradation is arrested if the enzyme for one step is lacking. As the sequence of the monosaccharides in the chains varies, this stepwise process would explain the varying lengths of the polysaccharide chains stored in these diseases.

In principle, the same methods were used to analyze the other enzyme defects; determination of the terminal residues invariably showed them to contain the bond for which the enzyme was lacking. From the nature of these enzyme defects, another property of the stored material found its explanation: a single defect leads to storage of chemically different materials. For example, Hurler patients accumulate dermatan sulfate as well as heparan sulfate. Both contain α-L-iduronic acid residues. Hence, the defect of an enzyme specifically directed at this residue causes accumulation of both types of polysaccharides containing it.

The result of these combined efforts is seen in Fig. 4.11 for chondroitined dermatan sulfate, Fig. 4.12 for heparan and keratan sulfate. The enzymes for which genetic blocks are known are noted. Table 4.7 shows the enzyme defects.

Consequences for Understanding of Genetic Heterogeneity. In Sect. 3.3.5, it was shown using muscular dystrophy as an example how genetic heterogeneity can be analyzed using genetic evidence - various modes of inheritance - as well as evidence from clinical characteristics such as age of onset, pattern of clinical manifestation, severity of symptoms, and other criteria. In the mucopolysaccharidoses, a striking interfamilial variability of all these indicators is found, whereas within the same family manifestations usually are similar. Hence, a subdivision into various genetic types seemed logical and was in fact carried out before the enzyme defects had been analyzed. How does this subdivision on the basis of "indirect" evidence from the phenotypes compare with the "direct" evidence from analysis of enzyme defects?

By and large, the correspondence is satisfactory (Tables 4.5 and 4.7). However, there are two exceptions:

1. From clinical evidence, Sanfilippo disease would have been regarded as a single entity. Yet, it was shown to consist of four different enzyme defects. This experience has been repeated in the analysis of many metabolic disorders. Different defects

Fig. 4.11, 4.12. In addition to the mucopolysaccharidoses described in the text, some other defects leading to mucolipidoses (Sandhoff's disease; M II gangliodosis) were also indicated. (From Kresse et al., Klin. Wochenschr., p. 870/71, 1981) ▷

Fig. 11. a Schematic representation of structure and catabolism of chondroitin sulfate. Sequential degradation starts from the non-reducing end *(on the left)*. Designation of diseases caused by the inactivity of an enzyme is given in brackets. *GlcUA*, glucuronic acid; *GalNAc*, N-acetylgalactosamine; *S*. SO_3H. **b** Schematic representation of structure and catabolism of dermatan sulfate. for details see legend to Fig. 4.11 a. *IdUA*, iduronic acid

Fig. 4.12. a Schematic representation of structure and catabolism of heparan sulfate. For details see legend to Fig. 4.11 a. *GlcN*, glucosamine; *GlcNAc*, N-acetylglucosamine; *IdUA*, iduronic acid. **b** Schematic representation of structure and catabolism of keratan sulfate. For details see legend to Fig. 4.11 a. *Gal*, galactose; *GlcNAc*, N-acetylglucosamine

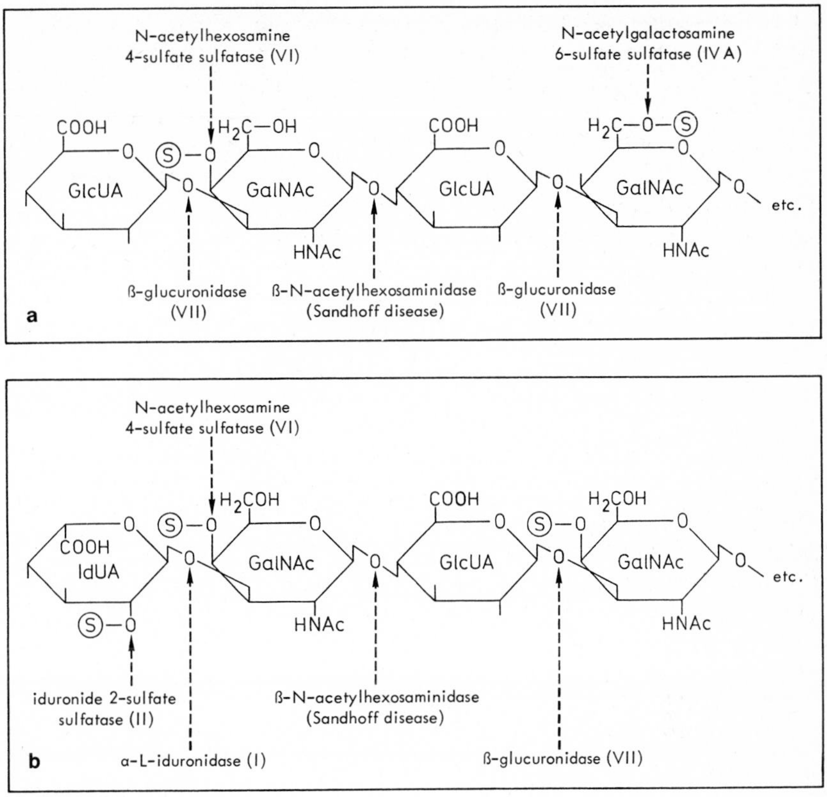

N-acetylhexosamine 4-sulfate sulfatase (VI)

N-acetylgalactosamine 6-sulfate sulfatase (IV A)

COOH
H_2C—OH
COOH
H_2C—O—Ⓢ
O
Ⓢ—O
O
O
O
GlcUA
O
GalNAc
O
GlcUA
O
GalNAc
O
etc.
HNAc
HNAc

ß-glucuronidase (VII)

ß-N-acetylhexosaminidase (Sandhoff disease)

ß-glucuronidase (VII)

a

N-acetylhexosamine 4-sulfate sulfatase (VI)

H_2COH
COOH
H_2COH
O
Ⓢ—O
O
O
Ⓢ—O
O
COOH
IdUA
O
GalNAc
O
GlcUA
O
GalNAc
O
etc.
Ⓢ—O
HNAc
HNAc

iduronide 2-sulfate sulfatase (II)

ß-N-acetylhexosaminidase (Sandhoff disease)

b

a-L-iduronidase (I)

ß-glucuronidase (VII)

Fig. 4.11 a, b

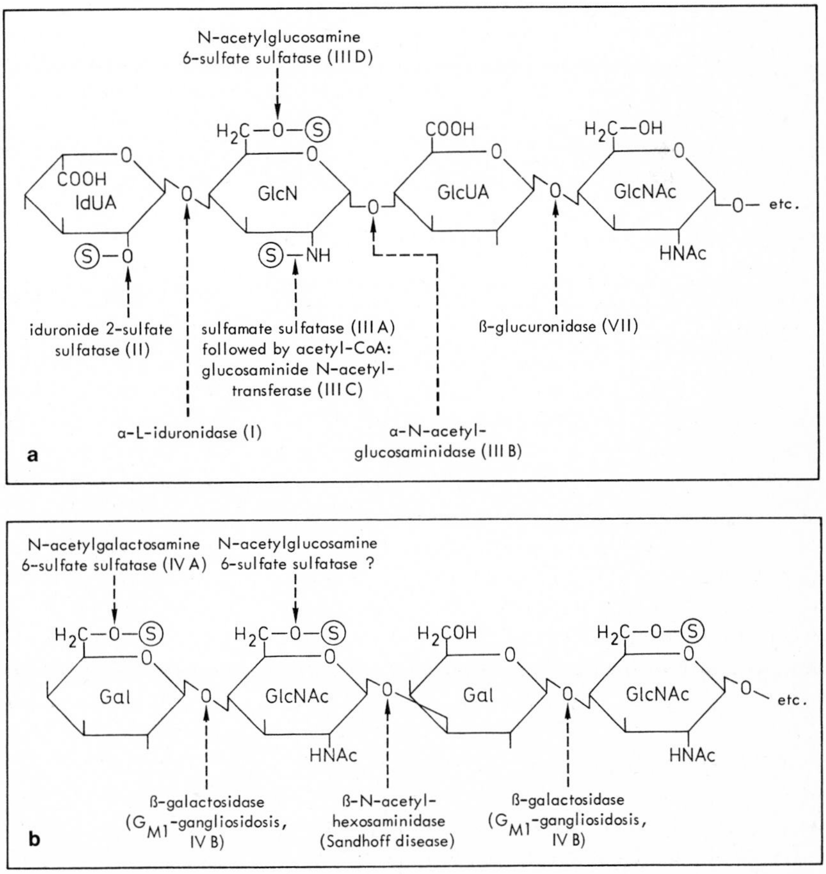

N-acetylglucosamine 6-sulfate sulfatase (III D)

H_2C—O—Ⓢ
COOH
H_2C—OH
O
O
O
O
COOH
IdUA
O
GlcN
O
GlcUA
O
GlcNAc
O— etc.
Ⓢ—O
Ⓢ—NH
HNAc

iduronide 2-sulfate sulfatase (II)

sulfamate sulfatase (III A) followed by acetyl-CoA: glucosaminide N-acetyl-transferase (III C)

ß-glucuronidase (VII)

a-L-iduronidase (I)

a-N-acetyl-glucosaminidase (III B)

a

N-acetylgalactosamine 6-sulfate sulfatase (IV A)

N-acetylglucosamine 6-sulfate sulfatase ?

H_2C—O—Ⓢ
H_2C—O—Ⓢ
H_2COH
H_2C—O—Ⓢ
O
O
O
O
Gal
O
GlcNAc
O
Gal
O
GlcNAc
O
etc.
HNAc
HNAc

ß-galactosidase (G_{M1}-gangliosidosis, IV B)

ß-N-acetyl-hexosaminidase (Sandhoff disease)

ß-galactosidase (G_{M1}-gangliosidosis, IV B)

b

Fig. 4.12 a, b

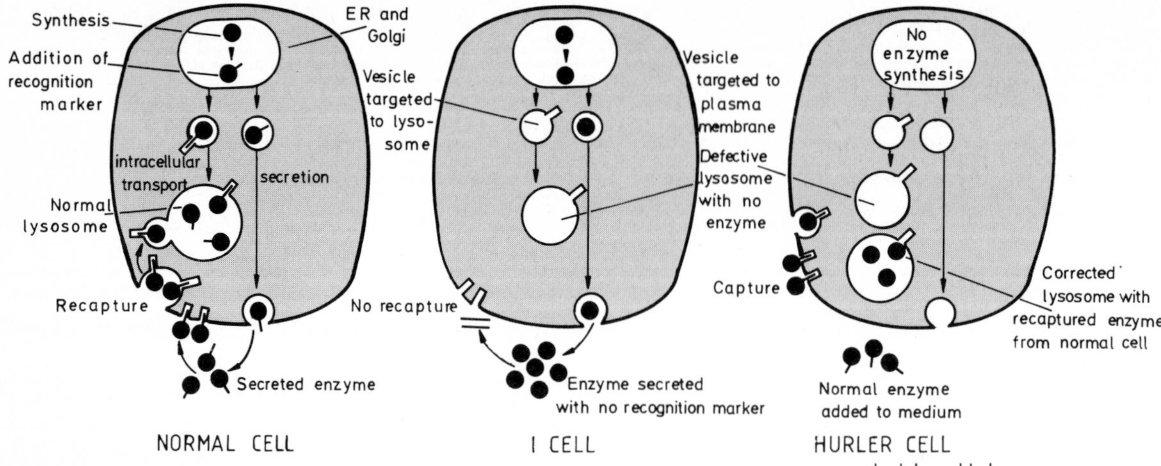

Fig. 4.13. Diagram illustrating secretion and recapture of lysosomal hydrolytic enzymes by normal and mutant cells grown in culture. Specific receptor proteins located at the plasma membranes of all three cells allow them to take up hydrolytic enzymes into their lysosomes. Hurler cells fail to make α-L-iduronidase but the defect can be corrected when the enzyme is added to the medium. In I-cell disease, all enzymes are present but a recognition marker for their uptake and transport to lysosomes is lacking. (From Alberts et al., 1983, modified)

within the same pathways may cause the same clinical picture. Other examples are the erythrocyte defects of glycolysis leading to nonspherocytic hemolytic anemia (Sect. 4.2.2.2).

2. On the other hand, the patterns of manifestation of the Hurler and Scheie types are very different, the course being much milder in Scheie's disease. Yet, the enzyme defect is identical. In analogy with erythrocyte enzyme defects, some explanations come to mind. The mutations may lead to enzymes with different degrees of residual activity, or the enzymes may consist of more than one polypeptide chain and the mutations may affect genes for different chains. So far, no differences in residual activity of α-L-iduronidase between Hurler and Scheie fibroblasts have been shown. Additional explanations are suggested by results of genetic analysis of hemoglobins (Sect. 4.3).

Studies that were successful in eliciting genetic heterogeneity within the gene locus as in G6PD deficiency have not been carried out in the mucopolysaccharidoses. There is no reason to assume that differences between single mutations will turn out to be less striking. The mutation present in double dose in a given patient affected with an autosomal recessive disease may have a common origin if the mutant allele comes from a common ancestor of both parents. This will be the case if the parents are relatives, for example, first cousins, or in genetic isolates (Sect. 6.4.1). In many other instances, however, the mutations found in the affected person will have a different origin. Hence, they will hardly ever be identical. If the term "homozygote" were to be confined to individuals carrying two absolutely identical mutations, many or even most patients with recessive diseases would not conform to the definition. They would be "compound heterozygotes" (Fig. 3.12). A number of patients have been observed with a clinical phenotype intermediate between the Hurler and Scheie types [1312]. Fibroblasts were shown to be devoid of the Hurler correcting factor. Some of these patients may have been indeed, compound heterozygotes for both the Hurler and the Scheie alleles. However, in at least four such families, consanguinity of parents was observed. Since an increase of consanguinity cannot be expected in families, where two different alleles segregate, occurrence of a third, intermediate allele is possible.

Differential Diagnosis and Treatment of Mucopolysaccharidoses. When a mucopolysaccharidosis is suspected on clinical grounds, the putative diagnosis should be checked by the demonstration of an excessive urinary excretion of sulfated glycosaminoglycans. Appropriate methods are now available [1312]. The final diagnosis, however, depends on demonstration of the enzyme defect, which can be studied in cultured fibroblasts, in leukocytes, and, for some enzymes, in the serum [1029]. Prenatal diagnosis is possible, the gene being expressed in amniotic cells (Sect. 9.1.1), either by demonstration of an excessive accumulation of radioactive glycosaminoglycans, or by the finding of an enzyme defect. But since the number of amniotic cells that can be cultured in the available time is restricted, the number of enzyme determinations that can be performed is limited. Therefore, every attempt should be made to arrive at an enzymatic diagnosis in the affected sib before prenatal testing. Then, only the specific enzyme deficient in this sib has to be looked for in the amniotic cells.

The finding that enzymes can be taken up by deficient cells, thus correcting deficiency, appears encouraging for attempts at enzyme therapy. So far, however, purified enzymes have not been available in sufficient amounts, and infusions of large amounts of plasma or leukocytes have led only to dubious, and in any case slight improvements. The uptake of lysosomal enzymes into a cell is a highly specific process in involving the presence of a particular recognition marker on the enzyme protein, which may be different for different tissues [1240]. Still, the approach appears promising.

Defect of a Recognition Marker for Lysosomal Hydrolases [1242]. In 1967, DeMars and Leroy described "remarkable cells" in cultures of skin fibroblasts from a patient thought to have Hurler disease. These cells were filled with acid phosphatase-positive inclusions that appeared dense under phase contrast microscopy. On the basis of the appearance of these inclusions, the disease was named I cell disease. Its clinical phenotype, while resembling the Hurler phenotype, is more severe; the mode of inheritance is autosomal recessive. Fibroblasts from such patients were deficient in β-hexosaminidase, arylsulfatase A, and β-glucuronidase, whereas these enzymes were found in increased concentration in the culture medium. Initially, a membrane defect was suspected; but lysosomes from affected patients were shown to take up and keep normal enzyme at a normal rate. However, hydrolases from I cells could not be endocytosed by normal cells. Therefore, the enzyme molecules themselves were altered. They were found to lack a recognition marker for endocytosis, i.e., mannose-6-phosphate. Normally, mannose-6-phosphate residues are added after synthesis of lysosomal enzymes. The sugar compound acts as a signal allowing lysosomal enzymes to bind to a mannose-6-phosphate receptor that directs lysosomal enzymes into the lysosome, where they are activated. Because of lack of this processing enzyme, most lysosomal enzymes in I cell disease lack mannose-6-phosphate. Instead of being taken up by the cell lysosomes via the specific receptor pathway the enzymes pass through the cell and are secreted into the plasma (Fig. 4.13). The multiple clinical defects in I cell disease can be explained by the single processing defect that fails to add the recognition marker mannose-6-phosphate to these enzymes.

Since the recognition site is common to several enzymes, I cell disease is a condition in which a single gene defect causes more than one enzyme deficiency. Such multiple enzyme effects might also have other reasons.

4.2.2.4 Enzyme Defects Involving More than One Enzyme

In the cases discussed so far, whether affecting energy supply of red blood cells or catabolism of glycosaminoglycans, one gene mutation has led to an alteration or deficiency of only one enzyme. Hence, these examples confirm the one gene one enzyme hypothesis. However, there are cases in which one mutation affects two enzymes. In some of these, activity of one enzyme may be impaired by the deficiency of the other. A possible instance may be the reduction of activity of glucose-6-phosphatase in cases of type III glycogen storage disease (23240), which is due to a deficiency of amylo-1,6-glucosidase, the debrancher enzyme. Steroids with cortisol-like effect normalized the activity of glucose-6-phosphatase in these cases; therefore, a structural defect of this enzyme is not likely [1199].

Other cases in which both enzymes have a structural mutation and enzyme activities in heterozygotes are reduced for both enzymes by about one-half cannot be explained in this way. Here, the most likely explanation is that these enzymes are polymers composed of genetically different polypeptide chains and that the enzymes showing the defect have one polypeptide chain in common.

Maple Syrup Urine Disease (Branched-Chain Ketoaciduria) [196]. One recessive disorder in which no less than three functionally related enzymes are deficient is maple syrup urine disease. This is a defect in degradation of the branched-chain amino acids leucine, isoleucine, and valine (Fig. 4.14). The genetic block is shown in Fig. 4.15. In the most frequent classic type of the disease, feeding difficulties, vomiting, hypertonicity, and a shrill cry are observed in the first week of life. Loss of tone and apnea (perhaps due to hypoglycemia) may intervene. Later, reflexes are lost, the child frequently has seizures, and may die in early infancy. Untreated children who survive suffer from severe mental retardation [182]. Apart from this classic type. "intermittent," "mild," and thiamin-responsive types have been described.

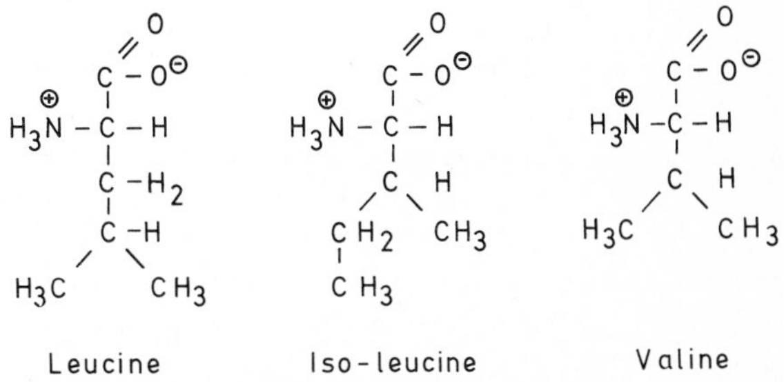

Leucine Iso-leucine Valine

Fig. 4.14. The branched-chain amino acids

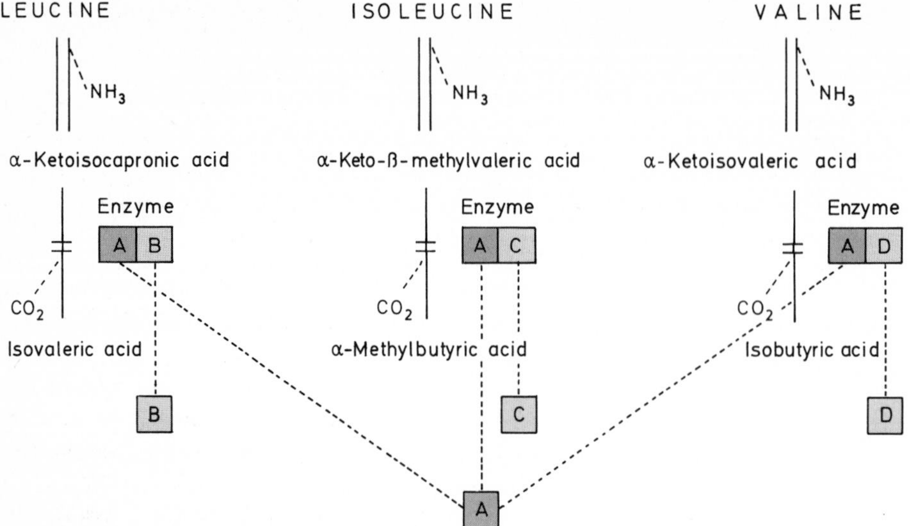

Fig. 4.15. Postulated enzyme systems for metabolism for branched-chain amino acids contain one polpeptide chain as a common component *(A)*. This polypeptide chain may be combined with three different polypeptide chains *(B, C, D)*, forming the three enzymes *(AB, AC,* and *AD)*. Alteration of this common polypeptide chain by mutation leads to genetic blocks in the metabolism of all three amino acids and to branched-chain ketoaciduria. Rare defects affecting polypeptide B and D have been detected

Analysis of the genetic block shows the transaminase activities that transform the three amino acids to the corresponding keto acids to be normal (Fig. 4.15). The anomaly is found in the next step: oxidative decarboxylation. For this step, three different multienzyme complexes seem to be responsible. It appears that one component is identical in these complexes and that this component is altered in the mutation causing this disease. In the intermittent type, episodes with ataxia and increased concentration of branched-chain amino acids and their keto acids are observed, especially during infectious diseases; between these episodes the blood values are normal, and the patients show neurological abnormalities. In one child with this condition, a residual enzyme activity of ~2%–10% was described. For related blocks see [1287].

Other Metabolic Defects Involving More Than One Enzyme [196]. Other metabolic defects in which one genetic block involves two enzymes of one metabolic pathway include orotic aciduria (25890), a deficiency in the formation of uridine, a ribonucleic acid precursor from orotic acid. The two genetic blocks are seen in Fig. 4.16. Enzyme activites in heterozygotes are reduced by about one-half for both enzymes. This rules out a secondary effect of the one enzyme on the activity of the other and argues in favor of one gene being involved in both reactions.

In at least one homozygous case, an altered electrophoretic mobility as well as a higher thermolability of the second enzyme, the decarboxylase, showed that the enzyme defect must be structural. It turned out that each reaction is performed by a discrete domain within the macromolecule, but all the domains are part of a continuous amino acid sequence [1145]. In Sandhoff gangliosidosis, on the other hand, the combined defects of hexosaminidase A and B have been traced back to a mutation affecting their common subunit, the β chain [1286].

A number of patients with simultaneous deficiencied of vitamin K-dependent coagulation factors II, VII, IX, and X have been observed. They are caused by a defect in post-transcriptional modification [1111 b].

A Fresh View on the One Gene-One Enzyme (or One Gene-One Polypeptide) Hypothesis [1088]. As already mentioned, single enzyme defects affecting more than one enzyme can often be traced back in humans to a common subunit in different enzymes. They are structural mutants.

Another important aspect of enzyme formation is post-transcriptional processing. Human sucrose-isomaltase, for example, comprises two polypeptide chains, each having a specific enzyme activity. The two chains are derived from proteolytic cleavage of a single precursor macromolecule [1297]. This proinsulin molecule must also be processed to form functionally active insulin.

More detailed evidence for the processing of peptide molecules is now available for a number of small peptides in the brain – the enkephalins and endorphins. In these cases, precursor molecules are tai-

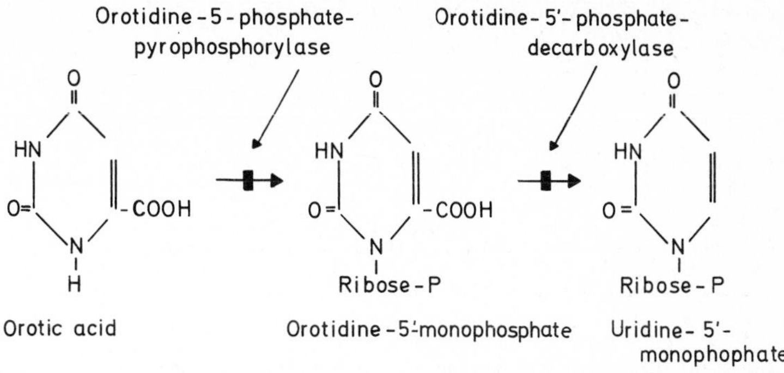

Fig. 4.16. Two genetic blocks in adjacend steps of the synthesis of the pyrimidine base uracil. Both blocks are seen in orotic aciduria

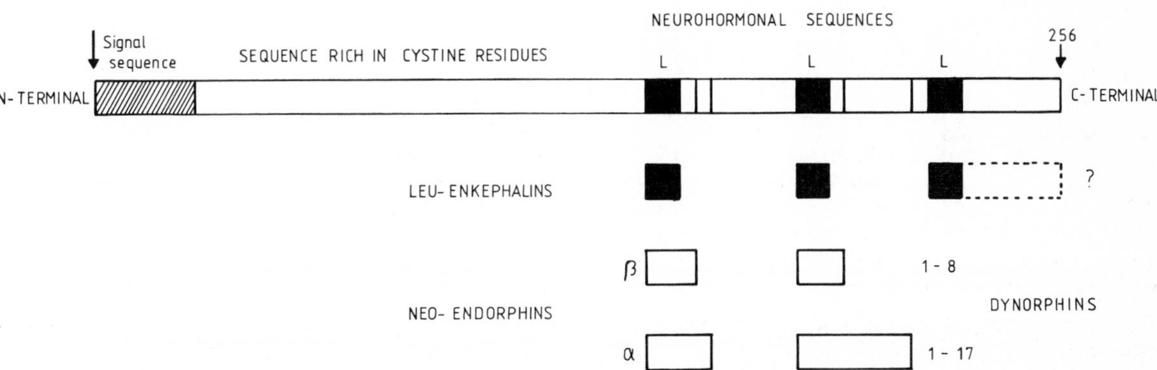

Fig. 4.17. Preprodynorphin (Protein). Leu-enkephalin: L = Tyr, Gly, Gly, Phe, Leu; neo-endorphin: L + Arg, Lys, Tyr, Pro; neo-endorphin: L + Arg, Lys, Tyr, Pro + Lys; dynorphin 1–8: L + Arg, Arg, Ile; dynorphin 1–17: L + Arg, Arg, Ile + Arg, Pro, Lys, Leu, Lys, Trp, Asp, Asn, Gln. The signal sequence at the N-terminal is followed by a sequence rich in cystine residues. The neurohormonal sequences follow close to the C terminal; they are spliced out of the preprodynorphin molecule. Since the endorphins and dynorphins are longer than the Leu-enkephalins, it is tempting to speculate that the splicing occurs in (at least) two steps: In a first step, the longer molecules are formed. Then, they are processed further to form the Leu-enkephalins. (From Frézal et al. 1983; modified)

lored according to cell type and the stage of cell development, one possible mechanism of differentiation during embryonic development. Figure 4.17 shows the preprodynorphin molecule and its processing into various types of enkephalins, endorphins, and dynorphins. The function of these proteins will be mentioned briefly in Sect. 8. Similar post-translational processing has been described for other hormones and neurohormones, especially those from the pituitary and the hypothalamus.

Functionally important modification of molecules occurs not only at the post-translational level, but also at levels closer to gene action; even the gene itself may be modified. Such modification was analyzed in the immune globulin genes (Sect. 4.4). These results do not detract from the heuristic value of the one gene-one *enzyme* hypothesis for most cases.

One special situation where several enzymes may be disturbed functionally by one mutation occurs when uptake, transport, or binding of cofactors is disturbed.

4.2.2.5 Influence of Cofactors on Enzyme Activity [182]

Enzyme Cofactors. Many enzymes need nonprotein molecules as cofactors for their activity. These cofactors may be simple ions, for example, Mg^{++}, or organic, nonprotein components. If the cofactor is a more complex nonprotein component, it is known as a coenzyme. Precursors of coenzymes have to be supplied by nutrition and are traditionally called vitamins. Often, a vitamin acts as a precursor for a coenzyme involved in many enzymatic reactions; nutritional deprivation of vitamins leads to a vitamin deficiency state.

Apart from exogenous deficiency, conditions showing decreased coenzyme function may also be due to genetic defects at various levels of vitamin uptake and utilization (Fig. 4.18). Every vitamin has to be taken up from the intestine, transported into the cells, and taken up into specific cell organelles. It must be converted into the coenzyme that, in turn, must combine with the apoenzyme to form a holoen-

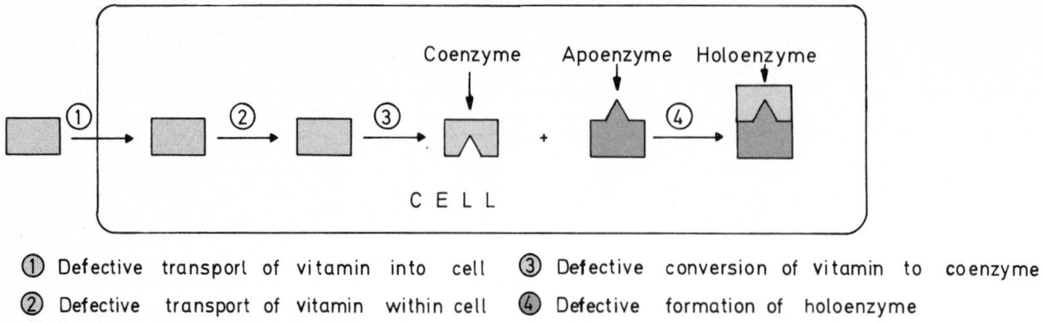

Coenzyme Apoenzyme Holoenzyme

CELL

① Defective transport of vitamin into cell ③ Defective conversion of vitamin to coenzyme

② Defective transport of vitamin within cell ④ Defective formation of holoenzyme

Fig. 4.18. Mutations may interfere with vitamin-catalyzed reactions at several levels, from vitamin transport into the cell to enzyme formation. (Adapted from Scriver and Rosenberg, 1973 [182])

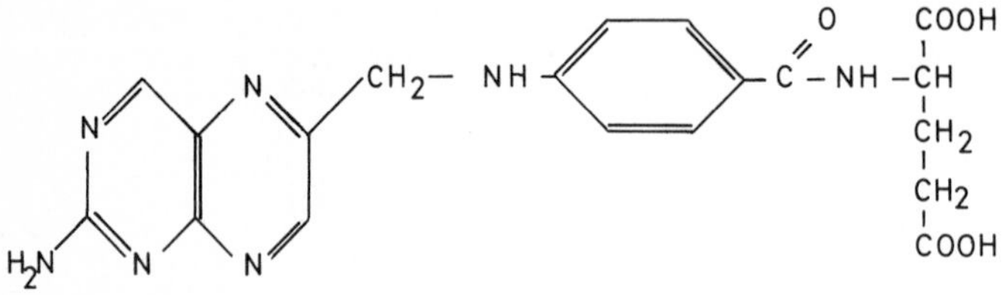

Fig. 4.19. Folic acid. From *left* to *right*, a pterin ring, β-aminobenzoic acid, and glutamic acid

zyme. Any of these steps may be impaired by genetic blocks. The exact mechanism of vitamin uptake is known mainly for vitamin B_{12} (cobalamin) and folic acid; for both substances, transport and coenzyme synthesis defects have been described.

Folic Acid Dependency (22903, 24930, 22905): *Deficiencies in Transport and Coenzyme Formation.* Folic acid is composed of three residues – a pterin ring, *p*-aminobenzoic acid, and glutamic acid (Fig. 4.19). The daily amount needed is normally present in a large number of foodstuffs. There are five coenzyme forms of folate, all of which are concerned with the transfer of 1-carbon units needed for DNA, RNA, methionine, glutamate, and serine synthesis. The principal steps of absorption and coenzyme formation are the following:

Step	Enzyme
1. Conversion of polyglutamyl to monoglutamyl form	Conjugase enzyme (intestinal mucosa; stomach; pancreas)
2. Absorption by active transport	Duodenum and jejunum (exact mechanisms?)
3. Transport to the tissues	
4. Conversion of folate to the coenzymes:	
a) Reduction of the pterin ring: formation of tetrahydrofolate	
b) Conversion into five different coenzymes	Five different enzyme reactions

Tetrahydrofolate (THF) is used for two different functions:

1. It serves as acceptor of the β-carbon atom of serine when it is cleaved to glycine. This carbon atom forms a methylene bridge between *N*-atoms 5 and 10 of THF to yield N^5, N^{10}-methyl THF, which is reduced to N^5-methyl THF.
2. Alternatively, THF may be converted to N^5, N^{10}-methyl THF, the precursor of the formyl coenzyme form of THF. These formyl or methyl coenzymes are required for a number of 1-carbon transfer reactions in purine and pyrimidine synthesis, methionine synthesis, and cyclic interconversion of the folate compounds themselves.

At least five inherited conditions have been described in which transport mechanisms or mechanisms of conversion into coenzymes are deficient (Table 4.8). In four of them, signs of marked central nervous system dysfunction including mental retardation were observed, and two showed megaloblastic anemia. Their most important feature is, however, that they can be treated succesfully provided that the condition has been diagnosed in time. In the intestinal absorption defect, for example, the folate requirement is normal; intramuscular instead of oral administration restores normality. In three of the four enzyme defects, increased amounts of folic acid intake improved the metabolic situation, but the diagnosis had been made too late to know whether the central nervous system improvement could have been prevented had folic acid been supplemented early enough.

Table 4.8. Inborn errors of folic acid metabolism. (Scriver and Rosenberg, 1973 [182])

Phase of metabolism affected	Nature of defect	Manifestation of defect			Folate requirement in vivo
		Serum folate concentration	Megaloblastic anemia	CNS dysfunction[a]	
Intestinal absorption	Undefined	Low	Yes	Yes	Normal
Tissue utilization	Formiminotransferase deficiency	High	No	Yes	Increased
	Cyclohydrolase deficiency	High	No	Yes	NR[b]
	Dihydrofolate reductase deficiency	Normal	Yes	No	Increased
	N^5,N^{10}-methyltetra-hydrofolate reductase deficiency	Low-normal	No	Yes	Increased

[a] Includes mental retardation, psychotic behavior, seizures, EEG abnormalities, cerebral cortical atrophy.
[b] NR, not reported.

Table 4.9. Inherited aminoacidopathies demonstrating vitamin B_6 deficiency. (Scriver and Rosenberg, 1973 [182])

Disorder	Main clinical symptoms	Apoenzyme affected
Infantile convulsions	Seizures	Glutamic acid decarboxylase
Pyridoxine-responsive anemia	Microcytic hypochromic anemia	Undefined
Cystathioninuria	Probably none	Cystathionase
Xanthurenic aciduria	Mental retardation (?)	Kynureninase
Homocystinuria	Ectopia lentis; thrombotic vascular disease; CNS dysfunction	Cystathionine synthase
Hyperoxaluria	Calcium oxalate nephrolithiasis; renal insufficiency	Glyoxylate: α-ketoglutarate carboligase

Deficiencies in coenzyme formation or uptake are expected to affect more than one enzyme at a time – in fact, all enzymes on which this coenzyme acts. On the other hand, deficiencies in the last step, the ability of the protein apoenzyme to form a holoenzyme by combining with the coenzyme, will affect only one enzyme. They are more similar to the usual enzyme defects described in the foregoing sections.

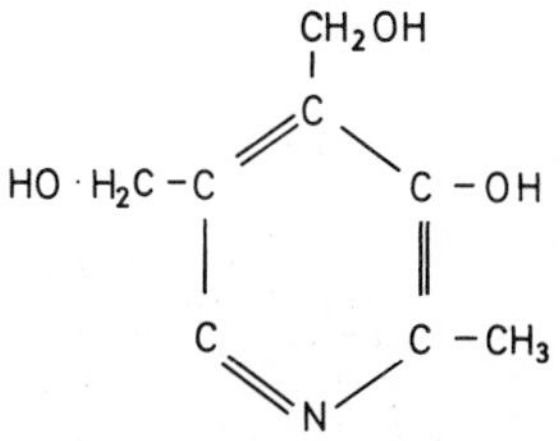

Fig. 4.20. Vitamin B_6 (Pyridoxine)

Pyridoxine (Vitamin B_6) Dependency (26610). Vitamin B_6 is a substituted pyridine ring that occurs in several natural forms in widely varying foodstuffs (Fig. 4.20). In the cell, these precursors are phosphorylated to pyridoxal-5′-phosphate or pyridoxamine-5′-phosphate by a specific kinase. These phosphorylated compounds act as coenzymes for a great many apoenzymes, which regulate the catabolism of amino acids, glycogen, and short-chain fatty acids. Six genetic disorders due to vitamin B_6 dependency are listed in Table 4.9. All of them require 5–50 times the physiologic doses of vitamin B_6 for biochemical and/or clinical improvement.

The initial cases of the condition listed at the top of Table 4.9 were two sibs who suffered from seizures not responsive to antiepileptic drugs. However, these children improved immediately after parenteral treatment with high doses of pyridoxine and needed a maintenance dose to keep them healthy [182]. In the meantime, a fair number of additional cases have been described, and the apoenzyme involved has been identified as glutamate decarboxylase.
Analogous results have been reported in cystathioninuria and xanthurenic aciduria. On the other hand, a mutation in coenzyme affinity does not explain the vitamin B_6 responsiveness in many patients with

homocystinuria (23620) due to cystathionine synthase deficiency; here, the exact type of interaction is still unexplored [182].

Apart from those metioned above, there are a number of other case reports in which high doses of a vitamin have improved the clinical and biochemical condition of patients. Further analysis of this disease group will be interesting theoretically, as it offers new insight into the mechanism of coenzyme binding and action. It is rewarding for medical practice, as well, since these conditions will respond most favorably to therapy with high doses of vitamins.

This general concept, which is well-founded for a few very rare inborn errors of metabolism, has been claimed to apply in a variety of more common diseases such as schizophrenia. An entirely new therapeutic concept known as "orthomolecular psychiatry" claims that schizophrenia is caused by this type of vitamin deficiency and responds favorably to high doses of niacin. Unfortunately, the scientific foundations for these claims are largely lacking. On the other hand, milder deficiencies of this sort may exist as genetic polymorphisms and may contribute to the etiology of common diseases.

4.2.2.6 X-Linked HPRT Deficiencies (30800) [1053]

Enzyme Defects as Tools for Some Basic Questions on Gene Action and Mutation. Some enzyme defects have proved useful as analytical tools for attacking problems of more general significance for our understanding of gene action and mutation. One group of defects has been utilized especially successfully: The hypoxanthine-guanine phosphoribosyl transferase (HPRT) deficiencies that are classified with the defects of purine metabolism (Fig. 4.21) [1294].

Lesch-Nyhan Syndrome [1293]. Lesch and Nyhan (1964) [1198] described a peculiar syndrome with athetosis, hyperreflexia, and a compulsive self-destructive behavior involving the chewing-off of lips or fingers [1164]. All patients show hyperuricemia and excessive excretion of uric acid in the urine, which may lead to uric acid nephrolithiasis with obstructive uropathy. Only boys are affected; the mode of inheritance is X-linked (Sect. 3.1.4). Heterozygotes can be identified but are not affected.

In 1967, Seegmiller et al. [1295] discovered almost complete deficiency of one enzyme of purine metabolism, hypoxanthine-guanine phosphoribosyl transferase (HPRT) in erythrocyte lysates from three patients and in cultured fibroblasts from another. The enzyme defect was then confirmed in other tissues such as liver, leukocytes, and brain. Different pathways lead to inosine-5-monophosphate in several steps. However, cells can also use preformed purine bases and nucleosides produced during the breakdown of nucleic acids. This „salvage" pathway involves the conversion of the free purine bases to their corresponding 5′-mononucleotides. There are two enzymes involved, one specific for hypoxanthine and guanine, the other for adenine (Fig. 4.21). When the first enzyme is deficient, hypoxanthine and guanine are not recycled but are converted into uric acid in large amounts. This leads to hpyeruricemia with kidney stone formation; it is, however, not known by which mechanism the CNS symptoms are produced. This enzyme defect, which can readily be examined in cultured fibroblasts, has been used to investigate a number of problems.

Molecular Heterogeneity. As evidenced by such usual characteristics as residual activity, Michaelis-Menten constant, thermolability, end-product inhibition by GMP or IMP, and others, the mutations observed in the various families are usually different. Sometimes, severe HPRT deficiency was observed even without the clinical symptoms of the Lesch-Nyhan syndrome. Moreover, partial HPRT deficiencies have been observed in some adult patients suffering from gout [1162, 1294]. The great majority of gout patients, however, have normal HPRT. The minority in which the abnormal enzyme is present show – like those with the severe defect – an X-linked mode of inheritance. This is additional evidence that the same locus is, indeed, involved.

Evidence for X-Inactivation. Some of the best confirming evidence for the Lyon hypothesis (Sect. 2.2.3.3) comes from analysis of enzyme activity in HPRT heterozygotes at the cell level [1164]. Moreover, these studies have provided some new insight into metabolic inter-relationships between the cells of an individual.

Metabolic Co-operation. Heterozygous carriers can be identified by studies on cultured skin fibroblasts. When fibroblasts are cloned and enzyme activity is measured by autoradiography of the uptake of ^{3}H-labeled hypoxanthine into the cell, about half the clones will show full HPRT activity, whereas the other half is deficient for this enzyme. However, in fibroblast cultures without cloning, the great majority of heterozygote cells showed some enzyme activity. Apparently, HPRT-deficient cells were corrected for their metabolic defect when they were in contact with normal cells [1090]. This was confirmed by artificial mixing of normal and defective cells in culture and is called „metabolic co-operation."

Three mechanisms for this co-operation are possible (Fig. 4.22):

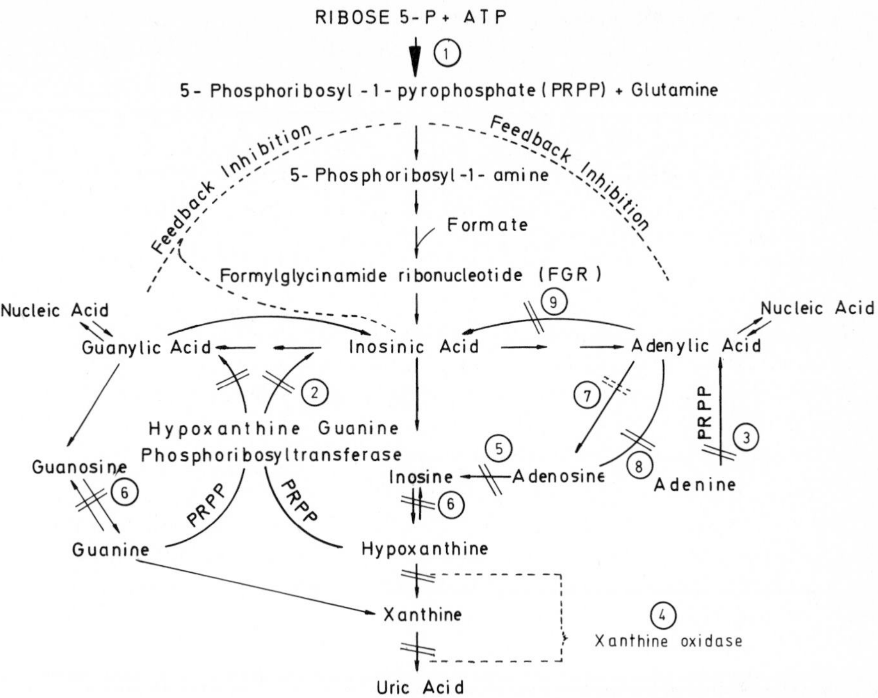

Fig. 4.21. Known enzyme defects in human purine metabolism. *(1)* Increased phosphoribosylpyrophosphate synthetase activity in patients with overproduction of uric acid leading to gout. *(2)* Gross deficiency of hypoxanthine-guanine phosphoribosyl transferase (HPRT) in children with Lesch-Nyhan disease and partial deficiency of the same enzyme in patients with overproduction of uric acid and gout. *(3)* Adenine phosphoribosyl transferase (APRT) deficiency in patients with kidney stones composed of 2-8 dioxyadenine that are often confused with uric acid stones. *(4)* Xanthine oxidase deficiency in patients with xanthinuria who are at increased risk for xanthine urinary calculi and, occasionally, for myalgia caused by xanthine crystals in the muscle. *(5)* Adenosine deaminase deficiency associated with a severe combined immunodeficiency state. *(6)* Purine nucleoside phosphorylase deficiency associated with isolated defect in T cells. *(7)* Purine 5'-nucleosidase activity is low in lymphocytes of patients with agammaglobulinemia that may be secondary to loss of B-cells. *(8)* Adenosine kinase deficiency has so far been developed only in the human lymphoblast cell lines. Its counterpart in patients is yet to be identified. *(9)* Myoadenylate deaminase deficiency is associated in some patients with development of weakness and muscle cramps after vigorous exercise and failure to show a rise in venous blood ammonia in response to muscle exercise. (From Seegmiller, 1983 [1294]). These enzyme defects provide excellent examples for the various, more and often less characteristic phenotypic consequences of different genetic blocks within the same functional network

1. Normal cells provide deficient cells with DNA or mRNA, thus enabling them to synthesize a functional enzyme protein.
2. The deficient cells receive preformed enzyme. This possibility is suggested by the observation of correcting factors in mucopolysaccharidoses (Sect. 4.2.2.3). Incubation of cultured. HPRT-deficient fibroblasts with ultrasound-minced material from normal cells show partial restoration of enzyme function.
3. Normal cells synthesize the nucleotide (the end product), which is transferred to the deficient cell. This mechanism is supported by the most evidence. After deficient fibroblasts are separated from normal cells they revert promptly to the mutant phenotype, although normal HPRT is stable for many hours under the experimental conditions. In another experiment, lymphocytes from a

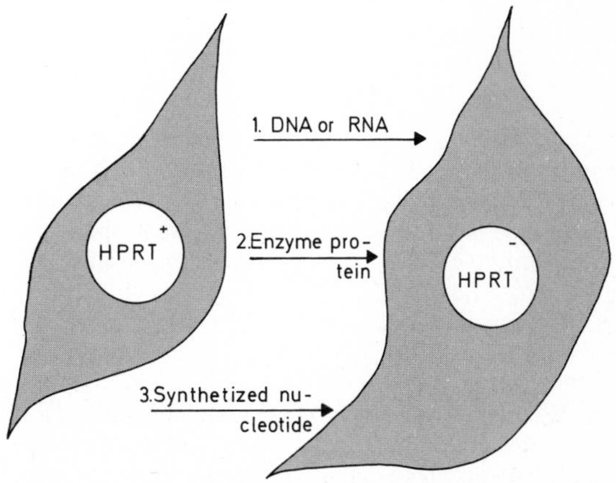

Fig. 4.22. Possibilities for metabolic co-operation between HPRT-active and HPRT-deficient cells in cell cultures from heterozygotes. See text for details

normal female were incubated with a ^{3}H hypoxanthine-containing medium. These cells were then mixed with untreated isolated lymphocytes from a male Lesch-Nyhan patient. After incubation, but before autoradiography, the male, HPRT-deficient cells were typed by staining the Y chromosome with quinacrine (Sect. 2.1.2.3). These cells showed labeling after autoradiography, indicating that the deficient cells had received nucleotides from normal cells. In a similar experiment, metabolic co-operation was also demonstrated between preincubated normal red cells and mutant lymphocytes and fibroblasts. The material transferred to the deficient cell appears to be inosine monophosphate or one of its derivatives; the cell membrane seems to play an active part in the transfer [1053].

Other Problems Examined with HPRT Deficiency. HPRT deficiency has proved to be a useful tool for investigations of the mutational process:

a) The possibility of identifying all hemizygotes and heterozygotes by enzyme determinations in the fibroblasts makes HPRT deficiency especially suited for comparison of spontaneous mutation rates in the two sexes (Sect. 5.15).
b) The gene is also expressed in amniotic cells. Therefore, HPRT deficiency can be diagnosed by amniocentesis – quite in contrast to other X-linked conditions such as hemophilia or Duchenne-type muscular dystrophy, where the biochemical defect is not expressed in amniotic cells.
c) Using an abnormal substrate, 8-azaguanine, a selective system for point mutations in fibroblast cultures has been developed that permits the examination of problems of spontaneous and induced mutations at the cellular level. Normal cells will metabolize 8-azaguanine by HPRT and will be killed. HPRT-deficient cells cannot metabolize this compound and survive.

Immune Deficiency Diseases Associated with Adenosine Deaminase and Nucleoside Phosphorylase Defects (Figure 4.21). Another defect of an enzyme, which is involved in nucleoside metabolism, leads to a different phenotype. This enzyme defect is all the more interesting because it involves a rare variant of an enzyme for which a genetic polymorphism is also known. Defects in one or more components of the immune system can result in increased susceptibility to microbial infections. The classic disease in this category is X-linked hypogammaglobulinemia, which is caused by a maturation defect of B-lymphocytes [1092, 1261]. B-lymphocytes are the production sites of humoral antibodies, and their absence causes failure of γ-globulin synthesis. T-lymphocytes are involved in cellular immunity and are intact in this disease.

Among the various immune deficiency diseases, one category has been classified as severe combined immune deficiency disease with increased susceptibility to infection with a broad spectrum of bacteria, viruses, and fungi. In many cases, there is defective function of both B- and T-lymphocytes. Other cases (see below) only have T cell dysfunction. The basic defect appears to be failure of differentiation of stem cells into more mature lymphoid cells [1274, 1267]. This group of diseases has a heterogeneous etiology since both X-linked and autosomal recessive cases have been identified. Further heterogeneity has been demonstrated in the autosomal recessive cases. Enzyme defects related to metabolism of purines, pyrimidines, and nucleic acids have been identified. Thus, combined immune deficiency can be caused by deficiency of adenosine deaminase (24275) or of nucleoside phosphorylase (16405) [1294].

Adenosine deaminase (ADA) catalyzes the irreversible deamination and hydrolysis of the purine nucleoside adenosine to inosine and ammonia. Nucleoside phosphorylase catalyzes the conversion of the purine nucleoside inosine to hypoxanthine and of guanosine to guanine. It has little activity in converting adenosine to adenine. These enzymes are key enzymes in DNA and RNA synthesis and breakdown.

Adenosine deaminase – located on chromosome 20 (Sect. 3.4) – exists as a polymorphic trait in the population as demonstrated by starch gel electrophoresis. The common allele is known as ADA^1 and a common polymorphic allele as ADA^2. ADA^2 has a gene frequency of about 0.05 in Western populations [1309]. Several other rare variants of ADA have been described [1294]. ADA deficiency is an autosomal recessive trait. Affected children lack any ADA activity in their red cells and other tissues [1098]. Their parents usually have intermediate amounts of the enzyme and are clinically normal. Some residual ADA activity can be demonstrated in affected patients [1294]. The defects appear to be a structural mutation affecting the ADA allele, which leads to almost complete loss of functional activity of the enzyme in affected homozygotes. Prenatal diagnosis of ADA deficiency has been achieved [1126].

Nucleoside phosphorylase (NP) is specified by a locus on chromosome 14 [1096]. No polymorphisms exist at this locus, but several rare variants have been demonstrated [1064]. In affected patients, there is complete absence of nucleoside phosphorylase activity while parents have the expected intermediate activity [1294]; inheritance is autosomal recessive. Using immunologic techniques, it has been demonstrated that at least two mutations – one with CRM and without CRM – (see Sect. 4.2.2.2) cause nucleoside phosphorylase deficiency [1256]. The defects presumably are structural mutations of the enzyme. Affected patients, as expected, may be compound heterozygotes for two mutations rather than true homozygotes.

Patients with ADA deficiency usually have severe B and T cell dysfunction [1294], while in NP deficiency B cell function and therefore immunoglobulin production appears to be intact. T-lymphocyte dysfunction as manifest by lymphopenia, failure of lymphocytes to respond to mitogens, and abnormal skin tests to a variety of antigen is striking in both ADA and NP deficiency.

The exact biochemical mechanism by which these enzyme deficiencies produce immunologic abnormalities is unknown. It has been speculated that ADA deficiency produces an accumulation of deoxy ATP, which inhibits pro-

duction of pyrimidine deoxyribonucleotides and therefore interferes with DNA synthesis, lymphocyte proliferation, and immune response. The mechanisms for NP deficiency may be similar [1294].
The treatment of choice for these defects appears to be bone marrow transplantation. The clinical manifestation of ADA deficiency in some cases has been improved by infusion of normal red cells as a source of the missing enzyme [1266].

4.2.2.7 *Phenylketonuria: Paradigm for Successful Treatment of a Metabolic Disease* [182; 203]

Metabolic Oligophrenia. Phenylketonuria (PKU) (26160) was first described by Følling (1934) [1080] in mentally retarded patients with a peculiar, "mousy" odor. The name was coined by Penrose (1935) [1262]. It is now one of the best-known inborn errors of metabolism in humans; the various aspects of this disease have repeatedly been reviewed [203, 1290; 1287]. Here, only three main aspects will be discussed: the introduction of a low-phenylalanine diet as the first successful approach to alter the phenotype of a genetic enzyme defect by suitable manipulation of the environment; the genetic heterogeneity of the condition as revealed by mass screening of newborns; and the problem of heterozygote detection and of possible phenotypic abnormalities in heterozygotes.

Enzyme Defect in PKU. L-Phenylalanine is an essential amino acid. However, only a fraction of the normal intake of L-phenylalanine can be utilized for protein synthesis; the major part is oxidized primarily to tyrosine but – to a much smaller extent – also to other metabolites, primarily phenylpyruvic acid. The parahydroxylation of phenylalanine to form tyrosine is a complex reaction. The hydroxylase consists of two protein components, one of which is labile and found only in the liver (and possibly, with lower activity, in the kidneys), the other being stable and found in many other tissues. This stable component contains a pteridine as cofactor.
PKU was shown to be caused by complete deficiency of hepatic phenylalanine hydroxylase (Fig. 4.23) (Udenfriend and Cooper, 1952 [1331]; Jervis, 1953 [1144]. The labile component of the enzyme system is affected. In more recent enzyme assays from liver biopsies, small residual activity (up to ≈6% of normal) is found in about half the cases with classic PKU. Other types of hyperphenylalaninemia with higher enzyme activities will be discussed below. The hydroxylase reaction is one step of the metabolic pathways of phenylalanine and tyrosine for which a number of genetic blocks are known (Fig. 4.24). The gene is located on chromosome 12 [1363a].

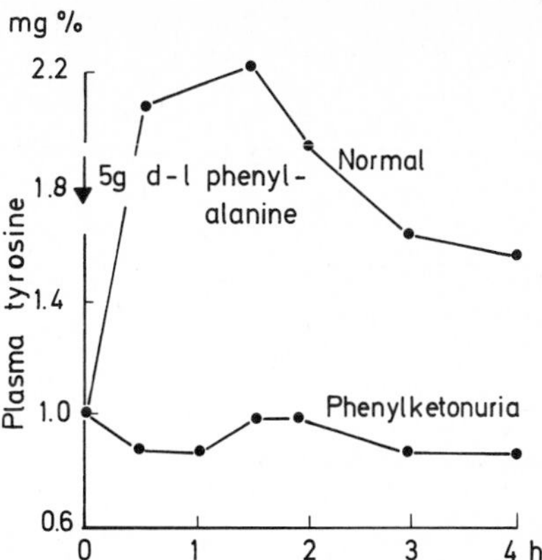

Fig. 4.23. Genetic block in phenylketonuria: Ingestion of 5 g d-L-phenylalanine leads to an increase of serum tyrosine in the normal but not in the phenylketonuric individual. (Adapted from Harris, 1959)

Dietary Treatment of PKU. Phenotypic damage due to a genetic block may be caused either by lack of a metabolite beyond the block or by accumulation of a metabolite ahead of the block. Examples for lack of a metabolite behind the block are albinism and cretinism with goiter (Fig. 4.24). In PKU, it soon became obvious that the abnormalities could hardly be due to tyrosine deficiency: tyrosine is usually available in sufficient amounts in food. On the other hand, the numerous metabolites found in the urine of PKU patients, along with a high increase in serum phenylalanine levels, indicated the opening of additional overflow pathways. This suggested that therapy consisting of reducing phenylalanine intake might be successful. Such an attempt was first made by Bickel et al. (1953) [1004]:

> On the assumption that the excessive concentration of phenylalanine (or perhaps of some break-down products) is responsible for the mental retardation found in this condition we decided to keep a girl, aged 2 years, with phenylketonuria on a diet low in phenylalanine. She was an idiot and unable to stand, walk, or talk; she showed no interest in her food or surroundings, and spent the time groaning, crying, and banging her head. The diet hat to be specially prepared, because a sufficiently low phenylalanine intake could only be attained by restricting practically all the nitrogen intake to a special casein (acid) hydrolysate ... This was treated with activated acid-washed charcoal, which removed phenylalanine and tyrosine. Tyrosine, tryptophane, and cystine were then added in suitable amounts ....

The child was at first treated in a hospital, so that careful observations could be made. During a 4-week preliminary period, when no phenylalanine was permitted, no definitive clinical change other than loss of weight was observed. The characteristic musty smell disappeared, the levels of phenylalanine

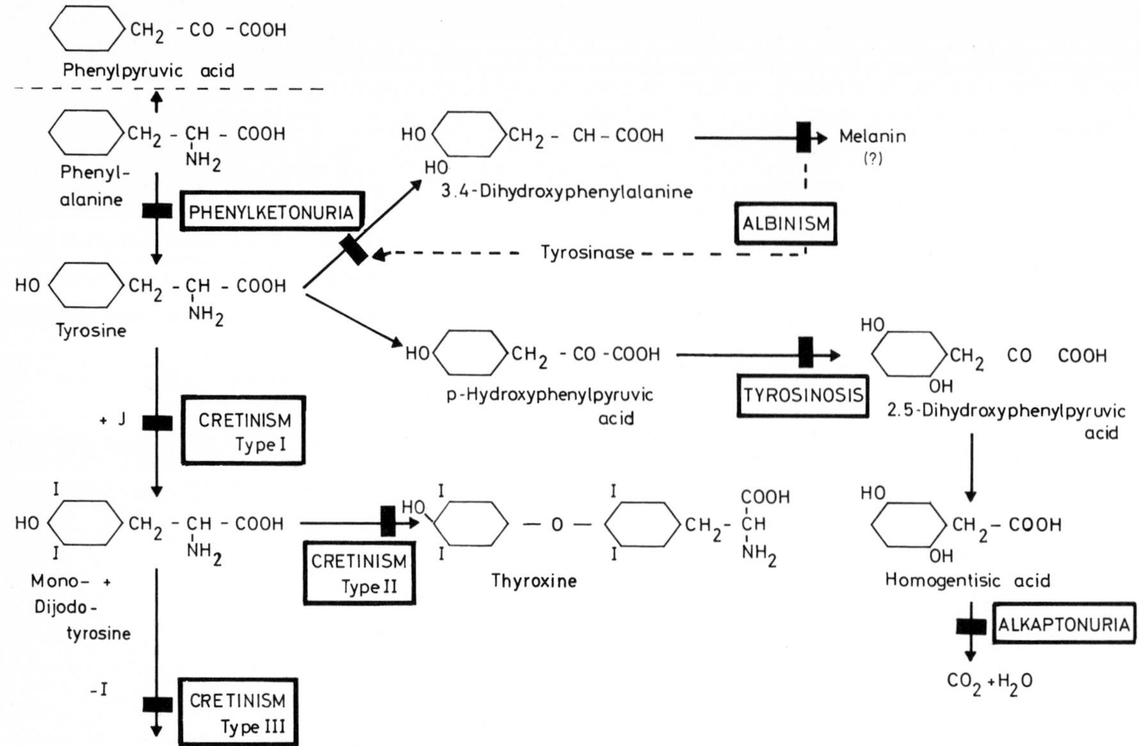

Fig. 4.24. Genetic blocks in metabolic pathways of some aromatic amino acids. The diagram is somewhat oversimplified. Genetic block leading to phenylketonuria, albinism, alcaptonuria, tyrosinosis, and three types of hereditary cretinism are included

in plasma and urine fell to normal, the excretion of phenylpyruvic acid decreased, and the ferric chloride reaction in the urine, indicating phenylpyruvic acid became negative (Fig. 4.25).

Subsequently, presumably as the result of tissue breakdown, the biochemical abnormalities returned to some extent, along with a generalized aminoaciduria. Phenylalanine was therefore added in small amounts in the form of whole milk, a daily intake of 0.3–0.5 g being found sufficient for normal weight gain, with greatly improved biochemical findings.

During continued outpatient treatment, a gradual improvement in the child's mental state took place within the next few months: She learned to crawl, her eyes became brighter, her hair grew darker, and she no longer banged her head or cried continuously.

To find out whether this improvement was real, an experiment was undertaken by adding 4 g phenylalanine/day to the diet. This resulted in a definite deterioration in the child's condition, which was soon reported by the mother. The study was repeated in the hospital under controlled conditions (Fig. 4.25) and again led to the expected biochemical and clinical changes. With this case, the beneficial effects of a low-phenylalanine diet were demonstrated. In the same paper, the authors stated that "... further controlled trials are being made, special attention being paid to the very young children, who are likely to benefit most."

The success of the dietary treatment was soon confirmed by other groups. Some doubts were raised, on the one hand by observations that some PKU patients do not become mentally defective even without any treatment, and on the other hand by failure of treatment in some cases. In spite of these complications, the evidence is now quite substantial that the diet does, indeed, have marked beneficial effect on the development of the PKU patient. There are, however, two qualifications:

1. To prevent brain damage, the diet should be started as soon as possible within the first few weeks of life.
2. The metabolic status of the child, and especially phenylalanine levels, should be monitored carefully.

Lifelong treatment is not necessary, since the adult brain appears to be resistant to the abnormal metabolite concentration found in PKU.

A number of women with treated PKU have had children. In spite of the fact that these children are only heterozygous, about 90% of them showed signs of severe mental retardation [1196]. Hence, intrauter-

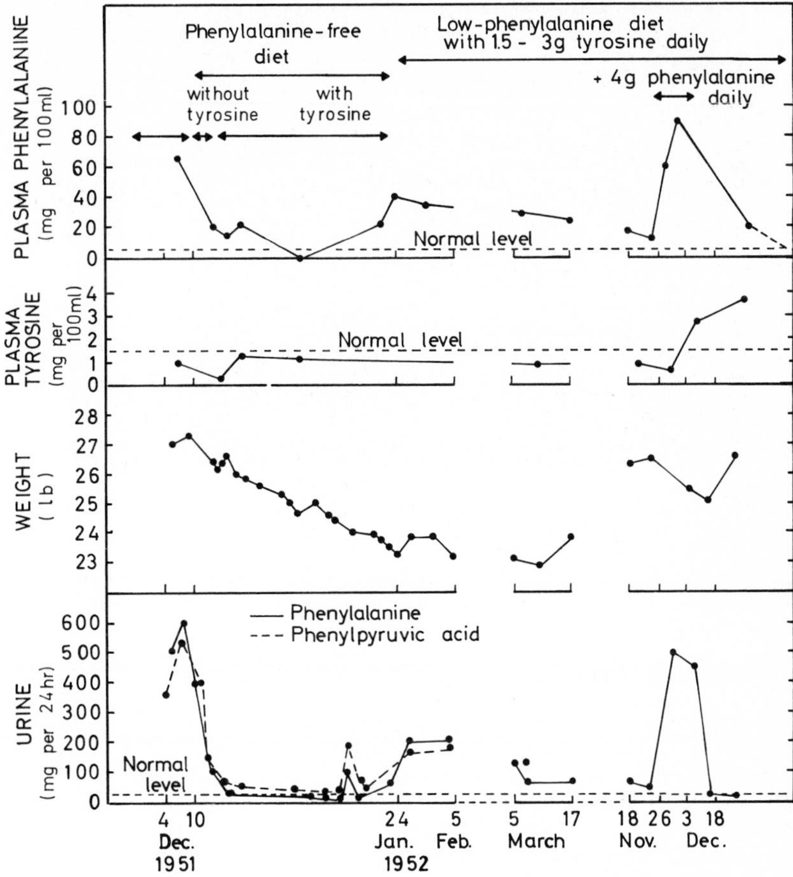

Fig. 4.25. First metabolic studies in a phenylketonuric child treated with a phenylalanine-free diet. Serum and urine phenylalanine were reduced quickly to normal values. Administration of 4 g phenylalanine per day led to a quick increase of phenylalanine level. (Bickel, 1954; Exp. Med. Surg. 12. 114–118)

ine hyperphenylalaninemia is harmful for the development of the fetus. A carefully controlled low-phenylalanine diet for all PKU patients right from the beginning of pregnancy appears mandatory to prevent this complication. This is a serious public health problem since careful search for the rare, previously treated PKU patient becomes necessary to avoid certain mental retardation in their offspring. Similar problems may arise with other treatable inborn errors of metabolism.

Genetic Heterogeneity of PKU. The possibility of successful treatment in early infancy before the appearance of clinical symptoms has led to the introduction of screening programs for newborns. In most Western industrialized countries, where PKU has a frequency ranging from 1:6000–1:20,000, practically all newborns are now being screened. A screening method should be easy and inexpensive. In most countries, the so-called Guthrie test is used [1116], which is based on the growth of phenylalanine requiring bacteria on a spot of blood containing phenylalanine above a certein threshold concentration (Fig. 4.26). Thus, only blood from infants with high phenylalanine levels will support growth of these bacteria.

When systematic newborn screening began, it was soon apparent that not every infant in which an unusually high blood phenylalanine level was found had PKU. Many had a less severe hyperphenylalaninemia not leading to clinical symptoms. In another group of children, the high blood level persisted for only a short time and normalized spontaneously. A whole range of biochemical – and genetic – heterogeneity was revealed. It is not known how many different types of mutations with varying effects on the phenotype exist. However, there is no a priori reason why the range of variability should be smaller than, for example, in the G6PD variants (Sect. 4.2.2.2). The results depend on the level of sophistication of the analysis. In a rare type, not phenylalanine hydroxylase but dihydropteridine reductase, an enzyme necessary for keeping the hydroxylase active, is deficient.

As in other enzyme defects, some or even many of the phenotypes may be due to compound heterozy-

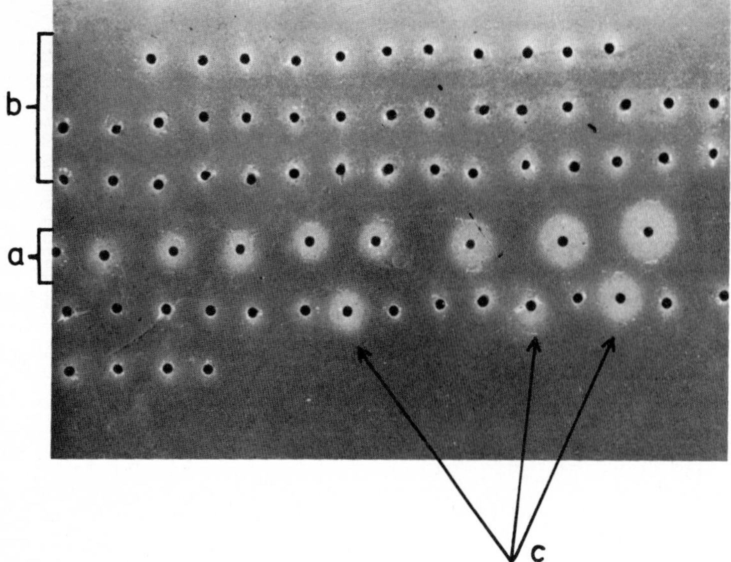

Fig. 4.26. A strain of bacillus subtilis requiring phenylalanine for growth is incubated in an agar plate. The strain is able to grow only if the phenylalanine level of the test blood is increased. This leads to a bacterial growth around the blood spot. The diameter of the growth area is related directly to blood phenylalanine level. *a*, standard phenylalanine concentrations from left to right; *b*, normal specimens; *c*, abnormal specimens between 6 and 12 mg/%

gosity for mutations with different molecular characteristics. From a practical point of view, it is important to note that a residual enzyme activity of ~10% seems to be sufficient for normal somatic and mental development without a special diet.

Just as for the G6PD variants, a more exact disentangling of the phenylalanine hydroxylase variants and the phenotypic consequences of their combination in homozygotes (or double heterozygotes) requires enzyme characterization in single heterozygotes. For the G6PD variants, this has been relatively easy technically, as the enzyme is present in blood cells and blood is easily available. Phenylalanine hydroxylase, on the other hand, is present only in liver cells so that needle biopsy of the liver is required. This procedure has a small morbidity and low mortality and therefore is not usually carried out for investigative purposes. Techniques may be found that may allow the "turning-on" of the gene for phenylketonuria in fibroblasts or lymphocytes so that the required characterization studies may be done in the future.

These screening programs, apart from their success in detection of PKU patients, have provided interesting information on the general frequency of PKU and other hyperphenylalaninemia genes in various human populations (Sect. 6.1). Many programs include screening for a varying number of other metabolic disease for which therapy is promising. However, the various problems, such as genetic heterogeneity and long-term follow-up, have not been solved for most other treatable inborn errors.

4.2.2.8 Heterozygote Detection

Heterozygote Detection for PKU and Hyperphenylalaninemia. As in other metabolic diseases, detection of heterozygotes is not only of theoretical interest but can be used in practice for genetic counseling of close relatives, for example, sibs of PKU patients. Such unaffected sibs have a ⅔ risk of being heterozygous. The ideal method for heterozygote detection would be direct enzyme assay, which at the present time would require liver biopsy. Other methods utilize procedures that specifically load this pathway; even if the reduced enzyme activity in the heterozygote is sufficient for metabolizing the amount of phenylalanine normally taken up with the food, it may be insufficient to cope with a high phenylalanine load. The first attempt to detect heterozygotes by a phenylalanine loading test was made by Hsia (1967) [1134], who administered L-phenylalanine to heterozygotes and followed the disappearance of this amino acid in the blood. They achieved a fairly good separation between heterozygotes and normals even though there was some overlap. Later, the method was refined by considering not only phenylalanine but also tyrosine levels, which were slightly decreased in heterozygotes.

Health Status of Heterozygotes. At first glance, heterozygous parents and sibs of PKU patients seem to be perfectly healthy. Penrose [1262], however, observed in a family of PKU patients a special type of mental illness of a depressive kind with onset at the age of about 50 in six relatives. He suspected that heterozygotes might run a higher risk for mental disease. In the past 50 years, however, the problem of possible anomalies and disease liabilities in PKU heterozygotes has received surprisingly little attention, and the few studies devoted to this problem often lack epidemiological sophistication. The problem has been reviewed recently [1340].

Recognizable peculiarities in PKU heterozygotes have been reported in four main fields: Deviations in I.Q. and other psychological tests; a higher risk for mental disease; EEG abnormalities; and disturbances of reproduction. The first three aspects will be discussed in greater detail in Sect. 8.2.3.2 on behavior genetics; suffice it to say at this point that the great majority of heterozygotes do *not* suffer from mental disease, but their risk of having a certain type of late-onset schizophrenia has been found to be significantly elevated. Moreover, a slight reduction of the average I.Q. - especially its verbal component - has been described, and minor EEG peculiarities appear to be more frequent. According to some data, which have been disputed, the risk for reproductive wastage in the form of abortions and stillbirths might be enhanced. Interestingly enough, a few reports point to an abnormal elevation of blood phenylalanine levels in stress situations, such as influenza with high fever [1008], or pregnancy. An interesting ecogenetic suggestion has been made in that aspartame - a high-phenylalanine Sweetening agent - might injure the fetuses of heterozygous women. (See Sect. 9.1 for discussion of maternal hyperphenylalanine). None of the above-cited data for PKU heterozygotes are proven. Phenylalanine blood levels were found to be increased in certain fraction of several hundred children with mild mental retardation and behavior problems [1125]. More work is indicated to substantiate these findings. If confirmed, such results would support the proposition that the reduced enzyme activity makes heterozygotes less able than other people to cope with certain stress situations; this subject could constitute a new chapter in ecogenetics (Sect. 4.5.2).

Heterozygote Detection in General. Neel (1949) [1236] was the first to recognize the general problem of heterozygote detection in medical genetics and to collect the scattered evidence available at that time. A more comprehensive report was given by Neel later (1953) [1237] and by Franceschetti and Klein (1954) [1084]. Since biochemical genetics has progressed rapidly, heterozygote tests for many diseases have become available, especially for those in which enzyme activities can be determined directly in fibroblasts or blood cells (Table 4.10). As a rule, heterozygotes have about half the enzyme activities of normal homozygotes. In cases in which the metabolic pathway but not the enzyme directly is tested (such as phenylketonuria), definite differences between mean values of normal and heterozygotes usually appear. However, as in phenylketonuria, the separation achieved between the two groups is in many conditions incomplete. Some individuals have borderline values. This holds true even when enzyme activities are measured directly. Considering the genetic heterogeneity between mutations at the same gene locus that lead to different types of changes of enzyme function, this result is not surprising. Apart from its value for theoretical understanding of enzyme function, diagnosis of heterozygotes is significant practically for two reasons.

First, it helps in genetic counseling of close relatives of patients with X-linked or autosomal-recessive diseases. Its practical value is revealed especially in X-linked diseases, since female heterozygotes run a risk of 50% for every son to be affected. In most

Table 4.10. Some principles of heterozygote detection

Principles	Examples
Unusual protein band in electrophoresis	Hemoglobinopathies
Reduced enzyme or protein activity	
in Erythrocytes	Galactosemia (gal-1-*p*-uridyl-transferase reduced)
Leukocytes	Glycogen storage disease type II (Pompe) (α-glucosidase reduced)
Fibroblasts and ammnion cells	Lesch-Nyhan syndrome (mosaics for defect of HPRT) and an increasing number of other enzyme defects
Clotting factors	Hemophilia, Prothrombin deficiency and stable factor (factor VII) deficiency
Cells from liver biopsy	Phenylketonuria (phenylalanine hydroxylase reduced)
Abnormal concentration of metabolites in blood	Phenylketonuria (abnormal increase of phenylalanine after loading)
Functional abnormality	
Enymatic reactions	Pseudocholinesterase variants with abnormal inhibition behavior
Structural elements	X-linked Duchenne-type muscular dystrophy (creatine phosphokinase (CPK) increased in serum)
Morphologic abnormality	X-linked ocular albinism (abnormal pigmentation pattern of retina)

autosomal-recessive diseases, heterozygote detection is less important provided that the possible heterozygote - in most cases a brother or sister of an affected homozygote - does not plan to marry a close relative such as a cousin. There is a risk for homozygous children only if both prospective parents are heterozygous, and most recessive conditions are so rare that the risk of a heterozygote to marry another one is low indeed (cf. the Hardy-Weinberg Law, Sect. 3.2).

A second application of methods for heterozygote detection consists in the screening of whole population groups. Such screening has sometimes been introduced in population groups where certain recessive genes are frequent. For example, about 8% of the American black population are heterozygous for the sickle cell gene, and about 3%-4% of Ashkenazi Jewish populations are heterozygous for the Tay-Sachs gene. However, well-meaning screening programs have repeatedly met with logistical and sociopsychological problems (Sect. 9.1.2).

The most frequent recessive disease in populations of northern and central European origin is cystic fibrosis with an incidence of about 1:2,000 and a heterozygote frequency of about 4%-5%. Here, heterozygote screening for genetic counseling of close realtives and particularly for detection of matings involving two heterozygotes (who are not aware of their carrier status) for intrauterine diagnosis of potentially affected offspring would be desirable. Unfortunately, however, the basic biochemical defect in this condition is unknown. Heterozygote detection and prenatal diagnosis have become much easier since the gene has been localized on chromosome 7, by a closely linked DNA marker [1167a; 1344a; 1357a].

Susceptibility to Common Diseases in Heterozygotes of Recessive Conditions. The lack of a reliable simple test for heterozygotes of cystic fibrosis is all the more regrettable since the solution to another problem depends on a reliable method for diagnosis of heterozygotes. In the early 1960s, it was claimed that heterozygotes for cystic fibrosis show an increased susceptibility to certain common diseases, such as gastric and duodenal ulceration and chronic bronchial infections [1170]. Subsequent studies failed to confirm this claim. The matter can only be settled by examination of heterozygotes in the age group affected by the diseases to be studied. Such testing requires a reliable heterozygote test.

In Sect. 3.7.4, α_1-antitrypsin deficiency was discussed as an example of a homozygous condition leading in many cases to increased susceptibility to common diseases - primarily chronic obstructive pulmonary disease. It was also shown that many heterozygotes, when exposed to environmental agents, such as tobacco smoking, might run an increased risk of developing chronic obstructive pulmonary disease. Sickle-cell heterozygous are healthy under normal conditions, but moderate hypoxia, such as an altitude above 2500 m, may cause in vivo sickling and splenic infarcts.

The PKU data have already been mentioned. In addition, there are many scattered reports [1340] for other diseases. For example, heterozygotes for various lipidoses were shown to have a minor decrease in (performance) I.Q. on average, combined with an increased incidence of personality disturbances. Cystinuria heterozygotes might run a slightly increased risk of kidney stones; galactokinase, deficiency heterozygotes may be susceptible to premature cataracts; and in some varieties of Wilson's disease, anomalies of kidney tubular function and minor neurological signs were described. Heterozygotes for xeroderma pigmentosum, a defect of DNA excision repair (Sect. 5.1.6.3), and for the chromosome instability syndromes often thought to be also caused by repair defects (Sect. 5.1.6.2) have been studied with great care for cancer risk. An increased risk of developing cancer at a relatively young age was found in Bloom's syndrome and in ataxia teleangiectasia (but not in Fanconi's anemia). Interestingly enough, the incidence of skin cancer was increased only in xeroderma pigmentosum heterozygotes living in the southern states of the United States of America, and not in those from other areas of the country [1316]. This again illustrates an ecogenetic problem: The high intensity of UV irradiation in the sunlight overloads the reduced capacity for excision repair, which can cope with this problem with lower UV irradiation exposure.

Detection and confirmation of increased risks for common diseases or slight physiologic abnormalities requires elaborate statistical designs, such as careful selection of controls, and above all a much larger number of biochemically and genetically well-defined heterozygotes than have so far been included in these studies. One difficulty with studies of this sort is the usually existing overlap in the laboratory values between heterozygotes and normals. In population studies, most values from the overlap zone will come from normals and not from heterozygotes since the total number of normals far exceeds the frequency of the heterozygotes. Studies on obligate heterozygotes (parents of homozygous children) are much more promising. Investigation of heterozygotes promises better insight into the genetic conditions underlying common diseases, as well as into many other problems. For example, an appreciable fraction of the genetic variability influencing performance and behavior as measured by I.Q. and personality tests could be caused simply by the high population frequency of heterozygotes for recessive diseases with well-known effects in homozygotes (for discussion, see Sect. 8.2.3.2). Our risk predictions for an increased population load due to mutagenic noxious agents such as ionizing radiation or chemical mutagens would have to be reassessed if slightly increased disease liabilities in heterozygotes were found to be the rule rather than the exception. Therefore, studies along these lines are urgently needed, particularly when unequivocal distinction between normals and heterozygotes is possible by a qualitative test system.

Heterozygote Testing in Hemophilia A (30670). In X-linked recessive conditions, heterozygote detection is of immediate importance for genetic counseling. One of the most frequent X-linked conditions in humans is hemophilia A (Sect. 3.1.4), which is caused by deficiency of one of the humoral factors needed for the first stage of blood coagulation (Fig. 4.27) – the antihemophilic globulin (AHF; factor VIII). Among normal individuals, activity of this protein in blood plasma ranges widely between about 40% and 300% of the population average (100%) [1277]. In hemophilic patients, factor VIII activity is severely reduced or even completely lacking. However, plasma of most patients contains an AHF-like substance that can be detected by anti-AHF antiserum. Apparently, these hemophilics produce an AHF protein that has lost its ability to take part in the blood clotting process but has maintained its immunologic identity as a cross-racting material (CRM); the concentration of such CRM is normal (see Sect. 4.2.2.2). In heterozygotes, due to X-inactivation half the cells capable of producing AHF will produce the active factor, which shows the blood-clotting activity as well as the specific immunologic properties of AHF. The remaining cells produce the CRM protein,

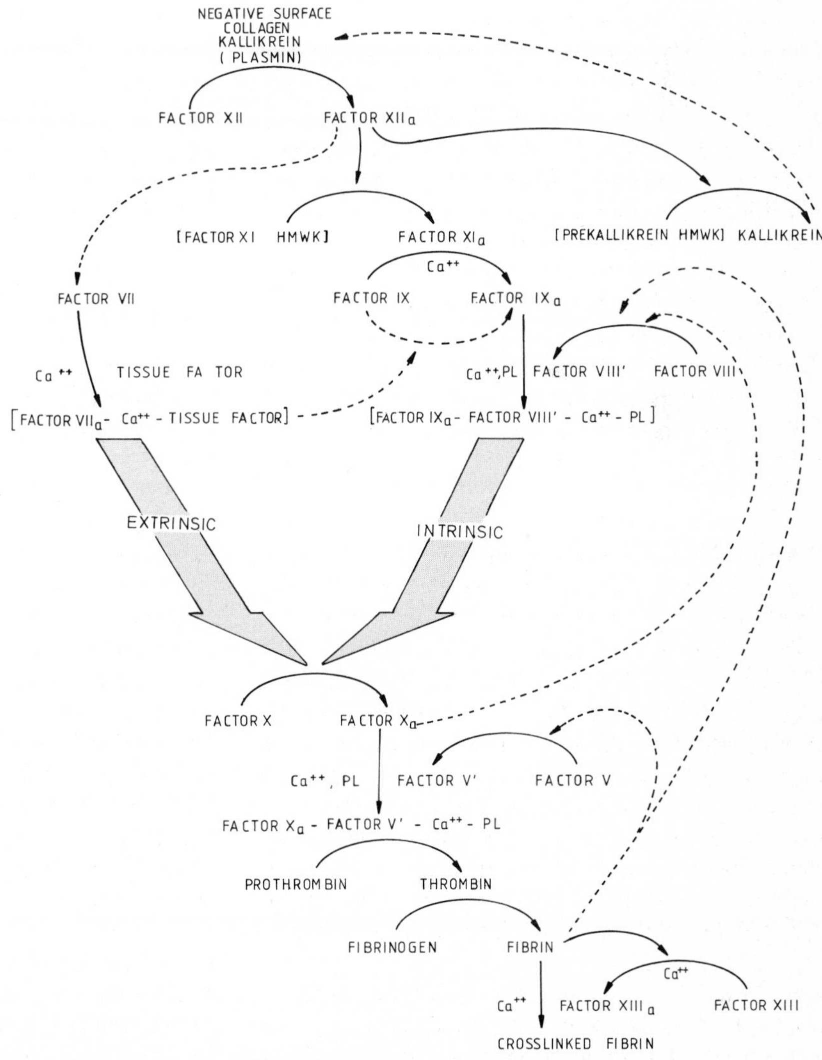

Fig. 4.27. The stages of blood coagulation. For almost all of them, genetic blocks are known. Factor VIII deficiency leads to hemophilia A, factor IX (Xmas factor) deficiency causes hemophilia B [207]

which shows the immunologic specificity but not the blood-clotting activity. Therefore, the mean clotting activity will be reduced to ~50% of normal; the immunologic test, on the other hand, will show a normal amount of protein, as it reacts with both active AHF and the CRM.

In earlier years, and before the immunologic test became available, heterozygote detection had to be based on AHF activity alone. However, while the mean activity was, indeed, much below the norm and ranged around 50%, there was still so much overlap between heterozygotes and normals that the method was of little use for genetic counseling. Considering the huge variability even in the normal range, this result was not surprising. The quantitative immunologic assay brought substantial improvement. However, before making use of this approach, the hemophilic patient(s) in the family should be tested for the CRM protein, since there is a variety of (severe) hemophilia A in which no such protein is produced by the mutant gene. It is a priori unlikely that *every* heterozygote can be detected since random X-inactivation will generate occasional females who will be entirely normal even with the combined immunologic and clotting assay. Moreover, because of some overlap with normals even with this technique, the final interpretation of test results depends upon the a priori expectation that the tested female will be a heterozygote. This probability is 50% for sisters of affected males but will drop in more remote female relatives. The test would be entirely unsuitable for population screening where with the very best technique about 5% of normals would be misidentified as hemophilia carriers.

More recently, improved biochemical methods, especially determination of the components of the factor VIII molecule, have been introduced into carrier detection, and more appropriate statistical methods which combine the probability derived from the pedigree with those obtained from the laboratory have been applied [1111a, b]. DNA variants in the factor VIII gene (Sect. 2.3.2.7) have opened up new ways not only for heterozygote detection, but also for prenatal diagnosis.

It may be mentioned in passing that in the entire field of blood coagulation and its disturbances, amazing progress has been made in recent years. Cooperation of various, genetically determined proteins in the process of blood coagulation and dissolution of blood clots provides an example for interaction of many genetic factors in a complex physiological function. This network has been disentangled largely with the aid of blood from patients with various genetic coagulation defects. Blood coagulation is not a simple and not even a branched chain reaction; in addition, feedback circles are involved. Figure 4.27 gives an impression. This topic has been discussed in more detail elsewhere [1111a, b, 1214].

Heterozygote Detection in Duchenne Muscular Dystrophy [1119, 1221]. Carrier detection is especially important for practical purposes in Duchenne muscular dystrophy (31020), since this disease is incurable and leads to prolonged suffering inevitably ending in early death. Prenatal diagnosis has been achieved only recently. Therefore, pregnancies in women having a high risk for giving birth to an affected boy are often terminated when the child is male. This, however, requires that risks and, therefore, the carrier status be determined as precisely as possible. To this end, a combination of statistical and biological techniques is required. Techniques are described in App. 8. With reference to biological methods, the most informative marker is the activity of the serum enzyme creatine phosphokinase (CK). Normally, this enzyme is found mainly in muscle, and its serum level is low. When muscle tissue is destroyd, however, more enzyme leaks into the blood, and the activity of this and other muscle enzymes increases. In male hemizygous Duchenne patients very high CK levels are found, except in the final stage of the disease when most of the muscular tissue has been destroyed. About 70% of obligate carriers (women having at least one affected son and at least one affected male on the female side of the family) have values clearly above the normal range. However, the diagnostic problem is complicated by the observation that separation of normals and carriers is better at younger age - primarily in children - and that CK values decrease with advancing age even in normals, but still more in heterozygotes. CK values are also lower during pregnancy.

In addition to CK determination, many other methods have been suggested, and successful application has been reported by some authors. Recently, impressive alterations in the thigh and calf muscles of carriers were detected by ultrasonographic or computerized ultrasound imaging techniques [1283]. Undoubtedly, DNA technology will help in the future in a more precise risk assessment by examination of linked DNA markers [1358] (Sect. 3.4.3). However, since many of the cases are sporadic and are often caused by new mutations in the germ cell of the mother (Sect. 5.1.3.4), linkage studies often are not feasible. Therefore, it is worthwhile to improve carrier detection methods based on phenotypes. Moser [1221] and Harper [1119] have given useful hints for counseling practice.

Problems with Heterozygote Detection. A frequent source of practical difficulty in heterozygote detection is the overlap in quantitative values when an enzyme assay or any kind of quantitative blood test is performed. The search for heterozygotes may be one for carriers of autosomal recessive or X-linked mutations, or for carriers of the gene for an autosomal dominant disease. In most conditions, clear statistical differences in the mean levels of the substance under assay can be found between heterozygotes and normals. However, there is usually significant overlap between the distributions, so that many normals have enzyme levels of the test substance that are consistent with heterozygosity. The reasons are usually not fully understood, but may relate to the presence of undetected "isoalleles" that each determine a unique range of different activity levels (Sect. 3.6). When using quantitative tests of this type for heterozygote detection it is essential for accurate assessment of the test results that the a priori (or Bayesian) likelihood of heterozygosity is considered.

Table 4.11 a. Prophobilinogen deaminase level in 217 normals and in 105 acute intermittent porphyria patients. (Modified from Bonalti-Pellié et al. 1984)

Units	Porphyria patients and obligate heterozygotes (%) (x)	Normal controls (%) (y)	Laboratory odds (likelihood ratio) in favor of heterozygosity (x):(y)
< 70[a]	20	0	Very high[a]
70–79	23.8	0.5	48 :1
80–89	22.9	0.9	25 :1
90–99	16.2	2.3	7 :1
100–109	9.5	3.7	2.6 :1
110–119	5.7	8.3	0.7 :1
120–129	1.9	14.3	0.13:1
> 129[b]	0	70	Unlikely[b]

Note that these values need to be determined for each laboratory separately in a sufficiently large sample size.

[a] Since 20% of heterozygotes and no normals were found in this range, the odds in favor of heterozygosity are very high. The lower the value, the higher the odds.

[b] None of the porphyria patients and 70% of normals had values above 129 units. With increasing levels, the risk of porphyria becomes increasingly unlikely.

Table 4.11 a shows specimen enzyme assay results in acute intermittent prophyria. In population screening in this example, where affected patients are heterozygotes, most of the suggestive results following enzyme assay would in fact be false positives since only one heterozygote would be expected in 10,000 persons of an *unselected* population – the population incidence. In a patient with a "negative" family history for porphyria who is suspected to have the disease because of suggestive clinical symptoms, the a priori expectation is less precise and can only be roughly estimated. In a sib of a definitely diagnosed patient, the a priori probability of heterozygosity is 50%. The probability of the test result being indicative of heterozygosity in these various instances can be calculated from the a priori odds and the degree of overlap between normals and heterozygotes, which needs to have been carefully defined (Tables 4.11 a, b). This latter requirement provides strong support for the existence of central reference laboratories where many normals and heterozygotes have been tested and genetic consultation is available.

In a study in which the levels of the enzyme porphobilinogen deaminase were investigated in acute intermittent prophyria, 30% of a control/normal population had values that overlapped those of a population of definite heterozygotes with acute intermittent prophyria (1) (Table 4.11). Only 20% of heterozygotes had enzyme values lower than those of any of the normals. Assuming that the same laboratory value of 95 units was found in different per-

Table 4.11 b. Different odds with same laboratory value (95 units) with various a priori probabilities of trait in acute intermittent prophyria. (Based on laboratory values from Bonaiti-Pellie et al., 1984)

Laboratory value (units)	A priori		Lab odds in favor of heterozygosity (See Table 4.11 a)	Joint odds in favor of heterozygosity	Final risk of heterozygosity[b]
	Odds[a]	Probability			
95	1/9,999 (e.g., population screening)	1/10,000	7:1	7:9999	0.0007 = 1/1500
95	1/99 (e.g., vagine clinical suspicion)	1/100	7:1	7:99	0.07
95	1/9 (e.g., clinical suspicion)	1/10	7:1	7:9	0.44
95	1:1 (e.g., sib or child of definitely diagnosed patient)	1/2	7:1	7:1	0.87

[a] odds = P:(1-p) where p = probability.

[b] Caluclated by multiplying the a priori odds and laboratory odds for the carrier state and noncarrier state to determine the joint odds for the carrier state as x/x+z where x is the joint odds for the carrier state and z the joint odds for the noncarrier state. Example: a priori odds 1:9; lab odds: 7:1; joint odds for carrier state to noncarrier state 7:9 derived from [(1 × 7):(9 × 1)]. Final risk 7/(7+9) = 7/16 = 44%.

sons (Table 4.11), according to these results the risks of carrying the gene for porphyria would be (a) 1:1500 for a member of the general population with no symptoms who is screened, (b) 7% for a person at 1% risk (i.e., clinical suspicion), (c) 44% for someone at 10% a priori risk (i.e., clinical suspicion), and (d) 87% for a first-degree relative at 50% risk. These data show the considerable ambiguity of a diagnosis with such a test result if the a priori expectation of the diagnosis cannot be defined clearly. Note that repeated testing will not necessarily be helpful in rendering these final risks more precise. With very low and high enzyme values the interpretation becomes easier if, as in this example, the exact range of enzyme levels on a large number of normals and heterozygotes is known.

In contrast to such uncertainties with quantitative testing, a definite diagnosis can be made regardless of the a priori probability if an "all or none" *qualitative* abnormality exists in heterozygotes that can be assessed biochemically or by DNA methodology.

4.2.2.9 Treatment of Inherited Metabolic Disease [1289; 1057; 1058]

General Principles. In earlier years, the conclusion that a certain character was inherited had the connotation that it could not be influenced by environmental manipulation. Hereditary diseases were therefore not considered to be amenable to treatment. These seemingly nihilistic attitudes have contributed to the stance of many doctors and behavioral scientists that genetics had little to contribute to their fields. The inborn errors of metabolism offer convincing examples that such beliefs are erroneous. Our ability to influence a disease or a behavioral abnormality often depends upon the depth of insight into mechanisms and not whether to etiology is genetic or nongenetic.

In principle, genetic traits can be influenced at all levels of gene action. In theory, the most thorough influence would be at the level of the genetic material: the DNA. Nonsexual transfer of DNA either with or without viral carriers was first demonstrated in bacteria and is now becoming possible in higher organisms, including human cells. For some enzyme defects, this approach may lead to useful therapeutic methods. Since the so-called methods of genetic engineering gained much attention among the general public, this approach with its false emphasis on cloning and the creation of human beings has frightened many people about the dangers of genetic research in general. In reality, gene therapy of a few Mendelian diseases may become possible and may at some time add a useful item to our therapeutic armamentarium. This topic will be dealt with in Sect. 9.2 on the genetic future of mankind.

In some cases, not the genes themselves but their immediate products, mRNA, may be supplied. So far no successful example of this approach is known. Much more advanced are the attempts at replacement therapy by enzymes or other proteins (see below). Often, the metabolic consequences of the genetic block may be influenced by suitable manipulation of the environment. The classic example, treatment of phenylketonuria by a phenylalanine-restricted diet, has been introduced in Sect. 4.2.2.7. In other cases, the clinical consequences are due not to accumulation of metabolites ahead of the block but to lack of a metabolite behind it. In these cases, substitution therapy may prove useful.

Finally, a great number of secondary consequences of a genetic disease may be successfully influenced, from disequilibrium of endocrine regulation caused by a block in hormone synthesis to blood transfusions in an inherited anemia. Figure 4.28 gives a synopsis of therapeutic possibilities. In the following, some examples will be discussed. For a more complete survey see [1058].

Substitution (Protein or Enzyme) Therapy. The classic example is hemophilia A. Factor VIII activity of 30%–40% of the normal average controls bleeding from wounds. This activity can be achieved by factor VIII infusions. Factor VIII concentrates are prepared from human blood. Thanks to these concentrates, home treatment and control of the bleeding episodes is now possible; hemophilia A patients may lead a nearly normal life. Problems are created by the large number of donors required to prepare sufficient amounts of this material [428]. Fortunately, the first and decisive steps toward production of pure factor VIII preparations by methods of genetic engineering – characterization and cloning of the factor VIII gene, and expression of plasmid-transferred genes in cell culture – have already been taken (Sect. 2.3.2.7).

Another example is substitution of pseudocholinesterase in patients with enzyme-deficient variants [1105]. Here, therapy is facilitated by two favorable conditions:

1. The enzyme deficiency is harmless under normal conditions; substitution is needed only for major surgery, requiring a muscle relaxant.
2. After injection of normal plasma, the activity decreases to one-half within 12 h. This means that suitable activity can be maintained by one injection for the duration of the operation.

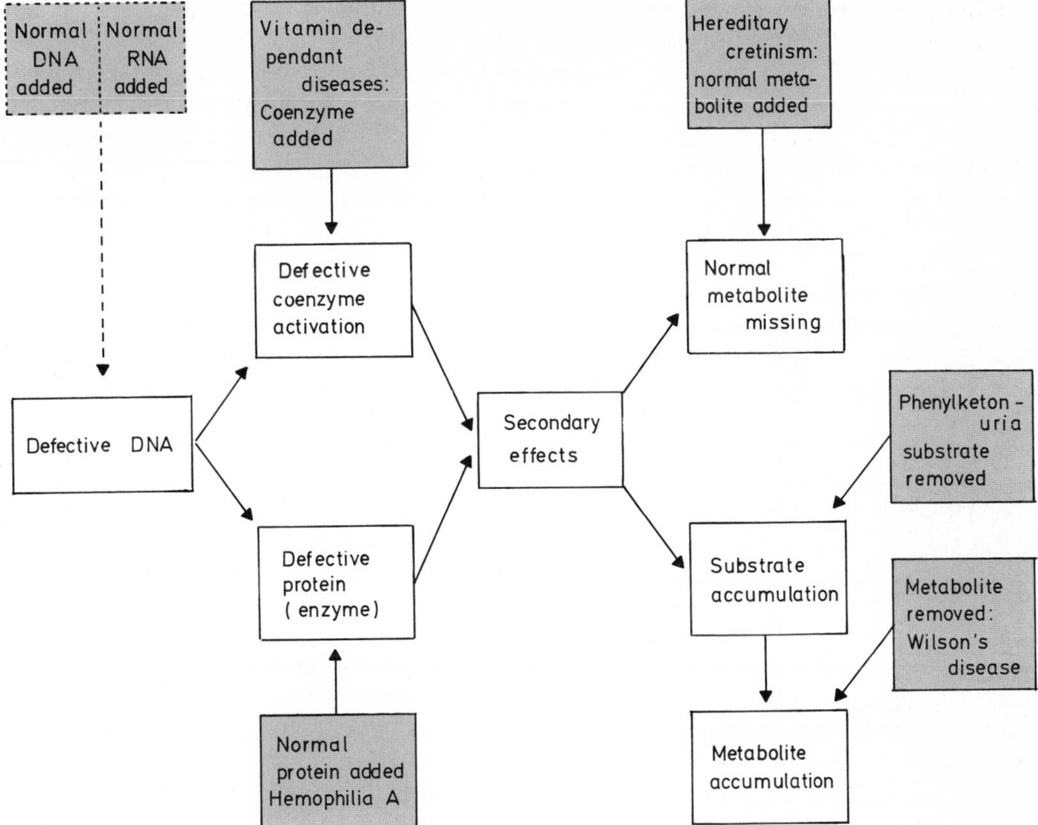

Fig. 4.28. Synopsis of therapeutic approaches to inherited metabolic diseases. Addition of normal DNA or RNA has so far not been successful. Normal protein or coenzyme (vitamin) may be added. The secondary effects of enzyme blocks may be removed by addition of a normal metabolite or removal of excess substrates or metabolites

The first therapeutic attempts were made using normal plasma as the source of enzyme, but it soon became apparent that purified enzyme preparations have distinctive advantages. As shown in Fig. 4.29, a suitable enzyme level is maintained for a sufficient time after injection and normalizes the duration of muscle relaxation.

Enzyme therapy is not necessarily required since prolonged apnea induced by succinylcholine in patients with pseudocholinesterase variants can be treated by external respiration provided by the anesthesiologist; but prolonged tracheal intubation often has some local after effects.

Enzyme therapy of no other genetically determined enzyme deficiency disease has yet reached the level of general clinical application, although encouraging studies are being carried out on conditions such as Gaucher's (23080; [1056]) disease where the missing glucocerebrosidase has been administered intravenously. Most enzyme defects would require lifelong corrective therapy. Here, further problems are encountered:

a) The enzyme is eliminated from the body within a relatively short time; continuous supplementation is needed.
b) Enzyme preparations may be recognized by the immune system as foreign proteins, and antibodies may make the injected material biologically ineffective.

The possibilities of overcoming these difficulties vary from one condition to another. To obtain enzyme preparations of human origin that are most desirable for therapy, cell cultures may be grown in large quantities. Human placentas are also often a good source of easily available tissues for enzyme preparation. The enzyme defect in the Hurler type of mucopolysaccharidosis (Sect. 4.2.2.3) has been corrected temporarily by transfusion of leukocytes [1168], and fibroblasts from close relatives (to minimize graft rejection) have been cultivated in vitro and then implanted in Hunter's disease [1052]. In both cases, a temporary decrease of storage was achieved but no clinical improvement. Recent developments in this field are discussed extensively elsewhere [1057, 1058]. Administration of the enzyme in

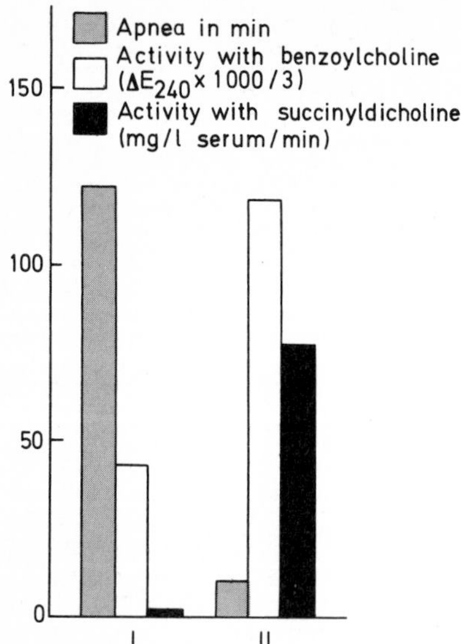

Fig. 4.29. Influence of an injection of a pseudocholinesterase preparation (1,3000 times enriched) on mean duration of apnea and pseudocholinesterase activity with two substrates (benzoylcholine and succinyldicholine) in three individuals with atypical enzymes before treatment *(I)* and 10 min after treatment *(II)*. After treatment, activity of the enzyme is increased. This leads to reduction of apnea time. (Schloot and Goedde, 1974 [196]). The height of the white and black columns reflect danges in activity

semipermeable micorcapsules may and in preventing access of possible antibodies to the enzyme while the substrate that is usually of smaller molecular weight may enter the membrane of the capsule [983]. These capsules may be as small as red blood cells.
Intravenous enzyme therapy - without capsules - may be possible with enzymes that can be taken up by the cells, such as the lysosomal enzymes involved in glycosaminoglycan catabolism (Sect. 4.2.2.3). Many enzymes, however, act only inside cells and cell compartments, and many not be amenable to such substitution. Hence, this approach has only limited applicability. Environmental manipulation by influencing metabolic consequences of enzyme defects appears to be a more successful alternative.

Environmental Manipulation: Removal of a Metabolite Ahead of the Block. The metabolite ahead of the block that acts as substrate for the deficient enzyme may be removed relatively easily, if it is not produced in the organism but is taken up as a normal nutrient. One example - phenylketonuria - has already been described. Another example is galactosemia due to deficiency of one of the three enzymes converting galactose into glucose. Here, removal of the substrate is easier, as galactose occurs almost exclusively in the milk. The problem becomes more difficult if the metabolite in question cannot be reduced without impairment of normal function.
In other instances, the substrate of the deficient enzyme is a metabolite normally formed in the organism. The mucopolysaccharidoses may again serve as one example: They are constantly synthesized and needed for many structural elements of the organism. In such conditions, complete removal of all symptoms by substrate reduction cannot be expected. However, a small reduction of synthesis may slow down clinical progression and in some cases may even help the organism to cope with the problem by using the residual activity of the enzyme or an alternative pathway. In other cases, the phenotypic ill-effects of an enzyme block are caused not by accumulation of a metabolite ahead of the block but by lack of a metabolite behind it.

Environmental Manipulation: Substitution of a Metabolite Behind the Enzyme Block. The best-known examples of such substitution therapy are disorders of hormone synthesis which have frequently been reviewed during recent years [171, 1288, 1243]. Other examples are the glycogen storage diseases types I and III (Fig. 4.2). Here, most of the clinical symptoms are caused not by the glycogen storage itself but by the failure of glycogen to be broken down to glucose, which leads to chronic hypoglycemia. Replacement therapy of blood glucose would meet with insuperable difficulties and, besides, would lead to still more glycogen storage. Therefore, surgical intervention to bypass the liver, with the blood stream coming from the intestine and containing the resorbed glucose, has been successfully tried. A shunt between the portal vein and the inferior vena cava causes most of the blood to bypass the liver and transport glucose directly to the heart muscle and to other organs. Definite improvements have been observed [1131].
Another example is orotic aciduria, which has been described in Sect. 4.2.2.4. Here, the excess of orotic acid seems to have no major ill-effects, but the deficiency of uridine compounds leads to impairment of nucleic acid synthesis and, specifically, to megaloblastic anemia and, additionally, to severe growth inhibition. Addition of uridine to the diet provides the missing metabolite and leads to amelioration of the clinical symptoms.

Elimination of the Metabolite Ahead of the Block and Substitution of the Metabolite Behind the Block. In one of the examples mentioned above, the glycogen storage diseases, "internal" substitution of the compound behind the block - glucose - by partial bypass of the liver helped also to reduce accumulation of the metabolite ahead of the block - glycogen. In

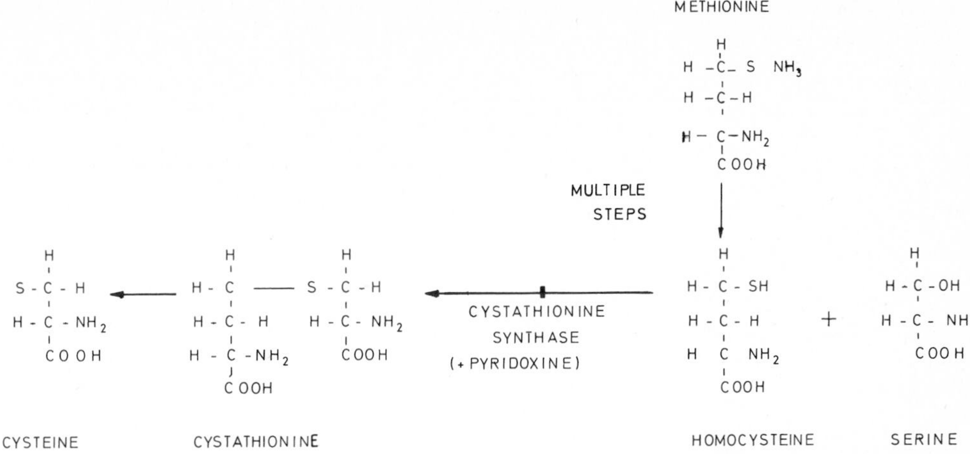

Fig. 4.30. Methionine pathway. Formation of cysteine via this pathway. Cystathionine synthase is inactive in homocystinuria. This leads to an increase of homocysteine and homocystine, on the one hand, and to a deficiency of cysteine, on the other

other diseases, clinical symptoms are caused by both mechanisms, and therapy should try to influence both. One example is homocystinuria (23620), which is caused by a defect of the enzyme cystathionine synthetase (Fig. 4.30). Homocysteine is formed from nutritional methionine. Therefore, the methionine supply should be reduced. However, since methionine is - like phenylalanine - an essential amino acid it cannot be eliminated altogether. Cysteine, on the other hand, is normally formed from methionine via the pathway shown in Fig. 4.30. Many of the numerous symptoms in homocystinuria are due to cysteine depletion; therefore, the diet should be enriched by cysteine. A different type of homocystinuria responds to pharmacologic doses of vitamin B_6 administration, which acts as coenzyme to cystathionine synthetase.

Treatment by Removing Secondary Effects of the Metabolic Defect. This is by far the largest group offering therapeutic possibilities in genetic defects. In contrast to the approaches discussed above, specific knowledge of the pathophysiologic and genetic mechanisms is not necessarily required. For example, we know next to nothing about the biochemical basis of developmental conditions, such as polydactyly or cleft lip and palate. Yet this does not prevent successful surgical correction. Very little is known regarding the biochemical basis of mental disease (Sect. 8.2.3.6). Still, drug therapy that was introduced on a purely empirical basis has proved to be fairly successful in managing patients with schizophrenia and the affective disorders. In all fields of medicine, most therapeutic measures - including the successful ones - are based on similar empirical evidence, whether the influences of genetic variability on the disease are large or small. On the whole, our present capacities of treating hereditary diseases, and alleviating human suffering are not too impressive at present [1045]. This, however, probably applies for much of our medical therapy.

Therapeutic intervention that requires specific knowledge of the pathophysiologic mechanism is the ultimate aim of most biomedical research. The group of adrenogenital syndromes due to enzymatic blocks in the synthesis of adrenal steroid hormones is one example. Cortisol (17-oxycorticosterone) cannot be formed; therefore, the normal feedback for inhibition of ACTH formation in the pituitary cannot work, and 17-ketosteroids are formed in large amounts from 17-oxyprogesterone. These, in turn, stimulate the development of sexual characters and lead to virilization of the female patients. Substitution of cortisol restores the feedback circle; ACTH and, consequently, 17-ketosteroid formation is reduced, and virilization is stopped (Fig. 4.31).

Dietary Treatment of Metabolic Diseases May Only Be the Extreme of a More General "Genetotrophic" Principle. In many metabolic diseases, the phenotypic consequences of an enzyme block can be avoided by a suitable change in nutrition. These conditions are called abnormal because they are rare. Had the great majority of our population one of these enzyme defects, we would have changed our eating habits accordingly, and what now appears to be a defect would be regarded as normal. One example is the intestinal lactose malabsorption found in most Orientals and Blacks and in many Europeans. Consumption of *large* amounts of milk and dairy products causes flatulence and bowel irritability in such lactase-deficient persons. Most persons of North-

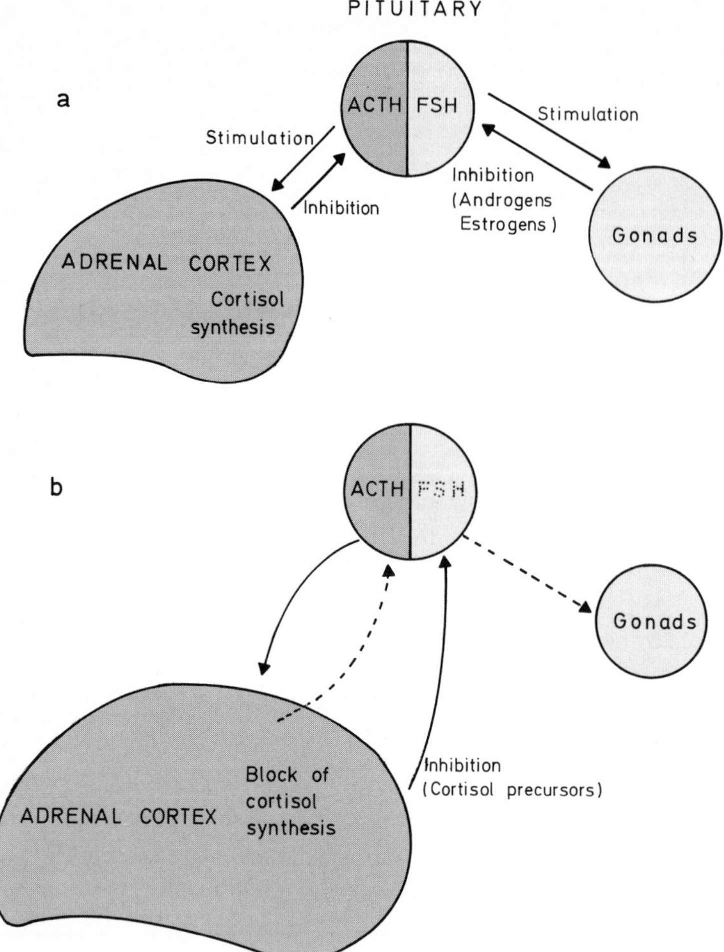

Fig. 4.31. **a** Negative feedback mechanism between pituitary and adrenal cortex. The adrenal cortex is stimulated by the pituitary hormone ACTH; the end-product of corticosterone synthesis, cortisol, inhibits ACTH formation. At the same time, the gonads are stimulated by FSH produced by the pituitary until the androgens (or estrogens) produced by the gonads inhibit FSH production. **b** In the adrenogenital syndrome (AGS), cortisol formation is inhibited by a genetic block. This has two effects on the pituitary: ACTH formation is not inhibited. Abnormally high ACTH formation leads to excessive formation of cortisol precursors, which inhibit FSH production due to their chemical similarity with androgens in the pituitary. Virilization in females results. Cortisol substitution restores the normal feedback cycle

western European ancestry have no such problems, as they have sufficient intestinal lactase for the breakdown of lactose (Sect. 7.3.1; [1924; 1922]).

In Sect. 4.2.2.5, diseases due to abnormal uptake, conversion, and utilization of coenzyme precursors (vitamins) were described. These conditions can usually be treated with unusually high doses of specific vitamins. From an evolutionary standpoint, however, even our normal dependency on vitamins may be regarded as a multiple genetic deficiency, since *Neurospora crassa* as well as *E. coli* bacteria are able to synthesize almost all vitamins. L-Ascorbic acid (vitamin C) serves as a potent reducing agent in mammalian metabolism and can be synthesized by all species with the exception of humans, higher primates, and guinea pigs. Humans need continuous "substitution therapy," which fortunately is supplied by normal nutrition. In exceptional situations, however, for example, when the food provided to sailors of a cargo ship on long voyages in past centuries did not contain enough vitamin C, scurvy developed.

Other pathways that have been lost during evolution are those needed for synthesis of the so-called essential amino acids. For some bacteria and fungi, these essential amino acids are not essential at all; they can be synthesized from simple nitrogen sources such as ammonia.

So far, we have mainly considered nutritional therapy of rare, exceptional genetic variants with extreme effects. However, even screening for phenylalanine

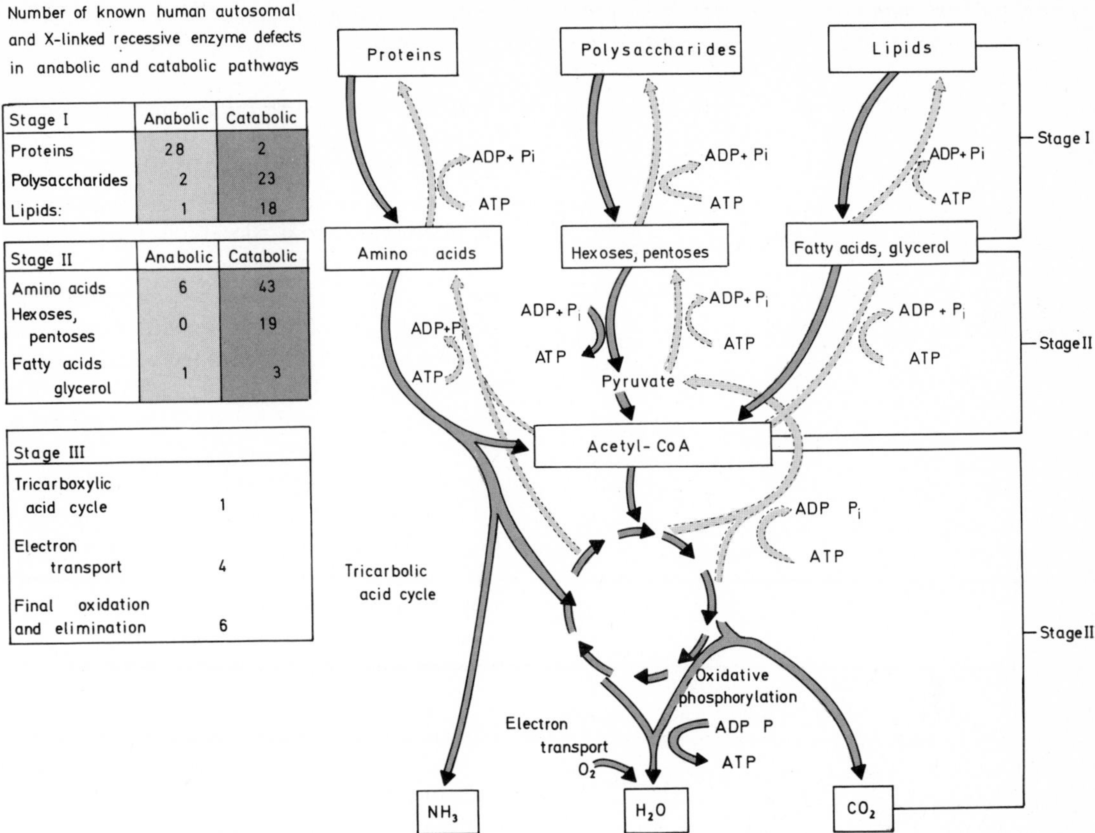

Fig. 4.32. The main catabolic *(dark arrows)* and anabolic *(light arrows)* pathways. Most human enzyme defects - excepting some defects of serum protein synthesis - are found in catabolic pathways. (Adapted from Lehninger, 1975 [120])

concentration in the serum has identified - apart from the extreme, classic PKU cases - those with mild hyperphenylalaninemia. Such persons do not need a special diet to keep their development within limits commonly regarded as "normal". However, there have been some data pointing to a shightly higher vulnerability of heterozygotes whose hydroxylase activity is reduced (Sect. 4.2.2.8). If this can be confirmed, it may be inferred that such irritability may depend on the amount of phenylalanine intake once the demand for protein synthesis has been satisfied.

In Sect. 6.1.2, the genetic polymorphisms will be discussed. We shall see that one-third of all blood enzymes occur in different molecular forms within the human population. Many of these molecular forms show differences in activity. This means that - apart from monozygotic twins - the metabolic pathways are utilized in a slightly different way in each individual, leading to "biochemical individuality," [225]. One aspect of this individuality is that nutritional requirements for optimum development may be slightly different for each individual. This "genetotrophic principle" is a part of the mutual adaptation between the individual, his peculiar genetic makeup, and his environment. This is one aspect of the more comprehensive field of ecogenetics that will be discussed in Sect. 4.5.2.

4.2.2.10 Enzyme Defects That Have Not Been Discovered

How Many Enzymes Are There and What Enzyme Defects Are Known? Some metabolic pathways have not yet been elucidated. Therefore, no one knows the exact number of enzymes in humans. Estimates run in the order of magnitude of at least 10,000. For approximately 200 enzymes, or about 2%, enzyme defects are known. What about the other 98%?

First, there are obviously a great number of inherited diseases that may be caused by an enzyme defect but have not been analyzed by appropriate techniques (Sect. 3.1.8). Most of the autosomal-recessive conditions listed in McKusick's catalogue may belong to this group [133].

Which Enzyme Defects Are Not Known? Figure 4.33 is an attempt to compare the major group of metabolic pathways [120] in which many enzyme defects are known and other groups in which few if any defects have been discovered. Well-known enzyme defects occur in:

a) Catabolic pathways of carbohydrates (i.e., glycolytic defects in hereditary hemolytic anemia)
b) Catabolic pathways of some amino acids (e.g., phenylketonuria)
c) Catabolic pathways in lysosomes for degradation of building material of cells and intracellular material (e.g., mucopolysaccharidoses)
d) Catabolic pathways for detoxification and excretion of internal metabolites such as ammonia (e.g., argininemia)
e) A few marginal reactions such as in the salvage pathway of nucleic acid metabolism (e.g., HPRT deficiency; Sect. 4.2.2.6)
f) Anabolic pathways for synthesis of biomolecules needed for special regulatory purposes (e.g., defects in thyroid hormone production)
g) Some pathways in transmembrane transport (e.g., cystinuria)
h) Some few DNA repair enzymes (e.g., xeroderma pigmentosum; Sect. 5.1.6.3)
i) Some metabolic steps in uptake and utilization of coenzyme precursors (e.g., vitamin D-resistant rickets)

We know few, if any, enzyme defects in:

a) Enzymes concerned with the processes of mitosis and meiosis
b) Enzymes needed for DNA and RNA synthesis, with the exception of a few repair enzymes
c) Enzymes concerned with protein biosynthesis
d) Energy supply sources, especially the cytochrome system
e) Enzymes for synthesis of most of the specialized compounds needed as neurotransmitters in the central and peripheral nervous system
f) Anabolic pathways in synthesis of many amino acids, pentoses, fats, and lipids
g) Anabolic enzymes for synthesis of tissue constituents such as sphingolipids, mucolipids, and mucopolysaccharides
h) The tricarboxylic acid cycle, which serves both catabolic and anabolic functions

In short, our knowledge of enzyme defects in humans is not only incomplete but also very much biased. Most enzyme defects affect enzymes concerned in the "house-keeping" functions of the cell. For most of the central building functions, no enzyme defects have become known so far. For catabolic pathways, and for biosynthesis of some specialized molecules such as hormones, the picture is more complete.

Why Do We Know So Little About Enzyme Defects of Central Building Functions? What are the reasons for this bias? Part of it is methodologic. Analysis of enzyme defects in humans depends vitally on availability of organ material. Blood cells are easily available, brain or even liver cells are not. The same problem must be faced in the analysis of genetic polymorphisms. Most polymorphisms detected so far affect blood constituents. If the blood – and not the brain – were our organ for thinking and feeling, our ignorance in the field of behavior genetics would have ended a long time ago (Sect. 6.2).

However, it is difficult to understand how these and other methodologic difficulties could explain the entire bias. The obvious alternative hypothesis is that deficiencies in these central building metabolic pathways would not be compatible with life; they would be lethal (Sect. 3.1.6). For example, it is difficult to imagine how a nearly complete defect of an essential DNA polymerase, which reduces or even abolishes DNA replication – and hence, cell division – could be compatible with life. The same argument holds true for the basic steps in the tricarboxylic acid cycle or the synthesis of vital metabolites.

For most enzymes, 50% of the normal activity is sufficient to maintain normal function, as evidenced by observations on heterozygotes for enzyme defects. Therefore, one might surmise that at least heterozygotes for such enzyme defects would be detected. However, such detection would require large-scale population studies of enzyme activities, which have not been carried out. Moreover, many enzymes show a marked interindividual variability in activity, which would make identification of heterozygotes difficult, if not impossible. This variability – and especially the observation that, in heterozygotes, ≈50% of the "normal" enzyme activity is sufficient for maintenance of function under normal living conditions – indicates an excellent "buffering" of the metabolism against intrinsic genetically determined weaknesses. Many functions are maintained by multiple pathways, and many mutations, even in the homozygous state, may not lead to inborn errors [820], or only in particular environmental conditions such as in the presence of a drug (Sect. 4.5).

The conclusion that there may be many recessive lethal mutations affecting essential pathways has far-reaching consequences for population genetics. There is no reason whatsoever to assume that the mutations affecting genes determining these "vital" enzymes are any less frequent than the mutations for which enzyme defects are known. Hence, all these mutations are expected to occur. They may lead occasionally to lethal homozygotes and hence would enhance the proportion of dead zygotes. One would expect this phenomenon to bring about an increase in the number of miscarriages under conditions favoring segregation of homozygotes in general, i. e., in consanguineous marriages (Sect. 6.3.1). This prediction, however, seems not to be borne out by experience. Probably, most of these zygotes die at so early a stage of development that a miscarriage is not realized and recorded.

4.2.2.11 Some General Conclusions Suggested by Analysis of Human Enzyme Defects

Detection of Enzyme Defects. In our consideration of human enzyme defects, several points recur. To be readily detected, an enzyme defect should be located in blood cells or should manifest itself in cultured fibroblasts. Moreover, it should lead to clear-cut clinical symptoms in affected individuals or should at least lead to alterations that will easily be detected by screening techniques, such as excretion of abnormal urinary metabolites. An inborn error with nonspecific symptoms, which is not accompanied by currently detectable biochemical disturbances, cannot be easily identified. Thus, even though several metabolic surveys have been carried out among patients with mental deficiencies, many more inborn errors than have been hitherto discovered might exist.

Elucidation of Metabolic Pathways by Utilization of Enzyme Defects. It is not so difficult to detect enzyme defects if the metabolic pathways are already known. In some cases, however, analysis of enzyme defects may provide a tool for elucidation of metabolic pathways that would otherwise be difficult to examine. The mucopolysaccharidoses are a cardinal example.

Characteristics of Mutations Leading to Enzyme Defects in Humans. In many of the enzyme defects analyzed so far, some residual activity of the enzyme could be observed. Moreover, qualitative changes in the enzyme protein are usually discovered, for example, cross-reacting material (CRM), change of kinetic characteristics, and many others. These findings point to qualitative changes in enzyme proteins due to mutations in structural genes and argue against a major share of regulatory mutations at all possible levels. Such mutations would be expected to cause quantitative changes only in enzyme activity. There is a high degree of genetic heterogeneity within a single gene locus, which adds to the heterogeneity between loci involved in the same pathways.

Mode of Inheritance: Heterozygotes. The mode of inheritance of enzyme defects is usually recessive - either autosomal or, in some cases, X-linked. Healthy heterozygotes have almost always enzyme activities of about half the population average. Hence, the human organism can work perfectly well with a single enzyme at half power. This reveals a remarkable amount of internal regulatory capacity within metabolic pathways. However, if the pathway is loaded with a substance requiring the deficient enzyme for its metabolism, its ability to cope with the metabolite is poorer than normal. Some observations raise the suspicion that this impairment may not be quite as unimportant for the health of heterozygotes as is usually assumed. It may contribute - possibly together with environmental stresses - to their susceptibility to common diseases, somatic or mental. Remarkably few systematic, large-scale, controlled investigations on the health status of heterozygotes for recessive diseases - especially during middle or advanced age - have been carried out. Like so many other shortcomings of our knowledge in human genetics, this lack of data may have sociologic reasons. Work on inborn errors is mostly being done by pediatricians or medical geneticists with a pediatric background, who usually are not interested in studies on epidemiology and population genetics. Conversely, population geneticists usually do not deal with biochemical refinements in field studies.
The observation that practically all enzyme defects are inherited as recessive traits inevitably raises the question as to the biochemical basis of dominant abnormalities. This problem will be discussed later (Sect. 4.6). First, the hemoglobin paradigm will be introduced. In this special case, many questions raised by the enzyme defects, and even the question as to possible mechanisms of Mendelian dominance, have been answered.

4.3 Man's Hemoglobin [119, 31,97 a]

The hemoglobin molecule can be studied with greater facility than any other human protein. Blood can easily be drawn from many individuals. Hemoglobin is the principal protein of red blood cells, and its extraction does not require complicated biochemical methods. It is therefore not surprising that we understand more about this protein than about all others. Genetically oriented studies on human hemoglobins have proceeded apace with the elucidation of the amino acid sequence and structure of the molecule. The hemoglobin system currently is a paradigm for our understanding of gene action at the molecular level. Hemoglobin research plays a role in human biochemical genetics similar to that of research with

Drosophila and phage in basic genetics. Most concepts derived from hemoglobin research apply readily to other proteins. In fact, many conceptual principles of human genetics could be taught by examples from the hemoglobin system.

4.3.1 History of Hemoglobin Research

Sickle Cell Anemia: a "Molecular" Disease. Work on human hemoglobin started with the investigation of a hereditary disease: sickle cell anemia. In 1910, Herrick [1121] observed a peculiar sickle-shaped abnormality of red cell structure in an anemic black student. It soon became apparent that this condition was fairly common among American blacks. Affected patients suffer from hemolytic anemia and recurrent episodes of abdominal and musculoskeletal pain. Taliaferro and Huck (1923) [1325] recognized that the condition is hereditary. It was shown by Neel (1949) [1235a] and, independently, by Beet (1949) [991] that patients with sickle cell anemia are homozygous for a gene that, in the heterozygous state, causes a much milder condition: sickle cell trait, which was found in about 8% of the American black population [1226].

The decisive step in the biochemical-genetic analysis of this disease was carried out by Pauling et al.[3] (1949) [1260] in a paper with the programmatic title, "Sickle Cell Anemia, a Molecular Disease." We cite from this paper:

"The evidence available at the time that our investigation was begun indicated that the process of sickling might be intimately associated with the state and the nature of the hemoglobin within the erythrocyte."

Therefore, the authors examined the hemoglobins of patients with sickle cell trait and sickle cell anemia, comparing them with the hemoglobin of normal individuals. In accord with the state of methodology for protein analysis at that time, these investigations were performed using the Tiselius zone electrophoresis (Fig. 4.33). The peaks in Fig. 4.33 represent the concentration gradients of hemoglobin in a suitable buffer solution; the positions of these peaks depend on the relative number of positive and negative charges in the protein molecule.

"The results indicate that a significant difference exists between the electrophoretic mobilities of hemoglobin derived from erythrocytes of normal individuals and from those of sickle cell anemic individuals."

In sickle cell trait, about 25%-40% of the hemoglobin turned out to be identical with that found in sickle cell anemia, whereas the remainder was indistinguishable from the normal. This result was compatible with the genetic data that sickle cell anemia represented the homozygous state of a gene for which carriers of the sickle cell trait are heterozygous.

"This investigation reveals, therefore, a clear case of a change produced in a protein molecule by an allelic change in a single gene involved in synthesis."

Single Amino Acid Substitution. In 1956, Ingram [1138], working in Cambridge in the laboratory where Perutz was pursuing his crystallographic work, Sanger had shown the amino acid sequence of insulin, and Crick and Watson had demonstrated the DNA model, discovered what precisely distinguished normal from sickle hemoglobin. Hydrolysis of the globin molecule with the protein-splitting enzyme trypsin yielded about 60 peptides, which were separated on paper in a two-dimensional array by electrophoresis in one direction and paper chromatography in the second. This "fingerprinting" method of protein analysis revealed that sickle cell hemoglobin was identical with the normal molecule in all peptides excepting one. Further analysis of this one differing peptide showed that sickle cell hemoglobin differed from normal hemoglobin in only one amino acid: glutamic acid was replaced by valine.

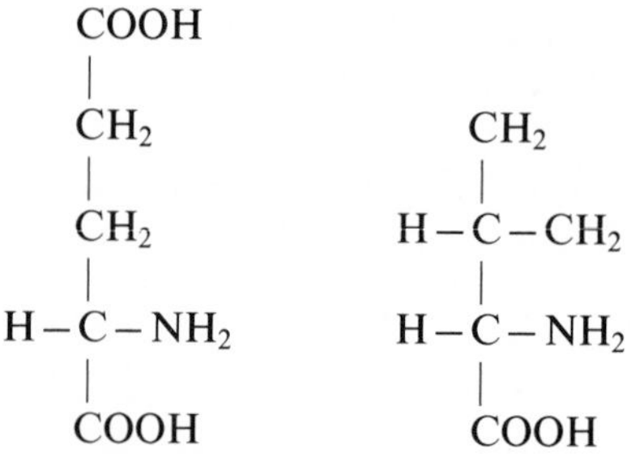

COOH	
|	
CH_2	CH_2
|	|
CH_2	$H-C-CH_2$
|	|
$H-C-NH_2$	$H-C-NH_2$
|	|
COOH	COOH
Glutamic acid	Valine

Glutamic acid has two COOH groups and one NH_2 group, whereas valine has only one COOH group. This charge difference explained the electrophoretic differences between normal and sickle hemoglobin.

Meanwhile, and especially after new methods of electrophoresis had replaced the cumbersome Tiselius electrophoresis, an increasing number of other hemoglobin variants were discovered. At present, over 400 of such variants are known [1194]. Further steps of great importance were the establishment and the elucidation of the full amino acid sequence

[3] Pauling, an outstanding chemist, heard from Castle, a renowned hematologist, (and son of one of the pioneers of mammalian genetics) about this disease and surmised that a defect of hemoglobin was likely to be its cause.

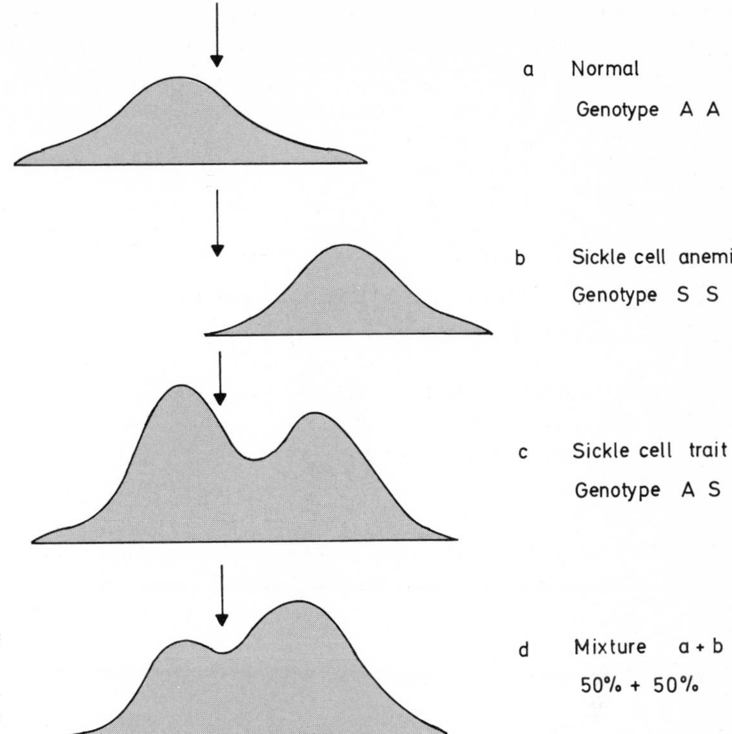

Fig. 4.33 a-d. Zone electrophoresis diagram of hemoglobins at pH = 6.9. **a** Normal homozygote *(AA)*; **b** patient with sickle cell anemia *(SS)*; **c** sickle cell trait *(AS)*; **d** mixture of equal parts of Hb A and Hb S. (Pauling et al., 1949 [1260]) The arrow indicates the starting point of electrophoresis

of the hemoglobin chains by Braunitzer et al. (1961) [1016], and of the three-dimensional structure of hemoglobin [1165, 1265]. Subsequent years have led to our understanding of structure-function relationships, and detection of different types of mutations, such as deletions and frameshifts. The isolation of the hemoglobin mRNA led to new insights into gene structure and function and opened new paths to the understanding of gene action.

Molecular work on the hemoglobins has proceeded at a rapid rate. The full DNA sequences of the various hemoglobin genes and their flanking sequences is now known, and the hemoglobin genes are better understood than any other mammalian genes. Mutations affecting the hemoglobins, particularly the thalassemias, have been elucidated and are models for the understanding of gene action at the molecular level. The genetics of hemoglobin as currently known will be described in the next section.

4.3.2 Genetics of Hemoglobins

Hemoglobin Molecules. Human hemoglobin consists of four globin chains. The general formula of the hemoglobin molecule is $\alpha_2 \beta_2$ signifying that of the four globin chains two are alike and each exist in duplicate [1348]. Most human hemoglobins have identical α chains while the non- α chains differ from each other. Each globin chain carries a heme group, a nonprotein molecule attached at a specific site of the globin molecule (Fig. 4.34). The four globin chains with their respective heme groups comprise the functional hemoglobin molecule that carries oxygen from the lungs to the tissues. A globin chain consists of a string of over 140 amino acids of specified structure (Fig. 4.35). The sequence of the various amino acids in a protein molecule, such as hemoglobin, is known as the primary structure. The spatial relationship between adjacent residues is known as the secondary structure and the three-dimensional arrangement of a protein subunit as the tertiary structure (Fig. 4.34). Quarternary structure refers to the arrangements of the four subunits into a functioning molecule.

The principal hemoglobin of children and adults is Hb A or adult hemoglobin ($\alpha_2 \beta_2$). The characteristic structure of HB A is the β-chain (Fig. 4.35). The α and β chains differ from each other in many amino acids. All adults carry a small amount (2%-3%) of Hb A_2 ($\alpha_2 \delta_2$). The characteristic δ chains differ in only ten amino acid positions from the β chain. A small amount (<1%) of fetal hemoglobin (Hb F: $\alpha_2 \gamma_2$) is also seen postnatally in all individuals (see below). The γ chain differs considerably from both the α and β chains. The α chains of Hb A, A_2, and Hb F are identical.

Several hemoglobins characteristic of embryonal and fetal development exist. ζ chains resemble α chains in their amino acid composition [1155] and

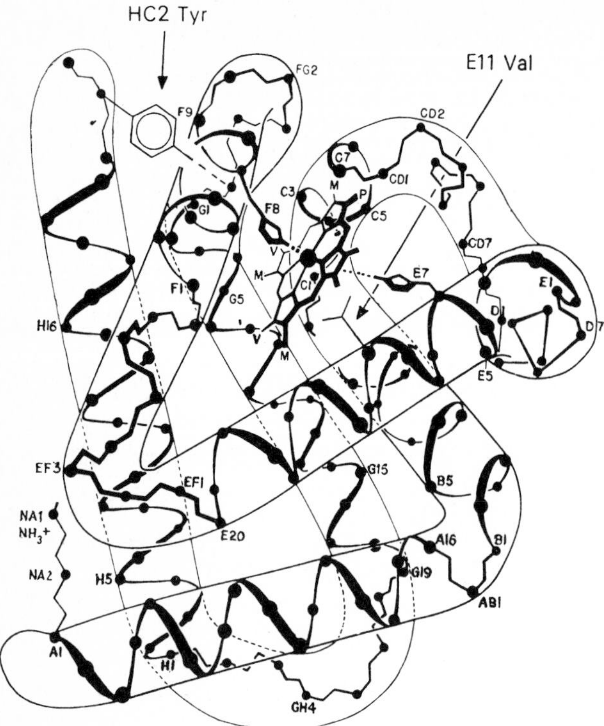

◁

Fig. 4.34. The diagram shows the three-dimensional structure (3° structure) of a typical globin chain, which consists of eight helical and six nonhelical segments. To allow comparisons of different globin chains, the helical segments are labeled *A* through *H* and nonhelical segments are denoted by two capital letters such as *CD, FG*, etc. The *black wavy line* indicates the spatial arrangement of the various amino acids (2° structure). Amino acids are numbered from the amino *(N)* terminus starting with *A1*. The *number* refers to a specific amino acid located at that position, which may differ in various globin chains. Structurally equivalent residues carry the same notation in all hemoglobins regardless of amino acid additions or deletions. Note the insertion of the nonprotein heme chain between *E7* and *F8*. Amino acid residues at *E7* (histidine), *E11* (valine), and *HC2* (tyrosine) are particularly important in the function of mammalian hemoglobins. The letters *M, V*, and *P* in the heme molecule refer to methyl, vinyl, and propionate side chains, respectively. (Perutz, 1976 [1265])

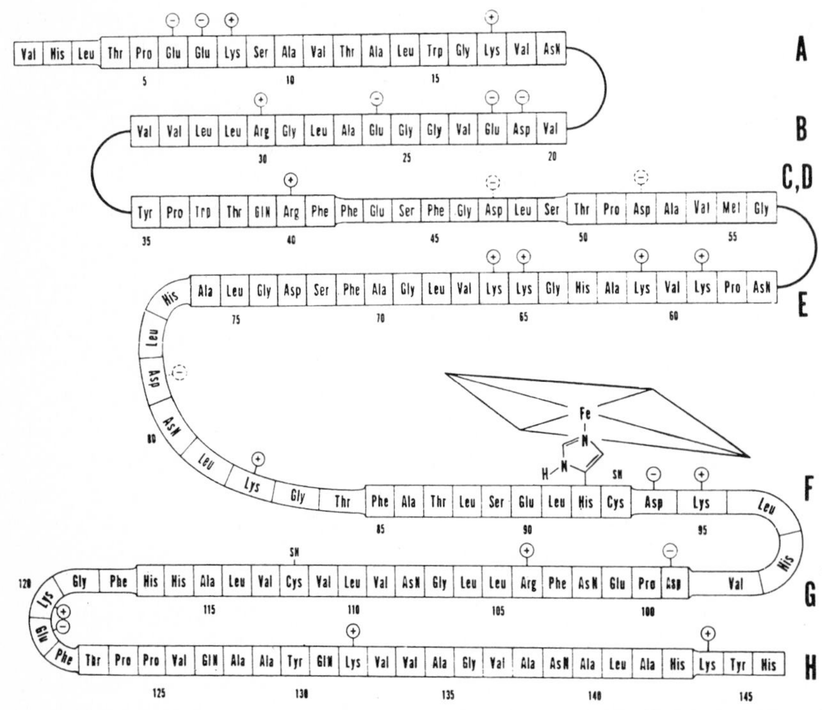

Fig. 4.35. The primary structure of amino acid sequence of the β chain of adult normal human hemoglobin (Hb A). Amino acids that are oriented in the characteristic folding of an α-helix are shown as *square boxes*. Nonhelical residues are depicted as *rectangles*. The site for heme attachment is shown. The specific amino acid sequence of the β-globin chain and its various features may be usefully compared with the three-dimensional molecular arrangement shown in Fig. 4.34

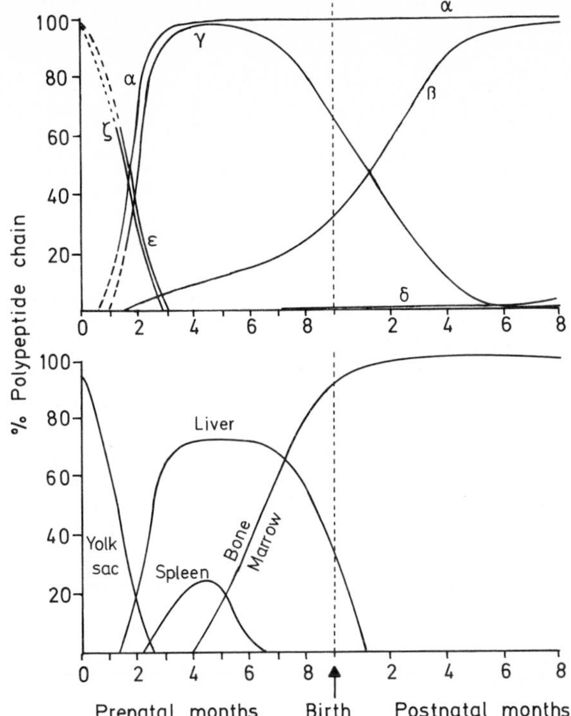

Fig. 4.36. Ontogeny of human hemoglobin chains before birth and in the first few months after birth. The *upper diagram* shows the characteristic developmental patterns of various globin chains. The *lower diagram* shows characteristic sites of erythropoiesis during development. There is remarkable similarity in the time sequences of the yolk sac and ε and ζ chain, hepatosplenic and γ chain, and bone marrow and β chain erythropoiesis. (Motulsky, A.G. in Cong. Malformations, Frazer, F.C., McKusick, V. (eds.), 1970 [1230])

ε chains have similarities to the β chains [1322]. The ζ chains probably are the earliest globin chains. The ζ and ε chains disappear after 8–10 weeks of embryonal life (Fig. 4.36) [1364]. The principal hemoglobin of fetal development is HB F ($\alpha_2\,\gamma_2$) with its characteristic γ chain. There are two types of γ chains: those with alanine at position 136 ($^{A}\gamma$) and those with glycine at that position ($^{G}\gamma$). A third type of γ chain with threonine instead of isoleucine at position 75 in the γ chain exists [1281, 1319]. Its frequency ranges between 10%–15% and it does not appear related to any discorder. Adult hemoglobin already can be demonstrated in fetuses at 6–8 weeks of age [1319, 1364].

While γ chain synthesis during fetal life largely occurs in liver and spleen, γ chains can also be produced by marrow erythropoietic cells. Conversely, while β chains in childhood and later are produced in the bone marrow, β chain production can also occur in extra marrow sites [1364]. The various normal hemoglobins are listed in Table 4.12.

Table 4.12. Human hemoglobins

Stage	Hemoglobin	Structure
Embryonic	Gower I	$\zeta_2\varepsilon_2$
	Gower II	$\alpha_2\varepsilon_2$
	Portland I	$\zeta_2\gamma_2$
Fetal	F	$\alpha_2{}^{G}\gamma_2$
		$\alpha_2{}^{A}\gamma_2$
Adult	A	$\alpha_2\beta_2$
	A_2	$\alpha_2\delta_2$

All of the normal human hemoglobins that have been investigated have an identical three-dimensional structure (Fig. 4.34), which is essential to allow the carriage of oxygen. All globin chains of the various hemoglobins had a common evolutionary origin and originated from each other by genetic duplication (see Sect. 7.2.3). The closer the resemblance between two chains, the more recent in evolutionary times duplicatory events occurred. Thus, the $^{A}\gamma$ and $^{G}\gamma$ chains with a single difference between them arose most recently while the β and α chain duplication had a more remote origin.

Hemoglobin Genes. The amino acid sequence of each of the globin chains is specified by a unique globin gene. A normal human therefore possesses at least one α, β, γ, δ, ε, and ζ gene in the haploid state or at least two of these genes in the diploid state. The gene for the α chain in most human populations exists in a duplicated state with no known differences between the two α genes [1350]. There are two γ genes – differing in the codon specifying position 136 $^{A}\gamma$ and $^{G}\gamma$. Some $^{A}\gamma$ genes have a variant codon specifying threonine rather than isoleucine at position 75 ($^{TA}\gamma$).

Genes also participate in synthesis of the nonprotein heme group by specifying a series of biosynthetic enzymes involved in heme synthesis.

The various globin genes with their respective globin chains and the various normal hemoglobins are shown in Table 4.12 and Fig. 4.37.

Extensive DNA sequence analysis has been carried out on all hemoglobin genes and their structure has been fully documented [981, 1041, 1200, 1273, 1304, 1314]. The human hemoglobin genes exist as two separate clusters of related multigene families, a frequent type of organization of mammalian genes (Fig. 4.38; 39). The α gene cluster is located on the short arm of chromosome 16 over a 25000 base pair (25 kb) region. The gamma-β-delta family is situated on the short arm of another chromosome, 11, across a 60 kb region. The genetic mechanisms which regulate coordinated gene function on the two different chromosomes to allow equal output of α and non-α

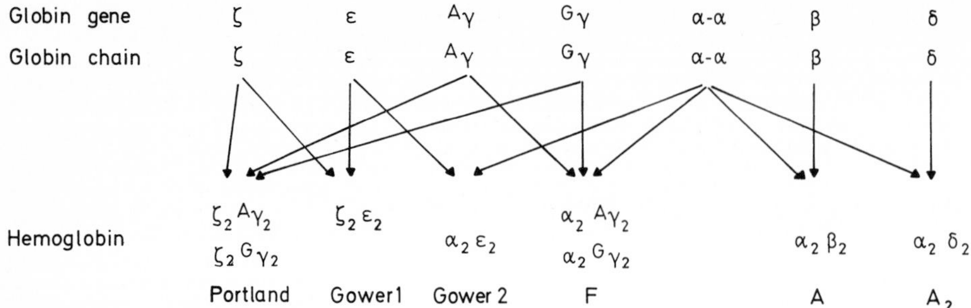

Fig. 4.37. Normal human globin genes. Single genes exist for Hb β, δ, ε and ζ. The genes for Hb α and Hb γ are duplicated. The products of the two hemoglobin γ genes (Hb $^{A}\gamma$ and Hb $^{G}\gamma$) differ from each other by a single amino acid residue, alanine (A) or glycine (G) at position 136. There are no known differences between the two Hb α genes. Tetrameric hemoglobin formation is shown in the lower portion of the figure

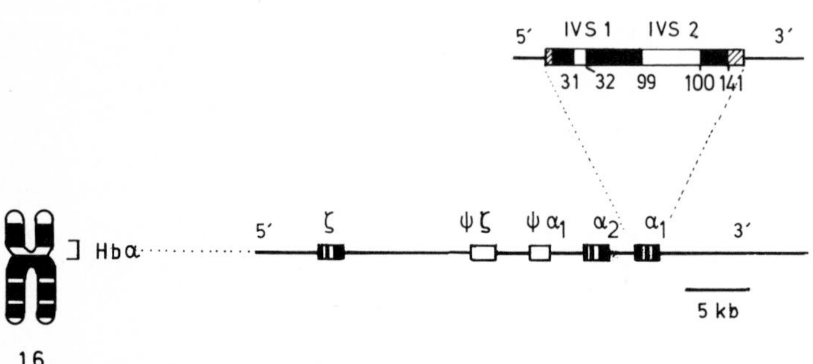

Fig. 4.38. Chromosomal location (16p) and organization of the human α globin gene cluster. ψ, pseudogene; *IVS*, Introns (intervening sequences, white boxes). 31,32,99 = number of base pairs in introns [972]

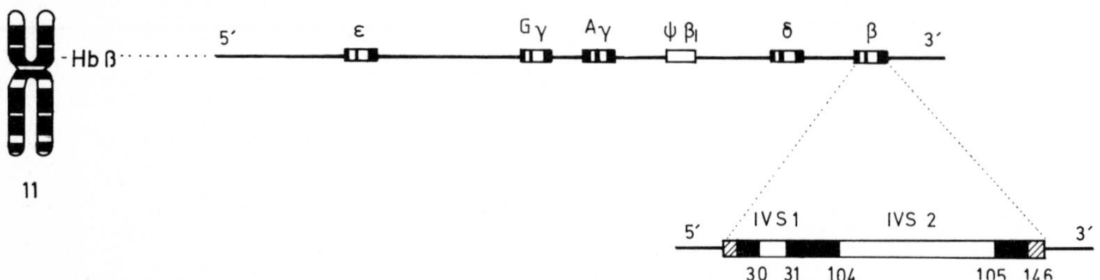

Fig. 4.39 Chromosomal location (11p) and organization of the human β globin gene cluster. Symbols and explanation identical as for Fig. 4.38 [972]

genes (such as β and gamma) remain unknown. The structural genes of the Hb α complex [from 5′ (upstream) to 3’ (downstream)] include: the embryonic ζ gene, a pseudogene for Hb ζ, a pseudogene for Hb α, and two identical α genes (Fig. 4.38). Similarly, the location of the various genes on the β cluster are: the embryonic epsilon gene, two fetal gamma genes (^{A}gamma and ^{G}gamma), the Hb α pseudogene, and a Hb delta and a Hb β gene (Fig. 4.39). The 5′ to 3′ arrangement of these genes is in the order of ontogenetic expression during development. Pseudogenes have DNA sequences that resemble those of their homologues. However, various mutational alterations have inactivated transcription, so that there is no functional expression. Pseudogenes are presumably duplication products that arose during evolution and were no longer required for normal function. The Hb delta gene whose gene product comprises only 2%–3% of total hemoglobin can be conceived of as a gene in transition to becoming a pseudogene.

All the globin genes have many functional similarities in organization. Three exons or coding sequences code for the unique amino acid sequence of each globin gene. Between exon 1 and exon 2 and between exon 2 and exon 3 there are unique inter-

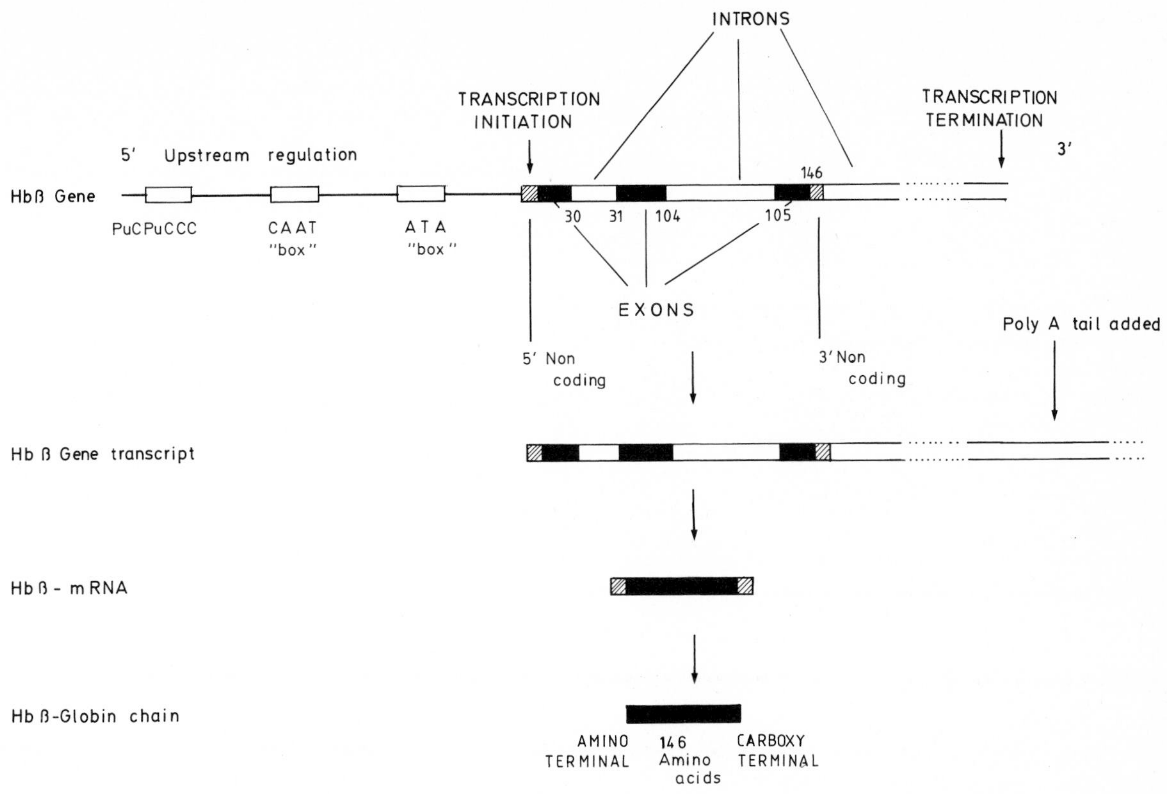

Fig. 4.40. Diagrammatic representation of Hb β gene, Hb β gene transcript, Hb β mRNA and Hb β globin chain. Note regulatory sequences, exons and introns. The poly A "tail" is added to the transcript and the introns are excised

vening sequences (IVS) or introns known as IVS-1 and IVS-2, respectively (Fig. 4.38, 39, 40). These introns are transcribed along with exons so that the initial gene transcripts reflect both coding and nocoding DNA sequences of the respective gene. Intervening sequences are excised during nuclear processing so that the terminus of exon 1 is spliced to exon 2 and the end of exon 2 to exon 3 to form functional mRNA that directs hemoglobin production on the ribosomes (Fig. 4.40). The two intervening sequences of the different genes on the γ-δ-β cluster are identical, but differ from shorter intervening sequences on the α cluster. The full details of splicing are not yet understood, but study of β-thalassemia mutations that interfere with normal excision and splicing (see below) has helped to elucidate this process. All introns start with GT (donor sites) and end with AG (acceptor sites) - these dinucleotides are part of so-called consensus sequences at the splicing sites. (See [1041 and 1238] for details.) Some details of the pathway leading from the gene to the hemoglobin molecule are shown in Figs. 4.40 and 41.

The biochemical evidence for nonlinkage of the Hb α and Hb β genes was preceded by genetic evidence that offspring of a mating of the double heterozygote for both an Hb α and Hb β mutation with a normal individual included four phenotypes: normal, Hb α^x, Hb β^x, and the double mutation (Fig. 4.42) [1014]. With close linkage of the Hb α and Hb β genes, both Hb α^x and Hb β^x mutations but not the parental types or normals would have been found among the offspring. Similarly, genetic evidence for close linkage between Hb δ and Hb β came from failure to find recombinants among children from matings between double hetrozygotes for both the Hb δ and Hb β mutations [1013] (Fig. 4.43). The existence of Hb Lepore, a δ-β fusion gene [1350], provided biochemical evidence for linkage of Hb δ and Hb genes on the same chromosome (see below). Linkage between γ and β genes was inferred by demonstrating that Hb Kenya was a γ-β fusion gene.

Promoters. Three different but similar sequences are located upstream (5′) of any gene and appear to be involved in the regulation of transcription (Fig. 4.40). These are also known as promoter regions [1041, 1238]. They include the TATA or ATA (Hogness) box 30 base pair proximal to the initiation site. This sequence serves as a promoter region for location of the precise site of transcription initiation. Another invariant sequence CAAT (at about minus 80 base pairs) is a recognition site for RNA polymerase. A

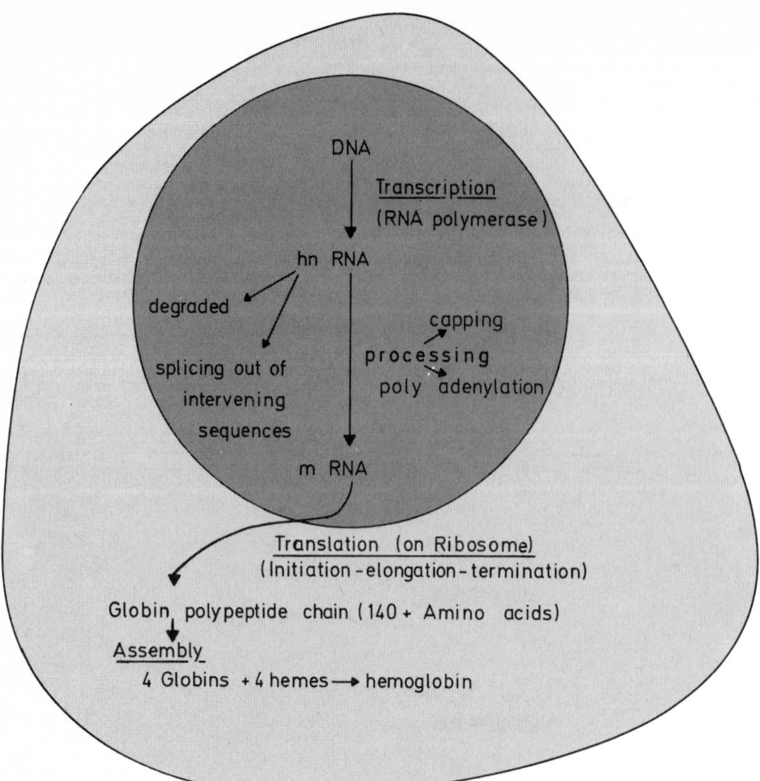

Fig. 4.41. Overview of protein synthesis with hemoglobin as a model. The nucleotides of the DNA hemoglobin gene are transcribed (transcription) by the enzyme RNA polymerase to form heterogeneous nuclear RNA (hn RNA). Intervening sequences not specifying structural information are spliced out. mRNA moves out of the nucleus *(dark grey)* into the cytoplasm *(ligth grey)* were globin synthesis or translation proceeds on the ribosomes by initiation, elongation, and termination. The globin polypeptide chain is formed and heme is inserted. Four globin chains form the functional hemoglobin molecule

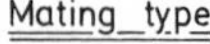

Genotype β^S][β^A α][$\alpha^{Hopkins\text{-}2}$ × β^A][β^A α][α

Phenotype Hb A; Hb S; Hb Hopkins-2; Hb Hopkins-2/S Hb A

Offspring

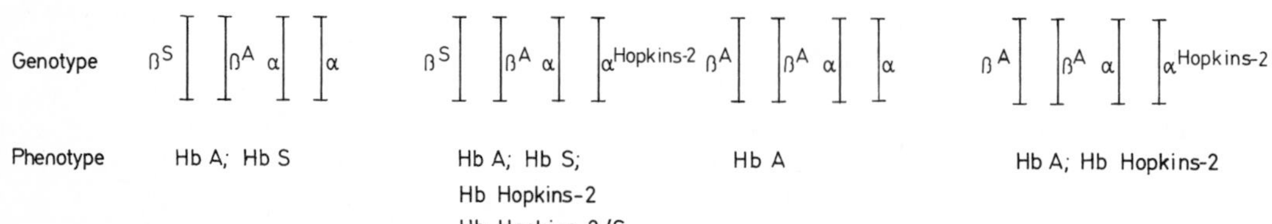

Fig. 4.42. Formal genetics of a mating between a double heterozygote for Hb α ($\alpha^{Hopkins\,2}$) and Hb β (Hb β^S) with a normal person. Since the genes for Hb α and Hb β are located on different chromosomes, free assortment of all chromosomes occurs and four classes of offspring in equal proportion were found: normal (Hb A); Hb A/S: sickle cell trait; Hb A/Hb Hopkins-2: Hopkins-2 trait and compound heterozygosity for Hb S and Hb Hopkins-2 identical to that of the affected parent. If the genes for Hb α and β were closely linked, the parental phenotypes would not be formed in the offspring except for possible recombination (Sect. 3.4). The closer the linkage, the less the chance of recombination (see Fig. 4.43)

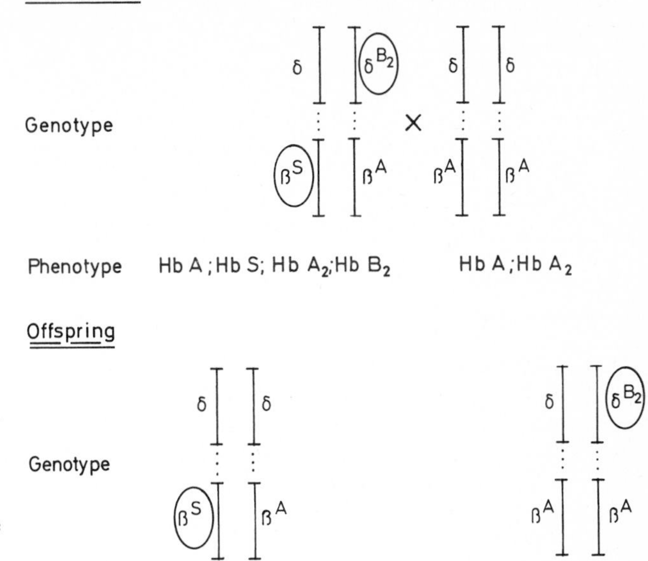

Fig. 4.43. Formal genetics of a mating between a double heterozygote for Hb β (Hb β^S) and Hb δ (Hb δ^{B_2}). The genes for Hb β and Hb δ apparently are closely linked on the same chromosome. All offspring either inherit the β^S or the δ^{B_2} abnormality. No normals or compound heterozygotes such as the parental type were seen among the children. These findings are consistent with close linkage of the two genes

third distal element is located at 80–100 base pairs and has the characteristic sequence PuCPuCCC (Pu, purine). Whether globin production requires other enhancers, i.e., genetic elements that influence efficience of transcription independent of position or orientation, remains unknown.

Downstream Sequences. – Transcription terminates about 1000 base pairs downstream from exon 3 of the β gene (Fig. 4.40). The highly conserved sequence AAU AAA provides the signal for endonucleolytic cleavage of RNA, which is followed by addition of the poly A tail of 220 residues. This poly A is not coded for by DNA at the globin gene site. The poly A nucleotides are required to stabilize mRNA, which carries the genetic information from the genes of the nucleus to the ribosomes where globin synthesis occurs by joining amino acids in their characteristic sequence (Fig. 4.41).

DNA Polymorphisms at the Globin Genes [972, 1253]. Gene mapping by restriction enzyme analysis of the $\gamma\delta\beta$ gene cluster led to the recognition of considerable variation in DNA sequence between different individuals (Fig. 4.44). All recognized DNA variants at the Hb β gene complex are single nucleotide substitutions and are symbolized as either present (+) or absent (−). Among 17 polymorphic sites at the Hb β cluster, 12 were located at flanking DNA, 3 within introns, 1 within a pseudogene, and 1 only within the coding (synonymous) portion of the Hb β gene. This distribution is not unexpected, since mutations affecting the coding regions would be more likely to cause harmful effects. Presumably, since much of the DNA between coding blocks is not expressed, sequence variation usually has no functional consequences. The various polymorphic sites are of ancient origin since they are found in all racial groups (Table 4.14). Some variants occur as polymorphisms only in blacks and not in other racial groups.

Two DNA polymorphisms at the Hb α locus show another frequent type of DNA variation – hypervariable regions (HVR) which consist of different numbers of randomly repeated DNA segments of 36 nucleotides, (Sect. 2.3.3.9).

A specific arrangement of polymorphisms at a gene cluster (or a gene locus) has been termed a haplotype. As an example, a given array of five polymorphisms might be symbolized as +-+-+ in an upstream (5′) to downstream (3′) direction. The term "framework" has been used to denote four basic haplotypes determined by minor variation at 5 sites (Table 4.13) of the β globin gene and represents the background upon which mutations leading to the hemoglobinopathies and thalassemias later occurred.

A remarkable feature of the DNA variants at the β gene cluster is their linkage disequilibrium. (Sect. 3.5.4) If there was free recombination over many generations one would expect random associations of any two polymorphic sites and a very large number of haplotypes (2^n where n is the number of polymorphisms; with 2^4 = there would be 16 expected haplotypes). Instead, only a few haplotypes have been found. For instance, strong linkage disequilib-

Table 4.13. β-globin framework sequences and their frequencies. (After [972 and 1253])

Designation	Codon 2	Intron 2				Frequency		
		Pos. 16	Pos. 74	Pos. 81	Pos. 666	Medit.	U.S. Blacks	S.E. Asians
1	CAC	C	G	C	T	53%	79%	18%
2	CAC	C	T	C	T	28%	12%	35%
3-Asians and Blacks	CAT[a]	G	T	C	C		9%	47%
3	CAT[a]	G	T	T	C	19%		

[a] Synonymous valine mutation.

rium for eight polymorphisms upstream from the delta gene exists (1-8 in Fig. 4.44) so that four haplotypes account for 94% of all chromosomes. Similarly, four haplotypes accounted for 90% of haplotypes for five polymorphisms (12-17 in Fig. 4.44), which are located over an 18 kb region downstream and including the β gene. Surprisingly, complete randomization was found when these upstream und downstream clusters of polymorphisms were compared. The most reasonable interpretation postulates a high recombination rate at a site separating these clusters - a recombinational hot spot; and a recombination has already been found in one family. The exact boundaries of this area of high recombination have not yet been defined.

Table 4.14. Frequency of DNA polymorphic sites in the β globin gene cluster in different groups. (From [972])

Polymorphisms	Greeks	American blacks	S.E. Asians
Taq I (1)[a]	1.00	0.88	1.00
Hine II (2)	0.46	0.10	0.72
Hind III (3)	0.52	0.41	0.27
Hind III (4)	0.30	0.16	0.04
Pvu II (5)	0.27		
Hinc II (6)	0.17	0.15	0.19
Hinc II (7)	0.48	0.76	0.27
Rsa I (8)	0.37	0.50	
Taq I (9)	0.68	0.53	
Hinf I (10)	0.97	0.70	0.98
Rsa I (11)			
HgiA (12)	0.80	0.96	0.44
Ava II (13)	0.80	0.96	0.44
Hpa I (14)	1.00	0.93	
Hind III (15)	0.72	0.63	
Bam HI (16)	0.70	0.90	
Rsa I (17)	0.37	0.10	

[a] The number in parentheses refer to the restriction enzymes in Fig. 4.44.

Hemoglobin Variants. Hemoglobin variants are caused by a variety of mutational events affecting a given hemoglobin gene. The most common hemoglobin variants are amino acid substitutions affecting a single amino acid of a globin chain. About 350 such substitutions have been described [119]. These substitutions are caused by replacements of a single nucleotide in a given codon triplet of the DNA, which changes the mRNA triplet to one that specifies a different amino acid such as GUA (valine) to GAA (glutamic acid) (see third line of Fig. 4.45). If the electric charge of the mutant amino acid is different, the variant hemoglobin can be recognized by its altered behavior on electrophoresis. Mutations that do not change electrophoretic charge are usually only detected if they affect hemoglobin function deleteriously and cause disease. Most hemoglobin mutations regardless of whether they do or do not affect electrophoretic charge have no effect on hemoglobin function and are compatible with normal health. In general, amino acid substitutions of the exterior of the hemoglobin chains cause fewer perturbations of function than those replacing amino acids in the chain interior or close to the insertion of the heme group. Substitutions affecting normal helical turns of the chain often cause hemoglobin instability. Amino acid replacements affecting subunit contacts often are associated with abnormalities in oxygen affinity [1320]. Most hemoglobin variants are rare. A few, such as Hb S, Hb C, and Hb E, have reached polymorphic higher frequencies and are further discussed in Sect. 6.2.1.6.

Polymorphisms at the nucleotide level in the coding area also exist. The genetic code is degenerate (Table 2.12), i.e. several code words can code for an identical amino acid (see Fig. 4.45). Considerations of two different amino acid substitutions at Hb β^{67} (Fig. 4.45) show that the original codon for valine at Hb β^{67} from which the mutation occurred must have been different in the two individuals where the mutation occurred and the new hemoglobin arose. The details are shown in Fig. 4.45. Different individuals may thus use different codons for the same amino acid.

Clinical Effects of Hemoglobin Variants. The results of compromised hemoglobin function can produce different types of disease. Four principal categories of hemoglobin diseases exist: (1) hemolytic anemia

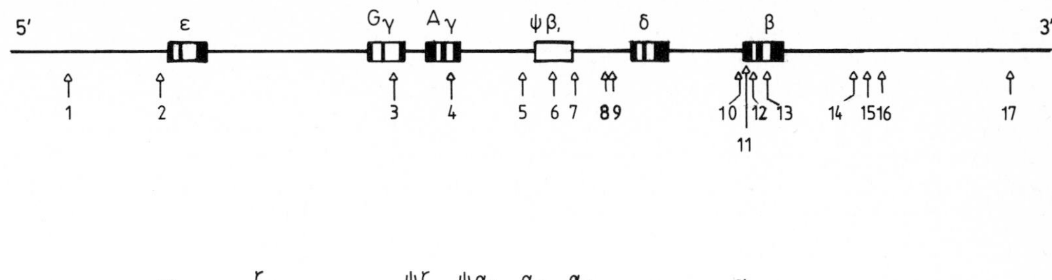

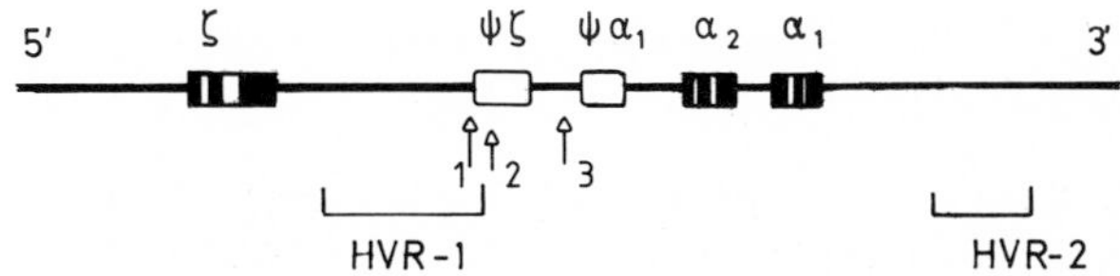

Fig. 4.44. Restriction enzyme polymorphisms at the Hb *β (top)* and Hb *α (below)* genes. The numbers refer to the sites where different restriction enzymes cut the DNA. HVR refers to hypervariable regions (minisatellite) (Table 4.14)

due to unstable hemoglobins, (2) methemoglobinemia due to more rapid hemoglobin oxidation, (3) erythrocytosis due to abnormal oxygen affinity, and (4) sickle cell disorders due to distortion of the red cell membrane by Hb S. In all cases except for the sickling disorders, heterozygotes for the various abnormal hemoglobins will be affected with the various diseases, i.e., the mutations manifest as autosomal dominants (see also Sect. 4.6).

Unstable Hemoglobins [31, 1355–1357]. Over 100 unstable hemoglobins have been described. Most affect the *β* chain. Many unstable hemoglobins have amino acid substitutions or deletions affecting the heme pocket of the globin chain. Clinical manifestations vary from mild instability, which is not clinically apparent, to severe instability, which causes increased blood destruction. Sulfonamides have produced more severe hemolysis in several unstable hemoglobins. The instability of these hemoglobins is often caused by premature dissociation of the heme from the globin chain. Such heme-depleted globin is precipitated as intracellular material known as Heinz bodies and interferes with cell membrane function. Heinz bodies may be removed ("pitted") by the spleen without destruction of the red cells that carry them. Ultimately, such red cells are removed prematurely by the reticuloendothelial system. In some unstable hemoglobins, splenectomy may ameliorate the severe hemolysis.

The diagnosis of unstable hemoglobins if not associated with electrophoretic mobility alterations is difficult and may require isolation of the precipitated globin chains for further analysis in specialized laboratories. The unstable hemoglobins contribute to the heterogeneous class of the congenital nonspherocytic hemolytic anemias. Such hemoglobins have been found as fresh mutations and identical unstable hemoglobins (i.e., Hb Köln, Hb Hammersmith) have been found several times as a new mutation in different individuals from different families [31].

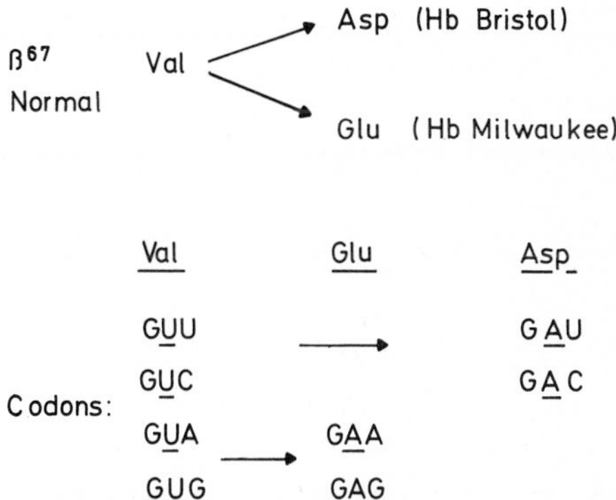

Fig. 4.45. Codon polymorphism. The usual amino acid at position 67 of the Hb *β* chain is valine. Hb Bristol and Hb Milwaukee are caused by different mutations at that site replacing the original valine with Glu (Hb Bristol) and Asp (Hb Milwaukee). The possible codon triplets for valine are shown in the *lower part* of the figure. A mutation from Val to Asp could only have arisen from GUU or GUC while that from Val to Gln could only have originated from GUA or GUG. Consequently, the original persons who underwent the two different mutations must have used different codons for specifying the normal valine at Hb β^{67}

Methemoglobinemia Due to Hb M [31]. Hb M is of historical interest in that the first globin abnormality was discovered as a dominant trait in a family with congenital cyanosis by Hörlein and Weber in 1948 [1130]. It is interesting that the first discovered human enzyme deficiency was the recessively inherited methemoglobin reductase deficiency, which also produces methemoglobinemia [1100]. Methemoglobinemia, therefore, can be caused both by a dominantly inherited globin abnormality as well as by a recessively inherited enzyme deficiency.

Five different mutations can produce Hb M. Methemoglobinemia is caused by the more rapid oxidation of divalent iron to trivalent iron (Table 4.15). Four Hb M's are caused by tyrosine replacements of the histidine residues that anchor the heme group in its characteristic pocket (Fig. 4.34) of the globin molecule and stabilize the heme iron in its oxidized form. The fifth mutation - Hb Milwaukee 1 - cannot yet be fully explained on molecular grounds. Patients with Hb M mutations of the α chain are cyanotic from birth. Those with Hb M mutation of the β chain do not develop severe cyanosis until 6 months of age, when the γ chains are replaced by β chains. Mild hemolysis is common in patients with Hb M.

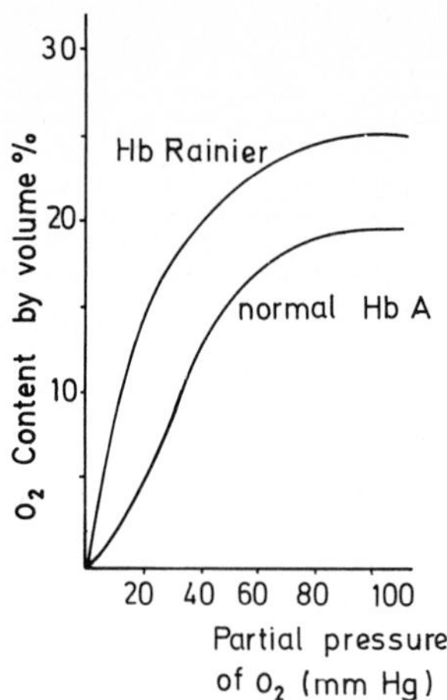

Fig. 4.46. O_2 dissociation curve of a hemoglobin with increased oxygen affinity. Note that the abnormal hemoglobin Rainier does not release oxygen at lower partial pressures of oxygen as readily as normal hemoglobin. Tissue hypoxia results and stimulates erythropoietin formation with resultant erythrocytosis

Table 4.15. Hb M

α chain	Hb M
M Boston	α^{58} His→Tyr
M Iwate	α^{87} His→Tyr
β chain	
M Saskatoon[a]	β^{63} His→Tyr
M Hyde Park	β^{92} His→Tyr
M Milwaukee 1	β^{67} Val→Glu

[a] Hörlein and Weber's classic Hb M [1130].

Erythrocytosis Due to Hemoglobins with Abnormal Oxygen Affinity [31, 992]. About 30 hemoglobins with increased oxygen affinity exist. Substitutions affect α_1 β_1 contact of the tetramer in 11 instances. Movement of the globin subunits during oxygenation occurs at this interchain contact. Stabilization of the oxy conformation or destabilization of the deoxy conformation by the mutation may result in increased oxygen affinity (Fig. 4.46). Most other hemoglobins with high O_2 affinity have substitutions of the COOH terminal of the β chain or at binding sites of diphosphoglycerate (DPG) regions, which normally are involved in maintenance of stability of the deoxy conformation.

The increased oxygen affinity reduces oxygen delivery to the tissues with resultant hypoxia (Fig. 4.46). Hypoxia leads to release of the hormone erythropoietin, which stimulates red cell production with resultant erythrocytosis. Patients with erythrocytosis due to abnormal hemoglobins are sometimes erroneously diagnosed as suffering from polycythemia vera. A dominant pattern of inheritance and the absence of splenomegaly, leukocytosis, and thrombocytosis differentiates erythrocytosis due to an abnormal hemoglobin from polycythemia vera. Occasional cases of this type of abnormal hemoglobin have occurred as fresh mutations.

Only three hemoglobins with reduced oxygen affinity have been detected [992]. With increased oxygen delivery to the tissues caused by the reduced affinity for hemoglobin, a lessened production of erythropoietin would be expected. The expected mild anemia has been demonstrated with two of these hemoglobins.

Sickle Cell Disorders [31, 1211, 1298]. Hb S is caused by the substitution of valine for glutamic acid in the sixth position of the β chain. Unlike all other substitutions, this particular mutation grossly affects the solubility and crystallization of this hemoglobin under conditions of hypoxia. Patients with sickle cell anemia inherit the abnormal gene from each of their parents and lack Hb A. With a relatively low degree of hypoxia, the Hb S of such patients polymerizes into filaments of high molecular weight that associate to form bundles of fibers. These abnormal hemoglobin crystals distort the red cell membrane to its characteristic sickling shape (Fig. 4.47). Some of these cells remain irreversibly sickled and are destroyed prematurely. Sickled cells increase blood viscosity and impede normal circulation in small blood vessels. The resultant hypoxia leads to more sickling with a vicious cycle of more stagnation and characteristic episodic sickle crises with abdominal and musculoskeletal pain. After several years, necrosis of poorly perfused tissues, such as in the spleen, often occurs and this organ atrophies.

Patients with sickle cell trait who have one normal (Hb β^A) and one abnormal gene (Hb β^S) only have 25%–40% Hb S. These individuals are clinically quite normal. Their red cells contain both Hb A and Hb S, have a normal red cell life span, and in vivo

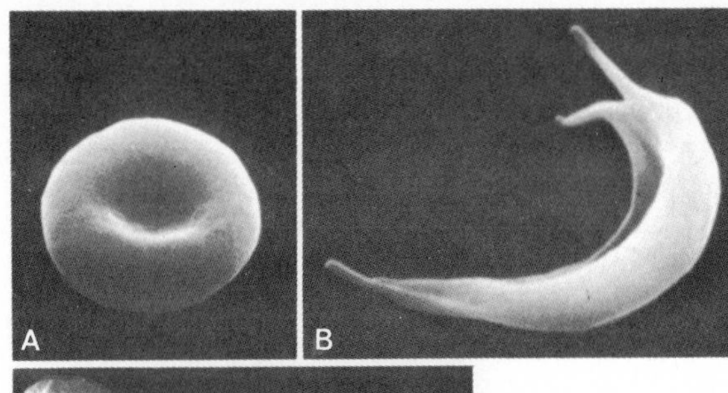

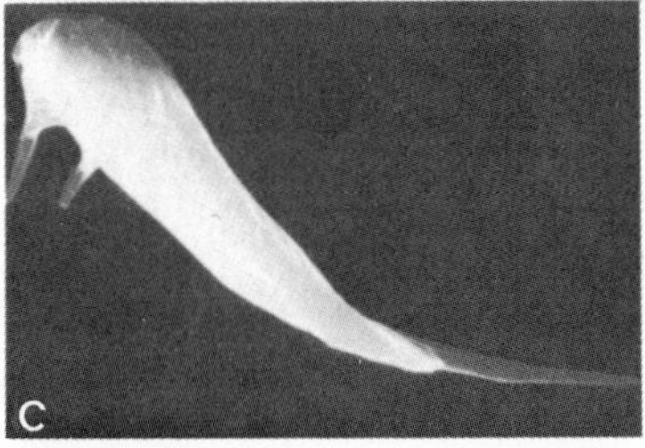

Fig. 4.47 A–C. Scanning electron micrographs of oxygenated (**A**) and deoxygenated (**B** and **C**) red cells from a patient with homozygous sickle cell anemia. Note the normal biconcave shape of the red cell without any Hb A and distortion under conditions of hypoxia. The appearance of cells such as depicted in **B** led to the term *sickle* cell since the red cell resembles a sickle – an agricultural tool. (Bunn et al., 1977 [31])

sickling only occurs under conditions of severe hypoxia, such as at atmospheric conditions over 3.000 m [1292].

Certain other hemoglobins when present together with Hb S decrease the extent of sickling. Hb F coexists with Hb S in red cells of patients with sickle cell anemia. Hb F reduces the gelling and crystallization of Hb S so that patients with sickle cell anemia and large amounts of Hb F have few or no symptoms of sickle cell anemia. Hb F in some of these instances is contributed by a gene for hereditary persistence of fetal hemoglobin (see below). In general, there is inverse correlation between the amount of Hb F and the severity of symptoms in sickle cell anemia. Any manipulation that would increase fetal hemoglobin production would therefore effect clinical improvement in sickle cell anemia [970]. The clinical manifestation of the thalassemias will be covered below.

4.3.3 Other Types of Hemoglobin Mutations
[1188; 1349]

Deletions. Deletions of whole hemoglobin genes have been identified. Deletion of Hb α genes causes α-thalassemia and deletion of both the Hb δ and Hb β genes causes hereditary persistence of fetal hemoglobin (HPFH) or Hb $\delta\beta$ thalassemia (see below).

The deletion of a single nucleotide triplet or codon would lead to a deletion of the amino acid specified by that codon. A deletion that will remove four codons or 12 nucleotides would cause deletions of four amino acids. Deletions of up to five amino acids corresponding to 15 nucleotides have been seen (see Table 4.16). It is likely that more extensive intraglobin deletions would be incompatible with formation of a viable hemoglobin molecule. Most deletions are either unstable or have increased O_2 affinity or both (Table 4.16). All deletions affect the Hb β gene. It is not quite clear why no Hb α deletions have been detected. Possibly, Hb α gene deletions are more deleterious during embryonic and fetal life where Hb β deletion would be less harmful (see Fig. 4.36).

If a deletion affects a number of nucleotides not divisible by 3, the continued reading of the code in triplets creates new sets of triplets that specify entirely different amino acids ("frameshift" mutants). A resultant globin structure can sometimes be identified. Hb Wayne (Fig. 4.48) appears to be caused by a deletion of a single nucleotide at the 139th codon near the terminus of the Hb α chain, which consists of 141 amino acids. The nucleotides of the termination codon at position 142 are read out of phase and the shifted reading frame continues until a new ter-

Table 4.16. Hemoglobin variants caused by deletions

Hb	Site of deletion	Amino acid residue(s) deleted	Properties
Leiden	β6 or 7	Glu	Unstable, ↑O_2 affinity
Lyon	β17–18	Lys, Val	↑O_2 affinity
Freiburg	β23	Val	↑O_2 affinity
Niteroi	β42–44 or β43–45	Phe, Glu, Ser	↓O_2 affinity, unstable
Tochigi	β56–59	Gly, Asn, Pro, Lys	Unstable
St. Antoine	β74–75	Gly, Leu	Unstable, normal O_2 affinity
Tours	β87	Thr	↑O_2 affinity, unstable
Gun Hill	β91–95 or β92–96 or β93–97	Leu, His, Cys, Asp, Lys	↑O_2 affinity, unstable
Leslie	β131	Gln	Unstable, normal O_2 affinity
Coventry	β141	Leu	Unstable
McKees Rock	β145–146	Tyr + His	↑O_2 affinity

mination codon (UAG) is encountered. Thus, a slightly elongated hemoglobin chain with five additional amino acids specified by nucleotides of the downstream flanking area of the α gene (see Figs. 4.38 and 4.48) results. Since the reading frame is shifted, the sequence of these amino acids will be different from the downstream amino acid sequence of the terminator mutations of the Hb α gene such as Hb Constant Spring (see Fig. 4.48) which is translated in phase.

It is not unexpected that the deletion characteristic of Hb Wayne was identified close to the end of the α chain. Any deletions of nucleotides that lead to "reading frame" shift errors at positions other than the end of the structural globin are unlikely to specify viable globin sequences. The resultant phenotypes would therefore be those of "thalassemia" without an identifiable gene product, e.g. β^{0} thalassemia (see below).

Deletions most likely occur because mispairing of homologous sequences of nucleotides takes place either during meiosis or during mitotic division in germ cell development. A consideration of nucleotide sequences around the areas of deletions for the various deletion mutants show expected homologies that would facilitate mispairing. Recombination or crossover events following mispairing then may lead to deletions of various sizes.

Fusion genes may be another result of mispairing. The homology of various globin genes may lead to mispairing between similar but not identical genes and nonhomologous crossover may lead to fusion genes that consist of the NH_2 terminal portion of one globin and the COOH terminal portion of another. "Hb Lepore" is a Hb δ-β fusion gene (Fig. 4.49), and several kinds of Hb Lepore with differing amounts of δ and β gene material depending upon the site of crossover exist (Fig. 4.49). In the various Hb Lepore, both the normal Hb δ and Hb β genes are deleted, and a new Hb δ-β fusion gene replaces them (Fig. 4.49). In Hb Kenya, misalignment between Hb$^{A}\gamma$ and Hb β occurs with crossover and resultant deletion of Hb$^{G}\gamma$, Hb$^{A}\gamma$, and Hb δ with a new chromosome consisting of HB$^{G}\gamma$ and a Hb$^{A}\gamma$-β fusion gene (Fig. 4.51).

Duplications. Duplications may affect whole genes, such as the duplications during evolution that led to the various globin chains (α, β, γ, δ, ε, ζ) (Sect. 7.2.3). The existence of two α globin genes and two γ globin genes ($^{A}\gamma$ and $^{G}\gamma$) on a single chromosome are examples of more recent evolutionary duplications. Intragenic duplications are known. In Hb α Grady residues 116–118 are duplicated [1136].

Duplications of one or two nucleotides may lead to frameshift mutations. Such frame shifts have been discovered near the terminus of the β chain [31]. Hb Tak is caused by duplication of the nucleotide AC following the 146th position and Hb Cranston owes its origin to duplication of the nucleotide AG following the 144th position of the β chain (Fig. 4.50). Hb Cranston has unique amino acids at positions 145 and 146. Hb Tak has a normal sequence up to and including position 146. The β chain normally has 146 amino acids. The frameshifts by insertion of the two nucleotides in both Hb Cranston and Hb Tak makes for an identical reading frame following positon 146. Both hemoglobins are elongated by the same amino acid sequence at the NH_2 terminal until a new stop codon (UAA) terminates the sequence at posi-

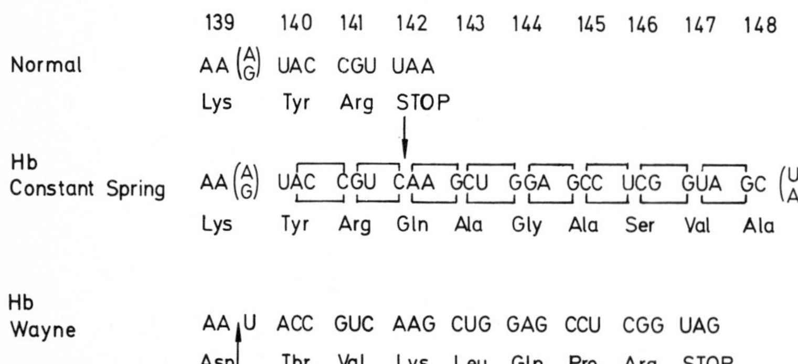

Fig. 4.48. 3′ end of Hb α gene. Note that in Hb Constant Spring, a mutation has changed the stop codon 142 UAA to CAA allowing translation of flanking nucleotides usually not expressed. The sequence of the first six nucleotides of a total of 30 additional nucleotides is shown. Hb Wayne is caused by deletion of the third nucleotide of the 139th codon. The first U nucleotide of the 140th codon is used as a third nucleotide of the 139th codon forming a new codon AAU, which specifies Asn. The resultant frameshift in reading the code results in hemoglobin Hb Wayne. The amino acid sequence of Hb Wayne can be predicted from the nucleotides of Hb Constant Spring by reading the shifted codes in multiples of three as shown by the *bracket above and below* the nucleotides of Hb Constant Spring. Hb Wayne has only five additional amino acids since a stop codon UAG is reached after translation of five codons

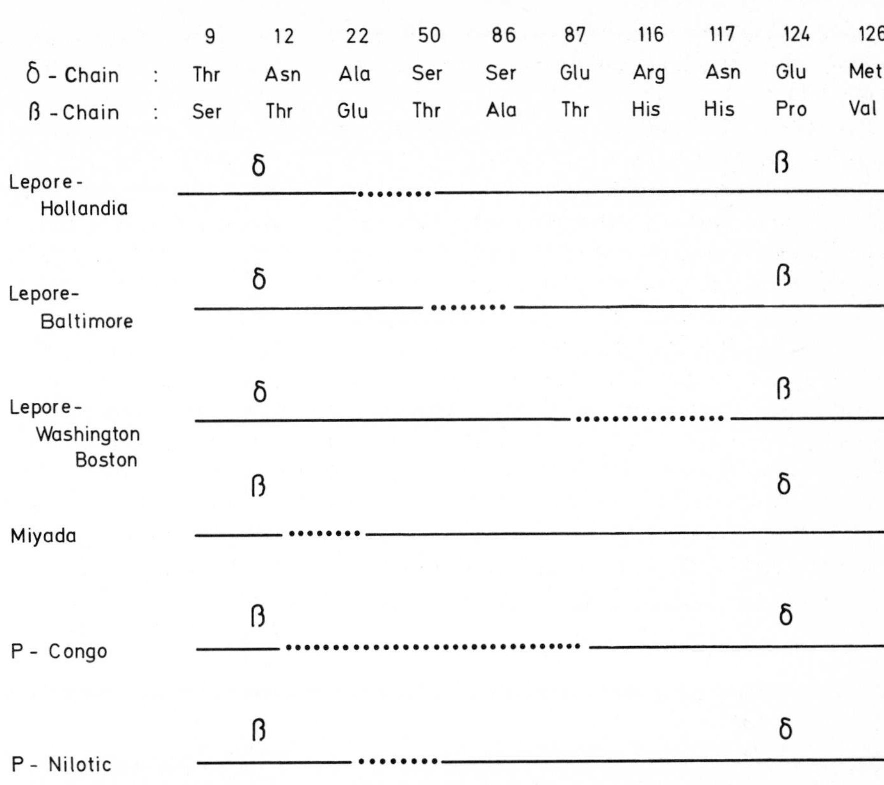

Fig. 4.49. $\delta\beta$ and $\beta\delta$ fusion genes. The ten differences between amino acids in the Hb γ and Hb β are shown in the *upper portion* of the figure. Otherwise, Hb δ and Hb β are identical in amino acid structure. Three different types of Hb Lepore have been found. In Hb Lepore Hollandia, the crossover between Hb δ and Hb β occurs between positions 22 and 50. The exact site of crossover is indeterminate since no differences between Hb δ and Hb β exist between these two positions. In Hb Lepore Baltimore, the crossover occurred between position 50 and 86 and in Hb Lepore Washington-Boston between positions 87 to 117. By similar reasoning, the sites of crossover are shown for various Hb β-δ or anti-Lepore hemoglobins. (Adapted from Forget [1082])

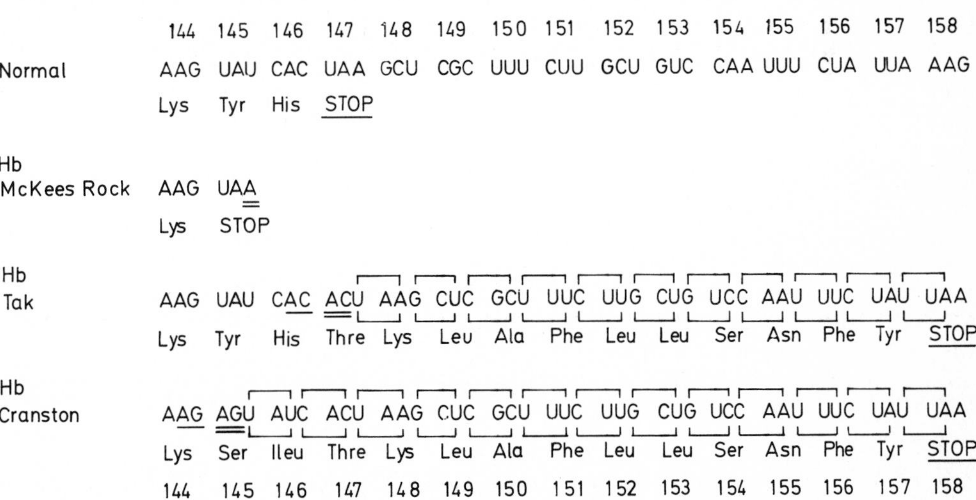

Fig. 4.50. 3′ end of hemoglobin β gene. The Hb β chain normally has 146 amino acids. Hb McKees Rock has 144 amino acids since a mutation has altered the codon UAU (Tyr) of the 145th position to the stop codon UAA. In Hb Tak and Hb Cranston the two last nucleotides AC at position 146 and the last two nucleotides AG of codon 144 have been duplicated respectively. The involved nucleotides are *singly* and *doubly underlined*. The resulting coding frameshift by two nucleotides makes for an identical amino acid sequence in Hb Tak and Hb Cranston starting with position 147 until the stop codon UAA at position 158 is reached. The *brackets* define the codon triplets of the normal sequence as shown in the upper part of the figure. The actual amino acid sequence of Hbs Tak and Cranston was determined and corresponded exactly to that of the actually determined structural nucleotides of the normal Hb β gene

tion 158. The new elongated sequence reflects the downstream flanking nucleotides of the β gene (Fig. 4.50).

Duplications of fewer than three nucleotides produce an out-of-phase reading frame and are not likely to give rise to a viable hemoglobin molecule if they occur at portions of the gene other than those specifying the terminus of the hemoglobin chains. Duplication gene products would also be expected from crossover events as the counterpart of fusion genes (Figs. 4.49 and 4.51). The resultant (δ, β-δ, β) gene product or Hb anti-Lepore has in fact been identified several times as Hb Miyada, P Congo, and P Nilotic (Fig. 4.49). The expected Hb anti-Kenya product (${}^{G}\gamma$, ${}^{A}\gamma$, δ, β-${}^{A}\gamma$, δ, β) (Fig. 4.51) has not yet been found.

Duplications presumably have the same origin as deletions and arose from mispairing followed by nonhomologous crossover as shown in Fig. 4.51.

4.3.4 Thalassemias and Related Conditions [31; 972; 1238; 1253; 222; 97a]

A variety of conditions are characterized by genetically determined diminished or absent synthesis of one or another of the hemoglobin chains. These diseases are known as the thalassemias. This term is derived from *thalassa*, the Greek word for the Mediterranean Sea, and was originally selected to describe the Mediterranean origin of many gene carriers of these conditions. While ethnologically and geographically incorrect, the term continues to be used widely. Thalassemias can be subdivided into α- and β-thalassemias. Different genetic mechanisms have been demonstrated to cause diminished production of globin chains and the etiology of the thalassemias is heterogeneous [1037].

Developments in understanding of the thalassemias at the molecular level have led to far better comprehension of the mutational lesions in this group of diseases than of any other mammalian mutations.

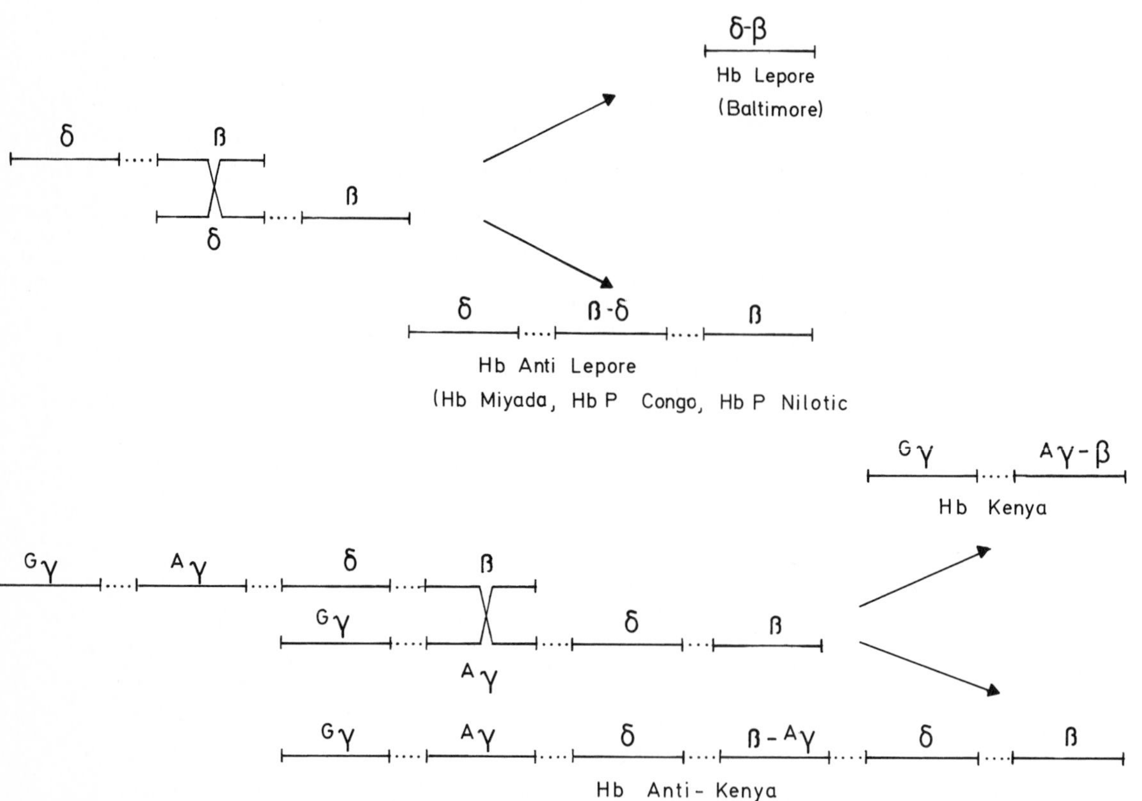

Fig. 4.51. Formation of hemoglobin fusion genes. Mispairing between the Hb δ and Hb β gene followed by recombination within the structural gene leads to a $\delta\beta$ fusion gene (Hb Lepore) with deletion of the normal Hb δ and Hb δ gene. The alternate product of such nonhomologous crossovers makes for creation of a β-δ fusion gene preceded by a normal Hbδ gene and followed by a normal Hb β gene. Such Hb anti-Lepore have also been discovered (Hb Miyada, Hb P, Hb Congo, see Fig. 4.49). Mispairing between the Hb β and Hb ${}^{A}\gamma$ genes followed by recombination has yielded a ${}^{A}\gamma$-β fusion gene known as Hb Kenya. The diagram indicates why the normal Hb ${}^{A}\gamma$, δ and β genes have been deleted and the Hb γ gene is preserved in cases of Hb Kenya. A postulated Hb anti-Kenya is pictured but has not yet been discovered

The elucidation of normal structure, function, and organization of the globin gene has been aided significantly by investigation of various thalassemia genes. It has become clear that mutational interference with the different steps involved in globin synthesis can reduce (β^+-thalassemias) or abolish globin production β^0-thalassemias [972, 1253, 1238, 4341] (Table 4.17; 4.19)

Transcription or Promoter Mutations. Thalassemia mutations that affect the noncoding 5′ upstream regions of the Hb β gene can be considered as regulatory mutations which affect gene transcription. Mutations at the more distal regulatory constant sequence PuCPuCCC and within the regulatory TATA box have been described (Table 4.17). These mutations diminish hemoglobin synthesis and manifest as relatively mild thalassemias. No mutations have yet been found at the CAAT box.

A RNA Cleavage Mutation. A single β^+-thalassemia mutation AATAAA→AACAA in the downstream flanking sequences of the Hb β gene has frequently been discovered among blacks, demonstrating that a downstream mutation can affect transcriptional efficiency. The apparent prevalence of the transcription mutation at the TATA box (see above) and the RNA cleavage mutation among blacks (Table 4.17) explains the rather mild nature of β-thalassemia in that racial group.

Terminator (Nonsense) and Frameshift Mutations. As explained above, mutations that lead to a terminator signal within exons coding for hemoglobin would produce a foreshortened nonfunctional globin chain and thus lead to β^0-thalassemia. Three such mutants have been found. One of these mutations is common in Mediterraneans (β^{39} C–T). A restriction enzyme (Mae 1) which recognizes this sequence has been identified and can be used for direct prenatal diagnosis of the β^{39} thalassemia mutation [1328 a].
Deletions or insertions of fewer or more than three bases produce frameshifts with garbled coding, causing effective termination of functional globin synthesis. Seven frameshift β^0-thalassemia mutations of this type have been identified in different populations. (Table 4.17)

RNA Processing Mutations. Processing of mRNA transcripts involves excision of intervening sequences with splicing of exons to make a functional mRNA molecule (Figs. 4.40, 4.52). Many different mutations affecting this process have been described. One set of mutations alters the dinucleotides GT (or AG) at the donor (or acceptor sites) of splice junctions. These dinucleotides are part of so-called consensus sequences that include several other nucleotides and are critical for splicing. If altered by single nucleotide alterations, splicing may be compromised to cause β^0-Thalassemia or β^+-thalassemia. So-called cryptic splice sites including the sequence GT and AG, which are probably never used during normal splicing, are sometimes activated by mutations elsewhere in the gene and cause interference with mRNA production.

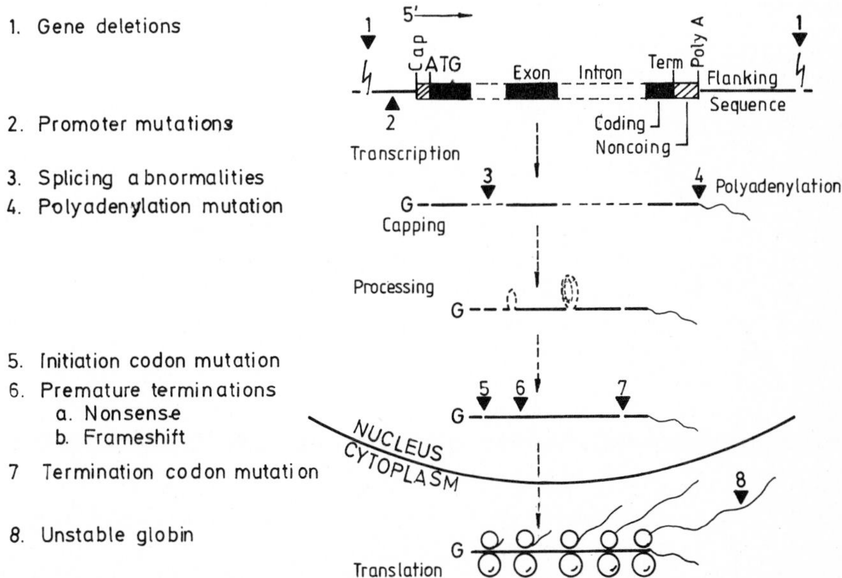

Fig. 4.52. Transcription and translation of hemoglobin genes. The numbers refer to the sites at which thalassemia mutations occur

Mutations in intervening sequences may create new splice sites within introns that compete with or retard normal processing thus causing thalassemia. Another class of mutations enhance existing cryptic sites. As an example, the common Hb E mutation activates such a site and is associated with decreased Hb β^E production - a type of thalassemia. The various splicing mutations are listed in Table 4.17. The location of all point mutations on the Hb β gene causing β-thalassemia is shown in Fig. 4.53

Deletion Mutations at the Hb β Globin Gene Cluster and Hereditary Persistence of Fetal Hemoglobin. Unlike the α-thalassemias (see below), most β-thalassemias are not caused by gene deletion. However, a 619-bp deletion extending from within intron 2 to past the end of the Hb β gene is the cause of over one-third of β-thalassemias among Asiatic Indians (Fig. 4.54, Table 4.18). Different rare Hb β deletions have also been described in a U.S. black and in a Dutch person. Several other more extensive deletions at the gamma-delta-β locus have also been detected. Their location and extent are shown in Fig. 4.54. None of these deletions can be recognized cytogenetically, since they are too small for microscopic detection. Hb Lepore and Hb Kenya are deletions which left parts of the functional genes intact and produced delta-β (Hb Lepore) and gamma-β (Hb Kenya) fusion genes. Several deletions have led to complete removal of all or almost all of the gene cluster with no synthesis of gamma, delta, and β chains. Conventionally, a functional distinction has been made between deletions that cause thalassemic (i.e., anemic) phenotypes (e.g., delta-β thalassemia) and deletions where fetal hemoglobin synthesis "compensates" for the absent delta and β loci (i.e., hereditary persistence of fetal hemoglobin; HPFH). This distinction is not absolute, since full compensation by gamma-chain production is not achieved in HPFH. The reason why some deletions activate the fetal hemoglobin gene remains unknown and is under active investigation.

Hereditary persistence of fetal hemoglobin can also be caused by nondeletion mutations. A point mutation has been found in the Greek variety of HPFH in the flanking sequence 5′ to the Aγ gene (−117) [1040]. A different point mutation at position 202 in a flanking region upstream of the Hb$^{G\gamma}$ has been detected among blacks with nondeletion HPFH [1093]. The sequences affected by these two point mutations are thought to represent key regulatory regions that are normally involved in postnatal suppression of gamma-chain synthesis. Studies directed at understanding the control of the Hb F switch have wide-ranging implications for treatment of thalassemia and sickle cell anemia, since increased Hb F production in these disorders would be of marked therapeutic benefit.

Heterocellular Hereditary Persistence of Fetal Hemoglobin [222]. A group of heterogenous genetic conditions is characterized by slight increase in fetal hemoglobin (2%–3%, but sometimes higher), which are unevenly distributed over the red cell population. This characteristic led to the designation of heterocellular HPFH. There is no anemia. When a gene for heterocellular HPFH interacts with Hb S or β-thalassemia, fetal hemoglobin levels are often higher than the usual 2%–3%. Particularly in sickle cell anemia, the resulting elevated Hb F levels may make for a mild clinical course. Heterocellular HPFH is transmitted as an autosomal dominant trait and is not associated with detectable deletions at the β gene cluster by restriction mapping. Its molecular basis remains unknown. There is some evidence that a gene that raises Hb F levels is not linked to the gamma-delta-β gene cluster [1012].

Clinical Implications. The β-thalassemias are widespread throughout the tropical and subtropical areas of the world and owe their frequency to a selective advantage vis-à-vis falciparum malaria [1227] (Sect. 6.2.1.6). β-Thalassemia heterozygotes have mild anemia (Table 4.20). Hb A_2 ($\alpha_2\delta_2$) is slightly increased. The red cells are smaller and less well filled with hemoglobin (MCH and MCV decreased) [31, 222]. Heterozygotes usually do not require any medical attention or treatment. The appearance of the red cells is shown in Fig. 4.55.

Severely affected β-thalassemia homozygotes have marked anemia requiring blood transfusions. Hb A is completely absent in β^O-thalassemia homozygotes and much decreased in β^+-thalassemia homozygotes. Most hemoglobin is of the fetal type. The disease is associated with growth failure and leads to death in adolescence or earlier. Homozygosity and compound heterozygosity for β^+/β^O thalassemia are severe hemoglobinopathies and are a serious public health problem in countries where these genes are common. The simultaneous presence of α-thalassemia (see below) ameliorates the clinical severity of homozygous β-thalassemia.

Hb S-β^+-thalassemia is common in black populations and resembles sickle cell anemia in its symptomatology. Hb E-β thalassemia is common in S.E. Asia and produces a severe anemia similar to homozygous β^O-thalassemia. This severity is explained by the fact that the Hb E mutation itself also causes mild thalassemia (see above).

More than 30 different point mutations causing β-thalassemia have now been identified (Fig. 4.53; Table 4.17), 33% are mutations that prevent transla-

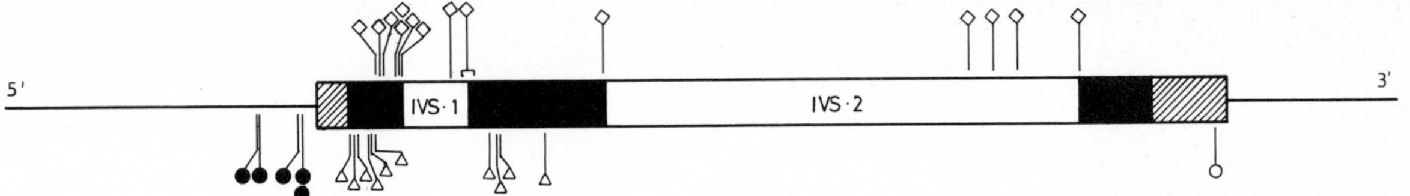

Fig. 4.53. Hb β globin gene and sites of mutations causing β-thalassemia. The various mutants are listed in Table 4.17

Table 4.17. β-thalassemia mutations. (After [972, 1253])

Position	Sequence change	Thalassemia type	Ethnic group
I. Transcriptional mutants			
a) Distal regulatory element			
−87	C–G	β^+	Mediterranean
−88	C–T	β^+	U.S. Black
b) TATA box			
−28	A–C	β^+	Kurdish
−28	A–G	β^+	Chinese
−29	A–G	β^+	U.S. black
II. RNA cleavage defect 1000 bp downstream from exon 3	T–C	β^+	U.S. black
III. Nonfunctional RNA			
a) Terminator defects			
Codon 17	A–T	β^0	Chinese
Codon 39	C–T	β^0	Mediterranean
Codon 15	G–A	β^0	Asiatic Indian
b) Frameshift mutations			
Codon 8	−2	β^0	Turkish
Codon 16	−1	β^0	Asiatic Indian
Codon 44	−1	β^0	Kurdish
Codon 8/9	+1	β^0	Asiatic Indian
Codon 41/42	−4	β^0	Asiatic Indian
Codon 6	−1	β^0	Mediterranean
Codon 71/72	+1	β^0	Chinese
IV. RNA processing mutants			
a) Splice junctions			
Donor sites			
IVS-1, pos 1	G–A	β^0	Mediterranean
IVS-1, pos 1	G–T	β^0	Asiatic Indian
IVS-2, pos 1	G–A	β^0	Mediterranean
IVS-1, pos 5	G–C	β^+	Asiatic Indian
IVS-1, pos 6	T–C	β^+	Mediterranean
Acceptor sites			
IVS-1	25 bp deletion	β^0	Asiatic Indian
IVS-1	A–G	β^0	U.S. Black
b) New splice sites			
New donor			
IVS-1, pos Gru	C–T	β^0	Chinese
IVS-2, pos 705	T–G	β^+	Mediterranean
IVS-1, pos 745	C–G	β^+	Mediterranean
IVS-1, pos 116	T–G	β^+	Mediterranean
New acceptor			
IVS-1, pos 110	G–A	β^+	Mediterranean
c) Enhanced cryptic sites			
Codon 24	T–A	β^+	U.S. black
Codon 26	G–A	β^E	Asians
Codon 27	G–T	$\beta^{Knossos}$	Mediterranean

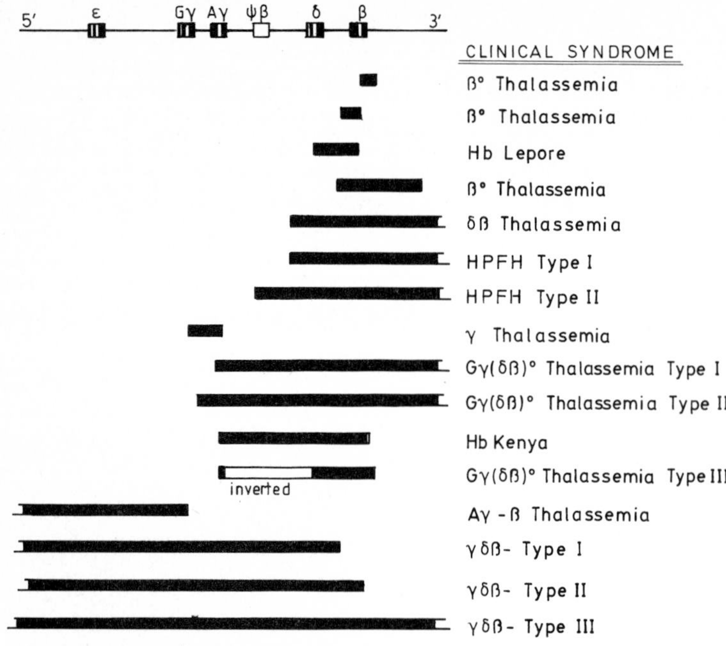

Fig. 4.54. Deletions at the $\gamma\delta\beta$ gene cluster. Most of these deletions are rare. *HPFH,* hereditary persistance of fetal hemoglobin

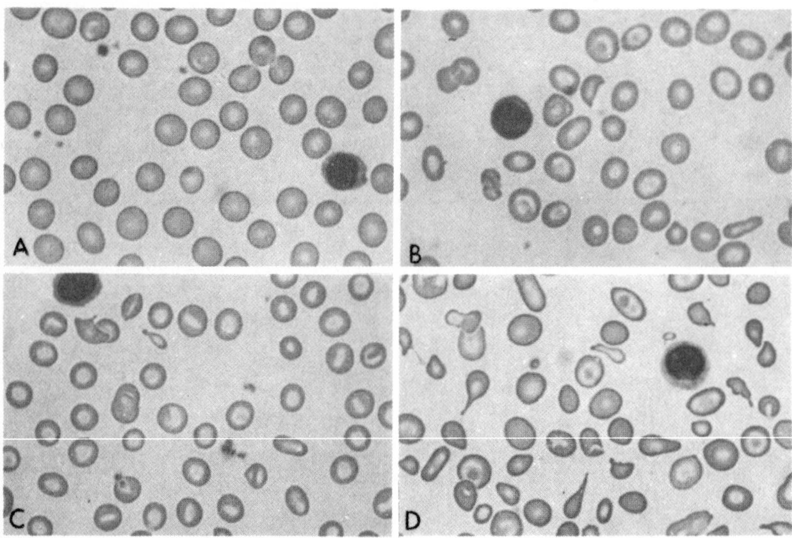

Fig. 4.55 A-D. Peripheral blood smears of a normal individual (**A**), of patients with heterozygous β-thalassemia (**B**), of heterozygous α-thal-1 (**C**), and of β-thalassemia major (**D**). (Bunn et al. 1977 [31])

tion of messenger RNA into protein by nonsense or frameshift mutations, while 47% affect RNA splicing, 16% affect transcription, and 1% affect the RNA cleavage mechanism. It is likely that many more thalassemia mutations exist. This remarkable heterogeneity of mutations at the β globin locus makes for the frequent finding of compound heterozygotes for β-thalassemia, i.e., affected patients are not homozygotes but have inherited a different thalassemia mutation from each parent. Such compound heterozygotes will be somewhat less frequent in population isolates, where a single thalassemia mutation may account for the majority of thalassemias. As an example, while the β^{39} nonsense mutation comprises about 27% of all β-thalassemia mutations in general Mediterranean populations (Table 4.19 a), it accounts for most of the β-thalassemia mutations in Sardinia. Since homozygosity for a given thalassemia mutation may range from only mild interference to complete absence of globin synthesis, and compound heterozygosity is frequent, a wide spectrum of different clinical severity will be encountered.

Table 4.18. α-Thalassemias caused by deletions

		Number of Hb α genes				Total number of active Hb α genes	Total number of deleted Hb α genes
		Maternal		Paternal			
		Active	Deleted	Active	Deleted		
Normal		2	–	2	–	4	–
Mild α-thalassemia		1	1	2	–	3	1
(α-thal-2)		2	–	1	1		
Severe α-thalassemia	a)	–	2	2	–		
(α-thal-1)	b)	2	–	–	2	2	2
	c)	1	1	1	1		
Hb H disease		–	2	1	1	1	3
(Hb H = β^4)		1	1	–	2		
Hydrops fetalis with Hb γ^4		–	2	–	2	–	4

Note that the deletions may be transmitted from mother or father or (except for mild α-thalassemia) from both parents. Note that different matings would give rise to the identical phenotypes of severe α-thalassemia. a) mother: severe α-thal, father: normal; b) mother: normal, father: severe α-thal; c) mother: mild α-thal, father: mild α-thal.

A compilation of the clinically important hemoglobinopathies is given in Table 4.19.

α-Thalassemia [1122, 1156a, 2322]: *Deletion α-Thalassemia.* Most α-thalassemias are caused by gene deletions. Gene conversion (see Sect. 2.3.4) and multiple crossover events occurred during evolution ("concerted evolution") and created a high degree of homology in the structural and flanking areas surrounding the two normal Hb α genes. Thus, sequence homology in that area allows incorrect chromosomal alignment, followed by recombination with subsequent deletions and duplications (Fig. 4.56). The crossover chromosomes bearing either a single Hb α-or a triple α locus ($\alpha\alpha\alpha$) have been observed. Malarial selection has amplified the frequency of the single Hb α gene ($-\alpha$) among tropical and subtropical populations, so that this variant is one of the most frequent thalassemias. The triple α chromosome seems to confer no (or slight) deleterious effects on its bearers and is much rarer.

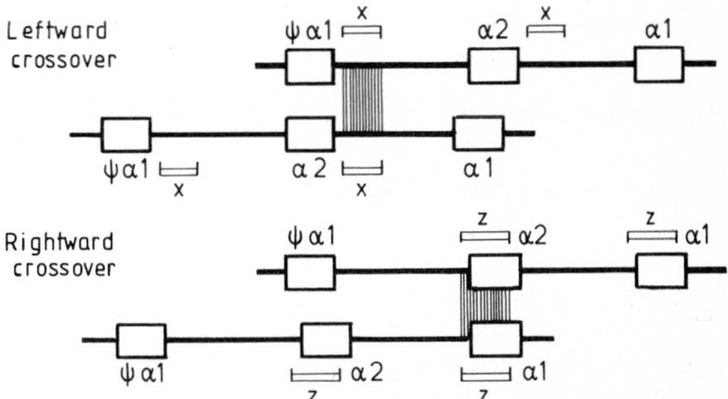

Fig. 4.56. Unique crossover at the X and Z homology boxes in the Hb α gene region. In the *leftward* crossover there is misalignment between the X boxes with crossover and creation of a single Hb α gene (4.2 kb deletion). In the *rightward* crossover there is misalignment between the Z boxes. Crossover inside the α gene causes an α_2-α_1 fusion gene with a 3.7 kb deletion. Note the formation of a triple α gene with these mechanisms. (From Kan 1985 [1156a])

Two types of deletion events cause the mild from of α-thalassemia (α-thal 2 or $-\alpha/\alpha\alpha$). Figure 4.56 shows the mechanisms of the deletion. The so-called leftward crossover creates a single Hb α gene by mispairing of a flanking sequence downstream from the pseudo α_1 gene and its homologue at the Hb α_2 locus. Recombination deletes 4.7 kb. Because the area of recombination is further upstream than that observed in the rightward deletion (see below), the term "leftward" is used. The so-called "rightward" crossover derives from misalignment of the Hb α_2 and Hb $\alpha 1$ genes, with crossover within these genes producing an Hb α_2-α_1 fusion gene and a 3.7-kb deletion. The rightward single α gene is the most common type of α-thalassemia in Africa and in the Mediterranean area, while in Asia both leftward and rightward crossovers have been found. The absence of one alpha gene in either type of mutation causes very mild thalassemia and is very common all over the world, reaching heterozygote frequencies of up to 33% in parts of Africa and in the Mediterranean areas.

Deletions that cause elimination of both α loci (--) are shown in Fig. 4.57. The resultant thalassemias are often designated as α-thal 1. These mutations are frequent in S. E. Asia, rare in Mediterraneans, and have never been observed in Africans.

All deletions are detectable postnatally and prenatally with restriction enzyme mapping techniques. Various phenotypes caused by deletions of one, two,

Table 4.19a. Most frequent β-thalassemias in different ethnic groups. (After [972])

Ethnic group	β-thal mutations	Type	Frequency
U.S. blacks	TATA box (−29)	β^+	39%
	Poly A site	β^+	26%
Mediterraneans	Intron 1 (pos 110)	β^+	35%
	β^{39} terminator	β^+	27%
Asiatic Indians	Intron 1 (pos 5)	β^+	36%
	Deletion (619 b.p.)	β^0	36%
Chinese	Frameshift (pos 71/72)	β^0	49%
	Intron 2 (pos 654)	β^0	38%

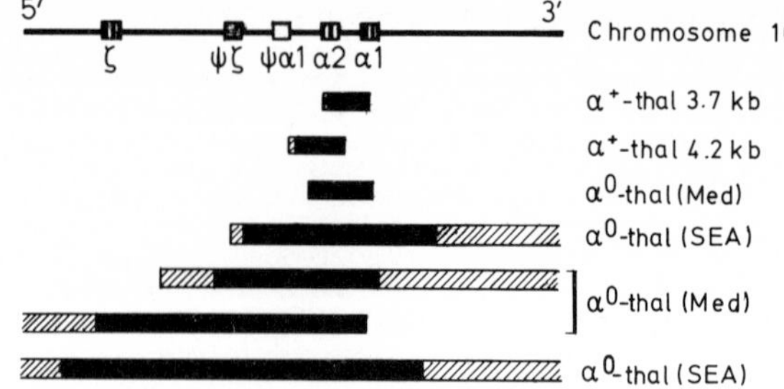

Fig. 4.57. Deletions at the Hb α gene. Most deletions are common (see text). *SEA*, South East Asia; *Med*, Mediterranean ■; ▨ extent of deletion not known

Table 4.19b. Nondeletion mutations in α-thalassemia. (After [972, 1156])

I. *RNA cleavage mutation*		
1. α_2 gene	AATAA*A*→AATAA*G*	Saudi Arabian
II. *Nonfunctional mRNA*		
1. α_2 initiation codon	AUG→ACG	Sardinian
2. α_1 gene	frameshift codon 14	Saudi Arabian
III. *RNA processing mutation*		
1. α_2 gene	IVS-1 5′ consensus: 5 B.P. deletion	Mediterranean
IV. *Chain termination mutations*		
1. Hb Constant Spring	UAA→CAA (Glu)	S.E. Asian
2. Hb Icaria	UAA→AAA (Lys)	Greek
3. Hb Seal Rock	UAA→GAA (Glu)	U.S. Black
4. Hb Koya Dora	UAA→UCA (Ser)	Asiatic India
V. *Impeded α-β dimer formations*		
1. Hb Quong-Sze (α_1)	CTG-CCG	Chinese

three, or four Hb α genes have been documented (Table 4.18) (Fig. 4.57). Absence of a single Hb α gene ($-\alpha/\alpha\alpha$) produces little or no hematologic impairment, since three genes remain active. Restriction enzyme analysis or biosynthetic assay for Hb α chain synthesis is required for detection of mild α-thalassemia due to deletion of a single Hb α gene. Deletion of all four Hb α genes (--/--) is a condition that is lethal perinatally and is known as hydrops fetalis, referring to the extensive edema of the stillborn infant. Most of the hemoglobin of such infants are subunits of four α chains (Hb α^4 or Hb Bart's). Survival of the fetus into late pregnancy is thought to be caused by the presence of functional Hb Portland ($\zeta^2\ \gamma^2$). The virtual absence of hydrops fetalis from African infants is related to the nonexistence of the chromosome bearing the two α gene (--) deletion in that population.

Deletion of two Hb α genes ($--/\alpha+$ or $--/\alpha\alpha$) produces mild anemia, while deletion of three Hb α genes ($-\alpha/-$) causes a more severe anemia characterized by production of Hb H – a Hb β^4 tetramer (Table 4.18). Hb H is formed because of the deficiency of Hb α chains, which does not affect Hb β synthesis.

Hb α Nondeletion Thalassemia [972, 1156]. Nondeletion mutations in α-thalassemia a similar to those detected in β-thalassemia would be expected. A variety of such mutations have in fact been found and are listed in Table 4.19b.

No regulatory mutations in the upstream region of the Hb α loci have been identified yet. However, a mutation characterized by absence of the initiation codon AUG has been detected. This mutation causes α-thalassemia by failure to produce functional mRNA since an AUG sequence that would specify initiation is only found at codon 32. An α globin chain that lacks the first 32 amino acids is grossly defective. A downstream mutation that produces decreased functional mRNA also exists.

Only one splicing mutation, consisting of a small 5-bp deletion that abolishes an acceptor site in IVS-1, has so far been detected.

Four different point mutations have altered the UAA termination codon of the α gene to one specifying an amino acid. As a result, the usually untranslated downstream sequence which specifies another 31 amino acids reads through until a new termination signal is reached. These extended abnormal hemoglobin mRNAs are unstable and only small amounts (5%) of these variants can be detected in the blood. Hb Constant Spring is the most common of these mutants. (See Fig. 4.48.)

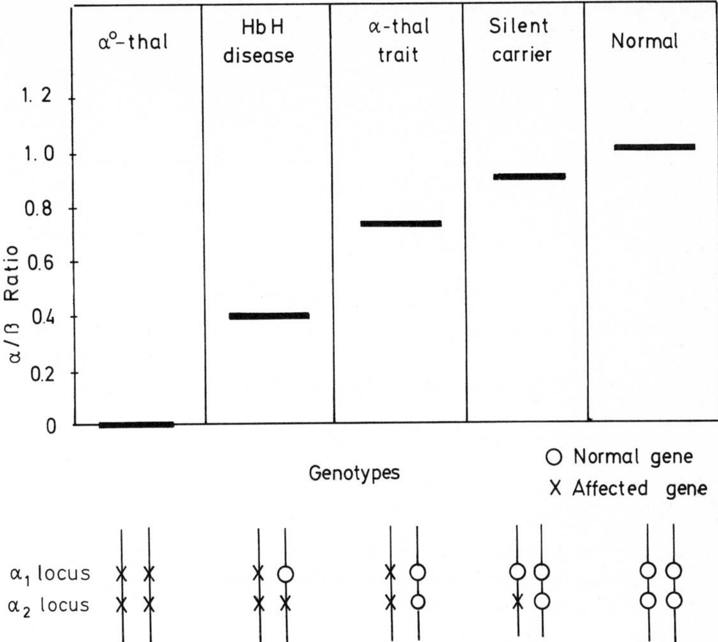

Fig. 4.58. Globin chain synthesis in the α-thalassemias. Globin chain synthesis is expressed as the ratio Hb α/Hb β chain synthesis which is close to 1 in normals, i.e., identical amounts of Hb α and Hb β chains are produced. "Silent" carriers are heterozygotes for α-thal-2 (mild α -thalassemia), which require biosynthetic methods for detection. (Kazazian et al., 1977 [1159])

In Hb Quong-Sze the mutation interferes with α_1/β_1 dimer formation and causes extensive instability. Other such variants probably exist.
No direct data are available regarding the frequency of nondeletion α-thalassemia; however, the nature of the allelic α-thalassemia genes in patients with Hb H (β^4) who carry at least one double deletion of Hb α $(--/\alpha\alpha)$ on one of their chromosomes has been analyzed. Both in Saudi Arabia and in China as many as 50% of such allelic α-thalassemia genes in HbH disease were nondeletion rather than deletion mutants.

4.3.5 Population Genetics of Hemoglobin Genes
(see [972]; Sect. 6.2.1.3)

The presence of a relatively large number of DNA polymorphisms at the Hb β locus has made it possible to study the mutational origin and spread of the different hemoglobinopathies. Taken at face value, the results appear to suggest that the Hb S mutation occurred at least three times [1233] in different geographic areas (Senegal, Benin, and Bantu) and then spread by malarial selection. Hb S in Mediterraneans appears to be of African origin, and both the Benin and Senegalese DNA haplotypes have been found in the Mediterranean area. Three chromosome backgrounds have been observed among Hb E chromosomes in southeast Asia. One of these is derived from crossover at the "hot spot" area (see above). Hb C has been largely observed in one haplotype (but see Sect. 6.2.1.6).
A specific β-thalassemia mutation has usually occurred upon a unique haplotype with subsequent expansion because of malarial selection. Among 31 thalassemia point mutations only 2 were found in more than one ethnic group. The association of a given mutation with a characteristic haplotype for DNA variants aided greatly in defining the nature of the various thalassemia mutations. Since different DNA haplotypes were selected for investigation, the chance that the thalassemia mutation under study was different from those previously investigated was high, and many different types of thalassemia were discovered [1253].
It can be concluded that all common thalassemias and hemoglobinopathies first occurred after human racial divergence and in most cases remained associated with the same chromosomal background on which the original mutation had occurred. When different haplotypes are associated with the same mutation, meiotic crossovers at the recombinational "hot spot" upstream of the β globin gene can be inferred in most instances.
Great interest attaches to the likely occurrence of interallelic gene conversion. Thus, the same Hb β^{39} mutation in Sardinia is associated with several ge-

Table 4.20. Clinically important hemoglobinopathies

	Disease	Genetics	Clinical severity
Sickle cell syndromes	Sickle cell anemia	Homozygote for Hb S	+ + +
	Sickle β-thal disease	Compound heterozygote for Hb S and β-thal	+ + to + + +
	Sickle-Hb C disease	Compound heterozygote for Hb S and Hb C	+ to + +
	Sickle cell trait	Heterozygote for Hb S	0
α-Thalassemias	Hydrops fetalis	4 Hb α deletions	Lethal
	Hb H disease	3 Hb α deletions (or 2 Hb α deletions and heterozygote for Hb CoSp) or point mutation	+ +
	α-thal-1 heterozygote	2 Hb α deletions or point mutation	+
	α-thal-2 heterozygote	1 Hb α deletion or point mutation	0
	Hb Constant Spring (CoSp) heterozygote	α Chain terminus mutant	+
β-Thalassemia	β^0 (Thalassemia major or Cooley's anemia)	Homozygote	+ + + +
	β^+-thal major (Cooley's anemia)	Homozygote	+ + + to + + + + [a]
	β^0/β^+ thalassemia	Compound heterozygote	+ + to + + +
	Hb Lepore heterozygote	δ-β fusion	+ (+ + + + for homozygotes)
	β^0, β^+, and $\delta\beta^0$-thal trait	Heterozygous	+
	Hb E-β-thal	Compound heterozygotes	+ + + +
Unstable hemoglobin diseases	Congenital nonspherocytic hemolytic anemia of Heinz body type	Heterozygous - dominant (many different varieties)	+ +
Hemoglobins with abnormal oxygen affinity	Familial erythrocytosis (high affinity)	Heterozygote - dominant (about 30 varieties)	+ +
M hemoglobin	Familial cyanosis (methemoglobinemia)	Heterozygote - dominant (5 varieties)	+ +

[a] Milder diseases in β-thal$^+$ homozygotes of African origin.

netic haplotypes that cannot simply be explained by recombination. Unidirectional transfer of sequence information (i.e., conversion) therefore is likely (Fig. 2.97). Similar suggestions have been made for analagous findings for Hb S, Hb E, and the β frameshift 6 mutation. Since there is strong evidence for gene conversion in globin gene evolution of nonhuman primates this mechanism appears to be the best explanation. A possible clue to the unknown mechanism of conversion has been suggested [972]. Spreading of a hemoglobin mutation from one framework to another has only been observed with four exon mutations (Hb S, Hb E, β^{39}, β^6 terminator). Six well-studied intronic thalassemia mutations have not been associated with gene conversion.

4.3.6 *Prenatal Diagnosis of Hemoglobinopathies* [966, 2269, 2322, 2361]

Prenatal diagnosis of the hemoglobinopathies is possible by several approaches: (a) Blood sampling by fetoscopy or placental puncture followed by biochemical characterization of the hemoglobin at the protein level; (b) amniocentesis at 15-16 weeks followed by DNA diagnosis of the amniotic cells which are of fetal origin; and (c) chorionic biopsy at 9-10 weeks of pregnancy, followed by DNA diagnosis of the chorionic villi removed (Sect. 9.1.1). Direct DNA diagnosis of fetal cells or tissue uses restriction enzymes that recognize unique mutations such as Hb S or oligonucleotide probes [1254; 2376] that hybridize to the thalassemia mutant to be detected (Fig. 4.59). Indirect methods of antenatal diagnosis utilize principles of genetic linkage by studying the fetus and the family with restriction fragment length polymorphisms (RFLP) closely linked to the thalassemia lesion that is being searched (Sect. 9.1.1).

All these methods have been used. Blood sampling by either fetoscopy or placental puncture has a relatively high fetal mortality (5% with experienced operators) but only requires expertise in methods of protein examination, since the studies are done on fetal blood. However, as the molecular methodology becomes simpler (Fig. 4.60), an increasing number of laboratories are introducing these techniques. Chorionic biopsy has the advantage over amniotic fluid puncture in that the procedure can be done earlier, more DNA is available, and results are rapidly available. Following amniocentesis, culture of several weeks' duration is required to obtain enough cells for DNA studies.

Direct DNA studies are always preferable to indirect DNA methods, which always require family studies. Thus, the direct oligonucleotide probe technique promises to be highly useful for thalassemias, particularly if nonradioactive methods for gene detection can be developed in the future. However, it must be known which one of the thalassemia mutants is suspected. The association of specific thalassemia mutations with certain DNA haplotypes may be helpful here. A preliminary study of cellular DNA of an affected family member is required to delineate the nature of the thalassemic lesion.

Hemoglobin as a Model System. Hemoglobin is the best analyzed genetic system in humans. Therefore, experiences with concepts developed in the course of its analysis may help toward better understanding of phenomena in other fields of human genetics. For example, when hereditary diseases with different phenotypes are found in different families, it is generally concluded that they are caused by mutations of different genes. The hemoglobin example shows that this must not necessarily be so. Methemoglobinemia, for example, is quite different phenotypically from hemolytic anemia or erythrocytosis due to a variant with abnormal oxygen affinity, yet allelic mutations are involved. The phenotype depends on the type of the molecular anomaly and the ways in which normal function is altered.

Another useful lesson is taught by the way in which the tetrameric structure of the hemoglobin molecule determines its function of oxygen binding capacity and how mutations may influence this function which may depend on interaction of the products of more than one gene. While the effects of most hemoglobin mutation are innocuous, the phenotypic effects of all pathologic hemoglobin variants are dominant with the exception of sickle cell anemia. This finding points to one mechanism for Mendelian dominance: disturbance of interaction between allelic products of alleles (see Sect. 4.6).

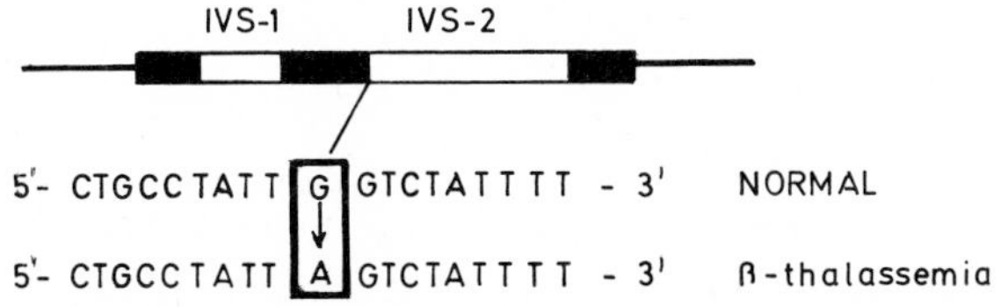

IVS-1 IVS-2

5'- CTGCCTATT G GTCTATTTT - 3' NORMAL

5'- CTGCCTATT A GTCTATTTT - 3' β-thalassemia

Fig. 4.59. Oligonucleotide probe (19-mer) for normal β globin gene differs from a probe for a β thalassemia gene with a G-A mutation in IVS-2 at one position (G-A). Under appropriate conditions of hybridization, the probe for this specific thalassemia mutation will only recognize the thalassemia gene and not the normal gene. Similarly, the "normal" probe will not hybridize with this thalassemia gene

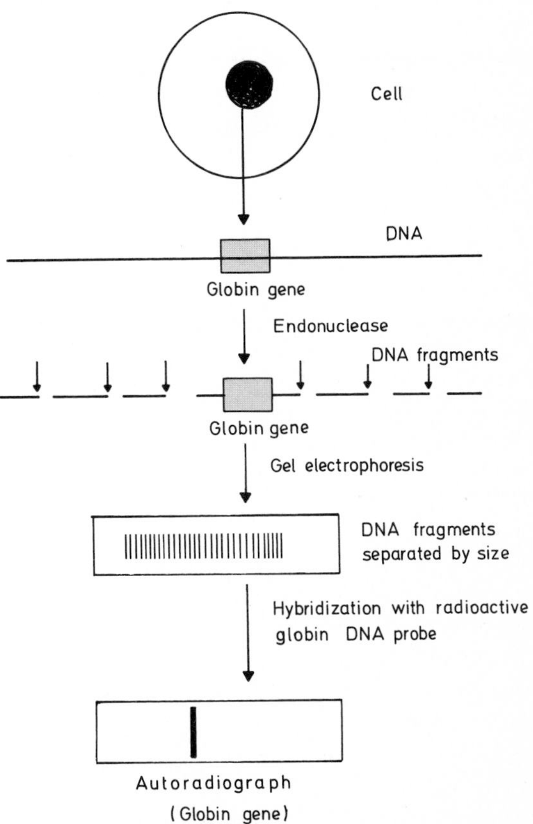

Fig. 4.60. Visualization of globin genes. DNA from any cell nucleus is chemically prepared. This DNA will contain the globin genes. Various restriction endonucleases will break up the DNA in many fragments. Different endonucleases recognize different sequences of nucleotides. The DNA fragments are separated on gel electrophoresis by size. A specific radioactive DNA probe for globin is prepared and reacted with the DNA fragments. Hybridization of the radioactive globin probe occurs with a globin gene and can be visualized following radioautography. (Adapted from Orkin et al., 1978 [1252])

Finally, the hemoglobins provide evidence for the action of many possible mechanisms of mutations. Such mutations may occur in the structural gene itself or in adjacent control areas. In most cases, they involve exchange of only one base but sometimes single bases or base sequences of various lengths may be deleted. Other types of mutations, such as frameshifts in reading the genetic code were somewhat unexpected to human and mammalian geneticists but might have been predicted by microbial geneticists who often encountered this type of mutation. As will be shown in Sect. 7, hemoglobin mutations have also significantly contributed to our understanding of the role of mutation in evolution.

4.4 Genetics of Antigen-Receptor/Antibody

Function and Formation of Antibodies. Living organisms are continuously challenged externally by bacteria or viruses and internally by cellular alterations that occasionally may lead to cancer. In the course of evolution, a complex defense system against such assaults has been developed, which includes both cellular and humoral factors. This defense system is called the immune system, and the mechanisms of its function are the object of immunology [100]. An oversimplified picture of the major components of immunologic defense is shown in Fig. 4.61, which also lists the main components for which genetic defects are known. Important structures within the immune system are lymphocyte receptors for antigens. Both types of lymphocytes, i.e., T- as well as B-lymphocytes, posess receptors which are similar in structure and encoded in the genome, although the genes of both types of receptors are different and located on separate chromosomes. The secreted B-cell receptors (antibodies) constitute the immunoglobulins. T-cell receptors are not secreted.

The specificities of the receptors are determined by the primary structure of the antigen-binding portion of the receptor. The antigen-binding portion of the molecule is encoded by a set of genes (variable part genes (V genes)) and during differentiation of the lymphocyte one of the V genes is randomly combined with the constant part gene (C gene). Thus, each differentiated lymphocyte has the ability to produce only one type of antigen-specific receptor;

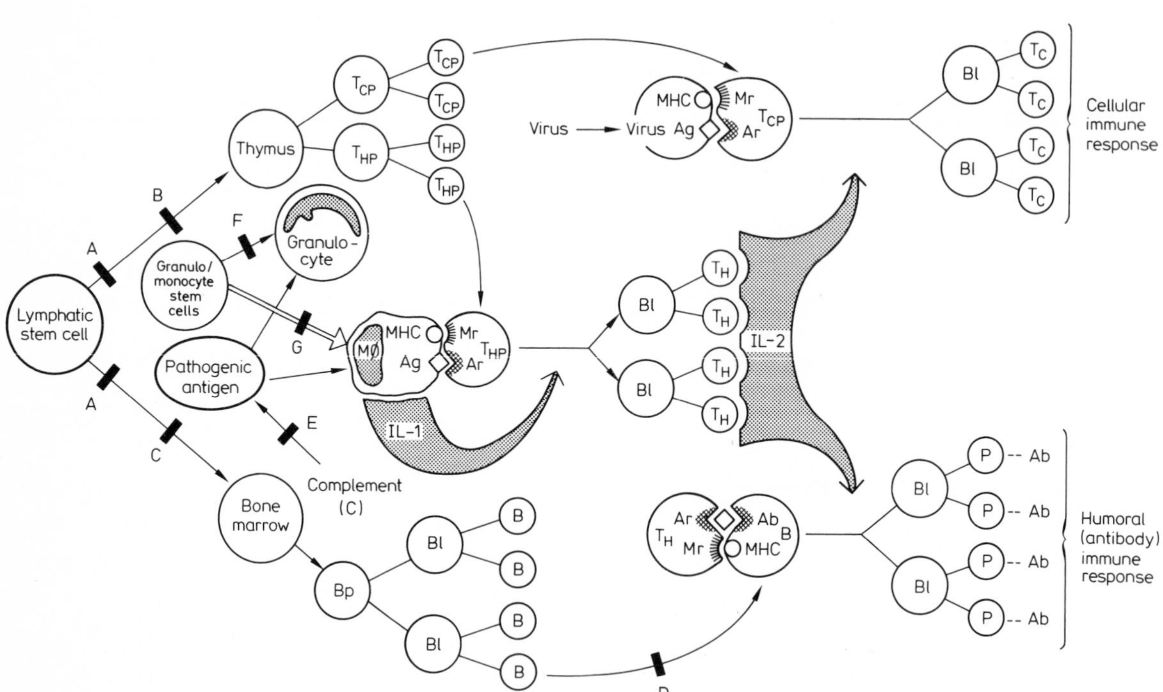

Fig. 4.61. The most important components and sequences of events of the immune system (simplified). Genetic deficiencies shown as black bars may impair the immune defense at various levels. *TCP,* cytotoxic T-cell precursor; *Tc,* cytotoxic T cell; *THP,* helper T-cell precursor; *TH,* helper T cell; *MØ,* macrophage; *Bp,* B-cell precursor; *B,* B cell; *P,* plasma cell; *Bl,* blast; *Ag,* antigen; *Ar,* antigen receptor; *Ab,* antibody (antigen receptor on B cells); *MHC,* major histocompatibility molecules; *Mr,* major histocompatibility molecule receptor; *IL-1,* Interleukin 1; *IL-2,* Interleukin 2; *A,* developmental abnormalities of lymphocytes; *B,* lymphopenic immune deficiencies and thymus hypoplasia; *C,* different types of B-cell defects; *D,* defects of specific immunoglobulins; *E,* defects of complement components; *F,* agranulocytosis; *G,* progressive granulomatosis of childhood

the lymphocyte population as a whole forms all the various receptors that the organism is able to produce. Exposure to a certain antigen leads to proliferation of those lymphocytes (clones) which possess a receptor fitting the antigen. The genetics of the B-cell receptor (antibody) molecule and its formation has been much better analyzed and is better known than the genetics and formation of the T-cell receptor. The underlying principles, however, appear to be similar.

Myeloma Proteins as Research Tools. The immunoglobulins (antibodies, secreted B-cell antigen receptor) of an average human being are a mixture of the gene products of very many different cell clones. At first glance, such heterogeneities seem to be an insuperable obstacle for any chemical analysis of antibodies that requires pure proteins. Quite as in many other situations, however, experiments of nature offered an opportunity to overcome this obstacle. Neoplasias originate from one single cell - in many cases by somatic mutation (Sect. 5.1.6). Hence, plasma cell tumors are expected to produce only one single antibody species in abundance, if differentiation of antibody-forming cells has occurred before the beginning of malignant cell growth. Such monoclonal proteins are, indeed, observed in mice and in humans suffering from myelomatosis - a not uncommon plasma cell tumor. Myeloma proteins can be isolated in sufficient amounts, purified, and their amino acid sequences determined. In this way, the structure of antibodies has been analyzed.

Classes of Immunoglobulins [1123, 1124]. Several classes of immunoglobulins can be distinguished: IgG, IgM, IgA, IgD and IgE. All these classes consist of several polypeptide chains of different lengths: the smaller light (L) chains and the longer heavy (H) chains. The H chains determine the class to which an immunoglobulin belongs. They may be of γ, μ, α, δ, or ε type. All classes of immunoglobulins utilize one of two L chains: κ or λ; both types of L chains may occur in all classes.
A common immunoglobulin is the IgG molecule; two chains are connected by S-S bridges with two L chains. The structure of the other classes is more complicated; one IgM molecule, for example, contains five subunits with two H chains each. During normal immunization, antibodies of the IgM class are formed first; they are then replaced by those of the IgG class without change of specificity. This switch occurs within the same cells (Fig. 4.62).

Constant and Variable Parts. It is a common property of all L and H chains, which distinguishes them from all other proteins known so far, that they consist of a *constant* and a *variable* part. The constant (C) part follows the rules that are familiar to us from other proteins; amino acid sequences are always identical for each type of chain, excepting a few positions for which genetic polymorphisms exist. These polymorphisms are usually recognized indirectly by the ability of some variants to inhibit agglutination of red blood cells by specific antibodies. They are called the GM or KM (Inv) groups and affect the heavy chain and light chain, respectively. The variable parts, on the other hand, have been different in their amino acid sequences in all human myeloma proteins analyzed so far. All variable parts of all light and heavy chains have a similar length of 107-120 amino acids. In the light chains, the constant part is similar in length to the variable part. In the heavy chains, the length of the constant part is an almost exact multiple of that in the light chain (Fig. 4.63). The constant parts of the γ_1 and α_1 chains are three times as long and those of the μ and ε chains are four times as long as those of the light chains. Moreover, all segments of the constant parts show some degree of homology with each other, i.e., their amino acid sequences, though different in many details, are much more similar than could be accounted for by chance.

Common Origin of the Genes for All Chains. The most obvious explanation for this similarity is a

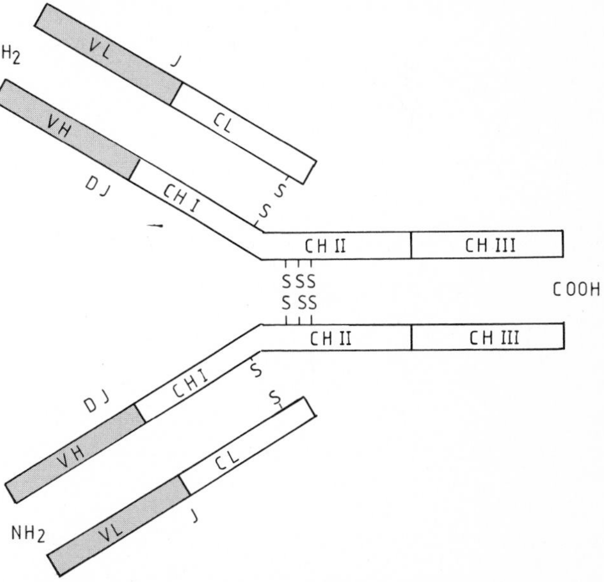

Fig. 4.62. Basic structure of an IgG molecule. - It is composed of two identical light chains and two identical heavy chains. Each chain consists of the amino-terminal V region and carboxy-terminal C region. The C region of the light chain has the approximate length of the V region; the C region at the IgG heavy chain is three times as long, consisting of three approximately structure-homologous subunits that were formed during evolution from one common gene

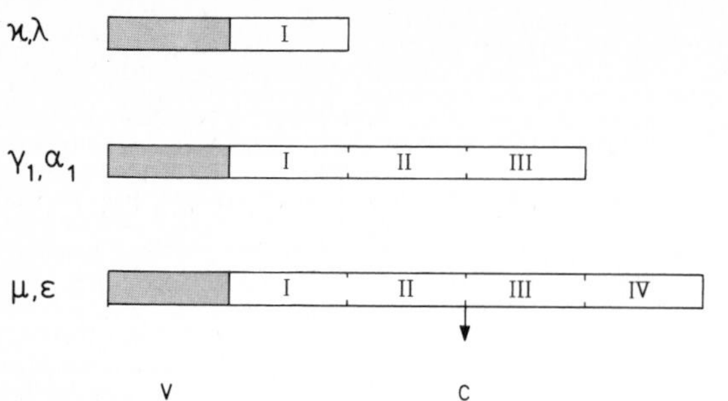

Fig. 4.63. The light chain *(top)* consists of a variable part *(V)* and one constant part, which may be a κ or a λ chain. Heavy chains (*center* and *below*) consist of a variable part *(V)* and a constant part consisting of a threefold or fourfold homologue of the basic antibody gene. The constant parts in γ_1 and α_1 chains, for example, which are components of the IgG and IgA classes of immunoglobulins, consist of threefold homologues of the basic gene. The μ and ε heavy chains consist of four-fold homologues. (Hilschmann et al., 1975 [1124])

common origin of all these segments in evolution; at the beginning, there was only one gene that determined a polypeptide chain, which had about the length of the constant part of a light chain. During evolution, this gene was repeatedly duplicated. Some of these duplications led to longer DNA stretches that determined polypeptide chains in which the same amino acid sequence was repeated three or even four times. These duplicated DNA stretches were now completely structure homologous but no longer position homologous. During the following millions of years, fixation of new mutations led to progressive diversification between these structure-homologous DNA stretches, causing their present-day differences in amino acid sequence.

The first duplication of a singular gene must have occurred by some chromosome rearrangement. Subsequent duplication can easily be produced by unequal crossing over after mispairing of structure-homologous but not position-homologous closely linked genes (Sect. 3.5.8). This seems to be the obvious mechanism for enhancement of the number of homologous stretches in the constant part of different heavy chain genes. Evolution of the various light chain and heavy chain genes requires additional steps of gene duplication and chromosomal rearrangement. The genes for light and heavy chains are not located close to each other on the same chromosome. Genetic polymorphisms of the light chain (the KM system) and of the heavy chain (the GM system) are not linked [78].

Genetic Determination of the Variable Chains. So far, we have only discussed the genetic determination of the constant parts, which can be explained satisfactorily using classic genetic principles. A simple genetic explanation of the variable parts of the immunoglobulin, however, is not possible. The fact that all their amino acid sequences have so far been found different can only be explained by assuming that each person harbors a great many plasma cell clones, each of which produces an immunoglobulin with a different variable part. This postulate immediately suggests that antibody specificity is located in the variable (V) parts. Two questions arise:

1. What genetic mechanisms determine the variable parts?
2. How do they cause antibody specificity?

Somatic Mutation or Selective Activation of Genes? Several hypotheses for the genetic determination of the variable parts have been brought forward. Two of these concepts were widely discussed: the "somatic mutation" hypothesis and the "selective gene activation" hypothesis. The somatic mutation hypothesis assumes that there is only one gene that undergoes many random mutations during proliferation of B-lymphocytes. Indeed, somatic mutations occur occasionally during proliferation of all cell types (Sect. 5.1.6). However, the hypothesis requires a specific mechanism by which the somatic mutation rate is enhanced specifically for this gene. Such mechanisms are conceivable; for example, the gene loci in question could be inaccessible to repair enzymes.

Somatic mutations are, of course, random in direction. Hence, this hypothesis predicts that the amino acid substitutions within variable chains from various antibodies as assessed by study of myeloma proteins be completely independent of each other. Of course, there may be sites at which no mutations are tolerated and which therefore are identical in all variable parts. Any other regularities, however, would be difficult to reconcile with this hypothesis. Such regularities have, indeed, been described. The variable parts known so far can be subdivided into a number of groups that have certain amino acid replacements in common, whereas other substitutions are different even within one group.

This finding suggested an alternative hypothesis for the genetic determination of the V parts: every individual could harbor a great number of genes that are arranged in a highly repetitive sequence. However, in every cell, only one of these genes could be active. This gene could be connected somehow with the gene for the constant part of the polypeptide chain, permitting continuous mRNA formation. If we assume that this gene sequence has been formed by repeated unequal crossing over followed in the course of millenia by random fixation of point mutations, then the regularities described above are easily explained. Mutations that are common to several polypeptide chains have been fixed before the genes for these chains were duplicated; mutations that are unique for one chain are of relatively recent origin.

Both hypotheses - the somatic mutation and the selective gene activation hypothesis - require one element of nonorthodox genetics. For accumulation of so many somatic mutations, cell-specific enhancement of the mutation rate or selection of mutants is necessary. Selective connection of one of the many variable genes with the constant gene required a linking mechanism that is unknown so far. The linking could not occur at the protein level and not even at the level of mRNA, as mRNA molecules already comprise the total information. Hence, it had to occur at the DNA level. Possible mechanisms were suggested by the discovery of restriction enzymes, i.e., enzymes able to cut DNA at specific base sequences (Sect. 2.3; 9.2). Cutting of DNA to enable it to rejoin in a new order before transcription could be one normal function of such enzymes. On the basis of amino acid sequence data, the controversy between the somatic mutation hypothesis and the multiple variable gene hypothesis could not be decided. A solution required direct investigation of the respective genes.

It was found when DNA technology (Sect. 2.3) became available. As it turned out, both parties - those who favored the "selective gene" hypothesis and the supporters of the "somatic mutation" hypothesis - were partially right. Figure 4.64 shows the structures of the mouse λ and κ light chain genes and the heavy chain genes. Many of these studies have been performed on mouse genes; all evidence points to very similar structures in humans.

All three types of genetic regions are, in principle, organized identically: They consist of genes for the constant parts, the variable parts, and a "joint region" that connects both. These are referred to below as C, V, and J, respectively, always using normal capital letters for the gene products and capital letters in *italics* for the gene segments, *C, V,* and *J*. As shown in Fig. 4.64, there are differences in detail. To start with the constant parts, each λ light chain gene sequence has two different *C* genes. Since the protein molecule has only one C region, the gene for this C region must be selected from 4 *C* genes (= two from each of two homologous chromosomes). Moreover, there are two *J* genes, each belonging to one *C* gene, and one *V* area. Moreover, at the left-hand side of the *V* segment, there is a short segment called *L* - the signal region where transcription starts. The light chain gene has one *C* region but five different *J* regions. There are numerous *V* regions; their number is now estimated at 90-300. In the heavy chain gene sequence, eight different *C* regions were found. The gene product Cμ is present in IgM proteins; the Cδ, γ1, γ2b and γ2a gene products are parts of IgD, IgG and IgA immunoglobulin. Moreover there are four different *J* regions. In distinction from the L chains, the H chains contain an additional amino acid sequence that is coded by a *D* region, this *D* region being present in twelve copies. Moreover, there are 100-200 L_H - V_H segments.

During differentiation of antibody-forming cells, *one L-V segment* is connected with one *J* segment (in H-chain-producing cell clones with one *D* and one *J* segment) and with one *C* segment. This leads to a great number of possible combinations; if the genome carries 2 Vλ, 3 Jλ, 300 Vκ and 4 Jκ segments this gives 1206 different light chains (2×3 plus 300×4). Likewise, if there exist 200 V_H, 12 *D* and 4 J_H segments, the maximum number of different V_H regions would be ($200 \times 12 \times 4$) = 9600. So much about verification of the gene selection hypothesis. Moreover, it was shown by direct examination that the gene region is, indeed, organized before and during B cell differentiation. The precise mechanism of this organization is now under investigation by several groups. Some aspects of this process are already known; for example, joining of ends is somewhat imprecise by some bases; this is an additional source of diversity. In other cases, this lack of precision leads to an out-of-phase reading frame that makes transcription impossible: Diversity is achieved at the expense of some waste. Sometimes, one or more nucleotides are inserted at the joints.

However, the above-mentioned sources of variation are not quite sufficient for creating antibody diversity. And indeed, the supporters of the somatic mutation hypothesis were also partially right: There is now conclusive evidence from comparison of homologous sequences of different origin that numerous somatic mutations occur. In almost all instances, they are simple base substitutions leading to the replacement of only one amino acid (See Sect. 5.1.4). They were observed not only in the *V* but also in the *J* and *D* segments.

In conclusion, the somatically generated diversity derives from four sources:

1. Combinational; *V* regions are present in multiple copies only one of which is connected with the corresponding *J, D* and *C* regions to form the functional gene.
2. and 3. Additional diversity occurs in the joint regions and may be named junctional site diversity and junctional insertion diversity: Joining ends are imprecise (2), and nucleotides may be inserted (3).

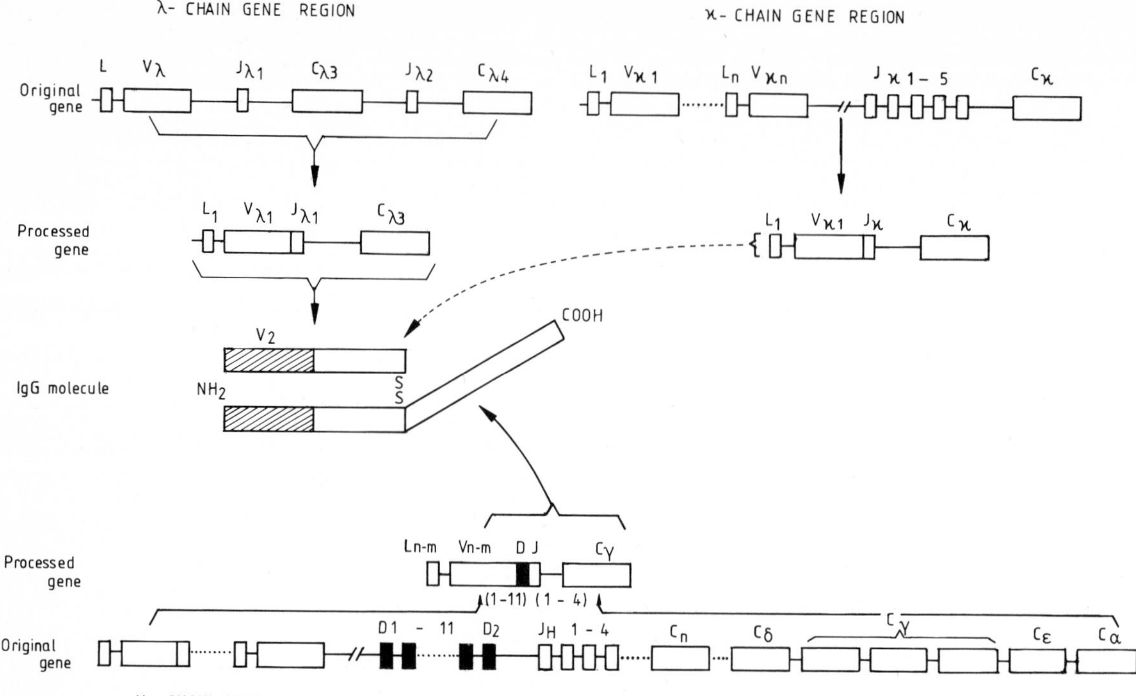

Fig. 4.64. Organization of immunoglobulin gene segments before and after somatic rearrangement. The rearranged state shown here is only one of many possible ones. The genes are rearranged before and during B cell maturation and determine, in the rearranged state, the IgG molecule (only one half of this molecule is shown). The *dotted line* from the processed κ gene to the IgG molecule means that either a λ chain or a κ chain is contributed to this molecule (see [1329 a])

4. Superimposed on these recombinational mechanisms of diversification is somatic mutation in the *V*, *J* and *D* regions.

In this way, it is easy to understand how the many thousands of different antibodies can be formed by one individual. Another result of interest for evolution biology might be mentioned here: As mentioned in Sect. 3.5.5, the HLA specificities are involved in the immune response, together with the immunoglobulins. It was therefore meaningful to look at possible similarities in the amino acid sequences of these proteins [1173, 1329 a]. As shown in Fig. 3.40, the HLA molecule consists of a heavy chain and a light chain. The heavy chain comprises the intracellular portion of the molecule and the region $\alpha 3$, $\alpha 2$, and $\alpha 1$. The regions of the HLA-B7 protein showed statistically significant homology in amino acid sequence with the *C* regions of IgG. Moreover, the two proteins have a number of additional peculiarities in common, such as matches of the three invariant residues of IgC domains - two cysteines and one tryptophane - a concentration of homologous residues around the cysteines and, in general, an amino acid sequence permitting folding into a three-dimensional structure very similar to that of the IgG molecule. All this evidence points strongly to a common origin of immunoglobulin genes and the HLA gene family in evolution. This topic will be taken up again in Sect. 7.

V Parts and the Specificity of Antibodies. As noted, antibody specificity is determined by the variable parts, which differ in amino acid sequences. Even within the variable parts, variability is higher in some regions than in others.

Antigens are bound to the combining site of an antibody molecule. If antibody specificity is, indeed, determined by the differences in amino acid sequences, then the combining sites should be found in the regions that are especially variable. In κ chains, these regions comprise positions 28-34, 50-56, and 91-96. A method for determining the spatial order of a molecule is X-ray crystallography. Such X-ray data are now available for the variable part of a κ chain (Fig. 4.65). The polypeptide chain forms two levels; neighboring segments in the same level run antiparallel and are connected by hydrogen bonds (grey lines in Fig. 4.65). There is evidence that the variable parts of the λ chain and of the H chains form a similar folding. In Fig. 4.65, the darkly shaded areas indicate the hypervariable regions. Re-

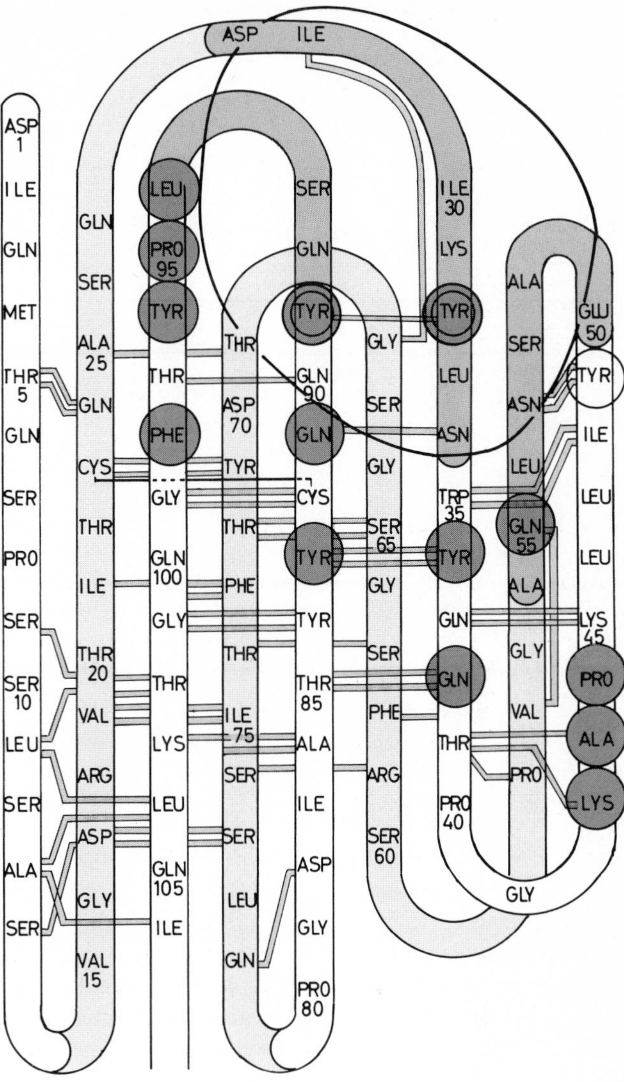

Fig. 4.65. Spatial reconstruction of the variable part of a κ chain from X-ray crystallographic data. The chain is arranged in two levels (lower level: *light shading*). Segments are connected by hydrogen bonds. Parts showing especially high interchain variability are designated by *dark shading*. All of them are located around a pocket that this chain forms together with another κ chain on top of it (shown by *heavy line on top* of the figure) and is probably the combining site with the antigen. Marked amino acids *(very dark shading)* give contacts in the dimer. (Adapted from Hilschmann et al., 1976 [1123])

gions that make contacts between the two monomers in the dimer are also marked. Two such variable parts form a pocket of 15 Å in diameter, with walls consisting of the hypervariable regions [1123]. It is likely that this pocket is part of the combining site with the antigen and that its shape determines antibody specificity. This shape, in turn, depends on the amino acid sequence of the hypervariable regions.

Since the early days of immunology, the relationship between antigen and antibody had often been compared to that between a key and a lock. The spatial model in Fig. 4.65 shows that this notion might be more than a metaphor.

4.5 Pharmacogenetics and Ecogenetics

4.5.1 Pharmacogenetics

The development of human biochemical genetics with its insights into genetically determined enzyme deficiencies gave rise to the field of pharmacogenetics. Garrod – the founder of human biochemical genetics [75] (Sect. 1.5) – and Haldane – the great British geneticist [87] – already hinted that biochemical individuality might explain unusual reactions to drugs and food. A series of abnormal and untoward drug reactions was shown to be caused by genetically determined variation of enzymes in the 1950s. G6PD deficiency (Sect. 4.2.2.2) explained hemolytic anemias caused by fava bean ingestion and by a variety of drugs in some individuals. Variation in the enzyme pseudocholinesterase [1195, 1152] was found to underlie prolonged apnea by suxamethoni-

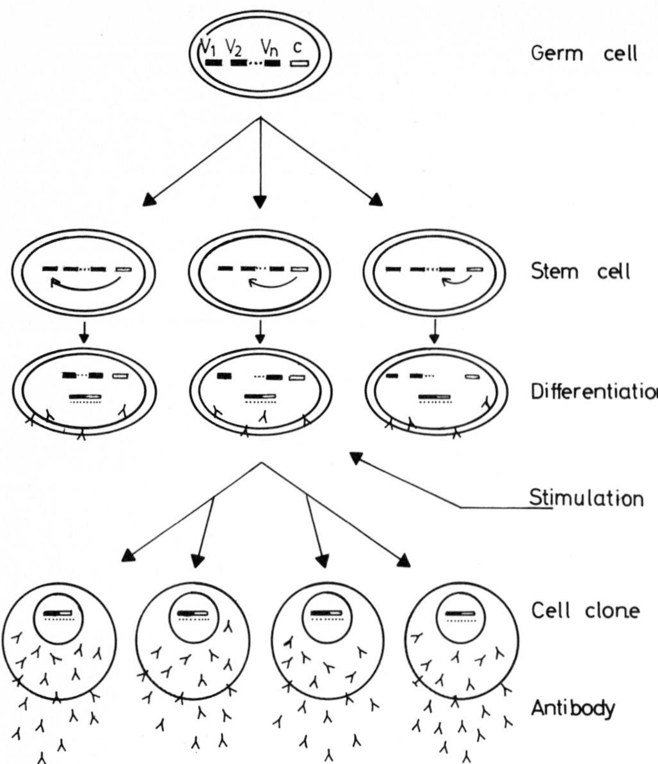

Fig. 4.66. During embryonic development and differentiation, several different stem cells for antibody production are formed. In each cell, only one antibody can be formed because the gene for the constant part *(white bar)* has been connected with only one gene for the variable part *(black bar)*. Stimulation by a specific antigen *(arrow from the right)* leads to proliferation of the stem cell clone capable of formation of the appropriate antibody and to increased production of this antibody. (Adapted from Hilschmann et al., 1976 [1123])

um - a drug widely used to relax muscles during surgery. Genetic differences in acetyltransferase activity explained marked variation in INH blood levels between individuals [1073].

It was therefore suggested by one of the authors (AGM) that abnormal responses to drugs sometimes might be caused by genetically determined enzyme deficiencies [1222]. The other author (FV) first introduced the term pharmacogenetics [1337].

G6PD System (30590). The G6PD system has already been discussed in Sect. 4.2. X-linkage of G6PD explained the preponderance of males with hemolytic drug reactions due to G6PD deficiency. G6PD levels intermediate between those of affected males and normals were often seen in heterozygous females but some obligatory heterozygous females had enzyme levels in the deficient range and many had enzyme levels within the normal range (see Fig. 4.6) [1234]. Heterozygote females with G6PD deficiency possess two red cell populations: normal and mutant cells. The ratio of normal to deficient red cells is most frequently 1:1 but may range from 1% normal and 99% mutant cells to 1% mutant and 99% normal cells in a few heterozygotes [999]. The frequency of females with G6PD-dependent drug reactions depends upon their population frequency (q^2: homozygotes; $2pq$: heterozygotes, where q is the frequency of affected males [1224]), and also on the degree of X inactivation as expressed by the ratio of normal and G6PD-deficient cells. Clinically affected females are the few homozygotes and usually only those heterozygotes who have a preponderance of mutant cells with low enzyme levels. Different possibly hemolytic drugs also vary in their potential for blood destruction.

Several common G6PD variants have been associated with hemolytic reaction (Table 4.4). In addition to drugs, hemolysis may be associated with bacterial and viral infection or may be encountered as neonatal jaundice, where the immature liver is unable to clear bilirubin - a metabolic product of hemoglobin released by hemolysis.

Hemolysis is particularly severe in the Mediterranean and Canton G6PD variants (Table 4.4) since in these conditions G6PD deficiency is associated with both decreased specific activity and molecular instability of the enzyme [1146]. In the common type of G6PD deficiency seen in persons of African origin, red cells younger than 60 days (red cell life is 120 days) have sufficient amounts of enzymes, and the molecular instability characteristic of the defect only affects the older red cells so that hemolysis is self-limited (Fig. 4.4). Increased G6PD levels in red cells following hemolysis will be seen since the older cells with deficient enzymes have been destroyed. Fatal hemolytic episodes are not usually observed in this milder type of G6PD deficiency. With more severe hemolysis, as occurs in the Mediterranean variety of G6PD deficiency, a lethal outcome may occur. The number of drugs that cause hemolysis is also larger in Mediterranean G6PD deficiency than in the African type. No detailed data on the spectrum of harmful drugs are available yet for most other G6PD variants.

Pseudocholinesterase Variation [1104]. The drug suxamethonium or succinyldicholine is commonly used as a muscular relaxant to facilitate surgical operations. Hydrolysis of the drug by the enzyme pseudocholinesterase occurs, and its normal action is brief. In occasional patients, the enzyme has poor affinity for the drug and such persons develop prolonged apnea due to depression of respiratory muscles. Under such circumstances, many hours of artificial respiration are necessary, unless the patient is administered the purified enzyme or plasma as a source of active pseudocholinesterase. The reason for the abnormal drug reactions are different mutations in the homozygous or compound heterozygous state affecting the active site of pseudocholinesterase, which no longer effectively can hydrolyze its substrate. The most common variant is the atypical allele (CHE_1^D). The allele specifying normal pseudocholinesterase is designated as CHE_1^U. About 3%-4% of the population of European origin are heterozygotes for normal and variant alleles (CHE_1^U/CHE_1^D); 1/3,500 persons are homozygotes (CHE_1^D/CHE_1^D) for the variant enzyme and are at risk for prolonged apnea when administered suxamethonium. The atypical enzyme is usually identified by failure to be inhibited by the inhibitor dibucaine in vitro. Abnormal homozygotes are relatively resistant to dibucaine, and heterozygotes have intermediate dibucaine inhibition (Table 4.21). Another allele of pseudocholinesterase is associated with complete absence of pseudocholinesterase activity (CHE_1^S) and homozygotes for this silent allele (CHE_1^S/CHE_1^S) are highly sensitive to suxamethonium since no pseudocholinesterase whatsoever exists in their plasma. This allele is common among Eskimos in Alaska. A third mutant allele can be differentiated by its inhibition behavior to fluoride and is known as the fluoride resistance allele (CHE_1^F). Laboratory tests for pseudocholinesterase variation are usually carried out with benzoylcholine as a substrate for the enzyme. Enzyme abnormalities in some patients with prolonged apnea can only be detected if succinyldicholine itself rather than benzoylcholine is employed as a substrate in laboratory testing [1103].

Acetyltransferase Variation [1062, 1204; 1271]. Several drugs are acetylated by the liver enzyme *N*-acetyltransferase. These drugs include INH, hydralazine, procainamide, phenelzine, dapsone, salicylazosulfapyridine, sulfamethazine, and nitrazepam. Human populations can be subdivided into distinct classes based on whether they do or do not acetylate a test drug such as INH or sulfamethazine, which is administered in vivo. About 50% of individuals of European and African origins are slow inactivators, whereas among Orientals, about 80-90% are fast inactivators. Family studies show that slow inactivators lack acetylation capacity, while rapid inactivators possess a single or double dose of the gene specifying the acetyltransferase enzyme. The exact biochemical mechanism responsible for the polymorphism affecting N-acetyltransferase is unknown. No in vitro tests for acetylation are yet available, and

Table 4.21. Pseudocholinesterase types and suxamethonium sensitivity

Genotype of clinical significance	Activity	Dibucaine number	Fluoride number	Phenotype frequency in European populations[a]	Suxa methonium sensitivity
$CHE_1^UCHE_1^U$	Normal	80	59	95%	None
$CHE_1^DCHE_1^D$	Moderately decreased	22	27	1: 3,200	+++
$CHE_1^SCHE_1^S$	Absent	0	0	1:170,000	++++
$CHE_1^FCHE_1^F$	Slightly decreased	66	35	1: 28,000	++
$CHE_1^DCHE_1^S$	Decreased	22	27	1: 11,000	+++
$CHE_1^DCHE_1^F$	Slightly decreased	49	33	1: 2,500	+++
$CHE_1^FCHE_1^S$	Slightly decreased	67	43	1: 33,000	++
$CHE_1^UCHE_1^D$	Slightly decreased	62	48	3.5%	(+)
$CHE_1^UCHE_1^F$	Slightly decreased	74	50	1.2%	(+)
$CHE_1^UCHE_1^S$	Slightly decreased	80	59	1:200	Unknown

[a] Based on the frequency of the heterozygous state of 3.5% for the atypical allele, 1.2% for the fluoride-resistant allele, and 0.5% for the silent allele. The frequencies of the homozygous and the heterozygous states were calculated by an expansion of the Hardy-Weinberg theorem for multiple alleles.
WHO From: Technical reports series No. 524. Geneva 1973 [969].

a drug that will be acetylated needs to be administered to assess the acetylation status of a patient. A standard dose of caffeine such as a cup of coffee may become useful for typing since an acetylated major urinary breakdown product of caffeine [5-*a*cetyl-a*m*ino-6-*f*ormyl-amino-3-methyl-*u*racil (AMFU)] is only excreted by rapid inactivators [1113]. Clinical consequences of the acetylation polymorphism relate to a higher frequency of INH polyneuropathy among "slow" inactivators. A higher frequency of lupus-like side effects are seen when slow inactivators are given hydralazine or procainamide. More hematologic side effects in slow inactivators are observed with dapsone and with salicylazosulfapyridine. Rapid inactivators may require higher doses of the various drugs to reach satisfactory therapeutic effects, and there are suggestions of increased hepatic toxicity of INH in rapid inactivators since acetylisoniazid is more hepatotoxic than INH [1280].

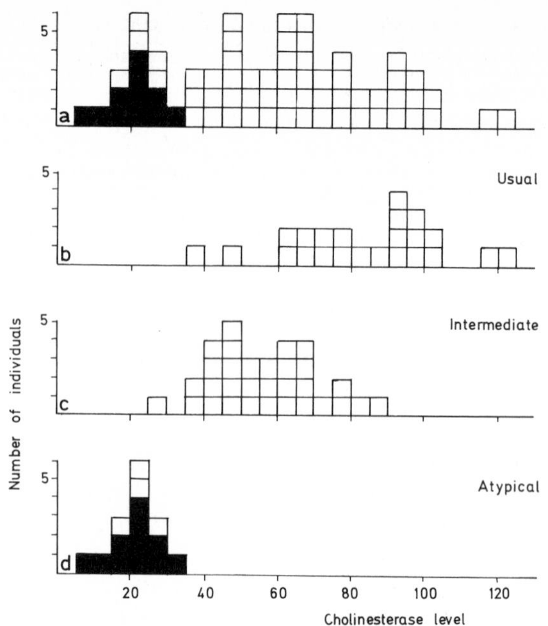

Fig. 4.67. a–d. Distributions of levels of cholinesterase activity in sera from 11 individuals found to be excessively sensitive to suxamethonium and 58 of their relatives (Harris et al., 1960). Each *square* represents one individual; the suxamethonium-sensitive propositi are marked in *black*. **a** Distribution of levels of serum cholinesterase activity. The activity levels were determined manometrically with acetylcholine as substrate. **b** Distribution of activity levels of those individuals classified as having the usual phenotype. **c** Distribution of activity levels of those individuals classified as having the intermediate phenotype. **d** Distribution of activity levels in those individuals classified as having the atypical phenotype. Note: distribution **a** represents the sum of distributions **b, c,** and **d**

Distribution Curves and Gene Action. Monogenic inheritance implies a qualitative difference between the mutant and the normal gene product. If measurements can be carried out at the level of the gene product and the results are plotted, the different genetic classes can usually be differentiated as separate modes in distribution curves. The G6PD levels of G6PD-deficient and normal males produces two modes of G6PD activity without overlap. All three genetic classes in pseudocholinesterase variation can be clearly distinguished without overlap by the inhibition measurements used for a study of this trait. In this system, the test method reflects the qualitative nature of the mutation more directly. However, if pseudocholinesterase blood levels rather than the inhibition behavior are measured, the distinction of the three classes is not readily possible, and overlap between normal homozygotes and heterozygotes, on the one hand, and between heterozygotes and abnormal homozygotes, on the other, occurs (Fig. 4.67).

These data show that a multimodal distribution curve usually can be interpreted to indicate operation of a single gene (Sect. 3.6.1). However, a unimodal curve may be encountered in a population as a result of variation at a single gene locus if the nature of the measurements used to reflect action of the mutant gene is more remote from primary gene action and is influenced by other environmental or genetic factors. Since unimodal distribution curves are usually interpreted as evidence of polygenic gene action (Sect. 3.6), one must be cautious in inferring such a mode of inheritance if data based only on frequency distribution are available.

Debrisoquine-Sparteine Polymorphism [971, 1072, 1305]. A common drug-related polymorphism observed among 3%–8% of Europeans and Africans makes for defective oxidation of a variety of drugs (Table 4.22) which require oxidative metabolism. This polymorphism was independently discovered in studies with the antihypertensive drug debrisoquine and the oxytocic and antiarrhythmic agent sparteine. Affected persons are homozygotes for the poor metabolizer (oxidizer) trait. Homozygotes for the rapid metabolizer trait cannot be clearly distinguished from heterozygotes. The biochemical nature of the polymorphism is unknown but there is good evidence from in vitro inhibition studies that a component of the P450 system of the liver (which is involved in metabolism of foreign substances) is the site of the difference [1050, 1219, 1258]. No in vitro test to assess oxidation status exists. For phenotype assignment, a test drug needs to be administered and urinary excretion of the metabolized and nonmetabolized test substance must be examined (i.e., 4-OH

Table 4.22. Sparteine/debrisoquine polymorphisms: adverse rections in poor metabolizers. (After [1071])

Drug	*Adverse reaction*
Debrisoquine	Hypotension
Sparteine	Exaggerated oxytocic action and cardiac depression
Phenacetin	Methemoglobinemia
Phenformin	Lactic acidosis
Perhexiline	Peripheral neuropathy and agranulocytosis
Captopril	Agranulocytosis
D Penicillamine	Proteinuria and thrombocytopenia
Nortryptyline	Postural hypotension
Guanoxan	Hypotension
Metiamide	Agranulocytosis
Encainid	Lack of drug action: Metabolized drug only is active
Beta blockers	
Propranolol	Brachycardia and hypotension
Metoprolol	Brachycardia and hypotension
Timolol	Brachycardia and hypotension
Alprenolol	Brachycardia and hypotension
Bufuralol	Brachycardia and hypotension

debrisoquine vs debrisoquine). The exact role of this polymorphism in side reactions to the widely used beta blocker drugs is under investigation [1051].

Mephenytoin Polymorphism [1353]. It has been shown that oxidation of the anticonvulsant mephenytoin is under control of a different gene and not related to the debrisoquine type of polymorphism. The relatively high frequency of side effects with this drug that were observed in the past presumably is accounted for by the failure of mephenytoin oxidation in a small percentage (2%–5%) of the population. Phenytoin (dilantin) may be oxidized by the same enzyme as is specified by this polymorphic gene.

Table 4.23. Monogenic pharmacogenetic traits

Enzymatic or metabolic abnormalities	Result and/or clinical abnormalities
A. *Well-established traits (see text)*	
a) *Common traits*	
Some G6PD variants	Hemolysis
N-Acetyltransferase polymorphisms	Reduced acetylation of several drugs (see text)
Poor oxidation (debrisoquine/ sparteine)	Untoward reactions to many drugs (See table 4.22)
b) *Rare traits*	
Pseudocholinesterase variants	Prolonged suxamethonium-induced apnea
Abnormal calcium metabolism	Malignant hyperthermia following inhalation anesthesia
Some unstable hemoglobins	Hemolysis
Various porphyrias	Several drugs precipitate symptoms of disease
Methemoglobin reductase deficiency	Cyanosis with some oxidizing drugs
B. *Traits whose clinical significance is less certain*	
Paraoxonase polymorphism	Low-activity carriers (~50%) more susceptible to parathione poisoning
Poor mephenytoin oxidation	Severe side effects from mephenytoin
Thiopurine methyl transferase polymorphism (cytosolic) [1354]	Lack of effective action of thiopurine drugs (e.g., mercaptopurine)
Catechol-*O*-methyl transferase polymorphism [2239, 1468]	Lack of effective action of *L*-dopa and α-methyl dopa
Epoxide hydrolase deficiency [1311]	Phenytoin hepatotoxicity

Other Monogenic Pharmacogenetic Traits. A variety of other traits that are inherited as simple Mendelian traits play a role in pharmacogenetics. These are listed in Tables 4.23.

Multifactorial Pharmacogenetics. A series of twin studies on the half-life of various drugs have emphasized the importance of genetic factors in drug metabolism. Whenever a drug has been given to identical twins and nonidentical twins [1335, 1270] much more similarity in half-life has been noted for the MZ twins. Heritability measurements based on such data have shown that the contribution of heredity to the total variation of drug halflife was high and sometimes reached 99% (Table 4.24).

When a drug is given to many members of a normal population, considerable variability in blood levels is noted in different persons. While a variety of factors affect such blood levels, differences in drug metabolism as assessed by half-life are usually major determinants. The half-life of a drug (or steady level of a drug) is a rather constant parameter in an individual and as suggested by the twin data is largely influenced by genetic factors. The underlying biochemical basis of the details of drug metabolism is as yet unknown. Variation in half-lives or response for most drugs can be plotted as a bell-shaped Gaussian distribution curve (Fig. 4.68). A certain number of persons at each end of the unimodal distribution curve will either have too large or an insufficient amount of drug following administration of an average dose. Toxicity due to excessive blood levels, on the one hand, or failure of the drug to exert its effects

Table 4.24 Studies in twins on drug elimination rate or under steady-state conditions. (Propping, 1978 [1270])

Drug	Authors, No. of twin pairs	Measured parameter	Range	r_{MZ}[a]	r_{DZ}[a]	h_2^2[d]
Antipyrine 18 mg/kg p.o. (single dose)	Vesell and Page (1968) 9 MZ, 9 DZ	Plasma half-life (h)	5.1 - 16.7	0.93	−0.03	0.99
Phenylbutazone 6 mg/kg p.o. (single dose)	Vesell and Page (1968) 7 MZ, 7 DZ	Plasma half-life (days)	1.2 - 7.3	0.98	0.45	0.99
Dicumarol 4 mg/kg p.o. (singel dose)	Vesell and Page (1968) 7 MZ, 7 DZ	Plasma half-life (h)	7.0 - 74.0	0.99	0.80	0.98
Halothane 3.4 mg I.V. (single dose)	Cascorbi et al. (1971) 5 MZ, 5 DZ	Urinary excretion of sodium trifluoroacetate in 24 h (% of injected dose)	2.7 - 11.4	0.71	0.54	0.63
Ethanol 0.5 g/kg p.o. (single dose)	Lüth (1939) 10 MZ, 10 DZ	β_{60} (mg/ml·h)[a]	0.051- 0.141	0.64	0.16	0.63
		EDR (mg/kg·h)[b]	50.00 -109.63	0.77	0.45	0.67
1 ml/kg p.o. (single dose)	Vesell et al. (1971) 7 MZ, 7 DZ	β_{60} (mg/ml·h)[a]	0.11 - 0.24	0.96	−0.38	0.98
1.2 ml/kg p.o. (single dose)	Kopun and Propping (1977) 19 MZ, 21 DZ	Absorption rate (mg/ml·30 min)	0.20 - 1.12	0.56	0.27	0.57
		β_{60} (mg/ml·h)[a]	0.073- 0.255	0.71	0.33	0.46
		EDR (mg/kg·h)[b]	57.6 -147.6	0.76	0.28	0.41
Diphenyl-hydantoin 100 mg I.V. (single dose)	Andreasen et al. (1973) 7 MZ, 7 DZ	Serum half-life (h)	7.7 - 25.5	0.92	0.14	0.85
Lithium 300 mg/12 h p.o. (for 7 days)	Dorus et al. (1975) 5 MZ, 5 DZ	Plasma concentration (mEq/l)	0.16 - 0.38	0.94	0.61	0.86
		Red blood cell concentration (mEq/l)	0.050- 0.102	0.98	0.71	0.83
		RBC/plasma concentration (each after 3 days of treatment)	0.18 - 0.56	0.84	0.62	0.92
Amobarbital 125 mg I.V. (single dose)	Endrenyi et al. (1976) 7 MZ, 7 DZ	Plasma clearance rate (ml/min)	16.0 - 67.2	0.87	0.55	0.83
		Weight-adjusted clearance (1/kg·h)	1.76 - 6.16	0.92	0.60	0.80
		Elimination rate constant (h^{-1})	2.09 - 8.17	0.93	0.03	0.91
Nortriptyline 0.6 mg/kg·d p.o. (for 8 days)	Alexanderson et al. (1969) 19 MZ, 20 DZ	Steady-state plasma level (ng/ml)	8 - 78	Published data do not allow calculation, but MZ twins are much more similar to one another than DZ		
Sodium salicylate 40 mg/kg I.V. (single dose)	Furst et al. (1977) 7 MZ, 7 DZ	Slope of serum salicylate decay (mg/dl·h)	0.64 - 1.02	0.64	0.32	0.86
Aspirin 65 mg/kg·d p.o. (for 3 days)	See Propping [1270]	Plateau serum salicylic acid (mg/dl)	11.9 - 36.4	0.90	0.33	0.98
		Salicylurate excretion rate (plateau) (mg/kg·h)	0.84 - 1.91	0.94	0.76	0.89

[a] β_{60} = Disappearance rate from blood.
[b] EDR = Ethanol degradation rate.
[c] r_{MZ}, r_{DZ} = intraclass correlation coefficient: in MZ and DZ twins, resp.
d h_2^2 (heritability) = $\frac{V_w(DZ) - V_w(MZ)}{V_w(DZ)}$
V_w = variance within twin pairs.

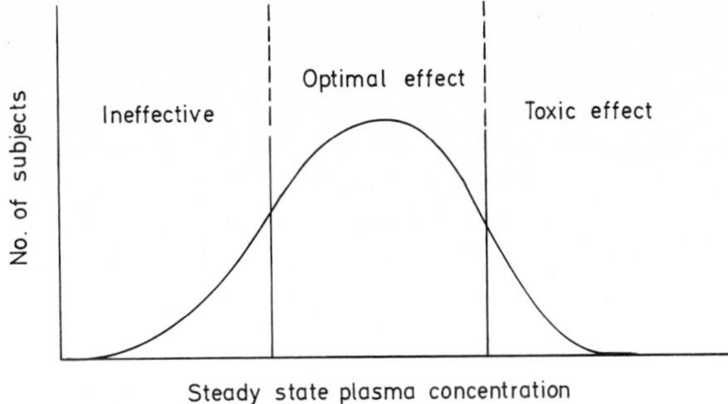

Fig. 4.68. Steady-state plasma concentration of a drug and biologic effect

because of the overly low blood level, on the other, would result. The demonstration that genetic factors play a role in the metabolism of most drugs moved pharmacogenetics from a field that dealt with a few unusual drug reactions to a discipline of central importance for pharmacology and therapeutics [1229].

Pharmacogenetic Variation at the Level of the Target Organ. Genetic variation at the level of the target organ has already been considered for G6PD deficiency of the red cell. The example of alcohol action on the brain will be discussed in Sect. 8.2.3.5. Furthermore, side effects of psychotropic drugs on the brain may have a genetic basis [1271]. For example, persons treated with phenothiazine occasionally develop parkinsonism; those having a relative with parkinsonism run a threefold risk [1271]. "Tardive dyskinesia" manifesting as abnormal, involuntary movements is not very rare among patients treated with psychopharmacologic agents; there is impressive familial aggregation. Since such therapy influences the action of the neurotransmitter dopamine (Sect. 8.2.3.6), it is interesting that an increase in number of dopamine receptors in the nucleus caudatus of rats has been described after administration of neuroleptics; this increase showed marked variation between strains. Schizophrenia - like psychosis - can be induced in humans by drugs such as LSD, metaphetamine, and even by alcohol abuse (= alcohol hallucinosis). The incidence of schizophrenia was found to be significantly increased among first-degree relatives of patients suffering from these complications.

A rare but dangerous complication of general anesthesia is malignant hyperthermia, often associated with increased muscular rigidity. In about half the patients, an incompletely autosomal dominant mode of inheritance has been observed. Many probands have minor muscular complaints, such as ptosis, strabismus, cramping, recurrent dislocations, or back trouble. Abnormal electromyograms and slight histological signs of myopathy have also been described. Slightly increased CPK values are common in these patients.

4.5.2 Ecogenetics [143, 969, 1228, 1250]

The concept of ecogenetics - first suggested by Brewer in 1971 [1017] - evolved historically from pharmacogenetics. Drugs are only a small fraction of environmental chemical agents to which mankind is exposed. Various other potentially toxic agents exist in the environment and may damage a fraction of the population who are genetically predisposed. Ecogenetics extends the central concept of genetically determined variable drug responses to other environmental agents. Since twin studies suggest that metabolism of all drugs is subject to genetic influences, it can be inferred that genetic control of biotransformation applies to all chemical agents. The field of human ecogenetics deals with variable responses of human individuals to environmental agents and attempts to explain why only a certain proportion of exposed human beings will be injured by harmful agents and how individuals differ in their adaptation to the environment. The working hypothesis of ecogenetics is the concept that an individual's internal biochemical makeup will often determine the response to an environmental agent, particularly in situations where it is already known that all human beings react unequally to that particular agent. Similar to the findings in pharmacogenetics, some ecogenetic reactions will be due to the presence of rare mutant genes and cause a grossly abnormal response or idiosyncratic reaction. In other instances, the variable response will be mediated by a polymorphic system, and a significant proportion (2%-50%) of a population will react differently. Most frequently, ecogenetic responses are likely to involve several genes and lead to unusual responses in a few individuals whose genetic makeup causes them to segregate at one and of the unimodal distribution curve.

Carcinogens. Recent data with bacterial test systems (see Sect. 5.2.2) suggest that most mutagenic substances are also carcinogenic. It is likely that pharmacogenetic principles apply to potentially carcinogenic chemicals. Genetic concepts provide a partial explanation for why most persons under equal exposure to a chemical or irritant do not develop cancer. Only individuals with variant metabolisms such as slow inactivators or those who transform a substance into a more powerful carcinogen are likely to responds with neoplasia. Genetic variation in repair enzymes (see Sect. 5.1.6.3) or in "immune surveillance" affecting mutant cells might be another mechanism that could give rise to cancer.

The enzyme system of arylhydrocarbon hydroxylase (AHH) is involved in the activation of polycyclic hydrocarbons into more potent carcinogenic agents. Arylhydrocarbon hydroxylase levels in man are under genetic control as assessed by twin and family studies. The exact mode of inheritance is not clear, but monogenic inheritance has been claimed for man [1160] and for an analogous enzyme system in mice [1235]. It is more likely, however, that inheritance is polygenic [1235, 1026]. In any case, it is conceivable that persons with high arylhydrocarbon hydroxylase activities may be at higher risk of cancer induced by polycyclic hydrocarbons, such as lung cancer associated with cigarette smoking [1161, 1068].

More recently, genetic variation in the activity of glutathione transferase has been described in human lymphocytes [1296]. The glutathione transferases play an important role in the biotransformation and detoxification of various endogenous and exogenous compounds. There was a clear trimodal distribution in a population sample of 248 individuals, 133 having low, 94 high and 21 very high activities. Data in eight families were compatible with a simple autosomal codominant mode of inheritance. It is well possible that individuals possessing a higher activity in the liver will be protected better against certain exogenous and endogenous reactive substances.

Much more work on the metabolism of carcinogenic substances needs to be done to demonstrate the likely role of human variation in environmental carcinogenesis. Variations in repair mechanisms may also play a role as suggested by the higher frequency of cancer in patients affected with genetic lesions of mutational repair (Fanconi's anemia, Bloom's syndrome, ataxia-teleangiectasia, xeroderma pigmentosum; Sect. 5.1.6). It is particularly noteworthy that the not infrequent heterozygotes for three of these conditions have a higher frequency of malignancy; but xeroderma pigmentosum heterozygotes only after extensive exposure to sun light (Sect. 4.2.2.8). Since many human cancer are thought to be related to environmental agents to which large portions of the populations are exposed, a genetic approach is likely to provide answers on why only some persons develop cancer with equal environmental exposure.

Recent studies [976] on the frequency of the oxidation polymorphism for the debrisoquine/sparteine class of drugs were done in 245lung cancer patients and 234 smoker controls (Fig. 4.69 a, b). It was hypothesized that rapid oxidizers might activate more carcinogenic agents from cigarette smoke. The results fitted this hypothesis and showed fewer poor metabolizers (i.e., those carrying the double dose of the polymorphis gene) among the cancer patients (1.9% vs 9% among controls). Moreover, the distribution of the test results among extensive metabolizers (i.e., "normal" homozygotes and heterozygotes for this gene) in lung cancer patients was shifted in a direction that suggested a higher frequency of homozygotes for rapid oxidation. Using the data for homozygosity among cancer patients, 77% can be calculated to be rapid oxidizer homozygotes and 22% heterozygotes as compared with 49% rapid oxidizer homozygotes and 42% heterozygotes among smoker controls (lung cancer patients: $q^2 = 0.016$; $2pq = 0.22$; $p^2 = 0.77$; controls: $q^2 = 0.09$; $2pq = 0.42$; $p^2 = 0.49$). While these data are suggestive of genetic differences for this polymorphism, it is conceivable that the test results (i.e., hydroxylation of debrisoquine in the liver and excretion in the urine) might be affected by the underlying cancer. Family studies need to be done to prove the claimed genotypic differences.

Bladder cancer can be related epidemiologically to industrial amines. Seven studies of the relationship of the acetylation polymorphism (see above) to bladder cancer have been done in 633 patients and appropriate controls [1268]. Counting all studies, the risk for bladder cancer among slow inactivators is 1.3 times that of controls suggesting that the aceylator type plays a small role in susceptibility to bladder cancer by carcinogenicity of nonacetylated amines.

A report of a much higher excretion of cotinine as compared with nicotine-1-*n*-oxide among smokers who developed bladder cancer needs validation, but is intriguing as it suggests possible abnormalities in nicotine metabolism in those at risk of developing bladder cancer [1110].

α_1-Antitrypsin Deficiency (10740). α_1-Antitrypsin deficiency is associated with the Z allele in the homozygous state and predisposes to early chronic obstructure pulmonary disease. Heterozygotes for this defect have somewhat impaired pulmonary function. It is possible that with smoking and polluted environments there is an increased frequency of chronic obstructive pulmonary disease among heterozygotes for this defect [1095; 653] (see Sect. 3.7).

Paraoxonase [1231]. Parathione is a widely used insecticide. The compound is metabolized to paraoxone by liver microsomes. Paraoxonase is further broken down by the serum enzyme paraoxonase. A definite bimodal distribution of widely varying paraoxonase levels exist in the European population, with 50% having low levels. Family studies have shown that those with low enzyme levels are homozygotes for a low-activity allele (gene frequency 0.7). Distinction between heterozygotes and homozygotes for the high activity allele is more difficult. No epidemiologic investigations are yet avail-

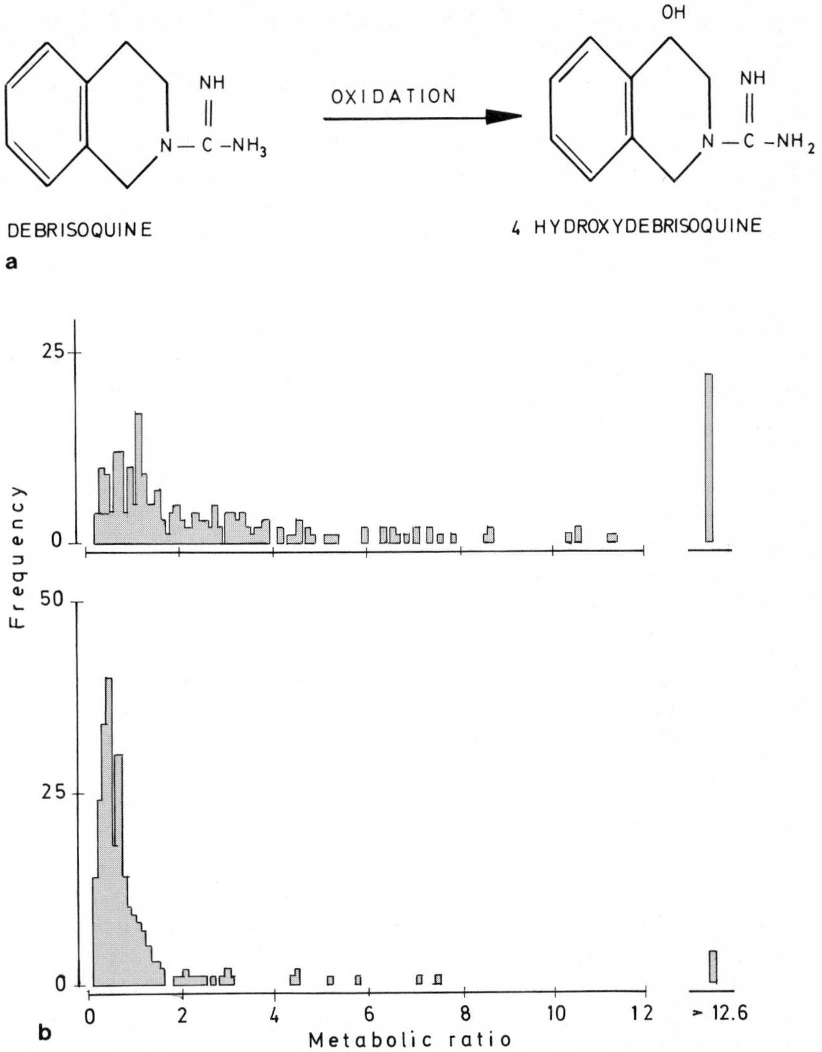

OH
NH
N — C –NH_3
OXIDATION
NH
N — C –NH_2
DEBRISOQUINE
a
4 HYDROXYDEBRISOQUINE
25
0
50
25
0
Frequency
0 2 4 6 8 10 12 > 12.6
b
Metabolic ratio

Fig. 4.69. **a** Oxidation of debrisoquine. **b** Distribution of metabolic ratios $\left(\text{urinary } \frac{\text{debrisoquine}}{\text{4-hydroxydebrisoquine}}\right)$ in smoking controls (**a**) and lung cancer patients (**b**). (Ayesh et al., Nature 312, 169, 1984, modified)

able regarding the significance of this polymorphism for those exposed to parathione. It could be expected that homozygotes would be at higher risk for poisoning with relatively low exposures. With more massive poisoning, the paraoxonase genotype presumably makes no difference for the development of symptoms.

Food. The best example of genetic difference in response to food is represented by adult hypolactasia. All human infants possess an intestinal enzyme lactase necessary for lactose absorption. In most human populations, the intestinal lactase disappears after weaning so that most human adults are lactose intolerant. A mutation makes for persistence of lactose absorption. This mutation presumably has a selective advantage in agricultural societies where cow milk is available for nutrition. In central and northern European populations, most persons possess this mutation in either single or double dose. Lactose intolerance is caused by lack of this gene and is therefore inherited as an autosomal-recessive trait. The mutation for persistence of lactose absorption is also common in some nomadic pastoralists in Africa and Arabia. Gene frequencies for persistence of lactase activity in some populations are discussed in Sect. 7.3.1. Persons with lactose intolerance tend to develop flatulence, intestinal discomfort, and diarrhea on exposure to milk and other lactose-containing foods [1285].

Celiac disease is a genetic condition associated with malabsorption of food due to dietary gluten sensitivity [1323]. When gluten is withdrawn from the diet, these patients lose their symptoms [1059].

Some, but not all G6PD-deficient persons develop hemolytic anemia on eating fava beans [1079]. Decreased urinary excretion of *d*-glutaric acid is a metabolic characteristic of patients with favism [1031] and may involve biotransformation to the toxic ingredient of the fava bean. The genetics of glutaric acid excretion is as yet unknown. One source of genetic variation in susceptibility to environmental influences that should be explored much more intensively is heterozygosity for autosomal recessive genes leading, in the homozygous state, to inherited metabolic disease. The scattered evidence available to date suggests more widespread susceptibilities than have hitherto become known (Sect. 4.2.2.8; [1340]). Moreover, phenotypic manifestations in the homozygotes and the nature of the enzyme defects give useful indications as to where disease liabilities of heterozygotes might be found.

Western diets with relatively high fat content are believed by most observers to cause the high frequency of coronary heart disease [1247]. Several autosomal dominant genes affecting lipid metabolism have been discovered: familial hypercholesterolemia, familial hypertriglyceridemia, and familial combined hyperlipidemia [1109] (see Sect. 3.8.14.2). It is unlikely that the frequency of familial hypercholesterolemia (about 1/500 in the population of the United States) has been much affected by Western diets, since the cholesterol levels of Japanese with familial hypercholesterolemia are very similar to those of affected persons in the United States and European populations [1206], while the mean cholesterol level of the normal Japanese population is significantly lower than that of Western populations. The genes for hereditary hypertriglyceridemia and combined hyperlipidemia [1109] do not express at all until the age of 25 years. It is probable that the frequency of these genetic conditions is affected by Western diets. Similarly, yet undefined genes (polygenes) that contribute to the determination of the cholesterol level and are unrelated to these major genes presumably are affected by diet to produce higher cholesterol levels in Western countries. The interaction of diet with all these genes is under intensive study (Section 3.8.14).

The gene for hemochromatosis is common in various European populations [1010]. About 1/500 of the population appear to be homozygotes and absorb iron at an increased rate, but only a fraction of homozygotes have symptoms. Fortification of bread with iron, as practiced in Sweden, has been recommended to prevent iron deficiency, which is a common condition in women and children. Such a practice would cause clinically apparent hemochromatosis to occur more frequently and earlier. The common heterozygotes (over 10% of the population) probably would not be harmed. Also, patients with thalassemia major who already are iron loaded would suffer. The practice has therefore been resisted in some countries, although undoubtedly iron deficiency would become less frequent than it is now. These considerations show the complex policy problems posed by the existence of genetic heterogeneity. What is helpful to one section of the population may be harmful to others. Furthermore, the exact scientific details of the extent of benefit and harm cannot always be fully specified. We can expect to face many of these problems as we learn more about genetic individuality.

4.6 Mechanisms of Autosomal Dominance

The basic mechanisms in autosomal-recessive diseases usually are enzyme deficiencies caused by structural mutations of the gene specifying the affected enzyme. The affected enzyme can often be shown to be structurally abnormal or unstable (see Sect. 4.2.2) [1069]. Heterozygotes usually have 50% of normal enzyme activity but are clinically unaffected, indicating that one-half the normal enzyme activity is compatible with normal function. In contrast, in autosomal-dominant inheritance, the heterozygote is clinically affected and the single dose of the mutant gene interferes with normal function.

The mechanisms for autosomal-dominant gene mutations are much more heterogeneous than those demonstrated for autosomal recessive traits. It has been suggested that autosomal-dominant diseases in general are more likely to be caused by mutations affecting structural proteins [132]. This concept has merit if one includes receptors and membranes as structural proteins, although other mechanisms also apply. In many autosomal-dominant diseases, the basic mechanism remains unknown. We have little insight as to how a single gene abnormality can produce the manifold manifestations of diseases such as neurofibromatosis or tuberous sclerosis. Similarly, we remain ignorant regarding the pathogenesis of diseases such as Huntington's chorea or polycystic kidneys. Better unterstanding of human developmental genetics may help to identify the defects in these conditions.

4.6.1 Abnormal Subunit Aggregations

Dysfibrinogenemias (13480) [1112]. A person with a dominant disease is a heterozygote. If such heterozygotes carry a protein mutation, there will be a mixture of normal and mutant molecules if the protein functions as a subunit aggregate. Presence of abnormal molecules in a mixture of normal and abnormal ones may interfere with the prcper formation of the aggregated proteins (Fig. 4.70). In some dysfibrinogenemias, various mutations involving the fibrinogen molecules may lead to a bleeding tendency. In some mutant forms of fibrinogen, the defects appeared to be at locations of the molecule that result in interference with fibrin molecule aggregation. In fibrinogen Detroit, an amino acid substitution at a site critical for conversion of fibrinogen to fibrin has been observed [1007] and was associated with severe bleeding. While the quantitative level of fibrinogen is normal in most abnormal fibrinogens, one fibrinogen defect was associated with lessened amounts of fibrinogen due to shortened molecular survival, presumably caused by molecular instability [1210]. Some genetically abnormal fibrinogens are associated with thrombosis. The critical mutational events making for increased clotting as contrasted with a bleeding tendency remain unknown. Most fibrinogen variants are not associated with clinical difficulties. More knowledge regarding the various amino acid substitutions in the dysfibrinogenemias is required for better understanding of structure-function relationship.

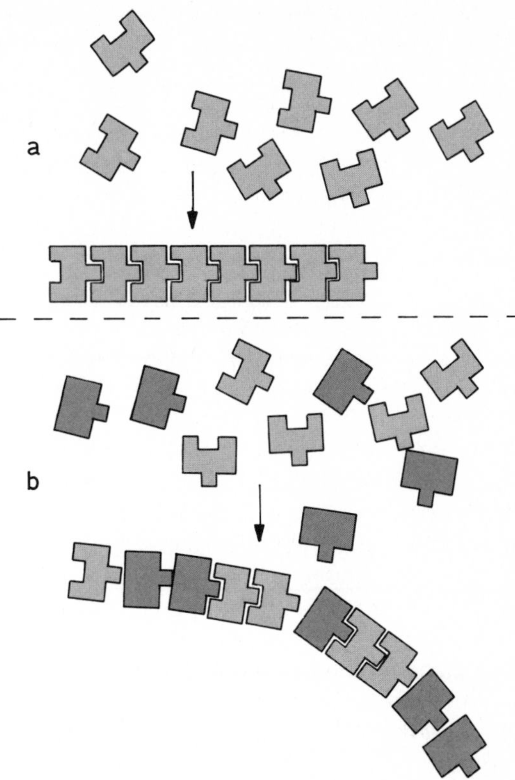

Fig. 4.70 a, b. Diagrammatic structure of a polypeptide chain in normals and heterozygotes. **a** The individual is homozygous and produces only normal polypeptides. **b** The individual is heterozygous. Normal and abnormal polypeptides are formed in equal amounts. The polypeptide chain cannot be properly assembled

4.6.2 Disturbance of Multimeric Protein Function by Abnormal Subunits

Hemoglobin Diseases. Somewhat analogous mechanisms in subunit formation appear to explain the various types of clinical abnormalities seen in the hemoglobin diseases. Since the functional hemoglobin molecule consists of four subunits produced under specification of two gene loci, heterozygotes will form hybrid molecules consisting of the normal and abnormal hemoglobin.[4] Depending upon the characteristics of the respective hemoglobin mutation, various manifestations such as methemoglobinemia, hemolytic anemia, or erythrocytosis may occur in heterozygotes (Sect. 4.3.2). Because of the high degree of co-operativeness among the four hemoglobin subunits, an abnormality in only one of the four subunits as encountered in a heterozygote for either an Hb α or Hb β mutation will lead to loss of normal function for the entire molecule. For example, some mutations lead to amino acid substitutions in regions of the α or β chains that are involved in contact between the four polypeptide chains within the tetramer. Such mutations may impair the heme-heme interaction necessary for oxygen exchange. This leads to a relative oxygen shortage in the tissues, which will be compensated by increasing the number of erythrocytes.

In these cases, human mutations are dominant because the resulting protein, to maintain normal function, has to undergo a functionally important interaction with a protein formed by the other allele. The amino acid substitution resultant from this mutation is located at a site necessary for this interaction. Functionally analogous amino acid substitutions could easily be responsible for other dominant mutations as well.

[4] Hybrid molecules such as Hb β^s β^A form but cannot be demonstrated with the usual methods of hemoglobin separation such as column chromatography and electrophoresis.

4.6.3 Abnormal Feedback Inhibition of Enzymes and Structurally Abnormal Enzymes

Porphyria (17600) [1282, 1217] - *Decreased Enzyme Activity.* In the various dominant porphyrias (Table 4.25), enzyme deficiencies affecting various steps of heme or porphyrin biosynthesis have been demonstrated. In each case, about 50% of the enzyme, as in the usual heterozygous situation, has been found. The pathophysiology of this group of

Table 4.25. Enzyme defects in the dominant hereditary porphyrias

Disease	Enzyme defect
Acute intermittent porphyria	Porphobihinogen diaminase
Variegate porphyria	Protoporphyrinogen oxidase
Hereditary coproporphyria	Coproporphyrinogen oxidase
Porphyria cutanea tarda	Uroporphyrinogen decarboxylase
Protoporphyria	Ferrochelatase

disease has been best studied in acute intermittent porphyria. Many patients with the characteristic enzyme deficiency of porphobilinogen deaminase do not have any symptoms of porphyria such as abdominal pain or neuropathy. Symptoms are usually associated with markedly elevated activity of the enzyme δ-aminolevulinic acid (ALA) synthetase, the first and rate-limiting enzyme of porphyrin synthesis. In fact, it had been thought at one time that the primary lesion of acute intermittent porphyria was a regulatory defect causing overproduction of this enzyme. ALA synthetase is induced by many drugs (for example, barbiturates, steroid hormones, and other chemicals) and normally is repressed by feedback inhibition from heme, the end product of the biosynthetic pathway which includes uroporphyrinogen synthetase. The diminished uroporphyrinogen synthetase activity in acute intermittent porphyria leads to less heme formation and to derepression of ALA synthetase with increased ALA formation. Half the normal amount of the enzyme is not sufficient to allow optimal function of the pathway, particularly when the pathway is stimulated by drugs such as barbiturates. In contrast to other enzyme deficiencies, the mutation in acute intermittent porphyria affects a critical rate-limiting enzyme in a tightly regulated biosynthetic pathway.

It is likely that other unidentified environmental chemicals and metabolites may also stimulate the pathway. Such a mechanism would explain why only some persons with the enzyme defect develop clinical symptoms. This principle of enzyme repression has been used for therapy. Hematin, as a source of heme, has been given to suppress ALA synthetase activity. A documented decline of ALA and porphobilinogen production has been associated with clinical improvements of acute intermittent porphyria.

Increased Enzyme Activity in Gout. Increased enzyme activity due to a structural lesion of an enzyme with increased specific activity has been demonstrated as a rare cause of dominant gout. Occasional patients have elevated amounts of phosphoribosylpyrophosphate synthetase activity [990]. The enzyme in these patients was shown to be structurally abnormal by electrophoretic and immunochemical analysis. These findings suggest that the primary defect in this condition affects the enzyme directly.

4.6.4 Receptor Mutations

Receptors. Receptors on the surface of cell membranes exist for many hormones, neurotransmitters, and drugs. Many mutations affecting such receptors that are proteins must exist [29], but so far only two groups of receptor mutations have been studied in detail.

One set are X-linked receptor defects that lead to androgen resistance by failure either to bind dihydrotestosterone to cell surfaces or inability to activate the nuclear binding sites of the hormone (see Sect. 4.7.5) The other set of receptor mutations affects cell binding by LDL cholesterol [29, 1107]. Cholesterol in the serum is carried mostly by a lipoprotein known as the low-density lipoprotein (LDL). A cell surface receptor (Fig. 4.71 a) exists in specialized structures known as coated pits on fibroblasts and lymphocytes (and by inference in liver cells), which binds LDL to the surface of the cell and transports the LDL-cholesterol complex into the cell by endocytosis. The LDL receptor binds only lipoproteins carrying apoprotein B and apoprotein E (B/E receptor). Such receptor-mediated endocytosis is a universal mechanism by which cells take up large molecules, each under the control of a highly specific receptor. Following the movement of LDL cholesterol into the cell cholesterol accumulation signals the cell to shut off further synthesis of LDL receptors the LDL protein is degraded by cellular lysosomes. The binding of LDL cholesterol and its transport into the cell further decreases cholesterol synthesis by diminution of the activity of the rate-limiting enzyme HMG CoA reductase. Cholesterol is esterified by the action of acyl-coA: cholesteryl acyltransferase (ACAT). The nature of the signals, which initiale these various pleiotropic reactions, is unknown.

Familial Hypercholesterolemia [1087, 4440] (Fig. 4.71). Familial hypercholesterolemia is caused by one of about ten different mutations at the same locus (chromosome 19) that affect LDL receptor function [913]. These mutations have been categorized in several classes as (a) no receptor synthesis, (b) defective transport to the cell surface after synthesis, (c) defective binding of LDL, (d) defective in-

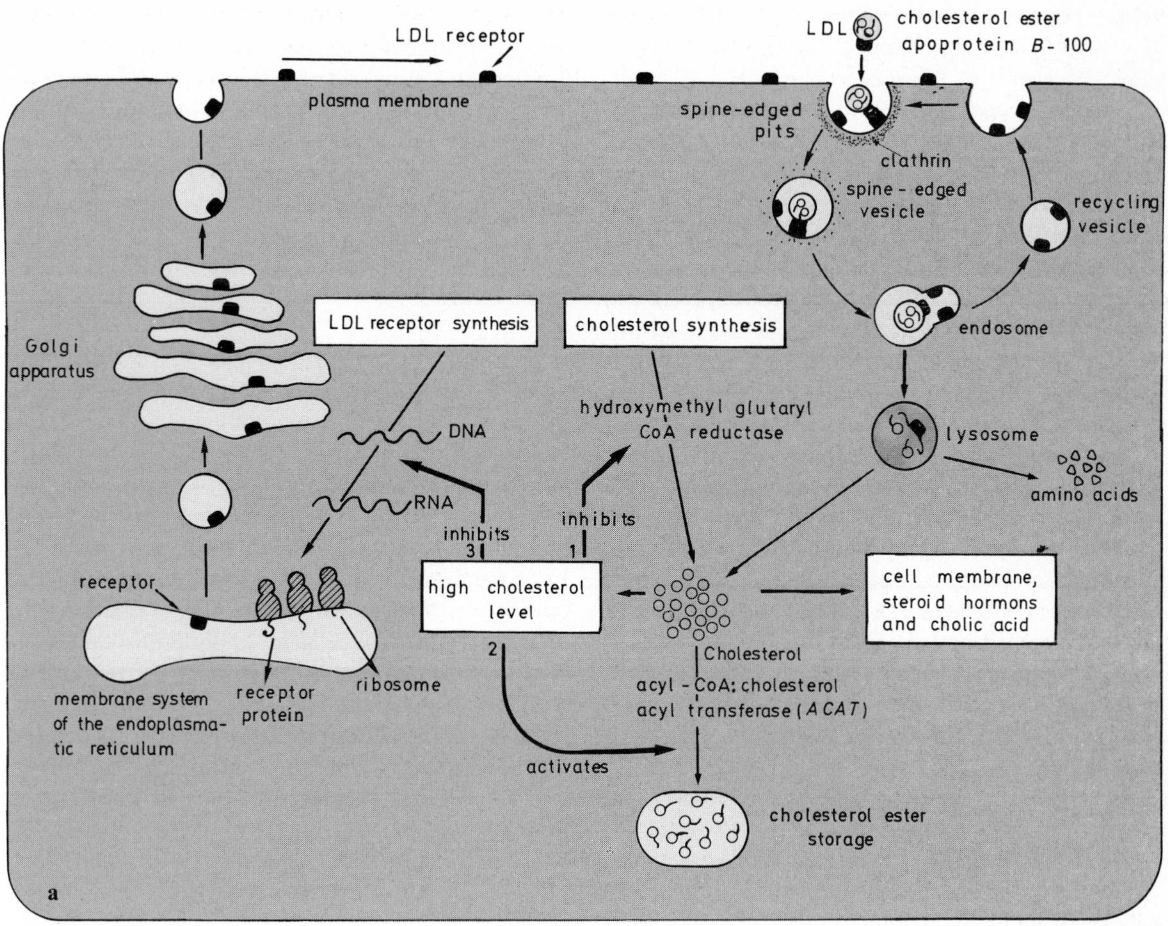

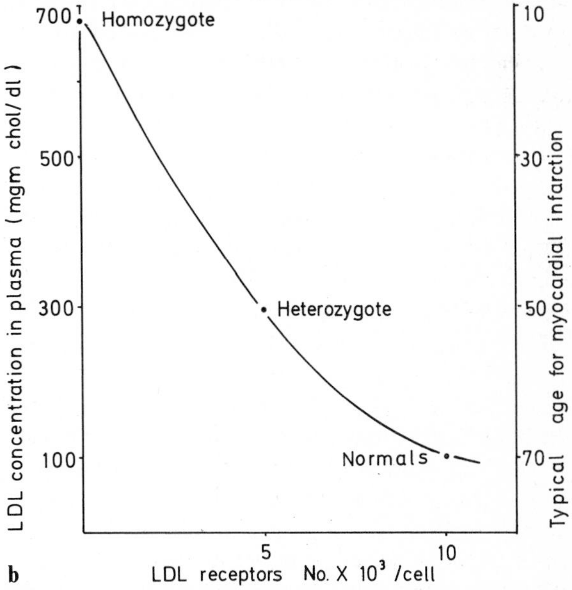

Fig. 4.71. a Cholesterol metabolism in the cell. Low-density lipoprotein (LDL) carries cholesterol *(above right)*. LDL is bound by a receptor in a spine-edged pit which is then transformed into a spine-edged vesicle. Several such vesicles fuse to form an endosome, where LDL separates from its receptor. The receptor returns to the cell membrane. LDL is taken up by a lysosome, where enzymes degrade apoprotein B-100 into its amino acids and open the ester bonds of cholesterol ester. Free cholesterol is used for production of cell membranes, steroid hormones or cholic acids. The cell regulates its cholesterol level, a high cholesterol level having three different effects: *1*, The enzyme HMG-CoA reductase, the limiting enzyme of cholesterol synthesis, is inhibited; *2*, the enzyme ACAT is activated, which esterifies cholesterol for storage together with fatty acids; *3*, synthesis of new receptors is inhibited by inhibition of transcription of the receptor gene. (From *Spektrum der Wissenschaft*, January 1985, modified) **b** Relation between LDL concentration and typical age for myocardial infarction due to coronary atherosclerosis as a function of the number of LDL receptors on fibroblasts from normal subjects and from persons with heterozygous and homozygous forms of familial hypercholesterolemia. The number of LDL receptors per cell was calculated from experiments in which maximal LDL binding was measured at 4 °C in actively growing fibroblasts deprived of LDL for 48 h (Goldstein and Brown, 1977 [1107])

ternalization, and (e) defective clustering in coated pits. As a result there may be complete absence or defective action of the receptor. About 1/500 persons in the general population are heterozygotes for familial hypercholesterolemia. They have one-half the number of normal LDL receptors and do not remove cholesterol at the normal rate from the circulation. Elevated serum LDL cholesterol levels result with development of atherosclerosis and premature heart attacks. About 50% of male heterozygotes will have overt clinical manifestation of coronary heart disease around the age of 50 years (Sect. 3.8.14). It has become possible to stimulate the heterozygote's normal LDL allele to increase the synthesis of LDL receptors by the oral administration of bile sequestrants (e.g., cholestyramine) which remove bile acids from the intestine [593]. This therapeutic maneuver, together with the drug mevinolin (an analog for the substrate of HMG CoA reductase) [722], which blocks cholesterol synthesis, allows normalization of cholesterol levels and the likely prevention of coronary heart disease. While bile sequestrants have been used for many years and appear safe, the long-term safety of mevinolin has not yet been demonstrated.

Homozygotes for thr receptor defect have no functional receptors and, because of their very high LDL levels, develop coronary heart disease and death in childhood or adolescence. Because of mutational heterogeneity many "homozygotes" are really compound heterozygotes for two different LDL receptor mutations. The severity of the clinical disease among true homozygotes and compound heterozygotes depends on the nature of the mutation they carry. Complete absence of LDL receptors causes more severe disease than lowered receptor activity due to defective LDL binding. Homozygotes do not respond to drug treatment and require other approaches such as portocaval shunts. A liver transplant has been done to furnish normal LDL receptors in one patient and has markedly reduced LDL cholesterol.

Dominant Hemolytic Anemia Due to Increased Red Cell Adenosine Deaminase (24275) [1334] - *a Receptor Defect?* A single kindred with 12 of 23 maternal relatives in several generations had severe hemolysis with excellent erythropoietic compensation and mild or no anemia. Affected individuals had 45-70-fold elevations of adenosine deaminase activity in the red cells. ATP levels were 47% those of appropriate controls. Other nucleotides were similarly decreased. The defect was inherited as an autosomal-dominant trait. The hemolytic syndrome presumably is caused by depletion of adenine nucleotides, which the nonnucleated red cells cannot synthesize. The basic defect, which makes for the remarkably high adenosine deaminase levels, remains unknown but may be a mutation of a postulated membrane receptor facilitating entry of adenosine into red cells. Inhibitors of adenosine transport into red cells also inhibit ADA activity. In analogy with the LDL receptor that modulates HMG CoA reductase activity, it is possible that a receptor involved in adenosine transport may modulate ADA activity. If this modulation is defective, ADA activity may be increased.

4.6.5 Membrane Defects

In some dominant diseases, the mutation seems to affect cellular membranes. One example is hereditary spherocytosis (18290), a common type of hemolytic anemia, in which the erythrocytes do not assume the disk-like, biconcave shape found normally but show a more spheroid form. There is a decrease of membrane surface area and diminshed membrane lipids and a specific increase of membrane permeability to sodium [1142]. Spherocytes are eliminated from circulation by the spleen at an increased rate. The exact membrane defect is unknown but probably affects the interaction of spectrin and other proteins to form a normally functioning cytosceleton. Different mutations appear to be involved.

4.6.6 Deposition of Abnormal Fibrillar Proteins: Hereditary Amyloidoses (10480-10525) [1102]

A group of diseases known as the amyloidoses is characterized by the deposition in many organs of different abnormal proteins that have been characterized physiochemically as "β-pleated sheet" fibrillar proteins. The nonhereditary amyloidoses occur as a complication of longstanding suppurative, inflammatory, or granulomatous infections. In other cases amyloidosis may be associated with various plasma cell dysplasias such as myelomatosis. In the plasma cell disorders, the fibrillar protein is derived from immunoglobulins, while the origin of the amyloidogenic protein is unknown in amyloidosis secondary to various chronic diseases.

Several clinically distinct types of inherited amyloidosis exist is autosomal dominant traits (Table 4.26). The most frequently recognized and best studied type is the so-called amyloid neuropathy type I (Portuguese or Andrade type) [1102]. Patients suffer from sensorimotor polyneuropathy and autonomous nervous system dysfunction. The amyloid material in this disease is unique and differs immunologically from the immunoglobulin-related amyloid fibrils seen in plasma cell diseases as well as from the

Table 4.26. Autosomal dominant amyloidoses. (Adapted from Glenner et al., 1978 [1102])

Generalized
- Neuropathic forms
  - Amyloid neuropathy type I (Portuguese or Andrade type)
  - Amyloid neuropathy type II (Indian type)
  - Amyloid neuropathy type III (Iowa type)
  - Amyloid neuropathy type IV (Finland type)
- Nonneuropathic forms
  - Visceral amyloidosis of Ostertag
  - Amyloidosis with deafness, urticaria, and limb pains
  - Amyloid cardiomyopathy

Localized amyloid deposits
- Cerebral arterial amyloidosis
- Lattice-like dystrophy of cornea
- Cutaneous lichenoid amyloidosis
- Medullary carcinoma of thyroid in multiple endocrine adenomatosis type 2 with amyloid deposits

amyloid fibrils of unknown protein origin in the acquired systemic amyloidoses. No immunologic or physicochemical studies of the amyloid fibrils have been carried out in the other clinical types of autosomal dominant amyloidoses.

The origin of the characteristic fibrillar protein in the hereditary amyloidoses is not clear. It is possible that the underlying mutation leads to production of an abnormal protein, which because of its altered characteristics takes on the characteristic β-pleated fibrillar structure of amyloid. Since the disease is a dominant trait, only 50% of this protein would be expected to be abnormal. The abnormal protein is deposited largely in neural tissues where it causes damage, leading to the various neuropathic types of hereditary amyloidosis. Recent studies on this type have shown that the amyloid fibril proteins in familial amyloid polyneuropathy are composed of variants of human prealbumin with single amino acid substitutions. Although the exact pathogenic role of prealbumin in the syndrome is not clear, use of DNA variants at the prealbumin locus allows preclinical and prenatal diagnosis of polyneuropathic amyloidosis.

Amyloidosis is also seen in an autosomal recessive disease, familial Mediterranean fever (FMF) (24910). Here the amyloid deposits are largely found in the kidney and resemble immunologically those seen in the acquired systemic amyloidoses. Amyloidosis in FMF appears secondary to the underlying disease.

4.6.7 Dominantly Inherited Tumor Diseases

Families showing a dominant mode of inheritance for malignant and nonmalignant tumors have been observed. Among these are well-known genetic diseases such as neurofibromatosis, hereditary polyposis, and retinoblastoma. In retinoblastoma (18020), the malignant eye tumor of infants and young children, a dominantly inherited and a non-hereditary form exist. The mechanism of dominant inheritance in retinoblastoma is quite unique and may apply to other tumors transmitted by this mode of inheritance. In contrast to sporadic retinoblastoma, patients with the hereditary variety (Sect. 5.1.6) inherit a single germinal mutation on chromosome 13 from one of their parents. This mutation alone does not cause retinoblastoma, but all retinal cells carry the mutation. The identical region of chromosome 13 transmitted by the other parent is then altered by a somatic mutation caused by a variety of somatic events in *one* of the many retinal cells (Fig. 5.37). The normal function of this chromosomal site is the prevention of tumor formation, which only occurs when both alleles are nonfunctional. This state in hereditary retinoblastoma is produced by a germinal mutation on one chromosome and a somatic mutation on the other.

Dominant transmission in these instances is inheritance of an otherwise silent germinal mutation in all retinal cells. However, the chance that an abnormal somatic mutation will occur to affect the same chromosomal region in one of the many retinal cells is about 90%. The resultant homozygous state in such a retinal cell that is now homozygous causes malignant proliferation and retinoblastoma. Dominant inheritance in these instances actually is transmission of a recessive germinal gene, which is rendered clinically apparent by a second somatic mutation affecting its allelic partner.

General Remarks. Table 4.27 lists various mechanisms of dominant disease. Even some enzyme deficiencies such as Cl inhibitor deficiency causing hereditary angioedema (10610) [1284] and antithrombin deficiency (10730) [1075] producing an increased tendency to venous thrombosis are inherited as autosomal dominant traits. It is not clear why disease develops in these heterozygous enzyme deficiencies. The many dominant anomalies, which are caused by disturbances in embryonic development, such as malformations of limbs (for example, brachydactyly; Sect. 3), remain largely unexplained. The same is true for complex syndromes, such as neurofibromatosis (16220) or osteogenesis imperfecta (16620). The examples given indicate that Mendelian dominance may result from a variety of different mecha-

Table 4.27. Some mechanisms of dominant disease

Mechanism	Example
Abnormal aggregation of protein subunits	Abnormal fibrinogens
Disturbance of multimeric protein function by abnormal subunits	Unstable hemoglobins
Diminished feedback inhibition by end product due to enzyme deficiency	Porphobilinogen deaminase deficiency in acute intermittent porphyria
Cell receptor defects	LDL cholesterol receptor defects in familial hypercholesterolemia
Cell membrane defects	Hereditary spherocytosis
Deposition of abnormal fibrillar protein	Hereditary amyloidosis (Portuguese type)
Somatic cell homazygosis plus recessive gene transmission	Retinoblastoma (See text)

nisms. Their elucidation will help to clarify many aspects of the genetic determination of structure and function. Scientifically, this promises to be much more interesting than the simple one-gene-one-enzyme relationship often found in recessive metabolic diseases. It is of particular interest that several intracellular enzymatic reactions can be disturbed by a single receptor mutation. Such biochemical pleiotropism offers an exciting model for the control of complex biochemical pathways by a single gene. The basic lesions in the common dominant diseases, offer exciting scientific challenges.

4.7 Genetics of Embryonic Development

Biochemical and molecular genetics have taught us much about the structure of genes, and the genetic control of enzymes and functional proteins. Information regarding the genetic basis of embryonic development is much less satisfactory. Consequently, enzyme defects are relatively well understood; the biochemical basis of morphologic abnormalities is in most cases obscure, despite successful attempts to elucidate biochemical mechanisms of dominant diseases (Sect. 4.6). The genetics of development is an unchartered area on the map of our knowledge of molecular genetic mechanisms.
As in other fields of molecular biology, developmental genetics largely utilizes experimental material other than humans because human experimentation is subject to obvious limitations. A textbook of human genetics cannot review the whole field. A broad outline will be sketched with indications where observations on humans may contribute some additional information. Developmental genetics is historically based on the classic developmental mechanics *(Entwicklungsmechanik)* and developmental physiology, which flourished in the first decades of this century, based on the work of Roux, Driesch, Spemann, Kühn, Waddington, and Hadorn.

4.7.1 Gene Activity in Early Development

It is useful to subdivide embryonic development into an early phase (comprising fertilization and the first cleaving divisions including gastrula formations) and a late phase, which comprises development of body shape and organ systems. Recent work on gene activity and regulation in early development, on more primitive organisms and later also on mammals, has yielded results that may apply to humans as well (Davidson, 1976 [49]; Magnuson and Epstein, 1981 [1207]).

a) The most fundamental genetic problem during embryonic development is differentiation: how is it possible that groups of cells assume different functions in spite of the fact that they have identical genomes? It is now known that older assertions according to which differentiation occurs mainly or even entirely at the gene level must be rejected: the genomes of all cells of an organism are – with exceptions – identical. In a classic experiment, it could be shown in frogs that an intestinal cell nucleus transplanted into an enucleated egg has all the genetic information to grow to a fully viable organism. Similar experiments have more recently been performed in mice. Hemoglobin genes have been successfully isolated from different types of nonerythrocytic tissues such as fibroblasts and lymphocytes (Sect. 4.3). Control of differentiation at the DNA level has been shown for the immunoglobulin genes (Sect. 4.4); it is likely that this might not be an absolute exception. As a rule, however, it occurs at the transcriptional level: differentiated cells produce different species of mRNA, some of which may overlap in various tissues. Regulation at other levels further away from primary gene action is possible. The exact mechanisms of this control in higher organisms are unknown; they may be related to chromosome structure as described in Sect. 2.3.
b) The number of diverse mRNA species that are translated in the early embryo is high; the pattern changes with stage of development. Hence, a large fraction of the genes seems to be necessary for early development.

c) In echinoderms, the first stages of development until gastrula or even postgastrula development are determined exclusively or predominantly by the maternal genome: the zygote receives a pool of maternal mRNA, which directs these early processes. Moreover, tRNA and ribosomes are also of maternal origin. Different parts of the paternal genome are switched on at slightly different times.

Investigations in the mouse gave a slightly different picture [1070, 1207]. In these studies, the advantages offered by the fact that many genetic variants, both normal and abnormal, are known have been used ingeniously. These variants have made it possible to combine genetic methods with other analytical techniques. At what time is the embryonic genome first expressed? Embryonic RNA synthesis starts as soon as the late two-cell stage. However, such analyses do not distinguish between maternal and paternal genomes, and RNA production does not mean that protein biosynthesis occurs as well. This problem can be solved, however, if the father can be distinguished by genetic markers from the mother. In this case, the appearance of paternal markers in the embryo indicates the latest possible stage at which embryonic gene products begin to be formed. (It does not exclude the possibility that the maternal genome is turned on earlier.) Such studies have shown that some paternal markers, such as HPRT (See Sect. 4.2.2.6) and the HY antigen, first became visible at the eight-cell stage. Others, such as β-galactosidase and β-glucosidase, can be found as soon as the four-cell stage, and at least one - the β_2 macroglobulin (see Fig. 3.39) - turns up even at the two-cell stage. Moreover, it has been shown that, for the X-linked HPRT gene, both parental genomes are active before X-inactivation (Sect. 2.2.3.3), since in female embryos, enzyme activity was twice that in males. One difficulty encountered in such experiments was sex determination, since the early embryo can be utilized either for karyotyping or for enzyme determination. This was overcome by the ingenious technique of creating artificial monozygotic twins: The two-cell stages are flushed out of the fallopian tubes, and the cells are then separated and grown in the culture medium to form two (genetically identical) embryos. Then, one embryo can be used for karyotyping and the other for biochemical studies.

Even if the embryonic genome starts to be expressed in the two-cell stage, this does not necessarily mean that its expression at this stage is necessary for normal development and function. In this connection, analysis of genetic variants leading to abnormalities in development proved useful. The dominant mutant "oligosyndactyly" in the heterozygote produces a limb abnormality and an anomaly of kidney formation. In the homozygote, it results in developmental arrest at the sixth cleavage division; the metaphase figures resemble those seen after treatment with a mitotic inhibitor such as colcemid. The abnormality of spindle formation leading to the accumulation of metaphases has been shown even in the blastocyst stage. Hence, some activity of the zygote's own genome is already necessary at this stage.

Larger Contribution of Maternal than Paternal Genotype to the Child's Phenotype? As already mentioned, the paternal genotype is switched on early in development. This does not necessarily mean, however, that maternal and paternal contributions to the development of the young embryo are in fact equal, especially since the young zygote receives a large amount of maternal RNA. There are indeed biological phenomena that suggest a larger contribution from the mother. For example: If in a crossing between horse and donkey the dam is a horse and the sire a donkey, a mule will result; if the dam is a donkey, the offspring is a hinny, which looks much more like a donkey than the mule.

In humans, sufficiently precise quantitative assessment is impossible for most normal characters in which children could be compared with their parents. Studies in this area require clearly quantifiable traits of high heritability and absolute constancy throughout life. Some family data in fact suggest children may be more similar to their mothers than to their fathers in the ridge patterns of fingers, palms, and toes [1169, 1279]. However, interpretation of such data is difficult.

Mouse Teratocarcinomas as Research Tools for Investigation of Early Development [1140]. Biochemical investigations at the cellular level on early stages of mammalian development are hampered by the scarcity of material. Many critical events take place in very small cell populations, and cells at different stages of differentiation are located close together. Recently, it has become possible to circumvent some of these difficulties by the use of mouse teratocarcinoma cells. Testicular or ovarian teratocarcinomata occur spontaneously or can be induced in a number of inbred mouse stains. They contain a wide variety of tissues corresponding to derivatives of the three embryonic germ layers. In addition, early embryo-like cells that are malignant show infiltrative growth. Their differentiated derivatives, however, are nonmalignant.

Teratocarcinoma-derived cells can be obtained in cell culture; here they may grow and differentiate into a number of tissues. A number of cell lines of early embryo-like cells have been established; when injected into mice, they may give rise to tumors containing a variety of differentiated cell types. Such cells show many similarities to normal embryonic cells and are, on the other hand, abundant. They can therefore be utilized as model systems for many aspects of differentiation. For example, they will mix with the normal embryonic cells when injected into a 4-day blastocyst, forming a genetic mosaic.

It is easy to obtain in culture large amounts of homogeneous teratocarcinoma-derived cells in a given status of differentiation. Such cell populations can be used for biochemical experiments, and the results can then be checked in the embryo. Such studies have given evidence, for example, of the common occurrence of most mRNA species in early embryonic cells and in precursors of blood cells (myeloblasts), while globin mRNA was found in myeloblasts but not in early embryonic cells. In other experiments, both X chromosomes were shown to be genetically active in clonal cultures of undifferentiated female cells; inactivation of one X occurs together with differentiation [885]. The system seems to be especially well-suited for investigation of the role of cell surface antigens in differentiation. It has been mentioned in the discussion on the major histocompatibility complex (MHC) in Sect. 3.5.5 that such antigens may have a function in cell differentiation. Early embryonic cells were shown to be completely devoid of the MHC H-2 antigens (the mouse counterparts of the human HLA specificities). Other antigens, however, are present on these cells, for example, an antigen called F-9. This antigen disappears with differentiation of early embryonic cells into fibroblasts, myoblasts, or other cell types that have H-2 antigens on their surface. In the adult animal, F-9 antigens are found only in cells of spermatogenesis. They seem to be related to the T/t gene complex. Some alleles of this system interfere with normal embryonic development, leading to arrest at specific stages of embryonic development.

The function of the F-9 antigen for morula formation was demonstrated by showing that morula formation that normally proceeds efficiently in tissue cultures could be prevented by a specific monovalent anti-F-9 antibody. Cleavage proceeded normally, indicating that no nonspecific toxic effect was involved. The antibody seems to loosen cellular interactions, thereby preventing morula compaction (and, as a consequence, blastocyst formation). In Sect. 4.2.2.6, metabolic co-operation in cultured fibroblasts between $HPRT^+$ and $HPRT^-$ cells was mentioned. This metabolic co-operation also occurs in early embryonic cells; it is inhibited by the anti-F-9 antibody.

The use of mouse teratocarcinomas for studying differentiation is one example of a widespread research strategy: studying a problem that seems to be too complex to be tractable exerpimentally in a simpler model system. It is possible that studies of this or similar systems may at some time offer an answer to the most intriguing questions of developmental genetics: Why do genetically identical cells become phenotypically different?

4.7.2 Later Phases of Embryonic Development, Phenocopies, Malformations

In later phases of embryonic development, differentiation of organ systems, extremities, head, and brain forms an organism. The occurrence of many inherited abnormalities indicates genetic control. These genetic mechanisms may be disturbed by a variety of external influences such as oxygen lack, ionizing radiation, infections with the rubella or cytomegalic inclusion viruses, or drugs, such as thalidomide and ethanol. Knowledge of such teratogenic agents is important for prevention of damage to the fetus. Details will not be given here since, apart from possible differential metabolism in the mothers and variable susceptibility of the fetus to teratogenic agents, the subject matter of human teratology, although clinically important in the differential diagnosis of birth defects, is outside the scope of this book.

Indications for Interaction Between Genetic and Nongenetic Factors in Malformation Production. Since Geoffrey St. Hilaire (1832–1836) produced malformations in chicken embryos by inhibiting gas exchange through covering the eggs with lacquer, many investigators have administered a variety of agents in efforts to disturb embryonic development. Most of these experiments were carried out in the hope of obtaining more information on the mechanisms of normal development. These hopes were fostered by the observation that sometimes phenotypes similar to those produced by gene mutation could be obtained (phenocopies; Goldschmidt, 1935 [1106]). In general, these studies have not fulfilled their expectations for an understanding of birth defects and will not be discussed further.

One aspect, however, is interesting: Some studies showed that genetic factors may be important even for induction of malformation by exogenous agents [1187]. In the chicken, hereditary malformations of the caudal part of the neural tube are not rare; "rumplessness" occurs as a dominant and as a recessive mutation; penetrance and expressivity depend strongly on the genetic background. Moreover, sporadic rumplessness is not very rare in chicken stocks that do not contain the major genes for this trait. The incidence differs from one stock to the other and is about 0.5% in jungle fowl, 1.6% in white leghorn, and 5.7% in silver-grey dorking. It is possible to induce phenocopies of rumplessness by a variety of chemicals such as insulin, boric acid, and pilocarpine. The higher the incidence of spontaneous defects in a given breed, the higher was the number of chemically induced phenocopies. Similar observations were made with other malformations in chicken.

In studies on mice, cortisone produced cleft lip and palate in 10% of the animals in a strain in which the spontaneous incidence for cleft lip and palate was 5%. In another strain with a spontaneous incidence of 0.2%, only 1.8% were observed after identical treatment with cortisone [1085].

Such clean experimental designs are impossible in humans. Hence, there is no proof that genetic variability exists for induction of human malformations by environmental agents. However, such a conclusion appears plausible.

For a better understanding of genetic mechanisms of normal development in humans, we are left with naturally occurring abnormalities. Here, two genetic situations proved to be especially revealing: chromosome aberrations and genetic anomalies of sex differentiation.

Before this evidence is discussed, however, the problem of gene action during development and differentiation will be considered from a more theoretical viewpoint. Some general rules for regulation of gene activity have been worked out mainly in bacteria and viruses. While these principles have turned out to be not directly applicable to explain regulation of gene activity in high organisms, their development has been of considerable historical significance and has had an impact on human genetics. The evidence cannot be reviewed here [117; 220], two classic examples will be shown, one for "negative," the other for "positive" control.

4.7.3 Gene Regulation in Bacteria and Eukaryotes

Negative and Positive Control. In bacteria, some genes are active only when their specific activity is required. Otherwise, they are inactive. Their activity is controlled by repressor and operator genes. The classical example is the lactose "operon" of *E. coli* [1141], in which three closely linked structural genes are under the common control of such control genes. In addition to such "negative control," positive control also occurs; in this case transcription requires a special protein in addition to an open operator.

Function of Regulatory Mechanisms. One important function for which bacteria have been selected during evolution is optimal utilization of varying energy sources for growth. Control systems, such as the lactose operon, fulfill this requirement. Bacteria use energy resources for producing lactase-cleaving enzymes only if lactose is available. Hence, the feedback circle of negative control helps to guide energy into the most useful direction.
cAMP is necessary for the metabolism of bacteria in a variety of ways; for example, it influences transcription. In higher organisms, including humans, it has an important additional function in hormone action. The action of most hormones of higher organisms act through the same effector substance, cAMP. Many hormones do not enter the cells in which they act. Consequently, these cells must have hormone receptors on their surface. Ligandreceptor interaction causes changes in the cell membrane that regulate the enzymes for cAMP formation and catabolism. cAMP, in turn, regulates transcription.
One important aspect of differentiation is the "decision" which of the hormone receptors will occur on a given cell type. Indeed, hormones have been shown to play an important role in embryonic development. One example, sex differentiation, will be described below in somewhat greater detail. In general, requirements for regulation of gene activity are much more complicated in higher organisms, including humans, than in bacteria. However, it can be expected from experience with evolutionary genetics that nature may have utilized similar principles for gene regulation as for differentiation. A stepwise adaptation to the more and more complex regulatory problems arising during evolution occurred. It will therefore be the logical course of research to follow evolution: investigating more and more complex biologic systems and adding more and more complexity to the interpetive models.
Research on regulatory mechanisms, such as on receptors, hormones and action of cAMP, and other chemical inducers of development in higher organisms, is being actively carried out in many fields. In this connection, some results from molecular biology suggest partial answers. In section 4.4, the genetics of antibody formation has been described. This is a first example for analysis of differentiation at the molecular level. It shows how various changes may occur at different levels – rearrangement of the DNA itself, at transcription, and in the molecular organization of the final product, i.e., the antibody molecule. For example, the soluble and the membrane-bound form of the IgM molecule differ by an additional amino acid sequence that enables the latter to be anchored in the cellular membrane. Both proteins are determined by the same gene complex; the primary transcript, however, is processed differently.
By the same token, one gene may even form two different proteins. The peptide hormone calcitonin of the rat is one example [968]. It is found in the thyroid and in the hpyothalamus where it occurs together with another polypeptide, the CGRP ("calcitonin gene-related product"). CGRP corresponds with calcitonin in 78 amino acids but has altogether 128 amino acids. Both molecules are determined by the same gene, whose primary transcript is processed differently: A complete exon is spliced out of the transcript that determines CGRP, and another part is added (Fig. 4.72). This process predominates in the hypothalamus but not in the thyroid, leading to different products of the same gene in different tissues. Determination of the growth hormone somatotropin follows similar rules. In the human pituitary, the normal hormone which consists of 191 amino acids occurs side by side with a second protein in which the 15 amino acids at positions 32–46 are lacking. This is precisely the beginning of the second exon; its first 45 nucleotides may function either as part of an exon, or as part of an intron.
Probably, such differences in the processing of the primary transcript are of more general significance for differentiation. But the question remains open by which mechanisms these differences in transcript processing are caused. A deeper understanding, may depend on elaborate theoretical concepts that are formulated as models. From such models, testable consequences can be derived, and suitable experiments may lead to modification of the model and to stepwise approximation toward the truth. Models of this sort should be treated as heuristic tools for directing further research.

Britten-Davidson Model. Britten and Davidson (1969) [1019] have proposed such a model for gene regulation in higher organisms. Based on the general features of the regulation models in microorganisms, it contains logical extensions for the more complex regulation requirements in differentiation.
According to this model, four kinds of genes are assumed: producer genes; receptor genes that are linked to producer genes and induce transcription under the influence of activator substances that are produced by integrator genes;

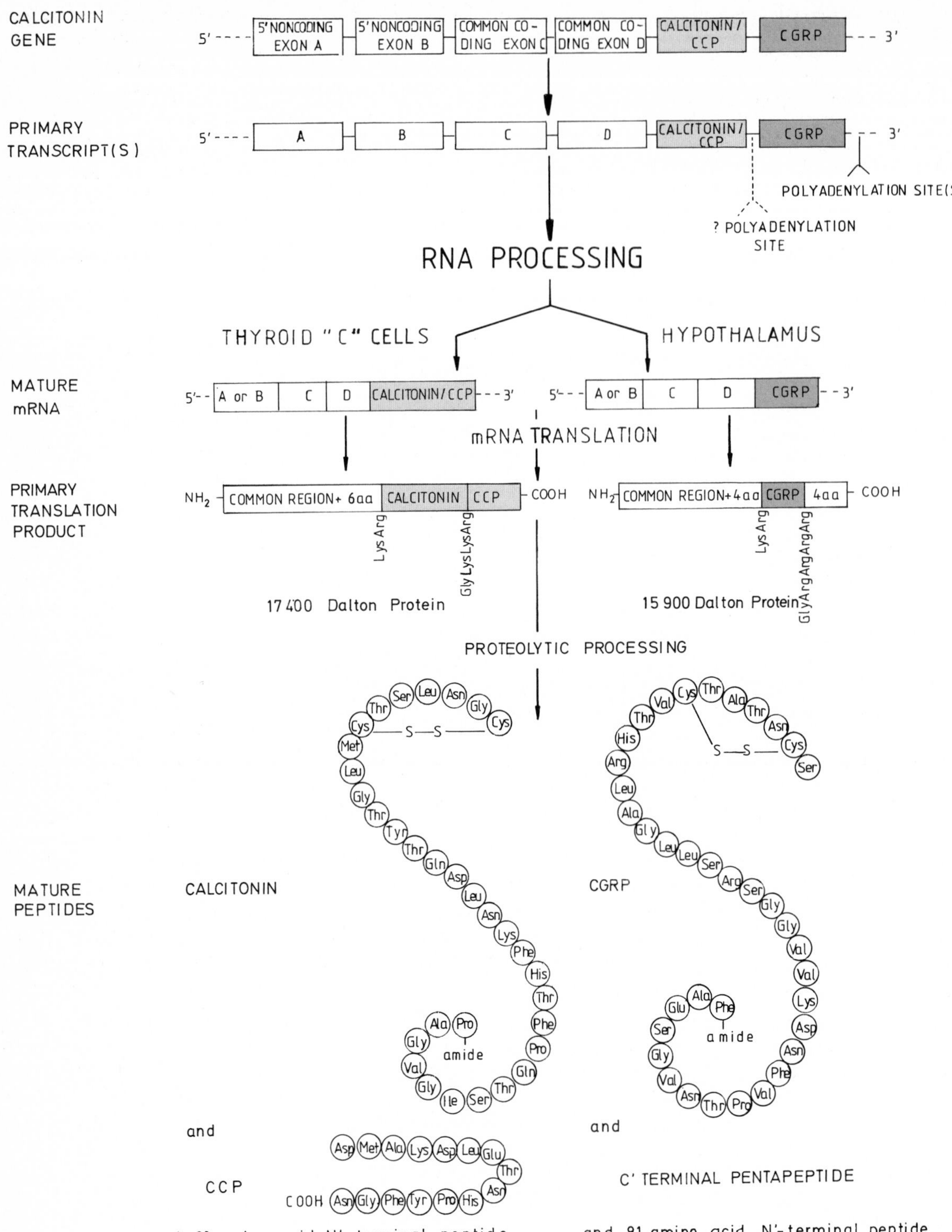

Fig. 4.72. Mechanism of tissue - specific expression of the calcitonin gene. The forked arrow indicates difference in mRNA processing between thyroid and hypothalamus. Some aspects of this pathway are still hypothetical but the genes and the end products (calcitonin and CGRP) are established. (From Amara et al., Nature 298, p. 243, 1982)

and sensor genes serving as binding sites for agents (hormones?) inducing specific activity patterns in the genome. This model was designed to be compatible with observations on embryonic development, for example, changes in state of differentiation that are induced by single genes and integrated activity of many different genes in a specific state of differentiation. The model allows for flexible interaction of genes at different levels of integration. During the 10 years since it has been proposed, little experimental evidence seems to have been assembled to test its elements.

Unfortunately, this model, like other attempts at explaining embryonic development, leaves open the most intriguing question: Why are different sets of genes activated in different cells, and why do these cells acquire different properties? This is the basic problem of differentiation. As mentioned, receptor molecules at the cell surface may be responsible for such differences, and the Britten-Davidson model additionally suggests differences in sensitivity of homologous sensor regions in different cells. But why do some cells have receptors that are lacking in others? And why are their genomes sensitive for different agents? This primary question of differentiation remains unanswered.

4.7.4 Genotype-Phenotype Relationships in Human Chromosome Aberrations [1176]

Most chromosome aberrations do not lead to defects of genes; in trisomies, for example, all genes of the trisomic chromosome are present, albeit in triplicate rather than in the normal duplicate number, and yet embryonic development is thoroughly disturbed. Hence, different mechanisms such as disturbances in regulation may be involved. Moreover, most clinical findings in patients with structural aberrations leading to loss of genetic material (for example: deletions, ring chromosomes) are not unlike those found in many trisomies. Many of such clinical signs are independent of the location of the structural defect. Therefore, it is not unlikely that the phenotypic abnormalities are produced by an imbalance of gene action during embryonic development rather than by specific gene loss. These observations suggest that analysis of abnormal development at different levels in zygotes with chromosome aberrations might be instructive for understanding of the normal regulatory processes.

4.7.4.1 Gene Dosage Effects in Trisomies and Gene Mapping

Early Attempts at Chromosome Mapping by Gene Dosage Effects. It is a general rule in autosomal-recessive enzyme defects that enzyme activities in heterozygotes are about halfway between those of the two homozygous phenotypes. Where enzyme activity in homozygotes is close to zero, about 50% of the normal enzyme activity is usually found in heterozygotes (Sect. 4.2.2.8). This means that, under normal conditions, enzyme activity directly reflects the amount of protein synthesized. No control or regulating mechanisms adjust protein synthesis to a "normal" level in such heterozygotes. It was therefore not unreasonable to argue that activities of enzymes that are determined by genes located on trisomic chromosomes or chromosome segments should occur in a ratio of 3:2 when compared with normal homozygotes. By similar reasoning, it was argued that measurements of enzyme activities in trisomics might offer evidence on whether the genes determining these enzymes are located on the trisomic chromosomes or chromosome segments.

In the 1960s, several instances of increased enzyme activity were, indeed, observed in trisomy 21, and some of them seemed to fit the expectation of an activity ratio of 3:2 between trisomics and normals. The use of these data for gene mapping became questionable when increased enzyme activity for G6PD, an X-linked enzyme, was found in the autosomal trisomy 21. Moreover, when more direct evidence on gene mapping became available from studies on cell hybridization between human and rodent cells (Sect. 3.4.3), gene assignments based on enzyme activity in trisomies turned out to be faulty. Variation in enzyme activities was found to be another aspect of unspecific disturbances due to genetic imbalance. More recently, however, an increasing number of enzymes for which genes had been localized by independent methods on chromosomes, did indeed show the expected gene dosage effects (Table 4.28). Moreover, when the amount of mRNA was assessed in fibroblast cell lines that were monosomic or trisomic for chromosome 21 the expected dosage effect was also found [1185]. Such studies might prove useful in the long run in helping to analyze chromosomal interaction in gene regulation.

So far, these gene dosage effects cannot yet be related to any of the clinical symptoms in Down's syndrome. Hence, these results - and also the gene dosage effects enumerated in Table 4.28 for genes located on other chromosomes - do not help to formulate a theoretical concept on the mechanisms of interaction between genes. In the future, they may be useful as building blocks for a theory, but at present we do not know exactly where they might fit in. At present, the only way seems to be to look for further evidence on phenotypic deviations in zygotes with chromosome aberrations.

Table 4.28. Autosomal gene dosage effects (Krone and Wolf, 1978 [1176][e]). Many more examples have been described in recent years. (See Sect. 3.4; Table A9.5)

Gene product		Chromosome or chromosome segment	Kind of anomaly	Kind of cells studied[b]	Ratio of abnormal/normal enzyme activity
Name	Code[a]				
Fumarate hydratase	FH (13685/86)	1 p2 or 3-1qter	Partial trisomy	F	1.57 1.6 (i)[c]
Acid phosphatase-1	ACP-1 (17150)	2p23-2pter	Partial monosomy	E	0.58
Acid phosphatase-1	ACP-1 (17150)	2p23-2pter	Partial trisomy	E	1.39[d]
Galactose-1-phosphate-uridyl transferase	GALT (23040)	3q21-3qter	Partial trisomy	E	1.44
Adenylate kinase-1	AK-1 (10300)	9q33-9qter	Partial trisomy	E	1.43
Glyceraldehyde-3-phosphate dehydrogenase	GAPD (13840)	12p12.2-12pter	Partial trisomy	E	1.47 1.37
Triosephosphate isomerase	TPI (19045)	12p12.2-12pter	Partial trisomy	E	1.86 2.20
Nucleoside phosphorylase	NP (16405)	14q11-14q21	Partial trisomies	E	1.49–1.73
Adenine-phosphoribosyl-transferase	APRT (10260)	16	Trisomy	A	1.69
Superoxide-dismutase-1	SOD-1 (14745)	21	Trisomy	E	1.45
Superoxide-dismutase-1	SOD-1 (14745)	21 21q22.1	Trisomy Partial trisomy	E E	1.56 1.75
Superoxide-dismutase-1	SOD-1 (14745)	21	Trisomy	P	1.56

[a] According to McKusick's catalogs (1978) [133].
[b] *A* amnion cells; *E* erythrocytes; *F* fibroblasts; *P* blood platelets.
[c] The normal value was taken to be equal to two-thirds of the sum of the specific activities contributed by two B alleles and one A allele.
[d] Immunologically ascertained.
[e] Note that gene dosage effects for these enzymes are in accord with gene assignments using other methods.

4.7.4.2 Other Biochemical Anomalies in Chromosome Aberrations

Fetal and Embryonic Hemoglobins in Trisomy 13. The normal pattern of hemoglobin synthesis has been described in Sect. 4.3.2.; ζ, ε and γ chains together with α chains are incorporated in the early embryonic hemoglobins Gower I and Gower II, respectively. During embryonic development, these hemoglobins are replaced by fetal hemoglobin (F), which in turn is almost entirely replaced by adult hemoglobin A in the early months of extrauterine life (Fig. 4.36).

In trisomies 18 and 21, the sequence and timing of the various developmental hemoglobins is normal. In trisomy 13, the level of Hb F is elevated at birth and decreases at a slower rate than normal with increasing age; small amounts of $Hb_{\gamma 4}$ (Barts) are present; a short time after birth even small amounts of Hb Gower II can still be found [1135]. These anomalies in the development of hemoglobins were observed in patients with trisomy 13 due to Robertsonian translocations as well. They seem to be associated with trisomy of a segment distal to 13–21.

The abnormality of hemoglobin synthesis in trisomy 13 appears to be an anomaly in the mechanism that controls the switchover from synthesis of the ε chain to the γ chain and the subsequent switchover from synthesis of the γ chain to that of the β and δ chains. The abnormality of hemoglobin synthesis in trisomy 13 was discovered at a time when the Jacob-Monod model of gene regulation (Sect. 4.7.3) had just been proposed and confirmed by experimental observations in bacteria. Therefore, it is not surprising that explanations in terms of this model were sought. Experimental evidence for or against any such hypotheses is not available; the persistence of the ε and γ chains in trisomy 13 is not accompanied by abnormal prolongation of the megaloblastic phase of hematopoiesis. This switch could be traced back to methylation and demethylation of Hb genes (Sect. 2.3.2.). It appears unconnected with a change of the tissue in which the synthesis proceeds.

The discovery of the anomaly in the synthesis of hemoglobin chains indicates ways in which progress in this field could be achieved: analaysis of changes in the switch-on and switch-off of the synthesis of specific proteins. Unfortunately, it is difficult to find other models for a similar analysis.

Trisomy 21 and DNA Repair. Another finding may possibly be related to abnormal embryonic development in Down's syndrome. Lymphocytes and fibroblasts from patients with trisomy 21 that are irradiated with X-rays or UV or both yield a higher rate of chromosome aberrations than normal lymphocytes. A 25%-30% reduction in UV-induced unscheduled DNA synthesis indicated that this finding is apparently caused by a defect in excision repair [1186] (Sect. 5.1.6).

4.7.4.3 Cellular Studies in Chromosome Aberrations

Cellular Phenotypes in Human Chromosome Aberrations. Biochemical investigations have not led to definite conclusions regarding the biochemical mechanisms of regulation in embryogenesis. Therefore, an alternative question could be posed: How are cells with chromosome aberrations different from normal cells? Comparative description of cellular phenotypes might give some clues as to the nature of developmental disturbance [1179, 1181-1183]. Cultured cells from a variety of chromosome aberrations - mostly from abortions - have been examined for cell cycle, cell morphology, histochemistry, and a number of immunologic and biochemical characters. A cell strain with trisomy 7, for example, had among other anomalies a reduced capacity for formation of histotypical structures, low collagen production, poor glycogen content and acid phosphatase activity. In the cell cycle, the G_2 period was found to be twice as long as in diploid cells [1181]; the S phase was shortened [1179]. A different cellular syndrome was described in a cell strain with trisomy 14. A low growth potential and inability to form histotypical structures were also found, but biochemical characteristics were different; for example, acid phosphatase was low, and there was a high concentration of polysaccharides measured by the PAS reaction [1182]. In trisomy 14 - and also in trisomy 9 - the cell cycle was normal; apart from trisomy 7 mentioned above, the G_2 phase was also prolonged in monosomy 21 and trisomy 21. In trisomy 21, some other growth parameters were determined, and a number of deviations from normal euploid cells have been reported (Table 4.29). It is especially interesting that the cellular phenotype in triploidy turned out to be almost normal so that no "cellular syndrome" could be established [1183]. Apparently, the malformations observed in triploidy cannot be related to detectable anomalies of the cells themselves and therefore appear to be produced at a different level of integration: Placental insufficiency [458].

Table 4.29. Studies on growth parameters in trisomy 21. (After Krone and Wolf, 1978 [1176])

Increased proportion of cells with intermediate DNA content in fibroblasts; suggestion of prolonged S phase
Retarded rate of DNA synthesis in fibroblasts
Significant increase of the population doubling time in fibroblasts; decrease of in vitro life span
Increased population doubling time in fibroblasts
Increase duration of G_2 and possibly of S phase in fibroblasts
Prolonged G_2 phase in fibroblasts
Decreased duration of cell cycle time in lymphocytes

Examination of Abortuses. As noted in Sect. 2.2.1, one characteristic feature of pregnancies carrying triploid fetuses is the hydatidiform mole of the placenta. The normal development of the placenta is arrested - especially when the origin of triploidy is paternal [1504] - before vascularization of the villi is completed, i.e., between the 21st and the 31st day of pregnancy. Therefore, development of the fetus often comes to an end at about the same time, as shown by histologic examination of triploid embryos [1094]. In this connection, we should remember that most triploid fetuses are aborted; survival up to the time of delivery is the exception. In trisomy 5, on the other hand, development of the amnion and of organs seem to be abnormal already at an earlier state of development [1180]. These examples show that better understanding of the pathogenesis of malformations could gradually emerge from careful histopathologic examination of embryos at different states of development and with different chromosomal aberrations [1011, 1180, 1302].

Aneuploid Mice as Models for Developmental Studies. Some naturally occurring populations of mice were found to have various Robertsonian translocations. In such mice, trisomies and monosomies of various chromosomes can be produced at will by suitable breeding. Here, the "tobacco mouse" has become especially famous [1115]. Monosomy of the smallest mouse chromosome, No. 19, leads to a delay in cell division starting 2 days after fertilization; death occurs just after blastula formation. Fusion with normal embryos resulted in the survival of monosomic cells in various tissues; therefore, this monosomy is not necessarily lethal in the single cell. The gene determining the enzyme aspartate aminotransferase-1, an enzyme involved in the synthesis of aspartate from glutamate, is located on chromosome 19. Assuming that a deficiency of this enzyme might somehow be involved in the lethal effects of monosomy 19, aspartate was added to the culture medium, resulting in survival for at least 2 more days [1070]. It might be possible to use trisomy of the mouse chromosome No. 16 as an experimental model for at least some aspects of the partiallly homologous chromosome 21.

Abnormal Phenotypes Due to Chromosome Aberration and Gene Regulation. Regulation of gene activity during embryonic development presupposes a certain equilibrium in quantity of the gene products that are produced by genes on different chromosomes. These gene products may act as enzymes or structural proteins, or they may have a regulating function, for example, as repressors of the function of other genes. It is understandable that discrepancies in the amount of genetic material will lead to disturbances in the interaction of genes and to regulatory defects in embryonic development. In this context, it is interesting that triploidy causes little or no defect at the cellular level; the developmental anomalies in triploid embryos seem to be caused by a specific anomaly of the placenta, the hydatidiform mole. This anomaly inhibits supply of oxygen and causes nonspecific malnutrition of the fetus. In triploidy, the *relative* amount of chromosome material is undisturbed. In trisomies, on the other hand, some chromosomal material is increased, whereas other material is present in normal amounts. If gene regulation requires interaction of gene products from different chromosomes – as implied, for example, by the above-mentioned Britten-Davidson model [1019] – developmental anomalies at the cellular level are to be expected in trisomies (and monosomies) but not in triploidy.
To examine the mechanisms by which these anomalies are produced, all steps in protein biosynthesis and metabolism need systematic comparison between normals and those affected with chromosomal aberrations. Such studies offer good opportunities for a better understanding of normal and abnormal embryonic development.

4.7.5 Sex Differentiation and Its Disturbances

Development of Sex Characteristics. One aspect of human development is relatively well understood: the development of sex characteristics. This field is an example of successful interplay between a well-founded theory, observations in humans, and animal experimentation. The theory was developed on the basis of intelligent inferences from a few scattered observations. It guided researchers in the design of animal experiments. These experiments helped to refine the theory. At the same time, additional observations in humans with genetic anomalies of sex differentiation could be fitted into the overall picture, helping to answer questions posed by theory and experiment and giving rise to new questions that could be attacked by new experiments.
Four levels of sexual development can be distinguished:

1. Determination of the chromosomal sex (46, XX or 46, XY)
2. Determination of gonadal sex (ovary or testicle)
3. Determination of phenotypic sex (female or male, internal and external sex characters)
4. Determination of psychological sex

The fourth level will be discussed in Sect. 8.2.3.3. The first one has been discussed in part in Sect. 2.2.3; as mentioned, analysis of autosomal chromosome aberrations, while of great *potential* significance for understanding the mechanisms of normal and disturbed development, have so far provided few *specific* clues as to developmental mechanisms. Analysis of numerical as well as structural chromosome aberrations involving the *sex chromosomes,* on the other hand, has yielded much valuable information not only on chromosomal sex (level 1) but also on determination of gonadal and phenotypic sex (levels 2 and 3), as discussed below and elsewhere [1068 a].
The gonadal anlage in the early embryo (up to the 5th or 6th week) shows no sex difference and does not contain germ cells. In humans, primordial germ cells become visible during the 3rd week of embryonal development in the ectoderm of the yolk sac. They then migrate under chemotactic influences into the gonads. This migration is independent of sex; in appropriate experimental systems female germ cells also migrate into male gonads and vice versa. The gonadal anlage may develop into either a testicle or an ovary (level 2). This direction normally depends on the presence of the Y chromosome: A male gonad develops if *one* Y chromosome is present – regardless of the number of X chromosomes. According to present evidence, this development is related to the H-Y antigen.

Y Chromosome and H-Y Antigen. Eichwald and Silmser (1955) [643] described an Y-determined transplantation antigen in the mouse (Sect. 3.1.4). Many years later, its significance for sex determination and specifically, for male gonad differentiation was recognized [1248, 1249, 1341, 1343; 1342]. This male-determining H-Y antigen has now been recognized in several species. It seems to be secreted by

the male primordial germ cells. As soon as these cells have reached the gonad anlage, testicular differentiation starts. This conclusion is based mainly on observations of cases with abnormal sexual development, such as freemartins (i.e., genetically female cows whose gonads are turned into testicles by a few cells transplanted to their gonads from male twins through common blood vessels). There is now experimental evidence that H-Y-receptors seem to be present on the gonadal cells of both sexes.
Incubation of minced testicular tissue leads to the assemblage of testicle-like structures by adding of H-Y antigen. Such an assemblage can also be induced in female gonadal cells. If the H-Y activity is removed by anti-H-Y antiserum, an ovary-like structure emerges.
Recent evidence from exceptional observations in humans, for example XX males, is compatible with the interpretation that the H-Y antigen may not be directly produced by the Y chromosome but that the Y chromosome may contain genetic information that controls a structural gene on an autosome, which, in turn, determines the H-Y antigen. This gene may be repressed in all individuals lacking the Y chromosome; it is derepressed by a factor normally produced by the Y. Hence, constitutive mutations may occasionally lead to individuals who have H-Y antigen and, therefore, male phenotypes in spite of the fact that there is no Y chromosome in any of their cells. It is possible, however, that a translocation of a segment of the H-Y determining part of the Y chromosome to an autosome has occurred. On the other hand, H-Y antigen has also been found in XO Turner females, albeit at a low titre, and in some XX males. This and other observation led to the conclusion that determination of the gonadal sex might be a threshold phenomenon: To turn a gonad anlage into a testicle, a certain minimum concentration of the H-Y substance might be necessary.
In recent years, additional observations in animals and humans have cast some doubts on the role of the H-Y antigen in testicle formation. For example, Hy antigen(s) leading to the transplantation reaction and giving a positive cytotoxicity test are probably not identical, since some animals are positive in one but negative in the other test system [1114]. Moreover, the cytotoxicity test is difficult to perform and is somewhat subjective [1038].
The third level, i.e., determination of the secondary sex characters, appears to be more complicated (and liable to error).

Development of Secondary Sexual Characters. Somatic sex differentiation follows the differentiation of gonads. The internal genital organs are formed from the Müllerian ducts and the Wolffian ducts, both descendants of the primordial kidney. In the female, the Müllerian duct develops into Fallopian tubes and the uterus; the Wolffian duct atrophies. In the male, a Wolffian duct forms the seminal ducts and seminal vesicles. Under the influence of the maternal HCG (human chorionic gonadotropin), the Leydig cells of the embryonic testicle produce the steroid hormones testosterone and 5-dihydrotestosterone. A hormone called müllerian inhibiting factor (MIF) is produced in Sertoli cells. These hormones meet bipotent anlagen for external and internal sex characters: primarily wolffian ducts, müllerian ducts, and the sinus urogenitalis. A normal male develops if all these elements act in time and in the right places. In their complete absence, on the other hand, female sexual characters will be formed. Hence, female development does not require any special promoting factors; it is "constitutive." Minor disturbances at several levels of this system lead to incomplete male development despite a male genotype (male pseudohermaphroditism); analysis of such anomalies has taught us much about the normal course of events. As Jost has put it: "Becoming a male is a prolonged, uneasy, and risky venture; it is a kind of struggle against inherent trends toward femaleness." At least 19 different gene defects (autosomal recessive or X-linked) are known that lead to a disturbance in the differentiation of internal and external male sexual characters. Anomalies occur, for example, in synthesis of androgens, lack of HCG, lack of HCG receptors in Leydig cells, and enzyme defects affecting the five enzymes involved in testosterone synthesis. Anomalies can also be caused by lack of sensitivity to testosterone or 5-dihydrotestosterone due to receptor defects in cells of wollfian ducts or the urogenital sinus.
The pathways to testosterone and their genetic blocks are shown in Fig. 4.73; the left-hand side of this figure shows the defects (F and G) leading to adrenogenital syndromes (Sect. 4.2.2.9) in females but not to pseudohermaphroditism in males. The blocks under A (congenital adrenal lipid hyperplasia) are not very well known; in addition to largely female external genitals, these males suffer from severe salt wastage. The same holds true for block B. The blocks on the right-hand side lead to male pseudohermaphroditism of various degrees without other manifestations of the adrenogenital syndrome.
An especially interesting enzyme defect not shown in Fig. 4.73 is the defect of 5 α-reductase, an autosomal recessive trait. This enzyme normally reduces testosterone in cells of the urogenital sinus to 5 α-dihydrotestosterone. If it is lacking, normal internal male sex organs (seminal vesicles; prostate) develop and the entire body, including muscular development, body hair, etc., is male except for the external

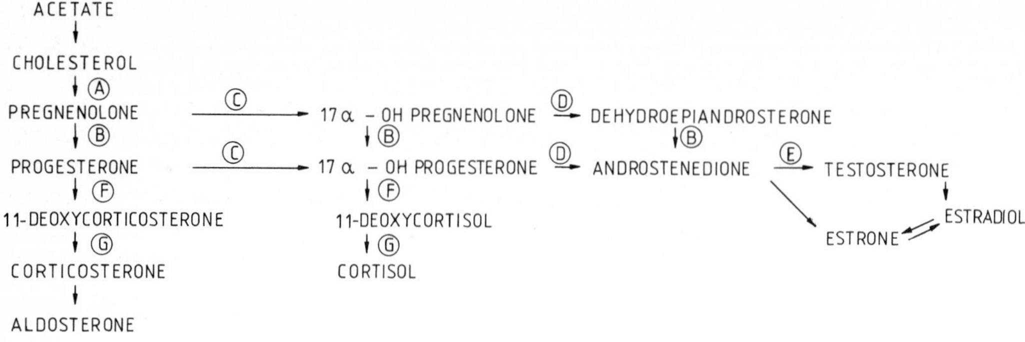

A = 20-hydroxylase, 22R-hydroxylase, and 20, 22-desmolase
B = 3ß-ol-dehydrogenase
C = 17α-hydroxylase
D = 17, 20-desmolase
E = 17-ketosteroid reductase
F = 21-hydroxylase
G = 11ß-hydroxylase

Fig. 4.73. Pathways leading to estradiol and aldosterone, and genetic blocks. (From Engel, 1982 [1068a])

genitalia, which upon superficial inspection are female. Hence the descriptive name of pseudovaginal perineoscrotal hypospadias (PPHS; 26460).

These enzyme defects are rare. Other more common syndromes exist in which androgens are normal, but the target tissue is completely or partially androgen resistant.

Testicular Feminization Syndrome (31370) [1114]. At birth, affected individuals appear to be normal females; the anomaly goes undetected during childhood, unless a testicle is discovered in an inguinal hernia. Affected persons with this syndrome have a male karyotype and male gonads; the term testicular feminization was coined by Morris (1953) [1220]. At puberty, primary amenorrhea and - in the majority of cases - absence or marked deficiency of axillary, pubic, and body hair are striking. In the adult, stature and proportions are normal for females, albeit legs are often somewhat longer. Breasts are well-developed. The mean body proportions conform more to the present-day ideal of female beauty than to the proportion of the average woman; it is therefore not surprising that affected patients have been found repeatedly among models.

The vagina is usually shorted and ends in a blind pouch. Instead of a uterus, some remnants of the Müllerian ducts are often present; instead of the Fallopian tube, a fibromuscular streak may be found. Testicles are located in the labia majora, the inguinal canal, or in the abdomen, and may show a normal or even increased number of the male hormone-producing Leydig cells. Usually there is no spermatogenesis. Occasionally, malignant tumors of the testicle develop.

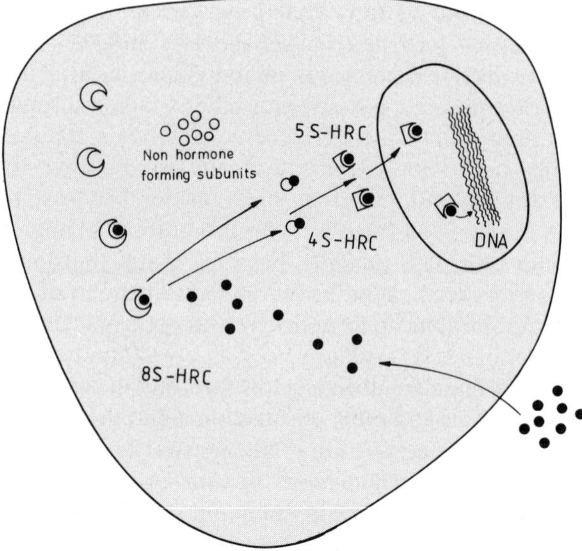

Fig. 4.74. Testosterone (T) and Dihydro-Testosterone (DHT) pathways inside the cell. - Three types of receptor molecules can be distinguished (by their sedimentation in the sucrose gradient) with defined ion concentration: 8 S; 5 and 4 S. After entering, the hormones are bound to the 8 S receptor, forming the 8 S-hormone-receptor complex (HRC). The 8 S-HRC dissociates into a 4 S-HRC and into some, - non-hormone forming - subunits. The 4 S complex is transformed into 5-HRC, which is then transported into the nucleus. Testicular feminization may be the result of various - complete and incomplete - blocks within this system. (From Engel, 1982 [1068a])

The patients secrete normal amounts of androgens, especially testosterone; normal male development would be expected. The most obvious explanation for the disease is an anomaly in end organ response to the hormone, which had been postulated for many years. More recently, an androgen receptor defect has been discovered [1359]. This receptor defect has been confirmed by a variety of studies: for example, testosterone treatment fails to exert normal response such as voice change, beard growth, clitoral hypertrophy, or (in affected mice) induction of alcohol dehydrogenase in the kidney. The psychological development of testicular feminization patients is entirely female; there is now evidence that interaction between androgens and androgen receptors influences brain development, EEG, and behavior (Sect. 8.2.3.3).

Many cases are sporadic, but whenever familial aggregation has been observed, transmission through normal women and additional cases among mother's "sisters" were detected. This suggested an X-linked mutation. However, as the patients cannot reproduce, the decisive criterion for X-linked inheritance - transmission from hemizygotes to all daughters but not to sons - could not be observed. Therefore, sex-limited manifestation of an autosomal-dominant gene could not be excluded. In 1971, however, a mutant with very similar phenotype was observed in the mouse [1205]. This mutant is clearly X-linked. The mammalian X chromosome remains stable through evolution; its genes are homologous in all mammalian species examined so far [156]. Moreover, testosterone receptors can also be demonstrated on fibroblasts, and heterozygotes were shown to have two populations of fibroblasts, one normal and the other lacking in testosterone-binding capacity, as expected from the Lyon hypothesis [985]. X-Linked inheritance can therefore be regarded as established. Interestingly enough, Patterson and Bonnier had concluded from pedigree information as long ago as in 1937 - at a time when assessment of the genotypic sex was impossible - that the patients are genotypic males and that the mode of inheritance is either X-linked or sex-limited. Sporadic cases may be caused by new mutations; the selective disadvantage of the gene is strong, the patients being infertile. Therefore, a high fraction of new mutants must be expected (Sect. 5.1.3).

Genetic Heterogeneity. Polypeptide hormones such as insulin (Sect. 3.8.14) are bound to membrane receptors of target cells. Steroid hormones such as testosterone, on the other hand, are bound to cytoplasmic receptors after having entered the cell by diffusion. Figure 4.74 shows the pathway of testosterone (T) and dihydrotestosterone (DHT) in the cell from its binding to the cytoplasmatic 8 S receptor to the 4 S receptor complex and to the 5 S receptor complex and its movement into the nucleus. Most mutations leading to testicular feminization appear to involve the 8 S receptor, but mutants affecting the 4 S- and 5 S complexes must be expected. And indeed, cases with testicular feminization and normal 8 S receptor have been observed. In addition, there are patients with incomplete testicular feminization and intersexual genitals; here, the 8 S receptor was found diminished but not completely lacking.

Only a few genetic disturbances of male development have been mentioned. Many others are known, which fit our theoretical concept of male differentiation and have increased our detailed knowledge on this process. Even in phenotypically apparently normal males, a number of genetic defects leading to male infertility occur - for example, by inhibiting maturation of germ cells into viable sperms [1032].

5 Mutation

5.1 Spontaneous Mutation

The most important feature of genes is their capacity to be reproduced identically from generation to generation. However, the evolution of life would never have been possible if no change in the genetic material ever had occurred. Since there is good evidence that all living beings on our planet have a common origin, genes as the carriers of genetic information must have the capacity for occasional alterations. Such changes are indeed observed; they are called mutations.

Animal and plant breeders for times immemorial observed that inherited characters might occasionally change and that such changes were transmitted to following generations. Darwin was very much interested in these observations. He called them "sports." One example at that time was the so-called Ancon sheep, a popular chondrodystrophic mutation since such sheep could not jump over fences.

The term mutation was introduced by de Vries (1901) [See 7] for sudden genetic changes in the plant *Oenothera lamarckiana*. In this species, de Vries observed for the first time a sudden and genetically stable change in one individual under experimentally controlled conditions. Later on, the relatively frequent occurrence of mutations in this plant was shown to be due to a particular karyotype: the chromosomes were connected end to end; they formed complicated patterns during pachytene, and centromeres were distributed in an orderly fashion during metaphase. This mechanism failed occasionally, producing what de Vries called "mutations."

5.1.1 Reappraisal of Genetic Variants That May Occur by New Mutation

In experimental genetics, the following types of mutations have been distinguished:

a) *Genome mutations* involve alterations in the number of chromosomes. Genome mutations are frequent in plants. Whole sets of chromosomes may be multiplied (polyploidy), or the number of single chromosomes may be increased (trisomy) or decreased (monosomy).
b) *Chromosome mutations* (Sect. 2.2). The detectable structure of chromosomes is changed, allowing microscopic detection. The total number of chromosomes is not altered.
c) *Gene mutations*. Here, no changes of chromosomes can be detected microscopically; the mutation can only be inferred from a change in the phenotype by genetic analysis (Section 3.6).

Analysis of human mutants, especially those occurring in the hemoglobin genes, at the protein and DNA levels has taught us much about the molecular nature of gene mutations. With these results and the development of banding techniques for chromosome analysis, especially high-resolution banding, the distinction of chromosome from gene mutation has become blurred. We know that at the molecular level deletions and insertions are possible and that unequal crossing-over might alter microstructure [748]. At the same time, the banding techniques have made it possible to detect previously invisible chromosomal rearrangements microscopically. However, chromosomal alterations detectable by banding involve structural alterations that differ by orders of magnitude from changes such as deletions of structural genes. The categorization into structural chromosomal aberrations and gene mutations is therefore useful for practical purposes.

Cells in Which Mutations May Occur. Apart from the *kind* of genetic damage, the *localization* of genetic alterations is of prime importance. Mutations may occur either in germ cells or in somatic cells. Germ cell mutations can be transmitted to individuals of the next generation and are usually found in all cells of the affected offspring. Somatic mutations are found only in the descendents of the mutant cell, making the individual a "mosaic." Phenotypic consequences will be observable only if the mutations happen to interfere with the specific functions of the affected cells.

Mutation Rates. One of the most frequently used parameters in mutation research is the *mutation rate,* which is defined in man as the probability with which a particular mutational event takes place per generation. It has become customary to define mutation rates as the rate at which a new mutation takes place in a fertilized germ cell per generation. It should be kept in mind that an individual is formed by two germ cells. For discussion on somatic cell mutation rates, see Sect. 5.1.6.

5.1.2 Genome and Chromosome Mutations in Humans

5.1.2.1 Mutation Rates

Methods Used. Estimation of mutation rates requires an assessment of the incidence of those cases with a trait or disease in the population where parents and other family members are unaffected (sporadic cases). Since the late 1950s, when the first chromosome aberrations in humans were discovered, it became obvious that the mutational events causing genome mutations must occur much more often than gene mutations leading to hereditary diseases. An exact assessment of chromosomal mutation rates became possible when Court Brown [42] initiated investigations on unselected population samples, such as on consecutive series of newborns. The calculation of the mutation rate is simple

$$\mu = \frac{\text{Number of sporadic cases with a certain anomaly}}{2 \times \text{numbers of individuals examined}}$$

This is the so-called direct method and can be applied to single gene-determined traits as well as to genome and chromosome mutations.

When based on series of newborns, this estimate should be qualified: it is confined to mutations whose carriers survive up to the time of birth. In man as well as in other mammals, however, the great majority of genome and chromosome mutations are lethal, leading to death of the zygote during embryonic life.

Incidence and Mutation Rates: Genome Mutations [1581]. Incidences of sex chromosome and autosomal abnormalities as estimated from seven studies on newborns have been collected in Tables 5.1 and 5.2. Many of these cases, with the exception of Turner's syndrome and some mosaics and translocations (Sect. 5.1.6) originated by nondisjunction during one of the two meiotic divisions in one of the parents' gonads. They are new mutants. Table 5.3 provides mutation rate estimates.

The lower values for trisomies 13 and 18 as compared with trisomy 21 are probably not real; more fetuses affected with trisomy 13 and 18 are already eliminated during embryonic life (for a discussion of the problem of chromosome aberration in abortions, cf. Sect. 2.2.4).

Table 5.1. Incidence of sex chromosome abnormalities in seven population samples. (Nielsen and Silessen, 1975 [1581])

Karyotype	Total		Rate/1,000	
47,XYY	28	35	0.81	1.02
47,XYY mosaics	7		0.20	
47,XXY	33	39	0.96	1.13
47,XXY mosaics	6		0.17	
♂ 46,XX	2		0.06	
♂ 45,X/46,XY	1		0.03	
46,X,inv(Y)	9		0.26	
45,X	2		0.10	0.39
45,X,mosaics	6		0.29	
47,XXX	20	24	0.98	1.18
47,XXX mosaics	4		0.20	
Population samples	♀ 20,370, ♂ 34,379, ♀ + ♂ 54,749			

The samples came from Edinburgh (U. K.), Ontario (Canada), Winnipeg (Canada), Boston (U.S.A.), Moscow (U.S.S.R.), and Århus (Denmark).

Table 5.2. Incidence of autosomal abnormalities (genome and chromosome mutations) in 54,749 newborns [1581]

Karyotype	Total	Rate/1,000
47,+13	3	0.05
47,+18	8	0.15
47,+21	63	1.15
47,+marker chromos.	12	0.22
47,+marker, mosaics	5	0.09
Deletions	5	0.09
Inversions	7	0.13
D/D translocations	43	0.79
D/G translocations	11	0.20
Reciprocal translocations	47	0.85
Unbalanced Y-autosomal translocations	2	0.04

Incidence and Mutation Rates: Chromosome Mutations. Incidences of structural autosomal chromosome abnormalities may be taken from Table 5.2. In the surveys on which these figures are based, conventional orcein staining was used. More recent studies with banding methods gave slightly higher frequencies, particularly of inversions but also of balanced reciprocal translocations; but the numbers are relatively small [1476, 1582]. The cases with deletions may be considered to be due to new mutation, giving a mutation rate estimate of 4.57×10^{-5}. For the different types of translocations, the table does not show how many have occurred de novo. Therefore, no mutation rate estimate can be derived from these data. Jacobs et al. [1500,1502] estimated the following mutation rates from more limited material: 1.9×10^{-4} - 2.2×10^{-4} for all balanced rearrangements (translocations and inversions) together 3.24×10^{-4} for unbalanced Robertsonion translocations, and 3.42×10^{-4} for unbalanced non-Robertsonian rearrangements. These mutation rates are defined for those "that survive long enough to give rise to clinically recognized pregnancies," i.e., *including* abortions. These data are of potential interest, since structural chromosome aberrations in newborns are likely candidates for future population monitoring programs for a potential increase of the mutation rate due to mutagenic agents such as ionizing radiation. It would be interesting to have mutation rates for *single,* specific translocations but here, only preliminary estimates based on insufficient data are available [1491, 1500].

Table 5.3. Mutation rates for genome mutations observed in newborns [1581]

Condition	Calculation	Mutation rate
Sex chromosome trisomies		
XXY including mosaics	$\frac{39}{2 \times 34{,}379}$	5.67×10^{-4}
XXX including mosaics	$\frac{24}{2 \times 20{,}370}$	5.89×10^{-4}
XXY and XXX together (X nondisjunction)		5.8×10^{-4}
XYY, including mosaics (Y nondisjunction)	$\frac{35}{2 \times 34{,}379}$	5.09×10^{-4}
Autosomal trisomies		
47,+21	$\frac{63}{2 \times 54{,}749}$	5.8×10^{-4}
47,+18	$\frac{8}{2 \times 54{,}749}$	7.3×10^{-5}
47,+13	$\frac{3}{2 \times 54{,}749}$	2.7×10^{-5}

Mutation rates for both genome and chromosome mutations are higher when calculated from amniocentesis data, indicating that fetal loss between the time of amniocentesis (16th-17th week of gestation) and birth is considerable [1665, 1496].

5.1.2.2 Nondisjunction and the Age of the Mother

Statistical Evidence. For Down's syndrome, an increase in risk with advancing age of the parents has been known for many years. Figure 5.1 shows the relative incidence in different age groups, compared with the population average. The risk does not change very much up to the age of 29 but rises steeply beginning with the age groups of 35-39. The population incidence of Down's syndrome is therefore expected to vary with changing age distribution of the parents. Table 5.4 gives an idea of the absolute risk in different age groups of women, as accepted until recently. Experience from amniocentesis in women above 35 indicates that these estimates for risks of trisomy 21 are too low (Table 5.4). The older statistical data have been collected before the chromosome era. Possible biases are: failure to diagnose the condition at birth or a higher risk for children with Down's syndrome to be stillborn. It is not very likely, however, that these biases are very important. The incidence figures from these old statistics agree too well with those derived from population cytogenetic studies (Table 5.2). The difference, which is still more marked for trisomies 13 and 18 [1496], is explained mainly, if not exclusively, by fetal death between amniocentesis ($\approx$ 16th-17th week of gestation) and birth. However, some recent data indicate

Table 5.4. Risk of Down's syndrome in relation to maternal age at delivery (live birth figures calculated for specific years) and at amniocentesis. (Modified from Harper, 1984)[a] [90]

Maternal age	At birth		Following amniocentesis			
	Down's syndrome		Down's syndrome		All chromosomal aberrations	
	Incidence per 1000	Risk (rounded)	Incidence per 1000	Risk (rounded)	Incidence per 1000	Risk (rounded)
30 years	1.4	1/700	NA	NA	NA	NA
35 years	2.2	1/450	3.9	1/250	8.7	1/115
37 years	4.0	1/250	6.4	1/150	12.1	1/80
39 years	6.5	1/150	10.4	1/100	18.4	1/50
41 years	12.5	1/80	16.9	1/60	29	1/35
43 years	20	1/50	27.4	1/35	45	1/20
All ages	1.5	1/650	NA	NA	NA	NA

NA, not available.
[a] Data from Sweden, Australia, and Wales (UK).

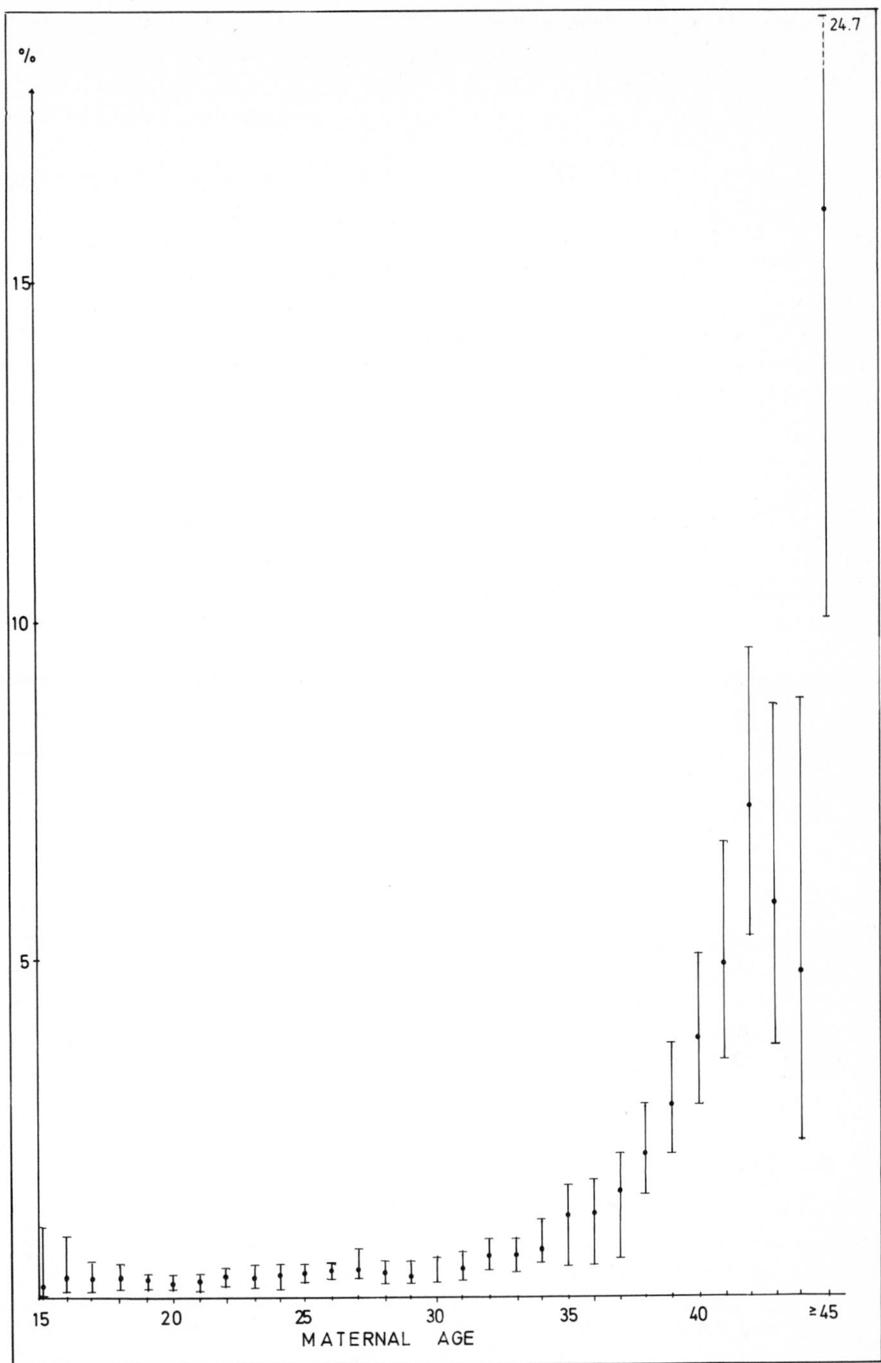

Fig. 5.1. Incidences and confidence intervals (95%) of Down's syndrome among newborns. (Data from Upstate New York, 1963-1974; Hook and Chambers, Birth Defects Orig. Art. Ser. Vol. 13, 1977) (933 cases among 1729909 live births)

a true increase in the risk of having a trisomic child for mothers 35 years old and older.

Some amniocentesis data seemed to show a certain leveling off of the increase in risk of trisomies in the oldest age group (mothers between 46 and 49). However, a very comprehensive study failed to confirm this effect [1494].

Ages of mothers and fathers in a society are obviously correlated: older wives tend to be married to older husbands.

Therefore, it is not immediately obvious whether an increase in risk with parental age can be traced to the fathers, the mothers, or both parents. In Down's syndrome, Penrose (1933) [1590] showed the increase in risk to be exclusively due to the mother's age. For

150 cases, in whom maternal as well as paternal age was known, he calculated the following partial correlation coefficients:

a) Partial correlation between maternal age and incidence of Down's syndrome, keeping paternal age constant, was $r = +0.221$.
b) Partial correlation between paternal age and Down's syndrome was $r = -0.011$.

The partial correlation techniques permits one to determine the correlations between two variables after the influence of the third had been eliminated. Penrose also analyzed the same data by a regression method with essentially the same result.

Is, Indeed, Only the Age of the Mother Involved? Maternal age effects have been reconfirmed and accepted for 50 years. Some years ago, however, evidence for an additional influence of paternal age has been brought forward [1646]. The data consisted of 224 Down's syndrome patients born in Denmark between 1960 and 1971, 176 of whom were cytogenetically identified as having trisomy 21. They were compared with a control of 6,053 randomly selected births from the same area and time period and with cases from the literature with the following results:

a) Men 55 years and older have a significantly increased risk that their newborn children will have Down's syndrome.
b) In almost all published larger series, cases occur where the age of the father greatly exceeds that of the mother.
c) When comparing patients and controls for separate maternal age groups with the partial correlation method, the proportion of fathers of children with Down's syndrome increases, although not significantly, with paternal age.

This problem has been reexamined frequently, with controversial results. For example, for Japan [1550], West Germany [1647], and Norway [1444] a small paternal age effect (in addition to the much stronger maternal age effect) was confirmed. For Sweden [1443], the State of New York [1493], and France [1606] no such effect could be demonstrated. These discrepancies led to a considerable amount of discussion on appropriate statistical methods. On the other hand, recent work with chromosomal markers has implicated the father as a source of the nondisjunction of chromosome 21 in about 20% of all cases of trisomy 21 [1503]. A study directed specifically at this group failed to show a paternal age effect [1495].

Higher Risk in Children of Very Young Mothers? It has been claimed that Down's syndrome also occurs more frequently in children of very young mothers. The evidence, however, is controversial. In the older literature, the most reliable series showed a decreased frequency for age groups under 20 in comparison to the 20-24 year age group [1532]. In some of the recent series from Canada, Sweden, and Denmark, incidence in the lowest age group is somewhat higher than in the next group [1527]. Figure 5.1 shows this increase especially for mothers 17-18 years or younger.

Penrose [1591] examined the absolute frequencies of children in relation to maternal ages. Figure 5.2 shows frequencies in different age groups for trisomies 21, 18, 13, XXY and XXX compared with the maternal age distributions in four representative populations. Contrary to the distribution in the general population, which is clearly unimodal, distribution among mothers of trisomy cases shows definite signs of bitangentiality. This points toward mixture of an age-dependent with an age-independent group, a conclusion supported by other evidence (Sect. 5.1.2.4).

Age-Specific Rates in Trisomies. Figure 5.3 shows the effect of maternal age on all clinically recognized pregnancies involving different trisomies (trisomy among liveborn and spontaneous abortions) assuming a spontaneous abortion rate of 15%. There is a strong maternal age effect, so that for women 42 years and older about one third of all clinically recognized conceptions are abnormal. Assuming that nondisjunction leads to equal numbers of monosomic and trisomic zygotes, it can be assumed that most oocytes among older women are aneuploid.

Maternal Age Effect in Other Trisomies. The relative incidences at various maternal ages are shown in Fig. 5.4 for trisomy 13 and 18 and in Fig. 5.5 for syndromes due to nondisjunction of the X chromosome (XXY and XXX). A maternal age effect for trisomy 9 may also be present, as one of the very few reported cases was diagnosed by amniocentesis in a 40-year old woman [1455]. Most trisomies detected in surveys of spontaneous abortions were also more frequent in children of older women (Table 5.5). However the effect for the large chromosomes is small and nonsignificant (it was slightly larger in some earlier series). Among the smaller chromosomes, which normally show a significant effect, it is small - albeit significant - for the most common one, trisomy 16. Some earlier studies had given no maternal age effect at all for this trisomy [212]. No significant effects were found for de novo chromosome rearrangements [1496]. A maternal age effect for 18p− unfortunately has not yet been reexamined.

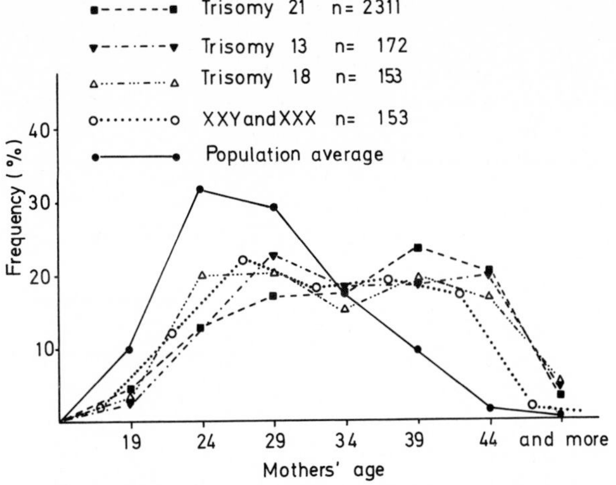

Fig. 5.2. Maternal age distribution of trisomies 21, 13, 18, and a combined sample of XXY, XXX compared with the population average in four representative populations. (Penrose, 1957 [1591]; Court Brown, 1967 [42]; Magenis)

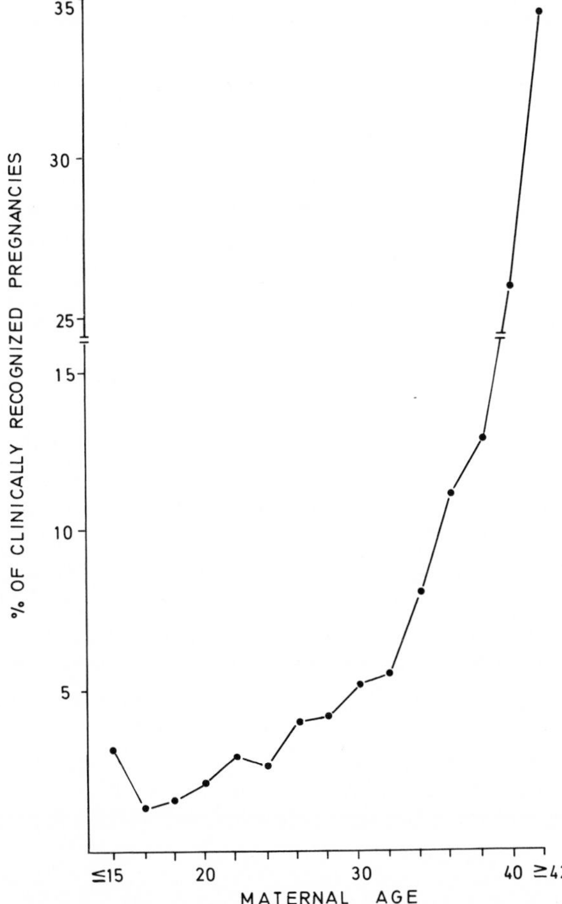

Fig. 5.3. Incidence of trisomy among all clinically recognized pregnancies, including spontaneous abortions, assuming a spontaneous abortion rate of 15%. (From Hassold and Jacobs, Ann. Rev. Genet. 18, p. 83, 1984 [1481])

Little or no paternal age effect was found for the XXY karyotype [1410].

5.1.2.3 In Which Sex and at Which Meiotic Division Does Nondisjunction Occur?

As explained in Sect. 2.2.1, trisomy is caused by meiotic nondisjunction. Two questions arise:

1. Does nondisjunction mainly occur during male or female meiosis?
2. Does it mainly occur in the first or in the second meiotic division?

Since the parental age effect as described above was shown to be exclusively – or according to controversial evidence predominantly – maternal in origin, it was tempting to conclude that most observed cases of nondisjunction occurred in the female germ line. Evidence from X-linked marker studies, however, showed that this is not true for the X chromosome.

Evidence for the X Chromosome from X-Linked Marker Studies. The principle of determining the origin of the trisomic germ cell is shown in Fig. 5.6. A deuteranomalous patient with Klinefelter's syndrome has a deuteranomalous mother and a father with normal color vision. The trisomic germ cell presumably originated in the mother who was homozygous for the deuteranomaly allele. It cannot be decided whether nondisjunction has occurred in the first or the second meiotic division. Such a distinction, however, is possible if the father has normal color vision and the mother is heterozygous as shown in Fig. 5.6. If the Klinefelter son is colorblind nondisjunction must have occurred during the second meiotic division in which sister chromatids of the same chromosomes are normally separated.
In principle, the same argument may be used by application of the X-linked blood group Xg. Race and Sanger (1969) [846], reviewing the evidence for nondisjunction in Klinefelter's (XXY) syndrome, arrived at an estimate of 40% occurring in the paternal germ lines, all in the first meiotic division. Fifty percent occurred in the first maternal meiotic division and 10% in the second maternal division. In all four cases of XXXY and XXXXY, the extra chromosomes came from the maternal germ line; it had originated during paternal meiosis in the XXYY cases.
Among XO patients whose Xg types contributed information to this analysis, about 74% possessed a maternal and about 26% a paternal X. If it is assumed that most XO cases are the result of chromosome loss during the early zygote stages and not of meiotic nondisjunction (Sect. 2.2.1), the findings

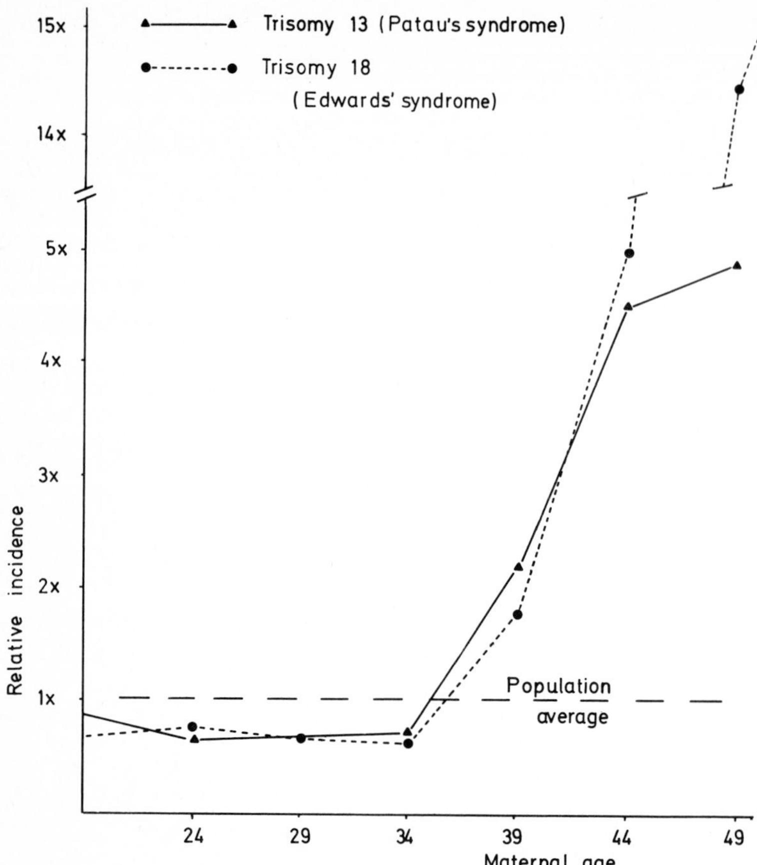

Fig. 5.4. Maternal age effect in trisomies 13 and 18. (Magenis et al., 1968 [see 212])

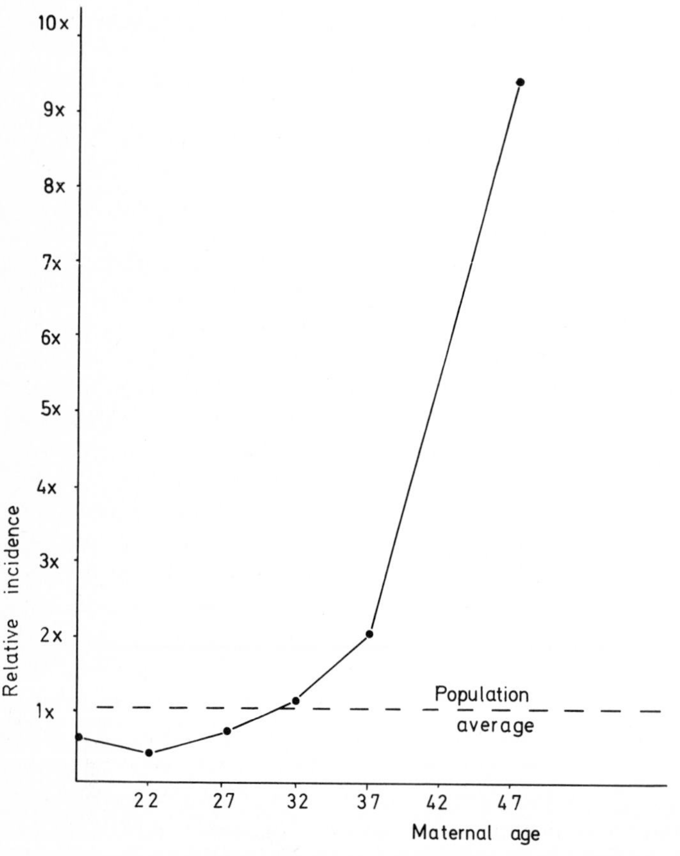

Fig. 5.5. Maternal age effect in syndromes due to nondisjunction of X chromosomes: XXY and XXX combined; 153 cases from Great Britain. (Court Brown et al., 1967 [42])

Table 5.5. Means and standard deviations of maternal ages in two series of spontaneous abortions (New York City, 372 abortions, 46, XX or XY, 190, 47, XX or XY; Hawaii, 418, 46, XX or XY; 172, 47, XX or XY). (From [1480])

46, XX or XY newborns	47, +2	3	4	5	6	7	8	9	10	11	12
25.8 25.0	28.3 ±6.2	25.5 ±6.4	27.9 ±6.2	22.0	24.0 ±4.2	30.2* ±6.7	24.6 ±4.9	29.0 ±6.9	31.8 ±7.7	25.0	26.6 ±7.0
46, XX or XY abortions	**13**	**14**	**15**	**16**	**17**	**18**	**19**	**20**	**21**	**22**	
26.7 ±6.2	29.2 ±5.6	25.6 ±7.4	31.0*** ±6.3	28.2* ±5.3	32.8 ±8.6	33.3*** ±5.6		33.8*** ±5.5	30.0** ±7.0	31.7*** ±5.3	

Figures at column heads indicate chromosome numbers.
Asterisks indicate statistically significant differences between normal and trisomic abortions: * $P=0.005$; ** $P=0.01$; *** $P=0.001$.

Table 5.6. Meiotic origin of extra chromosome 21 in Down's syndrome

	Total	Origin					
		Paternal			Maternal		
		First meiotic division	Second meiotic division	?	First meiotic division	Second meiotic division	?
Number	391	45	27	4	238	51	26
Percent age	100%	11.5%	6.9%	1%	60.9%	13%	6.6%
		19.4%			80.5%		
Percentage 1st meiotic division		72.4%			15.9% Paternal 84.1% Maternal		
Percentage 2nd meiotic division		27.6%			34.6% Paternal 65.4% Maternal		

concur with the results obtained in mice where the paternal X chromosome is particularly vulnerable a short time after fertilization (Sect. 5.2.1.3).
In principle, the same kind of argument may be used for identification of parental chromosomes by microscopically identifiable chromosome variants.

Direct Evidence from Chromosome Variants. Human chromosomes show individual variants that are constant over many generations and are also known as heteromorphisms. Their overall frequency varies between about 5% and 50%, depending on the methods used (Sect. 2.1.2.3) [441]. Fortunately for their utilization in assessing the origin of chromosomes in trisomies, these variants are especially frequent in the acrocentric chromosomes of the D and G groups. For the other chromosomes for which viable trisomies occur - X and 18 - little or no variability is known. Hence, the following discussion will be confined to trisomy 21.

The observed variants may be classified as elongations of the short arm (ph+), large satellites (ps+), double satellites (pss), and shortened short arms with or without satellites (ph−). Moreover, there are differences in fluorescence intensity of satellites and short arms and differences in the size of the heterochromatic block within the centromeric region that can be identified by C banding.

Utilization of Chromosomal Variants for Identification of Nondisjunction. Heteromorphisms can be utilized to trace a certain chromosome back to father or mother and to the first or second meiotic division (Fig. 5.7). In Fig. 5.7a, the trisomic child has three different chromosomes 21, a ph+ variant (a, hatched), a normal variant (b, black), and a ph− variant (c). Examinations of the parents shows the father to be homozygous for a and the mother to be heterozygous for b and c. We conclude immediately that nondisjunction must have occurred in the moth-

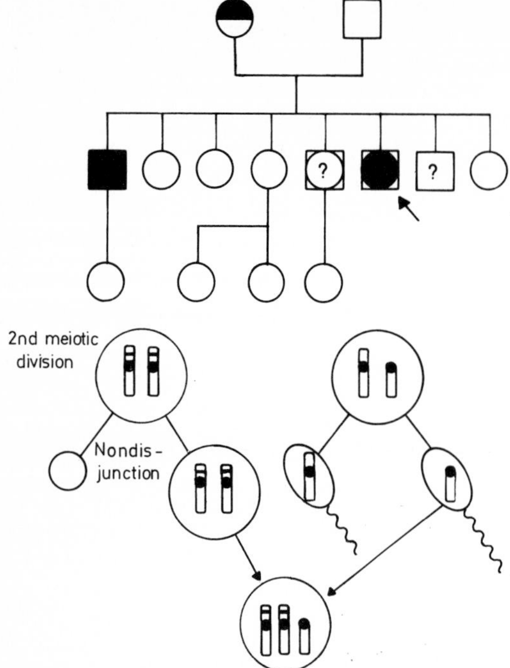

◁ **Fig. 5.6.** Determination of the origin of trisomic germ cells. A patient with Klinefelter's syndrome *(arrow)* is deuteranomalous. His father has normal color vision. The mother must be heterozygous because she has one deuteranomalous son. If the additional chromosome comes from the father, the Klinefelter son must be heterozygous or homozygous normal. The fact that he is deuteroanomalous shows that both his X chromosomes descended from one X of the mother (see Sect. 2.2.3.1). Moreover, nondisjunction must have occurred in the second meiotic division. (Lenz, 1964 [8])

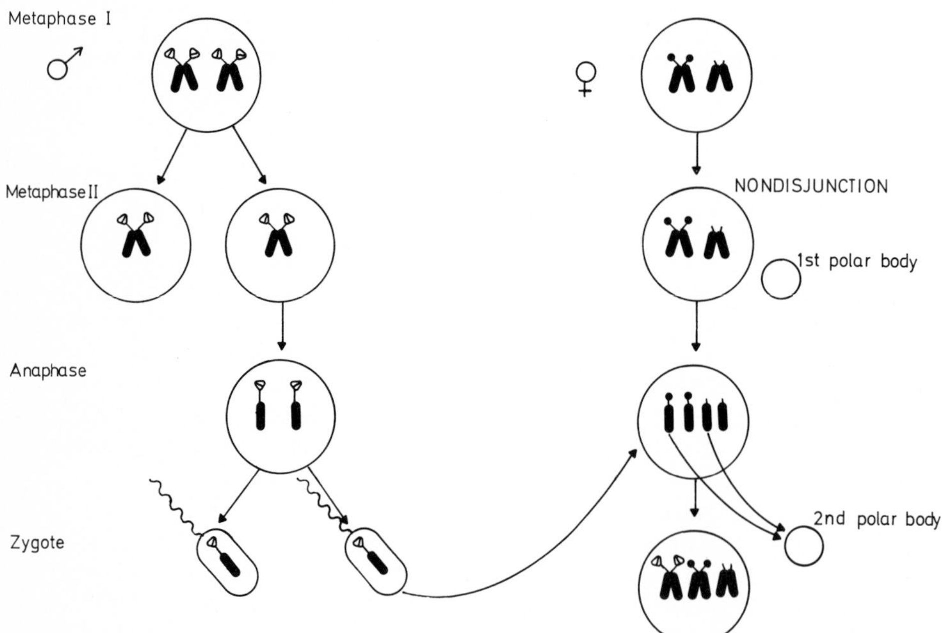

Fig. 5.7 a–c. Tracing back a certain chromosome to the father or the mother and to the first or the second meiotic division. **a** The trisomic child has three distinguishable chromosomes 21. Here, nondisjunction must have occurred in the first meiotic division of the mother. **b** Nondisjunction occurred in the first meiotic division. However, the mating is noninformative since nondisjunction in the second meiotic division and paternal nondisjunction could lead to the same result **c** Nondisjunction must have occurred in the maternal germ line and in the second meiotic division ▷

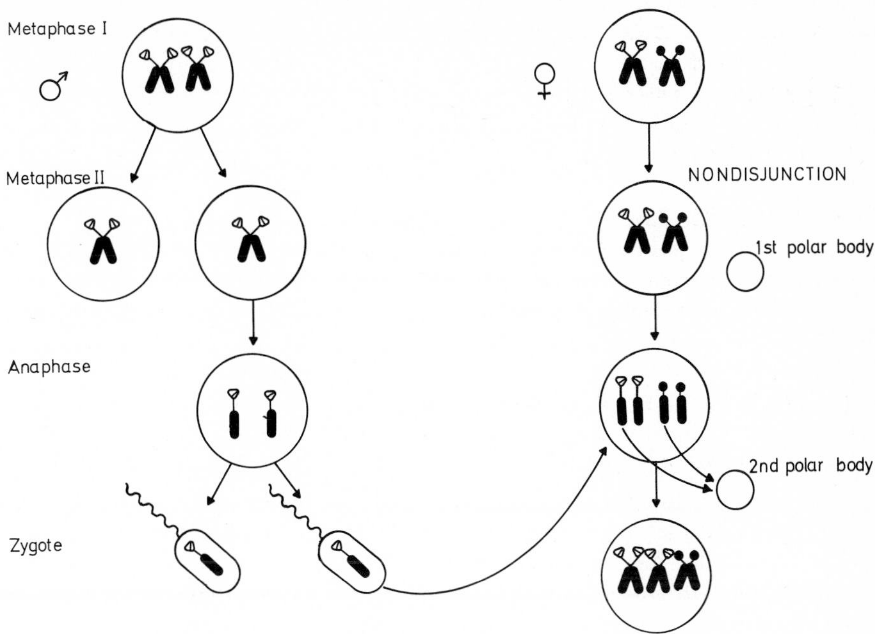
b) Two variants of the parents
Metaphase I
♂
♀
Metaphase II
NONDISJUNCTION
1st polar body
Anaphase
2nd polar body
Zygote

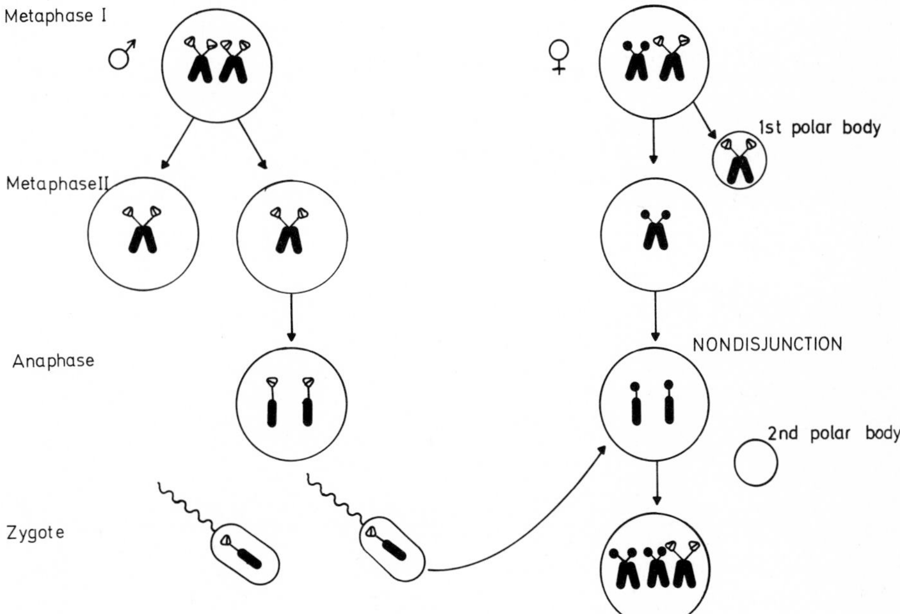
c) Nondisjunction in 2nd meiotic division of oogenesis
Two variants in the parents
Metaphase I
♂
♀
1st polar body
Metaphase II
Anaphase
NONDISJUNCTION
2nd polar body
Zygote

Fig. 5.7 b, c

er's germ cell, as the child has two maternal but only one paternal chromosome. Moreover, nondisjunction must have occurred in the first meiotic division since the child had both maternal chromosomes. Had nondisjunction occurred in the second division, the child would have received either two copies of b or two copies of c. Hence, nondisjunction could be traced back to the first meiotic division in the mother's ovary.

The situation is often different, however, as shown in Fig. 5.7 b. The mating is noninformative. Figure 5.7 c shows the same parental genotypes as Fig. 5.7 b. The extra chromosome is clearly maternal and could only have arisen by nondisjunction in the second meiotic division.

In the same way, the fraction of informative and noninformative matings for nondisjunction in the first and second meiotic divisions may be derived for all possible mating types [1526].

The results indicate that nondisjunction can occur in both the 1st and the 2nd meiotic divisions. It is of great interest that 20% of all cases of Down's syndrome are caused by paternal nondisjunction, with two thirds arising from a 1st division error and one third from a 2nd division error. Among the 80% of cases of maternal origin, the great majority resulted from 1st division nondisjunction.

These data ware corrected for an ascertainment error which makes it more likely that 2nd division errors are more frequently recognized [1504, 1526].

Information from Biochemical Variants Will Help to Enhance the Number of Informative Families. In principle, it is possible to use markers from all polymorphic systems. As shown above, this has already been done with color blindness and the Xg blood group polymorphism for the X chromosome. This approach has the advantage that alleles are usually well-defined and allele frequencies are known. Recently, a chromosome 21-specific DNA probe has been discovered that detects a restriction site polymorphism (RFLP). When 25 individuals from the London population were examined a frequency of the rarer allele of 0.38 and a heterozygote frequency of 0.47 were estimated [1433].

5.1.2.4 Nondisjunction, Chromosome Variants, and Satellite Association

The considerations reported in the foregoing paragraph were possible only because chromosome 21, like other acrocentric chromosomes, shows a high degree of variability, especially in the centromere region and in the short arms. An influence of this variability on the risk of nondisjunction has frequently been suspected and is all the more plausible since the short arm regions of acrocentric chromosomes are located closely together near the nucleolus. Even harsh methods of making metaphase preparations do not destroy completely their intimate relationship.

Satellite Association. The acrocentric chromosomes show a tendency to lie side by side on metaphase plates with their satellite regions facing each other [1446]. There is considerable interindividual variability in this phenomenon. It was, therefore, concluded that human beings with frequent satellite associations would manifest an increased probability of nondisjunction. The definition of a satellite association is shown in Fig. 5.8 [1712]. Satellite associations were studied in parents of 40 Down's syndrome cases in comparison with 81 controls (Table 5.7). There is a significant increase for chromosome 21 in mothers but not in fathers of the probands; a certain albeit smaller increase is also found for chromosome 14. In another study, significant heterogeneity of sat-

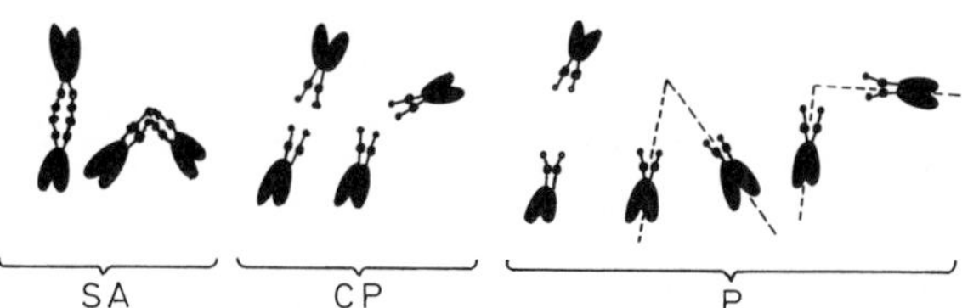

Fig. 5.8. Definition of a satellite association. *SA*, satellite association; *CP*, close proximity; *P*, proximity. (Zellweger et al., 1966 [1712])

Table 5.7. Satellite association index, 50 metaphases of each individual were evaluated (Mikkelsen)

Chromosome No.	Mothers of Down's syndrome ($n=36$) mean AI[a] $\pm\sigma$	Controls ($n=51$) mean AI $\pm\sigma$	t	P	Fathers of Down's syndrome ($n=27$) mean AI $\pm\sigma$	Controls ($n=30$) mean AI $\pm\sigma$	t	P
13	0.34 ± 0.019	0.34 ± 0.012	0.05	> 0.025	0.31 ± 0.020	0.36 ± 0.015	2.02	≃ 0.05
14	0.37 ± 0.015	0.32 ± 0.013	2.19	< 0.05	0.34 ± 0.019	0.34 ± 0.019	0.03	> 0.25
15	0.30 ± 0.017	0.29 ± 0.014	0.69	> 0.125	0.30 ± 0.018	0.31 ± 0.019	0.48	> 0.25
21	0.48 ± 0.018	0.34 ± 0.017	5.79	< 0.001	0.34 ± 0.025	0.33 ± 0.025	0.30	> 0.25
22	0.31 ± 0.015	0.31 ± 0.081	0.10	> 0.25	0.30 ± 0.024	0.29 ± 0.025	0.39	> 0.25

[a] *AI*, number of associated chromosomes of a specific type divided by the total number of chromosomes of their types.

ellite associations between different chromosomes was also reported in parents [1428]. These data support the hypothesis that an increased tendency to satellite association enhances the risk for nondisjunction.

In a recent study [1503], a slightly increased overall frequency of satellite associations in the parent in whose germ line nondisjunction occurred was observed, with no *specific* increase for the chromosomes involved in the event (e.g., chromosome 21). Nonetheless, the authors concluded that satellite association plays "an insignificant role" in the causation of nondisjunction, especially since trisomies are also common in nonacrocentric chromosomes, as shown by abortion studies, and parents having more than one trisomic child failed to show an increased frequency of satellite associations. In our opinion, the combined evidence does indicate a small influence of satellite association on nondisjunction risk.

Thyroid Disease and Antithyroid Antibodies. Altered thyroid function had long been suspected as a risk factor for nondisjunction. As early as 1921, Dollinger reported increased thyroid activity in mothers of Down's syndrome cases, and later on Ek (1959) [1441], investigating 41 mothers of Down's syndrome children, found their serum protein-bound iodine (PBI) to be significantly increased (7.1 μg%, compared with 5.9 μg% in the controls). These results were not confirmed later. This discrepancy may be due to a longer time interval between birth of the affected children and examination of the mother [1583]. Fialkow et al. (1967) [1447] found that mothers of children with Down's syndrome had a significantly higher frequency of thyroid autoantibodies than the controls (Fig. 5.9). This study was initiated by a clinical observation of one of us (A. G. M.): two girls with Turner's syndrome due to X isochromosomes also had Hashimoto's thyroiditis. It was difficult to believe that this finding was accidental.

In Fialkow's study, the frequency of thyroid autoantibodies was about the same in older compared to younger mothers. Because of the age-dependent increase of positive reactors with age in the control population, a significant difference between mothers of Down's syndrome children and controls was found only in the younger age group. Hence, there is good evidence that thyroid autoantibodies as a marker for autoimmunity enhance the risk for nondisjunction. Relative to other risk factors, this factor seems to be especially important in younger mothers where the age-dependent risk is low.

Do Thyroid Autoantibodies and Autoimmune Disease Also Enhance the Risk for Other Aneuploidies? Nondisjunction in acrocentric chromosomes is different from nondisjunction – and chromosome loss – in X chromosomes. Therefore, the results mentioned

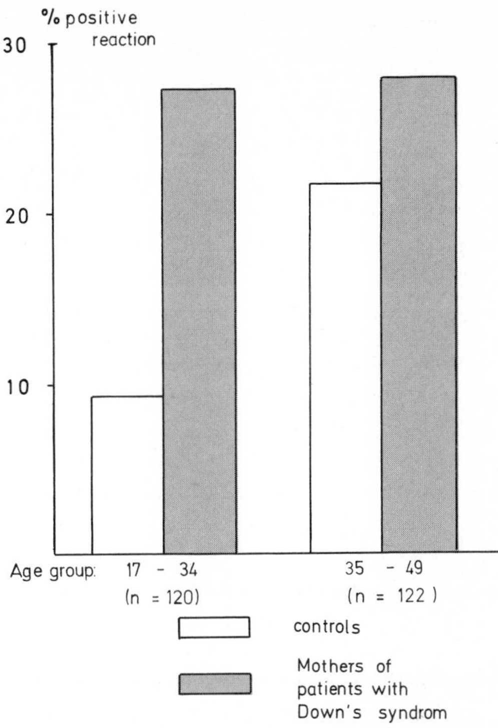

Fig. 5.9. Frequency of thyroid antibodies among mothers of children with Down's syndrome as compared with age-matched mothers of children not showing Down's syndrome. A significant increase was found in young but not in older mothers of Down's syndrome children. (Fialkow, 1967 [1448])

above do not necessarily imply that autoimmune thyroid disease will enhance the risk for X-chromosomal aneuploidy as well. However, they do provide a hint about where to look for risk factors. Some reports point to an increased frequency of autoantibodies in patients with gonadal dysgenesis (XO type) and in their parents. An increase of thyroid antibodies was even more pronounced in patients with mosaics and their mothers [1448]. Many cases of juvenile diabetes are now considered to be caused by an autoimmune mechanism (Sect. 3.7.3). A surprisingly high incidence of diabetes in close relatives of patients with the XO and XXY karyotypes, especially parents, was reported [1580a, 1583, 1687]. An influence of the *season* at which conception occurred on the incidence of nondisjunction has occasionally been suspected in the past. In a European collaborative study, it was shown that cases where nondisjunction occurred in the first meiotic division of the mother were conceived significantly more often in February, March, April, May, and October, and less often in June, July, August, November, and December [1507] (Fig. 5.10).

For the nondisjunction events occurring in the female 2nd division or in the male germ line, *no* con-

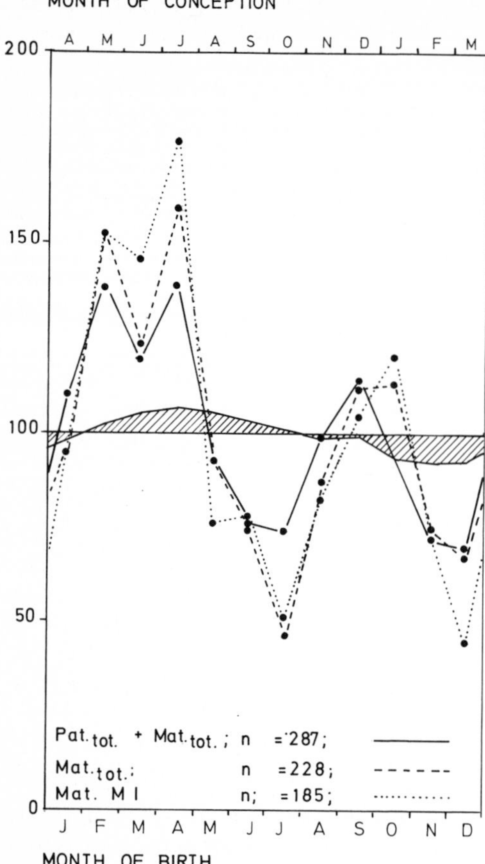

Fig. 5.10. Indexed (average = 100) monthly conception and birth frequency of cases with Down syndrome. The curves were corrected before comparison for the somewhat shorter duration of Down syndrome pregnancies. The *shaded area* is the standard curve of expected frequencies if Down's syndrome births would be randomly distributed among all births. Three curves are given: All D. S.-cases; only those in which the maternal origin of chromosome 21 could be established; and those in which nondisjunction had occurred in the first meiotic division (MI). (From Jongbloet [1507])

sistent trend could be discovered. The seasonal trend in female first divisions was interpreted – and had, in fact, been predicted – from an interesting hypothesis: Unlike human females, nonhuman primates and other mammals do not ovulate throughout the year; they have ovulatory and anovulatory seasons. If traces of this seasonality were maintained in our species, this would allow the following predictions:

1. Ovulation is delayed during the transition from anovulatory to ovulatory seasons and vice versa. In animal reproductive biology these transitions represent a "restoration phase," during which reproduction becomes functionally competent, and an "inhibition phase," during which reproduction decreases.
2. As in other animals preovulatory over-ripeness elicited by delayed ovulation leads to nondisjunction.

In humans, the spring months could represent a trace of the transition from anovulatory to ovulatory seasons; during the autumn, transition from the ovulatory to the anovulatory season could have occurred; and both phases could involve a higher nondisjunction risk. The hypothesis ingeniously connects evidence from general biology with statistical data from human cytogenetics.

Do Oral Contraceptives Enhance the Risk for Down's Syndrome? Some population studies failed to show the overall decrease of Down's syndrome expected from the decline in the frequency of women above 35 having children in recent, as compared with earlier times [1556]. The maternal age-specific incidence seems to have increased. This phenomenon could be due to more complete ascertainment of affected newborns, but a real increase is possible. What environmental factors have changed most drastically within the period during which the increase has been observed (about 1960–1975)? One factor has been the introduction of oral contraceptives. In Sect. 3.8.3, it was explained how the incidence of dizygotic twinning was influenced by such contraceptives. Could an increased risk for aneuploidies have a similar reason? Carr (1967) [1411] reported that women who became pregnant after they stopped taking the pill had an increased frequency of triploid spontaneous abortions. Other reports were contradictory; slight but not significant increases and negative results were published [1399, 1438]. In a recent study, 103 mothers of infants with Down's syndrome and an equal number of age-matched normal controls were compared for a history of usage of oral contraceptives. No evidence of increased pill use of the mothers of Down's syndrome patients either during pregnancy or in the year before conception was found [1505]. But the influence of contraceptives need not be direct. Probably, the decrease in the rate of dizygotic twinning (Sect. 3.8.3) was caused by a change in reproductive behavior concomitant with widespread usage of contraceptives. In earlier years, women having children when above the age of 35 were selected for fecundability (i.e., the ability to conceive). High fecundability, however, appears to be correlated with polyovulation – the prerequisite for dizygotic twinning. By the same token, it is conceivable that *lower* fecundability might be correlated with a higher risk for nondisjunction.

An influence of fecundability is also suggested by a recent study on mice [1403]. Unilateral ovariectomy resulted not only in an earlier cessation of reproduc-

tive life but also in an earlier rise of aneuploidy. Extrapolating to humans, the authors concluded that it is not chronological age but "distance in time from the approaching menopause" that is the decisive factor. It should be possible to test this hypothesis in retrospective studies of mothers of Down's syndrome patients.
Experiments with mice showed an increase of aneuploidies in metaphase II oocytes after treatment with norethisterone acetate only with very high doses and immediately after removal of the hormone. An optimist would conclude from this evidence that oral contraceptives do not enhance the risk of chromosomal anomalies. However, the data are not so clearly negative that a suspicion can be disregarded completely.

5.1.3 *Gene Mutation: Analysis at the Phenotype Level*

Almost all trisomies observed in the human population are caused by new mutations. In these cases, the patients are the only ones affected in their families; they are "sporadic" cases. If they have no children, the supernumerary chromosome will not be transmitted to the next generation. A person bearing a dominant or X-linked gene mutation will usually transmit this mutation to the next generation (Sect. 3.1). If the mutation does not interfere too much with health, reproduction will be almost normal, and pedigrees with many affected individuals may ensue. Almost every affected individual will have affected ancestors and relatives. If the condition prevents many of its bearers from having children, most of the newly mutant genes will be extinct soon after they have appeared, and a relatively large number of all observed cases will be sporadic and caused by a new mutation. Large-scale population surveys are required to estimate the mutation rate. Gene mutation rates may be analyzed at the qualitative-phenotypic-Mendelian, at the protein-enzyme, or at the DNA level.

5.1.3.1 *Methods for Estimating Mutation Rates*

In the following sections, various methods will be described for estimating mutation rates of rare hereditary diseases. In most cases, the methods are applied to conditions analyzed at the qualitative-phenotypic-Mendelian level implying that a simple mode of inheritance has been established, but a specific gene locus has not been identified.

Direct Method. The principle of the direct method has already been described (Sect. 5.1.2.1). Apart from genome mutations and structural chromosome aberrations, this approach can be utilized for dominant gene mutations. It simply requires determination of the *incidence* (frequency at birth) of a certain condition in the population and knowledge whether the case is, indeed, sporadic. All sporadic cases are assumed to be new mutations and the mutation rate can be calculated with ease (see formula in Sect. 5.1.2.1). In practice, application of this simple principle meets with difficulties and possibilities of error:

a) The most evident source of error is illegitimacy, a possibility that needs especially to be considered if the selective disadvantage of a trait is not obvious and if very few sporadic cases are observed among a majority of familial ones. On the other hand, if there is a strong selective disadvantage and if there are many sporadic cases along with some familial ones, an occasional case of illegitimacy does not disturb the estimate too much. Currently, with many genetic markers available, false paternity can usually be ruled out by appropriate tests.
b) A second possible source of error is the occurrence of phenotypically similar or identical cases that happen to be nonhereditary (phenocopies). The strictest genetic test for this bias is examination of offspring in such sporadic cases. If all such cases are mutants, a 1:1 segregation of affected and unaffected is expected. However, the test is insensitive if the total proportion of phenocopies is small. A preliminary hint is provided by considerations of genetic equilibrium. Selection against the condition should be strong enough to account for the fraction of sporadic cases among all cases [1667]. This principle leads to the general rule that the fraction of sporadic cases is roughly proportional to the selective disadvantage of the trait. In the extreme cases with no transmission to the next generation because of early death, all cases of autosomal-dominant mutation in the population will be sporadic. In intermediate cases, a variably large fraction of cases will be new mutants.
c) Often, different varieties of autosomal-dominant disease may exist. If all are lumped as one disease, the mutation rate may be spuriously elevated. Similarly, besides an autosomal-dominant type of disease, an autosomal-recessive variety may exist. Careful clinical and laboratory analysis of the phenotype, age of onset, natural history of the disease, and linkage analysis, using, for example, a DNA marker may help for discrimination. Consanguinity, if present, may provide a strong hint for autosomal recessive inheritance in populations where inbreeding is unusual.
d) Penetrance may be incomplete. If penetrance is not much below 100%, this bias can be corrected.

Mutation rate estimation can only be carried out after the proportion of sporadic cases that can be definitely considered as new mutants has been established satisfactorily and phenocopies have been ruled out. The most straightforward method is to compare the number of patients carrying new mutations with the total number of children born in the same year. This is equivalent to a determination of the incidence. This method is feasible mainly for conditions that can be identified in early infancy. Thus, acrocephalosyndactyly (Apert's syndrome) (10120) can be identified a birth in all affected infants, and the clinical status of the parents is clearly apparent for this defect. Based on the number of total births in the population, the mutation rate can be calculated. However, most diseases are not discovered at birth, and prevalence data ("cases at hand") alone are available. Thus, in hemophilia, the life span of hemophilia patients was one-third that of normals when a certain mutation study was carried out. Therefore, the prevalence figure had to be multiplied by a factor of 3 to obtain the incidence. Note that in this example the number of hemophilics was compared to the total population, while in the earlier example the number of cases of acrocephalosyndactyly per year was related to all births of that year alone.

The direct method, in spite of its simplicity, was introduced only after the concept of genetic equilibrium between mutation and selection had been established.

Danforth's Formula. Danforth [1429] first formulated this concept in 1921 and proposed its utilization for mutation rate estimation:

"It may be recalled that there is a considerable number of dominant traits which are slightly unfavorable ... The incidence of these traits is no doubt maintained in part by recurrent mutations. The frequency of such mutations could be estimated if the average number of generations through which they persist were known. In some of these there is evidence that the average duration is for only a few generations ... The rate of mutation must be such as to bring the incidence to its present value and to balance the adverse effect of selection."

He then proceeded to derive a formula to determine the average number of generations during which a mutation persists before elimination as a measure for the mutation rate. On Danforth's proposal, the concept of equilibrium between mutation and selection was clearly formulated. However, it was overlooked only to be rediscovered 15 years later by Haldane [1472].

Haldane's Indirect Method for Mutation Rate Estimation. Haldane explained the concept as follows:

"The sex-linked recessive condition haemophilia has been known for over a century. Since only a small minority of haemophiliacs live long enough to breed, and (as will be seen) over one-third of all haemophilia genes in new-born babies are in the X-chromosome of males, the condition would rapidly disappear unless haemophilia genes arose by mutation. The only alternatives would be that heterozygous females were more fertile than normal, or that in their meiosis the normal allelomorph ... was preferentially extruded into a polar body. Neither of these alternatives seems likely ...

We now assume, and will later attempt to show, that most large human populations are in approximate equilibrium as regards haemophilia, selection being balanced by mutation ... if x be the proportion of haemophilic males in the population, and f their effective fertility, that is to say their chance, compared with a normal male, of producing offspring, then in a large population of 2 N,(1 – f) xN haemophilia genes are effectively wiped out per generation. The same number must be replaced by mutation. But as each of the N females has two X-chromosomes per cell, and each of the N males one, the mean mutation rate per X-chromosomes is ⅓ (1-f) x, or if f is small, a little less than ⅓ x. Hence, we have to determine the frequency of hemophilia in males to arrive at the appropriate mutation rate."

Additionally, Haldane gave a formal treatment that led to the following results. The ratio of heterozygous females to hemophilic males is:

$$1+\frac{2f\mu+\nu}{2\mu+\nu}$$

Here, μ is the mutation rate in female germ cells, and ν in male. Of all cases of hemophilia, a fraction (1-f) μ (2 μ + ν) should be sons of nonheterozygous mothers, i. e., sporadic cases. In the same paper, Haldane showed that genetic equilibrium would indeed be re-established within a very short time after any perturbation.

Haldane's method is practical because information from only one generation is needed. This information, however, can be utilized in different ways. One useful extension concerns a separate assessment of mutations in male and female germ cells (Sect. 5.1.3.4).

Practical Problems in Application of the Indirect Method. As with the direct method, a number of practical problems arise when the indirect method is actually applied:

a) Illegitimacy is no problem since the estimate is based, in all cases, on conditions in the present-day population and not on previous generations.
b) Phenocopies and genetic heterogeneity raise problems identical to those encountered with the direct method.
c) Incomplete penetrance does not influence estimates of mutation rates provided that those gene carriers who fail to manifest the condition have no selective disadvantage compared with the population average.
d) A problem that is unique to the indirect method is the estimation of f (average fertility of patients relative to the population average). This problem is simple if $f=0$, i. e., the affected individuals do not reproduce at all. An example concerns the Duchenne-type of muscular dystrophy. The formula for the ratio of patients who are sons of normal mothers in $m=\frac{(1-f)\mu}{2\mu+\nu}$, which becomes ⅓ if $\mu=\nu$ and $f=0$. This means that one-third of the observed cases are due to new mutations if mutation rates in males and females are identical.

The problem is much more difficult if the fertility of the trait bearers is only slightly reduced. A useful approximation for f can be reached by comparing reproduction of patients with that of their unaffected sibs, if the patients' fertility is significantly subnormal. Otherwise, biases caused by family planning of sibs may be considerable. The best method is to determine the number of children in a random age group of the population, with follow-up to the end of their reproductive period in comparison with patients [1595, 1597]. Even this method may give distorted data if patients reproduce more frequently than in earlier times because of improved medical therapy. Under such conditions, the population is no longer in equilibrium and mutation rates will be underestimated.

These biases make all estimates of *f* unreliable. Therefore, the indirect method can be expected to give a general idea of the correct order of magnitude of the mutation rate only if the fertility of the patient (*f*) is markedly reduced.

The direct method was used in the great majority of estimates in Table 5.8. The indirect method was used mainly for X-linked recessive conditions. In hemophilia, fertility was markedly reduced at the time these data were collected. Fertility reaches zero in the Duchenne type of muscular dystrophy and in the two conditions in which the hemizygotes are lethal: incontinentia pigmenti and the orofaciodigital (OFD) syndrome. Therefore, these estimates can be regarded as fairly reliable.

Mutation Rates Cannot Be Estimated for Autosomal-Recessive Diseases. Obviously, the direct method cannot be used in fully recessive conditions because the mutation would most often occur in the germ cell of an individual who is mated with a normal homozygote, and the mating would therefore produce heterozygous and normal homozygous children. If methods to detect heterozygotes for various conditions were available, this problem might be obviated [1595], but since most methods for heterozygote detection involve some overlap with normals and because of the existence of silent alleles, there will still be problems. Currently, this method is not being utilized.

Theoretically, the indirect method could be used. With every homozygote who does not reproduce, two mutant genes are eliminated from the population, and the loss of these alleles would need to be compensated by mutation to achieve equilibrium. However, application of the method is subject to two conditions. The selective disadvantage must be confined to the homozygotes, and the heterozygous state must be selectively neutral. According to the Hardy-Weinberg Law (Sect. 3.2), the number of heterozygotes is $2pq$, that of affected homozygotes q^2. Hence, heterozygotes are much more frequent in comparison with affected homozygotes especially if the condition is rare. A very small selective disadvantage would therefore require a much higher mutation rate for maintenance of genetic equilibrium, whereas a small advantage would render mutations unnecessary to explain a genetic equilibrium with these concepts.

Furthermore, it is demonstrably wrong to assume genetic equilibrium for recessive mutations in present-day human populations. In the past, the human population was divided into many isolated groups, which showed different rates of population growth; most of these groups have begun mixing only in relatively recent times. Screening programs for rare inborn errors of metabolism show remarkable differences in incidence of recessive genes even between closely related populations (Sect. 6.1.3). The almost world-wide decrease in the number of consanguineous marriages has also contributed to the disturbance of any genetic equilibrium that may have existed in the past. At the moment, the number of cases of rare recessive disorders is below equilibrium value, and the increase to equilibrium is expected to be very slow [1473]; Sect. 6.3.1). Depending on more or less arbitrary assumptions, almost any mutation rate estimate can be calculated for a recessive condition using the same data. Hence, such estimates are guessing games without scientific value. Thus, until studies can be done with heterozygote states for autosomal-recessive traits using the principles of calculation for autosomal-dominant traits, mutation studies for autosomal-recessive traits should not be done at all. Such calculations should be confined to dominant and X-linked recessive conditions.

5.1.3.2 Results on Mutation Rates

Estimates Based on Population Surveys. Table 5.8 lists mutation rate estimates. The main criterion for inclusion in Table 5.8 was that the determination of incidence, especially of sporadic cases, be fairly reliable. Some remarks may be useful for some specific conditions.

Achondroplasia (10080) is a fairly well-defined condition characterized by shortening of the limbs, depressed nasal bridge, characteristic vertebral changes, and occasional internal hydrocephalus. Patients' reproduction is markedly reduced; therefore, the majority of all observed cases are sporadic and caused by new mutations. At least two superficially similar conditions that lead to death a short time after birth are not included in the estimates: achondrogenesis (20060, 20061, 20070) and thanatophoric dwarfism (27367). The rapid development of nosology in the skeletal dysplasias with many different subtypes, which may be confused with classic achondroplasia, makes earlier estimates of the mutation rates for this condition suspicious. Genetic heterogeneity is a general problem with most mutation rate estimates. If, on closer scrutiny, a condition turns out to consist of more than one genetic entity with a dominant mode of inheritance, then the mutation rate must be considered as a combined estimate for these genetic entities.

In view of the possibility to distinguish the various types of achondroplasia by clinical and radiographic criteria, a new look at the mutation rates in this condition and in some of its dominant genocopies carried out by an expert "dwarfologist" would be of great interest. Meanwhile, the three currently available estimates for achondroplasia are remarkably similar.

Table 5.8. Selected classic mutation rates for human genes

Trait	Population examined	Mutation rate	Number of mutants/10^6 gametes	Authors
A. Autosomal mutations				
Achondroplasia	Denmark	1×10^{-5}	10	Mørch, corrected by Slatis
	Northern Ireland	1.3×10^{-5}	13	Stevenson
	Four cities	1.4×10^{-5}	14	Gardner
	Germany (Reg. Bez. Münster)	$6\text{–}9 \times 10^{-6}$	6–9	Schiemann
Aniridia	Denmark	$2.9\text{–}5 \times 10^{-6}$	2.9–5	Møllenbach, corrected by Penrose
	Michigan (U.S.A.)	2.6×10^{-6}	2.6	Shaw et al.
Dystrophia myotonica	Northern Ireland	8×10^{-6}	8	Lynas
	Switzerland	1.1×10^{-5}	11	Klein, corrected by Todorov et al.
Retinoblastoma	England, Michigan (U.S.A.), Switzerland, Germany	$6\text{–}7 \times 10^{-6}$	6–7	Vogel
	Hungary	6×10^{-6}	6	Czeizel et al.
	The Netherlands	1.23×10^{-5}	12.3	Schappert-Kimmijser et al.
	Japan	8×10^{-6}	8	Matsunaga
	France	5×10^{-6}	5	Briart-Guillemot et al.
	New Zealand	$9.3\text{–}10.9 \times 10^{-6}$	~9–11	Fitzgerald et al.
Acrocephalosyndactyly (Apert's syndrome)	England	3×10^{-6}	3	Blank
	Germany (Reg. Bez. Münster)	4×10^{-6}	4	Tünte and Lenz
Osteogenesis imperfecta	Sweden	$0.7\text{–}1.3 \times 10^{-5}$	7–13	Smårs
	Germany (Reg. Bez. Münster)	1.0×10^{-5}	10	Schröder
Tuberous sclerosis (epiloia)	Oxford Regional Hospital Board Area (G.B.)	1.05×10^{-5}	10.5	Nevin and Pearce
	Chinese	6×10^{-6}	6	Singer
Neurofibromatosis	Michigan (U.S.A.)	1×10^{-4}	100	Crowe et al.
	Moscow (U.S.S.R.)	$4.4\text{–}4.9 \times 10^{-5}$	44–49	Sergeyev
Polyposis of intestines	Michigan (U.S.A.)	1.3×10^{-5}	13	Reed and Neel
Marfan's syndrome	Northern Ireland	$4.2\text{–}5.8 \times 10^{-6}$	4.2–5.8	Lynas
Polycystic disease of the kidneys	Denmark	$6.5\text{–}12 \times 10^{-5}$	65–120	Dalgaard
Diaphyseal aclasis (multiple exostoses)	Germany (Reg. Bez. Münster)	$6.3\text{–}9.1 \times 10^{-6}$	6.3–9.1	Murken
von Hippel-Lindau syndrome	Germany	1.8×10^{-7}	0.18	Burhorn
B. Sex-linked recessive mutations				
Hemophilia	Denmark	3.2×10^{-5}	32	Andreassen, corrected by Haldane
	Switzerland	2.2×10^{-5}	22	Vogel
	Germany (Reg. Bez. Münster)	2.3×10^{-5}	23	Reith
Hemophilia A	Germany (Hamburg)	5.7×10^{-5}	57	Bitter et al.
	Finland	3.2×10^{-5}	32	Ikkala
Hemophilia B	Germany (Hamburg)	3×10^{-6}	3	Bitter et al.
	Finland	2×10^{-6}	2	Ikkala
Duchenne-type muscular dystrophy	Utah (U.S.A.)	9.5×10^{-5}	95	Stephens and Tyler
	Northumberland and Durham (G.B.)	4.3×10^{-5}	43	Walton
	Südbaden (Germany)	4.8×10^{-5}	48	Becker and Lenz
	Northern Ireland	6.0×10^{-5}	60	Stevenson
	Leeds (G.B.)	4.7×10^{-5}	47	Blyth and Pugh
	Wisconsin (U.S.A.)	9.2×10^{-5}	92	Morton and Chung
	Bern (Switzerland)	7.3×10^{-5}	73	Moser et al.
	Fukuoko (Japan)	6.5×10^{-5}	65	Kuroiwa and Miyazaki
	Northeast England (G.B.)	10.5×10^{-5}	105	Gardner-Medwin
	Warsaw (Poland)	4.6×10^{-5}	46	Prot
	Venice (Italy)	$3.5\text{–}6.1 \times 10^{-5}$	35–61	Danieli et al.
Incontinentia pigmenti (Bloch-Sulzberger)	Germany (Reg. Bez. Münster)	$0.6\text{–}2.0 \times 10^{-5}$	6–20	Essig
Orofaciodigital syndrome (OFD)	Germany (Reg. Bez. Münster)	5×10^{-6}	5	Majewski

Source: Vogel and Rathenberg, 1975 [1683]. With addition of [1460, 1430, 1451].

In *aniridia* (Figs. 5.1 and 5.2) (10620), the irises are lacking. Affected persons may suffer additional visual handicaps such as nystagmus, cataracts, or glaucoma. The two estimates, which are based on sound population surveys, gave similar results. However, there are some pedigrees with atypical manifestations, indicating possible genetic heterogeneity [1388].

Myotonic dystrophy (16090) is a muscular disorder already mentioned in Sect. 1.7 in connection with allelic modification. Recent data [1406] suggest genetic heterogeneity.

Retinoblastoma (18020) is a malignant eye tumor that only affects children in the first years of their life. In any developed society, every patient will sooner or later be seen by a doctor, in most cases by an ophthalmologist. Hence, for ascertainment of all cases in a given population, only ophthalmologists and ophthalmologic departments need be screened. Series from a number of different populations are available. However, not all cases are due to germ cell mutations. Most unilateral, sporadic cases are nonhereditary and presumably caused by somatic mutations (Sect. 5.1.6). Only 10%–20% [1669, 1676] of such unilateral sporadic cases or even fewer [1402] are hereditary. The bilateral sporadic cases are all hereditary. Even so, the mutation rate estimates in Table 5.8, which were calculated on this basis, agree very well. The few cases that are due to 13q14 monosomy (Sect. 5.1.6) do not introduce a bias.

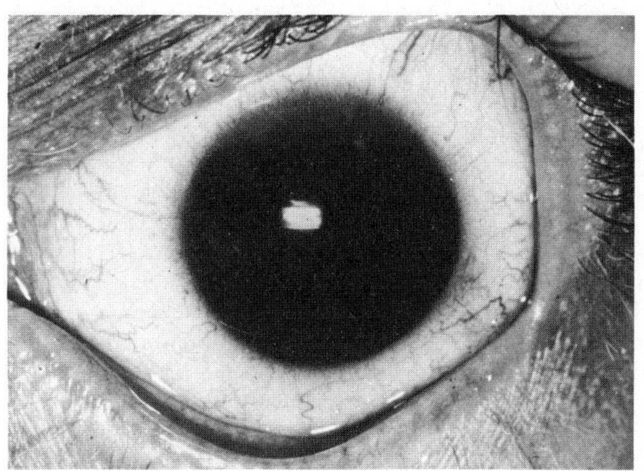

Fig. 5.11. Aniridia. In this case, the iris is totally absent. (Courtesy of Dr. W. Jaeger)

Acrocephalosyndactyly (Apert's syndrome) (10120) is a complex syndrome consisting of skull malformation and complete distal fusion of fingers with a tendency to fusion of the bones (Fig. 5.13). In a number of cases, additional malformations have been reported, and there is an increased risk for early death. While patients rarely have children, transmission has been observed at least three times. The conclusion that the condition is due to a dominant mutation is based on these findings and on the very strong paternal age effect (Sect. 5.1.3.3) [1394; 1591 a; 1689].

Osteogenesis imperfecta (16620) includes, apart from the increased fragility of bones, blue sclerae and frequently sensorineural hearing loss. Its extremely variable expressivity renders any mutation rate estimate relatively unreliable. Genetic heterogeneity and specifically the existence of recessive types adds to the difficulties.

Tuberous sclerosis (19110) is one of the first conditions for which a mutation rate estimate became available [1471]. However, it is not one of the best suited ones, as gene expression is quite variable.

Neurofibromatosis (16220) alwo shows a quite variable expressivity. The first mutation rate estimate is based on a very careful epidemiologic study in Michigan [1426]. The rate was estimated by both the direct and the indirect methods. The estimate given in Table 5.8 (1×10^{-4}) is the highest known so far for a human disorder. However, there seems to be no conclusive evidence for genetic heterogeneity apart from families with acoustic neurinoma ("central neurofibromatosis"). A more recent estimate from the Soviet Union [1617] gives a value somewhat more in line with other mutation rate estimates. Here, however, the incidence estimate was based on examination of 16-year-old prospective conscripts. At this age, some mild manifestations could still be overlooked.

With *Polyposis intestini* (17510), genetic heterogeneity poses a problem, as there is a least one other syndrome involving multiple colon polyps (Gardner's syndrome).

Marfans' syndrome (15470) has become complicated by the discovery of homocystinuria (23620), a condition that can easily be confused with Marfan's syndrome and has an autosomal-recessive mode of inheritance.

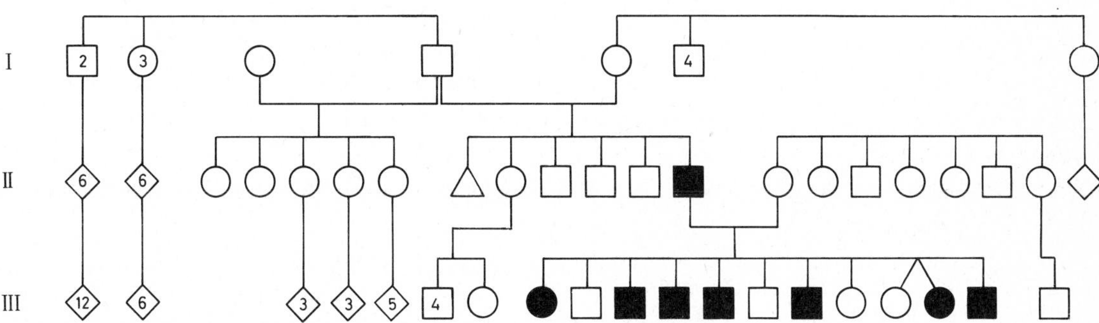

Fig. 5.12. Pedigree with new mutation to aniridia (Møllenbach, 1947 [1560])

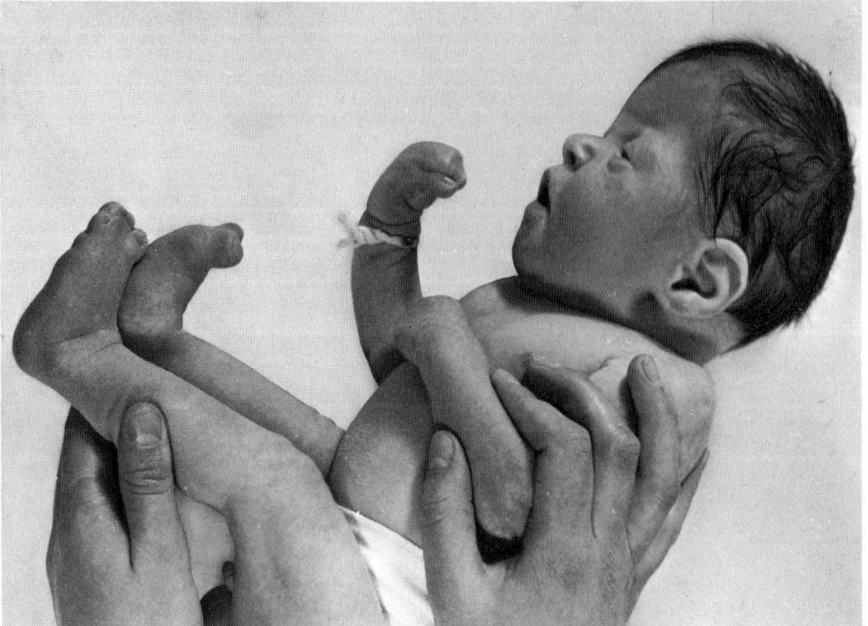

Fig. 5.13. Child with acrocephalosyndactyly (Apert's syndrome). Note syndactyly and deformed shape of head

Polycystic disease of the kidneys (17390) has, after neurofibromatosis, the second highest mutation rates calculated so far.

The estimate for *multiple exostoses* (diaphyseal aclasis) (13370) is based on seven sporadic cases in a relatively small population. Penetrance seems to be influenced by sex-limiting modifying genes (Sect. 3.1.7); the mutation rate estimate might not be too exact.

The *von Hippel-Lindau disease* (19330) mutation rate estimate is based on a survey carried out by Röhrborn and two graduate students [1407, 1587]. Although the results have not yet been published, the conclusions reveal that at least three conditions must be distinguished: isolated, unilateral angiomas of the retina, which are nonhereditary; isolated angiomas of the cerebellum, also nonhereditary; and the autosomal-dominant syndrome with often bilateral angiomatosis of retina, hemangiomas of the cerebellum, and other tumors of internal organs (kidney, etc.). Unfortunately, the mutation rate estimate is based on three cases with unaffected parents only. However, the population survey seems to be relatively reliable. The mutation rate is the lowest estimated so far for a classic dominant condition defined by a specific phenotype.

X-Linked Recessive Conditions. For *Hemophilia* (30670, 30690), estimates for different populations agree relatively well. The first estimates (for Denmark and Switzerland) included both hemophilia A and B; later, the two conditions were treated separately. The mutation rate for hemophilia A is about one order of magnitude higher than that for hemophilia B. The disorder is genetically interesting because biochemical analysis of hemophilia A and B has yielded information on the biological mechanism of mutation [1509, 1594]. Hence hemophilia is the only one of all the conditions for which mutation rates have been estimated in which analysis at the DNA (See Sect. 2.3.3.7) and protein-enzyme levels has been carried out.

For the *Duchenne-type of muscular dystrophy* (31020), at least 11 mutation rate estimates from different populations are available. As with retinoblastoma, the ascertainment problems could be overcome relatively easily. The diagnosis can be made without special difficulty [1221]. Application of the indirect method is obviously justified. The patients never have children. Hence, selection against this mutation is very strong. All ten estimates agree amazingly well in order of magnitude.

The *incontinentia pigmenti* (Bloch-Sulzberger syndrome) (30830) estimate is based on the hypothesis suggested by Lenz [759] and explained in Sect. 3.1.4 that the mode of inheritance of this disorder is sex-linked dominant with lethality of the male hemizygotes. This mode of inheritance is bound to lead to strong selection against the mutation.

The *orofaciodigital (OFD) syndrome* (31120) estimate is also based on the genetic hypothesis of lethality in hemizygotes and dominance in females and on many sporadic cases [674].

The two last-mentioned mutation rate estimates are relatively inexact because of the small number of cases.

Are These Mutations Representative of Comparable Mutations in the Human Genome? The mutation rate estimates in Table 5.8 are all of the order of magnitude between 10^{-4} and 10^{-6}. Taken at face value, these data might suggest that they represent the general order of magnitude for human mutation rates that results in more or less detrimental phenotypes with clear-cut dominant or X-linked recessive modes of inheritance. This conclusion, however, would be wrong. The disorders listed in Table 5.8 were selected on the basis of their suitability for a mutation rate estimate. This suitability depends on the ease with which a certain condition may be ascertained and

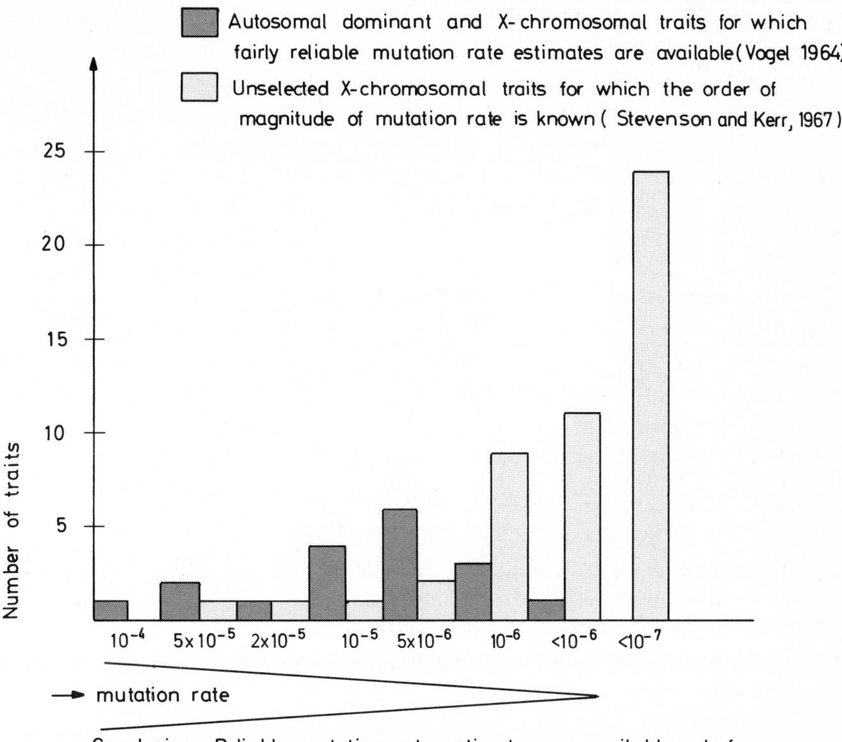

Fig. 5.14. Relatively well-established mutation rates for some autosomal-dominant and X-linked mutations (Table 5.10) in comparison with X-linked mutations for which only an order of magnitude is known. (Stevenson and Kerr, 1967 [1651]) Note much lower mutation rate of the second group of X-linked diseases

diagnosed and most particularly on its frequency in the population. In all cases examined so far on an epidemiologic scale, the surveyed population group has not been larger than about 10 million. It is necessary to use disorders that are relatively frequent to find a sufficient number of cases of a specific disorder to provide the basis for a reasonable acceptable mutation estimate in a population of this size.

This aspect was examined thoroughly by Stevenson and Kerr [1651] for X-linked defects. According to these authors, evidence of frequencies and mutation rates falls into three categories:

1. For a few common disorders, ad hoc studies have been carried out. Here, frequency estimates are relatively reliable.
2. With respect to uncommon disorders, the authors undertook to record all X-linked defects in 875,000 male newborns.
3. For very rare disorders, the only guide to frequency is the number of cases and affected families in the world literature.

Stevenson and Kerr [1651] analyzed 49 rare conditions. This number excludes polymorphisms whose frequency is evidently not dependent on an equilibrium between mutation and selection (color blindness, Xg blood group, G6PD variants). For the 49 diseases, the estimated aproximate order of magnitude of the mutation rates can be found in Fig. 5.14. The authors do not claim accuracy; they do, however, provide sufficient evidence to make this range of estimates plausible.

There is only one disorder, Duchenne-type muscular dystrophy, with a mutation rate higher than 5×10^{-5}. For 24 conditions, the estimated mutation rate is below 1×10^{-7}, and for another 11, it is estimated to range between 1×10^{-7} and 1×10^{-6}. This distribution makes it extremely difficult to calculate an average, especially since the list is by no means exhaustive. A number of other, mostly very rare X-linked defects could be added to the list. Obviously, the frequency of a specific disorder increases its probability of being known to us.

The authors' conclusion is acceptable as a first approximation. They arrive at an average mutation rate of 4×10^{-6} per gamete per generation for mutations on those loci on the X chromosome that lead to observable phenotypic deviations.

Since the distribution of mutation rates appears to be very much skewed toward lower rates, a consideration of the median value in these data may be useful. (The median of a distribution is the value that so divides the distribution that half lies above it and half below.) Cavalli-Sforza and Bodmer [36] calculated a median mutation rate of 1.6×10^{-7} from the same data. This suggests that mutation rates for many traits are very low indeed. Although specific data for autosomal-dominant mutations are not available, a similar rate may be assumed.

Do These Mutation Rates Comprise the Total Mutability of the Gene Loci Concerned? The foregoing does not imply, however, that the total mutability of the various gene loci has been established. The estimates involve only mutations that produce visible changes in the phenotype.
Two other types of mutations defy analysis:

a) Those leading to amino acid substitutions in a specific polypeptide chain that have no noticeable influence on the biologic function of the chain. From our experience with known polypeptides, especially the hemoglobin molecule, we can conclude that many mutations, perhaps even a majority, belong to this group.
b) Those affecting function to such a degree that they are lethal to the zygote during the course of embryonic development.
c) Most of the mutations affecting non-transcribed DNA segments, unless the required DNA techniques can be feasibly applied at the population level.

The large differences between mutation rates estimated from the phenotypes may have various causes: the most common explanation may be a higher degree of genetic heterogeneity for the apparently more frequent mutations. A second possible cause is that numerous mutations within a given gene may lead to the same mutant phenotype, whereas for some genes, very specific changes are needed to produce a certain phentotype and most mutations are either lethal or lead to quite different phenotypes [1571]. Thirdly, there could even be a genuine difference in mutation rate due to a difference either in number of nucleotides per gene or in mutation probability per nucleotide. The problem of total mutability of a gene cannot be solved by analysis at the phenotypic level (Sect. 3.6) but needs analysis at the DNA level. Not only the gene but the mutational site should be identified. This topic will be taken up again in Sect. 5.1.3.1.

In What Context Should Human Mutation Rates Involving Dominant or X-Linked Phenotypes Be Investigated? For assessment of mutation rates for a genetic disease, very thorough ad hoc studies are necessary. All persons who might possibly be affected by a designated disorder must be ascertained within a relatively large population as completely as possible, with special attention paid to "sporadic" cases. Those so ascertained and their families must be examined personally by an experienced investigator to establish the diagnosis and to exclude similar conditions that are genetically different. Such a study would require an unusual effort in time and manpower and would rarely, if ever, be justified simply to establish a mutation rate. Hence, this aspect might conveniently be included in more comprehensive epidemiologic studies designed, for example, to classify an ill-defined group of disorders into a number of genetic entities, to examine reproductive behavior of patients under the influence of modern medicine, or to set up a registration system as the basis for genetic counseling or population monitoring. In fact, most of the mutation rates in Table 5.8 have been estimated in the context of clinical population genetics studies in the 1940s and 1950s.
The activities of the centers mainly involved in such work have been described in Sect. 3.1.8. Since about 1960, these institutions have disappeared or have turned to other activities. At present, there is no center specializing in this type of genetic epidemiologic study, and only very few such studies are being carried out. In our opinion, the most plausible explanation for the abandoning of population studies is supplied by the sociology of science. In the late 1950s and early 1960s, the renaissance of biochemical genetics, cytogenetics, somatic cell genetics, and immunogenetics opened completely new prospects for genetic analysis in humans. Methods and concepts of molecular biology became available. Understandably enough, many research workers were fascinated by these new possibilities and turned away from relatively cumbersome and less satisfying tasks of ascertaining and examining cases in large populations. This trend was reinforced by the development of parts of human population genetics into a highly formalized and somewhat esoteric speciality, which appeared so remote from biology that its significance for a deeper understanding of general biologic problems seemed doubtful to many more biologically and clinically inclined workers.
This development, again, had sociologic consequences. Departments and research teams in medical genetics were founded in many countries, but they were no longer organized with epidemiologic work in mind. Their emphasis tended to be on cytogenetics and biochemistry and the other new laboratory methods. This necessarily funneled the work of new talent in the same direction, thus reinforcing the trend. However, problems such as incidence, mutation rate, genetic heterogeneity, and delineation of genetic from nonhereditary types of diseases are far from solved. In the face of increasing pollution of

our environment by potentially mutagenic chemical and physical agents (Sect. 5.2), knowledge about human mutations at all levels is more urgent than ever before. Although many geneticists speak out publicly about mutations, their pronouncements are based on the same, limited set of old data. Comprehensive knowledge of genetic heterogeneity and phenotypic delineation of disorders is particularly urgent now, since there is an increasing demand for genetic counseling and many new methods for improving genetic prediction have become available.
What can be done to correct this one-sidedness of development? Obviously, the remedy cannot be to abandon or even curb the new methodology in favor of old-fashioned population studies. Scientific progress depends on the quality of research workers, and qualified workers cannot be found for research that they regard as uninteresting. Moreover, the old studies, in spite of some merits, had indisputable weaknesses that impair their usefulness and should deter us from mere repetition. The time is ripe to plan studies in which the two approaches - analysis at the molecular or chromosomal level and analysis at the population level - are combined. For the Lesch-Nyhan syndrome, for example, probable mutations with different characteristics can now be observed in single cells in vitro (Sect. 5.1.5). Would it not be interesting to compare the spectrum of these mutations with the spectrum from a comprehensive population study? Similar comparative studies are conceivable in hemophilia as well as in other diseases of relatively high incidence. They would help not only the population geneticist to find better explanations for the phenomena observed in populations but also the biochemical and cell geneticists to improve their understanding of some findings observed in vitro. Last but not least, they could be of great help in management and genetic counseling of patients and their families.

Mutation Rates for Rare Enzyme Variants. In Sect. 6.1.2, the incidences of rare enzyme variants will be discussed. These incidences, together with the fact that, so far, no new mutations leading to such variants have been observed in persons not exposed to mutagenic agents, permit calculation of an upper limit for the mutation rate of $\mu = 2.24 \times 10^{-5}$ per gene per generation [1788]. This estimate leads to the conclusion that mutation rates for such genes are not higher than the mutation rates for dominant and X-linked diseases. It does not exclude the possibility that they may be appreciably lower; additional data need to be collected.

5.1.3.3 Mutation Rate and Age of the Father

One of Weinberg's Brilliant Ideas. In a paper entitled *Zur Vererbung des Zwergwuchses,*[1] Weinberg (1912) [1692] discussed the genetic basis of achondroplasia. The cases available to him consisted of the pedigrees published by Rischbieth and Barrington (1912) [1599]. Weinberg examined the possibility of a simple recessive mode of inheritance and rejected this hypothesis. He found that the data fitted somewhat better the assumption of a dihybrid recessive mode of inheritance. Then, he mentioned the opinion of Plate that the condition was dominantly inherited. Analysis of the evidence indicated that later-born siblings were more likely to be affected. Having made some remarks about possible biases, he continued:

> If a more exact analysis of birth order indeed confirmed a high incidence in the last-born children, this would point toward the formation of the "anlage" for dwarfism by mutation.[2]

About 30 years later, this prediction was confirmed by Mørch [795], who conducted an epidemiologic study of all achondroplastic dwarfs living in Denmark, including some recently deceased at the time of his investigations. He presented convincing evidence that sporadic cases are, indeed, due to new mutation. He also showed that average maternal as well as paternal age in these sporadic cases was significantly higher compared with the population average and that the maternal age effect was not due to an influence of birth order, which is correlated with paternal as well as maternal age. He was not able to determine whether the effect was due to maternal age, paternal age, or both.

Watson-Crick Model Stimulated New Research on Paternal and Maternal Age Influences. Meanwhile, Watson and Crick [1347] had published their model of DNA structure. In addition to explaining replication and information storage, the model suggested a convincing mechanism for spontaneous mutation: the insertion of a wrong base at replication. This mechanism required that the mutation process depend on replication. Moreover, certain investigations in microorganisms seemed to confirm that almost all mutations occur in dividing organisms [1668]. This concept gave a new impetus to statistical analysis of the effect of parental age on human mutation. The argument was set out by Penrose [1590 a] as follows:

> There are very few cell divisions in the female germ cell line but many in the male germ line since the spermatogo-

[1] *On the Heredity of Dwarfism.*
[2] Translation ours.

nia are continuously dividing. Thus the incidence of mutation due to failure to copy a gene at cell division would be unlikely to have any strong relation to maternal age; a marked increase of defects with this origin, however, would be seen at late paternal ages.

The predictions from this mutation mechanism can be compared with those derived from other plausible mechanisms [1590 a, 1668]. In Table 5.9, all five possibilities are suggested by one or another result in experimental genetics. The second model (mutations depend on cell division) was the one primarily considered by Penrose. It predicts an increase in mutation rates with age only in males and a higher mutation rate in germ cells of males as compared with females.

Cell Divisions During Germ Cell Development in the Two Human Sexes. To improve predictions of mutation rates beyond the qualitatively correct but very general statement of Penrose, the number of cells and cell divisions in the male and female germ line should be known. The evidence that is widely scattered in the literature of various fields has been collected, giving the following picture of early germ cell development, oogenesis, and spermatogenesis [1683].

Early Development. The human primordial germ cells emerge from the yolk sac 27 days after fertilization and begin to colonize the gonadal ridges. On day 46 of gestation, the indifferent gonad undergoes sex differentiation and becomes an ovary or a testicle.

Oogenesis. Oogenesis (Fig. 5.15) occurs only during fetal life and ceases by the time of birth. After sex differentiation, the ovarian stem cells rapidly increase in number by mitotic divisions. From the 2nd month of gestation onward, variable numbers of oocytes enter the prophase of meiosis; oogonia persisting beyond the age of 7 months undergo degeneration. Leptotene and zygotene stages (Sect. 2.1.2.4) are found between the 2nd and 7th month. Usually, after birth, all stem cells are utilized; oogonia have been transformed into oocytes or have degenerated.

The total population of germ cells in the embryo rises from 6×10^5 at 2 months to a maximum of 6.8×10^6 during the 5th month. The population then declines to about 2×10^6 at birth. On the plausible assumption of proliferation by dichotomous divisions, an oocyte will have passed through 22 divisions at the time of birth (Fig. 5.15). This figure is arrived at by finding the exponent of 2, which approximates the total number of oocytes, i.e., $2^{22} \approx 6.8 \times 10^6$. From birth up to mature age and fertilization, the cell will undergo only two meiotic divisions, irrespective of the age at which fertilization occurs.

Spermatogenesis. In spermatogenesis (Fig. 5.16), cell kinetics is different. At the same stage of embryonic life at which the primordial germ cells give rise to oogonia in the female, they become *gonocytes* in the male. From the early embryonic stage up to the age of puberty, the tubules continue to become populated by so-called Ad spermatogonia (d = dark), and at about age 16, spermatogenesis is fully established. The number of Ad spermatogonia can be estimated in three different ways: from volumetric data; from their average number per tubular cross section and the length of tubuli; and from the maximum amount of sperm produced per day. These estimates give values ranging from 4.3×10^8 to 6.4×10^8 per testis. An approximate estimate for both testicles is $\approx 1.2 \times 10^9$. This value can be reached by about 30 cell divisions.
In contrast to oocytes, however, these Ad spermatogonia undergo a continuous sequence of divisions. Of the two division products, one cell prepares for the next division into two Ad cells, whereas the other one divides, giving two Ap (p = pale) cells. These develop into B spermatogonia and spermatocytes, which then undergo meiotic divisions (Fig. 5.16). The timing of these cell divisions is well-known, partly from in vivo studies in young men. The division cycle of Ad spermatogonia lasts for about 16 days. This makes it possible to estimate the number of cell divisions according to age (Table 5.10).

If this calculation is approximately correct, the number of cell divisions a sperm will have undergone from early embryonic development up to the age of 28 is about *15 times greater than the number of divisions in the life history on an oocyte.* At later ages in males, this calculation would give still higher values. Such an extrapolation would be dangerous, as the

Table 5.9. Simple models for mutation and their statistical consequences

Model No. Mechanism:	1 Mutation depending on time only	2 Mutation depending on cell divisions only	3 Mutation during a certain time before puberty	4 Mutation after ceasing of divisions	5 Mutation in mature germ cells
Male germ cells	Linear increase of mutations with age; no sex difference	Increase of mutations with age; higher mutation rate in males	No increase with age; no sex difference	No increase with age; lower mutation rate in males	No increase with age; maybe somewhat higher mutation rate in males
Female germ cells	Linear increase of mutations with age	No increase with age; lower mutation rate in females	No increase with age	Increase with age; higher mutation rate in females	No increase with age; maybe somewhat lower rate in females

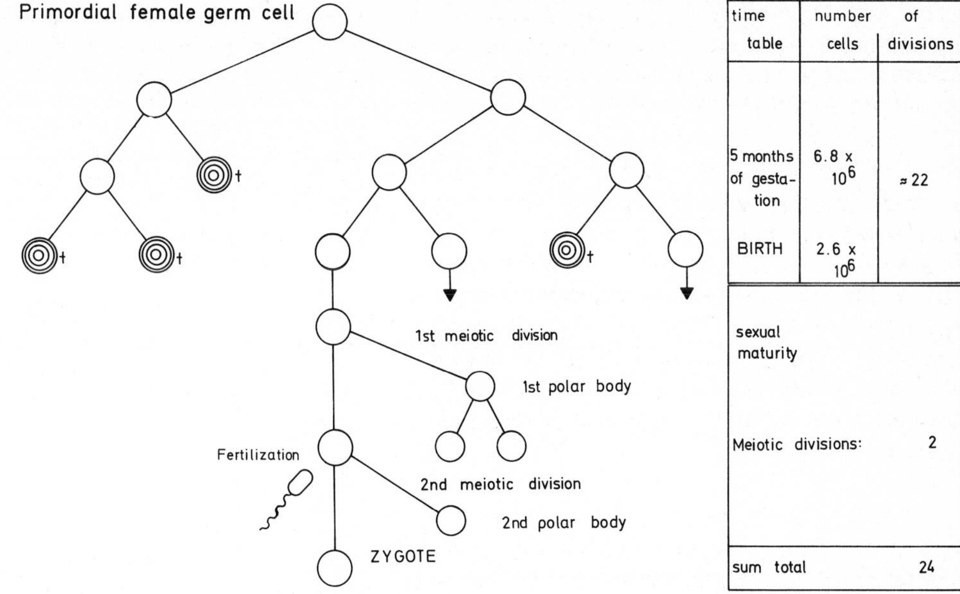

Fig. 5.15. Oogenesis in the human female. All cell divisions – excepting the two meiotic divisions – are already finished at the time of birth; ◎ cell atrophy

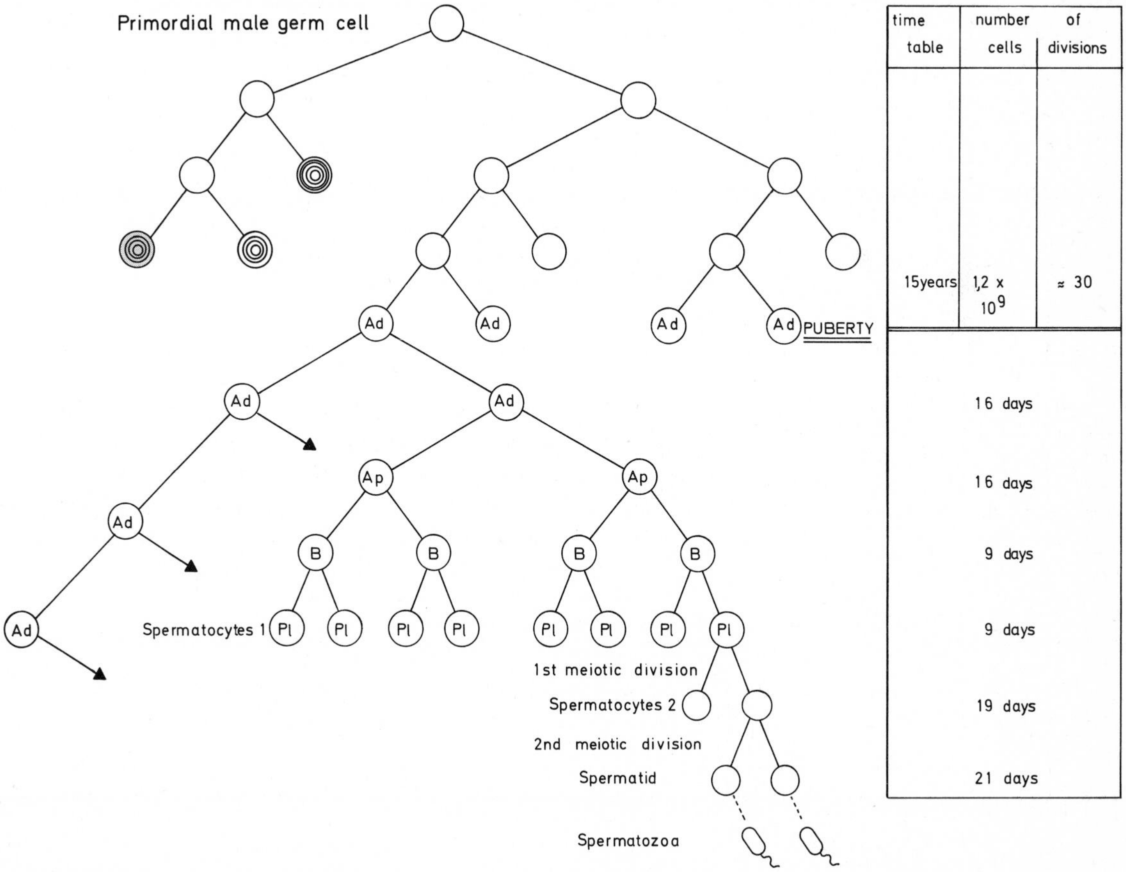

Primordial male germ cell
PUBERTY
Ad Ad Ad Ad
Ad Ad
Ap Ap
Ad
B B B B
Ad
Spermatocytes 1 Pl Pl Pl Pl Pl Pl Pl Pl
1st meiotic division
Spermatocytes 2
2nd meiotic division
Spermatid
Spermatozoa

time table	number cells	of divisions
15years	1,2 x 10^9	≈ 30
	16 days	
	16 days	
	9 days	
	9 days	
	19 days	
	21 days	

Fig. 5.16. Cell divisions during spermatogenesis. The overall number of cell divisions is much higher than in oogenesis. It increases with advancing age. *Ad (dark),* spermatogonia; *Ap, (pale)* spermatogonia; *B,* spermatogonia; *Pl,* spermatocytes; ○ cell atrophy

actual examinations on which the estimation is based were carried out in younger people. It is well-known, however, that sexual activity as measured by number of ejaculations decreases as early as during the 4th decade of life. Some data indicate that sperm counts at that age seem to increase a little, but undoubtedly spermatogenesis decreases with advancing age as shown in Fig. 5.17.

Figure 5.17 would also give the cumulative distribution of the number of spermatogonial divisions if slowing down of spermatogenesis were due to prolongation of the division cycles of Ad spermatogonia. Other mechanisms, however, are possible: for example, some Ad spermatogonia could degenerate, while others continued to divide at the same rate.

Table 5.10. Number of cell divisions in spermatogenesis (from embryonic development to meiosis)

From embryonic age to puberty	≈30
Ad-type spermatogonia (one division/ cycle = 16 days)	≈23/yr
Proliferation + maturation	4+2=6
Total	
At the age of 28	≈380 divisions
At the age of 35	≈540 divisions

Source: Vogel and Rathenberg, 1975 [1683].

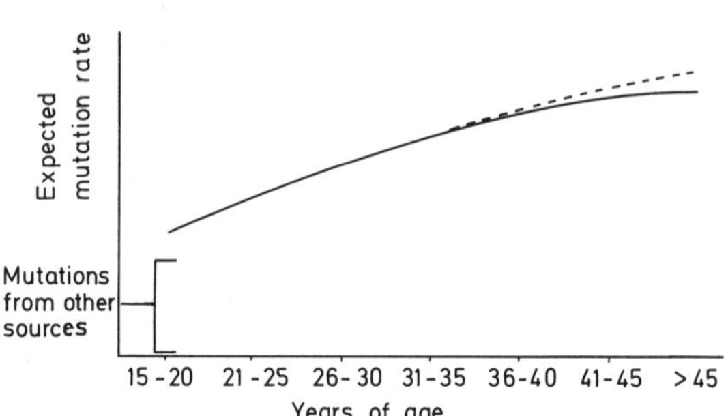

Fig. 5.17. Cumulative distribution of cell divisions in spermatogenesis and expected increase of the mutation rate with paternal age. The distribution curve is derived from Kinsey's data on number of ejaculations at various age groups; in the *dashed* curve, a possible higher sperm count in the ejaculate of older men is considered

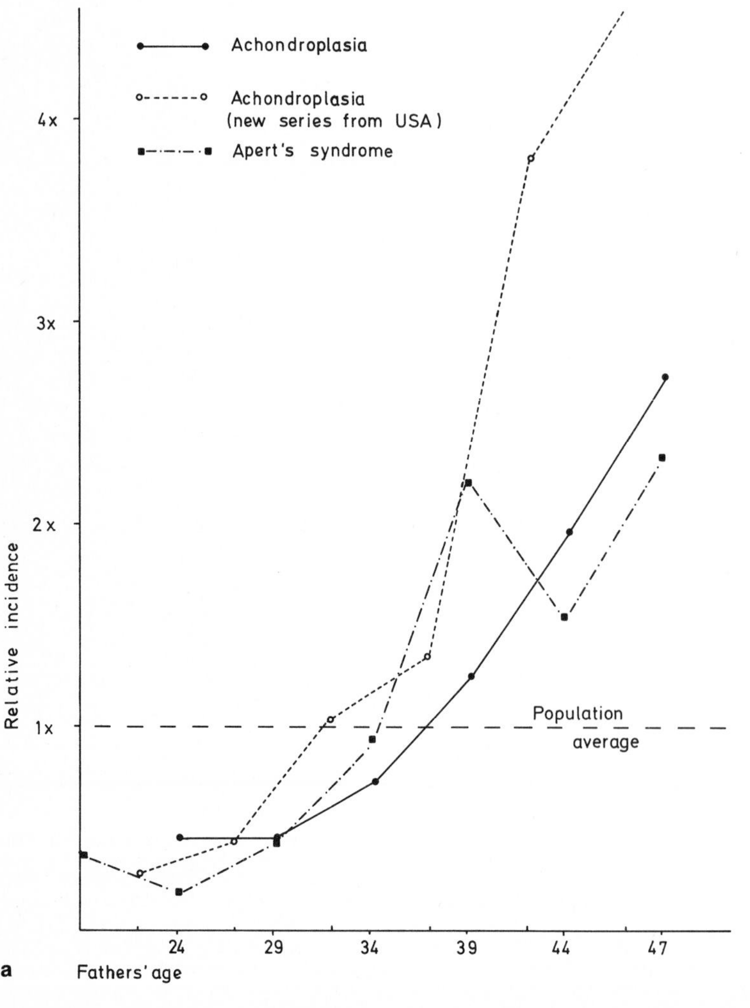

Fig. 5.18 a, b. Mutation rates (relative to the population averages) in various age groups of males and for a number of dominant conditions [1683]

Surprisingly enough, the changes in human spermatogenesis with advancing age have never been examined by histologic methods.[3]

Increase of Mutation Rate with Paternal Age. The foregoing diversion into histology was necessary to understand the meaning of increase of mutation rate with paternal age, which are actually observed. In Fig. 5.18, the relative mutation rates compared with the population average for different age groups of males are seen for a number of conditions; besides achondroplasia, these are acrocephalosyndactyly (Apert's syndrome) [1394]. Marfan's syndrome [1570], and myositis ossificans (a condition involving progressive ossification of the muscles) [1662] (Table 5.11). Especially interesting is the increase for maternal grandfathers of sporadic cases of hemophilia A [1486, 1598].

In all these series, there were some problems with adequate control samples from the general population, as many population statistics give the number of newborns in relation to maternal but not to paternal age. However, the effect is so pronounced that small-scale inconsistencies in the controls will not influence the outcome too much.

All the curves in Fig. 5.18 have two features in common:

1. The mutation rate in the oldest age group is several times - approximately five times - higher than in the youngest group.

[3] There is one exception: Von Winiwarter (1912), in his famous paper on the human chromosome number [543] examined spermatogenesis of four men: they were 21, 23, 25, and 41 years old. In the first three, he found all stages in all cross sections. The picture in the 41-year-old man was quite different. Whereas in some areas, spermatogenesis seemed to proceed normally, others only showed spermatogonia and spermatocytes, and some cross sections looked empty, not even spermatogonia being present. The proband, who was carefully examined, was described as completely healthy, and his sexual life was normal. Von Winiwarter was unable to find any traces of infection in his testis. He concluded that these changes signaled the normal aging process. If this single observation could be corroborated, the peculiar slope of the curve for mutation probability plotted against age may be compatible with the copy error hypothesis.

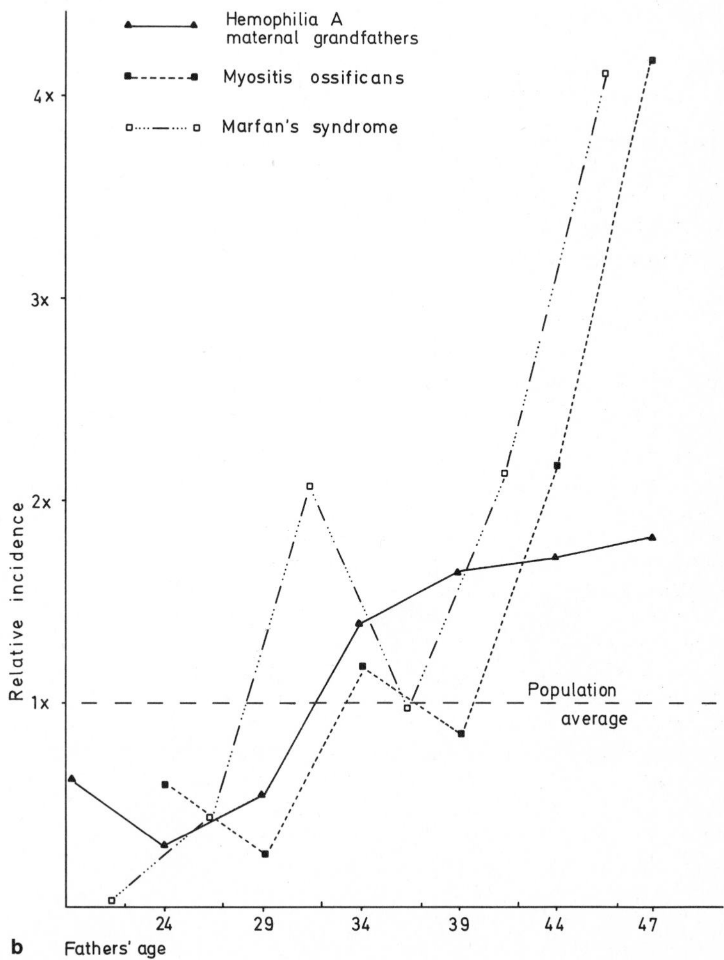

Fig. 5.18 b b

Table 5.11. Mean ages of parents in some dominant conditions[a]

Condition	Patients' parents			Controls' parents		
	Father	Mother	Difference	Father	Mother	Difference
Achondroplasia	38.7	32.1	6.6	33.3	28.6	4.7
	36.1	30.5	5.6	29.9	26.5	3.3
	37.7	34.3	3.4	30.9	28.6	2.3
	40.2	33.0	7.2	31.0	28.4	2.6
Acrocephalosyndactyly	36.9	31.8	5.1	31.0	28.0	3.0
	35.0	30.8	4.2	31.0	27.8	3.2
Myositis ossificans	36.6	30.4	6.2	32.0	?	?
Marfan's syndrome	36.6	29.3	7.3	29.9	26.5	3.3

[a] Adapted from Vogel and Rathenberg, 1975 [1683].

2. The slope of the increase curve has a tendency to become progressively steeper with advancing age. The first characteristic is compatible with the assumption that the increase is caused by accumulation of cell divisions. The second one, however, is not compatible with such an assumption; rather a flattening of the curve would be expected, at least if the division rate of Ad spermatogonia does indeed slow down with advancing age. This discrepancy has not been resolved.

Other Dominant Mutations for Which a Paternal Age Effect Is Possible. The paternal age may also give a clue to identifying dominant mutations as the agent in certain malformation syndromes that cause their carriers not to have children. Possible candidates are hydrocephalus without spina bifida and microphthalmus/anopthalmus [1590a]. Conditions known or assumed to be dominant in which there are suggestions for a paternal age effect are: the basal cell nevus syndrome, Waardenburg's syndrome, Crouzon's disease, the oculodentodigital syndrome, and Treacher Collins syndrome [1506].

Mutations Leading to Unstable Hemoglobins or Hemoglobin M and Paternal Age [1643]. As mentioned in Sect. 4.3.2, hemoglobins M (=methemoglobin) and unstable hemoglobins cause clinical syndromes transmitted as autosomal dominants. Stamatoyannopoulos collected world-wide information on pedigrees in which one of these hemoglobins occurred as a new mutation. In all, 50 cases were collected from 14 countries; they were born between 1922 and 1976. The overall average paternal age was 32.7 years, and the average maternal age, 28.3 years. In order to compare the parental ages of the probands with those of the general control population, the authors calculated for each year and each country the cumulative frequency distributions of the ages of all parents. The ages of each proband's father and mother were then expressed as percentiles of these distributions. The distribution of paternal age percentiles was shifted towards the upper end of the range; 11 of 50 paternal ages for the probands fell between the 90th and the 100th percentiles (Fig. 5.20). While this result was suggestive of a paternal age effect, the difference from the control distribution was not significant. Adequate paternity testing was impossible in the majority of cases. Hence, the series might comprise an appreciable number of "false mutants" due to wrong assumptions of paternity. More recently, a single new mutation of a single HbM was published; the father of the mutant child was 49, its mother was 37 years old [1586]. A new mutation for β-thalassemia has been reported and documented unusually carefully [1657, 1658]. Clinical and biochemical examinations of the child, who was 2 years old at the time of diagnosis, as well as of both parents and three siblings, left no reasonable doubt that it was indeed a new mutant; illegitimacy was excluded convincingly. The father was 45 and the mother 44 years old at the time of the proband's birth. In the same year (1965) in Switzerland, the average paternal age was 31 and the maternal age 28.2. A single case report cannot replace a statistical survey, yet this result points strongly toward a paternal age effect. However, the knowledge that advanced paternal ages are more likely for new mutations would lead to publication bias in such isolated cases.

Some Dominant Mutations Show Only a Small Paternal Age Effect. Penrose [1590] noted that not all dominant mutations show a strong paternal age effect. Among those for which the effect is much less is the well-analyzed example of bilateral retinoblastoma [1683]. Others include tuberous sclerosis, neurofibromatosis, and osteogenesis imperfecta. Figure 5.19 shows the data for these mutation rates. The increase with paternal age is not significant for the three last-mentioned conditions, which, however, show a significant birth order effect suggesting that paternal age does in fact play a role. Comparison between Fig. 5.19 and Fig. 5.18 points strongly toward heterogeneity of dominant mutations regarding paternal age effect. More than one type of mutation is likely to exist: one with a strong paternal age effect, the other with a weak or no effect.

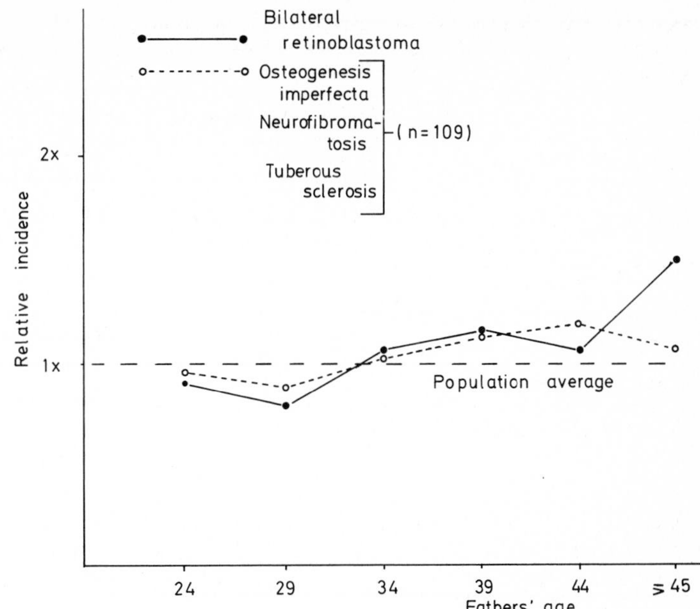

Fig. 5.19. Small paternal age effect for some mutations: bilateral retinoblastoma, tuberous sclerosis, neurofibromatosis, and osteogenesis imperfecta [1670]

The existence of a paternal age effect for lethal or semilethal mutation of X-linked genes has important consequences. One would expect to find such a paternal age effect frequently among the maternal grandfathers of cases of such X-linked diseases on the assumption that the mutation occurred frequently in the grandparental rather than the parental generation. Furthermore, if mutations were replication dependent, such mutations would be more often found in grandfathers rather than in grandmothers since in spermatogenesis there are many more cell divisions than in oogenesis. Therefore, this problem was examined in two series of 77 cases with hemophilia A in which a new mutation could be expected [1486, 1598]. A significant increase in the mutation rate with grandpaternal age was, indeed, observed (Fig. 5.18b), and confirmed in a later study. This is another argument in favor of a higher mutation rate in male germ cells (see below).

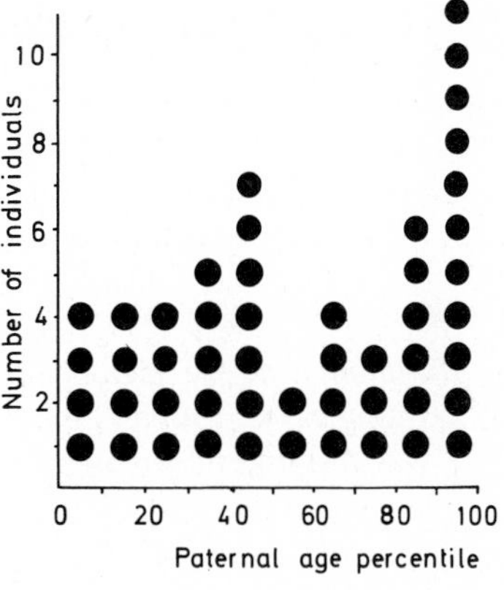

Fig. 5.20. Percentile distributions of ages of probands' fathers for new mutations of unstable hemoglobins or HbM. Expectation on absence of a paternal age effect would be five in each category. Note the high number of observation in the oldest group (⩽90% of the population). (From Stamatoyannopoulos et al. (1981) [1643])

Another X-Linked Disorder: Lesch-Nyhan Syndrome (30800). The X-linked condition Lesch-Nyhan syndrome is caused by a defect of the hypoxanthine-guanine-phosphoribosyltransferase (HPRT). This defect can be diagnosed not only in the male hemizygote, but also in cells of the female heterozygote that show lyonization (Sect. 4.2.2.6).

In a survey to be discussed below [1456], five cases were found in which the mother of a patient was heterozygous but the maternal grandmother was a normal homozygote. Hence, the mutation must have originated in the maternal grandfather's germ cell. The ages of both grandparents are shown in Table 5.12. The means are much higher than those of the average population in the United States. Hence, a paternal age effect of the order of magnitude found in the other conditions reviewed above is very likely.

5.1.3.4 Possible Sex Difference of Mutation Rates

If most mutations are dependent on cell divisions and DNA replication, one would expect not only an increase in the mutation rate with age in male germ

Table 5.12. Ages of maternal grandparents at birth of heterozygous daughters representing new mutation for HPRT defect[a]

	Grandfather	Grandmother
5 families	27	24
	35	35
	40	31
	38	32
	40	39
Mean ± SD	36.0 ± 5.43	32.2 ± 5.54
Population average (U.S.A.)	28.7 ± 6.8	25.9 ± 6.2

[a] Adapted from Francke et al., 1976 [1456].

cells but also an absolutely higher mutation rate in males than in females. The male germ cell undergoes a great many more cell divisions than does the female germ cell (Fig. 5.15 and 5.16).

This problem cannot be examined directly in dominant mutations of humans, as the individual mutation that shows up in the phenotype of a 'sporadic' case cannot be located either in the sperm or in the ovum. In X-linked mutations, however, a test is possible. For X-linked conditions - hemophilia A, the Lesch-Nyhan syndrome, and the Duchenne-type of muscular dystrophy - enough case material is available; all three have been analyzed, but with contradictory results.

Sex Difference in the Mutation Rate for Hemophilia A. Haldane [1474], analyzing cases from a Danish population survey [557], was the first to postulate a higher mutation rate in male as compared with female germ cells. According to Haldane [1472], of all bearers of an X-linked recessive condition, a fraction *(m)*:

$$m = \frac{(1-f)\mu}{2\mu + v}$$

must be sons of normal homozygous mothers, i.e., if there is a genetic equilibrium, their disorder must be due to new mutation in the mothers' germ cell. Mutations in mothers' germ cells are expected to lead to sporadic cases, i.e., to cases who are the only hemophilics in their sibships. But some isolated cases are to be expected for statistical reasons even if all mothers are heterozygous (Sect. 3.3.4; App. 3). Their fraction will be increased if, in addition, new mutants occur. This fraction m can be estimated from family data, or it is possible to test the "null hypothesis" that the data do not deviate significantly from expectation if *all* hemophilics are sons of heterozygous mothers. To avoid uncontrollable biases in the ascertainment of sibships with only one, and with more than one patient, however, the proportion of mutant males should be determined, whenever possible, in series of families that have been collected by complete (truncate) selection (Sect. 3.3.4). This means that within a predetermined time period all living hemophilics and their families within a defined population have been ascertained. In hemophilia and other sex-linked recessive diseases, such studies should be confined to male children and to families in which there is no pedigree evidence of heterozygosity in the mother.

Four series from the literature were found to meet the requirements necessary for sufficiently careful and complete ascertainment [1671]. Within these series, no surplus of sporadic cases was found over expectation if all the mothers were heterozygous. This means that the data fit the hypothesis that all or almost all new mutations occur in male germ cells (Table 5.13) unless the assumption of a genetic equilibrium between mutation and selection does not hold true, or the loss of alleles due to selection against hemophilics is offset by some other mechanism such as segregation distortion in favor of heterozygotes, or a heterozygote advantage. For hemophilia these possibilities were excluded. Had this study used data from more recent surveys, selection relaxation due to treatment of hemophilics with factor VIII preparations (Sect. 4.2.2.9) would have to be considered. The data are not compatible with a hypothesis of equal mutation rates in germ cells of both sexes. Additional evidence may be derived from heterozygote testing since a large proportion of heterozygotes can be detected by testing the ratio of factor VIII activity and hemophilia antigen. Therefore, a comparison of the results of such tests in obligatory heterozygotes and in mothers of sporadic cases should show more normal values among mothers of sporadic cases. Table 5.14 which summarizes several studies, however, shows few differences between obligatory heterozygotes and mothers of sporadic cases. This is another piece of evidence that almost all mothers of sporadic cases are heterozygotes. More recently, a maximum likelihood estimation [1699] using factor VIII data from obligate carriers in comparison with those from mothers of sporadic cases (probable carriers) gave an estimate of $\mu \approx 1/10\, v$; i.e., the mutation rate in male germ cells could be about ten times that in the female germ line [1701].

Hence, the evidence is quite good that most mutations in hemophilia do indeed occur in male germ cells. The only other explanations that female heterozygotes are much more fertile than normal women or that there is some segregation distortion in favor of the mutant gene have been excluded.

Likely Higher Mutation Rate in Male Germ Cells in the Lesch-Nyhan Syndrome [1456]. The advanced age of maternal grandfathers in this conditions has already been mentioned. The disorder is so severe in males that they never have children. Hence, their relative fertility *(f)* is 0. The formula is simplified to $m = 1/3$ if $\mu = \nu$. With equal mutation rates in the two sexes, one-third of all patients in one generation should have homozygous normal mothers, their disorders being due to a new mutation in their mothers' germ cell.

Epidemiologic surveys of all families within a specified population are not currently available. However, in this disease, heterozygotes can be diagnosed by laboratory examination. It was therefore worthwhile to determine whether of all known cases one-third would indeed have homozygous normal mothers. In a collaborative study [1456], 47 families were analyzed, 39 from the United States, 3 from Great Britain, and 1 each from Canada, Belgium, Germany, Ireland, and Switzerland. In 27 families, the only person affected was one male. In all these cases, the mothers were available for heterozygote testing, and only four were normal homozygotes. In the other 23 cases (as well as in all families containing more than one affected), the mothers were heterozygotes. Many fewer than one third of *all* mothers were normal homozygotes.

There are three hypotheses that would account for the high number of heterozygotes among the mothers:

1. Since families were not located by an epidemiologic survey, ascertainment could be biased in favor of families containing more than one case.
2. There could be a reproductive advantage of heterozygotes or segregation distortion.
3. There is a higher mutation rate in male germ cells. Both alternatives 1 and 2 appear unlikely. Hypothesis 3 of a higher mutation rate in the male germ cells is the most plausible explanation for these data.

Table 5.13. Expected and observed number of sibships with a single hemophilic patient among sibships with no other hemophilic relatives. Both hemophilia A and B included[a]

No. of brothers in sibship	No. of patients with hemophilia		
	1	>1	
2	20	18	
3	10	11	
4	5	6	
5	2	8	
6	–	1	
7	–	–	
8	–	1	
Observed	37	45	$\chi^2 = 0.104$
Expected	38.457	43.543	$P \geqq 0.05$

[a] Adapted from Vogel and Rathenberg, 1975 [1683].

The Marker (X) Syndrome. The marker (X) syndrome is probably the most common X-linked hereditary disease in humans (See Sect. 8.2.1.2), and a recent mutation rate estimate is the highest one for any "classical" phenotype in humans (Sect. 5.1.3.2). In the study in which this mutation rate was estimated [1619], segregation analysis of the pedigrees collected in a number of series failed to give any evidence for sporadic cases that could be traced back to new mutations in the germ cells of mothers. Hence, segregation analysis gave the same result for the marker (X) syndrome as had been obtained previously for hemophilia A (See Table 5.13). The authors concluded that *all* new mutations for the marker (X) syndrome occurred in male germ cells. In our opinion, this conclusion may be premature. There were inevitable biases in ascertainment of the pedigrees in which the analysis was performed; as mentioned in

Table 5.14. Diagnosis of carrier state for hemophilia A by determination of factor VIII activity and factor VIII antigen in obligatory carriers (mothers of more than one hemophilic son or with other hemophilic relatives) and probable carriers (mothers of only one hemophilic son and no other affected relatives). (From the literature; see also [1674])

Study	Obligatory carriers					Probable carriers				
	n	Mean activity %	Mean antigen %	Ratio[a] activity/antigen	Diagnosed %	n	Mean activity %	Mean antigen %	Ratio[a] activity/antigen	Diagnosed %
Hathaway et al. (1976)	32	72	126	0.61		12	99	129	0.65	
Biggs and Rizza (1976)	82	52.5	111.6	0.53		41	43.8	105.9	0.46	
Ratnoff and Jones (1977)	82				94.3[b]	39				85[b]

[a] The normal value for the ratio activity/antigen is 1.0. This ratio is reduced in both obligatory and probable carriers.
[b] On the basis of discriminant functions for activities and antigen content, the authors state that 94.3% of all obligatory and 85% of the probable carriers can be diagnosed with 95% confidence as carriers.

Table 5.15. Spontaneous single-locus mutations from wild-type (normal) in the mouse[a]

Sex	No. of gametes tested	Mutations	Frequency/locus
♂	649,227	36	7.9×10^{-6}
♀	202,812	7[b]	6.1×10^{-6}

[a] Adapted from Searle, 1972 [1616] and Russell et al., 1972 [1609].
[b] Includes a cluster of six. Alternative estimate based on one mutation (1.4×10^{-6}).

Sect. 5.1.3.2, the study was not based on an attempt at ascertaining all cases in a limited population. Moreover, inflation of the mutation rate estimate caused by partial compensation of gene loss due to increased fertility of mildly affected carriers in the past has not been excluded [1678]. Therefore, a *lower* mutation rate in female than in male germ cells, as estimated for hemophilia A ($\nu \approx 10\,\mu$) is likely [1682, 2230].

No Sex Difference in Mutation Rates in the Duchenne-Type of Muscular Dystrophy. A fourth disease for which sufficiently large and carefully examined population samples are available is the Duchenne-type of muscular dystrophy (31020). Here the relative fertility *(f)* of the patients is again 0. They never have children, and if the mutation rates in both sexes were equal, one-third of all patients are likely to be sons of homozygous normal mothers. The problem was analyzed on the basis of three comprehensive population samples. After correction for an obvious ascertainment bias, the data were found to agree remarkably well with the expectation $m = 1/3$. The mutation rate seemed to be identical in male and female germ cells. Since then, many more case series of this disease have been published, and methods for heterozygote testing are available (App. 8). While there have been claims both that the mutation rate in female germ cells is lower than in males [1404] and, on the other hand, that it is even much higher [1405], the bulk of the evidence now suggests *approximately* equal mutation rates in both sexes for this condition [1413, 1432]. According to a new assessment of all available data [1430], the mutation rate seems to be *slightly lower* in female than in male germ cells. Hence, Duchenne muscular dystrophy is a special case: It has an unusually high mutation rate, and an unusual sex distribution of mutation rates. Perhaps the gene is especially large and comprises several more or less structure-homologous pseudogenes. This might cause frequent unequal crossing over (Sect. 3.5.8), leading to mutant phenotypes [1700]. Since the gene is located on the X chromosome, crossing over in general (and also unequal crossing over) can only occur in the female germ line.

Indirect Evidence for a Higher Mutation Rate in Male Germ Cells. As noted above, the problem of differing mutation rates in male and female germ cells cannot be examined directly in autosomal-dominant anomalies. However, inferential evidence may be derived from the increase of the mutation rate with paternal age, for example, in achondroplasia. If the paternal age effect is caused exclusively by the father's age, which has been shown to be very likely [1591, 1596], the notion of equal mutation rates in the two sexes could not apply, even if all mutations in children of young fathers were due to mutation in female germ cells. If the mutation rates in young parents were equal in both sexes, the surplus of new mutations due to the paternal age effect would bring about a much higher mutation rate in male than in the female germ cells. For details of the argument see [1677].

Sex Difference in Mutation Rates of the Mouse. In humans, more or less indirect approaches must be used for these studies. These difficulties do not exist in laboratory animals. Therefore, one would expect this problem to have already been solved. However, the evidence is suggestive but not entirely conclusive. Only relatively scanty data from mice are available as a by-product of work on mutation induction with the multiple locus test. This method detects recessive new mutations in the F_1 by back-crossing with a test strain homozygous for seven recessive mutations (Sect. 5.2.1). Table 5.15 shows the data. The sex difference is not very impressive. The seven mutations observed in females include a cluster of six mutants apparently due to one single mutation in the very early development of the ovary. If this cluster is counted only once, the mutation rate in females becomes 1.4×10^{-6}, which is indeed much lower than the mutation rate in males. The hypothesis that the mutation rate in females is lower than in males is strengthened by the remarkably low yields of mutations obtained by irradiation of females at low dose rates [1616].

Statistical Results and Mutation Mechanism. These various results may be compared with the expectations derived from the five mechanisms listed in Table 5.9. Prepubertal mutations (model 3) and mutations occurring after cell divisions have been completed (model 4) can be excluded for the group of mutations with a strong paternal age effect. Time dependent mutations (model 1) are also very unlikely for this category. For the group with a lower paternal age effect, a linear increase with time remains a possibility, perhaps in combination with mutations in mature germ cells (model 5). Most of the data point to the cell division dependent model (2). The sex difference and the increase with paternal age are predicted by this model. However, there are two aspects that suggest caution in accepting this mechanism entirely:

1. Failure to detect evidence of a strong paternal age effect in some instances.
2. The slope of the age-dependent increase of mutations (in those diseases showing a strong paternal age effect) rises more steeply with advancing age.

As shown above, one would have expected a leveling-off of the increase in mutations in the higher age groups. The evidence, however, is not critical, as we know too little about the nature of changes in spermatogenesis with advancing age. Still, the data suggest that the mutation process is somehow related to DNA replication and cell division. Perhaps the statistical data should not be pushed too hard; more detailed evidence can be expected from analysis at the molecular level. Some authors believe that the arguments brought forward in this section are not valid and that mutation rates are, indeed equal in both sexes [1560].

5.1.3.5 Germ Cell and Somatic Cell Mosaics for Dominant or X-Linked Mutations

Pedigree Observations. If a mutation occurs during early development of the germ cells, a germinal mosaic can be created, a more or less extended sector of the gonad bearing the mutation. This situation is well-known from mutation work with *Drosophila melanogaster,* and a "cluster" of mutants due to mutation in early stages of oocyte development in the mouse was mentioned in Table 5.15. In humans, the chance of finding such clusters is very low; they can only be detected if a fairly large sector of the gonad is affected. The earlier in development the mutation occurs, the larger the involved gonadal sector. The proportion of affected germ cells is 1 if the first stem cell already bears the mutation; it is ½ if the mutation occurred at the first stem cell division, $(½)^2$ if it occurred at the second division; and in general $(\frac{1}{2})^n$ if the mutation occurred at the *n*'th division (Fig. 5.21). If the total number of cell divisions is of the order of magnitude estimated above (Sect. 5.1.3.3; Fig. 5.15 and 5.16) and the probability of mutation is equal for all cell divisions, then the fraction of new mutations that occur in clusters, revealing germinal mosaicisms, will be small but not negligible. If the mutation is dominant, occasional families in which both parents are normal but more than one child shows the mutant phenotype are expected.
Occasional pedigrees have been described in which such a cluster is a possibility, for example, a large pedigree with aniridia [1596] and a family with split hand and split feet [1542]. It is always very difficult to exclude the more trivial alternative of incomplete penetrance in one of the parents.

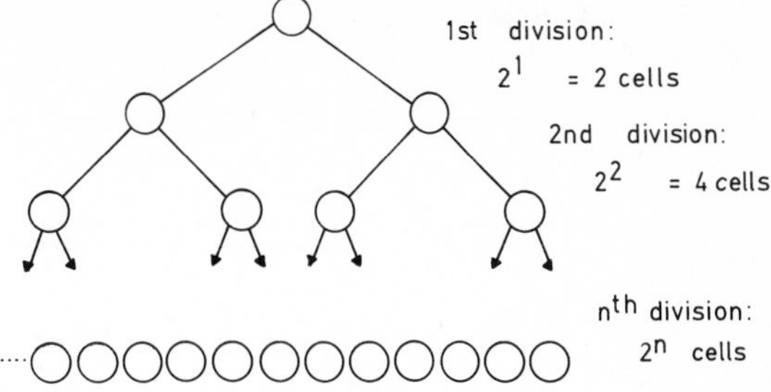

Fig. 5.21. Dependency of mutant germ cells on the stage of development of the germ cell at which the mutation occurred

Somatic Mosaicism for Dominant Mutations. Mosaicism caused by mutation occurs not only in germinal but also in somatic tissue. Such mosaicism might not only affect chromosome number, as described in Sect. 5.1.6, but applies to gene mutations as well. The pattern of phenotypic manifestation of gene mutations, however, makes detection of such mosaicism very difficult. Still, there is at least one observation. During a population survey on neurofibromatosis [1426], four individuals were observed in which the neurofibromata were confined to one sector of the body such as the extremities, the sacral area, and the back. In these four cases, the family history was negative. They produced a total of six children, none of whom was known to be affected. Hence, these four individuals probably represent somatic mutations affecting relatively early stages of development.

Half-Chromatid Mutations? Recently, Gartler and Francke (1975) [676] suggested a special mechanism for the production of mosaics for point mutations: half-chromatid mutations (Fig. 5.22). As mentioned above, many mutations seem to be due to a copy error in DNA replication. If such a copy error happens to occur in the last DNA replication cycle before germ cell formation, the resulting germ cell will contain a mismatching base pair, for example, AG instead of AT (Fig. 5.22). In the first cleavage division, A will pair with T and G will pair with C. Therefore, one of the two products of this division will have the base pair AT as before; the other one will contain the new base pair GC and will be mutant.
Incontinentia pigmenti may be an example. As mentioned in Sect. 3.1.4, the condition is probably caused by an X-linked dominant gene that is lethal in hemizygous males. A total of 593 cases have been described in females, six in males, who had normal

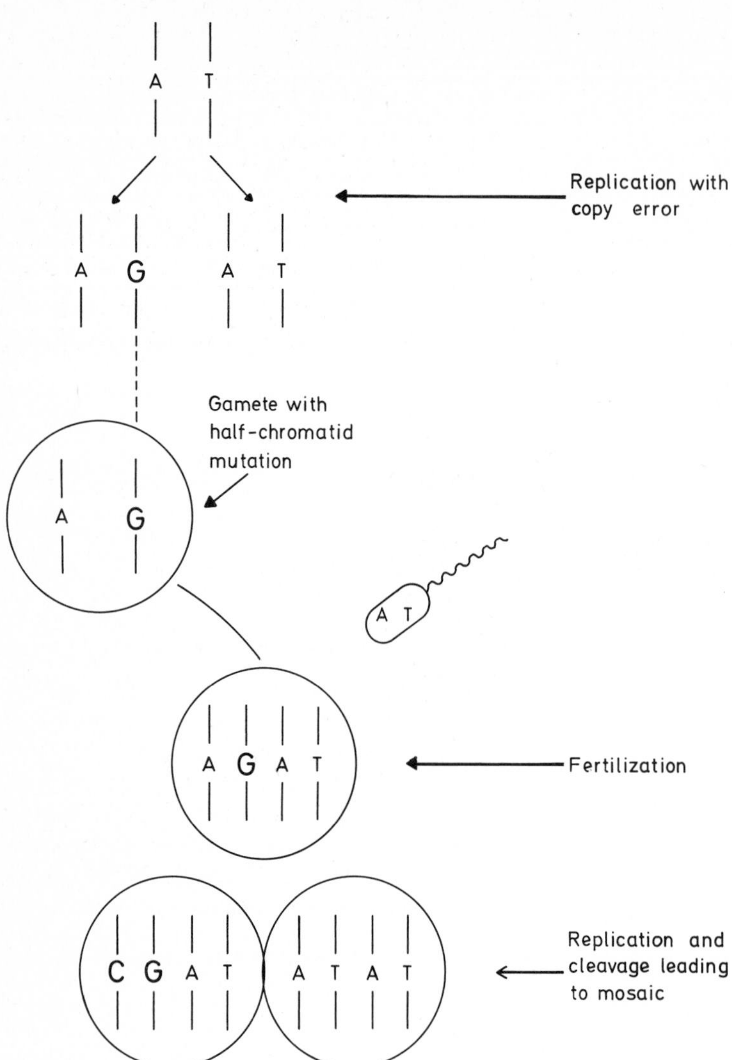

Fig. 5.22. Principle of a half-chromatid mutation: the base replacement in one nucleotide half-strand occurred during the last DNA replication before meiosis. In the first cleavage division, the mutant half-strand gives rise to a cell clone having a gene mutation; the normal half-strand produces a normal cell clone. The individual will be a 1:1 mosaic

XY karyotypes. The pattern of skin affection was similar in both sexes and resembled the mosaic pattern exhibited by the heterozygotes of some X-linked genes in mice, hamsters, and cats.
Moreover, the male cases were sporadic. The observation that the affected males are phenotypically so similar to the females who are known to be mosaics due to the Lyon effect (Sect. 2.2.3.3) makes it likely that males are also mosaics and that the mutation occurred in an early stage of embryonic development. Half-chromatid mutation is a good possibility [760].

5.1.4 Gene Mutation: Analysis at the Molecular Level

DNA and amino acid sequence analysis offer the chance for more specific insight into the mechanisms of mutation. Here, the human hemoglobin variants have proved to be especially informative (Sect. 4.3). Most observed mutations are due to replacement of one base by another; a minority are caused by deletions of some base pairs, by frameshift, by polypeptide chain elongation due to mutation of a terminator codon, or by recombinational events with mutation-like effects (Lepore-type mutations). At this level a number of questions may again be asked, some of which are familiar from our discussion of investigations at the phenotypic level.

5.1.4.1 Codon Mutation Rates

What is the probability that a particular codon will mutate in a specific direction, so that one amino acid will be replaced by another?

First Examination of This Problem. One of us (AGM) previously noted [1563]:

Ideally, such calculations could be done most directly by determining the frequency of a given variant in a population followed by testing the parents. In most cases, one of the parents will have transmitted the variant to a child. In families where both parents are normal (keeping the possibility of false paternity in mind), it is assumed that a new mutation has taken place.

At the time the following calculations were made, largescale electrophoretic population studies were not yet available. Therefore, the mutation rate had to be estimated in a more indirect way from the incidence of rare variants in the population (1:2,000), the fraction of possible variants that could be detected by electrophoresis, and the few observed mutations for one specific group of variants, methemoglobinemias due to the various types of Hb M. Considering all these factors, the order of magnitude of a codon mutation rate for a specific nucleotide substitution in hemoglobins was estimated to be:

$$\mu^n = 2.5 \times 10^{-9}$$

meaning the probability for a base to be replaced by a specific other base, leading to replacement of a specific amino acid by another one. The same indirect approach has been followed in a more recent study, in which new mutations for HbM and unstable hemoglobins identified as new mutants were collected [1643] (see also Sect. 5.1.3.3), and the number of cases observed in one country were compared with the total number of births in that country during one or two generations. The authors were aware that an estimate based on such slender evidence could give only a very crude approximation. It presupposes, for example, that in these populations all cases born during a certain time period had indeed come to their notice. It is not clear, on the other hand, whether all cases included in this study really are new mutants; paternity tests were performed in only 19 of the 55 observations. With all these reservations, the mutation rate for individual nucleotides was estimated at 5.3×10^{-9} for de novo unstable hemoglobins, 10.0×10^{-9} for α^M mutants, and 18.9×10^{-9} for β^M mutants, giving a nucleotide mutation rate of 7.4×10^{-9} for de novo β-chain mutations, based on unstable Hb and β^M mutants together. This estimate can then be extrapolated to the entire Hb β gene (excluding, of course, mutations of other molecular types and, especially, those occurring outside the transcribed DNA segment and leading to β-thalassemias). The result is: 8.6×10^{-6} per β-chain gene. To repeat once more: This is a very crude first approximation based on slim evidence with bold extrapolations.

By a quite different - and also indirect - approach, the codon mutation rate was estimated from the rate of evolution of globin pseudogenes, since in the absence of negative selection the codon replacement rate in evolution was shown to be equal to the mutation rate (see our discussions of Kimura's "neutral hypothesis," Sect. 7.2.3). The resulting estimate, 5×10^{-9}, is amazingly similar to the other estimates.

Estimate with More Direct Data. Data from investigations of a huge population sample of several hundred thousand people have become available from Japan and were used for a mutation rate estimate [1498, 1510, 1702, 1511]. Here, the incidence of rare variants was estimated to be 1:5,000, the fraction of new mutants among all variants being 1/12. This leads to a codon mutation rate of:

$$\mu^n = 5 \times 10^{-9}$$

Considering the very scanty evidence on which they are based, the four estimates agree remarkably well.

In a study designed mainly to utilize blood spots collected during newborn testing for inherited metabolic disease, especially PKU (Sect. 4.2.2.7), one new mutant, a Hb α variant, was discovered and confirmed by paternity testing among 25000 specimens tested for hemoglobin variants [1373]. Ideally, this approach should cover the mutations at ten parental gene loci (1 β, 2 α, and 2 γ from each parent), but not all mutations are ascertained by the electrophoretic method.

How Do These Codon Mutation Rates Compare with Estimates at the Phenotypic Level? Comparison of these estimates with the mutation rate estimates based on specific phenotypes (Table 5.8) show that the latter all approach a frequency of 1×10^{-5} within one order of magnitude. Yet, these rates concern the relatively frequent phenotypes that are well-suited for mutation rate calculations. As shown above, the average mutation rate for visible pathologic phenotypes is probably closer to 1×10^{-6} than to 1×10^{-7} per gamete or lower (Sect. 5.1.3.2). This value is about 40-400 times higher than the codon mutation rate estimates but, at first glance, it seems to compare well with the above-mentioned extrapolation for the Hb β gene. Obviously, this is due to the fact that the classic phenotypes may be the result of a large number of mutational events within the same gene. As known from examination of the polypeptide chains, there are longer and shorter genes. Furthermore, a 100-1000 times difference in order of magnitude between codon and gene mutation rates is a reasonable approximation to the theoretical expectation if we remember that some of the hemoglobin mutations do not lead to any obvious alteration in the phenotype. However, to make comparisons, a number of factors should be considered:

a) Some of the phenotypes in Table 5.8 may be the result of two or even more genetic entities.
b) Medical-genetic experience with the hemoglobin variants has shown that not all mutations within

the same gene must lead to identical or even similar phenotypes. Rather, the phenotype produced by the mutation depends on the special functional alteration of the protein concerned (Sect. 4.3). Mutation within hemoglobin genes may lead, for example, to hemolytic anemias, methemoglobinemia, erythrocytosis, or are not associated with any clinical symptoms at all. It would not be surprising if much of the difference in "classical" mutation rates between phenotypes were explained by the fact that within some genes very many or even almost all mutations lead to the same specific phenotype, whereas in other genes only one or a very few mutations produce the phenotype under investigation. The genes also have different lengths.

The increasing availability of DNA probes will soon help to solve these problems by Studies at the DNA level.

Relative Mutation Rates for the Different Molecular Types of Mutations in the Hemoglobin Genes. As noted above and explained in detail in Sect. 4.3, the mutations affecting the hemoglobin genes can be subdivided into various molecular types. On the assumption that the relative numbers of the different mutation types observed so far are more or less representative of their average mutation rate, it is possible to estimate relative mutation rates as shown in Fig. 5.23. These relative mutation rates give a crude approximation only. The main sources of bias are (1) the wide confidence limits of codon mutation rates for single amino acid replacements and (2) the biases involved in the ascertainment of the different types of human hemoglobin variants. Nevertheless, the relative order of magnitude for the different types of mutation may well be correct. For example, it is plausible that single amino acid replacements occur relatively frequently, that deletions are much rarer, and that chain elongations are very rare events.

At first glance, the most striking difference between the distribution of mutations for hemoglobins and the results from microbial genetics is the rarity of frameshift mutations in man. In bacteria the majority of mutations that can be analyzed biochemically are frameshifts [1698], and in bacteriophages frameshifts constitute the majority of replication-dependent mutations [53]. Besides, it is hardly conceivable that among the deletions only those affecting three or a multiple of three nucleotide pairs should occur. Obviously, deletions of a number of nucleotide pairs will often lead to frameshifts. In Fig. 5.23, the bias was corrected by multiplying the observed deletion rate by 3. The resulting value, however, is still much lower than the frequency of single amino acid substitutions. This paucity of deletions may partially be the result of a lower survival. On the other hand, the mutation spectrum observed in bacteria and phages is biased in the opposite direction. Many single amino acid substitutions will go unnoticed, as they do not lead to a detectable phenotypic effect [1698]. For example, only 10% of mutations known to occur in the histidine locus of *Salmonella* were recoverable as lethals; the remainder were all base substitutions

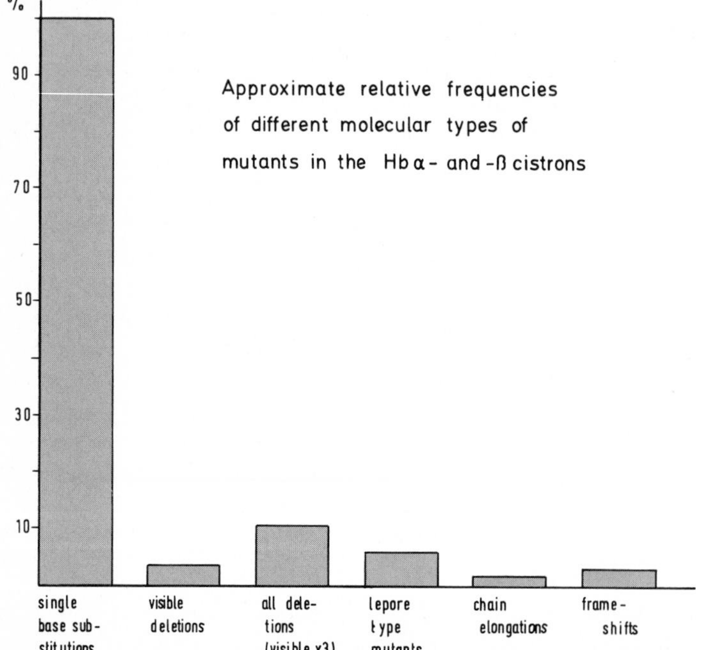

Fig. 5.23. Relative frequencies and presumable mutation rates for single base substitutions leading to amino acid replacements deletions at the molecular level, Lepore-type mutations (abnormal recombinations), frameshift mutations, and chain elongations. The number of splice mutations is of the same order of magnitude as the non-point mutations (See Sect. 4.3)

and either had no effect or only partially inactivated the gene loci and could not be recovered in the test [1374, 1512, 1695].

5.1.4.2 *Problem of the Total Mutation Rate Per Genome and Per Generation*

Requirements for an Estimate. Only mutation rates of single genes or codons have been considered so far. It would be tempting to extrapolate from these mutation rates to the total mutation rate per genome. This problem, however, requires information on a number of aspects:

a) More than a single codon mutation rate should be known.
b) There should be some information as to whether these codon mutation rates are representative of most codons within structural genes. For example, do mutations occur at random within these genes or are some sites more mutable than others?
c) Do mutations outside structural genes occur at the same rates as those inside structural genes? As mentioned (Sect. 2.3), the greatest part of human DNA does not consist of structural genes coding for polypeptide chains. As to the function of most of this DNA, however, only inferences are possible to date. Any attempt to estimate a genome mutation rate should take into account the possible role of DNA not coding for structural genes, of which we know so little.

Are the Known Hemoglobin Mutants Distributed at Random over the Length of the Hb α and β Genes? One of the most striking phenomena of spontaneous mutations in bacteriophages is the occurrence of "hot spots," i.e., the existence of points of an especially high mutability. This phenomenon was discovered by Benzer (1957) [1390] in the T4 phage of *E. coli.* If generalizations from the hemoglobin genes to the whole human genome are to be made, one must first ask whether such hot spots can also be observed in humans. The most direct approach would be to compare mutation rates of different sites within the genes. This, however, is not possible, as mutation rates are not available. Therefore, the distribution of the known mutants over the length of the α and β genes was examined. Here, deviations from randomness may be expected if some codons have a higher rate of mutations than others or if neighboring codons show a higher or lower tendency to mutate. The distribution of known mutations over the Hb α and β genes is seen in Fig. 5.24 (data from [119]). A slight deviation from randomness in the Hb α genes can plausibly be explained by assuming that special sites are so essential for function that mutations would not be compatible with life and therefore will not be recovered and counted as mutations.

There May Be Many New Mutations Per Genome and Generation, but a Reliable Estimate Is Not Yet Possible. The total amount of DNA in a haploid human cell is ~3.0–3.5×10^{-12} g, the haploid genome contains about 3–3.5×10^{9} nucleotide pairs. Since one codon comprises three nucleotide pairs, a maximum of about 1×10^{9} codons exists. Using the esti-

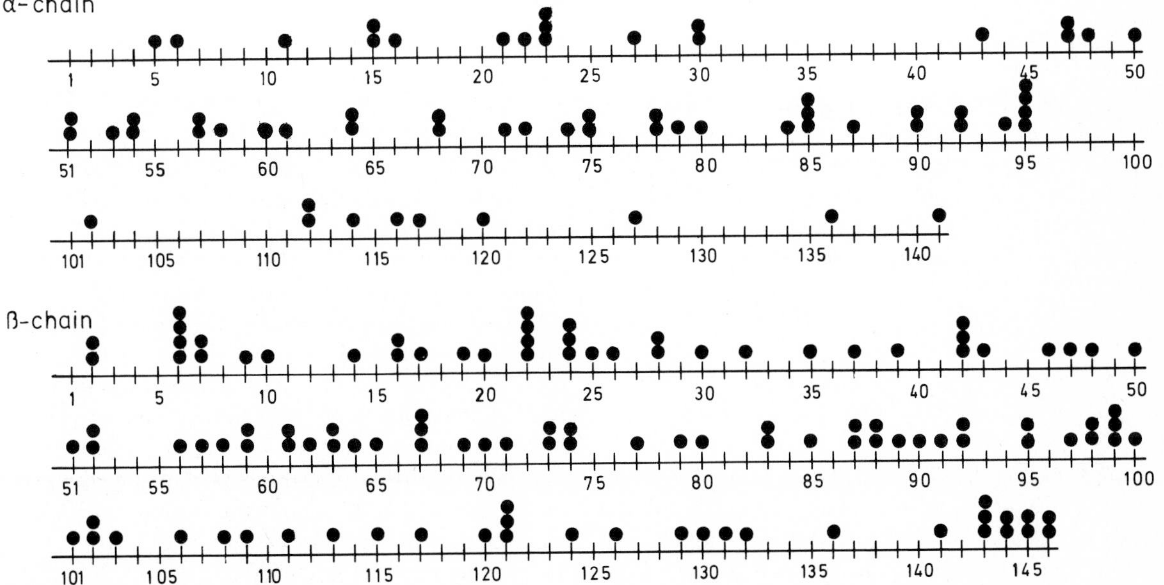

Fig. 5.24. Distribution of single amino acid replacements (due to single base substitutions) over the human hemoglobin α and β chains. This distribution is based on the hemoglobin variants observed in present-day human populations. (Lehmann and Huntsman, 1974 [119])

mate given above, a codon mutation rate of 2.5×10^{-9} for one specific base substitution may be assumed. This number must be multiplied by approximately 7 to arrive at the total mutability per codon, as expressed in amino acid substitutions: Any base can be replaced by three others, and approximately 75% of all base replacements, mainly those of the first two bases, lead to amino acid substitutions. This implies a total mutation rate of 1.75×10^{-8} per codon. This value multiplied by 1×10^{9} gives 17.5 mutations per haploid genome, or single germ cell, and 35 new mutations per individual (two germ cells) in the population.

This number is frighteningly high. It poses the question of how human beings continue to exist. Even considering that many of the known mutations in the hemoglobin genes have little or no ill-effect on their bearers, a mutation rate of this magnitude is hardly conceivable. The usual answer to this dilemma is that most of the human DNA does not consist of structural genes coding for specific proteins but has some other less specific, possibly regulatory function, or that large parts of it are really functionless: "junk," to cite Ohno [1588], or their function might be of such a general kind that most single base pair replacements do not really matter. Remember, for example, the "body guard" function assigned tentatively to the heterochromatin [385]. The amazing amount of restriction site polymorphism found in the nontranscribed DNA segments (Sect. 2.3.2.7) supports this assumption, since it could hardly have been maintained in the population if important functions were compromised. Neel [1578], however, remarks that this "silent" DNA can scarcely be immune to mutation and - sooner or later - to selection. He prefers to assume that mutation rates of different codons may be very different.

The sparse information available so far does not permit offering even a tentative solution to this problem. The aim of this section has been to pose rather than to answer questions [1510, 157, 1673]. Considering the more recent results on chromatin structure (Sect. 2.3), it can be hoped that some of these questions will soon be answered.

5.1.4.3 Mutations in Hemoglobin Genes and the Genetic Code

Amino Acid Replacements Indicate Corresponding Base Substitutions in the DNA. Since the genetic code became known in the early 1960s, analysis of amino acid replacements, especially in hemoglobins, has permitted identification of base substitutions within the RNA messenger and, as mRNA is complementary to the transcribing DNA strand, within the gene itself. This can be illustrated by the example of sickle cell hemoglobin:

a) Alteration of the biochemical phenotype: Hb A→Hb S
b) Amino acid replacement: glutamic acid→valine
c) Codons:

Glutamic acid		Valine	
mRNA	DNA	mRNA	DNA
GAA	CTT	GUA	CAT
GAG	CTC	GUG	CAC
		GUU	CAA
		GUC	CAG

d) Possible base replacements if only a single DNA base substitution has occurred:
CTT→CAT
CTC→CAC
e) Base substitution in the DNA:
T (thymine) →A (adenine)
f) Type of substitution: a purine base is substituted for a pyrimidine base. This type of substitution is called a "transversion." Substitution of a purine for another purine or of a pyrimidine for another pyrimidine is known as a "transition" [1457].

Transitions Are More Frequent Than Expected if Base Substitutions Occurred at Random. Any nucleotide base can undergo one transition and two transversions (Fig. 5.25). Therefore, if the direction of the mutation process were random, twice as many transversions as transitions would occur. Due to the degeneracy of the genetic code, however, not all base substitutions should manifest as amino acid replacements. These considerations allow the calculation of the expected amino acid substitutions. Table 5.16 shows the types of observed transitions and transversions. There are significantly more transitions than expected if the direction of mutation would be random [1681].

The result cannot be ascribed to a biased ascertainment of transitions due to the electrophoretic method which detects only amino acid substitutions associated with charge differences. In fact, among the total number of possible transversions, the proportion of the polar (and detectable) substitutions is higher than for transitions. A higher frequency of naturally occurring transitions is also suggested by amino acid replacements during evolution between proteins that are related by common descent as well as by comparison of related RNA species. Hence, it is quite likely that transitions occur more frequently among spontaneous mutations than would be expected, were the direction of the mutation process random. This result is not surprising, when the molecular mechanisms of induced mutations are considered.

Studies with many restriction enzymes, which were performed to detect polymorphic DNA sites in human single-copy DNA, showed an especially high efficiency in the detection of such sites of enzymes recognizing the CpG sequence [1382]. The cytosines in this sequence were often methylated; there is also evidence from other sources, for example bacteria, that methylated cytosine may be a hot spot for spontaneous mutations.

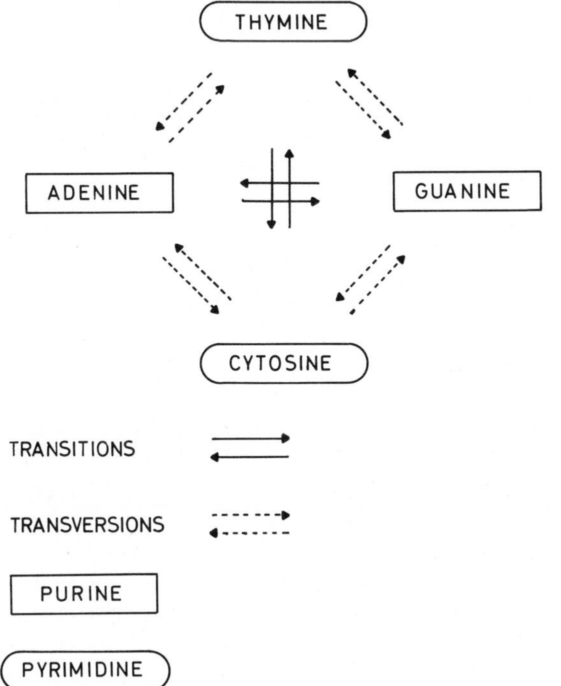

Fig. 5.25. Mechanism of mutations at the nucleotide level. Four transitions ⇆ and eight transversions ⇠⇢ are possible

5.1.4.4 Mutations in Microorganisms: Their Contribution to Understanding of Human Mutation

Mutations as Errors of DNA Replication. Data on humans point to a close relationship between mutation and cell division. Examination of this problem was triggered by the mechanism suggested by Watson and Crick (Fig. 5.26) [1347] and by early research in microorganisms, which suggested that many spontaneous mutations are, indeed, occurring during DNA replication, and erroneous introduction of the wrong nucleotide base will lead to a different base pair in future cell generations. Remarkably little new evidence on the mechanisms of *spontaneous* mutations has been accumulated in microorganisms in recent decades – in striking contrast to the huge amount of data now available on mechanisms of chemically induced mutation.

In view of the importance of spontaneous mutation for evolution (Sect. 7.2.3) and for all considerations on the biologic future of the human species, this lack of data is difficult to understand. The most obvious explanation is that evolutionary theorists are not usually concerned with planning of experiments in microbial genetics. Moreover, analyses of spontaneous mutation are in most cases much more time-con-

Table 5.16. Base substitutions in human hemoglobin variants. (Vogel and Kopun, 1977 [1681])

		All substitutions					Only nonfunctional substitutions		
		α chain	β chain	γ chain	δ chain	Sum total	α chain	β chain	Sum total
Transitions	A→G	1	10	0	0	11	0	0	0
	G→A	4	4	0	0	8	0	0	0
	T→C	7	16	4	1	28	7	8	15
	C→T	13	24	3	2	42	9	17	26
Sum total		25	54	7	3	89	16	25	41
Transversions	A→T	0	6	0	0	6	0	2	2
	T→A	2	4	0	0	6	2	3	5
	A→C	3	6	0	0	9	1	1	2
	C→A	4	5	1	0	10	2	2	4
	G→T	6	6	1	1	14	6	4	10
	T→G	4	10	1	0	15	2	6	8
	G→C	7	5	0	0	12	4	1	5
	C→G	9	6	0	0	15	6	3	9
Unspecified transversions		8	13	0	1	22	0	0	0
Sum total		43	61	3	2	109	23	22	45

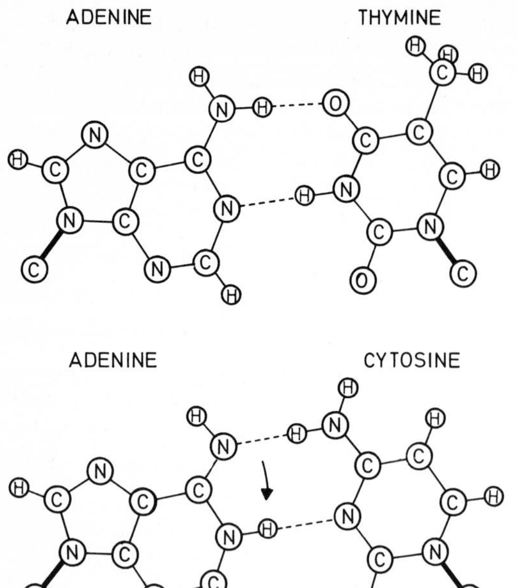

Fig. 5.26. Mechanism of point mutation by base replacement as suggested by the Watson-Crick model. A base may assume occasionally and for a short time a rare tautomeric configuration and pair with another base rather than with its usual partner, i.e. adenine with cytosine instead of with thymine. By the time of the next replication cycle, both bases will have attained their most probable configuration and will pair with their usual counterparts. Hence, the two double helices of the next generation are different; a point mutation has occurred [1347]

suming than are experiments with chemically or physically induced mutation. Yet, some data are available. As expected, amino acid substitutions due to spontaneous point mutation within the tryptophane synthetase A locus of *E. coli* are compatible with the exchange of one base, quite like the hemoglobin point mutations [1703].

Mutations in dividing and nondividing bacteriophage T4 rII mutants have been extensively examined [53]. The great majority of mutations has arisen by replication-dependent processes, and the majority are frameshifts. They are especially frequent in two hot spots; when these are excluded from consideration, the ratio of frameshifts to single base substitutions is reduced from 3.3 to 1.6. Evidence in favor of the replication dependence of mutations in bacteria has been presented by Kondo [1524].

We may conclude that in microorganisms many mutations, perhaps even the great majority, are replication dependent. Not all replication-dependent mutations, however, are caused by the exchange of one base: DNA replication also seems to enhance the risk of frameshift mutations. This indicates that the interpretation of paternal age effect and of the sex difference in human mutations (Sect. 5.1.3) by wrong base pairing during replication may be correct in many cases but does not necessarily explain all mutations.

For a long time, it appeared doubtful whether replication-independent spontaneous mutation occurred at all. Mutations were known to accumulate in nondividing gametes of *Drosophila*, in spores of *Neurospora*, and in the stationary phase of *E. coli*. Here, however, repair DNA synthesis may have complicated the matter. An unequivocal answer was finally provided by studies on bacteriophage T4 [53]. Neither repair nor replication occurs in free phage particles, yet certain rII mutants were observed to accumulate reverse mutations linearly with time. The mutation rate, however, was low compared to the rate during replication. Among the replication-independent mutations, frameshifts were rare, and transversions were probably more frequent than among replication-dependent mutations.

Mutator Genes. Demerec in 1937 [1435] described "unstable" genes in certain stocks of *Drosophila melanogaster.* Since that time, numerous examples of genetically determined, unusually high mutation rates have been observed in eukaryotes as well as in prokaryotes. Often, the enhanced mutability could be traced back to the influence of a "mutator gene." Analysis of the action of such mutator genes has provided valuable information on the interaction of various factors (polymerases, repair processes, etc.) [53, 1439, 1558] in the mutation process. For human point mutations in germ cells, no evidence for the actual occurrence of such mutator genes seemed to exist. Careful search for the extremely rare human families with two mutations would be of interest. For mutations in single cells, however, mutator genes have been identified (see below).

Mutation-Like Events Due to Extranuclear Entities Such as Viruses and Transposons? So far, spontaneous mutation has been discussed with reference to classical concepts, i.e., assuming a genetic change (base replacement; deletion; recombination) within nuclear DNA. However, in discussing more recent results on the structure of chromosomes and DNA in humans (Sect. 2.3), we also mentioned transposons – or "jumping genes." Their effects cannot be distinguished phenotypically from those of classical mutations. It is not known whether transposons play a role in the appearance and transmission of germ cell mutations in humans. But since they have been discovered not only in prokaryotes [520] but also in some eukaryotes (see Sect. 2.3.4), we regard their occurrence in our own species as likely.

Latent viruses can be transmitted vertically from generation to generation with no ill effects. They may, nonetheless, affect the physiology of their host. All King Edward potatoes carry the paracrinkle virus without pathologic lesions, but plants freed from the virus look different from the ordinary stocks and give a higher yield. In human in vitro systems certain pathogenic viruses, such as German measles virus, were shown to induce chromosomal aberrations [1386, 1479]. Such aberrations could also be induced in germ line chromosomes (1st meiotic division) of the male mouse [1655]. As long ago as 1963, Taylor [1654] discovered that the phage Mu of *E. coli* induces many gene mutations at various sites. Later, the Mu phages were identified as transposons. In some phages, most "spontaneous" mutations are, in fact, caused by transposons. More recently, it has been discovered that animal viruses such as SV 40 and polyoma virus are able to induce gene mutations in mammalian cells (Chinese hamster and mouse cell lines) [1463]. As already mentioned (Sect. 5.1.3.3), many mutation rates for dominant mutations increase strongly with paternal age. Some dominant mutations show a small parental age effect only – so small that it cannot be decided whether it is due to paternal, or maternal age, or both. In Fig. 5.19, the increased incidence of retinoblastoma has been shown as one example. Among the other dominant mutations for which a small increase only was shown are neurofibromatosis, tuberous sclerosis, and polyposis coli – all tumor diseases! Tumor induction in somatic tissues by viruses and the role of oncogenes will be mentioned in Sect. 5.1.6.5. It has been speculated, however, that for the above-mentioned tumor diseases transposable elements or viruses might play a role in producing mutationlike events in germ cells.

The interaction between mutation and viruses will again be mentioned in Sect. 5.1.6, where somatic mutation and cancer are dealt with.

5.1.5 Examination of Gene Mutations in Single Cells

With the success of genetic analysis in microorganisms, it appeared promising to study problems in human genetics in single cells. The development of this approach has been described in Sect. 4.2.2.1. Considering the low frequency of spontaneous mutations, and the technical obstacles to examining population samples of a sufficient size to establish even a crude order of magnitude at the level of the individual, such an approach would increase the resolving power of genetic analysis by several orders of magnitude.

Attempt at Examination of Mutations Occurring In Vivo. Atwood and Scheinberg [1378] worked out a technique that permitted removal by agglutination with an anti-A serum of all those blood cells from a blood sample that reacted with this anti-A serum, leaving only cells that did not react. In probands of blood group AB, some inagglutinable cells were found. These cells were interpreted to have lost the A but not the B antigen, suggesting that nonspecific antigen loss had not occurred. The relative frequency of these cells ranged in various individuals between 0.5 and 10.9/1,000 cells. These cells were interpreted to be somatic mutants. However, the very magnitude of the phenomenon made this interpretation unlikely. Moreover, additional immunologic observations with this system cast serious doubt on the mutational origin of the inagglutinable cells. This attempt remains interesting in spite of the fact that it was unsuccessful.

The problem of whether cells showing a biochemically or immunologically aberrant phenotype are indeed new mutants or are only the products of some secondary change not affecting the genetic material remains a central issue in mutation research on single cells. Recently, repeated attempts at showing hemoglobin variants in single red blood cells appear to have been successful [1642]

Examination of Mutant Cells In Vitro. The methods for cultivating normal human diploid cells in vitro have been described in Sect. 4.2.2.1 [1418]. One of the major difficulties involved in the study of mutation in such cells is the low mutation rate. To isolate the very few mutant cells from the great majority of normal ones, special selection methods are needed. The principle of such methods has been described in Sect. 4.2.2.6 for the Lesch-Nyhan syndrome (defect of the enzyme hypoxanthine-guanine phosphoribosyltransferase, HPRT). The cells are offered 8-azaguanine as a growth substance instead of hypoxanthine; 8-azaguanine when accepted by the normal enzyme will kill the cell. Only cells unable to metabolize this compound because of their enzyme defect in HPRT will survive.

Other selective systems are available, for example, in galactosemia, citrullinemia, and orotic aciduria. In a new and promising approach, red cells containing hemoglobin variants generated by presumed somatic mutation are being identified by use of specific antibodies against specific variant hemoglobins.

As noted above, a variant isolated in cell culture is not necessarily the product of a true genetic and transmissible alteration. At least two criteria are needed to confirm a mutation:

1. Demonstration of stability of the selected phenotype.
2. Positive results with the fluctuation test of Luria and Delbrück [1538], which is based on the principle that when a great number of cultures are compared, only a very few will have numerous mutant cells (mutation early in the development of the culture), whereas most cultures will contain very few mutants (mutation late in the development of the culture), and many others will have not even one mutation (Fig. 5.27).

Test methods for determining the spontaneous mutation rate are based on the fluctuation test; in Fig. 5.27 c, for example, every plate would be counted as one mutation if the colonies are clones from one cell each.

Table 5.17 shows a number of estimates for spontaneous mutation rates for the HPRT locus. At first glance, these mutation rates seem to be of the same order of magnitude as the germ cell mutation rates calculated at the phenotypic level for hereditary diseases (Table 5.8). These human mutation rates as ex-

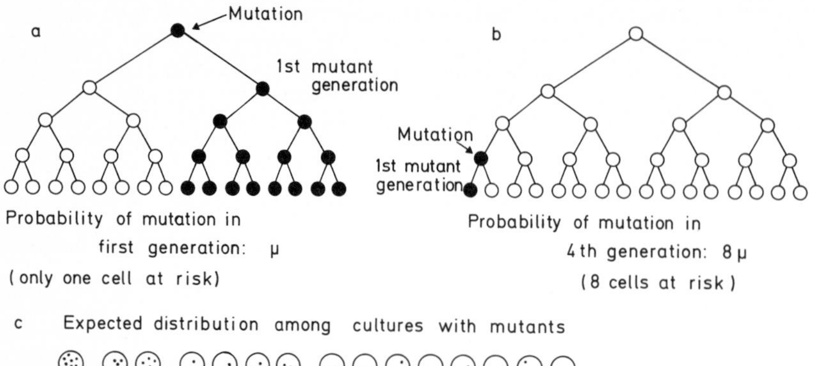

Fig. 5.27 a–c. Principle of the fluctuation test of Luria and Delbrück [1538]. If the mode of cell division is dichotomous, there are $2^1 = 2$ cells in the first generation after division has started; $2^2 = 4$ cells in the second; $2^3 = 8$ cells in the third generation, and so on. If the cell cultures to be examined all start with only one cell and if the mutation rates are identical in all cell generations, the relative number of cultures with 0, 1, 2 ... mutant cells can be calculated, depending on the cell generation in which the mutation has occurred. For example, if the mutation has occurred in the first division, half of all cells will be mutant (**a**). In this case, only one cell is at risk. On the other hand, if the mutation rate has occurred in the last division before the culture is examined, only one cell will be mutated (**b**). For such a mutation, however, 2^{n-1} cells are at risk, if n = number of cell generations. (**c**) Expected distribution among cultures

Table 5.17. Spontaneous mutation rate in human and Chinese hamster cells: resistance to 8-azaguanine. (Vogel and Rathenberg, 1975 [1683])

Cell line	Ploidy level	Rate
	A. Human	
D98[a]	Aneuploid	4.9×10^{-4}
L54[b]	Diploid	7.0×10^{-5}
Glen[b]	Aneuploid	7.0×10^{-5}
Fibroblasts[c]	Diploid	4.1×10^{-6}
	B. Chinese hamster	
237[b]	Hypodiploid	4.0×10^{-5}
V5[d]	Diploid	2.2×10^{-5}
V25[d]	Tetraploid	4.7×10^{-5}
V68[d]	Octaploid	1.9×10^{-5}
V79[e]		1.5×10^{-8}

[a] Data from Szybalski and Smith.
[b] Data from Shapiro et al.
[c] Data from De Mars and Held.
[d] Data from Harris.
[e] Chu et al.

plained in Sect. 5.1.3.2 represent a biased portion of traits with high mutation rates and are the end result of several dozen cell divisions. A comparison with somatic cell mutation rates, which are expressed as the number of mutations for a single cell division only, therefore is not appropriate.

Using the HPRT system, an elevation of the spontaneous mutation rate to $19–23 \times 10^{-6}$ mutants per cell per generation was described in fibroblasts from two patients with Bloom's syndrome, in comparison with a rate of $4.6–4.9 \times 10^{-6}$ in normal fibroblasts [1688]. Bloom's syndrome is a chromosome instability syndrome; hence, this gene is a human mutator gene. This higher mutation rate has also been demonstrated in vivo.

More recently, this increase in mutation rate has been confirmed for lymphocytes in Bloom's syndrome and for the other chromosome instability syndromes [1666; 1666a]. The figures are:

Normal control cells:	$2.0–\ 4.4 \times 10^{-4}$
Bloom's syndrome (7 patients):	$8.5–24.9 \times 10^{-4}$
Fanconi's anemia (2 patients):	$20.0–22.6 \times 10^{-4}$
Ataxia-teleangiectasia (1 patient):	8.5×10^{-4}

5.1.6 *Somatic Mutations*

Mutations may also occur in somatic cells. The effect of a somatic mutation is found in the descendants of the mutant cell, making the individual a mosaic. Mosaics are individuals with a mixed cell population. In the simplest situation, a normal cell population and a mutant cell population may coexist in a single individual. Often, such different populations may exist side by side. In fibroblasts – and possibly in other cell types – there is extensive mixing of genetically different cell groups since very small cell patches may be shown to exhibit different cell markers, such as different G6PD types.

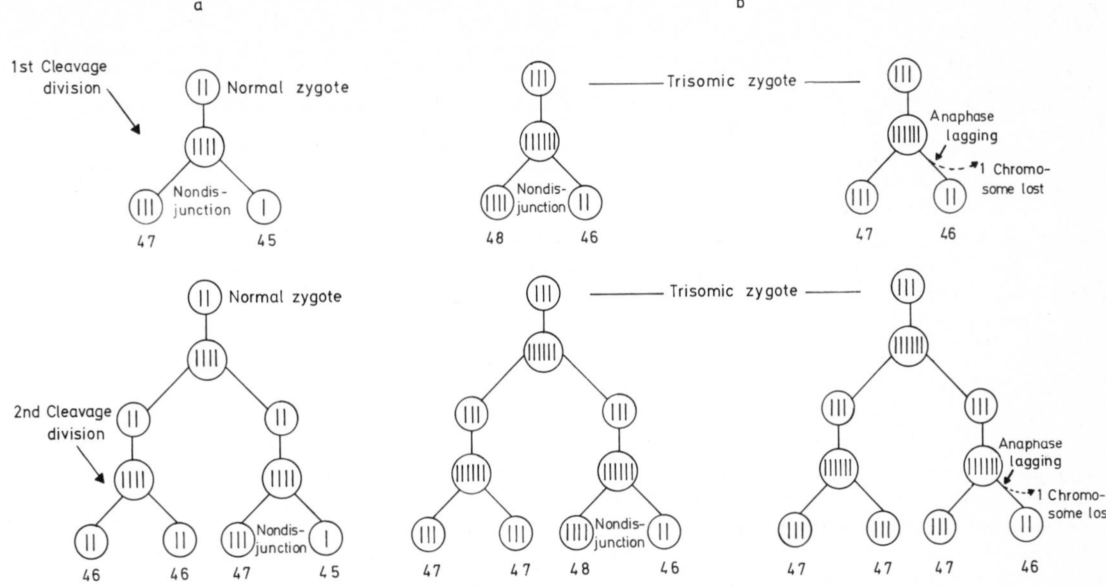

Fig. 5.28 a, b. Secondary nondisjunction and anaphase lagging as mechanisms for production of mosaics. **a** Normal zygote, secondary nondisjunction; **b** trisomic zygote, secondary nondisjunction, anaphase lagging

5.1.6.1 Formation of Mosaics for Genome Mutations

Mosaics for genome mutations are frequent. In Down's syndrome, for example, a ratio of 1 mosaic for 48 standard trisomy patients has been reported. An estimated population incidence of 1:650 for the Down's syndrome would thus result in a mosaic frequency of 1:31,000. Such mosaic cases also show dependence on maternal age but to a lesser degree than in simple trisomy 21 [483].

Mechanism of Mosaic Formation in Early Cleavage. An analysis of maternal age effects makes it possible to draw some conclusions regarding the origin of mosaics in Down's syndrome. A mosaic may be derived from a normal zygote. In such cases, nondisjunction would have to occur in an early (but not the first) cleavage division.[4] The monosomic product of the division is usually lost. Mosaicism can also be derived from a trisomic zygote. One cell strain would lose the extra chromosome by anaphase lagging, or nondisjunction may occur in a somatic cell (secondary nondisjunction; Fig. 5.28). The proportion of mosaics caused by each of these mechanisms can be estimated. With origin in a normal zygote no maternal age increase would be expected. With origin in a trisomic zygote, the increase in maternal age should be similar to that found for Down's syndrome in general. All mosaics represent a mixture of the two mechanisms; the average maternal age will depend on the fraction attributed to each cause. For 40 mosaics described in the literature, 20% were estimated as being derived from normal zygotes. From this calculation, a comparative estimate for the frequency of certain mitotic disturbances in normal and trisomic zygotes can be derived (Table 5.18). It follows that trisomic zygotes show an almost 40 times greater tendency toward anaphase lagging than normal cells, and nondisjunction is 70 times as frequent. These estimates, however, are only applicable to mosaics that evolve to a clinically recognizable Down's syndrome. This probability is much higher for zygotes originally trisomic than for those that were originally normal.

Mosaics with a small fraction of trisomic cells might occur at a later stage of development. They are often phenotypically normal or show only micromanifestations of Down's syndrome, such as abnormal dermatoglyphics. They may have children with Down's syndrome if a segment of the ovary or the testicle has the abnormal karyotype. The proportion of such minor mosaics among parents of children with Down's syndrome may be significant. The 1% recurrence risk of trisomic Down's syndrome conceivably could be related to gonadal trisomy 21 of this type.

[4] Nondisjunction in the first cleavage division would result in a trisomic and a monosomic division product. With loss of the monosomic cell, this could lead to standard trisomy.

Table 5.18. Estimated incidence of clinically recognizable mosaic mongolism arising from different types of zygotes and from different types of mitotic error. (Richards, 1969 [483])

Mitotic nondisjunction from normal zygote		Mitotic nondisjunction from trisomic zygote			Abnormal mitosis of zygote with structural anomaly involving chromosome 21 (unbalanced zygotes)[b]
Basis of estimate	Numbers	Basis for estimate	Mitotic non-disjunction	Anaphase lagging	
Relative incidence of mosaics from mitotic nondisjunction of a normal zygote among all cases of Down's syndrome	1/250[a]	Relative incidence of mosaics arising from nondisjunction or anaphase lagging in a zygote with trisomy 21	1/2,349	37/2,349	3/55
Incidence of Down's syndrome in population	1/650				
Estimated incidence of mitotic errors from normal zygotes	$1/250 \times 1/650 = 1/160{,}000$	Estimated incidence of mitotic errors from trisomy 21 zygote	1/2,300	1/60	1/18

[a] The analysis is based on 2,466 cases with Down's syndrome.
[b] The analysis is based on 250 observed mosaics. – G/G translocation seems to increase the risk for mitotic errors.

5.1.6.2 Hereditary Syndromes with Increased Chromosome Instability [1465; 1464; 1634]

Fanconi's Anemia (22765). Fanconi's anemia is a childhood panmyelopathy with bone marrow failure leading to a pancytopenia. Skeletal anomalies, especially of the thumb and of the radius, and hyperpigmentation are usually found; other malformations are frequent. The mode of inheritance is autosomal recessive. Genetic heterogeneity was suggested by analysis of age at onset [1638] and confirmed by mutual correction of chromosome instability after fusion of cells from patients with different clinical types [1707]. There is a more common type with onset in the first years of life, and a rarer one (see the description below) with juvenile onset. Complementation studies between cases according to different repair characteristics [1706] or different ethnic origin [1708] failed to identify additional genetic heterogeneity but recently, at least two different types have been identified by cell hybridization studies.

Schroeder et al., 1964 [1635] described two brothers with this disease, 21 and 18 years old. The parents and a younger brother (7 years) were healthy. The older brother showed, apart from normal karyotypes, metaphases with multiple chromosome aberrations, such as achromatic lesions (gaps), chromatid breaks; isochromatid breaks, acentric fragments, dicentric chromosomes, and chromatid interchanges; 19 of 39 metaphases showed at least one but in some cases multiple anomalies. Endoreduplication was seen in about 10% of all metaphases. The younger clinically unaffected brother showed a somewhat lower number of mitoses with chromosome aberrations but the same range of anomalies. Six years later, he developed clinical symptoms of the disease. He died of multiple hemorrhage at the age of 32. Autopsy revealed clinically unrecognized lung cancer [1634].

These were the first published cases of chromosome instability in a hereditary disease. The result was soon confirmed in other cases (Fig. 5.29).

Bloom's Syndrome (21090). Bloom's syndrome is a condition characterized by a low birth weight, stunted growth, sun-sensitivity of the skin and a facial butterfly-type lesion with telangiectasia. The mode of inheritance is autosomal recessive. Most families are of Ashkenazi Jewish origin. German et al. [1466], examining metaphases from blood cultures of seven patients, observed in six patients high frequencies (4%–27%) of cells with broken and sometimes rearranged chromosomes. Other cytogenetic anomalies described in Fanconi's anemia were present in Bloom's syndrome as well. However, the hallmark of Bloom's syndrome are symmetric quadriradial chromatid interchanges that are not seen in Fanconi's anemia. They presumably arose from chromatid exchanges between homologous chromosomes. In contrast, in Fanconi's anemia, asymmetric quadriradials are common, which are caused by random breaks of nonhomologous chromosomes. The frequency of sister chromatid exchanges (Sect. 2.1.2) in Bloom's syndrome was ten times higher than in normals or in patients with Fanconi's anemia. Although superficially somewhat similar, the basic mechanisms leading to Bloom's syndrome and Fanconi's anemia are quite different.

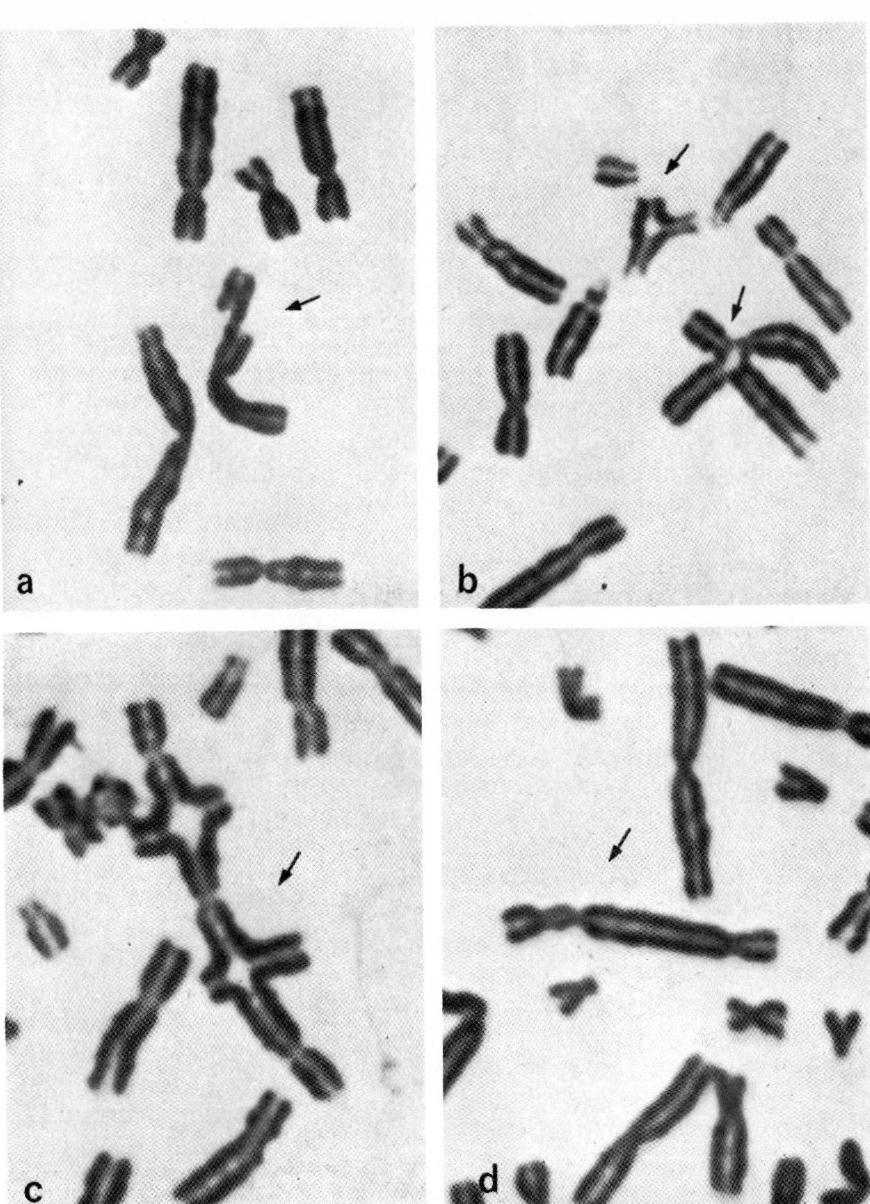

Fig. 5.29 a–d. Chromosomes from a patient with Fanconi's anemia. **a** Chromatin break; **b** two chromatin interchange figures with participation of nonhomologous chromosomes; **c** hexagonal interchange figure in which three chromosomes participate; **d** tricentric chromosomes. (Courtesy of Dr. T. M. Schroeder-Kurth)

Ataxia-Telangiectasia (20890) [1477]. The two constant clinical features of the ataxia-telangiectasia (Louis-Bar) syndrome are progressive cerebellar ataxia and oculocutaneous telangiectasia. Ataxia is usually recognized at the age of 12–14 months; the patient is confined to a wheelchair before adolescence. Various immune deficiencies have been reported that vary from patient to patient. The most common defect is a low level or complete absence of IgA. The mode of inheritance is autosomal recessive. Chromosome instability has been reported repeatedly; the number of breaks seems to be lower than in Fanconi's anemia and Bloom's syndrome [1396, 1469, 1485]. Breaks are apparently random. The level of chromosomal breaks often fluctuates. Chromosome analysis is complicated by the fact that phytohemagglutinin stimulation of lymphocytes is often impaired. Pseudodiploid clones are common, and a translocation involving the long arm of chromosome 14 is characteristic.

In all three conditions – Fanconi's anemia, Bloom's syndrome, and ataxia-telangiectasia – the increased chromosome instability is not an in vitro artifact but occurs in vivo as well. It is reasonable to assume that the clinical symptomatology of these conditions is directly related to chromosome instability. Moreover, chromosomes from patients with all three diseases show an increased sensitivity to chromosome breaking (=clastogenic) agents.

Chromosome Instability and Cancer. Patients with all three conditions have a strongly increased risk to develop malignant neoplasias. Many patients with Fanconi's anemia succumb during childhood and youth from bleeding or infections, but an increasing number of neoplasias are being reported [1465]. Up to 1981 45 cases had been collected; these included 22 acute leukemias, none of them lymphatic; 16 primary tumors of the liver; the rest carcinomas of other organs. However, these cases were observed only after steroid therapy had been introduced in the mid-1960s. It is not quite certain whether this is due to the life-prolonging effect of this therapy, or whether it *causes* cancer in these patients. A variety of malignant tumors were found in ataxia-telangiectasia [1465]. Among 108 patients, there were 48 with various non-Hodgkin lymphomas; 12 with Hodgkin's disease; 26 leukemias, mostly lymphatic; and 22 with other conditions (cancers of stomach, brain, ovary, skin, etc.). Hence, lymphatic neoplasias prevail. In Bloom's syndrome, of the 99 individuals known to suffer from this condition up to 1981, 23 developed at least one neoplasm. Considering the young age of these patients, a 100-fold increase of risk for neoplasia has been estimated. In distinction to ataxia-teleangiectasia, a great diversity of type and tissue distribution was observed.

It is reasonable to assume that the increased risk for development of neoplasias in these syndromes might be related directly to the increased rate of spontaneous chromosome breakage.

Such chromosome instability leads to many cells with various aneuploidies due to chromosome breakage. Most of these cells will die immediately, some will survive a few divisions. In an occasional cell, however, the structural defect provides a selective advantage in that the cell division rate is no longer inhibited. Such a cell will soon form a clone of genetically identical cells: the initial cancer cells. Due to their uninhibited growth, the abnormal cell clone will gradually replace the normal cells.

If such a cell clone contains a structurally abnormal chromosome, we should occasionally find specific chromosome aberrations in a certain fraction of cells of patients with one of the three syndromes with chromosome instability. Such cell clones have, indeed, been observed. Figure 5.30 shows a "marker chromosome," a 1p- that characterizes a clone from a patient with Fanconi's anemia that has been observed since 1974 [1636]. In Fig. 5.31 the fraction of metaphases with this marker chromosome are compared over several years. This clone probably had a

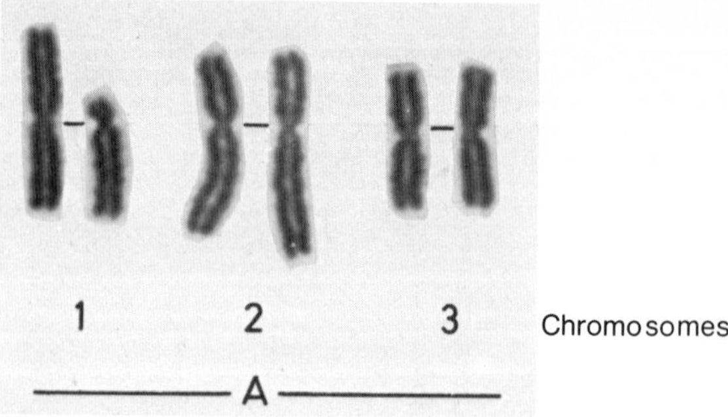

Fig. 5.30. Marker chromosome 1p- found in a cell clone from a patient with Fanconi's anemia. (Courtesy of Dr. T. M. Schroeder)

certain selective advantage which is, however, decreasing. Similar clones have been observed in the other two conditions as well. Full development of leukemia by gradual increase of a defined cell clone has been observed in ataxia-telangiectasia [1485]. Possible molecular mechanisms of malignant transformation in connection with chromosome breaks will be discussed in Sect. 5.1.6.5, on oncogenes.

5.1.6.3 Molecular Mechanisms of Chromosomal Instability and Tumor Formation Due to Somatic Mutation

Xeroderma Pigmentosum (27870–27875). Chromosome instability and the existence of marker chromosomes in the three syndromes with inherited chromosome instability suggests that repeated chromosome breakage may lead to cell clones that develop into malignancies. This raises the question of the molecular mechanism of chromosome instability. Another hereditary disease, xeroderma pigmentosum, has provided much information.

After exposure to ultraviolet light, the skin of the patients with xeroderma pigmentosum shows erythrema that is followed by atrophy and teleangiectasia (Sect. 3.1.3). Gradually, these areas become wartlike, and finally, skin cancer develops. It had been known from work in microorganisms that cells have an enzyme system that is able to repair defects of the DNA. Enzymatic repair of defects induced by ultraviolet light had been well analyzed in microorganisms at the molecular level. Xeroderma pigmentosum is characterized by an abnormally high sensitivity to ultraviolet light. Cleaver [1036, 1420] showed that this disease was caused by a defect in a DNA repair enzyme. Later, a number of different enzyme defects that lead to xeroderma pigmentosum-like phenotypes have been identified (Table 5.19).

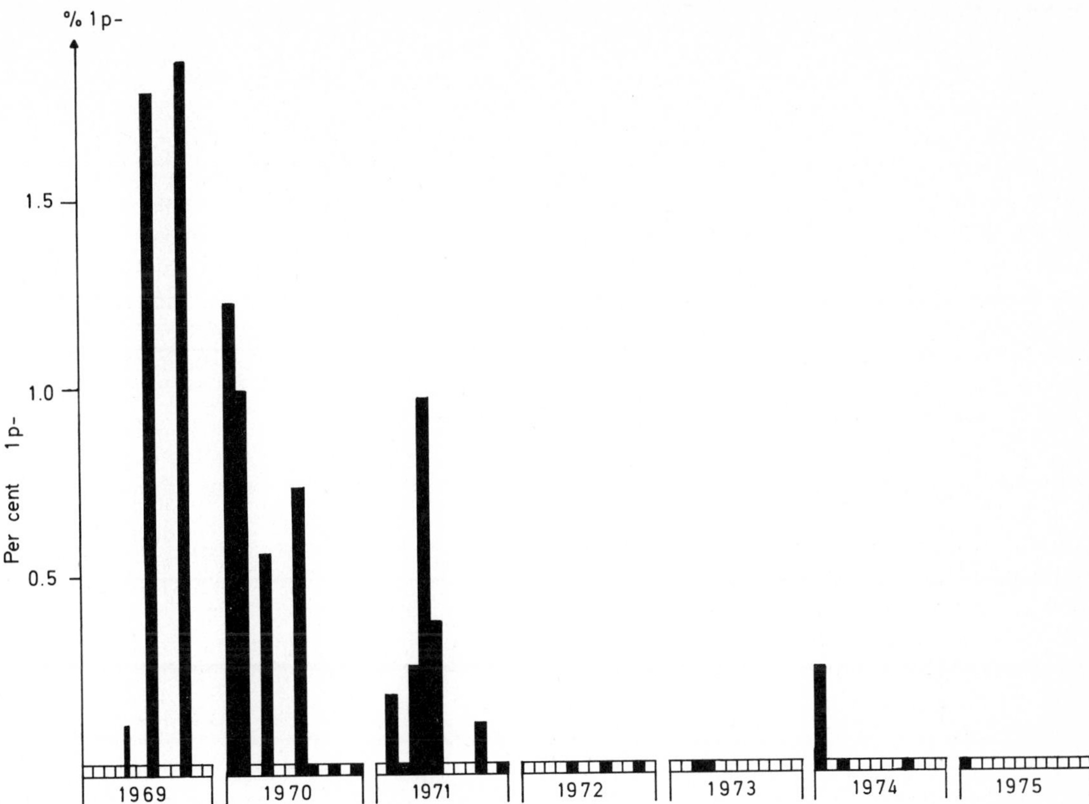

Fig. 5.31. Fraction of metaphases showing the 1p- marker chromosome of Fig. 5.30. Repeated examinations over many years. (Courtesy of Dr. T. M. Schroeder-Kurth).

Table 5.19. Distribution of complementation groups in XP studies [1450]

Country	Frequency of cases in the respective complementation groups								Number of patients
	A	B	C	D	E	F	G	Variant	
North America[a, b, c]	3	1	5	5	0	0	0	2	16
Europe and Great Britain[a, b, c]	10	0	14	8	2	0	2	5	41
Japan[d]	21	0	1	1	0	3	0	14	40
Egypt[b]	7	0	12	0	0	0	0	5	24
Germany	2	0	7	5	0	0	0	9	23
Number of analyzed cases	43	1	39	19	2	3	2	35	144

[a] Cleaver and Bootsma 1975; [b] Cleaver et al. 1981; [c] Kraemer 1980; [d] Takebe 1979.

Mechanisms of DNA Repair. Three main mechanisms of DNA repair have been analyzed in microorganisms: photoreactivation, excision repair, and postreplication repair (Fig. 5.32) [220, 1458].

1. Photoreactivation. Blue-violet light enhances the chance of survival of bacteria that had received prior ultraviolet light exposure. The main effect of ultraviolet irradiation is formation of thymine dimers from two neighboring thymines. Photoreactivation is due to an enzyme that restitutes these thymines by cleaving the dimers.
2. Excision repair. The second mechanism, excision repair, does not require light. In a first step, an endonuclease recognizes the dimers and cuts the affected DNA strand close to them, producing free ends. These free ends are recognized by an exonuclease that cuts off nucleotides, beginning at these free endings. Apart from the UV-induced dimers, up to 100 other nucleotides are removed. A polymerase induces resynthesis of the removed strand using the intact sister strand as template. Finally, the last gap between the resynthetized and the old strand is closed by a ligase.

3. Postreplication repair. If photoreactivation and excision repair are impossible, the damaged DNA strand cannot act as a template during replication since the dimer does not pair with any other bases; a gap remains in the newly synthetized complementary DNA strand. However, the genetic information distorted by dimer formation is available in a newly synthetized DNA strand along the old complementary strand. This information acts as a template for the production of another intact copy that replaces the damaged DNA strand. The exact mechanism of this replacement is still unknown; it seems to be similar to normal recombinational events. One normal DNA strand as a template is necessary for excision repair as well as for postreplication repair.

Enzyme Defects in Xeroderma Pigmentosum-Like Diseases (27870–27880). Cultured fibroblasts of patients with xeroderma pigmentosum (XP) were shown to have a reduced survival time after ultraviolet irradiation. Moreover, the survival of various UV-irradiated viruses grown in XP cells is less than in normal cells. This shows that the genetic defect of the host cell prevents it from correcting the defect in the virus genome. These results suggested a defect in one of the above-mentioned repair mechanisms. This was confirmed by direct investigation of these mechanisms in XP cells. Cells from different patients were found to be defective in their ability to perform excision repair. An early stage of excision repair, the ability to excise dimers, is deficient and this leads to a reduction of the insertion of new bases – repair replication (unscheduled DNA synthesis).

Genetic Heterogeneity [1420; 1421; 1450]. The enzyme system responsible for excision repair comprises a number of enzymes, and the clinical differences between the various XP patients suggest genetic heterogeneity – either by mutations within the genes for different polypeptide chains or at different sites within the same genes. One method for examining this problem is cell hybridization (Sect. 3.4.3) between fibroblasts from different pat-

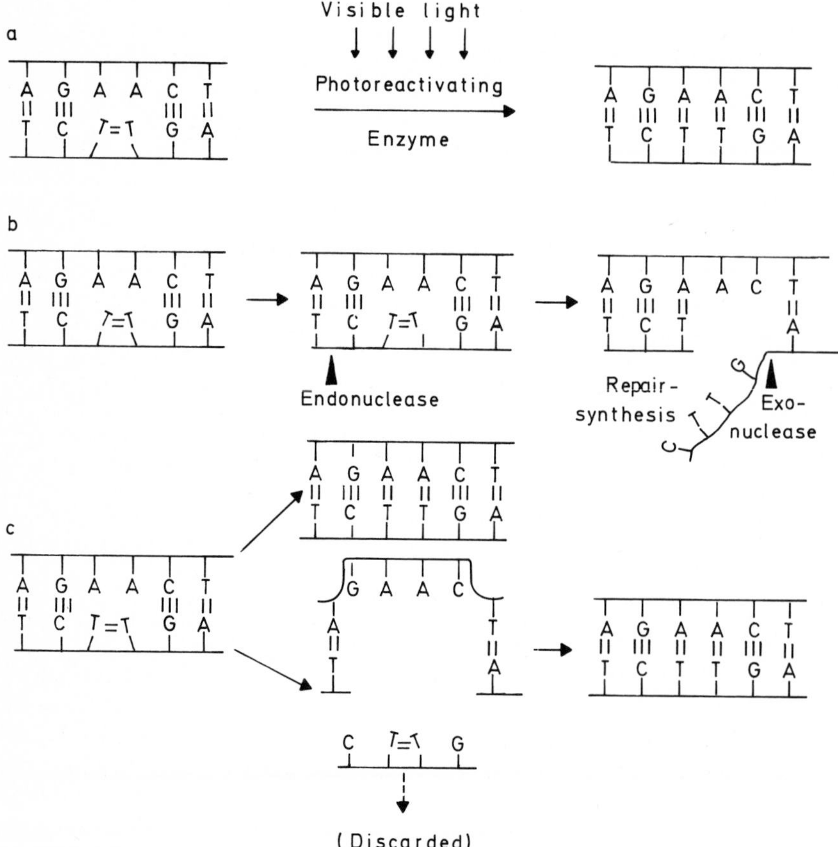

Fig. 5.32 a–c. Three types of DNA repair. **a** Photoreactivation, thymine dimers are reopened and A=T hydrogen bonds are re-established. **b** Excision repair, half-chromatid sequence containing the thymine dimer is excised, and a new half-chromatin is synthesized. **c** Postreplication (recombination) repair. A half-chromatid sequence is excised and repair occurs after replication, with assistance from the other division product

ients. The daughter cell of two fused cells will be able to carry out excision repair if their enzyme defects affect different loci. In this case, one genome provides one intact enzyme, the other genome produces the other enzyme, and the two defects are mutually compensated. If the enzyme defects are identical, even though different mutational sites are affected within the same gene, no such compensation is possible. In this way, at least eight complementation groups have been identified (Table 5.19). There are also clinical differences between the complementation groups: only patients of groups A, B, and D have additional neurological findings such as microcephaly, progressive mental deficiency, retarded growth and sexual development, deafness, ataxia, choreoathetosis, and areflexia (de Sanctis and Cacchione, 1932). Many XP patients with neurological manifestations do not show the full spectrum of symptoms. Even within the same complementation group, heterogeneity in the degree of neurological manifestation may be remarkable. In a number of patients diagnosed clinically as XP, excision repair has been found to be entirely normal. They are now classified as XP variant. In these patients, postreplication repair was shown to be deficient. Defects in photoreactivation have not yet been observed. Moreover, Table 5.19 shows a very unequal population distribution of these variants; for example, type A and the variant type are especially common in Japan, whereas type C, which is common in populations of European extraction, is rare in Japan.

Malignant Neoplasias in Patients with XP. Patients with XP sooner or later develop multiple malignant skin tumors. All cell types that are exposed to UV light may be involved. Basal and squamous cell carcinomas, malignant melanomas, keratoacanthomas, hemangiomas, and sarcomas may develop. Cancer formation can be prevented by minimizing exposure to UV irradiation with sun screens, ointments, or by avoidance of sunlight.

Increased Cancer Risk in Heterozygotes [1340]. The mode of inheritance of all three chromosome instability syndromes and of Xeroderma pigmentosum (XP) is autosomal recessive. Wherever enzyme defects have been identified, enzyme activity in heterozygotes is usually about half that found in normal homozygotes (Sect. 4.2.2.8). Therefore, it was reasonable to look for a possible increase of cancer risk among heterozygotes. The best evidence is now available for ataxia-telangiectasia (A-T) [1624, 1626]. The study was based on 27 families with 1639 close relatives of A.-T. patients. The data on cancer incidence in this group were compared with expectations calculated from the United States cause-, age-, time-, sex-, and race-specific mortality rates (Table 5.20). There is a definite increase of cancer mortality in the youngest age group of relatives (0–44 years) and in both sexes, in females more than males. In addition, a higher incidence of malignant neoplasms was also found in living relatives. There were many types of malignancies; especially, as in homozygotes, neoplasias of the lymphatic system, but also carcinomas of stomach and ovary.

Table 5.20. Observed and expected deaths from malignant neoplasms in heterozygotes [1624]

Age groups	Ataxia-teleangiectasia		Xeroderma pigmentosum	
	Observed	Expected	Observed	Expected
Male				
0–44	6	2.14		
45–74	21	19.02	38	35,4
75+	4	5.22		
All ages	31	26.38		
Female				
0–44	9	2.97		
45–74	23	18.34	30	33.2
75+	4	5.00		
All ages	36	26.31		
Both sexes and all ages	67	52.7	68	68.6

The incidence of homozygotes for A-T in the population was estimated at about 1:40000, corresponding to a heterozygote frequency of ≈1%. In this case, it was estimated that "A-T heterozygotes might comprise over 5% of all persons dying from a cancer before age 45 and about 2% of those dying from this cause between ages 45 and 75." In addition to the cancer risk, A-T heterozygotes might also have a somewhat increased susceptibility to diabetes, severe scoliosis, and neural tube defects. [1624]

Another study on cancer disposition was performed in close relatives of Xeroderma pigmentosum patients. There is no *general* increase in cancer mortality. However, an increased incidence of (non-fatal) nonmelanoma skin tumors was found in these heterozygotes [1316]. This study was based on 2597 close relatives of XP patients from 31 U.S. American families. Interestingly enough, an increase of skin cancers was found only for those living in the southern states of the United States, where exposure to sunlight is extensive. This, together with the negative result for all other cancers excepting skin carcinomas, suggests a specific defect in UV repair in epithelial cells manifesting only when the skin is exposed extensively to sunlight, that is to say, an ecogenetic phenomenon (Sect. 4.5.2).

In Fanconi's anemia, on the other hand, careful assessment of the available evidence failed to reveal any increased cancer risk of heterozygotes.

Molecular Mechanisms in Syndromes with Enhanced Chromosome Instability. Formation of thymine dimers only affects one of the two DNA sister strands. It therefore will not lead immediately to a chromosome gap or break. If the dimer cannot be excised, however, it cannot act as template in the next replication; the complementary DNA strand will be incomplete, and a break will become visible in the second replication cycle (Fig. 5.33). Hence, if interrup-

Table 5.21. Neoplasms with a known consistent chromosomal defect. (Adapted from [1704])

Disease	Chromosome defect	Breakpoints or deletion
Leukemias		
Chronic myelogenous leukemia	t (9; 22)	9q34.1 and 22q11.21
Acute nonlymphocytic leukemia		
M1 (Acute myelogenous leukemia)	t (9; 22)	9q34.1 and 22q11.21
M2 (Acute myelogenous leukemia)	t (8; 21)	8q22.1 and 21q22.3
M3 (Acute promyelocytic leukemia)	t (15; 17)	15q22 and 17q11.2
M4 (Acute myelo-monocytic leukemia	inv 16	p13.2 and q22
M4, M5 (Acute nonlymphocytic leukemia)	t (9; 11)	9p22 and 11q23
M1, M2, M4, M5, M6 (acute nonlymphocytic leukemia)	del 5q	5q22q23
	del 7q	7q33q36
	+ 8	
	+12	
Chronic lymphocytic leukemia	t (11; 14)	11q13 and 14q32
Acute lymphocytic leukemia		
L1-L2	t (9; 22)	9q34.1 and 22q11.21
L2	t (4; 11)	4q21 and 11q23
L3	t (8; 14)	8q24.13 and 14q32.33
Lymphomas		
Burkitt's, small noncleaved cell	t (8; 14)	8q24.13 and 14q32.33
Follicular small cleaved, mixed, and large cell	t (14; 18)	14q32.3 and 18q21.3
Small cell lymphocytic	+12	
Small cell lymphocytic, transformed to diffuse large cell	t (11; 14)	11q13 and 14q32
Carcinomas		
Neuroblastoma, disseminated	del 1p	1p31p36
small cell lung carcinoma	del 3p	3p14p23
Papillary cystadenocarcinoma of ovary	t (6; 14)	6q21 and 14q24
Constitutional retinoblastoma	del 13q	13q14.13
Retinoblastoma	del 13q	13q14
Aniridia; Wilm's tumor	del 11p	11p13
Wilm's tumor	del 11p	11p13
Benign solid tumors		
Mixed parotid gland tumor	t (3; 8)	3p25 and 8q21
Meningioma	−22	22

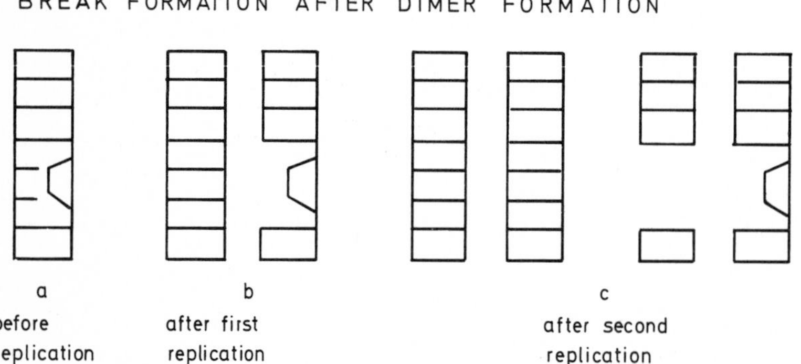

Fig. 5.33 a–c. Defect of excision repair. If thymine dimers formed by UV irradiation cannot be excised, they cannot act as templates in the next replication cycle. This leads to a chromosome break. **a** Dimer formation; **b** at first replication, no complementary bases are assembled opposite to the thymine dimer; **c** second replication: discontinuity of DNA structure in one division product is the result

tion of the DNA double helix is related to microscopically visible chromosome breakage we expect an increase in the number of chromosome breaks after irradiation of XP cells as compared with normal cells. This increase has, indeed, been described. In unirradiated XP cells, on the other hand, no chromosome instability has been observed in contrast to the three syndromes described before (Fanconi's anemia, Bloom's syndrome, and ataxia-telangiectasia), associated with spontaneously enhanced chromosome instability. Hence, the molecular defects are probably different. However, it is reasonable to assume that some of the mechanisms of DNA replication and repair might be involved in these conditions as well. Some observations seem to corroborate this conclusion.

Despite many efforts, however, the molecular basis of any of these syndromes has so far eluded deeper understanding. Many different mechanisms have been suggested, such as defects of various repair enzymes, reduced supply of enzymatic cofactors, an abnormality of transport of enzymes necessary for DNA replication such as topoisomerases through the nuclear membrane, or defects in energy supply. Some interesting details of possible significance have been observed, such as a retarded DNA chain growth in Bloom's syndrome. But the basic defects are still unknown. Their elucidation would be important, since these syndromes are key models for the elucidation of the molecular mechanisms of spontaneous mutations in general and of somatic mutations in their relationships to neoplasia in particular.

Chain of Events in the Formation of Malignant Neoplasias by Somatic Mutation [1633]. The chain of events that lead to the formation of neoplasias by somatic mutation is depicted in Fig. 5.34: The first step is damage to DNA. It might be caused either by internal factors, such as genetically defective replication or repair mechanisms, or by outside influences, such as ionizing radiation, chemical mutagens, or viruses. The DNA damage might lead to a complete breakdown of the replication mechanism and might

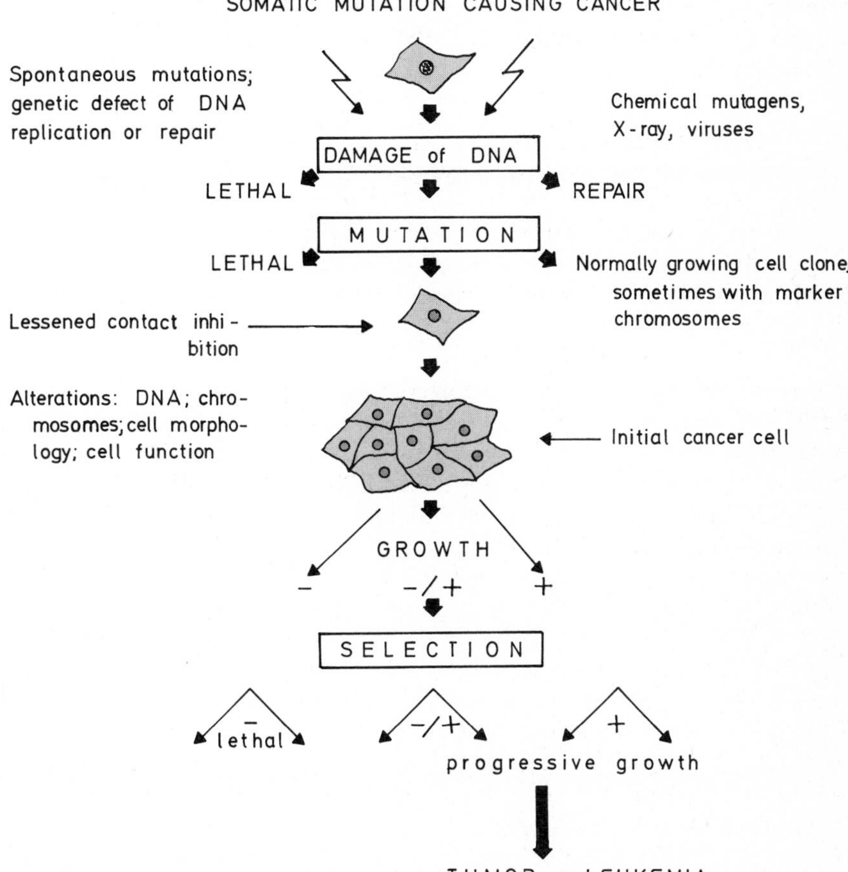

Fig. 5.34. Gradual development of a malignant tumor. A somatic mutation, such a chromosome break, may produce a cell clone with a selective advantage. This clone may develop gradually into a malignant tumor. (Adapted from Schroeder, 1972 [1633])

therefore be lethal. On the other hand, it might – and, in very many cases, will – be repaired. As a second possibility, a mutation might be formed. Here, it is of no principal importance which type of mutations is formed. For example, it may be a point mutation by one single base exchange, or a visible chromosome anomaly. This mutation may be lethal, leading to death of the affected cell clone due to a selective growth disadvantage in competition with normal cells, or it may show normal cell growth, and a marker chromosome may be the only indicator of the mutation. As a third possibility, the new cell clone may have a selective advantage due to a genetic deficiency in normal mechanisms of growth inhibition and regulation. In spite of varying fractions of cell death, cancer develops. Secondary genetic processes such as formation of additional aneuploidies may contribute to cell death or may occasionally lead to cell clones with stronger selective advantage. Growth of neoplasia therefore occurs unchecked until the affected individual dies from interference with normal functioning. This general formulation of chromosomal involvement in neoplastic origin was already developed in the early part of this century by Boveri (1914; see [77]) but had to await the development of better cytogenetic techniques in more recent years.

5.1.6.4 Other Observations Pointing to Somatic Mutation as a Mechanism in Carcinogenesis [1520]

History of the Somatic Mutation Hypothesis of Cancer. The hereditary syndromes with enhanced chromosome instability and with deficiencies of DNA replication and repair are remarkable, because they offer models for molecular mechanisms of somatic mutation and tumor formation. However, the hypothesis that cancer may be due to somatic mutation is much older. Von Hansemann (1890) [1475], on the basis of his studies on mitosis, postulated that the cell of the malignant tumor is a cell with an abnormal chromatin content [77]. Boveri (1914) specified this concept further, assuming unequal distribution of the chromosomes of a cell to its daughter cells. He emphasized, however, that the abnormal chromatin constitution as such, and not the mechanism of its production, is important. During the decades to follow, the somatic mutation hypothesis was elaborated by many authors and discussed from many aspects. Burnet (1957, 1974) [1408] most clearly formulated its most important consequences:

a) Neoplasias must be monoclonal, i.e., they must originate from one single cell.
b) Their incidence can be enhanced by chemical agents or viruses that react with DNA.
c) In a large population of proliferating cancer cells, additional mutations are to be expected in single cells. They will produce cells having additional selective advantages; subclones derived from these cells will rapidly overgrow the other tumor cells.
d) The mutation hypothesis also explains the increase in incidence of most cancers with age, if somatic mutation can – as a first approximation – be regarded as a time-dependent process. Besides, clones need a number of years to grow before they manifest clinically.

Induction of cancer by chemicals that react with DNA will not be discussed further here. Suffice it to mention that many chemical mutagens (Sect. 5.2.2) have indeed been shown to be carcinogenic as well. Malignant neoplasias usually show a variety of subclones with different karyotypes, indicating multiple anomalies of mitosis during tumor proliferation. Increase of neoplasia with age is a well-known general feature of cancer biology. Evidence from many different sources is now available to prove monoclonal origin of tumors. It is known, for example, that one B-lymphocyte produces only one specific type of γ-globulin light chain – a λ or κ chain. Different B-lymphocytes, however, synthesize light chains that differ in the "variable" part of the amino acid sequence (Sect. 4.4). All cells in myelomatosis – a malignant disease – on the other hand, produce light chains with identical variable parts. Uterine muscle cells of women who are heterozygous for two X-linked G6PD variants (Sect. 2.2.3.3) are mosaics of cells expressing one or the other of these variants, as expected from random inactivation of X chromosomes. Fibroid tumors of the uterus, on the other hand, always show the same variant in all tumor cells. Similar observations are available for other tumors [1519]. Most tumors are in fact monoclonal in origin. Some hereditary tumors such as neurofibromatosis have a multiclonal origin suggesting that the proliferation tendency is inherent in every cell [1449].

Virus Etiology Versus Somatic Mutation? There are now many observations, especially in animals, that tumors may be caused by viruses, and it is reasonable to assume that some human tumors have a viral origin as well. This hypothesis does not contradict the somatic mutation hypothesis. Viruses are often site-specific and may induce a mutational event in the chromosome. The course of tumor formation following viral damage may then be similar to that described for any kind of somatic mutation.

Neoplasias with Constant Chromosomal Aberrations. In a few neoplasias, a single specific chromosome aberration limited to the tumor tissue had been described many years ago. The classic example is the Philadelphia (Ph^1) chromosome that is almost regularly associated with chronic granulocytic leukemia [1584]. A translocation of much of the long arm of chromosome 22 to chromosome 9 is usually found [1478] (Fig. 5.35). Such patients are normal chromosomally in all tissues except for the hematopoietic system. In fact, the characteristic chromosome abnormality affects all blood cell precursors including megakaryocytic and erythropoietic cells. Biologic tagging techniques by G6PD variants has led to similar insights. The clinical and hematologic consequences of the translocation, however, only affects the granulocytic elements of the blood, showing that a given "mutation" while present in several cell types may only affect the growth pattern of a single differentiated tissue.

In some exceptional families, several members died from chronic granulocytic leukemia, and in one family [1488] some younger individuals showed the Ph^1 chromosome in their hematopoietic cells without clinical signs of leukemia. In this family, the susceptibility to breakage of chromosome 22 was inherited in an autosomal-dominant manner.

Meningiomas are tumors of the meningeal coverings of the brain. They are histologically benign and show in most cases hypodiploid cell lines in which one chromosome 22 is lacking or has a deletion. Subsequent steps in subclone formation affect other chromosomes [1709-1711, 1545].

With the advent of the new technique of high-resolution banding (Sect. 2.1.2.3), and improvement of methods for preparation of single cells from solid tumors for chromosome studies, such specific chromosome aberrations have been discovered for an increasing number of other neoplasias [1704]; in most cases either a band (or subband) deletion or a recip-

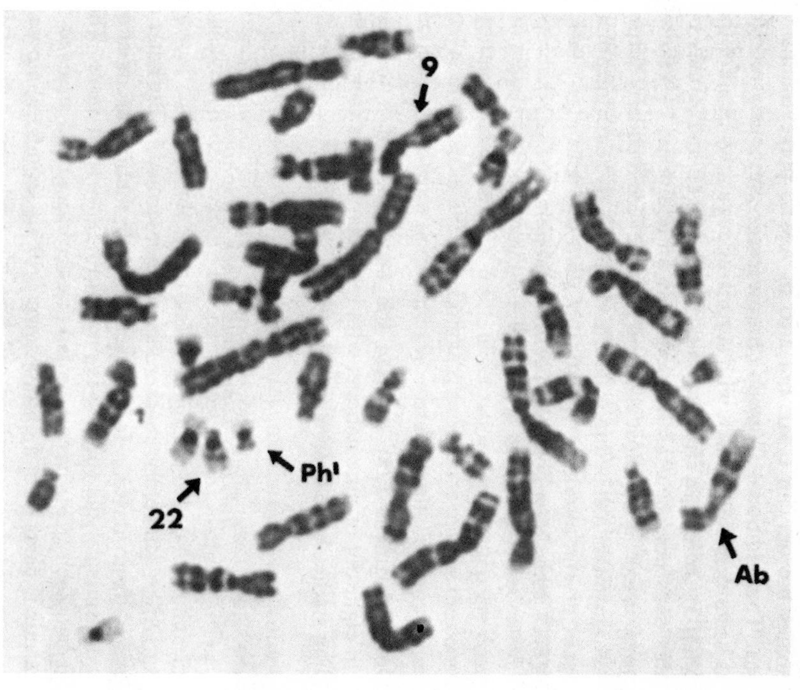

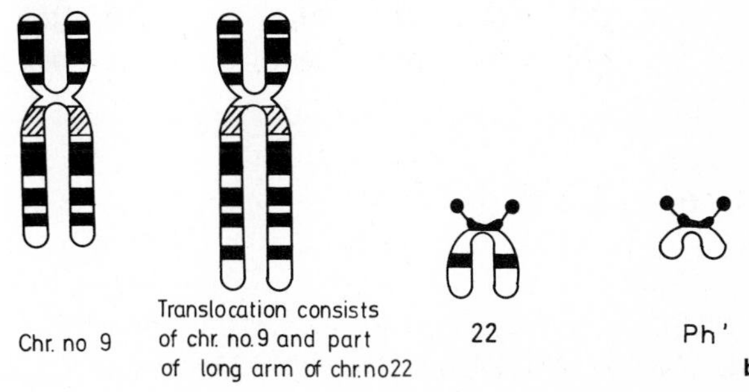

Fig. 5.35 a, b. Philadelphia (Ph^1) chromosome in a patient with chronic granulocytic leukemia. **a** Translocation between chromosome 22 and 9; Ab aberrant chromosome. **b** The participating chromosomes and the translocation products are shown diagrammatically. (Harnden, 1977 [1478])

rocal translocation between two chromosomes has been identified one of which is always the same in the same type of tumor and breaks are at a specific site, whereas the other one might vary. Table 5.21 shows neoplasias for which specific translocations and breakpoints have been identified. It is quite possible that this aberration is present in tumors where no such chromosome aberration has been discovered, but is so small that even high-resolution banding is inadequate for its detection. But why do only certain chromosomal aberrations lead to a selective advantage of the cell, and why is this advantage tissue-specific? In recent years, molecular biologists have found a partial answer to the first question (Sect. 5.1.6.6). But before these results are reported, the example of a solid tumor will be discussed.

Retinoblastoma (18020). Retinoblastoma is an ophthalmologic cancer of children and occurs as an inherited as well as a noninherited type [1669; 1676]. The inherited type shows an autosomal-dominant mode of inheritance with about 90% penetrance; the penetrance varies somewhat between families [1548]. Approximately 68% of all inherited cases are bilateral, the rest are unilateral. In some of the bilateral cases, more than one primary tumor in the second eye was observed [1519-1522]; in unilateral cases, inherited as well as noninherited, multiple origin of tumors is not usually observed.

Many cases of retinoblastoma are sporadic, i. e., they are the first affected in an otherwise healthy family. Within this group, only ~20%-25% are bilateral. Sporadic cases belong to either of two groups: dominant new mutations with 50% affected offspring and nonhereditary cases. All bilateral sporadic cases are new mutations; the segregation ratio among their offspring is not far below 50% [1628]. Of the unilateral sporadic cases, about 10%-12% are new mutations, the rest are nonhereditary cases. The consequences of this situation for genetic counseling have been described [1676] (see App. 8 for example).

Two Mutational Steps in the Inherited Type. In the inherited type, two mutational steps have been assumed. In a first step, a germ cell mutation leads to alterations of chromosome structure that enhance the risk for a somatic mutation to transform one or a few of the altered cells into retinoblastoma cells.

To produce noninherited retinoblastoma, two somatic steps have been regarded as necessary: The first step may turn the cell into a potential retinoblastoma cell; as a result, its state might now be identical with that of all cells in hereditary retinoblastoma. This means that the tumor develops only if a second step follows that might be identical with the one step necessary in the hereditary variety.

As long ago as 1963, a 13q deletion was described in all cells of a patient with bilateral retinoblastoma and (not very impressive) additional constitutional abnormalities [1531]. In recent years, and especially since high-resolution banding has been introduced, an increasing number of such patients have been discovered; in many of them only a very small segment of the long arm of chromosome 13 was deleted or involved in a reciprocal translocation (Fig. 5.36). By

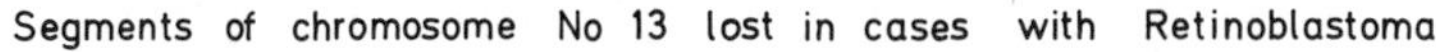

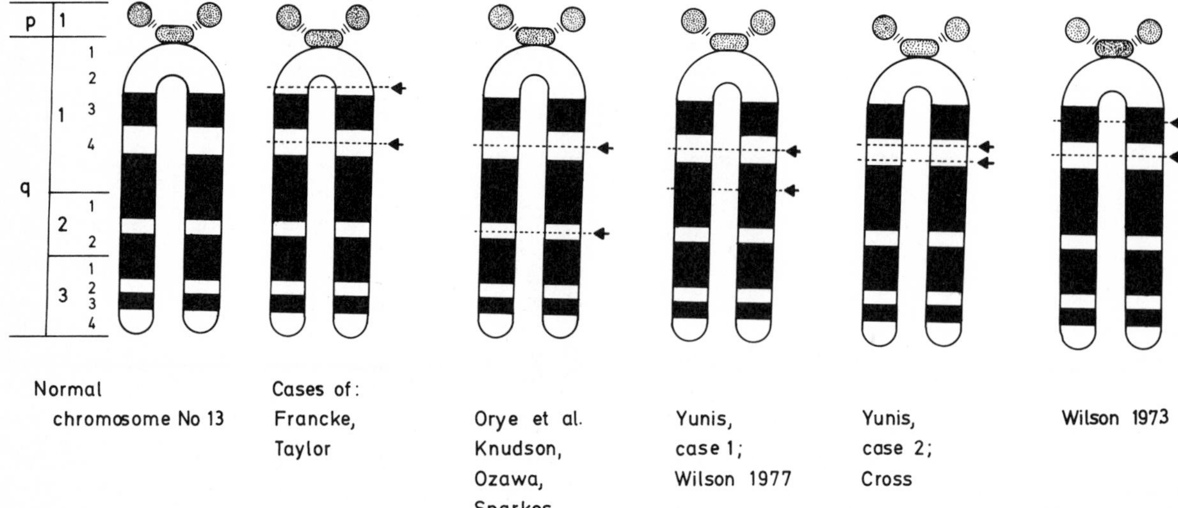

Fig. 5.36. Deletions in patients with retinoblastoma. (From Vogel, [1676])

comparison of many such observations, the deleted segment could be identified as 13q14; (or even 13q14.13; Fig. 5.36). The retinoblastoma "gene" (if we want to call this chromosome segment a gene), is closely linked to a gene for esterase D (ESD).

These observations raised another question: Might the second step – that leading to the development of a malignant cell clone – consist in deletion of a chromosome segment in the region 13q14? In this case, all tumor cells should show this deletion. For technical reasons, examination of tumor cell chromosomes had long been difficult; moreover, cells of solid tumors are intermingled intimately with cells from connective tissue. This might explain why an early study reporting 13q changes in tumor cells was not confirmed sooner and contradictory results were published. More recently, however, involvement of the 13q14 region has been confirmed in many cells from patients with the inherited as well as the noninherited variants [1380].

However, even if this result could be confirmed in most instances it would not necessarily solve the problem of manifestation in only a few of the cells containing the mutation. This problem has recently been solved in part by methods of molecular biology (Sect. 2.3) [1389, 1414, 1467], using DNA restriction fragment polymorphisms closely linked to the retinoblastoma gene. According to these studies, the decisive step occurred in the homologous chromosome carrying the normal allele. Sometimes, restriction markers from one parent were lacking in the tumor cells altogether, indicating that the chromosome from this parent had disappeared and been replaced by a second copy of the chromosome carrying the mutant retinoblastoma gene; in other cases, recombinational events involving only a part of the normal chromosome led to the same consequence: The mutation was made homozygous (Figs. 5.37, 5.38). The significance of this result might extend beyond the specific case in which it has been achieved: It suggests that events such as nondisjunction or recombination might be much more common in somatic tissue than hitherto expected. This might explain various phenotypic peculiarities and even diseases. More specifically, it might offer explanations for other, autosomal-dominant tumor diseases.

Genetic Syndromes Associated with Tumors. Osteosarcoma frequently occurs in patients with retinoblastoma. In such cases, as in retinoblastoma patients, homozygosity of the identical locus at chromosome 13 has also been demonstrated [1525]. Homozygosity of an identical site thus can cause either retinoblastoma or osteosarcoma. In Wilm's tumor, as predicted by Knudson, a site on chromosome 11, viz. 11p13 is rendered homozygous by similar mechanisms. Here an occasional autosomal dominant syndrome characterized by aniridia, genitourinary abnormalities (gonadoblastoma, ambiguous genitalia) and mental retardation (WAGR) is found. The same locus appears to be involved in the Beckwith-Wiedemann syndrome, which is associated with a variety of embryonal tumors (Table 5.23). The common mechanism of these various tumors also involves so-

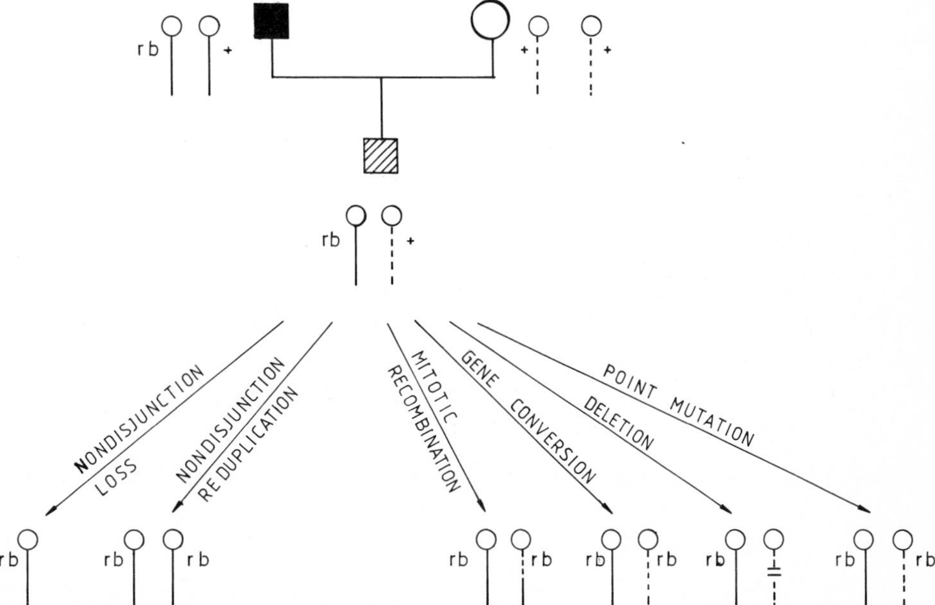

Fig. 5.37. Chromosome mechanisms which could lead to homozygosity of the retinoblastoma (Rb) allele in the somatic tissue of a heterozygote for this allele. (From Cavenee et al., 1983)

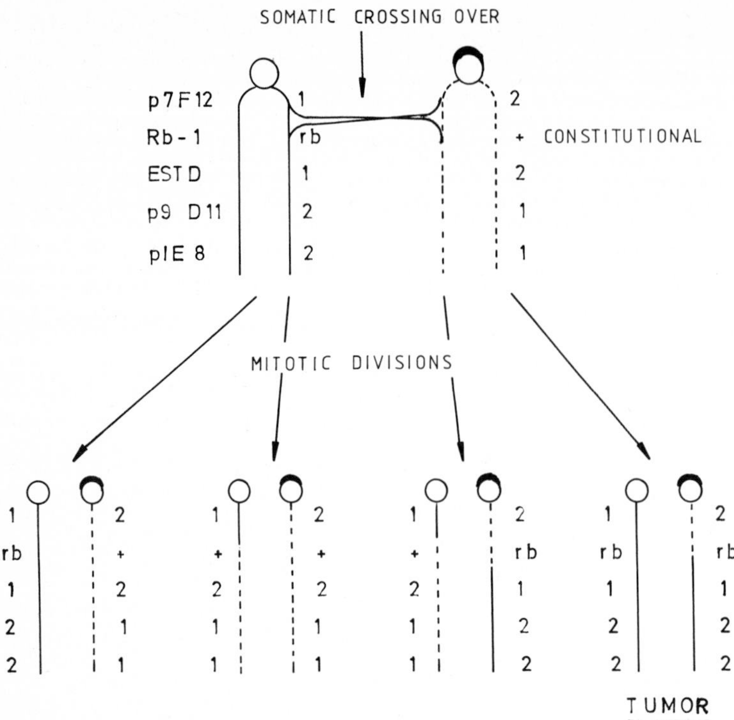

Fig. 5.38. Homozygosity of the Rb allele effected by a mitotic recombinational event (crossing over between two of the four chromatids in the G_2 phase). *ESTD,* esterase D locus; p7F12, p9D11, pIE8, DNA fragment length polymorphisms. The cap on the chromosome with the wild allele on the Rb locus represents a C-stained heterochromatic marker. (From Cavenee et al., 1983 [1414])

Table 5.22. Localization of proto-oncogenes on human chromosomes. (From [1694])

Proto-oncogene	Localisation	Proto-oncogene	Localisation
N-*ras*	1 cen-p21	*mos*	8q22
B-*lym*	1p32	*myc*	8q24
ski	lq12-qter	*abl*	9q34
N-*myc*	2p23-pter	H-*ras*-1	11p15
raf-1	3p25	*ets*	11q23-q24
raf-2	4	K-*ras*-2	12p12-pter
fms	5q34	*int*-1	12q14-qter
K-*ras*-1	6p23-ql2	*fos*	14q21-q31
myb	6q22-q24	*fes*	15q25-q26
erb-b	7	*erb*-A	17
met	7p11·4-7qter	*src*	20
		sis	22q12-q13
		H-*ras*-2	X

Table 5.23. Some genetic syndromes associated with tumors

Genetic syndrome	Tumor	Chromosome
Beckwith-Wiedemann syndrome	Hepatoblastoma Rhabdomyosarcoma Wilm's tumor Adrenal carcinoma	11p
Aniridia, genitourinary abnormalities, retardation, mental (AGR syndrome)	Wilm's tumor	11p13
Neurofibromatosis	Astrocytoma Sarcomatous transformation	
Multiple endocrine neoplasia (type II)	Medullary thyroid cancer	
Basal cell nevus syndrome	Medulloblastoma Astrocytoma Ovarian cell cancer Hamartoma	

matic development of homozygosity for a site on chromosome 11 [1525]. The allelic site presumably was inherited from one parent.

Still other tumors occur as noninherited, solitary evens as well as dominantly inherited types with multiple tumors. In skin tumors, this is a well-known pattern [1631]: Solitary tumors are usually noninherited, whereas multiple tumors of the same histologic type are dominantly inherited. Examples are neurofibromatosis versus single neurofibromas, multiple versus single lipomas, multiple versus sporadic cutaneous leiomyomas, multiple versus sporadic glomus tumors, or the basal cell nevus syndrome versus single basal cell naevi [1391].

Similarly, Knudson has adduced epidemiologic and statistical evidence that similar mechanisms may apply in a variety of other childhood tumors, such as neuroblastoma, and pheochromocytoma. In these conditions, both sporadic and hereditary cases are postulated. The hereditary cases are more often familial, are frequently bilateral, occur at an earlier age, and involve more than one site (i. e., both adrenals in neuroblastoma) [1519, 1522; 1549]. In the future, gene assignments for various genetic syndromes associated with tumors may be achieved by identification of chromosomal sites that undergo specific losses of constitutional heterozygosity during oncogenesis. Some examples of such genetic syndromes and the accompanying tumors are given in Table 5.23.

"Cancer families" are known for other malignant tumors. Carcinoma of the esophagus in many patients of a family with a special type of tylosis palmaris and plantaris has been discussed in Sect. 3.6 (Fig. 3.68). In this as well as in other examples of dominant inheritance of an increased liability for neoplasia, the tumors tend to occur earlier in life within these families than in the more common sporadic variety of the same cancer. There often is an increased tendency for multiple occurrences in the same individual. Again, these observations suggest a genetically determined and dominantly inherited increased probability for malignant transformation.

5.1.6.5 Oncogenes [1686; 1690; 1691; 1696]

Basic Principles. In recent years, research in molecular biology and especially the discovery of so-called oncogenes has helped our understanding of molecular mechanisms of cancer formation. Early studies with cell fusion (Sect. 3.4.3) suggested that mutations at certain gene loci might be important for malignant transformation, and treatment of hamster cells with polyoma viral DNA resulted in malignant transformation. On the other hand, the ability of some RNA viruses, the retroviruses, to induce tumors in animals had been known since the work of Ellermann and Bang (1908) and P. Rous (1911). These and later results caused many research workers to look for retroviruses that could be implicated in the causation of human tumors. Up to the end of the 1970s these studies failed to be convincing. In recent years, however, introduction of new methods from molecular biology has led to new and important insights.

The genome of a retrovirus consists of an one-stranded RNA. It comprises the following areas of information (from the 5' to the 3' end): 5'-regulating sequence; genes for proteins necessary for the internal structure; reverse transcriptase gene; genes for surface glycoproteins; 3'-regulating sequence. As soon as the virus particle enters the cell, the reverse transcriptase produces a double-stranded DNA copy of the one-stranded RNA. This DNA is then integrated into the chromosomal DNA of the cell; integration may occur at many sites of the host genome. This DNA induces the cell to produce new viral RNA as well as the proteins necessary for the synthesis of new viral particles.

In addition to this minimal information, the genomes of oncogenic retroviruses carry an additional gene, which is responsible specifically for malignant transformation of host cells. This gene is called a retroviral oncogene (v-onc). It was shown by DNA hydridization studies (Sect. 2.3) using DNA probes from v-onc genes that such genes are homologous with genes occurring at various sites of the host genome. But under normal conditions, they do not lead to malignant transformation. They are called proto-oncogenes or cellular oncogenes (c-onc). In a recent review [1696], many such c-onc genes are mentioned that could be localized on various human autosomes (Table 5.22; see also Tabl. A9.5). It is now assumed that these genes were integrated into the viral genome at some time during evolution. However, why they cause malignant transformation when carried into the cell by virus infection but not when transmitted as normal cell constituents is completely unknown. Three different c-*onc* genes code for three tyrosine-specific protein kinases, show homologies in the amino acid sequences of their gene products, and are located on human chromosomes 3, 15, and 20. Such protein kinases phosphorylate proteins and might, therefore, change their biological activity. This might lead to transformation, for example, by changing properties of the cell surface (contact inhibition?). Proto-oncogenes may code for normal growth factors or their receptors. Thus, the sis gene codes for one chain of the platelet derived growth factor and the erb-b encodes the receptor for the epidermal growth factor [1437]. Proto-oncogenes may thus be involved in the control of normal cell growth through the various growth factors and their receptors. It can readily be visualized how mutant proto-oncogenes (see below) might stimulate mitosis and cause cancerous growth. The platelet-derived growth factor is particularly interesting [1437]. It appears to be involved in stimulating the formation of atheromatous lesions - an event presumably mediated by increased synthesis of the normal growth factor. Its mutant counterpart sis is involved in neoplastic sarcoma formation. The relationship between the genes for normal growth factors and their mutations, i.e., oncogenes that induce tumors, is of great significance for developmental and cancer biology [1694]. There are structural differences between v-*onc* and the homologous c-*onc* genes; for example, the c-*onc* genes, like other eukaryotic genes, consist of exons and introns (Sect. 2.3), whereas the corresponding v-*onc* genes have maintained only the exons.

Cellular Transformation. In many cases, cellular oncogenes were discovered by direct gene transfer from transformed to normal cells. Treatment of preneoplastic mouse NIH/3T3 fibroblasts with DNA from tumorigenic cell lines leads, in a certain fraction of instances, to the expression of transformed tumorous phenotypes in the recipient cells. The first transforming gene characterized in NIH/3T3 transformants (c-Ha-ras 1) came from the human bladder carcinoma cell line EJ. The genetic lesion leading to activation (i.e., transforming ability) of the oncogene involved in this transformation was found by cloning of this gene to be a single point mutation, resulting in the replacement of just one amino acid in the gene-determined protein. The codon GGC was replaced by GTC; the transversion G→T led to replacement of glycine by valine in the resulting protein. However, a search by means of restriction enzyme analysis for this proto-oncogene in 29 human cancers failed to show additional instances; it appears to be rare in cancers. Interestingly, the viral counterpart of this oncogene in a mouse sarcoma virus showed a point mutation in the same position. Other gene activations by single point mutations have been discovered. Moreover, some oncogenes found in other tumors showed structural similarity with the C-Ha-ras1 gene. Possibly, mutant genes of the ras genes are involved in $\approx 15\%$ of human tumors.
A transforming gene with different structure (B-lym) was isolated from Burkitt lymphoma cell lines; its gene product is partially homologous with transferrin, the iron-transporting protein (see Sect. 6.1.2). Rearrangement of another gene - C-myc - was also found in Burkitt lymphoma cell lines. The fact that two different oncogenes were found in the same kind of tumor suggests the possibility that mutations at more than one gene locus are sometimes necessary for malignant transformation. Perhaps the different steps in carcinogenesis are represented by sequential activation of different proto-oncogenes, leading to qualitative and quantitative alterations in gene expression. The mechanisms for such an activation are now being investigated with great intensity, since they promise insights into molecular events leading to malignant transformation. In addition to the point mutations mentioned above, attachment of c-*onc* genes to strong promoter or enhancer regions of DNA, for example, by insertion of such regions close to c-onc genes proved to be successful in vitro. However, promoter (or enhancer) insertion close to an oncogene might be only *one* requirement for transformation that might be completed by changes in other proto-oncogenes.

Oncogenes Involved in Carcinogenesis Due to Chromosomal Rearrangements? These result on oncogenes – and especially the observation that their insertion close to strong promoters might lead to activation – suggest the question as to whether, in chromosomal rearrangements characteristic for certain neoplasias (Sect. 5.1.6.4), such rearrangements of proto-oncogenes in relation to their neighborhood of promoter/enhancer regions (and possibly other regulating genes) might be decisive. Therefore, many groups are now investigating oncogenes and their activities in tumors in relation to the localization in normal and rearranged chromosomes (Fig. 5.39). In Burkitt's lymphoma, for example, transcription of the c-myc gene may increase 20 fold [1704]. In mouse plasmocytomas a translocation similar to that leading in humans to the Burkitt lymphoma has been observed between the terminal segment of chromosome 15, which carries c-myc, and chromosome 12. The breakpoint coincides with the heavy-chain constant region gene (see Sect. 4.4). A similar situation appears to be present in human Burkitt's lymphoma. The human oncogene C-abl is located on the terminal band of the long arm of chromosome 9, the same band as is involved in the breakpoint of the 9:22 translocation in chronic myeloid leukemia. It is too early for final conclusions; but this preliminary evidence favors the hypothesis that activation of oncogenes might, indeed, be involved in carcinogenesis by chromosome rearrangements. Proto-oncogenes have also been found as *amplified* copies in tumor cells [1694]. Double minute chromosomes and homogeneous staining regions of chromosomes are cytologic counterparts of gene amplification in cancer cells. The myc oncogene involved in translocation of Burkitt's lymphoma is overexpressed in carcinoma of the lungs, colon

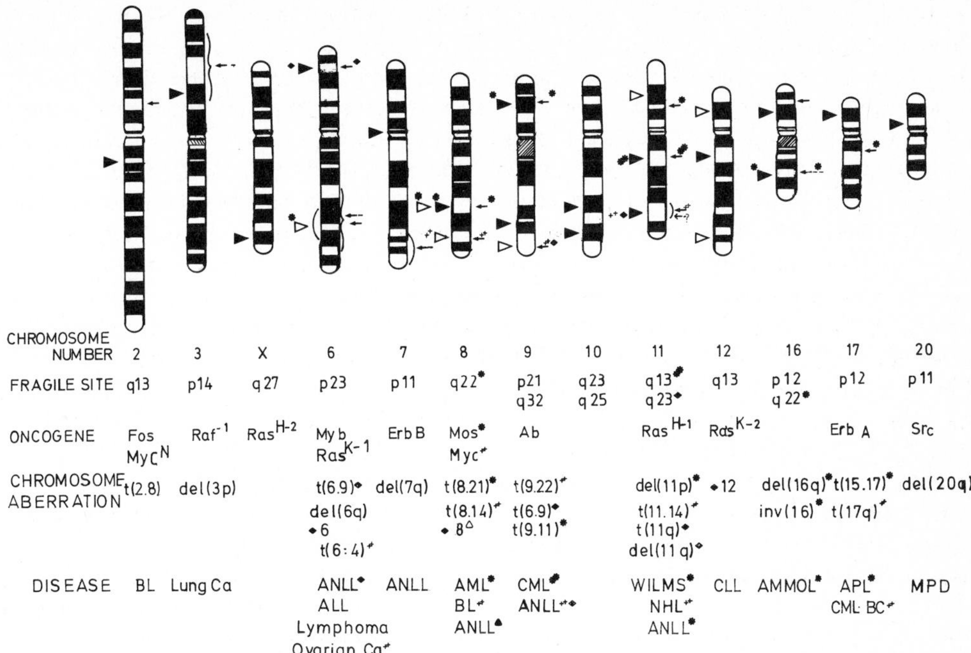

Fig. 5.39. Chromosomal fragile sites, localizations of oncogenes, and localization of chromosomal breaks in chromosomal aberrations leading to cancer. Diagram of chromosomes containing known fragile sites; the chromosome number, fragile site, oncogene, karyotypic aberrations and associated neoplastic diseases are indicated below the chromosome. The *arrowheads* to the left of each chromosome indicate the bands carrying the fragile site (▶) or cellular oncogene (▷); the *arrow(s)* to the right of a chromosome identify specific bands involved in consistent translocations (←) or deletions (---) observed in patients having the disorders listed. Additional symbols (*, #, +) link the aberration to a disease. *BL*, Burkitt's lymphoma; *ALL*, acute lymphoblastic leukaemia; *ANLL*, acute nonlymphocytic leukaemia; *AML*, acute myeloblastic leukaemia; *CML*, chronic myelogenous leukaemia; *AMMOL*, acute myelomonocytic leukaemia; *AMOL*, acute monoblastic leukaemia; *BC*, blast crisis; *NHL*, non-Hodgkin's lymphoma; *CLL*, chronic lymphocytic leukaemia; *APL*, acute promyelocytic leukaemia; *MPD*, myeloproliferative disorder. p indicates the short arm and q the long arm of the chromosome; *t*, del and inv indicate translocation, deletion and inversion [1705]

carcinoma, and promyelocytic leukemia. The N-myc gene is amplified in advanced stages of neuroblastoma and high levels of epidermal growth factor are expressed in squamous cell carcinomas – presumably as a result of amplification of the erb-b gene. Tumor *progression* may be related to activities of particular oncogenes, since certain oncogenes are found in the more aggressive forms of the tumor. If characteristic patterns can be identified, specific targets for therapy are available. Oncogene analogues or antioncogene antibodies, in contrast to current cancer chemotherapy, would not interfere with normal cell function but might arrest tumor growth.

Oncogenes – at least those known at present – certainly do not embody the *entire* answer to the question on mechanisms of carcinogenesis. In most human tumors, for example, activated oncogenes have *not* been found (which might well be because not all oncogenes are already known). Moreover, the question as to why infection by retroviruses harboring oncogenes normally leads to a malignancy in one specific tissue only is still completely unresolved. There must be strongly tissue-specific expression of such genes. Nevertheless, discovery of oncogenes has been an important step forward in our understanding of malignant transformation.

5.1.6.6 A Genetic View of Human Cancer

The various data regarding cancer in man allows some generalizations. Mendelian inheritance in human cancer is rare, but a variety of benign tumors particularly can be inherited as Mendelian traits. Neurofibromatosis, multiple polyposis, and the various multiple endocrine neoplasia syndromes are examples [1540, 1568]. Malignant transformation is common in most if not all of these conditions. The chance of malignant transformation may be related to the turnover time of the affected cell. With more frequent cell divisions, a malignant tumor cell is more likely to develop. It may be no coincidence that in the rapidly turning-over epithelium of the large bowel colon cancer invariably appears following benign hereditary polyposis.

Another tumor mechanism, i.e., homozygosity (as found in retinoblastoma, Wilm's tumor, and other embryonal tumors) produced by an inherited germinal mutation and followed by a somatic mutation or somatic rearrangement in a single somatic cell, may be more generally applicable to some other tumors (Sect. 5.1.6.4).

Chromosomal abnormalities are often seen in cancer cells but are rarely unique. They are usually considered to reflect abnormalities secondary to disordered growth. In addition to the translocation in chronic myelogenous leukemia, other nonrandom chromosomal changes (Table 5.21) were identified first with increasing frequency in various hematologic malignancies where detailed studies were more readily possible. In fact, Yunis has claimed that with high-resolution banding of chromosomes most tumors can be shown to have characteristic chromosomal defects [1705].

The association of tumors with the autosomal-recessive chromosomal breakage syndromes is certain and is particularly interesting in xeroderma pigmentosum where the basic lesions have been identified as a defect of DNA repair following UV light. Suggestions that the much more common heterozygotes for these syndromes may also suffer from increased frequency of cancer makes pathophysiologic sense but needs further documentation [1340]. The possibility that a significant fraction of human malignancy is caused by the carrier state for a variety of breakage syndromes would be of great public health importance for tumor prevention if tests for carrier detection for these defects would be fairly readily available.

Various genetic defects affecting chromosome breakage and repair, as well as lack of immune surveillance with environmental carcinogens are likely to produce neoplasias. Additional genetic factors bearing on environmental carcinogenesis relate to abnormal metabolism of the carcinogenic environmental agent. Arguing from twin studies in pharmacogenetics (Sect. 4.5.1), it is likely that metabolism of most foreign agents is under genetic control. Slow biotransformation or existence of enzyme systems that transform procarcinogens into more potent cancer-causing agents will cause a certain fraction of the population to be at higher risk for genetic reasons. Data on arylhydrocarbon hydroxylase levels, an enzyme that transforms polycyclic hydrocarbons into the more carcinogenic agents, and the debrisoquine-sparteine system in lung cancer in man suggest such mechanisms (Sect. 4.5.2). There is increasing evidence cited above that a large fraction of tumors owe their origin to somatic mutations. As an example, many of the chemicals that are mutagenic in bacterial test systems are also carcinogenic in animal systems. The data on the monoclonal origin of most tumors as shown by genetic tagging techniques using G6PD variants (see above) also point toward somatic mutation as the major source of neoplasia formation.

However, viral causation of human tumors is not excluded, and it is likely that some human tumors may be caused by viruses. Virus causation of human tumors is most certain in Burkitt's lymphoma, a tumor that is clearly of monoclonal origin. The fact that chromosome 14 and chromosome 8 are often involved in the development of Burkitt's lymphoma makes this tumor a model relating viral causation, monoclonal origin, oncogenes, and clonal chromosomal evolution.

Recent results on oncogenes have not only established a link between tumor cytogenetics and molecular biology by demonstrating that cancer genes might be located close to the very breakpoints involved in tumor-specific chromosome rearrangements. They also helped to bridge the gap between spontaneous and mutagen- and virus-induced somatic mutation by showing that oncogenes might be "activated," i.e., turned from normal and necessary to cancer-promoting genes either by classical point mutations or by rearrangements effected by passage through a virus.

Striking familial aggregation of tumors only occurs in the various hereditary tumor syndromes and occasionally in so-called cancer families, which are characterized by (a) increased occurrence of adenocarcinoma, primarily of colon and endometrium, (b) increased frequency of multiple primary malignant neoplasms, (c) earlier age of onset, and (d) vertical transmission compatible with autosomal-dominant inheritance. In many tumors, such as carcinoma of the breast and carcinoma of the stomach, a modest familial aggregation is found. These data are often interpreted as compatible with multifactorial inheritance, but the nature of the genetic predisposition is unknown.

The changing epidemiologic pattern of some tumors, such as the increasing frequency of lung cancer and the decreasing frequency of gastric cancer in the last generation, clearly suggests that genetic factors alone cannot explain the predisposition to these cancers. We know that cigarette smoking is related to lung cancer and suspect that disappearance of a carcinogen because of better food preservation has reduced the frequency of stomach cancer. The fact that only a certain fraction of heavy cigarette smokers develop lung cancer argues against a simple environmental explanation. It is most likely that specific environmental factors, such as the hydrocarbon of cigarette smoke and other irritants, interact with specific genetic factors, such as those affecting hydrocarbon metabolism, DNA repair, and immunologic surveillance in determining the establishment of a clinically significant lung cancer. In occasional cases, either the environmental factor alone or the genetic factor alone are likely to be all important. In most cases, interaction of both heredity and environment probably are operative. Since a large number of environmental agents and a number of genetic mechanisms appear involved in various cancers, no simple panacea to prevent all cancers is likely to be found. Hopefully, identification of genetically susceptible subgroups of the population by simple laboratory tests may become possible with better understanding of the various mechanisms underlying genetic susceptibility to various cancers.

5.1.6.7 Somatic Mutation and Aging

Aging and Death. Humans are the only living beings who know about their inevitable death. The difficulties in accepting this knowledge are epitomized by the enormous importance of death cults in cultural evolution - from burial rites of Neandertal men and highly sophisticated cultures dedicated almost exclusively to taking care of the deceased, such as the Egyptians, to the heaven-and-hell theories inherent in some present-day religions. In earlier periods, death generally came suddenly at a relatively young age. With the development of modern life styles, and especially modern medicine, we have gained decades of life span. But we are now confronted with the slow deterioration of our biological capacities during aging. An increasing fraction of the population, particularly in industrialized societies, consists of aged people; a large part of medical practice is now devoted to geriatric medicine. This has created many new social, biological, and medical problems.

Twin and family investigations in humans have shown a relatively strong genetic influence on life expectation [1508]. Certain cases of Alzheimer's disease, a type of senile dementia, show familial aggregation. Moreover, the onset and natural history of many physical and mental diseases is age dependent. In some diseases of the brain, characteristic age-dependent changes in specific groups of neurons have been observed, and are regarded by some research workers as simple models for normal aging.

Studies on Biological Mechanisms of Aging in Single Cells [1513]. The notion that aging and death might be reduced to properties of the single cell was first put forward by the zoologist A. Weismann about one hundred years ago. He suggested that the origin of "natural death" lies in the limitation of somatic cell's powers of reproduction. We read in one of his publications that "death takes place because a worn-out tissue cannot forever renew itself, and because a capacity for increase by means of cell division is not everlasting but finite" (Weismann, 1891 [1693]). Weismann was also the first to suggest that too long a survival of a multicellular organism beyond the period of reproduction and care for offspring would carry a selective disadvantage to the species.

Weismann's hypothesis of somatic cell deterioration seemed to be refuted by the claim of A. Carrell (1912) that somatic chicken cells could be cultured outside the donor indefinitely. For a long time, this "potential immortality" hypothesis was generally accepted and was among the most popular biologic "results." Failures to reproduce these results were explained by inappropriate culture conditions and other methodological factors, until the careful experiments of Hayflick in the early 1960s [1483, 1484] established that Carrell's claim had been wrong. It is now widely accepted that normal diploid, mammalian fibroblasts can undergo only a finite number of cell divisions; for human embryonic fibroblasts this number is about 50 ± 10. On the other hand, heteroploid, *transformed* cells generally appear to multiply indefinitely.

Several studies have sought to explore the link between aging of the organism as a whole and the limited number of cell divisions in somatic cells. If cells stop dividing in culture because they have "used up" an intrinsic maximum number of divisions, then their in vitro capacity for division should decrease with increasing age of the donor. In fact, this has repeatedly been demonstrated [1547, 1630]. Another implication of this hypothesis would be that cells of short-lived species would be able to undergo fewer divisions than those of long-lived ones. This was, in fact, confirmed when humans (maximum life span $\approx$ 110 years) were compared with mice (3½ years). Mouse fibroblasts can only undergo 14-28 divisions in vitro. However, when other species are included in the comparisons the association between species-specific life span and ability of cells to divide in vitro is blurred. Moreover, in humans, even cells taken from very old people have maintained the capacity of dividing about 20 times.

On the other hand, the significance of such cell studies for the problem of aging appeared to increase when it was discovered [1547] that cells affected by Werner's syndrome had a markedly reduced capacity for division in cell culture. The main clinical signs of this autosomal recessive syndrome are cataracts, subcutaneous calcifications, premature graying, premature arteriosclerosis, skin changes, diabetes mellitus, a higher incidence of malignant tumors, chromosome instability, and a prematurely

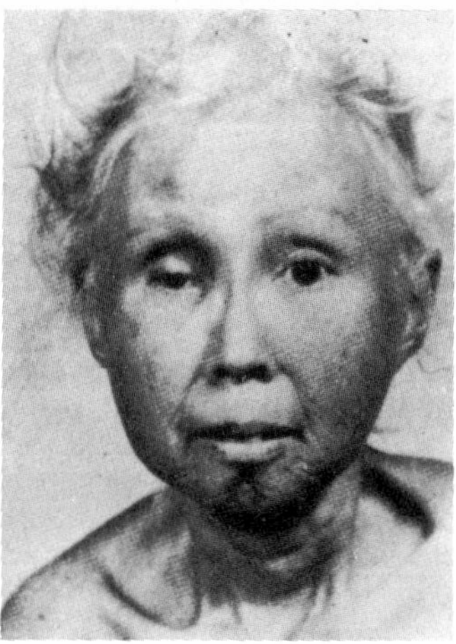

Fig. 5.40. A Japanese-American woman as a teenager and at 48 years of age. She had eight sibs, two of whom were also affected with Werner's syndrome. (From [1613])

aged face (Fig. 5.40). Life expectancy is considerably reduced [1442; 1613]. Werner's syndrome has sometimes been regarded as a genetic model of premature aging. However, while some of the clinical and pathological findings resemble those found during the normal aging process, many others do not.

Molecular and Chromosomal Mechanisms. Many studies are in progress on molecular and chromosomal mechanisms, but so far no coherent and generally accepted theory has emerged. With considerable oversimplification, two groups of hypotheses might be distinguished. One of them assumes that ceasing of cell divisions is somehow programmed by a biological mechanism of regulation. An important argument in favor of this hypothesis is the observation that transformed cells are not limited in their dividing capacity. In the other group of hypotheses it is assumed that cells lose their dividing capacity by accumulation of "errors" that prevent them from dividing when a certain threshold is exceeded. Such errors might occur somewhere at the translational or post-translational level; they might also occur at the DNA level in the form of somatic mutations. The last-mentioned hypothesis, which was first proposed by Szilard in 1959 [1627] and was extended by Burnet in 1974 [1409], is supported by evidence of an increasing frequency of defined somatic mutations with age in vitro [1459] and in vivo: The fraction of lymphocytes having a defect of HPRT that makes them resistant to 6-thioguanine increases with age [1559]. However, the somatic mutation theory does not explain all phenomena of cell aging; accumulation of errors at various post-translational levels is also likely [1490].
But how does the transformed cell escape the consequences of such an accumulation of errors, whether produced by somatic mutation or by another error-prone process? So far, this problem seems not to have been solved. Several possible mechanisms [1513] are under scrutiny, among them reactivation of normally suppressed correction mechanisms or more rapid selective elimination of cells harboring such errors. In conclusion, it is likely that somatic mutations contribute to the process of normal aging just as they are involved in cancer formation, which is also an age-dependent process. But how much they contribute and how they interact with other cellular processes is largely unknown.

5.2 Mutation Induction by Irradiation and Chemical Mutagens

Public Interest in Induced Mutation. In the preceding sections, spontaneous mutations were discussed. "Spontaneous" means that these mutations are unpredictable and without known cause, even though we know that some conditions - such as parental age - may enhance the probability of mutation. This probability increases under the influence of certain agents, such as energy-rich radiation and a number of chemicals. Since human beings in their normal environments are exposed to a variety of these agents, research on induced mutation is receiving more and more attention from the general public. Relatively large amounts of money have been allotted to this work, and scientists are expected to advise political authorities as to protective measures. In regard to radiation-induced mutations, an appropriate return from these investments has been forthcoming for a number of years. The World Health Organization, the International Commission on Radiation Protection (ICRP), the U.S. National Academy of Sciences, and other influential organizations have established expert groups and, with their help, have published estimates of genetic risks. There are still many gaps in our knowledge particularly regarding the effect of low-level radiation on humans but a fairly coherent picture of the radiation threat is now emerging. Relatively little is known about the possible impact of environmentally induced mutations by chemicals.

There is much confusion within the scientific community concerning the specific nature and extent of the chemical threat and the kinds of information and recommendations scientists should provide [1489, 212]. One reason is that the problem is, indeed, more complex than the problem of radiation-induced mutation. Another reason may be that the scientific community rewards success in research within a relatively narrow field, carried out with technically difficult methods. This requires ingenuity in recognizing and precisely formulating problems that by necessity are limited in scope and can be solved by suitable methods. Such an approach does not usually require experience in many different areas of science. Within the natural sciences, the barriers between specialties tend to impede free usage of the information available at different levels. Hence, scientists considered able to provide expert advice on the basis of their prominence within a given field frequently may not have a balanced view of a complex problem. They tend to see problems mainly from the viewpoint of their own specialty.

The traditional channels of scientific communication - scientific societies, congresses, journals - have failed so far to overcome these difficulties in the field of environmental mutagenesis. International organizations such as WHO have not worked as efficiently in the fields of chemically induced mutation as in the field of radiation. Some publications whose main emphasis is on research methods [1489, 212] have helped somewhat to clarify the situation. Perhaps the solution will lie in new institutions organizing scientific efforts at all relevant levels. Unfortunately, such institutions have so far been successful in moblilzing broad-scale scientific endeavors only under two conditions:

1. When the goal was clearly defined and purely technologic.
2. When there was a strong and immediate political motivation.

The examples that come to mind are the Manhattan Project in World War II, which led to the development of the atomic bomb, and the American and Russian projects to explore outer space.

In the following we shall give our view of the problems. This view may help others to realize the complexity of the matter and to participate in the definition of goals.

5.2.1 Radiation-Induced Mutation

5.2.1.1 Basic Facts and the Problems Posed by Them

Capacity of Energy-Rich Radiation to Induce Mutation. That energy-rich radiation can induce mutations had been suspected but was first proven by Muller (1927) [1567] in *Drosophila melanogaster* and by Stadler (1927-28) in barley after Mavor (1924) had shown before the induction of nondisjunction [1552]. A condition for Muller's discovery was the ingenious use of *Drosophila* mutants in developing test systems for counting mutations, especially X-linked lethals. Results of Muller's classic experiment are seen in Table 5.24. Here, dosis t_4 is twice as high as t_2. Doubling of the radiation dose approximately doubled the number of induced mutations. The scope of this experiment was too small, however, to detect spontaneous mutations among the controls.

Within the following two decades, classic radiation genetics developed. Its basic concepts were set out in the books of Lea and Catcheside (1942) [1528], Timoféeff-Ressovsky and Zimmer (1947) [1656], and Hollaender [97] (1954-1956).

Some Technical Remarks on Radiation. Two types of energy-rich radiation have to be considered: electromagnetic waves and corpuscular radiation. The biologic activity of electromagnetic waves in relation to their wavelength and energy is seen in Fig. 5.41. For a mutagenic effect, the energy should at least be sufficient to lift an electron from an inner to an outer shell, rendering the atom unstable and more prone to chemical reactions. UV radiation has this property and is therefore mutagenic, provided it reaches the DNA. Its best-known chemical reaction is the linking of two adjacent thymine molecules. This prevents them from pairing with adenine. Therefore, UV photons cause point mutation, but little in the way of structural defects. For the human germ cell, UV is not dangerous, being absorbed in the epidermis. There, however, it might induce somatic mutation - and cancer - of the skin (Sect. 5.1.6). Radiation photons of higher energy (X-rays, γ-rays) are able to push electrons out of the external shell so that the atom becomes a positive ion. These electrons may now

Table 5.24. Muller's classic experiment, which proved mutation induction by X-rays [1567]

Experiment	Number of chromosomes tested	Number of mutations observed		
		Lethals	Semilethal mutations	Visible mutations
Controls	198	0	0	0
X-rays (t_2)[a]	676	49	4	1
X-rays (t_4)[a]	772	89	12	3

[a] The dose t_4 was twice the dose t_2.

react with other atoms, converting them into negatively charged ions. Both types of ions - together with free radicals - form the material for secondary chemical reactions. Corpuscular radiation consists not of energy photons but of particles. They may be charged, like electrons and protons, or uncharged, like neutrons. Their physical effects depend on their kinetic energy. Ionization induced by neutrons is densely concentrated around the track of the particle, whereas electromagnetic waves (X and γ radiation) produce looser ionizations.

The biologic effects of all types of radiation depend on the locale of the source (inside or outside the body), type of radiation (electromagnetic waves, charged or uncharged particles), their energy, and properties (density, water content, etc.) of the absorbing material.

Irradiation of any kind has not only direct but also indirect effects. Energy-rich neutrons, for example, may be accepted into atom nuclei, or they transfer their kinetic energy to, say, hydrogen nuclei (protons). Thes protons are accelerated and undergo many secondary reactions with other molecules.

The radiation energy dose is usually measured in Grays (Gy): 1 Gy is equal to the energy dose produced when 1 J energy is transmitted by ionizing radiation to matter of 1 kg mass under constant and defined conditions. This corresponds to 100 rad (old nomenclature), and, as a rule, to 100 roentgen (R), but the last is defined with reference to the number of ionizations. Another important measure is the equivalent dose that measures the (negative) biological effect of a certain radiation dose. It is calculated by multiplying the energy dose by a factor that varies from one type of radiation to the other, depending mainly on the energy discharge; it is higher, for example, with densely than with loosely ionizing radiation. It is measured in J/kg (J = joule). An old measure is the rem (1 rem = 1/100 J/kg). In this work data from the older literature are often cited. We have used the new nomenclature: 1 Gy = 100 mGy = 100 rad = 100 R. This may not always satisfy the needs of radiation physicists; but for the purposes of this book, it may suffice.

Results and Concepts of Classic Radiation Genetics [1528, 1656, 97]. The most important results and concepts of classic radiation genetics may be summarized as follows:

a) To induce mutations in a certain cell, the radiation must reach this cell, for example a germ cell. This statement is not quite as self-evident as it may seem. Indirect influences, for example, by induction of a chemical compound that reaches the gonads via the circulation, are not impossible a priori; a limited amount of indirect action has indeed been demonstrated by more recent work. Nevertheless, for all practical purposes the statement is still a good approximation. For humans this principle is important, since the high absorption rate of some types of radiation (UV, or very low-energy X-rays) prevents them from being dangerous to the germ cells. They may still be dangerous to the individual by inducing somatic mutations and cancer.
b) Radiation does not create any new biologic phenomena; it only enhances the probabilities of various mutations and cellular events, which occur spontaneously from time to time. Radiation-induced mutations are not basically different from spontaneous mutations. This rule has also been confirmed for chemically induced mutations. However, not all types of spontaneous mutation are increased in the same proportion by all mutagenic agents. On the contrary, there are definite differences in the relative frequencies of different types of spontaneous as well as of radiation-induced and chemically induced mutations. The fact that any type of inducible mutation might also spontaneously occur creates a difficult statistical problem in trying to prove that an in-

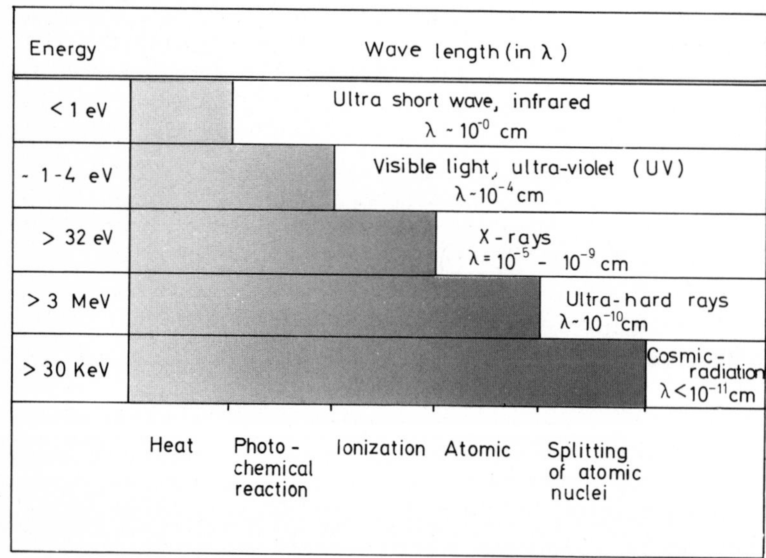

Fig. 5.41. Biologic activity of electromagnetic waves in relation to their wavelength and energy

crease in the mutation rate in a human population is caused by exposure to mutagenic agents. This may be explained by an example from human teratology: thalidomide, which was introduced as a sleep-inducing drug, turned out to be teratogenic when taken a few times or even once during early pregnancy. The pattern of malformations induced by the drug was strikingly characteristic, and the combination of short and malformed limbs (phocomelia) with ear, eye, and internal malformations had almost never been observed before. It was this specificity that first led doctors to realize that something new must have happened and then to look systematically for possible teratogenic agents. However, had the drug caused the same number of cases of cleft lip and palate or neural tube defects as it had of phocomelia, it very probably still would be regarded as a useful hypnotic agent with excellent indications for use in pregnancy.

These considerations suggest that any mutagenic agent inadvertently introduced into our environment and causing the same amount of damage as thalidomide did would surely be overlooked.

c) A third problem that has frequently been discussed is the dose factor in the mutation rate. For mutation requiring only one primary event, the dose-effect curve has been shown to be simply linear:

$$M = \mu + kD$$

(M = number of mutations; μ = spontaneous mutation rate; D = dose; k = mutation rate/dose unit). This linear dose effect relationship holds true as long as only a moderate fraction of all irradiated cells include an induced mutation. For higher mutation rates, there is a certain "saturation effect," which leads to flattening of the curve. Here, an exponential equation would describe the curve more properly. An example for *Drosophila melanogaster* is shown in Fig. 5.42. The radiation doses used are relatively high; for lower doses, the linearity is confirmed. Within this lower range, the evidence is not too conclusive, as the fitting of a dose-effect curve to data requires very large sample sizes.

These results were interpreted by the target theory. Every mutational even is produced by one hit in a susceptible structure, and the probability of this structure being struck increases linearly with increasing dosage. The saturation effect with very high doses (higher than those used in the experiments depicted in Fig. 5.42) is due to the same structure being struck more than once.

d) The theory predicted a nonlinear dose-effect relation for mutations needing more than one primary event. Many translocations require two breaks within a reasonably short distance and at approximately the same time for the chromosomes to rejoin after they are broken. However, the two breaks need not be produced by two separate hits. They may also be induced by one particle or photon, especially if the density of ionizations is very high, for example, if neutrons are used. In this case, a dose-effect curve could be expected in which a linear (one-hit) and quadratic (two-hit) component are mixed:

$$M = \mu + k_1 D + k_2 D2$$

Such dose-effect curves have indeed been observed in many experiments. Figure 5.43 shows an example for *Drosophila melanogaster.*

e) For single-hit events, the theory makes still another prediction of great practical importance. Obviously, the number of hits depends on the dose, not on the time within which the dose is applied. Hence, it should be unimportant whether the same dose is given within a very short time (high "dose rate") or extended over a much longer time (low "dose rate"). The first *Drosophila* experiments seemed to confirm this prediction (Fig. 5.44). More recent results from mice [1611], however, raise serious doubts as to their general applicability, and there now seems to be agreement that, in many cases, a certain dose applied at a lower dose rate will lead to a lower number of mutations than the same dose applied at a higher dose rate.

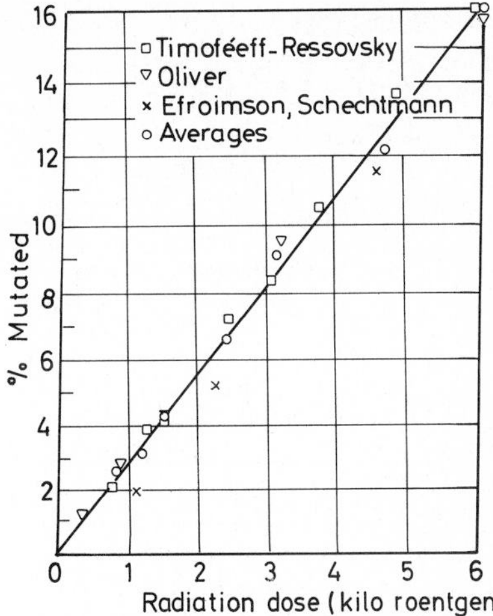

Fig. 5.42. Linear increase of the mutation rate for point mutations and single chromosome breaks in *Drosophila melanogaster.* Note linearity of effect. Data from various authors. (Timoféeff-Ressovsky and Zimmer, 1947 [1656])

Confirmation and Extension of These Results. A great number of studies using other materials have helped

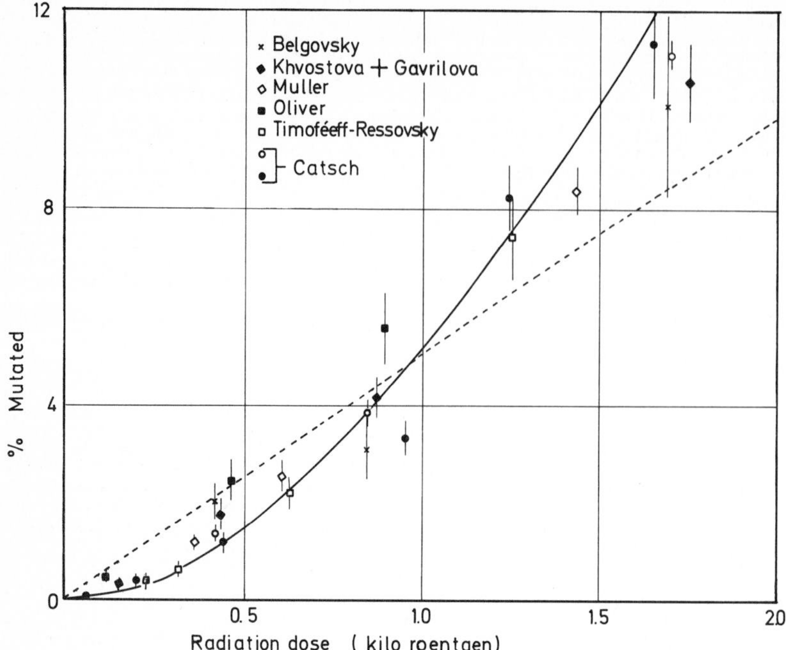

Fig. 5.43. Chromosome aberrations (two-break events) in relation to the radiation dose (in 10 Gy). The two-hit curve *(uninterrupted line)* fits the data better than the one-hit curve *(dotted line)*. Data from various authors. The figure gives experimental points and their standard deviations. (Timoféeff-Ressovsky and Zimmer, 1947 [1656])

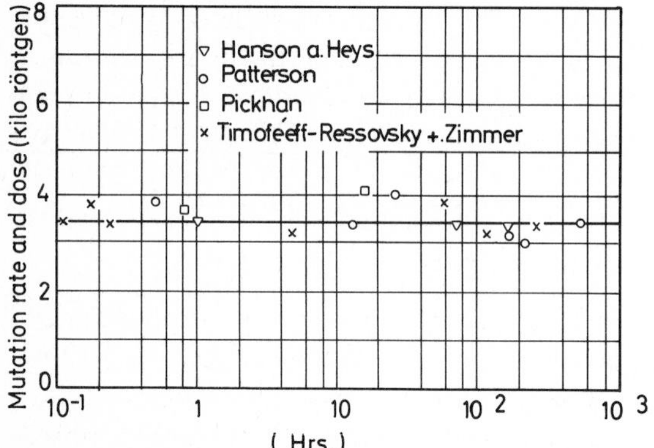

Fig. 5.44. Irradiation of *Drosophila melanogaster* with a radiation dosis of 35 Gy and different dose rates. There seems to be no difference in the mutation rate *(horizontal line)*. See, however, the text (Timoféeff-Ressovsky and Zimmer, 1947 [1656])

to confirm and partially modify these results. The developmental stage of the irradiated germ cell has been taken into account; different - for example, biochemical - phenotypes have been introduced, and radiation-induced mutations have been explained in terms of chemical reactions.

Influence of the Chemical Environment, Especially the O_2 Content of Irradiated Tissue. One of the secondary consequences of irradiation is the formation of highly reactive radicals such as peroxides. For their formation, oxygen is needed. Therefore, it is not surprising that high oxygen content of irradiated tissue enhances mutation induction. This effect is strong, for example, with X-rays. It is almost or totally absent with densely ionizing radiation, such as α particles.

Molecular Effects of Radiation [1660]. In classic radiation genetics, the mutagenic effects were defined at the phenotypic level. However, morphologic examination of chromosomes, especially in plants like tradescantia, was introduced at a relatively early stage, and it was shown that many of the mutational changes could be explained by induction of morphologically visible chromosome breaks and their sequelae, such as translocations.

For a long time, it was not even known whether ionizing radiation could induce gene mutations. Many investigators considered it plausible, but there remained the possibility that all radiation-induced mutations were basically small deletions or chromosome rearrangements. Recently, this matter was settled by using the φ X174 phage, which has only one DNA half-strand. Here, reverse mutations were induced that could not be explained by any other mechanism than a single point mutation. In bacteria, induction of gene mutations, among them transitions, was shown for the tryptophane locus of *E. coli* [1660].

The first eukaryotic system in which radiation-induced gene mutations were found was *Neurospora crassa* [1544, 1618]. Of the induced mutations, 42% were transitions, 37% insertions or deletions of single base pairs, and the rest were of different origin, among them probably transversions.

Basic Facts of Radiation Genetics Reconfirmed in Human Lymphocyte Chromosomes [1395, 1387]. Most of the work in classic radiation genetics has been done using organisms, such as *Drosophila*, which are only remotely related to humans. Irradiation experiments on chromosomes were carried out mainly on plants. Work on humans was begun soon after the method of chromosome preparations from lymphocytes had become available (Sect. 2.1.2.2). The technical details of diagnosis and classification of structural chromosome aberrations are discussed in Sect. 2.2.2. Dicentric chromosomes turned out to be an especially good indicator for chromosome breakage. Figure 5.45 shows the dose-effect curve. Its slope is more or less linear but has a tendency to become steeper with higher doses. Such dose-effect curves are produced if some of the primary events are one-hit and others are two-hit events. Formation of a dicentric chromosome requires two chromosome breaks in closely adjacent chromosomes. If our theoretical notions are correct, these breaks may be caused either by a single hit or by two hits. Hence, the dose-effect curve found does not contradict these notions. Fig. 5.46 also contains data for ring chromosomes, minutes, and fragments.

Figure 5.47 shows a decrease in the number of dicentric and ring chromosomes with the same dose (2 Gy) but decreasing dose rate, i.e., radiation given over a period of time at different rates. These data confirm the dose-rate effect that had been established for other organisms, such as the mouse. These few data show that basic phenomena from radiation genetics can, in principle, also be demonstrated with human chromosomes.

The human population is exposed to ionizing radiation from a number of sources. Geneticists are asked to advise the public on the extent of possible hazards. How can these questions be answered?

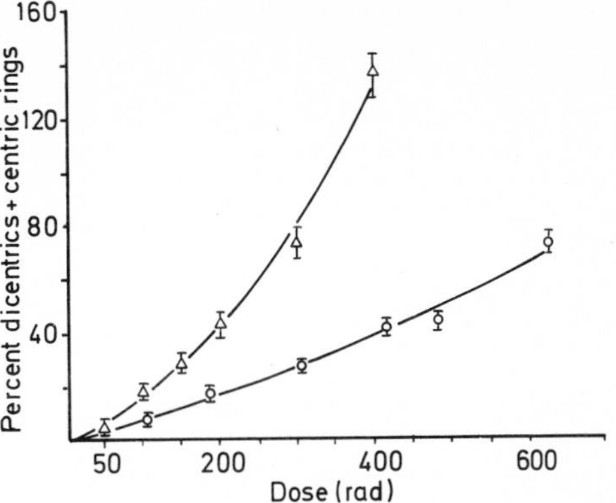

Fig. 5.45. Dose-effect curves for dicentric chromosomes and ring chromosomes after irradiation of human lymphocytes in vitro. Acute △ and chronic ○ irradiation. Standard errors are indicated at each point. (Adapted from Brewen and Luippold, [1400])

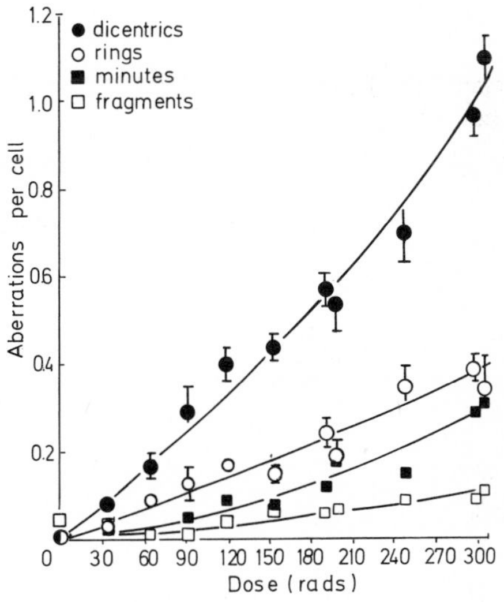

Fig. 5.46. Dose-effect curve from acute in vitro irradiation of human lymphocytes in vitro. Dicentrics, ring chromosomes, minutes (fragments containing a centromere), and fragments. (Bloom, 1972 [1395])

5.2.1.2 Problem of Estimating the Genetic Risk Due to Radiation and Other Environmental Mutagens

The problem of estimating the extent of risk to the human population and to all other mutagenic agents can be specified as follows:

a) In what way does the agent affect the genetic material?
b) How extensively is the human population exposed to this agent?
c) What is the probable increase of mutations compared with the "spontaneous" mutation rate?
d) What are the long-term consequences of this increase for the population?

These four questions should ideally be answered by the scientist. There is, however, a fifth question for which the scientist can only supply data. The answer should come from society as a whole.

e) How high an increase in mutational damage are we prepared to accept in exchange for the bene-

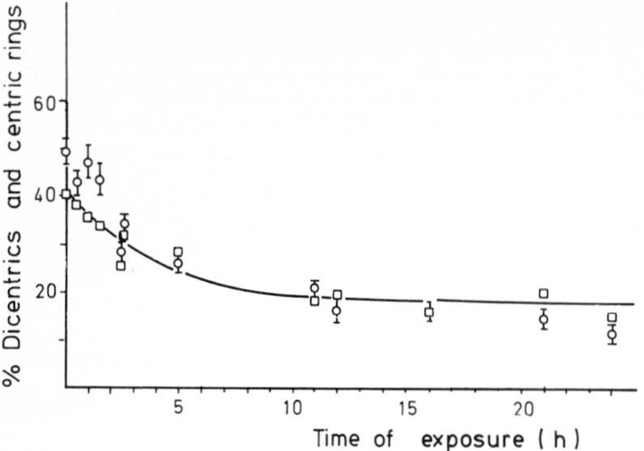

Fig. 5.47. Fraction of metaphases with dicentric chromosomes (□) and ring chromosomes (○) after in vitro irradiation with 2 Gy X-rays. Decrease of the effect with increasing exposure time and decreasing dose rate (= increasing exposure time). (Adapted from Brewen and Luippold, [1400])

fits present-day society enjoys from mutagenic agents - for example, diagnostic and therapeutic X-ray usage, nuclear energy, certain drugs, and amenities of all kinds?

Principles of Mutagenicity Testing. The question regarding effects on the genetic material is very complex. It can be subdivided into a number of more specific questions, such as:

1. What kinds of mutations are induced? Genome, chromosome, or gene mutations?
2. Are they induced mainly in germ cells or in somatic cells?
3. If they are induced in germ cells,
   a) What stage of germ cell development is primarily affected?
   b) Are the mutations transmitted to the next generation, or are they eliminated, for example, during meiosis?
   c) If they are transmitted, what phenotypic changes may be expected in offspring?
4. If they are induced in somatic cells,
   a) What cells are especially endangered?
   b) What are the consequences for the individual?

All these questions should be investigated by a comprehensive mutagenicity testing program. Which organisms should be tested in such a program? In theory, the answer is obvious. We are interested in humans; therefore, the best approach would be to examine these problems in humans. This, however, is not possible experimentally for ethical reasons. Only naturally occurring situations can be explored, and here the conditions are usually so complex that a clear-cut answer can not be expected. Therefore, an experimental animal has to be found for many studies. This animal should fulfill three main conditions:

1. It should be closely enough related to the human species for meaningful extrapolation. The inevitable differences between the experimental species and humans should be of such kind and order of magnitude that they can be accounted for either theoretically or experimentally when the comparison is made.
2. The generational turnover should be rapid so that genetic experiments can be carried out within a reasonable time.
3. It should be possible to keep test animals in sufficiently large numbers at a reasonable price.

The only animal that fulfills all these conditions is the mouse (*Mus musculus*). Therefore, radiation genetics of mammals - and mutation genetics in general - is largely that of the mouse. Other species, such as rats, Chinese hamsters, or marmoset monkeys, have only been used occasionally. Thus, the following sections will be devoted mainly to the mutation genetics of the mouse, with inferences to be drawn for human beings. When available, data from humans or from other species will be cited for comparison.

In Vivo Test Systems for Mutagenic Agents in Germ Cells of the Mouse. The test systems available for this animal can be subdivided into in vivo and in vitro systems and into systems for examining germ cells and somatic cells.

The available test systems for germ cell mutations are shown in Fig. 5.48. Assume that a mutation has been induced in spermatogonia. If this mutation is a chromosome mutation, it can be initially identified in mitotic spermatogonial divisions. During the first and second meiotic divisions, the influence of meiosis on induced aberrations may be observed. The haploid phase of spermatogenesis so far has been inaccessible (See, however, [431, 432]). The next mitotic divisions occur during early embryonic development of the F_1 generation. At this stage, it is possible to flush zygotes out of the Fallopian tube and to examine their chromosomes.

When the oocytes are prepared from the ovary, first meiotic divisions in the female germ cell can be observed. Depending on the time lapse between copulation and sacrifice of the animal, the stage of oocyte and early zygote can be examined. Cleavage stages may even be grown for a number of additional divisions in culture.

Implantation of the zygote in the uterus occurs on the 9th day of pregnancy; the zygote is then in the blastocyst stage. A few days afterward, examination of implanted embryos becomes possible. The so-called deciduomata indicate implantation sites of embryos that died a short time after implantation.

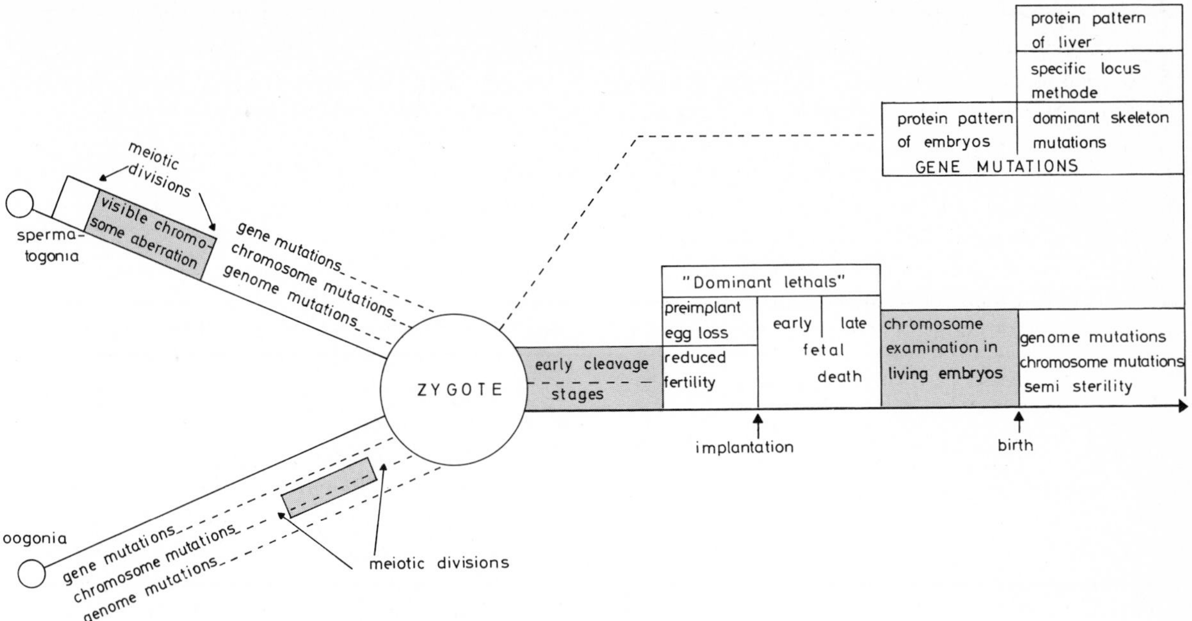

Fig. 5.48. In vivo test systems for examining the effects of a mutagenic agent in the mouse. For details, see the text

Also visible are some embryos that died at a later stage, as well as healthy embryos. Deciduomata, together with embryos that died later, represent postimplantation zygote loss. The preimplantation loss may be inferred from the difference between the number of corpora lutea in the ovaries and the total number of implantations. Preimplantation and postimplantation zygote loss, unless shown to be due to nongenetic factors, are usually attributed to "dominant lethals." In spite of more modern developments, the dominant lethal method is still one of the standard methods for mutagenicity testing [1603]. The surviving animals can be examined for chromosomal anomalies during late pregnancy.

Up to this stage of late pregnancy, all methods have been cytogenetic. Chromosome and genome mutations induced in early germ cell stages can now be followed through all stages of development to the newborn animal. There is no difficulty in examining mitotic chromosomes after birth in somatic tissue or even in following them up into the germ cells of the F_1 generation.

Examination of point mutations has to await birth. At birth, methods for a group of dominant and a group of recessive mutations are available. The dominant mutations are those affecting the skeleton. One method has been described in Sect. 3.6.2.5. Induced skeleton mutations showed an unexpectedly high degree of incomplete penetrance and variable expressivity.

Multiple Recessive Mutation Method. The method for assessing recessive mutations was one of the first to be developed for mouse radiation genetics. It permits detection of mutations at a small number of loci with great precision. Many of the basic results in this field have been achieved using this method. Wild-type animals, for example, males, are irradiated and mated with a test stock that is homozygous for seven autosomal-recessive mutations. When no mutations have been induced, the F_1 animals are all heterozygous and show the wild-type phenotype (Fig. 5.49). However, if one of the seven mutations has been induced in the paternal germ cell, the F_1 animal will be homozygous only for this mutation and will show the phenotype. The mutations in the test stock are selected in such a way that each homozygote can easily be identified. Figure 5.50, for example, shows a homozygote for one of these genes (piebald spotting).

Both methods used for screening of gene mutations – the skeleton method and the multiple loci method – permit analysis at the qualitative phenotypic level (Sect. 3.6) of gene action. Hence, the results, important as they are for a quantitative assessment of genetic effects in the mammalian organism, are not satisfactory for an analysis of the mechanisms of mutation induction. Besides, they only provide information for a limited number of gene loci. However, one of the results of this work suggests that the induced mutation rates of various genes might differ considerably.

When the methods of protein analysis by electrophoresis became available, a number of attempts were made to introduce them into mutagenicity testing. Theoretically, this approach is impeccable. Induced mutations can be identified at the molecular level, and electrophoretic systems are now available for a great number of loci (Sect. 6.1.2). The practical difficulty is simply the large number of examinations needed. Inexact as our estimates of codon mutation rates may be, they show an order of magnitude between 10^{-8} and 10^{-9} per codon or amino acid replacement (Sect. 5.1.4.1). Even if the spontaneous mutation rate per gene for electrophoretically detectable mutations were $\sim 10^{-6}$ to 10^{-7}, such a rate would require a huge amount of testing for the chance of discovering very few mutants. This problem is critical in devising ways of screening human populations for mutation rate increases. Is there any way to overcome this difficulty in animal experimentation – apart from setting up huge and expensive testing facilities?

A promising technique for the study of experimentally is the two-dimensional assay with electrofocussing in one direction and electrophoresis in the other that will also be discussed in Sect. 6.1.2 [1516] A highly constant pattern of some hundreds of protein spots can be obtained with liver extracts from inbred mice. Point mutation may lead to shifting of one "spot," which can easily be recognized (Fig. 5.51). To the best of our knowledge, the method has not yet been applied to radiation experiments.

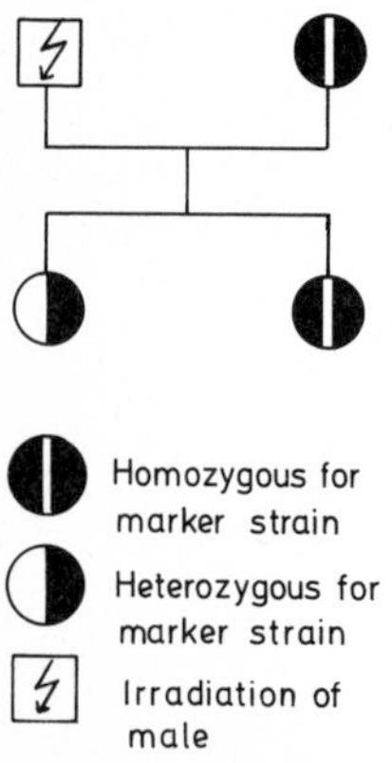

Homozygous for marker strain
Heterozygous for marker strain
Irradiation of male

Fig. 5.49. Principle of the multiple recessive mutation method. A wild-type male *(upper left)* is irradiated and mated with a female of the test stock *(upper right)*. If no mutation is induced, the offspring will be heterozygous for the test loci and, hence, phenotypically wild-type *(lower-left)*. If a mutation at one of the seven loci has been induced in one of the sperms, one animal in the offspring will be homozygous for this gene and show the mutant phenotype

Test Systems for Somatic Mutations. The simplest test system for somatic mutations is chromosome examination in lymphocyte cultures or in bone marrow. The lymphocyte system has frequently been used with humans. It is, in fact, the only one that permits examining high-risk groups of humans with a minimum of distress for genetic effects of exposure to mutagenic agents.

A still simpler test system examines chromosomes irradiated in vitro. In most cases, human lymphocyte cultures have been used. With this system, it was shown that the general rules of mutation induction as worked out in experimental genetics also apply for human chromosomes (Sect. 5.2.1.1). Assays for examining point mutations in single cells in vitro have been developed for some biochemical markers (Sect. 5.1.5), and the same methods are being utilized

Fig. 5.50. Mouse with the phenotype of one of the seven test loci *(piebald spotting)* together with the wild-type father, the test stock mother *(white)*, and the heterozygous wild-type sibs. (Courtesy of Dr. U. Ehling, Neuherberg)

for mutagenicity testing. Theoretically, this approach is very elegant. Its resolving power is several orders of magnitude higher than that of the in vivo methods for detection of gene mutation, since the experimental unit is not the individual but the single cell. Moreover, mutants can be cloned and subjected to all kinds of biochemical investigation. In practice, however, the method still meets with difficulties. For example, selective systems have become available for very few biochemical markers only; some of the variant cells may not be caused by mutation, and extrapolation from the very artificial in vitro conditions to the living body is always difficult.

5.2.1.3 Results of Radiation Mutagenicity Testing in Mammals [1377]

General Effects of Radiation on Mammalian Germ Cells. The germ cell development in both human sexes has been described in Sect. 5.1.3. It is similar in principle in all mammals; minor differences between species are important for planning of mutation experiments but will not be covered here. In the mouse, the following general effects of radiation on germ cell development have been observed.

Acute irradiation of the male germ cells with 2–4 Gy kills most spermatogonia, whereas more mature germ cells (spermatocytes and all postmeiotic cell stages) survive. Hence, there is little impairment of fertility during the first 6 weeks following irradiation. During this period, all germ cells that had already entered the spermatocyte stage will become mature sperms. A sterile period of 2–3 months follows depending on the radiation dose. After this time, fertility is resumed. Meanwhile, the testicular tubules have been repopulated, starting from a very small population of A spermatogonia.

Female animals, after a short lag phase, become permanently sterile even with low doses. When irradiated with 0.5 Gy of X-rays, female mice have three to four litters before they become sterile due to the destruction of oocytes. As they approach ovulation, oocytes become more resistant, and shortly before ovulation, their death rate does not differ from that of the controls. The high degree of sensitivity of oocytes makes mutation experiments difficult. These differences between the sexes and between the stages of germ cell development regarding radiation sensitivity should be kept in mind when the results of mutagenicity experiments are considered.

Chromosome Mutations in Male and Female Germ Cells of Mice [1454]. Spermatogonial mitoses show a high frequency of chromosome aberrations (breaks, reunion figures, dicentric chromosomes) after acute irradiation. Such aberrations can be detected when the first meiotic divisions are examined.

Experimental data from human males are available.[5] Testis biopsy material was obtained from nine human volunteers who had received testicular irradiation with high-dose X-rays (0.78, 2, or 6 Gy). The interval between irradiation and sampling varied,

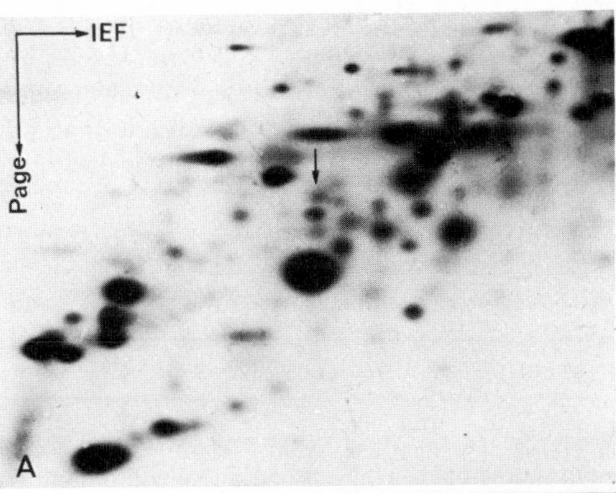

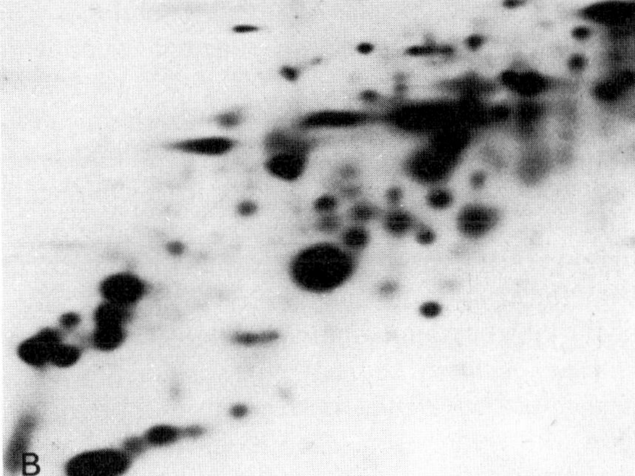

Fig. 5.51 A, B. Two-dimensional, electrophoretic separation of soluble proteins from fetal mouse liver; part of a protein pattern. The protein sample is taken from the supernatant after homogenization and centrifugation of single livers. Separation of proteins is performed on polyacrylamide by isoelectric focusing *(IEF)* in one dimension and by electrophoresis *(PAGE)* in the second. **A** Part of protein pattern of fetal liver after treating the sire with a mutagen (methylnitrosourea). A new protein spot is indicated by an *arrow*. **B** Normal protein pattern of a fetal liver from the same inbred mouse strain. (Courtesy of Dr. J. Klose, Berlin)

depending on the dose. For comparison, adult male individuals of various other animal species were irradiated with similar doses and under comparable conditions (Table 5.25). The doses used in these studies ranged between 100 rad and 600 rad. The increase in the yield of reciprocal translocation with increasing radiation dose was highest in the marmoset and second highest in humans. In various studies on the mouse, and also in rabbits, the effect

[5] Use of human beings for such experiments appears highly questionable. Although the investigators stated that they used prisoner volunteers only under conditions where these men will have no further children [1401], such assurances are difficult to adhere to and there is the further possible risk of carcinogenicity.

was about half that found in humans; guinea pigs and rhesus monkeys showed still lower values. On the other hand, a recent assessment of such studies yielded the somewhat surprising result that (almost?) no Robertsonian translocations were induced by ionizing radiation [1453].

Reciprocal translocations can also be induced by irration of female mice; in this case the details of the outcome vary depending on the experimental conditions (e.g., time interval between irradiation and ovulation).

Direct Evidence of the Outcome of Induced Chromosome Aberrations. There is strong selection against germ cells and zygotes containing chromosome aberrations. After irradiation of female mice with fairly high doses, for example, very few translocations were found among F_1 offspring. After irradiation with as much as 3 Gy, a total of eight translocations was recovered among 1735 progeny (female and male); about half as many as after irradiation of males [1377]. This selection has been examined in detail through various stages of embryonic development by direct examination of chromosomes in early embryonic stages after treatment by irradiation of female mice [1597a]. The decrease in number of zygotes with chromosome aberrations after treatment of oocytes during the preovulatory phase was examined through four stages: the second meiotic division; blastocysts; death during the embryonic age as evidenced by the dominant lethal test; and proportion of chromosome aberrations among living embryos in the late embryonic age. About 88% of all meiotic (II) cells showed induced numeric or structural chromosome aberrations. By late embryonic age all cells in which a chromosomal aberration had been identified were eliminated, and the number of embryos carrying an aberration was not higher than in the controls. The 1977 UNSCEAR report had estimated that a fraction of $\approx$6% of all fetuses with induced chromosome aberrations survived up to birth. The above-mentioned study showed that this was a very cautious estimate; probably the fraction of survivors is lower. The same was shown to be true for mutations induced by a strong chemical mutagen, the cytostatic drug Trenimon [1386a]. As already mentioned, the great majority of spontaneous chromosomal aberrations in humans are also eliminated before birth (Sect. 2.2.4). Therefore, it is reasonable to assume that this will occur with the great majority of all induced aberrations.

Radiation-Induced Genome and Chromosome Mutations: Sensitivity of Certain Cell Stages. By X-linked genetic markers, it is possible in the mouse to distinguish XO from XX animals and to ascertain whether the X chromosome in XO individuals is of paternal of maternal origin. With this method, it has been found that XO individuals occur spontaneously in a frequency of 0.1% to 1.7%, depending on the mouse strain examined. Usually, the single X chromosome is maternal. The paternal chromosome apparently is lost between impregnation and the first cleavage division [1609, 1612, 213].

During this relatively long period – about 4½ hrs – the paternal and maternal genomes do not become fused (Sect. 2.1.2.4); they form "pronuclei." It is at this time that the paternal X is at risk of being lost. This risk may be enhanced by radiation, especially if the animal is irradiated between impregnation and the first cleavage division, i.e., during the pronucleus phase. Radiation at this phase yields exclusively XO animals and no XXY types. This implicates X chromosome loss and not nondisjunction as the cause of the chromosome defect.

Nondisjunction seems to be enhanced by irradiation of male spermatocytes, mainly in the preleptotene stage, as evidenced by metaphase II examinations. Results with irradiation of oocytes are still controversial.

Loss of the X chromosome a short time after impregnation – and during the first cleavage divisions – is also frequent in humans. In fact, it seems ot be the most frequent cause for formation of XO zygotes. Moreover, the X chromosome lost is in most cases the paternal, as shown by marker studies with the Xg blood groups (Sect. 5.1.2). Hence, the similarity to spontaneous X chromosome loss in the mouse is close. It is therefore safe to conclude that the human preovulatory oocyte and the very early zygote may also be susceptible to chromosome loss and possibly to induction of nondisjunction and structural chromosome damage [71].

Radiation-Induced Gene Mutation in the Male Germ Line. Methods for chromosome examinations were not available to mammalian radiation genetics in its most fertile period during the 1950s. Therefore, emphasis was mainly on the induction of gene mutations. The most extensively used method was the multiple loci test for visible recessive mutations. The number of animals tested goes into the millions. As expected, the data show a marked increase in the number of mutations compared with the spontaneous mutation rate in both sexes. The mutation rate was estimated to be about 30×10^{-8} per roentgen per locus in the male and around 18.5×10^{-8} per roentgen per locus in the female. This induced mutation rate differs widely depending on treatment conditions. What are these conditions?

Table 5.25. Comparison of the slope of linear regression "b" estimated for various species of mammals for the induction of reciprocal translocations in spermatogonia. (Data from various authors [1377])

Animal	b: $10^{-4} \pm$ S.D. 10^{-4}
Rhesus monkey	0.86 ± 0.04
Mouse (various strains)	between 1.29 ± 0.02 and 2.90 ± 0.34
Rabbit	1.48 ± 0.13
Guinea pig	0.91 ± 0.18
Marmoset	7.44 ± 0.95
Man	3.40 ± 0.72

The linear regression coefficients (b) were estimated for all species using the data given by the authors on doses giving the peak yield of translocations.

Dose-Rate Effect. Contrary to the Drosophila data mentioned above, a dose administered at one time (a high dose rate) has a much stronger effect than the same dose distributed over a longer time (a low dose rate) [1611]. Figure 5.52 shows the dose-effect curves for a series of experiments on male mice with high and low dose rates. The dose-rate effect is obvious from a comparison of the upper two curves (acute X-ray radiation) with the lower two, which show the effect of chronic γ irradiation with extremely low dose rates. In spite of the huge number of examined animals, the 90% confidence intervals for all mutation frequencies were rather high. A dose applied at a high dose rate induced about three times as many mutations as the same dose at a low dose rate.

Such a dose rate effect was also shown in female germ cells; it is caused by repair processes that are more efficient when the damage is spread over a longer period. The lessened effect for single irradiation with 10 Gy is due to prior elimination of severely damaged cells. Such paradoxic dose-effect curves are not rare in mutation genetics; a mutagenic agent may damage a cell at two levels. It may impair it in a way that prevents it from further division, or it may leave the cell intact, damaging only the genetic material. In analyzing mutagenic activity at the phenotypic level, we consider only damage to the genetic material, tacitly assuming that the viability of the cell is not affected. If a relatively low dose affects only or mainly the genes, whereas a higher dose also damages cell viability, a paradoxic dose-effect curve may result.

Another factor that influences the mutation rate is the developmental stage of the germ cells at which irradiation occurs. In female mice, for example, irradiation administered more than 7 weeks before mating did *not* cause any increase in mutation rate. In

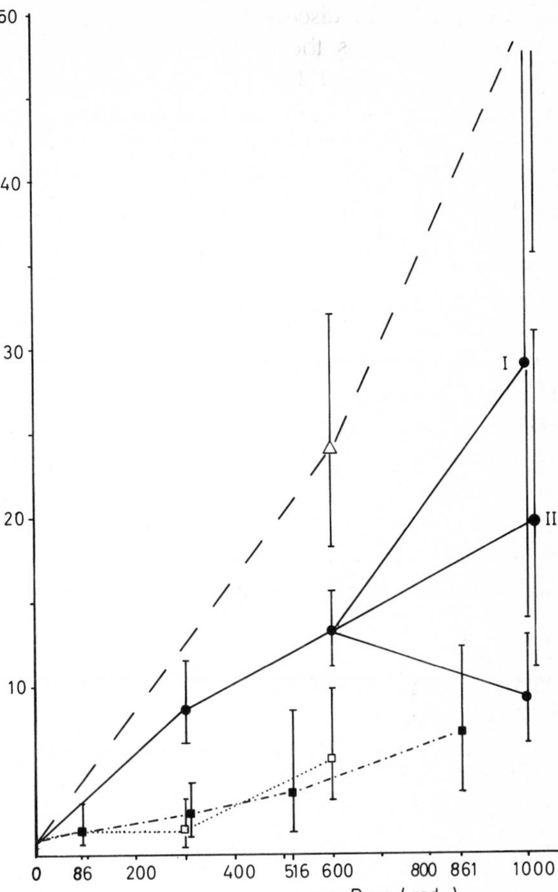

Fig. 5.52. Dose-effect curves and dose-rate effect after irradiation of male mice in the poststerile phase (irradiation of spermatogonia). Number of observed mutants per 100,000 gene loci tested. For all experimental points, the 90% confidence belts are included. ·-· irradiation with one acute dose. The three different points for 10 Gy indicate one acute dose (lowest point) and two experiments with fractionated application. △---△ two doses with an interval of 24 h. ■-·-·-■ chronic γ irradiation (0.9 mGy/min). □····□ chronic γ irradiation (0.1 mGy/min). (Adapted from W. L. Russell, 1972 [1609])

males, irradiation of more mature cells (spermatocytes) led to a higher mutation rate.

Ionizing radiation induces mainly structural chromosome damage and only relatively few point mutations [1607; 1608]. More thorough examination of many induced mutations at the seven loci of the mouse has shown that structural aberrations are also the most frequent cause of these mutations. Most of them were small deletions.

In addition to recessive mutations, dominant mutations – especially those affecting the skeleton or causing cataracts of the lenses – were also studied [212, 1377]. The extremely variable expressivity and markedly incomplete penetrance of such mutants has been discussed in Sect. 3.6.2.5.

Doubling Dose. In discussion on genetic radiation hazards for humans, the so-called doubling dose is frequently considered [1537]. The concept of a doubled mutation rate is entirely arbitrary and was selected as a convenient reference to find a dose of radiation that, if delivered to a human population, would double the natural mutation rate. *It is obvious from the foregoing discussions that there cannot be just one doubling dose.* The doubling dose will vary with the type of mutation, with the germ cell stage at which irradiation occurs, with the specific kind of radiation, and with the dose rate. The use of a single doubling dose for all types of exposure of humans is therefore meaningless. Doubling doses for specific situations, such as an acute exposure doubling dose or a chronic exposure doubling dose for radiation, is more meaningful. Data from mouse experiments suggest doubling doses of 18–52 rem for acute and ≈100 rem for chronic irradiation. It is reasonable to assume that doubling doses in humans may be of about the same order of magnitude, although data suggesting that humans are more radioresistant exist (see Sect. 5.2.1.5).

Population Experiments with Mice and Other Mammals. The experimental results discussed relate only to the first of the five questions of Sect. 5.2.1.2: In what way does radiation affect the genetic material? However, some animal experiments have been planned with the fourth question in mind: What are the long-term consequences for the population of a mutation rate increase due to irradiation?

Some experiments with long-term irradiation of artificial animal populations were continued over many generations. The radiation doses were of the order of magnitude of several hundred roentgen per generation; some were administered at high and others at low dose rates. The overall effects were surprisingly minor: Litter size was usually reduced, the number of dead implantations (dominant lethals) was increased, and in some experiments there were more sterile animals. Progeny after irradiation of many generations of ancestors in some cases even surpassed the nonirradiated controls in life span [213].

In one experiment with rats [1579, 1580], a behavioral characteristic – maze learning – was examined. Progeny of irradiated parents were on the average somewhat less "bright" than the controls.

The outcome of these population experiments can be interpreted optimistically or pessimistically. The optimist may conclude that even long-term irradiation with very high doses does not cause much genetic damage. Almost all induced mutations are eliminated during meiosis or lead to early zygote or embryonic death. Such effects, however, would not be very important for human populations. The pessimist, on the other hand, may object to extrapolation from multiparous animals to humans. He may argue that embryonic and newborn deaths found in the experiments would turn up as a high stillborn rate and as early death of malformed children. In general, such animal experiments may be criticized as not permitting reasonable extrapolation to the human species.

Most of these long-term population studies in mice date back to the late 1950s and early 1960s. Little if any work of this kind seems to be continuing since the explanatory power of these studies for the human situation is thought to be low.

One study was carried out in a rodent population exposed to a high level of natural irradiation in Kerala, southern India. On the average, γ irradiation was 16 mGy per year. This is 7.5 times higher than in a control area. There was no difference whatsoever as to skeletal abnormalities in comparison with animals from other areas.

Conclusions from Mouse Radiation Genetics for Genetic Hazards to Humans. In Sect. 5.2.1.2, a number of questions were formulated, which radiation genetic research in the mouse and in other mammals may help to answer.

1. What kind of mutations are induced? Ionizing radiation seems to induce mainly chromosome mutations. There is good evidence that genome mutations, especially aneuploidies, are also induced. Many of the induced mutations affect only a single functional gene; it has not yet been determined how many of them may be gene mutations in the strictest sense. All these findings probably can be extrapolated to humans.
2. Are mutations induced mainly in germ cells or in somatic cells? There is good evidence that mutations are induced in all exposed cells – germ cells as well as somatic cells. It may be assumed that answers to these two questions hold true for humans as well.

3a. What stage of germ cell development is primarily affected? In the male, all germ cell stages may be affected, but meiosis acts as a potent filter, especially for chromosome aberrations. Postmeiotic stages up to and including the time of fertilization seem to be more endangered than premeiotic germ cells. For this result, too, extrapolation to humans seems reasonable. In the female mouse, oocytes may be susceptible to mutation induction and to chromosome loss only when irradiated within the last 7 weeks before ovulation. Oocytes during the dictyotene stage of many years without cell divisions are resistant to mutation induction. This result from the multiple-loci experiments in mice has been largely confirmed by cytogenetic

evidence. An optimist would conclude that human oocytes will also be radiation resistant during most of their lifetime. A pessimist, on the other hand, would point to possible species-specific differences. We tend to hold the optimistic view.
The oocyte from some days before ovulation until several hours after impregnation is especially susceptible to chromosome loss - especially loss of the X chromosomes - as well as to induction of structural aberrations and possibly meiotic nondisjunction. There is no reason why this could not be true for the human oocyte as well. Careful radiation protection during the weeks around conception is the obvious countermeasure. Are oogonia also sensitive to mutation induction by radiation? Mammalian oogonia are only found during the embryonic age; therefore, the pregnant dams must be irradiated, and later on the daughters must be mated. So far, however, there is only limited experimental evidence that points to induction of translocation at about half the rate of spermatogonia. It is nevertheless worthwhile to look for further evidence since experiments with a chemical mutagen - the alkylating agent Trenimon - have shown a high sensitivity of oogonia.

3b. Are mutations transmitted to the next generation or are they eliminated? Information is available mainly for chromosome aberrations. Here, the meiotic division acts as a good filter. Many aberrations, however, pass this filter, and others are induced during and after meiosis in the female. At least 90% of the induced aberrations are eliminated during embryonic development - more than half of them already before implantation in the uterus, most of the remaining a short time after implantation. A minority, 5% or less, survive, giving rise to offspring with translocations or being aneuploid themselves. The high incidence of chromosome anomalies among spontaneous abortions in humans (Sect. 2.2.4) makes it very likely that most induced chromosome aberrations would be eliminated in the same way. More specifically, the experimental results suggest that in humans loss of zygotes before implantation may be at least as high as after implantation. A significant number of aberrations would survive, increasing the number of children with aneuploidies and balanced or nonbalanced chromosome aberrations.

3c. What phenotypic changes are to be expected in the offspring? Zygote loss before implantation in the uterus would probably go largely unnoticed, leading only to a short delay of menstruation. Zygote loss after implantation would show up mainly as spontaneous abortion, the distribution between early and late abortion possibly being similar to the distribution without radiation. Aneuploidies and unbalanced structural anomalies would lead to the well-known chromosomal syndromes. Balanced translocations would mostly be phenotypically unnoticeable in offspring of exposed parents but could lead to unbalanced zygotes in the following generations.
Dominant gene mutations may lead to a certain increase of the well-known dominant phenotypes. The frequency of such traits in the human population is maintained by an equilibrium between mutation and selection (Sect. 5.1.3.1). However, experience with skeletal mutants of the mouse suggests that other, and possibly much more frequent, dominant mutations may lead to less distinctive and more variable phenotypic changes, such as slight anomalies of the skeleton, connective tissue, or other organ systems. Here, our extrapolations are bound to be much more uncertain than with visible chromosome anomalies.

4. If mutations are induced in somatic cells, which cells are especially endangered, and what are the consequences for the individual? In principle, all types of somatic cells are at risk. However, it is safe to assume that the risk for cells that are frequently dividing is especially high. These cells may give rise to cell clones with selective advantages in comparison with other cells of the same type - and in the end, to malignant neoplasias. Such neoplasias caused by irradiation have actually been demonstrated in human populations (see below).

Altogether, radiation genetics of the mouse has provided results that are useful for estimation of the radiation risk to humans. They gave us a fairly good idea about induction and transmission of genome and chromosome mutations. As to gene mutations, the evidence is not quite as good, but it is known in principle that such mutations may be induced as well as transmitted.

Impact of Prenatal or Postnatal Mutational Damage on Man. In discussions of the impact of mutations of human populations, it is important to differentiate between prenatal and postnatal effects. Mutagenic agents that lead to early abortuses may not even be noticed and may be manifested by a slight delay of menstrual period. Their impact is zero. Production of abortuses during the first trimester will be recognized but have relatively little impact. Abortuses during the last trimester of pregnancy are rare, and stillbirths have a more significant impact on the family but cause less of a burden than a variety of genetic diseases and birth defects. Here again, those defects associated with relatively early mortality cause fewer mental, social, medical, and family problems than diseases associated with long term suffering for the patient and his family. Since most mutagenic agents will also be carcinogenic, the production of neoplasms will be a more direct and immediate effect.

5.2.1.4 Human Population Exposure to Ionizing Radiation

How strongly and how widely is the present-day human population exposed to ionizing radiation? This is the second question that needs to be answered by the scientist who wishes to estimate the extent of potential radiation damage to the human population, and it has frequently been examined. Only very vew aspects and a small selection of data will be discussed here. First the natural background radiation and then the increase due to modern civilization – including diagnostic medicine – will be considered.

Natural Background Radiation. All human beings are continuously exposed to natural sources of irradiation. The average dose per year for cosmic radiation depends on altitude above sea level and latitude. Terrestrial radiation is higher in areas with primitive rock than in alluvial lands. The total mean irradiation received by the gonads of human beings in low altitude areas within a 30-year period (one generation) is estimated to be between 3 and 4 rem (Table 5.26). More recent estimates are higher; in earlier reports, an additional radiation load due to radon in the air had not been given due consideration.

Additional Irradiation Due to Modern Civilization. Some irradiation estimates for Germany and the United States are listed in Table 5.6. Medical diagnostics and therapy constitute the major part. Within the large group of X-ray diagnostic measures, examination of the abdomen and the pelvis are the chief sources of relevant exposure. Improvement in X-ray technology as well as strict regulations for supervision have contributed much to minimize this load. However, some X-ray diagnostic measures will increase the probability of mutations; any X-ray exposure should be clearly indicated and be carried out with maximum shielding. The benefit to the individual should be clear and must be weighed against the potential damage to him or her and to future generations. By far the largest amounts of genetical-

Table 5.26. Average radiation load of human populations[a,b]

Sources	Europe		U.S.A.	
	mrem/year	$\frac{1}{100}$ J/kg/30 years	$\frac{1}{100}$ mJ/kg/year	$\frac{1}{100}$ J/kg/30 years
Natural background radiation				
Cosmic radiation	(30)–50 (–120)			
Mean terrestrial radiation	60			
Incorporation of radioactive elements	20			
	≈130	≈3.9	≈100	≈3.06
Additional radiation due to modern civilization				
Medical diagnostics and therapy				
1958	20	0.6	73	2.2
1971	50	1.5		
Occupational (without nuclear technology)	<1	<0.03	0.8	0.024
Nuclear technology	<1	<0.03	0.003	0.0001
Fallout (atomic bomb testing)	8	0.24		
Minor sources (television; watches)	<2	<0.06	2	0.06
Air traffic	<1	<0.03	0.4	0.120
Sum total additional irradiation	≈60	≈1.8	≈ 80	≈2.4
Sum total natural background and additial irradiation	190	5.7	180	5.4

[a] Adapted from Barthelmess, 1973 [1385].

[b] These figures do not include radiation received by the general population from nuclear power plants. Under circumstances of normal operations exposure of the general population is minimal. Workers in the nuclear power industry receive exposure of 0.006–0.008 J/kg/year.

ly relevant irradiation are to cancer patients needing radiation therapy of pelvic organs. Many such patients will be of postreproductive age or will die of their cancer within a relatively short time. However, if they do desire children, the results of radiation genetics research in mammals may help to minimize the genetic risk by advising them to avoid conception during and some weeks after therapy.

Nuclear power plants constitute a special problem that is being much discussed in all industrialized countries. Depending on the type of reactor and on other conditions, the actual irradiation varies and is said to be much lower than the allowed limit in most cases. This statement is true, however, only in the absence of irregular emissions due to technical failure, accidents or sabotage. The risk of such an incident occurring sooner or later cannot be estimated but is said to be very low. On the other hand, experts assert that the energy needs of future decades cannot be met without nuclear power and that the wellbeing, even the life, of future generations heavily depends on sufficient energy supply. The geneticist can only hope that a major effort will be made in the development of alternate technologies for clean energy.

Of all the sources contributing to exposure, the medical sector is the only one adding more than a trivial share to the whole load. Some of it is unavoidable. For example, hospital nurses administering a radium device to cancer patients will inevitably receive some irradiation. In many cases, however, radiation load can be kept at a minimum. In activities where heavier exposure cannot be avoided, only persons who are past child-bearing age, or who for other reasons are unlikely to reproduce, should be employed. It is an open question whether we will have to face increased exposure to radiation in the future. On the one hand, there is the indisputable increase due to professional use of radiation and radioactive substances and due to nuclear power plants. X-ray technology, on the other hand, is being so vastly improved, especially in medicine, that a certain decrease in exposure can be anticipated. As Table 5.26 shows, the average load from all sources of modern civilization, excluding nuclear power plants, has not quite reached the average natural background radiation.

5.2.1.5 How Much of an Increase in the Spontaneous Mutation Rate Must Be Anticipated?

How can the probable increase in spontaneous mutation rate be calculated? This is the third question to be answered if we are to get an estimate of the projected genetic damage due to radiation. We need information on three points:

1. How much irradiation will the average individual receive?
2. How many additional mutations per dose unit will be induced?
   When these two questions are answered, we shall know how much the mutation rate will increase relative to the spontaneous rate. To turn this relative estimate into an absolute one, another question must be answered.
3. How many mutations would occur "spontaneously," i.e., without additional irradiation due to modern civilization?

For the first two questions, answers may be derived from Sect. 5.2.1.3 and 5.2.1.4, together with some additional, direct observations in human beings. The third question turns out to be the most difficult and will need special discussion. The data in Table 5.26, indicate that the average irradiation per individual within a generation of 30 years will increase from 0.03–0.04 Gy ("natural" irradiation) to 0.07 Gy, i.e., that modern civilization will approximately double the background irradiation. Such a doubling must be carefully distinguished from the dose of irradiation which doubles the mutation rate. This dose is much higher than 0.03 Gy. This answers the first question. The answer to the second question is somewhat more complex and requires a more elaborate discussion.

How Many Additional Mutations Per Dose Will Be Induced? The various classes of mutations should be considered separately, and dose rate as well as sex should be taken into account (Sect. 5.2.1.3). Most humans are exposed to very low doses of chronic irradiation administered at a very low dose rate; very few are exposed to high doses at very high dose rates.

In the following, we shall first discuss direct evidence available for irradiated human populations. This evidence will then be compared with the data gained in experiments with mice and other mammals as discussed above. This will lead us to a quantitative estimate for various kinds of mutations relative to the spontaneous rate.

Phenotypic Characteristics in Irradiated Human Populations. Apart from ethically dubious experiments with human volunteers, genetic radiation effects can be assessed directly only when humans have been exposed to radiation either for therapeutic purposes or by accident. In therapy, dose estimations are usually fairly accurate, but the number of individuals is low, and they are selected for various diseases. Radiation is usually administered in a few very high doses during medical therapy or at a low dose rate during professional exposure. In the case of accidents, estimation of the dose may be very inaccurate, but dose and dose rate are usually high.

Survivors of Atomic Bombs in Hiroshima and Nagasaki [1639, 1640, 1573, 1576]. Following the explosions of the atomic bombs on Hiroshima and Nagasaki, genetic studies on the survivors were organized by American and Japanese scientists under the Atomic Bomb Casualty Commission (ABCC) as early as 1946 and have been continuing ever since, from 1975 onward under the auspices of the Radiation Effects Research Foundation (RERF). This has been the only attempt to organize an ongoing "big science" project in human genetics.

1. Are there any differences between children conceived after the parents had been exposed to the atomic bomb and children of unexposed parents?
2. If there are differences, how are they to be accounted for?

Data collection was facilitated by the food-rationing system in postwar Japan, which provided additional food for pregnant women. When applying for extra rations, the women were given a questionnaire regarding their past pregnancies, their exact location when the bomb was dropped, and any symptoms that might indicate radiation illness afterward. When the child was born, relevant data on delivery and health status of the newborn were added, and each child was examined by a doctor of the ABCC commission. About one-third were re-examined 9 months later. In the following years, the method of examination and follow-up was modified in various ways. The following parameters were considered:

a) Sex ratio (ratio between ♂ and ♀ at birth);
b) Congenital malformations;
c) Stillbirths;
d) Body weight at birth;
e) Death within the first 9 months of life and death during childhood and youth;
f) Anthropometric measurements;
g) Autopsy results.

The critical reader might ask why the two most important indicators of genetic damage - chromosome aberrations in somatic and germinal cells and dominant or X-linked recessive new mutations - were omitted. The answer is simple: The methods for chromosome examination in humans only became available 10-15 years later, and dominant or X-linked gene mutations are so rare that a sufficient number could not be expected in the examined population.
Every parent was categorized as belonging to one of five exposure groups, the lowest group consisting of those who were absent from the city at the time of the bombings; the highest group comprising persons who had suffered from symptoms of radiation sickness (loss of hair; petechiae; gingivitis) in the months afterward. Consanguinity of the parents and maternal age were considered to avoid biases due to these variables.

Shift in Sex Ratio Due to X-Linked Lethals? The results of these studies can be summarized very briefly. Regardless of the extent of irradiation exposure, no significant difference was found in the number of malformations, in stillbirth rate, or in any other of the examined parameters - with one possible exception: the sex ratio. If the mother's germ cell includes an X-linked recessive lethal gene, it will go unnoticed in a heterozygous daughter. A male zygote, however, will be hemizygous and will be killed. Hence, irradiation-induced X-linked recessive lethals will shift the sex ratio in favor of daughters. X-linked dominant lethals, on the other hand, when induced in female germ cells, will kill male and female zygotes alike. The sex ratio will not be influenced. When induced in the male, X-linked dominant mutations will kill only female offspring, as the X chromosome of fathers is transmitted only to daughters. X-linked recessive lethals induced in fathers will not influence the sex ratio of their children at all. Therefore, in contrast to the radiation effect on mothers, irradiation of fathers would shift the sex ratio in favor of boys.
The analysis was based on 46,752 children born within 10 years after exposure of the mother (27,057), father 7,525), or both parents (12,170). For all categories, the sex ratio showed shifts in the expected direction [1639]. There were more females after maternal and more males after paternal irradiation. When both parents were exposed, the maternal appeared to exceed the paternal effect. The results were not statistically significant using conventional tests. However, in 11 out of 12 comparisons, the effects were in the expected direction.

Support for the Sex Ratio Shift by Studies After Exposure to X-rays. A similar sex ratio shift was observed among children of parents exposed to X-ray therapy for treatment of ankylosing spondylitis, anal and vulval pruritus, and other conditions. Two studies, one from France [1530], the other from the Netherlands [1632], together comprised several hundred patients irradiated with high doses, sometimes several hundred roentgens. The sex ratio among children conceived after irradiation deviated in the expected direction, and the deviations were significant.
Therapeutic irradiation and irradiation by atomic bombs involve high doses and high dose rates. Pro-

fessionally exposed individuals, on the other hand, usually receive relatively low doses at very low dose rates. Do these doses also cause a shift in sex ratio? A study of radiologic technicians in Japan yielded a positive answer: Not only was the sex ratio altered but sterility appeared to be more frequent than in the general population [1653].

Second Study in Hiroshima and Nagasaki: No Sex Ratio Shift [1640]. Some years after the first report, additional data on pregnancies between 1956 and 1962 were published. The analysis proceeded in the same way, although ascertainment of data differed in some details from the first sample. The sample consisted of a slightly smaller number of children of irradiated partents. In this study, no differences in sex ratio could be demonstrated.

The whole set of sex ratio data from Hiroshima and Nagasaki may be interpreted in one of three ways: either the sex ratio deviations found in the earlier analysis were fortuitous, and the radiation doses received by the atomic bomb survivors did not induce enough X-linked lethals to bring about a sex ratio shift, or sex ratio is actually no indicator for X-linked mutations, the argument being too simplistic, or the effect in the earlier series was real, but in the 10–15 years between radiation exposure and the conception of children of the second series germ cells containing lethal mutations were eliminated. The results showing sex ratio shifts in medically irradiated individuals in Europe [1530, 1632] favor the latter explanation.

The fact remains, however, that sex ratio is an unsatisfactory measure of genetic effects, being influenced by many other variables, such as age of parents and general living conditions. The irradiation data, in spite of the great expenditure in time and manpower, have provided suggestive but not entirely convincing evidence for induction of X-linked lethals by irradiation.

A Reassessment Using Additional Data and Methods [1572, 1615, 1641](Table 5.27). A cytogenetic investigation of the children of survivors started in 1968. Hence, aberrations leading to severe malformations and early death could no longer be ascertained, since the children had died in the meantime. In the mid-1970s, studies were also begun using electrophoretic techniques for qualitative and quantitative variation (rare variants) of the blood. Moreover, data on "untoward pregnancy outcome," meaning a stillbirth and/or major congenital defect and/or death during the 1st postnatal week, – were reevaluated for 70,082 deliveries. In addition, death up to the age of 17 was considered. None of these parameters gave a statistically significant difference compared with the unexposed population. With these recent data, however, the evaluation strategy was changed. In the earlier studies, exposed individuals had been classified into subgroups according to their distance from the hypocenter at the moment of the detonation, and presence or absence of signs of radiation sickness. Then, deviation from the null hypothesis (no genetic effect at all) was tested. In the new studies, each exposed individual was assigned an individual dose estimate on the basis of better knowledge of the kind of radiation produced by the bombs, and taking account of all available information, such as distance from the hypocenter, sheltering, etc. The *average* body dose received by all those actually exposed (i.e., situated at 1600 m or less from the hypocenter) was somewhat more than 10 rad, the gonadal dose being about half this amount. However, the variation among exposed individuals was enormous, a few individuals having received doses of more than 100 rad. After individual calculation of exposures the authors no longer tested the null hypothesis. They argued that since it was known and generally accepted that ionizing radiation induces mutations any increase in effects that could reasonably be related to mutations, whether statistically significant or not, was probably caused by additional induced mutations. Hence, regressions of increase of ill effects on exposure were calculated. These regressions, together with certain assumptions on the share of new mutations explaining the "spontaneous" occurrence of the events under study, were used to esti-

Table 5.27. Genetic effects in children of atomic bomb survivors in Hiroshima and Nagasaki [1572, 1641, 1615]

	Exposed	Controls
a) Protein mutants (30 gene loci)		
Number of loci examined	543664	386706
Confirmed as new mutations	3	0
Mutation rate: 5.4×10^{-6}		
b) Gonosomal chromosome aberrations		
Number of children examined	5762	5058
Gonosomal aberrations (XXX, XXY, XYY)	16	13
c) Untoward pregnancy outcome		
Number of births examined	30625	39457
Untoward outcome	1498 (4.80%)	1846 (4.68%)
d) Death up to the age of 15		
Number of children	27925	22764
Deaths up to the age of 17	1833 (6.56%)	1398 (6.14%)

mate doubling doses. As an example, it was assumed that 1 in 400 newborns suffered from an untoward pregnancy outcome due to a fresh mutation. The estimated doubling doses were 535 rem for numerical sex chromosome anomalies, 69 rem for untoward pregnancy outcome, and 147 rem for childhood mortality. Moreover, a slight but statistically nonsignificant increase in rare electrophoretic variants was found; here no doubling dose estimate is possible, since there were no new mutants among the controls. Compared with doubling dose estimates from mouse data – 40 rem for acute and 100 rem for chronic irradiation (Sect. 5.2.1.3) – and considering that the human irradiation was acute, the estimates suggested that humans were more radioresistant than mice. However, the data are hardly comparable. Even the genetically best-defined parameter, frequency of chromosome aberrations, was obscured by the fact that newborns could not be examined, since cytogenetic methods were not available at that time. This excludes not only most structural aberrations but most autosomal trisomies as well. For example, there was no increase in Down's syndrome. The estimate is based entirely on X- chromosomal numerical aberrations, and for this category experimental evidence for an increase is slender even in the mouse, except for loss of the paternal X after irradiation in the pronucleus state of the fertilized oocytes. Untoward pregnancy outcome and death before the age of 17 depend on so many different environmental factors that any estimate of a genetic impact must be very crude indeed. This remains true even when factors of importance, such as consanguinity and socioeconomic status, are carefully controlled. The method used to search for electrophoretic mutants will miss deletions that are more likely to be induced by radiation than point mutations for which (almost) no differences were found.

Taken together, the studies on children of atomic bomb survivors in Hiroshima and Nagasaki show contrary to common opinion, that even an atomic holocaust does not lead to a genetic catastrophe among the progeny of the fairly heavily radiated survivors. On the contrary, genetic effects are hardly discernible in the first generation of offspring. We shall show that this will probably apply for future generations as well. The horror of such an atomic holocaust relates to the immediate and short-term effects on exposed individuals and, to a lesser extent, to mutations in somatic tissue manifesting in later life, for example as malignant neoplasias with a latent period of as long as 10–40 years.

Irradiation of Parents for Medical Reasons and Trisomy 21 in Children. Reports on the frequency of trisomy 21 (Down's syndrome) among children of mothers exposed to X-ray diagnostics or therapy are contradictory. Sigler et al. (1965) [1620] compared the radiation exposure of 216 mothers of children with Down's syndrome with that of 216 other mothers of the same age and origin. Exposure was clearly higher among the mothers of children with trisomy 21. Uchida et al. (1968) [1663] compared the frequency of aneuploidy among the children of radiation-exposed and unexposed women. Whereas the other examined parameters did not show differences, there were ten children with aneuploidy among the progeny of the irradiated women – eight cases with Down's syndrome and two with trisomy 18. Among the controls, only one child with Down's syndrome was found; the difference is statistically significant. However, the gonadal radiation doses received by these mothers were very low, between 0.007 and 0.126 Gy. There are additional studies suggesting a higher incidence of aneuploidy after radiation exposure. Meanwhile, the problem has often been reexamined (for a review, see [1377]); most studies failed to show a radiation effect. Some of the earliest studies were retrospective, i.e., radiation history was established after the children were born.

This mode of study can introduce a bias if the search for radiation exposure is – understandably enough – more thorough in mothers who have had a child with Down's syndrome than it is among control mothers. The study by Uchida, on the other hand, was prospective. The pregnancy outcome of mothers with and without a history of radiation exposure was compared. Studies yielding negative results were both retrospective and prospective.

From the mouse data discussed in Sect. 5.2.1.3, one would conclude that maternal irradiation in low doses considerably prior to conception should not induce a higher nondisjunction rate in the oocyte. Moreover, women in need of X-ray studies may differ from other women in their health status, which may affect their risk of nondisjunction. Hence, a spurious correlation is very well possible. Recently, evidence has been brought forward that irradiation of human lymphocytes in vitro with small doses may enhance the frequency of somatic nondisjunction [1664].

Higher Incidence of Structural Chromosome Anomalies and Down's Syndrome in Human Populations Exposed to High Background Radiation? Some areas in Brazil and in South India (Kerala) have a high background radiation, about 10–100 times the normal levels, due to a high content of radioactive elements such as thorium and radium in the monazite sands of the soil. In Brazil, a chromosome study was carried out in 12,000 chronically exposed inhabitants. There was a significant but marginal increase of chromosome-type aberrations such as deletions, dicentric chromosomes, and ring chromosomes in lymphocyte cultures [1381]. In a population of 12,918 individuals living in a high-irradiation area of Kerala, South India, 12 individuals with Down's syndrome, 12 others with severe mental deficiency and additional malformations, and 11 with severe mental deficiency but without additional malformations were found, as compared with no Down's syndrome cases, only one case with severe mental deficiency and malformations, and 2 patients with severe retardation without additional clinical signs observed among 5,938 controls. Moreover, chromosome counts in blood cultures from exposed individuals showed a slight increase in number of chromatid and chromosome aberrations [1523]. While the increase of chromosome aberrations in somatic cells, which was reported in both studies, is probably real, the higher incidence of Down's syndrome in the second study could easily be caused by an ascertainment bias, or by other differences between the test and control populations, since Down's syndrome was unusually rare in controls. Careful prospective studies are necessary.

In China, studies on a population of about 80,000 were performed in two high-radiation areas in Yangjiang county. In these areas, background radiation is about three times as high as in the control district, where it corresponds roughly to that found in most other populations of the world. As a rule, the families had lived for six or more generations in these regions; there was no major difference in general living conditions between high-irradiation and control areas. These studies failed to show any increase in cancer mortality or in the fraction of children suffering from specified genetic defects and diseases compared with controls. A slight increase in the number of Down's syndrome patients could be explained by a higher average maternal age in the exposed group. The fraction of chromosome abnormalities in lymphocytes was slightly but not significantly increased in exposed individuals [1416, 1427, 1436, 1487, 1535].

Evidence of Somatic Chromosome Mutations After Exposure to Radiation:

Medical Therapy. Tough et al. (1960) [1659] were the first to describe structural aberrations in the chromosomes of two patients who had been irradiated for treatment of ankylosing spondylitis. Since then, radiation sensitivity of human somatic chromosomes has often been examined [1387, 1395]. Some results may be mentioned:

a) Within a certain time after therapeutic irradiation, ~25%–35% of the cells (in most cases lymphocytes) contained structural chromosome aberrations.
b) With time elapsing between irradiation and examination, the number of cells showing dicentric chromosomes, ring chromosomes, and acentric fragments declines – mostly within the first 2 years after irradiation. Yet, after 10 years, such anomalies still were about four times more frequent than among the controls. The number of reciprocal translocations, on the other hand, was not much lower after 10 years than immediately after irradiation (Fig. 5.53). Following combined X-ray and radium therapy of gynecologic tumors, structural chromosome defects in the lymphocytes were visible as long as 25 years later. Similar aberrations were observed in patients treated with radioisotopes such as ^{131}I or ^{32}P.

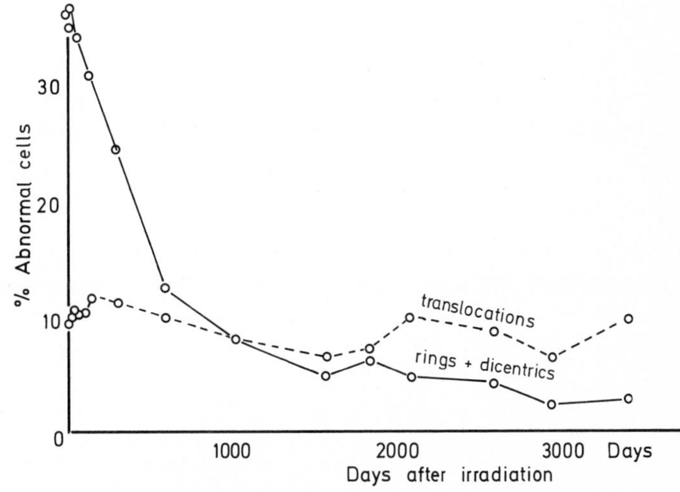

Fig. 5.53. After therapeutic radiation exposure of women with gynecologic tumors, the fraction of cells showing dicentric chromosomes and ring chromosomes decreases (O–O–O) in the years following exposure. The number of translocations, on the other hand, (O--O--O) remains more or less stable. (Bauchinger, 1968 [1387])

Professional Exposure. Chromosome aberrations in individuals professionally exposed to chronic irradiation have been frequently described. Cells with ring or dicentric chromosomes occur very rarely spontaneously (1/2,000 cells after 48 h of lymphocyte cultivation, 1/8,000 after 72 h). Therefore, they are good indicators of radiation exposure. However, simple chromosome breaks also increase in number [1387]. Exposed groups include workers painting phosphorescent materials on watch dials, reactor personnel, and persons involved in radiation accidents. It has even become possible to estimate the radiation doses an individual has received from the extent of cytogenetic changes. Such biologic dosimetry is a useful test in human radiobiology.

Atomic Bomb Survivors. Sasaki and Miyata [1614] examined 51 survivors of the atomic bomb about 22 years after exposure, comparing them with 11 untreated controls. No less than 83,506 cells were analyzed. In the irradiated sample, the rate of dicentric chromosomes and rings was 0.0027 per cell (201 in 73,996 cells). Among the controls, it was 0.0002 (2 in 9,510 cells). The number of cells with stable, symmetric translocations was also increased. There was a clear relationship between the distance from the hypocenter at the time of the bombing and the number of aberrant cells (Fig. 5.54).

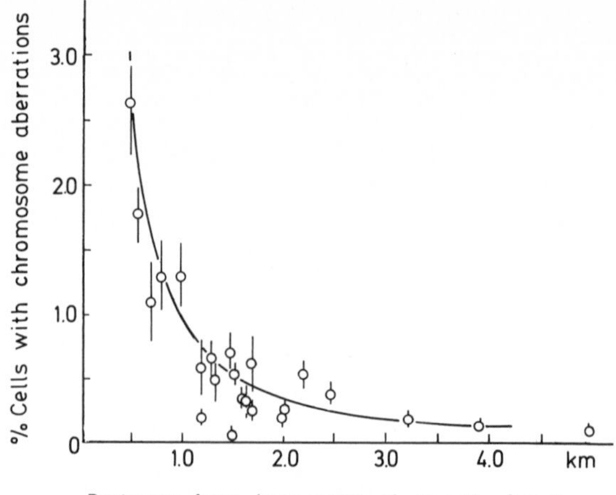

Fig. 5.54. Frequency of cells (lymphocytes) with chromosome aberrations in relation to the distance from the hypocenter at the time of atomic bombing. (Sasaki and Miyata, 1968 [1614])

Other studies with similar results were carried out by Bloom et al. [1395]. Here, four heavily exposed individuals who had received radiation doses of between 2.07 and 6.42 Gy were found to have developed clones with aberrant cytology. In Sect. 5.1.6, in the discussion on syndromes showing enhanced chromosome instability, it was explained that clone formation points to a selective advantage of the cells having this chromosome aberration. Cell clones having such an advantage may easily develop into malignant neoplasias.

Neoplastic Disease in Atomic Bomb Survivors. Follow-up studies among patients who have received extensive radiation and among survivors of the atomic bombs in Hiroshima and Nagasaki have shown an increased frequency of a variety of malignancies. These include leukemia [1393, 1557] (Fig. 5.55), cancer of the female breast [1377, 1555], cancer of the lung [1497], cancer of the thyroid [1434, 1557] and of the digestive system. These findings are compatible with the somatic mutation theory of cancer (Sect. 5.1.6) and therefore not entirely unexpected. The "latent" period between radiation exposure and clinical diagnosis of malignancy usually is very long (more than 10 years) particularly for solid tumors. Further study of low level radiation as possible precursors of human cancer is of great public health significance. It is not excluded that cellular repair mechanism may repair damage produced at low levels of radiation. Until this matter has been clarified by investigation, no threshold for somatic damage by irradiation should be assumed. The problems of carcinogenesis by somatic mutations induced by radiation are of immediate importance to the current generation in contrast to the induction of genetic mutations where effects may be postponed to the distant future.

There was some concern whether children exposed to the atomic bomb in utero would run a higher risk of developing malignancies, but the investigations carried out on this problem failed to produce any evidence of an increase [1499].

As mentioned in Sect. 5.1.6.7, some of the changes concomitant with normal ageing could be due to somatic mutation. Some animal experiments point to a certain shortening of life span after irradiation of the individual [1377].

Projected Additional Mutations Per Dose. How many additional mutations will be induced relative to the spontaneous mutation rate? Here, the information from animal genetics, especially mouse genetics, will serve as a guideline. In view of possible species differences in radiation susceptibility, we would certainly prefer to base all risk estimates on human data, but such data are so scanty that they do not permit any reasonable quantitative estimate. However, they can serve as a control and in some cases help to qualify estimates from animal data. We shall mainly consider the expected relative frequencies at the time of birth but shall occasionally make statements as to the expected zygote loss before birth.

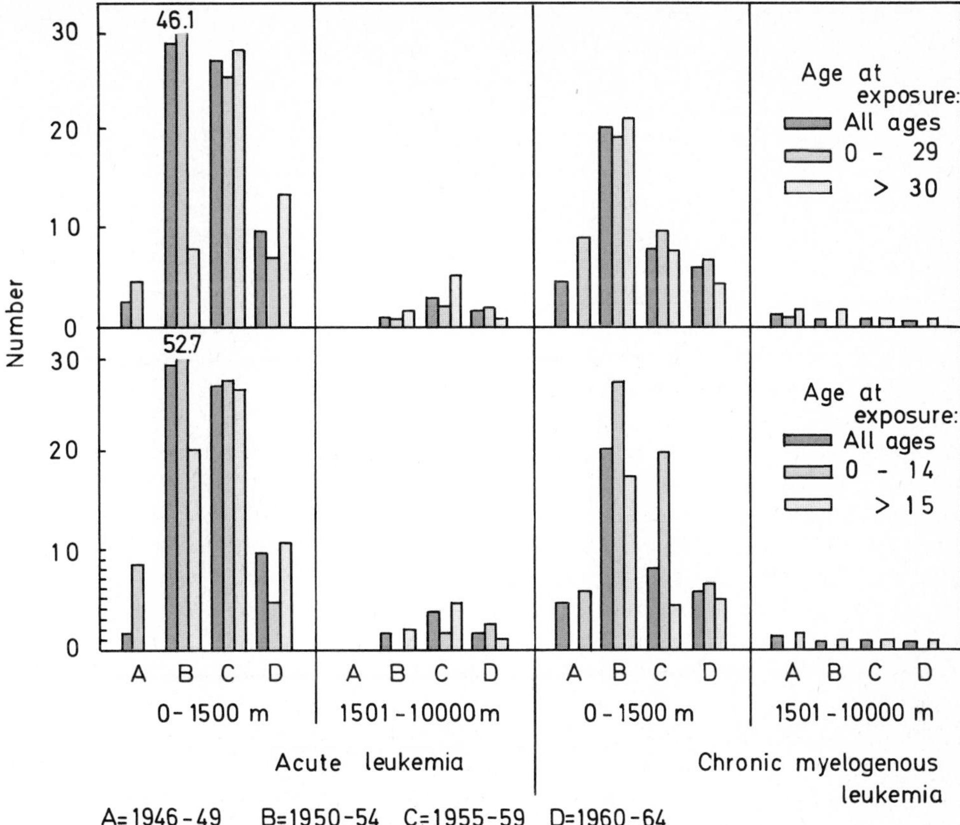

Fig. 5.55. Increase of leukemias after atomic bomb irradiation. Average annual rate per 100,000 population; based on 326 cases with acute or chronic leukemia who were at the time of the bombing 0–10,000 m away from the hypocenter. Subdivision according to age at the time of bombing (ATB). (Bizzozero et al., 1966 [1393])

The various groups of mutations will be considered separately.

Genome Mutations. A quantitative statement seems to be impossible. Too many factors are uncertain, for example: Does ionizing radiation enhance the nondisjunction rate when applied during the different stages of germ cell development, and if so, how much of an increase should be expected? Do the human data discussed above – a possible increase of X aneuploidy among children of atomic bomb survivors and a possible increase of autosomal trisomies, especially trisomy 21, many years after maternal X-ray diagnosis – establish that radiation induces nondisjunction, or is there some other possible explanation?

Unlike nondisjunction, chromosome loss caused by maternal irradiation of several weeks before or several hours after impregnation is well confirmed. The strong dose-rate effect indicates that the danger lies mainly in acute irradiation with high doses, whereas chronic exposure to very low dose rates may not increase the mutation rate. The quantitative effect in humans is difficult to predict, since most spontaneous XO zygotes in humans are aborted. Studies on patients with the XO karyotype clearly show that the majority of such patients owe their origin to loss of one sex chromosome by anaphase lagging or to mitotic nondisjunction during an early cleavage division. It is tempting to assume that aborted XO zygotes represent the XO products of meiotic nondisjunction since we are certain that nondisjunction affecting the X chromosomes occurs as evidenced by the existence of XXY and XXX types. It is not known whether chromosome loss after impregnation but before the first cleavage division leads to viable XO zygotes in humans, but their occurrence even without radiation during the cleavage divisions points to a higher general risk of chromosome loss during these divisions. Therefore, it is certainly advisable to regard the weeks around impregnation as especially susceptible to damage. Conception should be avoided at the time of and for several weeks after exposure to high doses of radiation with high dose rates.

Structural Chromosome Aberrations. Here, we are in a better situation. The most important group of structural aberrations permitting survival up to birth are the translocations. They may be balanced or unbalanced, leading either to an individual with a partial trisomy (or partial monosomy) syndrome or to a healthy translocation carrier. As examinations of newborns have shown, most translocation carriers are phenotypically healthy, and their translocations are balanced. However, carriers of de novo translocations that appear balanced may have a slightly increased risk of mental retardation (Sect. 2.2.2). The great majority of unbalanced zygotes die in the embryonic stage. Balanced translocations in humans cause segregation of unbalanced zygotes in the next generations. In experimental animals, the corresponding phenomenon is termed semisterility, as translocation carriers have a smaller number of healthy offspring than do healthy animals. The unbalanced zygotes die in utero. Lüning and Searle [1537] estimated the doubling dose (Sect. 5.2.1.3) for semisterility as 0.31 Gy after irradiation of the male mouse in the poststerile phase, i. e., of As spermatogonia. This means an increase of 0.34×10^{-2} per Gy per gamete. Compared with the mouse, direct data from human volunteers, and from the marmoset monkey show a two to three times higher susceptibility of spermatogonia for translocations visible at the first meiotic division. As noted, the elimination rate for translocations up to the time of birth is high in mice as well as in humans, but it is not known whether a higher elimination rate in humans would make up for the higher induction rate as shown by the above-cited studies. Therefore, it might be wise to assume a doubling dose for translocation induction of about 0.15 Gy in humans and an induction rate of 0.7×10^{-2} per Gy per gamete for translocations.

On the other hand, Robertsonian translocations are *not* (or almost not) induced [1453]. Postmeiotic male germ cells, especially spermatids, are more sensitive to translocation induction than are spermatogonia. Low dose rates, however, considerably reduce the yield of translocations from spermatogonia.

In female mice, the germ cells several weeks before conception and the oogonia of embryos appear to be radiation sensitive. For oogonia, an increase 0.16×10^{-2} per Gy per gamete has been estimated. This estimate is about half as high as that for spermatogonia.

For other structural aberrations, especially deletions, no reliable mouse data are available. However, the incidence of deletions in human newborns is small (Sect. 5.2.1, Table 5.3). Deletions invariably lead to severe malformation syndromes and usually to early death. As a first approximation, the increase in radiation-induced deletions may be estimated to be similar to that of translocations, since the basic event in both deletions and translocations is a chromosome break.

Point Mutations. Here, we are on safer ground, especially in regard to single autosomal-recessive mutations. Lüning and Searle (1971) [1537] arrived at a doubling dose of 0.32 Gy for acute irradiation of mouse spermatogonia and an estimate for induced mutations of 2.45×10^{-5} per Gy per roentgen per locus. The dose-rate effect is strong; for chronic irradiation with very low dose rates, a doubling dose of at least 1 Gy corresponding to an increase of $\approx 0.8 \times 10^{-5}$ Gy per locus, can be estimated.

The mutation rate in male postmeiotic germ cells is about twice as high as in spermatogonia, and there is little if any dose-rate effect. This would point to a doubling dose of 0.16 Gy, irrespective of the dose rate. This estimate may easily hold true in humans as well. Hence, the $\approx$6–8 weeks immediately preceding conception are especially dangerous. During this time acute irradiation and X-ray examination of pelvic organs, whereby the gonads are directly exposed, should be avoided. The danger of low-level chronic irradiation, on the other hand, will be very small, since individual germ cells remain in their most endangered stage of development for only a few weeks.

The sensitivity of oogonia to single locus mutation induction is still unknown. The oocyte up to about 7 weeks before ovulation seems to be resistant; no increase in mutation rates after acute or chronic irradiation has been reported. During the last 7 weeks before fertilization, the mutation rate has been estimated to be 1.85×10^{-5} per locus (irradiation with 2 Gy in one dose) and 18 per Gy per dose, when 2 Gy are given in 20 fractions [1541]. Considering a spontaneous mutation rate of 4.9×10^{-6} per locus (Sect. 5.2.1.3), this leads to a doubling dose of ~0,25 Gy for acute irradiation. With dose fractionation, the doubling dose would be much higher and, given the strong dose-rate effect, possibly still much higher for chronic irradiation.

The practical conclusion is that conception should be avoided within several weeks after irradiation. It is unknown whether the human oocyte like the oocyte of the mouse is resistant to induction of point mutations for most of its lifetime.

It is reasonable to assume that the effects are compared with those found in recessive mutations.

How Many of Naturally Occurring, "Spontaneous" Mutations Are Caused by Natural Background Radiation? The considerations on doubling doses and number of induced mutations per Grey per locus aid in estimating the fraction of naturally occurring mutations induced by natural background radiation.

This radiation is estimated to be between 0.03 and 0.04 Gy in 30 years (Table 5.26). The dose rate is extremely low. Therefore, an estimate of 1 Gy for the doubling dose may be reasonable for mutations in spermatogonia. Additional irradiation similar in amount to the background radiation would enhance the spontaneous rate by 3%-4%, and the same fraction of the spontaneous mutation rate might be caused by natural background radiation.

How Many Mutations Would Occur Spontaneously, i.e., Without Additional Irradiation Due to Modern Civilization? All the estimates derived above for point mutations are relative estimates. They were given either as doubling doses or as the number of new mutations per roentgen per locus. To convert them into absolute estimates, the number of loci at risk needs to be known. The problem of the total point mutation rate of the human genome was discussed in Sect. 5.1.4.2, by considering spontaneous mutation rates for single genes. There we saw that extrapolation from single mutation rates to the total human genome depends on assumptions as to the number of loci at risk, the number of mutations within one locus leading to a recognizable phenotype, and the effects, if any, of mutations in DNA sequences not coding for polypeptide chains. All these questions are so difficult to answer that a direct estimate of how much the point mutation rate might increase after a given radiation dose simply cannot be made.

Some of these difficulties can be circumvented if the question is put somewhat more simply: How much of an increase in genetically determined disease may be expected as a consequence of a given increase in radiation? To answer this question, we have to know the amount of genetically determined disease without additional irradiation, the proportion of genetic diseases maintained in the population by fresh mutation, and the increase in the mutation rate per radiation dose. Preliminary answers to the last question have been derived in the foregoing sections. Unfortunately we lack the required information regarding the frequency of genetically determined human disease.

There is only one group of genetic anomalies for which a relatively reliable absolute estimate is possible - the structural chromosome aberrations, mainly translocations. According to Jacobs (1972) [1501], 81 euploid autosomal structural aberrations were observed among 43,558 newborns, 20% of which were due to fresh mutation. Additionally, 21 aneuploid aberrations were observed in the same sample, one-third of which were due to new mutation. The doubling dose for translocation was 0.31 Gy for acute irradiation [1537] or three times as much 0.93 Gy as for chronic irradiation. This leads to the following estimate of the induced mutation rate per 0.01 Gy per gamete for euploid and aneuploid zygotes together:

$$\mu = \frac{\text{Number of individuals carrying a fresh mutation}}{2 \times \text{population examined}}$$

$$= \frac{23}{2 \times 43{,}558} \times \frac{1}{93} = 2.8 \times 10^{-6}.$$

Additional radiation of 0.01 Gy per generation would enhance the frequency of germ cells with new aberrations of this type by 2.8 per 1 million germ cells and, as two germ cells form an individual, by 5.6 individuals with such an aberration per 1 million. This estimate may possibly be too low by a factor of 2-4 if human spermatogonia are, indeed, more sensitive to radiation than are mouse spermatogonia.

For trisomies, a calculation determining the increase of trisomies per Gy is not possible. The incidence of trisomies is known and is exactly two times the mutation rate, as every case is a new mutant. Here, however, no estimate for the increase per Grey per gamete is available. On the basis of evidence from the mouse, however, there is good reason to assume that this increase might be smaller than for structural aberrations.

Estimates for gene mutations are much more difficult, since neither the amount of genetically determined disease in the absence of irradiation nor that fraction of the disease maintained in the population by fresh mutation is known.

Amount of Genetic Disease in the Population Is Not Known. It may come as a surprise for many readers that the incidence of hereditary disease in human populations has never been determined. Two serious attempts have been made at ascertaining all cases of hereditary diseases in well-defined populations, the first in Northern Ireland [1649], the second in British Columbia (Canada) [1661]. The data from these reports have been used by the U.N. as the basis for risk estimates. However, a critical study of the tables published with the reports shows that the authors did not carry out the examination and genetic classification of affected individuals personally but had to rely on diagnoses of a number of medical practitioners and sometimes even on their judgement as to modes of inheritance. Diagnoses of rare diseases are often wrong, given the unsatisfactory state in the classification of many genetic diseases and the poor general state of knowledge regarding genetic and environmental heterogeneity which often requires extensive clinical and laboratory assessment. Moreover, hereditary diseases tend to be rare, and diagnostic practice in medicine often does not (and must not) insist on finegrained analysis (See Sect. 3.8.14). The inconsistencies in these reports are obvious to clinical geneticists. Why do the U. N. reports rely so heavily on these data? The scientists serving on the committees that have to prepare these reports are mainly experimental geneticists without medical training.

Evaluation of Ad Hoc Studies Might Give an Idea of the Prevalence of Dominant and X-Linked Conditions. Attempts at determining overall frequencies of hereditary diseases in human populations have failed so far. For many diseases with dominant or X-- linked modes of inheritance, however, ad hoc studies have been carried out by physicians who were experts in the clinical manifestation of the disease and who used all available methods of ascertainment and checked the diagnosis of all cases. Some of these studies were received in Sect. 5.1.3 on spontaneous mutation rates. Examinations of these ad hoc studies should give an idea of the order of magnitude for the prevalence of these diseases. Table 5.28 does not list all dominant conditions known so far - there are several hundred (Sect. 3.1.8) - but it does comprise the most frequent ones [1412]. The relatively common disorders tend to have their onset in adult life. They are common because reproductive fitness is relatively high, the onset of the disease often occurring after the patient has already had children. The most frequent is monogenic familial hypercholesterolemia. The metabolic error in this disease has been discussed in Sect. 4.6.4. Heterozygous males have a risk of 50% for some manifestation of coronary disease by age 50 and a 50% risk of death from ischemic heart disease by age 60. Other diseases in this group are Huntington's chorea, neurofibromatosis, multiple exostoses, and polycystic kidney disease. The total incidence of autosomal dominant disorders - including those with low-incidence not listed in Table 5.28 - amounts to approximately 7/1,000.

A similar survey of ad hoc studies for X-linked recessive disorders gave an estimate of ~1/1,000, allowing for conditions not included in Table 5.28. It may be assumed that some dominant or X-linked conditions are not yet known. Hence, about 1% of the population will sooner or later develop a dominant or X-linked hereditary disease.

For autosomal-recessive disorders, striking ethnic and goegraphic variations are known, which will be discussed in Sect. 6.1.3. The frequency of autosomal-recessive diseases is only very remotely related to their mutation rate (Sect. 5.1.3). Therefore, predictions of future incidences with increases in mutation rates are impossible. Any increase in the mutation rate will only very slowly - after many, many generations - lead to an increase of homozygous, and affected, individuals. This topic will be discussed in Sect. 6.2.1.

Even for autosomal dominant and X-linked recessive conditions, however, frequency estimates can only be used to assess future incidence under the impact of induced mutations if the incidence is maintained by an equilibrium between mutations and selection. In this case, any increase in the mutation rate will ultimately result in a new equilibrium, at which the incidence will have been increased in the same proportion as the mutation rate. However, for some of the dominant conditions listed in Table 5.28, and especially for the most frequent ones, it is not known

Table 5.28. Estimated incidences at birth per 1,000 newborns for some of the most common dominant and X-linked diseases[a]

Incidence/live births (order of magnitude)	Diseases
2.0	Monogenic familial hypercholesterolemia
1.0	Dominant otosclerosis
0.8	Adult polycystic kidney disease, X-linked mental retardation
0.5	Multiple exostoses
0.4	Huntington's chorea
0.2	Hereditary spherocytosis, neurofibromatosis, Duchenne-type muscular dystrophy
0.1	Hereditary polyposis, dominant form of blindness, dominant form of early childhood onset deafness, dentinogenesis imperfecta, hemophilia A, dominant ichthyosis
0.04	Osteogenesis imperfecta, Marfan's syndrome
0.03	Hereditary retinoblastoma, hemophilia B
0.02	Achondroplasia, acute intermittent porphyria, X-linked deafness, ocular albinism, nystagmus
0.01	Tuberous sclerosis, Ehlers-Danlos syndrome, osteopetrosis tarda, variegate porphyria, cleft lip and/or palate with mucous pits of lip, X-linked imperforate anus, X-linked aqueductal stenosis, hypogammaglobulinemia, hypophosphatemic rickets, anhidrotic ectodermal dysplasia, amelogenesis imperfecta

[a] Adapted from Carter [1412].

whether they are, indeed, maintained by an equilibrium between mutation and selection. For Huntington's chorea, for example, Wendt and Drohm [941] were unable to find even a single unequivocal case of a new mutation among several thousand chorea patients. A selective advantage of the gene in earlier centuries, for example, by more uninhibited propagation in initial stages of the disease, is possible.

Frequency of New Mutations in Autosomal Dominant Disease and Biologic Fitness. Since individuals with dominant diseases that lower the chances of survival or marriage are less likely to reproduce, many cases of such a disease might be expected to present as new mutations with unaffected parents. The lower the biologic fitness the higher will be the proportion of cases due to fresh mutations. In a disease that leads to early death or no reproduction, most cases would be expected to be new mutants. The approximate proportions of parents affected by new mutations in autosomal dominant disorders are shown in Table 5.30. In diseases that do not interfere with reproduction because of late onset (such as polycystic kidney disease and Huntington's chorea), few cases or none at all are new mutants. Furthermore, such diseases will be more frequent in the population.
In spite of these strictures, it will be assumed in the following paragraph that all dominant mutations listed in Table 5.28 are maintained by an equilibrium between selection and mutation and that an increase in the mutation rate will lead sooner or later to a corresponding increase in their frequency. This assumption is the most cautious; it predicts the highest increase with induced mutations.

Educated Guesses as to Incomplete Penetrance and Multifactorial Diseases. In a great number of diseases, which are at least partially genetically determined, neither a chromosome aberration nor a simple mode of inheritance can be demonstrated. The genetic component can be inferred from a higher incidence among relatives of probands as compared with the general population, and from a higher concordance rate in monozygotic than in dizygotic twin pairs. Genetic analysis with such data can be performed only at the phenotypic-biometric-statistical level. Such biometric models are often used to compare the very general hypothesis of multifactorial inheritance with more specific genetic models, for example, autosomal dominant inheritance with reduced penetrance (Sect. 3.1.2). The three main categories within this group of disorders are congenital malformation, "common" diseases such as diabetes or hypertension, and mental disorders, such as schizophrenia and manic-depressive psychosis.
For most of these conditions, the multifactorial genetic model together with a threshold effect describes the pattern of inheritance fairly adequately

Table 5.30. Approximate proportions of patients affected by new mutations in autosomal dominant disorders. [Modified from Goldstein and Brown (1977)]

Disorder	Percentage
Apert's syndrome (acrocephalosyndactyly)	>95
Achondroplasia	80
Tuberous sclerosis	80
Neurofibromatosis	40
Marfan's syndrome	30
Myotonic dystrophy	25
Huntington's disease	1
Adult polycystic kidney	1
Familial hypercholesterolaemia	<1

Table 5.29. Estimated additional number of individuals with hereditary defects among 1 million newborns of the next generation and after a new equilibrium has been reached, per 0.01 Gy additional irradiation/generation [1377]

Type of mutation	Number of cases in the population	Number of additional cases/0.01 Gy in 1 generation	Increase per 0.01 Gy as % in first generation	Value of new equilibrium
Autosomal dominant and X-linked monogenic conditions	10000	15	0.15%	10100
Autosomal recessive diases (homozygotes)	1000–2500	very small		Very slow increase
Unbalanced translocations (and other structural chromosomal aberrations)	400	23	5.75%	430
Trisomies and X0	5000	Probably very small		
Congenital malformations and malformations manifesting later and constitutional diseases[a]	90000	4.5	Probably very small	

[a] Estimated on the basis of prevalence data in British Columbia [1661]. See, however, criticism of these data in the text.

(Sect. 3.6). However, careful analysis of phenotypes and pedigrees has in many instances succeeded in singling out rare types with simple modes of inheritance. Some examples, such as X-linked mental retardation, are listed in Table 5.28. The experience with radiation-induced dominant skeletal mutants of the mouse suggests that dominant major genes with very incomplete penetrance and extremely variable expressivity may be much more frequent than hitherto expected (Sect. 3.6.2.5). This category of conditions might easily be maintained by mutation and might therefore increase with a rising mutation rate.

On the other hand, studies of the disease associations with genetic polymorphisms, such as HLA antigens and blood groups (Sect. 3.7), have shown these polymorphisms to be involved in the genetic determination of some of these diseases. However, since the gene frequencies of these polymorphisms are not maintained by mutation but probably by balancing selection (Sect. 6.2.1.3), an increase in the mutation rate would not affect gene frequencies of polymorphisms and hence the frequencies of such multifactorial diseases (Sect. 3.6).

The impact of mutations on the frequency of birth defects is difficult to estimate. Very few such defects are clearly monogenic. However, the effect of major genes in some birth defects has not been ruled out. Similarly, few birth defects are entirely environmental in origin. Thus, striking differences in the population frequency of neural tube defects may be compatible with hitherto undetected environmental influences. In most birth defects, there is presumably interplay between multiple genetic and yet unknown environmental factors. Mutation effects would depend upon the nature of the underlying genetic variation. If much of the genetic variation predisposing to birth defects is caused by genetic polymorphisms maintained by selection, mutation would have little or no effect on frequencies. To complicate the matter further, it is conceivable in view of the rather low identical twin concordance rates that some complicated birth defects of organogenesis, such as congenital heart defects and others, may be caused by random or stochastic processes not influenced by either genetic or environmental factors [2348].

Table 5.29 shows an estimate of the additional number of individuals carrying hereditary defects per 0.01 Gy of additional irradiation. This estimate has been combined from expert estimates given by a task group of ICRP (International Commission on Radiation Protection) and by another expert group in the UNSCEAR report. Some of the forecasts have been modified somewhat in accordance with our own considerations given above. In agreement with the 1982 UNSCEAR report, the estimates given here are *lower* than those published earlier. This is because we have learned more about some problems. As an example, the risk for Robertsonian translocations was earlier assumed to be identical to that for reciprocal translocations. It is now known to be much lower, if it exists at all. This decrease of risk estimates shows that such committees were appropriately conservative in that they always assumed higher risks in doubtful cases.

On the other hand, the estimates relate only to hereditary defects; possible variation in the normal range has not been considered. For example, in recessive disease, the number of *homozygotes* will increase only very slowly. However, *if heterozygotes* for some recessive diseases such as phenylketonuria were proven to have a slightly depressed I. Q., an increase of average heterozygosity due to additional mutations could influence I. Q. variation in the normal population.

A superficial consideration of this table would create the impression that in the absence of ionizing radiation or other mutagenic agents, the incidence of genetic anomalies would remain the same over the generations. This assumption, is not warranted, however; for example, autosomal recessive diseases currently have an especially low incidence in industrialized countries, since the frequency of consanguineous marriages has decreased rapidly in the last few generations (Sect. 6.3.1.2). Hence, homozygote frequencies will increase very slowly, over hundreds of generations, until a new equilibrium is reached. The incidence of dominant and X-linked disorders, on the other hand, will be influenced negatively by the relaxation of natural selection due to improved therapy, as well as positively by increasingly effective genetic counseling and prenatal diagnosis. Even if an increase in genetic diseases and/or birth defects were found by careful monitoring, it would be difficult, if not impossible, to single out any one factor, such as irradiation, as the cause of such an increase.

For another, still lower estimate for dominant and X-linked mutations see [1417].

For improvement of these estimates, the following data are needed: (1) a more thorough knowledge of the genetic factors of multifactorial conditions and of dominants with low penetrance. (2) A much better knowledge of the incidence and prevalence of hereditary diseases; this knowledge can come only from carefully planned, large-scale epidemiologic studies in which the medicostatistical information is combined with ad hoc studies on single diseases; and (3) better knowledge of the interplay between mutation and natural selection, especially for those diseases in which genetic equilibrium between mutation and very strong selection is not immediately obvious.

In Sect. 5.2.1.2, four questions were asked:

1. In what way, if any, does the agent affect the genetic material?
2. How strongly is the human population exposed to this agent?
3. How high an increase in the "spontaneous" mutation rate must be anticipated?
4. What are the long-term consequences for the population?

A cumbersome chain of arguments has led us to question 3 and a short way into question 4. The main thrust of question 4, however, cannot be answered since we do not yet know the fraction of the total sum of mutations that are harmful. The population studies on irradiated mouse populations could lead us to rather optimistic conclusions as to consequences for the human population, but humans are not mice and extrapolation of these results could be misleading. The discussion of population genetics (Sect. 6) will show where answers regarding the long-term trends may eventually be found.
Under close scrutiny, the problem of the genetic risk to human populations by ionizing irradiation turned out to be unexpectedly complex. However, a trail through the wilderness of problems, though partially covered by weeds and interrupted occasionally by inaccessible ravines, has become visible in gross outline. The situation is quite different for the genetic risk due to chemical mutagens. Here, the problems are still more complex, and the scientific community has hardly begun to blaze a usable trail.

5.2.2 *Chemically Induced Mutations*

5.2.2.1 *Extent of the Problem*

History [1384, 1385]. Mutation induction by chemical compounds was suspected even in the early days of genetics. In his first publication on radiation-induced mutation, Muller (1927) [1566] wrote:

> It has been repeatedly reported that germinal changes, presumably mutational, could be induced by X or radium rays, but, as in the case of the similar published claims involving other agents (alcohol, lead, antibodies, etc.), the work has been done in such a way, that the meaning of the data, as analyzed from a modern genetic standpoint, has been highly disputatious at best; moreover, what were apparently the clearest cases have given negative or contrary results on repetition.

Following Muller's publication, a one-sided emphasis on radiation genetics developed, and for a long time there was little interest in any other mutagenic agents. In 1941, Muller again commented on the attempts to influence genetic material through chemicals (Cold Spring Harbor Symp.):

> But although many very drastic treatments (killing the majority of the organism treated) have been tried, none so far reported has met with marked success and has been tested and had its efficacy confirmed by independent workers ...
> In view of the high protection ordinarily afforded the genes by the cell which carries them, ... it is not to be expected that chemicals drastically affecting the mutation process while leaving the cell viable will readily be found by our rather hit-and-miss methods. But the search for such agents, as well as the study of the milder, "physiological" influences that may affect the mutation process, must continue, in the expectation that it still has great possibilities before it for the furtherance both of our understanding and of our control over the events within the gene ...

In 1942, Auerbach and Robson in Great Britain achieved unassailably positive results by producing mutations in *Drosophila* with nitrogen mustard (not a "mild" agent). For reasons of military secrecy, these results were not published until after the war (1946, 1947) [1379]. The motive of testing this particular substance for mutagenicity was the similarity between the skin lesions induced by nitrogen mustard and those induced by high acute doses of radiation.
Independently, Oehlkers (1943) [1585] in Germany had achieved positive results on *Oenothera,* particularly with urethane, a much-used drug, which was considered a good hypnotic for children and later proved carcinogenic. Oehlkers was exclusively interested in basic research. The publication gives no indication at all of any particular motive for choosing urethane as a test agent. Rapoport (1946) [1593] in Russia described the mutagenic action of carbonyl compounds.
After these first discoveries, the new field rapidly expanded, but there was little if any discussion on practical perspectives or concern with mankind's genetic health. The first moves in this direction can be found in a lecture by Lüers (1955) [1536] and in a survey article by Barthelmess (1956) [1383]. In 1961, Conen and Lansky [1422] described, for the first time, chromosome aberrations in the lymphocytes of individuals who had been treated with nitrogen mustard. Since the early 1960s - after a delay of about 20 years since chemical mutagens were first described - the scientific community gradually recognized the possible threat of chemical mutagens to the human population. Röhrborn (1965) [1600] was the first to summarize the evidence and to pose clearly the important questions. Meanwhile, the field has evolved rapidly, the Environmental Mutagen Society (EMS) was founded, and monographs were published [1489, 213; 1622]. Frequent conferences are held, which, however, are now leading to a certain consensus regarding the nature of the questions that should be answered [1375; 1376; 1425; 11; 22].

Mutagenic Compounds in the Human Environment. Mutagenic effects have been observed for a great many compounds and in a wide variety of test organisms. For most of these effects - such as genome mutations, chromosome breaks and rearrangements, or point mutations - the genetic significance is obvious. Others, like "stickiness" of chromosomes or chromosome "gaps" (Sect. 2.2.2) are difficult to evaluate. Some substances impair the function of the tubular apparatus necessary for spindle formation during mitosis. Altogether, the effects of chemicals on the genetic material are more varied than radiation effects.

Table 5.31 gives a selection of compounds found to be mutagenic, and depicts the mutational mechanisms. The strongest mutagens observed so far are alkylating agents, such as nitrogen mustard, ethyleneimine compounds, and methylsulfonic acid esters. Many of these compounds are used in treatment of malignancies or in conditions where an immune reaction or a cell proliferative process is to be inhibited. Other groups of mutagenic compounds used in therapy include antimetabolites of nucleic acids or folic acids and acridine compounds.

One of the most widespread substances to which man is exposed and which has been found to be mutagenic in microorganisms is caffeine. Nitrites, which are favored model substances for elucidating molecular mechanisms of point mutation [1457], are used as meat preservatives that, at the same time, give meat a fresh red color. These two examples suffice to indicate that the problem of chemical mutagenesis may be potentially serious.

Molecular Mechanisms of Chemical Mutagenesis [1458, 1602]. Like irradiation, chemical mutagens may cause disturbances of meiosis leading to nondisjunction, chromosome breaks, and point mutations. Little is known about the exact mechanisms of nondisjunction, since the events leading to nondisjunction take place at the chromosome level. Better information is available about the action of chemicals on the DNA molecule. Some point mutations can be explained in terms of molecular mechanisms, and it is likely that these mechanisms are important in chromosome break induction as well. The secondary reactions leading to chromosome reunions, on the other hand, are still insufficiently understood (Sect. 2.2.2).

It is convenient to distinguish two types of chemically induced DNA changes that interfere with the function of DNA as genetic material. These are: (1) inactivating alterations and (2) mutagenic alterations.

Inactivating alterations present DNA replication across the altered site, unless they have been eliminated by repair (Sect. 5.1.6). The most important types are depicted in Fig. 5.56. They frequently lead to cell death. The cell may survive, however, if, for example, the thymine dimer or an altered base is removed by a repair mechanism, if the chromosome breaks are restituted, and if the product of restitution does not disturb mitosis.

Mutagenic alterations are defined as changes that do not interfere with DNA replication but change the base sequence. Some mutagens, like BUdR, can only act on replicating DNA, as they have to be incorpo-

Table 5.31. Molecular mechanism of point mutations

Mutagen example	Mechanism
Base analogues: (5-bromuracil) incorporation during division	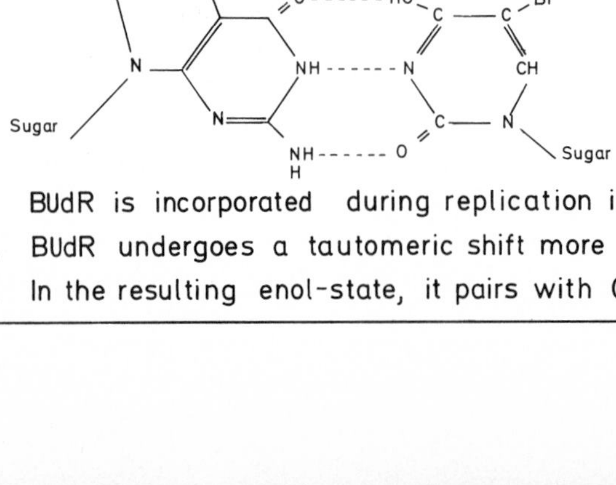 [Structural formulas: adenine–5-bromouracil pairing (keto form) and guanine–5-bromouracil pairing (enol form); labels: N, H, NH, O, C, Br, CH, HN, HO, Sugar] 1. BUdR is incorporated during replication instead of thymine 2. BUdR undergoes a tautomeric shift more frequently than thymine 3. In the resulting enol-state, it pairs with Guanine

Table 5.31 *(continued)*

Mutagen example	Mechanism
Nitrous acid: deamination of adenine and cytosine in resting DNA	Adenine to hypoxanthine 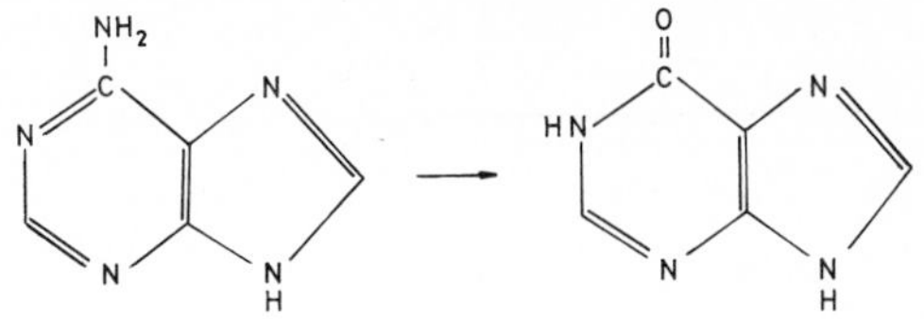 (NH$_2$ → O): Base pairing like guanine Cytosine to uracil  (NH$_2$ → O): Base pairing like thymine
Alkylating agents: methyl-methane sulfonate (MMS)	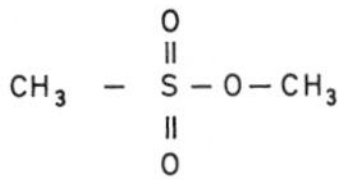 $CH_3-SO_2-O-CH_3$ Alkylation of guanosine at the 7 position with resulting tauto-meric shift (lower right) 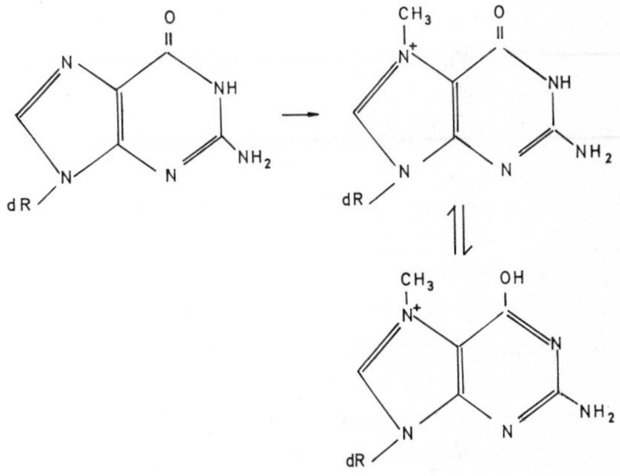 (guanosine, dR → N^+–CH_3 ⇌ OH tautomer, NH_2)
Hydroxylamines	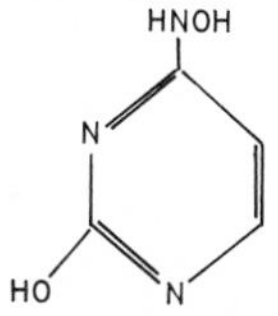 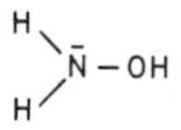 HNOH-substituted cytosine; $H_2\bar{N}-OH$ 1. Hydroxylamines react with cytosine, forming derivatives that are N-hydroxylated in 4-, 6-, or both positions 2. The derivatives are in a different tautomeric state than cytosine. They can pair with adenine instead of guanine
Acridine dyes (trypaflavine)	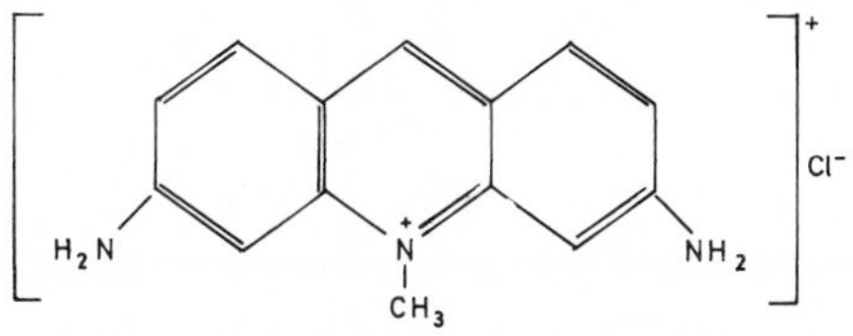  [H_2N–acridine(N^+–CH_3)–NH_2]$^+$ Cl^- 1. The acridine molecule intercalates in the DNA double helix 2. This stretches the transcribing DNA strand leading to a frameshift mutation

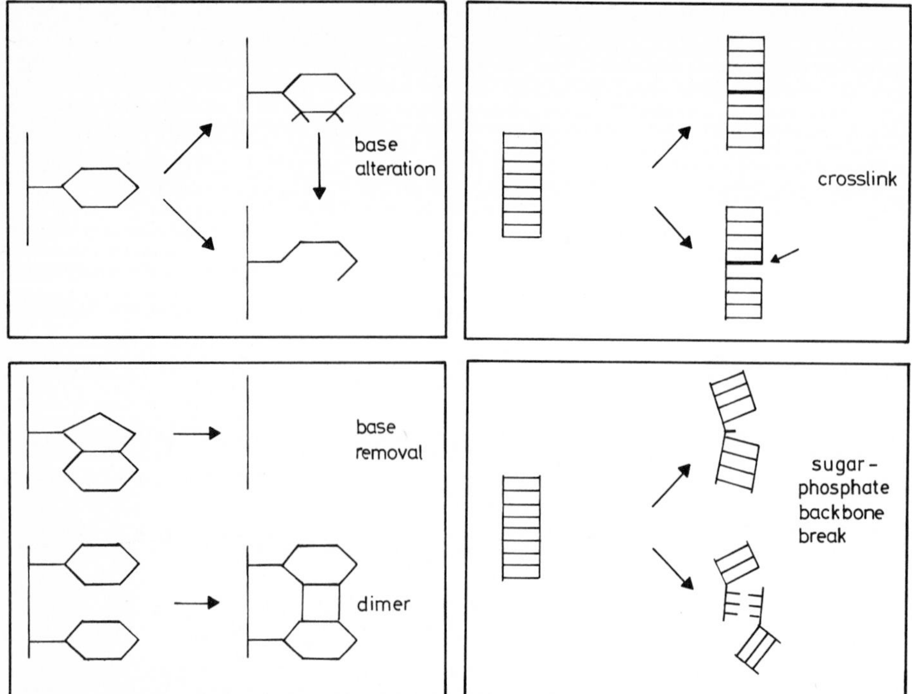

Fig. 5.56. The most important types of chemically induced inactivating DNA alterations. (Freese, 1971 [1458])

rated. Others, like nitric acid, may also alter resting DNA. Most chemicals that induce mutagenic alterations are also capable of inducing inactivating ones. Some inactivating agents, on the other hand, do not induce point mutations.

Note that point mutations caused by a change of one base affect only one of the DNA strands. Therefore, they are transmitted to one of the two division products only (Fig. 5.56). This is also to be expected in spontaneous mutations due to tautomeric shift of a normal base (Sect. 5.1.3.5). In fact, there is some evidence in humans that such mutations, if occurring in a germ cell during or after the last mitotic division, lead to a special kind of mosaics in which half the body cells contain the mutations [676]. Radiation-induced point mutations, on the other hand, are usually transmitted to both division products, indicating that the primary defect may be of the inactivating kind (Fig. 5.56), cell survival being secured by repair processes.

Other compounds do not affect the DNA itself but interfere with repair processes. The most widely used compound of this group is caffeine. This compound does not induce any chromosome abnormalities in mammalian in vivo-test systems but produces chromosome gaps and breaks in human lymphocyte cultures. Apparently, the culture conditions lead to considerable DNA damage, which normally is repaired but results in visible breaks in the presence of caffeine. The proof that caffeine is, indeed, a repair inhibitor comes from lower organisms [1419, 1604]. When administered in high doses together with an alkylating mutagen, caffeine enhanced the number of chromosome breaks and translocations in bone marrow cells of the Chinese hamster. Such synergistic effects of mutagenic agents may easily go undiscovered and may be important for humans exposed to a variety of potential mutagens.

Difficulties in Setting up Research Strategies for Discovering Genetic Threats Due to Chemical Mutagens. As shown in Sect. 5.2.1.2, assessment of genetic threats by ionizing radiation is a complex, multistep process. For assessing threats due to chemical mutagens, additional dimensions of complexity have to be considered.

a) In spite of the different sources of radiation (X- and γ-rays, neutrons, etc.), the overall biologic radiation effects are very similar. In contrast, there are innumerable chemical mutagens with different chemical constitutions. They interact in various ways with the DNA molecule. Their biologic action spectra are different, though partially overlapping.
b) Radiation is quenched in part by the tissue overlaying the irradiated cells, for example, germ cells, but its quality is not altered. Chemical mutagens, on the other hand, have to pass the metabolic system of the organism; they are converted into other compounds in most unpredictable

ways. Chemical substances may have lost their mutagenic capacity, or they may have gained new mutagenic capacities not present in the basic compound. Nonmutagenic chemicals that have been converted into mutagenic compounds are particularly dangerous since they may easily escape current screening techniques (Sect. 5.2.1.2). Cyclophosphamide, a much-used cytostatic drug, is an example. The compound as such is not mutagenic, but is converted into highly mutagenic compounds in the mammalian organism [212].

The formation of compounds with altered properties is, however, only one aspect of the complexities encountered. Others are uptake, distribution among the various organ systems, and excretion, or - if excretion is not possible - storage. All these pharmacokinetic aspects need to be considered for the compound itself and for its metabolites.

Extrapolation from Other Mammalian Species to Humans and from One Human Being to Another. Toxicologic studies are usually carried out in laboratory mammals such as mice or rats. In doing so, we take advantage of the phylogenetic relationship among all mammalian species, which postulates that many of the basic metabolic pathways are similar allowing cautious extrapolation from experimental animals to humans. However, this similarity is not at all pervasively present. While the broad general pathways of metabolism are similar between species, minor differences in metabolic pathways are common [1697]. Even within the human species, genetic polymorphisms affecting the metabolism of chemicals are found, the exploration of which is the field of pharmacogenetics (Sect. 4.5.1). Examples involving potential mutagens or carcinogens are the acetyltransferase polymorphism leading to quick or slow elimination of the antituberculosis drug isoniazid (INH) and genetic differences in arylhydrocarbon hydroxylase action affecting the metabolism of carcinogenic aromatic hydrocarbon compounds. Hence, toxicologic results in mammals other than man cannot be immediately extrapolated to humans.

Differences Between Chemical Mutagens and Ionizing Radiation and Between Different Classes of Chemical Mutagens in the Induction of Genome and Chromosome Mutations. So far, all our considerations have been theoretical, based on general experience with drug metabolism - or they have been indirect, based mainly on experimental results in molecular biology. Are there direct data showing differences between classes of mutagens in their effect on cells, particularly germ cells, in mammals? Such data do, indeed, exist. Two examples may suffice. The dominant lethal test has been described in Sect. 5.2.1.2. To examine dominant lethal induction at different stages of spermatogenesis, male mice are treated with the mutagen and are mated on three successive days with virginal females. The females are then sacrificed at about the 15th day of pregnancy, and the dead and living implantations are counted. All dead implantations are classified as dominant lethals. The experiment permits testing the relative sensitivity to the mutagen at different developmental stages of germ cells. Progeny of females mated immediately after exposure of the males have received paternal chromosomes treated as mature sperms. Progeny of matings about 50 days after exposure to the mutagen come from sperms treated as spermatogonia.

The pattern of germ cell sensitivity varies. Whereas an acridine compound *trypaflavine* affects all germ cell stages to about the same degree, *Trenimon,* an ethyleneimine compound, induces dominant lethals almost exclusively in postmeiotic germ cell stages; *mitomycin C* and ethyl-nitrosourea have their maximum effects before meiosis [1376]. The difference between Trenimon and trypaflavine can be explained. It is known from experiments in bacteria, that trypaflavine, being an acridine compound, induces mainly frameshifts (Table 5.31) that will pass the meiotic filter, whereas Trenimon causes numeric and structural chromosome aberrations that are efficiently screened by meiosis. For the difference between the effects of mitomycin C and Trenimon, no convincing explanation can be offered. The example shows the possible order of magnitude in cell stage specificity. Most alkylating substances exert their maximum mutagenic effect on male germ cells in postmeiotic germ cell stages. These substances are used in medicine mainly as cytostatics. Therefore, if a male patient wishes to have children, conception should be avoided throughout the first three months after the end of the treatment.

5.2.2.2 *Research Strategies for Assessment of Genetic Risks Due to Chemical Mutagens*

What Questions Should We Try to Answer? We have noted earlier that of the four questions posed - how does the agent affect the genetic material; how strongly is the human population exposed to the agent; how high an increase in the "spontaneous" mutation rate can we expect; and what are the long-term consequences of this increase - the fourth one cannot at this date be answered. This applies to any kind of mutation rate increase, independent of the source of mutation. Therefore, the question will not be discussed any further in this context, and we shall concentrate on questions 1, 2 and 3.

5.2.2.3 In What Way Do Chemical Mutagens Affect the Genetic Material?

Planning of Test Programs. It should be clear from the foregoing discussion that the first question - whether and how a specific agent affects the genetic material - cannot be answered in an abstract way. In some cases, the chemical constitution of the compound will help in formulating more specific hypotheses. An acridine compound, for example, can be expected to induce frameshift mutations primarily, and an ethyleneimine or nitrogen mustard compound may be assumed to have alkylating effects and to induce genome, chromosome, and gene mutations. Often, a substance may be degraded so quickly that there is very little danger. The chemical constitution of other compounds may give no hint at all and yet, the compound will be able to induce mutations by an unknown mechanism. Therefore, every new chemical to which mankind is exposed should be tested for mutagenicity.

The complexity of pharmacokinetics, the great number of substances already in use as pharmaceuticals, pesticides, cosmetics, and in many other contexts, as well as the enormous number of compounds that are constantly being introduced causes almost insuperable difficulties. However, two rather simple rules may help in working out reasonable solutions [1684]:

1. Determine precisely what you want to know. Keep extrapolations as meager as possible. Whenever forced to extrapolate, be aware of the problems introduced at every step of the extrapolation.
2. Plan your test program to answer specific questions.

Determine Precisely What You Want to Know. Extrapolations inherent in mutagenicity test programs are those from other test systems to humans, from one kind of mutation to another, from one tissue to another, and from high to low doses.

Experiments cannot be carried out on human germ cells. Therefore, extrapolations from other organisms are inevitable. These organisms should be as closely related to the human species as at all possible. For practical reasons, the laboratory mammals used in radiation genetics - mouse, Chinese hamster, rat - should be utilized. Results from more remotely related species such as bacteria or *Drosophila* will yield only slender evidence applicable to humans. Therefore, every opportunity to examine human population groups exposed to possible mutagens should be exploited. Examinations of somatic cells for chromosome aberrations, sister chromatid exchanges [1462], and as biochemical point mutations may give valuable clues.

The methods for mutation detection in mammals were reviewed in Sect. 5.2.1.2. Genome and chromosome mutations can now be detected in almost all stages of development. In radiation genetics, certain extrapolations to gene mutations are possible, since radiation induces chromosome as well as gene mutations, and the same stages of cell development are susceptible. For many chemical mutagens, however, this extrapolation is much more difficult. Some compounds may induce gene mutations but few, if any, chromosome mutations. Therefore, special search for gene mutations is more urgently required than in radiation genetics. Unfortunately, our spectrum of methods is less satisfactory than for genome and chromosome mutations. The multiple loci test, though applicable in theory, is much too time-consuming and expensive for general application. It is very useful, however, for testing the few substances for which faster but less reliable methods (see below) have raised suspicion of mutagenicity. The same applies for induction of dominant mutations leading to skeletal anomalies or cataracts [1440]. Methods for screening of mutations at the enzyme or protein level are still under development. Biochemical variation found in single cells cannot be identified with certainty as mutational in origin. A strong emphasis on further refinement of the methods for protein and enzyme screening might have been expected. The actual development, however, has been different. The so-called host-mediated assays seemed to promise an easier approach. A bacterial or fungal culture - the test organism - is grown in the peritoneal cavity of a mouse. The chemical is administered to the mouse by injection or feeding. It then passes through the animal's metabolism before coming into contact with the test organism that is screened for point mutations. At first glance, this method seems to be elegant. The metabolic changes a chemical might undergo in the mammalian organism are accounted for. However, the genetic material to be tested remains that of bacteria, and experience has shown that different bacteria may react differently to the same mutagens, some showing mutations whereas others do not [1592]. Another, similar approach has become popular in recent years. Most of the enzymes for drug metabolism are found in the microsomal fraction of liver cells. Therefore, the test chemical was exposed to microsomal preparations in vitro and then tested for mutagenicity in bacterial systems (Ames test). These methods may give some hints toward identification of potential mutagens, and especially the Ames test is valuable for screening of a great number of compounds.

Ames Test as a Screening Test for Carcinogens? There is now increasing evidence that cancer is due to somatic mutation - at least in many cases (Sect. 5.1.6). It is therefore not surprising that many mutagens are, at the same time, carcinogens and vice versa and that many of the compounds known as carcinogens have been identified as mutagens. At present, animal testing for carcinogenicity is time-consuming and expensive. It was, therefore, suggested to introduce a fast method for mutagenicity testing - and the above-mentioned Ames test is such as fast method - as a screening test for potential carcinogens. In a study on 300 different compounds, carcinogens and noncarcinogens, 157 of 175 carcinogens were shown to be mutagenic in this test [1553, 1554]. Only very few noncarcinogens showed mutagenic activity. This result has been discussed extensively during recent years; it is undisputed that some carcinogens may escape observations, as they do not act through DNA damage. In general, the problem of extrapolation from the bacterial to the human genome and from the in vitro assay to the conditions in vivo is similar to the problems encountered in testing for germ cell mutagenicity. The method may help to establish priorities in carcinogenicity testing of environmental chemicals, as many more compounds can be tested within a short time. However, it cannot replace the in vivo methods.

Metabolism. Another important factor affecting the probability of mutagenesis and carcinogenesis relates to genetically determined differences in the metabolism of foreign substances, including drugs and environmental agents (xenobiotics). Current data show that identical twins exhibited identical biotransformation rates of drugs in the face of significant variation in biotransformation in the general population (Sect. 4.5). Thus genetic factors appear to be of major importance in the breakdown of xenobiotic agents. As an example, careful work with mice and more limited work with men has shown that arylhydrocarbon hydroxylase plays an important role in the transformation of polycyclic hydrocarbons to epoxides. Epoxides are much more carcinogenic than hydrocarbons. More generally, a small proportion of the population who will be slow inactivators of a xenobiotic or who transform a xenobiotic to an active mutagenic agent will be at much higher risk than the general population at large. No test system with experimental animals can illuminate this important aspect of environmental mutagenesis. It may be useful, therefore, to incorporate human cell material from many different individuals (fetal livers from abortuses?) into the Ames test (see above) to find possible differences due to human variation in metabolism of xenobiotic substances. Methods of screening for mutants by study of variant enzymes and proteins in human populations will be a more direct way of finding mutation, but attempting to implicate a specific environmental deleterious agent will be more difficult to establish (see below).

Tissue Differences in Mutation? The *third extrapolation* is from one tissue to another. Chromosome mutations, for example, can easily be tested in the bone marrow of mice or Chinese hamsters, or in human lymphocyte cultures. It would be tempting to extrapolate from these somatic cells to germ cells. However, experience in radiation genetics has shown that sensitivity of germ cells may be very different from that of other cells (Sect. 5.2.1.3). For practical purposes, it is not so important whether this difference is caused by cell selection or a true difference in mutation induction. Moreover, the developmental stages of germ cells in both sexes differ in sensitivity. For all these reasons, mutagenicity testing of somatic cells does not provide the information needed for assessment of genetic risks. However, methods for cytogenetic in vivo testing are very valuable for risk assessments of somatic mutations, especially those possibly leading to cancer.
If genetic risk *for progeny* is at stake, germ cells of mammals should be tested. If a substance turns out to be mutagenic, possible differences in the metabolism of this compound between the test animal and humans, as well as within the human population, should be considered.

Extrapolation From High to Low Doses. The *fourth extrapolation,* which is frequently necessary, is from high to low doses of the chemical. Usually, this problem does not arise with pharmaceuticals, since drugs are often administered in fairly high doses. A major problem is encountered with so-called environmental mutagens, i.e., compounds to which many people are exposed in relatively low doses, but for a long time. Here, animal experiments should be planned to mimic the actual situation as closely as possible - for example, by exposure to low doses for 1-2 years. Mouse experiments of this type have, indeed, been set up for caffeine - fortunately for all coffee drinkers with negative results [212]. Naturally, however, such experiments are costly and time-consuming. Nevertheless, they should be planned for all chemicals to which human beings are exposed and which were shown to be mutagenic in higher doses. If a mutagenic effects is proven for reasonably high doses, i.e., for doses that do not kill the animal by other, toxic effects, it is wise to consider the compound as mutagenic at lower doses, unless the contrary has been proven.

Thresholds in Environmental Mutagenesis? In the interpretation of test results in chemical mutagenesis, the concept of a threshold needs consideration. It is usually assumed that the production of cancers by mutation is directly dependent upon dose. This concept implies that high doses are administered to a relatively small number of animals in a typical toxicity study. If cancers develop with high doses, these results are extrapolated to the low exposure levels usually operating in the human population, and it is assumed that the compound is carcinogenic. Such extrapolation appears reasonable if no threshold were to exist below which no cancers were likely to develop. The full and scientifically incontroversial proof that no thresholds exist would require the use of a forbiddingly large number of animals for each test substance. Since no studies of this sort are available, the no threshold model (i.e. linear extrapolation to low dosage) must be assumed for maximum protection of the human species. It should be clear, however, that extrapolations of this sort would be misleading if a threshold existed. Such a threshold is a biologic possibility in view of the presence of enzymes repairing mutational damage. All these considerations of carcinogenesis apply equally to induction of mutations in germ cells.

The complex and expensive test program required may sometimes be obviated:

Plan Your Test Program to Answer Specific Questions. Not every chemical has to be tested for all its possible genetic effects on germ cells and somatic cells. The test program depends on the purpose the compound is supposed to serve. Birth control pills, for example, to which many members of our species are exposed, should be tested carefully for any mutagenic activity in oogenesis, particularly during the "sensitive" period before and shortly after the time of impregnation. However, it would be trivial to test birth control pills for mutagenic activity in the male germ line.

Recently a consensus seems to have emerged among the many committees concerned with development of guidelines for testing newly introduced or widely used substances for mutation induction: First fast and inexpensive methods such as the Ames test for point mutations and the micronucleus test for chromosomal aberrations [212] are used. If these tests show mutagenicity, more "relevant" but time-consuming mammalian in vivo tests are applied [1425]. This procedure carries with it the risk of missing "false negatives" in the simple tests. This risk can be partially avoided if substances that are in especially wide use and/or suspected because of their chemical structure are tested in more "relevant" systems despite negative results in simple tests.

5.2.2.4 How Widely Is the Human Population Exposed to the Agent?

An Important, but Sometimes Neglected Question. The question of how extensively human populations are actually exposed to the agent is crucial for any estimate as to the genetic threat of chemical mutagens. These considerations have sometimes been neglected in discussions on chemical mutagenesis. Again, as with so many other problems, the most plausible explanation can be found by considering the sociology of science. Most research workers concerned with problems of chemical mutagenesis are experimental geneticists with experience in mutation research on some test system, for example, the mouse, human chromosomes, or bacteria. Understandably enough, their main concern is with the efficiency of test methods. Toxicologists employed by the pharmaceutic companies who take up these methods for practical use are usually not initiated along genetic lines. Epidemiologists, on the other hand, often have little interest in genetics and are not actively interested in problems of mutagenesis.

The problem of population exposure can be subdivided as follows:

a) How many people are exposed to a specified mutagenic agent?
b) What doses of this agent are used and for how long?
c) How do age and sex of the exposed population sample compare with the age- and sex-specific reproduction of the population?
d) To what extent, if any, do individuals exposed to the agent actually reproduce? For example, do they suffer from diseases that prevent them from having children?
e) Do the offspring show signs of genetic damage?

The first three questions can often be answered approximately by use of data available in statistical and medical publications.

Population Exposure to a Frequently Used Drug. INH (isoniazid) is a frequently used drug for therapy of tuberculosis - a disease that affects people of all age groups. Mutagenic activity of this drug has repeatedly been shown in bacteria using the host-mediated assay. In an in vitro polynucleotide assay, INH inhibited transcription [1514], and in Chinese hamster cells in vitro, it reduced postreplication repair [1515]. On the other hand, a comprehensive research program using many different mammalian in vivo test systems for chromosome aberrations failed to show any definite indication of increase in the spontaneous mutation rate [1605]. But induction of point mutations could not be tested in a mammalian in vivo system. Certainly a drug of this kind deserves a look from the standpoint of population exposure.

Such a study was carried out for the population of West Germany in the year 1970. It turned out that more than 35% of the total drug-exposed population were of reproductive age. Assuming that the reproduction of these tuberculosis patients would equal that of their age and sex mates in the normal population, it was estimated that 5,600 children per year should be expected from them. This means that every 162nd child was at risk as offspring of a marriage in which one parent had been treated for tuberculosis. If the compound were mutagenic, this proportion would point toward a small, but not negligible population load [1629].

Population Exposure to Highly Mutagenic Drugs. The situation is quite different for cytostatic drugs. Here, mutagenicity is undisputed. The analysis, however, cannot be sufficiently exact if only published or available statistical data are used. Life expectancy is usually very short at the beginning of therapy. Besides, the cancer patients treated with these drugs usually are in generally poor health, so that their reproduction cannot be assumed to equal that of the untreated population. An additional step specifically involving the number of children born to patients after the onset of treatment needs to be introduced into the analysis. This, together with all the steps set out earlier for INH, was done in another population study in West Germany [1680]. The results indicated that only 23 children per year could be expected from cytostatically treated patients after onset of therapy. Hence, cytostatic treatment would not have enhanced the overall mutation rate of the population to a noticeable extent. However, this conclusion is valid only as long as cytostatic therapy is limited to malignant diseases and to a few other rare conditions. The recent tendency to expand these indications and treat other patients with cytostatics could alter this picture.

Similar Studies Are Needed for Other Chemicals. The cited two studies from West Germany show that available information sometimes makes it possible to estimate the additional load of mutations imposed by specific compounds, provided the compound has been shown to be mutagenic for man. The results for INH, on the one hand, and cytostatic drugs, on the other, were essentially negative but for different reasons. The mutagenicity of INH in the mammalian test system used so far could not be confirmed. As for gene mutations, the matter has not been settled. Although cytostatic drugs are undisputed mutagens, reproduction by the exposed population is so low that there is virtually no genetic risk for the overall population.

In cases where data of the type used here indicate more exposure and the compounds involved are proven to be mutagenic in relevant in vivo experiments with mammals, the extent of the additional mutational load can be estimated by extrapolation from these experiments. The time has come to attempt similar estimates for all drugs or environmental chemicals, which have been shown to be mutagenic. In this way, it would be possible to achieve at least a minimum estimate of the additional genetic load due to chemically induced mutations. This information is urgently needed. In drawing possible conclusions, we should always keep in mind that while the exposure to any specific compound may be small, the total exposure to a great variety of possible mutagenic compounds could be appreciable.

In Sect. 5.2.1.4, the exposure of human populations to ionizing radiation was discussed. The data given were selected from a huge amount of information systematically collected over a long period of time in many countries and populations. Little work along such lines has been done so far to assess human exposure to chemical mutagens.

5.2.2.5 How High an Increase in the Spontaneous Mutation Rate Must Be Anticipated Due to Chemical Mutagens?

Chemically Induced as Compared to Radiation-Induced Mutations. In Sect. 5.2.1.5, no clearcut answer to the problem of increased mutations by ionizing radiation could be given. The expected order of magnitude could be determined with any confidence only for structural chromosome aberrations and - with some qualifications - for dominant and X-linked gene mutations. The uncertainty was due mainly to our ignorance of the "base line," i.e., the overall spontaneous mutation rate in humans. The same element of uncertainty would hamper any attempt to estimate the increase of mutations due to chemical mutagens. Unlike the situation in radiation genetics, however, other and quite essential information is also lacking. We know very little as to how large the populations exposed to known mutagens actually are. Our knowledge of the exact pattern of action of many of these compounds is fragmentary. We do not know enough about cell stage specificity, about mutation induced in germ cells and somatic cells, about induction of genome, chromosome, and gene mutations, or about pharmacokinetics. For a great number of compounds to which human beings are exposed, it is unknown whether they are mutagenic for mammals. The risk is somewhat higher when mutagenicity has been proven in other organisms, such as bacteria, bacteriophages, fungi, or *Drosophila,* but even chromosome breaks in human lymphocyte cultures do not prove mutagenic activity in the living organism, as evidenced in the case of caffeine. Sufficient in vivo studies on mammalian test systems are still lacking for many compounds.

As Grüneberg has put it [1470], the radiation risk resembles the risk we undergo when involved in automobile traffic: The enhanced rate can be estimated approximately, and adequate precautions are to a certain extent possible. The risk due to chemical mutagens, on the other hand, resembles the risk involved in a walk through the jungle at night. Here a crackling in the underbrush, there an unexplained sound may signal unknown hidden dangers.

Monitoring of Human Populations for Increased Mutation Rates. In view of all these difficulties, it is, indeed, a tempting idea to ask: Should we not try to monitor large populations for new mutations? Any increase would be observed directly, and we could try to relate it to radiation, chemical mutagens, or other possible causes.

There are various possibilities for selection of traits to be sampled in such a monitoring program. For example, it is possible to screen for some "sentinel" mutations - dominant mutations with specific phenotypes for which spontaneous mutation rates are fairly well-known (Table 5.8). However, the number of individuals to be surveyed is of the order of several million, and because mutations to a specific phenotype are rare, most of them pose diagnostic difficulties due to genetic heterogeneity and phenocopies that require a high level of medical expertise for diagnosis. While a good estimate of the *order of magnitude* of the mutation rate is feasible, it would be difficult to estimate an *increase* with confidence. An alternative would be to screen for genome and chromosome mutations. Technically, such as task would be much easier, as these conditions are more frequent, but obviously this approach does not provide any information about gene mutations.

The cleanest, but at the same time most ambitious, approach is examination of population samples to locate genetically determined protein and enzyme variants from blood and to establish for each variant whether it is transmitted from one of the parents, as is usually the case, or whether it represents a new mutation [1577, 1578]. We must be sure only to study rare variants and to omit the many polymorphic variants that are frequent.

There are many problems involved in studies of this type. The most critical one is statistical. Spontaneous mutation rates for single mutations discoverable at the protein level are very low (Sect. 5.1.4). Therefore, large sample sizes are needed to detect an appreciable increase in the mutation rate with certainty [1652, 1672; 1679].

The question may be put as follows: How precisely can a true trend (t)

$$t = \frac{\mu_2}{\mu_1}$$

between the mutations rates in two populations or time periods be estimated from the observed trend (t')

$$t' = \frac{x_2}{x_1}$$

if x_1 and x_2 are the observed numbers of new mutants in the two populations? An approximation can be found for the probability

$$P = 0.95$$

that t' is found between the following limits depending on the real trend t:

$$t' = t \pm 2t\sqrt{\frac{3.8}{x_1, x_2}}; \quad \frac{|t'-t|}{t} = 2\sqrt{\frac{3.8}{x_{1,2}}}$$

$x_{1,2}$ is the sample size of the two samples x_1 and x_2 together. Figure 5.57 shows the confidence limits for the apparent trend t' (ordinate) depending on the sizes of both samples together and the real trend (abscissa). For example, if the real trend t is 1.3, i.e., if the mutation rate in the second sample is 1.3 times the mutation rate in the first sample, 500 new mutants are needed to find an increase, and in this case, the apparent trend t' would be found with 95% probability between ~1.08 and ~1.52. These figures should be compared with the spontaneous human mutation rates of the possible human sentinel mutations (Sect. 5.1.3, Table 5.8) and of chromosome and genome mutations (Sect. 5.1.2, Table 5.3) to get some notion of the sheer size of the problem. Some data are available. Among 133, 478 alleles studied in individuals at the Galton Laboratory in London, 77 rare biochemical variants were found, and each biochemical variant was found to be transmitted from one parent or the other to the proband. No new mutations were detected. These data allow the calculation of a maximum mutation rate for these biochemical variants, which was 2.24×10^{-5} per gene per generation [1788].

Genome and chromosome mutations frequently occur spontaneously so that a monitoring program can remain within reasonable limits. To detect a significant increase of mutations for the usual dominant genetic diseases, whole populations of large countries would have to be screened thoroughly for decades. For mutations identifiable at the protein level, a large and extremely well-organized program would be successful if the number of genes available for screening were suitably enhanced [1577; 1575]. Two different approaches have been suggested. The first [1574, 1575] uses cord blood samples collected from placentas (infant blood) immediately after delivery. At the same time, blood samples are also collected from both parents. In these samples, as many electrophoretic polymorphisms are examined as possible. Quantitative variation in some enzymes is also assessed. Such a program needs a special organization for assuring the cooperation of parents and collecting blood. Therefore, this method is practicable wherever relatively small populations are exposed to a potential high risk, and a maximum amount of information must be collected from few

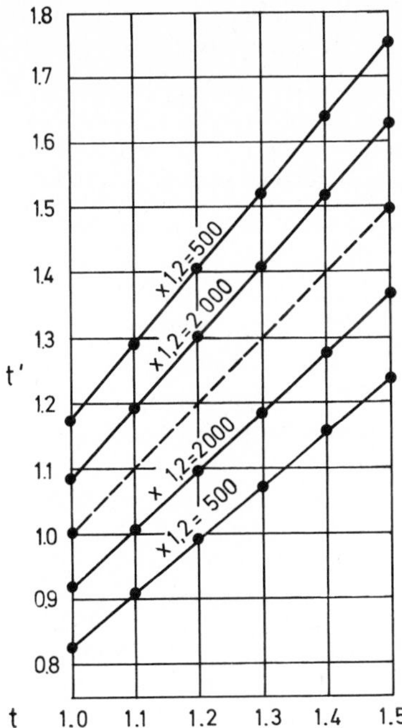

Fig. 5.57. Confidence limits ($P = 0.95$) for an apparent trend between two mutation frequencies (t', *ordinate*) depending on the real trend (t, *abscissa*) and the sizes of the two samples x_1 and x_2 together ($x_{1,2}$)

individuals. However, such programs are not feasible for continuous monitoring of populations consisting of millions of people. In this case, it should be asked whether relevant biologic specimens are available that have been collected for other purposes and could also be used for mutation monitoring. As mentioned in Sect. 4.2.2.7, newborns are screened for phenylketonuria (PKU) and other inherited metabolic diseases. For PKU, the Guthrie test is normally performed; for this test, blood spots dried on paper are used. These spots can be tested for hemoglobin variants [1679] and a number of other gene products [1373]. The methodology is fast and cheap, not only technically but also logistically. For example, the specimens are only numbered at the time of examination; names and addresses can easily be retrieved from the PKU screening registry in the few cases where a rare variant has been found and parental studies need to be done to determine whether one of the parents carries the variant, whether false paternity applies, or whether the variant indeed represents a new mutation. As mentioned in Sect. 5.1.4, one fresh Hb α mutation has been discovered in a pilot study on 25000 newborns.

Before such a program is introduced into large-scale population monitoring, co-operation within the scientific community needs to be organized in such a way that every variant that might be a new mutant is verified carefully – including parternity testing – by a laboratory experienced in human biochemical genetics. The paternity problem is particularly vexing. In Western countries, false paternity is universal in a few percent of family studies carried out for other programs. Mutations will be much rarer than false paternity. However, with an increasing number of genetic polymorphisms, paternity exclusion can be more frequently established. In the long run, human society would have information on mutation frequencies and their changes with changing environmental – as well as genetic – conditions. Statistical problems involved in ascertaining an increase of mutations by population monitoring are not trivial but appropriate methods have been suggested [1561]. However, even if an increase in mutations were to be demonstrated, the task to find the causative agent or agents would often be extremely difficult.

Present Attitudes of Societies Toward Mutagenicity Testing. Very little is being done at present in any society to estimate risks posed by chemical mutagens. There is, however, growing awareness that chemicals yet to be introduced – for example, as pharmaceuticals or pesticides – should be tested for mutagenicity. The problem is more difficult for chemicals already in use for a long time, but testing at least the most important of these compounds has been initiated. Hence, there is agreement *that* chemicals should be tested for mutagenicity. The confusion arises over the question of *how* they should be tested. The principles set out above, though not actually challenged, are far from being accepted within the scientific community. Moreover, the principle of minimal extrapolation requires mammalian in vivo test systems. These systems are available for all types of germ cell mutations and for most somatic mutations but in general require more time and more technical skill than methods using simpler test organisms like bacteria, fruit flies, or lymphocyte cultures. Therefore, it is always tempting for policy makers to fall back on the following argument. Since geneticists disagree in their recommendations, and since there is a correlation in mutation induction even between the most remotely related species, expensive mammalian in vivo systems need not be applied, and simple systems that are cheaper and quicker are sufficient. It is often overlooked that such a policy leaves all the important questions as to phase specificity, type of mutations, and most of the pharmacokinetic and pharmacogenetic problems unanswered – even if liver microsomes are used to allow for metabolic variation. Moreover, the differences usually found even among various "simple" systems make it more than likely that many mutagens would be overlooked even if the following two-step procedure were to be adopted:

1. Test a great number of compounds in a simple system.
2. Select those that are mutagenic and test them in the "relevant" mammalian in vivo systems.

Nevertheless, the sheer size of the problem forces compromises, and a first step involving quick testing of a great number of substances - for example, with microbial test systems involving mammalian liver microsomes - may be inevitable in some cases. However, the possibility of "false negatives" should not be overlooked. Compounds that are suspicious because of their chemical constitution, or those that are used by a great number of individuals of reproductive age ought to be tested with more relevant systems even if screening systems have given negative results.

In many countries, discussions among governmental agencies, scientists, and chemical or pharmaceutic companies are now in progress for the purpose of adopting guidelines for mutagenicity testing.

Medical and Social Significance of Different Types of Mutations. For a complete assessment of the impact of mutations on man, an analysis of the types of diseases produced by various kinds of mutations is required. All autosomal trisomies apart from trisomy 21 are lethal in infancy, and in fact most of them are aborted in the first trimester. The largest medical impact therefore concerns Down's syndrome, with a current incidence of 1-2/1,000. There is profound mental retardation that requires special care in the home or in institutions. The medical and personal impact of spontaneous abortions as the result of chromosome anomalies or lethal genes is much less severe. X-chromosomal aneuploidies are much more common. Their total impact on society largely comes from XXY (Klinefelter's) patients with a current frequency of 1/1,000 male births. The mean intelligence of such males is somewhat lower than that of the control population. There is some increased sociopathy, and infertility is the rule (Sect. 2.2.3.2). While there is a slightly increased load on medical facilities, the major impact of this condition occurs in the personal and social sphere, which is harder to quantitate. X0 females will usually be aborted. If there is postnatal survival, shortened stature and infertility constitute the main personal and medical impact. The XXX condition has no apparent medical implications but is associated with mental retardation.

Chromosome mutations, such as deletions and translocations, if unbalanced will usually lead to early fetal loss. The few postnatal survivors with such mutations will have serious malformations and may require extensive medical care but most will die early in life.

The total impact of point mutations is more difficult to assess. There are many autosomal-dominant and X-linked diseases clearly maintained in the population by mutational equilibrium (Sect. 5.1.3). However, the total frequency of all these diseases is about 1%. Many of these diseases are transmitted, such as Huntington's chorea, and few or no mutations have been observed. In others, such as achondroplasia, about 80% of all cases are new mutations. An increase in the mutation rate would lead to a substantial increase in that group of diseases where most cases now represent fresh mutations. A special group of dominant diseases are common, such as familial hypercholesterolemia and presumably familial hypertriglyceridemia and familial combined hyperlipidemia [1109]. It is unlikely that these conditions are maintained by mutational equilibrium, and selective factors probably have been prominent in making for relatively high frequencies. New mutations, therefore, would have a small impact.

New mutations for autosomal-recessive diseases usually manifest only in the homozygous state. The medical load will be minimal since it requires the mating of two heterozygotes, and the disease impact will be delayed over many generations. Furthermore, equilibrium for such recessive diseases no longer applies: This has led to a relatively low incidence in recent times.

The multifactorial group of diseases includes many birth defects, most common diseases of middle life, the common psychoses (schizophrenia and affective disorders), and many cases of mental retardation. The total medical and social impact of these diseases is far in excess of the strictly genetic diseases. The estimated impact of mutation depends upon the genetic model applicable to the hereditary causation of these disorders. If the underlying genetic causation is based on polymorphic gene variability, such as HLA and ABO in their contribution to disease (Sect. 3.7), mutation would have few effects because the main driving force of polymorphisms is selection and not mutation (Sect. 6.2.1). If these conditions were due to the action of several rare mutant genes or a few major genes, mutation would have a significant effect. In the almost complete absence of specific knowledge regarding the nature of most genetic systems underlying the etiology of common diseases, *no* predictions as to the effect of an increased mutation rate are possible.

These considerations show that serious and large-scale effects of mutations could potentially develop in diseases that are not usually considered as genetic diseases by the public or by medical geneticists. More genetic work on the specific genetic contribution to these diseases is therefore required.

6 Population Genetics

Population genetics deals with the consequences of Mendelian laws on the composition of the population with special reference to the effects of mutations, selection, migration, and chance fluctuation of gene frequencies. All these factors together determine genetic population structure. Knowledge of population genetics is useful for many purposes. For example, it provides a basis for an understanding of the epidemiology of genetic diseases and helps in the planning of measures for prevention of genetic damage by environmental agents. Another objective of studies in population genetics is better understanding of human evolution and prediction of future trends in the biologic evolution of mankind in the face of various environmental changes. Since the human population is much better described than any other species and much better records are available, there are many advantages to studying the population genetics of man.

The work of R. A. Fisher, J. B. S. Haldane, S. Wright, and their successors has provided an elaborate theoretical framework of population genetics. Empirical data on man and their interpretations, however, have lagged behind mathematic and theoretical considerations (Lewontin, 1977 [1808]). Several excellent expositions of population genetics exist (Li, 1955 [124]; Li, 1976 [1810]; Crow and Kimura, 1970 [45]; Cavalli-Sforza and Bodmer, 1971 [36]; Jacquard, 1974 [103]; Hartl, 1980 [93 a]; Ewens, 1980 [1757]). Our treatment of the subject therefore will not be exhaustive. Special attention will be paid to empirical data collected in human populations and their interpretation.

Work in human population genetics may conveniently be divided into two broad classes: description of populations and their genetic composition and studies designed to understand the causes for changes in the human gene pool. These two approaches are intimately connected. It is impossible to elaborate specific hypotheses and to design studies to test them unless certain underlying facts regarding population structure are known. Yet, there are so many different human populations and so many known genetic traits that the task of describing the genetic characteristics of all populations is formidable. A selection of the more important problems is necessary. What are the guidelines for such a selection?

In general, similar principles should apply for planning of scientific work in human population genetics as for formulating the design of laboratory investigations. Empirical studies without guidance by a specific hypothesis rarely lead to significant insights. The quality of scientific work generally depends on the profundity and specificity of the underlying hypotheses. One cannot expect, however, that all data collection in science will always be guided by a hypothesis. However, usually mere data-collecting activities are generally less satisfactory scientifically than projects that ask specific questions. Fortunately, descriptive data can often be obtained from a variety of sources:

a) Polymorphisms for DNA variants, enzymes and other proteins are continuously being described. Testing of population samples to determine gene frequencies will provide information regarding the gene frequencies of these newly discovered traits and allows testing for other genetic markers, thus providing data for assessment of population structure.
b) Population testing may be done for medical reasons. For example, newborn screening for phenylketonuria and sometimes for other rare recessive disorders is now performed as a routine procedure in many populations. These studies provide valuable information about population differences in gene frequencies. Family studies in cases of disputed paternity sometimes provide data of scientific interest.
c) Gene frequency data may be collected while testing a specific hypothesis in a population. Even if the hypothesis is refuted or if the outcome of the study is equivocal, the descriptive data may be useful.

The categorization of a human population using frequencies of genetic polymorphisms and genetic diseases is only the beginning step toward understanding the differences in gene frequencies between human populations. Explanatory hypotheses are required to explain the differences.

Consider the hypothesis that the high frequency of persistence of intestinal lactase activity in adult whites as contrasted with Mongoloids and blacks is caused by a selective advantage in a climate favoring the development of vitamin D deficiency and rickets, since lactose enhances calcium resorption from the intestine and calcium reduces the risk of developing rickets (Sect. 7.3.1). This hypothesis has a biologic basis, is specific, and can easily be refuted by showing that lactose does not increase calcium resorption. Hypotheses of this sort are highly desirable. Unfortunately, much work in human population genetics remains at the descriptive level.

6.1 Population Description

6.1.1 Hardy-Weinberg Law: Extended Consideration - Gene Frequencies

Hardy-Weinberg Law for Autosomal Genes [124]. The Hardy-Weinberg Law was discussed in Sect. 3.2. Let two alleles A_1, A_2 have the gene frequencies $A_1 = p$, $A_2 = q$; $p+q=1$, let mating be random. Then, the three phenotypes will occur in the following frequencies: $p^2 A_1A_1$, $2pq A_1A_2$, $q^2 A_2A_2$. This rule may be generalized as follows. If the gene frequencies of n alleles, $A_1, A_2 \ldots A_n$, are $p_1, p_2, \ldots p_n$ ($\sum p_i = 1$), and if the population breeds at random with respect to the gene locus A, then the phenotypes will occur according to chance combinations of these alleles in pairs:

$$(p_1A_1 + p_2A_2 + \ldots + p_nA_n)^2 =$$

$$\sum_{i=1} p_i^2 A_iA_i + \sum_{i<j} 2p_ip_j A_iA_j$$

In the absence of disturbing influences, the gene as well as genotype frequencies remain constant from generation to generation. For autosomal genes, this "Hardy-Weinberg equilibrium" is reached in the first generation of random mating. This is not so, however, for X-linked genes.

Hardy-Weinberg Law for X-Linked Genes. Let two alleles A_1 and A_2 have frequencies p_M and q_M ($p_M + q_M = 1$) in the male population; the phenotypes A_1 and A_2 are also found with frequencies p_M and q_M. In the female population, on the other hand, the genotypes A_1A_1, A_1A_2, and A_2A_2 occur in frequencies r, $2s$, and t ($r+2s+t=1$). Then the frequencies of the alleles A_1 and A_2 in the females are $p_F = r+s$ and $q_F = s+t$, respectively, and these women produce oocytes of the two types A_1 and A_2 also with frequencies p_F and q_F, respectively. Their male offspring are formed in the same proportion. Female offspring, on the other hand, are produced from a combination of the ($p_FA_1 + q_FA_2$) oocytes with the ($p_MA_1 + q_MA_2$) X-sperms of the men. Hence, the next generation will be composed as follows:

♂♂:
$p_FA_1 + q_FA_2$,

♀♀:
$p_Mp_FA_1A_1 + (p_Mq_F + p_Fq_M)A_1A_2 + q_Mq_FA_2A_2$

It follows that the frequencies of A_2 in the males and in the females of the next generation are:

$$q'_M = q_F, \quad q'_F = \tfrac{1}{2}(p_Mq_F + p_Fq_M) + q_Mq_F = \tfrac{1}{2}(q_M + q_F)$$

This means:

a) The gene frequency of A_2 in males of each generation is equal to the gene frequency of A_2 in females of the preceding generation.
b) The frequency of A_2 among females in each generation equals the mean of the frequencies of A_2 in males and females of the foregoing generation.
c) The equation:

   $$q'_M - q'_F = -\tfrac{1}{2}(q_M - q_F)$$

   holds.

This means that if the gene frequencies in males and females in one generation are not equal their difference is halved in the next generation. Besides, the sign of the difference changes: is q_F higher than q_M, then q'_M will be higher than q'_F. Both gene frequencies q_M and q_F converge to a common value $\hat{q}$; at the same time, the genotype distributions in the two sexes tend to the equilibrium states:

♂♂:
$(1-\hat{q})\, A_1 + \hat{q}A_2$,

♀♀:
$(1-\hat{q})^2\, A_1A_1 + 2\hat{q}(1-\hat{q})A_1A_2 + \hat{q}^2\, A_2A_2$

The mode of approximation to this equilibrium can be seen in Fig. 6.1. This example shows that establishment of a Hardy-Weinberg equilibrium after one generation of random mating is by no means self-evident.

The other limitations of the Hardy-Weinberg Law have been enumerated in Sect. 3.2; a great deal of population genetics can be regarded as the elaboration of this fundamental rule.

Gene Frequencies. Individuals found in a population can be classified on the basis of their phenotypes. In an increasing number of cases - especially in the genetic polymorphisms - these phenotypes give unequivocal evidence of the genotype. However, description of genetic variability in a population in terms of *genotypes* is usually clumsy. The Hardy-Weinberg Law makes it possible to describe this variability in terms of *gene frequencies.* This notation simplifies the matter; in a system with two alleles, a single number (p or q) comprises all the information needed. The estimation of gene frequencies has been described in principle in Sect. 3.2 and in App. 1; a detailed outline is given in Race and Sanger (1975) [166] or in Mourant et al. (1976) [144].

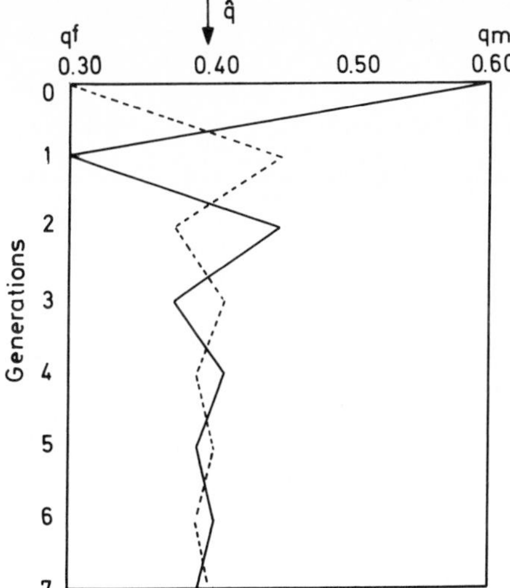

Fig. 6.1. Approximation of gene frequencies in males to equilibrium value, if the mode of inheritance is X-linked $q_M=0.6$; $q_F=0.3$; $\hat{q}=0.4$; ——— =, gene frequencies in males, q_M; ----- =, gene frequencies in females, q_F; $\hat{q}$ =, gene frequency at equilibrium

6.1.2 Genetic Polymorphisms

Definition and History. A polymorphism is a Mendelian or monogenic trait that exists in the population in at least two phenotypes (and presumably at least two genotypes), neither of which is rare - that is, neither of which occurs with a frequency of less than 1%-2%. Often we find more than two alleles, and more than two phenotypes, for a single locus. A polymorphism should be contrasted with a *rare* genetic variant. Rare genetic variants are arbitrarily defined as monogenic traits that occur in the population with a frequency of less than 1%-2% and usually at much lower frequencies. The first human polymorphism was the AB0 blood group discovered by Landsteiner (1900) [259]. Up to 1955, only polymorphisms for a number of other red cell surface antigens, i.e., the blood groups, were known. In 1955, Smithies [1307, 1308] described the method of starch gel electrophoresis, which permits separation of proteins on a fixed medium not only by charge but also by molecular size. The new method enabled Smithies to detect polymorphism of a serum protein: the hemoglobin-binding protein haptoglobin. This method proved to be useful, first for detection of serum protein polymorphisms and later - together with methods for specifically identifying enzyme activities - for detection of enzyme polymorphisms (Giblett, 1977) [78, 1769].

A great number of polymorphisms for other serum proteins - and later for enzymes of blood plasma, erythrocytes, or leukocytes - have been detected. Many polymorphisms are genetically straightforward with two alleles determining two variants of the same protein. Others are highly complex such as the major histocompatibility complex (MHC), with multiple, related loci in a complex system on human chromosome 6 (Sect. 3.5.5).

Present Situation. Table 6.1 shows the most important polymorphisms. Some polymorphisms exist in only one major race group. For a few polymorphisms, specific hypotheses as to their maintenance in human populations by natural selection have been brought forward. These hypotheses will be discussed in greater detail in Sect. 6.2.1.6 and 6.2.1.8.

Biochemical Individuality for Polymorphisms. Garrod (1902) [249] ended his paper on alcaptonuria (Sect. 1.5) with the sentence:

> If it be, indeed, the case that in alcaptonuria and the other conditions mentioned we are dealing with individualities of metabolism and not with the results of morbid processes the thought naturally presents itself that these are merely extreme examples of variations of chemical behavior which are probably everywhere present in minor degrees and that just as no two individuals of a species are absolutely identical in bodily structure neither are their chemical processes carried out on exactly the same lines.

This "biochemical individuality" is striking if all blood polymorphisms are considered. Let us, for example, select a northwestern European individual having the most frequent alleles of every polymorphism listed in Table 6.1. How many other individuals will have the same phenotype and genotype for all markers? This probability can be calculated by multiplying the relative frequencies of these phenotypes in the white population (Table 6.2), with the result of 3.1×10^{-5}; thus, among 100,000 male individuals, only two or three will have this phenotype, though this particular combination is the most frequent one. All other combinations are still rarer.

Table 6.1 does not even comprise the phenotypes of the major histocompatibility complex (MHC) (Sect. 3.5.5) or other less well-documented polymorphisms and many enzyme systems for which only rare variants have been observed. Had these been included, it could have been shown that every person on this planet (except for identical twins) is genetically unique. The physiologic function for only some of the polymorphic characters listed in Table 6.1 is known. The possible significance of these polymorphisms for predictions of health risks under varying environmental conditions has been discussed in Sect. 4.5.2.

What is the Proportion of Polymorphic Human Gene Loci? For what proportion of their gene loci are hu-

Table 6.1. Some important human polymorphisms

Name	Main alleles	Remarks
Erythrocyte surface antigens (blood groups)		
AB0	A_1, A_2, B, 0	For discussions of disease associations see Sect. 3.7.2; for natural selection see Sect. 6.2.1.8
ABH secretion	Se, se	Interaction with the Lewis system
Diego	Di^a, Di^b	Allele Di^a present only in Amerindians and Mongoloid populations (Sect. 7.3.1)
Duffy	Fy^a, Fy^b, Fy	Amorphic allele Fy common in negroids; discussion of selection in Sect. 7.3.1
Kell	K, k	Other, closely linked loci, e.g., Sutter (Js^a)
Kidd	Jk^a, Jk^b	Very few individuals with Jk (a-b-)
Lewis	Le^a, Le^b	Interaction with ABH secretor locus
Lutheran	Lu^a, Lu^b	
MNSs	MS, Ms, NS, Ns	There are some other, closely linked antigens: Hunter and Henshaw, especially in blacks.
P	P_1, P_2, p	Allele p is very rare
Rhesus	Gene complexes CDe, cde, cDE, C^wDe, cDe, cdE, CDE, and others in varying combinations	Discussion of maternal-fetal incompatibility in Sect. 3.5.4; structure of the gene complex and linkage disequilibrium in Sect. 3.5.5
Xg	Xg^a, Xg	X-linked
Serum protein groups		
α_1-antitrypsin (α_1-Protease inhibitor)	PI^{M_1}, PI^{M_2}, PI^{M_3}, PI^S, PI^Z	Numerous rare alleles. Discussions of α_1-antitrypsin deficiency, especially in homozygotes of the PIZ allele, in Sect. 3.7.4
Ceruloplasmin	CP^B, CP^A, CP^C	Most Europeans are homozygous CP^B/CP^B; blacks have a gene frequency of 0.06 for CP^A
Complement component-3	$C3^S$, $C3^F$	Apart from these two common alleles, there are a number of rare ones
Group-specific protein	GC^{1F}, GC^{1S}, GC^2	Special variants described, for example GC^{Chip} in Chippewa Indians, GC^{Ab} in Australian aborigines; subdivision of common alleles possible; discussion of natural selection in Sect. 7.3.1
Haptoglobin	HP^{1S}, HP^{1F}, HP^2	Many rare variants are known; genetic and nongenetic ahaptoglobinemia occurs
Immunoglobulins IGHG (gm)	$G1m^3$, $G3m^5$, $G1m^1$, $G1m^{1,2}$	This is a very complicated system with many rare haplotypes and specificities; cf. genetics of antibody formation, Sect. 4.4
IGKC (Km)	Km^1, Km^3	More alleles are known, but are usually not easily available
Properdin factor B (glycine-rich-β-glycoprotein)	BF^S, BF^F	Rare alleles are known
Transferrin	TF^{C_1}, TF^{C_2}, TF^{C_3}, TF^B, TF^D	Different D and B variants have been described, all of them are rare. D variants occur mainly in blacks
Red cell enzymes		
Acid phosphatase-1	$ACP1^A$, $ACP1^B$, $ACP1^C$	An additional allele $ACP1^R$ has been observed in Khoisanids
Adenosine deaminase	ADA^1, ADA^2	Rare alleles ADA^3 and ADA^4 have been described
Adenylate kinase-1	$AK1^1$, $AK1^2$	Some other, rarer alleles are known
Esterase D	ESD^1, ESD^2	Rare variants are also known

Table 6.1 *(continued)*

Name	Main alleles	Remarks
Peptidase-A	$PEPA^1$, $PEPA^2$	$PEPA^2$ has a gene frequency of about$_1$ 0.07 in blacks; whites have almost exclusively $PEPA^1$. Rare variants are known
Peptidase-D (Proline dipeptidase)	$PEPD^1$, $PEPD^2$, $PEPD^3$	$PEPD^3$ observed especially in blacks
Phosphoglucomutases		
PGM1	$PGM1^{a1}$, $PGM1^{a2}$, $PGM1^{a3}$, $PGM1^{a4}$	Rare alleles are known
PGM2	$PGM2^1$, $PGM2^2$	Allele $PGM2^2$ only in blacks, other alleles very rare
PGM3	$PGM3^1$, $PGM3^2$	Enzymes mostly in leukocytes, placenta and sperms. Linkage with the major histocompatibility complex (MHC); Sect. 3.5.5
Phosphogluconate dehydrogenase	PGD^A, PGD^B	Some other, rare alleles are known
Some other enzyme polymorphisms		
Alcohol dehydrogenase	$ADH3^1$, $ADH3^2$	ADH2 active in other organs, ADH3 active in the intestines
Cholinesterase (serum)-1	$CHE1^U$, $CHE1^D$, $CHE1^S$	Discussed in Sect. 4.5.1
Glutamic-pyruvic transaminase (alanine aminotransferase)	GPT^1, GPT^2	Activity discussed in Sect. 3.6.1.3
Liver acetyltransferase	Rapic and slow inactivators	Discussed in Sect. 4.5.1

mans polymorphic? Are polymorphic loci only a relatively small fraction of all human gene loci, or is the share of polymorphic loci appreciable? Blood groups can only be detected if an antibody is found against a certain antigen. However, the antibody could only be formed in an individual not carrying the same antigen. Hence, serologic detection of a gene locus usually presupposes genetic variability with respect to this gene locus - either a polymorphism or a rare variant. The detection of genetic variability proceeds differently for enzymes. The search for enzyme variability requires an ability to detect the enzyme readily in an electrophoretic medium. If there is polymorphism, abnormal electrophoretic mobility of the variant gene product often results. Detection of the normal (wild type) and variant enzyme can only be achieved by localization of the enzyme-specific biochemical reaction on the supporting medium of electrophoresis, for example starch gel. Such localization is achieved by developing a test system for enzyme assay that results in a colored end product allowing direct visualization on the gel. As an example, red cell acid phosphatase splits phenolphthalein phosphate at pH 6 into phenolphthalein and phosphate; the free phenolphthalein stains the starch gel only at those segments of the gel where the enzyme has migrated. Different patterns of enzyme staining were detected that could be related to major gene differences (Fig. 6.2).

When a number of enzymes were screened for variants by analogous methods, an unbiased estimate of the fraction of polymorphic enzymes became possi-

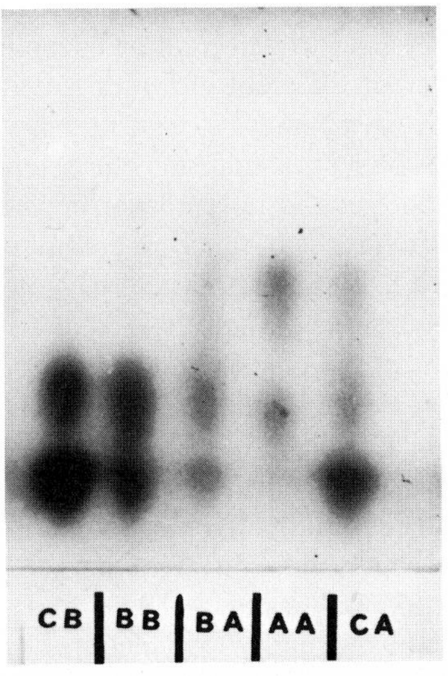

Fig. 6.2. The various electrophoretic types of erythrocyte acid phosphatase. Three alleles (A, B, C) produce six phenotypes (CC not shown). (Courtesy of Dr. J. Greiner)

Table 6.2. Phenotype of a male of European extraction having the most common phenotypes of the polymorphisms in Table 6.1

Phenotype	Approximate frequency
Blood groups[a]	
A1	35%
CdD. ee	33%
Fy (a+b+)	48%
Jk (a+b+)	50%
K-k+	91%
Lu (a-b+)	92%
MNSs	23%
P_1	79%
Se	78%
Xg (a+)	66%
Serum proteins and erythrocyte enzymes[b]	
ADA 1	87%
ACP1 AB	39%
AK1 1	93%
BF S	61%
C3 S	61%
GC 1S	34%
Gm (-1, -2, +3, +5)	45%
GPT 2-1	49%
HP 2-1	47%
km (-1)	87%
PGD A	96%
PGM1 a1	39%
Pi M1	55%
TF C1	57%

Indicence of this most common phenotype: 9.2×10^{-7}.
[a] Frequencies from Race and Sanger (1975).
[b] Frequencies from Hummel (1971, 1977, 1979).

ble. Such an estimation was carried out by Harris and Hopkinson in 1972 [1787]. They collected information from electrophoretic surveys of European populations on 71 different loci determining enzyme structure. The number of different enzymes tested was smaller, as one enzyme may consist of more than one polypeptide chain and hence may be determined by more than one gene. A polymorphism was registered if the gene frequency of the most common alleles did not exceed 0.99. Since the heterozygote frequency (2pq) is about twice the genotype frequency (q), about 2% of the population would be heterozygous for such alleles. According to this definition, 20 of the 71 (28.2%) loci showed polymorphisms. This figure does not comprise the total number of polymorphisms, as some variants cannot be identified by electrophoresis but require different methods of detection (e.g., pseudocholinesterase) (Sect. 4.5.1). Only electrophoretic polymorphisms are included in these calculations. Since in some cases the electrophoretic technique may not have been good enough to clearly show differences between variants, the figure of 28.2% is almost certainly an underestimate. For one of these enzymes – erythrocyte acid phosphatase with three common alleles – more than 50% of the population are heterozygous. For other loci, the degree of heterozygosity is much lower, and for all 71 loci the average fraction of heterozygous individuals was estimated as 6.7% per locus. It can be calculated from the genetic code that only about one-third of all possible base substitutions lead to polar amino acid replacements with an expected change in the electrophoretic mobility [1681]. However, there is no a priori reason to assume that mutations leading to nonpolar replacements have a lower probability of occurrence. Hence, the real average heterozygosity per locus may be about three times higher, amounting to something like 20%. Similar findings regarding the ubiquitous occurrence of genetic polymorphisms have been made in all biologic species examined (Fig. 6.3). This finding is therefore not unique to the human species where it was first discovered. The reasons for this remarkable genetic heterogeneity, which is not surprising in the human species with its highly differentiated individualization of physiognomic features, has elicited considerable controversy and led to theories of non-Darwinian evolution in which it has been claimed that most genes are neutral [1510, 1511] (See Sect. 7.2.3).

Rare Variants. The limiting frequency of the most common allele, and hence for a genetic polymorphism, was set at 0.99 in the study by Harris [1787]. However, for many enzymes, rare variants have become known. Often such rare variants are present in addition to polymorphisms; for many of the other loci, only rare and no common variants are known. Their incidence and distribution in European populations was examined by Harris et al. (1974) [1788]. The probability of detection of rare variants depends on sample size, and population samples of at least several hundred individuals need to be studied for an assessment of this problem. Such samples were available for 43 enzyme loci. Rare variants were defined as having a gene frequency of 0.005 or lower. At 22 of the 43 loci – 7 of 13 "polymorphic" loci; 15 of 30 "nonpolymorphic" ones – rare alleles were found (Fig. 6.4). There were 56 rare alleles in all, and of these the majority were extremely rare, 45 having gene frequencies of less than 0.001. The influence of sample size on the probability of variant detection was demonstrated by the fact that the mean sample size for those loci at which rare alleles were found was 4,023, while it was only 1,300 for loci at which no rare alleles could be detected. For some loci where this study detected no rare variants, such rare mutants have been found in other studies. It is reasonable to assume, therefore, that they may occur for all enzymes.

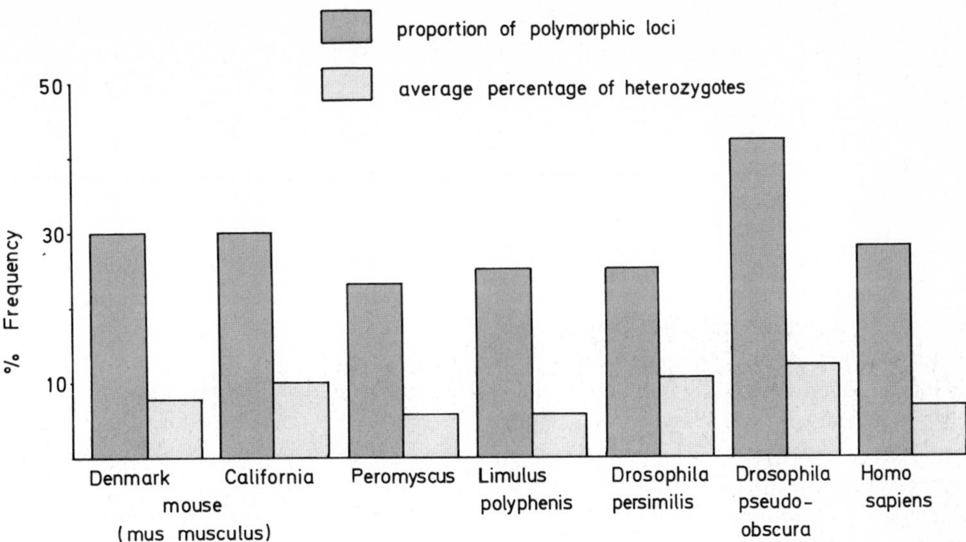

Fig. 6.3. Proportion of genetic enzyme polymorphism in different species, including man. (Adapted from Harris and Hopkinson, 1972 [1787])

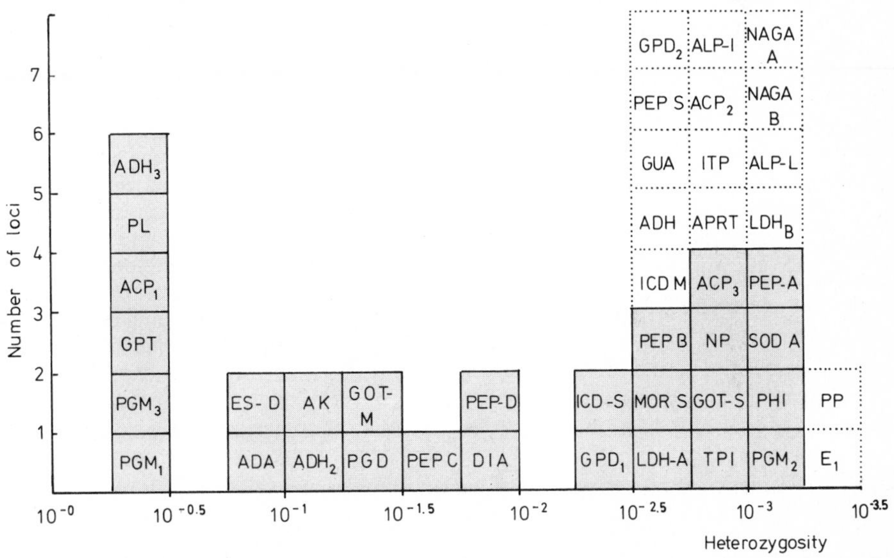

Fig. 6.4. Distribution of heterozygote frequencies for 43 human gene loci for which enzyme variants are available. The blocks enclosed within *dotted lines*-are overestimates representing the situation if one heterozygote had been detected in a sample where none were actually detected. (Harris et al., 1974 [1788])

Average heterozygosity per locus was estimated for rare alleles in a similar way as for polymorphic variants:

$$\frac{\text{Number of heterozygotes of rare alleles at all loci (204)}}{\text{Sum of individuals screened at each locus (115,755)}}$$

= 1.76 per 1,000 individuals.

One locus - that for placental alkaline phosphatase - had a particularly high level of heterozygosity attributable to rare alleles. When this locus was excluded from the calculations, the overall estimate was 1.14 heterozygotes per 1,000 individuals per locus.

In other populations, similar calculations have been carried out. For example, the frequency of rare variants of 12 glycolytic enzymes was estimated at about 2/1,000 from combined data of white, black, and Mongoloid populations [1891], and at about 3/1,000 from surveys of 17 enzymes and protein per loci in some Indian tribes of South America [1880].

In a Japanese sample, an incidence of 2/1,000 was found for 25 polypeptides, 23 of which were also included in the British sample of Harris et al. [1788]. For two loci, PGM_1 (phosphoglucomutase) and PHI (phosphohexose isomerase), higher frequencies were found in the Japanese as compared with the British population [1578].

Mutation Rate May Be Low. It is possible to estimate the average mutation rate, μ, for these rare variants [1788]. More precisely: A number can be calculated that is an upper confidence limit for μ. For this estimation, Harris et al. used 36 of the screened loci for which parental information was available. For these 36 loci, altogether 109,209 individuals were tested; 126 were carriers of a rare variant. Both parents could be tested in only 77 of the 116 cases. In all cases, one of the parents carried the rare variant; *there were no fresh mutations.* This subset of 77 carriers of rare genes corresponds to a reference sample of screened individuals with an expected size of

$$109{,}209 \times \frac{77}{126} = 66{,}739$$

It can therefore be stated that no fresh mutation occurred in $2 \times 66{,}739 = 133{,}478$ parental gametes, on the average. The probability of observing at least one new mutation in these gametes is given by $1 - e^{-n\mu}$, *where* n = the number of gametes and e = the base of natural logarithms. Solving the equation $1 - e^{-n\mu} = 0.95$ for μ therefore gives $\mu = 2.24 \times 10^{-5}$/gene/generation as an upper 95 per cent confidence limit for the actual mutation rate/gene/generation for electrophoretically detectable variants of these enzyme loci. This means: The calculated value overestimates the actual mutation rate with a probability of 95%. Using slightly different assumptions for these data and including other data from many laboratories that have failed to show a single instance of fresh mutations in biochemical systems, a lower mutation rate would be calculated. A higher average mutation rate than this value had also been made unlikely for a different class of mutations – those leading to dominant or X-linked hereditary disease – by direct assessment (Sect. 5.1.3). On the other hand, three new mutants have been discovered by electrophoretic screening of children born to atomic bomb survivers (Tab. 5.27), giving a mutation rate of 5.4×10^{-6}, and one Hb variant among loci by extended newborn screening [1679].

Distribution of Variants According to Their Frequencies. Figure 6.4 shows the distribution of enzyme variants according to their frequencies. There is a clear-cut dichotomy between common and rare variants. This dichotomy gave rise to speculations on the mechanisms of evolution (selection vs chance fluctuations of selectively neutral mutants). This problem will be discussed in Sect. 7.2.3.

Genetic Polymorphisms of Other (e.g., Structural) Proteins [1892, 1803]. On the basis of data showing a high frequency of genetic polymorphism in enzymes, it was initially concluded that most genes were highly polymorphic. Recent studies have shown, however, that this is not the case. In these studies, the method of two-dimensional electrophoresis illustrated in Figure 5.51 was mainly used [1892]. Table 6.3 shows results from one of these studies [1803]. Total cultivated fibroblasts, fibroblasts fractionated into pellets (= structure-bound proteins) and supernatant (soluble proteins), and soluble proteins from hair root cells were examined. The number of qualitative variants was low indeed. There was much more variation of a clear-cut, but only quantitative kind. Previous studies on mice had shown such quantitative variation to be largely genetical,

Table 6.3. Variation in protein patterns of human fibroblast cell lines and hair root cells as revealed by two-dimensional electrophoresis. (From [1803])

Source of protein	Pairs of patterns compared	Total number of spots (average)	Number and percentage of variant spots			
			Qualitative variants		Quantitative variants (average/comparison)	
			No.	%	No.	%
Total fibroblasts	6	659	4	1	46	6.92
Fractionation of fibroblasts by centrifugation:						
Supernatant	6	460	2	1	51	11.04
Pellet	4	502	0	0	17	3.45
Fractionation of hair root cells by centrifugation:						
Supernatant	6	289	3[a]	1	30	10.19

The preparations were made from cells taken from different individuals. Total fibroblasts and fibroblast fractionations were also from different individuals. Qualitative variants are given for all comparisons together, quantitative variants and total numbers of spots as averages of single comparisons.

[a] Three times the same protein.

as well. Moreover, variation - qualitative as well as quantitative - is wider in the soluble proteins of the supernatant. A similar lack of variation in tissue proteins has now been found not only in fibroblasts but also in other human tissues such as brain [1741] or lymphocytes [1782, 1783]. It is interpreted by the plausible assumption that structural proteins have to interact with many other proteins, and that this involves severe constraints on the number of mutations leading to conformational changes that can be tolerated [1992]. This hypothesis is supported by the observation that soluble proteins show more variation than structurally bound proteins. The higher degree of quantitative than of qualitative variation could possibly suggest a higher tolerance of nontranscribed DNA segments for mutations, since such mutations are often thought to influence rate of synthesis of proteins. (See, for example, Sect. 4.3.4 on the thalassemias.)

In phytohemagglutinin-stimulated lymphocytes, three genetic polymorphisms among about 200 different protein spots were discovered and confirmed by population and family studies [1782, 1783]. All three were found in the soluble fraction. In these lymphocytes, polymorphisms might be slightly more common than in fibroblasts. It has been concluded from these, as well as from an increasing number of studies on other species such as mice and *Drosophila*, that genetic polymorphisms might be rarer in structural proteins than in the soluble blood enzymes. However, premature generalizations should be avoided, since in a recent study on human blood platelet proteins that was planned mainly for assessing new mutations leading to protein variants, more genetic variability has been observed [1838]. Variability found in this study turned out to be comparable to that encountered in soluble plasma and erythrocyte proteins.

DNA Polymorphisms [328, 221, 60]. The large extent of polymorphisms in expressed gene products such as blood groups, tissue types, and blood proteins is vastly exceeded by genetic variation at the DNA level. Since a large fraction of the genome does not appear to be involved in the direct regulation or specification of gene products, mutations in these nonregulatory and noncoding regions of DNA appear to have no phenotypic effect and appear selectively neutral. Determination of DNA sequences in different individuals and the use of restriction enzymes in mapping the human genome has revealed tremendous variability at the level of DNA. When family studies were done, the different DNA variants were transmitted by Mendelian inheritance. An entirely new set of marker genes is therefore now available.

Types of DNA Polymorphisms. The most frequent type of DNA variant is discovered when a given restriction enzyme encounters a point mutation in its usual recognition sequence, no longer recognizes this site and therefore fails to cleave DNA at this site (Fig. 2.84). The DNA under examination can be probed by gene - specific or chromosome - specific or anonymous probes using one or another of the various restriction enzymes. The resultant pieces of DNA differ in fragment length and are known as restriction fragment length polymorphisms (RFLP). They can be recognized by their variable mobility after radioactive tagging on Southern blots (Fig. 6.5). It is likely that nonradioactive labeling of DNA fragments will become available in the future. Point mutations that change a nucleotide to another base in the noncoding region of DNA are very common. A few systematic studies on the extent of DNA variation were carried out by studying a given sequence of DNA with a large number of restriction (6-17) enzymes in a small sample of individuals (10-12). The results for the genomic regions studied so far (hemoglobin, albumin, and anonymous DNA segments from different chromosomes) [1143, 1742, 1959] indicate that the extent of nucleotide diversity is about an order of magnitude greater than that observed from data for structural genes that specify protein products. This means that on average 1/500 to 1/250 nucleotides will differ between random chromosomes (heterozygosity = 0.001-0.004). The enzymes Msp I and Taq I, which recognize the methylated dinucleotide EpG, are particularly useful for finding DNA variants. Most restriction enzyme fragment length polymorphisms are dimorphic in that only two "alleles" exist, as manifested by the presence (+) or absence (−) of the restriction site. The frequency of the polymorphic variant can vary from a few percent to a maximum of 50%.

Another type of DNA variant consists of different numbers of tandem repeat sequences sharing a common 10-15 base pair core ("minisatellites") [1795]. A chromosome area may bear a different number of such repeats. The production of such polymorphisms is facilitated by their sequence identity which causes deletions and duplications by unequal crossover (Fig. 6.6). The resultant restriction fragments will differ in size depending upon the number of repeats. Such hypervariable areas occur near the insulin gene and around the Hb β gene complex in chromosome 11 but are also common in other chromosomal areas. Since this type of DNA polymorphism manifests with several different numbers of repeats, allelic heterozygosity is common and homozygosity unusual. This property is a significant advantage for marker studies, since such variants are likely to be informative in almost all instances.

Using a minisatellite from an intron sequence of myoglobin, Jeffrey recently developed a probe that recognizes such minisatellites or hypervariable DNA [1795]. Many such hypervariable regions were detected at various human chromosomes. There was considerable heterozygosity reaching almost 100% in the small population that has been studied at this time. In an inbred pedigree of 54 individuals spanning 4 generations it was shown that segregation of

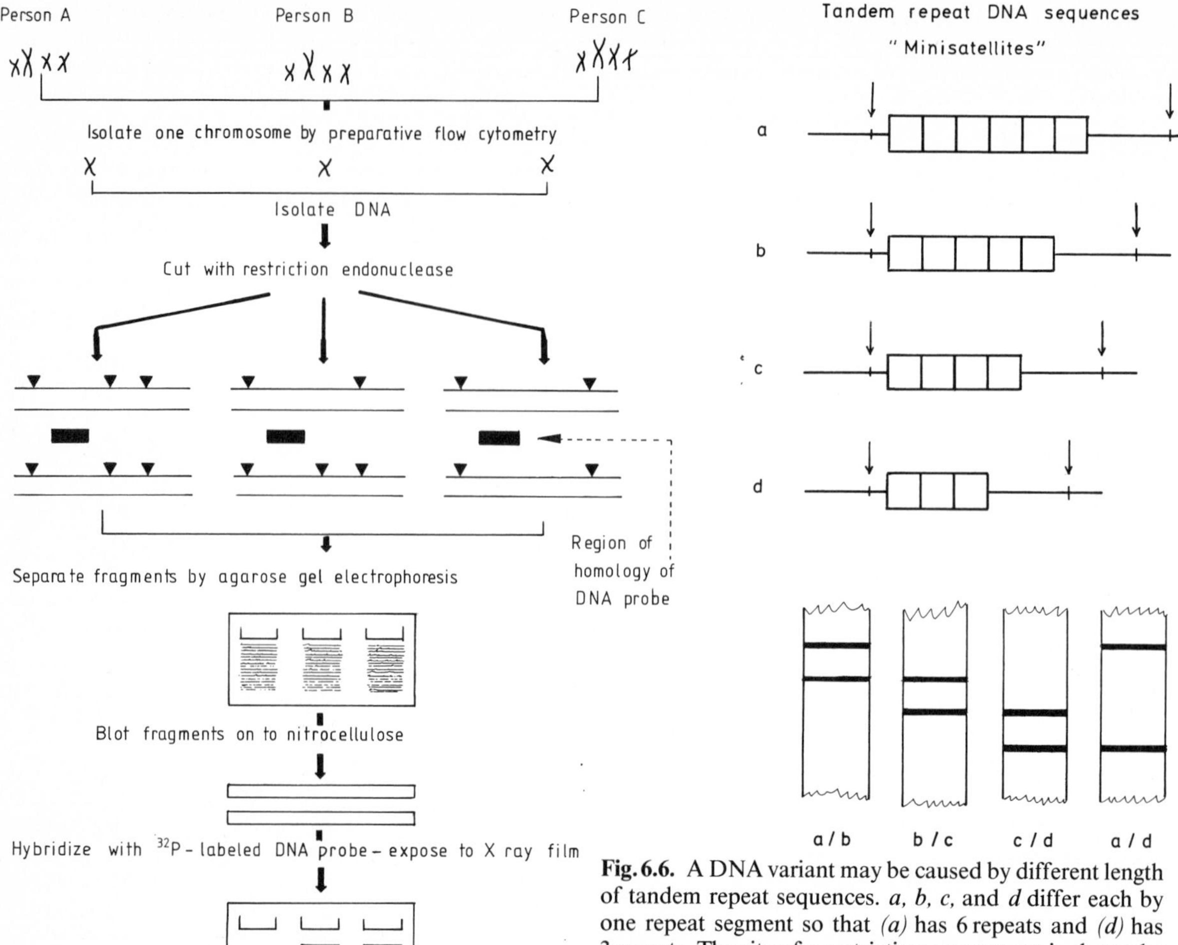

Fig. 6.5. Three individuals, two of them (*A, C*) homozygous for different restriction fragment length polymorphism *(RFLP)* haplotypes, the third (*B*) heterozygous. Principle of analysis and results of agar gel electrophoresis and Southern blotting

Fig. 6.6. A DNA variant may be caused by different length of tandem repeat sequences. *a, b, c,* and *d* differ each by one repeat segment so that *(a)* has 6 repeats and *(d)* has 3 repeats. The site of a restriction enzyme cut is shown by *arrows*. The cut sites are not altered. However, the size of the DNA fragments varies depending upon the number of repeats. The smallest *(d)* chromosome will move faster and the various DNA sizes can be distinguished on the Southern blot by mobility differences. Heterozygotes for various combinations are shown diagrammatically (*a/b; b/c;* etc.) [60].

the various heterozygote bands followed Mendelian laws. Most remarkably, every individual in this pedigree could be distinguished from every other person by this method of analysis. An unique DNA "fingerprint" (even for related persons) therefore will be available with further developments of this technology.

However, an occasional unique crossover or DNA slippage during replication might lead to differences in fragment length between parents and offspring – a process that would be expected to be more common than a point mutation. Among 27 individuals where 240 bands of this sort were studied a unique new band not detected in either parent was found once, giving a mutation rate of 1/240 or at least four orders of magnitude higher than the rate found with point mutations (Sect. 2.3.3.9).

Applications of DNA Marker Studies. The use of DNA markers extends the theoretical and practical applications of genetic linkage work considerably. In some years, many DNA markers that are evenly spread over the human genome are likely to be available. Such markers will definitely be useful for the preclinical and prenatal detection of various monogenic genetic diseases by linkage (Sect. 3.4) [946]. It may even become possible to get a better appreciation of the contribution of various genes to common diseases (that are not inherited as Mendelian traits) by attempting to detect major genic sites that are not yet identified by linkage. However, there will be difficult analytical problems in the differentiation of major and minor genes in polygenic conditions (Sect. 3.7). DNA variants can also serve as markers for tumor studies [1685] and should allow analysis of tumorigenesis in neoplasms such as retinoblastoma, Wilm's tumors, and others (Sect. 5.1.6). Prenatal diagnosis of the hereditary forms of those tumors also should become possible. Since the new polymorphic minisatellites appear to contain signals for recombination [1795], mechanisms of recombination can be studied. Paternity and maternity detection is foreseeable, as are forensic applications for identification of blood and seminal stains. The discovery of the different types of DNA polymorphisms is revolutionizing the field of human and medical genetics. A complete map of the human genome can be expected in the not too distant future. The utilization of this map for both practical and theoretical work will be an exciting development of the next few years.

Mitochondrial DNA Polymorphisms. Mitochondria are transmitted only from the mother to all her children; there is no diploidy, no meiosis, and no recombination. Polymorphisms of mitochondrial DNA (Sect. 2.3.4) are especially useful for population genetics, mainly for the analysis of relationships between population groups [1792] and population history: They do not appear to be subject to selection pressures. Therefore, comparison of maternally inherited mRNA restriction patterns between population groups gives an unbiased picture of the mutational history.

6.1.3 Hereditary Diseases

Dominant and X-Linked Recessive Diseases. Two subgroups of dominant and X-linked hereditary diseases may conveniently be distinguished from the standpoint of population genetics:

1. Those in whom reproduction is impaired considerably - either because affected persons are so severely sick that they often die early or because they are so severely disabled that they have little chance to marry and have children.
2. Those in whom reproduction is not impaired - either because the abnormality is trivial or because it manifests only after reproduction has been completed.

In the first subgroup of disabling genetic disease, the incidence and prevalence are mainly determined by the mutation rate: most mutations disappear after one or a few generations. This problem has been discussed in Sect. 5.1; data on mutation rates for such conditions (Sect. 5.1.3; Table 5.8) indicate similar mutation rates in all populations for which evidence is available. The sources of selection against these mutations are disease specific and have been more or less identical in all populations; therefore, the incidence is also similar. With different applications of successful treatments in different populations, variable incidences would be expected. For example, successful treatment of hemophilia will lead to an increase of affected patients.

The situation is quite different for conditions that do not impair reproduction. Here, incidence may definitely be unequal between populations, depending on such factors as population size, history, and breeding structure. These problems will be discussed in Sect. 6.4.2.

Autosomal Recessive Diseases. The various influences that have shaped the current frequencies of autosomal recessive diseases in human populations are largely unknown. How these frequencies will change under the influence of modern civilization, therefore, cannot be determined with certainty. The incidence of most autosomal recessive diseases is unknown. As a side-effect of the widespread newborn screening programs designed to secure early diagnoses of treatable metabolic defects, estimates for incidence of several inborn errors of metabolism have been collected recently [1881].

Screening centers that participated in a study designed to estimate the frequency of phenylketonuria were invited by the organizer to co-operate on the basis of acceptability of their techniques and the number of newborns tested. They were requested to report only if at least 70,000 newborns had been screened. Centers that were aware of regional differences within the population covered were asked to give separate figures for each region or population group. The data are given in Table 6.4.

Phenylketonuria (26160) *and Hyperphenylalaninemia*

a) In Europe, there is a higher incidence of PKU in eastern European populations than in western and southwestern populations. The difference between eastern and western Austria fits into this pattern.
b) The Scandinavian populations, and especially the Finns, show an exceptionally low PKU frequency. It is interesting that the Finnish population differs in other genetic aspects from the rest of the Europeans as well (Sect. 6.4.2); for example, a unique pattern of other inherited diseases and pericentric chromosome inversions unknown in other populations have been observed (Sect. 2.2.2).
c) High frequencies of PKU are found in the Republic of Ireland; differences within Great Britain, such as a high frequency in the Manchester area, may reflect migration from Ireland.
d) In the United States, Boston and Portland (Oregon) reported figures quite comparable to most frequencies in Europe. In Montreal, in the French-speaking part of Canada, there is a much lower frequency for PKU than in most European areas and the two centers in the United States, and a significantly lower rate than found in France where this population originated.
e) In Japan, the only Far East country screened for inborn errors, the frequency of PKU is especially low, compar-

Table 6.4. Frequency of PKU and hyperphenylalaninemia in some populations [1881]

Region	Frequency	
	PKU	Hyperphenyl-alaninemia
Warsaw, Poland	1 : 7782	1 : 16885
Prague, CSSR	1 : 6618	1 : 6303
East Germany	1 : 9329	1 : 52135
East Austria	1 : 8659	1 : 21982
West Austria	1 : 18809	1 : 18809
Switzerland	1 : 16644	1 : 24106
Evian, France	1 : 13715	1 : 13143
Hamburg	1 : 9081	1 : 61297
Münster, West Germany	1 : 10934	1 : 7997
Heidelberg	1 : 6178	1 : 14580
Denmark	1 : 11897	1 : 40790
Stockholm, Sweden	1 : 43226	1 : 22140
Finland	1 : 71111	1 : 71111
London	1 : 18292	1 : 50304
Liverpool, United Kingdom	1 : 10215	1 : 112362
Manchester	1 : 7707	1 : 80925
West Ireland, Eire	1 : 7924	1 : 68670
Eastern, Ireland, Eire	1 : 5343	1 : 32594
Boston, Mass., U.S.A.	1 : 13914	1 : 17006
Portland, Oregon, U.S.A.	1 : 11620	1 : 33700
Montreal, Canada	1 : 69442	?
Auckland, New Zealand	1 : 18168	1 : 95384
Sydney, Australia	1 : 9818	1 : 22091
Japan	1 : 210851	1 : 70284
Ashkenazi, Israel	1 : 180000	1 : 15000
Non-Ashkenazi, Israel	1 : 8649	1 : 7111

able only to the frequencies in Finland and among Ashkenazi Jews in Israel.

f) The frequencies of hyperphenylalaninemia (HPA) a condition not associated with mental retardation - vary greatly and independently of those for PKU. The PKU: HPA ratio also shows a surprisingly wide range, between 0.51 and 12.0 (Fig. 6.7).

Other Conditions. Other conditions included in the study were galactosemia due to transferase deficiency (23040) (Sect. 4.2), histidinemia (23580), maple sugar urine disease (24860) (Sect. 4.2.2.4), and homocystinuria (23620) (Sect. 4.2).

For galactosemia, most centers - Hamburg, Vienna, Auckland, Prague, Stockholm, Zürich - reported frequencies between about 1:30,000 and 1:65,000. Hence, the incidence in most populations of European origin seems to be similar. There was a significant difference between eastern and western Austria.

Histidinemia, in the centers with the best screening conditions, has an incidence between 1:13,000 and 1:19,000. Homocystinuria, leucinosis and - with the exception of Montreal - tyrosinosis (27670) are very rare, showing incidences between about 1:100,000 and 1:600,000; in Montreal, tyrosinosis reached a frequency of about 1:13,000, which could be traced to a specific French-Canadian isolate.

For conditions other than phenylketonuria screening is not nearly as complete - and the methods are not nearly as reliable. But methods are improving, and with extension of the programs the sample sizes will grow rapidly. Some apparent differences between populations may be due to small sample size, while at the same time some real differences may still remain hidden.

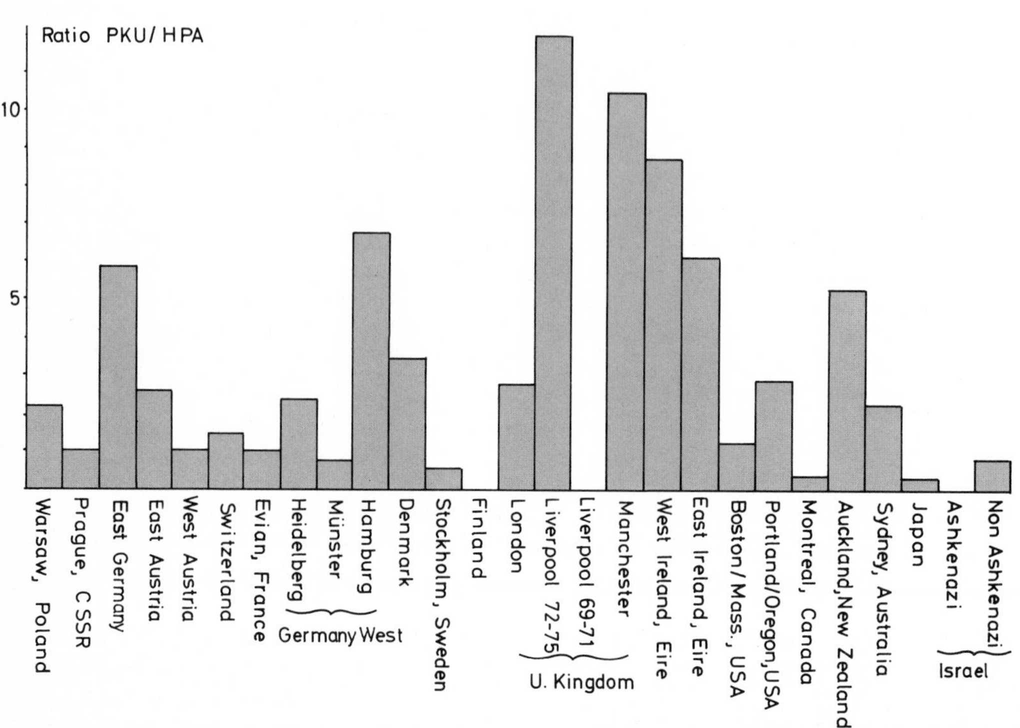

Fig. 6.7. Ratio of *phenylketonuria* to hyperphenylalaninemia in various populations. (Thalhammer et al., 1975 [1881])

Interestingly, PKU is more common in certain parts of East Germany (German Democratic Republic) than in West Germany. A study on the origin of families of PKU children in the Northwest of West Germany showed that a higher fraction of these families than expected by chance were refugees from the East [1761 a]. There appears to be a decline from the east to the (north-)west of continental Europe. Studies in molecular biology (RFLP haplotypes) will soon give new evidence on the history of these genes (1363 a).

High Frequencies of Recessive Diseases in Special Populations. A number of recessive diseases have attained high frequencies in special population groups. Tyrosinosis in the French-Canadian population has already been mentioned; others are Tay-Sachs (27280), Niemann-Pick (25720), and Gaucher's disease (23080) in the Ashkenazi Jewish population. In the Finnish population, a number of such diseases have attained high frequencies as a consequence of a distinct population structure and history (Sect. 6.4.2). Some diseases have been found only in smaller population groups (isolates) where they became frequent while being more or less unknown in any other group. Others - like thalassemia or sickle cell anemia - are frequent in certain geographic areas and in some racial groups. In general, recessive diseases unique to certain small populations owe their origin to genes that in the heterozygote state have no particular advantage, while those genes causing more common and widespread diseases such as thalassemia and sickle cell anemia have had a heterozygote selective advantage in the past. However, the occurrence of metabolically related gene mutations in Ashkenazi Jews (Tay-Sachs, Gaucher's., Niemann-Pick disease) might suggest selective factors in these instances. A full discussion of these problems will be given later (Sect. 6.4).

With progress of genetic counseling, including antenatal diagnosis, and of therapy of genetic diseases, more and more countries will monitor their population for genetic disease. Such programs will lead to more precise knowledge of incidence.

6.2 Systematic Changes in Gene Frequencies: Mutation and Selection

Gene frequencies remain unchanged in populations only in the absence of disturbing influences. (See Hardy-Weinberg-equilibrium; Sect. 3.2) The most important of these influences are:

a) New mutation
b) Natural selection
c) Migration
d) Chance fluctuations.

Spontaneous and induced mutations in humans have been dealt with in Sect. 5; these aspects will be extended below in the context of other influences. Selection, however, will be dealt with in detail. Another deviation from the Hardy-Weinberg equilibrium is assortative mating. A discussion of consanguinity and chance fluctuation of gene frequencies will lead us to a more general consideration of the breeding structure of human populations and set the stage for a better understanding of human evolution.

6.2.1 Natural Selection

Natural selection was recognized by Darwin as the main driving force of evolution. His evolutionary theory, which is based largely on the understanding of selection and its consequences for the origin of species, became the leading paradigm of biology in the nineteenth century. The concept of "fitness" is central for understanding this paradigm. In the early twentieth century, selection theory was given a sound mathematical foundation, and a number of empirical examples were analyzed.

6.2.1.1 Mathematical Selection Models: Darwinian Fitness

Scope of Mathematical Models in Selection Theory and Their Limitations [124]. In discussions on selection, we shall utilize mathematical models on a fairly large scale. In these models, a number of assumptions are made for some parameters - for example, gene frequencies and selective advantages or disadvantages of special genotypes. The consequences of these assumptions for the direction and extent of changes in gene frequencies over time will be examined. Such models help to understand the consequences of certain changes in these parameters by creating some order in the vast and at first glance unintelligible complexity of genetic differences among human populations.

Such order may be artificial: Certain aspects are singled out, while others are deliberately neglected. While the calculations become tractable, major errors may result. The most important oversimplifications are the following:

a) The population is considered to be infinitely large in size. Hence, gene frequency remains constant in the absence of other factors. No real population is infinitely large; on the contrary, many human breeding populations were very small throughout the long time periods most important for human evolution. Hence, all results derived from selection models have to be scrutinized critically in the light of the theory of small populations (Sect. 6.4) and our knowledge of human population size and mating structure. Unfortunately, this knowledge is relatively scanty, especially for earlier time periods. Conclusions in concrete cases are thus difficult if not impossible.

b) As a rule, selective advantages or disadvantages are assumed to be constant over long evolutionary periods. Closer scrutiny of their biologic mechanisms, however, will often show that they may have changed even within relatively short time periods.

c) The generations are assumed to be nonoverlapping. In real life, reproduction of a population is a continuous process; generations overlap strongly. This complexity, however, does not influence consequences derived from genetic models very much. For treatment of models with overlapping generations, see Jacquard (1974) [103].

Deterministic and Stochastic Models: Use of Computers. The various limitations apply mainly to models in which a functional relationship between parameters is assumed. For example, the change in gene frequency between generations is assumed to depend upon a certain mode of selection: The model is "deterministic." In reality, however, all parameters - gene frequencies, selection pressures, mutation rates - will show chance fluctuations because population size is not infinite. When computers became available, it was possible to include chance fluctuation in the calculations, thereby creating stochastic models. The change of gene frequency over the generations can now be simulated assuming a certain population size. The curve showing, for example, the change of gene frequencies over time does not give the "ideal" outcome but only one of many possible outcomes; it is not even known whether a vertain curve is a very likely one. Therefore, a single calculation for a certain set of parameters is not sufficient to obtain an unbiased impression of the consequences of certain assumptions; the same calculation has to be repeated several times. Such a method gives better information than a deterministic model; in addition to the main trend, the possible deviations from this trend caused by chance fluctuations can be demonstrated. In the following section, deterministic models will be used mainly, but occasionally stochastic models will also be considered.

How Should Models Be Used in Practice? Oversimplifications implied in all models, deterministic or stochastic, are necessary for recognition of the general laws of natural selection. In each concrete case, however, the fact that they are oversimplifications should be kept in mind. While the consequences inferred from such models are *formally* correct, the possibility that they may derive from aspects of the model that have no counterpart in the real world is often not considered. Much of human population genetics suffers from too uncritical interpretations of formal results deduced from oversimplified models.

The ideal sequence of events would be that:

a) A certain genetic situation requires theoretical understanding.
b) A hypothesis is formulated and a relatively simple model is constructed that includes the main parameter(s) of the hypothesis.
c) The consequences of this model for changes in gene and genotype frequencies over time or for differences among present-time populations are explored.
d) The result of this exploration is compared with the empirical evidence.
e) Discrepancies between the theoretically expected and the empirically observed results are noted. Critical interpretation may lead to rejection of the hypothesis, change in assumptions regarding important parameters, or refinement of the model.

Recognition that a concrete problem requires closer scrutiny by a special model may change with growth of further knowledge, as problems are not isolated but are parts of the whole context of a special field. This context may change; a problem formerly regarded as holding a key position may become less important, and a model designed to solve this problem, while not formally shown to be inadequate, may turn out to have trivial or unrealistic consequences.

We recognize that the ideal sequence of events in the study of problems in population genetics often is impossible because of technical and logistic limitations. We would like to stress, however, that we consider that hypothesis-oriented efforts in population genetics are ultimately of greater explanatory power for better understanding of population genetics than descriptive studies even if embellished by the most advanced and elegant statistical methods.

Concept of Darwinian Fitness. The central concept of selection theory is Darwinian fitness. Under given environmental conditions, not all individuals in a given population will perform equally well. These differences are partially caused by different genetic endowment of the individuals. This obviously has many medical, social, and ethical aspects. In connection with natural selection, however, only one aspect is relevant: different reproduction rates of individuals with different genotypes. Only reproductive differences can lead to a shift of gene frequencies over time, if population size is regarded as unlimited so that chance deviations can be neglected. This reproductive performance of a certain genotype in comparison with a norm is often called its (Darwinian) fitness. This notion of fitness can also be defined for a single allele, if it influences the reproduction of its carrier. Fitness of a certain genotype may be reduced or enhanced in two different ways:

1. The genetic constitution reduces the chance of a genotype to survive into reproductive age. This reduction in viability is frequent in hereditary diseases.
2. The genetic constitution reduces the chance of a genotype to produce offspring, i.e. fertility is diminished.

For the final result - a change in gene frequency - and, therefore, for the definition of Darwinian fitness, this subdivision is of little relevance.
From the point of view of population genetics, there is no difference between a gene that causes spontaneous miscarriage and one that causes sterility in an otherwise healthy person. Medically, and from the point of view of society, there are considerable differences between the impact of these two genes.

6.2.1.2 Selection Leading to Changes of Gene Frequencies in One Direction

Symbols Used. Fitness of a genotype is defined as its efficiency in producing offspring. It is measured not in absolute but in relative units, the fitness of the optimum genotype being taken as unity (1). Deviations from unity are denoted as s. For example, for a genotype having 80% fitness of the optimum genotype, $s=0.2$; this fitness is: $1-s=1-0.2=0.8$. To avoid confusion with the signs, it is sometimes desirable to have a direct measure for the fitness $1-s=w$. In the literature, s is sometimes defined with respect to the average, not to the optimum fitness of the population, $\bar{w}$. This convention has disadvantages since the fitness of a genotype varies with the genotype distribution in the population. We shall therefore define fitness with respect to an "optimum" genotype.

Elimination of Heterozygous "Dominant" Phenotypes. This case is frequent and simple: A mutation may change the phenotype of its carrier so thoroughly as to make reproduction impossible. All numeric and most structural chromosome aberrations and many dominant gene mutations (Sect. 5.1.3) have this effect. Whereas chromosome aberrations can be diagnosed directly, dominant new mutations can only be recognized by transmission of the condition to the offspring. Hence, nonreproductive dominant mutations are in principle unrecognizable. Still, extrapolation from a very few individuals who manage to reproduce, or a paternal age effect suggestive of a dominant mutation may make such mutation plausible even for anomalies preventing their carriers from reproduction. Some diseases and birth defects may be caused, in fact, by such dominant mutations that have not been recognized so far.

Partial Elimination of Autosomal Dominants. Most dominant diseases reduce the average reproduction of their bearers. In the absence of counteracting forces, such as mutation, such gene loss would lead to reduction of gene frequencies in every generation by a fraction, which depends on the selective disadvantage of the probands (Fig. 6.8). A genetic equilibrium between new mutation and selection can be assumed in almost every case. Let p be the frequency of the dominant allele A; q the frequency of the recessive normal allele a. Selection may be appreciable, and the mutation rate, μ, may be low (Sect. 5.3.1). Then, affected homozygotes AA can be neglected, as p^2 is very small. The genotype frequencies are:

Genotype	Fitness	Before selection	After selection
aa	$1(=w_{22})$	$q^2\approx 1-2p$	$\frac{1-2p}{1-2ps}$
Aa	$1-s(=w_{21})$	$2pq\approx 2p$	$\frac{(1-s)2p}{1-2ps}$

The loss of Aa individuals per generation is approximately $2ps$. Since only half of the genes are A, the loss of A genes is $\frac{1}{2}(2ps)=ps$. A genetic equilibrium exists if this loss is compensated for by mutation:

$$ps=\mu(1-p)\approx\mu$$

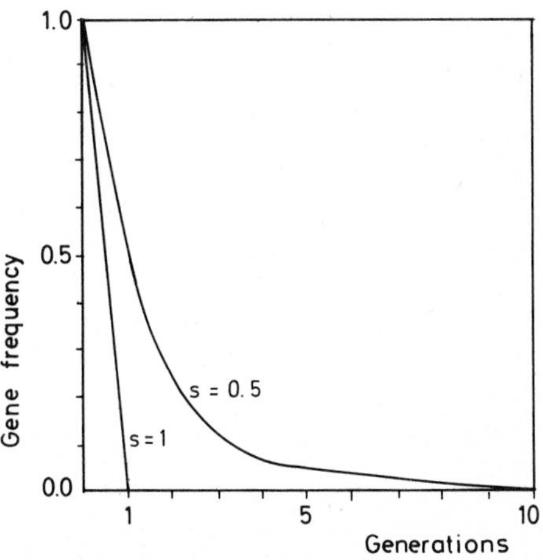

Fig. 6.8. Decrease of the gene frequency of a dominant gene in absence of new mutations with $s=1$ and $s=0.5$; s selection coefficient

This gives the equilibrium frequency of p:

$$\hat{p} = \mu/s \tag{6.1}$$

Two limiting cases are interesting:

1. No reproduction of heterozygotes occurs, i.e., $s=1$, $w_{21}=0$; in concordance with our conclusion from the last section, the frequency of the dominant allele (p) will be equal to the mutation rate (μ).
2. No selective disadvantage of the gene applies (i.e., $s=0$, $w_{21}=1$). Under such conditions, there is no equilibrium and the gene frequency of the dominant allele increases monotonously: $\mu/s \rightarrow \infty$.
3. Finally, an intermediary case with some disadvantage of the dominant gene, e.g., $s=1/3$, $w_{21}=2/3$; the final gene frequency of the dominant gene ($\hat{p}$) will be three times the mutation rate.

Most pathologic conditions have an appreciable selective disadvantage. Their population frequency is maintained by an equilibrium between mutation and selection. Here the prime mover is of course the mutation rate. The frequency of the condition increases until there are so many affected individuals that their selective disadvantage balances the mutation rate increase and a genetic equilibrium is reached. This equilibrium is stable. If the number of affected individuals shifts above the equilibrium value, the number of losses of the disadvantageous allele by selection will exceed the number of its production by mutation, and its frequency will be reduced in the next generation. On the other hand, if the number of affected individuals shifts below the equilibrium value, more disadvantageous genes will be produced by mutation than lost by selection, and the gene frequency increases until equilibrium is reached again.

Selection Relaxation. Equation (6.1) can be used to calculate the consequence of selection relaxation: Assume that medical therapy, by eliminating some of the phenotypic consequences of a dominant mutation, has succeeded in reducing the selective disadvantage of the allele from:

$$s_1 = 1/2 \quad \text{to} \quad s_2 = 1/4.$$

Let $\hat{p}_2$ be the equilibrium value for s_2. Then, Eq. (6.1) gives:

$$\hat{p}_1 = \frac{\mu}{1/2} = 2\mu; \quad \hat{p}_2 = \frac{\mu}{1/4} = 4\mu; \quad \hat{p}_2 = 2\hat{p}_1$$

Therefore, the new frequency will be twice that of the earlier frequency. Moreover, the new equilibrium will be reached within a few generations.

Selection Relaxation in Retinoblastoma. Retinoblastoma is a malignant eye tumor of young children. The great majority of all cases in the population are sporadic. However, familial cases are frequent and show autosomal dominant inheritance with about 90% penetrance. All bilateral but only about 10–12% of the unilateral sporadic cases are new mutations.

In earlier times, retinoblastoma was almost always fatal; the patients died during childhood, and none could ever reproduce. In 1865, A.v.Graefe introduced the enucleation of the diseased eye; more recently, additional therapeutic methods such as X-ray irradiation and light coagulation have become available, and currently, about 90% of the unilateral and 80% of the bilateral cases can be cured and will have children [1887]. Of all new mutations for hereditary cases 68% are bilateral and about 32% are unilateral. The changes in frequency of the condition can be predicted as follows:

Let X_o be the frequency of individuals in the population with hereditary retinoblastomas; since the condition is very rare, affected homozygotes can be neglected and the gene frequency of the normal allele may be assumed to be $q \approx 1$; therefore, X_o is nearly equal to the frequency of heterozygotes ($2pq \cong 2p$). Let $s_1 = 1$ before selection relaxation. After introduction of efficient therapy, selection against unilateral cases becomes $s_U = 0.1$ (only 10% of unilateral cases die in childhood; 90% survive to reproduce normally); selection against bilateral cases becomes $s_B = 0.2$ (80% survive to reproduce normally). This leads to the following overall estimate for s_2 (= remaining selection against the retinoblastoma allele after selection relaxation):

$$\begin{aligned} s_2 &= s_U \times 0.32 + s_B + 0.68 \\ &= 0.1 \times 0.32 + 0.2 \times 0.68 \\ &= 0.168 \end{aligned}$$

The new selection coefficient after selection relaxation is 16.8% of the coefficient prior to selection relaxation when s is 1. The frequency of heterozygotes in the nth generation, X_n, can be related to that in the $(n+1)$th generation, X_{n+1}, by the approximate formula:

$$X_{n+1} = X_n(1-s_2) + 2\mu = X_n(1-s_2) + X_o$$

From this recurrence formula, a general formula for X_n can be derived:

$$X_n = X_0[1 + (1-s_2) + (1-s_2)^2 + \ldots + (1-s_2)^n]$$

Thus, the new equilibrium value for the heterozygote frequency, X, is the sum of a geometric series with the initial term $X_0 = 2\mu$ and the factor $1-s_2$:

$$\hat{X} = \frac{X_o}{1-(1-s_2)} = \frac{X_o}{s_2} = \frac{X_o}{0.16\,s} = 5.95 \times X_o$$

This result can also be obtained from Eq. 6.1. It follows that inherited retinoblastoma will be about six times more common several generations after successful medical therapy has been introduced. Introducing actual estimates from present-day populations (see Table 5.8), $X_o = 2\mu = 1.2 \times 10^{-5}$:

$$\hat{X} = \frac{1.2 \times 10^{-5}}{0.16\ s} = 7.14 \times 10^{-5}$$

From this value, the overall frequency of *all* retinoblastoma cases – including the nonhereditary ones – can be estimated when the fraction of hereditary cases among all retinoblastoma cases in the population before selection relaxation is known. The overall incidence of retinoblastoma is $\approx 4 \times 10^{-5}$; it follows that $\approx 2.8 \times 10^{-5}$ are nonhereditary cases. From this, an incidence estimate at equilibrium after selection relaxation can be derived:

$7.14 \times 10^{-5} + 2.8 \times 10^{-5} = 9.94 \times 10^{-5}$; meaning that incidence increased from $\approx 1:25000$ to $\approx 1:10000$, i.e., by $\approx 150\%$. Moreover, while before selection relaxation $\approx 30\%$ of all cases were hereditary, $\approx 72\%$ will be hereditary in the new equilibrium state. This equilibrium will be established relatively quickly (Fig. 6.9): After nine generations the incidence of hereditary retinoblastoma is more than four times the value before selection relaxation started (equilibrium: $\hat{X} = 5.95 X_o$). In Fig. 6.9, calculations for two alternative assumptions are also given: $s_2 = 0.4$; and $s_2 = 0$, i.e., no selection. In the latter case, incidence increases in a linear fashion; there is no equilibrium. There is some hope that the prediction given here will not come true; artificial selection by genetic counseling and voluntary birth control on the part of gene carriers will partially replace natural selection.

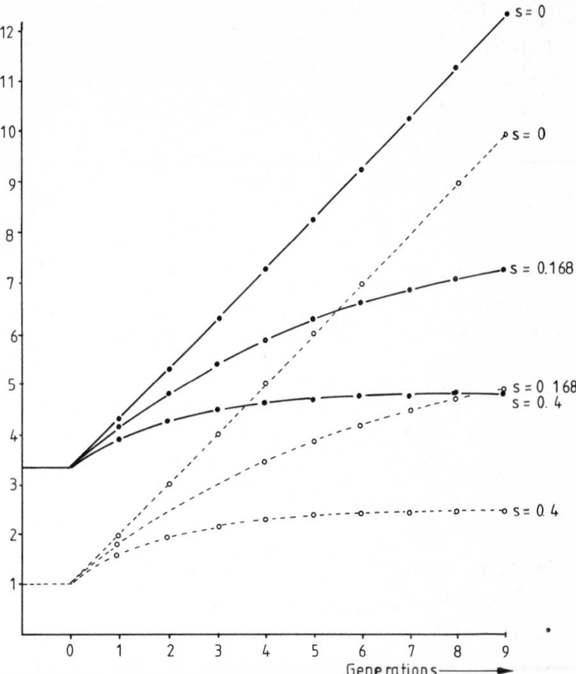

Fig. 6.9. Predicted increase of retinoblastoma in the population due to selection relaxation. *Ordinate:* 1 = incidence of hereditary retinoblastoma (--- = 2× mutation rate) with complete selection against retinoblastoma genes due to mortality of retinoblastoma patients. *Abscissa:* Number of generations. Three assumptions on the selection coefficient s: No selection at all (s = 0), weak selection (s = 0.168) and stronger selection (s = 0.4). (From Vogel 1979 [1887])

Selection by Complete Elimination of Homozygotes. In many autosomal recessive diseases, the homozygotes usually do not reproduce. Again, two alleles *A, a* with gene frequencies *p, q* are considered. This time, however, the homozygotes aa have a selective disadvantage:

$$s = 1; \quad w_{22} = 0$$

The frequency of zygotes is seen in the table. In general, the gene frequencies in two successive generations n and $n+1$ are related by:

$$q_{n+1} = \frac{q_n}{1+q_n} \qquad (6.2)$$

This formula reflects the recurrence relation between successive partial sums (= sum of the first n terms) of a harmonic series.[1] It leads to the general formula for the gene frequency q_n after n generations:

$$q_n = \frac{q_o}{1+nq_o} \qquad (6.3)$$

The change of gene frequency per generation is:

$$\Delta q_n = \frac{q_n}{1+q_n} - q_n = -\frac{q_n^2}{1+q_n} \qquad (6.4)$$

It follows from Eq. (6.3) that $q_n = q_o/2$, if $nq_o = 1$. Hence, the gene frequency is halved within $n = {}^1/q_o$ generations.

An important practical application will be considered. If all homozygotes of the most common recessive disease - cystic fibrosis - refrain from reproduction, how will the gene frequency be influenced?

Selection s = 1 against homozygotes

	AA	Aa	aa	Gene frequency of a
Before selection	p^2	$2pq$	q^2	q
After selection	$\frac{p^2}{p^2+2pq}$	$\frac{2pq}{p^2+2pq}$	0	$\frac{q}{1+q}$

The gene frequency is $q_o = 0.02$, corresponding to a homozygote frequency in a random mating population of $q_o^2 = 0.0004$. Of 10,000 individuals, 4 are affected.

After $n = 1/0.02 = 50$ generations, q_o is reduced from 0.02 to 0.01. Assuming a generation time of 30 years, this halving of the gene frequency will require 1,500 years. With rarer diseases such as galactosemia, $q_o = 0.005$, $q^2 = 1:40{,}000$; halving would need 200 generations, or 6,000 years.

Hence, the attempt to reduce the number of recessive genes by homozygotes refraining from reproduction is an extremely inefficient process. Moreover, new mutations are not even considered in this calculation.

Partial Elimination of Homozygotes. In some recessive diseases, the homozygotes are not completely unable to reproduce but there is significantly re-

[1] A harmonic series is one whose successive partial sums are the reciprocals of those of an arithmetic series:

$$u_o = \frac{c}{b}, \; u_1 = \frac{c}{b+k}, \; u_2 = \frac{c}{b+2k}, \text{ etc.}$$

In this special case, $b = 1$, $c = k = u_o$.

duced biological fitness. Let their reproduction be reduced by s ($1 > s > 0$). Gene frequencies before and after selection may be taken from Table 6.5. This gives the following frequency of the recessive allele in the next generation:

$$q_1 = \frac{pq + q^2(1-s)}{1-sq^2} = \frac{q(1-sq)}{1-sq^2} \tag{6.5}$$

The formula for the relationship between the frequencies of the recessive allele in two successive generations is:

$$q_{n+1} = \frac{q_n(1-sq_n)}{1-sq_n^2}.$$

This recurrence formula seems to have no general solution. The change per generation is:

$$\Delta q_n = q_{n+1} - q_n = -\frac{sq_n^2(1-q_n)}{1-sq_n^2}$$

Δq_n depends on the values of both q_n and $p_n = 1 - q_n$. It is small if one of these terms is small. For example, the following values can be calculated for $s = 0.2$:

q:	0.99	0.50	0.01
Δq:	−0.00244	−0.0263	−0.0000198

With very small q, Δq approximated $-sq^2$.

A Bit of Calculus. To determine the change in q over a greater number of generations, Δq can be replaced by:

$$\frac{dq}{dt} = -sq^2(1-q); \quad \frac{dq}{q^2(1-q)} = -sdt$$

Integrating both sides over n generations, we get:

$$\int_{q^o}^{q_n} \frac{dq}{q^2(1-q)} = -s\int_o^n dt = -sn$$

$$sn = \left[\frac{1}{q} + \log_e \frac{1-q}{q}\right]_{qo}^{qn} = \frac{1}{q_n} - \frac{1}{q_o} + \log_e \frac{1-q_n}{q_n} - \log_e \frac{1-q_o}{q_o}$$

$$= \frac{q_o - q_n}{q_o q_n} + \log_e \frac{q_o(1-q_n)}{q_n(1-q_o)}$$

Calculations for number of generations (n) needed to produce a certain change in q ($s = 0.01$ against the homozygotes) shows the following:

Decrease of q	*n generations*
0.9999–0.9990	230
0.9990–0.9900	232
0.9900–0.5000	559
0.5000–0.0200	5,189
0.0200–0.0100	5,070
0.0100–0.0010	90,231
0.0010–0.0001	900,230

In the meantime, other events may have occurred to disturb the genetic composition of the population much more.

Table 6.5. Partial elimination of homozygotes

	AA	Aa	aa	Total
Before selection	p^2	$2pq$	q^2	1
Fitness	1	1	$1-s$	
After selection	p^2	$2pq$	$q^2(1-s)$	$1-sq^2$

Table 6.6 Selection 2s against homozygotes and s against heterozygotes

	AA	Aa	aa	Total
Before selection	p^2	$2pq$	q^2	1
Fitness	1	$1-s$	$1-2s$	
After selection	p^2	$2pq(1-s)$	$q^2(1-2s)$	$1-2sq$

Gametic Selection. Up to now, selection has been assumed to act on the zygote. However, the gametes may already be affected. Because of their genetic make-up, some gametes may have a lower chance for fertilization than others.

If a mutation of any kind influences the probability of fertilization, a distortion of the segregation ratio for that mutation will result. Such distortions have been suggested to explain anomalous segregation ratios in Alport's syndrome and in Leber's optical atrophy (Sect. 3.1.5). For all other human conditions, conformity with Mendelian segregation ratios is usually taken for granted. Since large samples would be required to show small deviations from Mendelian proportions, the evidence is not critical. However, gross deviations are unlikely.

For gametes containing balanced and unbalanced translocations, segregation distortion is beyond any doubt, although the exact mechanism remains to be elucidated (Sect. 2.2.2). If the relative contributions of gametes A and a to the following generation are 1 and $(1-s)$, and their frequencies are p and q, their contribution to the next generation will be p and q $(1-s)$, respectively. Hence, the change per generation is:

$$\Delta q = \frac{q(1-s)}{1-sq} - q = \frac{-sq(1-q)}{1-sq}$$

All human examples mentioned before are not quite so simple. Since selection seems to work differently on the germ cells of the two sexes, the situation in conditions such as translocations is more complicated.

Selection against gametes is nearly identical formally with selection against homozygotes and intermediary heterozygotes (Table 6.6).

6.2.1.3 Selection Leading to a Genetic Equilibrium

So far, only those models of selection have been discussed that lead to an increase in the frequency of one allele at the expense of the other. An equilibrium, and hence a steady state of gene frequencies over generations, could only be achieved by introduction of an external force - mutation. However, there are modes of selection that, in themselves, lead to an equilibrium. A steady state with no systematic changes in gene frequencies may be created if selection favors heterozygotes.

Selection in Favor of Heterozygotes with Selective Disadvantage of Both Homozygotes. The biologic basis of this model - heterosis - has been known in experimental genetics for a long time. It has led to theoretical discussions as well as to many practical applications in plant breeding, most notably of corn [1744, 1755a, 1868, 1869].

Heterosis or hybrid vigor means the superiority of heterozygous genotypes with respect to one or more characters in comparison with the corresponding homozygotes. It is the phenotypic result of interactions between allelic genes [1754]. Heterosis due to the occurrence of mutations and gene combinations that convey a greater fitness on heterozygotes than is found in the corresponding homozygotes has been paid most attention in human genetics.

Heterozygote Advantage: Formal Consequences. Table 6.7 shows the genotypes before and after selection (s_1, s_2: selection against the phenotypes AA and aa). By using the relation $pq+q^2=q$, we get the change from one generation to the next:

$$\Delta q = \frac{q-s_2q^2}{1-s_1p^2-s_2q^2} - q = \frac{pq(s_1p-s_2q)}{1-s_1p^2-s_2q^2}$$

Figure 6.10 shows the relation for various values of q and Δq and for $s_1=0.15$, $s_2=0.35$; Δq may be positive or negative, depending on whether s_1p is larger or smaller than s_2q. If $s_1p=s_2q$, $\Delta q=0$. Solving for p or q gives the following equilibrium values:

$$\hat{p} = \frac{s_2}{s_1+s_2}; \quad \hat{q} = \frac{s_1}{s_1+s_2} \tag{6.6}$$

These equilibrium values depend only on s_1 and s_2; they are independent of the gene frequencies at the onset of selection. Moreover, the equilibrium is stable; Δq is positive if $q<\hat{q}$; it is negative if $q>\hat{q}$. If disturbed by accidental influences, the equilibrium has a tendency to re-establish itself.

The leading paradigm for heterozygote advantage in humans is a selective mechanism that has caused the high frequency of the sickle cell gene in some human populations. It will be described in more detail in Sect. 6.2.1.6.

6.2.1.4 Selection Leading to an Unstable Equilibrium

Selection Against Heterozygotes. A stable equilibrium may be established in the population if selection favors heterozygotes at the expense of the homozygotes. However, selection may also favor the homozygotes at the expense of the heterozygotes. Let 1, 1-s, 1 be the fitness of AA, Aa, and aa. Then the change in gene frequencies is seen in Table 6.8. This leads to:

Table 6.7 Selection in favor of heterozygotes

	AA	Aa	aa	Total
Before selection	p^2	$2pq$	q^2	1
Fitness	$1-s_1$	1	$1-s_2$	
After selection	$p^2(1-s_1)$	$2pq$	$q^2(1-s_2)$	$1-s_1p^2-s_2q^2$

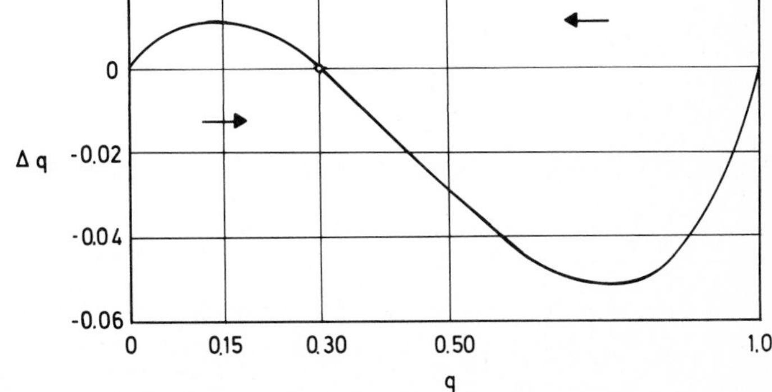

Fig. 6.10. The change of gene frequency Δq for various values of q, and for arbitrary selective disadvantages $s_1=0.15$ and $s_2=0.35$. s_1 and s_2 denote the selective disadvantages of the two homozygotes in comparison with the heterozygote. (Li, 1955 [124])

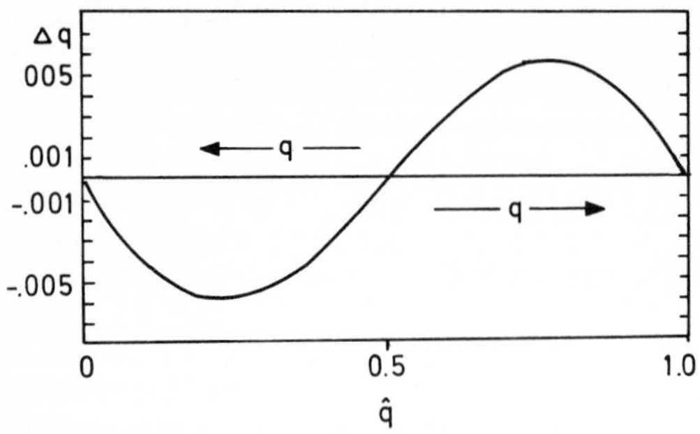

Fig. 6.11. Change of gene frequency Δq. Selection $s=0.50$ against the heterozygotes. Δa is negative if $q<\hat{q}$ and positive if $q>\hat{q}$. (Li, 1955 [124])

$$q'=\frac{pq(1-s)+q^2}{1-2spq}=\frac{q-spq}{1-2spq} \quad \text{and}$$

$$\Delta q=\frac{spq(2q-1)}{1-2spq}\approx 2spq(q-\tfrac{1}{2}), \qquad (6.7)$$

if s is small. Figure 6.11 shows Δq in dependence of q.

If $q=1/2$, $\Delta q=0$. This means that there is a genetic equilibrium, but this equilibrium is unstable. If q becomes higher than 1/2, Δq is positive. The gene frequency tends toward 1. If q is shifted below 1/2, on the other hand, Δq becomes negative: q tends toward 0. An unstable equilibrium alone cannot maintain a polymorphism in the population; even in the artificial case of two populations homozygous for two different alleles mixing in equal proportions, small shifts in gene frequencies by chance fluctuation would soon destroy the equilibrium state, causing the gene frequency to move either toward 1 or toward 0.

Still, such an unlikely event appears to have happened. In humans, two situations are known in which the only selective factor that can be analyzed so far is selection against heterozygotes. In one of these situations, the Rhesus system (11170) allelic frequencies conform to our definition of polymorphism. In pericentric inversions, heterozygotes have also been found in polymorphic frequencies and homozygotes have occasionally been observed.

Pericentric Inversions. In a heterozygote for a pericentric inversion, pairing of homologous chromosomes during meiosis may not proceed properly. The resulting unbalanced gametes may either be eliminated before fertilization or form lethal zygotes. On the other hand, no meiotic disturbance is expected in homozygotes; here, pairing of homologs will proceed properly. A high incidence of reproductive wastage has, indeed, been observed in the progeny of cases with some pericentric inversions [401 a]; adequate population genetic studies are lacking. They are all the more needed, as pericentric inversions have played a major role in human evolution, apparently providing a powerful mechanism for reproductive isolation. Pericentric inversions do not affect the health status of their heterozygous carriers but impair their reproductive capacity. This is a classic – and possibly the best – example of fertility influencing fitness, without any viability component. This may be the reason for its significance in evolution, especially speciation (Sect. 7.2.1).

Table 6.8. Selection against heterozygotes

	AA	Aa	aa	Total
Before selection	p^2	$2pq$	q^2	1
Fitness	1	$1-s$	1	
After selection	p^2	$2pq(1-s)$	q^2	$1-2spq$

Table 6.9. Mating types and classes of progeny against which selection due to Rh incompatibility works *(shaded areas)*

Fathers→ ↓Mothers	DD p^2	Dd $2pq$	dd q^2
DD p^2	p^4 DD	p^3q DD p^3q Dd	p^2q^2 Dd
Dd $2pq$	p^3q DD p^3q Dd	p^2q^2 DD $2p^2q^2$ Dd p^2q^2 dd	pq^3 Dd pq^3 dd
dd q^2	p^2q^2 Dd	pq^3 Dd pq^3 dd	q^4 dd

Selection Against Rh Heterozygotes. The unstable equilibrium in selection against heterozygotes was first discovered by Haldane (1942) [1777] for the special case of serologic mother-child incompatibility in the Rh factor. This situation is slightly more complicated than that considered for pericentric inversions. The danger of erythroblastosis and, hence, of selection against heterozygous children occurs if an Rh-negative mother bears a positive child. The Rh locus has a complex structure (Sect. 3.5.4). To understand the principle of selection, however, we need only consider the genes D (positive) and d (negative); this reduces the problem to that of a simple diallelic system. Erythroblastosis children occur in matings dd♀ × DD ♂ or dd♀ × Dd♂. Table 6.9 lists the mating types; the endangered children against whom selection works are indicated by shading. Combining all mating types leads to the formula for the change in the frequency of D:

$$\Delta q=\frac{p-\frac{1}{2}q^2(s_1p^2-s_2pq)}{1-q^2(s_1p^2+s_2pq)}-p=$$

$$\frac{p(p-\frac{1}{2})q^2(s_1p+s_2q)}{1-pq^2(s_1p+s_2q)}. \qquad (6.8)$$

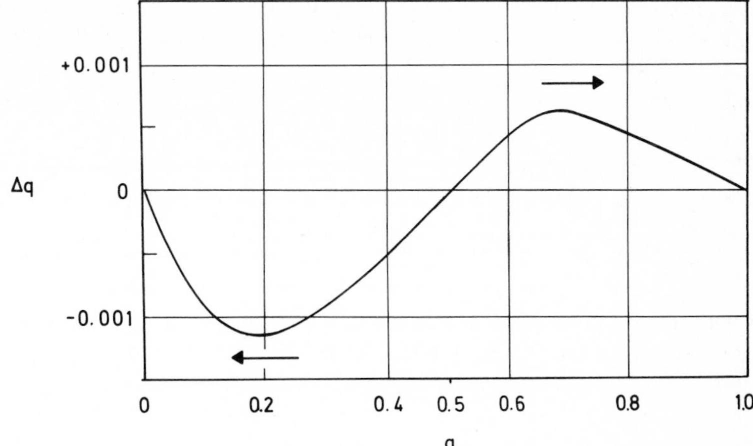

Fig. 6.12. Unstable equilibrium at $p=q=0.5$ in the special case of mother-child incompatibility (Rhesus factor); selection $s_1=0.05$ for heterozygous children of homozygous DD fathers and selection $s_2=1/2s_1$, for heterozygous children of heterozygous Dd fathers

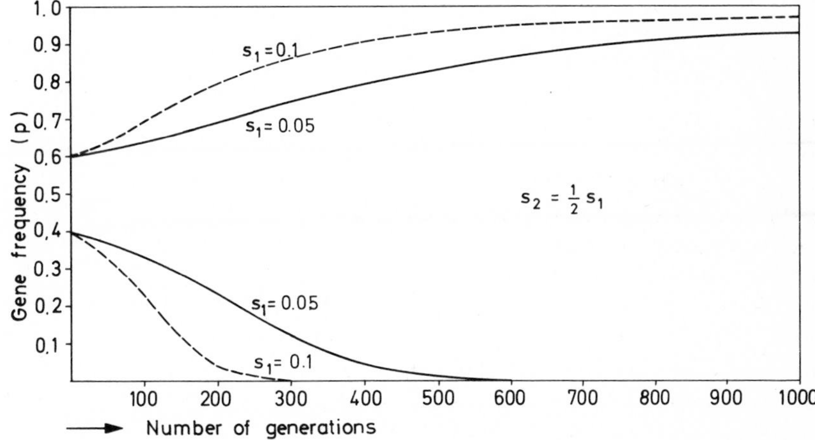

Fig. 6.13. Change in gene frequencies of one allele (p) under selection due to mother-child incompatibility. p decreases, if $p<0.5$ at the onset of selection; it increases, if $p>0.5$ at the onset of selection. (Vogel and Helmbold, 1972 [211])

Here, s_1 equals selection against children with dd mothers and DD fathers; s_2 equals selection against heterozygous children of dd mothers and Dd fathers. As the risk of immunization increases with the number of Dd children, it is lower when the father is Dd, because on the average only every second child will cause maternal immunization, whereas every child of a DD father is heterozygous and may immunize its mother; hence $s_2<s_1$.

It can easily be shown that $\Delta p=p'-p=0$ if and only if $p=1/2$. This means that Eq. (6.8) has the same equilibrium point as Eq. (6.7). Again, the equilibrium is unstable.

In current western European populations, d has a gene frequency $q=0.35$. It follows that its frequency will decrease, unless other selective mechanisms counteract this tendency. How rapidly will this decrease proceed? Figure 6.10 shows Δp for several generations with two assumptions on the initial gene frequencies p for D and q for d, and selection coefficients s_1 and s_2 in the range actually observed in humans. The number of affected children varied with the average number of pregnancies before prophylactic therapy of women at risk with anti-D antiserum was introduced [1740]. Erythroblastosis in about 5% of all Dd children of dd mothers is a reasonable estimate. In earlier times, when women had more pregnancies, this figure must have been somewhat higher.

Figure 6.13 shows that the change in gene frequencies under these conditions proceeds very slowly, the reason being that not all heterozygotes are subject to selection (Table 6.9). This may explain why the Rh polymorphism still exists. Other, still unknown modes of selection, even if only very slight, may have upset selection against heterozygotes. Random mating may not have been proceeding long enough for selection against heterozygotes to work, since the time when chance fluctuations in human populations consisting of small and relatively isolated subgroups created huge between-group differences in gene frequencies (Sect. 6.4.1). A full explanation of the population genetics of the Rh polymorphism remains an enigma in view of the high frequency of Rh system alleles in so many populations. Conceivably, other not yet understood functions were involved in selection.

AB0 Blood Group System. Serologic mother-child incompatibility also occurs in the AB0 blood group system. Contrary to Rh immunization, even the first child may be damaged [211]. On the average, erythroblastosis is milder than with Rh incompatibility. However, AB0 incompatibility may also lead to increase in the number of spontaneous miscarriages, although the evidence is controversial [211]. Selection seems to work mainly against A0 and B0 children of 0 mothers. Recurrent equations can be derived in the same was as shown above for the two-allele case of the Rh system – with the difference that selection is identical against heterozygous children of either homozygous or heterozygous fathers, since, unlike in Rh immunization, the first incompatible child may already be damaged. The following formula describes the change in gene frequency for allele A:

$$\Delta p = \frac{p - \frac{s}{2}pr^2}{1 - sr^2(1-r)} - p = \frac{spr^2(\frac{1}{2} - r)}{1 - sr^2(1-r)} \qquad (6.9)$$

An exactly analogous formula can be derived for Δq, the change in frequency for allele *B*. The formula for Δr, the change in frequency of allele 0, is slightly different:

$$\Delta r = \frac{r - \frac{s}{2}r^2(1-r)}{1 - sr^2(1-r)} - r = \frac{sr^2(1-r)(r-\frac{1}{2})}{1 - sr^2(1-r)} \qquad (6.10)$$

If $r = 0.5$, $\Delta p = \Delta q = \Delta r = 0$, independently of the ratio between *p* and *q*. This equilibrium is unstable in respect to *r*, and neutral in respect to *p* and *q*. Figure 6.14 shows the speed at which *r* approaches 1, and *p* and *q* approach 0. No other selective mechanism is at work, and initial gene frequencies corresponding to those found in western Europe are assumed (A: $p = 0.3$; B: $q = 0.1$; 0: $r = 0.6$). The selection coefficients are those suggested by the empirical evidence. Changes in gene frequencies are much more rapid than those found in the Rh case (Fig. 6.13).

6.2.1.5 Other Modes of Selection

Frequency-Dependent Selection [103, 1739]. So far, selection values have been treated as constants. However, these values may be functions of the genotype frequencies as well as of the population density. This type of selection is known as frequency- or density-dependent selection. More specifically, the correlation between the selective value of a genotype and its frequency may be negative, a genotype becoming more advantageous when its frequency declines. In Sect. 3.5.5, Batesian mimicry in *Papilio memnon* was described as an example for linkage disequilibrium of closely linkes loci due to a selective advantage of certain combinations (haplotypes) of alleles. The situation was compared with that encountered in the human major histocompatibility complex (MHC). The butterfly system can also be utilized to demonstrate frequency-dependent selection; the selective value of the mimic depends upon its frequency relative to the frequency of the animal mimicked or model. When the mimic is rare, predators encountering its visual pattern will find distaseful models more often than palatable mimics. The avoidance of the pattern shared by mimic and model will be strengthened; the mimic has a selective advantage. As soon

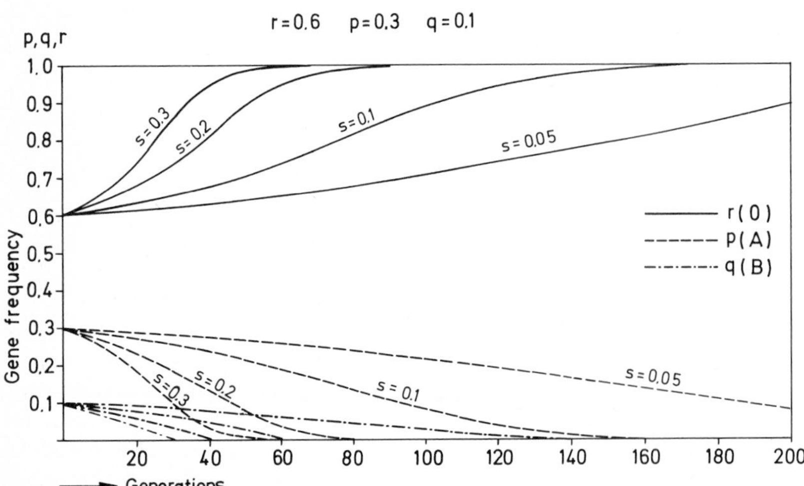

Fig. 6.14. Change of gene frequencies in the AB0 blood group system due to selection *s* against A0 children and B0 children of 0 mothers. *p* gene frequency of A; *q* gene frequency of B; *r* gene frequency of 0; *r* tends to 1; *p* and *q* tend to 0. (Vogel and Helmbold, 1972 [211])

as the mimic becomes too frequent, on the other hand, predators will more often find palatable mimics, and their avoidance will be weakened. If the mimics greatly outnumber the models, the predators may even learn to associate the pattern with palatability; the genotype now has a selective disadvantage.
Let us assume the following simple model of frequency-dependent selection in a dominant gene:

Phenotype	*Frequency*	*Fitness*
A	$1-q^2$	$w_1=1+s_1(1-q^2)$
a	q^2	$w_2=1+s_2q^2$

Then we have:

$$\Delta q = q' - q = \frac{pqw_1 + q^2w_2}{\bar{w}} - q = \frac{pq^2(w_2 - w_1)}{\bar{w}}$$

$$\bar{w} = (1-q^2)w_1 + q^2w_2 = 1 + s_1(1-q^2)^2 + s_2q^4$$

Therefore, the equilibrium condition is:

$$w_1 = w_2$$

or:

$$q^2(s_1+s_2) - s_1 = 0$$

This equation is solvable for the gene frequency, q, if and only if the selection coefficients s_1 and s_2 are both positive or both negative. The solution is:

$$\hat{q} = \sqrt{\frac{s_1}{s_1+s_2}}$$

For the equilibrium state, the fitness value will be:

$$\hat{w}_1 = \hat{w}_2 = \hat{\bar{w}} = 1 + \frac{s_1s_2}{s_1+s_2}$$

The equilibrium can be shown to be stable if s_1 and s_2 are both negative (and not less than -1). This means that a stable equilibrium exists when A has an advantage if it is rare, such as in Batesian mimicry.
At equilibrium, all the genotypes have the same fitness. If there is no dominance and the fitnesses of all three genotypes are different, calculation becomes more cumbersome [103]. However, it can be shown that in this case also frequency-dependent selection can yield stable polymorphisms when there is no heterozygote advantage. Polymorphisms can even be maintained by such a mechanism with a selective disadvantage of heterozygotes. Furthermore, in a particular situation, there may be more than one stable equilibrium. In humans, frequency-dependent selection is a plausible mechanism for mutual adaptation of a parasite to the human host and vice versa. The parasite - for example, a bacterium or virus - may become adapted to the commonest biochemical or immunologic variety of the host, and rarer varieties may gain a selective advantage [1779]. The parasite mimics the antigens of its host, either by acquiring the genetic ability to produce its antigens or by utilizing directly the host's membrane material for synthesis of its own membrane. Examples of the first - genetic - mechanism are bacteria that contain ABH-like antigen [211] (Sect. 6.2.1.8); the second mechanism is possibly utilized by some viruses. In both cases, the immune defense mechanism of the host will be deceived, and the parasite will be more successful than if it would have no antigen in common with the host. The selection will be frequency dependent, as the virus will adapt mainly to the most frequent genotype; the rarer ones will have an advantage.
One mechanism by which the host can defend itself against this "strategy" of the parasite is to create a highly polymorphic system with very many different antigenic patterns at its cell surface. This will prevent the parasite from gaining selective advantage by adapting to one specific pattern. Such a highly polymorphic system has developed for the major histocompatibility complex (MHC), which comprises the HLA loci together with a number of other gene loci involved in the immune response (Sect. 3.5.5).

Frequency-Dependent Selection in Combination with Linkage Disequilibrium. The second significant feature of the MHC complex - apart from the high degree of polymorphism - is the nonrandom distribution of allele combinations for loci forming this complex, especially the HLA loci. This *linkage disequilibrium* has been described in Sect. 3.5.4 and 3.5.5. The example of butterfly-wing patterns helped in understanding how not one allele at one locus but a certain combination of alleles at the various loci of a gene complex or supergene conferred a selective advantage to its carriers. The development of such a selective system depends on whether these alleles together are able to create an adaptive phenotypic pattern. Such groups of closely linked loci with related functions or supergenes may have been created during evolution when certain combinations of genes had a selective advantage. This advantage will usually be destroyed in the next generation by free recombination of the involved gene loci; it will be maintained if these loci are closely linked. Therefore, selection is expected to favor chromosome rearrangements leading to closer linkage between such loci. For the human MHC complex, these considerations lead to the hypothesis that certain genetically determined patterns of the immune response have been favored by natural selection.

Density-Dependent Selection [1739]. Selection may vary not only with the relative frequencies of genotypes within a population but with the absolute population size/habitat, i.e., the population density. It has been shown that density-dependent selection allows the maintenance of a balanced polymorphism under a wide range of conditions. Under some circumstances, a polymorphic population can support a larger number of individuals than a monomorphic one. Still more important, changes in population density may bring about *genetic revolutions* that mimic the effects of random genetic drift (Sect. 6.4). Changes in population density have been important in human evolution, for example, when neolithic agriculture techniques were acquired. It is therefore plausible to assume that density-dependent selection may have played an important role in human evolution.

Kin Selection. In recent years, another kind of selection has been discussed increasingly by evolutionary theorists: kin selection. As a rule, animals interact with each other and with the environment in social groups - families, hordes, gangs. Within such groups, behaviors are often observed that seem to contradict the expectation that each individual competes with all other members of the species for survival and reproduction, an expectation that is implicit in the earlier selection models. The biologic meaning, for example, of *altruistic* behavior - sometimes even at the expense of the individual's life - is obvious intuitively when we observe a mother defending her offspring against a predator. But sometimes, behaviors are observed where the biologic basis seems less clear; for example, when a lion sire takes over a harem and kills the young offspring of his predecessor. Theoretical interpretation of such social behaviors, altruistic and otherwise, remained an enigma for evolutionary theorists from Darwin onwards until in 1964 Hamilton [1784] developed a theory of kin selection. The full theory is intricate and will be described here in outline only. The basic idea, however, is simple and obvious: The individual fights for survival and transmission of its *genes*. Often, the interest of my genes will coincide with my interest as an individual; but if I have, for example, two children, then, sacrificing my life for their survival is equivalent to saving my own life at their expense, since each of them carries half my genes. If I have three or more children, such altruistic behavior is even superior, saving *more* of my genes than if I were to survive myself. The same reasoning applies for sibs, or for more remote relatives such as grandparents, uncles and nephews, and others. The lion, on the other hand, probably will own the harem for a limited - and often short - time span only. During this time, he is "interested" in reproducing as extensively as possible. Lionesses who are still nursing cubs from a different father do not ovulate and cannot become pregnant. Therefore, it is in the interest of the lion's genes to kill these cubs.

An increasingly important branch of science - sociobiology (Sect. 7.2.4) - is based on the theoretical concepts of Hamilton and his successors. The consequences have been popularized in discussions of the selfish gene and have led to heated discussions between biologists and social scientists and philosphers. Understanding of this requires some knowledge of the conceptual background. A simple formal treatment of altruistic behavior [1738] is given below. A gene A causes its bearers to act "altruistically." For example, a mother blackbird, by flying in front of the nose of a cat, diverts the cat from her young, but runs a risk of being caught. This may decrease personal fitness by an amount c, at the same time increasing the fitness of a relative by amount b. The gene A will have a selective advantage if $c/b < r$ where r is the "coefficient of relationship" between donor and recipient (Hamilton's law). r measures the probability that the gene of interest in the donor is found as a copy in the recipient; it depends only on the relative positions of "donor" and "recipient" in a pedigree. In the absence of inbreeding, values of r are:

Full sibs: 1/2 First cousins 1/8
Half sibs: 1/4 Parent-offspring 1/2

This means, r is identical with the fraction of genes two individuals have in common by descent (See Sect. 6.3.1).
This rule will now be derived for a simple diploid model with the following specifications:

1. A single locus, two alleles; A is an altruistic gene (gene frequency p), a is selfish (gene frequency q).
2. Panmictic mating, near infinite population size.
3. The gene codes for interaction between two kinds of relatives (e.g., sib-sib, cousin-cousin).
4. Since $r > c/b$ is the proposed condition for the spread of the A gene, the model will state the condition for the spread of the A gene as $c/b < ?$ and compare ? with r.

If the fitnesses of the three genotypes (AA, Aa, and aa) are w_{11}, w_{21}, w_{22}, respectively, the condition for the A gene to spread is [45]:

$$p(w_{11}-w_{21})+q(w_{21}-w_{22})>0$$

The fitnesses may be written as follows:

$$w_{11}=1+s,\ w_{21}=1+r,\ w_{22}=1+t$$

where r, s, and t refer to additional changes in fitnesses due to altruism. Now, define as follows:

N = population size (very large)
n = number of "altruistic acts" in a generation
b = increment to the fitness of an individual helped
c = decrement to the fitness of the individual helping
m = proportion of n acts which help AA individuals
e = proportion of n acts which help Aa individuals
x = proportion of n acts where the helper is AA

With these definitions, the fitnesses may be written as follows:

$$w_{11}=1+\frac{nbm-nxc}{p^2N}$$

$$w_{21}=1+\frac{nbe-nc(1-x)}{2pqN}$$

$$w_{22}=1+\frac{bn(1-m-e)}{q^2N}$$

These equations simply assign the c's (costs) and b's (benefits) to the various genotypes and then divide by the number of individuals of these genotypes present to get the average fitnesses. If the three equations are substituted in Eq. 1, the following is found:

$$c/b<\frac{2m+e-2p}{1+x-2p} \tag{6.11}$$

The < sign would of course be reversed if the cost c were really a gain, at the expense of the recipient.

In order to see whether Eq. 6.11 reduces to 1/2 in the cases of full sibs, as asserted above, expressions for m, e, and x are needed. They were calculated [1738] as:

$$m=(x+p)\left[\frac{1+p}{4}\right];\quad e=\frac{1+p-p^2-xp}{2}$$

If these are introduced into Eq. 6.11 it reduces to 1/2. Similar calculations are possible for other degrees of relationship. A more general and stringent treatment may be found, e.g., in Hamilton's papers.

Selection in Continuously Distributed, Multifactorially Determined Characters. So far, we have examined selection only for single genes. However, many normal and abnormal traits and diseases show a continuous distribution in the population and are determined by an unspecified number of different genes. In principle, the laws of selection theory apply for these characters as well. However, since genetic analysis has not yet penetrated to the Mendelian-phenotypic level (Sect. 3.6), biometric methods have to be used [63, 65, 124]. These methods have gained importance for animal and plant breeding. Some of these problems include:

a) Obviously, the change of a quantitative character in a population under the influence of natural selection is proportional to heritability (for the heritability concept, see Sect. 3.6.1.5).
b) The response of the character to selection depends on the strength of selection (Fig. 6.15 a and b).
c) It also depends on the degree of genetic variability in the population (Fig. 6.15). In the absence of genetic variability, selection is ineffective.

Figure 6.16 shows the effect of artificial selection over generations. The mean shifts toward the direction of selection but genetic variability decreases from generation to generation, until selection ceases to be effective.

In evolution, mutations causing slight, barely discernible shifts in multifactorial systems have presumably been of great importance. Under the influence of selection, these mutations have led to a slow and gradual shift of quantitative characters.

One genetic property that shows such a continuous distribution, and is therefore thought by many observers to be due to interaction of many genes, is the liability to common "multifactorial" diseases such as congenital malformations, common diseases, or mental disorders. Altering selective conditions – for example, by successful treatment of congenital heart disease or schizophrenia – will lead to a shift of the distribution curve of disease liability and, hence, to a higher incidence of these conditions. It is, however,

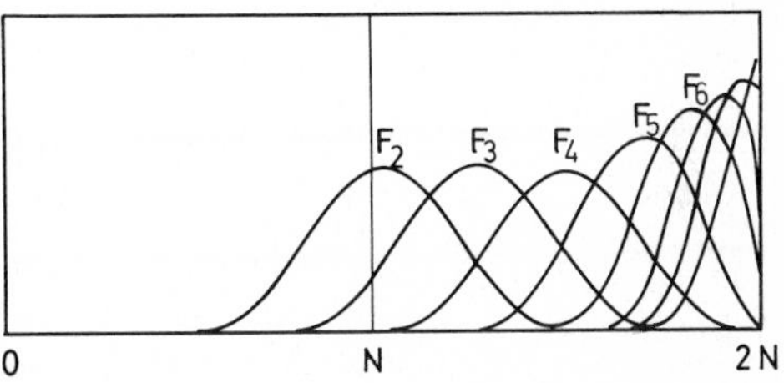

Fig. 6.16. Effect of selection over many generations on a continuously distributed character with multifactorial inheritance. With shifting of the mean, the variability gradually decreases. The same is true for the effect of selection in one generation. Ultimately, the population will be homogeneous genetically, and no further selection is possible [63]

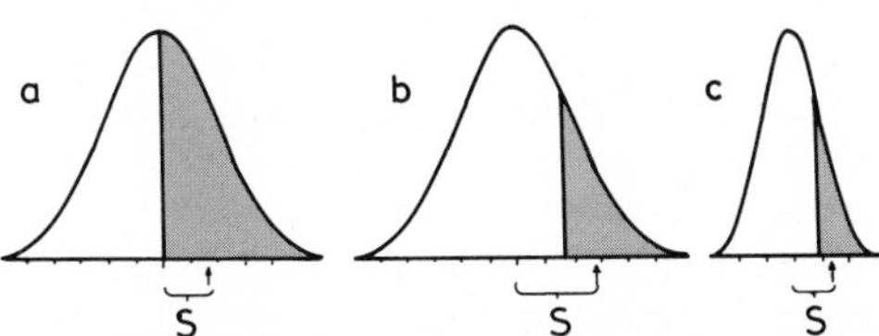

Fig. 6.15 a–c. Selection in continuously distributed multifactorial characters. **a** Selection by elimination of 50% of the population (those below the population mean, shown in white) from reproduction. In the F_1 generation, the mean is shifted by S. It can be shown that this shift must be 0.8 × the standard deviation (SD). **b** Selection by elimination of 80% of the population from reproduction. This leads to a shift of 1.4 × SD of the population mean in the next generation. **c** This shows a population with lower genetic variability. If 80% of the population are eliminated from reproduction as in b, the shift of the mean relative to the standard deviation is the same but absolutely it is much lower (in this case 1/2). (Falconer, 1960 [63])

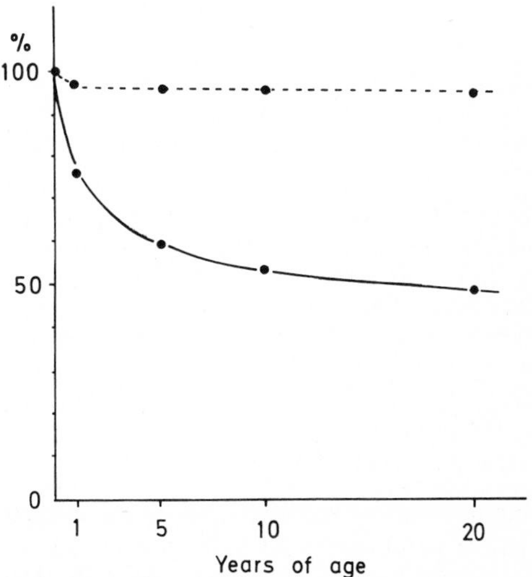

Fig. 6.17. Survival rate of newborn children up to the age of 20 in Prussia (central Europe) in the middle of the eighteenth century. Based on data from Süssmilch, 1776 [1875]). The *dotted line* gives survival rate for Berlin (1955) for comparison

difficult to estimate the extent of the shift. We do not know sufficiently well whether, and in which examples, the multifactorial genetic model gives an adequate description of reality.

6.2.1.6 Selection Due to Infectious Diseases [1831, 211]

In the foregoing sections, the most important mathematical selection models have been described, and some practical applications to situations occurring in human populations have been indicated. However, the most important probable source of natural selection has so far not been considered: selection due to genetic differences in susceptibility to infective agents.

Selection Due to Infectious Diseases in Historical Populations. Natural selection is especially effective when acting through differential mortality before the age of reproduction, i.e., during childhood and youth. The first reliable statistics on childhood mortality were available from Europe in the eighteenth century [1875]. The survival rate in children at birth up to the age of 20 is shown in Fig. 6.17 for Prussia, an average central European country, in the middle of the eighteenth century. More than half the newborns had died before reaching the age of 20. About one-quarter died within the 1st year of life. What were the causes of these early deaths? Statistics answer this question only partially, as not all diagnoses reported at that time can be identified with presently known diseases. However, the data leave little doubt that most children died from viral and microbial diseases, such as intestinal infections, smallpox, tuberculosis, and measles. Therefore, infectious diseases are good a priori candidates for possible selective mechanisms. Endemic infections that affect every generation would be more effective selective agents than epidemic diseases which only occur episodically such as plague.

History of Some Infectious Diseases. What are the infectious diseases that may have influenced gene frequencies of whole populations in the past? Four groups come to mind:

1. Acute infections that invaded whole countries, sweeping away large parts of the populations. Examples are plague, cholera, and smallpox.
2. Chronic infections such as tuberculosis, leprosy, and syphilis.
3. The heterogeneous group of intestinal infections that occurred in all age groups but were fatal mainly in infancy and early childhood.
4. Tropical diseases such as malaria.

The available information on the history of these infections is fragmentary and often misleading, as no bacteriologic evidence for historical epidemics is available, and the descriptions often do not permit a diagnosis. Some conclusions seem to be warranted, however [1791]:

a) Plague can be traced back to the end of the second or the beginning of the third century *B. C.* Good descriptions exist from Alexandria and Libya around the time of Christ's birth; during the Middle Ages, a number of "plagues" swept through Europe, killing a high fraction of the population. In more recent times, epidemics have come from the Near East (Turkey) and southern or southeastern Asia. Smallpox centers were in Africa and in Asia, especially India and China. In both countries, smallpox was known for thousands of years; India even has a smallpox goddess, Sitala.

b) Tuberculosis used to be endemic in large parts of the Old World and was introduced wherever white men came. Leprosy, to become endemic, needs special living conditions. This disease was frequent in Europe during the Middle Ages but disappeared with improved living conditions by the eighteenth century. At present, many millions are still affected, especially in India, southeastern Asia, Africa, and South America. Syphilis, according to one hypothesis, was a disease of the New World - Central and South America - and was introduced to Europe immediately after the discovery of America. Following epidemics in the late fifteenth and the sixteenth centuries, syphilis became an endemic disease [1728]. This hypothesis is far from being universally accepted, although it is in reasonable accord with many historical facts. Historical hypotheses of this kind are difficult to refute but still more difficult to confirm.

c) It is known that infant mortality due to intestinal infections leading to infant diarrhea was very high in Europe even up to 1900, and in large parts of Asia, Africa, and South America until much more recent times. It is reasonable to assume that such high infant mortality has always been experienced in human history. Beyond this statement, more specific hypotheses are impossible.

d) Whereas for tropical countries mortality statistics such as those mentioned for Prussia (Fig. 6.17) have only been available recently there is ample evidence that even after World War II, many of these countries still had a very high childhood mortality. This mortality was caused by a variety of conditions; infectious diseases such as (tropical) malaria and intestinal infections prevailed. If there were genetically conditioned susceptibilities to these infections, those individuals who were susceptible would die more often, whereas others who enjoyed a higher level of genetic resistance would have a better chance to survive and to transmit their genes to their children. Therefore, the genetic composition of the present-day human populations will be influenced strongly by such differences in resistance to infections. Recent decades have seen much progress in the analysis of mechanisms for such resistance and in studies of the genetic composition of human populations. In the following, two examples will mainly be analyzed:

1. Selection due to malaria in relation to the frequency of hemoglobin genes in populations of tropical countries.
2. Selection in relation to the AB0 blood groups.

With the progress of immunogenetics in recent years, other genetic systems will probably be included in such genetic population analysis soon.

Distribution of Sickle Cell Gene and Other Abnormal Hemoglobin Genes in the World Population. The leading paradigm for heterosis in humans is the selective mechanism causing the high frequency of the sickle cell gene in some human populations. The

molecular basis, genetic determination, and genotype-phenotype relationship of sickle cell hemoglobin has been described in Sect. 4.3 and 5.1.4.
Sickle cell hemoglobin (HbS) is produced by a single nucleotide substitution of the β-hemoglobin chain. Affected homozygotes suffer from a severe hemolytic anemia, whereas heterozygotes are clinically healthy under normal conditions.
From the standpoint of population genetics, the very unequal distribution in the populations of the world is the most striking feature of the sickle cell gene. Such a pattern is not, however, confined to this gene; a number of other hemoglobin variants such as Hb C, D, E, and the thalassemias show a similarly unequal distribution and are also polymorphic in some populations. Most frequent and widespread, however, is the sickle cell gene. Within a broad periequatorial belt from the Cameroons and the Congo to Tanzania, frequencies ranging between 25% to as high as 40% of heterozygosity for HbS have been found. The frequency decreases slightly toward the western part of Africa. In northern and southern Africa, it is much lower; in both these areas, HbS is found only sporadically in many populations. In the Mediterranean, HbS occurs especially in Sicily, Calabria, and some parts of Greece. On the Chalkidiki peninsula, heterozygote frequency may reach 30% (Fig. 6.18). The gene is relatively common in South Indian populations and has been found in Arabs. It is absent from native American populations and practically absent from all northern and northwestern European populations. Three explanations are possible for such an unequal distribution in principle.

1. The mutation rate may be different, either for exogenous reasons - for example, by differential exposure to a mutagenic influence (Sect. 5.2) - or for endogenous reasons, for example, unequal distribution of mutator genes (Sect. 5.1.4).
2. Selection may work differently under different environmental conditions.
3. Chance fluctuations of gene frequencies (genetic drift) may occur, especially if the effectively breeding populations are small (Sect. 6.4.1).

To begin with the last-mentioned possibility, it is very unlikely that differences between large population groups such as those observed for the frequency of HbS have come about by chance. Moreover, why should all known polymorphic hemoglobin variants (Hb, S, C, D, E and the thalassemias) be found almost exclusively in the tropical-subtropical belt? The drift hypothesis has, therefore, never been considered seriously to explain the distribution of these polymorphisms.

Differential Mutation Rates to HbS? The hypothesis of differential mutation rates has been seriously considered [1836]. To calculate the mutation rate to

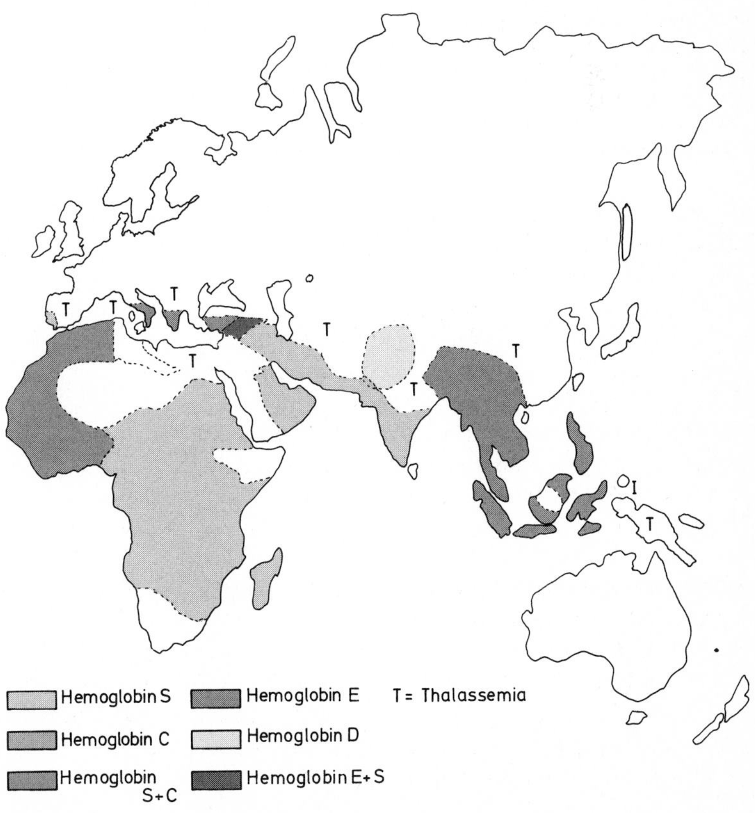

Fig. 6.18. Areas in which the sickle cell gene and genes for other abnormal hemoglobins are common. (See also [1790a, 1815a])

Hb S, Haldane's indirect method (Sect. 5.1.2) was used. First, the number of genes in nonreproducing affected homozygotes that are eliminated in each generation was estimated. Then, a mutation rate of equal order of magnitude was assumed by postulating genetic equilibrium between mutation and selection. This "mutation rate" turned out to be 10^{-2}(1:100) in some populations - an excessively high value. Moreover, the hypothesis of an unusually high mutation rate was refuted a short time later by direct examination [1805, 1882]:

Of 530 mothers who had at least one child with sickle cell anemia, 525 showed the sickle cell trait and were, therefore, considered heterozygotes. Of 483 fathers, 445 had the sickle cell trait. In such studies, only results from mothers can be considered, unless the "legal" fathers can be shown by comprehensive genetic marker studies to be the biologic fathers as well. Even the five mothers without the sickle cell trait, however, do not prove a new mutation; more probably, they were β-thalassemia carriers, and their children had sickle-β-thalassemia, which may mimic all clinical findings of sickle cell anemia. Anyway, these figures indicate that the maximum plausible mutation rate is not nearly high enough to explain the high population incidence.

All evidence points to very low mutation rates for single base substitutions in the DNA (Sect. 5.1.4), possibly ranging between 10^{-8} and 10^{-9}. This, together with the small overall population sizes in earlier centuries, suggested that ultimately all sickle cell genes could be traced back to very few single mutations or only one. More recently, linkage studies with 11 DNA polymorphisms in numerous blacks from the United States of America and Jamaica have confirmed this conclusion: Four clusters of haplotypes were discovered, none of which could be derived from any other by fewer than two crossover events. This could mean up to four mutations, or a still smaller number, if two crossing over events or gene conversion, which is likely (Sect. 2.3.4), are considered. These haplotypes were found with sickle cell genes from Mediterraneans indicating that sickle cell genes in Mediterranean populations are not due to a separate mutation but have been introduced by Africans [1721]. In view of the very low estimates of codon mutation rates (Sect. 5.1), we find the idea that ultimately all HbS alleles go back to just *one* mutation and that the present location of this allele in four RFLP haplotypes might be explained by very rare recombination or by gene conversion (see Sects. 2.3 and 4.3) an attractive one. The same reasoning might also apply for HbE.
But why did these genes become so frequent?
There must have been a selective advantage.

Malaria Hypothesis. Geographic distribution and genetic analysis offer two hints:

1. The selective advantage seems to be confined to tropical and subtropical areas.
2. Homozygotes suffer from a severe hemolytic anemia. Their reproductive fitness is estimated to be approximately 20%-25% of normal and under primitive living conditions approaches 0. The homozygous state for sickle cell anemia was unknown in central Africa until the last 2 decades because affected infants died very early. The high frequency of HbS could only be reached with a selective advantage of heterozygotes who are, at the same time, much more frequent than homozygotes.

The two criteria led to the hypothesis that heterozygotes are less susceptible to falciparum malaria than are normal homozygotes:

Beet (1946, 1947) [1724, 1725] observed in the Balovale district and in other parts of northern Rhodesia that during the dry season when malaria infectivity is in general lower, heterozygous AS children showed malaria parasites in their blood smears more rarely than normal homozygotes. The difference was not significant statistically, but splenomegaly was also less pronounced in heterozygotes. Beet appears to have been the first to suggest malaria as the principal selective agent. In 1951, the Lambotte- Legrands in Zaire had the impression that cerebral malaria was rarer among heterozygotes for sickle cell anemia.

Haldane (1949) [1778], noting a similar geographic distribution of thalassemia and malaria, suggested that thalassemia could be maintained in the population by a selective advantage of heterozygotes in the presence of malaria. For the sickle cell gene, this hypothesis was tested by Allison [1714-1717, 1719], who formulated it as follows:

1. The homozygous sickle-condition is virtually lethal in Africa.... The rate of elimination of the gene could not be compensated by recurrent mutations.
2. Balanced polymorphism has resulted because the sickle cell heterozygote is at an advantage, mainly as a consequence of protection against Falciparum malaria.
3. Malaria exerts its selective effect mainly through differential viability of subjects with and without the sickle cell gene between birth and reproductive age, and to a much lesser extent through differential fertility.
4. High frequencies of the sickle cell gene are found only in regions where Falciparum malaria is, or was until recently, endemic.
5. In most New World black populations. Frequencies of the sickle cell gene are lower than would be expected from dilution of the African gene pool by racial admixture. This is probably the result of elimination of sickle cell genes without counterbalancing heterozygote advantage.
6. In regions where two genes for abnormal hemoglobins coexist, and interact in such a way that individuals possessing both genes are at a disadvantage ... these genes will tend to be mutually exclusive in populations.

In the following sections, the evidence for the first four arguments of the "malaria hypothesis" in its specific form will be taken up. The evidence for points 5 and 6 will be deferred to later sections.

Evidence for the Malaria Hypothesis. Several lines of evidence will be discussed in detail, since this example illustrates the methodology available in humans for examination of hypotheses of natural selection. The evidence may be divided into two parts: (1) results from testing the proposed mechanism of selective advantage and (2) results from examining the proposed consequences for reproduction and population frequency.

1. The hypothesis predicts that young children (during the first 5 years of life) without the sickle cell trait, will be infected more massively, will become more severely sick, and will die more often from falciparum malaria than will sickle cell heterozygotes. The hypothesis does not predict that school children or adults will be more frequently or severely infected. In hyperendemic areas, immunization by frequent infection with malaria has developed early in life so that no differential mortality can be expected between heterozygotes and normal homozygotes in older children or adults.
   Table 6.10 shows the incidence of *Plasmodium falciparum* infections in sicklers and nonsicklers. For statistical evaluation, the method of Woolf (Sect. 3.7.2) was used; the weighted mean relative incidence shows that children without the sickle cell trait (normal homozygotes) incur a risk of 2.17 that of sickle cell heterozygotes to have a heavy falciparum infection. The table shows only cases with heavy infections, defined as showing more than 1,000 parasites per 1 μl blood. However, malaria can lead to a difference in fitness only if a higher percentage of affected nonsicklers die from malaria - or if the disease impairs their reproduction. Table 6.11 is a study of the number of fatal outcomes among normal homozygous and heterozygous children in several areas of Africa. With one exception, only normal homozygotes died of malaria. In view of the incidence of the sickle cell trait in these populations (8-29%), this finding cannot be a chance result. Hence, a higher malarial susceptibility during early childhood and, consequently, a higher death rate was demonstrated for normal homozygotes as compared with heterozygotes. The results of in vivo studies where volunteers with sickling trait were infected with malarial plasmodia are less clear-cut. In one study [1715], sickle cell heterozygotes showed a lower incidence of parasitemia than controls after experimental malaria inoculation; in other studies, this result was not confirmed.
2. In another set of investigations, the consequences of an increased disease susceptibility for the population were examined: (a) a higher nonsickler mortality during childhoods should result in a greater frequency of sicklers among adults com-

Table 6.10. Incidence of heavy *P. falciparum* infections in African children. (Allison, 1964 [1719])

Authors	Classification of infection	Sickle cell		Non-sickle cell		Relative incidence[a] (Woolf)	χ^2	Probability
		Heavy infections	Total	Heavy infections	Total			
Allison (1954a)	Group 2 or 3	4	43	70	247	3.86	6.16	$0.02 > P > 0.01$
Foy et al. (1955)	Heavy	21	241	38	241	1.96	5.44	$0.02 > P > 0.01$
Raper (1955)	>1,000 μl	35	191	374	1,009	2.63	23.74	$P < 0.001$
Colbourne and Edington (1956)	>1,000/μl	3	173	57	842	4.11	5.59	$0.02 > P > 0.01$
Colbourne and Edington (1956)	>1,000/μl	5	15	75	177	1.47	0.46	$P > 0.50$
Garlick (1960)	>1,000/μl	25	91	147	342	1.99	7.06	$0.01 > P > 0.001$
Allison and Clyde (1961)	>1,000/μl	36	136	152	407	1.66	5.27	$0.05 > P > 0.02$
Thompson (1962, 1963)	>5,630/μl	3	123	42	593	3.05	3.38	$0.10 > P > 0.05$

[a] Incidence of heavy *P. falciparum* infections in non-sickle cell trait groups relative to unity in corresponding sickle cell trait groups.
Weighted mean relative incidence = 2.17. Difference from unity $\chi^2 = 51.379$ for 1 d.f. $P < 0.001$.
Heterogeneity between groups $\chi^2 = 5.719$ for 7 d.f., $P > 0.5$.

pared with children of the same population group as shown in Table 6.12; (b) a higher mortality of nonsicklers should also lead to a greater number of surviving children from marriages in which sickle cell heterozygotes segregate. The evidence from one large study [1790, 1718] is shown in Table 6.13. Childhood mortality was highest in matings between two heterozygotes. This is not surprising, as one-fourth of them are expected to be homozygotes for the sickle cell allele and hence to suffer from sickle cell anemia. However, relatively fertile matings that, at the same time, show the lowest number of dead children were those between heterozygotes and normal homozygotes (AS × AA). This is to be expected if heterozygotes run a lower risk of dying during childhood.

From the difference between the frequency of the genotypes AA, AS, and SS in the adult population in the Musoma district and its Hardy-Weinberg expectation, Allison [1718] calculated the relative fitness of the different genotypes as compared with the population average:

$w_{AA} = 0.7961$; $w_{AS} = 1.000$; and $w_{SS} = 0.1698$

corresponding to:

s_1 (selection against AA) = 0.2039;

s_2 (selection against SS) = 0.8302.

Is the ratio of the selection coefficients sufficient to maintain a genetic equilibrium for the actually ob-

Table 6.11. Malarial mortality in Hb-S trait (AS) [1829]

	% AS (population frequency)	No. dead of malaria	Observed No. AS dead of malaria	Expected No. AS dead of malaria
Leopoldville, Congo (Zaire)	26	23	0	6
Luluabourg, Congo (Zaire)	29	21	1	6.1
Ibadan, Nigeria	24	27	0	6.5
Accra, Ghana	8	13	0	1
Kampala, Uganda	19	16	0	3
Total		100	1[a]	22.6

[a] $\chi^2 = 26.67$ $P < 0.001$.

Table 6.12. Comparison of prevalence of sickle cell heterozygotes among children and adults. (Allison, 1956 [1718])

Population	No. of children examined	Sickle cell heterozygotes %	No. of adults examined	Sickle cell heterozygotes %
Dar-es-Salaam	753	17.9	283	23.3
Congo; Baluba	147	16.3	775	23.5
Congo; Pygmies	119	22.7	327	28.1
Dakar	1,350	6.2	952	15.5
Ruanda-Urundi	516	14.2	928	13.2
Musoma, Tanzania	287	31.8	654	38.1
Mandingo, Gambia	211	9.0	713	11.5
Jola, Gambia	103	14.5	312	17.0
Fula, Gambia	69	17.3	127	18.9
Jolloff, Gambia	48	18.8	104	17.3

Table 6.13. Fertility and children's mortality in Africans living in the Musoma district. (Tanzania) [1718]

Mating type	Number of matings	All living children	Mean number of living children/mating	Deceased children	Mean number of deceased children/mating	% Deceased children	Total number of children (living or dead)	Total number of children/mating
AS × AS	18	44	2.44	35	1.94	44.3	79	4.39
AS × AA	84	221	2.63	121	1.44	35.3	342	4.07
AA × AA	74	172	2.32	115	1.55	40.1	287	3.88
All mating types	176	437	2.48	271	1.54	38.2	708	4.02

served frequency of the sickle cell gene? To test this, we return to Eq. (6.6) for equilibrium conditions:

$$\hat{q}=\frac{s_1}{s_1+s_2}=\frac{0.2039}{0.2039+0.8302}=0.1972$$

A gene frequency of this order of magnitude corresponding to a heterozygote frequency of 31.7% was, indeed, found in some populations. Hence, the data roughly agree with the predictions from the hypothesis of a balanced polymorphism.
In how many generations would such an equilibrium be established if this selection would start anew? Results of some model calculations [1872, 1848] are seen in Fig. 6.19. Forty generations correspond to about 1000 years if a generation time of 25 years is assumed. Since a millennium is not an unduly long time, the model is realistic from this standpoint as well. Some other results point to an additional loss of fertility due to placental malaria in infected homozygous normal women [1758]. The evidence, however, is not entirely convincing.

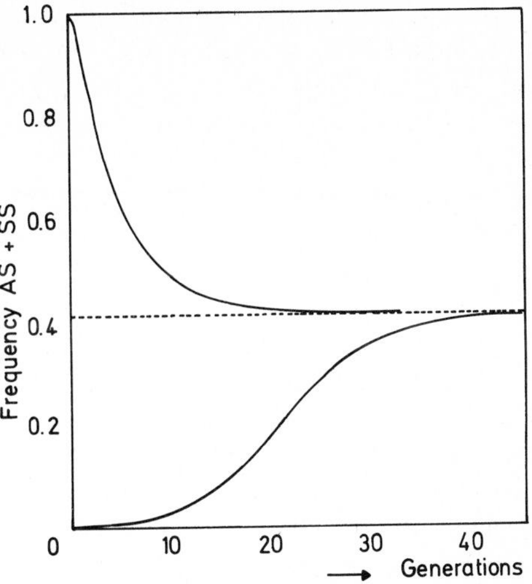

Fig. 6.19. Model calculation for the changes in the combined frequency of sickle cell homozygotes and heterozygotes for the sickle cell gene and the assumption about s_1 and s_2 set out in the text. Here, fitness of the heterozygote is assumed to be 1.26 times fitness of the normal homozygote. Fitness of the sickle cell homozygote is one-fourth that of the heterozygote. (Smith et al., 1954 [1872] With initial gene frequencies as different as 0 or 1, an identical equilibrium frequency will be reached

Some Other Aspects of the Malaria Hypothesis [1829, 1830]. In recent years, the malaria hypothesis has been examined from various points of view, the most important of which are:

a) It has been estimated [1814] that malaria contributed significantly to approximately 15% of childhood deaths in endemic areas with ubiquitous infection. Considering that childhood mortality approached or even exceeded 50%, this mortality level is enough to provide a high selective advantage for any gene offering at least partial protection.
b) Malaria is an ancient disease; it probably has existed for at least 2,000 years (80 generations) in Mediterranean areas. In some parts of Africa, malaria probably has been less ancient and presumably was introduced by slash and burn agriculture [1812], a practice that facilitated mosquito breeding in water ponds warmed by sunshine. Malaria - and, therefore, selection in favor of the sickle cell gene - depends on ecologic conditions.

What Will Happen if the Advantage of Heterozygotes Disappears? The advantage of sickle cell heterozygotes is only present in an environment in which many children die from falciparum malaria; it vanishes as soon as malaria is eradicated. There is little doubt that malaria has been reduced in many countries very much below its level of 20-30 years ago, although the condition is coming back in certain areas. This means that at present there is little if any heterozygote advantage left. The consequences for the gene frequency of the sickle cell gene are obvious. Selection against the sickle cell gene will lead to its gradual diminution provided only that selection against abnormal homozygotes (patients with sickle cell anemia) continues. The speed of this decline can be calculated from the formulae used in the discussion of complete selection against homozygotes. Since most New World black populations have been living for several generations in a malaria-free environment frequencies of the sickle cell gene are lower than would be expected from admixture of white genes (Sect. 6.3.4).

The expected reduction of the sickle cell gene was observed in blacks of malaria-free Curacao, whereas blacks in Surinam had higher HbS frequencies. Both groups indicated a similar origin in Africa [1797]. Only the Surinamese population was exposed to malaria, which persists to the present time. Results pointing in a similar direction were obtained in North American blacks in Georgia, when white admixture estimates (see Sect. 6.3.4) were calculated for a number of genes. The HbS trait turned out to give much higher admixture values than other markers.
Selection against HbS homozygotes in the absence of malaria would explain the results [1729, 1797, 1884].

Population Genetics of G6PD Variants and Falciparum Malaria [1829, 1830]. So far, only HbS has been discussed. However, some other genetic traits of red blood cells are also common in tropical and subtropical areas (Fig. 6.18). Examples are HbC in western Africa, HbD in parts of India, HbO in Arabia, HbE in South-East Asia, the various α- thalassemia and β-thalassemia variants that are observed all over the malaria-infested area, and some variants of the enzyme glucose-6-phosphate dehydrogenase (G6PD). It is plausible to conclude that malaria may be the major selective factor for these traits as well. The actual evidence, however, is not nearly as good as for the sickle cell gene. Here, it cannot be discussed in detail. Only the G6PD variants, thalassemia, and HbE will be considered.

Soon after the discovery of G6PD deficiency, it became apparent that G6PD deficiency was mainly found in populations originating from tropical and subtropical areas of the globe. The geographic distribution was similar to that of falciparum malaria and suggested that G6PD deficiency similar to the sickling trait owed its distribution to selection by this malarial organism. Malaria requires nonprotein glutathione (GSH) for growth. Since GSH is reduced in G6PD deficiency, proliferation of falciparum malaria might be curtailed in the enzyme deficiency. Lessened parasitization is associated with lower mortality. Microgeographic mapping of the G6PD frequencies in areas of the world where both high and low endemicity of falciparum malaria existed demonstrated that high gene frequencies were found where malaria was frequent and low gene frequencies were seen in areas with little or no malaria. The results from Sardinia were particularly impressive and showed high G6PD frequencies in the plains where malaria was endemic and low G6PD frequencies in the hills where malaria was absent. Hill and plains populations were similar when other genetic markers were studied.

The finding that various genetic types of G6PD deficiency of different mutational origin (such as the A^- type, the Mediterranean type, and various Asian types) reached high frequencies in different parts of the globe is strong presumptive evidence for selection. Evidence related to a good correlation of the frequencies of the sickling gene and that of the A^- variety of G6PD deficiency exists in African countries and between frequencies of β-thalassemia and the Mediterranean type of G6PD deficiency in Sardinia. Otherwise stated, both traits in a given area were either high or low in frequency, depending upon malarial prevalence in earlier generations. Since G6PD deficiency is X-linked and HbS and β-thalassemia are autosomal traits, no other explanation besides selection can adequately explain these correlations.

The dynamics of selection in an X-linked trait is complex since males have two (normal and mutant) and females have three genetic classes (normal, heterozygotes, and homozygotes). Various data and theoretical calculation make it likely that heterozygote females are most significantly protected in this system and have the optimal survival advantage. The advantage of G6PD-deficient males is counterbalanced by the disadvantage caused by hemolytic episodes. Heterozygote females have two red cell populations: a G6PD-deficient and a normal G6PD population (Sect. 2.2.3.3). The G6PD-deficient red cell population was demonstrated to show fewer malaria parasites than the normal population, thus providing direct evidence for protection of G6PD-deficient cells vis-a-vis falciparum malaria.

In Vitro Studies of Malarial Growth in Red Cells [1766, 1816, 1821, 1822, 1847]. The development of a methodology to grow malarial organisms in red cells in tissue culture provided a direct technique to study the ability of genetically abnormal cells to support malarial growth. However, owing to technical difficulties, work with this test system has not always provided definite answers.

Red cells from sickle cell trait heterozygotes were found to be a poor medium for falciparum malaria proliferation under conditions of hypoxia [1847]. Invasion of such cells also was somewhat reduced. Sickling itself was not required [1089, 1844]. G6PD deficient cells from males with either the African or the Mediterranean variety do not support falciparum growth as well as normal cells [1859, 1860], though some groups have been unable to find any differences [see 1845]. However, all groups agree that there is lessened proliferation of malarial organisms in G6PD heterozygote females, depending upon the number of deficient cells [1860]. It had already been shown earlier, by means of staining studies, that falciparum parasites were less frequent in the deficient cells of G6PD heterozygotes whose red cells are a mixture of deficient and normal cells [1823]. Somewhat low glutathione (GSH) levels were not further diminished in G6PD-deficient cells in culture, a somewhat surprising finding since GSH depletion was thought to be an important cause of the growth differential.

Heterozygotes for HbC and HbE appear to support malarial growth as well as normal cells, while the data from homozygotes for HbEE and HbCC are equivocal. Most studies have failed to show any proliferation differentials between normal and β-thalassemia cells except with additional oxidative stress [1861]. Among the α-thalassemias, only those with HbH disease, i. e., deletion of three alpha genes: α-/--, showed definitive decreased growth [1793]. The data in those with two abnormal α genes were less consistent, and normal growth was seen in those with a single Hbα deletion (α-thal 2). HbF appears to inhibit proliferation of falciparum malaria [1847a]. However, the various conditions characterized by hereditary persistence of fetal hemoglobin with high fetal hemoglobin levels in heterozygotes have not become as frequent as might be expected from the in vitro findings. Ovalocytes resist entry of falciparum organisms in vitro, and ovalocytosis is common in certain areas of New Guinea and Papua [1799]. However, cells with a rare and widespread blood group En (a-) resist invasion of falciparum malarial organisms in vitro [1846], but this blood group has not reached polymorphic frequencies in any population.

The various in vitro data suggest that a direct demonstration of diminished falciparum growth may be easy when the growth differential is relatively great, as with HbS heterozygotes. The occasionally observed proliferation differences in the somewhat rare homozygotes for HbC and E and in HbH disease are immaterial for population spread of these genes, but show that only fairly large genetic differences can be demonstrated by such laboratory studies. With more subtle differences, as in HbE and C heterozygotes, the methodology may not be sensitive enough to demonstrate small differences in growth. In any case, results of the in vitro methods are still too inconsistent to provide the decisive data that could prove or disprove the malarial hypothesis.

Ascertaining and Measuring of Selection in Humans. The section summarizes some research strategies for ascertaining and measuring selection in humans. Most of these strategies have been used for testing the malaria hypothesis for the sickle cell gene [1830]:

a) There are obvious geographic correlations between endemicity of the selecting disease and the protective gene.
b) Affliction with severe forms of the selecting disease and, hence, mortality is reduced in heterozygotes of the protective gene as compared with normal homozygotes.
c) There could also be increased fertility in A/S heterozygotes as compared with normal homozygotes [1829, 1830].
d) Mortality differences are expected to lead to age stratification in the population. If the selecting disease would selectively kill small children, there should be a relative increase in the frequency of the protective gene with age.
The most obvious way to measure selection in infancy is to compare gene frequencies in infants and in the adult population; this may be supplemented by comparison of effective fertility in families (Tables 6.11-6.13). If selection intensity is as strong as with the sickle cell trait, this approach may be successful. However, in human genetics, the expected selection intensities are generally much lower. For a recessive disease with 100% selection against the affected homozygotes and a selective disadvantage of 3% of normal homozygotes, the equilibrium value (Eq. 6.6) does not depend on gene frequency and may be calculated as follows:

$$\hat{q} = \frac{0.03}{0.03+1} = 0.0291; \quad q^2 = 0.00085$$

This implies a homozygote frequency of a little less than 1:1,000 - a frequency somewhat higher than that of cystic fibrosis in the western European population. For rarer recessive genes, the disadvantage of the normal homozygote (s_1) as compared to the heterozygote must be much lower to maintain a balanced system (Table 6.14). An enormous sample size is required to verify a selective disadvantage of this order of magnitude (0.5%-3.0%). If the rarer allele is moderately frequent, i.e., if a genetic polymorphism exists, the selective disadvantage of the normal homozygote is expected to be higher, and the required sample sizes will be more reasonable. Most studies on selection have been carried out with polymorphisms, such as blood groups. However, the results have turned out to be ambiguous [211]. This is not surprising since present day gene frequencies reflect selection processes of the past or may not be caused by selection at all. Infant mortality in general has been dramatically reduced. If the studies described in Tables 6.11-6.13 were to be repeated today in malaria-free areas, it is unlikely that childhood mortality differentials and difference in the number of surviving children would be observed. Hence, investigations of fertility and mortality differentials in human populations to assess selection may be impossible for practical reasons. It may be misleading to use present-day results for conclusions as to selection in the past when entirely different environmental conditions prevailed.

Table 6.14. s_1 (selective disadvantage of the normal homozygote) needed for maintaining a balanced system if $s_2=1$ (complete selection against the abnormal homozygote) under genetic equilibrium

q	q^2	s_1	Example
0.0291	0.000847	0.03	Cystic fibrosis
0.0109	0.000118	0.011	Phenylketonuria
0.00498	0.000025	0.005	Galactosemia

e) Difficulties of this type are inherent in most present-day studies on selection and can be circumvented by direct examination of a putative selective mechanism. However, to use this approach, a specific hypothesis regarding such a mechanism is needed. For the malaria hypotheses, this was not too difficult. The geographic distribution of the sickle cell gene showed striking similarity with the distribution of falciparum malaria, and the *Plasmodium* was known to specifically attack red blood cells.

In general, formulation of such a causal hypothesis requires knowledge of the physiologic function of the gene concerned. Once a reasonable hypothesis on a mechanism is available, the task of testing for selection is much simpler. It has even been asserted that no case of balanced polymorphism in any species has ever been discovered without knowledge, or at least a plausible hypothesis, of the biologic mechanisms by which selection works (B. Clarke, personal communication).

6.2.1.7 *Natural Selection and Population History: HbE and β-Thalassemia*[2]

It has often been discussed how genetic data, for example, gene frequencies of genetic polymorphisms can be utilized to derive conclusions as to population history and population affinities. The following, well-analyzed example demonstrates how various techniques of population genetics and data from history and linguistics may be combined to answer such questions. This example will be used also to examine the problem of two different alleles under selection and to demonstrate the usefulness of computer simulation of population processes for population genetic analysis.

Interaction of Two Abnormal Hemoglobin Genes in a Population. In one of his predictions, Allison concluded that in regions where two genes for abnormal hemoglobins coexist and interact in such a way that individuals possessing both genes are at a disadvantage these genes will tend to be mutually exclusive in populations [1718]. This problem will be illustrated by the interaction of HbE and thalassemia genes in southeastern Asia [1760]. The homozygous state of thalassemia has been described in Sect. 4.3. HbE disease is a hemolytic anemia that is much milder than that found in HbS homozygotes. β-Thalassemia major (Cooley's anemia) is associated with severe hemolysis and reduced hemoglobin synthesis. Most compound heterozygotes for both β-thalassemia and HbE (thalassemia-HbE disease) suffer from a marked chronic anemia approaching the severity of β-thalassemia major. The genes for the Hbβ-chains variants such as HbE and the β-thalassemias are so closely linked (Sect. 4.3.4) in trans-position that they can be treated as alleles.

[2] This section is not absolutely necessary for understanding further sections.

Distribution of HbE and Thalassemia. The distribution of HbE in southeastern Asia has a center of maximal frequency in the Khmer-speaking population of northern Cambodia and the adjacent areas of northeastern Thailand; here, the gene frequency may reach 0.3, corresponding to a heterozygote frequency of 42%, one of the highest hemoglobinopathy frequencies ever attained. In other parts of Thailand, on the Malaysian peninsula, and in Indonesia, the frequency is much lower. HbE has also been found in China, Assam, and in Bengal (Fig. 6.20). The total number of carriers of this gene may be around 20 million, β-thalassemia alleles occur in the same areas; they are, however, much more widespread.

HbE and Malaria. Once the relationship between HbS and falciparum malaria had been worked out, a similar mechanism for maintenance of other hemoglobin polymorphisms was considered plausible. Attempts to test this hypothesis directly met with difficulties, for example, the problems of field study in populations not benefitting from public health and medical supervision, as well as the presence of other protective genetic mechanisms such as G6PD deficiency and thalassemia in the same populations. A protective effect of the HbE allele in heterozygotes and homozygotes was, however, strongly inferred. Comparisons between the geographical distributions of HbE and malaria had to take into account that in mainland southeastern Asia the main vector is *Anopheles minimus,* a forest mosquito that is abundant in hilly and mountainous areas. This fact causes the distribution of malaria to be opposite that of Mediterranean countries, where swamp and brackish water mosquitos are most important. In southeastern Asia, malaria is a disease of the hills and forests. Indeed, it is in these areas that frequency of HbE tends to be highest [1762].

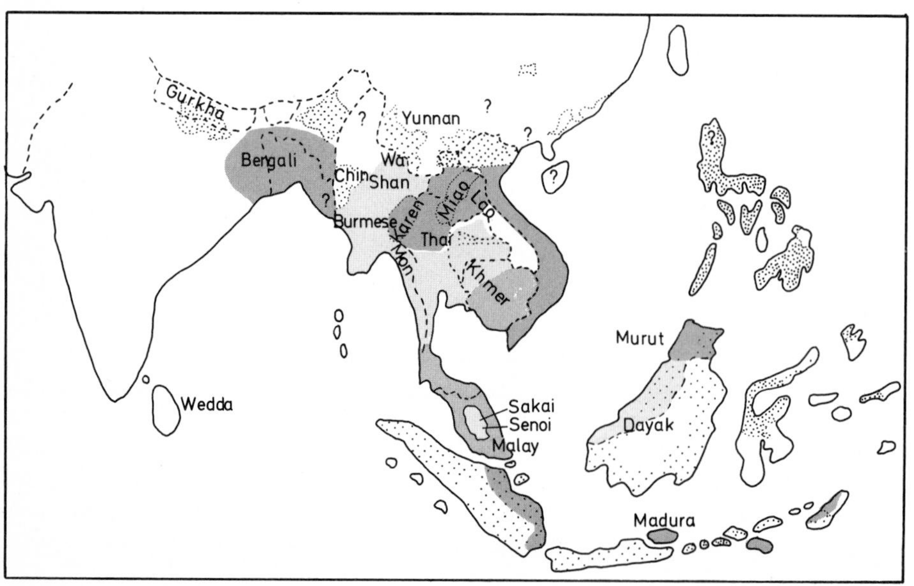

Fig. 6.20. Populations in which HbβE has been observed. (Flatz, 1967 [1760] *Shaded areas,* high frequency; *lightly shaded areas,* moderate frequency; *stippled areas,* occasional observations

Fitness of the Genotypes Involving HbE and Thalassemia: Problem of a Genetic Equilibrium. What are the conditions for change in gene frequencies and for genetic equilibria in such a system of three alleles (HbβA, HbβE, HbβT)? To answer this question, the fitness values (i.e., the selective advantages or disadvantages) of the various genotypes had to be estimated. On the basis of gene frequencies in the nuclear Khmer group and the clinical manifestations in HbE homocygotes, the following values were estimated:

HbβE/HbβE:	$w_{EE}=0.7$ to 0.8	$s_{EE}=0.2$ to 0.3
HbβE/HbβA:	$w_{AE}=1.05$ to 1.2	$s_{AE}=-0.2$ to -0.05
HbβA/HbβA:	$w_{AA}=0.9$ to 0.95	$s_{AA}=$ 0.05 to 0.1
HbβT/HbβT:	$w_{TT}=0$	$s_{TT}=$ 1
HbβT/HbβA:	$w_{AT}=1.05$ to 1.2	$s_{AT}=-0.2$ to -0.05
HbβT/HbβE:	$w_{ET}=0.2$ to 0.5	$s_{ET}=0.5$ to 0.8

where w is the fitness of a certain genotype compared with average fitness in the population and s is the selective disadvantage (definition: $s=1-w$). Thus, in the presence of a selective advantage, as in s_{AE}, the s value will be negative. For simplification, the letters A, E, and T have been used to refer to the HbβA, HbβE, and Hbβ thalassemia genes.

Is a stable genetic equilibrium under these conditions possible? Contrary to the two-allele system described in Eq.(6.6), selective advantage of heterozygotes in a triallelic system does not necessarily lead to a stable genetic equilibrium. The requirements for a stable equilibrium must meet the following four conditions:

1) $s_{AA}>0,\ s_{EE}>0,\ s_{TT}>0$
2) $s_{AA}\,s_{EE}-s^2_{AE}>0$
3) $s_{AA}\,s_{TT}-s^2_{AT}>0$
4) $s_{EE}\,s_{TT}-s^2_{ET}>0$

The derivation of these formulas, which cannot be given here, was developed by Penrose et al. [1848].

On the basis of the fitness data for the different genotypes, the following conclusions can be drawn: requirement 1, selective disadvantage of the homozygotes, is met; requirements 2 and 3 may be met with by some reasonable combination of data; requirement 4, on the other hand, i.e. that the product of selective values of the two homozygotes is greater than the squared selective value of the compound heterozygote will be fulfilled only in unlikely combinations of the selection coefficients. The selective disadvantage of the compound heterozygote as compared with the homozygote E/E is very probably too severe to permit the establishment of an equilibrium, let alone a stable equilibrium.

This conclusion is largely independent of heterozygote advantages that may vary depending on malaria exposure and is based mainly on the inherited disease states associated with the genotypes HbβE/E, HbβT/T, and HbβE/T. The fitness of these two homozygote states and that of the compound heterozygote, however, is influenced very little by environmental factors.

What do we expect of the distribution of HbβE and HbβT gene frequencies, q_E and q_T, in various population groups when there is a stable or semistable equilibrium, as compared with an unstable equilibrium or no equilibrium? A stable equilibrium will result in a clustering of the distribution points, which represents the value pairs (p_E, q_T) in a two dimensional co-ordinate system around a certain equilibrium point. If the equilibrium is only semistable, the clustering effect is lessened: After a disturbance of the equilibrium, the distribution points do not necessarily return to the same equilibrium point as before but – as Penrose et al. (1956) [1848] have shown – to some point lying on the straight line connecting the "unopposed" equilibrium points of HbβE and HbβT (where only one of the alleles exists).

The distribution actually found [1760] suggests that no stable or semistable equilibrium obtains. This means that the two alleles HbβE and HbβT have a tendency to reduce one another below equilibrium frequency. This reciprocal effect is caused by the strong selective disadvantage of the compound heterozygote.

Population Dynamics of HbβE and HbβT. If the population is not at equilibrium, how quickly and in which direction are the gene frequencies expected to change? Or – if we look at the problem from the point of view of population history – how was the present distribution of gene frequencies attained?

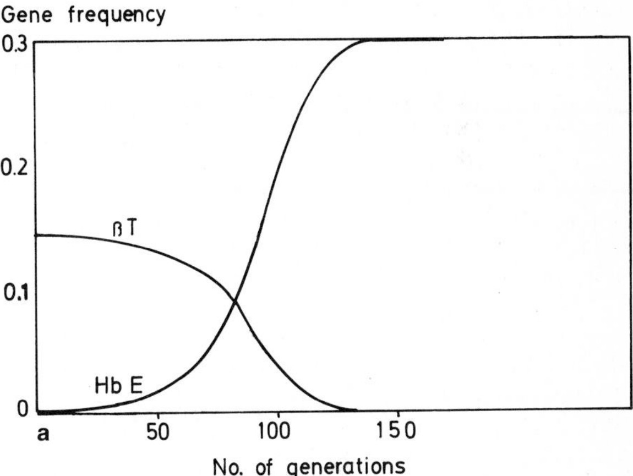

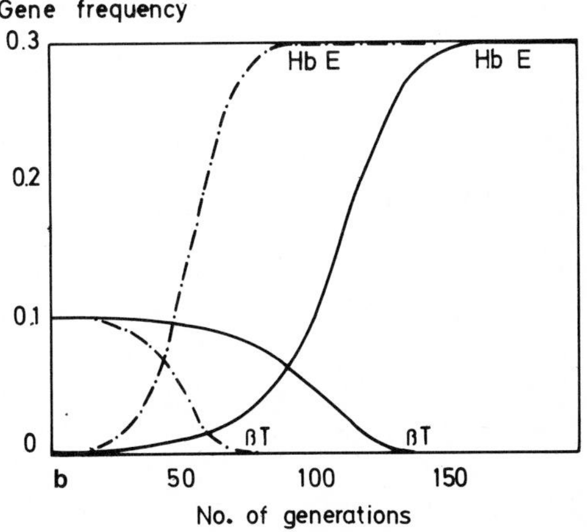

Fig. 6.21. a Elimination of HbβT by HbβE; high fitness of HbβA/E and HbβA/T heterozygotes; simulation of conditions with high selective pressure. Assumed fitness values: A/A 1.0; A/E 1.225; A/T 1.2; E/E 0.7; E/T 0.25; T/T 0. Note that HbE replaces thalassemia. **b** Elimination of HbβT by HbβE; HbβA/T fitness lower than **a.** Fitness values: A/A 1.0; A/E 1.15; A/T 1.125; E/E 0.8; E/T 0.25; T/T 0.0 -·-·-·- A/A 1.0; A/E 1.225; A/T 1.125; E/E 0.7; E/T 0.25; T/T 0.0. (Flatz, 1967 [1760])

In the foregoing sections, formulas for change in gene frequency from one generation to the next (Δq) were derived for special cases. In a similar manner, it is possible to derive relevant equations to deal with the change in gene frequencies between generations for a three-allele situation. In this way, the speed of change in gene frequencies under different selection pressures can be examined. In precomputer times, the amount of numeric calculation would have been prohibitive; now, it is an easy problem.

Figures 6.21-6.24 show some possibilities. In Fig. 6.21, for example, gene E is introduced into a population with high T frequency. Both A/E and A/T heterozygotes have a high selective advantage; gene E replaces T. If the selection advantage of the A/E heterozygotes is lower, there exists a critical value of this advantage below which the gene E can no more replace T. And even if gene E is able to replace T, the speed of elimination depends strongly on the initial frequency of E. Under similar fitness conditions but with high initial frequency of E, gene T fails to replace E (Fig. 6.22). If the fitness of A/E heterozygotes is very much lower than that of A/T heterozygotes, T may be able to replace E (Fig. 6.23). If both genes E and T start with low frequencies (i.e., both alleles are introduced anew in the population), E has a good chance to establish itself at the expense of T, even if only A/T heterozygotes and not A/E heterozygotes have a selective advantage over A/A (Fig. 6.24).

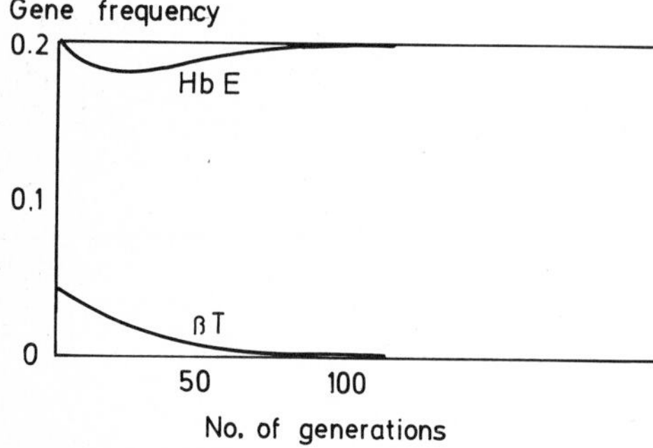

Fig. 6.22. Failure of HbβT under fitness conditions similar to those of Figure 6.19 but with high initial HbβE frequency. A/A 1.0; A/E 1.1; A/T 1.125; E/E 0.7; E/T 0.4; T/T 0.0 (Flatz, 1967 [1760])

Selection Relaxation. Malaria may be eradicated in Southeastern Asia in the future. Hence, the selective advantage of A/T and A/E heterozygotes no longer exists, but the disadvantages of E/E and T/T homozygotes as well as of E/T heterozygotes still prevail. What will be the consequences for the frequencies of the E and T genes?

Figure 6.25 shows two situations, the first with very high E and very low T frequencies (as in northeastern Thailand), the second with more similar E and T frequencies (as in central Thailand). In both cases, the decline of both genes will be fairly rapid, especially at the beginning. For gene E, this decline reaches an order of magnitude that can be ascertained if population samples of reasonable size from two different populations are compared.

Implications of These Results for the Population History of Southeastern Asia. Present-day gene frequencies are the result not only of selection pressures but also of the demographic history of populations. More specifically, if an ecologic situation has led to different ways of adaptation in two different populations - for example, adaptation to malaria in one population by HbβE, the other by thalassemia - and if the two mechanisms are to a certain degree mutually exclusive, a comparison of gene frequencies with the known facts of populations history may give some clues as to the genetic relationship of these populations.

In southeastern Asia, there is little evidence of a neolithic culture comparable to that found in Europe. Three stages of social development and ecologic situations can be discerned:

1. Hunter and gatherer groups. Judging from similar groups that exist today, the habitat of these people was in forested areas in the hills and mountains. If it is assumed that the distribution of malaria was similar to current conditions, these hunter and gatherer populations must have been exposed to intense malaria pressure. Nevertheless, conditions for diffusion of a gene carrying heterozygote advantage were unfavorable because of the small size of the breeding population and the few possibilities for gene diffusion among them; the ultimate fate of even favorable mutations would in most cases be extinction (Sect. 6.4).
2. At about 1,000 B.C., with the introduction of rice cultivation in irrigated fields, a social organization at the vil-

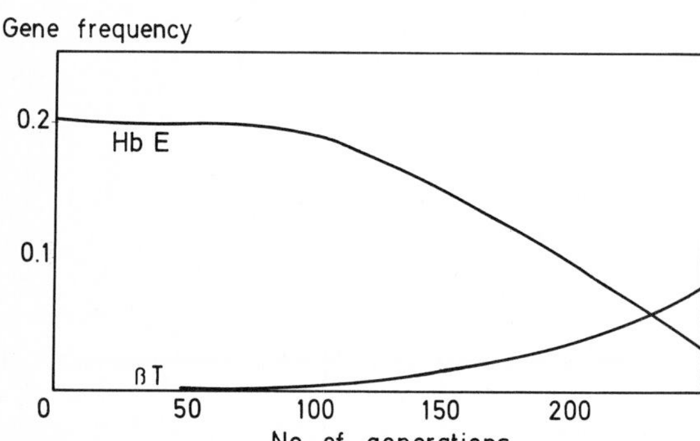

Fig. 6.23. Elimination of HbβE by HbβT; HbβE/E fitness high, HbβA/E fitness low. Fitness values: A/A 1.0; A/E 1.033; A/T 1.125; E/E 0.0; E/T 0.25; T/T 0 (Flatz, 1967 [1760])

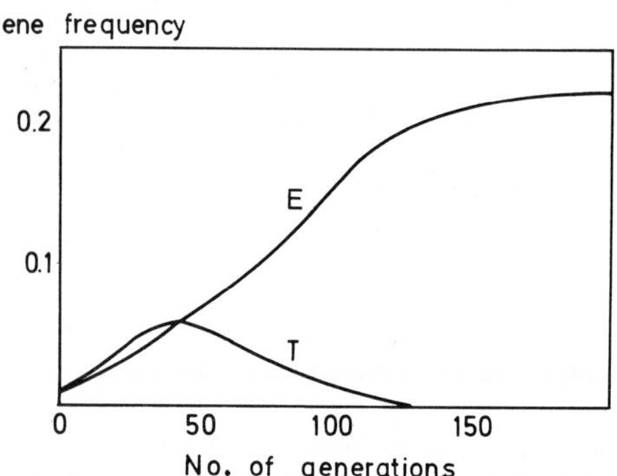

Fig. 6.24. Both HbβE and HbβT start with low frequency; rise of HbβE to equilibrium frequency, initial rise and subsequent elimination of HbβT. A/A 1.0; A/E 1.0; A/T 1.125; E/E 0.75; E/T 0.25; T/T 0 (Flatz, 1967 [1760])

lage and district level appeared. Most of the known settlements of this area were located at the margin of valleys. In such a society, conditions for diffusion of a gene maintained by protection against malaria were most favorable. The time available for the HbβT and HbβE genes to become frequent (about 3,000 years or 120 generations assuming a generation time of about 25 years) was sufficient for attaining the present-day frequencies as evidenced by Fig. 6.24.

3. At present, the majority of the population in most southeast Asian countries resides in the great river basins and deltas, which were mostly uninhabitable in prehistoric times; social and political development during historical times permitted organized cultivation in the lowlands and let to a continued migration of people into the plains. Due to the special ecologic requirements of the vector, *Anopheles minimus*, malaria is rare in the plains. Hence, the move to the relatively malaria-free plains is believed to have caused a considerable relaxation of selection against the HbβA homozygote and a diminishing advantage of the HbβE and HbβT heterozygotes. The two genes HbβT and HbβE are, indeed, less frequent in the plains than in the adjacent hill areas.

Comparison with HbβS in Western Africa. In western Africa, the principal malaria vectors are mosquitoes requiring open spaces and stagnant water for their propagation. The sickle cell gene was probably introduced to western Africa in the neolithic period concomitantly with improved agricultural methods (slash and burn farming). This development opened wide spaces for the malaria mosquitoes, leading to high endemicity, which, in turn, set the stage for the spread of the sickle cell gene and establishment of its polymorphism.

Similar trends with the introduction of agriculture in two different populations in the presence of different ecologic requirements of the mosquitoes led to the establishment of the HbβS polymorphism in the plains in Africa and of the HbβE polymorphism in the hills of southeastern Asia.

Hemoglobin βE in the Austroasiatic (Mon-Khmer) Language Group. The Austroasiatic language group now comprises the Khmer (Cambodia), tribal languages in Vietnam, Mon in Lower Burma, western and norther Thailand, tribal languages in Thailand, Burma, and southern China, and several languages in Assam and Bengal. Historical and linguistic evidence suggests that the entire area of mainland southeastern Asia, with the exception of southern Malaya and parts of Vietnam, was inhabited by Austroasiatic people up to the fifth or sixth century A. D. when large-scale migration began.

In Fig. 6.26, the areas of Austroasiatic languages, past and present, are compared with the areas in which the HbβE

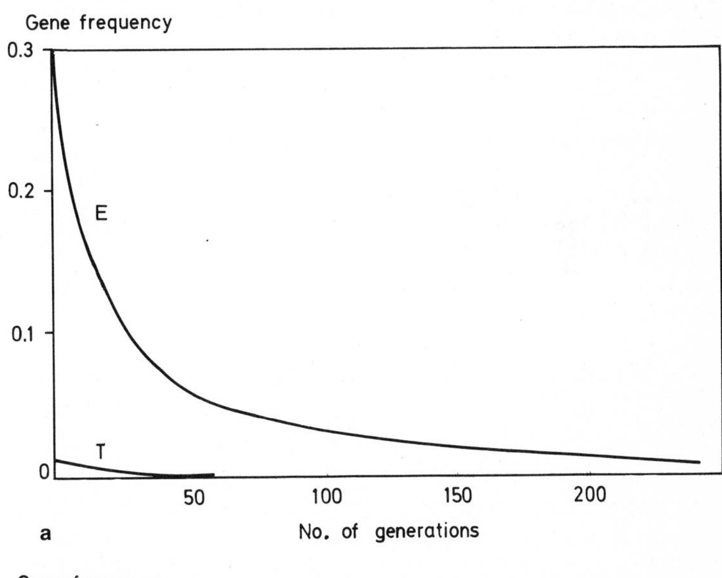

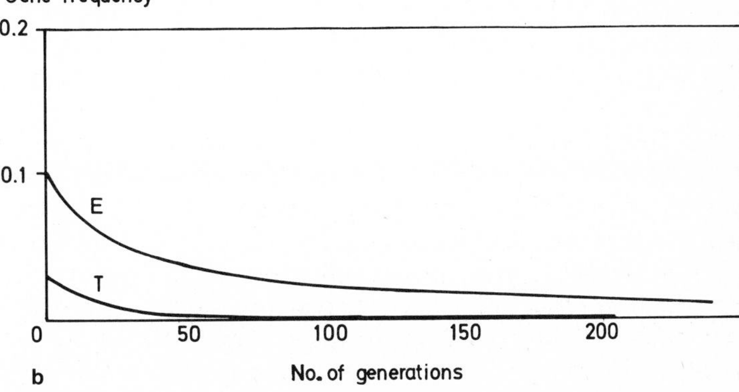

Fig. 6.25. **a** Simulation of conditions in northeastern Thailand with assumed selection relaxation. A/A = A/E = A/T = 1.0; E/E 0.7; E/T 0.25; T/T 0. **b** Selection relaxation under conditions found in river basin areas (e. g., central Thailand). A/A 1.0; A/E 1.0; A/T = 1.0; E/E 0.7; E/T 0.25; T/T 0

gene is polymorphic. The congruence is evident; the most likely explanation is a concomitant diffusion process: HbβE may have emerged in an original Austroasiatic group, and both HbβE and Austroasiatic language and culture may gradually have diffused throughout mainland southeastern Asia. The dynamic models (Fig. 6.21) indicate that this diffusion could have occurred in a population with preexisting high thalassemia frequency, as the HbβE gene will replace thalassemia genes under many selective conditions.

Prediction Confirmed. As noted, other Austroasiatic groups are living outside the southeast Asian mainland, for example in Assam. If the hypothesis of the history of this gene is correct, and if malaria is, indeed, the ecologic factor keeping the gene in high frequency, populations of the Austroasiatic language group outside southeastern Asia who lived for a long time in a malaria-infested area should have high HbβE frequencies.

Two population groups were compared in Assam; The Khasi, an Austroasiatic group that has maintained its ethnic identity, and the Ahom, a group of Mongoloid origin which immigrated from Thailand more recently - in the thirteenth century. Both populations live in areas in which malaria has been endemic. Table 6.15 shows the results. The frequencies of HbβE are very high in both populations. The prediction derived from the hypothesis that HbβE should be common in a population originating from the Austroasiatic group and subject to malaria selection was confirmed for the Khasi. Occasionally, however, the HbβE gene may be introduced into a neighboring population of non-Austroasiatic origin. If favored by selection, it may even attain a relatively high frequency. This was observed in the Ahom [1752].

When we discussed research strategies in human population genetics, it was suggested that such studies were most meaningful if guided by specific hypotheses. Investigations on HbE in Austroasiatic population groups of Assam are examples based on such hypotheses of population studies.

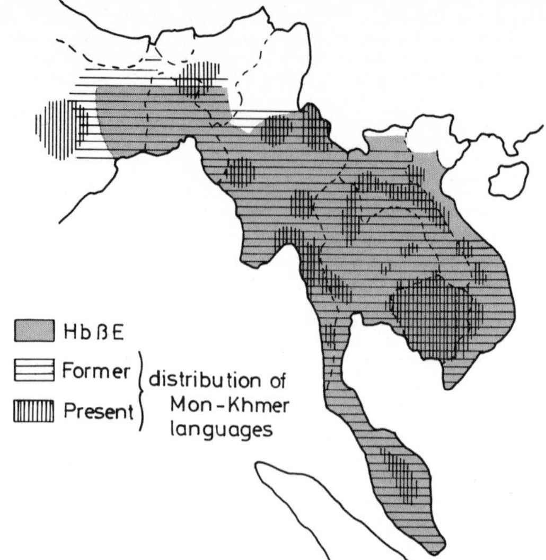

Fig. 6.26. Polymorphic distribution of HbβE in mainland southeastern Asia ▭; former ▭ and present ▭ distribution of Austroasiatic (Mon-Khmer) languages. (Flatz, 1967 [1760])

Table 6.15. Hemoglobin E in samples of the Khasi (Bhoi) and the Ahom population of Assam. (Das et al., 1971 [1752])

Group	N	Hemoglobin type			Percent HbβE carriers AE, E	Frequency HbβE
		A	AE	E		
Khasi	80	47	31	2	41.3	0.2188
Ahom	82	34	37	11	58.5	0.3598

Difference of HbβE genes $\chi^2 = 7.82$ $P < 0.01$.

Some General Conclusions from the Studies on HbE and Thalassemia. The studies on HbE and thalassemia in southeastern Asia lead to more general conclusions in the interpretation of population differences in gene frequencies. They show how these differences in gene frequencies may be determined by both population history and natural section.

Comparing populations exposed to the same relevant ecologic agent - in this case malaria - we find a definite genetic difference that is caused by different population histories. In one population, adaptation to the agent was achieved by the HbβE gene, in the others by β-thalassemia genes; both adaptations were shown to be mutually exclusive up to a certain degree. Since Hb E/E homozygotes are less severely affected than the Hb T/T homozygotes, adaptation through the HbβE gene offers antimalaria protection at a lower price and will tend to replace adaptation afforded by the β-thalassemia gene. These results will be obtained in the long run in spite of a counteracting force, the partial mutual exclusiveness of the two genes that is caused by strong selection against the compound heterozygote.

Comparing populations that had been exposed to malaria in different degrees, we also find definite genetic differences. In populations of the hilly areas, HbβE and thalassemia tend to be more frequent than in the great plains, where there has been less malaria. This is expected but provides little information about the ethnic relationship between populations of hills and plains. Resolution of this ambiguity required knowledge of the selective agent - in this case, malaria.

The literature in anthropology and human genetics abounds with comparisons between different population groups using certain genetic markers. However, critical analyses of such differences in terms of selection, on the one hand, and population history, on the other, are usually lacking.

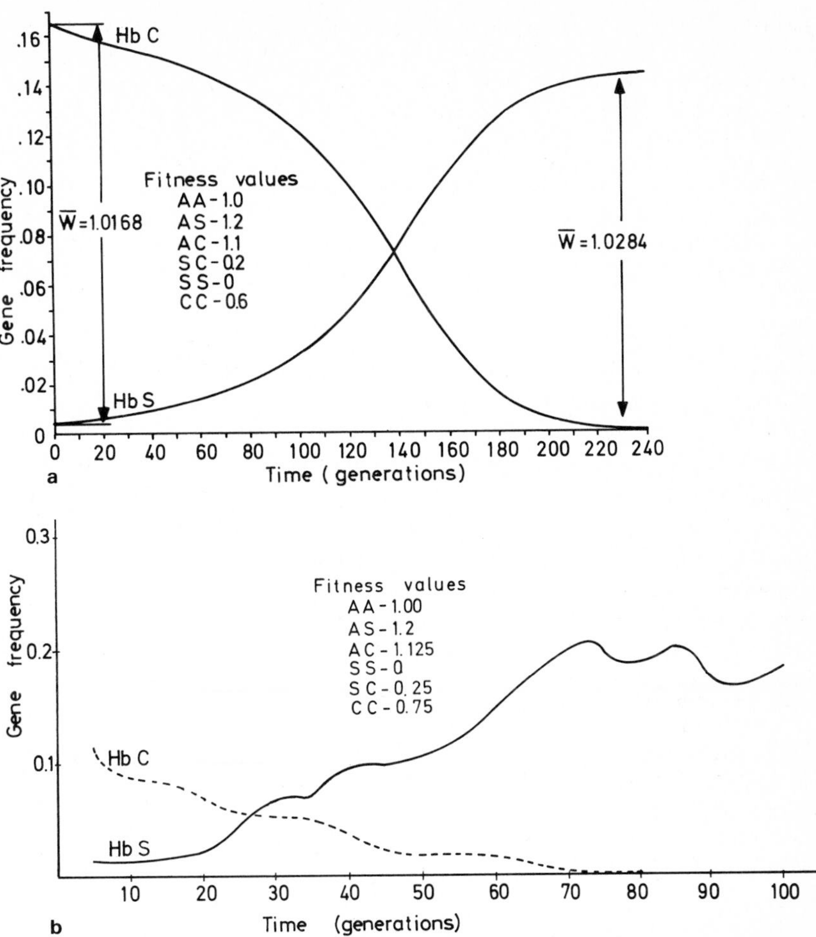

Fig. 6.27. a Replacement of HbβC by HbβS due to higher selective advantage of the heterozygote HbβA/S as compared with the heterozygote HbβA/C. **b** Computer simulation of the model shown in Fig. 6.25a, but assuming an effectively breeding population of N=1,000 and allowing for chance fluctuation of gene frequencies. (Note slight differences in fitness of AC, SC and CC as compared with 6.25a). (Adapted from Livingstone, 1967 [1815; 126])

Similar Studies on Sickle Cell Polymorphism in Africa: a Stochastic Model for Replacement of One Allele by Another [126]. In similar studies, the population history of western Africa has been analyzed together with exposure to malaria and the frequencies of the genes for HbβS and HbβC. The situation is similar to that encountered for HbβE and thalassemia in southeastern Asia: there are two alleles, HbβS and *HbβC,* offering protection against malaria – with different fitness values w_i of homozygotes and heterozygotes and with strong selection against the double heterozygote. Figure 6.27a shows changes of gene frequencies concomitant with replacement of allele HbβC by HbβS, the decisive factor here being the higher selective advantage of the heterozygote for HbβS as compared with the heterozygote for HbβC. The selection model is deterministic, like the model used for HbβE and thalassemia. Population size is assumed to be infinite. In Fiq. 6.27b, on the other hand, the size of the effectively breeding population (Sect. 6.4.1) was assumed to be 1,000, and the resulting chance fluctuations were allowed for. This model is stochastic. The general tendency is the same as in Fig. 6.27a; however, chance fluctuations are obvious.

In western Africa, as in southeastern Asia, research on abnormal hemoglobins has contributed to our knowledge of population history. There are other areas in the world in which such studies could help in a similar way – in which population history is complicated and further studies of G6PD and hemoglobin variants are required. One such area would be India, especially in the south and east.

Several types of G6PD deficiency co-exist in polymorphic frequencies in the Philippines and in Thailand, unlike the presence of only two principal G6PD variants (A^- and A^+) in African populations. Presumably, population mixture of groups who originally only carried a single G6PD mutation brought about this situation. More studies on the relevant variants and on population history in these parts of Asia are required for a detailed analysis of this problem.

6.2.1.8 Selection in the AB0 Blood Group Systems and in Other Polymorphisms

AB0 Blood Groups and Disease. Apart from the interaction between hemoglobin and G6PD variants and malaria, no other human example could be analyzed as thoroughly. However, it may be useful to discuss some difficulties in arriving at a clear picture in another, much more complicated and controversial example: the AB0 blood groups. As noted above, one aspect of selection in this system is generally accepted, though there is no agreement as to the extent of selection: serologic mother-child incompatibility. However, such incompatibility leads to an unstable equilibrium and to slow changes in gene frequencies (Sect. 6.2.1.4). In the absence of other modes of selection, the polymorphism would slowly disappear. Contrary to this prediction, the AB0 polymorphism is present in almost all human populations. This finding suggests other selective factors. Do we have positive evidence for such selection?

AB0 Blood Groups and Infections Disease. Widespread disease associations have been reported for the AB0 blood (Sect. 3.7.2). For example, carriers of type A are more susceptible to a number of malignant tumors and some other diseases, whereas those with type 0 show a higher susceptibility to gastric and duodenal ulcers. Moreover, rheumatic fever for which immune mechanisms are undisputed [211] also shows a blood group association; the risk of being affected is lower for group 0 than for A, B, and AB. While possibly leading to a higher average chance of group 0 carriers to survive to a more advanced age, these associations presumably had little if any influence on natural selection, as most of them affect individuals of middle and older age, i.e., after reproduction. However, they show a fundamental influence of the AB0 antigens on body physiology. More specifically, the data on rheumatic fever suggest that this influence may have something to do with the immune response [1789]. Even the associations of A with cancer and of 0 with peptic ulcers may be due to differences in immunologic response.

Infectious diseases specifically challenge the immune response of the organism. If this response is influenced by the AB0 blood group, selection by differential susceptibility to infections could cause differential mortality in childhood and youth.

Distribution of AB0 Alleles in the World Population. Figures 6.28–6.30 show the distribution of the alleles A, B, and 0 [144]. This distribution suggests an influence of natural selection. Had it been caused by chance fluctuation of gene frequencies, all possible combinations of gene frequencies, all possible combinations of gene frequencies for the three alleles should have appeared. This, however, is not the case. Only a limited number of the possible combinations are observed [1732].

Certain hints regarding the kind of selection can be derived from the distribution of allele 0. This allele is usually frequent in populations that have been living for a long time in relative isolation, such as aborigines of Australia and Polynesia, the Arctic, and northern Siberia. Also within Europe, certain isolated population groups usually have high frequencies, e.g., Irish, Basques, Icelanders, inhabitants of Corsica and Sardinia, and the Walsers of Switzerland. An especially high frequency of allele 0 is found among the Indians of Central and South America setting them apart from other populations. The differences in other polymorphisms, for example Rh, argue against the hypothesis that all these areas were once inhabited by a homogeneous population with high 0 frequency. The data are suggestive of natural selection. What kind of selection might have led to an increase in gene frequencies in relatively isolated areas – or, conversely, to a decrease in areas in the mainstream of world "traffic"? A group of plausible candidates are infectious diseases, especially the great epidemics of the past.

In Sect. 6.2.1.6 the following groups of infectious diseases have been mentioned as possibly important for natural selection:

a) Epidemics such as plague, cholera, and smallpox.
b) Chronic infections, for example, tuberculosis and syphilis.
c) Intestinal infections mainly in children.
d) Tropical diseases of children and young adults.

Unlike the polymorphic hemoglobin variants, which are confined to tropical countries, the AB0 polymorphism is found all over the world. Therefore, tropical diseases are unlikely to play a role in selection. Three aspects could be used for constructing a testable hypothesis:

1. The population of Central and South America was almost completely isolated before Columbus' discovery and may have had a special group of infections not shared by the rest of the world population, e.g., syphilis and, possibly, related treponemal diseases. Blood group 0 is extremely frequent in these populations. Is there any evidence for an advantage of group 0 toward infection with treponema pallidum, the microorganism causing syphilis?

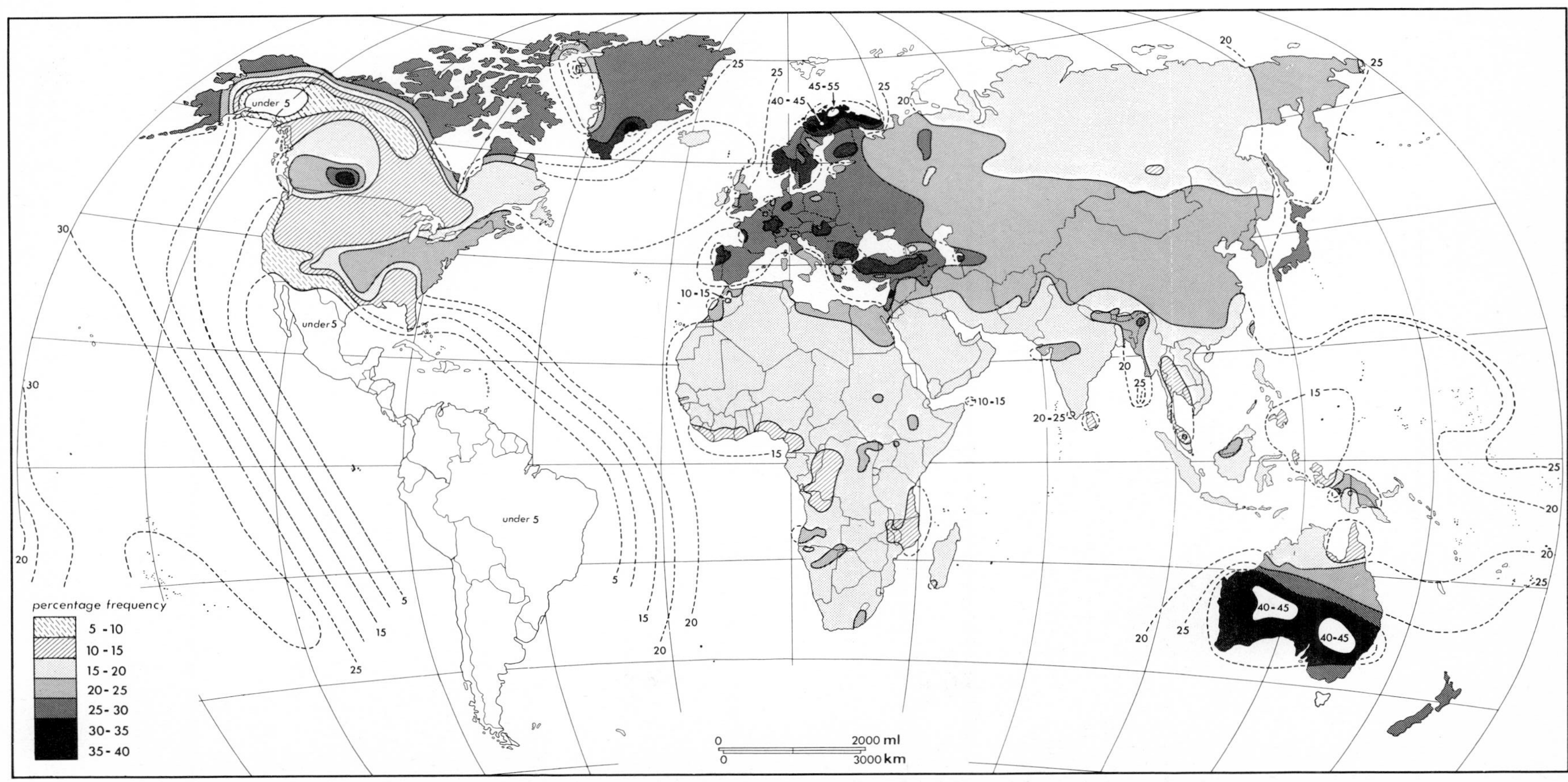

Fig. 6.28. Frequency distribution of allele A in the aboriginal populations of the world. (Mourant, 1974 [144])

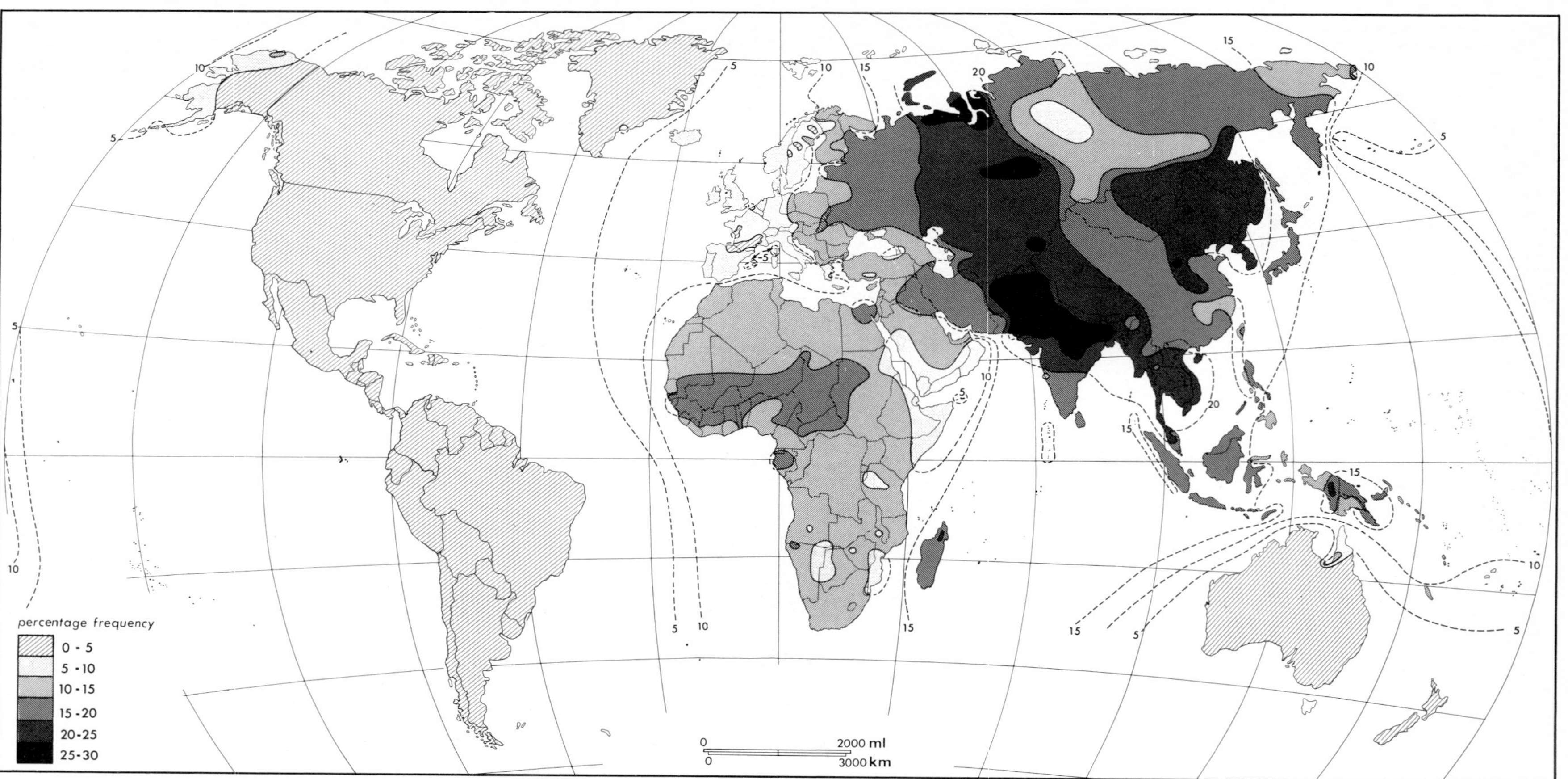

Fig. 6.29. Frequency distribution of allele B in the aboriginal populations of the world. (Mourant, 1974 [144])

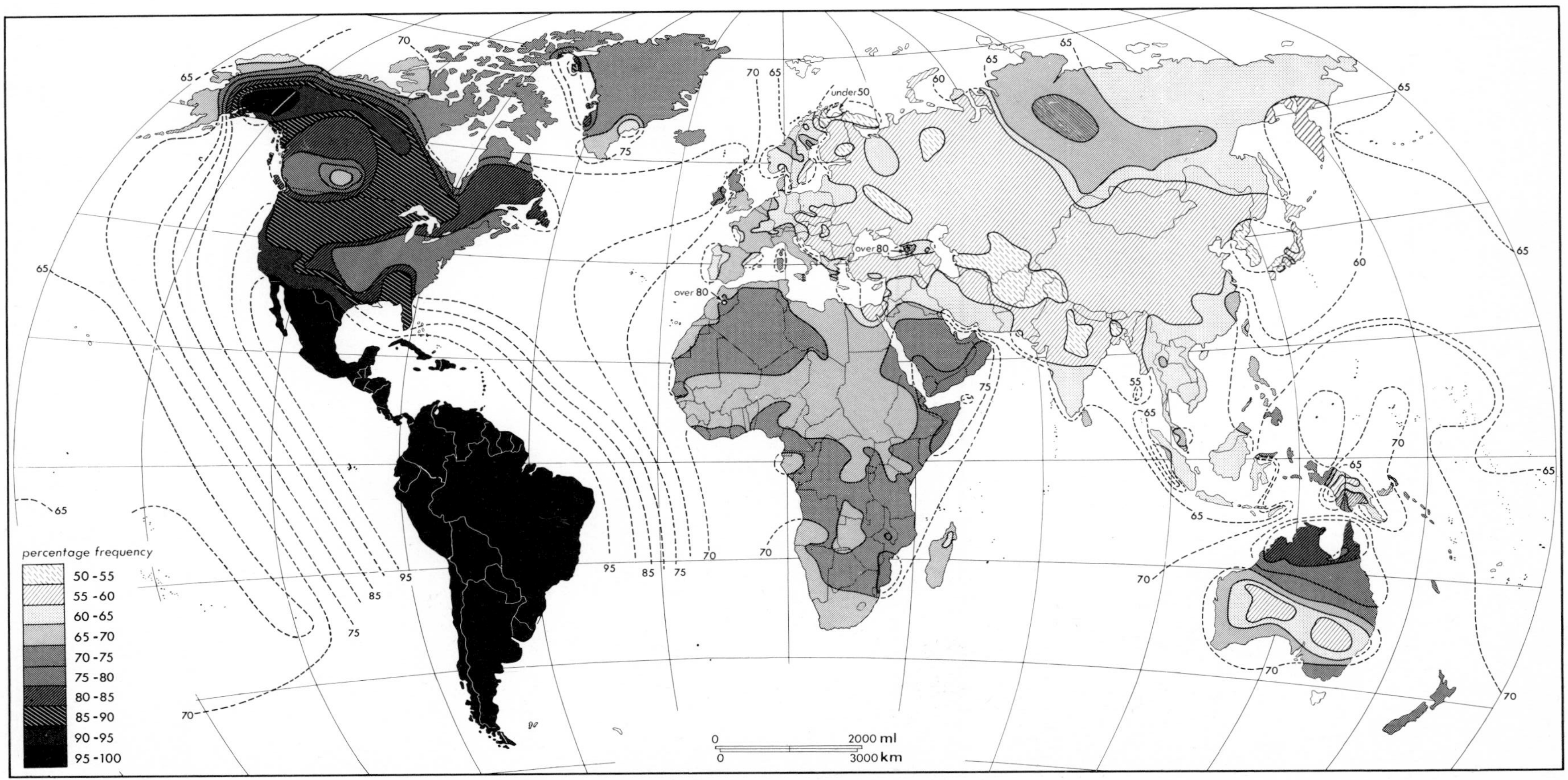

Fig. 6.30. Frequency distribution of allele 0 in the aboriginal populations of the world. (Mourant, 1974 [144])

2. Plague repeatedly devastated Europe, mainly affecting densely populated areas. Marginal and partially isolated populations that may have been affected to a lesser degree usually show high group 0 frequencies. Is there any evidence for a disadvantage of blood group 0 in coping with the plague bacillus?
3. Smallpox was not eradicated until the mid 1970s. Up to that time, it had been frequent in many countries. Therefore, modern statistics as to frequency and death rate are available, especially for Africa and the Indian subcontinent. AB0 distributions are fairly well-known for these areas. The blood group with the higher susceptibility should be rarer in areas with high smallpox rates. Is there any evidence for this prediction?

Syphilis and Blood Group 0. We are faced with the problem to judge the influence of an infection on differential biologic fitness in a population before 1492. However, just as for the example of sickle cell protection vis-à-vis malaria (Sect. 6.2.1.6), the most convincing indirect evidence would be that possession of group 0 conveys an advantage for coping with this infection. Such a hypothesis can no longer be examined, since syphilis is treated with penicillin so successfully that individual differences in outcome due to different immune responses no longer apply. In the 1920s, however, penicillin therapy was not available, and at that time comprehensive data on blood groups and syphilis were collected that could be analyzed with the following results [211]:

a) There was no association between the risk of new syphilis infection and AB0 blood groups.
b) However, after the customary therapy of that time - neosalvarsan - individuals with group 0 had a much better chance to become seronegative than did those with the other blood groups (Fig. 6.31).
c) Tertiary syphilis, such as general paralysis, is less frequent in blood group 0 than in the other AB0 blood groups (Fig. 6.32).

Hence, the combined data suggest an advantage of group 0 in the immune response to syphilis. Thus, the prediction derived from the hypothesis could be verified. If syphilis influenced reproduction, this effect resulted mainly through infection of the fetus by a syphilitic mother. Such an infection is known to lead to late fetal death frequently. Hence, the evidence allows the tentative hypothesis that the high frequency of group 0 in Central and South American Indians could be due to selection by syphilis and related treponema infections *if* these infections did come from America, *if* they were widespread in earlier times, *if* they did influence survival of children or fertility of their mothers, and *if* no other selective agents were responsible.

Cholera and Blood Group 0 [1770]. A clear association of AB0 type and a lethal endemic infectious disease has come from recent large-scale studies dealing with cholera in Bangladesh. While patients with diarrhea due to rotavirus,

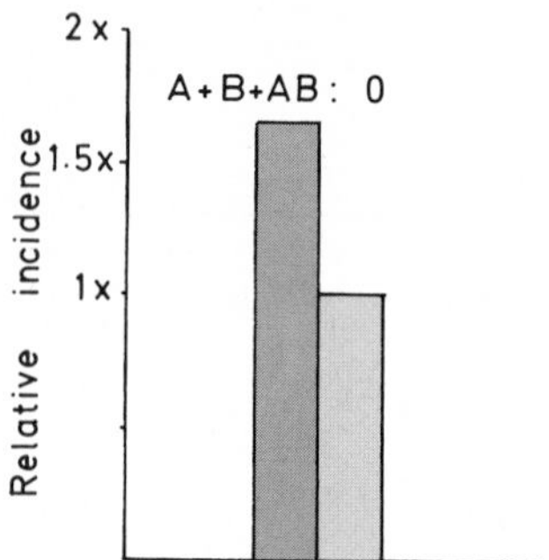

Fig. 6.32. Relative incidence of teriary syphilis in relation to the AB0 blood groups. (Vogel and Helmbold, 1972 [211])

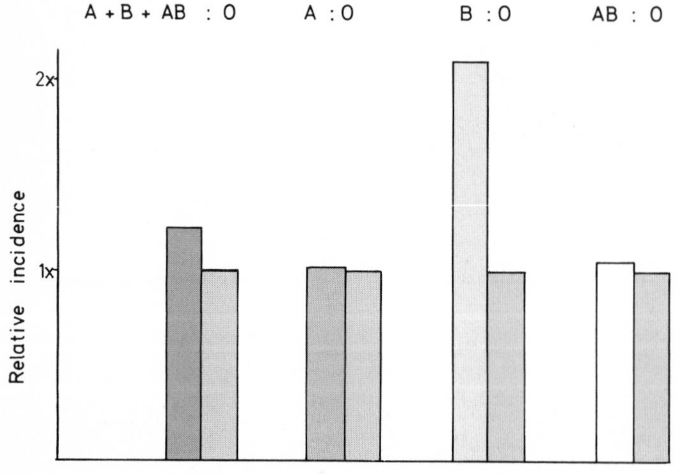

Fig. 6.31. Relative incidence of various blood groups for positive serum-Wassermann reaction after neosalvarsan treatment of syphilis. (Vogel and Helmbold, 1972 [211])

shigella, toxicogenic *E. coli,* or nontoxicogenic cholera had blood group 0 frequencies similar to controls (about 30%), patients infected with toxicogenic (0I) vibrio cholerae had a blood group 0 frequency of 57%. This difference was highly significant statistically. Among family members infected with the toxicogenic cholera strain there was a statistically significant tendency for the frequency of group 0 to increase with increasing severity of diarrhea. Severe diarrheal epidemics had been described in the past in this area, and were most probably caused by cholera. The low frequency of blood group 0 in this region might have been induced by a higher susceptibility to cholera with resulting death of blood group 0 carriers. The mechanism of the interaction remains obscure.

Plague and Blood Group 0. Was plague related to the distribution of Group 0 in Europe? Unlike for syphilis and Cholera, no blood group data on plague are available. The disease is very rare now; it only occurs in areas that are unaccessible to research workers. Therefore, indirect evidence has to be examined.

In Sect. 6.2.1.5 on frequency-dependent selection, it was mentioned that parasites may adapt to their hosts by producing surface antigens common to those of the host, thereby deceiving the host's immune response. Such evidence for vertebrates and their parasites has been collected [1739, 1751]. It has been known since the late 1950s that humans share ABH-like antigens with a great number of bacteria, especially those of the *E. coli* group. Even the "normal" anti-A and anti-B isoantibodies are thought to be immune antibodies against ubiquitous infections with intestinal germs. It was therefore of interest to investigate whether *Pasteurella pestis* had ABH-like antigens [1849, 1850]. H antigen, which is much more abundant in the human group 0 than in other blood groups was indeed detected. This finding suggests a poorer immune response against *P. pestis* in group 0 individuals and agrees with the assumption that 0 individuals may have had a selective disadvantage due to a higher death rate from plague. Needless to say, this finding does not prove the hypothesis.

Does a Common Blood Group Antigen of the Microorganism Impair the Immune Reaction of the Host? E. coli and Infectious Diarrheas. The next step for testing this hypothesis would be to examine whether communality of a blood group antigen impairs the immune response of the human host. As noted, such a study cannot be done with plague, since no plague patients are available for examination. However, this approach has become possible for another group of diseases that are caused by bacteria whose ability to synthesize ABH antigens is well-known – the *E. coli* group. In the 1950s and early 1960s, central Europe was swept by a succession of infective waves of infant diarrhea. The causative organisms were identified as *E. coli* that, by serologic examination of their antigen profiles, could be subdivided into various substrains. The outcome, unlike in earlier times, was rarely fatal because of therapy with antibiotics, plasma and fluid infusions.

An Austrian pediatrician, Kircher (1961, 1964) [1800, 1801], observed a more severe course of infant diarrhea in patients of group A as compared with those of other blood groups. This topic was restudied, drawing on comprehensive material that had been collected over many years, and could be compared with suitable controls [1889] (Sect. 3.7.2). Definite heterogeneity was found; in some years, A patients were frequently affected; in other years, 0 carriers were common (Fig. 6.33). From the case histories, a number of clinical criteria for severity of the disease could be derived: As an example, infusions given to the more severely afflicted infants (Table 6.16). In the years with higher incidence for group A, there was a more severe couse of the disease. In the years where group 0 had a higher incidence, infants of this

Table 6.16. Course of infant diarrhea in 396 patients with positive *E. coli* results. (Vogel 1964 [1889])

Variable compared	1956 and 1960–1963		1957–1959	
	A	0	A	0
Patients with plasma infusions (%)	56.6	39.4	41.9	52.3
Loss of body weight (g)	161.0	137.1	154.0	158.6
Average frequency of stools	6.10	5.42	5.95	6.0
Highest body temperature (°C)	38.40	38.18	38.51	38.60
Time in hospital (days)	26.54	26.13	23.67	28.10
Gain of body weight (g)	577.8	549.0	506.3	585.6

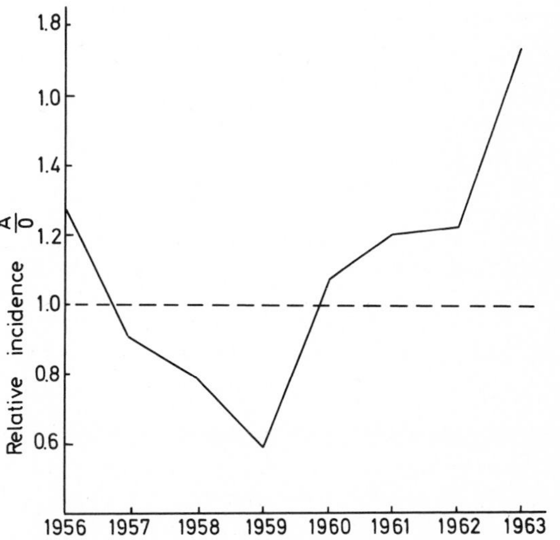

Fig. 6.33. Relative incidence of blood group A as compared with 0 among 1,200 children with infant diarrhea in Heidelberg, 1956–1963 [211]

blood group were somewhat more severely affected. This tendency was especially pronounced in that part of the data in which specific *E. coli* strains could be identified.

Studies on various serologically identifiable *E. coli* strains during these years suggested that the observable differences were probably related to corresponding variations in strains of *E. coli*.

Blood group associations with infant diarrhea have been shown in other series as well [1858, 1873], and the antibody titers against *E. coli* 0 86 were found to be higher in persons of groups A, B, and AB as compared with 0, indicating that these individuals had more serious infections [1756] (Fig. 6.34). The involved *E. coli* strain is known to have B and also A antigen.

These results make it likely that the proposed mechanism - antigen communality between parasite and host - may indeed lead to a more severe infection if the host is human and the antigen is part of the ABH system. By analogy, the H antigen of the plague bacillus could have led to more severe disease among carriers of group 0. Thus, selection against this allele would occur.

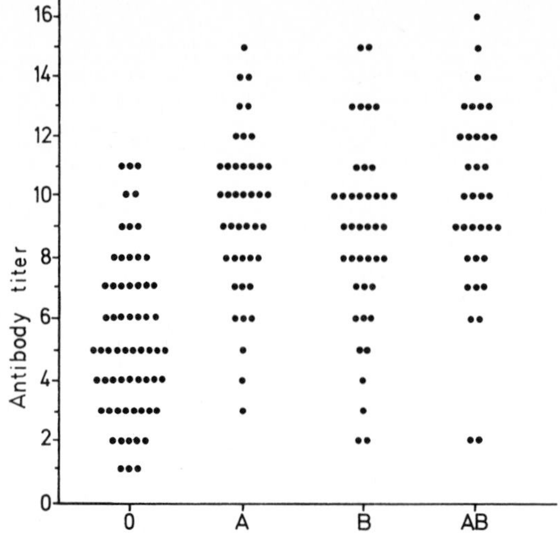

Fig. 6.34. Antibody titers in the AB0 system against *E. coli* 086 in individuals with various blood groups. (Eichner et al., 1963 [1756])

AB0 Blood Groups and Smallpox. The question of a possible blood group association with smallpox is still more controversial than the associations discussed so far but is cited here because the experimental rationale may be used as a model for future research of interaction between a virus and a human host.

After the hypothesis was elaborated that the human AB0 distribution might be related to great epidemics and that antigen communalities could be the decisive variable, vaccinia virus was examined for ABH activity. For technical reasons, these investigations were carried out not with the smallpox (variola) but with the closely related vaccinia virus. Strong A activity was found [1849, 1890]. This suggested an obvious immunologic mechanism for a disease association. During viremia with smallpox, a virus having A antigen will be partially inactivated by the anti-A antibodies present only in group B and 0 individuals but not in A and AB individuals who lack anti-A. Therefore, a more severe couse of smallpox is expected in patients with groups A and AB. As smallpox affects children and is often fatal such a blood group difference would have a strong impact on selection.

The result on A-like antigen in the virus has been challenged [1786]. The A antigen was said to be derived not from the virus but from the medium on which the virus was grown. At the time of these studies, the possibility that viruses could take up material of the host into their own capsid was still unknown. Evidence for such a mechanism has now been put forward to explain the different clinical reactions to infections with the hepatitis B virus [1818]. This virus appears to take up serum protein - especially gammaglobulin - components and to transfer them to the new host, whose immune reaction may depend at least in part on the similarity of these components with his own genetically determined proteins.

Association Studies on Smallpox Patients Gave Contradictory Results. The prediction that individuals of groups A and AB would be more frequently and severely afflicted with smallpox was tested repeatedly with conflicting results. In one study of 986 fresh and former smallpox cases, the relative incidence was shown to be much higher in persons of groups A or AB as compared with B and 0 [1888]. The same tendency was apparent when severity of clinical symptoms and mortality were considered [1888] (Fig. 6.35). Furthermore, individuals who had survived earlier smallpox epidemics showed a slight excess of groups B and 0 - an indication of the high death rates of A and AB. Among the survivors, severe scarring was, again, more frequent in groups A and AB. This study was carried out in Indian villages during a smallpox epidemic; the sibs of the affected probands who stayed healthy despite similar exposure to the infection were used as controls. Almost none of the individuals, affected or healthy, had ever been vaccinated. This research design maximizes differences in blood group distribution between patients and controls, while reducing possible errors due to population stratification as much as possible.

No study with such a design has ever again been carried out. However, two studies on hospital populations in Indian cities [1755, 1874] and one study from Brazil with a milder variety of smallpox [1804] failed to confirm this association. There ist one likely explanation for these discrepancies: The study mentioned above on smallpox in Indian villages [1888] was performed mainly in children. For the two Indian hospital studies, no information on the age distribution of the patients was published, but there is some background information suggesting that most patients might have been adults. As mentioned above (Sect. 6.2.1.6), the selective advantage of HbS heterozygotes is present only in young children; surviving adults are highly immunized irrespective of their HbS types. The same might easily have been true in smallpox: Up to the time of systematic eradication, this infection was almost endemic in large parts of India. High titres of hemagglutination-inhibiting variola antibodies were indeed found in

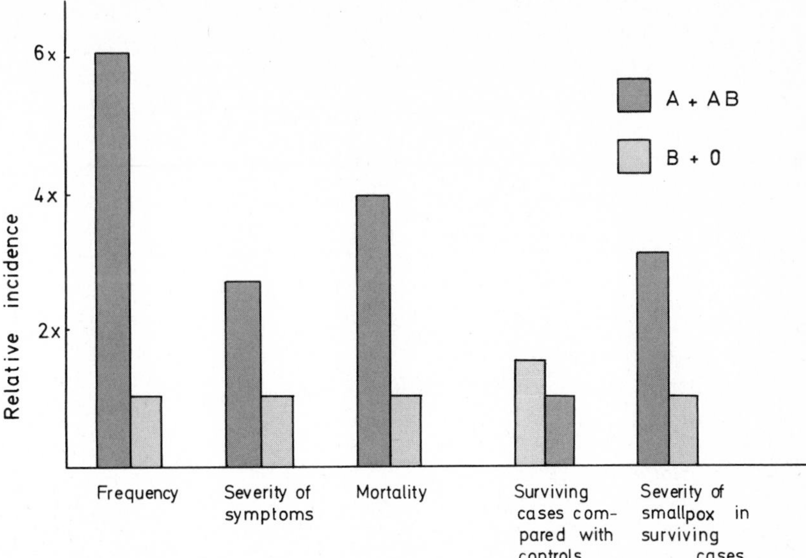

Fig. 6.35. Relative incidence of smallpox in blood group A+AB and B+0 patients. From *left* to *right:* incidence (437 fresh cases *vs* 428 unaffected control siblings); 300 cases with severe *vs* 137 cases with mild symptoms; 225 patients who died from smallpox *vs* 212 survivors; blood group distribution among 428 surviving patients in comparison with 324 controls; severe *vs* mild smallpox scars among 548 smallpox survivors [211]

many individuals who lived in these areas but had no history of clinically discernible affliction with smallpox and had never been vaccinated [1737]. However, the evidence remains contradictory, with little chance to settle the matter by direct examination, since smallpox has been almost entirely eradicated.

Blood Group A and Smallpox in the World Population. As noted, smallpox was frequent in many populations up to the early 1970s. Therefore, comprehensive population statistics are available. If smallpox has been an important selective agent against allele A, a negative correlation between A frequency and smallpox frequency or mortality must be expected. This negative correlation was found for the Indian subcontinent [1727] and for Africa [1885], when the frequency of type A in various subpopulations was compared with the occurrence of smallpox.

	Number of population groups	No. of individuals included in the calculations	Spearman Rank correlation
India and Pakistan (mortality)	18	87,153	$\rho = -0.634$, $P < 0.01$
Africa (morbidity)	27	195,313	$\rho = -0.499$, $P < 0.01$

The correlations are in the direction expected if selection of smallpox were in fact important but, of course, they do not prove it.

Distribution of AB0 Blood Group Genes in the World Population and Selection by Infectious Diseases. What aspects of the world distribution of AB0 alleles might be explained by these selective mechanisms, and which ones elude explanation? The following highly tentative conclusions could be drawn:

a) The extremely high 0 frequency in Central and South America could be due to advantage of 0 in the presence of syphilis.
b) The higher frequency of 0 in marginal European populations could be caused by lower selection against 0 carriers by plague and cholera.
c) The relatively low A frequencies in central and southern Asia could be due to selection by smallpox. In the same areas, gene 0 is also not so frequent, the gap being filled by allele B. This advantage of group B could be due to long-standing selection against type A by smallpox as well as against type 0 by plague amd cholera.

In view of the strong and long-standing selection against gene A, it is justifiable to ask why this blood group is still present in these populations. It is conceivable that another, still unknown, selective advantage of A exists. There is conflicting evidence as to associations with other viruses [211]. If many viruses are able to incorporate host-surface antigens, frequency-dependent selection by common virus diseases may be common. If frequency-dependent selection favors in general the rarer gene, this mechanism could easily be the major force maintaining the AB0 polymorphism in human populations.

One aspect of the world AB0 distribution cannot be explained satisfactorily. Why is allele B so frequent in central and southern Asia and why is it so rare in most other areas? Long-standing selection against type A as well as 0 may be part of the story. Interaction of the human host with intestinal germs – and possibly foods containing ABH-like antigens – is one of the major unknown factors in ABH selection [1843].

The experiences with HbβE in Thailand (Sect. 6.2.1.7) suggest that such frequency clines in population are difficult to interpret. They might reflect population history and gene diffusion, but they may also indicate clines in selection intensities.

Lesson of Studies on AB0 Blood Group Selection for Research on Natural Selection in Human Populations. In spite of their shortcomings, studies on selection and the AB0 blood groups have been described extensively for their possible significance:

a) It is an oversimplification to treat selection as constant over long time periods in the same population. For the main selective factor of the hemoglobin variants, *Plasmodium falciparum,* this oversimplification might hold true for many centuries or even several thousand years, since malaria remained endemic as long as the ecologic conditions for the mosquito vector did not change. Many other infections, on the other hand, come and go as epidemics. Here, selection may change even over short time periods. Sometimes cataclysmic events may result that will long be remembered in history, such as the plague epidemic of the Middle Ages. In other cases, some infections may not be recorded, such as infantile diarrheas. The variety of selective agents and their change over time is an almost all-pervading element.
b) In such a situation, genetic variability in itself may be an advantage for the species. If one epidemic kills almost all individuals carrying one genetic variant only, many of those having other variants and who therefore are less susceptible survive. The next epidemic may wipe out this variant but favor the first. This may lead to a dynamic situation in which gene frequencies oscillate over time, depending on the prevalent selective agents. Data on blood group determination on bones from the fifteenth to the seventeenth centuries were interpreted as evidence for such oscillations [1798]. In view of the difficulties of accurate AB0 blood group measurements in ancient bones and mummies due to bacterial cross reactions, these results must be considered with caution. If genetic variability in itself is an advantage, no single optimum genotype exists.
c) The fact that an unstable polymorphic situation conveys advantages for survival of the species does not account for maintenance of such a situation. Considering the small population sizes in isolated groups of earlier human history, one would expect many more populations to have become monomorphic in the course of time. To maintain a polymorphism, a stabilizing element is required. It is suggested that this stabilizing element might have been frequency-dependent selection (Sect. 6.2.1.5).

Genetic Susceptibilities and Infectious Disease. In the foregoing sections, many examples of genetic selection vis-à-vis infectious disease have been discussed. Future work in this field will most likely be successful if concerned with diseases causing maximum selection. Endemic diseases, because they act at all times, will be more effective agents than epidemic diseases, which act episodically. Diseases that affect a large portion of the population will be more effective as selective agents than those limited to restricted segments of the population. Diseases that kill children will be more effective selective agents than those compatible with survival or those largely affecting adults, particularly after their reproductive period.

Natural Selection by Infectious Agents Is Likely for the MHC Polymorphism. Apart from the AB0 blood groups, associations with infectious diseases have been claimed for a number of other genetic polymorphisms. The best a priori candidates for such associations might be the polymorphisms of the major histocompatibility complex (MHC) and especially the HLA genes. At present, information on associations of HLA types with major infectious diseases is scarce. Scientists interested in this genetic system are for the most part working in countries where the major epidemics are eradicated or have lost much of their importance for public health. However, studies on HLA and disease have shown the MHC to be a major component in the genetic variability of immune response in humans (Sect. 3.5.5). Therefore, associations with epidemic diseases and a strong influence of natural selection on gene frequencies can safely be predicted. In fact, HLA associations with leprosy [1772, 1883], typhoid [1884], malaria [1852], as well as immune responsiveness to streptococcal [1773] and tetanus [1866] antigens have been suggested. It is possible that AB0 selection will be regarded in the future as a minor factor in comparison with selection due to components of the MHC locus.

Does Genetic Liability to Atopic Diseases Lead to an Increased Resistance to Helminth Infestation [1774]? One of the main health risks for children in tropical countries is the almost ubiquitous infestation with intestinal worms; mainly tapeworms, ascaris, and hookworms. Hookworms can cause severe anemias, which in combination with other infections may contribute to premature death. Characteristic clinical signs of worm infestation are an increase in the level or eosinophilic granulocytes, and an increased IgE blood level. Such findings are also observed in atopic diseases such as asthma, hay fever, and atopic dermatitis. There is good evidence that atopic diseases have a multifactorial genetic basis, and that a gene or genes influencing IgE levels are involved (Sect. 3.6.2.7). Atopic diseases are common in present-day populations even though some manifestations of such diseases may impair health significantly. It is therefore conceivable that genotypes associated with atopic manifestations had a selective advantage in earlier times. Studies therefore were carried out to assess whether atopic genotypes in Papua New Guinea [1774] had a selective advantage in relation to helminthic infestation. In 500 villagers, skin tests for immediate hypersensitivity responses against a number of allergens were performed. Based on these tests, 10% of these villagers were identified as atopic. Moreover, a clinical diagnosis of asthma was made on the basis of clinical evidence and pulmonary function tests on all inpatients of a local district hospital; in 24 patients, asthma was diagnosed. Stool samples from these patients and also from 50 nonasthmatic villagers diagnosed as atopic and 139 non-atopic villager controls were examined for hookworm egg counts. The result is seen in Fig. 6.36. Average egg counts were found to be lowest in asthmatic patients, higher in the nonasthmatic, atopic individuals, and highest among the controls. Possible biases, such as different distribution of atopic and nonatopic individuals among different villages, were carefully excluded. The study confirms the hypothesis proposed by the authors: The concomitants of atopy, such as IgE elevation, apparently afford relative protection against hookworm infestation. There is also some limited clinical evidence for a relative protective effect of IgE levels mitigating the severity of worm infestation [1851].

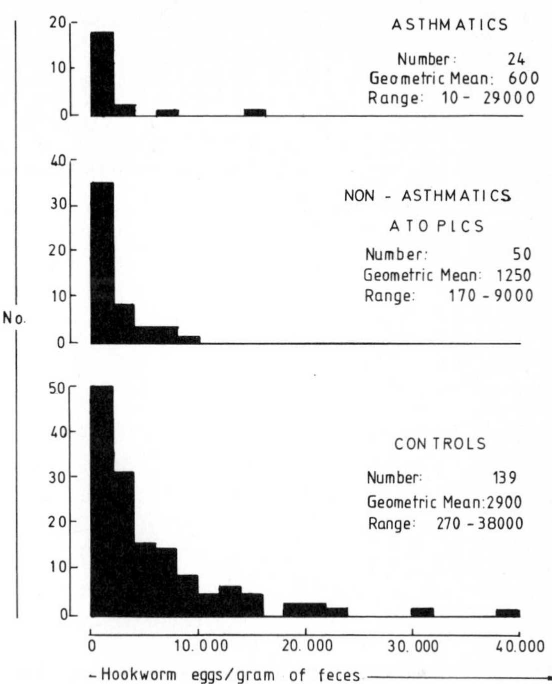

Fig. 6.36. Distribution of the numbers of hookworm eggs/g faeces in three samples of a heavily infested population in Papua/New Guinea. Note the enormous differences in infestation range between asthmatics, non-asthmatic atopics, and controls. (From Grove and Forbes [1774])

6.3 Deviations from Random Mating

In previous considerations, random mating has been presupposed, and Hardy-Weinberg proportions have been assumed to hold true. However, such assumptions are an abstraction. In modern outbreeding populations, mating may approximate randomness for some genetic traits, such as blood groups and enzyme types, but will certainly be nonrandom for some traits and certain hereditary diseases, such as congenital deafness. Because of their need for special schools and professional training, deaf individuals form social groups with intense in-group contacts but remain partially isolated from the outside world. Naturally, so-called assortative mating between deaf partners will occur frequently. If both partners carry the same type of recessive gene for deafness, all their children will be deaf. 'Assortative matings' are less conspicuous but much more common with regard to psychologically or socially significant aspects of life, such as social status, income, range of interest, education, or intelligence (Fig. 6.37). Human populations, far from mating at random, comprise a complex and ever-changing system of more or less isolated subgroups. These subgroups may be called "isolates," if they are well-delimited, mating being more or less confined to members of one group. They have been called "demes" if they only represent groups within which the probability of mating is enhanced as compared to matings with outside individuals [103]. There is no sharp line between isolates and demes.

One type of assortative mating is mating between relatives. As relatives share some ot their genes by common descent, consanguineous matings influence the incidence of some inherited diseases. Comparison of the progeny of consanguineous with those of nonconsanguineous marriages will uncover the manifestation of recessive genes by an increased frequency of specific diseases and will give evidence regarding the role of recessive genes in morbidity and mortality of many diseases where recessive genes may play a subsidiary role. These studies also provide data to assess the concept of genetic load (see Sect. 6.3.2).

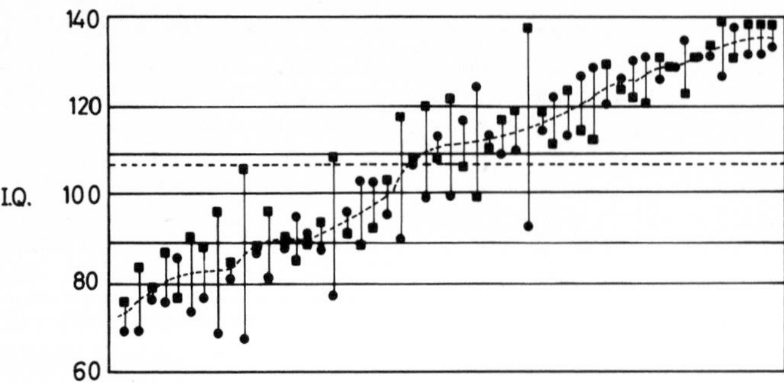

Fig. 6.37. Assortative mating as to intelligence quotient (I. Q.) in a sample of married couples in the United States. ■ husband; ● wife; *stippled line*, mean of the couples. (Outhit, 1933; Schwidetzky, Das Menschenbild der Biologie 1959)

Another aspect is the widespread tendency to prefer marriages within the same subgroup, which, in the long run, leads to genetic differences between such subgroups. Measures of "population distance" have been developed to assess such differences. Population structure and genetic composition of the population are also influenced by migration of individuals between subpopulations. Migration counteracts the effects of isolation; it is currently of increasing importance.

6.3.1 Consanguineous Matings

6.3.1.1 Inbreeding Coefficient [103]

All Human Beings Are Relatives. Relatives are defined as individuals who have a certain portion of their genes in common by descent. If we take this definition literally, all human beings are relatives. We have common ancestors. The progenitors might even have been one single couple (see Sect. 7.2.1). Then, why are our genes so different? For the simple reason that our common descent dates back thousands of generations. During this long time period, many intervening mutations have caused genetic variability. Obviously, it would be meaningless operationally to treat all of mankind as relatives, since no conclusions could be drawn from this - albeit formally correct - assumption. To the contrary, a main point of interest motivating us to measure consanguinity concerns these intervening mutations and the effect of consanguinity on their phenotypic manifestation. However, in measuring consanguinity, we should always keep in mind that it is merely a matter of practical convenience how many generations we go back.

Degrees of Relationship Normally Considered. Most frequently, lines of descent are studied for only three generations. This restriction means that parents, grandparents, and great-grandparents but no more distant ancestors are taken into account. The most remote relatives considered in describing the degree of consanguinity between two individuals are therefore second cousins. This convention was initially established for reasons of convenience. For Catholics, a special dispensation is required for a marriage between second cousins or closer relatives, and the church registers for these dispensations are an easy source of information for frequencies of consanguineous marriages in Catholic populations. Types of consanguineous marriages usually found within this range of relatives are shown in Fig. 6.38.
The convention to limit assessment of consanguinity to more narrow relationship is also theoretically reasonable. Beyond this range, the coefficient of inbreeding of an individual increases only very slowly with the number of additional consanguineous marriages in his or her ancestry.

Two Useful Measures: Coefficient of Kinship and Inbreeding Coefficient [130]; [1819]. In a population, various types of consanguineous matings may occur: those between second or first cousins, between uncle and niece, occasionally even between brothers and sisters or between fathers and daughters. It is, of course, possible to describe the frequencies of all such matings, and such data might offer interesting insights from a sociologic point of view. Only one aspect is interesting to the geneticist: How closely are the spouses related - what is their share of common genes? Whether we intend to compare individuals by the degree to which they are inbred or to describe population groups by the average degree of inbreeding of their members, we need a measure for this share. If possible, a single number would simplify our task in the same way as the notion of "gene frequency" simplifies a description of a population in terms of genotypes. Several measures for the degree of inbreeding have been proposed; the choice among them is largely arbitrary. The "inbreeding coefficient" (Wright, 1922; [1885]) has proved to be the most useful. It is closely related to the "coefficient of kinsip" (Malécot, 1948; [130]). These coefficients are defined as follows:

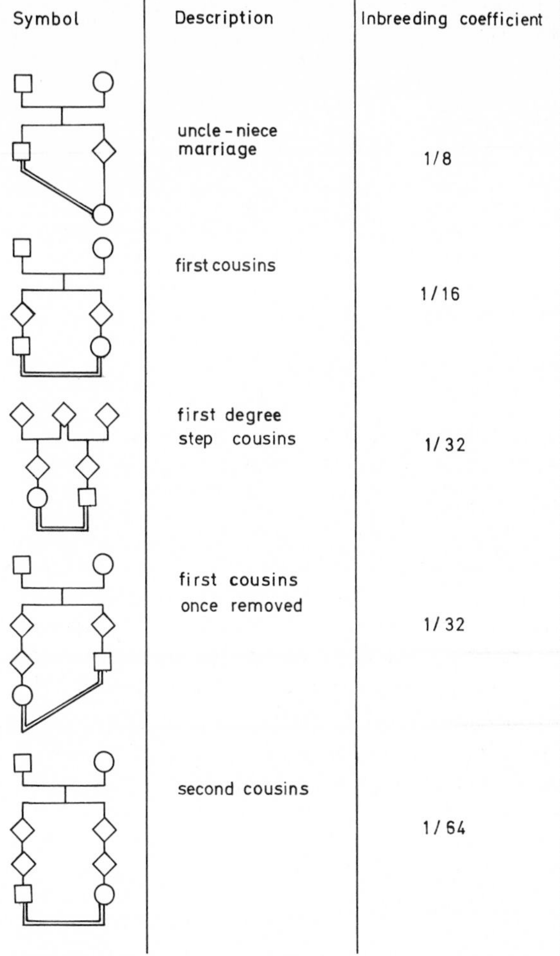

Fig. 6.38. The most important types of consanguineous marriages

a) The coefficient of kinship, Φ_{AB}, ot two individuals, A and B, is the probability that a gene taken at random from A is identical by common descent with a gene at the same locus taken at random from B.
b) The inbreeding coefficient, F, of an individual is equal to the coefficient of kinship, Φ, of his father and mother.

The distinction between the two coefficients is that a coefficient of kinship applies to two individuals who may have common ancestors. An inbreeding coefficient applies to one individual and measures the degree of relationship between his parents and, hence, the resemblance between the two genes received from his parents at each locus. In fact, *the inbreeding coefficient is equal to the probability that the two genes, which the individual has at a given locus, are identical by descent.*

Coefficient of Inbreeding and the Hardy-Weinberg Law. Let us consider an autosomal gene pair A, a (gene frequencies p, q). In a randomly mating population, the three genotypes will occur in the proportions $p^2:2pq:q^2$. If the genotype contains N such gene pairs with gene frequencies p_i, q_i ($i=1,2,\ldots, N$), the degree of heterozygosity under random mating will be:

$$2\sum_{i=1}^{N}\frac{p_iq_i}{N}$$

and of homozygosity:

$$\sum_{i=1}^{N}\frac{p_i^2+q_i^2}{N}$$

the sum of both being 1. This degree of heterozygosity indicates the fraction of autosomal genes with two alleles, for which an individual is, on the average, heterozygous. For a single gene, it indicates the probability for an individual to be heterozygous.
In a consanguineous mating (Fig. 6.38) a pair of alternate alleles A and a are considered. An oocyte may contain the gene a. If the mating is random, the probability of this oocyte to be fertilized by a sperm with a is p and by a sperm with A is q. If the parents are related, they have a certain fraction of genes in common by descent; accordingly, p is increased to $(p+Fq)$ and q is reduced to $(q-Fq)$, and correspondingly for oocytes containing allele A. Here, the value that corresponds to F in the former delineation may be called F'. If the mode of inheritance is autosomal, the two parents will show the same distribution of genes A and a. Therefore, it must hold that $pq(1-F)=qp(1-F')$ and hence $F=F'$.
It can be shown that F is equal to the inbreeding coefficient as defined above. This means that the genotypes of children having an inbreeding coefficient F do not occur in Hardy-Weinberg proportions but in the proportions:

$$\begin{array}{ccccc} AA & : & Aa & : & aa \\ (p^2+Fpq) & : & 2(1-F)pq & : & (q^2+Fpq) \end{array}$$

For the child, his or her own degree of heterozygosity is diminished, on the average by a factor F. Or, to put it differently, F is the probability that the two homologous chromosomes carry two genes that are derived from the same ancestral gene at a randomly chosen gene locus.

Calculation of the Inbreeding Coefficient F. In most practically occurring situations in human genetics, actual calculation of Φ or F is not necessary, as the coefficients for the degrees of consanguinity occurring in human populations are known as shown in Fig. 6.38. An occasional pedigree might require individual calculation. This is quite different from the situation in animal breeding, where very complicated relationships between mates may be encountered. For their assessment, Wright proposed the method of path coefficients [124], [961]. The pedigrees of the two mates are drawn, and all their common ancestors are marked. Then, one of the least remote common ancestors is selected, and the two mates are connected by all possible pathways, which:

a) Lead to this common ancestor;
b) Consist of "steps" (one step being defined as the connection between an individual and one of his parents);
c) Do not lead to one person more than once.

The other common ancestors are treated in the same way. The number of steps in each path is counted. For one ancestor, x paths with $m_1 \ldots m_x$ steps may exist, giving for t common ancestors $\left(\sum_{i=1}^{t} x_i\right) = r$ paths. Then:

$$F=\tfrac{1}{2}(2^{-m_1}+2^{-m_2}+\ldots+2^{-m_r})=\frac{1}{2}\sum_{i=1}^{r}2^{-m_i} \qquad (6.12)^{3}$$

The following simple consideration may help to understand this formula. A child shares with each of his parents 1/2 of its genes, with a grandparent 1/4, with a great-grandparent 1/8, etc. If a path of a steps from the mother to an ancestor, she has a fraction of $(1/2)^a=2^{-a}$ genes in common with that ancestor. If a path of b steps leads from the father to this ancestor, he shares 2^{-b} genes with this ancestor. This means that father and mother have $2^{-a}\times 2^{-b}=2^{-m}$ genes ($m=a+b$) in common. This number divided by 2 gives the probability that a random gene of the mother is identical by descent with a random gene of the father.[4]

[3] If some of the common ancestors come from a consanguineous mating, the terms have to be corrected considering their inbreeding coefficients.

[4] For a more rigorous derivation, and for other methods of calculation, cf. Li (1955) [124], Jacquard (1974) [103] and, Kempthorne (1957) [108], who gives a useful matrix method.

Examples. Figure 6.40 shows a first-cousin marriage. The path over the common grandfather of the couple has four steps, and the same is true for the path over the common grandmother. Inserting in Eq. (6.12) gives:

$$F=\tfrac{1}{2}(2^{-4}+2^{-4})=\tfrac{1}{2}\left(\tfrac{1}{16}+\tfrac{1}{16}\right)=\tfrac{1}{16}.$$

Figure 6.41 shows a case of father-daughter incest. There is only one path and it consists of one step:

$$F=\tfrac{1}{2}2^{-1}=\tfrac{1}{2}\cdot\tfrac{1}{2}=\tfrac{1}{4}.$$

The third example is of brother-sister incest (Fig. 6.42):

$$F=\tfrac{1}{2}(2^{-2}+2^{-2})=\tfrac{1}{2}\left(\tfrac{1}{4}+\tfrac{1}{4}\right)=\tfrac{1}{4}.$$

As mentioned above, for humans a useful convention limits these calculations to three generations back. The convention is sound quantitatively. For example, if a common ancestor is found five generations back, the corresponding path has ten steps; therefore, it contributes to F only $1/2\times 2^{-10}=1/2{,}048$.

Inbreeding Coefficient of a Population. Frequently, we are interested in an index that measures mean consanguinity in a population, considering all types of consanguineous marriages together. If the calculation is confined to the last three generations, the resulting coefficient K of the "apparent consanguinity" is given by:

$$K=\sum F_i\,M_i$$

Summation goes over the various types of consanguineous marriages, with F_i and M_i being the inbreeding coefficient and the relative frequency of

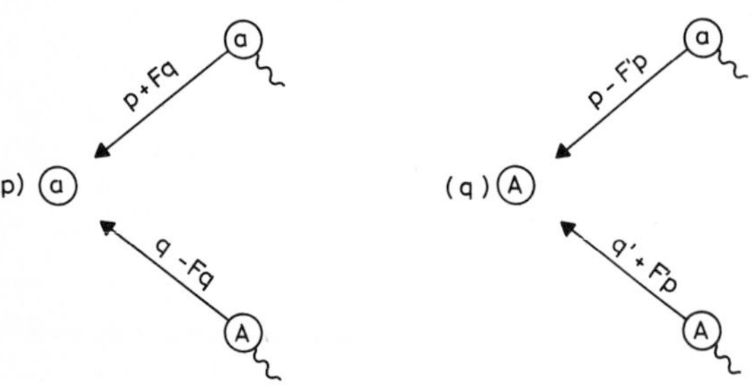

Fig. 6.39. An oocyte may contain the allele a. With random mating, probabilities for this oocyte to be fertilized by a sperm with allele a $=p$, with allele A $=q$. In a consanguineous mating, these probabilities are $(p+Fq)$ or $(q-Fq)$. *Left,* fertilization of oocyte with allele *a*; *right,* fertilization of oocyte with allele A. (Ludwig, 1944 [1819])

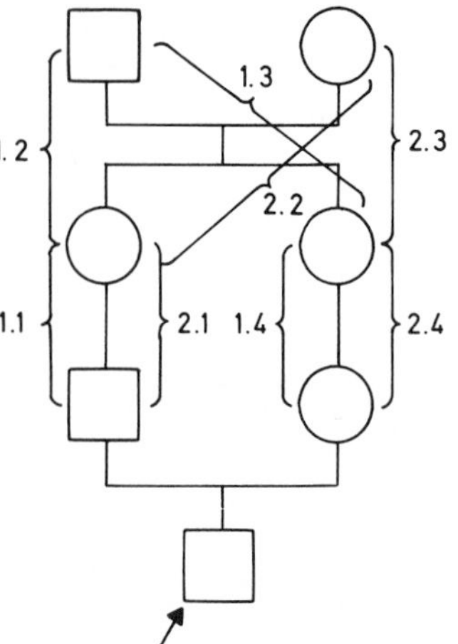

Fig. 6.40. First-cousin marriage. Calculation of F by the method of path coefficients. Four steps (1.1–1.4) connect the proband's father and his mother via the common grandfather of the parents. Four other steps (2.1–2.4) connect the father and the mother via the common grandmother. $F=\frac{1}{16}$

the i-th type of consanguineous marriage. K is frequently referred to simply as the mean of F or as F of the population. In comparing different populations for this parameter, one should keep in mind that the convention of including three ancestral generations has often not been followed. Besides, in almost all population studies, the coefficients of kinship (according to the definition on p. 483) of all couples and not the inbreeding coefficients of all individuals have been calculated. This kinship coefficient gives an unbiased estimate of the mean inbreeding coefficient of the individuals only if inbreeding does not influence reproduction.

6.3.1.2 Inbreeding, Isolate Size, and Inherited Disease

Frequency of Children with Recessive and Multifactorial Diseases in Consanguineous Matings Compared with Nonconsanguineous Matings. Let the allele that, in the homozygous state, leads to a recessive disease have a gene frequency q. Then, the phenotype frequency in the random mating population will be q^2; in the population of individuals with the inbreeding coefficient F, it will be $q^2 + Fpq$. With decreasing q, the ratio Fpq/q^2 increases: *The lower the gene (and genotype) frequency, the higher the frequency of consanguineous marriages among parents of the affected*

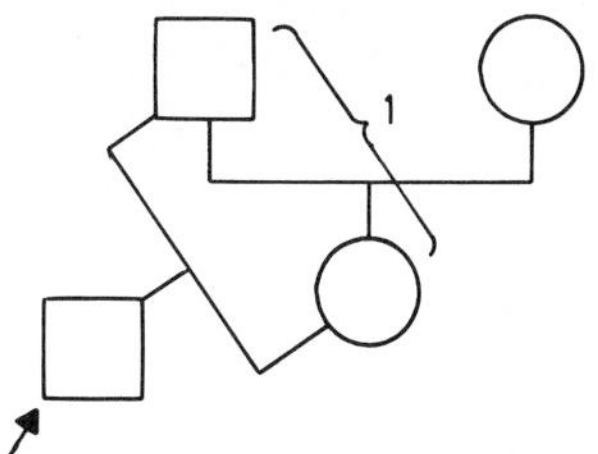

Fig. 6.41. Father-daughter incest. There is only one path

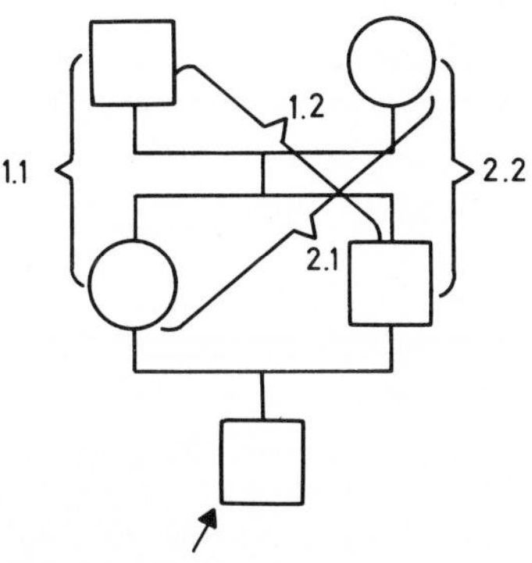

Fig. 6.42. Brother-sister incest. There are two paths with two steps each

homozygotes. This holds true not only for recessive diseases but also for multifactorial characters (Sect. 3.6). Among individuals with inbreeding coefficient F, the variance of a normally distributed liability with heritability $h^2 = 1$ is

$$V_F = V_0(1+F)$$

V_0 being the variance in a noninbred population. However, with increase of the variance, the relative number of individuals beyond the threshold will also increase. Hence, children from consanguineous matings will have a slightly higher risk of being affected with a multifactorial threshold character than children from nonrelated parents (Fig. 3.63).
For recessive genes, the argument can be reversed theoretically. If more consanguineous marriages are found than are expected on the basis of the population incidence of a recessive condition, not one but several recessive genes with correspondingly lower gene frequencies may be involved, i.e., genetic heterogeneity may exist. Under the assumption that these genes have equal frequencies, even their number has been estimated. In practice, this approach will almost always be futile for the following reasons:

a) Due to the decrease of inbreeding in modern populations, the number of homozygotes for recessive diseases has steeply declined (see below).
b) The decrease of inbreeding has mainly taken place in big cities and densely populated areas. Many of the recessive diseases come from remote rural areas where consanguineous marriages are in general more frequent.
c) Hence, a high ratio of consanguineous to nonconsanguineous matings may be encountered simply due to population heterogeneity, even if there is only one recessive gene.

These problems will be discussed in connection with the genetic basis of deaf mutism (App. 3).

Calculation of Isolate Size from the Frequency of Consanguineous Marriages. If matings occur at random, and the population is infinite in size, the probability of a marriage being consanguineous converges to 0. This argument can be reversed as follows.
If matings within isolates occur at random, how large must the isolates be to account for the fraction of consanguineous matings actually found? This is the basic idea of the estimation of isolate size (Dahlberg, 1929 [1750]). Let n be the number of individuals in an isolate, and $n/2$ the marriageable individuals for each sex. Let b be the average number of children per marriage. Then, the average number of first cousins of the other sex for every individual is $b(b-1)$. If they have the same chance of being married as every other individual within the population, this gives:

$$c = \frac{b(b-1)}{n/2}$$

Table 6.17. Estimate of isolate size from different types of consanguineous marriages. (Adapted from Morton, 1955 [1826])

Populations	Frequency of consanguineous marriages (actual numbers)				Estimate of isolate size		
	All marriages	First cousin	Second cousin	Uncle-niece, auntnephew	First cousins	Second cousins	Uncle-niece, auntnephew
Japan (Hiroshima-Nagasaki)	66,417	2,683	1,116	1	100	950	130,000
Austria (Orel, 1932)	117,431	840	309	63	560	6,100	3,700
England (Bell, 1940)	59,551	340	67	4	700	14,000	30,000
Brazil (Freire-Maia, 1952)	43,082	2,185	1,124	22	80	610	3,900

for the fraction of consanguineous matings. Solving for n, the isolate size, we get:

$$n=\frac{2b(b-1)}{c} \qquad (6.13)$$

The same calculation can be carried out for other degrees of relationship. These and similar calculations have been widely used in the literature to assess the breeding structure of human populations.

However, consanguineous matings do not occur at random; they are influenced by many, and varying, social and sociopsychological influences. The weakness of the method can most easily be shown by demonstrating that estimates of isolate sizes based on various classes of consanguineous marriages give quite different results in the same populations (Table 6.17).

Inbreeding Coefficient F in Various Population Groups. In Table 6.18 (see p. 488, 489), the frequencies of consanguineous matings in various populations have been collected. Table 6.18 lists first-cousin marriages (1-C), and the F values calculated on the basis of the available data.[5] The data in Table 6.18 were collected by the genealogic method, i.e., pedigrees of couples of consanguineous marriages were assessed. Depending on the method of ascertainment, F may be more or less underestimated because:

a) The data from Catholic countries are usually based on dispensation registers. However, probably not all Catholics who planned to marry a close relative have actually asked for dispensation, and especially in cities the priests may not know the couples well enough to realize that they are related.
b) In some studies, the investigators relied on the statements of the families who, in many cases, concealed consanguinity.

One example: On the Japanese island Hosojima, the seven members of the island council knew of 19 of 45 marriages in which the spouses were related. A check of the Koseki, a Japanese population register, brought this number to 25, which was again enhanced to 29 by careful analysis of pedigrees [1794].

In almost all European countries and in the United States, the inbreeding coefficients are very low; high coefficients are usually found in some small communities and in religious, geographic, and ethnic isolates. In South America, which has been very well examined, the mean inbreeding coefficient appears to be about twice or three times that in Europe [1763, 1764]. High values of F are also found in Japan. The highest values were described for parts of southern India, especially the state of Andhra Pradesh, among the Nubic tribes in Egypt, and among the Fulbe in Guinea.

Decline of Consanguinity in Industrial Countries. In the industrialized countries of the West, a decline in consanguineous marriages has been observed since the beginning of this century. The trend began in the highly industrialized areas and big cities and is now spreading into the more remote rural provinces. France has been especially carefully examined; Figs. 6.43 and 6.44 show the frequencies of apparently consanguineous marriages between 1926–1930 and between 1956–1958 [103]. In the first period, the mean coefficient of inbreeding (more exactly, the coefficient of apparent consanguinity; see above) was 86.1×10^{-5}; it had decreased to 23×10^{-5} by 1956–1958. The decline is usually explained by the higher mobility of a population in an industrialized society and the possibility of choice among more individuals of the other sex. This explanation is corroborated by studies in which the distances between the birth places of spouses were shown to grow over

[5] For Israel, only frequencies of first cousin, uncle-niece, and aunt-nephew marriages are given, with the result that the inbreeding coefficient may be underestimated.

time. The phenomenon is usually referred to as the "breaking up" of isolates. More recently, the tendency toward a lower consanguinity rate has been strengthened by the reduction in number of children per marriage, which reduces the number of eligible cousins.

These and other considerations raise the question of the degree to which consanguineous marriages represent an otherwise unbiased sample of all marriages. The question becomes important when consanguineous marriages are utilized to estimate the "genetic load" (Sect. 6.3.2) due to genes that are lethal or detrimental in the homozygous state.

Social and Psychological Influences on the Frequency of Consanguineous Marriages. Some biases are obvious from inspection of Figs. 6.43 and 6.44; usually, the quality of obstetric and pediatric care is better in developed provinces of France with low consanguinity rates. Even within these provinces, the rates are lower in towns than in villages. Selective migration of healthier people from rural into urban areas may further add a spurious trend toward higher infant and perinatal mortality in consanguineous marriages.

However, the bias may be much more subtle. In a German study, for example, individuals married to a close relative were shown to differ psychologically from the average population [1896]. As an example, males living in consanguineous marriages suffered from more difficulties in establishing interpersonal contacts than other men and therefore selected a relative rather than a nonrelative as a spouse. On the other hand, completely different sociopsychological conditions for consanguineous marriages may exist; in populations of southern India with very high consanguinity rates, a marriage between daughter and mother's brother is the most preferred marriage type socially.

In India as well as in Japan until very recently, marriages are arranged by the families of the two spouses. Economic reasons and the advantage of knowing the spouse very well are the most prominent factors in such arrangements. Personality factors such as those described in the above-mentioned German study seem to be of minor importance. For this and other reasons, it is reasonable to conclude that in Japan consanguineous marriages are a less biased sample of all marriages than in European countries. Still, some bias might be left. In a study of consanguinity carried out on the Japanese island Hirado [1877, 1878, 1879], for example, there was a tendency for older brothers to marry a first cousin more frequently. Moreover, it was the older brothers who inherited the land owned by the families and hence tended to stay more often in their villages, while many of the younger brothers emigrated to other areas. Such biases must be considered, when the influence of parental consanguinity on the health status of their progeny is assessed.

Influence of the Decline of Consanguinity on the Incidence of Recessive Diseases. Let us assume a population with average inbreeding coefficient F in which an equilibrium between mutation rate μ and the selection coefficient s has been established at a gene frequency q. Now, the degree of inbreeding is reduced within a short time period, say, one generation, from F_1 to F_2. Consequently, the number of homozygotes will drop from q^2+F_1pq to q^2+F_2pq. This change will disturb the genetic equilibrium, selection now being insufficient to eliminate the number of genes produced by new mutations since there are fewer homozygotes. For example, F_1 for European populations has ranged between about $0.003-0.005$. Let us also assume a mutation rate of 10^{-5} and selection coefficient (s) of 0.5 against the homozygotes. The equilibrium frequency is $\hat{q}=2.6-3.2\times10^{-3}$. In this case, complete cessation of inbreeding leads to a drop of the homozygote frequency as shown in Fig. 6.45. Moreover, 175-185 generations (about 4,500 years) are required for the gene to reach one-half the way to the new equilibrium frequency.

The decline of inbreeding is one of the reasons why "modern" societies are now enjoying - and will enjoy for a long time to come - an unusually low incidence of recessive disorders. The other, related reason is that, due to stochastic processes in small populations (Sect. 6.4), genes for recessive diseases have attained unequal frequencies in various population groups. With increasing intermarriage between these groups, gene frequencies will level out, and populations with high gene frequencies and, hence, high homozygote frequencies will disappear.

6.3.2 Concept of Genetic Load

6.3.2.1 Theory

Estimation of the Overall Number of Recessive Genes in the Human Population [1856, 1871]. Homozygotes in general and especially those for rare diseases are more frequent among children from consanguineous marriages than in the general population. This fact can be used to estimate the number of such recessive genes in each individual of the population.

For example, a gene taken at random from an individual has a probability of 1/2 to be identical by descent with one of the two genes at the same locus of

Table 6.18. Frequency of consanguineous marriages and inbreeding coefficients F ($\times 10^{-5}$) in various countries. (v. Fumetti [1767])

Country/region/diocese	Method of ascertainment[a]	Time	Population N	% First cousin marriages	% Consang. marriages (All types together)	F ($\times 10^5$)	Time	Population	% First cousin marriages	% Consang. marriages	F
Europe											
Belgium	DA	1918–59	2,404,027	0.49	1.47	50	1955–59	300,592	0.22	0.97	29
C.S.S.R. diocese Brno	DA	1930–66	230,988	?	0.93	28	1960–66	?	?	0.20	8
Germany Bavaria, Württemberg	DA+CR	1848–1922	16,182	0.50	1.18	44	–	–	–	–	–
5 localities near Tübingen	CR+MR	about 1920	453	4.91	20.00	472	–	–	–	–	–
Archdiocese Cologne	DA	1898–1943	192,980	0.37	0.93	35	–	–	–	–	–
Diocese Münster and Osnabrück	DA	–	–	–	–	–	1946–51	119,899	0.18	0.59	19
France	DA	1926–58	6,061,000	0.52	1.36	49	1956–58	530,000	0.22	0.67	23
Ireland	DA	–	–	–	–	–	1959–68	190,547	0.13	0.53	16
Italy	DA	1911–60	13,687,897	1.33	3.00	118	1956–60	1,646,612	0.77	1.90	70
Austria: archdiocese Vienna	DA	1901/02 1914/14 1929/30	117,294	0.67	1.28	60	–	–	–	–	–
Switzerland 4 mountain villages	A+CR	about 1920	538	2.79	32.71	509	–	–	–	–	–
Spain (total)	DA	1930	17,000	2.00	5.34	203	–	–	–	–	–
D. Ciudad Rodrigo	DA	1940–64	11,394	2.18	9.41	254	1960–64	2,069	2.10	10.63	275
North America											
Canada Catholic pop.	DA	–	–	–	–	–	1959	51,729	0.37	1.51	45
French population	DA	1885–95 1915–25 1945–65	149,992	1.03	4.17	180	1955–65	50,128	0.37	2.10	90
U.S.A. Catholic pop.	DA	–	–	–	–	–	about 1958	133,228	0.08	0.11	8
Mormons	FB	1930–50	132,524	0.04	?	?	–	–	–	–	–
Mormons, in 9 rural parishes	FB	–	–	–	–	–	1950	625	1.44	9.92	189
Central and South America											
Argentina 12 D[b]	DA	–	–	–	–	–	1956/57	51,391	0.75	1.12	58
Brazil 72 D	DA	–	–	–	–	–	1956/57	212,090	2.63	4.82	225
Brazil 95 D	DA	–	–	–	–	–	1965/67	198,088	2.14	4.00	176
Bolivia 5 D	DA	–	–	–	–	–	1956/57	4,130	0.32	0.63	28
Chile 8 D	DA	–	–	–	–	–	1956/57	28,596	0.80	1.31	74
Chile, province Valparaiso	DA	1917–66	195,721	0.60	0.98	50	1957–66	51,828	0.41	0.68	35
Costa Rica 1 D	DA	–	–	–	–	–	1954	3,833	0.94	3.39	114
Ecuador 3 D	DA	–	–	–	–	–	1956/57	3,954	2.17	6.27	229
El Salvador 2 D	DA	–	–	–	–	–	1956/57	2,494	1.04	4.85	142

Table 6.18 *(continued)*

Honduras 3 D	DA	–	–	–	–	–	1956/57	3,759	0.56	3.43	110
Columbia 13 D	DA	–	–	–	–	–	1956/57	34,470	1.25	2.95	119
Cuba 3 D	DA	–	–	–	–	–	1956/57	2,277	0.53	0.83	54
Mexico 10 D	DA	–	–	–	–	–	1956/57	28,292	0.17	1.27	31
Panama 1 D	DA	–	–	–	–	–	1956/57	350	0	0	–
Peru 3 D	DA	–	–	–	–	–	1956/57	565	2.12	4.07	279
Uruguay 3 D	DA	–	–	–	–	–	1956/57	8,822	0.85	1.43	65
Venezuela 4 D	DA	–	–	–	–	–	1956/57	2,931	1.60	4.46	191
Asia											
Japan total	I+FB	–	–	–	–	–	about 1950	213,148	5.39	8.16	400
Hirado[c]	I+FB	1880–1964	10,403	5.58	14.72	461	1960–64	853	1.76	10.79	200
India Bombay[d]	I	–	–	–	–	–	about 1955	3,520	9.29	11.39	612
Southern India	I	–	–	–	–	–	after 1950	26,042	24.58	39.37	2,835
Israel	I	–	–	–	–	–	1955–57	11,424	5.22	9.68	387
Africa											
Egypt, nubic populations	I	–	–	–	–	–	1967/68	1,782	45.29	75.76	3,335
Guinea, Fulbe	I	–	–	–	–	–	about 1960	1,280	15.21	29.91	819

[a] *DA* dispensation archives; *A*, ancestry tables; *I*, interview; *MR*, marriage registers; *FB*, family books; *CR*, church registers.
[b] Cipher + D shows the number of dioceses.
[c] The figures for Hirado relate to the time of marriage for those marriages of which at least one spouse was still alive at the time of examination.
[d] Inbreeding coefficient was calculated only from uncle-niece and aunt-nephew matings; therefore, it might be too low by 1/4 to 1/3.

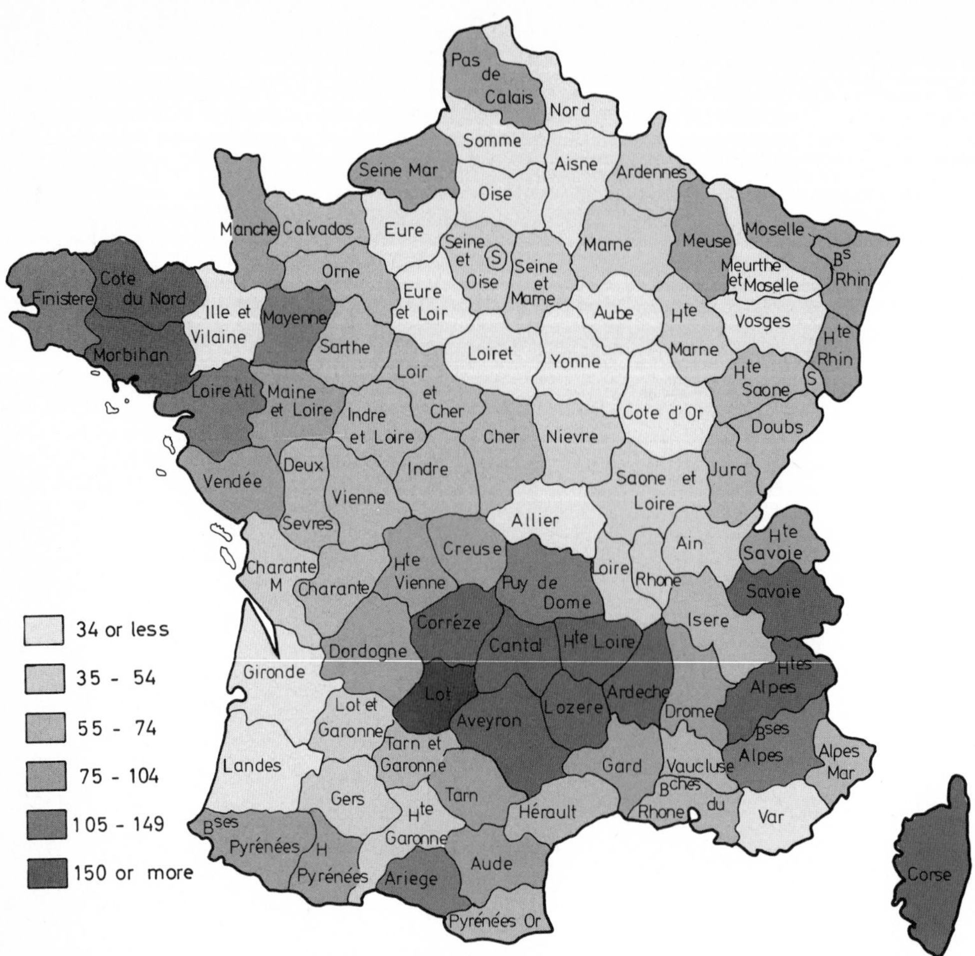

Fig. 6.43. Coefficient of apparent consanguinity in France 1926–1930. (Jacquard, 1974 [103]) The figures $\times 10^{-5}$ refer to F

the individual's sib. If one of the sibs in a brother-sister mating is a carrier of a gene leading to a recessive disease in the homozygous state, the other sib also carries the gene with a probability of 1/2, and every child of this mating has a risk of 1/4 to be affected with the disease. Hence, the probability that such a mating produces at least one affected child is $1-(3/4)^s$, s being the number of children per mating. Thus, when an otherwise unselected series of brother-sister matings is examined with respect to the incidence of recessive diseases among their progeny, the average number of individuals carrying such a recessive gene in the general population can be estimated. The same argument holds true, for example, for father-daughter matings. These matings are proscribed by law and custom and are rare. Moreover, nobody would regard individuals involved in such a mating as an unbiased population sample (see Sect. 6.3.2.4).

Intuitive Background: Our Load of Mutations. H.J. Muller, the famous geneticist, had been concerned since his teens with the idea that the human species might be in danger of deteriorating biologically, that the evolutionary system might collapse sooner or later, and that our species might finally be submerged in an ocean of suffering from disease, mental defects, miscarriage, and other catastrophes.

We saw at the beginning of this book (Sect. 1.8) that this concern was shared by many scientists at the beginning of the century, and was in fact, the motive behind the work of F. Galton and the eugenics movement.

Muller's arguments were comprehensively set out in his paper *Our Load of Mutations* in 1950 [1835]. His most important theses may be formulated as follows:

a) A large share of all human zygotes are killed or prevented from reproduction by mutations.
b) The overall mutation rate per individual, i.e., the total number of new mutations contained in both

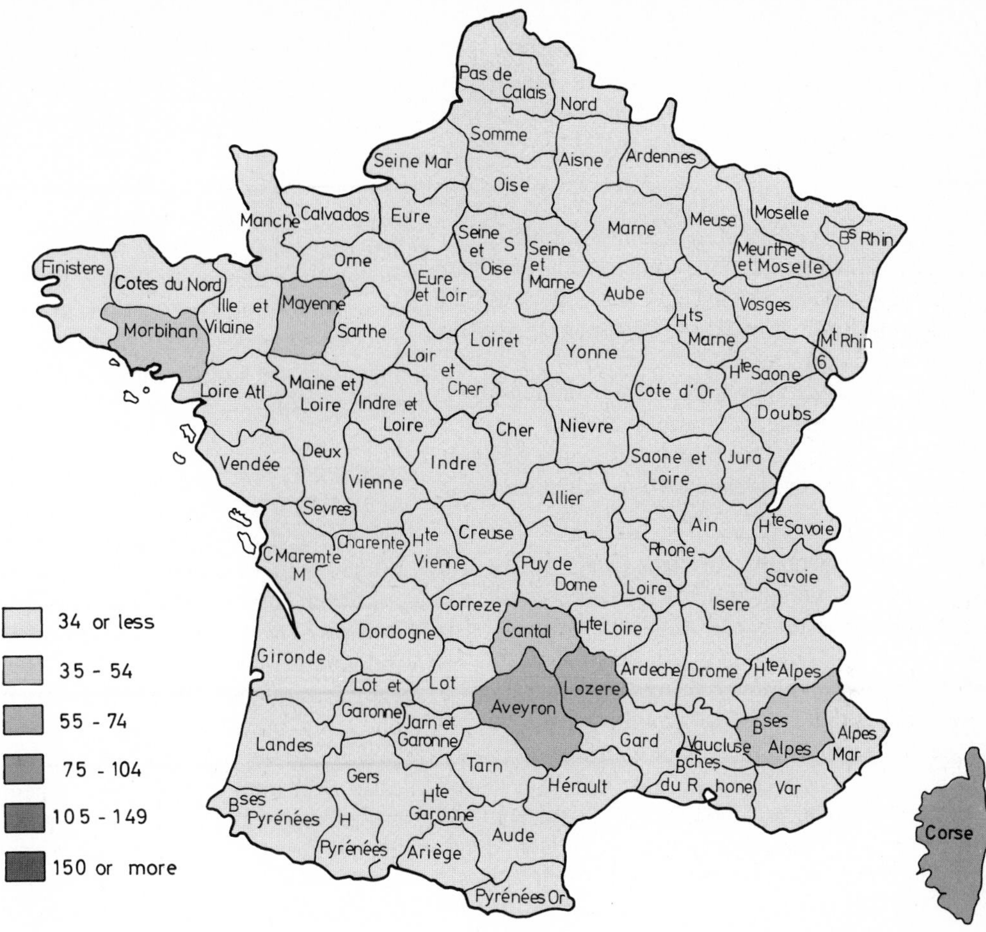

Fig. 6.44. Coefficient of apparent consanguinity in France 1956–1958. (Jacquard, 1974 [103])

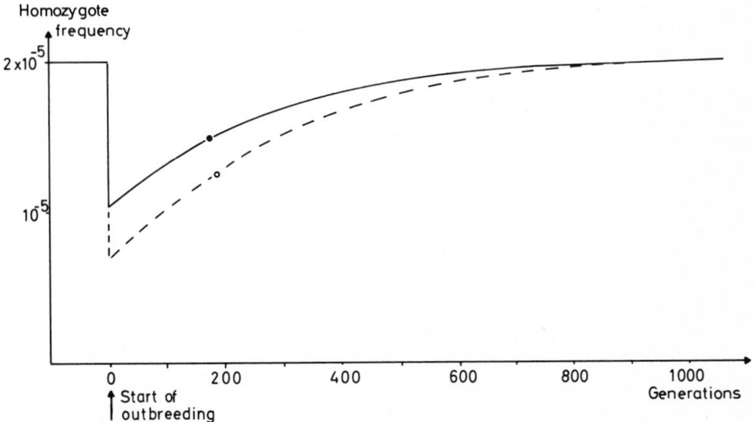

Fig. 6.45. Drop of the frequency of recessive homozygotes in a population with long-lasting inbreeding by complete cessation of inbreeding and very gradual increase to the old value as the result of an excess of freshly produced compared with selectively eliminated mutants. Mutation rate $\mu = 10^{-5}$; selection coefficients for the recessive homozygotes $s = 0.5$; inbreeding coefficients $F = 0.003$ (—) and $F = 0.005$ (- - -). At the points ● and ○, respectively, the gene frequency q reaches one-half of the new equilibrium frequency. The values of the gene frequency c were calculated by solving the equation:

$$q^2 + Fq(1-q) = \mu/s$$

for q, and then stepwise for each new generation by the recurrence formula:

$$q' = (q^* - sq^{*2})/(1 - sq^{*2})$$

where $q^* = q + \mu p$

germ cells that form this individual, is between one mutation per two to ten germ cells.

c) Every individual is heterozygous for several genes that would kill him if homozygous. These genes are usually deleterious even in the heterozygous state.
d) Natural selection has relaxed; hence, the number of deleterious genes in the human population increases dangerously; they may reach a critical threshold above which the whole genetic system may break down, leading to disappearance of the human species.
e) The danger becomes more acute by increased exposure to ionizing radiation.
f) We should try to curb this dangerous development by regulating human reproduction artificially.

Since Muller brought forward these theses, our knowledge of human genetics has improved, and some of his questions can now be answered fairly precisely [1886]. One aspect will be singled out: the assertion that every human being is heterozygous for several genes that would kill him if homozygous but are even deleterious in the heterozygous state.

Effect of Variation on Fitness. A more formalized and more balanced concept was created by Haldane in several papers, especially the one titled *The Effect of Variation on Fitness* [1775, 1780]. He defined the fitness of a genotype as the average number of progeny left by an individual of that genotype and explained that the average fitness of a species will almost always be close to unity, as otherwise the population would increase very rapidly.

But in any species some genotypes have a fitness less than unity, ranging to zero in the case of lethal genes and genes causing complete sterility. So it is clear that the fitness of the standard type containing no deleterious genes must exceed unity.
It is at once clear that in equilibrium such abnormal genes are wiped out by natural selection at exactly the same rate as they are produced by mutation. It does not matter whether the gene is lethal or almost harmless. In the first case, every individual carrying it, or if it is recessive, every individual homozygous for it, is wiped out. In the second, the viability or fertility of such individuals may only be reduced by one thousandth. In either case, however, the loss of fitness to the species depends entirely on the mutation rate and not at all on the effect of the gene upon the fitness of the individual carrying it.

The alert reader may already have discovered that this statement is in essence a generalization of Haldane's argument used to derive the indirect method for estimating human mutation rates (Sect. 5.1.3.1). Having made some preliminary calculations on the extent of this overall mutation rate in *Drosophila,* he continued:

This may be taken as a rough estimate of the price the species pays for the variability which is probably a prerequisite for evolution.

In a later paper, Haldane [1780] examined the reduction of fitness needed to permit natural selection to carry out an adaptive gene substitution in evolution.
On the basis of Muller's and Haldane's contributions, the concept of genetic load was utilized for testing on human populations in another famous paper *An Estimate of the Mutational Damage in Man from Data on Consanguineous Marriages* by Morton et al. (1956) [1827].

Definition of the Genetic Load [1827]. Morton, Crow and Muller [1827] distinguish between the total damage due to disadvantageous mutations present in the human genome and the expressed damage. Both are described as lethal equivalents. A lethal equivalent is a group of mutations that, distributed among various individuals, causes one death, on the average, for genetic reasons. Such mutations may be, for example, one lethal mutation leading to death in all cases or two mutations each leading to death in 50% of the cases. The total damage per gamete was defined as the average number of lethal equivalents in the zygote, if the zygote were formed by doubling all chromosomes of the gamete. The expressed damage per gamete is the average number of lethal equivalents of this gamete that would be manifested if combined in the zygote with another gamete according to the mating system actually prevailing in the population.

The total genetic damage may be estimated as follows. Let us consider one gene locus. The probability for a given zygote to survive deleterious effects of mutations at this locus is given by:

$1 - qFs$	$-q^2(1-F)\,s$	$-2q(1-q)(1-F)\,sh$	
Probability of death due to homozygosity from consanguinity	Probability of death due to homozygosity not from consanguinity	Probability of death in a heterozygote	(6.14)

Here, s is the probability for a zygote homozygous for the mutation to die; h is a measure for the dominance of this mutation ($h=0$ if the gene is completely recessive, $h=1$ if the gene leads to death in the heterozygote as often as in the homozygote); F is the inbreeding coefficient.
Another assumption is that genetic and environmental causes of death act independently. With this condition, the fraction of survivors may be expressed as follows:

$$S=\prod_{i,j}(1-x_i)[1-q_jFs_j-q_j^2(1-F)s_j -2q_j(1-q_j)(1-F)s_jh_j] \qquad (6.15)$$

Here, x_i is the probability for a certain environmental cause of death. The product comprises all x_i and all q_j, the gene frequencies of deleterious mutations. The number of these mutations as well as the number of environmental factors x_i can be assumed to be large; the single probabilities, however, are small. Therefore this expression is approximated by the following:

$$S = 1 - \Sigma x - \Sigma Fqs - (1-F)\Sigma q^2 s - 2(1-F)\Sigma q(1-q)sh$$

This, in turn, may be approximated by:

$$S = e^{-(A+BF)} \text{ or } -\log_e S = A + BF \qquad (6.16)$$

where

$$A = \Sigma x + \Sigma q^2 s + 2\Sigma q(1-q)\, sh$$

$$B = \Sigma\, qs - \Sigma\, q^2 s - 2\, \Sigma q(1-q)\, sh$$

The summation includes all environmental factors and, respectively, all loci with mutant alleles.

In a randomly mating population ($F=0$), the expressed genetic damage – together with the environmental damage – is represented by A. B, on the other hand, is a measure for the hidden genetic damage that could manifest itself only with complete homozygosity ($F=1$). The total genetic damage is expressed by $\Sigma\, qs$, which is the sum of B and the genetic component of A and, hence, lies between B and $B+A$.

It is possible to estimate A and B using weighted regression coefficients of $\log_e S$ (S=fraction of survivors) on F. Considering the low degree of inbreeding normally found in human populations and the low death rate among offspring of unrelated couples, the following simplified formula gives a satisfactory approximation:

$$S = 1 - A - BF \qquad (6.17)$$

The actual calculation may proceed as follows:

$$S_1 = 1 - A, \quad S_2 = 1 - A - FB, \quad S_1 - S_2 = FB;$$

$$A = 1 - S_1; \quad B = \frac{S_1 - S_2}{F},$$

S_1 = number of survivors in nonconsanguineous matings, and S_2 = number of survivors in consanguineous matings.

The number of lethal equivalents is estimated from the difference between consanguineous and nonconsanguineous marriages in the number of children stillborn and deceased before reproductive age.

An Example. The three authors [1827] utilized some data from France for a preliminary calculation (Table 6.19). For stillbirth and death during childhood and youth (before reproduction) together, a value of B between 1.5 and 2.5 was calculated; $A+B$ was not very much higher. The ratio B/A, which was to play a major role in later discussions, varies between 15.06 and 24.41. This would mean that the average gamete carries a number of deleterious genes that, if distributed in single individuals and made homozygous, would kill 1.5–2.5 individuals before reproductive age. The total genetic damage is 1.5–2.5 lethal equivalents per gamete; 3–5 lethal equivalents per zygote. In this calculation, miscarriages and deaths at a later age (for example, during the reproductive period) are not included. Hence, this approach underestimates the real damage. *Apparently, every human being is heterozygous for several genes that would be deleterious in the homozygous state.*

The authors cautiously admit that the differences encountered between consanguineous and nonconsanguineous marriages may in part not be biological; only the pregnancy outcome of consanguineous marriages was determined by direct interview; socioeconomic differences between rural and urban populations might add another bias toward higher mortality among children from consanguineous marriages. We shall see later that this caution was only too well justified.

Estimate of the Expressed Genetic Damage. As a next point, the authors concluded that the same genes may have a certain disadvantage even in the heterozygous state, i.e., that their "dominance" h is greater than 0. According to Eq. (6.15), the total probability of a particular mutant to be eliminated under the naturally occurring breeding structure is approximately $z \times s$, where $z = F + q + h$ (notations as above). The number of expressed lethal equivalents per gamete can be shown to be equal to the total number of lethal equivalents multiplied by the harmonic mean of the values of z for the particular mutants. For determining h, data on human beings were not available. Therefore, data from *Drosophila* were used, which gave values of h for 16 autosomal lethals with a mean of about 0.04. Considering the fact that the more deleterious mutants might be rarer in

Table 6.19. Number of stillbirths and childhood mortality in consanguineous compared with nonconsanguineous marriages in France. (Adapted from Morton et al., 1956 [1827])

	First-cousin marriages	1½ cousin marriages	Second-cousin marriages	No consanguinity $F=0$
District Morbihan				
Stillbirths and neonatal death	51/461 (0.111)	3/78 (0.038)	23/309 (0.076)	72/1628 (0.044)
Death during early childhood	64/410 (0.156)	17/75 (0.227)	32/286 (0.112)	138/1556 (0.089)
District Loir-et-Cher				
Stillbirths and neonatal death	18/282 (0.064)	6/105 (0.057)	11/240 (0.046)	36/1117 (0.032)
Death during early childhood	32/264 (0.121)	1/99 (0.010)	17/229 (0.074)	60/1081 (0.056)

natural populations, and assuming that most of the adverse effects are produced in heterozygotes (due to their higher frequency), the harmonic mean of z for all deleterious genes was estimated to be 0.02. With 1.5–2.5 as the total number of lethal equivalents per gamete, this corresponds to 3%–5% of expressed lethality per gamete or 6%–10% per zygote.

Estimate of the Overall Mutation Rate of Detrimental Mutations. As noted, Haldane (1935) [1472] had already postulated a genetic equilibrium between selection and mutation. Over a sufficiently long time, the number of new mutations would be equal to the number of detrimental alleles per generation that are lost due to lethality. Hence, the mutation rate was also estimated to be $\approx 0.03–0.05$ per gamete per generation. The authors assumed that one-half to two-thirds of the real genetic damage could not be discovered by analysis of stillbirths and infant mortality; for example, early embryonic death could not be detected. Taking this into account, a total mutation rate of 0.06–0.15 per gamete was calculated [1827], a value in line with Muller's estimate in his paper *Our Load of Mutations* [1835]. The reader should keep in mind, however, that this estimate rests on two assumptions:

1. That the higher incidence of stillbirth and neonatal death in offspring from consanguineous as compared with nonconsanguineous matings – as analyzed in their paper and leading to a high B/A ratio – is a true biologic consanguinity effect.
2. That lethal and detrimental genes reduce the fitness of the heterozygotes as well.

Much of the criticism of conclusions from the theory of genetic load is centered around these two assumptions.

Impact of the Genetic Load Concept on Human Population Genetics. The overall picture revealed by this concept is rather gloomy. Everyone is heterozygous for a number of genes that lead to genetically determined death not only in the homozygous state, especially in a consanguineous marriage, but in heterozygotes as well. There is a constant influx of new mutations at a high rate and with deleterious effects. Due to the adverse effects of these mutations, virtually everyone is less healthy and carries more defects than were he free from these mutations.

This concept had a strong impact on theoretical thinking and the planning of research in human population genetics. Partially, this effect may have been due to its intrinsic appeal, as research along these lines promised to give an overall view of problems crucial for the future of our species. The scientific reputation of the team that came forward with this concept may have contributed strongly to its success: Muller, the Nobel laureate who, out of deep concern for the future of our species, had left his fruit flies to help save mankind; Crow, the population geneticist of high reputation who guaranteed the solidity of the approach; and Morton, the brilliant young man whose imagination broke open the way toward a bright scientific future.

Discussions and Controversies Concerning the Load Concept. The concept of genetic load has been discussed extensively by population geneticists [1809, 1863]. On the one hand, it has been asserted that investigation of the outcome of consanguineous as compared with nonconsanguineous matings could contribute to the problem of whether detrimental mutations ("mutational load") or balanced polymorphisms due to heterozygote advantage ("segregational load") contribute more to the genetic load of the human species [1745–1747]. On the other hand, it was shown that the genetic load concept might lead in some cases to absurd consequences [1809]. Presently, many geneticists seem to share the feeling that definition of genetic advantages and disadvantages in the basic mathematic model is too static, and, therefore, the concept should be applied with great caution. For a recent, somewhat more realistic variant of the concept see [1748].

6.3.2.2 Practical Applications of the Theory

Many attempts have been made to estimate the actual genetic load in human populations. Most of the older studies are based on the theory of genetic load and may be considered as practical applications of this theory. Some ot the more recent studies, however, are based on more direct medical evidence. They will be discussed in Sect. 6.3.2.4.

Attempts to Assess the Genetic Load by Consanguinity Studies. The effect of parental consanguinity on stillbirth frequency and childhood mortality has been examined in many studies. The most comprehensive and for many reasons most reliable set of data is from Japan [198, 1839, 1840, 1876].

Table 6.20 gives an overview of these studies up to 1972. The sizes of the noninbred control samples are not given; they are usually larger than the inbred samples. The B/A ratio shows variations between +15.2 and −5. A negative value means that childhood mortality was even lower in consanguineous than in nonconsanguineous matings. Formally, such a result would establish a "negative genetic load," which is nonsense biologically.

Most consanguinity studies on genetic load yielded B/A values between 5.7 and 7.8; a simple unweighted average gives a B/A ratio of 6.7. This result is probably exaggerated by a lower socioeconomic status of the consanguineous couples. Even though consanguineous marriages are socially much more accepted – and also more frequent – in Japan than in other, especially Christian, societies, socioeconomic biases are all-pervading and, moreover, variable in direction [1879].

Table 6.20. Results of various studies on effect of inbreeding on death during infancy, childhood, and young adulthood among Japanese. (Schull and Neel, 1972 [1879])

Investigator and Locale	*A*	*B*	*B/A*	Size of inbred sample	Ascertainment
	(For definition see p. 493)				
Watanabe					
Fukushima prefecture	0.0881	0.5157	5.8	4,594	Through child surviving to high school
Tanaka, Kishimoto					
Shizuoka	0.1253	0.7191	5.7	2,205	Through child surviving to elementary school
Schull et al.					
Nagasaki prefecture (Kuroshima)	0.0927	1.4074	15.2	223	Koseki and Catholic church records, followed (average) 15 years, deaths before age 20
Schull and Neel					
Hiroshima prefecture	0.0875	0.5317	6.1	1.697	Pregnancy registration at 5th month followed (average) to 8 years
Nagasaki prefecture	0.0986	0.1060	1.1	2,608	
Schull and Neel					
Kure	0.0929	0.0405	0.4	564	Pregnancy registration at 5th month, followed (average) to 15 years
Yanase					
Fukuoka prefecture					
Hs	0.0962	1.2535	13.0	277	Household survey, deaths before 6 years
Hi	0.1292	0.3308	2.6	304	
Ta-Ko	0.0916	0.9884	10.8	301	
Fujiki et al.					
Yamaguchi prefecture					
Mis	0.1222	0.3287	2.7	497	Koseki records plus household interviews, followed (average) to midchildhood
Nuw	0.1985	−0.8107	−4.1	234	
Kur	0.1936	−0.9608	−5.0	79	
Nagano					
Fukuoka prefecture (Fukuoka City)	0.0873	0.6765	7.8	5,953	Through elementary and junior high school followed through age 12
Schull et al.					
Nagasaki prefecture (Hirado)	0.1157	0.7703	6.7	6,626	Household survey, nonaccidental deaths largely through age 20
Freire-Maia et al.					
Japanese immigrants in Bauru state of São Paulo, Brazil	0.1378	0.6995	5.1	105	Household survey, subjects followed through age 21
Average	0.1036	0.6700	6.7	–	–

This effect is still stronger in studies carried out in South America, France, the United States, India, and Africa (Table 6.21). For easier comparison, calculation of the *B/A* ratios was restricted here to first-cousin marriages; hence, the data are not strictly comparable with the Japanese data of Table 6.20. However, the variability seems to be still larger than that between the various Japanese series. This is not surprising, as in European countries and the United States, consanguineous marriages have become very rare (Table 6.18). In countries with a strong Christian tradition, there is social pressure against such unions, and consanguineous couples differ in social and even psychological aspects from the average population, as explained above.

One way to eliminate at least a part of these biases is to use the children of brothers and sisters of consanguineously married persons as controls. In a study along these lines carried out in the Vosges mountains of France, [1768], 189 consanguineous marriages were compared with 646 control marriages. The difference in perinatal death rate was small and not significant for first-cousin marriages; it was negligible for the more remote degrees of consanguinity (Table 6.22). The difference in the number of sterile couples is significant between consanguineous and nonconsanguineous marriages. This result may, but need not necessarily, point to a higher intrauterine death rate. Moreover, we should always keep in mind that, in the B/A ratio, A contains not only the genetic component of mortality in a random mating population but all environmental components as well.[6]

Table 6.21. Mortality effects from inbreeding studies. (Fraser and Mayo, 1974 [1763])

Reference	Population studied		B/A from marriages of first cousins compared with unrelated couples
	Racial classification	Location	
Neel (1963)[a]	Blacks	Brazil	6.9
Neel (1963)[a]	Blacks	Brazil	7.6
Freire-Maia (1963)	Blacks	Brazil	9
Freire-Maia and Azevedo (1971)	Blacks	Brazil	3
Neel (1963)	Africans	Tanganyika	−1.0
Neel (1963)	Whites	Brazil	−0.0
Neel (1963)	Whites	Brazil	−0.6
Freire-Maia (1963)	Whites	Brazil	1.0
Freire-Maia and Azevedo (1971)	Whites	Brazil	3
Freire-Maia et al. (1963)	Mixed	Brasil	16.5
Neel (1963)	Whites	U.S.A.	7.2
Neel (1963)	Whites	Chicago, U.S.A.	6.6
Neel (1963)	Whites	Morbihan France	20.2
Neel (1963)	Whites	Loir-et-Cher, France	13.1
Neel (1963)	Whites	N. Sweden	−3.0
Kumar et al. (1967)		Kerala	20
Roberts (1969)[b]	Indians	Kerala	14.8

[a] Neel (1963) summarizes the work of many authors.

[b] Roberts modified the results of Kumar et al. to take account of consanguinity other than first cousin relationships.

Altogether, the data in Tables 6.20 and 6.21 are rather disappointing. Considering the wide spread of B/A ratios, we cannot even conclude with confidence that inbreeding as such enhances the risk of stillbirth and child mortality, though this conclusion is plausible. One conclusion, however, is obvious. The high B/A ratios detected in the data analyzed by Morton et al. [1827] were not found in any other studies.[7] They were most likely caused by ascertainment biases or by socioeconomic differences between consanguineous and nonconsanguineous matings.

Recessive Diseases and Congenital Malformations in the Offspring of Consanguineous Marriages. So far, the discussion has been very abstract. If children from consanguineous matings are stillborn more often or tend to die during infancy and childhood, the question comes up immediately: Why do they die? Do they suffer from known recessive diseases or from multifactorial threshold conditions such as malformations?

Here the most comprehensive sets of data come again from Japan (Table 6.23 [1876]). This table shows the incidence of major malformation in one of the Japanese series. There is a slightly significant difference between consanguineous and nonconsanguineous matings. In the same large cohort of newborn Japanese infants, the overall frequency of death together with major congenital defects was 4.3% in the control children and 6.2% in the offspring of first cousins. Most of the anomalies in the Japanese data were malformations – sometimes complex in nature – for which a recessive mode of inheritance had never been established. Even from the most detailed survey [198], the approximate number of identifiable diseases with a confirmed autosomal-recessive mode of inheritance cannot be established. This means that the huge amount of work

[6] Apart from the biases, the B/A ratio is also sensitive to details of statistical evaluation [1763].

[7] The French data in Table 6.21 refer to this material.

Table 6.22. The effects of inbreeding, Vosges, (France), 1968. (Jaquard, 1974 [103])

	First-cousin marriages		Second-cousin marriages		All couples	
	Consanguineous couples	Control couples	Consanguineous couples	Control couples	Consanguineous couples	Control couples
Mean number of children per family	4.2	5.1	4.8	4.8	4.5	4.8
Percent of sterile couples	–	–	–	–	6.9	4.6
Perinatal deaths (%)	11.1	9.0	8.0	7.9	8.9	8.5

Table 6.23. Distribution of children with severe congenital anomalies by city of origin and parents' degree of consanguinity. (Schull, 1958 [1876])

City	[a]	First-cousin marriages	First-cousin once removed marriages	Marriage of 2nd degree cousins	Parents unrelated	Total
Hiroshima	*n*	936	313	384	26,012	27,645
	m	17	2	4	293	316
	p	0.0182	0.0064	0.014	0.0113	0.0114
Kure	*n*	318	113	140	7,544	8,115
	m	4	2	1	58	565
	p	0.0126	0.0177	0.0071	0.0077	0.0080
Nagasaki	*n*	1,592	412	637	30,240	32,881
	m	27	4	8	300	339
	p	0.0170	0.0097	0.0126	0.0099	0.0103
Total	*n*	2,846	838	1,161	63,796	68,641
	m	48	8	13	651	720
	p	0.0169	0.0095	0.0112	0.0102	0.0105

Analysis[b]

	χ^2	DF	P
Cities	7.269	2	$0.02 < P < 0.05$
Consanguinous marriages *vs* normal marriages	11.775	3	$0.001 < P < 0.01$
Interaction	2.535	6	$0.75 < P < 0.90$

[a] *n*, number of children; *m*, number odf children with severe congenital anomalies; $p = \frac{m}{n}$.

[b] Roy and Kastenbaum's method (1956) is used; see Schull (1958). There is a weakly significant difference between consanguineous and normal marriages in regard to frequency of major malformations.

invested in studies on consanguinity effects has not provided information that would enable us to allot at least part of the inbreeding effects to a clearly defined group of genes – those leading to recessively inherited diseases in the homozygous state.

Other Parameters Showing an Inbreeding Effect: Cognitive Abilities. In the course of several studies, additional parameters were examined for consanguinity effects such as anthropologic measurements, dental characteristics, blood pressure, co-ordination, visual and hearing acuity, intelligence, school performance [1840, 1879, 198, 1765]. On the whole and neglecting some inconsistencies among the various sets of data, there was usually a slight lowering of performance with inbreeding. This was especially interesting – and only partially accounted for by socioeconomic differences – for intelligence and school performance. After exclusion of socioeconomic factors by appropriate statistical techniques, there remained a decrease of about 6 points in the I.Q. value per 10% of F (= 1.6 times the inbreeding coefficient in first-cousin matings)[8] for the verbal and performance part of a standard intelligence test (Wechsler Intelligence Scale for Children, WISC). School performance was comparably lower [198].

The conclusion that inbreeding reduces average cognitive performance was corroborated by a study on 3,203 Arab school children in Israel. In this Arab population, the rate of first-cousin marriages is about 34%; in about 4%, double first-cousin marriages are found. Socioeconomic conditions were considered carefully; they were practically identical in the inbred group and the noninbred controls. The average performance in three different intelligence tests was significantly lower in the children from first-cousin and especially from double first-cousin matings than in the control group. Average school performance in four major subjects showed the same difference [1723].

Overall Estimate of Zygote Loss Due to Parental Consanguinity. Consanguinity data on intelligence or anthropological measurements are interesting from many points of view. However, they do not contribute to our knowledge of the influence of parental consanguinity on mortality of the zygotes before reproductive age. This is the only important parameter for the problem of "lethal equivalents." Tables 6.20 and 6.21 give some data confined to stillborns, i.e., conceptuses that were lost after the 7th

[8] For comparison: a child from a first-cousin marriage has $F = 1/16 = 6.25\%$.

month of pregnancy, and to those who died during early childhood. The data do not include loss of conceptuses due to miscarriages before the 7th month or those who survived from about the age of 8 up to adulthood. The latter period may safely be neglected, as mortality within this group is very low in most modern societies. Only very scanty data are available regarding spontaneous abortion. Schull and Neel (1972) [1879] attempted an estimate, which is shown in Fig. 6.46.

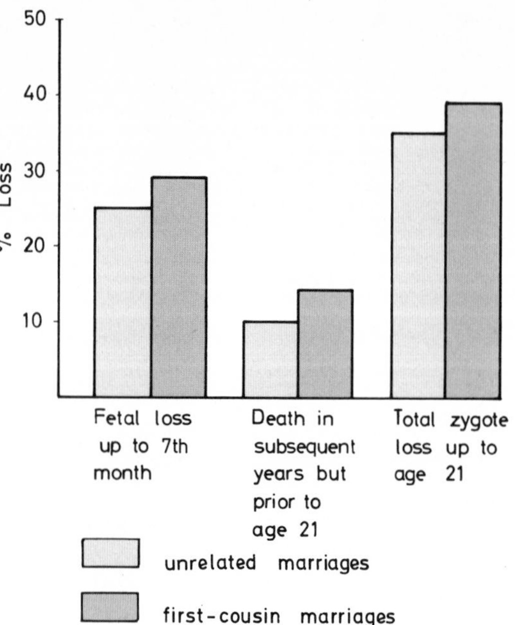

Fig. 6.46. Loss of zygotes in relation to parental consanguinity by miscarriage and by early death (8th month of pregnancy to the age of 21). (Schull and Neel, 1972 [1879])

6.3.2.3 *Critical Evaluation*

Theoretical Interpretation. Taken at face value, all these data could be interpreted in terms of lethal equivalents, as proposed by Morton et al. [1827], and their number could be estimated. However, the very fact of the variability of *B/A* values among the various studies (Tables 6.20 and 6.21) should make us skeptical. The genetic model underlying the analysis of these data assumes that the effect of lethal genes are independent of each other. Such an assumption is certainly an oversimplification that soon leads to conceptual difficulties which might be overcome by adoption of other genetic models such as assuming a certain threshold for the tolerable number of homozygote genes compatible with survival. However, even this adjustment does not remove the main difficulty inherent in the interpretation of such data: the lack of specificity of the phenotypic differences between offspring of consanguineous and nonconsanguineous marriages.

Medical Evidence. Let us look at stillbirth frequency and neonatal death – the most often used parameters. Both are known from medical information to be caused sometimes by genetic factors. In the majority of cases, however, no hereditary factors can be implicated. The placenta may detach itself prematurely, or the child is strangled by the umbilical cord, or its position is such that delivery cannot proceed properly, or delivery lasts so long that the child is suffocated, and so on. Death in infacny or early childhood may be due to infection, malnutrition, or many other reasons. For some fatalities a genetic component may often be plausible, for others occasionally, for still others not at all. Moreover, pregnancy supervision, improved obstetrical techniques – for example, frequent cesarean section – and better medical care for the newborn have succeeded in reducing the perinatal death rate to 1%–3% or less in all developed countries. This frequency is less than 10% of perinatal mortality around 1900 and much less than in the French rural samples described in Table 6.19. One might argue that many children suffer from a gene mutation that would be lethal under more primitive living conditions. However, such children do not appear to suffer any ill-effects following the critical perinatal period.

One example that has been analyzed fairly well is pyloric stenosis, which is more frequent in male infants. The recurrence risk is about 2%–6% for first-degree relatives of male probands and 10%–20% for first-degree relatives of female probands. All data point to multifactorial inheritance in combination with a threshold effect (Sect. 3.6.2); hence, a certain increase with parental consanguinity should be expected. Patients have a hypertrophic pyloric muscle that prevents the stomach from emptying its contents into the duodenum, and in earlier times, they often died as infants. Now, the surgeon cuts the pyloric muscle, which allows normal gastric emptying. This simple operation removes all ill-effects, and the children grow up to become normal adults. There are suggestions that they have well-developed muscles and score above-average success in athletics.[9]

If the effects of many other lethal equivalents causing early death are somewhat similar to this example, then "so what"? Let us live happily ever after with our lethal equivalents. But other causes of early death may not be quite as trivial. For example, the studies in Japan provide hints that one reason for the increased death rate of infants and children from consanguineous marriages may be a higher susceptibility to infections [198]. The reader is invited to con-

[9] C.O. Carter, personal communication.

sult textbooks on obstetrics and pediatrics for causes of stillbirth and neonatal and infant death. Some are genetic and many can now be defined exactly. Other genetic conditions will certainly be discovered by future research in medical genetics; large-scale consanguinity studies have so far contributed little to the identification of genetic causes of neonatal deaths.

6.3.2.4 More Direct Approaches for Estimating the Number of Recessive Genes per Individual

In recent years, more direct approaches to the estimation of the average number of recessive genes per individual have been used. These studies were limited to genes leading to abnormal phenotypes in the homozygote. This means that mainly that part of the overall "genetic load" was considered, which is relatively well-defined from the genetic and medical viewpoint. One approach is the study of children from incestuous matings between close relatives. Here the highest inbreeding coefficients of all human matings are observed ($F=1/4$ in father-daughter and brother-sister matings). Hence, if human beings are, indeed, often heterozygous for deleterious recessive genes, a high fraction of homozygous and, therefore, severely handicapped children should be detected among the offspring of such matings.

Studies on Children from Incestuous Matings [1713, 1734, 1867]. Four studies have been published on children from matings between first-degree relatives (Table 6.24). All investigations show that a high proportion of children from such matings are severely defective. Taken at face value, this result would suggest that the number of recessive genes carried in the heterozygous state by the individuals involved in these matings is very high indeed. Moreover, diseases with a known autosomal recessive mode of inheritance such as Sanfilippo disease, a mucopolysaccharidosis (Sect. 4.2.2.3), homocystinuria (Sect. 4.2), cystic fibrosis, or deaf mutism (see, however, App. 3) were found among these children.

If it is assumed that all severely defective children of these series are homozygotes for recessive genes, the number of such genes can be estimated [1856]. In the three series of Table 6.24, 85 of 190 surviving children (44%) are affected with a disease that could be caused by homozygosity of a recessive gene. If a man is heterozygous for one recessive gene (Aa), his daughter (or sister) has a chance of 1/2 to be heterozygous as well, and there is a risk of $1/2\times1/4=1/8$ that the child will be homozygous for the anomaly. If the man is heterozygous for n recessive genes, his child from a father-daughter or brother-sister mating has a chance of $(7/8)^n$ not to be homozygous and, therefore, to be unaffected. This opens the possibility to estimate n, the number of recessive genes for which an individual is heterozygous. In the data of Table 6.24, $(7/8)^n=1-0.44=0.56$; $(7/8)^4=0.586$; $(7/8)^5=0.513$. Therefore, the number of recessive genes per individual would be between 4 and 5, well in line with Muller's estimate.

Most of the defective children, however, suffered from more uncharacteristic abnormalities such as "uncomplicated idiocy." Moreover, the study comprising the most comprehensive data [1867] showed clearly that many of the adults involved in these incestuous matings were mentally subnormal. Therefore, it is extremely difficult to disentangle the contribution of autosomal recessive defects that were uncovered by close consanguinity, on the one hand, and the contribution of otherwise defective parental genotypes and adverse environmental conditions, on the other. How many children might have suffered, for example, from the fetal alcohol syndrome [1824]? Still, it is reasonable to assume that homozygosity may contribute a significant fraction of these defects. This conclusion would mean that, in the parent population, many individuals may be heterozygous for such genes.

Consanguinity in Parents of Severely Mentally Retarded Children. If heterozygosity for genes leading to severe mental retardation is relatively common, homozygotes of such genes should be found relatively often among severely mentally retarded individuals and one would expect an increase of parental consanguinity in this group. Moreover, sibs of such patients should fall clearly into two groups, normal and severely mentally retarded. This prediction is confirmed by some recent studies. In a study of 904 families of mental retardates form Israel [1743], the fraction of homozygotes was estimated to be 18% among severe retardates with nonconsanguineous parents and no affected sibs and $\approx 75\%$ among severe retardates with no affected sibs and first-cousin parents. Most of these probands had no metabolic diseases. In another study [1828] on 703 probands and their families in Hawaii, a certain increase of parental consanguinity, especially in a group called "biologic" and comprising mainly severely retarded patients, was observed. This, again, pointed to a certain fraction of cases with simple recessive inheritance.

Alternative Approach for Estimating the Average Number of Recessive Genes in Humans. With the increasing utilization of screening programs for inherited metabolic disease, it is now feasible to estimate the fraction of heterozygotes for such diseases ($2pq$)

directly from the homozygote frequency (q^2). Harris (1975) [93] based such an estimate on data from a comprehensive screening program in Massachusetts [1806]. From the data in Table 6.25, it can be concluded that, on the average, 11% of individuals in the parent population may be heterozygous for one of the 14 conditions screened in this population.

It is tempting to expand this consideration. Let us assume that there are genes for another 100 recessive diseases in the same population and that each of them has a homozygote frequency of 1:1,000,000 (gene frequency $q = 1/1{,}000$; heterozygote frequency $2pq \approx 1/500$). Neglecting multiple heterozygotes, this would add another 20% to the number of heterozygotes, giving an overall frequency of 11% + 20% = 31% heterozygosity for anyone of 114 (14 + 100) recessive genes in the general population. Since there are almost 1300 recessive diseases and the frequency of some of these is higher than 1 in 1 million, the figure of 31% is a minimum estimate of heterozygosity. As noted, a number of recessive genes leading to unspecified severe mental retardation seem to exist. All these genes together might even be fairly common. No quantitative estimates for the exact frequency of such genes are possible with current methodology.

Consanguinity Effects and the Level of Genetic Analysis. Analyses in population genetics will turn out to be the more satisfying the closer the objects of analy-

Table 6.24. Children from matings between first-degree relatives (father-daughter and brother-sister matings)

Authors	Country	No. of children	No. of severely defectives	Remarks
Adams and Neel (1967) [1713]	U.S.A.	18	6	In a control series of the same size that was matched for age, race, weight, stature, and social status, only one defective child was found
Carter (1967) [1734]	U.K.	13	8	
Seemanova (1971) [1867]	C.S.S.R.	161 (138 surviving)	60 of 138 survivors; (40 severely mentally defective)	In a control series of 95 children from the mothers involved in the incestuous matings with nonconsanguineous spouses, only five severe defects were found; no child was severely mentally retarded
Baird and McGillivray (1982) [1722]	Canada	21	9	

Table 6.25. Incidence of certain metabolic disorders among newborn infants in Massachusetts. (Levy, 1973) [1806]

Disorder	Total screened	No. detected	Incidence	Estimated frequency of heterozygotes (per 1,000)
Phenylketonuria	1,012,017	66	1:15,000	16
Cystinuria	350,176	23	1:15,000	16
Hartnup disease	350,176	22	1:16,000	16
Histidinemia	350,176	20	1:17,500	15
Argininosuccinic acidemia	350,176	5	1:70,000	8
Galactosemia	588,827	5	1:118,000	6
Cystathioninemia	350,176	3	1:117,000	6
Maple syrup urine disease	872,660	5	1:175,000	5
Homocystinuria	480,271	3	1:160,000	5
Hyperglycinemia (nonketotic)	350,176	2	1:175,000	5
Propionic acidemia (ketotic hyperglycinemia)	350,176	1	<1:350,000	3
Hyperlysinemia	350,176	1	<1:350,000	3
Vit. D dependent rickets (with hyperaminoaciduria)	350,176	1	<1:350,000	3
Fanconi syndrome	350,176	1	<1:350,000	3
				110 = 11%

sis are to gene action. One of the reasons for the satisfying results of analysis of natural selection on the hemoglobin variants is that these variants were analyzed at the direct level of gene action (Sect. 4.3). This permitted an incisive scrutiny of the mechanism of selection.

So far, the possible genetic damage by lethal or detrimental mutations as revealed by offspring of consanguineous as compared to nonconsanguineous marriages has mainly been analyzed at the biometric-phenotypic level: Genetic variability seems to be present, but simple modes of inheritance or specific genes have not been identified. As a consequence, data have been collected and possible consanguinity effects have been revealed. Interpretations, however, in terms of genetic mechanisms are extremely difficult and, for many questions, controversial. To overcome these difficulties, elaborate statistical techniques have been applied but mainly with the result that socioeconomic variability was shown to obscure most of the biologic effects, making specific conclusions regarding genetic mechanisms hazardous if not impossible. We shall see in Sect. 8 that essentially the same is true for many aspects of human behavior genetics. Analysis at the phenotypic-biometric level leads to ambiguous results.

More satisfactory results can be expected from studies on the frequency of heterozygosity for well-defined, recessive diseases, such as the metabolic defects included in population screening programs. These diseases have been analyzed at the single-gene level and often at the level of enzyme activity. Therefore, estimation of average heterozygosity per individual may lead to fairly clearcut results.

In studies on incestuous matings, many of the observed phenotypes are not so well-defined, but it is reasonable to assume that for some a single recessive mode of inheritance may also apply. The studies on the outcome of consanguineous marriages show again that a global, all-comprehensive approach with broadly defined phenotypes leads to less satisfactory results than an analysis of specific, well-defined traits or diseases.

Despite the mentioned difficulties, it is possible to ask a number of specific questions in consanguinity studies. What is the consequence of maternal (or paternal) inbreeding for the incidence of nondisjunction and of numeric chromosome abnormalities?

Effect of Long-Standing Inbreeding. What is the consequence of long-lasting inbreeding – as in South Indian populations, where uncle-niece unions with $F=1/8$ are socially preferred – for the frequency of recessive genes in the population? To what degree are these genes continuously being eliminated? Any hidden deleterious genes can be exposed by inbreeding only if the ancestors of individuals who undertake a consanguineous marriage had mated at random for many generations. A past history of many generations' inbreeding would have "cleaned" the gene pool from deleterious mutations a long time ago. Assuming a constant overall mutation rate, the overall selection against deleterious genes could adapt itself to this mutation rate, but the gene frequencies at equilibrium between mutation and selection would be much lower than in a population with a long history of random mating. Hence, in a population with a history of inbreeding, comparison of progeny from consanguineous and nonconsanguineous matings is expected to exhibit a smaller B/A value than in a population with a past history of more or less random mating.

This expectation has recently been confirmed in studies on fetal development, incidence of malformations, and reproductive wastage (abortions, stillbirths, death during first year of life) among the offspring of more than 20000 women in South India [1854, 1855]. As mentioned above, consanguineous marriages are very common in Dravidian speaking South Indian populations, and have been for a long time. In the sample examined for the study on reproductive wastage, almost 47% of women living in rural areas and 29% of those living in towns were married to a close relative; in at least 80% of cases to a maternal uncle ($F=1/8$). Among more than 70000 pregnancies, there was a marked difference between women living in rural and urban areas, rural women having suffered a much higher incidence of fetal loss and infant death during the first year of life. But there was only a marginal difference between consanguineously and nonconsanguineously married women either in the rural or in the urban subsamples, and no consistent increase with degree of consanguinity. In the study on fetal development and malformation, which was performed using a prospective research design in the same sample of mothers, more than 14000 pregnancies could be followed. No increased incidence of congenital malformations in comparison with the controls could be observed in children of consanguineous parents. There was no influence of consanguinity either on gestational age or on birth weight or body length.

The entire body of data appears to confirm the expectation that long-standing, high inbreeding depletes the gene pool of deleterious genes. However, in view of the difficulties discussed above in disentangling biological from socioeconomic effects, even this conclusion should be regarded with caution, although it is plausible.

This notion explains why, in theory, a comparison of two populations, one having a long history of close inbreeding and the other a history of nearly random mating, would be interesting. However, the two populations should live under otherwise similar conditions and should have a similar anthropological background. Populations with a long history of close inbreeding exist, for example, in some areas of southern India, where the uncle-niece union is the socially preferred one, and first-cousin marriages are also very frequent. In other South Indian populations, inbreeding has been so much lower that they could serve as suitable controls. Again, however, the problem for such a research project would be the suitable selection of gene-defined phenotypic characters. The aspects of consanguinity for genetic counseling will be discussed in Sect. 9.1.

6.3.3 Differentiation Between Population Subgroups: Genetic Distance

Real Mating Structure of Human Populations. The usual assumption that in human populations random matings prevail is an abstraction. Even beyond the circle of immediate relatives, the choice of mates is not random at all. It depends on distance of birth places, limitations due to language, race, or social class, religion, and other factors. At present, this nonrandomness has a tendency to decrease strongly in many societies. Throughout history, however, nonrandom matings have been much more frequent than currently. A restricted choice of mates was especially operative during the thousands of generations in prehistory during which our ancestors lived in small groups as hunters and gatherers. Studies of present-day populations still living under comparable conditions are especially well-suited to give us an impression of population structure at that time.

Population groups living in relative isolation from each other will gradually become different genetically - either under the influence of differing selection pressure or due to different adaptations to the same selective agent, or simply due to chance fluctuations of gene frequencies (Sect. 6.4). The ultimate goal of population genetics is a causal analysis of such population differences. In most cases, however, such a goal will be impossible. In the absence of more specific evidence, it is often useful to begin with the assumption that genetic similarities between populations are caused by common descent. Hence, the more similar two populations are, the closer their genetic relationship will be. This means that their separation is assumed to date back only a short time, if they are similar, and a longer time if they are less similar. Hence, we draw conclusions from the present-day genetic composition of population groups as to their history. This concept is in principle the method underlying any classification of human beings into subtypes such as races.

In Sect. 6.2.1.7, a well-analyzed example of this kind was shown: the spread of Hb E in the Austro-Asiatic language group of southeastern Asia. Here the mutation to the HbβE allele became frequent under the influence of a selective advantage due to malaria. It "migrated" with subpopulations of this group into various regions, and a high frequency of Hb E is now an indicator not only of continuing selection due to malaria but also of descent from a certain population group. Thus, the Hbβ E gene is an indicator of population history.

Population History or Selection? At the same time, the Southeast Asian example showed how ambiguous the distribution of such a genetic trait may turn out to be. The Hbβ E gene is usually rarer in populations of the great southeast Asian river plains than in the more malarious hills. This differential distribution, however, has nothing to do with a separation in remote history but is simply a consequence of less intense selection pressure due to malaria in relatively recent times.

Genetic similarities between populations are ambiguous. They may, but need not necessarily, reflect a common history. They may result from a parallel development under the influence of similar selection pressures. In the southeast Asian example, only a single gene has been used as an indicator of similarity between populations. One could argue that the greater the number of different genes included in such considerations the less likely it will be that all of them were subject to similar selection pressure. Therefore, similarity between populations in many different genes argues very much in favor of their relationship by common descent. Indeed, research work describing populations in terms of their overall similarities and differences has usually tried to include as many inherited characters as possible and has used methods of multivariate statistics. Nevertheless, conclusions from such studies should always be considered with the reservation that usually nothing is known of the selection pressures that have been involved.

Methods for Determining Genetic Distances. A number of excellent reviews on methods for assessing genetic distances between populations are available [36, 103, 1893]. For continuously distributed characters such as anthropologic measurements, the generalized distance D^2 of Mahalanobis is often used; a simplified version is the index $C_H{}^2$ by Penrose. For alternatively distributed traits such as genetic polymorphisms with simple modes of inheritance, the arc and chord measure of Cavalli-Sforza and Edwards (1964) [1735] has become popular.

6.3.4 Gene Flow

Apart from selection (discussed above) and chance fluctuations of gene frequencies (to be examined below), the composition of a population is also influenced by gene flow. The term "migration" is often used to denote gene flow from one population to another.

Effects of Migration on Gene Frequencies [124]. The effect of migration on gene frequency will be examined using a somewhat oversimplified model. A large population may be subdivided into many subgroups. The average gene frequency may be $\bar{q}$; each subgroup exchanges a fraction m of its genes with a random sample of the whole population every generation; q may be the gene frequency in the first generation in the subgroup to be considered. Then, gene frequency in this subgroup in the next generation will be:

$$q' = (1-m)q + m\bar{q} = \bar{q} - m(q-\bar{q})$$

$$\Delta q = q' - q = -m(q-\bar{q})$$

Δq is proportional to the deviation of the subgroup gene frequency q from the overall average ($\bar{q}$), as well as to m. In the long run, and in absence of other factors such as differential selection between subgroups, differences will level out, and all subgroups will have a common gene frequency, $\bar{q}$.

This model is not realistic, as immigrants will often come mainly from neighboring subgroups. If neighbors tend to deviate from the average population mean in the same direction as the "receptor" subgroup, the speed of the leveling-out between subgroups is reduced. For actual calculations, it is more realistic to regard $\bar{q}$ not as the overall population average but as the average of actual immigrants.

Migration and Selection. If the population subgroups are subject to different forces of selection, the process of leveling-out may be counteracted by selection. Three different situations may be distinguished.[10]

1. If rate of migration and intensity of selection are of the same order of magnitude, the gene frequencies of subgroups may remain very different from each other.
2. If selection intensity is much larger than immigration rate, the subgroup gene frequency will be determined largely by selection, with only a weak diluting effect due to migration.
3. On the other hand, if the proportion of immigrants is much larger than selection intensity, the effects of migration will override that of selection.

In any case, a stable genetic equilibrium between selection, on the one hand, and migration, on the other, may ensue. This situation is somewhat similar to that of the equilibrium between selection and mutation (Sect. 5.2).

Measuring the Admixture of Genes to a Population Subgroup. The proportion of genes a population has received from another by gene flow is often investigated.

Let q_a be the frequency of a gene in the "pure" ancestor population and q_n that of the same gene in the present-day hybrid population for which admixture of foreign genes is assumed. The frequency of this gene in the "donor" population may be called q_c. Then, the fraction m of genes in the present-day hybrid population that originated in the donor population can be estimated as follows:

$$q_n = mq_c + (1-m)q_a$$

and therefore:

$$m = \frac{q_n - q_a}{q_c - q_a}$$

The (large-sample) variance of m can be calculated as follows:

$$V_m = \frac{1}{(q_c - q_a)^2}[V_{qn} + m^2 V_{qc} + (1-m)^2 V_{qa}]$$

Rationale for Measuring the Admixture of Genes to a Population Subgroup. In recent years, the problem of how many genes from whites (and other groups) may be present in the United States black populations has received much attention. Such an estimation, while simple in principle, depends on a number of conditions that are difficult to meet [1857]:

a) The exact ethnic composition of the ancestral populations and the gene frequencies of the genes used for these estimates should be known.
b) There should be no systematic change in gene frequency between ancient and modern generations within either of the two populations for the gene or genes to be included in such a study. Such systematic changes may be caused by natural selection. In the American black populations, for example, the sickle cell gene is relatively common. This gene has attained its present frequency in Africa by selection due to falciparum malaria (Sect. 6.2.1.6), which does not occur in North America. Hence, the sickle cell gene must have been subject to negative selection due to segregation of homozygotes with sickle cell anemia in the United States; an estimate based on this gene would overrate white admixture.

However, this argument can be reversed: If an estimate of admixture is based on one or ideally many genes that fulfill the preconditions, the difference between this admixture estimate and the estimate from the gene under selection may be used to indicate selection and to measure selection intensity.

Estimates for Admixture of Genes from Whites to American Black Populations. The American black population is descended from slaves who were imported from West Africa (Nigeria, Senegal, Gambia, Ivory Coast, Liberia, and other areas). Gene frequencies of most genetic markers show definite variation in these ancestral populations. The same is true for the American white population, which originated from immigration of various northern, western, central, and southern European groups. Moreover, those genes that entered the gene pool of blacks may not be an unbiased and random sample of the genes of all North American whites. It is possible that certain segments of the white population were more extensively engaged in outbreeding than others. Still, care-

[10] For mathematical treatment of these problems, see Li [124] and, at a more sophisticated formal level, Jacquard [103].

ful assessment of possible biases may help to estimate the correct order of magnitude of admixture [1857].

Estimates for admixture based on blood group systems and serum factors thought to be under little or no differential influence by selection (Rhesus: Duffy; Gm serum groups) range between $m=0.04$ and $m=0.30$ for different black subpopulations. Estimates from southern rural areas give lower values than those from metropolitan cities such as Baltimore or New York City, which usually ranged above 0.2.

One example may show the method of estimation. The allele Fy^a of the Duffy blood group system has a gene frequency $q_c=0.43$ in American whites. In the population of western Africa, its present frequency q_a is less than 0.03; in most subgroups, Fy^a is completely absent. It may be assumed that its frequency might have been very low at the time when slaves were taken. In a present-day population of blacks in Oakland, Calif. ($n=3{,}146$), the Fy^a gene was found with a gene frequency of $q_n=0.0941\pm0.0038$; the corresponding value in the white population ($n=5046$) is $q_c=0.4286\pm0.0058$; q_a (gene frequency in the African population) may be assumed to be 0.0. The formula given above leads to the following estimate of admixture:

$$m=\frac{q_n}{q_c}=\frac{0.0941}{0.4286}=0.2195$$

If q_n is assumed to be 0.02, this estimate would be 0.181. Hence, the admixture of genes from whites as estimated from the Duffy blood groups accounts for 18%–22% of the gene pool of the black population in Oakland (Calif.). An estimate of admixture for the same population from AB0 blood groups leads to a similar result ($m=0.20$).

Evidence for Selection. As noted, admixture estimates for m calculated for genes that had been subject to selection in Africa can be used to examine whether and in which direction a change of the selection pattern has occurred in their new habitat. A number of studies on American blacks have consistently indicated higher estimates for three genetic markers: the sickle cell gene (Hbβ S); the allele for the African variant of G6PD (Gd^{A-}), and the haptoglobin allele Hp^1. As noted in Sect. 6.2, Hbβ S and Gd^{A-} in Africa are subject to selection by falciparum malaria; haptoglobin is a hemoglobin transport protein. The values for admixture were consistently higher than those derived from Duffy or AB0 blood groups; they ranged between about 0.49 (Gd^{A-} in Seattle in the northwest) and about 0.17 (Gd^{A-} in Memphis in the south). These results point to selection against these genes in the United States, a country without malaria. Because of lack of necessary data, the exact extent of this selection cannot be determined. It has been pointed out that because of the many biases in such studies other approaches should be used to confirm selective factors for genes that do not conform to the migration estimates by this method.

6.4 Chance Fluctuations of Gene Frequencies

6.4.1 Genetic Drift

Deterministic and Stochastic Models. So far, all our discussions have been based on Mendelian segregation ratios and the Hardy-Weinberg Law. Parameters, such as mutation rates, selective values, or inbreeding coefficients were treated as constants that have certain relationships with each other. The models were deterministic. In real life, however, these parameters are statistical variables that are subject to chance fluctuations. The processes examined in population genetics, such as changes in gene or genotype frequencies under the influence of these parameters, are not strictly deterministic; they are stochastic or random processes.

For infinitely large populations, chance fluctuations become negligible, and the stochastic results are approximately equal to those calculated from the deterministic models. Utilization of deterministic models is justified if we are interested in learning how, in principle, a certain parameter changes the genetic constitution of a population.

In one example – the dynamics of the alleles Hbβ S and Hbβ C in western Africa (Sect. 6.2.1.7; Fig. 6.27) – we explicitly considered fluctuations of gene frequencies under selection pressure in a finite population, but even if we compare the observed population frequencies of a blood group system with their Hardy-Weinberg expectations statistically, we implicitly consider chance fluctuations as well. Even if all available individuals could be blood-typed, there still would be chance deviations.

During most of human evolution, population size was relatively small, and the human species was split into several small breeding groups. Such isolates have been frequent up to recent times, and some are still in existence. Therefore, some of the theoretical consequences need to be considered.

Island Model. Let us look at an extreme example. On 160 solitary islands in the Pacific, one married couple each settles. All 320 individuals have blood group MN. Every pair has a son and a daughter who, in an incestuous mating, become the ancestors of each island population. No MN genotype has any selective advantage or disadvantage. Several hundred years later, we return to the islands

to examine the MN blood groups of their populations. What shall we find?
The founder population was identical genetically with respect to this character. Hence, our first assumptions will be that the same holds true for their progeny, as there has been no selection and - let us also assume this - no new mutation. Upon somewhat closer scrutiny, however, this assumption turns out to be wrong. Since, in all cases, both parents are MN, their children are expected in a proportion:

$$1/4\,\mathrm{MM}+1/2\,\mathrm{MN}+1/4\,\mathrm{NN}$$

This leads to the following probabilities for the genotypes of the sib pairs assumed to be the ancestors of the island populations:

$$\mathrm{MM}\times\mathrm{MM}=\tfrac{1}{4}\times\tfrac{1}{4}=\tfrac{1}{16}$$

$$\mathrm{MN}\times\mathrm{MN}=\tfrac{1}{2}\times\tfrac{1}{2}=\tfrac{1}{4}$$

$$\mathrm{NN}\times\mathrm{NN}=\tfrac{1}{4}\times\tfrac{1}{4}=\tfrac{1}{16}$$

$$\mathrm{MM}\times\mathrm{MN}=2\times\tfrac{1}{4}\times\tfrac{1}{2}=\tfrac{1}{4}$$

$$\mathrm{MM}\times\mathrm{NN}=2\times\tfrac{1}{4}\times\tfrac{1}{4}=\tfrac{1}{8}$$

$$\mathrm{NN}\times\mathrm{MN}=2\times\tfrac{1}{4}\times\tfrac{1}{2}=\tfrac{1}{4}$$

It follows that in the first generation, on the average, ten islands will contain only MM individuals, another ten islands will have only NN individuals. On 40 islands, both children are MN. The other 100 islands contain two genotypes each: 40 MM and MN, 40 MN and NN, and 20 MM and NN.
The genotype frequencies of subsequent generations obviously will depend on this distribution. This consequence is most obvious for the ten islands with the MM × MM pair and the ten islands with an NN × NN pair. Their populations must be purely MM and NN, respectively. The other allele has been lost by chance - without any selective disadvantage. We may call these processes random fixation and random extinction, respectively.
M and N are common alleles. However, one of the few individuals founding such an island population may be heterozygous for a rare allele. This allele will have a good chance to become frequent in subsequent generations, unless eliminated by random extinction (see below). Such "founder effects" are common in human populations. For example, the dominant condition - porphyria variegata (17620) - is common in the Afrikaans-speaking white population of South Africa and was traced back to an early immigrant, one of the founders of this population group.

A More General Case. Returning to our island example, random extinction and, correspondingly, random fixation of an allele may occur not only in the first generation - as described above - but in subsequent generations as well. Its probability will increase with decreasing size, N, of the effectively breeding population.[11] Moreover, random extinction and fixation are only extreme cases; in less extreme cases, both alleles are maintained, but their frequencies fluctuate at random.
This fluctuation is frequently called "genetic drift." In the following, genetic drift will be examined in a slightly more formal way [124].

Let us assume an effectively breeding population of N diploid individuals. This population may be visualized as the result of a random sample of $2N$ gametes from the preceding generation. In this generation, let q be the gene frequency of the allele a, and $p=1-q$ that of the allele A. The number of genes a in the present generation of N individuals will follow a binomial distribution: $(p+q)^{2N}$. This means: the $2N+1$ possible values of the frequency q of a in this generation are:

$$0, \frac{1}{2N}, \frac{2}{2N}, \frac{3}{2N}, \ldots, \frac{i}{2N}, \ldots, \frac{2N-1}{2N} \tag{6.18}$$

and the probability for q to assume a certain value $q_j=j/2N$ is:

$$\binom{2N}{j} p^{2N-j} q^{j} = \binom{2N}{2Nq_j} p^{2Np_j} q^{2Nq_j} \tag{6.19}$$

Now, let $\delta q = q_i - q$ be the chance deviation of q from one generation to the next.[12] Then it follows from the above distribution that:

$$\sigma^2_{\delta q} = \frac{q(1-q)}{2N} \tag{6.20}$$

Hence, this variance - which describes the extent of the fluctuation in q from one generation to the next - depends inversely on N, the effective population size. One example:

$N=50;\quad q=0.5$

We get:

$$\sigma^2_{\delta q} = \sqrt{\frac{0.5\times 0.5}{2\times 50}} = 0.05$$

The probabilities with which the various values of q occur are given in Table 6.26.
This distribution may still be viewed from another angle. Let us assume a large number of loci, all with a gene frequency $q=0.5$ in the parent generation. Then, some of them will assume a higher and others a lower frequency in the next generation, according to the distribution set out in Table 6.26.

Decay of Variability [124]. The island example showed that a given allele may disappear by chance from the population; in this case, the partner allele is fixed. As seen from Eq. (6.19) and Table 6.26, this process has a certain - albeit usually small - probability in a finite population. Once fixation has occurred, however, it cannot be reversed. The probability of fixation (i.e., of q becoming 0 or 1) tends to 1

[11] Here the capital letter N is used, as n comprises the total population including the aged, the children, and the nonfertile.

[12] In contrast, Δq denotes the systematic deviation as caused, for example, by natural selection.

Table 6.26. Probability distribution of gene frequency q in $N=50$ children if $q=0.5$ in the parent's generation (Li, 1955 [124])

$q=$	<0.35	0.35 –0.40	0.40 –0.45	0.45 –0.50	0.50 –0.55	0.55 –0.60	0.60 –0.65	>0.65	Sum total
Prob.	0.002	0.021	0.136	0.314	0.341	0.136	0.021	0.002	1,000

with increasing number of generations. Hence, in the long run, a population group will sooner or later become completely homozygous even in the absence of selection, if neither mutation nor migration disturb the process. This phenomenon is called "decay of variability."

Let K be the rate of fixation or extinction of alleles per generation, $K/2$ being either the fixation or the extinction rate. Then it can be shown (cf. Jacquard [103]) that $K=1/2\,N$. Hence, in one generation, $1/4\,N$ of all such alleles will be extinct, $1/4\,N$ will be fixed.

We shall return to this discussion, and extend it, in the context of molecular evolution (Sect. 7.2.3).

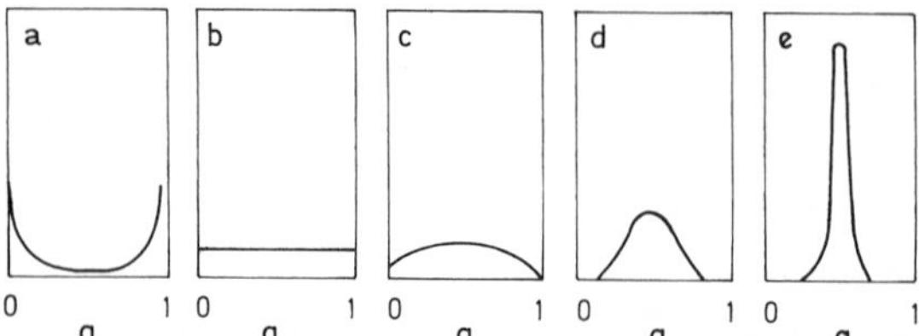

Fig. 6.47 a–e. Distribution of gene frequency q in small populations in relation to the effectively breeding population size N. Mutation rates μ and back-mutation rates ν are assumed to be identical. **a** $N\mu$ very small; **b** $4N\mu=4N\nu=1$; **c** $4N\mu=4N\nu=1.5$; **d** $4N\mu=4N\nu=10$; **e** $4N\mu=4N\nu=20$. With small population size (**a**), many populations will be homozygous for one of the two alleles ($q=0$ or $q=1$); with very large population size (**e**), q, clusters around 0.5. (Li, 1955 [124])

6.4.2 Genetic Drift in Co-Operation with Mutation and Selection

Mutation. Let us imagine a large population that is composed of a great number of small or moderately sized subpopulations. The distribution of gene frequencies q in these subpopulations depends on the size of the effectively breeding population (N), the mutation rate μ (A→a), and the back mutation rate ν (a→A). If the mutation rate is constant and identical in both directions (A→a; a→A), the distributions in Fig. 6.47 are possible, depending on N. The mean of q is identical in all cases (in our example: $q=0.5$). The variance, however, increases with decreasing N. With small N, the marginal classes close to $q=0$ and $q=1$ are strongly occupied, pointing to a high rate of random extinction and random fixation. With high N, on the other hand, the distribution clusters closely around the mean.

Fate of a New Mutation. In the foregoing paragraph, mutations occurring with a constant, recurrent rate have been considered. What is the fate of one single new mutation in a population? The mutation is present in one sperm or one oocyte. The zygote that is formed from these two germ cells will be heterozygous. Each child of the first carrier has a chance of 1/2 to carry this mutation. If he has two children, there is a risk of $(1/2)^2=1/4$ for the mutation to be lost by chance after one generation. On the other hand, the chance that the number of new alleles will be two in the next generation will also be 1/4. Fisher (1930) [1759] calculated the risk for extinction of such a new allele under the assumption that the number of children in sibships follows a Poisson distribution with mean 2 (Fig. 6.48). If the mutation is neutral, i.e., if it has neither a selective advantage nor a disadvantage, the risk of its disappearing from a population sooner or later is overwhelmingly high. This remains true even if the mutation has a small selective advantage. Such a risk holds, however, only for an infinitely large population; chance fluctuations in a small population may still lead to fixation of this mutation.

Selection. Co-operation between genetic drift and selection may be more complicated, as there are many different modes of selection (Sect. 6.2.1). In Fig. 6.49, the homozygotes AA are assumed to have fitness 1, the heterozygotes Aa, $1-s$, the homozygotes aa, $1-2s$. In an infinitely large population and in the absence of mutations, the gene frequency q will approach 0. For a great number of small subpopulations, q will most often become 0. In some cases, however, it will assume a higher value, and in a few cases fixations will occur at $q=1$ – in an opposite direction to the selection pressure.

Quite a different distribution of q among subpopulations is found if heterozygotes have a selective advantage (Fig. 6.50). If the disadvantage of both homozygotes is identical, the values of q will cluster around $q=0.5$, but fixation or extinction will still have a small probability of occurring that, again, de-

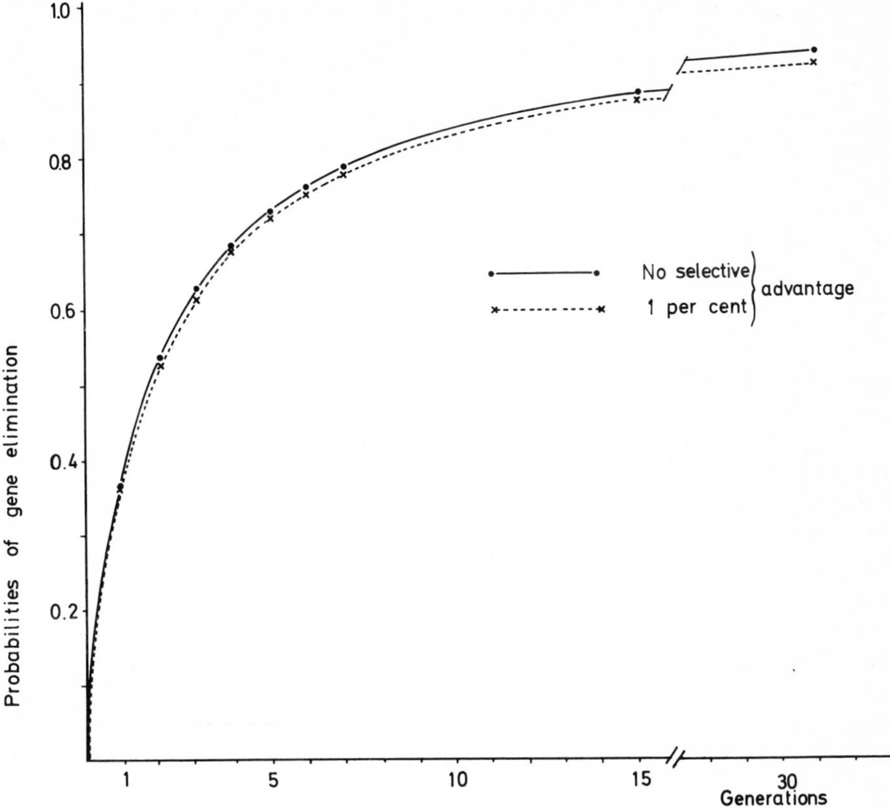

Fig. 6.48. Increase of the probability that a new mutation will disappear from the population. With 1% selective advantage, this probability is almost identical to neither a selective advantage nor a disadvantage. Sibship size is assumed to follow a Poisson distribution with a mean of 2. (Fisher, 1930 [1759])

pends on the effective breeding population size, N. Table 6.27 contrasts the patterns of spread of new genes by drift and selection.

Mutation and Selection Together. Let us now turn to the distribution in subpopulations of gene frequencies of a completely recessive gene with a mutation rate μ, a selective disadvantage s of homozygotes aa, and, again, an effectively breeding population of size N. Let us also assume that q is small so that back mutations a→A can be neglected. Then, the distribution of q for the extreme case $s=1$ (complete selection against the homozygotes aa) and for different values of N is given in Fig. 6.48. The mutation rate is taken as $\mu=10^{-5}$, in accordance with the order of magnitude of the human mutation rates for some visible inherited anomalies (Sect. 5.1.3). It turns out that, even with a moderate population size of $N=10^4$, the majority of subpopulations will not harbor the recessive gene at all. In a few populations, however, the gene will be much more frequent than in the general population. The affected homozygotes will still show much greater frequency differences, since their number corresponds to the square of the gene frequency.

Table 6.27. Patterns of spread of new genes by drift and selection. (After [1821])

Drift	Selection
Spread will depend on migration of people more than on environment	Spread will depend on environment more than on migration of people
Increase in gene frequency can be very fast if population is small	Increase in gene frequency is essentially independent of population size
Gene spreading may be advantageous or neutral	Gene spreads only if advantageous in a particular environment
Mutant gene will be "identical by descent" wherever it has spread	Different mutant genes at the same locus or at different loci that give similar phenotypes may be selected for (evolutionary convergence)

In principle, the same applies for dominant mutations, i.e., if the heterozygote Aa has a selective disadvantage. However, because of this selection against heterozygotes, an increase of gene frequency

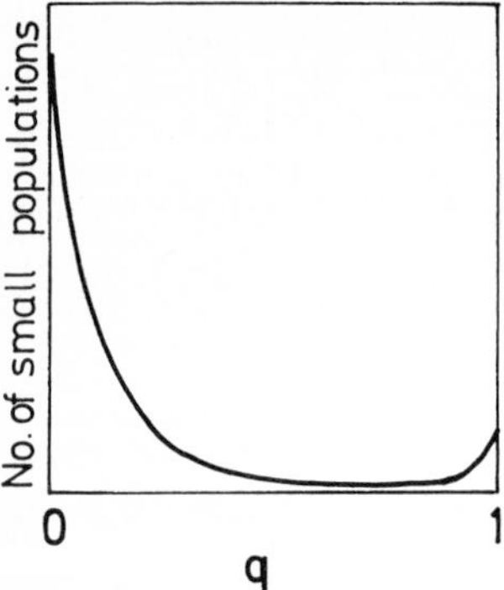

Fig. 6.49. Distribution of gene frequency q for gene a in small populations with $4Ns=5$ (s, selective disadvantage of heterozygotes Aa; $2s$, selective disadvantage of homozygotes aa). In most populations, the gene a will not be present at all. In a few populations, it will attain a moderate or high gene frequency, and in some, it will even replace the other allele. See text for other details. (Li, 1955 [124])

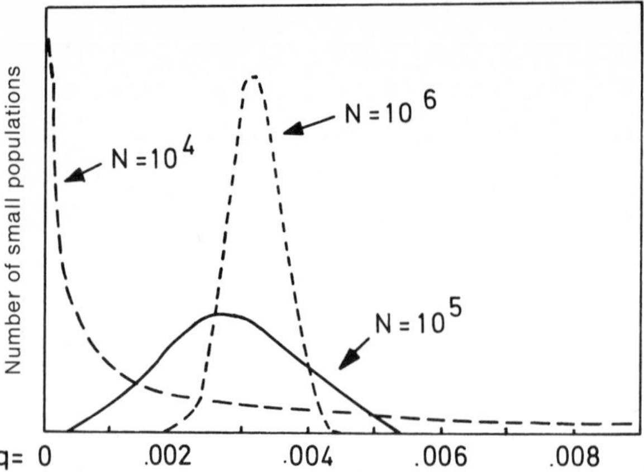

Fig. 6.51. Distributions of gene frequencies q for gene a in relation on the size N of the effectively breeding population with selection $s=1$ against the homozygotes and a mutation rate $\mu=10^{-5}$. The distribution of gene frequencies q depends critically on the size of the effectively breeding population N. (Li, 1955 [124])

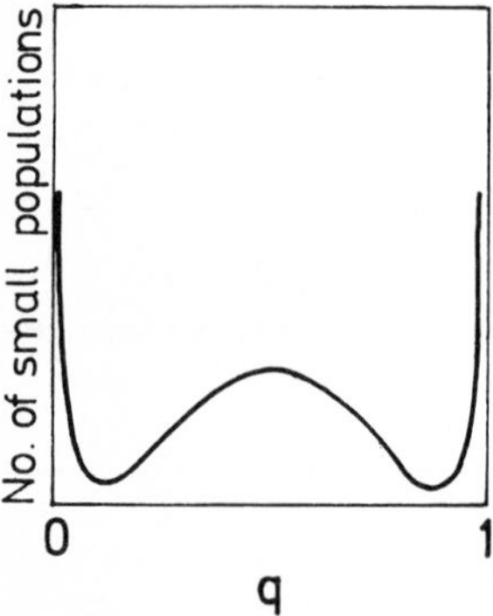

Fig. 6.50. Distribution of gene frequency q for gene a in small populations with $4Ns=10$ and selective disadvantage s of both homozygotes compared with the heterozygotes. The distribution shows three maxima: in many populations, gene frequencies cluster around 0.5; however, there are other populations in which either the allele a or the allele A are completely lacking. (Li, 1955 [124])

by chance in spite of negative selection pressure becomes less likely. Therefore, dominant disorders will become frequent in this way only if they cause very little selective disadvantage. With this mode of inheritance, the increase of individuals with aberrant phenotypes corresponds only to the gene frequency, not to the frequency of homozygotes.

Rare Inherited Diseases in Human Isolates. These considerations explain why hereditary diseases - especially recessives - may occasionally become frequent in small populations that have been living in relative isolation for a long time. Either the allele was introduced by a founder (founder effect) or it happened to be produced by a new mutation. Usually this distinction cannot be made. In any case, such a gene had a chance to become frequent even in spite of selection against it.

This is one reason why isolates provide much information on rare recessive diseases. Furthermore, consanguineous matings have become less frequent in most modern nonisolate populations. The incidence of homozygotes for rare recessive diseases is therefore reduced below the equilibrium value (Sect. 6.3.1). In relatively isolated - and in many cases rural - populations, the traditional ways of choosing a husband or wife and, hence, the old level of inbreeding tend to be maintained. Consequently, there is no overall drop in homozygote frequencies, and the average frequency of recessive homozygotes tends to be higher in isolates than in the general population. This fact, together with the unequal distribution of gene frequencies, makes isolates well-suited for the discovery of hitherto unknown recessive conditions. Another pssible factor may be selective migration. For a relatively long time, the generally fitter and more active elements have migrated from the isolates into cities and industrial centers. It has been shown, on the other hand, that heterozygotes for some severe recessive abnormalities might in some cases suffer from minor symptoms (Sect. 4.2.2.8) and might be underrepresented in this migration group.

On the other hand, however, investigation of isolates has brought a certain one-sidedness into our knowledge of inherited diseases. The method might be compared with the microscope: It enables us to examine special parts of a structure very exactly, whereas other parts escape observation completely. There is only one difference. We can direct the mi-

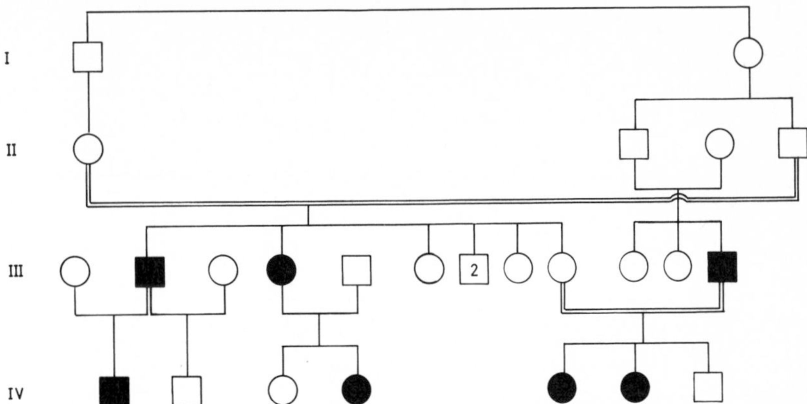

Fig. 6.52. Pedigree from the island Mljet with consanguineous matings and cases with Mal de Meleda. This recessive disease attained a high incidence in the population of this island but is almost entirely unknown elsewhere. (Courtesy of Dr. U. W. Schnyder)

croscope to the areas in which we are interested. In isolate studies, on the other hand, chance directs the microscope.

Example: Mal de Meleda. Mal de Meleda is an autosomal recessive condition categorized in the group of palmoplantar hyperkeratoses; yet, unlike the most common dominant type (Unna-Thost), the keratotic changes are not confined to the palms and soles but may affect other parts of hands and feet. The anomaly was discovered about 150 years ago on the small island of Mljet (Meleda) off the Yugoslavian coast. The population numbers only a few thousand individuals; in 1930, no fewer than 14 of 93 marriages were between first cousins [1731]. In the 1960s, many cases carrying the anomaly were still alive. A typical pedigree is seen in Fig. 6.52. Due to the high frequency of heterozygotes and the great number of consanguineous matings, marriages between affected homozygotes and heterozygotes are not rare, and the phenomenon of "pseudodominance" (Sect. 3.1.3) can be observed. Outside Mljet, similar cases have been observed only rarely.

There are many other examples of recessive diseases that have been discovered or described in detail in isolates.

One of the first was classic myoclonus epilepsy described by Lundborg [1820] (25480) as early as 1913 in the Swedish province Blekinge. Other examples include the juvenile type of amaurotic idiocy (20420), which was described by Sjögren [1870] in a Swedish isolate, Friedreich's ataxia (22930) and a special type of dwarfism investigated by Hanhart in isolated valleys of the Swiss alps [1785]; Werner's syndrome (27770) (Sect. 5.1.6.7) in Sardinia [1736]; the Ellis-van Crefeld syndrome (chondroectodermal dysplasia; 22550) in the Amish, a religious sect living in Pennsylvania [1825]; and tyrosinemia (27670) among French Canadians [1726]. Diseases that are fairly common might show additional clustering due to genetic drift and founder effects in isolated population groups. Examples are the high frequency of cystic fibrosis in a small area of Brittany, France (1:377 births; about six to eight times as common as in northwestern European populations) [1730] and the same disease (incidence 1:569) in the Ohio Amish [1802].

A special problem is posed by the high incidence of three lipid storage diseases due to defects of different lysosomal hydrolases (catabolic enzymes) among the Ashkenazi Jews of Eastern Europe [1771, 1837]. The conditions are: The infant form of Tay-Sachs disease (gangliosidosis G_{M2}; 27280), Niemann-Pick disease (sphingomyelin lipidosis; 25720), and the adult form (type I) of Gaucher disease (glucosylceramide disease; 23080) [1264]. On the one hand, some conditions in the history of this population seem to favor the hypothesis of genetic drift: During long periods of its history, this population lived as a religious minority in relative isolation, and some claim that the ancestral population of current Ashkenazi Jews was fewer than 10000 people at the beginning of the ninth century. Upon closer scrutiny of history and demography, however, this evidence becomes more dubious [Neel: in 1771]: The population was subdivided into many, sometimes widely scattered, isolates; the population size was very large for drift to be effective, at least in certain periods, and there was some dilution of the gene pool by admixture. On the other hand, even if all these arguments could be explained away the fact would remain that no less than *three* pathogenetically and biochemically similar genes have spread in the same population. In our opinion, this can only be explained by some probably very specific selective advantage of heterozygotes under the special living conditions of this population in the past. Tuberculosis has been suggested as a selective agent, but the fraction of the Ashkenazy Jewish population that appears to be resistant to tuberculosis is much larger than the 4% who are carriers for Tay-Sachs disease. Neel further objects that other urban populations were also infested with tuberculosis, as well, and these genes have nonetheless not become common. The problem of cooperation between selection and drift will be taken up again in Sect. 7.2.3.

"Rare Flora in Rare Soil": Hereditary Diseases in Finland [1842]. With the growth of the world population during recent centuries, some populations that began as relatively small and marginal isolates have developed into nations comprising several millions of people. When, due to favorable geographic and political conditions, this population growth occured

Fig. 6.53. Origin of patients with three recessive diseases in Finland. *Left:* Congenital chloride diarrhoea; greatgrandparents of 11 evident and 3 probable sibships (64). *Middle:* Congenital nephrotic syndrome of Finnish type; 60 grandparents of 57 sibships. *Right:* Cornea plana congenita recessiva. Grandparents of 32 sibships. (Norio et al., 1972 [1842]) (*Stippled areas* refer to pre-world war II borders).

without much disturbance due to migration and admixture by other populations, the random sample of recessive genes present by chance in the founders will be represented in the current population. Such a nation will display a unique collection of recessive diseases. Genes that are relatively common elsewhere will be lacking, and other genes will be found that are unknown in other areas. Since in historical times population migration and gene flow were intensive, especially in populations of European origin for which most information is available, examples of undisturbed growth of a relatively isolated population are rare. The best is the Finnish-speaking[13] population of Finland. Several favorable features are: an undisturbed, "traditional" population structure; an advanced level of medicine permitting reliable diagnosis of rare diseases, and excellent church records, which serve as a reliable population register for roughly the last ten generations.

Population History of Finland. Most ancestors of the current Finnish population immigrated during the first millennium from the Baltic region over many centuries. They were, like the Estonians, descendants of a common basic population, the Baltic Finns, who are also members of a special language group. It is possible that their immigration came to an end long before historical times, i. e., 1,000 A. D. They settled in the southwest of the country; the scarcity of burial grounds and other remnants indicates that the number of early settlers was probably very limited. In the following centuries, settlement spread slowly toward the north and east. The total population is estimated to have been 400,000 during the seventeenth century; it reached 1.6 million in 1850, and 4.6 million in 1970. Up to recent times the population has been largely rural; this has changed now, as with industrialization many families have been moving into the big cities. However, there still is little migration between rural areas.

The slow immigration of a limited number of settlers and the relatively independent growth of subpopulations with little gene flow between them afford the best conditions for founder effects and for subsequent shifts of gene frequencies within these subpopulations by genetic drift.

First-cousin marriages were prohibited by law prior to 1872, but more remote, multiple consanguinity between spouses has been, and is still, very common. We might expect to find the following:

a) A relatively high incidence in certain subpopulations of recessive disorders that are otherwise rare or unknown, along with lower frequencies in adjacent Finnish subpopulations due to relatively recent migrations.
b) Very low frequencies of these disorders in the big cities, which are melting pots for the overall Finnish gene pool.
c) Very low incidence or absence of some recessive diseases known in other, non-Finnish populations.

This is exactly what has been found.

Recessive Disorders in Finland. Table 6.28 shows a number of recessive diseases that have been found fairly frequently in Finland but only rarely or not at all outside this country. Other disorders listed in Table 6.28 have been observed in Finland as well as in other countries. Table 6.28 also shows disorders that occur more commonly in other populations but are rare or nonexistent in Finland. There can be no doubt that Finland presents, indeed, a number of otherwise very rare or even unknown disorders, whereas some diseases prevalent elsewhere are miss-

[13] Finland also has a Swedish-speaking population group.

Table 6.28. Recessive diseases in Finland. (Norio et al., 1973 [1842])

Rare recessive diseases that are relatively common in Finland	Recessive diseases that are about as common in Finland as elsewhere	Recessive diseases otherwise relatively common that are rare in Finland
Congenital nephrotic syndrome (25630) (Finnish type)	Fructose intolerance, Tyrosinemia	Galactosemia
Aspartyl glycosaminuria (AGU) (20840)	Mucopolysaccharidosis I (Hurler)	Hepatic glycogenoses
Early infantile ceroidlipofuscinosis	Mucolipidosis II (I cell disease)	Cystinosis
Progressive dementia with lipomembranous polycystic osteodysplasia	Adrenogenital syndromes (for example, 21-hydroxylase deficiency)	Maple syrup urine disease
Dystrophia retinae pigmentosa-dysacusis syndrome	Polycystic disease of kidney (perinatal form)	Phenylketonuria
Cornea plana congenita	Profound childhood deafness	Homocystinuria
Åland eye disease (X-linked)	Xeroderma pigmentosum	Cystic fibrosis of the pancreas
Lysinuric protein intolerance (LPI)		Tay-Sachs disease
Mulbrey dwarfism		Gaucher's disease
Cartilage-hair hypoplasia		
Diastrophic dwarfism		
Congenital chloride diarrhea		
Selective malabsorption of vitamin B12		

ing. Most conspicuous among the latter is phenylketonuria, which, in a comprehensive newborn-screening program, has been searched very carefully (Sect. 6.1.3, Table 6.4).

Attempts to locate the origin of specific "Finnish" disorders led to interesting results. They turned out to have their centers in limited geographic areas, with some scattering outside these areas, and to be virtually absent in the native population of the big cities. Figure 6.53 shows the birthplaces of ancestors (grandparents or great-grandparents) of patients with three of the recessive diseases given in Table 6.28. As noted above, dominant anomalies may also show founder and drift effects, provided there is no strong selection against these genes. A Finnish example is amyloidosis with corneal lattice-like dystrophy and cranial neuropathy (see also Sect. 4.6). The corneal lattice-like dystrophy manifests itself at the age of 20–35 years but affects visual acuity only slowly and to a moderate degree. Symptoms of amyloidosis only become visible in later life. Therefore, the disease does not impair reproduction very much. In Fig. 6.54, 207 patients are indicated by the birth place of the affected parent.

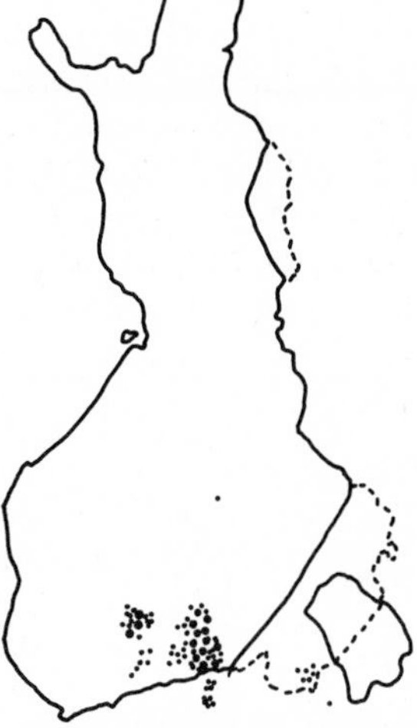

Fig. 6.54. A dominant gene in Finland: familial amyloidosis with corneal lattice dystrophy and cranial neuropathy, 207 patients are indicated by the birth place of the affected parents. *Big dot,* ten parents; *small dot,* one parent. (Norio et al., 1972 [1842])

Conclusions from the Experience in Finland for Research in Population Genetics of Rare Disorders. A population history and breeding structure such as that found in Finland still exists in many populations of the Old World. However, most of them lack the other conditions that are so convenient in Finland: the exact reports on family histories, the excellent medical facilities, and last but not least, research workers determined to exploit these possibilities. Most other populations having such medical facilities and research workers, such as those of the United States and most western and central European countries, have been subject to large-scale outbreeding by population mixing. Therefore, the opportunities for discovering new recessive conditions are unfavorable. In Sect. 3.1.8, it was stated that not all of the defects leading to hereditary disease in humans are as yet known and that many more could be discovered in countries that still have a traditional population structure. The considerations of this section have shown why this is so, and the example of Finland has proved that this statement is not merely a theoretical speculation.

7 Human Evolution

7.1 Paleoanthropologic Evidence

Population Genetics Helps to Understand Evolution. The concepts developed in population genetics and the examples from human genetics will help in understanding human evolution and the resulting genetic differences between humans and other mammals and especially between our closet relatives, the great apes. These concepts will also improve our understanding of genetic variability within and between present-day human populations. We shall discuss the evidence in three parts: In the very short introductory section, we shall sketch the evidence on human evolution from paleonthropology. This survey will lead us to a more elaborate discussion of the genetic mechanisms of human evolution. Finally, we shall discuss the genetic variability between present-day human population groups in the light of the genetic evidence.

Evidence from Paleoanthropology [186, 1976, 1987]. Evolution of the higher primates including *homo* is now fairly well-known in outline. We know the most important anatomic features of the main ancestral forms and partially their geographic distribution, and we can reconstruct their way of living. There are many differences in opinion between the various paleoanthropologists about details of human evolution and the position of single specimens and populations within this context. However, most specialists agreed up to recently on the following conclusions: *Homo,* as well as the great apes go back to a common ancestral population, the *Dryopithecinae* that lived in Africa about 15–20 million years ago. The branch that was to evolve to *Homo* split off from the *Dryopithecus* population, forming the *Ramapithecines.* The role of this group as a human ancestor has become dubious in recent years. The *Australopithecines,* a prehuman population that lived for a few million years, appeared in Africa after the *Ramapithecines.* The evidence points to three subpopulations in the *Australopithecus,* a robust type with especially large mandibles and teeth that lived mainly as vegetarians *(Australopithecus robustus),* a more gracile form *(Australopithecus africanus),*

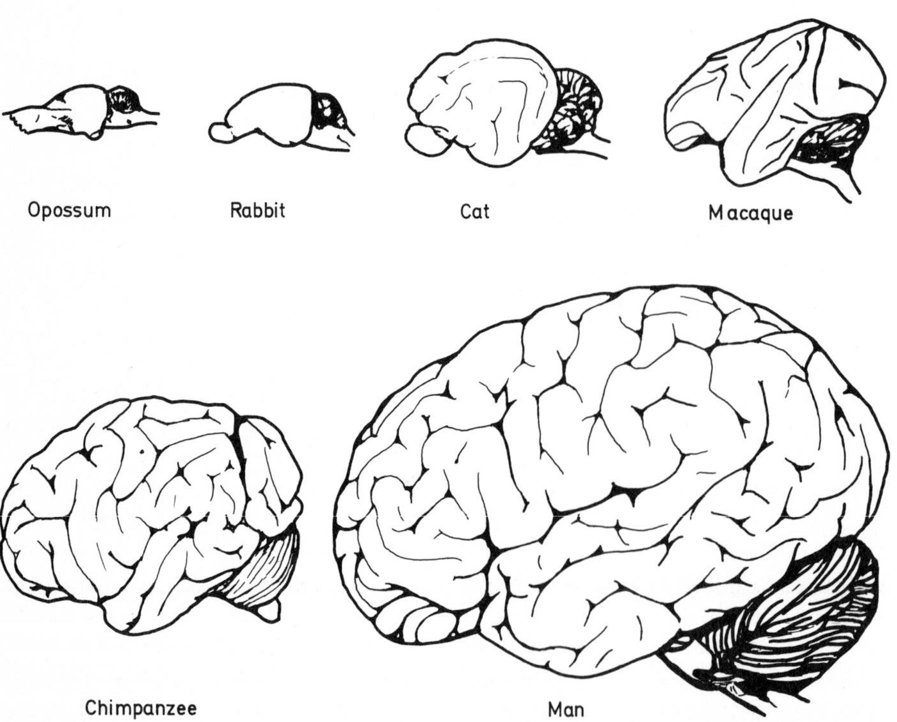

Fig. 7.1. The brain in various species of mammals. Note the developmental differences in size and structure of the cerebral cortex

and a third form that was characterized by a slightly larger brain size, gradually acquired the ability to make tools, and finally developed into *Homo erectus*, a variety that is classified as human from all available evidence. *Homo erectus* managed to leave the boundaries of Africa; important fragments have been found in Java, China, and Germany (Heidelberg). Attempts are often made to give a precise estimate of the time of emergence of the human species. The species probably developed gradually to its specifically human characteristics (see below) over many millions of years. There appears to be general agreement that the human species existed as long as 2-3 million years ago. With further development of mental capacities, the species conquered a progressively wider habitat. Another subtype, *Neandertal* man, turned up mainly in northwestern Europe during the ice ages, and finally, *Homo sapiens* took over, conquering the whole globe. The pedigree of *Homo sapiens* has been worked out in painstaking analysis by physical anthropologists using skeletal materials. Thus, the morphologic evidence will not be discussed here. Suffice it to say that the most important single feature was the development of the human brain during this period (Fig. 7.1). This trend is reflected in a crude way in the increase of skull capacity (Fig. 7.2).

At the same time, biologic evolution is supplemented more and more by cultural evolution until, at present, cultural conditions have become the main driving force for the biologic changes within our species that have to be expected in future. Table 7.1 contrasts biologic and cultural evolution in humans. Table 7.2 provides a synthetic overview of human evolution.

7.2 Genetic Mechanisms of Evolution of the Human Species

Genetic mechanisms involved in the development of the human species can be analyzed mainly by comparing present-day man with their closest phylogenetic relatives, the great apes. Our goal is twofold:

1. Establishment of the degree of relationship between species and construction of a phylogenetic tree that shows in which order these species have developed from common ancestral populations.
2. Analysis of the genetic mechanisms of evolution and speciation.

Construction of phylogenetic trees requires evidence from skeletal fragments of supposed ancestors as well as evidence from comparative anatomy and comparative genetics. For an analysis of the mechanisms of evolution, these skeletal remains are of no value. We have to rely on a comparison of genetic differences between present-day species. Here, intraspecies variability might help to understand interspecies differences somewhat better.

In the following, the available evidence will be discussed at four levels:

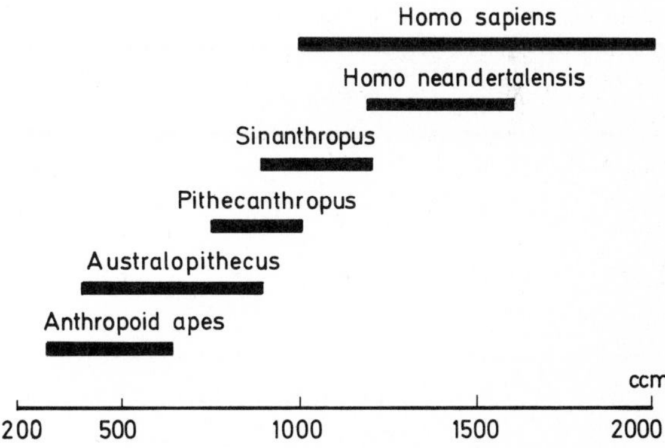

Fig. 7.2. Estimated interindividual variability of skull capacity in populations at different stages of human phylogeny. (Adapted from Heberer, in [8])

Table 7.1. Comparison of biologic and cultural evolution [1968]

	Biologic evolution	Cultural evolution
Mediated by	Genes	Ideas
Rate of change	Slow	Rapid and exponential
Agents of change	Random variation (mutations) and selection	Usually purposeful; directional variation and selection
Nature of new variant	Often harmful	Often beneficial
Transmission	Parents to offspring	Wide dissemination by many means
Nature of transmission	Simple	May be highly complex
Distribution in nature	All forms of life	Unique to man
Interaction	Man's biology requires cultural evolution	Human culture required biologic evolution to achieve the human brain
Complexity achieved by	Rare formation of new genes by chromosomal duplication	Frequent formation of new ideas and technologies

Note that the evolutionary destiny of all species was dependent upon interaction of genetic constitution and environment beyond its control. Man alone has the means to control both his environment and, to some extent, his genetic constitution.

Table 7.2. Evolution of man [1968]

Mean brain volume (cc)	Time scale		Tool use	Life style	Arts and language
	Years ago	Generations ago			
400-550	1.7 million	85000	Simplest stone & bone	Hunting & gathering	
900	600000	30000	More refined stone tools	Similar	
1300	50000	2500	Stone axes	Still hunters	Cave painting Early languages
	30000	1500			
	10000	500	Metal tools	Agriculture	Hieroglyphic, Iconic written languages
	8000	400			
	600	300	More complex tools & vehicles for transportation	Cities & agriculture	Alphabetized languages
	3500	175			
	300	15	Complex machinery	Industrialized centers	Printing
	30	1	Nuclear energy use	Atomic age	Radio, TV
	20		Computers	Postindustrial Age	

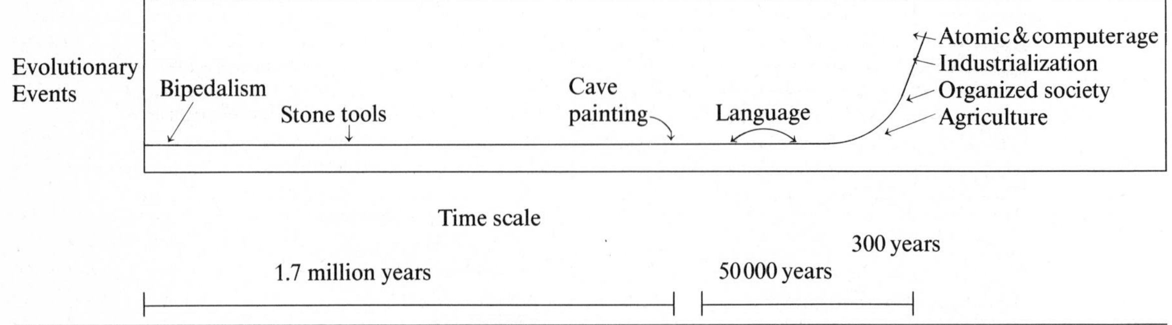

1. Chromosomes
2. Satellite DNA
3. Amino acid sequences of specific proteins
4. Behavior.

7.2.1 Chromosome Evolution and Speciation

Chromosome Number of Humans and Closely Related Nonhuman Primates [1912]. The following species will specifically be described: chimpanzee *(Pan troglodytes),* pygmy chimpanzee *(Pan paniscus),* gorilla *(Gorilla gorilla),* and orang-utan *(Pongo pygmaeus).* The chromosome number of all four species was soon established as 48; the main difference between *Homo* and both chimpanzees species was a fourth pair of acrocentric chromosomes of the D group (Fig. 7.3). In the other two species, gorilla and orang-utan, more acrocentric chromosomes were found. There was a strong overall similarity between the chromosomes of *Homo* and *Pan* that confirmed the evidence from morphologic and biochemical research (Sect. 7.2.3) that *Pan* is our closest living relative.

Comparison of Chromosome Structure with Banding Methods. Comparison of karyotypes between two species should help in reconstructing the number and kind of chromosome rearrangements that have occurred since these species have been separated in evolution. This reconstruction became possible when banding methods were introduced in 1970 (Sect. 2.1.2.2). Pericentric inversions were soon established as the mechanisms responsible for most of the species differences between man and apes [1929]. The difference in chromosome number could be explained by joining of two different acrocentric chromosomes about the length of D chromosomes to form one large submetacentric chromosome – the human No. 2. Such joining of different chromosomes is well-known from the present-day human population: the prevalent mechanism is centric fusion that implies loss of the short arms of a chromosome. It is therefore not surprising that centric fusion was assumed to be the mechanism for these species differences as well. However, a more detailed analysis of banding patterns showed that short-arm material is, indeed, present. The human No. 2 was produced by a telomeric fusion [1912].

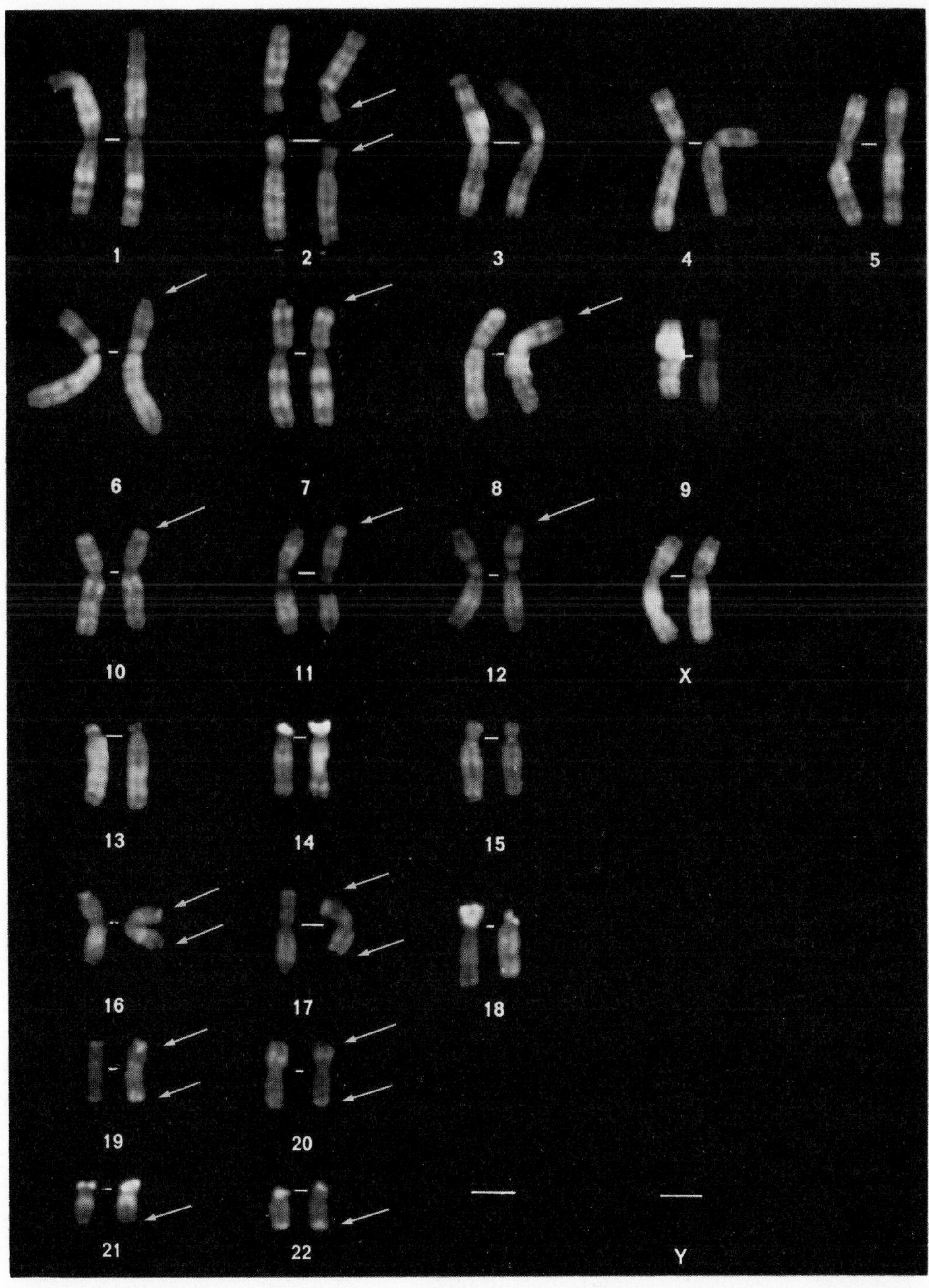

Fig. 7.3. Karyotype of a chimpanzee *(Pan troglodytes),* Q banding. A counterpart of the human chromosome 2 is obviously lacking. Instead, two additional pairs of acrocentric chromosomes are present. *Arrows* indicate terminal Q bands not present in humans (and orang-utan). (Courtesy of Dr. B. Dutrillaux, Paris). Numbers correspond to human chromosome numbers

Such a fusion chromosome has two centromeres; this should lead to mitotic complications known from dicentric chromosomes produced by interchanges, such as after radiation-induced chromosome breakage (Sect. 5.2.1.1). Such problems were avoided by only one of the centromeres carrying on its normal mitotic function. This suppression of one centromere has occasionally been observed in present-day chromosome aberrations as well.

The goal of a detailed analysis of all discernible rearrangements by which the species differ from each other and from man using all available banding techniques was achieved largely by Dutrillaux (1975) [1912].

Example. The homologues of the human No. 2 are seen in Fig. 7.4. *Pongo* and *Gorilla* differ by an inversion in 2 q, *Gorilla* and *Pan* by another inversion in 2 p. A telomeric fusion must have occurred between *Pan* and *Homo*. In some preparations, the human No. 2 shows a secondary constriction at the fusion point (2 qh). Very rarely, an endoreduplication of the segments corresponding to the former 2 q chromosome is observed and indicates some independence between the fusion partners. On the basis of these results, the evolution of No. 2 can be reconstructed.

Comparison of Overall Karyotypes of the Five Species. Species differences can be used to reconstruct evolution of all single chromosomes in the same way as shown above for No. 2. Apart from the one telomeric fusion and the pericentric inversions, a number of paracentric inversions have been observed. Table 7.3 shows the number and kind of chromosome rearrangements by which the species differ. As expected, the two species of chimpanzees show the highest degree of similarity. *Homo* is most closely related to chimpanzees; his most remote relative is *Pongo*. This could be expected from morphologic evidence.

Inconsistencies and Their Possible Explanation by an Unorthodox Principle. The interrelationship of the various rearrangements in the different species and their phylogency could easily be established for every single chromosome. An unexpected difficulty arose, however, when these chromosome phylogenies were combined to give an overall phylogeny of these species. *Pan troglodytes* and *Gorilla* have three inversions in common (5, 12, and 17), suggesting a common ancestor not shared by *Homo*, *Homo* and *Pan*, on the other hand, have two inversions (2, 7) in common that are not found in *Gorilla*. This finding suggests a common ancestor of these two species not shared by *Gorilla*. How can this dilemma be resolved? Dutrillaux (1975) [1912] suggested three possibilities (Fig. 7.5):

1. The common ancestral population may have shown a chromosome polymorphism for these rearrangements. However, this polymorphism must have existed for a very long time, probably over millions of years. Moreover, we know that pericentric inversion of the size observed here may lead to meiotic difficulties and, hence, to a reproductive disadvantage. Selection against heterozygotes that is involved here does not lead to a

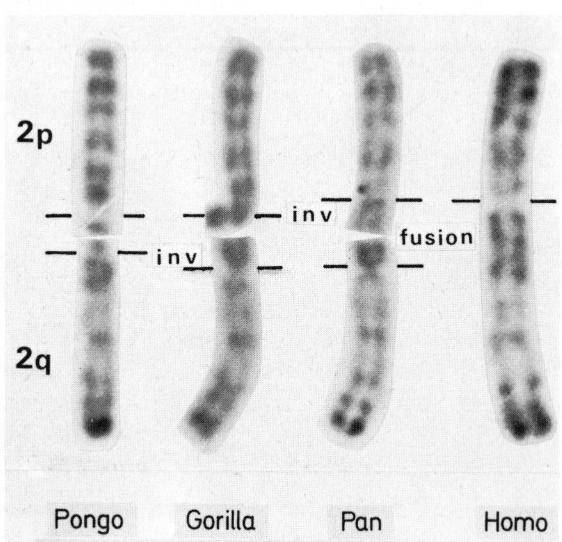

Fig. 7.4. Phylogenetic development of human chromosome 2 through some inversions and one telomeric fusion. G banding. *Pongo* and *Gorilla* differ by an inversion *(inv)* in 2 q; *Gorilla* and *Pan* differ by an inversion in 2 p. *Homo* differs from all the three other species by a telomeric fusion of two chromosomes. (Courtesy of Dr. B. Dutrillaux, Paris)

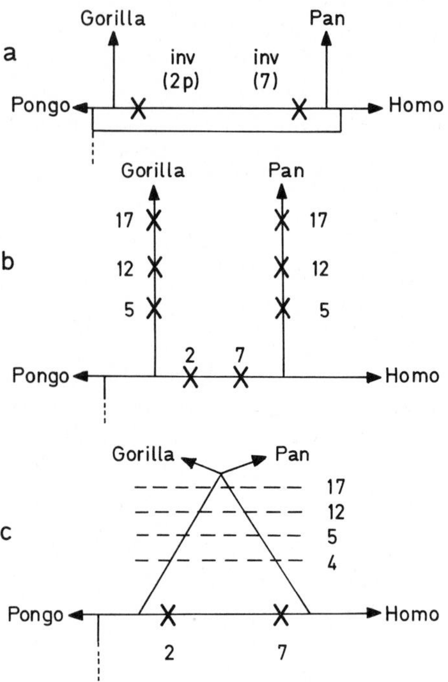

Fig. 7.5 a–c. Three hypotheses to explain the inconsistency in the phylogenetic tree of *Homo sapiens*, *Pan troglodytes*, and *Gorilla*. **a** Heterozygosity of the population for two inversions. **b** Parallel fixation of three inversions in the two phyla of *Gorilla* and *Pan*. **c** Separation of *Gorilla* and *Pan* from the human phylum; repeated hybridization between the ancestral populations of *Gorilla* and *Pan*, which makes the establishment of three chromosomal rearrangements in the hybrid population possible. Later, definite separation of the two phyla. This hypothesis is considered most likely. For details, see text. (Dutrillaux, 1975 [1912])

Table 7.3. Comparison of structural differences[a] between [81] the human karyotype and the karyotype of the great apes[b]. Numbers in parentheses indicate likely but not confirmed events

Comparison	Pericentric inversions	Paracentric inversions	Telomeric fusions	Some other rearrangements
H. sapiens-P. troglodytes	6	0	1	
H. sapiens-P. paniscus	6	(1)	1	
H. sapiens-G. gorilla	8	2	1	1 translocation
H. sapiens-P. pygmaeus	7	3	1	1 complex change
P. paniscus-P. troglodytes	0	(1)	0	
P. paniscus-G. gorilla	6	2(+1)	0	1 translocation
P. paniscus-P. pygmaeus	9	3(+1)	0	1 complex change
P. troglodytes-G. gorilla	6	2	0	1 translocation
P. troglodytes-P. pygmaeus	9	3	0	1 complex change
G. gorilla-P. pygmaeus	10	3	0	1 translocation, 1 complex change

[a] Additions of bands or heterochromatic material are not taken into account.
[b] Condensed from Dutrillaux (1975) [1912].

stable genetic equilibrium (Sect. 6.2.1.4). On the other hand, a high frequency of pericentric inversions, of almost polymorphic proportions, has been described in Finland (Sect. 2.2.2).

2. A second explanation would be independent fixation of three identical rearrangements in the phyla of *Gorilla* and *Pan troglodytes*. This event, however, is a very unlikely coincidence. Therefore, Dutrillaux regarded a third possibility as most likely.
3. Recurrent hybridization between the ancestral populations of *Gorilla* and *Pan* after separation from human ancestors. It is possible that a separation occurred between the ancestral *Gorilla* and the common ancestry of chimpanzee and man, possibly supported by ecologic isolation. Much later, the chimpanzee and human phyla separated; the *Pan* phylum moved into the same environment in which the protogorilla was already living (possibly the tropical rain forest). Here, hybridization could occur repeatedly, until the two species finally developed reproductive barriers, thereby completing the speciation process.

Another species difference that is hard to explain is the presence or absence of certain chromosome segments.

Presence and Absence of Certain Segments. In addition to the rearrangements described above, some of these primate species regularly show certain chromosome segments that are lacking in others:

a) Terminal Q bands: After quinacrine staining, small Q bands are observed at the ends of many chromosomes in *Pan* and *Gorilla*; they are in most cases located in the same chromosome arms in both species. They are lacking in *Pongo* as well as in *Homo*. There are two possibilities: Either they appeared as a fresh mutation in a common ancestor of *Pan* and *Gorilla* or they existed in a primitive ancestor and disappeared during the evolution of *Pongo* and *Homo*. Both events are difficult to understand on the basis of classic concepts: duplication or deletion of single chromosome segments as isolated and random events.
b) Heterochromatic regions are seen at the short arm of some acrocentric chromosomes. Their number diminishes in the following order *Pongo* → *Gorilla* → *Homo* → *Pan*. Presumably, such heterochromatic material is formed as a new mutation from time to time in immediate proximity to the centromeres of acrocentric chromosomes and is then distributed by random chromosome rearrangement to other chromosome parts. The secondary constriction of the chromosome 9 of *Homo* (Sect. 2.1.2) might be such material. The No. 9 of humans and of *Pan* also comprise a heterochromatic block close to the centromere.
c) Variations also exist in the occurrence of T bands. Repeated de novo synthesis of some T band material at the ends of the chromosome and – in some cases – secondary distribution to other chromosome parts by rearrangements is one possible explanation.
d) *Gorilla* and *Homo* show additional Q bands close to the centromeres of chromosomes 3 and 4; only *Gorilla* shows a similar band at No. 9, *Homo* on No. 13, and *Pan* only on No. 3. They are lacking completely in *Pongo* – and in the gibbon.

All three observations – terminal Q and T bands, heterochromatic regions, and juxtacentromeric Q bands – show that the karyotype differences between the five closely related species do not only consist in rearrangements of genetic material that can be explained using classical principles. An additional mechanism seems to be de novo synthesis as well as loss of chromosome material. Studies at the

DNA level have helped to understand the nature of such newly synthesized material somewhat better.

Recently, a tentative phylogenetic tree of the primates from the prosomians to man has been established by study of the karyotypes of more than 60 species of primates using almost all available banding techniques [1913]. The entire euchromatic material, i.e. the nonvariable R and Q bands (Sect. 2.1.2.1) appears to be identical in all species of *monkeys, apes,* and in *humans.* Quantitative and qualitative variations all involve heterochromatin. The types of chromosome rearrangements reconstructed from species differences in chromosome structure vary from one subgroup to the next; for instance, Robertsonian translocations (centric fusions) prevail among the *Lemuridae*; chromosome fissions are frequent among the *Cercopithecinae* but were not found elsewhere; and pericentric inversions were common in the evolution of *pongidae,* including *homo sapiens.*

Chromosome Rearrangements in Evolution and in the Current Population. There is one major difference between breakage points and chromosome rearrangements in evolution and in the current population. The most frequent chromosome rearrangements currently observed are centric fusions, i.e., stable connection between acrocentric chromosomes of the D and G groups with loss of short-arm material (Sect. 2.2.2). Surprisingly enough, not even one of these centric fusions has been fixed during evolution of the five species of pongidae, including homo sapiens. One explanation would be a selective disadvantage due to formation of aneuploidies, for example, zygotes with trisomy for the long arm of 21 q, which would produce Down's syndrome (Sect. 2.2.2), or those leading to miscarriage. To the best of our knowledge, however, not all centric fusions lead to a selective disadvantage. (For a more detailed discussion, see Sect. 2.2.2.2.)
Could the high frequency of centric fusions and the resulting zygote loss be a relatively recent genetic adaptation to the special conditions of childbearing and development in humans?

Selective Advantage of High Rate of Spontaneous Miscarriage in Humans [1937]? About 5%–7% of all recognized conceptions in humans are chromosomally abnormal (Sect. 2.2.4); most are lethal. They lead to miscarriage or – in exceptional cases – to delivery of a severely malformed child that would have no chance for survival under primitive living conditions. In the surviving infants – largely with X-chromosomal aneuploidies – fertility is considerably reduced (Sect. 2.2.3.2). Loss of additional zygotes before implantation goes unnoticed in most cases. At first glance, this considerable loss, caused by chromosomal aberrations, indicates considerable impairment in the reproductive fitness of our species. Upon closer scrutiny, however, the problem takes on quite a different aspect and led to the following hypothesis: In humans, offspring requires a high amount of parental care for an extended time period. Hence, there must be an optimum interval between live births to maximize the probability that a large proportion of the offspring will survive to reproduction. Any mechanism that tends to reduce the number of live births from the maximal to the optimal and the birth interval from the shortest to the best, without placing the mother's life into jeopardy, may have been of a selective advantage. Early miscarriage due to chromosomal aberrations may have fitted these requirements well under the primitive living conditions of our ancestors since mothers were not exposed to the risks of childbirth as frequently. Furthermore, the average breast-feeding period per child was extended in such instances and therefore may have protected an infant from the hazards of malnutrition and intestinal infections. The overall number of infants and young children per mother is clearly reduced by a high frequency of spontaneous abortions.
The higher incidence of centric fusions might also be caused by selection. The mechanism may be related to some function of the nucleolar region since the chromosomal areas involved in centric fusion are concentrated in the nucleolar region. This hypothesis would predict that centric fusions are rarer in other higher primates than in humans. Because of the small population size of all nonhuman primates examined so far, no incidence data for centric fusions – and for chromosome aberrations in general – are available. Trisomies, on the other hand – the most potent source of reproductive wastage due to chromosome anomalies – do occur in nonhuman primates, as shown by the occurrence of trisomy 21 in chimpanzees [1951]. These few observations do not permit any conclusions as to incidence; considering the relatively small number of chimpanzees under surveillance, trisomy 21 may not be much rarer in chimpanzees than in humans.

Homologies of Chromosomes and Chromosomal Segments Between Humans and Other, More Remotely Related Species. Homologies in chromosome structure and order of genes can be found not only between primates – including homo sapiens – but between more remotely related species as well. The mouse, the most commonly used mammalian model for many genetic problems, has been well studied in this respect, as the linkage maps of humans and mice are relatively well known (Sect. 3.4.3). And indeed, there is a remarkable tendency for linked human loci to be also linked in the mouse [1910, 1949]. These homologies are so strong as to suggest that various linkage groups have been kept together in different species by natural selection. This conclusion implies a functional significance of sequence and ordering of the genetic material beyond the level of single genes (see Sects. 2.3, and 3.5.5). The X chromosome has been conserved almost entirely through mammalian evolution [156]; at least ten X-linked loci have been found on the X chromosomes of mouse and man [1910, 1932]. Ohno [156] discussed the hypothesis that this might be caused by phenomena such as X inactivation and gene dosage compensation.
For the entire genome, the average length of the chromosome segments conserved since divergence of the ancestors of humans and mice was estimated by comparison of the two linkage maps to be ≈ 8.1 cM [1960]. This estimate was used to assess the number of chromosomal rearrangements fixed in the two ancient populations since this split has occurred.

How Can a Chromosome Rearrangement Be Fixed in a Population? As shown in Sect. 6.4.2, the overwhelming majority of all new mutations in a population will disappear. Such loss applies not only to selectively neutral mutations but even to those with a small selective advantage. Most chromosome rearrangements, such as pericentric inversions, will often have a selective disadvantage due to meiotic difficulties. On the other hand, it was shown by Kimura (1968) [1510] that the rate of fixation of a near-neutral mutation only depends on its mutation rate. Mutation rates for pericentric inversions are unknown so far. In the absence of this knowledge and of reliable information on the extent of the selective disadvantage - which may vary from one rearrangement to the other - no predictions as to probability of fixation can be made.

Yet, we have to keep in mind one special feature of selection against inversions. Such selection only works against heterozygotes. Inversion homozygotes have normal fertility, regardless of the site of the inversion, since pairing of homologous chromosomes at meiosis becomes normal. Do we know a genetic situation in which the "dangerous" state of heterozygosity is overcome quickly - even within two generations? Such a situation may occur when a brother and a sister inherit the same rearrangement from one parent and produce homozygous offspring in a brother-sister mating. Within this homozygous group, fertility would now again be normal, whereas matings within the general population would only produce heterozygous offspring with reduced fertility. Hence, this mechanism would build an effective reproductive barrier, providing the best conditions for gradual establishment of a new species (Fig. 7.6) [1947].

Present-day primates often live in small groups. Similar conditions might have been true for our prehuman ancestors. Therefore, the conditions for brother-sister matings or matings between other close relatives were favorable. If the "incest taboo" that prevents mother-son or brother-sister matings had already been observed in these ancestral groups, one or a few more generations of heterozygotes may have occurred before the two homozygotes that were to form the ancestral couple would be produced. Moreover, homozygosity of a pericentric inversion has been observed in the present-day human population in a child from a father-daughter mating [1902]. Are new primate species founded by one couple? More specifically: Do all human beings have in common one ancestral couple? The myth of Adam and Eve as the ancestral couple of mankind thus may even have a scientific basis.

A comparative study on 1511 species, representing 225 genera of vertebrates, has shown a strong correlation between speed of chromosome evolution and speciation, both processes being very fast, for example, in primates [1905]. The authors put forward evidence that population subdivision into small demes has probably been the decisive factor. This conclusion is in full accord with the hypothesis discussed above.

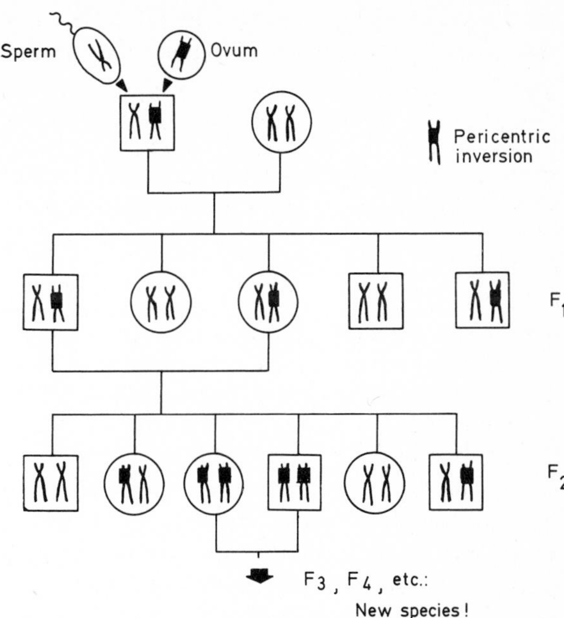

Fig. 7.6. Possible establishment of a new species by close inbreeding. A pericentric inversion is formed in a germ cell of one individual, giving rise to one heterozygous carrier in the next generation. This carrier may have a number of heterozygous offspring who may mate with each other and produce homozygous progeny

7.2.2 Comparison of Satellite DNA in Higher Primates

Human Satellite DNA. The establishment of a pericentric inversion in a population does not necessarily require mating between two pericentric inversion heterozygotes. A slow increase in the frequency of such an inversion in a limited population by chance or by drift (Sect. 6.4.2) - even in the face of small selective disadvantage - would be another possibility. However, such an explanation would meet with more difficulties in explaining another, recently discovered phenomenon: species differences in satellite DNA.

Human satellite DNA has been described in Sect. 2.3.1.1, where it was explained that the term satellite relates to observations of DNA centrifugation in a cesium chloride density gradient, which in addition to the main DNA peak show smaller peaks that are characteristic for each species (Fig. 2.80). Satellite DNA consists of relatively short, highly repetitive DNA sequences; their biological function is unknown, but they may influence crossing over during meiosis [437]. In humans, four satellite DNA fractions, SAT I-IV, have been distinguished, isolated, and characterized; they make up about 4% of the human DNA or ⅕ to ⅙ of the entire highly repetitive DNA. These four satellite fractions were transcribed to yield radioactive complementary cRNAs, which were hybridized in situ to metaphase chromosomes of humans and great apes (Fig. 7.7) to assess their evolutionary origin. These studies show how incomplete evidence may be used for the construction of a hypothesis which fits well into an established theoretical framework but has to be modified when more complete data become available.

As mentioned in Sect. 2.3.1.1, distribution of various satellite fractions on human chromosomes is unequal. Comparison with the apes showed that there is less homology between species for *distribution* of SAT DNA on chromosomes than for the banding patterns, though a certain degree of homology is observed. Human SAT III turned out to be very similar with chimpanzee SAT A; moreover, both fractions hybridized with chromosomes from *Gorilla* and *Pongo* [1938]. It was concluded that *Homo* shares SAT III with the three other species through a common ancestral form. On the other hand, SAT III did not hybridize with the DNAs of any of the lower primates tested. Hence, SAT III arose in evolution at a time subsequent to the divergence of the common ancestor of *Homo, Pan, Gorilla,* and *Pongo* from the other primates.

Human satellites I and II, on the other hand, did not hybridize in these studies with chimpanzee chromosomes or chimpanzee DNA. It was concluded that these fractions might be younger than SAT III, and might have appeared *after* the human ancestor had separated from the ancestors of the great apes. This conclusion was confirmed by hybridization characteristics.

This conclusion, however, had to be modified when SAT IV was included in these studies and when the various species were examined more completely [1928]. Figure 7.8 shows that *Homo* shares SAT I, III, and IV, but not II, with *Pan*. All four fractions are shared with *Gorilla,* and at least I, II, and III with *Pongo*. Hence, the picture drawn above does not hold up entirely against this new evidence. It now appears "that the fundamental sequences of all four satellites were present in the common ancestor of the species, ... but possibly only in one or a few copies per chromosome. The subsequent amplification of these sequences may have occurred after speciation, and, although the majority of amplification occurred at homologous sites in different species, sufficient differences exist to provide further evidence for the independent nature of this amplification event or events" (Fig. 7.8).

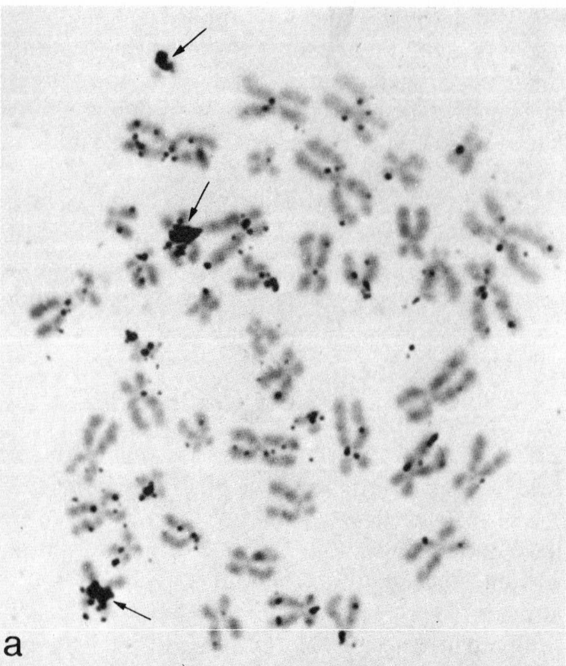

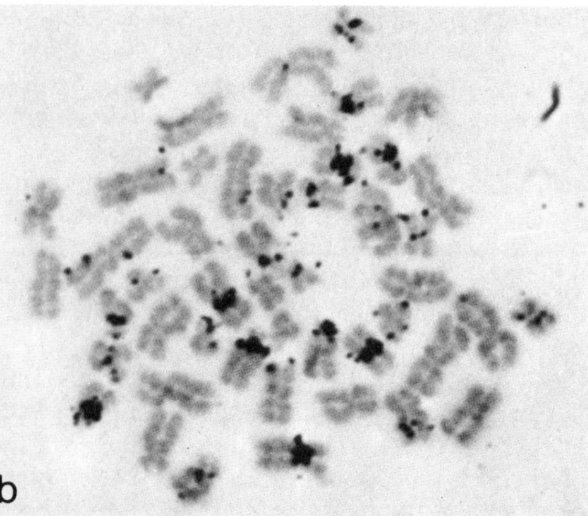

Fig. 7.7. **a** Autoradiographs of chromosomes of human male and **b** chimpanzee showing the locations of in vitro hybridized human satellite III RNA. In man, concentrations are found mainly in the centromeric heterochromatin of chromosome 9 *(arrow)* and the Y chromosomes *(arrow)*, with minor concentrations in other areas. In the chimpanzee, about five chromosome pairs show high concentrations of satellite DNA A, which is sufficiently related to human satellite III to form heteroduplexes. (Jones, 1976 [1938])

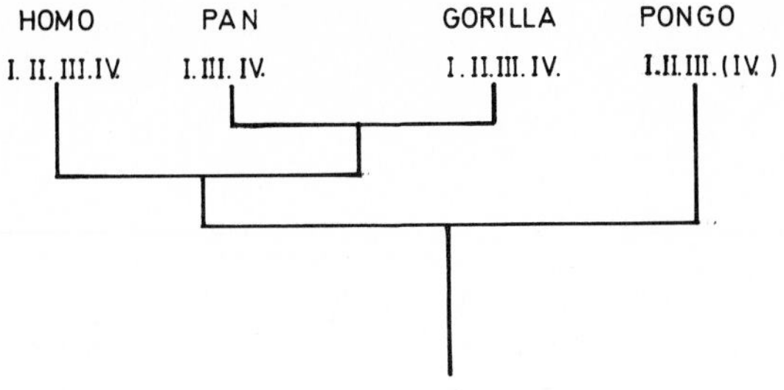

Fig. 7.8. Phylogeny of the Hominidae showing contemporary distribution of sequences homologous to the four human satellite DNA species I–IV. For explanation see the text

Comparison with Chromosome Evolution (Sect. 7.1.2). Differences between the karyotypes of *Homo* and the great apes were found in heterochromatin. They partially involved the centromeric regions. Additional telomeric regions showed species differences in Q and T bands that do not contain any of the satellite fractions identified so far, but still unidentified satellite fractions are likely to exist. As mentioned, euchromatic chromosome bands, which are thought to contain most structural genes (Sect. 2.3), appear to be identical in the primate species examined so far (Sect. 7.2.1). Variation is found in the satellite DNA and heterochromatin fractions. This points to a possible role of these fractions for evolution of specific human characters.

Species-specific satellite DNA fractions are known notably in the higher primates but in other species as well. Their possible biological function has been discussed in Sect. 2.3; their significance for evolution is still obscure.

7.2.3 ***Protein Evolution*** [1988]

Protein Sequences [51, 81]. One of the major achievements of biochemistry has been the determination of amino acid sequences within proteins. When the first sequences became known in the late 1950s and the early 1960s, homologies of sequences between homologous proteins of different species soon became apparent. Similarly within the same species homologies between different, but functionally related proteins were found. The sequences were usually identical at some positions; they showed differences in others. At that time, it was already known from study of some human hemoglobin variants that point mutations usually lead to the replacement of one single amino acid in a polypeptide chain. When the genetic code was deciphered, such replacements were shown to be caused by substitution of only one base in the transcribing DNA strand. Determination of biologic relationships between species by comparing the number of differences in amino acid sequences of their homologous proteins followed. Thus, phylogenetic trees were constructed that could be compared with those derived from classic paleontologic and morphologic evidence. The methods of tree construction are given in [51, 1919, 1921, 1954].

Phylogenetic Tree for Hemoglobin Genes [1991]. Figure 7.9 shows a phylogenetic tree for a number of vertebrate species, including man, chimpanzee, and gorilla using amino acid sequences of myoglobin and the Hb α and Hb β genes. The time scale comes from paleontologic evidence. This tree shows that there is only one base substitution for these genes between man and chimpanzee and three between man and gorilla.

Similar trees can be constructed for other proteins and – by combining the evidence – for all known proteins together. They can be extended beyond the vertebrates by including primitive animals and – for ubiquitous proteins such as histones or cytochrome C – even plants, fungi, and microorganisms.

Rates of Evolution for Different Proteins. The number of necessary mutations for a given number of steps of speciation can be compared for different proteins. Some proteins have turned out to evolve at a much higher rate than others (Table 7.4). The histones, for example, have been amazingly stable, whereas the fibrinopeptides evolved at a much faster rate. Remembering that only a minute fraction of all mutations are fixed during evolution – about 1 in 3.5 million on the average [1941], – we can discount differential mutation rates as a possible explanation; rather, the restraints imposed upon the amino acid sequence by protein function seem to be important. Fibrinopeptides, for example, are released in the process of fibrin formation from fibrinogen; its function is not very specific (covering of the fibrin surface required for formation of the fibrin network). This might explain the high rate of evolution. It is understandable, on the other hand, that histones are very much restricted in their conformation. They occur in the chromatin in intimate spatial relationship with DNA and therefore might be considered to assume a critical functional role that could be impaired by even small molecular changes.

Gene Duplications. As noted in Sect. 4.3, the molecule of hemoglobin A consists of two α and two β chains, Hb F has γ chains and Hb A_2 has δ chains instead of the β chains. All four types of chains have many homologous amino acids in common. The most obvious explanation is that all these genes – together with the gene for the myoglobin chain – are

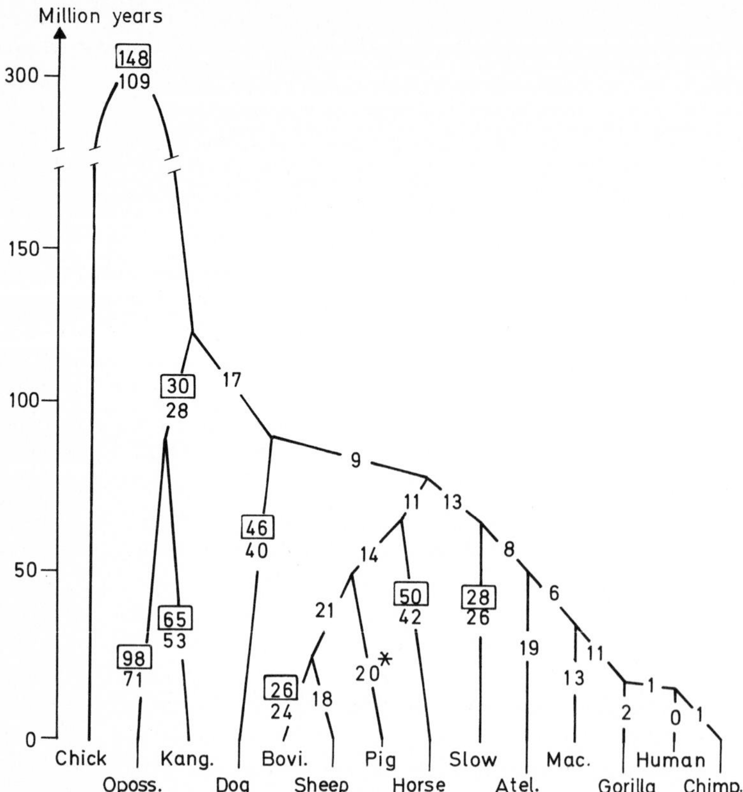

Fig. 7.9. Phylogenetic tree for myoglobin and the Hb α and β genes. The *ordinate* gives the approximate time of the splits between two phyla, on the basis of paleontologic evidence. The *numbers* refer to the number of nucleotide replacements between species. Numbers in □ refer to substitutions corrected for multiple-step mutations. The myoglobin sequence of the pig is still incomplete so that the number of replacements is slightly underestimated. (Goodman, 1976, [81])

Table 7.4. Rates of mutation acceptance in evolution [51]

Proteins	PAMs[a] per 100 million years
Fibrinopeptides	90
Growth hormone	37
Pancreatic ribonuclease	33
Immunoglobulins	32
κ chain C region	39
κ chain V regions	33
γ chain C regions	31
λ chain C region	27
Lactalbumin	25
Hemoglobin chains	14
Myoglobin	13
Pancreatic secretory trypsin inhibitor	11
Animal lysozyme	10
Gastrin	8
Melanotropin β	7
Myelin membrane encephalitogenic protein	7
Trypsinogen	5
Insulin	4
Cytochrome C	3
Glyceraldehyde 3-PO_4 dehydrogenase	2
Histone IV	0.06

[a] 1 PAM = 1 accepted point mutation/100 amino acid residues in 100 million years. An accepted point mutation in a protein is a replacement of one amino acid by another.

derived from one ancestral hemoglobin chain. Stepwise, functional differentiation required duplication of these genes so that one copy could maintain the original function whereas the duplicated one was free to acquire a new function. Figure 7.10 shows the steps of duplication for the hemoglobin genes, together with the evolutionary level at which they occurred and the approximate time scale.

Duplication of genetic material – either of individual genes or short chromosomal parts or of the whole genome (polyploidization) – have been of major importance in evolution. Apparently, polyploidizations have not occurred any more in the evolution of mammals [157]; minor duplications – and presumably also deficiencies – have been frequent.

Evolution of Genes for Protein Domains. So far, we have only considered changes in amino acid sequences. However, proteins have a characteristic three-dimensional configuration. This conformation is usually formed by a succession of two or more "domains," i.e., sequences with a molecular weight of 20000, within which there are many contacts, whereas much fewer contacts are found *between* do-

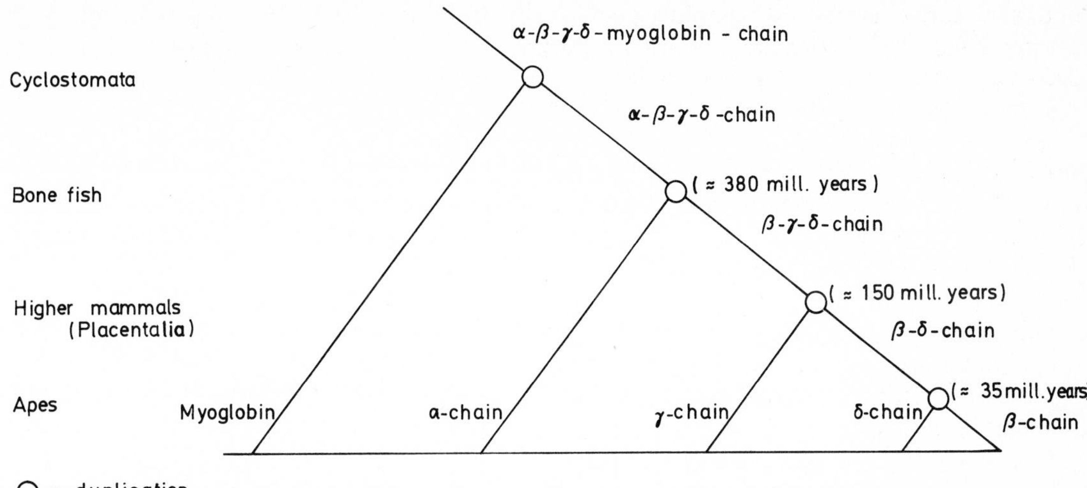

Fig. 7.10. Steps of duplication for the hemoglobin genes and evolutionary stage at which they occurred. Additional duplications have led to the hemoglobin chains of early embryonic age, the ς and ε chains. Moreover, there are, in humans, two γ and two α chains. Information on the exact time of gene duplication for these chains is not available

mains of one protein. By comparison of domains from different proteins, conformational similarities were shown to be more widespread than anticipated from amino acid sequences. Protein domains may be very similar in conformation in the absence of similarity in amino acid sequence. In the course of evolution, mutations were accepted only when the resultant amino acid replacement did not disturb conformation [1979]. As suggested by model calculations, only about 200-500 such domains may have been the basic units from which the great number of different proteins that occur in nature have been found. As mentioned in Sect. 2.3.3, genes of higher organisms consist of several exons (=transcribed DNA sequences) separated by introns (=nontranscribed sequences). Often, a single exon appears to comprise the DNA sequences for determination of such a protein domain.

The factor VIII C gene, for example (Sect. 2.3.3.7), can be traced back to three domains A, B, and C. The A domain consists of 330 amino acids, the B domain has 980, and the C domain 150 amino acids. These domains are arranged in the order A1-A2-B-A3-C1-C2. The A domains show an unexpected but "significant homology to ceruloplasmin," a copper-binding serum protein which also consists of three A domains, but lacks the B and C domains (see Sect. 2.3.7).

Advantageous or Neutral Mutations? Why are certain amino acids in a sequence replaced by others in the long term? There are two possibilities: Either they improve the functional condition of the molecule, which makes for a selective advantage, or they are replaced - although they are selectively neutral or even very slightly deleterious - by genetic drift or random fixation (Sect. 6.4). The latter possibility has been stressed by Kimura [1941, 1942].

In the introduction to his monograph [1941], Kimura writes:

"The neutral theory asserts that the great majority of evolutionary changes at the molecular level, as revealed by comparative studies of protein and DNA sequences, are caused not by Darwinian selection but by random drift or selectively neutral or nearly neutral mutants. The theory does not deny the role of natural selection in determining the course of adaptive evolution, but it assumes that only a minute fraction of DNA changes in evolution are adaptive in nature, while the great majority of phenotypically silent molecular substitutions exert no significant influence on survival and reproduction. ...

The neutral theory also asserts that most of the intraspecific variability at the molecular level, such as is manifested by protein polymorphism, is essentially neutral so that most polymorphic alleles are maintained in the species by mutational input and random extinction. ... (It) regards protein and DNA polymorphisms as a transient phase of molecular evolution and rejects the notion that the majority of such polymorphisms are adaptive and maintained in the species by some form of balancing selection."

On the other hand, Kimura clearly differentiates between positive (adaptive) selection, which he thinks is very rare *at the molecular level,* and negative selection, by which disadvantageous mutations are being eliminated in great numbers. He even estimates that only $\approx 10\%$ of all new mutations might be neutral (or even slightly deleterious); $\approx 90\%$ are outright deleterious and have no chance of being fixed.

This "neutral" hypothesis, if applicable to a significant amount of genetic change in time, would have important consequences for our understanding of genetic differences between and also within species that must have originated during evolution. Therefore, it is not surprising that it has led to heated discussions among population geneticists. These discussions were beset with misunderstandings, many of which were clarified when Kimura recently summarized his views [1941].

To understand this hypothesis, two scientific developments should be visualized:

1. In the 1960s, an enormous amount of genetic variability became apparent at the protein and therefore at the DNA level. Methods for determining amino acid sequences showed differences between homologous proteins in different species as well as between related proteins in the same species. Moreover, analysis of the genetic code opened up new sources of variation for analysis. Among other aspects, the enormous amount of DNA available in the eukaryotic cell (Sect. 2.3.1.1) raised the question of its function. The large amount of DNA and its considerable variability induced some scientists to question whether natural selection as assumed by the neo-Darwinian theory of evolution accounted for these findings or whether, at the molecular level, random processes were more important. If selection were the critical factor, as assumed by the conventional synthetic theory, the number of DNA sites subject to selection would be enormous.
2. At the same time, Kimura had developed mathematical diffusion models that allow answers to such questions as: "What is the probability that a single mutant which appears in a finite sized population will eventually spread throughout the whole population?" (i.e., the probability of gene fixation, Sects. 6.4.1 and 6.4.2).

In a classic paper, he formulated the problem and its solution as follow:

"Consider a population of size *N*. ... If we look sufficiently long into the future, the population of genes at a particular locus will all be descended from a single allele in the present generation. ... This is the result of the inexorable process of random genetic drift.[1] If, in the present generation, an allele A_1 exists in frequency p, the probability is ... p that the lucky allele from which the whole population of genes is descended is A_1 rather than some other allele.

Now, if mutation occurs at a rate μ per gene per generation, then the number of new mutants at this locus in the present generation is $2N\mu$.[2] Furthermore, the probability that a particular gene will eventually be fixed in the population is $1/2N$. So the probability of a mutant gene rising in this generation and eventually being incorporated in the population is

$$2\,N\mu \times \frac{1}{2N} = \mu$$

The rate of neutral gene substitution is identical with the mutation rate – irrespective of population size."

More recently, Kimura [1941] has amended this statement somewhat by remarking that a mutant that is selectively disadvantageous in a large population may be neutral in a small one, so that the rate of substitution would in fact be more rapid in small populations.

In his model calculations, Kimura treated random mutation as a time-dependent process. Even in his recent review [1941] he assumes that random fixation of mutations, only depending on the mutation rate, would predict accumulation of genetic differences at a linear time scale, regardless of species, generation time and other parameters. It would be limited only by the constraints imposed by the functional requirements of the genes and their products, the proteins: "Negative selection" eliminated base substitutions and amino acid replacements *not* compatible with normal function. Differences in the speed of evolution between different proteins (Table 7.4) can be explained by this negative selection, in agreement with conventional theory. Only the different sequences *maintained* in the course of evolution cannot, in his opinion, be explained by corresponding differences in *positive* selection.

The value of the "neutral hypothesis" for an explanation of aspects of evolution can be tested by examining whether its predictions can be verified or refuted by actual data. Such predictions have been made at two levels: amino acid sequences and replacements in proteins; base sequences and substitutions in the DNA.

As mentioned, one prediction has been the linear time dependence of the amino acid replacement rate – the *evolutionary clock*. Its establishment required measurements of the time when two branches in a phylogenetic tree have split, using an independent time scale such as paleontologic evidence. Discussions in the literature on this prediction have often dealt with this evidence. For example, some species of fish have been living in the deep ocean from time immemorial, and the ecological conditions must have been very similar if not identical throughout, but nonetheless, protein "evolution" proceeded at the same speed. The α- and β-hemoglobin chains in nonhuman mammals have diverged since their separation by the same amount as those in humans and fish. On the other hand, more detailed examinations

---

[1] This statement follows from the rules of random fixation; Sect. 6.4.

[2] Where N = size of populations.

of some parts of the "phylogenetic tree" revealed deviations: protein evolution of primates, for example, has been slower than predicted by the "clock."

Arguments Against General Applicability of the Neutral Hypothesis. Much of the discussion about the neutral hypothesis drew on arguments from theoretical population genetics. Suffice it to say that, depending on specific assumptions and the parameters of the genetic models used, conclusions other than those derived by Kimura are possible (see, for example, Ewens [1757]). Ideally, the theory and the propositions derived from it should be tested by empirical evidence.
The following statements are undisputed:

1. Many forms of mutation lead to genetic defects. They have a strong selective disadvantage, however, and will soon be eliminated from the population.
2. Other mutations are subject to special modes of selection that maintain genetic polymorphisms, either by heterosis (Sect. 6.2.1.3) or by frequency-dependent selection (Sect. 6.2.1.5).
3. Many amino acids found in proteins would, under present-day conditions, have no measurable selective advantage in comparison with the amino acids that they replaced during evolution probably many millions of years ago.

It is conceivable that an undetermined fraction of mutations had a slight selective advantage at the time of fixation. Some recent evidence points to such influences:

1. Once the function of a certain gene product has been established, selection will mainly tend to preserve its functional characteristics, or even adapt the molecule a little better to its function. A substitution of one amino acid for another with similar conformational and biochemical properties will therefore tend to be maintained [1509, 1907]. There ist, indeed, a strong correlation between biochemical similarity of amino acids and probability of substitution.
2. Once a mutant is formed, for example by duplication, selection is expected to adapt it to its new function; the frequency of amino acid replacement will presumably increase. This was asserted for the hemoglobin genes, which show increased amino acid replacement after exhibiting duplications [1927]. This result has been disputed, mainly by questioning the paleontologic evidence on which it was based and the method of tree construction [1941].
3. The neutrality hypothesis postulates that many, if not most, of the present-day polymorphisms in the human population are not maintained by selection but are neutral alleles on the way to fixation by drift.

At first glance, the distribution of rare and common electrophoretic variants seemed to contradict this postulate (see Sect. 6.1.1; Fig. 6.4). The observed distribution is strongly bimodal; there is a group with relatively high (intermediary) gene frequencies, which are thought to be maintained by heterozygote advantage or frequency-dependent selection, and a group with low gene frequencies that could comprise genes maintained without selective advantage or disadvantage by genetic drift. The neutral hypothesis would predict a relatively high frequency of different rare and very common (gene frequency $\geqslant 0.9$) variants, and a lower frequency of variants with intermediary gene frequencies; this is apparently at variance with the distribution actually found. But it is very difficult to distinguish whether such a distribution has been caused mainly by genetic drift of neutral alleles, by a mixture of various modes of selection, or by both.

4. This hypothesis can also be discussed at the level of DNA. For example, base substitutions not leading to amino acid replacements - especially those affecting the third base of a codon - were found to be more common than those causing such replacements; and DNA sequences outside transcribed regions turned out to be especially variable. This also applies to variation *within* the human species; there are many DNA restriction site polymorphisms (Sect. 2.3.3.9 and 6.1.2). According to a recent estimate [328], an average heterozygosity/codon might be about ten times as high in noncoding DNA sequences of the human genome, as in coding sequences. Moreover, the base substitution rate in functionally inert pseudogenes, e.g., an Hb α pseudogene of the mouse (Sect. 4.3), appears to be higher than in their active counterparts [1953]. On the other hand, a comparison for the mRNA of the hemoglobin-β chains for humans, mouse, and rabbit *failed* to show the randomness expected from the neutral hypothesis; on the contrary, the pattern of base substitutions was decidedly nonrandom [1918].

As shown by these and other arguments, much of the evidence brought forward for and against the "neutral hypothesis" is ambiguous. Until these questions are settled some plausible inferences can be attempted.

"Genetic Sufficiency" [1992, 1993]. Let us assume that environmental conditions change in such a way that functional adaptation of a certain polypeptide becomes less efficient. Then, if a mutation occurs that

meets the new demand in a more efficient manner, its carriers will have a selective advantage. The new mutation does not necessarily improve the polypeptide to its optimal conceivable state. Only some improvement has occurred. Moreover, a number of different mutations might have brought about this improvement; nature has a number of options by which a given demand can be met, not always in an optimum fashion but often to an adequate degree. The actual substitution selected depends on the availability of mutants in the population at the time when the demand arises. Availability, in turn, depends on the mutation rate (in addition to genetic drift). Indeed, there is evidence that base transitions are more frequent among mutations that were fixed during evolution than are transversions; transitions also seem to be more frequent among fresh mutations (Sect. 5.1.4) [1985]. Hence, there is an element of randomness within the boundaries imposed by functional demands and selection.

One example familiar to human geneticists may help to demonstrate the meaning of "genetic sufficiency." When malaria became widespread in populations of tropical countries, an improved resistance against this disease became useful to the population. Genetic adaptations soon occurred in all exposed populations. The precise mode of adaptation, however, was different (Sect. 6.2.1.6). In Africa, HbS and HbC were selected, while in the Austroasiatic population it was HbE, and in various populations different thalassemias and various G6PD deficiencies became frequent. The adaptive value of these mutations was by no means identical; HbE, for example, offered protection from malaria at a much lower price than β-thalassemia, as HbE homozygotes have a much milder form of anemia than β-thalassemia homozygotes (Sect. 6.2.1.7). Nevertheless, both adaptations were sufficient, since the population survived. Obviously, the adaptations were dependent on the kind of mutation that happened to be present and could therefore be favored by selection.

This concept of "evolutionary sufficiency" has been brought forward by a molecular biologist (Zuckerkandl). However, it is very similar to the result of the population geneticist Ewens [1757] that various, not necessarily "optimal", combinations of genes might meet a special demand from the environment. Kimura's ideas, as developed in his recent summary [1941], are not far from this concept. The main difference seems to be that authors such as Ewens and Zuckerkandl think that *positive selection,* i.e., selection in favor of slightly advantageous substitutions, may have a somewhat greater importance than Kimura believes *in addition to random processes.* But it appears to be agreed that in a constant environment most selection is of a negative kind, i.e., tends to *preserve* a certain function by elimination of deleterious mutants.

Limitations of Present Knowledge of Natural Selection and Neutral Substitutions in the Evolution of Proteins. As already mentioned, most research workers agree that natural selection has been responsible for substitution of some amino acids in proteins and for some of the genetic protein polymorphisms found in the human population. On the other hand, a portion of the observed variation between species and within the human population is probably caused by random drift; here, selective advantages or disadvantages may be trivially smaller or even completely lacking. However, current evidence is insufficient to decide what fraction of all genetic variability is caused by selection and what fraction by random processes. In this context, the amount of genetic polymorphism in the human population should be remembered: The human genome might comprise about 50000–100000 structural genes [1943] that transcribe for proteins. Several hundred of these are known, and up to 30% of these may be polymorphic.

These polymorphisms were largely determined in the blood. As explained elsewhere (Sect. 6.1.2), the amount of polymorphism in other, less accessible tissues might be much lower, but there are probably hundreds or even thousands of polymorphic loci, only a minute fraction of which are known so far. Moreover, the physiological function of many polymorphic enzymes is completely unknown. However, inferences regarding natural selection are much more likely to be correct if founded on knowledge about the physiological function of the polymorphism under study.

In most proteins a characteristic function depends crucially on a few amino acid positions. Functional restrictions are of such a general kind that they may be met with by many different amino acids; for example, maintenance of three-dimensional conformation. Here, genetic drift may shift bases freely, and this may lead to polymorphisms at the protein level. These polymorphisms may even bring about small functional differences that do not or only trivially influence fitness (Sect. 6.2.1.1) of their carriers, or only trivially, and natural selection. When the ecological conditions change, they might provide a reservoir for fast adaptation. On the other hand, the fact that selective influences are still unknown for most polymorphisms does not mean that selection has been absent. On the contrary, it is difficult to detect selection – especially in human populations, where modern civilization has changed living conditions considerably within a very few centuries, having eliminated potentially important selective agents such as infectious diseases and - for large parts of the world population - malnutrition from the environment. The discovery of selective mechanisms requires specific, functionally founded hypotheses. This does not mean, on the other hand, that all functional differences found between polymorphic variants must at some time have influenced fitness. On the other hand, it would be difficult to explain without such assumption that, as a rule, the rare variants of polymorphic enzymes show lower activities (Sect. 6.1.2). The undisputed fact that, for most human polymorphisms, selective influences are still unknown, does not mean that selection has been absent; rather, it testifies to our inability to propose and test well-founded hypotheses on selective mechanisms. The neutral hypothesis, when applied to the study of human polymorphisms, might even have a counterproductive effect if it discourages the search for sources of natural selection.

Evolutionary Clock and Mutation. As already noted, the evolutionary clock can be explained if mutations are time dependent irrespective of species and if they are fixed at random. We have seen in Sect. 5.1.3 that human mutation rates for some mutations seem to be more frequent in male than in female germ cells, that some mutation rates increase with age of the father, and that many mutations seem to be related to DNA replication. Even the fact of the vastly different generation times found in various animals makes the hypothesis that mutation is simply time-dependent seem highly unlikely.

For a more realistic approximation to the mutation rate, it would be desirable to construct a timetable based on the number of DNA replication cycles/time unit in the various species, but the necessary information on germ cell kinetics is not available [1985]. If a clock-like regularity of base substitutions could indeed be shown (which is, in our opinion, dubious) it could certainly not be used as argument in favor of random fixation of mutations: According to the rule derived by Kimura himself (see above), the rate of fixation only depends on the mutation rate, which cannot be assumed to be merely time dependent in all species.

A possible way out of these difficulties might be the assumption that mutations that are slightly disadvantageous in large populations are neutral in small ones and therefore have a higher probability of being fixed: Species with large body size (such as elephants) have a longer generation time on average (and, probably, also a lower number of DNA replication cycles/*time* unit) but a smaller population size than species with small body size (such as mice). One could also argue that the mutation rate/time unit does *not* only depend on the number of replication cycles/time unit but has been adapted to an optimum rate by natural selection. But why, if most mutations are neutral anyway?

In conclusion, Kimura, by proposing his "neutral theory," has certainly made an important point by stressing that evolution at the molecular level has some aspects not revealed when studied at the level of phenotypes. There is little doubt that random (or near-random) processes at the molecular level were much more important that most biologists have thought before. It is often observed that the explanatory power of new theoretical concepts is overrated by their originators. But, as Popper says, science can proceed only by boldly advancing hypotheses and then submitting them to severe tests.

Evolution by Reshuffling of Exons. Discovery of the exon-intron structure of genes (Sects. 2.3 and 4.3) has opened up a new path for our understanding of protein evolution: Exons may be separated from each other and rearranged in a new order, or some exons only of a gene may be transcribed in one species, whereas the entire set is used in another. As already mentioned (Sect. 4.2.2.4), such differences in gene usage have been observed even between different tissues of the same individuals; this may be one mechanism of differentiation.

Some sets of exons are utilized for the construction of different proteins. As an example, the low density lipoprotein receptor has homology with eight exons that code for a precursor of the epidermal growth factor. This and other work suggests that functional proteins are mosaics of simpler structures that are shuffled together (see Gilbert, 1925 a). The complexity of proteins appears to be derived from the combinational assembly of a relatively reduced number of smaller genes that specify exon structure.

Comparison of the Protein Data with Data from Chromosome Evolution and Satellite DNA. The data on evolution of proteins show amazingly little differences between *Homo* and the higher primates, chimpanzee and gorilla. For practical purposes, these proteins can be regarded as identical. In the hemoglobin molecule, for example, these species differences are functionally less important than rare variants in human populations that are fully compatible with life but may lead to mild hemolytic anemia. This extremely slow evolution can be explained by the assumption that the function of these proteins remained largely identical. However, even if we look at karyotypes, we find only a few chromosomal rearrangements, mainly pericentric inversions. Similar rearrangements are not so rare in the present-day human population, and they do not influence the phenotype at all. They might explain reproductive barriers that once were an important condition for speciation; they do not explain the genetic mechanisms that created the specific human phenotype. Little is known about the functions of additional R and T bands and species differences in heterochromatic material and satellite DNA. However, centromeric heterochromatin shows much variability within current human populations. An effect of these heteromorphisms on human phenotypes such as behavior has been claimed but is not generally accepted [1914].

Thus, we are left with the conclusion that genes important for human evolution during the phase of human brain development are completely unknown. Since most human DNA does not code for proteins and might either be "junk" or of great importance for regulation of gene activity, especially during embryonic development (Sect. 4.8), relevant changes might have occurred within this nonstructural DNA [1993]. Such alterations might have occurred within the nontranscribed parts of those sections of DNA separating structural genes, that are postulated to have regulatory functions. It is conceivable that DNA sequences not important in structural gene function may somehow be required for development and, hence, that changes within those DNA species may be especially efficient in bringing about improvements in brain function. This concept, however, is very speculative and general. To formulate more specific hypotheses, more needs to be known regarding genetic determination of embryonic development and regarding genes affecting intraspecies variation in human behavior (Sect. 8). Even disregarding any phenotypic effects and only considering the analyzed genetic phenomena such as chromosome rearrangements, addition or reduction of chromosome material, satellite DNA and amino acid sequences of proteins many aspects are still poorly understood. For example, how were chromosome rearrangements fixed in populations? Are the mechanisms identical to those applying to amino acid replacements? What single events formed satellite DNA species? Do such events have specific significance for speciation or for regulation of gene functions?

7.2.4 Restriction Fragment Length Polymorphisms (RFLPs) and Evolution

Restriction site polymorphisms have been described in sections 2.3.2.7 and 6.1.2. As mentioned in sect. 7.2.3, neutrality of base substitutions is more plausible for RFLPs situated outside coding DNA sequences than for polymorphisms leading to changes in proteins. Therefore, comparisons, based on nuclear and mitochondrial DNA, between humans and higher apes and between human populations have been performed. Almost identical polymorphic sites were shown to be present within the region of the albumin genes, in humans as well as in chimpanzees, for example [1958].

Origin of Various Coding Genes Identical in Repetitive Oligomer Sequences? The discovery of an increasing number of repetitive DNA families (Sect. 2.3.1.1) has led to speculations and studies on the origin of the genetic code and of coding sequences. Ohno, for example, found out by comparison of many DNA sequences that the same sequences, often about 10 bases but sometimes up to 28 bases long, are found in many and quite different genes, and in different reading frames [1967]. Probably, repetitive DNA sequences have been adapted to various quite different functional demands during evolution.

7.2.5 Behavior

Man the Toolmaker. In the face of our deficient knowledge of the genetic determinations of human behavior, it seems premature to speculate regarding the nature of the genetic changes that have brought about the last steps in the development of our species. It is, however, possible to develop hypotheses on the nature of selective pressures under which these changes have occurred, since the living conditions of *Australopithecinae* from which Homo emerged can be reconstructed [1968]. These creatures lived mainly in the open savanna; their food consisted of vegetables and of meat. They hunted big game, such as antelopes, as evidenced by their tools that were partially made from animal bones. For a long time, production of these tools was regarded as their most important achievement, and the bipedal gait was considered necessary for the freeing of hands for manipulation of objects. Meanwhile, we have learned from observation of apes and monkeys that quadrupedy does not prevent an animal from manipulating, as quadrupedal animals use their hands extensively while sitting [1982]. Still, toolmaking is one important feature that has to be considered in a larger context.

Social Structure of Early Prehuman and Human Groups. The context within which all activities of these early prehuman forms must be considered is the social group structure. Hunting big game with primitive tools requires co-operation of a fair number of individuals. Such co-operation has to be planned carefully and requires exchange of information over a certain distance. At a time when the male group is hunting, the females have to keep their children safe from predatory animals. All these activities require a fair group size of between 20 and 50 individuals.

Functioning of such a group could be improved by two conditions. The first was leadership. It is easily conceivable that hunting was planned by one male who convinced other group members to follow his advice. The second conditon was language. The task of a game-hunting group must have been much easier when its members were able to inform and direct one another by acoustic signals.
These simple considerations give an idea of the selection pressures favoring evoluation of the two most prominent features that distinguish humans from all other animals: intelligence that enables them to develop abstract concepts and to plan for the future and, closely connected with it, language.
The selective value of a genotype depends on higher reproductive rate of individuals with this genotype as compared with others. It is, however, easily understandable that dominant males in such a group - those who planned hunting and decided in general where the group would move to find favorable living conditions - also had easiest access to women and would father a higher fraction of children than other males. Given the special requirements of this way of life, these dominant males would, at the same time, also be the most intelligent and, more specifically, those most proficient in mastering language. In a very primitive present-day South American Indian group, the Xavantes, for example, the dominant male was, indeed, shown to be the father of many of the children [1966]. The same tendency for headmen to father more children than non-headmen was found in another tribe, the Yanomama [1963].

Precursors of Language and Cultural Tradition in Apes and Monkeys. In an effort to better understand evolution of social structure and cultural tradition, such as toolmaking and language development, many studies have been performed in higher primates, such as chimpanzees, gorillas, baboons, and rhesus monkeys. The social structures of primate groups show vast differences - from baboon species that live in groups of hundreds to the male orangutan who lives alone in his territory and mates with females that happen to inhabit an overlapping territory. The group might be open or closed; one male might be the "pasha," or more subgroups consisting of one male and a varying number of females may be present, or there may even be sexual promiscuity. The groups tend to be larger in the open savanna and half-desert than in the tropical rain forest, but even this rule has exceptions. These observations do not permit direct conclusions as to the structure of early human groups. At least three kinds of observations might lend themselves to extrapolation [1982]:

1. Most groups show "social stratification." Some animals hold a dominant position in the hierarchy, others are of lower rank. Such stratification is found - in one or the other form - in present-day primitive human populations and might have been one feature of early prehuman groups as well.
2. Monkeys show "cultural" transmission of behavior. New behaviors, such as bathing in the sea or dipping a potato in seawater to improve its taste by salt, are invented by single individuals and eventually taken over by the group. Whether a certain behavior will be accepted depends on conditions that appear very human to us. A new behavior has a higher chance to be accepted by the group if the inventor is high-ranking than if he is low-ranking. Younger group members take over new behaviors much easier than older ones.
3. While so far no primate has ever learned to vocalize sounds resembling a human language, some chimpanzees could be trained to convey detailed information by giving hand signals using a symbolic sign language similar to that used by the deaf or using computer consoles with symbols.

Others were able to figure out ways of producing tools from parts given to them separately, to understand and use concepts such as shape vs color, or to recognize themselves in a mirror and produce a picture of themselves by combining parts of a puzzle. When challenged sufficiently by the experimental design, apes display amazing intellectual capacities. The question of why they apparently do not utilize more of these capacities in their daily life in the wild state has puzzled many research workers. Probably, they use such capacities not for developing techniques or solving puzzles, but to achieve success within the social structure of their group, thereby improving their chance for reproduction [1983, 1984]. In primitive human societies the achieving of headman status is an analogy. The abilities developed under the selection pressure of social life in a primate group probably served as preadaptation for human cultural development.

Behavioral Characteristics of Humans in Common with Other Species. So far, behavioral characteristics by which humans differ from other species have been examined. However, there are also a great many characteristics in which humans and other animals are similar. For example, we are frightened like animals if our life is threatened, we are sexually aroused if appropriate stimuli are offered, and an assailant may enrage us so that we will react by aggressive fighting.

The question of whether aggression is an innate human characteristic has led to many discussions. According to Lorenz (1966) [127], "there is an internal urge to attack. An individual does not simply wait to be provoked, but, if actual attack has not been possible for some time, this urge to fight builds up until the individual actively seeks the opportunity to indulge in fighting." Many other scientists, however, believe that fighting derives principally from the situation, but they do not deny that the emotional status of the animal is important as well. In Sect. 8.1.2, we shall see that the ease by which aggressive behavior in the mouse is elicited may vary between inbred strains and that this variation is correlated with the amount of the neurotransmitter epinephrine.

According to Lorenz, humans, unlike other animals engaging in intraspecies fighting, tend to carry on until destruction of the enemy; the inhibition to kill other members of the species is overruled mainly by development of weapons, especially those acting at longer distance so that eye-to-eye contact is avoided. According to this concept, in a society that does not permit outlets for aggression from time to time, aggressive potentials will build up, and the society may even be destroyed in a final outbreak.

Most social scientists, and many biologists, feel that this view oversimplifies the situation. To avoid destructive intraspecific aggression, it seems important to look for different means of education apart from providing harmless outlets such as athletics, as proposed by Lorenz. Study of aggressive behavior in animals and possibly in nonhuman primates might help to understand possible biologic aspects of the human situation somewhat better.

Some other reactions seem to be inborn in human beings as well. For example, even children that are born blind and deaf will smile when they feel that somebody caresses them. The habit of recognizing somebody by raising the eyebrows has been observed in all human cultures and might well be an inborn behavior pattern.

Possibly, intensified exchange of ideas and concepts between human behavior genetics and ethology will help in the future to better understand the genetic determination of human behavior. These and similar observations and speculations have gained much popularity in recent years, in the United States as "sociobiology" and in Europe as "human ethology."

Human Sociobiology [129a]. Sociobiology is the biologic and evolutionary study of all forms of social behavior. The term was coined by Wilson in 1975 in his discription of existing knowledge of this area [1989]. However, the field was not entirely new. Ethology and behavioral ecology were areas of study investigating social behavior in animals. Investigators and theorists in sociobiology are distinguished by invoking genetic and evolutionary principles to understand and predict such behavior. While only a few years old, the field has elicited much attention and can be considered as the beginning of a new "paradigm" in a sense of Kuhn's models of scientific advances. The central theorem of sociobiology states that each individual can be expected to behave so as to maximize its inclusive biologic fitness [1899, 1989]. As an example, the Darwinian theory of natural selection had difficulties explaining how altruistic behavior could survive in evolution since it diminished the fitness of the "altruist." Such altruistic behavior, however, could be explained if the principle of kin or group selection is invoked (Sect. 6.2.1.5). Under such selection, altruism can evolve if the advantages to survival of the group or the kinship outweigh the negative effects of selection acting on individuals. As Haldane put it succinctly many years ago [1930], he was prepared to lay down his life for two brothers and eight cousins since he shared one-half of his genes with his brothers and one-eighth of his genes with his cousins! In general, however, kin selection will be a weak selective agent in comparison to individual selection and can be invoked only under certain extreme conditions [1978].

Sociobiologists compare the behavior of hundreds of species in the light of evolutionary principles and hope to shed light on some new aspects of behavior that could not be fully understood before. Using such methods, sociobiology has provided novel explanations for certain genetically fixed behavior patterns in many animals. First, claimed insights had been achieved without attempts to understand the underlying neurobiologic mechanisms involved in such behaviors. This has changed more recently [129a]. But the particular genes involved in social behavior remain largely hypothetical. Nevertheless, the existence of inherited genetic behavior patterns, such as those determining navigation in migratory birds, cannot be contested.

The possible implications of sociobiologic theory for human behavior have elicited considerable interest. By extending their reasoning to the human species, sociobiologists are attempting to interpret human emotions, human sexuality, aggression, and social status by evolutionary principles [1898, 1989]. It has been suggested that a human biogram exists, a pattern of potentials and constraints built into these species. Genes set limits within which cultures can develop. Facial expressions conveying various emotional feelings appear very similar across all human cultures. Sexuality is considered the device of natural selection to ensure pair bondings. Polygyny (mating of one male with many females) is claimed to have a physical basis by having conferred a natural advantage on the species. The forms of polygyny - polygamy, mistresses, multiple marriages, etc. - may vary in different cultures. Wilson [1990] postulates a physical basis for sacralization and for inherited mythopoietic tendencies and therefore limits to "scientific enlightenment" as a basis for social cohesion. Sociobiologists feel that many more of the recognized human constants are physically rather than socially determined. These include male dominance, sexual division of labor, prolonged maternal care, and extended socialization of the young.

Some ethologists became especially interested in the biological laws influencing individual development during infancy, childhood, and youth. Here, the phenomenon of "imprinting," discovered by K. Lorenz in the wild goose as long ago as 1935 [1948] and analyzed later in many other animal species, has influenced thinking on cognitive and emotional learning. The young gosling invariably follows the animal first seen after hatching, even if this is a human being; most often it is the mother. There are, in animals, many other behavior patterns that can be learned only in a specific phase of individual development, and must be learned then. It is a matter of controversy whether imprinting occurs in humans at all but there can be little doubt that human interaction is required, for example, for learning to speak and for social and emotional learning [1934].

Wilson admits that the preprogramming of the human brain is much more nonspecific when compared with other species [1990] and therefore allows for much more plasticity of human behavior. Nevertheless, he interprets the evidence of little variation between human cultures to indicate that traits, such as incest taboos, use of body ornaments, and elaborate kinship rules, are biologically derived.

Sociobiologists see their field as an antithesis to the environmentalism of the social sciences, such as social anthropology and sociology, which often assume human social and cultural life to be entirely culturally determined and constrained only by the most rudimentary biologic drives. *The major substantial criticism of sociobiology is the lack of direct evidence for the operation of genetic factors influencing most human behaviors claimed to be under genetic control.* Despite this current lack of direct evidence, it appears highly likely that certain aspects of human behavior have been programmed genetically by natural selection. *It is improbable that the human species is entirely autonomous in its behavior and that genetic determinants of the central nervous system and their influence on social behavior have been entirely over-ridden by cultural factors* (Sect. 8). The human species and its brain are a continuum in the evolutionary chain. The complete independence from biologic constraints of traits mediated by the central nervous system is therefore unlikely.

Sociobiology has been heatedly condemned by scientists and others who deny that there are any relevant constraints placed on social processes by human biology [1898]. These critics see sociobiology as another manifestation of social Darwinism used by privileged members of the ruling classes to justify the current status quo of Western societies by "biologizing" the raison d'être of fundamentally unjust and sexist behavior. Critics of sociobiology are intensely conscious of the misuse of past pseudogenetic theories that were used to justify discrimination and social injustice (Chap. 1).

These matters will not be resolved by further polemics nor by the kind of evidence likely to be brought forward by the current school of sociobiologists. Genetically oriented experimental designs in families that attempt to dissect human behavioral patterns into their subcomponents with attention to their biologic substrates will be required to decide the extent of biologic programming of social behavior in the human species.

It is questionable, however, whether such designs will be sufficient to solve the problem: Most conclusions and concepts of sociobiology have been derived from comparisons between humans and other species, or are based on behavioral *similarities* between various human populations in spite of differences in cultural patterns. In contrast to this, classical genetic analysis utilizes *differences* between individual members of the same populations as analytical tools for elucidating basic mechanisms. This difference in approaches between ethology and genetics should be kept in mind in all discussions on genetic determination and evolution of behavior patterns. In this way, misunderstandings in discussions between ethologists and geneticists could be avoided.

7.2.6 *Investigation of Current "Primitive" Populations*

Most approaches to the study of human evolution are based on indirect evidence. Conclusions are derived from skeetal findings, from chromosomes and proteins, or from comparative observations of different species. A slightly more direct approach exists. Despite the worldwide "progress" of modern civilization, a few human populations are still living as hunters and gatherers, i.e., under conditions not too different from those in which our ancestors existed for the longest time periods during human evolution. In recent years, such populations have increasingly been studied by human geneticists and anthropologists in the hope that such investigations would provide information regarding environmental forces that shaped our genetic makeup. The results have helped to better understand human evolution.

Problems for Which Primitive Populations Could Provide Evidence. There are a number of questions for which study of primitive populations may suggest answers [1962]:

a) Size of population groups and isolation: As explained in Sect. 6.4, size and isolation of effectively breeding population groups are the major factors for the extent of chance fluctuations of gene frequencies, formation of subgroups such as races, and finally speciation. On the other hand, these parameters are especially liable to differences in environmental conditions.
b) Population control: There is evidence that overall population size of human populations was fairly constant over long periods and an equilibrium between population size and the ecologic conditions, especially food supply, existed. Modern civilization has deeply disturbed this equilibrium, making the present-day "population explosion" one of the most dangerous threats for the future of the human species. While study of primitive populations cannot provide meaningful clues for the survival of modern civilization, it is interesting to observe how primitive populations manage to adjust population size to their ecologic conditions.

c) Natural selection due to differential fertility: As explained in Sect. 6.2.1, natural selection implies that various genes within the gene pool of a population have different chances to reach the gene pool of the next generation. These changes depend on mortality and/or fertility of their bearers. Differential mortality cannot be readily studied in primitive populations. Therefore, observations of differential fertility are of special interest.
d) Disease patterns: Historical evidence has shown conclusively that, during recent centuries, diseases – especially those caused by infectious agents – have played a major role in mortality during infancy and childhood. Presumably, therefore, some infectious diseases have played a major role in natural selection. This raises the question of whether and to what extent genetic selection by resistance to infectious disease might apply to primitive hunters and gatherers as well.
e) Selection relaxation: Natural selection has undoubtedly relaxed for many traits that were harmful under primitive conditions. Are there genetic differences between present-day primitive and civilized populations that suggest selection relaxation?

Populations in Which These Problems Have Been Investigated. Investigations on South American Indians living in the jungles of Brazil and Venezuela – the Xavantes, Yanomama, and Makiritare – proved to be informative. At the time of these studies – in the 1960s and early 1970s – these tribes were among the least acculturated in South America. Still, they depart in many ways from the strict hunter-gatherer way of life prevalent during much of human evolution. Unfortunately, the remaining true hunter-gatherers are either all greatly disturbed by modern life styles or are so reduced in numbers and withdrawn to such inaccessible areas that appropriate study of a sufficient number of individuals appears impossible. However, the tribes studied were in their general way of life and breeding structure much closer to hunter-gatherers than to modern man. They live in primitive villages that serve as bases for their hunting and gathering expeditions. These villages are usually abandoned after a number of years. Primitive agriculture (manioc, squash, sweet potatoes, cooking bananas, and maize) provides a smaller share of the required food in the Xavantes and a much larger fraction in the Yanomama [1906].
A few of the more important results of these studies will be mentioned.

Size of Population Groups and Isolation. The village is the most important unit; population size of a village ranges between ~40–50 and ~150–200 [1906]. There is a strong tendency for marriage within the village. If the population becomes too large, their social rules become increasingly endangered, and part of the community consisting of several families may split off. Obviously, such small population size – together with the inherent tendency to isolation – favors the creation of many subpopulations with different gene pools and therefore rapid evolution.

Population Control. The maximum human birth rate is much higher than that required for maintenance of a constant population size. In civilized human populations of past centuries, a high mortality in infancy and youth and during reproductive age was the rule. The most important causes of death were infectious diseases, malnutrition, and demise of women related to pregnancy and delivery.
Among the primitive Indians under study, these causes of death turned out to be less important; the number of children per woman was limited, and an effective livebirth rate of approximately one child every 4 or 5 years was maintained by a variety of measures, such as intercourse taboos, prolonged lactation (the children were usually weaned at an age of about 3 years), abortion, and infanticide.
Infanticide is practiced especially when a child is grossly defective or when several births follow each other closely [1962]. Female newborns are killed more often than males. The health of the surviving children, however, seems to be excellent and remains well up to the age of about 40. The death rate of the population below 40 years of age appeared to be lower than that of present-day civilized populations of developing countries such as India – or, presumably, of the western European population of 200 or more years ago. On the other hand, individuals above the age of 40 have been observed only rarely. Their causes of death are not evident. Death by warfare, by common intratribal man-to-man fights or by pneumonia are plausible possibilities. Among the younger age groups, however, these tribes maintain a health standard lost by our not so distant ancestors as a tribute to continuous settlement and agriculture and regained – and partially surpassed – only recently as a result of modern hygiene and medicine.

Natural Selection Due to Differential Fertility. As explained in Sect. 7.1, the most important aspect of human evolution has been the improvement of innate mental capacities. Presumably, changes in brain size have been accompanied by alteration in structure and function of the human brain. Such improvement in mental activities requires a reproductive advantage in favor of individuals bearing genes for such abilities. While our knowledge of the genetic basis for such behavioral variability is limited (Sect. 8), it is reasonable to assume that genes for such capacities are found in a higher proportion among individuals holding a leading position in the social hierarchy of their village since they are able to plan hunting trips, provide for food supply, or to settle controversies between members of the community.

Indeed, village headmen were shown to have several wifes and a disproportionately great number of children [1963]. The effect of their polygamy was all the more striking, as newborn girls were much more endangered by infanticide than boys, creating in the Yanomama a sex ratio of 128♂/100♀ during the age interval 0–14. Together with the polygamy of high-ranking men, this can only mean that some males will be barred absolutely from reproduction.

Thus, in a Xavantes group, 16 of 37 married men were polygamous; 65 of the 89 surviving children came from these polygamous marriages. The headman had married no fewer than five times – more than any other member of the group. These five unions resulted in 23 surviving children, approximately one-fourth of all children of the group [1966].

If the high reproduction of socially high-ranking males has been a general feature of primitive human populations and if mental ability leading to high social rank has at least a partial genetic basis, a plausible mechanism for relatively quick evolution of this specifically human characteristic exists.

Balance by Disease [1962, 1965]. The population up to the age of 40 was usually in excellent health. At the same time, serum gamma globulin levels were about twice as high as in civilized populations. Hence, newborns are expected to have a high level of transplacentally acquired antibodies. To quote Neel:

From the first months of their lifes, these infants are in an intimate contact with their environment that would horrify a modern mother – or physician. They nurse at sticky breasts, at which the young mammalian pets of the village have also suckled, and soon are crawling on the feces-contaminated soil and chewing on an unbelievable variety of objects. Our thesis is that the high level of maternally derived antibody, early exposure to pathogens, the prolonged period of lactation and the generally excellent nutritional status of the child make it possible for him to achieve a *relatively* smooth transition from passive to active immunity to many of the agents of disease to which he is exposed.

On the other hand, an epidemic from outside may have catastrophic consequences – not so much because the individuals cannot overcome it but because village life comes to a standstill, when almost everyone is sick, as was observed when a measles epidemic was introduced to a Yanomama community.

Can These Observations on a Few Indian Tribes Be Generalized? Various, more or less "primitive" populations that are the main objects of ethnologic research show an enormous range of variation in most aspects of social life and culture. Unfortunately, studies on the medical genetics and even simpler demography of such populations are much scantier. The reason can probably be found in the sociology of science: Most ethnologists and social anthropologists engaged in field work with such populations are usually trained in linguistics, sociology, and other social sciences and are not oriented toward biology and medicine. Some even share "antibiologic" prejudices characteristic of some scholars in the humanities and social sciences.

In view of scanty information, it may be argued that some of the biologic features cited above are specific for the populations examined and that extrapolations to our primitive ancestors are unwarranted. Interdisciplinary research on the few remaining primitive populations of the world is sorely needed to find out which aspects of their biology are unique for certain cultures and natural environments and which can be generalized. Such interdisciplinary research is all the more urgent since primitive populations are now rapidly disappearing as a consequence of improved traffic and an increasing opening-up of remote areas. The studies on South American Indians cited above are paradigmatic for this interdisciplinary approach.

Relaxation of Selection. Comparisons between primitive and civilized populations can also be used to test the hypothesis that modern civilization has led to relaxation of selection of genes necessary for survival in an adverse environment [1969–1974].

Probably the best example is X-linked red-green color blindness because diagnosis is fairly quick and easy by standardized color plates. For a more exact diagnosis, however, testing by anomaloscopy is required, which is very difficult to perform in the field. A survey of 13 samples of primitive populations (for example, Eskimos, Australian aborigines, Fiji islanders, North and South American Indians) with altogether 7,712 individuals revealed a pooled prevalence of 2% for all types of red-green colorblindness together, whereas in 99 samples of civilized populations with 436,853 individuals, about 5% color vision defects were found. Populations somewhat removed from hunters and gatherers in time or in habitat usually show intermediate rates. For the survival of a hunter and gatherer, color vision provides obvious advantages in recognition of game or other food as well as of dangerous animals such as snakes. Considering the relatively short time elapsed since the onset of neolithic culture and introduction of agriculture, the increase of color blindness is surprisingly high, suggesting either a considerable mutation rate [1974], possibly due to frequent unequal crossing over [825a], or a temporary selective advantage of colorblind individuals. Diagnostic errors, especially in studies of primitive populations, cannot be entirely excluded.

Similar comparisons have been carried out for visual acuity [1969], hearing acuity [1970], deformed nasal septa [1972, 1973], small tear ducts, breast cancer, and spina bifida occulta [1974], always with the result that slight deficiencies are more common among civilized than among primitive populations. While some of the data are open to criticsm, the overall argument that civilized populations differ from primitive ones in such characters and that the differences are due to relaxed selection is probably correct.

7.3 Genetics of Group Differences Within the Human Species

7.3.1 Races

Race Classification. All humans who are living at present belong to one species: their matings have fertile offspring. It is impossible to state with certainty whether any of the ancient human types, such as Neandertal man, were members of the species *Homo sapiens.*

The species *Homo sapiens* is divided into populations that are commonly called races. *A race is a large population of individuals who have a significant fraction of their genes in common and can be distinguished from other races by their common gene pool.* In earlier times, members of a race often lived together under similar sociocultural conditions. The connotation of "race" as a broad population group merges without sharp limits into that of smaller units such as "demes." Race classification and race history have been one of the major fields of research within classic anthropology in the nineteenth and especially the early twentieth century. The classifications were based on visual impression and on statistical distributions of physical traits. With the development of human genetics in recent decades, such categorizations have been suplemented by evidence based on frequencies of genetic polymorphisms. Classifications by various authors differ somewhat in detail [41]; subdivision into the three main races, Negroids, Mongoloids, and Caucasoids, is undisputed. Two smaller groups, Khoisanids or Capoids (Bushmen and Hottentotts) and Australoids (Australian aborigines and Negritos) are often added.

Genetic Differences Between Races. The definition of race used here is a genetic one, and it would be desirable to base race classification on genetically well-defined characters that have been analyzed at the gene level. Several groups of such characters can be distinguished.

Many genes are common to all human beings, possibly with small, quantitative differences. For example, everyone has the genes that determine the enzymes needed for various basic metabolic processes. Exceptional individuals with rare mutations affect-

Table 7.5. Estimated gene frequencies of Atlantic and Pacific populations. (Morton et al., 1967 [1955])

System	Gene	Atlantic[a]		Pacific[a]	
		A_1	A_2	P_1	P_2
Se	Se	0.513	0.570	0.500	0.500
	se	0.487	0.430	0.477	0.500
Le	Le	0.816	0.660	0.758	0.545
	le	0.184	0.340	0.242	0.455
Lu	Lu^b	0.975	0.964	1.000	1.000
	Lu^a	0.025	0.036	0	0
IGHG (gm)	$G1m^1$	0.196	0.204	0.674	0.775
	$G1m^{1,2}$	0.021	0.092	0.226	0.123
IGK (km)	$G1m^{1,3,5}$	0	0	0.100	0.102
	$G1m^5$	0.783	0.704	0	0
	Km^2	0.907	0.900	0.698	0.693
	Km^3	0.093	0.100	0.302	0.307
PTC	t	0.548	0.506	0.292	0.207
	T	0.452	0.494	0.708	0.793
ABO	A_1	0.182	0.236	0.170	0.256
	A_2	0.065	0.068	0	0
	B	0.101	0.066	0.110	0.139
	O	0.652	0.630	0.720	0.605
MNSs	MS	0.247	0.270	0.103	0.236
	Ms	0.283	0.310	0.473	0.462
	NS	0.080	0.076	0.062	0.107
	Ns	0.390	0.344	0.362	0.195
P	P_2+p	0.510	0.458	0.751	0.570
	P_1	0.490	0.542	0.249	0.430
K	k	0.961	0.952	1.000	1.000
	K	0.039	0.048	0	0
Fy	$Fy+Fy^b$	0.575	0.603	0.164	0.318
	Fy^a	0.425	0.397	0.836	0.682
Di	Di^b	1.000	1.000	0.985	0.982
	Di^a	0	0	0.015	0.018
Rh	r	0.348	0.404	0.092	0.053
	r'	0.009	0.008	0	0
	r''	0	0.005	0	0.033
	r_y	0	0.001	0	0
	R_o	0.035	0.053	0	0
	R_1	0.498	0.415	0.665	0.602
	R_2	0.099	0.110	0.179	0.308
	R_z	0.011	0.004	0.064	0.004
HP	HP^1	0.319	0.384	0.275	0.397
	HP^2	0.681	0.616	0.725	0.603
TF	C	0.993	0.994	0.978	0.996
	B	0.007	0.006	0.006	0.002
	D	0	0	0.016	0.002
G6PD	+	0.978	1.000	1.000	0.970
	–	0.022	0	0	0.030
Jk	Jk^a	0.512	0.523	0.742	0.310
	Jk^b	0.488	0.477	0.258	0.690

[a] One of the Atlantic *(A)* populations is English, one of the Pacific *(P)* ones is Japanese; the others were picked rather haphazardly from published evidence ([144]).

ing those genes suffer from inborn errors of metabolism. Many of these genes are more or less identical in humans and other living beings.

Other genetically determined characters are the common heritage of all or almost all members of a single race; they are lacking in individuals of other races. The number of such characters seems to be fairly small; genetically they are not well-defined. One example is the eyelid fold of Monogoloids.

Another group of genetic characters occurs – apart from a few exceptions – only in one of the three main races but is absent in the other two. This group comprises a number of well-defined genetic polymorphisms (Table 7.5). One example is the Diego blood factor [1944–1946]. This blood group was discovered in 1953 in Venezuela in four generations of a family and was absent from most whites. Investigations in American Indian populations showed phenotype frequencies of the Diego factor between 0.025 and 0.48. White and black populations did not show this allele at all. Populations of Mongoloid extraction such as Japanese and Chinese, on the other hand, often had this allele, albeit with a lower average frequency than South American Indians. This finding confirmed the assertion of classic anthropology that American Indians were part of the main Mongoloid race.

Yet another class of genetic characters is more frequent in some populations than in others. It comprises genetic characters and alleles that are observed in all human races but with different frequencies. Examples are most genetic polymorphisms and genes determining continuously varying characters, such as stature, body proportions, and physiologic functions. Genetic polymorphisms are increasingly used for the genetic characterization of different populations. Over the last few years, many different markers have become available for such studies. A highly refined categorization of populations by clear-cut genetic criteria is now possible and summarized in Mourant's monumental compilation [144]. Tracing anthropologic affinities and the possible role of selection in determining some gene frequencies has become possible. The overall results clearly show overlaps in gene frequencies between populations and illustrate the difficulties of fixed racial classifications. While differences, for instance, between a Swedish and Korean population are very clear-cut using all markers, no such differences can often be found between neighboring populations. It has been pointed out that differences between members within each racial group exceed those between each of the principal races (Mongoloid, blacks and whites).

How Did the Genetic Race Differences Evolve? The main factor in evolution of phenotypes – and in the formation of races specifically – *is natural selection in adaptation to different environmental conditions.* For selection to be effective in producing genetic dif-

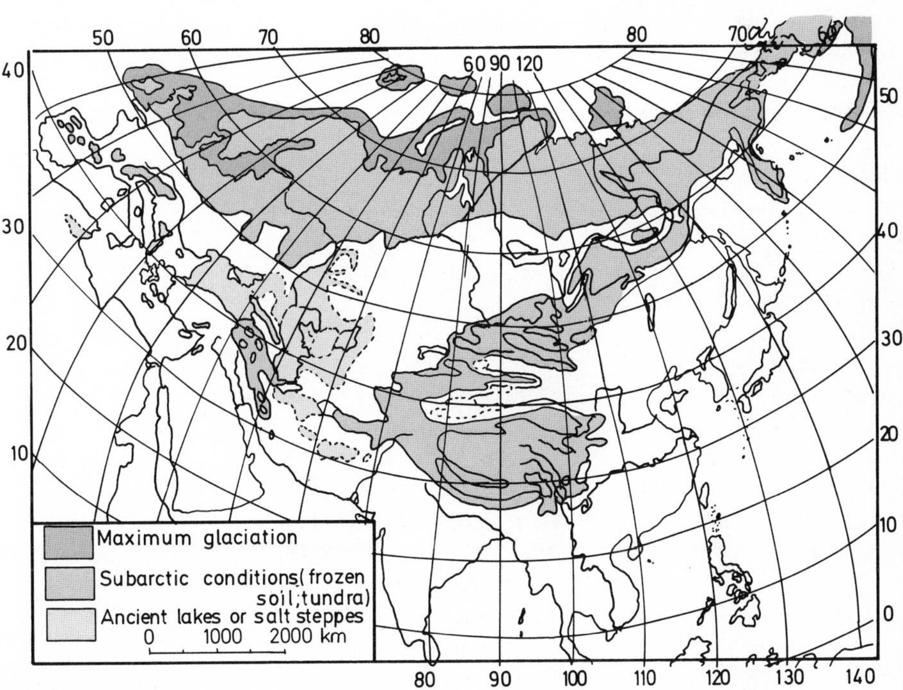

Fig. 7.11. The Eurasian continent about 100,000 years ago. Three habitats are almost completely separated from each other by the Himalayan and Altai mountains together with their glacier areas. (Von Eickstedt, 1934 [57])

ferences, such as those existing between the main races, considerable reproductive isolation between subpopulations is required. Is there a period in early history, during which the human population was subdivided into three more or less isolated subpopulations?

During the most recent glacial period, i.e., about 100,000 years ago, much of the earth's expanse was covered by ice (Fig. 7.11). The Himalaya and Altai mountains together with their glacier areas separated the Eurasian continent into three areas, providing the conditions for separate evolution of whites in the west, Mongoloids in the east, and blacks in the south. Wherever the present-day living areas of the three main races do not conform with the areas in which they evolved, migrations can explain the discrepancies [57].

Genetic Differences That Can Be Explained by Specific Selective Mechanisms: Skin Pigmentation and Ultraviolet Light. The most conspicuous difference between the main races is the difference in skin pigmentation. Most present-day subhuman primates are darkly pigmented, and it is reasonable to assume that in ancient human populations dark pigmentation was also prominent, particularly since the earlier humans originate from Africa. Why then are whites and Mongoloids so lightly pigmented?

According to a plausible hypothesis, there has been an adaptation to low ultraviolet (UV) irradiation in the habitats of these two races. UV light is necessary for conversion of provitamin D to vitamin D in the human skin (Fig. 7.12). Vitamin D is needed for calcification of bones; its deficiency leads to rickets. One of the most dangerous features of rickets is pelvic deformation which impairs normal childbirth, often leading to the death of mother and child under primitive living conditions. This effect obviously makes for strong selection pressures. Figure 7.13 shows the degree of skin pigmentation and intensity of UV light irradiation in various areas of the world [1986].

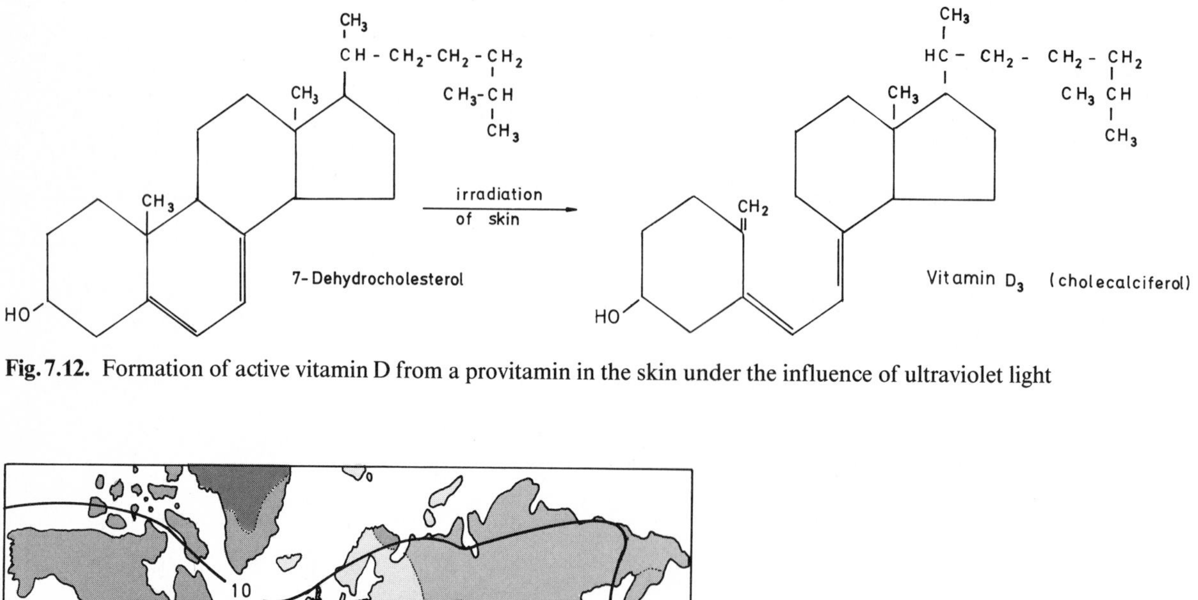

7-Dehydrocholesterol —irradiation of skin→ Vitamin D_3 (cholecalciferol)

Fig. 7.12. Formation of active vitamin D from a provitamin in the skin under the influence of ultraviolet light

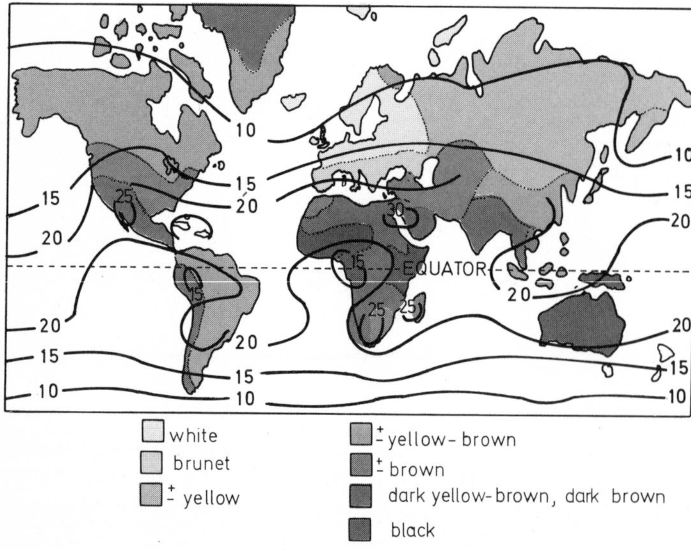

Fig. 7.13. Intensity of ultraviolet light and average skin pigmentation in various areas and indigenous populations of the world. Numbers denote mean intensities of global solar irradiation on a horizontal plane at the surface of the Earth (mW cm^{-2} averaged over 24 h, for the whole year). (Adapted from Walter, 1970 [1986] and Mourant [1957])

The hypothesis implies that UV radiation can penetrate easier into a lightly pigmented skin and, hence, that identical UV doses will lead to more vitamin D formation in lightly pigmented skin. Experiments with saddle pigs were informative. These animals are darkly pigmented in the middle of their bodies, whereas the cranial and caudal regions show little if any pigmentation. Vitamin D formation after UV irradiation in vitro was higher in unpigmented than in pigmented skin of the same animals [1901] (Fig. 7.14). Two human populations are apparent exceptions of the geographic correlation with skin pigmentation: Eskimos and African Pygmies. Both – and especially the latter – are darkly pigmented despite the fact that UV irradiation in arctic regions and also on the grounds of the tropical rain forest is scanty. Eskimos seem to get abundant Vitamin D from fish and seal liver, the pygmies possibly from the insect larvae that form part of their nutrition [1917].

Frequency of the Fy^- Allele in Blacks. The Fy^- allele of the Duffy blood group system was found to be frequent among blacks but is very rare or does not occur at all in Mongoloids and whites. Blacks with this allele have a complete resistance against the infective agent of tertiary malaria, *Plasmodium vivax* [1952]. Tertiary malaria alone is hardly if ever fatal. Hence, a selective advantage is not immediately obvious. However, under primitive living conditions and in a population that is exposed to multiple infective agents and parasites, a malaria infection can be a severe health hazard indeed.

The discovery that the Duffy blood group is involved in receptor activity for *Plasmodium vivax* is of great paradigmatic importance because it illustrates the biologic significance of a previously discovered red cell polymorphism whose function was unknown, as is the case with most polymorphisms. Since practically all Africans are Duffy negative, this example also demonstrates how an allele that is usually observed in polymorphic frequencies spreads throughout the entire population because of its selective advantage.

An alternate hypothesis has been recently elaborated [1817]. It has been suggested that the pre-existing high frequencies of the Duffy negative allele prevented vivax malaria from becoming endemic in West Africa. It is argued that vivax malaria originated in a primate ancestor and failed to spread through Africa because of the existence of the Duffy negative allele.

Lactose Absorption and Malabsorption. Lactose is the only nutritionally important carbohydrate in milk (Fig. 7.15). To be absorbed by the small intestine, lactose has first to be hydrolyzed by a specific enzyme, lactase, that is located in the brush border of the intestinal epithelial cells. The milk of almost all mammals contains lactose; lactase activity is high during the newborn and suckling period in all populations and races but declines at the time of weaning. Afterward, lactase activities are maintained at low levels, usually at less than 10% of the activity in the newborn.

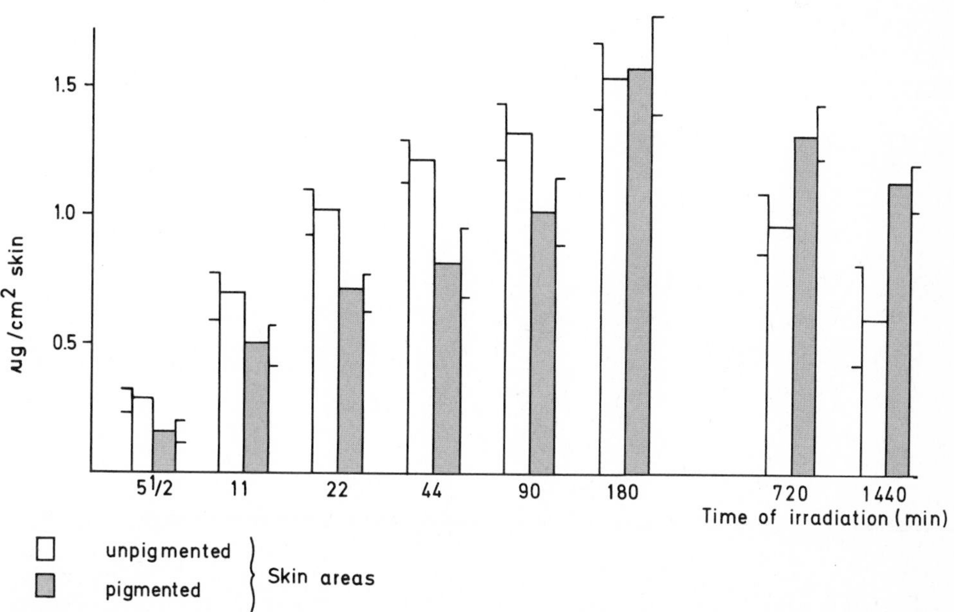

Fig. 7.14. Formation of vitamin D (mg/cm² skin: *ordinate*) in saddle pigs after irradiation with UV light (S 300; 50 cm distance). The *abscissa* shows the time of irradiation. After irradiation up to 90 min, vitamin D formation is lower in pigmented as compared with unpigmented skin. Synthesis in pigmented skin is greatly delayed and higher than in unpigmented skin only after 180 min. The *brackets* refer to standard deviations of the means. (Bekemeier, 1969 [1901])

Up to a few years ago, humans were thought to represent an exception: They "normally" seemed to maintain their lactase activity into adult age. Such persons with high lactase activity can tolerate large amounts of lactose; after a lactose load, they show a considerable rise in blood concentration of glucose and galactose, the sugars that constitute lactose.

Lactose Malabsorption [1924]. In persons with low lactase activity, there is little or no increase of blood glucose after milk ingestion. Many develop clinical symptoms of intolerance after consuming 25-50 g of lactose (1 liter of cow's milk contains 45-50 g of lactose). These symptoms are watery, acid diarrhea, colicky abdominal pain, and flatulence. Smaller amounts of milk and milk products in which much of the lactose has been removed by fermentation, such as yoghurt or whey, are tolerated without untoward effects. Comparative studies of lactose tolerance in American blacks and whites showed that lactose intolerance due to lactose malabsorption was much more frequent in blacks [1900]. Since that time, studies from many populations have become available [1924] (Fig. 7. 16). The most reliable method to determine whether intestinal lactase is present is enzyme determination in intestinal biopsies. Such a method is not suitable for population or family studies. Therefore, standard lactose tolerance tests have been designed that measure the rise of H_2 in the exhaled air after ingestion of a standard oral dose of lactose [1936].

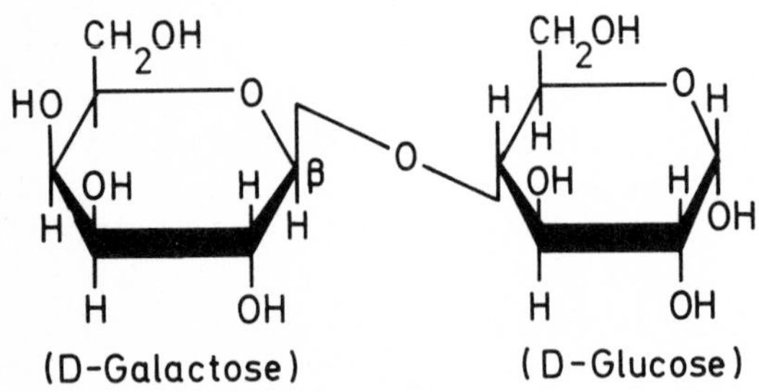

(D-Galactose) (D-Glucose)

Fig. 7.15. The disaccharide lactose

Persistence of lactase activity in older children and adults or lactose tolerance is very rare or absent in most Mongoloid populations including American Indians and Eskimos. A similarly low incidence of lactose tolerance is observed in most Arab and Jewish populations and in the populations of tropical Africa, Australian aborigines, and Melanesians. A consistent high prevalence of persistent lactase activity (over 75%) is found only in northern and central Europe and in their descendants in other continents. In addition, high frequencies of lactose tolerance were reported in a few groups of nomadic pastoralists in Africa. Intermediate frequencies (30%-70%) were observed in Spain, Italy, and Greece. Southern Asia shows a high variability; possibly, the trait was introduced by migration. The American black population shows somewhat higher values than African blacks.

What Is Normal? What Is Abnormal? In conclusion, most human populations show a decrease of lactase activity after weaning; this is the common pattern in

Fig. 7.16. Persistence of lactase absorption *(PLA)* in the world population. (Courtesy of G. Flatz)

two of the three major races; a high incidence of persistent activity is found only in whites, and even here, not in all populations. Hence, the loss of this unique activity following weaning together with its consequence, lactose malabsorption, is the "normal" condition in humans as it is in other mammals.
It is an interesting cultural phenomenon that scientists considered lactose tolerance, i.e., persistence of lactase activity, normal since this trait was most common in populations of northern European origin where this work was done. This conceptual bias even had economic consequences. In an effort to improve protein intake of children in African and Asian countries, powdered milk was distributed in large amounts on the not unreasonable hypothesis that what is good for European children must also be good for children in developing countries. These programs may need to be reconsidered in the light of our knowledge of population distributions. Yet, small amounts of milk and milk products are tolerated; therefore, suitably organized milk and dairy programs might still be useful. It is therefore unwise to prohibit any consumption of lactose-containing foods by lactose-intolerant populations who otherwise suffer from protein deficiency.

Enzyme Induction or Genetic Variability? There are two possible biochemical explanations for lactose malabsorption.

1. Lactose malabsorption could be a consequence of the low lactose intake of most individuals after weaning. We know that the activity of many enzymes can be enhanced by the substrate (substrate-specific induction) [1903]. This hypothesis first received wide recognition but was subsequently tested in animals and in humans with negative results [1924].
2. Some observations on a few isolated families [1915] suggested a genetic basis, and more specifically, an autosomal-recessive mode of inheritance of malabsorption.

Autosomal-recessive inheritance of lactose malabsorption was demonstrated by a large-scale study of all relevant mating types in Finland (Sahi, 1974 [1975]) and confirmed in many other populations. Lactose "absorbers" are either homozygotes or heterozygotes for the lactose absorption gene, malabsorbers lack this gene.

Multiple Allelism? Low lactase activities of the recessive type are not attained at the same age in different populations. In Thailand and in the Bantu, all children above the age of 4 years show no blood glucose increase after lactose loading. In black American children, the proportion of lactose malabsorbers increases to 14 years, and in Finland, full expression may be delayed up to the time between the 15th and 20th year. This phenotypic variability might be due to multiple alleles or to differences in the amount and type of milk consumed during childhood and requires further study.

Genetic Mechanism. Residual lactase activities seem to be also present in adult malabsorbers. So far, it is not known whether the lactase formed by malabsorbers is different in its protein structure from that formed by absorbers. The switchover from high to low lactase activity at the time of weaning in malabsorbers is somewhat similar to that from hemoglobin γ chain to β chain production and, hence, from Hb F to Hb A formation in hemoglobin synthesis; persistence of lactase activity can be compared with the persistence of fetal hemoglobin (Sect. 4.3).

Natural Selection. The predominance of lactose malabsorption in adults of most human populations together with its regular occurrence in other mammals suggests that the gene responsible for persistence of lactase activity occurred by mutation sometime in human evolution and that a selective advantage has caused the high frequencies in some populations. What is the nature of this advantage? Two main hypotheses have been brought forward [1924]:

1. A culture historical hypothesis
2. A hypothesis that advocates improved calcium absorption by lactose [1923].

According to the cultural hypothesis, domestication of milk animals in the neolithic period (about 9,000 years ago) resulted in a selective advantage of individuals who could now satisfy a higher fraction of their nutritional requirements mainly protein by milk consumption. Indeed, there are some populations with milk consumers the cattle-breeding tribes mentioned before for which this hypothesis could apply. Its general validity, however, is not certain. For example, there is no parallelism between the milking habit and the prevalence of lactose absorption. Large populations in Africa and Asia are milk consumers but have a very low incidence of lactose absorption. Yet, there are always some individual absorbers; hence, the gene was present and could have been favored by selection. The highest frequency of this gene in Europe is found in southern Scandinavia (0.7–0.75), where breeding of milking animals was introduced in more recent times because of unfavorable conditions in the postglacial period. Moreover, at a time when artificial refrigeration of milk or production of dried milk had not been invented, milk turned sour within a short time, and malabsorbers probably discovered easily that sour milk which is low in lactose content causes less di-

gestive symptoms. A specific advantage of milk in the environmental conditions of northern Europe is therefore an alternative worthwhile considering.
In connection with the selective advantage of lightly pigmented skin, vitamin D deficiency in northern areas due to diminished UV irradiation and severe selective disadvantage of rickets by pelvic deformation of childbearing women were mentioned. It has now been suggested that lactose can replace vitamin D, improving the uptake of calcium [1923]. Late rickets and osteomalacia occur at an age at which lactose malabsorption has already developed. The crucial problem of this hypothesis is the mechanism of a possible protective effect of high lactose absorption against rickets. Is there specific enhancement of calcium absorption with hydrolysis of lactose? Animal experiments can only give ambiguous results, since adult animals are lactose malabsorbers. Recently, however, it has been shown by studies in human subjects that calcium absorption is, indeed, enhanced by lactose absorption [1908].
Regardless of the ultimate fate of the calcium hypothesis, the problem as posed is the type of heuristic hypothesis needed in human population genetics. The hypothesis is specific, provides a mechanism by which selection can work, and suggests experiments by which it could be tested.

Vitamin D and GC Serum Groups. A genetic polymorphism of an immunologically defined fraction of a human β_2 serum protein has been known since 1959 [1935]: many alleles are known but most populations are polymorphic only for two of them, GC^1 and GC^2 Australian aborigines have a third allele, GC^{Abo} and the Chippewa Indians have a fourth one, GC^{Chip} [1940]. When the first data on gene frequencies became available, the allele GC^2 turned out to have low frequencies in areas of high aridity. This finding was explained when the function of the GC proteins was discovered. They are carrier proteins of vitamin D [1909].
A more recent survey suggests a relationship between sunshine and the GC polymorphism; in the aboriginal habitats of the world population, high frequencies of GC^2 were found in most populations that had been living for a long time in areas with low sunlight intensity [1957].
This geographic distribution suggests a selective advantage due to a more efficient transport of vitamin D, especially if the supply is limited and, hence, a lower incidence of rickets in individuals either heterozygous or homozygous or both for the GC^2 allele. The exact mechanism remains to be elucidated.

Possible Selective Mechanisms for Other Racial Characteristics. Apart from the examples mentioned in the foregoing paragraphs and those used in the section on population genetics, little is known about selective advantages or disadvantages of racial characters. It is reasonable to assume that the small and heavyset stature of Eskimos and their relatively thick subcutaneous fat has advantages in a cold climate and the broad and deep thorax of South American Indians living in the high Andes relates to respiratory adaptation to life at high altitudes.
Members of the different racial groups show differences in susceptibility to common diseases under the living conditions of the United States and northwestern Europe and other affluent countries. American blacks, for example, are more susceptible to high blood pressure than whites. Some groups of Asian Indians, for example, in Trinidad, were shown to have higher frequencies of diabetes mellitus than other population groups. Better understanding of such racial differences will undoubtedly be forthcoming as soon as investigators with a medical background and intimate knowledge of diseases and their causes become interested in population genetics.
Hypotheses such as the "thrifty genotype" concept [1961] and the rapid mobilization of lipids have been intertwined to explain diabetes and atherosclerosis. It is assumed that under conditions of starvation, the diabetic genotype allows more efficient mobilization of carbohydrates while atherosclerosis-favoring genes have been explained as permitting more rapid mobilization of fat.
Such selective mechanisms acting in the past when starvation was common over many generations have been suggested to explain the high frequency of diabetes and atherosclerosis in current times. Unfortunately, neither of these hypotheses can be clearly reconciled with current pathophysiologic concepts of carbohydrate and lipid metabolism.

7.3.2 Future of Human Races: Race-Crossing

Will Races Disappear? The one important condition for the formation of races has been isolation. Within their old habitats in Asia, Africa, and Europe, this isolation is still largely in existence. In the newly conquered areas of settlement, such as North and South America, huge "melting pots" were formed that comprised elements of all three race groups. In spite of social customs, which keep intermarriage at a low level, there is little doubt that race mixture will increase, leading sooner or later to a hybrid population.
How long the main races in the Old World will maintain their separate status depends on political developments and cannot be predicted. It is possible that in the long run race-specific differences will disappear by interbreeding.

Interracial Crosses in Hawaii [1955]. What are the genetic consequences of race mixture? At a time when biologists did not think in terms of variabilities within populations but thought of human variability in terms of racial "types," race mixture was often regarded as disruptive and leading to disharmonious phenotypes. It was agreed that by long selection certain combinations of genes were "coadapted". If such coadaptation were broken up by race mixture, disharmonious phenotypes would result. It therefore came as a surprise that hybrid populations, such as those between Hottentots and whites in southwestern Africa, turned out to be fully viable and quite normal in health [1916]. Beyond this statement, however, the numerous old studies could not answer the question of possible genetic effects in interracial crosses. Another consideration predicted beneficial effects of race-crossing. Since "hybrid vigor" had been shown in lower forms of life, it was suggested that racial hybrids might be particularly healthy. Some answers came from a carefully planned study in Hawaii [1955].

The population of Hawaii is composed mainly of Hawaiian (Polynesian), white, Chinese, and Japanese elements, with smaller additions of Koreans, Filippinos, Puerto Ricans, and others. Population statistics and medical facilities are reliable, and the environmental conditions under which the racial groups are living are not too different and, in any case, strongly overlapping. Intermarriage between racial groups is rather common. The trend toward intermarriage is relatively recent so that the results in the generation following the start of hybridization could be observed.

Scope of the Study and Material. The scope of the study was defined as follows:

a) What are the genetic effects of out-crossing in man on the first generation of hybrids?
b) Do human populations represent coadapted genetic combinations that are disrupted after the first generation of out-crossing?

The major part of the data consisted of 172,448 live birth certificates and 6,879 still birth certifcates for all births registered between 1948–1958. These data comprise entries on racial affiliation of the parents. This affiliation does not mean that the parents were members of "pure races;" they contained varying admixtures from other race groups that were estimated using data from genetic polymorphisms. The population of parents was 62.7% Pacific (Hawaiian and Mongoloid); the rest was mainly white. Additional information came from records on maternal stature and weight in one clinic.

The analysis was based on a stepwise regression considering environmental factors such as socioeconomic differences and medical care. In such an analysis the question as to how differences between certain categories can be explained by concomitant variables and how much must be attributed to interracial cross can be examined.

Results and Interpretation. The main result of this study can be summarized very briefly. There were no obvious ill-effects of out-breeding either on early or late fetal deaths or on postnatal infant deaths. Birth weight and maternal weight and stature were not significantly related to maternal hybridity. A maternal effect on birth complications could be expected when the mothers were Japanese or Chinese and the fathers whites, because Japanese and Chinese women are smaller. No such effect, however, was found. As noted in Sect. 3.8.3, the rate of dizygotic twinning is much lower in Mongoloids than in white women and is probably caused by a difference in frequency of polyovulation, i.e., by a maternal factor. This conclusion was corroborated by study of interracial crosses; the DZ twinning rate only depended on the race of the mother, irrespective of the father's race and hybridity of the children. Mothers who were hybrids between whites and Pacific races had a low dizygotic twin frequency closely resembling that of Pacific mothers. If the genetic disposition to polyovulation would be a threshold character determined by additive gene action, one would have expected these women to show an intermediate DZ frequency. As it stands, the result suggests participation of recessive genes in polyovulation. The data on congenital malformation were not quite as reliable as the mortality data. Taken at face value, there was no difference in the overall incidence of major malformations between racial groups, whereas frequencies of certain categories of malformations varied between these groups. Spina bifida, for example, was more frequent in Japanese of Hawaii than reported in the Japanese mainland but still remained significantly less frequent than in whites. Genetic as well as environmental factors may be important. The occurrence of severe clubfoot appeared to be high in Hawaiians and low in other Pacific groups. The incidence of congenital malformations was not influenced by out-breeding.

If anything, there was a small nonsignificant overall advantage of hybrid children that could be due to a lower risk of recessive detrimental genes to become homozygous.

Questions Not Answered by the Hawaii Study. The two questions posed by the Hawaii study can be answered as follows:

1. The genetic effects of out-crossing in man on first generation hybrids do not lead to any ill-effects manifested by perinatal and infant death or major congenital malformations as assessed by vital statistics.
2. There is no evidence for coadapted genetic combinations that are disrupted after the first generation of out-crossing when newborns are assessed indirectly by scruting of birth and death certificates.

These conclusions, however, do not answer all possible questions. In the older anthropologic literature, disharmonic combinations between jaws and teeth were claimed on the basis of inadequate evidence, an assertion that was severely criticized. Race-crossing in rabbits between races of extreme size did not show any such disharmonies [1980]. Some of the older authors specifically mentioned emotional difficulties of race hybrids by such disharmonic gene combinations; such speculations were usually based on a biologically wrong understanding of race where races were considered specific types rather than population groups with different gene frequencies. The role of the environment was generally neglected. There is no evidence to suggest that racial mixture has any deleterious genetic consequences. Less homozygosity with beneficial consequences for the incidence of many recessive diseases is certain. Whether hybrid vigor and greater physical and mental health will result cannot be answered with certainty. Additional studies of physical and mental health of hybrids as compared with their ancestral populations would be required under carefully controlled environmental conditions for a scientific assessment of the possible genetic effects of race crossing.

8 Genetics and Human Behavior

Scope and Conceptual Difficulties of Human Behavior Genetics. A survey of the genetic aspects of evolution has shown great similarity between human beings and higher nonhuman primates in chromosomes, proteins, and in many other genetically determined characters. The human species differs substantially in only one characteristic: language and abstract thinking. No other species can look into the past nor into the future! The analysis of the essential differences between man and other species must therefore be directed to the brain - the organ of thought and language. These characteristics have enabled our species to supplement biologic evolution with "cultural evolution," with all its consequences for creation of human civilizations and for alteration in life style (Table 7.1). The uniqueness of the human brain that allowed these developments is part of our genetic heritage. All experiments with the most advanced species of nonhuman primates - chimpanzees - have shown that attempts at raising such animals similar to and together with human children failed to develop spoken language. Although cognitive functions of chimpanzees were found to be more advanced than previously assumed, even these animals never reach the level of conceptualizing that older children do.

On the other hand, development of typical human behavior in children requires interaction with other human individuals and with the environment at large, including sensory stimuli and opportunities for motor behavior. Sensory and motor deprivations - especially if occurring at critical periods of childhood - lead to deficiencies that may even be detectable by changes in the histology of nerve cells and their interconnections [55]. Human genetics is concerned mainly with the analysis of genetic mechanisms leading to phenotypic differences between members of our species. The fact that behavioral characteristics can develop only in intimate and continuous interaction with the environment makes genetic analysis conceptually very difficult. Fingers, for example, are formed during a brief period of embryonic development. The trait of brachydactyly or short fingers persists all through later life regardless of environmental changes. The usual dominant mode of inheritance of this condition can easily be traced over many generations (Sect. 3.1.2). Most severe enzyme deficiencies will also lead to an abnormal phenotype under all conditions, but specific manipulation of the environment may help to alleviate the symptoms. We are learning now that many simple genetic traits require certain environmental factors for their phenotypic expression such as in pharmacogenetics (see Sect. 4.5.1). Most behavioral characteristics, on the other hand - with the exception of severe mental deficiency - need to be considered in the context of a certain environment. From birth on, the individual is not only shaped and modified by the environment - he or she in turn actively manipulates the environment and creates a complex interaction of many components. Analysis of genetic variability in behavior between individuals is difficult since the genetic aspect is only one of many parts in this complicated system.

Practical Difficulties. In addition to these conceptual problems, there are also practical obstacles that impede scientific progress in the field. Studies of genetic variability usually require examination of sizeable groups of individuals. Such examinations are feasible if the material to be studied can be readily obtained such as blood or even a skin biopsy. This is the main reason why genetic variability of erythrocyte enzymes (Sect. 4.2.2.2) and hemoglobin variants is so well understood (Sect. 4.3). Investigations of human *brain* function, on the other hand, have to use more indirect methods since human brain material is available rarely except from autopsies.

Importance of the Field. We feel that human behavior genetics promises to be the most interesting and important branch of human genetics with possibly far-ranging consequences. Elucidation of genetic variability will allow better understanding of different human behaviors and emotions. It is probably no mere concidence that work in human genetics began with a problem of behavior genetics - the work of F. Galton on the frequency of high performance among relatives of outstanding men (Sect. 1.3).

Later, the mainstream of research moved away from behavior genetics, as other fields became more accessible to genetic research methods and concepts.

However, some research in human behavior genetics has always been carried out. This work was usually influenced by concepts of quantitative genetics developed with quite different problems in mind, such as animal breeding (Sect. 3.6). The development of the twin method seemed especially well suited for analysis of genotype-environmental interaction and also encouraged such research (Sect. 3.8). However, conscious or subconscious prejudice often biased the results of these studies.

Paradigms of Mendel and Galton in Behavior Genetics. We believe that concepts and methods in human genetics have reached a phase in which they can be applied more frequently to problems of behavior genetics. Research that will be most fruitful and least controversial is likely to be guided by the paradigm of Mendel - the gene concept - and its subsequent extensions, to the level of molecular biology. Most research in human behavior genetics, on the other hand, has been guided by the paradigma of F. Galton using biometric methods in the quantitative analysis of relationships between behavioral phenotypes. This approach, too, has gradually been extended by various methodologies and has reached a considerable degree of sophistication. However, the passionate polemics of recent years concerning the genetics of intelligence suggest that the Galtonian paradigm may have reached the limits of its explanatory power. Such biometric approaches can never provide an explanation of *genetic mechanisms*. For practical reasons, the clean experimental designs needed to obtain unequivocal answers to these problems by the Galtonian approach simply cannot be achieved in humans.

We predict that progress in our understanding of the genetic variability underlying the interindividual differences in behavior will depend on our ability to apply concepts and methods of the central paradigm of genetics, i.e., analysis of gene action, to research on these problems. This does not mean, however, that the methods of quantitative genetics now utilized in human behavior genetics will soon become obsolete. They may retain an important auxiliary function in data analysis. Decisive progress, however, is likely to come from Mendelian analysis of biologic parameters of gene action in the central nervous system.

8.1 Animal Models

Even if the "jump" from the mental capacity of animals that are closest to us in evolution seems to be enormous, many basic principles of the action of brain and nervous system are still identical in different species, although the degree of complexity is lower in animals. A frequently used research strategy followed successfully in genetics (Sect. 4.1) analyzes simpler systems to gradually develop concepts and methods that will finally allow to attack the more complex situations successfully. This strategy is now being pursued for investigation of the nervous system. It is not our intention to survey modern brain research. We shall confine ourselves to a few examples that - by analogy or by contrast - may help to elucidate the situation in humans.

8.1.1 Research in Insects

Dialects in the Language of Bees [2044, 2047]. Insects, especially the colony-forming species, may exhibit complex and meaningful sequences of movement. Honeybees, after having found food in a close-by field with suitable flowers, inform other working bees of their hive about their finding by a circular dance. However, if this field is farther away, they perform another type of dance that indicates the direction in which food can be found. These behavior patterns are tightly fixed genetically, but provide no clues as to genetic mechanisms. Experiments have shown that bees are able to improve their capacity of spatial orientation by learning and that learning ability in itself is a genetic character in which genetic stocks may differ [2111, 2112].[1] At the level of formal genetics, a genetic mechanism could be elucidated for another behavior character of honeybees.

American Foul Brood: A Problem of Hive Hygiene [2185]. There are differences in nest-cleaning behavior between breeding stocks of bees. These differences were discovered in the process of selective breeding for a character that seemed to be void of any behavioral aspect: resistance against a larvae-killing infection, American foulbrood, caused by a bacterial organism. Two stocks were compared, a resistant and a susceptible one. The worker bees from the resistant stock immediately opened the combs containing affected larvae and removed them, whereas the bees from the susceptible stock failed to do so. When the two stocks were crossed, none of the F_1 generation cleaned their hives. When back-crossed with the resistant stock, however, four types of progeny were observed in a ratio of 1:1:1:1. The first group behaved quite like the resistant bees: they opened the comb cells containing affected larvae and removed the dead. The second group opened the combs but did not remove the dead. The third group did not open the combs but removed the dead if the combs were opened artificially. Finally, the fourth group neither opened the cells nor removed dead larvae; lhese animals behaved similar to the susceptible stock. This behavior has been explained by assuming a recessive mode of inheritance for two different genes: one gene controls the open-

[1] The application of the terms language and dialect to the dancing patterns of bees is semantically inappropriate since such behavioral patterns are quite different from human language and dialects. Their neurophysiology is therefore likely to be very different. We have used these terms (following v. Frisch) [2044] in the absence of a simple descriptive term for this behavior.

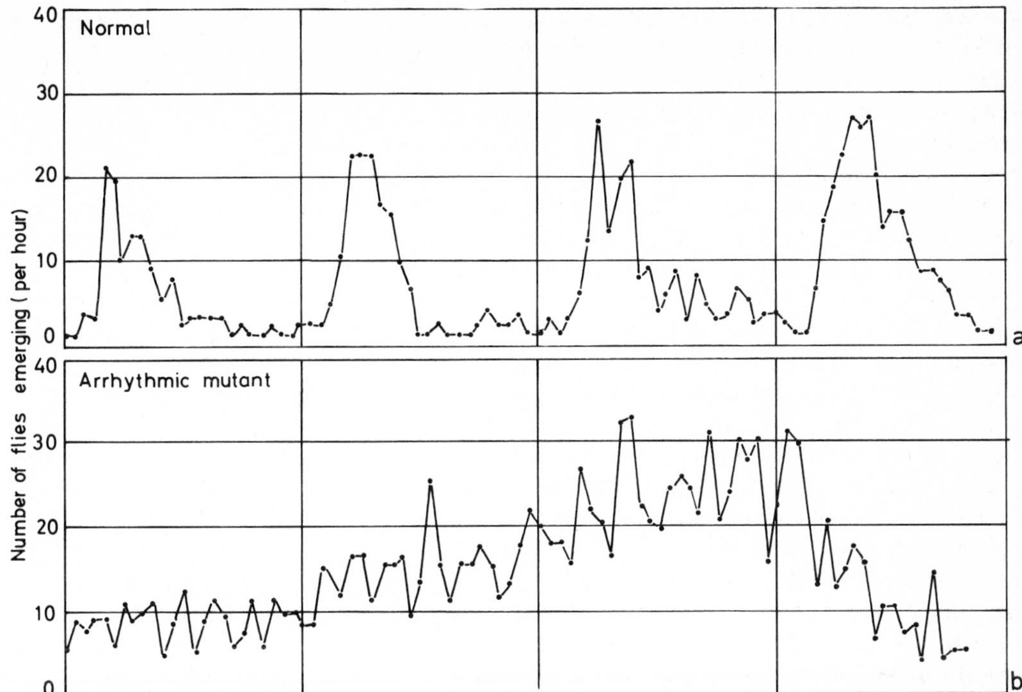

Fig. 8.1 a, b. Number of flies eclosing from the pupae over 4 days. The *ordinate* shows the number of flies eclosing, the abscissa the time of the day. **a** The normal wild-type usually ecloses early in the morning. **b** A mutant in which the normal day-night rhythm has been lost: the flies will eclose at about identical rates throughout the day. (Adapted from Benzer, 1972 [2004])

ing of comb cells, the other one the removing of afflicted larvae.

This example shows how a rather complicated genetically fixed behavioral pattern may be controlled by differences in single genes. However, we do not know the reason why the bees behave in this way. The steps of analysis from the observable phenotypic-Mendelian level (Sect. 3.6.1) to the physiologic mechanisms of gene action remain to be carried out. Do the resistant bees smell the dead larvae? Do they see a slight discoloration of the comb? Or what stimulus triggers their behavior?

Comprehensive experiments on the genetic mechanisms of genetically determined behavior characteristics were performed in another insect more familiar to the geneticist: *Drosophila melanogaster.*

"Genetic Dissection of Behavior" in Drosophila [2004]. A great many mutants are known. Most are characterized by morphologic and color criteria. Sturtevant showed as early as 1915 [2212] that the phenotype of the X-linked mutation white-eyed influenced choice of mates for pairing. Other mutations specifically influence courtship and mating behavior or motor activity. One mutant does not fly in spite of anatomically normal wings, and others lack the day-night rhythm. Fig. 8.1 shows the number of flies eclosing from the pupae over 4 days. Normal flies mostly eclose at dawn when the air is moist and cool and the fly has time to unfold its wings and harden its cuticle before there is much risk of desiccation or

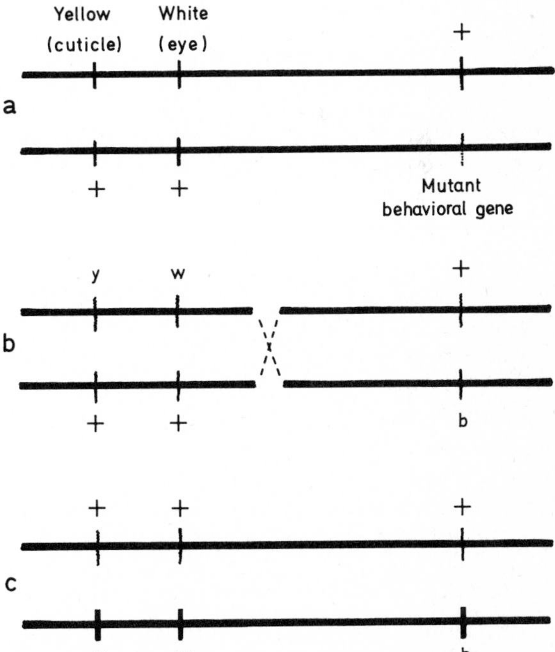

Fig. 8.2 a–c. X chromosomes tagged with the X-linked recessive mutations white and yellow also carry the behavioral mutation to be examined. **a** The mutant, behavioral gene *(b)* is located on a homologous X chromosome. **b** Crossing over. **c** The behavioral gene *(b)* is now located on the X that carries the genes y and w. (Adapted from Benzer, 1972 [2004])

from predators. This behavior, like other characteristics such as activity of the fly, is controlled by an "internal clock" independent of environmental influences; for example, this clock even works in complete darkness.

A certain mutant, however, has lost this internal clock. The flies eclose at all times of the day (Fig. 8.1b). The morning maximum has completely disappeared. Genetic analysis has shown his mutation to be X-linked.

What is the mechanism of this aberrant behavior? To answer this question, an ingenious genetic technique was developed: the X-linked mutant gene is brought by recombination to the chromosome carrying the mutations white and yellow that serve as phenotypic markers (Figs. 8.2 and 8.3). A male showing both mutants is mated with a female having a ring X chromosome (Fig. 8.3). As explained in Sect. 2.2.2, ring chromosomes are not always stable; they may be disrupted during mitosis, and the fragments may be lost. Mosaic zygotes, such as X ring X/XO or XY/XO are formed. Contrary to humans, XO zygotes are male in *Drosophila*. The exact kind of the mosaics, however, can be recognized by inspection; body parts with only the paternal X chromosome show the white and yellow phenotype and are hemizygous for the behavior mutation. In body parts that have maintained the ring X, the dominant wild alleles for the three mutations will determine the phenotype (Fig. 8.4).

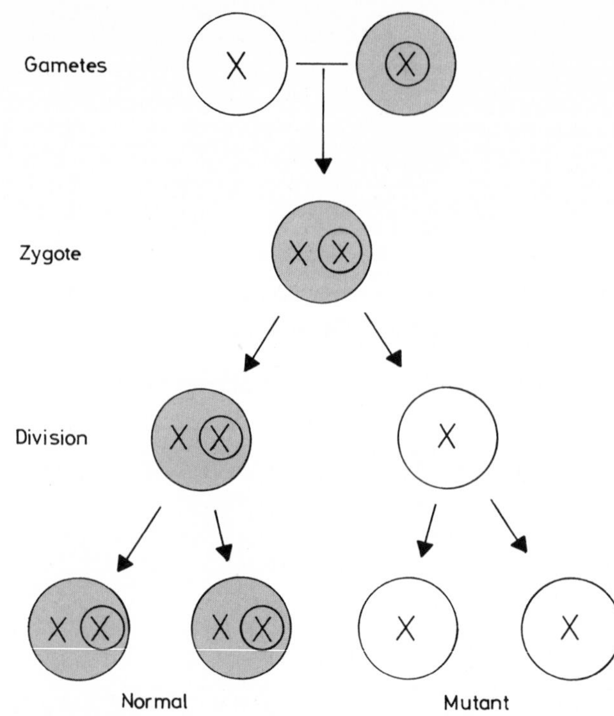

Fig. 8.3. Mating of a male having the X chromosome with the two markers white and yellow and the behavioral mutation with a female carrying an X ring chromosome shown as a circle around X in the *shaded* figure. The ring chromosome may be lost in one cell division, leading to a cell clone with only the normal X chromosome. (Adapted from Benzer, 1972 [2004])

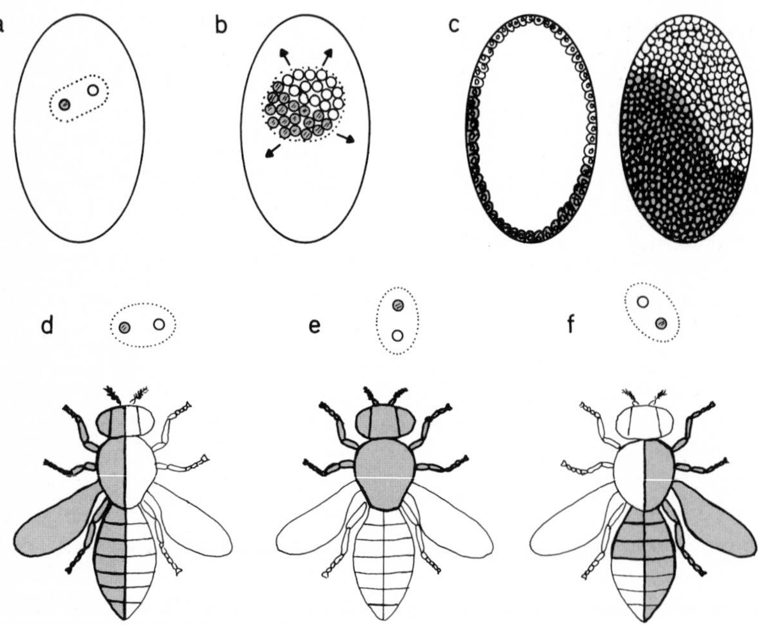

Fig. 8.4 a–f. Different types of mosaics in offspring from matings between male carriers of the X chromosome, which carries the white and yellow mutants and the behavioral mutant, and female ring chromosome carriers. **a–c** Different stages of embryonic development. **d–f** Position of the two cell types with and without X ring chromosome and mosaic patterns of animals. (Adapted from Benzer, 1972 [2004])

The behavioral phenotypes of a great number of mosaics with different distributions of the two types of cells can be compared. Such analysis is possible in this insect by a convenient property of embryonic development: there is little migration and mixing of cells. Cells maintain their relative position within the body. Using this technique, the internal clock was found to be most closely associated with the head of the insect. Among flies with a mosaic head, some showed the normal rhythm and others the abnormal one. A few flies, however, showed a unique rhythm, suggesting that each side of the brain produced its rhythm independently, and that the fly responded to both of them.
The method has been refined by including an acid phosphatase-deficient X-linked mutant. The enzyme is normally present in all cells, including nerve cells, and can be made visible by a histochemical staining technique. Using this enzyme mutant as a marker in the same way as the external markers *white* and *yellow,* it may even be possible to localize an aberrant behavior pattern to certain groups of nerve cells. It was shown, for example, that the photoreceptor cells of the eye come from a different area of the embryonic blastoderm than do the neurons of the lamina to which they project. The photoreceptor cells arise in the eye, while the laminar neurons come from the brain. Therefore, some mosaic flies have a normal retina and a mutant lamina and vice versa. Both defects cause an identical disturbance in phototaxis - a nice example of genetic heterogeneity (Sect. 3.3.5).
Most behavior patterns in *Drosophila* analyzed in this way are genetically predetermined in all details. While simplifying genetic analysis, this high degree of genetic determination makes comparison with humans difficult, as our most important behavioral characteristic is the ability to learn from experience. However, a limited ability to learn was also shown in *Drosophila*. This property may open up new paths for genetic analysis of learning capacity.

Mouse Mutants Affecting Embryonic Development of the Brain. Analysis of deviating phenotypic patterns in mouse mutants was initiated to elucidate mechanisms of normal embryonic development in the central nervous system. One example is the reeler [2198], a mutation leading to severe difficulties to maintain body balance. The reeler brain develops normally except for the cerebellar cortex and hippocampal formation, which show disorganization of cell alignment and of intracortical synaptic connections. These regions are the only parts in the mouse brain where cells destined to establish synaptic contacts with each other become initially transposed. This and other evidence points to a failure in a specific recognition mechanism that normally allows patterned cell alignment in the developing brain. Other mouse mutants have provided material for a continued analysis of brain development in the mouse. The use of such experiments of nature for an analysis of normal biologic function is turning out to be a useful investigative tool. It is likely that similar mechanisms apply to brain development of man.

What Can We Learn from Drosophila and Mouse Experiments for Genetic Analysis of Human Behavior? These experiments, interesting as they are, do not help much in our attempts to assess the role of genetic factors in human behavioral variability. Their design is mostly based on peculiarities of *Drosophila* not shared by humans. For example, mosaics can neither be constructed at will nor can they be tagged by suitable mutations that alter the phenotype; embryonic development is too complicated to make this approach feasible; most importantly, variations in human behavior are never fixed as suitably as in insects. Looked at from this viewpoint, the questions posed by *Drosophila* experiments are not so much genetic as embryologic and - in part - neuroanatomic and neurophysiologic. These studies can be compared, for example, with experiments in cats in which certain parts of the brain or of the peripheral nerves are destroyed, and the functional consequences are analyzed to find out how these parts cooperate in normal function. The study of neurological symptoms carried out by neurologists to arrive at brain localization of a given lesion is a human counterpart of this work in *Drosophila*. The *Drosophila* mutants utilized in these experiments have been produced artificially by chemical mutagens. While they might also occur spontaneously, their fitness will in most cases be low. For example, the mutant with loss of the internal clock will hatch all through the day. Thus, most animals will soon be dried out or be devoured by predators. These mutants are analogous to rare hereditary diseases whose mechanism is unknown and shed less light on genes influencing behavioral variability within the "normal" range. However, deleterious mutants, even those affecting the function of certain nerve cell groups, are known in humans. Various methods for analysis of deviant gene actions are in use, and this research approach has little to learn from work with *Drosophila*.
Still, a more general lesson can be learned. As other sources of phenotypic variation, behavioral variability is likely to be at least partially determined by genes. To analyze the possible genetic mechanisms affecting behavioral variability, we should analyze the pathways in which gene differences influence behavior. Contrary to *Drosophila,* a straightforward one-to-one relationship between gene and phenotype cannot be expected, however, and a more indirect approach is necessary. Do we have any chance to find such genes? This topic will be taken up in Sect. 8.2.3.4. First, we ask whether experiments with mammals can give us additional clues. The animal

that has been examined most extensively is the mouse.

8.1.2 Behavioral Genetic Experiments in the Mouse

Three main approaches have been used in the mouse:

1. Known mutants have been examined for special behavioral characteristics [2067]. Similar to the described *Drosophila* mutants, mouse mutants have low fitness and are therefore rare in natural populations. They have little or no bearing on the naturally occuring genetic variability of behavior and might be compared to severe hereditary diseases in humans. Their study contributes little to understanding the genetic basis of the range of normal human behavior.
2. Various inbred strains have been compared for behavioral characteristics, such as temperature preferences, geotactic reaction as measured by the behavior in climbing slopes of different angles, explorative behavior, general motor activity, ability to find a way through a maze, and emotionality as measured by the frequency of defecation. Frequently, inbred strains show between-strain differences in such behavioral characteristics. Such characters can then be analyzed genetically by between-strain crossings.
3. In randomly mating populations, artificial selection can be used to demonstrate genetic variability for quantitatively varying characteristics. As predicted (Sect. 6.2.1.5), populations usually respond to selection by shifting of the mean in the direction desired by the experimenter and by diminution of the variance of the trait under selection. After a variable number of generations, the mean reaches a plateau. Now, the remaining variance at that point is exclusively environmental, and continuing response to selection would be possible only if new genetic variability were created by mutation.

These approaches show that genetic variability for measured behavioral characteristics exist. However, we remain ignorant regarding the genes involved. More specific conclusions may be possible if animals performing differently in the "tests" are compared in physiologic variables that could possibly be involved. Here, selection methods may be especially useful. If selection for a behavioral trait results in differences in other behavioral or in anatomical, physiological or biochemical characters, such differences may suggest a working hypothesis for a causal relationship [2126].

One Example for a Single-Gene Abnormality: the Obese Mouse [2047]. *Inferences to Human Obesity.* Mice homozygous for the obese gene ob/ob, if reared under standard conditions, eat heavily, become obese, and are relatively inactive. Increased food intake by these animals appeared to be caused by a failure of a satiation mechanism.
From an analogy with rats that were made hyperphagic by hypothalamic lesions, it was concluded that the primary defect might be located in the hypothalamus; histologic examinations, however, showed no anomaly.
For many cases of obesity in humans, failure of a satiation mechanism could be an explanation. Observations in MZ twins, adoption studies and accumulation of obese individuals in families point to genetic factors, even if a simple mode of inheritance has not been found so far. But the hypothesis of familial eating habits as the main cause of human obesity can hardly be excluded.
Several single gene mutations with different mechanisms may cause obesity in the mouse. Any one or a combination of these or other mechanisms may explain obesity in man. Just as in other common diseases, heterogeneity with several monogenic varieties, multifactorial types, and strictly environmental variants is likely to be operative. A genetic analysis of human obesity of unselected families without any clues as to mechanisms will therefore start with a "bouillabaise" of obesity variants. Biometric analysis of such heterogeneous data will lead to the discovery of heritability for human obesity but give no insight into the causes in individual families.

Genetic Differences in Alcohol Uptake. Another behavior extensively studied in mice and rats is the tendency to prefer alcohol and the degree of susceptibility to the narcotic activity of this substance. Here, no single gene has been identified so far, but after the pioneering research of Williams et al. (1949) [2243], comparison of inbred strains and selection experiments have shown much of the variability underlying alcohol preference to be genetic in origin [2191]. Differences in alcohol preference between various inbred mouse strains were shown when the animals were allowed to choose between water and alcohol. Animals of the C57Bl strain consumed, on the average, about two-thirds of their total daily liquid intake from a 10% alcohol solution. Some other strains, notably the DBA/2 strain, avoided alcohol almost completely, and still others showed an intermediate behavior (Fig. 8.5) [2182].
Strain differences were also found for susceptibility to the effects of alcohol. The sleeping time after intraperitoneal injection of an anesthetic dose of alcohol was found to be short in the C57Bl strain – the strain with the highest alcohol preference. It was much longer in the DBA/2 strain that avoided alcohol and turned out to be still longer in the BALB/c strain. Still more interesting were the effects of alcohol on spontaneous locomotor activity. The activity of C57Bl mice was reduced significantly; that of BALB/C mice was unchanged, and the activity of the C3H strain was enhanced after alcohol exposure [2127].
Therefore, the strain differences in response of the brain to alcohol are not just quantitative, but the effect of the drug may go in opposite directions. The data regarding consumption of alcohol and of alcoholism (Sect. 8.2.3.5) [2163] showed analogous findings in humans.
Quantitative differences in the parameters tested, for example, sleeping time, could be due either to a different rate

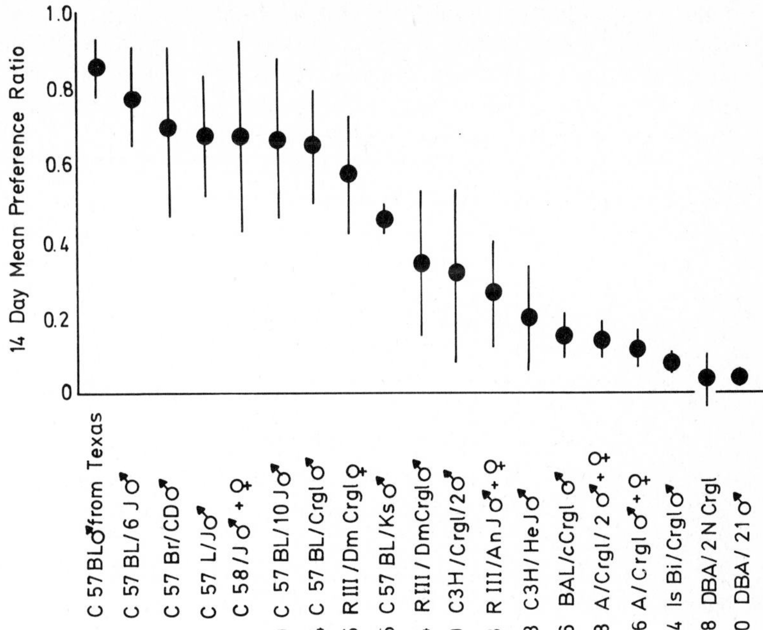

Fig. 8.5. Differences in alcohol preference in many inbred mouse strains. (Preference ratio = Alcohol/Water) ● mean; the lines refer to $\pm 1\sigma$. (Rogers et al., 1963 [2182])

in metabolism of alcohol or to a difference in the susceptibility of the brain. A comparison of blood and brain tissue alcohol levels at various times after injection and the rates of elimination were found to be practically identical between the strains showing the greatest difference in sleeping time - C57Bl and BALB/C. Therefore, the differences must lie in the susceptibility of the brain to alcohol. Here again, similar data in humans are of interest [2161]. No data regarding the biochemical mechanisms of this strain difference are available. However, alcohol preference also depends on hepatic metabolism, and here, strain differences in alcohol dehydrogenase and aldehyde dehydrogenase have been detected [2194].

A genetic analysis by crossing inbred strains with high and low alcohol preference (C57Bl/6 and DBA/2) gave preliminary evidence that two loci or two independent clusters of closely linked loci may control the major part of the variance between the strains. The low preference in the DBA strain is thought to be due to a higher acetaldehyde accumulation, as acetaldehyde dehydrogenase is less active [2046].

To extrapolate these results to human problems of alcoholism, correlations of alcohol preference with other behavioral characteristics should be investigated. Alcoholism in humans is most likely the result of complex interaction between postulated genetic differences affecting liver, brain, or other tissues and behavior patterns influenced by the social environment. Aspects of this problem will be discussed in Sect. 8.2.3.5.

Emotionality in Rats and Alcohol. The Maudsley reactive and nonreactive strains (MR and MNR) of rats have been developed by selection. Selection was carried out for high and low frequency of defecation following exposure of animals to a novel, mildly stressful situation as a measure of emotional reactivity. The two strains were then compared for preference of a 5% alcohol solution and total alcohol intake. The nonreactive strain showed significantly higher mean blood levels than the reactive strain [2007]. A naive, anthropomorphic point of view would have expected the opposite. Ready extrapolation from rat experiments to man in this instance became untenable when alcohol preference in mice as well as in rats turned out to be conditioned largely by the caloric content and not by the psychopharmacologic action of alcohol [2182, 2115].

Learning Ability. One of the most important abilities of humans is the ability to learn from experience. The psychology of learning is a well-developed speciality of modern psychology. Three examples will show attempts at analysis of the problem of interaction of genetic and environmental variability.

1. *Simple Mode of Inheritance for Conditioned Avoidance Learning.* A special aspect of learning ability is avoidance learning. Animals are brought into a cage with a grill bottom through which an electric current may be applied. This electric shock causes a slight pain which can be avoided by jumping into another compartment of the cage. When the animal has learned this reaction, the current is announced by a flash of light, and the animal learns quickly to jump into the other compartment as soon as it sees the flash. The number of flashes necessary for learning this conditioned avoidance reaction is measured.

   Examination of several inbred mouse strains showed definite strain differences, and extensive breeding experiments using the strains with the most extreme reactions were interpreted to show this difference to be caused by a simple, monogenic mode of inheritance. Exchange of the newborns of the two strains immediately after birth did not change strain-specific learning ability. This behavior pattern could not even be altered by implantation of fertilized oocytes into the uteri of animals from the other strain. Both results - the simple mode of inheritance as well as the failure of early maternal environment to change the strain-specific behavior pattern - shows that the strain difference in avoidance learning must be genetically determined [2013]. For a morphological correlate, see below.

This finding does not mean, however, that no environmental influences could ever modify the genetically determined learning ability. Such modifications are possible and were analyzed in studies of maze learning.

2. *Heredity and Environment in Maze Learning.* To get to a food source, rats are made to pass through a maze. According to the speed with which individual rats achieve this goal, and by counting the number of mistakes made, "bright" and "dull" rats can be distinguished from each other. In a genetically heterogeneous population, selective breeding leads to two populations with almost no overlap after about seven generations. The dullest rat of the bright stock is still brighter than the brightest animal of the dull stock. This result points to appreciable genetic variability of maze-learning ability in the base population.

One of the many modifications of this experiment proved to be of special importance for formulating hypotheses on learning ability in humans [2023]. Following establishment of the bright and dull rat colonies, three groups were formed from the young of the two stocks. One group was reared as usual. The second group was reared from birth with restricted opportunities for cognitive and explorative experiences. The third group was reared in a specially enriched environment. Their cage wall was illustrated with modernistic designs, and the cages contained ramps, mirrors, balls, and tunnels. Figure 8.6 shows the results. The difference in maze running between the two stocks was seen only in the groups that came from the "normal" or usual environment. All deprived rats from both strains showed a high number of errors, and all rats from both strains under the "enriched" environmental conditions performed almost equally well.

It would, of course, be premature to conclude that in humans a suitable environment can level out all inherited differences in learning ability. Moreover, maze learning in animals is different from learning ability in humans, as many experiments have shown.

However, the experiment shows that in the rat deprivation at a young age makes for disadvantage in later performance. On the other hand, an environment rich in opportunities for varied experiences may improve learning ability. This conclusion corroborates similar hypotheses derived from experiences with humans [168].

3. *Psychosexual Behavior Also Has to Be Learned.* Two inbred stocks of guinea pigs were compared

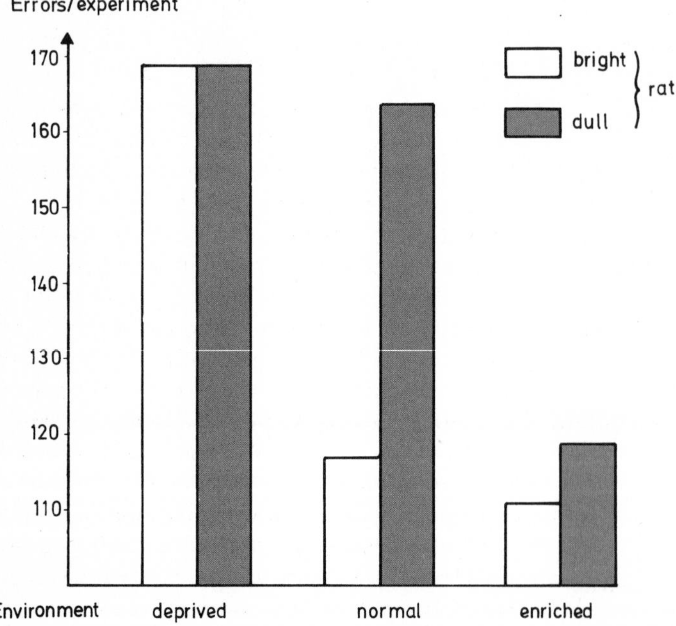

Fig. 8.6. Result of a maze running experiment with bright and dull rats, in a deprived, normal, and enriched environment. There is a definite difference of performance in normal environment. The difference disappears almost entirely under deprived as well as under enriched conditions. (Data from Cooper and Zubek, [2023])

regarding a sequence of activities involved in sexual behavior. Both inbred stocks showed lower sexual acitivty than the heterozygous stock; there were definite interstock differences. One stock, for example, showed more courtship behavior, whereas the other stock proceeded to copulating behavior with few preliminaries. To examine the influence of learning on sexual behavior, males that had been kept isolated from the age of 25 days were compared with others who grew up with a group of females. At the age of 77 days, test animals were placed together with a female in heat. In one of the stocks, only 6% of the isolated males achieved ejaculation compared with 84% of the controls. Similar results were obtained with the other stock. This experiment shows the importance of environment for development of normal, adult sexual behavior. The genetic component, on the other hand, is obvious from the strain difference [2223, 2224].

Attempts at Elucidating the Biologic Mechanisms of Behavior Differences. Animal experiments can be tools for elucidating the biologic mechanisms of interindividual differences in behavior [2021, 2181].

A difference in two-way avoidance learning in the mouse, transmitted by a simple Mendelian mode of inheritance, has been mentioned above. The mouse strains used in these experiments underwent thorough morphological study of their brains, with particular emphasis on the hippocampal areas, since earlier work had suggested a relationship of these areas with certain aspects of learning. The extent of the mossy fiber terminal field, reflecting the number of synapses of the dentohippocampal circuitry in the basal parts of the hippocampal pyramid cell layer, was found to be strongly correlated with poor acquisition of two-way avoidance. Animals with extended mossy fibers and, therefore, many synapses showed poorer performance than those in which this special field was less well developed. This correlation between a behavioral characteristic and a specific feature of brain morphology could be coincidental, reflecting two different features of two inbred mouse strains. However, a causal relationship was strongly suggested by two additional experiments: Rats selectively bred for differential avoidance performance and genetically heterogeneous mice tested individually for avoidance performance and subsequently sacrificed for brain study showed the same correlation [2207, 2208]. Subsequent biochemical studies showed, for example, an influence of thyroxin on avoidance learning as well as on mossy fiber projection [2117a]. These investigations not only contribute to our understanding of the biological basis of behavior modification; but also raise questions regarding the concept of "learning": Intuitively, we would have expected a correlation in the opposite direction, i.e., many synapses making for good learning ability. However, at the phenomenological level, there are other kinds of behavior modification or learning that show only loose correlations or even no correlation at all with two-way avoidance learning. Extrapolations from animal experiments to humans - especially in the field of neurobiology and behavior - should be made with caution. Yet, such experiments may ultimately help to elucidate aspects of learning in humans. Other attempts have been made, to relate biochemical variables to behavior.

Differences were found, for example, in the endocrine system of "emotional" and "nonemotional" rats. Emotional males had larger adrenals and thyroid glands. Endocrine function in man is one of the most powerful influences on the brain.

Attempts at relating behavioral traits to biochemical features of brain function are of great interest. Here, neurotransmitters have recently received much interest. Neurotransmitters are compounds that transmit impulses from one neuron to the other through *synapses* (Sect. 8.2.3.6). Their variation in conditions with normal and psychotic human behavior has attracted much attention in recent years. Early studies had shown cholinesterase levels to be higher in some brain areas of "bright" as compared with "dull" rats [2107]. In more recent experiments [2021, 2022], two related sublines of one inbred mouse stock, BALB/cJ and BALB/cN, differed in their fighting behavior. Males of the BALB/cJ subline would immediately attack other males that were put into their cage, whereas BALB/cN males remained peaceful. This behavior difference was not caused by a higher general activity of the first subline, as the sublines did not differ in their general motor activity, but was specific for "fighting." Cross-rearing experiments showed that this type of aggressiveness was not caused by maternal environmental influence. A number of enzymes involved in the metabolism of an important group of neurotransmitters, the catecholamines (Sect. 8.2.3.6), were studied in the adrenals and the brain of these animals. The levels of three enzmyes, tyrosine hydroxylase, dopamine-β-hydroxylase, and phenylethanolamine *N*-methyltransferase were twofold higher in adrenals of the BALB/cJ, the fighting subline (Table 8.1). The difference in the brain was in the same direction but smaller and not significant.

Breeding experiments between the two sublines were interpreted to show a difference in one single gene pair; nonfighting was dominant over fighting. Enzyme levels of all three enzymes segregated with the behavior. The data point to a lessened degradation of these enzymes as the most likely cause for their elevated enzmye activity. Remembering the many alterations that mutations may cause in proteins and their possible consequences for enzmye function, a number of explanations for these findings can be suggested. For example, the three enzymes may share a common polypeptide chain changed by mutation, a degrading enzyme may be

Table 8.1. Levels of biosynthetic enzymes involved in catecholamine metabolism in adrenal glands of BALB/c sublines [2022]

Subline	Behavioral phenotype	N	Tyrosine hydroxylase	Dopamine-β-hydroxylase	Phenylethanol-amine-N-methyltransferase
BALB/cJ	Fighter (F)	9	8.87 ± 1.19	30.23 ± 2.38	0.371 ± 0.014
BALB/cN	Nonfighter (NF)	9	4.51 ± 0.46	17.33 ± 1.18	0.193 ± 0.011
F_1 (F × NF)		12	6.05 ± 0.40	23.82 ± 1.47	0.276 ± 0.016

Activity of each enzyme is expressed in nanomoles of product formed per hour per adrenal pair. In each case, differences between high and low parents and between F_1 and either parent were statistically significant.

reduced in activity, the three strucutral genes may be closely linked, or the mutation of a regulatory gene was involved that influenced all three structural genes. It is also possible that the mode of inheritance may not be monogenic. In any case, the correlation between enzmye levels and aggressive behavior must not mean that high enzyme levels - and consequently a high production of catecholamines - *cause* this behavior. However, since other experiments also showed this possible role for catecholamines in aggressive behavior, a causal relationship is at least likely.

Behavioral differences between inbred mouse strains such as duration of catalepsy were also detected after administration of haloperidol, a compound that binds to receptors for the neurotransmitter dopamine. The expected differences in drug binding were in fact demonstrated; they suggest genetic differences in regulation of the number of dopamine receptors in the stem ganglia [2192].

Possible Significance of Experiments with Mice and Other Mammals for Behavioral Genetic Analysis in Humans. Animal experiments, especially with mammals, are suitable for formulating hypotheses that can be examined in humans. On the other hand, hypotheses suggested by certain observations in humans can be put to a more stringent test in animal experiments. The examples given above have already indicated two different approaches to the problem of genetic determination of behavior.

1. Comparisons between inbred strains and selection experiments were primarily designed to demonstrate that genetic variability exists. The experiments are usually designed to permit quantification of the extent of genetic variability. This approach frequently also allows analysis of the interaction of genetic with environmental variability.
2. Very soon, however, this kind of analysis becomes unsatisfactory, as the answers are not specific. To gain insight into the mechanisms of genetically determined behavioral variations morphological, physiological (usually neurophysiologic) or biochemical characters have to be investigated. A satisfactory state of this analysis is reached when such variation can be traced to single gene mutations.

8.2 Behavioral Genetics in Humans

Normal and Abnormal Behavior. Normal behavior is largely studied by psychologists. Abnormal behavior is mostly studied by psychiatrists, while research into the biologic causes of major psychoses is often done by neuroscientists. The definition of the normal range of behavior is difficult and strongly depends upon a given society's definition of "normal." However, transcultural studies of recent years have clearly indicated that the major psychoses, such as schizophrenia, occur in all societies and are not produced by artificial labeling as claimed by some social psychiatrists.

Results of genetic analysis for many traits have shown that abnormal function is often caused by single genetic defects, including single gene mutations. Variability within the normal range rarely has been traced to differences in a single allelic series and presumably owes its origin to contributions from several genes intracting with environmental factors. By analogy, gross abnormal behavior, therefore, is more likely to be caused by single gene defects, while normal behavior has a more complex genetic and environmental determination. As a consequence, the elucidation of the genetics of abnormal behavior is a simpler task than genetic understanding of behavior in the normal range.

Our treatment of human behavior genetics will approach the subject similar to our analysis of gene action elsewhere in this book (Sect. 4). More descriptive phenomenologic studies will be covered first, with a few glimpses toward a more causal analysis at a deeper genetic level.

Observation and Measurement of Human Behavior. Human behavior can be observed by outsiders or reported following introspection or both methods may be applied. The outside observer can note, for example, how quickly an individual moves - whether (s)he prefers to be motorically active or rather passive - or whether (s)he likes to engage in social contacts. On a larger scale, we may be interested to see how an individual passes through the stages of life - how childhood, adolescence and young adulthood are experienced, how and why occupations are selected and whether a person leads a reasonably happy and successful life. In all these aspects, we can compare one individual with others. Furthermore, to simplify the task of measurement and comparison, we may wish to standardize our procedures of observation. For example, certain abilities or interests can be examined and compared as comprehensively as possible within a reasonable time by specifically constructed test systems. Such *psychological tests* should have a high *reliability;* repeated application of the same or analogous test to the same individual should lead to similar results. The reliability of a test can be assessed relatively easily. Much more difficult to determine is its *validity,* i.e., the question of whether the test measures anything of relevance for real life. Intelligence tests, for example, should be able to predict a person's performance in a situation in which, according to the conventions of our society, intelligence is needed; for example, success in school or university or performance in an occupation.

Psychologists interested in test construction have primarily been successful in measuring sensorimotor skills and the group of abilities that are usually called "intelligence." So-called intelligence tests have been applied on a very large scale, and these studies - especially when interpreted in genetic terms - have raised a considerable uproar. However, such observations and measurements are only one side of the matter. Data obtained from introspection may be just as important. We wish to know how a person feels. Does (s)he feel relaxed or excited, elated or depressed - in brief: What is his or her feeling tone? Does the person feel the urge to take an active part in life or to let things just happen? What attitudes does a person have toward other people - toward family, colleagues, occupational life, cultural activities? All these matters are often lumped together under the label "personality."

Strictly speaking, the psychologist has only inside knowledge of one person - himself. However, human beings communicate with each other - by nonverbal means such as physiognomic expression, movements of their bodies, especially their hands, and above all, by that natural instrument that is unique to our species, language. The investigator uses these means of communication to gather more indirect information also of the feelings and thoughts of other people. The simplest method is the *interview.* This interview may be structured if the investigator has prepared in advance a number of topics to be explored. Other more refined methods are *personality questionnaires* with queries on feelings, attitudes, and opinions. Most questionnaires are constructed so that the proband has several choices. The number of statements in one or the other direction to various topics are then compared with standard populations. Widely used questionnaires are the Minnesota Multiphasic Personality Inventory (MMPI) and the 16 Personality Factors (16PF) test. The difficulty with this approach is whether the proband is indeed "honest" and express his or her true opinion. Thus, the questionnaires usually contain internal controls.

For a certain time, some psychologists had hoped to gain deeper insight into personality by recording the reactions to tasks that had no obvious relationship with normal life but were believed to evoke responses that revealed subconscious feelings, anxieties and desires. The bestknown example for such a *projective test* is the Rorschach test. Many psychologists have become disappointed with this approach; in earlier years, a number of twin studies with projective tests were carried out but the results suffered from the ambiguities of the test method.

All the above-mentioned aspects of human behavior show variations between individuals, which can be examined by genetic approaches.

8.2.1 Investigations with Classic Phenomenologic Methods

8.2.1.1 Reappraisal of Classic Methods

(See also Sect. 3)

Family Investigations. The most straighforward - and most unsophisticated - approach for assessing a genetic contribution to the variability of a certain trait is the comparison of this trait between biologically related individuals. If an all-or-none or alternative characteristic is genetically determined, it will usually be more frequent among relatives. With increasing biologic relationship, the frequency in relatives will be higher. Such alternative individual characteristics are various diseases as contrasted with general health, and mental retardation in contrast to normal intelligence. If the character is defined on a graded or quantitative scale, the similarity between relatives is, on the average, expected to grow with increasing biologic relationship.

One goal of biometric research is to establish the expectation for relatives of probands to develop the same conditions. Calculation of such *empirical risk figures* has been explained in Sect. 3.3.6, using schizophrenia as an example.
In quantitative traits, the similarities between relatives are usually expressed and measured as correlation coefficients. Such analysis are strictly empirical and require no genetic concepts. The conceptual difficulties begin if we try to interpret empirical risk figures or correlations between relatives in terms of genetic variability. Theoretically, certain correlation coefficients can be derived from assumptions regarding the degree of relationship between two persons and the degree of dominance of these genes. The theoretical expectations can then be compared with those empirically observed, and from this comparison, the "heritability," i.e., the fraction of variability contributed by genes, can be estimated. Heritability and methods for calculation – and the assumptions normally inherent in such calculations – have been described in Sect. 3.6.1.
Theoretically, heritability calculations can also be carried out for alternatively varying characteristics, such as diseases. For these calculations, information on the occurrence of the test trait in relatives and in the general population is needed. Here, too, a number of assumptions regarding the biologic meaning of such a calculation need to be made. Worse still, most of these assumption cannot be tested with most empirical data. The problem has been discussed in Sect. 3.6.2.
One objection to such heritability calculations comes to mind immediately. In most cases, relatives not only share some of their genes but live under similar environmental conditions as well. In heritability estimates, however, it is assumed that genetic and environmental variance components are independent of each other. Attempts at correcting for this bias by introducing estimates of environmental parameters regarded as relevant for the trait in question are of dubious value. The classic approach to overcome this difficulty is the comparison of monozygotic and dizygotic twins.

Twin Method (cf. Sect. 3.8). Monozygotic (MZ) twins are, as a rule, genetically identical. Therefore, differences found between them must be nongenetic. Dizygotic (DZ) twins are no more similar genetically than normal sibs. Like MZ twins, however, they are usually reared together; therefore, the influence of environmental factors may be assumed to be similar in first approximation. Hence, if MZ twins are more similar than DZ twins in a quantitatively varying trait or if they are more frequently concordant in an alternately varying character, genetic components in the variability of this character are assumed. This assumption is subject to a number of qualifications that grow out of the observation that twins – and especially MZ twins – are in many aspects not a random population sample. These problems have been explained in detail in Sect. 3.8.4. For planning of research in behavior genetics, it needs to be remembered that the special twin situation, especially in childhood and youth, leads to certain deviations from average development. More generally, the environmental differences to which twins – and to a slightly lesser degree other sibs as well – are exposed can hardly be regarded as representative for such differences in the general population. Usually, such differences are much smaller in relatives.

MZ Twins Reared Apart: Studies on Adopted and Foster Children [152, 2119]. To overcome these difficulties, two approaches were chosen especially for problems in behavior genetics: first, the comparison of *monozygotic twins who had been separated* in infancy or early childhood and reared apart. This design theoretically avoids biases due to a common environment and interaction between the twins in a very elegant way. Secondly, in a comparison of adopted children (or of children living in foster homes) with their *biologic* parents, common environmental factors usually acting on parents and their children no longer apply, and the parent-child relationship is reduced to its biologic components. Influences of a common environment are excluded.
Unfortunately, in reality matters are somewhat different. Placing of separated MZ twins or of adopted children never occurs at random; selective placement based on socioeconomic status and behavioral characteristics is commonly carried out by social agencies; couples adopting children are a biased sample of all couples; foster homes may offer aberrant living conditions. The ideal research design of a randomized environment cannot be achieved in humans.

8.2.1.2 Mental Retardation and Deficiency

Definition. There are many definitions of mental retardation and deficiency. A useful definition that does not depend on tests or school results is that given in the British Wood Report (1929). A mentally deficient individual was defined as "one who by reason of incomplete mental development is incapable of independent social adaptation." Table 8.2 gives the usual nomenclature.
The last two columns of Table 8.2 list quantitative measures of the degree of mental subnormality. The "intelligence quotient" (I.Q.) as devised by Binet and Simon (1907), and modified later, was primarily designed as an aid to teachers in Paris in the allocation of pupils to their

Table 8.2. Traditional nomenclature for the mentally subnormal

Degree of defect	British	American	French	Approximate Binet intelligence level	
				I.Q. (children)	Mental age in years (adults)
Mild (high-grade)	Feeble-minded	Moron	Débile	50–69	7–10
Severe (medium or low-grade)	Imbecile	Imbecile	Imbécile	20–49	3– 6
Severe (low-grade)	Idiot	Idiot	Idiot	0–19	0– 2
All grades[a]	Mentally defective, mentally handicapped, mentally subnormal	Feeble-minded, mentally retarded	Arrieré, oligophrénie	0–70	0–10

[a] The terms *Schwachsinn, Geistesschwäche,* and *Oligophrenie* are used in Germany.

correct grades in school [2157]. The children were classified in years and months of mental age. By comparing mental age with the chronologic age, it was immediately apparent whether a child was advanced or retarded. For example, if a child was 12 years old and his performance was equal to that of the average value of all 9-year-olds, the I.Q. was 9/12=75. For individuals above the age of 14–16 years, this formulations is meaningless. Therefore, their I.Q. is defined by comparison with a standardized sample of their age peers. Even this formulation is only a crude approximation, as the mental capacity of a subnormal individual is qualitatively different from that of a normal individual.

Incidence of Mental Subnormality. Incidence figures of mental subnormality vary widely, depending on the definition. If the definitions of Table 8.2 are adopted, i.e., if an I.Q. of 69 or a mental age of 7–10 is taken as the lower limit of normalcy, incidences cluster around ~2%–3% of the population. The great majority of mentally subnormal persons can be classified in the mild (high-grade) group; only about 0.25% of the overall population are categorized as severely retarded (I.Q. < 50). Among severe cases, boys are over-represented.

The frequency of detectable mental retardation increases after the age 6–7 years due to school attendance where mental retardation will be more likely recognized because of learning difficulties. The detected population frequency again decreases after school age since many individuals unable to succeed in school can achieve a satisfactory social adaptation. These data emphasize the importance of school attendance for different definitions of mental retardation.

Two Biologic Groups. Mentally subnormal individuals can be divided into two classes, for which the terms "pathologic" and "subcultural" became most popular [2116]. The pathologic group comprises a mixture of cases with various genetic and nongenetic causes. It was once believed that many cases falling into this category were environmentally caused, as parents were usually normal. We know today that this is not entirely true. While this group comprises exogenous cases, for example, those due to brain injury or to infectious disease, such as meningitis or encephalitis, many cases are due to genetic causes. Down's syndrome has long been the most frequent single diagnosis in this pathologic group (Sect. 2.2.2). Most inherited and predominantly autosomal-recessive inborn errors of metabolism that cause mental deficiency are also included in this class (Table 8.4), which also contains some autosomal-dominant and X-linked recessive conditions. Segregation analysis (Sect. 3.3) gave hints that among cases of this class currently diagnosed as nonspecific mental retardation further monogenic autosomal-recessive conditions might be hidden.

X-Linked Mental Retardation. The higher incidence of boys in this group was explained, when a number of X-linked conditions were discovered [2151]. From a population study in British Columbia (Canada), prevalence of all types of X-linked mental retardation together has been estimated at ≈1.8/1000 males [2220]; studies in populations of mental retardates confirm this order of magnitude. Hence, X-linked mental retardation is at least as common in the male population as Down's syndrome.

The best-known and most common single type (one in 2000–4000 males [1619]) is the Martin-Bell syndrome. Among males with this syndrome, a specific anomaly of the X chromosome is found: In ≈2%–35% of their X chromosomes in lymphocyte cultures a "fragile site" is seen at the end of the long arm (Xq 28). The same "marker X" can also be found in some of the female carriers; it "can readily be seen in ... carrier females under the age of about 25, but to demonstrate it in older women is difficult unless they are intellectually impaired" [2220]. Even among younger carriers, mar(X)-positive ones are much more common among mental retardates than among normals.

In addition to the mar(X), many male patients show some characteristic physical features, including the finding of large testicles, as large or lop ears, and prominent forehead and jaw (Fig. 8.7; Table 8.3). At birth, some have a large head and a slightly increased birthweight. The I.Q. may be as low as ≈30, but in most instances is in the 50–60 range. As a rule, ability measured by the verbal part of standard I.Q. tests is depressed more than in the performance part. Speech has been reported as "repetitive," and stuttering appears to be common. However, there is wide variability within and especially between families in clinical signs [2201] as well as in the frequency with which the marker can be detected. In an increasing number of pedigrees, the mutant gene must have been transmitted by a clinically and cytogenetically unaffected male, and segregation analysis [1619] has revealed ≈80% penetrance of mental retardation in hemizygous males. Moreover, cases and pedigrees have been observed with mental retardation and macro-orchidism, but without confirmed mar(X), and in many instances the physical features are not nearly as characteristic as in the patient show in Fig. 8.7a. Among female heterozygotes, phenotypic variability is still wider. According to one study [1619], "mental impairment" was seen in about 30%. There has been some discussion on the mental status of the 70% without outright mental impairment. In one study [2155] most subjects had a below average I.Q.

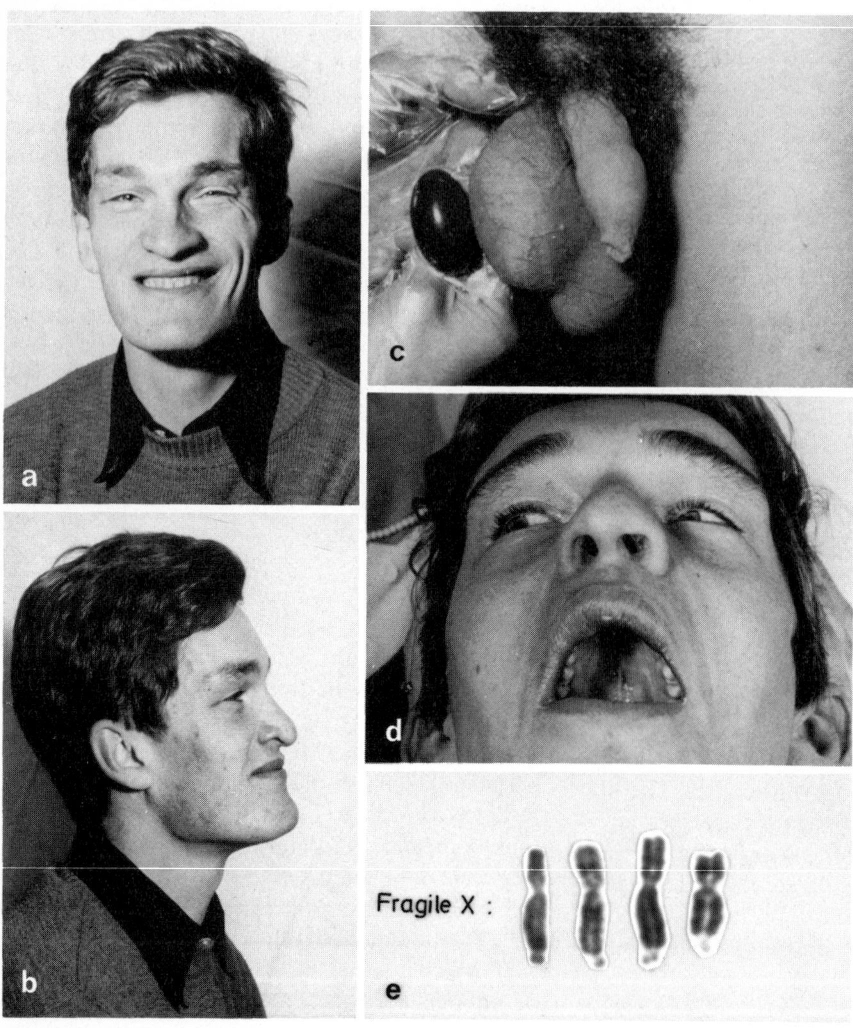

Fig. 8.7 a–e. Patient with X-linked mental retardation with macro-orchidism, typical face, and high-arched palate. In **d** the size of the testicle is compared with a model indicating average size. The marker X could be demonstrated in 35% of the cells. Note, however, that clinical features are not always as characteristics as in this case! **e** The marker (X) chromosome. (From Tariverdian and Weck [2213])

This condition raised interesting questions in population genetics. If the indirect method of estimating mutation rates (Sect. 5.1.3.1) is applied, assuming an equilibrium between loss of alleles due to reduced fertility of mentally retarded patients, an unusually high mutation rate results (Sect. 5.1.3.2), which would be confined entirely to mutations in male germ cells. Therefore, the alternative possibility was discussed that the loss of genes due to impaired fertility in moderately or severely retarded gene carriers might be compensated in part by increased fertility of gene carriers - especially females - in the borderline or slightly subnormal range [2230]. There are many reports suggesting above-average reproduction in persons mentally borderline or slightly subnormal, especially in the period between 1850 and 1950.

In this syndrome, the fraction of mitoses showing this characteristic X chromosome can be enhanced by folic acid deprivation in cell culture. This has led to speculations on possible pathogenetic mechanisms but a study on folic acid pathways in mar(X) cells failed to reveal a special defect [2237]. On the other hand, therapeutic attempts with folic acid were reported to have limited success [2202], and substitution therapy during embryonic development has even been proposed. Moreover, heritable fragile sites have been observed in other chromosomes. Most of them are also folic acid sensitive. Heterozygosity of such sites appears to be more common in mental retardates than in the general population of newborns [192].

Table 8.3. Clinical features of mar(X) mental retardation (Martin-Bell syndrome). (Modified from [2220])

Intelligence	I.Q. range 30-65, sometimes borderline normal or even normal. Occasional hyperactivity or autism in childhood; generally friendly, shy, nonaggressive in teenager years; speech anomaly
Growth	Birth weight normal; usually bigger than normal sibs; head circumference above 50th, sometimes above 97th percentile
Facies	Prominent forehead and jaws, long face and big ears
Testicles	May be 3-4 ml in childhood (normal 2 ml); postpuberal boys 30-60 ml (normal < 25 ml)
Occasional features	Epilepsy; increased reflexes in lower limbs; gynecomastia, striae, fine skin; thickening of scrotal sac

In addition to the Martin-Bell [mar(X)] syndrome, several other types of X-linked mental retardation exist. Opitz and Sutherland, in an attempt at nosological classification, enumerated no less than 17 entities, some so rare or so little known that only one or a few pedigrees have been published so far. Some relatively well-defined ones are a form with hypotonia, muscular weakness, and cerebellar signs (Allen-Herndon-Dudley syndrome; 30960; [1997]); and the Renpenning syndrome (30950; [2178]), with a small skull, generally small growth, and a characteristic facies different from that in the Martin-Bell syndrome. The Juberg-Marsidi syndrome (30959; [2089]) is characterized by deafness, microcephaly, vesicourethral reflux, and blepharophimpsis or ptosis. For other types, see [2151]).

A different pattern is observed for the mild or high-grade group. As a rule, there are far fewer cases of dramatic exogenous origin even though exogenous causes do exist now and then. Few neurological abnormalities or other clinical findings are noted. Instead, a distinct pattern of familial recurrence is observed. Whereas patients with severe mental retardation usually have normal parents and only occasionally affected sibs, the fraction of affected relatives in the group of *mild* mental retardation is high (Fig. 8.8).

Empirical Risk Figures [2251]. Severe mental retardation is a typical example of a mixed etiologic category. For an increasing number of cases within this class, genetic analysis at the phenotypic Mendelian level has become possible, and many mutations at individual gene loci have been detected. Other cases

Table 8.4. Degree of mental retardation according to chromosomal status. (Sutherland, 1976) [2203])

		Normal karyotype	Down's syndrome	Other abnormality	Total
Borderline	(I.Q. 68-85)	7			7
Mild	(I.Q. 52-67)	65	9	2	76
Moderate	(I.Q. 36-51)	116	29	3	148
Severe	(I.Q. 20-35)	195	31	7	233
Profound	(I.Q. < 20)	115	4	5	124

The data have been collected in an Australian institution for the mentally retarded. The 588 in-patients of this hospital were all karyotyped.

were shown to be caused by a chromosomal anomaly. For the unclassified rest, empirical risk figures of recurrence have been calculated that are, however, of limited practical value. Genetic counseling requires careful analysis of the single case and must include a consideration of exogenous brain damage. A relatively large unclassified core remains where genetic counseling is difficult.

The calculation of empirical risk figures is more useful for mild mental retardation, especially when the minority of cases with known exogenous causes have been excluded. Table 8.5 shows empirical risk figures derived from the famous Colchester Survey of Penrose [2156]. The difference in familial occurrence of mental retardation between severely and mildly affected probands, the latter groups showing higher familial aggregation, is noteworthy. Other studies usually gave similar figures [2251, 2070]. Comparison of data from various authors is difficult as criteria of diagnosis vary. The data on familial aggregation and the absence of clear segregation of normals and mentally retarded in the many families studied is compatible with multifactorial inheritance. If a scale, such as I.Q., is used with continuously varying test results, no threshold exists. However, it may be useful to establish artificial thresholds to delineate that fraction of the population unable to benefit from regular schools.

Twin Studies. A way to estimate heritability of a character is the comparison of MZ and DZ twins. Three twin studies are described in Fig. 8.9. Considering that in Smith's series discordance in two pairs of MZ twins could be related to exogenous causes, concordance of mental subnormality for which no exogenous causes (early brain damage or infection) can be found, approaches 100%, and hence heritability seems indeed to be 100%.

Somewhat closer scrutiny of the data casts some doubt on this interpretation. First of all, the twins included in these studies lived with their families; even if the parents were not outright subnormal, they were in most cases within the lower range of the intelligence distribution. A certain amount of deprivation from stimuli necessary for normal intellectual development may safely be assumed to be responsible for part of the mental retardation in these MZ twins. This assumption is corroborated by another interesting observation in the families of Juda's twin series: among the fathers of retarded twins, one-fourth were recorded as retarded and among the mothers of retarded twins, more than one-third. The significance of the mother for mental development of the infant and child, especially in the first three years of life, is now an undisputed finding of developmental psychology. Hence, the data suggests an additional maternal influence, even if other explanations (X-linked genes) are possible.

Juda [2090–2092] collected her data by contacting the families of all children in special schools for retarded children in southern Germany. Among 18183 students, she ascertained 488 twin children, i.e., one twin per 37.3 students. The frequency of twin births in Germany was at that time about 1 twin birth in 84 deliveries, i.e., 1 twin individual in 42 mothers. Due to the higher infant mortality of twin children, the expectation to find a twin was calculated to be about 1 twin per 60 children in the population. Hence, twins were over-represented in her data. This result agrees with the general experience that development of twins is slower [2078, 2079] and mental subnormality more frequent than in the general population. The influence of this bias on concordance rate cannot be predicted. In any case, these considerations show that interpretation of empirical risk figures and twin concordances as well as herita-

Table 8.5. Prevalence of mental retardation among parents and sibs of mentally retarded probands. (Penrose [2157])

Degree of mental retardation of proband	*n*	Above average	Borderline and mildly retarded	Severely retarded	Total retarded
		Parents			
Borderline or mildly retarded (627 probands)	1,254	0.32%	27.59%	0.24%	27.83%
Severely retarded (653 probands)	1,306	0.53%	15.00%	0.08%	15.08%
All degrees	2,560	0.43%	21.17%	0.16%	21.33%
		Sibs			
Borderline or mildly retarded	2,321	1.21%	19.52%	2.50%	22.02%
Severely retarded	2,549	1.57%	12.24%	4.28%	16.52%
All degrees	4,870	1.40%	15.71%	3.43%	19.14%

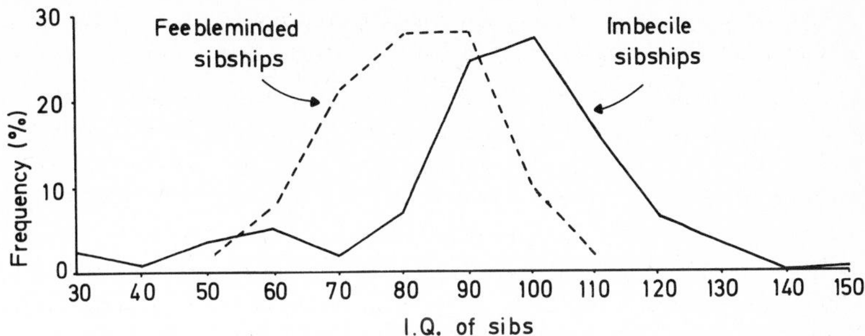

Fig. 8.8. Distribution of I.Q. in sibships with mildly affected (high-grade mental retardation; feeble-minded) and severely affected probands (imbecile; low-grade mentally retarded probands). Feeble-minded sibships show a nearly normal distribution, which is shifted toward slightly reduced I.Q. values pointing to a multifactorial basis. The sibships of severely mentally retarded probands show a bimodal distribution. There is a minority of mentally retarded sibs, but the I.Q. distribution of most siblings equals that of the average population (the mean is close to I.Q. = 100). These data point to discrete, major causes, genetic or nongenetic, for severe mental retardation. (Fraser Roberts, 1952)

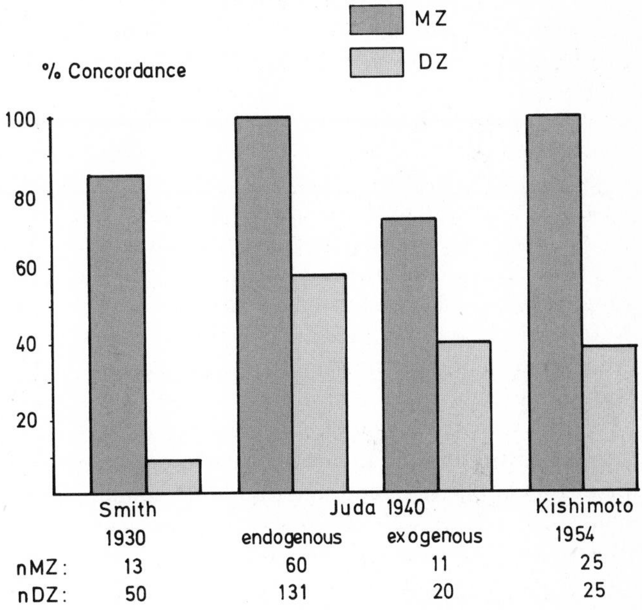

Fig. 8.9. Concordance in MZ and DZ twins in twin studies on mild (high-grade) mental retardation. *nMZ,* No. of MZ twins; *nDZ,* No. of DZ twins. (Zerbin-Rüdin, 1967 [2251])

bility estimates derived from such data may be ambiguous for two reasons: genetic and environmental influences are correlated, and the development of twins is different from that of singletons. As mentioned repeatedly, the genetic model of multifactorial inheritance describes the genetic situation in a preliminary manner only (Sect. 3.6). More recent data indicate that specific genetic causes may soon be identified for some of the genetic variability in the subnormal range: for example chromosomal aberrations, heterozygosity for inborn errors such as phenylketonuria, or the mar(X) syndrome (see above) [1994].

8.2.1.3 Intelligence and Performance in the Normal and Superior Ranges

Superior Achievement. So far, we have concentrated on the lower end of the variation, i.e., those people who, because of their low intellectual endowment, have difficulties in adapting to the demands of society. F. Galton first reported on individuals whose achievement is regarded as superior by the standards of their society in his classic paper on "Hereditary Talent and Character" (1865) [248] and in the subsequent monograph on *Hereditary Genius* (1869). This work established one of the two paradigms on which human genetics of founded. He showed that men regarded as "eminent" in British society had many times more male close relatives in the eminent group than could be expected if the distribution of high achievement would be random.

Since that time, repeated attempts have been made to document the inheritance of genius or of special talents. For example, the pedigrees of famous artists and scientists, such as Bach, Darwin and Galton, or Bernoulli have been reported as evidence for hereditary talent, and comprehensive statistics have been published [2093]. In all such cases, there is confounding of heredity and environment, and no specific statements regarding the role of genetic factors can be made.

Variability in the Normal Range: Nature of Intelligence. Many studies have been performed to determine the relative contributions of heredity and environment to behavior within the normal range. Success in life and contribution to human society clearly depend on a variety of factors that can roughly be classified as intelligence and personality. Individual differences in intelligence have for a long time been a principal field of research in psychology. More recently, however, these studies have been criticized fiercely. To understand this controversy somewhat better, we should consider the history of intelligence testing and some modern concepts of intelligence [168].

Intelligence and Intelligence Testing. Mental subnormality and eminent performance have been defined using social criteria: subnormality as making an individual incapable of independent social adaptation (Sect. 8.2.1.2) and eminence as being recognized by a group of present-day professional colleagues as one of the leading figures in the field [2093]. The intelligence quotient or the mental age had at first been introduced and utilized as a device for school assignment and for classifying the degrees of mental subnormality. Thus, the I.Q. represented a criterion for delimiting the normal range from mild subnormality in a more clearly defined but somewhat more arbitrary way than independent social adaptation! Indeed, Binet introduced these measurements for the purpose of such classifications. It is said that Binet himself was not happy with the direction I.Q. research had subsequently taken [2221]. Thus, emphasis shifted to classification of normal individuals according to their I.Q. An important event that precipitated this development was World War I, which afforded American psychologists the opportunity to test large numbers of army recruits.

Such extensive testing required selection of tasks and questions that would force respondents to give scorable right or wrong answers. Because testers were not communicating directly with individual testees, the reasons for the wrong answers could not be analyzed. Qualitative differences would have to be ignored. Scores became all important [2221].

After World War I, interest in quantitative measurement continued, and the tests became more sophisticated. The modern intelligence test comprises items that examine ability to handle words and verbal concepts, abstractions, mathematical tasks, spatial visualization, and memory. Such tests have turned out to be useful in practice; despite many criticisms about their meaning, operationally they give fairly reliable predictions about performance in school, college, and university.

Certain professional occupations such as architecture, engineering, science, and medicine require a minimum I.Q. above the average for success, while other occupations make less of an intellectual demand on their practitioners. Failure to reach an above-average I.Q. result necessary for occupational success in the professions has predictive value in establishing that such persons are not likely to succeed in professional schools and universities preparing for these professions. However, cognitive ability as assessed by I.Q. tests is certainly not the only criterion for professional success.

The older theoretical work was mainly centered around the problem of whether intelligence is one basic ability that influences all single tasks, different as they may be, or whether different tasks demand different abilities. One method for examining such problems is factor analysis. The correlations between single score results were examined for correlation clusters that were believed to point to such basic abilities. The results were consistent in showing relatively high correlations between single items, pointing to a common factor of "general intelligence" (g) that influences all test scores (Spearman). There is some agreement that in addition to this g factor, specific abilities are necessary for performing verbal and mathematical tasks, and for space perception.

Growing Uneasiness over Intelligence Testing Among Psychologists. In spite of its undisputed success in predicting school achievement, the method of measuring intelligence and the I.Q. concept has met with growing uneasiness among psychologists. It is asserted, for example, that the test measures the ability for "puzzle solving," i.e., for solving problems that are uninteresting by themselves but ask for certain specific formal skills. These skills are stressed in all school systems [2143]. They are also needed in many professional fields, for example, in all engineering fields. The highest puzzle-solving skills are necessary for solution of problems in the physical, natural, and social sciences. Th. Kuhn [257] has even described most science ("normal science") as consisting of puzzle solving (see Introduction).

Many other problems in daily life, however, require "intelligent performance in natural situations," which may be defined as "responding appropriately in terms of one's long-range and short-range goals, given the actual facts of the situation as one discovers them" [168]. Such abilities are tapped less well by current I.Q. tests. African natives who never experienced any Western contacts will thus react inappropriately to many test items [2056].

New Approaches for a Better Understanding of Human Intelligence [168]. There are a number of attempts to gain a better understanding of how intelligent behavior develops and how individual differences may come about. For example, what are the basic cognitive processes that enable us to cope with our environment in an intelligent way? How important are basic processes, such as short- or long-term memory or complex phenomena, such as language? Can we learn about these processes by constructing computer programs for solving problems? Can we gain more relevant information by observing individual behavior in natural environments [2017]? Many investigators are now emphasizing the interactions of internal and external factors in the development of the individual; here the observations of Piaget on how children gradually acquire logical concepts are influencing many pschologists. These attempts do not challenge the practical usefulness of intelligence testing for prediction of success in educational institutions; they do challenge its relevance for comparison of groups.

Family and Twin Studies for Assessing the Genetic Contribution to Normal Variability of Intelligence. When methods for measuring intelligence became available, pioneers in this field attempted to determine whether and to what degree normal variability was influenced by heredity or environment. The *Zeitgeist* before and after World War I was very much influenced by the eugenics movement (Sect. 1.8), and F. Galton had impressed on many scientists the idea that intelligence could be measured and statistical comparison between close relatives could help to solve the age-old problem of how nature and nurture co-operate in creating intelligence. When Siemens [869, 870] showed in 1924 first how MZ can be readily distinguished from DZ twins (Sect. 3.8), the twin method was quickly adopted and became the most cherished tool for this kind of research.

Success in Schools. The most readily available material were school grades; from the point of view of intelligence testing, they are a good indicator since they usually show a fairly high correlation with I.Q. test results. Moreover, school grades might even have higher validity since teachers will observe their students for a long time [2096]. Many studies show resemblances in school grades between students, their parents, and sibs and are generally consistent with a model of multifactorial inheritance. However, the data could also be explained in a different way. Parents who were themselves successful at school provide more help for their children - either directly, by advising them in their homework and offering rewards for success in school, or indirectly, by providing more opportunities for exercising cognitive abilities.

Twin studies may help to distinguish environmental from genetic influences but suffer from the difficulties discussed in Sect. 3.8. Figure 8.10 shows the differences in school grades in 60 MZ and 41 DZ twin pairs from Germany [2045]. The differences in MZ are only half as large as in the DZ twins. Taken at face value, these results point to an appreciable genetic component in school performance. However, even at this relatively low level of sophistication, other results cast some doubt on this explanation. In a Finnish study, for example, a strong difference in school performance between DZ and MZ twins was shown in male but not in female twins [2096]. Do the teachers treat girls differently from boys?

Data such as school grades are not sufficiently critical. At the time when these studies were carried out, twins were usually kept together in the same class. Fooling the teacher about their identity was a popular trick among MZ twins. Shall we really believe that their grades were always assigned independently? More objective methods are obviously necessary.

Intelligence Tests in Families and Twins. Genetic studies with intelligence tests have been carried out on a very large scale. Bouchard and McGue (1981) [2006] collected 111 investigations comprising 526 samples of individuals with different degrees of relationship, among them 47 series with parent-child comparisons, 71 such comparisons between sibs, 41 series with DZ twins, and 37 series with MZ twins. These results will not be discussed in every de-

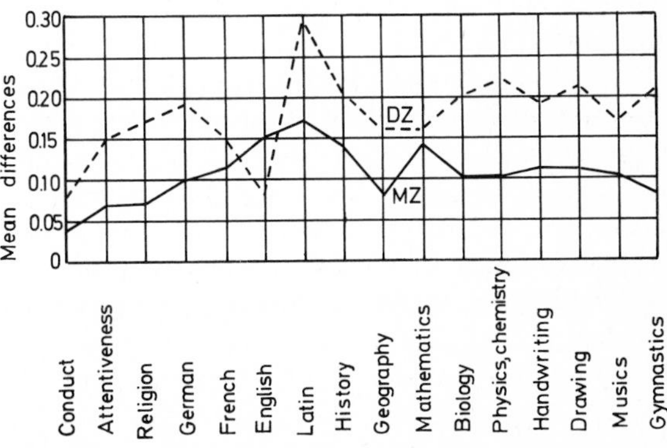

Fig. 8.10. Average differences in school grades among 60 MZ (32 ♂, 28 ♀) and 41 DZ (20 ♂, 21 ♀) DZ twins from Germany. (Frischeisen-Köhler, [2096])

tail, as critical analysis of recent years has uncovered that there are errors in reporting, and many of the data were biased, mainly in favor of a high genetic component of the described variability [104, 2214]. The lowest correlations were obtained in unrelated individuals and the highest in identical twins. The rank order (from lowest to highest) of these data was as follows: unrelated individuals < parents-adopted children < parents-children = sibs < DZ twins < MZ twins. This result is, of course, compatible with the hypothesis that the variability measured by the I.Q. test score has a strong genetic component. The question is whether other hypotheses can be convincingly excluded. Before discussing various kinds of evidence, we should examine the distribution of actual differences between MZ and DZ twins usually found in such studies. Figure 8.11 shows the pooled actual differences in I.Q. points from three studies. Twin pairs showing no difference are certainly more frequent among MZ than among DZ pairs. However, the number of MZ pairs that are quite different is by no means negligible. This finding shows that genetic factors cannot be exclusively responsible for the variability found in the population.

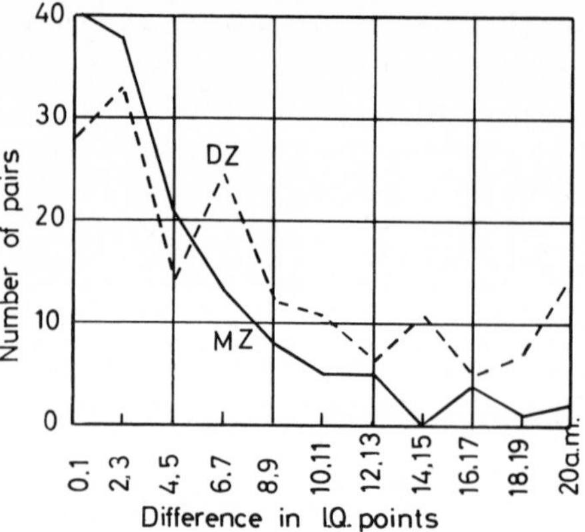

Fig. 8.11. Distribution of differences in I.Q. points between MZ and DZ twins from three series. Note that some MZ pairs show very large differences. (Stocks, 1933 [2209])

Heritability Estimates. The concept of heritability has been introduced in Sect. 3.6.1, using another quantitative and measurable trait - stature - as an example. We made the following statements:

a) The theoretical correlations between relatives can be derived from Mendel's laws and the Hardy-Weinberg Law under certain assumptions. Heritability can be estimated by comparing these theoretical correlations with those empirically observed. Hence, the heritability concept has a foundation in Mendelian genetics.
b) Heritability estimation is genetic analysis at the phenotypic-statistical level. The genes determining the measurable trait cannot be identified, and it is not possible to make any specific inferences regarding their number mode of inheritance, or mode of action.
c) Even an exact quantitative assessment of heritability requires a number of assumptions that in most cases can not be tested in human data, such as random mating with respect to the trait examined, and absence of correlation or interaction between genetic and environmental influences. The methods may be refined by considering assortative mating or by including environmental - for example, socioeconomic - scores in the calculation. Assessment of such scores, however, creates new problems. In App. 6, the problem of estimating heritability from twin data will be discussed with the following results.

d) An unbiased estimate of heritability from twin data is not possible theoretically. Three different ways of estimation were proposed: a comparison of MZ with DZ twins (h_1^2), which may overestimate or underestimate heritability depending on some assumptions; a comparison of MZ twins with control pairs taken from the twin sample and matched by age and sex (h_2^2), which overestimates heritability since the correlation of environmental influences on MZ twins is neglected; and, as a third possibility, calculating of h_3^2 from the intraclass correlation coefficient. This last method is in most cases the least adequate. It may overestimate or underestimate heritability depending on assumptions regarding common environmental factors in MZ and DZ twins, biases due to the age distribution in the twin sample, differences in socioeconomic background of MZ and DZ twins, and others. This least adequate method has almost exclusively been applied to heritability estimates from twin studies.
c) Another possible shortcoming of the twin method is that twins are treated as an otherwise unbiased sample of the population for which conclusions have to be drawn. However, twin pregnancies as well as the twin situtï¿½ion during childhood and youth create special conditions that might bias the results. With these limitations, what conclusions can be drawn from existing data?

Twin Study on Swedish Conscripts. Husén (1959) [721] examined all male Swedish twin pairs born between 1928 and 1933 who were inducted for military service between 1948 and 1952. 215 MZ pairs and 416 DZ pairs could be analyzed. The Swedish conscription test series, which comprises a number of tasks common in standard intelligence tests (such as finding of synonyms, distinction of concepts, and complementing of matrices, similar to the Raven test) was used. In the whole series, the intraclass correlation coefficients were reported to be 0.90 in MZ and 0.70 in DZ

Table 8.6. Heritability estimates and intraclass correlations in MZ and DZ twins reared together and reared apart

Reference	Psychological test variable (P)	No. of pairs (n)	S.D. of $2n$ individuals (σ)	Intra-pair correlation (r_P)	Intra-pair variance (V_P^w)	Estimates				
						$\hat{h}_0^2$	$\hat{h}_1^2$	$\hat{h}_3^2$	$\hat{r}_E^{(1)}$	$\hat{r}_E^{(2)}$
Husén (1959) year groups 1949-52	I.Q. by I-test	MZT 215 DZT 416	30.48 31.90	0.894 0.703	98.11 302.32	— —	0.675 ±0.039	0.382 ±0.057	— —	— —
Newman, Freeman, and Holzinger	Binet Mental Age	MZS 19 MZT 50 DZT 50	23.55 29.5 35.4	0.637 0.922 0.831	201.1 67.9[a] 211.8[a]	0.637 ±0.136	0.679 ±0.099	0.182 ±0.097	0.785 ±0.100	0.662 ±0.171
(1937) [90]	Binet I.Q.	MZS 19 MZT 50 DZT 50	13.00 17.3 15.7	0.670 0.910 0.640	55.84 26.9[a] 88.7[a]	0.670 ±0.126	0.697 ±0.093	0.540 ±0.174	0.727 ±0.128	0.518 ±0.244
	Otis score	MZT 50 DZT 50	20.7 21.3	0.947 0.800	22.7[a] 90.7[a]	— —	0.750 ±0.077	0.294 ±0.106	— —	— —
	Otis I.Q.	MZS 19 MZT 50 DZT 50	13.58 16.0 15.8	0.727 0.922 0.621	50.42 20.00[a] 94.6[a]	0.727 ±0.108	0.789 ±0.065	0.602 ±0.179	0.714 ±0.137	0.603 ±0.201
	Stanford Educat. Age	MZS 19 MZT 50 DZT 50	23.47 30.5 32.3	0.502 0.955 0.883	274.1 41.9[a] 122.1[a]	0.502 ±0.172	0.657 ±0.106	0.144 ±0.067	0.910 ±0.040	0.847 ±0.077
Shields (1962) [1651]	Dominoes Intell.-Test	MZS 37 MZT 34	9.02 8.33	0.758 0.735	19.68 18.40	0.758 ±0.070	— —	— —	−0.095 ±0.454	0.065 ±0.352
	Mill Hill Vocabul. scale	MZS 38 MZT 36	5.75 3.98	0.741 0.742	8.566 4.097	0.741 ±0.073	— —	— —	0.004 ±0.403	0.522 ±0.176

[a] Recalculated by the formula $V_P^w = \sigma^2(1 - r_p)$.

Explanations

MZS = Monozygotic twins separated;

MZT = Monozygotic twins } brought up together.
DZT = Dizygotic twins }

$\hat{h}_0^2 = r_{P,\mathrm{MZS}}$ (unbiased estimate of h^2 if $r_{E,\mathrm{MZS}} = 0$).

$\hat{h}_1^2 = 1 - V_P^w(\mathrm{MZT})/V_P^w(\mathrm{DZT})$, S.E. ($\hat{h}_1^2$) calculated by (A 6.14).

$\hat{h}_3^2 = 2(r_{P,\mathrm{MZT}} - r_{P,\mathrm{DZT}})$, S.E. ($\hat{h}_3^2$) calculated by (A 6.16).

$\hat{r}_E^{(1)} = (r_{\mathrm{MZT}} - r_{\mathrm{MZS}})/(1 - r_{\mathrm{MZS}})$ (estimate of $r_{E,\mathrm{MZT}}$ if $r_{E,\mathrm{MZS}} = 0$, cf. A 6.2).

$$\mathrm{S.E.}(\hat{r}_E^{(1)}) = \frac{1 - r_{P,\mathrm{MZT}}}{1 - r_{P,\mathrm{MZS}}} \sqrt{\frac{(1 + r_{P,\mathrm{MZT}})^2}{n_{\mathrm{MZT}}} + \frac{(1 + r_{P,\mathrm{MZS}})^2}{n_{\mathrm{MZS}}}}$$; (according to Newman, Freeman, and Holzinger (1937)).

$\hat{r}_E^{(2)} = 1 - V_P^w(\mathrm{MZT})/V_P^w(\mathrm{MZS})$ (estimate of $r_{E,\mathrm{MZT}}$ if $r_{E,\mathrm{MZS}} = 0$, cf. A 6.9).

S.E. ($\hat{r}_E^{(2)}$) calculated by (A 6.14).

pairs. The test-retest reliability values for the same test were 0.92-0.93. This result means that the difference between MZ twins in hardly larger than that between two successive examinations of the same individuals. Husén's paper does not provide heritability estimates, which we therefore supplied (Table 8.6). At first glance, the values seem to be very high. However, the correlations between DZ twins are relatively high as well. This finding results in a large discrepancy between h_1^2, on the one hand, and h_2^2 and h_3^2, on the other, as defined in Sect. 3.8 and App. 6. Most correlations between same-sexed sibs that have been reported in the literature are much lower. In another Swedish twin series by Husén [2079] that comprised 268 MZ and 360 DZ pairs of school age and in which a series of school performance tests was used, correlations between MZ and DZ of the same order of magnitude as in the conscript study were found.

Twin Performance on I.Q. Test Is Lower Than That of Singletons. Both Swedish studies showed still an-

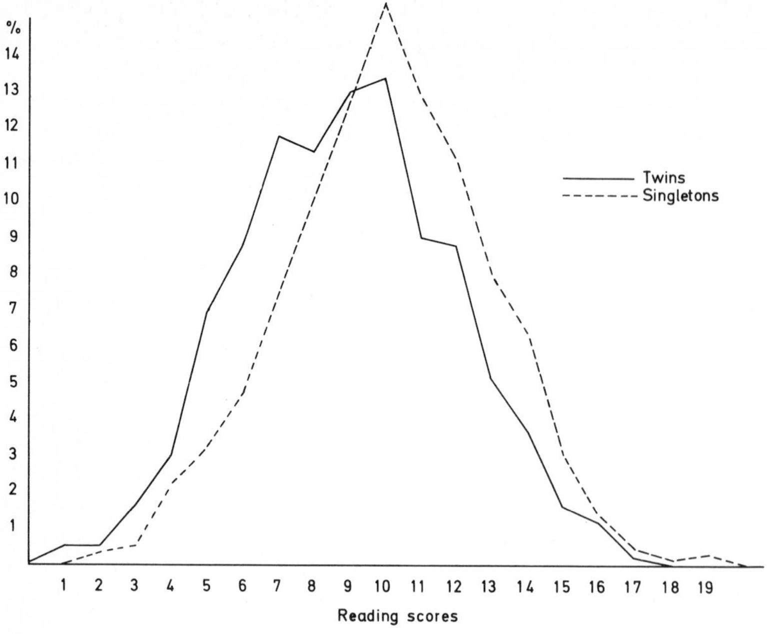

Fig. 8.12. An example for the difference in cognitive test scores between twins and non-twins: distribution of a score for reading ability in a series of 12 to 13-year-old Swedish school children (girls). (Husén, 1960 [2079])

other interesting result. The average test scores of twins were significantly lower than the scores of non-twins; among the conscripts, this difference amounted to about one-fourth of one standard deviation; in the school children who were about 12–13 years old, this difference tended to be even larger. The distribution of test scores in girls for one item, reading ability, is shown in Fig. 8.12. In Sect. 8.2.1.2 we mentioned that twins were found in unexpectedly high frequency among children attending special schools for backward children. This higher frequency is explained by a lower mean I. Q. level of twins. In part, this is caused by biologic influences, especially the intrauterine "transfusion syndrome" of MZ twins (Sect. 3.8.4), since it can also be found when the co-twin had died [2141]. However, in Husén's series, DZ twins performed, on the average, still more poorly than MZ twins, contrary to our expectations if the reason were predominantly biologic in origin. A socioeconomic alternative could be that the incidence of DZ births but not of MZ births increases with the age of the mothers and parity. Probably, not only somewhat older mothers but also mothers with higher fecundability have a higher chance to give birth to DZ twins (Sect. 3.8.3). It is well-known that at the time when Husén's twins were born, women from higher socioeconomic strata used to have fewer children than women from lower classes. These women obviously practiced less birth control and relied more on their "natural fecundability." Therefore, they had a higher chance to have DZ twins. Hence, DZ twins tended, on the average, to come from a somewhat lower socioeconomic group than MZ twins and the non-twin population. However, whatever the reason, individuals from a lower social stratum tend to perform more poorly in I. Q. and cognitive tests.

Is there an environmental explanation for the twin correlations? DZ twins, being of the same age, normally spend more time together and tend to be exposed to more similar environmental influences than sibs of different ages. Therefore, if they are more similar in their test results, these environmental influences should be of importance. However, we know from a number of studies [27] that MZ twins tend to have much closer relationships than DZ twins: "they are often eager to do the same things, to help each other, not to compete with each other, and to be similar in most aspects." If MZ twins really behave so differently from DZ twins and if differences in behavior between DZ twins and the genetically equivalent sibs of unequal ages cause remarkable differences in their test score correlations, should smaller difference in test scores of MZ in comparison with DZ twins not also be related to their peculiar situation? This question could be answered by finding an experimental design that separates the genetic influence from the special effect of the twin situation.

MZ Twins Reared Apart. In theory, the ideal probands for such studies are MZ twins who were separated, if possible, immediately after birth, and have been reared in different environments. The first observation was published in 1922 by Popenoe [2159] and in 1925 by H. J. Muller [2139]. In spite of their different environments, Jessie and Bessie were very similar in intelligence, both achieving above-average scores. Their feeling tones and temperaments, on the other hand, differed and this could plausibly be explained by their life histories.

It is obviously very difficult to find such twin pairs, but three series of cases are available in which an attempt to ascertain relatively unbiased twin series was successful: Newman et al. (1937) [152] in the United States; Shields (1962) [2195] in Great Britain; and Juel-Nielsen (1965) [2095] in Denmark.[2]

Newman et al. compared 19 MZ pairs, who had often been separated during infancy, but never after the age of 6, with 50 MZ and 50 DZ pairs reared together. The ages at the time of examination ranged between 11 and 59 years. Every pair was given a careful biologic and psychologic examination. The results for heritability estimates are compared with those of one other study in Table 8.6. They were recalculated from the originals. In all three series, the heritabilities are high, though not as high as for MZ twins reared together. What are the reasons for the differences? Is it possible to correlate them with known environmental factors?

In Newman's study, the differences between the conditions in which the twins had been brought up were estimated by five observers and classified by separate scoring scales with a maximum of 50 points each for educational, socioeconomic, and physical (i.e., health) advantages separately. The twin who stayed in school longer and had, in general, the better educational score also tended to have better test scores. The correlation between education and I.Q. scores (+0.79) was significant. A lower correlation (+0.51) was found with the social scale and a still lower one (+0.3) with the health score. One case report may illustrate possible environmental differences. Alice and Olive were born in London and were separated at the age of 18 months. Alice's foster parents lived in London; Olive was adopted by relatives who lived in a small Canadian town. The separation lasted up to the age of 18; examination took place about 1 year later. Alice's foster parents were classified as lower middle class; they had four daughters of their own who were much older than Alice. She went to school up to the age of 14, took an 18-month business training course, and started working in business offices. Her parents were not able to care much for her; due to the events of World War I, the quality of her school education was rather poor. Olive, on the other hand, grew up as the only child in a family that was better off than Alice's. She was spoiled by her parents; she attended grade school, took a 2-year commercial course equivalent to high school, and worked as an office clerk like her sister. There was a significant I.Q. difference, Alice having a value of 84.9, whereas Olive scored 96.9. Hence, the sister that had lower educational opportunities had a definitely lower score. Alice and Olive were very similar in temperament, although Olive was more active and domineering. In a separate chapter of their book, Newman et al. explain part of the I.Q. difference by the fact that the test was designed for American students and was therefore "not quite fair for an English girl."

In the second study (Shields, 1962 [2195]), only two short tests, one verbal and one nonverbal, were used.[3] The calculation of heritability (App.6) was impaired by lack of control DZ pairs; the other data have been recalculated.

The 12 twin pairs observed by Juel-Nielsen (1965) [2095] were between 22 and 77 years of age at the time of investigation. Their ages at separation ranged between 1 day and 5¾ years. They were extensively tested using the Wechsler-Bellevue (W-B) intelligence scale, form I, and Ravens' progressive matrices. For both tests, a marked resemblance between the test scores achieved by the twin partners was found. The correlations coefficients between the co-twins' scores for W-B values were as follows: total I.Q. 0.62; verbal I.Q. 0.78; performance I.Q. 0.49. Raven raw scores: $r = 0.79$. As explained in App.6, these intraclass correlation coefficients are heritability estimates and, more specifically, estimates of h_3^2. Intelligence testing was repeated in nine pairs; the time between the two examinations varied, an average time lag being 12 months, with a minimum of 6 months for all tests. The correlation between the first and second test of one individual, i.e., the test-retest reliability, was smaller than the difference between co-twins. This means that the study revealed real differences in test intelligence between twins of MZ pairs.

Due to an unlucky shortage of information on DZ controls – such controls are scanty in the Shields study and absent in the Juel-Nielsen study – heritability was estimated mainly from the intraclass correlation coefficients (h_3^2). As explained in App.6, this may introduce a bias in favor of a spuriously high h^2 estimate if the MZ sample comprises pairs of different ages and if the character to be tested is age-dependent. In theory, intelligence tests are standardized for age, and the test scores should be age independent. In practice, however, this is not quite true; besides, the age differences within all three samples of MZ twins reared apart were unusually high.

In all three studies, intelligence testing was carried out only as part of a more comprehensive assessment of personality development in relation to environmental differences. In general, striking similarities despite these differences are stressed by the authors; the observed differences could often plausibly be accounted for by corresponding differences in educational opportunities and environmental influences in general.

[2] In the older literature, a fourth series, that of C. Burt, is usually cited. There is now convincing evidence, however, indicating that Burt's data, which were published very incompletely, had been fabricated by the author.

[3] Synonyms section (set A) of the Mill Hill vocabulary scale and the (nonverbal) Dominoes test.

Overall Results of Studies on MZ Pairs Reared Apart. Altogether, MZ twins reared apart show a remarkable degree of similarity of intellectual performance. This similarity is found not only during childhood and youth but in mature life as well. It is greater than normally observed between sibs and even between DZ twins reared together. Differences in social status during childhood, educational opportunities, and later experiences influenced intellectual abilities up to a certain degree and, in some cases, created appreciable differences. On the whole, however, the impression of similarities prevails.

In view of some recent criticism of twin studies assessing intellectual abilities, it should be remembered that the authors of all three studies were keenly aware of the limitations of their studies as well as of the shortcomings of their own data. Moreover, in their papers, we find no preoccupation whatsoever with heritability measurements nor with a numerical estimate of the relative contributions of heredity and environment on the variance of test scores. The estimates in Table 8.6 are ours, not theirs.

In our opinion, these studies provide some evidence in favor of a view that within our present-day white populations and under the environmental conditions under which children have grown up in the first half of our century in societies, such as that of Denmark, Great Britain, or the white, English-speaking population of the United States, genetic variability is responsible for an appreciable part of the variance in test performance on the usual intelligence tests. We make these statements realizing a variety of criticisms (See also [2214, 2038]). These include:

a) The sampling process by which the twin pairs of the first two studies [152, 2195] were ascertained may have created a bias in favor of more similarity between the twins and may, on the other hand, have induced them to exaggerate the degree of separation in the information provided to the investigators.
b) Some of the twin pairs included in these studies were separated so late that common influences during infancy may have helped to shape their attitudes.
c) The homes in which they grew up were, on the average, more similar than random homes.
d) Some of the twins may have met during childhood, for example, at school, so that the known mechanisms of mutual identification may occasionally have come into play.

In view of these criticisms, some observers still deny any heritability of I.Q. [104]. After looking strictly at single studies of this sort, such skepticism may be warranted. We are impressed, however, by many lines of evidence suggesting that some determinants of I.Q. may have a genetic basis. It is quite clear that the available evidence does not permit to assess the exact degree to which the variance of intellectual abilities in these populations is determined by genetic as opposed to nongenetic influences. One argument against generalization of such results from twin studies always remains that twins are not a random sample from the general population. Therefore, other approaches were needed.

Studies on Adopted and Foster Children. The obvious alternative to get around the problem of a correlation between genetic and environmental variances are studies on foster children and on adopted children. For children living in foster homes, environmental influences are supposedly randomized, leaving only the genetic variability of parents. In adopted children and children living with foster parents, the genetic influence of the biologic parents can be compared with the environmental influence of the adopting parents [2119].

The evidence from studies on adopted and foster children with foster parents was reviewed by Loehlin in 1980 [2119]. The earliest studies were performed as long ago as in the 1920s and 1930s. When the data were pooled they gave substantially lower correlations in I.Q. between adopted children their and adoptive parents than between biological children and the same parents. Since these studies were criticized for various reasons new studies were carried out in the 1970s. In three American studies, more than 500 adoptive families with nearly 800 adopted and about 550 biological children were included. All three yielded substantially the same result: Correlations between adopted children and adoptive parents were lower than those between these parents and their biological children (Table 8.7). In general, the correlations were somewhat lower than in earlier studies; this was explained convincingly by a more restricted range of I.Q. values in foster families.

In one study, adopted children could be compared with their biological mothers, from whom they had been separated as newborns. The correlation was surprisingly high (0.32), pointing to a substantial genetic contribution.

Studies on adopted and foster children are subject to a number of unavoidable biases; on the one hand, adoption definitely is not carried out at random. Adoption agencies have the understandable tendency to place children in "suitable" homes. Therefore, a genetic correlation between foster children and foster parents is expected. On the other hand, some of these children have spent the first year(s) of their lives with their biologic mothers or in foster homes, reducing the environmental influence of the foster parents. Moreover, couples willing to adopt children are not an unbiased sample of all parents in the population; their mean I.Q. is not only higher, but there is also a smaller variance [2038]. This may lead to a

Table 8.7. Parent-child and sib-sib correlations of I.Q. in three recent adoption studies. (From [2119])

Study	Adoptive children		Biological children		Biological mothers and their children adopted by other parents	Sibs in adoptive families	
	Fathers	Mothers	Fathers	Mothers		Biological-adoptive	Biological-biological
Minnesota I	0.15	0.23	0.39	0.35		0.39[a], 0.30[b]	0.42
Minnesota II (Both: Scarr and Weinberg [2190])	0.16	0.09	0.40	0.41		−0.03	0.35
Texas (Loehlin [2119])	0.17	0.19	0.42	0.23	0.32	0.22[a], 0.29[b]	0.35

[a] Adopted-adopted, [b] correlation between adopted and biological "sibs".

lower correlation as a statistical artifact [104, 2119].

On the whole, the limited evidence derived from all adoption studies may be summarized as follows. Genetic variability, as well as environmental factors, influence the intellectual development of the child. Due to unavoidable biases in the examined series, however, the relative contribution of these influences cannot be estimated.

For studies carried out with so much expenditure in time and money, this conclusion is certainly a disappointingly general result. Based on current knowledge, one might have guessed at such a conclusion before. These investigations do not give answers to more specific questions. They are valuable, however, in providing possibly useful information on social and psychological aspects of adoption.

In recent years two different attempts have been made to overcome some of the methodological problems and to elucidate the nature-nurture problem in I.Q. research by applying sophisticated statistical techniques and by including relatives of many degrees and with different degrees of similarity of the environment in the models: The Birmingham school used analysis of variance [2087], and the Hawaii school applied the principles of path analysis [139, 140] (see also Appendix 7). Both attempts have been criticized severely [2214] because of the many assumptions that had to be made by each school.

What Does the Available Evidence Prove Regarding Genetic Variability of Intelligence in the Normal Range? The answer is short: very little. With some stretching of the evidence for biases in sampling, statistical evaluation, reporting, and interpretation of ambiguous results, and with some not too implausible assumptions regarding environmental influences of biologic and adoptive parents or the way by which MZ twins influence one another, the point has been made that genetic variability does not influence I.Q. performance at all [104], i.e., that heritability is zero, or that it is very low [2214].

This conclusion does not mean that a zero heritability is the most likely answer. Interpretation of the same data with different assumptions on biases has led other authors to the conclusion that heritability of I.Q. performance might be as high as 0.8 [2085, 2086]. Most scientists, if asked for an educated guess, will probably settle on values somewhere in-between, more because they dislike extreme points of view than because of a strong conviction in favor of any positive evidence.

If the same data lend themselves to such diametrically opposite interpretations, the actual evidence must be very "soft". As the foregoing discussion has shown, clean experimental designs in humans are in most cases impossible. Genetic and environmental influences in families are correlated; attempts to separate both factors by examining adopted children suffer from nonrandomness of adoption. MZ twins living together influence one another in unpredictable ways that change with the cultural setting. MZ twins reared apart are often educated in homes with similar backgrounds. MZ twins tend to show differences in development due to peculiarities of twin pregnancy.

The entire approach of measuring intellectual ability and determining the genetic and environmental contributions to its variance leaves us unsatisfied. Why?

As geneticists, we are ultimately interested in an analysis at the individual gene level. The absence of such data highlights the contrast between the two paradigms of human genetics: the biometric paradigm founded by F. Galton (1865, Sect. 1.3) and the gene concept of G. Mendel (1865, Sect. 1.4). These two paradigms have been discussed in Sect. 3.6.1. The genetic analysis of intelligence had long been regarded by many research workers as the field in which Galton's paradigm scored its most impressive victories, while analysis on the lines suggested by Mendel seemed to be doomed to failure. The discussion of recent years that started with Jensen's work on group differences and educability [2085] has exposed the weaknesses of the biometric paradigm so mercilessly that one can hardly imagine how it should survive. On the other hand, research on genetic variability in other, more easily accessible fields of human genetics, as well as in population genetics of other species, has revealed an amazing amount of genetic variability in populations (Sect. 6.1). For example, no less than one-third of blood enzymes examined so far show genetic polymorphisms, and usually the normal variants show slight functional differences within the normal range [1787]. Studies on the genetic basis of common disease, as well as the recent progress in pharmaco- and ecogenetics, are now increasingly revealing influences of such normal genetic variability on the health status of the individual under varying environmental conditions. It is our contention that genetic variability of biologic factors influencing intelligence and other aspects of human behavior is likely to be quite as extensive. However, phenotypic expression may be more complex than in somatic traits. The ambiguity of test results in revealing such genetic variability may - at least in part - be caused by the insufficiency of test methods rather than by weak genetic influences on intelligence. But can a Mendelian approach replace current efforts? This problem will be deferred to a later section.

Paradigm Clash Also in Psychology [2152]. The phenomenologic and biometric approach that is associated with Galton has by no means been undisputed among psychologists concerned with problems of genetic variability of human behavior in general and, more specifically, intelligence. On the contrary, its shortcomings have been explained for decades, and repeated attempts were made at identifying specific determinants of intelligence [2062]. Such attempts have remained only partially successful, and considerably more work with attention to such components using a variety of approaches is sorely needed.

Research on I. Q. and Politics. Adherents of a paradigm frequently form a group that also shares certain convictions that are not immediately connected with the scientific content of this paradigm. Sometimes, they even try to separate socially from members of concurring paradigm groups. From more general experience in social psychology, it is not very surprising that such tendencies lead to identification of the own group with the "good guys," whereas members of the other group are condemned as the "bad guys" - reactionary, politically authoritarian, antisocial, and altogether wicked. This attitude leads to proposals to curb certain kinds of research - and, at a slightly less "academic" level - to threatening certain investigators physically or even destroying their research materials. In spite of the fact that we agree scientifically with much of the criticism of the biometric approach, we wish to stress that we are fiercely opposed to such politics. The work in theoretical thinking and empirical analysis that has been done in recent years within the context of the biometric paradigm [2086] has been useful. It explored the inherent possibilities of this approach and - largely inadvertently - exposed its limitations.

8.2.1.4 Special Cognitive Abilities and Personality

Special Cognitive Abilities. Most of the debate regarding intelligence is centered around the overall outcome of the intelligence test - the I. Q. However, most modern tests consist of a number of subtests. Scores of these subtests are positively correlated in the same individuals. However, this correlation is far from complete; the subtests measure in part independent abilities. For example, the Wechsler test consists of a verbal and a performance subtest. One special ability that shows a relatively low correlation with the other scores is the ability to visualize spatial relationships of objects (Sect. 8.2.2; Fig. 8.19). It is not surprising that twin and family studies have also been carried out for such subtests [2235]. So far, however, no consistent differences in heritability between such subtests have emerged. The hope that this approach would reveal more basic abilities that might not be subject to the problems of heredity-environment interaction has not been fulfilled.

Intelligence Is Not Everything. Almost since the time when the I. Q. test became popular, it was obvious that for achievement in school, college, and universities, and still more so in professional fields, other qualities were required in addition to intellectual ability. Terman et al. [2215, 2216] performed a longitudinal study on individuals who had been selected at school age as especially gifted. Many of them were unusually successful professionally but some were not; in the unsuccessful group, personal instability was much more frequent than in the successful group.

Twin Data for Temperament, Sensory and Motor Functions, and Personality. Abundant literature is available on twin studies for temperament, sensory and motor functions, and personality. Most methods that psychologists use have also been tried on twins to find genetic variability in these parameters. Almost all these studies had an ever-recurring result: MZ were more similar than DZ twins [3, 27, 2047, 2122, 2226, 2227]. A number of attempts were made to classify the observed variability. The bestknown attempts are those of Cattell and his school [2014-2016] and of Eysenck [2033-2035, 2036,2037] using the method of factor analysis. From correlation clusters of a great number of single test scores – such as questions in a personality questionnaire – Cattell isolated a number of factors that were interpreted as basic dimensions of personality. Eysenck [2033, 2034, 2035], also on the basis of personality questionnaires, distinguished three main dimensions of personality: extraversion, neuroticism, and psychoticism. Such studies were useful for conceptualization in psychology, and Eysenck's studies helped to focus attention to the relation between behavior and brain function. So far, however, no definite conclusions on genetic mechanisms have emerged.

Longitudinal Study of Twins. The most comprehensive picture of the action of different environments on genetically identical individuals is the prospective study of the life histories of MZ as compared with DZ twins. Obviously, the investigator who plans such studies must begin early in his or her career and must occupy himself or herself with this project again and again over an entire lifetime. Most investigators are not prepared to do so. However, one twin study has now been carried out for over 40 years by the same investigator. This investigation is, indeed, producing more satisfying results than most other studies of the more conventional, cross-sectional type [2062, 2063].

Gottschaldt and his co-workers organized two holiday camps for twins on the island of Norderney (Germany) in 1936 and 1937. In these camps, 136 twin pairs between 4 and 18 years of age were examined for several weeks. The program consisted of a great number of systematic observations and tests that included not only the usual intelligence tests but also tests of sensory and motor ability and behavior in more complex situations, such as searching for hidden objects. The investigators took great care to avoid the typical "test situation" and to create a socially relaxed climate between twins and examiners. For every individual, a diary was kept. This approach comes close to the controlled observation recently suggested by the ethnologist Charlesworth [2017] for research on intelligence. It also enabled the investigators to assign scores for qualities such as drive, sensitivity, and feeling tone (Fig. 8.13).

Data obtained by comparison of the ability to learn by experience were of interest. Pairs of colored figures that occurred in changing orders had to be connected by pencil. After 30 sec, the experiment was stopped, and the number of connected pairs was counted. The experiment was carried out twice a day for 5 weeks. It produced typical learning curves with an initial increase and a plateau that was reached after individually different times; this speed in increase of performance was characteristically different among MZ and DZ twins (Fig. 8.14). The test

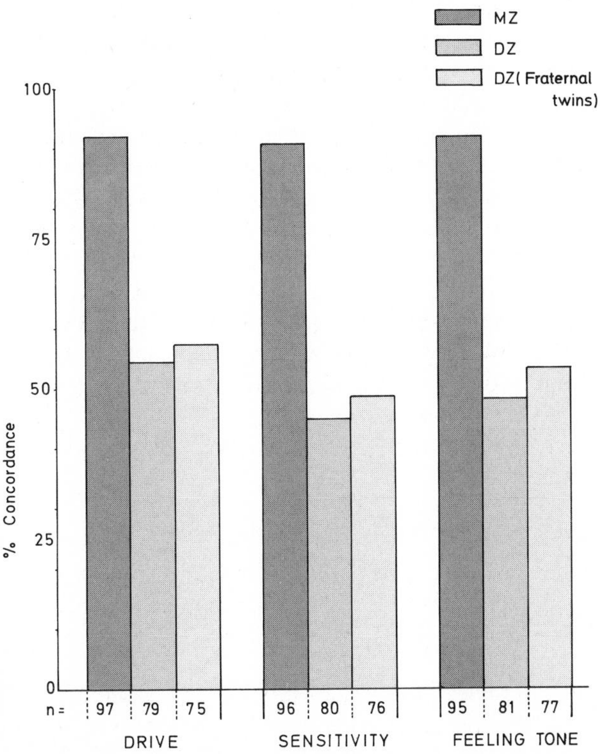

Fig. 8.13. Concordance and discordance in ratings for drive, sensitivity, and feeling tone in juvenile MZ pairs, DZ twin pairs of same sex, and DZ pairs of different sex (Fraternal twins), (observations in holiday camps). (Geyer and Gottschaldt, 1939 [2003])

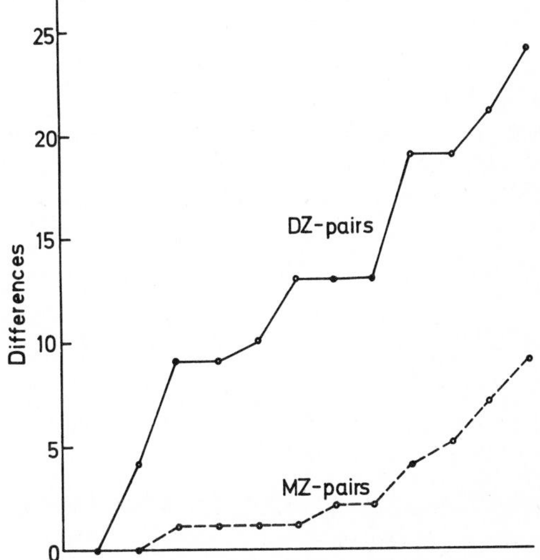

Fig. 8.14. Difference (in minutes) between MZ and DZ twin pairs in learning speed (reaching of a certain learning level in a 50-min experiment; each point represents one twin pair; see text for further details). (Wilde, cited in Gottschaldt, 1939 [2062])

results confirmed in all cases the greater similarity of MZ as compared with DZ twins; detailed results have been published.

The investigators utilized many tests of their own, making comparability of their data difficult with other studies. Undoubtedly, this approach gave a much more comprehensive, if less well-quantitated picture of the development of MZ as compared with DZ twins. Unfortunately, such results do not permit disentanglement of the effects of the special environmental conditions from genetic effects.

However, some of these twins could be re-examined three times, for the last time in the early 1970s. Meanwhile, World War II brought thorough changes in life style. Hence, the situation was especially well-suited for following how MZ twins would behave under such conditions.

Table 8.8 gives an overview of the results of three examinations, carried out in the 1930ies, the late 1940s, and again in the 1960s. Two groups of abilities usually considered important as the basis of formal intelligence, i.e. capacity of information uptake and abstract thinking, remained concordant well into middle age. Certain pairs became more different, others more similar. A certain concordance up to the 20ies can be found for temperament; here definite differences became apparent in middle age, and they could plausibly be correlated with life histories and life experience. For the other three characters – behavior control and level of aspirations, mental attitudes, and range of interest – the striking concordances that were seen in childhood tended to disappear already in the 20s and still more when the probands had reached middle age.[4]

Hence, examination of MZ twins in adult age revealed striking differences. These differences did not become visible in abilities normally described as "intelligence" but in mental attitudes and life aspirations, in behavioral control, and to a certain degree even in temperament. These discordances were confirmed partially in another longitudinal twin study in Germany that was carried out at about the same time and on twins that were somewhat older in 1945 [918]. In this study, three classes of MZ twins could be distinguished tentatively on the basis of their life experiences: (1) A class of especially gifted and successful twins who were relatively discordant for personality and life experiences; apparently such individuals had a number of choices between different life styles. (2) The largest group with about average intelligence and performance remained relatively concordant for personality and life experiences up to middle age and in some cases up to old age. Their choices in life were more restricted by convention. (3) A class of twins with below-average success turned out to be rather discordant for personality and life experiences. In this class, the course of their lives seems to be largely determined by accidental circumstances.

One could argue that these two studies give an exaggerated picture of the possible differences in psychological development of MZ twins and, hence, of the plasticity of personality characteristics by differing environmental influences. World War II and the following years brought to this study population changes in life style, such as displacement from homes and occupations, interruption of education, and loss of close relatives. Whether and how a normal life could be re-established was often a matter of luck. Very probably, higher concordances between twins would have been maintained had the living conditions been more normal. However, these results show how much personal development may be and will be modeled by environment, even if the genetic background is identical. More recently, a longitudinal study on somatic and psychological development of twins has been started in Louisville (USA).

[4] This description is based on a preliminary and relatively short and general report of Gottschaldt (1968) [2063].

Table 8.8. Results from a longitudinal study of 20 MZ twin pairs. (Gottschaldt, 1968 [2063])

Mean age	First examination (1937) 11.7 years	Second examination (1950/51) 23.3 years (20.3–31)	Third examination (1968) 41.5 years (34.9–46)
Capacity of information uptake	++	++	++
Abstract thinking	++	++	++
Mental attitudes (range of interest; appraisal of own situation)	++	(+)	–
"Vitality"	++	+	–
Activity	++	+	–
Mental responsiveness	++	++	–
Control of behavior	++	+	–

++, strong concordance; +, weaker concordance; (+), doubtful concordance; –, no concordance.

The twins are being examined at relatively short time intervals from birth onward [2244]. During the first 2 years of their lives monozygotic twins, but to a surprisingly high degree also dizygotic twins, were found to be highly concordant for mental development as well as for the spurts and lags in development.

Possible Consequences for Educational Policy. Educational thinking is traditionally centered around children of school age and young people. However, twin investigations have shown increasing differences in mental attitudes, behavior control, and aspirations between MZ twins in mature life. These findings suggest that there are great potentials for mental and emotional development that our society does not utilize sufficiently. Would it not be a good idea to spend more money and ingenuity on educational opportunities to adults? Could this help individuals to achieve more fulfillment and happiness? Could the society as a whole overcome some of its future problems more efficiently by such approaches? Recent trends in adult education, which started in the United States to occupy the educational establishment in the face of declining enrollments of young people because of the falling birth rate, will be observed with much interest.

Genetics of Scent Differences ("Olfactogenetics") [989]. An interesting aspect of human individuality concerns differences in odor. It has been stated that police dogs are unable to differentiate the scent of identical twins. Presumably, the characteristic scent of an individual is under the control of genes that determine the excretion of chemical substances from the skin. These substances give off a distinct scent for each individual. Scents are undoubtedly also influenced by diet and by the bacterial composition of the skin. However, it appears likely that the microbial flora of the skin depends to a large extent on the genetically determined biochemical makeup of the various compounds excreted in dermal sweat. It has been shown in mice that the ability to sniff differences in odor depends upon the *H2* locus - the analog of the HLA complex in man [989]. In these ingenious experiments, mice that were genetically identical except for differences at the *H2* locus were able to detect scent differences between each other. These studies also showed that mice preferred mating partners that differed at the *H2* locus, which is an interesting evolutionary strategy to promote genetic diversity. It appears that the *H2* locus (and possibly the human *HLA* locus) is a major genetic determinant that imparts a characteristic smell to each individual. The pathway from genotype to phenotype is unknown. HLA studies in various types of poorly delineated human anosmias (not associated with nervous system diseases) such as anosmia for isobutyric acid (20700) and anosmia for cyanide (30430) would be of interest. It has also been claimed that the supposed presence or absence of odor of urinary asparague metabolites is caused by the fact that some people are unable to smell a urinary metabolite that is always present after eating asparagus [1202]. It was previously thought that there was a polymorphism in urinary excretion of a metabolite (10840).

Sensory Human Genetics and Behavior. The genetics of sensory perception in humans is in its infancy. However, a good case can be made for the idea that human beings perceive their environment slightly differently depending upon their genetic makeup. Tasting of phenylthiocarbamide (PTC) is a genetic trait; some individuals cannot taste this substance. Color vision varies in that depending upon the specific color vision defect some persons see colors differently or not at all. Tone deafness exists, and some persons are unable for genetic reasons to recognize different tones, so that they have poor musical ability [2099]. The counterpart - high musical ability - seems to aggregate in families. Information processing in the brain varies. Dyslexia is a genetic trait and presumably represents a defective ability to process words by the central nervous system [2150].

All these examples and the above-mentioned differences in perceiving scents presumably are only the tip of the iceberg, and many more polymorphic differences in sensory perception (taste, smell, vision, hearing) undoubtedly exist. Each person therefore probably perceives his or her environment somewhat differently; each of us probably has his and her different external world. Our reactions to this uniquely perceived environment therefore may differ, and variability in behavior may result. The marked similarity of even trivial behavioral characteristics observed in identical twins reared apart (Lykken and Boudard, work in progress, 1985) may be a partial consequence of their perceiving the environment in an identical manner.

8.2.1.5 "Abnormal" and Socially Deviant Behavior

Criminality. Since J. Lange (1929) [118] published his monograph on *Crime as Fate,* a number of studies have been published in which MZ and DZ twins were compared for concordance in criminal behavior [2047]. The relevant information is collected in Fig. 8.15. The most important conclusions may be summarized as follows:

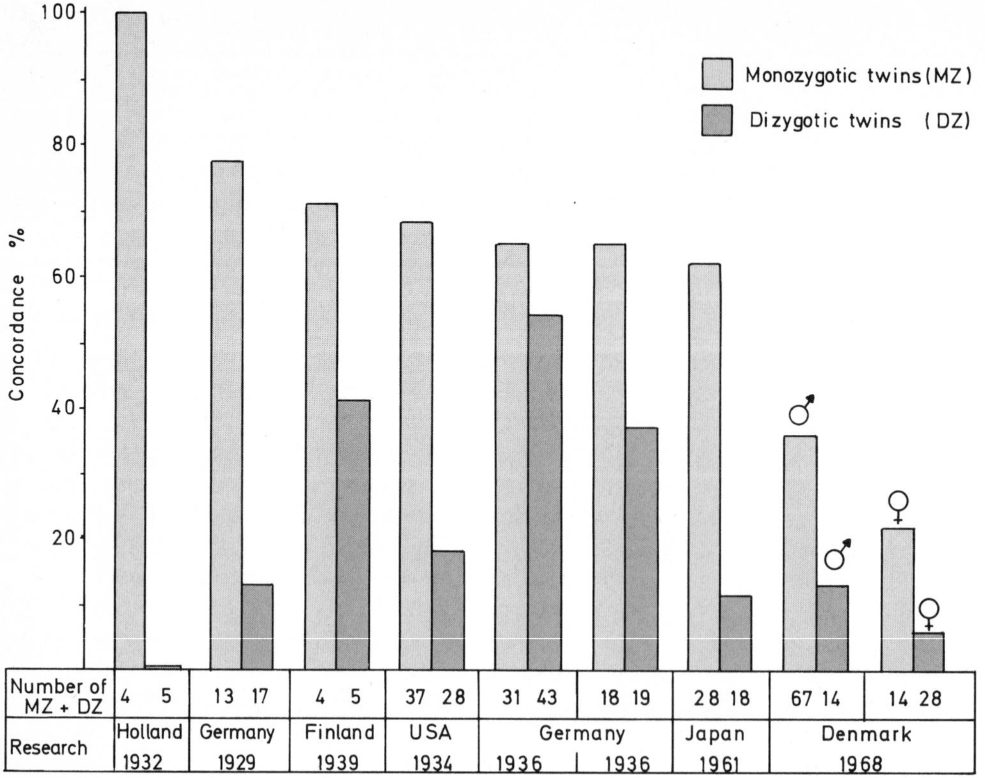

Fig. 8.15. Concordance rates of MZ and DZ twins for criminality – as defined by court conviction – in studies from various populations. ([72] and some additions)

a) Concordance of MZ as compared with DZ twins is higher in all series and in many of them considerably higher.
b) Absolute concordance of MZ twins varies between series. There is a tendency for a higher concordance in the older investigations.

The main reason for these differences in concordance rates is the spectrum of crimes covered by these investigations. For example, the original series of Lange was selected for severe and repeated criminality, whereas the more recent series, for example those from Denmark, tend to refer to all types of criminality, including occasional and smaller offenses.

Taken at face value, these results suggest that the liability to become a convicted offender depends strongly on the genetic endowment of an individual with particularly striking effects for severe and repeated criminality. This conclusion, if proven to be correct, might elicit two different responses by society: to isolate law offenders as biologically deviant or to consider them as sick and to attempt quasi-medical therapy. However, we must be cautious not to explain MZ concordances too readily by genetic influences before enlistment into criminal behavior by social interaction between members of these twin pairs has been excluded as an alternative interpretation. On the other hand, should such social interaction be the only important factor? This possibility seems to be unlikely as well. Again, we are left with the ambiguity of twin data.

Studies on adoptees are of some interest [2025, 2080, 2206]. One study started with adoptees who were at least 18 years old (the oldest was 47) and had been adopted out following conviction of their mothers for felony, prostitution, larceny, and other offenses; sufficient information on fathers was not available. With three exceptions, they were adopted by unrelated persons. A control group of adoptees was matched for sex, race, and age at the time of adoption. The most important results are given in Table 8.9. There is a significantly higher risk of arrests, incarceration, and psychiatric hospital records among the proband group in comparison with controls. Structured psychiatric interviews, while showing no other difference between the probands and the control groups, led to the diagnosis of "antisocial personality" in 6 of 46 cases of the proband group but in only one questionable case of the 46 controls. On the other hand, detailed case histories gave evidence for an influence of environmental factors on the manifestation of antisocial behavior. Five of the six antisocial probands spent over 12 months in or-

Table 8.9. Criminality in adoptive children, records follow-up. (Crowe, 1974 [2025])

	Probands,[a]	Controls,	
	$N=37$	$N=37$	
Arrest records			
Adult arrest	7	2	
Adult conviction	7	1	$P=0.03$[b]
	$N=42$	$N=42$	
Incarceration			
Juvenile[c]	3	0	
Adult	4	0	
Either	6	0	$P=0.01$
	$N=42$	$N=42$	
Psychiatric hospital records			
Number hospitalized	7	1	
Outpatient only	1	1	
Total number seen	8	2	$P=0.04$
Number with both psychiatric and arrest records	6	0	$P=0.01$

[a] Mothers convicted of felony, prostitution, larceny and other offenses.
[b] Probabilities are Fisher's exact, one-tailed.
[c] One juvenile was actually ascertained through hospital records and was sent to the training school shortly thereafter.

phanages and temporary foster homes before adoption and were placed up for adoption when they were over 1 year of age. Most of the other probands had been placed earlier. Socioeconomic status of the adoptive homes did not correlate with the outcome. However, among the six homes in which the antisocial probands were placed, two were broken homes. In the control group, adverse environmental influences, such as late adoption, were also found in some cases, but there they did not result in abnormal behavior. On the other hand, in five of the six cases diagnosed as antisocial personality, the alleged father was said to also have had records of offenses.

This study succeeded in a relatively clean separation of genetic and environmental factors. The results point to genetic influences. However, the development of a manifestly deviant personality also required unfavorable environmental conditions. The first year of life already seems to be important, in accordance with other results from child psychology. On the other hand, the majority of children overcome unfavorable environmental conditions without recognizable ill-effects; only a presumably genetically predisposed group reacted by developing a deviant personality. It is comforting that a majority of adoptees developed normally.

In another study [2080], psychiatric records of biologic relatives of psychopathic[5] adoptees revealed a higher incidence of psychopathy as compared with that of adoptive relatives and controls. In a third study [2206], biologic relatives of criminal adoptees were found to have a higher rate of criminality than the adoptive relatives and biologic or adoptive relatives of control adoptees. However, criminality in an adoptee correlated independently with criminality in biologic as well as adoptive parents. This study thus also gave evidence for genetic as well as environmental influences.

Homosexuality. Kallmann (1953 [2098]) examined concordance of homosexual behavior in 95 male twin pairs, 44 MZ and 51 DZ. He classified the degree of homosexuality using a scale of six grades according to Kinsey. Grades 1-4 indicate a low degree, 5-6 a high degree of homosexual behavior. The probands were taken from court files; hence, they were selected for conflict with the law. Table 8.10 shows a very high concordance among MZ twins. There is not even one pair in which the other twin is completely free from homosexual tendencies. Of the 51 DZ pairs, on the other hand, 38 proved to be completely discordant in spite of the fact that low-grade homosexual behavior (grades 1-4) is fairly common in the American population (Kinsey).

One obvious argument against the genetic interpretation of the high MZ concordance is the possibility that one twin may have seduced the other one to homosexual activities. This, however, does not seem to be so: all twins denied emphatically any sexual relationship with their co-twin. It did happen occasionally that a twin introduced his co-twin to another sex partner. As a rule, however, homosexual behavior is reported to have developed independently. Concordance of MZ twins was much more specific than indicated in Table 8.10 by concordance figures. Most pairs played similarly active or passive roles in homosexual relationships. Later, other authors described discordant MZ pairs, and the literature is now giving the impression that concordance for homosexuality in MZ twins is higher than in DZ twins but by no means complete [2073, 2068 a].

Neuroses. A review of the many studies on twins' families and adoptees in which aspects of neurotic or otherwise abnormal behavior were investigated will not be attempted [2003, 2066, 2081, 2083, 2153, 2187, 2199, 2204]. Especially interesting are intensive parallel studies on MZ twins with psychological and psychoanalytic methods - in cases in which either one or both twins had shown neurotic behavior [2068 a]. Concordance as to neurotic symptoms was, in general, higher among MZ as compared with DZ

[5] Psychopathy defined according to the American definition of the term which is more restrictive than the European definition, comprising mainly antisocial behavior.

twins. Within these limits, however, development of abnormal behavior, and especially the specific symptomatology of a neurosis, depended on the familial situation that might be quite different even for MZ.

The most thoroughly studied examples of the influence of different social interaction patterns within a family on personal development and mental disease in genetically identical individuals are a group of monozygotic quadruplets with schizophrenia that will be discussed here. Other genetic aspects of psychoses will be deferred to Sect. 8.2.3.7.

The Genain Quadruplets [2183].[6] The monozygotic quadruplets Nora, Iris, Myra, and Hester had a somewhat primitive, narcistic, and in general severely psychopathic father and an overprotective mother. The father managed only with difficulty to earn enough money for the family. The quadruplets grew up under socially very restricted conditions; the parents did everything to isolate the four girls from other children, with the result that they could make no friends at school; they shunned their age mates and were shunned by them. Right from infancy, they developed differently. Hester, for example, was and remained the weakest. This was also reflected in their school grades. Here, Nora was the best, Myra the second, and Hester dropped out after the 11th grade. After they had left high school, one girl after the other broke down and developed schizophrenia. Finally, Myra recovered almost completely, Nora managed to achieve a marginal adjustment, Iris was still hospitalized at the end of the study (her clinical condition fluctuated), and Hester was hospitalized with a "deteriorated hebephrenic psychosis.". Hence, this genetically identical group was concordant for schizophrenia but discordant for course and outcome.

The girls were repeatedly and carefully observed at the National Institute of Mental Health, Bethesda, Md., where the parents could also be observed in a prolonged study. The differences in outcome could plausibly be related to environmental factors that were known from other studies to influence manifestation. There was, for example, an obvious relationship between the social interactions of the twins with each other and with their parents and the course of the disease. For example, the weakness of Hester and of Iris had resulted in formation of two groups of two each, the good ones, Nora and Myra, and the bad ones, Iris and Hester. This group formation could be followed through the life histories and was still present in the hospital, as evidenced by the number and kinds of contacts the mother established with her daughters during the observation period (Fig. 8.16), as well as by the number and intensity of contacts between the quadruplets themselves (Fig. 8.17 and 8.18). The mother had the most positive contacts (equalitarian; expression of affection) with the most self-reliable daughter, Myra, whereas negative contacts, such as ignorance, punitiveness, use of fear to control behavior, occurred most often with the sickest daughter, Hester. The intensity of contacts between the quadruplets themselves shows the same pattern: the two stronger ones communicated with each other, leaving the weaker to themselves.

Other aspects in the development of the four girls are their identification with father and mother and their different abilities to establish contacts with the outer world. The documentation, which is infinitely richer than can be shown here, demonstrates how the common genetic disposition of the four girls was stressed by sometimes similar but predominantly different influences from the environment, especially the relations with parents, sibs, and - less importantly - other persons, to produce mental illness of the same general kind but with different course and outcome. This case history may help the reader to visualize specific aspects of human behavior subsumed under the abstract terms used to describe genetic and environmental

[6] This section is not absolutely necessary for understanding the entire book.

Table 8.10. Homosexuality in mono- and dizygotic twins. (Kallmann, 1953 [2098])

No.	Concordance (severity 5-6)	Concordance (severity 1-4)	Discordance	Concordance
MZ: 44	31	13		100%
DZ: 51	2	11	38	25%

Note that ascertainment only included homosexual twin index cases of severity 5-6. See text for definition of severity.

Table 8.11. I.Q. in patients with abnormal numbers of sex chromosomes

I.Q.	XXY	XXXY	XXXXY	XYY	XXYY	XXX	XXXX XXXXX
- 19	–	1	4	–	–	1	–
20- 39	–	1	16	–	4	4	1
40- 59	5	5	6	1	12	16	7
60- 79	32	5	3	10	8	14	5
80- 99	23	–	–	8	2	1	2
100-119	12	–	–	1	1	–	1
120-	–	–	–	–	–	–	–
Sum total	72	12	29	20	27	36	16

Source: Moor (1967), Pena (1974), Lenz (1976) [121].

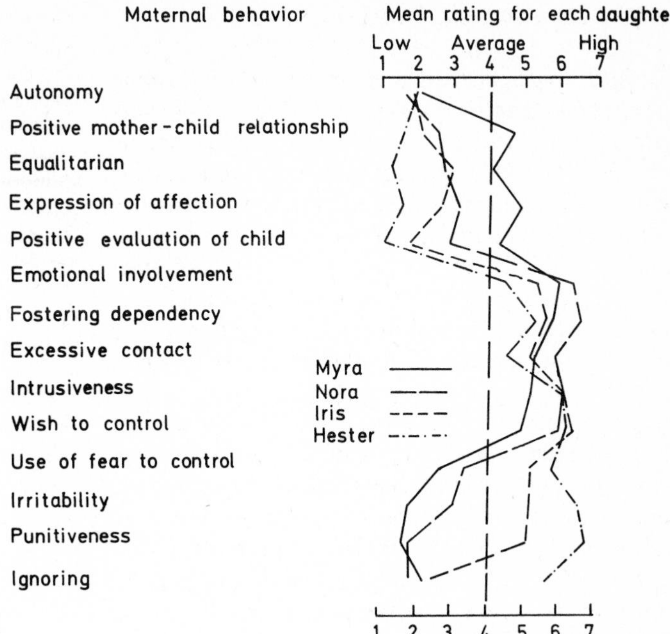

Fig. 8.16. Number and kind of social contacts between the mother and the Genain quadruplets. (Rosenthal et al., 1963 [2183])

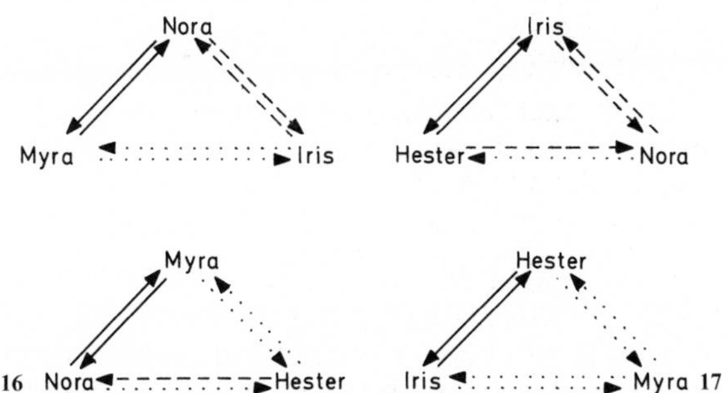

Figs. 8.17, 8.18. Intensity of social contacts between the Genain quadruplets. (Rosenthal et al. 1963 [2183])

variance or interaction between genotype and environment.

8.2.2 Chromosome Aberrations and Psychological Abnormalities

The studies discussed in Sect. 8.2.1 were carried out using classic methods of comparison between twins and other relatives. In such investigations, genetic and environmental influences can only be separated with difficulty, and never unambiguously, even if twins reared apart or adoptees are examined. The genetic analysis remains at the phenotypic-biometric level; no individual gene action can be identified and the biologic mechanisms of the genetic variability that causes behavioral variation remains unexplored. This approach has been compared with the attempt to elucidate the way a clock works by the statistics of the movements of its hands [2039]. To solve the riddle of the clock's mechanism, one must open it and examine its clockwork. How can this be done?

Human Chromosome Aberrations and Behavior: Possibilities and Limitations. Patients with chromosome aberrations usually show, along with many other findings (Sect. 2.2.2), behavioral abnormalities, which may be mild or severe and may affect all carriers of an aberration or only some of them. For our analysis they offer the unique opportunity to relate the behavioral abnormality to an independently ascertained, undisputable, and clearly defined genetic cause. Unfortunately, however, opportunities for a more incisive analysis are exhausted with this statement at the present state of our knowledge. Chromosome aberrations influence embryonic development in multiple and ill-defined ways. These influences cannot be related to functional anomalies of single genes; disturbances of regulatory processes are the most likely explanation. Unfortunately, this explanation says little since regulatory mechanisms remain largely unknown (Sect. 4.7). There is hope that, as soon as regulatory mechanisms and the ways in which they are disturbed by chromosome aberrations are better known, we may better understand behavioral abnormalities caused by chromosome aberrations [2205]. Meanwhile, much work can be done at intermediate levels of analysis. For example, the influence of brain morphology, of endocrine abnormality, and of more indirect social and psychologic influences on behavior can be examined.

8.2.2.1 Autosomal Aberrations

Down's Syndrome. Most unbalanced autosomal aberrations lead to multiple and severe malformations (Sect. 2.2.2) that also affect the brain, causing severe mental deficiency. Down's syndrome is the most common of these abnormalities observed after birth. The range of I.Q. of Down's patients is between 20-60. The median I.Q. is 40-50. Many such persons can be educated to read and write, and a few can learn to make a marginal social adjustment. The vast majority are unable to function on their own in society.

In general, there seems to be no difference between the intelligence test profiles of Down's syndrome cases and of other individuals with a comparable degree of mental retardation. Down's syndrome does not lead to any specific defects, for example, in linguistic abilities, psychomotor coordination, perceptual speed, or spatial ability [2027]. Penrose (1962) [2157] described the patients as "cheerful and friendly personalities. Their capacities for imitation and their memories for people, for music and for complex situations may be found to range far beyond their other abilities." Adult patients, however, are sometimes rather morose. Intellectual abilities tend to decrease in older patients [2106]. One - a son of teachers - has even written an autobiography [2077], which is interesting for our understanding of the inner world of such children. Some situations are described fairly vividly and even with a certain sense of humor, but even the slightest attempts at abstraction are lacking. The father in his two social roles as father and as teacher is not clearly identified as the same person.

Social Problems. Patients with Down's syndrome are now surviving longer than they used to when infections could not be controlled. Many of them reach adulthood, and they should be educated to achieve a certain degree of self-reliance within protected surroundings. When given the chance to select a partner of the other sex, they may even develop stable and satisfying relationships, usually within an institutional framework.

8.2.2.2 Aberrations of the X Chromosome

Numeric and structural aberrations of the X and Y chromosomes generally lead to much milder disturbances in embryonic development than do autosomal aberrations (Sect. 2.2.3). Many somatic abnormalities found in these syndromes are related to abnormal sexual development. The psychologic disturbances are less overwhelmingly severe and may be sometimes specific.

Klinefelter's Syndrome. The standard karyotype of the Klinefelter syndrome is XXY; other karyotypes as well as mosaics occur (Sect. 2.2.3.1). Adult patients are on the average about 6 cm taller than their normal brothers; especially their legs are longer in relation to overall stature. This growth abnormality can already be observed during childhood; at the age of puberty, the subnormal sexual development becomes obvious: the testicles are small, and there is aspermy so that these patients are infertile. They are, however, capable of sexual intercourse. Many of their psychological symptoms can be explained by their diminished androgen production which is normally required for the expression of male-specific psychologic development.

The patients may show slightly reduced intelligence with special difficulties in the ability to learn how to read and write. Their vitality and ability to establish social contacts is often reduced. Table 8.11 shows some data on I.Q. values in patients with Klinefelter's syndrome as compared with other anomalies of the X and Y chromosomes. The mean is shifted below the norm; data from unselected probands range between 88-96. However, I.Q. values well above average are not rare. On the other hand, Klinefelter's syndrome has been found more often in a series of mildly mentally subnormals. The literature reports do not point unanimously to a well-defined, specific defect of mental abilities; according to one report, 7 of 30 patients showed severe dyslexia, and in three others, residues of such dyslexia were found [235].

School problems are more frequent than expected from intellectual ability and seem to be caused by behavioral problems. Adult patients usually hold unskilled jobs; success in higher professional careers has been reported but seems to be the exception.

Psychiatric reports on the personalities of Klinefelter patients point to a variety of deviations from the norm [235]. It has even been asserted that all patients are abnormal in one way or the other. Their behavior has been described as passive-aggressive, withdrawn, self-contented, and mother-dependent, but they have also been characterized as quiet and law-abiding citizens who do not attract attention. In 32 juvenile and 52 adult patients who were not inmates of psychiatric institutions, their development can be described as follows: Erections, ejaculations, and masturbation occur; interest in girls is sometimes present. Sexual daydreaming seems to be fairly frequent, whereas actual sexual relationships are rare. The libido is reduced in most cases and may be lacking completely, but in some patients it is normal. Some patients have no erections and no ejaculations; other do have intercourse but in most cases only rarely. If sexual activity is present, it usually fades out early - at about the age of 40. On the other hand, fairly many patients live in stable marriages.

The typical Klinefelter patient seems to be an unhappy person who has difficulties in coping with the normal demands of life in school and in the job market; his ability to establish normal social and sexual relationships is reduced. Sometimes, he revolts against this situation with aggressive outbreaks. With this background, it is not surprising that the patients are more often found among law-offenders than in the general population [2140, 2144, 2219]. Table 8.12 shows their frequencies in two series of law-offenders [2247]. Criminal patterns and activities are not specific; however, white-collar crimes are almost completely lacking, probably due to reduced average intelligence. In a recent study with small but unbiased material, Klinefelter patients turned out to be no more frequent among law-offenders than XY men of the same intelligence and educational achievement [2247].

Klinefelter Variants. Variants of Klinefelter's syndrome with more than two X chromosomes or with more than one Y have been discussed in Sect. 2.2.3.2. Many of the patients described in the literature were ascertained in institutions for the mentally retarded. Table 8.11 lists some I.Q. data: the severity of mental deficiency increases with the number of X chromosomes. The data on psychiatric results in mosaic Klinefelter patients are scanty. There seem to be no conspicuous differences compared with the standard karyotype.

Therapy and Prevention. Therapy with male hormones leads to increased virilization and to improvement of libido [235]. Part of the favorable psychological effect may be due to this virilization, but an additional, direct effect on brain function is possible. Symptoms that can be described as "climacteric," such as moodiness, nervousness, and psychasthenia, seem to disappear. However, in some cases, hormonal therapy seems to enhance restlessness and aggressive tendencies. When a personality has developed over many years under so abnormal internal conditions, a simple series of hormone injections cannot normalize long-established psychological and social behavioral patterns. Psychotherapy coupled with treatment may be useful in many cases. If a psychiatrist or psychotherapist comes across a patient with the described psychological syndrome, the possibility of Klinefelter's syndrome should be considered and ruled out by a chromosome examination.

Turner's Syndrome. Clinical and chromosomal results have been described in Sect. 2.2.3.2; the standard type shows the karyotype XO; however, many mosaics and structural variations are observed.
The distribution of the I.Q. in 72 patients from the literature is shown in Table 8.13 as an example. These patients are often successful in school [235]. In one study of 126 patients, 2 achieved a university degree, 10 passed a European school comparable to an American high school, 93 went to elementary school, 21 of them with difficulties, and 21 attended a special school for the mentally backward. The discrepancy between measured intelligence and success at school may have various reasons:

a) The I.Q. distribution in Table 8.13 may be biased if, in some cases, the impression of a lower intelligence led to referral for psychological examination.
b) Turner patients are dainty little girls with pleasant personalities who are well-behaved and industrious. This may induce teachers to overestimate their intelligence and school achievement. This bias should be contrasted to the image of a Klinefelter boy with the same I.Q.: a somewhat flabby boy of above-average stature with little social contact in the class room, somewhat morose, apparently not working very hard, and tending to occasional aggressive outbursts.
c) Moreover, the intellectual abilities of Turner patients are unequally developed as discussed below.

Table 8.12. Prevalence of the XXY Klinefelter's syndrome among immates of institutions for criminals, juvenile penitentiaries, etc.

Author	Kind of institution	Prevalence of XXY
Tsuboi (1970) [2219]	Two institutions for criminal psychopaths (Denmark)	5/480 ≈ 1%
Murken (1973) [2140]	Four institutions for criminals and aggressive psychopaths (Germany)	7/728 ≈ 1%

Note that the frequency is seven to ten times higher than in the general population (see Table 5.1).

Table 8.13. Distribution of I.Q. in 72 patients with Turner's syndrome. (Züblin, 1969 [235])

I.Q.	<51	51-70	71-90	91-110	111-130
No. of patients	3	19	28	14	8

Intelligence Defect in Turner's Syndrome. Shaffer (1962) [2193] observed in 20 patients with Turner's syndrome (age between 5.8 and 30.9 years; mean 15.9) a discrepancy between verbal and performance I.Q. in the Wechsler Intelligence Scales for Adults and Children, respectively (WAIS; WISC). Verbal abilities were within the normal range while performance was impaired. At a somewhat more anecdotal level,

There were several bits of information to suggest that the test deficiencies ... had their counterparts in every-day behavior. Almost without exception, the subjects reported they had great difficulties in understanding mathematics, especially algebra. One girl had an extremely poor sense of direction and frequently became lost. Another went through an elaborate ritual when putting away kitchen utensils since any departure from this procedure left her thoroughly confused.

Shaffer pointed to the similarity of these results with those often found in cases with certain types of organic brain damage.

This specific defect was confirmed and characterized more precisely [1995, 2133, 2134, 235]. The items in which the patients score especially low are concerned with perceptual organization (block design and object assembly, Fig. 8.19). Dyscalculia may also be present but is usually not quite so severe. The space-form blindness also leads to difficulties with right-left directional discrimination. The patients did very poorly, for example, in a road map test that requires orientation to right and left (Fig. 8.20) [1996]. Money [2133] has suggested a functional defect of the parietal lobe, perhaps involving the nondominant more than the dominant hemisphere.

Other symptoms described frequently are a general retardation of psychological development with psychic infantilism also among adults. Often, interests are confined to the most banal matters; the patients are dependent in their social relationships. Among the more intelligent patients, a compensatory activity, especially in sports and school, has been described. There seems to be little or no antisocial behavior in contrast to XXY and XYY patients. The sexual drive is usually underdeveloped; however, patients are often married. By and large, they seem to suffer more from their small stature than from their sexual underdevelopment. Estrogen treatment usually leads to a better development of secondary sexual characters, to menstrual spotting, and in some cases to improved sexual responsiveness. It does not influence the space-form blindness; other effects of treatment on personality are doubtful and, in any case, irregular.

Triple-X Syndrome. This syndrome has been described in Sect. 2.2.3.1. Many women with the karyotype XXX are normally developed and have children. Twelve of 119 patients described in the literature had epileptic seizures; in a home for epileptic patients, 2 of 209 patients had an XXX karyotype [2075]. Their intelligence tends to be well below average (Table 8.11); they are more frequent in institutions for the mentally retarded than in the general population. Somatic symptoms lead to medical examination only occasionally, and many of the described patients have been ascertained by screening populations of mental institutions. It cannot be decided definitely how much the XXX karyotype enhances the liability to psychoses but some authors estimate that the rate of schizophrenia-like psychoses might be threefold increased [2164]. In recent years, chromosome studies on newborns have been carried out in many countries showing a frequency of about 1:1000 female births. A prospective study of the XXX individuals is urgently needed.

EEG Abnormalities. Epilepsy seems to be more frequent in XXX women than in the general population. The same may possibly be true for XXY men. Abnormalities of the EEG have often been described in some patients with all three syndromes

Fig. 8.19. Block design; an item of the Wechsler intelligence scale

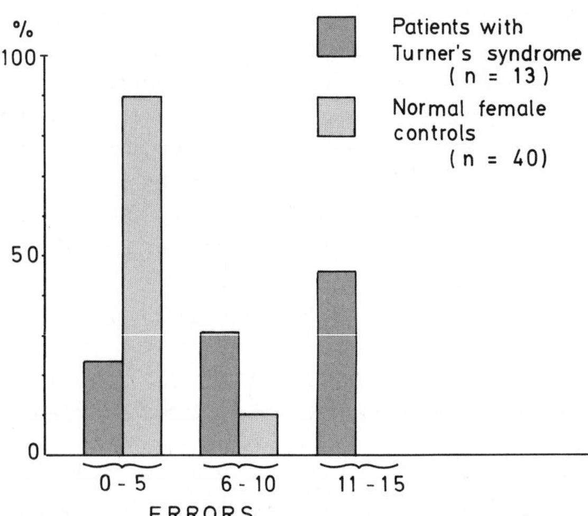

Fig. 8.20. Road map test; number of errors in a series of 13 Turner patients *(left)* in comparison with 40 nurses *(right)* [2193]

(XXX, XXY, and XO). The EEG in the same patient may vary at different times between normal and abnormal. The abnormalities are usually of a diffuse type. Similar EEGs are often found in individuals with behavioral disorders, especially psychopathy [235]. However, most studies were not carried out on unbiased samples of patients but on patients living in institutions and other biased samples. Therefore, all these descriptions should be regarded as preliminary.

8.2.2.3 XYY Syndrome

Somatic Symptoms. For a description of the XYY syndrome, cf. Sect. 2.2.3. The mean stature of these patients is appreciably taller than in the population from which they come. Many of them show normal sexual development and are fertile. The distribution of I. Q. is shifted to a lower average; some of the patients show average intelligence, but the mean I. Q. ranges between 80–88 (Table 8.12).

Higher Prevalence Among Criminals. The XYY syndrome has become widely known since Jacobs et al. (1965) [394] carried out a survey of patients who were mentally subnormal and under surveillance in a special institution because of "dangerous, violent, or criminal propensities." Among 196 probands, 12 had an abnormal karyotype; seven with XYY and one with XXYY. This frequency was much higher than expected; however, the authors stated that they could not decide whether these men had been institutionalized mainly because of mental subnormality, aggressive behavior, or some combination of these factors. These results were soon confirmed in a number of studies from institutions for mentally subnormal men with behavior problems, especially on examination of especially tall inmates. On the basis of such evidence, it was concluded that their antisocial behavior was caused by the extra Y chromosome and that they were genetically predisposed to criminality. The explanation seemed simple. Normal men are more aggressive than normal women; normal men have one Y chromosome, women do not. Hence, if someone has two Y chromosomes, he will be twice as aggressive as normal men; his aggressivity may fall outside the socially acceptable range, and he may commit acts of violence. He is a "supermale."

At about the same time, R. Speck murdered eight Chicago nurses, and it was concluded on the basis of physical evidence (tall stature, facial acne, mild mental retardation, etc.) that he might have an XYY karyotype. He actually turned out to have a normal XY karyotype [2005].

The "murderer chromosome" was born. Discussions started on "whether society is justified in restricting an XYYs freedom before he violates the law. The XYY individual is a perpetual threat since at any time he may face a situation in which he will be unable to control his behavior" (see 2005). At the same time, during the trial of Daniel Hugon who had murdered an elderly prostitute in Paris, the defense attorney claimed that his client was not legally responsible for his act, as he had an extra Y chromosome. He was given a reduced sentence, presumably because of his anomaly. Other similar cases were soon to follow.

Gradually, however, some pertinent questions were asked: above all, how frequent is the XYY karyotype in the general population of nonconvicts? Studies on incidence among male newborns showed a frequency of around 1:1100, similar to that of Klinefelter's syndrome (Sect. 5.1.2). Even in the absence of reliable prevalence studies among the male adult population – they are still lacking – it was fair to conclude that the prevalence might not be too different from the incidence at birth, i. e., that there was no preferential mortality. This, however, could only mean that the great majority of all XYY men do not get into conflict with the law.

Another question was whether the nature of their crimes revealed a certain pattern and, more specifically, whether acts of violence and – as had been suggested – sexual aggression prevailed. This was not the case: the pattern was quite similar to that of other groups of law-offenders with the same degree of intelligence impairment and, more specifically, to that found in patients with Klinefelter's syndrome who were also more often detected among law-offenders than in the general population [2005, 2140, 2219]. Offenses against property were most frequent; socalled white-collar crimes were lacking, presumably due to the lower average intelligence of the probands. These results detracted from the romantic picture of the savage supermale. This image was completely shattered when somebody asked how XYY men behave when institutionalized. Are they more aggressive than other men detained in the same institutions? They turned out to be, in fact, more agreeable; they had, on the average, better relationships with supervisory personnel [2210]. Many more psychological and psychiatric studies were carried out. While varying in details, they gave an overall picture not usually different from other, chromosomally normal inmates of the same institutions in the same range of intelligence.

All these results suggested alternative explanations for the undisputed higher frequency of XYY probands in institutions for law-offenders, especially those where criminals suffering from mental subnormality, psychopathy, and behavior problems were incancerated.

Intellectual Dysfunction or Simply Stature? There are many studies on convicted and imprisoned law-offenders. Their mean I.Q. is generally lower. Intellectually subnormal people are involved more often in criminal activities, or they run a higher risk to be found out; if they are found out, they cannot hire good lawyers and have a lower chance to escape imprisonment. Is the supposedly higher crime rate of XYY men only a result of their reduced average intelligence?

Support for this contention came from studies that showed convicted XYY offenders to come often but less frequently than other criminals from broken families and from families with a poor socioeconomic background [2118, 2144]. It seemed even possible that their tall stature influenced the rate of detention. A huge and strong man may arouse in many a judge or jury the feeling that it might be safer to detain him in jail.

Studies on Unbiased Samples. To really establish the range of variation in phenotypic manifestation, to determine the intervening variable between the abnormal karyotype and criminal behavior, and to explore the influence of socioeconomic and educational differences, studies on unbiased samples of cases were needed.

a) From chromosome studies in newborn (Sect. 5.1.2.1), a fair number of XYY boys are available. In the long run, careful surveillance of their development should give the most reliable information. From such studies, interesting information is now available that will be discussed below.
b) In theory, chromosome examination and, at the same time, assessment of behavior characteristics in a sufficiently large, unselected population of adult males should help to answer our questions. Considering the incidence at birth (1:1000), such a study would need an extremely large sample size. However, XYY males are prominent by their stature. Therefore, most of them will be found in the upper range of stature, thereby reducing the screening effort.

Such a study has been carried out in Denmark [2247]. The population from which the sample was drawn consisted of all male Danish citizens born between January 1, 1944 and December 31, 1947 to women who were residents of Copenhagen: 31437 males. Their present addresses were ascertained by the Danish National Register. These men were at least 26 years old when the study began; the military draft boards provided information on stature of 28884 men; most of the others had died; a few had emigrated or were not available for other reasons.

A cutoff point of 184 cm (≈6 ft, ½ inch) was used in establishing the tall group in which the search for sex chromosome anomalies was to be conducted; the resulting sample consisted of 4591 men. After preparation of the general public through the mass media, these men were contacted, and if possible, a buccal smear for X and Y chromatin determination (Sect. 2.1.2.3) and blood for karyotyping including Q banding was taken. The investigators succeeded in securing co-operation from 4139 men (90.8% of the sample); the rest declined co-operation, had emigrated, or could not be found.

For the entire study group, limited but reliable data on the pertinent behavioral characteristics were available: Convictions for criminal offenses were documented in police registries; the military draft boards give all draftees a screening test for cognitive abilities, the so-called BPP. This test covers only a limited number of cognitive dimensions. However, the educational level was used as an additional index for intellectual achievement. In Denmark, examinations are given at the end of the 9th, 10th, and 13th years of schooling. The index was constructed simply from the number of examinations a man had passed, ranging from none to three.

Results of the Study. In the population sample, 12 XYYs and 16 XXYs were identified. A search in the penal registers showed that 5 of the 12 XYYs (41.7%), 3 of the XXYs (18.8%), and 9.3% of all the XY men had been convicted of one or more criminal offenses (Table 8.14). The difference between XYY and XY is significant; the difference between XXY and XY is not. Both groups with abnormal karyotypes (XYY and XXY) showed a marked reduction in the indices for intellectual achievement - BPP and educational index - despite the fact that the socioeconomic status of their parents was identical to that in the control sample.

Three hypotheses on intervening variables between abnormal karyotype and criminality were examined: Is it their aggressivity that brings these men into conflict with the law, is it their reduced intelligence, or simply their tall stature? A first answer was provided by the nature of the crimes of the five convicted XYY men. Only one of them had committed an act of aggression against other people. The other offenses were, for example, thefts, arson, burglary, and procuring for prostitution. With one exception (imprisonment for somewhat less than 1 year), all penalties were mild, indicating small offenses. The one act of aggression was an isolated occurrence in a long criminal career of the only man who could be regarded as a "typical" criminal. The range of offenses of the three XXY men was similar to the control population. Thus, in both XXY and XYY men,

Table 8.14. Criminality rates and mean values for background variables of XY, XYY, and XXY. Significance level pertains to comparison with the control group (XY) using a two-sided test. For criminality rate, an exact binomial test was used; for all other variables a *t* test was used. (Witkin et al., 1976 [2247])

Group	Criminality		Army selection test			Educational index[a]			Parental SES[b]		
	Rate (%)	N	Mean	SD	N	Mean	SD	N	Mean	SD	N
XY	9.3	4,096	43.7	11.4	3,759	1.55	1.18	4,084	3.7	1.7	4,058
XYY	41.7[c]	12	29.7[d]	8.2	12	0.58[c]	0.86	12	3.2	1.5	12
XXY	18.8	16	28.4[d]	14.1	16	0.81[e]	0.88	16	4.2	1.8	16

[a] Refers to educational achievement by type of school attended successfully (maximum: 3).
[b] Socioeconomic Status.
[c] $P<0.01$.
[d] $P<0.001$.
[e] $P<0.05$.
S.D., Standard deviation; *N*, Number of men.

the nature of the crimes committed resembled that of the control population.

Together with the other results mentioned, these data, in spite of the small sample size of XYY men, permit the conclusion that the aggression hypothesis is most probably wrong. Is the higher liability of the probands to become law-offenders caused by their lower intellectual capability?

The data showed, first, that normal XY men with no records of crimes had a mean intelligence score (not to be confounded with the I.Q.) of 44.5, whereas those with one or more such crimes had a score of 35.5. The educational index showed a similar trend for criminality (1.62 for noncriminals, 0.74 for criminals). Can the criminality rate of XYY and XXY men be accounted for completely by their intellectual dysfunction? To examine this question, their criminal rates were compared with that of the control group after statistical adjustment of the background variables, such as intellectual functioning, socioeconomic conditions of the parents, and stature. The analysis consisted of three stages. The first step established the probability that an XY male with a particular set of values for the background variables is a criminal; the second step established for each XYY or XXY men the probability that he could be a criminal if he were an XY man with the same background-variable values; in a third step, the observed frequency of criminals in the proband group was compared with the frequency predicted in the second step. This analysis gave the following results:

a) Background variables, i.e., lower intelligence and the socioeconomic status of the parents, account for some of the difference in the criminality between the XYY and XY groups. However, an elevated crime rate among the XYYs remains even after these adjustments have been made. It is significant at the 5% level.
b) XXY men are not significantly different in criminality from the XY control group after adjustment for background variables.

Hence, in the XYY state, intellectual functioning seems to be an important mediating variable between the abnormal karyotype and the above-average liability to criminal behavior. However, this behavior is not explained completely in this way. Either, the indices used (BPP test and educational index) only gave an incomplete account of the cognitive defect, or an additional "personality" factor was involved. We consider the latter hypothesis more likely; psychological examination of unselected XYY probands will help to answer this question. Abnormal EEG results in an unselected series of XYY patients suggest an "organic" component.

The aggression hypothesis can now be rejected with confidence; the hypothesis that stature enhances the risk of XYY men to be convicted could also not be confirmed. XYY probands without a criminal record even tended to be a little bit taller than those with such a record.

This investigation together with many other studies led to the following conclusions:

a) There can be little doubt that men with the chromosome constitution XYY run a higher relative risk than normal XY men to show antisocial behavior and to get into conflict with the law.
b) Part of this risk may be traced back to their impaired intellectual function.
c) However, many of them seem to be afflicted with an additional, more specific personality disturbance that may lead to difficulties in adjustment to the social environment. Probably, conviction for criminal acts is only the "tip of the iceberg;" social difficulties not leading to conflict with the law may be much more widespread.

The conclusion that XYY men are different psychologically is confirmed by a psychological double-blind study on seven randomly chosen young XYY men in comparison with 28 XY men from the same French population [2148]. It was possible to distinguish the XYY men from the XY controls on the basis of psychological results. Examination of 14 XYY males (i. e. including some other cases in addition to the randomly chosen ones) by a number of psychological tests (Rorschach; Thematic Apperception Test; interview) uncovered increased impulsivity after emotional stimulation, predominant responses to immediate gratification, and lack of emotional control. In some of the probands, this lability was checked by rigid self-control. Defense mechanisms against anxiety were poor, and the concept of self was weak, easily fragmented, and often infantile. Such personality characteristics could, indeed, constitute an increased risk for antisocial behavior. Contrary to most other studies, including the Danish study described above, the average I. Q. in this series was not subnormal.

EEG studies on eight XYY men (from the Danish study described above) revealed that slowing of the α-frequency in the resting (awake) EEG appears to be a fairly constant finding in this group. The *maximum* α-frequency in these subjects was *lower* than the *minimum* frequency in 16 XY controls [2234 a]. This result, together with others, led to the (admittedly very general) hypothesis that "neural factors" contribute to behavioral difficulties in some XYY men [2076].

Social and Therapeutic Consequences. The evidence shows that the legal consequences for prevention of crimes by XYY men considered in the heydays of the aggression hypothesis would have no basis at all. Still, problems remain. If the XYY status is discovered in a newborn study, should parents be informed? In view of the publicity the condition has gained, could such information have the effect of a self-fulfilling prophecy in that parents would treat their boy differently, and could this enhance his tendency for deviating behavior? We are in favor of providing all information; however, great care is needed in conveying the facts to the parents in a form that causes as little embarrassment as possible and, above all, no damage. The parents should understand that their boy might possibly need some more special help during his education than an XY boy but that, given a stable environment and the same amount of parental protection that other boys enjoy, normal social adjustment will be the most likely outcome. No effective somatic therapy is known; the blood androgen levels have normal mean values with a little higher than normal variance.

An ongoing project in Denmark shows how many of these problems could be solved [2146, 2147]. This study started from the 11 148 consecutive newborns in whom chromosomes were studied (Sect. 5.1.2.1). In this series, 25 aneuploidies involving the sex chromosomes were discovered. As of 1981 [2147], these patients were followed up to the age of 7-10 years, examined four times during this period, and compared with their 36 sibs and with two larger groups of Danish children. Among them were 8 with XXY Klinefelter's syndrome (including 2 mosaics); 5 with XYY (1 mosaic); 7 girls with triple-X, and 4 girls and 1 boy with a 45,X cell line. The parents of 23 of these children were informed of the anomaly, but not until the children were 1-2 years old; the parents of the two remaining children did not receive this information until the children were already 3 and 5 years old. Motor coordination, intelligence, psychological aspects of brain dysfunction, school achievement, and behavioral problems received special attention. The children showed slight but definite deviations from normal behavior in early life. For example, the mothers of each of 16 children reported that during the first 12-18 months the child had been quiet and easy to take care of, and had slept a lot - much more than the sibs or other children. During later development motor coordination was found to be poor in many of the children with additional X and Y chromosomes, especially in XXX girls. In the opinion of the authors, "the clumsiness in many of the children ... will most probably tend to worsen their tendency to isolate themselves ... it would be desirable to stimulate such children to a better gross motor function by gymnastics and other types of sports. The decrease of fine motor coordination might cause writing difficulties ..., and special teacher attention should also be paid to this aspect."

A decrease in *verbal* intelligence level was observed in triple X girls, in XXY boys and, to a lesser degree, also in XYY boys. This delay in intellectual development was reflected in diminished school achievement; in addition, the teachers reported lower levels of activity, endurance, and self-confidence, among other things. In conclusion, children with extra X or Y chromosomes need special stimulation, support, and teaching efforts when learning to read and write. "The children should be stimulated to group activities inside as well as outside school in order to overcome problems in their relationship with other children. This may also help to increase their self-confidence, drive, and activity and stimulate them to be more independent." Androgen therapy in XXY boys from the age of 12 onwards was recommended. However, therapeutic measures, and especially a positive attitude of parents, were found to be helpful; there were several examples of passive XXX or XXY children with delayed development who showed marked improvement after appropriate treatment (see also [2170; 2171]. This information should be given to prospective parents when a sex chromosomal aberration is found by prenatal examination and the couple considers pregnancy termination.

In the few XO girls studied no major delay in general development was observed; but the well-known weakness in space perception (see above) was found in psychological tests carried out even in childhood. In another series of 24 patients between ages 9 months to 18 years this was again confirmed; in addition, mild feeding problems and sleeping difficulties and, in some cases, slightly retarded motor development were described. Social skills, however, were excellent [2018].

The results of the Danish studies on the psychological development of children with gonosomal aneuploidies (XXY, XXX, XYY) are confirmed in principle by investigations in other series, e.g., the results from Edinburgh and Colorado [2174–2176].

Inevitably, with the widespread use of antenatal diagnostics, XYY as well as XXY and XXX karyotypes will be discovered by amniocentesis. Parents usually should be fully informed of the findings. The implications of the anomaly should be fully explored. The option of abortion as a possible choice needs careful discussion; such genetic counseling should be nondirective and the decision should be left to the parents. On the other hand, the attitude and feelings of the genetic counselor regarding the impact of these states are likely to differ depending upon individual interpretations of the facts cited above. A completely objective and neutral stance is therefore difficult and practically not achievable. Some parents at the current state of knowledge would not hesitate to proceed with the pregnancy while others would prefer an abortion.

What Can We Learn from the XYY Story About the Attitude of Scientists in the Face of a Problem of Great Public Concern? The surprisingly high incidence of XYY men in certain groups of convicts coincided with a growing public concern about increasing violence. In this situation, some scientists reacted almost as irrationally as representatives of public opinion - mainly the mass media. Premature conclusions and sweeping generalizations were made on the basis of results that obviously came from biased data. The principal reason for such misinterpretation might have been that cytogeneticists who primarily worked on the XYY problem were influenced by their experience in clinical cytogenetics of a fairly straightforward relationship between cause - an abnormality of the karyotype - and effect - a certain phenotype. These workers were simply not sensitized to the careful consideration of interaction between different influences that shape psychological development and fate of a human being and are usually studied by behavioral scientists, such as psychologists, social anthropologists, or sociologists. This trend played into the hands of the media who preferred simplistic interpretations to complex explanations.

The discussion on intelligence (Sect. 8.2.1.3) has made it abundantly clear that we do not consider the biometric analysis of complex phenotypic relationships as a suitable approach to analysis of genetic mechanisms. However, once a genetic mechanism has been established with other, more incisive methods, the combined sophistication of social science, epidemiology, and statistics is needed to assess the phenotypic impact of this genetic mechanism.

Chromosome Aberrations and Behavior: Some General Conclusions. The chromosome aberrations, especially those involving X and Y chromosomes, provide a model to show how genetic variability and environment may interact in producing a psychological phenotype and which intervening variables should mainly be considered: abnormalities in brain physiology or biochemistry and in the endocrine system. The example of the sex chromosome aberrations has also indicated a research strategy: first identify a variant genotype, then explore its influence on the phenotype, considering concomitantly the intra- and interindividual differences of the environment. This strategy is opposite to the usual approach, which starts with the phenotype. However, it has already led to success in the genetic analysis of common diseases (Sect. 3.7). On the other hand, the disturbance in embryonic development and physiology caused by chromosome aberrations is poorly understood. We would be happier if this strategy would be feasible for genetic variability at single gene loci with known physiological mechanisms.

Brain Morphology in Chromosomal Aberrations. Since chromosomal aberrations often lead to reduced intellectual performance and to increased behavioral difficulties, and since the EEG in these disorders points to anomalies in brain development and maturation, we would expect morphological investigations to provide evidence regarding the mechanisms by which such aberrations impair brain function. The literature, however, is scanty, even though increasing numbers of affected fetuses have become available [2064] following abortion of affected fetuses diagnosed by prenatal diagnosis. In a series of 274 fetuses and children with trisomies 13, 18, and 21 and some other aberrations, gross neuropathological changes were not detectable in more than two thirds of the cases. The most common findings were holoprosencephaly/arhinencephaly, mainly in trisomy 13 but also in one case of trisomy 18; corpus callosum defects, mainly found in trisomy 18 but sometimes also in trisomies 13 and 21; and neurons that had not reached their normal location during embryonic development (= heterotypias and microdysplasias), mainly in trisomy 21 but also in cases with other chromosomal syndromes. It should be noted that such anomalies may occur at lower frequencies in fetuses and infants without chromosomal aberration; for example, in about one third of cases with holoprosencephaly, *no* chromosomal aberration is found. Moreover, cerebellar nerve cell heterotopias are frequently found in chromosomally normal newborns, mostly prematures, and disappear with increasing age. Recently, complex dystopias of cortical neurons including misdirection of pyramidal cells by 180° have been described in a case with partial deletion/duplication 17p, and lacking brain convolutions [2004a]. There are some scattered reports on sex chromosomal aberrations; for example on cortical dysplasia, disturbances of cell migration, and delayed cellular maturation in a 45, X case and similar anomalies [2009] or abnormal cortical convolutions of the frontal cerebral cortex [2001] in two XYY patients. The significance of these findings remains unclear until the frequency of such anomalies in individuals *not* suffering from a chromosomal aberration is firmly determined. In our opinion, the application of refined neuropathological methods to fetuses and newborns with chromosomal aberrations - always keeping in mind variability in the normal range - might open new approaches to the study of biological mechanisms affecting behavior-genetic variation. For example, - is there a morphological substrate for the deficiency in space perception in Turner's syndrome? On the basis of other evidence, such as the results of aphasia studies, this has been tentatively localized to the parieto-occipital region of the cerebral cortex.

8.2.3 Suggested New Approaches to Human Behavior Genetics

Most material covered in Sect. 8.2.3 relates to possible applications of genetic concepts to behavioral phenomena. These genetic concepts have been developed in the foregoing sections, especially in Sect. 3 (Formal Genetics) and Sect. 4 (Gene Action). So far, their explanatory power has not been utilized fully for behavior genetic problems. Therefore, we advocate their full application. This does not mean, however, that these concepts will prove to be sufficient in explaining genetic determination of human behavior, its development, and its variability as satisfactorily as, for example, a point mutation in the structural gene of the β hemoglobin locus explains sickle cell anemia.

We discussed in the introduction of this book that the genetic approach to biologic phenomena is "reductionistic:" genetic analysis is regarded successful if an inherited difference can be traced to a difference in the condition of a gene. In most cases - for example, when we want to understand an enzyme deficiency - this approach leads to satisfactory results. Its limitations become obvious when we try to analyze genetic variability in normal embryonic development and its deviations that lead to congenital malformations. Here, a complex system of feedback mechanisms apparently regulates the activity of genes in the various cell groups and in different phases of development. The development of human behavior is never finished and proceeds through our lifetime. The feedback mechanisms involved appear to be more complex than those presumably operative during embryonic somatic development. It is possible that for some mental functions new principles for gene action have appeared during evolution. Thus, our current genetic concepts that have proved to be so successful in other fields may only have limited success in behavior genetic analysis. Still, we have to start with these concepts. This approach is the only way to find out its limits, and it may hint at possible new concepts, which may help to overcome current constraints.

8.2.3.1 Genetic Variability That Could Influence Human Behavior

Few actual data exist in these areas. We describe possible approaches in some detail because we feel that this field should be further explored [1968] (Fig. 8.21).

General Metabolism. With gross disturbance of metabolism, conscience will be blurred, and mental processes will come to a halt. Nongenetic examples are hepatic failure leading to liver coma or renal failure causing uremia. The mechanisms of both conditions are obvious, at least in gross outline. Due to the failure to metabolize or to excrete certain compounds properly, their concentration increases, they interact with the normal metabolic processes in the brain, disturbing these processes, and if the disturbance is severe enough, the resulting intoxication abolishes mental functioning.

Brain function may be altered by intoxication not only when a toxic agent is produced by metabolic processes within the body but also when it is taken up from the outside. The alteration of conscience and feeling tone that follows ingestion of naturally occurring toxic foods has been observed fairly early in development of the human species, as witnessed by the fact that some of these effects have been known in almost all primitive populations. In modern societies, drug addiction is one of the most threatening social problems. Modern medicine, on the other hand, has learned to utilize drug action affecting the nervous system in a great variety of ways - from general anesthesia to psychopharmacologic agents that help to manipulate the psychic state of patients with mental diseases. Before drugs can enter the brain and alter its functions, they have to be metabolized with changes in chemical composition and modification of their action. These metabolic changes are mediated mainly by various enzymes. Enzymes, in turn, may show genetic variability that may influence activity, substrate specificity, and other characteristics (Sect. 4.2). The ensuing differences in drug action as well as in side-effects are the object of pharmacogenetics (Sect. 4.5.1). In principle, genetic variability may be found for enzymes that metabolize compounds of our normal food intake needed for certain metabolic processes. In some cases, the genetic alteration of the enzyme molecule may be so profound that its activity is totally or almost totally abolished. Examples involving develop-

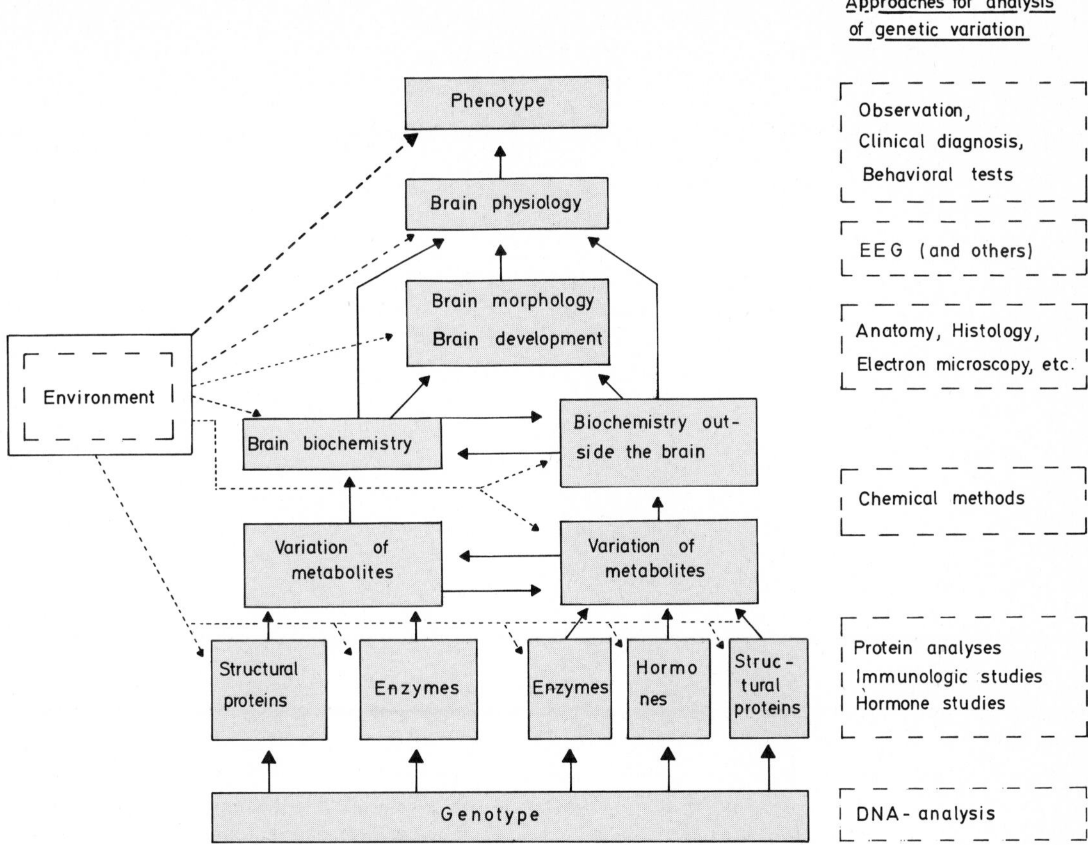

Fig. 8.21. Levels at which genetic variability of brain function could be investigated. ☐ Level at which genetic variability might occur; ⬚ method of investigation

ment and proper function of the brain are the phenylalanine hydroxylase defect in phenylketonuria, which leads to a flooding of the brain with phenylalanine and other abnormal toxic metabolites. The chronic states of ammonia intoxication that ensue from enzyme defects in the urea cycle are another example. Many inborn errors of metabolism are associated with mental retardation by analogous mechanisms.

Variability of Hormones. Many of the genetic regulatory processes in the human organism are mediated by *hormones* (Sect. 4.7). We know a great many qualitative and quantitative genetic defects of hormonal function. Most of them also affect human well-being and behavior. Obvious examples are defects of thyroid function that lead to hypothyroidism with its lack of alertness. Sexual hormones also have a profound effect on embryonic and adolescent development and on the mental state and behavior of the adult. In looking for genetic mechanisms that could influence human behavior and could therefore be screened for appropriate genetic variability, endocrine glands would be major candidates (Fig. 8.22).

Genetic Variability Within the Brain. Genetic differences in quantitative hormone production or in the structure of hormone molecules are only one possible source of genetic variability of hormone function. A second source is their effect on the target organs. Here, *hormone receptors* are important mediators. Some of the genetic defects that are now described as receptor diseases (Sect. 4.6.4) have implications for brain function. There are suggestions that genetic variability of hormone receptors within the brain may affect brain function and, hence, produce genetic variability in behavior.

This is only one example of how genetic variability in structure and function of the brain itself may have an impact on behavioral variability. To find other possible mechanisms, we have to survey all levels at which brain structure and function can be examined - from anatomy and histology to electrophysiology and to the basic biochemical processes that are involved in excitation and inhibition of nerve cells. In all these fields, a huge amount of information is available, but surprisingly little has been done to screen this information for possible genetic variability. In part, this may have technical reasons. Assess-

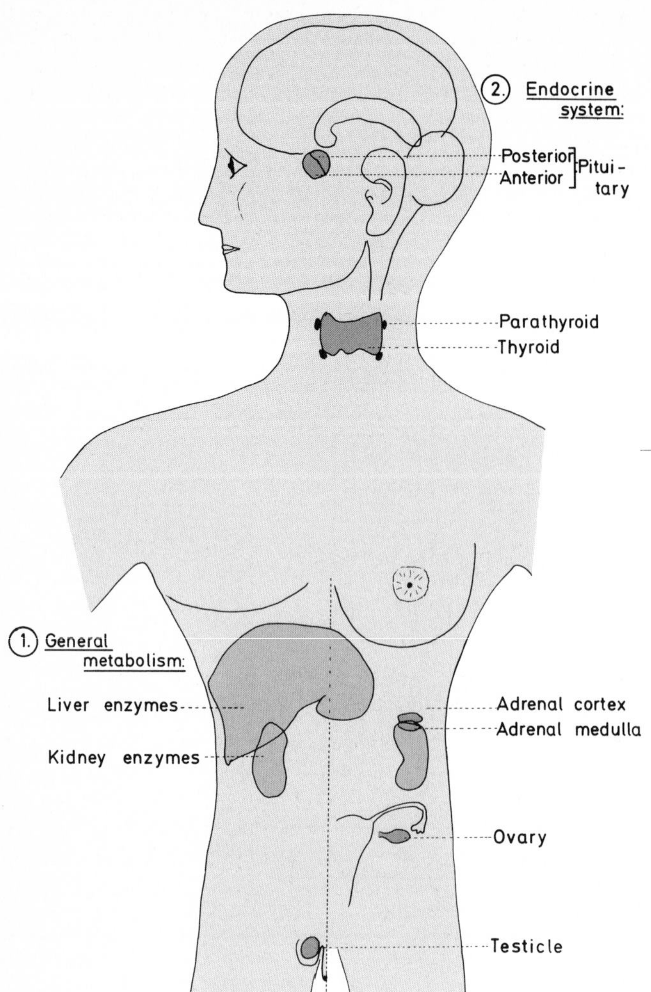

Fig. 8.22. Genetic variability outside the brain that may influence human behavior. In general, metabolic influences of liver and kidney enzymes are especially important. Other influences may come from endocrine glands, such as the anterior or posterior pituitary, the adrenals, gonads, thyroids, and parathyroids

ment of genetic variability normally requires examination of fairly large series of individuals. This is technically easy, for example, for traits affecting blood. If brain samples could be taken as easily as blood samples, we would already know more about genetic variability that influences human behavior. It should be quite clear, however, that even complete understanding of genetic polymorphisms affecting the brain is not likely to explain human behavior since the determinations of behavior is so much more complex than, for example, red cell function. We therefore need to look for alternative strategies.

a) The first possibility is to look for genetic variability in a physiologic parameter that can be measured directly by a noninvasive technique. Here, the genetic analysis of the normal electroencephalogram (EEG) has so far offered the best opportunities [921]. An obvious disadvantage of this approach is that the relationship between gene action and physiologic phenotype is less direct and therefore more subject to distortion by intervening variables than if we would, for example, examine an enzyme.
b) We can look for genetic variability that is expressed not only in the brain but in other tissues as well - preferably various types of blood cells that might express genes active in the brain. This approach is limited by the biochemical differentiation during embryonic development that reduces the possibilities of extrapolation from one tissue to the other. It is not unreasonable to assume that differentiation and functional specialization may be especially well-developed in the central nervous system, making extrapolations from other tissues especially hazardous. Still, this approach has led to some promising results, such as by examination of platelet and serum enzymes in affective disorders. Under some conditions, genes not expressed in an accessible tissue might even be activated in other tissue systems.
c) The genetic variability in its relationship to behavior can be examined in the brains of experimental animals. This approach relies on the well-

known homology in many physiologic processes between humans and other mammals that has its foundation in their common phylogenetic origin. Animals as model systems for genetic analysis in situations where experimentation in humans can obviously not be done for ethical reasons are in widespread use, for example, in research on induced mutation [2142]. For behavioral genetics, the most obvious advantage is direct access to the brain. The obvious disadvantages are the likely existence of species differences and the unique role of the human brain. Hence, this approach can provide us with models for the analysis of mechanisms by which differences in brain physiology may lead to behavioral differences. These models may give hints about where such differences might also be found in humans. In the following discussion, we shall rely heavily on animal experimentation.

Experimental animal populations often are derived from inbred strains and are radically different from human populations in breeding structure. Animals of an inbred strain are genetically identical. They may be compared to human MZ twins. Results achieved in one inbred strain should be compared with those achieved in other strains and even in natural populations.

These approaches are recommended on the assumption that important aspects of genetic variability of brain function that lead to behavioral differences between human beings can be analyzed if the well-established principles of genetics - i.e., the gene concept, gene-determined enzymes and receptors, regulation of gene action, widespread genetic variability at the protein level - are applied to brain function. This is not to deny that understanding of complex intelligent behavior will probably require additional principles of information processing and organization. Figure 8.20 gives an overview of the levels at which genetic variability of brain function may cause corresponding variability of behavior.

8.2.3.2 Genetic Variability Outside the Brain That Influences Human Behavior

Enzyme Defects Leading to Mental Deficiency (Fig. 8.22). Many of the known enzyme defects in humans lead, apart from their other widely different phenotypic effects, to mental deficiency. The human brain is especially susceptible to various kinds of metabolic disturbance (Table 8.15). For our understanding of normal brain function, however, these observations have so far contributed very little. It is, for example, plausible that nerve cells cannot work properly if they are stuffed with insufficiently degraded macromolecules, such as those found in the mucopolysaccharidoses [1240] or mucolipidoses. For understanding of normal function, however, this is not very revealing. In general, the brain's development and function appears to be especially sensitive to changes in the biochemical environment. Many different inborn errors will tend to increase or decrease a variety of biochemical substances that will often cause mental retardation. A comparison of inborn errors that do not produce mental retardation with those that do may provide some clues but will not give us an explanation of normal brain function.

Table 8.15. Types of selected inherited metabolic diseases leading to mental deficiency

Kind of defect	Metabolic disease (examples)
Amino acid metabolism	Phenylketonuria Maple sugar urine disease Hyperammonemias
Carbohydrate metabolism	Galactosemia
Endocrine disorders	Types of cretinism with goitre
Disorders in binding to cofactors (vitamins)	Pyridoxine dependency
Lysosomal diseases	Mucopolysaccharidoses Gangliosidoses

Genetically determined metabolic anomalies may not only lead to mental retardation but, in some cases, may contribute to susceptibility for mental diseases labeled as "schizophrenia." Examples are the adult type of metachromatic leukodystrophy, various types of porphyria, and even Huntington's chorea [2164]. The topic will be taken up again in Sect. 8.2.3.7.

Self Mutilating Behavior in the Lesch-Nyhan Syndrome: Uric Acid. The Lesch-Nyhan syndrome was described in Sect. 4.2.2.6 as caused by a defect of the enzyme hypoxanthine-guanine phosphoribosyltransferase (HPRT) with failure in recycling hypoxanthine into guanine synthesis. Instead, high amounts are converted into uric acid and excreted. Apart from hyperexcitability of the nervous system that leads to hypereflexia, and almost continuous movements, patients suffer from a compulsive tendency to self-destruction. In spite of pain, they bite their fingers and lips and mutilate themselves. This destructive tendency has no counterpart in "normal" or "abnormal" human psychology; still, we know of other types of compulsive neuroses. Some patients, for example, feel compelled to wash their hands over and over again. Moreover, schizophrenic psychoses seem to occur among relatives of patients with compulsion neurosis more often than in the general population [2187]. Possibly, analysis of the specific damage caused by this enzyme defect in the brain could be somewhat instructive regarding neuronal mechanisms of compulsive behavior. Moreover, it would be interesting to explore whether heterozygotes of the HPRT defect show behavioral peculiarities, especially when they reach middle or advanced age.

A positive correlation between blood uric acid level and I.Q. has repeatedly been asserted and is usually backed by an impressive list of outstanding men in history who suffered from gout. If uric acid slightly enhances neuronal excitability, this nonspecific stimulation could conceivably have a positive effect on intelligence and performance. On the other hand, a recent twin study failed to show a correlation between uric acid level and intelligence (Propping, personal communication).

Heterozygotes of Recessive Disorders. Enzyme activity in heterozygotes of inherited metabolic diseases is usually only half the normal activity. Therefore, we would not be surprised if phenotypic abnormalities were detected, at least when the specific metabolic pathways are placed under stress, or possibly with advancing age. However, systematic studies on such heterozygotes, especially on their mental status and performance, are remarkably scanty (see also Sect. 4.2.2.8). The widely scattered evidence has recently been reviewed [1340]. The most comprehensive studies have been performed in phenylketonuria. Despite deficiencies of epidemiological methods used in some of these studies, some conclusions are likely: (a) On average, PKU heterozygotes show a slight reduction of a few points in I.Q. when compared with adequate controls. The verbal part of the test appears to be reduced more than the performance portion. (b) Some studies point to a slightly higher risk of psychotic disorder, the psychoses having a late onset and a benign course and involving depressive symptoms. (c) Some heterozygotes have shown an increased level of cortical irritability as evidenced by the EEG. (d) These slight anomalies could be caused, at least in part, by an increased level of intracellular phenylalanine and tyrosine, since these amino acids were found to be increased in the (more easily accessible) lymphocytes.

Other autosomal recessive diseases for which slight phenotypic deviations have been described include several lipidoses, such as the late infantile form of metachromatic leukodystrophy (25010), globoid cell leukodystrophy (Krabbe) (24520), and a few heterozygotes for Sandhoff's disease (26880), Niemann-Pick disease (25720), and Wolman's disease (27800). Here, the slight deficits were found mainly in the performance I.Q., especially in subtests for space perception. Moreover, scores in personality questionnaires testing for psychosomatic disorder; depression; and emotional lability were increased. Again, accumulation of abnormal metabolites could be a causative factor, since some accumulation of mucolipids is found not only in homozygotes but also in heterozygotes (Fig. 8.23).

In two studies on autosomal recessive microcephaly (25120) from different parts of the world, Canada and the Soviet Union, about one third of normocephalic heterozygotes were found to be mentally subnormal. According to Quazi and Reed [2172], who carried out the Canadian study, such heterozygotes may constitute an appreciable fraction of all individuals suffering from mental subnormality "of unknown origin." According to these authors, homozygotes occur in a frequency of about 1:40000; heterozygote frequency is about 1:100 (Sect. 3.2.1). If one third of these are mentally subnormal, this would mean that about one third of 1% "of the population at large is mentally retarded because they carry one gene for microcephaly. The prevalence of mental retardation in the United States and England using an I.Q. of 69 as a threshold has been estimated to be about three percent" [2157]. Thus about one of nine mentally retarded individuals from the general population could be a heterozygote for the gene of microcephaly. Of course, this estimate can only be regarded as a crude approximation; in other populations, recessive microcephaly appears to be much rarer, and moreover, there may be genetic heterogeneity. In principle, however, this argument is plausible. It could even be generalized: When one third of such heterozygotes are mentally defective, the other two thirds will probably have an I.Q. in the lower normal range. Moreover, about 1 in 50 individuals in our population are heterozygous for PKU (assuming a homozygote frequency of $\approx$ 1:10000), and heterozygotes for other recessive diseases for which slight

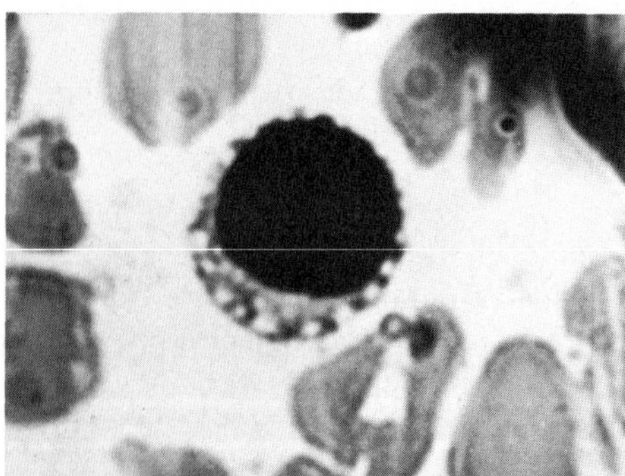

Fig. 8.23. Lymphocyte of a heterozygote for juvenile amaurotic idiocy. The cell shows vacuole-like structures that indicate lipid storage [1060]

reduction in I.Q. have been described do also occur. It should be remembered that many types of autosomal or X-linked recessive mental retardation have been only incompletely characterized so far. If the rather frequent types of X-linked mental retardation are added (keeping in mind the observation that $\approx$ 20% of male and $\approx$ 70% of female carriers of this gene are not frankly retarded but may have a below-average I.Q.) (Sect. 8.2.1.2), it could be argued that a significant proportion of the "normal" genetic variability of I.Q. in the lower range might be caused by heterozygosity of autosomal or X-linked recessive diseases.

8.2.3.3 Hormone Action

How Do Hormones Act? As explained in Sect. 4.7, hormones usually act on special cells that have receptors by which the hormones are bound. This triggers synthesis of specific proteins within these cells; here, cyclic AMP acts as a mediator – as the "second messenger." Recently, some "receptor diseases" have become known; the best-analyzed examples are familial hypercholesterolemia and testicular feminization (Sect. 4.7.5). In the latter condition, a deficiency of androgen receptors causes a female phenotype in individuals having an XY karyotype and therefore testicles.

Hormone receptors are also present on cells of the CNS; their development and function can be influenced at least at three levels: by hormone-induced metabolic processes in other tissues that influence brain function indirectly, by quantitatively or qualitatively abnormal hormone supply, and by individual differences in receptors. For instance, the effect of thyroxin on mental and neural activity, which is so impressive when hypothyroid patients respond to thyroid therapy, probably is an indirect one since thyroxine enhances the basal metabolic rate in all tissues except the brain [120].

Sexual hormones, on the other hand, appear to have a direct influence on brain development [138]. This influence begins already in embryonic age, as witnessed by observations on girls who had been exposed to the masculinizing effect of synthetic progestins administered to the mother as explained below.

Tomboyism in Girls Prenatally Exposed to Masculinizing Compounds [138]. Tomboyism can be defined as follows:

a) Tomboyish girls like to join boys in outdoor sports, especially ball games, and prefer toys that boys usually play with.
b) Self-assertiveness in competition for position in the dominance hierarchy of childhood is strong enough to permit successful rivalry with boys. This, however, does not mean aggressiveness. Such girls often establish a position of leadership among younger children.
c) Tomboyish girls generally prefer slacks and shorts though they do not have an aversion to dressing up in a feminine fashion for special occasions.
d) They usually do not play with dolls. Later in childhood, there is no great enthusiasm for baby-sitting.
e) The tomboyish girl reaches the boyfriend stage later than most of her compeers. Romance and marriage are given second place to achievement and career. There is, however, no tendency to lesbianism.

While such behavior may develop in response to environmental influences, evidence has been brought forward that tomboyish behavior might be primed by hormone action during embryonic development. Progestins are steroids related in chemical structure to androgens that can substitute for progesterone in their pregnancy-preserving function. When first introduced in therapy, their masculinizing effect was unknown. In the 1950s, some mothers gave birth to otherwise normal daughters with masculinization of the clitoris. Such children grew up as girls; the masculinizing effect ceased at birth.

Another group of masculinized girls are those with adrenogenital syndromes mainly due to the autosomal-recessive defect of 21-hydroxylase, one of the enzymes needed for synthesis of cortisol (Sect. 4.2). The condition is treated by cortisol substitution that, by reducing the ACTH production of the pituitary, curbs synthesis of cortisol precursors with androgen-like effect. Hence, exposure is limited to the prenatal period as encountered in daughters of progestin-treated mothers.

In Table 8.16, daughters of progestin treated mothers and patients with adrenogenital syndrome are compared for tomboyish behavior with testicular feminization patients and with Turner patients who do not

Table 8.16. Tomboyism in various genetic conditions among girls. (Money and Ehrhardt, 1972 [138])

	Daughters of progestin-treated mothers	Adrenogenital syndrome	Turner's syndrome	Testicular feminization
Tomboyish behavior	9	11	0	(1?)
Behavior conforming to traditional sex role	1	4	15	9
Sum total	10	15	15	10

develop gonads and are, therefore, not exposed to gonadal hormones during embryonic life. At the time of examination, the oldest girl of the progestin-exposed group was 16; 15 Turner patients ranged between 8 and 16½ years. Tomboyism as described above was present in nine of the ten girls with the progestin-induced syndrome, in 11 of the 15 girls with adrenogenital syndrome, but only in a very few of the Turner and testicular feminization patients. The attitudes of all these girls toward their future roles as married women and mothers were modified accordingly.

Interpretation of these results was corroborated in experiments with female rhesus monkeys who had been artificially androgenized during embryonic development and, during childhood, showed behavior that resembled that of their *male* age mates [138]. These data favor the view that sex-specific brain patterns influence the attitudes of children and adolescents in accepting their sex roles.

Testicular Feminization. The testicular feminization syndrome (31370) is caused by androgen insensitivity. If the brain also lacks functional androgen receptors, one would expect in these individuals "typically female" attitudes toward marriage and motherhood. Such tendencies were, indeed, found in most of the ten patients that could be examined [138]. However, the interpretation is more ambiguous than of data in prenatally masculinized girls. Patients with testicular feminization are almost perfect girls externally and grow up as such; therefore, their identification with the female role could also be explained in a straightforward psychological manner.

After these first reports had appeared, the problem was reexamined repeatedly both in animals and in human beings, not always with the same results (see, e.g., [2030, 2128]. However, it is our impression from the bulk of the literature that prenatal androgen exposure in some way influences brain development.

Homosexuality and Hormones. The twin studies discussed in Sect. 8.2.1.5 suggested a genetic component in homosexual behavior of males. It has been suggested that the distortion of the sexual release scheme of male homosexuals may be caused by an abnormally low androgen effect. This problem was repeatedly investigated by measuring urinary excretion or serum levels of steroid hormones [138]. Different proportions of various selected metabolites of sex hormones were claimed to be detectable in urine; however, the studies were poorly controlled for age, amount of sexual activity, or general health status. Moreover, the degree of variation of these metabolites in the urine of controls is considerable and strongly overlaps that of homosexuals. Even males awaiting estrogen therapy and sex-assignment surgery had normal androgen levels [2028].

These results show that there is no simple relationship between androgen production and homosexuality. But they do not exclude the possibility of minor differences in androgen receptors in their relationship to homosexuality. Here, much work must still be done. New and independent evidence for sex-specific differences in brain function comes from studies with a noninvasive method that enables us to follow up the functional state of the central nervous system – the electroencephalogram (EEG).

8.2.3.4 Brain Physiology: Genetics of the EEG

Direct investigation of genetic variability in brain function in humans is fraught with technical difficulties. Therefore, more indirect approaches are necessary. Several were mentioned: direct examination of brain physiology, investigation of genetic variability outside the brain that might be related to brain function, and animal experimentation. For studies of brain physiology, the electroencephalogram (EEG) has mainly been utilized.

Human EEG. The most important features of the human EEG have been described in Sect. 3.6.1 when criteria for a simple autosomal-dominant mode of inheritance in a continuously varying character were discussed using the low-voltage EEG as an example. It was mentioned that the resting EEG develops during childhood and youth from irregular forms with relatively slow waves to the final pattern, which is reached at about 19 yrs. at the latest. This adult pattern is dominated by α waves (frequencies 8–13/sec) with variable admixture of β waves ($\geqq$14/sec) and ϑ waves (4–7/sec); usually, α waves are most prominent over the occipital region of the brain. Under normal conditions, and in healthy individuals, there is little variation between leads from the same individual taken at different times; variation between individuals, on the other hand, is considerable.

Twin Studies. The EEG is a highly complex trait that varies in many dimensions, e.g., distribution of frequencies and amplitudes in one lead, variations between leads from the various parts of the head surface, and form of the waves. To get an overall impression of the role of genetic factors in this variability, it was reasonable to begin with comparison of MZ and DZ twins. As the EEG shows a characteristic development from infancy to adult age, with varying speed between individuals, twins in their 1st and 2nd decades of life, together with a few young adults, were the most appropriate probands. In the absence of disturbing factors, such as severe fatigue, brain disease such as epilepsy, or brain tumors, and severe metabolic anomalies, the brain wave pattern under standard conditions (relaxed, with closed eyes) is practically completely genetically determined. This conclusion also applies to the speed of brain maturation as evidenced by EEG development [2228].

It could be argued that this concordance was caused by the common environment of the twins. EEG studies carried out on 8 Danish pairs reared apart [2095] and in adult age at the time of their examination showed the same degree of EEG concordance [2095]. This strong similarity is maintained up to old age. Only those EEG abnormalities that are caused by disease such as small strokes might be discordant [2074].

The adult EEG pattern is even identical in twin pairs with quite different emotional histories, for example, when one suffered from a severe neurosis. Organic brain diseases, on the other hand, such as epilepsy, may cause clear-cut and permanent differences even between MZ twins.

As already mentioned, the development of the EEG during childhood shows marked interindividual variability and is determined by genetic factors, as evidenced by twin studies. Psychological maturation also varies between individuals. Moreover, in children regarded by psychologists as immature or showing behavioral abnormalities, irregular EEG patterns were often observed. Such data suggested the hypothesis that the pattern of EEG maturation could be related to differences of psychological maturation in the normal range, as measured by developmental tests and performance. This expectation was confirmed for the α EEG frequency [2233]. Hence, EEG frequency measures part of the genetic variability influencing individual differences in normal psychological development. This relationship deserves to be studied in much greater detail.

Family Studies. The result of twin studies encouraged attempts to find more clear-cut evidence for genetic mechanisms. If monogenic inheritance is expected, alternatively distributed characters should be chosen for study. Such alternately distributed characters do, indeed, occur, the most evident being the so-called low-voltage EEG with little or no occipital α activity. The mode of inheritance turned out to be autosomal-dominant (Sect. 3.6.1; Fig. 3.54), with little overlap between the two phenotypic classes.

Another hereditary EEG type is dominated by monomorphic α waves. The average α rhythm has a maximum over occipital parts of the cerebral cortex and is more irregular and mixed with other waves over frontoprecentral areas. In the monomorphic α wave type, very regular α waves of high amplitudes are usually seen over the whole cerebral cortex. Twin and family studies leave no doubt that this type is hereditary, as is the low-voltage EEG. Family data point to simple dominance [921]. However, delineation from the more average EEG is more difficult than with the low-voltage EEG.

In a certain way, this EEG type can be regarded as the "counter-type" of the low-voltage EEG; while the low-voltage EEG has a weak α rhythm, this rhythm seems to be especially strong in the monomorphic α EEG.

Some other traits with simple dominant modes of inheritance could also be identified, for example, a variant in which the occipital α waves are replaced by 16-19/sec waves that show the general characteristics of α waves, such as blocking after eye opening and other stimuli. In other families, genetic variants of the fast component, the β waves, were discovered. In distinction to the α waves, β waves are in most cases concentrated above the frontal and precentral parts of the brain; in some families, these β waves form characteristic spindle-like groups. Two autosomal-dominant types were distinguished [921].

Sex Difference in EEG Patterns. Most EEGs with β waves show a definite accumulation in certain families; the data do not conform to simple monogenic inheritance. They are most easily explained by multifactorial inheritance in combination with threshold effects (Sect. 3.6.2).

Furthermore, the prevalence of EEGs with (mainly diffuse) β waves increases with age, and there is a definite sex difference. In all age groups - with the exception of childhood - the prevalence of EEGs with β waves is higher in females than in males. This sex difference is often associated with few α waves in occipital leads [2043].

How Is the EEG Produced in the Brain? Hereditary variation in EEG patterns points to differences in physiologic function of the human brain. For understanding the nature of these differences and their possible influence on behavior, it is necessary to know how the EEG is produced. Here, results from experimental neurophysiology suggest a relatively clear-cut picture [1998]. The EEG waves, especially α waves, are formed by interaction of neurophysiologic processes at several, at least three to four levels (Fig. 8.24). The EEG "battery" is located in the cerebral cortex; here, groups of neurons discharge in rhythmic order. However, their activity is co-ordinated by a "pacemaker" (more precisely, a group of interrelated pacemakers) in the thalamus. The thalamus, in turn, is influenced in its activity by input from brain structures at lower levels. The ascending reticular activating system (ARAS) is located in the reticular formation, mainly in the pons and medulla oblongata. The ARAS has a leading function, for example, in sleeping and dreaming. In the waking state, it maintains a level of "tonic arousal," which is influenced by input of specific, centripetal pathways, for example, sensory stimulation. A high level of arousal causes EEG desynchronization. Other influences on the EEG come from the limbic system, a functional unit comprising the hippocampus, amygdala, mamillary bodies, and connecting structures. The limbic system is involved in emotionality, activi-

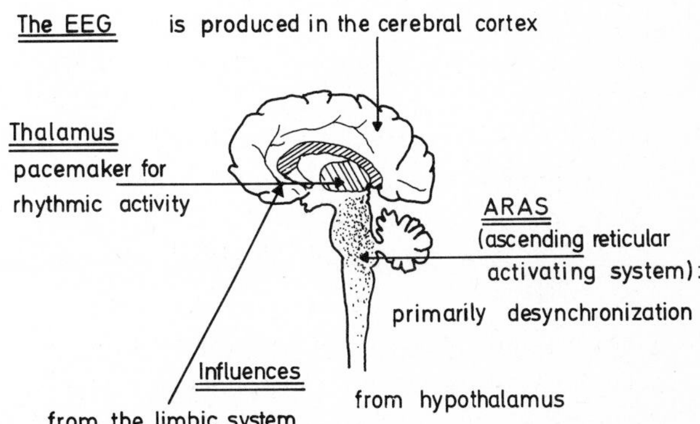

Fig. 8.24. Oversimplified drawing of the human brain showing the structures involved in production of the EEG

ty, and motivation. Neurophysiologic studies have also provided some hints as to the physiologic function of α activity [1998]: the α rhythm seems to modulate and selectively amplify afferent stimuli.

Influences of Inherited EEG Variations on Personality. Personality and performance of an individual depends on the way in which its brain is active spontaneously and handles information. Individual differences in such neurophysiological parameters will therefore result in psychological differences. What influences on 'personality' and intellectual performance can be expected in individuals carrying the genetic EEG variants described above, considering the neurophysiologic results outlined in the foregoing paragraph? Probands with monomorphic α waves might be expected to be "strong modulators and amplifiers," whereas probands with a low-voltage EEG might be weak modulators and amplifiers. β waves, especially those of the diffuse variant (which seems to be multifactorially inherited) are generally assumed to be the result of a high level of tonic arousal in the ARAS. Therefore, in probands with this EEG variant, disturbances in the modulating function of α activity are expected.
Indeed, comparative studies on 298 probands with various EEG variants had results that were compatible with these expectations [2231-2234]:

a) The monomorphic α tends to be on the average, sthenic, stable, and reliable. The probands are more likely to show signs of high spontaneous activity and toughmindedness; precision of work, especially under stress conditions, and short-term memory are especially good. On the other hand, information processing is not very quick.
b) The low-voltage category shows little spontaneous activity, the probands tend to be group-oriented and extraverted. Spatial orientation is especially good. Recently, the expected difference in information processing between groups (a) and (b) has been shown directly by studying averaged evoked EEG potentials [2230a].
c) The group with diffuse β waves shows a high error rate in spite of low working speed in tests measuring concentration and precision. Stress resistance seems to be low. The disturbance of the α mechanism by high tonic arousal leads to a disturbance in intelligence test performance, especially in tasks measuring the ability for spatial orientation.

An oversimplified picture of the proposed relationship between EEG variation and personality is given in Table 8.17, which also lists some additional EEG variants. For example, probands with fast occipital α variants (16-19 c/sec) seem to excel in abstract thinking and in motor skills. Probably, they are able to process information rapidly. The EEG literature has reports on possible positive correlation between α frequency and intellectual performance [see 2231].
Another, very rare EEG variant not shown in Table 8.18 differs from superficially similar types by a number of characteristics; for example, the 4-5 cps waves are blocked by eye opening and are immediately replaced by α waves after minute disturbances. The genetic basis is not quite clear; two concordant MZ pairs and a small number of families with more than one affected member point to genetic factors, but most probands are the only affected persons in otherwise normal families. Many probands with this EEG variant show emotional disturbances and abnormalities in the autonomic nervous system; among psychiatric patients, the variant is much more common than in the general population [2138]. An anomaly in the limbic system may explain these results.

Association Between α Waves and Spatial Ability. The mean performance of women is poorer than that of men in test items requiring visual orientation in space (Sect. 8.2.2.2). Figure 8.19 showed one standard task designed to test spatial ability. It was shown to be especially poor in patients with Turner's syndrome. As parieto-occipital parts of the cortex are involved in visual perception, a relationship of occipital α-activity to performance in tests for spatial perception is possible. Some studies suggested a decrease of spatial ability with decreasing α- and increasing β-activity as expected from the sex difference in both measures. However, the evidence is ambiguous.

Averaged Evoked EEG Potentials. The brain reacts to a stimulus by producing a characteristic wave. This reaction, however, is so small that it is lost in the "noise" produced by the resting EEG. This problem can be overcome by repeating stimuli (e.g., light flashes and sounds) many times and adding up the reactions. In this way, background noise will level out and the *averaged evoked potential (AEP)* will remain. Its characteristic form was found to be concordant in MZ twins [2010]; and some of its properties were claimed to be correlated with measures for intelligence [2069], personality characteristics, and susceptibility to mental disease [2011]. However, there are many controversies regarding details of methods, artifacts, epidemiological basis of samples examined, and results. Recently the correlation with intelligence could not be confirmed in a comprehensive study [Vogel et al., in press].

8.2.3.5 Genetic Aspects of Alcoholism

Alcoholism is defined as a dependence on alcohol, which leads to social disability. Alcoholism strongly

Table 8.17. Hereditary EEG variants, their genetic basis, and their psychologic consequences [2231]

EEG type and mode of inheritance	Genetic variation of EEG	Functional consequences	Psychological consequences
Monomorphic α waves (probably autosomal-dominant)	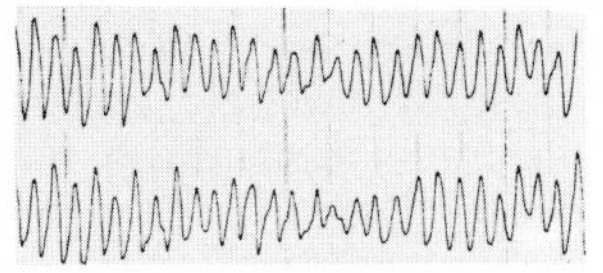	Strong selection and amplification	Sthenic, stable, resistant to stress
Fast α variants (16–19/s) (autosomal-dominant)	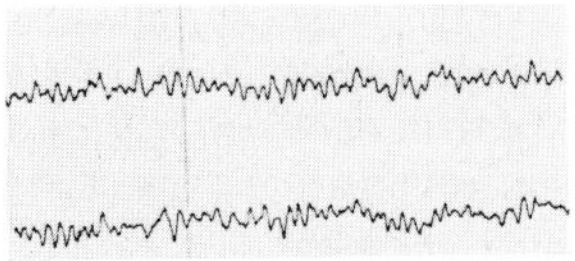	Fast information processing	Good intelligence and good motor skill
Low-voltage EEG (autosomal-dominant)	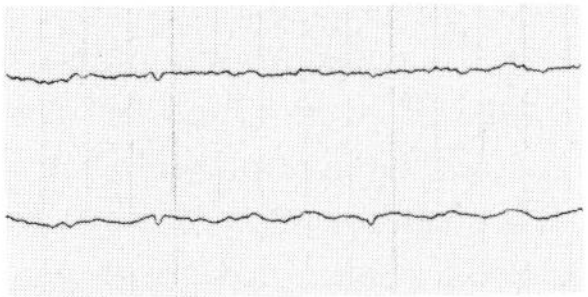	Weak amplification	Relaxed, little activity, conformistic
Low-voltage borderline (mixed group)	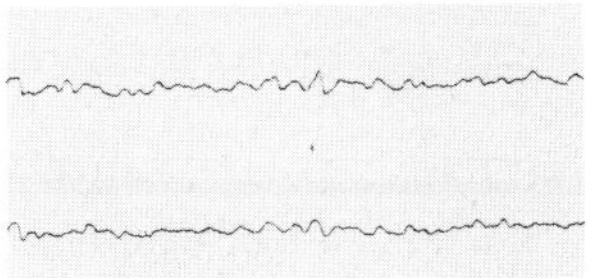	Weak amplification; disturbed	(Mixed group)
Frontoprecentral β groups (autosomal-dominant)	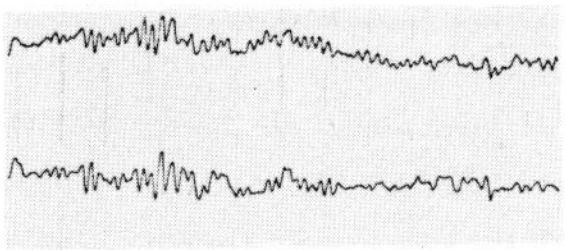	?	(Inconspicuous; undisturbed)
Diffuse β waves (polygenic)	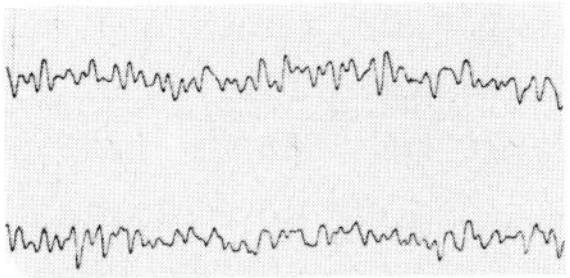	Disturbed information processing due to high tonic arousal	Tense, disturbed spatial orientation, susceptible to stress
Normal EEG (polygenic)	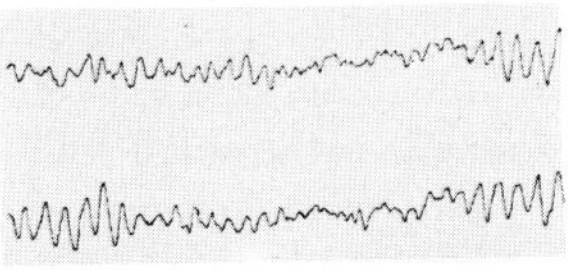		

depends on the environment. In a society where alcohol-containing drinks are not available, nobody can become an alcoholic. Even within Western societies where opportunity is abundant and social pressure tends to favor alcohol consumption, individuals who become alcohol-dependent often suffer from personality disturbances for which no psychological or social explanation can be found. There are individual differences in alcohol susceptibility, and the possible genetic basis of such differences can be studied.

Animal Models. It is suggested by experiments with mice and rats that genetic variability may influence the susceptibility to alcoholism (Sect. 8.1.2). Definite differences in alcohol preference were found between inbred strains [2182]. These differences were shown to be associated with differences not only in alcohol metabolism but with quantitative and qualitative differences in response of the brain to alcohol as well. These results suggest that genetic differences between humans should also be sought at two levels – in alcohol metabolism and in influence on brain physiology.

Studies with Classic Methods: Twin and Adoption Studies. In a study on drinking habits in 174 twin pairs in Sweden, (among them 48 MZ pairs) a higher concordance of MZ as compared with DZ pairs was found, which also extended to the drinking pattern (steady vs. "spree" drinking; amount of alcohol uptake) [2097]. The investigator had to rely on historical data; therefore, the study may not be fully reli-

Table 8.18. Comparison of drinking problems and patterns in two adoptive groups (%) [2057, 2058]

	Probands[a] ($N=55$)	Controls ($N=78$)
Hallucinations[b]	6	0
Lost control[c]	35	17
Amnesia	53	41
Tremor	24	22
Morning drinking[c]	29	11
Delirium tremens	6	1
Rum fits	2	0
Social disapproval	6	8
Marital trouble	18	9
Job trouble	7	3
Drunken driving arrests	7	4
Police trouble, other	15	8
Treated for drinking, any[b]	9	1
Hospitalized for drinking	7	0
Drinking pattern		
Moderate drinker	51	45
Heavy drinker	22	36
Problem drinker	9	14
Alcoholic[c]	18	5

[a] Adult male adoptees: At least one biologic parent hospitalized for alcoholism.
[b] $P<0.05$.
[c] $P<0.02$.

able. Some confirming information, however, is available for MZ twins reared apart [2195].

Studies with Classic Methods: Family, Twin, and Adoption Studies. Studies in families of alcoholic probands have often been performed. Some years ago, Cotton [2024] reviewed no less than 27 studies published in English, in which the families of altogether 6251 alcoholics and 4083 nonalcoholics were compared. Despite the fact that these studies are not of equal caliber since controls were not always investigated –, some general conclusions are possible: Almost one third of all alcoholics had at least one affected parent; in the majority of cases (25%; range of variation between various studies: 2.5%–50%), the father was an alcoholic. In most studies, relatives of female alcoholics were affected more often than those of affected males. This could indicate that women become alcoholic more often for endogenous reasons, whereas in men the immediate cause of alcoholism is related more frequently to nonfamilial environmental factors. Among family members of patients with schizophrenia and affective disorders alcoholism was much rarer than in families of alcoholics. Hence, an increased susceptibility to alcoholism is not a general feature of patients with major psychiatric disorders. No specific personality type among alcoholics could be discovered.

Åmark [1997] in Sweden studied the relatives of 645 alcoholic probands. In addition to alcoholism, he also found an increased risk of "psychopathy," "psychogenic psychoses," and criminality, but not of mental retardation, epilepsy, or endogenous psychoses. In this connection, it should be remembered that in Scandinavian countries, excessive alcohol consumption is shunned more strongly by society than in countries such as Italy or France, where alcoholic drinks are a part of daily life.

Of course, aggregation of alcoholics in certain families alone cannot be accepted as indicating a genetic susceptibility. Therefore, family investigations were supplemented by twin and adoption studies. Three comprehensive twin studies are available; two from Sweden [2097; 2088] and one from Finland [2154]. In addition to a higher concordance of MZ than of DZ twins, these studies led to some slightly more specific conclusions. In the Finnish study, for example, frequency of alcoholic excesses and amount of alcohol ingested were definitely more often concordant in MZ than in DZ twins; loss of control showed higher concordance up to the age of 30 but not later in life, and when social consequences were considered no closer similarity was observed between MZ than between DZ twins. Contrary to expectation, comparison of life histories of discordant MZ pairs failed to give any relevant information on specific environmental factors.

The shortcomings of the twin method have often been discussed (Sect. 3.8.4). They can be avoided by adoption studies. A number of such studies have been performed [2229], for example in Denmark.

In Table 8.18, 55 Danish adoptees in the age range from 20 to 40, of whom at least one biological parent had been hospitalized for alcoholism, were compared with 78 controls [2057, 2058]. The data came from a pool of 5,483 adoption cases originally established for the study of schizophrenia. No difference was found between the adopting families of probands and controls as to drinking habits or socioeconomic background. There was a strong dissimilarity between the two groups as to alcoholism; a corresponding difference emerged in the frequency of divorce, whereas other types of psychopathology were similar. The two groups differed only as to alcoholism defined as social and job problems, loss of control, withdrawal hallucinations, and psychiatric treatment. Heavy and even problem drinkers without these symptoms were found among the controls in about the same frequency. This study was supplemented by one in which 20 sons of alcoholics who were adopted by other families were compared with 30 of their biologic brothers who had not been adopted and grew up in the homes of their biologic parents [2057].As expected, the socioeconomic conditions in these homes were, on the average, poorer than in the families of the adoptive parents. The result of the study, however, was unexpected: Among the nonadopted brothers, the fraction of alcoholics was about the same as in the adopted group. Had the environment played an important role in the causation of alcoholism, one would have expected many more alcoholics in the nonadopted group. Therefore, these data suggest that in a society in which alcoholic drinks are universally available genetic factors largely determine whether an individual becomes an alcoholic.

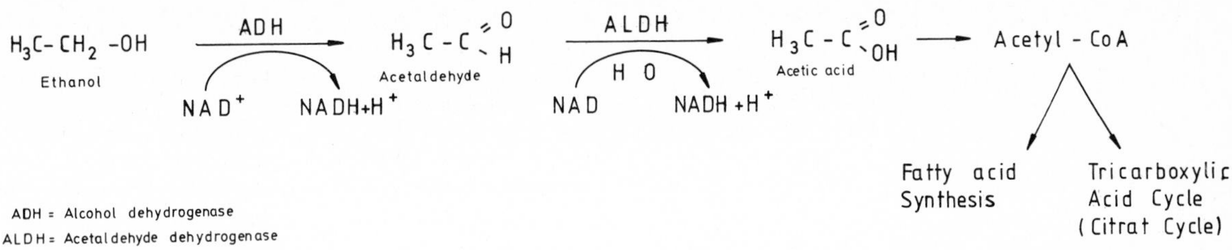

H_3C-CH_2-OH (Ethanol) —ADH (NAD^+ → $NADH+H^+$)→ $H_3C-C(=O)-H$ (Acetaldehyde) —ALDH (H_2O; NAD → $NADH+H^+$)→ $H_3C-C(=O)-OH$ (Acetic acid) → Acetyl - CoA → Fatty acid Synthesis; Tricarboxylic Acid Cycle (Citrat Cycle)

ADH = Alcohol dehydrogenase
ALDH = Acetaldehyde dehydrogenase

Fig. 8.25. Alcohol degradation in two successive steps, the first step controlled by alcohol dehydrogenase, the second by acetaldehyde dehydrogenase

In a similar study, daughters of alcoholics who gave them for adoption were compared with their sisters who had stayed with their alcoholic parents. Despite the pronounced difference in familial environment alcoholism was about equally frequent in both groups, and more common than in the female population [2059].

Genetic Variability of Alcohol Metabolism. If there is a genetic variability in susceptibility to alcoholism, what is the mechanism? The answer can be sought at two levels: Alcohol metabolism and alcohol action on the brain.

A strong genetic influence on alcohol metabolism was shown by twin studies [2162]. Figure 8.25 shows the two most important steps of ethanol oxidation. The two most important enzymes are alcohol dehydrogenase (ADH) and aldehyde dehydrogenase (AlDH). Both are localized in the liver. ADH is determined by three autosomal gene loci (ADH_1, ADH_2, ADH_3). The genes ADH_1 and ADH_3 are active mainly during fetal life; in adults, ADH_2 is responsible for most of the activity in liver and kidney. In 5%–20% of individuals of European extraction, but in 90% of Japanese, an atypical variant was discovered. Since at physiological pH the atypical enzyme shows much more activity than the more common one, it was suggested that alcohol oxidation proceded much faster in carriers of the atypical enzyme. Moreover, it was known that many Japanese show the phenomenon of "flushing" after intake of relatively small amounts of alcohol; i.e., the face reddens, pulse rate increases, and the individual feels poorly. A similar effect can also be produced in carriers of the more common ADH variant when they take the drug disulfiram (Antabus) with alcohol. Since this drug is known to enhance the level of acetaldehyde by inhibiting the enzyme aldehyde dehydrogenase (ALDH) (Fig. 8.25) the elevated acetaldehyde levels were assumed to be the cause of "flushing."

This enzyme also shows a genetic polymorphism in the Japanese population, allele frequencies each being around 50%, whereas in people of European extraction the deviant enzyme is rare. The common Japanese polymorphism is associated with decreased ALDH activity and appears to account for the frequent flushing phenomenon. The combination of faster production of aldehyde combined with diminished breakdown of this compound presumably accounts for the flushing.

Has this genetic difference in alcohol metabolism something to do with susceptibility to alcoholism? As mentioned above, flushing causes not only an increased pulse rate but is associated with definite discomfort. This discomfort may prevent carriers of the ALDH variant from drinking too much and might therefore protect them against becoming alcoholics. And indeed, this variant was shown to be much less common among Japanese alcoholics than in the general population [2250].

Reaction of the Brain to Alcohol as Measured by the EEG. Hints for the genetic determination of alcoholism were found, when the reaction of the brain to alcohol was examined using the EEG [2161, 1270]. In a twin sample, measurements of EEG characteristics, such as amplitudes and distribution of frequencies, gave high heritabilities: The EEG's of MZ twins even had a tendency to become still more similar after alcohol. Still more interesting, however, were the differences in reaction when the resting EEG was considered. Individuals with a prominent and stable α rhythm in the resting stage showed relatively little change after alcohol intake (Fig. 8.26). On the other hand, persons whose EEG at rest showed less well-developed α waves showed a stronger response to alcohol intake: their α waves were much more prominent and stable after alcohol than in the resting EEG (Fig. 8.27). These reactions were highly concordant in MZ twins but in some cases discordant in DZ twins.

Some other twins showed qualitatively deviating reactions, for example, β waves. Individuals with the low-voltage EEG, a variant with a single dominant mode of inheritance (Sect. 3.6.1), did not react to alcohol by increased α wave production. Hence, there are strong genetically determined differences between individuals in the reaction of their brains to alcohol. These differences were not caused by a corre-

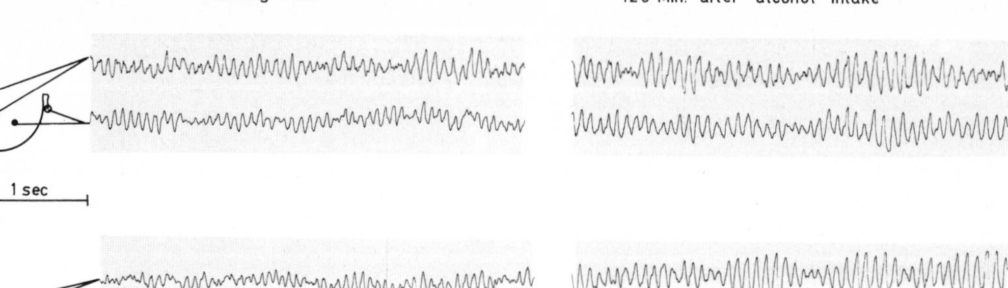

Fig. 8.26. A male adult MZ twin pair with well-developed occipital α rhythm. Loading with 1.2 g/kg ethanol leads to a relatively small increase of α activity 120 min after ethanol intake. (Propping, 1977 [2161])

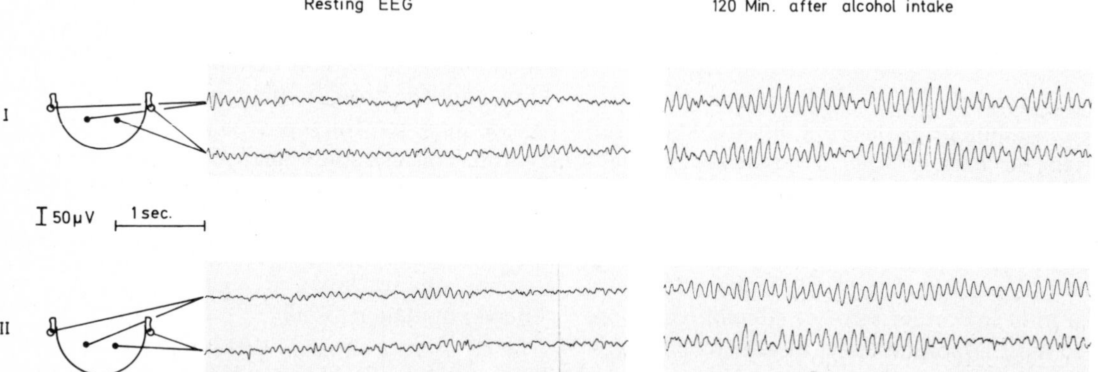

Fig. 8.27. A male adult MZ twin pair with relatively poorly expressed α waves in the resting EEG; 120 min after intake of 1.2 g/kg ethanol, the α rhythm is strikingly enhanced. (Propping, 1977 [2161])

sponding variability in alcohol metabolism. There was no relation between alcohol blood level and EEG reactivity. Still more important, the aclohol-induced modification of the EEG pattern persists for a relatively long time after most of the drug has been metabolized. The response of the brain itself appears to be subject to a genetically determined difference.

Can Our Knowledge on Neurophysiologic Mechanisms of the EEG Explain the Differential Reaction to Alcohol? The mechanisms of EEG production, which have been explained in Sect. 8.2.3.4, might help to understand brain function as altered by alcohol. The pacemaker in the thalamus is only influenced slightly in individuals with well-developed α rhythm. Its synchronizing function is improved, on the other hand, in subjects in whom the normal spontaneous α rhythm shows tendencies to desynchronization. This tendency may be caused by a stronger desynchronizing input from the reticular system. Is there any evidence that could connect these two bits of information, suggesting a specific hypothesis for a genetic component of alcoholism?

It has been described before that some alcoholics tend to have a poor α rhythm [2161]. More specifically, their α rhythm has often been described to be similar so that depicted for twins showing a strong alcohol reaction (Fig. 8.27). It was not quite clear, however, whether this EEG pattern was one of the causes of alcoholism or a consequence of alcohol action on the brain either directly or via liver damage. On the other hand, meditation techniques, such as transcendental meditation, were shown to enhance α activity. The subjective result of meditation is that of relaxation and peacefulness. The same result can sometimes be achieved by direct attempts at enhancing α activity by biofeedback. One can feed the EEG into a device that emits a sound as long as α waves are produced. The proband is asked to try to maintain this sound as long as possible. In this way, α activity can be enhanced for a limited time. Often, the probands describe their feeling tone as relaxed, bordering on happiness; their condition seems to be similar to that achieved by meditation [16].

These results suggest a neurophysiologic and genetic hypothesis for some cases of alcoholism. Susceptibility to alcoholism is high in persons who normally suffer from a relatively high level of tonic arousal, or a weak resistance of the thalamic pacemaker. This high arousal level is attenuated by alcohol, and the person feels better. Therefore, alcohol consumption is positively reinforced and may result in alcoholism. This hypothesis was confirmed by EEG studies on nonalcoholic family members of alcoholics [2167; 2158]. Some of them showed similar EEG patterns; therefore, the EEG could not be the effect of chronic alcohol abuse. In one study, however, the reduction of average α-activity was found only in female and not in male alcoholics. At the same time, these women were found by independent psychiatric evidence to be affected more often regardless of external circumstances, whereas most of the men had become alcoholics as a result of social pressure and other external factors [2167].

This example shows how genetic susceptibility may lead to different phenotypic effects, depending on the sociocultural conditions of the environment. In a Western society with its strong emphasis on social drinking, the genetically susceptible individual is endangered to become an alcoholic. In a Buddhist setting, the same person would probably become one especially dedicated to meditation.

The arousal-attenuating effect on the reticular system is not the only alcohol effect on the brain. The twin studies suggest that genetic variability may exist for other aspects of brain function as well. This example has been described in detail, since it is one of the first instances in which neurophysiologic data and concepts have been utilized to develop a genetic hypothesis. Neurophysiology is a highly developed field of science that has so far been little influenced by ideas and concepts of human genetics. Human geneticists have only rarely tried to include concepts of neurophysiology in their thinking. As occurs so often in the development of science, the main reason for this barrier is the compartmentalization of science in that neurophysiology specifically and physiology in general have been less influenced by genetics than most other fields in biology. The reason may be that physiology is less reductionistic and deals more with interpretation of integrated systems and feedback processes than other areas in biology [267].

8.2.3.6 Brain Physiology: Genetic Variability of Neurotransmitters

Analysis at the Biochemical Level Is Needed: The Synapsis. In the long run, genetic analysis at the EEG level, while being conceptually more satisfactory than analysis of behavioral phenotypes, will have limited success. Resolution at the level of enzymes and proteins will have to be achieved. Where could we find genetic variability of enzymes and proteins that may influence brain function? The main functional components of the nervous system are neurons [55]. They are single cells with one nucleus, one long branch that is called neurite or axon and functions as the effector organ of the neuron, and a number of elaborately branched dendrites that establish contacts with other nerve cells via the so-called synapses. Figure 8.28 shows the main organelles of a synapsis. There is complete separation between the presynaptic terminal and the postsynpatic membrane; the two are separated by a narrow synaptic cleft. When the nerve impulse reaches the presynaptic terminal, transmission across the synapsis is effected not by an electric mechanism but by chemical means. The specific transmitter substances are prepacked in quanta of some thousands of molecules in the vesicles of the presynaptic terminals. The arriving impulse causes one or a very few vesicles to liberate their quanta of transmitter into the synaptic cleft. Thus, the transmitter can act on specific receptor sites of the postsynaptic membrane. This reaction causes Na^+ ions to diffuse across the membrane and bring about a change in electric potential. There are two types of synpases: excitatory and inhibitory. When a neuron has received a sufficient number of impulses from excitatory synapses, the axon "fires," i.e., it releases an impulse. An inhibitory synapsis, on the other hand, may cause an inhibitory hyperpolarization of the postsynaptic membrane, which prevents depolarization from reaching the critical level above which the neuron will fire. In this way, an excitatory impulse can be transmitted to an ever-increasing number of excitatory nerve cells; this 'chain reaction' is prevented from evolving into an "explosion" by intercalation of inhibitory nerve cells [55].

This chain of events suggests a number of possibilities for genetic variability. For example, the enzymes for synthesis or break down of neurotransmitter molecules may have different activities, the membrane may show structural differences that influence its permeability for neurotransmitter or enzyme molecules, three might be differences in receptors, or outside regulating processes may influence the function of the synapsis at various levels. The most straightforward possibility is quantitative variation in the amount of transmitter molecules present. There are, indeed, some results in mental diseases that suggest abnormalities in neurotransmitter function.

Chemical Types of Neurotransmitters (Fig. 8.29). Several compounds are used in the brain as neurotransmitters; synapses are specialized for one type. The best-known examples are norepinephrine (adrenergic synapses) and acetylcholine (cholinergic synapses). The reason that their analysis has so far been most successful is technical: both can be studied in cells of the peripheral nervous system. For example, neurons of the sympathetic nervous system are adrenergic, neurons of the parasympathetic nervous system are cholinergic. In the brain, however, these two types of synapses together constitute only a minority of all neurons; a variety of amino acids (histamine, glutamic acid, aspartic acid, glycine, and possibly others) act as neurotransmitters. For their activity, not only synthesis but also inactivation is important. Figure 8.29 shows the main groups.

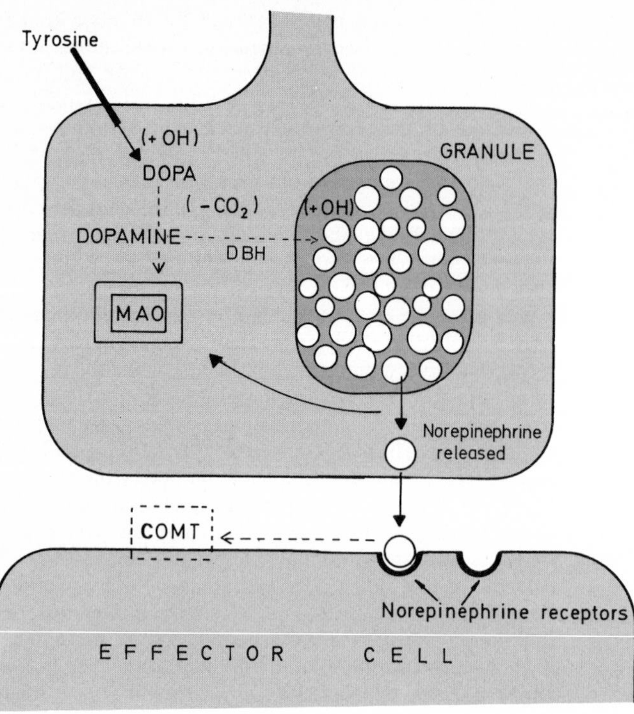

Fig. 8.28. Schematic representation of an adrenergic synapsis with its most important organelles. Norepinephrine is synthesized from tyrosine, stored in granules, released into the synaptic cleft, and bound to receptors of the effector cell on the postsynaptic membrane (see text for further explanation) (*MAO*, monoamine oxidase; *COMT*, catecholamine o-methyltransferase); *DBH*, dopamin β-hydroxylase

Two classes of substances have recently attracted much attention as they appear to be altered in affective disorders as well as in schizophrenic psychoses: the catecholamines norepinephrine and epinephrine and their precursors, and the indoleamines, especially 5-hydroxtryptamine (serotonin). In the following, we mainly confine ourselves to one group, the catecholamines.

Catecholamines. Epinephrine and norepinephrine are formed from tyrosine. The function of an adrenergic synapsis is shown in Fig. 8.28. This example will be used to demonstrate some possibilities for genetic variability and, at the same time, for experimental approaches to analyze this variability. In the context of this book, we can only give an oversimplified picture. Norepinephrine, when not used as a neurotransmitter, or after having performed this function, has to be inactivated. Here, two enzymes have often been examined: catecholamine-*O*-methyltransferase (COMT) and monoaminooxidase (MAO). The concentration of norepinephrine may be enhanced or lowered – or the synapses may even be depleted of this neurotransmitter – either by altering synthesis or degradation of this substance.

Genetic analysis of variability in these and other enzymes is difficult as the human brain is not directly accessible to our analysis. Two ways to overcome this difficulty are:

1. Experiments with animals.
2. Examination of these enzymes in other, more easily accessible tissues.

Animal Experiments on Genetic Variability in Catecholamine Metabolism [2002, 2021, 2022]. The enzymes tyrosine hydroxylase, dopamine-β-hydroxylase, and phenylethanolamine-*N*-methyltransferase were shown to have about twofold greater activities in the adrenals of the inbred BALB/c than in the BALB/cN mouse line. In F_1, F_2 and back-cross progenies, single genes were found to control these enzyme activities was detected, suggesting that either the structural genes for these enzymes are closely linked or that they are under a common regulatory genetic control.

It has been shown that conversion of norepinephrine to epinephrine can be altered by other influences, for example, adrenal steroids, hypophysectomy, or cold stress. The breakdown time of the enzyme involved also shows interstrain differences. Moreover, the mechanism of control may vary even between different mouse strains [2002].

Another source of genetic variability is that of cAMP, which has been found to act as a second messenger for various hormones and neurotransmitters [120, 220]. cAMP content was found to be different in the brains of four inbred mouse strains.

These experiments suggest complex regulation of the quantity of norepinephrine in adrenergic synapses of the brain; it may be recalled that the resulting differences in activity show a correlation with differences in aggressive behavior (Sect. 8.1.2). In view of these complexities, it seems to be a difficult undertaking to examine the same enzymes in humans and to draw conclusions from activity determinations to possible differences in neurotransmitter function in the human brain. Still, some indications exist

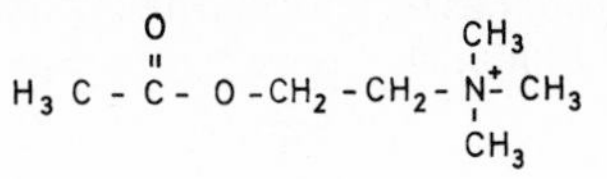 $H_3C-C(=O)-O-CH_2-CH_2-N^+(CH_3)_3$	Acetylcholine
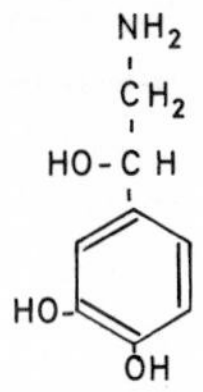 $NH_2-CH_2-CH(OH)-C_6H_3(OH)_2$	Norepinephrine
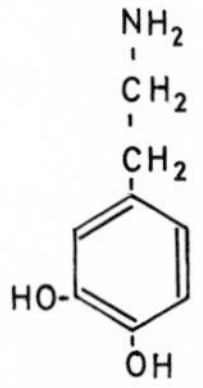 $NH_2-CH_2-CH_2-C_6H_3(OH)_2$	Dopamine
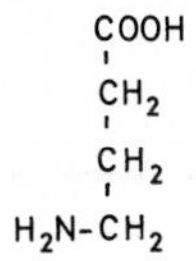 $COOH-CH_2-CH_2-CH_2-NH_2$	(GABA) Gamma-aminobutyric acid
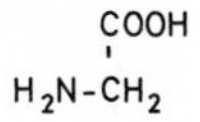 $COOH-CH_2-NH_2$	Glycine
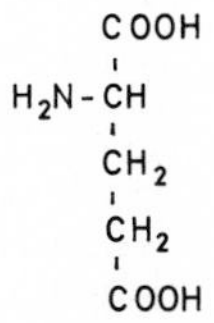 $COOH-CH(NH_2)-CH_2-CH_2-COOH$	L-glutamic acid
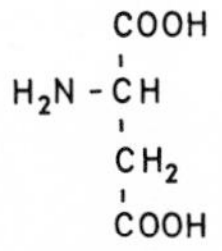 $COOH-CH(NH_2)-CH_2-COOH$	L-aspartic acid
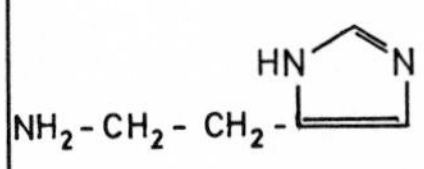 $NH_2-CH_2-CH_2-$ (imidazole ring: HN, N)	Histamine
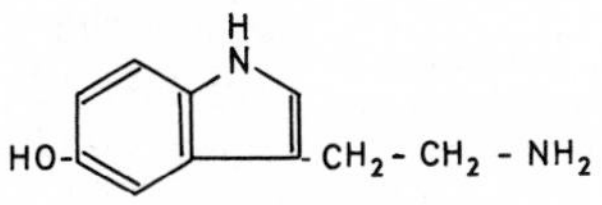 HO-(indole ring: H, N)$-CH_2-CH_2-NH_2$	Serotonin

Fig. 8.29. The main groups of neurotransmitters

that the approach might lead to detection of genetic variability that might help to understand the genetics of normal and deviant behavior.

Psychotropic Drugs [2169]. Research on mental disease was encouraged by the observation that the symptoms of affective and mental disorders can be influenced by psychopharmacologic agents. These drugs were shown to influence synaptic transmitter function, especially the function of norepinephrine. It was observed, for example, that some patients with depression responded better to monoaminooxidase (MAO) inhibitors and others to tricyclic antidepressants, such as imipramine. Moreover, relatives, when suffering from depressions, responded positively to the same drug as the proband. This familial tendency to respond to one class of drugs and not to the other suggests genetic factors. Both drugs have an influence on the function of norepinephrine on adrenergic synapses; MAO inhibitors reduce the degradation of epinephrine, thereby enhancing the available amount of this chemical at the synapsis. The tricyclic antidepressants, such as imipramine, reduce reabsorption of epinephrine into the neuron from which it had been released, thereby enhancing the available amount of epinephrine for neurotransmission. Conceivably, the interfamilial differences in therapeutic efficiency of those drugs could point to different genetic anomalies, at the synapsis in these families. Such specific conclusions, however, are made difficult by the observation that genetic differences between humans exist in metabolism of these drugs and, hence, in their steady-state blood levels. The differences between families might therefore be caused by genetic differences in drug metabolism rather than in target organs. Such metabolic differences have been examined especially carefully for the tricyclic antidepressant nortriptyline, which only differs slightly from imipramine [2169].

In the interpretation of genetic differences of psychopharmacologic reactions both drug metabolism and the target organ - mainly the brain - need always be considered. Experiments are urgently needed in which biosynthesis and blood level are held constant so that effects at the brain level can be studied. Such investigations in humans are necessary not only for a deeper understanding of the genetic basis of affective and other mental diseases but for rational drug therapy as well.

8.2.3.7 Affective Disorders and Schizophrenia

Genetic Investigations in Affective Disorders and Schizophrenia. The affective disorders and schizophrenia are diseases with clearly distinguishable clinical signs in most cases. Genetic studies in these conditions have a long history. After many case reports in the pre-Mendelian area, the classic paper of Rüdin (1916) [2186], which owes its statistical sophistication to collaboration with Weinberg of "Hardy-Weinberg" fame, became a paradigmatic model for the phenotypic-biometric approach in such disorders. Following this paradigm, many family and twin studies were carried out. These studies clearly established that genetic variability plays a major part in the causation of affective diseases and schizophrenia (Fig. 8.30 and 8.31); moreover, this work contributed much to the increasing sophistication of statistical methodology for establishing empirical risk figures needed for genetic counseling [744]. These empirical results, however, left an increasing number of research workers dissatisfied, and there is a long list of - mostly futile - attempts at moving genetic analysis to a level closer to gene action to find the biologic basis for these disorders.

Affective disorders include manic-depressive or bipolar disease and unipolar depression. Another large group of common psychoses is usually classified as schizophrenia. Affective disorders are mainly characterized by cyclic anomalies of feeling tone - depression or mania - whereas in schizophrenia, anomalies of thought patterns and loss of contact with reality are the main symptoms.

Twin and Family Studies in Affective Disorders [2252] [2051]. In older studies, the group of affective disorders was usually regarded as one entity. Twin and family studies started from unselected series of patients or from affected twins to establish empirical risk figures for the various degrees of family relationship (Sect. 3.3.6). Figure 8.30 shows the reported twin series. Obviously, concordance in MZ twins is much higher than in DZ twins, pointing - if the straightforward interpretation is accepted - to a genetic contribution. It is particularly important that the concordance rate of 12 MZ pairs reared apart was 67% and thus of the same order of magnitude as concordance rates of MZ twins reared together [2252]. Concordance even in MZ twins is far from complete. This finding proves the importance of environmental factors. Unfortunately, the concordance rates in Fig. 8.30 were calculated without age correction. Hence, it is possible that some discordant pairs sooner or later will become concordant.

Bipolar and Unipolar Types: Empirical Risk Figures. The twin data as well as the older family studies did not separate bipolar patients, i.e., those with manic as well as depressive phases, from unipolar cases, i.e., those suffering only from depression. Leonhard [2114] first suggested that these two disorders may be genetically distinct. Subsequent studies have con-

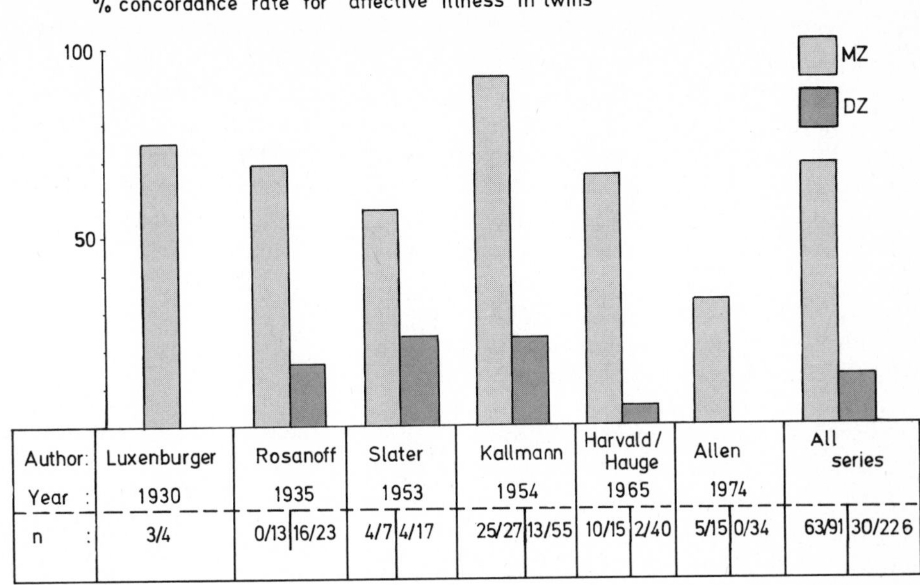

Fig. 8.30. Concordance and discordance for affective illness in MZ and DZ twins. The *figs.* 3/4 etc. refer to the total number of twin pairs and the number of concordant pairs

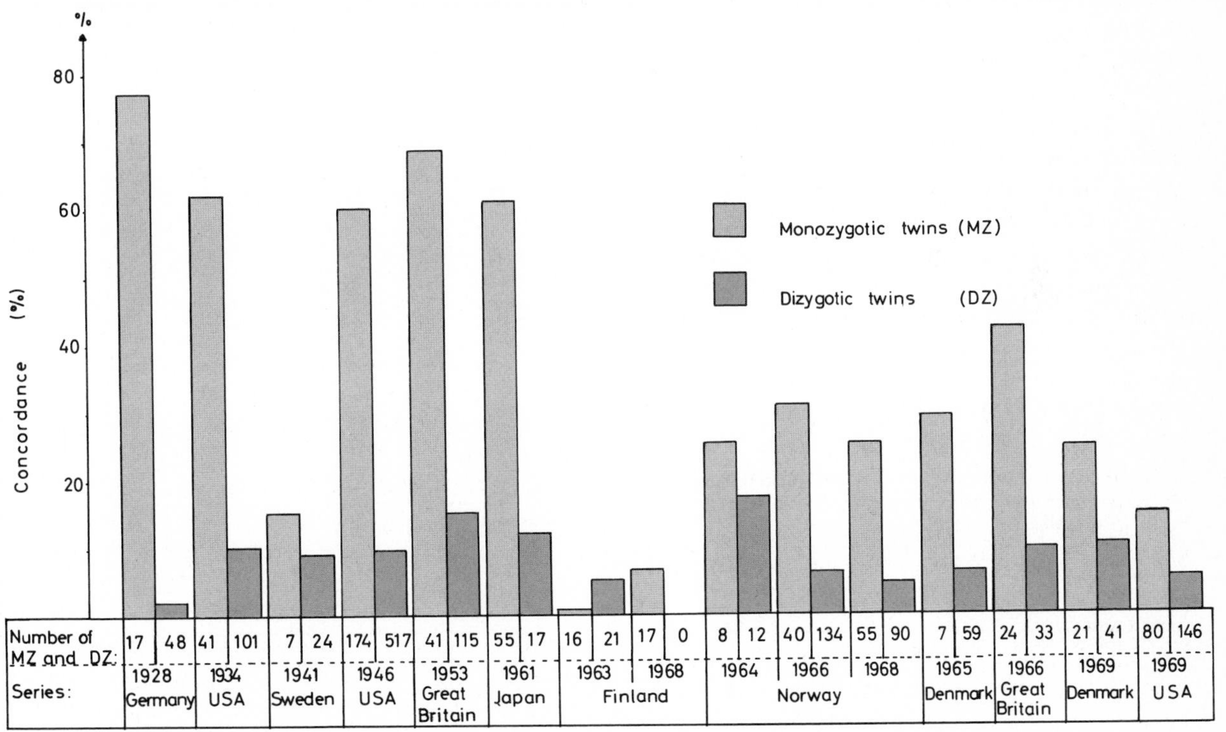

Fig. 8.31. Concordance and discordance for schizophrenia in MZ and DZ twins from various populations

firmed consistently that bipolar patients have more bipolar relatives than unipolar patients. However, the prevalence of unipolar depression is much greater among relatives of bipolar patients than in the general population. These conclusions were confirmed by a comparative study of biological and adoptive parents of bipolar patients [2131]. The degree of psychopathology, especially affective disorders, in the biological parents of these adoptees was found to be similar to that of the parents of nonadopted bipolar patients, whereas the rate of psychiatric disorder in the adoptive parents of manic-depressive patients was similar to that of the adoptive parents of unaffected controls.

An adoption study on depressive disorders mainly of the unipolar type had a quite different result [2105]. The amount of registered psychiatric illness in *adoptive* fathers was about fivefold that in the controls (adoptive fathers of carefully matched, psychiatrically normal controls). On the other hand, a threefold increase in psychiatric illness was found only in biological mothers of female patients. These data point to a major contribution of familial, exogenous factors in unipolar, affective disease. Moreover, some other psychiatric disturbances, such as a personality type characterized by unusual mood swings, mild to moderate depressions, alcoholism, and acute nonrecurrent psychosis, have been observed more frequently among relatives of affective disorder patients. Another interesting aspect is a sex difference. Female first-degree relatives of bipolar probands are 1.5–2 times more frequently affected than male first-degree relatives. No consistent sex difference has been shown among first-degree relatives of unipolar probands. Table 8.19 shows the most important empirical risk figures.

In recent years, attempts have been made to subdivide affective disorders still further into biologically more homogeneous subgroups. In a remarkable study, Angst et al. [2000] distinguished three subgroups of bipolar disease, depending on whether depressive or manic phases predominated or both occurred with about equal frequencies. In the latter group, the highest risk for close relatives of being affected was found. Moreover, the genetic risk increases with the number of affected relatives in the family.

Simple Modes of Inheritance. Taken at face value, these empirical risk figures are not compatible with a simple mode of inheritance. However, some observations fairly consistently suggest participation of major genes in at least some cases. Some pedigress point to an autosomal-dominant mode of inheritance. Other predigrees suggest X-linked dominant

Table 8.19. Empirical risk figures for affective disorders (six series). (Gershon et al., 1976 [2051])

a) Influence of age at onset in proband

Type of disease	Age at onset	Number of first-degree relatives	Affected (%)
Bipolar	<40	561	19.9
	>40	276	11.2
Unipolar	<40	886	16.7
	>40	933	9.5

b) Influence of sex of proband (bipolar probands)

Sex of proband	Sibs				Children			
	n[a]		Affected (%)		n[a]		Affected (%)	
	♂	♀	♂	♀	♂	♀	♂	♀
♂	146.9	135.1	12.3	15.5	115.9	122.2	8.6	21.3
♀	136.9	137.3	11.0	19.7	179.0	167.8	13.4	16.7

c) Influence of sex of proband (unipolar probands)

	Sibs				Children			
	n[a]		Affected 6%)		n[a]		Affected (%)	
	♂	♀	♂	♀	♂	♀	♂	♀
♂	296.6	307.0	16.2	12.1	305.5	335.8	10.5	11.0
♀	743.9	789.1	7.8	13.6	717.4	755.2	7.8	15.2

[a] Correction for age may lead to decimals in numbers of relatives at risk.

inheritance; such cases seem to be characterized by relatively early onset and good response to lithium treatment [2245]. In such families, severe as well as mild, questionable, bipolar and unipolar psychoses were observed. Delineation of an X-linked dominant from an autosomal-dominant mode of inheritance is difficult (see Sect. 3.1.4), especially if a clinically indistinguishable autosomal-dominant type exists.

An argument in favor of X-linked inheritance would be linkage with an X-linked marker, such as red-green color-blindness. Evidence both in favor of linkage with this marker and against such linkage [2053] has been brought forward. Some doubt was raised by the fact that linkage was also asserted with the *Xg* locus despite its location far away from the red-green color blindness loci (3.4.3). A more recent study, however, pointed to close linkage with color vision defects in some pedigress (recombination fraction between 0 and 0.05), whereas there was no linkage with *Xg* [2180]. The fraction of families with this X-linked type appears to be small, since father-to-son transmission in a random series of families was not found to be reduced [2000].

There is no evidence whatsoever for linkage with X-linked markers in families where only *unipolar psychoses* are found. Furthermore, in this group, women are more often affected than men, age at onset is usually later, and in a large subgroup, depressions occur in phases of hormonal instability, such as pregnancy, after delivery, and especially during menopause. In conclusion, the family data on affective disorders, besides confirming the hypothesis of a strong genetic contribution, point to genetic heterogeneity and to different biologic mechanism.

Twin and Family Studies in Schizophrenia [173, 2061, 2060, 2252]. The available twin studies on schizophrenia are summarized in Fig. 8.31. While showing in general higher concordance in MZ as compared with DZ twins, these data indicate a considerable degree of statistical heterogeneity for concordance in MZ pairs. More recent studies usually showed lower concordances than older studies. In part, this discrepancy is caused by the mode of ascertainment. The older studies were based on "limited representative" sampling (Sect. 3.8.6). Probands were an unselected series of patients and the authors ascertained whether these probands had twins and whether these twins were affected or unaffected. The more recent Scandinavian series, on the other hand, were based on "unlimited representative" sampling [2042, 2108, 2217]. First, all twin pairs in the population were ascertained. Then, all pairs in which at least one twin was schizophrenic were included in the studies. In contrast to "limited representative" sampling, this method also includes probands who are less severely affected and therefore would have little or no chance to be included in a series largely based on hospital admissions. Therefore, lower concordance rates are not surprising. Lower concordance rates for MZ twins with milder clinical symptomatology have repeatedly been reported in other studies [2236].

Table 8.20. Environmental factors in 16 discordant MZ twins discordant for schizophrenia. (Tienari, 1963 [2217])

	Twins with schizophrenia	Non-psychotic twins
First born	10	6
Lower birth weight	8	6
More difficult delivery	5	2
Bigger during childhood	7	9
Dominating during childhood	1	14
More lively during childhood	6	8
"Speaker" during childhood	4	12
More adjusted during childhood	8	–
More sensitive during childhood	5	–
More timid during childhood	8	4
Better in school	3	10
Started working earlier	1	6
Left parental home earlier	–	6
Married earlier	1	10
Unmarried (twin married)	7	–
Lower social status	6	–
More withdrawn	7	–
More often sick during childhood	3	2
Stronger during childhood	7	9

In some studies, analysis of discordant MZ pairs revealed differences in the premorbid personality during childhood and youth [2217, 2218]. Table 8.20 shows some of the environmental factors studied in a Finnish twin series. The description of the identical Genain quadruplets (Sect. 8.2.1.5) illustrates how differences in health and in attitudes of parents and sibs, especially the co-twin, combined with external stress lead to a severe psychosis in some cases, whereas others are more mildly affected or even healthy in spite of their identical genetic constitution. One argument against a genetic interpretation of concordance in MZ twins reared together is based on their close social interaction, which would reinforce schizophrenic symptomatology in the second twin if one were affected (Sect. 3.8.4). However, there were reports on at least 12 twin pairs reared apart in which at least one partner became schizophrenic; in seven pairs the other twin was also schizophrenic [2061], a figure well in line with that reported for MZ twins reared together (Fig. 8.31). Another argument would be that a schizophrenic parent would provide especially poor conditions for mental development of the child and would therefore enhance his or her risk of becoming schizophrenic too. This question can be answered by a comparative study of children of discordant MZ. If the presence or absence of a schizophrenic parent were a significant factor, the children of the clinically healthy twin would be affected less often than children of the twin with schizophrenia. Such a study in Denmark failed to show this difference [2040], thus supporting the genetic interpretation.

Analysis of interaction patterns between parents and twins showed that as a rule parents do not spontaneously *act* in a more similar way toward MZ than toward DZ twins, but their *reactions* to MZ twins are more similar, since these twins *act* similarly [2102]. Moreover,

Empirical Risk Figures. Empirical risk figures for schizophrenia suggest the following conclusions (see also Table 3.7, Sect. 3.3.6).

a) First-degree relatives of schizophrenics have a 10–20 times increased risk to develop the disease when compared with the general population. The actual absolute risk is about 10%–15%.
b) The risk is higher in relatives of probands with catatonic and hebephrenic symptomatology than in relatives of probands with paranoid or so-called simple schizophrenia.
c) Within families, there is a correlation of clinical subtypes. Relatives of catatonics have a higher risk to become catatonic than relatives of the paranoid type. However, the risk of relatives of catatonics and hebephrenics to become paranoid is also enhanced when compared with the population average, and vice versa.
d) The risk for sibs of affected probands is lowest (about 10%) when both parents of the proband are unaffected; it is higher with one (actual risk: 15%) and still higher with two affected parents (actual risk, 40%–60%).
e) Among relatives of schizophrenics, personality types are often observed who, while not outright abnormal, deviate from the normal and are often classified as schizoid [19].

A characteristic example of such risk figures is given in Table 3.17, Sect. 3.3.6. Familial aggregation of a psychiatric disorder does not necessarily mean that genetic factors must be involved. Pathogenic familial environments could lead to a similar result. Social psychiatrists are actively searching for such family factors. The twin results provide more critical evidence in favor of genetic susceptibility. However, the usefulness of twin data, especially in behavior genetics, can also be disputed (see Sect. 3.8.4). The best way to separate the contribution of genetic and environmental influences is an examination of adopted children in comparison with their biologic and adopted parents.

Adoption Studies in Schizophrenia. The first adoption study on schizophrenia was published by Heston (1966) [2072]. He examined adults who were adopted early in life and whose mothers suffered from schizophrenia. The probands had been separated from their biologic mothers, had no contact with them, and did not live with maternal relatives.

Five of the 47 offspring in this study were affected with schizophrenia. This incidence is similar to that of children who lived with their schizophrenic parents. On the other hand, none of 50 individuals in a matched control group of adopted persons with non-schizophrenic parents became schizophrenic. Half of the children of schizophrenic women who were not affected with overt schizophrenia showed a significant excess of psychosocial disability; the other half were notably successful adults.

More elaborate studies were carried out by a team of American and Danish workers in Denmark. In that country, records of good quality on adoption as well as on schizophrenia are available. It was shown that the incidence of schizophrenia in biologic relatives of schizophrenic adoptees was about three times as high as in their adoptive relatives; that adoptive parents of schizophrenic probands did not have schizophrenia more often than adoptive parents of non-schizophrenic children (they showed other types of psychological abnormality more often); and that children of schizophrenic parents had the same risk for schizophrenia regardless of whether they were educated by their parents or were adopted early in life. Intrauterine or perinatal influences from the biologic mothers of schizophrenic patients were excluded by examining adoptees who were the paternal half-sibs of schizophrenics and therefore only showed common heredity with the father. In this group, schizophrenia was also more frequent than among the controls (adoptees without schizophrenic half-sibs or other relatives) [2104, 2184, 2241, 2242].

Biologic Hypotheses in Schizophrenia. The yield of biologic research in schizophrenia has been low despite all research efforts that have been carried out for many decades. Interpretations are difficult because there is no qualitative or quantitative measure by which schizophrenia can be diagnosed; diagnostic standards will differ with different psychiatrists and in various countries.

All data from twin, family, and adoption studies in schizophrenia can be most easily explained by the model of multifactorial inheritance in combination with a threshold effect. This formulation does not exclude the possibility that in some families major genes may contribute to the genetic disposition since genetic heterogeneity between families is likely. At this preliminary level of genetic analysis, the situation is described by a *diathesis-stress hypothesis* [19, 2060]: stress situations that are overcome by most individuals and may lead to "neurotic" symptoms in others can trigger a schizophrenic psychosis in genetically predisposed persons. This explanation, however, is not quite satisfactory. The question is: Which genetically determined physiologic changes enhance the risk of an individual to become schizophrenic?

Again, as in affective disorders, the current hypotheses are centered around anomalies of neurotransmitter metabolism [2200]. For example, methionine may lead to acute psychotic reactions in chronic schizophrenics. On the other hand, the psychotomimetic drug mescaline is a methylated derivative of dopamine, a norepinephrine precursor. It is possible that some schizophrenics have an abnormally high methylation capacity for dopamine and that methionine being a general source of methyl groups enhances this effect. Other candidates for a possible endogenous psychotoxin are methylated derivatives of serotonin (5-hydroxytryptamine). Indole metabolism in general and especially serotonin is now at the center of much work in this area. For example, brain uptake of the serotonin precursor, tryptophane, could be reduced; this would also be compatible with the methionine effect, as methionine blocks tryptophane uptake competitively. Another possibility is an imbalance between an overactive brain dopamine system and an underactive serotonin system. This would be similar to Parkinson's disease, where an imbalance between an overactive acetylcholine and an underactive dopamine system exists (cf. the list of neurotransmitters in Fig. 8.29). Combination of tryptophane loading with a MAO inhibitor - to prevent serotonin degradation - had indeed led to behavioral improvement of some schizophrenics.

These examples show in which direction concepts and experimentation are currently moving. It needs to be stressed, however, that we have no conclusive evidence about biologic mechanisms of schizophrenia. Moreover, other hypotheses are now being discussed, for example, those involving membrane properties and abnormal immune processes. The twin and family data discussed above strongly suggest a multifactorial genetic model and genetic heterogeneity. Hence, it is possible that in many cases not one major biochemical abnormality but a summation of a few or even several minor abnormalities, together with external stress, may push the individual beyond the threshold and into psychosis. In other cases, one major abnormality may be decisive. The history of somatic theories in schizophrenia is not encouraging [2103]. In such a situation, it might be useful to ask a seemingly simple question: What is schizophrenia? Does it really exist as a disease unit?

"Schizophrenia" in the Light of Human Genetics. It has already been mentioned that the diagnosis of schizophrenia might be difficult and that it often depends on somewhat arbitrary criteria. In Sect. 3.8.14 the contribution of human genetics to a theory of disease was discussed in more general terms and it was mentioned that the concept of a disease determined by a single, main cause originated in the last decades of the nineteenth century, when medical bacteriology was the leading biomedical science. This concept showed its explanatory power when the causative organisms of tuberculosis and of syphilis were discovered. The success of specific therapies with chemotherapeutic agents or antibiotics would have been impossible without this concept of disease.

The diagnosis of dementia praecox by Kraepelin (later termed schizophrenia by E. Bleuler) was conceived in the light of such disease models by combining a set of clinical signs with the gradual deterioration observed in longitudinal studies of patients. This diagnostic concept implicitly implied one major common cause. Human geneticists are intuitively sympathetic to such an idea because their ideal disease concept is provided by Garrod's "inborn errors of metabolism" [75] or, even more specifically, by the hemoglobinopathies, where *one* specific mutation determines *one* protein anomaly, leading to *one* characteristic disease (Sect. 4.3.2). Thoughtful psychiatrists such as K. Jaspers or E. Bleuler realized early that too direct an application of this model to this group of diseases might lead to wrong conclusions. During the following decades, however, Kraepelin's disease concept turned out to be remarkably viable, even surviving the discovery that many cases did *not* show deterioration and that much of the observed deterioration (*not* all of it) had been an artifact caused by long-term hospitalization [19]. Survival of this diagnostic concept was achieved - at least in part - by an interesting strategy: Whenever symptoms characteristic of schizophrenia were observed in association with findings that suggested organic disease, the diagnosis of schizophrenia was withheld; often it was replaced by diagnoses such as "schizophrenic reaction." When all patients with schizophrenic symptoms who showed signs of a specific organic disease in addition were excluded a disease group remained for which specific causative factors could not be found.

Schizophrenia-like symptoms have been described more often than could be expected by mere chance in a large number of organic brain diseases [2164]. Moreover, there are many reports of brain atrophy in chronic schizophrenics. It is true that in many of these cases no genetic disposition for "true" schizophrenia could be shown by family studies and the natural history of the condition was that of the underlying disease and not that of schizophrenia. It is difficult, on the other hand, to escape the conclusion "... that had the organic diagnosis not been reached independently of the psychiatric symptomatology, most of the cases would have been regarded as indubitably schizophrenic" [2101].

Table 8.21. Genetic disorder in which an increased risk of developing a schizophrenia-like psychosis is probable. (Abridged from [2164])

Condition	Reported findings
XXY (Klinefelter)	The rate of schizophrenia-like psychoses is probably increased by a factor of 3
XXX	The rate of schizophrenia-like psychoses is probably increased by a factor of 3
$18q^-$ or r(18)	Moderate mental retardation, poor speed development, psychotic episodes (schizophrenia-like or of manic-depressive type) in childhood or adulthood
Huntington's disease	Schizophrenia-like psychoses in the initial phases of the disease
Acute intermittent porphyria	Various psychiatric symptoms including "schizophrenia" are frequently reported
Porphyria variegata	One case report of schizophrenia-like psychosis; serine loading produced psychotic symptoms
Metachromatic leukodystrophy, adult type	Numerous reports on psychoses presenting as "schizophrenia"; cases were detected by chance or by systematic screening of psychotics
Familial basal ganglia calcification	Concordant schizophrenia-like psychosis in a pair of identical twins, and familial occurrence

There are also a number of well-defined genetic conditions in which schizophrenia-like psychoses appear to be more common than could be expected by chance (Table 8.21). Such psychoses have also been described in a number of other conditions, in which an increased risk is possible but evidence is so scanty that no conclusion can be drawn. Examples are 45,XO and XYY karyotypes; adult types of various lipidoses; congenital adrenal hyperplasia; homocystinuria; Wilson's disease, and several others [2164]. Some of these conditions, when compared with pathogenetic hypotheses discussed on the basis of other evidence, even suggest plausible biological mechanims: for example those leading to a reduced supply of folic acid or influencing metabolism and function of sulfated amino acids such as methionine. Such patients may have an abnormally high capacity for methylation of dopamine; and methionine, being a general source for methyl groups, might enhance this effect.

Most of the patients with a diagnosis of schizophrenia cannot be classified as having clear-cut genetic or nongenetic organic disease. However, in many cases slight functional deviations have been described, such as reduction of MAO and other enzymes, a "choppy" rhythm together with a certain reduction of α-activity in the EEG, or slight deviations in visually or acoustically evoked EEG potentials. However, none of these deviations can be found in all schizophrenics, and there is always the question of whether they are related to the causation of the disease. An additional complication is introduced by the fact that most patients observed for the first time have been treated with psychotropic drugs, which may have changed many of the parameters of possible interest.

These small deviations, however, suggest still another question: Some of the conditions listed in Table 8.21 have an autosomal recessive mode of inheritance; in a dominant condition, such as porphyria, clinical signs are seen only under special stress conditions. Is it not possible that heterozygotes – for metachromatic leukodystrophy or homocystinuria, for example – have an increased liability of becoming schizophrenic, especially if this genetic "weakness" combines with other such liabilities, or if somatic or psychological stress factors are added? A possible slight increase of the risk for psychoses in PKU heterozygotes has been mentioned in Sect. 8.2.1.3. Results from family and twin studies in schizophrenia are compatible with the model of multifactorial inheritance (Sect. 3.6), including environmental stress factors. Such a genetic model only serves the description of observations at a preliminary level: It poses questions, rather than answering them. In schizophrenia, answers will be easier if the monocausal disease concept suggested by Kraepelin and his successors is replaced by a multicausal concept. The human brain is a complex system, in which a great number of structural and biochemical subsystems interact. Its reaction to exogenous and endogenous stress factors depends on the individual, genetically determined variation within these subsystems, on the life history of the individual, and on the kind and localization of stress. Many combinations of extrinsic and intrinsic stress factors may lead to the same, or a similar, end result: Apparently, the brain has only a limited number of ways of reacting to such stress, the reaction depending on the specific subsystems toward which the stress is directed and their intrinsic susceptibilities.

This leads from the question of *causes* to the question of pathogenesis. Which functional *mechanisms* of the brain are altered in schizophrenia, and how are the clinical signs and symptoms produced by these alterations? So far, no coherent theory has won the sympathy of the various experts. However, there is some evidence that the common final pathway in which all etiological factors come together could have something to do with attentional dysfunction or "faulty filtering of information" (Erlenmeyer-Kimling; see also [2124]). Once abnormal functioning has begun there will be an inherent tendency for self-perpetuation. Evolution has provided the human brain with the ability to "learn" (Sect. 7.2.4). This means that patterns of function change the structure of connections between neurons in a way that repetition of this function will become easier. Under normal conditions, this ability has selective advantages; it helps the individual to cope with the demands from a wide variety of environments. But, like many other biological adaptations, it might become disastrous under special conditions, i.e., when the functional pattern offered for "learning" is counterproductive.

A similar mechanism is well established for another brain disease (or group of brain diseases): Epilepsy. Every epileptic seizure can be considered to be helping to prepare for the next. Apparently, the epileptic seizure is another of the very few ways of our brain to react to many different stimuli. Most of the principles discussed at some length in this section on schizophrenia could also have been derived using the etiology and pathogenesis of epilepsy as an example [1999].

Research Strategies for Further Elucidation of the Genetic Basis of Schizophrenia. As explained above, the disease concept of "schizophrenia" is phenomenological; unlike tuberculosis or phenylketonuria it does not imply a single major cause. This does not necessarily detract from its value for diagnostics as a basis for therapy. Moreover, analysis of familial aggregation, in addition to its usefulness for genetic counseling, offers hints for a better understanding of causes. For further elucidation, the following strategies are suggested:

1. Studies on "multiplex families" (families with several affected members) offer the best chance of discovering major genes and, hence, single, major biochemical abnormalities.
2. Long-term prospective studies on high-risk children (e.g., children of schizophrenic parents) promise valuable information. Such studies are now being performed by a number of groups [2032, 2049, 2084, 2130].
3. Genetic variability in the normal range of parameters thought to be of possible importance should be studied.
4. In some families, an unfortunate combination of slight quantitative variation in biochemical and structural variables might have caused the disease in approximately the same way as has been described for squinting (Sect. 3.6.2.7).

Neurotransmitter Enzymes and Genetic Variability of Normal Behavior. Neurotransmitter enzymes do not only show deviations in affective or psychotic disorders; there is also an appreciable range of variability between normal individuals, and twin investigations showed that this variability seem to be largely genetically determined [2246, 2029]. For one enzyme, dopamine-β-hydroxylase, family studies have led to the detection of one genetic variant with zero activity in the serum [2238, 2240]. However, correlations with behavioral parameters or with the function of the autonomic nervous system - the sympathic synapses are adrenergic - have so far not been reported.

Neurotransmitters, like many other biologically active molecules, must react with receptors - mostly at the surfaces of cells - before they can exert their biological functions. Like enzymes, these receptors may show genetic variability. So far, most of the receptor diseases that have become known occur when receptor activity is reduced nearly to zero (Sect. 4.6.4), but genetic variability in the normal range has been established by a twin study for adrenergic receptors on human platelets [2165] and for the LDL receptor. Moreover, receptors might be examined like enzymes by various physicochemical separation methods, and such studies may uncover new, and functionally important genetic variability.

Other Possible Parameters of Brain Biochemistry That Could Influence Brain Function and Behavior. The brain has a very high demand for energy. It utilizes about 20% of the O_2 uptake. Moreover, it is very susceptible to O_2 deprivation; a few minutes are sufficient to cause irreversible brain damage. It was therefore sensible to look for genetic variability of glycolytic enzymes in human brain tissue [2149]. However, no polymorphism in glycolytic enzymes was found.

Critical Assessment of the Attempts to Relate Behavioral Variability to Biochemical Differences in Brain Function. In spite of suggestive hypotheses, attempts at explaining abnormal or normal behavior in terms of a genetically determined biochemical mechanism of deviant brain function have failed so far. There is not even a single model in which the chain of events from the genetic variant to the behavioral trait could be elucidated. Still, this approach appears worthwhile following. Considering the current stage of research a breakthrough is most likely to come in the field of affective disorders.

We should keep in mind, however, that norepinephrine, dopamine, and serotonin are only a fraction of all neurotransmitters; little if any genetic research has been done on some of the other neurotransmitters shown in Fig. 8.29. However, neurotransmitters may only be one aspect of brain function in which individual differences can be expected. So far, other approaches such as distribution of neurotransmitters in various regions of the brain, individual differences in growth of the brain, number of nerve cells, number of connections between nerve cells, or myelinization have not been studied at all. Besides, it was shown that even the development of synaptic connections between nerve cells can proceed properly only when the nerve cells are functioning. For example, the number of apical dendrites in the optical region of mouse brains is reduced if the animals are raised in darkness [55, 2225]. The occipital α rhythm of the human EEG is, on the average, less well-developed in adults who were born blind [16]. As noted above, interindividual variability of the EEG, and especially the development of the α rhythm, is exclusively genetically determined [921] under normal conditions. However, the brain structures that produce the EEG of the optical cortex can develop properly only if they receive adequate sensory input. The genetic program of brain development can be realized only in interaction with the environment.

Comparing our insight into genetic variability in function of the brain influencing human behavior with our knowledge of the genetic endowment of the red blood cell (Sect. 4.2, 4.3), we realize how fragmentary and almost nonexisting is our information regarding behavior. Whereas in a hemoglobin gene, we can follow the influence of a well-defined alteration in the DNA sequence step-by-step to the altered phenotype, in behavioral genetics we are left with measurements and comparisons of phenotypes that are far away from gene action. We have only taken the first steps in the analysis of the intervening variables. Are the mechanisms more complicated? Is their development in the individual less strictly programmed and less autonomous? Will interaction with the environment turn out to be more essential for proper development of brain function and even of brain structure? Such finding would not be surprising, since the human brain is the latest and most complex product of evolution. The few available results endorse this view [55].

At the beginning of this section, we noted that the field of human behavioral genetics is conceptually the most important area of human genetics. At the same time, however, our knowledge is still most unsatisfactory and fragmentary in this field, and our theoretical framework is least elaborate. In such a situation, scientific hypotheses find little objective foundation. Scientists are human beings with prejudices and emotions; they are influenced by their personal biases stronger than in fields with a more elaborate theoretical framework and a sounder empirical foundation. In behavioral genetics, one specific problem has raised the most bitter controversies: the problem of possible, genetically determined, intellectual differences between ethnic groups.

8.2.4 Differences in I. Q. and Achievement Between Ethnic Groups

Group Differences in Behavioral Traits? The human population is subdivided into subpopulations. They are called races if they have a certain amount of their genes in common in which they differ from other subpopulations. The term "ethnic group" is often used when historical and cultural aspects are included in the criteria of classification. However, genetically related groups tend also to have common cultural traditions and social systems; the two concepts "race" and "ethnic group" overlap strongly (Sect. 7.3.1). The genetic composition of races differs because they have developed under reproductive isolation, which might have produced chance fluctuations of gene frequencies and especially random fixation of alleles. Moreover, there may have been different selective conditions. Nowadays, migration with its accompanying gene flow tends to diminish group differences. It has already been shown in Sect. 7.3.1 that genetic susceptibilities to disease, such as diabetes, and the ability to digest certain foods, such as lactose, show different distributions in racial groups. It is therefore conceivable that the various environments in which human groups lived in the past made different demands on their behavior, thereby selecting for different combinations of genes that influence behavior. Chance fluctuation sometimes may also have affected such genes. The existence of some genetic group differences in behavior would therefore not be surprising.

Beyond this very general statement, however, more specific predictions are hazardous. We hardly know any human genes that influence behavior in the normal range; we know very little about the special abilities human groups needed to cope with the different microenvironments to which they were exposed in the past; we have even less knowledge about everyday living conditions of our remote ancestor. Did people living in cold areas with long winters need a better ability to plan in advance for food supply? Did hunters and gatherers in the tropical rain forest need more alertness and versatility in coping with sudden dangers? Were people who lived in open savannas and half-deserts accustomed to larger social groups, while the rain forest rather favored smaller band size as in subhuman primates? We simply do not know. Ethnologists have observed enormous, and in some cases extreme differences in behavior patterns in the same race and between groups living under conditions that appeared very similar. These differences may, in turn, influence the genetic composition of such populations, especially for genes affecting behavioral characters. Therefore, inferences from present-day primitive populations as to the behavioral patterns of our ancestors who supposedly lived under similar conditions are very hazardous. Nevertheless, such inferences are often made but depend significantly on whether the current primitive population selected for study is peaceful or aggressive, sexually restrained or uninhibited, co-operative or selfish. These very differences strongly indicate that simplistic genetic interpretations cannot apply.

To compare behavioral characteristics that could show genetic variability between races or ethnic groups within present-day "civilized" populations, comparisons should be confined to groups living under identical conditions, such as family patterns, education, chances to enter into various occupational careers, and other such functions. In conceptually simpler situations, for example, in animal breeding, we would not dream of drawing conclusions for selective breeding of animal stocks unless we had carefully kept the environment constant. In humans, on the other hand, comparable conditions hardly, if ever, exist. This difficulty makes all judgments ambiguous.

Two group differences have mainly been documented: The higher average intelligence and intellectual achievements of Ashkenazi Jews as compared with the gentile European and North American populations in which they live and the lower average I. Q. of the American black population as compared with the white and Oriental population groups of the United States. The Jewish-gentile difference may have contributed to - but certainly was not the only reason for - the antisemitic movements in many European countries that led to the genocide of most of the central and eastern European Jewish population by the Nazi government. The black-white difference is now providing racists with pseudoscientific arguments for discrimination of black population groups.

Intelligence and Achievement of Ashkenazi Jews. Ashkenazi Jews of Europe lived for many hundreds of years under conditions of severe discrimination. They were confined to restricted quarters within the cities, which were called ghettos; they were not allowed to own property and were barred from many occupations. The situation changed in the nineteenth and early twentieth centuries when civil rights were achieved in western European countries. Social discrimination of various degrees continued beyond this time. Nevertheless, many occupational careers had now opened up with the result that Jews were soon frequently found in many professions requiring high intellectual abilities.

In 1907, for example, about 1% of the German population was Jewish, but 6% of physicians and 15% of lawyers were Jewish [2113]. Among university professors, the fraction of Jews (including those baptized - the usual requirement for such a position) in 1900/1910 was: in the law faculties, 14.2%; in the arts and sciences faculties, 12%; and in the medical faculties, 16.8%. During the winter term 1924/1925, Jews were about six times more frequent among university students than in the general population. Comparable and more recent data are available from the United States. For example, 27% of the Americans who received Nobel prizes from 1901 to 1965 were of Jewish origin, while Jews comprise only about 3% of the American population (Zuckerman, 1977) [2065].

Most observers will agree that Marx, Freud, and Einstein were among those who most profoundly influenced civilization in the last century. All three were Jews of Ashkenazi origin.

In comparing test scores between ethnic groups carried out in the United States and Canada, the mean of Jewish subjects (almost entirely of Ashkenazi origin) was 5-10 I.Q. points higher than that of non-Jewish whites - especially in the verbal parts of I.Q. tests [2120].

What are the reasons for this unusually high performance of a relatively small ethnic group? Here, cultural explanations undoubtedly could account for some or even all of the difference. Socioeconomically, the group lived for centuries under conditions in which only intellectual performance could secure survival. A high emphasis on intellectual ability was characteristic of the cultural climate. High aspirations, encouragement of superior performance, and an intellectually rich environment favor intellectual development. In recent decades, the relatively smaller number of children in Jewish families might provide an additional explanation as high performance of children from small sibships, and especially for first-borns, has often been observed [2249]. The challenge of the "marginal outsider" to higher performance in most societies may have contributed to the results.

On the other hand, genetic explanations cannot be entirely dismissed. Selection in favor of intellectual capability with the Jewish communities favored the "scholars," i.e., those who were especially able to interpret traditonal texts such as the Talmud. These scholars were maintained by their communities and were given the opportunity to marry the most wealthy girls. Since they lived under more favorable economic conditions than the majority of Jews who were rather poorly off in Poland and Russia, the mortality of their infants may have been lower than that of the general Jewish population.

In fact, data from Poland from the middle of the eighteenth century suggest that the poorer Jewish families had 1.2-2.4 *surviving* children per family while the more prominent Jews had 4-9 children who reached adulthood. Another conceivable genetic factor is selective survival under conditions of persecution over the centuries. Possibly, the more nimble and smarter young adults might have been able to escape violent death more readily and therefore transmitted their genes to their descendants. The effect of such "I.Q.-dependent" mortality can be significant [2135, 2136].

In the absence of specific knowledge of the genetic mechanisms that may underlie individual differences in intellectual performance, we have no way to decide whether genetic factors have contributed to the intellectual excellence of the Jewish group. The means to a solution of this problem do not exist, and there is no way of tackling this question unambiguously at the present state of knowledge in human behavior genetics.

Difference in Mean I.Q. Between Ethnic Groups in the United States, Especially Between Blacks and Whites. Similar difficulties apply to a problem that has aroused an unusual degree of public controversy: the difference in average I.Q. between ethnic groups in the United States, especially blacks and whites [2196, 2188a]. The reader who has followed our considerations on human behavioral genetics, and especially the discussion of the heritability concept and its application to I.Q. test performance, should by now be able to provide his own answers. There are enormous difficulties in setting up an unassailable experimental design to assess the causation of a variable, such as I.Q., that is determined by

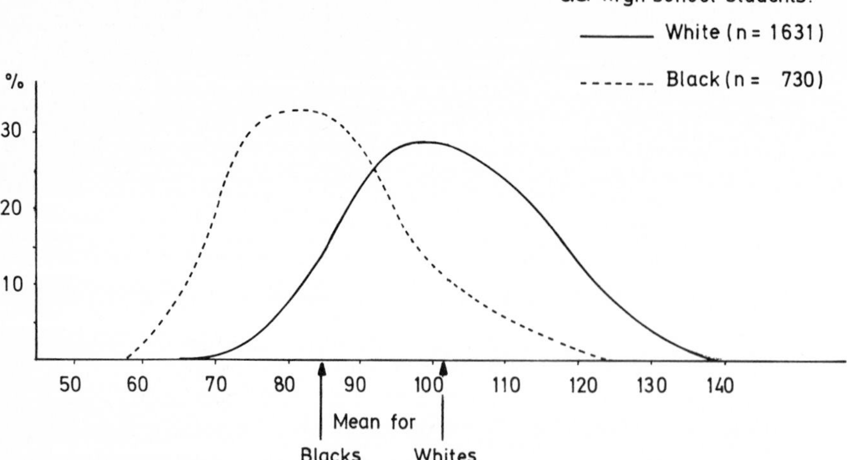

Fig. 8.32. Distribution of I.Q. in a series of white and black American high school students. (Data from Roland and Swan, 1965). See H. Walter, Grundriß der Anthropologie. München: BLV 1970. In other series some blacks were observed whose test results equaled those of the best whites

a complex interaction of a genetic disposition with many different influences from the physical and sociocultural environment. The problem is even more complex than that posed by I.Q. differences in Jews since the environmental conditions under which American blacks have been living in the past are much more different than those between Jewish and non-Jewish whites.

However, the problem has attracted much attention. Therefore, we shall at least briefly describe the most essential facts and some arguments in favor and some against genetic interpretations [2085, 2086, 2120, 2188 a].

Since intelligence testing was first carried out in U.S. Army conscripts of World War I, American blacks have fairly consistently shown lower average I.Q. values than white Americans. Figure 8.32 shows one characteristic result. Certain findings seem to be undisputed:

a) There is a mean difference of about 15 I.Q. points in most comparisons between American blacks and whites.
b) The distributions of blacks and whites overlap strongly; differences between individuals of each group may be much larger than the between-group differences of mean values. Some blacks reach highest levels of performance; the black population as a whole comprises the full range of human talents [2085].
c) There is considerable variability of means and distributions between subgroups within the white and black populations, depending on whether the samples come from the south or the north, from urban or rural areas, from children or adults, ect. [2120].

Explanation: Genetic or Socioeconomic? Two groups of explanations have been offered for the black-white difference: genetic and socioeconomic. In the following tabulation some of the arguments will be opposed to each other.

Genetic explanation	Nongenetic explanation
I.Q. has a high heritability within the white population [2085, 2086]	The within-population heritability estimates are of highly dubious scientific value [2214], (see Sect. 3.6, 3.8, App. 6) The data are often based on politically biased scientific evidence and some data were fraudulent [2039, 2054, 2055, 2117]
Cautious extrapolation from heritability of in-group differences to that between groups is possible [2086]	Extrapolation from heritability of in-group differences to between-group differences requires too many untestable assumptions and is scientifically unsound (see Sect. 3.8.10 on stature)
Sociocultural differences can only explain a minor part of the differences: a) Other, equally discriminated groups such as American Indians fail to show an I.Q. difference of similar order	The sociocultural and educational differences between whites and blacks are fully sufficient to explain existing differences
b) Differences are especially pronounced in so-called culture-free tests	Culture-free tests are not really culture free. Performance depends on:

Genetic explanation	Nongenetic explanation
c) Differences in standard intelligence tests are more pronounced in performance than in verbal tests	a) intellectual training b) active interest in solving puzzles that are not related to every-day problems
The I.Q. differences are not removed when tests using the special language of American blacks are applied	I.Q. test are based on the language of the white middle-class population. Tests using the language of American blacks show improved results
Contrary to their poor test performance (and their correspondingly poor school success), black children often show a remarkable degree of practical intelligence in daily-life situations. This contrast can be explained by two different levels of intellectual performance: level 1, the ability to process information in a simple, straightforward way that is sufficient for most purposes of daily life, and level 2, the ability to process and rearrange information in a more complex way that is necessary for abstract thinking. The deficiency of most blacks relates to level 2, not level 1 [2086]	The difference between practical and theoretical intelligence in blacks can be explained sociopsychologically: a) Blacks consider abstract thinking as less interesting b) Blacks experience school and its values as imposed from the outside, in which they do not actively take part c) Black children are not motivated and supported by their parents to strive for intellectual achievements
Only a small part of the test difference can be removed when the tests are given by black instructors	I.Q. tests have in most cases been given by white instructors with diminished motivation on the part of black children
Even if the average I.Q. of black children can be boosted by adoption (see opposite) into especially favorable white homes, there remains an I.Q. difference among these children in the following rank order (from high to low): white children of biologic parents/white adopted children/adopted black children with a higher share of white genes/adopted blacks with a lower share of white genes [2190]	Black children who have been adopted by white parents show I.Q. values that are above the average of the white population. This shows the powerful influence of the environment, which is favorable in families who adopt children [2190]. The I.Q. difference between adopted black children with a higher and those with a lower share of white genes can be explained fully by specified environmental differences [2190]
The U.S. Army refused a much higher fraction of black than of white army conscripts because of poor performance in intelligence tests. The black fathers of German children (see opposite) were selected for higher I.Q.	There is no I.Q. difference between German children whose fathers were black American soldiers and appropriate German controls [2120]
The differences in white admixture between the examined children [2188a, 2189] are so small that the expected I.Q. correlation could only be shown in a much larger sample. The correlation of skin color and black marker genes was so small in this study that lack of correlation between I.Q. and white admixture is difficult to interpret	There is no correlation between I.Q. and white admixture in black children [2188a, 2189]
Those not considering genetic reasons for the I.Q. difference between American blacks and whites are so preoccupied by their liberal ideology that they have lost their ability to think clearly and to face facts	Those considering genetic reasons for the I.Q. differences between American blacks and whites are racist reactionaries who consciously or unconsciously want to discriminate against ethnic minorities to maintain the privileges of their own class or ethnic group. Moreover, they are poor scientists

These arguments show how little even very sophisticated argumentation based on biometric studies may help in gaining decisive evidence as long as nothing is known regarding the biologic mechanisms involved. Given the present state of our knowledge of the biologic mechanisms that underlie genetic differences in intelligence within the normal range, attempts at elucidating the possible genetic reasons for group differences in intellectual performance - especially the I.Q. difference between American blacks and whites - appear futile.

Is All Research That Has Been Done in This Field Scientifically and Socially Worthless? We do not consider the work that has been done so far in this field as scientifically worthless. Studies on heritability have helped to expose the severe inherent limitations of this concept, especially in its application to human populations. Moreover, once well-defined genetic variability at the physiologic and biochemical level can be established, many of the methods and considerations will find useful applications. Experiences with numeric chromosome aberrations, for example, the XYY karyotype, have already shown the extent to which biologic studies can be improved by statistical, biometric, and epidemiologic sophistication.

Moreover, many results are interesting in their own right, even if the main purpose - to collect evidence for the causal explanation of group differences - cannot be achieved. It is, for example, useful to know how much I.Q. and school performance of adopted children can be boosted under the favorable conditions of adoption. The concept of two levels of information processing, even if basically incorrect in biologic terms, may help to develop teaching strategies better adapted to different requirements of various children than those used by our present-day school systems. In these and other ways, some of these attempts that seem to be futile when considered from the standpoint of a genetic analysis may gain new significance.

If Genetic Group Differences Did, Indeed, Exist, Would They Suggest Any Consequences in Social Policy? Let us assume for the sake of argument that a part of the I.Q. difference between blacks and whites is, indeed, genetically determined. What conclusions would we have to draw for social policy?

All considerations of this problem have to start with the contention that the individual and not the ethnic group should be the target of social policy. It would certainly be the goal of a society to create conditions in which every individual has a fair chance to develop his or her abilities and fit into society in a way that secures the highest possible degree of personal fulfillment - in a manner that, at the same time, best serves the requirements of society as a whole. One way of achieving this goal is to offer adequate facilities for education. This condition includes teaching methods that have an optimum effect on individually different learning abilities and motivations, regardless of whether and to what degree any differences in abilities are genetically determined. All experiences of modern genetics teach us that the phenotype is the result of an interaction between genotype and environment and that specific genotypes need appropriate, and in many cases also specific, environmental conditions for optimal development. It is the task of behavioral genetics to define these conditions and to provide advice regarding development of individual-oriented educational strategies that will best fit the genetically determined strengths and weaknesses of the child. Current educational approaches to dyslexia, which frequently is genetically determined [2071] are an example of such an approach.

Whether programs of compensatory education for children who show certain difficulties should be part of such strategies cannot be answered fully and will require future experience. These decisions have nothing whatsoever to do with the question of whether and to what degree genetic variability influences these deficiencies, not to mention the question of whether the individual belongs to a minority that differs in its population average from the majority. Should we push a slightly mentally subnormal Jewish child through all levels of higher education only because such a student belongs to a subpopulation with a higher average level of achievement? This is obviously absurd. Logically, it would be just as absurd to deny a black child compensatory education that could increase his or her chance of success in life because the child belongs to a minority in which a higher proportion of children are in need of compensatory measures. Nevertheless, racial sensibilities are often aggrieved when such steps are taken.

We realize that emphasis on optimum educational opportunities and consideration of the individual's abilities regardless of ethnic identification, while theoretically based on sound principles, may practically cause many problems at the present time. What teaching methods should a teacher use in a multiracial or multiethnic class with a wide range of abilities? Should the teacher give most attention to the lowest ability group? To the average? We have no ready answers. Is equal representation of all ethnic minorities desirable if some minorities excel in certain areas? Should there be different standards for minority representation for various occupational groupings? In our opinion, the ideal society would provide each individual, regardless of race or ethnic origin, with opportunities for maximum development of his or her abilities. It is conceivable that such a scheme may lead to over-representation or under-representation of some racial or ethnic groups, whether for genetic or cultural reasons.

In Sect. 3.8.10, the increase of stature during the recent century has been described. Stature has a high heritability under most environmental conditions. Despite high heritability, better nutrition has brought about a very significant increase in the population average.

This conclusion does not mean, however, that every environmental change will have effects; environmental alterations must be appropriate. Compensatory programs may fail, not because of high heritability of the characters they are meant to influence but because of their inadequacy to compensate for what is lacking. The geneticist has no reason whatsoever to discourage any such programs; he should strongly encourage all attempts by social and behavioral scientists to explore the specific conditions that cause individuals to do less well than others and to try to change these conditions.

Intermarriage. Marriage is a matter not of ethnic groups but of individuals. Individual members of two different ethnic groups might be much more similar genetically, i.e., they might have many more genes in common than two random individuals of a single ethnic group. This conclusion also applies for genes that might influence behavioral variability. We interpret all evidence on racial admixture to indicate that no biologic ill-effects for the children from racial intermarriage have been observed (Sect. 7.3.2). In modern societies in which spouses are selected individually and marriages are not arranged by families, there is strong assortative mating for behavioral phenotypes, such as intelligence (Sect. 6.3; Fig. 6.35). Such assortative mating will remain, regardless of whether intermarriage between groups will become more frequent or not.

Unlike the old national states of Europe, the societies of two world powers - the United States and the Soviet Union - consist of one ethnic majority and several strong minority groups. The fact that such minority groups exist creates tensions and conflicts. One could suggest that these conflicts could be solved most easily if the minority groups would be absorbed by the majority. It could be strongly argued, however, that the richness of a country's cultural heritage could be preserved best if minority groups were to keep their biologic and cultural identity. Such an advantage may be worth the sacrifice of living with the ensuing tensions and conflicts.

9 Practical Applications of Human Genetics and the Biological Future of Mankind

The living conditions of human populations have altered in recent generations and are still changing at an increasing rate in most countries. From the viewpoint of human health and life expectancy, the most spectacular changes have occurred due to progress in hygiene and medicine. These developments must influence human fertility and mortality and, hence, the genetic composition of future generations. Moreover, improving knowledge in human genetics has led to practical applications at an increasing rate - especially in genetic counseling and genetic screening. These approaches are promoted not primarily to improve future generations genetically but to avoid unnecessary hardships for present-day families. However, the widespread utilization of genetic counseling and genetic screening will also influence the genetic composition of future generations. More recently, molecular biology has provided increasingly efficient techniques for genetic diagnosis and, probably, therapy. Human geneticists need to consider whether these influences will be beneficial or not. What will be the impact of all these new developments on the human species? These problems will be considered in the following section.

9.1 Applications of Human Genetics

9.1.1 Genetic Counseling [71, 90, 101, 129, 136, 149, 205, 2258, 2293, 2323 a, 2351]

Genetic counseling has become an important area of applied human genetics, and an increasing number of patients request advice or are referred by their physicians for counseling about the diagnosis, impact, and recurrence risks of genetic diseases. As the public media and the medical literature disseminate more news about genetics, public and medical interest in genetic disease is growing further. What is genetic counseling? Genetic counseling refers to the totality of activities that (a) establish the diagnosis; (b) assess the recurrence risk; (c) communicate to the patient and family the chance of recurrence; (d) provide information and sympathetic counsel regarding the many problems raised by the disease and its natural history, including the potential medical, economic. psychological, and social burdens; and (e) provide information regarding the reproductive options to be taken including prenatal diagnosis, and refer the patients to the appropriate specialists.

The range of problems seen during genetic counseling covers a wide area. There is some variation because local expertise for certain diseases differs, but many conditions are usually encountered. Generally, only 30%-50% of the patients and families turn out to have classical genetic illnesses, such as monogenic diseases or chromosomal aberrations. Many consultations deal with various birth defects, mental retardation, delayed development, dysmorphic-looking children, short stature, and similar problems, which may or may not have a genetic etiology.

Genetic counseling is usually done by specialist physicians, and many doctors all over the world are now specializing in medical genetics. A new profession of "genetic associates" has also emerged in the United States. Genetic associates usually are women who have been trained in specialized post-baccalaureate 2-year university programs. They usually work with physicians in medical genetics clinics and carry out much of the pre- and postclinic visiting for information gathering, counseling, and follow-up. Their participation in genetic service activities has made it possible to use medical manpower more efficiently.

Genetic counseling is actually a medically oriented, rather than a "eugenic" endeavor. Most observers consider it inappropriate to advise couples regarding reproductive decisions that are based on eugenic considerations, even though the outcome of a given decision may worsen the genetic load by adding harmful genes that would otherwise be eliminated. Couples asking for advice are encouraged to make those reproductive decisions which are most appropriate for them regardless of possible deleterious effects on the gene pool of the population. This practice places genetic counseling squarely within the framework of medical practice, where the individual and his family rather than the population is the focus of advice and treatment. Fortunately, the course of action selected by most couples (i.e., limitation of reproduction when there is a high risk) coincides with a favorable impact on the population gene pool (see Sect. 6.3).

Diagnosis. An accurate diagnosis of a genetic disease using all the modalities of modern medicine is essential. Accurate diagnosis is emphasized, since similar phenotypes may sometimes have different modes of inheritance or may not be inherited at all. The family history will often be important since a clearcut pattern of inheritance such as in autosomal dominant traits will sometimes provide the basis for counseling when an definitive diagnosis may not be clear. Previous medical and hospital records are often helpful in making a precise diagnosis. Since many genetic diseases are associated with somewhat characteristic facial features an inspection of photographs of family members may be helpful. Chromosomal examinations are frequently required in the diagnosis of complex birth defects (see Sect. 2.2.2). Since many genetic diseases are rare, even specialists in a given field of medicine may have difficulty in arriving at an accurate diagnosis. The new specialty of clinical genetics occupies itself with diagnosis as well as with the management of genetic and related diseases. Even specialists in this field cannot be equally knowledgeable about all genetic diseases in every area of medicine, but need to be aware of various up-to-date monographs and compendia [13, 85, 133, 187, 231] so as to establish the appropriate diagnosis. The *Catalogue of Mendelian Traits in Man* by McKusick [133] is helpful, but often needs to be supplemented with other work for non-Mendelian diseases. A good library and the knowledge of how to consult the current literature in clinical genetics is essential. Because of the rapid expansion of knowledge, an examination of the journal literature, as against textbooks and monographs, is more important in clinical genetics than in most fields of medicine.

Parents whose infants are stillborn or die in the neonatal period often request genetic counseling regarding recurrence risks. Usually little or no information is available regarding the specific abnormality of the stillbirth, since no pathological or other diagnostic study was done. It has been recommended that as a minimum a gross autopsy, photography, radiography, and bacterial cultures should be performed in all cases of stillbirth or early neonatal death to allow a diagnosis, since this is needed for genetic counseling [71]. Chromosomal studies and histopathology are not likely to provide diagnostic information if the gross autopsy findings are normal. However. such studies are usually indicated if multiple anomalies are found on autopsy.

A definitive diagnosis often cannot be arrived at even by experienced specialists. The reason is the enormous complexity of development and its possible perturbation by known and mostly unknown genetic, epigenetic, and environmental factors. Fewer diagnostic uncertainties occur with monogenic diseases than with various birth defects. However, even in this area the growth of the McKusick catalogue over the years (i.e., from 866 definitive loci in 1971 to 1826 definitive loci in 1985) attests to the rapid expansion of knowledge in this field.
To keep up with this explosive advance in our knowledge, several research groups have placed information about clinical findings in genetic diseases and birth defects on computers and worked out programs to allow diagnosis using this information. While this approach has been useful to some physicians, many experienced clinical geneticists have not been convinced of the utility of current computerized diagnosis compared with conventional practice. Many clinical findings are subtle and are difficult to translate into words. Further developments are expected in this area, however, and are ultimately likely to be useful for clinical work.

Recurrence Risks. Genetic risks in Mendelian diseases are clearly defined and depend upon the specific mode of inheritance (Table 9.1). The actual clinical risks to the patient, particularly in autosomal-dominant inheritance, depend upon variable penetrance and expression and late onset of many disorders. Patients are more interested in the actual recurrence risk of the clinical symptomatology than the formal genetic risks alone. In diseases with lowered penetrance, the actual recurrence risk will be lower than the formal risk of genetic transmission. As an example, an offspring's risk for an autosomal dominant disease with 70% penetrance will be 35% rather than 50% ($0.5 \times 0.7 = 0.35$). With late onset of the disease, the risk becomes smaller as a person remains unaffected beyond the age at which the disease first becomes manifest. As an example, in Huntington's disease (14310), excellent data regarding the age of onset in affected patients exist and can be used to assess the specific risks more precisely [878] (see Fig. 3.4 and 9.1, App. 8). With increasing understanding of the human gene map, an increasing number of genetic diseases can now be diagnosed by study of the segregation of closely linked marker genes. DNA variation - particularly in noncoding DNA - appears to be ubiquitous and DNA variants will almost always be found with gene-specific probes (Sect. 2.3). This gene-specific approach requires molecular knowledge of the mutant gene that causes the disease (Table 9.2a). Linkages of disease loci with "anonymous" DNA markers have also been discovered and may be clinically useful (Table 9.2b). An attempt to link an "anonymous" DNA marker to a disease gene does not require the nature of the mutation of the genetic disease to

Table 9.1. Risks for rare Mendelian disorders in families of affected patients

Mode of inheritance	First-degree relatives at risk	Risk	Other relatives at risk	Risk
Autosomal-dominant	Sibs, parents, children (both sexes)	50%	Uncles, aunts, nephews, nieces, first cousins	25% 12.5%
Autosomal-recessive	Sibs (both sexes), children	25% Negligible[a]	Uncles, aunts, nephews, nieces, cousins	Negligible
X-linked recessive	Brothers, sisters as carriers	50%[b] 50%[b]	Maternal uncle, maternal aunt as carrier	50%[b] 50%[b]

[a] Risks for children of patients affected with common autosomal-recessive diseases depend upon gene frequency [highest risk (4%) for sickle cell anemia (0.08 × 0.5)].

[b] Risks are negligible when the disease in an affected patients is caused by a new mutation. Recent work suggests that the proportion of new mutations except in Duchenne muscular dystrophy may be much less than the expected 33% in X-linked lethal diseases [2396].

(The risks for X-linked recessive disease apply only if the disease is familial and not if the mother's carrier status is caused by a new mutation).

Table 9.2. Diagnosis and prenatal detection of genetic diseases by study of closely linked DNA (and other) markers

A) *Detection with gene-specific probes*[a]
- Sickle cell anemia (AR)
- Alpha and beta thalassemia (AR)
- Hemophilia A (XLR)
- Hemophilia B (XLR)
- Lesch-Nyhan syndrome (XLR)
- Familial hypercholesterolemia (AD)
- Phenylketonuria (AR)
- Alpha-antitrypsin deficiency (AR)
- Antithrombin III deficiency (AR)
- Growth hormone deficiency (AR)

B) *Detection by "anonymous" DNA variants linked to disease gene*[b]
- Huntington's disease (AD)
- Duchenne and Becker muscular dystrophy (XLR)
- Myotonic dystrophy (AD)
- Hemophilia A (XLR)
- X-linked mental retardation (XLI)
- Retinoschisis (XLR)
- Polycystic kidneys (AD)

AD, autosomal dominant; *AR*, autosomal recessive; *XLR*, X-linked recessive; *XLI*, X-linked.

[a] Crossovers are highly unlikely, since DNA variants detected by gene-specific probes are tightly linked and extremely close to the mutational site within or around the affected gene.

[b] Crossovers between DNA variant and disease depend upon the distance of the DNA marker from the disease locus (see Sect. 3.4).

known. However, unless the disease locus has been assigned to a specific chromosomal site and a DNA marker from that chromosomal site can be found, the chance of finding such a linkage is low. As the human gene map becomes increasingly saturated with various DNA markers this approach promises to be increasingly successful.

The linked marker approach for diagnosis with both gene-specific and chromosome-specific DNA probes requires that a DNA variant exists in the family under study and that a sufficient number of family members are available to allow an unequivocal assignment of the *cis* or *trans* relationship of the disease locus to the marker gene and its normal allele. There may be difficulties. As an example, it has been shown that the family structure in Huntington's disease [2307] will not allow preclinical diagnosis with linkage methodology in most cases, since the relevant family members are often dead or are too young to be affected and therefore cannot be used for marker-disease gene assignments.

Diagnosis using the linkage approach is only indicated when an diagnosis cannot be made or is not certain by conventional methods. Thus, the DNA approach via linkage is invaluable for prenatal diagnosis, for preclinical diagnosis of late onset conditions, or for carrier detection of X-linked diseases and to a lesser extent for carrier detection of autosomal recessive disorders. In contrast, DNA analysis is not used for the postnatal diagnosis of diseases such as sickle cell anemia, hemophilia, or phenylketonuria where appropriate biochemical tests for the expressed gene products rather than for the gene can be done on the patient's blood and require no family study. Occasionally, a direct DNA diagnosis is possible when the nature of the mutation is known. Because of mutational heterogeneity (i.e., different mutations causing the same disease), this approach has limitations unless the exact mutation being searched for can be specified. We know already that many different mutations occur in hemophilia A and in familial hypercholesterolemia [685] but not in sickle cell anemia and alpha antitrypsin deficiency [43].

Genetic advice in multifactorial conditions, such as in birth defects, common diseases of middle life, and major psychoses, lacks the precision possible with counseling involving Mendelian genes since the number of genes and their relative contributions are usually unknown. Empirical risk figures based on the frequency of recurrence of the disease in many affected families need to be used for counseling. These recurrence risks are usually lower than those in the Mendelian diseases and range between 3%-5% for many common birth defects, such as the neural tube defects and cleft lip and palate. Risks for first-degree relatives (sibs, parents, and children) for the more common diseases of middle life such as hypertension, schizophrenia, and affective disorders range between 10%-15%. Careful distinction of the rare monogenic variety of a disease that appears multifactorial must always be kept in mind. Thus, rare patients with gout may have an X-linked disease due to HPRT (hypoxanthine-guanine phosphoribosyl-transferase) deficiency (30800) or their gout may be caused by the autosomal-dominant PRPP (phosphoribosylpyrophosphate synthetase) deficiency (13894). Among male patients with coronary heart disease below the age of 60 years, about 5% have familial hypercholesterolemia - an autosomal-dominant trait (14440) (see Sect. 4.6.4).

Transmitted chromosome abnormality, such as translocations, do not segregate by Mendelian ratios, and counseling must be based on empirically derived risk figures (Sect. 3.3.6; App. 3).

The quotation of absolute recurrence risk in percent is more meaningful to a family than the communication of relative risks where the frequency of the disease is compared with that of the general population. A 100-fold increase for a condition that occurs in the population with a frequency of 1:100,000 carries an actual risk of only 1:1,000 - a negligible recurrence risk. In Mendelian conditions, recurrence risks are fixed regardless of whether several or no affected children preceded. Chance has no memory! In multifactorial diseases such as congenital heart disease or cleft lip and palate, if two or more first-degree relatives are affected in a given family, more disease-producing genes are operative in that family, and the risk for future offspring becomes higher than the usual 3-5% [71]. Detailed discussions of the approaches to genetic counseling and risk data for many different types of diseases can be found in recent books [71, 91].

Communication. The meaning of genetic risks must be conveyed in terms understandable to patients. The probability that 3%-4% of all children of normal parents will develop serious birth defects, genetic diseases, or mental retardation should be communicated as a measure against which additional risk can be gauged. There may be problems in communicating the extent of uncertainty. For example, with a sporadic case of an undiagnosable birth defect the risk might be zero if the disease is nongenetic, 2%-3% if there is a multifactorial etiology, and 25% if it is caused by an autosomal recessive trait. An integrated empirical risk based on the probability of the various possibilities is often given as an empirical risk. Such a risk in this example might be 5% on the assumption that monogenic recessive varieties of this birth defect tend to be rare. However, many counselees prefer to be told about the full extent of uncertainty rather than being offered a single risk figure [2265]. The burden of the disease must be clearly explained. Very severe but invariably fatal conditions in early life carry a less severe burden to the family than those associated with chronic crippling disease. Various reproductive options and alternatives must be discussed. Since problems under discussion may be complex and prove to be emotionally difficult to the patient, it may be necessary to have several counseling sessions. In any case, the counselor should provide a written summary in lay language.

Consanguinity. First cousins and more remote relatives who contemplate marriage occasionally ask for advice about the risks of having children with inherited diseases. There are laws which make marriages between cousins illegal in over half of the United States. Consanguinity definitely increases the risks of disease caused by homozygosity for recessive genes (Sect. 6.3), but the absolute risks are relatively low. It has been estimated that the rate of various diseases, birth defects, and mental retardation among offspring of first cousin matings is at most twice the background rate faced by any couple. Thus, the chance that a child from such a mating will be normal is around 93%-95%. These risks are still lower for more remote consanguinity and are difficult to discriminate from the population background rate for such disorders. There are no additional risks for offspring of a normal person married to an unrelated person when one partner has consanguineous parents. On the other hand, the risks are considerable for children of incestuous matings involving first-degree relatives, such as sib-sib and father-daughter matings (see Sect. 6.3.2.4). There is an almost 50% risk that the child will be affected by a severe abnormality, childhood death, or mental retardation (Sect. 6.3.2.4). It is therefore advisable that children of incestuous matings who are to be placed for adoption be observed for about 6 months before the adoption is finalized. By that time, most of the potential defects should have become apparent.

It is remarkable that where defects have been found in offspring of consanguineous matings these have mostly taken the form of nonspecific congenital malformations, childhood death, and mental retardation rather than well-defined autosomal recessive diseases. However, detailed studies searching for the many different recessive inborn errors of metabolism have not been done, and it is likely that a significant proportion of childhood deaths involved unrecognized inborn errors.

It is possible that in societies where inbreeding has been practiced for many generations (such as in India) the risks for offspring of consanguineous matings are lower, since selection against homozygous gene combination would have removed many such genes over the generations (Sect. 6.3).

Heterozygote Detection. Heterozygote detection is particularly important in sisters of boys affected with X-linked recessive diseases, such as hemophilia (30670) and Duchenne-type muscular dystrophy (31020). Regardless of their husband's genetic constitution, there is a 50% risk that the sons of female heterozygotes will be affected with these diseases. In contrast, autosomal-recessive diseases become evident only when *both* parents are heterozygotes; a heterozygote sib of an affected patient must mate with another heterozygote for the disease to occur. The chance that an unrelated mate of a person who is a carrier of an autosomal-recessive disease will also be a carrier is usually quite low.

Specialized laboratory tests for carrier detection (such as creatine phosphokinase enzyme assays for Duchenne-type muscular dystrophy and a combined assay of antihemophilic globulin clotting and antigenic activity for hemophilia A) (30670) may be helpful [2302, 2254].

Such tests, however, must be carefully standardized on normal subjects and obligate heterozygotes before applying them for individual carrier identification [2254]. The detection of heterozygotes is accurate and simple in the hemoglobinopathies. An increasing number of heterozygote states for various autosomal enzyme deficiencies such as the hexosaminidase deficiency of Tay-Sachs disease (23070) can also be recognized [2320].

If there is overlap in laboratory results between normals with a low value and carriers with a high value, the significance of an identical laboratory result in various individuals may differ depending upon the a priori probability of the tested person being a carrier. Tests that are excellent for carrier detection in sisters of males affected with X-linked diseases may give too many "false positives" in screening studies of extended kindreds or particularly in the general population where the probability that the tested subject ist a carrier is slight [2300, 2349]. As an example, 5% of the normal female population would be identified as

Table 9.3. Genetic diseases: carrier detection[a] advisable for reproductive decisions

Disorder	Mode of inheritance	Method of carrier diagnosis	Preventive measures in carriers	Population affected
Duchenne-type muscular dystrophy	X-linked	CPK level DNA variants	Amniocentesis	All
Hemophilia	X-linked	AHG level and AHG cross-reactive material, DNA variants	Amniocentesis	All
Lesch-Nyhan syndrome	X-linked	HPRT assay DNA variants	Amniocentesis: HGPRT assay	All
Translocation Down's syndrome	Empirical recurrence risks apply	Chromosomal tests for balanced carrier	Amniocentesis: chromosomal study	All
Sickle cell anemia	Autosomal-recessive	Hemoglobin electrophoresis	Genetic counseling, intrauterine diagnosis	Blacks
β-Thalassemia major	Autosomal-recessive	Red cell abnormalities, Hb A-2 increased	Genetic counseling, intrauterine diagnosis	Mediterranean and tropical populations
Tay-Sachs disease	Autosomal-recessice	Hexosaminidase A assay	Amniocentesis, hexosaminidase assay	Ashkenazi Jews

[a] Limited to translocation Down's syndrome, X-linked, and *frequent* autosomal-recessive diseases. Detection of carriers for many rare autosomal inborn errors associated with enzyme deficiency is possible, but the risk for normal sibs of affected patients to have affected offspring is very small since the frequency of the carrier state for such inborn errors is very low in the population.

carriers for hemophilia using the same standards that identify sisters of hemophilic boys as heterozygotes with a high probability. Some conditions where carrier detection tests should be done are listed in Table 9.3.

In some of these situations, additional statistical techniques can occasionally be helpful for refinement of a genetic prognosis. For example: A woman's brother is affected with an X-linked recessive condition; a maternal uncle was also affected. She has therefore a 50% risk to be heterozygous. Assume that she has already two nonaffected sons and that a test for heterozygote detection is not available. The information that her two sons are normal reduces her chance of her being a carrier. Alternately, such a woman may have a negative result for a test that detects 90% of heterozygotes. In this case, her risk of being a carrier is very low. Appendix 8 and Murphy and Chase's book [149] deal with the statistical principles of calculating the exact recurrence risks in such situations.

The increasing availability of DNA markers will make carrier diagnosis in X-linked diseases more effective (Table 9.2). Already several probes (both gene-specific and X-chromosome-specific) exist which recognize DNA variants linked to the hemophilia A and B genes. Using several hemophilia A probes, all females are heterozygous for one or the other of these variants. Carrier diagnosis therefore usually will be possible in families where the hemophilia A gene is segregating. The situation is similar for Duchenne muscular dystrophy where DNA variants that are physically close to the Duchenne muscular dystrophy gene are available. The information from DNA markers can be combined with CK tests and pedigree information (see Appendix 8 for an example with detailed calculations). Extensive work with X-linked mental retardation is being carried out, which should provide the appropriate probes for carrier detection of this common condition.

Reproductive Options and Alternatives. If a couple decides that the risks for further reproduction are too high, several options besides contraception should be discussed. Adoption is becoming less practicable because fewer babies are available. Sterilization of either husband or wife may be considered, but it must be emphasized that this is an irreversible procedure. Thus, sterilization is undesirable for prevention of autosomal-recessive conditions, because remarriage to a non-carrier after possible divorce or death could eliminate the genetic risks almost entirely. Artificial insemination by a donor other than the husband may be acceptable to rare couples to prevent autosomal recessive disease or autosomal-dominant disease contributed by the husband.

Detection of Genetic Diseases in Relatives. Optimal genetic counseling in some diseases should include the testing of relatives at risk (Table 9.4). In some conditions, the detection of latent disease in relatives may be life-saving if followed by suitable therapy. A sib of a patient with Wilson's disease has a 25% chance of being affected but may be too young to exhibit overt symptoms. Sibs of patients with hereditary polyposis (17510) [2391] have a 50% chance of being affected and therefore carry the certain risk of malignant transformation in one of the many polyps in this condition. In general, vigorous attempts should be made to examine relatives when a genetic condition causes serious preventable or treatable diseases. A case can also be made for early detection of diseases, such as polycystic kidneys (17390) [2276]; [2369 a] to allow those affected better reproductive decisions, choice of life style and appropriate occupations, and better preparation for ultimate renal transplantation or dialysis. Possible carriers for serious X-linked diseases (such as hemophilia and Duchenne-type muscular dystrophy) and for chromosomal carrier status (such as Down's syndrome associated with translocation) should be searched for in families for prevention by antenatal diagnosis (see below).

Directive vs Nondirective Genetic Counseling. After genetic advice that includes an estimate of the recurrence risk (see below) has been given, parents need to decide whether to have further children or not. Many physicians are paternalistically inclined and are accustomed to giving specific directive advice and counsel for or against future pregnancies. In the practice of medical genetics, however, a fairly strong tradition of nondirectiveness has developed. Some of this nondirectiveness may have sociological reasons. In the United States of America, when genetic counseling was first started some 30 years ago it was usually carried out by nonphysician geneticists, who lacked the medical profession's tradition of dispensing directive advice. However, the nondirectiveness of genetic counseling fits in well with recent trends towards increasing patient autonomy. Since each family is unique and reactions to risks vary, nondirectiveness fosters mature decision making. However, absolutely neutral advice is rarely possible or even desirable. The person or family requesting advice usually want and need more than a computer-like professional who only dispenses facts. The counselor may emphasize the more positive or the more frightening aspects of a given disease. These feelings will tend to affect the counseling process di-

Table 9.4. Treatable and preventable *adult* genetic diseases with autosomal-dominant inheritance for which search in family members of affected patients is mandatory

Disorder	Method of diagnosis	Treatment	Advantages of early diagnosis and treatment
Hemochromatosis (autosomal recessive)	Transferrin saturation, ferritin levels, liver biopsy most reliable	Venesection	Prevents liver, heart, and pancreatic disease
Hereditary spherocytosis	Incubated osmotic fragility test	Splenectomy	Prevents anemia and gall stones; protects against splenic rupture
Hereditary polyposis	Coloscopy	Colectomy	Prevents colon cancer
Gardner's syndrome	Coloscopy, Benign cysts, lipomas, fibromas on physical examination	Colectomy	Prevents colon cancer
Familial hyperparathyroidism	Serum calcium, phosphorus, parathyroid hormone	Surgery	Prevents renal damage and other complications of hypercalcemia
Multiple endocrine adenomatosis	Serum calcium, phosphorus, blood sugar, gastrointestinal, and skull X-ray	Surgery	Prevents complications of hyperparathyroidism, hypoglycemia, peptic ulcer
Medullary thyroid carcinoma pheochromocytoma syndromes	Calcitonin, measurement of blood pressure	Surgery	Prevents thyroid carcinoma and complications of hypertension
Familial hypercholesterolemia	Serum cholesterol, LDL receptor	Diet, drugs?	Prevents premature coronary heart disease
Malignant hyperthermia	Serum creatine phosphokinase	Avoid general anesthesia	Prevents fatalities during general anesthesia
Acute intermittent porphyria	Phorphobilinogen deaminase in red cells	Avoid precipitating drugs	Prevents abdominal and neurological symptoms

rectly or indirectly, by nonverbal clues. A cup may be half full or half empty, and the positive or the negative aspect of such facts may be stressed more vigorously. Nor do all couples have the necessary educational background and social and emotional maturity to make fully informed decisions. A hasty assessment of the family in these matters, however, may lead to a more paternalistic stance in advice giving than is desirable. Nevertheless, many couples expect the medical geneticist, who has the required knowledge and experience, to aid them in making their decisions. "What would you do if you were in my position?" is a frequent question from counselees, regardless of background. However, since a couple's situation, religion, and cultural background may be quite different from that of the counselor, the counselor's choice for his/her own circumstances may not necessarily be appropriate. Reproductive decisions will differ between couples even if the genetic facts and the disease burden are identical. In some diseases, such as Huntington's chorea, the prognosis is so dismal that many counselors take a fairly directive stance and urge abstinence from reproduction for those at 50% risk. Furthermore, cultural traditions vary in different countries. We have the impression that counseling tends to be more directive in central Europe and the eastern socialist countries than in the English-speaking world.

Assessment of Genetic Counseling and Psychosocial Aspects [2289, 2332, 2382]. Genetic counseling is a relatively new field, and its practice has not been standardized. Most professionals engaged in genetic counseling agree that counselees should achieve a good understanding of the medical significance and social impact of the disease, to allow them to make appropriate reproductive decisions. Some observers have measured the success of genetic counseling by subsequent reproductive behavior. If more couples with a high risk (>10%) were deterred from reproduction than couples with a low risk, genetic counseling was considered to have been successful. This result has in fact been noted in several studies [2271]. Such a narrow end-point is considered an inadequate objective of genetic counseling. It would be better to know whether full sharing of information and comprehension of the disease and its recurrence risks has been achieved and whether all needs for information and psychological and social support have been met.

Various studies of genetic counseling agree that many patients after counseling are confused about recurrence risks and do not fully understand the nature of the disease. The largest study was carried out by a group of sociologists in the late 1970s in 47 genetic counseling clinics in the United States, involving 205 counselors and over 1000 female counselees [2382]. Many different conditions were included and both counselors and counselees were questioned about their experiences and assessment of the counseling process. The results showed that counselors tended to emphasize recurrence rates during the counseling sessions while counselees often were interested in causation, prognosis, and treatment of the disease - an area which according to the counselees often was not discussed as fully as desired. While both counselors and counselees generally were more interested in the medical and genetic aspects of the consultation, counselees occasionally had psychosocial concerns which were not addressed by the professionals.

In this study, 54% of counselees given a risk and 40% of counselees given a diagnosis were unable to report these data shortly after counseling. This failure of learning occurred independently of whether MDs, PhDs, or genetic associates had done the counseling and was unrelated to the experience of the counselor. Counselors with many years of experience had no better results than more recent graduates. In several other studies the results if comprehension of the risks were substantially better but by no means perfect [2289]. Usually, but not always, education was correlated with a better level of understanding.

Genetic counseling services have been used more extensively by families with good educational backgrounds than by more disadvantaged population groups. Couples who are motivated to learn about the disease and its recurrence risks are more likely to be affected in their reproductive decisions by the information provided than those who have been referred and are not always certain about the purpose of the genetic consultation. Thus, self-referred patients also tended to have better comprehension of genetic counseling information.

In another study perception of counseling information was investigated [2334, 2335]. Perception of recurrence rates was often not utilized by the counselees in the probabilistic sense represented by the figures given. Percentage risks were more frequently perceived as binary, i.e., even with lower risks it was believed that the disease either would or would not occur, with all the attendant fears of recurrence. Parents were then overwhelmed by multiple uncertainties such as how to make reproductive choices, how others would react to their decision, what it would mean to have an affected child, and whether they would be able to fulfill their role as parents. Such perceptions appeared more important for decision-making than the actual facts of diagnosis, prognosis, and risk. These data show that there often is a discrepancy between the mental set of the scientifically oriented counselor and the thought processes of the counselees, who find it difficult to deal with probabilistic information. Bridging this gap is a real challenge.

Genetic counseling - as currently practiced in most countries - places less specific emphasis on emotional aspects as compared to "counseling" activities in other areas such as psychological and marriage counseling. Some observers have recommended that more attention be given to the psychodynamic aspects of genetic disease [2325, 2270, 2287]. It is our experience, however, that psychologically oriented genetic counseling, in which a significant amount of time is spent on depth psychology and psychodynamics, is rarely required. If there are deep psychological problems, referral to a psychiatrist or psychotherapist is the most appropriate course of action.

Extensive psychological counseling by genetic counselors does not appear indicated and would require more extensive training in clinical psychology than is currently provided. An empathic and understanding approach to families with an awareness of the many social and psychological aspects of the disease and support in these matters needs to be encouraged. Genetic counseling is more than mere diagnosis, risk assessment, and "cold" dispensation of information.

There are imperfections in the genetic counseling process as currently practiced. Nevertheless, most educated counselees who receive definite information about the matters troubling them usually appear to be satisfied, and the majority of counselees with low risks are relieved to find that their actual risks are much lower than they had feared.

The interaction of patients and professionals in any encounter has many variables, and scientific study of this process is difficult. Nevertheless, genetic counseling as a new field demands further investigation of the process, its psychosocial effects, and outcomes, so that optimum results can be worked out. Controlled studies comparing patients who received counseling with those with a similar disease who did not would be particularly interesting.

Prenatal Diagnosis [2312, 2357, 2373, 2286]. The field of prenatal diagnosis has grown rapidly in recent years and has altered the practice of genetic counseling. Specific information regarding the possibility of antenatal diagnosis is now usually provided in genetic counseling situations. Prenatal diagnosis substitutes definite information for a probability of recurrence - a much more acceptable outcome to most individuals.

Antenatal diagnosis includes a variety of techniques (Table 9.5), among which amniocentesis and ultrasonography are most frequently used.

Amniocentesis (Fig. 9.2). Amniocentesis is carried out at the beginning of the second trimester of pregnancy (15-17th week of pregnancy) by transabdominal puncture. The procedure has proved to be safe in the hands of trained obstetricians, but is not 100% harmless. There is a slight risk of fetal loss (0.5%). Infection and hematomas are much rarer still, and other obstetric complications are even rarer. The procedure is done in an outpatient setting in conjunction with ultrasonography, a procedure that decreases the failure ratio and diminishes the occurrence of blood-stained fluid and fetomaternal hemorrhage. A chromosomal study requiring culture of the aspirated amniotic cells of fetal origin is usually carried out and the results are obtained 2-3 weeks later.

In addition to chromosomal aberrations, many enzyme deficiencies can be detected in amniotic cells (Table 9.6). Since individual enzyme deficiencies are

Table 9.5. Prenatal diagnosis for genetic diseases and birth defects

Amniocentesis or chorionic villus sampling	
	Chromosomal disorders
	Fetal sexing
	Inborn errors of metabolism
	All disorders detectable with DNA methods
	Open neural tube defects (amniocentesis only)
Ultrasonography	
	Neural tube defects
	Structural malformations
Fetoscopy	
	Blood aspiration
	Coagulopathies
	Immunodeficiencies
	Marker (X) syndrome
	Inspection
	Gross visible malformation (limb defects)
	Biopsies
	Epidermolysis bullosa and others
	Liver biopsy
Maternal blood screening	
	Neural tube defects

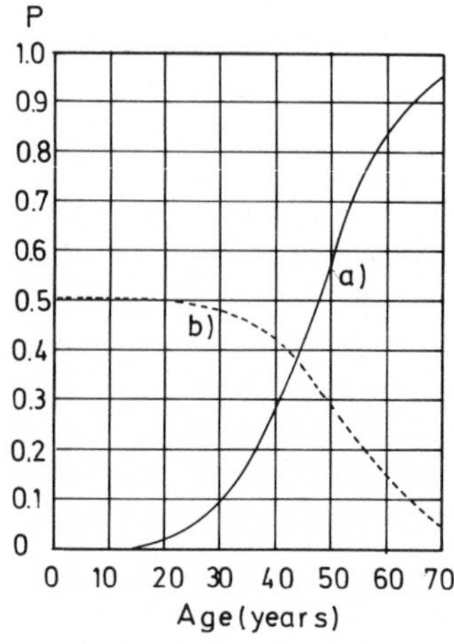

Fig. 9.1. Cumulative age at onset (—) and risk of becoming affected (- - -) for a so far unaffected son or daughter of a patient with Huntington's disease [2308]

Table 9.6. Prenatal diagnosis of inherited inborn errors of metabolism. (Modified from [2308])

Disorder	Usual enzyme deficiency	Comments
Acid phosphatase deficiency (lysosomal)	Acid phosphatase	
Adenosine deaminase deficiency (combined immuno-deficiency)	Adenosine deaminase	
Adrenogenital syndrome	21-Hydroxylase	Amniotic fluid analysis; also indirectly by HLA linkage
Adrenoleucodystrophy	Long-chain fatty acid defect	
Argininosuccinic aciduria	Argininosuccinase	Argininosuccinic acid also raised in amniotic fluid
Citrullinaemia	Argininosuccinate synthetase	
Cystinosis	Unknown	Accumulation of intracellular ^{35}S-labeled cystine
Fabry's disease	α-Galactosidase	X-linked; variable expression
Farber's disease	Ceramidase	
Fucosidosis	α-L-Fucosidase	
Galactosaemia (classic)	Galactose-1-phosphate uridyl transferase	Treatment available
Galactosaemia (galactokinase deficiency)	Galactokinase	Relatively benign and treatable disorder

Table 9.6 *(continued)*

Disorder	Usual enzyme deficiency	Comments
Gaucher's disease	Glucocerebrosidase	Heterogeneous
Generalized gangliosidosis	β-Galactosidase	
Glucose-6-phosphate dehydrogenase deficiency	G-6-PD	Often mild. Many enzyme variants. X-linked
Glutaric aciduria	Glutaryl-CoA- dehydrogenase	
Glycogenosis Type I (von Gierke's disease)	Glucose-6-phosphatase	Fetal liver biopsy possible
Glycogenosis Type II (Pompe's disease)	α-1,4-Glucosidase	Heterogeneous
Glycogenosis Type III	Amylo 1-6-glucosidase	
Glycogenosis Type IV (Andersen's disease)	Brancher enzyme	
Homocystinuria	Cystathionine synthetase	Heterogeneous
Hyperammonemia, X-linked	Ornithine carbamyl transferase	X-linked; variable expression in female. Fetal liver biopsy
Hypercholesterolemia, familial	Low-density lipoprotein receptors	Homozygote only detectable; detection of heterozygote likely with gene probe
Hypophosphatasia	Alkaline phosphatase	Only severe infantile type detectable
'I' cell disease (mucolipidosis II)	? Lysosomal membrane defect	Increase in multiple lysosomal enzymes
Krabbe's disease	β-Galactosidase	
Lesch-Nyhan syndrome	Hypoxanthine-guanine phosphoribosyl-transferase	X-linked recessive. Heterogeneous
Mannosidosis	α-Mannosidase	
Maple syrup urine disease	α-Ketoacid decarboxylase	
Menkes disease	Defective copper metabolism	X-linked. Abnormal copper uptake
Metachromatic leukodystrophy	Arylsulfatase A	Heterogeneous
Methylmalonic aciduria	Methylmalonyl CoA mutase	Methylmalonic acid detectable in amniotic fluid. May be treatable *in utero*. Heterogeneous
Mucopolysaccharidosis I (Hurler's syndrome)	α-L-Iduronidase	MPS IS (Scheie syndrome) has same enzyme deficiency. Amniotic fluid MPS levels useful in Types I, II and III
Mucopolysaccharidosis II (Hunter's syndrome)	Iduronate sulfatase	X-linked; enzymatic diagnosis possible from amniotic fluid as well as cells
Mucopolysaccharidosis III A (Sanfilippo A syndrome)	Heparan sulfate sulfatase	
Mucopolysaccharidosis III B (Sanfilippo B syndrome)	α-*N*-Acetylhexosaminidase	Carrier detection feasible on serum
Mucopolysaccharidosis IV (Morquio syndrome)	Chondroitin sulfate sulfatase	
Mucopolysaccharidosis VI (Maroteaux-Lamy syndrome)	Arylsulfatase B	
Niemann-Pick disease	Sphingomyelinase	Heterogeneous
Phenylketonuria (classic)	Phenylalanine hydroxylase	Treatable DNA probe applicable to some families
Phenylketonuria (Dihydropteridine reductase type)	As indicated	Severe and difficult to treat
Porphyria, acute intermittent	Porphobilinogen deaminase	Autosomal dominant
Porphyria, congenital erythropoietic	Uroporphyrinogen cosynthetase	
Propionic acidaemia	Propionyl-CoA-carboxylase	Also directly detectable from amniotic fluid
Refsum's disease	Phytanic acid oxidase	Possible. Not actually confirmed
Sandhoff's disease	β-N-Acetyl hexosaminidase (A and B)	
Tay-Sachs disease	β-N-Acetyl hexosaminidase A	Carrier detection and high risk population screening feasible
Wolman's disease	Acid lipase	
Xeroderma pigmentosum	DNA repair enzyme (endonuclease)	Heterogeneous

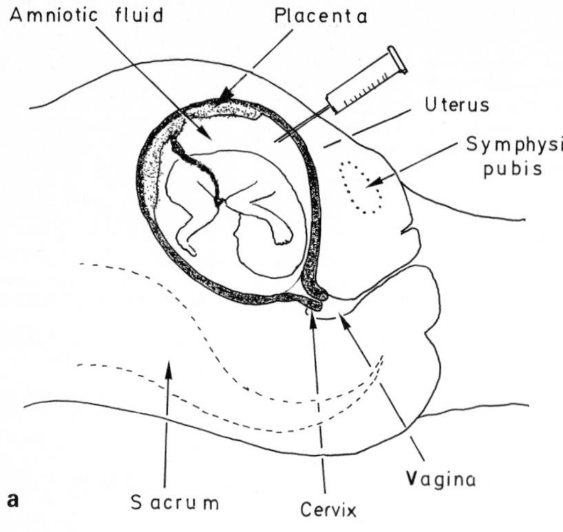

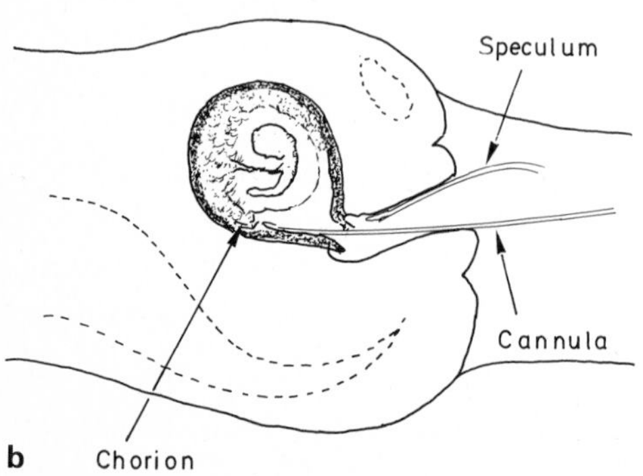

Fig. 9.2. a Amniocentesis. Puncture of the amniotic cavity through the abdominal wall. **b** Chorion villus sampling. The uterus is entered through vagina and portio vaginalis uteri. (Emery 1984 modified)

Table 9.7. Ultrasound in prenatal diagnosis

Obstetric indications	
Accurate gestational dating	
Multiple pregnancy	
Placental localization	
CNS disorders	*Skeletal defects*
Anencephaly	Severe bone dysplasias
Hydrocephaly	Congenital types of osteogenesis imperfecta
Encephalocele	Limb defects
Meningomyelocele	*Chest*
Spina bifida	Diaphragmatic hernia
Holoprosencephaly	Intrathoracic cysts
Microcephaly	Pulmonary hypoplasia
Abdominal/gastrointestinal	Small chest wall (various skeletal syndromes)
Gastroschisis	*Renal/GU*
Omphalocele	Renal agenesis
Duodenal atresia	Polycystic kidney (infantile)
Esophageal atresia	Severe obstructive uropathy
Various fetal tumors	

rare and the technical problems of assay are considerable, specialized reference laboratories ideally carry out the appropriate testing. In view of the rarity of most inborn errors, biochemical testing in prenatal diagnosis (unlike the search for chromosomal errors in older mothers) is never done routinely but only on specific indications in high-risk pregnancies (e.g., previously affected child).

Chorionic Villus Sampling [2267, 2318, 2323, 2346, 2379, 2399]. During the past few years, chorionic villus sampling has been introduced. This procedure obtains chorionic tissue of fetal trophoblastic origin under ultrasound guidance for cytogenetic, biochemical, or DNA testing. The procedure can be carried out between the 8th and 10th weeks of pregnancy and therefore has psychological advantages for the patient compared with amniocentesis, which is done at the 15-17th week of pregnancy. Cytogenetic results are available immediately or within one day. The specific risks of the procedure have not been fully established and extensive clinical trials to specify these risks are under way. The problems of abortion risk and diagnostic errors due to chromosomal mosaicism in the trophoblast are now under scruting.

Ultrasonography [2268]. Wide use of noninvasive ultrasound examination of fetuses is now allowing prenatal diagnosis of a variety of fetal anomalies. Table 9.7 lists the various diseases that can be detected. Ultrasonography has changed rapidly in recent years. Equipment such as real-time scanners has markedly improved and allows better discrimination of fetal morphology. However, the training and experience of ultrasound operators in fetal diagnosis vary considerably.

While all current studies indicate that ultrasound is harmless to the developing fetus, the indiscriminate application of the procedure in all pregnancies is the cause of some concern in the absence of absolute proof of its innocence. In some countries, however, repeated ultrasonographic examination has become part of routine pregnancy surveillance, and many instances of previously unsuspected malformations have been diagnosed. Various authoritative bodies (National Institutes of Health, USA; World Health Organization) suggest caution and recommend the use of ultrasonography only when definite maternal or fetal indications exist. The increasing technical perfection of ultrasonography complements other forms of prenatal diagnosis, particularly in the detection of neural tube defects.

Fetoscopy [2342, 2345, 2374]. Fetoscopy with small fiberoptic instruments allows entry into the amniotic cavity and is usually carried out between the 18th and 22nd weeks of pregnancy. Even in experienced hands this procedure carries a 5%–10% abortion rate. Inspection of the fetus to detect defects has limitations because of the restricted field of vision. Sampling of fetal blood under direct vision is possible and has been used in the diagnosis of the thalassemias as well as in the hemophilias and the mar (X) syndrome. Any genetic condition that manifests in fetal blood can be diagnosed. Fetal skin biopsies may be carried out and even fetal liver biopsies have been done to diagnose diseases that are only expressed in the liver.

Fetal Blood Sampling. Placental punctures have been done frequently to obtain fetal blood in the hemoglobinopathies, but because of a relatively high rate of fetal hemorrhage and need to repeat, such testing is used less frequently.

Maternal Blood Sampling [216, 2274, 2298, 2342]. Maternal blood sampling by venepuncture for alpha fetoprotein elevations as a screening procedure to detect neural tube defects and some other fetal anomalies (Table 9.8) has been carried out in many centers. The decreasing incidence of neural tube defects in most countries and their increasing detectability by ultrasonography has made definite recommendations regarding the universal adoption of such screening difficult. Alpha fetoprotein screening may also be useful in detection of Down's syndrome, since fetuses with trisomy 21 have a lower AFP level than normal fetuses (e.g., median: ~0.7 multiple of the median) [2261, 2275]. It has been shown that if amniocentesis for chromosomal study were to be offered to all women whose blood AFP level were at or below a specified level (i.e., 0.5 multiple of the median level) an additional 20%–40% of cases of Down's syndrome would be detected over those detected with current methods using amniocentesis at specified maternal ages. However, for each case of trisomy 21 so detected 150–200 additional amniocenteses on normal fetuses would have to be performed.

Other methods, such as the detection of fetal cells in the maternal circulation, are under investigation. The successful application of such a technique would be useful in screening of maternal blood for fetal cells with chromosomal and biochemical aberrations. However, even though some fetal cells exist in the maternal circulation, many technical difficulties need to be overcome before this procedure could be applied routinely.

Table 9.8. Abnormalities that raise and decrease amniotic fluid alpha-fetoprotein levels

Increased AFP	Lowered AFP
Neural tube defects	Down's syndrome
Spontaneous intrauterine death	
Omphalocele	
Gastroschisis	
Nephrosis (Finnish type)	
Sacrococcygeal teratoma	
Bladder extrophy	
Some skin defects	
Meckel's syndrome	

Indications for Prenatal Diagnosis. With newer modalities of prenatal diagnosis more indications exist, and an increasing number of fetal conditions can be diagnosed. Chorionic villus sampling is used for the same indications as amniocentesis and has the advantage of being available significantly earlier in pregnancy.

1. *Maternal Age:* Amniocentesis is most frequently performed to rule out Down syndrome and other chromosomal aberrations in women of "advanced" maternal age. In most countries, this age has been somewhat arbitrarily selected as 35 years, where the risk for Down's syndrome at birth is about 1/400, rising to 1/100 by age 40 and to 1/40 by age 44 (Table 5.4). The incidence of Down's syndrome and other chromosomal aberrations is significantly higher at amniocentesis than at birth, since many aneuploidies are aborted prior to birth (Table 5.4).
2. *Previous Aneuploidy:* A previous child with Down's syndrome or other autosomal trisomy slightly increases the risk of recurrence. The risk for Down's syndrome is about 1/250 for those < 35 years of age and is probably about twice the age-specific risk for those > 35 years.
3. *Parental Chromosomal Rearrangements* [310, 334]. The carrier status for translocations or pericentric inversion gives an increased risk of unbalanced, abnormal fetuses (e.g., translocation Down syndrome) (Sect. 2.2.2). The risks do not correspond to those expected from chromosomal segregation, but are based on empirical data, presumably because of selection against unbalanced gametes. The risk for translocation trisomy 21 is about 15% when the mother is a carrier and only 3% if the father is the carrier (t14q21 and t21q22q). In reciprocal translocations the risks of future affected offspring are significantly higher (~20%) if ascertainment occurs via an affected live offspring as against ascertainment by recurrent abortions (5% risk) (for details see Sect. 2.2.2.2). The more extensive unbalanced duplication/deletion (3–6 chromosome bands out of 200 total) are associated with lower recurrence risks (9%–16%) than those with duplications/deletions affecting only 1–2 bands (34%). Presumably, larger defects are often not viable and have already aborted spontaneously prior to amniocentesis.
4. *Risk for X-Linked Disorders:* Sex determination to detect male fetuses who are at 50% risk for serious X-linked disease is occasionally done. Here, chorionic villus biopsy early in pregnancy is more acceptable to most parents than amniocentesis. Increased availability of DNA methodology will

decrease the need to use male sex as a diagnostic criterion.

5. *Marker X Syndrome:* Diagnosis of this common mental retardation syndrome (Sect. 8.2.1.2) requires the aspiration of fetal blood by fetoscopy since the defect cannot yet be properly diagnosed on amniotic cells.
6. *Hemoglobinopathies* [2345]. Fetoscopy is used to obtain fetal blood, followed by assay of hemoglobin production to detect the various thalassemias. This approach is being replaced by DNA diagnosis using fetal cells that are obtained by chorionic villus sampling. Direct detection of the mutation in sickle cell anemia is possible (see Sect. 4.3) while the linkage approach with DNA variants is utilized for diagnosis of the thalassemias.
7. *Inborn Errors of Metabolism:* Enzyme assay of fetal cells needs to be done. The list of such conditions is large (Table 9.6). Diagnosis by the linkage approach with DNA variants is possible in classic phenylketonuria where the enzyme is not expressed in amniotic cells.
8. *Various Genetic Diseases by DNA Linkage Diagnosis (cf. Hemophilia A):* A list of such diseases is given in Table 9.2 and discussed in Sect. 2.3. This list will be growing rapidly in the next few years.
9. *Neural Tube Defects.* Amniocentesis (*not* chorionic biopsy) for amniotic fluid AFP is usually carried out in high-risk women such as those with previously affected children or following a confirmed maternal high blood AFP level. Ultrasound is very useful and always should be done in these circumstances.

Prenatal diagnosis is widely used in the developed countries, although a significant fraction of pregnant women above 35 years of age are not examined in many areas. Extensive public information via women's magazines in Denmark has led to the highest acceptance rate, with 80% of eligible women being studied. Often, the availability of prenatal diagnosis has encouraged parents to start a pregnancy in circumstances where fear of an affected infant previously deterred childbearing. While abortion for fetal indications has become accepted in many countries, a significant proportion of the population in the United States and elsewhere has strong feelings against termination of pregnancy for religious and other reasons. Some antiabortionists are particularly concerned about the making of value judgments regarding the continuation of fetal life as is done with genetic disorders. They feel that such practices are the beginning of the "slippery slope" that ultimately will lead to rejection of relatively minor defects in the search for the "perfect" baby and lead to a resurgence of eugenic and racist schemes. Fears have also been expressed that society will be less inclined to pay large sums of money to take care of children with genetic diseases where abortion could have avoided the birth of the disabled child. However, since most disabilities cannot be prevented at present, such trends have not emerged. Furthermore, the fact that most societies in recent years have given better financial and social support to handicapped people argues against these fears [2357].

9.1.2 Genetic Screening [2256, 2344, 2350]

With better understanding of various genetic diseases, public health applications have developed. It is reasoned that all members of a population at risk should be screened for a given defect if treatment or preventive measures are possible. Similarly, screening for certain genetic carrier states has been recommended to allow genetic counseling or intrauterine diagnosis before a sick person has been born. These programs are distinct from the usual retrospective genetic counseling where patients and families ask the advice of genetic counselors because a person with a genetic disease exists in the family.

Phenylketonuria Screening: Prevention of Mental Retardation [2256, 2333]. Phenylketonuria (26160) is one of the most common inborn errors of metabolism in populations of European origin with a frequency of about 1/10,000 births. The condition is an autosomal-recessive trait and is caused by a mutation affecting the enzyme phenylalanine hydroxylase, which fails to metabolize phenylalanine. The resultant buildup of metabolites damages the developing brain and leads to profound mental retardation. About 1% of residents of institutions for the mentally retarded have phenylketonuria. If diagnosed shortly after birth, an appropriate diet that restricts phenylalanine can prevent mental retardation. The disease can be diagnosed by simple and inexpensive blood tests obtained by puncture of the infant's heel before leaving the hospital (Sect. 4.2.2.7). Most developed countries have introduced phenylketonuria screening as a routine test for all newborns. A positive screening test does not necessarily mean that the baby has phenylketonuria because variants of hyperphenylalaninemia exist that may not cause mental retardation. Follow-up by a team of experienced biochemists and pediatricians is therefore required to assure that treatment is only administered if necessary since a phenylalanine-restricted diet may be injurious.

Furthermore, classical phenylketonuria must be distinguished by appropriate tests from defects that cause malignant hyperphenylalanemia, such as dihydropteridine reductase deficiency and errors in biopterin synthesis [1290]. Affected patients with these defects are not clinically influenced by phenylalanine-restricted diets.

The success of the PKU program has raised new problems, in that girls with PKU who had effective treatment during childhood and are no longer on diets are now becoming pregnant. The pregnant woman's phenylalanine levels are high, so that the developing fetus (who will be an obligatory heterozygote for PKU) is injured by the high phenylalanine levels of the mother. Multiple abortions, microcephaly with mental retardation, cardiac defects, and intrauterine growth retardation have invariably been found. Reinstatement of a phenylalanine-restricted diet before pregnancy is started should prevent these abnormalities. In view of logistic difficulties in identifying all mothers who have had PKU and the occurrence of unplanned pregnancies, it will be difficult to ascertain all women who need renewed phenylalanine-restricted diets.

Many other treatable inborn errors of metabolism can be tested on blood obtained for phenylketonuria testing. All of these conditions, such as maple syrup urine disease (24860), homocystinuria (23620), and galactosemia (23040) are much rarer than phenylketonuria [2333]. Pilot investigations involving testing for these conditions are now being widely carried out. The problem of "false positives" and the required follow-up study for each of these defects increase the logistical difficulties with each additional disease that is being tested for.

Cost-benefit calculations have sometimes been carried out to justify the cost of screening programs for inborn errors of metabolism. Such calculations contrast the financial costs of caring for a diseased child with the expenses of the screening program. Humanitarian aspects cannot be readily quantified and by necessity must be neglected in cost-benefit analyses. Even economic considerations are often treated simplistically in such analyses. For example, the disappearance of all cases of phenylketonuria will reduce the number of patients in hospitals for the mentally retarded by 1%. It is quite unlikely that such a change will have any but the most trivial effect on the budgets of such institutions. The full cost of all personnel, such as professionals who are paid from other funds but spend much of their time on screening programs, are often not taken into account [2316, 2359]. It is therefore considered dangerous to base the initiation of screening programs on cost-benefit analyses alone.

Screening Mothers-at-Risk for Chromosomal Malformations. The existence of a maternal age effect for many chromosomal defects makes it desirable to search for abnormalities such as Down's syndrome or trisomy 21 in older mothers. Currently, women over the age of 35 are advised to have amniocentesis, since the risks for trisomy 21 rise steeply with increased maternal age (Sect. 5.1.2.2). Because of small family size, such as in the United States population, most cases of trisomy 21 will be born to younger women [2313], and the population impact of amniocentesis is less than ideal. The procedure, however, is highly recommended for the individual pregnant woman over the age of 35.

Screening for Autosomal-Recessive Traits. Certain heterozygote traits are common in some populations. Screening of these populations for the carrier state would identify carrier couples who will have a 25% chance of affected children.

Following screening, carriers should be counseled regarding the genetic and medical risks of the disease and provided with information regarding the various reproductive alternatives. These would include: (I) avoidance of mating with another carrier, (2) avoidance of child-bearing if married to another heterozygote, and (3) amniocentesis and possible abortion if married to another heterozygote for conditions where techniques for intrauterine detection are available. Matings of heterozygotes with persons who are not carriers would have no untoward medical or genetic consequences. It is not difficult to see that the options of avoiding a mate who happens to carry the identical genetic trait and preclusion of child-bearing are not very popular.

Therefore, screening programs have been most successful in relatively well-informed populations for traits where intrauterine diagnosis was available. The carrier state for Tay-Sachs disease (27280) occurs at a frequency of about 4% among Ashkenazi Jewish populations. The hexosaminidase deficiency characteristic of the defect can be readily detected in serum, and heterozygotes can be identified. Fetuses with Tay-Sachs disease can be identified by assay of enzyme in fetal cells aspirated by amniocentesis and grown in tissue culture. Programs for Tay-Sachs screening have been carried out in many metropolitan cities of the United States, and many fetuses with Tay-Sachs diseases have been identified and aborted [2320, 2321]. In families with previous cases of Tay-Sachs disease, the program has been life-giving by allowing couples to have healthy children as assured by intrauterine diagnosis. Without amniocentesis, such couples would practice contraception. The programs have required a fairly high level of publicity to attract those at risk for screening. Even with the relatively high heterozygote frequency of 4%, only 1 in 2000 of Ashkenazi Jewish children will be affected ($0.04 \times 0.04 \times 0.25$) and only as a result of matings where both parents are of Ashkenazi ancestry. Most obstetricians are not acquainted with the disease and therefore do not have their Jewish patients tested for the trait. One community has refused to initiate screening on the grounds that the frequency of the disease did not justify potential emotional disturbances in those identified as carriers. In general, however, with adequate education, public response to the programs has been good.

Sickle cell screening has been less successful until recently [2256, 2288, 2347]. Black populations in which the sickle cell trait is common are often less well educated, and the purpose of a genetically oriented screening program has not always been made clear to the potential screenees. The distinction between the innocuous sickle cell trait and sickle cell anemia particularly has often not been apparent to those who were at risk. Occasional sickle cell trait carriers have been discriminated against in occupations, life insurance, and even in marital choice. These consequences illustrate the importance of extensive public education before screening programs can be initiated.

While the sickle cell trait can easily be tested for, intrauterine diagnosis for sickle cell anemia was not available until recently. Such antenatal diagnosis is now feasible by fetoscopy and more readily by direct DNA study [2322]. However, the disease is less serious than Tay-Sachs disease and beta thalassemia major (see below). There is considerable variability in clinical expression, and some patients are not very sick. Intrauterine diagnosis therefore is less often requested for sickle cell anemia than for the more severe genetic diseases.

Beta thalassemia screening for identification of couples for antenatal diagnosis has been very successful in several Mediterranean areas, such as Cyprus, Ferrara area of Italy, Sardinia, and Greece, where the frequency of the condition has fallen dramatically since the late 1970s when fetal diagnosis (on blood obtained during fetoscopy) and selective abortion was initiated [966, 2269]. Beta thalassemia trait screening is more difficult, since in contrast to sickle cell trait screening no single test is available. Nevertheless, the results in Mediterranean countries demonstrate that screening and an antenatal diagnostic program is possible and can have a great public health impact.

Screening for sickle cell anemia at birth has been advocated to identify affected children and at the same time to detect couples at high risk of howing more children with sickle cell anemia. There is some controversy about such programs, since no specific therapy exists for the disease except for early treatment of infections. Counseling of heterozygotes in such instances is usually not done. Follow-up studies are scarce.

Screening of newborns for congenital hypothyroidism has become well established and is based in assay of blood thyroxine (T4) followed by measurement of thyroid-stimulating hormone (TSH) by radioimmunoassay if T4 levels are elevated. Treatment is highly effective and prevents mental retardation and other signs and symptoms of hypothyroidism. The frequency of congenital hypothyroidism is about 1/4000 - two to three times more common than phenylketonuria. The etiology of congenital hypothyroidism is usually non-Mendelian and often nongenetic. Currently, phenylketonuria and congenital hypothyroidism can be recommended unequivocally for routine screening of all newborns.

Screening for Neural Tube Defects [216, 2260, 2274, 2298]. Neural tube defects have a multifactorial etiology. Genetic factors appear to apply but their specific nature is unknown. Their frequency ranges from 1/200 in Southeast England to 1/1000-1/1500 in Germany. The lower figures also apply to the United States. Rates have been decreasing in recent years. Population screening for these defects can be done by maternal alpha fetoprotein (AFP) determinations in serum at 16-18 weeks of pregnancy. Most mothers who are carrying a fetus with an open neural tube defect will have a markedly elevated AFP level (Tabl. 9.8; see [71]). Such elevated levels, however, can be caused by many other fetal diseases and by multiple pregnancy, intrauterine death, or underestimation of gestational age. There is an overlap between normals and affected subjects. The programs will detect 80%-90% of open neural tube defects. A blood AFP test is repeated if AFP levels are elevated and ultrasound is done to rule out twins, to confirm gestational age, and to find indications of neural tube defect. If the elevated blood AFP level is confirmed, amniocentesis for amniotic fluid AFP is done. Such screening programs are complex and somewhat controversial, since most women with elevated AFP levels will not have an affected fetus and there is no need for them to undergo repeated testing with consequent anxiety. The lower the frequency of neural tube defects in a given population, the higher will be the number of "false-positive" AFP results without neural tube defects. Nevertheless, those who have directed such programs feel that AFP maternal screening is valuable in populations where the frequency of neural tube defects is not rarer than 1/1000, and if a quality laboratory and trained personnel are available to allow the extensive counseling and ultrasound studies that are required [2260].

Extensive Screening of All Newborns for Many Polymorphisms in Future Times? So far, screening programs have been introduced for certain inherited diseases. However, as noted in discussions on the genetic basis of common disease (Sect. 3.8.13), on genetic polymorphisms in relation to disease (Sect. 3.7), and on pharmacogenetics and ecogenetics (Sect. 4.5), some "normal" genes whose products can be recognized may influence susceptibility to common disease, interacting sometimes with specific environmental conditions. This raises the question of whether it could be useful in the future to screen every newborn for many of these polymorphisms to allow an individual prognosis regarding disease risks. Preventive measures, such as avoidance of certain food stuffs, tobacco smoking, alcohol, other drugs, or occupational exposure to specific environmental conditions, such as dust or chemicals, might be recommended depending upon the specific results. It thus might become possible to reduce the risks for certain diseases to which certain genotypes are more susceptible.

Such screening would have significant societal impact. How would a society react to the knowledge that some of its future members will have a good chance to reach an advanced age in reasonably good health, whereas others would need lifelong protection from certain ubiquitous environmental influences and still others might have a relatively restricted life expectancy under the best conditions? Who in a society should have access to this information? How could confidentiality be ensured in storage by computers? Will the solidarity within human communities overcome realization of this kind of genetic inequality, or will there be strife between groups with different genotypes? How will detailed knowledge of one's liabilities influence individual happiness and fulfillment?

These are only some of the ethical problems raised by the development of human genetics and specifically by the prospects of genetic screening. Mankind does not yet appear to be spiritually prepared for such developments.

9.2 Genetic Manipulation

Human beings have been manipulating the genes of plants and animals by domestication for several thousand years. The development of agriculture therefore is a form of genetic engineering. The breeding of different types of dogs is an example of manipulating genes affecting behavior [72].

We are constantly and increasingly manipulating the genetic constitution of the human species indirectly, by manipulating the human environment and by treating diseases that have genetic determinants. Both therapy and public health measures affect the human gene pool by preserving harmful genes that in the absence of these measures would have been eliminated. For example, it is very likely that genes involved in predisposition to various infections will increase in frequency because of the wide use of antibiotics in the last two generations. Previously, such genes disappeared with the death of the affected patients. Another example: Since marital partners tend to resemble each other in intelligence by assortative mating, the distribution of genes affecting intelligence (see Sect. 8.2.1.3 for evidence of the existence of such genes) will tend to concentrate a larger share of high intelligence genes among the offspring of gifted couples. However, if genetic manipulation is discussed, the public usually considers something quite different, such as the creation of a human being in the laboratory by genetic specifications and similar bizarre scenarios. Such speculations have been occasionally voiced since the foundation of modern genetics. However, they have found a quasi-scientific basis only since the Watson-Crick model of DNA inaugurated a new period of genetic research. In the early 1960s, several symposia became important as a prototype of these discussions. The most notable was a symposium. *Man and His Future* (1963) [308], where prominent scientists discussed the prospects of genetic manipulation without restraint. The rapid developments of the last few years in molecular biology have again led to many discussions regarding genetic engineering. Although we are far from being able to modify human genes, there is a great deal of concern among the public about "mad scientists" "tampering" with the human gene pool to modify human characteristics. Since the genetic determination of complex human traits, such as personality, intelligence, stature, remains poorly understood and is affected by many genes, such traits could not be manipulated. Furthermore, the current methodology of genetic engineering is not sufficiently advanced nor safe enough to attempt any human genetic manipulation. However, it is conceivable that in the distant future we will understand more about human genetics and will have developed technology to influence some human traits directly. It is therefore important to deal with these issues.

It may be useful to subdivide the proposed methods of genetic manipulation into two groups: more conservative steps that make use of well-established biologic principles needing only some technical improvements, and the more revolutionary approaches requiring major breakthroughs in molecular biology.

"Conservative" Approach: Germinal Choice [2358]. H.J. Muller was the main promoter of germinal choice. He repeatedly suggested that prospective parents should not only rely on their own germ cells

but should choose freely between the germ cells of many individuals, selecting the future phenotypes of their children by knowledge about the personality and achievements of the individuals from whom the germ cells would be taken. According to Muller,

> a choice is not a real one unless it is a multiple choice, one carried out with maximum foreknowledge of the possibilities entailed and hampered as little as possible by irrational restrictions ... Moreover ..., the final decision regarding the selection to be made should be the prerogative of the couple concerned. These conditions can be fulfilled only after plentiful banks of germinal material have been established, representing those who have proved to be most outstanding in regard to valuable characteristics of mind, heart and body ... Catalogued records should be maintained, giving the results of diverse physical and mental tests and observations of all the donors, together with relevant facts about their lives, and about their relatives ... The germinal material used should preferably have been preserved for at least twenty years ... [obviously because the performance of a donor can only be finally assessed after this time]. Such an undertaking by a couple would assume the character of an eminently moral act, a social service that was in itself rewarding ...

From a technical point of view, this proposal can already be carried out now; storage of human sperm is possible. In fact, artificial insemination with stored sperm is utilized on a large scale in cattle breeding. In humans, artificial insemination - albeit in most cases with fresh semen obtained from healthy donors - is being carried out fairly extensively in women who cannot conceive because of their husbands' sterility. Genetic investigation of the donor usually is not performed by physicians involved in artifical insemination. However, with increasing emphasis on human genetics by the public and the involved couples, guidelines for genetic screening of artificial insemination donors have been published [2295] and are being implemented in some centers. Such genetic screening is largely based on a careful family history questionnaire. Chromosome testing is not usually done, and heterozygote testing is only done for high-frequency traits in certain ethnic groups (i.e., Tay-Sachs screening in Ashkenazi Jews; sickle cell testing in blacks, etc.). While there are few problems with defining rejection criteria for a family history of clear-cut monogenic diseases, it is more difficult to establish such criteria for the more common multifactorial and polygenic diseases. The distress caused by a rejection for genetic reasons when the potential donor considers the establishment of his own family is one result of genetic screening in this area.

Sperm banks have been established in some metropolitan cities of the United States, particularly for deposit of sperm of men who undergo vasectomy but want to ensure the possibility of having a child in case they change their minds. Such sperm banks are also useful to allow future offspring of men with neoplastic diseases who undergo treatment with high doses of cytostatic agents or irradiation and will be sterile or genetically damaged following such treatments. A sperm bank in California has been established by a millionaire industrialist, which solicits sperm samples from proven high achievers in the hard sciences (such as Nobel Prize winners and those with similar distinction) for artificial insemination of self-selected volunteer women, and several babies have been born. Based on the heritability of contribution of genes to intelligence (see Sect. 8.2.1.3), it is hoped that the donors' children will have an increased likelihood of intellectual distinction - a reasonable premise. It is, of course, highly unlikely that such a child will have the exact genetic and environmental determinants which made the father an exceptionally creative person. This sperm bank has rightly received considerable unfavorable publicity in the media. A commercial sperm bank in the USA circulates to doctors a listing of the ethnic background and social and professional characteristics of potential donors for selection by couples. The desired sperm sample is then shipped to the doctor of the infertile couple for insemination. Artificial inovulation is also possible. Oocytes are gathered from human ovaries during laparoscopy and fertilized in vitro. Artificial fertilization in vitro is more difficult. Extensive efforts in recent years have led to successful in vitro fertilization of human eggs [2369]. Meanwhile, many such babies have been born. Multiple eggs are collected by laparoscopy following medicinal and hormonal stimulation of the ovary and are fertilized with the husband's sperm in vitro. After several cell divisions have occurred, several eggs are placed in the uterus in the hope that at least one fertilized egg will implant and grow to

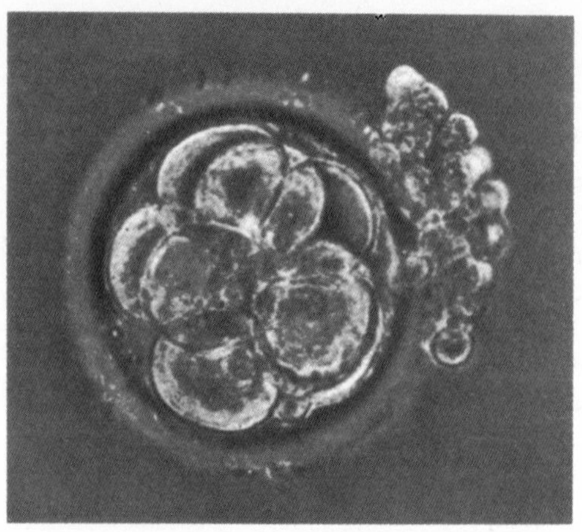

Fig. 9.3. Human oocyte after fertilization in vitro and cultivation in Ham's F10 medium. (Edwards, 1970; in Klingmüller, 1976 [2327])

maturity [2259]. Because of this methodology, multiple births often occur. Other fertilized eggs are frozen away for future use [2389]. A number of clinics in different parts of the world now offer this technology to couples where the female partner has so far been infertile because of blocked fallopian tubes. The procedure is technically demanding and requires meticulous attention to methodologic detail. Often no pregnancies result, but success rates have been improving.

This technique opens up the possibility of in vitro fertilization of sperms and eggs from any human source. Some pregnancies under such a scheme have been carried by "surrogate" mothers rather than by the biological mother [2299]. There has been much concern about such practices and their possible abuse as well as about the legal status of frozen fertilized eggs, their use, and their disposal [2257, 2259]. Some observers have criticized the practice of implanting multiple zygotes into one uterus, a practice meant to keep the chance for success reasonably high, since at present each single zygote runs a high risk of early death. On the other hand, this practice has led to a relatively high incidence of multiple births, with all the additional risks for the children (Sect. 3.8.4.). Human reproduction can now be manipulated in many ways at different levels. It is hard to predict how this will influence the attitudes of future generations.

Do We Have to Face Large-Scale Attempts at Breeding Human Beings by Such Methods? It was Muller's intention not only to prevent rare hereditary diseases by occasional artificial insemination but to achieve an improved human quality by large-scale selective breeding. Do we have to face such attempts in the future? And would they be successful?

We cannot imagine that selective breeding will ever become popular in an open democratic society. It will probably remain restricted to a small minority of the population. It has been suggested that a dictatorship might set out to breed nuclear physicists, for example, by utilizing the sperm of Nobel prize winners in this field. It is likely that many children produced in this way will perform above average in the sciences and some may even be outstanding if provided with the appropriate environment. Still, the danger that such a scheme will be attempted is relatively small. Even dictatorships usually have more immediate concerns; they will probably not invest their - necessarily limited - resources in such an undertaking that would not yield results for a least 20-30 years.

Molecular Biology and Speculations on Genetic Manipulation. Speculations regarding possible future methods of genetic engineering are based on the following results of molecular biology:

a) The mutagenic activity of certain chemical (Sect. 5.2.2) may be utilized to induce specific mutations at well-defined gene loci.
b) In microorganisms, DNA can be incorporated by extrasexual means (transformation or transduction). Similar attempts may be made in eukaryotes, including humans.
c) Defective genes may be replaced by using viral genes as vectors.
d) Artifically synthesized genes may be included into the human genome.

Induction of Specific Mutations. In the majority of gene mutations, one base of the DNA sequence is replaced by another base. Such nucleotide substitution may lead to an amino acid replacement in a specific protein that may become functionally inadequate (Sect. 5.1.4). Some chemical mutagens, on the other hand, selectively attack specific bases and induce such point mutations. Only a few years ago, attempts at attacking single, specific sites by a mutagen seemed to be doomed to failure by the sheer size of the problem; Far too many identical sites are present in the human genome. More recently, however, single sites have been attacked in viruses, restriction endonucleases (Sect. 2.3.) being used first at a specific site, followed by enzymatic removal of some bases, and finally attack on one base only, by a mutagen that reacts with this base only in a single-copy sequence [117]. Hence, the problem of localized mutagenesis is now accessible in principle, even if the practical problems are still formidable.

Gene Transfer and Expression in Eukaryotes. In microorganism, two principal methods are known for introducing foreign genetic material into a cell. In transformation, pure DNA may under certain insufficiently defined conditions enter a microbial cell and be integrated into the genetic material. In transduction, a bacteriophage incorporates a particle of the bacterial genome. When released from the host and infecting another bacterium, the bacterial material carried by the virus is transferred to the new host, where it will be genetically active. Transformation experiments played an important role in the history of genetics; they established that DNA is the genetically active material (Avery et al., 1944) [220].

In eukaryotes, transformation and transduction by DNA and also expression of transferred genes have repeatedly been reported. Early examples were plants [2315] and cultured animal cells. Prokaryotic gene expression in eukaryotic cells has been achieved in an increasing number of instances. The prokaryotic DNA came in most cases from viruses but sometimes from bacteria. A famous example was the transfer of the galactose operon of *E. coli* to human fibroblasts in 1971 [2341]. In humans, galactose is metabolized via the same pathway as in *E. coli,* and deficiency mutants for the three enzymes involved are known. The most common is galactosemia (23040), a defect of P-gal-uridyltransferase. Incubation of such cells in vitro with lambda phages carrying *E. coli* gal operon led to transferase production in these cells.

In subsequent years, replication of this result proved to be difficult and progress was slow. More recently, however, the introduction of new DNA techniques (Sect. 2.3.) has brought much faster progress, and now, practical application of gene transfer for somatic gene therapy appears to be within our reach.

Artificial Genes [114]. The in vitro synthesis of genes is one of the most spectacular success stories of molecular biology. Khorana's group synthesized the gene for alanine transfer RNA in vitro. The exact nucleotide sequence had been known beforehand, and the gene was synthesized by techniques of organic chemistry starting with single fragments. Several years after the gene had been synthesized, Khorana's group managed to induce this gene to function in tRNA synthesis. Hence, not only the gene itself but all surrounding regulating regions necessary for activity were produced (Fig. 9.4).

Until recently, production of an artificial gene by chemical methods was a gigantic undertaking. Currently, synthesis of artificial oligo-nucleotides has become commonplace. The technology for construction of human genes is therefore at hand.

These developments open the prospect that in the future it might become possible to synthesize genes with any desired nucleotide sequence and information content. Using the restriction enzymes mentioned before, these genes would then be incorporated at will into any genome.

Prospects of Human Gene Therapy. The state of the art in 1984 has been reviewed by French-Anderson [2296]. Two different goals for gene therapy have to be distinguished clearly: Correction of a genetic defect in somatic cells, and in germ cells or very early zygotes. So far, the former has aroused little if any concern, whereas the latter is rejected – or at least, considered very sceptically – by most observers.

But even somatic gene therapy should be tried only if certain conditions are met. They are not different from the conditions for any medical therapy. Following French-Anderson [2296], we shall discuss them briefly under three headings: Delivery, expression, and safety.

Delivery and Expression. At present, the only human cells that can be used for gene transfer are those from bone marrow or fibroblasts. No other cells can

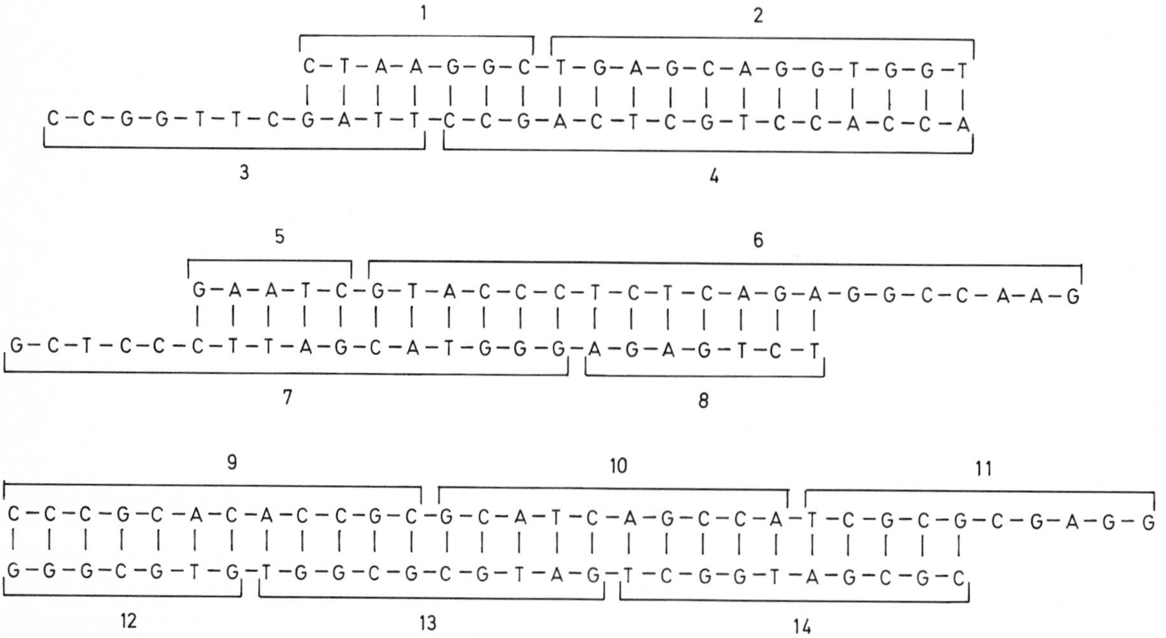

```
                        1                          2
              C-T-A-A-G-G-C-T-G-A-G-C-A-G-G-T-G-G-T
              | | | | | | | | | | | | | | | | | | |
C-C-G-G-T-T-C-G-A-T-T-C-C-G-A-C-T-C-G-T-C-C-A-C-C-A
            3                          4

            5                    6
      G-A-A-T-C-G-T-A-C-C-C-T-C-T-C-A-G-A-G-G-C-C-A-A-G
      | | | | | | | | | | | | | | | | | |
G-C-T-C-C-C-T-T-A-G-C-A-T-G-G-G-A-G-A-G-T-C-T
                  7                  8

        9                      10                      11
C-C-C-G-C-A-C-A-C-C-G-C-G-C-A-T-C-A-G-C-C-A-T-C-G-C-G-C-G-A-G-G
| | | | | | | | | | | | | | | | | | | | | | | | | | |
G-G-G-C-G-T-G-T-G-G-C-G-C-G-T-A-G-T-C-G-G-T-A-G-C-G-C
     12                  13                   14
```

Fig. 9.4. Composition of the tRNA gene for alanine from artificially synthesized DNA fragments (Numbers 1–14). The gaps are closed by enzymatically introduced nucleotides in combination with a ligase. (Khorana, [114])

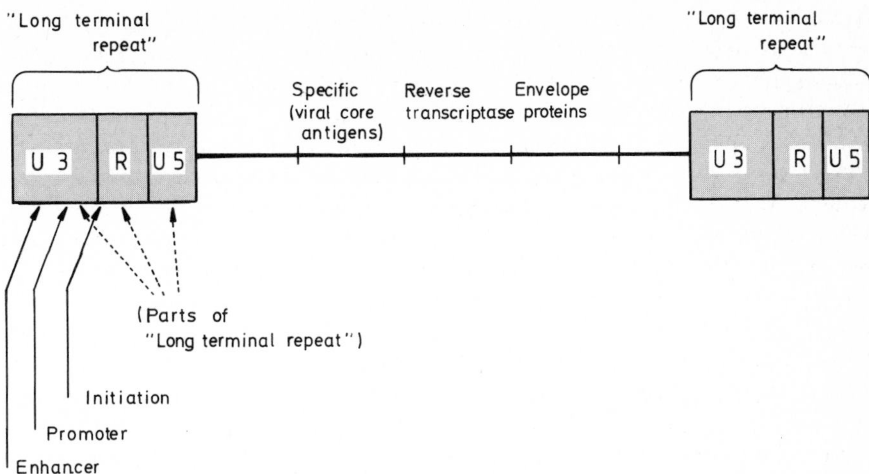

Fig. 9.5. The Maloney murine leukemia virus genotype, a retrovirus (RNA virus containing a gene for the enzyme reverse transcriptase) whose DNA transcript inserts in the mammalian genome, and can be used as vector for transporting mammalian genes into a cell. (From French Anderson [2296], modified)

be extracted from the body, grown in culture to allow gene transfer, and reimplanted into the patient. Four methods are available for transferring cloned genes into such cells: (a) Viral (RNA or retroviruses and DNA viruses) (Fig. 9.5); (b) chemically mediated e.g., with calcium phosphate; (c) fusion techniques involving fusion of DNA-loaded liposomes, red blood cell ghosts, or protoplasts, to cells; and (d) physical (microinjection or electroporation). Fusion techniques are currently not very well developed. DNA microinjection has been used for many experiments in developmental biology of other vertebrates [535], but the amount of material to be injected is normally very large and can hardly be controlled. At present, retrovirus-based vectors appear to be the most promising: Up to 100% of the target cells can be infected, and the DNA can integrate as a single copy at a single (albeit random) site. Moreover, the structure of the inserted DNA sequence is known. Figure 9.5 shows one of the system in use, the Maloney murine leukemia virus (MoMLV) system. Figure 9.6 shows the principle of the experimental approach.

A functioning gene for neomycin resistance has been introduced into hematopoetic cells of adult mice [2402], and the human gene for the enzyme hypoxanthine-guanine phosphoribosyltransferase (HPRT) (see Sect. 4.2.2.6.) was transferred and brought to expression in a HPRT-deficient cell line [2343]. This experiment will possibly open up a way for gene therapy of the Lesch-Nyhan syndrome (30800; see Sect. 4.2.2.6) which has so far eluded other therapeutic attempts. According to [2296], this disease could be an especially promising candidate, because the gene seems to work on an "always on" basis; even a

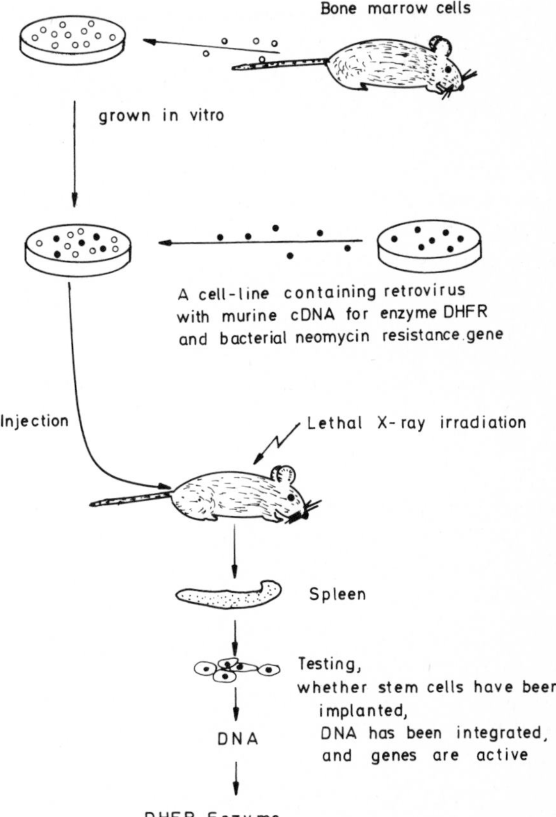

Fig. 9.6. Principle of gene transfer by a retrovirus with inserted cDNA for murine genes into a receptor animal, after its own bone marrow has been destroyed by X-rays. *DHFR*, dihydrofelate reductase. (Principle from French Anderson [2296])

small amount of enzyme production could be beneficial; and a slight overproduction of enzyme would not do much harm. It is doubtful, on the other hand, whether the severe behavioral abnormalities in this syndrome could be influenced by the introduction of HPRT genes into bone marrow cells, since they are caused by the enzyme defect in brain cells. On the other hand, metabolic cooperation has been observed in vitro: HPRT-producing cells can feed HPRT-negative ones the enzyme or its precursors (Sect. 4.2.2.6). This might occur in vivo, improving the function of neuronal cells. A still better candidate is adenine deaminase (ADA) deficiency leading to an immune defect (Sect. 4.2.2.6; see [2363]). Here, the functional defect appears to be present mainly or exclusively in cells that occur in the bone marrow and can therefore be treated in vitro and reimplanted.

Safety. Before such techniques are introduced into the therapy of any human disease, safety must be guaranteed. For example, human oncogenes (Sect. 5.1.6.5) are, in part, structure-homologous with retroviruses. There must be safeguards against production of malignancies by infection with such a virus that might have been modified, e.g., by recombination. Risks are also imaginable for other delivery systems. Experiments at several levels - human bone marrow in vitro; mice and primates in vivo - will help to eliminate any such risks as far as possible. In very severe and, so far, untreatable diseases, safety requirements are less stringent than in milder diseases or in those for which other therapies are available. Hence, the situation is not so different from that encountered in other, more traditional, fields of medical therapy.

Gene Transfer into Oocytes or Early Zygotes. The general public is concerned not so much about gene transfer into somatic cells, but much more about "genetic manipulation" of germ cells and early zygotes, with the goal of changing the genetic constitution of future generations. Irresponsible speculations by some molecular biologists and journalists have led to public concern regarding the motivations and intentions of scientists. The ambitious molecular biologist planning outrageous experiments with human beings is feared by many individuals.
Meanwhile, "gene therapy" at the germ cell level has been performed in the mouse: Injection of growth-hormone genes from rats into mouse fertilized eggs resulted in dramatic growth of some of the treated animals: 21 mice developed from these eggs; 7 carried the fusion gene, and 6 of these grew much larger than their littermates [2365]. In a later experiment of this group [2304], dwarf mice of the strain little (lit) were used. In this strain, dwarfism is caused by an isolated growth hormone (GH) deficiency; this is an animal model for autosomal-recessive GH deficiency type I in humans (26240). The same rat genes that had been utilized in the experiments mentioned above were used for gene transfer experiments. Of 41 animals that developed from microinjected eggs, 7 carried the foreign gene and 6 showed a remarkable increase in body size; growth was not only restored but phenotypes were changed from dwarf to giant.

Therapy of Fertilized Eggs in Humans? It is apparent from these results that gene therapy has been successful in some of the treated animals. In the majority it was unsuccessful. However, we can anticipate that the efficiency of gene transfer may be improved with time. More disturbing is the excessive effect of the treatment in the animals in which it worked. Imagine human beings reaching a height of, say, three meters after such a therapy! Obviously, the newly inserted gene has not been subject to the normal regulation. It is possible that the use of regulatory sequences might help in solving these problems. However, it has proved impossible so far to target genes to their usual chromosomal location. Insertion occurs at random and in a few cases has produced serious mutational damage in the recipient mice by interfering with normal gene production at the site of insertion.
Medical geneticists therefore agree that gene therapy techniques should not be used with human fertilized eggs. The risks are too high, and will continue so in the foreseeable future. Moreover, other means are available for achieving the same goal. Pituitary dwarfism, for example, can be treated efficiently by human growth hormone, which is available from genes cloned in bacteria. Still more important, there is no medical indication for such a therapy: Most metabolic diseases are inherited in a recessive fashion. Among children of a heterozygous couple there is a 1:3 segregation ratio between affected and unaffected children. Of course, the early zygote immediately after fertilization (which would have to occur in vitro) would have to be treated. But before treatment, homozygosity of the mutant allele must be established. This cannot be performed at the gene-product level, since the genes are not yet active; it has to be done at the DNA level. However, the cell will not survive the diagnostic procedure unless it is separated into two and the diagnosis is made on one cleavage product while the other is kept for gene transfer. A very complicated and dangerous procedure! Provided the homozygous state could, indeed, be diagnosed at such an early stage, would it not be preferable to discard the zygote and encourage the couple to try again?

In conclusion, we believe that there is no reason for gene therapy in oocytes or early zygotes. We should refrain from attempting it.

Public Reaction to New Achievements and Prospects of Molecular Biology. The achievements and – still more – the prospects of molecular biology in the recent decade have aroused extremely strong reactions in the general public, and especially among the opinion-forming elites (theologians, philosophers, journalists). Biologists and medical geneticists have often been shocked by fierce attacks; they often feel that their benevolent intentions are sadly misinterpreted and that an image has been set up of the ambitious and ruthless scientist who will soon start manipulating human populations for sinister purposes unless he ist stopped by an alert public. In this connection, however, we should not forget that scientists were among the first to "ring the alarm" when the possibilities for experimental recombination of DNA by restriction endonucleases became apparent. Still more, this first alarm proved to be unnecessary and could have been avoided had the protagonists acted in a more circumspect way.

The fundamental scientists who were involved in early research with restriction enzymes became concerned about hypothetical dangers of the new scientific developments that allowed splitting of genes at random, joining genes from different organisms across species, and using the ubiquitous *E. coli* organism for gene transfer. They called a conference to discuss hypothetical dangers of spread of uncontrollable infections and of cancer using these new methods [2264]. No microbiologists or epidermiologists experienced in human infectious disease or in human cancer were consulted or invited to the widely publicized conference. The public became greatly alarmed, and laws were rapidly passed in the United States to regulate work with the new techniques. It soon became clear, however, that the envisaged dangers were largely hypothetical. The organisms used were so enfeebled that the kind of epidemic infections feared could never be established. It was also soon realized that cross-species DNA transfers had occurred in nature for millenia. Using the newer experimental laboratory data and the experience of a century of background in clinical microbiology and epidemiology, most informed and knowledgeable observers regarded the early fears as unjustified. The unnecessary alarms regarding "recombinant DNA research" are an excellent example of failure of communication between molecular biologists who knew little about the realities of cancer and infectious disease and the relevant medical and biologic scientific community. Nevertheless, a few eminent scientists continue to be worried about possible dangers of cross-species DNA transfers.

Further Speculations on Gene Manipulation. Starting from the above-mentioned results and prospects, some biologists have speculated on much more ambitious goals for gene manipulation: In their opinion, human beings with novel capacities should be created. If, for example, the skin of the head or back could be replaced by a tissue containing chlorophyll, this could give the individual the capacity for photosynthesis – a possible partial solution for the food shortage problem in an increasingly overpopulated world.

Cloning of frogs has been achieved by introducing an intestinal frog cell nucleus into an enucleated frog egg. The genetic information of the nucleus was able to specify the normal development of a frog under appropriate conditions. Some scientists and the public have been fascinated by the possibility of cloning a human being by analogous principles. Research work toward this end, however, is not being carried out to our knowledge. Nevertheless, this does not discourage some scientists from embarking on such phantasies under circumstances where wide publicity is guaranteed, and a science writer in 1978 claimed in a book that the cloning of a human being was achieved [2375]. No proof was given, but this scenario helped to sell his book and the media were full of stories about the implications. It has been fantasized that cloning of human beings would make it possible to duplicate outstanding and creative human beings such as Einstein or Mozart. However, it is apparent that the genetic material of an Einstein alone will not guarantee another Einstein. Others have suggested that dictatorships would clone groups of military scientists or brute soldiers in the service of the state. If such a feat were ever to become possible, it is unlikely that a country would embark on such an undertaking. Since clones would take a generation to reach their full potential, politicians and statesmen would be more likely to look for more rapid ways of ensuring political and military success.

Other scenarios have implied the creation of subhuman creatures following fusion of cells with chromosomes from human and subhuman primates and subsequent insertion into an enucleated egg. Such humanoids have been envisaged to carry out dull and repetitive tasks of no interest to normal humans. Again, we are far from being able to carry out such schemes.

Along more "conventional" lines, Lederberg [2330] proposed manipulation of the central nervous system by as yet nonexistent chemical or growth factors to improve the efficiency of the human brain. With such schemes, the genetic material would of course not be changed – a process that was termed "euphenics", in contrast to eugenics.

Brainstorming of this sort and discussion of "far-out" scenarios [2303] are useful to recognize possible novel possibilities for research. Maybe more importantly, they may alert us as to possible abuses of science. It is unfortunate and dangerous for the public's understanding of science that the media often leave the impression that these novel reproductive schemes were seriously planned or are already being carried out by scientists [2329].

The Need for a Dialogue on Ethical Tissues. Many results of molecular biology are already being applied in various branches of medical genetics, from cytogenetic diagnosis to genetic counseling using restriction fragment length polymorphisms (RFLPs; see, e.g., [694]); gene transfer to somatic cells is just around the corner. The more pretentious goals of some visionaries have also been mentioned and their list should be extended; it includes the creation of human beings according to genetic specifications, improving the efficiency of human brains, and increasing health status and life span. Compared with our present-day knowledge of molecular biology of higher organisms, and our ignorance of the genetics of much of the normal morphological variation in humans, these proposals are somewhat analogous to the idea that a boy who has just been given his first electronic set for Christmas, could successfully improve the latest IBM computer. One could also argue that even if the technical conditions for a few of these phantasies do become available their practical application will probably remain impossible for sociological reasons.

However, the possibility of misuse on a much smaller scale cannot be entirely dismissed. We should therefore be glad that these problems are widely discussed, even though much of the discussion is uninformed as far as the technical facts and concepts are concerned. We should enter the public dialogue wherever we regard this as meaningful and should attempt to enhance its standard of scientific and ethical sophistication. We can learn from the questions that are being asked and become more critical toward our own goals. And in the long run, we may convince our partners in this dialogue that the ethical problems posed by the new methods are often quite different from those initially posed. One example: In Huntington's disease (14310) prenatal diagnosis is possible in some families by use of a DNA marker. This method will sometimes lead to the diagnosis of Huntington's disease in a parent who previously was only known to be at 50% risk. Some medical geneticists argue that the information should be given provided that the clients have been informed before of the possible consequences and insist on this information. Others maintain that this information should be withheld, since the clients might not be able to cope without severe trauma or eben suicide even if they requested such information.

In their attempts to help patients and their families, medical geneticists face ethical dilemmas which may be difficult to resolve. A continuing public dialogue on ethical issues is necessary. Both authors of this book have been involved repeatedly in such dialogues [2258, 2354, 2396a].

9.3 Biologic Future of Mankind

Human Evolution Is Not Finished [2395]. Evolution of the human species is not confined to the past. The mechanisms that bring about changes of gene frequencies from one generation to the next are still in operation. Knowledge of these mechanisms should help to predict future trends in the genetic composition of human populations. To understand such predictions and to make proper use of them, we should keep in mind the limitations that are inherent in all such attempts.

a) We can only extrapolate from trends that are already visible at the present time; human history, however, has often been shaped by unpredicted events. Such an assessment is also true for the biologic history of our species, which is now inextricably intertwined with cultural, social, and political history as well as with the future development of human biology and medicine with the potential to actively influence the future evolution of mankind.
b) It is a well-documented fact that the visible signs of revolutionary future scientific developments that will be easily recognized in retrospect are often not clear to contemporary scientists. It would be presumptuous for the present authors to claim an exception. Future colleagues may cite this book as an example of how important trends have been overlooked.
c) All predictions made here assume that modern civilization will continue to exist and that within modern societies concern for health and medicine will maintain its prominent position. In this context, we assume that preventive medicine will continue to gain in importance.

If we assume that attitudes toward these problems will be increasingly rational, there is little doubt that concepts of human and medical genetics will be utilized at an ever-growing scale. One could argue that this assumption is unrealistically optimistic, as it presupposes that human societies will learn to keep their social and technologic structures from destruction. Thus, it is by no means impossible that a large fraction of the human species will sooner or later be doomed to destruction by an atomic holocaust. We are also intensely conscious that many of the concerns expressed here have a much lower priority in the developing countries of the world where problems of overpopulation, undernutrition, and infectious diseases are the key problems needing solution.

The following extrapolations are subject to the conditions of human survival and the fact that social progress desired by both "capitalistic" and "socialistic" societies will continue.

Main Forces Determining Evolution. The principal mechanisms that determine evolution have been discussed in the section on population genetics (Sect. 6). Their influence on the development of present-day human populations has been covered in Sect. 7. What can be predicted for the future?

Genetic Drift. Chance fluctuations of gene frequencies may lead to appreciable genetic differences between subpopulations, provided that these populations are almost completely isolated from each other, keeping gene flow between them very low. This effect becomes stronger with smaller size of an effectively breeding population and even leads to lessened variability due to random loss of alleles for a predictable proportion of genes (Sect. 6.4).

In present-day human populations, we are observing a strong overall tendency toward breaking down of isolates and increasing intermarriage between different populations. There seems to be no reason to assume that this trend will reverse in the foreseeable future and that new isolates of a small size will be formed. Hence, chance fluctuations, contrary to their indisputable importance for human evolution in the past, will become less significant in the future. If this trend in the human breeding structure persists, new species of man will not develop, since speciation always needs reproductive isolation of a population subgroup (Sect. 7.2). The creation of a human-like new species is sometimes discussed by science fiction writers. Currently existing genetic variability coupled with selective breeding of man could not lead to such a new species. One would have to create novel genetic combinations by presently nonexistent techniques and enforce selective breeding. These possibilities, therefore, are extremely remote. For future shifts of gene frequencies within our species, two factors causing systematic changes in gene and genotype frequencies remain: mutation and selection.

Mutation. Our current knowledge regarding spontaneous and induced mutations has been discussed in Sect. 5, and the problem of a genetic load due to mutations has been reviewed in Sect. 6.3.2. We can safely assume that practically all chromosome aberrations and many gene mutations are unfavorable for the individual as well as for the population; most chromosome aberrations kill the zygote during embryonic development; a minority survive up to birth or still longer, but affected patients suffer from severe malformations. Gene mutations will often lead to inherited disease with simple modes of inheritance or to defects in multifactorial genetic systems (Sect. 3.6.2). A very large fraction of point mutations leads to changes in amino acid sequences of proteins and causes no apparent functional deficiency as indicated by hemoglobin variants. In comparison to *all* mutations, the fraction of advantageous mutations constitutes at best a small minority.

The combined evidence justifies the conclusion that influences enhancing the overall mutation rate will be unfavorable. The extent of the impact of various mutations on health has been discussed in Sect. 5.2 and suggests surprising differences depending on mutational mechanisms and phenotypic manifestation.

Trends in Spontaneous Mutations Rates: Chromosome Mutations. The mutation rates of numeric chromosome mutations increase with the age of the mother. Therefore, every change in maternal age will lead to a corresponding alteration in the overall incidence of such chromosome mutations. In many modern populations, there is a trend toward a decrease in the number of children per family and a concentration of childbirths within the age groups with the lowest risk: women in their 20s. It has been calculated that in Western countries and in Japan this trend should have reduced the number of children with Down's syndrome by about 25%–40% [2338, 2371, 2394]. Will this trend continue? Some more recent figures show that the tendency of many modern women to postpone childbearing to a somewhat higher age could easily lead to a reversion of this trend.

On the other hand, antenatal diagnosis (Sect. 9.1.1) is most efficient for early recognition of chromosome abnormalities. Many countries offer this diagnostic tool to all women above the age of 35. If all pregnant women above 35 years would use this procedure, a significant reduction of Down's syndrome, depending on the age distribution of mothers would be expected. With enhanced safety of antenatal diagnosis for mother and child, this procedure may become almost routine for most pregnancies in developed countries. Under such conditions, anomalies due to numeric or structural chromosome aberrations will become completely avoidable.

Gene Mutations. The mutation rate for many gene mutations increases with the age of the father [2367] (Sect. 5.1.3.3). Therefore, any trend in the age distribution of fathers will influence the mutation rate accordingly. The changes will not be quite as large for autosomal-dominant rare conditions with a paternal age effect as for numeric chromosome aberrations, the paternal age effect on dominant and X-linked mutation rates being smaller than the effect of maternal age on incidence of numeric chromosomal anomalies. The total medical impact of the paternal age effect probably is quite low and the actual absolute risk of an older father to have a child affected with a dominant mutation is almost negligible.

Ionizing Radiation and Chemical Mutagens. It was shown in Sect. 5.2.1.5 that any conceivable increase of radiation exposure will probably enhance the mutation rate by a few percent. In view of the fluctuation of the "spontaneous" mutation rate, for example, due to changing age distributions of parents, any increase due to radiation will probably not even be recognizable except with refined epidemiologic techniques. Still, the effect will exist. Therefore, one of the major goals of preventive medicine in the future is to keep irradiation as low as possible. At present, the main source of radiation exposure is diagnostic medicine; here, improvement of technology can do a lot. As to occupational and general background irradiation, it can only be hoped that the technology for securing the long-term energy demands of the world by sources other than nuclear energy will be improved in the future so that nuclear energy can be de-emphasized in the long run.

Too little is known regarding the exposure of our population to chemical mutagens (Sect. 5.2.2) to venture any predictions. The general statement can be made that we shall probably have to live with a certain number of chemically induced mutations, as human society will probably not be prepared to forego the immediate advantages of some chemicals in exchange for long-term avoidance of small and undefined genetic damage. Current, often hysterical reactions to possible health dangers of chemicals will hopefully be replaced by more appropriate responses once we have learnt more about chemical mutagenesis and carcinogenesis.

In conclusion, a certain, presumably small increase of the natural mutation rate will have to be faced. This increase will lead to a corresponding increase of numeric and structural chromosome aberrations and dominant or X-linked hereditary disease (Sect. 5.2.2). Whether this increase will be offset by a decrease due to a shifting parental age distribution cannot be predicted. Since cancer is often caused by somatic mutations possibly induced by environmental agents (5.1.6), increases in neoplastic diseases are possible. The decrease in the frequency of gastric cancer for unknown causes and the anticipated decline of lung cancer with less cigarette smoking indicate that overall beneficial trends may compensate for certain unfavorable trends.

Selection: Dominant and X-Linked Diseases. It is a widespread contention that natural selection has relaxed due to modern medicine. This statement, however, is only partially true. No therapy, for example, has so far succeeded in preventing miscarriages caused by chromosome aberrations. Patients with Down's and Klinefelter's syndrome still do not reproduce. Natural selection has not changed for these conditions.

Selection did relax for some pathologic traits with autosomal-dominant or X-linked recessive modes of inheritance. These are genetic diseases that have been maintained so far by genetic equilibrium between mutation and selection. One example is hemophilia A, for which substitution therapy with factor VIII now enables patients to lead an almost normal life. Life expectancy and the chance to have children is vastly improved. For such conditions, an appreciable increase of incidence within only a few generations can be quantitatively predicted when the presentday frequency, the selection coefficients before and after selection relaxation, and the mode of inheritance are known; for retinoblastoma [2393], such a calculation has been given in Sect. 6.2.1.2 (Fig. 6.9).

There are many other dominant and X-linked conditions, however, for which no satisfactory therapy is available, and natural selection still acts with full strength. Examples are neurofibromatosis, tuberous sclerosis, or the Duchenne-type of muscular dystrophy.

Undoubtedly, the future may bring therapeutic progress for these diseases as well, which will lead to selection relaxation. On the other hand, the entire group of dominant diseases is an easy 'target' for genetic counseling, and X-linked conditions can be avoided by prenatal diagnosis using DNA polymorphisms. We therefore have reason to hope that a growing fraction of patients affected with such disorders will voluntarily refrain from reproduction. Under such circumstances, artificial selection will replace natural selection, and keep population incidence close to the mutation rate.

Figure 9.7 illustrates this trend in a more anecdotal way: a man in the grandparents' generation who was blind due to congenital cataract had seven living children, four of

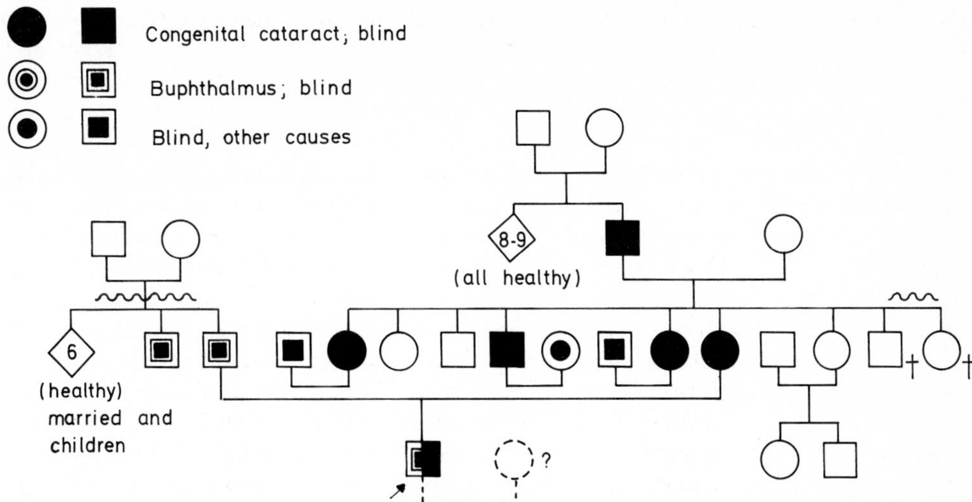

Fig. 9.7. Pedigree with congenital cateract, demonstrating assortative mating for blindness with different causes, and voluntary restriction of reproduction in these marriages, The grandfather of the family was blind due to congenital cataract; nevertheless, he had nine children; four were also blind. All are married to blind persons (assortative mating!). Only one of these couples had one single son, who being blind, asked for genetic counseling (Observation F.V.)

whom were affected and blind. All four were married to spouses who were blind for other reasons, demonstrating assortative mating. It is remarkable, however, that three of these couples voluntarily - and without genetic counseling - refrained from having children. The fourth couple had just one son, who asked for genetic counseling even before marriage. Apparently, the mutation that had multiplied no less than fourfold at a time when contraceptive measures were not yet in general use was wiped out within two more generations by voluntary decision of its carriers. Every medical geneticist with experience in genetic counseling knows of such examples.

Natural Selection: Recessive Diseases. The most conspicuous success in therapy of genetic diseases was achieved in recessive enzyme defects (Sect. 4.2.2.7). Treatment causes individuals affected with some of these diseases to grow up healthy so that they are able to have children. Moreover, it can be predicted that if gene therapy will be successful in the foreseeable future, it will be so for some of these enzyme defects. However, reproduction of abnormal homozygotes leads only to a very slow increase of gene frequency (Sect. 6.2.1.2), which gives little cause for concern.

Still, an increase of recessive diseases has to be faced for another reason: at present, most populations are not in equilibrium for recessive genes. Breakdown of isolates and steep decrease in the number of consanguineous marriages have created a situation in which the overall number of homozygotes is far below the expected equilibrium value. In the absence of other factors, such as changes in known selective advantages of heterozygotes, this trend should lead to a slow increase of homozygotes over hundreds of future generations (Sect. 6.3.1.2). Since 100 generations correspond to a time period of 2500-3000 years, such an increase need not worry us now. Living conditions are likely to change in an entirely unpredictable manner in such a long time period.

In conclusion, we expect a very slow increase in the incidence of recessive diseases above their present-day frequency in randomly mating populations of the industrialized countries in Europe and in the United States. However, this prediction does not take into account artificial selection by genetic counseling, antenatal diagnosis, and possibly screening programs for heterozygotes. If antenatal diagnosis becomes a routine procedure, the most frequent recessive diseases, for example, the β thalassemias in Mediterranean and S. E. Asian populations or cystic fibrosis in the northwestern European and North American whites, are likely to be included for routine intrauterine diagnosis. The rarer defects are unlikely to be screened.

Taking all evidence for genetically well-defined abnormalities, such as chromosome aberrations, dominant and X-linked diseases, and autosomal-recessive diseases together, we can predict fairly safely that modern civilization will *not* bring about a large increase in the incidence of these abnormalities. Whether there will be a small and very gradual increase, or whether the overall incidence will stay more or less at its present rate, or whether even a decrease may occur depends largely on many developments in our society. Shall we succeed in keeping population exposure to mutagenic agents within reasonable limits as explained earlier? How effectively will artificial selection through genetic counseling and antenatal diagnoses reduce genetic damage? And to a much smaller degree, how long will society be prepared to pay for the relative "luxury" of working out schemes for prevention and treatment of rare diseases? We cannot make any accurate predictions at this time.

Gradual Loss of Functions That Are Now Being Maintained by Multifactorial Genetic Systems. Apart from the well-defined genetic defects discussed so far, genetic variability within the population also exists for functional systems that depend on a complex but ordered interplay of various genes during embryonic development. Examples are the heart, the eyes, or the immune system. During evolution, these systems have developed under constant and intensive selection pressure. As soon as this selection pressure abates, mutations will accumulate that lead to small functional infirmities, and over very long evolutionary periods, these systems will slowly but inevitably "fall apart". Among animals, the best known examples are found in species that have lived for many generations in the absolute darkness of caves or at great oceanic depths where an intact visual system brings no selective advantage. Typically, the variability of the eyes first increases, individuals with small defects becoming more frequent. Later on, most animals will have more or less severely defective eyes, and finally, an eye-less species results. In a civilized society, minor defects of the visual system fail to carry any selective disadvantage. This trend has already led to considerable increase of variability on visual acuity; conditions such as myopia, hyperopia, or astigmatism are probably more frequent than in primitive populations that have lived up to more recent times under conditions of stronger selection (Sect. 7.2.5). Will this process continue? One could argue that such trend will be bearable; there will just be more people who use glasses. Their „defect" provides a living for other professions, such as ophthalmologists, optometrists, and opticians, and creates the spectacle-making industry.

A slow deterioration of the immune system would be more dangerous. As explained in Sect. 6.2.1.6, mortality during infancy and childhood has been high up to very recent times with infections being the principal contributory factor. Under the influence of strong selection, a complex system for recognition and elimination of infective agents has evolved (Sect. 4.4). We know a great number of genetic defects which impair the efficiency of this system. Formerly, such defects usually led to death of the individual from infection; today, with antibiotic therapy, many of these patients survive and will have children. For extreme defects, we can hope for therapies such as bone marrow transplantation, which substitute normally functioning immune cells. However, mutations may change protein molecules in a much more subtle way, only slightly impairing their function. Studies on the hemoglobin molecule have shown that such mutations are probably very frequent. Will relaxation of selection against such variants lead to a slow deterioration of the whole system? Shall our descendants slowly become more susceptible to infections of all kinds that need to be counteracted by a more elaborate combination of antibiotics and immune therapy? Possibly still more important, will this deterioration of the immune defense system enhance the frequency of various types of cancer, since potentially malignant cell clones may often be kept from proliferation by so-called immune surveillance?

It is also possible that the relaxation of selection against multifactorial traits may lead to their higher frequency. All trends in the past, for instance, made for fairly strong selection against cleft lip and palate. Nowadays, because of effective surgical therapy, such children survive more readily and have children. The frequency of this trait in the next few generations is therefore likely to increase. The exact extent of such increase is difficult to predict. Similar considerations apply to congenital heart defects.

Increase in the Number of Intellectually Subnormals? As noted above, conscious planning of reproduction is one of the main factors to be considered in every estimate. This factor, however, may have favorable as well as unfavorable consequences. In populations, one subgroup with a reduced ability to plan for their own future exists - the mildly mentally subnormal. Indeed, their mean reproduction exceeds the population average. While environmental factors are well known that may help to bring about mental subnormality, there is evidence to suggest that genetic variability is important especially for the mildly retarded group (Sect. 8.2.1.2). Therefore, a continuing higher reproduction of mildly retarded persons might enhance their population frequency. This conclusion has been challenged by Penrose [2157] by the argument that in families with mildly subnormal individuals a certain fraction of the more severely handicapped will segregate who are not able to have children at all. This counterbalancing effect might keep the gene frequency low even in the face of higher reproduction of the married subnormals. Comprehensive studies of Reed and Reed [167] (1965) seemed to support this argument: the number of unmarried and childless retarded family members was, indeed, increased in families of the mentally subnormal. We cannot be sure, however, whether this compensation will be sufficient in the future; too much depends on unpredictable social conditions: mentally subnormal women can be taught to take birth control pills regularly and might even be more reliable than women of average or above-average intelligence.

Yet, it is possible that genes for mental subnormality will increase; even though the number of mentally subnormal individuals may be higher, the mean ability of the rest of the population will not be affected. There is strong assortative mating for intelligence in general and especially for mental subnormality. This trend may create social problems, as modern societies with increasing demands for technical skills will provide progressively fewer jobs for mildly retarded persons.

A Favorable Selective Trend: Abandoning of Genetic Adaptations with Otherwise Unfavorable Effects. So far, we mainly considered unfavorable selective trends, the only benefical trend being artificial selection by genetic counseling. Another favorable trend, discussed below, will probably lead to genetic improvement much quicker than most of the unfavorable trends can lead to genetic deterioration.

Some anomalies and diseases of the erythrocytes, most notably hemoglobin variants, attained a high incidence in tropical countries despite the fact that homozygotes suffer from severe blood diseases such as sickle cell anemia or thalassemia (Sect. 4.3). However, heterozygotes have a selective advantage vis-à-vis falciparum malaria, which has been endemic in these areas until recent times (Sect. 6.2.1.6). Unless malaria will reappear in the future, selection against the affected homozygotes will reduce the gene frequencies.

Here, nature has evolved a compromise: for the advantage on increased resistance against malaria, the price of many cases with unherited disease had to be paid. As soon as malaria has been eradicated, this compromise is no longer needed and the harmful gene gradually disappears (Sect. 6.2.1.6).

A similar mechanism may exist for the ABO blood groups (Sect. 6.2.1.8). It is possible that this genetic polymorphism arose as an adaptation to multiple and varying infective agents and that selection has been frequency dependent. The price to be paid has been zygote loss due to serologic mother-child incompatibility. If selection due to infections would disappear and only selection by incompatibility would remain, a gradual and slow elimination of the rarer alleles, A and B, and fixation of the most frequent one. O would ensue. Similar "compromises" are likely for other genetic systems as well.

Table 9.9. Favorable and unfavorable trends (without genetic engineering) affecting genetic composition of future human populations

Unfavorable trends	Probable significance
Mutation rate increase due to ionizing radiation	Not very significant
Mutation rate increase due to chemical mutagens	Unknown
Higher reproduction of patients with inherited diseases	Probably not very significant
Increase of recessive diseases (new equilibrium)	Insignificant in the next few centuries
Deterioration of normal functions due to "selection relaxation"	May be significant in the long run
Favorable trends	
Elimination of genetic adaptations with otherwise unfavorable effects (infections or malnutrition)	Possibly significant
Decrease of mutation rates for chromosome aberrations and point mutations due to decreasing parental age	Significant
Voluntarily decreased reproduction in families with genetic diseases	Unknown; will become more important in the future
Genetic counseling including antenatal diagnosis	Important, even after a short time

The Human Species in the Future. From these considerations, the following picture of the future genetic composition of the human species emerges: its overall genetic composition will be similar to that encountered at present. A tendency toward diminution of racial and ethnic differences will probably continue. Genetic defects might be slightly more or less frequent than at present but will be under effective control by counseling and antenatal diagnosis. Living cases with autosomal chromosome aberrations may become rarities. Diseases caused by polygenic factors are likely to increase because of medical and surgical treatments and other cultural factors leading to relaxation of selection. Unfortunately, the extent of this increase cannot yet be accurately assessed because of our lack of knowledge regarding the specific genetic contribution in these disease. In Table 9.9, the favorable and unfavorable trends are compared.

Appendix 1

Methods for the Estimation of Gene Frequencies

In the context of this book only the principles of gene frequency estimation can be shown (Sect. 3.2). For details, see Race and Sanger [166]; Mourant [144]: and others. We will start with the simplest example.

One Gene Pair: Three Genotypes Can Be Identified in the Phenotype. Here, every single allele (*M* or *N*) can be identified, and the gene frequency can be established by direct counting. The *MN* blood types may be used as examples:

$$\hat{p} = \frac{2\overline{M} + \overline{MN}}{2(\overline{M} + \overline{MN} + \overline{N})} = \frac{\overline{M} + \frac{1}{2}\overline{MN}}{\overline{M} + \overline{MN} + \overline{N}}; \quad \hat{q} = 1 - \hat{p} \qquad \text{(A 1.1)}$$

The calculation of variance *V* is also straightforward:

$$V = \frac{\hat{p}\hat{q}}{2(\overline{M} + \overline{MN} + \overline{N})} \qquad \text{(A 1.2)}$$

The gene frequencies $\hat{p}$ and $\hat{q}$ can now be used to test whether the phenotype frequencies agree with their expectations according to the Hardy-Weinberg Law. Using the following formula, the explicit calculation of these expectations can be spared:

$$\chi^2 = \frac{(\overline{M} + \overline{MN} + \overline{N})(\overline{MN}^2 - 4 \times \overline{M} \times \overline{N})^2}{(2\overline{M} + \overline{MN})^2(\overline{MN} + 2\overline{N})^2}$$

In principle the same counting method can be used when there are more than two alleles and when all genotypes can be distinguished phenotypically, e.g., in polymorphisms such as acid erythrocyte phosphatases.

One Gene Pair; Only Two Genotypes Can Be Identified by the Phenotypes. The problem becomes slightly more complicated if one of the two alleles in dominant, i.e. if the heterozygote is identical phenotypically with one of the two homozygotes. In this case, the homozygote of the recessive gene provides the information necessary for gene frequency calculations. The frequency of the homozygotes is q^2. The Diego blood factor (Sect. 7.3.1.) may serve as example. There are two classes in American Indian or Mongoloid populations: Those showing a positive agglutination reaction with anti-Di^a serum, and others who fail to show this reaction. Family studies have shown the negative type to be recessive.

$$q\,(\text{Frequency of the gene } Di^a) = \sqrt{\frac{\text{Di(a}-)}{Di(a+) + Di(a-)}}$$

$$V = \frac{1 - q^2}{4[Di(a+) + Di(a-)]}$$

There is no degree of freedom left to test for a Hardy-Weinberg equilibrium.

If an anti-Di^b serum is available, the heterozygotes can be identified, and gene frequency can be calculated by the gene counting method as shown above for the MN blood group.

More Than Two Alleles: Not All Genotypes Can Be Distinguished in the Phenotype. A special case, the AB0 blood group, has already been discussed in Sect. 3.2.2.

Maximum Likelihood Principle of Estimation. We are confronted with the general problem that a parameter is not known a priori but must be estimated from empirical data. According to R. A. Fisher, an estimate should fulfill the following conditions:

a) It should be consistent. This means that the estimate converges stochastically on the parameter with an increasing number of observations.
b) The estimate should be sufficient. It must be impossible to extract additional knowledge about the parameter by calculating other statistics from the data.
c) The estimate should be efficient. It should extract the maximum possible amount of information from the data. The variance should be as small as possible.

Generally, the best solution of the estimation problem is the maximum likelihood principle established by R. A. Fisher. First, a simple example:
The probability for the occurrence of n_1 events, each of which has the probability *p*, and $n - n_1$ events, each of which has the probability $1 - p$, is:

$$L = \frac{n!}{n_1!(n - n_1)!} p^{n_1}(1-p)^{(n-n_1)}$$

according to the binomial distribution. To find the value of *p* for which this probability has a maximum, the first derivative of *L* with respect to *p* must be 0. For the sake of convenience, instead of *L* its logarithm is usually maximized:

$$\log L = \log \frac{n!}{n_1!(n-n_1)!} + n_1 \log p + (n - n_1) \log(1-p)$$

$$\frac{d(\log L)}{dp} = \frac{n_1}{p} - \frac{n - n_1}{1-p} = \frac{n_1 - np}{p(1-p)}$$

$$\frac{d(\log L)}{dp} = 0; \text{ hence: } p = \frac{n_1}{n}$$

a result which is intuitively obvious. It means: $n_1 = np$, i.e., for the binomial distribution the most likely value of the parameter is one for which the expectation is identical to the observed value. For reasons that cannot be explained here, the (large sample) variance of this estimate for *p* can be calculated from the negative reciprocal of the second derivative by inserting the maximum likelihood estimate of p. In this case:

$$\frac{d^2(\log L)}{dp^2} = -\frac{n_1}{p^2} - \frac{n-n_1}{(1-p)^2};$$

$$\frac{1}{s^2} = I = \frac{n^2}{n_1} + \frac{n^2}{n-n_1}, \quad s^2 = \frac{n_1(n-n_1)}{n^3}$$

This expression for s^2 would have resulted from a more conventional way of calculation as well: The general formula for the variance of the binomial distribution:

$$V = \frac{p(1-p)}{n}; \quad \text{inserting of } p = \frac{n_1}{n}; \quad 1-p = \frac{n-n_1}{n}$$

gives the same result.

In the following, the formulation will be somewhat more general.[1] Let x be a random variable the distribution of which depends only on p. This may be symbolized by the expression $f(x; p)$ for the density function of x. Let us assume that from the universe of all x, a sample of size n is drawn with x-values $x_1, x_2 \ldots x_n$. Then, the probability for every possible sample can be written as follows:

$$L = f(x_1; p)\, f(x_2; p)\, f(x_3; p) \ldots f(x_n; p) = \prod_{i=1}^{n} f(x_1; p)$$

By inserting the particular, observed sample values into this expression and regarding it as a function of p, one obtains the likelihood function of the sample. The maximum likelihood estimate is found by solving the equation:

$$\frac{d(\log L)}{dp} = 0$$

for p. The variance of this estimate is obtained by calculating the second derivative and taking the negative reciprocal of its expectation (general symbol: E)

$$s^2 = -1/E\left(\frac{d^2 \log L}{dp^2}\right)$$

$1/s^2$ is also called the information of p or *Ipp*.

The simple gene counting method as explained above with the *MN* blood types does indeed gives the maximum likelihood estimate. The calculation becomes somewhat more complicated when more than two alleles are present and not all genotypes can be identified in the phenotype as in the AB0 system. Here, many authors have given rules for maximum likelihood estimation.

However, the Bernstein method, with corrections, has turned out to be practically equivalent. Therefore, formulas derived from maximum likelihood equations can be used for calculating the variances of the Bernstein estimates:

$$Vp = \frac{p}{8n}\left(4 - 3p + \frac{pr}{pq + r}\right)$$

$$Vq = \frac{q}{8n}\left(4 - 3q + \frac{qr}{pq + r}\right)$$

$$Vr = Vp + Vq - \frac{pq}{4n}\left(4 - \frac{pq}{pq + r}\right)$$

Here, n means sample size (all four blood groups together).

[1] Cf. Neel and Schull (1954) [150].

Calculation of AB0 Gene Frequencies by the Bernstein Method.

Bernstein, in his investigations on the genetic basis of the AB0 system (Sect. 3.2), also derived a method for estimating AB0 gene frequencies. Later, he refined this method by first estimating provisional frequencies p', q', r' and then correcting them to calculate the definite gene frequencies p, q, r:

$$p' = 1 - \sqrt{(\bar{B} + \bar{O})/n} \qquad p = p'(1 + D/2)$$

$$q' = 1 - \sqrt{(\bar{A} + \bar{O})/n} \qquad q = q'(1 + D/2)$$

$$r' = \sqrt{\bar{O}/n} \qquad r = (r' + D/2)(1 + D/2)$$

D is the difference between 1 and the sum $p' + q' + r'$. Estimates using this improved Bernstein method were shown to be practically identical with the maximum likelihood estimate.

Example: Estimation of Gene Frequencies by Gene Counting.

Race and Sanger [166] gave the following phenotype frequencies for individuals from London, Oxford, and Cambridge:

$\overline{M}$	$\overline{MN}$	$\overline{N}$	Sum total
363	634	282	1279

The gene frequency p of allele M, and q of allele N, respectively therefore is according to Eq. A.1.1:

$$p = \frac{363 + \frac{1}{2} \times 634}{1279} = 0.5317$$

$$q = \frac{282 + \frac{1}{2} \times 634}{1279} = 0.4683$$

It follows: $p^2 = 0.2827$; $2pq = 0.4980$; $q^2 = 0.2193$.

To calculate the expected genotype frequencies (E), these figures have to be multiplied by the number of individuals, 1279:

$$E(\overline{M}) = 361.6$$

$$E(\overline{MN}) = 636.9$$

$$E(\overline{N}) = 280.5$$

These expectations are now compared with the observed frequencies:

$$X_1^2 = 1279 \times \frac{[634^2 - (4 \times 363 \times 282)]^2}{[(2 \times 363) + 634]^2 [634 + (2 \times 282)]^2} = 0.027;$$

$P \gg 0.05$

There is no statistically significant difference between observed and expected gene frequencies.

Example: Estimate of AB0 Gene Frequencies [711]. In 21,104 individuals from Berlin, the blood group distribution found was:

$\overline{\text{A}} = 9123$

$\overline{\text{B}} = 2987$

$\overline{\text{O}} = 7725$

$\overline{\text{AB}} = 1269$

By the improved Bernstein method, this leads to the following results. (For details see Sect. 3.2.2)

$p = 0.287685 \pm 0.002411$

$q = 0.106555 \pm 0.001545$

$r = 0.605760 \pm 0.002601$

The maximum likelihood method was shown to lead to exactly the same results [711].
The following maximum likelihood variances were calculated:

$Vp = 0.000005811$

$Vq = 0.000002386$

$Vr = 0.000006763.$

The standard deviations listed above are the square roots of these variances.
From these gene frequencies, expectations for genotype frequencies can now be calculated in the same way as shown above for the MN blood types and compared with the observed frequencies by a conventional χ^2 test.

Still more complicated problems are encountered for the Rh blood groups and in general for all systems for which many different combinations of antigens are inherited together. Here, computer programs have been published or mentioned in publications. For the Rh system, the following publications may be used [585, 586]. A number of authors have published rules for the calculation of gene and haplotype frequencies for the HLA system [554, 738, 779, 805, 962]. The ALLTYPE system is also useful [789].
Inadequate sampling, however, will not be compensated by computer processing. All methods mentioned so far are based on the assumption that sampling of individuals has been independent, i.e., that the selection of any person does not enhance or diminish the chance for any other person in the population to be selected. This rule is violated, for example, when relatives are accepted into the sample. Samples containing relatives are not necessarily useless for gene frequency calculations. But such inclusion must be noted, together with the degree of the relationships, and special statistical methods must be used for analysis [211].

Appendix 2

Testing for Segregation Ratios: Frequent Traits, No Ascertainment Bias, Dominance [876, 877]

Segregation analysis is straightforward if the mode of inheritance is codominant, so that every genotype corresponds to a distinct phenotype, and if the families to be analyzed have been sampled from the population irrespective of genotypes of family members. In this case, the number of individuals in each genotypic class needs to be compared with its expectation from Mendel's law by the χ^2 test as shown in Sect. 3.3.3., Table 3.7.

With dominance, segregation analysis is more complicated than with codominant inheritance: In matings involving phenotypes $\overline{A}$ and $\overline{a}$, two mating types each are involved. For example, in mating type $\overline{A} \times \overline{a}$ the two mating types AA × aa and Aa × aa are involved. The mating $\overline{A} \times \overline{A}$ comprises the mating types AA × AA, AA × Aa, and Aa × Aa. A useful method for segregation analysis has been given by C.A.B. Smith [876].

Mating Type $\overline{A} \times \overline{a}$. In this group, two genetic types of families are represented: AA × aa, and Aa × aa. The first type yields only Aa children with phenotype $\overline{A}$, the second type gives Aa and aa children in the expected ratio of 1:1. For illustration, blood group data are used (Tables A 2.1, A 2.2). The observed results can be arranged as in Table A 2.1 the numbers in column r and row s representing the number of families with s examined children, r of which have the phenotype $\overline{a}$.[1] The number of families with at least one recessive child is:

s	1	2	3	4	5	6	total
n_s	2	12	7	0	1	1	23

In these families, the number of recessive-type children is expected to follow a "truncate binomial distribution." Twochild families, for example, having 0, 1 and 2 recessive children are expected in the ratio 1:2:1; however, the class with 0 recessive has been excluded. Therefore, there is a probability of ⅔ that such a family will have one recessive child and a probability of ⅓ that it will have two recessive children. The expected number of recessive children in a family of two is accordingly:

$$1 \times \tfrac{2}{3} + 2 \times \tfrac{1}{3} = \tfrac{4}{3} = a_2$$

and the variance:

$$1^2 \times \tfrac{2}{3} + 2^2 \times \tfrac{1}{3} - a_2^2 = \tfrac{2}{9} = b_2$$

In principle the same argument can be used for families having 3,4 ... etc. children, and $a_3, a_4 \ldots$ and $b_3, b_4 \ldots$ can be calculated.

In general, the probability that a family of s children, with at least one recessive, has exactly r recessives is:

$$P(s;r) = \binom{s}{r} (\tfrac{1}{2})^s / [1 - (\tfrac{1}{2})^s] = \binom{s}{r} 2^{-s} / (1 - 2^{-s})$$

(cf. the binomial distribution, Sect. 3.3.2). It follows that:

$$a_s = \frac{s}{2(1-2^{-s})}; \quad b_s = \frac{s}{2(1-2^{-s})} \left[\frac{1+s}{2} - \frac{s}{2(1-2^{-s})} \right] \qquad \text{(A 2.1)}$$

The total number of expected recessive children in the sample will be:

$$E_1 = 2a_1 + 12a_2 + 7a_3 + a_5 + a_6 = 35.623$$

using the values in Table A 2.3 (from Smith). From a similar combination of b values, the variance can be calculated:

$$V_1 = 8.555$$

[1] Data in Tables A 2.1, A 2.2 for P, p blood types from Lang-Brown et al. (1953) [876].

Table A 2.1. Families of type $\overline{A} \times \overline{a}$

r = No. of children of recessive type $\overline{a}$	s = No. of children/family tested						Total
	1	2	3	4	5	6	
0	7	5	3	2	0	0	17
1	2	7	2	0	0	0	11
2	0	5	5	0	0	0	10
3	0	0	0	0	1	1	2
Totals	9	17	10	2	1	1	40
Total (excluding $r=0$)	2	12	7	0	1	1	

Table A 2.2. Families of type $\overline{A} \times \overline{A}$

r = No. of children of recessive type $\overline{a}$	s = No. of children/family tested							Total
	1	2	3	4	5	6	8	
0	20	15	10	3	1	1	1	51
1	2	5	1	1	0	0	0	9
2	0	0	0	0	0	1	0	1
Total	22	20	11	4	1	2	1	61
Total (excluding $r=0$)	2	5	1	1	0	1	0	10

In general, the values will be:

$$E_1 = \Sigma n_s a_s; \qquad (A2.2)$$

$$V_1 = \Sigma n_s b_s \qquad (A2.3)$$

The observed number of recessives in Table A2.1 is:

$$O_1 = 11 \times 1 + 10 \times 2 + 2 \times 3 = 37$$

The agreement can be checked by comparing the difference $O_1 - E_1 = 1.377$ with its standard error $s_1 = \sqrt{V_1} = 2.925$. Since $\chi^2 = \frac{(O_1 - E_1)^2}{V_1} = 0.222$, there is an excellent fit.

A similar method can be applied to families of mating type $\bar{A} \times \bar{A}$ (Table A2.2) with the one difference that in matings Aa × Aa, children of types $\bar{A}$ and $\bar{a}$ are expected in a ratio of 3:1. Expections and variances are A_s and B_s, and can also be taken from Table A2.3. In the ten families with at least one recessive child, 11 such children were observed, compared with an expectation E_2 of 12.3 and a variance V_2 of 2.069:

$$\chi^2 = \frac{(O_2 - E_2)^2}{V_2} = 0.817$$

Again, the observed values fit the expectation excellently. So far gene frequencies have not been used. The observed numbers of families with at least one recessive child must be compared with the expected numbers of such families as calculated from the total number of families in the sample. For this purpose, reliable gene frequency estimates are needed; these should come from a large sample of unrelated individuals. Families of mating type $\bar{A} \times \bar{a}$, for example, may be of type AA × aa, or Aa × aa. They are expected to occur with the frequency $2 \times p^2 \times q^2 + 2 \times 2pq \times q^2 = 2p^2q^2 + 4pq^3$. At the same time, this value is the probability that a randomly selected family will be of mating type $\bar{A} \times \bar{a}$. A family can have recessive children only if the mating type is, in fact, Aa × aa. Even in that case, the probability of having at least one recessive among s children is $1-(\frac{1}{2})^s$, for there is the probability $(\frac{1}{2})^s$ that only dominant children will segregate.
From this reasoning, the probability π that a family of mating type $\bar{A} \times \bar{a}$ with s children will produce at least one recessive child can be deduced:

$$\pi = \frac{4pq^3[1-(\frac{1}{2})^s]}{2p^2q^2 + 4pq^3} = ud_s \qquad (A2.4)$$

where $u = \frac{2q}{1+q}$, $d_s = 1-(\frac{1}{2})^s$

The values of d_s are tabulated in Table A2.3, and when the gene frequencies are known, ud_s can easily be calculated. When, in the entire series of families, there are n_s families of size s ($s = 1,2,..$), then the expectation and variance of the number of families with at least one recessive child are:

$$E_3 = \Sigma n_s ud_s; \quad V_3 = \Sigma n_s ud_s(1 - ud_s) \qquad (A2.5)$$

For example, gene frequency q of the recessive allele p in the P system is 0.51. Therefore, $u = 0.675$. The further calculation is carried out in Table A2.4.
In principle, the same calculation can be carried out for the $\bar{A} \times \bar{A}$ families with u^2 and $D_s = 1-(\frac{3}{4})^s$ instead of u and d_s, respectively; the appropriate D_s values are again tabulated in Table A2.3.

Table A 2.3. Expectations a_s, variances b_s, and d_s values for mating type $\bar{A} \times \bar{a}$; expectations A_s, variances B_s, and D_s values for mating type $\bar{A} \times \bar{A}$

s	a_s	b_s	d_s	A_s	B_s	D_s
1	1.000	0.000	0.500	1.000	0.000	0.250
2	1.333	0.222	0.750	1.143	0.122	0.438
3	1.714	0.490	0.875	1.297	0.263	0.578
4	2.133	0.782	0.938	1.463	0.420	0.684
5	2.581	1.082	0.969	1.639	0.592	0.763
6	3.048	1.379	0.984	1.825	0.776	0.822
7	3.528	1.667	0.992	2.020	0.970	0.867
8	4.016	1.945	0.996	2.223	1.172	0.900
9	4.509	2.215	0.998	2.433	1.380	0.925
10	5.005	2.478	0.999	2.649	1.592	0.944
11	5.503	2.737	1.000	2.871	1.805	0.958
12	6.001	2.992	1.000	3.098	2.020	0.968
13	6.501	3.245	1.000	3.329	2.234	0.976
14	7.000	3.497	1.000	3.563	2.446	0.982
15	7.500	3.748	1.000	3.801	2.658	0.987
16	8.000	3.999	1.000	4.040	2.867	0.990

Smith, C.A.B.: Ann. Hum. Genet. *20*, 265 [876].

Table A 2.4. Expected and observed frequencies of families with recessive children; mating type $\bar{A} \times \bar{a}$

No. of children s	Total number of families n_s	No. with at least one recessive child m_s	Expected $n_s ud_s$	V
1	9	2	3.038	2.012
2	17	12	8.606	4.249
3	10	7	5.906	2.418
4	2	0	1.266	0.465
5	1	1	0.654	0.226
6	1	1	0.664	0.223
Total	40	23	$E_3 = 20.134$	$V_3 = 9.593$

Table A 2.5. Expected and observed frequencies of families with recessive children; mating type $\bar{A} \times \bar{A}$

No. of children s	Total numbers of families N_s	No. with at least one recessive child M_s	Expected $N_s UD_s$	V
1	22	2	2.508	2.222
2	20	5	3.995	3.197
3	11	1	2.899	2.135
4	4	1	1.248	0.858
5	1	0	0.348	0.227
6	2	1	0.750	0.469
8	1	0	0.410	0.242
Total	61	10	$E_4 = 12.158$	$V_4 = 9.350$

An overview of all comparisons is given in Table A2.6, together with a summation of all χ^2 comparisons. The first two χ^2 values test for fit with the expected segregation ratios, whereas with the last two the observed frequencies of the different mating types are compared with their expectations on the basis of the Hardy-Weinberg Law and the gene frequencies used. This clear separation makes the method easily understandable.

Table A 2.6. χ^2 comparisons

Test	Mating	χ^2	d.f.[a]
No. of recessive children, given m_s	$\bar{A}\times\bar{a}$	0.222	1
No. of recessive children, given M_s	$\bar{A}\times\bar{A}$	0.817	1
$\sum m_s$, given n_s	$\bar{A}\times\bar{a}$	0.856	1
$\sum M_s$, given N_s	$\bar{A}\times\bar{A}$	0.498	1
Total		2.393	4

[a] Degrees of freedom.

Appendix 3

Formulas and Tables for the Correction of Ascertainment Biases, Testing and Estimating of Segregation Ratios: Other Statistical Problems and a Calculated Example

Whereas in App. 2 a method for testing segregation ratios in common traits (such as genetic polymorphisms) has been described, App. 3 contains methods for segregation analysis in rare traits (such as genetic disease), including corrections for ascertainment biases. As explained in Sect. 3.3, segregation analysis may proceed in two different ways: testing of empirical data for correspondence with a given theoretical segregation ratio and estimation of segregation ratios.

In both cases, the biases introduced by the way in which the data had been ascertained must be corrected. As discussed in Sect. 3.3, two different modes of ascertainment have to be distinguished: Single selection ($k=0$) and complete or truncate selection ($k=1$). With single selection ($k=0$), every family has been ascertained by one single proband. Examples are family surveys starting with a series of hospital patients. With complete or truncate selection ($k=1$), all affected individuals in the population have been ascertained. Correction of segregation ratios is necessary because sibships not comprising at least one affected sib just by chance despite heterozygosity of one (in a dominant or X-linked recessive condition) or both parents (in a recessive condition) have not been ascertained.

In the following, some of the available methods for testing agreements of a segregation ratio with its theoretical expectation, as well as methods for estimating segregation ratios are given. In principle, we followed suggestions given by Kaelin (1955; [729]). The method recommended here only requires a pocket or desk calculator, preferentially programmable unless the use of the tables of Kaelin is preferred (which are also printed on p. 557 in the first edition of this book). First, the methods for segregation analysis will be described. Then, some problems posed by genetic heterogeneity and by admixture of sporadic cases are discussed, and a method for examination of birth order effects is given. Then, the methods will be applied to a published population survey on deaf mutism in Northern Ireland. Finally, the somewhat more complicated ascertainment problems encountered with translocation families (and with other families comprising more than one sibship) will be discussed.

At first glance, recommendation of a "do it yourself" method without using one of the existing, mostly very sophisticated computer programs for segration analysis (e.g. the program SEGRAN of Morton, 1969, cf. also Yee et al., 1969) might look somewhat oldfashioned; however, the investigator who takes the trouble to go through such an analysis himself is rewarded by the big advantage of being able to evaluate critically the results in the light of the peculiarities and possible shortcomings of his data. Readers who have access to a personal computer (PC) and have knowledge of the computer language BASIC may easily write a program according to the methods described in the following.[1]

Testing of Empirical Family Data for Correspondence with a Given Segregation Ratio. In this approach, the observed numbers of affected sibs r in sibships of size s are compared with their expectations $E_s(r)$. These expectations were calculated from the formulas $\frac{sp}{1-q^s}$ (complete or truncate selection, $k=1$), $(s-1)p+1$ (single selection, $k=0$) by entering a few program steps into the pocket calculator or using the tables A3.1 and A3.2 in the first edition. Here, s = number of children/sibship; n_s = number of sibships of size s; r = number of affected sibs; p = segregation ratio to be tested; $q=1-p$. To calculate the expectation $\Sigma E_s(r)$ for the entire set of sibships to be analyzed, the appropriate expectations $E_s(r)$ must be summed up. For example, if the family data comprise five sibships of size 6 with two affected sibs and one sibship of size 8 with three affected sibs, if these sibships have one proband each (single selection, $k=0$), and if the expected segregation ratio is 0.25 (recessive inheritance), the expectation for the entire number of affected sibs is obtained as follows:

$$\Sigma E_s(r) = 5 \times 2.25 + 1 \times 2.75 = 14.00$$

[1] Such a BASIC program for the IBM-PC as well as one for the computer HP 9826 is available on request from the first author (F. V.) if an empty 5.25" floppy disk is enclosed.

Table A3.1. Marriages U × U assumed to be heterozygous Aa × Aa: testing of segregation for deaf mutism according to the a priori method

Sib-ship-size s	Number of sibships	Number of affected		V
		Expected	Observed	
2	35	35 × 1.143 = 40.005	40	4.270
3	39	39 × 1.297 = 50.583	50	10.257
4	34	34 × 1.463 = 49.742	41	14.280
5	35	35 × 1.639 = 57.365	48	20.720
6	49	49 × 1.825 = 89.425	68	38.024
7	34	34 × 2.020 = 68.680	57	32.980
8	33	33 × 2.223 = 73.359	69	38.676
9	15	15 × 2.433 = 36.495	32	20.700
10	6	6 × 2.649 = 15.894	10	9.552
11	3	3 × 2.871 = 8.613	6	5.415
12	4	4 × 3.098 = 12.392	10	8.080
13	1	1 × 3.329 = 3.329	1	2.234
	288	505.882	432	205.188

The observed number of affected sibs is: $5 \times 2 + 1 \times 3 = 13$. These two figures can now be compared with each other using the formula $\chi = \frac{O-E}{V}$; the variance is calculated from the formula

$$V_s(r) = (s-1)pq \text{ if } k=0,\ spq\,\frac{1-q^s-spq^{s-1}}{(1-q^s)^2} \text{ if } k=1$$

for the variance in the single sibships:

$$V = \Sigma V_s(r)\ 5 \times 0.9375 + 1 \times 1.3125 = 6.0000$$
$$\sqrt{V} = 2.4494.$$

$$\chi = \frac{1.0000}{2.4494} = 0.4082; \quad P > 0.05$$

A calculated example is given in the genetic analysis of deaf-mutism. (Table A3.1)

Estimating a Segregation Ratio from a Family Material. The testing method described above only answers the question whether a given set of data agrees with the expectations derived from a specified genetic hypothesis. More often, however, such a hypothesis is not obvious. It is therefore advisable to *estimate* the segregation ratio. The first estimation methods described were Weinberg's "Geschwistermethode" (sib method) and "Probandenmethode" (proband method). The sib method is appropriate if all affected sibs are also probands, i.e. if $k=1$. The unaffected and affected sibs are counted, starting from every affected sib. For example, a sibship may comprise six sibs, three affected and three unaffected. The sib method gives the following result: 3 × 2 sibs are counted as affected, 3 × 3 sibs are counted as unaffected. The estimated segregation ratio

$$\hat{p} = \frac{3 \times 2}{3 \times 2 + 3 \times 3} = \frac{6}{15} = 0.4$$

If not all affected sibs are ascertained as probands, the above-mentioned procedure is performed starting only with the probands; it is then called the proband method. If every sibship has been ascertained by only one proband, the enumeration is carried out only once. For the above-mentioned sibship, this means: $\hat{p} = 2/5 = 0.4$. This is the case $k=0$. For one single sibship, the two estimates for $k=1$ and $k=0$ are identical. They may be different, however, if many sibships of different sizes are included. In this case, the estimate for $k=1$ gives the highest value for p, that for $k=0$ gives the lowest value.
More recently, more sophisticated estimation procedures have been developed.
In the following, we describe the method given by Finney [663] in the version of Kaelin [729]. For every sibship, a weighted score is calculated:

$$W_s Y_s = \frac{r-e}{pq} - B_s \qquad (A3.1)$$

Here, s and r are the number of all sibs and the number of affected sibs respectively, and:

$$e=1, \quad B_s=0, \quad W_s = \frac{s-1}{pq} \quad \text{for } k=0,$$

$$e=0, \quad B_s = \frac{s^2pq^{s-2}}{(1-q^s)^2} \quad W_s = \frac{s}{pq}\,\frac{1-q^s-spq^{s-1}}{(1-q^s)^2} \quad \text{for}$$
$k=1$

The weighted scores W_sY_s as well as the weights W_s are separately summed up for all sibships. That value of p for which the quotient of the score sum and the weight sum, $\frac{\Sigma W_sY_s}{\Sigma W_s}$, equals p, and is an estimate, $\hat{p}$, of the true segregation ratio. Except for single selection ($k=0$), the value $\hat{p}$ can only be calculated iteratively. One starts with a first approximation p_1 of $\hat{p}$, for which one may take the estimate of Weinberg's proband method (see above), calculates a new approximation $p_2 = \Sigma W_sY_s/\Sigma W_s$ using p_1, and replaces p_1 by p_2. This procedure is repeated until p_2 no longer differs from p_1. The calculation described is simplified in the following way. If p_2 is higher than p_1 (meaning that $p_1 < \hat{p}$), the calculation is repeated with higher values of p_1 until p_2 becomes lower than p_1; conversely, if p_2 is initially lower than p_1 (meaning that $p_1 > p$), p_1 is diminished until p_2 becomes higher than p_1. If p is included in this way between two values $p_1 < p_1$, then it can be calculated by linear interpolation. $\hat{p}$ results from the intersection of the two straight lines:

$$y=x, \quad \text{and} \quad y=\frac{\overline{p_2}-\underline{p_2}}{\overline{p_1}-\underline{p_1}}(x-\overline{p_1})+\overline{p_2}$$

The right sides of the two equations are equated and the resulting equation is solved for x, giving $\hat{p}$. The variance is calculated as follows:

$$\frac{1}{V(\hat{p})}=W(\hat{p})=\frac{(\hat{p}-\underline{p_1})\overline{W}+(\overline{p_1}-\hat{p})\underline{W}}{\overline{p_1}-\underline{p_1}} \qquad \text{(A 3.2)}$$

(Linear interpolation between the weights $\underline{W}$ and $\overline{W}$ corresponding to $\underline{p_1}$ and $\overline{p_1}$). The procedure will be demonstrated below using a practical example.
For $k=0$, the final estimate $\hat{p}$ is equal to:

$$\frac{\text{Number of affected sibs (without probands)}}{\text{Number of all sibs (without probands)}}$$

(Weinberg's proband method) and is reached already in the first cycle of iteration.
Let us again consider our example, the sibship with $s=6$ children, $r=3$ of whom are affected. With complete selection ($k=1$), the following score is calculated starting with a provisional estimate $p_1=0.45$:

$$W_sY_s=\frac{3}{0.45\times 0.55}-1.568=12.1212-1.568$$
$$=10.5532$$

$$\hat{p}_1=\frac{W_sY_s}{W_s}=\frac{10.5532}{21.448}=0.4920.$$

Here, the numerical values for B_s and W_s are calculated by Eq. A 3.1. As the estimated $\hat{p}_1$ is higher than the provisional value, 0.45, the calculation is repeated with $p_2=0.5$.

$$\hat{p}_2=\frac{W_sY_s}{W_s}=\frac{10.839}{22.059}=0.4913.$$

The true value of $\hat{p}$ is between these two estimates; it can be found by interpolation.
So far, only the two limiting cases, $k=1$ (complete selection) and $k=0$ (single selection) have been treated explicitly. However, methods have also been developed for incomplete multiple selection, i.e. for any number of probands within a sibship. Morton et al. [800; 802; 954; 963] have refined the method, taking into account the number of ascertainments per proband. In the course of a population study, probands may be ascertained not just once but several times. Theoretically, such multiple ascertainment actually permits estimating the real prevalence of the condition in a population when ascertainment has been incomplete. Assume for simplicity that two rounds of ascertainments are made, that the probability of ascertainment in each round is π, and that the chances of a person being ascertained in the first and the second rounds are independent of each other. Then, the probability of being ascertained twice is π^2 and of being ascertained once (in either the first or the second round) $2\pi(1-\pi)$. The ratio

$$\frac{\text{twice ascertained}}{\text{once ascertained}}=\frac{\pi^2}{2\pi(1-\pi)}$$

for which observed values may be obtained, allows the calculation of π.

This calculation, however, rests on a very crucial condition. Different ascertainments of a proband must be independent of each other. It has been explained in Sect. 3.3.4 that even single ascertainments of different probands in one family are almost never independent. It is clear from all medical and epidemiologic experience that – whatever practical ways are chosen to collect family material – different ascertainments of probands will never be independent. To take two extreme examples, a medical doctor suffering from a hereditary disease will easily be ascertained several times in different hospitals that he consulted because they specialize in his disease, whereas an unskilled worker living in a remote village will most likely not be ascertained by any kind of survey.
We therefore consider these refined methods of analysis to be inadequate for most family studies of humans. It is even our contention that methods considering multiple or proband selection should not be used because ascertainment of probands within the same family is not independent, as explained in Section 3.3.4. We even consider it dangerous to apply these methods to series of families for which independent ascertainment has not been strictly controlled. In spite of all refinements of statistical analysis offered in the meantime, we believe that Kaelin (1955) [729] and C. A. B. Smith (1959) [878] are correct in the following recommendation:

In practice a geneticist investigating a rare condition is in a position of considerable difficulty. He can only make precise statements about segregation ratios if he knows exactly what are the statistical properties of his method of collecting the data. However, if the condition is rare, he will wish to collect as many cases as he can through inquiries to hospitals, doctors, etc.; and this will not in general constitute a well-defined sampling scheme from the statistical point of view. Thus it seems practically inevitable that there should be some doubt as to the exact value of p in such cases. It is usual to assume that the situation will be intermediate between the two cases ... of truncate and (single selection), so the simplest method of testing would seem to be to show that the number of affected is not greater than might be expected on the hypothesis of (complete selection) and not less than might be expected on the hypothesis of (single selection) (Smith) [878].

In our opinion, the only exception from this rule is complete ascertainment of all families with probands in one population, with complete or truncate selection when families are ascertained through the children's generation, and no ascertainment bias at all for families ascertained in the parent's generation. Therefore, epidemiological studies on rare hereditary diseases should be based, wherever possible, on a full ascertainment of all cases in a defined population and for a given time period.

The usefulness of statistical correction methods for poor primary data should not be overestimated. Even the best cook is unable to prepare a delicious roast hare from a dead cat. (By the way, the one big secret of French cuisine is that it insists on utilizing only the very best ingredients.)

Other Statistical Problems Connected with the Testing of Genetic Hypotheses

Genetic Heterogeneity. The methods described above give satisfying results when the condition to be examined is homogeneous genetically and the mode of inheritance is simple. Under these conditions, the test method shows agreement between expectation and observation, and the estimation yields a $\hat{p}$ close to the expected segregation ratio (25% in the case of recessive inheritance). Even in this case, the question remains unresolved as to whether one ore more genes with the same mode of inheritance are involved, i.e., whether genetic heterogeneity exists (Sect. 3.3). Genetic considerations can help to decide this question if matings between two homozygotes occur. Here, all children should be affected when the same locus is involved in both parents.

In recessive inheritance, one further argument in favor of genetic heterogeneity can be derived from the relative frequencies of consanguineous matings (especially first-cousin matings) among the parents of probands and in the general population (Sect. 6.3.1). This ratio increases with decreasing gene frequency. Hence, if consanguinity is much higher than expected on the basis of gene frequency calculated from observed prevalence and under the assumption of one recessive gene, it may be concluded that more than one recessive gene exists. The frequencies of these genes would be lower, explaining the higher consanguinity rate.

This argument, however, though sound theoretically, should be used with great restraint in practice, as it assumes random distribution of consanguineous matings in the population. However, families with recessive diseases will frequently come from semi-isolates, among whom the proportion of consanguineous matings is generally higher than in the entire population. As a rule, therefore, this reasoning should be used only together with other corroborating evidence.

Admixture of Sporadic Cases. Another problem arises when the estimated segregation ratio turns out to be lower than expected. Here, admixture of sporadic cases (nonhereditary or dominant new mutations) is usually the most obvious explanation. In this case, the number of sibships with just one affected sibling should be increased above its expectation (E), which is, according to the binomial distribution:

$$E = \sum_{s=2}^{\infty} E_s = \sum_{s=2}^{\infty} n_s \frac{spq^{s-1}}{1-q^s} \quad \text{(complete selection)} \tag{A3.3}$$

$$E = \sum_{s=2}^{\infty} E_s = \sum_{s=2}^{\infty} n_s q^{s-1} \quad \text{(single selection)} \tag{A3.4}$$

where n_s is the number of sibships with size s. This expectation can be calculated by inserting the estimate $\hat{p}$ for p, and can then be compared with the observed frequency O by calculating:

$$\chi = \frac{O-E}{\sqrt{V}}$$

which is distributed approximately normally:

$$V = \sum_{s=2}^{\infty} V_s \text{ with } V_s = \frac{E_s(n_s - E_s)}{n_s} \tag{A3.5}$$

An example is given at the end of this appendix.

This problem has no practical importance in conditions with complete dominance, as the sporadic cases (nonhereditary or new mutants) cases can easily be discovered because both parents are unaffected. It may come up in cases of incomplete penetrance. Here, the method can help to discriminate between cases in which the gene was present in one of the parents but did not manifest itself and the real sporadic cases with genotypically normal parents, i.e., new mutations. It is especially important, however, when we have reason to suspect recessive inheritance and the estimated $\hat{p}$ is lower than 0.25. In such a case, when the analysis shows that the number of sibships with just one unaffected sibling has increased, we may want to have an estimate of the undisturbed segregation ratio. There are two ways to obtain such estimates. First, analysis can be limited to sibships with at least two affected.

As a second possibility, analysis may be limited to sibships whose parents are consanguineous, as most of these patients will be real recessives. The number of sporadic cases can be estimated, again using the binomial distribution (example below).

Birth Order and Maternal Age. The hypothesis of a simple mode of inheritance also predicts that the sequence of affected and unaffected sibs is random and that there is no influence of paternal or maternal age or of birth order. The most general tests for randomness of sequences are based on the theory of runs and were adapted for use in human pedigree data. Older assertions that in Fanconi's anemia sibs

Table A 3.2. Calculation of birth order effects. Mean *(black figures)* and variance *(ordinary figures)* of 6 A in complete sibship

s \ r	1	2	3	4	5	6	7	8	9	10	11	12	13	14	15	16	17	18	19
2	**9**	**18**	–	–	–	–	–	–	–	–	–	–	–	–	–	–	–	–	–
	9	0																	
3	**12**	**24**	**36**	–	–	–	–	–	–	–	–	–	–	–	–	–	–	–	–
	24	24	0																
4	**15**	**30**	**45**	**60**	–	–	–	–	–	–	–	–	–	–	–	–	–	–	–
	45	60	45	0															
5	**18**	**36**	**54**	**72**	**90**	–	–	–	–	–	–	–	–	–	–	–	–	–	–
	72	108	108	72	0														
6	**21**	**42**	**63**	**84**	**105**	**126**	–	–	–	–	–	–	–	–	–	–	–	–	–
	105	168	189	168	105	0													
7	**24**	**48**	**72**	**96**	**120**	**144**	**168**	–	–	–	–	–	–	–	–	–	–	–	–
	144	240	288	288	240	144	0												
8	**27**	**54**	**81**	**108**	**135**	**162**	**189**	**216**	–	–	–	–	–	–	–	–	–	–	–
	189	324	405	432	405	324	189	0											
9	**30**	**60**	**90**	**120**	**150**	**180**	**210**	**240**	**270**	–	–	–	–	–	–	–	–	–	–
	240	420	540	600	600	540	420	240	0										
10	**33**	**66**	**99**	**132**	**165**	**198**	**231**	**264**	**297**	**330**	–	–	–	–	–	–	–	–	–
	297	528	693	792	825	792	693	528	297	0									
11	**36**	**72**	**108**	**144**	**180**	**216**	**252**	**288**	**324**	**360**	**396**	–	–	–	–	–	–	–	–
	360	648	864	1008	1080	1080	1008	864	648	360	0								
12	**39**	**78**	**117**	**156**	**195**	**234**	**273**	**312**	**351**	**390**	**429**	**468**	–	–	–	–	–	–	–
	429	780	1053	1248	1365	1404	1365	1248	1053	780	429	0							
13	**42**	**84**	**126**	**168**	**210**	**252**	**294**	**336**	**378**	**420**	**462**	**504**	**546**	–	–	–	–	–	–
	504	924	1260	1512	1680	1764	1764	1680	1512	1260	924	504	0						
14	**45**	**90**	**135**	**180**	**225**	**270**	**315**	**360**	**405**	**450**	**495**	**540**	**585**	**630**	–	–	–	–	–
	585	1080	1485	1800	2025	2160	2205	2160	2025	1800	1485	1080	585	0					
15	**48**	**96**	**144**	**192**	**240**	**288**	**336**	**384**	**432**	**480**	**528**	**576**	**624**	**672**	**720**	–	–	–	–
	672	1248	1728	2112	2400	2592	2688	2688	2592	2400	2112	1728	1248	672	0				
16	**51**	**102**	**153**	**204**	**255**	**306**	**357**	**408**	**459**	**510**	**561**	**612**	**663**	**714**	**765**	**816**	–	–	–
	765	1428	1989	2448	2805	3060	3213	3264	3213	3060	2805	2448	1989	1428	765	0			
17	**54**	**108**	**162**	**216**	**270**	**324**	**378**	**432**	**486**	**540**	**594**	**648**	**702**	**756**	**810**	**864**	**918**	–	–
	864	1620	2268	2808	3240	3564	3780	3888	3888	3780	3564	3240	2808	2268	1620	864	0		
18	**57**	**114**	**171**	**228**	**285**	**342**	**399**	**456**	**513**	**570**	**627**	**684**	**741**	**798**	**855**	**912**	**969**	**1026**	–
	969	1824	2565	3192	3705	4104	4389	4560	4617	4560	4389	4104	3705	3192	2565	1824	969	0	
19	**60**	**120**	**180**	**240**	**300**	**360**	**420**	**480**	**540**	**600**	**660**	**720**	**780**	**840**	**900**	**960**	**1020**	**1080**	**1140**
	1080	2040	2880	3600	4200	4680	5040	5280	5400	5400	5280	5040	4680	4200	3600	2880	2040	1080	0
20	**63**	**126**	**189**	**252**	**315**	**378**	**441**	**504**	**567**	**630**	**693**	**756**	**819**	**882**	**945**	**1008**	**1071**	**1134**	**1197**
	1197	2268	3213	4032	4725	5292	5733	6048	6237	6300	6237	6048	5733	5292	4725	4032	3213	2268	1197

Source: Haldane, J. B. S., Smith, C. A. B.: Ann. Eugen. *14* (1947–1949) [698].

affected with a recessive blood disease (Sect. 5.1.6) occurred in clusters within their sibships have recently been refuted by a test based on the theory of runs [895].

Examples of paternal and maternal age effects are given in Sect. 5; a very useful approach is the examination of birth order. Obviously, birth order is correlated with paternal and maternal age and can be examined in the family data itself, without recourse to a control population. A useful test [698] has been tabulated: A = sum of birth orders of affected sibs, s = number of all sibs, and r = number of affected sibs in the sibship. Then, the expectation and variance of $6A$ (which is easier to tabulate than for A), when all siblings can be classified, is:

$$E(6A)=3r(s+1) \quad V(6A)=3r(s+1)(s-r)$$

The formulas are more complicated when classification is incomplete (cf. the original). Expectations and variances are tabulated in Table A3.2.

Practical Example of Segregation Analysis Using a Large Sample: Profound Deafness

Some of the methods described in Sect. 3.3.6 and in App. 3 will now be applied to one specific example. To demonstrate as many methods as possible, we shall proceed in a somewhat roundabout way.

A. C. Stevenson and E. A. Cheeseman (1955) [899] collected all cases of profound deafness in Northern Ireland. At the time of their study, 613 deaf mutes were living who had been born deaf or had lost their hearing capacity very early in life. The authors had additional data on 85 deaf mutes who were no longer living.

Ascertainment of Families. The profoundly deaf are usually known to welfare officers or school medical officers. Therefore, since all doctors in Northern Ireland were contacted, ascertainment could be assumed to be reasonably complete. Although a few affected were ascertained indirectly through affected relatives, even in these cases notes of one kind or another were found in the files. Hence, all affected deaf individuals can be regarded as probands according to the definition given above. Since persons were ascertained in more than one way, complete or truncate selection can be assumed.

The pedigree data were collected by personal visits of one of the authors or by a specifically trained field worker and supplemented as far as possible by physical examinations and objective data. Here, records from special schools about affected persons in families, consanguinity of parents, and other such data proved to be helpful. Prevalence was found to be 45/100,000 inhabitants.

Clinical Aspects. Clinical examination of the patients was carried out to enlarge knowledge of pathogenesis and symptoms, and to rule out the environmental agents that may cause deaf mutism in early childhood, such as rubella, erythroblastosis, ototoxic drugs, perinatal trauma, encephalitis, meningitis, and otitis media. Such cases had to be eliminated from the sample. Frequently, however, a full etiologic diagnosis was not possible from clinical and audiometric evidence. Together with use of case histories, it was possible to eliminate 183 living and 2 deceased patients who were not included in the genetic analysis.

Genetic Analysis. The following three types of families were analyzed separately:

1. Parents HD × HD
2. Parents HD × U
3. Parents U × U

(HD, hereditary deaf; U, unaffected)

Data for the third group (U × U) are shown in Table A3.1. Previous experience and the high number of unaffected siblings suggest an autosomal-recessive mode of inheritance. Therefore, the test method described above is used, and the observed frequencies are compared with their expectations for complete selection. The result, however, is not compatible with the genetic hypothesis. There is a highly significant deficit of affected patients ($\chi^2=26.60$ with 1 d.f.).

This result shows the disadvantage of the test method in comparison with the estimation method. The estimation would have provided an estimate of the segregation ratio in these families, whereas the test method simply says "no."

Therefore, the estimation method is used with $p=0.20$; $q=0.80$ as preliminary values (cf. Eq. A3.1).

$$\begin{aligned}\Sigma W_s Y_s &= \frac{\Sigma r}{pq} - \Sigma B_s = \frac{432}{0.2\times 0.8}\\ &\quad -35\times 6.173-39\times 6.047-34\times 5.875\\ &\quad -35\times 5.664-49\times 5.417-34\times 5.142\\ &\quad -33\times 4.845-15\times 4.532-6\times 4.211\\ &\quad -3\times 3.887-4\times 3.566-1\times 3.251\\ &= 1127.554\end{aligned}$$

$$\begin{aligned}\Sigma W_s &= 35\times 3.858+39\times 8.188+34\times 12.967\\ &\quad +35\times 18.163+49\times 23.739+34\times 29.651\\ &\quad +33\times 35.856+15\times 42.308+6\times 48.962\\ &\quad +3\times 55.775+4\times 62.706+1\times 69.721\\ &= 6301.800\end{aligned}$$

It follows:

$$\hat{p}=\frac{1127.554}{6301.800}=0.17893$$

This value is much lower than the preliminary value 0.20; therefore, the calculation is repeated with $p=0.15$, $q=0.85$.

Result:

$\Sigma Y_s W_s = 1290.674$; $\Sigma W_s = 7197.210$;

$$\hat{p}=\frac{\Sigma Y_s W_s}{\Sigma W_s}=0.17933$$

Interpolation:

$$y = x;\ y = \frac{0.17893 - 0.17933}{0.20 - 0.15}(x\text{-}0.15) + 0.17933$$

It follows by equating the right-hand sides:

$$x = \hat{p} = 0.1791$$

This value can now be used to calculate the variance from Eq. A3.2:

$$1/V = W$$

$$= \frac{(0.20 - 0.1791) \times 7197.210 + (0.1791 - 0.15) \times 6301.800}{0.05}$$

$$= 6676.081$$

$$V = 0.00014979; \quad s = \sqrt{V} = 0.01224$$

This estimate is: $\hat{p} = 0.1791 \pm 0.01224$. This value is, again, significantly different from the expectation of recessive inheritance. Besides, it is known that in a fair number of cases the condition has an exogenous cause. It is possible that the authors did not succeed in eliminating all these cases from their data. They should show up as sporadic cases, enhancing the number of families with just one patient (Table A3.3).
The expectation of this number calculated from Eq. A3.3. is 181.56; 198 were found; the variance is obtained from formula A3.5 as 59.052. The comparison gives $\chi = (198 - 181.56)/\sqrt{59.052} = 2.139$, $P < 0.05$ (one-tailed test): There are, indeed, too many sporadic cases i.e. setting $r_{min} = 2$. Now, the following formulas for B_s and W_s are used in connection with Eq. A3.1:

For $k = 0$: the formulas for $r_{min} = 1$, $k = 1$ with s replaced by $s - 1$

$$\text{for } k = 1\text{:}\ B_s = \frac{s(s-1)pq^{s-3}(sp + q^s - 1)}{(1 - q^s - spq^{s-1})^2}, \qquad \text{(A3.6)}$$

$$W_s = \frac{s}{pq}\,\frac{(1 - q^{s-1})^2 - (s-1)^2 p^2 q^{s-2}}{(1 - q^s - spq^{s-1})^2}$$

Therefore, the estimation of p must be repeated using only the sibships with at least two affected siblings (90 sibships with 234 affected siblings). Our preliminary estimate of $\hat{p}$ is 0.25. The calculation gives:

$$\frac{\Sigma Y_s W_s}{\Sigma W_s} = \frac{385.993}{1428.131} = 0.27028$$

Therefore, the calculation is repeated with $\hat{p} = 0.30$, giving:

$$\frac{\Sigma Y_s W_s}{\Sigma W_s} = \frac{389.955}{1439.769} = 0.27084$$

Interpolation leads to the final estimate. Standard deviation can be calculated from Eq. A3.2:

$$\hat{p} = 0.27051 \pm 0.02642$$

Now, the estimate fits quite well with the expected segregation ratio of 0.25. On this basis, it is possible to estimate the number of sporadic cases that are not inherited within the families having more than one child but only one affected.
Of all $n = \Sigma n_s$ families, $\Sigma n_{s,h}$ will have inherited cases. $n_{s,h}$ can be estimated from the binomial distribution using the estimate $\hat{p}$ and the number of families with more than one affected child $n_{s,r>1}$, applying the following formulas:

Table A 3.3. Examination of whether sibships with one affected sib are more frequent than expected

Size of sibship s	Number of sibships with 1	2	3	4	5	>1 affected children	Total number of sibships n_s
2	30	5				5	35
3	30	7	2			9	39
4	29	3	2			5	34
5	26	5	4			9	35
6	35	9	5			14	49
7	18	11	3	2		16	34
8	15	9	3	3	3	18	33
9	7	4	1	1	2	8	15
10	4	1		1		2	6
11	1	1	1			2	3
12	2		1		1	2	4
13	1					0	1
Total	198	55	22	7	6	90	288

$$n_{s,r>1} = \frac{1 - q^s - spq^{s-1}}{1 - q^s} \times n_{s,h}$$

or:

$$n_{s,h} = n_{s,r>1}\frac{1 - q^s}{1 - q^s - spq^{s-1}}$$

In our case:

$$\hat{p} = 0.27051, \quad \hat{q} = 0.72949, \quad \Sigma n_{s,h} = 180.03$$

It could be argued that in this case use of theoretical segregation ratio, 0.25, would be preferable. There are arguments against and in favor of this possibility. The difference is small, however. With the theoretical segregation ratio, $\Sigma n_{s,h} = 193.08$ families would be obtained. Instead, 288 families were found. This means that there must be, on the average, 107.97 nonhereditary cases among the sporadic ones. Additionally, there are 21 cases for which this information is not available, as they are single children. If we assume that the same fraction is nonhereditary, then an additional $21 \times 108/432 = 5.25$ cases will be nonhereditary, giving a total of 113.22 nonhereditary cases (=24.99%) of $432 + 21 = 453$ cases, both parents of which were unaffected.
An alternative way to examine the problem of nonhereditary cases would be to estimate the segregation ratio only among children from consanguineous marriages (Table A3.4). Here, the estimate is:

$$\hat{p} = 0.269 \pm 0.038$$

i.e., it is identical with the result from families having more than one affected child.
So far only the mating type, where both parents are normal (U × U) has been analyzed. Now, the matings between two affected (HD × HD) will be examined (Table A3.5). Among deaf mutes, this type is relatively frequent, because assortative mating is favored by the educational system,

Table A 3.4. Sibships from consanguineous marriages

Size of sib-ship s	Number of sibships with					Total number of sibships n_s	Number of children		
	1	2	3	4	5		Affected	Healthy	Total
	affected children								
1	4					4	4	0	4
2	4	1				5	6	4	10
3	3	1				4	5	7	12
4	1	1				2	3	5	8
5	1	1	1			3	6	9	15
6		2	1			3	7	11	18
7	1	4		1		6	13	29	42
8	1	2			1	4	10	22	32
9		1	1			2	5	13	18
10	1			1		2	5	15	20
12	1					1	1	11	12
Total	17	13	3	2	1	36	65	126	191

Table A 3.5. Sibships from marriages between two deaf partners with at least one affected child

Size of sib-ship s	Number of sibships with					Total number of sibships n_s	Number of children		
	1	2	3	4	5		Affected	Healthy	Total
	affected children								
1	2					2		2	2
2	3	1				4	3	5	8
3	1		1			2	2	4	6
4				1		1		4	4
7					1	1	2	5	7
9					1	1	4	5	9
Total	6	1	1	1	2	11	11	25	36

which creates a "social isolate" for deaf mutes. If deaf mutism were always caused by the same recessive gene, then all children from these matings should be affected. This is the case in five sibships only, whereas in six matings affected as well as unaffected children were observed. Besides, the authors observed no less than 21 matings between deaf partners whose children (a total of 53) were all unaffected. Considering only the matings with exclusively unaffected children, the conclusion would be that genetic heterogeneity exists, with a number of different recessive genes. In view of the numerous nongenetic cases in matings U × U, however, a second possibility offers itself. Does at least one of the parents suffer from a nonhereditary type? This hypothesis can be tested partially by looking for hints as to a genetic cause in these families, such as consanguinity or affection of other relatives - for example, parents' sibs. Such hints were, indeed, discovered in the families of both spouses in 12 cases and in families of one spouse in five cases. Hence, in many of these families, the fact that children are unaffected appears due to genetic heterogeneity and not because one parent had exogenous deafness.

Table A 3.6. Marriages deaf mute × unaffected with at least one deaf child

s	Number of sibships with r deaf-mutes					n_s	Number of sibs		
	1	2	3	4	5		Deaf mute	Healthy	Total
2	2					2	2	2	4
3		1				1	2	1	3
4			2			2	6	2	8
7			1			1	3	4	7
	2	1	3			6	13	9	22

The amount of information supplied by the mating type HD × HD, however, is not yet exhausted. There remains the group with affected as well as unaffected children - no less than six marriages with 11 healthy and 14 deaf children. They can be explained neither by genetic heterogeneity nor by exogenous factors. The most obvious explanation would be that, apart from the recessive types, there are dominant types as well. However, when dominants do occur, then some of the sibships HD × HD with only affected sibs could belong to the dominant type, and the lack of unaffected siblings could be a chance result.

Now, the third mating type, HD × U, will be examined (Table A3.6). There were 45 matings with at least one child. In 39 of them, only healthy children were observed, 102 in all. Obviously, these matings do not contradict the hypothesis that the affected parent suffers from a recessive type. There were, however, six marriages with at least one affected child. According to Eqs. A3.1 and A3.2 for $k=1$, $\hat{p}=0.548\pm0.119$ is the estimate for the segregation ratio. Obviously, this points to matings between heterozygotes and homozygotes. But there are still two possibilities. Either these mutations are dominant - in which case the affected parent is heterozygous and the unaffected homozygous - or the mutation is recessive - in which case the unaffected parent is heterozygous and the affected parent is the homozygous one. These six matings do not permit a decision between the two possibilities. At first glance, it is unlikely that 6 of 45 affected would end up marrying a heterozygote just by chance. This argument, however, neglects the high amount of assortative mating in the families of deaf mutes, which naturally leads not only to marriages between affected but also between affected and heterozygotes, for example, sibs and other near relatives of deaf mutes.

Population Genetics of Deaf Mutism. The authors compared the reproduction rate of the patients with that of the contemporary general population (which, by the way, was very high). The proportion of those married as well as the average number of children were somewhat diminished but not very much. This result supports the conclusion that among the "nonhereditary cases" some new mutants of dominant mutations may be hidden. Their number, however, cannot even be estimated approximately, as there are reasons to believe that the selective disadvantage of profound deafness may have altered due to changing social conditions. Besides, the dominant types involved are too ill-defined. Therefore, a mutation rate estimate for these dominant types based on these data is not feasible. For the recessive types, no mutation rate estimates are possible (Sect. 5.1.3.1).

As indicated in Table A3.5, many of the U × U matings with affected children were consanguineous. Theoretically, this observation can be utilized for further analysis of genetic heterogeneity. Let q be the gene frequency of a recessive allele and c the frequency of first-cousin marriages. Then, the relative frequency g of first-cousin marriages among parents of homozygotes depends on q. It is given by the formula (Sect. 6.3.1.2):

$$g = \frac{c\frac{1}{16}q(1+15q)}{(1-c)q^2 + c\frac{1}{16}q(1+15q)} = \frac{c1+15q)}{16q + c(1-q)}$$

In this sample, $q^2 = 0.00027$. With $c = 0.1\%$, g would be 0.47%, assuming only one recessive gene. For $c = 1\%$, g would be 4.57%. The real proportion c for the contemporary Northern Irish population was estimated by the authors to be between 0.1% and 0.4%. The 6.8% consanguineous matings that were detected among the families analyzed here are not compatible with the assumption of only one recessive gene; this result supports the assumption of genetic heterogeneity.

However, this argument by itself should be regarded with caution. Isolates within the population may cause the same effect, and the social group made up of deaf and their families has many properties of a social isolate. It is needless to add that an estimation of the number of recessive genes involved - though possible formally if identical frequencies of these genes are assumed - would require so many untested, and untestable, assumptions that such a calculation is simply not warranted.

Summary. The sample comes from a survey exhaustively covering all congenital and early-onset deafness in Northern Ireland.

On the basis of clinical examinations, a few cases were diagnosed as exogenous and eliminated from further analysis. Not all patients who owe their deafness to exogenous causes, however, can be identified by either case histories or symptoms. Such cases were analyzed together with patients having genetic deafness.

The progeny of the different mating types were analyzed statistically. Matings between unaffected (U × U) yielded a significantly lower estimate of the segregation ratio p than the expected 0.25 in recessive inheritance. This underestimate was shown by further analysis to be caused by admixture of sporadic cases. Restriction of the statistical analysis to sibships with more than one affected sib or to children from consanguineous marriages gave estimates fitting the expectation of autosomal-recessive inheritance. The fraction of sporadic cases in all shibships with only one affected child and unaffected parents was estimated with the result that many of these sporadic cases were obviously nonhereditary in origin, confirming medical experience with exogenous factors. However, some of them may have been due to dominant mutations, especially since matings of the deaf with normals or with each other produced a significant minority of cases that fitted best the criteria of dominant inheritance.

Within the autosomal-recessive group, genetic heterogeneity was proven by the observation that in the majority of matings between two affected deaf persons, (HD × HD), all the children had normal hearing capacity. This conclusion was corroborated by the observation that there are relatively many consanguineous marriages among the unaffected parents of the deaf mutes. In our opinion, no further conclusions can be drawn from this study.

Why has this example been discussed at such length?:

To show that segregation analysis using the methods recommended here can be done by the research worker himself without serious difficulties. For the few calculations, an electronic desk calculator is sufficient, but when a larger computer is available, programming is straightforward. In most cases, the research worker and his team have spent many months or even years ascertaining and examining families. It is essential to set aside an adequate time period for statistical analysis. This has one great advantage. The analysis is carried out stepwise, and every step can be considered critically in the light of relevant information on sampling process, clinical results, and background knowledge of the population. Nobody knows the data better than those who have collected them. Therefore, the original research workers are best suited for a critical assessment.

Furthermore, the prospect that the sample will be analyzed by the team itself will lead to more care in the planning and conduct of the survey. Here, advice by a statistically experienced co-worker indeed is needed. Sampling should be planned with the intended statistical analysis already in mind and should be carried out according to well-defined, strict rules which are not altered in the course of the study. Most statistical errors in human genetics - and generally in the life sciences - are caused not by inadequate statistical techniques but by use of inappropriate methods based on improper assumptions. The interface between sample and analysis is critical, and no one knows this area better than the research worker himself.

The data on deafness analyzed above have also been subjected to a statistical analysis by a group using more sophisticated statistical methods. From this analysis, conclusions were drawn about a number of other parameters, for example, mutation rates of dominant and recessive genes and the number of recessive genes involved. We consider such conclusions to be hazardous, since they are based on untested assumptions about the data.

The same group offers a program package "segregation analysis" including prevalence estimates, separation of modes of inheritance, number of recessive genes, and mutation rates [800]. It is, of course, comfortable to have the data analyzed in this statistically impeccable way. However, one should remain critically aware of the properties and sometimes inevitable shortcomings of one's own data and not accept an elegant appearing computer analysis as being necessarily biologically meaningful.

Genetic Heterogeneity in Deaf Mutism. Since Stevenson and Cheeseman's survey was published, much work has been done on deaf mutism, and both genetic heterogeneity and a variety of dominant types have been found. A number of genetic types of deafness can be identified on the basis of additional clinically or biochemically recognizable symptoms [825, 669]. There is little doubt that these conditions comprise only a small part of the genetic heterogeneity. As the authors [825] write: "The interaction of literally hundreds - perhaps thousands - of genes is required for the development of the normal human ear. Consequently, defects in any of the great many genes can lead to hearing loss."

Correction for Ascertainment Biases, in Families with More than One Sibship, and with Various Kinds of Probands. Often, families are observed that consist of more than one sibship. Moreover, various kinds of probands might exist. For example, families with reciprocal translocations might be ascertained through an unbalanced translocation carrier, in most cases a child with multiple malformations. Or the proband might be a balanced translocation carrier; he or she might have been ascertained in the course of chromosome screening such as carried out in adult normal populations, in populations of newborns, in mental retardates, in any individuals carrying certain malformations, or in a study of spontaneous abortions. If ascertained by way of abortions, families will normally be examined only if at least two abortions have occurred. Furthermore, there are differences depending on whether the analysis is based on single published families or on families collected within the framework of a collaborative study. Such collaborative studies are preferable, since there is less danger of a combination of "interesting cases." Schäfer [501 a], in his study on segregation of translocations, discussed these problems and gave suggestions for correcting the most important biases.

When a collection of cases has been obtained the type of ascertainment has to be determined. The majority of published cases are usually ascertained via a child with an unbalanced translocation. At first glance, statistical correction according to the single selection model ($k=0$) in the sibship of this child-proband seems to be appropriate. One could argue that ascertainment and/or publication of the family also depends on the clinical status of other relatives; it is therefore prudent to repeat the calculation with a model of truncate selection; the true segregation ratio might be closer to the single selection result. However, this is true only if analysis is based on families ascertained via a clinically affected individual. In future, more and more family studies will be based on probands from long-term, complete collections of all malformed newborns in entire populations, as is already being done in Hungary, for example [616; 617]. In such cases, correction according to the model of truncate selection would be adequate.

As a rule, however, such pedigrees will consist of more than one sibship. In these cases, additional - secondary or even tertiary - probands occur. All individuals whose presence has induced the investigator to extend his studies to a further generation or sibship must be regarded as probands. This will be explained with reference to a model pedigree (Fig. A 3.1; Table A 3.7; [501 a]).

The family was discovered through III, 16, a carrier. She would not have been karyotyped, had she not suffered from multiple abortions. Since an indication for karyotyping is normally assumed when at least two abortions have occurred, two abortions are considered as probands among her children, and eliminated from the risk calculation for abortions. III, 16 is a "secondary proband." Moreover, if her mother (II, 5) were to have a normal karyotype,

Table A 3.7. Data for risk calculation for the family (Fig. A 3.1)

Sibships at risk	No. of children (excluding probands + unclassified persons)	Probands	Unbalanced cases	Balanced carriers	Abortions	Normals
II, 2-5	2	1	0	2	0	0
III, 2-9	5	0	0	3	0	2
III, 11-16	5	1	0	3	0	2
IV, 1, 2	1	0	0	0	1	0
Iv, 3, 4	2	0	1	0	0	1
Iv, 5-7	3	0	1	1	0	1
IV, 8-12	3	2	0	0	3	0

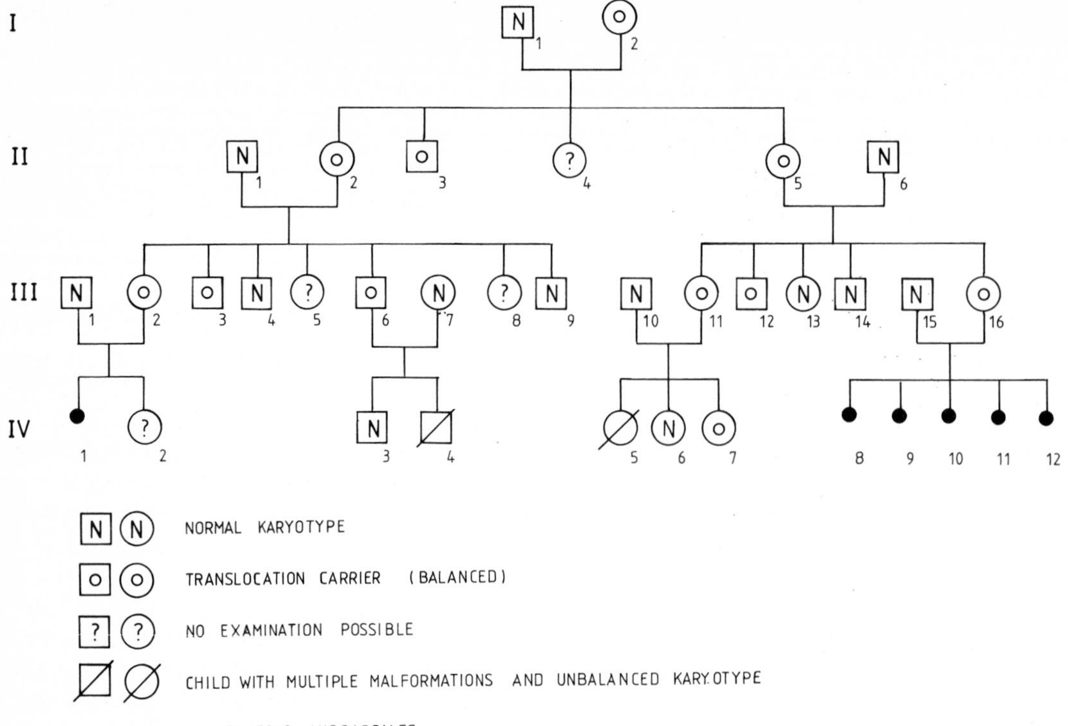

Fig. A3.1. Model pedigree with translocations. For analysis see Table A3.7. (From Schaefer, 1984 [501 a])

III, 16 would be regarded as de novo translocation, and the sibship to which II, 5 belongs would not have been examined. Therefore, II, 5 is another secondary proband and must be eliminated from risk calculations in her sibship. II, 4 was not karyotyped; since she has no children it is unknown whether she is a carrier. Therefore, she too must be excluded from risk calculations. II, 11, on the other hand, is a carrier, being a sister of the proband III, 16, and would have been examined in any case, irrespective of her children; therefore, she does *not* count as a proband; she must be *included* in the risk calculation for children of translocation carriers. She has three children who would have been examined anyway, regardless of their phenotypes. Therefore, no correction is necessary; they can all be included in the risk calculation.

The next step of the analysis must be carried out carefully: If generation II would have been examined in any case, then II, 5 is the only proband in this generation, and all other sibs can be used for risk calculation (excepting, of course, II, 4). Moreover, the sibships III, 2 to III, 9; IV, 1, 2; and IV, 3, 4 have been ascertained through an affected parent; therefore, no correction is necessary. However, if, for example, sibship IV, 3, 4 has only been examined because III, 11 has told the investigator that her cousin also has a malformed child (and if this sibship would not have been examined otherwise), the unbalanced child IV, 4 is a (tertiary) proband and has to be eliminated from risk estimation. This example shows how important explicit reporting of the ascertainment process is. In the following analysis it is assumed that the sibships on the left side of Fig. A3.1 have, indeed been ascertained through an affected parent. The result can be seen in Table A3.7. The following risk estimates can be derived:

a) For unbalanced patients: 2/21
b) For abortions: 4/21
c) For balanced carriers: 9/21
d) For normal offspring: 6/21

For these estimates, the single cases from all sibships have simply been pooled (= precumulation). This procedure could be criticized on the grounds that larger sibships carry too much weight in relation to smaller ones. It is also possible to make risk estimates for all sibships separately and then to pool them (= postcumulation). However, most studies in the literature on translocations, including that of Schäfer, have been performed with precumulation. This procedure appears to be justified especially if the sibships belong to larger pedigrees, since it can safely be assumed that within such a pedigree, the real risks will be identical in all sibships. On the other hand, it is necessary to perform such risk calculations separately for families in which at least one unbalanced patient has been observed as proband, those who have been ascertained through abortions, and those who were ascertained by a balanced carriers, since only some of all unbalanced translocations can lead to unbalanced offspring; many unbalanced zygotes cannot survive the early zygote stage. Basically the same rules for risk calculation should be applied to large pedigrees with (in most cases) autosomal dominant or X-linked diseases.

Appendix 4

Multifactorial Inheritance and Major Genes

The study of segregation ratios is relatively straightforward when the analysis has advanced to the qualitative phenotypic level (Sect. 3.6.1.3), i.e., when a simple Mendelian mode of inheritance of clearly distinguishable phenotypes can be anticipated. For many human characters, however, such an analysis is not yet possible; they must be studied at the quantitative phenotypic-biometric level (Sect. 3.6.1.4). Normal traits, such as stature or I.Q., and physiological and biochemical characteristics, such as serum cholesterol levels, fall into this category, but so also do most common diseases. Some approaches to analysis of such traits have been described in Sect. 3.6.1. The heritability concept was explained, and strategies for a stepwise dissection of characters and traits subsumed by the model of multifactorial inheritance with and without threshold were suggested. Among these strategies we discussed searching for phenotypic subclasses, physiological components, or associations with genetic polymorphisms.

In recent years, however, several authors have proposed statistical methods for more rigorous testing of multifactorial vs single-gene models and for identifying effects of major genes in the presence of a multifactorial background [139, 140, 646, 647]. In principle, these methods consist of two steps. In a first step, certain assumptions are made regarding the mode of inheritance of a certain condition, and the consequences of these assumptions for frequency (in alternately distributed characters) or distribution (in continuously distributed characters) among certain groups of relatives are calculated. In this way, a tentative "model" of a particular mode of inheritance is created. Then, the goodness of fit of a set of empirical data is tested by a statistical method against the expectations derived from a given model. Hence, this approach to analysis is not different in principle from that described in Sects. 3.3.3 and 3.3.4 for testing whether family data fit the expectations of simple Mendelian inheritance. Often, several alternative models are constructed and are compared with the actual data.

Models cannot be constructed without simplifying assumptions. This is inevitable and does not cause much harm, provided that all assumptions are set out clearly. However, the fact that a data set fits expectations based on a certain model does not prove that this model describes the real situation adequately. Other plausible models must be excluded. Very often, such exclusion is simply not possible for models normally encountered in human genetics, e.g., multifactorial inheritance vs autosomal dominant inheritance with incomplete penetrance. Geneticists who usually work with simple Mendelian models are "spoiled"; there are few situations that convincingly mimic a monogenic mode of inheritance without additional assumptions. As a rule, they are on solid ground. This is not so, however, in the morass of multifactorial models.

Two models of practical importance for analysis of the genetic background of common diseases will be compared below: A multifactorial model with a threshold and the model of a simple dominant mode of inheritance with incomplete penetrance. On the whole we shall follow the analysis given by Krüger [746], since this author has clearly set out the assumptions and implications of the models. Some other, related approaches will be mentioned briefly. (For a more general discussion, see Sect. 3.6.2.2.)

Simple Diallelic Mode of Inheritance with Incomplete Penetrance. The penetrances in genotypes AA and Aa are w_1 and w_2 respectively, while genotype aa is assumed to be always unaffected. Then, the incidence of the trait in the population is:

$$P=p^2\,w_1+2pqw_2=p^2(w_1-2w_2)+2pw_2$$

(p = frequency of the allele A). It is realistic to simplify this model for practical calculations by assuming $w_1=1$ (complete penetrance in AA homozygotes).

Multifactorial Inheritance with Threshold Effect. The phenotypic value of the *disease liability* [654] may be called x. It can be divided into components as shown in Sect. 3.6.1. It is assumed that the environmental value E is not correlated with the genotypic value G:

$$V_x=V_G+V_E$$

and that the genotypic value contains no epistatic component. Then, G is the sum of the contributions of independently acting gene pairs, and its distribution in the population is approximated by a normal distribution when the number of gene pairs increases. Therefore, it is realistic to assume that G is normally distributed in the population and that the environmental deviation E has a normal distribution. Under these conditions the phenotypic value x will be normally distributed as well.

Being a hypothetical variate, the liability can be defined in such a way that x, G, and E each has the mean 0, and x has the variance 1. Then, the *threshold* is uniquely determined by the population incidence P as that point that subdivides the standard normal distribution (normal distribution with mean 0 and variance 1) into two parts with frequencies $1-P$ and P. Consider two relatives of defined degree of relationship taken randomly from the population. The pair of their liabilities (x_1, x_2) is a random variable that has a bivariate normal distribution. When the correlation coefficient of the two liabilities, $r_{x_1x_2}$, is given, this distribution is completely specified, and the probability that each one singly is or both together are affected can be calculated. Under the above conditions, the following relation exists among the coefficients of the phenotypic, genotypic, and environmental correlations of the liabilities of the two relatives [488]:

$$r_{x_1x_2} = r_{G_1G_2}H^2 + r_{E_1E_2}E^2$$

where $H^2 = V_G/V_x = V_G$ is the heritability (in the broad sense) and $E^2 = 1 - H^2 = V_E$. The correlation between the environmental components E_1, E_2 of the two relatives cannot in most cases be determined. Therefore, it will be assumed to be 0. Furthermore, only the special case $H^2 = h^2$ (i.e., $V_G = V_A$) will be examined – in agreement with the experience from quantitative genetics that the nonadditive component $H^2 - h^2$ is usually very small. Then the following equation holds true:

$$r_{x_1x_2} = r \times h^2$$

where $r = r_{G_1G_2}$ has a fixed value depending only on the type of relationship, and the model depends only on the parameters h^2 and P. The additional consideration of an environmental component E (equivalent to $h^2 < 1$) removes the unrealistic assumption of a sharp threshold, which is replaced by a "threshold region" the extent of which is given by V_E. Within this threshold region, the probability of manifestation of the anomaly is assumed to increase continuously from 0 to 1.

Comparison of the Single Locus and the Multifactorial Models. In the following, the models will be compared for a number of values of population incidences P, for a number of penetrance values w for the diallelic model, and for different assumptions regarding h^2 for the multifactorial model. The calculation is straightforward for the diallelic model when single selection is assumed for ascertainment (Sect. 3.3). For the multifactorial model, $r = h^2/2$ for parents, siblings, and children, $r = h^2$ for monozygotic twins. From this and the bivariate normal distribution of the liabilities of the two relatives I_1 and I_2 the conditional probability, Q, can be obtained that I_2 also shows the anomaly, when I_1 (the proband) is affected. Q is the ratio of the probability that both relatives are affected to the probability P_1 that I_1 is affected. Q is represented by the volume of the darkgrey shaded region under the normal probability density surface in Fig. A 4.1, while the lightgrey shaded regions represent the probabilities of the events I_1 affected, I_2 normal and I_1 normal, I_2 affected, respectively. Figure A 4.1 a shows the case of two unrelated individuals. Their risks are independent of each other: $Q = P$. This is expressed by the central symmetry of the density surface. Figure A 4.1 b shows the joint liability distribution for first-degree relatives; in this special case, $h^2 = 1$ (and thus $r = ½$) is assumed. A consequence is that the observation that I_1 is affected increases the risk of the relative I_2 to be affected: $Q > P$. The volumes of the shaded regions under the density surface can be calculated by numeric integration, the details of which cannot be described here.[1]

Figures A 4.2 and A 4.3 show the comparisons. The following notations were used: Q_1 = incidence of the anomaly in children or parents of probands; Q_2 = incidence in siblings or dizygotic twins of probands, Q_1 and Q_2 without taking into account parental phenotypes; Q_3 = incidence in monozygotic twins of probands; $Q_{1.1}$ = incidence in siblings of probands with two unaffected parents, $Q_{1.2}$ incidence in siblings of probands one of whose parents was affected, $Q_{2.2}$ incidence in siblings of probands both of whose parents were affected.

Figures A 4.2 and A 4.3 are very similar. They show the incidences in the two models for children (or parents) and for sibs, irrespective of the parental mating types. The curves reveal a certain overlap for high P (incidence = 0.2%–0.5% and higher) between dominant inheritance with low penetrance and multifactorial inheritance with high heritability. On the other hand, separation of the two models for an anomaly with low incidence is quite good. In monozygotic twins (figure not shown) the diallelic model can be simulated throughout by the multifactorial

[1] The tetrachoric functions of K. Pearson, which were used for the calculation by some authors, have disadvantages. For a full discussion of the problem [cf. 746].

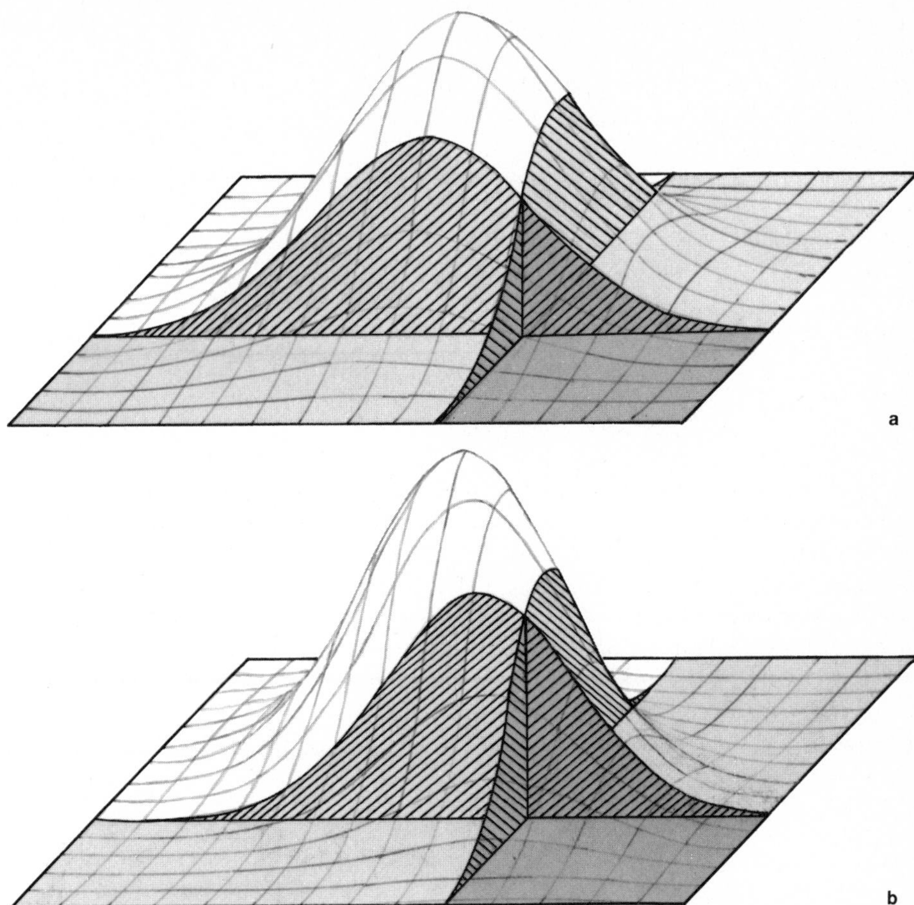

Fig. A4.1 a, b. Density surface of the bivariate normal distribution of the disease liabilities of two individuals. The two thresholds are indicated by the *hatched* areas; the *darkly shaded* areas in the anterior right corner indicate probability Q that both individuals are affected. **a** Two uncorrelated individuals in a randomly mating population. **b** Two first-degree relatives. In **b** the *darkly* shaded area is much larger than in **a**, indicating the increased risk of a relative to be affected if the proband suffers from the disease

model. The latter, however, can be simulated by the diallelic model only for high h^2 values, not for low h^2 values.

Hence, simulation is possible for monozygotic and dizygotic twins separately, but it is also possible for MZ and DZ simultaneously. To examine this problem, the ratio:

$$R_1 = \frac{Q_3}{Q_2}$$

of incidences in MZ and DZ twins was examined (Fig. A 4.4). Again, not every multifactorial model can be simulated, but the models with high heritabilities can be separated from the single locus model. For the latter model, the upper limit of R_1 (for varying values of penetrance) tends monotonically to 4 when the population incidence approximates 0 (and practically equals 4 for $P \leqq 0.01\%$). This is a confirmation of the twin criterion of Penrose [837]: if the concordance of monozygotic twins is more than four times as high as among dizygotic twins, the single locus model can be excluded in favor of the multifactorial model. On the other hand, a value $R_1 <$ 4 does not rule out a multifactorial model.

Figures A 4.5–A 4.7 show the incidences $Q_{1,1}$, $Q_{2,1}$, and $Q_{2,2}$ in siblings and parents for a given mating type unaffected × unaffected, affected × unaffected, affected × affected. There is no overlap in $Q_{1,1}$ in families with two unaffected parents. The incidences for multifactorial inheritance, even with high heritabilities, are always lower than for dominance with incomplete penetrance, even when the penetrance is very low. For $Q_{2,1}$ (Fig. A 4.6), on the other hand, there is considerable overlap: in families with one affected parent, the multifactorial model can be distinguished from the single locus model only if heritability is very low. $Q_{2,2}$ (mating type plus × plus) shows in principle identical findings.

Up to now, the analysis has neglected dominance effects. All examinations of the multifactorial model

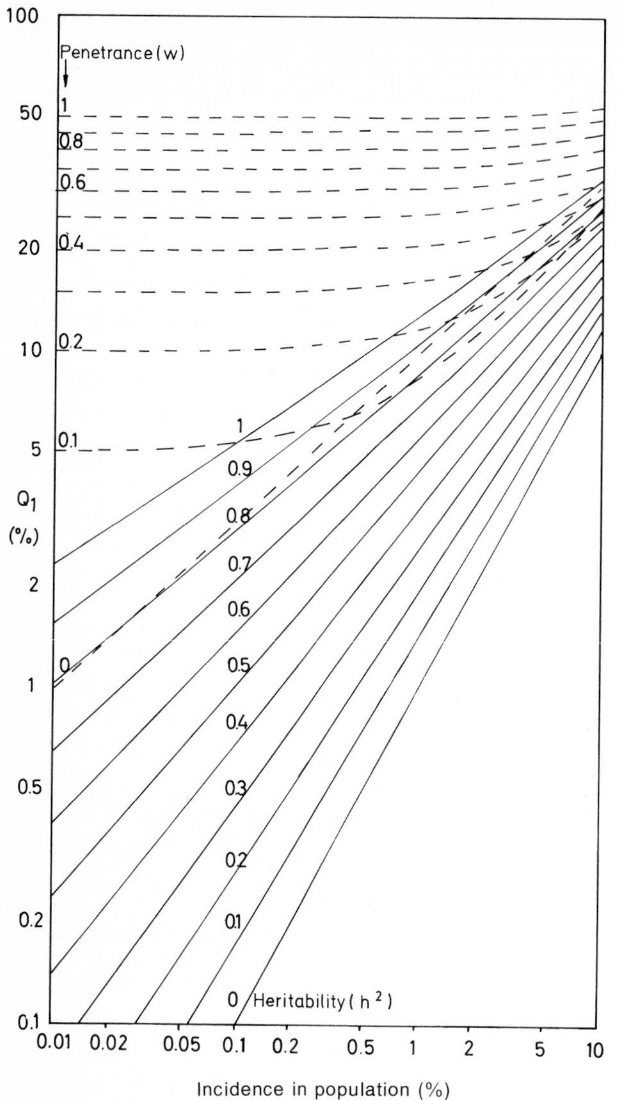

Fig. A4.2. Frequency of the character in children (or parents) of probands (Q_1) in the diallelic model (- - -) and in the multifactorial model (—) [746]

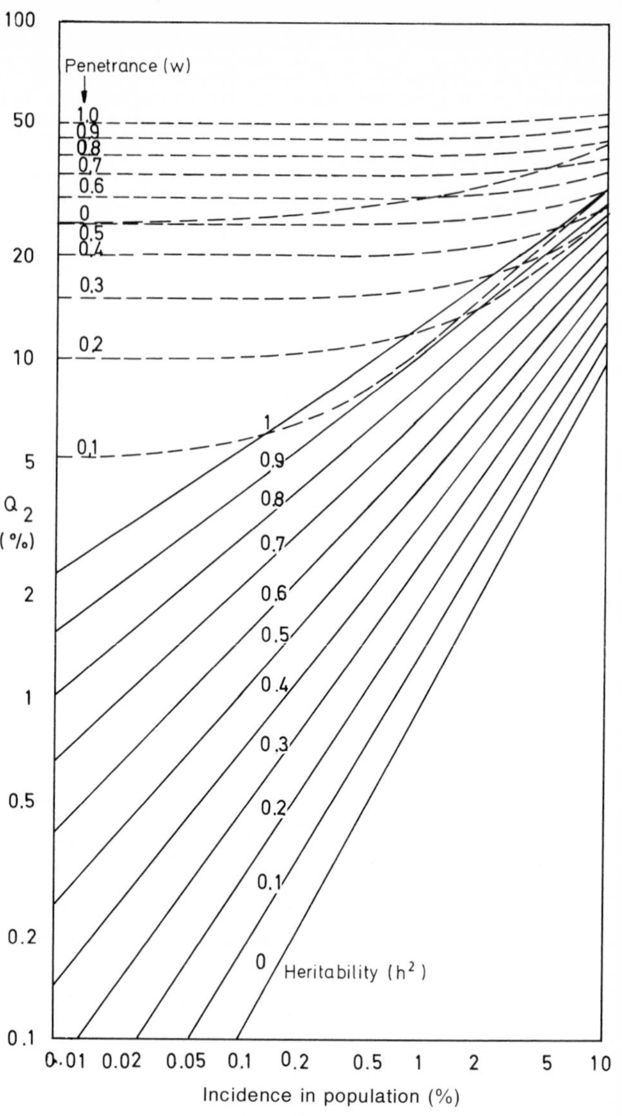

Fig. A4.3. Frequency of the character in sibs of probands (Q_2) in the diallelic model (- - -) and in the multifactorial model [746]

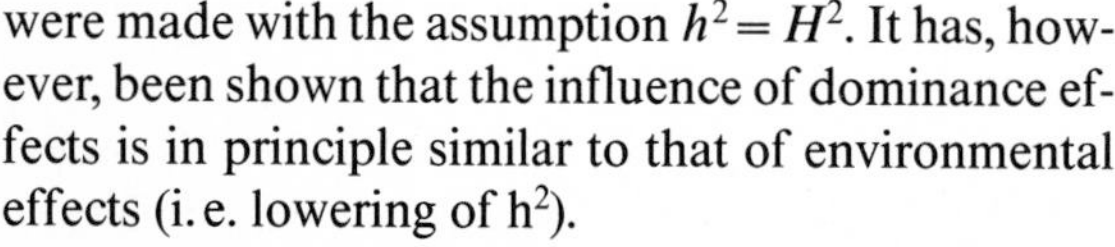

were made with the assumption $h^2 = H^2$. It has, however, been shown that the influence of dominance effects is in principle similar to that of environmental effects (i.e. lowering of h^2).

In sum, there are always areas of overlap. Considering the fact that the multifactorial model is an abstraction and that the data usually available for such an analysis are subject to appreciable sampling errors, the result cannot be considered altogether satisfying. A ratio:

$$R_2 = \frac{Q_{2,1}}{Q_{1,1}}$$

i.e. between the expected frequencies among children of one affected parent ($Q_{2,1}$) and those of two unaffected parents ($Q_{1,1}$) as proposed earlier as criterion might be thought possibly to give a better separation. But as Fig. A 4.8 shows, for high P, the overlap is still appreciable, although for lower P, the separation is indeed much better. The multifactorial model can be distinguished from the single locus model if $R_2 \geqq 2.5$. If the incidence among sibs of probands with one affected parent is 2.5 times higher than among sibs with two unaffected parents, then the diallelic model is highly unlikely. This criterion can be compared with Penrose's twin criterion [837]; here, too, no definite conclusion is possible if the value $R_2 < 2.5$, - quite as in a situation with $R_1 < 4$. It is interesting that both criteria, which had been established largely on an intuitive basis, have been corroborated by the analysis.

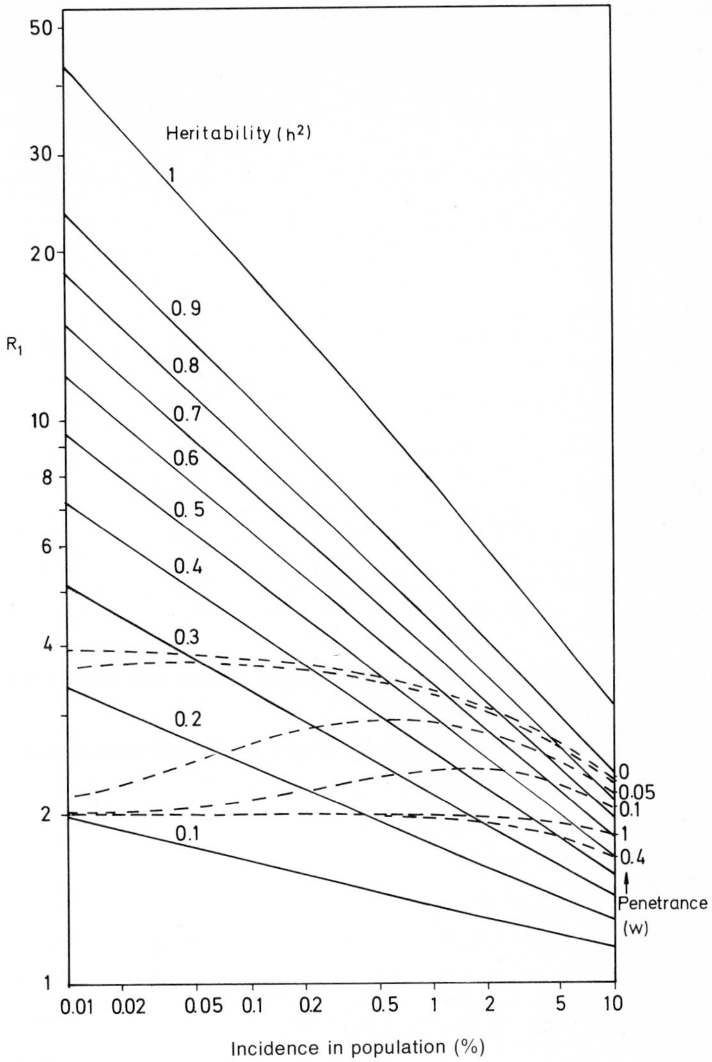

Fig. A4.4. Frequency of the character among monozygotic and dizygotic twins of probands ($R_1 = Q_3/Q_2$) in the diallelic model (- - -) and in the multifactorial model (—) [746]

An important parameter that needs to be specified is h^2. Therefore, it should be estimated beforehand. In practice, this is possible in two independent ways. First, by using concordance rates in MZ twins and secondly by comparing the frequency Q among relatives of probands with the population incidence P. The first method gives H^2, not h^2, but we can reasonably hope that the difference is negligible. This aspect is discussed in Sect. 3.8; App. 6. The second method depends on properties of the multifactorial model that need not be realistic and cannot very well be controlled.

Falconer (1965, 1967) [654, 655] has given the principle, which is formally analogous to the treatment of a selection experiment in quantitative genetics. Let G be the mean liability in the population, A the mean liability of the affected persons, and R that of the relatives (with given degree of relationship) of the affected persons. Then, the ratio of the differences $R-G$ ("response") and $A-G$ ("selection differential") is equal to the coefficient of regression b for the liability in the relatives on the liability in the probands:

$$b = \frac{R-G}{A-G}$$

Under the above assumptions about the liability (normal distribution with mean 0 and variance 1), the left side is equal to the correlation coefficient of the liabilities:

$$b = r = h^2$$

being the coefficient of relationship (for example, $b = \frac{1}{2}$ for first-degree relatives). The denominator of the right side can be calculated from the (standard) normal distribution, using the distance between the population mean and the threshold value corresponding to the population incidence, P, of the

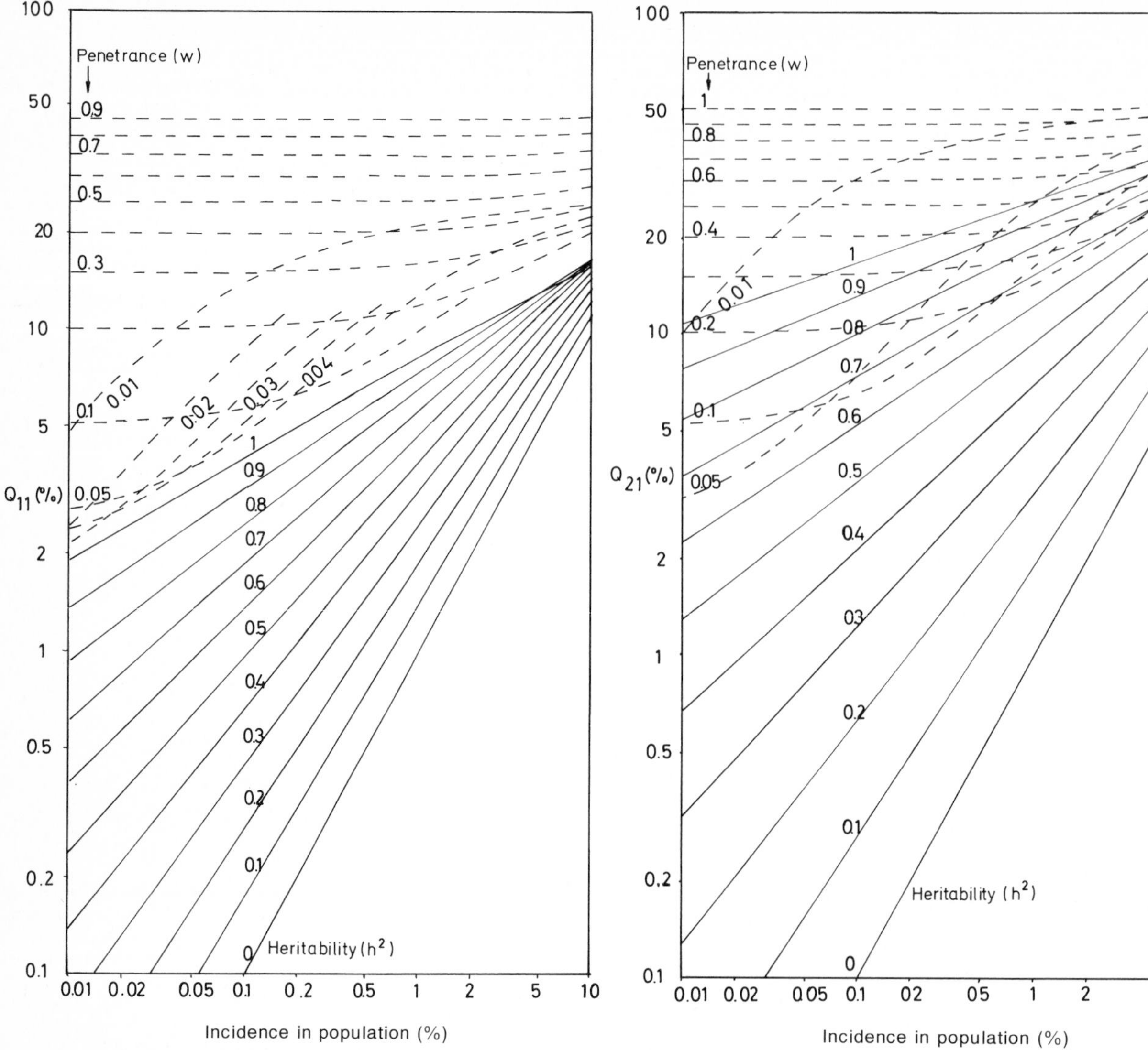

Fig. A4.5. Frequency of the character among sibs of the probands with parents of type unaffected × unaffected ($Q_{1,1}$) in the diallelic model (---) and the multifactorial model (—) [746]

Fig. A4.6. Frequency of the character among sibs of probands with parents of type affected × unaffected ($Q_{2,1}$) in the diallelic model (---) and the multifactorial model (—) [746]

anomaly. Falconer proposed that the numerator be calculated as the difference between the threshold value corresponding to the incidence, Q, of the anomaly in the relatives and the threshold value for the population. An improved nomogram (Fig. A 4.9), with the correlation coefficient r of liabilities of proband and relative instead of h^2, was published by Ch. Smith [880]. This nomogram also comprises negative values for h^2 (below and on right-hand side of the zero line in Fig. A 4.9). These negative values are meaningless biologically but may occur as a consequence of small sample size, and may be used when several samples are combined for a compound estimate.

Comparison of Expectations Derived from These Models with Sets of Actual Data. The nomograms A4,1–A4 can be used for comparing actual family and twin data with expectations based on the two models. Often, such an assessment provides an intuitive impression as to which model may be closer to the truth. However, such comparisons are unsatisfactory if performed for various degrees of relationship separately: Even if all these comparisons fail to give criteria for deciding against one of the two models, the entire *pattern* of frequencies among the various categories of relatives may still provide a clue against one alternative. Moreover, it has so far been assumed implicitly that the population incidence p is known. This, however, is almost never the case; as a rule, p must be estimated from a population sample.[2]

[2] Footnote see p. 665

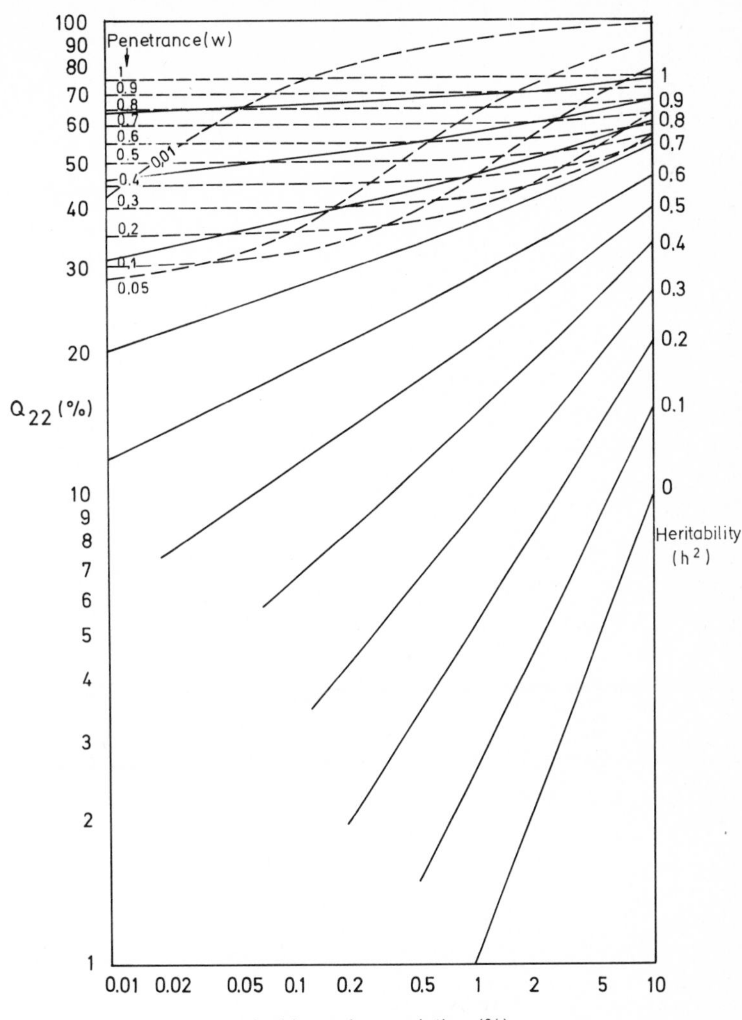

Fig. A4.7. Frequency of the character among sibs of probands with parents of type affected × affected ($Q_{2,2}$) in the diallelic model (- - -) and the multifactorial model (——) [746]

A method which allows comparison of the entire pattern with the models is based on the maximum likelihood principle. This method will be described below for a comparison of the two models discussed above. The incidences of the character to be examined in the various categories of relatives of probands may be named $Q_1, Q_2, Q_3 \ldots Q_u$ (in our case: u = 6). For each incidence Q_i ($i = 1 \ldots$ u), an observed value $\tilde{Q}_i = k_i/n_i$ is available that has been determined from a sample of n_i relatives of a certain degree, k_i of which are affected by the disease to be examined. Moreover, an observed value for incidence in the general population $Q_0 = k_0/n_0$ for the estimation of P may be available from examination of a population sample of n_0 individuals. If we assume in addition that the samples for estimation of the incidences in relatives Q_i comprise only one relative of a certain type for each proband, the probability for all these observed values together is given by the formula

$$\mathrm{w} = \prod_{i-0}^{u} \binom{n_i}{n_k} Q^{kj} (1 - Q_i)^{n_i - k_i} \qquad (1)$$

Here, Q_i denotes the (unknown) expectation of the "real" frequency in the ith category of relatives. Coming back to the two models described above, Q_i is a function of the model parameters either of the simple dominant model with incomplete penetrance, or of the multifactorial/threshold model (parameters: $\tau_1 = \mathrm{P}$, $\tau_2 = \mathrm{w} =$ penetrance for single inheritance; $\tau_1 = \mathrm{P}$, $\tau_2 = \mathrm{h}^2$ for the multifactorial model). In this case, Eq. 1 becomes the *likelihood function* of the observed values $\tilde{Q}_1 \ldots \tilde{Q}_u$, assuming hypothesis that the empirically found pattern of incidences among these various categories of relatives has been caused by the mode of inheritance assumed in one of these two models. For each of the two models those two values of the parameters τ_1 and τ_2 are determined for which the likelihood function has a maximum. These *maximum likelihood (ML) estimates* of τ_1 and τ_2 may be denoted as: $\hat{\tau}_1$ and $\hat{\tau}_2$, and $\hat{Q}_i = Q_i(\hat{\tau}_1, \hat{\tau}_2)$ may be the corresponding ML estimates of

[2] In practice, estimation of *p*, which is required for estimating h^2, e.g., by using table A 4.9, often poses difficult problems. If possible, estimates of *P* should be made in the same populations from which families are ascertained for analysis, since many multifactorial conditions, such as birth defects or common diseases, show wide differences between populations in incidence or prevalence. Moreover, incidence estimates from prevalence data might pose problems if the condition causes early mortality or if age at onset is variable.

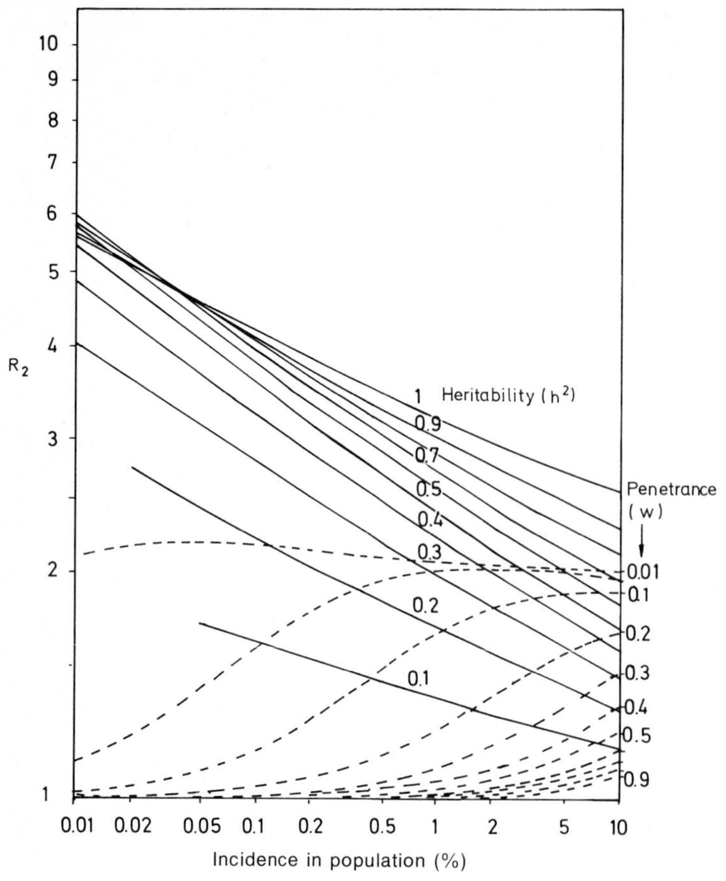

Fig. A4.8. Ratio of frequencies in sibs of probands with parents of types affected × unaffected and unaffected × unaffected ($R_2 = Q_{2,1}/Q_{1,1}$) in the diallelic model (- - -) and the multifactorial model (—) [746]

Q_i ($\hat{Q}_0 = \hat{t}_i$). These ML estimates $\hat{Q}_i$ represent the best estimates for the frequency of the disease in question among the various degrees of relatives included in the study if the model applies. They must now be compared with the frequencies actually observed. The precision with which the model describes the observations can be tested by the expression:

$$\chi^2 = \sum_{i=0}^{u} \frac{(k_i - n_i\hat{Q}_i)^2}{n_i\hat{Q}_i(1-\hat{Q}_i)} = \sum_{i=0}^{u} \frac{(\tilde{Q}_i - \hat{Q}_i)^2\, n_i}{\hat{Q}_i\,(1-\hat{Q}_i)} \quad (2)$$

When the null hypothesis holds true, i.e., when the distribution of incidences in the various categories of relatives is indeed caused by the mode of inheritance assumed by the model, this expression is distributed approximately like χ^2, with $u-1$ degrees of freedom. Hence, a value of expression (2) found in a χ^2 table to be higher than that given for must be rejected. When both models are tested in this way, four outcomes are possible:

1. There is no difference between the empiricial data and either model; no model can be excluded.
2. Only the single-gene model can be excluded.
3. Only the multifactorial model can be excluded.
4. Both models are excluded. One should remember that even alternatives 2 and 3 do not *prove* either of the two models; many other models might explain the data equally well. In practice, the expression

$$L = \sum_{i=0}^{u} [k_i \ln Q_i + (n_i - k_i)(\ln(1-Q_i)] \quad (3)$$

is often used, which is[3] the natural logarithm of Eq. 1. Instead of Eq. 2, the expression

$$\chi^2 = 2 \sum_{i=0}^{u} \left[k_i \ln\frac{\tilde{Q}_i}{\hat{Q}_i} + (n_i - k_i)\ln\frac{1-\tilde{Q}_i}{1-\hat{Q}_i}\right] \quad (4)$$

can be used for testing, since it is also distributed in a similar way to χ^2 with $u-1$ degrees of freedom.

The methods of computing the ML estimates of the parameters $\hat{t}_1$ and $\hat{t}_2$, i.e. the values of these parameters for which the logarithmic likelihood function (Eq. 3) has a maximum cannot be described here; they are standard methods for the mathematician [746, 804].

How Should We Proceed in Practice? The analysis of common disease in practice is difficult and usually requires the help of a genetical statistical expert. Simple Mendelian modes of inheritance may be difficult to detect, because of the high frequency of the trait and diminished penetrance. However, even in the most expert hands the distinction between a monogenic trait of low penetrance and a multifactorial trait is frequently impossible.

Careful consideration of heterogeneity is usually essential. A given disease diagnosis in this group of

[3] Excepting an additive constant that does not depend on the model parameters.

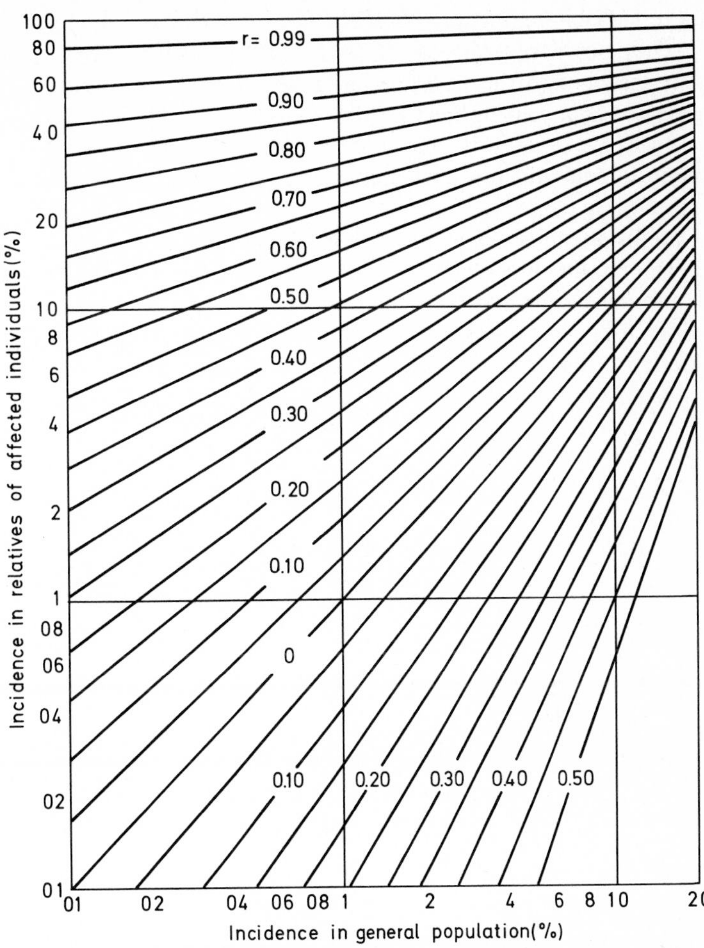

Fig. A4.9. Table for estimation of h^2 from the frequency (in %) of a character in the general population and first-degree relatives of probands. For example, if a disease has an incidence of 0.2% in the general population and of 4% among children of the probands, then this result conforms to a correlation coefficient of $r = 0.40$ and, as $h^2 = 2 \times r$, to a heritability $h^2 = 0.8$. Generally, the formula $h^2 = r/R$ applies. Here R is a measure for the relationship that can be derived from the formulas for h^2 in Sect. 3.6.1.5. For example $R = ½$ in first-degree relatives, ¼ in second-degree relatives, and 1 in MZ twins. (Smith [875 a])

common conditions is often a blanket term for several diseases with different genetic and nongenetic causes. Furthermore, gene-environment interaction is often operative and difficult to estimate. A very large monogenic pedigree may be easy to elucidate genetically, but the results obtained from such a kindred may not apply to other persons and their families. Genetic analysis requires that the data are standardized for differences in age of onset, sex differences, and similar confounding variables before detailed analysis is performed. Such analysis is best performed by comparing the investigator's data against various genetic models. Computer programs for such models exist, which determine the likelihood that the data will fit one or other of the possible modes of inheritance. Such models may include simple dominant inheritance, recessive inheritance, intermediate inheritance, polygenic inheritance, major gene inheritance superimposed on polygenic inheritance, nongenetic familial aggregation, etc. However, even with such sophisticated data processing, it is necessary to be cautious about the finality of the conclusions. Sometimes, some small changes in the data change the conclusions. One must therefore be tentative about accepting a model of single gene inheritance from such analyses. The failure to find monogenic inheritance, on the other hand, does not necessarily mean that no major gene is involved. It is therefore essential that the investigator be aware of the biologic, biochemical, and pathophysiological background of the disease under study. The application of the most advanced biologic methodology applicable to the disease under study therefore will often reveal heterogeneity and will usually be closer to the action of genes than use of a blanket type of diagnosis. Of course, for many diseases we lack the necessary biological approaches and can only use phenotypes based on current concepts. Generally, our insight into the transmission genetics of such diseases (e.g., schizophrenia) will be scanty although we may be convinced from other evidence that genetic factors and not environmental agents play a key etiologic role.

Careful attempts to identify minor deviations from normal in the direction of the disease (as shown for squinting, atopy) may be useful for analysis and provide clues not available when the overt phenotype alone is considered. For these various reasons, the understanding of these diseases is unlikely to come from genetic techniques alone, but requires meaningful collaboration between geneticists, statisticians, and the appropriate specialists for the disease.

Model testing has been performed for several common diseases, such as epilepsy [611, 793], red blood cell catechol-O-methyltransferase activity [666], hypercholesterolemia [794], coronary heart disease [847], and hyperlipidemia [954], but see also [819–822] for some caveats.

Appendix 5

Diagnosis of Zygosity

Exclusion of Monozygosity by Genetic Polymorphisms. It is obvious that there will be no differences in sex or in any genetic marker between MZ twins. Hence, in the absence of laboratory error, any difference in genetic markers (blood, serum, or isoenzyme groups) in a twin pair proves its dizygotic status. Moreover, chance may cause identical marker status in dizygotic twins even if differences are likely as inferred from the parental genotype. For example, the father of a twin pair may have blood type M, the mother MN. Then, the dizygotic twin pair may have the blood types M, M (probability ¼) or MN, MN (probability also ¼) or M, and MN (probability ½). Hence, in half of all such cases, we expect the twins to have identical blood types; in the other half, their blood types will be different. By testing for additional genetic markers, it is possible to enhance the probability that a DZ pair differs in at least one of them. From this knowledge, it is possible to calculate the "inverse" probability for a concordant twin pair to be dizygotic, using the mathematic principle of conditional probabilities established by Bayes already in 1793. This principle proved to be useful not only in diagnosis of twin zygosity but in genetic counseling as well (App. 8) [881].

Bayes Principle of Conditional Probabilities. Consider one twin pair for which the probability of monozygosity or dizygosity has to be determined. More precisely, our question is: What proportion of all twin pairs with the same combination of genetic markers in the twins themselves and their parents are expected to be dizygotic? Or, to put it somewhat differently: How often would the twin pair be misclassified, if monozygosity were assumed in every such case? The general formula of Bayes is:

$$P(A_1/B) = \frac{P(A_1) \times P(B/A_1)}{P(A_2) \times P(B/A_2) + P(A_1) \times P(B/A_1)} \quad \text{(A5.1)},$$

where A_1 and B are different events, and A_2 means the event "not-A_1."

In our case, $P(A_1/B)$ may be the probability of monozygosity among twin pairs with identical blood types. Then, $1 - P(A_1/B)$ is the probability for the twin pair to be dizygotic or the probability of error when the twin pair is classified as monozygotic. $P(A_1)$ is the a priori probability for MZ twins among all twins in the population. It is about 30% in European populations. $P(A_2)$ is the a priori probability for a twin pair to be dizygotic. $P(A_2) = 1 - P(A_1) = 0.7$; Eq. A5.1 may be simplified as follows:

$$P(A_1/B) = \frac{1}{1 + (Q \times L)} \quad \text{(A5.2)}$$

Here, Q is the ratio DZ/MZ in the population (if 30% of all twin pairs are MZ, then $Q = 2.33$). L is the likelihood ratio of conditional probabilities of DZ and MZ twins to be identical in a given combination of genetic markers. Its value can be calculated by multiplication from the L_i of the various marker systems used:

$$L = L_1 \times L_2 \times \ldots \times L_n \quad \text{(A5.3)}$$

One Example. Tables A5.1 and A5.2 show one example. From the list of genetic markers, it turns out that some are not informative, parents and children being identical genetically (IGHG (Gm), IGHG (Km), HP, AK). For most other markers, the mating types and, hence, the expected segregation ratios among the children are obvious. For example, in the GC system, the father is homozygous 2–2, the mother is heterozygous 2–1. Therefore, the expect-

Table A5.1. Example of a zygosity diagnosis

	Father	Mother	Both twins
Sex	♂	♀	♂
Blood groups	A_2	0	A_2
	MS/Ms	MS/Ms	MS/MS
	Kk	kk	Kk
	Fy(a+b+)	Fy(a−b+)	Fy(a−b+)
	R_1 r	R_2 r	rr
Serum proteins	G1m(−1)	G1m(−1,−2)	G1m(−1,−2)
	Km(−1)	Km(−1)	Km(−1)
	HP 2−2	HP 2−2	HP 2−2
	GC 2−2	GC 2−1	GC 2−1
Isoenzymes	ACP B	ACP AB	ACP AB
	PGM1 2−1	PGM1 2−1	PGM1 1−1
	AK1 1−1	AK1 1−1	AK1 1−1

Table A5.2. Calculations for the data from Table A5.1

	P_{DZ}	P_{MZ}
A priori probability	0.70	0.30
Conditional probabilities[a]		
Sex	0.50	1.00
AB0	0.50–1.00	1.00
MNSS	0.25	1.00
Kell (K)	0.50	1.00
Duffy	0.50	1.00
Rhesus	0.25	1.00
GC	0.50	1.00
ACP	0.50	1.00
PGM1	0.25	1.00

[a] Conditional probability of the second twin being phenotypically identical with the first for given phenotype of the first twin.

Table A5.3. Random mating table for a system of two alleles

Mating	Frequency	Children		
		AA	Aa	aa
AA × AA	p^4	p^4	–	–
AA × Aa	$4p^3q$	$2p^3q$	$2p^3q$	–
AA × aa	$2p^2q^2$	–	$2p^2q^2$	–
Aa × Aa	$4p^2q^2$	p^2q^2	$2p^2q^2$	p^2q^2
Aa × aa	$4pq^3$	–	$2pq^3$	$2pq^3$
aa × aa	q^4	–	–	q^4
Total	1	p^2	$2pq$	q^2

Source: Maynard Smith and Penrose (1955) [881]

ed segregation ratio among the children is 1:1, and if twin 1 has type 2−1 and the twins are dizygotic, then the probability for twin 2 to be 2−1 as well is 0.50. For the AB0 groups, the situation is not as obvious since the father (phenotype A_2) may have the genotypes A_2A_2 or A_20. If he is A_2A_2, both twins must have the phenotype A_2 even if they are dizygous. If he is A_20, the probability of the second twin to be A_2 too is 0.50. Sometimes, the genotype of the parent may be established, for example, by another child showing group 0. Otherwise, the AB0 blood groups may be regarded as uninformative and omitted.[1] For the other systems, the calculation proceeds as follows (Eq. A5.3):

$$L = \frac{0.7}{0.3} \times 0.5 \times 0.25 \times 0.5 \times 0.5 \times 0.25 \times 0.5 \times 0.5 \times 0.25$$

A priori sex MNSs K Fy Rh GC ACP PGM1 $= 0.0011$

$P(A_1/B) = 0.9989$

Hence, the probability that the twin pair is dizygotic in spite of its concordance in all informative marker systems is extremely low. For all practical purposes, monozygosity can be assumed. Inclusion of additional Mendelian marker genes will increase the probability that a pair of twins are monozygotic. The use of HLA markers is a particularly powerful tool in view of the great variability of this system. Note that in the example, the denominators in Eq. A5.3 have always been 1, i.e., MZ twins are always identical in their markers. In principle, the method also permits the inclusion of quantitative traits in which MZ twins may be different but are, on the average, more similar than DZ twins. In this case, the denominator in Eq. A5.3 differs from 1. In practice, however, this extension of the method is of little importance.

[1] The two possible genotypes are to be expected in the proportions p_2: $2r$ where p_2 and r are the gene frequencies of the alleles A_2 and 0 in the population. Then the conditional probability for twin 2 to have the same phenotype as twin 1 if this twin has type A_2 can be deduced as

$$(p_2 \times 1 + 2r \times \tfrac{1}{4})/(p_2 \times 1 + 2r \times \tfrac{1}{2}) = \frac{p_2 + \tfrac{1}{2}r}{p_2 + r}$$

However, the distribution of genotypes is biased by the exclusion of those unambiguously identified by another 0 child. This, again, depends on distribution of sibship sizes in the population – an information that is usually not available.

Genotype of the Parents May Not Be Known. In the above-mentioned example, genetic markers were known not only in the twin pair but in the parents as well. In many cases, however, the parents will not be available for examination. Under these circumstances, the known gene frequencies of the marker systems in the populations can be used for the calculation. The rules have been set out by Maynard Smith and Penrose (1955) [881]. The conditional probability $P_{i,\,DZ}$ of twin 2 having the same phenotype as twin 1 if this twin has phenotype i is calculated from the frequencies of mating types in the population (Table A5.3) and from the relative number of children with different genotypes expected from these matings (Tables A5.4 and A5.5). Tables A5.6–A5.21 provide values for $P_{i,\,DZ}$ and the commonly used genetic polymorphisms. They are based mostly on northwestern European gene frequencies.

Methods from Classic Anthropology. A fairly reliable method for distinguishing MZ from DZ twins had already been available before most genetic polymorphisms were known. It was established by Siemens (1924) [869, 870] and is based on comparison in a great number of visible physical characters and on

Table A5.4. Sib-sib frequencies for a system of two alleles

Sib	Genotypic				Sib	Phenotypic		
	AA	Aa	aa	Total		$\bar{A}$	$\bar{a}$	Total
AA	$\frac{1}{4}p^2(1+p)^2$	$\frac{1}{2}p^2q(1+p)$	$\frac{1}{4}p^2q^2$	p^2	$\bar{A}$	$p(1+q)-\frac{1}{4}pq^2(3+q)$	$\frac{1}{4}pq^2(3+q)$	$p(1+q)$
Aa	$\frac{1}{2}p^2q(1+p)$	$pq(1+pq)$	$\frac{1}{2}pq^2(1+q)$	$2pq$	$\bar{a}$	$\frac{1}{4}pq^2(3+q)$	$\frac{1}{2}q^2(1+q)^2$	q^2
aa	$\frac{1}{4}p^2q^2$	$\frac{1}{2}pq^2(1+q)$	$\frac{1}{4}q^2(1+q)^2$	q^2				
Total	p^2	$2pq$	q^2	1	Total	$p(1+q)$	q^2	1

Source: Maynard Smith and Penrose (1955) [882].

Table A5.5. Relative chances in favor of dizygotic twin pairs in a system of two alleles

Genotypic		Phenotypic	
Twin pair both	Relative chance in favor of dizygotic twins	Twin pair both	Relative chance in favor of dizygotic twins
AA	$\frac{1}{4}(1+p)^2$	$\bar{A}$	$1-\frac{1}{4}q^2(3+q)/(1+q)$
Aa	$\frac{1}{2}(1+pq)$		
aa	$\frac{1}{4}(1+q)^2$	$\bar{a}$	$\frac{1}{4}(1+q)^2$

Source: Maynard Smith and Penrose (1955) [882].

Table A5.6. AB0 system, according to Ikin et al. (1939); 3459 persons from England. (Vogel and Helmbold, 1972 [211])

Gene frequencies

A_1 0.208959 B 0.061166
A_2 0.069649 0 0.660226

Phenotype of twins	Frequency	L = Likelihood ratio DZ/MZ	Log $(1/L)$
0	0.435898	0.68909	0.16173
A_1	0.348692	0.64697	0.18912
A_2	0.096819	0.48236	0.31663
B	0.084508	0.47407	0.32415
A_1B	0.025562	0.32392	0.48956
A_2B	0.008520	0.28483	0.54541

Table A5.7. MNSs system, according to Race and Sanger (1962); 1419 persons from England. (Vogel and Helmbold, 1972 [211])

Gene frequencies

MS 0.247172 NS 0.080208
Ms 0.283131 Ns 0.389489

Phenotype of twins	Frequency	L = Likelihood ratio DZ/MZ	Log $(1/L)$
MS	0.201058	0.51614	0.28723
Ms	0.080163	0.41161	0.38552
MNS	0.277611	0.50438	0.29724
MNs	0.220553	0.47329	0.32487
NS	0.068914	0.41379	0.38322
Ns	0.151702	0.48267	0.31635

Table A5.8. Rhesus system, designation of phenotypes according to reactions with the antisera $C+W^W$, c, D, E, C^W, e, according to Race et al. (1948); 2000 persons from England

Gene frequencies

CDE 0.0024 cDe 0.0257
CDe 0.4076 cdE 0.0119
Cde 0.0098 cde 0.3886
cDE 0.1411 C^WDe 0.0129

Phenotype of twins	Frequency	L = Likelihood ratio DZ/MZ	Log $(1/L)$
− + − − − +	0.151010	0.4821	0.31691
− + + − − +	0.020635	0.3683	0.43384
− + − + − +	0.009249	0.3524	0.45292
− + − + − −	0.000142	0.2560	0.59178
− + + + − −	0.023267	0.3319	0.47901
− + + + − +	0.117527	0.4179	0.37890
+ + − − − +	0.007617	0.3515	0.45407
+ + + − − +	0.338241	0.5400	0.26757
+ + + − + +	0.010689	0.3595	0.44433
+ + − + − +	0.000233	0.2555	0.59264
+ + + + − +	0.129480	0.4241	0.37250
+ + + + − −	0.000734	0.2890	0.53905
+ + + + + +	0.003947	0.2925	0.53393
+ − − − − +	0.000096	0.2549	0.59359
+ − + − − +	0.174127	0.5021	0.29919
+ − + − + +	0.010935	0.3651	0.43756
+ − + + − +	0.002004	0.3555	0.44922
+ − + + − −	0.000006	0.2512	0.59998
+ − + + + +	0.000062	0.2538	0.59544

Table A5.9. P system, according to Henningsen (1949, 1952); 2345 persons from Sweden

Gene frequencies

P_1 0.5401
P_2 0.4599

Phenotype of twins	Frequency	L = Likelihood ratio DZ/MZ	Log $(1/L)$
P_1	0.78849	0.8747	0.05815
P_2	0.21151	0.5328	0.27341

Table A5.10. Kell system, according to Race and Sanger (1962); 1108 persons from England. (Vogel and Helmbold, 1972 [211])

Gene frequencies
K 0.0457
k 0.9543

Phenotype of twins	Frequency	L = Likelihood ratio DZ/MZ	Log $(1/L)$
K+	0.08931	0.5393	0.26814
K−	0.91069	0.9548	0.02008

Table A5.11. Duffy system, according to Race and Sanger (1962); 1944 persons from England

Gene frequencies
Fy^a 0.4213
Fy^b 0.5787

Phenotype of twins	Frequency	L = Likelihood ratio MZ/DZ	Log $(1/L)$
Fy(a+b−)	0.1775	0.5050	0.2967
Fy(a+b+)	0.4876	0.3326	0.4781
Fy(a−b+)	0.3349	0.6231	0.2055

Table A5.12. Kidd system, according to Race and Sanger (1962); 4275 persons from the whole world

Gene frequencies
Jk^a 0.5142
Jk^b 0.4858

Phenotype of twins	Frequency	L = Likelihood ratio DZ/MZ	Log $(1/L)$
Jk(a+b−)	0.2644	0.5732	0.2417
Jk(a+b+)	0.4996	0.6249	0.2042
Jk(a−b+)	0.2360	0.5519	0.2581

Table A5.13. Lutheran system, according to Race and Sanger (1962); 1373 persons from England

Gene frequencies
Lu^a 0.039
Lu^b 0.961

Phenotype of twins	Frequency	L = Likelihood ratio DZ/MZ	Log $(1/L)$
Lu(a+b−)	0.0015	0.2699	0.5688
Lu(a+b+)	0.0750	0.5187	0.2851
Lu(a−b+)	0.9235	0.9614	0.0171

Table A5.14. Secretors and nonsecretors, according to Clarke et al. (1959) and McConnell (1960); 1118 persons from Liverpool

Gene frequencies
Se 0.5233
se 0.4767

Phenotype of twins	Frequency	L = Likelihood ratio DZ/MZ	Log $(1/L)$
Secretors	0.77276	0.8662	0.06236
Nonsecretors	0.22724	0.5452	0.26348

Table A5.15. Hp system

Gene frequencies
HP^1 0.396
HP^2 0.604

Phenotype of twins	Frequency	L = Likelihood ratio DZ/MZ	Log $(1/L)$
HP 1−1	0.1568	0.4872	0.31229
HP 2−1	0.4784	0.6196	0.20789
HP 2−2	0.3648	0.6432	0.19165

Table A5.16. GC system, according to Baitsch et al. (1964); 678 persons from southwestern Germany

Gene frequencies
GC^1 0.7367
GC^2 0.2633

Phenotype of twins	Frequency	L = Likelihood ratio DZ/MZ	Log $(1/L)$
GC 1−1	0.54273	0.7540	0.12261
GC 2−1	0.38795	0.5970	0.22404
GC 2−2	0.06933	0.3990	0.39905

Table A5.17. IGHG (Gm) system, according to Ritter (1968); 1234 persons from southwestern Germany

Gene frequencies
$G1M1^1$ 0.2787
$G1M1^{not\ 1}$ 0.7213

Phenotype of twins	Frequency	L = Likelihood ratio DZ/MZ	Log $(1/L)$
G1M1 (1)	0.47937	0.7188	0.14339
G1M1 (−1)	0.52027	0.7407	0.13035

anthropologic measurements. The following characters have proved useful: color, form, and density of hair; shape and proportions of face; detailed structure of facial regions, such as the eye, including eyebrows, color and structure of the iris; details of the nose and mouth region; chin, ears, shape of hands and feet; dermatoglyphics; color and structure of the skin, including freckling. Various anthropometric measurements of the body, head and face are also helpful. The anthropologic literature has lists of informative characters. The experienced investigator bases his diagnosis not so much on comparison of single characters as on the whole "gestalt." It could be shown that in comparison of anthropologic and serologic methods no divergence between the two approaches were found [893].

This does not mean, however, that zygosity diagnosis based on physical characters is always easy. Due to

Table A5.18. IGKC (Km) system, according to Ritter (1968); 1234 persons from southwestern Germany

Gene frequencies
Km^1 0.0710
$Km^{not\ 1}$ 0.9290

Phenotype of twins	Frequency	L = Likelihood ratio DZ/MZ	Log $(1/L)$
Km (1)	0.13696	0.5605	0.25139
Km (−1)	0.86304	0.9303	0.03140

Table A5.19. AK system, according to Wille and Ritter (1968); 108 persons from southwestern Germany

Gene frequencies
$AK1^1$ 0.972
$AK1^2$ 0.028

Phenotype of twins	Frequency	L = Likelihood ratio DZ/MZ	Log $(1/L)$
AK1 1−1	0.9448	0.9722	0.01225
AK1 2−1	0.0544	0.5136	0.28937
AK1 2−2	0.0008	0.2642	0.57807

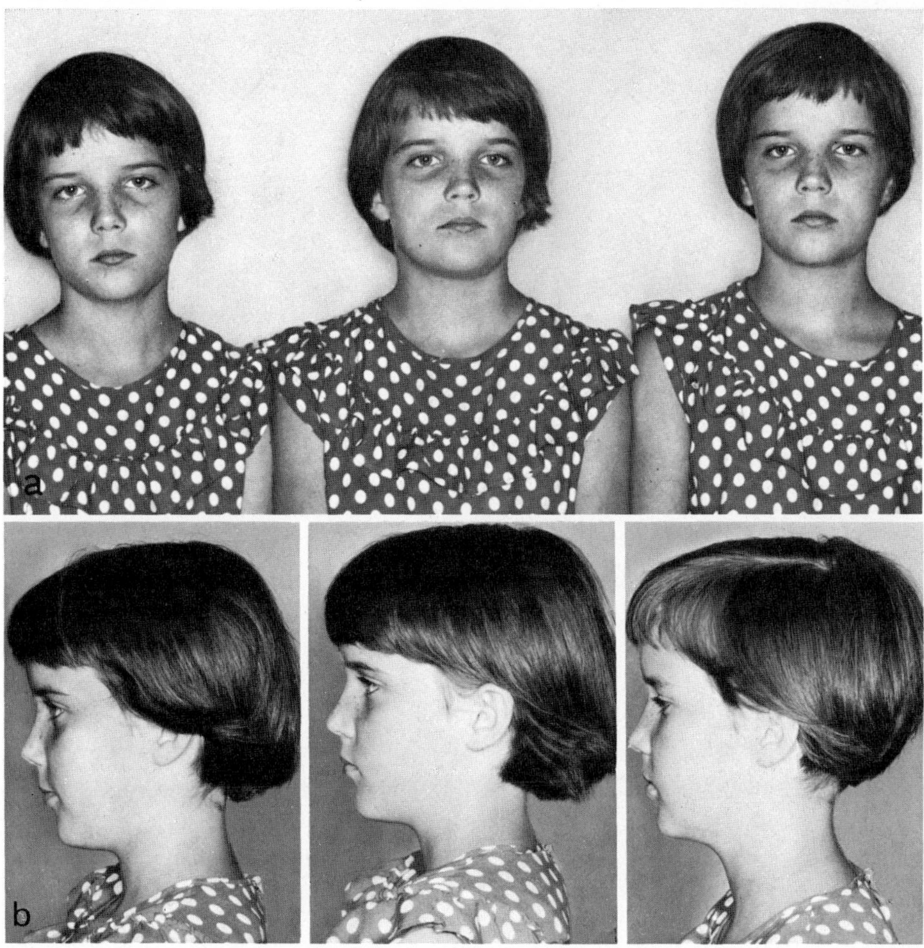

Fig. A5.1 a, b. MZ triplets at the age of 10

Table A5.20. ACP1 system, according to Fuhrmann et al. (1968); 528 persons from southwestern Germany

Gene frequencies

A 0.320
B 0.626
C 0.054

Phenotype of twins	Frequency	L = Likelihood ratio DZ/MZ	Log (1/L)
A	0.1024	0.4356	0.36091
BA	0.4006	0.5867	0.23161
CA	0.0346	0.3521	0.45328
B	0.3919	0.6610	0.17982
CB	0.0676	0.4369	0.35962
C	0.0029	0.2777	0.55638

Table A5.21. PGM1 system, according to Wille et al. (1968); 68 persons from southwestern Germany

Gene frequencies

PGM_1^1 0.754
PGM_1^2 0.246

Phenotype of twins	Frequency	L = Likelihood ratio DZ/MZ	Log (1/L)
PGM1 1−1	0.5685	0.7691	0.11400
PGM1 2−1	0.3710	0.5927	0.22713
PGM1 2−2	0.0605	0.3881	0.41102

For Tables 5.6 to 5.21, various European sources have been used.

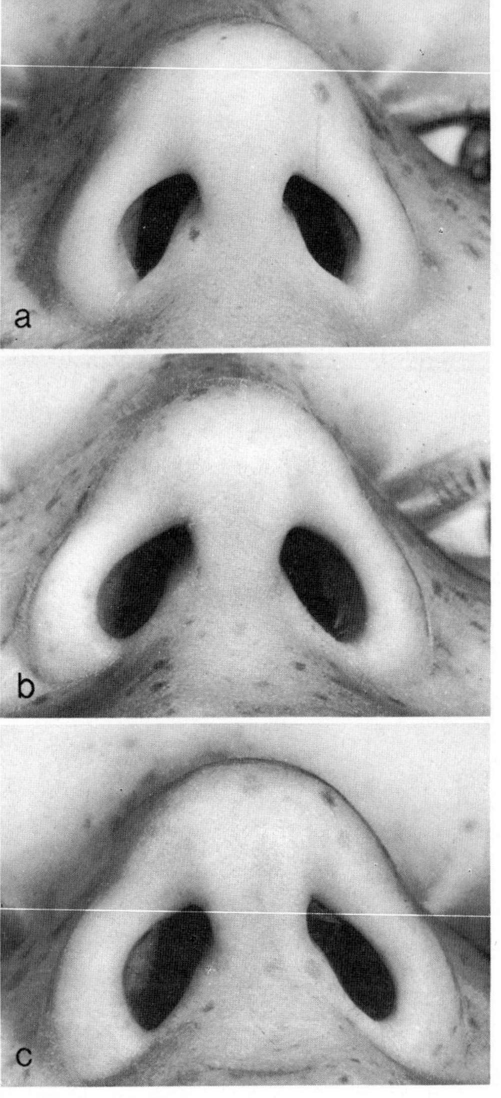

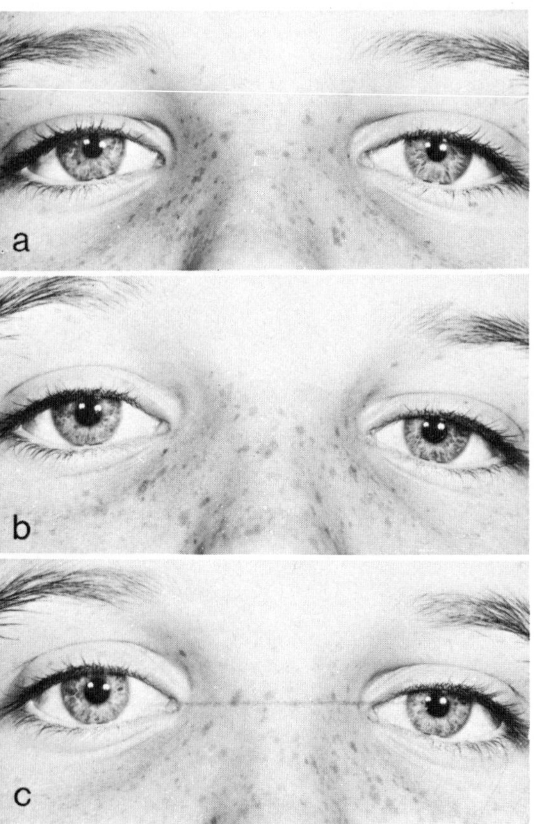

Fig. A5.2 *(left)*, **A5.3** *(right)*. Physiognomic details of the MZ triplets shown in Fig. A5.1

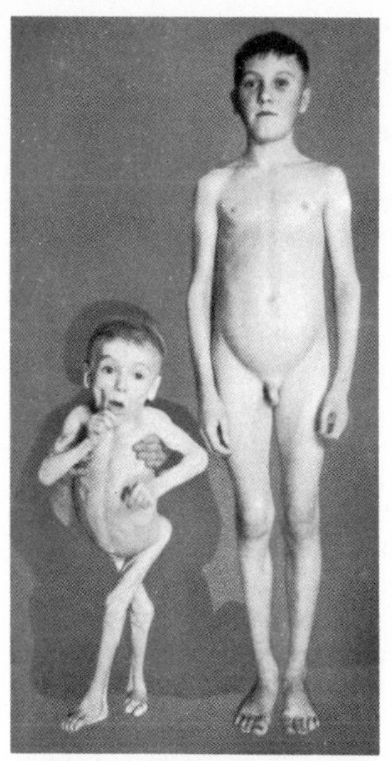

differing living conditions, MZ twins may sometimes look so different that the layman would not even identify them as sibs and only painstaking anthropologic examination will identify them as monozygotic. On the other hand, DZ twins, like other sibs, may occasionally look very similar. Figures A5.1–A5.5 give an impression of the degree of similarity or – in some cases – dissimilarity found in MZ twins.

How Should We Proceed in Practice? From the foregoing discussion, it seems as if a comprehensive study of genetic marker systems would be the most appropriate and also a sufficient method for reliable

◁ **Fig. A5.4.** MZ twins at the age of 10. Discordant dwarfism. The second born, dwarfed twin never learned to walk or talk. He showed an undiagnosed developmental disturbance of the skeleton and died a short time after examination. The twin brother showed normal stature but a bilateral coloboma of the iris, which was absent in the malformed twin. (Grebe, 1959 [687])

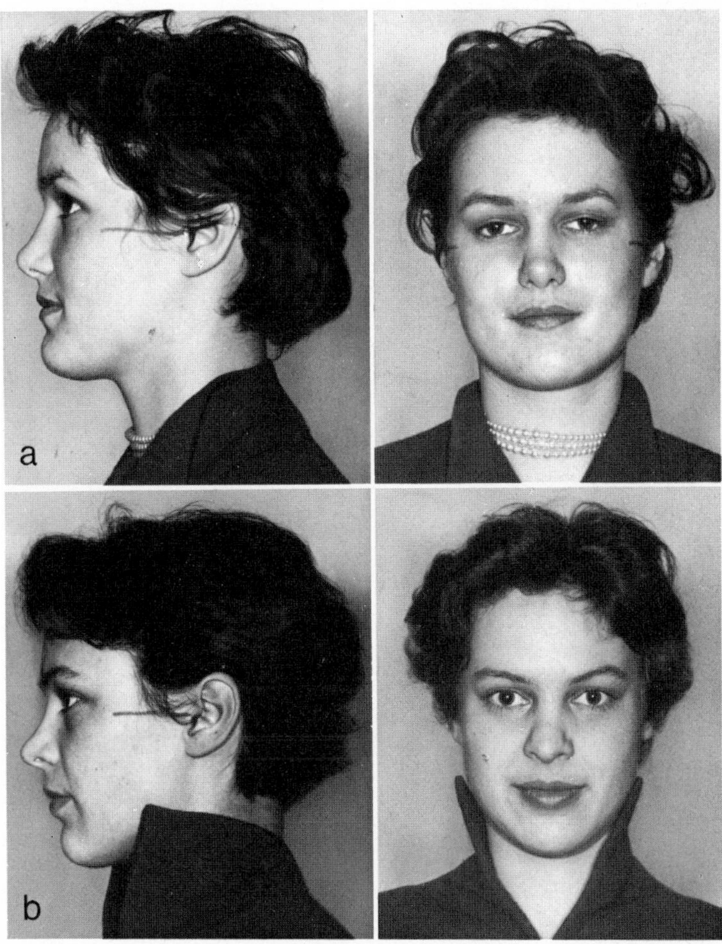

Fig. A5.5 a, b. DZ twins at the age of 19. Note the conspicuous similarity

zygosity diagnosis. This conclusion, however, needs some qualification. On the one hand, serology is not immune to errors. An error in determination of only one system in only one of the two twins will lead to misclassification of an MZ pair as dizygotic. The investigator should therefore trust his eyes. Wherever possible, he should add (and document) a physiognomic comparison of the twin pair and should insist on repetition of serologic examinations whenever a twin pair is regarded as monozygotic in spite of discordance in a marker system. Apart from the possibility of laboratory errors, comprehensive study of serologic and enzyme systems has two other disadvantages: it is expensive and requires a genetic marker laboratory. The anthropologic comparison, on the other hand, is much less time-consuming and, above all, very much less costly. Therefore, research workers involved in twin research should familiarize themselves with this method or should cooperate with a physical anthropologist experienced in this technique. They will learn to classify unambiguously most twin pairs. As a rough rule, twins whose physical appearance led to identity mix-ups by teachers and others are very often MZ in type. Serologic examination may be then added in cases in which a physical comparison fails to give clear-cut results. Such a procedure is advisable especially in studies of large twin series of several hundred pairs. In studies of single twin pairs or small series with rare diseases, examination of genetic markers will normally not meet with such logistical difficulties. Anthropologic diagnosis requires more experience and judgment; it is therefore more subjective. Serologic diagnosis, on the other hand, is more objective, but laboratory errors are always possible.

Ideally, twin examination should also include a registration of placentation and embryonic membranes: DZ twins have two placentas, two amnions, and two chorions, whereas MZ twins may have one chorion or even one amnion and one placenta. As mentioned in Sect. 3.8.4, presence of only one chorion may be important for intrauterine development of MZ twins. In practice, however, reliable information is only rarely available, and DZ twins may have a single fused placenta with resembles that of MZ twins. Hence, an attempt of including such data into zygosity diagnosis is often futile and may even be misleading.

Appendix 6

Heritability Estimates from Twin Data

The concept of heritability has been introduced in Sect. 3.6.1.5 and in Sect. 3.6.2. Methods were described for heritability estimates in threshold characters. The criterion was the ratio between incidence in close relatives of affected probands and in the general population. In a continuously distributed character, such as stature, heritability was estimated from comparison between parents and children.

Twin data can be utilized as an alternative way to get heritability estimates. In Sect. 3.6.1.5, heritability was defined as

$$h^2 = \frac{V_A}{V_P}$$

i.e. the ratio between the additive genetic variance (V_A) and the total phenotypic variance (V_P). It was also noted that, in human genetics, h^2 is often called "heritability in the narrow sense," and is contrasted with $H^2 = V_G/V_P$ (heritability in the broad sense; also called degree of genetic determination), where V_G is the total genetic variance including dominance, epistasis, and interaction variance.

In this appendix, estimation of heritability, h^2, will be restricted to twin data. It is not possible to estimate V_D, the dominance variance from twin data alone. Furthermore, V_D is usually negligible compared to V_A (Falconer [63]). Thus, the error in assuming that the total genetic variance (V_G) is identical to the additive variance (V_A) is likely to be small. Therefore, the assumption of $V_G = V_A$ will be made and the definition will be

$$h^2 = \frac{V_G}{V_P} \qquad \text{(A6.1)}$$

As shown in Sect. 3.6.1,

$$V_P = V_A + V_D + V_E + V_I + V_M + \mathrm{Cov}_{GE}$$

Here, V_E = environmental variance, V_I = variance due to interaction between heredity and environment, and V_M = the variance between repeated measurements of the same character, representing either truly different values - such as blood pressure on different days - or, in a constant character, measurement errors. Cov_{GE} is the covariance between the genetic and environmental components of the phenotypic value. Heritability estimates from twin data require that V_I and Cov_{GE} be 0. This assumption is not realistic in most cases, especially in behavioral genetics, but estimation of these parameters, if they deviate from 0, poses almost unsuperable practical difficulties. The measurement variance, V_M, however, needs to be considered but is often disregarded in many twin studies. Equation A6.1 becomes:

$$h^2 = \frac{V_G}{V_G + V_E + V_M}$$

It might be biologically meaningful to offer an alternative definition

$$h'^2 = \frac{V_G}{V_G + V_E}$$

Here, only the constant part of the phenotypic variance, not the part that varies from day to day, is considered.

If $\mathrm{Cov}_{GE} = 0$, the following relation exists between the correlation coefficient $r_{P_1P_2}$ of the phenotypic values P_1, P_2 of two relatives, the correlation coefficient of their genotypic values G_1, G_2, and the correlation coefficients of their environmental values E_1, E_2:

$$r_{P_1P_2} = r_{G_1G_2} h^2 + r_{E_1E_2} E^2$$

with

$$E^2 = \frac{V_E}{V_P}$$

In analogy with heritability, E^2 might be called the "environmentability." The correlation coefficients can be defined as intrapair correlations if the two relatives are twins:

$$r_{P,\mathrm{MZ}} = h^2 + r_{E,\mathrm{MZ}} E^2 \quad \text{for MZ twins} \qquad \text{(A6.2)}$$

$$r_{P,\mathrm{DZ}} = r_{G,\mathrm{DZ}} h^2 + r_{E,\mathrm{DZ}} E^2 \quad \text{for DZ twins} \qquad \text{(A6.3)}$$

Here, the theoretical genotypic correlation $r_{G,\mathrm{MZ}}$, which is 1 for MZ twins, (Sect. 3.6.1.5), has been used.

If the environmental correlation between the twins of an MZ pair is assumed to be identical with the environmental correlation between the twins of a DZ pair, it follows that:

$$h^2 = \frac{r_{P,\mathrm{MZ}} - r_{P,\mathrm{DZ}}}{1 - r_{G,\mathrm{DZ}}} \qquad \text{(A6.4)}$$

This expression is known as the H index (Holzinger). The phenotypic intrapair correlation coefficient is given by:

$$r_{P,\text{twins}} = \frac{V_P^B}{V_P^B + V_P^W}$$

where V_P^B = phenotypic variance between pairs and V_P^W = phenotypic variance within pairs.
The variance components V_P^B and V_P^W can be estimated from the phenotypic values p_{i1}, p_{i2} ($i=1, 2, \ldots, n$) observed on n twin pairs:

$$V_P^W = DQ_W, \quad V_P^B = (DQ_B - DQ_W)/2 \tag{A6.5}$$

Here, DQ_W and DQ_B are the within-pairs and between-pairs mean squares:

$$DQ_W = \frac{1}{2n}\sum_{i=1}^{n}(p_{i1}-p_{i2})^2 \tag{A6.6}$$

$$DQ_B = \frac{2}{n-1}\sum_{i=1}^{n}(\bar{p}_{i.}-\bar{p}_{..})^2 \tag{A6.7}$$

with

$\bar{p}_{i.} = (p_{i1}+p_{i2})/2$ = mean phenotypic value for i-th pair

$\bar{p}_{..} = \frac{1}{n}\sum_{i=1}^{n}\bar{p}_{i.}$ = total mean of all measurements in the twin sample

For the actual calculation, the following formula may be used instead of Eq. A6.7:

$$DQ_B = \frac{1}{2(n-1)}\left[\sum_{i=1}^{n} y_i^2 - \frac{1}{n}\left(\sum_{i=1}^{n} y_i\right)^2\right] \tag{A6.7'}$$

with $y_i = p_{i1} + p_{i2}$ ($1 = 1, 2, \ldots, n$).

The variance component V_P^W may be decomposed in the following way:

$$V_P^W = V_G(1-r_{G,\text{twins}}) + V_E(1-r_{E,\text{twins}}) + V_M \tag{A6.8}$$

This equation can be applied to MZ pairs, DZ pairs, or unrelated control pairs from the general population (CP).

$$V_P^W(\text{MZ}) = V_E(1-r_{E,\text{MZ}}) + V_M \tag{A6.9}$$

$$V_P^W(\text{DZ}) = V_G(1-r_{G,\text{DZ}}) + V_E(1-r_{E,\text{DZ}}) + V_M \tag{A6.10}$$

$$V_P^W(\text{CP}) = V_G + V_E + V_P = V_M \tag{A6.11}$$

For further analysis, two steps are suggested. First, one should examine whether h^2 deviates significantly from 0. Then, h^2 should be estimated.

Testing of the Null Hypothesis ($h^2=0$). Under the assumption $r_{E,\text{MZ}} = r_{E,\text{DZ}}$ it follows from Eqs. A6.2 and A6.3 that the hypothesis $h^2=0$ is equivalent to the hypothesis:

$$r_{P,\text{MZ}} = r_{P,\text{DZ}}$$

For testing the latter hypothesis, the fact is used that under this hypothesis

$$z = 1/2\log_e\frac{1+\hat{r}_{P,\text{MZ}}}{1-\hat{r}_{P,\text{MZ}}} - 1/2\log_e\frac{1+\hat{r}_{P,\text{DZ}}}{1-\hat{r}_{P,\text{DZ}}}$$

has approximately a normal distribution with mean 0 and variance

$$\frac{1}{n_{\text{MZ}}-3/2} + \frac{1}{n_{\text{DZ}}-3/2}$$

Here, $\hat{r}_{P,\text{MZ}}$ and $\hat{r}_{P,\text{DZ}}$ are estimates for $r_{P,\text{MZ}}$ and $r_{P,\text{DZ}}$ from n_{MZ} MZ pairs and n_{DZ} DZ pairs.

Estimates for h^2 (and h'^2):

$$h_1^2 = \frac{V_P^W(\text{DZ}) - V_P^W(\text{MZ})}{V_P^W(\text{DZ}) - V_M} \tag{A6.12}$$

$$h_2^2 = \frac{V_P^W(\text{CP}) - V_P^W(\text{MZ})}{V_P^W(\text{CP}) - V_M} \tag{A6.13}$$

To estimate h^2 (or h'^2), the intrapair variances are replaced by their estimates from the analysis of variance. An exact formula for the standard error of these two estimates of h^2 does not exist. If V_M may be neglected compared to V_P^W, the following formula holds approximately:

$$\text{S.E.}^2 = 2F^2\frac{n_2^2(n_1-1)(n_1+n_2-4)}{n_1^2(n_2-3)^2(n_2-5)} \tag{A6.14}$$

where F means the observed value of the ratio V_P^W (MZ)/V_P^W (DZ) or V_P^W (MZ)/V_P^W (CP), respectively, and n_1, n_2 are the number of pairs from which variances in numerator and denominator have been estimated. Apart from sampling errors, the estimates by Eqs. A6.12 and A6.13 are biased, as will now be explained in detail:
Equation A6.12: If environmental correlations between MZ and DZ are assumed to be identical: $r_{E,\text{MZ}} = r_{E,\text{DZ}}$, then it follows from Eqs. A6.9 and A6.10:

$$h_1^2 = \frac{V_G(1-r_{G,\text{DZ}})}{V_G(1-r_{G,\text{DZ}}) + V_E(1-r_{E,\text{DZ}})}$$

$$= h'^2\frac{1-r_{G,\text{DZ}}}{1-r_{G,\text{DZ}} + (1-h'^2)(r_{G,\text{DZ}} - r_{E,\text{DZ}})}$$

Hence, $h_1^2 \lesseqgtr h'^2$, if $r_{E,\text{DZ}} \lesseqgtr r_{G,\text{DZ}}$.
Therefore h_1^2 overestimates h^2, if $r_{E,\text{DZ}} > r_{G,\text{DZ}}$, because always $h^2 \leqq h'^2$. For other cases, no prediction of the bias of h_1 in estimating h^2 is possible.

Equation A6.13: It follows from Eqs. A6.9 and A6.11 that:

$$h_2^2 = \frac{V_G + V_E r_{E,\text{MZ}}}{V_G + V_E} = h'^2 + (1 - h'^2) r_{E,\text{MZ}} \geqq h'^2 \geqq h^2$$

(Here, $r_{E,\text{MZ}} \geqslant 0$ is assumed). This means that h_2^2 will usually overestimate h^2.

In these two estimates for h^2, intrapair variances have exclusively been used. Frequently, h^2 is also estimated from *intrapair correlation coefficients,* which have been calculated utilizing the variance of the whole sample of MZ or DZ twins, i.e., also the variance between twin pairs. From Eq. A6.4, the following estimation formula can be derived:

$$h_3^2 = 2(r_{P,\text{MZ}} - r_{P,\text{DZ}}) \qquad \text{(A6.15)}$$

This formula rests on the assumption $r_{G,\text{DZ}} = 1/2$, which holds true only if mating is random and there is neither dominance nor epistasis. Practically , the condition is at best only approximately fulfilled; moreover, it is unknown whether $r_{G,\text{DZ}} > 1/2$, or $r_{G,\text{DZ}} < 1/2$. Therefore, the bias in estimating h^2 from Eq. A6.15 cannot be predicted.

A correction is possible, if $r_{G,\text{DZ}} > 1/2$ due to assortative mating and there is neither dominance nor epistasis. The standard error of the estimate h_3^2 can be calculated only very approximately:

$$\text{S.E.}^2(h_3^2) \approx 4\left[\frac{(1 - r_{P,\text{MZ}}^2)^2}{n_{\text{MZ}}} + \frac{(1 - r_{P,\text{DZ}}^2)^2}{n_{\text{DZ}}}\right] \qquad \text{(A6.16)}$$

Comments on These Methods of Heritability Estimation. The above considerations show that an unbiased estimate of h^2 from twin data is impossible even if such components as covariance between heredity and environment (Cov_{GE}) as well as the interaction term V_I are neglected and if the very unlikely assumption is made that the environmental correlations $r_{\text{E,MZ}}$ and $r_{\text{E,DZ}}$, i.e., the common environmental influences affecting MZ pairs and DZ pairs, are identical. Even with these oversimplifying assumptions, there remain systematic errors that can only be partially controlled.

An empirical way to overcome this difficulty partially is to calculate alternative estimates from the same data to see how well they coincide. The three alternative estimates proposed above can be characterized as follows: h_1^2 is calculated from the classic comparison between MZ and DZ twins. The bias of this estimate includes the genotypic correlation between sibs, $r_{G,\text{DZ}}$. This value is ½ with random mating. With regard to many characters in which heritability estimates are used, e.g., intelligence quotient or stature, mating is however known to be assortative. The direction and the degree of bias depends on the difference of the genotypic and environmental correlations between the sibs, which is usually unknown. Therefore, the heritability estimate based on control persons (h_2^2) may be a useful second choice, even though by depending on environmental correlations $r_{\text{E,MZ}}$ and $r_{\text{E,DZ}}$ this estimate overestimates h^2 systematically.

The additional comparison of control pairs was proposed by Vogel and Wendt in 1956 [926] but, apparently, has never been used since then. A similar procedure has now again been suggested by Kamin (1974) [104]. Control pairs from twin samples can easily be matched for age and sex, thus eliminating variance components contained in most twin samples but without significance for the problem to be examined.

This nuisance variance is the main argument against utilizing the estimate from the intraclass correlation coefficients (h_3^2), which contains these additional variance components unless the twin samples are very homogeneous, for example, a single cohort of army conscripts. The problem is taken up in Sect. 8.2.1.3.

Heritability of I. Q. as an Example. The twin sample consists of 50 German adult male twin pairs between 23 and 30 years, 25 MZ and 25 DZ. The twins were conscripts and were, therefore, unbiased regarding socioeconomic status, education, and test intelligence. In this sample, the I-S-T (Intelligence Structure Test according to Amthauer) was administered [2234]. Here, only the total intelligence score, corrected for age, is considered; this score is proportional to I. Q. To construct a set of control pairs with intrapair age differences as small as possible, the following procedure was used. All twin pairs, regardless of zygosity, were arranged in ascending order of age. In this arrangement, the first and second pair, the third and fourth pair and so on were combined as quadruples that were transformed into two new pairs, each by exchanging one co-twin between the original pairs; the co-twins to be exchanged were selected at random. The observed value pairs of total intelligence score (p_{i1}, p_{i2}) as well as the quantities Σy_i, Σy_i^2, and $\Sigma(p_{i1} - p_{i2})^2$ derived from them (for notations see the foregoing text) are given in Table A6.1 for the twin pairs and in Table A6.2 for the control pairs.

From these tables one calculates:

a) *For MZ pairs* According to Eq.

$$\hat{V}_P^W = 1{,}005/50 = 20.100 \qquad \text{(A6.6)}$$

$$DQ_B = (1{,}017{,}589 - 5{,}009^2/25)/48 = 291.370 \qquad \text{(A6.7)}$$

$$\hat{V}_P^B = (291.370 - 20.100)/2 = 135.635 \qquad \text{(A6.5)}$$

$$\hat{r}_{P,\text{MZ}} = \frac{135.635}{135.635 + 20.100} = 0.871$$

Confidence limits (99%) for $\frac{1}{2}\log_e \frac{1+r_{P,\mathrm{MZ}}}{1-r_{P,\mathrm{MZ}}}$:

$$\tfrac{1}{2}\log_e \frac{1+0.871}{1-0.871} \pm 2.58 \frac{1}{\sqrt{25-1.5}} = 0.805 \text{ and } 1.869$$

to which correspond

$$\frac{e^{2\times 0.805}-1}{e^{2\times 0.805}+1} = 0.667, \quad \frac{e^{2\times 1.869}-1}{e^{2\times 1.869}+1} = 0.954$$

as confidence limits for $r_{P,\mathrm{MZ}}$.

b) *For DZ pairs:*

$\hat{V}_P^W \; = 1{,}682/52 = 31.308$

$DQ_B \; = (1{,}048{,}268 - 5{,}180^2/26)/50 = 325.052$

$\hat{V}_P^B \; = (325.052 - 31.308)/2 = 146.872$

$$\hat{r}_{P,\mathrm{DZ}} = \frac{146.872}{146.872 + 31.308} = 0.824$$

Confidence limits (99%) for $\frac{1}{2}\log_e \frac{1+r_{P,\mathrm{DZ}}}{1-r_{P,\mathrm{DZ}}}$:

$$\tfrac{1}{2}\log_e \frac{1+0.824}{1-0.824} \pm 2.58 \times \frac{1}{\sqrt{26-1.5}} = 0.648 \quad \text{and}$$

1.690

with corresponding confidence limits

0.570 and 0.934

for $r_{P,\mathrm{DZ}}$.

c) *For the control pairs:*

$\hat{V}_P^W \; = 16{,}659/100 = 166.590$

$DQ_B \; = (2{,}024{,}211 - 10{,}023^2/50)/98 = 153.066$

$\hat{V}_P^B \; = (153.066 - 166.590)/2 = -6.762$

$$\hat{r}_{P,\mathrm{CP}} = \frac{-6{,}762}{166.590 - 6.762} = -0.042$$

The above confidence limits for $r_{P,\mathrm{MZ}}$ and $r_{P,\mathrm{DZ}}$ show that the intrapair correlation coefficient of I.Q. deviates significantly from 0 ($P<0.01$) in both types of twins. This means that twins, regardless of their zygosity, are more similar in I.Q. than two unrelated persons. This result - although expected if I.Q. has a genetic basis - does not yet exclude the possibility of a purely nongenetic explanation because twins also partially share their environment. To examine this possibility, we test hypothesis $r_{P,\mathrm{MZ}} = r_{P,\mathrm{DZ}}$ (null hypothesis).

$$z = \tfrac{1}{2}\log_e \frac{1+\hat{r}_{P,\mathrm{MZ}}}{1-\hat{r}_{P,\mathrm{MZ}}} - \tfrac{1}{2}\log_e \frac{1+\hat{r}_{P,\mathrm{DZ}}}{1-\hat{r}_{P,\mathrm{DZ}}} = 0.168$$

$$\operatorname{var} z = \frac{1}{25-1.5} + \frac{1}{26-1.5} = 0.0834$$

$$z/\sqrt{\operatorname{var} z} = 0.582$$

Under the null hypothesis, the probability of a z value as extreme as or more extreme than the value found is greater than 10%. This means that from the comparison of the intrapair correlation coefficients in our two twin series the hypothesis $h^2=0$, i.e., there is no genetic contribution to the variation of I.Q. in the population, cannot be rejected. Consequently, the estimate for h^2 according to Eq. A6.15:

$$h_3^2 = 2(\hat{r}_{P,\mathrm{MZ}} - \hat{r}_{P,\mathrm{DZ}}) = 2 \times (0.871 - 0.824) = 0.094,$$

also cannot be regarded as significantly different from 0; this is confirmed by considering the standard error of h_3^2: its approximate value, calculated by Eq. A6.16 is 0.159. Now it is possible that the difference of the intrapair correlations of I.Q. in MZ and DZ twin pairs is biased, the between-pairs variance of I.Q. being, for some unknown reason, smaller in the DZ twins than in the MZ twins. In this case, only the two estimates of h^2 based solely on within-pair variances have to be used. But even if there is no such bias - as obviously in our data - the calculation of these other estimates of h^2 (Eqs. A6.12 and A6.13) is strongly recommended. In our case

$$h_1^2 = \frac{V_P^W(\mathrm{DZ}) - V_P^W(\mathrm{MZ})}{V_P^W(\mathrm{DZ})} = \frac{31.308 - 20.100}{31.308} = 0.358$$

$$h_2^2 = \frac{V_P^W(\mathrm{CP}) - V_P^W(\mathrm{MZ})}{V_P^W(\mathrm{CP})} = \frac{166.590 - 20.100}{166.590} = 0.879$$ [1]

Standard errors for these estimates can be calculated by Eq. A6.14:

$\mathrm{SE}(h_1^2) = 0.301, \quad \mathrm{SE}(h_2^2) = 0.045$

giving (nearly) 95% confidence intervals

-0.23 to 0.95 for h_1^2

and

0.79 to 0.97 for h_2^2

The two intervals overlap partially, but the subinterval common to both is remote from h_3^2 by more than two times the standard error of the latter estimate, the three estimates for h^2 thus appearing incompatible. On the other hand, h_2^2 presumably overestimates h^2: the assumption inherent in this estimation - that environmental intrapair differences are identical for twins and unrelated control pairs - is incorrect. Thus, the differences between the three estimates may be explained by this bias combined with sampling errors that, due to the small size of our twin series, are large. In any case, these results are hardly compatible with the high values of heritability re-

[1] The subtracend V_M in the denominators representing the variance between repeated measurements of I.Q. has been omitted; here the test reliability of I-S-T may be inserted.

Table A 6.1. Total intelligence scores observed in twin pairs

MZ pairs				DZ pairs			
Pair No.				Pair No.			
(i)	p_{i1}	p_{i2}	$\lvert p_{i1}-p_{i2}\rvert$	(i)	p_{i1}	p_{i2}	$\lvert p_{i1}-p_{i2}\rvert$
1	107	105	2	1	86	98	12
2	88	80	8	2	112	100	12
3	89	102	13	3	89	84	5
4	96	110	14	4	125	128	3
5	84	84	0	5	105	99	6
6	100	89	11	6	90	84	6
7	87	78	9	7	103	98	5
8	79	87	8	8	91	102	11
9	96	97	1	9	94	84	10
10	111	113	2	10	97	107	10
11	114	114	0	11	112	109	3
12	106	111	5	12	106	110	4
13	114	113	1	13	90	85	5
14	120	117	3	14	98	100	2
15	110	107	3	15	116	104	12
16	87	87	0	16	78	79	1
17	92	93	1	17	104	115	11
18	103	101	2	18	95	113	18
19	107	99	8	19	113	115	2
20	83	84	1	20	84	83	1
21	99	105	6	21	110	109	1
22	86	95	9	22	98	93	5
23	107	101	6	23	77	85	8
24	122	117	5	24	76	86	10
25	118	115	3	25	117	117	0
				26	117	110	7
Σy_i		5,009					5,180
Σy_i^2		1,017,589					1,048,268
$\Sigma(p_{i1}-p_{i2})^2$		1,005					1,628

Table A 6.2. Total intelligence scores observed in control pairs (See p. 678 for choice of control pairs)

Pair No. (i)	p_{i1}	p_{i2}	$\lvert p_{i1}-p_{i2}\rvert$	Pair No. (i)	p_{i1}	p_{i2}	$\lvert p_{i1}-p_{i2}\rvert$
1	89	88	1	26	113	78	35
2	102	80	22	27	94	89	5
3	106	114	8	28	84	84	0
4	114	110	4	29	90	100	10
5	96	114	18	30	112	84	28
6	97	113	16	31	110	109	1
7	125	86	39	32	112	107	5
8	128	95	33	33	103	76	27
9	77	83	6	34	101	86	15
10	85	84	1	35	99	117	18
11	78	101	23	36	120	105	15
12	107	79	28	37	96	87	9
13	104	93	11	38	110	87	23
14	92	115	23	39	122	105	17
15	84	99	15	40	117	99	18
16	107	83	24	41	117	113	4
17	103	110	7	42	111	117	6
18	98	109	11	43	98	86	12
19	91	97	6	44	93	98	5
20	102	107	5	45	84	106	22
21	98	115	17	46	84	111	27
22	118	100	18	47	107	104	3
23	95	100	5	48	116	105	11
24	113	89	24	49	90	110	20
25	87	115	28	50	117	85	32
Σy_i							10,023
Σy_i^2							2,024,211
$\Sigma(p_{i1}-p_{i2})^2$							16,659

ported for I.Q. by some authors [2086]. This example illustrates the problems in estimating heritability in human twin data and should make one cautious of accepting such data for definitive scientific conclusions.

For those interested in the use of more refined quantitative-genetic models for the study of quantitative traits in twins the introduction by Eaves [634] is recommended. It also mentions the enormous sample sizes needed for such an analysis.

Appendix 7

Method of Path Coefficients

Basic Concept. The concept of heritability as described is based on correlations between relatives. The first correlations were calculated by biometricians. Later, Fisher showed that certain theoretical correlations were expected as a consequence of Mendel's laws. The meaning of correlations, however, is ambiguous. If A and B are correlated, then A may be the partial cause of B, B may be the partial cause of A, or A and B may have a common partial cause, C. In genetics, however, the sequence of events is often unambiguous: a correlation between parent and child is caused by the fact that the child inherits half its genes from the parent and that the child lives in an environment that has largely been created by the parents. S. Wright (1921; 1931) [961] designed a statistical technique to take account of this sequential relation.

"A qualitative interpretation of a system of variables ... is conveniently represented by a diagram in which arrows are used to indicate which variables are to be treated as functions of which others ... Unanalyzed correlations may be represented by two-headed arrows to indicate connection through common factors (Fig. A7.1).

It is convenient to measure each variable in terms of its standard deviation. Letting

$$x_0 = \frac{X_0 - \bar{X}_0}{\sigma_0} \quad \text{etc.}^1$$

we can write the best linear expression for deviations of a given variable in terms of those from which arrows are drawn to it in the form:

$$x_0 = p_{02}x_2 + p_{03}x_3$$

The coefficients p_{02}, p_{03}, etc., are abstract numbers which I have called path coefficients, related numerically to the concrete partial regression coefficients in the same way that the correlation coefficient is related to total regression. They differ from correlation coefficients, however, in having direction ... For any two variables of such a system, the correlation can

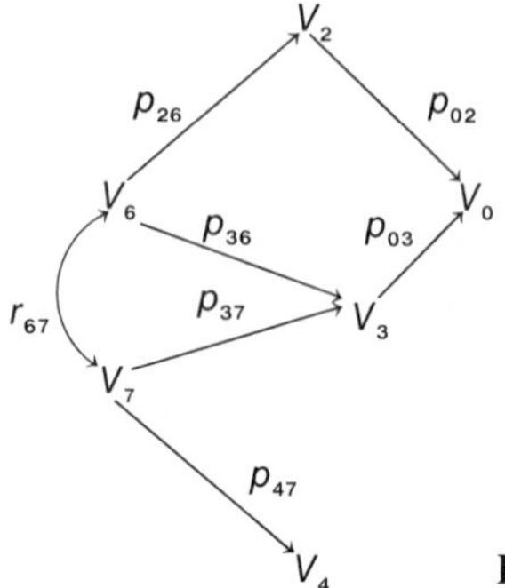

Fig. A7.1

be analyzed into contributions tracing through the represented factors of either one. Letting s stand for the factors of x_0 and t for those of x_1,

$$r_{01} = \Sigma p_{0s} r_{1s} = \Sigma p_{1t} r_{0t}$$

By further analysis of the correlation terms, this leads to the easily remembered principle that any correlation can be analyzed into contributions from all of the paths through the diagram (direct or through common factors) by which the two variables are connected, and that each of these contributions is a product of the coefficients pertaining to the elementary paths. One of these elementary paths in each case may be an unanalyzed bidirectional one, measured by a correlation coefficient."

The foregoing may be elucidated by an example (Fig. A7.1).

The following formulas hold true[2]:

$$\begin{aligned} r_{04} &= p_{02}r_{24} + p_{03}r_{34} \\ &= p_{47}r_{07} \\ &= p_{02}p_{26}r_{67}p_{47} + p_{03}p_{36}r_{67}p_{47} + p_{03}p_{37}p_{47} \end{aligned}$$

Here p_{ij} is the path coefficient from the variable v_j to the variable V_i and r_{ij} the coefficient of correlation between V_i and V_j $(i, j = 0, 2, 3, 4, 6, 7)$.

---

1 X_0, X_2, X_3 mean variable, $\bar{X}_0, \bar{X}_2, \bar{X}_3, \ldots$ their means, $\sigma_0, \sigma_2, \sigma_3, \ldots$ their standard deviations, and $x_0, x_2, x_3, \ldots$ the corresponding standardized variables.

2 The derivation of these formulas follows from the theory of partial correlations and regressions; for an elementary explanation, cf. Li, *Population Genetics*, Chap. 12 [124].

Application to Data on Human I. Q. Wright used data collected by Burks (see also Sect. 8.2.1.3) on correlation of the intelligence quotient (I. Q.) between parents and adopted children, on the one hand, and biologic children, on the other, to exemplify the approach through path coefficients. The two groups of parents were thought to be closely similar. Apart from determining I. Q. values for parents and children, Burks had constructed for each family a "cultural index" measuring the quality of home environment. For both groups of parents, Burks then calculated the correlations between the child's I. Q. and cultural index, r_{CE}, and between the child's and the midparent's I. Q., r_{CP}. Additionally, the correlation, r_{EP}, between the midparent's I. Q. and environment as measured by the cultural index was calculated, but only for the biologic families.
Now, we follow again Wright's analysis using the principle of path coefficients (see also Fig. A7.2):

Adopted children:[3]

$r_{CE} = p_{CE} = +0.29$

$r_{CP} = p_{CE} r_{EP} = +0.23$

$p_{CE}^2 + p_{CH}^2 = 1.00$[4]

Own children:

$r_{EP} = +0.86$

$r_{CE} = p_{CE} + p_{CH} r_{HE} = 0.49$

$p_{CP} = p_{CE} r_{EP} + p_{CH} r_{HP} = +0.61$

$p_{CE} = \frac{0.29}{0.96} p_{CH}$

$p_{CE}^2 + p_{CH}^2 + 2 p_{CE}\, p_{CH} r_{HE} = 1.00$[4]

"If I. Q. of the foster parents is related to child's I. Q. only through correlation with home environment, the parent-offspring correlation should be the product of the two intermediary coefficients. This leads to a value of the correlation between midparent and environment (+0.79) closely similar to the observed in the control data (+0.86). This indicates that there was no influence of the parents other than through the home environment as actually measured [in the biologic families]. There was only 9 per cent determination of variance by home environment ($=0.29^2$) leaving a residuum of 91 percent determination and a path coefficient of about 0.96. How far this traces only to child's heredity and how far to unmeasured

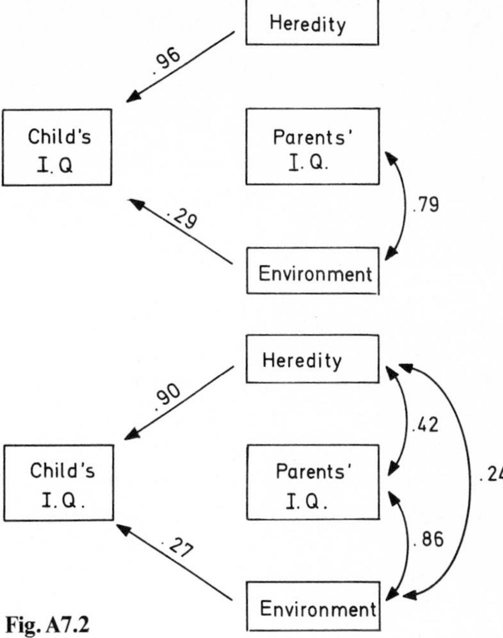

Fig. A7.2

environmental factors the data give no answer. But since home environment is presumably much the most important environmental factor ... one may surmise that the residual group is largely hereditary.
In the other groups, the situation is more complex. We can at once write three equations representing analysis of the three known correlations "(r_{EP}, r_{CE} and r_{CP})." If we assume that the only factor of child's I. Q. apart from the home environment as measured is heredity, we can write a fourth equation expressing complete determination. But there are five coefficients to be determined. No solution is possible and no quantitative interpretation is possible from the data of the control group .... In the present case, however, we have another resource. The control group of parents was carefully selected for comparability with the foster group. Presumably home environment has closely similar effects in the two cases. We should be able to borrow the environmental coefficients from the foster data."
The path coefficient p_{CE}, however, cannot directly be transferred, because it is influenced in the control data by the correlation between heredity and environment. However, "it may be deduced that the ratio $p_{CE} : p_{CH}$ should be the same in the two cases, giving a fifth equation." These equations were solved, giving the results of Fig. A7.2.
Generally, the number of equations must at least equal the number of paths or correlation coefficients to be determined. It can be shown mathematically that path coefficients are equivalent to standardized regression coefficients, i. e., to regression coefficients

[3] In these formulas r always means a correlation coefficient and p a path coefficient. The indices refer to the variables set into relation: p = midparent's I. Q., C = child's I. Q., E = environment (cultural index).

[4] This is the so-called theorem of complete determination by the causes; it follows from the fact that the correlation of a variable with itself equals 1.

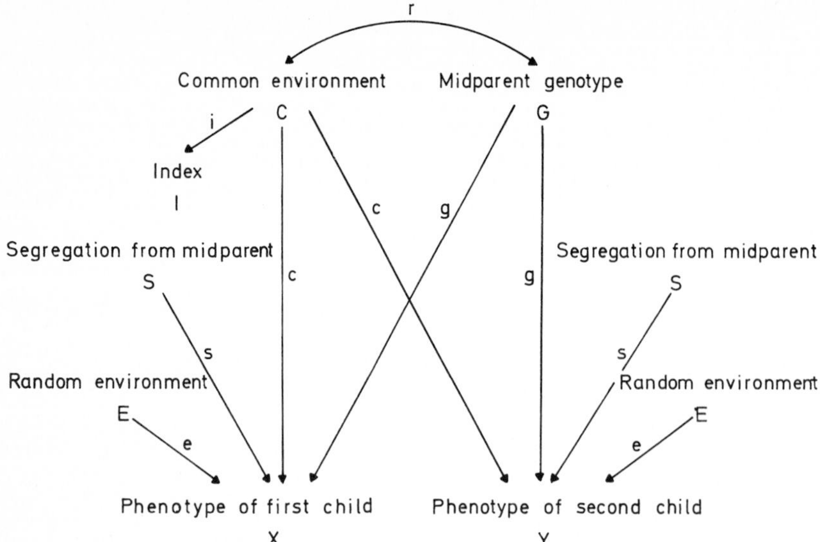

Fig. A7.3. Path diagram for correlation between sibs [803]

that refer to standardized variables rather than to variables in original physical units. It is therefore a matter of taste whether the analysis is carried out using the standard partial regression or partial correlation methods or whether path coefficients are used. The advantage of the latter method is that the diagram provides a straightforward and intuitively convincing way to visualize the components that influence the character.

More Recent Developments of the Method. In recent years, N. E. Morton and his collaborators [803] have refined this approach by introducing a number of additional correlations. Figure A7.3 shows a path diagram for correlation between siblings. Here, the phenotypes of two children, X and Y, are assumed to be "resultants of four additive factors: common environment (C), random environment (E), midparent genotype (G) and segregation from the midparent genotype (S). Midparent genotype and common environment are correlated (r) and common environment is imperfectly but linearly measured by an index (I) based on such variables as socioeconomic status, income, parental education, neighbourhood, school-district, and cultural and mental advantages of the home."

The method has been used by Morton and his colleagues during recent years in the hope that it would provide a solution to the question of whether the average I. Q. differences between American blacks and whites is due to genetic or sociocultural factors. The authors came to the conclusion that it is largely sociocultural (5th Int. Congr. Hum. Genet. 1976).

While this conclusion may well be true, the evidence on which it is based - and especially the method of path coefficients and its application to these problems - has been severely critized for methodologic reasons (Goldberger, [683]; Taylor [2214]). It is a major drawback of the method that it does not provide enough equations for estimation of all relevant parameters. Therefore, some of them must be assumed to be 0. Moreover, operationalization of the assumed factors is open to criticism; socioeconomic status and education, for example, which are used to estimate the influence "common environment," may have genotypic components. We are doubtful whether this alternative method of biometric analysis will contribute much to our understanding of genetic influences on heredity-environment interaction in complex situations as encountered in human behavioral genetics. As Wright puts it, he himself "has never made the preposterous claim that the theory of path coefficients provides a general formula for the deduction of causal relations." On the contrary, the prior knowledge of the causal relations is a prerequisite for the applicability of the method, and the results obtained by its use depend on the correctness of the assumed causal relations placed into the equations.

Appendix 8

Genetic Counseling: Use of Conditional Probabilities

A Problem in the Estimation of Genetic Risks. As noted in Sect. 9.2.1, estimation of a genetic risk is based either on segregation ratios in Mendelian diseases or on empirical risk figures when the mode of inheritance is complex. Utilization of such figures for estimating the specific risk for a certain proband or family is straightforward if no additional information is available. For example, every future child of an affected member of the big Farabee family with brachydactyly (Sect. 3.1.2, Fig. 3.2) will have a risk of 50% to have brachydactyly. However, there are many situations in which additional information can be included in the risk estimate.

Example: Inherited and Sporadic Retinoblastoma. As noted in Sect. 5.1.6, retinoblastoma (R), a malignant eye tumor of young children, may occur either as a dominant disease with about 90% penetrance or as a nonhereditary condition, probably due to somatic mutation. In the latter case, both parents – and all other family members – are unaffected, and the genetic risk for children is not higher than the incidence in the general population – about 1:15,000–1:25,000. Moreover, a somatic mutation will always lead to unilateral retinoblastoma. However, among all sporadic unilateral cases, about 10% are caused by mutation in the germ cell of one parent. Hence, parents and other family members are also unaffected; however, the risk for every child of the individual affected with sporadic unilateral retinoblastoma to carry the tumor is now about 45% (90% penetrance with a 50% segregation ratio). Consider the following situation: A sporadic, unilateral case asks about the genetic risks for his children. If no other information is available, this risk amounts to $0.9 \times 0\%$ (for the noninherited fraction) $+ 0.1 \times 45\%$ (for the germ cell mutations) = 4,5%. The situation becomes more complicated, however, if our proband already has one or more healthy children. Had his disease been caused by a dominant mutation, each of these children would have had a risk of 45% to be affected. Therefore, the fact that they are not affected, increases his chance to have the noninherited form of the disease and, hence decreases the risk for his future children. How can this risk be calculated?

Probability of Being a Hereditary Case [2393]. As noted, the prior probability of our proband being a hereditary case is $P(H)=0.1$. If he is a hereditary case, then the conditional probability that his first child would be unaffected (event U), i.e., the probability that it is unaffected despite the fact that the proband carries the gene, is $P(U/H)=0.55$. On the other hand, the prior probability that the proband is a nonhereditary case is $P(\text{not } H)=0.9$. In this case, the conditional probability that his child would be unaffected would be $P(U/\text{not } H)=1$, as there is (almost) no risk. From these considerations, the formula for his posterior probability of being a hereditary case can be derived:

$$P(H/U) = \frac{P(H) \times P(U/H)}{P(H) \times P(U/H) + P(\text{not } H) \times P(U/\text{not } H)} \tag{A8.1}$$

Inserting the figure from our example leads to:

$$P(H/U) = \frac{0.1 \times 0.55}{0.1 \times 0.55 + 0.9 \times 1} = \frac{0.055}{0.955} = 0.0625$$

Hence, one unaffected child has reduced the probability of our proband being a hereditary case from 0.1 to 0.058. The risk for his next child to be affected is now:

$$R_2 = 0.058 \times 0.45 = 0.0261$$

or a reduction from 4% down to 2.6%. With two nonaffected children, the conditional probability $P(U/H)$ becomes $0.55^2 = 0.3025$. Inserting into Eq. A8.1 leads to $P(H/U)=0.0325$, $R_2=0.015$. For n children, $P(U/H)$ becomes 0.55^n. The principle can easily be grasped by considering Fig. A8.1.

A Convenient Notation System and a Form of Graphic Representation. Murphy [149] suggested a clear and convenient system of notation that makes the calculation described above more evident, especially for professional workers who have difficulties with abstract mathematical concepts. A table is constructed in which the stepwise calculation is visualized. Table A8.1 shows the above calculation for retinoblastoma. From the a posteriori probability of our client to be a hereditary case (0.58), the probability that the first child will be affected can now be calcu-

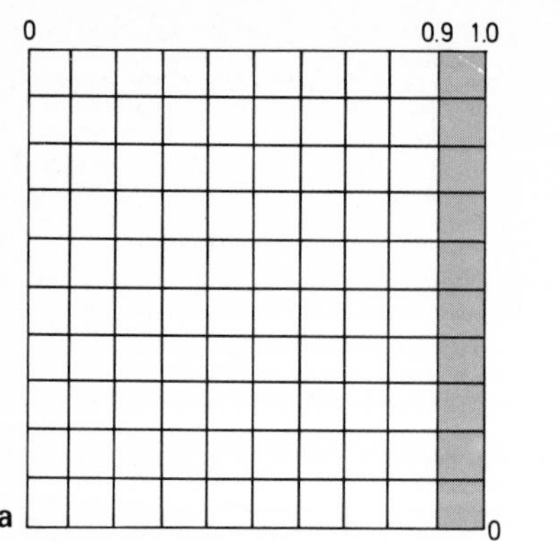

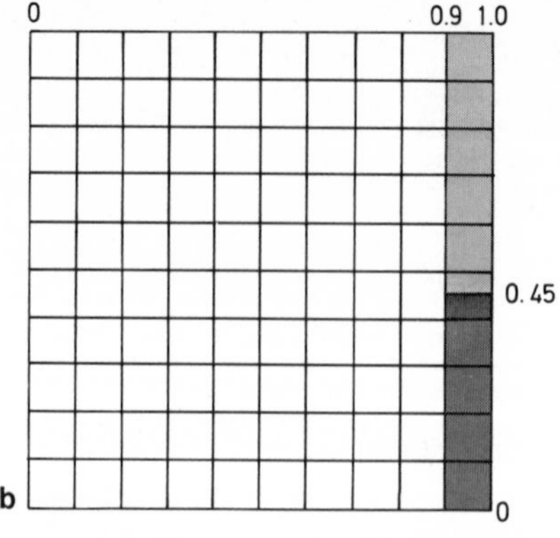

Fig. A8.1 a, b. Graphic representation of the risk calculation for children of parents with unilateral retinoblastoma. **a** The lightly shaded area represents the parents with hereditary retinoblastoma (≈10% of all unilateral sporadic cases). **b** After the first child is born, 45% of these parents (=4,5% of *all* parents) are revealed as carriers of the hereditary form. Only $\frac{5.5}{95.5}$% of the rest (=first child normal) will be carriers of the genes for the hereditary form

Table A8.1. Probability calculation for retinoblastoma

	Client is a hereditary case	Client is a nonhereditary case
a) The a priori probability (the client's chance for belonging to either group)	0.1	0.9
b) Conditional probability (that the first child will not be affected, given the group to which the parent belongs)	0.55	1.0
c) Combined probability (the chance that both a and b will occur)	$0.1 \times 0.55 = 0.055$	$0.9 \times 1.0 = 0.9$
d) The a posteriori probability (the client has a normal child and is/ is not a hereditary case	$\frac{0.055}{0.055+0.9} = 0.058$	$\frac{0.9}{0.055+0.9} = 0.942$

lated as shown in the previous paragraph: $0.58 \times 0.45 = 0.0261$ or 2%–3%.

The principle of calculation can also be shown graphically (Fig. A8.1). In this square, the *white area* represents the nonhereditary group, the *lightly shaded area* the hereditary cases. After the first child has outlived the danger period (the first few years of life), the parents of affected children, namely 45% of the hereditary cases (Fig. A8.1, *dark area*), are eliminated from the group of sporadic unilateral retinoblastoma patients as a whole. These 45% have been clearly established as hereditary, with the corresponding consequence of a 45% risk for subsequent children. The risk for clients with healthy children must now be calculated on the basis of the total area *exclusive* of the *dark portion*. The hereditary cases are now no longer represented by 10/100 squares but only by 5.5/95.5.

Example: Huntington's Chorea. A healthy man at the age of 35 comes for genetic counseling; his father and his grandmother are affected with Huntington's chorea, and he is concerned about the risk for himself and for his future children. Huntington's chorea is an autosomal-dominant disease with full penetrance; age at onset, however, varies between about 20 and 70 years of age (Sect. 3.1.2; Fig. 3.4). If the proband showed signs of the disease, the problem would be simple. Each of his children would have a risk of 50%. If he has not yet reached the age of manifestation, the problem is also simple: he has a risk of 50%; the risk for his children is 50% of 50% or 25%. In fact, however, he has already passed through part of the manifestation period without being affected. This fact increases his chance of being homozygous for the normal allele and of remaining unaffected.

How does this situation influence the risk for his children? At the age of 35, about 30% of all heterozygotes have already shown clinical signs of the disease. This leads to the following calculation:

	Client heterozygous	Client homozygous normal
Prior probability	0.5	0.5
Conditional probability	0.7	1.0
Joint probability	$0.5 \times 0.7 = 0.35$	0.5
Posterior probability	$\frac{0.35}{0.5+0.35} = 0.412$	$\frac{0.5}{0.05+0.35} = 0.588$
Risk for child	$0.412 \times 0.5 = 0.206$	

Hence, the risk for the child has been reduced from 0.25 to 0.206.

Such calculations can be performed in many other specific situations in autosomal dominant and recessive diseases (for a detailed discussion, see [71]).

Heterozygotes of X-Linked Recessive Diseases. This type of calculation has its most important practical application in counseling of women who are at risk of being heterozygous for an X-chromosomal recessive trait and therefore having affected sons. Let us look at the pedigree in Fig. A8.2. We are virtually certain that Alma is heterozygous. Her daughter, Barbara, therefore, has a prior probability of 50% of being heterozygous also. This means a risk of $0.5 \times 0.5 = 0.25$ for any son that he will manifest the trait. If there is no further information the above values most form the basis of any counseling.

The situation is different in the pedigree of Fig. A8.3. In this case, Barbara already has a normal son. The conditional probability of having a normal son even though she is heterozygous is 0.5. The calculation proceeds as follows:

	Barbara carrier	Barbara not carrier
Prior probability	0.5	0.5
Conditional probability	0.5	1.0
Joint probability	$0.5 \times 0.5 = 0.25$	$0.5 \times 1.0 = 0.5$
Posterior probability	$\frac{0.25}{0.5+0.25} = 0.333$	$\frac{0.05}{0.5+0.25} = 0.667$
Risk that son will be affected $0.333 \times 0.5 = 0.167$		

The calculation is performed in a similar way if the pedigree is more complicated, i.e., if Barbara has a daughter and the daughter wants to know the risk for her sons, etc. In this case, the posterior probability of Barbara would be used to calculate the prior probability of her daughter. For a number of specific examples, see [71].

A basically *new* situation occurs when the carrier of the disease concerned is a sporadic case (Fig. A8.4). In this case, he may either be a new mutant, when, his mother is homozygous normal and there is no increased risk for sons of his sisters; or the mother is heterozygous, in which case his sisters also have a prior probability 0.5 of being heterozygous. As mentioned in Sect. 5.1.3.4, the fraction of new mutants among the bearers of a (rare) X-linked recessive condition is

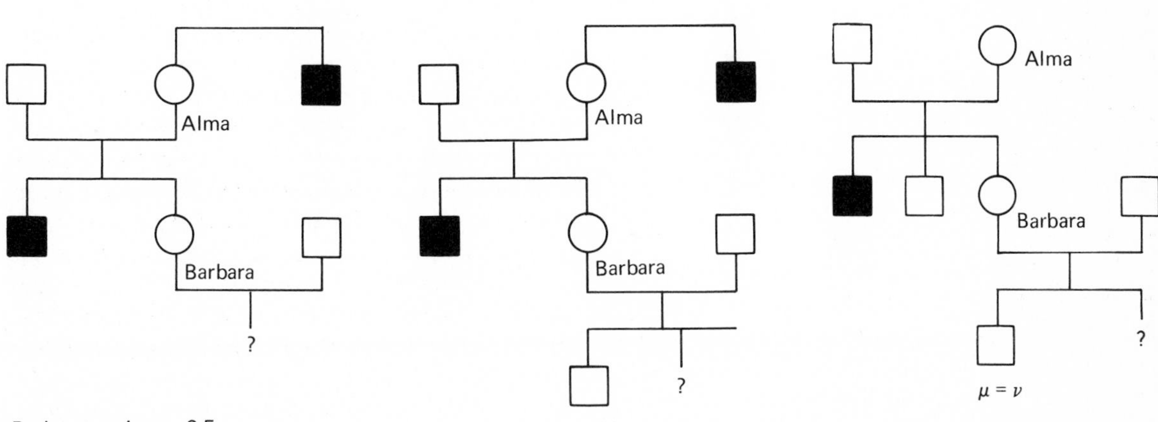

Fig. A8.2 **Fig. A8.3** **Fig. A8.4**

$$m = \frac{(1-f)\ \mu}{2\mu + \nu}$$

(f = relative fertility of trait bearers in relation to the general population; μ = mutation rate in female germ cells; ν = mutation rate in male germ cells). When mutation rates are equal in the two sexes, and when f = 0, the formula is reduced to m = ⅓. This means that the mother has a prior probability of two thirds of being heterozygous. This leads to the following risk calculation for Barbara's son:

	Barbara carrier	Barbara not carrier
Prior probability	$\frac{2}{3} \times \frac{1}{2} = \frac{1}{3}$	$\frac{2}{3}$
Conditional probability	$\frac{1}{2}$	1
Joint probability	$\frac{1}{3} \times \frac{1}{2} = \frac{1}{6}$	$\frac{2}{3}$
Posterior probability	$\frac{\frac{1}{6}}{\frac{1}{6}+\frac{2}{3}} = \frac{1}{5}$	$\frac{\frac{2}{3}}{\frac{1}{6}+\frac{2}{3}} = \frac{4}{5}$

Risk that son will be affected: $\frac{1}{5} \times \frac{1}{2} = \frac{1}{10}$

This simple calculation holds true only when the two conditions mentioned above (μ = ν; f = 0) are fulfilled and when there is a genetic equilibrium between mutation and selection. This appears to be true in Duchenne muscular dystrophy; the most common X-linked recessive disease in most centers. For other mutations such as hemophilia A and HPRT deficiency, mutation rates appear to be much higher in male than in female germ cells (Sect. 5.1.3.4). Here, the fraction m must be calculated on the basis of empirical evidence, ν = 10 μ being a useful approximation since the male mutation rate is about 10 × as high as that in females. When no specific data are available, assuming a prior probability 1 for Alma – and ½ for Barbara (Fig. A8.4) might be a good choice (it overestimates the risk slightly).

We shall now give only one slightly more complicated example; for other examples see [149; 71; 205; 90]. Figure A8.5 shows the pedigree. Barbara has an affected and a normal brother; but she also has a sister, Bettina, who is the mother of two normal sons. Bettina is either a normal homozygote (½ if Alma is heterozygous), in which case normal sons are to be expected, or she is heterozygous, in which case the conditional probability of having two normal sons is ¼. This aspect ($\frac{1}{2} + \frac{1}{4} \times \frac{1}{2} = \frac{5}{8}$) is included in the calculation of the conditional probability for Alma, in which her conditional probability of having an unaffected son if she is heterozygous is also considered. The evaluation goes as follows (again for μ = ν; f = 0).

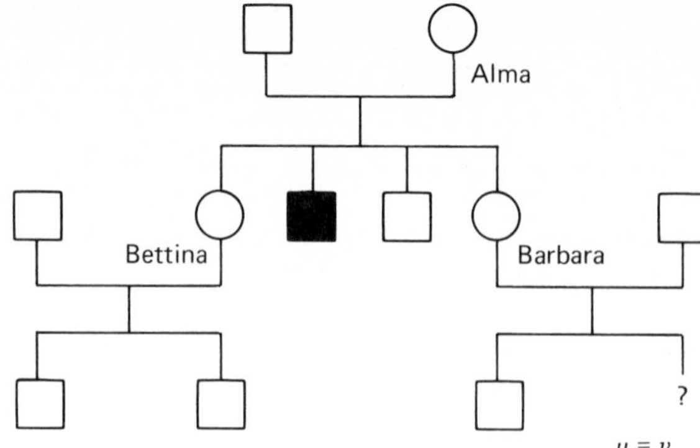

Fig. A8.5

	Alma carrier	Alma not carrier
Prior probability	$\frac{2}{3}$	$\frac{1}{3}$
Conditional probability	$\frac{1}{2} \times \frac{5}{8} = \frac{5}{16}$	1
Joint probability	$\frac{2}{3} \times \frac{5}{16} = \frac{5}{24}$	$\frac{1}{3} \times 1 = \frac{8}{24}$
Posterior probability	$\frac{5}{13}$	$\frac{8}{13}$

	Barbara carrier	Barbara not carrier
Prior probability	$\frac{5}{13} \times \frac{1}{2} = \frac{5}{26}$	$\frac{21}{26}$
Conditional probability	$\frac{1}{2}$	1
Joint probability	$\frac{5}{26} \times \frac{1}{2} = \frac{5}{52}$	$\frac{21}{26} \times 1 = \frac{42}{52}$
Posterior probability	$\frac{5}{47}$	$\frac{42}{47}$
Risk for Barbara's son:	$\frac{5}{47} \times \frac{1}{2} = \frac{5}{94}$	

These various calculations give the risk of the carrier state based on the pedigree information alone. In actual practice, additional information based on biochemical studies and DNA studies should be included to refine the risk. In Duchenne muscular dystrophy, CK values in heterozygotes are sometimes increased, but a clear-cut separation of heterozygotes and normal homozygotes based on CK values is impossible (for discussion of heterozygote tests in this disease, see Sect. 4.2.2.8). Similar considerations apply for the diagnosis of hemophilia A carriers using factor VIII and immunological assays. It is desirable to provide an optimum estimate of the genetic probability of a woman to be heterozygous for either one of these diseases by including the risk of the carrier state as determined by specific laboratory tests (see below).

Ideally, a central reference laboratory in a given region should perform the appropriate tests on a series of obligatory carriers as well as on normal women. These data and the degree of overlap between normals and carriers (Table A8.2 for muscular dystro-

Table A8.2. Likelihood of being a carrier for Duchenne muscular dystrophy at different levels of serum creatine kinase (CK). [Data from Sibert et al. (1979) Arch Dis Child 54: 534–537]

CK (iu/l)	Likelihood ratio (odds)	CK	Likelihood ratio (odds)
< 40	0.12	120–	12.79
40–	0.12	130–	25.12
50–	0.16	140–	49.02
60–	0.27	150–	94.34
70–	0.46	160–	180.9
80–	0.86	170–	342.5
90–	1.67	180–	641.0
100–	3.28	190+	<1000
110–	6.49		

These figures are based on an upper limit (95% of normal for adult females of 100 IU/l. The likelihood ratio represents the odds that a given CK value for an adult nonpregnant woman came from the obligatory carrier population as against the control population (i. e., at a CK of 100 IU/l the odds of the carrier state are 3.28:1. Note that equal numbers of normals and obligatory carriers are tested for construction of a table of this kind. The actual risks depend on the genetic risks as calculated from the pedigree (see text).

phy) are used to determine the odds (or likelihood ratio) that a given woman is or is not a carrier. However, *this information alone is not sufficient for counseling,* since the *final risk depends strongly upon the probability of the carrier state as established from the specific pedigree information* as explained above. This two-step procedure is essential and often not appreciated by laboratory-oriented personnel who use *only* the laboratory information for risk counseling. Thus, when the probability of being a carrier for an X-linked disease is low on the basis of family history (such as in a male patient's remote female relative), a somewhat elevated laboratory value makes the carrier state less probable than the identical value in a woman with a high probability of being a carrier (see below).

For practical purposes, the probability of the carrier state based on the pedigree information is first determined and is expressed in odds or likelihood ratio [odds $= p:(1-p)$ where p is probability. Example: if $p = 1/4$, the odds are $0.25:0.75 = 1:3$]. From the empirical information of CK values in carriers and normals (such as in Table A8.2), the odds of the carrier state are noted for the laboratory value obtained. The odds from the pedigree and laboratory information are multiplied and transformed into the actual probability or risk.

Example: Sister of male patient with Duchenne muscular dystrophy with affected uncle (Fig. A8.2). Sister's mean CK value following three determinations: 100 IU/l (95% of normal for adult females).

	Carrier	Not carrier
Pedigree odds	1	1
Lab odds*	3.28	1 (From Table A8.2)
Joint odds	3.28	1

* Odds based on actual laboratory findings

Probability of carrier state $= a/a+b = 3.28/3.28+1 = 3.28/4.28 = 0.77 = 77\%$.
Where a = odds of carrier state
b = odds of noncarrier state
The odds that this woman is a carrier have been increased from 50% to 77% by the CK value. Note that the probability of a woman being a carrier with the same laboratory value of 100 would only be 14% if her genetic risk (as calculated from the pedigree information) were 1:20. This is still a much higher risk than if there were no family history whatsoever. In that case, the risk of the carrier state is 1/2000 (carrier frequency in general population), and the actual risk (still based on a laboratory value of 100) is calculated to be 0.0016 or 1/628.

Quick Method by Dichotomizing Normals and Carriers by Laboratory Tests. Since complete information on laboratory values in known carriers as compared with the normal population often is not available, a less desirable "short cut" uses the information that about one-third of carriers of Duchenne muscular dystrophy have "normal" CK values (i. e., below 2 standard deviations from the population mean) or similar information in hemophilia. For instance, if normal values (regardless of the specific level of CK) are found, the risk for a carrier with a pedigree probability of 1/2 (Fig. A8.2) becomes 1/4 or 25% (Fig. A8.6):

	Carrier	Not carrier
Probability from pedigree information	1/2	1/2
Probability from normal CK value	1/3	1
Joint probability	1/6	1/2=3/6
Relative risk	1	3
Final risk	1/4=25%	3/4

Alternately the final risk of the carrier state can be calculated as the posterior probability: $1/6/1/6+3/6 = 1/4$ (see also Fig. A8.6).
The actual risk may be lower if the CK values are in the lower values of the normal range and higher for CK levels at the higher range. At a certain CK level (where the odds of the carrier/noncarrier are 1:1), CK values will not influence the risks of being a carrier.

Use of RFLP Information. The final assessment can be further refined by utilizing the three restriction fragment length (RFLP), polymorphisms that exist 5′ and 3′ to the gene of Duchenne muscular dystrophy. It is known that 45% of British women are heterozygous for at least one of these loci. Since each of these DNA markers are fairly far removed from the gene (13–20 cM), crossovers between the marker and the muscular dystrophy gene will be rela-

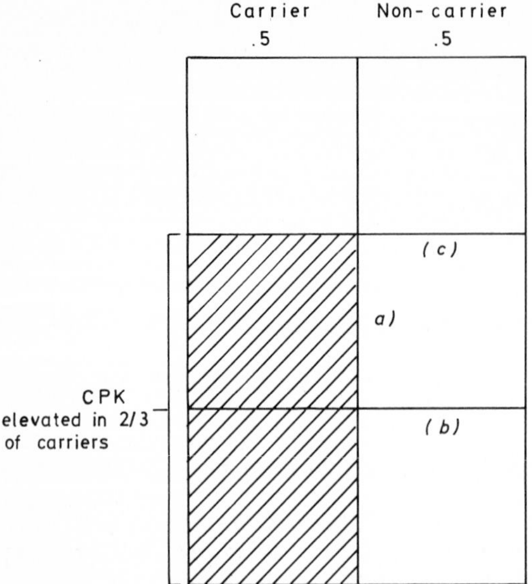

Fig. A8.6. Use of CK levels in Duchenne muscular dystrophy. *Example:* Pedigree information gives risk of carrier states as 0.5 and risk of noncarrier state as 0.5. The total probability "space" (see diagram) is divided into two equal sized portions by line *(a)*. Since two thirds of carriers have an elevated CK value, the probability space is divided by lines *(b)* and *(c)* into boxes of 1/3 size of the total. The carriers who have an abnormal CK level are indicated by the crosshatched area (=2/3 of carriers). Among those with normal CK levels 1 *(white box)* out of a total of 4 white boxes will be a carrier

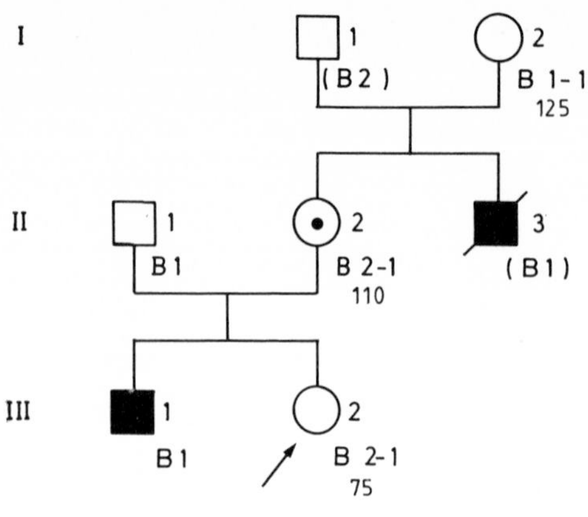

Fig. A8.7. A pedigree that is informative for an X-linked RFLP (RC8) in Duchenne muscular dystrophy. B1 refers to the more common allele and B2 to the less common allele at this locus. (B2) and (B1) indicate inferred genotypes. The numbers below the symbols refer to the CK levels. The B1 allele in the patient III-1 was transmitted from the carrier mother (II-2) who had obtained this allele from her mother (I-2). III-2, who is at 50% risk of being a carrier, obtained the normal B2 allele from her mother and a normal B1 allele from her father. III-2 therefore did not inherit the muscular dystrophy allele unless a crossover transferred the mutant gene to the chromosome bearing the B2 allele (15% chance)

tively frequent. However, adding the RFLP data to the pedigree and CK information will often provide further useful information that reduces the risk.

An example of use of all information is shown on Fig. A8.7. Subject III-2 has a brother and deceased uncle with Duchenne muscular dystrophy. She wants to know whether she is a carrier. Her CK value is 75 IU/liter, while those of her mother and maternal grandmother are 110 IU/liter and 125 IU/liter. Her pedigree risk of being a carrier is 50:50. Her odds of being a carrier based on the CK value are 0.46:1 (Table A8.1). First, the probability of the carrier state are calculated for pedigree and CK information as follows:

	Carrier	Not carrier
Probability from pedigree information	1/2	1/2
Probability from CK information	0.46	1 (From Table A8.1)
Joint probability	0.23	0.5

Final probability = 0.23/0.73 = 0.31 = 31%.

The RFLP data (RC8) in her family (Fig. A8.7) suggest that she inherited the normal X chromosome (B2) from her mother, although a crossover could have produced the observed pattern. Knowing the distance of the marker from the gene, her chance of being a carrier based on the RFLP information alone is about 15%, i.e., chance of a crossover.

Combining all information reduces her risk to 1.4% (0.31 = 0.15 = 0.014) and her risk of having an affected son to 0.7%. Sometimes the use of DNA variants can yield a great deal of information, as shown in Fig. 0.00, where there is a high probability that a woman who is at 50% risk of being a carrier (III-2) is not a heterozygote.

RFLPs for hemophilia A and B are also becoming available. It appears that practically all women are heterozygous for a gene-specific or X chromosomal segment specific probe for hemophilia A. If the family structure is otherwise favorable, RFLP carrier diagnosis in hemophilia A therefore will usually be possible with these tightly linked markers (Figs. A8.8, A8.9).

Recurrence Risks for Children of Unaffected Gene Carriers in Autosomal Dominant Inheritance. Occasionally, a clinically unaffected person who has relatives affected with an autosomal dominant disease of lowered penetrance wants advice regarding the disease risk for his or her offspring. Regardless of the exact penetrance, it has been shown that the risk for children of a person at 50% risk is never more than 9% [2366]. The reason for this is that the unaffected parent is unlikely to carry the gene in diseases with high penetrance. Conversely, in diseases with low penetrance, even though the parent is a gene carrier the chance of a child being clinically affected is small.

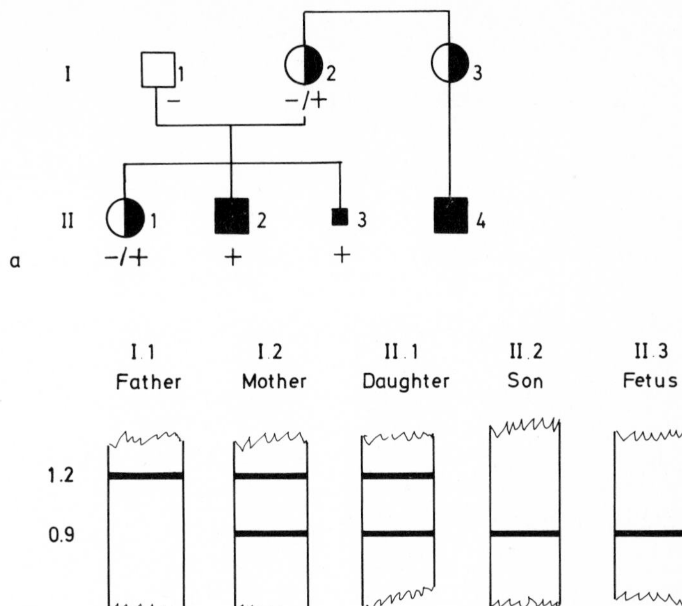

Fig. A8.8. a, b. Principle of heterozygote identification and prenatal diagnosis in hemophilia A. The mother (I,1) is double heterozygous for the hemophilia allele and the RFLP marker +. The father (I,1) is healthy and has the marker −. Since the affected son, II,2, has inherited the hemophilia allele as well as the marker + , from his mother the marker + must be on the same chromosome as the hemophilia allele (= coupling, Sect. 3.4). Since the daughter, II,1, is heterozygous for −/+, she must have inherited the chromosome containing + and the hemophilia gene from her mother; she must be heterozygous for hemophilia. The fetus II,2 is + positive; he must have inherited the same chromosome, and be affected with hemophilia

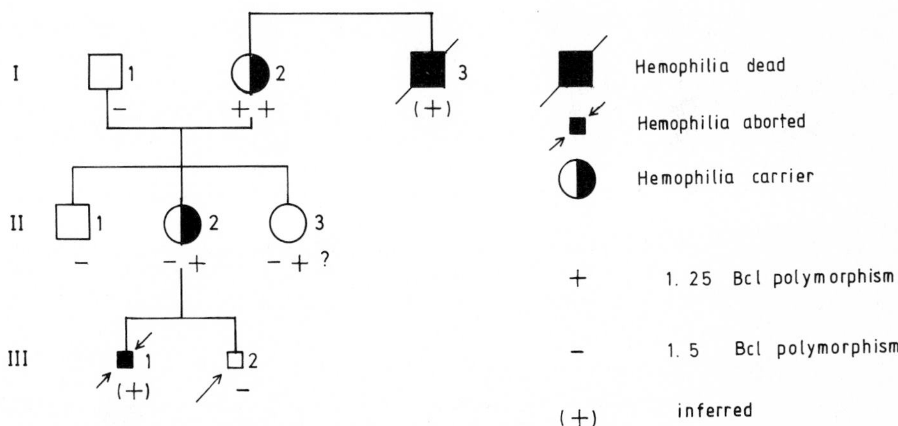

Fig. A8.9. Hemophilia A pedigree (Din et al., 1985). Note that hemophilia A could be ruled out in a fetus (III-2) by antenatal diagnosis using a factor VIII probe and restriction enzyme Bgl without actual testing the DNA phenotypes of the two hemophiliacs (III-1 and I-3). The presence of the DNA variant in the fetus (III-2) identical to that of the normal maternal grandfather (I-1) indicated that the fetus was normal. Note that the sister's (II, 3) carrier state cannot be determined using the DNA variant

Appendix 9

Examples for Linkage Calculations

In Sect. 3.4.2, methods for examining pedigree material for linkage have been described. It was explained that the method of lod scores is the method of choice; the derivation of this method from pedigree probabilities was shown. The tables and formulas for actual calculations can be found in various standard references [882, 796, 797, 798].

In the following two examples for linkage calculations will be given: one establishing moderately close linkage for two X-linked conditions, and another one showing close linkage for two autosomal traits.

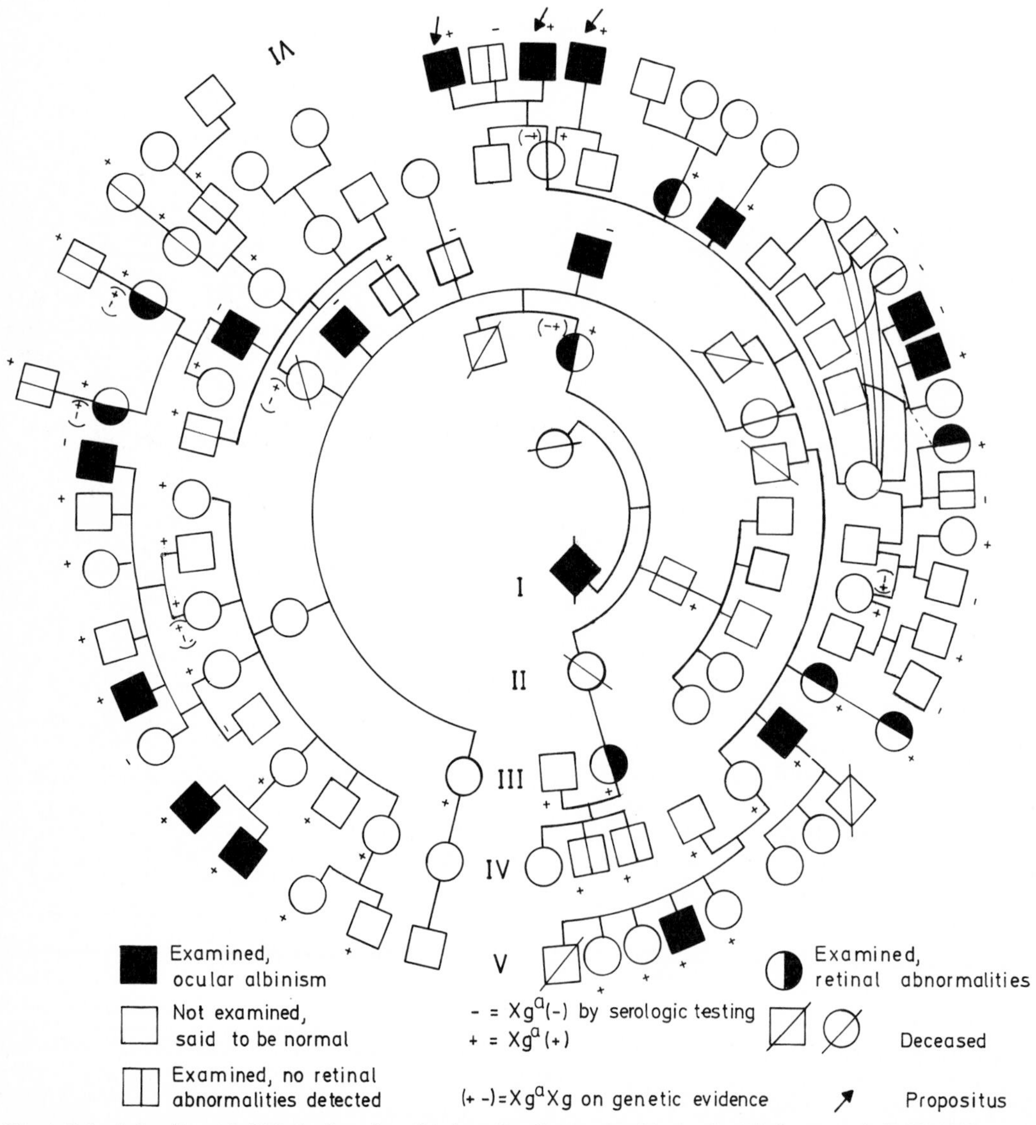

Fig. A.9.1. A family with X-linked ocular albinism (family M of Table A9.1). (Fialkow et al. 1967 [661])

Table A9.1. Lod scores for linkage of ocular albinism and Xg blood group [661]

Family	Mother	Children	Recombination fraction, Θ								
			0.05	0.10	0.15	0.20	0.25	0.30	0.35	0.40	0.45
M (Fig. A9.1)	II-3	$Z_1 3:1$, $e_1 3:1$	−0.483	−0.241	−0.126	−0.064	−0.030	−0.012	−0.003	−0.001	0.000
	III-10	$Z_1 2:0$, $e_1 1:1$	0.154	0.131	0.107	0.085	0.062	0.041	0.024	0.011	0.003
	IV-12	$Z_1 3:1$, $e_1 3:1$	−0.483	−0.241	−0.126	−0.064	−0.030	−0.012	−0.003	−0.001	0.000
	IV-19	$Z_1 4:0$, $e_1 3:1$	0.795	0.708	0.614	0.513	0.407	0.297	0.190	0.094	0.025
	IV-30	$Z_1 2:0$, $e_1 1:1$	0.154	0.131	0.107	0.085	0.062	0.041	0.024	0.011	0.003
	IV-31	$Z_1 1:1$, $e_1 2:0$	−0.584	−0.340	−0.215	−0.138	−0.087	−0.052	−0.028	−0.012	−0.003
	V-34	1 nonrecombinant	0.279	0.255	0.230	0.204	0.176	0.146	0.114	0.079	0.041
	V-35	1 nonrecombinant	0.279	0.255	0.230	0.204	0.176	0.146	0.114	0.079	0.041
W (Fig. A9.2)	IV-2	3 nonrecombinants	0.837	0.765	0.690	0.612	0.528	0.438	0.342	0.237	0.123
	IV-4	1 recombinant	−1.000	−0.699	−0.523	−0.398	−0.301	−0.222	−0.155	−0.097	−0.046
L (Fig. A9.3)	III-2	1 nonrecombinant	0.279	0.255	0.230	0.204	0.176	0.146	0.114	0.079	0.041
Sum of lod scores			0.277	0.979	1.218	1.243	1.139	0.957	0.733	0.479	0.041
Antilog (relative probability of Θ)			1.687	9.528	16.52	17.5	13.77	9,057	5.408	3.013	1.690

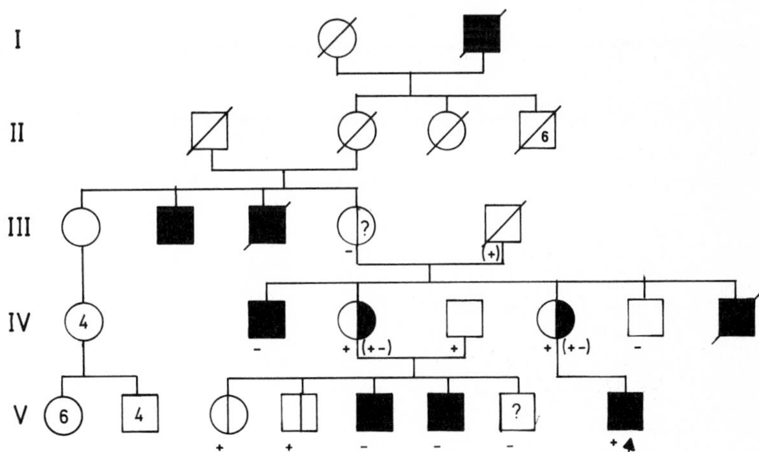

Fig. A9.2. A family with X-linked ocular albinism (family W of Table A9.1; symbols as in Fig. A9.1). (Fialkow et al., 1967 [661])

Example: Linkage Between the X-Linked Genes for Ocular Albinism and the Xg Blood Group [661]. Figures A9.1–A9.3 show three large pedigrees in which the X-linked genes for ocular albinism and for the Xg blood groups are segregating; 25 sons of 11 women known to be heterozygous for both ocular albinism and the Xg type give information for linkage. Table A9.1 classifies the sibships of the three pedigrees according to their types in the tables [882, 796, 797, 798] and the distribution of lod scores. The likelihood ratio $P(\Theta)$ of the pedigree for the recombination fraction Θ is calculated as follows:

$$\log P(\Theta)=z(\Theta)$$

$$P(\Theta)=10z(\Theta)$$

where $z(\Theta)$ is the tabulated lod score.

In Fig. A9.4, the lod scores are plotted with interpolation. The probability of the pedigree for linkage

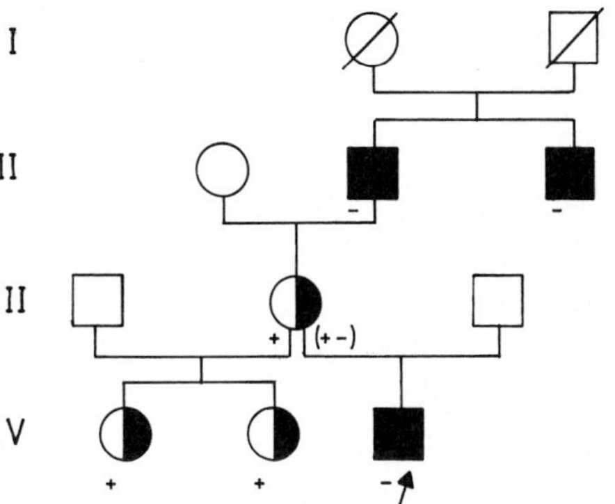

Fig. A9.3. A family with X-linked ocular albinism (family L of Table A9.1; symbols as in Fig. A9.1). (Fialkow et al., 1967 [423])

($\Theta = 0{,}175$) is 18 times higher than the probability for no linkage.

So far, the 11 sibships have been evaluated independently; the fact that they belong to three large pedigrees has been neglected. Inclusion of this aspect may help to extract considerably more information. It requires setting-up of individual equations for the three pedigrees. Computer programs for evaluations of such large pedigrees are available [831 a, b, c]. The results of such a computerized evaluation are shown in Table A9.2. In Fig. A9.5, the distribution of lod scores for different recombination fractions is compared for this computerized evaluation and for separate evaluation of 11 sibships. Again, the most likely recombination fraction Θ is 0.175; however, the pedigree probability for linkage is now 470 times that for no linkage. As expected, computer analysis has succeeded in extracting considerably more information from the data.

In 1980, Coneally and Rivas [612 a] gave some useful rules for linkage studies. As a rule, a *main locus* such as a rare disease with an autosomal dominant mode of inheritance is selected and tested for linkage by studying several *test loci,* i. e., frequent genetic polymorphisms. One parent should be heterozygous for both the test locus and the main locus. Hence, the loci to be studied should be chosen so as to maximize the occurrence of informative matings. Most heterozygotes for the test locus can be expected if frequencies of two codominant marker alleles are equal or when a recessive allele has a slightly higher frequency than its dominant allele.

Linkage studies are laborious and expensive. Sometimes, time and money can be saved if the analysis is initiated with a first screening step using preexisting information for selection of test loci and a second step of more detailed analysis. If two test loci are known to be relatively closely linked, only one should be examined. HLA testing, for example, is time-consuming and expensive. However, the *Bf* locus is closely linked with HLA, and can be tested much more easily. Therefore, this locus might be used for screening.

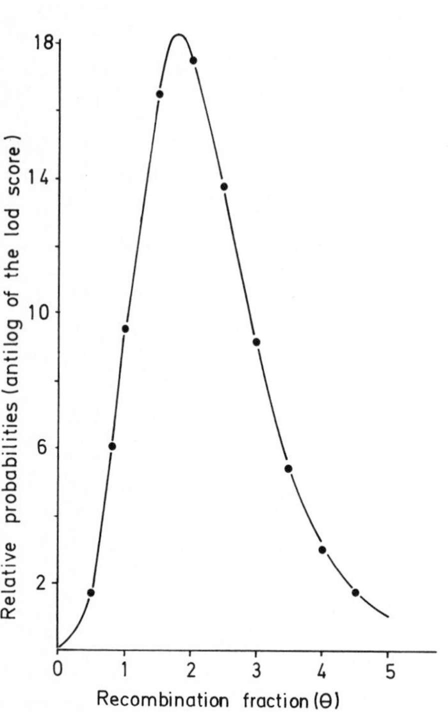

Fig. A9.4. Relative probability for linkage between gene loci for ocular albinism and Xg blood groups based on lod scores (Table A9.1). Note that the relative probability of linkage ($\Theta = 0.175$) is 18 times that of no linkage. (Fialkow et al., 1967 [661])

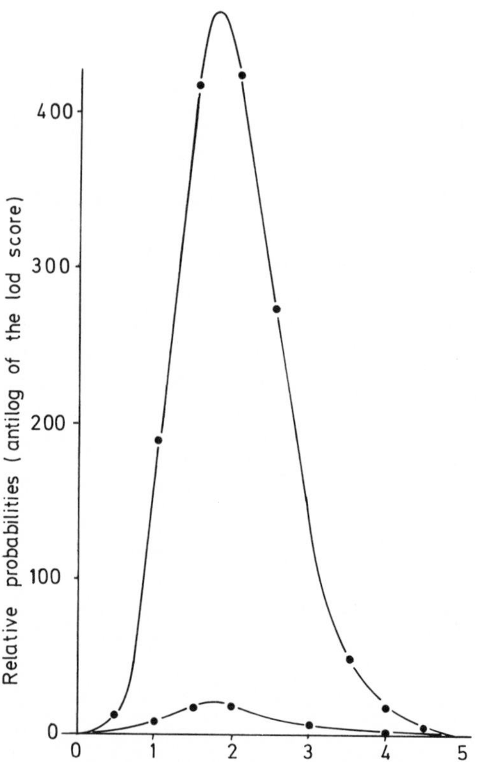

Fig. A9.5. Relative probability for linkage between gene loci for ocular albinism and the Xg blood group. Data obtained with use of a computer *(upper curve)* and with the conventional lod score method *(lower curve).* The *lower curve* is identical to the data in Fig. A9.4 but drawn on a different scale. Note that the computer has extracted considerably more information from the data. The recombination fraction remains 0.175, but the relative probability is 470 times that of no linkage. (Fialkow et al., 1967 [661])

Table A9.2. Lod scores (computer analysis)-linkage of ocular albinism and Xg blood groups

Family	Recombination fraction, Θ								
	0.05	0.10	0.15	0.20	0.25	0.30	0.35	0.40	0.45
M (Fig. A9.2)	0.8960	1.8371	2.1277	2.1369	1.9811	1.7139	1.3662	0.9579	0.5008
W (Fig. A9.2)	−0.1637	0.0668	0.1685	0.2144	0.2272	0.2165	0.1869	0.1406	0.0784
L (Fig. A9.3)	0.4264	0.3759	0.3256	0.2760	0.2272	0.1796	0.1331	0.0878	0.0436
Sum of lod scores	1.1587	2.2798	2.6218	2.6273	2.4355	2.1100	1.6862	1.1863	0.6228
Antilog (relative probability of Θ)	144.1	190.5	418.6	423.9	272.6	128.8	48.6	15.4	4.2

Table A9.3. Pedigree with dentinogenesis imperfecta and GC blood types

Θ_M	Θ_F									
	0.05	0.10	0.15	0.20	0.25	0.30	0.35	0.40	0.45	0.50
0.05	5.7385	7.0780	7.6460	7.8825	7.9238	7.8286	7.6263	7.3316	6.9506	6.4831
0.10	5.4992	6.8370	7.4048	7.6418	7.6841	7.5906	7.3908	7.0997	6.7240	6.2639
0.15	4.9709	6.3028	6.8690	7.1058	7.1488	7.0569	6.8596	6.5723	6.2019	5.7492
0.20	4.3034	5.6209	6.1826	6.4177	6.4605	6.3697	6.1747	5.8911	5.5262	5.0811
0.25	3.5484	4.8381	5.3897	5.6203	5.6613	5.5704	5.3772	5.0971	4.7376	4.3001
0.30	2.7353	3.9794	4.5123	4.7332	4.7690	4.6761	4.4831	4.2057	3.8512	3.4214
0.35	1.8921	3.0723	3.5733	3.7754	3.7997	3.6999	3.5039	3.2269	2.8764	2.4540
0.40	1.0653	2.1707	2.6243	2.7935	2.7940	2.6774	2.4705	2.1884	1.8390	1.4252
0.45	0.3170	1.3584	1.7647	1.8954	1.8641	1.7217	1.4957	1.2033	0.8581	0.4764
0.50	0.2835	0.7003	1.0796	1.1946	1.1545	1.0094	0.7890	0.5196	0.2383	0.0000

Table A9.4. Part of Table A.9.3, with finer subdivisions

Θ_M	Θ_F										
	0.20	0.21	0.22	0.23	0.24	0.25	0.26	0.27	0.28	0.29	0.30
0.01	7.1367	7.1579	7.1721	7.1797	7.1813	7.1771	7.1676	7.1529	7.1334	7.1092	7.0807
0.02	7.5746	7.5958	7.6100	7.6177	7.6193	7.6152	7.6057	7.5911	7.5716	7.5475	7.5190
0.03	7.7649	7.7862	7.8005	7.8082	7.8098	7.8058	7.7963	7.7818	7.7624	7.7384	7.7100
0.04	7.8524	7.8737	7.8880	7.8957	7.8974	7.8934	7.8840	7.8695	7.8502	7.8263	7.7979
0.05	7.8825	7.9038	7.9182	7.9260	7.9277	7.9238	7.9144	7.9000	7.8807	7.8569	7.8286
0.06	7.8757	7.8970	7.9114	7.9193	7.9211	7.9171	7.9079	7.8935	7.8743	7.8505	7.8223
0.07	7.8427	7.8642	7.8786	7.8865	7.8883	7.8844	7.8752	7.8609	7.8418	7.8181	7.7900
0.08	7.7902	7.8117	7.8261	7.8341	7.8360	7.8321	7.8230	7.8087	7.7897	7.7660	7.7380

With an increasing number of gene loci that can be included in a linkage study, the chances of detecting linkage increase. For example, if loci are uniformly distributed over the genome, and if $\Theta = 0.4$ is assumed as the maximum recombination fraction detectable by linkage studies, the probability of detecting at least one linkage is 0.35 with 20 marker loci, and 0.5 for 30 markers. Additional information might be derived from preexisting knowledge on assignment to specific chromosomes. For example, a linkage study with a rare skin anomaly (acrokeratoelastoidosis) brought positive but statistically nonsignificant lod scores for two markers; the rest of the markers gave negative lod scores. Normally, the study would have been regarded as inconclusive. However, since these two marker loci had been assigned before to chromosome No. 2, the main locus could also be assigned tentatively to this chromosome, and linkage with the marker loci could be assumed [688 a].

As mentioned in Sect. 3.4.2, the actual map distance of two loci can be derived from the estimate of Θ by using Fig. 3.26. With increasing density of the human gene map, however, linkages of three and more gene loci are being discovered in increasing number. The estimates of Θ between pairs of such loci have to be

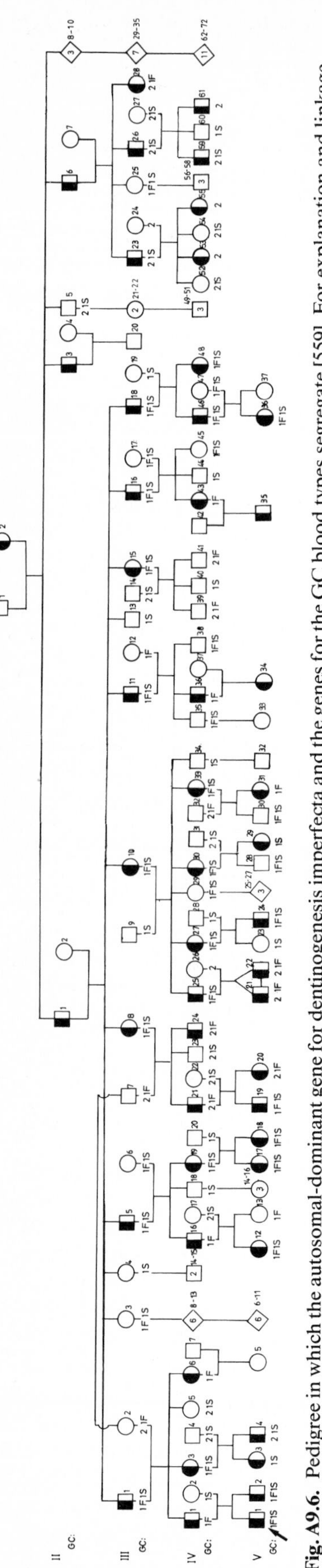

Fig. A9.6. Pedigree in which the autosomal-dominant gene for dentinogenesis imperfecta and the genes for the GC blood types segregate [559]. For explanation and linkage analysis see the text

combined for optimum mapping of the entire region. Rules and formulas for such multipoint mapping can be found in [612 a].

In recent years, an increasing number of restriction site polymorphisms (Sect. 2.3.3.9) have been introduced in linkage analysis. This analyis therefore assumcs an increasing importance in human and medical genetics. With these polymorphisms, multipoint mapping will become the rule rather than the exception.

Figure 9.6 shows a large pedigree in which an autosomal dominant condition, dentinogenesis imperfecta (12550), segregates together with the GC protein types [559]. There are three alleles, GC1S, GC1F, and GC2. Full evaluation of the pedigree implies calculation of probably genotypes for individuals who could not be typed for the marker (I, 1, 2; II, 1, 2) on the basis of gene frequencies. Then, the full pedigree can be scored. Scoring is difficult, but fortunately a computer program (LIPED) is available and a detailed working prescription can be obtained from the author, Dr. J. Ott[1] [831 a–c].

Table A9.3 gives the lod scores for Θ_M (recombination fraction in males) and Θ_F (recombination fraction in females). The maximum lod score (7.9238) is found at $\Theta_M = 0.05$ for males and $\Theta_F = 0.25$ for females. In Table A9.4, a more detailed calculation has been performed for the "critical" area of Table A9.3; the best estimates are $\Theta_M = 0.05$ and $\Theta_F = 0.24$; lod score $z = 7.9277$.

[1] Present address: Dr. J. Ott, Statistisches Amt der Stadt Zürich, Napfgasse 6, CH-8001 Zürich.

Table A9.5. The human gene map (McKusick, August 1, 1985, with additions). Human genes have been mapped using the following techniques:

1. F - Study of traits in families, e.g., AB0 blood groups and nail-patella locus
   Fc - Chromosome variant used, e.g., Duffy blood groups and uncoiler - 1
2. S - Segregation of cellular traits and chromosomes in clones from somatic cell hybrids
3. M - Microcell mediated gene transfer (MMGT), e.g., a collagen gene (COL1A1) to chromosome 17
4. C - Chromosome-mediated gene transfer (CMGT), e.g., cotransfer of galactokinase and thymidine kinase.
5. R - Irradiation of cells followed by "rescue" through fusion with nonirradiated (nonhuman) cell (Goss-Harris method of radiation-induced gene segregation); e.g., order of genes on Xq
6. A - In situ DNA-RNA or DNA-DNA annealing ("hybridization"), e.g., ribosomal RNA genes to acrocentric chromosomes; kappa light chain genes to chromose 2
7. HS - DNA/cDNA molecular hybridization in solution ("Cot analysis"), e.g., assignment of Hb beta to chr. 11 in derivative hybrid cells
8. RE - Restriction endonuclease techniques, e.g., fine structure map of non-alpha globin (NAG) region on (beta-globin cluster, HBBC) 11p; physical linkage of 3 fibrinogen genes (on 4q) and APOA1 and APOC3 (on 11p).
   a. Combined with somatic cell hybridization, e.g., NAG (HBBC) to 11p
   b. Combined with chromosome sorting, e.g., insulin to 11p.
   Includes Lebo's adaptation (dual laser chromosome sorting and spot blot DNA analysis), e.g., MGP to 11q
9. D - Deletion mapping (concurrence of chromosomal deletion and phenotypic evidence of hemizygosity), trisomy mapping (presence of three alleles in the case of a highly polymorphic locus), or gene dosage effects (correlation of triplicate state of part or all of a chromosome with 50% more gene product). Examples: acid phosphates-1 to chromosome 2; glutathione reductase to chromosome 8. Includes DNA dosage, e.g., fibrinogen loci to 4q2
10. AAS - Deductions from the amino acid sequence of proteins; e.g., linkage of delta and beta hemoglobin loci from study of hemoglobin Lepore. (Includes deductions of hybrid protein structure by monoclonal antibodies, e.g., close linkage of MN and Ss from study of Lepore-like MNSs blood group antigen)
11. LD - Linkage disequilibrium, e.g., beta and delta globin genes (HBB, HBD)
12. V - Induction of microscopically evident chromosomal change by adenovirus (probably represents change comparable to "puffing" in insects; accompanied by activation of kinases), e.g., adenovirus 12 changes on chr. 1 and 17
13. Ch - Chromosomal change associated with particular phenotype and not proved to represent linkage (Fc), deletion (D) or virus effect (V); e.g., loss of 13q14 band in some cases of retinoblastoma. ("Fragile sites", observed in cultured cells with or without folate-deficient medium or BrdV treatment, fall into this class of method, e.g., fragile site at Xq27 in one form of X-linked mental retardation. Fragile sites are useful as markers in family linkage studies, e.g., FS16q22 and haptoglobin
14. OT - Ovarian teratoma (centromere mapping), e.g., PGM3 and centromere of chr. 6
15. EM - Exclusion mapping, i.e., narrowing the possible location of loci by exclusion of parts of the map by deletion mapping, extended to include negative lod scores from families with marker chromosomes and negative lod scores with other assigned loci; e.g., support for assignment of MNSs to 4q
16. H - Based on presumed homology; e.g., assignment of LDHC to 12p

The certainty with which assignment of loci to chromosomes or the linkage between two loci has been established has been graded into the following classes:

C = Confirmed - observed in at least two laboratories or in several families (not used in the following lists but can be assumed when one of the other symbols is not given)
P = Provisional - based on evidence from one laboratory or one family
I = Inconsistent - results of different laboratories disagree
L = Limbo group - evidence not as strong as provisional, but included for heuristic reasons. (Same as "tentative" in other terminology)

Chromosome No. 1

1p36.3	RNA, U1 small nuclear (18068) REa, A
distal to 1p36.13	(P) Adenovirus-12 chromosome modification sites-1A (10292) V
1p (distal to Rh?)	(L) Retinitis pigmentosa-1 (18010) F
?1p (?linked to Rh)	(L) Malignant melanoma, cutaneous (15560) F (lod 2.0, theta 0.30)
1p (12cM distal to PGD?)	(P) Dombrock blood group (11060) F (linked to PGD only in female)
1p36.13-1pter	Glucose dehydrogenase (13809) S, F
1p36.13-1pter	Enolase-1 (17243) S, F, R
1p36.13-1pter	6-Phosphogluconate dehydrogenase (17220) F, S
1p36.12	(P) Atrial natriuretic factor (10878) REa, A, H
1p34-1q36	Oncogene src-2 (19013) A, REb

Table A 9.5 *(continued)*

1p32-1pter (?1p34)	(P)Neuroblastoma (25670) Ch
(1p32-1pter)	Elliptocytosis-1 (13050) F
1p32-1p36 (?1p36.11)*	Rhesus blood group (11170) F, D - Order: cen--E-C-D--1pter
linked to Rh (theta 0.044)	(P)Erythrokeratodermia variabilis (13320) F
?linked to Rh	(L)Lutheran inhibitor, InLu (11115) F
1p32-1pter	UDP galactose-4-epimerase (23035) S, LD
1p32	(P)Oncogene BLYM1: chicken bursal lymphoma (16483) A
1p34	Alpha-L-fucosidase (23000) S, F, R
?linked closely to FUCA1	(L)Alpha-L-fucosidase regulator, FUCT (13683) LD
(1p32-1p34)	Scianna blood group (11175) F
(1p32-1p34)	(P)Radin blood group (11162) F - ? same as Scianna
cen-PCM1-about	(P)C8 (beta polypeptide) (12096) F — not closely linked
8cM-C8-UMPK	(P)C8 (alpha-gamma polypeptide) (12095) F — not closely linked
1p32 (distal to PGM1)	Uridine monophosphate kinase (19171) S, R
1p34	Adenylate kinase-2, mitochondrial (10302) S, F, R
?distal to 1p22.1	(P)Transformation suppressor-1 (19019) S, H
1p22.1	Phosphoglucomutase-1 (17190) F, S, R
1pter-1p23	(L)Alcohol dehydrogenase (10372) S - ?control of expression (M. Smith, '84)
1p22.1 to distal 1p21	Nerve growth factor, beta- (16203) REa, H, A
1q21-1q23	Apolipoprotein A-II (10767) REa
1p21-1qter	(P)Renin (17982) REa
1p	Ferritin, heavy chain (13477) REa, REb - presumably pseudogene (also X et al.)
1p13 or 1q25	(L)Aspermiogenesis factor (10842) Ch
1p21	Amylase, pancreatic (10465) F, A, REa — Multiple genes
1p21	Amylase, salivary (10470) F, A, REa — Multiple genes
1p22 (?1p11-1p13,A)	Oncogene NRAS (16479) REa, A (?same as NCF)
1qh	Satellite DNA III (DIZI) (12637) A
1q21; ?linked to Duffy)	(L) Elliptocytosis-2 (13060) F — ?same locus
1q22-1q25	(P) Alpha-spectrin (18286) REa, A — ?same locus
1q (close to Fy)	Cataract, zonular pulverulent (11620) F
1q12-1qter	(P)Oncogene Sloan-Kettering, SK, chicken virus (16478) REa (=D1S3)
1p21-1qter	Actin, skeletal muscle alpha chains (10261) REa
(I)1q42; 1q21-1q31	Acid beta-glucosidase (glucocerebrosidase) (23080) S
Prox. 1p or 1q	(P)C-reactive protein (12326) REa — probably close
	(P)Amyloid P (10477) REa — probably close
1q12-q21 (distal to 1qh) (about 15 cM)	Duffy blood group (11070) F, Fc
	Charcot-Marie-Tooth disease, slow nerve conduction type (11820) F
1q2	(P)Histone Cluster B (H3, H4) (14278, 14275) REa, A
1q21-1q23	Uridyl diphosphate glucose pyrophosphorylase-1 (19175) S, R
1q22 (or 1q21)	(P)Adenovirus-12 chromosome modification site-1B (10294) V (class 1, U2 snRNA pseudogenes at this site)
(I)1q25; 1q42	Peptidase C (17000) S, R
1q42.1	Fumarate hydratase (13685) S, R, D
1q32.1-1q42	Guanylate kinase-1 (13927) S, D — Genetic independence unproven
	Guanylate kinase-2 (13928) S, D — Genetic independence unproven
1q42-1q43	5S ribosomal RNA gene(s) (18042) A — same site
1q42	(P)Adenovirus-12 chromosome modification site-1C (10293) V — same site
1q23.1-1q23.9	Antithrombin III (10730) F, D, A, REa (about 10cM from Fy)
	(P)5-Methyltetrahydrofolate: L-homocysteine S-methyltransferase (tetrahydropteroyl-glutamate methyltransferase) (15657) S
1p22.1-qter	(P)Succinate dehydrogenase (1 of 2 polypeptides) (18547) S
1cen-1q32	(P)Phosphofructokinase, muscle type (23280) S
1q	(P)Xeroderma pigmentosum A (27870) S
	(P)Uroporphyrinogen decarboxylase (17610) S
1q31-1q321 (at junction)	(L)Microcephaly, true (25120) Ch
	(P)Complement component Clq, B chain (12057) REa
	(P)Cystic fibrosis antigen (21970) S

In addition: 2 anonymous DNA segments, 5 O'Farrell protein spots, and 2 antigens defined by monoclonal antibodies MSK1, 15801; MKS2, 15802).

Table A 9.5 *(continued)*

Chromosome No. 2

	(P)Apoliporotein B (20010) REa
?2p	(P)Carbamoylphosphate synthetase I (mitochondrial CPS) (23730) REa
2p23-2p24	Oncogene NMYC (16484) REa, A
2p	Aryl hydrocarbon hydroxylase (10833) S
(2p23)	(L)Aniridia (10620) F
2p25.1-2pter	(L)Coloboma of iris (12020) Ch
(2p)	(L)Acrokeratoelastoidosis (10185) F (?linked to ACP1, Jk, IGKC)
(I)2p25; 2p23	Acid phosphatase-1 (17150) D, S
?2p23 (close to ACP1?)	Proopiomelanocortin (17683) REa
2p23	Malate dehydrogenase, soluble (15420) S
2p11-2p22	(P)Galactose enzyme activator (13703) S
	(P)Immunoglobulin heavy chain attachment site (14710) S
	(P)Acetylcholinesterase regulator, or derepressor (10068) D
2p23-2q32?	Adenosine deaminase complexing protein-2 (10272) S
2p11 (distal to IGK)	(P)Leu-2 T-cell antigen (T8 lymphocyte antigen) (18691) REa, A
	(P)Uridyl diphosphate glucose pyrophosphorylase-2 (19176) S
2p12 (2p11.2 by high resolution *in situ* mapping) Order: cen-V-J-C-ter.	Immunoglobulin kappa Light Chain Gene Family (Km; Inv) A, REa
	Variable region of kappa light chain (many genes) (14698)
	J region of kappa light chain (several genes) (14697)
	Constant region of kappa light chain (14720)
about 23cM	(P)Kidd blood group (11100) F
?linked to Jk (lod 2.15, theta 0.14 male, 0.27 female)	(L)Optic atrophy, Kjer type (16550) F
linked to Jk (lod 3.8 at theta 0.29)	(L)Colton blood group (11045) F (chr. 7 suggested by dosage effect in monosomy 7, *Hum. Genet. 62*:40, 1982)
2q14.3-2q21.3	(L)Gardner syndrome (17530) Ch
2q32-2qter	Ribulose 5-phosphate 3-epimerase (18048) S
2q32-2qter	Isocitrate dehydrogenase, soluble (14770) S
	Fibronectin-1 (13560) S (prob. structural gene; see chr. 8, 11)
2p25	(L)Cataract, anterior polar (11565) Fc (see 14q24)
2p23-qter	(P)Interferon, bet-3 (see 14764) S
2q36-2q37	Glucagon (13803) REa, A
2q31-2qter	(P)Elastin (13016) A
2q24.3-2q31	(P)Collagen V, alpha-2 chain (12019) REa, A
2q31-2q323(2q24.3-2q31)	Collagen III (alpha-1) (12018) REa, A

In addition: 2 anonymous DNA segments (HGM 7) and 1 surface polypeptide (18561).

Chromosome No. 3

	(P)Dihydrofolate reductase pseudogene-4 (see 12606) REa
3p25	(P)Oncogene RAF1 (16476) REa
3p21-3cen	Beta-galactosidase-1 (23050) S
?3p21.1 (or 7p13)	(L)Greig craniopolysyndactyly syndrome (17570) Ch
3p21	Aminoacylase-1 (10462) S
3p14-3p23	Small-cell cancer of lung (18228) Ch, D
3p14.2	(L)Renal cell carcinoma (14470) Fc
3p13-3q12	Glutathione peroxidase-1 (23170) S
3cen-3q21	(P)Orotate phosphoribosyltransferase/OMP decarboxylase (UMP synthase) (25892) (S)
	(P)Melanoma-associated antigen p97 (15575) S ┐ ?related or identical
3q23-3qter	Transferrin receptor (19001) S, H, REa ┘
	(P)Ceruloplasmin (11770) F, H ┐ (m, F)*
3q21-3q25	Transferrin (19000) S, H, REa ┘
	(P)Pseudocholinesterase-1 (17740) F ┐ (m, F)*
3q28	Somatostatin (18245) REa ┘
3cen-3q13	(P)Alpha-2HS-glycoprotein (13868) F**, S, D
	(P)Temperature sensitive (tsAF8) complement (11695) S
	(P, I) Herpes virus sensitivity (14245) S (see chr. 11)

In addition: 4 anonymous DNA segments (HGM 7).

* See Section I, B for definition.

** Probable order = cen-TF-CHE1-A2HS.

Table A 9.5 *(continued)*

Chromosome No. 4

about 15cM	(L)Alpha-fucosidase-2 (13682) F (linked to PLG; lod 7.37, male theta 0.12)
	(L)Plasminogen (17335) F (far from MN; lod 2.35, male theta 0.30, with GC)
4p14-4qter	(P)Formaldehyde dehydrogenase (13649) S
4p12-4q12	Peptidase S (17025) S, EM
4pter-4q21	(P)Phosphoribosylpyrophosphate amidotransferase (17245) S
4p (far from MN, GC)	Huntington disease (14310) Fd
	(P)D4S10 (G8 DNA segment linked to HD, q.v.) REa
(eq; about 25 cM from MNSs)	Stoltzfus blood group (11180) F
(4q; linked to MNSs)	Sclerotylosis (18160) F
about 11cM	Dentinogenesis imperfecta-1 (12549) F
4q11-4q13	Group-specific component (13920) F, Fc, H, D
?linked to GC	(L)Von Recklinghausen neurofibromatosis (16220) F
4q28-4q31 (by EM)	MN blood group (11130) F, Fc, AAS (male lod 3.79, theta 0.32, GC)
	Ss blood group (11174) F, Fc, AAS
(Linked to MN)	(P)Anterior segment mesenchymal dysgenesis (10725) F
4q23-4q27	(L)Rieger syndrome (18050) Ch
4p12-4q13	Phosphoglucomutase-2 (17200) S
4q11-4q13	Albumin (10360) F (linked to GC), A, REa — 5′-ALB-3′--5′-AFB-3′
4q11-4q13	Alpha-fetoprotein (10415) H, A
	Fibrinogen gene cluster (FGC)
4q2(4q29-4q31)	Fibrinogen, alpha chain (13482) RE, REa, H, D
4q2	Fibrinogen, beta chain (13483) RE, REa, D
(Linked to MNSs), 4q2	Fibrinogen, gamma chain (13485) F, REa, H, RE, D
4q21-4qter	(P)Aspartylglucosaminidase (20840) S
	(P)Temperature sensitivity complementation, ts13 (18732) S
	(P)Quinoid dihydropteridine reductase (26163) S
	(P)Leukocyte antigen group five (LAG5) (15145) S
4q21-4qter	(P)Alcohol dehydrogenase, class I (10370) REa (cluster of ADH1, ADH2, ADH3 loci determining alpha, beta, and gamma chains)
	(P)Alcohol dehydrogenase, class III (ADH Ch1; ADHX) (10371) S
4q21-4qter	Epidermal growth factor (13153) REa, H
4q26-4q28	(P)T-cell growth factor (interleukin-2) (14768) REa
	(P)Oncogene RAE2 (see 16476) REa - processed pseudogene

In addition: 3 anonymous DNA segments and 1 antigen defined by a monoclonal (HGM 7).
Possible order: PLG - CEM - GC - DGI - Sa - MN - FGG

Chromosome No. 5

5pter-5q1	Leucyl-tRNA synthetase (15135), S, H
5q11-5q13 (not q12)	Arylsulfatase B (25320) S
5p	(L) Antiviral state repressor requlator (10747) (S) - not in MIM
5p	(P) Interferon, beta-2 (see 14757) S
5q11-5q22 (5q23)	Dihydrofolate reductase (12606) S, REa, H (to other chrs. with amplification)
5q13	Hexosaminidase B (14265) S, Ch
5q15-5qter	Diphtheria toxin sensitivity (12615) S
5q34	Oncogene FMS (McDonough feline sarcoma) (16477) REa
5q35	Chromate resistance (sulfate transport) (11884) S
5q31-5q35	Emetine resistance (ribosomal protein S14) (13062) S
5q	(L) Macrocytic anemia, refractory (15355) Ch
	(P) Beta-adrenergic receptor (10969) S
	(P) Glucocorticoid receptor, lymphocyte (23157) S
	Histocompatibility: type II antigens, gamma chain of (14279) S, REb
	(P) 3-hydroxy-3-methylglutaryl coenzyme A (HMG COA) reductase (14291) REa

In addition: 3 anonymous DNA segments (HGM 7) and 1 surface polypeptide (18561).

Table A 9.5 *(continued)*

Chromosome Nr. 6		
6pter-6p23	(P)Clotting factor XII (Hageman factor) (23400) D	
10cM distal to F13A	(P)Malic enzyme mitochondrial (ME2) (15427) F	
6q22 (distal to HLA)	(P)Clotting factor XIII, A component (13457) F (male theta, 17 vs. HLA)	
6p21.3	Major histocompatibility complex (MHC) F, S, A, RE, Ch***	
No crossover with BP	Complement component-2 (21700) F, LD - 2% recomb. with HLA-B*	
	Complement component-4F, or C4B (12028) F, H - 10kb from C4S	
	Complement component-4S, or C4A (12081) F, H (on HLA-B side of C4B)	
	(L)Neuraminidase-1 (16205) H	
No crossover with C2	Properdin factor B (Bf) (13847) F - less than 1kb from C2, 30kb from C4	
Linked to C2, C4, BF	Con. adrenal hyperplasia: 21-hydroxylase deficiency; P450(C21) (20191) F	
Near HLA-A end	(P)Mixed lymphocyte reaction, weak (15786) F	2 loci, A and B** only B active
Near HLA-D	Primed lymphocyte test-1 (17668) F	
	(L)Pagweed sensitivity (17945) F	
In A/B segment	(P)Blastogenic response to synthetic polypeptides (14695, 14696) F	
In B/D segment	(P)Blastogenic response to synthetic polypeptides (14681, 14682) F	
	(P)Immune suppression (IS) (14685) H	
6p23-6q12 (prob. close to HBA)	Phosphoglycerate kinase, testicular: PGKA or PGK2 (17227) REa, H (homologous to T region gene of mouse)	
Close to HLA-A	Hemochromatosis (23520) LD, F	
	(P)T-complex locus TCP-1 (18698) REa, H	
	(P)Neutrophil differentiation factor (2070) LD	
	(L)F9 embryonic antigen (13701) H	
Linked to C3BR	(L)Complement component-4 binding protein (12083) H (in MHC in mouse, not closely linked in man.	
Not close to HLA	(P)Complement component C3b, receptor for: C3BR (12062) S	closely linked to C4BP; location on 6 in limbo
	(P)Complement component C3d, receptor for: C3DR (12065) S	
	(P)Complement factor H (13437) F	
	(P)Atrial septal defect, secundum type (10880) F	
6p21.3-6p21.2 (about 3cM from HLA)	Glyoxalase: (13875) F, S	
(p(?linked to MHC)	(L)Corticosterone side-chain isomerase (12255) H	
6p23-6q12	Prolactin (17676) REa (?between 6cen and (GLOI)	
6p23-6q12	(P)Insulin-like DNA sequence (14749) REa	
6p23-6q12	(P)Initiator methionine tRNA (18062) REa (2 of 12+ genes on 6)	
6q12	Malic enzyme, soluble (MEI) (15425) S	
?between GLOI and PGM3	Spinocerebellar ataxia-1 (16440) F - 15% male recomb./HLA	
6q12	Phosphoglucomutase-3 (17210) S, F, OT	
6q12-6q21	(P)Chorionic gonadotropin, alpha chain (shared with LH, FSH, TSH) (11885) REa	
6q21-6q22 (6q22-6q24)	Oncogene: avian myeloblastosis virus, MYB (18999) S, A	
6q23-6q12 (6p11-6p12,A)	Oncogene: Kirsten rat sarcoma virus-1, KRAS1 (19011) S - pseudogene	
	(P)Oncogene: yes-2 (16489) REa	
6q21	(P)BKM (banded krait minisatellite) DNA (related to heterogametic sex) (10978) A	
6q21	Superoxide dismutase-2, mitochondrial (14746) S25	
	(L)Insulin dependent diabetes mellitus (22219) F, LD	
	(L)Manic-depressive illnes (12548) F	
	(L)Pepsinogen (16970) F, H	
	(L)Paget disease of bone (16725) F	
	(L)Laryngeal adductor paralysis (15027) F	
	(L)Plasminogen activator tissue (17337) S	
	Baboon M7 virus replication, GEVI (19018) S	
	(P)Monkey RBC receptor (15805) S	
	(I) Adenosine deaminase complexing protein-1 (10271) S	
	(P)Argininosuccinate synthetase pseudogene-2 (10784) REa (others on 8 or more other chrom. incl. X and perhaps Y)	
	(P)Dihydrofolate reductase pseudogene-2 (see 12606) S	

Table A 9.5 *(continued)*

Chromosome No. 7	
7pter-7q22	(P)Actin, cytoskeletal beta (10263) REa about 20 pseudogenes
7p14-7pter	(P)Gamma-glutamylcyclotransferase (13717) S
7p11-7q11	(P)Asparagine synthetase (10837) S
?7p	(L)Goldenhar syndrome (14140) Ch
7p21-7q22	(P)Argininosuccinate lyase (20790) S
7pter-7q22	Phosphoserine phosphatase (17248) S
7pter-7q22	(P)T-cell antigen receptor, gamma subunit (18697) REa
7p14-7cen	Biliverdin reductase (10975) S
	Uridine phosphorylase (19173) S
?7p13 (or 3p21.1)	(L)Greig - polysyndactyly craniofacial dysmorphism syndrome (17570) Ch
7p22-7q22	Malate dehydrogenase, mitochondrial (15410) S
7p13-7p22 (?p21-p22)	Epidermal growth factor receptor (same as S7) (13155) S
7pter-7q22	Oncogene ERBB (19014) REa (same as EGFR?; similar sequence)
7p11.4-7qter	(P)Oncogene MET (16486) REa
7ren-7q22,	Beta-glucuronidase (25322) S, D
7q21.3-7q22.1 (A)	Collagen I (alpha-2) (12016) S, REa, D
	(P)Hydroxyacyl CoA dehydrogenase (14345) S
7q22-7qter	Neutrophil migration (NM; granulocyte glycoprotein; GP130; formerly neutrophil chemotactic response, NCR) (16282) D
7q22-7qter	(P)Trypsin-1 (27600) REA ⎤ both serine proteases
7q22-7qter	(P)Carboxypeptidase A (11485) REa ⎦
7q22-7qter	(P)Actin, cytoskeletal beta, pseudogene-5 (10264) (about 20 in all: 1 on X chr.; 2 on chr. 5; 3 on chr. 18; 4 on chr. 5; etc.)
7q21-7q36	Histone Cluster A (H1, H2A, H2B) (14271, 14272, 14276) A
	(P)Nonhistone chromosomal protein-1 (11887) S
7q35	T-cell antigen receptor, beta subunit (18693) REa, A
	(P)Growth rate controlling factor-1 (13922) S
	(L)Diaphorase-2 (12587) S
	(L)Colton blood group (11045) D (but see chr. 2)
	(L)Elastin (13016) REa

In addition: 5 anonymous DNA segments, 1 surface antigen (18552), and 2 O'Farrell protein spots (HGM 7).

Chromosome No. 8	
	(L)Fibronectin (13560) S (?concerned with expression on cell surface)
8p11	(P)Spherocytosis (18290) F, Ch (linkage with t8;12 and t3;8; see chr. 14) Glutathione
8p21.1	reductase (13830) S, D
?8q	(P)Clotting factor VII expression (13445) D
8q23.3 (8q24.1)	Langer Gledion syndrome (15023) Ch
?8q24.1	(L)Trichorhinophalangeal syndrome; type I (19035) Ch ⎤ ?deletion of both
	(L)Multiple exostoses (13370) Ch ⎦ loci in LGS
?8q13-8qter (EM)	(L)Glutamate-pyruvate transaminase, ?soluble liver (13822) S, EM (see Chr. 16)
8q21.1-8qter	(P)Glycine auxotroph B, complementation of hamater; GLYB (13848) S
8q22	Oncogene: Moloney murine sarcoma virus, MOS (19006) S
8q24.3	(L)Neoplastic lymphoproliferation-1 (16184) Ch
8q24(cen-5′-3′-ter)	Oncogene MYC: avian myelocytomatosis virus (19008) A
8q24(distal to MYC)	Thyroglobulin (18845) S, A
	(P)Carbonic anhydrase II (11481) REa (2 forms of CA linked in Old World monkey and mouse)

ANLL, M2 -t(8;21) (q22.1;q22.3) Burkitt and ALL, L3-t (8;14) (q24.13;q32.33). Yunis, *Science: 221*:227, 1983. Burkitt also t(8;2) (q24;p12), t8;22) (q24;q11)

Table A 9.5 *(continued)*

Chromosome No. 9

Location	Locus	Remarks
	(P)Aldolase B; fructose-1-phosphate aldolase (22960) REb	
	(P)Coproporphyrinogen oxidase (12130) S	
	(P)Temperature sensitivity complementation, H142 (18729) S	
9pter-9q12	(P)Relaxin, H1 (17973) S	cluster
9pter-9q12	(P)Relaxin, H2 (17974) S	
9pter-9q12	(P)Methylthioadenosine phosphorylase (15654) S	
9q13-9p24	Adenylate kinase-3, mitochondrial (10303) S, D	
9p13-9p22	Aconitase, soluble (10088) S	
9p13-9p21	Galactose-1-phosphate uridyltransferase (23040) S, D, F	
9p21(distal to IFL)	Interferon-F (beta/or fibroblast interferon) (14764) REa, A	very closely linked (Fd, LD) IFF duplicate in some
9pter-9p13	Leukocyte interferon gene family (Interferon-L or alpha) (14766) REa, A	
9qh (9q12)	(P)Cytoplasmic membrane DNA (12633) A	
9cen-9q34	(P)Polylpolyglutamate synthetase (13651) S	
9cen-9qter	(P)Glucose-regulated protein, GRP78 (13812) REa	
9q14	Argininosuccinate synthetase (21570) S, D (14 pseudogenes on 11 chrs.)	Linked to AB0 definitely or provisionally. Linkage group assigned to No. 9 by assignment of AK1
(9q14)	AB0 blood group (11030)F	(Su, *AJHG 36:*954, '84)
9q14	Adenylate kinase-1, soluble **(10300) F, S, D	m*.
	Nail-patella syndrome (16120) F	
	Delta-aminolevulinate dehydratase (12527) F, S	
	Orosomucoid (13860) F	
	(L)Dopamine-beta-hydroxylase (22336) F (lod 2.32, RF 0.0 with AB0)	
	(L)Waardenburg syndrome-1 (19350) F	
	(L)Multiple exostoses (13370)	
	***Oncogene: Abelson strain, murine leukemia virus (18998) REa	
	(P)Immunoglobulin epsilon heavy chain pseudogene: IGE2 (14721) A	

In addition: 2 anonymous DNA segments and 4 O'Farrell protein spots (HGM 7).
** Proximal to break in Ph^1 rearrangement (9q32.1). *** Distal to 9q34.1.

Chromosome No. 10

Location	Locus
10p14-10p15	(P)Interleukin-2 (T-cell growth factor) receptor (14768) REa, A
10pter-10p11.1	Phosphofructokinase, platelet type (17184) S
10p11.2	Hexokinase-1 (14260) S, D
?10q11-10q24 (not on 10p)	Adenosine kinase (10275) S, D, EM
10q11.1-10q24	Inorganic pyrophosphatase (PP) (17903) S, D
10q25.3-q26.1	Glutamate oxaloacetate transaminase, soluble (13818) S, D, H
	(P)Glutamate-gamma-semialdehyde synthetase (13825) S
	(P)Polykaryocytosis promoter (FUSE) (17475) S
	(P)External membrane protein-130 (13371) S
?10q24-10q25	(P)Lysosomal acid lipase-A (27800) S, H
	(I) Chorionic gonadotropin (11885) S (also see chr. 18; not structural locus)
	(P)Oligomycin resistance (mitochondrial ATPase, ATPM) (16436) S
10q25.3-10q26.1	(P)Phosphoglycerate mutase A (17225) D, H
	(P)Glycoprotein neuraminidase (25655) S
10q24-10qter	(P)Urokinase (19184) REa, A
	(P)Ornithine aminotransferase (25887) S
	(P)Sphingolipid activator protein-1 (24990) S

In addition: 2 anonymous DNA segments and 2 O'Farrell protein spots (HGM 7).

Table A9.5 *(continued)*

Chromosome No. 11	
11p	(P)Tyrosine hydroxylase (19129) REa, A
11pter-11q12	(P)Cathepsin D (11684) S
11p15(11p14.1-11p14.2*)	Insulin (17673) S, A, REb
11p11-11p15	Insulin-like growth factor II, or somatomedin A (14747) REa-separate gene for variant (14741)
11p13-11p15	(L)Beckwith-Wiedemann syndrome (13065) Ch (duplication)
11p15.1-11p15.5 (11p14.1-11p14.2*)	Oncogene HRAS1 (Harvey rat sarcoma-1) (19002) S (pseudogene HRAS2 on X) (distal to insulin and beta-globin)
11p15(11p11.22*) (polarity: centromere-- 5′-HBE-3′---5′-HBG-3′ --////-5′-HBB-3′;; same as ontogenetic order)	Non-Alpha Globin Cluster (NAG; hemoglobin beta cluster; HBBC)** — S, HS, REa
	Hemoglobin epsilon (14210) AAS
	Hemoglobin gamma 136 glycine (14225)
	Hemoglobin gamma 136 alanine (14220)
	Hb gamma regulator (14227) RE
	Hemoglobin delta (14200) AAS
	Hemoglobin beta (14190) LD, AAS, F
	Hpa I recognition polymorphism (14302) RE
	(Cluster: S, HS, REa)
about 15cM from HBB	F-cell production (heterocellular hereditary persistence of fetal hemoglobin) (14247) F — Order: Cen--PTH-7cM-HBBC-8cM-HRAS1-4cM-INS--ter. HBBC-11cM-INS-pter. (Fearon et al., *Am. J. Hum. Genet. 36:*329, 1984)
Proximal to HBBC (11p11.21*)	Parathyroid hormone (16845) REa
11p13-11p15	Calcitonin (11413) REa, A - also calcitonin gene related peptide, CGRP
11p13	Wilms tumor/aniridia/gonadoblastoma/retardation (WAGR) complex (19407) Ch
11p13	(L)Aniridia-2 (10621) Ch
11p1305-11p1306	Catalase (11550) S, D (distal to WAGR; about 10 cM from NAG)
11p1203-11p1208	Lactate dehydrogenase A (15000) S
11p11-11q12	Acid phosphatase-2 (17165) S
	(P) BVIX (BALB virus induction, xenotropic) (11399) S
	(P) Herpes virus sensitivity (14246) S (see chr. 3)
11q13	(P)B-cell leukemia-1; BCL1 (15140) RE
11q13	Apolipoprotein Cluster I
11q13	Apoliporprotein A-I (10768) REa, RE — diagram of region. Protter et al., DNA, 3: 449, 1984.
	Apolipoprotein C-III (10774) REa, RE (2.6kb 3′ to APOAI), F
(11p13)	Apolipoprotein A-IV (10769) F
11q13-11q14	(P)BKM (banded krait minor satellite) DNA (related to heterogaemtic sex) (10978) A.
11q13-11q22	Esterase-A4 (13322) S
11p13-11q22	Glutathione S-transferase-3 (formerly called GST1) (13837) S
11q13-11pter	(P)Neutral alpha-glucosidase AB (10416) S
11q13-11pter	(P)Muscle glycogen phosphorylase (23260) REb
11q23.2-11qter	Porphobilinogen (PBG) deaminase (uroporphyrinogen I synthase) (17600) S
11q23-11q24	(P)Oncogene ETC (16474) REa
	(L)Fibronectin (?13560) S (?fibrillar morphology of cell FN)
	(P)Leu-7 (HNK-1) antigen of natural killer lymphocytes (15129) S
	(L)Congenital glaucoma-1 (23130) Ch
	(P)Collagenase (recessive epidermolysis bullosa dystrophica) (22660) S
	(P)2′,5′-oligoisoadenylate synthetase (16435) S
	(P)Perritin, heavy chain (13477)
	(P)Thy-1 T-cell antigen (18823) REa, H
11q23-11qter	(P)T3 T-cell antigen receptor, delta chain (18679) REa

In addition: 13 anonymous DNA segments, 5 surface antigens (e.g., 15125, 15126, 15127, 18554), 29 antigens, not necessarily all separate and different, defined by monoclonal antibodies (e.g., 14304, 15806, 15807), and 1 O'Farrell protein spot (HGM 7). * By *in situ* hybridization to meiotic pachytene chrs. (Chaganti et al., *Somat. Cell Molec. Genet. 11:* 197-202, 1985). ** See *Mendelian Inheritance in Man,* 6th ed., 1983, for amino acid substitutions in gamma, delta and beta variants and for molecular changes in deletion and nondeletion forms of beta-thalassemia.

Table A 9.5 *(continued)*

Chromosome No. 12

(P) Salivary Protein Complex (SPC)

Proline tich parotid salivary protein *(Pr)* (16879) ┐
Parotid acidic protein *(Pa)* (16873) ┘ (c, F, LD)
Double band parotid salivary protein *(Db)* (16877) ─ (?, m, F, LD)
Parotid salivary glycoprotein (Gl) (16884) - Linkage disequilibrium
Salivary protein Pe (18097) - prob. closely linked to Pa Probable order: Pa - Pr - Db - Gl
PIF (parotid isoelectric focusing) protein (16872) (Ps and PIF closer to Gl and Db than Pr or Pa)
Parotid middle band protein (Pm) (16878) ┐
Parotid size variant (Ps) (16881) ┘ linked to above five loci
Parotid basic protein (Pb) (16875) ┐
Post-parotid basic protein (PPb) (16876) ┘ linkage unknown
CON1 (16887) - closely linked to Ps
CON2 (16888) - closely linked to PmF
Sequence (AJHG *36*:52, 1984): Ps-Pr-Pm-Gl-Db.
Salivary protein Po (18099) - prob. closely linked to CON2
Parotid proline-rich protein Pc (16871) - linked to Ps

	(P)Methioninyl-tRNA synthetase (15656) S
	(P)Alpha-glycerophosphate dehydrogenase (13842) S
12pter-12p12	(P)von Willebrand factor (19340) A
12pter-12q12	Branched chain amino acid aminotransferase-1 (11352) S
12p	(P)Aromatic alpha-keto acid reductase (10792) S
12pter-12q14	(P)Oncogene INT1 (murine mammary cancer virus) (16482) REa
12p13	* Glyceraldehyde-3-phosphate dehydrogenase (13840) S, D, R
12p13	Triosephosphate isomerase-1 (19045) S, D, R
	(L)Triosephosphate isomerase-2 (19046) S
12q13.1-12q13.2	Collagen II (alpha-1) (12014) REa, A
?12p12.1, ??12q24.4 (12p11.1-12.1)	Oncogene ras2 (KRAS2): Kirsten rat sarcoma virus (19007) S
12p121-12p122	Lactate dehydrogenase B (15010) S, D
	(L)Lactate dehydrogenase C (15015) H
12p11-12p12	Enolase-2 (13136) S
12p11-12qter	Citrate synthase, mitochondrial (11895) S
12q12-12q14	Serine hydroxymethyltransferase (13845) S, R
	(L)Human leukocyte antigen: MIC3 (?distinct from S8) (14303) S
12q21	Peptidase B (16990) S
12q21-12q24.2	(L)Familial intestinal neurofibromatosis (16222) Ch
12q22-12qter	Insulin-like growth factor I, or somatomedin C (14744) REa (2nd separate IGF2 gene, 00000)
12q24.1	Interferon, gamma or immune type (14757) S, A (3 introns; IFF, IFL none)
	Alpha-2-macroglobulin (10395) REa
12q24.1	(P)Phenylalanine hydroxlase (26160) REa, A
	(P)Elastase-1 (13012) REa
	(P)T4/LEU 3 (18694) REa

Also: 3 DNA segments, 1 surface antigen (18556), 3 antigens defined by monoclonals, 3 O'Farrell protein spots (HGM 7).
* Probable order: 12pter-TPI-GAPD-LDHB-ENO2-cen-SHMT-PEPB-12qter.

Chromosome No. 13

13p12	Ribosomal RNA (18045) A
4cM (from ESD	Wilson disease (27790) F
13q14 (proximal to RB1)	Esterase D (13328) S, F, D
13q14	Retinoblastoma-1 (18020) Ch
13q14	(L) X-ray sensitivity (19437) Ch
	(L) Dubin-Johnson syndrome (23750) LD (with factor VII deficiency)
13q34	Clotting factor VII (22750) D
13q34	Clotting factor X (22760) D
?13q34	(L) Carotid body tumor (16800) F (?linked to factors VII and X)
13q14-13q31	(L) Letterer-Siwe disease (24640) Ch
	(P) UV-damage, excision repair of (Xp complementation group I) (19206) S
	(P) Collagen IV alpha-1 chain (12013) S

In addition: 7 anonymous DNA segments (HGM 7).

Table A 9.5 *(continued)*

Location	Gene / trait	
Chromosome No. 14		
4p12	Ribosomal RNA (18045) A	
	(P)Temperature sensitivity complementation, K12 (18731) S	
	Lentil agglutinin binding (LCH) (15102) S	
14q11.2	(L)T-cell lymphoma/leukemia (18696) Ch	
14q112	T-cell antigen receptor, alpha subunit (18688) H, REa, A	
14q12.00-14q13.105	Nucleoside phosphorylase (16405) S, D	
14q21-14q31	(P)Oncogene FOS: FBJ murine osteosarcoma virus (16481) REa, A	
14q21-14qter	Tryptophanyl-tRNA synthetase (19105) S	
	(P)External membrane protein-195 (13374) S	
14q32	Creatine kinase, brain type (12328) S	
14q22-14qter	(P)Phosphoribosylglycineamide formyltransferase (17246) S	?1 multifunctional protein involved in *de novo* purine synthesis
	(P)Phosphoribosyl formylglycinamidine synthetase (10258) S	
	(P)Esterase activator (13325) S	
?linked to IGHC	(L)Spherocytosis, Denver type (18290) F	
?14q23-14q24.2	(L)Holt-Oram syndrome (14290) Ch	
14q24	(L)Cataract, anterior polar (11565) Fc (see 2p25)	
14q24.3-q32.2	Protease inhibitor (PI; alpha-1-antitrypsin) (10740) F, S	
14q32.1-14q32.2	DNA segment D14S1 (10775) REa, A (2 other anonymous DNA segments - HGM7)	
About 20cM apart	Immunoglobulin Heavy Chain Gene Cluster REa, A	
	Variable region genes (about 250) (14707)	
	D (for diversity) region genes (many) (14691)	
	J (for joining) region genes (more than 4) (14701)	
	Constant region of heavy chain of IgM1 (14702)	
	Constant region of heavy chain of IgM2 (14703)	
	Constant region of heavy chain of IgD (14717)	
	Constant region of heavy chain of IgG2 (14711)	5′-G2-17kb-G4-3′
(14q32.33-14qter)	Constant region of heavy chain of IgG4 (14713)	
	Constant region of heavy chain of IgG3 (14712)	close linkage known from Lepore-like myeloma protein.
Prob. orientation cen-PI-D14S1-IGHC--IGHV-ter (3′centromeric 5′, telomeric; IgM telomeric to IgG).	Constant region of heavy chain of IgG1 (14710)	
	Constant region of heavy chain of IgE (14718)	
	Constant region of heavy chain of IgEP1 (14713)	---IGEP2 on chr. 9
	Constant region of heavy chain of IgA1 (14690)	
	Constant region of heavy chain of IgA2 (14700)	
Chromosome No. 15		
15p12	Ritosomal RNA (18045) A	
15q11-15q13	Alpha-mannosidase-A, cytoplasmic (15458) S, D	
15q11	Prader-Willi syndrome (17627) Ch	
?near centromere	(P) Dyslexia-1 (12770) FC	
15q11-15qter	(P) Coronavirus 229E sensitivity (12246) S	
15q13-15q15	Beta-2-microglobulin regulator (10971) D	
15q21-15q22	Beta-2-microglobulin (10970) S, D, H (on 15q+ in APL)	
15q22-15qter	Mannosephosphate isomerase (15455) S	
15q22-15qter	Pyruvate kinase-3 (PKM2) (17905) S, D	
15q22	Hexosaminidase-A (27280) S (on 15q+ in APL)	
15q21-15qter	Isocitrate dehydrogenase, mitochondrial (14765) S	
	(P) BVIN (BALB virus induction, N-tropic) (11398) S	
	(P) Neutral alpha-glucosidase C (10418) S	
15pter-15q21	Sorbitol dehydrogenase (18250) (S, H	
15q25-15q26(?15q26)	Oncogene: feline sarcoma virus, FES (19003) S, A*	
15q11-15qter	(P) Actin, cardiac alpha (ACTC) (10254) REa	
	(P) Attached cell antigen 28.3.7 (MIC7) (10899) S	
	(P) Deoxin-inducible P1-450 (12401) S	

In addition: 1 anonymous DNA segment, 2 antigens defined by monoclonal antibodies, and 2 O'Farrell protein spots (HGM 7). * Far from breakpoint in acute promyelocytic leukemia (APL): t(15;17)(q22;q21).

Table A 9.5 *(continued)*

Chromosome No. 16

Location	Locus
	(P)Polycystic kidney disease (17390) F, S
5cm from GPT1	(P)Macular dystrophy, atypical vitelliform (15184) F (on 16 if GPT1 is)
?16p11-16pter	(I) Glutamate pyruvate transaminase, ?soluble red cell (13820) S (see chr. 8) (c, F)
	(P)Epidermolysis bullosa, Ogna type (EBS1) (13195) F
	(P)Cystathionase (21950) S
16p12-16pter	(L)Hb H mental retardation syndrome (14175) F
	Alpha Globin Gene Cluster****
16p12-16pter	Hemoglobin zeta (14231) RE — RE, A, D ?polarity vis-a-vis centromere
16p12-16pter	Hemoglobin alpha (1, 2, or 3 loci) (14180) S, HS
	(L)Transcobalamin II (27535) H - linked to HBA in mouse (19% recombination)
16p21	Phosphoglycolate phosphatase (17228) S
16q12-16q21	Diaphorase-4 (12586) S
16p12-16q22	Glutamate oxaloacetic transaminase, mitochondrial (13815) S
Order: pter-PGP-0.25-16qh-0.17-GOT2-0.08-HP-qter (Jeremiah et al., *Ann. Hum. Genet. 46:* 145, 1982)	
16q12-16q22***	Adenine phosphoribosyltransferase (10260) S, D
?linked to HP	(L)Congenital cataract (11559)
16q22	Haptoglobin (14010) FC
16q22	Lecithin-cholesterol acyltransferase (24590) F, LD
16q22	Metallothionein I and II Clusters (15635, 15632) REa, A
	(P) Thymotrypsinogen B (11889) REa - 19% AA homology to HP
	(P) Thymidine kinase, mitochondrial (18825) S
	(P) Lysosomal acid lipase-B (24798) S
	(P) Growth rate controlling factor-2 (13923) S
	(P) Esterase-B3 (13329) S
	(P) Glyoxalase II (hydroxyacyl glutathione hydrolase; BACH) (13867) S
	(P) Nonhistone chromosomal protein-2 (NHCP-2) (11888) S
	(P) Antiviral state repressor regulator (interferon regulator) (10747) D
	(P) Vesicular stomatitis virus defective interfering particle repressor (12526) S
	(L)Transcobalamin II (27535) H (of perhaps chr. 17)

In addition: 2 anonymous DNA segments (HGM 7).

*** Distal to GOT2 and DIA4.

**** See Antonarakis, Kazazian and Orkin (*Hum. Genet. 69:* 1-14, 1985) for a tabulation of deletion and nondeletion formes of alpha-thalassemia.

Chromosome No. 17

Location	Locus
	(P)Aldehyde dehydrogenase-3 (10066) S
17p13.3	(P)Miller-Dieker lissencephaly syndrome (24720) Ch
17pter-17p11	Myosin, Heavy Chain (16073) REa (a myosin gene cluster also on 7)
	MYHCA = myosin, cardiac heavy chain, alpha (adult) (16071) A
	MYHCB = myosin heavy chain, beta (fetal) (16076) A (5′ to MYHCA)
	MYHSA1 = myosin heavy chain, adult-1 (16073) REa
	MYHSA2 = myosin heavy chain, adult-2 (16074) REa
	MYHSE1 = myosin heavy chain, embryonic-1 (16072) REa
17q11-17q22	Homeo box-1 (14295) REb, A, H
17q11-17q22	Homeo box-2 (14296) REb, A, H
17q21	RNA, U2 small nuclear (A)
17q210-17q220	Galactokinase (23020) S, Ch, R, C
17q210-17q220	Growth Hormone/Placental Lactogen Gene Cluster S, REa, A, C — 5′
(17q22-17q24 by *in situ*	GHN = Growth hormone, normal (13925) S, REa, A
hybridization. By CMGT,	CSL = Chorionic somatomammotropin-like (15020) S, REa, A
order - cen-GALK-GH-	CSA = Chorionic somatomammotropin A (15020) S, REa, A — about 45 kb
(TK-COL1A1)	GHV = Growth hormone variant (13925) S, REa, A
	CSB = Chorionic somatomammotropin B (15020) S, REa, A
17q21.13-17q22.05	— 3′
closetto TK than GALK)	Collagen I alpha-1 polypeptide (12015) S, M, A, REa
17p11-17q21(17q21.33-17q22)	Oncogene: avian erythroblastic leukemia virus, ERBA (19012) REa (prox. to APL
17q210-17q220	break at 17q21.1)
17q210-17q220	Adenovirus-12 chromosome modification site-17 (10297) V
17q23-17q25	Thymidine kinase-1 (18830) S, Ch, R, C*

Table A 9.5 *(continued)*

17q23-17qter	Acid alpha-glucosidase (23230) S, A
	Peptidase E (17020) S
	(P)Aldolase C (10387) REb
	(L)Creatine kinase, brain type (12328) S

In addition: 1 anonymous DNA segment, 1 antigen defined by a monoclonal antibody, and 1 surface antigen (18557).
* Moves to 15q+ in acute promyelocytic leukemia (AP): t(15;17)(q22;q21).

Chromosome No. 18

18q122	(Subband critical to trisomy 18 phenotype)]
18q23	Peptidase A (16980) S, D
18p11	(I) Human chorionic gonadotropin, alpha subunit (11885) S, A (see chr. 10)
18q21	(P) Oncogene B-cell leukemia/lymphoma 2, BCL2 (15143) Ch, RE
18q24.3	(L) Neoplastic lymphoproliferation-2 (16185) Ch - t(14;18) in non-Hodgkin lymphoma
	(P) Oncogene: endogenous retrovirus-1, ERV1 (13115) REa
	(P) Asparaginyl-tRNA synthetase (10841) REa
	Thymidylate synthase (18835) S
	(L) Diastrophic dysplasia (22260) Ch
	(P) Prealbumin (transthyretin) (17630) REa
	(P) Oncogene YES-1 (16488) REa
	(P) Dihydrofolate reductase pseudogene (see 12606) REa (shows absence-presence polymorphism)

In addition: 3 anonymous DNA segments (HGM 7).

Chromosome No. 19

19p13.2-19p13.3	(P)Insulin receptor (14741) REa, A
19p13.3-19q13	Lysosomal alpha-D-mannosidase-B (24850) S
19pcen-19q13	Glucosephosphate isomerase (17240) S, D
19q	Polio virus sensitivity (17385) S
19q	(L)Echo 11 sensitivity (12915) S
19p13.2-19cen	Peptidase D (prolidase) (17010) S, F, H (closely linked to APOC2)
	(P)Beta-glucuronidase, mouse, modifier of (23161) S
	(P)Baboon M7 virus receptor (?same as virus RD114 receptor) (10919) S
	(P)Coxsackie B3 virus susceptibility (12005) S
	(P)Branched chain amino acid transaminase-2 (11353) S
19pter-19q13	(P)Lysosomal DNA-ase (12635) S
19pter-19p13	Familial hypercholesterolemia (LDL receptor) (14389) F, S (about 20 cM from C3)
	LW (Landsteiner-Wiener) blood group (11125) F (very close to C3, Lu)
19pter-19p13.2 (?order: FHC-C3-Le-DM-Se-Lu)	Complement component-3 (12070) S (Le about 7 cM in males vs. C3 RFLP)
	Lewis blood group (11110) F (linked to C3)
Linked to DM?	(P)Neurofibromatosis (16220) F
Linked to PEPD (about 5cM)	Myotonic dystrophy (16090) F (in C3 linkage group; SRO for C3, 19pter-19q13.2)
	Secretor (18210) F
about 13cM from C3	Apolipoprotein Cluster II
	Apolipoprotein E (20776) F, REa (?close to Se)-v,F Order suggested by Eiberg et al., 1983): Le-C3-DM-(Se-PEPD)-Lu: FHC-C3-APOE/APOC2. APOC1 6 a, b downstream from APOE.
	Apolipoprotein C-II (20775) REa, F, LD
	Apolipoprotein C-I (10771) REa, RE (second APOCI gene on 19)
	Lutheran blood group (11120) ⎤ (v, F)
	(P)Bombay phenotype (Bh) 21110) ⎦
19q13.32-19qter	Ferritin, light chain (13479) S, A
19q13.32-19qter	Chorionic Gonadotropin, Beta Chain (at least 8 genes) (11886) REa, H
	Luteinizing hormone, beta-chain (15278) RE
	(P)Elongation factor-2 (13061) S

Beta chains of FSH, TSH linked to LHB? TSHB and LHB unlinked in mouse. CGB, LHB, FSHB, TSHB = ?glycopeptide hormone gene cluster.

Table A 9.5 *(continued)*

19q13.1-19pter	Phenobarbital-inducible cytochrome P450 (12396) REa, A
	(P)Creatine kinase miracle type (12331) REa
	(P)Complementation of CHO DNA repair defect UV20 (12638) S - ?form of xeroderma pigmentosum (?27870)
	(P)Complementation of CHO DNA-repair defect EM9 (12634) - ?Bloom syndrome locus (21090)

In addition: 3 anonymous DNA segments HGM 7).

Chromosome No. 20

20p	Inosine triphosphatase (14752) S
20p12.2	(L) Multiple endocrine neoplasia, types 2 and 3 (17140, 16230) Ch
20q131-20qter	Adenosine deaminase (10270) S, D, REa — functionally related
20cen-20q131	S-adenosylhomocysteine hydrolase (18096) S — functionally related
	Growth hormone releasing factor (somatocrinin) (13919) REa, REb
	(P) Vasopressin (19/q234) REa
	(P) Desmosterol-to-cholesterol enzyme (DCE) (12565) F
20q12-20q13	Protooncogene *src* (Rous sarcoma) (19009) REa, A
	(P) Neuraminidase and beta-galactosidase expression (25654) S

In addition: 4 anonymous DNA segments (HGM 7).

Chromosome No. 21

21p12	Ribosomal RNA (18045) A
21q21-21qter	Antiviral protein (interferon receptor) (10745) S, D
21q22.1	Superoxide dismutase-1 (soluble) (14745) S, D
21q22.1	Phosphoribosylglycinamide synthetase (13844) S, H — ?1 multifunctional protein involved in *de novo* purine synthesis
	(P) Phosphoribosylaminoimidazole synthetase (17244) S — ?1 multifunctional protein involved in *de novo* purine synthesis
21q22	Phosphofructokinase, liver type (17186) S, D
21q11-qter	(P) Primary thrombocytosis (18795) Ch
	(L) Beta-amino acids, renal transport of (10966) D
	(L) Beta-adrenergic stimulation, response to (10967) D
	(L) 5-hydroxytryptamine oxygenase regulator (00000) D - not in MIM
21q21-21q22.1	Cystathionine beta-synthase (23620) S, D

In addition: 9 anonymous DNA segments, 1 surface antigen (18559) and 2 antigens defined by monoclonals (HGM 7). Determinants crucial to the Down syndrome phenotype located 21q22-21qter. 21q22 = critical band for Down syndrome also breakpoint in AML-M2, t(8;21)(q22;q22).

Chromosome No. 22

22p12	Ribosomal RNA (18045) A
22pter-22q11	(L) Cat eye syndrome (11547) Ch
22pter-22q11(Ph[1])	(P) Alpha-L-iduronidase (25280) S, D
22q1	DiGeorge syndrome (18840) Ch
?22q11.12(on Ph1)	Immunoglobulin Lambda Light Chain Gene Family REa, A — order 5′ to 3′: cen-V-C-ter
	Variable region of lambda light chains (many genes) (14724)
6 or more constant region genes in tandem, each with J gene	J region of lambda light chains (several genes) (14723)
	Constant region of lambda light chains (several genes; (14722)
22q11.3 (distal to lambda)	Chronic myeloid leukemia (15141) Ch- $Ph^1 = t(9; 22)$ (q34.1;q11.21)
22q12	(L) Ewing sarcoma (13345) Ch - t(11;22)(q24;q12)
22q13.31-22qter	** Arylsulfatase A (25010) S
22q13.31-22qter	NADH-diaphorase-1 (25080) S
22q11-22q13	** Aconitase, mitochondrial (10085) S
22q11-22q13	(P) Myoglobin (16000) REa
22q13	** N-acetyl-alpha-D-galactosaminidase (alpha-galactosidase B) (10417) S
22q13-22qter	Beta-galactosidase-2 (protective protein) (10968) S
22q11-22qter	Oncogene SIS: simian sarcoma virus (19004) REa (same as platelet derived growth factor)

In addition: 7 anonymous DNA segments and 1 surface antigen (18558) (HGM 7).
** Distal to the break point that creates Philadelphia chromosome.

Table A 9.5 *(continued)*

A. *Regional assignments on the X chromosome by study of cell hybrids and other methods as indicated.*

Location	Locus
Xp22.32	Xg blood group (31470) F — related or identical
Xp22.32	MIC2X, or monoclonal antibody 12E7 (31347)
Xp22.32	Steroid sulfatase (ichthyosis) (30810) F — escape lyonization
(L)Xp22	Aicardi syndrome (30405) Ch (balanced translocation)
(L)Xp22	Hypomagnesemia, X-linked primary (30760) Ch (?balanced translocation)
(P)Xp22.32	Chondrodysplasia punctata, X-linked (30295) Ch
(L)Xp21-Xp22	Gonadal dysgenesis, XY female type (30610) F, Chr (?deficiency of gonad-specific H-Y receptor) — May be the same
(P)Xp223	H-Y regulator (or repressor) (30697) Ch (See II, B for reference to structural H-Y locus on Xp.)
Xp21-Xp223	Polymorphic lambda RC8 probe (DXS9) about 10 cM from DMD, 15cM from RS, linked to STS, prob. linked to Menkes syndrome (*Hum. Genet. 65:* 53, 1983)
Xp21	Duchenne muscular dystrophy (31020) Ch, F
Xp21	Becker muscular dystrophy (31010) Fd, ?Ch (?allelic to DMD)
Xp21.1 (prox. to DMD)	Ornithine transcarbamylase (31125) Ch
Xp11.2-Xp21	(L) Glycerol kinase deficiency (30703) Ch
Xp11.2-Xp21	(L) Primary adrenal hypoplasia (30020) Ch
Xp11-Xq11 (Xq25)	Actin (30002) REa - ?pseudogene
Xp11-Xq13	Menkes syndrome (30940) Fd, Fc, H (close to Tfm & Pgk-1 in mouse) - 5cM from cen.
(P)Near centromere	Alport-syndrome-like nephritis (30105) Fd
Xp110-Xp113	Polymorphic DNA probe L1.28 (DXS7) about 15 cM from BMD (proximal to Menkes) S
(P)Xp11-Xq13	
close to DXS7	Testicular feminization (androgen resistance; DHT-receptor defect) (31370) S
	Norrie disease (31060) Fd
?linked to DXS	(P) Retinis pigmentosa with metallic sheen fundus reflex in carrier (30320) Fd
3cM prox. to DXS7 (Xp21)	Retinitis pigmentosa, X-linked (31260) Fd, Ch***** (?allelic with 30320)
Xp21	Chronic granulomatous disease (30640) Ch*****
(L)Xq112-Xq211(Xq13)	X chromosome controlling element (X-inactivation center) (31467).
(P)Xq13	Aarskog-Scott syndrome (30540) Ch (balanced translocation)
(L)Xq12	Anhidrotic ectodermal dysplasia (30510) Ch (balaned translocation)
(L)Xq13	Sideroblastic anemia (30130) Ch (somatic cell chromosome rearrangement)
Xq13	Phosphoglycerate kinase (PGK) (31180) S****
Near PGK	Monoamine oxidase A (30985) S
Xq13-q21 (linked to DXYSI)	Chorioderemia (30310) Fd
Xq22	Alpha-galactosidase A (Fabry disease) (30150) S, R
Xq22-Xq26	Phosphoribosylpyrophosphate synthetase (31185) S, R
Xq26-Xq27	Hypoxanthine-guanine phosphoribosyltransferase (30800) S, M, C, R** - about 35 cM from FS
Xq26-Xq27	Hunter syndrome (mucopolysaccharidosis II) (30990) Ch (balanced translocation)
near HPRT (Xq13-Xq27)	Temperature sensitivity, mouse and hamster, complement; BA2R (31365) S
Xq27 (distal to HPRT)	Hemophilia B (factor IX) (30690) REa, A, F, D - proximal but very close to FS
Xq27.3*	Fragile site Xq27.3 (at interface between Xq27 and Xq28) (30955) Ch***
Xq28	Glucose-6-phosphate dehydrogenase; G6PD (30590) S, R
Xq28 (distal to G6PD)	Goeminne TKCR syndrome (31430) Ch (balanced translocation)

* A second X chromosome fragile site is located at Xq26.

** Three HPRT pseudogenes located on chromosomes 3, 5, 11 (Caskey, 1984).

*** Siniscalco hypothesis (1984): 1)Xq27.3 = "hotspot" for meiotic recombination. 2)Fragile site at that location = small chromosomal aberration resulting from anomalous recombination. 3)Recombination is suppressed at that location in females heterozygous for the fragile site.

**** Radiation-induced segregation (Goss-Harris) supported order: PGK-GALA-PRPS-HPRT-SAX-G6PD.

***** Interstitial deletion of Xp21 resulted in make with RP, DMD, and CGD with McLeod red cell phenotype (Francke et al. *Am. J. Hum. Genet., 37:* 250-267, 1985).

References

References to cited authors are numbered consecutively in the text and are listed alphabetically for each chapter. All references of more comprehensive interest such as to various books, monographs, and general reviews are presented as a separate list. A number of additional references of general interest not cited specifically in the text are included in the list.

Textbooks, Handbooks, Monographs and Reviews of More General Interest

1. Hirschhorn K (eds) 1970-1986 Advances in human genetics, Vol 1-15. Harris H, New York, Plenum
2. Alberts B, Bray D, Lewis J, Raff M, Roberts K, Watson JD (1983) Molecular biology of the cell. Garland Publ, New York London
3. Anastasi A (1958) Differential psychology, 3rd ed. McMillan, New York
4. Anderson TW (1958) An introduction to multivariate statistical analysis. Wiley, New York London Sidney
5. Applebaum EG, Firestein SK (1983) A genetic counseling casebook. The Free Press, New York
6. Baraitser M (1982) The genetics of neurological disorders. Oxford Univ Press, Oxford etc
7. Barthelmess A (1952) Vererbungswissenschaft. Alber, Freiburg München
8. Becker PE (ed) Humangenetik, ein kurzes Handbuch in fünf Bänden, Vol I/1 (1968), Vol I/2 (1969), Vol I/3 (1975), Vol I/4 (1972), Vol II (1964), Vol II-I/I (1964) Vol III/2 (1972), Vol III/3 (1976), Vol IV (1964), Vol V/I (1966), Vol V/2 (1967). Thieme, Stuttgart
9. Becker PE (1977) Topics in human genetics: Vol III, Myotonia congenita and syndromes associated with myotonia. Becker PE, Lenz W, Vogel F, Wendt GG (eds). Thieme, Stuttgart
10. Berg K (1979) Genetic damage in man caused by environmental agents. Academic Press, New York etc
11. Berg K (1979) Inherited variation in susceptibility and resistance to environmental agents. In: Berg K (ed) Genetic Damage in Man Caused by Environmental. Academic Press, New York
12. Berg K (ed) (1985) Medical Genetics; present, past, future. Alan R Liss, New York
13. Bergsma D (1979) Birth Defects Compendium, 2nd ed, An encyclopedic guide to birth defects. Alan R Liss, New York
14. Bickel H, Guthrie R, Hammersen G (eds) (1980) Neonatal screening for inborn errors of metabolism. Springer Verlag, Berlin Heidelberg New York
15. Bickel H, Hudson FP, Woolf LI (eds) (1971) Phenylketonuria and some inborn errors of amino acid metabolism. Biochemistry, genetics, diagnosis, therapy. Thieme, Stuttgart
16. Birbaumer N (1975) Physiologische Psychologie. Springer, Berlin Heidelberg New York
17. Birth Defects (1972) Part XV: The cardiovascular system. Williams & Wilkins, Baltimore 8: (5)
18. Birth Defects (1974) Part XVI: Urinary system and others. Williams & Wilkins, Baltimore 10: (4)
19. Bleuler M (1972) Die schizophrenen Geistesstörungen. Thieme, Stuttgart
20. Bodmer WF, Cavalli-Sforza LL (1976) Genetics, evolution, and man. Freeman, San Francisco
21. Bond DJ, Chandley AC (1983) Aneuploidy. Oxford University Press, Oxford etc
22. Bora KC, Douglas GR, Nestmann ER (eds) (1982) Chemical Mutagenesis, Human Population Monitoring and Genetic Risk Assessment. Progress in Mutation Research, Vol 3. Elsevier Biomedical Press, Amsterdam etc
23. Borgaonkar DS (1980) Chromosomal variation in man. A catalogue of chromosomal variants and anomalies, 2nd ed. Liss, New York
24. Boveri T (1889) Ein geschlechtlich erzeugter Organismus ohne mütterliche Eigenschaften. Sitzungsber Morphol Physiol (Munich) 5
25. Boveri T (1888) Zellstudien II. Die Befruchtung und Teilung des Eies von Ascaris megalocephala. Z Naturwiss 22: 685-882
26. Bowman JE (ed) (1983) Distribution and Evolution of Hemoglobin and Globin Loci. Elsevier, New York
27. Bracken H von (1969) Humangenetische Psychologie. In: Becker PE (ed) Humangenetik, ein kurzes Handbuch. Vol I/2, pp 409-562. Thieme, Stuttgart
28. Brock DJH (1982) Early Diagnosis of Fetal Defects. Churchill Livingstone, Edinburgh
29. Brown MS, Goldstein JL (1976) New directions in human biochemical genetics: understanding the manifestations of receptor deficiency states. Prog Med Genet [New Series] 1: 103-119
30. Bulmer MG (1970) The biology of twinning in man. Clarendon, Oxford

30a. Bunge M (1967) Scientific Research Vols I, II. Springer-Verlag, Berlin Heidelberg New York

31. Bunn HF, Forget BS, Ranney HM (1977) Human hemoglobins. Saunders, Philadelphia London Toronto
32. Burgio GR, Fraccaro M, Tiepolo L, Wolf U (eds) (1981) Trisomy 21. Springer Verlag, Berlin Heidelberg New York
33. Butterworth T, Ladda RL (1981) Clinical genodermatology, Vol 1, 2. Praeger, New York
34. Calabrese EJ (1984) Ecogenetics. Genetic Variation in Susceptibility to Environmental Agents. John Wiley and Sons, New York
35. Carter CO, Fairbank TJ (1974) The genetics of locomotor disorders. Oxford University Press, London
36. Cavalli-Sforza LL, Bodmer WF (1971) The genetics of human populations. Freeman, San Francisco

36a. Scarr S (1981) Race, social class, and individual differences in I.Q. Erlbaum, Hillsdale

37. Cherfas J (1982) Man-made life. Blackwell, Oxford
38. CIBA Found Symp (1963) Man and his future. Wolstenholme GEW (ed). Churchill, London

39. CIBA Found Symp [new series] (1976) Embryogenesis in mammals. Elsevier Excerpta Medica, North-Holland 40
40. Cohen BH, Lilienfeld AM, Huang PC (1978) Genetic issues in public health and medicine. Thomas, Springfield Ill
41. Coon CS (1965) The living races of man. Knopf, New York
42. Court Brown WM (1967) Human population cytogenetics. North-Holland, Amsterdam
43. Cox DW, Woo SLC, Manfield T (1985) DNA restriction fragments associated with α_1-antitrypsin indicate a single origin for deficiency allele PI Z. Nature 316: 79-81
44. Creutzfeldt W, Köbberling J, Neel JV (eds) (1976) The genetics of diabetes mellitus. Springer, Berlin Heidelberg New York
45. Crow JF, Kimura M (1970) An introduction to population genetics theory. Harper & Row, New York
46. de la Cruz FF, Gerald PS (eds) (1981) Trisomy 21 (Down syndrome). University Park Press, Baltimore
47. Current developments in anthropological genetics (1982) Mielke JH, Crawford MH (eds) Theory and Methods, Vol 1. Crawford MH, Mielke JH (eds) Ecology and population structure, Vol 2. Plenum Publ Corp, New York
48. Dausset J, Svejgaard A (eds) (1977) HLA and disease. Munksgaard, Copenhagen
49. Davidson EH (1976) Gene activity in early development, 2nd ed. Academic Press, New York San Francisco London
50. Davies KE (1981) The application of DNA recombinant technology to the analysis of the human genome and genetic disease. Hum Genet 58: 351
51. Dayhoff MO (1972) Atlas of protein sequence and structure. Vol 5. National Biomedical Research Foundation (Suppl 1973), Washington (DC)
52. Der Kaloustian VM, Kurban AK (1979) Genetic diseases of the skin. Springer, Berlin Heidelberg New York
53. Drake JW (1969) The molecular basis of mutation. Holden Day, San Francisco
54. Dutrillaux B (1975) Sur la nature et l'origine des chromosomes humaines. Monographies des Annales de Génétique. L'expansion scientifique, Paris
55. Eccles JC (1970) Facing reality. Springer, New York Heidelberg
56. Ehrman L, Parsons PA (1976) The genetics of behavior. Sinauer, Sunderland (Mass.)
57. Eickstedt E von (1934) Rassenkunde und Rassengeschichte der Menschheit. Enke, Stuttgart
58. Emery AEH (1976) Methodology in medical genetics. Churchill Livingstone, Edinburgh London New York
59. Emery AEH (1973) Antenatal diagnosis of genetic disease. Williams & Wilkins, Baltimore
60. Emery AEH (1984) An Introduction to Recombinant DNA. John Wiley and Sons, Chichester
61. Emery AEH, Rimoin DL (1983) Principles and practice of medical genetics, 2 vols. Churchill Livingstone, Edinburgh etc
62. Efroimson VP (1964) Introduction to medical genetics (in Russian). Goverment Medical Publishers, Moscow
63. Falconer DS (1981) Introduction to quantitative genetics. 2nd ed. Oliver & Boyd, Edinburgh London
64. Ferguson-Smith MA (ed) (1983) Early prenatal diagnosis. Churchill Livingstone, London
65. Fisher RA (1930) The genetical theory of natural selection. Oxford University Press, Oxford
66. Fox SW (ed) (1984) Chemical and Biological bases. Plenum, New York
67. Fraser GR (1976) The causes of profound deafness in childhood. The Johns Hopkins University Press, Baltimore/London
68. Fraser GR, Friedmann AI (1967) The causes of blindness in childhood. The Johns Hopkins Press, Baltimore
69. Fraumeni JF jr (ed) (1975) Persons at High Risk of Cancer. Academic Press, New York etc
70. Friedberg EC (1984) DNA repair. Freeman WH, New York
71. Fuhrmann W, Vogel F (1983) Genetic counseling, 3rd ed. Heidelberg Science Library 10. Springer, New York
72. Fuller JL, Thompson WR (1978) Foundations of behavior genetics. Mosby, St Louis
73. Galjaard H (1980) Genetic metabolic diseases. Early diagnosis and prenatal analysis. Elsevier/North Holland, Amsterdam etc
74. Gardner LI (ed) (1975) Endocrine and genetic diseases of childhood and adolescence. Saunders WB, Philadelphia etc
75. Garrod AE (1963) Inborn errors of metabolism. Henry Frowde, London (1923). Reprinted by Oxford University Press London
76. Gedda L (ed) (1961-1962) De genetica medica, Vols I-VI. Orrizonte Medico, Roma
77. German J (1974) Chromosomes and cancer. John Wiley & Sons, New York London Sidney Toronto
78. Giblett ER (1969) Genetic markers in human blood. Blackwell, Oxford Edinburgh
79. Giblett ER, Polmar SH (1979) Inherited immunodeficiency diseases: relationship to lymphocyte metabolic dysfunction. Prog Med Genet N.S.3, 177-219
80. Goldschmidt RB (1955) Theoretical genetics. University of California Press, Berkeley
81. Goodman M, Tashian RE (ed) (1976) Molecular anthropology. Plenum, New York
82. Goodman RM, Gorlin RJ (1977) Atlas of the face in genetic disorders. 2nd ed. The CV Mosby Comp, St Louis
83. Gorlin RJ, Pindborg JJ, Cohen MM jr (eds) (1976) Syndromes of the head and neck (2nd ed). McGraw Hill Book Co, New York etc
84. Gottron HA, Schnyder UW (eds) (1966) Handbuch der Haut- und Geschlechtskrankheiten. Supplement VII: Vererbung von Hautkrankheiten. Springer, Berlin Heidelberg New York
85. de Grouchy J, Turleau C (1982) Clinical Atlas of Human Chromosomes. John Wiley, New York
86. de Grouchy J, Tourleau C (1982) Atlas des maladies chromosomiques, 2nd ed. Expansion Scientifique Française, Paris
87. Haldane JBS (1954) The biochemistry of genetics. Allan & Unwin, London
88. Hamerton JL (1971) Human cytogenetics. Vol I, II. Academic Press, New York London
89. Harper PS (1984) Genetics of muscular dystrophies. In: Progress in Medical Genetics, Vol 6. Saunders, Philadelphia

90. Harper PS (1984) Practical Genetic Counseling, 2nd ed. Wright, Bristol
91. Harper PS (1985) The genetics of muscular dystrophies. Prog Med Genet N.S. 6: 53-90
92. Harris H, Hopkinson DA (1976) Handbook of enzyme electrophoresis in human genetics. North Holland, Amsterdam
93. Harris H (1980) The principles of human biochemical genetics, 4th ed. North-Holland, Amsterdam Oxford
93a. Hartl DL (1980) Principles of population genetics. Sinauer Associates, Inc, Sunderland Massachusetts
94. Hayden MR (1981) Huntington's Chorea. Springer, Berlin Heidelberg New York
95. Hirsch J (1970) Behavior-genetic analysis and its biosocial consequences. Semin Psychiat 2: 89-105
96. Holborow E (1977) Immunology in medicine. Academic Press, New York
97. Hollaender A (ed) (1954-56) Radiation biology. 4 Vols. Mc Graw-Hill, New York
97a. Honig GR, Adams JG III (1986) Human Hemoglobin Genetics. Springer, Vienna
98. Hsia YE, Hirschhorn K, Silverberg RL, Godmilow L (eds) (1979) Counseling and Genetics. Alan R Liss, Inc, New York
99. Human Gene Therapy - A Background Paper (1984) Washington DC. US Congress, Office of Technology Assessment, OTA-BP-BA-32, December
100. Humphrey JH, White RG (1970) Immunology for students of medicine, 3rd ed. Blackwell, Oxford Edinburgh
101. International Directory of Genetic Services (1983) National Foundation March of Dimes, New York
102. Ionasescu V, Zellweger H (1983) Genetic in Neurology. Raven Press, New York
103. Jacquard A (1974) The genetic structure of populations. Springer, New York Heidelberg Berlin
104. Kamin LJ (1974) The science and politics of I.Q. Erlbaum, Potomac
105. Keats BJB (1981) Linkage and chromosome mapping in man. Population Genetics Laboratory, University of Hawaii, Honolulu
106. Keats BJB, Morton NE, Rao DC, Williams WR (1979) A source book for linkage in man. The Johns Hopkins University Press, Baltimore London
107. Kelly TE (1980) Clinical Genetics and Genetic Counseling. Yearbook Medical Publ, Chicago
108. Kempthorne O (1957) An introduction to genetic statistics. John Wiley & Sons, New York
109. Kessler S (1979) Genetic counseling, Psychological dimensions. Academic Press, New York
110. Kety SS, Rowland LP, Sidman RL, Matthyse SW (1983) Genetics of Neurobiological and Psychiatric Disorders. Raven Press, New York
111. King M-C, Lee GM, Spinner NB, Thomson G, Wrensch MR (1984) Genetic epidemiology. Ann Rev Public Health 5: 1-52
112. Kinsey AC, Pomeroy WB, Martin CE (1948) Sexual behavior in the human male. Saunders, Philadelphia London
113. Klein J (1975) Biology of the mouse histocompatibility-2 complex. Springer, Berlin Heidelberg New York
114. Klingmüller W (1976) Genmanipulation und Gentherapie. Springer, Berlin Heidelberg New York
115. Koller S (1983) Risikofaktoren der Schwangerschaft. Springer, Berlin
116. Konigsmark BW, Gorlin RJ (1976) Genetic and metabolic deafness. Saunders, Philadelphia London Toronto
117. Knippers R (1985) Molekulare Genetik, 4. Aufl. Thieme Verlag, Stuttgart New York
118. Lange J (1929) Verbrechen als Schicksal. Thieme, Leipzig
119. Lehmann H, Huntsman RG (1974) Man's hemoglobins. North-Holland, Amsterdam Oxford
120. Lehninger AL (1975) Biochemistry, 2nd ed. Worth, New York
121. Lenz W (1983) Medizinische Genetik, 6th ed. Thieme, Stuttgart
122. Lewin B (1983) Genes. J Wiley & Sons, New York
123. Lewontin RC (1974) The genetic basis of evolutionary change. Columbia University Press, New York London
124. Li CC (1955) Population genetics. University of Chicago Press, Chicago
125. Lima-de-Faria A (1983) Molecular evolution and organization of the genome. Elsevier, Amsterdam etc
126. Livingstone FB (1967) Abnormal hemoglobins in human populations. Aldine Chicago (see also 1815 a)
127. Lorenz K (1966) On aggression. Methuen, London
128. Lorenz K (1978) Vergleichende Verhaltensforschung. Springer, Wien New York
129. Lubs HA, de la Cruz F (eds) (1977) Genetic counseling. Raven, New York
129a. Lumbsden CJ, Wilson EO (1981) Genes, mind, and culture. The coevolutionary process. Harvard University Press, Cambridge Massachusetts and London England
130. Malécot G (1948) Les mathématiques de l'hérédité. Masson, Paris
131. Marois M (ed) (1982) Prevention of Physical and Mental Congenital Defects. Part A: The Scope of the Problem. Progress in Clinical and Biological Research, Vol 163 A. Alan R Liss Inc, New York 1985. Proc Conference of the Institut de la Vie, Strasbourg
131a. Marois M (ed) (1985) Prevention of Physical and Mental Congenital Defects. Part B: Epidemiology, Early Detection and Therapy, and Environmental Factors. Progress in Clinical and Biological Research, Vol 163 B. Alan R Liss Inc, New York
131b. Marois M (ed) (1985) Prevention of Physical and Mental Congenital Defects. Part C: Basic and Medical Science, Education, and Future Strategies. Progress in Clinical and Biological Research, Vol 163 C. Alan R Liss Inc, New York
131c. Mayr E (1982) The growth of biological thought Cambridge, Mass: Harvard Univ. Press
132. McKusick VA (1972) Heritable disorders of connective tissue, 4th ed. Mosby, St Louis
133. McKusick VA (1983) Mendelian inheritance in man, 6th ed. The Johns Hopkins University Press, Baltimore
134. McKusick VA, Clairborne R (1973) Medical genetics. HP Publ Comp, New York
135. Mielke JH, Crawford MH (1980, 1982) Current developments in Anthropological genetics Vol 1,2. Plenum Press, New York London

136. Milunsky A (1975) The prevention of genetic disease and mental retardation. Saunders, Philadelphia London Toronto
137. Milunsky A (1979) Genetic disorders and the fetus. Plenum Press, New York London
138. Money J, Ehrhardt AA (1972) Man and woman, boy and girl. Johns Hopkins University press, Baltimore
139. Morton NE (1982) Outline of genetic epidemiology. S Karger, Basel etc
140. Morton NE, Chung CS (eds) (1978) Genetic epidemiology. Academic Press, New York
141. Motulsky AG (1985) Hereditary syndromes involving multiple organ systems. In: Wyngaarden JB, Smith LH (eds) Cecil Textbook of Medicine, pp 1172-1173. Saunders, Philadelphia
142. Motulsky AG (1984) Medical genetics. JAMA 252: 2205-2208. Also published in French (1985) JAMA Edition Francaise 10: 30-34
143. Motulsky AG, Vogel F, Buselmaier W, Reichert W, Kellermann G, Berg P (eds) (1978) Human genetic variation in response to medical and environmental agents: pharmacogenetics and ecogenetics. International Titisee Conference. Titisee, 13-15 October 1977. Hum Genet [Suppl 1]
144. Mourant AE, Kopeć AC, Domaniewska-Sobczak K (1976) The distribution of the human blood groups and other polymorphisms, 2nd ed. Oxford University Press, London
145. Mourant AE, Kopeć AC, Domaniewska-Sobczak K (1978) Blood groups and diseases. Oxford University Press, London
146. Mulvihill JJ, Miller RW, Fraumeni JF (eds) (1977) Genetics of human cancer. Process in Cancer Research and Therapy. Vol 3. Raven, New York
147. Murken J-D, Stengel-Rutkowski S (1978) Pränatale Diagnose. Enke, Stuttgart
148. Murphy EA (1982) Biostatistics in Medicine. The Johns Hopkins University Press, Baltimore
149. Murphy EA, Chase GA (1975) Principles of genetic counseling. Year Book Medical Publ, Chicago
150. Neel JV, Schull WJ (1954) Human heredity. University of Chicago Press, Chicago
151. Nei M (1975) Molecular population genetics and evolution. North Holland, Amsterdam New York
152. Newman HH, Freeman FN, Holzinger KJ (1937) Twins: A study of heredity and environment. (4th impression 1968). University of Chicago Press, Chicago
153. Nora JJ, Fraser FC (1974) Medical Genetics: Principles and Practices. Lea & Febiger, Philadelphia
154. Nora JJ, Nora AH (1978) Genetics and counseling in cardiovascular diseases. Thomas, Springfield
155. Nyhan EL, Sakati NO (1976) Genetic and malformation syndromes in clinical medicine. Year Book Med Publ, Chicago
156. Ohno S (1967) Sex chromosomes and sex-linked genes. Springer, Berlin Heidelberg New York
157. Ohno S (1970) Evolution by gene duplication. Springer, Heidelberg New York
158. Old RW, Primrose SB (1981) Principles of Gene Manipulation. Studies in Microbiology, Vol 2. Blackwell Scientific Publications, Oxford etc
159. Omenn GS, Gelboin HV (eds) (1984) Genetic Variability in Responses to Chemical Exposure. Cold Spring Harbor Laboratory
160. Penrose LS, Smith GF (1966) Down's anomaly. Churchill, London
161. Persaud TVN (ed) (1979) Advances in the study of birth defects, Vol 1-4. MTP Press, Lancaster (continued)
162. Pontecorvo G (959) Trends in genetic analysis. Columbia University Press, New York
163. Progress in medical genetics, Vol I-X. Steinberg AG, Bearn AG (eds) (1961-1974) Grune & Stratton, New York
164. Progress in medical genetics (new series), (-1985) Steinberg AG, Bearn AG, Motulsky AG, Childs B (eds) Saunders, Philadelphia
165. Puck TT, Kao F-T (1982) Somatic cell genetics and its application to medicine. Ann Rev Genet 16: 225-271
166. Race RR, Sanger R (1975) Blood groups in man, 6th ed. Blackwell, Oxford
167. Reed EW, Reed SC (1965) Mental retardation: a family study. Saunders, Philadelphia London
168. Resnick LB (ed) (1976) The nature of intelligence. Erlbaum, Hillsdale
169. Rieger R, Michaelis A, Green MM (1968) A glossary of genetics and cytogenetics. Springer, Berlin Heidelberg New York
170. Rimoin DL (1975) The chondrodystrophies. Adv Hum Genet 5: 1-118
171. Rimoin DL, Schimke RN (1971) Genetic disorders of the endocrine glands. Mosby, St Louis
172. Rodeck CH, Nicolaides KH (eds) (1984) Prenatal Diagnosis. John Wiley & Sons, Chicester New York
173. Rosenthal D, Kety SS (1968) The transmission of schizophrenia. Pergamon, Oxford
174. Rothschild H (ed) (1981) Biocultural aspects of disease. Academic Press, New York etc
175. Rowland LP (1977) Pathogenesis of human muscular dystrophies. Experta Medica, Amsterdam/Oxford
176. Russell PJ (1980) Lecture notes on genetics. Blackwell, Oxford etc
177. Rutter WJ (1984) Molecular genetics and individuality. In: Fox SW (ed) Individuality and Determinism. Chemical and Biological Bases. Plenum, New York, pp 61-71
178. Salmon MA (1978) Developmental defects and syndromes. HM + M Publishers, Aylesbury
179. Salzano FM (ed) (1975) The role of natural selection in human evolution. North-Holland/American Elsevier, Amsterdam New York
180. Sandberg A (ed) (1983) Cytogenetics of the mammalian X chromosome, Vol 1,2. A Liss, New York
181. Sankaranarayanan K (1982) Genetic effects of ionizing radiation in multicellular eukaryotes and the assessment of genetic radiation hazards in man. Elsevier, Amsterdam
182. Scriver CR, Rosenberg LE (1973) Amino acid metabolism and its disorders. Saunders, Philadelphia
183. Sedano HO, Sauk JJ, Gorlin RJ (1977) Oral manifestations of inherited disorders. Butterworths, Boston London
184. Setlow JK, Hollaender A (eds) (1979-1983) Genetic engeneering, Vol 1 1979, Vol 2 1980, Vol 3 1981, Vol 4 1982, Vol 5 1983. Plenum Press, New York London
185. Siegel S (1956) Nonparametric statistics for the behavioral sciences. McGraw-Hill, New York

186. Simons EL (1972) Primate evolution. An introduction to man's place in nature. Macmillan, New York
187. Smith DW (1982) Recognizable patterns of human malformation. 3rd ed. WB Saunders, Philadelphia
188. Snell GD, Dausset J, Nathenson S (1977) Histocompatibility. Academic Press, New York
189. Sorsby A (1970) Ophthalmic genetic, 2nd ed. Butterworths, London
190. Sparkes RS, Comings DE, Fox CF (1977) Molecular human cytogenetics. Academic Press, Ney York San Francisco London
191. Spranger JW, Langer LO, Wiedemann H-R (1974) Bone dysplasias. Fischer, Stuttgart
192. Sutherland GR, Hecht F (1985) Fragile sites on human chromosomes. Oxford Univ Press, Oxford etc
193. Svejgaard A, Hauge M, Jersild C, Platz P, Ryder LP, Staub Nielsen L, Thomsen M (1979) The HLA system. An introductory survey. Monogr Hum Genet 7 Karger, Basel
194. Schimke RT (ed) (1982) Gene amplification. Cold Spring Harbor Lab
195. Schinzel A (1984) Catalogue of unbalanced chromosome aberrations in man. De Gruyter, Berlin New York
196. Schloot W, Goedde HW (1974) Biochemische Genetik des Menschen. In: Vogel F (ed) Handbuch der Allgemeinen Pathologie. Vol IX: Erbgefüge, pp 325-494. Springer, Heidelberg
197. Schreier K (1979) Die angeborenen Stoffwechselanomalien, 2 Aufl. Thieme, Stuttgart
198. Schull WJ, Neel JV (1965) The effect of inbreeding on Japanese children. Harper & Row, New York
199. Schulman JD, Simpson JL (eds) (1981) Genetic Diseases in Pregnancy. Academic press, New York
200. Schulz-Schaeffer J (1980) Cytogenetics. Plants, animals, humans. Springer, New York Heidelberg Berlin
201. Schwarzacher HG, Wolf U (1974) Methods in human cytogenetics. Springer, New York Heidelberg Berlin
201a. Schwarzacher HG (1976) Chromosomes. Hdb d Mikrosk Anat d Menschen I/3. Springer Verlag, Berlin Heidelberg New York
202. Stamatoyannopoulos G, Nienhuis AW (eds) (1985) Experimental approaches for the study of hemoglobin switching. Alan R Liss, New York
203. Stanbury JB, Wyngaarden JB, Fredrickson DS, Goldstein JL, Brown MS (1983) The metabolic basis of inherited disease, 5th ed. McGraw-Hill, New York Toronto London
204. Stern C (1973) Principles of human genetics, 3rd ed. Freeman, San Francisco
205. Stevenson AC, Davison BCC, Oakes MW (1976) Genetic counselling, 2d ed. Lippincott, Philadelphia
206. Stewart RE, Prescott GH (1976) Oral facial genetics. The CV Mosby Comp, Saint Louis
207. Therman E (1986) Human chromosomes. 2nd ed. Springer, New York Heidelberg Berlin
207a. Tiwari JL, Terasaki PI (1985) HLA and disease associations. Springer-Verlag, New York
208. Tsuang MT, Vandermey R (1980) Genes and the mind - Inheritance of mental illness. Oxford University Press, Oxford
209. Vogel F (ed) (1972) Spontaneous mutation. International Titisee Workshop. Hum Genet 16: 1-180
210. Vogel F (ed) (1974) Handbuch der allgemeinen Pathologie: Vol IX, Erbgefüge. Springer, Berlin Heidelberg New York
211. Vogel F, Helmbold W (1972) Blutgruppen - Populationsgenetik und Statistik. Humangenetik, ein kurzes Handbuch. Becker PE (ed), Vol I/4, Thieme, Stuttgart, pp 129-557.
212. Vogel F, Röhrborn G (eds) (1970) Chemical mutagenesis in mammals and man. Springer, Berlin Heidelberg New York
213. Vogel F, Röhrborn G, Schleiermacher E, Schroeder TM (1969) Strahlengenetik der Säuger. Fortschr Allg Klin Humangenetik
214. Waardenburg PJ, Franceschetti A, Klein D (1961, 1963) Genetics and ophthalmology, Vol I, II. Blackwell, Oxford
215. Wald NJ (ed) (1984) Antenatal and neonatal screening. Oxford University Press, Oxford New York Tokyo
216. Wald NJ, Cuckle HS (1984) Neural tube defects: screening and biochemical diagnosis. Rodeck CH, Nicolaides KH (eds) prenatal Diagnosis. John Wiley & Sons, Chichester New York
217. Wallace B (1981) Basic population genetics. Columbia University Press, New York
218. Warkany J, Lemire RJ, Cohen MM (1981) Mental retardation and congenital malformations of the Central Nervous System. Year Book Med Publ, Chicago London
219. Washburn SL (1982) Human evolution. Perspect Biol Med 25: 582-602
220. Watson JD (1976) Molecular biology of the gene, 3rd ed. Benjamin, New York
221. Weatherall DJ (1985) The New Genetics and Clinical Practice. Oxford Univ. Press
222. Weatherall DJ, Clegg JB (1981) The Thalassemia Syndromes, 3rd ed. Blackwell, Oxford
223. White MJD (1973) Animal cytology and evolution, 3rd ed. Cambridge University press, New York
224. Wiedemann H-R, Grosse F-R, Dibborn H (1982) Das charakteristische Syndrom. Schattauer, Stuttgart New York
225. Williams RJ (1956) Biochemical individuality. John Wiley & Sons, New York
226. Wolpert L (1984) DNA and its message. Lancet 2: 853-856
227. Wood C, Trounson A (1984) Clinical in vitro fertilization. Springer, Berlin etc.
228. Work TS, Burden RH (eds) (1983) Laboratory techniques in biochemistry and molecular biology. Elsevier, Amsterdam New York Oxford
229. Wyngaarden JB, Smith LH (eds) (1985) Cecil Textbook of Medicine. Saunders, Philadelphia
230. Yunis JJ (ed) (1969) Biochemical methods in red cell genetics. Academic Press, New York London
231. Yunis JJ (ed) (1979) Chromosomal Syndromes. Academic Press, New York
232. Zacharov AF, Benusch VA, Kuleshov NP, Baranowskaya LI (in Russian) (1982) Human Chromosomes (Atlas). Meditsina, Moscow
233. Zaleski MB, Dubiski S, Niles EG, Cunningham RK (1983) Immunogenetics. Pitman, Boston etc
234. Zerbin-Rüdin E (1967) Hirnatrophische Prozesse. In: Becker PE (ed) Humangenetik, ein kurzes Handbuch, Vol II/2, pp 84-157. Thieme Verlag, Stuttgart

235. Züblin W (1969) Chromosomale Aberrationen und Psyche. Karger, Basel New York

References to Introduction and Chapter 1

236. Allen GE (1975) Genetics, eugenics and class struggle. Genetics 79: 29-45
236a. Becker PE Wege ins dritte Reich. Wissenschaftler und Ideologen vor Hitler (To be printed)
237. Cairns J, Stent GS, Watson JD (1966) Phage and the origin of molecular biology. Cold Spring Harbor Lab, New York
238. Committee for the Study of Inborn Errors of Metabolism, Assembly of Life Sciences NRC. Genetic screening. Programs, principles, and research. National Academy of Sciences 1975, Washington (DC)
239. Baltzer F (1962) Theodor Boveri - Leben und Werk. Wissenschaftliche Verlagsgesellschaft, Stuttgart
240. Bernstein F (1924) Ergebnisse einer biostatistischen zusammenfassenden Betrachtung über die erblichen Blutstrukturen des Menschen. Klin Wochenschr 3: 1495-1497
241. Bodmer WF (ed) (1978) The HLA system. Br Med Bull 34: (3) 213-319
242. Brock DJH (1977) Biochemical and cytological methods in the diagnosis of neural tube defects. Prog Med Genet 2: 1-40
243. Bunge M (1967) Scientific research. I. The search for system. II. The search for truth. Springer, Berlin Heidelberg New York
244. Capelle W (1953) Die Vorsokratiker. Kröner, Stuttgart
244a. Cremer T (1986) Von der Zellenlehre zur Chromosomentheorie. Springer, Berlin etc
245. Dungern E von, Hirszfeld L (1911) On the group-specific structures of the blood. III. Z Immunitätsforsch 8: 526-562
246. Dunn LC (1962) Cross currents in the history of human genetics. Am J Hum Genet 14: 1-13
247. Ephrussi B, Weiss MC (1965) Interspecific hybridization of somatic cells. Proc Natl Acad Sci USA 53: 1040
247a. Falk R (1984) The gene in search of an identity. Hum Genet 68, 195-204
248. Galton F (1865) Hereditary talent and character. Macmillan's Magazine 12: 157
249. Garrod AE (1902) The incidence of alcaptonuria: A study in chemical individuality. Lancet II: 1616-1620
250. Graham LR (1977) Political ideology and genetic theory: Russia and Germany in the 1920's. Hastings Cent Rep 7: 30-39
251. Haller M (1963) Eugenics; hereditarian attitudes in American thought. New Brunswick (NJ)
252. Hardy GH (1908) Mendelian proportions in a mixed population. Science 28: 49-50
253. Harris H (1969) Enzyme and protein polymorphism in human populations. Br Med Bull 25: 5
254. Harris H, Watkins JF (1965) Hybrid cells from mouse and man: artificial heterokaryons of mammalian cells from different species. Nature 205: 640
255. Joravsky D (1970) The Lysenko affair. Harvard University Press, Boston
256. Kevles DJ (1985) In the Name of Eugenics. Genetics and the Uses of Human Heredity. Alfred A Knopf, New York
257. Kuhn TS (1962) The structure of scientific revolutions. University of Chicago Press, Chicago
258. Lakatos I, Musgrave A (1970) Criticism and the growth of knowledge. Cambridge University Press, New York
259. Landsteiner K (1900) Zur Kenntnis der antifermentativen, lytischen und agglutinierenden Wirkungen des Blutserums und der Lymphe. Zentralbl Bakteriol 27: 357-362
260. Landsteiner K, Wiener AS (1940) An agglutinable factor in human blood recognized by immune sera for rhesus blood. Proc Soc Exp Biol 43: 223
261. Levine P, Burnham L, Katzin EM, Vogel P (1941) The role of isoimmunization in the pathogenesis of erythroblastosis fetalis. Am J Obstet Gynecol 42: 925-937
262. Levine P, Stetson RE (1939) An unusual case of intragroup agglutination. J Am Med Assoc 113: 126-127
263. Ludmerer K (1972) Genetics and American society. Johns Hopkins University Press, Baltimore
264. McKusick VA (1975) The growth and development of human genetics as a clinical discipline. Am J Hum Genet 27: 261-273
265. Medvedev Z (1977) Soviet genetics: new controversy. Nature 268: 285-287
266. Mendel GJ (1865) Versuche über Pflanzenhybriden. Verhandlungen des Naturforschenden Vereins (Brünn)
267. Mohr H (1977) The structure and significance of science. Springer, New York Heidelberg Berlin
268. Motulsky AG (1959) Joseph Adams (1756-1818). Arch Intern Med 104: 490-496
269. Motulsky AG (1972) History and current status of pharmacogenetics. In: Human Genetics: Proceedings of the 4th International Congress of Human Genetics. Paris, September 1971. pp 381-390. Excerpta Medica, Amsterdam
270. Motulsky AG (1978) The genetics of common diseases. In: Morton NE, Chung CS (eds) Genetic epidemiology, pp 541-548. Academic Press, New York
270a. Motulsky AG (1984) Genetic epidemiology. Genet Epidemiol 1: 143-144
271. Motulsky AG (1977) Ecogenetics: genetic variation in susceptibility to environmental agents. In: Human Genetics: Proceedings of the 5th International Congress of Human Genetics. Mexico City, 10-15 October 1976, pp 375-385. Excerpta Medica, Amsterdam
272. Motulsky AG (1978) Presidential address: Medical and human genetics 1977: trends and directions. Am J Hum Genet 30: 123-131
272a. Müller-Hill B (1984) Tödliche Wissenschaft. Rowohlt, Hamburg
273. Nachtsheim H (1952) Für und wider die Sterilisierung aus eugenischer Indikation. Thieme, Stuttgart
274. Nasse CF (1820) Von einer erblichen Neigung zu tödlichen Blutungen, p 385. Horns Archiv
275. Neel JV (1966) Between two worlds. Am J Hum Genet 18: 3-20
276. Penrose LS (1967) Presidential address - the influence of the English tradition in human genetics. In:

Crow JF, Neel JV (eds) Proceedings of the 3rd International Congress of Human Genetics, pp 13-25. Johns Hopkins University Press, Baltimore

277. Ploetz A (1895) Die Tüchtigkeit unserer Rasse und der Schutz der Schwachen: Ein Versuch über Rassenhygiene und ihr Verhältnis zu den humanen Idealen, besonders zum Sozialismus. S Fischer, Berlin
278. Pollack W, Gorman JG, Freda VJ (1969) Prevention of Rh hemolytic disease. In: Brown E, Moore CV (eds) Progress in Hematology. VI. Vol VI, pp 121-147. Heinemann Medical Books, London
279. Popper KR (1970) Normal science and its dangers. In: Lakatos I, Musgrave A (eds) Criticism and the growth of knowledge, pp 51-58. Cambridge University Press, New York
280. Popper KR (1934) Logik der Forschung. Mohr, Tübingen (4rd ed. 1971)
281. Popper KR (1963) Conjectures and refutations. Rutledge & Kregan Paul, London
282. Reitlinger G (1961) The final solution. Barnes, New York
283. Rosenberg CE (1976) No other gods. On science and American social thought. Johns Hopkins University Press, Baltimore
284. Rüdin E (1916) Studien über Vererbung und Entstehung geistiger Störungen. I. Zur Vererbung und Neuentstehung der Dementia praecox. Springer, Berlin
285. Stanbury JB (1974) Inborn errors of the thyroid. Progr Med Genet 10: 55-80
286. Verschuer O von (1937) Was kann der Historiker, der Genealoge und der Statistiker zur Erforschung des biologischen Problems der Judenfrage beitragen? Forschungen zur Judenfrage 2: 216-222
287. Watkins JWN (1970) Against "normal science". In: Lakatos I, Musgrave A (eds) Criticism and the growth of knowledge, pp 25-38. Cambridge University Press, New York
288. Weinberg W (1901) Beiträge zur Physiologie und Pathologie der Mehrlingsgeburten beim Menschen. Arch Ges Physiol 88: 346-430
289. Weinberg W (1908) Über den Nachweis der Vererbung beim Menschen. Jahreshefte des Vereins für vaterländische Naturkunde in Württemberg 64: 368-382
290. Weinberg W (1912) Weitere Beiträge zur Theorie der Vererbung. IV. Über Methode und Fehlerquellen der Untersuchung auf Mendelsche Zahlen beim Menschen. Arch Rass Ges Biol 9: 165-174
291. Zimmerman D (1973) Rh. The intimate history of a disease and its conquest, pp 371. Macmillan, New York

References to Chapter 2

292. Aller V, Albisqueta JA, Perez A, Martin MA, Goday C, Del Mazo J (1975) A case of trisomy 8 mossaicism 47, XY, +8/46, XX. Clin Genet 7: 232-237
293. Anderson S, Bankier AT, Barrel BG, de Bruyn MHL, Coulson AR, Dromin J, Eperon IC, Nierlich DP, Rue BA, Sanger F, Schreier PH, Smith AJH, Staden R, Young IG (1981) Science and organization of the human mitochondrial genome. Nature 290: 457-465
294. Angell RR, Hitken RJ, van Look PFA, Lumsden MA, Templeton AA (1983) Chromosome abnormalities in human embryos after in vitro fertilization. Nature 303: 336-338
295. Appelhans H, Vosberg H-P (1984) Characterization of a human genomic DNA fragment coding for a myosin heavy chain. Hum Genet
296. Arber W (1979) Promotion and limitation of genetic exchange. Science 205: 361-365
297. Arnold J (1879) Beobachtungen über Kernteilungen in den Zellen der Geschwülste. Virchows Arch [Pathol Anat] 78: 279
297.a Balkan W, Martin RH (1983) Chromosome segregation into the spermatozoa of two men heterozygous for different regional translocations. Hum Genet 63: 345-348
298. Barr ML, Bertram LF (1949) A morphological distinction between neurones of the male and the female and the behavior of the nucleolar satellite during accelerated nucleoprotein synthesis. Nature 163: 676-677
299. Berg P (1981) Dissections and reconstructions of genes and chromosomes. Science 213: 296-303
300. Berger R, Tonati G, Derre J, Ortiz MA, Martinelli J (1974) Cri du chat syndrome with maternal insertional translocation. Clin Genet 5: 428-432
301. Bergsma D (ed) (1974) Urinary system and others. Birth Defects 10: (4) Part XVI. Williams & Wilkins, Baltimore
302. Beutler E (1963) Autosomal inactivation. Lancet I: 1242
303. Beutler E (1964) Gene inactivation: The distribution of gene products among populations of cells in heterozygous humans. Cold Spring Harbor Symp Quant Biol 29: 261
304. Beutler E, Yeh M, Fairbanks VF (1962) The normal human female as a mosaic of X chromosome activity: Studies using the gene for G-6-PD as a marker. Proc Natl Acad Sci USA 48: 9
305. Blanc H, Chen K-H, D'Amore MA, Wallace DC (1983) Amino acid change associated with the major polymorphic Hinc II site of Oriental and Caucasian Mitochondrial DNAs. Americ J Hum Genet 35: 167-176
306. Bochkov NP, Kuleshov NP, Chebotarev AN, Alekhin VI, Midian SA (1974) Population cytogenetic investigation of newborns in Moscow. Hum Genet 22: 139-152
307. Bodmer WF (1981) The William Allan Memorial Award Adress: Gene clusters, genome organization and complex phenotypes. When the sequence is known, what will it mean? Amer J Hum Genet 33: 664-682
308. Böök JA, Santesson B (1960) Malformation syndrome in man associated with triploidy (69 chromosomes). Lancet I: 858-859
309. Born G, Grützner P, Hemminger HJ (1976) Evidenz für eine Mosaikstruktur der Netzhaut bei Konduktorinnen für Dichromasie. Hum Genet 32: 189-196
310. Boué J, Barichard F, Deluchat C, Der Sarkissian H, Galano P, Boué A (1981) Diagnostic prenatal des anomalies de la structure chromosomique. 226 observations. La Nouvelle Presse Medicale 10: 3299-3301

311. Bridges CB (1916) Non-disjunction as proof on the chromosome theory of heredity. Genetics I: 1-52; 107-163
312. Brisson N, Verma DPS (1982) Soybean leghemoglobin gene family: Normal, pseudo, and truncated genes. Proc Natl Acad Sci USA 79: 4055-4059
313. Brown SW (1966) Heterochromatin. Science 151: 417-435
314. Brownlee GG, Rizza A (1984) Clotting factor VIII cloned. Nature 312: 307
315. Bühler EM (1980) A synapsis of the human Y chromosome. Hum Genet 55: 145-175
315a. Burgoyne PS (1986) Mammalian X and Y crossover. Nature 319: 258-259
316. Carr DH (1970c) Chromosome studies in selected spontaneous abortions. I. Conception after oral contraceptives. Can Med Assoc J 103: 343-348
317. Carr DH (1971) Chromosomes and abortion. Adv Hum Genet 2: 201-257
318. Carr DH (1967) Chromosome anomalies as a cause of spontaneous abortion. Am J Obstet Gynecol 97: 283
319. Carter CO, Hamerton JL, Polani PE, Gunalp A, Weller SDV (1960) Chromosome translocation as a cause of familial mongolism. Lancet II: 678-680
320. Caspersson T, de la Chapelle S, Foley GE, Kudynowski J, Modest EJ, Simonsson E, Wagh V, Zech L (1968) Chemical differentiation along metaphase chromosomes. Exp Cell Res 49: 219
321. Cattanach BM (1975) Control of chromosome inactivation. Ann Rev Genet 9: 1-18
322. Chadefaux B, Allord D, Rethoré MO, Raoul O, Poissonier M, Gilgenkrantz S, Cheruy C, Jérôme H (1984) Assignment of human phosphoribosylglycinamide synthetase locus to region 21q 22.1. Hum Genet 66, 190-192
323. Chapelle A de la, Schröder J, Stensand K, Fellman J, Herva R, Saarni M, Auttolainen I, Tallila L, Tervilae L, Husa L, Tallquist G, Robson EB, Cook PJL, Sanger R (1974) Pericentric inversion of human chromosomes 9 and 10. Am J Hum Genet 26: 746-765
324. Clendenin TM, Benirschke K (1963) Chromosome studies on spontaneous abortions. Lab Invest 12: 1281-1292
325. Cohen MM, Shaw MW (1964) Effects of mitomycin C on human chromosomes. J Cell Biol 23: 386-395
326. Cohen MM (1971) The chromosomal constitution of 165 human translocations involving D group chromosomes identified by autoradiography. Ann Génét (Paris) 14: 87-96
327. Collman RD, Stoller A (1963) A life table for mongols in Victoria, Australia. J Ment Defic Res 7: 53
328. Cooper DN, Schmidtke J (1984) DNA restriction fragment length polymorphism and heterozygosity in the human genome. Hum Genet 66: 1-16
328a. Cooper DN, Schmidtke J (1986) Diagnosis of genetic disease using recombinant DNA. Hum Genet 73: 1-11
329. Creasy MR, Crolla JA, Alberman ED (1976) A cytogenetic study of human spontaneous abortions using banding techniques. Hum Genet 31: 177-196
330. Cremer C, Gray JW, Ropers H-H (1982) Flow cytometric characterization of a Chinese hamster x man hybrid cell line retaining the human Y chromosome. Hum Genet 60: 262-266
331. Davidson WM, Smith DR (1954) The nuclear sex of leucocytes. In: Overzier (ed) Intersexuality, pp 72-85. Academic Press, New York
332. Davies KE (1981) The applications of DNA recombinant technology to the analysis of the human genome and genetic disease. Hum Genet 58: 351-357
333. Davies KE, Young BD, Elles RG, Hill ME, Williamson R (1981) Cloning of a representative genomic library of the human X chromosome after sorting by flow cytometry. Nature 293: 374-375
334. Davis JR, Rogers BB, Hageman RM, Thies CA, Veomett IC (1985) Balanced reciprocal translocations: risk factors for aneuploid segregant viability. Clin Genet 27: 1-19
335. Delhanty JA, Ellis JR, Rowley PT (1961) Triploid cells in a human embryo. Lancet I: 1286
336. Denaro M, Blanc H, Johnson MJ, Chen KH, Wilmsen E, Cavalli-Sforza LL, Wallace DC (1981) Ethnic variation in Hpα I endonuclease cleavage patterns of human mitochondrial DNA. Proc Natl Acad Sci USA 78: 5768-5772
337. Drets ME, Shaw MW (1971) Specific banding patterns of human chromosomes. Proc Natl Acad Sci USA 68: 2073
338. Ducos J, Marty Y, Sanger R, Race RR (1971) Xg and X chromosome inactivation. Lancet II: 219-220
339. Dutrillaux B (1973) Nouveau système de marquage chromosomique: Les bandes T. Chromosoma 41: 395
340. Dutrillaux B, Laurent C, Robert JM, Lejeune J (1973) Inversion péricentrique, inv (10), chez la mère et aneusomie de recombinaison, inv (10), rec (10), chez son fils. Cytogenet Cell Genet 12: 245-253
341. Dutrillaux B, Lejeune J (1975) New techniques in the study of human chromosomes: Methods and applications. Adv Hum Genet 5: 119-156
342. Dutrillaux B, Viegas-Péguignot E, Aurias A, Mouthuy M, Prieur M (1981) Non random position of metaphasic chromosomes: A study of radiation induced and constitutional chromosome rearrangements. Hum Genet 59: 208-210
343. Edwards JH, Harnden DG, Cameron AH, Crosse VM, Wolff OH (1960) A new trisomic syndrome. Lancet I: 787
344. Engel J, Gunning P, Kedes L (1982) Human actin proteins as encoded by a multigene family. In: Pearson ML, Epstein HF (eds) Muscle development, molecular and cellular control, pp 107-117. Cold Spring Harbor Lab Cold Spring Harbor, New York
345. Epstein CJ (1969) Mammalian oocytes: X chromosome activity. Science 163: 1078
346. Evans HJ (1977) Chromosome anomalies among livebirth. J med Genet 14: 309-314
347. Fincham JRS, Sastry GRK (1974) Controlling elements in maize. Ann Rev Genet 8: 15-50
348. Flemming W (1882) Beiträge zur Kenntnis der Zelle und ihrer Lebenserscheinungen. III. Arch Mikr Anat 20: 1
349. Flemming W (1897) Über die Chromosomenzahl beim Menschen. Anat Anz 14: 171
350. Ford CE (1969) Mosaics and chimaeras. Br Med Bull 25: 104-109
351. Ford CE, Hamerton JL (1956) The chromosomes of man. Nature 178: 1020-1023
352. Ford CE, Miller OJ, Polani PE, Almeida JC de,

Briggs JH (1959) A sex-chromosome anomaly in a case of gonadal dysgenesis (Turner's syndrome). Lancet I: 711-713
353. Fraccaro M, Kaijser K, Lindsten J (1959) Chromosome complement in gonadal dysgenesis (Turner's syndrome). Lancet I: 886
354. Fraccaro M, Kaijser K, Lindsten J (1960) Chromosomal abnormalities in father and mongoloid child. Lancet I: 724-727
355. Fraccaro M, Lindsten J, Ford CE, Iselius L (in cooperation with many other scientists) (1980) The 11q; 22q translocation: A European collaborative analysis of 43 classes. Hum Genet 56: 21-51
356. Gartler SM, Chen S-H, Fialkow PJ, Giblett ER, Singh S (1972) X-chromosome inactivation in cells from an individual heterozygous for two X-linked genes. Nature 236: 149
357. Gartler SM, Riggs AD (1983) Mammalian X-chromosome inactivation. Ann Rev Genet 17: 155-190
358. Gartler SM, Sparkes RS (1963) The Lyon-Beutler hypothesis and isochromosome X patients with the Turner Syndrome. Lancet II: 411
358a. Geissler E (1984) Movable DNA elements and evolution. In: Geissler E, Scheler W (eds) Darwin today. VIII. Kühlungsborner Kolloquium. Akademie-Verlag, Berlin
359. German J, Archibald R, Bloom D (1965) Chromosomal breakage in a rare and probably genetically determined syndrome of man. Science 148: 506
360. Giles RE, Blanc H, Cann HM, Wallace DL (1980) Maternal inheritance of human mitochondrial DNA. Proc Natl Acad Sci USA 77: 6715-6719
361. Gitschier J, Wood WI, Goralka TM, Wion KL, Chen EY, Eaton DH, Vehar GA, Capon DJ, Lawn RM (1984) Characterization of the human factor VIII gene. Nature 312: 326-330
362. Gitschier J, Wood WI, Tuddenham EGD, Shuman MA, Goralka TM, Chen EY, Lawn RM (1985) Detection and sequence of mutations in the factor VIII gene of haemophiliacs. Nature 315: 427-430
363. Goodpasture C, Bloom SE (1975) Visualization of nucleolar organizer regions in mammalian chromosomes using silver staining. Chromosoma 53: 37-50
364. Gray JW, Langlois RG, Carrano AV, Burkhart-Schulte K, Van Dilla MA (1979) High resolution chromosome analysis: One and two parameter flow cytometry. Chromosoma 73: 9-27
365. Green MM (1980) Transposable elements in Drosophila and other diptera. Ann Rev Genet 14: 109-120
366. Grossman L, Moldave K (1980) Methods in Enzymology, Vol 65. Part I: Nucleic Acids. Academic Press, New York
367. Grouchy J de, Lamy M, Thieffry S, Arthuis M, Salmon C (1963) Dysmorphie complexe avec oligophrénie. Délétion des bras courts d'un chromosome 17-18. CR Acd Sci (Paris) 256: 1028-1029
368. Grüneberg H (1966) The case for somatic crossing over in the mouse. Genet Res 7: 58-75
369. Gusella JF, Wexler NS, Coneally PM, Nyler SL, Anderson MA, Tanzi RE, Watkins PC, Ottina K, Wallace MR, Sakaguchi AY, Young AB, Shoulson I, Bonilla E, Martin JB (1983) A polymorphic DNA marker genetically linked to Huntington's disease. Nature 306: 234-238
370. Habedank M, Rodewald A (1982) Moderate Down's syndrome in three siblings having partial Trisomy 21q22.2 and therefore no SOD-1 excess. Hum Genet 60: 74-77
371. Hagemeijer A, Smit EME (1977) Partial trisomy 21. Further evidence that trisomy of band 21q22 is essential for Down's phenotype. Hum Genet 38: 15-23
372. Haldane JBS (1936) A search for incomplete sex linkage in man. Ann Eugen 7: 28-57
373. Hamerton JL (1968) Robertsonian translocation in man. Evidence for prezygotic selection. Cytogenetics 7: 260-276
374. Hamerton JL, Ray M, Abbot J, Williamson C, Durcasse GC (1972) Chromosome studies in a neonatal population. Can Med Assoc J 106: 776-779
375. Hanauer A, Levin M, Heilig R, Daegelen D, Kahn A, Mandel JL (1983) Isolation and characterization of DNA clones for human skeletal muscle alpha-actin. Nucleic Acid Res 11: 3503-3516
376. Harnden DG, Lindsten JE, Buckton K, Klinger HP (1981) An international system for human cytogenetic nomenclature. High resolution Banding. Birth Defects: Original Article Series, Vol XVII, No 5
377. Harper ME, Ullrich A, Saunders GF (1981) Localization of the human insulin gene to the distal end of the short arm of chromosome 11. Proc Natl Acad Sci USA 78: 4458-4460
378. Hecht F, Jacky PB, Sutherland GR (1982) The fragile X chromosome. Amer J Hum Genet 11: 489-495
379. Heberer G (1940) Die Chromosomenverhältnisse des Menschen. In: Just G (ed) Handbuch der Erbbiologie des Menschen, Vol I, pp 2-30. Springer, Berlin
380. Heitz E (1928) Das Heterochromatin der Mouse. I. Pringsheims Jb wiss Botanik 69: 762-818
381. Hindley J (1983) DNA sequencing. In: Work TS, Burden RH (eds) Laboratory techniques in biochemistry and molecular biology. Elsevier, Amsterdam New York Oxford
382. Hoo JJ, Förster C, Kindermann I, Zabel B, Hansen S (1974) Supernumerary small ring chromosome. Human Genet 25: 17-28
383. Hooft C, Coetsier H, Oyre E (1968) Syndrome de Turner et inversion péricentrique probable du chromosome 2, 45, X, 2. (p+, qt). Ann Génét 11: 181-183
384. Hsu TC (1952) Mammalian chromosomes in vitro. I. The karyotype of man. J Hered 43: 167
385. Hsu TC (1975) A possible function of constituitive heterochromatin: The bodyguard hypothesis. Genetics (Suppl) 79: 137-150
386. Hsu TC, Pomerat CM (1953) Mammalian chromosomes in vitro. II. A method for spreading the chromosomes of cells in tissue culture. J Hered 44: 23-29
387. Hultén M, Lindsten J (1973) Cytogenetic aspects of human male meiosis. Adv Hum Genet 4: 327-387
388. Humphries SE, Whittall R, Minty A, Buckingham M, Williamson R (1981) There are approximately 20 actin genes in the human genome. Nucleic Acid Res 9: 4895-4908
389. Iselius L, Lindsten J, Aurias A, Fraccaro M and many other authors (1983) The 11q;22q translocation: A collaborative study of 20 new cases and analysis of 110 families. Hum Genet 64: 343-355

390. Itakura K (1985) Antisense RNA sequences. First Intern Symp on the Role of Recombinant DNA in Genetics (RL Teplitz et al eds) Crete, (In the Press 1986)
390 a. Jacobs PA (1977) Human chromosome heteromorphisms (variants). Progress in Med Genet NS Vol II, 251-274
391. Jacob F, Brenner A, Cuzin F (1963) On the regulation of DNA replication. Cold Spring Harbor Symp Quant Biol 28: 329-348
392. Jacobs PA (1977) Human chromosome heteromorphisms (variants). Progr in Med Genet (New series) 2: 251-274
393. Jacobs PA, Baikie AG, MacGregor TN, Harnden DG (1959) Evidence for the existence of the human "superfemale." Lancet II: 423-425
394. Jacobs PA, Brunton M, Melville MM, Brittain RP, McClermont WF (1965) Aggressive behavior, mental subnormality and the XYY male. Nature 208: 1351-1352
395. Jacobs PA, Strong JA (1959) A case of human intersexuality having a possible XXY sex-determining mechanism. Nature 183: 302-303
396. Jacobsen P, Mikkelsen M, Rosleff F (1974) The trisomy 8 syndrome: Report of two further cases. Ann Génét (Paris) 17: 87-94
397. Jalbert P, Sele B (1979) Factors predisposing to adjacent-2 and 3: 1 disjunctions: Study of 161 human reciprocal translocations. J Med Genet 16: 467-478
398. Jeffreys AJ, Wilson V, Thein SL (1985) Hypervariable "minisatellite" regions in human DNA. Nature 314: 67-73
399. Johannisson R, Gropp A, Winking H, Coerdt W, Rehder H, Schwinger E (1983) Down's syndrome in the male. Reproductive pathology and meiotic studies. Hum Genet 63: 132-138
400. Jones RS, Potter SS (1985) L1 sequences in HeLa extrachromosomal circular DNA: Evidence for circularization by homologous recombination. Proc Nat Acad Sci USA 82: 1989-1993
401. Joseph JL, Brasch JM, Smyth DR (1982) Patterns of exchange induced by mitomycin C in C-bands of human chromosomes. II. High frequency of Y-Y exchange in XYY cells. Hum Genet 62: 346-348
401 a. Kaiser P (1984) Pericentric inversions. Problems and significance for clinical genetics. Hum Genet 68: 1-47
402. Kakati S, Nihill M, Sinah A (1973) An attempt to establish trisomy 8 syndrome. Hum Genet 19: 293-300
403. Khalili K, Salas C, Weinemann R (1983) Isolation and characterization of human actin genes in phage Lambda vectors. Gene 21: 9-17
404. Kirsch-Volders M, Hens L, Susanne C (1980) Telomere and centromere association tendencies in the human male metaphase complement. Hum Genet 54: 69-74
405. Kjessler B (1966) Karyotype, meiosis and spermatogenesis in a sample of men attending an infertility clinic. Monogr Hum Genet 2 Karger, Basel
406. Koske-Westphal Th, Passarge E (1974) Die Chromosomen des Menschen und ihre Untersuchung in somatischen Zellen. In: Vogel F (ed) Handbuch der allgemeinen Pathologie. Vol IX: Erbgefüge, pp 261-323. Springer, Berlin Heidelberg New York
407. Koskull H von, Aula P (1974) Inherited (13, 14) translocation and reproduction. Hum Genet 24: 85-91
408. Kunze J, Tolksdorf M, Wiedemann H-R (1975) Cat eye syndrome. Hum Genet 26: 271-289
409. Kurilo LF (1981) Oogenesis in antenatal development in man. Hum Genet 57: 86-92
410. Kurnit DM (1979) Satellite DNA and heterochromatin variants: The case for unequal mitotic crossing over. Hum Genet 47: 169-186
411. Latt SA (1973) Microfluorometric detection of deoxyribonucleic acid replication in human metaphase chromosomes. Proc Natl Acad Sci USA 70: 3395-3399
412. Latt SA, Schreck RR, Laveday KS, Dougherty CP, Schuler CF (1980) Sister chromatid exchanges. Adv in Hum Genet 10: 267-331
413. Lauritsen JG (1977) Genetic aspects of spontaneous abortion. University of Aarhus: Laegeforeninges
414. Lauritsen JG, Bolund L, Friedrich U, Therkelsen AL (1979) Origin of triploidy in spontaneous abortuses. Ann Hum Genet 43: 1-6
415. Lejeune J (1968) De la duplication de structures circulaires. Ann Génét. (Paris) 11: 71-77
416. Lejeune J, Berger R (1965) Sur deux observations familiales de translocations complexes. Ann Génét (Paris) 8: 21-30
417. Lejeune J, Gautier M, Turpin MR (1959) Étude des chromosomes somatiques de neuf enfants mongoliens. CR Acad Sci (Paris) 248: 1721-1722
418. Lejeune J, Lafourcade J, Berger R, Vialatte J, Roeswillwald M, Seringe P, Turpin R (1963) Trois cas de délétion partielle du bras court d'un chromosome 5. CR Acad Sci (Paris) 257: 3098-3102
419. Leonard C, Hazael-Massieux P, Bocquet L, Larget-Piet L, Boué J (1975) Inversion péricentrique inv (2) (p11q13) dans les familles non apparentées. Hum Genet 28: 121-128
420. Lifschytz E, Lindsley DL (1972) The role of X-chromosome inactivation during spermatogenesis. Proc Natl Acad Sci USA 69: 182-186
421. Lilley DMJ, Pardon JF (1979) Structure and function of chromatin. Ann Rev Genet 13: 197-233
422. Lubs HA, Ruddle FH (1970) Applications of quantitative karyotypy to chromosome variation in 4400 consecutive newborns. In: Jacobs PA, Price WH, Law P (eds) Human population cytogenetics, pp 120-142. Pfizer Medical Monographs 5. University of Edinburgh Press, Edinburgh
423. Lyon MF (1971) Possible mechanisms of Y chromosome inactivation. Nature New Biol 232: 229
424. Lyon MH (1968) Chromosomal and subchromosomal inactivation. Ann Rev Genet 2: 31-52
425. Lyon MF (1961) Gene action in the X-chromosome of the mouse. Nature 190: 372-373
426. Lyon MF (1961) Gene action in the X-chromosome of mammals including man. Proceedings 2nd International Conference of Human Genetics. Rome, August 1961, pp 1228-1229
427. Madan K (1983) Balanced structural changes involving the human X: Effect on sexual phenotype. Hum Genet 63: 216-221
428. Maddox J (1984) Who will clone a chromosome? Nature 312: 306
429. Mager DL, Henthorn PS (1984) Identification of a retroviruslike repetitive element in human DNA. Proc Nat Acad Sci USA 81: 7510-7514
430. Martin GM (1978) The pathobiology of aging. University of Washington Medicine 5: 3-10

431. Martin RH, Balkan W, Burns K, Lin CC (1982) Direct chromosomal analysis of human spermatozoa. Amer J Hum Genet 34: 459-468
432. Martin RH, Balkan W, Burns K, Rademaker AW, Lin CC, Rudd NL (1983) The chromosome constitution of 1000 human spermatozoa. Hum Genet 63: 305-309
433. McClintock B (1956) Controlling elements and the gene. Cold Spring Harbor Symp Quant Biol 21: 197-216
434. Merril CR, Harrington MG (1985) The search for mitochrondrial inheritance of human diseases. TIG 1: 140-144
435. Metz CW (1916) Chromosome studies on the diptera II. The paired association of the chromosomes in the diptera and its significance. J Exp Zool 21: 213
436. Mikelsaar A-V, Schmid M, Krone W, Schwarzacher HG, Schnedl W (1977) Frequency of Ag-stained nucleolus organizer regions in the acrocentric chromosomes of man. Hum Genet 31: 13-17
437. Miklos GLG, John B (1979) Heterochromatin and satellite DNA in man: properties and prospects. Amer J Hum Genet 31: 264-280
439. Miller OJ (1985) Dosage compensation in mammals: Why does a gene on the inactive X yield less product than one on the active X? Hum Genet 69: 97-101
440. Miller OL, Beatty BR (1969) Visualization of molecular genes. Science 164: 955-957
441. Miller OJ, Miller DA, Warburton D (1973) Application of the new staining techniques to the study of human chromosomes. Prog Med Genet 9: 1-48
442. Miller OJ, Schreck RR, Beiser SM, Erlanger BF (1973) Immunofluorescent studies of chromosome banding with antinucleotide antibodies. In: Nobel Symposium 23: Chromosome identification, pp 43-48. Academic Press, New York, London
443. Mikkelsen M (1971) Down's syndrome. Current stage of cytogenetic research. Hum Genet 12: 1-28
444. Mikkelsen M, Hansson A, Jacobsen P, Hobolth N (1975) Translocation (13q21q). Four generation family study with analysis of satellite associations, fluorescent markers, and prenatal diagnosis. Hum Genet 27: 303-307
445. Mikkelsen M, Stene J (1970) Genetic counseling in Down's syndrome. Hum Hered 20: 457-464
446. Milunsky A (1973) The prenatal diagnosis of hereditary disorders. Thomas, Springfield
447. Moorhead PS, Nowell PC, Mellman WJ, Battips DM, Hungerford DA (1960) Chromosome preparations of leukocytes cultured from human peripheral blood. Exp Cell Res 20: 613-616
448. Morgan TH (1910) Sex-limited inheritance in drosophila. Science 32, 120-122
449. Morton NE, Jacobs PA, Frackiewicz A (1975) The effect of structural aberrations of the chromosomes on reproductive fitness in man. 1. Methodology. Clin Genet 8: 159-168
450. Morton NE, Lindsten J, Iselius L, Yee S (1982) Data and theory for a revised chiasma map of man. Hum Genet 62: 266-270
451. Moser H, Emery AEH (1974) The manifesting carrier in Duchenne muscular dystrophy. Clin Genet 5: 271-284
453. Muller HJ (1941) Induced mutations in drosophila. In: Genes and chromosomes, structure and organization. Cold Spring Harbor Symp on Quant Biol Vol IX, 151-167
454. Mullinger AM, Johnson RT (1980) Packing DNA into chromosomes. J Cell Sci 46: 61-86
455. Nagakome Y, Iimura K, Tangiuchi K (1973) Points of exchange in a human no. 5 ring chromosome. Cytogenet Cell Genet 12: 35-39
456. Natarajan AT, Zwanenburg TSB (1982) Mechanisms for chromosomal aberrations in mammalian cells. Mutation Research 95: 1-6
457. Nathans D (1980) Restriction endonucleases, Simian Virus 40, and the new genetics. Science 206: 903-909
458. Niebuhr E (1974) Triploidy in man. Hum Genet 21: 103-125
459. Niebuhr E (1972) Localization of the deleted segment in the cri-du-chat syndrome. Hum Genet 16: 357-358
460. Noel B, Quack B, Rethoré MO (1976) Partial deletions and trisomies of chromosome 13; mapping of bands associated with particular malformations. Clin Genet 9: 593-602
461. Nowakowski H, Lenz W, Parada J (1958) Diskrepanz zwischen Chromatinbefund und chromosomalem Geschlecht beim Klinefelter-Syndrom. Klin Wochenschr 36: 683-684
462. Nowakowski H, Lenz W, Parada J (1959) Diskrepanz zwischen Chromatinbefund und genetischem Geschlecht beim Klinefelter-Syndrom. Acta Endocrinol 30: 296-320
463. Ohno S, Kaplan WD, Kinosita R (1959) Formation of the sex chromatin by a single X chromosome in liver cells of Rattus norvegicus. Exp Cell Res 18: 415-418
464. Ohno S, Makino S (1961) The single-X nature of sex chronatin in man. Lancet I: 78-79
465. Øster J, Mikkelsen M, Nielsen A (1964) The mortality and causes of death in patients with Down's syndrome (mongolism). International Copenhagen Congress of Scientific Study of Mental Retardation 1: 231
466. Pachmann U, Rigler R (1972) Quantum yield of acridines interacting with DNA of defined base sequence. Exp Cell Res 72: 602
467. Painter TS (1923) Studies in mammalian spermatogenesis. II. The spermatogenesis of man. J Exp Zool 37: 291-321
468. Paris Conference 1971: Standardization in human cytogenetics. Cytogenetics 11: 313-362 (1972)
469. Passarge E (1979) Emil Heitz and the concept of heterochromatin: Longitudinal chromosome differentiation was recognized fifty years ago. Am J Hum Genet 31: 106-115
470. Passarge E, Fries E (1973) X-chromosome inactivation in X-linked hypohidrotic ectodermal dysplasia. Nature New Biology 245: 58-59
471. Patau K (1960) The identification of individual chromosomes, especially in man. Am J Hum Genet 12: 250-276
472. Patau K, Smith DW, Therman E, Inhorn SL, Wagner HP (1960) Multiple congenital anomaly caused by an extra chromosome. Lancet I: 790-793
473. Pawlowitzki IH (1972) Frequency of chromosome abnormalities in abortions. Hum Genet 16: 131-136

474. Pearson PL, Bobrow M (1970) Fluorescent staining of the Y chromosome in meiotic stages of the human male. J Reprod Fertil 22: 177-179
475. Penrose LS, Delhanty JDA (1961) Triploid cell cultures from a macerated foetus. Lancet I: 1261
476. Pirastu M (1985) The oligonucleotide technique for antenatal diagnosis of β-thalassemia in Italy. In: Teplitz RL et al. (eds) First Intern Symp on the Role of Recombinant DNA in Genetics. Crete (In the press 1986)
477. Polani PE (1962) Sex chromosome abnomalies in man. In: Hamerton JL (ed) Chromosomes in medicine, pp 73-139. Heinemann, London
478. Polani PE, Bishop PMF, Lennox B, Ferguson-Smith MA, Stewart JSS, Prader A (1958) Color vision studies in the X-chromosome constitution of patients with Klinefelter's syndrome. Nature 182: 1092-1093
479. Polani BE, Briggs JH, Ford CE, Clarke CM, Berg JM (1960) A mongol child with 46 chromosomes. Lancet I: 721-724
480. Quie PG, White JG, Holmes B, Good RA (1967) In vivo bactericidal capacity of human polymorphonuclear leucocytes: Diminished activity in chronic granulomatous disease of childhood. J Clin Invest 46: 668-679
481. Rao PN, Johnson RT, Sperling K (1982) Premature chromosome condensation. Application in basic, clinical and mutation research. Academic Press, New York etc
482. Rappold GA, Vosberg H-P (1984) Chromosomal localization of a human myosin heavy-chain gene by in situ hybridization. Hum Genet 65: 195-197
483. Richards BW (1969) Mosaic mongolism. J Ment Defic Res 13: 66-83
484. Rigler R (1966) Microfluoreometric characterization of intracellular nucleic acids and nucleoproteins by acridine orange. Acta Physiol Scand 67: [Suppl 267], 1
485. Röhme D, Heneen WK (1982) Banding patterns in prematurely condensed chromosomes and the underlying structure of the chromosomes, pp 131-157. In: Rao PN, Johnson RT, Sperling K (eds) Premature chromosome condensation. Academic Press, New York etc
486. Ropers HH, Migl B, Zimmer J, Müller CR (1981 a) Steroid sulfatase activity in cultured fibroblasts of XX males. Cytogenet Cell Genet 30: 168-173
487. Ropers HH, Migl B, Zimmer J, Fraccaro M, Maraschio PP, Westerveld A (1981 b) Activity of steroid sulfatase in fibroblasts with numerical and structural X chromosome aberrations. Hum Genet 57: 345-356
488. Ropers HH, Wienker TF, Grimm T, Schroetter K, Bender K (1977) Evidence for preferential X-chromosome inactivation in a family with Fabry disease. Am J Hum Gen 29: 361-370
488 a. Rouyer F, Simmler MC, Johnsson C, Vergnaud G, Cooke HJ, Weissenbach J (1986) A gradient of sex linkage in the pseudoautosomal region of the human sex chromosomes. Nature 319: 291-295
489. Rudok E, Jacobs PA, Yanaginachi R (1978) Direct analysis of the chromosome constitution of human spermatozoa. Nature 274: 911-913
490. Ruzicka F (1973) Über die Ultrastruktur menschlicher Metaphase-Chromosomen. Hum Genet 17: 137-144
491. Sachs L (1953/54) Sex-linkage and the sex chromosome in man. Ann Hum Genet 18: 255-261
492. Southern EM (1975) Detection of specific sequences among DNA fragments separated by gel electrophoresis. J Mol Biol 98: 503-517
493. Setlow JK, Hollaender A (1979) Genetic engineering: Principles and methods, Vol 1. Plenum Press, New York
494. Shmookler Reis RJ, Lumpkin CK, McGill JR, Riabowol KT, Goldstein S (1983) Extrachromosomal circular copies of an "inter-Alu" unstable sequence in human DNA are amplified during in vitro and in vivo ageing. Nature 301: 394-398
495. Smith DW, Patau K, Therman E, Inhorn SL (1960) A new autosomal trisomy syndrome: Multiple congenital anomalies caused by an extra chromosome. J Pediatr 57: 338-345
496. Smith HO (1979) Nucleotide sequence specificity of restriction endonucleases. Science 205: 455-462
497. Solari AJ (1980) Synaptonemal complexes and associated structures in microspread spermatocytes. Chromosome 81: 315-337
498. Sperling K (1984) Frequency and origin of chromosome abnormalities in man. In: Obe B (ed) Mutation in man, pp 128-146 Springer, Berlin Heidelberg New York
499. Sperling K (1984) Genetische Sektion - Anatomie der menschlichen Gene. In: (Passarge E ed) pp 73-100 Verlag Chemie, Darmstadt
500. Sperling K, Rao PN (1974) The phenomenon of premature chromosome condensation: Its relevance to basic and applied research. Humangenetik 23: 235-258
501. Summitt RL, Martens PR, Wilroy RS (1973) X-autosome translocation in normal mother and effectively 21-monosomic daughter. J Pediatr 84: 539-546
501 a. Schaefer MSD (1983) Segregation and Pathologie autosomaler familiärer Translokationen beim Menschen. Diss Univ Kaiserslautern
502. Schemp W, Meer B (1983) Cytologic evidence for three human X-chromosomal segments escaping inactivation. Hum Genet 63: 171-174
503. Schinzel A (1979) Autosomale Chromosomenaberrationen. Archiv für Genetik 52: 1-204
504. Schleiermacher E, Schliebitz U, Steffens C (1974) Brother and sister with trisomy 10p: A new syndrome. Hum Genet 23: 163-172
505. Schmid CW, Jelinek WR (1982) The Alu family of dispersed repititive sequences. Science 216: 1065-1070
506. Schmidt ER (1985) Sequenzen von DNA. In: Blin N, Trendelenburg MF, Schmidt ER (eds) Molekular- und Zellbiologie, pp 35-51. Springer Verlag, Berlin Heidelberg New York
507. Schmidtke J, Cooper DN (1983) A list of cloned DNA sequences. Hum Genet 65: 19-26
508. Schmidtke J, Cooper DN (1984) A list of cloned human DNA sequences - Supplement. Hum Genet 67: 111-114
509. Schmidtke J, Epplen JT (1980) Sequence organization of animal nuclear DNA. Hum Genet 55: 1-18
510. Schnedl W (1971) Banding pattern of human chromosomes. Nature 233: 93
511. Schnedl W (1978) Structure and variability of human chromosomes analyzed by recent techniques. Hum Genet 41: 1-10

512. Schnedl W (1974) Banding patterns in human chromosomes visualized by Giemsa staining after various pretreatments. In: Schwarzacher HG, Wolf U (eds) Methods in human cytogenetics, pp 95-116. Springer, New York Heidelberg Berlin
513. Schneider EL, Epstein CJ (1972) Replication rate and life span of cultured fibroblasts in Down's syndrome. Proc Soc Exp Biol Med 141: 1092-1094
514. Schneiderman LJ, Smith CAB (1962) Non-random distribution of certain homologous pairs of normal human chromosomes in metaphase. Nature 195: 1229-1230
515. Schroeder TM, Anschütz F, Knopp A (1964) Spontane Chromosomenaberrationen bei familiärer Panmyelopathie. Hum Genet 1: 194-196
516. Schwarzacher HG (1974) Fluorescence microscopy of chromosomes and interphase nuclei. In: Schwarzacher HG, Wolf U (eds) Methods in human cytogenetics, pp 83-93. Springer, New York Heidelberg Berlin
517. Schwarzacher HG (1970) Die Ergebnisse elektronenmikroskopischer Untersuchungen an somatischen Chromosomen des Menschen. Hum Genet 10: 195-208
518. Schwarzacher HG, Wachtler F (1983) Nucleolus organizer regions and nucleoli. Hum Genet 63: 89-99
519. Simpson JL (1976) Disorders of sexual differentiation. Etiology and clinical delineation. Academic Press, New York
520. Starlinger P (1980) IS elements and transposons. Plasmid 3: 241-259
521. Stern C (1936) Somatic crossing-over and segregation in Drosophila melanogaster. Genetics 21: 625-730
522. Stern C (1959) The chromosomes of man. J Med Educ 34: 301-314
523. Strayer D, Heintz N, Roeder R, Gillespie D (1983) Three organizations of human DNA. Proc Nat Acad Sci USA 80: 4770-4774
524. Taylor AL (1963) Bacteriophage-induced mutation in Escherichia coli. Proc Natl Acad Sci USA 50: 1043-1051
525. Taylor JH (1960) Asynchronous duplication of chromosomes in cultured cells of Chinese hamster. J Biophys Biochem Cytolog 7: 455-464
526. Taylor KM, Wolfinger HL, Brown M-G, Chadwick DL (1975) Origin of a small metacentric chromosome. Familial and cytogenetic evidence. Clin Genet 8: 364-369
527. Teplitz RL (1985) The use of synthetic oligonucleotides in and prenatal diagnosis of genetic disease. In: Teplitz RL et al (eds) First Intern Symp on the Role of Recombinant DNA in Genetics. Crete (In the press 1986)
528. Therman E, Meyer-Kuhn E (1981) Mitotic crossing-over and segregation in man. Hum Genet 59: 93-100
529. Therman E, Patau K (1974) Abnormal X chromosomes in man. Origin, behavior and effects. Hum Genet 25: 1-16
530. Therman E, Sarto GE, Palmer CG, Kallio H, Denniston C (1979) Position of the human X inactication center on Xp. Hum Genet 50: 59-64
531. Tode JJ, Knopf JL, Wozney JM, Sutzman LA, Bueker JL, Pittman DD, Kaufman RJ, Brown E, Shoemaker Ch, Orr EC, Amphlett GW, Foster WB, Coe ML, Knutson GJ, Fass DN, Hewick RM (1984) Molecular cloning of a cDNA encoding human antihaemophilic factor. Nature 312: 342-347
532. Tjio HJ, Levan A (1956) The chromosome numbers of man. Hereditas 42: 1-6
533. Tjio HJ, Puck TT (1958) The somatic chromosomes of man. Proc Natl Acad Sci USA 44: 1229-1237
535. Trendelenburg MF (1983) Progress in visualization of eukaryotic gene transcription. Hum Genet 63: 197-215
536. Vehar GA, Keyt B, Eaton D, Rodriguez H, O'Brian DP, Rotblat F, Oppermann H, Keck R, Wood WI, Harkins RN, Tuddenham EGD, Lawn RM, Capon DJ (1984) Structure of human factor VIII. Nature 312: 337-342
537. Waardenburg PJ (1932) Das menschliche Auge und seine Erbanlagen. Bibliogr Genet 7
538. Wachtel S (ed) Errors of sex determination (Proc of the Kroc Foundation Conf.) Hum Genet 58: 1-127
539. Wallace RB (1985) The use of synthetic DNA hybridization probes as tools for genetic analysis. In: Teplitz RL et al (eds) First Intern Symp on the Role of Recombinant DNA in Genetics. Crete (In the press 1986)
540. Weisblum B, de Haseth PL (1972) Quinacrine, a chromosome stain specific for deoxyadenylate-deoxy-thymidinate-rich regions of DNA. Proc Natl Acad Sci USA 69: 629
541. Wilson GN, Lynne Szura L, Rushford C, Jackson D, Erickson J (1982) Structure and variation of human ribosomal DNA: The external transcribed spacer and adjacent regions. Am J Hum Genet 34: 32-49
542. Windhorst DB, Holmes B, Good RA (1967) A newly defined X-linked trait in man with demonstration of the Lyon effect in carrier females. Lancet I: 737-739
543. Winiwarter H von (1912) Études sur la spermatogenèse humaine. I. Cellule de Sertoli. II. Hétérochromosome et mitoses de l'epithelium seminal. Arch Biol (Liege) 27: 91-189
544. Wolf U, Reinwein H, Porsch R, Schröter R, Baitsch H (1965) Defizienz an den kurzen Armen eines Chromosoms Nr 4. Hum Genet 1: 397-413
545. Wolff E de, Schärer K, Lejeune J (1962) Contribution a l'étude des jumeaux mongoliens. Un cas de monozygotisme hétérocaryote. Helv Paediatr Acta 17: 301-328
546. Wood WI, Capon DJ, Simonsen CC, Eaton DL, Gitschier J, Keyt B, Seeburg PH, Smith DH, Hollingshead P, Wion KL, Delwart E, Tuddenham EGD, Vehar GA, Lawn RM (1984) Expression of active human factor VIII from recombinant DNA clones. Nature 312: 330-337
547. Wollenberg C, Kiefaber MP, Zang KD (1982) Quantitative studies on the arrangement of human metaphase chromosomes VIII. Localization of homologous chromosomes. Hum Genet 60: 239-248
548. Wyman AR, White R (1980) A highly polymorphic locus in human DNA. Proc Natl Acad Sci USA 77: 6754-6758
549. Yuncken C (1968) Meiosis in the human female. Cytogenetics 7: 234-238
550. Yunis JJ (1981) Mid-prophase human chromosomes. The attainment of 2000 bands. Hum Genet 56: 293-298

551. Zankl H, Zang KD (1979) Quantitative studies on the arrangement of human metaphase chromosomes. VII. The association pattern of acrocentric chromosomes in carriers of Robertsonian translocations and in their relatives with normal karyotypes. Hum Genet 52: 119-125

References to Chapter 3 and Appendices Nos. 1, 2, 3, 4, 5, 6, 7, 9

552. Aird I, Bentall HH, Roberts JAF (1953) A relationship between cancer of stomach and the AB0 groups. Br Med J I: 799-801
553. Albert ED, Baur MP, Mayr WR (eds) Histocompatibility Testing 1984. Springer-Verlag, Berlin Heidelberg New York
554. Allen F, Amos DB, Bathelor R, Bodmer W, Ceppelini R, Dausset J, Engelfriet C, Jeannet M, Kissmeyer-Nielsen F, Morris P, Payne R, Terasaki P, van Rood JJ, Walford R, Zmijewski C, Albert E, Mattuiz P, Mickey MR, Piazza A (1970) Joint report of 4th Intern Histocompatibility Workshop. Histocompatibility testing (Terasaki P ed), Munksgaard, Kopenhagen 17-47.
555. Allison AC, Blumberg BS (1958) Dominance and recessivity in medical genetics. Am J Med 25: 933-941
556. Amos DB, Kostyu DD (1980) HLA-A central immunological agency in man. Adv in Hum Genet 10: 137-208
557. Andreassen M (1943) Haemofili i Danmark. Opera ex Domo Biologiae Hereditariae Humanae Universitatis Hafniensis Munksgaard Vol, 6 Copenhagen
558. Antonarakis SE, Copeland KL, Carpenter RJ jr, Carta CA, Hoyer LW, Caskey CT, Toole JJ, Kazazian HH, Jr (1985) Prenatal diagnosis of haemophilia A by factor VIII gene analysis. Lancet 1: 1407-1409
559. Ball SP, Cook PJL, Mars M, Buckton KE (1982) Linkage between dentinogenesis imperfacta and GC. Ann Hum Genet 46: 35-40
560. Baralle FE, Shoulders CC (1984) Lipoprotein genes and hyperlipidemia. Schweiz Med Wschr 114: 1351-1358
561. Bateson W, Punnett RG (1908) Saunders: Confirmations and extensions of Mendel's principles in other animals and plants. Report to the Evolution Committee of the Royal Society (London)
562. Bauer KH (1927) Homoiotransplantation von Epidermis bei eineiigen Zwillingen. Beitr Klin Chir 141: 442-447
563. Becker PE (1964) Myopathien. In: Becker PE (ed) Humangenetik, ein kurzes Handbuch, Vol III/I, pp 411-550. Thieme, Stuttgart
564. Becker PE (1953) Dystrophia musculorum progressiva. Thieme, Stuttgart
565. Becker PE (1972) Neues zur Genetik und Klassifikation der Muskeldystrophien. Hum Genet 17: 1-22
565a. Bell J (1934) Huntington's chorea. Treasury of Human Inheritance 4
566. Bell J (1947) Dystrophia myotonica and allied diseases. Treasury of Human Inheritance 4: Teil V
567. Benirschke K, Kim CK (1973) Multiple pregnancy. N Engl J Med 288: 1276-1284; 1329-1336
568. Bennett T (1975) The T-locus of the mouse. Cell 6: 441-454
569. Benzer S (1957) The elementary units of heredity. In: McElroy WD, Glass B (eds) The chemical basis of heredity, John Hopkins University Press, Baltimore pp 70-93.
570. Berg K (1983) Genetics of coronary heart disease. Progr Med Genet N.S. 5: 35-90
571. Bernstein F (1931) Zur Grundlegung der Chromosomentheorie der Vererbung beim Menschen. Z Induktive Abstammungs-Vererbungslehre 57: 113-138
572. Bernstein F (1930) Fortgesetzte Untersuchungen aus der Theorie der Blutgruppen. Z. Induktive Abstammungs-Vererbungslehre 56: 233-237
573. Bernstein F (1930) Über die Erblichkeit der Blutgruppen. Z. Induktiven Abstammungs-Vererbungslehre 54: 400
574. Bernstein F (1925) Zusammenfassende Betrachtungen über die erblichen Blutstrukturen des Menschen. Z. Induktive Abstammungs-Vererbungslehre 37: 237
575. Berwick DM, Cretin S, Keeler EB (1980) Cholesterol, Children, and Heart Disease. An Analysis of Alternatives. Oxford New York
576. Bhende YM, Deshpande CK, Bhata HM, Sanger R, Race RR, Morgan WTJ, Watkins WM (1952) A "new" blood-group character related to the AB0 system. Lancet I: 903-904
577. Bieber FR, Nance WE, Morton CC, Brown JA, Redwine FO, Jordan RL, Mohanakumar T (1981) Genetic studies of an acardiac monster: Evidence of polar body twinning in man. Science 213: 775-777
578. Bilheimer DW, Goldstein JL, Grundy SM, Starzl TE, Brown MS (1984) Liver transplantation to provide low-density-lipoprotein receptors and low plasma cholesterol in a child with homozygous familial hypercholesterolemia. N Engl J Med 311: 1658-1664
579. Birch-Jensen A (1949) Congenital deformities of the upper extremities. Opera ex Domo Biologiae Hereditariae Humanae Universitatis Hafniensis Munksgaard, Kopenhagen 19:
580. Blackburn H (1979) Diet and mass hyperlipidemia: a public health view. In: Levy R, Rifkind B, Dennis B, Ernst N (eds) Nutrition, Lipids, and Coronary Heart Disease. Raven Press, New York
581. Bodmer WF (1972) Population genetics of the HL-A system: Retrospect and prospect. In: Dausset J, Colombani J (eds) Histocompatibility testing, Munksgaard, Copenhagen pp 611-617.
582. Bodmer WF, Bodmer JG (1978) Evolution and function of the HLA system. Br Med Bull 34: 390-316
583. Boman H, Ott J, Hazzard WR, Albers JJ, Cooper MN, Motulsky AG (1978) Familial hyperlipidemia in 95 randomly ascertained hyperlipidemic men. Clin Genet 13: 108
584. Botstein D, White RL, Skolnick M, Davis RW (1980) Construction of a genetic linkage map in man using restriction fragment length polymorphisms. Am J Hum Genet 32: 314-331
584a. Bowman BH, Kurosky A (1982) Haptoglobin: The evolutionary product of duplication, unequal crossing over, and point mutation. In: Harris H, Hirschhorn K (eds) Advances in Human Genetics Vol 12. Plenum Press, New York and London, pp 189-261
585. Boyd WC (1955) Maximum likelihood estimation of Rh gene frequencies in Pacific populations. Nature 176: 648

586. Boyd WC (1955) Simple maximum likelihood method for calculating Rh gene frequencies in Pacific populations. Am J Phys Anthrop 13: 447-453

587. Bracken H von (1934) Mutual intimacy in twins: Types of structure in pairs of identical and fraternal twins. Character and Personality 2: 293-309

588. Bridges CB (1936) The bar "gene", a duplication. Science 83: 210

589. Bridges RA, Berendes H, Good RA (1959) A fatal granulomatous disease of childhood. Am J Dis Child 97: 387

590. Brönnestam R (1973 a) Studies on the C3 polymorphism. Relation between phenotype and quantitative immunochemical measurements. Hum Hered 23: 128-134

591. Brönnestam R (1973 b) Studies on the C3 polymorphism. Relationship between C3 phenotype and rheumatoid arthritis. Hum Hered 23: 206-213

592. Brown MS, Goldstein JL (1979) Abetalipoproteinemia. In: Goodman RM, Motulsky AG (eds) Genetic diseases among Ashkenazi Jews. Raven, New York

593. Brown MS, Goldstein JL (1984) How LDL receptors influence cholesterol and atherosclerosis. Sci Am 251: 58-66

594. Brunzell JD, Schrott HG, Motulsky AG, Bierman EL (1976) Myocardial infarction in the familial forms of hypertriglyceridemia. Metabolism 25: 313-320

595. Burke W, Hornung S, Copeland BR, Furlong CE, Motulsky AG (1984) Red cell sodium-lithium countertransport in hypertension. In: Villarreal H, Sambhi MP (eds) Topics in Pathophysiology of Hypertension. Martinus Nijhoff Boston, pp 88-99

596. Burke W, Motulsky AG (1985) Genetics of hypertension. Pract Cardiol 11: 159-173

597. Burke W, Motulsky AG (1985) Hypertension - some unanswered questions. JAMA 253: 2260-2261

598. Cagianit B, Rhyner K, Furrer W, Schnebli HP (1981) Thiosulfate-sulphur transferase (rhodanase) deficiency in Leber's hereditary optic atrophy. Lancet II: 981-982

599. Carney RG jr (1976) Incontinentia pigmenti: A world statistical analysis. Arch Dermatol 112: 535-542

600. Carter CO (1969) Genetics of common disorders. Brit Med Bull 25: 52-57

601. Carter CO (1961) The inheritance of congenital pyloric stenosis. Br Med Bull 17: 251-254

602. Carter CO (1976) Genetics of common single malformation. Br Med Bull 32: 21-26

603. Carter CO (1977) Genetics of common diseases. In: Gene-environment interaction in common diseases. Japan Medical Research Foundation (ed), pp 108-117. University of Tokyo Press, Tokyo

604. Caskey CT (1985) New aid to human gene mapping. Nature 314: 19

605. Ceppellini R, Dunn LC, Turri M (1955) An interaction between alleles at the Rh locus in man which weakens the reactivity of the Rh_0 factor (D^u). Proc Natl Acad Sci USA 41: 283-288

606. Ceppellini RL, Siniscalco M (1955) Una nuova ipotesi genetica per il sistema Lewis-secretore e suoi reflessi nei reguardi di alcune evidenze di linkage con altri loci. Riv Ist sieroter Ital 30: 431-445

607. Ceppellini R, Siniscalo M, Smith CAB (1955/56) The estimation of gene frequencies in a random-mating population. Ann Hum Genet 20: 97

608. Chakravartti MR, Vogel F (1973) Topics in human genetics. Vol I: A twin study on leprosy. Thieme, Stuttgart

609. Chown B, Lewis M, Hiroko K (1957) A 'new' Kell blood group phenotype. Nature 180: 711

610. Chung CS, Morton NE (1959) Discrimination of genetic entities in muscular dystrophy. Am J Hum Genet 11: 339-359

611. Cloninger CR, Rice J, Reich Th, McGriffin P (1982) Genetic analysis of seizure disorders as multidimensional threshold characters, In: Anderson VE (ed) Genetic basis of the epilepsies. Raven Press, New York pp 291-309.

612. Cohen-Hagenauer O, Robbins E, Messart C, Busson M, Deschamps I, Hors J, Lalouel J-M, Dausset J, Cohen D (1985) A systematic study of HLA class II-β DNA restriction fragments in insulin-dependent diabetes mellitus. Proc Natl Acad Sci USA 82: 3335-3339

612a. Coneally PM, Rivas ML (1980) Linkage analysis in man. Adv in Hum Genet 10: 209-266

612b. Cook PJL, Robson EB, Buckton KE, Jacobs PA, Polani PE (1974) Segregation of genetic markers in families with chromosome polymorphisms and structural rearrangements involving chromosome 1. Ann Hum Genet 37: 261-274

613. Carney G, Seedburgh D, Thompson B, Campbell DM, Mac Gillivray I, Timlin D (1979) Maternal Height and Twinning. Ann Hum Genet 43: 55-59

614. Cudworth AG, Wolf E (1982) The genetic susceptibility to type I (insulin-dependent) diabetes mellitus. Clin Endocr Metab 11: 389-408 1

615. Cuénot L (1905) Les races pures et leur combinaisons chez les souris. Arch Zool Exp Génét 3: 123-32

616. Czeizel A (1978) The Hungarian congenital malformation monitoring system. Acta Paediat Academ Scientiarum Hungaricae 19: 225-238

617. Czeizel A, Kiss P, Oszotovics M et al (1978) Nationwide investigations of multiple malformations. Acta Paediat Academ Scientiarum Hungaricae 19: 275-280

618. Czeizel A, Pazonyi I, Métreki J, Tomka M (1979) The first five years of the Budapest twin register, 1970-1974. Acta Genet Med Gemellol 28: 73-76

619. Czeizel A, Tusnády G, Vaczó G, Vizkelety T (1975) The mechanism of genetic predisposition in congenital dislocation of the hip. J Med Genet 12: 121-124

620. Dalgaard OZ (1957) Bilateral polycystic disease of the kidneys. Opera ex Domo. Biologiae Hereditariae Humanae Universitatis Hafniensis Munksgaard, Copenhagen 38: 255 pp.

621. Das HK, Biro PA, Duceman B, Sood AK, Shyma E, Reddy P, Lawrance S, Pan J, Weissman SM (1983) Molecular studies of the genes of the human major histocompatibility In: Caskey CT, White RL (eds) Banbury Report 14. Cold Spring Harbor Lab p 41-51.

622. Dausset J, Colombani J (1972) Histocompatibility testing 1972. Munksgaard, Copenhagen

623. Dausset J, Contu L (1980) The MHC and immune response in man. Progr Immunol 4: 513-529

624. Davies KE, Young BD, Elles RG, Hill ME, Williamson R (1981) Cloning of a representative genomic

library of the human X chromosome after sorting by flow cytometry. Nature 293: 374-376

625. Davis SH, Gavin J, Goldsmith KLG, Graham JB, Hamper J, Hardisty RM, Harris JB, Holman CA, Ingram GIC, Jones TG, McAfee LA, McKusick VA, O'Brien JR, Race RR, Sanger R, Tippett P (1963) The linkage relations of hemophilia A and hemophilia B (christmas disease) to the Xg blood group system. Am J Hum Genet 15: 481-492
626. Deeb SS, Motulsky AG, Albers JJ (1985) A partial cDNA clone for human apolipoprotein B. Proc Natl Acad Sci USA 82: 4983-4986
627. Degnbol B, Green A (1978) Diabetes mellitus among first- and second-degree relatives of early onset diabetics. Ann Hum Genet 42: 25-47
628. Demenais F, Bonaïti C, Briard M-L, Feingold J, Frézal J (1979) Congenital glaucoma: Genetic models. Hum Genet 46: 305-317
629. Donahue RP, Bias WB, Renwick JH, McKusick VA (1968) Probable assignment of the Duffy bloodgroup locus to chromosome 1 in man. Proc Natl Acad Sci USA 61: 949
630. Dorn H (1959) Xeroderma pigmentosum. Acta Genet Med Gemellol (Roma) 8: 395-408
631. Dungern E von, Hirzfeld L (1911) Über gruppenspezifische Strukturen des Blutes. III. Z. Immunitätsforsch 8: 526-562
632. Dunsford I, Bowley CC, Hutchinson AM, Thompson JS, Sanger R, Race RR (1953) A human bloodgroup chimera. Br Med J II: 81
633. Dupont B, Smithwick EM, Oberfield SE, Lee TD, Levine LS (1977) Close genetic linkage between HLA and congenital adrenal hyperplasia (21-hydroxylase deficiency). Lancet II: 1309-1312
634. Eaves LJ (1982) The utility of twins, In: Anderson VE et al (eds) Genetic basis of the epilepsies. Raven Press, New York pp 249-276.
635. Eder HA, Gidez LJ (1982) The clinical significance of the plasma high density lipoproteins. Med Clin North Am 66: 431-440
636. Edwards JH (1960) The simulation of mendelism. Acta Genet (Basel) 10: 63-70
637. Edwards JH (1965) The meaning of the associations between blood groups and disease. Am J Hum Genet 29: 77
638. Edwards JA, Gale RP (1972) Camptobrachydactyly: A new autosomal sominant trait with two probable homozygotes. Amer J Hum Genet 24: 464-474
639. Egger J, Wilson J (1983) Mitochondrial inheritance in a mitochondrially mediated disease. New England Journal Med 309: 142-146
640. Ehling UH (1966) Dominant mutations of the skeleton in offspring of X-irratiated male mice. Genetics 54: 1381-1389
641. Ehling UH (1970) Evaluation of presumed dominant skeletal mutation. In: Vogel F, Röhrborn G (eds) Chemical mutagenesis in mammals and man, Springer, Berlin Heidelberg New York pp 162-166.
642. Ehling UH, Randolph ML (1962) Skeletal abnormalities in the F_1 generation of mice exposed to ionizing radiations. Genetics 47: 1543-1555
643. Eichwald EJ, Silmser CR (1955) Communication. Transplant Bull 2: 148-149
644. Ellison RT III, Kohler PF, Curd JG, Judson FN, Reller LB (1983) Prevalence of congenital or acquired complement deficiency in patients with sporadic meningococcal disease. N Engl J Med 308: 913-916
645. Elsas LJ, Endo F, Strumlauf E, Elders J, Priest JH (1985) Leprechaunism: an inherited defect in a high-affinity insulin receptor. Am J Hum Genet 37: 73-88
646. Elston RC (1979) Major locus analysis for quantitative traits. Amer J Hum Genet 31: 655-661
647. Elston RC (1981) Segregation analysis. Adv in Hum Genet 11: 63-120
648. Elston RC, Glassman E (1967) An approach to the problem of whether clustering of functionally related genes occurs in higher organisms. Genet Res (London) 9: 141-147
649. Elston JH, Lange K (1975) The prior probability of autosomal linkage. Ann Hum Genet 38: 341-350
650. Epstein FH (1976) Genetics of ischaemic heart disease. Postgrad Med J 52: 477-480
651. Eriksson S (1965) Studies in alpha1-antitrypsin deficiency. Acta Med Scand 177: 175
652. Erlich H, Stetler D, Grumet C (1983) Restriction length polymorphism analysis of HLA-typed families using cloned HLA probes, In: Caskey CT, White RL (eds) Banbury Report 14. Cold Spring Harbor Lab pp 327-334.
653. Fagerhol MK, Cox DW (1981) The Pi polymorphism: Genetic, biochemical and clinical aspects of human $alpha_1$-antitrypsin. Adv in Hum Genet 11: 1-62
654. Falconer DS (1965) The inheritance of liability to certain diseases, estimated from the incidence among relatives. Ann Hum Genet 29: 51
655. Falconer DS (1967) The inheritance of liability to diseases with variable age of onset with particular reference to diabetes mellitus. Ann Hum Genet 31: 1-20
656. Farabee (1905) Inheritance of digital malformations in man. Papers of the Peabody Museum for American Archeology and Ethnology. (Harvard University) 3: 69
657. Farhud DB, Ananthakrishnan R, Walter H (1972) Association between the C3 phenotypes and various diseases. Hum Genet 17: 57-60
658. Farquhar JW, Maccoby N, Wood PD, Alexander JK, Breitrose H, Brown BW jr, Haskell WL, McAlister AL, Meyer AJ, Nash JD, Stern MP (1977) Community education for cardiovascular health. Lancet 1: 1192-1195
659. Feller W (1950) An introduction to probability theory and its applications. John Wiley & Sons, New York
660. Ferns GAA, Stocks J, Ritchie C, Galton DJ (1985) Genetic polymorphisms of apolipoprotein C-III and insulin in survivors of myocardial infarction. Lancet 2: 300-303
661. Fialkow PJ, Giblett ER, Motulsky AG (1967) Measurable linkage between ocular albinism and Xg. Am J Hum Genet 19: 63-69
662. Finkelstein S, Walford RL, Myers LW, Ellison GW (1974) HL-A antigens and hypersensitivity to brain tissue in multiple sclerosis. Lancet I: 736
663. Finney DJ (1947/9) The truncated binomial distribution. Ann Eugen 14: 319-328

664. Fisher RA (1918) The correlation between relatives on the supposition of Mendelian inheritance. Trans R Soc (Edinburgh) 52: 399-433
665. Fisher RA, Race RR (1946) Rh gene frequencies in Britain. Nature 157: 8
666. Floderus Y, Iselius L, Lindsten J, Wetterburg L (1982) Evidence for a major locus as well as a multifactorial component in the regulation of human red blood cell catechol-O-methyltransferase activity. Hum Hered 32: 76-79
667. Folstein SE, Phillips JA III, Meyers DA, Chase GA, Abbott MH, Franz ML, Waber PG, Kazazian HH Jr, Conneally PM, Hobbs W, Tanzi R, Faryniarz A, Gibbons K, Gussella J (1985) Huntington's disease: two families with differing clinical features show linkage to the G8 probe. Science 229: 776-779
668. Franceschetti A, Klein D (1957) Two families with parents of different types of red-green blindness Acta Genet (Basel) 7: 255
669. Fraser GR (1976) The causes of profound deafness in childhood. Johns Hopkins University Press, Baltimore London
670. Fraser GR, Friedman AI (1967) The causes of blindness in childhood. Johns Hopkins University Press, Baltimore
671. Fredrickson DS, Goldstein JL, Brown MS (1978) The familial hyperlipoproteinemias. In: Stanbury JB, Wyngaarden JB, Fredrickson DS (eds) The Metabolic Basis of Inherited Disease, 4th ed. McGraw-Hill, New York, pp 604-655
672. Fuhrmann W (1963) Das Syndrom der erblichen Nephropathie mit Innenohrschwerhörigkeit (Alport-Syndrom). Dtsch Med Wochenschr 88: 525-532
673. Fuhrmann W (1974) Die formale Genetik des Menschen. In: Vogel F (ed) Handbuch der Allgemeinen Pathologie. Vol IX, Erbgefüge, Springer, Berlin Heidelberg New York pp 147-259.
674. Fuhrmann W, Stahl A, Schroeder TM (1966) Das oro-facio-digitale Syndrom. Humangenetik 2: 133-164
675. Galton F (1876) The history of twins as a criterium of the relative powers of nature and nurture. J Anthropol Inst
676. Gartler SM, Francke U (1975) Half chromatid mutations: Transmission in humans? Am J Hum Genet 27: 218-223
677. Gaul LE (1953) Heredity of multiple benign cystic epithelioma. Arch Dermatol Syph (Chicago) 68: 517
678. Gelb AF, Klein E, Lieberman J (1977) Pulmonary function in nonsmoking subjects with $alpha_1$ antitrypsin deficiency (MZ phenotype). Am J Med 62: 93
679. Gerard G, Vitrac D, Le Pendu J, Muller A, Oriol R (1982) H-deficient blood groups (Bombay) of Reunion island. Amer J Hum Genet 34: 937-947
680. Gesell A, Thompson H (1929) Learning and growth in identical infant twins: An experimental study by the method of co-twin control. Genet Psychol Monogr 6: 5-124
681. Geyer E (1940) Ein Zwillingspärchen mit zwei Vätern. Arch Rassenbiol 34: 226-236
682. Giblett ER, Klebanoff SJ, Pincus SH, Swanson J, Park BH, McCullough J (1971) Kell phenotypes in chronic granulomatous disease: a potential transfusion hazard. Lancet I: 1235
683. Goldberger AS (1978) Pitfalls in the resolution of I.Q. inheritance, pp 195-222. In: Morton NE, Chung CS (eds) Genetic epidemiology. Academic Press, New York etc
684. Goldstein JL, Brown MS (1977) Atherosclerosis: the low-density lipoprotein receptor hypothesis. Metabolism 26: 1257-1275
685. Goldstein JL, Brown MS (1979) LDL receptor defect in familial hypercholesterolemia. Med Clin North Am 66: 335-362
686. Goldstein JL, Hazzard WR, Schrott HG, Bierman EL, Motulsky AG (1973) Hyperlipidemia in coronary heart disease. II. Genetic analysis of lipid levels in 176 families and delineation of a new inherited disorder. J Clin Invest 54: 1544-1568
687. Grebe H (1959) Erblicher Zwergwuchs. Ergeb Inn Med Kinderheilkd 12: 343-427
688. Greiner J, Schleiermacher E, Smith T, Lenhard V, Vogel F (1978) The HLA system and leprosy in Thailand. Hum Genet 42: 201-213
688 a. Greiner J, Krüger J, Palden L, Jung EG, Vogel F (1983) A linkage study of acrokeratoelastoidosis. Possible mapping to chromosome 2. Hum Genet 63: 222-227
689. Grosse-Wilde H, Bertrams J, Schuppien W, Netzel B, Ruppelt W, Kuwert EK (1977) HLA-D typing in 111 multiple sclerosis patients: Distribution of four HLA-D alleles. Immunogenetics 4: 481-488
690. Grüneberg H (1952) Quasi-continuous variations in the mouse. Symp Genet 3: 215-227
691. Grüneberg H (1952) Genetical studies in the skeleton of the mouse. IV. Quasi-continuous variations. J Genet 51: 95-114
692. Grzeschik K-H (1973) Utilization of somatic cell hybrids for genetic studies in man. Hum Genet 19: 1-40
693. Gusella JF, Tanzi RE, Anderson MA, Hobbs W, Gibbons K, Raschtchian R, Gilliam TC, Wallace MR, Wexler NS, Conneally PM (1984) DNA markers for nervous system diseases. Science 225: 1320-1326
694. Gusella JF, Wexler NS, Conneally PM, Naylor SL, Anderson MA, Tanzi RE, Watkins PC, Ottina C, Wallace MR, Sakaguchi AY, Young AB, Shoulson I, Bonilla E, Martin JB (1983) A polymorphic DNA marker genetically linked to Huntington's disease. Nature 306: 234-238
695. Haase FH (1938/39) Die Übersterblichkeit der Knaben als Folge recessiver geschlechtsgebundener Anlagen. Z Menschl Vererbungs-Konstitutionslehre 22: 105-126
696. Hadorn E (1955) Developmental genetics and lethal factors. John Wiley & Sons, London New York
697. Haldane JBS (1941) The relative importance of principal and modifying genes in determining some human diseases. J Genet 41: 149-157
698. Haldane JBS, Smith CAB (1947/9) A simple exact test for birth order effect. Ann Eugen 14: 116-122
699. Haldane JBS, Smith CAB (1947) A new estimate of the linkage between the genes for colour blindness and haemophilia in man. Ann Eugen 14: 10-31
700. Hall JG, Dorst JP, Taybi H, Scott CI, Langer LO, McKusick VA (1969) Two probable cases of homozygosity for the achondroplasia gene. Birth Defects 4: 24-34
701. Hamby RI (1981) Hereditary aspects of coronary artery disease. Am Heart J 101: 639-649

702. Harris H (1970) Cell fusion. Clarendon, Oxford

702a. Harris H (1948) A sex-limiting modifying gene in diaphyseal aclasis (multiple exostoses). Ann Eugen 14: 165-170

703. Harris H, Smith CAB (1948) The sib-sib age of onset correlation among individuals suffering from a hereditary syndrome produced by more than one gene. Ann Eugen 14: 309-318

704. Hatzold O (1966) Die Sexualproportion der Geborenen und ihre Schwankungen als präkonzeptionelles Wahrscheinlichkeitsproblem. Dtsch Akad f Bevölkerungswissenschaft, Reihe B, Vol 5

705. Hauge M, Harvald B, Fischer M, Gottlieb-Jensen K, Juel-Nielsen N, Raebild I, Shapiro R, Videbech T (1968) The Danish twin register. Acta Genet Med Gemellol (Roma) 18: 315-332

706. Havel RJ (1982) Familial dysbetalipoproteinemia. Med Clin North Am 66: 441-454

707. Haverkamp-Begemann N, Lookeren-Campagne van A (1952) Homozygous form of Pelger Huet's nuclear anomaly in man. Acta Haematol (Basel) 7: 295-303

708. Haws DV, McKusick VA (1963) Farabee's brachydactylous kindred revisited. Johns Hopkins Med J 113: 20-30

709. Hegeman JP, Mash AJ, Spivey BE (1974) Genetic analysis of human visual parameters in populations with varying incidences of strabism. Am J Hum Genet 26: 549-562

710. Helmbold W (1959) Über den möglichen Aufbau des Rh-Genkomplexes. Blut 5: 141-148

711. Helmbold W, Prokop O (1958) Die Bestimmung der AB0-Genfrequenzen mittels der maximum-likelihood-Methode und anderer Verfahren anhand forensischer Blutgruppenbestimmungen in Berlin. Blut 4: 190-201

712. Hershon K, Brunzell J, Albers JJ, Haas L, Motulsky A (1981) Hyper-apo-B-lipoproteinemia with variable lipid phenotype (familial combined hyperlipidemia). Arteriosclerosis 1, 380a

712a. Hirszfeld L (1928) Konstitutionsserologie und Blutgruppenforschung. Springer, Berlin

713. Hitman GA, Jowett NI, Williams LG, Humphries S, Winter RM, Galton DJ, (1984) Polymorphisms in the 5'-flanking region of the insulin gene and non-insulin-dependent diabetes. Clin Sci 66: 383-388

714. Hitzig WH, Seger RA (1983) Chronic granulanatous disease, a heterogenous syndrome. Hum Genet 64: 207-215

715. Holmes B, Quie PG, Windhorst DB, Good RA (1966) Fatal granulomatous disease of childhood: an inborn abnormality of phagocytic function. Lancet I: 1225

716. Howell-Evans W, McConnell RB, Clarke CA, Sheppard PM (1958) Carcinoma of the oesophagus with keratosis palmaris and plantaris (tylosis). Q J Med (new series) 27: 413-429

717. Hrubec Z, Robinette CD (1984) The study of human twins in medical research. N Engl J Med 310: 435-441

718. Hully SB, Rosenman RH, Bawol RD, Brand RJ (1980) Epidemiology as a guide to clinical decisions. The association between triglyceride and coronary heart disease. N Engl J Med 302: 1383-1389

719. Humphries SE, Kessling AM, Horsthemke B, Donald JA, Seed M, Jowett N, Holm M, Galton DJ, Wynn V, Williamson R (1985) A common DNA polymorphism of the low-density lipoprotein (LDL) receptor gene and its use in diagnosis. Lancet 1: 1003-1005

720. Hultén M (1974) Chiasma distribution at diakinesis in the normal human male. Hereditas 76: 55-78

721. Husén T (1959) Psychological twin research. Almquist & Wiksell, Stockholm

722. Illingworth DR, Sexton GJ (1984) Hypocholesterolemic effects of mevinolin in patients with heterozygous familial hypercholesterolemia. J Clin Invest 74: 1972-1978

723. Inkeles S, Eisenberg D (1981) Hyperlipidemia and coronary atherosclerosis: a review. Medicine 60: 110-123

724. Iselius L, Lalouel JM (1982) Complex segregation analysis of hyperalphalipoproteinemia. Metabolism 31: 521-523

725. Jackson JF, Currier RD, Terasaki PI, Morton NE (1977) Spinocerebellar ataxia and HLA linkage. Risk prediction by HLA typing. N Engl J Med 296: 1138-1141

726. Johannsen W 1909 Elemente der exakten Erblichkeitslehre. G Fischer (3rd ed 1926), Jena

727. Johnson BC, Epstein FH, Kjelsberg MO (1965) Distribution and familial studies of blood pressure and serum cholesterol levels in a total community - Tecumseh, Michigan. J Chron Dis 18: 147

728. Jörgensen G (1974) Erbfaktoren bei häufigen Krankheiten. In: Vogel F (ed) Handbuch der allgemeinen Pathologie: Vol IX, Erbgefüge. Springer, Berlin Heidelberg New York, pp 581-665

729. Kaelin A (1955) Statistische Prüf- und Schätzverfahren für die relative Häufigkeit von Merkmalsträgern bei einem der Auslese unterworfenen Merkmal mit Anwendung auf das Retinagliom. Arch Julius Klaus Stiftung Vererbungsforsch 30: 263-485

729a. Kämpe O, Larhammer D, Wiwan K, Scheuning L, Claesson L, Gustafsson K, Pääbo S, Hyldig-Nielsen JJ, Rask L, Peterson PA (1983) Molecular analysis of MHC antigens. In: Möller E, Möller G (eds) Genetics of the immune response. Plenum, 001 York, pp 61-79

730. Kagan A, Rhoads GC, Zeegan PD, Nichaman MZ (1971) Coronary heart disease among men of Japanese ancestry in Hawaii: the Honolulu heart study. Isr J Med Sci 7: 1573

731. Kallmann FJ (1938) The genetics of schizophrenia. Augustin, New York

732. Kallmann FJ (1946) The genetic theory of schizophrenia. An analysis of 691 schizophrenie twin index families. Am J Psychiatry 103: 3

733. Kan YW, Dozy AM (1978) Antenatal diagnosis of sicklecell anaemia by DNA analysis of amniotic-fluid cells. Lancet II: 910-912

734. Kannell WB (1976) Some lessons in cardiovascular epidemiolog from Framingham. Am J Cardiol 37: 268-282

735. Keats BJB, Morton NE, Rao DC, Williams W (1979) A source book for linkage in man. Johns Hopkins Univ. Press, Baltimore

736. Keats JB, Morton NE, Rao DC (1981) Reduction of physical assigment to a standard lod table: Chromosome 1. Hum Genet 56: 353-359

737. Kelley WN, Greene ML, Rosenbloom FM, Henderson JF, Seegmiller JE (1969) Hypoxanthine-guanine phosphoribosyltransferase deficiency in gout. Ann Intern Med 70: 155

738. Kiekebusch-Müller BD, Arnold H, Mayr WR (1980) Ein Algorithmus zur Berechnung der HLA-Haplotypfrequenzen (an algorhithm for calculating HLA haplotype frequencies). Anthrop Anzeiger 38: 1-10

739. Kissmeyer-Nielsen F (ed) (1975) Histocompatibility testing 1975. Munksgaard, Copenhagen

740. Klein J (1974) Genetic polymorphism of the histocompatibility 2 loci of the mouse. Ann Rev Genet 8: 63-77

741. Klein J (1975) Biology of the mouse histocompatibility-2 complex. Springer, Berlin Heidelberg New York

742. Kloimwieder R (1942) Die Intelligenz in ihren Beziehungen zur Vererbung, Umwelt und Übung. Z Menschl Vererbungs-Konstitutionslehre 25: 582-617

743. Knowler WC, Pettit DJ, Vasquez B, Rotwein PS, Andreone TL, Permutt MA (1984) Polymorphism in the 5'flanking region of the human insulin gene. J Clin Invest 74: 2129-2135

744. Koller S (1940) Methodik der menschlichen Erbforschung II. Die Erbstatistik in der Familie. In: Just G, Bauer KH, Hanhart E, Lange J (eds) Handbuch der Erbbiologie des Menschen, Vol II, Methodik, Genetik der Gesamtperson. Springer, Berlin, pp 261-284

745. Kravitz K, Skolnick M, Edwards C, Cartwright G, Amos B, Carmelli D, Baty B (1978) Pedigree analysis of the linkage between HLA and hemochromatosis. In: Morton NE, Chung CS (eds) Genetic epidemiology. Academic Press, New York, pp 241-246

746. Krüger J (1973) Zur Unterscheidung zwischen multifaktoriellem Erbgang mit Schwellenwerteffekt und einfachem diallelem Erbgang. Hum Genet 17: 181-252

747. Krüger J, Propping P (1976) Rückgang der Zwillingsgeburten in Deutschland. Dtsch Med Wochenschr 101: 475-480

748. Krüger J, Vogel F (1975) Population genetics of unequal crossing over. J Mol Evol 4: 201-247

749. Kueppers F (1975) α_1-Antitrypsin. In: Becker PE (ed) Humangenetik, ein kurzes Handbuch, Vol I/3. Thieme, Stuttgart, pp 35-49

750. Kunkel LM, Tantravahi U, Eisenhard M, Latt S (1978) Regional localization on the human X of DNA segments cloned from flow sorted chromosomes. Nucleic Acid Res 10: 1557-1578

751. Kurachi K, Chandra T, Degen SJ, White TT, Marchioro TL, Woo SLC, Davie EW (1981) Cloning and sequence of cDNA coding for alpha1-antitrypsin. Proc Natl Acad Sci USA 78: 6826-6830

752. Lalouel JM, Morton NE, MacLean CJ, Jackson J (1977) Recurrence risks in complex inheritance with special regard to pyloric stenosis. J Med Genet 14: 408-414

753. Landsteiner K, Wiener AS (1940) An agglutinable factor in human blood recognized by immune sera for rhesus blood. Proc Soc Exp Biol Med 43: 223

754. Lange K, Boehnke M (1982) How many polymorphic genes will it take to span the human genome? Amer J Hum Genet 34: 842-845

754a. Lange K, Page BM, Elston RC (1975) Age trends in human chiasma frequencies and recombination fractions. I. Chiasma frequencies. Am J Hum Genet 27: 410-418

755. Laurell C-B, Eriksson S (1963) The electrophoretic α_1-globulin pattern of serum α_1-antitrypsin deficiency. Scand J Clin Lab Invest 15: 132

756. Lemser H (1938) Zur Erb- und Rassenpathologie des Diabetes mellitus. Arch Rassenbiol 32: 481

757. Lenz F (1923) Die Übersterblichkeit der Knaben im Lichte der Erblichkeitslehre. Arch Hyg 93: 126-150

757a. Lenmark A (1985) Molecular biology of type 1 (insulin-dependent) diabetes mellitus. Diabetes 28: 195-203

758. Lenz W (1959) Ursachen des gesteigerten Wachstums der heutigen Jugend. In: Akzeleration und Ernährung. Fettlösliche Wirkstoffe. Vol 4

759. Lenz W (1961) Zur Genetik der Incontinentia pigmenti. Ann Paediatr (Basel) 196: 141

760. Lenz W (1975) Half chromatid mutations may explain incontinentia pigmenti in males. Am J Hum Genet 27: 690-691

761. Lenz W (1973) Vererbung und Umwelt bei der Entstehung von Mißbildungen. Humanbiologie 121: 132-145

762. Levine P, Stetson RE (1939) An unusual case of intragroup agglutination. J Am Med Assoc 113: 126-127

763. Levy RI (1981) Declining mortality in coronary heart disease. Arteriosclerosis 1: 312-325

764. Lewis EB (1951) Pseudoallelism and gene evolution. Cold Spring Harbor Symp Quant Biol 16: 159-174

765. Lieberman R, Humphrey W (1972) Association of the H-2 types with genetic control of immune responsiveness to (γ2a) allotypes in the mouse. J Exp Med 136: 1222-1230

766. Lilly F (1966) The inheritance of susceptibility to the Gross leukemia virus in mice. Genetics 53: 529-539

767. Lundsgaard R (1944) Leber's disease. A genealogic, genetic and clinical study of 101 cases of retrobulbar optic neuritis in 20 Danish families. Acta Ophthalmol [Supp] (Kbh) 30

768. Luxenburger H (1935) Untersuchungen an schizophrenen Zwillingen und ihren Geschwistern zur Prüfung der Realität von Manifestationsschwankungen. Z Gesamte Neurol Psychiat 154: 351-394

769. Luxenburger H (1940) Zwillingsforschung als Methode der Erbforschung beim Menschen. In: Just G, Bauer KH, Hanhart E, Lange J (eds) Handbuch der Erbbiologie des Menschen, Vol II, Methodik, Genetik der Gesamtperson. Springer, Berlin, pp 213-248

770. Maartmenn-Moe K, Magnus P, Golden W, Berg K (1981) Genetics of the low density lipoprotein receptor: III. Evidence for multiple normal alleles at the low density lipoprotein receptor locus. Clin Genet 20: 113-129

771. Maartmann-Moe K, Magnus P, Børresen A-L, Berg K (1981) Low density lipoprotein receptor activity in cultured fibroblasts from subjects with or without ischemic heart disease (in the absence of familial hypercholesterolemia). Clin Genet 20: 337-346

772. Madlener M (1928) Eine Bluterfamilie. Arch Rassenbiol 20: 390-394

773. Mann JD, Cahan A, Gelb A, Fisher N, Hamper J, Tippett P, Sanger R, Race RR (1962) A sex-linked blood group. Lancet I: 8
774. Maroteaux P (1974) Les maladies osseuses de l'enfant. Flammarion, Paris
775. Marsh DG, Meyers DA, Bias WB (1981) The epidemiology and genetics of atopic allergy. New Engl J Med 305: 1551-1559
776. Marsh WL (1978) Chronic granulomatous disease, Kx antigen and the Kell blood groups. In: Brewer GJ (ed) Progress in clinical and biological research. The red cell. Vol 21. Liss, New York, pp 493-507
777. Marsh WL, Øyen R, Nichols ME, Allen FH (1975) Chronic granulomatous disease and the Kell blood groups. Br J Haematol 29: 247
778. Massart C, Busson M, Deschamps I, Hors J, Lalouel J-M, Dausset J, Cohen D (1985) A systematic study of HLA class II-β DNA restriction fragments in insulin-dependent diabetes mellitus. Proc Natl Acad Sci USA 82: 3335-3339
779. Matthiuz PL, Ihde D, Piazza A, Ceppelini R, Bodmer WF (1970) New Approaches to the population genetics and segregation analysis of the HL-A system. In: Terasaki P (ed) Histocompatibility Testing. Munksgaard, Kopenhagen, pp 193-205
780. Maynard-Smith S, Penrose LS, Smith CAB (1961) Mathematical tables for research workers in human genetics. J & A Churchill, London
781. McArthur N (1953) Statistics in twin birth in Italy. 1949 and 1950. Ann Eugen 17: 249
782. McClintock B (1944) The relation of homozygous deficiencies to mutations and allelic series in Maize. Genetics 29: 478-502
783. McDevitt HO et al (1972) Genetic control of the immune response, mapping of the Ir-1 locus. J Exp Med 135: 1259-1278
784. McKusick VA (1984) The human gene map 15 november 1983. The morbid anatomy of the human genome. Clin Genet 25: 89-123
785. McMichael A, McDewitt H (1977) The association between the HLA system and disease. Prog Med Genet [New Series] 2: 39-100
786. Menzel H-J, Kladetzky R-G, Assmann G (1983) Apolipoprotein E polymorphism and coronary artery disease. Arteriosclerosis 3: 310-315
787. Métneki J, Czeizel A (1980) Contraceptive pills and twins. Acta Genet Med Gemellol 29: 233-236
788. Migeon BR, Miller SC (1968) Human-mouse somatic cell hybrids with single human chromosome (group E): Link with thymidine kinase activity. Science 162: 1005-1006
789. Miki C, Yee S, Yasuda N, Morton NE (1969) Alltype. In: Morton NE (ed) A genetics program library. The Univ of Hawaii Press, Honolulu
790. Milch RA (1959) A preliminary note of 47 cases of alcaptonuria occurring in 7 interrelated Dominican families, with an additional comment on two previously reported pedigrees. Acta Genet (Basel) 9: 123-126
791. Mohr J (1954) A study of linkage in man. Opera ex Domo Biologiae Hereditariae Humanae Universitatis Hafniensis 33: Munksgaard, Copenhagen
792. Mohr OL, Wriedt C (1919) A new type of hereditary brachyphalangy in man. Carnegie Inst (Wash) Publ Nr 295: 1-64
793. Moll PP (1982) Alternative genetic models: An application to epilepsy. In: Anderson VE et al (eds) Genetic basis of the epilepsies. Raven Press, New York, pp 277-289
794. Moll PP, Berry TD, Weidman WH, Ellefson R, Gordson H, Kottke BA (1984) Detection of genetic heterogeneity among pedigrees through complex segregation analysis: An application to hypercholesterolemia. Amer J Hum Genet 36: 197-211
795. Mørch ET (1941) Chondrodystrophic dwarfs in Denmark. Opera ex Domo Biologiae Hereditariae Humanae Universitatis Hafniensis 3: Munksgaard, Copenhagen
796. Morton NE (1956) The detection and estimation of linkage between the genes for elliptocytosis and the Rh blood type. Am J Hum Genet 8: 80-96
797. Morton NE (1957) Further scoring types in sequential tests, with a critical review of autosomal and partial sex linkage in man. Am J Hum Genet 9: 55-75
798. Morton NE (1955) Morton NE (1955) Sequential tests for the detection of linkage. Am J Hum Genet 7: 277-318
799. Morton NE (1962) Segregation and linkage. In: Burdette WJ (ed) Methodology in human genetics. Holden Day, San Francisco, pp 17-52
800. Morton NE (1969) Segregation analysis. In: Morton NE (ed) Computer application in genetics. Univ of Hawaii press, Honolulu, pp 129-139
801. Morton NE (1976) Genetic markers in atherosclerosis: a review. J Med Genet 13: 81-90
802. Morton NE (1983) Outline of genetic epidemiology. S Karger, Basel etc
803. Morton NE, MacLean CJ (1974) Analysis of family resemblance III. Complex segregation of quantitative traits. Am J Hum Genet 26: 489-503
804. Morton NE, Yee S, Elston RC, Lew R (1970) Discontinuity and quasi-continuity: Alternative hypotheses of multifactorial inheritance. Clin Genet 1: 81-93
805. Morton NE, Simpson SP, Lew R, Yee S (1983) Estimation of haplotype frequencies. Tissue Antigens 22: 257-262
806. Motulsky AG (1976) The genetic hyperlipidemias. N Engl J Med 294: 823-827
807. Motulsky AG (1977) The George M Kober lecture. A genetical view of modern medicine. Trans Assoc Am Physicians 40: 76-90
808. Motulsky AG (1978) The genetics of common diseases. In: Morton NE, Chung CS (eds) Genetic epidemiology. Academic Press, New York, pp 541-548
809. Motulsky AG (1979) The HLA complex and disease. Some interpretations and new data in cardiomyopathy. N Engl J Med 300: 918-919
810. Motulsky AG (1982) Genetic approaches to common diseases. In: Bonné-Tamir B (ed) Human Genetics, Part B: Medical Aspects. Alan R Liss, New York, pp 89-95
811. Motulsky AG (1980) Approaches to the genetics of common diseases. In: Rotter JI, Samloff IM, Rimoin DL (eds) The Genetics and Heterogeneity of Common Gastrointestinal Disorders. Academic Press, New York, pp 3-10
812. Motulsky AG (1984) Editorial: Genetic epidemiology. Gen Epidemiol 1: 143-144
813. Motulsky AG (1984) Genetic research in coronary heart disease. In: Rao DC, Elston RC, Kuller LH,

Feinleib M, Carter C, Havlik R (eds) Genetic Epidemiology of Coronary Heart Disease: Past, Present, and Future. Alan R Liss, New York, pp 541-548

814. Motulsky AG (1984) The 'new genetics' in blood and cardiovascular research: applications to prevention and treatment. Circulation 70 (Suppl III): III-26-III-30
815. Motulsky AG, Boman H (1975) Genetics and atherosclerosis. In: Schettler G, Weizel A (eds) Atherosclerosis, Vol III. Springer, Berlin, pp 438-444
816. Motulsky AG, Boman H (1975) Screening for the hyperlipidemias. In: Milunsky A (ed) The prevention of genetic disease and mental retardation. Saunders, Philadelphia, pp 306-316
817. Mühlmann WE (1930) Ein ungewöhnlicher Stammbaum über Taubstummheit. Arch Rassenbiol 22: 181-183
818. Munro A, Waldmann H (1978) The major histocompatibility system and the immune response. Br Med Bull 34: 253-258
819. Murphy EA (1980) The quantiatative genetics of disease. Am J Med Genet 7: 103-113
820. Murphy EA (1981) The genetic dynamics of disease. Am J Med Genet 8: 35-52
821. Murphy EA (1981) Only authorized persons admitted: The quantitative genetics of health and disease. Johns Hopkins Med J 148: 114-122
822. Murphy EA (1982) Muddling, meddling, and modeling. In: Anderson VE et al (eds) Genetic basis of the epilepsies. Raven Press, New York, pp 333-348
823. Nachtsheim H (1950) The Pelger anomaly in man and rabbit. A Mendelian character of the nuclei of the leucocytes. J Hered 41: 131-137
824. Naeslund J (1956) Metodiken rid den första läsundervisningen. En översikt och experimentella bidrag. Svenska, Uppsala
825. Nance WE, McConnell FE (1973) Status and prospects of research in hereditary deafness. Adv Hum Genet 4: 173-250

825 a. Nathans J, Piantanida TP, Eddy RL, Shows TB, Hogness DS (1986) Molecular genetics of inherited variation in human color vision. Science 232, 203-210

826. Neel JV, Fajons SS, Conn JW, Davidson RT (1965) Diabetes mellitus. In: Neel JV, Shaw MW, Schull WJ (eds) Genetics and the epidemiology of chronic disease. Govt Print Office, Washington, pp 105-132
827. Neufeld HN, Goldbourt U (1983) Coronary heart disease: genetic aspects. Circulation 67: 943
828. Nicholas JW, Jenkins WJ, Marsh WL (1957) Human blood chimeras. A study of surviving twins. Br Med J I: 1458
829. Nora JJ (1980) Identifying the child at risk for coronary heart disease as an adult: a strategy for prevention. J Pediatr 97: 706-714
830. Nora JJ, Lortscher RH, Spangler RD, Nora AH, Kimberling WJ (1980) Genetic-epidemiologic study of early-onset ischemic heart disease. Circulation 62: 503-508
831. Olefsky JM (1985) Diabetes mellitus. In: Wyngaarden JB, Smith LH Jr (eds) Cecil Textbook of Medicine, 17th ed. WB Saunders, Philadelphia, pp 1320-1341

831 a. Ott J (1974) Estimation of the recombination fraction in human pedigrees: Efficient computation of the likelihood for human linkage studies. Am J Hum Genet 26: 588-597

831 b. Ott J (1976) A computer program for linkage analysis of general human pedigree. Am J Hum Genet 28: 528-529

831 c. Ott J (1977) Counting methods (EM algorithm) in human pedigree analysis. Linkage and segregation analysis. Ann Hum Genet 40: 443-454

832. Pauli RM (1983) Editorial comment: Dominance and homozygosity in man. Amer J Med Genet 16: 455-458
833. Pearson K (1904) On the generalized theory of alternative inheritance with special references to Mendel's law. Philos Trans R Soc A 203: 53-86
834. Pearson RJC (1964) Blood groups and disease. Br Med J I: 840
835. Penrose LS (1938) (Colchester survey) A clinical and genetic study of 1280 cases of mental defect. HMSO, London, Spec Rep Ser Med Res Counc 229
836. Penrose LS (1947/49) The problem of anticipation in pedigrees of dystrophia myotonica. Ann Eugen 14: 125-132
837. Penrose LS (1953) The genetical background of common diseases. Acta Genet (Basel) 4: 257-265
838. Petrányi GG, Irángi P, Hollán SR (1974) Relations of HL-A and Rh systems to immune reactivity. Vox Sang 27: 470-482
839. Pious D, Erlich H, Gladstone P, Levine F (1983) Analysis of the HLA regions using somatic cell mutants. In: Caskey CT, White RL (eds) Banbury Report 14. Cold Spring Harbor Laboratory, pp 61-68
840. Pola V, Svojitka J (1957) Klassische Hämophilie bei Frauen. Folia Haematol (Leipz) 75: 43-51
841. Poll H (1914) Über Zwillingsforschung als Hilfsmittel menschlicher Erbkunde. Z Ethnol 46: 87-108
842. Propping P, Krüger J (1976) Über die Häufigkeit von Zwillingsgeburten. Dtsch Med Wochenschr 101: 506-512
843. Propping P, Vogel F (1976) Twin studies in medical genetics. Acta Genet Med Gemellol (Roma) 25: 249-258
844. Propping P, Voigtländer V (1983) Was ist gesichert in der Genetik der Atopien? Allergologie 6: 160-168
845. Puska P, Tuomilehto J, Salonen J et al (1979) Changes in coronary risk factors during a comprehensive five-year community programme to control cardiovascular diseases (North Karelia Project). Br Med J 2: 1173-1178
846. Race RR, Sanger R (1969) Xg and sex chromosome abnormalities. Br Med Bull 25: 99-103
847. Rao CD, Elston RC, Kuller LH, Feinleib M, Carter C, Havlik R (eds) (1984) Genetic Epidemiology of Coronary Heart Disease. Past, Present, and Future. Alan R Liss, New York
848. Rappold G, Cremer T, Cremer C, Back W, Bogenberger J, Cooke HJ (1984) Chromosome assignment of two cloned DNA probes hybridizing predominantly to human sex chromosomes. Hum Genet 65: 257-261
849. Rath B (1938) Rotgrünblindheit in der Calmbacher Blutersippe. Nachweis des Faktorenaustausches beim Menschen. Arch Rassenbiol 32: 397-407

850. Reed TE, Neel JV (1959) Huntington's chorea in Michigan. 2. Selection and mutation. Am J Hum Genet 11: 107
851. Rees A, Shoulders CC, Stocks J, Galton DJ, Baralle FE (1983) DNA polymorphism adjacent to human apoprotein A-1 gene: relation to hypertriglyceridemia. Lancet 1: 444-446
852. Reich T, Rice J, Cloninger CR, Wette R, James J (1979) The use of multiple thresholds and segregation analysis in analyzing the phenotype heterogeneity of multifactorial traits. Ann Hum Genet 42: 371-388
853. Reid DH, Parsons PH (1963) Sex of parents and variation of recombination with age in the mouse. Heredity 18: 107
854. Renwick JH (1956/7) Nail-patella syndrome: Evidence for modification by alleles at the main locus. Ann Hum Genet 21: 159-169
855. Renwick JH (1969) Progress in mapping human autosomes. Br Med Bull 25: 65
856. Richter S (1967) Zur Heredität des Strabismus concomitans. Humangenetik 3: 235-243
857. Richter S (1966) Untersuchungen über die Heredität des Strabismus concomitans. Sammlung zwangloser Abhandlungen auf dem Gebiet der Augenheilkunde. Thieme, Leipzig
857a. Robbins DC, Blix PM, Rubenstein AH, Kanazawa Y, Kosaka K, Tager HS (1981) A human proinsulin variant at arginine 65. Nature 291: 679-681
858. Robertson FW (1981) The genetic component in coronary heart disease - review. Genet Res Camb 37: 1-16
859. Rose NR, Vladutiu AO, David CS, Shreffler DC (1973) Autoimmune murine thyroiditis. V. Genetic influence on the disease in BSVS and BRVR mice. Clin Exp Immunol 15: (2) 281-287
860. Rosenfield RE, Allen FH, Rubinstein P (1973) Genetic model for the Rh blood group system. Proc Natl Acad Sci USA 70: 1303-1307
861. Rotter JI, Rimoin DL (1981) The genetics of the glucose intolerance disorders. Am J Med 79: 116-126
862. Rüdiger HW, Dreyer M (1983) Pathogenetic mechanisms of hereditary diabetes mellitus. Hum Genet 63: 100-106
863. Rushton WAH (1975) Visual pigments and color blindness. Sci Am 232 (3): 64-75
864. Scott J, Knott TJ, Priestley LM, Robertson ME, Mann DV, Kostner G, Miller GJ, Miller NE (1985) High-density lipoprotein composition is altered by a common DNA polymorphism adjacent to apoprotein AII gene in man. Lancet 1: 771-773
865. Selby PB, Selby PR (1978) Gamma-ray-induced dominant mutations that cause skeletal abnormalities in mice. II. Description of proved mutations. Mutat Res 51: 199-236
866. Sheppard PM (1975) Natural selection and heredity, 4th ed. Hutchinson, London
867. Shoelson S, Haneda M, Blix P, Nanjo A, Sanke T, Inouye K, Steiner D, Rubenstein A, Tager H (1983) Three mutant insulins in man. Nature 302: 540-543
868. Shows TB, Sakaguchi AY, Naylor SL (1982) Mapping the human genome, cloned genes, DNA polymorphisms and inherited disease. Adv Hum Genet 12: 341-452
869. Siemens HW (1924) Die Zwillingspathologie. Springer, Berlin
870. Siemens HW (1924) Die Leistungsfähigkeit der zwillingspathologischen Arbeitsmethode. Z Induktive Abstammungs-Vererbungslehre 33: 348
871. Siemens HW (1925) Über einen, in der menschlichen Pathologie noch nicht beobachteten Vererbungsmodus: Dominant geschlechtsgebundene Vererbung. Arch Rassenbiol 17: 47-61
872a. Simon M, Alexandre JL, Bourel M, LeMarec B, Scordia C (1977) Heredity of idiopathic haemochromatosis: a study of 106 families. Clin Genet 11: 327-241
872b. Simon M, Bourel M, Genetet B, Fauchet R (1977) Idiopathic hemochromatosis: demonstration of recessive transmission and early detection by family HLA typing. N Engl J Med 297: 1017-1021
873. Simonds B (1963) Tuberculosis in twins. Putnam, London
874. Sing CF, Davignon J (1985) Role of the apolipoprotein E polymorphism in determining normal plasma lipid and lipoprotein variation. Am J Hum Genet 37
875. Slack J (1979) Inheritance of familial hypercholesterolemia. In: Paoletti R, Grott AM Jr (eds) Atherosclerosis Reviews, Vol 5. Raven Press, New York, pp 35-66
875a. Smith Ch (1971) Recurrence risks for multifactorial inheritance. Am J Hum Genet 23: 578-588
876. Smith CAB (1956) A test for segregation ratios in family data. Ann Hum Genet 20: 257
877. Smith CAB (1956/7) Counting methods in genetical statistics. Ann Hum Genet 21: 254-276
878. Smith CAB (1959) A note on the effect of ascertainment on segregation ratios. Ann Hum Genet 23: 311-323
879. Smith CAB (1970) A note on testing the Hardy-Weinberg law. Ann Hum Genet 33: 377
880. Smith CH (1970) Heritability of liability and concordance in monozygous twins. Ann Hum Genet 34: 85
881. Smith SM, Penrose LS (1955) Monozygotic and dizygotic twin diagnosis. Ann Hum Genet 19: 273-289
882. Smith SM, Penrose LS, Smith CAB (1961) Mathematical tables for research workers in human genetics. Churchill, London
883. Smithies O (1964) Chromosomal rearrangements and protein structure. Cold Spring Harbor Symp Quant Biol 29: 309
884. Smithies O, Connell GE, Dixon GH (1962) Chromosomal rearrangements and the evolution of haptoglobin genes. Nature 196: 232
885. Snyder LF, Doan CA (1944) Is the homozygous form of multiple teleangiectasia lethal? J Lab Clin Med 29: 1211-1216
886. Sørensen H, Dissing J (1975a) Association between the $C3^F$ gene and atherosclerotic vascular diseases. Hum Hered 25: 279-283
887. Southern EM (1982) Application of DNA analysis to mapping the human genome. Cytogenet Cell Genet 32: 52-57
888. Svejgaard A, Jersild C, Staub Nielsen L, Bodmer WF (1974) HL-A antigens and disease. Statistical and genetical considerations. Tissue Antigens 4: 95-105
889. Svejgaard A, Hauge M, Jersild C, Platz P, Ryder LP, Staub Nielsen L, Thomsen M (1979) The HLA system. An introductory survey. 2nd ed. Karger, Basel New York

890. Swanson J, Park B, McCullough J (1972) Kell phenotypes in families of patients with X-linked chronic granulomatous disease. 12th Congress of the International Society for Blood Transfusion. Washington (Abstracts) p 26
891. Schaefer EJ (1984) Clinical, biochemical, and genetic features in familial disorders of high density lipoprotein deficiency. Arteriosclerosis 4: 303-322
892. Schepank H (1974) Erb- und Umweltfaktoren bei Neurosen. Psychiatry Series 11: 1-221
893. Schiff F, v Verschuer O (1933) Serologische Untersuchungen an Zwillingen. II. Mitt Z Morphol Anthropol 32: 244-249
894. Schnyder UW (1955) Neurodermitis und Allergie des Respirationstraktes. Dermatologica 110: 289
895. Schroeder T-M, Tilgen D, Krüger J, Vogel F (1976) Formal genetics of Fanconi's anemia. Hum Genet 32: 257-288
895a. Schrott HG, Karp L, Omenn GS (1973) Prenatal prediction in myotonic dystrophy: guidelines for genetic counseling. Clin Genet 4: 38-45
896. Steinberg AG (1965) Evidence for a mutation or crossing over at the Rh-locus. Vox Sang 10: 721
897. Stern C (1957) The problem of complete Y-linkage in man. Am J Hum Genet 9: 147-165
898. Stern C, Walls GL (1957) The Cunier pedigree of "color blindness". Am J Hum Genet 9: 249-273
899. Stevenson AC, Cheeseman EA (1955/56) Hereditary deaf mutism with particular reference to Northern Ireland. Ann Hum Genet 20: 177-231
900. Stevens WL (1950) Statistical analysis of the AB0 blood groups. Hum Biol 22: 191-217
901. Stocks P, Barrington A (1925) Hereditary disorders of bone development. Treasury of Human Inheritance 3, Part 1
902. Stone NJ (1979) Genetic hyperlipidemia and atherosclerosis. Artery 5: 377-397
903. Strasser G (1982) Coronary risk factors revisited. World Health Forum 3: 85-88
904. Sturtevant AH (1925) The effects of unequal crossing over at the bar locus in drosophila. Genetics 10: 117
904a. Svejgaard A, Platz P, Ryder LP (1983) HLA and disease 1982. In: Möller G (ed) HLA and disease susceptibility (Immunological reviews, vol 70). Munksgaard, Copenhagen, pp 193-218
905. Tattersall R (1976) The inheritance of maturity-onset type diabetes in young people. In: Creutzfeldt E, Köbberling J, Neel JV (eds) The genetics of diabetes mellitus. Springer Verlag, Berlin Heidelberg New York, pp 88-95
906. Tattersall RB, Pyke DA (1972) Diabetes in identical twins. Lancet II: 1120-1125
907. ten Kate LP, Boman H, Daiger SP, Motulsky AG (1982) Familial aggregation of coronary heart disease and its relation to known genetic risk factors. Am J Cardiol 50: 945-953
908. ten Kate LP, Boman H, Daiger SP, Motulsky AG (1984) Increased frequency of coronary heart disease in relatives of wives of myocardial infarct survivors: assortative mating for lifestyle and risk factors. Am J Cardiol 53: 399-403
910. Terasaki PI (1980) Histocompatibility testing 1980. UCLA Tissue Typing Laboratory, Los Angeles
911. Terasaki PI, McClelland JD (1964) Microdroplet assay of human serum cytotoxins. Nature 204: 998-1000
912. Timoféef-Ressovsky NW (1931) Gerichtetes Variieren in der phänotypischen Manifestierung einiger Generationen von Drosophila funebris. Naturwissenschaften 19: 493-497
913. Tolleshaug H, Goldstein JL, Schneider WJ, Brown MS (1982) Posttranslational processing of the LDL receptor and its genetic disruption in familial hypercholesterolemia. Cell 30: 715-724
914. Trevor-Roper PD (1952) Marriage of two complete albinos with normally pigmented offspring. Br J Ophthalmol 36: 107
915. Utermann G (1983) Coronary heart disease. In: Emery AEH, Rimoin DL (eds) Principles and practice of medical genetics. Churchill Livingstone, Edinburgh etc, pp 956-978
916. Utermann G, Hardewig A, Zimmer F (1984) Apolipoprotein E phenotypes in patients with myocardial infarction. Hum Genet 65: 237-241
917. Utermann G, Kindermann I, Kaffarnik H, Steinmetz A (1984) Apolipoprotein E phenotypes and hyperlipidemia. Hum Genet 65: 232-236
918. Verschuer O von (1954) Wirksame Faktoren im Leben des Menschen. Steiner, Wiesbaden
919. Verschuer O von (1958) Die Zwillingsforschung im Dienste der inneren Medizin. Verh Dtsch Ges Inn Med 64: 262-273
920. Vogel F (1969) Does the human X chromosome show evidence for clustering of genes with related functions? J Génét Hum 17: 475-477
921. Vogel F (1970) The genetic basis of the normal human electroencephalogram (EEG). Hum Genet 10: 91-114
922. Vogel F (1957) Methoden zur Prüfung der Reihenfolge von Merkmalsträgern und Gesunden in Geschwisterschaften. Z Menschl Vererbungs-Konstitutionslehre 34: 194-204
923. Vogel F (1982) Die Bedeutung der Humangenetik für eine Theorie der Krankheit. Verh Dtsch Ges Path 66: 1-15
924. Vogel F, Dorn H (1964) Erbliche Hautkrankheiten. In: Becker PE (ed) Humangenetik, ein kurzes Handbuch, Vol IV. Thieme, Stuttgart, pp 346-535
925. Vogel F, Krüger J (1967) Multifactorial determination of genetic affections. Proceedings of the 3rd International Congress of Human Genetics. Johns Hopkins University Press, Baltimore, pp 437-445
926. Vogel F, Wendt GG (1956) Zwillingsuntersuchung über die Erblichkeit einiger anthropologischer Maße und Konstitutionsindices. Z Menschl Vererbungs-Konstitutionslehre 33: 425-446
927. Waardenburg PJ (1957) The twin study method in wider perspective. Acta Genet (Basel) 7: 10-20
928. Waardenburg PJ (1953) Zum Kapitel des außerokularen erblichen Nystagmus. Acta Genet (Basel) 4: 298-312
929. Wahlund S (1928) Zusammensetzung von Populationen und Korrelationserscheinungen vom Standpunkt der Vererbungslehre aus betrachtet. Hereditas 11: 65-105
930. Walter H (1976) Körperbauform und Klima. Kritische Überlegungen zur Übertragbarkeit der Bergmann'schen Regel auf den Menschen. Z Morphol Anthropol 67: 241-263

931. Watkins Winifred M (1966) Blood-group substances. Science 152: 172-181
932. Weicker H (1959) Die genetischen Grundlagen der Fanconi-Anämie. Schweiz Med Wochenschr 89: 1081
933. Weight M, Cortese C, Sule U, Miller NE, Lewis B (1982) Heritability of the low density lipoprotein receptor activity of human blood mononuclear cells: studies in normolipidaemic adult male twins. Clin Sci 62: 397-401
934. Weinberg W (1902) Beiträge zur Physiologie und Pathologie der Mehrlingsgeburten beim Menschen und Probleme der Mehrlingsgeburtenstatistik. Z Geburtshilfe Gynäkol 47: 12
935. Weinberg W (1909) Der Einfluß von Alter und Geburtenzahl der Mutter auf die Häufigkeit der ein- und zweieiigen Zwillingsgeburten. Z Geburtshilfe Gynäkol 65: 318-324
936. Weinberg W (1912) Methoden und Fehlerquellen der Untersuchung auf Mendelsche Zahlen beim Menschen. Arch Rassenbiol 9: 165-174
937. Weiner W, Lewis HBM, Moores P, Sanger R, Race RR (1957) A gene, y, modifying the blood group antigen A. Vox Sang 2: 25-37
938. Weiss MC, Green H (1967) Human-mouse hybrid cell lines containing partial complements of human chromosomes and functioning human genes. Proc Natl Acad Sci USA 58: 1104-1111
939. Weitkamp LR (1972) Human autosomal linkage groups. Proceedings of the 4th International Congress of Human Genetics. Paris, 1971. Excerpta Medica, Amsterdam, pp 445-460
940. Welander L (1957) Homozygous appearance of distal myopathy. Acta Genet (Basel) 7: 321-325
941. Wendt GG, Drohm D (1972) Fortschritte der Allgemeinen und Klinischen Humangenetik. Vol IV: Die Huntingtonsche Chorea. Thieme, Stuttgart
942. Werdelin O (1982) Immune response genes. Allergy 37: 451-461
943. Wettke-Schäfer R, Kantner G (1983) X-linked dominant inherited diseases with lethality in hemizygous males. Hum Genet 64: 1-23
944. White R, Leppert M, Bishop DT, Barker D, Berkowitz J, Brown C, Callahan P, Holm T, Jerominski L (1985) Construction of linkage maps with DNA markers for human chromosomes. Nature 313: 101-105
945. White R (1985) Mapping human chromosomes in genetic diseases. First Intern Symp on the Role of Recombinant DNA in Genetics (Teplitz RL et al, eds). Crete, (In the press, 1986)
946. White R, Leppard M, Bishop DT, Barker D, Berkowitz J, Brown C, Callahan P, Holm T, Jerominski L (1985) Constitution of linkage maps with DNA markers for human chromosomes. Nature 313: 101-105
947. Whittingham S, Mathews JD, Schanfield MS, Tait BD, Mackay IR (1981) Interaction of HLA and Gm in autoimmune chronic active hepatitis. Clin Exp Immunol 43: 80-86
948. Wieland W (1975) Diagnose. Überlegungen zur Medizintheorie. De Gruyter, Berlin New York
949. Wieland W (1983) Systematische Bemerkungen zum Diagnosebegriff. Münstersche Beiträge zur Geschichte und Theorie der Medizin 20: 17-34
950. Wiener AS (1943) Additional variants of the Rh type demonstrable with a special human anti-Rh-serum. J Immunol 47: 461-465
951. Wiener AS (1941) Hemolytic reactions following transfusion of blood of the homologous group II. Arch Pathol 32: 227-250
952. Wiener AS, di Diego N, Sokol S (1953) Studies on the heredity of the human blood groups. I. The MN types. Acta Genet Med Gemellol (Rome) 2: 391-398
953. Williams DL (1985) Molecular biology in arteriosclerosis research. Arteriosclerosis 5: 213-227
954. Williams WR, Lalouel JM (1982) Complex segregation analysis of hyperlipidemia in a Seattle sample. Hum Hered 32: 24-26
955. Wilson SR (1973) The correlation between relatives under the multifactorial model with assortative mating I. The multifactorial model with assortative mating. Ann Hum Genet 37: 189-204
956. Wilson SR (1973) The correlation between relatives under the multifactorial model with assortative mating. II. The correlation between relatives in the equilibrium position. Ann Hum Genet 37: 205-215
957. Wimer BM, Marsh WL, Taswell HF, Galey WR (1977) Haematological changes associated with the McLeod phenotype of the Kell blood group system. Br J Haematol 36: 219
958. Winters RW, Graham JB, Williams TF, McFalls VC, Burnett CH (1957) A genetic study of familial hypophosphatemia and vitamin D-resistant rickets. Trans Assoc Am Physicians 70: 234-242
959. Woolf B (1955) On estimating the relation between blood group and disease. Ann Hum Genet 19: 251-253
960. Wright S (1934) The results of crosses between inbred strains of guinea pigs differing in number of digits. Genetics 19: 537-551
961. Wright S (1931) Evolution in Mendelian populations. Genetics 16: 97-159
962. Yasuda N, Tsuji K (1975) A counting method of maximum likelihood for estimating haplotype frequency in the HL-A system. Jap J Hum Genet 20: 1
963. Yee S, Lew R, Morton NE (1969) A general program for segregation analysis. In: Morton NE (ed) A genetics program Library. The Univ of Hawaii Press, Honolulu
964. Yokoyama S (1985) DNA polymorphism and the susceptibility to diabetes. Am J Med Genet 21: 649-654

References to Chapter 4

965. Agarwal DP, Goedde HW (1984) Alkoholmetabolisierende Enzyme: Alkoholunverträglichkeit und Alkoholkrankheit. In: Zang KD (ed) Klinische Genetik des Alkoholismus. Verlag W Kohlhammer, Stuttgart, pp 65-89
966. Alter BP (1985) Antenatal diagnosis of thalassemia: a review. Ann NY Acad Sci 445: 393-407
967. Alter BP, Nathan DG (1978) Antenatal diagnosis of haematological disorders - '1978'. Clin Haematol 7: 195-216
968. Amara SG, Jonas V, Rosenfeld MG, Ong ES, Evans RM (1982) Alternative RNA processing in calcitonin gene expression generates mRNAs en-

coding different polypeptide products. Nature 298: 240-244
968a. Amrhein JA, Meyer WJ III, Jones HW, Migeon, CJ (1976) Androgen insensitivity in man: Evidence for genetic heterogeneity Proc Natl Acad Sci USA 73: 891-894
969. Anonymous (1973) Pharmacogenetics. Report of a WHO scientific group. WHO technical Reports Series N. 524. Geneva
970. Anonymous (1978) Editorial: Fetal haemoglobin in sickle-cell anaemia and thalassaemia - a clue to therapy? Lancet I: 971-972
971. Anonymous (1984) Report on the Workshop "Molecular Basis of Polymorphic Drug Oxidation in Man", Otzenhausen, 1983. Eur J Clin Pharmacol 27: 253-257
972. Antonarakis SE, Kazazian HH, Jr, Orkin SH (1985) DNA polymorphism and molecular pathology of the human globin gene clusters. Hum Genet 69: 1-14
973. Anton-Lamprecht I, Schnyder UW (1974) Ultrastructure of inborn errors of keratinization. VI. Inherited ichthyoses - a model system for heterogeneities in Keratinization disturbances. Arch Dermatol Forsch 250: 207-227
974. Anton-Lamprecht I, Hashimoto I (1976) Epidermolysis bullosa dystrophica dominans (Pasini) - A primary structural defect of the anchoring fibrils. Hum Genet 32: 69-76
975. Atlas SA, Vesell ES, Nebert DW (1976) Genetic control of interindividual variations in the inducibility of aryl hydrocarbon hydroxylase in cultured human lymphocytes. Cancer Res 36: 4619
976. Ayesh R, Idle JR, Ritchie JC, Crothers MJ, Hetzel MR (1984) Metabolic oxidation phenotypes as markers for susceptibility to lung cancer. Nature 312: 169-170
977. Bach G, Friedman R, Weissman B, Neufeld EF (1972) The defect in the Hurler and Scheie syndromes: deficiency of α-L-iduronidase. Proc Natl Acad Sci USA 69: 2048
978. Bangham AD (1968) Membrane models with phospholipids. Progr Biophys 18: 29-95
979. Bank A (1978) Critical review. The thalassemia syndromes. Blood 51: 369-384
980. Bank A, Mears JG, Ramirez F (1979) Organization of the human globin genes in normal and thalassemia cells. In: Stamatoyannopoulos G, Nienhuis A (eds) Cellular and molecular regulation of hemoglobin switching. Grune & Stratton, New York, pp 521-539
981. Baralle FE, Shoulders CC, Proudfoot NJ (1980) The primary structure of the human epsilon-globin gene. Cell 21: 621-626
982. Barranger JA (1984) Marrow transplantation in genetic disease. N Engl J Med 311: 1629-1631
983. Bartoldi F, Giovenco S, Sostia W, Maiconi W, Morisi F, Pittalis F, Prospen G, Spotorno G (1977) Biomedical application of fibre entrapped enzymes. Pharmacol Res Comm 9: 521-546
984. Bartholomé K, Lutz P, Bickel H (1975) Determination of phenylalanine hydroxylase activity in patients with phenylketonuria and hyperphenylalaninemia. Pediatr Res 9: 899-903
985. Bauknecht T (1977) Studies on steroid hormone receptors (5α-Dihydrotestosterone, Estradiol, and Dexamethasone) in cultured human fibroblasts and amniotic fluid cells. Hum Genet 39: 321-328
986. Beadle GW (1945) Biochemical genetics. Chem Rev 37: 15-96
987. Beadle GW, Ephrussi B (1936) The differentiation of eye pigments in drosophila as studied by transplantations. Genetics 21: 225-247
988. Beadle GW, Tatum EL (1941) Genetic control of biochemical reactions in neurospora. Proc Natl Acad Sci USA 27: 499-506
989. Beauchamp GK, Yamazaki K, Boyse EA (1985) The chemosensory recognition of genetic individuality. Sci Am 253: 86-92
990. Becker MA, Kostel PJ, Meyer LJ, Seegmiller JE (1973) Human phosphoribosylpyrophosphate synthetase: increased enzyme specific activity in a family with gout and excessive purine synthesis. Proc Natl Acad Sci USA 70: 2749
991. Beet EA (1949) The genetics of the sickle cell trait in a Bantu tribe. Ann Eugen 14: 279
992. Bellingham AJ (1976) Haemoglobins with altered oxygen affinity. Br Med Bull 32: 234-238
993. Benöhr HChr, Waller HD (1973) Eigenschaften der Glutathionreductase von Erythrocyten Gesunder und Enzymmangelträger. Klin Wochenschr 51: 1177
994. Benöhr HC, Waller HD (1975) Metabolism in haemolytic states. Clin Haematol 4: 45-62
995. Bercroft DM, Phillips LI (1965) Hereditary orotic aciduria and megaloblastic anemia: A second case with response to uridine. Br Med J I: 547
996. Betke K, Beutler E, Brewer GJ, Kirkman HN, Luzzato L, Motulsky AG, Ramot B, Siniscalco M (1967) Standardization of procedures for the study of glucose-6-phosphate dehydrogenase. WHO Tech Rep Ser No 366
997. Beutler E (1969) Glutathione reductase stimulation in normal subjects by riboflavin supplementation. J Clin Invest 48: 1957
998. Beutler E (1969) Effect of flavin compounds on glutathione reductase activity: in vivo and in vitro studies. J Clin Invest 48: 1957-1966
999. Beutler E (1969) G-6-PD activity of individual erythrocytes and X-chromosomal inactivation. In: Yunis JJ (ed) Biochemical methods in red cell genetics. Academic Press, New York, pp 95-113
1000. Beutler E (1975) Red cell metabolism. A manual of biochemical methods. Grune & Stratton, New York London
1001. Beutler E (1978) Hemolytic anemia in disorders of red cell metabolism. Plenum, New York London
1002. Beutler E (1983) Glucose-6-phosphate dehydrogenase deficiency. In: Stanbury JB, Wyngaarden JB, Fredrickson DS Goldstein J, Brown M, (eds) The metabolic basis of inherited disease, 5th ed. McGraw-Hill, New York, pp 1629-1653
1003. Beutler E (1979) Review: Red cell enzyme defects as nondiseases and as diseases. Blood 54: 1-7
1004. Bickel H (1953) Influence of phenylalanine intake on phenylketonuria. Lancet II: 812
1005. Bienzle U (1981) Glucose-6-phosphate dehydrogenase deficiency. Clin Haematol 10: 785-799
1006. Bird AP (1984) DNA methylation - how important in gene control? Nature 307: 503-504
1007. Blombäck M, Blombäck B, Mammen EF, Prasad AS (1968) Fibrinogen Detroit - a molecular defect in the

N-terminal disulphide knot of human fibrinogen? Nature 218: 134

1008. Blyumina MG (1981) Blood serum phenylalanine level in heterozygotes for the phenylketonuria gene under conditions of intensified protein catabolism. Genetika (Moskva) 17: 910-914
1009. Boivin P, Gerland C (1965) La synthese du glutathion au cours de l'anémie hemolytique congenitale avec deficit en glutathion reduit. Deficit congenital en glutathion-synthetase erythrocytaire? Nouv Rev Fr Hematol 5: 606
1010. Bothwell TH, Charlton RW, Motulsky AG (1983) Idiopathic hemochromatosis. In: Stanbury JB, Wyngaarden JB, Fredrickson DS, Goldstein JL, Brown MS (eds) The Metabolic Basis of Inherited Disease, 5th Ed. McGraw-Hill, New York, pp 1269-1298
1011. Boué J, Boué A, Philippe E, Giraud A, Deluchat C (1973) Phenotypes of karyotyped abortuses. Bull Eur Soc Hum Genet 32
1012. Boyer SH, Dover GJ, Serjeant GR, Smith KD, Antonarakis SE, Embury SH, Margolet L, Noyes AN, Boyer ML, Bias WB (1984) Production of F cells in sickle cell anemia: regulation by a genetic locus or loci separate from the β-globin gene cluster. Blood 64: 1053-1058
1013. Boyer SH, Rucknagel DL, Weatherall DJ, Watson-Williams EJ (1963) Further evidence for linkage between the β and δ loci governing human hemoglobins and the population dynamics of linked genes. Am J Hum Genet 15: 438-448
1014. Bradley TB, Boyer SH, Allen FH (1961) Hopkins-2-hemoglobin: a revised pedigree with data on blood and serum groups. Bull Johns Hopkins Hosp 108: 75-79
1015. Brady RD, Koloday EH (1973) Disorders of ganglioside metabolism. Prog Med Genet 8: 225-242
1016. Braunitzer G, Hilschmann N, Rudloff V, Hilse K, Liebold B, Müller R (1961) The haemoglobin particles. Chemical and genetic aspects of their structure. Nature 190: 480
1017. Brewer GJ (1971) Annotation: Human ecology, an expanding role for the human geneticist. Am J Hum Genet 23: 92-94
1018. Brewer GJ (1980) Inherited erythrocyte metabolic and membrane disorders. Med Clin North America 64: 579-596
1019. Britten RJ, Davidson EH (1969) Gene regulation for higher cells. A theory. Science 165: 349-357
1020. Bross K, Dittes H, Krone W, Schmid M, Vogel W (1973) Biochemical and cytogenetic studies on the nucleolus organizing regions (NOR) of man. I. Comparison of trisomy 21 with balanced translocations t(DqGq). Hum Genet 20: 223-229
1021. Bross K, Krone W (1972) On the number of ribosomal RNA genes in man. Hum Genet 14: 137
1022. Bross K, Krone W (1973) Ribosomal cistrons and acrocentric chromosomes in man. Hum Genet 18: 71-75
1023. Brown MS, Goldstein JL (1974) Expression of the familial hypercholesterolemia gene in heterozygotes: mechanism for a dominant disorder in man. Science 185: 61-63
1024. Buckley RH, Gilbertsen RB, Schiff RI, Ferreira E, Sanal SO, Waldmann TA (1976) Heterogeneity of lymphocyte subpopulations in severe combined immunodeficiency: evidence against a stem cell defect. J Clin Invest 58: 130-136
1025. Burke BE, Shotton DM (1983) Erythrocyte membrane skeleton abnormalities in hereditary spherocytosis. Br J Haematol 54: 173-187
1026. Bürki K, Liebelt AG, Bresnick E (1975) Simple vs. complex inheritance of inducible aryl hydrocarbon hydroxylase in mouse tissues. Biochem Genet 13: 417-433
1027. Butenandt A (1953) Biochemie der Gene und Genwirkungen. Naturwissenschaften 40: 91-100
1028. Butterfield DA, Chesnut DB, Roses D, Appel SH (1974) Electron spin resonance studies of erythrocytes from patients with myotonic muscular dystrophy. Proc Natl Acad Sci USA 71: 909-913
1029. Cantz M, Gehler J (1976) The mucopolysaccharidoses: Inborn errors of glycosaminoglycan catabolism. Hum Genet 32: 233-255
1030. Carson PE, Flanagan CL, Ickes CE, Alving AS (1956) Enzymatic deficiency in primaquine-sensitive erythrocytes. Science 124: 484-485
1031. Cassimos C, Malaka-Zafiriu K, Tsiures J (1974) Variations in salycilamide glucuronide formation in normal and in G-6-PD deficient children. J Pediatr 84: 110-111
1032. Chandley AC (1981) The chromosomal basis of human infertility. Brit Med Bull 35: 181-186
1033. Chakravarti A, Buetow KH, Antonarakis SE, Waber PG, Boehm CD, Kazazian HH (1984) Nonuniform recombination within the human β-globin cluster. Am J Hum Genet 36: 1239-1258
1034. Childs B, Zinkham W (1958) A genetic study of a defect in glutathione metabolism of the erythrocyte. Johns Hopkins Med J 102: 21-37
1035. Cleaver JE (1972) Xeroderma pigmentosum: Variants with normal DNA repair and normal sensitivity to ultraviolet light. J Invest Dermatol 58: 124-128
1036. Cleaver JE, Bootsma D (1975) Xeroderma Pigmentosum: Biochemical and Genetic Characteristics. Ann Rev Genet 9: 19-38
1037. Clegg JB, Weatherall DJ (1976) Molecular basis of thalassaemia. Br Med Bull 32: 262-269
1038. Cleve H (1981) H-Y antigen and sex determination. The Tenth Arne Tiselius Memorial lecture. In: Peeters H (ed) Protides of the biological fluids. Vol 29. Pergamon Press, Oxford New York, pp 3-12
1039. Cohen AS (1972) Inherited systemic amyloidosis. In: Stanbury JB, Wyngaarden JB, Fredrickson DS (eds) The metabolic basis of inherited disease, 3rd ed. McGraw-Hill, New York, pp 1273-1294
1040. Collins FS, Metherall JE, Yamakawa M, Pan J, Weissman SM, Forget BG (1985) A point mutation in the Aγ-globin gene promoter in Greek hereditary persistence of fetal haemoglobin. Nature 313: 325-326
1041. Collins FS, Weissman SM (1984) The Molecular Genetics of Human Hemoglobins Proc Nucleic Acid Res Mol Biol 31: 315-462
1042. Comings DE (1972) The structure and function of chromatin. Adv Hum Genet 3: 237-431
1043. Conner BJ, Reyes AA, Morin C, Itakura K, Teplitz RL, Wallace RB (1983) Detection of sickle β^s-globin allele by hybridization with synthetic oligonucleotides. Proc Natl Acad Sci USA 80: 278

1044. Cori GT, Cori CF (1952) Glucose-6-phosphatase of the liver in glycogen storage disease. J Biol Chem 199: 661

1045. Costa T, Scriver CR, Childs B (1985) The effect of Mendelian disease on human health: a measurement. Am J Med Genet 21: 231-242

1046. Cotte J, Kissin C, Mathieu M, Poncet J, Monnet P, Salle B, German D (1968) Observations on a case of partial deficiency of erythrocyte ATPase. Rev Fr Etudes Clin Biol 13: 284

1047. Courtney M, Jaliat S, Tessier L-H, Benavente A, Crystal RG, Lecocq J-P (1985) Synthesis in E coli of α_1-antitrypsin variants of therapeutic potential for emphysema and thrombosis. Nature 313: 149-151

1048. Cox RP, Krauss MR, Balis ME, Dancis J (1970) Evidence for transfer of enzyme product as the basis of metabolic cooperation between tissue culture fibroblasts of Lesch-Nyhan disease and normal cells. Proc Natl Acad Sci USA 67: 1573-1579

1049. Dacie JV (1953) Mollison PL, Richardson N, Selwyn JG, Shapiro L (1953) Atypical congenital haemolytic anemia. Q J Med 22: 79

1050. Dayer P, Balant L, Fabre J (1984) The genetic control of drug oxidation in the liver. Internat J Clin Pharmacol 3: 421-425

1051. Dayer P, Balant L, Kupfer A, Courvoisier F, Fabre J (1983) Contribution of the genetic status of oxidative metabolism to variability in the plasma concentrations of beta-adrenergic blocking agents. Eur J Clin Pharmacol 24: 787-799

1052. Dean MF, Muir H, Benson PF, Button LR, Boylston A, Mowbray J (1976) Enzyme replacement therapy by fibroblast transplantation in a case of Hunter syndrome. Nature 261: 323-326

1053. De Bruyn CHMM (1976) Hypoxanthine-Guanine phosphoribosyl transferase deficiency. Hum Genet 31: 127-150

1054. Deisseroth A, Nienhuis A, Lawrence J, Giles R, Turner P, Ruddle FH (1978) Chromosomal localization on human β globin gene on human chromosome 11 in somatic cell hybrids. Proc Natl Acad Sci USA 75: 1457-1460

1055. Deisseroth A, Nienhuis A, Turner P, Velez R, Anderson WF, Ruddle F, Lawrence J, Creagan R, Kucherlapati R (1977) Localization of the human α-globin structural gene to chromosome 16 in somatic cell hybrids by molecular hybridization assay. Cell 12: 205-218

1056. Desnick RJ (1979) Prospects for enzyme therapy in the lysosomal storage diseases of Ashkenazi Jews pp 253-270 In: Goodman RM, Motulsky AG (eds) Genetic diseases among Ashkenazi Jews. Raven Press, New York

1057. Desnick RJ (ed) (1980) Enzyme therapy in genetic diseases. A Liss, New York

1058. Desnick RJ, Grabowski GA (1981) Advances in the treatment of inherited metabolic diseases. Adv Hum Genet 11: 281-369

1059. Dicke WK, Weijers HA, van de Kamer JH (1953) Coeliac disease: the presence in wheat of a factor having a deleterious effect in cases of coeliac disease. Acta Paediatr 42: 34-42

1060. Diebold K, Häfner H, Vogel F, Schalt E (1968) Die myoklonischen Varianten der familiären amaurotischen Idiotie. Hum Genet 5: 119-164

1061. Dittes H, Krone W, Bross K, Schmid M, Vogel W (1975) Biochemical and cytogenetic studies on the nucleolus organizing regions (NOR) of man. II. A family with the 15/21 translocation. Hum Genet 26: 47-59

1062. Drayer DE, Reidenberg MM (1977) Clinical consequences of polymorphic acetylation of basic drugs. Clin Pharmacol Ther 22: 251-258

1063. Dugaiczyk A, Woo SLC, Lai EC, Mace ML jr, McReynolds L, O'Malley BW (1978) The natural ovalbumin gene contains seven intervening sequences. Nature 274: 328-333

1064. Edwards YH, Hopkinson DA, Harris H (1971) Inherited variants of human nucleoside phosphorylase. Ann Hum Genet 34: 395-408

1065. Eichelbaum M, Bertilsson L, Sawe J, Zekorn C (1982) Polymorphic oxidation of sparteine and debrisoquine: related pharmacogenetic entities. Clin Pharmacol Therap 31: 184-186

1066. Elston RC, Graham JB, Miller CH, Reisner HM, Bouma BM (1976) Probabilistic classification of hemophilia A carriers by discriminant analysis. Thromb Res 8: 683-695

1067. Emery AEH (1980) Duchenne muscular dystrophy. Genetic aspects, carrier detection and antenatal diagnosis. Br Med Bull 36: 117-122

1068. Emery AEH, Anand R, Danford N, Duncan W, Paton L (1978) Aryl-hydrocarbon-hydroxylase inducibility in patients with cancer. Lancet I: 470-472

1068a. Engel W (1982) Geschlechtsdifferenzierung und ihre Störungen. Verh Dtsch Ges Path 66: 329-343

1069. Epstein CJ (1977) Inferring from modes of inheritance to the mechanisms of genetic disease. In: Rowland LP (ed) Pathogenesis of human muscular dystrophies. Excerpta Medica, Amsterdam, pp 9-22

1070. Epstein CJ (1985) Mouse monosomies and trisomies as experimental systems for studying mammalian aneuploidy. TIG 1: 129-134

1071. Evans DAP (1986) Pharmacogenetics. In: King RA, Rotter JI, Motulsky AG (eds) The Genetic Basis of Common Disease. McGraw-Hill, New York (in press)

1072. Evans DAP, Harmer D, Downham DY, Whibley EJ, Idle JR, Ritchie J, Smith RL (1983) The genetic control of sparteine and debrisoquine metabolism in man with new methods of analysing bimodal distribution. J Med Genet 20: 321-329

1073. Evans DAP, Manley K, McKusick VA (1960) Genetic control of isoniazid metabolism in man. Br Med J II: 485

1074. Evans HJ (1977) Facts and fancies relating to chromosome structure in man. Adv Hum Genet 8: 347-438

1075. Filip DJ, Eckstein JD, Veltkamp JJ (1976) Hereditary antithrombin III deficiency and thromboembolic disease. Am J Hematol 2: 343-349

1076. Fisch RO, Simes LK, Torres F, Anderson JA (1965) Studies on families of phenylketonurics. Am J Dis Child 109: 427-431

1077. Flatz G (1971) Population study of erythrocyte glutathione reductase activity. II. Hematological data of subjects with low enzyme activity and stimulation characteristics in their families. Hum Genet 11: 278-285

1078. Flatz G (1971) Population study of erythrocyte glutathione reductase activity I. Stimulation of the enzyme by flavin adenine dinucleotide and by riboflavin substitution. Hum Genet 11: 269-277
1079. Flatz G, Xirotiris N (1976) Glukose-6-phosphat-Dehydrogenase. In: Becker PE (ed) Humangenetik, ein kurzes Handbuch Vol I/3. Thieme, Stuttgart, pp 494-535.
1080. Fölling A (1934) Über Ausscheidung von Phenylbrenztraubensäure in den Harn als Stoffwechselanomalie in Verbindung mit Imbezillität. Hoppe Seylers Z Physiol Chem 227: 169
1081. Forbes GB (1953) Glycogen storage disease. J Pediatr 42: 645
1082. Forget BG (1978) Molecular lesions in thalassemia. Trends in Biochemical Science Vol 3, 86-90
1083. Forget BG, Wilson JT, Wilson LB, Cavallesco C, Reddy VB, de Riel JK, Biro AP, Ghosh PK, Weissman SM (1979) Globin mRNA structure: general features and sequence homology. In: Stamatoyannopoulos G, Nienhuis A (eds) Cellular and molecular regulation of hemoglobin switching. Grune & Stratton, New York, pp 569-593
1084. Franceschetti A, Klein D (1954) Le dépistage des hétérozygotes. In: Gedda L (ed) Genetica medica. Orrizonte Midico, Rome
1085. Fraser FC, Walker BE, Trasler DG (1957) Experimental production of congenital cleft palate: Genetic and environmental factors. Pediatrics [Suppl] 19: 782
1086. Fratantoni JC, Hall CW, Neufeld EF (1968) Hunter and Hurler syndromes. Mutual correction of the defect in cultured fibroblasts. Science 162: 570-572
1087. Fredrickson DS, Goldstein JL, Brown MS (1978) The familial hyperlipoproteinemias. In: Stanbury JB, Wyngaarden JB, Fredrickson DS (eds) The metabolic basis of inherited disease, 4th ed, pp 604-655. McGraw-Hill, New York
1088. Frézal J, Munnich A, Mitchell G (1983) One gene, several messages. From multifunctional proteins to endogenous opiates. Hum Genet 64: 311-314
1089. Friedman MJ (1978) Erythrocytic mechanism of sickle cell resistance to malaria. Proc Natl Acad Sci USA 75: 1994-1997
1090. Friedmann T, Seegmiller JE, Subak-Sharpe JH (1968) Metabolic cooperation between genetically marked human fibroblasts in tissue culture. Nature 220: 272-274
1091. Garrod AE (1963) Inborn errors of metabolism. London 1923, reprinted by Oxford Univ Press
1092. Geha RS, Rosen RS, Merler E (1973) Identification and characterization of subpopulations of lymphocytes in human peripheral blood after fractionation on discontinuous gradients of albumin: the cellular defect in X-linked agammaglobulinemia. J Clin Invest 52: 1726-1734
1093. Gelinas R, Endlich B, Pfeiffer C, Yagi M, Stamatoyannopoulos G (1985) G to A substitution in the distal CCAAT box of the Aγ-globin gene in Greek hereditary persistence of fetal hemoglobin. Nature 313: 323-325
1094. Geisler M, Kleinebrecht J, Degenhardt K-H (1972) Histologische Analysen an triploiden Spontanaborten. Hum Genet 16: 283-294
1095. Gelb AF, Klein E, Lieberman J (1977) pulmonary function in nonsmoking subjects with alpha 1-antitrypsine deficiency (MZ phenotype). Am J Med 62: 93
1096. George DL, Francke U (1976) Gene dosage effect: regional mapping of human nucleoside phosphorylase on chromosome 14. Science 194: 851-852
1097. Gerhard DS, Kidd KK, Kidd JR, Egeland JA (1984) Identification of a recent recombination event within the human β-globin gene cluster. Proc Natl Acad Sci USA 81: 7875-7879
1098. Giblett ER, Anderson JE, Cohen F, Pollara B, Neuwissen HJ (1972) Adenosine deaminase deficiency in two patients with severely impaired cellular immunity. Lancet II: 1067-1069
1099. Gibson QH (1948) The reduction of methaemoglobin in red blood cells and studies on the cause of idiopathic methaemoglobinaemia. Biochem J 42: 13-23
1100. Gibson QH, Harrison DC (1947) Familial idiopathic methemoglobinemia. Lancet II: 941-943
1101. Gilbert W (1978) Why genes in pieces? Nature 271: 501
1102. Glenner GG, Ignaczak TF, Page DL (1978) The inherited systemic amyloidoses and localized amyloid deposits. In: Stanbury JB, Wyngaarden JB, Fredrickson DS (eds) The metabolic basis of inherited disease, 4th ed. McGraw-Hill, New York, pp 1308-1339 (Summary in 5th ed. pp 1468-1469)
1103. Goedde HW, Agarwal DP (1978) Pseudocholinesterase variation. International Titisee Conference. Titisee, 13-15 October 1977. Hum Genet [Suppl 1], 45-56
1104. Goedde HS, Altland K (1971) Suxamethonium sensitivity. Ann NY Acad Sci 179: 695-670
1105. Goedde HW, Altland K, Scholler KL (1967) Therapie der durch genetisch bedingte Pseudocholinesterase-Varianten verursachten verlängerten Apnoe nach Succinylcholin. Med Klin 62: 1631-1635
1106. Goldschmidt RB (1935) Gen und Außeneigenschaft (Untersuchungen an Drosophila) I. und II. Mitt Z Vererbungslehre 10: 74-98
1107. Goldstein, JL, Brown, MS (1977) The low-density lipoprotein pathway and its relation to atherosclerosis. Ann. Rev. Biochem. 46, 897-930
1108. Goldstein JL, Brown MS, Stone NJ (1977) Genetics of the LDL receptor: evidence that the mutations affecting binding and internalization are allelic. Cell 12: 629-641
1109. Goldstein JL, Hazzard WR, Schrott HG, Bierman EL, Motulsky AG (1973) Hyperlipidemia in coronary heart disease. II. Genetic analysis of lipid levels in 176 families and delineation of a new inherited disorder. J Clin Invest 54: 1544-1568
1110. Gorrod JW, Jenner P, Keysell GR, Mikhael BR (1974) Oxidations metabolism of nicotine by cigarette smokers with cancer of the urinary bladder. J Natl Cancer Inst 52: 1421-1424
1111 a. Graham JB (1979) Genotype assignment (carrier detection) in the haemophilias. Clin Haematol 8: 115-145
1111 b. Graham JB, Barrow ES, Reisner HM, Edgell CJS (1983) The genetics of blood coagulation. Adv Hum Genet 13: 1-81
1112. Gralnick HR, Finlayson JS (1972) Congenital dysfibrinogenemias. Ann Intern Med 77: 471-473

1113. Grant DM, Tang BK, Kalow W (1982) Polymorphic N-acetylation of a caffeine metabolite in man. Clin Pharmacol Therap 33: 355-359
1114. Griffin JE, Wilson JD (1980) The syndromes of androgen resistance. N Engl J Med 302: 198-209
1115. Gropp A, Kolbus U, Giers D (1975) Systematic approach to the study of trisomy in the mouse. II. Cytogenet Cell Genet 14: 42-62
1116. Guthrie R, Susi A (1963) A simple phenylalanine method for detecting phenylketonuria in large populations of newborns. Pediatrics 32: 338
1117. Haldane JBS (1954) The Biochemistry of Genetics. London
1118. Hanel HK, Cohn J, Harvald B (1971) Adenosine-triphosphate deficiency in a family with nonspherocytic anaemia. Hum Hered 21: 313-319
1119. Harper PS (1986) Carrier detection in Duchenne muscular dystrophy: a critical assessment. Prog Med Genet N.S.
1120. Harvald B, Hanel HK, Squires R, Trap-Jensen J (1964) Adenosine-triphosphatase deficiency in patients with nonspherocytic hemolytic anemia. Lancet II: 18
1121. Herrick JB (1910) Peculiar elongated and sickle-shaped red blood corpuscles in a case of severe anemia. Arch Intern Med 6: 517
1122. Higgs DR, Hill AVS, Micholls R, Goodbourn SEY, Ayyub H, Teal H, Clegg JB, Weatherall DJ (1985) Molecular rearrangements of the human α-gene cluster. Ann NY Acad Sci 445: 45-56
1123. Hilschmann N, Kratzin H, Altevogt P, Ruban E, Kortt A, Staroscik C, Scholz R, Palm W, Barnikol H-U, Barnikol-Watanabe S, Bertram J, Horn J, Engelhard M, Schneider M, Dreher W (1976) Evolutionary origin of antibody specificity. In: Goodman M, Tashian RE (eds) Molecular anthropology. Plenum Press, New York London, pp 369-386
1124. Hilschmann N, Watanabe S, Barnikol HU, Laure CJ, Bertram J, Horn J, Engelhard M, Schneider M, Dreker L (1975) Die Rolle der Evolution im Immunsystem. Nova Acta Leopoldina 42: 189-222
1125. Hirsch W, Mex A, Vogel F (1967) Metabolic traits in mentally retarded children as compared with normal populations: Phenylalanine and tyrosine in serum and urine. J Ment Defic Res 11: 212-227
1126. Hirschhorn R, Beratis N, Rosen FS, Parkman R, Stern R, Polmar S (1975) Adenosine deaminase deficiency in a child diagnosed prenatally. Lancet I: 73-75
1127. Hirschhorn R, Martiniuk F, Rosen FS (1979) Adenosine deaminase activity in normal tissues and tissues from a child with severe combined immunodeficiency and adenosine deaminase deficiency. Clin Immunol Immunopathol 14, 107-120
1128. Hirschhorn R, Weissmann G (1976) Genetic disorders of Lysosomes. Prog Med Genet [New Series] 1: 49-101
1129. Hoof F van, Hers HG (1964) Ultrastructure of hepatic cells in Hurler's disease (gargoylism). CR Acad Sci [D] (Paris) 259: 1281
1130. Hörlein H, Weber G (1948) Über chronische familiäre Methämoglobinämie und eine neue Modifikation des Methämoglobins. Dtsch Med Wochenschr 72: 476
1131. Howell RR (1972) Genetic disease: The present status of treatment. Hosp Pract 7: 75-84
1132. Howell RR, Stevenson RE (1971) The offspring of phenylketonuric women. Soc Biol 18: 519-529
1133. Howell DR, Williams JC (1983) The glycogen stoarage diseases. In: Stanbury et al (eds) The metabolic basis of inherited disease. 5th ed. McGraw Hill, New York etc, pp 141-166
1134. Hsia DY-Y (1957) The laboratory detection of heterozygotes. Am J Hum Genet 9: 97-116
1135. Huehns ER, Dance N, Beaven GH, Hecht F, Motulsky AG (1964) Human embryonic hemoglobins. Cold Spring Harbor Symp Quant Biol 29: 327-331
1136. Huisman THJ, Wilson JB, Gravely M, Hubbard M (1974) Hemoglobin Grady: The first example of a variant with elongated chains due to an insertion of residues. Proc Natl Acad Sci USA 71: 3270-3273
1137. Huisman TJH, Wrightstone RN, Wilson JB, Schroeder WA, Kendall AG (1972) Hemoglobin Kenya, the product of fusion of γ and β polypeptide chains. Arch Biochem Biophys 153: 850-853
1138. Ingram VM (1956) A specific chemical difference between the globins of normal human and sickle cell anaemia haemoglobin Nature 178: 792
1139. Ingram VM (1957) Gene mutations in human haemoglobin: The chemical difference between normal and sickle cell hemoglobin. Nature 180: 325-328
1140. Jacob F (1978) Mouse teratocarcinoma and mouse embryo. The Leuwenhoek lecture 1977. Proc R Soc Lond [Biol] 201: 249-270
1141. Jacob F, Monod J (1961) On the regulation of gene activity Cold Spring Harbor Symp Quant Biol 26: 193-209
1142. Jandl JH, Cooper RA (1978) Hereditary spherocytosis. In: Stanbury JB, Wyngaarden JB, Fredrickson DS (eds) The metabolic basis of inherited disease, 4th ed. McGraw-Hill, New York, pp 1396-1409
1143. Jeffreys AJ (1979) DNA sequence variants in Gγ-, Aγ-, δ- and β-globin genes of man. Cell 18: 1
1144. Jervis GA (1953) Phenylpyruvic oligophrenia: Deficiency of phenylalanine oxidizing system. Proc Soc Exp Biol Med 82: 514-515
1145. Jones ME (1980) Pyrinidine nucleotide biosynthesis in animals: Genes, enzymes, and regulation of UMP biosynthesis. Ann Rev Biochem 49: 253-279
1146. Kahn A (1978) G6PD variants. International Titisee Conference. Titisee, 13-15 October 1977. Hum Genet [Suppl 1], 37-44
1147. Kahn A, Etiemble J, Meienhofer MC, Boivin P (1975) Erythrocyte phosphofructokinase deficiency associated with an unstable variant of muscle phosphofructokinase. Clin Chim Acta 61: 415
1148. Kahn A, Kaplan J-C, Dreyfus J-C (1979) Advances in hereditary red cell enzyme anomalies. Hum Genet 50: 1-27
1149. Kahn A, Marie J, Galand C, Boivin P (1975) Molecular mechanism of erythrocyte pyruvate kinase deficiency. Hum Genet 29: 271-280
1150. Kalckar HM (1957) Biochemical mutations in man and microorganisms. Science 125: 105-108
1151. Kalow W (1982) The metabolism of xenobiotics in different populations. Can J Physiol Pharmacol 60: 1-12
1152. Kalow W, Staron N (1957) On distribution and inheritance of human serum cholinesterase, as indicated by dibucaine numbers. Can J Biochem 35: 1305

1153. Kalter H, Warkany J (1983) Congenital malformations. Etiologic factors and their role in prevention. N Engl J Med 308: 424–431
1154. Kalter H, Warkany J (1983) Congenital malformations. N Engl J Med 308: 491–497
1155. Kamuzora H, Lehmann H (1975) Human embryonic haemoglobins including a comparison by homology of the human ζ and α chains. Nature 256: 511–513
1156a. Kan YW (1985) Molecular pathology of α-thalassemia. Ann NY Acad Sci 445: 28–35
1156b. Kan YW, Chang JC, Poon R (1979) Nucleotide sequences of the untranslated 5′ and 3′ regions of human α-, β- and γ-globin mRNAs. In: Stamatoyannopoulos G, Nienhuis A (eds) Cellular and molecular regulation of hemoglobin switching. Grune & Stratton, New York, pp 595–606
1157. Kan YW, Dozy AM (1978) Polymorphism of DNA sequence adjacent to the human β globin structural gene. Its relation to the sickle mutation. Proc Natl Acad Sci USA 75: 5631–5635
1158. Kan YW, Dozy AM, Trecartin R, Todd D (1977) Identification of a non-deletion defect in α-thalassemia. N Engl J Med 297: 1081–1084
1158a. Kappas A, Sassa S, Anderson KE (1983) The porphyrias. In: Stanbury JB, Wyngaarden JB, Fredrickson DS, Goldstein JL, Brown MS (eds). The metabolic Basis of inherited Disease 5th ed. McGraw-Hill, New York
1159. Kazazian HH, Cho S, Phillips JA III (1977) The mutational basis of the thalassemia syndromes. Prog Med Genet 2: 165–204
1160. Kellermann G, Luyten-Kellerman M, Shaw CR (1973) Genetic variation of aryl hydrocarbon hydroxylase in human lymphocytes. Am J Hum Genet 25: 327–331
1161. Kellermann G, Shaw CR, Luyten-Kellerman M (1973) Aryl hydrocarbon hydroxylase inducibility and bronchogenic carcinoma. N Engl J Med 289: 934
1162. Kelley WN, Greene ML, Rosenbloom FM, Henderson JF, Seegmiller JE (1969) Hypoxanthine-guanine phosphoribosyl transferase in gout. Ann Intern Med 70: 155–206
1163. Kelly WN, Wyngaarden JB (1970) Studies on the purine phosphoribosyltransferase enzymes in fibroblasts from patients with the Lesch-Nyhan syndrome. Clin Res 18: 394
1164. Kelly WJ, Wyngaarden JB (1983) Clinical syndromes associated with HPRT deficiency. In: Stanbury JB, Wyngaarden JB, Fredrickson DS, Goldstein JL, Brown MS (eds) The metabolic basis of inherited disease, 5th ed. McGraw-Hill, New York, pp 1115–1143
1165. Kendrew JC, Dickerson RE, Strandberg BE, Hart RG, Davies DR, Phillips DC, Shore VC (1960) Structure of myoglobin – a three-dimensional Fourier synthesis at 2 Å. Resolution. Nature 185: 422–427
1166. Kirkman HN (1972) Enzyme defects. Prog Med Genet 8: 125–168
1167. Kirkman HN, Hendrickson EM (1963) Sex-linked electrophoretic difference in glucose-6-phosphate dehydrogenase. Am J Hum Genet 15: 240
1167a. Knowlton RG (et al.) (1985) A polymorphic DNA marker linked to cystic fibrosis is located on chromosome 7. Nature 318, 380–382
1168. Knudson AG, DiFerrante A, Curts FJ (1971) The effect of Leucocyte transfusion in a child with MPS type I. Proc Natl Acad Sci 68: 1738–1741
1169. Knussmann R (1973) Unterschiede zwischen Mutter-Kind- und Vater-Kind-Korrelationen im Hautleistensystem des Menschen. Hum Genet 19: 145–154
1170. Koch E, Bohn H, Koch F (1964) Mucoviscidosis. Schattauer, Stuttgart
1171. Korenberg J, Therman E, Denniston C (1978) Hot spots and functional organization of human chromosomes. Hum Genet 43: 13–22
1172. Kornberg RD (1977) Structure of chromatin. Ann Rev Biochem 46: 931–945
1173. Krangel MS, Orr HT, Strominger JL (1980) Structure, function and biosynthesis of the major human histocompatibility antigens (HLA-A and HLA-B). Scand J Immunol 11: 561–571
1174. Kresse H, Cantz M, von Figura K, Glössl J, Paschke E (1981) The mucopolysaccharidoses: Biochemistry and clinical symptoms. Klin Wschr 59: 867–876
1175. Kresse A, Neufeld EF (1972) The Sanfilippo A corrective factor. Purification and mode of action. J Biol Chem 247: 2164
1176. Krone W, Wolf U (1978) Chromosomes and protein variation. In: Brock DSH, Mayo O (eds) The Biochemical Genetics of Man, 2nd ed. Academic Press, New York London
1177. Krooth RS, Weinberg AN (1961) Studies on cell lines developed from the tissues of patients with galactosemia. J Exp Med 133: 1155–1171
1178. Kühn A (1961) Grundriß der Vererbungslehre. Quelle & Meyer, Heidelberg
1179. Kukharenko VI, Kuliev AM, Grinberg KN, Terskikh VV (1974) Cell cycles in human diploid and aneuploid strains. Hum Genet 24: 285–296
1180. Kulazenko VP (1974) Morphogenetic disturbances in a spontaneous abortus with trisomy. 13 Hum Genet 25: 53–59
1181. Kuliev AM, Kukharenko VI, Grinberg KN, Vasileysky SS, Terskikh VV, Stephanova LG (1973) Morphological, autoradiographic, immunochemical and cytochemical investigation of a cell strain with trisomy 7 from a spontaneous abortion. Hum Genet 17: 285–296
1182. Kuliev AM, Kukharenko VI, Grinberg KN, Terskikh VV, Tamarkina AD, Begomazov EA, Vasileysky SS (1974) Investigation of a cell strain with trisomy 14 from a spontaneously aborted human fetus. Hum Genet 21: 1–12
1183. Kuliev AM, Kukharenko VI, Grinberg KN, Mikhailov AT, Tamarkina AD (1975) Human triploid cell strain. Phenotype on cellular level. Hum Genet 30: 127–134
1184. Kupfer A, Preisig R (1984) Pharmacogenetics of mephenytoin: a new drug hydroxylation polymorphism in man. Eur J Clin Pharmacol 26: 753–759
1185. Kurnit DM (1979) Down syndrome: gene dosage at the transcriptional level in skin fibroblasts. Proc Natl Acad Sci USA 76, 2372–2375
1186. Lambert B, Hansson K, Bui TH, Funes-Carvicto F, Lindsten J, Holmberg M, Strausmanis R (1976) DNA repair and frequency of X-ray and u.v.-light induced chromosome. Aberrations in leukocytes from patients with Down's syndrome. Ann Hum Genet 39: 293–302

1187. Landauer W (1957) Phenocopies and genotype with special reference to sporadically occurring developmental variants. Am Naturalist 91: 79-90
1188. Lang A, Lorkin PA (1976) Genetics of human haemoglobins. Br Med Bull 32: 239-245
1189. Lawn RM, Efstratiadis A, O'Connell C, Maniatis T (1980) The nucleotide sequence of the human beta-globin gene. Cell 21: 647-651
1190. Layzer RB, Rowland LP, Bank WJ (1969) Physical and kinetic properties of human phosphofructokinase from skeletal muscle and erythrocytes. J Biol Chem 244: 3823
1191. Layzer RB, Rowland LP, Ramey HM (1967) Muscle phosphofructokinase deficiency. Arch Neurol 17: 512
1191a. Leder, P (1978) Discontinuous genes. N Engl J Med 298, 1079-1081
1192. Leder P (1982) The genetics of antibody diversity. Sci Am 246: 102-115
1193. Leder P, Tilghman S, Tiemeier D, Kunkel D, Seidman JG (1979) The organization of mouse β-globin genes. In: Stamatoyannopoulos G, Nienhuis A (ed) Cellular and molecular regulation of hemoglobin switching. Grune & Stratton, New York, pp 493-500
1194. Lehmann H, Kynoch PAM (1976) Human haemoglobin variants and their characteristics. North-Holland, Amsterdam
1195. Lehmann H, Ryan E (1956) The familial incidence of low pseudocholinesterase level. Lancet II: 124
1196. Lenke RR, Levy HL (1980) Maternal phenylketonuria and hyperphenylalaninemia: An international survey of the outcome of untreated and treated pregnancies. New England J Med 303: 1202
1197. Leroy JG (1983) The oligosaccharidoses (formerly mucolipidoses). In: Emery AEH, Rimoin DL (eds) Principles and Practice of Medical Genetics, Vol 2. Churchill Livingstone, Edinburgh etc, pp 1348-1365
1198. Lesch M, Nyhan WL (1964) A familial disorder of uric acid metabolism and central nervous system function. Am J Med 36: 561
1199. Levin S, Moses SW, Chayoth R, Jagoda N, Steinitz K, Levinson G (1967) Glycogen storage disease in Israel. Isr J Med Sci 3: 397-410
1200. Liebhaber SA, Goossens MJ, Kan YW (1980) Cloning and complete sequence of human 5'-alpha-globin gene. Proc Natl Acad Sci USA 77: 7054-7058
1201. Liebhaber SA, Rappaport EF, Cash FE, Ballas SK, Schwartz E, Surrey S (1984) Hb I mutation encoded at both globin loci on the same chromosome: Concerted evolution in the human genome. Science 226: 1449-1451
1202. Lison M, Blondheim SH, Melmed RN (1980) A polymorphism of the ability to smell urinary metabolites of asparagus. Br Med J 281: 1676-1678
1203. Löhr GW (1969) Genetische Enzymdefekte der Hexokinase und der Transport-Adenosin-Triphosphat-Phosphohydrolase der Erythrozyten. Folia Haematol 91: 28
1204. Lunde PKM, Frislid K, Hansteen V (1977) Disease and acetylation polymorphism. Clin Pharmacokin 2: 182-197
1205. Lyon WF, Hawkes SG (1970) An X-linked gene for testicular feminization of the mouse. Nature 227: 1217-1219
1206. Mabuchi H, Haba T, Ueda K, Ueda R, Tatami R, Ito S, Kametani T, Koizumi J, Miyamoto S, Ohta M, Takeda R, Takegoshi T, Takeshita H (1977) Serum lipids and coronary heart disease in heterozygous familial hypercholesterolemia in the Hokuriku district of Japan. Atherosclerosis 28: 417-423
1207. Magnuson T, Epstein CJ (1981) Genetic control of very early mammalian development. Biol Rev 56: 369-408
1208. Mars R de (1964) Some studies of enzymes in cultivated human cells. In: Metabolic control mechanisms in animal cells. Natl Cancer Inst Monogr 13: 181-193
1209. Martin GR, Epstein CJ, Travis B, Tucker G, Yatziv S, Martin DW jr, Clift S, Cohen S (1978) X-chromosome inactivation during differentiation of female teratocarcinoma stem cells in vitro. Nature 271: 329-333
1210. Martinez J, Holburn RR, Shapiro S, Erslev AJ (1974) Fibrinogen Philadelphia: a hereditary hypodysfibrinogenemia characterized by fibrinogen hypercatabolism. J Clin Invest 53: 600
1211. May A, Huehns ER (1976) The mechanism and prevention of sickling. Br Med Bull 32: 223-233
1212. McDevitt H, Bodmer W (1974) HLA immune-response genes and disease. Lancet I: 1269-1274
1213. McLaren A, Simpson E, Tomonari K, Chandler P, Hogg H (1984) Male sexual differentiation in mice lacking H-Y antigen. Nature 312: 552-555
1214. McKee PA (1983) Hemostatis and disorders of blood coagulation. In: Stanbury JB, Wyngaarden JB, Fredrickson DS, Goldstein JL, Brown MS (eds) The metabolic basis of inherited disease, 5th ed. McGraw-Hill Book Co, New York etc, pp 1531-1560.
1215. McPherson E, Taylor CA Jr (1982) The genetics of malignant hyperthermia: evidence for heterogeneity. Am J Med Genet 11: 273-285
1216. Mears JG, Ramirez F, Leibowitz D, Nakamura F, Bloom F, Konotey-Ahulu F, Bank A (1978) Changes in restricted human cellular DNA fragments containing globin gene sequences in thalassemias and related disorders. Proc Natl Acad Sci USA 75: 1222-1226
1217. Meyer RA, Schmid R (1978) The porphyrias. In: Stanbury JB, Wyngaarden JB, Fredrickson DS (eds) The metabolic basis of inherited disease, 4th ed. McGraw-Hill, New York, pp 1166-1220
1218. Miller M, Opheim KE, Raisys VA, Motulsky AG (1984) Theophylline metabolism: variation and genetics. Clin Pharmacol Therap 35: 170-182
1219. Minder EI, Meier PJ, Muller HK, Minder C, Meyer-UA (1984) Bufuralol metabolism in human liver: a sensitive probe for the debrisoquine-type polymorphism of drug oxidation. Eur J Clin Invest 14: 184-189
1220. Morris JM (1953) The syndrome of testicular feminization in male pseudohermaphrodites. Am J Obstet Gynecol 65: 1192
1221. Moser H (1984) Duchenne muscular dystrophy: Pathogenetic aspects and genetic prevention. Hum Genet 66: 17-40
1222. Motulsky AG (1957) Drug reactions, enzymes and biochemical genetics. J Am Med Assoc 165: 835-837

1223. Motulsky AG (1964) Current concepts of the genetics of the thalassemias. Cold Spring Harbor Symp Quant Biol 29: 399-413
1224. Motulsky AG (1965) Theoretical and clinical problems of glucose-6-phosphate dehydrogenase deficiency. In: Jonxis JHP (ed) Abnormal haemoglobins in Africa. Blackwell, Oxford, pp 143-196 a
1225. Motulsky AG (1972) Hemolysis in glucose-6-phosphate dehydrogenase deficiency. Fed Proc 31: 1286-1292
1226. Motulsky AG (1973) Frequency of sickling disorders in US blacks. N Engl J Med 288: 31-33
1227. Motulsky AG (1975) Glucose-6-phosphate dehydrogenase and abnormal hemoglobin polymorphisms - evidence regarding malarial selection. In: Salzano FM (ed) The role of natural selection in human evolution. North-Holland, Amsterdam, pp 271-291
1228. Motulsky AG (1977) Ecogenetics: genetic variation in susceptibility to environmental agents. In: Human genetics. proceedings of the 5th International Congress of Human Genetics. Mexico City, 10-15 October 1976. Excerpta Medica, Amsterdam, pp 375-385
1229. Motulsky AG (1978) Multifactorial inheritance and heritability in pharmacogenetics. International Titisee Conference. Titisee, 13-15 October 1977. Hum Genet [Suppl 1], 7-12
1230. Motulsky AG (1970) Biochemical genetics of hemoglobins and enzymes as a model for birth defect research. In: Frazer FC, McKusick VA (eds) Congenital malformations. Excerpta Medica, Amsterdam, p 199
1231. Mueller RF, Hornung S, Furlong CE, Anderson J, Giblett ER, Motulsky AG (1983) Plasma paraoxonase polymorphism: a new enzyme assay, population, family, biochemical, and linkage studies. Am J Hum Genet 35: 393-408
1232. Nagel RL, Fabry ME, Pagnier J, Zouhoun I, Wajcman H, Baudin V, Labie D (1985) Hematologically and genetically distinct forms of sickle cell anemia in Africa. The Senegal type and the Benin type. N Engl J Med 312: 880-884
1233. Nagel RL, Labie D (1985) The consequences and implications of the multicentric origin of the Hb S gene. In: Stamatoyannopoulos G, Nienhuis A (eds) Experimental Approaches for the Study of Hemoglobin Switching. Alan R Liss, New York, pp 93-103
1234. Nance WE (1975) Genetic studies of human serum and erythrocyte polymorphisms. University of Wisconsin: Ph D Dissertation 1967. Quoted in: Harris H: The principles of human biochemical genetics. North-Holland, Amsterdam, p 163
1235. Nebert DW, Goujon FM, Gielen JE (1972) Aryl hydrocarbon hydroxylase induction by polycyclic hydrocarbons: Simple autosomal dominant trait in the mouse. Nature New Biol 236: 107
1235 a. Neel JV (1949) The inheritance of sickle cell anemia. Science 110: 64
1236. Neel JV (1949-1950) The detection of the genetic carriers of hereditary disease. Am J Hum Genet 1/2: 19-36
1237. Neel JV (1953) The detection of the genetic carriers of inherited disease. In: Sorsby A (ed) Clinical genetics. Mosby, St Louis, p 27
1238. Nienhuis AW, Anagnou NP, Ley TJ (1984) Advances in thalassemia research. Blood 63: 738-758
1239. Nelson TE, Flewellen EH (1983) The malignant hyperthermia syndrome. N Engl J Med 309: 416-418
1240. Neufeld EF (1974) The biochemical basis for mucopolysaccharidoses and mucolipidoses. Prog Med Genet 10: 81-101
1241. Neufeld EF, Barton RW (1972) Genetic disorders of mucopolysaccharide metabolism. In: Gaull GE (ed) Biology of brain dysfunction, pp 1-30. Plenum, New York
1242. Neufeld EF, McKusick VA (1983) Disorders of lysosomal enzyme synthesis and localization: I-cell disease and Pseudo-Hurler polydystrophy. In: Stanbury JB et al (eds) The metabolic basis of inherited disease. 5th ed. McGraw-Hill Book Co, New York etc, pp 778-787
1243. New SI, Levine LS (1973) Congenital adrenal hypoplasia. Adv Hum Genet 4: 251-326
1244. Ng WG, Rae TF, Donnell GN (1983) Disorders of carbohydrate metabolism. In: Emery AEH, Rimoin DL (eds) Principles and practice of medical genetics. Churchill Livingstone, Edinburgh etc, pp 1267-1285
1245. Niebuhr E (1974) Triploidy in man. Hum Genet 21: 103-125
1246. Nienhuis AW (1978) Mapping the human genome. N Engl J Med 299: 195-196
1247. Norum KR (1978) Genetic and nongenetic hyperlipidemia and Western diet. International Titisee Conference. Titisee, 13-15 October 1977. Hum Genet [Suppl 1], 125-130
1248. Ohno S (1976) A hormone-like action of H-Y antigen and gonadal development of XY/XX mosaic males and hermaphrodites. Hum Genet 35: 21-25
1249. Ohno S (1976) Major regulatory genes for mammalian development. Cell 7: 315-321
1250. Omenn GS, Motulsky AG (1978) "Ecogenetics": genetic variation in susceptibility to environmental agents. In: Cohen BH, Lilienfeld AM, Huang PC (eds) Genetic issues in public health and medicine, pp 83-111. Thomas, Springfield
1252. Orkin SH, Alter BP, Itay C, Mahoney MJ, Lazarus H, Hobbins JC, Nathan DG (1978) Application of endonuclease mapping to the analysis and prenatal diagnosis of thalassemias caused by globin-gene deletion. N Engl J Med 299: 166-172
1253. Orkin SH, Kazazian HH Jr (1984) The mutation and polymorphism of the human β-globin gene and its surrounding area. Ann Rev Genet 18: 131-171
1254. Orkin SH, Markham AF, Kazazian HH (1983) Direct detection of the common Mediterranean thalassemia gene with synthetic DNA probes. An alternative approach for prenatal diagnosis. J Clin Invest 71: 775
1255. Ortigoza-Ferado J, Richter RJ, Hornung SK, Motulsky AG, Furlong CE (1984) paraoxonase hydrolysis in human serum mediated by a genetically variable arylesterase and albumin. Am J Hum Genet 36: 295-305
1256. Osborne WRA, Chen SH, Giblett ER, Biggar WD, Ammann AJ, Scott CR (1977) Purine nucleoside phosphorylase deficiency: evidence for molecular heterogeneity in two families with enzyme deficient members. J Clin Invest 60: 741-746

1257. Ottolenghi S, Comi P, Giglioni B, Tolstoshev P, Lanyon WG, Mitchell GJ, Williamson R, Russo G, Musumeci S, Schiliro G, Tsistrakis GA, Carache S, Wood WG, Clegg JB, Weatherall DJ (1976) $\delta\beta$ thalassemia is due to a gene deletion. Cell 9: 71
1258. Otton SV, Inaba T, Kalow W (1983) Inhibition of sparteine oxidation in human liver by tricyclic antidepressants and other drugs. Life Sciences 32: 795-800
1259. Paigen B, Gurtoo HG, Minowada J et al (1977) Questionable relation of aryl hydrocarbon hydroxylase to lung-cancer risk. N Engl J Med 297: 346-350
1260. Pauling L, Itano HA, Singer SJ, Wells IC (1949) Sickle cell anemia: a molecular disease. Science 110: 543
1261. Pearl ER, Vogler LB, Okos AJ, Crist WM, Lawton AR III, Cooper MD (1979) B-lymphocyte precursors in human bone marrow: an analysis of normal individuals and patients with antibody deficiency states. J Immunol 120, 1169-1175
1262. Penrose LS (1935) Inheritance of phenylpyruvic amentia (Phenylketonuria). Lancet II: 192-194
1263. Penrose LS, Quastel JH (1937) Metabolic studies in phenylketonuria. Biochem J 31: 266-271
1264. Percy AK (1983) The gangliosidoses and related lipid storage diseases. In: Emery AEH, Rimoin DL (eds) Principles and practice of medical genetics, Vol 2. Churchill Livingstone, Edinburgh etc, pp 1366-1388
1265. Perutz MF (1976) Structure and mechanism of haemoglobin. Brit Med Bull 32: 195'-208
1266. Polmar SH, Stern RC, Schwartz AL, Wetzler EM, Chase PA, Hirschhorn R (1976) Enzyme replacement therapy for adenosine deaminase deficiency and severe combined immunodeficiency. N Engl J Med 295: 1337-1343
1267. Polmar SH, Wetzler EM, Stern RC, Martin DW Jr (1978) Evidence for the role of ribonucleotide reductase inhibition in adenosine deaminase deficiency. Pediatr Res
1268. Price Evans DA (1984) Survey of the human acetylator polymorphysm in spontaneous disorders. J Med Genet 21: 243-253
1269. Prins HK, Oort M, Loos JA, Zürcher C, Beckers T (1966) Congenital nonspherocytic hemolytic anemia associated with glutathione deficiency of the erythrocytes. Blood 27: 145
1270. Propping P (1978) Pharmacogenetics Rev Physiol Biochem Pharmacol 83: 124-173
1271. Propping P (1984) Genetic aspects of neurotoxicity. In: Blum K, Manzo L (eds) Neurotoxicology. Marcel Dekker Inc, New York pp 203-218
1272. Proudfoot NJ, Brownlee GG (1976) Nucleotide sequences of globin messenger RNA. Br Med Bull 32: 251-256
1273. Proudfoot NJ, Gil A, Maniatis T (1982) The structure of the human zeta globin gene and a closely linked nearly identical pseudogene. Cell 31: 553
1274. Pyke KW, Dosch H-M, Ipp MM, Gelfand EW (1975) Demonstration of an intrathymic defect in a case of severe combined immunodeficiency disease. N Engl J Med 295: 424-428
1276. Raghuram TC, Koshakji RP, Wilkinson GR, Wood AJJ (1984) Polymorphic ability to metabolize propanolol alters 4-hydroxy propanolol levels but not beta-blockade. Clin Phamacol Therap 36: 51-56
1277. Ratnoff OD (1978) Hereditary disorders of hemostasis. In: Stanbury JB, Wyngaarden JB, Fredrickson DS (eds) The metabolic basis of inherited disease, 4th ed. McGraw-Hill, New York, pp 1755-1791
1278. Ratnoff OD, Jones PK (1977) The laboratory diagnosis of the carrier state for classic hemophilia. Ann Intern Med 86: 521
1279. Reed T, Young RS (1982) Maternal effects in dermatoglyphics: Similarities from twin studies among palmar, plantar and fingertip variables. Am J Hum Genet 34: 349-352
1280. Reidenberg MM, Drayer DE (1978) Aromatic amines and hydrazines, drug acetylation, and lupus erythematosus. International Titisee Conference. Titisee, 13-15 October 1977. Hum Genet [Supp 1], 57-64
1281. Ricco G, Mazza U, Turi RM, Pich PG, Camaschella C, Saglio G, Benini LF (1976) Significance of a new type of human fetal hemoglobin carrying a replacement isoleucine-threonine at position 74 (E19) of the γ chain. Hum Genet 32: 305-313
1282. Romeo G (1977) Analytical review. Enzymatic defects of hereditary porphyrias: an explanation of dominance at the molecular level. Hum Genet 39: 261-276
1283. Rott HD, Mulz D (1983) Duchenne's muscular dystrophy: Carrier detection by muscle ultrasound. J Génét hum 31: 63-65
1284. Ruddy S, Austen KF (1978) Inherited abnormalities of the complement system. In: The metabolic basis of inherited disease, 4th ed, Stanbury JB, Wyngaarden JB, Fredrickson DS (eds). McGraw-Hill, New York, pp 1737-1754
1285. Sahi T (1978) Intestinal lactase polymorphisms and dairy foods. International Titisee Conference. Titisee, 13-15 October 1977. Hum Genet [Suppl 1], 115-124
1286. Sandhoff K, Christomanou H (1979) Biochemistry and genetics of gangliosidoses. Hum Genet 50: 107-143
1287. Scott CR (1983) Disorders of amino acid metabolism. In: Emery AEH, Rimoin DL (eds) Principles and Practice of Medical Genetics. Churchill Livingstone, Edinburgh etc, pp 1241-1266
1288. Scott CI (1973) The genetics of short stature. Prog Med Genet 8: 243-299
1289. Scriver CR (1969) Treatment of inherited disease: Realized and potential. Med Clin North Am 53: 941-963
1290. Scriver CR, Clow CL (1980) Phenylketonuria and other phenylalanine hydroxylase mutants in man. Ann Rev Genet 14: 179-202
1291. Scriver CR, Mackenzie S, Clow CL, Delvin E (1972) Thiamine-responsive maple syrup urine disease. Lancet I: 310-312
1292. Sears DA (1978) The morbidity of sickle cell trait. A review of the literature. Am J Med 64: 1021-1036
1293. Seegmiller JE (1980) Diseases of purine and pyrimidine metabolism. In: Bondy PK, Rosenberg LE (eds) Metabolic control and disease, 8th ed. WB Saunders Cop, Philadelphia, pp 777-937
1294. Seegmiller JE (1983) Disorders of purine and pyrimidine metabolism. In: Emery AEH, Rimoin DL (eds)

Principles and Practice of Medical Genetics. Churchill Livingstone, Edinburth etc, pp 1286-1305
1295. Seegmiller JE, Rosenbaum FM, Kelly WN (1967) An enzyme defect associated with a sex-linked human neurological disorder and excessive purine synthesis. Science 155: 1682
1296. Seidegaard J, Pero RW (1985) The hereditary transmission of high glutathione transferase activity towards trans-stilbene oxide in human mononuclear leukocytes. Hum Genet 69, 66-68
1297. Semenza G (1981) Intestinal oligo- and disaccharides. In: Randle PJ, Steiner DF, Whelan WJ (eds) Carbohydrate metabolism and its disorders, Vol 3. Academic Press, London, pp 425-479
1298. Serjeant GR (1974) The clinical features of sickle cell disease. Clinical Studies. IV. North Holland, Amsterdam
1299. Sibert JR, Harper PS, Thompson RJ et al (1979) Carrier detection in Duchenne muscular dystrophy. Evidence from a study of obligatory carriers and mothers of isolated cases. Arch Dis Child 54: 534-537
1300. Simpson JL (1983) Disorders of gonads and internal reproductive ducts. In: Emery AEH, Rimoin DL (eds) Principles and practice of medical genetics. Churchill Livingstone, Edinburgh, pp 1227-1240
1301. Siniscalco M (1979) Approaches to Human Linkage. Progress in Medical Genetics. NS Vol 3. Saunders, Philadelphia, pp 221-307
1302. Singh RP, Carr DH (1967) Anatomic findings in human abortions of known chromosomal constitution. Obstet Gynecol 29: 806
1303. Sinnot EW, Dunn LC, Dobzhansky T (1958) Principles of genetics, 5th ed. McGraw-Hill, New York
1304. Slighton JL, Blechl AE, Smithies O (1980) Human fetal G-gamma- and A-gamma-globin genes: complete nucleotide sequences suggest that DNA can be exchanged in these duplicated genes. Cell 21: 627-638
1305. Sloan TP, Mahgoub A, Lancaster R, Idle JR, Smith RL (1978) Polymorphism of carbon oxidation of drugs and clinical implications. Br Med J 2: 655-657
1306. Sly WS, Achard DT, Kaplan A (1977) Correction of enzyme deficient fibroblasts: Evidence for a new type of pinocytosis receptor which mediates uptake of lysosomal enzymes. (Abstract) Clin Res 25: 471 A
1307. Smithies O (1955) Grouped variations in the occurrence of new protein components in normal human serum. Nature 175: 307-308
1308. Smithies O (1955) Zone electrophoresis in starch gels: Group variations in the serum proteins of normal human adults. Biochem J 61: 629-641
1309. Spencer N, Hopkinson DA, Harris H (1968) Adenosine deaminase polymorphism in man. Ann Hum Genet 32: 9-14
1310. Sperling O, Boer P, Eilam G, de Vries A (1972) Evidence for molecular alteration of erythrocyte hypoxanthine-guanine phosphoribosyl transferase in a gouty family with partial deficiency of the enzyme. Rev Eur Etudes Clin Biol 17: 72-75
1311. Spielberg SP, Gordon GB, Blake DA, Goldstein DA, Herlong HF (1981) Predisposition to phenytoin hepatotoxicity assessed in vitro. N Engl J Med 305: 722-727
1312. Spranger J (1972) The systemic mucopolysaccharidoses. Ergeb Inn Med Kinderheilk [NF] 32: 165
1313. Spranger J (1983) The mucopolysaccharidoses. In: Emery AEH, Rimoin DL (eds) Principles and Practice of Medical Genetics. Churchill Livingstone, Edinburgh, pp 1339-1347
1314. Spritz RA, DeRiel JK, Forget BG, Weissman SM (1980) Complete nucleotide sequence of the human delta-globin gene. Cell 21: 639-646
1315. Suskind SR, Yanofsky C, Bonner DM (1955) Allelic strains of neurospora lacking tryptophan synthetase: A preliminary immunochemical characterization. Proc Nat Acad Sci USA 41: 577
1316. Swift M, Chase Ch (1979) Cancer in families with Xeroderma pigmentosum. J Natl Cancer Inst 62: 1415-1421
1317. Schmid C, Deininger PL (1975) Sequence organization of the human genome. Cell 6: 345-358
1318. Schnyder UW (1976) Hereditäre Epidermolysen: Klassifikation, Erbprognose und Therapie. Fortschr Prakt Dermatol 8: 1-8
1319. Schroeder WA, Huisman THJ (1978) Human gamma chains: structural features. In: Stamatoyannopoulos G, Nienhuis A (eds) Cellular and molecular regulation of hemoglobin switching. Grune & Stratton, New York, pp 29-45
1320. Stamatoyannopoulos G (1972) The molecular basis of hemoglobin disease. Ann Rev Genet 6: 47
1321. Stamatoyannopoulos G, Nienhuis AW (eds) (1985) Experimental Approaches for the Study of Hemoglobin Switching. Alan R Liss, New York
1322. Steinheider G, Melderis H, Ostertag W (1975) Embryonic ε chains of mice and rabbits. Nature 275: 714-716
1323. Stokes PL, Asquith P, Cooke WT (1973) Genetics of coeliac disease. Clin Gastroenterol 2: 547-556
1324. Studencki AB, Conner BJ, Impraim CC, Teplitz RL, Wallace RB (1985) Discrimination among the human β^A, β^S, and β^C-globin genes using allelespecific oligonucleotide hybridization probes. Am J Hum Genet 37: 42-51
1324a. Takizawa T, Huang I-Y, Ikuta T, Yoshida A (1986) Human glucose-6-phoshate dehydrogenase: primary structure and cDNA cloning. Proc Natl Acad Sci USA, in press
1325. Taliaferro WH, Huck JG (1923) The inheritance of sickle cell anaemia in man. Genetics 8: 594
1326. Tang BK, Grant DM, Kalow W (1983) Isolation and identification of 5-acetylamino-6-formylamino-3-methyluracil as a major metabolite of caffeine in man. Drug Metab Dispos 11: 218-220
1327. Tarni S, Kono N, Nasu T, Nishikawa M (1969) Enzymatic basis for the coexistence of myopathy and hemolytic disease in inherited muscle phosphofructokinase deficiency. Biochem Biophys Res Commun 34: 77
1328. Thalhammer O, Havelec L, Knoll E, Wehle E (1977) Intellectual level (I.Q.) in heterozygotes for phenylketonuria (PKU). Hum Genet 38: 285-288
1328a. Thein SL, Wainscoat JS, Lynch JR, Weatherall DJ, Sampietro M, Fiorelli G (1985) Direct detection of β*39 thalassaemic mutation with Mae 1. Lancet 1: 1095
1329. Tilghman SM, Tiemeyer DC, Seirman JG, Peterlin BM, Sullivan M, Maizel JV, Leder P (1978) Intervening sequence of DNA identified in the structural portion of a mouse β-globin gene. Proc Natl Acad Sci USA 75: 725-729

1329a. Tonegawa S (1983) Somatic generation of antibody diversity. Nature 302: 575-581

1330. Toniolo D, Persico MG, Battistuzzi G, Luzzatto L (1984) Partial purification and characterization of the messenger RNA for human glucose-6-phosphate dehydrogenase. Mol Biol Med 2: 89-103

1331. Udenfriend S, Cooper JR (1952) The enzymatic conversion of phenylalanine to tyrosine. J Biol Chem 194: 503

1332. Utermann G, Langenbeck U, Beisiegel U, Weber W (1980) Genetics of the apolipoprotein E system in man. Amer J Hum Genet 31: 339-347

1333. Valentine WN, Fink K, Paglia DE, Harris SR, Adams WS (1974) Hereditary hemolytic anemia with human erythrocyte pyrimidine S′ nucleotidase deficiency. J Clin Invest 54: 866-879

1334. Valentine WN, Paglia DE, Tartaglia AP, Gilsanz F (1977) Hereditary hemolytic anemia with increased adenosine triphosphate. Science 195: 783-784

1335. Vesell ES (1973) Advances in pharmacogenetics. In: Prog Med Genet 9: 291-367

1336. Vischer TL (1983) Pharmacogenetics in therapy with gold and other slow-acting anti-rheumatic drugs. Rheumatology 8: 220-228

1337. Vogel F (1959) Moderne Probleme der Humangenetik. Ergeb Inn Med Kinderheilk 12: 52-125

1338. Vogel F (1964) Eine vorläufige Abschätzung der Anzahl menschlicher Gene. Z Menschl Vererbungs-Konstitutionslehre 37: 291-299

1339. Vogel F (1964) preliminary estimate of the number of human genes. Nature 201: 847

1340. Vogel F (1984) Relevant deviations in heterozygotes of autosomal-recessive diseases. Clin Genet 25: 381-415

1341. Wachtel SS (1977) H-Y antigen and the genetics of sex determination. Science 198: 797-799

1342. Wachtel S (ed) (1981) Errors of sex determination. (Proc of the Kroc Foundation Conf.) Hum Genet 58: 1-127

1343. Wachtel SS, Ohno S, Loo GC, Boyse EA (1975) Possible role of H-Y antigen in primary sex determination. Nature 257: 235-236

1344. Wainscoat JS, Thein SL, Higgs DR, Bell JI, Weatherall DJ, Al-Awamy BH, Serjeant GR (1985) A genetic marker for elevated levels of haemoglobin F in homozygous sickle cell disease. Br J Haematol 60: 261-268

1344a. Wainwright BJ (et al.) (1985) Localization of cystic fibrosis locus on human chromosome 7cen-p22 Nature 318, 384-385

1345. Waller HD, Benöhr AChr (1976) Enzymdefekte in Glykolyse und Nucleotidstoffwechsel roter Blutzellen bei nichtspherocytären hämolytischen Anämien. Klin Wochenschr 54: 803-850

1346. Wang T, Roden DM, Wolfenden HT, Woolsey RL, Wood AJJ, Wilkinson GR (1984) Influence of genetic polymorphism on the metabolism and disposition of encainide in man. J Pharmacol Exp Therap 228: 605-611

1347. Watson JD, Crick FHC (1953) The structure of DNA. Cold Spring Harbor Symp Quant Biol 18: 123-132

1348. Weatherall DJ (ed) (1976) Haemoglobin: Structure, function and synthesis. Br Med Bull 32: (3) 193-287

1349. Weatherall DJ, Clegg JB (1976) Molecular genetics of human hemoglobin. Ann Rev Genet 10: 157-178

1350. Weatherall DJ, Clegg JB (1981) The thalassaemia syndromes, 3rd ed. Blackwell, Oxford

1351. Weatherall DJ, Wainscoat JS, Thein SL, Old JM, Wood WG, Higgs DR, Clegg JB (1985) Genetic and molecular analysis of mild forms of homozygous β-thalassemia. Ann NY Acad Sci 445: 68-80

1352. Weatherall DJ, Wood WG, Jones RW, Clegg JB (1985) The developmental genetics of human hemoglobin. In: Experimental Approaches for the Study of Hemoglobin Switching. Stamatoyannopoulos G, Nienhuis AW (eds). Alan R Liss, New York, pp 3-25

1353. Wedlund PJ, Aslanian WS, McAllister CB, Wilkinson GR, Branch RA (1984) Mephenytoin hydroxylation deficiency in Caucasians: frequency of a new oxidative drug metabolism polymorphism. Clin Pharmacol Therap 36: 773-780

1354. Weinshilboum RM, Sladek SL (1980) Mercaptopurine pharmacogenetics: monogenic inheritance of erythrocyte thiopurine methyltransferase activity. Am J Hum Genet 32: 651-662

1355. White JM (1974) The unstable haemoglobin disorders. Clin Haematol 3: 333-356

1356. White JM (1976) The unstable haemoglobins. Br Med Bull 32: 219-222

1357. White JM, Dacie JV (1971) The unstable hemoglobins - molecular and clinical features. Prog Hematol 7: 69-109

1357a. White R (et al.) (1985) A closely linked genetic marker for cystic fibrosis Nature 318, 382-384

1358. Wieacker P, Davies K, Pearson P, Ropers HH (1983) Carrier detection in Duchenne muscular dystrophy by use of cloned DNA sequences. Lancet I: 1325-1326

1359. Wilkins L (1950) The diagnosis and treatment of endocrine disorders in childhood and adolescence. Thomas, Springfield

1360. Williamson R (1976) Direct measurement of the number of globin genes. Br Med Bull 32: 246-250

1361. Wilson JM, Young AB, Kelley WN (1983) Hypoxanthine-guanine phosphoribosyltransferase deficiency. N Engl J Med 309: 900-910

1362. Wolf U, Engel W (1972) Gene activation during early development of mammals. Hum Genet 15: 99-118

1363. Wolfe LC, John KM, Falcone JC, Byrne AM, Lux SE (1982) A genetic defect in the binding of protein 4.1 to spectrin in a kindred with hereditary spherocytosis. N Engl J Med 307: 1367-1374

1363a. Woo SLC, Güttler F, Ledley FD, Lidsky AS, Kwok SCM, DiLella AG, Robson KJH (1985) The human poenylalanine hydroxylase gene. In (K. Berg, ed.) Medical genetics Past, present, future. New York: A. R. Liss pp 123-135

1364. Wood WG, Clegg JB, Weatherall DJ (1977) Developmental biology of human hemoglobins. In: Progress in hematology, Brown EB (ed) Vol X. Grune & Stratton, New York, pp 43-90

1365. Worthy TE, Grobner W, Kelley WN (1974) Hereditary orotic aciduria: Evidence for a structural gene mutation. Proc Natl Acad Sci USA 71: 3031-3035

1366. Yoshida A (1968) Subunit structure of human glucose-6-phosphate dehydrogenase. Biochem Genet 2: 237

1367. Yoshida A (1982) Molecular basis of difference in alcohol metabolism between Orientals and Caucasians. Jpn J Hum Genet 27: 55-70

1368. Yoshida A, Beutler E (1983) G-6-PD variants: another up-date. Ann Hum Genet 47: 25-38

1369. Van Zeeland AA, Van Diggelen MCE, Simons JWIM (1972) The role of metabolic cooperation in selection of hypoxanthine guanine phosphoribosyl transferase (HGPRT) deficient mutants from diploid mammalian cell strains. Mutation Res 14: 355-363

1370. Zenzes MT, Wolf U, Günther E, Engel W (1978) Studies on the function of H-Y antigen: dissociation and reorganization experiments on rat gonadal tissue. Cytogenet Cell Genet 20: 365-372

1371. Zimmermann U, Riemann F, Pilwat G (1976) Enzyme loading of electrically homogeneous human red blood cell ghosts by dielectric breakdown. Biochim Biophys Acta 436: 460-474

References to Chapter 5

1372. Abrahamson S, Bender MA, Conger AD, Wolff S (1973) Uniformity of radiation induced mutation rates among different species. Nature 245: 460-462

1373. Altland K, Kaempfer M, Forssbohm M, Werner W (1982) Monitoring for changing mutation rates using blood samples submitted for PKU screening. In: Human Genetics. Part A: The unfolding genome. A Liss, New York, pp 277-287

1374. Ames BN, McCann J, Yamasaki E (1975) Methods for detecting carcinogens and mutagens with the Salmonella/mammalian microsome mutagenicity test Mutation Res. 31, 347-364

1375. Anonymous (1982) Identifying and estimating the genetic impact of chemical environmental mutagens. The National Academy Press, Washington

1376. Anonymous (1983) International Comission for Protection against environmental mutagens and carcinogens. Committee 4 report. Mutation Res 115: 255-291

1377. Anonymous (1982) Ionizing radiation: Sources and biological effects. United Nations Scientific Committee on the Effects of Atomic Radiation. United Nations, New York

1378. Atwood KC, Scheinberg SL (1958) Somatic variation in human erythrocyte antigens. Sympos Genet Approches on Somatic Cell Variation. Gatlinburg, April 2-5, 1958, J of Cellular and Comparative Physiology Vol 52, Suppl 1, pp 97-123

1379. Auerbach C, Robson JM (1946) Chemical production of mutations. Nature 157: 302

1380. Balaban G, Gilbert F, Nichols W, Madans AT, Shields J (1982) Abnormalities of chromosome No 13 in retinoblastomas from individuals with normal constitutional karyotype. Cancer Genet Cytogenet 6: 213-221

1381. Barcinski MA, Abreu MC, de Almeida JCC, Naya JM, Fonseca LG, Castro LE (1975) Cytogenetic investigation in a Brazilian population living in an area of high natural radioactivity. Am J Hum Genet 27: 802-806

1382. Barker D, Schafer M, White R (1984) Restriction sites containing CpG show a higher frequency of polymorphism in human DNA. Cell 36: 131-138

1383. Barthelmess A (1956) Mutagene Arzneimittel. Arzneimittelforsch 6: 157

1384. Barthelmess A (1970) Mutagenic substances in the human environment. In: Vogel F, Röhrborn G (eds) Chemical mutagenesis in mammals and man. Springer, Berlin Heidelberg New York, pp 69-147

1385. Barthelmess A (1973) Erbgefahren im Zivilisationsmilieu. Goldmann, München

1386. Bartsch HD (1970) Virus-induced chromosome alterations in mammals and man. In: Vogel F, Röhrborn G (eds) Chemical Mutagenesis in Mammals and Man. Springer Verlag, Berlin Heidelberg New York, pp 420-432

1386a. Basler A, Buselmaier D, Röhrborn G (1976) Elimination of spontaneous and chemically induced chromosome aberrations in mice during early embryogenesis. Hum Genet 33, 121-130

1387. Bauchinger M (1968) Chromosomenaberrationen und ihre zeitliche Veränderung nach Radium-Röntgentherapie gynäkologischer Tumoren. Strahlentherapie 135: 553-564

1388. Behnke H, Holtermann W (1961) Häufigkeit, Vererbung und klinische Ausprägung der Aniridie in Schleswig-Holstein. Proceedings of the 2nd International Congress on Human Genetics. Roma, pp 1879-1883

1389. Benedict WF, Murphree AL, Banerjee A, Spina CA, Sparkes MC, Sparkes RS (1983) Patient with 13 chromosome deletion: Evidence that the retinoblastoma gene is a recessive cancer gene. Science 219: 973-975

1390. Benzer S (1957) The elementary units of heredity. In: McElroy W, Glass B (eds) The chemical basis of heredity, pp 70-93. Johns Hopkins University press, Baltimore

1391. Berendes U (1974) Multiple tumors of the skin: Clinical, histopathological, and genetic features. Hum Genet 22: 181-210

1392. Berg K (ed) (1979) Genetic damage in man caused by environmental agents. Academic Press, New York etc

1393. Bizzozero OJ, Johnson KG, Ciocco A (1966) Radiation-related leukemia in Hiroshima and Nagasaki. 1946-1964. New Engl J Med 274: 1095

1394. Blank C (1960) Apert's syndrome (A type of acrocephalosyndactyly). Observations on a British series of 39 cases. Ann Hum Genet 24: 151-164

1395. Bloom AB (1972) Induced chromosome aberrations in man. Adv Hum Genet 3: 99-172

1396. Bochkov NP, Lopukhin YM, Kulsehov NP, Kovalchuk LV (1974) Cytogenetic study of patients with ataxia-teleangiectasia. Hum Genet 24: 115-128

1397. Bora KG, Douglas GR, Nestmann ER (eds) (1982) Chemical mutagenesis, human population monitoring and genetic risk assessment. Progr in Mut Res Vol 3. Elsevier, Amsterdam

1398. Borberg A (1951) Clinical and genetic Investigations into tuberous sclerosis and Recklinghausen's neurofibromatosis. Munksgaard, Copenhagen

1399. Boué A, Boué J (1973) Études chromosomiques et anatomiques des grosseuses suivant l'arrêt de contraceptifs stéroides. J Gynecol Obstet Biol Reprod (Paris) 2: 141-154

1400. Brewen JG, Luippold HE (1971) Radiation-induced human chromosome aberrations. In vitro dose-rate studies. Mutat Res 12: 305-314

1401. Brewen JG, Preston RJ (1974) Cytogenetic effects of environmental mutagens in mammalian cells and the extrapolation to man. Mutat Res 26: 297-305

1402. Briard-Guillemot ML, Bonaïti-Pellié, C, Feingold T, Frézal T (1974) Étude génétique de rétinoblastome. Hum Genet 24: 271

1403. Brook JD, Gosden RG, Chandley AC (1984) Maternal ageing and aneuploid embryos - evidence from the mouse that biological and not chronological age is important influence. Hum Genet 66: 41-45

1404. Bucher K, Ioanescu V, Hansen J (1980) Frequency of new mutants among boys with Duchenne muscular dystrophy. Am J Med Genet 7: 27-34

1405. Bundey S (1981) A genetic study of Duchenne muscular dystrophy in the West Midlands. J of Med Genet 18: 1-7

1406. Bundey S (1982) Clinical evidence for heterogeneity in myotonic dystrophy. J Med Genet 19: 341-348

1407. Burhorn D (1970) Klinisch-genetische Analyse des von Hippel-Lindau-Syndroms, ausgehend von Patienten mit Angiomatosis retinae. MD Dissertation, University of Heidelberg

1408. Burnet M (1974) The biology of cancer. In: Chromosomes and cancer German J (ed). Wiley & Sons, New York London Sidney Toronto

1409. Burnet FM (1974) Intrinsic mutagenesis: A genetic approach to ageing. Wiley & Sons, New York

1410. Carothers AD, Collyer S, DeMey R, Johnston I (1984) An aetiological study of 290 XXY males, with special reference to the role of paternal age. Hum Genet 68: 248-253

1411. Carr DH (1967) Chromosomes after oral contraceptives. Lancet II: 830-831

1412. Carter CO (1977) Monogenic disorders. J Med Genet 14: 316-320

1413. Caskey CT, Nussbaum RL, Cohan LC, Pollack L (1980) Sporadic occurrence of Duchenne muscular dystrophy. Evidence for a new mutation. Clin Genet 18: 329-341

1414. Cavenee WK, Dryja TP, Phillips RA, Benedict WF, Godbout R, Gallie BL, Murphree AL, Strong LC, White RL (1983) Expression of recessive alleles by chromosomal mechanisms in retinoblastoma. Nature 305: 779-784

1416. Chen Dequing et al. (1982) Cytogenetic investigation in a population living in the high background radiation area. Chin J of Radiolog Med and Protection 2: 61-63

1417. Childs JD (1981) The effect of a change in mutation rate on the incidence of dominant and X-linked recessive disorders in man. Mutation Res 83: 145-158

1418. Chu EHY, Powell SS (1976) Selective systems in somatic cell genetics. Adv Hum Genet 7: 189-258

1419. Clarke AM (1967) Caffeine- and amino acid-effects upon try^+ revertant yield in UV-irradiated hor^+ and hor^- mutants of E Coli B/r. Mol Gen Genet 99: 97-108

1420. Cleaver JE (1972) Xeroderma pigmentosum. Variants with normal DNA repair and normal sensitivity to ultraviolet light. J Invest Dermatol 58: 124-128

1421. Cleaver JE, Bootsma D (1975) Xeroderma pigmentosum: Biochemical and genetic characteristics. Ann Rev Genet 9: 19-38

1422. Conen PE, Lansky GS (1961) Chromosome damage during nitrogen mustard therapy. Br Med J I: 1055-1057

1423. Croce CC (1985) Chromosomal translocations, oncogenes, and B-cell tumors. Hosp Pract 20: 41-48

1424. Croce CM, Klein G (1985) Chromosome translocations and human cancer. Sci Am 252: 54-60

1425. Crow JF (1983) Chemical mutagen testing: A committee report. Environmental Mutagenesis 5: 255-261

1426. Crowe FW, Schull WJ, Neel JV (1956) A clinical, pathological, and genetic study of multiple neurofibromatosis. Thomas, Springfield

1427. Cui Yanwei (1982) Heredity diseases and congenital malformation survey in high background radiation area. Chinese J of Radiol Med and Protection 2: 55-57

1428. Curtis D (1974) Acrocentric associations in Mongol populations. Hum Genet 22: 17-22

1429. Danforth GH (1921) The frequency of mutation and the incidence of hereditary traits in man. Eugenics, genetics and the family. Scientific Papers of the 2nd International Congress of Eugenics. Vol 1, New York, pp 120-128

1430. Danieli GA, Barbujani G (1984) Duchenne muscular dystrophy. Frequency of sporadic cases. Hum Genet 67: 252-256

1431. Danieli GA, Mostacciuolo ML, Pilotto G, Angelini C, Bonfante A (1980) Duchenne muscular dystrophy. Data from family studies. Hum Genet 54: 63-68

1432. Davie AM, Emery AEH (1978) Estimation of proportion of new mutants among cases of Duchenne muscular dystrophy. J Med Genet 15: 339-345

1433. Davies KE, Harper K, Bonthron D, Krumlauf R, Polkey A, Pembrey MW, Williamson R (1984) Use of a chromosome 21 cloned DNA probe for the analysis of nondisjunction in Down Syndrome. Hum Genet 66: 54-56

1434. DeGroot LJ, Paloyan E (1973) Thyroid carcinoma and radiation. A Chicago endemic. J Am Med Assoc 225: 487-491

1435. Demerec M (1937) Frequency of spontaneous mutations in certain stocks of Drosophila melanogaster. Genetics 22: 469

1436. Deng Shaozhuang et al (1982) Birth survey in high background radiation area. Chin J Radiol Med Protection 2: 60

1437. Deuel TF, Huang JS (1984) Roles of growth factor activities in oncogenesis. Blood 64: 951-958

1438. Dhadial RK, Machim AM, Tait SM (1970) Chromosomal anomalies in spontaneously aborted human fetuses. Lancet II: 20-21

1439. Drake JW (ed) (1973) Proceedings of an International Workshop on the Genetic Control of Mutation. Genetics [Suppl] 73

1439 a. Duckworth-Rysiecki G, Cornish K, Clarke CA, Buchwald M (1985) Identification of two complementation groups in Fanconi anemia. Somatic Cell and Mol Genet 11, 35-41

1440. Ehling UH (1982) Risk estimate based on germ cell mutations in mice. In: Sugimura T, Kondo S, Takebe H (eds) Environmental mutagens and carcinogens. Proc 3rd Intern Conf on Environmental Mutagens. Univ of Tokyo Press, Tokyo and A Liss, New York, pp 709-719

1441. Ek J (1959) Thyroid function in mothers of mongoloid infants. Acta Paediatr 48: 33-42

1442. Epstein CJ, Martin GM, Schultz AL, Motulsky AG (1966) Werner's syndrome: A review of its symptomatology, Natural history, pathological features, genetics and relationship to the natural aging process. Medicine 45: 177-221
1443. Erickson JD (1979) Paternal age and Down syndrome. Amer J Hum Genet 31: 489-497
1444. Erickson JD, Bjerkedal T (1981) Down syndrome associated with father's age in Norway. J of Med Genet 18: 22-28
1445. Evans JA, Hunter AGW, Hamerton JL (1978) Down syndrome and recent geographic trends in Manitoba. J Med Genet 15: 43-47
1446. Ferguson-Smith MA, Handmaker SD (1961) Observations on the satellited human chromosomes. Lancet I: 638-640
1447. Fialkow PJ (1967) Autoantibodies and chromosomal aberration. Lancet I: 1106
1448. Fialkow PJ (1967) Thyroid antibodies, Down's syndrome and maternal age. Nature 214: 1253-1254
1449. Fialkow PJ (1977) Clonal origin and stem cell evolution of human tumors. In: Genetics of Human Cancer. Mulvihill JJ, Miller RW, Fraumeni JF Jr (eds). Raven, New York, pp 439-453
1450. Fischer E, Thielmann HW, Neundörfer B, Rentsch FJ, Edler L, Jung EG (1982) Xeroderma pigmentosum patients from Germany: Clinical symptoms and DNA repair characteristics. Arch Dermatol Res 274: 229-247
1451. Fitzgerald PH, Stewart J, Suckling RD (1983) Retinoblastoma mutation rate in New Zealand and support for the two-hit model. Hum Genet 64: 128-130
1452. Ford CE (1970) Cytogenetics and sex determination in man and mammals. J Biosoc Sci [Suppl] 2: 7-30
1453. Ford CE, Evans EP, Searle AG (1978) Failure of irradiation to induce Robertsonian translocations in germ cells of male mice. In: Conference on Mutations: Their origin, nature and potential relevance to genetic risk in man. Jahreskonferenz 1977, Zentrallaboratorium für Mutagenitätsprüfungen. H Boldt Verlag, Boppard, pp 102-108
1454. Ford CE, Searle AG, Evans EP, West BJ (1969) Differential transmission of translocations induced in spermatogonia of mice by X-irradiation. Cytogenetics 8: 447-470
1455. Francke U, Benirschke K, Jones OW (1975) Prenatal diagnosis of trisomy 9. Hum Genet 29: 243-250
1456. Francke U, Felsenstein J, Gartler SM, Migeon BR, Dancis J, Seegmiller JE, Bakay F, Nyhan WL (1976) The occurrence of new mutants in the X-linked recessive Lesch-Nyhan disease. Am J Hum Genet 28: 123-137
1457. Freese E (1963) Molecular mechanism of mutations. In: Molecular genetics Taylor JH (ed). Plenum, New York, pp 207-269
1458. Freese E (1971) Molecular mechanisms of mutations. In: Chemical mutagens Hollaender A (ed), Vol 1. Plenum, New York, pp 1-56
1459. Fulder SJ, Holliday R (1975) A rapid rise in cell variants during the senescence of populations of human fibroblasts. Cell 6: 67-73
1460. Gardner RJM (1977) A new estimate of the achondroplasia mutation rate. Clin Genet 11: 31-38
1461. Gebhardt E (1974) Antimutagens. Data and problems (review). Hum Genet 24: 1-32
1462. Gebhart E (1981) Sister chromatid exchange (SCE) and structural chromosome aberration in mutagenicity testing. Hum Genet 58: 235-254
1463. Geissler E, Theile M (1983) Virus-induced gene mutations of eukaryotic cells. Hum Genet 63: 1-12
1464. German J (ed) (1983) Chromosome mutation and neoplasia. AR Liss, New York
1465. German J (1983) Neoplasia and chromosome-breakage syndromes. In: Chromosome mutation and neoplasia German J (ed). AR Liss, New York, pp 97-134
1466. German J, Bloom D, Archibald R (1965) Chromosome breakage in a rare and probably genetically determined syndrome of man. Science 148: 506-507
1467. Gilbert F (1983) Retinoblastoma and recessive alleles in tumorigenesis. Nature 305: 761-762
1468. Glatt H, Oesch F (1984) Variations in epoxide hydrolast activities in human liver and blood. In: CHS Banbury Report 16. Genetic Variability in Responses to Chemical Exposure, Omenn GS, Gelboin HV (eds). Cold Spring Harbor Laboratory, pp 189-201
1469. Gropp A, Flatz G (1967) Chromosome breakage and blastic transformation of lymphocytes in ataxia-teleangiectasia. Hum Genet 5: 77-79
1470. Grüneberg H (1970) Das Problem der Mutationsbelastung. In: Genetik und Gesellschaft Wendt GG (ed). Wissenschaftliche Verlagsanstalt, Stuttgart, pp 72-77
1471. Gunther M, Penrose LS (1935) The genetics of epiloia. J Genet 31: 413-430
1472. Haldane JBS (1935) The rate of spontaneous mutation of a human gene. J Genet 31: 317-326
1473. Haldane JBS (1939) The spread of harmful autosomal recessive genes in human populations. Ann Eugen 9: 232-237
1474. Haldane JBS (1947) The mutation rate of the gene for hemophilia, and its segregation ratios in males and females. Ann Eugen 13: 262-271
1475. Hansemann D von (1890) Über asymetrische Zellteilung in Epithelkrebsen und deren biologische Bedeutung. Virchow's Arch Pathol Anat 119: 299-326
1476. Hansteen I-L, Varslot K, Steen-Johnsen J, Langard S (1982) Cytogenetic screening of a newborn population. Clin Genet 21: 309-314
1477. Harnden DG (1974) Ataxia teleangiectasia syndrome: Cytogenetic and cancer aspects. In: Chromosomes and cancer. German J (ed). Raven, New York, pp 87-104
1478. Harnden DG (1977) Cytogenetics of human neoplasia. In: Genetics of human cancer. Mulvihill JJ, Miller RW, Fraumeni JF, Jr (eds). Raven, New York, pp 87-104
1479. Harnden DG (1974) Viruses, chromosomes, and tumors: The interaction between viruses and chromosomes. In: German J (ed) Chromosomes and Cancer. J Wiley & Sons, New York, pp 151-190
1480. Hassold T, Jacobs P, Kline J, Stein Z, Warburton D (1981) Effect of maternal age on autosomal trisomies. Ann Hum Genet 44: 29-36
1481. Hassold TJ, Jacobs PA (1984) Trisomy in man. Ann Rev Genet 18: 69-77
1482. Hayes A, Costa T, Scriver CR, Childs B (1985) The effect of Mendelian disease on human health. II. Response to treatment. Am J Med Genet 21: 243-255

1483. Hayflik L (1965) The limited in vitro lifetime of diploid cell strains. Exp Cell Res 37: 614-636

1484. Hayflik L, Moorhead PS (1961) The serial cultivation of human diploid cell strains. Exp Cell Res 25: 585-621

1485. Hecht F, Kaiser McCaw B (1977) Chromosome instability syndromes. In: Genetics of human cancer. Mulvihill JJ, Miller RW, Fraumeni JF Jr (eds). Raven, New York, pp 105-123.

1486. Herrmann J (1966) Der Einfluß des Zeugungsalters auf die Mutation zu Hämophilie A. Hum Genet 3: 1-16

1487. High Background Radiation Research Group, China (1980) Health survey in high background radiation area in China. Science 209: 877-880

1488. Hirschhorn K (1968) Cytogenetic alterations in leukemia. In: Perspectives in leukemia. Dameshek W, Dutcher RM (eds). Grune & Stratton, New York, pp 113-122.

1489. Hollaender A (ed) (1973) Chemical mutagens. Principles and methods for their detection. Plenum, New York. Vol I/II 1971; Vol III 1973

1490. Holliday R, Kirkwood TBL (1981) Predictions of the somatic mutation and mortalisation theories of cellular ageing are contrary to experimental observations. J Theor Biol 93: 627-642

1491. Hook EB (1981) Unbalanced Robertsonian translocations associated with Down's syndrome or Patau's syndrome: Chromosome subtype, proportion inherited, mutation rates and sex ratio. Hum Genet 59: 235-239

1492. Hook EB, Cross PK (1981) Temporal increase in the rate of Down syndrome livebriths to older mothers in New York State. J Med Genet 18: 29-30

1493. Hook EB, Cross PK (1982) Paternal age and Down's syndrome genotypes diagnosed prenatally: No association in New York State data. Hum Genet 62: 167-174

1494. Hook EB, Cross PK, Regal RR (1984) The frequency of 47,+21, 47,+18 and 47,+13 at the uppermost extremes of maternal ages: results on 56094 fetuses studied prenatally and comparisons with data on livebirths. Hum Genet 68: 211-220

1495. Hook EB, Regal RR (1984) A search for a paternal age-effect upon cases of 47,+21 in which the extra chromosome is of paternal origin. Am J Hum Genet 36: 413-421

1496. Hook EB, Schreinemachers DM, Willey AM, Cross PK (1983) Rates of mutant structural chromosome rearrangements in human fetuses: Data from prenatal cytogenetic studies and associations with maternal age and parental mutagen exposure. Amer J Hum Genet 35: 96-109

1497. Ishimaru T, Cihak RW, Land CE, Steer A, Yamada A (1975) Lung cancer at autopsy in A-bomb survivors and controls, Hiroshima and Nagasaki, 1961-1970. II. Smoking, occupation and A-bomb exposure. Cancer 36: 1723-1728

1498. Iuchi I (1968) Abnormal hemoglobin in Japan: Biochemical and epidemiologic characters of abnormal hemoglobin in Japan. Acta Haematol Jpn 31: 842-851

1499. Jablon S, Kato H (1970) Childhood cancer in relation to prenatal exposure to A-bomb radiation. Lancet II: 1000

1500. Jacobs PA (1981) Mutation rates of structural chromosome rearrangements in man. Amer J Hum Genet 33: 44-45

1501. Jacobs PA, Frackiewitz A, Law P (1972) Incidence and mutation rates of structural rearrangements of the autosomes in man. Ann Hum Genet 35: 301-319

1502. Jacobs PA, Funkhauser J, Matsuura J (1981) In: Hook EB, Porter IH (eds) Population and biological aspects of human mutation. Academic Press, New York London, pp 133-145

1503. Jacobs PA, Mayer M (1981) The origin of human trisomy: A study of heteromorphisms and satellite associations. Ann Hum Genet 45: 357-365

1504. Jacobs PA, Morton NE (1977) Origin of human trisomies and polyploids. Hum Hered 27: 59-72

1505. Janerich DT, Flink EM, Keogh MD (1976) Down's syndrome and oral contraceptive usage. Br J Obstet Gynaecol 83: 617-620

1506. Jones KL, Smith DW, Harvey MAS, Hall BD, Quan L (1975) Older paternal age and fresh gene mutation: Data on additional disorders. J Pediatr 86: 84-88

1507. Jongbloet PH, Mulder A, Hamers AJ (1982) Seasonality of pre-ovulatory nondisjunction and the aetiology of Down syndrome. A European collaborative study. Hum Genet 62: 134-138

1508. Kemp T (1940) Altern und Lebensdauer. In: Handbuch der Erbbiologie des Menschen. Just G (ed). Springer Verlag, Berlin, pp 408-424

1509. Kerr CB (1965) Genetics of human blood coagulation. J Med Genet 2: 254

1510. Kimura M (1968) Evolutionary rate at the molecular level. Nature 217: 624-626

1511. Kimura M, Ohta T (1973) Mutation and evolution at the molecular level. Genetics [Suppl] 73: 19-35

1512. King JL (1971) The role of mutation in evolution. Proceedings of the 6th Berkeley Symposium on Mathematical Statistics and Probability. University of California Press, Berkeley, pp 69-100.

1513. Kirkwood TBL, Cremer Th (1982) Cytogerontology since 1881: A reappraisal of August Weisman and a review of modern progress. Hum Genet 60: 101-121

1514. Klamerth OL (1976) Inhibition of transcription by isonicotinic and hydrazide. Mutat Res 35: 53-64

1515. Klamerth OL (1978) Inhibition of post-replication repair by isonicotinic and hydrazide. Mutat Res 50: 251-261

1516. Klose J (1975) Protein mapping by combined isoelectric focusing and electrophoresis of mouse tissue. Hum Genet 26: 231-243

1517. Klose J, Blohm J, Gerner I (1977) The use of isoelectric focusing and electrophoresis to obtain highly complex-protein patterns of mouse embryos. In: Methods in prenatal toxicology. Neubert D, Merker H-J, Kwasrigroch TE (eds) Thieme, Stuttgart, pp 303-313

1518. Knudson AG (1971) Mutation and cancer: Statistical study of retinoblastoma. Proc Natl Acad Sci USA 68: 820-823

1519. Knudson AG (1973) Mutation and human cancer. Adv Canc Res 17: 317-352

1520. Knudson AG (1977) Genetics and etiology of human cancer. Adv Hum Genet 8: 1-66

1521. Knudson AG, Hethcote HW, Brown BW (1975) Mutation and childhood cancer. A probabilistic model for the incidence of retinoblastoma. Proc Natl Acad Sci USA 72: 5116-5120

1522. Knudson AG, Strong LC (1972) Mutation and cancer: Neuroblastoma and pheochromocytoma. Am J Hum Genet 24: 514-532
1523. Kochupillai N, Verma JC, Grewal MS, Ramalingaswami V (1976) Down's syndrome and related abnormalities in an area of high background radiation in coastal Kerala. Nature 262: 60-61
1524. Kondo S (1973) Evidence that mutations are induced by error in repair and replication. In: Workshop on the Genetic Control of Mutation. Genetics [Suppl] 73: 109-122
1525. Koufos A, Hansen MF, Copeland NG, Jenkins NA, Lampkin BC, Cavenee WK (1985) Loss of heterozygosity in three embryonal tumours suggests a common pathogenetic mechanism. Nature 316: 330-334
1526. Langenbeck U, Hansmann I, Hinney B, Hönig V (1976) On the origin of the supernumerary chromosome in autosomal trisomies - with special reference to Down's Syndrome. Hum Genet 33: 89-102
1527. Lawry RB, Jones DC, Renwick DHG, Trimble BK (1976) Down syndrome in British Columbia, 1972-1973: Incidence and mean maternal age. Teratology 14: 29-34
1528. Lea DE, Catcheside DG (1942) The mechanism of the induction by radiation of chromosome aberrations in Tradescantia. J Genet 44: 216-245
1529. LeBeau M, Rowley JD (1984) Heritable fragile sites in cancer. Nature 308: 607-608
1530. Lejeune J, Turpin R, Rethoré MO (1960) Les enfants nés de parents irradiés (Cas particuliers de la sex-ratio). 9th International Congress Radiology 23.-30.7. 1959, München, pp 1089-1096
1531. Lele KP, Penrose LS, Stallard HB (1963) Chromosome deletion in a case of retinoblastoma. Ann Hum Genet 27: 171
1532. Lenz W (1959) Die Abhängigkeit der Mißbildungen vom Alter der Eltern. Verh Dtsch Ges Inn Med 64. Kongr Bergmann Verlag, München, pp 74-88.
1533. Lindgren D (1972) The temperature influence on the spontaneous mutation rate. I. Literature review. Hereditas 70: 165-178
1534. Lindsten J, Marsk L, Berglund K, Iselius L, Ryman N, Annerén G, Kjessler B, Mitelman F, Nordenson I, Wahlström J, Vejlens L (1981) Incidence of Down's syndrome in Sweden during the years 1968-1977. In: Burgio et al (eds) Trisomy 21. Springer, Berlin Heidelberg New York, pp 195-210
1535. Lu Bingxin et al (1982) Survey of hereditary ophthalmopathies and congenital ophthalmic malformations in high background areas. Chin J Radiol Med and Protection 2: 58-59
1536. Lüers H (1955) Zur Frage der Erbschädigung durch tumortherapeutische Cytostatica. Z Krebsforsch 60: 528
1537. Lüning KG, Searle AG (1970) Estimates of the genetic risks from ionizing irradiation. Mutation Res 12: 291-304
1538. Luria SE, Delbrück M (1943) Mutations of bacteria from virus sensitivity to virus resistance. Genetics 28: 491
1539. Lynas MA (1956/57) Dystrophia myotonica with special reference to Northern Ireland. Ann Hum Genet 21: 318-351
1540. Lynch HT (1976) Miscellaneous problems, cancer, and genetics. In: Lynch HT (ed) Cancer genetics. Thomas, Springfield
1541. Lyon MF, Philipps RJS (1975) Specific locus mutation rates after repeated small radiation doses to mouse oocytes. Mutation Res 30: 375-382
1542. Mackenzie HJ, Penrose LS (1951) Two pedigrees of ectrodactyly. Ann Eugen 16: 88
1543. Magenis RE, Overton KM, Chamberlin J, Brady T, Lovrien E (1977) Parental origin of the extra chromosome in Down's syndrome. Hum Genet 37: 7-16
1544. Malling HV, DeSerres FJ (1973) Genetic alterations at the molecular level in X-ray induced ad-3B mutants of Neurospora crassa. Radiat Res 53: 77-87
1545. Mark J (1974) Cytogenetics of the human meningioma. In: German J (ed) Chromosomes and cancer. J Wiley & Sons, New York, pp 497-517
1546. Marsden CD (1982) Neurotransmitters and CNS disease: Basal ganglia disease. The Lancet II: 1141-1146
1547. Martin CM, Sprague CA, Epstein CJ (1970) Replicative lifespan of cultivated human cells: Effect of donor's age, tissue and genotype. Lab Invest 23: 86-92
1548. Matsunaga E (1976) Hereditary retinoblastoma: Penetrance, expressivity and age of onset. Hum Genet 33: 1-15
1549. Matsunaga E (1981) Genetics of Wilm's tumor. Hum Genet 57: 231-246
1550. Matsunaga E, Tonomura A, Oishi H, Kikuchi Y (1978) Reexamination of paternal age effect in Down's syndrome. Hum Genet 40: 259-268
1551. Mattei JF, Mattei MG, Ayme S, Siraud F (1979) Origin of the extra chromosome in trisomy 21. Hum Genet 46: 107-110
1552. Mavor JW (1924) The production of nondisjunction by X-rays. J Exp Zool 39: 381-432
1553. McCann J, Ames BN (1976) Detection of carcinogens as mutagens in the Salmonella/microsome test: Assay of 300 chemicals: Discussion. Proc Natl Acad Sci USA 73: 950-954
1554. McCann J, Choi E, Yamasaki E, Ames BN (1975) Detection of carcinogens as mutagens in the Salmonella/microsome test: Assay of 300 chemicals. Proc Natl Acad Sci USA 72: 5133-5139
1555. McGregor DH et al (1977) Breast cancer incidence among atomic bomb survivors, Hiroshima and Nagasaki, 1950-1969. J Natl Cancer Inst 59: 799-811
1556. Mikkelsen M, Fischer G, Stene J, Stene E, Petersen E (1976) Incidence study of Down's syndrome in Copenhagen, 1960-1971: With chromosome investigation. Ann Hum Genet 40: 177-182
1557. Miller RW (1969) Delayed radiation effects in atomic bomb survivors. Science 166: 569
1558. Mohn G, Würgler FE (1972) Mutator genes in different species. Hum Genet 16: 49-58
1559. Morley A, Cox S, Holliday R (1982) Human lymphocytes resistant to 6-thioguanine resistance increase with age. Mech Ageing Dev 19: 21-26
1560. Morton NE, Lalouel JM (1978) Genetic counseling in sex linkage. Birth Defects Conference. San Francisco, 11-14 June 1978
1561. Morton NE, Lindsten J (1976) Surveillance of Down's syndrome as a paradigm of population monitoring. Hum Hered 26: 360-371

1562. Møllenbach CJ (1974) Medfødte defekter i ojets indre hinder klinik og arvelighedsforhold. Munksgaard, Kopenhagen

1563. Motulsky AG (1968) Some evolutionary implications of biochemical variants in man. Proceedings of the 8th International Congress of the Anthropology and Ethnology Society. September 1968, Tokyo

1564. Motulsky AG (1982) Interspecies and human genetic variation, problems of risk assessment in chemical mutagenesis and carcinogenesis. In: Chemical Mutagenesis, Human Population Monitoring and Genetic Risk Assessment. (Progress in Mutation Research, Vol3), Bora KC, Douglas GR, Nestmann ER (eds) Elsevier Biomedical Press, pp 75-83

1565. Motulsky AG (1984) Environmental mutagenesis and disease in human populations. In: Mutation, Cancer, and Malformation. Chu EHY, Generoso WM (eds). Plenum New York, pp 1-11

1566. Muller HJ (1927) Artificial transmutation of the gene. Science 66: 84-87

1567. Muller HJ (1955) Artificial transmutation of genes. In: Great experiments in biology. Gabriel ML, Fogel S (eds). Prentice Hall, Englewood Chiffs (Reprint of the 1927 paper), pp 260-266

1568. Mulvihill JJ, Miller RW, Fraumeni JF Jr (eds) (1977) Genetics of human cancer. Raven, New York

1569. Murdoch JL, Walker BA, Hall JG, Abbey H, Smith KK, McKusick VA (1970) Achondroplasia - A genetic and statistical survey. Ann Hum Genet 33: 227

1570. Murdoch J, Walker BA, McKusick VA (1972) Parental age effects on the occurrence of new mutations for the Marfan syndrome. Ann Hum Genet 35: 331-336

1571. Neel JV (1957) Some problems in the estimation of spontaneous mutation rates in animals and man: Effects of radiation on human heredity. WHO, Geneva, pp 139-150

1572. Neel JV (1981) Genetic effects of atomic bombs. Science 213: 1206

1573. Neel JV, Kato H, Schull WJ (1974) Mortality in the children of atomic bomb survivors and controls. Genetics 76: 311-326

1574. Neel JV, Mohrenweiser H, Satoh C, Hamilton B (1979) A consideration of two biochemical approaches to monitoring human populations for a change in germ cell mutation rates. In: Berg K (ed) Genetic damage in man caused by environmental agents. Academic Press, New York etc, pp 29-47

1575. Neel JV, Mohrenweiser HW, Meisler MH (1980) Rate of spontaneous mutation of human loci encoding protein structure. Proc Nat Acad Sci USA 77: 6037-6041

1576. Neel JV, Schull WJ et al (1956) The effect of exposure to the atomic bombs on pregnancy termination in Hiroshima and Nagasaki. Natl Acad Sci Natl Res Counc Publ, Washington (DC), 461

1577. Neel JV, Tiffany TO, Anderson NG (1973) Approaches to monitoring human populations for mutation rates and genetic disease. Chemical mutagens. Hollaender A (ed), Vol 3. Plenum, New York, pp 105-150

1578. Neel JV, Ueda N, Satoh C, Ferrell RE, Tanis RJ, Hamilton HB (1978) The frequency in Japanese of genetic variants of 22 proteins. V. Summary and comparison with data on Caucasians from the British isles. Ann Hum Genet 41: 429-441

1579. Newcombe HB (1965) The study of mutation and selection in human populations. Eugen Rev 57: 109-125

1580. Newcombe HB, McGregor F (1964) Learning ability and physical wellbeing in offspring from rat populations irradiated over many generations. Genetics 50: 1065-1081

1580a. Nielsen J (1966) Diabetes mellitus in parents of patients with Klinefelters' syndrome. Lancet I: 1376

1581. Nielsen J, Sillesen I (1975) Incidence of chromosome aberration among 11, 148 newborn children. Hum Genet 30: 1-12

1582. Nielsen J, Wohlert M, Faaborg-Andersen J, Hansen K, Hvidman L, Krag-Olsen B, Moulvad I, Videbech P (1982) Incidence of chromosome abnormalities in newborn children. Comparison between incidences in 1969-1974 and 1980-1982 in the same area. Hum Genet 61: 98-101

1583. Nilsson C, Hansson A, Nilsson G (1975) Influence of thyroid hormones to satellite association in man and the origin of chromosome abnormalities. Hereditas 80: 157-166

1584. Nowell PC, Hungerford DA (1960) A minute chromosome in human chronic granulocytic leukemia. Science 132: 1497

1585. Oehlkers F (1943) Die Auslösung von Chromosomenmutationen in der Meiosis durch Einwirkung von Chemikalien. Z Induktiven Abstammungs-Vererbungslehre 81: 313-341

1586. Oehme R, Kohne E, Kleihauer E, Horst J (1983) HbM Milwaukee: Direct detection of the β-globin gene mutation of an afflicted family. Hum Genet 64: 376-379

1587. Oertelt R (1970) Klinisch-genetische Analyse des von Hippel-Lindau-Syndroms: Ausgehend von den Anginoblastomen des Kleinhirns. University of Heidelberg: MD Dissertation

1588. Ohno S (1972) Gene duplication, mutation load, and mammalian genetic regulatory systems. J Med Genet 9: 254

1589. Pearson M, Rowley JD (1985) The relation of oncogenesis and cytogenetics in leukemia and lymphoma. Ann Rev Med 36: 471-483

1590. Penrose LS (1933) The relative effects of paternal and maternal age in mongolism. J Genet 27: 219-224

1590a. Penrose LS (1955) Parental age and mutation. Lancet II: 312

1591. Penrose LS (1957) Parental age in achondroplasia and mongolism. Am J Hum Genet 9: 167-169

1591a. Pfeiffer RA (1964) Dominant erbliche Akrocephalosyndaktylie. Z Kinderheilkd 90: 301

1592. Propping P, Buselmaier W, Röhrborn G (1973) Kritische Betrachtung über die intraanimale Kultur von Mikroorganismen, eine Methode zum Nachweis chemisch induzierter Mutationen. Arzneim Forsch 6: 746-749

1593. Rapoport IA (1946) Carbonyl compounds and the chemical mechanism of mutation. CR Acad Sci USSR 54: 65

1594. Ratnoff OD, Bennett B (1973) The genetics of hereditary disorders of blood coagulation. Science 179: 1291-1298

1595. Reed TE (1959) The definition of relative fitness of individuals with specific genetic traits. Am J Hum Genet 11: 137

1596. Reed TE, Falls HF (1955) A pedigree of aniridia with a discussion of germinal mosaicism in man. Am J Hum Genet 7: 28-38

1597. Reed TE, Neel JV (1955) A genetic study of multiple polyposis of the colon (with an appendix deriving a method for estimating relative fitness). Am J Hum Genet 7: 236-263

1597 a. Reichert W, Buselmaier W, Vogel F (1984) Elimination of X-ray-induced chromosomal aberrations in the progeny of female mice. Mutation Res. 139, 87-94

1598. Reith W (1970) Mutationen zu Hämophilie A. Häufigkeit im Regierungsbezirk Münster und Abhängigkeit vom Zeugungsalter. University of Münster: MD Dissertation

1599. Rischbieth H, Barrington A (1912) Treasury of human inheritance. Parts VII and VIII, Section XV A: Dwarfism. University of London, Dulau London, pp 355-573

1600. Röhrborn G (1965) Über mögliche mutagene Nebenwirkungen von Arzneimitteln beim Menschen. Hum Genet 1: 205-231

1601. Röhrborn G, Berrang H (1967) Dominant lethals in young female mice. Mutat Res 4: 231-233

1602. Röhrborn G (1970) Biochemical mechanisms of mutation. In: Chemical mutagenesis in mammals and man. Vogel F, Röhrborn G (eds). Springer, Berlin Heidelberg New York, pp 1-15

1603. Röhrborn G (1970) The dominant lethals: Method and cytogenetic examination of early cleavage stages. In: Chemical mutagenesis in mammals and man. Vogel F, Röhrborn G (eds). Springer, Berlin Heidelberg New York, pp 148-155

1604. Röhrborn G, Buckel U (1976) Investigation on the frequency of chromosome aberrations in bone marrow cells of Chinese hamsters after simultaneous application of caffeine and cyclophosphamide. Hum Genet 33: 113-119

1605. Röhrborn G et al (1978) A correlated study of the cytogenetic effect of INH on cell systems of mammals and man conducted by thirteen laboratories. Hum Genet 42: 1-60

1606. Roth MP, Feingold J, Baumgarten A, Bigel P, Stoll C (1983) Reexamination of paternal age effect in Down's syndrome. Hum Genet 63: 149-152

1607. Russell LB, de Hamer DL, Montgomery CS (1973) Analysis of the c-locus region by means of complementation testing and biochemical studies. Biol Div Ann Prog Rep ORNL-4915: 101-103

1608. Russell LB, de Hamer DL, Montgomery CS (1974) Analysis of 30 c-locus lethals by viability of biochemical studies. Biol Div Ann Prog Rep ORNL-4993: 119-120

1609. Russell WL, Kelly EM, Hunsicker PR et al (1972) Effect of radiation dose-rate on the induction of X-chromosome loss in female mice. In: United Nations. Report of the United Nations Science Committee on the Effect of Atomic Radiations. Ionizing radiation: levels and effects. Vol II, Effect. New York

1610. Russell WL (1965) Effect of the interval between irradiation and conception on mutation frequency in female mice. Proc Natl Acad Sci USA 54: 1552-1557

1611. Russell WL, Russell LB, Kelly EM (1958) Radiation dose rate and mutation frequency. Science 128: 1546-1550

1612. Russell LB, Saylors CL (1963) The relative sensitivity of various germ cell stages of the mouse to radiation-induced nondisjunction, chromosome losses and deficiency. In: Sobels FH (ed) Repair from genetic damage and differential radiosensitivity in germ cells. Pergamon, Oxford pp 313-340

1613. Salk D (1982) Werner's syndrome: A review of recent research with an analysis of connective tissue metabolism, growth control of cultures cells and chromosomal aberrations. Hum Genet 62: 1-15

1614. Sasaki MS, Miyata H (1968) Biological dosimetry in atomic bomb survivors. Nature 220: 1189-1193

1615. Satoh C, Awa AA, Neel JV, Schull WJ, Kato H, Hamilton HB, Otake M, Goriki K (1982) Genetic effects of atomic bombs. In: Human Genetics, Part A: The Unfolding Genome. Bonné-Tamir B (ed). A Liss, New York, pp 267-276

1616. Searle AG (1972) Spontaneous frequencies of point mutations in mice. Hum Genet 16: 33-38

1617. Sergeyev AS (1975) On mutation rate of neurofibromatosis. Hum Genet 28: 129-138

1618. de Serres FJ, Malling HV (1969) Identification of the gene alteration in specific locus mutants at the molecular level. Jpn J Genet 44: 106-113

1619. Sherman SL, Morton NE, Jacobs PA, Turner G (1984) The marker (X) syndrome: A cytogenetic and genetic analysis. Ann Hum Genet 48: 21-37

1620. Sigler AT, Lilienfeld AM, Cohen B-H, Westlake JE (1965) Radiation exposure in parents with mongolism (Down's syndrome). Johns Hopkins. Med J 117: 374

1621. Sillence DO (1983) Disorders of bone density, volume and numeralization. In: Principles and practice of medical genetics (Emery, AEH, Rimoin DL, eds.), Churchill Livingstone, Edinburgh etc. pp 736-751.

1622. Sugimura T, Kondo S, Takebe H (eds) (1982) Environmental mutagens and carcinogens. Proc 3rd Intern Conf Environmental Mutagens. University of Tokyo Press, A. Liss, Tokyo and New York

1623. Sutherland GR (1982) Heritable fragile sites on human chromosomes. VIII. Preliminary population cytogenetic data on the folic acid sensitive fragile sites. Am J Hum Genet 34: 452-458

1624. Swift M (1982) Disease predipostion of ataxia-teleangiectasia heterozygotes. In: Ataxia teleangiectasia - a cellular and molecular link between cancer, neuropathology and immune deficiency, Bridges A, Harnden DG (eds). John Wiley & Sons, New York

1625. Swift M, Chase Ch (1979) Cancer in families with xeroderma pigmentosum. J Natl Cancer Inst 62: 1415-1421

1626. Swift M, Sholman L, Perry M, Chase Ch (1976) Malignant neoplasms in the families of patients with Ataxia-teleangiectasia. Cancer Res 36: 209-215

1627. Szilard L (1959) On the nature of the ageing process. Proc Natl Acad Sci USA 45: 30-45

1628. Schappert-Kimmijser J, Hemmes GD, Nijland R (1966) The heredity of retinoblastoma. In: 2nd Congress of European Society of Ophthalmology. Vienna, 1964. Ophthalmologica 151: 197-213

1629. Schmidt H (1973) Wahrscheinliche genetische Belastung der Bevölkerung mit INH (Isonikotinsäure-Hydrazid). Hum Genet 20: 31-45
1630. Schneider EL, Mitsui Y (1976) The relationship between in vitro cellular ageing and in vivo human age. Proc Nat Acad Sci USA 73: 3584-3588
1631. Schnyder UW (1966) Tumoren der Haut in genetischer Sicht. Praxis 55: 1478-1482
1632. Scholte PJL, Sobels FH (1964) Sex ratio shift among progeny from patients having received therapeutic X-radiation. Am J Hum Genet 16: 26-37
1633. Schroeder TM (1972) Genetische Faktoren der Krebsentstehung. Fortschr Med 16: 603-608
1634. Schroeder TM (1982) Genetically determined chromosome instability syndromes. Cytogenet Cell Genetics 33: 119-132
1635. Schroeder TM, Anschütz F, Knopp A (1964) Spontane Chromosomenaberrationen bei familiärer Panmyelopathie. Hum Genet 1: 194-196
1636. Schroeder TM, Drings P, Beilner P, Buchinger G (1976) Clinical and cytogenetic observations during a six-year period in an adult with Fanconi's anaemia. Blut 34: 119-132
1637. Schroeder TM, Kurth R (1971) Analytical review. Spontaneous chromosomal breakage and high incidence of leukemia in inherited disease. Blood 37: 96
1638. Schroeder TM, Tilgen D, Krüger J, Vogel F (1976) Formal genetics of Fanconi's anemia. Hum Genet 32: 257-288
1639. Schull WJ, Neel JV (1958) Radiation and the sex ratio in man. Science 128: 343-348
1640. Schull WJ, Neel JV, Hashizume A (1968) Some further observations on the sex ratio among infants born to survivors of the atomic bombings of Hiroshima and Nagasaki. Am J Hum Genet 18: 328-338
1641. Schull WJ, Otake M, Neel JV (1981) Genetic effects of the atomic bombs: A reappraisal. Science 213: 1220-1227
1642. Stamatoyannopoulos G (1979) Possibilities for demonstrating point mutations in somatic cells, as illustrated by studies of mutant hemoglobins. In: Berg K (ed) Genetic damage in man caused by environmental agents, Academic press, New York pp 49-62.
1643. Stamatoyannopoulos G, Nute PE, Miller M (1981) De novo mutations producing instable hemoglobins or hemoglobin M. I. Establishment of a depository and use of data for an association of de novo mutation with advanced parental age. Hum Genet 58: 396-404
1644. Stamatoyannopoulos G, Nute PE (1982) De novo mutations producing unstable Hbs or Hbs M. II. Direct estimates of minimum nucleotide mutation rates in man. Hum Genet 60: 181-188
1645. Starlinger P, Saedler H (1972) Insertion mutations in microorganisms. Biochemie 54: 177-185
1646. Stene J, Fischer G, Stene E, Mikkelsen M, Petersen E (1977) Paternal age effect in Down's syndrome. Ann Hum Genet 40: 299-306
1647. Stene J, Stene E, Stengel-Rutkowski S, Murken JD (1981) Paternal age and Down's syndrome. Data from prenatal diagnoses (DFG). Hum Genet 59: 119-124
1648. Stevenson AC (1957) Achondroplasia: An account of the condition in Northern Ireland. Am J Hum Genet 9: 81-91
1649. Stevenson AC (1959) The load of heredity defects in human populations. Radiat Res [Suppl 1], 306-325
1650. Stevenson AC, Bobrow M (1967) Determinants of sex proportions in man, with consideration of the evidence concerning a contribution from X-linked mutations to intrauterine death. J Med Genet 4: 190-221
1651. Stevenson AC, Kerr CB (1967) On the distribution of frequencies of mutation in genes determining harmful traits in man. Mutat Res 4: 339-352
1652. Strobel D, Vogel F (1958) Ein statistischer Gesichtspunkt für das Planen von Untersuchungen über Änderungen der Mutationsrate beim Menschen. Acta Genet Stat Med 8: 274-286
1653. Tanaka K, Ohkura K (1958) Evidence for genetic effects of radiation on offspring of radiologic technicians. Jpn J Hum Genet 3: 135-145
1654. Taylor AM (1963) Bacteriophage - induced mutation in Escherichia coli. Proc Nat Acad Sci USA 50: 1043-1051
1655. Thadani MA, Polasa H (1979) Cytogenetic effects of inactivated influenza virus on male germ cells of mice. Hum Genet 51: 253-258
1656. Timoféeff-Ressovsky NW, Zimmer KG (1947) Das Trefferprinzip in der Biologie. Leipzig
1657. Tönz O, Glatthaar BE, Winterhalter KH, Ritter H (1973) New mutation in a Swiss girl leading to clinical and biochemical β-thalassemia minor. Hum Genet 20: 321-327
1658. Tönz O, Winterhalter KH, Glatthaar BE (1973) New mutation leading to β-thalassaemia minor. Nature 241: 127
1659. Tough IS, Buckton KE, Baikie AG, Court Brown WM (1960) X-ray induced chromosome damage in man. Lancet 1960/II: 849-851
1660. Traut H (1976) Effects of ionizing radiation on DNA. In: Molecular biology, biochemistry and biophysics. Hüttermann J, Köhnlein W, Téoule R (eds), Vol XXVII, Springer, Berlin Heidelberg New York pp 335-347.
1661. Trimble BK, Doughty JH (1974) The amount of hereditary disease in human populations. Ann Hum Genet 38: 199-223
1662. Tünte W, Becker PE, v Knorre G (1967) Zur Genetik der Myositis ossificans progressiva. Hum Genet 4: 320-351
1663. Uchida IA, Holunga R, Lawler C (1968) Maternal radiation and chromosomal aberrations. Lancet II: 1045-1049
1664. Uchida IA, Lee CPV, Byrnes EM (1975) Chromosome aberrations induced in vitro by low doses of radiation: Nondisjunction in lymphocytes of young adults. Am J Hum Genet 27: 419-429
1665. Van Dyke DL, Weiss L, Roberson JR, Babu VR (1983) The frequency and mutation rate of balanced autosomal rearrangements in man estimated from prenatal genetic studies for advanced maternal age. Am J Hum Genet 35: 301-308
1666. Vijayalaxmi, Evans HJ, Ray JH, German J (1983) Bloom's syndrome: Evidence for an increased mutation frequency in vivo. Science 221: 851-853
1666a. Vijayalakshmi, Wunder E, Schroeder TM (1985) Spontaneous 6-thioguanine-resistant lymphocytes in Fanconi anemia patients and their heterozygous parents. Hum Genet 70: 264-270

1667. Vogel F (1954) Über Genetik und Mutationsrate des Retinoblastoms (Glioma retinae). Z Menschl Vererbungs-Konstitutionslehre 32: 308-336

1668. Vogel F (1956) Über die Prüfung von Modellvorstellungen zur spontanen Mutabilität an menschlichem Material. Z Menschl Vererbungs-Konstitutionslehre 33: 470-491

1669. Vogel F (1957) Neue Untersuchungen zur Genetik des Retinoblastoms (Glioma retinae). Z Menschl Vererbungs-Konstitutionslehre 34: 205-236

1670. Vogel F (1958) Gedanken über den Mechanismus einiger spontaner Mutationen beim Menschen. Z Menschl Vererbungs-Konstitutionslehre 34: 389-399

1671. Vogel F (1965) Sind die Mutationsraten für die X-chromosomal rezessiven Hämophilieformen in Keimzellen von Frauen niedriger als in Keimzellen von Männern? Hum Genet 1: 253-263

1672. Vogel F (1970) Monitoring of human populations. In: Chemical mutagenesis in mammals and man. Vogel F, Röhrborn G (eds), Springer, Berlin Heidelberg New York pp 445-452.

1673. Vogel F (1970) Spontaneous mutation in man. In: Chemical mutagenesis in mammals and man. Vogel F, Röhrborn G (eds), Springer, Berlin Heidelberg New York pp 16-68.

1674. Vogel F (1975) Mutations in man. Approaches to an evaluation of the genetic load due to mutagenic agents in the human population. Mutation Res 29: 263-269

1675. Vogel F (1977) A probable sex difference in some mutation rates. Am J Hum Genet 29: 312-319

1676. Vogel F (1979) Genetics of Retinoblastoma. Hum Genet 52: 1-54

1677. Vogel F (1983) Mutation in Man. In: Emery AEH, Rimoin DL (eds) Principles and Practice of Medical Genetics, Churchill Livingstone, Edinburgh etc. pp 26-48.

1678. Vogel F (1984) Mutation and selection in the marker (X) syndrome. Ann Hum Genet 48: 327-332

1679. Vogel F, Altland K (1982) Utilization of material from PKU-screening programs for mutation screening. In: Bora KC (ed) Proceedings of an International Symposium on chemical mutagenesis, human population monitoring and genetic risk assessment. Progr in Mut Res Vol 3, Elsevier, Amsterdam pp 143-157.

1680. Vogel F, Jäger P (1969) The genetic load of a human population due to cytostatic agents. Humangenetik 7: 287-304

1681. Vogel F, Kopun M (1977) Higher frequencies of transitions among point mutations. J Mol Evol 9: 159-180

1682. Vogel F, Krüger J, Brøndum Nielsen K, Fryns JP, Schindler D, Schinzel A, Schmidt A, Schwinger E (1985) Recurrent mutation pressure does not explain the prevalence of the marker (X) syndrome. Hum Genet 71: 1-6

1683. Vogel F, Rathenberg R (1975) Spontaneous mutation in man. Adv Hum Genet 5: 223-318

1684. Vogel F, Röhrborn G, Hansmann I (1974) Die Testung von Fremdstoffen auf Mutagenität. Arzneim Forsch 24: 1665-1677

1685. Vogelstein B, Fearon ER, Hamilton SR, Feinberg AP (1985) Use of restriction fragment length polymorphisms to determine the clonal origin of human tumors. Science 227: 642-645

1686. Vogt PK (1983) Onkogene. Verh Ges Dtsch Naturf und Ärzte, Wiss. Verlagsgesellschaft, Stuttgart 235-247.

1687. Wais S, Salvati E (1966) Klinefelter's syndrome and diabetes mellitus. Lancet II: 747-748

1688. Warren ST, Schulz RA, Chang CC, Wade MH, Troske JE (1981) Elevated spontaneous mutation rate in Bloom syndrome fibroblasts. Proc Nat Acad Sci USA 78: 3133-3137

1689. Weech AA (1927) Combined acrocephaly and syndactylism occurring in mother and daughter. A case report. Johns Hopkins Med J 40: 73

1690. Weinberg RA (1983) A molecular basis of cancer. Sci Am, Nov. 1983, 126-142

1691. Weinberg RA (1984) Ras oncogenes and the molecular mechanisms of carcinogenesis. Blood 64: 1143-1145

1692. Weinberg W (1912) Zur Vererbung des Zwergwuchses. Arch Rassen- und Gesellschafts Biol 9: 710-718

1693. Weismann A (1891) Essays upon heredity and kindred biological problems. Vol I. Claredon Press (1st ed 1889, 2nd ed 1891) Oxford

1694. Weiss RA, Marshall CJ (1984) Oncogenes. Lancet 2: 1138-1142

1695. Whitfield HJ Jr, Martin RG, Ames B (1966) Classification of aminotransferase (C gene) mutants in the histidine operon. J Mol Biol 21: 335-355

1696. Willecke K, Schäfer R (1984) Human Oncogenes. Hum Genet 66: 132-142

1697. Williams RT (1959) Detoxication mechanisms, 2nd ed Wiley & Sons, New York

1698. Winkler U (1972) Spontaneous mutations in bacteria and phages. Hum Genet 16: 19-26

1699. Winter RM (1980) Estimation of male to female ratio of mutation rates from carrier detection tests in X-linked disorders. Am J Hum Genet 32: 582-588

1700. Winter RM, Pembrey ME (1982) Does unequal crossing over contribute to the mutation rate in Duchenne muscular dystrophy? Amer J Med Genet 12: 437-441

1701. Winter RM, Tuddenham EGD, Goldman E, Matthews KB (1983) A maximum likelihood estimate of the sex ratio of mutation rates in hemophilia A. Hum Genet 64: 156-159

1702. Yanase T, Hanada M, Seita M, Ohya T, Imamura T, Fujimura T, Kawasaki K, Yamaoka K (1968) Molecular basis of morbidity - from a series of studies of hemoglobinopathies in Western Japan. Jpn J Hum Genet 13: 40-53

1703. Yanofsky C, Ito J, Horn V (1966) Amino acid replacements and the genetic code. Cold Spring Harbor Symp Quant Biol 31: 151-162

1704. Yunis JJ (1983) The chromosomal basis of human neoplasia. Science 221: 227-236

1705. Yunis JJ, Soreng AL (1984) Constitutive fragile sites and cancer. Science 226: 1199-1204

1706. Zakrzewski S, Koch M, Sperling K (1983) Complementation studies between Fanconi's anemia cells with different DNA repair characteristics. Hum Genet 64: 55-57

1707. Zakrzewski S, Sperling K (1980) Genetic heterogeneity of Fanconi's anemia demonstrated by somatic cell hybrids. Hum Genet 56: 81-84
1708. Zakrzewski S, Sperling K (1982) Analysis of heterogeneity in Fanconi's anemia patients of different ethnic origin. Hum Genet 62: 321-323
1709. Zankl H, Zang KD (1971) Cytological and cytogenetical studies on brain tumors. III. Ph^1-like chromosomes in human meningiomas. Hum Genet 12: 42-49
1710. Zankl H, Zang KD (1972) Cytological and cytogenetical studies on brain tumors. IV. Identification of the missing G chromosome in human meningiomas as no 22 by fluorescence technique. Hum Genet 14: 167-169
1711. Zankl H, Zang KD (1974) Quantitative studies on the arrangement of human metaphase chromosomes II. The association frequency of human acrocentric marker chromosomes. Hum Genet 23: 259-265
1712. Zellweger H, Abbo G, Cuany R (1966) Satellite association and translocation mongolism. J Med Genet 3: 186-189

References to Chapter 6

1713. Adams MS, Neel JV (1967) Children of incest. Pediatrics 40: 55-62
1714. Allison AC (1954) Protection afforded by sickle-cell trait against subtertian malarial infection. Br Med J I: 290
1715. Allison AC (1954) The distribution of the sickle-cell trait in East Africa and elsewhere, and its apparent relationship to the incidence of subtertian malaria. Trans R Soc Trop Med Hyg 48: 312
1716. Allison AC (1954) Notes on sickle-cell polymorphism. Ann Hum Genet 19: 39
1717. Allison AC (1955) Aspects of polymorphism in man. Cold Spring Harbor Symp Quant Biol 20: 239
1718. Allison AC (1956) The sickle and hemoglobin C-genes in some African populations. Ann Hum Genet 21: 678
1719. Allison AC (1964) Polymorphism and natural selection in human populations. Cold Spring Harbor Symp Quant Biol 24: 137-149
1720. Anonymous (1983) WHO Working Group (including A Motulsky and 13 others): Community control of hereditary anaemias. Bull WHO 61: 63-80 (1983). Also published in French, Bull WHO 61: 277-297
1721. Antonarakis SE, Boehm CD, Serjeant GR, Theisen CE, Dover GJ, Kazazian HH (1984) Origin of the β^s-globin gene in Blacks: The contribution of recurrent mutation or gene conversion or both. Proc Natl Acad Sci USA 81: 853-856
1722. Baird PA, McGillivray B (1982) Children of incest. J Pediatr 101: 854-857
1723. Bashi J (1977) Effects of inbreeding on cognitive performance. Nature 266: 440-442
1724. Beet EA (1946) Sickle-cell disease in the Balovale district of North Rhodesia. East Afr Med J 23: 75
1725. Beet EA (1947) Sickle-cell disease in Northern Rhodesia. East Afr Med J 24: 212-222
1726. Bergeron P, Laberge C, Grenier A (1974) Hereditary tyrosinemia in the province of Quebec: Prevalence at birth and geographic distribution. Clinical Genetics 5: 157-162
1727. Bernhard W (1966) Über die Beziehung zwischen AB0-Blutgruppen und Pockensterblichkeit in Indien und Pakistan. Homo 17: 111
1728. Bloch J (1901) Der Ursprung der Syphilis. Fischer, Jena
1729. Blumberg BS, Hesser JE (1971) Loci differently affected by selection in two American black populations. Proc Natl Acad Sci USA 68: 2554
1730. Bois E, Feingold J, Demenais F, Runavot Y, Jehanne M, Toidic L (1978) Cluster of cystic fibrosis cases in a limited area of Brittany (France). Clin Genet 14: 73-76
1731. Bosnjakovic S (1938) Vererbungsverhältnise bei der sogenannten Krankheit von Mljet. Acta Derm Venereol (Stockh) 19: 88
1732. Brues AM (1954) Selection and polymorphism in the AB0 blood groups. Am J Phys Anthropol 12: 559-598
1733. Cann RL, Brown WM, Wilson AC (1984) Polymorphic sites and the mechanism of evolution in human mitochondrial DNA. Genetics 106: 479-499
1734. Carter CO (1967) Risk of offspring of incest. Lancet I: 436
1735. Cavalli-Sforza LL, Edwards AWF (1967) Phylogenetic analysis: Models and estimation procedures. Evolution 21: 550-570
1736. Cerimele D, Cottoni F, Scappaticci S, Rabbiosi G, Sanna E, Zei G, Fraccaro M (1982) High prevalence of Werner's syndrome in Sardinia. Description of six patients and estimate of the gene frequency. Hum Genet 62: 25-30
1737. Chakravartti MR, Vogel F (1971) Haemaglutination-inhibiting variola antibodies in blood serum of former smallpox-patients, their healthy siblings and unvaccinated controls in other areas. Humangenetik 11: 336-338
1738. Charnov E (1977) An elementary treatment of kin selection. J Theor Biol 66: 541-550
1739. Clarke B (1975) Frequency-dependent and density-dependent natural selection. In: The role of natural selection in human evolution. Salzano F (ed), North Holland, American Elsevier, Amsterdam, New York pp 187-200.
1740. Clarke LA (1974) Rh haemolytic disease. Original papers commentaries. Medical and Technical Publ Comp, Newcastle
1741. Comings DE (1982) Two-dimensional gel electrophoresis of human brain proteins. III. Genetic and non-genetic variations in 145 brains. Clin Chem 28: 798-804
1742. Cooper DN, Smith BA, Cooke HJ, Niemann S, Schmidtke J (1985) An estimate of unique DNA sequence heterozygosity in the human genome. Hum Genet 69: 201-205
1743. Costeff H, Cohen BE, Weller L, Rahman D (1977) Consanguinity analysis in Israeli mental retardates. Am J Hum Genet 29: 339-349
1744. Crabb AR (1947) The hybrid-corn makers. New Brunswick
1745. Crow JF (1958) Some possibilities for measuring selection intensities in man. Hum Biol 30: 1-13
1746. Crow JF (1963) 2. The concept of genetic load: A reply. Am J Hum Genet 15: 310-315
1747. Crow JF (1970) Genetic loads and the cost of natural selection. In: Mathematical topics in population ge-

netics. Kojima K (ed), Springer, Berlin Heidelberg New York, pp 128-177.

1748. Crow JF, Denniston C (1981) The mutation component of genetic damage. Science 212: 888-893

1749. Cruz-Coke R (1982) Nomogram for estimating specific consanguinity risk. J Med Genet 19: 216-217

1750. Dahlberg G (1929) Inbreeding in man. Genetics 14: 421-454

1751. Damian RT (1964) Molecular mimicry: antigen sharing by parasite and host and its consequences. Am Naturalist 98: 129-150

1752. Das BM, Chakravartti MR, Delbrück H, Flatz G (1971) High prevalence of Haemoglobin E in two populations in Assam. Hum Genet 12: 264-266

1753. Das BM, Deka R (1975) Predominance of the haemoglobin E gene in a Mongoloid population in Assam (India). Hum Genet 30: 187-191

1754. Dobzhansky T (1952) Nature and origin of heterosis. In: Heterosis. Gowen JW (ed), Iowa State College Press, p 218.

1755. Downie HW, Meiklejohn G, Vincent LSt, Rao AR, Sundara Babu BV, Kempe CH (1966) Smallpox frequency and severity in relation to A, B and 0 blood groups. Bull WHO 33: 623

1755a. East EM, Jones DF (1919) Inbreeding and outbreeding. Lippincot, London, Philadelphia

1756. Eichner ER, Finn R, Krevans JR (1963) Relationship between serum antibody levels and the AB0 blood group polymorphism. Nature 198: 164

1757. Ewens WJ (1980) Mathematical population genetics. Springer, Berlin Heidelberg New York

1758. Firschein IL (1961) Population dynamics of the sickle cell trait in the Black Caribs of British Honduras. Am J Hum Genet 13: 233

1759. Fisher RA (1930) The distribution of gene ratios for rare mutations. Proc R Soc Edinb 50: 205-220

1760. Flatz G (1967) Haemoglobin E: Distribution and population dynamics. Hum Genet 3: 189-234

1761. Flatz G (1976) Populationsgenetik der Hämoglobinanomalien. Humangenetik, ein kurzes Handbuch. Becker PE (ed), Vol III/3, Thieme, Stuttgart pp 557-579.

1762. Flatz G, Pik C, Sundharayati B (1964) Malaria and haemoglobin E in Thailand. Lancet II: 385

1762a. Flatz G, Oelbe M, Herrmann H (1983) Ethnic distribution of phenylketonuria in the North German population. Hum Genet 65: 396-399

1763. Fraser GR, Mayo O (1974) Genetic load in man. (Review). Hum Genet 23: 83-110

1764. Freire-Maia N, Azevedô JBC (1971) The inbreeding load in Brazilian White and Negro populations as estimated with sib and cousin controls. Am J Hum Genet 23: 1-7

1765. Freire-Maia N et al (1983) Inbreeding studies in Brasilian schoolchildren. Am J Med Genet 16: 331-355

1766. Friedman MJ, Trager W (1981) The biochemistry of resistance to malaria. Sci Am 244: 154-164

1767. v Fumetti C (1976) Inzuchtkoeffizienten und Häufigkeiten konsanguiner Ehen. Biolog Diplomarbeit, Heidelberg

1768. Georges A, Jacquard A (1968) Effects de la consanguinité sur la mortalité infantile. Resultats d'une observation dans le département des Vosges. Population 23: 1055-1064

1769. Giblett ER (1977) Genetic polymorphisms in human blood. Ann Rev Genet 11: 13-28

1770. Glass RI, Holmgren J, Haley CE, Khan MR, Svennerholm A-M, Stoll BJ, Hossain KMB, Black RE, Yunus M, Barua D (1985) Predisposition for cholera of individuals with 0 blood group. Am J Epidemiol 121: 791-796

1771. Goodman RM, Motulsky AG (eds). (1979) Genetic diseases among Ashkenazi Jews. Raven Press, New York

1772. Greiner J, Schleiermacher E, Smith T, Lenhard V, Vogel F (1978) The HLA system and leprosy in Thailand. Hum Genet 42: 201-213

1773. Greenberg LJ, Gray ED, Yunis EJ (1975) Association of HL-A 5 and immune responsiveness in vitro to streptococcal antigens. J Exp Med 141: 935-943

1774. Grove DI, Forbes IJ (1975) Increased resistance to helminth infestation in an atopic population. Med J Australia 1: 336-338

1775. Haldane JBS (1937) The effect of variation on fitness. Am Naturalist 71: 337-349

1776. Haldane JBS (1939) The spread of harmful autosomal recessive genes in human populations. Ann Eugen 9: 232-237

1777. Haldane JBS (1942) Selection against heterozygotes in man. Ann Eugen 11: 333

1778. Haldane JBS (1949) The rate of mutations of human genes. Proceedings of the 7th International Congress on Genetics. Hereditas [Suppl] 35: 267

1779. Haldane JBS (1955) On the biochemistry of heterosis, and the stabilization of polymorphism. Proc R Soc (London) [Biol] 144: 217-220

1780. Haldane JBS (1957) The cost of natural selection. J Genet 55: 511-524

1781. Haldane JBS, Moshinsky P (1939) Inbreeding in Mendelian populations with special reference to human cousin marriage. Ann Eugen 9: 321-340

1782. Hamaguchi H, Yamada M, Shibasaki M, Mukai R, Yabe T, Kondo I (1982) Genetic analysis of human lymphocyte proteins by two-dimensional gel electrophoresis: 3. Frequent occurrence of genetic variants in some abundant polypeptides of PHA-stimulated peripheral blood lymphocytes. Hum Genet 62: 142-147

1783. Hamaguchi H, Ohta A, Mukai R, Yabe T, Yamada M (1981) Genetic analysis of human lymphocyte proteins by two-dimensional gel electrophoresis. I. Detection of genetic variant proteins in PHA-stimulated peripheral blood lymphocytes. Hum Genet 59: 215-220

1784. Hamilton WD (1964) The genetical evolution of social behavior. I. J Theoret Biol 7: 1-16

1785. Hanhart E (1955) Zur mendelistischen Auswertung einer 33 Jahre langen Erforschung von Isolaten. Navant'Anni dell Leggi Mendeliane, Ed Orrizante Medico, Roma pp 397-415.

1786. Harris R, Harrison GA, Rondle CJM (1963) Vaccinia virus and human blood group A substance. Acta Genet (Basel) 13: 44

1787. Harris H, Hopkinson DA (1972) Average heterozygosity per locus in man: an estimate based on the incidence of enzyme polymorphisms. Ann Hum Genet 36: 9-20

1788. Harris H, Hopkinson A, Robson EB (1974) The incidence of rare alleles determining electrophoretic variants: Data on 43 enzyme loci in man. Ann Hum Genet 37: 237-253

1789. Helmbold W (1959) Über den Zusammenhang zwischen AB0-Blutgruppen und Krankheit. Betrachtungen zur Ursache der AB0-Frequenzverschiebung bei Patienten mit Carcinoma ventriculi, carcinoma genitalis und ulcus pepticum. Blut 5: 7-22
1790. Hiernaux J (1952) La génétique de la sicklémie et l'interèt anthropologique de sa fréquence en Afrique noir. Annal Mus Congo Belge, Science de l'homme, Anthropologie 2, Tervuren, 42 pp
1790a. Hill AVS, Wainscoat JS The evolution of the α- and β-Globin gene clusters in human populations. Hum Genet (In the press 1986)
1791. Hirsch A (1881) Handbuch der historisch-geographischen Pathologie, Part 1: Infektionskrankheiten, 2nd ed, Enke, Stuttgart
1792. Horai S, Gojobori T, Matsunaga E (1984) Mitochondrial DNA polymorphism in Japanese. Hum Genet 68: 324-332
1793. Ifediba TC, Stern A, Ibrahim A, Rieder RF (1985) Plasmodium falciparum in vitro: diminished growth in hemoglobin H disease erythrocytes. Blood 65: 452-455
1794. Ishikumi N, Nemoto H, Neel JV, Drew AL, Yanase T, Matsumoto YS (1960) Hosojima. Am J Hum Genet 12: 67-75
1795. Jeffreys AJ, Wilson V, Thein SL (1985) Hypervariable 'minisatellite' regions in human DNA. Nature 314: 67-73
1796. Jeffreys AJ, Wilson V, Thein SL (1985) Individual-specific 'fingerprints' of human DNA. Nature 316: 76-79
1797. Jonxis JHP (1959) The frequency of haemoglobin S and C carriers in Curacao and Surinam. Symposium on Abnormal Hemoglobins. Blackwell, Oxford
1798. Kellermann G (1972) Further studies on the AB0 typing of ancient bones. Hum Genet 14: 232-236
1799. Kidson C, Lamont G, Saul A, Nurse GT (1981) Ovalocytic erythrocytes from Melanesians are resistant to invasion by malaria parasites in culture. Proc Natl Acad Sci USA 78: 5829-5832
1800. Kircher W (1961) Untersuchungen über den Zusammenhang von Dyspepsieverlauf und AB0-Blutgruppenzugehörigkeit. Monatsschr Kinderheilkd 109: 369
1801. Kircher W (1964) Weitere Untersuchungen über den Zusammenhang zwischen Verlauf und Häufigkeit der Säuglingsenteritis and AB0-Blutgruppenzugehörigkeit. Monatsschr Kinderheilkd 112: 415
1802. Klinger KW (1983) Cystic fibrosis in the Ohio Amish: Gene frequency and founder effect. Hum Genet 65: 94-98
1803. Klose J, Willers I, Singh S, Goedde HW (1983) Two-dimensional electrophoresis of soluble and structure-bound proteins from cultured human fibroblasts and hair root cells: Qualitative and quantitative variation. Hum Genet 63: 262-267
1804. Krieger H, Vicente AT (1969) Smallpox and the AB0 system in Southern Brazil. Hum Hered 19: 654
1805. Lambotte-Legrand J, Lambotte-Legrand C (1955) Anémie drépanocytaire et homozygotisme (à propos de 300 cas). Ann Soc Belge Méd Trop 35: 47
1806. Levy HL (1973) Genetic screening. Adv Hum Genet 4: 1-104
1807. Lewontin RC (1967) An estimate of average heterozygosity in man. Am J Hum Genet 19: 681-685
1808. Lewontin RC (1977) Population genetics. Proceedings of the 5th International Congress on Human Genetics 1976. Armendares S, Lisker R (eds), Excerpta Medica, Amsterdam, Oxford pp 13-18.
1809. Li CC (1963) 3. The way the load works. Am J Hum Genet 15: 316-321
1810. Li CC (1976) First course in population genetics. Boxwood, Pacific Grove
1811. Livingstone FB (1957) Sickling and malaria. Brit Med J I: 762
1812. Livingstone FB (1958) Anthropological implications of sickle cell gene distributions in West Africa. Am Anthropologist 60: 533-562
1813. Livingstone FB (1962) The origin of the sickle cell gene. Conference on African Historical Anthropology. Northwestern University, Chicago
1814. Livingstone FB (1971) Malaria and human polymorphisms. Ann Rev Genet 5: 33-64
1815. Livingstone FB (1973) Data on the abnormal hemoglobins and glucose-6-phosphate dehydrogenase deficiency in human populations, 1967-1973. University of Michigan, Ann Arbor
1815a. Livingstone FB (1985) Frequencies of Hemoglobin Variants. Thalassemia, The Glucose-6-Phosphate Dehydrogenase Deficiency, G6PD Variants, and Ovalocytosis in Human Populations. Oxford, New York
1816. Livingstone FB (1983) The malaria hypothesis. In: Distribution and Evolution of Hemoglobin and Globin Loci, Bowman JE (ed), Elsevier, New York pp 15-44.
1817. Livingstone FB (1984) The Duffy blood groups, vivax malaria, and malaria selection in human populations: a review. Hum Biol 56: 413-425
1818. London WT, Sutnick AI, Millman I, Coyne V, Blumberg BS, Vierucci A (1972) Australia antigen and hepatitis: Recent observations on the serum protein polymorphism, infectious agents hypotheses. Canad Med Ass J 106, Special Issue, 480-485
1819. Ludwig W (1944) Über Inzucht und Verwandtschaft. Z Menschl Vererbungs Konstitutionslehre 28: 278-312
1820. Lundborg H (1913) Medizinisch-biologische Familienforschungen innerhalb eines 2232-köpfigen Bauerngeschlechtes in Schweden (Provinz Blekinge). 2 vols Fischer, Jena
1821. Luzzatto L (1979) Genetics of red cells and susceptibility to malaria. Blood 54: 961
1822. Luzzatto L, Sodeinde O, Martini G (1983) Genetic variation in the host and adaptive phenomena in Plasmodium falciparum infection. In: Malaria and the Red Cell, Evered D, Whelan J (eds). Pittman, London pp 159-173
1823. Luzzatto L, Usanga EA, Reddy S (1969) Glucose-6-phosphate dehydrogenase deficient red cells resistance to infection by malarial parasites. Science 164: 839
1824. Majewski F (1980) Untersuchungen zur Alkoholembryopathie. Thieme Verlag, Stuttgart
1825. McKusick VA, Egeland JA, Eldridge R, Krusen DR (1964) Dwarfism in the Amish. I. The Ellis-van Creveld syndrome. Bull Johns Hopkins Hosp 115: 306

1826. Morton NE (1955) Nonrandomness in consanguineous marriage. Ann Hum Genet 20: 116-124

1827. Morton NE, Crow JF, Muller HJ (1956) An estimate of the mutational damage in man from data on consanguineous marriages. Proc Natl Acad Sci USA 42: 855-863

1828. Morton NE, Matsuura J, Bart R, Lew R (1978) Genetic epidemiology of an institutionalized cohort of mental retardates. Clin Genet 13: 449-461

1829. Motulsky AG (1964) Hereditary red cell traits and malaria. Am J Trop Med Hyg 13 (1), Part 2, 147-158

1830. Motulsky AG (1975) Glucose-6-phosphate dehydrogenase and abnormal hemoglobin polymorphisms - evidence regarding malarial selection. In: The role of natural selection in human evolution. Salzano F (ed), North-Holland, Amsterdam pp 271-291.

1831. Motulsky AG (1960) Metabolic polymorphism and the role of infectious disease in human evolution. Hum Biol 32: 28-62

1832. Motulsky AG (1979) Possible selective effects of urbanization on Ashkenzai Jewish populations. In: Genetic Diseases Among Ashkenazi Jews. Goodman RM, Motulsky AG (eds). Raven Press 1979, New York pp 201-212

1833. Motulsky AG (1980) Ashkenazi Jewish gene pools: admixture, drift and selection. In: Population Structure and Genetic Disorders. Eriksson AW, Forsius H, Nevanlinna HR, Workman PL, Norio RK (eds). Academic Press London: pp 353-365

1834. Motulsky AG, Murray JC (1983) Conference summary: current concepts of hemoglobin genetics. In: Distribution and Evolution of Hemoglobin and Globin Loci, Bowman JE (ed), Elsevier, New York pp 345-355.

1835. Muller HJ (1950) Our load of mutation. Am J Hum Genet 2: 111-176

1836. Neel JV (1951) The population genetics of two inherited blood dyscrasias in man. Cold Spring Harbor Symp Quant Biol 15: 141

1837. Neel JV (1979) History and the Tay-Sachs allele. In: Goodman RM, Motulsky AG (eds): Genetic diseases among Ashkenazi Jews, Raven Press, New York pp 285-299.

1838. Neel JV, Rosenblum BB, Sing CF, Skolnick MM, Hanash SM, Sternberg S (1984) Adapting two-dimensional gel electrophoresis to the study of human germline mutation rates. In: Two dimensional gel electrophoresis of proteins, Academic Press, New York pp 259-306.

1839. Neel JV, Schull WJ (1962) The effect of inbreeding on mortality and morbidity in two Japanese cities. Proc Natl Acad Sci USA 48: 573

1840. Neel JV, Schull WJ, Kimura T, Yanijawa Y, Yamamoto M, Nakajima A (1970) The effect of parental consanguinity and inbreeding in Hirado, Japan. III. Vision and hearing. Hum Hered 20: 129-155

1841. Nei M, Li WH (1979) Mathematical model for studying genetic variation in terms of restriction endonucleases. Proc Natl Acad Sci USA 76: 5269-5273

1842. Norio R, Nevalinna HR, Perheentupa J (1973) Hereditary diseases in Finland: Rare flora in rare soil. Ann Clin Res 5: 109-141

1843. Otten CM (1967) On pestilence, diet, natural selection, and the distribution of microbial and human blood group antigens and antibodies. Curr Anthropol 8: 209

1844. Pasval G, Weatherall DJ, Wilson JM (1978) Cellular mechanism for the protective effect of haemoglobin S against falciparum malaria. Nature 274: 701-703

1845. Pasvol G (1982) The interaction of malaria parasites with red blood cells. Br Med Bull 38: 133-140

1846. Pasvol G, Wainscoat JS, Weatherall DJ (1982) Erythrocytes deficient in glycophorin resist invasion by the malarial parasite Plamodium falciparum. Nature 297: 64-44

1847. Pasvol G, Weatherall DJ, Wilson RJM (1978) Cellular mechanism for the protective effect of haemoglobin S against P. falciparum malaria. Nature 274: 701-703

1847a. Pasvol G, Weatherall DJ, Wilson RJM (1977) Effects of foetal haemoglobin on susceptibility of red cells to Plasmodium falciparum. Nature 270: 171-173

1848. Penrose LS, Smith SM, Sprott DA (1956) On the stability of allelic systems, with special reference to haemoglobins A, S, and C. Ann Hum Genet 21: 90-93

1849. Pettenkofer HJ, Bickerich R (1960) Über Antigen-Gemeinschaften zwischen den menschlichen Blutgruppen und gemeingefährlichen Krankheiten. Zentralbl Bakteriol I Abt 179

1850. Pettenkofer HJ, Stöss B, Helmbold W, Vogel F (1962) Alleged causes of the present-day world distribution of the human AB0 blood groups. Nature 193: 444

1851. Phills JA, Harrold J, Whiteman GV, Perelmutter L (1972) Pulmonary infiltrates, asthma and eosinophilia due to *ascaris suum* infestation in man. New Engl J Med 286: 965-970

1852. Piazza A (1973) et al In: Histocompatibility Testing 1972. (Report of an International Workshop and Conference held at Evian, 23-27 may 1972). Dausset J, Colombani J (eds), Munksgaard, Copenhagen pp 73-84.

1853. Povey S, Hopkinson DA (1981) The use of polymorphic enzyme markers of human blood cells in genetics. Clin Haematol 10: 161-184

1854. Rao PSSS, Inboraj SG (1979) Trends in human reproductive wastage in relation to long-term practice of inbreeding. Ann Hum Genet 42: 401-413

1855. Rao PSS, Inboraj SG (1980) Inbreeding effects on fetal growth and development. J of Med Genet 17: 27-33

1856. Reed SC (1954) A test for heterozygous deleterious recessives. J Hered 45: 17-18

1857. Reed TE (1969) Caucasian genes in American negroes. Science 165: 762-768

1858. Robinson MG, Tolchin D, Halpern C (1971) Enteric bacterial agents and the AB0 blood groups. Am J Hum Genet 23: 135

1859. Roth EF Jr, Raventos-Suarez C, Rinaldi A, Nagel RL (1983) Glucose-6-phosphate dehydrogenase deficiency inhibits in vitro growth of Plasmodium falciparum. Proc Natl Acad Sci USA 80: 298

1860. Roth EF Jr, Raventos-Suarez C, Rinaldi A, Nagel RL (1983) The effect of X chromosome inactivation on the inhibition of Plasmodium falciparum malaria growth by glucose-6-phosphate dehydrogenase-deficient red cells. Blood 62: 866

1861. Roth E Jr, Raventos-Suarez C, Gilbert H, Stump D, Tanowitz H, Rowin KS, Nagel RL (1984) Oxidative stress and falciparum malaria: a critical review of the evidence. In: Malaria and the Red Cell, Alan R. Liss, New York pp 35-43.
1862. Salzano FM (ed) (1975) The role of natural selection in human evolution. North Holland, Amsterdam, Oxford
1863. Sanghvi LD (1963) The concept of genetic load: A critique. Am J Hum Genet 15: 298-309
1864. Sanghvi LD, Balakrishnan V (1972) Comparison of different measures of genetic distance between human populations. In: The assessment of population affinities in man. Weiner JS, Huizinga J (eds), Clarendon Oxford pp 25-36.
1865. Sangvichien S (1966) A preliminary report on non-, metrical characteristics of neolithic skeletons found at Bankhoro, Kanchanaburi. J Siam Soc (Bangkok) 54: 1
1866. Sasazuki T, Kohno Y, Iwamoto I, Tanimura M, Naito S (1978) Association between an HLA haplotype and low responsiveness to tetanus toxoid in man. Nature 272: 359-361
1867. Seemanová E (1971) A study of children of incestuous matings. Hum Hered 21: 108-128
1868. Shull GH (1908) Composition of a field of maize. Rep Am Breeders Assoc 4: 296-301
1869. Shull GH (1911) Experiments with maize. Bot Gaz 52: 480
1870. Sjögren T (1931) Die juvenile amaurotische Idiotie. Hereditas 14: 197-425
1871. Slatis HM (1954) A method of estimating the frequency of abnormal autosomal recessive genes in man. Am J Hum Genet 6: 412-418
1872. Smith SM (1954) Notes on sickle-cell polymorphism. Ann Hum Genet 19: 51
1873. Socha W, Bilinska M, Kaczera Z, Pajdak E, Stankiewicz D (1969) Escherichia Coli and AB0 blood groups. Folia Biol (Kraków) 17, (4)
1874. Sukamaran PK, Master HR, Undesia JV, Balakrishnan B, Sanghvi LD (1966) AB0 blood groups in active cases of smallpox. Indian J Med Sci 20: 119
1875. Süssmilch (1786) Die göttliche Ordnung, 9th ed Teil II. Berlin
1876. Schull WJ (1958) Empirical risks in consanguineous marriages: Sex ratio, malformation, and viability. Am J Hum Genet 10: 294-343
1877. Schull WJ, Furusho T, Yamamoto M et al (1970) The effects of parental consanguinity and inbreeding in Hirado, Japan. IV. Fertility and reproductive compensation. Hum Genet 9: 294-315
1878. Schull WJ, Nagano H, Yamamoto M, Komatsu I (1970) The effect of parental consanguinity and inbreeding in Hirado, Japan. I. Stillbirth and prereproductive mortality. Am J Hum Genet 22: 239-262
1879. Schull WJ, Neel JV (1972) The effect of parental consanguinity and inbreeding in Hirado, Japan. V. Summary and interpretation. Am J Hum Genet 24: 425-453
1880. Tanis RJ, Neel JV, Dovey H, Morrow M (1973) The genetic structure of a tribal population, the Yamomama Indians. IX. Gene frequencies for 17 serum protein and erythrocyte enzyme systems in the Yamomama and five neighboring tribes; nine new variants. Am J Hum Genet 25: 655-676
1881. Thalhammer O (coordinator) (1975) Frequency of inborn errors of metabolism, especially PKU, in some representative newborn screening centers around the world. A collaborative study. Hum Genet 30: 273-286
1882. Vandepitte JM, Zuelzer WW, Neel JV, Colaert J (1955) Evidence concerning the inadequacy of mutation as an explanation of the frequency of the sickle-cell gene in the Belgian Congo. Blood 10: 341
1883. de Vries RRP, Fat RFMLA, Nijenhuis LE, van Rood JJ (1976) HLA-linked genetic control of host response to mycobacterium leprae. Lancet II: 1328-1330
1884. de Vries RRP, van Rood JJ (1977) Abstract Tissue Antigens 10: 212
1885. Vogel F (1970) Anthropological implications of the relationship between AB0 blood groups and infections. Proceedings of the 8th International Congress of Anthropologic and Ethnologic Sciences. Tokyo, 1968. Vol I, p 365,
1886. Vogel F (1979) 'Our load of mutation': reappraisal of an old problem. Proc Roy Soc London B 205: 77-90
1887. Vogel F (1979) Genetics of retinoblastoma. Hum Genet 52: 1-54
1888. Vogel F, Chakravartti MR (1966) AB0 blood groups and smallpox in a rural population of West Bengal and Bihar (India). Hum Genet 3: 166-180
1889. Vogel F, Dehnert J, Helmbold W (1964) Über Beziehungen zwischen AB0-Blutgruppen und der Säuglingsdyspepsie. Hum Genet 1: 31-57
1890. Vogel F, Pettenkofer HJ, Helmbold W (1960) Über die Populationsgenetik der AB0-Blutgruppen. II. Genhäufigkeit und epidemische Erkrankungen. Acta Genet (Basel) 10: 267-294
1891. Wade Cohen PT, Omenn GS, Motulsky AG, Chen S-H, Giblett ER (1973) Restricted variation in the glycolytic enzymes of human brain and erythrocytes. Nature 241: 229
1892. Walton KE, Steyer D, Gruenstein EI (1979) Genetic polymorphism in normal human fibroblasts as analyzed by two-dimensional polyacrylamide gel electrophoresis. J Biol Chem 254: 7951-7960
1893. Weiner JS, Huizinga J (ed) (1972) The assessment of population affinities in man. Clarendon, Oxford
1894. Workman PL, Blumberg BS, Cooper AJ (1963) Selection, gene migration and polymorphic stability in U.S. White and Negro population. Am J Hum Genet 15: 429
1895. Wright S (1922) Coefficients of inbreeding and relationship. Am Naturalist 56: 330-338
1896. Zerbin-Rüdin E (1960) Vorläufiger Bericht über den Gesundheitszustand von Kindern aus nahen Blutsverwandtenehen. Z Menschl Vererbungs-, Konstitutionslehre 35: 233-302
1897. Zuckerkandl E (1976) Evolutionary processes and evolutionary noise at the molecular level. J Mol Evol 7: 167-183

References to Chapter 7

1898. Anonymous (1976) Sociobiology Study Group of Science for the People; Sociobiology: Another biological determinism. Bioscience 26: 182-190
1899. Barash DP (1977) Sociobiology and behavior. Elsevier, New York

1900. Bayless TM, Rosenzweig NS (1966) A racial difference in incidence of lactase deficiency. A survey of milk intolerance and lactase deficiency in healthy adult males. J Am Med Assoc 197: 968-972
1901. Bekemeier H (1969) Evolution der Hautfarbe und kutane Vitamin D-Photosynthese. Dtsch Med Wochenschr 94: 185-189
1902. Betz A, Turleau L, de Grouchy J (1974) Hétérozygotie et homozygotie pour une inversion péricentrique du 3 humain. Ann Génét 17: 77
1903. Bolin TD, Davis AE (1969) Asian lactose intolerance and its relation to intake of lactose. Nature 222: 382-383
1904. Brown WM, Prager EM, Wang A, Wilson AC (1982) Mitochondrial DNA sequence of primates: Tempo and mode of evolution. J Mol Evol 18: 225-239
1905. Bush GL, Case SM, Wilson AC, Patton JL (1977) Rapid speciation and chromosomal evolution in mammals. Proc Natl Acad Sci USA 74: 3942-3946
1906. Chagnon NA (1968) Yanomamö, the fierce people. Holt, Rinehart & Winston, New York
1907. Clarke B (1970) Selective constraints on amino-acid substitutions during the evolution of proteins. Nature 228: 159-160
1908. Cochet B, Jung A, Griessen M, Bartholdi P, Schaller Ph, Donath A (1983) Effects of lactose on intestinal calcium absorption in normal and lactase-deficient subjects. Gastroenterology 84: 935-940
1909. Daiger EI, Schonfield MS, Cavalli-Sforza LL (1975) Group-specific component (Gc) proteins bind vitamin D and 25-hydroxyvitamin D. Proc Natl Acad Sci USA 72: 2076-2080
1910. Dalton DP, Edwards JH, Evans EP, Lyon MF, Parkinson SP, Peters J, Searle AG (1981) Chromosome maps of man and mouse. Clin Genet 20: 407-415
1911. Dayhoff ME (ed) (1978) Atlas of Protein Sequence and Structure. Vol 5, Suppl 3. National Biomedical Research Foundation, Washington DC
1912. Dutrillaux B (1975) Sur la nature et l'origine des chromosomes humains. L'expansion scientifique, Paris
1913. Dutrillaux B (1979) Chromosomal evolution in primates. Tentative phylogeny from Microcebus Murinus (Prosimian) to man. Hum Genet 48: 251-314
1914. Erdtmann B (1982) Aspects of evaluation, significance and evolution of human C-band heteromorphism. Hum Genet 61: 281-294
1915. Ferguson A, Maxwell J (1967) Genetic aetiology of lactose intolerance. Lancet II: 188-190
1916. Fischer E (1913) Die Rehobother Bastards und das Bastardierungsproblem beim Menschen. G. Fischer, Jena
1917. Fischer E (1961) Über das Fehlen von Rachitis bei Twiden (Bambuti) im Kongourwald. Z Morphol Anthrop 51: 119-136
1918. Fitch WM (1980) Estimating the total number of nucleotide substitutions since the common ancestor of a pair of homologous genes: Comparison of several methods and three beta hemoglobin messenger RNA's. J Mol Evol 16: 153-209
1919. Fitch WM, Farris JS (1974) Evolutionary trees with minimum nucleotide replacements from amino acid sequences. J Mol Evol 3: 263
1920. Fitch WM, Langley CH (1976) Evolutionary rates in proteins: Neutral mutations and the molecular clock. In: Molecular anthropology. Goodman M, Tashian RE (eds), Plenum, New York pp 197-219.
1921. Fitch WM, Margoliash E (1967) The construction of phylogenetic trees. A generally applicable method utilizing estimates of the mutation distance obtained from cytochrome C sequences. Science 155: 279
1922. Flatz G, Howell JN, Doench J, Flatz SD (1982) Distribution of physiological adult lactase phenotypes, lactose absorber and malabsorber, in Germany. Hum Genet 62: 152-157
1923. Flatz G, Rotthauwe HW (1973) Lactose nutrition and natural selection. Lancet II: 76-77
1924. Flatz G, Rotthauwe HW (1977) The human lactase polymorphism. Physiology and genetics of lactose absorption and malabsorption. Prog Med Genet N.F. 2: 205-250
1925. Flatz G, Saengudom Ch, Sanguanbhokai T (1969) Lactose intolerance in Thailand. Nature 221: 758-759
1925a. Gilbert W (1985) Genes-in-pieces revisited. Science 228, 823-824
1926. Goodman M (1976) Toward a genealogical description of the primates. In: Molecular anthropology. Goodman M, Tashian RE (eds), Plenum, New York pp 321-353.
1927. Goodman M, Moore GW, Matsuda G (1975) Darwinian evolution in the genealogy of haemoglobin. Nature 253: 603-608
1928. Gosden JR, Mitchell AR, Seuanez HN, Gosden CM (1977) The distribution of sequences complementary to human satellite DNAs I, II and IV in the chromosomes of chimpanzee (Pan troglodytes), Gorilla (Gorilla gorilla) and Orang Utan (Pongo pygmaeus). Chromosoma 63: 253-271
1929. de Grouchy J, Turleau C, Raudin M, Klein M (1972) Evolutions caryotypiques de l'homme et du chimpanzé. Etude comparative des topographies des bandes après dénaturation ménagée. Ann Génét 15: 79
1930. Haldane JBS (1975) Quoted by: Smith JM The theory of evolution, 3rd ed Penguin Books, London p 180.
1931. Hamilton WD (1964) The genetical theory of social behavior: I. and II. J Theor Biol 7: 1-52
1932. Happle R, Phillips RJS, Roessner A, Junemann G (1983) Homologous genes for X-linked achondroplasia punctata in man and mouse. Hum Genet 63: 24-27
1933. Harris H, Hopkinson DA, Edwards YH (1977) Polymorphism and the subunit structure of enzymes: a contribution to the neutralist-selectionist controversy. Proc Natl Acad Sci USA 74: 698-701
1934. Hassenstein B (1973) Verhaltensbiologie des Kindes. Piper, München, Zürich
1935. Hirschfeld J (1959) Immuno-electrophoretic demonstration of qualitative differences in human sera and their relation to the haptoglobins. Acta Pathol Microbiol Scand 47: 160-168
1936. Howell JN, Schockenhoff T, Flatz G (1981) Population screening for the human adult lactase phenotypes with a multiple breaths version of the breath hydrogen test. Hum Genet 57: 276-278
1937. Jacobs PA (1975) The load due to chromosome abnormalities in man. In: The role of natural selection in human evolution. Salzano F (ed), North-Holland, Amsterdam, Oxford pp 337-352.

1938. Jones KW (1976) Comparative aspects of DNA in higher primates. In: Molecular anthropology. Goodman M, Tashian RE (eds), Plenum, New York pp 357–368.

1939. Jones KW (1978) Speculations on the functions of satellite DNA in evolution. Z Morphol Anthropol 69: 143–171

1940. Jörgensen G, Ritter H, Vogel F (1975) Gc-Polymorphismus ("gruppenspezifische Komponente"). In: Humangenetik, ein kurzes Handbuch. Becker PE (ed), Vol I/3, Thieme, Stuttgart pp 105–134.

1941. Kimura M (1983) The neutral theory of molecular evolution. Cambridge University Press, Cambridge etc.

1942. Kimura M (1982) The neutral theory as a basis for understanding the mechanism of evolution and variation at the molecular level. In: Kimura M (ed) Molecular evolution, protein polymorphism and the neutral theory, Japan Sci Soc Press Tokyo/Springer, Berlin, pp 3–56.

1943. King JL, Jukes TH (1969) Non-Darwinian evolution. Science 164: 788–798

1944. Layrisse M (1958) Anthropological considerations of the Diego (Di^a) antigen. Am J Physical Anthropol 16: 173–186

1945. Layrisse M, Arends T (1957) The Diego system – steps in the investigation of a new blood group system. Further studies. Blood 12: 115–122

1946. Layrisse M, Arends T, Dominguez Sisco R (1955) Nuevo gropo sanguineo encontrado en descendientes de Indios. Acta Med Venez 3: 132–138

1947. Lejeune J (1968) Adam and Eve ou le monogénisme. Nouv Rév Théol 90: 191

1948. Lorenz K (1935) Der Kumpan in der Umwelt des Vogels. J Ornithol 83: 137–213, 289–413

1949. Lundin L-G (1979) Evolutionary conservation of large chromosomal segments reflected in mammalian gene maps. Clin Genet 16: 71–81

1950. Mayr E (1967) Artbegriff und Evolution. Parey, Hamburg und Berlin

1950a. Mayr E (1985) The Growth of Biological Thought Harvard Univ Press, Cambridge Mass

1951. McClure H, Belden KH, Pieper WA (1969) Autosomal trisomy in a chimpanzee. Resemblance to Down's syndrome. Science 165: 1010–1011

1952. Miller LH, Mason SJ, Clyde OF, McGinniss MH (1976) The resistance factor to *plasmodium vivax* in blacks. N Engl J Med 295: 302

1953. Miyata T, Yasunaga T (1981) Rapidly evolving mouse α-globin related pseudogene and its evolutionary history. Proc Natl Acad Sci USA 78: 450–453

1954. Moore GW (1976) Proof for the maximum parsimony ("Red King") algorithm. In: Molecular anthropology. Goodman M, Tashian RE (eds), Plenum, New York pp 117–137.

1955. Morton NE, Chung CS, Mi MP (1967) Genetics of interracial crosses in Hawaii. Monogr Hum Genet 3

1956. Motulsky AG (1968) Human Genetics, society, and medicine. J Hered 59: 329–336

1957. Mourant AE, Tills D, Domaniewska-Sobczak K (1976) Sunshine and the geographical distribution of the alleles of the Gc system of plasma proteins. Hum Genet 33: 307–314

1958. Murray JC, Demopulos CM, Lawn RM, Motulsky AG (1983) Molecular genetics of human serum albumin: restriction enzyme fragment length polymorphisms and analbuminemia. Proc Natl Acad Sci USA 80: 5951–5955

1959. Murray JC, Mills KA, Demopulos CM, Hornung S, Motulsky AG (1984) Linkage disequilibrium and evolutionary relationships of DNA variants (RFLPs) at the serum albumin locus. Proc Natl Acad Sci USA 81: 3486–3490

1960. Nadeau JH, Taylor BA (1984) Length in chromosomal segments conserved since divergence of man and mouse. Proc Natl Acad Sci USA 81: 814–818

1961. Neel JV (1962) Diabetes mellitus: A "thrifty" genotype rendered detrimental by "progress"? Am J Hum Genet 14: 353–362

1962. Neel JV (1970) Lessons from a "primitive" people. Science 170: 815–822

1963. Neel JV (1980) On being headman. Perspec Biol Med 23: 277–294

1964. Neel JV (1982) The wonder of our presence here: a commentary on the evolution and maintenance of human diversity. Perspec Biol Med 25: 518–558

1965. Neel JV, Centerwall WR, Chagnon NA, Casey HL (1970) Notes on the effect of measles and measles vaccine in a virgin-soil population of South American Indians. Am J Epidemiol 91: 418–429

1966. Neel JV, Salzano FM, Junqueira PC, Keiter F, Maybury-Lewis D (1964) Studies on the Xavante indians of the Brazilian Mato Grosso. Am J Hum Genet 16: 52–140

1967. Ohno S (1985) Frequent recurrence of certain base oligomers reflects the ultimate derivation of all structural genes from oligomeric repeats. First Intern Symp on the Role of Recombinant DNA in Genetics (Teplitz RL et al, eds) Crete, in the press 1986

1968. Omenn GS, Motulsky AG (1972) Biochemical genetics and the evolution of human behavior. In: Genetics, environment and behavior. Ehrmann L, Omenn GS, Caspari E (eds), Academic Press, New York, London pp 129–179.

1969. Post RH (1962) Population differences in vision acuity. Eugen Q 9: 189–212

1970. Post RH (1964) Hearing acuity variation among negroes and whites. Eugen Q 11: 65–81

1971. Post RH (1965) Notes on relaxed selection in man. Anthropol Anz 29: 186–195

1972. Post RH (1966) Deformed nasal septa and relaxed selection. Eugen Q 13: 101–112

1973. Post RH (1969) Deformed nasal septa and relaxed selection II. Soc Biol 16: 179–196

1974. Post RH (1971) Possible cases of relaxed selection in civilized populations. Hum Genet 13: 253–284

1975. Sahi T (1974) The inheritance of selective adult-type lactose malabsorption. Scand J Gastroenterolog 9 [Suppl 30], 1–73

1976. Simons EL (1976) The fossil record of primate phylogeny. In: Molecular anthropology. Goodman M, Tashian RE (eds), Plenum, New York pp 35–62.

1977. Simpson GG (1951) Zeitmaße und Ablaufformen der Evolution. Musterschmidt, Göttingen

1978. Smith M (1976) Commentary: Group Selection. Q Rev Biol 51: 277–283

1979. Schulz GE (1981) Protein-Differenzierung: Entwicklung neuartiger Proteine im Laufe der Evolution. Angewandte Chemie 93: 143–151

1980. Stengel H (1958) Gibt es eine "getrennte Vererbung von Zahn und Kiefer" bei der Kreuzung extrem großer Kaninchenrassen? Ein experimenteller Beitrag zum sogenannten "Disharmonieproblem". Z Tierzücht Züchtungsbiol 72: 255–286

1981. Tinbergen N (1968) On war and peace in animals and men. Science 1960: 1411–1418

1982. Vogel Ch (1975) Neue Aspekte zur Evolution des Menschen. Nova Acta Leopold. N.F. (Halle) 42: 253–269

1983. Vogel Ch (1975) Prädispositionen bzw. Präadaptionen der Primaten-Evolution im Hinblick auf die Hominisation. In: Hominisation und Verhalten, Kurth G, Eibl-Eibesfeld I (eds), G. Fischer, Stuttgart pp 1–31.

1984. Vogel Ch (1983) Personelle Identität und kognitiv-intellektuelle Leistungsfähigkeit im sozialen Feld nicht-menschlicher Primaten. Veröff Joachim Jungius-Ges Wiss Hamburg 50: 23–39

1985. Vogel F, Kopun M, Rathenberg R (1976) Mutation and molecular evolution. In: Molecular anthropology. Goodman M, Tashian RE (eds), Plenum, New York pp 13–33.

1986. Walter H (1970) Grundriß der Anthropologie. BLV, München

1987. Walker A (1976) Splitting times among hominids deduced from the fossil record. In: Molecular anthropology. Goodman M, Tashian RE (eds), Plenum, New York pp 63–77.

1988. Wilson AC, Carlson SS, White TJ (1977) Biochemical evolution. Ann Rev Biochem 46: 573–639

1989. Wilson EO (1975) Sociobiology: the new synthesis. Belknap Press of Harvard University, Cambridge (MA)

1990. Wilson EO (1978) On human nature. Harvard University Press, Boston

1991. Zuckerkandl E (1965) The evolution of hemoglobin. Sci Am 212 (5): 110–118

1992. Zuckerkandl E (1976) Evolutionary processes and evolutionary noise at the molecular level. II. A selectionist model for random fixations in proteins. J Mol Evol 7: 269–311

1993. Zuckerkandl E (1978) Molecular evolution as a pathway to man. Z Morphol Anthropol 69: 117–142

References to Chapter 8

1994. Åkesson HO (1984) Intelligence and polygenic inheritance. A dogma to be reexamined. Acta Pediatrica Scand 73: 13–17

1995. Alexander D, Money J (1965) Reading ability, object constancy, and Turner's syndrome. Percept Mot Skills 20: 981–984

1996. Alexander D, Walker AT, Money J (1964) Studies in direction sense: I. Turner's syndrome. Arch Gen Psychiatr 10: 337–339

1997. Allen W, Herndon CN, Dudley EC (1944) Some examples of the inheritance of mental deficiency: Apparently sex-linked idiocy and microcephaly. Am J Ment Def 48: 325–334

1998. Andersen P, Andersson SA (1968) Physiological basis of the alpha rhythm. Appleton-Century Crofts, New York

1999. Anderson VE, Hauser WA, Penry JK, Sing CF (eds) (1982) Genetic basis of the epilepsies. Raven Press, New York

2000. Angst J, Frey R, Lohmeyer B, Zerbin-Rüdin E (1980) Bipolar manic-depressive psychoses: Results of a genetic investigation. Hum Genet 55: 237–254

2001. Austin GE, Sparkes RS (1980) Abnormal cerebral cortical convolutions in an XYY fetus. Hum Genet 56: 173–175

2002. Barchas JD, Ciaranello RD, Dominic JA, Deguchi I, Orenberg EK, Renson J, Kessler S (1974) Genetic differences in mechanisms involving neuroregulators. J Psychiatr Res 11: 347–360

2003. Becker PE (1958) Die Neurosen im Lichte der Genetik. Dtsch Med Wochenschr 83: 612–616

2004. Benzer S (1973) Genetic dissection of behavior. Sci Am 222: 24–37

2004 a. Bordarier C, Robain O, Rethoré O, Dulac O, Dhellemes C, Inverted neurons in Agyria – A Golgi study of a case with abnormal chromosome 17

2005. Borgaonkar DS, Shah SA (1974) The XYY chromosome male – or syndrome? Prog Med Genet 10: 135–222

2006. Bouchard TJ, McGue M (1981) Familial studies of intelligence: A review. Science 212: 1055–1059

2007. Brewster DJ (1972) Ethanol preference in strains of rats selectively bred for behavioral characteristics. Nature and nurture in alcoholism. Ann N Y Acad Sci 197: 49–53

2008. Brown WT, Jenkins EC, Friedman E, Brooks J, Cohen IL, Duncan C, Hill AL, Malik MN, Morris V, Wolf E, Wisniewski K, French JH (1984) Folic acid therapy in the fragile X-syndrome. Am J Med Genet 17: 277–288

2009. Brun A, Gustavson K-H (1982) Letter to the editor. Hum Genet 60: 298

2010. Buchsbaum MS (1974) Average evoked response and stimulus intensity in identical and fraternal twins. Physiol Psychol 2 (3A): 365–370

2011. Buchsbaum MS (1975) Average evoked response augmenting/reducing in schizophrenia and affective disorders. In: Freedman DX (ed) The biology of the major psychoses: A comparative analysis, Raven Press, New York pp 129–142.

2012. Buchsbaum MS, Coursey RD, Murphy DL (1976) The biochemical high-risk paradigm: Behavioral and familial correlates of low platelet monoamine oxidase activity. Science 194: 339–341

2013. Buselmaier W, Vierling Th, Balzereit W, Schwegler H (1981) Genetic analysis of avoidance learning by means of different psychological test systems with inbred mice as model organisms. Psychol Res 43: 317–333

2014. Cattell RB (1955) The inheritance of personality. Am J Hum Genet 7: 122

2015. Cattell RB (1964) Personality and social psychology. Knapp, San Diego

2016. Cattell RB (1965) Methodological and conceptional advances in evaluating heredity and environmental influences and their interaction. In: Methods and goals in human behavior genetics. Vandenberg SG (ed), Academic Press, New York p 95

2017. Charlesworth WR (1976) Human intelligence as adaptation: An ethological approach. In: The nature of intelligence. Resnick LB (ed), Erlbaum, Hillsdale pp 147–168.

2018. Chen H, Faigenbaum D, Weiss H (1981) Psychological aspects of patients with the Ulrich-Turner syndrome. Amer J Med Genet 8: 191-203
2019. Childs B (1972) Genetic analysis of human behavior. Ann Rev Med 23: 373-406
2020. Childs B, Finucci JM, Preston MS, Pulver AE (1976) Human behavior genetics. Adv Hum Genet 7: 57-97
2021. Ciaranello RD, Hoffman HF, Shire JGCh, Axelrod J (1974) Genetic regulation of the catecholamine biosynthetic enzymes. J Biol Chem 249: 4528-4536
2022. Ciaranello RD, Lipsky A, Axelrod J (1974) Association between fighting behavior and catecholamine biosynthetic enzyme activity in two inbred mouse sublines. Proc Natl Acad Sci USA 71: 3006-3008
2023. Cooper RM, Zubek JP (1958) Effects of enriched and restricted early environments on the learning ability of light and dull rats. Can J Psychol 12: 159-164
2024. Cotton NS (1979) The familial incidence of alcoholism. J Stud Alcohol 40: 89
2025. Crowe RR (1974) An adoption study of antisocial personality. Arch Gen Psychiatry 31: 785-791
2026. Crowe RR (1975) Adoption studies in psychiatry. Biol Psychiatry 10: 353-371
2027. Dingman HF (1968) Psychological test patterns in Down's syndrome. In: Progress in human behavior genetics. Vandenberg SG (ed), Johns Hopkins Press, Baltimore pp 19-26.
2028. Dörner G (1976) Hormones and brain differentiation. Elsevier, Amsterdam
2029. Dunnette J, Weinshilboum R (1982) Family studies of plasma dopamine-β-hydroxylase thermal stability. Amer J Hum Genet 34: 84-99
2030. Ehrhardt AA, Meyer-Mahlburg HFL (1979) Prenatal sex hormones and the developing brain: Effects on psychosexual differentiation and cognitive function. Ann Rev Med 30: 417-430
2031. Ehrman S, Omenn GS, Caspari E (1972) Genetics, environment, and behavior. Academic Press, New York, London
2032. Erlenmeyer-Kimling L, Bernblatt B, Fluss J (1979) High-risk research in schizophrenia. Psychiatr Annals 9: 38-51
2033. Eysenck HJ (1947) Dimensions of personality. Routledge & Kegan Paul, London
2034. Eysenck HJ (1952) The scientific study of personality. Routledge & Kegan Paul, London
2035. Eysenck HJ (1960) Levels of personality, constitutional factors, and social influences: An experimental approach. Int J Soc Psychiatry 6: 12
2036. Eysenck HJ (1980) Intelligenz-Struktur und Messung. Springer, Berlin Heidelberg New York
2037. Eysenck HJ (ed) (1982) A model for intelligence. Springer, Berlin Heidelberg New York
2038. Farber SL (1981) Identical twins reared apart. Basic Books Inc., New York
2039. Feldman MW, Lewontin RC (1975) The heritability hangup. Science 190: 1163-1168
2040. Fischer M (1971) Psychoses in the offspring of schizophrenic monozygotic twins and their normal cotwins. Brit J Psychiatr 118: 43-52
2041. Fischer M (1972) Umweltfaktoren bei der Schizophrenie. Intrapaarvergleiche bei eineiigen Zwillingen. Nervenarzt 43: 230-238
2042. Fischer M, Harvald B, Hauge M (1969) A Danish twin study of schizophrenia. Brit J Psychiat 115: 981-990
2043. Friedl W (1977) Untersuchungen des Ruhe-EEG normaler weiblicher und männlicher junger Erwachsener mit Hilfe der elektronischen EEG-Analyse. M. D. Dissertation, University of Heidelberg
2044. Frisch K von (1950) Bees, their vision, chemical senses, and language. Cornell University Press, Ithaca
2045. Frischeisen-Köhler I (1930) Untersuchungen an Schulzeugnissen von Zwillingen. Z Angew Psychol 37: 385
2046. Fuller JL, Collins RL (1972) Ethanol consumption and preference in mice: A genetic analysis. In: Nature and nurture in alcoholism. Ann NY Acad Sci 197: 42-48
2047. Fuller JL, Thompson WRT (1960) Behavior genetics. John Wiley & Sons, New York
2048. Garmezy N (1974) Children at risk: The search for the antecedents of Schizophrenia. Part I. Conceptual models and research methods. NIMH Schizophrenia Bull 8: 14-90
2049. Garmezy N (1974) Children at risk: The search for the antecedents of schizophrenia. Part II: Ongoing research programs, issues and intervention. NIMH Schizophrenia Bull 9: 55-125
2050. Gershon ES (1982) Genetic studies of affective disorders and schizophrenia. Human Genetics, Part A: The Unfolding Genome, A. R. Liss Inc., New York pp 417-432.
2051. Gershon ES, Bunny WE, Leckman JF, van Eerdewegh M, de Bauche BA (1976) The inheritance of affective disorders: A Review of data and of hypotheses. Behav Genet 6: 227-261
2052. Gershon ES, Targum SD, Kessler LR, Mazure CM, Bunney WE (1978) Genetic studies and biologic strategies in the affective disorders. Prog Med Genet N S Vol II, 101-166
2053. Gershon ES, Targum SD, Matthyse S, Bunney WE Jr (1979) Color blindness not closely linked to bipolar illness. Report of a new pedigree series. Arch Gen Psychiatry 36: 1423-1430
2054. Goldberger A (1976) Jensen on Burks. Educ Psychol 12: 64-78
2055. Goldberger AS (1977) Models and methods in the I. Q. debate. Part I. Social Systems. Res. Inst. (SSRI) Workshop Series 7710, University of Wisconsin
2056. Goodnow JJ (1976) The nature of intelligent behavior: Question raised by cross-cultural studies. In: The nature of intelligence. Resnick LB (ed), Erlbaum, Hillsdale pp 169-188
2057. Goodwin D (1976) Is alcoholism hereditary? Oxford University Press, New York
2058. Goodwin DW, Schulsinger F, Hermansen L, Gruze SB, Winokur G (1973) Alcohol problems in adoptees raised apart from alcoholic biological parents. Arch Gen Psychiatry 28: 238-243
2059. Goodwin DW, Schulsinger F, Knop J, Mednick S, Gruze SB (1977) Psychopathology in adopted and nonadopted daughters of alcoholics. Arch gen Psychiatr 34: 1005
2060. Gottesman II, Shields J (1972) Schizophrenia and genetics. Academic Press, New York, London
2061. Gottesmann II, Shields J (1982) Schizophrenia, the epigenetic puzzle. Cambridge Univ. Press, Cambridge etc.
2062. Gottschaldt K (1939) Erbpsychologie der Elementarfunktionen der Begabung. In: Handbuch der Erb-

biologie des Menschen. Just G (ed), Vol V/1, Springer, Berlin pp 445-537.

2063. Gottschaldt K (1968) Begabung und Vererbung. Phänogenetische Befunde zum Begabungsproblem. In: Begabung und Lernen. Roth H (ed), pp 129-150. Deutscher Bildungsrat: Gutachten und Studien der Bildungskommission, 4. Klett, Stuttgart
2064. Gulotta F, Rehder H, Gropp A (1981) Descriptive neuropathology of chromosomal disorders in man. Hum Genet 57: 337-344
2065. Haag E v d (1969) The Jewish mystique. Stein & Day, New York
2066. Hallgren B (1960) Nocturnal enuresis in twins. Acta Psychiatr Scand 35: 73-90
2067. Hawkins JD (1970) Single gene substitutions and behavior. In: Contributions to behavior genetic analysis. The mouse as a prototype. Appleton-Century-Crofts, New York pp 139-159.
2068. Heath AC, Berg K, Eaves LJ, Solaas MH, Corey LA, Sundet J, Magnus P, Nance WE (1985) Education policy and the heritability of educational attainment. Nature 314: 734-736
2068a. Heigl-Evers A, Schepank H (Eds) (1980, 1982) Ursprünge seelisch bedingter Krankheiten. Vols I, II. Vandenhoeck & Rupprecht, Göttingen
2069. Hendrickson AE and DE (1982) The psychophysiology of intelligence. Part I, II. In: Eysenck HJ (ed) A model for intelligence, Springer, Berlin Heidelberg New York pp 151-230.
2070. Herbst DS, Baird PA (1982) Sib risk for nonspecific mental retardation in British Columbia. Am J Med Genet 13: 197-208
2071. Herschel M (1978) Dyslexia revisited. Hum Genet 40: 115-134
2072. Heston LL (1966) Psychiatric disorders in foster home reared children of schizophrenic mothers. Br J Psychiatry 112: 819-825
2073. Heston LL, Shields J (1968) Homosexuality in twins. Arch General Psychiat 18: 149-160
2074. Heuschert D (1963) EEG-Untersuchungen an eineiigen Zwillingen im höheren Lebensalter. Z Menschl Vererbungs- Konstitutionslehre 37: 128
2075. Hiroshi T, Akio A, Shingi T, Eiji I (1968) Sex chromosomes of Japanese epileptics. Lancet I: 478
2076. Hook EB (1979) Extra sex chromosomes and human behavior: The nature of the evidence regarding XYY, XXY, XXYY and XXX genotypes. In: Genetic mechanisms of sexual development, Vallet HL, Porter IH (eds), Academic Press, New York pp 437-461.
2077. Hunt, Nigel (1966) The world of Nigel Hunt - Diary of a mongoloid youth. Darwen Finlay-Son Lt., London
2078. Husén T (1953) Twillingstudier. Almquist & Wiksell, Stockholm
2079. Husén T (1960) Abilities of twins. Scand J Psychol 1: 125-135
2080. Hutchings B, Mednick S (1975) Registered criminality in the adoptive and biological parents of registered male criminal adoptees. In: Genetic research in psychiatry. Fieve RR, Rosenthal D, Brill H (eds), Johns Hopkins University Press, Baltimore pp 105-116.
2081. Ihda S (1961) A study of neurosis by twin method. Psychiatr Neurol Jpn 63: 681-892
2082. Inborn alcoholism? (1985) Lancet 1: 1427-1428
2083. Inoue E (1965) Similar and dissimilar manifestations of obsessive-compulsive neurosis in monozygotic twins. Am J Psychiatry 121: 1171-1175
2084. Itil TM, Hsu W, Saletu B, Mednik S (1974) Computer EEG and auditory evoked potential investigations in children at high risk for schizophrenia. Am J Psychiat 131: 892-900
2085. Jensen AR (1969) How much can we boost I.Q and scholastic achievement? Harvard Educ Rev 39: 1-123
2086. Jensen AR (1973) Educability and group differences. Methuen, London
2087. Jinks JL, Fulker DW (1970) Comparison of the biometrical genetical MAVA, and classical approaches to the analysis of human behavior. Psychol Bull 73: 311-349
2088. Jonsson AE, Nilsson T (1968) Alkoholkonsumtion hos monozygota og dizygota tvillingspar. Nord Hyg 49: 21
2089. Juberg RL, Marsidi I (1980) A new form of X-linked mental retardation with growth retardation, deafness and microgenitalism. Amer J Hum Genet 32: 714-722
2090. Juda A (1939) Neue psychiatrisch-genealogische Untersuchungen an Hilfsschulzwillingen und ihren Familien. I. Die Zwillingsprobanden und ihre Partner. Z Ges Neurol Psychiatr 166: 365-452
2091. Juda A (1940) Neue psychiatrisch-genealogische Untersuchungen an Hilfsschulzwillingen und ihren Familien. II. Die Kollateralen. Z Ges Neurol Psychiatr 168: 448-491
2092. Juda A (1940) Neue psychiatrisch-genealogische Untersuchungen an Hilfsschulzwillingen und ihren Familien. III. Aszendenz und Deszendenz. Z Ges Neurol Psychiat 168: 804-826
2093. Juda A (1953) Höchstbegabung, ihre Erbverhältnisse sowie ihre Beziehungen zu psychischen Anomalien. Urban & Schwarzenberg, München, Berlin
2094. Juel-Nielsen N, Harvard B (1958) The electroencephalogram in monovular twins brought up apart. Acta Genet (Basel) 9: 57-64
2095. Juel-Nielsen N (1965) Individual and environment. A psychiatric-psychological investigation of monozygotic twins reared apart. Acta Psychiatr Scand [Supp] 183
2096. Just G (1970) Erbpsychologie der Schulbegabung. In: Handbuch der Erbbiologie des Menschen. Just G (ed), Vol V/1. Springer, Berlin, pp 538-591.
2097. Kaij L (1960) Alcoholism in twins. Almquist & Wiksell, Stockholm
2098. Kallmann FJ (1953) Heredity in health and mental disorder. Norton, New York
2099. Kalmus H, Fry DB (1980) On tune deafness (dysmelodia): frequency, development, genetics and musical background. Ann Hum Genet 43: 369-382
2100. Karopka RJ, Benzer S (1971) Clock mutants of Drosophila melanogaster. Proc Natl Acad Sci USA 68: 2112-2116
2101. Kay DWK (1963) Late paraphrenia and its bearing on the aetiology of schizophrenia. Acta Psychiat Scand 39: 159-169
2102. Kendler KS (1983) Overview: A current perspective on twin studies of schizophrenia. Am J Psychiat 140: 1413-1425

2103. Kety SS (1959) Biochemical theories of schizophrenia. Science 129: 1528-1532

2104. Kety SS, Rosenthal D, Wender PH, Schulsinger F, Jacobsen B (1975) Mental illness in the biological and adoptive families of adopted individuals who have become schizophrenic. A preliminary report based upon psychiatric interviews. In: Genetic research in psychiatry. (ed): Fiere RR, Rosenthal D, Brill H, Johns Hopkins Press, Baltimore p 147

2105. Knorring AL v, Cloninger R, Behman M, Sigvardsson S (1983) An adoption study of depressive disorders and substance abuse. Arch Gen Psychiatr 40: 943-950

2106. König K (1959) Der Mongolismus. Hippokrates, Stuttgart

2107. Krech D, Rosenzweig MR, Bennet EL, Kraeckel BA (1954) Enzyme concentrations in the brain and adjustive behavioral patterns. Science 120: 994-996

2108. Kringlen E (1967) Heredity and environment in the functional psychoses. Heinemann, London

2109. Krüger J, Propping P (1976) Rückgang der Zwillingsgeburten in Deutschland. Dtsch Med Wochenschr 101: 475-480

2110. Kuhlo W, Heintel H, Vogel F (1969) The 4-5/sec. rhythm. EEG Clin Neurophysiol 26: 613-619

2111. Lauer J, Lindauer M (1971) Genetisch fixierte Lerndispositionen bei der Honigbiene. Inf. Org. 1, Akademie der Wissenschaft und Literatur. Mainz

2112. Lauer J, Lindauer M (1973) Die Beteiligung von Lernprozessen bei der Orientierung. Fortschr Zool 21: 349-370

2113. Lenz F (1932) Menschliche Auslese und Rassenhygiene. In: Bauer E, Fischer E, Lenz F: Menschliche Erblichkeitslehre und Rassenhygiene. Lehmann, München

2114. Leonhard K (1957) Aufteilung der endogenen Psychosen, 1st ed Akademie-Verlag, Berlin

2115. Lester D, Freed EX (1972) A rat model of alcoholism. In: Nature and nurture in alcoholism. Ann N Y Acad Sci 197: 54-59

2116. Lewis EO (1933) Types of mental deficiency and their social significance. J Ment Sci 79: 298

2117. Lewontin RC (1975) Genetic aspects of intelligence. Ann Rev Genet 9: 387-405

2117a. Lipp HP, Schwegler H, Driscoll P (1984) Postnatal modification of hippocampal circuity alters avoidance learning in adult rats. Science 225, 80-82

2118. Little AJ (1974) Psychological characteristics and patterns of crime among males with an XYY sex chromosome complement in a maximum security hospital. B.A. Sp. Hon. Thesis (Quoted from Borgaonkar and Shah), Sheffield University

2119. Loehlin JC (1980) Recent adoption studies of I.Q. Hum Genet 55: 297-302

2120. Loehlin JC, Lindzey G, Spuhler JN (1975) Race differences in intelligence. Freeman, San Francisco

2121. Lubs HA (1983) X-linked mental retardation and the marker X. In: Emery AEH, Rimoin DL (eds) Principles and Practice of Medical Genetics, Churchill Livingstone, Edinburgh etc. pp 216-223.

2122. Manosevitz M, Lindzey G, (1969) Thiessen DD Behavioral genetics: methods and research. Appleton-Century-Crofts, New York

2123. Marsden CD (1982) Neurotransmitters and DNS disease: Basal ganglion disease. Lancet II: 1141-1146

2124. Matthyse S, Spring BJ, Sugorman J (eds) (1978) Attention and information processing in schizophrenia. J Psychiat Res 14: 1-331

2125. Maubach M, Diebold K, Friedl W, Propping P (1982) MAO-Aktivität in Thrombocyten von affektpsychotischen Patienten und ihren Verwandten ersten Grades. In: Beckmann H (ed) Biologische Psychiatrie, Thieme Verlag, Stuttgart-New York pp 182-188.

2126. McClearn GE (1972) Genetics as a tool in alcohol research. In: Nature and nurture in alcoholism. Ann N Y Acad Sci 197: 26-31

2127. McClearn GE, Rodgers DA (1972) Differences in alcohol preference among inbred strains of mice. Q J Stud Alcohol 20: 691-695

2128. McGuire LS, Ryan KO, Omenn GS (1975) Congenital adrenal hyperplasia. II. Cognitive and behavioral studies. Behav Genet 5: 175-188

2129. Mendels J, Stern S, Frazer A (1976) Biochemistry of depression. Dis Nerv Syst 37: 4-36

2130. Mednik SA, Mura E, Schulsinger F, Mednik B (1973) Perinatal conditions and infant development in children with schizophrenic parents. Soc Biol 20: 111-112

2131. Mendlewicz J, Rainer JD (1977) Adoption study supporting genetic transmissions in manic-depressive illness. Nature 268: 327-329

2132. Migeon B, DerKaloustian VM, Nyhan WL, Young WJ, Childs B (1968) X-linked hypoxanthine-guanine phosphoribosyl transferase deficiency: Heterozygote has two clonal populations. Science 160: 425-427

2133. Money J (1968) Cognitive deficits in Turner's syndrome. In: Progress in human behavior genetics. Vandenberg S (ed), Johns Hopkins Press, Baltimore pp 27-30.

2134. Money J, Alexander D (1966) Turner's syndrome: Further demonstration of the presence of specific congnitional deficiencies. J Med Genet 3: 47

2135. Motulsky AG (1979) Possible selective effects of urbanization on Ashkenazi Jewish populations. In: Genetic Diseases among Ashkenazi Jews. Goodman RM, Motulsky AG (eds). Raven, New York pp 201-212

2136. Motulsky AG Ashkenazi Jewish gene pools: admixture, drift and selection. In: Population genetic studies on isolates. Sigrid Juselius Symposium VII. Academic Press, London pp 353-365

2137. Motulsky AG (1981) Some research approaches in psychiatric genetics. In: Genetic Strategies in Psychobiology and Psychiatry. Gershon EL, Matthysse S, Breakefield XO, Ciranello RD (eds) The Boxwood Press, Pacific Grove, CA pp 423-428

2138. Müller-Küppers M, Vogel F (1965) Über die Persönlichkeitsstruktur von Trägern einer seltenen erblichen EEG-Variante. Jahrb Psycholog Psychother Med Anthropol 12: 75-101

2139. Muller HJ (1925) Mental traits and heredity. J Hered 16: 433-448

2140. Murken JD (1973) The XYY-syndrome and Klinefelter's syndrome. Topics Human Genetics. Thieme, Stuttgart

2141. Myrianthopoulos NC, Nichols PL, Broman SH (1976) Intellectual development of twins - comparison with singletons. Acta Genet Med Gemellol (Roma) 25: 376-380

2142. Nachtsheim H (1959) Probleme vergleichender Genetik bei Säugern. Naturwissenschaften 20: 565-573
2143. Neisser U (1976) Academic and artificial intelligence. In: The Nature of Intelligence. (ed): Resnick LB, Erlbaum, Hillsdale pp 135-144.
2144. Nielsen J (1970) Criminality among patients with Klinefelter's syndrome. Br J Psychiatry 117: 365-369
2145. Nielsen J, Hreidersson AB, Christensen KR (1973) D/D translocations in patients with mental illness. Hereditas 75: 131-135
2146. Nielsen J, Sillesen I, Sørensen AM, Sørensen K (1979) Follow-up until age 4 to 8 of 25 unselected children with sex chromosome abnormalities, compared with sibs and controls. Birth Defects: Original Article Series, Vol XY, pp 15-73
2147. Nielsen J, Sørensen AM, Sørensen K (1981) Mental development of unselected children with sex chromosome abnormalities. Hum Genet 59: 324-332
2148. Noel B, Duport JP, Revil D, Dussuyer I, Quack B (1974) The XYY syndrome: Reality or myth? Clin Genet 5: 387-394
2149. Omenn GS, Wade Cohen PT, Motulsky AG (1977) Genetic variation in glycolytic enzymes in human brains. International Congress of Human Genetics. Experta Medica, Paris, 1977. International Congress Series 233: 135
2150. Omenn GS, Weber BA (1978) Dyslexia: search for phenotypic and genetic heterogeneity. Am J Med Genet 1: 333-354
2151. Opitz JM, Sutherland GR (eds) (1984) Conference report: International workshop on the fragile X and X-linked mental retardation. Am J Med Genet 17: 5-385
2152. Overton WF (1973) On the assumptive base of the nature-nurture controversy: Additive versus interactive conceptions. Hum Dev 16: 74-89
2153. Parker N (1964) Twins: A psychiatric study of a neurotic group. Med J Aust 51: 735-742
2154. Partanen J, Bruun K, Markkanen T (1966) Inheritance of drinking behavior. The Finnish Foundation for Alcohol Studies, Vol 14
2155. Paul J, Froster-Iskenius U, Moje W, Schwinger E (1984) Heterozygous female carriers of the marker-X-chromosome: I. Q. estimation and replication status of fra (X) (q). Hum Genet 66: 344-346
2156. Penrose LS (1938) (Colchester survey) A clinical and genetic study of 1280 cases of mental defect. Spec Rep Ser Med Res Counc (London) 229. His Maj Stat Off London
2157. Penrose LS (1962) The biology of mental defect, 3rd ed Grune & Stratton, New York
2158. Pollock VE, Volavka J, Goodwin DW, Mednik SA, Gabrielli WF, Knop J, Schulsinger F (1983) The EEG after alcohol administration in men at risk for alcoholism. Arch of Gen Psychiat 40: 857, 861
2159. Popenoe P (1922) Twins reared apart. J Hered 5: 142-144
2160. Praag van HM (1982) Neutrotransmitters and CNS disease: Depression. The Lancet II, pp 1259-1264
2161. Propping P (1977) Genetic control of ethanol action on the central nervous system. Hum Genet 35: 309-334
2162. Propping P (1978) Alcohol and alcoholism. In: Human genetic variation in response to medical and environmental agents: Pharmacogenetics and ecogenetics. (ed): Motulsky AG et al. Hum Genet Suppl 1: 91-99
2163. Propping P (1980) Genetic aspects of alcohol action on the electroencephalogram (EEG). In: Biological research in alcoholism. Begleiter H, Kissin (eds). Plenum, New York, pp 589-602
2164. Propping P (1983) Genetic disorders presenting as "schizophrenia". Karl Bonhoeffer's early view of the psychoses in the light of medical genetics. Hum Genet 65: 1-10
2165. Propping P, Friedl W (1983) Genetic control of adrenergic receptors on human platelets. A twin study. Hum Genet 64: 105-109
2166. Propping P, Friedl W (1983) Platelet MAO activity and high risk for psychopathology in a German population. Mod Probl Pharmacopsychiat 19: 304-314
2167. Propping P, Krüger J, Mark N (1981) Genetic disposition to alcoholism. An EEG study in alcoholics and their relatives. Hum Genet 59: 51-59
2168. Propping P, Friedl W, Nebel B, Feige A (1979) Plasma DBH platelet MAO and proteins of red blood cell membranes in individual with variants of the normal EEG. Neuropsychobiology 169, 5, 309-316
2169. Propping P, Kopun M (1973) Pharmacogenetic aspects of psychoreactive drugs: facts and fancy. Hum Genet 20: 291-320
2170. Puck MH (1981) Some considerations bearing on the doctrine of self-fulfilling prophecy in sex chromosome aneuploidy. Am J Med Genet 9: 129-137
2171. Puck MH, Bender BG, Borelli JB, Salbenblatt JA, Robinson A (1983) Parents' adaptation to early diagnosis of sex chromosome anomalies. Am J of Med Genet 16: 71-79
2172. Quazi RH, Reed TE (1975) A possible major contribution to mental retardation in the general population by the gene for microcephaly. Clin Genet 7: 85-90
2173. Rao DC, Morton NE, Elston RC, Yee S (1977) Causal analysis of academic performance. Behav Genet 7: 147-159
2174. Ratcliffe SG (1982) Speed and learning disorders in children with sex-chromosome abnormalities, Developm Med and Child Neurolog 24: 80-84
2175. Ratcliffe SG, Field MAS (1982) Emotional disorders in XYY children: Four case reports. J Child Psychol Psychiatr 23: 401-406
2176. Ratcliffe SG, Tierney I, Nshaho J, Smith L, Springbett A, Callan S (1982) The Edinburgh study of growth and development of children with sex-chromosome abnormalities. Birth Defects, Original Article Series 18 (4): 41-60
2177. Reed TE, Kalant H, Gibbins RJ, Kapur BM, Rankin JG (1976) Alcohol and aldehyde metabolism in Caucasians, Chinese and Amerinds. Can Med Assoc J 115: 851-855
2178. Renpenning H, Gerrard JW, Zaleski WA, Tabata T (1962) Familial sex-linked mental retardation. Canad Med Ass J 87: 954-956
2179. Risch N, Baron M (1982) X-linked and genetic heterogeneity in bipolar-related major affective illness: Reanalysis of linkage data. Ann Hum Genet 46: 153-166

2181. Rodgers DA (1970) Mechanism-specific behavior: An experimental alternative. In: Contributions to behavior-genetic analysis. The mouse as a prototype. Appleton-Century-Croft, New York pp 207–218.

2182. Rodgers DA, McClearn GE, Bennett EL, Herbst M (1963) Alcohol preference as a function of its caloric utility in mice. J Comp Phys Psychol 56: 666–672

2183. Rosenthal D (ed) (1963) The Genain quadruplets. Basic Books, New York, London

2184. Rosenthal D, Wender PH, Kety SS, Schulsinger F, Welner J, Østergaard L (1968) Schizophrenics' offspring reared in adoptive homes. In: The transmission of schizophrenia. Rosenthal D, Kety S (eds), Pergamon, Oxford p 377.

2185. Rothenbuhler N (1964) Behavior genetics of nest cleaning in honey bees. 4. Responses of F_1 and backcross generations to disease-killed brood. Am Zool 4: 111–123

2186. Rüdin E (1916) Studien über Vererbung und Entstehung geistiger Störungen. Springer, Berlin

2187. Rüdin E (1953) Ein Beitrag zur Frage der Zwangskrankheit, insbesondere ihrer hereditären Beziehungen. Arch Psychiatr Z Neurol 191: 14–54

2188. Saunders JB (1982) Alcoholism: new evidence for a genetic contribution. Br Med J 284: 1137–1138

2188a. Scarr S (1981) Race, social class, and individual differences in I.Q. Erlbaum, Hillsdale

2189. Scarr S, Pakstis AJ, Katz SH, Barker UB (1977) Absence of a relationship between degree of white ancestry and intellectual skills within a black population. Hum Genet 39: 69–86

2190. Scarr S, Weinberg RA (1976) I.Q. test performance of black children adopted by white families. Am Psychol 31: 726–739

2191. Seixas FA, Omenn GS, Burk ED, Eggleston S (eds) (1972) Nature and nurture in alcoholism. Ann NY Acad Sci 197

2192. Severson JA, Randall PK, Finch CE (1981) Genotypic influences on striatal dopaminergic regulation in mice. Brain Res 210: 201–215

2193. Shaffer JW (1962) A specific cognitive deficit observed in gonadal aplasia (Turner's syndrome). J Clin Psychol 18: 403–406

2194. Sheppard JR, Albersheim B, McClearn GE (1970) Aldehyde dehydrogenase and ethanol preference in mice. J Biol Chem 245: 2876–2882

2195. Shields J (1962) Monozygotic twins brought up apart and brought up together. Oxford University Press, London

2196. Shuey AM (1966) The testing of negro intelligence, 2nd ed Social Science Press, New York

2197. Siervogel RM, Weinshilboum R, Wilson AF, Elston RC (1984) Major gene model for the inheritance of catechol-O-methyltransferase activity in five large families. Am J Med Genet 19: 315–323

2198. Sidman RL, Greene MC (1970) Nervous new mutant mouse with cerebellar disease. In: Les mutants pathologiques chez l'animal. C.N.R.S. Paris p 69–79

2199. Slater E (1964) Genetic factors in neurosis. Brit J Psychol 55: 265–269

2200. Smythies JR (1976) Recent progress in schizophrenia research. Lancet II: 136–139

2201. Soudek D, Partington MW, Lawson JS (1984) The fragile X syndrome I: Familial variation in the proportion of lymphocytes with the fragile site in mamales. Am J Med Genet 17: 241–252

2202. Sutherland GR (1983) The fragile X chromosome. Intern Rev Cytol 81: 107–143

2203. Sutherland GR, Murch AR, Gardiner AJ, Carter RF, Wiseman C (1976) Cytogenetic survey of a hospital for the mentally retarded. Hum Genet 34: 231–245

2204. Schepank H (1974) Erb- und Umweltsfaktoren bei Neurosen. Springer, Berlin Heidelberg New York

2205. Schmid W, Nielsen J (eds) (1981) Human Behavior and Genetics. Amsterdam, Elsevier North Holland, New York, Oxford

2206. Schulsinger F (1972) Psychopathy, heredity, and environment. Int J Ment Health 1: 190–206

2207. Schwegler H, Lipp H-P (1983) Hereditary covariations of neuronal circuity and behavior: Correlations between the proportions of hippocampal synaptic fields in the regio inferior and two-way avoidance in mice and rats. Behavioral Brain Res 7: 1–38

2208. Schwegler H, Lipp H-P, Van der Loos H, Buselmaier W (1981) Individual hippocampal mossy fiber distribution in mice correlates with two-way avoidance performance. Science 214: 817–819

2209. Stocks P (1930) A biometric investigation of twins and their brothers and sisters. Ann Eugen 4: 49–108

2210. Street DRK, Watson RA (1969) Patients with chromosome abnormalities in Rampton Hospital. In: Criminological implications of chromosome abnormalities. West DJ (ed), Cropwood Round Table Conference,Institute of Criminology, University of Cambridge pp 61–67.

2211. Strömgren E (1967) Neurosen und Psychopathien. In: Humangenetik, ein kurzes Handbuch. Becker PE (ed), Vol V/2, Thieme, Stuttgart pp 578–598.

2212. Sturtevant AH (1915) Experiments of sex recognition and the problem of sexual selection in Drosophila. J Anim Behav 5: 351–366

2213. Tariverdian G, Weck B (1982) Nonspecific X-linked mental retardation. A review. Hum Genet 62: 95–109

2214. Taylor HF (1980) The I.Q. game. A methodological inquiry into the heredity-environment controversy. The Harvester Press, Brighton

2215. Terman LM, Merrill MA (1937) Measuring of intelligence. Houghton Mifflin, Boston

2216. Terman LM, Oden MH (1959) Genetic studies of genius: Vol V, The gifted group at midlife. Stanford University Press, Stanford (Cal)

2217. Tienari P (1963) Psychiatric illnesses in identical twins. Acta Psychiatr Scand [Suppl] 171

2218. Tienari P (1971) Schizophrenia and monozygotic twins. Psychiatria Fennica 1971: 97–104. Helsinki University General Hospital, Helsinki

2219. Tsuboi T (1970) Crimino-biologic study of patients with the XYY syndrome and Klinefelter's syndrome. Hum Genet 10: 68–84

2220. Turner G, Jacobs PA (1984) Mental retardation and the fragile X. Adv Hum Genet 13

2221. Tyler LE (1976) The intelligence we test. – An evolving concept. In: The nature of intelligence. (ed): Resnick LB, Erlbaum, Hillsdale pp 13–26.

2222. Usdin E, Mandell AJ (eds) (1978) Biochemistry of mental disorders. Dekker, New York, Basel

2223. Valenstein ES, Riss W, Young WC (1954) Sex drive in genetically heterogeneous and highly inbred stains of male guinea pigs. J Comp Physiol Psychol 47: 162–165

2224. Valenstein ES, Riss W, Young WC (1955) Experimental and genetic factors in the organization of sexual behavior in male guinea pigs. J Comp Physiol Psychol 48: 397-403
2225. Valverde F (1967) Apical dendritic spines of the visual cortex and light deprivation in the mouse. Exp Brain Res 3: 337-352
2226. Vandenberg SG (ed) (1965) Methods and Goals in human Behavior Genetics. Academic Press, New York
2227. Vandenberg SG (1968) Progress in human behavior genetics. Johns Hopkins Press, Baltimore
2228. Vogel F (1958) Über die Erblichkeit des normalen Elektroencephalogramms. Thieme, Stuttgart
2229. Vogel F (1981) Humangenetische Aspekte der Sucht. Dtsch Med Wschr 106: 711-714
2230. Vogel F (1984) Mutation and selection in the marker (X) syndrome. A hypothesis. Ann Hum Genet 48, 327-332
2230 a. Vogel F, Krüger J, Höpp HP, Schalt E, Schnobel R (1986) Visually and auditory evoked EEG potentials in carriers of four hereditary EEG variants. Human Neurobiology 5, 49-58
2231. Vogel F, Schalt E (1979) The electroencephalogram (EEG) as a research tool in human behavior genetics: Psychological examinations in healthy males with various inherited EEG variants. III. Interpretation of the results. Hum Genet 47: 81-111
2232. Vogel F, Schalt E, Krüger J (1979) The electroencephalogram (EEG) as a research tool in human behavior genetics: Psychological examinations in healthy males with various inherited EEG variants. II. Results. Hum Genet 47: 47-80
2233. Vogel F, Schalt E, Krüger J, Klarich G (1982) Relationship between behavioral maturation measured by the "Baum" test and EEG frequency. A pilot study on monozygotic and dizygotic twins. Hum Genet 62: 60-65
2234. Vogel F, Schalt E, Krüger J, Propping P (1979) The electroencephalogram (EEG) as a research tool in human behavior genetics: Psychological examinations in healthy males with various inherited EEG variants. I. Rationale of the study; material; methods; heritability of test parameters. Hum Genet 47: 1-45
2234a. Volavka Jan, Mednick SA, Rasmussen L, Sergeant J (1977) EEG Spectra in XYY and XXY Men. Electroencephalography and Clinical Neurophysiology, 43: 798-801. Elsevier/North-Holland, Scientific Publishers, Ltd
2235. DeVries JC, Vandenberg SG, McClearn GE (1976) Genetics of specific cognitive abilities. Ann Rev Genet 10: 179-207
2236. Wahl OF (1976) Monozygotic twins discordant for schizophrenia: A review. Psychol Bull 83: 91-106
2237. Wang JCC, Erbe RW (1984) Folate metabolism in cells from fragile X syndrome patients and carriers. Am J Med Genet 17: 303-310
2238. Weinshilboum RM (1978) Human biochemical genetics of plasma dopamine-β-hydroxylase and erythrocyte catechol-0-methyltransferase. Human genetic variation in response to medical and environmental agents: Pharmacogenetics and ecogenetics (ed): Motulsky AG et al. Hum Genet Suppl 1: 101-111
2239. Weinshilboum RM (1983) Biochemical genetics of catecholamines in humans. Mayo Clin Proc 58: 319-330
2240. Weinshilboum RM, Schrott HG, Raymond FA, Weidman WH, Elveback LR (1975) Inheritance of very low serum dopamine-β-hydroxylase. Am J Hum Genet 27: 573-585
2241. Wender PH, Rosenthal D, Kety SS (1968) A psychiatric assessment of the adoptive parents of schizophrenics. In: The transmission of schizophrenia. (ed): Rosenthal D, Kety SS, Pergamon, Oxford p 235.
2242. Wender PH, Rosenthal D, Kety SS, Schulsinger E, Welner J (1976) Crosstesting. A research strategy for clarifying the role of genetic and environmental factors in the etiology of schizophrenia. Arch Gen Psychiatr 30: 121
2243. Williams RJ, Berry LJ, Beerstecher E (1949) Biochemical individuality. III. Genetotrophic factors in the etiology of alcoholism. Arch Biochem 23: 275-290
2244. Wilson RS (1972) Twins: Early mental development. Science 175: 914-917
2245. Winokur G, Tanna VL (1969) Possible role of X-linked dominant factor in manic-depressive disease. Dis Nerv Syst 30: 89-94
2246. Winter H, Herschel M, Propping P, Friedl W, Vogel F (1978) A twin study on three enzymes (DBH, COMT, MAO) of Catecholamine metabolism. Psychpharmacology 57: 63-69
2247. Witkin HA, Mednick SA, Schulsinger F, Bakkestrøm E, Christiansen KO, Goodenough DR, Hirschhorn K, Lundsteen C, Owen DR, Philip J, Rubin DB, Stocking M (1976) Criminality in XYY and XXY men. Science 193: 547-555
2248. Young W, Goy RW, Phoenix CH (1964) Hormones and sexual behavior. Science 143: 212-218
2249. Zajonc RB (1976) Family configuration and intelligence. Science 192: 227-236
2250. Zang KD (ed) (1984) Klinische Genetik des Alkoholismus. W. Kohlhammer Verlag, Stuttgart, Berlin, Köln, Mainz
2251. Zerbin-Rüdin E (1967) Idiopathischer Schwachsinn. In: Humangenetik, ein kurzes Handbuch. Becker PE (ed), Vol V/2, Thieme, Stuttgart pp 158-205
2252. Zerbin-Rüdin E (1967) Endogene Psychosen. In: Humangenetik, ein kurzes Handbuch. Becker PE (ed), Band V/2, Thieme, Stuttgart pp 446-573

References to Chapter 9 and Appendix 8

2253. Adinolfi A, Adinolfi M, Lessof MH (1975) Review: Alphafeto-protein during development and in disease. J Med Genet 12: 138-151
2254. Anonymous (1977) Methods for the detection of haemophilia carriers: a memorandum. WHO Bull 55: 675-702
2255. Anonymous (1973) Law and ethics of A.I.D. and embryo transfer. Ciba Found Symp (new series) 17
2256. Anonymous (1975) National Research Council. Committee for the Study of Inborn Errors of Metabolism. Genetic screening: programs, principles, and research. National Academy of Sciences, Washington (DC)
2257. Anonymous (1985) Embryo research. Lancet 1: 255-256

2258. Anonymous (1983) President's Commission for the Study of Ethical Problems in Medicine and Biomedical and Behavioral Research: Screening and Counseling for Genetic Conditions. The Ethical, Social, and Legal Implications of Genetic Screening, Counseling, and Education Program. US Government Printing Office, Washington, DC
2259. Anonymous (1984) Report of the Committee on Inquiry into Human Fertilisation and Embryology. Her Majesty's Stationery Office, London
2260. Anonymous (1985) Maternal serum alpha-fetoprotein screening for neural tube defects. Results of a consensus meeting. Pren Diag 5: 77-83
2261. Anonymous (1985) Low maternal serum alphafetoprotein and Down syndrome. Lancet 1: 259-260
2262. Applebaum EG, Firestein SK (1983) A Genetic Counseling Casebook. The Free Press, New York
2263. Bakker E, Hofker MH, Goor N, Mandel JL, Wrogemann K, Davies KE, Kunkel LM, Willard HF, Fenton WA, Sandkuyl L, Majoor-Krakauer D, Essen AJV, Jahoda MGJ, Sachs ES, van Ommen GJB. Pearson PL (1985) Prenatal diagnosis and carrier detection of Duchenne muscular dystrophy with closely linked RFLPs. Lancet 1: 655-658
2264. Berg P, Baltimore D, Boyer HW, Cohen SN, Davis RW, Hogness DS, Nathans D, Roblin RO, Watson JD, Weissman S, Zinder ND (1974) Potential biohazards of recombinant DNA molecules. Science 185: 303
2265. Bloch EV, DiSalvo M, Hall BD, Epstein CJ (1979) Alternative ways of presenting empiric risks. In: Risk, Communication, and Decision Making in Genetic Counseling. Epstein CJ, Curry CJR, Packman S, Sherman S, Hall BD (eds), Birth Defects: Orig Art Ser XV/5C. Alan R. Liss, New York
2266. Bonaiti-Pellié C, Phung L, Nordmann Y (1984) Recurrence risk estimation of acute intermittent porphyria based on analysis of porphobilinogen deaminase activity: A Bayesian approach. Am J Med Genet 19: 755-762
2267. Brambati B, Simoni G, Danesino C, Oldrini A, Ferrazzi E, Romitti L, Terzoli G, Rossella F, Ferrari M, Fraccaro M (1985) First trimester fetal diagnosis of genetic disorders: clinical evaluation of 250 cases. J Med Genet 22: 92-99
2268. Campbell S, Pearce JM (1983) Ultrasound visualization of structural anomalies. Br Med Bull 39: 322
2269. Cao A, Cossu P, Falchi AM, Monni G, Pirastu M, Rosatelli C, Scalas MT, Tuveri T (1985) Antenatal diagnosis of thalassemia major in Sardinia. Ann N Y Acad Sci 445: 380-302
2270. Capron AM, Lappe M, Murray RF, Powledge TM, Twiss SB, Bergsma D (eds) (1979) Genetic counselling: facts, values, and norms. Birth Defects: Orig Art Ser XV (2). Alan R. Liss, New York
2271. Carter CO, Fraser Roberts JA, Evans KA, Buck AR (1971) Genetic clinic: a follow-up. Lancet 1: 281
2272. Chargaff E (1976) On the dangers of genetic meddling. Science 192: 938
2273. Cote GB (1982) Odds in genetic counseling. J Med Genet 19: 455-457
2274. Crandall BF, Robertson RD, Lebherz TB, King W, Schroth PC (1983) Maternal serum α-fetoprotein screening for the detection of neural tube defects. West J Med 138: 531-534
2275. Cuckle HS, Wald NJ, Lindenbaum RH (1984) Maternal serum alpha-fetoprotein measurement: a screening test for Down syndrome. Lancet 1: 926-929
2276. Dalgaard OZ (1957) Bilateral polycystic disease of the kidneys. A follow-up of two-hundred and eighty-four patients and their families. Acta Med Scand 328 [suppl]
2277. Davis BD (1977) The recombinant DNA scenarios: Andromeda strain, chimera, and golem. Sci Am 65: 547-555
2278. Di Lonardo AM, Orrego C, Darlu P, King M-C, Baur M (1984) Human genetics and human rights. Identifying the families of kidnapped children. Am J Foren Med Pathol 5: 339-347
2279. Din N, Schwartz M, Kruse T, Vestergaard SR, Ahrens P, Caput D, Hartog K, Quiroga M (1985) Factor VIII gene specific probe for prenatal diagnosis of haemophilia A. Lancet 1: 1446-1447
2280. Dworkin RB, Omenn GS (1985) Legal aspects of human genetics. Ann Rev Public Health 6: 107-130
2281. Edwards RG, Fowler RE (1970) The genetics of preimplantation human development. Mod Trends Hum Genet Vol 1: 181-213
2282. Edwards RG, Steptoe PC (1973) Biological aspects of embryo transfer. pp 11-18. In: Law and ethics of A. I. D. and embryo transfer. Ciba Found Symp (new series) 17: 11-18
2283. Elles RG, Williamson R, Niazi M, Coleman DV, Horwell D (1983) Absence of maternal contamination of chorionic villi used for fetal gene analysis. N Engl J Med 308: 1433
2284. Emery AEH, Pullen I (1984) Psychologic aspects of genetic counseling. Academic Press
2285. Epstein CJ (1975) Genetic counseling: present status and future prospects. In: Early diagnosis and prevention of genetic disease. Went L, Vermeij-Keers C, Linden AGJM van der (eds), University of Leiden Press, Leiden pp 110-131.
2286. Epstein CJ, Cox DR, Schonberg SA, Hogge WA (1983) Recent developments in the prenatal diagnosis of genetic diseases and birth defects. Ann Rev Genet 17: 49-83
2287. Epstein CJ, Curry CJR, Packman S, Sherman S, Hall BD (1979) Risk, Communication, and Decision Making in Genetic Counseling. Birth Defects: Orig Art Ser XV (5C), Alan R. Liss, New York
2288. Erbe RW (1975) Screening for the hemoglobinopathies. In: The prevention of genetic disease and mental retardation. Milunsky A (ed), Saunders, Philadephia pp 204-220.
2289. Evers-Kiebooms G, Berghe H van den (1979) Impact of genetic counseling: a review of published follow-up studies. Clin Genet 15: 465-474
2290. Falk R, Motulsky AG, Vogel F, Weingart P (1985) Historische und ethische Aspekte der Humangenetik. Ein Interdisziplinares Kolloquium im Wissenschaftskolleg, in Wissenschaftskolleg - Institute for Advanced Study - Berlin Yearbook 1983/84. Siedler Verlag, Berlin pp 75-121
2291. Farrow MG, Juberg RC (1969) Genetics and laws prohibiting marriage in the United States. JAMA 209: 534
2292. Ferguson-Smith MA (1984) Prenatal diagnosis of chromosome anomalies: who is at risk? In: Prenatal

Diagnosis, Rodeck CH, Nicolaides KH (eds) John Wiley & Sons, Chichester, New York

2293. Fletcher JC, Berg K, Tranøy KE (1985) Ethical aspects of medical genetics. A proposal for guidelines in genetic counseling, prenatal diagnosis and screening. Clin Genet 27: 199-205
2294. Fraser FC (1974) Genetic counseling. Am J Hum Genet 26: 636-659
2295. Fraser FC, Forse RA (1981) On genetic screening of donors for artificial insemination. Am J Med Genet 10: 399-405
2296. French-Anderson W (1984) Prospects for human gene therapy. Science 226: 401-409
2297. Fuhrmann W (1971) Arteriosklerose; Erkrankungen der Koronargefässe. In: Humangenetik: Ein kurzes Handbuch in fünf Bänden, Vol III/2. Becker PE (ed) Thieme, Stuttgart p 508
2298. Fuhrmann W, Weitzel HK (1985) Maternal serum alpha-fetoprotein screening for neural tube defects. Report of a combined study in Germany and short overview on screening populations with low birth prevalence of neural tube defects. Hum Genet 69: 47-61
2299. Furrow BR (1984) Surrogate motherhood: A new option for parenting. Law, Medicine and Health Care, p 106
2300. Galen RS, Gambino SR (1975) Beyond normality: The predictive value and efficiency of medical diagnoses. John Wiley & Sons, New York
2301. Gitschier J, Drayna D, Tuddenham EGD, White RL, Lawn RM (1985) Genetic mapping and diagnosis of haemophilia A achieved through a BclI polymorphism in the factor VIII gene. Nature 314: 738-740
2302. Graham JB (1977) Genetic counseling in classic hemophilia A. N Engl J Med 296: 996
2303. Haldane JBS (1963) Biological possibilities for the human species in the next ten thousand years. In: Man and his future. Wolstenholme G (ed), Churchill London pp 337-361.
2304. Hammer RE, Palmiter RD, Brinster RL (1984) Partial correction of murine hereditary growth disorder by germlike incorporation of a new gene. Nature 311: 65-67
2305. Hammer RE, Pursel VG, Rexroad CE Jr, Wall RJ, Bolt DJ, Ebert KM, Palmiter RD, Brinster RL (1985) Production of transgenic rabbits, sheep and pigs by microinjection. Nature 315: 680-683
2306. Harper PS, O'Brien T, Murray JM et al (1983) The use of linked DNA polymorphisms for genotype prediction in families with Duchenne muscular dystrophy. J Med Genet 20: 252-254
2307. Harper PS, Sarfarazi M (1985) Genetic prediction and family structure in Huntington's disease. Br Med J 290: 1929-1931
2308. Harper PS, Shaw D, Williams H (1984) Prenatal diagnosis and the muscular dystrophies. In: Prenatal Diagnosis, Rodeck CH, Nicolaides KH (eds) John Wiley & Sons, Chichester, New York
2309. Harris H (1975) Prenatal diagnosis and selective abortion. Harvard University Press, Cambridge (MA)
2310. Hellerman JG, Cone RC, Potts JT, Rich A, Mulligan RC, Kronenberg HM (1984) Secretion of human parathyroid hormone from rat pituitary cells infected with a recombinant retrovirus encoding preproparathyrpoid hormone. Proc Natl Acad Sci USA 81: 5340-5344
2311. Hess D (1972) Transformationen an höheren Organismen. Naturwissenschaften 59: 348-355
2312. Holtzman NA, Leonard CO, Farfel MR (1981) Issues in antenatal and neonatal screening and surveillance for hereditary and congenital disorders. Ann Rev Public Health 2: 219-251
2313. Holmes LB (1978) Genetic counseling for the older pregnant woman: new data and questions. N Engl J Med 298: 1419-1421
2314. Horst J, Kluge F, Bayreuther K, Gerok W (1975) Gene transfer to human cells: Transducing phage λ plac gene expression in GM_1-gangliosidosis fibroblasts. Proc Natl Acad Sci USA 72: 3531-3535
2315. Horst J, Stanbro H, Merril CR (1980) On procaryotic gene expression in eucaryotic systems. Hum Genet 54: 289-302
2316. Inman RP (1978) On the benefits and costs of genetic screening. Am J Hum Genet 30: 219-223
2317. Itakura K, Hirose T, Crea R, Riggs AD, Heyneker HL, Bolivar F, Boyer HW (1977) Expression in *Echerichia coli* of a chemically synthesized gene for the hormone somatostatin. Science 198: 1056-1063
2318. Jackson LG (1985) First-trimester diagnosis of fetal genetic disorders. Hosp Pract 20: 39-48
2319. Jeanpierre M, Junien C (1984) DNA analysis as clinical investigation: when and how. Ann Genet 27: 134-147
2320. Kaback MM, Zeiger RS, Reynolds LW, Sonneborn M (1974) Approaches to the control and prevention of Tay-Sachs disease. Prog Med Genet 10: 103-134
2321. Kaback MM (ed) (1977) Tay-Sachs disease: screening and prevention. Alan R. Liss, New York
2322. Kazazian HH Jr, Boehm CD, Dowling CE (1985) Prenatal diagnosis of hemoglobinopathies by DNA analysis. Ann N Y Acad Sci 445: 337-368
2323. Kazy Z, Rozovsky IS, Bakharev VA (1982) Chorion biopsy in early pregnancy: a method for early prenatal diagnosis for inherited disorders. Pren Diag 2: 39-45
2323a. Kelly P (1977) Dealing with Dilemma. A Manual for Genetic Couselors. Springer, New York
2324. Kessler S (1979) The genetic counselor as psychotherapist. In: Capron AM, Lappe M, Murray RF, Powledge TM, Twiss SB, Bergsma D (eds): Genetic counselling: facts, values, and norms. Birth Defects: Orig Art Ser XV (2). Alan R. Liss, New York
2325. Kessler S (1980) The psychological paradigm shift in genetic counseling. Soc Biol 27: 167-185
2326. Kingston HM, Sarfarazi M, Newcombe RG, Willis N, Harper PS (1985) Carrier detection in Becker muscular dystrophy using creatine kinase estimation and DNA analysis. Clin Genet 27: 383-391
2327. Klingmüller W (1976) Genmanipulation und Gentherapie. Springer, Berlin Heidelberg New York
2328. Koch M, Fuhrmann W (1985) Sibs of probands with neural tube defects - a study in the Federal Republic of Germany. Hum Genet 70: 74-79
2329. Lawless EW (1977) Technology and social shock. Rutgers University Press, New Brunswick (N.J.)
2330. Lederberg J (1963) Biological future of man. In: Man and his future. Wolstenholme G (ed), Churchill, London pp 263-273.

2331. LeMeur M, Gerlinger P, Benoist C, Mathis D (1985) Correcting an immuneresponse deficiency by creating E_α gene transgenic mice. Nature 316: 38-42
2332. Leonard C, Chase G, Childs B (1972) Genetic counseling: a consumer's view. N Engl J Med 287: 433
2333. Levy HL (1974) Genetic screening. Adv Hum Genet 4: 1
2334. Lippman-Hand A, Fraser FC (1979) Genetic counseling - the postcounseling period: I. Parents' perceptions of uncertainty. Am J Med Genet 4: 51-71
2335. Lippman-Hand A, Fraser FC (1979) Genetic counseling - the postcounseling period: II. Making reproductive choices. Am J Med Genet 4: 73-87
2336. MacSorley K (1964) An investigation into the fertility rates of mentally ill patients. Ann Hum Genet 27: 247
2337. Marx JL (1985) Making mutant mice by gene transfer. Science 228: 1516-1517
2338. Matsunaga E (1965) Measures affecting population trends and possible genetic consequences. United Nations World Populations Conference. Belgrad, August-September
2339. McLaren A (1973) Biological aspects of A. I. D. In: Law and ethics of A. I. D. and embryo transfer. Ciba Found Symp (new series) 17: 3-9
2340. McLaren A (1985) Prenatal diagnosis before implantation: opportunities and problems. Pren Diag 5: 85-90
2341. Merrill CR, Geier MR, Petricciani JC (1971) Bacterial virus gene expression in human cells. Nature 233: 398-400
2342. Mibashan RS, Rodeck CH (1984) Haemophilia and other genetic defects of haemostasis. In: Prenatal Diagnosis, Rodeck CH, Nicolaides KH (eds) John Wiley & Sons, Chichester, New York
2343. Miller AD, Eckner RJ, Jolly JD, Friedmann I, Verma IM (1984) Expression of a retrovirus encoding human HPRT in mice. Science 223: 630-632
2344. Milunsky A (1975) The prevention of genetic disease and mental retardation. Saunders, Philadelphia
2345. Modell B (1984) Haemoglobinopathies - diagnosis by fetal blood sampling. In: Prenatal Diagnosis, Rodeck CH, Nicolaides KH (eds) John Wiley & Sons, Chichester, New York
2346. Modell B (1985) Chorionic villus sampling. Evaluation, safety and efficacy. Lancet 1: 737-740
2347. Motulsky AG (1973) Screening for sickle-cell hemoglobinopathy and thalassemia. Isr J Med Sci 9: 1341-1349
2348. Motulsky AG (1974) Brave new world? Current approaches to prevention, treatment, and research of genetic diseases raise ethical issues. Science 185: 683-663
2349. Motulsky AG (1975) Family detection of genetic disease. In: Early diagnosis and prevention of genetic disease. Went L, Vermeij-Keers C, van der Linden AGJM (eds), University of Leiden Press, Leiden pp 101-110.
2350. Motulsky AG (1975) Problems of screening for genetic disease. In: Early diagnosis and prevention of genetic disease. Went L, Vermeij-Keers C, van der Linden AGJM (eds), University of Leiden Press, Leiden pp 132-140.
2351. Motulsky AG (1979) Genetic counseling. In: Textbook of medicine, 15th ed. Beeson PB, McDermott W, Wyngaarden JB (eds), Saunders, Philadelphia
2352. Motulsky AG (1977) A genetical view of modern medicine. The Kober lecture. Transact Ass Am Phys 40: 76-90
2353. Motulsky AG (1982) Genetic counseling. In: Cecil Textbook of Medicine, 16th ed, Wyngaarden JB, Smith LH Jr (eds), Saunders, Philadelphia pp 23-26
2354. Motulsky AG (1983) Impact of genetic manipulation on society and medicine. Science 219: 135-140
2355. Motulsky AG (1984) Genetic engineering, medicine and medical genetics. Biomedicine and Pharmacotherapy 38: 185-186
2356. Motulsky AG, Fraser GR (1980) Effects of antenatal diagnosis and selective abortion on frequencies of genetic disorders. Clin Obstet Gynecol 7: 121-134
2357. Motulsky AG, Murray J (1983) Will prenatal diagnosis with selective abortion affect society's attitude toward the handicapped? In: Research Ethics, Berg K, Tranoy KE (eds) Alan R Liss, New York, pp 277-291
2358. Muller HJ (1963) Genetic progress by voluntarily conducted germinal choice. In: Man and his future. Wolstenholme G (ed), Churchill, London pp 247-262.
2359. Nelson WB, Swint JM, Caskey CT (1978) An economic evaluation of a genetic screening program for Tay-Sachs disease. Am J Hum Genet 30: 160-166
2360. Nyhan WL (1985) Neonatal screening for inherited disease. N Engl J Med 313: 43-44
2361. Old JM, Weatherall DJ, Wart RHT, Petrou M, Modell B, Rodeck CH, Warren R, Morsman JM (1985) First trimester diagnosis of the hemoglobin disorders. Ann N Y Acad Sci 445: 349-356
2362. Omenn GS (1982) Predictive identification of hypersusceptible individuals. J Occup Med 24: 369-374
2363. Orkin SH (1985) Molecular biology of β-thalassemia. In: First International symposium on the role of recombinant DNA in genetics. Teplitz RL, Loukopoulos D (eds). Crete, (in the press 1986)
2364. Ottman R, Pike MC, King M-C, Henderson BE (1983) Practical guide for estimating risk for familial breast cancer. Lancet 2: 556-558
2365. Palmiter RD, Brinster RL, Hammer RE, Trumbauer ME, Rosenfeld MG, Birnberg NC, Evans RM (1982) Dramatic growth of mice that develop from eggs microinjected with metallothionein-growth hormone fusion genes. Nature 300: 611-615
2366. Pauli RM, Motulsky AG (1981) Risk counselling in autosomal dominant disorders with undetermined penetrance. J Med Genet 18: 340-343
2367. Penrose LS (1955) Parental age and mutation. Lancet II: 3/2
2368. Perry TB, Fraser FC (1973) Variability of serum creatine phosphokinase activity in normal women and carriers of the gene for muscular dystrophy. Neurology 23: 1316
2369. Plachot M, Mandelbaum J (1984) La fecondation in vitro: 5 ans, bientôt l'age de raison. Ann Genet 27: 133
2369 a. Reeders ST, et al. (1986) Two genetic markers closely linked to adult polycystic kidney disease on chromosome 16. Brit Med J 292, 851-853
2370. Reilly P (1975) Genetic screening legislation. Adv Hum Genet 5: 319-376

2371. Richards BW (1967) Mongolism: The effect of trend in age at child birth and chromosomal type. J Ment Subnormality 13: 3
2372. Robertson FW, Cumming AM (1985) Effects of apoprotein E polymorphism on serum lipoprotein concentration. Arteriosclerosis 5: 283-292
2373. Robinson A (1985) Prenatal diagnosis by amniocentesis. Ann Rev Med 36: 13-16
2374. Rodeck CH (1984) Obstetric techniques in prenatal diagnosis. In: Prenatal Diagnosis, Rodeck CH, Nicolaides KH (eds). John Wiley & Sons, Chichester, New York
2375. Rorvik DM (1978) In his image: The cloning of a man. Lippincott, Philadelphia
2376. Rosatelli C, Falchi AM, Tuveri T, Scalas MT, DiTucci A, Monni G, Cao A (1985) Prenatal diagnosis of beta-thalassaemia with the synthetic-oligomer technique. Lancet 1: 241-243
2377. Scriver CR (1980) Predictive medicine: a goal for genetic screening. In: Neonatal Screening for Inborn Errors of Metabolism, Bickel H, Guthrie R, Hammersen G (eds). Springer, Berlin
2378. Silvestroni E, Bianco I (1975) Screening for microcytemia in Italy: analysis of data collected in the past 30 years. Am J Hum Genet 27: 198-212
2379. Simoni G, Brambati B, Danesino C, Terzoli GL, Romitti L, Rossella F, Fraccaro M (1984) Diagnostic application of first trimester trophoblast sampling - 100 pregnancies. Hum Genet 66: 252-259
2379 a. Simoni G, Gimelli G, Cuoco C, Romitti L, Terzoli G, Guerneri S, Rosella F, Pescetto L, Pezzolo A, Porta S, Brambati B, Porro E, Fraccaro M (1986) First trimester fetal karyotyping: One thousand diagnoses Hum Genet 72, 203-209
2380. Sinsheimer RL (1977) Recombinant DNA. Ann Rev Biochem 46: 415-438
2381. Sinsheimer R (1977) An evolutionary perspective for genetic engineering. New Scientist 20: 150
2382. Sorenson JR, Swazey JP, Scotch NA (eds) (1981) Reproductive Pasts Reproductive Futures. Genetic Couselling and its Effectiveness. Birth Defects: Orig Art Ser XVII (4). Alan R. Liss, New York
2383. Szybalska EH, Szybalski W (1962) Genetics of human cell lines. IV. DNA-mediated heritable transformation of a biochemical trait. Proc Natl Acad Sci USA 48: 2026-2034
2384. Schrott HG, Karp L, Omenn GS (1973) Prenatal prediction in myotonic dystrophy: guidelines for genetic counseling. Clin Genet 4: 38-45
2385. Starlinger P (1984) Medizinische Gentechnologie: Möglichkeiten und Grenzen. Deutsches Ärzteblatt 81: 2091-2098
2386. Steptoe PC, Edwards RG (1976) Reimplantation of a human embryo with subsequent tubal pregnancy. Lancet I: 880-882
2387. Steptoe PC, Edwards RG (1978) Birth after the reimplantation of a human embryo. Lancet II: 366
2388. Terheggen HG, Lowenthal A, Lavinha F, Colombo JP, Rogers S (1975) Unsuccessful trial of gene replacement in arginase deficiency. Z Kinderheilkd 119: 1-3
2389. Trounson A, Mohr L (1983) Human pregnancy following cryopreservation, thawing and transfer of an eight-cell embryo. Nature 305: 707-709
2390. Ullrich A, Shine J, Chirgwin J, Pictet R, Tischer E, Rutter WJ, Goodman HM (1977) Rat insulin genes: Construction of plasmids containing the coding sequences. Science 196: 1313-1319
2391. Veal AM (1965) Intestinal polyposis. Eugenics laboratory memoirs XL. Cambridge University Press, London
2392. Villa-Komaroff L, Efstratiadis A, Broome S, Lomedico P, Tizzard R, Naber SP, Chick WL, Gilbert W (1978) A bacterial clone synthesizing proinsulin. Proc Natl Acad Sci USA 75: 3727-3731
2393. Vogel F (1957) Die eugenische Beratung beim Retinoblastom (Glioma retinae). Acta Genet 7: 565-572
2394. Vogel F (1967) Wie stark ist die theoretische Häufigkeit von Trisomie-Syndromen durch Verschiebungen im Altersaufbau der Mütter zurückgegangen? Zoologische Beiträge 13: 451-462
2395. Vogel F (1973) Der Fortschritt als Gefahr und Chance für die genetische Beschaffenheit des Menschen. Klin Wochenschr 51: 575-585
2396. Vogel F (1977) A probable sex difference in some mutation rates. Am J Hum Genet 29: 312-319
2396a. Vogel F (1985) New DNA techniques - chances and risks for mankind. First Int Symp on the Role of Recombinant DNA in Genetics. Crete (In the press, 1986)
2397. Vosberg HP (1977) Molecular cloning of DNA. An introduction into techniques and problems. Hum Genet 40: 1-72
2398. Wald NJ, Cuckle HS (1984) Open neural tube defects. In: Antenatal and Neonatal Screening. Wald NJ (ed), London, Oxford
2399. Ward RHT (1984) First trimester chorionic villus sampling. In: Prenatal Diagnosis, Rodeck CH, Nicolaides KH (eds) John Wiley & Sons, Chichester, New York
2400. Weatherall D (1984) Gene transfection. A new step nearer gene therapy? Nature 310: 451-452
2401. Wendt GG, Landzettel HJ, Unterreiner I (1959) Das Erkrankungsalter bei der Huntingtonschen Chorea. Acta Genet (Basel) 9: 18
2402. Williams DA, Lemischka IR, Nathan DG, Mulligan RC (1984) Introduction of new genetic material into pluripotent haemotapretic stem cells of the mouse. Nature 310: 476-480
2403. Wood L, Trounson A (1984) Clinical in vitro fertilization. Springer, Berlin Heidelberg New York

Subject Index

Springer

W. Fuhrmann, F. Vogel

Genetic Counseling

Translated from the German by S. Kurth-Scherer

3rd edition 1983. 50 figures. X, 188 pages. ISBN 3-540-90715-7

Contents: Appearances Deceive. – Recording a Family Medical History or Pedigree. – The Autosomal Dominant Mode of Inheritance. – New Mutations and Nonhereditary Cases (Phenocopies and Somatic Mutations). – The Autosomal Recessive Mode of Inheritance and Tests for the Detection of Heterozygotes. – Sex-Linked Modes of Inheritance. – How Can Additional Information Be Incorporated into Genetic Risk Calculations? – Chromosal Aberrations. – Malformations Not Subject to a Simple Mode of Inheritance. – Prenatal Diagnosis of Genetic Abnormalities and Diseases. – Other Diseases Without a Simple Mode of Inheritance. – Mental Retardation and Mental Illness. – Genetic Prognosis for a Consanguineous Marriage. – Exposure to Mutagenic Noxes. – Teratogenic Effects During Early Pregnancy. – Psychological and Social Considerations. – References. – Index.

"... In summary the new edition of this now established introduction to genetic counselling can be happily recommended to the clinician wishing to gain an insight into the general principles involved...."

The Journal of Inherited Metabolic Disease

E. Therman

Human Chromosomes

Structure, Behavior, Effects

2nd edition. 1986. 87 figures. XVI, 313 pages.
ISBN 3-540-96173-9

"... Technical standards of production are excellent, as one might expect from an author who is a master of the craft, the photographic illustrations are superb... Although the text is very simply written, it contains hardly any ambiguous, misleading or controversial points, even for a critical reader; a really notable achievement. [This book] is intended as an introductory text for students of cytogenetics and for biologists interested in human cytogenetics. As such it is thoroughly recommended."

Heredity

Springer-Verlag
Berlin Heidelberg New York
London Paris Tokyo

J. L. Tiwari, P. I. Terasaki

HLA and Disease Associations

1985. 23 figures. XXIII, 472 pages. ISBN 3-540-96081-3

Contents: HLA Nomenclature. – The HLA Complex. – The Data and Statistical Analysis. – Mechanisms of HLA and Disease Associations. – Overview. – Rheumatology and Joint Diseases. – Dermatology. – Neurology. – Endocrinology. – Gastroenterology. – Ophthalmology. – Malignancy. – Allergy. – Urogenital Diseases. – Cardiovascular Diseases. – Connective Tissue Diseases. – Pulmonary Diseases. – Infectious Diseases. – Musculoskeletal Diseases. – Psychiatric Diseases. – Hematologic Diseases. – Complement and Immune Deficiencies. – Immune Response. – Miscellaneous Diseases. – Index.

Tumour Aneuploidy

Editors: T. Büchner, C. D. Bloomfield, W. Hiddemann, D. K. Hossfeld, J. Schumann

1985. 49 figures, 56 tables. VIII, 141 pages. ISBN 3-540-15376-4

Contents: Incidence and Prognostic Significance of DNA Aneuploidy in Childhood Acute Lymphoblastic Leukemia. – DNA Aneuploidy in Childhood Acute Lymphoblastic Leukemia as Detected by Flow Cytometry: Relation to Phenotype, Presentation Features and Prognosis. – Aneuploidy in Acute Myeloid Leukemia. – Incidence and Heterogeneity of DNA Aneuploidies in Solid Tumors and Acute Leukemias. – Chromosome Findings in Secondary Acute Leukemias. – Cytogenetic Studies in Acute Leukemia and in the Blast Phase of the Chronic Myelocytic Leukemia. – DNA Aneuploidy – A Common Cell Marker in Human Malignancies and Its Correlation to Grade, Stage and Prognosis. – The Clinical Usefulness of Chromosome Abnormalities in Acute Leukemia. – Chromosomal Heterogeneity in Solid Tumors. – Flow Cytometry as a Diagnostic and Prognostic Tool in Cancer Medicine. – Multiparameter in Flow Cytometry for Determination of Ploidy, Proliferation and Differentiation in Acute Leukemia: Treatment Effects and Prognostic Value. – Subject Index.

Current Topics in Microbiology and Immunology Volume 114

1985. 39 figures. VII, 255 pages. ISBN 3-540-15103-6

Contents: *W. Zillig, R. Schnabel, K. O. Stetter:* Archaebacteria and the Origin of Eukaryotic Cytoplasm. – *H.-P. Vosberg:* DNA Topoisomerases: Enzymes That Control DNA Conformation. – *D. P. Nayak, T. M. Chambers, R. K. Akkina:* Defective-Interfering (DI) RNAs of Influenza Viruses: Origin, Structure, Expression, and Interference. – *R. W. Compans, D. H. L. Bishop:* Biochemistry of Arenaviruses. – *P. M. Colman, C. W. Ward:* Structure and Diversity of the Influenza Virus Neuraminidase.